ENCYCLOPEDIA OF

BIOPROCESS TECHNOLOGY: FERMENTATION, BIOCATALYSIS, AND BIOSEPARATION

VOLUME 2

WILEY BIOTECHNOLOGY ENCYCLOPEDIAS

Encyclopedia of Bioprocess Technology: Fermentation, Biocatalysis, and Bioseparation
Edited by Michael C. Flickinger and Stephen W. Drew

Encyclopedia of Molecular Biology
Edited by Thomas E. Creighton

Encyclopedia of Cell Technology
Edited by Raymond E. Spier

Encyclopedia of Ethical, Legal, and Policy Issues in Biotechnology
Edited by Thomas J. Murray and Maxwell J. Mehlman

ENCYCLOPEDIA OF BIOPROCESS TECHNOLOGY: FERMENTATION, BIOCATALYSIS, AND BIOSEPARATION
EDITORIAL BOARD

ENCYCLOPEDIA OF BIOPROCESS TECHNOLOGY: FERMENTATION, BIOCATALYSIS, AND BIOSEPARATION

VOLUME 2

Michael C. Flickinger
University of Minnesota
St. Paul, Minnesota

Stephen W. Drew
Merck and Co., Inc.
Rahway, New Jersey

A Wiley-Interscience Publication
John Wiley & Sons, Inc.
New York / Chichester / Weinheim / Brisbane / Singapore / Toronto

21434743

This book is printed on acid-free paper. ♾

Published simultaneously in Canada.

For ordering and customer service, call 1-800-CALL-WILEY.

Library of Congress Cataloging-in-Publication Data:

Flickinger, Michael C.
The encyclopedia of bioprocess technology : fermentation, biocatalysis, and bioseparation / Michael C. Flickinger, Stephen W. Drew.
p. cm.
Includes index.
ISBN 0-471-13822-3 (alk. paper)
1. Biochemical engineering--Encyclopedias. I. Drew, Stephen W., 1945- . II. Title.
TP248.3.F57 1999
660.6′03--dc21 99-11576
CIP

Printed in the United States of America.

10 9 8 7 6 5 4 3 2 1

CHROMATOGRAPHY, SIZE EXCLUSION

Charles M. Roth
Martin L. Yarmush
Massachusetts General Hospital, Shriners Burns Hospital, and Harvard Medical School
Boston, Massachusetts

KEY WORDS

Desalting
Excluded volume
Large scale
Packing materials
Purification
Resolution

OUTLINE

INTRODUCTION

Size-exclusion chromatography (SEC) is a widely used means for separation of macromolecules of different sizes. Porath and Flodin in 1959 first demonstrated the use of SEC, or gel filtration chromatography, for separations by resolving glucose from two size fractions of dextrans (1). In addition, they applied the technique to desalting of serum proteins. Remarkably, these two applications—crude separations and desalting—remain the primary uses of SEC today. A wide range of materials, including sugars, proteins, and nucleic acids, have been analyzed or separated using this technique. In addition, SEC is used extensively in the characterization of polymers. However, in this article, the emphases are on analysis and on separations involving biological macromolecules.

Basic Principle

The essential principle of size exclusion chromatography is partitioning of species on the basis of preferential sieving into the pores of support particles. Sample is applied to the top of a bed packed with porous stationary phase particles. Elution results from flow of a fixed composition buffer through the bed (isocratic elution), in contrast with other modes of chromatography in which the buffer composition is varied (gradient elution). Molecules that are larger than the pores of the support access only the interstitial space between particles and therefore pass through the column quickly, whereas smaller particles access the volume within the pores and are retained for a longer time by the column (Fig. 1). Therefore, a support must be selected with a pore size distribution so that there exists a differential partitioning of the molecules to be separated.

Lexicon

In any chromatographic process, sample is injected into the column and eluted by continuous flow of a buffer through the column. The time between injection and the peak of any eluted solute is termed its retention time (t_R). An equivalent measure of retention is the retention volume (V_R), which, for the usual case of constant flow rate, is merely the product of the flow rate and retention time. The terms *elution time* and *elution volume* are used interchangeably with *retention time* and *retention volume*. For solute molecules larger than the size of the pores, the elution volume represents the volume of the column between the support particles plus that of extracolumn tubing between the injection loop and the column and from the column to the detector. This is known as the void volume (V_0), and it is frequently estimated using large, inert tracer solutes. The pore volume (V_p) is the volume within the particles occupied by fluid and, in principle, accessed entirely by very small particles, for which the retention volume is maximized:

$$V_{R,max} = V_0 + V_p \tag{1}$$

In general, the elution volume is intermediate between the void volume and the sum of the void and pore volumes and thus defines a distribution coefficient (K_d) through the relation

$$V_R = V_0 + K_d V_p \tag{2}$$

By definition, the distribution coefficient assumes values only between zero and unity. The other relevant volume is that of the support particles or chromatographic matrix (V_m). In principle, one can calculate the pore volume based on the amount and density of the support through the relation

$$V_p = \frac{W_r \rho_{gel}}{1 + W_r \rho_{H_2O}} (V_t - V_0) \tag{3}$$

where V_t is the total volume (sum of void, pore, and matrix volumes, the first corrected for the volume of extracolumn tubing), ρ represents mass density, and W_r is the water

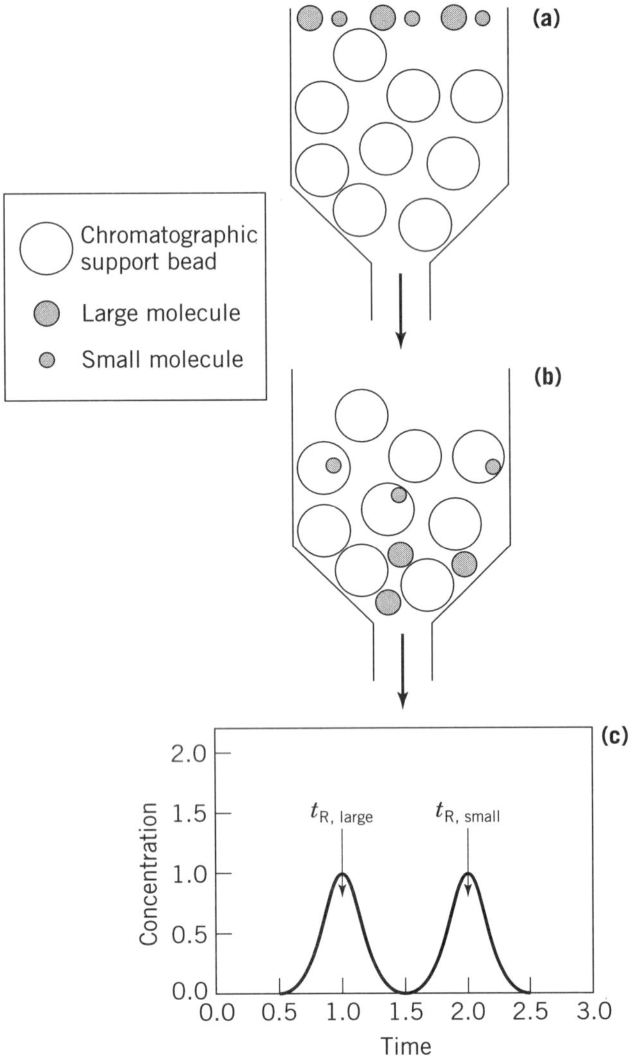

Figure 1. Principle of SEC. (**a**) A mixture of large and small particles is introduced together into the top of the column. (**b**) As the particles move down the column, the small particles are able to access the volume within the pores of the particles. Therefore, the larger particles are eluted from the column before the smaller ones. (**c**) Schematic chromatogram showing the elution profiles of the large and small particles.

regain of gel (grams of water per gram of dry gel divided by the density of water). Experimentally, V_p can be measured as the retention volume of a small molecule that does not interact with the support. With determination of V_0 and V_p, the distribution coefficient of any solute can be determined from its retention and equation 2; conversely, the retention time of a solute could be predicted from a model for its distribution coefficient (see "Process Variables") and from the column parameters.

The typical chromatographic peak is Gaussian in shape. The quality of separation of two (or more) species in SEC is characterized by differences in their retention times and by certain measures of the distribution of material in each peak about the mean. The resolution (R_s) is a measure of separation of two bands, defined by

$$R_s = \frac{t_{R2} - t_{R1}}{\frac{1}{2}(w_2 - w_1)} \qquad (4)$$

where w_i is the baseline width of peak i, related to the standard deviation of the Gaussian peak through $w_i = 4\sigma_i$, and t_{Ri} is the retention time of the ith peak. Another measure of separation is the selectivity, α, between two bands. It is defined by

$$\alpha = \frac{k'_2}{k'_1} \qquad (5)$$

where k′ of species i is its retention factor (or capacity factor). The retention factor is a relative retention time defined as

$$k' = \frac{t_R - t_0}{t_0} = \frac{V_R - V_0}{V_0} \qquad (6)$$

For non-Gaussian bands, t_R and σ can be determined via statistical analysis of the chromatogram from the first and second moments, respectively, of a trace.

MATERIALS

Equipment

A typical laboratory setup for SEC is depicted in Figure 2. The equipment required includes solvent and reservoir, valves and tubing, pump, packed column, detector, fraction collector, and recording device. For the most part, this equipment is universal to the various modes of chromatography. The precolumn setup for SEC, especially at low pressure, is relatively straightforward. Solvent delivery is simpler in SEC than for other modes of chromatography because there is no need for gradient formation and the only pump requirement is for stable flow. Peristaltic pumps are adequate for low-pressure work, whereas piston or diaphragm pumps are used for high pressures (2).

The most important aspect of the equipment is the column packing. The gel must be packed uniformly throughout the column to avoid channeling or other heterogeneous effects that act to blur separation of the solutes. Frequently, the gel is supplied as a powder or as a suspension in a storage solvent. In the latter case, this solvent is filtered off, and in both cases the gel is then suspended in a packing solvent, which is usually an aqueous buffer with some NaCl added. A stable, uniform packing is attained by degassing the gel slurry before packing, pouring it at once, and then packing at a flow rate greater than used in normal operation.

Detection of proteins and nucleic acids is most commonly accomplished with UV absorbance detectors, although more sensitive techniques such as light-scattering and mass spectrometry are also possible (3). The detection limit (C^*) is dictated by the absorbance sensitivity (A^*) of the instrument, the extinction coefficient (ϵ) of the solute, and the path length (L) of the detector, according to

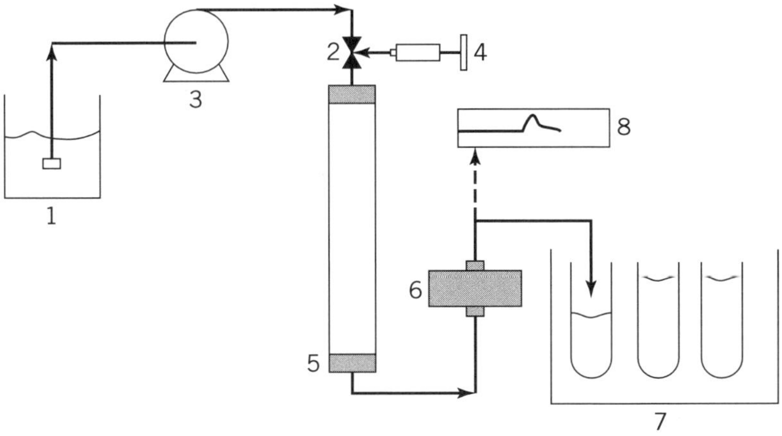

Figure 2. Laboratory SEC setup. Solvent (*1*) is pumped through a three-way valve (*2*) by a peristaltic pump (*3*). The desired volume of sample is introduced by syringe (*4*) through one port of the valve while the solvent flow is temporarily blocked. The valve is repositioned to allow elution of the sample though the packed column (*5*). The eluting solution is monitored by a detector (*6*) and collected in a fraction collector (*7*). A permanent record of the chromatogram is obtained from a recorder (*8*) or computer.

$$C^* > \frac{A^*}{\epsilon l} \tag{7}$$

The sensitivity of absorbance detectors is usually about 5 $\times$ 10^{-4}, and their path lengths can be up to 1 cm. Typical proteins have extinction coefficients on the order of 1 cm^2/mg; therefore, the detection limit is on the order of 1 μg/mL. Nucleic acids exhibit strong absorbance, peaking around 260 nm, with extinction coefficients on the order of 25 to 50 cm^2/mg; consequently, lower concentrations down to 20 ng/mL may be detected. Carbohydrates, on the other hand, exhibit no significant absorbance in UV or visible light; refractive index detectors are often used for their quantitation.

Packing Materials

A wide variety of packing materials are available for use in SEC. The choice of gel for a particular application depends on a number of factors, including materials compatibility, operating conditions, and particularly the sizes of the molecules to be separated. Manufacturers typically provide a good deal of information about their packing materials, including fractionation range, maximum flow rate or pressure drop, and stable operating ranges of pH and temperature.

Properties. The key feature of the packing material with respect to a particular separation is the pore size distribution, because this determines the range of particles that may be fractionated. Very small pores are required for desalting applications, so that all macromolecules are excluded and low molecular weight species are retained. Materials possessing much larger pores are used for analysis and purification of biological macromolecules. Because the pore sizes are normally inferred from the sieving of test solutes, the useful molecular weight fractionation range, rather than the actual pore size, is generally reported. It is important to note that different types of molecules exhibit different fractionation ranges; therefore, a fractionation range reported using protein standards will not be identical to that of, for example, nucleic acids. Some manufacturers report two fractionation ranges: one using protein standards and another using nucleic acids or dextrans. Usually, these two values will be sufficient to at least select an appropriate gel for the application of interest.

Once a fractionation range is determined, a material appropriate for the solutes of interest must be identified. Although the gel materials used in SEC have been developed to exhibit minimal adsorption, biological macromolecules inevitably interact to some extent with these materials. Proteins in particular will adsorb to a variety of surface functionalities. A general rule is that the gel material should be hydrophilic and uncharged. However, most hydrophilic materials exhibit at least a weak charge. Undesirable electrostatic interactions caused by this charge can be minimized by eluting with a buffer of moderate ionic strength, 0.3 to 0.5 M. In addition, the column must be operated under conditions at which the material is stable. Packing materials can deteriorate when subjected to extremes of pH, ionic strength, flow rate, and temperature. For some materials, enzymatic degradation is also possible in the event of bacterial contamination.

The size of the particles and their degree of compressibility determine the pressure drop across a bed and limit the flow rate that may be used. Particles ranging from 5 to 600 μm are available (4,5). Some SEC media, particularly cross-linked materials, are rigid over their normal operating range (Table 1). For these rigid materials, the pressure drop (Δp) across a gel of length (L) is inversely proportional to the square of the particle diameter (d_p) and proportional to velocity (v_0) (flow rate divided by cross-sectional column area), in accordance with Darcy's law:

$$\Delta p = \frac{\mu v_0 L}{k d_p^2} \tag{8}$$

where k is a constant and μ is the fluid viscosity. However, as the particles compress, the pressure drop increases dramatically, effectively setting an upper limit on the maximum attainable velocity. From the standpoint of materials selection, manufacturers supply information regarding particle size, fraction range, and maximum allowable flow.

Available Materials. Materials comprised of dextran, agarose, acrylamide, or combinations have traditionally been the workhorses of SEC. However, novel polymeric materials and polymer-coated silica media have been de-

Table 1. Packing Materials for SEC

Name	Manufacturer	Composition	MW fraction range (proteins)	DNA exclusion limit (bp)	Bead size (μm)	pH range	Maximum back pressure (MPa)	Maximum flow rate (cm/h)
Superose 12 prep grade	Pharmacia	Highly cross-linked agarose	1×10^3–3×10^5	150	20–40	3–12	0.7	30
Superose 6 prep grade	Pharmacia	Highly cross-linked agarose	5×10^3–5×10^6	450	20–40	3–12	0.4	30
Sephacryl S-100 HR	Pharmacia	Dextran/bisacrylamide	1×10^3–1×10^5	—	25–75	3–11	0.2	20–39
Sephacryl S-200 HR	Pharmacia	Dextran/bisacrylamide	5×10^3–2.5×10^5	118	25–75	3–11	0.2	20–39
Sephacryl S-300 HR	Pharmacia	Dextran/bisacrylamide	1×10^4–1.5×10^6	118	25–75	3–11	0.2	24–48
Sephacryl S-400 HR	Pharmacia	Dextran/bisacrylamide	2×10^4–8×10^6	271	25–75	3–11	0.2	33–63
Sephacryl S-500 HR	Pharmacia	Dextran/bisacrylamide	4×10^4–2×10^7[c]	1,078	25–75	3–11	0.2	24–48
Sephacryl S-1000 SF	Pharmacia	Dextran/bisacrylamide	5×10^5–1×10^8[e]	20,000	40–105	3–11	—	40
Sepharose 6B	Pharmacia	Agarose	1×10^4–1×10^6	194	45–165	4–9	2.8	14[a]
Sepharose CL-6B	Pharmacia	Cross-linked agarose	1×10^4–1×10^6	194	45–165	3–13	2.8	30[a]
Sepharose 4B	Pharmacia	Agarose	6×10^4–2×10^7	872	45–165	4–9	0.9	11.5[a]
Sepharose CL-4B	Pharmacia	Cross-linked agarose	6×10^4–2×10^7	872	45–165	3–13	1.7	26[a]
Sepharose 2B	Pharmacia	Agarose	7×10^4–4×10^7	1,353	60–200	4–9	0.6	10[a]
Sepharose CL-2B	Pharmacia	Cross-linked agarose	7×10^4–4×10^7	1,353	60–200	3–13	0.9	15[a]
Sephadex G-10	Pharmacia	Cross-linked dextran	<700	—	40–120	2–13	DL	2–5
Sephadex G-25	Pharmacia	Cross-linked Dextran	1×10^3–5×10^3	—	20–150	2–13	DL	2–5
Sephadex G-50	Pharmacia	Cross-linked dextran	1×10^3–3×10^4	—	20–80	2–10	DL	2–5
Sephadex G-75	Pharmacia	Cross-linked dextran	3×10^3–7×10^4	—	10–120	2–10	0.016[b]	18
Sephadex G-100	Pharmacia	Cross-linked dextran	4×10^3–1.5×10^5	—	10–120	2–10	0.0096[b]	12–50
Sephadex G-150	Pharmacia	Cross-linked dextran	5×10^3–3×10^5	—	10–120	2–10	0.0036[b]	6–23
Sephadex G-200	Pharmacia	Cross-linked dextran	5×10^3–6×10^5	—	10–120	2–10	0.0016[b]	3–12
Superdex 75 HR 10/30	Pharmacia	Dextran/Cross-linked agarose	3×10^3–7×10^4		13–15	3–12	1.8	110
Superdex 200 HR 10/30	Pharmacia	Dextran/cross-linked agarose	1×10^4–6×10^5	200	13–15	3–12	1.5	76
Bio-Gel P-2	Bio-Rad	Polyacrylamide	1×10^2–1.8×10^3	—	45–90	2–10	—	5–10[d]
Bio-Gel P-4	Bio-Rad	Polyacrylamide	8×10^2–4×10^3	—	45–180	2–10	—	10–20[d]
Bio-Gel P-6	Bio-Rad	Polyacrylamide	1×10^3–6×10^3	5	45–180	2–10	—	10–20[d]
Bio-Gel P-10	Bio-Rad	Polyacrylamide	1.5×10^3–2×10^4	—	45–180	2–10	—	10–20[d]
Bio-Gel P-30	Bio-Rad	Polyacrylamide	2.5×10^3–4×10^4	20	45–180	2–10	—	6–13[d]
Bio-Gel-P-60	Bio-Rad	Polyacrylamide	3×10^3–6×10^4	55	45–180	2–10	—	3–6[d]
Bio-Gel P-100	Bio-Rad	Polyacrylamide	5×10^3–1×10^5	—	45–180	2–10	—	3–6[d]
BioGel A 0.5m	Bio-Rad	Agarose	$<1 \times 10^4$–5×10^5	—	38–300	4–13	—	7–20[e]
BioGel A 1.5m	Bio-Rad	Agarose	$<1 \times 10^4$–$1.5^{1}0^{6}$	—	38–300	4–13	—	7–20[e]
BioGel A 5m	Bio-Rad	Agarose	1×10^4–5×10^6	—	38–300	4–13	—	7–20[e]
BioGel A 15m	Bio-Rad	Agarose	4×10^4–1.5×10^7	—	38–300	4–13	—	7–20[e]
BioGel A 50 m	Bio-Rad	Agarose	1×10^5–5×10^7	350	75–300	4–13	—	5–25[e]
Macro-Prep SE 100/40	Bio-Rad	Cross-linked Agarose	5×10^3–1×10^5	—	40	1–14	4	240
Macro-Prep SE 1000/40	Bio-Rad	Cross-linked Agarose	1×10^4–1×10^6	—	40	1–14	3	240
Hydropore-5-SEC	Varian/Rainin	Polymer-coated silica	5×10^3–1×10^6	—	5	—	—	—
Toyopearl HW-40	TosoHaas	Poly(ethylene glycol/ methacrylate)	1×10^2–1×10^4	—	20–60	2–12	0.3	DL
Toyopearl HW-50	TosoHaas	Poly(ethylene glycol/ methacrylate)	5×10^2–8×10^4	—	20–60	2–12	0.3	DL
Toyopearl HW-55	TosoHaas	Poly(ethylene glycol/ methacrylate)	1×10^3–2×10^5	—	20–60	2–12	0.3	DL
Toyopearl HW-65	TosoHaas	Poly(ethylene glycol/ methacrylate)	4×10^4–5×10^6	—	20–60	2–12	0.3	DL
Toyopearl HW-75	TosoHaas	Poly(ethylene glycol/ methacrylate)	$5 \times 10^5 \times 10^7$	—	30–60	2–12	0.3	DL
TSK-GEL G2000SW	TosoHaas	Bonded silica	5×10^3–1.5×10^5	55	4–13	2.5–7.5	1–12	DL
TSK-GEL G3000SW	TosoHaas	Bonded silica	1×10^4–5×10^5	110	4–10	2.5–7.5	1–12	DL
TSK-GEL G4000SW	TosoHaas	Bonded silica	2×10^4–1×10^7	375	8–17	2.5–7.5	1–3.5	DL
TSK-GEL G1000PW	TosoHaas	Poly(ethylene glycol/ methacrylate)	$<2 \times 10^3$	—	10	2–12	2	DL
TSK-GEL G2000PW	TosoHaas	Poly(ethylene glycol methacrylate)	$<5 \times 10^3$	—	10–20	2–12	2–4	DL
TSK-GEL G2500PW	TosoHaas	Poly(ethylene glycol/ methacrylate)	$<8 \times 10^3$	—	6–20	2–12	2–4	DL

Table 1. Packing Materials for SEC *(continued)*

Name	Manufacturer	Composition	MW fraction range (proteins)	DNA exclusion limit (bp)	Bead size (μm)	pH range	Maximum back pressure (MPa)	Maximum flow rate (cm/h)
TSK-GEL G3000PW	TosoHaas	Poly(ethylene glycol/ methacrylate)	5×10^2–8×10^5	—	6–20	2–12	2–4	DL
TSK-GEL G4000PW	TosoHaas	Poly(ethylene glycol/ methacrylate)	1×10^4–1.5×10^6	—	10–22	2–12	1–2	DL
TSK-GEL G5000PW	TosoHaas	Poly(ethylene glycol/ methacrylate)	$<1 \times 10^7$	1,500	10–22	2–12	1–2	DL
TSK-GEL G6000PW	TosoHaas	Poly(ethylene glycol/ methacrylate)	$<2 \times 10^8$	—	13–25	2–12	1–2	DL
TSK-GEL GMPW	TosoHaas	Poly(ethylene glycol/ methacrylate)	$<2 \times 10^8$	—	13–17	2–12	1–2	DL
TSK-GEL G-Oligo-PW	TosoHaas	Poly(ethylene glycol/ methacrylate)	$<3 \times 10^3$	—	6	2–12	4	DL
TSK-GEL G-DNA-PW	TosoHaas	Poly(ethylene glycol/ methacrylate)	$<2 \times 10^8$	7,000	10	2–12	2	DL

Note: Properties described are as stated by the manufacturers. In many cases, several sizes of particles are available for a given material; these are indicated as a range. DL indicates that the material obeys Darcy's law (i.e., flow rate proportional to pressure drop per bed length) over the normal operating range.
[a]Determined using distilled water at room temperature on a 2.5- by 30-cm column.
[b]Determined using distilled water at room temperature on a 2.6- by 30-cm column.
[c]The indicated fractionation range was determined for dextrans rather than proteins.
[d]Typical flow rates determined using a 1.5- by 70-cm column.
[e]Typical flow rates determined using a 1.5- by 20-cm column.

veloped. Silica possesses excellent mechanical strength, but it is prone to adsorption of biomolecules. Materials are available from a number of manufacturers, and many are summarized in Table 1. Although this list is not exhaustive, it represents many of the commonly used materials, and it is notable that these have not changed significantly over the past 15 years (5–7).

Dextrans are used in a wide variety of protein and nucleic acid separations. They are polysaccharides that are made into a porous gel by cross-linking with epichlorohydrin (8). These materials are quite stable for pH > 2 and for temperatures up to at least 120 °C and can be used with either aqueous or organic solvents. Dextran gels with small pore sizes exhibit low compressibility, but versions with larger pore sizes are quite compressible (9). As a result, some dextran gels are limited to flow rates that can be less than 1 mL/min. Because dextrans are polysaccharides, they are susceptible to bacterial degradation; however, because of their thermal stability, they can be autoclaved. A variation of the traditional dextran gel is provided by cross-linking dextran with *N,N′*-methylenebisacrylamide. This provides a material that is substantially more rigid but has a smaller stable pH range and is more susceptible to adsorption of proteins. The latter obstacle can often be overcome by eluting with a higher ionic strength solvent.

Acrylamide gels are formed by cross-linking acrylamide monomer with *N,N′*-methylenebisacrylamide. The resulting matrix is quite stable over the pH range from 2 to 10 and to high temperatures and is not prone to bacterial degradation. It is less adsorptive than the corresponding dextran material. Consequently, polyacrylamide gels are commonly used for protein separations. Their main disadvantage is that they are incompatible with organic solvents.

A copolymer of D-galactose and 3,6-anhydro-L-galactose is used to form an agarose gel. The composition and degree of cross-linking have a profound effect on the properties of agarose-based media. With less cross-linking, the gels are typically stable only in the pH range of 4 to 9, are compressible at flow rates as low as 1 mL/min, and must be sterilized chemically, but they exhibit little adsorption of macromolecules and broad fractionation ranges that make them much more suitable than acrylamide or dextran for the separation of large proteins and nucleic acids. With a higher degree of cross-linking, the material is stable over a wider range of pH and flow rate, and it can be autoclaved; however hydrophobic interactions may occur between proteins and the support, and the available fractionation range is reduced.

A number of variations on traditional materials have been established. Generally, these are produced to combine the desirable sieving properties of one material with the mechanical or chemical stability of another. The three most commonly used materials—dextran, acrylamide, and agarose—have been combined in all possible pairs. The dextran–bisacrylamide material (Sephacryl) mentioned earlier is one such example. Another is a composite of dextran and cross-linked agarose, covalently bound together to form a material (Superdex, [9]) that has the fractionation range of a dextran support with the rigidity of agarose. The resulting material is autoclavable and stable in the range 3 < pH < 12 and to a variety of strong solvents. Acrylamide has also been entrapped in an agarose gel to form the commercial Ultrogel matrix (7).

Other supports, based on materials such as silica and hydrophilic polymers, have been produced for SEC (5). Although porous silica is inherently rigid and can be manufactured in a variety of pore sizes, it exhibits marked adsorption of proteins and is unstable at basic pH. Other polymer gels exist commercially, such as Toyopearl and Spheron, which are based on polyethers and poly(2-hydroxyethylmethacrylate), respectively. The latter is a material approved by the Food and Drug Administration and can thus be used to produce biocompatible supports.

APPLICATIONS

A number of applications exist for SEC in the research laboratory and in industrial processes. These can be divided into the two categories of sample analysis and purification. The former is most often performed on small columns with high resolution, and the latter is a bulk operation in which the separations are usually fairly crude. A summary of some examples of SEC applications is provided in Table 2.

Analysis

At the analytical level, SEC is used as a tool in characterizing the size, shape, and interactions of biological macromolecules. For example, SEC has been used to study the kinetics and equilibrium dimerization of proteins (10,11). Likewise, it can be used to study the hybridization of nucleic acids (12). Similar applications have been made to reversible interactions between macromolecules and low molecular weight species (13,14). It has also been used to identify and characterize molten globule states in studies of protein unfolding (15). In addition, it is an excellent tool for identification and separation of small molecules that are conjugated to biomolecules, such as probes attached to antibodies or drugs conjugated to targeting molecules (16–18).

SEC is often used for molecular weight determination, although this is an approximate method. Molecules that are similar to the solute of interest but with known molecular weight are used to create a calibration curve of retention time versus molecular weight. The unknown molecular weight of the solute is then estimated from its retention time. The degree of success of this approach is highly dependent on how similar the unknown solute is to the molecular weight standards. That is, a calibration curve prepared with globular proteins will give an inaccurate molecular weight for an oligonucleotide or even for a glycoprotein. Nonetheless, for similar molecules, molecular weights can be successfully estimated. For example, the retention times of a number of globular proteins correlate well with log(MW), as shown in Figure 3. Under the conditions of Figure 3, the retention time of α_1-acid glycoprotein is 8.70, from which an apparent molecular weight of 57 kDa is calculated, significantly greater than its actual molecular weight of 40 kDa.

For molecules of similar chemistry, such as globular proteins, the deviation of the observed elution volume from the calibration curve can be an indication of molecular shape. Deviation from spherical to ellipsoidal shape results in increased molecular friction, and thus the molecule sieves as if its apparent molecular weight is greater than its actual value. Although exact molecular weight determination is complicated by the shape and chemical nature of different protein molecules, changes in a particular structure can be discerned with high resolution. For instance, the extent of modification of a protein by a dye or other ligand can be monitored by the relatively small decrease in retention time if the same column, buffer, and operating conditions are used (16).

Separations

By design, SEC separates molecules based on their size, so it is not possible to achieve well-resolved separations of biomolecules from a mixture of similar species using this technique. Although fair discrimination of molecules according to molecular weight was seen in the previous section, small differences in retention time are insufficient for resolution in purification (see "Dependence of Separation Parameters on Operating Variables"). Nonetheless, crude fractionation of a particular size range is possible with

Table 2. Examples of SEC Applications

Analysis
Molecular weight estimation
Identification and analysis of protein dimers
Analysis of reversible protein–small molecule interactions
Studies of protein unfolding
Identification of products in molecular conjugates
Evaluation of the stability of macromolecular complexes
Separations
Desalting
Crude protein separations
Purification of molecular conjugates
Removal of endotoxins and viruses from pharmaceutical formulations

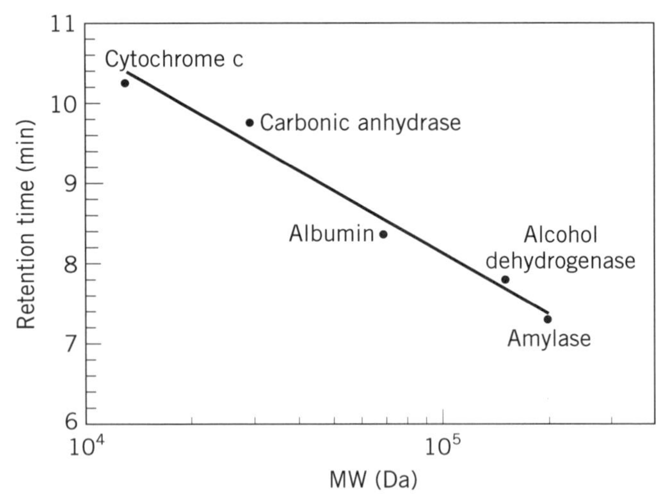

Figure 3. Calibration curve of an analytical SEC column. The elution volumes of several globular proteins were determined on a TSK-GEL G3000SW$_{XL}$ column, 7.8 mm i.d. and 30 cm length. Twenty microliter aliquots (concentration approx. 1 mg/mL) were injected and elution was performed in a buffer of 0.1 M KH_2PO_4 + 0.1 M Na_2SO_4, pH 7. The eluate was monitored by UV absorbance detection at 280 nm.

SEC. Thus, it is best suited either for purification of one molecule of interest from a small set of dissimilar contaminants in the laboratory or as an early or late step in conjunction with other steps in an industrial process.

SEC finds many laboratory applications in manipulations of biological molecules. For example, unreacted species in molecular conjugates are often removed via SEC (18). Cleanup of materials involved in the synthesis of peptides and oligonucleotides is also performed frequently via SEC (19). It is also used for removal of aggregates resulting from a variety of biochemical preparations. For example, an intermediate or final step in purification of a monoclonal antibody is often removal of dimers and aggregates that have been generated as a result of prior processing. In the production of recombinant proteins, a certain fraction may be clipped by proteolysis; these smaller molecules can be separated by SEC (20).

The most common application of SEC at both laboratory and process scales is desalting or buffer exchange. Many processing steps (e.g., extraction, crystallization, other modes of chromatography) involve the use of a solvent that is undesired for future processing or formulation steps. Very small pore SEC columns are used to replace the undesired buffer by the desired one. Usually, a wide disparity in molecular sizes exists between the biological solute and the buffer salts, so short columns can be used, limiting the dilution to twofold or less (21).

PROCESS VARIABLES

Selection of a support material for an application of interest is only the first step in performing an analysis or separation. The amount of sample to be analyzed or purified must be considered. The column length and total amount of material to be injected into the column must be determined. Finally, a flow rate must be found that will provide adequate throughput and resolution. In the process of designing a configuration, convective dispersion, transport of the solutes to and within the particles, and partitioning of the solutes between the mobile and gel phases must all be considered. For these reasons, a substantial body of literature has been devoted to the engineering issues associated with SEC. The most general and applicable models are discussed in this section.

Plate Theory

An operational description of chromatography can be obtained from plate theory, which is a phenomenological description of zone spreading in chromatography, first proposed by Martin and Synge in 1941 (22). The essential concept is that as the solute traverses the column, it undergoes repeated equilibration with the stationary phase, so that each equilibration defines one stage, or plate. The column length over which equilibration occurs is known as the height of an equivalent theoretical plate (HETP), or simply plate height. The development of a solute peak zone is shown schematically in Figure 4. In each stage, fluid in the mobile phase is mixed, and partitioning occurs between the stationary and mobile phases. A mass balance on a particular stage takes the form

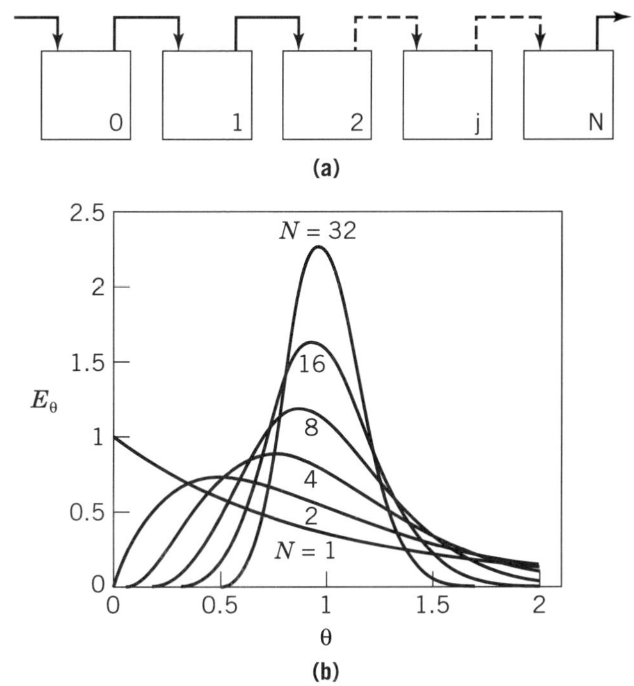

Figure 4. (**a**) Plate theory. The progression of solute through the column can be viewed as passing through a number of stages. In each stage, fluid in the mobile phase is mixed and partitioning occurs between the stationary and mobile phases. (**b**) Residence time distribution from N successive plates. The residence time distribution E, which is the output from a pulse input, in terms of the dimensionless variable $\theta = t/\tau$, is shown as a function of θ for various numbers of plates (N). It can be seen that as the number of plates exceeds approximately 10, the residence time distribution begins to assume a Gaussian shape, which is enhanced as N increases further.

$$Q(C_{j-1} - C_j) = v_m \frac{dC_j}{dt} + v_p \frac{d\bar{C}_{pj}}{dt}, \tag{9}$$

where Q is the flow rate, C_j is the concentration of solute in the jth stage, v_m is the mobile phase volume per stage, v_p is the pore volume per stage, and $\bar{C}_{pj}$ is the average pore concentration in the jth stage. For linear chromatography, the average pore concentration is proportional to the bulk concentration, and equation 9 becomes directly analogous to the mass balance on stirred tanks in series (23).

The input to a chromatography column is essentially a pulse; consequently, the output from a series of N stages is analogous to the residence time distribution (E) from N tanks in series. The output from a single stage is an exponential distribution, but, as the number of plates increases, the residence time distribution takes on a Gaussian shape, as shown in Figure 4. It can be shown that the variance σ^2 of the output peak is dependent only on the retention time and number of stages through

$$\sigma^2 = t_R^2/N \tag{10}$$

Therefore, the number of plates (N) in chromatography is defined as

$$N = t_R^2/\sigma^2 \quad (11)$$

and the plate height (H) as the length of the column (L) divided by the number of plates

$$H = L/N = L\sigma^2/t_R^2 \quad (12)$$

Relation of Size to Sieving

Because the purpose of SEC is to separate molecules on the basis of their size, interest developed from the time of its introduction to understand quantitatively the relationship between the size of a solute molecule and its retention volume. This is of importance in the use of SEC for molecular weight determination, in its interpretation for applications involving monitoring changes in molecular structure, and in the selection of SEC media and operating conditions to perform a separation.

Volume Exclusion. The primary mechanism of sieving is based on the excluded volume of particles in the pores. At equilibrium, the solute molecules will attain a concentration within the pore based on the volume accessible to the center of the molecule. When the pore size is of the same order as the solute molecular size, the accessible volume becomes significantly less than the pore volume (Fig. 5). As a result of the excluded volume effect, even a monodisperse pore size distribution will produce a continuum of distribution coefficients with increasing molecular size.

To develop this concept into a quantitative model, a geometry is usually assumed. An early result was that of Laurent and Killander (24), who used the expression of Ogston (25) for the available volume in a network of rigid rods to develop an expression for the distribution coefficient:

$$K_d = \exp[-\pi L(r_s + r_r)^2] \quad (13)$$

where L is the total length of rods per volume, r_s is the radius of the (assumed) spherical particles, and r_r is the radius of the rods. The parameters L and r_r are generally not known and must be fit to experimental data. Other expressions were derived for cylindrical, conical, and slitlike pores (26). The problem of ascribing a particular pore geometry was circumvented by Ackers (27), who represented a pore by the maximum characteristic radius particle that could pass through it and postulated that the distribution of pore sizes is distributed normally. For this model, the distribution coefficient takes the form

$$K_d = \text{erfc}\left[\frac{a - a_0}{b_0}\right] \quad (14)$$

where a is the radius of the solute particle, a_0 is effectively an average pore radius, and b_0 is a measure of the variance of the distribution. These parameters must still be determined for each gel from sieving experiments on at least two (preferably more) solutes.

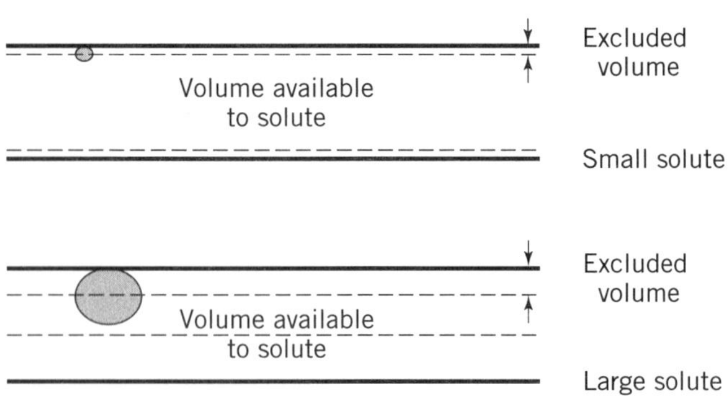

Figure 5. Volume exclusion of solutes in pores. Solutes assume concentrations within pores of varying sizes according to the volumes accessible by the centers of its molecules.

Rates of Pore Diffusion. The models based on volume exclusion (see previous section) are entirely thermodynamic; that is, they do not include any transport limitations and do not predict a dependence of distribution coefficient on flow rate. In practice, a slight dependence is observed, which led to the development of theories that account for transport of solute molecules into the pores. An approach to accounting for diffusion is to model the distribution of molecules between the mobile and gel phases using the one-dimensional (1-D) unsteady-state diffusion equation (28), which has the solution

$$C_g = C_m \,\text{erfc}\left[\frac{x}{2(Dt_d)}^{1/2}\right] \quad (15)$$

where C_g and C_m are concentrations in the gel and mobile phases, respectively, x is distance into the 1-D gel phase, t_d is the diffusion time, and D is the diffusion coefficient of the solute. With t_d inversely proportional to the mobile phase velocity (v_0) and b representing the power to which diffusion coefficient is inversely proportional to molecular weight (M), the distribution coefficient can be determined to be (28)

$$K_d = \frac{k}{(\pi v_0 M^b)^{1/2}}(1 - e^{-v_0 M^b/k^2}) + \text{erfc}\left[\frac{(v_0 M^b)^{1/2}}{k}\right] \quad (16)$$

where k is a grouping of constants fit to experimental data.

Purely Empirical Correlations. The equations initially used to correlate SEC data were essentially empirical. The most commonly used forms are

$$V_R = A - B \log(M) \quad (17)$$

and

$$\log(V_R) = A' - B' \log(M) \quad (18)$$

where M is the molecular weight of the solute and A, B, A', and B' are empirical constants for the particular column configuration. As shown in Figure 3 and discussed in "Analysis", very good correlations of data for similar species are possible, but a priori prediction or extrapolation beyond the measured range of molecular weights or to molecules of different chemistry is problematic. Note that the more sophisticated theoretical models discussed earlier also make use of essentially empirical constants that require calibration of the column using proteins (or other molecules) of known molecular weight. Calibration of the column must be repeated if the pH, ionic strength, or temperature is changed significantly or if the column is repacked. Furthermore, the calibration standards must be well characterized and of a similar nature to the solutes of interest. Nonetheless, the use of theoretical models attaches a physical significance to the parameter's fit from experimental data and provides some insight into the relative contributions of the volume-exclusion versus pore-diffusion mechanisms (28).

Transport Issues

The previous section dealt with the retention time, which mostly results from the equilibrium partitioning of species into the gel support. For separations, it is necessary to consider that chromatography is conducted in a packed bed, under flow, sometimes at high pressures. The configurational aspects of the chromatography column lead to band broadening, or spreading of the eluted solute band. The main contributions to band broadening are eddy diffusion, axial dispersion, and mass transfer resistance at the surface of the particles and within the pores. Eddy diffusion results from the tortuosity of extraparticle paths that solute particles take from the top to the bottom of the column. Axial dispersion is essentially molecular diffusion along the length of the column. The mass transfer resistance at the surface arises from the boundary layer, or film, through which solute must pass to reach the gel particle, and the mass transfer resistance of the pores is due to diffusion of solute into them and is enhanced by their nonuniformity.

A chromatography column is a packed bed of gel particles and is subject to the same types of scaling analysis as for packed beds in other chemical processes. Most of the resistance to mass transfer is at the level of the particle; consequently, the particle diameter is the characteristic length for scaling. Hence, a reduced velocity or Péclet number, *(Pe)* can be defined in terms of the particle diameter as

$$Pe = \frac{d_p u}{D_0} \tag{19}$$

where u is the interstitial velocity of the fluid and D_0 is the bulk solute diffusion coefficient. Likewise, the reduced plate height (h) is

$$h = \frac{H}{d_p} \tag{20}$$

where H is the plate height defined in equation 12.

The relation of interest is that between plate height and flow rate (or velocity) or, equivalently, between reduced plate height and Péclet number. This is obtained from solution to the governing differential mass balance equations in lesser or greater complexity (29–31).

Dependence of Separation Parameters on Operating Variables

Analysis of mass transport in chromatography columns has made it possible to make quantitative predictions regarding plate heights as a function of flow rate in SEC (32). The original analysis of mass transport was conducted by van Deemter (29). His equation has served as the basis for all subsequent analyses:

$$h = 2\lambda + \frac{2\gamma}{Pe} + CPe \tag{21}$$

where λ, γ, and C are constants accounting for eddy dispersion, axial diffusion, and mass transfer resistance at the particle surface, respectively. This equation predicts a minimum plate height with respect to increasing fluid velocity. More detailed accounting of the mass transfer contributions, including that caused by intraparticle diffusion, have produced alternative expressions for the plate height, for example, the expanded form of Horváth and Lin (33):

$$h = \frac{2\gamma}{Pe} + \frac{2\lambda}{1 + \omega Pe^{-1/3}} + \frac{\kappa k_0^2}{(1 + k_0)^2} Pe^{2/3} + \frac{\theta k_0}{30(1 + k_0)^2} Pe \tag{22}$$

where θ is a geometric factor for a given particle, λ, γ, ω, and κ are parameters dependent on the details of the particular column packing, and k_0 is the ratio of intraparticle to interstitial volumes available to the solute of interest. The third and fourth terms represent mass transfer resistances at the particle surface and within the pores, respectively.

For liquid chromatography involving macromolecules, the contributions from axial dispersion and resistance at the particle surface are small, and the largest contribution is from pore diffusion. Consequently, the minimum observed by van Deemter for GLC (29) is typically not observed in SEC of biomolecules, and the plate height is roughly proportional to velocity (32,34,35). Plate height also depends on particle size:

$$H \sim \frac{u d_p^2}{D_0} \tag{23}$$

From this approximate equation, two practical points can be inferred. First, because plate height is a measure of zone spreading and should generally be minimized, smaller particles are preferred because of the strong dependence of plate height on diameter. However, pressure drop is inversely proportional to the square of particle diameter, so the mechanical properties of the material limit the extent to which particle size can be reduced. Second, lower velocities, that is, slower flow rates, will result in less zone spreading. This advantage may be offset where the speed of a separation is important. Furthermore, the flow rate cannot be reduced too drastically or the effects of axial diffusion will become limiting.

An important measure of the separation is the resolution between two or more solutes to be separated. The dependence of the resolution on column variables can be determined from the equations already developed. Combining equations 4, 6, and 12 gives an expression for resolution in terms of capacity factors and plate heights (36):

$$R_s = \frac{L^{1/2}}{2} \frac{k_2' - k_1'}{H_2^{1/2}(1 + k_2) + H_1^{1/2}(1 + k_1')} \quad (24)$$

Employing the approximation that the plate heights of the two solutes are equal, an expression for resolution as a function of retention, selectivity, and column efficiency can be derived by combining the definition of selectivity (equation 5) with equation 24:

$$R_s = \frac{1}{2}\left(\frac{\alpha - 1}{\alpha + 1}\right) \frac{\bar{k}'}{1 + \bar{k}'} \frac{L^{1/2}}{H^{1/2}} \quad (25)$$

where $\bar{k}' = (k_1' + k_2')/2$ is the average retention time of the solutes.

Equation 25 expresses resolution as a function of three terms representing selectivity, retention, and zone spreading, respectively. Selectivity is the most important determinant of resolution, up to a point. An increase in α from 1.1 to 1.2 results in an 83% increase in resolution, all other factors being equal; however, an increase from 2.1 to 2.2 results in only a 6% further increase in resolution. Selection of a column best suited to fractionate the species of interest is thus critical to obtaining high resolution. Because the capacity factor is greater than or equal to unity, the middle term takes values only between 0.5 and 1; nonetheless, some improvement in resolution is seen to be possible by increasing the retention time. Because of the weak dependence of resolution on length, significant increases in column length are often required to effect a modest improvement in resolution. The inverse relationship between resolution and plate height suggests that minimizing zone spreading by the methods discussed earlier—reducing particle diameter or flow rate—will concomitantly improve resolution. This is true, subject to the same limitations on flow rate and pressure drop. Furthermore, if speed of separation is important, a trade-off exists.

In addition to the particle size and composition, column length, and operating flow rate, the SEC practitioner must choose column diameter, sample volume, and buffer composition. Column diameter does not appear explicitly in the equations describing resolution, but they are derived with the assumption of a uniform packing, which is increasingly difficult to achieve as the diameter-to-length ratio increases. Therefore, columns used for fractionation are prepared with a length-to-diameter ratio of about 10:1. But in industrial scale desalting applications, where the selectivity is great and resolution not an issue, much wider columns can be used. Sample volumes are usually chosen to be less than 5% of the total column volume, because values greater than a couple of percent tend to produce significant increases in zone spreading. In principle, the solute molecules should not interact with the stationary phase in SEC, but in practice biological molecules exhibit some interaction with a variety of materials, including gels used for SEC (37). Variation of buffer composition can be used to modulate adsorptive interactions. For example, adsorption is often mediated by charge interactions between solute and sorbent and can be reduced by increasing the ionic strength of the buffer, which provides double-layer screening of the electrostatic interactions.

CONSIDERATIONS FOR LARGE SCALE

SEC is used in large-scale industrial processes for the production of proteins for agricultural and pharmaceutical use. At this scale, economic considerations become imperative, and in particular it is important for the chromatographic operation to achieve a high level of productivity and throughput while maintaining high resolution.

General Strategies

For large-scale operation, the throughput required necessitates reuse of chromatography columns in a repeated-batch operation. Usually, the primary goal is to maximize productivity, which is defined as the amount of purified material per column area per time. This can be achieved by maximizing flow rate and minimizing the amount of time during which the column is not actively performing a separation. From a productivity standpoint, the time between elution of the valuable solute and all others is wasted time. One route to improving productivity is to reduce the resolution of the solutes to the minimum necessary (Fig. 6). Likewise, the time after elution of all solutes before the next injection can be used more efficiently by applying the next sample before the column volume of the present run is finished (21). These general guidelines govern both desalting and fractionation; further optimization depends on the particular application.

Another strategy to maximize throughput is to maximize the sample size. However, dispersion effects are enhanced for large sample volumes. Consequently, for fractionation applications, it is advisable to use sample sizes less than 3% of the total column volume (38). Although it is often desirable to have highly concentrated samples, higher concentrations result in higher viscosities, which result in slower diffusion into and equilibration with the gel. The requirement of sample equilibration with the gel also limits the flow rate that may be applied.

For desalting, on the other hand, larger sample volumes can be applied, because the sample elutes in the void volume and the salts are retained in the pore. Because of this large discrepancy in elution volumes, the increased dispersion resulting from large sample sizes is not much of an issue, and sample sizes as great as 25 to 30% of the column volume may be used. Furthermore, small molecules (e.g., ions) equilibrate quickly with the pores because of their small excluded volume and high diffusion coefficients, and macromolecules are completely excluded, so high flow rates may be used without compromising the separation.

Scale-Up Guidelines

The principle of scale-up is to determine the values of new process variables that will allow a desired change in one

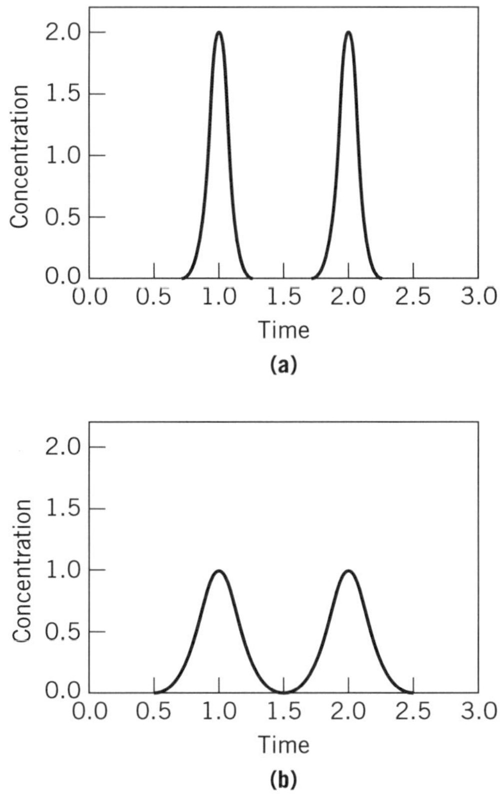

Figure 6. Optimization of resolution for large-scale SEC. (**a**) A small injection of solutes at low flow rate results in well-resolved bands. (**b**) The productivity is increased by increasing flow rate, increasing sample size, or decreasing column length to the point where the solutes are just resolved.

process variable while keeping others constant. Generally, it is desired to increase the throughput while maintaining the same resolution, but sometimes it is necessary to change resolution while keeping throughput constant. In any case, the key is to have a description of the relationship among process variables. It is here that the analysis of the preceding section is most useful.

Scaling rules for isocratic elution chromatography, which includes both SEC and adsorption chromatography, have been described by Wankat and Koo (36). The starting point is the definition of scaling parameters describing the ratios of process variables in the new and old designs, that is,

$$\delta = \frac{d_{\mathrm{p,new}}}{d_{\mathrm{p,old}}}$$
$$\lambda = \frac{L_{\mathrm{new}}}{L_{\mathrm{old}}}$$
$$\rho = \frac{D_{\mathrm{new}}}{D_{\mathrm{old}}}$$
$$\pi = \frac{\Delta p_{\mathrm{new}}}{\Delta p_{\mathrm{old}}}$$
$$\theta = \frac{Q_{\mathrm{new}}}{Q_{\mathrm{old}}} \qquad (26)$$

where D is the column diameter and Q is the buffer flow rate. The capacity factor in SEC is a thermodynamic quantity representing the partitioning of solutes in the pores of the support particles and therefore should not vary with column configuration or process variables. Therefore, the description of column resolution in equation 24 can be used for both the new and old designs to define a ratio of design resolutions, Ψ (36):

$$\Psi \equiv \frac{R_{\mathrm{s,new}}}{R_{\mathrm{s,old}}} = \left(\frac{L_{\mathrm{new}}}{L_{\mathrm{old}}}\right)^{1/2} \frac{[H_2^{1/2}(1 + k_2') + H_1^{1/2}(1 + k_1')]_{\mathrm{old}}}{[H_2^{1/2}(1 + k_2') + H_1^{1/2}(1 + k_1')]_{\mathrm{new}}} \qquad (27)$$

The difference between plate heights of the different solutes is usually ignored (see equation 25) and equation 27 is greatly simplified, that is,

$$\Psi = \lambda^{1/2}\left(\frac{H_{\mathrm{old}}}{H_{\mathrm{new}}}\right)^{1/2} \qquad (28)$$

As discussed in "Dependence of Separation Parameters on Operating Variables", transport resistances at the particle surface and within the pores usually dominate, with the result that the plate height can be represented by a scaling of the form equation 21. Substituting in equation 28 gives a simple design equation:

$$\Psi = \left(\frac{\lambda}{\theta}\right)^{1/2} \frac{\rho}{\delta} \qquad (29)$$

Note that more complete plate height expressions (derived from either theoretical or empirical correlations) may be substituted into equation 28 to develop more sophisticated design equations.

Equations such as equation 29 can be used to derive new sizes and operating conditions for a given change in any of the others. As a simple example, suppose that a design involving a reduction in particle size is desired, while maintaining constant column dimensions and throughput. In this case, λ, ρ, and θ are equal to unity, and the resolution is inversely proportional to the particle size

$$\Psi = \frac{1}{\delta} \qquad (30)$$

Alternatively, the objective of reducing particle size may be to allow a greater flow rate while maintaining constant resolution. In this case, Ψ, ρ, and λ are equal to unity, and the flow rate scales as the inverse square of particle diameter

$$\theta = \frac{1}{\delta^2} \qquad (31)$$

However, this is only valid in the regime where the particles are rigid, and it may not be feasible because of the large increase in pressure drop that would result.

The application of scaling rules to the interrelationship of process variables in equation 24 is an idealization that is subject to a number of constraints. Arbitrary column diameters and lengths are not possible, and certain length-to-diameter ratios (on the order of 10:1) are desirable to achieve a uniform packing. Also, only certain particle sizes are available, and the range in which they are manufactured is dependent to a large degree on their resistance to flow (i.e., pressure drop). As discussed in "Packing Materials", many packings are rigid only within a certain range of pressure drops. If a new column design takes the pressure drop out of this range, the performance may be drastically altered. For rigid particles, the pressure drop is described by equation 8; therefore, its scaling is

$$\pi = \frac{\theta\lambda}{\rho^2\delta^2} \tag{32}$$

For compressible particles, the same procedure may be used with a correlation that describes the relationship between pressure drop and flow rate (39).

SUMMARY

SEC has been applied extensively to biological macromolecules in the laboratory and has found utility as a step in the production of proteins at large scale as well. Further growth of the biotechnology industry and the ease and applicability of the technique ensure that SEC will be used widely in the future as well. Many packing materials have been developed to enable the chromatographer to perform an analysis or separation quickly, efficiently, and unambiguously through improvements in the inertness and mechanical stability of the support. Engineering analysis has enabled quantitative descriptions to be made regarding the influence of mass transfer resistances to the efficiency (plate height) of the column. These can be incorporated into simple design equations for the development and optimization of column configurations and operating conditions for a number of industrially important separations.

BIBLIOGRAPHY

1. J. Porath and F. Flodin, *Nature* **183**, 1657–1659 (1959).
2. H. Colin, in G. Ganetsos and P.S. Barker eds., *Preparative and Production Scale Chromatography*, Dekker, New York, 1993, pp. 11–45.
3. H.G. Barth, B.E. Boyes, and C. Jackson, *Anal. Chem.* **68**, 445R–466R (1996).
4. C.V. Uglea, *Liquid Chromatography of Oligomers*, Dekker, New York, 1996, pp. 269–274.
5. R. Eksteen, in E.D. Katz ed., *High Performance Liquid Chromatography: Principles and Methods in Biotechnology*, Wiley, New York, 1996, pp. 144–145.
6. M.L. Yarmush, K. Antonsen, and D.M. Yarmush, in C.L. Cooney, A.E. Humphrey eds. *The Principles of Biotechnology: Engineering Considerations*, Pergamon Press, Oxford, UK, 1985, pp. 489–505.
7. J.-C. Janson, in G. Subramanian, ed., *Process Scale Liquid Chromatography*, VCH, Weinheim, Germany, 1995, pp. 81–98.
8. H. Determann, *Gel Chromatography*, Springer-Verlag, New York, 1969.
9. Pharmacia Biotech catalog, 1995, pp. 308–322.
10. C. De Felice, K. Hayakawa, T. Watanabe, and T. Tanaka, *J. Chromatogr.* **645**, 101–105 (1993).
11. T.W. Patapoff, R.J. Mrsny, and W.A. Lee, *Anal. Biochem.* **212**, 71–78 (1993).
12. M.K. Dewanjee, A.K. Ghafouripour, M. Kapadvanjwala, and A.T. Samy, *Biotechniques* **16**, 844–850 (1994).
13. J.P. Hummel and W.J. Dreyer, *Biochim. Biophys. Acta* **63**, 532–534 (1962).
14. S. Kadokura, T. Miyamoto, and H. Inagaki, *Biopolymers* **20**, 1113–1122 (1981).
15. V.N. Uversky, *Biochemistry* **32**, 13288–13298 (1993).
16. X.-M. Lu, A.J. Fischman, S.L. Jyawook, K. Hendricks, R.G. Tompkins, and M.L. Yarmush, *J. Nucl. Med.* **35**, 269–275 (1994).
17. M.S. Wadhwa, D.L. Knoell, A.P. Young, and K.G. Rice, *Bioconjugate Chem.* **6**, 283–291 (1995).
18. S.S. Wong, *Chemistry of Protein Conjugation and Cross-Linking*, CRC Press, Boca Raton, Fla., 1991, p. 213.
19. R.K. Mishra, C. Moreau, C. Ramazeilles, S. Moreau, J. Bonnet, and J.Y. Toulmé, *Biochim. Biophys. Acta* **1264**, 229–239 (1995).
20. V.B. Lawlis and H. Heinsohn, *LC-GC* **11**, 720–729 (1993).
21. J.-C. Janson and T. Pettersson, in G. Ganetsos and P.S. Barker eds., *Preparative and Production Scale Chromatography*, Dekker, New York, 1993, pp. 559–590.
22. A.J.P. Martin and R.L.M. Synge, *Biochem. J.* **35**, 1358–1368 (1941).
23. O. Levenspiel, *Chemical Reaction Engineering*, Wiley, New York, 1972, pp. 1–293.
24. T.C. Laurent and J. Killander, *J. Chromatogr.* **14**, 317–330 (1964).
25. A.G. Ogston, *Trans. Faraday Soc.* **54**, 1754 (1958).
26. P.G. Squire, *Arch. Biochem. Biophys.* **107**, 471–478 (1964).
27. G.K. Ackers, *J. Biol. Chem.* **242**, 3237 (1967).
28. W.W. Yau and C.P. Malone, *Polym. Lett.* **5**, 663–669 (1967).
29. J.J. van Deemter, F.J. Zuiderweg, and A. Klinkenberg, *Chem. Eng. Sci.* **5**, 271–289 (1956).
30. J.F.G. Reis, E.N. Lightfoot, P.T. Noble, and A.S. Chiang, *Sep. Sci. Technol.* **14**, 367–394 (1979).
31. A.M. Lenhoff, *J. Chromatogr.* **384**, 285–299 (1984).
32. A.M. Athalye, S.J. Gibbs, and E.N. Lightfoot, *J. Chromatogr.* **589**, 71–85 (1992).
33. C. Horváth and H.-J. Lin, *J. Chromatogr.* **149**, 43–70 (1978).
34. F.E. Regnier and K.M. Gooding, *Anal. Biochem.* **103**, 1–25 (1980).
35. H.W. Blanch and D.S. Clark, *Biochemical Engineering*, Dekker, New York, 1996, p. 584.
36. P.L. Wankat and Y.-M. Koo, *AIChE J.* **34**, 1006–1019 (1980).
37. D. Corradini, R. Filippetti, and C. Corradini, *J. Liq. Chromatogr.* **16**, 3393–3408 (1993).
38. R.K. Scopes, *Protein Purification: Principles and Practice*, Springer-Verlag, New York, 1984, pp. 238–250.
39. A.W. Mohammed, D.G. Stevenson, and P.C. Wankat, *Ind. Eng. Chem. Res.* **31**, 549–561 (1992).

See also ADSORPTION, PROTEINS WITH SYNTHETIC MATERIALS; CHROMATOGRAPHY, ION EXCHANGE; MEMBRANE CHROMATOGRAPHY.

CITRIC ACID, PROCESSES

M.R.V. Krishnan
Anna University
Chennai, India

KEY WORDS

Aspergillus niger
Enzymc CoA
Fermentation
Molasses
Mold
pH
Solid substrate fermentation
Submerged culture
Surface culture
Yeast

OUTLINE

INTRODUCTION

Citric acid ($C_6H_8O_7$) is a white or translucent solid with a molecular weight of 192.12. It occurs as a natural constituent in citron, lemon, lime, pineapple, pear, peach, and similar fruits (Table 1). It is also found in animal tissues. The popular forms of citric acid are the anhydrous forms, the monohydrate and sodium salts of the acid.

Table 1. Citric Acid Contents of Various Fruits and Vegetables

Fruit/vegetable	Citric acid (wt %)
Lemon	4.0–8.0
Grape	1.2–2.1
Tangerine	0.9–1.2
Orange	0.6–1.0
Currant	
Black	1.5–3.0
Red	0.7–1.3
Raspberry	1.0–1.3
Strawberry	0.6–0.8
Apple	0.008
Tomato	0.25
Potato	0.35–0.5
Asparagus	0.08–0.2
Turnip	0.05–1.1
Pea	0.05
Corn	0.02
Lettuce	0.16
Eggplant	0.01

Citric acid was first isolated by Scheele in 1784 when he crystallized it from lemon juice. It was later synthesized (via symmetrical dicholoroacetone) from glycerol by Grimocex and Adam in 1880. It was first made in crystallized form from lemon juice. Calcium citrate was precipitated through the reaction of hot lemon juice with calcium carbonate, followed by decomposition of the product with sulphuric acid. Two hundred years later, this process is still being used. However, raw material is the principal limiting factor for production on a large scale. Thirty tons of lemon are needed to produce 1 ton of citric acid. The first commercial citric acid was prepared in 1860 in England from calcium citrate imported from Italy and Sicily. By 1880, France, Germany, and the United States had begun manufacturing citric acid using similar methods. In 1913, recovery of citric acid from calcium citrate began in Italy, and by 1922, Italy produced 90% of the world's supply.

MICROBIAL PRODUCTION

In 1893, Wehmer recognized citric acid as a microbial metabolite (1) (Table 2). Although his attempt to develop a commercial fermentation process with a species of *Penicillium* was unsuccessful, he established the basis for the subsequent complete reorganization of the citric acid industry. It was later confirmed that the use of *Aspergillus niger* (Fig. 1) favors sucrose fermentation to citric acid.

In 1923, a plant was started in New York to make citric acid by a fungal fermentation technique developed by Currie (2). This production effectively broke the power of the Italian cartel. In this process, *A. niger* was grown on the surface of a shallow pan of sugar medium. This surface fermentation process was subsequently used in England, The Netherlands, Belgium, Germany, Switzerland, Argentina, and the former Soviet Union.

A new fungal fermentation process was introduced successfully in the United States in 1952. In this process,

Table 2. Chronology of Events in Citric Acid Production

Authors (ref.)	Investigation/remarks
Wehmer 1893 (cited in 1)	Demonstrated citric acid to be a metabolic product of certain molds.
Currie 1917 (2)	First chose *Aspergillus niger* for deriving citric acid.
Doelger & Prescott 1934 (3)	Fermentation above 30 °C decreased citric acid yield and increased accumulation of oxalic acid.
Cahn 1935 (4)	First to describe solid-state fermentation for production of citric acid.
Mazzadroli 1938 (5)	Potassium ferrocyanide could be added to eliminate excess Fe^{2+}.
Szucs 1944 (6)	Citric acid synthesis took place only after total assimilation of phosphates.
Moyer 1953 (7)	Suggested use of methanol as stimulant for increasing acid yield.
Usami & Takatomi 1958 (1)	Spores produced under submerged fermentation were poor, and spores from surface culture were better acid formers.
Usami et al. 1960 (9)	Increase in aeration resulted in higher yields and reduction in cycle time.
Kovats 1960 (10)	Demonstrated lower concentration of sugar leads to lower yields of citric acid and accumulation of oxalic acid.
Schweiger 1961 (11)	Iron concentration above 0.2 ppm affected yield of acid. However, addition of copper 0.1 to 500 ppm at the time of inoculation or during the first 50 h of fermentation countered the deleterious effect of iron.
Naguchi & Bando 1960 (12)	Concentration of ammonium nitrate greater than 0.25% lead to accumulation of oxalic acid.
Sanchez-Marroquin et al. 1963 (13)	Higher yields of citric acid occurred in a simple medium rather than a complex medium.
Clark 1962 (14)	Ferrocyanide addition on the acid yield 10–200 μg/mL was tolerated during growth. But less than 20 μg/mL helped acid production.
Millis et al. 1963 (15)	Addition of vegetable oils, fatty acids, etc., increased acid yield.
Czech. Acad. Sci. 1964 (16)	Mycelial digests from *A. terreus* or *A. Niger* stimulated acid production by 60–70% when added to the medium.
Leopold 1965 (17)	Addition of pressed baker's yeasts to culture medium increased acid yield from 61.5 to 72.1%.
Bruchmann 1966 (18)	Addition of mild oxidizing agents stimulated acid production.
Sussman & Halvorson 1966 (19)	Spore viability varies with age.
Wendel 1967 (20)	In the initial stages, citric acid diffused into the medium, which leads to a stratification on the mycelium. This stratification inhibited fungal metabolism, leading to decreased acid yields. This stratification can be eliminated by stirring.
Tabuchi et al. 1969 (21)	Production of citric acid from *n*-paraffin using yeast, (*Candida*) medium composition. Hydrocarbon, 40–60 gs; NH_4Cl_2, 2 gs; KH_2PO_4, 0.5 g; $MgSO_4$, 0.5 g; cornsteep liquor, 1 g; $CaCo_3$, 30 g/L.
Khan et al. 1970 (22)	Phosphates promoted more growth at less acid yields.
Fedoseev et al. 1970 (23)	Copper sulfate at 4.7 mg/100 g molasses resulted in better conversion of sugar to citric acid.
Kyowa Fermentation Industry 1970 (24)	Citric acid production from dodecane (or) C_{12}–C_{14}. Using arthrobacterium in aqueous medium, inoculum 5% air at 3 vvm[a] at 28 °C, yielded 28 mg/mL.
Sanchez-Marroquin et al. 1970 (25)	The use of ion-exchange resin for reduction of metal content was better than chemical treatment.
Fukuda et al. 1970 (26)	Use of *Corynebacterium* on *n*-paraffins; yield 41.4 mg/mL; culture period 64 h, 32 °C.
Leopold 1971 (17)	Glycerol addition increased acid yield by 30%.
Dhankar et al. 1972 (27)	Acid yield improved by addition of peanut oil to molasses medium.
Kumamoto & Okamura 1972 (28)	Mn^{2+}, Ba^{2+}, Al^{3+} had an effect on fungal morphology.
Sardinas 1972 (29)	Patented a process for citric acid production involving *B. licheniformis*.
Chaudhary et al. 1972 (30)	A low pH in molasses medium was inhibitory to growth.
Ohmori & Ikeno 1973 (31)	Bacterial production of citric acid from media containing isocitric acid.
Dhankar et al. 1974 (32)	Sodium nitrate at a concentration of 0.4% was superior to ammonium nitrate.
Zhuravskii, 1974 (33)	Patented continuous multistage process for citric acid production.
Halama 1974 (34)	Chemicals such as pentachloralphenolate, semicarbozone, tetracycline, etc., used for control of bacterial growth.
Ohtsuka et al. 1975 (35)	Addition of malic hydrazide increased yield from 30 to 70%.
Wold & Suzuki 1973 (36)	The rate of acid production increased when AMP was added to the culture medium.
Wold & Suzuki 1976 (37)	Zn^{2+} regulated growth and production.
Brezhnoi et al. 1976 (38)	Continuous production of citric acid from molasses.
Choudhary 1978 (39)	A 3-day-old slant culture was as good as 7- to 8-day-old spore.
Krishnan & Natarajan 1987 (40)	Liquid–liquid extraction of citric acid.
Krishnan & Annadurai 1992 (41)	Precipitation of citric acid.
Krishnan et al. 1996 (42)	Reduction in cycle time for citric acid production.

[a] vvm, volume air per volume medium per minute.

A. niger was allowed to grow in the entire volume of the culture solution in large deep tanks. This process was introduced in Mexico in 1959 and in Israel in 1961.

The earliest attempt to produce citric acid by the submerged growth technique was that by Amelung, who aerated the culture by bubbling air through the solution. Numerous studies, subsequent to the work of Amelung, resulted in the development of two approaches to the control of citric acid production in submerged cultures. In the first approach, Perquin produced citric acid by inducing the establishment of a deficiency of one of the major nutrients, such as phosphate. In another approach, Johnson et al. studied the effect of manganese and iron and related the concentration of these to the amount of inhibition of citric acid accumulation.

Further development of methods for the control of iron and manganese levels in the culture solution led to the popularity of the submerged fermentation process, now used in the United States, Mexico, and Israel. Excess iron may be precipitated by ferrocyanide. The removal of iron and the presence of specific enzyme inhibitors that control the destruction of citric acid form the basis for the successful submerged fermentation process. Other mineral nutrients required for the submerged fermentation are the same as those for surface fermentation.

The most interesting change in citric acid manufacture was done by Miles Laboratories. They changed the raw materials from molasses to glucose, doubling the production. The company patented a process in which starchy material is first converted to sugar by enzymatic action and then used as the raw material for citric acid production. They also claim that with the addition of various phenols and other compounds to check the growth of *A. niger*, it is possible to use less highly purified starting material. This is similar to the use of the lower alcohols, which has been known for many years and is being investigated now as an additive to blackstrap molasses.

STORAGE OF CITRIC ACID

The popular forms of citric acid are the anhydrous forms, the monohydrate form and sodium salts of the acid. Both citric acid and citric monohydrate are available in variety of sieve sizes. They are conventionally available as granular, fine granular, and powder forms.

Crystalline anhydrous citric acid can be stored in dry form without difficulty, although highly humid conditions and elevated temperatures should be avoided to prevent caking. The product should be stored in tight containers to prevent exposure to moist air. Several granulations are commercially available with larger particle sizes having less tendency toward caking. Materials packed with desiccants are also available. Solutions of citric acid are corrosive to normal concrete, aluminium, carbon steel, copper alloys and should not be used with nylon, polycarbonates, polyamides, polyimides, or acrylics. Recommended materials of construction for pipes, tanks, and pumps handling citric acid solutions are 316 stainless steel, fiberglass-reinforced polyester, polyethylene, polypropylene, and polyvinyl chloride. Sodium and potassium citrates are not as corrosive as citric acid but they should be handled in the same type of equipment as citric acid.

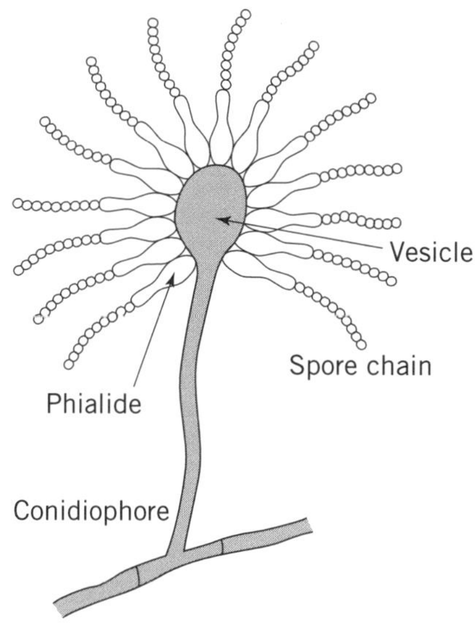

Figure 1. *Aspergillus niger*.

PRODUCT APPLICATIONS

The food and pharmaceutical industries use citric acid extensively because of its high solubility, pleasant sour taste, very low toxicity, and ready assimilability (Table 3). Citric acid also finds application in some cosmetic preparations, metal cleaning, electropickling, copper plating, secondary oil recovery, and other industrial uses (Table 4).

Foods and Beverages

Citric acid is widely used in foods, beverages, and jams and jellies owing to its ability

Table 3. Global Production Pattern of Citric Acid

Place	Production (%)
Western Europe	41
North America	28
South and Central America	11
Far East, Australia, and New Zealand	11
Other	9

Table 4. World Consumption Pattern of Citric Acid

Uses	Consumption (%)
Food and beverages	60
Pharmaceuticals	12
Detergents	12
Metal cleaning	6
Textile dyes	5
Cosmetics	3
Others	2

1. To maintain desired pH levels
2. To impart refreshing and tingling taste
3. To act as a flavor-enhancing agent
4. To act as a color stabilizer

In candy manufacture, citric acid helps to enhance the flavor and taste of berries and other ingredients. As a very dependable agent for maintenance of pH, citric acid finds wide application in the production of jams, jellies, preserves, soft drinks, syrup, and soft drink tablets. For pH control and color stabilization, citric acid is used in fruit and vegetable juices. The pH prevents juice spoilage, and natural taste and flavor are greatly enhanced. Lowering of pH inactivates certain oxidative enzymes, thus improving the keeping qualities of frozen fruits, peaches, apricots, plums, pears, and cherries against flavor or color spoilage. Citric acid also combines with trace metals, helping avoid the undesirable oxidation changes.

Pharmaceuticals

Citric acid's wide application in the manufacture of drugs and pharmaceuticals is for enhancing the taste and flavor. It is used as an anticoagulant for blood. Powders and tablets owe their effervescent property to citric acid. The free acid and its sodium and potassium salts are used as mild acidulants in astringent preparations. Several salts of citric acid are used in the manufacture of antianemic tonics, vitamins, liver tonics, antipyretic agents, cough syrups, antidysentery, and antidiarrheal agents. It is also used in inducing fertility.

Cosmetics

Hair rinses and hair setting agents derive benefit from citric acid. It is also used in astringent lotions, bleaching lotions, and others.

Metallurgy

Citric acid finds extensive use in metallurgical application and as a sequestering agent for such metals as iron, copper, zinc, nickel, cobalt, chromium, and manganese. The acid and ammonium salts are added for scale removal in boilers; cleaning of reactors could be done very well using the acid. Citric acid is widely used in electropickling of copper and its alloys and in copper coating. Any iron plugging in the oil-flow line could be removed by the acid. Tanning liquors, bottle washing compounds, and diazo printing paper all contain citric acid.

Industrial Uses

It is used in the manufacture of several esters, linoleum, inks, silvering compounds, and fabric dyes (Table 5). Citric acid enjoys a superior status above phosphates in the manufacturing of detergents.

Other Applications

Some of the algicides, pesticides, fish preservation, chemicals, and insecticides are based on citric acid. Preservation of latex from certain natural plants involves the use of citric acid.

Table 5. Some Leading Manufacturers and Suppliers of Citric Acid

Manufacturers
Miles Laboratories Inc., Elkhart, Indiana
Joh. A. Benckshiser Gmbh, Ludwingshafen/Rhein, Germany
Rhone-Poulenc S.A., France
John & E. Sturge Ltd., Birmingham, England
Cargill Inc., Minneapolis, Minnesota
San Fu Chemical Co., Ltd., China
Boehringer Ingelheim KG, Rhein, Germany
Pfizer Inc., New Jersey
Fermenta Products, Quimicos, SA, Brazil
Showa Chemical Co., Pvt Ltd., Osaka, Japan
Suppliers
Lurgi AG. Frankfurt am Main, Germany
Snamprogetty SPA, San Donato, Milan, Italy
Mannesman Anlagenbau, Dusseldorf, Germany
Boehringer Ingelheim KG, Burgerstrasse, Germany
Vogelbusch GmbH, Austria

THE TECHNOLOGY OF CITRIC ACID PRODUCTION

Citric acid production centers around one strain, *Aspergillus niger*, which is still extensively used. In the late 1970s, processes involving *Candida* yeast became commercial. *Candida guillermondi* has unique benefits over *A. niger* in that trace metal removal is not essential. It can also tolerate higher pHs (3.5 to 5) and sugar concentrations for acid production. The rate of acid production is also faster. As in the case of *A. niger*, nitrogen limitation triggers acid production. Certain bacteria also produce citric acid on a variety of substrates.

Citric acid can be manufactured commercially by three methods:

1. Surface culture technique
2. Submerged culture technique
3. Solid substrate technique

Commercially used substrates are citrus fruits (not adapted presently because the fruits are costly and seasonal), sucrose, glucose, molasses cane juice, and certain petroleum fractions.

Shallow pan reactors, stirred tank fermenters, and airlift fermenters are commonly used depending on the process requirements. Fermenters work in a semibatch mode or a continuous mode. It will be of interest to state here that yields of product vary with the pretreatment to which the media are subjected. For example, yields for pretreatment with suitable exchange resins is 98% in the case of sucrose and 75% for molasses; ferrocyanide-treated molasses yields 68% as contrasted to the untreated molasses, for which the yield is 62%.

Citrus Fruits and Pineapples

Lemon peels and the white layer are removed and separated for the recovery of oil and pectin, respectively. The remaining portion is pulped and filtered. The filtrate is likely to contain pectin and albumin, which are removed by spontaneous fermentation. In the case of pineapples, the waste and low quality fruits are crushed and filtered. The lemon juice and the pineapple juice, recovered separately, could contain 4% and 0.75% citric acid, respectively. Calcium carbonate or hydroxide is added to the individual juice to the required amount, and the temperature is maintained between 55 and 95 °C. Any oxalate present is filtered preferentially before citric acid. The calcium citrate cake obtained is washed and reacted with a small amount of sulfuric acid. The resultant slurry is filtered to recover citric acid solution from the precipitated calcium sulfate. The clear filtrate is subjected to decolorization before recovering the citric acid crystals by evaporation. This process is on the decline, as stated earlier.

Cane Molasses and Beet Molasses

Blackstrap molasses, about 40 °Bé with a sugar content of 52 to 57% sugar, or beet molasses, 41 °Bé containing 48 to 52% sugar, could be used for citric acid production (Table 6). However, cane molasses should be purified before subjecting it to fermentation. Alternate methods for molasses purification are discussed below:

1. *Sulfuric acid treatment*. The pH of the molasses (10% total reducing sugar) is adjusted to 3.0 by adding 0.1 N sulphuric acid. This is allowed to stand for 1.5 h and then centrifuged at 3,000 rpm for 15 min. The supernatant is collected and used.
2. *Potassium ferrocyanide treatment*. The molasses (10% TRS) is heated to 85 °C for 30 minutes and centrifuged at 3,000 rpm for 15 min; 100 mL of the supernatant is collected, and 0.5 mL of K_4 $Fe(CN)_6$ (10% solution) is added. The pH is adjusted to 6.5.
3. *Tricalcium phosphate treatment*. The pH of the molasses (10% TRS) was adjusted to 7.0 by the addition of 0.1 N NaOH and treated with 2% (w/v) tricalcium phosphate followed by heating at 105 °C for 5 minutes. The mixture was cooled and centrifuged at 3,000 rpm for 15 minutes. The supernatant was used.
4. *Tricalcium phosphate with hydrochloric acid treatment*. The TCP-treated liquor was adjusted to pH 2.0 by the addition of 0.1 N Hcl followed by vigorous shaking. The mixture was allowed to stand for 6 h. The supernatant was used after centrifugation at 3,000 rpm for 20 min.
5. *Bentonite treatment*. The pH of molasses (10% TRS) was adjusted to 7.0. Bentonite 2% (w/v) was added and kept in a boiling water bath for 30 min. The solution was then centrifuged at 3,000 rpm for 15 min. The supernatant was collected and used.

Table 6. Analyses of Cane and Sugar Beet Molasses

Component	Beet (%)	Cane (%)
Water	20.0	16.5
Sugar	62.0	53.0
Nonsugars	10.0	19.0
Ash	8.0	11.5

Molasses Medium

Beet molasses is the most widely used raw material in the United States and Europe. Latin American and Caribbean plants use sugar as raw material (3 tons of sucrose per ton of citric acid). Smaller plants in the Caribbean use citric wastes from citrus fruits. The production rates are Argentina, 2,000 tons per annum, Mexico, 1,000 tons per annum; and Uruguay, 500 tons per annum.

The beet molasses or pretreated cane molasses is taken in a mixing vessel where dilute sulphuric acid is added to register a pH of 5.5 of 6.5. Phosphorous, potassium, and nitrogen are added as nutrients to the required amount for growth and citric acid production. This mixture is sterilized with live steam. After this, a requisite amount of sterile water is added to give a sugar percentage of 15 to 20%. The feed is admitted into shallow aluminium pans arranged in convenient stacks in a sterile room. Each tray is shallow (about 75-mm deep) but big enough to hold solutions up to 500 L. The temperature (28 to 32 °C) and humidity (40 to 60%) are controlled for maximum yield. This sugar medium is inoculated with the spores of a selected strain of *A. niger*. In about 8 to 10 days, the fermentation process is over (pH being about 2). The acid might be contaminated with oxalic acid and gluconic acid. The tray contents are sent out for recovery of the acid, and the trays and the chamber are sterilized with steam. Dilute formic acid or sulfur dioxide could also be used for positive sterilization.

The recovery consists of adding hydrated lime (1 part of lime to every 2 parts of liquor) to the fermenter broth kept at 95 °C. The precipitated calcium citrate could be reacted with enough dilute sulfuric acid when citric acid remains in solution, leaving behind calcium sulfate. Based on the sugar content of the raw materials, the yield could be 35 to 65% by weight of sugar. In this process, pH maintenance is difficult.

Submerged Fermentation

The majority of industries have adopted this process. Conventionally, the growth phase and the production phase are two distinct regions. Two different media are used, one for growth and the other for production (Table 7). After 3 to 4 days of lush growth, the mycelia are separated from the growth medium and added to the production fermenter under stirred conditions. In the production fermenter, the pH and temperature are maintained around 3 and 30 °C, respectively. Air is sparged without interruption into the solution at about 1 to 1.2 volume air per volume medium per minute (vvm). In about 4 to 5 days, the production becomes nearly complete with about 65 to 70% of sugar being converted to the final product.

The submerged process is admirably suited for a flexible operation. It is possible to introduce certain process accel-

Table 7. Medium for Citric Acid

Component	Sporulation (g/L)	Production (g/L)
Sucrose	140	140
Bactoagar	20	0.0
Ammonium nitrate	2.5	2.5
Potassium hydrogen phosphate	1.0	2.5
Magnesium sulphate heptahydrate	0.25	0.25
Copper ion	0.0048	0.00006
Zinc ion	0.0038	0.00025
Ferrous ion	0.0022	0.0013
Manganous ion	0.001	0.001

erators or nutrients at desired and convenient time intervals; activated carbon or ascorbic acid could be used as the accelerator. The flexibility of such a fermenter is so good that sugar concentration and pH could be maintained at desired values. The acid recovery is through the calcium citrate process. For every ton of citric acid produced, one needs about 3,650 kg of molasses, 5 to 14 kgs of nutrient, 650 kg of sulfuric acid, and 450 kg of lime. Although one might use several organisms (*A. niger* is still the most sought after), the yield depends mainly on the nutrient used, presence of trace metals, and above all, the pH and air supply.

In the new process by Miles Laboratories mentioned previously, sucrose has been replaced with glucose for acid production. Starch is converted to sugar through enzymatic reaction, and the resulting sugar is the raw material for acid production. Specific chemicals control the growth of the mold at the expense of citrate production. An interesting point is that their starting material is not of high purity. They use very large fermentation vats into which sucrose solution is pumped, and fungal spores are added to help the growth and production cycles. After the cycle is over, the vats are charged with fresh medium.

Solid-Substrate Fermentation

In the solid-substrate fermentation process, the mold *A. niger* ferments molasses or sucrose adsorbed on beet or cane pulp. Carriers such as beet or cane pulp offer large areas for microbial contact with the medium and ensure good aeration. Bagasse is preferred over beet pulp because of its larger surface area and lower cost. In addition, bagasse can be used several times over. Bagasse is broken down into fine pieces, 1 to 2 mm long, sterilized, and soaked in molasses or sucrose solution before being inoculated with the selected strain. The citric acid yields are about 45% of the sugar content of molasses and 55% of sucrose. This process is labor intensive and needs more floor space, plus there is the likelihood of contamination. It is difficult to maintain pH. The process requires good aeration with controlled humidities.

Relative Merits between Submerged and Surface Fermentation

The starting sugar cane concentration and the yield percentage are more or less the same for submerged and surface fermentation. However, the surface process is highly labor intensive and not easily amenable for pH or tem-

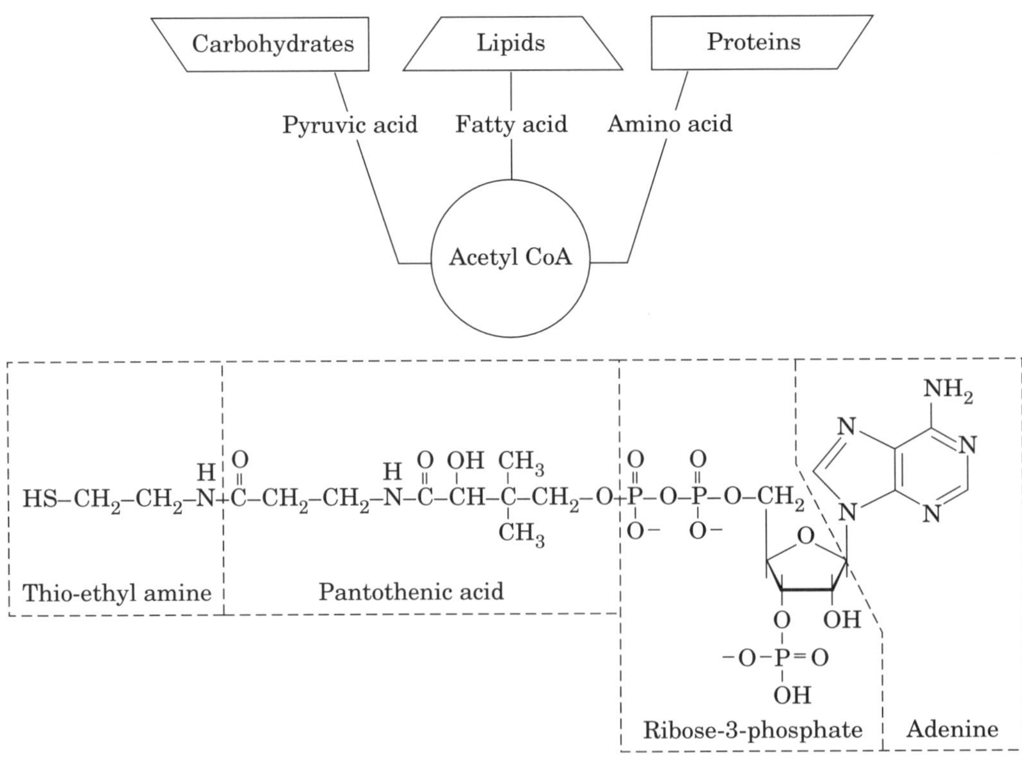

Figure 2. Enzyme CoA.

perature control. Flexibility of operation is higher in the submerged process because addition or removal of chemicals is easier. The risk of contamination is less in the submerged process. The surface process needs a larger working area, and the chance of pollution (caused by spores) is also high. Oxalic acid formation is higher in surface fermentation.

One advantage of the surface process is that it is less energy intensive. Although automation of the submerged process may be easier, in the event of contamination, the losses suffered will be enormous.

BIOCHEMISTRY OF CITRIC ACID PRODUCTION

Successful production of citric acid depends on several factors. External factors can be controlled directly, and biochemical factors, the various biological reactions that go

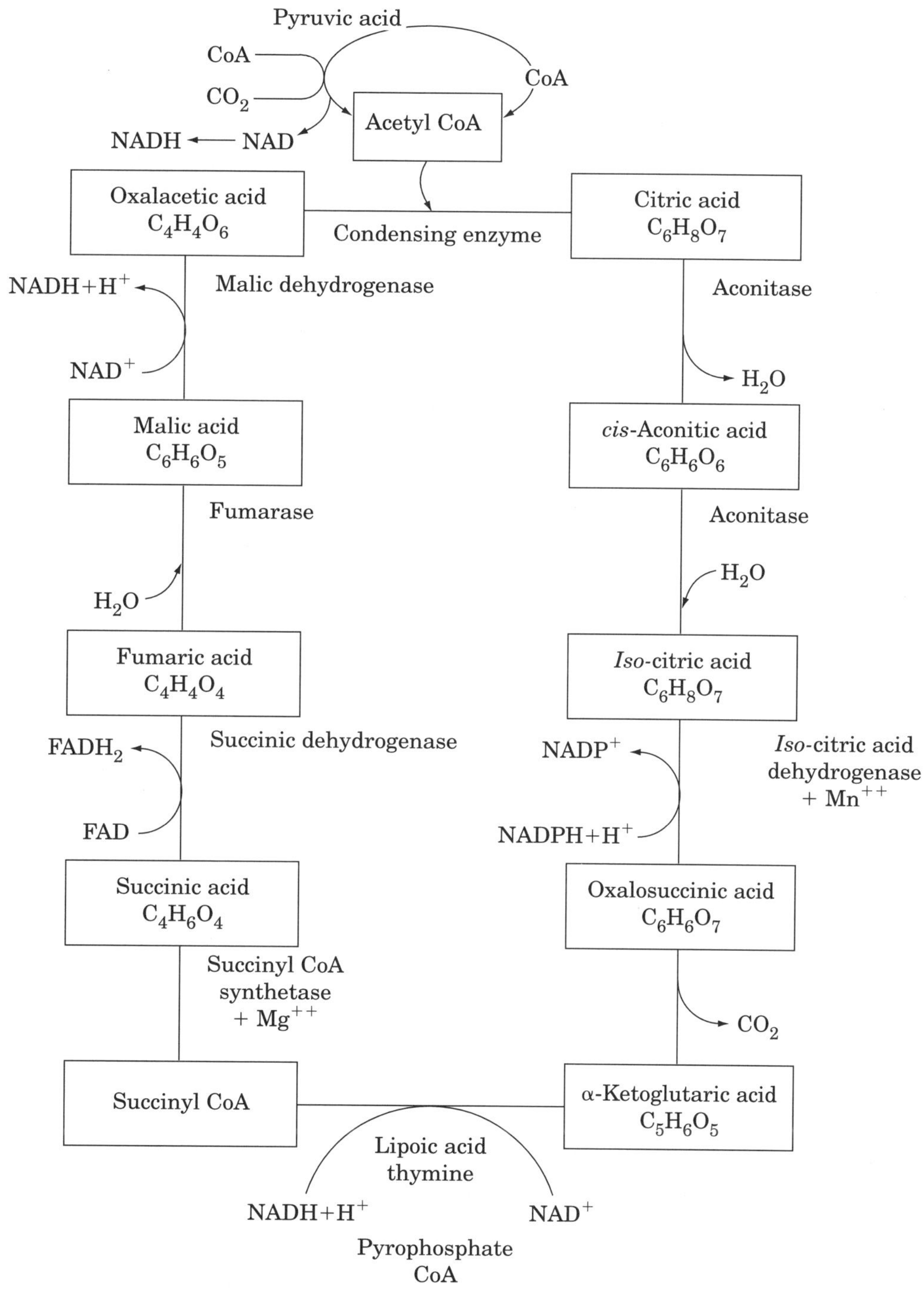

Figure 3. Tricarboxylic acid-cycle.

on inside the cell, can be controlled indirectly by controlling the external factors.

External Factors

- The type of sugar used and sugar levels
- The pH in the milieu
- The temperature of production
- The presence or absence of trace metals such as Fe, Cu, Zn, Mn
- The presence or absence of phosphates and nitrates in the medium
- The oxygen concentration in the environment

Biochemical Factors

- Breakdown of sugars to pyruvic acid and acetyl CoA (Fig. 2)
- Formation of oxalacetic acid from pyruvic acid and carbon dioxide
- Promotion of the enzymatic reaction necessary for citric acid accumulation
- Suppression of the enzymatic reaction that represses citric acid production
- Oxygen concentration in the medium and cell (citric acid production needs a copious supply of oxygen)
- Favoring citric acid accumulation at the expense of cell growth

Based on the external and biochemical factors, one could draw the following conclusions:

1. Glycolytic and pyruvate enzymes should be activated for the citrate synthesis by maintenance of high sugar levels.
2. Repression of enzymes in the tricarboxylic acid (TCA) cycle would inhibit citrate production.
3. Manganese deficiency should be ensured for NH_4^{2+} generation, which would counter the inhibition of phosphofructokinase by citrate.
4. A copious oxygen supply will overcome possible cell starvation, even under emergent conditions.
5. Ideal pH maintenance encourages citrate production rather than oxalic or gluconic acid formation.

$$\begin{array}{l}CH_3\\ |\\ C{=}O\\ |\\ COOH\end{array} + C_OA + NADP + ADP + P_i \xrightarrow[\text{Pyr. decarboxylase}]{\text{Pyr. dehydrogenase}} CH_3COC_OA + CO_2 + NADH + ATP$$

$$CH_3COSC_OA + \begin{array}{l}COOH\\ |\\ CO\\ |\\ CH_2\\ |\\ COOH\end{array} + H_2O \overset{\text{condensing}}{\underset{\text{enzyme}}{\longleftrightarrow}} C_6H_8O_7 + SC_OA$$

$$C_6H_8O_7 \overset{\text{Aconitase}}{\longleftrightarrow} C_6H_6O_6 + H_2O$$

$$C_6H_6O_6 + H_2O \overset{\text{Aconitase}}{\longleftrightarrow} C_6H_8O_7$$

$$C_6H_8O_7 + NADP + ADP + P_i \xrightarrow[\text{decarboxylase}]{\text{Isocit. dehydrogenase}} C_6H_6O_7 + H_2O + NADPH_2 + ATP$$

$$C_6H_6O_7 \overset{\text{Oxalo succinic}}{\underset{\text{decarboxylase}}{\longleftrightarrow}} C_5H_6O_5 + CO_2 + H_2O$$

$$C_5H_6O_5 + NAD + ADP + P_i + H_2O\ SC_OA \xrightarrow[\text{decarboxylase}]{\text{a-ket. glut. dehydro}} \text{succinyl } SC_OA + CO_2 + NADH_2 + H_2O +$$

$$\text{succinyl } SC_OA + H_2O + ADP + P_i \xrightarrow[\text{kinase}]{\text{Succinyl thio}} C_4H_6O_4 + ATP + SC_OA$$

$$C_4H_6O_4 + FAD + ADP + P_i \xrightarrow[\text{dehydrogenase}]{\text{Succinic}} C_4H_4O_4 + FADH_2 + H_2O + ATP$$

$$C_4H_4O_4 + H_2O \overset{\text{Fumarase}}{\longleftrightarrow} C_4H_6O_5$$

$$C_4H_6O_5 + NAD + P_i + ADP \xrightarrow[\text{dehydrogenase}]{\text{Malic}} C_4H_4O_5 + H_2O + NADH_2 + ATP$$

The Krebs cycle

In most cells, catabolism of simple sugar is the major source of energy. It starts with the glycolytic pathway. Glucose is broken down into two molecules of three-carbon compounds called pyruvic acid. The cell gains two ATP molecules for each molecule of glucose entering the glycolytic pathway. The ATP molecules are storehouses of energy, releasing energy necessary for cell work. The catabolic reaction also gives rise to NADH, which is useful for several anabolic reactions.

Depending on the type of cell and availability of oxygen, the pyruvic acid produced earlier can undergo two different types of reactions. When anaerobic conditions prevail, pyruvic acid is converted to lactic acid or ethyl alcohol. Under aerobic conditions, the pyruvic acid is converted to carbon dioxide and acetyl CoA (Fig. 2). This acetyl CoA is further degraded in the (Fig. 3) TCA cycle or election transport system to produce energy for cell work. Several other substrates exist that can give rise to acetyl CoA. For example, proteins and lipids can be degraded to provide acetyl CoA. Thus, cells can degrade several chemical compounds to acetyl CoA and subject them to the TCA cycle and electron transport chain for obtaining energy.

The TCA cycle together with the electron transport system constitutes the cells' primary metabolic furnace. These two combine to burn acetyl CoA in oxygen to liberate carbon dioxide, water, and useful energy. This energy from the TCA cycle and electron transport chain is captured and taken up in ATP for storage of energy.

During oxidation of one molecule of acetyl CoA (by 0the TCA cycle occurring in the mitochondrial matrix), one molecule of flavoprotein (FP or FAB) and three molecules of nicotinamide adenine dinucleotide (NAD) are reduced. These reduced coenzymes are oxidized by molecular oxygen by way of a system of enzymes and coenzymes called the respiratory chain or electron transport system, occurring in the inner mitochondrial membrane. During this oxidation process, enormous amounts of energy are released, some of which is utilized by the inner membrane subunits.

The TCA cycle occurs in the following ways:

- The pyruvic acid is decarboxylated; the acetyl group combines with CoA to give acetyl CoA. The H and NAD combine to give NADH. Energy transfer occurs, yielding ATP.
- Active acetyl combines with oxalacetic acid to form citric acid. The released SCoA can now combine with more CH_3CO.
- Citric acid looses water to give rise to aconitic acid.
- Isocitric acid results as *cis*-aconitic acid gets rehydrated.
- Isocitric acid loses hydrogen to yield oxalosuccinic acid. The H combines with NADP to form NADPH. Energy transport takes place, yielding ATP. Loss of CO_2 also occurs.
- Decarboxylation of oxalosuccinic acid takes place to yield α-ketoglutaric acid. Hydrogen reacts with NAD to form $NADH_2$. Energy is released to convert ADP to ATP.
- SCoA combines with the succinyl group of α-ketoglutaric acid and gets converted to succinic acid, yielding SCoA and ATP.
- Dehydrogenation of succinic acid leads to fumaric acid. Hydrogen joins FAD to form $FADH_2$ and energy. This leads to formation of ATP.
- Malic acid is formed as fumaric acid gets hydrated.
- Malic acid is dehydrogenated to form oxalacetic acid. Oxalacetic acid combines with active acetyl to form citric acid and the cycle continues. The released hydrogen forms $NADH_2$ from NAD and energy released gives rise to ATP.

Several yeasts and bacteria can produce citric acid from specific substrates:

Citric Acid Production by Yeasts

Candida lipolytica
Candida tropicalis
Candida zylenoides
Candida fibrae
Candida intermedia
Candida parpsilosis
Candida petrophylum
Candida subtropicalis
Candida oleophila
Candida hitachinica
Candida citra
Candida guillermondi
Candida sucrosa

A variety of substrates can be utilized by the *Candida* species. They could utilize glucose, acetic acid, calcium acetate, hydrocarbons, molasses, alcohols, fatty acids, and natural oils (such as coconut oil). For example, *Candida* can utilize glucose or molasses as indicated in the list. *C. lipolytica* can use *n*-paraffins, n-alkanes, and alkenes.

From Glucose	From Molasses
Glucose, 10–18%	Sugar cane/beet molasses, 5–250 g/L
NH_4Cl	$(NH_4)_2 SO_4$, NH_4Cl, NH_4NO_3
KH_2PO_4	0.3–1.5 VVM, 6 days
$MgSO_4 \cdot 7H_2O$	5.5–6.50 pH for growth
22–30 °C, 3–6 days	2.8–4 pH for production
	3,154 kg sugar → 1,119 kg monocitrate

Citric Acid Production by Bacteria

Certain bacteria are preferred for producing citric acid because of their low doubling time. Examples include:

Bacillus licheniformis, *Bacillus subtilis*, *Brevibacterium flavum*	They utilize glucose or isocitric acid or hydrocarbons
Arthrobacter paraffinens can utilize dodecane (C_{12}–C_{14}) at 28 °C over a period of 72 h; yield 28 mg/mL	*Bacillus licheniformis* can utilize glucose medium: 30–37 °C 36–120 h; pH 7; nitrogen salts act as nutrients; yield 42 g/L

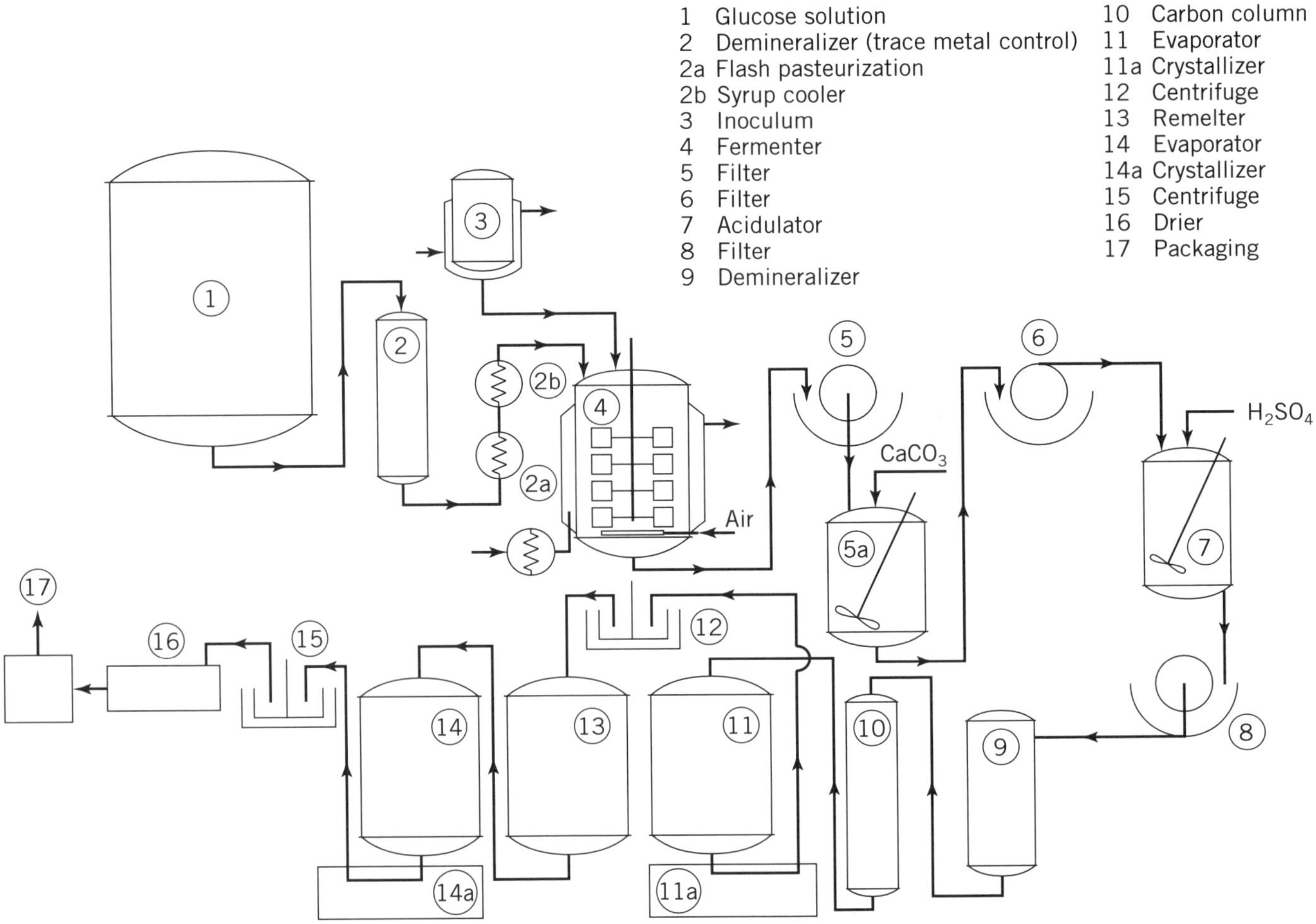

Figure 4. Citric acid production and separation flowsheet.

Recovery of Citric Acid

Two important routes are followed. One is the precipitation as calcium citrate (Fig. 4). The second is the liquid–liquid extraction technique as discussed by Krishnan and Natarajan (40). Solvents such as esters, ketones, amines, and alcohols have been successfully used for the recovery, as outlined by Rieger and Kioustelidis (43).

CONCLUSIONS

Citric acid enjoys a unique place in several important everyday applications. The capabilities of *A. niger* have been almost totally exploited by using several strains of its culture and their mutants. The application of *A. niger* for citric acid production has been stretched nearly to the limit. It is the wishful thinking of the author that some more work could be undertaken in producing such stains that could utilize unclarified raw materials in shorter periods. No effort should be spared in developing newer and cheaper chemicals (promoters) that could accelerate the production rate.

Care has been taken to cover the entire world literature on citric acid production. However, some references might have been missed more because of specific constraints rather than because of oversight or willful omissions. I should like to end by saying that some more work on the application of bacterial culture would be rewarding. It may not be right to say that the last word has been spoken on citric acid production.

Dissolve away difficulties,
Filter out gloom
And from the mother liquor of life
Gather crystals of (citric acid) joy.

Adapted from Krishnan.

BIBLIOGRAPHY

1. S.C. Prescott and C.G. Dunn, *Industrial Microbiology*, McGraw-Hill, New York, 1959.
2. J.N. Currie, *J. Biol. Chem.* **31**, 15–37 (1917).
3. W.P. Doelger and S.C. Prescott, *Ind. Eng. Chem.* **26**, 1142 (1934).
4. F.J. Cahn, *Ind. Eng. Chem.* **27**, 201–203 (1935).
5. Fr. Pat. 833,631 (Oct. 26, 1938), G. Mazzadroli.
6. U.S. Pat. 2,353,771 (July 18, 1944) J. Szucs.
7. A.J. Moyer, *Appl. Microbiol.* 1–7 (1953).

8. S. Usami and N. Takatomi, *Kogyo Kagaku Zasshi* **61**, 1494–1497 (1958).
9. S. Usami, H. Suzuki, and N. Takatomi, *Kogyo Kagaku Zasshi* **63**, 1766–1768 (1960).
10. J. Kovats, *Acta Microbiol. Pol.* **9**, 275–287 (1960).
11. U.S. Pat. 297,084 (Jan. 31, 1961), L.B. Schweiger.
12. Y. Naguchi and Y. Bando, *Hakko Kogaku Zasshi* **38**, 485–488 (1960).
13. A. Sanchez-Marroquin et al., *Rev. Soc. Quim. Mex.* **7**, 191–198 (1963).
14. D.S. Clark, *Biotechnol. Bioeng.* **4**, 17–21 (1962).
15. N.F. Millis, B.H. Trumphy, and B.M. Palmer, *J. Gen. Microbiol.* **30**, 365–379 (1963).
16. Fr. Pat. 1,355,922 (March 20, 1964), Czech. Acad. Sci.
17. Czech. Pat. 142,338 (Aug. 15, 1971), J. Leopold.
18. E.E. Bruchmann, *Naturwissenschaften* **53**, 226–227 (1966).
19. A.S. Sussman and H.O. Halvorson, *Spores: Their Dormancy and Germination*, Academic Press, New York, 1966.
20. D.G. Wendel, *Zucker* **20**, 238–246 (1967).
21. T. Tabuchi, M. Tanaka, and M. Aee, *Nippon Nogei Kagaku Kaishi* **443**, 154–158 (1969).
22. M.A.A. Khan, M.M. Hussain, M.A. Khalique, and M.A. Rahman, Pak. *J. Sci. Ind. Res.* **13**, 439–444 (1970).
23. V.F. Fedoseev et al., *Khlebopek. Konditer. Promst.* **14**, 33–35 (1970).
24. Br. Pat. 1,187,610 (April 8, 1970), Kyowa Fermentation Industry.
25. A. Sanchez-Marroquin et al., *Appl. Microbiol.* **20**, 888 (1970).
26. Ger. Pat. 2,003,221 (Dec. 3, 1970), H. Fukuda, T. Suzuki, Y. Sumino, and S. Akiyama.
27. H.S. Dhankar, M.Sc. Thesis, Haryana Agric. Univ., Hissar, India, 1972.
28. Jpn. Pat. 7,234,955 (Sept. 2, 1972), H. Kumamoto and T. Okamura.
29. Fr. Pat. 2,113,668 (July, 1972), J.L. Sardinas.
30. K. Chaudhary et al., *J. Res. Haryana Agric. Univ.* **1**, 48 (1972).
31. Jpn. Pat. 7,443,157 (Nov. 19, 1973), I. Ohmori and Y. Ikeno.
32. H.S. Dhankar, S. Ethiraj, and S.R. Vyas, *Ind. J. Technol.* **12**, 316–317 (1974).
33. U.S.S.R. Pat. 432,186 (June 15, 1974), G.I. Zhuravskii et al.
34. D. Halama, *Biotechnol. Bioeng. Symp.* **4**, 891–898, 1974.
35. Jpn. Pat. 7,530,158 (Sept. 29, 1975), M. Ohtsuka, Y. Ohzaki, and H. Aramiya.
36. W.S.M. Wold and I. Suzuki, *Biochem. Biophys. Res. Commun.* **50**, 237–244 (1973).
37. W.S.M. Wold and I. Suzuki, *Can. J. Microbiol.* **22**, 1083–1092 (1976).
38. U.S.S.R. Pat. 510,509 (April 15, 1976), Y.D. Brehznoi, L.A. Novikova, Y.L. Ignatov, and Y.E. Kachanov.
39. K. Chaudhary et al., *J. Ferment. Technol. Jpn.* **56**, 554 (1978).
40. M.R.V. Krishnan and M. Natarajan, M. Tech. Thesis Anna Univ. Madras, India, 1987.
41. M.R.V. Krishnan and G. Annadurai, M. Tech. Thesis Anna Univ., Madras, India, 1992.
42. M.R.V. Krishnan, et al. *Bioprocess Eng.* **15**, 323, 1996.
43. Ger. Pat. 2,355,039 (May 7, 1975), M. Rieger and J. Kioustelidis.

ADDITIONAL READING

A.Q. Chaudhary and S.J. Pirt, *J. Gen. Microbiol.* **43**, 71–81 (1966).
U.S. Pat. 3,118,821 (Jan. 21, 1964), D.S. Clark.
Ger. Pat. 2,108,094 (Sept. 2, 1971), H. Fukuda, T. Suzuki, S. Akiyama, and Y. Sumino.
Can. Pat. 973,824 (Sept. 2, 1975), H. Fukuda, T. Suzuki, S. Akiyama, and Y. Sumino.

CLEANING, CLEANING VALIDATION

GAIL SOFER
BioReliance
Rockville, Maryland

KEY WORDS

Assays
Cleaning
Cleaning methods
Equipment
Multiproduct
Nucleic acids
Proteins
Unit operations
Validation
Viruses

OUTLINE

INTRODUCTION

The importance of adequate cleaning in a biological process cannot be overemphasized. The required stringency of cleaning and cleaning validation is best evaluated by performing a risk analysis. The risks associated with the use of improperly cleaned equipment and materials (such as filters and chromatography media) vary with the product source, intended use of end product, the position of the equipment and materials in the processing train, the phase of development of the process, and the type of unit operation. A further consideration is whether the equipment or facility is used for more than one product. Robust cleaning methods, suitable cleaning agents, and cleaning validation ensure that a product with the desired properties is manufactured without contamination from processing additives, well-defined impurities, unexpected contaminants, and previously run product.

PRODUCT SOURCES

For certain complex product sources such as plasma, the potential safety risk from inadequate cleaning is clearly very high. The impurities in the source material are not fully characterized, making evaluation of cleaning efficiency more difficult. For other products, such as secreted products from a mammalian cell culture such as chinese hamster ovary (CHO) cells, the impurities are clearly defined, cleaning efficacy is more readily assessed, and the risks are fewer than with products derived from multiple sources such as blood donors. Robust cleaning procedures are also required for synthetic processes such as those used to produce oligonucleotides used for antisense therapy and synthetic peptides, but the cleaning methods and assessment of their efficacy may be even simpler.

INTENDED USE OF PRODUCT

If the end product is intended for use as a therapeutic agent delivered in a large dose for an extended time in a healthy individual, the criticality of cleaning is obvious. For a product intended for in vitro diagnostic use, the potential contamination that can arise from inadequate cleaning can result in an incorrect diagnosis and even result in a dangerous treatment for a patient. For industrial enzymes, contamination from previously run product may or may not have a detrimental effect on product performance, but even in this case the limits of cleanliness should be defined.

POSITION IN PROCESSING TRAIN

A typical processing scheme will consist of several unit operations (Table 1). The closer one gets to final product, the more critical the cleaning becomes. For example, it has been suggested by some regulators during oral presentations that firms using contract manufacturers for formulation and filling consider supplying their own dedicated equipment to ensure no risk of contamination from other products caused by insufficient cleaning. But even upstream in fermentation or synthetic processes, the lack of adequate cleaning can lead to contamination of the next batch with unknown impurities that adulterate the product in such a way that its safety, potency, and efficacy are compromised.

Table 1. Unit Operations in a Typical Biotechnology Process

Cell culture or fermentation
Isolation
Clarification
Purification
Formulation or final filling

PHASE OF PRODUCT DEVELOPMENT

From research to full-scale manufacturing, proper cleaning is essential for reproducible processes. Not surprisingly, there is a great rush to get a product into the clinic to demonstrate efficacy. In this rush, cleaning is sometimes overlooked, resulting in laboratory data that are unreliable. For example, when a researcher uses a chromatography column for developing a purification process and fails to adequately clean precipitated impurities, the purification process in the next run is likely to yield a product that does not represent what is intended. Good manufacturing practices (GMP) are required for all clinical trials (1). This requirement is also applied in many nations other than the United States. Although cleaning validation might not be complete at this stage, routine cleaning is essential, and it is advantageous to have a cleaning validation plan.

Cleaning and cleaning validation are essential in the manufacture of marketed products and are part of good manufacturing practices. For therapeutics, the legal requirements are given in the U.S. Code of Federal Regulations in the form of general guidance (21 CFR 211.67 and 21 CFR 600.11).

MULTIPRODUCT VERSUS DEDICATED EQUIPMENT AND FACILITIES

In the past, most facilities used for manufacturing therapeutics were dedicated to one product or a family of products from the same source. For complex biologicals, such as plasma products and traditional vaccines, this is still the case. For these biologics, starting materials, and even the final products, may not be sufficiently well defined to evaluate the effectiveness of the cleaning regimes in removing all traces of previously run product and processing materials. Advances in analytical technologies, however, are allowing firms to better assess cleaning effectiveness, especially for well-characterized products. Products that are considered well characterized include therapeutic DNA plasmids, therapeutic synthetic peptides of 40 or fewer amino acids, monoclonal antibodies for in vivo use, and therapeutic recombinant DNA-derived products (2).

Today, there is a trend to reduce processing costs by employing contract manufacturers to produce clinical trial materials, thereby avoiding the construction of costly facilities for products that may not ultimately obtain regulatory approval. There are also many firms that campaign products (produce one at a time) to fully utilize space, equipment, and personnel. Regulators carefully scrutinize cleaning and cleaning validation in such facilities to ensure that there is no cross-contamination of products. Cleaning validation using sensitive, specific, and nonspecific analytical methods is essential for such operations (see "Methods for Evaluating Cleanliness").

UNIT OPERATIONS

General Considerations

Each unit operation has somewhat different cleaning requirements, which as previously stated are dependent on the nature of the product source and its intended use, its position in the process train, and whether it is used for more than one product. Cleaning is not an issue to be addressed after a process is finalized; it should be designed into the process. In fact, the selection of the most suitable equipment and materials may be dependent on their ability to be cleaned. Records kept during research and development may be valuable in the design of a cleaning process. For example, a researcher may find that a standard cleaning procedure recommended by an equipment vendor is not sufficiently harsh for a particular product feedstream.

Although each feedstream and each unit operation have unique features that must be considered before the most efficient cleaning protocol can be designed, there are many commonalities that can used to minimize development time. References such as regulatory guidelines and PDA's book on cleaning should be consulted (3). Keeping up with the latest technical and regulatory publications and comments at meetings and workshops is essential.

In the design of the cleaning regime for each unit operation, critical factors to consider include the type of cleaning method, compatibility, and detection methods. Cleaning may be automated or manual. Automated cleaning can provide more consistent results. Manual cleaning, on the other hand, is clearly much more difficult to validate, and operators must be certified to perform the task. Cleaning-in-place (CIP) is preferred, but for some components, cleaning-out-of-place (COP) is necessary. Cleaning may be accomplished by chemical or physical methods, or, most likely, by a combination of both.

Chemical Methods

When chemical methods are used, the contact time, temperature, and cleaning agent concentration must be specified. The order of adding cleaning agents and rinsing agents may be critical. Additionally, if the chemical agent is flowing through the equipment, then flow must be specified. It is essential to ensure that the cleaning agent is compatible not only with the specified equipment, but also with any ancillary components that might be exposed to the agent. The nature of the soil must also be considered. Whereas hot alkali and detergents might work for one unit operation, for another they may actually exacerbate the cleaning problem by causing deterioration of the equipment or aggregation of impurities. Organic solvents used for removal of lipids may extract potentially harmful chemicals from some equipment components in a given unit operation. If a packed chromatography column containing a soluble hydrophobic protein is cleaned by treatment with a salt solution, the protein is likely to precipitate and actually leave behind a larger residue. For filters and chromatography media, compatibility with the cleaning agent is of particular concern. For example, when incompatible cleaning conditions are used, an immunoadsorbent might leach significant amounts of antibody, leading to reduced capacity and hence inconsistent results in the next batch. Table 2 shows some common cleaning agents and their mechanisms.

Physical Methods

Physical methods include high velocity flow, jet sprays, and agitation. If physical methods are used, it is essential to ensure compatibility with the equipment. Physical methods commonly used for cleaning tanks and fermentation vessels (e.g., spray balls) are clearly not suitable for any packed chromatography columns or thin piping.

Combining Methods

A combination of physical and chemical methods is usually used. Mechanical forces remove gross soils and deliver cleaning solutions to large areas, but chemicals (such as alkali, acid, and detergents) are generally required as well. According to Rohsner and Serve, cleaning processes require energy as well as chemical and physical reactions (4). Cleaning mechanisms are further addressed by LeBlanc et al. (5).

Some development work is necessary to put into place the most efficient and cost-effective cleaning methods for each unit operation. It is important to recognize that the nature of the soil may change somewhat upon scale-up of a process, necessitating a modification of the cleaning procedure. This is most common in the earlier processing steps such as fermentation. Ranges of acceptable conditions should be established and validated to accommodate the worst-case soil.

Fermentation and Cell Culture. At this early step in the process, the nature of the soil to be cleaned is not well defined, but it may contain substances such as lipids; soluble and precipitated proteins; endotoxins (if Gram-negative bacterial fermentation is used); small, soluble

Table 2. Common Cleaning Mechanisms and Cleaning Agents

Mechanism	Agent
Dissolve	Water, alkali, acid
Saponify	Alkali
Degradation of proteins	Alkali, acid
Wetting	Surfactants
Emulsification, suspension	Phosphates, surfactants
Sequester	Chelating agents

highly charged molecules; large DNA; particulate cell debris; and even whole cells. The amount of each substance to be removed depends on the fermentation or cell culture process. Filamentous fungi generally require harsh cleaning, vessels used for yeasts and bacteria are more easily cleaned, and animal cell cultures in which the products are secreted require even less stringent conditions. Assessing the cleanliness at this early stage requires the use of nonspecific assays such as total organic carbon, and in some cases specific assays are also used (see "Methods for Evaluating Cleanliness").

If the product is secreted and very few cells die during the process, the substances to be removed from the bioreactor may be limited, and relatively mild cleaning procedures might be very effective. However, if the cells die and lyse at unknown rates in a process that is not well controlled, there will be more debris and released cellular components to remove. The heat used to kill cells may also cause aggregation and precipitation, making the cleaning more difficult. The type of cell culture media (i.e., serum, serum-free, defined) must also be considered.

In addition to designing the cleaning procedure to accommodate the type of cells or microorganisms, it is necessary to evaluate the bioreactor design and ensure that the CIP mechanisms are suitable. CIP system design and operation for bioreactors have been addressed by Chisti and Moo-Young (6).

Cell Removal and Clarification. Cell removal and clarification most commonly use filtration and centrifugation operations. Techniques such as expanded bed absorption, two-phase separations, and precipitation are also used. Each one of these methods may require its own unique cleaning procedures. Cleaning must ensure removal of cells, cell membranes, cell walls, nucleic acids, proteins, lipids, etc.

Filtration. In filtration operations, one must consider the membranes as well as the hardware. Chemical cleaning is most common, and the cleaning agents must be compatible with both the membrane and the wetted components of the hardware. Filter housing designs that should be considered for ease of cleaning are discussed in Ref. 3. In addition to the piping and membranes, wetted components may include spacers, gaskets, valves, and seals. Adhesives may also be present. The cleaning agent should not damage any of these over time. Some membranes are incompatible with cleaning agents such as sodium hydroxide; some may be cleaned with mild detergents and then sanitized with bleach. Furthermore, some filter materials have pressure and temperature requirements that must be considered when designing the cleaning protocol. And as with other unit operations, the number of uses of wetted components may need to be defined to ensure that with extended exposure to cleaning agents extractables are not released and that the membrane performance is not changed.

The stringency of cleaning required is dependent on the level of gel polarization and membrane fouling. Brose and Waibel have addressed the adsorption of proteins to microfiltration capsules (7).

The cleanliness of tangential flow membranes is often determined by measuring the flux of clean water through the system at a standard transmembrane pressure. Guidance and cautions for performing this measurement have been described by Michaels (8). Other methods of assessing cleanliness include total organic carbon, physical inspection, and comparing the specifications of a water-for-injection (WFI) rinse.

Centrifugation. Because centrifuges represent a significant capital cost, they are frequently used for more than one product. Cleaning validation is essential to prevent cross-contamination. Although continuous desludging disk centrifuges can be designed so that they can be cleaned in place, those that do not desludge continuously cannot. The bowl in tubular bowl centrifuges must usually be cleaned out of place after disassembly (3). Aerosols generated during centrifugation may require further evaluation of the equipment area to be cleaned. For low-volume products, cleaning can be minimized by using disposable self-contained tubes. This approach may be preferable, for example, when centrifugation is used to prepare multiple products over a relatively short period of time, such as for monoclonal antibodies used for in vitro diagnostics.

Phase Partitioning and Precipitation. With the exception of aerosol generation, phase partitioning and precipitation present the same general cleaning issues as found in tanks and centrifuges. Equipment design plays a major role in determining how well these pieces of equipment can be cleaned.

Expanded Bed Adsorption and Big Beads. Expanded bed adsorption and the use of large (on the order of 300 μm) beads are two techniques in which product isolation is achieved by binding to chromatographic media. Unlike standard chromatography (see next section), particulates and even whole cells are applied to the chromatography media. Cleaning is essential to ensure that the equipment does not retain soils from previous runs and that media can be reused. Chromatography media, whether used for isolation or purification, have a high degree of surface area that allows for adsorption and even entrapment of impurities. Even though chromatography media are dedicated to one product, cleaning must address the specific feedstream—both impurities and product—and its interaction with the media. Furthermore, the equipment must be designed to ensure that cleaning procedures will remove all traces of previously run product and that no residual live organisms remain.

In a study to evaluate the effectiveness of a cleaning and sanitizing method, a challenge test was performed by adding an *Escherichia coli* homogenate containing approximately 10^9 colony forming units (cfu) per milliliter to two STREAMLINE expanded bed columns. One column contained 250 mL of STREAMLINE SP and the other 250 mL of STREAMLINE DEAE. Sodium chloride cleaning solutions of increasing concentrations from 0.1 to 0.5 M were added, and the reduction of living microorganisms measured. As seen in Figure 1, the relative reduction in living cells is approximately 10^5. The increase in the number of cells in the eluate from the DEAE illustrates the necessity to understand the nature of the impurity to be removed and its interaction with the component being cleaned. In this case, the increase occurs when the increased sodium

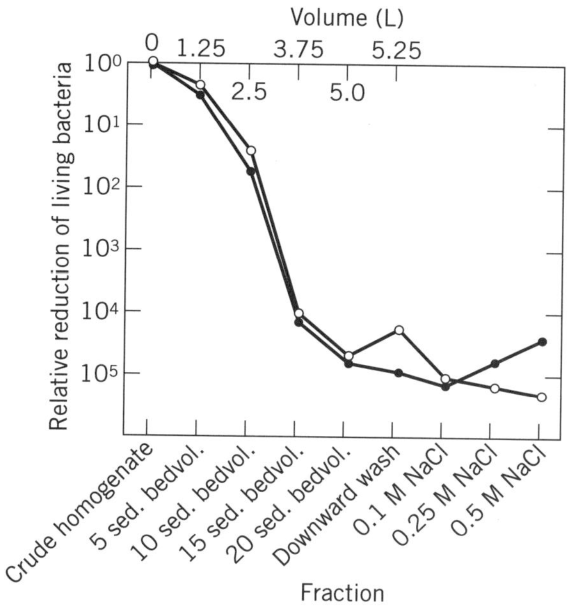

Figure 1. Relative reduction of living *E. coli* cells for STREAMLINE DEAE (●) and SP (○) in relation to the volume of the washing solution. *Source*: With kind permission from Kluwer; Ref. 9.

chloride concentration releases negatively charged *E. coli* cells from the positively charged DEAE (9).

Chromatography. A robust chromatographic process provides the purity required for proteins, peptides, oligonucleotides, plasmids, viral vectors, etc. But it is essential to select equipment and media that can be cleaned in place.

Most of today's chromatographic media can tolerate rather harsh conditions, such as exposure to sodium hydroxide. In affinity chromatography, the ligand usually determines the cleaning solutions that can be used. Protein A, for example, has been shown to withstand even sodium hydroxide (10). On the other hand, with immobilized monoclonal antibodies, it is often necessary to protect the column by using only clarified, partially purified feed. The cost of preparing a monoclonal antibody for use as an immunoadsorbent used in the purification of a therapeutic product is extraordinarily high, because the immobilized antibody must have the same purity as the therapeutic itself (11). One example of the highly successful use of such a column is described by Feldman et al. for the purification of factor VIII from human plasma (12).

As with other unit operations, specific impurities must be addressed in designing the cleaning regimes to ensure that there is no residue carried over from batch to batch of product. Clearly, efficient cleaning is even more important the closer one gets to final product purity. All cleaning procedures, contact time, and temperature must be specified. To maintain column lifetime, the least harsh method that provides a safety window should be implemented.

From a typical bacterial fermentation, one might have to remove soluble and precipitated proteins, hydrophobic proteins and lipids, nucleic acids, viruses, and endotoxins.

Proteins and lipids. Soluble proteins are the easiest to remove. For ion exchange, gel filtration, and most affinity chromatography media, a high salt wash (e.g., 1 to 2 M NaCl) is most common. For hydrophobic interaction chromatography (HIC), low-ionic-strength buffers or water are usually sufficient. In some cases, a detergent must be used, but this approach is more costly and requires validation of detergent removal.

Precipitated proteins frequently require up to 1 M sodium hydroxide or 1 M acetic acid. This is often followed by a water rinse and then a salt wash (except for HIC). Like precipitated proteins, hydrophobic proteins usually require sodium hydroxide. Other approaches include the use of ethanol, detergents, and a combination of cleaning agents. For example, ethanol (20%) in 1 M sodium hydroxide has been used at 40 °C for 30 to 60 minutes to remove host cell proteins from an ion exchanger.

For lipids, very little information is available. Commonly used solubilizing agents may destroy some chromatography media. Detergents and organic solvents (i.e., ethanol and acetonitrile) have been used with some success, but it is necessary to be aware that explosion-proof areas may be required for organic solvents. For those already working in such an area, this does not present a problem, but for firms used to working only in aqueous conditions, this may increase costs significantly. Furthermore, solvent disposal or recycling must be implemented. Removal of lipids before chromatography is strongly recommended to alleviate these concerns and added expenses.

Nucleic Acids. Nucleic acids bind tightly to anion exchangers. Some earlier work to evaluate their removal was performed by loading radiolabeled calf thymus DNA onto a DEAE Sepharose® Fast Flow column. Combinations of sodium chloride and sodium hydroxide were insufficient to totally remove all of the DNA. More recently, the ability of different cleaning agents to remove nucleic acids from a quaternary amine anion exchanger was tested using three different samples (13). The data from two of the samples are shown in Table 3 and 4. It is interesting to observe that for the monoclonal antibody (Table 4), DNase was required to thoroughly remove all traces of nucleic acids.

Viruses. Regulatory agencies have expressed concern about the retention of viruses on columns. As stated at the International Conference on Harmonization Guideline on Viral Safety, "Assurance should be provided that any virus potentially retained by the production system would be adequately destroyed or removed prior to reuse of the system. For example, such evidence may be provided by demon-

Table 3. Mass Balance of Calf Thymus DNA Clearance from Q-Sepharose® Fast Flow

Fraction tested	μg DNA	% of total DNA
Starting material	699.75	100
Buffer A wash	0.79	0.11
100% B eluate	253.80	36.27
Acid wash	0.64	0.09
1 M NaOH/1 M NaCl	449.60	64.25
Water wash	3.92	0.56
Total DNA eluted	708.75	101.2

Table 4. Mass Balance of Total DNA Clearance from a Monoclonal Antibody on Q-Sepharose Fast Flow

Fraction tested	μg DNA	% of total DNA
Total DNa in MAb	481.50	100
Flow through	0.31	0.06
Wash	0.12	0.02
Peak 1	10.00	2.08
Peak 2	270.68	56.21
100% B wash	6.08	1.26
Acid wash	4.00	0.83
1 M NaOH/1 M NaCl	64.00	13.29
2 M NaOH	3.21	0.67
3 M NaCl	0.65	0.13
Water wash	0.29	0.06
Subtotal	359.34	74.61
Dnase 1 treatment	125.00	25.96
Total DNA eluted	485.9	100.6

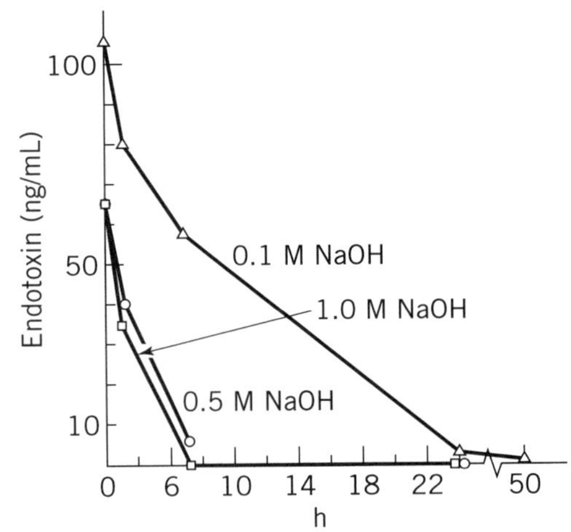

Figure 2. Inactivation of *E. coli* endotoxin by sodium hydroxide.

strating that the cleaning and regeneration procedures do inactivate or remove virus" (14). Table 5 shows work from Q-One Biotech summarizing the inactivation of eight different viruses by 0.1 M and 0.5 M NaOH. The kinetics of inactivation, which must be taken into account for specifying contact time, are also shown.

Endotoxins. Like nucleic acids, endotoxins are highly negatively charged and may bind tenaciously to anion exchangers. Sodium hydroxide has been shown to be very effective in inactivating endotoxins. Figure 2 shows the inactivation of *E. coli* endotoxin in solution at room temperature. Although a contact time of up to 5 hours with 1 M NaOH is indicated here to inactivate endotoxin, the added action of flow may enhance removal. In a study on a chromatography system challenged with a solution containing 1,200 endotoxin units (EU) per milliliter, it was found that circulating 1 M NaOH for 60 minutes was sufficient to inactivate endotoxin in the both the system and the packed column of a quaternary amine anion exchanger that was placed in line. In addition to the *Limulus* amebocyte lysate (LAL) test for endotoxin, a total organic carbon assay was used to analyze the rinse fluids for any carbon-containing residues remaining from the endotoxin. No residue was detected (15).

Combining Cleaning and Sanitization Methods. The ability to remove microorganisms from chromatography media is an issue that should be addressed in development. Depending on the microorganisms, they may produce endotoxins, enterotoxins, spores, and proteases—all of which could adulterate a product. Sanitization of chromatography columns is readily accomplished with sodium hydroxide (16). Combining cleaning and sanitization is the most cost-effective approach. In many cases, addition of sodium chloride to a sodium hydroxide solution provides an efficient, simple method that is easy to validate for removal of bioburden (tested by routine monitoring) and impurities from the feedstream. The importance of compatibility of the wetted materials with such combinations should be considered at the design phase.

Formulation and Final Filling. Cleaning issues in formulation and filling operations are addressed in depth elsewhere. Some relevant texts include those by Olson and Groves (17) and Cole (18). A recent publication that ad-

Table 5. Virus Titles and Inactivation values (log_{10})

		HIV	BVD	CPV	BHV	POL	SV-40	MLV	ADV
0.1M NaOH	Spike	2.0×10^{8}	9.5×10^{5}	2.0×10^{9}	6.9×10^{9}	7.1×10^{8}	1.7×10^{8}	2.6×10^{5}	2.2×10^{8}
	t = 0 min	[b]*5.9×10^{2}*	2.7×10^{7}	1.9×10^{3}	1.2×10^{2}	3.5×10^{4}	1.5×10^{5}	*3.7×10^{1}*	1.7×10^{2}
	t = 10 min	*5.7×10^{2}*	2.7×10^{5}	2.4×10^{3}	1.5×10^{1}	2.7×10^{3}	3.6×10^{5}	*3.8×10^{1}*	6.0×10^{1}
	t = 20 min	*5.8×10^{2}*	1.5×10^{4}	9.6×10^{2}	4.5×10^{1}	2.0×10^{4}	4.7×10^{4}	*4.0×10^{1}*	6.3×10^{1}
	t = 60 min	*5.8×10^{2}*	2.7×10^{4}	5.0×10^{3}	4.5×10^{1}	2.1×10^{3}	2.0×10^{4}	*4.3×10^{1}*	*2.9×10^{1}*
Inactivation (log_{10})		>3.5	2.5	5.6	8.2	5.5	3.9	>3.8	>6.9
0.5M NaOH	Spike	2.0×10^{6}	9.5×10^{6}	2.0×10^{9}	6.9×10^{9}	7.1×10^{8}	1.7×10^{8}	2.6×10^{5}	2.2×10^{8}
	t = 0 min	*5.7×10^{2}*	1.9×10^{4}	9.4×10^{2}	*5.9×10^{1}*	1.1×10^{5}	1.5×10^{5}	*6.3×10^{1}*	9.4×10^{1}
	t = 10 min	*5.6×10^{2}*	1.3×10^{2}	1.2×10^{3}	*5.9×10^{1}*	1.1×10^{5}	1.7×10^{3}	*4.7×10^{1}*	7.5×10^{1}
	t = 20 min	*5.6×10^{2}*	*1.7×10^{2}*	1.5×10^{3}	*5.9×10^{1}*	2.0×10^{4}	8.4×10^{3}	*4.7×10^{1}*	*2.0×10^{1}*
	t = 60 min	*6.7×10^{2}*	*1.7×10^{2}*	1.5×10^{3}	*5.9×10^{1}*	6.2×10^{3}	1.0×10^{2}	*5.5×10^{1}*	*2.2×10^{1}*
Inactivation (log_{10})		>3.5	>4.7	6.1	>8.1	5.1	6.2	>3.7	>7.0

Notes: Virus titers expressed in $TCID_{50}$ units for all viruses except BHV and MLV (expressed in pfu). Italicized titers: No virus was detected in these samples and values listed are theoretical minimum detectable titers.

HIV = human immunodeficiency virus; BVDA = bovine viral diarrhea virus; CPV = canine parvovirus; BHV = bovine herpes virus; POL = poliovirus; SV-40 = simian virus-40; MLV = murine leukemia virus; ADV = adenovirus.

dresses reducing pyrogens in cleanroom wiping materials provides valuable information on cleaning (19).

METHODS FOR EVALUATING CLEANLINESS

Several methods for evaluating cleanliness were discussed previously in conjunction with the different unit operations. There are three basic methods to evaluate cleanliness: visual inspection, assaying rinse water, and assaying surfaces by swabbing. The nature of the unit operation and equipment design frequently determine which method will be most appropriate. A combination of methods is usually necessary for validation. Visual inspection may be quite useful as a qualitative assessment of cleanliness for equipment such as tanks. As noted by several regulators, "you can always see the ring around the bathtub." However, more sensitive, quantitative methods are also required. For multiproduct facilities, it is absolutely necessary to use product-specific assays. Product-specific assays are also used for dedicated facilities.

Rinse water can be collected after cleaning and examined for residues. In many cases, the quality of the rinse water is compared to WFI that has not been run through the component being tested. Other tests commonly run on rinse water (3) include LAL for endotoxin, total organic carbon (TOC) (20,21), protein assays, sodium dodecylsulfate–polyacrylamide gel electrophoresis (SDS-PAGE), spectrophotometric analysis, and product-specific assays such as ELISAs.

Swabbing is performed by using non-carbon-containing materials that do not interfere with subsequent testing but from which swabbed soils can be extracted. The swabs themselves should not generate particles, background carbon, or nonvolatile residues. Swabbing of surfaces allows evaluation of the cleanliness of areas in which non-water-soluble residues might remain. However, there may be components or areas of systems that are hard to reach by swabbing. In this case, a soil can be applied to pieces of the equipment, called coupons, that are cut out or provided by a vendor. The soil is allowed to dry, and then the coupon is swabbed after cleaning. Coupons can be useful in designing the cleaning protocol. The coupon can be placed inside equipment and cleaned during a normal cleaning cycle. Sometimes, the coupon and the soil are heated to simulate a difficult cleaning problem. Cleaning methods can be tested on the baked-on soil. After the swabbing is performed, the swab is extracted, and the extract is analyzed by assays such as TOC, HPLC, and ELISA.

A survey of five U.S. biotechnology companies revealed which assays are commonly performed and how they are used at different stages of manufacturing (22). The report presents the purpose of each test, detection limits, acceptance criteria, and methodology in a concise tabular form.

SETTING LIMITS

How clean is clean? As analytical methods become more sensitive, the likelihood of finding something is increased. Attempts to define acceptance limits of cleanliness have been made by Fourman and Mullen (23) and Zeller (24) and discussed in Reference 3. Fourman and Mullen suggested that "No more than 10 ppm of any product should appear in another product" and "No more than 0.001 dose of any product will appear in the maximum daily dose of another product." But clearly, a risk analysis should be performed to assess the potential hazard for each product.

Whereas for some therapeutics these low levels suggested by Fourman and Mullen may be accepted by regulatory authorities, for many multiproduct facilities, "not detectable" may be required. On the other hand, for diagnostic products, the real issue should be whether the product is consistent and whether the carryover affects its accuracy.

The acceptance criteria for upstream unit operations such as fermentation may not be as stringent as those for final purification and processing steps. Typical acceptance criteria for CIP of fermenters has been discussed by Naglak et al. (25).

In addition to setting specifications for residual impurities derived from the product feedstream and final product, it is necessary to consider the removal of cleaning agents. Analytical strategies for organic cleaning agent residues have been addressed by Gavlick et al. (26). The questions: "What is the level of detergent residue that would be acceptable to the Food and Drug Administration (FDA)? What is the basis for arriving at this level, if any?" were answered by the FDA's Office of Compliance as follows: "FDA has repeatedly stated that it is the firm's responsibility to establish acceptance limits and be prepared to provide the basis for those limits to FDA. Thus, there is no fixed standard for levels of detergent residue. Any residues must not adversely alter drug product safety, efficacy, quality, or stability" (27).

CLEANING VALIDATION

Although it is not a regulatory requirement, a master plan for cleaning validation can be quite valuable. The master plan will provide a time frame for performing the work, prioritizing tasks, designating responsibilities, describing protocols, and listing all facility, equipment, and process validation requirements. Although the time frame will most likely change, the plan will ultimately save time and may prevent regulatory delays by ensuring that no critical items are overlooked. A final validation report should be signed by management.

Agalloco described a series of questions that provide insight into cleaning validation (28): What items are being removed? How are samples taken? What analytical tests are utilized? When shall cleaning validation be performed? Which physical parameters shall be evaluated? How much residual will be permitted? These questions should be addressed during the preparation of the master plan.

Two cleaning validation guides written by the U.S. FDA are very useful (29,30). Although the intent of these guides is to provide FDA investigators with information that results in consistency in inspections, there are many relevant points that should be considered by any firm that has to perform cleaning validation. For example, in the mid-

Atlantic version, it is stated that operators performing cleaning operations should be aware of problems and have special training, and the length of time between the end of processing and each cleaning step should be controlled. The documents further address issues such as equipment design, documentation, analytical methods, sampling, and establishment of limits. Although no one approach is required, the guidance provided in these documents should be followed or a reason why it was not adhered to provided.

One of the more difficult techniques to validate is swabbing. Some of the variables include operator swabbing technique, extraction efficiency from the swab, and recovery of residue from different wetted materials. In spite of its difficulties, this technique provides valuable information that cannot be obtained from visual inspection and rinse water analysis. Total protein analysis and total carbon analysis of swab samples for the cleaning validation of bioprocess equipment have been addressed by Strege et al. (31,32). Use of this technique for validating detergent residues has also been presented (33).

Most cleaning validation is performed at pilot or full scale, but for certain unit operations such as chromatography, the parameters that are to be validated can often be established with scaled-down models using production feedstream. Seely et al. scaled down a chromatography system and used cleaning conditions that were harsher than those used in normal production to determine if residues remained on the packed column (34). At the same time, production columns were also evaluated for cleaning efficacy by collecting relevant product fractions from blank runs and assaying them for UV280 absorbing material and endotoxin. In addition, reversed-phase HPLC, ELISA, and silver-stained SDS-PAGE assays were performed. No endotoxin or UV280 absorbing material was found. A trace amount of the product was found, but the level (about 0.3% of the amount normally collected) was thought not to represent a significant cycle-to-cycle carryover risk to the final product. Furthermore, the eluate after storage was evaluated by TOC for evidence of further cleaning effects of the storage solution.

Once cleaning is validated, it is much more convenient to monitor key process conditions such as cleaning agent concentration, pH, conductivity, temperature, contact time, and flow if applicable. Revalidation of cleaning may have to be performed when there is a change in a process or equipment, after maintenance or shutdown periods, or after area contamination. Each firm should establish the frequency of cleaning revalidation.

CASE STUDIES

Some case studies further illustrate the critical issues in cleaning and cleaning validation.

Validation of CIP of a Bioreactor

In a 1994 paper, Geigert et al. presented a case study on the CIP of a bioreactor (35). They addressed five key questions: What components need to be cleaned from the bioreactor? What is a scientifically sound and appropriate sampling plan? How will the test samples be handled prior to analysis? Are the test methods chosen scientifically sound and appropriate for the intended purpose? What is clean?

The components to be cleaned included active protein, host cell components, media ingredients, and cleaning agents. The sampling plan included swabbing of a few select surface sites, performing a worst-case cleaning challenge using artificial soil or reducing cleaning parameters, and evaluating liquid from the rinse. Both final rinse and earlier rinse samples were taken to show that the cleaning process was effective. Samples were evaluated for their stability during storage and handling. As an example, it was noted that hydrophobic proteins may adhere to certain plastic storage containers.

Several assays, both specific and nonspecific, were evaluated. Three assays were selected: SDS-PAGE, a bioassay, and pH. TOC was not yet set up at the facility; SDS-PAGE and the bioassay had already been validated by quality control and only needed to be further validated for the limit of detection for the cleaning studies. The absence of cleaning agent was determined by pH, and the absence of final product by the bioassay. Four protein products were analyzed by SDS-PAGE with silver staining. The limit of detection varied from 1 ppm to 10 ppm. The bioassay was performed for the three products that might be campaigned in a single bioreactor. Two products were from yeast; one was from *E. coli*. The two yeast products could be detected at levels as low as 4 ppb, but the *E. coli* product, which was very hydrophobic, aggregated. This study illustrates that the test procedures to evaluate cleanliness are critical in supporting the validation work. By using validated test methods, the firm was able to provide a high degree of assurance that a 1,600-L bioreactor could be cleaned sufficiently between product campaigns, thereby avoiding the additional costs of purchasing costly dedicated equipment.

Changeover of Equipment from Development to a Dedicated Unit

In this example, a firm had used a pilot-scale chromatography unit for development work in the purification of monoclonal antibodies (IgA and IgM) (36). To avoid the expense of purchasing a new unit, the firm decided to use this system for the manufacture of clinical trial material. This required extensive cleaning validation. Because the unit contained tubings and threaded fittings, it was decided that replacing low-cost wetted components was less expensive than validating their cleaning. After the new components were installed, the cleaning of the system was validated by demonstrating absence of carryover on previously exposed sites and by using new surface sites to establish a baseline for future cleaning validation. Samples were taken by swabbing and collecting rinse water, and samples were analyzed by both TOC and ELISAs. The IgM ELISA has a limit of detection of 25 ng/mL, and the IgA ELISA has a limit of 4 ng/mL. The results showed that the 16 used surfaces sampled were below the detection levels of the ELISAs.

CIP Validation in a Multiproduct Facility

A multiproduct biopharmaceutical contract manufacturing facility must address cleaning and cleaning validation to satisfy its own requirements as well as the requirements of their customers and regulators. Sherwood has presented a description of CIP experiences in such a facility (37). The facility manufacturers monoclonal antibodies and recombinant proteins by mammalian cell culture. The author points out that "Each company must determine their own cleaning policies and acceptance criteria based upon knowledge of the equipment and processes employed."

Visual inspection, final rinse water, and swabbing of surface samples were used to evaluate CIP of fermenters and purification holding vessels. Because visual inspection is subjective, two observers are always used. For the rinse water samples, sample containers are single use, and sample storage temperatures and duration were validated. The following tests were performed on rinse water: pH and conductivity, nitrate and protein concentration, bioburden, TOC, endotoxin, and an enzyme immunoassay (EIA). The EIA was only performed for the purification holding vessels. In general, nonspecific assays are used for screening both product and manufacturing materials in fermentation, whereas in the purification area more specific assays are used. For the swab samples, TOC and protein concentration were performed for the fermentation, and in the purification area, the EIA was added.

Acceptance criteria are more difficult to establish in such a facility because the final product dose is not known. Acceptance limits have to be based on analytical method performance criteria, pharmacopeial requirements, and historical data.

The usefulness of performing cleaning validation was illustrated by the performance qualification of the CIP of the fermenter vessel. In one run, a sight glass swab showed a high TOC result. After investigation, a manual cleaning of the sight glass was implemented.

For the initial performance qualification, three consecutive CIP runs were monitored and found to meet acceptance criteria. On an annual basis, one CIP run is fully monitored. The author noted that ongoing monitoring in such a contract manufacturing facility may also require final rinse and swab sampling.

SUMMARY

Cleaning and cleaning validation require sound scientific judgment. Cleaning is not just an afterthought, but should be designed into a process. Today there is a trend to use contract manufacturing, and most of these facilities will manufacture more than one product. Cleaning validation is particularly critical for such facilities. Regulatory agencies are spending more time investigating cleaning and cleaning validation. Ultimately, adequate cleaning leads to enhanced product quality, fewer batch failures, and hence better profitability.

BIBLIOGRAPHY

1. *Guidelines on the Preparation of Investigational New Drug Products*, U.S. Food and Drug Administration, Rockville, Md., March 1991.
2. Elimination of Establishment License Application for Specified Biotechnology and Specified Synthetic Biological Products. *Fed. Regist.* **61**, 24227–24233 (1996).
3. PDA Biotechnology Cleaning Validation Subcommittee, *Cleaning and Cleaning Validation: A Biotechnology Perspective*, PDA, Bethesda, Md., 1996.
4. D. Rohsner and W. Serve, *Pharm. Eng.* **15**, 20–28 (1995).
5. D.A. LeBlanc, D.D. Danforth, and J.M. Smith, *Pharm. Technol.* **17**, 84–92 (1993).
6. Y. Chisti and M. Moo-Young, *J. Ind. Microbiol.* **13**, 201–207 (1994).
7. D.J. Brose and P. Waibel, *Pharm. Technol.* **20**, 48–52 (1996).
8. S.L. Michaels, *BioPharm* **7**, 38–45 (1994).
9. R. Hjorth, S. Kåmpe, and M. Carlsson, *Bioseparation* **5**, 217–233 (1995).
10. G. Hale, A. Drumm, P. Harrison, and J. Phillips, *J. Immunol. Methods* **171**, 15–21 (1994).
11. Points to Consider in the Manufacture and Testing of Monoclonal Antibody Products for Human Use, U.S. Food and Drug Administration, Rockville, Md., 1994.
12. F. Feldman, S. Chandra, M. Hrinda, A. Schreiber, in *Quality Assurance in Transfusion Medicine, Vol II. Methodological Advances and Clinical Aspects*, CRC Press, Boca Raton, Fla., 1993, pp. 259–284.
13. Y. Dasarathy, *BioPharm* **9**, 41–44 (1996).
14. International Conference on Harmonization; Guidance on Viral Safety Evaluation of Biotechnology Products Derived from Cell Lines of Human or Animal Origin, *Fed. Regist.* **63**, 51074–51084 (1998).
15. Pharmacia Biotech, *Depyrogenation of a BioProcess™ Modular System and a Packed BPG Column with Sodium Hydroxide*, Technical Note 216, Research Report, 1996.
16. G. Sofer and L. Hagel, *Handbook of Process Chromatography: A Guide to Optimization, Scale up, and Validation*, Academic Press, London, 1997.
17. W.P. Olson and M.J. Groves, *Aseptic Pharmaceutical Processing*, Interpharm Press, Prairie View, Ill., 1987.
18. G.C. Cole, *Pharmaceutical Production Facilities Design and Applications*, Ellis Horwood, New York, 1990.
19. D.W. Cooper, *Pharm. Eng.* **16**, 51–61 (1996).
20. R. Baffi, G. Dolch, R. Garnick, Y.F. Huang, B. Mar, D. Matsuhiro, B. Niepelt, C. Parra, and M. Stephan, *J. Parenteral Sci. Technol.* **45**, 13–19 (1991).
21. K.M. Jenkins, A.J. Vanderwielen, J.A. Armstrong, L.M. Leonard, G.P. Murphy, and N.A. Piros, *PDA J. Pharm. Sci. Technol.* **50**, 6–15 (1996).
22. P.R. McArthur and M. Vasilevsky, *Pharm. Eng.* **15**, 24–31 (1995).
23. G.L. Fourman and M.V. Mullen, *Pharm. Technol.* **17**, 54–60 (1993).
24. A.O. Zeller, *Pharm. Technol.* **17**, 70–80 (1993).
25. T.J. Naglak, M.G. Keith, and D.R. Omstead, *BioPharm* **7**, 28–36 (1994).
26. W.K. Gavlick, L.A. Ohlemeier, and H.J. Kaiser, *Pharm. Technol.* **19**, 136–144 (1995).
27. Division of Manufacturing and Product Quality, HFD-320, Office of Compliance, Center for Drug Evaluation and Research, *Human Drug CGMP Notes* **3**, 1995.
28. J. Agalloco, *J. Parenteral Sci. Technol.* **46**, 163–168 (1992).
29. Mid-Atlantic Region Inspection Guide, Cleaning Validation, U.S. Food and Drug Administration, Rockville, Md., July 28, 1992.

30. Division of Field Investigations, Guide to Inspections of Validation of Cleaning Processes, U.S. Food and Drug Administration, Rockville, Md., July 1993.
31. M.A. Strege, J.J. Dougherty, R. Green, and A.L. Lagu, *BioPharm.* **7**, 40–42 (1994).
32. M.A. Strege, T.L. Stinger, B.T. Farrell, and A.L. Lagu, *BioPharm* **9**, 42–45 (1996).
33. J.M. Smith, *Pharm. Technol.* **16**, 60–66 (1992).
34. R.J. Seely, H.D. Wight, H.H. Fry, S.R. Rudge, and G.F. Slaff, *BioPharm* **7**, 41–48 (1994).
35. J. Giegert, R. Klinke, K. Carter, and A. Vahratian, *PDA J. Pharm. Sci. Technol.* **48**, 236–240 (1994).
36. P. Gariepy, R. Moore, and G. Sofer, *Biopharm* **11**, 32–36 (1998).
37. D. Sherwood, *PDA Fourth Int. Congress*, Vienna, Austria, February 19–23, 1996.

See also ASTM STANDARDS FOR BIOTECHNOLOGY; CHROMATOGRAPHY, COMPUTER-AIDED DESIGN; GOOD MANUFACTURING PRACTICE (GMP) AND GOOD INDUSTRIAL LARGE SCALE PRACTICE (GLSP); PILOT PLANTS, DESIGN AND OPERATION; PROCESS VALIDATION.

CLOSTRIDIA, SOLVENT FORMATION

PALMER ROGERS
University of Minnesota
Minneapolis, Minnesota

KEY WORDS

Biochemistry
Clostridium
Clostridium genetics
Physiology
Solvent fermentations
Solvent production
Strain development

OUTLINE

INTRODUCTION

The general biology of the genus *Clostridium* has been treated recently in detail by Hippe et al. (1). More than 100 species of bacteria fall into this group, probably because only four simple criteria need be met for membership: *(1)* ability to form heat-resistant endospores, *(2)* obligate requirement for anaerobic energy metabolism; *(3)* inability to reduce sulfate as a final electron acceptor, and *(4)* a gram-positive type of cell wall. The extreme heterogeneity of this huge "genus" has become more and more evident over the past 20 years. First, the GC content of the DNA of more than 100 species of *Clostridium* ranges between 21 and 54 mol %. Second, the 16S rRNA gene sequences have been determined for most of these species together with their close relatives. A recent analysis of these data indicates that the genus *Clostridium* is in fact mixed in with some of the closely related bacteria to form several very divergent "clusters", providing a strong basis for a taxonomic revision (2).

The largest rRNA homology group, which is newly designated cluster I (2), contains all the butanol-producing species of *Clostridium* (Table 1) (2–7). Fifty-five strains classified as *Clostridium acetobutylicum* in national collections were recently examined by means of DNA–DNA hybridization (3) and a combination of DNA fingerprinting and 16S RNA gene sequence analysis (4). The results show that these strains fall into four taxonomic groups. These taxa, or species, are listed in Table 1, together with three other butanol-producing species. Also listed in Table 1 are five ethanol-producing clostridia, which are considered important because of their potential to produce ethanol from fermentation of specific substrates such as crystalline cellulose or at a high temperature. Reclassification suggests that these "clostridia" fall into three or four distinct phylogenetic clusters (2). *C. thermocellum* is the most thoroughly studied ethanol-producing cellulolytic clostridial species. It is grouped with seven other cellulolytic species in cluster III (2) or group E (5). *C. thermohydrosulfuricum* strains have recently been divided into two groups and reclassified as *Thermoanaerobacter thermohydrosulfuricus* or *T. ethanolicus* (6) and placed in cluster V (2) or group A (5). *C. thermosaccharolyticum* most likely belongs in the new genus *Thermoanaerobacterium* (6) in cluster VII (2) or group C (5). The phylogenetic position of *C. saccharolyticum* has not been determined (7). Table 1 also includes strain numbers and culture collection numbers that appear most frequently in the literature dealing with solvent production by these clostridia.

The major role of the solvent- and acid-forming saccharolytic clostridia in nature is the degradation of organic materials, namely the biopolymers starch, cellulose, hemicelluloses, and proteins. Utilization of these polymers requires that these bacteria produce and secrete enzymes such as amylases, cellulases, proteinases, and xylanases. Some of these enzymes are of potential importance to commercial processes (8). The anaerobic fermentation of sug-

Table 1. Clostridial Species Important for Solventogenic Fermentations

Former name(s)	Proposed (new) name	Ref.
Ethanol-producing species		
C. thermocellum LQRI, ATCC 35609, DSM 2360	No change (cluster III)	2,5
C. thermohydrosulfuricum E100-69^T, ATCC 35045, DSM 567	*Thermoanaerobacter thermohydrosulfuricus* (cluster V)	2,5,6
C. thermohydrosulfuricum, Thermoanaerobacter ethanolicus JW200^T, 39E, ATCC 33223	*Thermoanaerobacter ethanolicus* (cluster V)	2,5,6
C. thermosaccharolyticum NCA 3814, ATCC 7956	Misclassified, belongs in genus *Thermoanaerobacterium* (cluster VII)	2,5,6
C. saccharolyticum NRCC 2533, ATCC 35040	No change	7
Butanol-producing species[a]		
C. acetobutylicum ATCC 824, 4259 Weizmann strain; DSM 792, and 8 strains	No change, "taxon I"	3,4
C. acetobutylicum NRRL B643, NCP 262, and four NCP strains	*C* (new species) "taxon II"	3,4
"*C. saccharoperbutylacetonium*" NI-4, NI-504, ATCC 27021	No change, "taxon III"	3,4
C. beijerinckii, C. acetobutylicum ATCC 25732^T, NRRL B 592, NCIMB 9362^T, 8052^T, and 19 strains	*C. beijerinckii*, "taxon IV"	4
C. puniceum NCIMB 11596	No change; closely related to taxa II and III	4
C. aurantibutyricum ATCC 17777	No change	4
C. tetanomorphum MG-1, ATCC 49273, DSM 4474	No change	4

Note: Culture collections: ATCC, American Type Culture Collection, Rockville, Md.; DSM, Deutsche Sammlung von Mikroorganismen und Zellkulturen GmbH, Göttingen, Germany; NCIMB, National Collection of Industrial and Marine Bacteria, Aberdeen, Scotland; NRRL, Northern Utilization and Development Division, Peoria, Ill.; NI, Nagao Institute, Tokyo, Japan; NRCC, National Research Council, Ottawa; NCA, National Canner's Association, Washington, D.C.; NCP, Department of Microbiology, University of Capetown, Rondebosch, South Africa.
[a]All butanol-producing species fall into cluster I(2).

ars from digestion of polymers or other sources is called a saccharolytic fermentation. The major products that accumulate are organic acids, neutral solvents, CO_2, and H_2. The solvents and acids are listed in Table 2 for typical fermentations by 11 solvent-producing *Clostridium* species (9–19). Production of CO_2 and H_2 gases is always high in these fermentations (see Table 2 note *a*); however, the exact amount produced is not listed for each species. In a natural setting these clostridia occur in conjunction with a consortium of other anaerobic bacteria, which utilize these fermentation products as rapidly as they are formed for their own energy and growth needs. In this way most of the organic starting materials are mineralized. Only a few isolated strains of *Clostridium* produce significant amounts of solvents and hence are potential candidates for industrial fermentations.

The reason for the high level of interest in basic research and development of fermentation technology for solvent production by the clostridia is straightforward. First, although there is no immediate threat that the world's petroleum reserves will be used up, a severe dwindling of these nonrenewable resources, which serve as the major source of fuels and chemicals, will most likely occur in the mid-twenty-first century (20). Second, it is now technologically possible to produce essentially all commodity chemicals from renewable biomass feedstocks such as starch and cellulose. Third, from about 1915 until the mid-1940s, fermentations employing *Clostridium acetobutylicum* made a significant impact on the commercial solvents industry. Typically, more than 1000 kg of butanol, 500 kg of acetone, and 175 kg of ethanol plus other utilizable byproducts were produced in 90,000-L fermentors in North America and elsewhere (21).

From about 1925 to 1980 research on the physiology and biochemistry of the clostridial fermentations was centered on the unraveling of the metabolic pathways essential to our current understanding of the biology of solvent formation (8). Over the past 10 years, basic research on clostridial solventogenesis has made great strides in the following areas: development of genetic methods applied to the solventogenic *Clostridium* species, regulation of pathways of solvent production, degradation and use of biopolymers, and relation of solvent production to developmental processes such as sporulation. This summary treatment of the biology of the solventogenic clostridia will emphasize our current understanding of the clostridial fermentation strategies, physiologic control and genetic mechanisms regulating solvent production, and future directions for development of new solvent-producing strains. The advantages and problems inherent in the biology of these bacteria relevant to their use in commercial solvent production also are presented.

Table 2. Ethanol and Butanol Fermentations by *Clostridium* Species

Organism	Substrates utilized	
	Complex polymers	Sugars
	Ethanol fermentations	
C. thermocellum	Cellulose, xylan	Cellobiose, few hexoses and pentoses
C. thermohydrosulfuricum (T. ethanolicus)	Starch, pectin, salicin, xylan	Many dissacharides, hexoses, pentoses, etc.
C. thermosaccharolyticum	Starch, xylan	Dissacharides, hexoses, and pentoses
C. saccharolyticum		Dissacharides, hexoses, and pentoses
	Butanol fermentations	
C. acetobutylicum taxon I (corn-starch-utilizing strains)	Starch, xylan	Cellobiose, lactose, some hexoses and pentoses
C. acetobutylicum taxon II		
C. saccharoperbutylacetonicum taxon III (molasses-utilizing strains)	Starch	Sucrose, cellobiose, some hexoses and pentoses
C. beijerinckii taxon IV	Starch	Sucrose, cellobiose, some hexoses and pentoses
C. aurantibutyricum	Starch	Sucrose, hexoses, and pentoses
C. puniceum	Starch, pectin	Hexoses, pentoses
C. tetanomorphum		Cellobiose, some hexoses and pentoses

[a]The gases CO_2 and H_2 are also produced in these fermentations. Production per 100 mmol of hexose CO_2, 150–220 mmol; H_2, 50–100 mmol.
[b]N, not reported.
[c]Amount of butyrate is variable depending on conditions.
[d]Acetone only is produced.
[e]Either isopropanol or acetone is produced, depending on the strain.
[f]Both isopropanol and acetone are produced.

PHYSIOLOGY AND BIOCHEMISTRY OF THE SOLVENT FERMENTATIONS

Clostridium's General Fermentation Strategy

Thauer et al. (22) outlined the general energy metabolism of the chemotrophic anaerobic bacteria, which include the clostridia. The clostridia depend almost exclusively on the fructose biphosphate pathway (Embden–Meyerhof pathway) for conversion of one hexose to two pyruvates with the net production of two ATPs and two NADHs (Fig. 1, pathway 1). In the fermentation of pentoses, the intermediates from 3 mol of pentose-5-phosphate, are one glyceraldehyde 3-phosphate and two fructose 6-phosphates. A combination of the enzymes transaldolase and transketolase is used (Fig. 1, pathway 6). The sugar-phosphates enter the fructose biphosphate pathway and have the capability of producing five ATPs and five NADHs per 3 mol of pentose fermented. The phosphorolytic 3-2 cleavage of xylulose-phosphate yielding acetyl-phosphate is not found in the clostridia or in other obligate anaerobes. Also, the clostridia do not have the enzyme pathway for oxidation of glucose 6-phosphate to NADPH, pentose 5-phosphate, and CO_2. However, high activities of glucose 6-phosphate dehydrogenase and of 6-phosphogluconate dehydrogenase were found to be induced in sporulating cells of *C. thermosaccharolyticum* (12). Possibly this pathway (Fig. 1, pathway 7) is induced during sporulation to provide extra NADPH for biosynthesis and has been overlooked in other clostridia. Yet, NADPH–ferredoxin oxidoreductase (Fig. 1, reaction 4), which is ubiquitous in all clostridia investigated (23), probably provides the NADPH for biosynthesis during both vegetative growth and sporulation. Most pyruvate produced from sugars during the clostridial fermentation is cleaved by pyruvate–ferredoxin oxidoreductase in a coenzyme A (CoA) dependent reaction yielding CO_2, acetyl CoA, and reduced ferredoxin (Fd_{red}), as shown in Figure 1, reaction 2. Acetyl CoA is central to all clostridial fermentations. The thioester bond of acetyl CoA is a very high energy bond; thus it is an important source of ATP in most clostridial fermentations, because for every mole of butyrate or acetate produced, a mole of ATP is formed (Fig. 2, reactions 2 and 6). Acetyl CoA is also the most important precursor of all alcohols and organic acids synthesized by these organisms: it thus serves as a major intermediate in these fermentations (Fig. 1). The NADH, which is formed during oxidation of glucose, is thus reoxidized and recycled so that the fermentation can continue.

The second key element in the clostridial fermentation strategy is the system that manages electron flux diagrammed in Figure 1, reactions 3–5. This system consists of ferredoxin and/or other iron–sulfur proteins that can accept or donate electrons at a very low potential ($E^{0\prime} = -0.41$ V) close to the hydrogen electrode; a ferredoxin–NADH oxidoreductase for passing electrons between NADH and Fd; a ferredoxin–NADPH oxidoreductase for controlled production of NADPH needed for biosynthesis of bacterial mass; and one or more hydrogenases permitting the use of protons as final electron acceptors, producing hydrogen gas, typical of most clostridial fermentations.

The electron distribution system functions in coordination with the branched fermentation pathways typical of clostridia, which eventually yield from two to five major acids and solvents (Table 2). The amounts of reduced versus neutral and oxidized products always are balanced with the amounts of H_2 and ATP produced, and there is a

Major fermentation products (mmol/100 mmol hexose)[a]

Ethanol	Butanol	Acetone isopropanol	Acetate	Butyrate	Lactate	Ref.
Ethanol fermentations						
85			31		26	9
194			11		2	10
109			75	2[c]	38	11,12
180			19		4	13
Butanol fermentations						
6	55	28[d]	11	9		14
5	46	24[d]	3	8		15
N[b]	43	26[e]	3	4		16
N	42	25[f]	N	N		17
2	68	15[d]	16	5.5		18
43	47	0	23	5		19

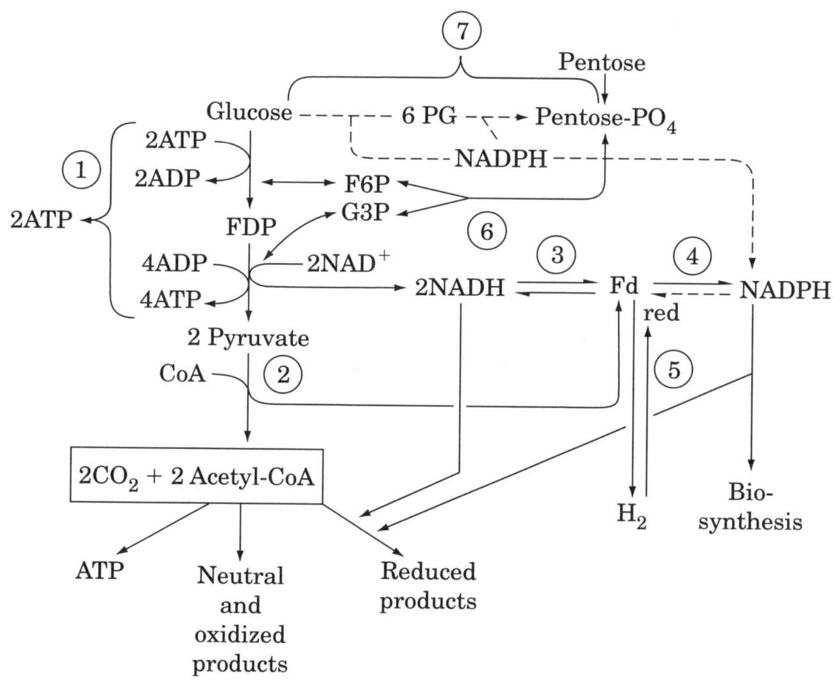

Figure 1. The clostridial fermentation strategy. Numbers indicate either pathways or enzymes as follows: 1, fructose biphosphate (Embden–Meyerhof) pathway; 2, pyruvate–ferredoxin oxidoreductase; 3, NADH–ferredoxin oxidoreductase and ferredoxin–NAD oxidoreductase; 4, NADPH–ferredoxin oxidoreductase; 5, hydrogenase; 6, transketolase + transaldolase; and 7, oxidative pentose phosphate pathway. *Source*: Reproduced with permission from Ref. 24, p. 6.

great deal of natural variation dependent on species and fermentation conditions (24). Figure 2 diagrams an example of this principle, in which *C. acetobutylicum* produces five organic fermentation products from acetyl CoA by utilizing three branch points in the fermentation pathways. Certainly, this variety of pathways available to the clostridia provides physiologic versatility and thus adaptability to different growth conditions. Presently, research is focused on elucidating the physiologic signals and details of molecular sensor and responder systems that control the direction of flow toward specific products in these fermentations (see "Regulation of Solvent Production"). The usefulness of these organisms will depend on directing these fermentations toward one or perhaps two major products. Another consequence of the biology of *Clostridium* (and perhaps all solvent-forming microbes) is that the yield of

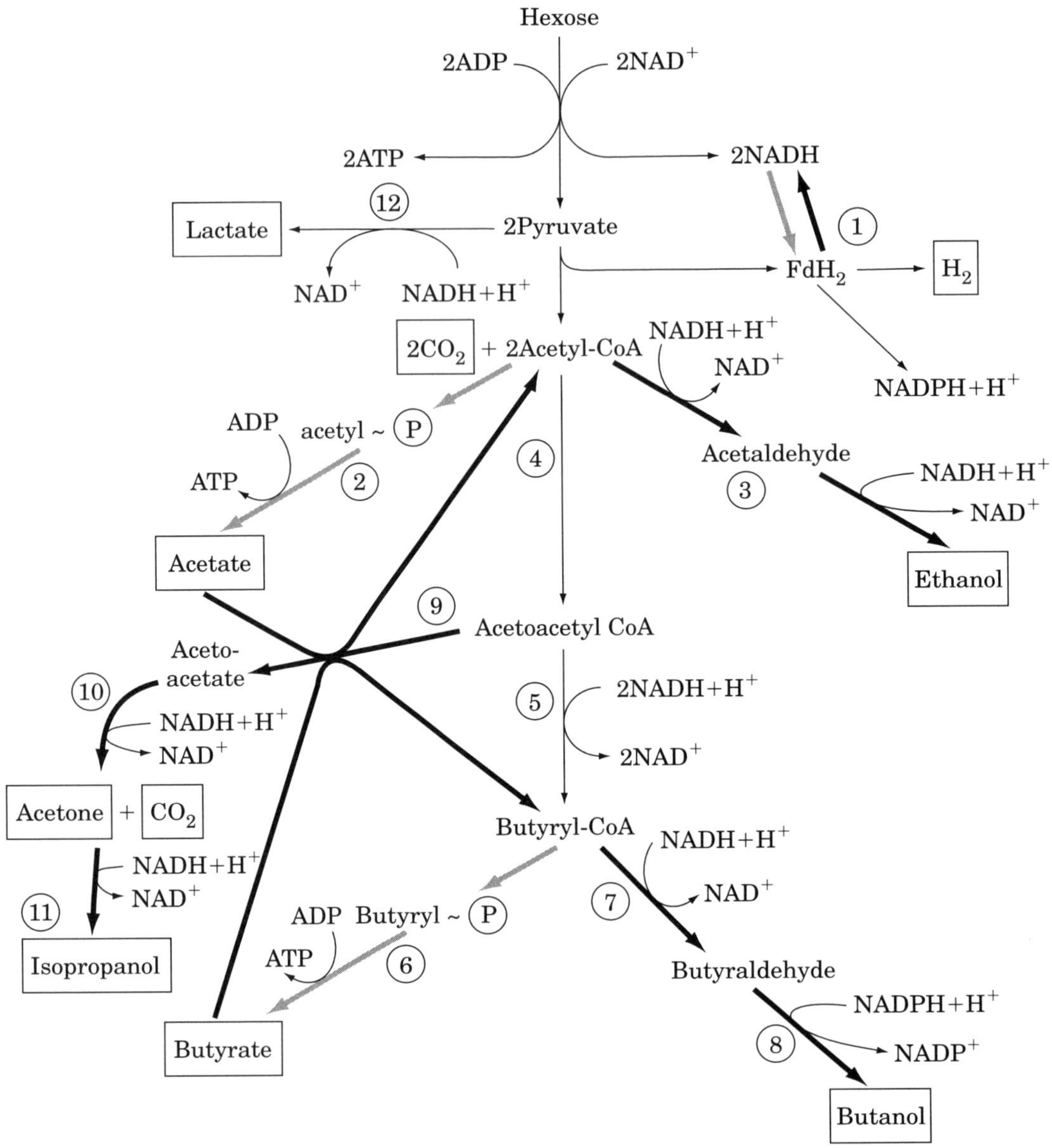

Figure 2. The butanol fermentations. Numbers indicate enzymes as follows: 1, three enzymes that exchange electrons with reduced ferredoxin (FdH_2), NAD^+: and $NADP^+$: ferredoxin oxidoreductases, and hydrogenase; 2, phosphotransacetylase and acetate kinase; 3, acetaldehyde dehydrogenase and ethanol dehydrogenase; 4, acetyl CoA acetyltransferase (or thiolase); 5, three enzymes producing butyryl CoA; 6, phosphotransbutyrylase and butyrate kinase; 7, butyraldehyde dehydrogenase; 8, butanol dehydrogenase; 9, acetoacetyl CoA: acyl CoA transferase; 10, acetoacetate decarboxylase; 11, isopropanol dehydrogenase; 12, lactate dehydrogenase. Thick gray arrows indicate reactions that predominate during acidogenesis; thick black arrows indicate reactions that predominate during solventogenesis. *Source*: Reproduced with permission from Ref. 25, p. 26.

organic solvents and acids formed is limited to only two-thirds of the substrate sugars fermented. As diagrammed in Figures 2 and 3, all fermented hexoses are funneled into CO_2, which is expelled, and acetyl CoA, which is the precursor of solvents and organic acids. Thus, only two-thirds of the substrate carbon is recovered as useful organic products. Exceptions to this one-third loss of productivity are the fermentations of sugars to produce two lactic acids by *C. thermolacticum* and the fermentation of sugars to produce three acetic acids by *C. thermoaceticum* (25).

Ethanol Fermentations

Tables 1 and 2 list three thermophilic clostridia, *C. thermohydrosulfuricum, C. thermocellum*, and *C. thermosaccharolyticum*, that produce significant amounts of ethanol together with acetate and some lactate during growth at 55–70 °C (26). *C. saccharolyticum*, first described in 1982, is a mesophile (37 °C optimum for growth) that grows on 20 different sugars and sugar alcohols including cellobiose, but it does not utilize starch or cellulose (7). In contrast,

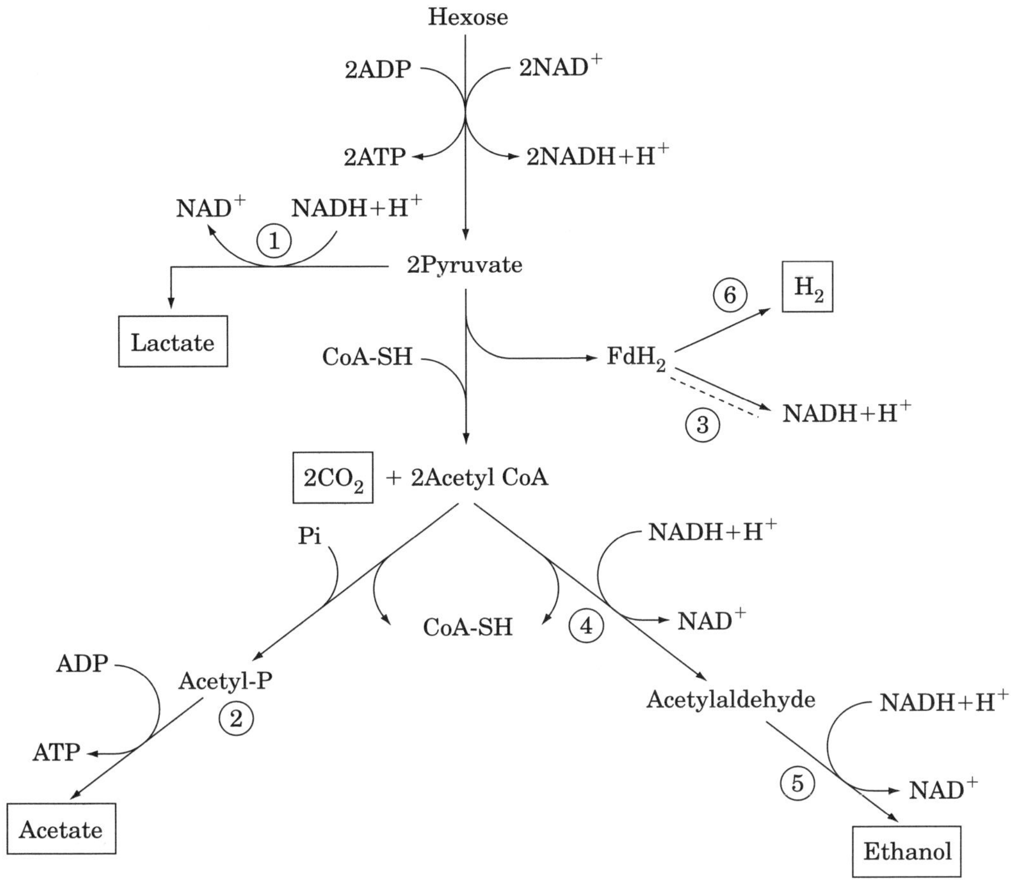

Figure 3. Clostridial ethanol fermentations. Numbers indicate enzymes as follows: 1, lactate dehydrogenase; 2, phosphotransacetylase + acetate kinase; 3, NADH: ferredoxin oxidoreductase and ferredoxin: NAD oxidoreductase; 4, acetaldehyde dehydrogenase; 5, ethanol dehydrogenase, and 6, hydrogenase. *Source*: Reproduced with permission from Ref. 25, p. 35.

C. thermocellum, which will utilize crystalline cellulose and hemicelluloses as feedstocks, is an excellent candidate for future bioconversion of cellulosic biomass to solvents. All these thermophilic clostridia utilize starch but are more limited in their ability to ferment sugars than is *C. saccharolyticum*. During the ethanolic fermentation of hexoses, the bulk of the pyruvate is converted to acetyl CoA, reduced ferredoxin (FdH_2) and CO_2 (Fig. 3). Depending on fermentation conditions, some of the pyruvate is diverted to lactate by lactate dehydrogenase. In one branch of the major fermentation pathway, acetyl CoA is reduced to acetaldehyde and then to ethanol by means of two NAD^+-linked dehydrogenases (Fig. 3, reactions 4 and 5). Acetate is formed via a second branch yielding a stoichiometric amount of ATP as in other clostridial fermentations (Fig. 3, reaction 2). Using this three-branched pathway, *C. thermocellum* produces about 0.6–1.0 mol ethanol/mol hexose, whereas the *C. thermohydrosulfuricum* and *C. saccharolyticum* fermentations yield 1.8–1.9 mol ethanol/mol hexose, which makes the latter two species attractive ethanol production organisms (Table 2).

The biochemical basis for the different product ratios produced by different species using the same branched pathway (Fig. 3) remains unknown, although the problem has been studied indirectly. Batch cultures of *C. thermosaccharolyticum* subjected to slow feeding of glucose show restricted growth, producing excess ethanol (1.0 mol/mol glucose) and little acetate or butyrate. On the other hand, rapidly growing cultures of this bacterium produce mostly acetate, lactate, and butyrate (12). Batch fermentation of cellulose by *C. thermocellum* yielded ethanol as the major product for the first two days, while lactate and acetate accumulated from days 2–5. Maintaining the pH at 7.2 rather than 6.5 favored a high ethanol/lactate ratio (9). Thus, in batch fermentation, the amount of ethanol observed depends on the pH and the time of sampling. Referring to Figure 3, it is clear that when these clostridia produce more than 1 mol of ethanol/mol of hexose fermented, a significant transfer of electrons must occur from FdH_2 to NAD^+ catalyzed by Fd:NAD^+ oxidoreductase. The role of this electron-balancing function was made clear when strain LQRI of *C. thermocellum*, which produced exactly 1:1 ethanol–acetate from cellobiose, was found to be devoid of Fd:NAD^+ oxidoreductase (27). Both strain AS39 of *C. thermocellum* and *Thermoanaerobacter brockii* contained this enzyme and could produce excess ethanol and low H_2 in experimental fermentations. Because *T. brockii* is in the same genus as *C. thermohydrosulfuricum* (5,6), it is likely that the high ethanol production by this clostridium is a result of the same electron-balancing pathways (see Table 2).

Because of the effective degradation of various crystalline cellulose sources by *C. thermocellum*, the cellulose enzyme complex has been extensively studied, as discussed later in "Biology of Substrate Utilization." To take advantage of this capability, the approach has been to develop mutants or devise both batch and continuous culture fermentation conditions that will increase the relative yield of ethanol to other products and increase tolerance of the organism to produce ethanol at 15 g/L rather than the normal 5 g/L (9). A second approach, which uses the principle of symbiotic coculture, is also available. Coculture fermentations have been developed at the experimental level to exploit the advantageous properties of two thermophilic clostridia to produce ethanol from biomass (26,28). A metabolic explanation for a stable communal coculture of *C. thermocellum* and *C. thermohydrosulfuricum* fermenting cellulose to ethanol has been presented. *C. thermohydrosulfuricum* depends on the products cellobiose and glucose formed from degradation of cellulose by *C. thermocellum*. Glucose utilization by *C. thermocellum* is inhibited by cellobiose, and both glucose and cellobiose are more rapidly utilized by *C. thermohydrosulfuricum*, which then allows some growth of *C. thermocellum*; however, the major fermentation, which is carried out by *C. thermohydrosulfuricum*, yields a high ratio of ethanol to acetate plus some lactate. Recent example cocultures of strains of these two bacteria fermenting 10 g of cellulose per liter yielded an end-product ethanol to acetate ratio of 10.5 to 4.3 in 19 hours (28). Since the thermophilic clostridia also secrete potent amylases and pullulunases, starch biomass is clearly an important feedstock for ethanol production catalyzed by these organisms, as discussed later (see "Development of New Solvent-Producing Strains").

Recently, 10 or more mesophilic cellulose-fermenting clostridia have been isolated, which degrade cellulose and ferment cellobiose to a mixture of acetate, ethanol, and other products (29). *C. saccharolyticum*, which was originally isolated from a cellulose-enriched sewage sludge, has been grown on cellulose in stable coculture with "*Bacteroides cellosolvens*" (30). Since *C. saccharolyticum* ferments cellobiose primarily to ethanol at 25 °C (Table 2), cellulose fermentation employing this strain in coculture with a mesophilic cellulolytic clostridium may form the future basis for direct bioconversion of cellulose to ethanol.

Butanol Fermentations

Very few species of the butyric acid clostridia are capable of producing significant amounts of butanol. The seven organisms listed in Table 1 have quite distinctive properties. *C. acetobutylicum*, taxon I, is the amylolytic type strain, ATCC 824^T or DMS 792; and phylogenetically it is only distantly related to the saccharolytic strains belonging to taxonomic groups II, III, and IV (4). *C. acetobutylicum*, taxon II (strains NCP 262 or NRRLB 643), contain a very small genome of 2.9 Mbp (versus 4.0 Mbp for strain ATCC 824 or 6.5 Mbp for a strain of *C. beijerinckii*, strain NCIMB 8052T) (31). Taxonomic group III, *C. saccharoperbutylacetonicum*, strain N1-4, is closely related to the other saccharolytic strains (4). In addition to *n*-butanol, all four of these strains produce butyrate, acetate, ethanol, and acetone (Table 2). Alternatively, some *C. beijerinckii* strains typically produce isopropanol instead of acetone. Even though these four organisms carry out basically the same fermentation, they are quite distinct biologically as to rapid growth and fermentation with a high butanol yield on different substrates in industrial-scale batches. The taxonomic group I strains utilize maize or potato starch effectively, whereas organisms of taxa II and III were selected to carry out a saccharolytic fermentation with a molasses substrate. *C. puniceum* carries out the same amylolytic fermentation, producing mostly butanol under the proper conditions (18). *C. puniceum* has the added ability to produce pectinolytic enzymes, which would increase access to plant starch important for some starting substrates. Taxonomic group IV, *C. beijerinckii* (NRRL B592) and *C. aurantibutyricum*, which ferment starch and a variety of sugars to butanol, ethanol, and isopropanol, have been the subjects of research but not commercial fermentations. *C. tetanomorphum* differs from the other organisms in that, during a saccharolytic fermentation, it produces equal amounts of butanol and ethanol but not acetone or isopropanol. Both the more recently described strains, *C. tetanomorphum* and *C. puniceum*, differ from the other five organisms in that they produce significant butanol and ethanol throughout exponential growth rather than only during the later stages of batch fermentations (18,19).

The earlier studies up to 1988 on the biochemistry, genetics, and development of butanol fermentation by the clostridia have been reviewed (24,32,33). Recent advances in the molecular biology, regulation of the fermentation, and other new developments are reviewed by Rogers and Gottschalk (25) and in a review of a meeting on solventogenic clostridia (34). The major enzymatic pathways for the products of butanol fermentation are now known, and the physiologic controls of the switch from an acidogenic to a solventogenic fermentation observed in some of the organisms have been identified. Figure 2 depicts the five major pathways of this fermentation. A characteristic shift in the fermentation pattern was known for some time by microbiologists running industrial fermentations employing strains of "*C. acetobutylicum*" (21,24). In a typical experimental batch culture fermenting glucose, there is an early accumulation of acids (acidogenic phase) followed by butanol formation (solventogenic phase) and a reutilization of butyrate and acetate (Fig. 4). *C. acetobutylicum* growing exponentially on sugars or starch at a pH 5.6 or greater produces butyrate, acetate, CO_2, and H_2 as major fermentation products (Fig. 2, thick gray arrows). During this acidogenic fermentation phase, the bacteria are running a basic butyrate–acetate fermentation. To maintain electron balance, for every mole of acetate formed, an extra mole of H_2 is formed, similar to the case of ethanolic fermentation. Therefore, about 2.4 mol of H_2 is produced from a fermentation of 1 mol of hexose that produces 0.4 mol of acetate and 0.8 mol of butyrate. The extra 0.4 mol of H_2 comes from transfer of electrons from NADH through NADH:Fd oxidoreductase and hydrogenase to H_2 (Fig. 2, reaction 1).

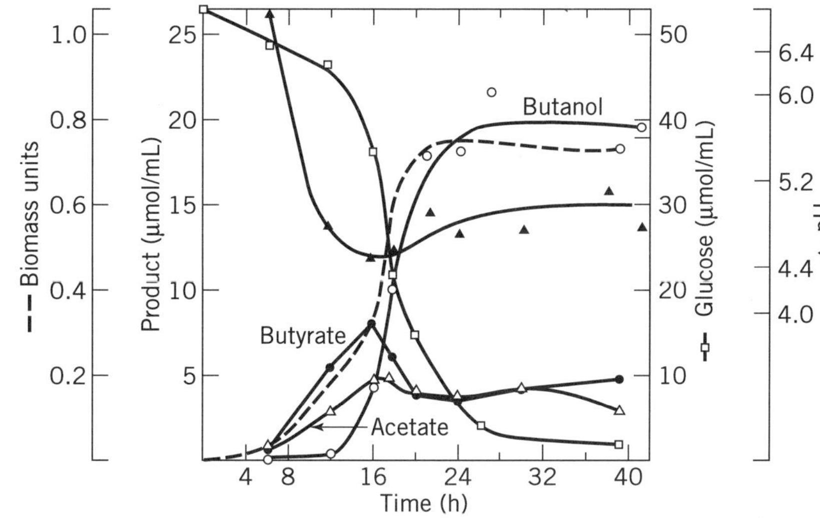

Figure 4. The switch from acidogenic to solventogenic fermentation by *C. acetobutylicum*, strain NRRL B643, in batch culture. *C. acetobutylicum* was grown in a yeast extract, amino acid medium at 37 °C with glucose as an energy source. Samples withdrawn at times shown were analyzed for biomass (dashed line), glucose (open squares), and pH (solid triangles); products in the broth were determined using gas chromatography: acetate (open triangles), butyrate (solid circles), and butanol (open circles). Acetone and ethanol determinations are omitted. *Source*: Reproduced with permission from Ref. 24, p. 6.

With accumulation of acids in a batch culture, the pH drops to pH 4.0–4.5 (depending on the strain), the growth becomes linear, then stops, and the classical switch in the fermentation occurs. The triggering of solvent formation requires the induction of new enzyme pathways in the cells (Fig. 2, thick black arrows) catalyzing formation of butanol, acetone, and ethanol. The connection among the processes of organic acid accumulation, drop in pH, slowing of growth, and induction of solvent-forming enzymes is addressed in the next section. During this phase, there is a net uptake of butyrate and acetate from the fermentation beers back into the cells (Fig. 4) and their recycling and conversion to butanol or ethanol (Fig. 2, arrow, reaction 9). Finally, acetone or isopropanol is produced as a result of decarboxylation of acetoacetate (Fig. 2, reactions 10 and 11).

Depending on the clostridial strain, one or two CoA-dependent aldehyde dehydrogenases and two or three alcohol dehydrogenases are needed for butanol, ethanol, and isopropanol formation, and some of these enzymes are induced 40 to 200-fold at the shift (Fig. 2, reactions 3, 7, 8, and 11). Three additional changes occur as a result of the shift from acidogenesis to solventogenesis:

There is a lower ATP yield per mol of glucose fermented because net accumulation of acetate and butyrate allows additional production from acetyl phosphate and butyryl phosphate during exponential growth (Fig. 2, reactions 2 and 6).

Operation of the alcohol pathways requires 2 mol additional NADH per mol of alcohol produced.

The amount of H_2 made is reduced to about 1.4 mol of H_2 per mol of glucose fermented, since electrons must be shunted from FdH_2 to NADH (Fig. 2, reaction 1).

These changes in electron flux may be involved in regulatory signals for promoting the induction of the shift. This is discussed later ("Regulation of Solvent Production").

Acetone and eventually isopropanol production requires two enzymes: acetoacetyl–CoA transferase (ACT) and acetoacetate decarboxylase (Fig. 2, reactions 9 and 10). In *C. acetobutylicum* and *C. beijerinckii* strains, these enzymes are induced 50- to 100-fold during the shift to solventogenesis. The production of acetone by many of the solvent-forming clostridia is biologically interesting because the pathway neither produces extra ATP nor acts as an electron sink. Even a reduction of acetone to isopropanol by *C. beijerinckii* strains is oddly inefficient, requiring only one NADH, while reduction of acetoacetyl CoA to butyryl CoA recycles two NADHs (Fig. 2, cf. pathways 9–11 and 5). The best guess is that the physiologic role of this pathway is recycling of accumulated organic acids for detoxification and pH control. Thus, after induction of these enzymes, acetate and butyrate are taken up by the bacteria, the ACT enzyme transfers CoA from acetoacetyl–CoA to these acids to form their acyl–CoA intermediates (Fig. 2, reaction 9), and then they are funneled through the alcohol pathways to butanol and ethanol. The decarboxylase pulls the reaction, forming CO_2 and acetone (Fig. 2, reaction 10), or is reduced to isopropanol (Fig. 2, reaction 11) and excreted. During the solvent production phase, reduction in concentration of acids in the medium occurs early, while acetone continues to accumulate long after, when there is no net increase in acids. It appears that in these clostridia, acid production and acid recycling occur throughout solvent formation. High levels of the butyrate pathway enzymes are present during this period, and added [^{14}C]butyrate is rapidly recycled to [^{14}C]butanol continuously. In contrast, *C. tetanomorphum*, which produces both butanol and butyrate simultaneously throughout the fermentation of sugars, contains no ACT enzyme, produces no acetone, and does not recycle [^{14}C]butyrate to [^{14}C]butanol at any time (25).

Other Solvent Products

The butanol-fermenting clostridia can be forced to produce several alternative solvents by growth under special conditions with specific carbon sources. For example, growth of *C. beijerinckii* in a chemostat culture at pH 6.5–6 with glycerol as a carbon source yielded 61% conversion to 1,3-propanediol (35). Glycerol fermentation by 17 of 21 solvent-producing clostridia produced 1,3-propanediol. *C.*

acetobutylicum and *C. beijerinckii* grown in defined media with rhamnose, with a gas phase (CO_2) plus 0.2% $NaHCO_3$, yielded 1,2-propanediol, propionic acid, and propanol as major fermentation products. Three strains of *C. thermosaccharolyticum* ferment glucose yielding up to 7.9 g/L of *R*(–)-1,2-propanediol with a best yield ratio of 0.27 g per gram of glucose. *C. sphenoides* normally produces lactate, acetate, and ethanol from fermentation of glucose. During phosphate limitation, this organism yields up to 34 mM D(–)-1,2-propanediol per 100 mM glucose fermented. New fermentation pathways are probably induced under these conditions. However, the biochemistry has not been carefully studied (see Ref. 25 for specific references).

Another strategy for unusual solvent production uses the principle of enzyme recruitment. It has been demonstrated that *C. acetobutylicum*, grown until it has shifted into the solventogenic phase, can convert several organic acids to their analogous alcohols when added to the growth medium through the recycle mechanism outlined earlier. For example, propionate, valerate, and 4-hydroxybutyrate are taken up by this bacterium and converted to *n*-propanol, *n*-pentanol, and 1,4-butanediol, respectively (25).

REGULATION OF SOLVENT PRODUCTION

Physiology of Control of Solvent Formation

To obtain a high yield of a specific product in a fermentation for solvents, it is often necessary simply to adjust the growth medium, carbon source, and other conditions that will ensure maximum amounts of the desired solvent and minimize production of other products. However, research on the cellular regulatory mechanisms that govern the fermentation not only will allow more effective control of culture conditions but will open the door to development of new highly productive strains. Because many clostridial fermentations typically depend on branched fermentation pathways, analysis of the control elements becomes particularly important. Because most of the recent research is centered on the *C. acetobutylicum* butanol fermentation, this work is emphasized here, although a few observations from the clostridial ethanol fermentations are mentioned. The pathways, products, and characteristics of the acidogenic and the solventogenic fermentations of *C. acetobutylicum* were presented in the preceding section. The culture conditions that promote the shift away from acid production and favor optimal butanol and acetone production are listed in Table 3. These conditions are about the same for batch or continuous cultures of *C. acetobutylicum* (24,32,33) and of *C. puniceum* (18). The conditions that favor butanol and ethanol production without acetone (the alcoholic fermentation) by continuous cultures of *C. acetobutylicum* (36) are also listed in Table 3. The internal status of the bacterium has been measured for each of the three different fermentations (36–38). During the acidogenic fermentation in batch or continuous cultures, the following conditions are required: adequate glucose, a pH_o greater than pH 6.0; and a low level of organic acids. The intracellular pH (about pH_i 6.8) is maintained close to neutral either by excretion of H_2 through the action of hydrogenase or by hydrolysis of ATP and pumping H^+ out of the cell. At a pH_i 6.8 the cell maintains a low positive ΔpH and a very low level of undissociated butyric acid. ATP is produced and utilized at a high rate in this fermentation. The low NADH/NAD ratio is probably a result of the high NADH:Fd oxidoreductase activity, whereby excess electrons are dumped onto FdH_2 and exit through hydrogenase as molecular hydrogen. H_2 is produced in excess in the acidogenic fermentation (Table 3 and Fig. 2, reaction 1, gray arrow).

The shift to a solventogenic fermentation requires an adequate carbon source (glucose), an acid pH_o (pH 4.8–4.4), and high concentrations of butyrate and acetate in batch cultures. Continuous chemostat cultures limited for PO_4 or nitrogen source can be forced to carry out a solventogenic fermentation either by adding a high concentration of organic acids or by setting the pH_o to pH 4.8–4.3, depending on the strain of clostridium (36,37). The intracellular pH (pH_i 5.9) is more acid and results in a greater than 10-fold higher concentration of undissociated butyric and acetic acids in the cell. Also a very high positive ΔpH of 1.1–1.5 is maintained by these cells. Curiously, this high ΔpH persists even when solvent-producing cells are treated with *N,N'*-dicyclohexylcarbodiimide (an ATPase inhibitor) or after CO gassing (a hydrogenase inhibitor) (36). A high concentration of ATP is maintained, and turnover of ATP is slowed in the cells, because growth is slower during solvent production. The NADH/NAD ratio is maintained similar to acidogenic cells, even though production of butanol results in new sites for oxidation of NADH. This is due to a net transfer of electrons from FdH_2 as mentioned earlier. This physiologic state is somehow essential for the induction of the enzymes required for solvent fermentation (Fig. 2, reactions 7–10).

The alcoholic fermentation, in which high amounts of butanol and ethanol are produced without acetone, has been studied only recently (36). The alcoholic fermentation in continuous cultures requires high glycerol and limited glucose, at a neutral pH. Table 3 lists three inhibitors of hydrogenase activity that, when added to continuous cultures with adequate glucose, also force cells to produce butanol, butyrate, and ethanol, but no acetone (37). These conditions cause an overload of NADH in the cell, resulting in a high NADH/NAD ratio and various levels of ATP (Table 3). The high level of NADH leading to butanol and ethanol formation is accompanied by increased activities of NADH-dependent alcohol and aldehyde dehydrogenases, which differ from the NADPH-dependent dehydrogenases that are expressed at high levels during solventogenesis in the bacterium. For example, the NADH-dependent butyraldehyde dehydrogenase activity is 100-fold higher in alcohologenic cells, while the NADPH-dependent enzyme activity is induced in solventogenic cells. This suggests that different sets of genes are expressed in these three fermentation conditions (36). The inversion of the transmembrane pH gradient ($-\Delta pH$ 0.3) observed during the alcoholic fermentation at neutral pH (Table 3) is apparently due to the low rate of H_2 production. But this inversion does not seem to create an energy crisis for the bacteria, since a similar proton motive force (pmf) was measured for both alcohologenic and acidogenic cells (36). An increase in the electrical potential ($\Delta\Psi$) compensates for the low chem-

Table 3. Conditions for Three Fermentations by *Clostridium acetobutylicum*

	Acidogenic	Solventogenic	Alcoholic
Intracellular status			
$[H^+]$	pH_i 6.8 ΔpH 0.3–0.6	pH_i 5.9 ΔpH 1.1–1.5	pH_i 6.2 ΔpH − 0.3
Undissociated butyrate, acetate	Low concentration	High concentration	Low concentration
ATP (μmol/g cell mass)[a]	Low (0.62), High turnover	High (1.6), Slow turnover	Variable (0.3–1.3), slow turnover
NADH/NAD (ratio)[a]	Low (0.14)	Low (0.1–0.2)	High (0.5–0.8)
Enzymes	Acid pathways (2 and 6 in Fig. 2) NADH:Fd oxidoreductase very high (gray arrow at 1, Fig. 2)	Solvent pathways, induced: butanol (7 and 8 in Fig. 2), acetone (9 and 10 in Fig. 2)	Alcohol pathways, induced: NADH-dependent alcohol and aldehyde dehydrogenases, Fd:NAD $(P)^+$ reductases
Major products	Butyrate, acetate, excess H_2	Butanol, acetone, low H_2	Butanol, ethanol, very low H_2
Fermentation conditions	Adequate glucose: pH_o 6.5–6.2; low $[\text{organic acid}]_o$	Adequate glucose: (1) pH_o 4.8–4.4, high $[\text{organic acids}]_o$; (2) Limited PO_4 either neutral pH at high $[\text{organic acids}]_o$ or pH 4.3 at low $[\text{organic acids}]_o$	(1) Glycerol and low glucose, pH_o 6.5; (2) Adequate glucose, pH_o 6.5–5.6 + neutral red (1 μM) or + CO gassing or + methyl viologen (4 μM) or limited Fe

[a]Data from Ref. 21.

ical potential (ΔpH). This may be related to the distribution of K^+ across the membrane.

Thus, in *C. acetobutylicum* each of the three fermentation conditions produces a distinct intracellular state, which in turn causes the bacterium to develop the enzyme network needed for each type of fermentation. The current view of physiologic control of *C. acetobutylicum* metabolism is a dual signal model in which a high internal ATP and/or butyric acid level generates one signal and a high NADH level provides the other (37). But how these internal physiologic signals are sensed by molecular sensors in the cell and transferred to molecular responders that specifically regulate gene expression by *C. acetobutylicum* is still unknown.

Genetics of Solvent Formation

Mutations That Affect Solvent Production and Its Regulation. Isolation and characterization of mutants of the solventogenic clostridia has provided valuable information on gene function and regulation of the solvent-producing pathways and of the linkage of this process with developmental changes in the bacterial cell (39). During industrial production of acetone and butanol by *C. acetobutylicum*, the early brewmasters recognized that high solvent-producing strains produced a high percentage of spores. Thus, the high solvent-producing strains were selected by subjecting cultures to cycles of sporulation, heat activation, and outgrowth (32). Mutants that fail to sporulate often also fail to produce solvents. These mutant cells tend to accumulate when *C. acetobutylicum* and related clostridia are repeatedly subcultured in batch cultures or grown in continuous culture, rather than starting cultures from fresh heat-activated spores. When the culture loses the capacity to produce solvents and form spores, this change is called "strain degeneration." Some strains, such as *C. beijerinckii* (NCIMB 8052), tend to degenerate more rapidly than others. In contrast, *C. acetobutylicum* (ATCC 824) will often give rise to asporogenous cells that continue to form acids or solvents (40). Recent studies on the phenomenon of clostridial strain degeneration have been reviewed (41). No satisfactory molecular explanation for the permanent degeneration of clostridial cells was available until just recently. In *C. acetobutylicum*, strains ATCC 824 and ATCC 4259, a large 210-kb plasmid (pWEIZ or pSOL-1) has been demonstrated that carries four solvent-forming genes of the *sol* operon, *ctfA*, *ctfB*, *adc*, and *aad* (see Fig. 5) (42). Degenerate strains that produce no butanol or acetone have lost the pSOL-1 plasmid. Since degenerate strains lacking pSOL-1 are also unable to sporulate or carry out any of the coordinate developmental functions, it is likely that major regulatory genes controlling development are located on this plasmid as well. A similar link between the induction of endospore formation and a fermentation shift occurs in *C. thermosaccharolyticum*. Exponentially growing cells produce acetate, butyrate, and lactate. A change in culture conditions produces elongate forms that ferment sugars to primarily ethanol, CO_2, and H_2 before the cells proceed to sporulate. Asporogenous mutants were selected that remained in the intermediate stage and continued to produce ethanol (43).

Mutants of solvent-forming clostridia that have defects in the pathways leading to acid production have been isolated to bias the carbon flow toward solvent production. Following mutagenesis, pathway mutants are often selected through the use of suicide substrates, which are toxic analogs of normal end products formed during the fermentation. For example, allyl alcohol acts as a suicide substrate when oxidized to the toxic aldehyde, acrolein, by the activity of alcohol dehydrogenases. Members of one set of allyl alcohol resistant mutants of *C. acetobutylicum* are

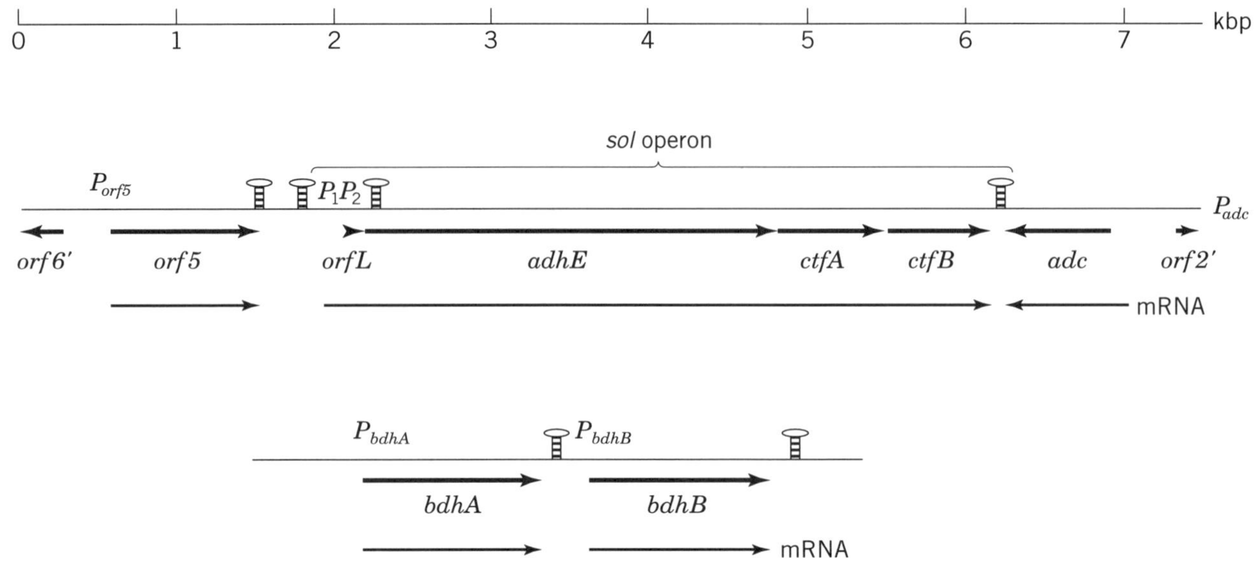

Figure 5. Organization of chromosomal gene regions encoding solventogenic enzymes in *C. acetobutylicum*, strains DSM 792 and ATTC 824. Arrows and arrowheads represent lengths, locations, and orientation of genes; lines with arrows indicate primary mRNA transcripts. The *sol* operon is marked off by a dashed line. A prime at the end of an open reading frame (ORF) indicates that the ORF is truncated. Promoter positions are indicated by P_{adc}, P_1, P_2, P_{bdhA}, P_{bdhB}, and P_{orf5}; possible stem–loop structures are indicated by hairpin symbols. *Source*: Reproduced with permission from Ref. 51, p. 253.

defective in butanol and ethanol production because of altered alcohol dehydrogenase enzymes. However, these mutants contain normal butyraldehyde dehydrogenase activity and produce significant amounts of butyraldehyde (15) (see Fig. 2, reactions 7 and 8). Fluoroacetate-resistant mutants of *C. thermosaccharolyticum* were found to be deficient in acetate kinase or phosphotransacetylase enzyme activities and thus produce less acetate and more ethanol than the parental strains (see Fig. 3, reaction 2) (44). An asporogenous ethanol-tolerant mutant of *C. thermocellum* (LD 1) was selected by means of fluoropyruvate. When fermenting cellobiose, this mutant strain produced a high ratio of ethanol to acetic acid, in contrast to the parental strains, which produce mostly lactic acid and acetic acid (9). Mutants of *C. acetobutylicum* (B643) and *C. beijerinckii* (NCIMB 8052) that produce excess butanol and acetone and more importantly also recycle virtually all acetate and butyrate have been isolated (15,16). These super recycling strains may be used in future production methods covered in a later section "Development of New Solvent-Producing Strains").

Conjugative Transposons and Transfer of Plasmid Vectors. Conjugative transposons are a class of mobile genetic elements usually found as residents in the chromosomes or plasmids of gram-positive bacteria (see reviews, Refs. 45,46). Originally discovered in *Enterococcus faecalis*, these elements have a broad host range and are important tools for genetic analysis. These mobile elements can integrate almost at random into the chromosome of the recipient bacterium, therewith creating an insertion that disrupts the sequence of the DNA information in a specific gene. Thus, the transposon mutagenesis differs from mutagenesis due to physical or chemical means in that only one specific alteration in the host cell DNA occurs, rather than a possibility of two or more random changes in the DNA that obscure the alteration in the phenotype and gene under study. Genetic techniques ensure that only one copy of the transposon has been integrated into the cell for each mutational event. The gram-positive bacterium *Enterococcus faecalis* is the usual donor in matings with *Clostridium*, yielding a transfer frequency of 10^{-3} to 10^{-4} per recipient. The transferred transposon, Tn916 (16.4 kbp), Tn925 (14.5 kbp), or Tn1545 (23.5 kbp), carries an antibiotic resistance marker such as tetracycline resistance (*tetM*) that allows selection of the transconjugants. Because the genes required for the conjugation and DNA transfer are located within the transposon, a transposon carried either on a plasmid or on the chromosome of the donor will be transferred. A number of specific Tn916-induced mutants have been identified in *C. acetobutylicum*. Both structural gene mutants in the fermentation pathway and regulatory mutants preventing solventogenesis, sporulation, and granulose formation have been selected (45,46).

Development of methods for transfer of plasmid vectors into *Clostridium* is vital to the application of genetics to further our understanding of the molecular basis of regulation of solvent formation. A summary of these transformation methods and the clostridial cloning vectors used with *C. perfringens* and *C. acetobutylicum* is available (48). There are two effective methods of transformation of *C. acetobutylicum*. Transformation of protoplasts requires conversion of cells to wall-less protoplasts followed by uptake of plasmid or bacteriophage DNA induced by polyethylene glycol (PEG), regeneration back to rod-shaped cells, and selection of transformed cells. Using this procedure with strain N1-4 of *C. saccharoperbutylacetonicum*, 10^4–10^5 transformants per microgram of plasmid or phage DNA are obtained. Transformation of whole cells requires suspension of washed cells together with plasmid DNA, electroporation in a cuvette at high voltage (5.0 kV/cm) using a BioRad Gene Pulser, and, after a recovery period of 2–4 hours, selection for transformants. When *C. beijerinckii* (NCIMB 8052) is used, about 10^2–10^4 transformants are obtained per microgram of DNA. Basically, the short pulse of electric current punches temporary holes in the cell membrane, permitting the plasmid DNA to enter the cell. To be useful and identifiable, the plasmid DNA must be able to survive and replicate along with cell division of the host bacterium. In addition, there must be selectable markers such as the *ermC* gene, for antibiotic (erythromycin) resistance, so that colonies of transformed cells can be identified.

Vectors for *C. acetobutylicum* are chimeric cloning vectors that contain an origin of replication that operates in *Clostridium* and a second origin of replication that permits the plasmid to replicate in *E. coli*. Thus, specific pieces of *Clostridium* DNA can be ligated (cloned) into specific sites of this vector, replicated and multiplied in *E. coli*, extracted, and then transferred to mutant *Clostridium* cells to study their effects on the physiology of the host cell. This technique has permitted identification of the roles of suspected regulatory genes and overexpression of specific fermentation pathways in *Clostridium acetobutylicum* (49).

Unfortunately, the successful transformation process is strain specific, as is plasmid survival. Originally, transformation procedures were successful only when *Clostridium* strains in taxa III and IV (see Table 1) were used. Later, electroporative transformation of *C. acetobutylicum* (ATCC 824, taxon I) with plasmid pIM13 from *Bacillus subtilis* was successful. But when a shuttle plasmid was prepared by ligating pIM13 together with *E. coli* plasmids, no transformants were seen. It was discovered that strain ATCC 824 contains a restriction endonuclease (*Cac* 824 I) that cut these shuttle vectors in more than 40 sites, thus destroying the plasmid in the cell. The strain 8052 (taxon IV) does not carry this restriction endonuclease, which explains why the same shuttle plasmids survive in these bacteria (49). The clever solution to survival of these shuttle vectors in the *C. acetobutylicum* (ATCC 824) host requires replication of the shuttle plasmid in *E. coli* in the presence of a second plasmid that expresses a specific transmethylase. This Φ3TI transmethylase adds a methyl group to a cytosine base in the sensitive sites of the shuttle vector DNA. This protects the plasmid DNA, and when extracted and electrotransferred into *C. acetobutylicum* (ATCC 824), the methylated plasmid survives at a high frequency (10^5 transformants/μg DNA) and replicates (49). To date, the important production strains of *C. acetobutylicum*, NRRL B643 or NCP 262 (taxon II), have not been successfully transformed by means of the methods and vectors available. Perhaps there are additional restriction endonucleases that act as barriers to vector survival in these strains.

Cloning, Genetic Organization, and Expression of Solvent and Acid Pathway Genes. Application of molecular genetic techniques together with basic enzymology to the steps in the *C. acetobutylicum* fermentation pathway has revealed a great deal about how this network is organized and regulated. Most of the genes coding for the enzymes involved in acid and solvent fermentation have been cloned in *E. coli*, since this organism expresses many foreign proteins from plasmid-borne genes on specially constructed expression vectors (50). Sequencing of the DNA regions coding for the solvent genes together with the adjacent upstream promoter regions has revealed the basic organization of these genes (Fig. 5, Ref. 51). In strains ATCC 824 and DMS 792, acetoacetate decarboxylase is encoded by a gene (*adc*) in a single operon. Adjacent is the convergently transcribed *sol* operon, which encodes an aldehyde/alcohol dehydrogenase (*adhE*), and the two subunits of acyl CoA transferase (*ctfA* and *ctfB*). The promoters P, P_2, and P*adc*, all have high homology to typical gram-negative and gram-positive bacterial promoter sequences and control expression of these genes. A typical terminator sequence (Fig. 5, hairpin symbols) between the two regions stops transcription from both directions. This entire region of the DNA that includes the *sol* operon and other linked genes such as *adc* is carried on the 210 kbp plasmid (pSOL-1) in *C. acetobutylicum* (ATCC 824), along with other yet undetermined regulatory genes. The remaining genes coding for solvent- and acid-forming enzymes appear to be carried on the chromosome in this strain.

At present, the distribution of the solvent genes in the other six groups of butanol-producing clostridia is not known. The DNA sequences reveal that these solvent genes as well as the genes encoding for butyrate production are transcribed by the RNA polymerase with sigma H (σ^H) that carries out expression of many housekeeping genes in gram-positive bacteria. Alternative sigma factors do not appear to play a role in regulation of solventogenesis in *C. acetobutylicum* (51).

The expression of a synthetic acetone operon carried on a shuttle plasmid (pFNK6) and transferred back into *C. acetobutylicum* (ATCC 824) reveals some additional properties of regulation of solvent production (49). The synthetic acetone operon was constructed from the *adc*, *ctfa*, and *ctfb* genes arranged to transcribe in the same direction from the P*adc* promoter (see Fig. 5). In cell extracts from strain ATCC 824 (pFNK6), the activities of the two acetone enzymes were more than fourfold higher than in cell extracts of strain ATCC 824. The course of the batch fermentation by ATCC 824 (pFNK6) differed from ATCC 824, with almost double the yield of both butanol and acetone and the complete disappearance of butyrate and acetate from the broth by 20 h (Fig. 6). The increase in butanol is consistent with the view that diversion of acetoacetyl CoA to acetone formation and away from butyryl CoA might cause a temporary high rise in NADH/NAD ratio and trigger early induction of the enzymes for butanol synthesis (see Fig. 2, reactions 5, 9, and 10, and 7 and 8). Two "degenerate" mutants of ATCC 824, namely M5 and DG1, which had lost the pSOL plasmid and produce none of the enzymes encoded by the *sol* operon, were transformed with plasmid pCAAD containing the *adcE* gene or with pFNK6 containing the synthetic acetone operon D (42). The degenerate mutants with pCAAD produced significant butanol and butyric acid but no acetone. The same degenerate mutants with pFNK6 produced butyric acid and acetone but

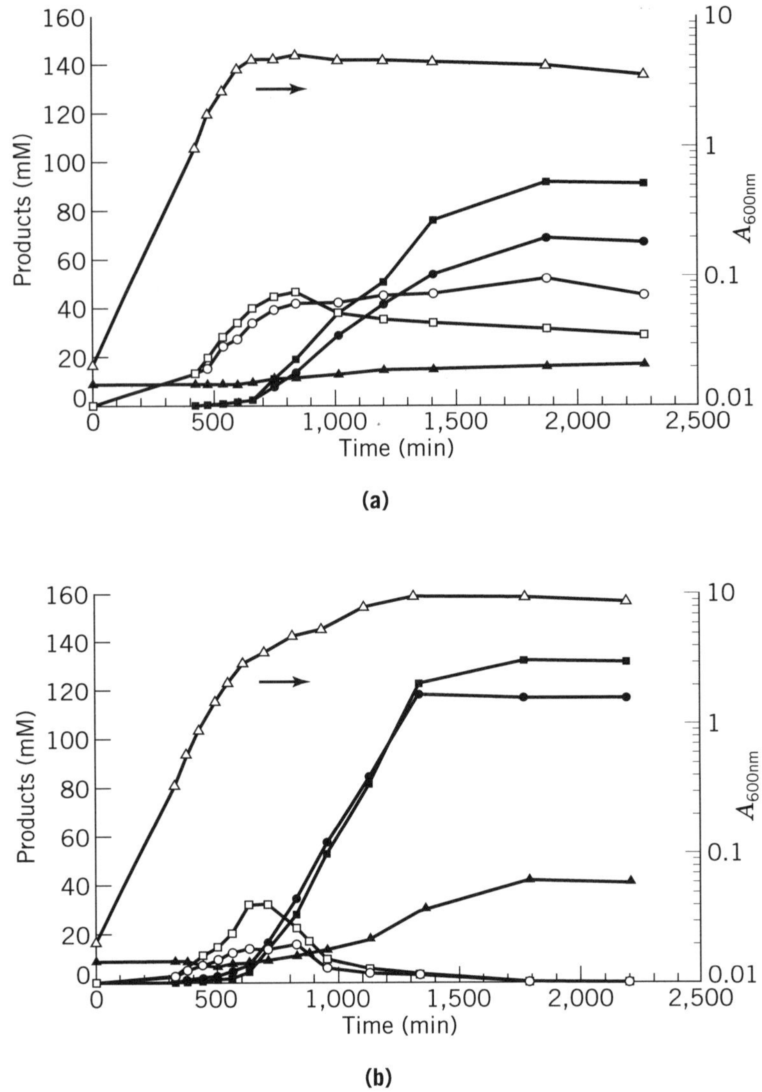

Figure 6. Optical density and product profiles in pH 4.5 batch fermentations for *C. acetobutylicum* strains: (**a**) ATCC 824 and (**b**) ATCC 824 (pFNK6). Open triangles, optical density at 600 nm; solid squares, butanol; solid circles, acetone; solid triangles, ethanol; open squares, butyrate; open circles, acetate. *Source*: Reproduced with permission from Ref. 49, p. 62.

no butanol. The butyraldehyde dehydrogenase (BYDH) activity was produced only in late growth phases in M5 (pCAAD), as is typical for strain 824 (50). M5 cells alone produced no butanol or BYDH activity. Thus the regulatory signals for the *sol* operon must operate "in trans" and be able to control multiple copies of the plasmid-borne genes for solvent production.

Chromosome Mapping and Gene Sequencing. When gene transfer methods and a number of cloned genes became available, construction of a genetic map of the *C. beijerinckii* chromosome was begun to expand genetic analysis of a solventogenic clostridia. About 32 genes have been positioned relative to one another on the 6.5-Mbp circular chromosome map (31). The process of mapping depended on preparation of the genomic DNA by lysis of cells within agar blocks, treatment with exonucleases that cut this DNA in just a few places, and then pulsed-field gel electrophoresis. The large chromosome pieces displayed on the gel are than probed with cloned genes to position each gene. As indicated earlier ("Butanol Fermentations"), this method has revealed wide divergence in strains of *C. acetobutylicum* (4). Other differences in arrangement of solvent genes and regulatory genes may emerge later.

The complete genomic DNA sequences of a number of Eubacteria and of Archaea, such as *Methanococcus jannaschii* are now available (52). The complete DNA sequences of the genome and the pSOL-1 plasmid of *C. acetobutylicum* (ATCC 824) are currently being determined (53). These sequences should provide valuable information on the genetic basis for expressing the proteins and regulatory networks essential for cellular vegetative and developmental changes carried out by this organism. Future successful strain development employing *Clostridium* will depend on complete analysis of the genetic potential of these organisms.

Insertional Mutagenesis with Integrational Plasmids. Recently, it has been possible to develop gene transfer involving homologous recombination in the solvent-producing strains *C. beijerinckii* and *C. acetobutylicum* (ATCC 824). Conjugation procedures with *C. beijerinckii* and electrotransformation with *C. acetobutylicum* were used to obtain transconjugants or transformed cells at high efficiency, leading to the detection of the integration of plasmids that lack a replication function active in *Clostridium* (Rep^-). Transformants or transconjugants arise by a Campbell-type recombination event between the chromosomal segment in the plasmid and the same region in the bacterial chromosome (31). This insertional event results in an altered or inactivated gene. Nonreplicative (R^-) plasmid constructs containing parts of the acetate kinase gene (*ack*) or butyrate kinase (*buk*) gene fragments were integrated into normal regions of the chromosome of *C. acetobutylicum*. The altered strains produced more butanol during fermentation of sugars (54).

Is Solvent Production a Cellular Stress Response?

Endospore Formation and Solventogenesis. The ultimate response of a number of bacteria to physical or chemical stress or starvation is formation of stress-resistant, metabolically inactive resting cells. Species of the gram-positive *Bacillus* and *Clostridium* differentiate into highly resistant endospores. The biology of endospore formation, including the molecular events that take place during this developmental process, has been studied primarily in *B. subtilis* by many microbiologists over the past 30 years (55). The physiologic changes inside the cell that signal the shift of *C. acetobutylicum* to solventogenesis are most likely either a shift in the redox state of elements in the electron transfer system, such as NADH or reduced Fd, or an increase in the concentration of butyric acid or H^+, or maybe both, as suggested earlier (see "Physiology of Control of Solvent Formation"). In this bacterium, the shift to solventogenesis is associated with other rather massive changes, which include induction of endospore formation and must involve the induction and activation of complex regulatory networks, proteins, and pathways (25). What are the molecules that receive the physiologic signals, and how are all these events triggered in this bacterium? As mutant studies cited earlier ("Genetics of Solvent Formation") indicate, initiation of solvent production and sporulation are tightly coupled, although clearly solventogenesis is not a prerequisite to sporulation in *C. acetobutylicum, C. beijerinckii*, and *C. thermosaccharolyticum*. Hence there must be both separate and common control mechanisms. Two lines of recent research support the view that these bacteria are regulated by two classes of the molecular signaling systems that normally operate in bacteria for stress control and adaptation to environmental change.

First, in many bacteria, coordinate transcription of sets of genes is regulated by alternative sigma factors that confer new promoter specificities on RNA polymerase. Examples are heat shock, or stationary phase in *E. coli*, sporulation in *Bacillus* and streptomycetes, and flagella synthesis in many bacteria. Development of endospores in *Bacillus* requires a temporal sequencing of gene expression regulated by four sigma factors (σ^F, σ^E, σ^G, and σ^K) and a protease (*spo* II GA) that activates the σ^E protein. All the genes encoding these sigma factors and the protease have been identified in *C. acetobutylicum* by cloning and sequencing (56). Thus, even though the physiologic signals for sporulation in *Bacillus* (nutrient starvation) are different from those for *Clostridium* (blocking of growth in the presence of excess nutrients), they utilize the same general regulatory features for development. As pointed out above, σ^H the sigma factor that directs RNA polymerase to general cellular promoters in *C. acetobutylicum*, similar to σ^A in other bacteria, apparently also directs the expression of solvent and acid enzyme genes in this bacterium. Therefore, it is clear that the genes concerned with solventogenesis (*sol* genes) are not part of the sporulation cascade. Indeed, a different mechanism coordinates sporulation and solventogenesis, along with other key physiological and morphological changes that occur during the metabolic "switch."

The second major signal system required for endospore formation is the phosphorelay pathway essential for initiation of sporulation (57). In *B. subtilis* a number of response regulator proteins respond to environmental, cell cycle, and metabolic signals, resulting in the phosphorylation of the SpoOA transcription factor. The SpoOA~P controls the transcription of numerous genes by activating or inhibiting the levels of other transcription regulators, which in turn regulate transcription of proteins such as sigma factors required to enter the sporulation cycle. The SpoOA protein acts as a transcriptional regulator both as a repressor of transcription and as an activator by binding to DNA sequences called the "OA boxes" (5-TGNCGAA-3) (57). Only the phosphorylated form, SpoOA~P, is active. PCR primers of portions of SpoAO averaged from sequences of four bacilli, were used to isolate SpoOA segments from the DNA of 11 other species of *Clostridium* and *Bacillus* including *C. acetobutylicum, C. beijerinckii*, and *C. pasteuranium*. The complete sequence of the *spoOA* gene from *C. beijerinckii* has been determined and, as in the case of two *Bacillus* strains, the promoter region of the gene has a σ^H promoter with an OA box (58). This is consistent with the autoregulation of *spoOA* by SpoOA~P, first shown in *B. subtilis*. Further, possible "OA box motifs" were found in all the promoter regions for the solvent genes *adc, ctfAB, adhE, bdhA*, and *bdhB* (see Fig. 5) and for the genes *ptb-butk* of *C. acetobutylicum* (31). Expression of these genes is either activated or repressed during the switch to solventogenesis.

Thus, in *C. acetobutylicum* (ATCC 824), direct control by SpoOA is likely. This view was tested in *C. beijerinckii* (NCIMB 8052) by means of the insertional mutagenesis technique described earlier ("Genetics of Solvent Formation"). An R^- (unable to replicate) integrational plasmid containing a DNA segment in the *spoOA* coding region was prepared and transferred into the cell. Recombination occurred by the Campbell-type mechanism, yielding an insert of the plasmid with its antibiotic resistance marker (Em^R) within the chromosomal *spoOA* gene inactivating its expression. The $SpoOA^-$ strain was asporogenic, unable to produce granulose, and unable to produce solvents (31). This indicates that all three of these "switch" responses are most certainly regulated by the phosphorelay system.

Heat Shock Response System in Clostridium. Exposure to a heat shock of all organisms, from bacteria to humans, results in the rapid burst of synthesis of from 15 to 40 proteins. This transient upshift in synthesis of this set of heat shock proteins is also triggered by exposure to toxic chemicals, virus infection, ethanol, and other agents. The major proteins induced include two important categories. First to be induced are the chaperone proteins such as GroES, GroEL, DnaK, DnaJ, and GrpE, which are required for folding newly synthesized proteins and preventing aggregation of denatured proteins destined to be renatured or digested by proteases. Second, a series of ATP-dependent proteases (e.g., Lon, ClpP, ClpX) is induced, since such enzymes must play a role in protein processing or in digesting damaged proteins. There are also a number of sigma factors and other proteins induced. These heat shock proteins function as well in cells growing in normal conditions by "chaperoning" the proper folding of many proteins and digesting abnormal or denatured proteins. Thus the general function of this stress-induced family of proteins is to ensure a high yield of correctly folded proteins even in the presence of protein-denaturing physical and chemical shocks (for general review, see Ref. 59).

Since solvent production in *C. acetobutylicum* and other clostridia is often enhanced by stress conditions such as accumulation of toxic acids that inhibit growth as outlined earlier, it is important to ask how regulation of this heat shock system is related to induction of solventogenesis. The two major chaperon families (the GroE and the Dna K-J groups), as well as 10 or so other heat shock proteins, are transiently induced when *C. acetobutylicum* is upshifted from 30 °C to 42 °C for 5 min. (60). When the same organism shifts to solvent production by a drop in pH at a high butyric acid concentration, the same transient induction of heat shock proteins occurs. Thus, this stress response is coordinated with the induction of the expression of genes encoding solvent-forming enzymes. In the case of *E. coli*, the expression of the heat shock genes depends on a special sigma factor, σ^{32}, which forces cellular RNA polymerase to recognize specific promoter sequences of the heat shock gene family. However, the regulatory mechanism for heat shock gene expression in *C. acetobutylicum* appears to be different. No evidence for an alternative sigma factor has been found, and the promoter sequences of the *groE* and *dnaK* operons conform to the common promoter region for gram-positive bacteria. However, in both *C. acetobutylicum* and *B. subtilis* there is a conserved inverted repeat (IR) of 9 bp separated by a 9-bp spacer within the promoter region of the *dnaK* and *groE* operons. This IR has been identified in 27 bacterial species (both gram-negative and gram-positive), always occurring only before the *dnaK* and *groE* operons. There is evidence in *B. subtilis* that this IR is a site that interacts with a regulatory protein (ORF 39) that may act alternatively as an activator and a repressor (61). A similar protein (ORF A) identified in *C. acetobutylicum* is a candidate for the same regulatory function.

How the mechanisms regulating heat shock and general stress response in the clostridia may be connected with the *SpoOA* phosphorelay system involved in the initiation of endospore formation and solventogenesis is still not known. In any case, the stress response in gram-positive bacteria is designed to combat a variety of stress situations. In summary, the overall stress response of the solvent-forming *C. acetobutylicum* and perhaps other solvent-producing clostridia appears to involve three response systems. First, heat shock genes are induced, expressing a 10- to 20-fold increase in chaperone proteins protecting vital cellular enzymes and other proteins against denaturing effects of internal butyric acid. Second, solvent genes are induced, expressing a 40- to 100-fold increase in butanol and acetone pathway enzymes and resulting in recycling of butyric and acetic acids, thus detoxifying the environment. Finally, the first two stress responses permit the cell time to induce the complex machinery required for differentiation to endosporulation, which is the ultimate stress response available to all the clostridia.

DEVELOPMENT OF NEW SOLVENT-PRODUCING STRAINS

The Product Tolerance Problem

When a *Clostridium*-based fermentation is contemplated for a commercial solvent production system, the factors such as rate and efficiency of substrate conversion to solvents must be considered. But the major problem to be faced for all clostridial fermentations is extraction of low concentrations of solvents from water. It has been calculated that when distillation is used in the recovery of ethanol at less than 6–8%, or *n*-butanol at less than 2–3% from fermentation beers, the energy required exceeds the net energy recovery in the final products obtained in the commercial process. Yeasts and strains of *Zymomonas mobilis* convert sugars to an ethanol concentration high enough (10–20%) to permit separation of the solvent from the fermentation beer on an energy budget sufficiently low to allow profitable industrial production. In contrast, cocultures of *C. thermocellum* and *C. thermohydrosulfuricum* fermenting liquefied wood to ethanol produce a low final yield of 2% ethanol, a concentration above which growth of both these bacteria is inhibited. Likewise, the butanol–acetone fermentation by *C. acetobutylicum* stops when only 2% (w/v) solvents is reached, primarily owing to the toxicity of butanol.

There has been only limited research on the biological basis of low tolerance of clostridia to solvents involving strains of three organisms: *C. thermocellum, C. thermohydrosulfuricum*, and *C. acetobutylicum* (24). Growth of *C. thermocellum* at low concentration of ethanol results in an increase in the ratio of unsaturated/saturated fatty acids in the cytoplasmic membrane consistent with work with other bacteria. A mutant strain, C9, adapted to growth at higher levels of ethanol (2%) exhibited even greater increase in the amount of unsaturated fatty acids. The view is that ethanol enters the cell, causing a decrease in membrane fluidity, and the bacteria adapt by altering the fatty acid content to increase fluidity. The low ethanol tolerance of *C. thermocellum* (5 g/L) has been ascribed to specific inhibition of a number of enzymes required for glycolysis. Glucose fermentation by *C. thermohydrosulfuricum* (strain 39E) was inhibited by 0.5–2% ethanol. A mutant strain (39EA) grew in 8% ethanol at 45 °C and 3.3% ethanol at 68 °C. Unfortunately, the mutant strain produced a 1:1 ratio of ethanol to lactate, while the parent strain (39E) produces an ethanol-to-lactate ratio of 8:1. The ethanol-tolerant mutant also, incomprehensibly, lost the ability to utilize starch for growth (see review, Ref. 24).

The studies with *C. acetobutylicum* on product tolerance reveal that the effect of butanol appears to be opposite of ethanol in that butanol increases the cell membrane fluidity (24). Growth of *C. acetobutylicum* in 1% butanol increases the ratio of saturated to unsaturated fatty acids in the membrane. Thus, this adaptation is similar to fatty acid changes in mesophilic bacteria in response to maximum growth temperatures. Growth of this organism is inhibited at 11, 51, and 43 g/L of acetone, ethanol, and butanol, respectively, while as little as 6–8 g/L of butyrate and acetate inhibit growth. Thus, the view that the metabolic shift of these clostridia from an acidogenic to a solventogenic metabolism with recycling of organic acids is a detoxification stratagem is consistent with these observations. However, the fermentation is finally restricted by a low concentration of the major solvent, butanol.

Preliminary experimental approaches to improving butanol tolerance have been reported. Physiologic replacement of normal membrane fatty acids with long-chain fatty acids such as oleic acid (C_{18}) and elaidic acid (C_{20}) in cell membranes of *C. acetobutylicum* has been partially successful. Adding oleic acid or elaidic acid to growing cells in a biotin-free medium results in bacteria twice as tolerant to butanol inhibition. Butanol-resistant mutants selected from *C. acetobutylicum* (ATCC 824) will ferment corn starch to yield 7.9 g per liter of butanol (24). These butanol-tolerant mutants yielded 14 g per liter of butanol during fermentation and tolerated 16 g per liter of butanol. Autolysin-resistant mutants of strain NCP 262 are somewhat more butanol tolerant but produce only 20 g per liter of total solvents during fermentation. Clearly, the mechanism of solvent tolerance has not been elucidated. Thus, the low solvent yields characteristic of all *Clostridium* fermentations remain a major stumbling block for commercialization.

Biochemical engineers have developed new designs for continuous fermentation combined with product recovery technologies to cope with solvent intolerance inherent in clostridial fermentations. Application of these techniques for the acetone–butanol fermentation was reviewed by Maddox et al. (62). A combination of the new separation technologies and specialized mutant strains may be the best approach for future solvent production systems. For example, a pervaporation technology was combined with a fed-batch fermentation employing the superrecycling mutant B18 of *C. acetobutylicum* (NRRL 643) (15). Continuous removal of acetone and butanol through silicone tubing maintained the level of butanol in the broth below 4.5 g/L while hourly glucose consumption was held at about 2 g/L for 80 h. There was little accumulation of organic acids, and a high solvent yield (63). A long-term (2,000 h) continuous cultivation of *C. beijerinckii* (NRRLB 592) in a two-stage chemostat with on-line solvent removal by membrane evaporation produced solvents at a rate of 9.27 g/L/h without degeneration (64). The substitution of the high solvent, superrecycling *C. beijerinckii* strain BA101 should improve yield and productivity (16). The combined application of either mutant or genetically altered clostridia with integrated fermentation and product recovery technologies may successfully sidestep the low product tolerance characteristic of the clostridia.

Biology of Substrate Utilization

Solventogenic clostridia have the potential for converting carbohydrate substrates from excess agricultural crops or forest and agricultural wastes into solvents and fuels. The carbohydrate feedstocks for these fermentations are primarily starch, cellulose, xylans (or hemicelluloses), and lactose. Three recent reviews present the details of the enzymology, and genetics of utilization of these carbohydrates by the clostridia (62,65,66). Most of the solvent-producing saccharolytic clostridia utilize starch as a carbon source, reflecting their synthesis of extracellular and cell-bound amylolytic enzymes. Each species carries an array of three or four of the following enzymes: α-amylase, which hydrolyzes α-1,4-glucosidic linkages of amylose at random; β-amylase, which hydrolyzes α-1,4-glucosidic linkages from the ends of amalose, releasing disaccharides; glucoamylase, a hydrolyzer like β-amylase, but one releasing glucose; and pullulanase, which hydrolyzes α-1,6-glucosidic linkages or branch points of amylopectin. Some thermophilic clostridia, such as *C. thermohydrosulfuricum*, produce a novel double-headed amylopullulanase that has both α-amylase and pullulanase activity. Strain development for utilizing specific starch-containing feedstocks will require selection of strains that contain the best array of these enzymes for the effective utilization of the selected carbon source. In addition, *C. puniceum* produces pectinolytic enzymes that seem to aid in exposing the starch feedstock to the amylolytic enzymes (8). Because most of the genes involved in expressing the enzymes for starch utilization have been cloned, it is now possible to genetically engineer solventogenic clostridial strains for each starch source.

The thermophilic *C. thermocellum* has been the central object of research on bacterial utilization of cellulose as a carbon source for solvent production (66). There are also seven or so mesophilic cellulolytic clostridia that are closely related but with distinctly different arrangements of cellulases (29). Although as noted earlier ("Ethanol Fermentation"), *C. thermocellum* has major drawbacks as a candidate for commercial production of ethanol, extensive study of the genetics and biochemistry of the extracellular cellulose complex indicates that it may play a role in new strain development. During batch growth of *C. thermocellum* in media with insoluble cellulose, most of the cellulose activity is located in large-cell-associated multiprotein aggregates (cellulosomes) that also bind to cellulose fibers. Clusters of cellulosomes are observed, and the basic subunit in *C. thermocellum* strain YS as a 2.1-MDa complex including 14 different polypeptides. In other strains, cellulosomes up to 3.5 MDa consisting of 50 polypeptides have been described. Multiple forms of β-1,4-endogluconases and exogluconases plus a 210-kDa scaffolding protein are organized in the cellulosome structure.

Other separate components in *C. thermocellum* include two xylanases and two β-glucosidases. Many of the genes for these proteins have been cloned, partially sequenced, and also expressed in *E. coli*. Thus, in future strain development, many or all of the component parts of the cellulosome from *C. thermocellum* could be genetically transferred and expressed in other solventogenic clostridia such as *C. acetobutylicum*. This later organism already produces xylanases and can grow slowly on xylans as a source of fermentable sugars. Six different strains of *C. acetobutylicum* contain both extracellular and cell bound β-1,4-exogluconases; but cannot utilize cellulose as a carbon source (65). The genetic technology is now available to develop a cellulolytic *C. acetobutylicum* that will convert cellulose to acetone and butanol.

Metabolic Pathway Inactivation and Amplification

As noted earlier ("Genetics of Solvent Formation"), the genetic methods for DNA transfer and insertional mutagenesis now available for some of the clostridial strains that produce solvents together with recent advances in cloning and sequencing of many of the important genes encoding solventogenesis and pathways of substrate utilization promise a new era of strain development. The best experimental examples for changing the ratios of fermentation products come from studies with *C. acetobutylicum*. Normally, the bacteria carrying out this fermentation produce five different organic products (as well as H_2 and CO_2) in different amounts, depending on growth conditions (see earlier: "Butanol Fermentations" and "Physiology of Control of Solvent Formation"). Ideally, during solvent production, a mutant that produced only acetone, ethanol, and butanol, with no residual organic acids, would be desirable. By transferring a plasmid carrying the three genes required for acetone production to a normal *C. acetobutylicum* (ATCC 824), one can produce an excess of the two acetone enzymes in the cell (49). This amplifies the capability to recycle the organic acids and produce a high yield of both butanol and acetone (see Fig. 6). The same result is obtained with a mutant strain, B18, of strain NRRL 643, selected for low acid production following mutagenesis (15).

The second approach to directing the flow of carbon into specific products in the *C. acetobutylicum* fermentation is "gene knockout" by insertional mutagenesis. Recent experiments indicate that inactivation of acetyl kinase or of butyrate kinase prevents acetate or butyrate production, respectively, and permits an earlier and increasing flux toward butanol. Since autolysin, an enzyme produced by *C. acetobutylicum*, causes a late growth phase self-lysis, inactivation of this enzyme might also prolong solvent production. Autolysin-deficient mutants of *C. acetobutylicum* strain P262 show increased cell stability and a higher butanol tolerance.

Another problem facing the use of *C. acetobutylicum* for a continuous fermentation is the natural coupling of solventogenesis with development of endospores. The normal biology of the organism has coordinated these and other processes to adapt to organic acid stress and other stress situations. In *C. acetobutylicum* (ATCC 824), the loss of the large 210-kb pSOL-1 plasmid causes loss of the *sol* operon as well as certain unknown genes required for sporulation and granulose formation (42). Thus, in strain ATTC 824, loss of the pSOL-1 plasmid may serve as an advantage for development of a continuously fermenting, solvent-producing strain. By transferring the genes of the *sol* operon and the *adc* gene on a plasmid back into a pSol-1-negative strain, it is possible to induce solvent formation uncoupled from spore formation (42). In a number of strains of *Clostridium* as well as the butanol-producing strains, the initiation of sporulation and solvent formation is controlled by a complex phosphorelay mechanism as in *B. subtilis*, as noted earlier ("Is Solvent Production a Cellular Response?"). Further study of this regulatory network in the clostridia may reveal ways to upregulate solvent formation and also block initiation of endospore formation.

Since microbiologists estimate that only about 0.1% or less of all the bacteria in nature have ever been successfully isolated, there must be a large untapped natural pool of uncharacterized clostridia with valuable capabilities worthy of the attention of industrial microbiologists. For instance, in 1979 only five clostridial thermophilic bacteria were known. By 1995 more than 16 clostridial thermophiles had been purified (26). Recent reclassification demonstrates that *Thermoanerobium* sp. and *Thermoanaerobacter* sp. are close relatives to the clostridia, which expands the number of candidates for new strain development to more than 25. Thus, just considering growth temperatures in the range of 55–70 °C as a desirable property, many natural species are only recently isolated from nature. Perhaps a program of genetic engineering should be preceded by the application of new enrichment techniques to a search for natural species with new pathways, increased solvent production, or increased tolerance to chemical or physical stress.

ACKNOWLEDGMENT

The work reported from our laboratory was supported by grant DE-FG02-86ER-13512 from the U.S. Department of Energy.

BIBLIOGRAPHY

1. H. Hippe, J.R. Andreesen, and G. Gottschalk, in A. Balows, H.G. Trüper, M. Dworkin, W. Harder, and K.-H. Schleifer eds., *The Prokaryotes*, 2nd ed., Springer-Verlag, New York, 1991, pp. 1800–1866.
2. M.D. Collins, P.A. Lawson, A. Willems, J.J. Cordoba, J. Fernandez-Garayzabal, P. Garcia, J. Cai, H. Hippe, and J.A.E. Farrow, *Int. J. Syst. Bacteriol.* **44**, 812–826 (1994).
3. J.L. Johnson and J.-S. Chen, *FEMS Microbiol. Rev.* **17**, 223–240 (1995).
4. S. Keis, C.F. Bennett, V.K. Ward, and D.T. Jones, *Int. J. Syst. Bacteriol.* **45**, 693–705 (1995).
5. F.A. Rainey, N.L. Ward, H.W. Morgan, R. Toalster, and E. Stackebrandt, *J. Bacteriol.* **175**, 4772–4779 (1993).
6. Y.-E. Lee, M.K. Jain, C. Lee, S.E. Lowe, and J.G. Zeikus, *Int. J. Syst. Bacteriol.* **43**, 41–51 (1993).
7. W.D. Murray, A.W. Kahn, and L. Van den Berg, *Int. J. Syst. Bacteriol.* **32**, 132–135 (1982).
8. J.G. Morris, in D.R. Woods ed., *The Clostridia and Biotechnology*, Butterworth-Heinemann, Boston, Mass., 1993, pp. 1–24.

9. P. Tailliez, H. Girard, J. Millet, and P. Beguin, *Appl. Environ. Microbio.* **55**, 207–211 (1989).
10. J.G. Zeikus, *Annu. Rev. Microbiol.* **34**, 423–464 (1980).
11. C.K. Lee and J. Ordal, *J. Bacteriol.* **94**, 530–536 (1967).
12. E.J. Hsu and Z.J. Ordal, *J. Bacteriol.* **102**, 369–376 (1970).
13. W.D. Murray and A.W. Kahn, *Can. J. Microbiol.* **29**, 342–347 (1983).
14. H. Bahl and G. Gottschalk, *Biotechnol. Bioeng. Symp.* **14**, 215–223 (1984).
15. P. Rogers and N. Palosaari, *Appl. Environ. Microbiol.* **53**, 2761–2766 (1987).
16. J. Formanek, R. Machi, and H.P. Blaschek, *Appl. Environ. Microbiol.* **63**, 2306–2310 (1997).
17. H.A. George, J.L. Johnson, W.E.C. Moore, L.V. Holdeman, and J.S. Chen, *Appl. Environ. Microbiol.* **45**, 1160–1163 (1983).
18. R.A. Holt, A.J. Carins, and J.G. Morris, *Appl. Microbiol. Biotechnol.* **27**, 319–324 (1988).
19. M. Gottwald, H. Hippe, and G. Gottschalk, *Appl. Environ. Microbiol.* **48**, 573–576 (1984).
20. U.S. Congress, Office of Technology Assessment, *Commercial Biotechnology: An International Analysis*, OTA-BA-218, Government Printing Office, Washington, D.C., 1984, pp. 237–250.
21. S.C. Prescott and C.G. Dunn, *Industrial Microbiology*, 3rd ed., McGraw-Hill, New York, 1959, pp. 240–284.
22. R.K. Thauer, K. Jungermann, and K. Decker, *Bacteriol. Rev.* **41**, 100–180 (1977).
23. H. Petitdemange, C. Cherrier, G. Raval, and R. Gay, *Biochim. Biophys. Acta* **421**, 334–347 (1976).
24. P. Rogers, *Adv. Appl. Microbiol.* **31**, 1–60 (1986).
25. P. Rogers and G. Gottschalk, in D.R. Woods ed. *The Clostridia and Biotechnology*, Butterworth-Heinemann, Boston, Mass., 1993, pp. 25–50.
26. F. Canganella and J. Wiegel, in D.R. Woods ed., *The Clostridia and Biotechnology*, Butterworth-Heinemann, Boston, Mass., 1993, pp. 393–430.
27. R. Lamed and J.G. Zeikus, *J. Bacteriol.* **144**, 569–578 (1980).
28. Y. Mori, *Appl. Environ. Microbiol.* **56**, 37–42 (1990).
29. L.B. Leshine, *Annu. Rev. Microbiol.* **49**, 399–426 (1995).
30. W.D. Murray, *Appl. Environ. Microbiol.* **51**, 710–714 (1986).
31. S.R. Wilkinson, D.I. Young, G. Morris, and M. Young, *FEMS Microbiol. Rev.* **17**, 275–285 (1995).
32. D.T. Jones and D.R. Woods, *Microbiol. Rev.* **50**, 484–524 (1986).
33. H. Bahl and G. Gottschalk, in H.J. Rehm and G. Reed eds., *Biotechnology*, vol. 6B, VCH, Weinheim, Germany, 1988, pp. 1–30.
34. P. Dürre, N.P. Minton, E.T. Papoutsakis, and D.R. Woods eds., Papers presented at a meeting on solventogenic *Clostridia*. *FEMS Microbiol. Rev.* **17**, 221–364 (1995).
35. C.W. Forsberg, *Appl. Environ. Microbiol.* **53**, 639–643 (1987).
36. L. Girbal, C. Croux, I. Vasconcelos, and P. Soucaille, *FEMS Microbiol. Rev.* **17**, 287–297 (1995).
37. H. Grupe and G. Gottschalk, *Appl. Environ. Microbiol.* **58**, 3896–3902 (1992).
38. C.L. Meyer and E.T. Papoutsakis, *Appl. Microbiol. Biotechnol.* **30**, 450–459 (1989).
39. D.T. Jones, in D.R. Woods ed., *The Clostridia and Biotechnology*, Butterworth-Heinemann, Boston, Mass., 1993, pp. 78–98.
40. R.C. Woolley and J.G. Morris, *J. Appl. Bacteriol.* **69**, 718–728 (1990).
41. E.R. Kashket and Z.-Y. Cao, *FEMS Microbiol. Rev.* **17**, 307–315 (1995).
42. E. Cornillot, R.V. Nair, E.T. Papoutsakis, and P. Soucaille, *J. Bacteriol.* **179**, 5442–5447 (1997).
43. S.L. Landuyt, E.J. Hsu, and M. Lu, *Ann. N.Y. Acad. Sci.* **413**, 473–478 (1983).
44. D.M. Rothstein, *J. Bacteriol.* **165**, 319–320 (1986).
45. P. Dürre, in D.R. Woods ed., *The Clostridia and Biotechnology*, Butterworth-Heinemann, Boston, Mass., 1993, pp. 228–246.
46. D.M. Mattsson and P. Rogers, *J. Ind. Microbiol.* **13**, 258–268 (1994).
47. M. Young, in D.R. Woods ed., *The Clostridia and Biotechnology*, Butterworth-Heinemann, Boston, Mass., 1993, pp. 99–118.
48. N.P. Minton, J.K. Brehm, T.-J. Swinfield, S.W. Whelan, M.L. Mauchline, N. Bodsworth, and J.D. Oultrum, in D.R. Woods ed., *The Clostridia and Biotechnology*, Butterworth-Heinemann, Boston, Mass., 1993, pp. 119–150.
49. L.D. Mermelstein, N.E. Welker, D.J. Petersen, G.N. Bennett, and E.T. Papoutsakis, *Ann. N.Y. Acad. Sci.* **721**, 54–68 (1994).
50. E.T. Papoutsakis and G.N. Bennett, in D.R. Woods ed., *The Clostridia and Biotechnology*, Butterworth-Heinemann, Boston, Mass., 1993, pp. 157–200.
51. P. Dürre, R.-J. Fischer, A. Kuhn, K. Lorenz, W. Schreiber, B. Stürzenhofecker, S. Ullmann, K. Winzer, and U. Sauer, *FEMS Microbiol. Rev.* **17**, 251–262 (1995).
52. C.J. Bult, O. White, G.J. Olsen, and 37 other authors, *Science*, **273**, 1058–1073 (1996).
53. D. Holzman, *Am. Soc. Microbiol. News* **62**, 8–9 (1996).
54. E.M. Green, Z.L. Boynton, L.M. Harris, F.B. Rudolf, E.T. Papoutsakis, and G.N. Bennett, *Microbiology* **142**, 2079–2086 (1996).
55. P.J. Piggot, C.P. Moran, Jr., and P. Youngman, *Regulation of Bacterial Differentiation*, American Society for Microbiology, Washington, D.C. 1994.
56. U. Sauer, J.D. Santangelo, A. Trever, M. Bucholz, and P. Dürre, *FEMS Microbiol. Rev.* **17**, 331–340 (1995).
57. J.A. Hoch, *Annu. Rev. Microbiol.* **47**, 441–465 (1993).
58. D.P. Brown, L. Ganova-Raevn, B.D. Green, S.R. Wilkinson, M. Young, and P. Youngman, *Mol. Microbiol.* **14**, 411–426 (1994).
59. C.A. Gross, in F.C. Neidhardt ed., *Escherichia coli and Salmonella Cellular and Molecular Biology*, vol. 1, 2nd ed. ASM Press, Washington, D.C., 1996, pp. 1382–1399.
60. H. Bahl, H. Müller, S. Behrens, H. Joseph, and F. Narberhaus, *FEMS Microbiol. Rev.* **17**, 341–348 (1995).
61. M. Hecker, W. Schumann, and U. Volker, *Mol. Microbiol.* **19**, 417–428 (1996).
62. I.S. Maddox, N. Qureshi, and N.A. Gutierrez, in D.R. Woods ed., *The Clostridia and Biotechnology*, Butterworth-Heinemann, Boston, Mass., 1993, pp. 343–370.
63. Q. Geng and C.-H. Park, *Biotechnol. Bioeng.* **43**, 978–985 (1994).
64. J.R. Gapes, D. Nimcevic, and A. Friedl, *Appl. Environ. Microbiol.* **62**, 3210–3219 (1996).
65. K. Bronnenmeier, and W.L. Staudenbauer, in D.R. Woods ed., *The Clostridia and Biotechnology*, Butterworth-Heinemann, Boston, Mass., 1993, pp. 261–310.
66. G.P. Hazlewood and H.J. Gilbert, in D.R. Woods ed., *The Clostridia and Biotechnology*, Butterworth-Heinemann, Boston, Mass., 1993, pp. 311–342.

See also ANAEROBES; CULTURE COLLECTIONS.

COAGULATION FACTORS, THERAPEUTIC

David L. Aronson
Consultant
Bethesda, Maryland

KEY WORDS

Antithrombin
CPMP
Factor VIII
Food and Drug Administration
Gene therapy
Hemophilia
Recombinant coagulants
Transgenics
von Willebrand factor

OUTLINE

INTRODUCTION

For the past 40 years, therapeutic preparations of plasma coagulation factors have been derived from plasma. The plasma fractionation industry developed during and after the second World War and was often a partnership between a governmental or not-for-profit organization that obtained the blood from volunteer donors and the nascent commercial plasma industry. Since the beginning of blood banking and the production of plasma products, the regulatory oversight has been done by the national organizations responsible for biologics control. The authority and the procedures of these organizations are different than those for pharmaceutical regulation.

Regulation of the plasma products in the United States was based on the Public Health Service Act of 1902. This act was promulgated to regulate the production of vaccines and therapeutic serums. The latter phrase was construed to include blood and plasma products when this industry developed later on. Compliance was monitored by the Public Health Service Laboratories, later to become the National Institutes of Health. In 1972, the regulation of U.S. biologics was moved to the Food and Drug Administration (FDA).

One of the major efforts in any of the biological control agencies was the development of standards and assays for impure and poorly characterized natural products. This major component of biologics regulation is seen in the name of the former U.S. control organization, Division of Biologics Standards at NIH, and in the U.K., the National Institute of Biological Standards and Control (NIBSC).

Although regulation has properly belonged to national regulatory bodies, there are other inputs to the regulatory process from supranational organizations, such as the European Community and the International Conference on Harmonisation (ICH). In the United States, the FDA requires that an advisory committee votes for approval of licensure. These committees include representatives of advocacy groups that have voting membership in the FDA blood product advisory committee (BPAC). The recommendations of professional organizations are also quoted in regulatory documents or regulatory decisions. The Scientific and Standardization Committee of the International Society of Thrombosis and Haemostasis has provided recommendations for the hemophilias and thrombotic disease. In addition, civil liability law has an impact on all therapeutic producers, particularly in the United States.

Regulatory documents of the FDA are published in the Code of Federal Regulations (CFR), and updated guidance documents can be obtained from the FDA World Wide Web site (*http://www.fda.gov*). European documents are found in the European Pharmacopeia, and guidance documents are obtained from the Committee on Proprietary Medicinal Products (CPMP).

Blood Coagulation

Blood coagulation results from a series of proteolytic reactions that first activate a procoagulant protein, and then the resulting coagulant is inactivated either proteolytically or by interaction with an inhibitor (1) (Fig. 1). Although presented as the coagulation cascade (Fig. 2), implying amplification, the complexity results in many points where there is sensitive biochemical control. Individual components are identified by roman numerals, and the activated form is indicated by a lowercase "a."

Factor IX, the defective protein in hemophilia B, is the proenzyme of a serine protease that is cleaved to activated factor IX (IXa). Factor VIII, the defective or deficient pro-

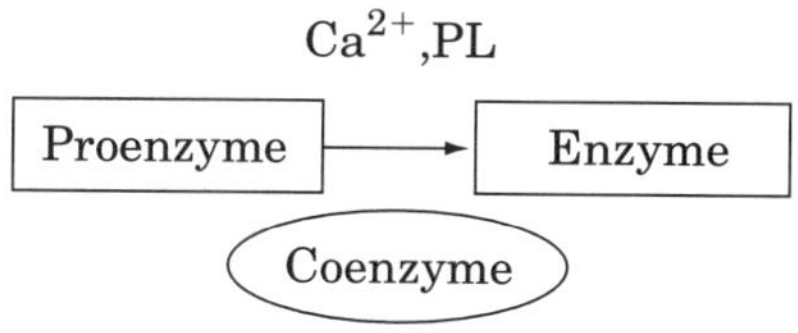

Figure 1. Prototype coagulation step.

tein in hemophilia A, is a protein cofactor that circulates bound to a large polymeric protein, von Willebrand's factor, which is defective in von Willebrand's disease. In the bound state the factor VIII remains in the plasma; in the presence of small amounts of thrombin, it is released from the von Willebrand factor and activated. In the presence of factors IXa and VIIIa there is activation of the downstream enzymes in the coagulation cascade (Fig. 2). Absence or defects of either factor IX or factor VIII leads to inability to form fibrin clots. Absence of the carrier protein, von Willebrand's factor, leads to low circulating levels of factor VIII as well as decreased recruitment of platelets to damaged endothelium and leads to a hemorrhagic state.

The factor VIII gene product is ~250,000 kDa derived from a gene of 186 kb in 26 exons. The gene product is glycosylated intracellularly, and the tyrosine at position 1680 must be sulfated. During passage through the Golgi apparatus there is excision of much of the central portion of the protein (B-domain), resulting in a heavy and light chain held together by Ca^{2+}. This is the form found in the circulation.

Genetic Hemorrhagic Disorders

Hemophilia A and hemophilia B are the result of mutation of genes located on the X chromosome. As such they are sex-linked, and sons of carrier mothers have a 50% likelihood of carrying the defective gene and presenting with the clinical phenotype of abnormal hemorrhage. This leads to crippling and death in those severely afflicted.

Treatment of these diseases began with plasma and evolved to more purified and concentrated plasma products (2). Adequate treatment for those with severe hemophilia involves lifetime intravenous administration given approximately weekly. Although the treatment is effective, there are two major safety issues with plasma products: blood-borne viral infections and the development of antibodies that inhibit the therapeutic effect. Hemophilia A and B were one of the early targets of the biotechnology industry.

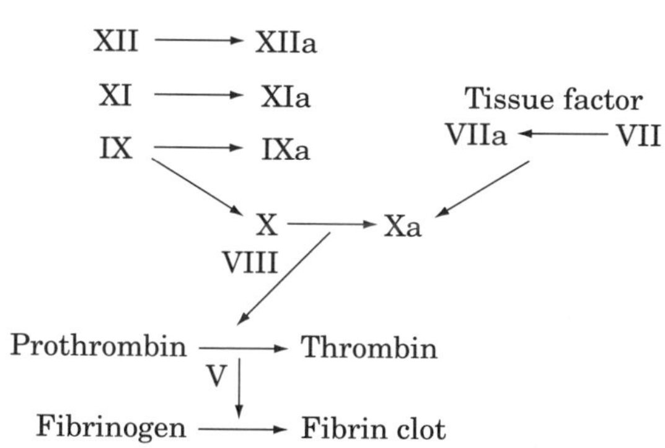

Figure 2. Coagulation sequence.

The patient population is small, with 16,000 patients with hemophilia A (deficiency of factor VIII) and hemophilia B (deficiency of coagulation factor IX) in the United States. Of these, about 8,000 have severe disease and need frequent treatment. Treatment with plasma products temporarily improved the well-being of this patient population until the HIV epidemic; 6,600 U.S. patients were infected (3). Patients with severe hemophilia A need frequent and lifelong treatment; this group uses ~50,000 to 100,000 factor VIII units annually.

Viral infection from plasma products has been the major problem in the treatment of the congenital bleeding disorders. Infection with most blood-borne viruses has been minimized during the past 10 years by viral inactivation. The most common procedure is the destruction of the lipid-coated viruses (HIV, hepatitis B and C) by extraction with a solvent-detergent mixture. Another procedure is the use of heating either in solution or in the dry state. Many regulatory authorities are demanding two viral inactivation and removal steps for recombinant and plasma-derived factor VIII and factor IX.

The second problem for the treatment of hemophilia A is that 15% of patients will develop inhibiting antibodies to factor VIII and become refractory to standard treatment. Although most patients who develop antibodies may be genetically predisposed, in other cases this is because of neoantigenicity induced in the factor VIII during production and viral inactivation (4).

A third hemorrhagic disease, von Willebrand's disease, is the result of heterogeneous mutations of the gene for a polymeric carrier protein of factor VIII, von Willebrand's factor.

REGULATORY ISSUES FOR RECOMBINANT COAGULATION FACTORS

Statutory drug law is based on the principles of safety and efficacy. Regulatory oversight of pharmaceutical production is done in large part through good manufacturing practices (GMPs) and good laboratory practices (GLPs). This is a body of regulation and policy that can change rapidly. In addition, in the regulation of biologics there is a need for product consistency whether the product is derived from animals or from cells. The particular origin of biologics leads to special needs in the control of the starting materials, including cells, genes, etc. This is best exemplified in the regulations for the establishment of intensively characterized master cell banks for cell-derived products (5). This cell bank will be expanded in number under standard conditions to produce a consistent product over decades.

The advent of the biotechnology products for coagulation deficiencies (i.e., cell-derived rather than plasma-derived coagulation factors) raised few unique regulatory issues. Traditional biologics regulation had already dealt with cell-derived products, including the potentially dangerous live virus vaccines. A regulatory framework for the characterization and control of cell lines had already been constructed, and in 1984, a workshop sponsored by the Bureau of Biologics dealt with many of the residual issues. This was followed by a WHO report in 1987 (6,7). This was followed by the development of *Points-to Consider* for products derived from mammalian cells (8) and the EEC equivalent (9). The *Points-to-Consider* are not legal regulatory documents, but they inform those who are dealing with the agency of the current views of the regulatory authority. There is an effort to have as much commonality as possible between the international regulatory agencies, and extensive efforts to harmonize the guidelines are ongoing by the ICH (10). Of particular concern originally was the quality and quantity of residual DNA in the final product (11); a fear that has diminished with time.

The FDA does not produce specific documents on coagulation or fibrinolytic products. In contrast, the European Community has specific and detailed regulatory documents for the coagulation products (12). The basis for regulation of recombinant coagulation factors is similar to plasma-derived coagulation factors: the assessment of safety, efficacy and consistency. More particularly, there is intensive and extensive evaluation of the cell production systems, the control and standardization of the final product, and clinical effectiveness and safety. The main point of the original regulation for cell-derived coagulation factors was that the products should be identical with plasma-derived products. More recently, products with similar but not identical structures have been received by regulatory authorities with little additional concern.

The GLPs and GMPs are in principle identical around the world; however, there can be significant differences in detail. There has been rapid change in pharmaceutical science during the past decade, including filling procedures, air and water systems, and analytical methods. Approved production facilities of today may not meet GMPs in 5 years. In addition, regulatory policy can change without any formal notification or publication. For these reasons, it is important that the manufacturer maintain constant communication with the regulatory agency from inception to final approval.

Many facilities for recombinant products will be designed and maintained as single product manufacturing areas. This will include raw materials, cell lines, expansion of cell lines, fermenters, purification equipment, and the filling line. It will only be for the postfilling operations that there will be contact with other products. Some useful products may have such small demand that a few lots will meet needs for several years. Such a case may be recombinant activated factor VII (VIIa) for use in hemophilics with inhibitors not responsive to other modes of therapy. In this case, the plant must have a validated procedure for producing a different product.

Cellular Issues in Recombinant Coagulation Products

All cells used for the production of recombinant coagulation factors must either meet the *Points-to-Consider* documents by the FDA or have information indicating that there is no compromise to the safety and efficacy of the product if some of the points are modified. The relevant *Points-to-Consider* are those involved with production of the master cell banks (5) and the *Points-to-Consider* for products derived from mammalian cells (8). The European equivalents are CPMP *Note for Guidance: Production and Quality Control for Medicinal Products Derived by Recombinant DNA Technology* (9) and the ICH documents on the analysis of the expression construct (10).

The cells must not only synthesize the desired protein, but efforts must be made to simplify the purification process and minimize the presence of adventitious infectious agents by simplification of the growth medium. In particular, the use of fetal calf serum is avoided if at all possible to minimize a significant source of microbial and antigenic contamination. It is likely that regulatory agencies in the future will require more effort to develop culture systems free of serum or other animal protein.

During the production time, there must be no evidence of either microbial contamination or of change in the cell or the cell product. Both the U.S. and European documents require characterization of the vector to demonstrate the stability of the inserted gene over the time of a production run or longer. The development of PCR technology has allowed the characterization of cDNA and RNA. Frequent sampling and testing of the conditioned media during production and the final purified product are a requirement.

The cells must be capable of appropriate post-translational modifications. In the case of factor IX, there must be carboxylation of the 10 to 11 glutamic acids so that the molecule has full function. This has been accomplished by inclusion of a gene for a protease to remove the propeptide and allow carboxylation of the glutamic acids at the NH_2 terminal of the factor IX molecule. Other significant post-translation modifications include the formation of a β-OH aspartate residue, sulfation, phosphorylation, and glycosylation. In addition, there must not be the production of any protease that will degrade or activate factor IX.

A similar constraint is true for cells producing factor VIII. This is a heavily glycosylated molecule, and each production system could yield different products, some of which might be immunogenic. Although this is a concern, no such immunogenic derivatives have been found as yet in factor VIII produced by recombinant technology.

Von Willebrand's factor must be produced in cells allowing polymerization of the monomer and secretion of these virus-sized protein molecules into the media. The cell culture system must also be able to sulfate a specific tyrosine residue to allow binding to factor VIII and thrombin activation of the factor VIII.

Purification Issues in Recombinant Coagulation Products

Purification of coagulation products revolves around three issues: the removal of foreign allergens, the removal of DNA, and the removal of adventitious agents originating from the host cell and the media. The issue of allergens is of importance because the recipients need lifelong therapy. In contrast to the older plasma products, the purification process using mouse immunoglobulin affinity columns may be a new hazard, and at this time, monoclonal antibody purification is the universal purification method for factor VIII. The monoclonal antibodies do not have to meet the full requirements for a therapeutic monoclonal antibody but must have been examined for murine viruses.

There are no absolute requirements or regulations for protein purity within the U.S. or European regulations.

However, the manufacturer is obligated to have satisfactory test systems to detect nontherapeutic proteins derived either from the cell source or the purification procedure (e.g., leached monoclonal antibodies). The ultimate test will be in recipients of the product (see "Clinical Trials").

The factor IX product being developed not only is grown in media without protein but the purification steps do not involve monoclonal antibodies. This strategy simplifies the purification process and is also deemed desirable in terms of the introduction of viruses either in tissue culture or in the final product (13). Although this is apparently the first coagulation product from cells grown in the absence of protein in the media and purified without the use of monoclonal antibodies, it may set a new standard.

The same procedures for the removal and inactivation of viruses that are required for cell-derived products are required for the plasma products. Although the use of cell-derived products in contrast to plasma-derived products will only rarely contain viruses transmissible to humans, the newer knowledge of virology and the public perception demands extreme measures. The manufacturer must show that the purification methodology has the capability to remove or inactivate viruses of different types. However, the European guidelines are much more specific than the comparable U.S. documents (14).

One of the first issues for cell derived products was the establishment of limits for DNA. This was set at 10 pg/dose for all products in the original FDA documents (10), but the current WHO level of 100 pg/dose (7) seems to be the actual figure used by the FDA. Products used for the treatment of hemophilias are infused 1,000 times in a lifetime; however, there is no change in the dosage for products used a few times as opposed to 1,000 times in a lifetime. The requirements for DNA removal revolve around the theoretical ability of cell-derived DNA to be incorporated into somatic cells in the recipient (11).

Final Product

The structure of the gene product must be extensively studied. This will include extensive peptide mapping, molecular weight analysis by electrophoretic assessment, and perhaps mass spectrometry. In some cases such as factor IX, the complete amino acid structure is feasible, whereas for factor VIII the total amino acid sequence is more onerous than useful in the light of the established gene structure. Disulfide bonding should be characterized and shown to be identical to plasma-derived coagulants. As noted earlier, there are specific post-translational modifications for given products important for their function, such as glutamyl carboxylation of factor IX, that should be characterized.

The final product will include degradation products of the gene product. These should be identified, analyzed for their sequence, and discriminated from other protein contaminants.

A specific issue for the coagulation factors is that they are in the appropriate state (i.e., nonactivated for factor VIII or IX). For von Willebrand's factor, the polymeric forms must be functional. As noted earlier, the circulating form of the coagulant proteins should be the proenzyme; factor IXa can induce thrombus formation. In the case of the cofactor, factor VIII, the uncleaved, nonactivated form is necessary; the presence of the activated form yields inaccurate potency measurements and is rapidly cleared. Thus, the purification system should remove the activated forms and be done in a manner that does not produce these forms.

Preclinical Testing

In general, the preclinical testing of recombinant products is similar to plasma-derived products and includes routine pyrogen, safety, and potency testing. Specific and accurate potency measurements for labeling and dosing must be available.

Potency Measurement

The assumed efficacy of the final product is assessed by the potency assay, the results of which yield the labeled value for treatment. The standard assay is a bioassay with intrinsic variability (2,15). The unit definition for coagulation products is the functional activity found in 1 mL of normal plasma containing a 10th volume of anticoagulant. This operational definition is translated into WHO standards for coagulants by the preparation of plasma or plasma fractions that originally were compared to normal plasmas. All WHO coagulation standards have been prepared by the NIBSC in the United Kingdom. These are calibrated in multicenter studies organized by the Scientific and Standardization Committee of the International Society on Thrombosis and Haemostasis. New standards are compared to previous WHO standards and also to plasma standards. As both factor VIII and factor IX therapeutic products become more purified, there are inconsistencies with older standards (15). A further complication is the development of what are apparently more accurate quantitative potency tests based on better biochemical knowledge but that can show very substantial differences with the standard reference tests (16–18). The European Pharmacopia designates a reference methodology for the measurement of coagulation factors. In general, this seems to be the assay used by the control laboratory rather than that with which there is the most experience. At this writing, the reference methodology for factor VIII in the European Pharmacopiea is the chromogenic assay. In contrast, the FDA has no designated reference method for coagulation factors but does distribute national references. There is a consistent difference in the pharmacokinetics when assessed by different methods (17,18).

The development of variant proteins such as the B-domain deleted factor VIII aggravates the difference seen between the different methods of measuring potency. The specific activity of the B-domainless variant is reported as being substantially higher than plasma-derived factor VIII (17). The in vitro potency assays are approximately equivalent on a molar basis. Whether they are the same therapeutic quantity must be answered.

A recombinant von Willebrand factor is in development (19). Von Willebrand's factor has several different functional attributes of importance to its efficacy. These include the support of platelet aggregation, binding to collagen, and the binding to factor VIII. At this time there is no consensus as to the appropriate potency test for a therapeutic product to treat von Willebrand's disease. The WHO standard for von Willebrand's factor is plasma and not a purified preparation.

Animal Models

Animal models add useful information before clinical testing of the products. A dog model of hemophilia A has been used extensively in the preclinical testing of factor VIII. One of the most immediate questions was the comparison of the pharmacokinetics of recombinant factor VIII with the plasma product. If the proposed product was more rapidly removed from the circulation than the plasma product, this would raise serious doubts about the efficacy of the product. Experiments in the dog hemophilia model showed normal pharmacokinetics and rapid binding of factor VIII to canine von Willebrand's factor (20). Various approaches have been and are being assessed to determine abnormal immunogenicity with animal models: induction of immune tolerance in newborn animals in addition to laboratory comparison of the epitopes of the plasma and recombinant product (21).

Thrombogenicity of the factor IX preparations has been assessed by both laboratory testing and assessment of the ability to form clots in animals (22). Although this concern arises from thrombosis associated with the use of older, less-purified products, there is still concern that even the purified products and recombinant factor IX can produce thrombosis. Animal pharmacokinetics comparing plasma-derived and recombinant factor IX are a necessary prelude to clinical studies. Dog models of hemophilia B are available, but the animal pharmacokinetic studies are generally performed in smaller laboratory animals. Because there is endogenous factor IX coagulant activity in the samples, the plasmas are assayed for the quantity of human factor IX antigen present over time (23).

Animal models for coagulation factors have been developed over the years for specific known adverse effects. In addition, animal testing is of value for identifying unsuspected effects.

Clinical Trial Information

Clinical trials are the final step in the assessment of the efficacy and safety of a new therapeutic product. The FDA regulations outline the information needed to begin human studies and the type of studies needed to establish safety and efficacy (24). European regulations are similar. Efficacy of treatment of the hemophilias has never been established in a placebo-controlled study because observational studies coupled with historical data clearly established the effectiveness originally of both factor VIII and factor IX replacement therapy for preventing or treating hemorrhagic episodes. This type of data was acceptable to the regulatory authorities around the world in the mid-60s and was consistent with the biologics definition of efficacy (25) based on the pharmacologic and pharmacokinetic studies. Current policy would, however, require comparison with an approved efficacious product for the factor VIII and factor IX products. The FDA does not prescribe an exact protocol but describes the general types of clinical trials that could give efficacy data (26). In contrast, the EEC guidelines describe in detail the clinical trials to be undertaken (12). Many of these details are derived from articles promulgated by a nonregulatory professional organization, the International Society on Thrombosis and Haemostasis (ISTH) (27,28).

In addition to establishing efficacy, clinical trials are designed to detect adverse events. The most common concern is for known or unknown infectious agents. Recommendations of the ISTH serve as guidelines for some of the European community policies.

The original protocols for recombinant factor VIII started with pharmacokinetic trials in older patients with frequent need for treatment. The strategy was to keep these patients solely on recombinant factor VIII for several years. This would give ample opportunity to test any abnormal immunogenicity of the product. In addition, there would be interim pharmacokinetic studies as the most accurate way of assessing low-level inhibitors. A trial was also done in previously untreated patients (PUPs) independent of any request from a regulatory agency (29). This study yielded valuable results on the treatment of hemophilia in very young patients without the confounding of immunologic or viral infection by other products. This set a precedent for the evaluation of all future recombinant products and has been incorporated into the EEC requirements for all recombinant factor VIII and factor IX. This will probably also be required by the FDA in the future. Although the immunogenicity of recombinant factor VIII and plasma factor VIII is a major concern, it is not clear what information comes from the routine PUP studies, because any excess immunogenicity should be observed in older patients. There are no specific European clinical guidelines for recombinant coagulation factors. There are the general clinical guidelines (12) for the coagulation products, which specifically exclude recombinant products (6,8). Based on the historical knowledge and precedent, there is no requirement for a blinded efficacy trial. Efficacy assessment is based on the comparison of pharmacokinetics with an approved product. For factor VIII, the requirements are to study 12 pretreated (PTP) subjects over 1.5 days with eight sampling times. A minimum of three lots are to studied. These subjects should be maintained solely on the experimental product for a further 3 to 6 months and then have a repeat pharmacokinetic study. Before approval, a study of pharmacokinetics in 20 PUPs must be started using the same protocol as in the other patients. There is a requirement for at least 10 surgical procedures in a minimum of 5 patients and clinical assessment of 30 immunocompetent patients after licensure.

In addition, information should be submitted on 30 patients who have been treated on at least 10 days with at least 6 months of follow-up. The factor VIII inhibitor titer must be repeated every 3 months. After approval, there is a periodic update on the laboratory parameters of at least 50 patients treated for at least 2 years.

With plasma-derived factor IX, the major adverse event other than viral disease transmission was thrombosis associated mostly with the earlier less-purified plasma fractions containing large amounts of the other vitamin K coagulants. The clinical trials for a recombinant factor IX must test for evidence of in vivo coagulation with currently available tests for thrombin–antithrombin complex and prothrombin fragment 1.2. In addition, there will probably be required testing for the enzymatically active form of factor IX (IXa). Although the FDA does not specifically deal with this issue, this still must be addressed for U.S. licensure. The European guidance documents call for tests for in vivo activation of coagulation.

PUP studies will be required by the FDA and by the European Community for factor VIII and factor IX products. The European community requirements for plasma-derived factor IX require pharmacokinetic studies in 10 subjects with 10 sampling times. Treatment with the same product should be maintained for 6 months, and repeat pharmacokinetic studies should be performed in 3 to 6 months. Current published clinical experience with recombinant factor IX are limited to abstracts indicating pharmacokinetics and clinical effectiveness similar to plasma-derived factor IX (30). These studies are not inconsistent with the European guidelines and were accepted by the BPAC (December 19, 1996).

Von Willebrand's factor is a new molecule to deal with, and there are no national regulatory comments addressing the therapeutic use of this particular protein. Although the pharmacokinetic analysis of the clearance of the coagulation activity of factors VIII and IX is accepted throughout the world, there is no accepted or validated functional assay for von Willebrand's factor. Should the functional assay be based on the ristocetin cofactor activity or the ability to bind to platelets or the content of multimers or the content of intact monomers or the ability to bind factor VIII?

One can anticipate that the regulatory agencies will use the SSC/ISTH guidelines for appropriate clinical studies (31). The current FDA policy is to request comparative clinical data with another product, but there is no product licensed in the United States for the treatment of von Willebrand's disease.

Although von Willebrand's disease is a common genetic deficiency, the clinical phenotype is extremely variable, and for the most part, the hemorrhages are less severe and less frequent than in hemophilia A or B. This leads to difficulty in recruiting patients into a structured protocol. Heretofore, the clinical trials of von Willebrand's factor have been uncontrolled, and current treatment strategies are based on overall experience. The surrogate markers that have been used traditionally are the level of factor VIII, the shortening of the bleeding time, and the functional level of the von Willebrand factor as assessed by the ability to support platelet aggregation in the presence of ristocetin. Inhibitor development is an issue in von Willebrand's disease, and this will have to be assessed. With recombinant von Willebrand's factor, the factor VIII level will be endogenously regulated and will increase to peak levels 10 to 20 hours after infusion. This is in contrast to the current practice of infusing patients with plasma products containing von Willebrand's factor and factor VIII with instantaneous increase in the factor VIII level.

Two recombinant factor VIII products have been licensed in the United States and in Europe. Recombinant factor IX is in clinical trials and the license was submitted to the FDA in 1996. Recombinant factor VIIa, a product proposed for use in factor VIII and IX patients with inhibitors, is being reviewed by the FDA (1997) and has been licensed in Europe and Canada for the treatment of patients who acquire inhibitors to factor VIII. A recombinant von Willebrand factor is undergoing preclinical development.

Tissue Plasminogen Activator (Alteplase)—Regulatory Issues

The in vivo formation of blood clots is usually a process limited in time and space. The formation of thrombi, pathologic blood clots, are the result of subacute or chronic disease processes resulting in blockage of blood vessels. The development of fibrinolytic agents, enzyme that digest fibrin clots, was the first major advance in the treatment of thrombotic diseases in 50 years. The biologics used for this purpose are activators of the endogenous proenzyme, plasminogen. Tissue plasminogen activator (tPA) (alteplase) was the first fibrinolytic agent produced by recombinant technology.

The regulatory issues for the licensure of tPA are documented in contemporary news reports (31) and in a recent book by Beebe and Murano (32).

Production of tPA

One technical issue was emphasized: the effect on product of altering culture conditions. Early production of tPA in roller bottles was used for the initial clinical investigation, including the initial pharmacokinetic studies. When there was change to fluid-phase culture, there was a significant decrease in the pharmacokinetics, leading to the need for an increase in the clinical dosage. It was originally proposed that this was due to the chain structure of the tPA but that has been ruled out (33). Despite investigation of the carbohydrates and other possible structural changes, the post-translational change resulting in the pharmacokinetic change is still a mystery.

CLINICAL TRIALS

Gene Therapy in Hemorrhagic Diseases

Gene therapy is the procedure in which a new gene is transplanted into the body of a host to improve its health. Recombinant proteins are produced by placing the desired gene in a cell system and producing the desired gene product in vitro. Ex vivo gene therapy is accomplished by transplanting cells containing the desired gene into the host; these cells will then propagate and produce the desired biologic effect. In vivo gene therapy injects a gene construct that will enter the cell of the host and produce the desired protein.

Hemophilias A and B are prime targets for gene therapy. The ideal of a one-time or infrequent treatment as opposed to weekly intravenous injections is attractive, but despite substantial advances there are many hurdles to overcome (34,35). Gene therapy procedures are changing rapidly, and this is noted in the regulatory documents with the suggestion that there be constant contact between the regulators and developers of any product proposed for gene therapy (36,37). Many of the molecular biology issues are covered by the regulatory documents on recombinant products (5,7–10).

Vectors

Vectors and gene product are characterized in the same manner as for cells producing rDNA proteins (8–10). For smaller viral vectors this would include sequencing the entire viral sequence; for larger viral vectors this may not be feasible. The gene product should be completely characterized, however, because the vectors used for in vivo gene transfer will not be expressed in the same cell, and many of the post-translational changes may differ. As with other cell-derived products, the development and characterization of master cell banks will be required (5,7). In addition, there must be establishment of a vector bank, that is, the final product for most of the proposed vectors is the result of the interaction of a cell and the vector. The vector bank should be tested for the presence of adventitious agents. Stability of the host cell–vector system must be established.

The most promising vectors under study for treatment of the hemophilias include replication of defective retroviral and adenoviral vectors, adeno-associated virus, and nonviral vectors. Retrovirus vectors have promise for in vitro exchange with a variety of cells, including fibroblasts, endothelium, and myoblasts. The nonviral vectors would seem to raise fewer issues albeit much lower in vivo expression. The strategy issues to be resolved include choice of vector and in vitro or in vivo gene transfer.

Production Issues

There are different regulatory issues and pathways for the licensure of vectors used in vitro and in vivo. Products for in vivo transduction will be produced purified and packaged in the final form for administration to the patients in a facility meeting all the standards for biologics production (38). For vectors where the transduction will be done in vitro, the producing institution will be responsible for the production of the vector and will have to meet all GMPs; however, the actual product administered will be done after expansion of the patient's transduced cells and will be a service for an individual patient similar to treatment with blood.

Current standards for the production of pharmaceutical products encourage the development of single product production facilities. However, the target group for gene therapy in the hemophilias is small, and only a few production runs may be needed if the hope of long-lasting or permanent effectiveness is achieved. This would lead to a multiuse production area with stringent validation of the facility after changeover from one product to another to prevent cross-contamination.

There are no specific guidelines as yet for the degree of purification for a gene therapy product; rather, it is left to the producer to show adequate purification and justify the adequateness of the procedure. The consistency of the purification method must be demonstrated. Published purification methods for viral vectors involves density gradient ultracentrifugation; it is unlikely that this will be implemented as a production method. Published purification procedures give little information on the degree of purity achieved. Because the desired outcome with gene therapy would be one-time or infrequent exposure to the agent, the presence of small amounts of cell-derived or process-derived protein is less of an issue than for recombinant proteins administered frequently.

The fidelity of reproduction of the vector is an issue because an error in replication could yield a toxic gene product. Where possible the correct nucleotide sequence should be verified.

Lot release testing suggested by the FDA includes (35):

1. Test for RNA or DNA content
2. Test for homogeneity of size and structure of RNA/DNA
3. Test for contamination with RNA or host DNA
4. Test for noninfectious viruses
5. Tests for toxic materials
6. Identity tests by restriction enzyme mapping
7. Sterility including testing for adventitious agents and replication competent viruses
8. Potency

The only test that would be unique for the coagulation products would be the potency test. It will be the responsibility of the manufacturer to design and validate a potency test during the development phase. It is likely that the standard coagulation tests of conditioned media produced by cells treated with a given amount of vector would be adequate for the validation of a lot. Whole animal methods could be developed but could be anticipated to be less quantitative and too time consuming to be useful for lot release.

Possible Adverse Effects

Viral mutagenesis and tumorigenicity are of concern even in replication of incompetent vectors because the genes could be inserted into chromosomes and disrupt growth regulation genes. Identification of the insertion mode and testing for replication competent viruses is recommended for the master cell bank, the working cell bank, and lot release testing on the adenoviral vector products.

The FDA and EEC guidelines (36,37) specifically are concerned with the ability of replication-incompetent viral genomes to form recombinants with cellular viral components. Strategies suggested include separation of the structural and enzymatic genes, removal of complementary viral sequences from the packaging cell lines, and prevention of infection of the packaging cell line with wild-type viruses that could lead to the formation of replication-competent recombinant viruses.

Specific issues with viral vectors are the pathology induced by the viral components. As many pathologic viral sequences as possible should be removed. The induction of virus antibodies preventing retreatment is an issue that can be modulated by minimization of the viral structure and perhaps regulation of the administered dose.

Animal Studies

Preclinical animal testing will be of importance not just for determining efficacy but for supporting safety and studying tissue distribution. All viral vectors will have to be examined by appropriate routes of administration and dose effects in the most sensitive animal models. Species will depend on sensitivity to particular viral toxicity. There is no requirement or indication that regulatory authorities will require primate studies.

Studies are mandated of the vector distribution in animals. A specific issue will be whether there is uptake into germ-line cells. Although this can be studied in animals by study of ovarian and testicular cells, it is not clear whether this will be a reliable indicator of human cell or organ location. Although there is public and political concern about germ-line modification, the FDA does not have any specific prohibition of a product with germ-line modification. The organ of synthesis for both factor VIII and IX is the liver, and this is the site of uptake of the majority of adenoviral vectors administered intravenously.

The cell or organ location is of importance to the final gene product; post-translation modification will be affected by the cell type transduced. There is at this time, however, no set experimental protocol for such studies, and the manufacturer will be expected to use all appropriate tools and develop specific tools as necessary for a given product.

Retroviral transduction and implantation of the ex vivo transduced cells should give an identifiable localized expression of the gene product. There may, however, be specific cell–cell interactions.

Presence of functional clotting should ensure adequate in vivo function. Human factor IX and factor VIII have been maintained in the therapeutic range for up to a half a year in various animal models with various vectors (35). The most promising are the ex vivo retrovirus-transformed fibroblasts or muscle cells and the adenovirus vectors with in vivo transfer.

As noted above, measurable shifts in clearance occur with recombinant products. Although the steady-state levels are measurable in both animals and humans, it will be difficult to measure clearance in patients, although this might be a regulatory issue.

A particular problem with factor VIII may be the need to have the gene product secrete directly into the plasma compartment. In the absence of the stabilizing effect of von Willebrand's factor, it is possible that factor VIII will be proteolyzed to an inactive form before reaching the plasma compartment where it functions.

Persistence of the transgene depends on absence of antibody to the vector-infected cell or the gene product. The antibody response to the vector and the toxicity of the vector can be limited by introducing lower quantities of the viral DNA. More recent experience indicates that immunity to the adenovirus may be circumvented.

Adverse effects of the vector and the implanted gene may be detected before patient exposure only in animal models, and at this time there have been no unexpected events. As expected, there are reports of liver damage with the use of adenoviral vectors, which can be minimized by dose adjustment (39). Adverse effects of the vector and the implanted gene may not be detected in an animal model.

Clinical Trial Issues

The initial decision to attempt gene therapy in hemophilia will be difficult and controversial and will be reviewed by regulatory groups, patient advocates, lawyers, and politicians. In the United States, any clinical use directly or indirectly involved with government funding formerly was approved by the NIH's Recombinant DNA Advisory Committee (RAC) and subsequently by the Office of Recombinant DNA Activities after approval of the the originating institution's internal review board. The results of these reviews were placed in the public domain, and it was understood that all adverse effects would be open information. In addition, the RAC and the FDA often had parallel review processes for other projects. In 1996, the role of the RAC was made more limited, and the FDA assumed all regulatory review and approval responsibility.

Certain principles in regard to informed consent have already been established by the RAC and the FDA. These include many of the usual points such as the trial goal and known risks. In addition, there must be clear discussion of the methodology and the duration of the therapy and of follow-up. One specific point mentioned is the provision for autopsy. The patient must have a commitment and understanding of the trial and the importance of each patient's value to the study as well as the advantages he or she may achieve.

The hemophilia patient who would benefit most would be the patient without access to adequate means of alternative treatment. Because 80% of the hemophilic population in the world does not have access to treatment, one can argue that this is a potential target group. Outside the United States, there has been an effort to treat human factor IX deficiency by gene therapy in China (40).

A second group to be considered for early treatment with gene therapy might be patients with inhibitors (antibodies) to factor VIII. These patients are already undergoing frequent exposure to induce tolerance. Treatment with gene therapy would simply be a modification of tolerance therapy, which could have great benefit for patients with inhibitors. Assessment of therapy would be more difficult if there is no measurable plasma factor VIII. Alternatives could be evidence of a decrease in the inhibitor titer, similar to that seen in classic tolerance regimens, or evidence that there is long-term production of gene product or mRNA. A third group would be those without adequate venous access for administration of current products.

The group that might benefit the most would be the very young or newborn hemophiliacs who, in principle, would not have any of the damage caused by lack of factor VIII or factor IX. It can be anticipated that gene therapy treatment of this group will not be approved in the absence of substantial long-term safety data from older patients.

There will be a regulatory requirement for long-term follow-up: how long and for what are not discussed in the *Points-to-Consider*. The obvious points are evidence of functional expression, unusual immunologic outcomes, viral shedding if a viral vector is used, and cell trafficking. Semen samples will be collected to assess germ-line effects. The patients currently undergoing gene therapy have been informed of the importance of autopsy information.

It is easy to state that the clinical trial must be designed to assess clinical safety and efficacy; it is far more difficult at this time to anticipate with precision the scientific or regulatory requirements, such as the number of patients to be included, the duration before licensure by a regulatory agency, and the statistical design and the number of patients in a monogenic disease where the assessment of bleeding reduction can be done with retrospective data.

The first major effectiveness issue will be the expression level. Clinical efficacy initially will be a plasma level deemed sufficient to minimize hemorrhage, 5 to 10% of the normal plasma level. The need for other replacement therapy will be the ultimate test of efficacy. The initial trials will begin with a dosage based on the animal data or some fraction thereof, and there may well need to be adjustment. Expression should be seen in days to weeks; duration of expression must be monitored frequently. There are no historical data to support toxicity of supranormal levels of factor VIII in the patient with hemophilia. However, nonhemophiliac subjects with supranormal levels of factor VIII have increased incidence of venous thrombosis (40,41).

The immunologic issues, inhibitor formation, and immune complex disease will have to be studied. The worst possible outcome would be the induction of chronic immune complex disease, which has never been reported in the 30-year history of the treatment of hemophilia despite studies indicating increased levels of immune complexes. The other side of the issue is that for years large daily doses of factor VIII have been used to develop immune tolerance in patients with factor VIII inhibitors, and no chronic immune complex disease has been reported. If recipients of gene therapy do develop immune complex disease, it would be observed within days to months and perhaps should be part of the normal treatment protocol as well as play a role in the investigative use of gene therapy in hemophilia.

Although hemophilia patients with inhibitors are routinely treated with factor VIII, these are patients with no intrinsic synthesis of the factor VIII. A concern raised by the FDA is that the cells producing the gene product could be destroyed by an immune cytolysis. This potential outcome would argue against the use of known inhibitor patients for the initial trials. Patients without hemophilia A who develop antibodies to factor VIII have not been reported to have any organ damage despite the presence of cells producing normal amounts of factor VIII.

The development of inhibitors usually occurs with the first 20 treatments, but the current European guidelines request prolonged inhibitor follow-up, and it is likely that long-term follow-up will be requested in gene therapy. Follow-up for the development of antibodies to the vector may also be suggested. If therapy is by a modified viral vector, there may be limited antibody formation. Methods should be developed to detect antibodies to this vector in the presence of wild-type virus infection.

Surveillance of the impact of the vector will be mandatory. The older hemophilia patient population is infected with a variety of chronic viral diseases, and many already have liver disease. Adenoviral vectors targeted to the liver will at least temporarily induce some inflammatory reaction. Other unknown adverse effects resulting from the vector may occur with this novel and dramatic therapeutic approach.

Gene therapy science is moving rapidly, and as old problems are solved, new areas of concern arise. The developers of any such product must be aware of all the issues. Preclinical and clinical studies must be rigorous; at this time, the regulatory guidelines are appropriately rigorous. The best updated source of information on the regulatory issues may be found at the FDA's Center for Biologics Evaluation and Review (CBER) internet site (*www.CBER.FDA.gov*).

TRANSGENIC ANIMAL PRODUCTS

The term *transgenic animal* is used for animals with either somatic cell DNA alterations or germ-line alterations. The potential value of transgenic animals for the production of therapeutic proteins would involve the production of animal strains with germ-line DNA alterations, a regulatory base exists both in the United States and Europe (42,43). The use of transgenic animals to produce proteins secreted into milk is an attractive alternative to the large investment in a sophisticated plant to produce therapeutic proteins. At this time, there is little evidence of commercial interest in this approach. Many coagulation proteins have been produced in the milk of transgenic animals (44).

The regulation in the United States of transgenic products would involve not only the FDA drug or biologics regulatory groups but in addition the veterinary medicine and food regulators of the FDA in addition to the Department of Agriculture. The FDA regulatory issues are spelled out in Reference 42.

Many of the issues of blood-derived products would be involved, including screening for viruses, assessing the health of the donor, etc. The possibility of adapting unknown animal viruses to a human host will not be a trivial issue and will have to be addressed by the producer.

The actual gene and the regulatory element characterization are covered by the *Points-to-Consider* for cell-derived products (8). The regulatory elements are of particular interest, and extensive characterization of the enhancers, promoters, suppressors, and presence of dominant control regions should be reported.

There must be a full record of all the procedures in obtaining the ova, methods for in vitro fertilization, and introduction of the altered DNA. After homologous recombination, it is important to show loss of function of the wild-type gene.

Both the history of the donors of the gametes and the recipient (foster parent) animals need to be documented in detail. Intensive veterinary evaluation of health with particular emphasis on diseases related to the species and the breed will be required.

The founder animal must be studied for the presence of the introduced gene and for the expression of the gene product in addition to the absence of functional product from the inactivated wild-type gene. The characterization of the gene expression should be evaluated as a function of age, season, etc. The acceptable range for the gene product should be established to allow for all the systematic variations. Anatomical localization and function of the transgene should be evaluated, and that the transgene is at the desired location must be verified. There is particular concern about the post-translational modifications in a species and at a tissue site different from the normal human product, which could lead to functional and immulogic problems.

The maintenance of a given transgenic line is presumed desirable, and methods similar to the master working cell bank for cell-derived products is suggested.

In addition to study of the transgene in the founder, there must be extensive study of the transgene stability in the succeeding generations until stability of the gene and the gene expression are demonstrated. In the founder, there may well be multiple gene copies inserted which over succeeding generations will be rearranged or deleted. It is desirable to demonstrate at the founder stage that there is integration at a single chromosomal site. If this is not possible, studies of the chromosomal integration must be done in the later generations. In addition, the RNA transcript should be examined for quality and quantity and identification of the tissues or cells producing the transcript.

The animals must be kept isolated from sources of known and unknown infections during their productive life. Animals would only be acceptable in the absence of disease and with a consistent level of production of the gene product.

ACKNOWLEDGMENTS

The author gratefully acknowledges the helpful guidance and education on gene therapy and international regulatory insights and issues by Terry Snape, Sheila Connelly, Milt Mozen, and Arthur Thompson.

BIBLIOGRAPHY

1. Y. Nemerson, in W.J. Williams, E. Beutler, A.J. Erslev, M.A. Lichtman eds., *Hematology*, McGraw-Hill, New York, 1990, pp. 1295–1304.
2. D.L. Aronson, *Transfusion* **30**, 748–758 (1990).
3. B.L. Evatt, *Thromb. Haemostasis* **74**, 36–39 (1995).
4. K. Peerlick, J. Arnout, J.G. Gilles, J.M. Saint-Remy, J. Vermylen, *Thromb. Haemostasis* **69**, 115–118 (1993).
5. *Points-to-Consider in the Characterization of Cell Lines Used to Produce Biologics*. Center of Biologics Evaluation and Review, U.S. Food and Drug Administration, Bethesda, Md., 1993.
6. H. Hopps, J.C. Petricciani, *Abnormal Cells, New Products and Risks*. Tissue Culture Association, Gaithersburg, Md., 1985.
7. *Acceptability of Cell Substrates for Production of Biologicals*. World Health Organization Tech. Report Ser. **747**, 1 (1987).
8. *Points-to-Consider for Products Derived from Mammalian Cells*. Center of Biologics Evaluation and Review, U.S. Food and Drug Administration, Bethesda, Md., 1987.
9. Committee for Proprietary Medicinal Products, *Note for Guidance: Production and Quality Control of Medicinal Products Derived by Recombinant DNA Technology*, 1995 (*http://www.eurdra.org/emea.html*).
10. *Quality of Biotechnological Products: Analysis of the Expression Construct in Cells Used for Production of R-DNA Derived Protein Products*, International Conference on Harmonization (*http://ifpma.org/ich/*).
11. J. Petricciani, F. Horaud, *Biologicals* **23**, 233–238 (1995).
12. European Agency for the Evaluation of Medicinal Products, Human Medicines Evaluation Unit, *Note for Guidance to Assess Efficacy and Safety of Human Plasma Derived Factor VIII: C and Factor IX: C Products in Clinical Trials in Haemophiliacs before and after Authorisation*, 1996 (*http://www.eurdra.org/emea.html*).
13. S. Harrison, B. Clancy, S. Brodeur, P. Oakes, D. Miller, D. Drapeau, M. Hamilton, T. Charlebois, M. Leonard, M. McCarthy, R. Zollner, S.R. Adamson, *Thromb. Haemostasis* **73**, 1235 (1995).
14. Committee for Proprietary Medicinal Products, *Note for Guidance on Viral Validation Studies: The Design, Contribution, and Interpretation of Studies Validating the Inactivation and Removal of Viruses*, 1996, (*http://www.eurdra.org/emea.html*).
15. A.B. Heath and T.W. Barrowcliffe, *Thromb. Haemostasis* **68**, 155–159 (1992).
16. G. Kemball-Cook, J.E. Tubbs, N.J. Dawson, T.W. Barrowcliffe, *Br. J. Haematol.* **84**, 273–278 (1993).
17. K. Fijnvandaat, E. Berntorp, J.W. ten Cate, H. Johnson, M. Peters, G. Savidge, L. Tengborn, J. Spira, C. Stahl, *Thromb. Haemostasis* **77**, 298–302 (1997).
18. C.A. Lee, T. Barrowcliffe, G. Bray, E. Gomperts, A. Hubbard, G. Kemball-Cook, P. Lilley, D. Owens, I. Von tilberg, J. Pasi, *Thromb. Haemostasis* **76**, 950–956 (1996).
19. H.P. Schwarz, P.L. Turecek, L. Pichler, A. Mitterer, W. Mundt, E. Dorner, J. Roussi, L. Drouet, *Thromb. Haemostasis* **78**, 571–576 (1997).
20. A.R. Giles, S. Tinlin, H. Hoogedorn, M. Fournel, P. Ng, N. Pancham, *Blood* **72**, 335– (1988).
21. I.R. MacGregor, L.E. McLauglin, M.C. MacGregor, C.V. Prowse, D.S. Pepper, *Vox Sanguinis* **69**, 319–327 (1995).
22. I.R. MacGregor, J.M. Ferguson, L.E. McLaughlin, T. Burnouf, C.V. Prowse, *Thromb. Haemostasis* **66**, 609–613 (1991).
23. J.C. Keith, T.J. Ferranti, B. Misra, T. Frederick, B. Rup, K. McCarthy, R. Faulkner, L. Bush, R.G. Schaub, *Thromb. Haemostasis* **73**, 101–105 (1994).
24. Investigational New Drug Applications, *Code of Federal Regulations*, Title 21, Part 312.
25. *Code of Federal Regulations*, Title 21, Part 601.25.
26. *Code of Federal Regulations*, Title 21, Part 314.126.
27. M. Morfini, M. Lee, A. Messori, *Thromb. Haemostasis* **66**, 384–386 (1991).
28. P.M. Mannucci, M. Colombo, *Thromb. Haemostasis* **61**, 532–534 (1989).
29. J.M. Lusher, S. Arkin, C.F. Abilgaard, R.S. Schwartz, *N. Engl. J. Med* **328**, 453–459 (1993).
30. G. White, J. Lusher, A. Shapiro, K. Tubridy, S. Courter, *Blood* **88**, 1296 (1996).
31. B. Brody, *Ethical Issues in Drug Testing, Approval, and Pricing*, Oxford Univ. Press, Oxford, U.K., 1995.
32. D. Beebe, G. Murano, in A.S. Lubiniecki, S.A. Vargo eds., *Tissue Plasminogen Activator: Regulatory Issues in Regulatory Practice for Biopharmaceutical Production*, Wiley-Liss, New York, 1994, pp. 281–294.
33. J.J. Emeis, C.M. van den Hoogen, D. Jense, *Thromb. Haemostasis* **54**, 661–664 (1985).
34. Thompson A., *Thromb. Haemostasis* **74**, 45–51 (1995).
35. S. Connelly, M. Kaleko, *Thromb. Haemostasis* **78**, 31–41 (1997).
36. *Points to Consider in Human Somatic Cell and Gene Therapy*, Center of Biologics Evaluation and Review, U.S. Food and Drug Administration, Bethesda, Md., 1996.

37. Committee for Proprietary Medicinal Products, *Note for Guidance: Gene Therapy Products—Quality Aspects in the Production of Vectors and Genetically Modified Somatic Cells* (III/5863/93 Final) (*http://www.eurdra.org/emea.html*).
38. Current Good Manufacturing Practices for Finished Pharmaceuticals, *Code of Federal Regulations*, Title 21, Part 211.
39. S. Connelly, J.M. Gardner, R.M. Lyons, A. McClelland, M. Kaleko, *Blood* **87**, 4671–4677 (1996).
40. D.-R. Lu, J.-M. Zhou, B. Zheng, X.F. Qui, J.L. Xue, J.M. Wang, P.L. Meng, F.L. Han, B.H. Ming, X.P. Wang, *Sci. China [B Series]* **36**, 1342–1351 (1993).
41. T. Koster, A.D. Blann, E. Briet, J.P. Vandenbroucke, and F.R. Rosendaal, *Lancet* **345**, 152–155 (1995).
42. *Points-to-Consider in the Manufacture and Testing of Therapeutic Products for Human Use Derived from Transgenic Animals*, Center of Biologics Evaluation and Review, U.S. Food and Drug Administration, Bethesda, Md.
43. Committee for Proprietary Medicinal Products, *Note for Guidance: Use of Transgenic Animals in the Manufacture of Biological Medicinal Products for Human Use*, 3612/93 (*http://www.eurdra.org/emea.html*).
44. W.N. Drohan, *Thromb. Haemostasis* **78**, 543–547 (1997).

COFACTOR REGENERATION, NICOTINAMIDE COENZYMES

JOERG PETERS
Bayer AG
Wuppertal, Germany

KEY WORDS

Coenzyme recycling
Enzyme-coupled approach
Immobilization
Inhibition
Reaction kinetics
Reaction mechanisms
Reactor systems
Regeneration methods
Substrate-coupled approach
Total turnover number

OUTLINE

INTRODUCTION

Today, there are about 3500 known enzymes (1), of which about 70% are cofactor dependent and about 15% are commercially available. Cofactors such as adenosine triphosphate (ATP), nicotinamide adenine dinucleotide (NADPH), acyl coenzyme A, *S*-adenosylmethionine (SAM), sugar nucleotides, and 3′-phosphoadenosine-5′-phosphosulfate (PAPS) are involved in stoichiometric amounts in cofactor-dependent enzymatic reactions (2). The very high cost of these cofactors prohibits their use as stoichiometric reagents. For economic reasons, regeneration of the cofactors from their reaction products is an essential requirement in making use of cofactor-dependent enzymes in preparative synthesis of organic compounds.

Using coenzyme recycling, the expensive cofactor is needed only in catalytic amount, so there is a drastic reduction in cost of the respective reactions. Additionally, coenzyme regeneration can accomplish three major objectives. First, it can influence the position of equilibrium of a reaction. A thermodynamically unfavorable reaction can be driven by coupling with a favorable cofactor-regeneration reaction. Second, regeneration affords only catalytic amounts of coenzyme, preventing the accumulation of cofactor by-product, which may inhibit the reaction of interest. Third, coenzyme regeneration can simplify the downstream processing of the reaction mixture.

The following chapter deals with methods for the regeneration of nicotinamide coenzymes, because this type of cofactor is the most significant. Hitherto, more than 650 oxidoreductases are known (3), of which about 90 are commercially available. For the majority of these redox enzymes, about 80% and 10% require nicotinamide adenine dinucleotide (NADH) or its respective phosphate (NADPH) respectively.

REGENERATION OF NICOTINAMIDE COENZYMES

Introduction

Nicotinamide adenine dinucleotide (NAD) and the analogous 2′-phosphate (NADP) are involved in redox reactions catalyzed by alcohol dehydrogenases. Two electrons and a proton (hydride) are transferred from the reduced coenzyme NADPH to the carbonyl compound in a stoichiometric reaction. Using coenzyme recycling, the expensive cofactor is needed only in catalytic amounts, leading to a drastic reduction in the cost of dehydrogenase-catalyzed reactions. This chapter discusses methods for regeneration of nicotinamide coenzymes.

Cost of Coenzymes. Nicotinamide adenine dinucleotides (NADH, NADPH) are expensive compounds. In Table 1 the actual prices per mole are listed and compared with the costs in the year 1976. If these coenzymes were used in stoichiometric amounts, enzymatic reductions with dehydrogenases would be prohibitively expensive. Therefore, the application of dehydrogenases on the industrial scale requires effective methods for coenzyme regeneration. The economic barrier to large-scale reactions posed by cofactor costs has been recognized for many years. A range of publications address this topic (2,5,6).

The turnover number (TN) is defined as the number of moles of product formed per mole of cofactor or enzyme per unit time. This number is indicative of the productivity of

Table 1. Prices of Nicotinamide Coenzymes

Coenzymes	Price in dollars per mole (depending on purity, 1976)[a]	Price in dollars per mole (mean catalog prices, 1998)
NAD^+	1.722 to 49.550	710
NADH	6.030 to 52.766	3
$NADP^+$	20.667 to 168.604	25
NADPH	177.032 to 535.197	215

[a]from Ref. 4.

the process. The total turnover number (TTN) is defined as the total number of moles of product formed per mole of coenzyme or enzyme during the course of the entire reaction. Hence, the TTN reflects the loss of coenzyme and enzyme throughout the process, and thus it is a useful estimate of the operational cost of the coenzyme or the enzyme. In order to overcome economic limitations, total turnover numbers of greater than 100 are desirable. For a batch synthesis, the cost of the coenzyme or enzyme is calculated on the basis of its initial cost. In contrast, for a continuous process, the cost may be calculated on the basis of the activity loss over the entire process time. When estimating the cost of a given reaction per unit time, the rate of the reaction must also be considered. The regeneration cost is defined as the cost of the components (enzymes, reagents, and coenzymes) required to regenerate one mole of coenzyme per unit time (7).

The highest TTN can be achieved as follows:

- The substrate concentration is as high as possible, determined by the substrate solubility, product inhibition, and enzyme stability.
- The concentration of coenzyme and enzyme to achieve acceptable reaction rates is as low as possible.
- The coenzymes and enzymes are highly stable.

Thus, both the initial costs of the cofactor and the enzyme, and the efficiency of their regeneration and utilization, determine their contribution to the cost of the product.

Origin of Coenzymes. Nicotinamide adenine dinucleotide (NAD) is presently isolated from yeast (8), and its 2′-phosphate NADP is normally synthesized by enzymatic phosphorylation of NAD using NAD-kinase (EC 2.7.1.23) and ATP (9,10). NADPH can be prepared from NADP by three different methods: chemical (11), enzymatical (12,13), or microbial reduction (14). Additionally, a combined chemical and enzymatic route for the synthesis of NADP has been developed that may be useful for the preparation of NADP analogs (15–17).

Requirements of an Economical Coenzyme Regeneration System. Chenault and Whitesides (4) compiled the following criteria for an ideal coenzyme regeneration system:

- Compatibility with the synthetic reaction
- Commercial availability of inexpensive and stable enzymes with high specific activity
- Simple and cheap auxiliary substrates; the corresponding products should not interfere with the isolation of the product of interest. They should not influence the stability of enzymes and coenzymes.
- High turnover numbers (TN) and total turnover numbers (TTN).
- The equilibrium constants of the coupled coenzyme regeneration system should be as high as possible.
- The analysis of the yield should not be effected by the substrates or products of the coenzyme regeneration.

As a rule of thumb, 10^3–10^4 cycles are sufficient for laboratory-scale redox reactions, whereas at least 10^5–10^6 cycles are desirable for industrial-scale synthesis. The exact turnover number required depends on the initial cost of the coenzyme and the value of the product of interest. A high turnover number requires a high selectivity for the formation of enzymatically active coenzyme. 1,4-dihydro-NAD^+, or $NADP^+$, is the only enzymatically active form; 1,2- and 1,6-dihydro species are not active. For instance, if 50% of the original coenzyme activity is to remain after 1,000 cycles, the regeneration reaction must be 99.3% regioselective. In the case of industrial applications of dehydrogenases, 10^6 turnovers are desired, resulting in a requirement for the regioselectivity of 99.99993%. Chemical, electrochemical, photochemical, or enzymatic methods (see later) show different selectivities in the formation of 1,4-dihydro $NADP^+$. Electro- and photochemical methods especially suffer from poor regioselectivities. Thus, the requirement for very high regioselectivity necessitates enzymatic catalysis, particularly in the recycling of NADH and NADPH (2,6,18).

Stability of Nicotinamide Coenzymes. *Nicotinamide coenzymes contain three labile bonds that may be the target of general or specific acid-or base-catalyzed hydrolysis*:

- the N-glycosidic bond between nicotinamide and ribose
- The phosphodiester bond between the two riboses
- The bond between ribose and phosphate (NADP)

Oppenheimer (19) and Chenault and Whitesides (6) compiled data on the influence of different buffers on the stability, and the possible products from the hydrolysis of NADPH. Under basic conditions, the reduced nicotinamide coenzymes are stable; under acidic conditions, they are unstable. In contrast, the oxidized coenzymes are stable under acidic conditions, and labile under basic conditions (19,20). As a compromise, NAD and NADH are used at pH 7–7.5, whereas NADP and NADPH are used at pH 8–8.5. The mechanisms of the decomposition of NADPH were studied in great detail (20–22). The N-glycosidic bond between nicotinamide and ribose is the most labile bond of the molecule. Base-catalyzed hydrolysis of this bond results in the removal of the nicotinamide ring and, thus, the coenzymatic activity is destroyed. On the other hand, the general acid-catalyzed protonation of the nicotinamide ring at position C_5 is followed by a rapid rearrangement to a cyclic ether product. Additionally, the 2′-phosphate group of NADPH catalyzes this decomposition intramolecularly. The half lifes for NADH and NADPH in 0.1 M phosphate (pH 7, 25 °C) are 27 h and 13 h, respectively (20). Organic buffers such as imidazole, Tris, Hepes, and triethanolamine, partially stabilize reduced coenzymes, resulting in a half life that is three times as long. The major pathway for the decomposition of $NADP^+$ is nucleophilic addition at C_4 of the nicotinamide ring, yielding a 1,4-dihydropyridine structure (23–25). In Tris-buffered solution at pH 7–8, $NADP^+$ is stable over at least 14 days (Peters, unpublished data). There is a strong dependence of the stability of the oxidized coenzymes on the concentration of the buffer (Peters, unpublished data). At low molarity (50 mM), the half-life in Tris buffer is lower compared with the half-life in buffer of high molarity (500 mM). Poly(ethylene)-glycol-coupled $NADP^+$ is very stable at pH 8 in Tris buffer of any molarity. Half-lifes of more than 240 days have been reported (Peters, unpublished data).

Principles for the Regeneration of Nicotinamide Coenzymes

Nonenzymatic Regeneration. Nonenzymatic methods for coenzyme regeneration can be divided in chemical, electrochemical, and photochemical methods. Chemical reduction of $NADP^+$ using a reducing agent like sodium dithionite ($Na_2S_2O_4$) suffers from very low total turnover number of less than 100 (26). Moreover, this agent can deactivate enzymes, presumably by modification of thiol groups in the protein (27). The major disadvantages of most electrochemical and photochemical regeneration methods (28–31) are low regioselectivity, leading to coenzyme inactivation (as already discussed) occurrence of side reactions, and low total turnover numbers of less than 1,000 (2,6). However, Simon and coworkers (32) stated that electro- and photochemical methods should not be underestimated. Schummer and coworkers (33) reported the synthesis of polyfunctional (*R*)-2-hydroxycarbonic acids on the preparative scale with resting cells of *Proteus vulgaris* by using electrochemical regeneration. Several groups (34–37) reported on the electrochemical regeneration of NADPH by using a bipyridine rhodium-(I) complex as an electron-transfer agent and formate as an electron donor. The bipyridine-rhodium complex can be coupled to poly(ethylene) glycol and thereby retained in an enzyme-membrane reactor. Dicosimo and coworkers (38) described the electrochemical regeneration of NADPH using the dye methyl viologen and flavoenzymes. However, enzyme denaturation on the surface of the electrode or enzyme deactivation by redox agents may occur.

In summary, enzymatic methods meet most of the requirements for an ideal coenzyme regeneration system as described earlier. The difficulties encountered with enzymatic coenzyme regeneration are mainly competitive or mixed inhibition by the substrate or product (39,40). In the following sections, the two general concepts for coenzyme regeneration using enzymes are presented (18). Focus will then turn to the most attractive methods for the recycling of reduced and oxidized nicotinamide coenzymes.

Substrate-Coupled Regeneration. In the substrate-coupled approach, both the main reaction and the regeneration of the coenzyme are catalyzed by a single enzyme (Fig. 1). The auxiliary substrate (S'_{red}) of the coenzyme regeneration reaction serves as the hydride donor. In order to shift the equilibrium of the reaction to the desired position, the hydride donor, usually a low-cost alcohol, is applied in large excess. However, yield of the system is limited by equilibrium of the coupled system. For the reduction of acetophenone using 2-propanol as auxiliary substrate, the equilibrium was reached at 60% yield (41).

In the case of the reduction of 2-acetyl-naphthaline using 2-propanol as hydride donor, the equilibrium could be shifted by the use of cyclodextrins, and thereby the total yield was increased to 93% of 1-(2-naphthyl)-ethanol (42). In the continuous synthesis, yields of more than 80% were obtained, and total turnover numbers for the coenzyme were in the range of 10^4. In this application the native coenzyme was not recovered from the product solution but was directly used in substoichiometric amounts. The total turnover number for the enzyme was greater than 10^6, and the productivity was 2.5 times as high as the comparable system with enzyme-coupled cofactor regeneration (formate dehydrogenase system). This demonstrates the potential of substrate-coupled systems for coenzyme regeneration.

Due to the lack of a well-established method for NADPH regeneration, the substrate-coupling method is mainly used for this purpose. However, there are some drawbacks of this method:

- The auxiliary substrate has to be present in a large excess, which may lead to complications in product recovery, or which may cause enzyme deactivation or inhibition.
- Enzyme deactivation may also occur as the result of the accumulation of highly reactive carbonyl species like acetaldehyde or cyclohexenone.

A special enzyme reactor employing a membrane that is only permeable to gases (comparable to Gore-Tex) may avoid most of these drawbacks. This type of reactor was successfully applied to improve cofactor recycling via the yeast alcohol dehydrogenase–ethanol system in the stereoselective reduction catalyzed by lactate dehydrogenase (43). The reaction scheme is shown in Figure 2.

Enzyme-Coupled Regeneration. In the case of the enzyme-coupled approach to NAD regeneration (Fig. 3), the reduction of the main substrate and the coenzyme re-

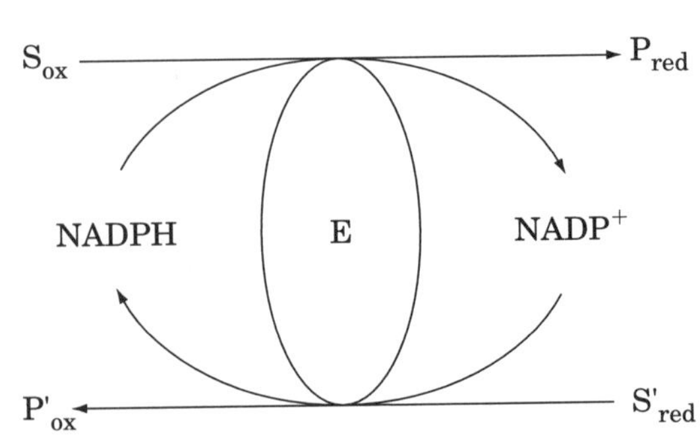

Figure 1.

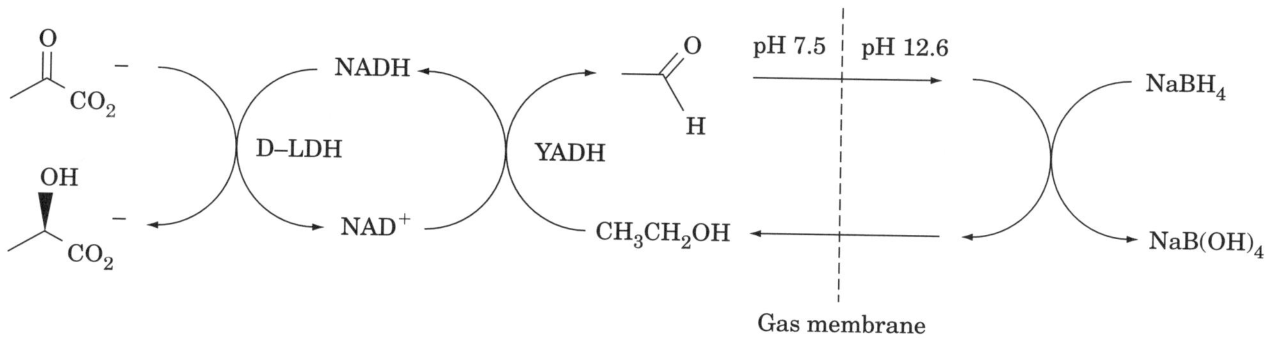

Figure 2.

generation reaction are catalyzed by two different enzymes (44). The enzymes employed should have sufficiently different substrate specificities to avoid interferences. The enzyme-coupled cofactor recycling has the following major advantages:

- The position of the equilibrium of a redox reaction can be influenced. Thermodynamically unfavorable reactions can be driven by coupling with a favorable cofactor regeneration reaction.
- High turnover numbers can be achieved: >600,000 (45).
- The activity ratio has to be optimized for each pair of enzymes to ensure neither lack nor excess of reduction equivalents (46).

However, there are also some disadvantages to this type of cofactor-recycling method:

- Cross inhibition between the main substrate and the coenzyme regeneration system may occur (46).
- A second enzyme must be used, generating additional costs.
- The characteristics of the second enzyme, such as pH and temperature optimum, must be compatible with the main reaction.

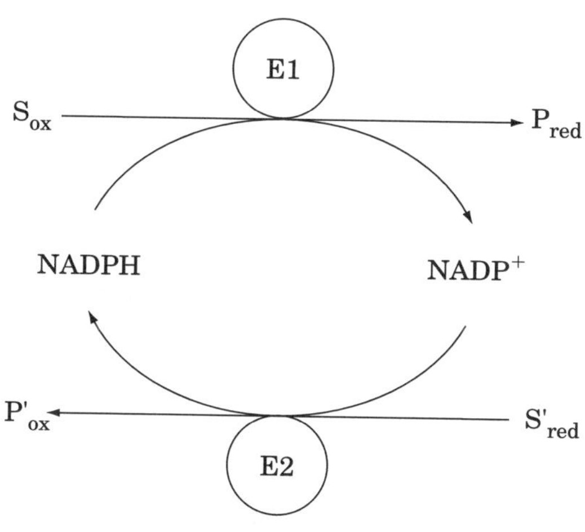

Figure 3.

Survey of Coenzyme Regeneration Methods and Their Assessment

Different methods for the regeneration of NADPH have been discussed in great detail by several authors (2,6,7,18,20,32,47–52). In this chapter, only the most useful methods are discussed and assessed with respect to their applicability in batch and continuous synthesis. Furthermore, recently developed new methods for nicotinamide coenzyme recycling are highlighted.

Regeneration of NADPH. In Table 2, the major disadvantages of a selection of available cofactor-recycling systems are listed, and each method is assessed with respect to its applicability in batch or continuous processes.

For the regeneration of native and poly(ethylene) glycol-bound NADH (PEG-NADH), the formate dehydrogenase (FDH) from *Candida boidinii* (53) has been widely and successfully applied (45,54–57). This enzyme catalyzes the oxidation of inexpensive formate (HCOOH) to carbon dioxide (CO_2). The reverse reaction is hardly detectable (53). Therefore, the FDH reaction can shift the equilibrium of unfavorable reactions to the product side. Both the auxiliary substrate and the coproduct are innocuous to enzymes and are easily removed from the reaction. Therefore, product isolation is easy. FDH has a broad pH optimum for activity, so it can be easily implemented in coupled enzymatic synthesis. The major disadvantages are the low specific activity of 4 U/mg and product inhibition by NADH. However, these drawbacks can be circumvented or minimized: the former by using immobilized or membrane-retained systems (54,58); the latter by adjusting the optimal ratio of enzymatic activities for the main reaction and the coenzyme-recycling reaction (46). Overall, the formate–FDH system is the most economical method for regeneration of NADH for batch and continuous processes. The total turnover numbers (TTN) range typically from 10^3 to 10^5 (45,56,58). Unfortunately, the *C. boidinii* formate dehydrogenase does not accept $NADP^+$.

Although several NADP-dependent formate dehydrogenases have been discovered (59–63), only the recently described mutant formate dehydrogenase from *Pseudomonas* sp. 101 (64,65), using $NADP^+$ as coenzyme, was applied in batch and continuous synthesis. Acetophenone was reduced by the NADPH-dependent alcohol dehydrogenase from *Lactobacillus* sp. (EC 1.1.1.99) with this enzyme for the regeneration of NADPH, at a space–time yield

Table 2. Methods for In Situ Regeneration of NADPH

Methods	Major disadvantages	Assessment
Glucose/glucose dehydrogenase	Not applicable in the enzyme-membrane reactor K_m value for PEG-$NADP^+$ 1,000 times that of native $NADP^+$ Complication of product isolation caused by gluconate High amounts of K^+ or Na^+ necessary to stabilize enzyme	Useful only for lab-scale reactions in batch reactors with native NADH or NADPH, but not for PEG-NAD(P)H
Glucose-6-P/glucose-6P dehydrogenase	High costs for glucose-6-phosphate (~5.500 $/mol) Coenzyme deactivation by 6-phosphogluconate No efficient usage of the coenzyme Complication of product recovery caused by 6-phosphogluconate	Useful only for lab-scale reactions, regeneration of NADH and NADPH
Glucose-6-sulfate/glucose-6-P dehydrogenase	Complication of product recovery caused by 6-sulfogluconate Substrate is accepted only by glucose-6-phosphate dehydrogenase from yeast	Circumvents some limitations of the former method; useful for preparative applications in continuous reactors; regeneration of NADH and NADPH
Ethanol/alcohol and aldehyde dehydrogenase	Relatively complicate and unstable multienzyme system with low TTN Oxygen sensitivity of alcohol and aldehyde dehydrogenases Low reactivity of the aldehyde dehydrogenase with $NADP^+$ Only activated carbonyl substrates are reduced at good yields, which may cause enzyme deactivation	Useful only for lab-scale reactions; regeneration of NADH and NADPH
Formate/formate: NAD^+ dehydrogenase	Low specific activity (3 U/mg) Deactivated by autooxidation Product inhibition caused by PEG-NADH	Simple and stable system for regeneration of PEG-NADH in batch or continuous mode of operation up to the process scale; very good compatibility with other enzymes
Formate/formate: $NADP^+$ dehydrogenase	Enzyme not commercially available yet	Simple and stable system for regeneration of NADPH in batch or continuous mode of operation; very good compatibility with other enzymes
Sec. alcohol/*Th. brockii* alcohol dehydrogenase	Deactivation by autooxidation possible Relatively expensive enzyme (58 cents/unit)	Simple and stable system for regeneration of PEG-NADPH in batch or continuous mode of operation; very good compatibility with other enzymes

of 8 g/(L day) and a conversion rate of 90–95% (66,67). The new FDH is quite stable, showing no loss in activity over one year at 4 °C (67). Unfortunately, this enzyme is not yet commercially available. A further disadvantage is the relatively high K_m value of the mutant FDH for NADP (320 μM). Additional work is required to engineer FDH mutants with improved kinetic properties. In conclusion, the new formate–NADP dehydrogenase (EC 1.2.1.2) has a large potential for use as a regeneration system for NADPH.

The secondary alcohol dehydrogenases from *Thermoanaerobium brockii* (TBADH) accept a broad range of secondary alcohols and $NADP^+$ as coenzyme (68). The enzyme is commercially available and contains a very low NADP-2′-phosphatase activity (Peters, unpublished data) (NADP(H) is hydrolyzed at the 2′-ribose phosphate bond at a rate of 0.1% per day in the continuous synthesis of NADPH (69).) This enzyme has been applied in the reduction of numerous carbonyl compounds using the substrate-coupled approach for coenzyme recycling (70–75). Since this enzyme tolerates up to 30% isopropanol, the substrate-coupled regeneration of NADPH is highly favorable. The TBADH-enzyme family is also an excellent system for the regeneration of native and poly(ethylene) glycol-bound NADPH (Peters, unpublished data; 69). In a model system, glutamate dehydrogenase (GluDH) was employed for the synthesis of glutamate from α-ketoglutarate. A complete set of kinetic data was recorded and the effectivity of the regeneration of PEG-NADPH by the TBADH was studied in a batch reactor. The optimal activity ratio of GluDH/(GluDH + TBADH) was 0.6. At this ratio, 95% of the total coenzyme is reduced. The conversion of α-ketoglutarate is affected negatively only at a TBADH portion of less than 20%. For comparison, in the leucine dehydrogenase–FDH system studied by Wichmann and coworkers (55), only 10% PEG-NADH is present at the optimal activity ratio of 1:1. The reason for this different behavior is the product inhibition of FDH by PEG-NADH. In contrast, TBADH shows only slight product inhibitions by 2-butanone and PEG-NADPH. The loss of coenzyme in the enzyme-membrane reactor is a function of the absolute coenzyme concentration (76) and, therefore should be as low as possible. The K_m value of the TBADH for native and

PEG-bound NADPH is 10 μM. Hence, only 0.2–0.3 mM NADPH is usually sufficient to ensure a maximum usage of the catalytic activities in the reactor.

Overall, the TBADH is recommended for the recycling of NADPH for the following reasons. First, the enzyme is commercially available and is stable in a polypropylene enzyme-membrane reactor. Second, the specific activity is almost 10 times as high as that of formate dehydrogenase. Favorably, the K_m values for native and PEG-bound NADPH are low (10 μM), and the product inhibitions by 2-butanone and PEG-NADPH are also low. Third, the TBADH is compatible with high amounts of organic solvent (e.g., 30% isopropanol) and is stable at elevated temperatures.

Regeneration of $NADP^+$. Oxidized nicotinamide coenzymes are used for the synthesis of carbonyl compounds from the corresponding racemic hydroxy compounds. For their regeneration, both enzymatic and nonenzymatic methods have been reported (for review see Ref. 7). Enzymatic methods seem to be preferred because of their compatibility with biological systems and their simplicity. The regeneration of oxidized nicotinamide coenzymes is somewhat problematic due to thermodynamics that are often unfavorable. Additionally, product inhibition by the reduced coenzyme may cause problems (see later).

In Table 3, the major disadvantages of various available cofactor-recycling systems are listed, and each method is assessed with respect to its applicability in batch or continuous processes. The alcohol dehydrogenase from *Pseudomonas* sp. which is not yet commercially available, has potential for the recycling of NAD^+ (77,78). A range of simple ketones can be used as hydrogen acceptors, which are reduced at high velocity (up to 270 U/mg). However, the kinetic limitations of this enzyme must be explored in more detail.

For recycling of $NADP^+$, the secondary alcohol dehydrogenase from *Thermoanaerobium brockii* (TBADH) is problematic due to the low product-inhibition constant K_{ip} for $NADP^+$ of 50 μM (Peters, unpublished data; 69). The K_m value of NADPH is 10 μM. Therefore, the oxidation of alcohols producing the reduced coenzyme is the favorable reaction.

Acetaldehyde and the commercially available alcohol dehydrogenase from baker's yeast (YADH) have also been used to regenerate NAD^+ from NADH (79). The total turnover numbers were 10^3 to 10^4. However, deactivation of the enzyme by the highly reactive acetaldehyde and the self-condensation of acetaldehyde outweigh the advantages of the inexpensive enzyme and the volatility of the reagents involved.

The most widely applied method for regeneration of oxidized nicotinamide coenzymes involves glutamate dehydrogenase (GluDH), which catalyzes the thermodynamically favorable reductive amination of inexpensive α-ketoglutarate to give L-glutamate (40,80). Both the substrate and the product are innocuous to enzymes. The commercially available and inexpensive GluDH accepts NADH, NADPH, and their respective poly(ethylene) glycol-bound coenzymes (Peters, unpublished data) and has a reasonably high specific activity of 40 U/mg protein. The products of the reductive amination of α-ketoglutarate, PEG-$NADP^+$ and L-glutamate, do not strongly inhibit the reaction (Peters, unpublished data). The ratio of the product inhibition constant to the Michaelis–Menten constant (K_{ip}/K_m) will determine the efficacy of the reaction (40). At a K_{ip}/K_m ratio greater than 1, the reaction may proceed to acceptable conversions. If this ratio is less than 1, the reaction can never proceed efficiently. The K_{ip}/K_m ratio for the substrate and product of the reductive amination of α-ketoglutarate to L-glutamate is 10.5. For the PEG-NADPH and PEG-$NADP^+$, the K_{ip}/K_m ratio is about 30 (Peters, unpublished data). Hence, product inhibition by L-glutamate and PEG-$NADP^+$ does not play a significant role in the efficacy of the regeneration reaction. The disadvantage of this regeneration method is that L-glutamate may complicate product isolation from the reaction mixture. This must be addressed case by case.

Pyruvate and commercially available and inexpensive lactate dehydrogenase (LDH) have been used for recycling of NAD^+ (6,81). The LDH is stable and has a high specific activity of about 1,000 U/mg protein. However, the redox potential is less favorable and, in contrast to GluDH, LDH does not accept NADPH. A further disadvantage is that pyruvate tends to polymerize in solution and reacts with NAD^+ in a process catalyzed by LDH. Despite these draw-

Table 3. Methods for In Situ Regeneration of $NADP^+$

Methods	Major disadvantages	Assessment
Alcohol/*Pseudomonas* sp. alcohol dehydrogenase	Enzyme not commercially available	
Sec. alcohol/*Thermoanaerobium brockii* dehydrogenase	Strong product inhibition by PEG-$NADP^+$; K_{ip} (NADP), 50 μM; K_m (NADPH), 10 μM	Not useful due to the severe product inhibition by PEG-$NADP^+$
Acetaldehyde/yeast alcohol dehydrogenase	Self-condensation of acetaldehyde Enzyme deactivation by highly reactive acetaldehyde	Not useful for the regeneration of NAD^+ due to side reactions and deactivation by acetaldehyde
α-Ketoglutarate glutamate dehydrogenase	Glutamate may complicate product isolation	Simple and stable system for regeneration of PEG-NADPH, good compatibility with other enzymes
Pyruvate/lactate dehydrogenase	Pyruvate tends to polymerize in solution and reacts with NAD^+ in a process catalyzed by lactate dehydrogenase	Simple and stable system for regeneration of NAD^+; side reactions may happen

backs, the pyruvate–LDH system has been applied in enzymatic oxidation reactions of 10–100 mmol of material.

Immobilized Coenzymes and Reactor Concepts

Enzyme-catalyzed reactions are often accompanied by the following disadvantages:

- Enzymes may not be sufficiently stable under the reaction conditions employed. Some lose their catalytic activity due to autooxidation, others due to self-digestion and/or denaturation by the solvent, and others due to mechanical shear forces.
- Isolated enzymes are water soluble and must be used repeatedly due to the economics of the processes for which they are used (82). They must be separated from the reaction mixture without destroying their catalytic activity.
- The productivity, expressed as the space–time yield, is often low due to the limited tolerance of enzymes to high concentrations of substrate(s) and product(s).

Some of these problems may be overcome by "immobilization" of the enzyme. However, a special problem is encountered with NADPH-dependent dehydrogenases because the coenzyme has to associate and dissociate freely to and from the active site(s) of the enzyme(s). Therefore, coimmobilization of the dehydrogenase and the cofactor is necessary to make the system practical (83). Either dehydrogenase, coenzyme, or both may decompose, leading to the replacement of the immobilized system.

If the K_m value for the coenzyme is in the micromolar range and the substrate-coupled approach for coenzyme regeneration is employed (as discussed earlier), a column reactor can be used with a small amount of free coenzyme dissolved in the mobile phase (72). This situation is particularly efficient if the reaction mechanism is an ordered bi-bi- or Theorell–Chance mechanism where the coenzyme is bound first to the active site, followed by the substrate. These mechanisms were found for a broad range of alcohol dehydrogenases.

The immobilization techniques most widely used for coenzyme-dependent dehydrogenases are shown in Figure 4. Immobilization techniques for coenzymes may be divided into coupling and entrapment methods.

Coupling Methods. For dehydrogenases, coupling methods can be further divided into methods for coupling the coenzyme on cross-linked enzymes or methods for coupling the enzyme and coenzyme on a carrier.

Coupling of the Coenzyme to Cross-Linked Enzymes. Enzymes can be attached to each other by covalent bonds, termed *cross-linking*. Bifunctional reagents such as glutardialdehyde, dimethyl adipimidate and dimethyl suberimidate can be used. The coenzyme can be bound to cross-linked enzymes using a spacer long enough so that the coenzyme has access to the active sites of both enzymes, thereby transferring the hydride. These requirements are very difficult to meet in practice. Lortie and coworkers (84) reported on the coimmobilization of horse liver alcohol dehydrogenase (HLADH), albumin, and NAD^+ using glutaraldehyde as the cross-linking agent. Long-chain aldehydes were prepared using this immobilized HLADH/NAD^+ system in a packed-bed reactor (glass column). The conversions were performed in organic solvent (hexane) saturated with buffer. Because proteins and coenzymes are insoluble in the water immiscible hexane, they remained in the aqueous phase. The long-chain alcohols (substrate) diffused into the aqueous phase, underwent enzymatic conversion, and the product (aldehyde) finally diffused back into the organic phase. This system was limited by the mass transfer between the organic phase and the water-containing porous particles. Lortie and coworkers (84) generalized this result to all types of fixed-bed enzyme reactors in biphasic systems. Importantly, the regeneration of the coenzyme was not a rate-limiting step.

Coupling of the Coenzyme on a Carrier. Another option for immobilizing coenzymes is to bind both the dehydrogenase(s) and the coenzyme to an insoluble carrier such as cellulose, dextrans, starch, chitin, agarose, or synthetic copolymers (85,86). The activation of the hydroxyl groups of polysaccharides is usually done by reaction with cyanogen bromide, leading to the formation of reactive imidocarbonates. Synthetic polymers can be activated using partial hydrolysis of the acetate groups followed by activation with epichlorhydrin. A range of preactivated polysaccharides and synthetic polymers (Biosynth, Eupergit, etc.) may be obtained from commercial sources. The enzyme is coupled via its amino groups. Also, in this case, a spacer long enough to allow the coenzyme to have access to the active sites of the enzyme(s) must be used. Horse liver alcohol dehydrogenase (HLADH) and an NADH analogue, N^6-[(6-aminohexyl)carbamoyl-methyl]-NADH, have been successfully coimmobilized to Sepharose 4B (83). It was shown that the coenzyme could be regenerated at a cycling rate of 3,400 cycles/h using the substrate-coupled approach. Both thermal and storage stability of the HLADH were increased substantially. It was not necessary to add soluble coenzyme to the substrate solution to maintain activity.

Entrapment Methods. Polymer-linked coenzyme derivatives are often introduced in enzyme reactors with coenzyme-dependent processes, because they can be retained upon ultrafiltration or dialysis and are recycled (Fig. 5) (35,39,45,47,56,57,76,87–93). Small substrate and product molecules can freely pass through the membrane (e.g., polyamide, polyethersulfone), but large enzymes and the polymer-linked coenzyme cannot. Synthetic membranes of defined pore size covering the range between 500 and 300,000 Da are commercially available at a reasonable cost. Additionally, a variety of synthetic membrane shapes, such as membrane discs or hollow fibers, are on the market. A simple form of a membrane reactor that does not require any special equipment is also shown in Figure 4. In this simple technique, termed *membrane-enclosed enzymatic catalysis* (MEEC), the enzymes and polymer-bound coenzyme are entrapped in a dialysis bag, which is mounted on a gently rotating magnetic stirring bar (81).

One modification of the enzyme membrane-reactor concept makes use of charged ultrafiltration membranes in order to retain the native coenzyme in the reactor (94–97). The application of charged ultrafiltration membranes in membrane reactors is limited to the production of noncharged products. In contrast, nanofiltration membranes

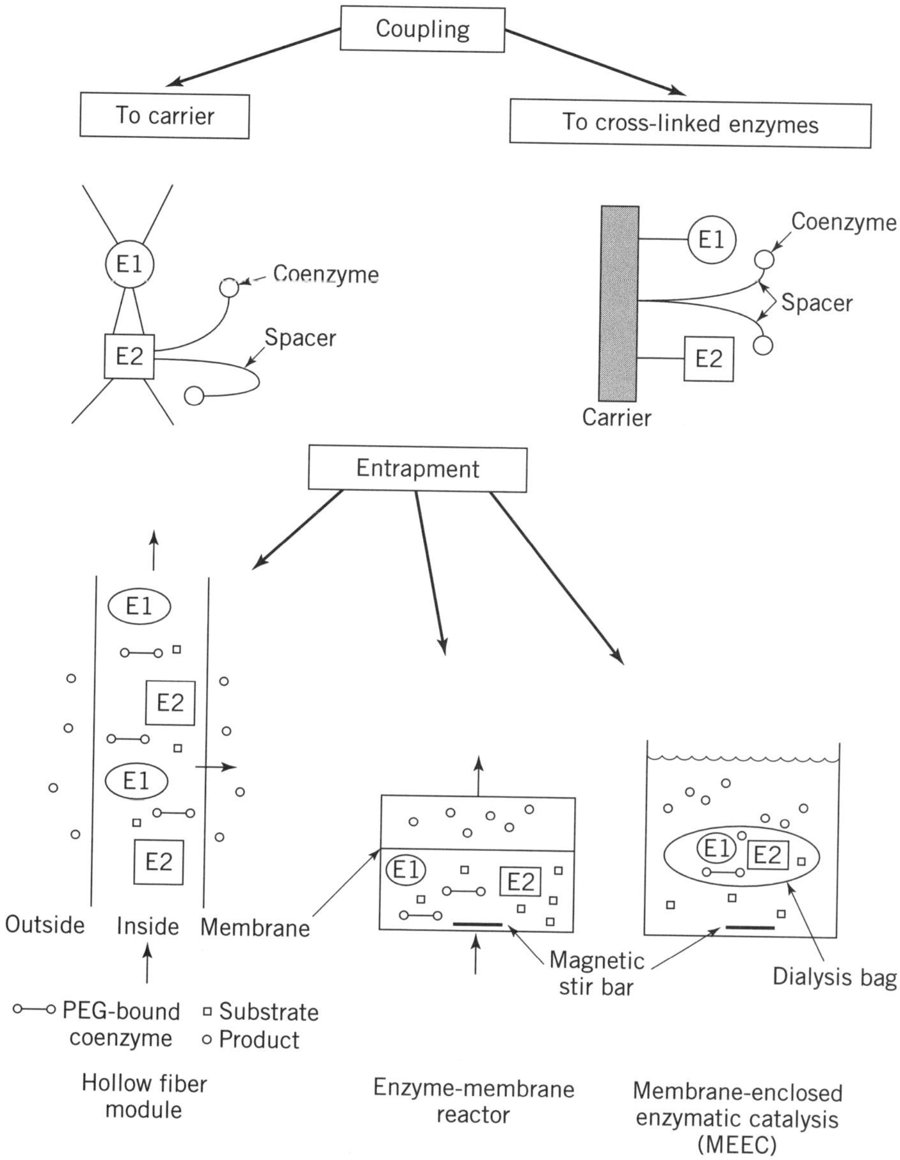

Figure 4.

can be used to retain the coenzyme in the reactor, whereas the reaction product(s) may freely pass the membrane (98). The separation of coenzyme and reaction product is due to the differences in the molecular weight, without a negative net charge of the membrane. For a recent overview of the applications of dehydrogenases using the enzyme membrane reactor technology, the reader is referred to Ref. 99.

Many reports on the preparation of polymer-linked coenzyme derivatives are available (100–109). The dextran-bound coenzymes have been demonstrated to the serious problem of protein leakage, due to the instability of the isourea linkage between coenzyme derivatives and dextran (110). However, more stable dextran derivatives can be synthesized by using bromohydroxypropyl derivatives of dextrans (111). These derivatives are used mostly in enzyme electrodes or as stationary phases in affinity chromatography (112).

The polyethyleneimine and polylysine derivatives of coenzymes (109) have the disadvantage of having electrostatic interactions between the multicharged polymers and enzymes, thus interfering with their effectiveness as active coenzymes. The relative reaction rates compared to native NAD^+ were 2–25% (polyethyleneimine) and 2–60% (polylysine), depending on the enzyme tested. The synthesis of polyacryl copolymers has the advantage that parameters like solubility, charge and the functional group can be varied just by exchange of the monomer (106). However, polyacryl-bound NAD^+ showed only 30% of the V_{max} value of the native coenzyme (113).

Another approach was established by Bückmann and coworkers using poly(ethylene) glycol derivatives with molecular weights of 3,000 to 20,000 kDa and modified coenzymes. The yield of the coupling of *N*(1)-(2-aminoethyl)-NAD^+ with the carboxy-PEG derivative was 90% (103). The alkylation of NADPH with 1,4-bis-(2,3-epoxypropoxy)-butane was investigated by Stein (114). This method allows the direct alkylation of NADPH in a one-step reaction, whereas all other described alkylation methods are mul-

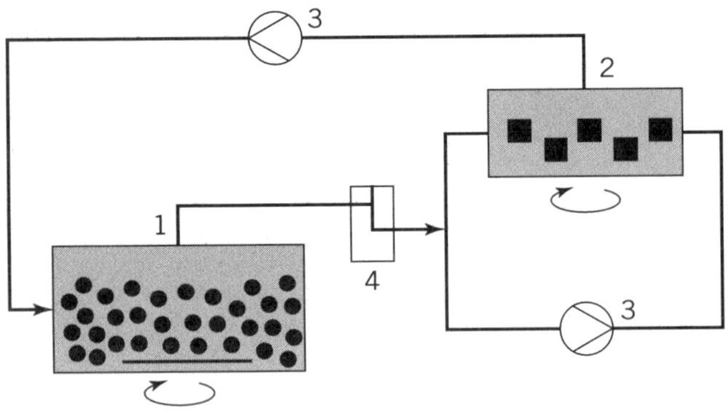

Figure 5.

tistep reactions starting from the oxidized coenzymes. Moreover, the coenzymatic activities of the NADPH derivatives were almost identical to the native coenzymes. The NADPH derivatives carry a hydrophilic spacer and a terminal epoxy group, which can be coupled easily to PEG-NH_2, resulting in a stable, covalent bond between PEG and the coenzyme derivative. For most of the enzymes tested, the coenzymatic activity of PEG-NADH is comparable to the native coenzyme (88,103,115). However, there are also exceptions (105). The PEG–coenzyme derivatives have been successfully applied in the above-mentioned enzyme membrane reactor (35,39,45,56, 57,76,87–93). PEG-NADH, as well as other PEG-derivatives, are commercially available from spin-offs of German research centers in Braunschweig and Jülich (ASA Spezialenzyme GmbH; Jülich Enzyme Products).

Modern reactor concepts for the application of dehydrogenases and native coenzymes in organic solvents have been worked out by several groups (89,116–118). These concepts circumvent the immobilization of the coenzyme by using hydrophobic microfiltration membranes and two-phase systems, an emulsion membrane reactor or a differential circulation reactor with continuous extraction, respectively.

Conclusions

Enzymatic reductions catalyzed by NADP-dependent dehydrogenases require stoichiometric amounts of nicotinamide coenzymes, which are very expensive compounds. The application of dehydrogenases on a preparative scale therefore requires effective methods for the regeneration of these coenzymes.

Today, coenzyme regeneration of both NADH and NADPH are very straightforward and the cofactor is no longer the dominant cost in preparative reductions. Nicotinamide coenzymes are labile compounds, but if they are used under optimal conditions, the half-life can be extended to more than 14 days. Coupling of NADPH to poly(ethylene) glycols stabilizes the coenzymes, by which means the half-life can be extended to more than 240 days. Enzymatic methods for coenzyme regeneration are preferable to nonenzymatic methods because they meet most of the requirements for an ideal recycling system. Two different principles can be distinguished: the substrate-coupled and the enzyme-coupled approaches. The enzyme-coupled method is advantageous because even unfavorable reactions can be driven to the product side by coupling with a favorable cofactor-regeneration reaction.

The most economical method for PEG-NADH recycling is the formate–formate dehydrogenase system, which has been applied up to the commercial-scale process.

For the regeneration of (PEG-NADPH, the alcohol–alcohol dehydrogenase from the *Thermoanaerobium brockii* system or the formate–formate:NADP dehydrogenase system are recommended.

The α-ketoglutarate–glutamate dehydrogenase system catalyzes the thermodynamically favorable reductive amination of α-ketoglutarate. Both NAD^+ and $NADP^+$, as well as PEG-NAD$(P)^+$, can be regenerated effectively.

The use of the enzyme membrane reactor technology in conjunction with poly(ethylene) glycol-linked coenzymes made possible the scale-up of enzymatic processes to preparative scale (91,119,120). Among others, this technology was successfully applied for the synthesis of different L-amino acids and a range of chiral hydroxy compounds of high value.

BIBLIOGRAPHY

1. H. Bielka, B.L. Horecker, W.B. Jakoby, P. Karlson, B. Keil, C. Lièbecq, B. Lindberg, and E.C. Webb eds., *Enzyme Nomenclature*, Academic Press Inc., New York, 1992.
2. H.K. Chenault, E.S. Simon, and G.M. Whitesides, *Biotechnol. Genet Eng Rev.* **6**, 221–267 (1988).
3. H.G. Davies, R.H. Green, D.R. Kelly, and S.M. Roberts, *Biotransformations in Preparative Organic Chemistry: the Use of Isolated Enzymes and Whole Cell Systems in Synthesis*, Academic, London, 1989.
4. J.B. Jones and J.F. Beck, in J.B. Jones, C.J. Sih, and D. Perlman eds., *Applications of Biochemical Systems in Organic Chemistry*, Wiley, New York, 1976, pp. 107–402.
5. I. Willner and D. Mandler, *Enzyme Microb. Technol.* **11**, 467 (1989).
6. H.K. Chenault and G.M. Whitesides, *Appl. Biochem. Biotechnol.* **14**, 147–197 (1987).
7. C.H. Wong and G.M. Whitesides, in J.E. Baldwin, and P.D. Magnus eds., *Tetrahedron Organic Chemistry Series: Enzymes in Synthetic Organic Chemistry*, Pergamon, Oxford, 1994, pp. 1–370.
8. T. Sakai, T. Uchida, and I. Chibata, *Agric. Biol. Chem.* **37**, 1049–1956 (1973).
9. K. Murata, J. Kato, and I. Chibata, *Biotechnol. Bioeng.* **21**, 887 (1979).
10. T. Hayashi, Y. Tanaka, and K. Kawashima, *Biotechnol. Bioeng.* **21**, 1019 (1979).
11. A.L. Lehninger, *Methods Enzymol.* **3**, 885–887 (1957).
12. G.W. Rafter and S.P. Colowick, *Methods Enzymol.* **3**, 887–889 (1957).
13. S. Suye and S. Yokoyama, *Enzyme Microb. Technol.* **7**, 418 (1985).
14. S.Y. Eguchi, N. Nishio, and S. Nagai, *Agric. Biol. Chem.* **47**, 2941–2943 (1983).
15. A. Traub, E. Kaufmann, and Y. Teitz, *Anal. Biochem.* **28**, 469–476 (1969).
16. A.J. Kornberg, *J. Biol. Chem.* **182**, 779–793 (1950).

17. D.R. Walt, M.A. Findeis, V.M. Rios-Mercadillo, J. Auge, and G.M. Whitesides, *J. Am. Chem. Soc.* **106**, 234–239 (1984).
18. W. Hummel and M.R. Kula, *Eur. J. Biochem.* **184**, 1–13 (1989).
19. N.J. Oppenheimer, in J. Everse, B. Anderson, and K.S. You eds., *The Pyridine Nucleotide Coenzymes*, Academic, New York, 1982, pp. 51–89.
20. C.H. Wong and G.M. Whitesides, *J. Am. Chem. Soc.* **103**, 4890–4899 (1981).
21. S.P. Colowick, N.O. Kaplan, and M.M. Ciotti, *J Biol Chem.* **191**, 447–459 (1951).
22. N.J. Oppenheimer and N.O. Kaplan, *Biochemistry* **13**, 4675–4685 (1974).
23. S.L. Johnson and K.W. Smith, *Biochemistry* **15**, 553–559 (1976).
24. J.F. Biellman, C. Lapinte, E. Haid, and G. Wiemann, *Biochemistry* **18**, 1212–1217 (1979).
25. J. Everse, E.C. Zoll, L. Kahan, and N.O. Kaplan, *Bioorg. Chem.* **1**, 207–233 (1971).
26. J.B. Jones, D.W. Sneddon, W. Higgins, and A.J. Lewis, *J. Chem. Soc. Chem. Commun.* 856–857 (1972).
27. R. Ramio and E.M. Lilius, *Enzymologia* **40**, 360–368 (1971).
28. D. Mandler and I. Willner, *J. Chem. Soc. Perkin Trans.* **II**, 805–811 (1986).
29. J.B. Jones and K.E. Taylor, *Can. J. Chem.* **54**, 2969–2973 (1976).
30. M.D. Legoy, V. Laretta-Garde, J.M. LeMoullec, F. Ergan, and D. Thomas, *Biochimie (Paris)* **62**, 341–345 (1980).
31. M. Julliard, J. Le Petit, and P. Ritz, *Biotechnol. Bioeng.* **28**, 1774 (1986).
32. H. Simon, H. Günther, J. Bader, and S. Neumann, *Ciba Found. Symp.* **111**, 97–111 (1985).
33. A. Schummer, H. Yu, and H. Simon, *Tetrahedron* **47**, 9019–9034, (1991).
34. R. Ruppert, S. Herrmann, and E. Steckhan, *Tetrahedron Lett.* **28**, 6583 (1987).
35. E. Steckhan, S. Herrmann, R. Ruppert, J. Thömes, and C. Wandrey, *Angew. Chem., Int. Ed. Engl.* **102**, 445–447 (1990).
36. D. Westerhausen, S. Herrmann, W. Hummel, and E. Steckhan, *Angew. Chem.* **104**, 1496–1498 (1992).
37. R. Wienkamp and E. Steckhan, *Angew Chem. Int. Ed. Engl.* **21**, 782–783 (1982).
38. R. Dicosimo, C.H. Wong, L. Daniels, and G.M. Whitesides, *J. Org. Chem.* **46**, 4622–4623 (1981).
39. M.R. Kula, R. Wichmann, U. Oden, and C. Wandrey, *Biochimie (Paris)* **62**, 523–536 (1980).
40. L.G. Lee and G.M. Whitesides, *J. Org. Chem.* **51**, 25–36 (1986).
41. J. Peters, T. Minuth, and M.R. Kula, *Tetrahedron Asymm.* **4**, 1683–1692 (1993).
42. T. Zelinski, Ph.D. Thesis, University of Düsseldorf, Germany, 1995.
43. P. van Eikeren, D.J. Brose, D.C. Muchmore, and J.B. West, *Ann. N.Y. Acad. Sci.* **613**, 796–801 (1990).
44. H.R. Levy, F.A. Loewus, and B. Vennesland, *J. Am. Chem. Soc.* **79**, 2949–2953 (1957).
45. E. Schmidt, W. Hummel, H. Schütte, and M.R. Kula, in H. Chmiel, W.P. Hammes, and J.E. Bailey eds., *Biochemical Engineering—A Challenge for Interdisciplinary Cooperation*, Gustav Fischer, Stuttgart, 1987, pp. 487–490.
46. J. Peters, Ph.D. Thesis, University of Düsseldorf, Germany, 1993.
47. S.S. Wang and C.K. King, in T.K. Ghose, A. Fiechter, and N. Blakebrough eds., *Advances in Biochemical Engineering*, Springer, Berlin, 1979, pp. 119–146.
48. J. Bader, H. Günther, S. Nagata, H.J. Schuetz, M.L. Link, and H. Simon, *J. Biotechnol.* **1**, 95–109 (1984).
49. C.H. Wong, D.G. Drueckhammer, and H.M. Sweers, *J. Am. Chem. Soc.* **107**, 4028–4031 (1985).
50. L.G. Lee and G.M. Whitesides, *J. Am. Chem. Soc.* **107**, 6999 (1985).
51. L. Poppe and L. Novak, *Selective Biocatalysis. A Synthetic Approach*, VCH Verlagsgesellschaft mbH, Weinheim, 1992.
52. K. Faber, *Biotransformations in Organic Chemistry*, Springer, 1995.
53. H. Schütte, J. Flossdorf, H. Sahm, and M.R. Kula, *Eur. J. Biochem.* **62**, 151–160 (1976).
54. Z. Shaked and G.M. Whitesides, *J. Am. Chem. Soc.* **102**, 7105–7107 (1980).
55. R. Wichmann, C. Wandrey, A.F. Bückmann, and M.R. Kula, *J. Biotechnol.* **23**, 2789–2802 (1981).
56. E. Schmidt, D. Vasic-Racki, and C. Wandrey, *Appl. Microbiol. Biotechnol.* **26**, 42–48 (1987).
57. D. Vasic-Racki, M. Jonas, C. Wandrey, W. Hummel, and M.R. Kula, *Appl. Microbiol. Biotechnol.* **31**, 215–222 (1989).
58. W. Hummel, H. Schütte, E. Schmidt, C. Wandrey, and M.R. Kula, *Appl. Microbiol. Biotechnol.* **26**, 409–416 (1987).
59. I. Yamamoto, T. Saiki, S.M. Liu, and L.G. Ljundahl, *J. Biol. Chem.* **258**, 1826–1832 (1983).
60. J.R. Andreesen, *J. Bacteriol.* **120**, 6–14 (1974).
61. L.R. Krumholz, R.L. Crawford, M.E. Hemling, and M.P. Bryant, *J. Bacteriol.* **169**, 1886–1890 (1987).
62. L.G. Ljundahl and J.R. Andreesen, *FEBS Letts.* **54**, 279–282 (1975).
63. S.W. Tannenbaum, *Biochim. Biophys. Acta* **21**, 335–342 (1956).
64. V. Tishkov, A. Galskin, L.B. Kulakova, and A. Egorov, *Enzyme Engineering XII*, Deauville, France, 1993.
65. V.I. Tishkov, A.G. Galkin, G.N. Marchenko, Y.D. Tsygankov, and H.M. Ergorov, *Biotechnol. Appl. Biochem.* **18**, 201–207 (1993).
66. K. Seelbach, Diplomarbeit, University of Bonn, Germany, 1994.
67. K. Seelbach, B. Riebel, W. Hummel, M.R. Kula, V.I. Tishkov, A.M. Egorov, C. Wandrey, and U. Kragl, *Tetrahedron Lett.* **37**, 1377–1380 (1996).
68. R.J. Lamed and J.G. Zeikus, *Biochem. J.* **195**, 183–190 (1981).
69. J. Peters and M.R. Kula, *Biotechnol. Appl. Biochem.* **13**, 363–370 (1991).
70. E. Keinan, K.K. Seth, R.J. Lamed, R. Ghirlando, and S.P. Singh, *Biocatalysis* **3**, 57–71 (1990).
71. E. Keinan, S.C. Sinha, and A. Sinha-Bagchi, *J. Org. Chem.* **57**, 3631–3636 (1992).
72. E. Keinan, K.K. Seth, and R.J. Lamed, *J. Am. Chem. Soc.* **108**, 3474–3480 (1986).
73. E. Keinan, E.K. Hafeli, K.K. Seth, and R.J. Lamed, *J. Am. Chem. Soc.* **108**, 162–169 (1986).
74. E. Keinan, K.K. Seth, and R.J. Lamed, *Ann. N.Y. Acad. Sci.* **501**, 130–150 (1987).
75. R.J. Lamed, E. Keinan, and J.G. Zeikus, *Enzyme Microb. Technol.* **3**, 144–148 (1981).

76. M.R. Kula, and C. Wandrey, in S.P. Colowick and N.O. Kaplan eds., *Methods in Enzymology Part C*, Academic, New York, 1987, pp. 9–21.
77. C.W. Bradshaw, H. Fu, G.J. Shen, and C.H. Wong, *J. Org. Chem.* **57**, 1526–1532 (1992).
78. G.J. Shen, Y.F. Wang, and C.H. Wong, *J. Chem. Soc. Chem. Commun.* **9**, 677–679 (1990).
79. G.L. Lemiere, J.A. Lepoivre, and F.C. Alderweireldt, *Tetrahedron Lett.* **26**, 4527–4528 (1985).
80. G. Carrea, R. Bovara, P. Cremonesi, and R. Lodi, *Biotechnol. Bioeng.* **26**, 560–563 (1984).
81. M.D. Bednarski, H.K. Chenault, E.S. Simon, and G.M. Whitesides, *J. Am. Chem. Soc.* **109**, 1283–1285 (1987).
82. C.J. Suckling and K.E. Suckling, *Chem. Soc. Rev.* **3**, 387–406 (1974).
83. S. Gestrelius, M.O. Mansson, and K. Mosbach, *Eur. J. Biochem.* **57**, 529–535 (1975).
84. R. Lortie, I. Villaume, M.D. Legoy, and D. Thomas, *Biotechnol. Bioeng.* **33**, 229–232 (1989).
85. P.W. Carr, and L.D. Bowers, *Immobilized Enzymes in Analytical and Clinical Chemistry*, Wiley, New York, 1980.
86. C.Y. Lee, and A.F. Chen, in J. Everse, B. Anderson, and K.S. You eds., *The pyridine Nucleotide Coenzymes*, Academic, New York, 1982, pp. 189–224.
87. K. Hosono, S. Kajiwara, Y. Yamazaki, and H. Maeda, *J. Biotechnol.* **14**, 149–156 (1990).
88. N. Katayama, I. Urabe, and H. Okada, *Eur. J. Biochem.* **132**, 403–409 (1983).
89. S. Kise and M. Hayashida, *J. Biotechnol.* **14**, 221 (1990).
90. German Patent 3937892 C2 (November 15, 1989), U. Kragl, J. Peters, C. Wandrey, and M.R. Kula (to the Regents of the Forschungszentrum Jülich GmbH); Swiss Patent CH679676A5 (November 5, 1990), U. Kragl, J. Peters, C. Wandrey, and M.R. Kula (to the Regents of the Forschungszentrum Jülich GmbH).
91. W. Leuchtenberger, M. Karrenbauer, and U. Plöcker, *Ann. N.Y. Acad. Sci.* **434**, 78–86 (1984).
92. C. Wandrey and R. Wichmann, in A.I. Laskin ed., *Enzymes and Immobilized Cells in Biotechnology*, Benjamin/Cummings, Menlo Park, Calif., 1985, pp. 177–208.
93. C. Wandrey, R. Wichmann, M.R. Kula, and A.F. Bückmann, *Die Umschau.* **3**, 88–91 (1984).
94. M. Ikemi and Y. Ishimatsu, *Biotechnol. Bioeng.* **36**, 155–165 (1990).
95. M. Ikemi, N. Koizumi, and Y. Ishimatsu, *Biotechnol. Bioeng.* **36**, 149–154 (1990).
96. B. Nidetzky, D. Haltrich, and K.D. Kulbe, *Chemtech.* 31–36 (1996).
97. T.R. Röthig, K.D. Kulbe, F. Bückmann, and G. Carrea, *Biotechnol Lett.* **12**, 353–356 (1990).
98. K. Seelbach and U. Kragl, *Enzyme Microb. Technol.* **20**, 389–392 (1997).
99. J. Peters, in H.J. Rehm, and G. Reed eds., *Biotechnology*, Wiley-VCH, Weinheim, 1998, pp. 391–473.
100. A.F. Bückmann and G. Carrea, *Adv. Biochem. Eng. Biotechnol.* **39**, 98–152 (1989).
101. A.F. Bückmann, *Biocatalysis* **1**, 173–186 (1987).
102. A.F. Bückmann, *Heterocycles* **27**, 1623–1628 (1988).
103. A.F. Bückmann, M.R. Kula, R. Wichmann, and C. Wandrey, *J. Appl. Biochem.* **3**, 301–315 (1981).
104. A.F. Bückmann, M. Morr, and G. Johansson, *Makromol. Chem.* **182**, 1379–1384 (1981).
105. A.F. Bückmann, M. Morr, and M.R. Kula, *Biotechnol. Appl. Biochem.* **9**, 258–268 (1987).
106. C.W. Fuller and H.J. Bright, *J. Biol. Chem.* **252**, 6631–6639 (1977).
107. P.O. Larsson and K. Mosbach, *FEBS Letts.* **46**, 119–122 (1974).
108. K. Okuda, I. Urabe, and H. Okada, *Eur. J. Biochem.* **151**, 33–38 (1985).
109. P. Zapelli, A. Rossodivita, and L. Re, *Eur. J. Biochem.* **54**, 475–482 (1975).
110. R. Axen, J. Porath, and S. Ernback, *Nature (London)* **214**, 1302–1304 (1967).
111. C.Y. Lee, A. Leigh-Brown, C.H. Langley, and D. Charles, *J. Solid-Phase Biochem.* **2**, 213–224 (1978).
112. C.R. Lowe and K. Mosbach, *Eur. J. Biochem.* **49**, 511–520 (1974).
113. C.W. Fuller, J.R. Rubin, and H.J. Bright, *Eur. J. Biochem.* **103**, 421–430 (1980).
114. A. Stein, Ph.D. Thesis, University of Münster, Germany, 1990.
115. N. Katayama, K. Hayakawa, I. Urabe, and H. Okada, *Enzyme Microb. Technol.* **6**, 538–542 (1984).
116. W. Kruse, W. Hummel, and U. Kragl, *Recl. Trav. Chim Pays-Bas.* **115**, 239–243 (1996).
117. A. Liese, M. Karutz, J. Kamphuis, C. Wandrey, and U. Kragl, *Biotechnol. Bioeng.* **51**, 544–550 (1996).
118. A. Liese, T. Zelinski, M.R. Kula, H. Kierkels, M. Karutz, B. Schulze, U. Kragl, and C. Wandrey, *J. Mol. Cat. B.* **4**, 91–99 (1998).
119. A. Bommarius, M. Schwarm, K. Stingl, M. Kottenhahn, K. Huthmacher, and K.H. Drauz, *Tetrahedron Asymm.* **6**, 2851–2888 (1995).
120. U. Kragl, D. Vasic-Racki, and C. Wandrey, *Bioprocess Eng.* **14**, 291–297 (1996).

CONDUCTIVITY

ALESSANDRO D'APRANO
University of Rome, La Sapienza
Rome, Italy

KEY WORDS

Carrier mechanism
Channel mechanism
Electrolytes
Electrolytic conductors
Electronic conductors
Ionic transport
Membrane
Molar conductivity
Resistance
Solid-state polymer electrolytes

OUTLINE

INTRODUCTION

Transport properties of electrolyte solutions (i.e., *conductance, diffusion, viscosity, thermal diffusion*, etc.) play a relevant role in basic and applied researches because they control most processes occurring in many fields of pure and applied science, such as chemistry, biology, medicine, colloids science, engineering, and materials science. In this article, electrolytic conductance will be emphasized, considering in particular the physical phenomena on which this quantity is based, the experimental apparatus used for its measurement, and some of the theories proposed to explain the variation of conductance with the electrolyte concentration. It will also be shown that conductance is a macroscopic property yielding both kinetic information in the form of ionic mobilities and thermodynamic information in the form of thermodynamic equilibrium constants for the formation of ion pairs. Conductance behavior also yields information on electrolytic systems at the microscopic level when the experimental data are analyzed with the theoretical equations based on models in which ion–solvent and ion–ion interactions are included.

CONDUCTION OF ELECTRICITY THROUGH MATTER

The transport of electricity through matter is effected by the movement of charge carriers. Depending on the nature of the charge carriers involved in the process, two basic classes of conductors can be distinguished. In the first class (*electronic conductors*), electricity is carried by electrons; in the second class (*electrolytic conductors*), electric current is carried by ions. Electronic conductors include metals and semiconductors.

The properties of the electronic conductors follow solid-state band theory. The rigorous treatment of electronic conductors is beyond the scope of this article, but to provide some background information for the reader, the mechanism of electrical conduction in metals can be summarized as follows. The energy levels of isolated atoms have descrete values. The filling of these energy levels by electrons is governed by the laws of quantum mechanics. As atoms approach one another to form a crystal lattice, the original energy levels combine to form energy bands. In metals, the conductivity bands correspond to the highest, partially occupied, energy levels of the metal atom in the ground state and contain a sufficient number of loosely bound electrons in random motion. Under the influence of an external electric field, the original random motions of electrons become oriented, and an electron flux (*electrical current*) runs along the conductor.

Quite different is the mechanism of conduction in electrolytic conductors. In electrolyte solutions, ions and solvent molecules are in rapid thermal motion, and any net ionic migration is essentially a small perturbation along the direction of the electric field superimposed on the former. Because ionic velocities are only of the order of 10^{-6} s^{-1} for an electrical field of unity (1 V/cm^{-1}), the solvent is usually approximated as a continuum in which the ions are assumed to move at a constant velocity, which is the time average equal to the sum of their components of motion in the direction of the applied field divided by the time of observation. Electrolytic conductance is, in last analysis, a measure of the steady-state flux achieved between the effects of an external electrical force and those of the internal hydrodynamic resistance to ionic motion. As such, it is an important research tool with which molecular interactions may be conveniently investigated.

In many technological processes (i.e., *electrodialysis, ion-selective electrode production, environmental and waste applications*, etc.) as well as in some biological processes (i.e., *permeation of ions through cellular membrane*), an artificial or natural membrane is interposed between two electrolyte solutions. From a physical point of view, the *membrane* is a potential barrier for the ionic flux; thus, the mechanism of conduction in these systems differs from that occurring in the simple electrolytic systems. Membranes can be divided in two main types: membranes that have a preferential permeability for a particular ion and membranes that are totally unpermeable to any kind of ions. The membranes of the first type are used in the technological applications where ion selectivity is the main factor of the process. The ionic conduction in these electrolyte membrane systems is only allowed for a particular ionic species. Quite different is the conduction mechanism for the systems containing a membrane of the second type. In these systems, the flux of ions (*hydrophilic*) is hindered by the hydrophobic character of the membrane (e.g., *lipidic bilayers*). To overcome this handicap, two alternative mechanisms are possible (*carrier and channel mechanisms*). Ionic transport mediated by carrier is obtained by using macrocyclic compounds (1,2) (e.g., natural antibiotics such as valinomycin and enniatin B or synthetic macrocyclic compounds such as crown ethers, cryptands, etc.). These carrier compounds form intrusion complexes with alkali metal ions. In these complexes, the central hydrophilic ion is hosted into a cavity containing oxygen atoms, and the exterior of the complex is hydrophobic. Such a feature, decreasing the energy barrier imposed by the membrane to the ionic flux, allows for the transport of the hosted ionic species across the hydrophobic membrane. It has been observed that the conductance of lipidic bilayers (10^{-10} S/cm) linearly increases with the carrier concentration. Such a result suggests that current-carrying species are 1:1 complexes between alkali metal cation and macrocyclic molecules. A transport mechanism consistent with the experimental results is depicted in Figure 1. According to this mechanism, the ionic transport across the membrane occurs in four distinct steps: (*1*) Association of ion M^+ and carrier S in the interface electrolyte solution–membrane (rate constant K_R); (*2*) Translocation of the complex M^+S to the opposite side of the membrane (rate con-

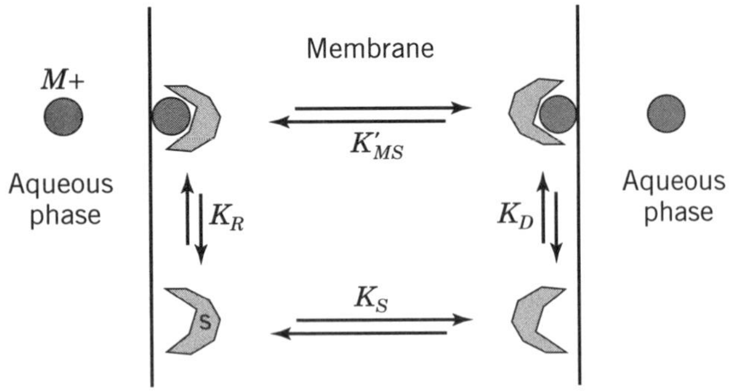

Figure 1. Carrier mechanism.

stant K'_{MS}); (*3*) Release of the ion to the solution (rate constant K_D); and (*4*) Back transport of the unloaded carrier (rate constant K_S). The kinetics of this carrier mechanism for a given system can be analyzed by studying the electric conductance behavior (3).

Different from the ionic transport mediated by carrier compounds is the channel mechanism. A channel consists of one or several binding sites arranged in a transmembrane sequence, and it is accessible from both sides at the same time. A number of simple *molecules* (e.g., the linear pentadecapeptide gramicidin A [4]) introduced into a biological or artificial membrane form channels permeable to cations. A detailed model of the gramicidin A channel was proposed by Urry (5). Information on the mechanism of ion transport through channels has been obtained in experiments (3) in which small amounts of gramicidin A were added to a planar bilayer membrane. Under this condition, the membrane current under a constant applied voltage shows discrete fluctuations with more or less the same amplitude. The experiment has been explained assuming that the single current fluctuation results from the formation and disappearance of a single channel permeable to ions.

In addition to electronic and electrolytic conductors, there is a special class of *solid-state polymer electrolytes* (6) that have recently attracted considerable worldwide interest for their prospective applications in batteries, sensors, and other practical devices. Polymer electrolytes are simply formed by dissolving an inorganic salt in suitable polymeric hosts. These host polymers contain electronegative heteroatoms (*oxygen*, *nitrogen*, *sulfur*, or *phosphorus*) that interact with cations by donor–acceptor bonds to form polymer–cation complexes. Poly(ethylene oxide) (PEO) and poly(propylene oxide) (PPO) are the most-used polymers that form solid-state polymer electrolytes.

```
  PEO:          PPO:
-CH-CH-O-    -CH-CH-O-
                |
                CH
```

The mechanism of ionic transport in such systems was initially attributed to ion hopping between fixed coordination sites of polymer. Recently, the percolation theory (7) formally describes these ion conductors as a random mixture of conductive islands interconnected by a nonconductive matrix in terms of ion size, charge, and interactions with other ions and the host matrix.

As stated before for the electronic conductors, a detailed description of the conductivity properties of polymer electrolytes and the transmembrane ionic transport is beyond the scope of this article, which focuses mainly on electrolyte conductance, namely, *conductance in liquid electrolyte solutions*.

ELECTROLYTIC CONDUCTANCE

In an electrolytic conductor, the relationship between the steady-state flux (J) of ions and the driving forces of the electric field (X) can be represented by the power series

$$J = A + BX + CX^2 + DX^3 + \ \ldots \qquad (1)$$

where A, B, C, and so forth are constants, and the units of J and X are $cm^3/s/K$ and volt/cm, respectively. For small applied fields, the terms containing the powers of X greater than unity can be neglected, and equation 1 therefore becomes

$$J = A + BX \qquad (2)$$

The constant A in equation 2 must be zero because the ionic flux vanishes when $X = 0$. Hence, for small electric fields, the ionic flux is linearly related to the driving force

$$J = BX \qquad (3)$$

where B is a constant specific of the system under investigation. The ionic flux J, representing the number of moles of ions crossing a unit area per second, can be easily converted into the charge flux or the current density i (number of electrical charge crossing unit area per second) by the relationship

$$i = JzF = zFBX \qquad (4)$$

In the preceding equation, z is the charge number, i has units of ampere (A) per cm^{-2}, and F is the faraday equal to $F = eN$ (e is the electron charge equal to $e = 1.602177 \times 10^{-19}$ coulomb, and N is the Avogadro's number equal to $N = 6.02214 \times 10^{23}\ mol^{-1}$). For a given electrolytic system, zFB is constant. Setting this constant equal to χ, equation 4 becomes

$$i = \chi X \qquad (5)$$

that can be rearranged in the form

$$\chi = i/X \qquad (6)$$

where χ is the *electrolytic conductance* as of a solution of known concentration (also referred to as the *specific conductance* in older literature) and has units of siemen (S) cm^{-1}. Table 1 compares representative values of the specific conductance of electrolytic solutions with those of metallic conductors and nonelectrolytic systems.

Table 1. Representative Values of Electrolytic Conductance at 25 °C

Substance	Type of conductor	Electrolytic conductance (S/cm)
Acetonitrile	Nonelectrolyte	1×10^{-8}
Water	Nonelectrolyte	2×10^{-7}
0.1 M KCl	Electrolytic	1×10^{-2}
4 M H_2SO_4	Electrolytic	8×10^{-1}
Iron	Metallic	1×10^{5}
Copper	Metallic	6×10^{5}

By definition, the electric field X is the ratio of the potential difference ΔV between two regions of the electrolyte and the distance l between them.

$$X = \Delta V/l \quad (7)$$

On the other hand, the density current (i) is the ratio of the total current (I) and the area (A) of the electrodes

$$i = I/A \quad (8)$$

Thus, combining equations 7 and 8 with equation 6 one has

$$\Delta V = I(l/\chi A) \quad (9)$$

According to Ohm's law, the quantity $(l/\chi A)$ must be a resistance, that is,

$$R = (1/\chi)(l/A) \quad (10)$$

The modern units for resistance is reciprocal siemen (S^{-1}), and the older unit of ohm (Ω) is no longer used. Equation 10 indicates that electrolytic conductors, like metallic conductors are ohmic conductors (i.e., they obey the Ohm's laws) and can therefore be represented in an electrical circuit as a resistor.

Measurements of Electrolytic Conductance

In conductance experiments, the electrolyte is placed in a conductivity cell of controlled geometry, and the resistances (R) of solutions of known concentration are measured. The cells are generally constructed in Pyrex glass with sealed platinum electrodes. Other noble or passive metals, such as gold, silver, titanium, stainless steel, and tungsten, as well as graphite can be used as electrodes for technical applications of conductivity (i.e., routine or in situ industrial measurements). Cells designs are available for a variety of research and industrial applications. The different types of conductance cells have in common an electrode compartment, shown schematically in Figure 2. The geometry factor, l/A (cm^{-1}), known as the cell constant, κ, is generally determined by calibrating the cell with a solution of known conductivity. Aqueous potassium chloride solutions are usually used for this purpose (8) because the specific conductance of this electrolyte at different concentrations and temperatures has been determined with a very high degree of accuracy. Cells are designed with various cell constants so that the resistance of different electrolytic systems ranges between 1,000 and 30,000 S^{-1} in order to avoid errors arising from high density current and insulation linkage. Table 2 summarizes recommended cell constants for various conductance ranges.

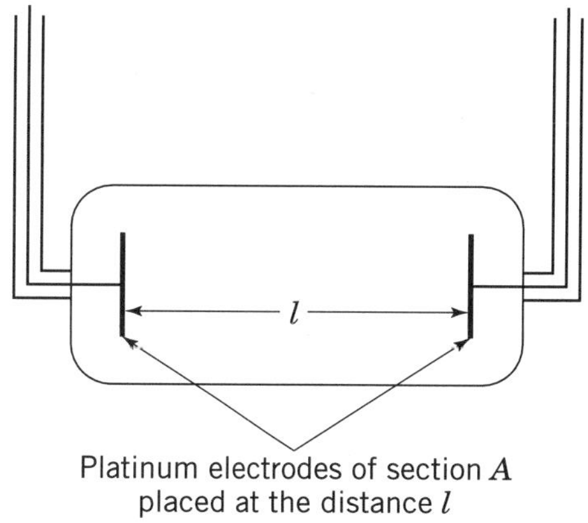

Figure 2. Schematic representation of conductometric cell.

The electrical resistance of electrolyte solutions cannot be measured with a traditional bridge circuit (e.g., a Wheatstone bridge) operating with direct current (dc) because this would give rise to polarization at the electrodes and in some cases to the electrolysis of the solution. Polarization produces undesirable changes in the electrolyte concentration during measurements that must be avoided because the resistance of the electrolytic system depends on ionic concentration. The simplest way to achieve this requirement is to use alternating current (ac) in the frequency range between 500 and 20,000 Hz. This rapid reversible change in applied potential effectively eliminates polarization effects.

The theory and design of ac bridges suitable for electrolyte conductance measurements were discussed by Jones and Josephs (9) and by Shedlovsky (10). A crude representation of the different ac bridges commercially available can be made in reference to the Kohlraush's bridge circuit, schematically depicted in Figure 3. The conductivity cell forms one arm of the bridge and a standard variable resistance box (R) forms another arm of the bridge. The other two arms of the bridge consist of two closely equal resistances, R_a and R_b. With ac bridge impedances are determined. The impedance Z is defined by the relation

$$Z = (R + X_R)^{1/2} \quad (11)$$

where R is the resistance and X_R the reactance given by

Table 2. Recommended Cell Constants for Various Conductance Ranges

Conductance Range (S cm)	Cell constant (cm^{-1})
$(0.5–20) \times 10^{-6}$	0.01
$(0.1–20) \times 10^{-5}$	0.1
$(0.2–20) \times 10^{-4}$	1.0
$(0.1–20) \times 10^{-3}$	10.0
$(0.1–20) \times 10^{-2}$	50.0

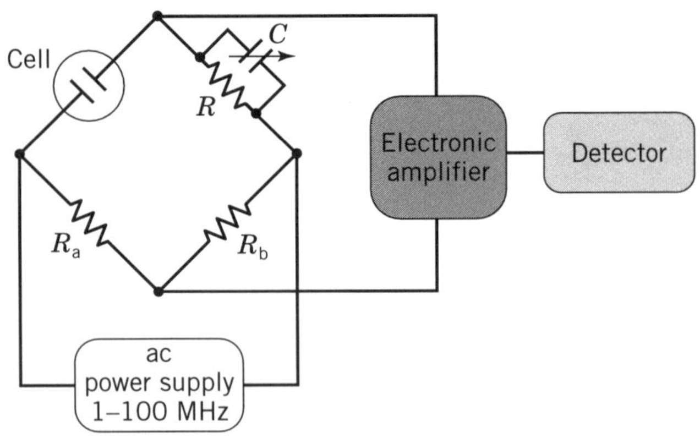

Figure 3. Schematic representation of a conductance apparatus based on Kohlraush's bridge circuit.

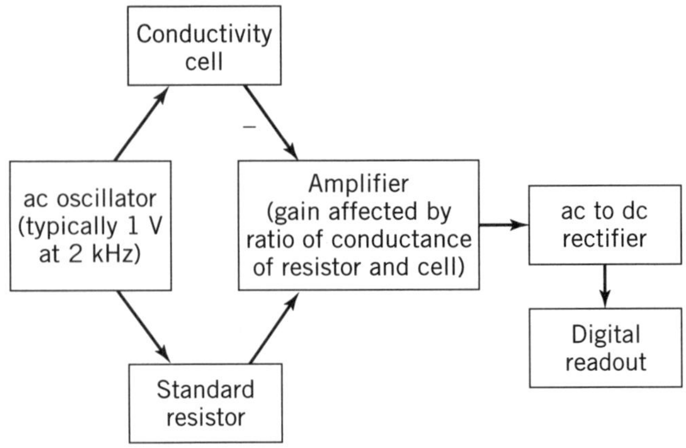

Figure 4. Block diagram of a typical digital conductivity meter.

$$X_R = X_L - X_C \quad (12)$$

where $X_L = 2\pi fL$ is the inductive reactance and $X_C = 1/(2\pi fC)$ is the capacitive reactance. In the preceding equations, R, X_L, and X_C are in reciprocal siemen (S^{-1}), f is the frequency in hertz, L is the inductance in henry, and C is the capacitance in faraday. Because the resistances used in the bridges for conductometric measurements are generally noninductance, capacities and resistances are the variables determined by balancing the bridge. Thus, only a standard variable condenser box (C) needs to be placed in parallel to the resistance box (R). For the measurement, both C and R are adjusted until zero signal from the electronic amplifier is detected. For this null condition, the following relations for the reactance capacities $X_C = 1/(2\pi fC)$ as well as for the resistances (R) can be written

$$\begin{aligned}[1/(2\pi fC)]_{cell}/[1/(2\pi fC)]_a &= [1/(2\pi fC)]/[1/(2\pi fC)]_b \\ &= [1/C)]_{cell}/[1/C)]_a \\ &= [1/C)]/[1/C]_b \\ R_{cell}/R_a &= R/R_b \end{aligned} \quad (13)$$

Because $[1/C)]_a = [1/C]_b$ and $R_a = R_b$, one obtains

$$\begin{aligned} C_{cell} &= C \\ R_{cell} &= R \end{aligned} \quad (14)$$

The ac bridge circuit is, in other words, a simple RC circuit allowing us to measure the individual impedance contributions from the resistance and the capacitance reactance. The block diagram for typical digital instruments used for routine resistance measurements is shown in Figure 4.

The conductance of electrolytic solutions changes by as much as 2% per degree Kelvin (Fig. 5). Thus, it is very important to have accurate temperature control within ±0.001 K for high-precision measurements. Although the thermostatic bath may have any convenient and appropriate design, mineral oil rather than water is recommended for use in the bath. This is because the high dielectric constant and conductivity of water compared with mineral oil may result in leakage of electrical energy from the cell sufficient to introduce errors of appreciable magnitude. For routine or industrial applications such as in situ conductivity measurements, temperature control need not be very precise, and temperature regulation within ±0.1 K is sufficient in most cases. Cells are available with built-in temperature sensors and associated instrumentation that automatically convert conductance values to a reference temperature, generally 25 °C. This feature on commercial bridges is referred to as automatic temperature compensation of conductivity.

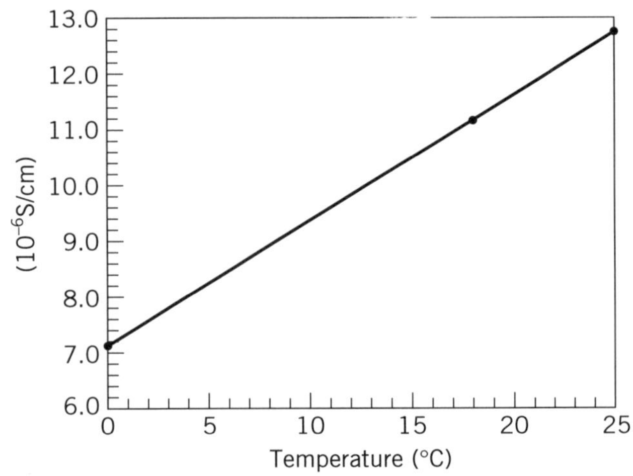

Figure 5. Dependence of the specific conductance of aqueous KCl solution (0.1 mol/L) with temperature.

For high-precision and high-accuracy conductivity measurements of dilute electrolyte solutions, the conductance of the solvent, χ_o, must be subtracted from the *apparent* (i.e., *experimental*) conductance, χ_a, as measured in the cell to obtain the real conductance, χ, of ionic species in solution.

$$\chi = \chi_a - \chi_o \quad (15)$$

The electrolytic conductance of commonly studied solvents is summarized in Table 3. Because conductivity measurements can be extremely accurate, by comparing measured

Table 3. Specific Conductance of Common Solvents at 25 °C

Solvent	Specific conductance (χ_o[S/cm])
Water	$1\text{–}2 \times 10^{-7}$
Methanol	$1\text{–}2 \times 10^{-8}$
Glycerol	$1\text{–}3 \times 10^{-8}$
Ethylene glycol	$6\text{–}8 \times 10^{-8}$
Acetonitrile	$3\text{–}4 \times 10^{-8}$
Propylene carbonate	$1\text{–}2 \times 10^{-6}$
Nitromethane	$1\text{–}3 \times 10^{-8}$
N-N-Dimethylformamide	$2\text{–}4 \times 10^{-8}$

conductances to those determined for highly purified solvents (see Table 3) the purities of solvents can be easily determined.

Molar, Equivalent, and Ionic Conductivity

The current density (i) across unit area per second can be expressed in terms of ionic velocity (v_i) by

$$i = F\Sigma_i c_i z_i v_i \tag{16}$$

where c_i, z_i, and v_i are the number of moles per cubic centimeter, the charge number, and the velocity in cm/s of the nth ionic species. Dividing equation 16 for the strength of electrical field (X) gives

$$i/X = (F/X)\Sigma_i c_i z_i v_i \tag{17}$$

Substituting equation 6 in equation 17 and setting the mobility of the nth ions $U_i = (v_i/X)$ one obtains

$$\chi = F\Sigma_i c_i z_i U_i \tag{18}$$

Restricting the treatment to 1:1 symmetrical electrolytes for which $z^+ = z^- = z$ and $c^+ = c^- = c$, equation 18 becomes

$$\chi = Fzc(U^+ + U^-) \tag{19}$$

Equation 19 shows that, contrary to electronic conductors, the specific conductance of electrolyte solutions depends on the ionic concentration. Thus, in order to normalize the electrolytic conductance for the concentration, a new quantity, the molar conductivity (Λ_m) is defined as

$$\Lambda_m = 10^3\chi/c_m \tag{20}$$

where c_m is the concentration (mol dm^{-3} units) of the electrolyte. The units of Λ_m are S cm^2 mol^{-1}. Molar conductivities taking into account only the molar concentration of the electrolyte do not always normalize the conductance, as in the cases where multiply charged ions are involved. To account for unsymmetrical electrolytes, such as $MgCl_2$, normalization is achieved by considering equivalent concentrations, $c = c_m/z$. In terms of equivalent concentrations, the normalized quantity is the *equivalent conductance* Λ (S cm^2 eq^{-1}), defined similarly to equation 20 by

$$\Lambda = 10^3\chi/c \tag{21}$$

Figure 6 shows that the equivalent conductance for a typical electrolyte (KCl in water) decreases as the ionic concentration increases. Thus, for meaningful comparisons of conductivity differences for different electrolytes, it is necessary to define a common reference state for equivalent conductances. The reference state is that of infinite dilution where the equivalent conductance at zero electrolyte concentration is referred to as the *limiting molar conductivity* and given the symbol Λ_o. As will be shown in the next section, the dependence of conductance on ionic concentration results from interionic interactions. Because at infinite dilution such interactions are negligible, only ionic velocity is affected by the external electrical field. Under these circumstances, Λ_o represents the conductivity of the electrolyte solution in its ideal state.

Substituting equation 21 in equation 19 gives

$$\Lambda = F(U^+ + U^-) = \lambda^+ + \lambda^- \tag{22}$$

where λ^+ and λ^- are the ionic equivalent conductance of the cation and anion at the given concentration. At infinite dilution, equation 22 for the limiting equivalent conductance becomes

$$\Lambda_o = \lambda_o^+ + \lambda_o^- \tag{23}$$

An important implication of this equation is that at infinite dilution, molar conductivities are *additive*. This property, discovered by Kohlrausch, is known as the *Kohlrausch law of independent ionic conductivity*. Thus, the limiting equivalent conductance for any electrolytic system can be obtained from the sum of the contributions of the individual ions. Limiting equivalent conductivity for a large variety of ionic species in different solvents as a function of temperature has been measured by many researchers. Table

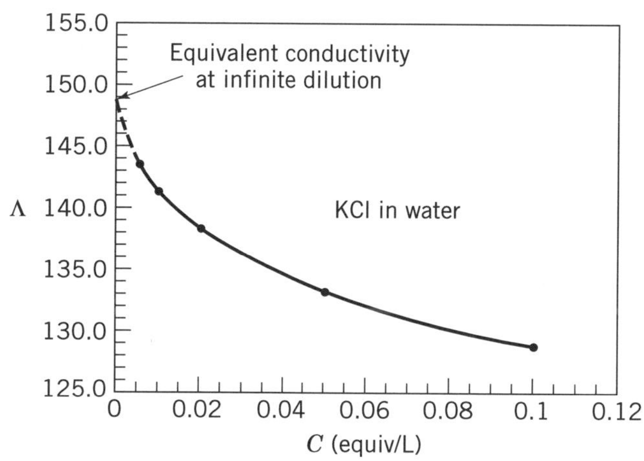

Figure 6. Dependence of Λ(S cm^2 equiv^{-1}) with electrolyte concentration C (equiv/L).

4 reports individual ionic λ_o values for selected ions in water at 25 °C.

THEORIES OF ELECTROLYTIC CONDUCTANCE

The search for a mathematical expression able to theoretically predict the concentration dependence of the molar conductivity has been the object of numerous scientists since the time of Arrhenius (1883). An ideal theory of conductance would predict Λ from fused salt systems (*solute mole fraction equal to unity*) to infinite dilution (*solvent mole fraction equal to unity*). In spite of this intense effort, the different conductance equations so far proposed take into account only the influence of the ion–ion and ion–solvent interactions on the ionic mobility and the concentration of free ionic species in solution. Thus, they are limited to concentrations up to where the effect of short range three-ion interactions (i.e., *triple ions formation*) becomes significant.

Debye–Onsager–Fuoss Theory

Because of the electrical charge on the ions, the ionic distribution in an electrolyte solution is modulated by the ion–ion interaction field superimposed to the ionic thermal motion. In order to quantitatively express the contribution of electrostatic interactions on ionic distribution, it is necessary to choose a model to represent the system. Although the model must include all the relevant properties of solute and solvent, the problem is to select a sufficiently simple model that is amenable to available mathematical treatments. All models for conductance behavior, though different in some details, are essentially based on the *primitive model* proposed by Debye and Hückel (11) in which

1. The electrolyte is completely dissociated at all the concentrations.
2. The solution is considered to be a collection of point charges (ions) in a homogeneous medium of viscosity (η) and relative permittivity (ϵ) (hydrodynamic continuum).
3. The Coulombic field surrounding the ions has spherical symmetry.
4. Only ion–ion long-range electrostatic interactions are considered.
5. The electrostatic potential energy of each ion is smaller than the thermal energy, that is, $|z_i e\psi| \ll kT$.

Table 4. Equivalent Ionic Conductivity of Selected Ions at Infinite Dilution in Water at 25 °C

Cations	λ_0^+ (S cm^2 equiv^{-1})	Anions	λ_0^- (S cm^2 equiv^{-1})
H^+	349.8	OH^-	198.30
Li^+	38.7	Cl^-	76.4
Na^+	50.1	I^-	76.8
K^+	73.5	Br^-	78.1
Rb^+	77.8	NO_3^-	71.4
Cs^+	77.3	ClO_4^-	67.3
Ag^+	61.9	ClO_3^-	64.0
NH_4^+	73.5	HCO_3^-	44.5
$(n\text{-}Bu)_4N^+$	19.5	$CH_3CO_2^-$	40.9
Ca^{2+}	59.5	$HC_2O_4^-$	40.2
Ba^{2+}	63.6	$C_6H_5CO_2^-$	32.4
Mg^{2+}	53.1	Picrate$^-$	30.4
Zn^{2+}	52.8	CO_3^{2-}	69.3
Cu^{2+}	53.6	$C_2O_4^{2-}$	74.2
Pb^{2+}	69.5	SO_4^{2-}	80.0
Fe^{2+}	54.0	$Fe(CN)_6^{3-}$	101.0
Fe^{3+}	68.0	$Co(CN)_6^{3-}$	98.9
$Co(NH_3)_6^{3+}$	102.3	$Fe(CN)_6^{4-}$	110.5

In the Debye–Hückel theory, the central ion is considered, from a time-averaged point of view, at rest. The ion is surrounded by a symmetrical ionic cloud localized at the distance (*Debye distance*) given by

$$x^{-1} = [(\varepsilon T/\Sigma z_i c_i)(1{,}000\ \kappa/8\pi e^2 N]^{1/2} \tag{24}$$

where z_i and c_i are the charge number and the concentration of the nth ion, respectively, ϵ is the static dielectric constant of the solvent, T is the temperature (Kelvin), κ is the Boltzman constant (1.38 10^{-6} erg/degree), e is the electron charge (4.80 10^{-10} absolute electrostatic unity [e.s.u.]), and N is the Avogadro's number. To the extent that one seeks to interpret the equilibrium properties of electrolytic solutions (i.e., *activity coefficients*), this picture of a static central ion surrounded by a ionic cloud with spherical symmetry is quite reasonable. Ionic interactions yielding the formation of spherically symmetrical ion clouds in the absence of external fields change from spherical to axial symmetry (*egg shape*) when a uniform external field (i.e., electric field as in the conductance or chemical potential field as in the diffusion) is applied to the electrolyte solution (Fig. 7).

Onsager (12) accounts for this asymmetrical distribution by considering the mutual interactions between an ion and its asymmetrical cloud. In the mathematical treatment of the migration of ions, he assumed that the drift velocity of an ion, besides the electric force arising from the externally applied field, is also determined by two other extra forces (i.e., *relaxation* and *electrophoretic forces*). The basic conductivity equation obtained by Onsager is

$$\Lambda = \Lambda_{0^\circ} - S\sqrt{c} \tag{25}$$

where Λ_0 is the limiting conductance due to the unper-

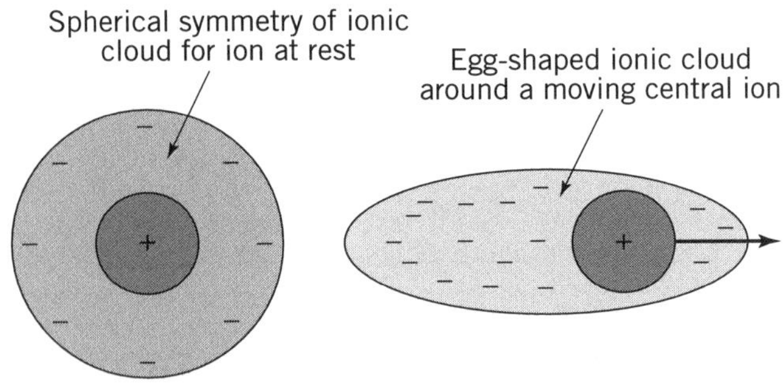

Figure 7. Ionic cloud for central ion at rest and in movement.

turbed ion movement in the external field at infinite dilution, and S is a constant that depends on absolute temperature, valence type of solute, dielectric constant, and viscosity of solvent. S includes the relaxation and the electrophoretic effect calculated via two particle correlation functions, assuming ions to be point charges and truncating series terms at the term $c^{1/2}$. It is important to realize that the constant S (and the constant E to be described in the next paragraph) does not account for finite size of ions and their solvation shells. For solvents of high dielectric constants ($\varepsilon > 45$), equation 25 can be considered an acceptable extrapolation formula, and we refer to it as the *limiting law*. However, for solvents of lower dielectric constant, deviations from equation 25 become large even in the case of completely dissociated electrolytes. The reasons for this drastic limitation arises from the crude assumption of the Debye–Hückel physical model, which represents the ions as point charges.

A further development of the conductance theory was made by Fuoss and Onsager (13), who treated the ions as finite-size particles. The resulting conductance equation derived by Fuoss and Onsager is

$$\Lambda = \Lambda^{\circ} - S\sqrt{c} + Ec \log c + J_a c \tag{26}$$

where Λ_0 and S have the same meaning as in equation 25, E is a new constant dependent on the same variables as S, and J_a is a parameter that according to the new model depends on the cation–anion distance a, which can be used as a variable in fitting data to equation 26 or simply estimated as the sum of crystal radii of cation and anion; a is therefore referred to as the *distance of closest approach*. The logarithmic term in equation 26 derives from the integration of the higher terms of the continuity equation. Although Equation 26 is based on a more realistic model (*rigid charged spheres in a continuum*) it does have limitations because it often fails to accurately describe conductance behavior in many solvents, particularly solvents with low dielectric constants. However, for solutions containing salts with large ions, such as tetraalchylammoniun salts in high dielectric constant solvents, equation 26 does in fact reproduce the experimental data with acceptable accuracy. The problems encountered with equation 26 arise when electrolytes with small ions in solvents of moderate dielectric constants are involved, and where ion pair formation (Fig. 8) becomes significant. Ion pairs behaving as a dipole in the external electric field do not contribute to the conductance current.

In 1926, J. Bjerrum (14) developed a theory of ionic association. This theory was based on the observation that for certain systems the ratio of measured activity coefficients to the stoichiometric concentration was less than the activity coefficients calculated by the Debye–Hückel theory. To account for such discrepancies, Bjerrum postulated that some ions associated to ion pairs that as dipoles are no longer included in the ionic atmosphere. In order to obtain the degree of association of ions into neutral ion pairs, Bjerrum calculated the probability function [$W(r)$] of finding an anion in the spherical shell surrounding the reference cation. By applying a mathematical treatment analogous to that used by Debye– Hückel in the derivation of the activity coefficients, he found an analytical form for $W(r)$ that after a sharp minimum, diverged to infinity (Fig. 9). In order to mathematically treat this function, Bjerrum arbitrarily cut the integral off at a certain critical distance between the cation and the anion at which the electrostatic attractive force is balanced by the mean force corresponding to the thermal motion. Such a distance, in which the ion is at minimum energy, depends on the valence type (z), electron charge (e), dielectric constant of solvent (ε), Kelvin temperature (T), and Boltzman constant (k). For uni-univalent electrolytes, this critical distance (*Bjerrum distance*) is represented by the equation

$$R = e^2/2\varepsilon kT \tag{27}$$

Integration of the probability function from the distance of closest approach (a) of the ions to R allows one to calculate the fraction of ions present in solution in the associated state. Thus, the degree of ion-pair association and the ion-pair association constant can be calculated.

Taking ion pair formation into account, Fuoss (15) extended the previous conductivity equation 26 to

$$\Lambda = \Lambda^{\circ} - S\sqrt{c\gamma} + Ec\gamma \log c\gamma + jc\gamma - K_a c\gamma f_{\pm}^2 \Lambda \tag{28}$$

where $f_{\pm}^2$ is the mean ionic activity coefficient calculated by the Debye–Hückel theory, and γ is the fraction of the electrolyte that exists as unassociated ions. In terms of γ, the thermodynamic equilibrium constant for the formation of ion pairs is determined by the mass equation

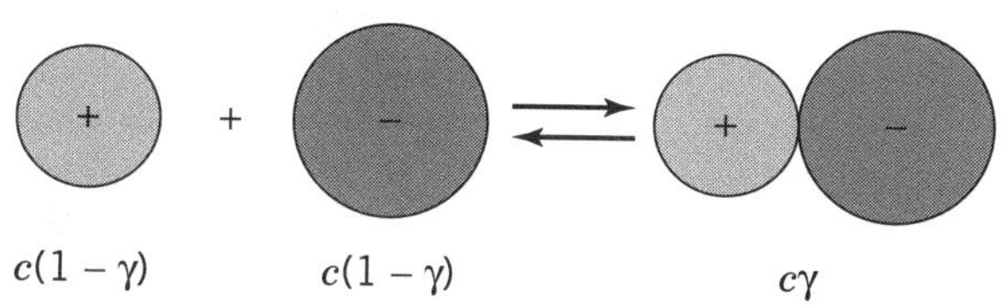

Figure 8. Schematic representation for ion-pair formation.

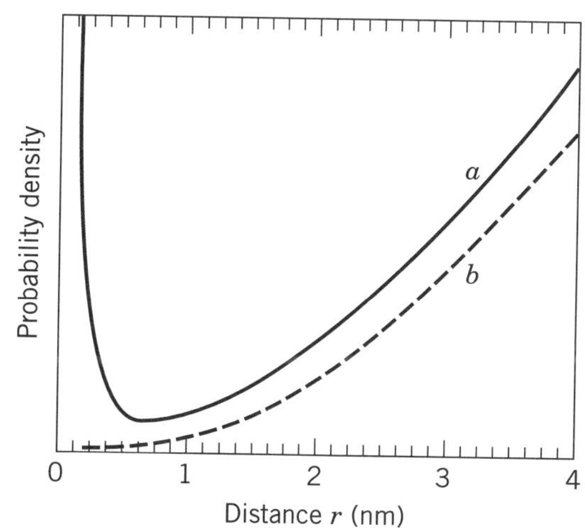

Figure 9. Bjerrum's curves of probability. (**a**) Two ions of opposite charges; (**b**) uncharged particles.

$$K_a = (1 - \gamma)/c\gamma^2 f^2 \quad (29)$$

Subsequently, Quint and Viallard (16) and Lee and Wheaton (17) extended the above equation to mixed electrolytes and for unsymmetrical electrolytes (e.g., $MgCl_2$). After the publication of equation 28, the theoretical treatment of the conductance of electrolyte solutions seemed to be completely solved in term of the electrostatic theory. However, numerous accurate experiments studies during the past 25 years have shown that equation 28 still has severe limitations. In particular, the systematic investigation on the conductivities of electrolytes in different solvents carried out by many researchers confirmed the limitation of equation 28 (or similar conductivity equations derived by Pitts [18], Kraeft [19], and Fuoss et al. [20]). Part of these limitations were eliminated by refining the mathematical treatment of the electrophoretic and relaxation terms [21]. Nevertheless, the most important limitation of the theory was the small concentration range over which equation 28 can be applied.

Several mathematical improvements were made to push the concentration limit of the conductivity equation to higher values. Fuoss and Hsia (22), by a new integration of the equations of continuity and motion with retention of terms of order $c^{3/2}$, presented in 1967 the empirical $\Lambda(c)$ function

$$\Lambda + \Lambda^o - S\sqrt{c} + Ec\gamma \log c + Ac + Bc^{3/2} \quad (30)$$

This equation is valid for 1:1 electrolytes up to the concentration $C_{max} = 3\ 10^{-7}\ \varepsilon^3$ mol dm^3, above which ion pair formation becomes statistically undefinable due to the presence of triple ions (23). Thereafter, Fernandez-Prini (24) expanded the function of the Fuoss–Hsia equation using, as suggested by Justice (25), the Bjerrum radius (R) as the lower limit in the integration instead of the contact distances a. The modified conductivity equation derived by Justice and Fernandez-Prini is

$$\Lambda = \Lambda_o - S\sqrt{c\gamma} + Ec\gamma \log c\gamma + J_1(R)c\gamma + J_2(R)c^{3/2}\gamma^{3/2} - K_a c\gamma f_\pm^2 \Lambda \quad (31)$$

It turned out that equation 32 was also not free from theoretical and experimental problems. The most serious was the prediction that a given electrolyte in different solvents of the same dielectric constants should have the same association to ion pairs.

Many violations of the failure of this *isodielectric rule* (26) show that the primitive model can be considered an oversimplified physical model adequate to describe the conductometric behavior of electrolytic systems, in which long-range ion–ion interactions play the main role, but it completely ignores additional effects caused by anion–cation, anion–solvent, and cation–solvent short-range interactions. Recently, Fuoss (27) proposed a new model for ions in solution, based on the concept of the Gurney cosphere (28) surrounding each ion. This *chemical model* takes into consideration short-range specific ion–solvent interactions (*solvation*) related to the polarizability, dipole moment, and shape and size of ions and solvent molecules, both of which are explicitly ignored in the primitive model. This new model involves the dual roles played by the ions involved in ion-pair formation. As long as the distance between the two ions is greater than the diameter of the Gurney cosphere, an ion involved in pairing is an anonymous member of an enormous population; once inside the cosphere, the ions behave like particles and move from site to site by interchanges with solvent molecules. Because of this duality, the long-range interactions (*relaxation field* and *electrophoresis*) can be determined by using continuum electrostatic and hydrodynamic theories, whereas short-range interactions can be determined by the thermodynamic changes of the free energy of the pairing process involving both solvent separate ion pairs and contact ion pairs. For symmetrical electrolytes, the conductance equation based on the chemical model can be written in the following symbolic form

$$\Lambda = \gamma\{[\Lambda_o(1 + (\Delta E/E)] + \Delta\Lambda^v + \Delta\Lambda^H + \Delta\Lambda^{el}]\} \quad (32)$$

In this equation, where γ is the fraction of solute that contributes to the conductance current, $(\Delta E/E)$ is the part of the relaxation field generated by purely electrostatic interactions between ions, $\Delta\Lambda^v$ is the part of the relaxation field caused by the change in the electrophoretic current caused by perturbation potential, $\Delta\Lambda^H$ is the Sanding–Feistel (29) hydrodynamic–hydrodynamic interactions term, $\Delta\Lambda^{el}$ is the electrophoretic countercurrent, and Λ_o is the limiting conductance. Starting from this chemical model proposed by Fuoss, Lee and Wheaton, in a series of papers (30), derived a conductivity equation for the general case where any number of ionic species of any valence type may be presented in solutions. Thus, not only 1:1 electrolytes but also asymmetrical and mixed electrolytes can be analyzed.

BIBLIOGRAPHY

1. B. Sesta and A. D'Aprano, *J. Solution Chem.* **18**, 1163–1171 (1989).
2. A.Y. Ovvhinnikov, V.T. Ivanov, and A.M. Skrob, *Membrane-Active Complexones*, Elsevier, Amsterdam, 1974.
3. R. Benz and P. Lauger, *J. Membr. Biol.* **27**, 171–191 (1976), P. Lauger, *Science* **178**, 24–30 (1972).
4. S.B. Hladky and D.A. Haydon, *Biochim. Biophys. Acta* **274**, 294–312 (1972).
5. C.W. Urry, *Proc. Natl. Acad. Sci. USA* **68**, 672–676 (1971).
6. J.R. MacCallum and C.A. Vincent eds., *Polymer Electrolyte Reviews*, Elsevier, London, 1987.
7. M.A. Ratner and D.F. Shriver, *Chem. Rev.* **88**, 109–112 (1988).
8. J.E. Lind Jr., J.J. Zwolenik, and R.M. Fuoss, *J. Am. Chem. Soc.* **81**, 1557–1559 (1959).
9. G. Jones and R.C. Josephs, *J. Am. Chem. Soc.* **50**, 1049–1092 (1928).
10. T. Shedlovsky, *J. Am. Chem. Soc.* **52**, 1806–1811 (1930).
11. P. Debye and E. Hückel, *Z. Phys.* **24**, 305–325 (1923).
12. L. Onsager, *Z. Phys.* **28**, 277–284 (1927).
13. R.M. Fuoss and L. Onsager, *Proc. Natl. Acad. Sci. USA* **41**, 1010–1025 (1955).
14. J. Bjerrum, *Kgl. Danske Videnskab. Selskab.* **7**, 9–16 (1926).
15. R.M. Fuoss, *J. Am. Chem Soc.* **79**, 3301–3303 (1957); R.M. Fuoss and F. Accascina, *Electrolytic Conductance*, Interscience Publishers, New York, 1959.

16. J. Quint and A. Viallard, *J. Chim. Phys.* **72**, 335–342 (1975).
17. W.H. Lee and R.J. Wheaton, *J. Chim. Phys.* **74**, 689–692 (1977).
18. E. Pitts, *Proc. R. Soc.* **217**, 43–48 (1953).
19. W. Kraeft, *Z. Phys. Chem.* **237**, 289–295 (1968).
20. R.M. Fuoss, L. Onsager, and J. Shinner, *J. Phys. Chem.* **69**, 2581–2594 (1965).
21. R.M. Fuoss, *J. Phys. Chem.* **63**, 633–636 (1959); R.M. Fuoss and L. Onsager, *J. Phys. Chem.* **67**, 628–632 (1963); R.M. Fuoss and L. Onsager, *J. Phys. Chem.* **68**, 1–8 (1964); M.S. Chen, Ph.D. Thesis, Yale Univ., New Haven, Conn., 1969.
22. R.M. Fuoss and K.L. Hsia, *Proc. Natl. Acad. Sci. U.S.A.* **57**, 1550–1601 (1975).
23. R.M. Fuoss, *J. Am Chem. Soc.* **56**, 1857–1859 (1934).
24. R. Fernandez-Prini and J.C. Justice, *Pure Appl. Chem.* **56**, 541–563 (1984).
25. J. C. Justice, *J. Chim. Phys. Phys.-Chim. Biol.* **65**, 353–362 (1968).
26. A. D'Aprano and R. M. Fuoss, *J. Solution Chem.* **3**, 45–55 (1974); A. D'Aprano, I. D. Donato, and E. Caponetti, *J. Solution Chem.* **3**, 371–376 (1974); A. D'Aprano, M. Goffredi, and R. Triolo, *J. Chem. Soc. Faraday Trans.* 1 **71**, 1188–1191 (1975).
27. R.M. Fuoss, *Proc. Natl. Acad. Sci. USA* **77**, 34–46 (1980); R.M. Fuoss, *J. Solution Chem.* **15**, 231–236 (1986); A. D'Aprano, F. Accascina, and R.M. Fuoss, *J. Solution Chem.* **19**, 65–78 (1990).
28. G. Gurney, *Ionic Process in Solution*, Dover, New York, 1953.
29. R. Sanding and R. Feistel, *J. Solution Chem.* **8**, 141–145 (1979).
30. W.H. Lee and R.J. Wheaton, *J. Chem. Soc., Faraday Trans.* 2 **74**, 743–766 (1978); W.H. Lee and R.J. Wheaton, *J. Chem. Soc., Faraday Trans.* 2 **74**, 1456–1482 (1978); W.H. Lee and R.J. Wheaton, *J. Chem. Soc., Faraday Trans.* 2 **75**, 1128–1145 (1979).

See also PROCESS MONITORING.

CORROSION, MICROBIAL

PATRIZIA PEREGO
BRUNO FABIANO
University of Genoa
Genoa, Italy

KEY WORDS

Biofouling
Cathodic reaction
Caulobacter
Chemio-lithotrophic microorganisms
Depassivation of metal surfaces
Desulfotomaculum
Desulphovibrio
Gallionella
Sphaerotilus natans
Thiobacillus

OUTLINE

INTRODUCTION

In its extensive meaning, biological corrosion, or biodeterioraton, refers to any corrosion caused by biological activity. Indeed, the original meaning of the term *corrosion*, as derived from *rodere*, the Latin root for "act of gnawing," includes biological activity. Microbiological corrosion refers, by definition, to the degradation of metallic structures due to the activity of a wide variety of microorganisms, usually embedded in a gel matrix, the biofilm, attached to the metal surface.

Before further discussion on this subject, a short review of the milestones in its historical development is presented. Evidence of anaerobic corrosion was observed as early as 1840, when Mallet noted sulfide corrosion products typical of anaerobic process on an ancient anchor on the bottom of the Seine River (1).

The phenomenon of microbially influenced corrosion (MIC) was reported by the end of the nineteenth century by Garret (2), who postulated that corrosion of lead in water was caused by metabolites produced by bacterial degradation of organic matter. He suggested that biooxidation of nitrite to nitrate enhanced the corrosive action of water.

Sulfate-reducing bacteria were discovered in 1895 by Beijerinck. In 1910 Gaines (3) indicated that iron bacteria and sulfur bacteria were responsible for the corrosion of buried ferrous metals.

The formation of deposits in waterpipes by iron bacteria (in particular aerobic species *Sphaerotilus natans, Caulobacter*, and *Gallionella*) was investigated by Harder (4). In 1921 Grant et al. (5) produced evidence that the rate of pitting of brass was highly increased by ammonia, which could be produced by bacterial activity.

In 1934, Von Wolzogen Kuhr and Van Der Vlugt (6) proposed a theory on the direct role of sulfate-reducing bacteria in the corrosion mechanism by their utilization of cathodic hydrogen and stimulation of the anodic dissolution of iron.

In 1953 Uhlig reported on the role of microorganisms, such as slime-forming bacteria, fungi, algae diatoms, and protozoa, attributing their corrosive action to the formation and conservation of differential aeration and concentration cells (7).

Several reviews dealing with MIC have been published since 1971 (8–11), with a particular effort in research during the so-called energy crisis. In fact, marine fouling on offshore platforms and biofouling in industrial cooling systems, as well as the problem of oil and gas reservoir contamination by sulfide, enhanced the need for a deeper insight into the role of, and the mechanism associated with, bacterial corrosion of metal. The adoption of accelerated tests, designed ad hoc according to newly developed measurement techniques, can address the right selection of materials, surface treatments, and welding filler materials, as well as the optimization of chemical and mechanical protective or preventive measures.

MICROORGANISMS IMPLICATED IN BIOLOGICAL CORROSION

To indicate proper defensive means against corrosion of metal materials, an accurate study and a deeper understanding of the mechanisms of action of the microorganisms involved are undoubtedly of the utmost importance. An essential role in the corrosion process is often played by the cell structures, for instance, the capsular material, essentially of a polysaccaridic nature, which has a protective function and allows surface adhesion. It is in this way that microorganisms create stratified deposits under which the cell's metabolic activity takes place. The capsular material contains substances in such concentrations as to actually exert a corrosive action on metals and, further, to ensure suitable conditions for development of some anaerobic microorganisms.

Microorganisms from a wide range of genera and species are involved in corrosion problems. They may be classified in three general groups: bacteria, fungi, and algae.

Bacteria

The bacteria associated with metal corrosion are unicellular, with a rigid cell wall, and divide by binary fission. They can be either autotrophic or heterotrophic, and either aerobic or anaerobic. There are three basic shapes of unicellular bacteria: spherical, rodlike, and curved or spirilloid. They vary considerably in size. Some iron bacteria are filamentous, obtaining the energy for carbon dioxide fixation by oxidation of ferrous ions to ferric ions with the consequent precipitation of ferric hydroxide. They are responsible for the formation of tubercles in potable-water-transmission pipelines. The taxonomy and physiology of these bacteria have been reviewed by Pringsheim (12) and Stokes (13).

The most important bacteria associated with the corrosion process are those whose metabolism is closely related to the well-known sulfur cycle reproduced in Figure 1. Aerobic bacteria of the *Thiobacillus* genus, known as sulfur-oxidizing bacteria, undertake the upper part of the cycle, (i.e., the oxidation of sulfur to sulfuric acid), and the central part of the cycle (i.e., reduction of sulfate to sulfide) is carried out by anaerobic bacteria (*Desulphovibrio* and *Desulfotomaculum* genus), indicated as sulfate-reducing bacteria (SRB). A list of the SRB and their main characteristics can be found in Table 1 (modified from Ref. 14). A list of other bacteria relevant to biological corrosion is found in Table 2.

Fungi

The fungi associated with corroding metal are generally filamentous or yeastlike. Unlike some bacteria, fungi require oxygen for growth, and obtain energy for growth through aerobic oxidation of an organic substrate, which releases organic acid metabolites as the end product, or through anaerobic fermentation. The fungi can utilize hydrocarbons as a source of carbon; they may be active in largely hydrocarbon environments and have proven to be the agents responsible for the corrosion of aluminium aircraft fuel tanks in the presence of condensed water.

Algae

Algae are a heterogeneous group of chlorophyll-containing plants found in seawaters and freshwaters. They are autotrophic organisms, and algae are commonly found in cooling water systems. Their role is similar to that of the

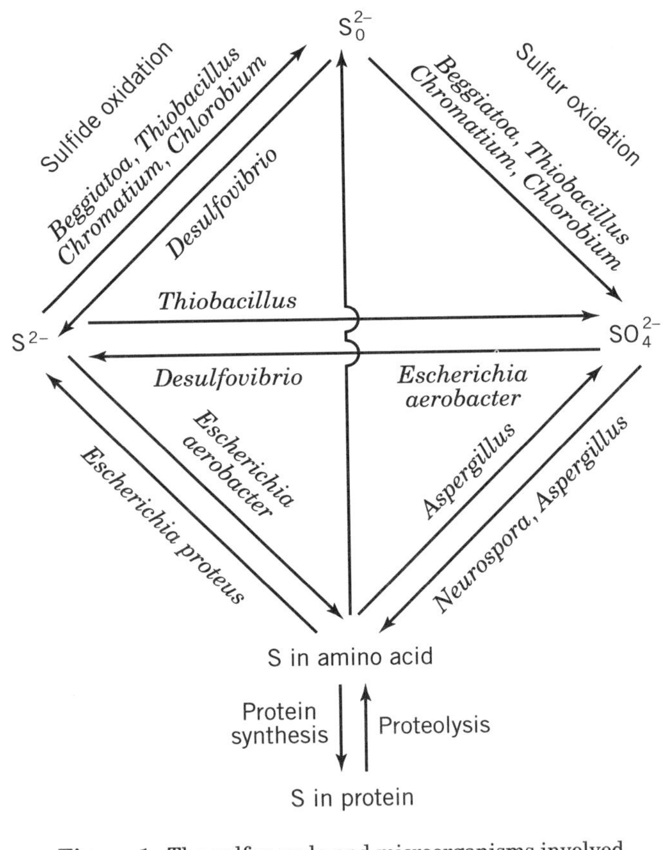

Figure 1. The sulfur cycle and microorganisms involved.

Table 1. List of Sulfate-Reducing Bacteria

Genus or species	pH	Temperature (°C)	Characteristics	Metal influenced	Mechanism
Desulfovibrio	4–8	10–40	Anaerobic	Iron and steel, stainless steel, aluminum, zinc, copper alloys	H_2 utilization to reduce SO_4^{2-} to S^{2-} and H_2S; promotion of sulfide film formation; no spore formation
Dv. desulfphuricans					Growth on pyruvate or choline in sulfate-free media
Dv. vulgaris					
Dv. salexigens					Obligate saltwater species (Cl^-)
Dv. gigas					
Dv. africanus					
Desulfotomaculum	6–8	10–40	Anaerobic	Iron and steel, stainless steel	Reduction of SO_4^{2-} to S^{2-} and H_2S; spore formation
Dt. nigrificans		55 optimum			Hydrogenase activity variable; not coupled to sulfate reduction; growth on pyruvate in sulfate-free media
Dt. orientis		10–40			Hydrogenase apparently absent
Dt. rumnis		10–40			Growth on pyruvate in sulfate-free media
Desulfomonas		10–40	Anaerobic	Iron and steel	Reduction of SO_4^{2-} to S^{2-} and H_2S

Note: From Ref. 15.

fungi: By adhering to the metal surface they encourage the formation of differential aeration and concentrate cells, thereby accelerating already-existing corrosion processes. The corrosive effect of algae attached to metal surfaces was investigated by Terry and Edyvean (15), who demonstrated that microfouling organisms readily attach to unprotected steels, cathodically protected steels, and painted steels within 100 days of exposure, causing alternating pH values night and day and enhancing pitting corrosion. Evidence of siliceous skeletons of diatoms causing defects in electroplated coatings of metals was shown by Ebneth and Ritterskamp (16).

MICROBIOLOGICAL CORROSION FEATURES

Metal corrosion in an aqueous environment is recognised as an electrochemical phenomenon where the anode reaction takes place when part of the substratum (metal) is oxidized and transferred into solution. The balancing cathode reaction consists of the contemporaneous reduction of components in the corrosive environment to fully preserve electroneutrality. Metal ions in solution might subsequently precipitate as free insoluble products, or may firmly adhere to the metal surface, where they can have a protective effect.

In favorable circumstances and under natural conditions, these films may have a protective function; a typical example of this function was seen in archaeological findings consisting of iron handiwork, which is well preserved even though found in corrosive ground. These well-preserved items were covered with a slightly compact, tightly adherent, thin, black film of ferric and ferrous phosphate.

When the metal is submersed in water, wet soil, or any other aqueous system, the initial reaction is metal dissolution to cation, which involves freeing of electrons:

$$Fe \rightarrow Fe^{++} + 2e^-$$

For the corrosion to continue, these electrons need to be consumed by reduction reactions that take place on the same metal on the so-called cathodic sites. Generally, the oxygen reduction to hydroxyl ion is the most common cathode reaction in a neutral environment:

$$O_2 + 2H_2O + 4e^- \rightarrow 4OH^-$$

In a neutral environment, the overall reaction is formation, and possibly subsequent precipitation, of insoluble products by the reaction of the ferrous ions with the hydroxyl ion (and, often, oxygen) in solution. As a rule, reactions occur evenly over the metal surface. Anode sites may be determined by either heterogeneous material (e.g., stress areas, inclusions) or preexisting, unevenly distributed oxide films. The overall corrosion rate may depend on quite a number of factors: Anode and cathode reactions, however, must remain balanced in the process to fully maintain electroneutrality. Lacking oxygen, the cathode reaction that may replace the preceding reaction in aqueous ambient in sustaining the corrosion process is reduction of the ion H^+, as follows:

$$2H^+ + 2e^- \rightarrow H_2$$

The reaction, however, is quickly checked in neutral solutions so that the corrosion process is rapidly stopped. At lower pH values, hydrogen evolution usually prevails, in both the presence and absence of oxygen. Besides these basic reactions, other factors affect the whole process such as, in particular, the metal surface microgeometric characteristics and the ambient oxygen concentration variation that creates differential aeration cells. Cathode sites are, in this case, the highest oxygenation areas. Similar concentrations of cells may occur on metal in the case of con-

Table 2. List of Other Bacteria Relevant to Microbial Corrosion

Genus or species	pH	Temperature (°C)	Characteristics	Metal influenced	Mechanism
Thiobacillus thiooxidans	0.5–8	20–40	Aerobic	Iron and steel, stainless steel	Sulfur and sulfide oxidation to H_2SO_4; protective coating damage
Thiobacillus ferrooxidans	1–7			Iron and steel	Fe^{2+} oxidation to Fe^{3+}
Gallionella	7–10			Iron and steel, stainless steel	Fe^{2+} oxidation to Fe^{3+}; manganous oxidation to manganic; tubercule formation promotion
Sphaerotilus	7–10	20–40	Aerobic	Iron and steel, stainless steel	Fe^{2+} oxidation to Fe^{2+} oxidation to Fe^{3+}; manganous oxidation to manganic; tubercule formation promotion
S. natans				Aluminium alloys	
Pseudomonas	4–9	20–40	Aerobic	Iron and steel, stainless steel	Some strains reduce Fe^{3+} to Fe^{2+}
P. aeruginosa	4–8			Aluminium alloys	

tact with soils or solutions where the concentration of aggressive ions is not homogeneous.

Many articles (10,11,17,18) on this subject have originally assumed that in order to participate in the corrosion process, microorganisms ought to be in a vegetative phase and produce oxydizing agents through their metabolism, or else directly intervene in one or both of the metal's electrochemical reactions. It has always been known that all bacterial colonies (not only the ones consisting of vegetative cells) may be responsible for differential aeration cell formation on metal surfaces. Microhabitats generated by these colonies often create anaerobic conditions that stimulate, in turn, the activity of SRB.

There are several ways of classifying MIC, such as on the basis of the metabolism of implicated microorganisms, as follows (10,11,19):

- Absorption of nutrients by microbial growth in adhesion to the metal surface
- Evolution of corrosive metabolites or end products of fermentative growth
- Sulfuric acid production
- Interference in the cathodic process under anaerobic conditions by obligate anaerobes

Another method of classification is on the basis of the mechanism involved (20):

- Corrosion by SRB
- Corrosion by acid producers
- Corrosion due to heavy microbial growth and deposits
- Corrosion associated with the rupture of protective coatings

Even though such classifications are useful, the mechanism and means of recognition of MIC are discussed, in view of industrial application, with reference to aerobic or anaerobic environmental conditions. Some practical examples are reported as well to provide the reader with the visual appearance of different MIC phenomena and practical ways of identification.

Bacterial Corrosion under Aerobic Conditions

Many aerobic microorganisms are capable of affecting the corrosion process. Some microorganisms are responsible for surface cell-deposit formation (bioscaling phenomena). In a marine environment, the biofouling phenomenon is mostly due to cirripeds, mussels, and so on. The corrosion process, in such cases, consists of formation of a differential-aeration cell due to oxygen assumption by the microbial colonies, tending to exhaust oxygen concentration under such mass. Poorly aerated surfaces behave as anodes, while better aerated surfaces, farther from the deposit, provide the cathode balancing of the reaction. The simple mechanism of aerobic corrosion is schematized in Figure 2 (after Ref. 10).

The formation of an occluded area on a metal surface is a mechanism potentially involved in MIC. It is based on the observation that when microorganisms form colonies on the surface of a metal, they do not constitute uniform layers but local colonization sites related to such metallurgical features as roughness, inclusions, surface charge, and superficial defects. The development of an occluded cell formed by sticky polymers is reproduced in Figure 3; biological and nonbiological (e.g., metals and chloride) species tend to aggregate to the colony site.

Microbiological corrosion on a reactor made of steel, with no addition of particular alloy elements and belonging to the FE33 commercial category, can be so marked under aerobic conditions as to be detectable by simple optical microscopy. A case study, taken from a reactor employed for yeast biomass industrial production and showing uniform generalized corrosion, is shown in Figure 4. The corroded surface, examined under a scanning electron microscope and reproduced in Figure 5, shows how the difference in molar volumes existing between oxide flakes and metal causes the process of spalling. Spalling of the corrosion products and superficial pitting due to the action of localized cellular deposits are shown in Figure 6.

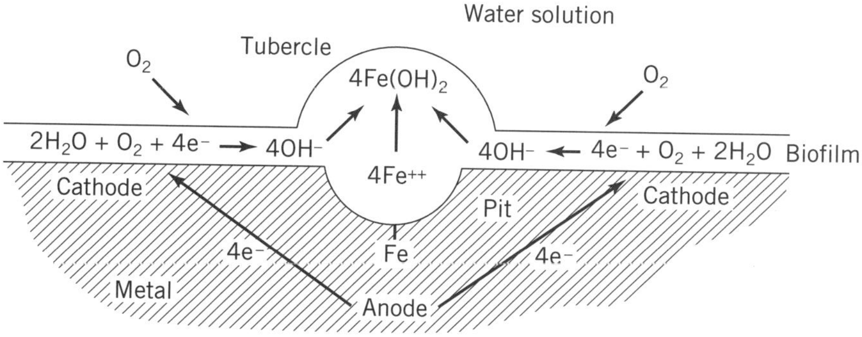

Figure 2. Aerobic corrosion mechanism by differential aeration cell (after Ref. 10)

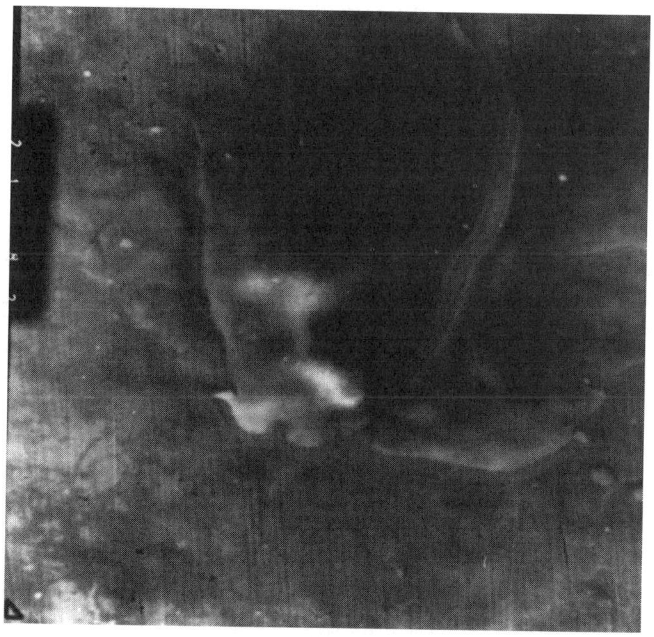

Figure 3. AISI 304 (500×): development of occluded cell under aerobic conditions.

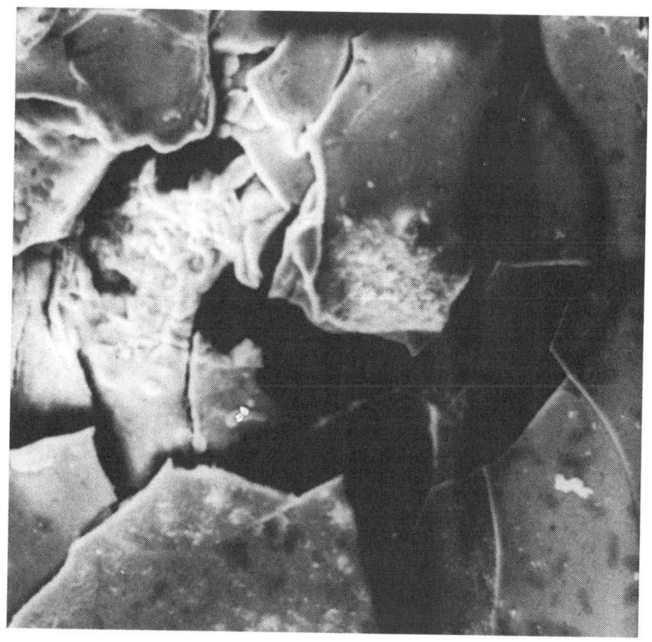

Figure 5. FE33 (500×): morphology of aerobic corrosion products showing spalling phenomenon.

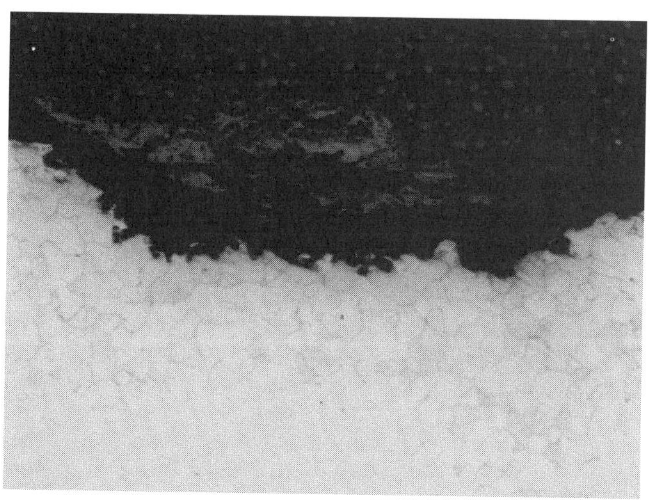

Figure 4. FE33 (120×): uniform generalized aerobic corrosion phenomena.

One particular type of aerobic corrosion is due to microorganisms belonging to the *Thiobacillus* genus. These are chemio-lithotrophic microorganisms that draw the energy needed for their vital processes from oxidation of inorganic substances. The nutrition functioning of these microorganisms, in particular *Thiobacillus ferrooxidans*, was investigated; various models have been proposed, among which one of the most feasible is that iron as a complex is absorbed through the microorganism's cell wall and donates the electron to the respiratory tract. The energy liberated during the subsequent metabolic phases is employed to form ATP, and the ferric iron and water formed are expelled from the cell, forming colloidal ferric hydroxide (21). So little energy is made available through the ferrous iron oxidation reaction (about 6.6 kcal/g-atom) that the organism is compelled to oxidize vast quantities of iron to obtain what it needs to grow.

Another type of microorganism, *Thiobacillus thiooxidans*, is capable of producing sulfuric acid. The relevant

Figure 6. FE33 (300×): spalling of aerobic corrosion products and superficial pitting due to localized cellular deposit.

mechanism has been studied in detail by Booth (22) and Purkiss (23): The generally accepted path involves the following reactions that may depend on the substrate and the environment:

$$2H_2S + 2O_2 \rightarrow H_2S_2O_3 + H_2O$$

$$5Na_2S_2O_3 + 4O_2 + H_2O \rightarrow 5Na_2SO_4 + H_2SO_4 + 4S$$

$$4S + 6O_2 + 4H_2O \rightarrow 4H_2SO_4$$

Bacterial Corrosion under Anaerobic Conditions

According to the previously mentioned considerations concerning electrochemical reactions, a corrosion process involving ferrous metals would seem unlikely in a near pH-neutral, oxygen-deficient environment, because neither of the two common cathode reactions can take place under these conditions at a sustained rate. This, however, is not always true, and when a corrosion phenomenon takes place in such conditions it is much more serious than under aerobic conditions.

The main factor responsible for this problem is the activity of SRB as well as their sulfured metabolic products. Buried metals corroded by this sort of process are one example of the severe economic losses and damage caused by this type of corrosion.

The process of anaerobic corrosion is characterized by a coating on corroding steel consisting of strongly reduced black sulfide–containing corrosion products. Graphitization takes place in cast iron, and little superficial evidence of corrosion can be noticed; the iron migrates from the material, leaving a soft residue, mainly carbon-made. As with bacterial involvement in the corrosion process, the explanation proposed by Wolzogen Kuhr and Van der Vlugt (6) is based on the assumption that cathode hydrogen may be utilized for sulfate reduction; the equations describing the process are as follows:

$$4Fe \rightarrow 4Fe^{2+} + 8e^- \quad \text{anode reaction}$$

$$8H_2O \rightarrow 8H^+ + 8OH^- \quad \text{water dissociation}$$

$$8H^+ + 8e^- \rightarrow 8H \quad \text{cathode reaction (*)}$$

$$SO_4^{2-} + 8H \rightarrow S^{2-} + 4H_2O \quad \text{cathode depolarization (*)}$$

$$Fe^{2+} + S^{2-} \rightarrow FeS \quad \text{corrosion product}$$

$$3Fe^{2+} + 6\,OH^- \rightarrow 3Fe(OH)_2 \quad \text{corrosion product}$$

The overall reaction can be written as

$$4Fe + SO_4^{2-} + 4H_2O \rightarrow 3Fe(OH)_2 + FeS + 2OH^-$$

Through the reactions marked by an asterisk, a biological action develops, bringing about hydrogen activation that, in turn, causes sulfate reduction. Several research projects carried out in recent years seem to support this assumption (24,25); lately, however, some studies point to an alternative process involving sulfide and the iron sulfide film (26,27).

Laboratory studies employing cultures of SRB are unlikely to produce heavily-pitted corrosion, and normal iron oxide and hydroxide corrosion products are nearly always absent. A similar phenomenon also occurs in the presence of elementary sulfur, which often forms, under natural conditions, in the periphery of pitted areas; it very seldom does so under laboratory conditions.

A combination of these and other observations related by King and Miller (27) has led to a more critical approach to the classic theory. In view of the particularly complex problems involved, we find it advisable to analyze the classic theory and cathode depolarization alternative mechanism separately.

Classic Theory. The Classic theory is based on the fact that some SRB strains possess the hydrogenase enzyme, which is capable of catalyzing both the reversible reactions involving hydrogen; that is,

$$H \leftrightarrow 2H \leftrightarrow 2H^+ + 2e^-$$

as well as sulfide reduction using molecular hydrogen, according to following:

$$SO_4^{2-} + 4H_2 \rightarrow S^{2-} + 4H_2O$$

Hydrogenase-positive strains of alophile, thermophile, and mesophile bacteria were shown by Booth and Tiller (28) to be capable of cathode depolarization, whereas hydrogenase-negative bacteria proved completely inactive in this process. Thus, strains such as *Desulphovibrio desulphuricans* and *Desulphovibrio vulgaris*, both hydrogenase positive, use the cathode hydrogen for substrate reduction, whereas hydrogenase-negative strains, such as *Desulfotomaculum orientis*, were found to play no role at all in this mechanism.

Among other hypotheses, a relation has been supposed to exist between the microorganism hydrogenase coefficient and the corrosion rate. In this manner, the steel corrosion rate would be related to hydrogenase activity, even if only in the limited field of experimental conditions. However, subsequent experiments that employed active cell cultures of several different strains of SRB indicated that the corrosion rate did not depend on the hydrogenase activity. And, further, hydrogenase-negative bacteria proved to be at least as aggressive as the *Desulphovibrio* species. This clearly indicates that other factors must be involved.

Cathode Depolarization Alternative Mechanism. Studies on cathode depolarization always refer to the acknowledged property of sulfides, both in solution and in solid state, (for instance, elemental sulfur) to act as corrosive agents. Further, it might be noted that the mild steel corrosion rate increases by up to 230% when in contact with iron sulfide. Serious corrosions in acidic environments were likewise found to be imputable to the presence of sulfured compounds and to elemental sulfur, with the corrosion rate varying from 5.3 mm per annum for iron sulfide to 1,100 mm per annum for sulfur.

Laboratory studies on the influence of ferrous iron on anaerobic mild steel corrosion by active SRB cultures hinted that a protective film was formed on the metal surface at low iron-ion concentration. However, the presence in solution of sufficiently high iron concentrations caused all of the bacterially produced sulfide to precipitate as iron sulfide, so that no protective layer was formed in the medium on the metal and the corrosion rate increased considerably (29). Electrochemical studies showed that substantial and energetic cathode depolarizations took place, even if the ferrous iron concentration was subsequently reduced.

In a previous study (30) associated with bacterial growth in a medium containing 0.5% ferrous iron, the corrosion rate of mild steel was shown to be an approximately linear function of time and independent of the hydrogenase activity of bacteria. The results of further tests clearly showed that the highest contribution to the corrosion process is provided by the ferrous sulfide mechanism. A comparison between mild steel corrosion by means of chemically prepared and biogenic ferrous sulfide indicated that at high ferrous iron concentrations most of the corrosion in bacterial cultures may be ascribed to biogenic ferrous sulfide. King et al. (31) showed that corrosiveness of chemically prepared ferrous sulfide is a stoichiometric function of sulfur, and that stimulation of corrosion due to iron sulfide decreases with time.

High corrosion rates in free bacterial systems may only be obtained by means of a continuous iron sulfide supply. The efficiency decrease of depolarization due to iron sulfide as a function of time without the presence of bacteria is ascribed to the hydrogen atomic bond of the iron sulfide structure. However, the iron sulfide activity may be renewed in the presence of SRB because of their capability of using the hydrogen contained in the film. In this way an explanation can be provided for this alternative mechanism, which is considered responsible for the high corrosion rate observed in the ground.

Under anaerobic conditions, too, the mechanism of formation of occluded areas on the metal surface is often associated with microbiologically induced corrosion. In connection with the development of occluded cells, microorganisms create conditions under the colony very different from those on the surrounding metal, leading to the formation of crevices (as shown in Fig. 7) as well as of ion concentration cells. This hypothesis is supported, in the presented case study, by energy dispersion spectrometer EDAX™ investigation revealing the presence of a series of deposits containing Fe, Cr, Ni, Mo, Ca, and S.

In parallel with the previously mentioned phenomenon, it may be noted that the microbiological community usually remains fixed to the colonization site, thus fixing the anodic site. The resulting pitting corrosion is characterized by the formation of small anodes and corresponding large cathodes, which allow the formation of pits in the metal matrix. Figure 8 shows incipient pitting triggered by microbiological activity on AISI 316. Incipient pitting engendered by microbiological activity on AISI 347 is shown in Figure 9; a detail of pits in the growing phase on the same metal is shown in Figure 10.

An example of intergranular corrosion, usually taking place on the border of uniform corroded areas, is reproduced in Figure 11. EDAX™ investigation (Fig. 12) on the selected corroded point revealed the presence of the following elements (% w/w): S: 1.58, Ca: 0.17, Cr: 34.7, Fe: 48.4, and Ni: 15.15. In the surrounding area the following elements were present (% w/w): Nb: 2.3, Cr: 14.7, Fe: 70.7, and Ni: 12.3.

DETECTION AND MONITORING OF MIC

In the recent past, the most important developments related to MIC research were provided by a multidisciplinary

Figure 7. AISI 304 (1,000×): crevice corrosion due to the formation of occluded area.

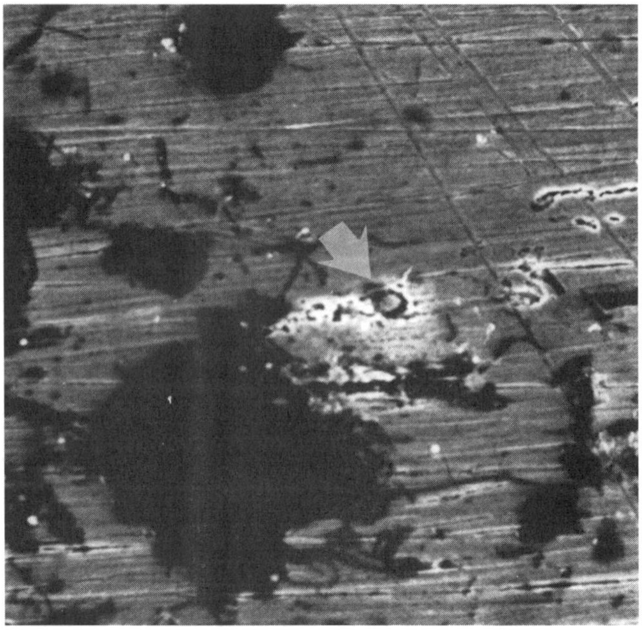

Figure 8. AISI 316 (1,000×): anaerobic incipient localized pitting corrosion.

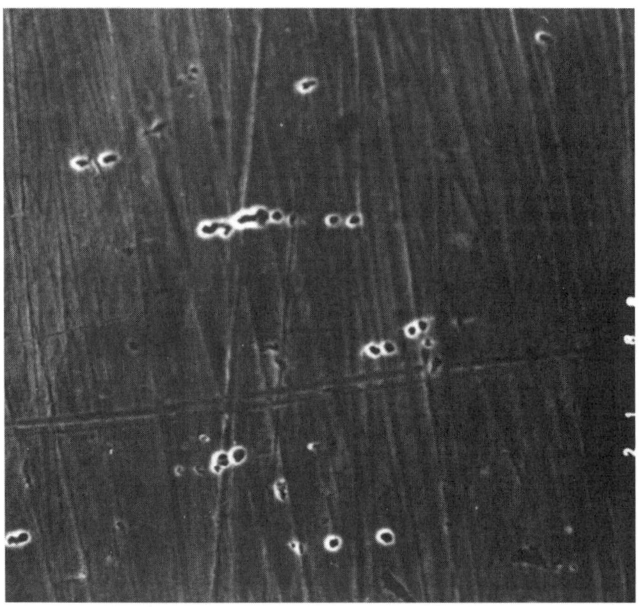

Figure 9. AISI 347 (1,000×): anaerobic incipient pitting corrosion.

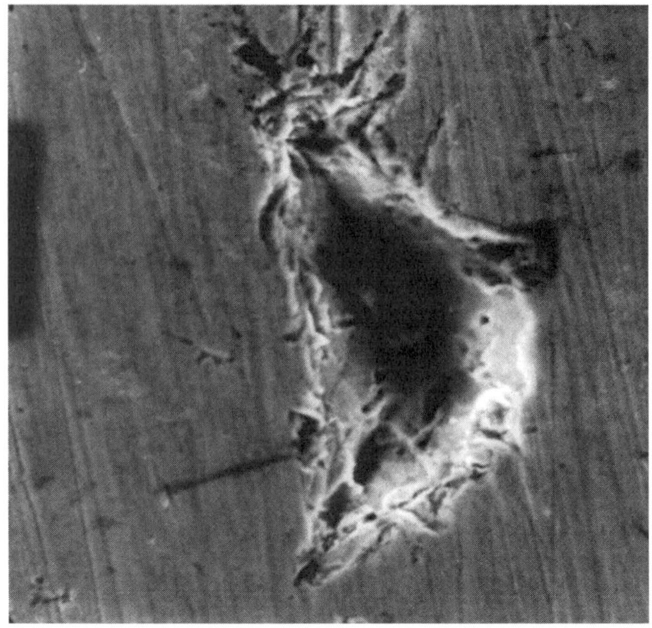

Figure 10. AISI 347 (1,500×): particular of anaerobic pitting in the growing phase.

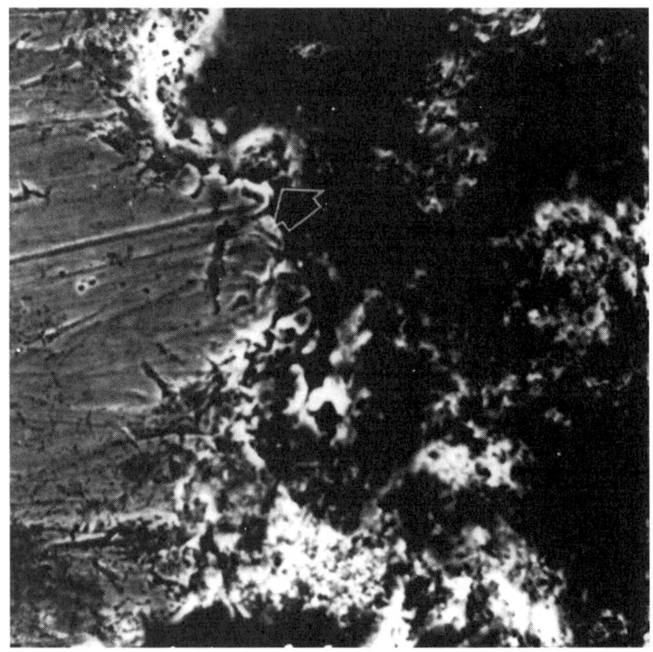

Figure 11. AISI 347 (1,000×): intergranular anaerobic corrosion on the border of uniform corroded area.

approach combining nondestructive analytical and electrochemical investigation methods. Indeed, appropriate techniques for detection and monitoring of the phenomenon are essential to dealing successfully with the nature of mutual interactions and to setting up effective control methods. Evidence of MIC lies in the conspicuous presence of microorganisms in those areas where the integrity of the material has deteriorated markedly. However, the electrochemistry of MIC has not been clearly defined because of the limits of the techniques employed in detection and monitoring (32). Some of the most important developments in MIC research over the past years were the result of a combination of nondestructive analytical and electrochemical methods aimed at characterizing microbial biofilms and corrosion reactions (33). In the following, the main traditional and innovative techniques are briefly presented and discussed.

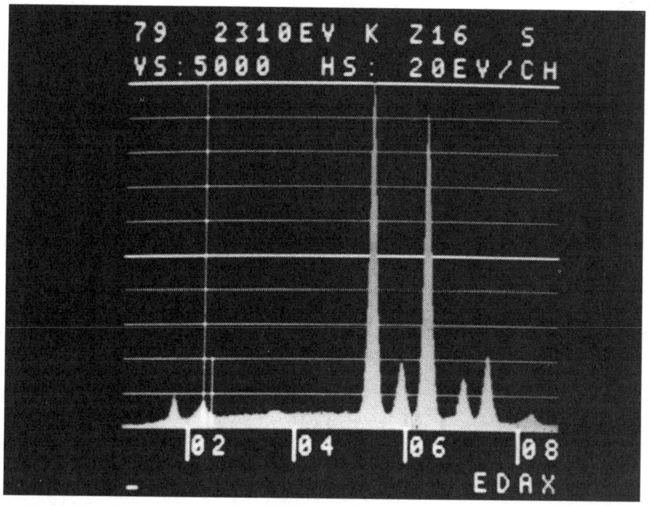

Figure 12. EDAX™ investigation in correspondence to the corroded point shown in Figure 11.

Microscopic Analysis

The identification of bacteria in pitted areas by means of scanning electron microscopy (SEM) was performed by several researchers (34–37), even though microbially induced pitting on stainless steel, the degree of interaction on this material being limited to a few surface monolayers, is not easily revealed. Moreover, it requires large amounts of corrosion products.

Energy-dispersive X-ray spectroscopy is widely applied in analyzing pit composition and is commonly employed for failure analysis of corroded metals. Recently developed (38) atomic force microscopy was applied to investigate sample topography in corroded pipework, allowing the analysis of profiles, depth, and size of pits, even in the presence of a nonconductive sample. In situ biofilm formation occurring under MIC conditions may be monitored by confocal laser microscopy (39); it allows the attainment of three-dimensional biofilm structure and has the potential of becoming an important method in the study of MIC kinetics.

Spectroscopic Analysis

Fourier transform infrared spectroscopy has been applied to monitor copper corrosion under biofilm (40,41), adopting Ge (an IR transparent internal reflection element) as substratum to be coated by a thin Cu film. The technique is severely limited by IR absorption by water, as well as by its inability to distinguish dead cells from live ones. X-ray photoelectron spectroscopy (XPS) and Auger electron spectroscopy (AES) may provide meaningful information about corroded surface composition and chemical states, as binding energies of the photoelectrons and Auger electrons are atom specific and changes in the binding energies of electrons are related to electron distribution around the atom. The limitations of these techniques, recently applied to study MIC (42–44), are connected with the kinetic energies of the Auger electrons and photoelectrons, which permit characterization of only few monolayers of the top surface, as well as with the need for cooling equipment to prevent bioproduct evaporation and decomposition under operative conditions (ultrahigh vacuum $<10^{-9}$ torr). XPS can provide composition and valence states of surface elements, as well as explore the mechanicistic nature of MIC over a relatively large area; Auger electron spectroscopy may be used to analyze smaller areas, obtaining chemical information of pits (31).

Extended X-ray absorption fine structure (EXAFS) may be employed to characterize enzymatic corrosion products created by bacterial consortium on metal surfaces (45,46), exploiting the oscillation of X-ray absorption spectrum in the range of 1 keV above the K-absorption edge (47). Limitations are related to in situ monitoring, as the X-ray energy may be too high.

Electrochemical Techniques

Conventional electrochemical techniques for detection and monitoring of MIC have been used for a long time and include corrosion and pitting potential measurements, linear polarization resistance (48), and electrochemical direct current (DC) methods (49,50); these techniques provide evidence of MIC and partial mechanistic information on loss of corrosion resistance and on electrochemical interactions (51), even if DC techniques require high applied currents and therefore can alter biofilms. Large polarization techniques have limited applications in MIC studies because of the destructive nature of applied currents to microbial biofilms. Consequently, electrochemistry of the biofilm–metal interface can be better investigated by means of techniques requiring small or zero applied potential.

Time-dependent changes in open circuit potential (OCP) were used to prove 316L stainless steel oxidation under a photosynthetic biofilm producing oxygen under illumination (52). Combined OCP and redox potential measurements of the bacterial medium permit accurate estimation of the corrosion tendency of a metal in microbially active environments (53), as well as of changes occurring in the passive films. Time-dependent changes of galvanic current between identical electrodes in dual cells demonstrated the effects of biofilms on induction of corrosion (54,55).

The concentric electrodes system, consisting of a small area anode surrounded by a large-area annular cathode, allowing maintenance of a uniform current profile with reduced overpotential, was successfully applied to monitor pitting commencement and development in SRB-containing anaerobic artificial seawater (56). Because the passive film has a blocking effect on anodic dissolution, this test apparatus, if employed with stainless steel, can result in overpotential to the anode so much higher than the redox potential as to impair bacterial growth.

The scanning electrode technique (SVET) has provided a nondestructive, on-line method to define the magnitude and sign of current densities in solution over freely corroding metals; in particular, it was applied to locate the presence and position of anodic electrochemical activity on the surface of carbon-steel samples by analysis of current maps and OCP (57), demonstrating the possibility of obtaining qualitative and quantitative information regarding localized corrosion influenced by bacteria.

Electrochemical impedance spectroscopy (EIS) was used to investigate the impedance of metal specimens exposed to microbial biofilms, in conjunction with biofilm analyses. It consists of the measurement of amplitude and phase shift of sinusoidal currents resulting from the application of small-amplitude sinusoidal signals to metal samples over a large frequency range (10 kHz to 5 mHz) (58). EIS can provide polarization resistance values (inversely proportional to corrosion rate), double-layer capacitance, and solution resistance so as to evaluate localized corrosion and three-dimensional surface deposits (51).

Future studies of MIC electrochemistry will focus on the combined use of different techniques, includings, confocal laser microscopy to characterize biofilm structure, EIS to investigate initiation of pits too small to be observed microscopically, and XPS and AES to analyze the surface chemistry and structure resulting from the corrosive phenomenon.

PROTECTIVE AND PREVENTIVE MEASURES

Replacement costs, maintenance shutdown, and loss of profits caused by microbiologically influenced corrosion of metal structures are problems in many industries. MIC, once recognized, is a question of optimizing prevention measures and minimizing maintenance and countermeasure efforts. Studies of MIC have addressed both understanding the complexities of microbial interactions with metal substrata and setting up protective and preventive measures. In particular, the involvement of microbial biofilms was amply investigated in the corrosion of several metals, including aluminium (58), copper (59), mild steel, and stainless steel (60). No universal approach to prevent MIC, has been developed so far, however, and the problem is tackled more from the corrosion than from the microbiological viewpoint.

Cathodic Protection

Cathodic protection principles and theory have been discussed extensively in the literature (61,62). Two techniques are commonly adopted: sacrificial anodes usually made of magnesium, zinc, or aluminium alloys, which corrode preferentially to provide structure protection, or impressed current from a rectifier via an inert anode. The cathodic protection method is used extensively to protect structures buried in soil or in marine environments.

In the absence of SRB, protection of steel is achieved at a potential depression of -0.85 V versus $Cu/CuSO_4$. In the presence of a soil environment presenting risk of microbial corrosion, a further -0.1-V potential depression is required to ensure the protection of the structure (63,64). The achievement of this value allows the suppression of SRB growth as well, even if, in the case of bare steel in direct contact with soil, the formation of iron sulfide can lead to extraordinary current requirements. On the other hand, far lower potential values must be avoided, because hydrogen cracking and alkaline blistering of protective coatings can occur.

In the presence of cell suspensions of sulfate reducers, a linear relationship between the hydrogenase-enzyme microbial activity and the additional current requirement to maintain different cathodic potential was verified (63). A focal conclusion is that the corrosion rate is enhanced in the case of an incompletely protected sample because of the accelerating effect of bacteria (65).

Protective Coatings

Protective coatings provide a barrier, impermeable to soluble aggressive species, between the metal and the surrounding environment. The range of available protective coatings is wide, and includes products based on coal tar, asphalt bitumens, epoxy resins and coal tar epoxies, thermoplastic polymers, synthetic polymeric coatings, and cementitious materials. Paints are also used as coating materials. Many of these coatings, being of organic nature, may biodeteriorate. Moreover, they usually cannot prevent gas diffusion and, for instance, allow hydrogen sulfide to pass the coating layer and reach the metal.

In marine environments, metallic coatings, especially aluminium and zinc, are often employed. Protection of steel by these metals is partly due to both their action as physical barriers and their anodic effect on the basic material.

In terms of industrial applications, when contamination connected with corrosion products must absolutely be avoided (e.g., in the food and pharmaceutical industries), the inner sides of plant sections or apparatuses should not be lined with coatings or resins (e.g., epoxy resins). In fact, the resulting unevenness (pores, microcrevices, blisters) might cause an extremely nonuniform, seriously corrosive process under the coating, due to an unfavorable relation between anodic and cathodic zones (66). Figure 13, referring to a FE33 steel sample from a reactor employed in yeast-biomass production, clearly demonstrates how crevice or surface unevenness of the protective paint leads to corrosion under the coating—a process detectable only under an optical microscope.

Figure 13. FE33 painted (300×): start of aerobic MIC below the paint due to protective coating unevenness.

Inhibitors and Biocide

In closed systems it is possible to use corrosion inhibitors and biocides to stop or limit the biological activity. The use of such chemicals has limitations because biocides are potential hazards to health and environment. Organic biocides can take part in complex and uncontrolled reactions, and completely unexpected compounds can be formed. Oxidizing biocides, for example, hydrogen peroxide, which cause death of cells by attacking cell enzymes, can be used in cooling systems and drinking water mains. Oxygen-based biocides (e.g., ozone) are used to prevent bacterial hydrogen-sulfide production. Nonoxidizing biocides, which cause cell-wall rupture, blockage of enzyme systems, or denaturation of proteins, include heavy-metal organic compounds, organic copper salts, organic sulfur compounds, amines, chlorinated phenols, and quaternary ammonium salts. The treatment is ineffective if the bacteria have previously formed nodules as their habitat, so in this case a preventive mechanical removal of nodules is necessary.

Careful and well-documentated observations by plant operating engineers have recognized the efficacy of regular scraping in pipeline corrosion control; experimental runs showed that well-established consortia of SRB are deeply disturbed by the mechanical treatment, with reestablishment only after 1–2 weeks (67). On the basis of these remarks, microbial corrosion in pipelines can be effectively prevented or controlled by combining mechanical scraping to remove protective nodules with proper dose treatment using a penetrating biocide (68).

Remedial action in the case of SRB-induced corrosion is usually approached by a combined action consisting of the application of a biocide and a corrosion inhibitor. Compounds such as molybdate show a double effect, inhibiting both SRB and the corrosion (69). It was observed (70) that 18 mM 2-*n*-Heptyl-4-hydroxyquinoline *N*-oxide inhibited respiration in *Hemophilus parainfluenza* by blocking electron transport between cytochromes b and c_1. Inhibition of the periplasmic hydrogenase enzyme in *Desulfovibrio vulgaris* was verified by both carbon monoxide and copper II chloride (71). In case of *Desulfovibrio desulphuricans*, characterized by a cytoplasmatic hydrogenase, only carbon monoxide was found effective in inhibiting the hydrogenase, and copper II chloride had no effect (72). Sodium molybdate evidenced inhibiting action on stainless steel corrosion by making passive the anodic reaction of an active corrosion cell and inhibition of the growth of a culture of *D. vulgaris* and unidentified *Vibrio* sp. (73). The main conclusion is that viable bacteria are necessary only in the early stages of pit formation, and not for the maintenance of galvanic current; in any case, the effective treatment of actively pitting systems requires the adoption of both a corrosion inhibitor and a biocide.

Alternative Materials

Solution of the problem of MIC based on safety considerations (74) may consist of the application of the substitution principle, in which a safer material is used in place of a more hazardous one, thus decreasing the need for added protective equipment and also decreasing plant cost and complexity. The use of common steel should be avoided in the construction of industrial plants for bioleaching of mine drainage. Biologic desulfurization of pit coal and the use of stainless steel, such as AISI 304, is compulsory to avoid safety problems connected with possible metallic structure failures (75). A number of resistant alloys providing adequate corrosion strength are available (e.g., all experimental runs on higher nickel–chromium alloys and titanium highlight their resistance to MIC), although their cost makes their use uneconomical in most situations. Steel and iron will be replaced extensively by nonmetallic materials, such as composite fiber reinforced resin systems, due to the availability of reliable and long-lasting materials. Other plastic materials, such as PVC and polyethylene, must be used with care, because they may contain plasticizers susceptible to biodeterioration, resulting in progressive embrittlement of the polymer. Cement products, including spun concrete and reinforced or prestressed concrete, are widely used for buried piping; however, their selection must be site specific, because they may have problems in some heavily contaminated soils and waters (63,76).

CONCLUSIONS

Microbiological corrosion is an interesting but costly and frustrating problem more widespread than one would expect, and it is often associated with the more common problem of biofouling. Microorganisms play a very complex role, and several mechanisms were suggested to highlight their involvement in the corrosion process. In many instances their role consists of the formation of differential aeration and concentration cells, in the production of aggressive metabolites, or in the acceleration of an existing corrosion process. Microbial involvement in the process can be indicated by localized corrosion associated with tubercles, discolorations, slime, anaerobic metabolism odors, sludges, and underdeposit corrosion phenomena.

The majority of these problems can be ascribed to SRB. The knowledge of aspects of SRB physiology and of their ability to grow on hydrogen, as well as the role of enzyme hydrogenase, has shown the most probable mechanism of corrosion due to these organisms.

From a practical industrial viewpoint, avoidance or minimization of microbial corrosion problems can be faced first by optimal selection of materials and, subsequently, by interposition of barrier systems or modification to the surrounding environment. The combined use of chemical inhibitors and biocides represents the most common solution to dealing with internal problems in process systems.

MIC problems will be positively contained and safely controlled in industrial situations only if designers and plant operators properly consider projects, design features, operating and maintenance procedures, and system monitoring.

BIBLIOGRAPHY

1. R. Mallet, *Br. Assoc. Adv. Sci. Rpt.* **9**, 221 (1940).
2. J.H. Garrett, *The Action of Water on Lead*, H.K. Lewwis, London, 1891.

3. R.H. Gaines, *J. Eng. Ind. Chem.* **2**, 128 (1910).
4. E.C. Harder, *Iron Depositing Bacteria and Their Geologic Relations*, Governmental Printing Office, Washington, D.C., 1919.
5. R. Grant, E. Bate, and W.H. Myers, Inst. Eng. Australia Sydney Div., Paper no. 8, 1921.
6. C.A.H. von Wolzogen Kuhr and L.S. van der Vlugt, *Water* (den Haag) **18**, 147 (1934).
7. H.H. Uhlig, *Corrosion Handbook*, 4th ed., Wiley, New York, 1953.
8. J.D.A. Miller ed., *Microbial aspects of Metallurgy*, MTP, Aylesbury, 1971.
9. G.H. Booth, *Microbial Corrosion*. M and B Monograph CE/1, Mills & Boon, London, 1971.
10. W.P. Iverson, in M.G. Fontana and R.W. Staehle, eds., *Advances in Corrosion Science and Technology*, vol. 2, Plenum Press, London, 1972, pp. 70–94.
11. J.D.A. Miller and R.A. King, in D.W. Lovelock and R.J. Gilbert, eds., *Microbial Aspects of the Deterioration of Material*, Academic Press, London, 1975, p. 83.
12. E.C. Pringsheim, *Biol. Rev. Cambridge Phil. Soc.* **24** 200 (1949).
13. J.S. Stokes, *J. Bact.* **67**, 278 (1954).
14. J.D.A. Miller and A.K. Tiller, in J.D.A. Miller, ed., *Microbial Aspects of Metallurgy*, MTP, Aylesbury, 1971, p. 61.
15. L.A. Terry and G.J. Edyvean, *Bot. Mar.* **24**, 177–183 (1981).
16. H. Ebneth, D. Rempel, and W. Hoyer, *Galvanotechnik* **63**, 27 (1972).
17. W.P. Iverson, *Science* **151**, 986 (1966).
18. G.H. Booth, A.W. Cooper, and P.M. Cooper, *Chem. Ind.* **86**, 2084 (1967).
19. D.J. Crombie, G.J. Moody, and J.D.R. Thomas, *Chem Ind.* **21**, 500 (1980).
20. J. Chantereau, *Corrosion Bacterienne. Bactéries de la Corrosion*, 2nd edition, Technique et Documentation ed., Paris, 1980.
21. G. Rossi, *Resoconti dell' Associazione Mineraria Sarda* **79**, 1–21 (1974).
22. G.H. Booth, *Microbial Corrosion*, MandB Monograph CE/1, Mills & Boon, London, 1971.
23. B.E. Purkiss, in J.D.A. Miller, ed., *Aspects of Metallurgy*, MTP, Aylesbury, 1971, p. 107.
24. J. Horvath, *Acta Chem. Acad. Sci.* **25**, 65 (1960).
25. G.H. Booth and F. Wormwell, in *1st Proc. Int. Congr. Metallic Corrosion*, Butterworth, London, 1962, p. 341.
26. G.H. Booth, L. Elford, and D.S. Wakerely, *Br. Corr. J.* **3**, 242 (1968).
27. R.A. King and J.D.A. Miller, *Nature* **233**, 491 (1971).
28. G.H. Booth and A.K. Tiller, *Trans Faraday Soc.* **58**, 2510 (1962).
29. R.A. King, J.D.A. Miller, and D.S. Wakerley, *Br. Corr. J.* **8**, 89 (1973).
30. G.H. Booth, P.M. Cooper, and D.S. Warkerley, *Br. Corr. J.* **1**, 345 (1966).
31. R.A. King, J.D.A. Miller, and J.S. Smith, *Br. Corr. J.* **8**, 137 (1973).
32. G. Chen, R.J. Palmer, and D.C. White, *Biodegradation* **8**, 189–200 (1997).
33. M.J. Franklin and D.C. White, *Current Opinion in Biotechnology* **2**, 450–456 (1991).
34. D.E. Nivens, P.D. Nichols, J.M. Henson, G.G. Geesey, and D.C. White, *Corrosion* **42**, 204–210 (1986).
35. T.R. Jack, G. Van Boven, and M. Wilmott, *Material Performance* **35**, 39–45 (1996).
36. W. Lee and W.G. Charaklis, *Corrosion* **49**, 186–99 (1992).
37. B.J. Little, P. Wagner, S.M. Gerchakov, M. Walch, and R. Mitchell, *Corrosion*, **42**, 533 (1986).
38. A. Steele, D.T. Goddard, and I.B. Beech, in *International Biodeterioration and Biodegradation*, Elsevier, London, 1994, pp. 35–46.
39. J.R. Lawrence, G.M. Wolfaardt, and D.R. Korber, *Appl. Environ. Microbiol.* **60**, 1166–1173 (1994).
40. P.J. Bremer and G.G. Geesey, *Appl. Environ. Microbiol.* **57**, 1956–1962 (1991).
41. G.G. Geesey and P.J. Bremer, *Mar. Technol. Soc. J.* **24**, 36–43 (1990).
42. P.J. Bremer, G.G. Geesey, B. Drake, J.G. Jolley, and M.R. Hankins, *Surface Interface Anal.* **17**, 767–772 (1991).
43. G. Chen, R.A. Sadowski, and C.R. Clayton, in *CORROSION / 95*, paper 217, NACE National Association of Corrosion Engineers, Houston, Tex., 1995.
44. G. Chen, S. Kagwade, G.E. French, C.R. Clayton, T.E. Ford, and R. Mitchell, *Corrosion* **52**, 891–899 (1996).
45. A.J. Davenport, J.A. Bardwell, and C.M. Vitus, *J. Electrochem. Soc.* **142**, 721–724 (1995).
46. A.J. Davenport and M. Sansone, *J. Electrochem. Soc.* **142**, 725–730 (1995).
47. L.C. Feldman and J.W. Mayer, in *Fundamentals of Surface and Thin Film Analysis*, North-Holland, New York, 1986, pp. 201–204.
48. J.S. Luo, X. Campaignolle, D.C. White, and I. Vance, in *CORROSION / 93*, paper 301, NACE National Association of Corrosion Engineers, Houston, Tex., 1993.
49. S.C. Dexter, D.J. Duquette, O.W. Siebert, and H.A. Videla, *Corrosion* **47**, 308, (1991).
50. B.J. Little, P. Wagner, and F. Mansfeld, *Electrochimica Acta* **37**, 2185–2194 (1991).
51. F. Mansfeld and B.J. Little, *Corrosion Sci.* **32**, 247–272 (1991).
52. N.J.E. Dowling, J. Guezennec, J. Bullen, B.J. Little, and D.C. White, *Biofouling* **5**, 315–322 (1992).
53. X. Zhang, R.A. Buchanan, E.E. Stansbury, and N.J.E. Dowling, in *CORROSION / 89*, paper 512, NACE National Association of Corrosion Engineers, Houston, Tex., 1989.
54. N.J.E. Dowling, J. Guezennec, M.L. Lemoine, A. Tunlid, and D.C. White, *Corrosion* **44**, 869–874 (1988).
55. S.M. Gerchakov, B.J. Little, and P. Wagner, *Corrosion* **42**, 689 (1986).
56. P. Angell, J.-S. Luo, and D.C. White, *Corrosion Sci.* **37**, 1085–1096 (1995).
57. M.J. Franklin, D.C. White, and H.S. Isaacs, *Corrosion Sci.* **32**, 945–952 (1991).
58. E.S. Ayllon and B.M. Rosales, *Corrosion* **44**, 638–644 (1988).
59. P. Angell, Ph.D., Thesis, University of Surrey, Surrey, OK, 1992.
60. B. Little, P. Wagner, and F. Mansfeld, *Int. Mat. Rev.* **36**, 253–272 (1991).
61. J.H. Morgan, *Cathodic Protection*, Leonard Hill, London, 1959.
62. L.L. Shrier ed, *Corrosion*, vol. 1, 2nd ed., Newnes Butterworth, London, 1976.
63. G.H. Booth and A.K. Tiller, *Corrosion Sci.* **8**, 583 (1968).
64. J. Horvath and M. Novak, *Corrosion Sci.* **8**, 538 (1968).

65. J. Guezennec, N.J.E. Dowling, J. Bullen, and D.C. White, *Biofouling* **8**, 133–136 (1994).
66. P. Perego, B. Fabiano, A. Ferraiolo, and R. Pastorino, *Chem. Biochem. Eng. Q.* **11**, 101–106 (1997).
67. J.W. Costerton and G.G. Geesey, in *Biologically Induced Corrosion NACE 8*, NACE National Association of Corrosion Engineers, Houston, Tex., 1986, pp. 223–232.
68. G.N. Kirby, *Chem. Eng. Prog.* **6**, 29–38 (1995).
69. J.S. Luo, X. Campaignolle, J. Bullen, M.W. Mittleman, D.C. White, and J.F. Zibrida, in *CORROSION/92*, paper 186, NACE National Association of Corrosion Engineers, Houston, Tex., 1992.
70. D.C. White and L. Smith, *J. Biol. Chem.* **239**, 3956–3963 (1964).
71. R.M. Fitz and H. Cypionka, *Arch. Microbiol.* **155**, 444–448 (1991).
72. R.M. Fitz and H. Cypionka, *Arch. Microbiol.* **152**, 369–376 (1989).
73. P. Angell and D.C. White, *J. Ind. Microbiol.* **15**, 329–332 (1995).
74. T. Kletz, *Plant Design for Safety. A User-Friendly Approach*, Hemisphere Publishing, New York, 1991, pp. 53–66.
75. P. Perego, B. Fabiano, R. Pastorino, and G. Randi, *Bioprocess Eng.* **17**, 103–109 (1997).
76. P. Perego, Ph.D. Thesis, University of Milan, Milan, Italy, 1986.

CORYNEBACTERIA, BREVIBACTERIA

P. A. Lessard
S. Guillouet
L. B. Willis
A. J. Sinskey
Massachusetts Institute of Technology
Cambridge, Massachusetts

KEY WORDS

Corynebacterium
Brevibacterium
Fermentation
Glutamic acid
Lysine
Threonine
Isoleucine
Biochemical engineering
Metabolic engineering

OUTLINE

INTRODUCTION

Corynebacteria and brevibacteria are Gram-positive, aerobic actinomycetes. The primary industrial interest in corynebacteria and brevibacteria relates to the production of amino acids by fermentation. In 1956, a researcher at Kyowa Hakko Kogyo Co. in Japan isolated soil bacteria capable of producing large amounts of glutamic acid (1). Until that time, glutamic acid was produced commercially from *konbu* a type of seaweed used in popular Japanese cuisine. The discovery of these bacteria proved to be a major innovation in the industry at the time, and today these bacteria account for worldwide production of more than 800,000 metric tons of glutamic acid (primarily for use as a flavor enhancer) and 330,000 metric tons of lysine (for feed additives) annually.

SPECIES RELATIONSHIPS AMONG GLUTAMIC ACID–PRODUCING CORYNEBACTERIA

When first identified, the species we now call *Corynebacterium glutamicum* was classified as *Micrococcus glutamicus*. Several "species" of glutamic acid–producing bacteria were similarly identified by a number of groups, and classified into genera such as *Brevibacterium, Corynebacterium, Arthrobacter*, and *Microbacterium*. During the subsequent four decades, considerable effort has been spent on reconciling the phylogeny of these bacteria. Nonetheless, the scientific literature regarding these bacteria is still laden with inconsistencies in the nomenclature, and an opportunity exists to establish conventions for species designation among these bacteria.

Systematics among the glutamic acid–producing bacteria have evolved from the classical morphological and physiological analyses to include molecular genetic characterizations. It can be argued that morphological analyses caused much of the confusion encountered in the literature because corynebacteria and brevibacteria belong to a group of bacteria, frequently called the "coryneform" bacteria, whose common characteristic appears to be pleomorphism. Abe and coworkers (2) identified several characteristics that are useful for distinguishing the corynebacteria and brevibacteria from sister clades. In addition to their prolific glutamate production under aerobic conditions, they require biotin, and individual bacteria are nonmotile, nonsporulating, and occur as short clubs or rods in V-shaped aggregates of two or more. Chemotaxonomic information has permitted a more methodical classification of these bacteria (3). Based on the presence of relatively short-chain mycolic acids (containing 22–36 carbon atoms) in the cell walls, and the types of lipids that comprise their cell membranes, these bacteria cluster with members of the genera *Rhodococcus, Mycobacterium*, and *Nocardia* into an informal collection known as the CMN group, which includes many medically significant pathogens.

In recent years, molecular systematics have been used to more precisely define the glutamic acid–producing bacteria. Chromosomal DNA of the glutamic acid–producing bacteria comprises approximately 52–56 mol % G + C. Comparison of 16S and 23S ribosomal RNA structures has revealed that many of the glutamic acid–producing species that have traditionally been classified into the different genera *Corynebacterium* and *Brevibacterium* (e.g., *B. divaricatum, B. flavum, B. lactofermentum, C. glutamicum*, and *C. lilium*) may all actually be members of the same species, *Corynebacterium glutamicum* (4). In the same studies, other species within the genus *Brevibacterium* (e.g., *B. linens*) were found to be much more distantly related, containing a higher G + C content in their genomic DNA, showing greater divergence among their ribosomal RNA structures, and lacking the mycolic acids that are characteristic of the CMN group. For this reason, in this article we shall refer to all of the glutamic acid–producing bacteria simply as corynebacteria, with emphasis on the well-characterized *C. glutamicum*.

COMMERCIAL INTERESTS IN CORYNEBACTERIA

The main industrial interest in corynebacteria concerns the production of amino acids. The twenty amino acids can be commercially manufacture and are used in a variety of products including food additives, animal feed supplements, cosmetics, and medicines, as well as some industrial applications. Currently, the highest revenues are from the animal feed industries and human food uses. The overall global market for amino acid animal feed additives is estimated to be worth more than $2 billion, and totals about 700,000 metric tons of material. Lysine and methionine account for an overwhelming majority of the market, which also includes lower-volume products such as threonine (Table 1). The market for monosodium glutamate alone, used in the human food industry as a flavor enhancer, is also estimated to be $2 billion. The worldwide market growth rate is around 8–12% annually for amino acids (5), but individual amino acids could experience different spurts of growth. For example, the price of cysteine doubled during 1996 due to increased demand that sparked a worldwide shortage (6). In 1990, the world production of threonine was 1,000 metric tons per year, but already new projects will give BioKyowa, for example, a total capacity of 10,000 metric tons annually for tryptophan and threonine in 1998 (7).

In 1990, aspartic acid was produced at around 10,000 metric tons per year, mainly for the production of aspartame. Most of the world's aspartic acid still goes into aspartame, but Donlar Corp. and Rohm & Haas believe the emerging application for polyaspartate, a newer specialty polymer with a wide range of applications, will drive new production of aspartic acid (8).

Table 1. Industrial Production of Amino Acids

Amino acid	Worldwide production (ton/year)	Production process
L-Glutamic acid	>800,000	Production by corynebacteria
L-Lysine	330,000	Production by corynebacteria
DL-Methionine	350,000	Chemical synthesis
L-Aspartic acid	10,000[a]	Bioconversion by *E. coli*
L-Phenylalanine	5,000[a]	Production by corynebacteria
L-Threonine	1,000[a]	Production by corynebacteria and *E. coli*

[a]Production data from 1990; all other data from 1997.

Amino acids are produced industrially by four different processes: production in reactors by bacteria (glutamate, lysine, phenylalanine, threonine, tryptophan, valine), bioconversion (aspartic acid), chemical synthesis (methionine), and extraction from natural materials (cysteine). Corynebacteria in particular are involved in the production of the two most important amino acids, glutamic acid and lysine (Table 1).

Corynebacteria are able to accumulate 8–10 times as much intracellular amino acid, such as glutamic acid, as accumulated by Gram-negative bacteria such as *Escherichia coli* (9). In addition, the amino acid biosynthetic pathways in corynebacteria are usually less strongly regulated than those in *E. coli*, for instance (Fig. 1). *E. coli* has three isoenzymes for aspartokinase and as many as five for acetohydroxy acid synthase (AHAS). The abundance of isoenzymes complicates the prospects of using genetic engineering to improve amino acid production in *E. coli*. These observations explain why corynebacteria are preferred for the production of the major amino acids.

Other industrial applications for corynebacteria concern the production of vaccines from *Corynebacterium tetani* and *C. diphtheriae* (10), the production of L-malic acid from fumaric acid by the fumarase of immobilized *Brevibacterium ammoniagenes* (11) and the production of L-ascorbic acid by 2,5-diketo-D-gluconic acid reductase of *Corynebacterium* sp. (11). Corynebacteria may also be used to produce recombinant proteins such as gamma interferon, proteases, and cellulase (12).

Corynebacteria, and specifically *Brevibacterium linens*, are involved in the cheese ripening process. Although the lactic acid bacteria play a key role in cheesemaking by catabolizing nitrogen compounds, the roles of other bacteria, such as brevibacteria, are also important. Possessing powerful and varied enzymatic systems, they contribute to the aging of ripened cheeses by producing unique flavors (13). Flavored compounds produced by these bacteria, including peptides, amino acids, and sulfured compounds, have found applications in the food industry as additives in cheese-flavored products (sauces, crackers, etc.) (14).

Not surprisingly, the *Corynebacterium* strains that are used in industrial fermentations are not the original iso-

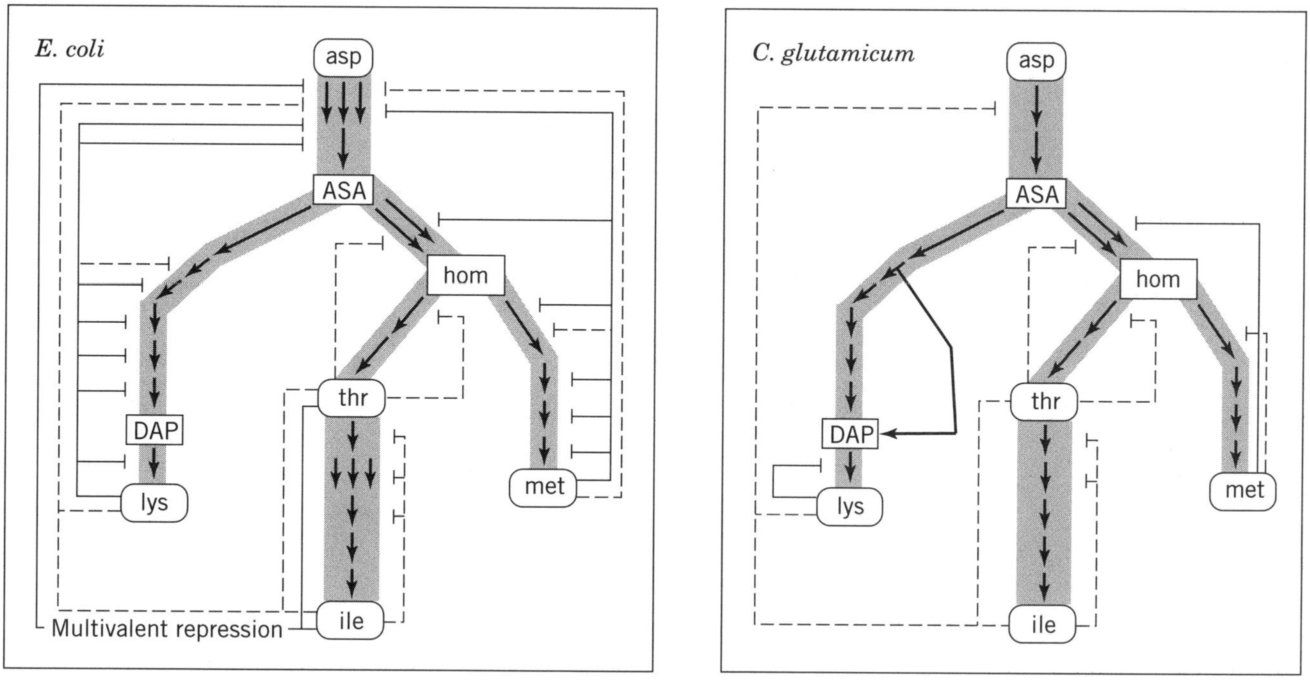

Figure 1. Summary of feedback inhibition and genetic repression operating in pathways for synthesis of aspartate derived amino acids in *E. coli* and *Corynebacterium*. Dashed lines denote feedback inhibition, solid lines denote genetic repression. Arrows indicate individual, enzymatically catalyzed steps. Steps that are catalyzed by multiple isozymes in *E. coli* are indicated by more than one arrow. Note that control of these pathways in *E. coli* is much more complex than that known in *Corynebacterium*. *E. coli* lacks the parallel pathways leading to lysine biosynthesis (described later in the text). More detailed depictions of the enzymes and intermediates involved in these pathways are depicted later in this chapter. ASA, aspartate semialdehyde; hom, homoserine; DAP, *meso*-diaminopimelate.

lates that were identified for their prolific amino acid production. A great deal of research has focused on improving the properties of the production strains. Glutamate excretion from the cells appears to be the major limitation to yield in glutamate production. Because excretion can be enhanced by manipulating the culture conditions, biochemical engineering and reactor design have been the primary routes toward improving glutamate yield. In lysine fermentations, however, the primary bottleneck appears to be in the rate of synthesis of the amino acid. For this reason, improvements in lysine production have come mainly from metabolic engineering. Following, we discuss both aspects, describing first biochemical engineering of the glutamate production process, and second, metabolic engineering of the pathways leading to lysine and other aspartate-derived amino acids.

BIOCHEMICAL ENGINEERING FOR IMPROVED AMINO ACID PRODUCTION

General Characteristics of Industrial Amino Acid Fermentations

Corynebacteria are mainly involved in industrial production of amino acids by fermentation processes. The volume of industrial fermentors for amino acid production may reach up to 100 m^3 in order to support the high demand of the market. Some production facilities produce up to 100,000 metric tons per year of amino acids. These large cultures are usually realized in fed-batch mode, in order to reach the highest productivities. Although continuous cultures also lead to high productivities, they are rarely used in industry due to the higher risks of bacterial contamination. The industrial raw materials are complex media such as cane or beet molasses, corn steep liquor, or starch hydrolysates. The carbon sources are thus mainly sucrose (from beet molasses), or sucrose, glucose, and fructose (from cane molasses). The nitrogen sources are ammonium salts provided by the medium in the first stage of the cultures, and later during the fermentation as ammonium hydroxide, which is used to regulate pH in the reactor.

Glutamic Acid Fermentation

Glutamic acid fermentation provides a good example of the development of a process in order to answer industrial problems. Glutamic acid is produced industrially by fed-batch fermentations of 25–40 h. The final concentration of glutamic acid ranges between 90 and 120 g/L with a biomass production of 15–25 g/L. Depending on the raw materials used, the conversion yields vary between 50 and 60%. Glutamic acid production is an aerobic process. Therefore it is very important to control the dissolved oxygen at its optimal level. Oxygen limitation leads to a drastic decrease in glutamic acid titer, and to significant production of succinic and lactic acids. Moreover, an ammonium limitation with sufficient oxygen supply results in a deviation of the fermentation process to a production of 2-oxoglutarate. In industrial-scale processes, automatic addition of gaseous ammonia avoids ammonium limitation and enables the pH regulation of the broth without the harmful effects of liquid ammonia and the undesirable dilution of the fermentation liquid.

Corynebacteria usually excrete little or no glutamic acid into the medium. Therefore, different processes were developed to increase the excretion of this amino acid. The first industrial process was based on biotin limitation. Takinami et al. (15) reported that in media with high biotin concentrations, *B. lactofermentum* did not produce glutamic acid. But when the biotin concentration was low (<3 μg/L), glutamic acid was excreted in the external medium (Fig. 2). Glutamic acid fermentation by biotin limitation occurs in two phases (Fig. 3). Cultures begin with a relatively high concentration of biotin to permit cell growth. When the biotin becomes limiting, bacterial growth slows down and glutamic acid is excreted. Using biotin limitation to enhance glutamate, industrial production has given glutamate titers up to 90 g/L.

The process of biotin limitation introduces different problems on an industrial scale: it does not allow the use of complex media that often contain high biotin concentrations, and it necessitates a precise and tedious monitoring of the biotin concentration. These problems were partly solved by the use of antibiotics or surfactants. The addition of sublethal concentrations of penicillin to a culture of *C. glutamicum* led to excretion of glutamic acid (16). This process was used in industry but was quickly replaced by a process involving surfactants. Addition of either penicillin or surfactants leads to an excretion of glutamic acid. Current modern production processes are fed-batch fermentations consisting of an initial period of rapid growth of the bacteria followed by a phase of extensive glutamate excretion, which is induced by the addition of surfactants such as amine surfactant (17) or Tween (18) (Fig. 4). Glutamate production via surfactant addition may reach titers of 120 g/L in industrial media.

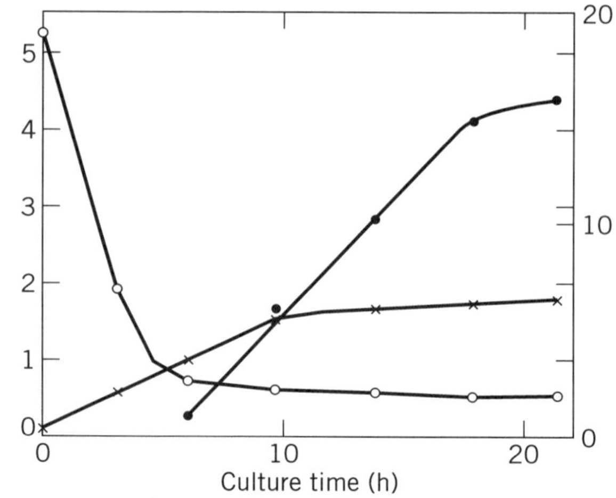

Figure 3. Production of glutamic acid by biotin limitation process. Closed circles, glutamic acid accumulated (g/L); x, dry cell concentration (g/L). *Source*: Adapted from Ref. 15.

The development of the fermentations that rely on penicillin or surfactants led to benefits for the amino acid industries. These processes have the main advantage of allowing the use of cheap, biotin-rich materials, such as molasses. Nevertheless, the monitoring of these fermentations remains delicate due to the sensitivity of the bacteria to surfactants. The ratio of surfactant to biomass concentration, and the physiological state of the cells when the surfactants are added, are critical determinants for production of high levels of glutamic acid (19).

The high levels of glutamic acid reached in industrial processes are usually explained by the fact that the excretion of glutamate leads to a reduction in the internal pool

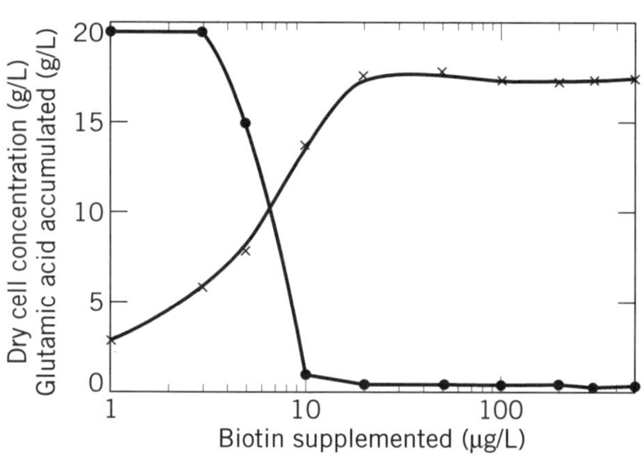

Figure 2. Relationship between the level of external biotin and the accumulation of glutamic acid in *Brevibacterium lactofermentum*. Closed circles, glutamic acid; open circles, dry cell concentration. *Source*: Adapted from Ref. 15.

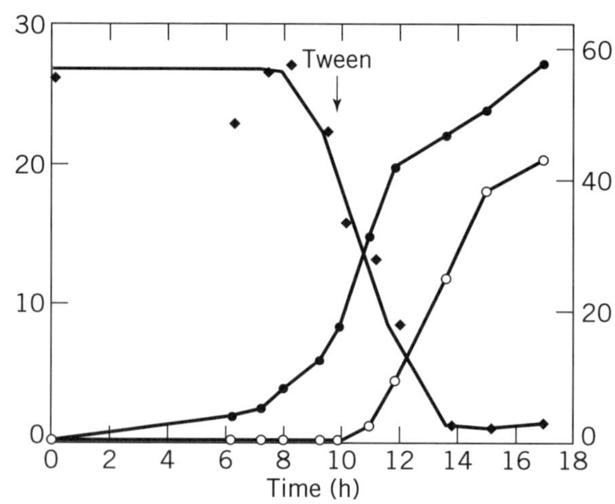

Figure 4. Production of glutamate by fed-batch culture of *Corynebacterium glutamicum* with addition of Tween. Closed diamond, glucose (g/L); open circles, glutamate (g/L). *Source*: Adapted from Ref. 18.

of this amino acid, which in turn relieves feedback control mechanisms. Regarding the regulatory aspects, 2-oxoglutarate dehydrogenase, which converts 2-oxoglutarate to succinyl-CoA at the branch point where glutamate formation diverges from the TCA cycle, is a key enzyme for glutamic acid biosynthesis (Fig. 5). The properties of the enzyme must be favorable for deviation of the carbon flux towards the glutamic acid pathway in order to prevent further oxidation through the TCA cycle. It has been shown that in overproducing strains, the V_{max} of the glutamate dehydrogenase is 150 times higher than that of 2-oxoglutarate dehydrogenase. Also, the intracellular 2-oxoglutarate concentration is high enough to direct the flux to glutamic acid biosynthesis, even with a 70-fold lower K_m value of 2-oxoglutarate dehydrogenase for 2-oxoglutarate (20). Recently, Kawahara et al. (21) showed that the activity of 2-oxoglutarate dehydrogenase was reduced in cells cultured under glutamate-production conditions (biotin limitation and addition of penicillin or surfactant) as compared to those in glutamate nonproductive conditions. In contrast, glutamate dehydrogenase activity, which draws carbon away from the TCA cycle into glutamate, was unchanged under these conditions. Thus, a high flow of carbon to the TCA cycle could be easily converted to glutamate after exhaustion of biotin, or addition of antibiotic or surfactants.

The second aspect of regulatory mechanisms of glutamic acid biosynthesis concerns the carbon dioxide fixation to form oxaloacetate. The efficiency of glutamic acid biosynthesis depends on the supply of oxaloacetate to the first part of the TCA cycle (Fig. 5). Anaplerotic pathways,

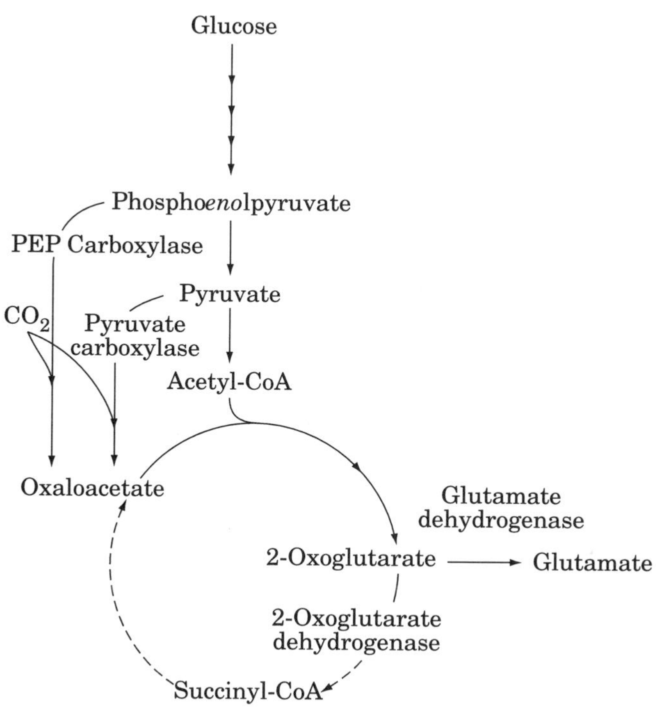

Figure 5. Simplified scheme of glutamic acid biosynthesis pathway through anaplerotic pathways as oxaloacetate generating system.

such as phosphoenolpyruvate carboxylase and pyruvate carboxylase, may then play an important role in the production of glutamic acid. In the future, increases in carbon dioxide fixation might improve the production yield of the strains.

The mechanism of glutamate excretion into the medium has long been a subject of controversy in the literature. Three different mechanisms have been proposed to describe glutamate efflux. The "leak" model proposes that glutamate is excreted because the cell membrane is physically altered after different treatments, such as biotin limitation, addition of sublethal antibiotic concentration, or addition of surfactants. The second hypothesis is based on a general model for many efflux processes in bacteria, namely, that inversion of the influx system of glutamate is induced by changes in chemical potential or regulation, or by uncoupling due to changes in the membrane structure. Finally, the last hypothesis proposes the existence of a specific glutamate carrier analogous to the recently characterized lysine transport system. These different hypotheses are discussed in a recent review (22).

Momose and Takagi (23) isolated mutant strains of *C. lactofermentum* sensitive to temperature and able to excrete glutamic acid after a thermal upshock from 30 °C to 37 °C in biotin-rich media. One of their mutants produced 20 g/L glutamic acid from molasses. In the future, with such strains it could be possible to produce glutamate from biotin-rich materials without the use of expensive additives, such as antibiotics or surfactants, just by simple control of the temperature.

METABOLIC ENGINEERING OF CORYNEBACTERIA

Through classical mutagenesis and selection, it has been possible to identify *Corynebacterium* mutants that overproduce lysine. By 1958, Kinoshita and colleagues (24) had learned that homoserine auxotrophs secreted higher amounts of lysine than did wild-type strains. Mutants with altered feedback inhibition in the lysine biosynthetic pathway have also been found to overproduce the amino acid. Thus, resistance to the lysine analog 2-aminoethyl-L-cysteine (AEC) has been a hallmark of the lysine overproduction phenotype. Other mutants that overproduce lysine have been found to contain lesions in pathways, such as those for leucine and alanine biosynthesis, which compete for the pyruvate that is required for lysine biosynthesis. Lysine production by *Corynebacterium* strains that are resistant to one or more lysine analogs and that carry auxotrophies for one or more additional amino acids has been described previously (25). The introduction of molecular genetics coupled with a greater understanding of how *Corynebacterium* regulates lysine synthesis has made it possible to enhance lysine productivity even further through metabolic engineering.

Metabolic engineering is an emerging field of research that combines the analysis of complex biochemical networks with biochemistry, enzymology, and molecular biology. Through metabolic engineering it is possible, at least in principle, to identify processes within the cell that limit overall lysine productivity, for example, and introduce specific changes into the cell to alleviate these limitations. Metabolic engineering requires both a detailed understanding of biosynthetic pathways and their regulation, and the ability to alter specific steps in these pathways. The most direct method for making these alterations is genetic manipulation. In this way, *E. coli* strains, with their well-characterized physiology and wide array of molecular genetic resources, have been engineered to produce large amounts of threonine, valine, or tryptophan. Metabolic engineering in *Corynebacterium*, however, cannot draw on the same diversity of resources, and although the physiology of these bacteria is becoming ever more understood, methods for introducing changes into these strains at the genetic level are not as well developed as they are for *E. coli*.

Genetic Tools for Manipulating *Corynebacterium*

To make the genetic changes necessary for metabolic engineering in *Corynebacterium*, researchers need to be able to identify and clone the genes that are involved in the target pathway. They also need methods for altering these genes to affect the regulation or level of expression of the enzymes they encode, and for subsequently reintroducing the altered genes into *Corynebacterium* to monitor their effects on amino acid biosynthesis. Therefore, metabolic engineers must have at their disposal an array of plasmids that can replicate in both *Corynebacterium* and other, more easily manipulated hosts, such as *E. coli*. Also required are a collection of selectable markers encoding, for example, antibiotic resistance, well-characterized transcriptional promoters that permit regulation of the altered genes, and efficient transformation or conjugation systems that allow the plasmids to be inserted into the target *Corynebacterium* strain.

Plasmids. Several different plasmids have been isolated and developed for the introduction and expression of genes in *Corynebacterium* (26). The majority of these were originally identified as small (3–5 kbp), cryptic plasmids from *C. glutamicum, C. callunae*, and *C. lactofermentum*. They fall into four compatibility groups, exemplified by the plasmids pCC1, pBL1, pHM1519, and pGA1. Shuttle vectors, plasmids that are capable of replicating in both *Corynebacterium* and *E. coli*, have been developed from these cryptic plasmids by incorporating elements from known *E. coli* plasmids (particularly the ColE1 origin of replication from pBR322 or pUC18), as well as antibiotic resistance markers. A fifth class of plasmids that is very useful for manipulating *Corynebacterium* is based on pNG2, a plasmid originally isolated from *Corynebacterium diphtheriae* (27). This plasmid and its derivatives replicate efficiently in many species of corynebacteria, as well as in *E. coli*. Since the sole origin of replication in pNG2 (an element of only 1.8 kbp) functions in both the Gram-positive and Gram-negative host, there is no need to add an additional ColE1-type element to it. As a result, pNG2 derivatives (e.g., pEP2) are much smaller than other *Corynebacterium* shuttle vectors and are therefore more easily manipulated.

Selectable Markers. Several genes conferring antibiotic resistance have proven useful for plasmid selection and other recombinant DNA work in corynebacteria. These include the kanamycin resistance determinant from Tn*903*, a hygromycin resistance marker isolated from *Streptomyces hygroscopicus*, a tetracycline resistance gene from *Streptococcus faecalis*, a bleomycin resistance gene from Tn*5*, and a chloramphenicol resistance marker from *Streptomyces acrimycini*. Work in our laboratory has shown that this latter gene does not confer chloramphenicol resistance to *Corynebacterium* unless its native promoter is substituted by a promoter that is known to work in *Corynebacterium*. The β-lactamase gene that is employed in many *E. coli* plasmids such as pBR322 does not confer ampicillin resistance in *Corynebacterium*.

Transformation Systems. Several methods have been devised for introducing foreign DNA into *Corynebacterium*. The earliest method to be employed routinely was based on protocols that had been successful for other Gram-positive species involving incubation of spheroplasts in the presence of DNA and polyethylene glycol (28). While useful, these methods were generally inefficient, often yielding fewer than 10^5 transformants per milligram of DNA. Electroporation of *Corynebacterium* spheroplasts has proven to be a much more efficient and reliable means of transformation. Spheroplasts are generated by growing the cells in rich media containing glycine and/or low concentrations of other inhibitors of cell wall biosynthesis, such as isonicotinic acid hydrazide (isoniazid), ampicillin, penicillin G, or Tween-80. The spheroplasts are then washed in low-salt buffers containing glycerol, concentrated, and mixed with DNA before being subjected to electroporation. Efficiencies as high as 10^7 transformants per microgram of plasmid DNA have been reported with this protocol.

A third method for DNA transfer into corynebacteria involves transconjugation. This method takes advantage of the promiscuity of *E. coli* strains carrying derivatives of the plasmid RP4. In *E. coli*, RP4 encodes many functions that mediate the conjugal transfer of plasmids from the host strain to other recipient strains of *E. coli*, or even to other species. These "*tra* functions" mediate pilus formation and plasmid transfer. RP4 also carries an origin of transfer, *oriT*, a *cis*-acting element that is recognized by the transfer apparatus that allows the plasmid to be conducted through the pilus and into the recipient strain. From this system Simon et al. (29) have developed a useful transconjugation tool that allows the transfer of plasmids from *E. coli* to *Corynebacterium*. They relocated the *tra* functions from RP4 into the *E. coli* chromosome in a strain called S17-1. Plasmids carrying the RP4 *oriT* can be mobilized from S17-1 into other recipients very efficiently. Although this method has proven useful for introducing replicating plasmids into *Corynebacterium*, it has proven even more useful for generating gene disruptions. This is accomplished by introducing a selectable marker into a clone of the *Corynebacterium* gene that is targeted for disruption. This construct is then ligated into an *E. coli* plasmid that carries the RP4 *oriT* but lacks an origin to support replication in *Corynebacterium*. S17-1 carrying this plasmid is then incubated with the recipient strain and the mixture is later transferred to a selective medium. Because the plasmid that was introduced is unable to replicate in corynebacteria, transconjugants that express the selectable marker are most likely to have undergone a cross-over recombination within the genomic DNA.

Restriction-Deficient Strains. Regardless of the transformation system used, there is clear precedent in the literature that corynebacteria are able to recognize *E. coli*–derived DNA as foreign and will most often degrade it. This ability has been attributed to the *Corynebacterium* restriction and modification system. To overcome this system, some transformation and transconjugation protocols call for briefly heating the recipient strain prior to transformation. The heat treatment presumably inactivates the enzymes responsible for the restriction system, allowing the introduced DNA to become established before the enzymes are turned over. Another strategy for improving the efficiency of DNA transfer has been to isolate *Corynebacterium* mutants that are deficient in the restriction system. These strains will incorporate plasmids that had been propagated in *E. coli* with almost the same efficiency as plasmids that had been propagated in *Corynebacterium*. In an alternate strategy used to circumvent the restriction system in *Corynebacterium*, Leblon and coworkers (30) developed an "integron" system for gene disruption. Integrons are DNA molecules that have the same restriction/modification properties as the target host's DNA, carry DNA that is homologous to a portion of the host genome (i.e., a region of the genome that is to be disrupted), and are unable to replicate in the host cell. A cloned gene from *Corynebacterium* is first interrupted with a selectable marker in a plasmid that is propagated in one *Corynebacterium* strain. This construct is then excised from the corynebacterial plasmid and self-ligated to form a nonreplicating circular molecule. This "integron" is then electroporated into the restrictive host. Modification of the DNA allows the integron to elude the host restriction system, and recombination into the host genome permits expression of the selectable marker.

Promoters. Reliable transcriptional promoters are required for efficient expression of foreign genes in *Corynebacterium*. For certain experiments, there is also a need for regulated promoters whose activity can be induced under specific culture conditions. Promoters such as the *fda*, *thrC*, and *hom* promoters derived from *Corynebacterium* genes have proven useful for heterologous gene expression. Inducible promoters from *E. coli*, such as P_{lac}, P_{tac}, and P_{trc}, which are induced by isopropylthiogalactopyranoside (IPTG) when the *lac* repressor (*lacI*) is present; P_{trp}, which responds to the inducer indole acrylic acid when the *trp* repressor (*trpR*) is present; and lambda P_L, which is repressed in the presence of the temperature-sensitive lambda repressor (cI857), have all been used to modulate gene expression in *Corynebacterium*. There is also intense interest in identifying developmentally regulated promoters from *Corynebacterium*, such as promoters that are activated during late exponential or stationary phases of culture, although this work has proceeded slowly.

Gene Identification. With all other genetic tools in place, there still remains the challenge of identifying relevant genes from *Corynebacterium*. In *E. coli*, some of the resources that have been used to isolate genes are transducing phage, transposable elements, genetic maps of the *E. coli* chromosome from transduction and transconjugation experiments, and more recently, complete physical and sequence maps of the chromosome. To date, the most successful method for identifying and recovering genes from *Corynebacterium* has been to use *Corynebacterium* genomic DNA to complement known auxotrophs of *E. coli*. In this exercise, libraries of plasmids carrying fragments of the *Corynebacterium* genome are introduced into *E. coli* strains that are deficient in a particular enzyme or function. Transformants that no longer display the auxotrophy (e.g., homoserine deficiency) are likely to carry the complementing gene from *Corynebacterium*. This strategy has led to isolation of numerous *Corynebacterium* genes, including several from the pathways responsible for synthesis of aspartate-derived and aromatic amino acids, intermediary metabolism, and other cellular processes. One limitation to this strategy is that not all genes from *Corynebacterium* will be expressed in the *E. coli* host. Thus, although a gene may be represented in the plasmid library, it may be unable to complement the *E. coli* mutation and therefore would not be recovered during selection. Overcoming this limitation, a smaller number of genes have been identified with a similar strategy in which a plasmid library from wild-type *Corynebacterium* was used to directly complement mutations in other *Corynebacterium* strains. Although this strategy avoids the concern of insufficient gene expression in the auxotrophic host, its utility is limited by poor plasmid transformation efficiency in the auxotrophs. Still other genes have been identified by hybridization with nucleic acid probes based upon homologous genes from other species, and direct amplification of genes using the polymerase chain reaction and degenerate oligonucleotide primers.

Transposable Elements. Transposable elements are extremely powerful tools in gene identification because they couple mutagenesis with gene recovery. Unlike classical mutagenesis techniques, which generate point mutations or small deletions within a gene, when transposable elements insert within a gene they form large disruptions, thereby "tagging" the altered gene for easier identification. A number of transposable elements have been found to transpose in *Corynebacterium*. Transposons found in the plasmids pTP10 of *C. xerosis* and pNG2 of *C. diphtheriae* have been shown to transpose in *C. glutamicum* and confer resistance to erythromycin. A group from the Mitsubishi Chemical Company in Japan developed a series of artificial transposons from an insertion sequence, IS*31831*, that they discovered in *C. glutamicum* (31). After inserting a selectable marker between the inverted repeats of IS*31831*, these researchers were able to introduce the resulting transposon into *C. glutamicum* strains on an *E. coli* plasmid (unable to replicate in *Corynebacterium*) via electroporation. They found that the selectable marker had inserted into the genome of the target cell at a frequency of approximately 4×10^4 mutants/μg DNA. The use of such transposons to generate *Corynebacterium* auxotrophs has led to the isolation of several genes responsible for amino acid biosynthesis, as well as other functions in corynebacteria.

Transducing Phage. Transducing phage have been used in other systems for mapping genetic loci and for isolating genes. In 1976, researchers at Ajinomoto Co. in Japan surveyed 150 strains of characterized and uncharacterized glutamic acid–producing coryneform bacteria to identify phage that might be useful for transduction (32). Of 24 different phage isolates recovered from this screen, only three were able to transduce a *trp* marker from a trp^+ donor to a trp^- recipient with any appreciable frequency, although even this efficiency was only 10^{-7} or less. These researchers were able to improve transduction efficiency slightly by including 4 mM cyclic adenosine monophosphate (cAMP) or 1.2 M magnesium chloride. Several different researchers have attempted to develop reliable transduction methods by isolating corynephages from sources such as contaminated industrial fermentations, soil, and animal waste. Although many phage have been isolated and characterized, few have been associated with transduction, and an opportunity still exists to develop a reliable, high-efficiency transduction system for general use with the glutamic acid–producing bacteria.

Metabolic Engineering for Aspartate-Derived Amino Acids

Classical mutagenesis and selection has given rise to amino acid–overproducing strains that carry mutations that alleviate feedback inhibition in critical regulatory enzymes, eliminate reactions that compete for substrates, alter control of gene expression, or somehow modify other aspects of the cell's physiology. Metabolic engineering of the pathways for the aspartate family of amino acids has followed very similar themes. Aspartate is derived from oxaloacetate, a tricarboxylic acid cycle (TCA) intermediate. All of a cell's lysine, methionine, threonine, and isoleucine are produced from aspartate (Fig. 1). *Corynebacterium* modulates the partitioning of precursors into pathways specific for the individual amino acids in this family at several levels, including feedback sensitivity of several key enzymes, differences in the specific activities of competing enzymes at nodes where the pathways for two amino acids diverge, and genetic repression. Although it is complex, regulation of the aspartate pathways in *Corynebacterium* does not appear to be as complex as that in *E. coli* (Fig. 1).

Lysine. Although there appears to be some level of transcriptional regulation of the operon that encodes diaminopimelate decarboxylase, the enzyme that carries out the final step in lysine biosynthesis (Fig. 6), the main control in lysine biosynthesis in *Corynebacterium* is exercised by aspartokinase (AK), which catalyzes the first common step toward the synthesis of all of the aspartate-derived amino acids. Many AEC-resistant strains of *Corynebacterium* have been shown to produce an aspartokinase that is insensitive to the feedback inhibition by lysine and threonine that is displayed by the wild-type enzyme. Sequence analysis of cloned aspartokinase genes from AEC-resistant *Corynebacterium* strains has revealed the molecular basis

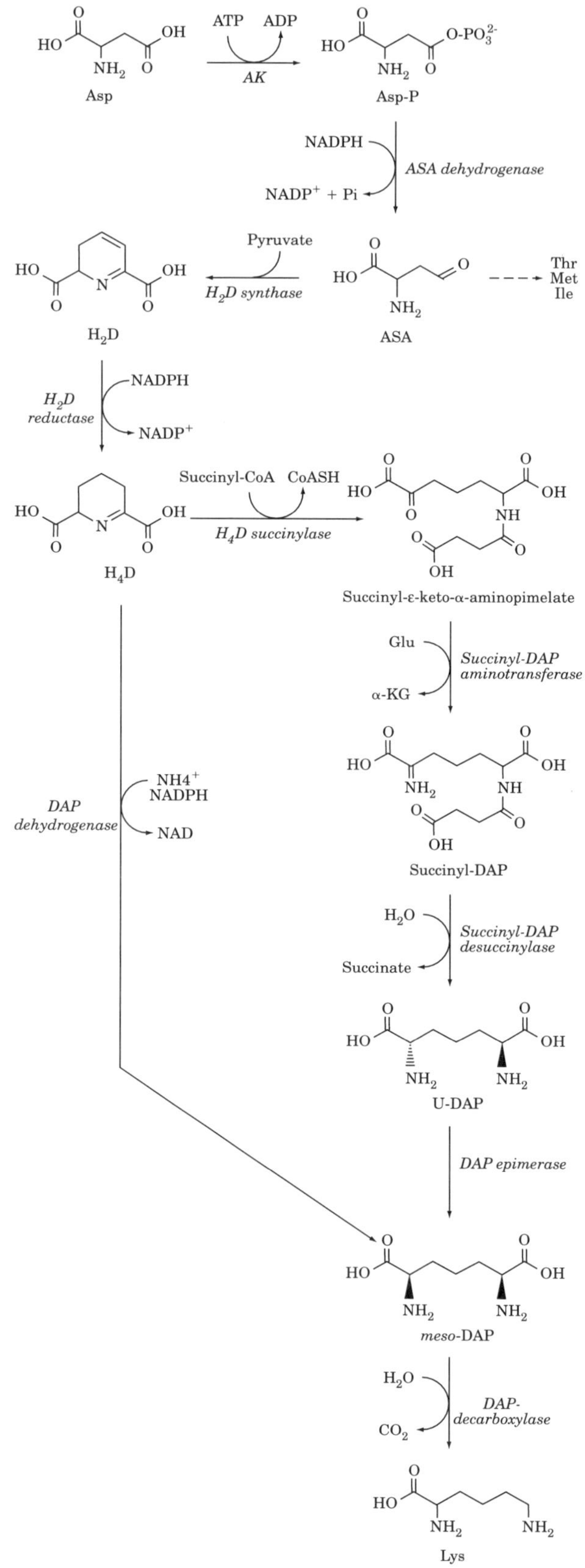

Figure 6. Lysine synthesis in *Corynebacterium glutamicum*. AK, aspartokinase; ASA, aspartate semialdehyde; H_2D, dihydrodipicolinate; H_4D, tetrahydrodipicolinate; DAP, diaminopimelate; α-KG, α-ketoglutarate.

of lysine overproduction in these strains. In *Corynebacterium*, aspartokinase is synthesized as two polypeptides, AKα and AKβ, which are encoded by overlapping reading frames and assemble as an $\alpha_2\beta_2$ heterotetramer. Feedback resistance is due to mutations that lead to amino acid substitutions within the AKβ subunit: specifically, substitution of tyrosine in place of serine at position 301, aspartate in place of glycine at position 345, or glycine in place of arginine at position 349 (22). From this it is postulated that the β-subunit of aspartokinase mediates regulation by lysine and threonine, although it is not clear whether these amino acids are directly involved in binding of either lysine, threonine, or both. Regardless, these mutations alone allow *Corynebacterium* to accumulate and secrete high levels of lysine.

However, it is clear that feedback inhibition of aspartokinase is not the only limitation to lysine overproduction. In AEC-resistant lysine overproducers, amplifying expression of the deregulated aspartokinase gene through recombinant DNA techniques further increases the synthesis of lysine, indicating that the specific activity of this enzyme is also limiting. A further limitation in lysine production, especially at the industrial scale, comes at the point of lysine secretion (33). Wild-type *Corynebacterium* cannot secrete significant amounts of lysine. Many lysine-overproducing strains have been found to possess enhanced lysine export systems in addition to deregulated aspartokinase activity. The lysine export mechanism apparently involves symport of two hydroxyl ions and can proceed against a lysine gradient and against the membrane potential (33). At this time, the molecular basis for increased lysine export during overproduction is not yet fully understood.

Unlike *E. coli, C. glutamicum* possesses two biochemical pathways for synthesizing lysine or, more precisely, for synthesizing *meso*-diaminopimelate (*meso*-DAP), the immediate precursor to lysine. The first pathway (the succinylase pathway) is initiated by the enzyme tetrahydrodipicolinate succinylase and involves a total of four enzymatic steps to produce *meso*-DAP. The second pathway (the dehydrogenase pathway) converts tetrahydrodipicolinate to *meso*-DAP in a single step through the action of DAP dehydrogenase. Because these two pathways incorporate nitrogen via different substrates (the succinylase pathway uses glutamate while the dehydrogenase pathway uses ammonium), the operation of these two mechanisms allows *C. glutamicum* to grow in the presence of different nitrogen sources. *meso*-DAP is an important cell wall component in corynebacteria, providing the cross-linking of the peptidoglycan material in this structure. Strains of *C. glutamicum* that are deficient in the succinylase pathway were found to have altered morphology and cell wall rigidity (34). Deficiency in the dehydrogenase pathway reduces productivity during lysine fermentations without significantly altering growth rates (35). The dehydrogenase enzyme has a lower affinity for the common substrate than does the succinylase, and it is constitutively expressed. The succinylase pathway, on the other hand, is energetically more expensive. ^{13}C-nuclear magnetic resonance studies on carbon flux (36) have confirmed that during normal growth the majority of the *meso*-DAP is produced via the succinylase pathway, and that during conditions of high lysine production the dehydrogenase pathway accounts for the bulk of this flux.

Threonine. Aspartate semialdehyde is the last common intermediate in the pathways for the synthesis of lysine and the other aspartate-derived amino acids (Fig. 7). Given

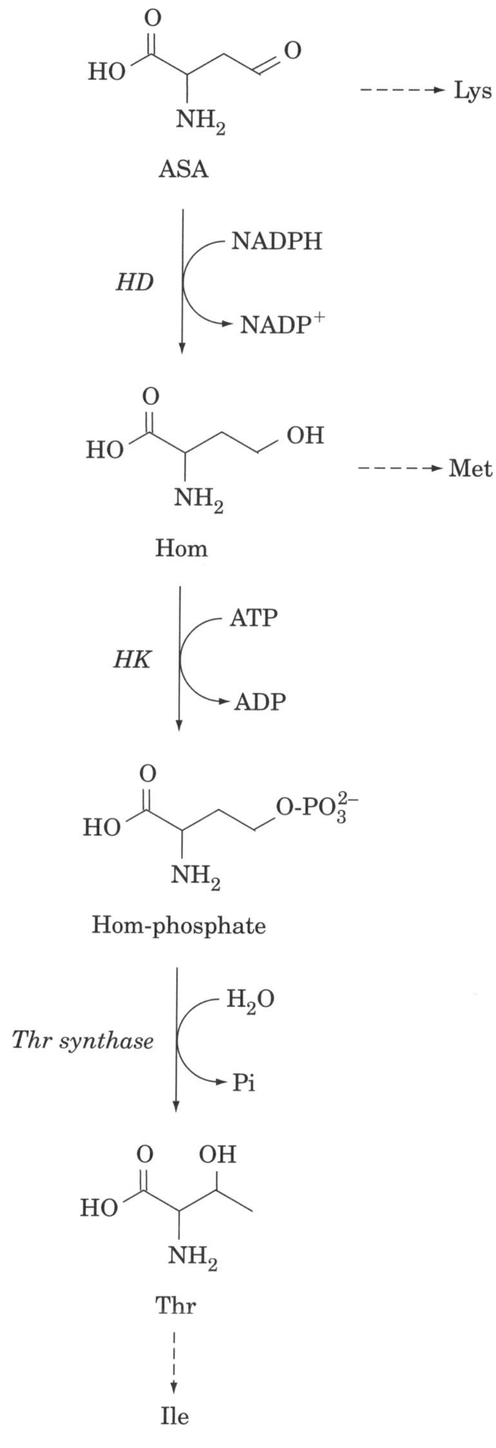

Figure 7. Threonine synthesis in *Corynebacterium glutamicum*. ASA, aspartate semialdehyde; Hom, homoserine; HD, homoserine dehydrogenase; HK, homoserine kinase.

the fact that lysine, methionine, threonine, and isoleucine are all derived from aspartate, it would seem to follow that deregulation of aspartokinase should boost synthesis of these other products along with lysine. The observation that the first committed step toward threonine, methionine, and isoleucine is catalyzed by an enzyme (homoserine dehydrogenase) with a stronger affinity for aspartate semialdehyde than the enzyme (dihydropicolinate synthase) that leads to lysine would seem to support this notion. However, when feedback inhibition of aspartokinase is relieved, the predominant product to accumulate is lysine because, unlike the pathway for lysine biosynthesis, the threonine and methionine biosynthetic pathway contains additional regulatory mechanisms. Homoserine dehydrogenase itself is inhibited by accumulation of threonine, and expression of the homoserine dehydrogenase gene is repressed by methionine. The result is that, even though homoserine dehydrogenase can outcompete dihydrodipicolinate synthase for aspartate semialdehyde, accumulation of the products of this branch in the pathway is quickly squelched. Lysine synthesis, which has almost no subsequent direct regulation, then resumes unhindered.

Because of its strong feedback regulation, homoserine dehydrogenase is an obvious target for metabolic engineering to improve threonine biosynthesis in *Corynebacterium.* Mutations near the carboxy terminus of this enzyme have been found to confer insensitivity to feedback regulation. When the deregulated homoserine dehydrogenase gene, hom^{dr}, and copies of *thrB*, which encodes homoserine kinase (the next enzyme in the pathway toward threonine), were integrated into the genomes of AEC-resistant strains of *Corynebacterium*, threonine accumulated at the expense of lysine (22,23). Inefficient export of the excess threonine in these cells triggered feedback inhibition of homoserine kinase, causing homoserine to accumulate as well. Concomitant overexpression of *thrB* from an inducible promoter was found to partially compensate for the deficiencies in threonine export by allowing more homoserine to be converted to threonine, but this effort led to the discovery that excess threonine was rapidly lost as the degradation products isoleucine and glycine. Disruption of the gene encoding threonine dehydratase, the first enzyme in the pathway from threonine to isoleucine, permitted more threonine to accumulate. However, threonine was still lost as glycine through two additional pathways, the first involving threonine dehydrogenase, and the second involving serine hydroxymethyl transferase, which is also active in serine catabolism (22). The difficulties encountered in development of *Corynebacterium* strains that overproduce threonine illustrates the complex and elegant regulation of this pathway.

Isoleucine. Overproduction of isoleucine has been accomplished by introducing excess threonine dehydratase (encoded by *ilvA*) into threonine production strains (37) (Fig. 8). Thus, combining a deregulated aspartokinase, deregulated homoserine dehydrogenase, overexpressed homoserine kinase, and overexpressed threonine dehydratase will result in abnormally high levels of isoleucine accumulation. Threonine dehydratase is normally feedback-inhibited by isoleucine, so it is not surprising that

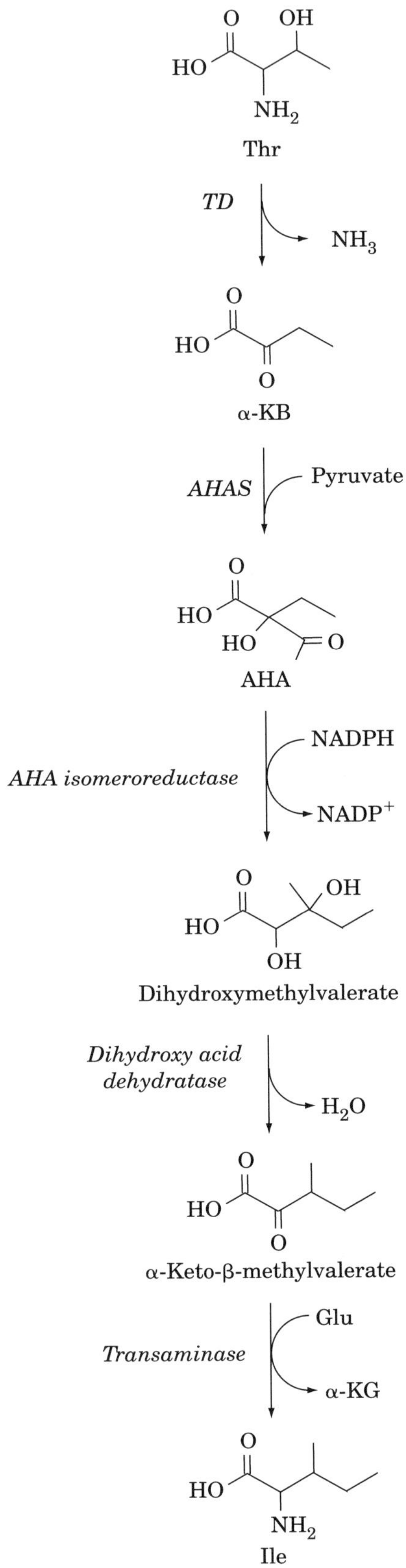

Figure 8. Isoleucine synthesis in *Corynebacterium glutamicum*. TD, threonine dehydratase; α-KB, α-ketobutyrate; AHA, acetohydroxy acid; AHAS, acetohydroxy acid synthase; α-KG, α-ketoglutarate.

mutant derivatives of threonine dehydratase with reduced sensitivity to isoleucine provided an additional dividend in this isoleucine production system (38). Despite these gains, it appears that amino acid export is once again a serious limitation to the effectiveness of amino acid production (33). An increase in threonine dehydratase activity causes α-ketobutyrate to accumulate. This induces the activity of the next enzyme in the isoleucine pathway, acetohydroxy acid synthase (AHAS), which might be expected to promote the conversion of threonine to isoleucine. Indeed, the addition of α-ketobutyrate alone to the medium enhanced flow through the latter parts of this pathway. However, when threonine dehydratase is overexpressed in the threonine production system, α-ketobutyrate is excreted into the medium. AHAS is also involved in valine and leucine synthesis and is itself feedback inhibited by the branched-chain amino acids. It appears that under isoleucine fermentation conditions a concomitant accumulation of valine may inhibit AHAS, thereby overcoming the stimulatory effects of α-ketobutyrate accumulation (37). Observations of intracellular concentrations of isoleucine suggest that isoleucine export is also a major rate-limiting step in isoleucine overproduction (39).

PERSPECTIVES

Considerable progress has been made in the production of amino acids by corynebacteria since the isolation of the first strain of *C. glutamicum*. In only four decades, the field has advanced from classical genetic manipulation of strains with uncharacterized lesions, through elegant biochemical analyses of enzymes from wild-type and overproducing strains, and on to the current molecular approaches. Aided by information gathered from sophisticated metabolic flux analyses and a greater understanding of biochemical pathways and cellular processes, the field has begun to apply a more rational strategy based on metabolic engineering of improved strains. These new strategies, facilitated by the development of improved genetic tools for the manipulation of corynebacteria, should permit the unraveling of complicated regulatory pathways. Using these methods, researchers have overcome many obstacles to amino acid overproduction and gained valuable insights that have benefited the broader scientific community.

Challenges to further enhancements in amino acid production remain in the areas of amino acid transport and global regulatory mechanisms (40), particularly those relating to central carbon metabolism. There is also opportunity for further improvement of large-scale fermentation techniques. The availability of the complete genomic sequence of closely related Gram-positive bacterial strains, including *Mycobacterium tuberculosis* and *M. leprae*, will undoubtedly aid in the isolation of homologous genes from corynebacteria. As the analysis of bacterial genomes and proteomes continues to mature, functional genomics and high-throughput chip-based technologies for the visualization of gene expression and cellular protein content hold out the promise for synthesis of an accurate and detailed picture of cellular metabolism. The study of the glutamic acid–producing corynebacteria will likely lead to continuous gains in productivity while offering new insights into microbial physiology.

BIBLIOGRAPHY

1. S. Udaka, *J. Bacteriol.* **79**, 754–755 (1960).
2. S. Abe, K.-I. Takayama, and S. Kinoshita, *J. Gen. Appl. Microbiol.* **13**, 279–301 (1967).
3. W. Liebl, in A. Balows, H.G. Trueper, M. Dworkin, W. Harder, and K.H. Schleifer eds., *The Prokaryotes,* Springer-Verlag, Berlin, 1991, pp. 1157–1171.
4. W. Liebl, M. Ehrmann, W. Ludwig, and K.H. Schleifer, *Int. J. Syst. Bacteriol.* **41**, 255–260 (1991).
5. S. Muirheid, *Feedstuffs* **69**, 1 (1997).
6. *Medical & Healthcare Marketplace Guide*: 1997–1998, 13th ed., Dorland's Biomedical, Philadelphia, Pa., 1997, pp. III–237.
7. M. Lerner, *Chemical Market Reporter* **253**, January 26, 1998, p. 4.
8. M. Lerner, *Chemical Marketing Reporter* **250**, September 16, 1996, p. 18.
9. J.C. Measures, *Nature* **257**, 398–400 (1975).
10. A.J. Makoff, I.G. Charles, and N.F. Fairweather, in Y. Murooka and T. Imanaka eds., *Recombinant Microbes for Industrial and Agricultural Applications*, Marcel Dekker, New York, 1994, pp. 205–232.
11. M. Ikemi, in Y. Murooka and T. Imanaka eds., *Recombinant Microbes for Industrial and Agricultural Applications,* Marcel Dekker, New York, 1994, pp. 797–814.
12. H. Billman-Jacobe, L. Wang, A. Kortt, D. Stewart, and A. Radford, *Appl. Environ. Microbiol.* **61**, 1610–1613 (1995).
13. M.E. Johnson and J.L. Steele, in M.P. Doyle, L.R. Beuchat and T.J. Montville, eds., *Food Microbiology: Fundamentals and Frontiers*, ASM Press, Washington D.C. 1997, pp. 581–594.
14. J.Y. Leveau, *Industries Alimentaires Agricoles* March 102 (1992).
15. K. Takinami, Y. Yamada, and H. Okada, *Agric. Biol. Chem.* **30**, 674–682 (1966).
16. T.D. Nunheimer, J. Birnbaum, E.D. Ihnen, and A.L. Demain, *Appl. Microbiol.* **20**, 215–217 (1970).
17. M. Marquet, J.L. Uribelarrea, A. Huchenq, G. Laneelle, and G. Goma, *Appl. Microbiol. Biotechnol.* **25**, 220–223 (1986).
18. D. Leyval, F. Debay, J.M. Engasser, and J.L. Goergen, *J. Microbiol. Methods* **29**, 121–127 (1997).
19. F. Debay, D. Leyval, J.M. Engasser, and J.L. Goergen, *Récents Progrès en Génie des Procédés* **57**, 43–48 (1997).
20. I. Shiio and K. Ujigawa, *Agric. Biol. Chem.* **42**, 1897–1904 (1978).
21. Y. Kawahara, K. Takahashi-Tuke, E. Shimizu, T. Nakamatsu and S. Nakamori, *Biosci. Biotech. Biochem.* **61**, 1109–1112 (1997).
22. M.S.M. Jetten and A.J. Sinskey, *Crit. Rev. Biotechnol.* **15**, 73–103 (1995).
23. H. Momose and T. Takagi, *Agric. Biol. Chem.* **42**, 1911–1917 (1978).
24. S. Konishita, K. Nakayama, and S. Kitada, *J. Gen. Appl. Microbiol.* **4**, 128 (1958).
25. K. Nakayama, in ed Blanch, S. Drew and Wang, eds., *Comprehensive Biotechnology,* vol. 3, 1985, pp. 607–620.

26. H. Sonnen, G. Thierbach, S. Kautz, J. Kalinowski, J. Schneider, A. Pühler, and H.J. Kutzner, *Gene* **107**, 69–74 (1991).
27. T.M. Serwold-Davis, N. Groman, and M. Rabin, *Proc. Natl. Acad. Sci. USA* **84**, 4964–4968 (1987).
28. M. Yoshihama, K. Higashiro, E.A. Rao, M. Akedo, W.G. Shanabruch, M.T. Follettie, G.C. Walker, and A.J. Sinskey, *J. Bacteriol.* **162**, 591–597 (1985).
29. R. Simon, U. Preifer, and A. Pühler, *Bio/Technology* **1**, 784–791 (1983).
30. O. Reyes, A. Guyonvarch, C. Bonamy, V. Salti, F. David, and G. Leblon, *Gene* **107**, 61–68 (1991).
31. A.A. Vertes, Y. Asai, M. Inui, M. Kobayashi, Y. Kurusu, and H. Yukawa, *Mol. Gen. Genet.* **245**, 397–405 (1994).
32. H. Momose, S. Miyashiro, and M. Oba, *J. Gen. Appl. Microbiol.* **22**, 119–129 (1976).
33. H. Sahm, L. Eggeling, B. Eikmanns, and R. Kramer, *FEMS Microbiol. Rev.* **16**, 243–252 (1995).
34. A. Wehrmann, B. Phillipp, H. Sahm, and L. Eggeling, *J. Bacteriol.* **180**, 3159–3165 (1998).
35. B. Schrumpf, L. Eggeling, and H. Sahm, *Appl. Microbiol. Biotechnol.* **37**, 566–571 (1992).
36. C.A. Shaw-Reid, Ph.D. Thesis, Massachusetts Institute of Technology, Cambridge, Mass., 1997.
37. G.E. Colón, T.T. Nguyen, M.S.M. Jetten, A.J. Sinskey, and G. Stephanopoulos, *Appl. Microbiol. Biotechnol.* **43**, 482–488 (1995).
38. S. Morbach, H. Sahm, and L. Eggeling, *Appl. Environ. Microbiol.* **61**, 4315–4320 (1995).
39. S. Morbach, H. Sahm, and L. Eggeling, *Appl. Environ. Microbiol.* **62**, 4345–4351 (1996).
40. R. Krämer, *J. Biotechnol.* **45**, 1–21 (1996).

CRYSTALLIZATION, BULK, MACROMOLECULES

Ana Paulina Barba de la Rosa
Alexander McPherson
University of California, Irvine
Irvine, California

KEY WORDS

Bulk purification
Crystals
Drug delivery
Drug design
Protein engineering
Proteins
X-ray diffraction

OUTLINE

INTRODUCTION

There are three principle areas of modern biotechnology where crystallization plays important roles. These are (*1*) The crystallization of proteins as well as viruses, nucleic acids, and macromolecular complexes for X-ray diffraction analysis; (*2*) the crystallization of specific proteins as a part of their formulation in a therapeutic compound, agricultural chemical, or drug; and (*3*) the crystallization of enzymes, and in some cases other proteins, as a means of purification from bulk processes. Although the objectives in each of these situations may be quite different, with the exception of scale, the underlying principles and approaches are fundamentally the same.

In cases of crystallization for X-ray analysis, one is usually dealing with quite pure, often exceptionally pure molecules, and the goal is to grow only a few large, perfect crystals like those shown in Figure 1. The X-ray analysis is a singular event or study, confined to the research laboratory, and the product is basic scientific knowledge. The crystals themselves have no material or medicinal value, but serve as intermediaries in the crystallography process. They provide X-ray diffraction patterns that in turn serve as raw data allowing the visualization of the macromolecules composing the crystals. Thus, they serve as an essential component in the elucidation of molecular structure.

Macromolecular structure is of formidable value in biotechnology because it is the essential knowledge required to apply the technique of rational drug design in the creation and discovery of new drugs and pharmaceutical products. It is the basis of powerful approaches now being applied in small, emerging biotechnology companies as well as major pharmaceutical companies to identify lead compounds to treat a host of human ailments, veterinary problems, and crop diseases in agriculture. The underlying hypothesis is that if the structure of the active site of a salient enzyme in a metabolic or regulatory pathway is known, then chemical compounds, such as drugs, can be rationally designed to inhibit or otherwise affect the behavior of that enzyme.

A second approach, of equal importance to biotechnology, that also requires X-ray quality crystals is the genetic engineering of proteins. Although recombinant DNA techniques provide the essential synthetic role that permits modification of the proteins, crystallography provides the analytical function. It serves as the structural guide for the introduction of intelligent and purposeful changes, in the place of random and chance amino acid substitutions. Crystallography allows direct visualization of the structural changes introduced and offers new directions for chemical or physical enhancements.

Bulk crystallization, along with the detailed consideration of crystallization processes and products, has long been a staple in the pharmaceutical and chemical industry (1), where it has been applied to conventional small mol-

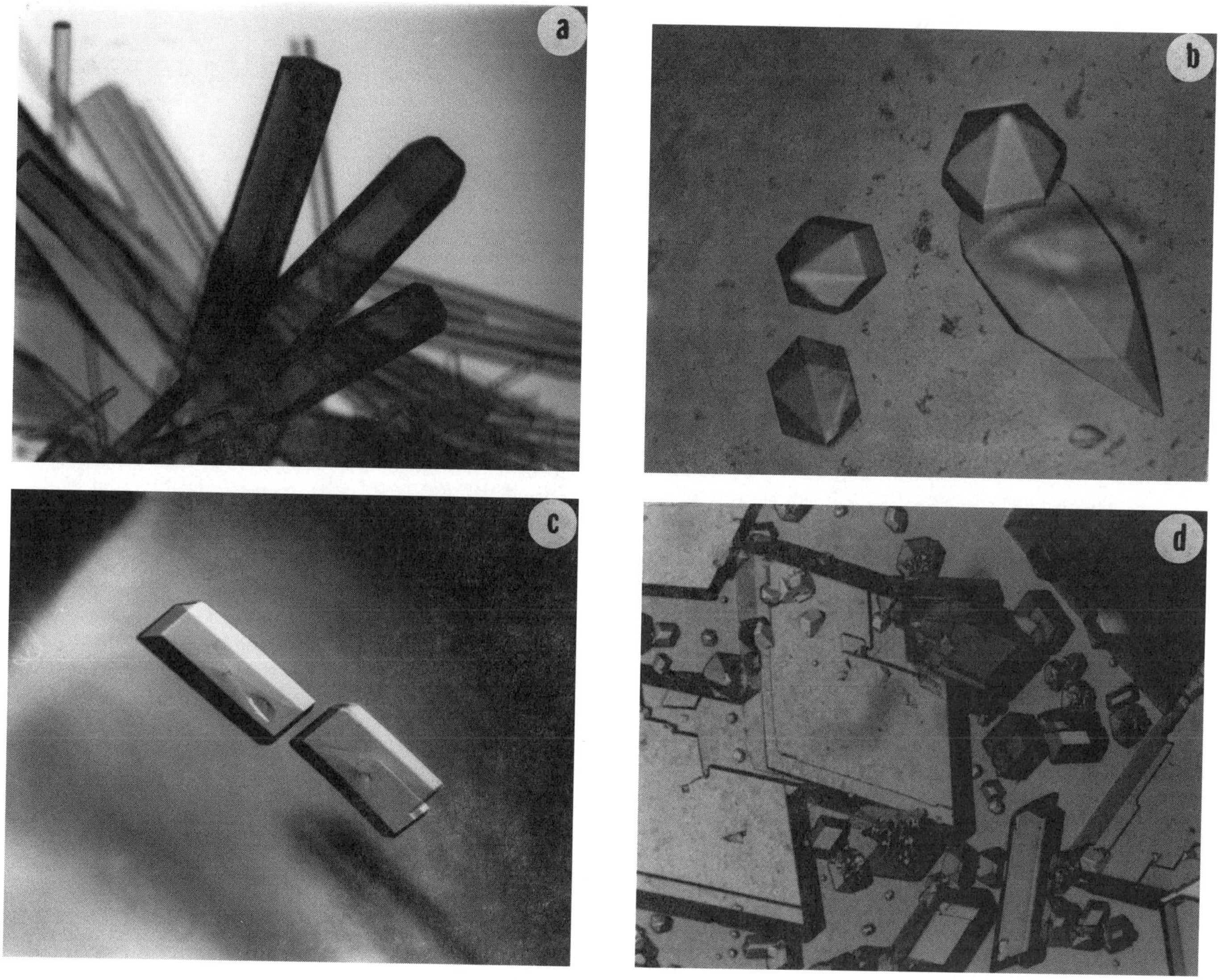

Figure 1. The crystals shown here were grown for X-ray diffraction studies that require a relatively small number of crystals, but of large size and high quality. (**a**) Beef liver catalase, a detoxifying enzyme; (**b**) α_1-acid glycoprotein, a human serum protein; (**c**) Fab fragments of an antiviral mouse antibody; and (**d**) canavalin, the major storage protein from legumes.

ecules and even more complex biological molecules such as glycosides, peptides, and antibiotics. A few examples from the literature where the crystallization procedures are described in some detail are for L-alanine (2), fructosylxyloside from *Scopulariopsis brevicaulis* (3), cephalosporin (4), and a flavanone, hesperidin from orange peels (5). Here, it has long been recognized that manipulation of crystalline formulations could dramatically affect solubility, shelf life, impurity content, thermal stability, release, efficacy, or a host of other commercially significant properties. Much research has, therefore, been devoted to the crystallization of commercially important conventional, small molecules during the past hundred years.

Only more recently has it become evident that crystalline formulations hold substantial promise for the administration and delivery of biological macromolecules as well (6,7). The first major example was that of insulin, a small protein of striking physiological importance to nearly all animals and crucial to the treatment of diabetes. In this instance, it was found that injection of crystalline protein resulted in a more prolonged, slower release of the protein drug and a significant improvement in its efficacy. This timed-delivery feature, first realized for insulin, now appears possible for other protein drugs.

In the coming years, a majority of biotechnology products will likely be proteins, particularly genetically modified proteins designed for introduction into living organisms, including humans. Their delivery and release will be a major problem, and their formulation as crystals will serve a major role in solving those problems. Here, unlike the growth of protein crystals for X-ray diffraction analysis, the crystals are in fact the product, and their production must be reproducible. The physical properties of the crystals, such as size, morphology, solubility, and impurity levels will be rigorously specified. They will be produced, under strict regulatory control, as is any drug. Thus, the objectives are quite different than for crystals produced for diffraction analysis.

In terms of pharmaceutically important proteins, natural insulin from animal tissues was certainly the first to be produced in bulk quantities by crystallization (8), and recombinant insulin followed (9). Others, however, are now known, including a large-scale facility for the production of crystalline Fab fragments from monoclonal antibodies (10).

The third area for crystal growth, the downstream processing of proteins produced in large fermentations for the purpose of purification, has associated with it yet another set of objectives and problems. Crystallization is one of the oldest chemical purification technologies and one that has been seriously underutilized to this point in biotechnology. Here, the goal is to promote the crystallization of macromolecules from extremely large volumes, commonly reaching several thousands of liters, of rather impure solutions or culture broths. The ideal is to achieve a very high degree of purification in a single, economical step, and at the same time generate a concentrated, stable, and active product that appeals in its form to a broad customer base. Although certain crystal sizes, morphologies, size distribution, and other features of the crystals may be preferred, these are not the most important parameters. Often crystals such as those shown in Figure 2 are quite acceptable. The major considerations are yield, reproducibility, degree of purification, and the ability to crystallize the material in spite of very high and variable contaminant levels.

Major difficulties accompany the crystallization of proteins in downstream processing and purification that are not seen in X-ray diffraction or formulation crystal growth. For example, processes such as vapor diffusion (discussed later), which is applicable to microliter volumes, are completely inapplicable to thousand liter volumes. Exotic precipitants may not be economical, sensitive procedures such as column chromatography may be out of the question, and speed is often of the essence. Waste product disposal can be a significant consideration. The process must be carefully conceived and developed to be as economical and efficient as possible and still yield the desired crystalline product at the required purity.

There are many examples in the literature of crystallization used for the purification of proteins from bulk fermentations, and these provide useful guides to the design of new procedures for other macromolecules. In one of the earliest reports, Fukumoto et al. (11) reported the purification of lipase from a strain of *Aspergillus niger* by fractionation with ammonium sulfate and its ultimate crystallization from acetone. Tsujisaka et al. (12) in 1973 purified the same enzyme from *Geotrichum candidum* using salt fractionation, DEAE Sephadex, and gel chromotography. The protein was crystallized from its aqueous solution. Buckel et al. (13) in 1981 and Mack et al. (14) describe a scheme based on salt fractionations and chromatography for the production of glutaconate CoA-transferase from *Acidaminococcus fermentans*.

In the 1990s, the appearance of reports of bulk crystallizations accelerated, including those on procedures for protein iminopeptidase from *Bacillus coagulans* (15), chymosin mutants from *Trichoderma reesei* (16), and pyroglutamyl peptidase from *Bacillus amyloliquefaciens* (17). Although only abstracts are currently available, some additional enzymes have also been crystallized for commercial production, these include cellulase (18,19), subtilisin protease and glucose isomerase (18), and alcohol oxidase (20).

THE UNIQUENESS OF MACROMOLECULAR CRYSTALS

Macromolecules, being unique in their properties, both in terms of size and complexity, give rise to crystals that are unique as well (21,22). We cannot, therefore, expect that

Figure 2. Needle crystals of α-amylase (**a**) and platelike microcrystals and needles of pineapple bromelin (**b**) are entirely suitable for downstream processing of enzymes produced in bulk for industrial purposes but are not of sufficient size nor of useful morphology for analytical studies.

macromolecular crystals will necessarily develop according to precisely the same mechanisms and principles as do conventional crystals (23). Indeed, recent investigations using advanced physical techniques such as inelastic light scattering, interferometry, and atomic force microscopy (22) have yielded evidence of new mechanisms and properties not before encountered with conventional crystals.

Macromolecular crystals are relatively small in comparison with conventional crystals, rarely exceeding 1 mm on an edge and generally smaller. Because only one stereoisomer of a biological macromolecule naturally exists, they do not form crystals possessing inversion symmetry and, therefore, generally exhibit simple shapes that lack the polyhedral character of small molecule crystals. They are extremely fragile, often crushing at the touch; degrade outside a narrow temperature, ionic strength, or pH range; generally exhibit weak optical properties (24); and diffract X-rays to resolutions far short of the theoretical limit (25). The reason for most of these character deficiencies is that macromolecular crystals incorporate large amounts of solvent in their lattices, ranging from about 30% at the lower limit to 90% or more in the most extreme cases (26,27). Proteins also have, as individual molecules, an array of water molecules surrounding them that are relatively tightly bound both in solution and in the crystal (28).

There are two other crucial differences between macromolecular and conventional crystal growth that have important practical consequences. The first is that macromolecular crystals are usually nucleated at extremely high levels of supersaturation, often several hundred to a thousand percent. Small molecule crystals, on the other hand, usually nucleate at only a few percent supersaturation. Virtually every quantitative aspect of crystal growth is a direct function of supersaturation (29). Although high supersaturation may be essential to promote nucleation, it is far from ideal for growth, and the many problems observed for macromolecular crystals attest to this. Furthermore, supersaturated macromolecular solutions, in addition to crystal nuclei, produce alternate solid states that we refer to collectively as amorphous precipitates. Unlike conventional systems, competition exists at both the nucleation and growth stages between crystals and precipitate. This is particularly acute because competition is promoted by high levels of supersaturation. Because amorphous precipitates are kinetically favored, though of less-favored energy states, they tend to dominate the equilibration process and often inhibit or preclude crystal formation.

Given the complexities that beset macromolecules, can we reasonably expect their crystallization to resemble that of conventional molecules? Evidence to this point suggests that the answer is in principle yes, but in practice no. It appears that the fundamental mechanisms and pathways of macromolecular crystal growth are the same as for conventional crystals (30,31) but that the magnitudes of the underlying kinetic and thermodynamic parameters that govern the process differ dramatically.

METHODS FOR GROWTH OF PROTEIN CRYSTALS

The growth of macromolecular crystals has been studied for more than 150 years, initially as a purification tool, and more recently for other purposes. It has been the topic of several texts and reviews and, with the increasing importance of protein crystallography, has become a major factor in the biotechnology industry. Because this body of literature exists, no attempt will be made here to exhaustively review all methods and aspects of macromolecular crystal growth. An effort will be made, however, to describe the underlying principles, present the most commonly employed laboratory approaches, summarized in Table 1, and suggest ways in which the field can be addressed by novice crystal growers.

The strategy used to bring about crystallization (22,32–37) is to direct a macromolecular solution slowly toward a state of supersaturation by modifying the properties of the solvent or the solute. This is generally achieved by altering the concentration of a precipitating agent, such as salt, or by altering some physical property, such as pH. This is illustrated by a version of the classical phase diagram for supersaturation, modified somewhat for the case of proteins in Figure 3.

In concentrated solutions, macromolecules, unlike most conventional molecules, may aggregate as an amorphous precipitate, a result to be avoided. It is usually indicative that supersaturation has proceeded too far or been reached too swiftly. One would like to slowly and marginally exceed the equilibrium point of saturation and thereby allow the macromolecules reasonable opportunity to assemble into a crystalline lattice.

The traditional procedure, illustrated by Figure 4, for inducing proteins to separate from solution as a solid phase is to gradually increase the level of saturation of a salt. Classically, the salt was ammonium sulfate, but others are also in common use (see, for example, Table 2 and the databases described in Refs. 26 and 27). Although the protein may separate as an amorphous precipitate, with appropriate care, manipulation of salt concentration is an effective means for growing protein crystals. This approach has, in fact, probably yielded more varieties of protein crystals than any other (22,26,27).

The accepted explanation for this salting-out phenomenon is that the salt ions and macromolecules compete for chemical interactions, principally hydrogen bonds, with the solvent molecules, that is, water. This occurs because both salt ions and protein molecules depend on hydration layers to maintain solubility. When competition between

Table 1. Methods for Attaining a Solubility Minimum

Bulk crystallization
Batch method in vials
Evaporation
Bulk dialysis
Concentration dialysis
Microdialysis
Liquid bridge
Free-interface diffusion
Vapor diffusion on plates or slides
Vapor diffusion in hanging drops
Sequential extraction
pH-Induced crystallization
Temperature-induced crystallization
Crystallization by effector addition

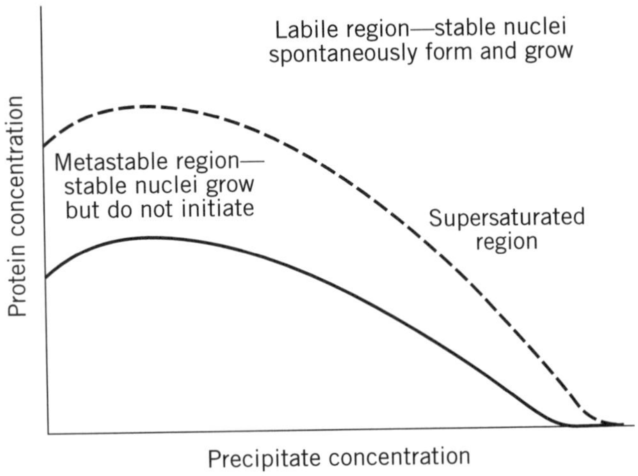

Figure 3. A phase diagram for a hypothetical protein showing its solubility as a function of precipitant concentration. The solid line represents the maximum solubility, or saturation curve, for the protein. Note that the protein is less soluble at very low and very high concentrations of the precipitant, corresponding to the salting-in and salting-out regions, respectively. The supersaturated region lies above the maximum solubility curve, and is, in turn, demarcated by a boundary discriminating the metastable region of supersaturation from the labile region. In the labile region, crystal nuclei can both spontaneously form and grow, whereas in the metastable region they can only grow.

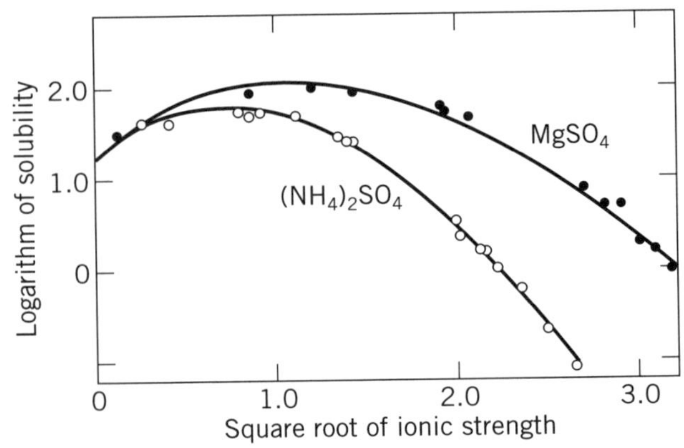

Figure 4. The solubility of the protein enolase at constant pH and temperature as a function of ionic strength produced by two separate salts. The two ends of the curves correspond to the salting-in and the salting-out regions of the solubility diagram.

ions and proteins becomes sufficiently intense, the protein molecules begin to self-associate in order to satisfy, by intermolecular interactions, their electrostatic requirements. Thus, dehydration, or the elimination and perturbation of solvent layers around protein molecules, reduces solubility. For a specific protein, the solubility minima are usually quite dependent on the pH, temperature, chemical composition of the precipitant, and properties of both the protein and the solvent (for more extensive discussions, see Refs. 38–42).

In the low ionic strength range, as seen in Figure 4, a phenomenon known as salting-in occurs in which the solubility of the protein increases as the ionic strength rises from zero. Protein solubility at low ionic strength is diminished by the removal of ions essential for satisfying the electrostatic requirements of the molecules. As ions are removed, and in this region of low ionic strength cations are most important (38–41), the protein molecules again balance their electrostatic requirements by forming interactions among themselves.

The salting-in effect can be applied in the opposite sense, that is, reducing ionic strength can serve as a crystallization tool. In practice, a protein is optimally solubilized at moderate ionic strength and then extensively dialyzed against distilled water. Many proteins, such as catalase, concanavalin B, immunoglobulins, and plant seed proteins, have been crystallized by this means (26,27,32,33).

Proteins may be forced from solution, at constant pH and temperature, by increases in ionic strength, or they can be crystallized or precipitated at constant ionic strength by changes in pH or temperature (38–41). This is illustrated by Figures 5 and 6. Reduction in solubility occurs because the charge distribution on a macromolecule, its surface features, or its conformation may change as a function of these variables (39). By virtue of its ability to inhabit a range of states, a protein may exhibit a number

Table 2. Common Precipitants Used in Macromolecular Crystallization

Salts	Organic solvents	Polymers
Ammonium or sodium sulfate	Ethanol	Poly(ethylene glycol) 1,000, 3,350, 6,000, 20,000
Lithium sulfate	Isopropanol	Jeffamine T
Lithium chloride	1,3-Propanediol	Polyamine
Sodium or ammonium citrate	2-Methyl-2,4-pentanediol	
Sodium or potassium phosphate	Dioxane	
Sodium or potassium or ammonium chloride	Acetone	
Sodium or ammonium acetate	Butanol	
Magnesium or calcium sulfate	Acetonitrile	
Cetyltriethyl ammonium salts	Dimethyl sulfoxide	
Calcium chloride	2,5-Hexandediol	
Ammonium nitrate	Methanol	
Sodium formate	1,3-Butyrolactone	
	Ethylene glycol 400	

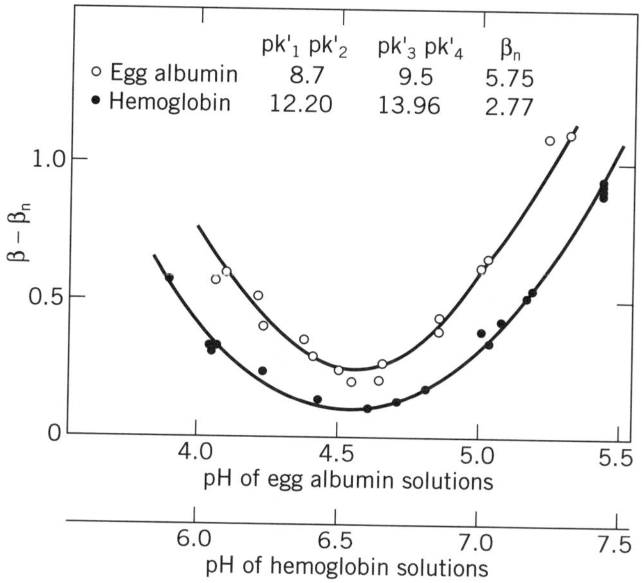

Figure 5. The solubility behavior of two proteins, hemoglobin and egg ovalbumin, at constant ionic strength as a function of pH. The solubility minima are often good targets for the crystallization of proteins by variation of pH.

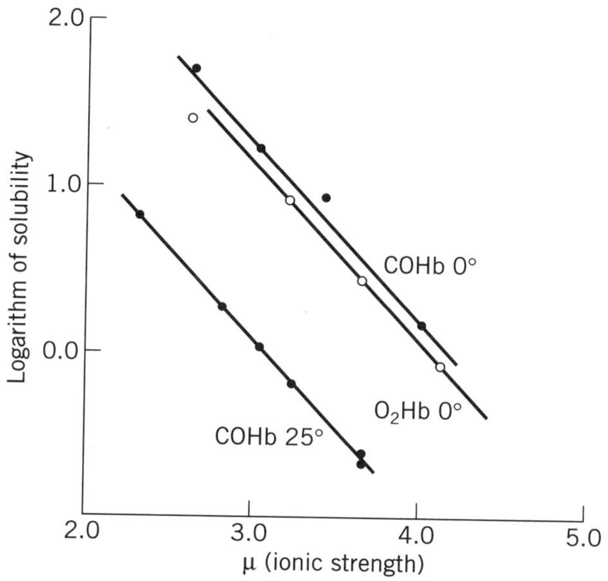

Figure 6. The solubility of hemoglobin in concentrated phosphate buffer at various temperatures.

of environmentally dependent solubility minima, and each of these minima may afford the opportunity for crystal growth.

Similar effects may be achieved as well by the slow addition to the mother liquor of some relatively mild alcohols or other organic solvents such as ethanol, acetone, or methylpentanediol. The only essential requirement for the precipitant is that at the specific temperature and pH of the experiment, the additive not adversely affect the structure and integrity of the protein. This is a stringent requirement and deserves some consideration. An organic solvent will compete, to some extent like salt, for water molecules. It also reduces the dielectric shielding capacity of intervening solvent. Reduction of bulk dielectric increases the effective strength of the electrostatic forces responsible for one protein molecule associating with another.

Polymers such as polyethylene glycol also serve to dehydrate macromolecules in aqueous solution, as do salts, and they furthermore alter the dielectric properties in a manner similar to organic solvents. They produce, however, an additional important effect. Polyethylene glycol perturbs the natural structure of the solvent and creates a more complex network having both water and itself as structural elements. A consequence of this restructuring of solvent is that macromolecules, particularly proteins, tend to be excluded, and phase separation is promoted (43,44). This is illustrated for several proteins in Figure 7.

Combinations of approaches using both precipitating agents, such as those seen in Table 2, and variation of physical parameters are frequently effective in promoting the crystallization of macromolecules. Increasing the concentration of a precipitating agent to a point just below supersaturation and then adjusting the pH or temperature to further reduce the solubility of the protein is a common approach. Modification of pH can be brought about quite well using the vapor diffusion technique (see later section) with volatile acids and bases such as acetic acid and ammonium hydroxide.

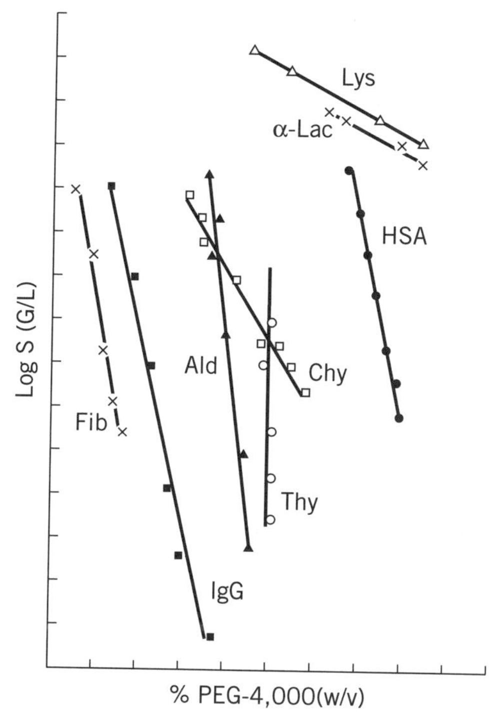

Figure 7. Solubility of various proteins in PEG-4,000. Measurements were made in 0.05 M potassium phosphate, pH 7.0, containing 0.1 M KCl. Human fibrinogen (×), human γ-globulin (■), lysozyme (△), α-lactalbumin (+), chymotrypsin (□), aldolase (▲), thyroglobulin (○), human serum albumin (•).

Because of the complexity of interactions between solute and solvent and the shifting character of the protein (45), the methods of crystallization must usually be applied over a broad set of conditions with the objective of discovering the particular minimum (or minima) that yield crystals. In practice, one determines the precipitation points of the protein at sequential pH values with a given precipitant, repeats the procedure at different temperatures, and then examines the effects of different precipitating agents and their concentrations.

METHODS FOR PROMOTING SUPERSATURATION

The oldest technique used for the crystallization of proteins and nucleic acids (see Table 1), with a history of more than 150 years (46), is what is commonly called batch crystallization (32,33,38,47–49). It is the simplest, most thoroughly tested, and when optimized, the most reliable of all crystallization methods. Lysozyme, for example, is most generally crystallized by the batch method. It relies on the mixing of a protein solution at undersaturated concentration with precipitant components. The latter alter the electrolyte properties of the solution to produce a mother liquor supersaturated with respect to protein. From such a solution the solid state may be expected to eventually form, and if the conditions and physical parameters have been chosen correctly, this solid state will be crystalline rather than amorphous. Alternatively, the physical state of an undersaturated protein solution may be altered by, for example, lowering or raising the temperature. The supersaturated solution produced as a result may yield crystals.

The batch method is attractive because of its inherent simplicity and reproducibility. It requires nothing more than the combination of two or more solutions into one and a period of time until nucleation commences and growth follows. The concept is appealing because it can be carried out with an absolute minimum of human intervention. Because the apparatus is nothing more than a container for a small quantity of fluid, which can be varied as circumstances require from a few microliters to hundreds of milliliters, it is fully amenable to interface with both simple and complex observational systems designed to monitor and record the process of protein crystal growth. Thus, the technique is very useful, for example, for inelastic light scattering studies of nucleation (50–57), or time-lapse photography investigations of protein crystal growth (58,59). It is particularly attractive as well for macromolecular crystallization experiments in space.

Batch crystallization requires no complex device to produce equilibration, either through the liquid or the vapor phase. It relies strictly on the energetics of the system—the activation barriers to nucleation—to regulate the process. Although it has not, in recent years, been extensively used to screen for protein crystallization conditions, this reflects more a prejudice of investigators and the popularity of other methods rather than any deficiency in the batch method itself. Because it was commonly used in the past with fairly large volumes of protein solution, investigators have mistakenly come to believe that it is inappropriate for small samples, which is untrue. It can even be carried out using microliter amounts in glass capillaries or under oil. It can be a useful screening technique as well as a good method for the preparation of large quantities of protein crystals once optimal values for parameters are established.

There are a number of ingenuous devices, procedures, and methods for dynamically altering the degree of saturation of a protein solution. Generally, these rely on the slow increase in concentration of some precipitant such as salt or polyethylene glycol. These same approaches can of course also be used for salting-in, modification of pH, and the introduction of materials that alter protein solubility. These techniques have been extensively reviewed elsewhere (32–36,60–63), and only a selection will be dealt with here.

Dialysis is familiar to nearly all biochemists as a means of modifying the components of a protein solution (32,33). As illustrated in Figure 8, the solution containing the protein is enclosed in a membrane casing or in a container having a semipermeable membrane partition. The membrane selectively allows the passage of small molecules and ions, but pore size prevents passage of the much larger protein molecules. The vessel or dialysis tube containing the protein is submerged in a larger volume of liquid that has the desired properties of pH, ionic strength, ligands, and so forth. With successive changes of the exterior solution and concomitant diffusion of small molecules and ions across the semipermeable membrane, the protein solution gradually acquires defined properties.

The dialysis approach, in some manifestation, has been used to crystallize many proteins (27,33). It was usually applied on a large scale, and only when substantial

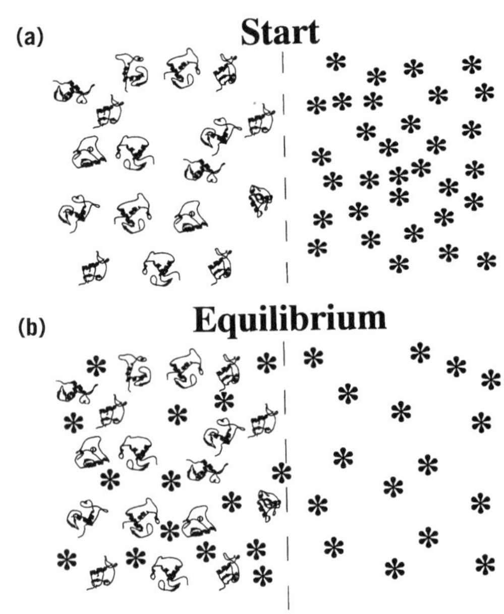

Figure 8. A schematic description of the dialysis process. (top) Protein molecules are separated from conventional small molecules or ions by a semipermeable membrane that prevents the macromolecules from passing. (bottom) The small molecules or ions have passed through the membrane and at equilibrium are equally distributed on its two sides.

amounts of protein were available. It has the advantage that liquid–liquid diffusion through a semipermeable membrane exposes the protein to a continuum of potential crystal-producing conditions without physically perturbing the mother liquor. Diffusion through the membrane is slow and controlled. Because the rate of exchange of substituents in the mother liquor is a function of the concentration gradient across the membrane, the nearer the system to equilibrium, the more slowly it changes.

This method has been adapted to much smaller amounts of protein by crystallographers who use almost exclusively microtechniques. These typically involve no more than 5 to 50 μl of protein solution in each trial. First described by Zeppenzauer et al. and Zeppenzauer (64,65) and subsequently modified and refined by numerous others, the method confines a protein solution within a glass capillary or the microcavity of a small plexiglass button. The cavity of the button or the ends of the microcapillary tube are closed off by a dialysis membrane. The whole arrangement, charged with protein solution, is then submerged in a much larger volume of an exterior liquid that is held within a closed vessel such as a test tube or vial. The microdyalysis method is shown schematically in Figure 9. If the exterior solution is at an ionic strength or pH that induces the mother liquor to become supersaturated, crystals may grow. If not, the exterior solution may be exchanged for another and the experiment repeated.

The dialysis buttons, seen in Figure 10, are particularly ingenious. Not only are they compact and easy to examine with a dissecting microscope, but they have a shallow groove about their waist. A section of moistened dialysis membrane is placed over the mother-liquor-filled cavity, and it is held securely in place by simply slipping a rubber O-ring over the top of the button and seating it in the groove. These buttons are in wide use and have proven quite successful.

The technique of liquid–liquid diffusion (Figs. 11 and 12), also known as free-interface diffusion, for the crystallization of macromolecules (25–27,62,63,66) is in principle and practice simple and straightforward. It relies on the direct interdiffusion of a precipitant solution into a protein solution across their common liquid–liquid interface. When the two solutions are juxtaposed, precipitant gradients form in the region of the interface and produce local supersaturation. This promotes the formation of crystal nuclei and subsequent growth, often to sizes useful for analysis.

The fundamental idea of liquid–liquid diffusion of a solvent or solute into a macromolecular solution is similar to dialysis. The method is particularly applicable in microgravity because the necessity of a membrane is eliminated and the two liquids can be maintained relatively stable with respect to one another for long periods of time. Because the diffusion rate of macromolecules is so much less than that of the small molecules of most precipitants, there is relatively little penetration of protein into the precipitant during the course of most experiments.

The technique of liquid–liquid diffusion has been applied in laboratories on earth to grow macromolecular crystals, and with some success. A search of the National Institute of Standards and Technology (NIST) database (26,27) shows 35 entries using this technique with another 193 ascribed to various kinds of dialysis. Direct liquid diffusion has not, however, been as extensively used in conventional laboratories as have some other techniques, such as batch or vapor diffusion, because of some inherent disadvantages. Significantly, most of these methodological deficiencies are relieved or eliminated in the absence of gravity.

On earth, density-driven convective currents produce immediate and continuous turbulence at the liquid–liquid interface. Transient mixing, if layering of solutions is used, or the necessity of a membrane, as in dialysis, lessens the methods' effectiveness. Sedimentation produces a particularly deleterious effect with the free-interface method because nuclei and small growing crystals, under the influence of gravity, fall from the liquid–liquid interface. Thus, as a crystal forms, gravity increasingly pulls it downward from the region of the solution where growth would most favorably continue. Gravity essentially removes growing crystals from an optimal growth environment and thereby imposes a self-limiting character on the process, in turn restricting crystal size.

Currently, the most widely used method for bringing about supersaturation in microdrops of mother liquid is vapor diffusion (26,27) (Fig. 13). This approach (32,33,35, 60,62,63,67) has provided a diversity of applications and may be divided into those procedures that use a "sitting

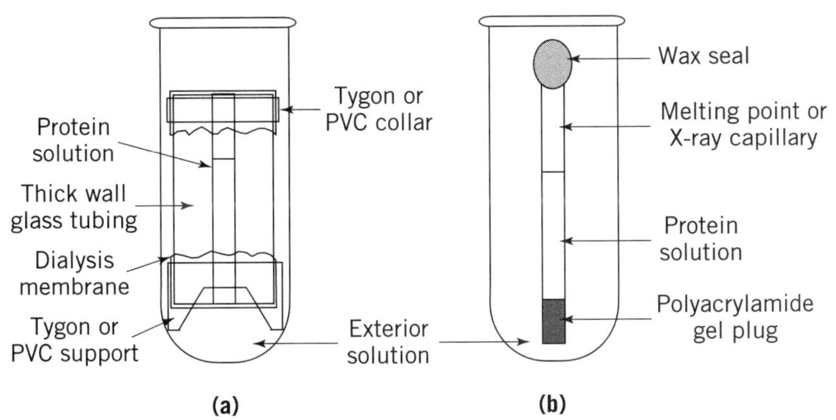

Figure 9. A microdialysis cell (**a**), for crystallization of small protein samples, is closed at one end by a section of dialysis membrane held in place by a Tygon ring that also serves as a support. (**b**) A similar arrangement using a thin glass capillary, but the membrane has been replaced by a polyacrylamide plug that is permeable to small molecules and ions.

Figure 10. The thin glass capillaries shown here may be used for either batch crystallization or crystallization by dialysis with volumes in the microliter range. The dialysis buttons seen both face on and on edge can be closed by dialysis membrane held in place by the rubber O-rings. Once filled with protein solution they can be submerged in a larger volume of solution contained, for example, in the small weighing bottle.

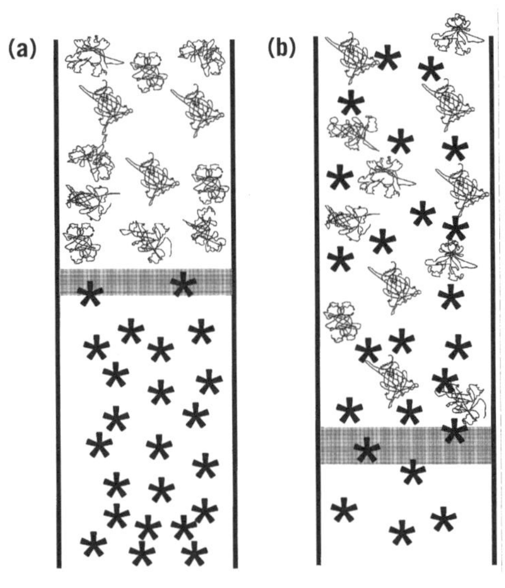

Figure 11. In free-interface diffusion the protein solution is (left) initially layered atop a precipitant solution, such as a highly concentrated salt. Diffusion of the small molecules into the protein solution (right) induces crystallization to occur.

drop" and those that use a "hanging drop." In either form, the method relies on the transport of water or some volatile agent between a microdrop of mother liquor, generally 2 to 25 μL volume, and a much larger reservoir solution of 0.75 to 25 mL in volume. Through the vapor phase, the droplet and reservoir come to equilibrium. Because the reservoir is of much larger volume, the final conditions at equilibrium are essentially those of the initial reservoir state. A variety of devices currently in use for protein crystal growth by vapor diffusion are shown in Figure 14.

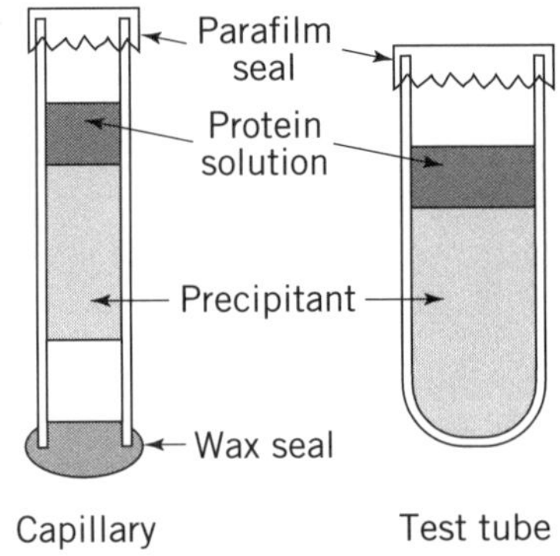

Figure 12. Diagram illustrating the free-interface diffusion technique for the crystallization of microliter amounts of protein solution on the left and milliliter amounts on the right. The method relies on the direct diffusion of precipitant and small molecules across the liquid–liquid interface.

In the vapor phase, water can be removed slowly from droplets of mother liquor, pH may be changed, or volatile components such as ethanol may be gradually introduced. As with those methods already described, the procedure may be carried out to further advantage at a number of different temperatures.

According to one popular procedure (32,33,67) (Fig. 15), droplets of 10 to 20 μL are placed in the nine wells of siliconized glass depression plates. The samples are then sealed in transparent containers, such as Pyrex dishes or plastic boxes that also hold reservoirs of 20 to 50 mL of the precipitating solution. The plates bearing the protein or nucleic acid samples are supported above the reservoirs by the inverted halves of disposable Petri dishes. Through the vapor phase, the concentration of salt or organic solvent in the reservoir equilibrates with that in the sample. In the case of salt precipitation, the droplet of mother liquor must initially contain a level of precipitant lower than the reservoir, and equilibration proceeds by diffusion of water from the droplet into the reservoir. This holds true as well for nonvolatile organic solvents, such as methylpentanediol and for polyethylene glycol. In the case of volatile precipitants, none need be added initially to the microdroplet, because distillation and equilibration proceed in the opposite direction.

The method described here has the advantage that it requires only microliter amounts of material and is well suited for screening large numbers of conditions. A disadvantage is that all the samples in a single container must be equilibrated against the identical reservoir solution. It does, however, permit some flexibility in varying conditions once the samples have been dispensed, by modification of the concentration or pH of precipitants in the reservoir. When clear plastic boxes are used, large numbers of samples can be quickly inspected under a dissecting microscope and conveniently stored.

The disadvantage of identical reservoir conditions throughout a single container was overcome by the introduction of plastic plates specifically designed for protein crystallization. One of these, sponsored by the American

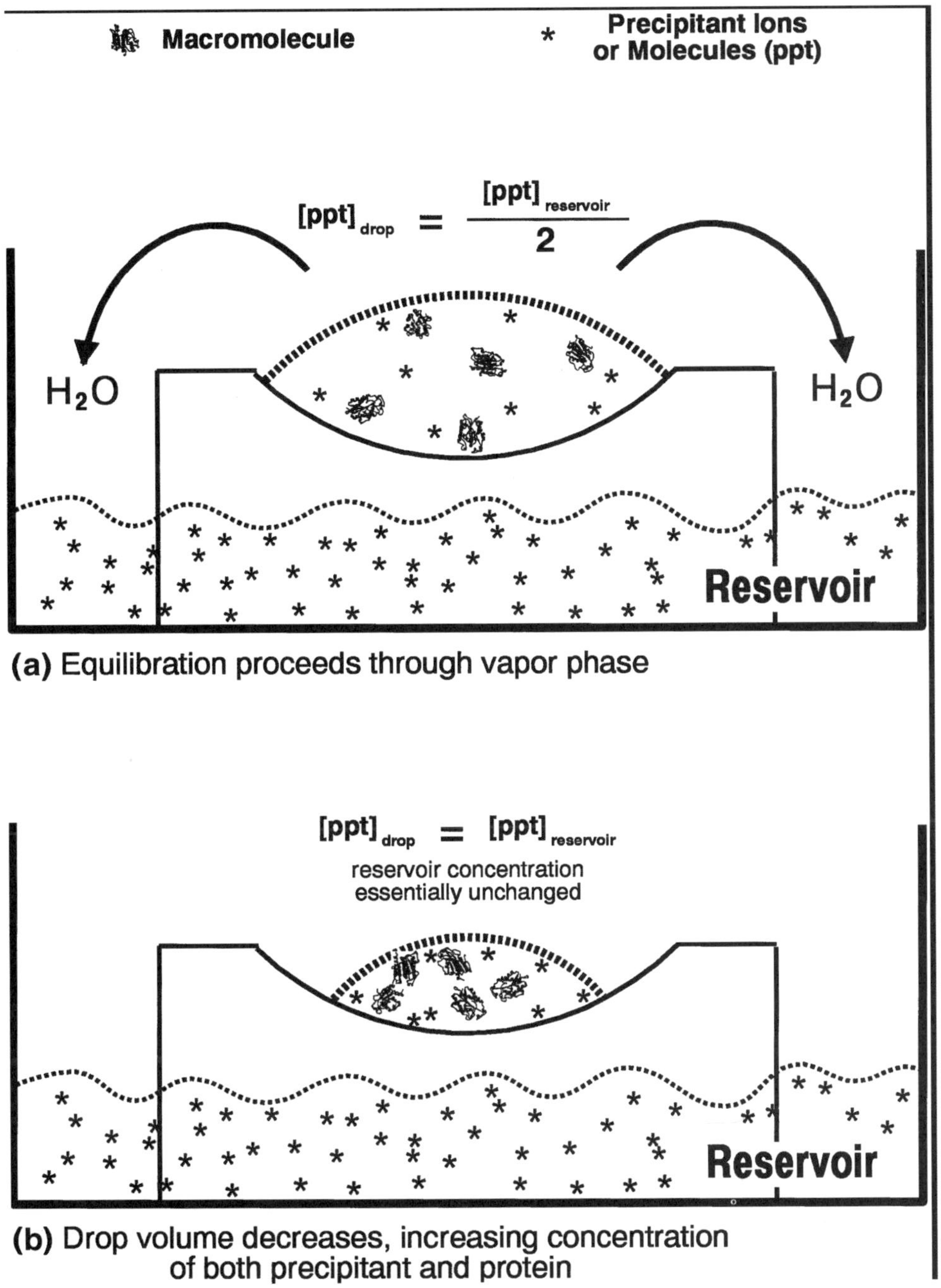

Figure 13. In vapor diffusion, a small drop of mother liquor containing protein and a low concentration of precipitant is, in (**a**), initially exposed to a much larger reservoir solution containing a high concentration of precipitant. At equilibrium in (**b**), through loss of water and reduction in volume, the mother liquor achieves a precipitant concentration approximately equal to that in the reservoir, thereby promoting crystallization of the protein to occur.

Crystallographic Association, is a plastic plate having accommodation for 15 protein samples. Each chamber has a separate reservoir compartment and the mother liquor microdroplet may be either suspended from the underside of a glass cover slip, as in the hanging drop method described later, or sandwiched between two glass cover slips. Sealing of the chambers from air requires silicon grease or oil between cover slips and the plastic rims of the chambers. With these plates, the optical properties are very good but equilibration tends to be slow, and the setup is quite tedious.

A second crystallization plate (68), also seen in Figure 14, is supplied by Charles Supper. With these plates, the drop sits atop a clear support post that protrudes upward from a circular moat containing the reservoir solution. The chambers can be rapidly and conveniently sealed from air by transparent plastic tape pressed over the upper surface of the plate after the reservoirs have been filled and the drops of mother liquor dispensed. Equilibration, as with the other plate, is through the vapor phase. Although the optical properties are less favorable with these devices, they are convenient and compact and can be used for rapid screens of crystallization conditions.

The hanging drop procedure (Fig. 16) also uses vapor phase equilibration, but with this approach a microdroplet of mother liquor (as small as 2 μL) is suspended from the underside of a microscope cover slip, which is then placed over a small well containing a milliliter or less of the pre-

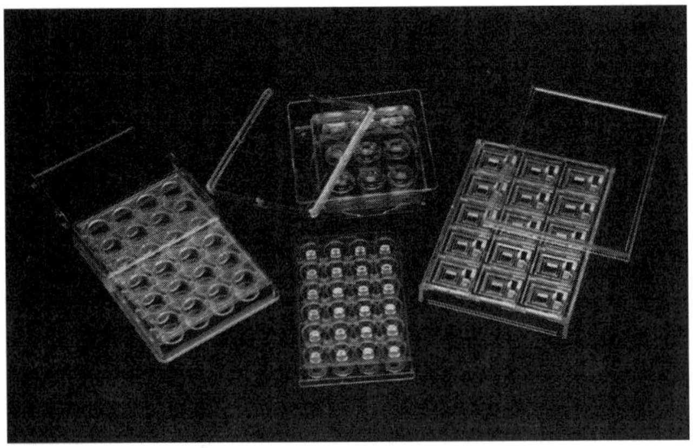

Figure 14. Four examples of trays, plates, or combinations used for crystallization by vapor diffusion. All are commercially available and can be used for either the hanging-drop or the sitting-drop approach.

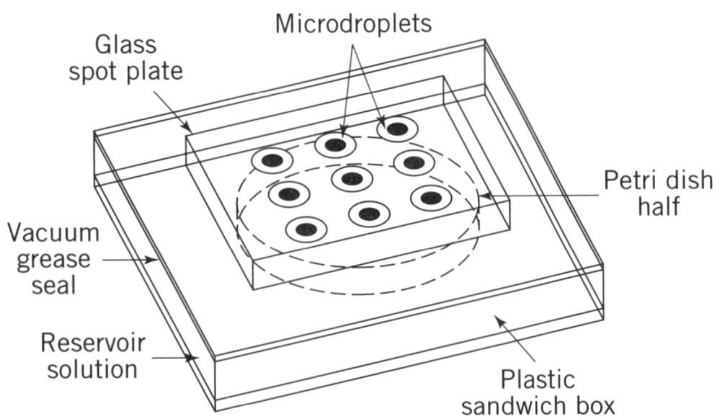

Figure 15. Schematic diagram of a nine-well depression spot plate enclosed in a plastic sandwich box used for the crystallization of macromolecules by the vapor diffusion method. Microdrops of 10 to 25 μL can be placed in the depressions with 25-mL reservoirs of precipitating solution in the bottom of the box. Equilibration proceeds through the vapor phase.

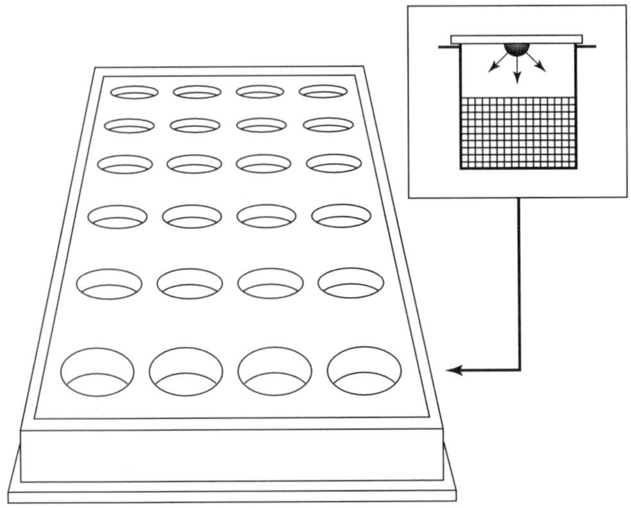

Figure 16. The standard configuration for hanging drop protein crystallization experiments is shown here. A tissue culture plate provides 24 wells for reservoirs of about 0.5 to 0.75 mL. Each well can be covered by a glass slip with a drop of protein solution hanging from its underside. The drop of 2 to 20 μL equilibrates with the reservoir solution over time through the vapor phase, causing precipitant concentrations to increase in the drop and induce crystallization.

cipitating solution. The wells are most conveniently supplied by disposable plastic tissue culture plates that have 24 wells with rims that permit sealing by application of silicone vacuum grease around the circumference. The hanging drop technique can be used both for the optimization of conditions and for the growth of large single crystals.

Although the principle of equilibration with both the sitting drop and the hanging drop is essentially the same, they frequently do not yield the same results even though the reservoir and protein solutions may be identical. Presumably because of geometrical differences in the apparatus used to achieve equilibration, the path to supersaturation is different, even though the end point may be the same. In some cases there are striking differences in the degree of reproducibility, final crystal size, morphology, or time required for growth. These observations emphasize the point that in crystal growth the route leading to supersaturation and the kinetics of the process may be as important as the end point achieved. There are nearly limitless variations on these themes for liquid–liquid or vapor diffusion, ranging from the very simple to the unnecessarily complex. You need only use a bit of imagination to devise a novel approach for yourself.

With some exceptions, the methodologies used for protein crystallization, most of those described above, have been devised for the growth of crystals from very small volumes of mother liquor. Indeed, reducing the amount of material needed has been a principal objective in X-ray crystallography laboratories for some time. Crystals can now be obtained from only 2 to 4 μL of solution, and entire screening matrixes (see later discussion) can be explored using only 2 to 5 mg of protein or nucleic acid.

Two major applications of protein crystallization, for formulations, and for downstream processing of macromolecules do not, however, benefit from reduction in scale. On the contrary, they profit from very large volumes and high yields of the crystalline product. As a consequence, the microtechniques described earlier may represent model approaches, but they are, in themselves, quite inappropriate to bulk processes.

Large-scale, industrial crystallizations are, at least currently, limited to methods based essentially on batch crystallization in combination with manipulation of pH and temperature. In addition, many chemical compounds that might be used as precipitants or additives in laboratory settings cannot be used in many cases in bulk processes because of expense, hazard, or the creation of unacceptable by-products (7). Although this might appear severely limiting, remarkable success has been achieved using only these simple tools. The large number of biological macromolecules in crystalline form and produced in kilogram (or larger) amounts available on the commercial market attests to this.

The restriction on techniques is compensated to a great extent in industrial processes by other factors. For example, the objective protein is usually in relatively high concentration and comprises a significant fraction of the total protein present. It is usually produced by genetically engineered or overproducing strains of simple microorganisms; hence, it does not suffer from many of the sources of microheterogeneity, seen in Table 3, that afflict many natural products. In addition, crystals produced in large-scale processes are usually very small crystals of marginal perfection that would be unsuitable for analytical work. Such crystals may, however, be entirely appropriate for the intended purpose of purification and product formulation.

THE CRYSTALLIZATION OF PROTEINS BY VARIATION OF PH OR TEMPERATURE

One of the most useful techniques for producing a supersaturated solution is adjustment of pH to values where the protein is substantially less soluble. This may be done in the presence of a variety of precipitants so that a spectrum of possibilities are created whereby crystals might form. The gradual alteration of pH is particularly subtle because it may be accomplished by a variety of gentle approaches that do not otherwise perturb the system. Although microdialysis is probably equally suitable, more success with pH has been achieved by the vapor diffusion method, using "sitting" microdroplets on spot plates or in one of the plastic plates available for protein crystallization. The ambient salt, effector, or buffer conditions are established before dispensing the microdroplets in the depressions on the plate. The pH is then slowly raised or lowered by adding a small amount of volatile acid or base to the reservoir. Diffusion of the acid or base then occurs from reservoir to sample, just as for a volatile precipitant.

Table 3. Sources of Microheterogeneity

Presence, absence, or variation in a bound prosthetic group, substrate, coenzyme, inhibitor, or metal ion
Variation in the length or composition of the carbohydrate moiety of a glycoprotein
Proteolytic modification of the protein during the course of isolation or crystallization
Oxidation of sulfhydryl groups during isolation
Reaction with heavy metal ions during isolation or storage
Presence, absence, or variation in post-translational side-chain modifications such as methylation, amidiation, and phosphorylation
Microheterogeneity in the amino or carboxy terminus or modification of termini
Variation in the aggregation or oligomer state of the protein caused by association or dissociation
Conformational instability caused by the dynamic nature of the molecule
Microheterogeneity caused by the dynamic nature of the molecule
Partial denaturation of sample
Genetically different animals, plants, or microorganisms that make up the source of protein preparations
Bound lipid, nucleic acid, or carbohydrate material or substances such as detergents used in the isolation

If the pH is to be raised, for example, a small drop of concentrated ammonium hydroxide can be added to the reservoir; a drop of acetic acid may be used to lower it. The pH can also be gradually lowered over a period of days by simply placing a tiny chip of dry ice in the reservoir. The liberated CO_2 diffuses and dissolves in the mother liquor to form weak carbonic acid. When a specific pH end point is required, the mother liquor may be buffered at that point and then moved significantly away by addition of acid or base. The microdroplets of mother liquor may then be returned to the buffer point by addition of an appropriate volatile acid or base to the reservoir.

As shown in Figure 4, individual proteins display distinct solubilities at varying levels of salt concentration and can be selectively separated from solution as amorphous precipitate or crystals. At very low ionic strengths where salting-in occurs, the solubility of the protein increases with the ionic strength because of a decrease in its activity coefficient. The salting-out effect, which Hofmeister (39,41) suggested to be caused by a reduction of the chemical activity of water by salt, occurs as the ionic strength increases. Because ionic strength is a function of the second power of the valence, multivalent ions are most efficient. In the salting-out region of a protein's solubility profile, the logarithm of protein solubility as a function of the ionic strength is linear and can be expressed by

$$\log S = \beta - K_s(I/2)$$

where I is the ionic strength in moles per kilogram of water, S is the solubility of the protein in grams per kilogram of water, and K_s is a constant. K_s serves as a measure of the slope of the solubility extrapolated to zero ionic strength and is the logarithm of the solubility at that point (39).

In principle, the precipitation or crystallization of a protein is effected by manipulating the factors that appear in the above equation. One can crystallize a macromolecule by increasing the ionic strength at constant pH and temperature, a K_s fractionation procedure, or it can be brought about at constant ionic strength by manipulating the pH and temperature, which in most cases strongly influence solubility.

With all techniques, the objective is to produce a supersaturated protein solution by very gradual and nonperturbing adjustment of conditions. An example is shown in Figure 17. In practice, a solubility minimum is determined for the protein at some specific pH, temperature, and ionic strength. The protein solution is set initially at the requisite salt concentration for this solubility minimum, but either the pH or temperature is set initially to maintain protein solubility. The temperature is then slowly relaxed to minimize protein solubility, or the pH is gradually equilibrated via the liquid or vapor phase to an appropriate value.

In the case of temperature shifts, the protein is generally raised to a higher temperature, as, for example, with squash globulin, insulin, or excelsin (41,42,50,52,69,70), a

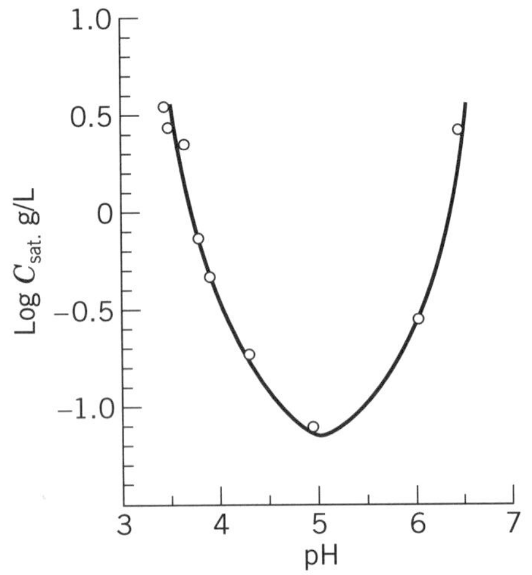

Figure 17. The solubility of insulin falls dramatically in the immediate neighborhood of pH 5, providing a ready means for creation of a supersaturated solution. The protein is simply dissolved to high concentration at pH 7 or higher, and the pH subsequently lowered through addition of H_3O^+ by vapor diffusion or dialysis.

process that increases solubility. The solution is then cooled slowly to where the protein is no longer soluble. This can be achieved by placing the warm solution in an insulated or thermally controlled container or by placing the sample in a Dewar flask. The approach to supersaturation thus occurs at an almost imperceptible rate. Exactly the same technique can be applied if the protein is more soluble at cold than warm temperature. Under these circumstances, the protein solution is cooled under established ionic strength conditions and then allowed to warm slowly. This approach was used, for example, to grow crystals of DNase by Kunitz (71) and is the basis of Jakoby's sequential extraction procedure using decreasing concentrations of ammonium sulfate (72).

The use of pH gradients in space and time has been successful in many investigations (26,27); they can be used in association with direct liquid dialysis or with vapor diffusion methods as well. They can, furthermore, be used in conjunction with a host of different precipitating agents. Temperature can be used in an analogous manner (73).

Proteins and nucleic acids are structurally flexible and are believed to alternate between several conformational states. In particular, they may assume substantially different conformations when they have bound coenzyme, substrate, or some other ligand. As a consequence, a protein with bound effector may exhibit different solubility properties than the nonliganded form. In addition, if multiple conformational states are available, the presence of effector may be used to select for only one of these, thereby imposing a degree of conformity of structure that would otherwise be absent.

The addition of ligands can sometimes be used to induce supersaturation and crystallization in those cases where its binding to the protein produces solubility differences under a given set of ambient conditions. The effector may be slowly and gently combined with the protein, for example, by dialysis, so that a resulting complex becomes supersaturated. The addition of ligands, substrates, and other small molecules has seen widespread use in protein crystallography, because it provides useful alternatives if the apoprotein itself cannot be crystallized.

No matter what the ultimate objective of macromolecular crystallization is, it is first necessary to (*1*) identify a set of physical and chemical conditions that yield some kind of crystals, (*2*) optimize those conditions, and (*3*) reproducibly apply the conditions according to reliable procedures that will consistently produce the requisite crystals. Thus, the initial step consists of screening large numbers of possible precipitants, buffers, effectors, pHs, temperatures, etc. The second is a careful screening using a finely sampled matrix of conditions focused on those that yielded crystals, and the third is on devising reproducible procedures for application. Clearly, the first stage is the greatest hurdle that must be overcome, but it is not necessarily the step demanding the most effort. Table 4 is included here, not to alarm or intimidate the novice, but to alert those interested in macromolecular crystal growth of the complexities and the many factors that may have an impact on a crystal growth investigation. Fortunately, only a few of these factors may be simultaneously operative for a particular problem; the trick is to discover which they are for your specific case.

How does one begin the process, that is, how do you discover those initial conditions that yield crystals and provide starting points? Over the years, a vast number of reports and observations entered the literature, without much form or even clear purpose (46). Attempts were made to sort and classify (48,74); to shape ad hoc experiments into more rational approaches (26,32,34,55,75); to gather,

Table 4. Factors That Do or Could Affect Protein Crystal Growth

pH and buffer
Ionic strength
Temperature and temperature fluctuations
Concentration and nature of precipitant
Concentration of macromolecule
Purity of macromolecules (see Table 3)
Additives, effectors, and ligands
Organism source of macromolecule
Substrates, coenzymes, inhibitors
Reducing or oxidizing environment
Metal and other specific ions
Rate of equilibration and rate of growth
Surfactants or detergents
Gravity, convection, and sedimentation
Vibrations and sound
Volume of crystallization sample
Presence of amorphous or particulate material
Surfaces of crystallization vessels
Proteolysis
Contamination by microbes
Pressure
Electric and magnetic fields
Handling by investigator and cleanliness
Viscosity of mother liquor
Heterogeneous or epitaxial nucleating agents

collate, and systematize the experiences of thousands of investigators into rationally constructed databases (27,37); and finally to extract from those bases arrays of conditions most likely to yield protein and nucleic acid crystals.

These sets of conditions form the foundation for the crystal screens (57,76–82), now commercially available (Hampton Research), that have virtually revolutionized the initial search for crystallization conditions. Although some important macromolecules continue to slip through these still imperfect nets, the number of successes is impressive. Improved screening techniques alone are to a great extent responsible for the explosion of new crystals, grown not only by crystallographers, but by the myriad of biochemists and molecular biologists whose need for structure is acute. The search for novel, and perhaps even better, precipitants and conditions for crystal growth continues, and we can reasonably expect even more useful developments in this area.

With arrays of high-probability conditions that need be tested against no more than a few milligrams of protein or nucleic acid, attention has focused on the macromolecules themselves. Three principles have clearly come forward: purity, stability, and solubility. Failure in crystallization can probably be attributed, in general, to one or more of these.

Homogeneity has always been recognized as important, though not always essential. The trick is to recognize when it is the crucial factor and direct one's efforts there. Important advances have been made with regard to purification. Not only are there now more powerful techniques and instruments available (36,83), but crystallography has benefited to an extraordinary degree from the refinements of recombinant DNA technology and the development of high-yield expression systems (84–86). With the introduction of endogenous purification tools such as the histidine tag or glutathione sulfur transferase (GST) chimers (87), even greater availability of ultrapure crystallizable protein seems on the horizon. Recombinant systems provide us not only with pure, native proteins, but a way to introduce mutations and thereby create almost limitless alternatives should the native molecule prove unyielding. They offer systematic approaches as well to the intelligent engineering of solubility properties and stability.

Chemical and conformational stability of the target macromolecule, too, has long been recognized and given serious attention. Ligands are used to lock discrete conformations, disulfides introduced by mutation or chemical modifications are made to reduce mobility (88), proteins and ribosomes from thermophilic organisms are used for their natural stability, and proteolytic or mutational truncations are made to remove naturally flexible domains or elements (36). We are only beginning to understand how to manipulate conformation to enhance crystallization properties (89), and this area also promises much for the future, particularly for challenging topics such as antibodies, multiple domain enzymes, and multimolecular complexes.

Perhaps the principle least recognized but now receiving the most attention is solubility, meaning here not only dissolution of macromolecules in a liquid, but in a disaggregated, monodisperse molecular form. Indeed, we have come to recognize aggregation and polydispersity as a principal reason why many macromolecules cannot be crystallized. This includes not only membrane and lipophilic proteins (90) but macromolecules that have a strong propensity to aggregate, such as antibodies (91), hormone polypeptides, and proteins whose purpose is ordered aggregation, such as those from viral capsids.

Extending observations that nonionic detergents promote the growth of hydrophilic (92) as well as membrane proteins (90,93), light-scattering studies have confirmed the salient importance of monodispersity in macromolecular crystallization (50,53,78,94,95). The current consensus among those involved in crystallization research is that if by some means monodispersity can be ensured, then there is better than an 80% chance that crystals will be obtained. To promote monodispersity, the common approaches are to incorporate detergents in crystallization trials or to crystallize at low protein concentrations, at higher temperatures, or in the presence of chaotropic agents. This is an area that continues to be under active investigation.

Additional factors have emerged from recent crystal growth research and from the empirical results generated by the current revolution in crystallography. Among them are the importance of surfaces and their promotion of heterogeneous nucleation (96–100), the value of unique environments such as microgravity (101–103), gels (104–106), thin capillaries (107), the broad range of polymeric precipitants useful in growing crystals (108,109), and new approaches to solving the problem of persistent microcrystals (110).

BIBLIOGRAPHY

1. A.S. Myerson, *Handbook of Industrial Crystallization*, Butterworth-Heinemann Series in Chemical Engineering, Stoneham, Mass., 1993.
2. S. Takamatsu and D.D.Y. Ryu, *Biotechnol. Bioeng.* **32**, 184–192 (1988).
3. H. Takeda, T. Hidano, and S. Kinoshita, *J. Ferment. Bioeng.* **80**, 492–498 (1995).
4. M.J. Akers, and T. Miyakawa, *Pharm. Res.* **11**, S133, PT6012 (1994).
5. N. Inaba, S. Ayano, Y. Ozaki, M. Mitake, H. Maeda, and Y. Ifuku, *J. Jpn. Soc. Food Sci. Technol.* **40**, 833–840 (1993).
6. B. Cudney and S. Patel, *Am. Biotechnol. Lab.* **12**, 42 (1994).
7. M.D. Weiss, *Gen. Eng. News* **8**, 15 (1995).
8. M.M. Long, L.S. DeLucas, C. Smith, M. Carson, K. Moore, M.D. Harrington, D.J. Pillion, S.P. Bishop, W.M. Rosenblum, R.J. Naumann, A. Chait, J. Prahl, and C.E. Bugg, *Microgravity Sci. Technol.* **7**, 196–202 (1994).
9. M. Butler, *Biotechnology Series Animal Cell Technology: Principles and Products*. Taylor and Francis Inc., Philadelphia, Pa., Open University Press, 1991.
10. W. Schiweck and A. Skerra, *Proteins: Struct. Funct. Genet.* **23**, 561–565 (1995).
11. J. Fukumoto, M. Iwai, and Y. Tsujisaka, *J. Gen. Appl. Microbiol.* **9**, 353–361 (1963).
12. Y. Tsujisaka, M. Iwai, and Y. Tominaga, *Agr. Biol. Chem.* **37**, 1457–1464 (1973).

13. W. Buckel, U. Dorn, and R. Semmler, *Eur. J. Biochem.* **118**, 315–321 (1981).
14. M. Mack, K. Bendrat, O. Zelder, E. Eckel, D. Linder, and W. Buckel, *Eur. J. Biochem.* **226**, 41–51 (1994).
15. A. Kitazono, T. Yoshimoto, and D. Tsuru, *J. Bacteriol.* **174**, 7919–7925 (1992).
16. J.E. Pitts, J.M. Uusitalo, D. Mantafounis, P.G. Nugent, D.D. Quinn, P. Orprayoon, and M.E. Penttila, *J. Biotechnol.* **28**, 69–83 (1993).
17. T. Yoshimoto, T. Shimoda, A. Kitazono, T. Kabashima, K. Ito, and D. Tsuru, *J. Biochem.* **113**, 67–73 (1993).
18. H. Gros, *Abstracts of Papers, American Chemical Society* **21**, Biot 114 (1996).
19. M.H. Heng, N.T. Becker, E.L. Braunstein, R.C. Fewkes, and H. Gros, *Abstracts of Papers, American Chemical Society* **21**, Biot 177 (1996).
20. R.G. Harrison and L.P. Nelles, *16-Fermentations* **113**, 585 (1990).
21. A. McPherson, *Sci. Am.* **260**, 62–69 (1989).
22. A. McPherson, A.J. Malkin, and Y.G. Kuznetsov, *Structure* **3**, 759–768 (1995).
23. A.J. Malkin, Y.G. Kuznetsov, T.A. Land, J.J. DeYoreo, and A. McPherson, *Nat. Struct. Biol.* **2**, 956–959 (1995).
24. E.A. Wood, *Crystals and Light: An Introduction to Optica Crystallography*, 2nd ed., Dover, New York, 1977.
25. J.M. Cowley, *Diffraction Physics*, 2nd ed., North Holland, Amsterdam, 1984.
26. G.L. Gilliland, *J. Cryst. Growth* **90**, 51–59 (1988).
27. G.L. Gilliland, M. Tung, D.M. Blakeselee, and J.E. Ladner, *Acta. Crystallogr.* **D50**, 408–413 (1994).
28. M. Frey, *Acta Crystallogr.* **D50**, 663–666 (1994).
29. A.A. Chernov, *Modern Crystallography III: Crystal Growth*, Springer-Verlag, New York, 1984.
30. H.E. Buckley, *Crystal Growth*, Wiley, London, 1951.
31. W.K. Burton, N. Cabrera, and F.C. Frank, *Philos. Trans. R. Soc. London, Ser. A* **243**, 299 (1951).
32. A. McPherson, in D. Glick ed., *Methods of Biochemical Analysis*, vol. 23, Academic Press, New York, 1976, pp. 249–345.
33. A. McPherson, *The Preparation and Analysis of Protein Crystals*, Wiley, New York, 1982.
34. P.C. Weber, *Adv. Protein Chem.* **41**, 1–36 (1991).
35. B. Lorber and R. Giege, *Crystallization of Nucleic Acids and Proteins: A Practical Approach*, Oxford Univ. Press, Oxford, U.K., 1992.
36. R. Giege, B. Lorber, and A. Theobald-Dietrich, *Acta Crystallogr.* **D50**, 339–350 (1994).
37. A.M. Hassell, T.J. Harrocks, E.H. Dashman, and A. Mistry, *Acta. Crystallogr.* **D50**, 459–465 (1994).
38. J.H. Northrop, M. Kunitz, and R.M. Herriott, *Crystalline Enzymes*, Columbia Univ. Press, New York, 1948.
39. E.J. Cohn and J.D. Ferry, in E.J. Cohn and J.T. Edsall eds., *Proteins, Amino Acids, and Peptides*, Reinhold, New York, 1950.
40. R. Czok and T. Bucher, *Adv. Protein Chem.* **15**, 315–415 (1960).
41. F. Hofmeister and Z. Hoppe-Seylers, *Physiol. Chem.* **14**, 165 (1890).
42. S.N. Timasheff and T. Arakawa, *J. Cryst. Growth* **90**, 39–46 (1988).
43. D.H. Atha and K.C. Ingham, *J. Biol. Chem.* **256**, 12108–12117 (1981).
44. J.C. Lee and L.L.Y. Lee, *J. Biol. Chem.* **256**, 625–631 (1981).
45. T. Arakawa and S.N. Timasheff, *Methods Enzymol.* **114**, 49–77 (1985).
46. A. McPherson, *J. Cryst. Growth* **110**, 1–10 (1991).
47. K. Bailey, *Nature*, **145**, 934 (1940).
48. K. Bailey, *Trans. Faraday Soc.* **38**, 186 (1942).
49. J.B. Sumner and G.F. Somers, *The Enzymes*, Academic Press, New York, 1943.
50. A.J. Malkin, J. Cheung, and A. McPherson, *J. Cryst. Growth*, **126**, 544–554 (1993).
51. Z. Kam, A.B. Shore, and G. Feher, *J. Mol. Biol.* **23**, 539–555 (1978).
52. G. Feher and Z. Kam, *Methods Enzymol.* **114**, 77–112 (1985).
53. E.T. Baldwin, K.V. Crumly, and C.W. Carter, *Biophys. J.* **49**, 47–48 (1986).
54. S.D. Durbin and G. Feher, *J. Cryst. Growth* **76**, 383–392 (1986).
55. C.W. Carter, E.T. Baldwin, and L. Frick, *J. Cryst. Growth* **90**, 60–73 (1988).
56. L. Betts, L. Frick, R. Wolfenden, and C.W. Carter, *J. Biol. Chem.* **264**, 6737–6740 (1989).
57. W. Kadima, A. McPherson, M.F. Dunn, and F.A. Jurnak, *Biophysical J.* **57**, 125–132 (1990).
58. S. Koszelak and A. McPherson, *J. Cryst. Growth* **90**, 340–343 (1989).
59. S. Koszelak, D. Martin, J. Ng, and A. McPherson, *J. Cryst. Growth*, **10**, 177–181 (1991).
60. T.L. Blundell and L.N. Johnson, *Protein Crystallography*, Academic Press, New York, 1976.
61. A. McPherson, *Methods Enzymol.* **114**, 112–120, (1985).
62. A. McPherson, in H. Michel ed., *Crystallization of Membrane Proteins*. CRC Press, Boca Raton, Fla., 1991, pp. 1–52.
63. A. McPherson, *Eur. J. Biochem.* **189**, 249–345 (1990).
64. M. Zeppenzauer, H. Eklund, and E. Zeppenzauer, *Arch. Biochem. Biophys.* **126**, 564 (1968).
65. M. Zeppenzauer, *Methods Enzymol.* **22**, 253 (1971).
66. B.A. Weber and P.E. Goodkin, *Arch. Biochem. Biophys.* **141**, 489–498 (1970).
67. A. Hampel, M. Labanauskas, P.G. Conners, L. Kirkegard, U.L. RajBhandary, P.B. Sigler, and R.M. Bock, *Science* **162**, 1384 (1968).
68. D. Morris, C.Y. Kim, and A. McPherson, *Biotechniques* **7**, 522–527 (1989).
69. G.E. Rindone, *Materials Processing in the Reduced Gravity Environment of Space*, North-Holland, New York, 1982.
70. A.J. Malkin and A. McPherson, *J. Cryst. Growth* **133**, 29–37 (1993).
71. M. Kunitz, *J. Gen. Physiol.* **35**, 423 (1952).
72. W.B. Jakoby, *Anal. Biochem.* **26**, 295 (1968).
73. F. Rosenberger and E.J. Meehan, *J. Cryst. Growth* **90**, 74–78 (1988).
74. E.T. Reichert and A.P. Brown, *The Differentiation and Specificity of Corresponding Proteins and Other Vital Substances in Relation to Biological Classification and Organic Evolution: The Crystallography of Hemoglobins*, Carnegie Ins., Washington, D.C., 1909.
75. C.W. Carter, Jr. and C.W. Cater, *J. Biol. Chem.* **254**, 12219–12223 (1979).
76. A. McPherson, *J. Cryst. Growth* **122**, 161–167 (1992).
77. R. Cudney, S. Patel, K. Weisgraber, Y. Newhouse, and A. McPherson, *Acta Crystallogr.* **D50**, 414–423 (1994).

78. A. D'Arcy, *Acta Crystallogr.* **D50**, 469–471 (1994).
79. P.D. Shaw Stewart and M. Khimasia, *Acta Crystallogr.* **D50**, 441–442 (1994).
80. E.A. Stura, A.C. Satterthwait, J.C. Calvo, D.C. Kaslow, and I.A. Wilson, *Acta Crystallogr.* **D50**, 448–455 (1994).
81. L.J. Harris, E. Skaletsky, and A. McPherson, *Proteins: Struct., Funct., Genet.* **23**, 285–289 (1995).
82. J. Jancarik and S.H. Kim, *J. Appl. Crystallogr.* **24**, 409–411 (1991).
83. R.F. Boyer, *Modern Experimental Biochemistry*, Benjamin/ Cummings, New York, 1993.
84. B. Nilsson, G. Forsberg, T. Moks, M. Hartmanis, and M. Uhlén. *Curr. Opin. Struct. Biol.* **2**, 569–575 (1992).
85. H. Ago, N. Habuka, J. Kataoka, M. Furuno, H. Tsuge, M. Noma, M. Miyano, B.C. Wang, and N.H. Xuong, *Acta Crystallogr.* **D50**, 355–360 (1994).
86. A.R. Rees, M.J.E. Sternberg, and R. Wetzel, *Protein Engineering: A Practical Approach*, IRL Press, Oxford Univ. Press, New York, 1992.
87. G.G. Privé, G.E. Verner, C. Weitzman, K.H. Zen, D. Eisenberg, and H.R. Kaback, *Acta Crystallogr.* **D50**, 375–379 (1994).
88. I. Rayment, W.R. Ryppniewski, K. Schmidtbase, R. Smith, *Science* **261**, 50–58 (1993).
89. F. Dyda, A.B. Hickman, T.M. Jenkins, A. Engelman, R. Craigie, and D.R. Davies, *Science* **266**, 1981–1986 (1981).
90. H. Michel, in H. Michel ed., *Crystallization of Membrane Proteins*, CRC Press, Orlando, Fla., 1990, pp. 73–88.
91. L.J. Harris, S.B. Larson, K.W. Hassel, J. Day, A. Greenwood, and A. McPherson, *Nature* **360**, 369–372 (1992).
92. A. McPherson, S. Koszelak, H. Axelrod, J. Day, R. Williams, M. McGrath, L. Robinson, and D. Cascio, *J. Biol. Chem.* **261**, 1969–1975 (1986).
93. R.M. Garavito, in H. Michel ed., *Crystallization of Membrane Proteins*, CRC Press, Orlando, Fla., 1990, pp. 89–106.
94. V. Mikol, E. Hirsh, and R. Giege, *J. Mol. Biol.* **213**, 187–195 (1990).
95. S. Veesler, S. Marcq, S. Lafont, J.P. Astier, and R. Boistelle, *Acta Crystallogr.* **D50**, 355–360 (1994).
96. W.J. Ray, Jr. C.E. Bracker, *J. Cryst. Growth* **76**, 562–576 (1986).
97. A. McPherson and P. Shlichta, *Science* **239**, 385–387 (1988).
98. C.J. Leung, B.T. Nail, and G.D. Brayer, *J. Mol. Biol.* **206**, 783 (1989).
99. S.P. Wood, R.W. Janes, E. Sweeney, and R.A. Palmer, *J. Cryst. Growth* **122**, 204–207 (1992).
100. W. Littke and C. John, *Science* **225**, 203–204 (1984).
101. L.J. DeLucas, C.D. Smith, H.W. Smith, V.K. Senagdi, S.E. Senadhi, S.E. Ealick, C.E. Bugg, D.C. Carter, R.S. Snyder, P.C. Weber, F.R. Salemme, D.H. Ohlendorf, H.M. Einspahr, L. Clancy, M.A. Navia, B.M. Mckeever, T.L. Nagabhushan, G. Nelson, Y.S. Babu, A. McPherson, S. Koszelak, D. Stammers, K. Powell, and G. Darby, *Science* **246**, 651–654 (1989).
102. J. Day and A. McPherson, *Protein Sci.* **1**, 254–1268 (1992).
103. S. Koszelak, J. Day, C. Leja, R. Cudney, and A. McPherson, *Biophysical J.* **69**, 13–19 (1995).
104. S. Patel, R. Cudney, and A. McPherson, *Acta. Crystallogr.* **D50**, 479–483 (1994).
105. M.C. Robert, Y. Bernard, and F. Lefaucheux, *Acta Crystallogr.* **D50**, 496–503 (1994).
106. K.J. Thiessen, *Acta Crystallogr.* **D50**, 491–495 (1994).
107. J.M. Garcia-Ruiz and A. Moreno, *Acta Crystallogr.* **D50**, 484–490 (1994).
108. A.M. Brzozowski and S.P. Tolley, *Acta Crystallogr.* **D50**, 466–468 (1994).
109. S. Patel, B. Cudney,and A. McPherson, *Biochem. Biophys. Res. Comm.* **207**, 819–828 (1995).
110. A. McPherson, *J. Appl. Crystallogr.* **28**, 362–365 (1995).

CRYSTALLIZATION, PROTEINS, KINETICS

Michael R. Johns
University of Queensland
Brisbane, Australia

KEY WORDS

Crystal growth kinetics
Nucleation
Osmotic second virial coefficient
Precipitation
Protein crystallization
Protein solubility
Purification

OUTLINE

INTRODUCTION

Crystallization is a mass transfer separation process commonly used in many chemical industries. It permits highly selective purification of a material from an impure mixture, and the crystalline state of the final product is frequently advantageous. Whereas bulk crystallization has been applied for the purification of many small organic biological products, including antibiotics and amino and organic acids, and has played an important role in chiral resolutions of organic materials, until recently it has been applied only rarely to the bulk purification of proteins. It is well known that proteins, despite their large size, can be crystallized. Nevertheless, a survey of the vast protein crystallization literature, almost entirely devoted to producing crystals for X-ray studies, reveals that protein crystallization has been perceived generally to be tedious, difficult, and disrupted by the presence of impurities. These perceptions have perhaps arisen because of the unique requirements of protein X-ray crystallography, which remains the primary method for the determination of protein structure. The requirements for X-ray crystallography and bulk protein crystallization are contrasted in Table 1. The significant differences highlight the fact that many of the well-known issues that bedevil the production of protein crystals for X-ray crystallography are not germane to the bulk crystallization of proteins for manufacturing uses. In fact, the bulk crystallization of many proteins is found to be reasonably straightforward, similar in nature to the crystallization of inorganic molecules (1), and capable of achieving high degrees of purification at good yield and large throughput. This article will focus on existing knowledge relevant to the bulk crystallization of proteins from solution and largely bypass the more specialized field of protein crystallization for X-ray crystallography, for which excellent reviews already exist (2,3). Useful general reference books on bulk crystallization of materials include those by Randolph and Larson (4) and Mullin (5).

THE PROTEIN CRYSTAL

In addition to the beauty of the crystalline form (Fig. 1), protein crystals share many features in common with smaller inorganic and organic molecule crystals. They consist of highly ordered molecular arrays with distinct crystal faces. Proteins usually exhibit polymorphism, leading to several crystallographic forms, depending on various factors including the nature of the precipitant or the temperature (Table 2). Obtaining the most appropriate crystal form is often important for process manufacture, because "chunkier" crystal forms are easier to wash, store, and transport without excessive crystal damage and fragmentation. Needle forms, which are difficult to process, occur commonly in industrial practice. Transition boundaries between different polymorphic forms must be established experimentally and may be affected by many factors, including temperature, pH, and precipitant type and concentration. A common example is the tetragonal-orthorhombic transition for lysozyme, which occurs at about 25 °C (6). Further, the crystals may undergo significant habit change in the presence of impurities or under varying conditions (7) and exhibit dendritic morphologies during very high crystal growth rates.

A significant feature of all protein crystals is their large degree of incorporated solvent, which is typically 50% v/v (8), but varies generally between 30 and 80%. This has the practical result of making protein crystals unstable when exposed to air, resulting in their rapid collapse to an amorphous state. Consequently, protein crystals are usually maintained as a liquid crystal slurry.

SOLUBILITY

Crystallization processes depend on the production of a supersaturated solution that leads to the formation of the crystalline solid phase, consisting primarily of solute. Consequently, an accurate protein solubility data set determined for the conditions chosen for the crystallization process is essential. Two processes are fundamental to crystallization: nucleation and subsequent crystal growth by addition of additional solute to the nucleated crystal. The extent of supersaturation is the primary determinant of both the yield of crystalline solute from the solution and the rate of crystal growth.

Nucleation

The influence of solubility on the crystallization process is illustrated by Figure 2. Under any set of conditions, there is a unique protein solubility given by the solubility curve. This curve represents the thermodynamic boundary at which there is stable coexistence of solid and dissolved protein forms. Below the solubility line, solution conditions thermodynamically favor the maintenance of a single, homogeneous liquid phase, and protein crystals added to this solution will dissolve. Solution conditions that lie above the solubility curve represent supersaturated conditions under which the formation of a protein-rich solid phase and a saturated liquid phase are favored. The solubility data represent the lower limit of the crystallization process in that once the protein concentration in solution declines to this concentration, the system is at equilibrium and no further crystallization is possible.

Metastable Region. The supersaturation region can be further divided into a metastable region and a nucleation

Table 1. The Conflicting Crystallization Needs of X-ray Crystallographers and Bulk Protein Manufacturers

	X-ray crystallography	Bulk manufacture
Aim of crystallization	Protein structure determination	Protein purification
Protein used	Usually novel	Usually well characterized
Protein quantity available	Usually limited (mg)	Large (grams or more)
Number of crystals desired	One	Unlimited
Crystal size	Generally >300 μm	25 μm or more
Lattice order	Very highly ordered	Ordered
Solution purity	Preferably free of impurities and protein heterogeneity	Usually contains many impurities

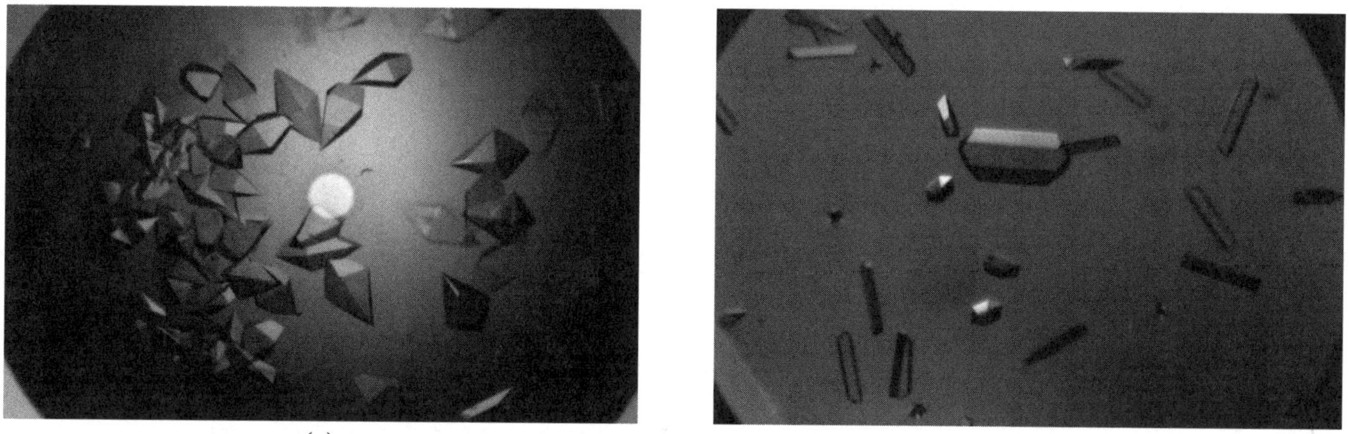

(a) (b)

Figure 1. (**a**) Crystals of the commercial sweetener protein, thaumatin, produced from 0.5 M sodium/potassium tartrate. *Source*: Courtesy of W.W. Wilson and A. Arabshahi, Mississippi State Univ. (**b**) Crystals of subtilisin. *Source*: Courtesy of Richard Bott, Genencor International, Inc.

Table 2. Polymorphic Forms of Proteins

Protein	Crystal form	Condition	Reference
Lysozyme	Tetragonal, orthorhombic	Temperature	6
Lysozyme	Monoclinic; triclinic; tetragonal	Nature of salt	9
Cytochrome c_2	Hexagonal; triclinic; tetragonal	Ionic strength, pH	10
Ovalbumin	Monoclinic; triclinic	pH	11

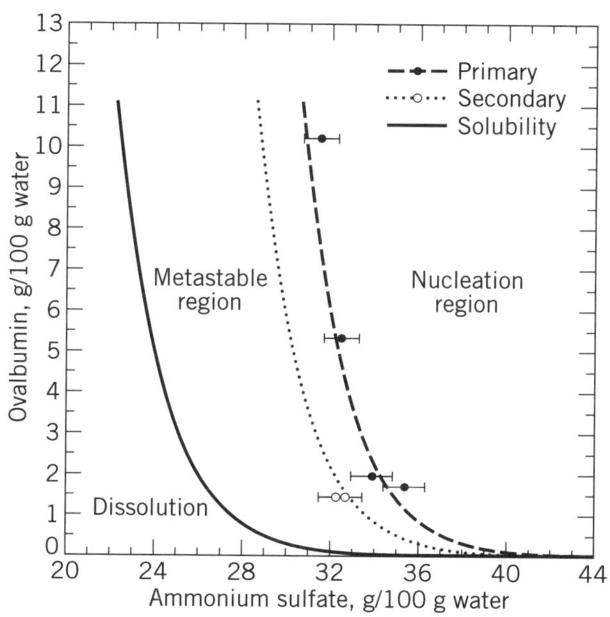

Figure 2. Solubility and nucleation data for ovalbumin at pH 5.2 and 30 °C (11).

region. The metastable region represents a region of low supersaturation in which nucleation does not occur spontaneously within times useful for crystallization processes, despite the existence of supersaturation. However, if crystals are already present, crystal growth will occur. This is the preferred region for the operation of industrial crystallization processes, since it provides the greatest degree of control over the crystal number in the crystallizer and, hence, crystal size distribution and quality. The width of the metastable region is of great significance for crystallization processes and appears to be reasonable generous for proteins (Table 3), although the available data are extremely limited.

When crystallization processes are operated in the metastable region, the crystallization process must be initiated by providing crystal nuclei to the crystallizer (seeding), so that crystallization can begin. The number, mean size, and size distribution of these seeds are variables that permit the quality of the final crystal product and the process crystallization time to be manipulated (Fig. 3).

Nucleation Region. In some crystallization processes, seeding is not preferred, usually because of requirements

Table 3. Typical Size and Metastable Regions for Various Molecular Species

Molecular category	Size of metastable region (c/c*)
Ionic, inorganic salts	≅1.1 or less
Small organics (e.g., sugars)	≅5–30
Proteins	≅1–20
Large proteins (MW > 200 kDa)	≅1.1 or less

Note: c, concentration; c*, concentration at solubility limit under same conditions.

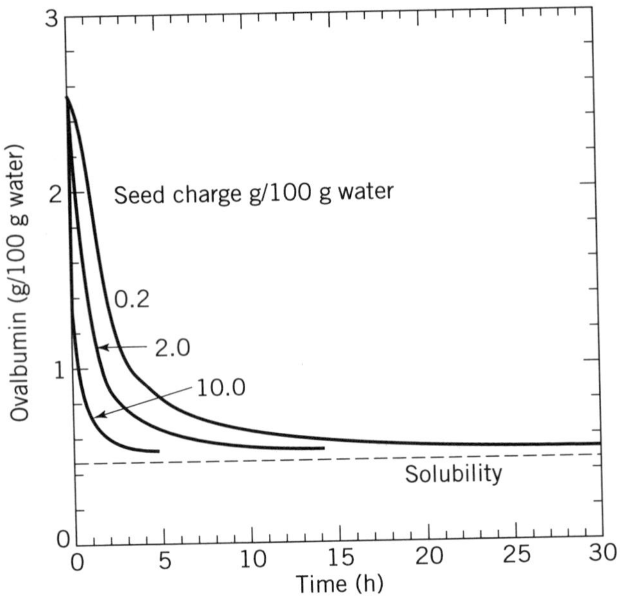

Figure 3. The effect of initial crystal seed charge on the rate of crystallization (given as decrease in ovalbumin concentration in solution) during ovalbumin crystallization from aqueous ammonium sulfate (27% w/v) at pH 4.9 and 30 °C (11).

for aseptic operation (e.g., pharmaceutical preparations) or when a very small (<30 μm diameter) final mean crystal size is desired. In these cases, batch crystallization processes must be initiated in the nucleation region. In this region of higher supersaturation, spontaneous production of large numbers of tiny crystal nuclei will occur after an initial induction time. The control of such processes can be difficult, because rates of crystal nucleation and growth are usually very high at the large supersaturation typical of this region. The nucleation region can be further subdivided into a primary nucleation region, in which nucleation will occur spontaneously from the supersaturated solution (heterogeneous nucleation if active foreign nuclei, such as dust, are present, or homogeneous nucleation if they are not) or secondary region (at lower supersaturation) if crystals are already present. In industrial crystallization, secondary nucleation predominates. However, the threshold between these regions is fuzzy.

Precipitation. At very high supersaturation, conditions will favor the production of a disordered solid phase (precipitation), which usually forms rapidly on nucleation. Such operation generates amorphous protein precipitates, which lack the purity and structural properties of protein crystals. Precipitation processes have been widely used for protein recovery and are covered elsewhere in this encyclopedia. Recent dynamic light scattering (DLS) studies of supersaturated protein solutions have suggested that there is a significant difference in the manner in which homogeneous nucleation occurs for crystallization and precipitation processes (12).

Factors Affecting Protein Solubility

Proteins vary widely in their solubility in aqueous solutions. Most commercially manufactured commodity proteins exhibit high solubilities (40 to 100 mg/mL) in aqueous buffers, but more esoteric proteins can have unusual behavior. For example, some proteins (prolamins) are more soluble in aqueous ethanol than in water (13). Accurate measurement of protein solubility is required for proper design of crystallization processes. This is frequently a tedious process, made difficult by the challenge of obtaining sufficient quantities of pure protein material and the common contamination of the material with heterogeneous forms of the protein, whose effect on the solubility is unclear. For real crystallization processes, the effect of impurities present, often in significant concentration, on protein solubility must not be overlooked. Issues and protocols for protein solubility determination have been discussed by Ries-Kautt and Ducruix (9). Recently, a useful apparatus for the rapid measurement of protein solubility (Fig. 4) using only small quantities of protein has been developed in which solutions near the solubility limit are equilibrated with small volume crystal beds of the protein (14). Protein solubility is affected by many factors, and the major factors are discussed below.

Solution pH. The pH of solution usually has a marked influence on protein solubility. The solubility often exhibits a minimum at the isoelectric point (IEP), with a sharp increase in solubility within small pH changes either side of the IEP (Fig. 5). In concentrated protein solutions, typical of those necessary for protein crystallization, this generalization is fallible. For example, glucose isomerase has an IEP of 4.0, but has minimum solubility in ammonium sulfate at about pH 7 (15).

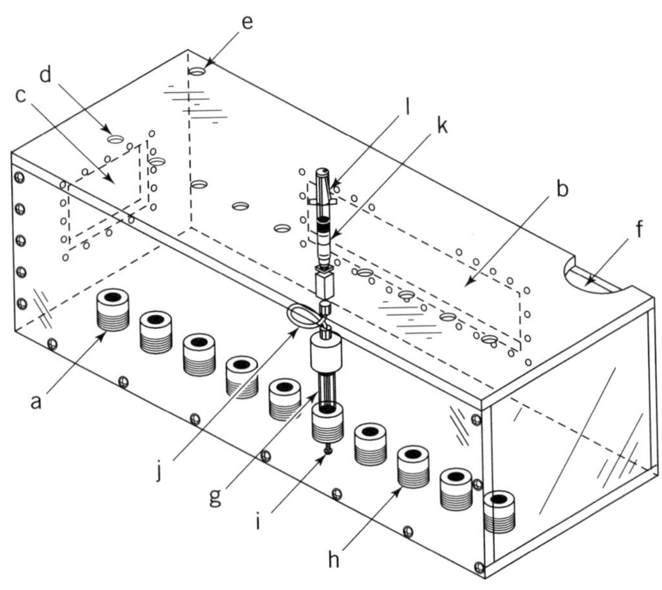

Figure 4. A temperature-controlled apparatus for the rapid determination of protein solubility using small amounts of material. a, column mounts; b, cooling slot; c, circulation pump slot; d, holes for reservoir access; e, temperature probe access; f, heater port; g, crystal column holder; h, base column mount; i, eluent tube; j, tube from column to reservoir; k, reservoir syringe; l, syringe pump. *Source*: Reprinted from Ref. 14 with kind permission of Elsevier Science-NL, Amsterdam.

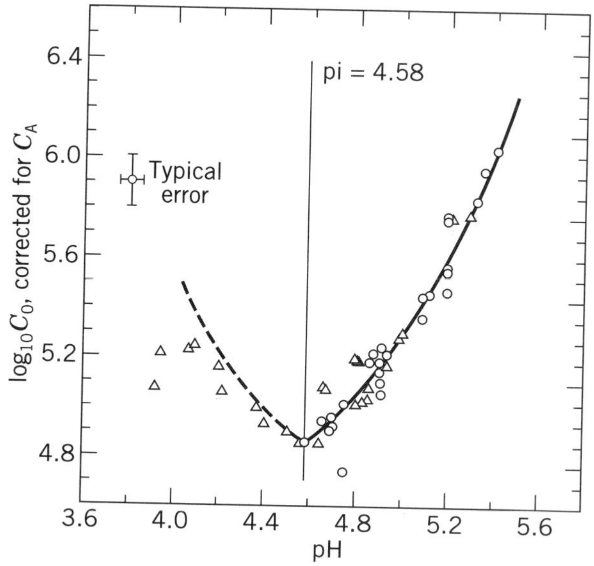

Figure 5. The effect of pH on ovalbumin solubility (C_o) near the isoelectric (pI) point at 18 and 30 °C. *Source*: Reprinted with permission from Ref. 51. Copyright © 1996 American Chemical Society.

Temperature. Proteins exhibit a wide range of solubility responses to temperature. The solubility of some proteins is largely unaffected by temperature; others, such as glucose isomerase (15) and lysozyme (16), show marked increase with temperature; and others exhibit a retrograde response. The effect of temperature and salt concentration on lysozyme solubility is presented in Figure 6. The two common crystal forms of lysozyme have been noted to have a markedly different solubility response to temperature (6).

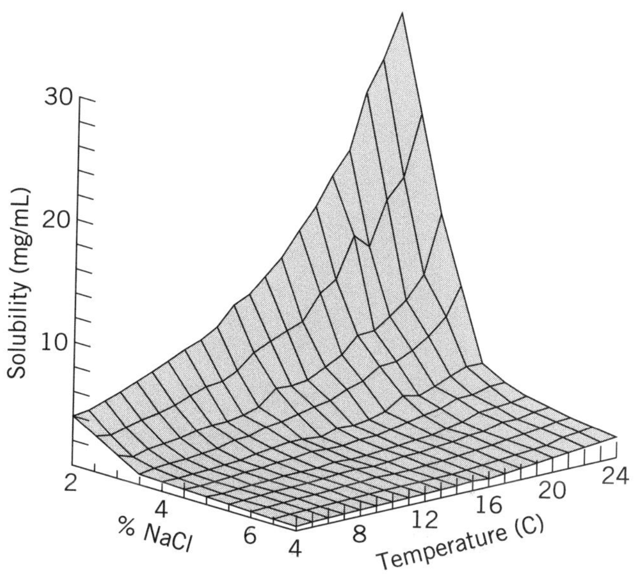

Figure 6. The solubility of tetragonal lysozyme as a function of salt concentration and temperature at pH 4.6 in 0.1 M sodium acetate. *Source*: Reprinted from Ref. 16 with kind permission of Elsevier Science-NL, Amsterdam.

Ionic Strength. Increasing ionic strength decreases globular protein solubility, although diverse effects occur, including regions of low ionic strength ($I < 0.15$ M) at which salting-in phenomena are commonly observed. In some cases, a solubility minimum with concentration has been observed for some protein–salt combinations, particularly for divalent salts (15,17,18). Protein solubility during precipitation processes has often been analyzed by the Cohn plot in which protein solubility is graphed as a logarithmic function of ionic strength. It has been frequently noted, however, that protein solubility in crystal suspensions is lower than for amorphous precipitate suspensions and that the Cohn relationship exhibits poor fit for the former (6,19).

Salt Type. The varying ability of different salts to alter protein solubility is well known and is described by the Hofmeister series (lyotropic series), which ranks ions according to their salting-in or salting-out effect. However, care is needed in applying the series, as shown for lysozyme (20).

Nonelectrolytes. A wide variety of nonelectrolytes can influence protein solubility, usually in a complex manner. Organic solvents (ethanol, acetone), polyethylene glycol (PEG), sugars, and even amino acids have been used to salt-out proteins. Where waste disposal of inorganic salt-rich mother liquors is difficult or expensive, nonelectrolyte nonsolvents may be a good alternative. The complex behavior of these systems is illustrated well by ethanol, which salts-out at low temperatures and low concentrations, but salts-in at all concentrations as temperature rises. These effects do not depend on the dielectric constant (polarity) of the solvent, but on the unusual properties of ethanol and water mixtures (13).

PREDICTING CONDITIONS FOR CRYSTALLIZATION

Obtaining the appropriate conditions for protein crystallization is more challenging than for many small organic and inorganic molecules, because relatively little is known about protein solubilities, often only small quantities of the protein are available early in process development, protein crystallization even under optimal conditions can be slow (days, months), and the use of a precipitant, rather than temperature, to create supersaturation adds an extra degree of freedom in locating conditions for good crystallization. Because X-ray crystallography has until recently been the sole method for protein structure determination, immense effort has been put into the development of protocols for quickly finding suitable conditions for protein crystallization at the microscale. Typical techniques used are well known and include dialysis, vapor diffusion, and batch crystallization methods. Various statistical search strategies, such as sparse matrix screens using commercially available solution kits (21), and robotic technology for repetitive microscale tests (22) are used to reduce the amount of material and time required and optimize the likelihood of finding a suitable set of conditions.

This body of work exists as an accessible database (23) and serves as an excellent resource for quickly narrowing down suitable systems for the bulk crystallization of proteins. Unfortunately, the transient solution environment typical of microscale systems means that additional tests are required to define crystallization conditions exactly. Process engineering issues can then be applied as an additional set of criteria. Precipitants used for the crystallization will need to be available in bulk at low cost and, for some sites, waste disposal concerns may impose constraints on the use of concentrated salt solutions. In this vein, the use of biodegradable alternative precipitants has been investigated.

Both DLS (12,17) and fluorescence spectroscopy techniques (24,25) have been applied recently to attempts to differentiate conditions leading to precipitation, or crystallization. Although changes in solute structure are seen, the predictive ability of these techniques remains elusive. An alternative approach for predicting optimal solution conditions for protein crystallization is to measure the osmotic second virial coefficient (B_{22}) of the protein in the crystallization mixture, but at dilute (<8 g/L) protein concentrations (26). It was discovered that for the proteins tested (Table 4) there exists a crystallization slot characterized by B_{22} values in the range of -1 to -8×10^{-4} mol/mL/g^2. At more negative values, the solution was found to promote the formation of amorphous protein aggregates, whereas at positive values, the protein preferred to remain in solution. Although independent experimental validation of this approach has not yet been published, subsequent work by Wilson to extend the method to more proteins and by others investigating the application of lattice model (8) or colloidal theory (27) to describe protein interaction in concentrated solutions has supported the validity of this approach. Measurements of B_{22} may provide a faster approach to optimal conditions than do microscale tests, which require several days for the result to be known.

Table 4. Values of B_{22} Corresponding to Crystallization Conditions

Protein	Precipitant and pH conditions[a]	$B_{22} \times 10^{-4}$ mol/mL/g^2
Lysozyme	2% NaCl, pH 4.6	−3.0
Canavalin	0.7% NaCl, pH 7	−0.8
Concanavalin A	1 M $(NH_4)_2SO_4$, pH 7	−2.5
Concanavalin A	0.1 M NaCl, pH 6	−1.9
BSA	52% satd $(NH_4)_2SO_4$, pH 6.2	−2.0
Ovostatin	7.5% PEG 8000, pH 7.5, 20°C	−7.1
Ribonuclease A	50% n-propanol, pH 5, 24°C	−4.1
α-Chymotrypsin	10% PEG 3350, pH 4.6	−8.4
STMV	12.5% SAS, pH 6.5	−1.8
Ovalbumin	43% SAS, pH 5.4	−6.1

Note: BSA = bovine serum albumin; SAS = saturated ammonium sulfate. [a] −298 K unless stated otherwise; original buffer conditions given in reference. *Source*: Ref. 26.

KINETIC STUDIES

The knowledge of rates of protein crystal growth is of primary importance in industrial crystallization processes, but except for lysozyme, few systematic studies have been published. Most kinetic studies have been performed using static crystal measurements, the most common method of which involves nucleating several crystals in a growth cell located on a microscope and following the increase in crystal dimension with time as supersaturated solution is pumped gently around the crystals (28,29). These are useful for fundamental studies of growth mechanisms, but have a number of disadvantages for process design:

- The growth rate is typically expressed as a face (e.g., two-dimensional increase) growth rate whose applicability to three-dimensional crystal growth in suspended crystal slurries in industrial crystallizers, in which the hydrodynamic conditions may be vastly different, has not been validated.
- Because crystals grow at different rates, a large number of crystals must be followed to obtain a reasonable approximation of mean population growth rates.

Nevertheless, the method has benefits in that data can be obtained on individual crystals. Few kinetic studies of protein crystallization have been performed in bulk crystallizers in which the growth of a large crystal population can be measured under conditions more typical of industrial conditions (11,30).

Factors Affecting Rates of Protein Crystal Growth

Supersaturation. Chief among the factors determining the rate of crystal growth is the degree of supersaturation, which defines the driving force for crystallization. Supersaturation (s) can be defined a number of ways and is typically defined in the manner most convenient for the user. The relative supersation (σ) = s/c^* is commonly used. In industrial practice, it is convenient to express the relationship between the growth rate (G, μm/h) and supersaturation (s) as a power law:

$$G = ks^{\mathrm{n}}$$

where k is the nth-order growth rate constant and n is the kinetics order. From the studies performed on protein growth kinetics, a range of dependencies of growth rate on supersaturation have been reported, but most have revealed a second-order dependence (Table 5). This may be interpreted as surface integration kinetics control of the rate of crystal growth. In isolated cases, a first-order dependence was measured, for example, for concanavalin A (Table 5), which is more indicative of a diffusion-controlled growth mechanism. The growth rates of protein crystals and their dependence on supersaturation are contrasted in Figure 7. Protein crystal growth rate appears to be independent of crystal size (31,32).

In the case of lysozyme, which has been extensively studied, the usual growth rates experienced in static cell studies suggest that about two to three layers of molecule are added to the crystal face every second, amounting to a

Table 5. The Relationship between Protein Growth Rate and Supersaturation

Protein	Kinetics order, n
Concanavalin A[a], porcine pancreatic α-amylase[b]	1
Insulin[c], ovalbumin[d], canavalin[e], lysozyme[f]	2

[a]Ref. 7.
[b]Ref. 33.
[c]Ref. 30.
[d]Ref. 31.
[e]Recalculated from Ref. 28.
[f]Ref. 29.
Source: Ref. 31.

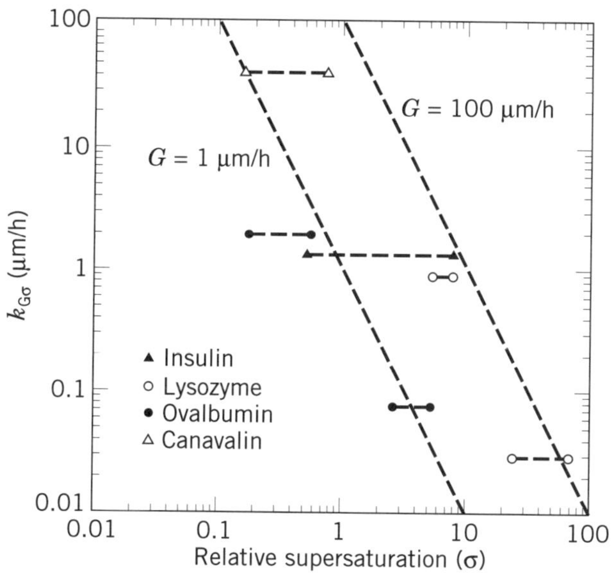

Figure 7. Values of the second-order growth rate constant, $k_{G\sigma}$, against relative supersaturation for various proteins. Most average crystal growth rates, G (expressed on a volume equivalent diameter basis), fall between 1 and 100 μm/h over the range of supersaturation commonly used.

face growth rate of 0.1 to 1 nm/s (0.4–4 μm/h) (34). Recent work has proposed that protein crystal growth occurs by the addition of oligomers of lysozyme into the crystal, rather than by monomer incorporation (35). There is a large body of supporting evidence for this view, both with lysozyme and for other proteins (1,7), although this finding remains controversial (36).

A comparison of protein growth rates with those of other commonly crystallized materials reveals that proteins are slower to crystallize relative to most smaller molecules for a given relative supersaturation (Fig. 8). Nevertheless, known protein crystal growth kinetics are comparable to those of aluminum trihydroxide, which is crystallized industrially on a very large scale. Furthermore, the metastable zone available for protein crystallization (e.g., the values of supersaturation in which it is convenient to operate a crystallizer) is many fold more than that possible in many small molecule crystallizations (Table 3).

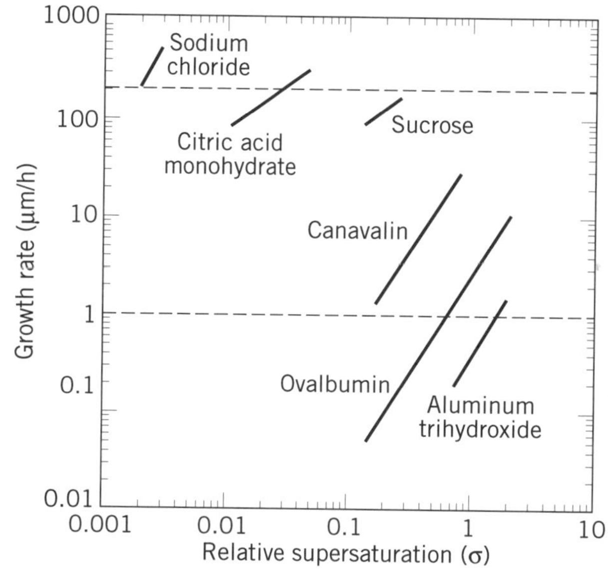

Figure 8. Comparison of average crystal growth rates for various values of supersaturation for industrially crystallized small molecules and the proteins ovalbumin and canavalin.

Most commercial protein crystallization processes are batch operations. During isothermal batch crystallizations, the supersaturation will fall rapidly from an initially high value to much lower values, especially because supersaturation is typically achieved by the addition of precipitant. In these cases, it may be uneconomical to wait for the system to reach equilibrium before terminating the crystallization. Although low supersaturation is undesirable because it will result in long crystallization times, excessive supersaturation is also undesirable and generally results in any of a number of inconvenient features, including greater inclusion of impurities in the crystal lattice, unusual and dendritic crystal forms that are hard to process (Fig. 9), and amorphous precipitation. Where the solubility of the protein being crystallized exhibits sensitivity to pH, temperature, or other factors, the supersatu-

Figure 9. Rosette of α-amylase crystals. *Source*: Courtesy of Richard Bott, Genencor International, Inc.

ration may be maintained at a moderate value by changing these factors as crystallization occurs, provided that there are no deleterious results such as decreased rate, change in crystal form, denaturation, etc. This can only be determined by thorough solubility studies.

Temperature and pH. The effect of temperature and pH on the kinetics of protein crystallization have been little studied. There are frequent comments in the literature that different pH and temperature conditions, or even changes in the primary structure of the protein, altered the rate of crystallization or made crystallization easier. In most cases, there are insufficient data to confirm that the experimental variable (pH, temp, etc.) had a direct influence on the rate of crystallization, rather than an indirect influence through some other effect, such as an altered protein solubility, which in turn would result in increased supersaturation for the same initial protein concentration and hence a faster crystallization rate.

The pH has been observed to have a marked effect on the crystallization rate, independent of its effect on solubility, in the case of ovalbumin (11). An increase in pH at values up to 1 pH unit higher than the isoelectric point (pH 4.6) resulted in an increase in the rate of crystallization of up to 10-fold. Whether this trend continued at larger deviations from the IEP was not studied. Reports on the effect of temperature are rare, but the probable dependence of the rate of protein crystallization on kinetic mechanisms implies that the rate should increase with temperature at a given supersaturation. Although the results above are too limited to generalize to all commercially produced proteins, it is interesting to note that conditions that have been observed to enhance crystallization rate (increased temperature, pH values away from the IEP) also correspond to conditions of greater protein solubility. Consequently, a compromise is necessary between high crystallization rates and the need for high recovery of protein as crystal.

Impurities. The presence of some impurities in a crystallizing solution is known to result in depressed rates of crystal growth for many molecules. In most cases, such poisoning is difficult to prove unequivocally. The impurities likely to be present in protein solutions undergoing crystallization can be arbitrarily divided into three categories:

- Nonprotein impurities
- Protein contaminants not derived from the product protein
- Heterogeneous forms of the product protein

The term *pure* typically means crystals devoid of the latter two categories, because large quantities of the precipitant inevitably will be present in the solvent channels within the protein crystal. In fact, it is exactly this property that makes protein crystals amenable to X-ray crystallography.

Nonprotein Impurities. The rate of protein crystallization appears to be remarkably unaffected by the presence of small compounds, such as salts, precipitants, sugars, and pigments, compared to small molecule crystallization processes (1), although they may have a pronounced effect on solubility of the protein (15). This resilience may be because of the comparatively small number of contact points between adjacent protein molecules in the crystal, which makes disruption of the ordered lattice by small molecules more unlikely compared to the tightly packed lattices of salts and other small molecules. This is convenient for protein bioprocessing. However, polysaccharides derived from microbial cell wall or components of the media are commonly present in fermentation broths and have been implicated as causing difficulties in bulk crystallization processes, particularly affecting the crystal morphology and properties of the final product (37). This problem was resolved by the use of carbohydrate-hydrolyzing enzymes.

Contaminating Proteins. The presence of contaminating proteins has often been considered a primary reason for not adopting crystallization as a protein purification process. In fact, in the few cases studied to date, the presence of contaminating proteins had little or no deleterious effect on the rate of crystallization. The rate of crystallization of ovalbumin was unaffected by the presence of up to 0.22-mol fraction of total protein as lysozyme and conalbumin (11); similarly, the rate of lysozyme crystallization was unaffected by the presence of up to 5% w/w ovalbumin (36). However, the literature on this issue is scarce.

Heterogeneous Forms of the Product Protein. An important issue is the effect of microheterogeneous forms of the product protein on rates of crystallization. There is mounting evidence that even small changes in the primary sequence of a polypeptide may lead to large alterations in its crystallizability. Unfortunately, the term *crystallizability* is vague and may refer to an improved propensity for nucleation, altered solubility under the conditions used for crystallization, or an actual enhancement in the rate of growth of the crystal face for a given supersaturation. Discerning between these effects requires extensive characterization of protein behavior, which typically is not performed. Nevertheless, numerous reports of protein variants that have either improved crystallizability or that inhibit crystallization of the preferred protein exist. Thus, variants of recombinant human insulin were generated by site-directed mutagenesis that exhibited improved crystallizability under physiological conditions (38), whereas for ovalbumin, the subtilisin-treated variant, plakalbumin (which possesses 6 fewer amino acid residues out of the original 385), is reported to exhibit greatly improved crystallizability (39). This opens the possibility of engineering protein variants with improved crystallization characteristics to improve properties germane to their final use or to reduce processing cost. Some forms of heterogeneity, however, may also be responsible for poisoning rates of protein crystallization. This has been reported for lysozyme in mixtures of that protein from different sources (40). This adverse effect of variant forms is more likely, because they would share a high degree of homology with the native protein and might be expected to integrate into the lattice in such a way as to poison or disrupt further growth. The ability of crystallization process to select for crystals of only one microheterogeneous form of a protein is unclear.

Other Factors. The effect of a number of other operating variables on the rate of protein crystallization have not been investigated. It would be expected that the precipitant used, or its concentration, would alter the rate of crystallization. However, data are scarce. Certainly, the effect of salt concentration (ammonium sulfate) on the rate of ovalbumin crystallization was negligible, independent of its effect on protein solubility, which is large (11). Agitation, or a convective environment, has been argued to have a deleterious effect on crystal growth in small-scale studies performed on microscope cells (29), although other studies have contradicted these findings (41,42). Nevertheless, some agitation is essential in industrial crystallizers to prevent the crystals settling in the vessel. A more serious effect of agitation is crystal breakage. Protein crystals appear to be relatively fragile, probably because of the small density of intermolecular contact points between neighboring molecules in the crystal lattice. Consequently, rough handling of the crystals should be minimized. High pressure (up to 3,000 bar) was reported to increase the crystallization rate of glucose isomerase relative to that of solutions at atmospheric pressure (43); however, it is more likely that the enhanced rates observed were the result of higher rates of nucleation at high pressure, which would generate larger crystal surface areas for growth (44). This latter effect would be advantageous for the production of large numbers of small (<10 μm) protein crystals in the absence of seeding, because induction times for nucleation were also reduced by high pressure.

Growth Dispersion

Growth rate dispersion (GRD) refers to the existence of a population of crystal growth rates under identical crystallization conditions. The primary cause of GRD remains unclear, but it is now accepted that crystals, under conditions of constant supersaturation, will grow at different, but constant, growth rates. The most significant effect of GRD is to broaden the crystal size distribution with time, which is generally undesirable for most industrial applications. The existence of GRD in protein crystallization is largely unstudied, but appears to be small relative to crystallizations that exhibit the effect strongly, such as sugars.

PURIFICATION BY CRYSTALLIZATION

Crystallization is capable of highly selective purification of most compounds from impure mixtures. Protein crystals sought for X-ray crystallography need to be highly pure, because the crystal is required for structure determination. This requirement has commonly led to the comment that protein solutions must be highly pure to attain crystallization. However, for bulk crystallization purposes, this requirement is not relevant. Because of the highly open structure of protein crystals, small impurities, such as salts, sugars, etc., are likely to be present in the final crystals to a similar degree as in solution, although protein crystals appear to exclude the pigments and colorants found routinely in fermentation broths (37). Successful protein crystallization has been routinely obtained in the presence of other contaminating proteins with degrees of purification similar to those achieved by ion exchange. Most of the carryover of contaminating protein is in the liquid adhering to the exterior surface of the crystals. Studies showed that ovalbumin crystals (MW 45,000 Da) produced in bulk crystallization were substantially free of both smaller and larger protein contaminants, which comprised 22 mol % of the total protein initially present in solution (11), after a single wash of the crystals (Fig. 10). Similar results have been observed for lysozyme (36).

PROCESS APPLICATIONS OF PROTEIN CRYSTALLIZATION

Despite the ubiquity of crystallization for the structural determination of proteins, actual process applications of crystallization as a technology for the purification and recovery of proteins remained relatively rare until the last few years. An increasing number of proteins are now produced in multikilogram or multiton quantities by crystallization from clarified fermentation broth. These include insulin, glucose isomerase, asparaginase, subtilisin, lipases, thermolysin, and penicillin acylase (45–47). The first major use of crystallization was in the purification of insulin, which was crystallized as a zinc complex from salt solution. This technology was developed in the 1950s and simulated the natural process found for storage of insulin as zinc hexameric crystalline arrays in the pancreas (30). More recently, the ability of insulin to be crystallized has been used in slow release forms to achieve a more sustained titer of the hormone in the blood (38), due presumably to the regulation of insulin adsorption by slow dissolution kinetics at physiological pH.

The salient features of protein crystallization processes reported in the patent literature are listed in Table 6, al-

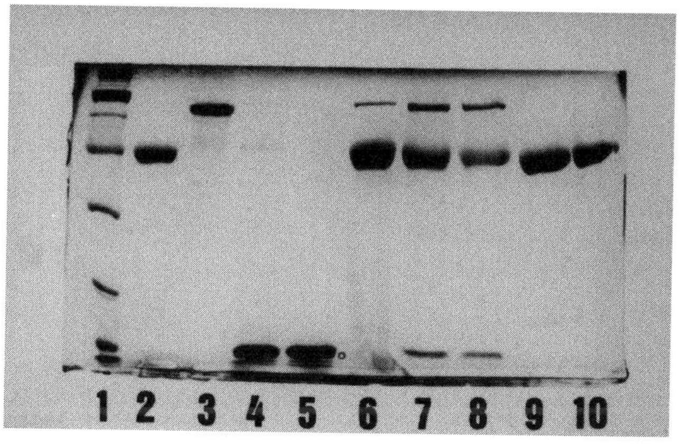

Figure 10. Crystallization of ovalbumin from a solution mixture of ovalbumin (0.784 mol fraction of protein), lysozyme (0.182), and conalbumin (0.034) results in high purity crystals by overloaded SDS-PAGE analysis with Coomassie blue R-250 staining. Lanes: 1, molecular weight; 2, ovalbumin; 3, conalbumin; 4 and 5, lysozyme; 6, initial solution; 7 and 8, final mother liquor; 9 and 10, washed ovalbumin crystals separated from the mother liquor (11).

Table 6. Selected Protein Crystallization Processes from Patent Literature

Protein	Conditions	Yield (%)	Time (h)	Company	Reference
Subtilisin	pH 5.2–5.8; 0.5M NaCl	84	24–48	Genencor	18
Glucose isomerase	1.5% $MgSO_4$	97	3	Stabra AG	47
Peroxidase	pH 4.0; 1.5% $CaCl_2$; 20% PEG 4,000	90	4	Novo-Nordisk	48
Alkaline protease	pH 7.5; 0.5M NaCl	76	4	Solvay	37
Amylase	pH 7.6; 8% NaCl	91.5	24	Solvay	49

though the process may not be used to produce the protein in commercial practice. In the majority of cases, halide salts are preferred for producing the supersaturation needed to obtain crystallization, although a recent report suggests that the use of PEG may greatly enhance the process at lower salt contents (48).

Crystallization is usually performed using clarified fermentation broth that has been concentrated normally by ultrafiltration so that supersaturation is attained on addition of precipitant. The crystallization pH is generally adjusted to values at or near the isoelectric point of the protein to take advantage of the lower solubility at these pH values. Alkaline proteases, such as those from *Bacillus* species (e.g., subtilisin) and lysozyme, are exceptions. These enzymes are typically crystallized at lower pH values to minimize autolysis during the process (46). Process temperatures vary widely. Temperature change plays a minor role compared to precipitant addition for obtaining supersaturation. This is usually the result of the constraints imposed on process temperature by protein stability requirements or the relative insensitivity of protein solubility to temperature, particularly at high salt concentrations. Typically, the crystals produced in commercial crystallization processes are small (30 to 100 μm in diameter) to aid their more rapid dissolution, in either product applications or subsequent purification steps. By the end of the crystallization process, the crystal content of the liquor can be up to 20 to 30% v/v (18).

The crystalline product can be separated from the mother liquor by settling of the denser crystal phase and decantation of the mother liquor from the crystals, or by usual solid–liquid separation technologies such as microfiltration, pressure filtration, or centrifugation. Washing to remove the mother liquor from the crystals improves crystal purity and the overall purification ability of the process. Generally, the crystals must be maintained as a concentrated slurry in a suitable salt solution to avoid redissolution or drying out and reversion to an amorphous state.

A novel approach to protein crystallization, which overcomes the problem of maintaining the crystal form, is the subsequent cross-linking of the crystals by, for example, glutaraldehyde, which renders the crystals insoluble in aqueous solutions. This process has been used to develop mechanically robust and stable enzyme crystals called CLECs (cross-linked enzyme crystals) suitable for repeated use and easy separation from the reaction liquor in organic syntheses (45,50). A photograph of CLECs is given in Figure 11. A further advantage of CLECs is their enhanced stability relative to the noncrystalline form and even to the un-cross-linked crystal form of the enzyme under adverse conditions, such as low water environments,

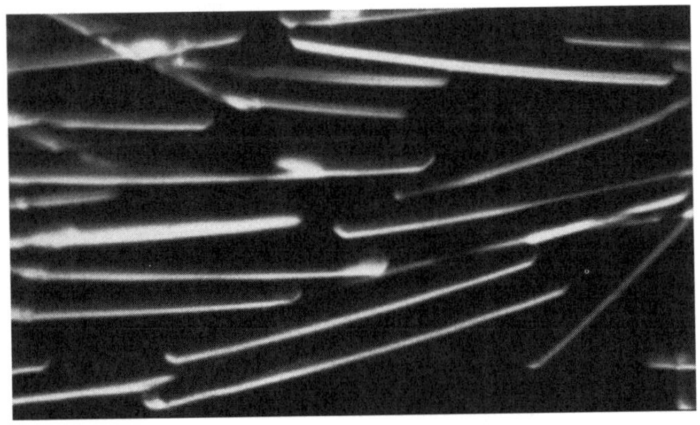

Figure 11. Cross-linked enzyme crystals (CLECs) of thermolysin (100 μm length). *Source*: Courtesy of Alex Margolin, Altus Biologics Inc. Reprinted from Ref. 45 with kind permission of Elsevier Science-NL, Amsterdam.

elevated temperatures or the presence of proteases, which makes them of value in the synthesis of many chirally pure drugs and chemical intermediates. A recent example (50) comprising a lipase revealed the half-life of the soluble enzyme to be 5 h compared to more than 13 days for the cross-linked crystal form.

CONCLUSION

In addition to its critical importance for protein structure determination by X-ray crystallography, protein crystallization is now emerging as a useful bulk technology for protein purification. It is particularly well suited to commodity proteins where bioprocessing costs must be contained in the face of demands for improved product quality. Protein crystallization offers potentially high degrees of purification at any scale in reasonable process times and cost. Nevertheless, the gaps in knowledge concerning the application of protein crystallization at process scale remain large, despite its successful use in the production of commercially significant proteins. Protein crystals also offer exciting prospects as novel forms of catalysts in controlled-release formulations and as biomaterials. The sensitivity of protein crystallizability to even single amino acid changes in the polypeptide chain further suggests that fruitful results can be obtained using tools of modern molecular biology and protein engineering, already well established in the production of commodity proteins. Perhaps what is needed most, however, is the realization that crystallization, long overlooked in production processes for proteins and far more elegant than its wilder sibling—precipitation—deserves to be considered alongside the many other operations commonly selected for protein purification.

ACKNOWLEDGMENTS

I would like to thank Professor Ted White and Dr. Russell Judge for their participation in much of the work described in this article and for their reading of the drafts. I also give my heartfelt thanks to Todd Becker, John Kan, Alfred Gaertner, Rick Bott, and all the team at the California Technology Center of Genencor International, Inc., Palo Alto, Calif., for their hospitality during my stay with them and for their insight on protein crystallization and proteins. Finally, I acknowledge the many useful discussions with researchers in the field, especially Marc Pusey (NASA, Huntsville, Ala.), Bill Wilson (Mississippi State Univ.), Harvey Blanch (Univ of California, Berkeley), Kris Berglund (Michigan State Univ.), Chuck Glatz (Iowa State Univ.), and many others. Thank you.

BIBLIOGRAPHY

1. F. Rosenberger, *J. Cryst. Growth* **76**, 618–636 (1986).
2. R. Giegé, B. Lorber, and A. Théobald-Dietrich, *Acta Crystallogr.* **D50**, 339–350 (1994).
3. A. McPherson, *Eur. J. Biochem.* **189**, 1–23 (1990).
4. A.D. Randolph and M.A. Larson, *Theory of Particulate Processes. Analysis and Techniques of Continuous Crystallization* 2nd ed, Academic Press, San Diego, 1988.
5. J.W. Mullin, *Crystallization* 3rd ed., Butterworth-Heinemann, Oxford, U.K., 1993.
6. F. Ewing, E. Forsythe, and M.L. Pusey, *Acta Crystallogr.* **D50**, 424–428 (1994).
7. S. Moré and W. Saenger, *J. Cryst. Growth* **153**, 35–41 (1995).
8. C. Haas and J. Drenth, *J. Cryst. Growth*, **154**, 126–135 (1995).
9. M. Ries-Kautt and A. Ducruix, in A. Ducruix and R. Giegé eds., *Crystallization of Nucleic Acids and Proteins*, Oxford Univ. Press, Oxford, U.K., 1992, pp. 195–218.
10. H.L. Axelrod, G. Feher, J.P. Allen, A.J. Chirino, M.W. Day, B.T. Hsu, and D.C. Rees, *Acta Crystallogr.* **D50**, 596–602 (1994).
11. R. Judge, M. Johns, and E. White, *Biotechnol. Bioeng.* **48**, 316–323 (1995).
12. A.J. Malkin and A. McPherson, *Acta Crystallogr.* **D50**, 385–395 (1994).
13. F. Franks, in F. Franks ed., *Protein Biotechnology*, Humana Press, Totowa, N.J. 1993, pp. 133–189.
14. M. Pusey and S. Munson, *J. Cryst. Growth* **113**, 385–389 (1991).
15. K. Visuri, in *10th Int. Conf. of Enzyme Engineering*, Kashikojima, Japan, 1989.
16. E. Cacioppo and M.L. Pusey, *J. Cryst. Growth* **114**, 286–292 (1991).
17. Y. Georgalis and W. Saenger, *Adv. Colloid Interface Sci.* **46**, 165–183 (1993).
18. U.S. Pat. 5,041,377 (August 20, 1991), T. Becker and V.B. Lawlis (to Genencor International Inc.).
19. C.W. Bunn, P.C. Moews, and M.E. Baumber, *Proc. R. Soc. London, Ser. B* **178**, 245–258 (1971).
20. M. Ries-Kautt and A. Ducruix, *J. Biol. Chem.* **264**, 745 (1989).
21. J. Jancarik and S.H. Kim, *J. Appl. Crystallogr.* **24**, 409–411 (1991).
22. H.A. Kelders, K.H. Kalk, P. Gros, and W.G.J. Hol, *Protein Eng.* **1**, 301–303 (1987).
23. G.L. Gilliland, *J. Cryst. Growth* **90**, 51–59 (1988).
24. M. Jullien and M.-P. Crosio, *J. Cryst. Growth* **110**, 182–187 (1991).
25. B. Pan and K. Berglund, *J. Cryst. Growth* **130**, 587 (1993).
26. A. George and W.W. Wilson, *Acta Crystallogr.* **D50**, 361–365 (1994).
27. D. Rosenbaum, P.C. Zamora, and C.F. Zukoski, *Phys. Rev. Lett.* **76**, 150–153 (1996).
28. R. DeMattei and R. Feigelson, *J. Cryst. Growth* **97**, 333–336 (1989).
29. E. Forsythe, F. Ewing, and M. Pusey, *Acta Crystallogr.* **D50**, 614–619 (1994).
30. J. Schlichtkrull, *Acta Chem. Scand.* **11**, 439–460 (1957).
31. R.A. Judge, Ph.D. Thesis, University of Queensland, Brisbane, Australia, 1995.
32. M. Pusey and R. Naumann, *J. Cryst. Growth* **76**, 593–599 (1986).
33. R. Boistelle, J.P. Astier, G. Marchis-Mouren, V. Desseaux, and R. Haser, *J. Cryst. Growth* **123**, 109–120 (1992).
34. N. Niimura, Y. Minezaki, M. Ataka, and T. Katsura, *J. Cryst. Growth* **154**, 136–144 (1995).
35. A. Nadarajah, E. Forsythe, and M. Pusey, *J. Cryst. Growth* **151**, 163–172 (1995).
36. M. Skouri, B. Lorber, R. Giegé, J.-P. Munch, and J. Candau, *J. Cryst. Growth* **152**, 209–220 (1995).
37. U.S. Pat. 5,439,817 (August 8, 1995), J.K. Shetty, C.P. Patel, and M.A. Nicholson (to Solvay Enzymes, Inc.).
38. J. Markussen, P. Hougaard, U. Ribel, A.R. Sorensen, and E. Sorensen, *Protein Eng.* **1**, 205–213 (1987).
39. M. Miller, J. Weinstein, and A. Wlodawer, *J. Biol. Chem.* **258**, 5864–5866 (1983).
40. C. Abergel, M.P. Nesa, and J.C. Fontecilla-Camps, *J. Cryst. Growth* **110**, 11–18 (1991).
41. M.L. Grant and D.A. Saville, *J. Cryst. Growth* **153**, 42–54 (1995).
42. H. Lin, F. Rosenberger, J. Alexander, and A. Nadarajah, *J. Cryst. Growth* **151**, 153–162 (1995).
43. K. Visuri, E. Kaipainen, J. Kivimaki, H. Niemi, M. Leisola, and S. Polasaari, *Biotechnology* **8**, 547–549 (1990).
44. M. Saikumar, C. Glatz, and M. Larson, *J. Cryst. Growth* **151**, 173–179 (1995).
45. A.L. Margolin, *Trends Biotechnol.* **14**, 223–230 (1996).
46. U.S. Pat. 5,256,557 (October 26, 1993), J.K. Shetty, C.P. Patel, and M.A. Nicholson (to Solvay Enzymes, Inc.).
47. U.S. Pat. 5,120,650 (June 9, 1992), K. Visuri (to Stabra AG).
48. World. Pat. WO 95/01989 (January 19, 1995), B.M. Nilsson, M.A. Laustsen, and A. Rancke-Madsen (to Novo-Nordisk A/S).
49. U.S. Pat. 5,281,526 (January 25, 1994), I.C. Good, C.Y. Kim, J.K. Shetty, and K.M. Sproat (to Solvay Enzymes, Inc.).
50. J. Lalonde, C. Govardhan, N. Khalaf, A. Martinez, K. Visuri, and A. Margolin, *J. Am. Chem. Soc.* **117**, 6845–6852 (1995).
51. R. Judge, M. Johns, and E. White, *J. Chem. Eng. Data* **41**, 422–424 (1996).

See also PROTEIN PURIFICATION, AQUEOUS LIQUID EXTRACTION.

CULTURE COLLECTIONS

PHILIP PACKER
ALAN DOYLE
Centre for Applied Microbiology and Research
Salisbury, Wilts
U.K.

KEY WORDS

Culture collection
Culture supply
Gene banks
Identification
Quality Control
Resource centers

OUTLINE

INTRODUCTION

The Increased Importance of Microorganisms and Cell Lines

Microorganisms have a significant effect on human life, whether as a result of the diseases or benefits they cause humans or to their food source (plants and animals) and through their effect on the environment (food spoilage, water pollution). In addition, microorganisms and cell lines are used increasingly within biotechnology and are therefore an integral part of the infrastructure of microbiology and fundamental in underpinning scientific research. They are used in crop improvement, food production, pharmaceutical production, and treatment of waste. Industry has the capacity to screen isolates for particular biochemical properties and metabolite production, to evaluate and develop commercial kits and rapid detection kits for microbially produced compounds, and to evaluate organisms' capacity for use in bioremediation and as biocontrol agents.

With the decline in the use of animals in scientific experimentation over recent years, both for economic and ethical reasons, many modern standards and specifications require the use of microorganisms and cell lines in the testing of products and quality control procedures. Culture collections, therefore, supply microorganisms and cell lines for the British, European, and U.S. testing standards, disinfection testing, the assessment of mutagenicity against a range of microorganisms, toxicity testing, NAMAS accreditation, quality accreditation, water and environmental reference strains, and biological and biomedical research.

The supplier is responsible for adequate quality control procedures to ensure the authenticity of these cultures. Failure to do so may result in potential loss incurred as a result of contamination, work on the wrong organism or biochemical product, or invalid manufacturing quality control procedures.

Preservation of Existing Biodiversity

Over recent years, there has been increased concern regarding the conservation of biodiversity and the wealth of genetic information still to be harnessed. This matter has been found to be so important that it has been taken up governmentally at an international level. The Convention on Biological Diversity was signed by the governments of 155 countries and the European Community at the Earth Summit in Rio de Janeiro in 1992. The objectives of the convention were the conservation of biological diversity, the sustainable use of its components, and the fair and equitable sharing of benefits arising from the use of biological resources. The consequences of the introduction of this complex convention has been to generate interest in the scientific aspects of diversity and to begin to establish a number of major programs and projects aimed at increasing knowledge and improving conservation procedures, particularly in situ methodologies.

At a legal level, the common inheritance principle regarding ownership of biological resources has changed to sovereign rights, and access to genetic resources will be controlled by the contracting parties and subject to national legislation. Scientifically, countries are required to establish research and training programs, develop guidelines and procedures for conservation (ex situ and in situ), introduce procedures for assessing and monitoring environmental impact, establish emergency response procedures, and establish partnerships that support capacity building and technology transfer in developing countries.

These regulations are far from being neatly juxtaposed to the Budapest Treaty legislating the patenting of microorganisms used in the industrial microbiology industry because the patent system limits the access to biological resources. The role of the Budapest Treaty is discussed later.

Conservation of microorganisms is obviously a very important component of the biological diversity initiative because they are the most biochemically adaptable organisms on the planet on which they are ubiquitous, occupying every ecological niche. Thus, culture collections have a primary role in conserving microorganisms and cells.

THE DEVELOPMENT OF CULTURE COLLECTIONS AS A RESOURCE

Since Krael established the first culture collection in Prague in 1890 (1), many others have been set up, not only to conserve living examples of microbes (e.g., bacteria, fungi, protozoa, algae, viruses), but also cell lines (human, animal, and plant cells) and DNA probes. Although the preservation of genetic diversity will fundamentally benefit humankind anyway, it is not known which materials may be useful or, indeed, for what purpose they may be required in the future. It is also impossible to predict changes in special habitats or in the general environment, nor what effect such changes may have on the bacterial genome or chromosomal elements. Therefore, some cultures that are of no importance today may be of significance in the future, and it is therefore of fundamental importance to preserve today's biodiversity.

There are three main types of culture collections, the classification of which largely depends on the size of the collection, its documentation, and the service provided. The first type is one where conservation of a small collection is a secondary activity of the laboratory. As such culture collections evolve and increase in size, a second type that is larger and more specialized will arise. These collections often result from specific research programs that are vulnerable when projects come to an end or individual senior scientists retire. Such collections need correct documentation to be a valuable resource and a greater effort to maintain the resource successfully.

The third type of culture collection is the nationally and internationally recognized service-supply culture collections where conservation is the primary function. In order that such service-culture collections can become a recognized international resource, three principal elements must be satisfied. First, the culture collection has to be preserved in a viable state. Second, the history, original source, and properties of the culture must be documented. Third, the cultures must be available to the wider scientific community.

With the increasing demand on such culture collections for authenticated, reliable biological material, the World Federation for Culture Collections (WFCC) has published recommendations for good practice in culture collections so that new and existing collections have guidance and approved standards of operation (2).

Service-supply culture collections exist primarily to maintain and supply on request cataloged, authenticated cultures and to give expert advice on cultivation and preservation to scientists, technologists, industrialists, and teachers. The primary task of the staff therefore is curatorial, thus distinguishing it from secondary collections. Through continuity of work, these culture collections become expert in documentation and preservation techniques and in the knowledge of the behavioral properties of the holdings.

For a service collection to be a truly valuable resource center, it must keep abreast of the developing science it serves. There must be an active accessioning policy, ongoing reassessment of preservation procedures, awareness of taxonomic changes, identification of new requirements, and an original research program. Recently, as a consequence of this, other services have become equally important, including cryopreservation, bulk supply of cultures for screening for products and quality control, patent and safe deposit facilities, identification of cultures, and the delivery of training courses.

OPERATION OF CULTURE COLLECTIONS

It is important that all culture collections operate good practice in all aspects of their function, from microbiology to dispatch of samples. The various activities of culture collections and their successful management are discussed in this section.

Organization, Funding, and Objectives

It is of fundamental importance that the establishment that houses a culture collection realizes its importance, accepts the responsibility inherent in maintaining a public service to appropriate standards, and is committed to its long-term maintenance. In the case of existing collections, this aspect should be continually clarified with the director of the parent institute, its scientific council, senior university officials, governing board, or other appropriate authorities.

Intrinsic to such a commitment is appropriate funding. The operation of such culture collections is a long-term commitment and requires core-funding from the overseeing parent organization. Support solely in the form of short-term contracts is totally inappropriate, because necessary levels of funding need to be guaranteed for the future activities of the culture collection, including the services being planned and the standards users would expect.

All culture collections should have a mission statement and long- and short-term objectives relating to the scope of the holdings and to the range of services that it will offer. If these statements are not agreed upon by the parent institute and other funding bodies, then the culture collection will not be able to operate effectively.

A considerable degree of income generation is possible from culture collection activities, varying from as little as 25% to as much as 85%. This can depend on a mixture of sources of income, culture sales, and contract services.

Holdings

The diversity of microorganisms and cells and their number should be continually addressed by a culture collection because appropriate funding will be needed to maintain the collection. Microorganisms that require particular containment facilities are those that are potentially pathogenic to humans, animals, or plants or those that are toxic.

When a new culture collection is being established, it often relates to the interests of the parent organization. However, it is advisable to aim to complement existing collections rather than duplicating existing ones. It is necessary, therefore, to have a clearly defined accession policy on which new strains are to be added to the collection. Lack of such a policy will result in many unsolicited strains entering a collection uncritically without due regard to collection objectives, storage capacity, personnel, and financial resources. Conversely, the range of a collection should not be restricted as to limit the effectiveness of the service provided to the scientific community.

New accessions may arise as a result of any of the following: (*1*) From the collection's own research; (*2*) From samples sent to the collection for identification; (*3*) From active solicitation by the collection; (*4*) They may arise unsolicited; (*5*) As a formal patent deposit under the Budapest Treaty.

Staff

The effective curation, management, and staffing of a culture collection is a demanding task. Routine accessions, preservation, maintenance, viability checking, culture supply, and additional services are all extremely time consuming. The staff have to have considerable knowledge not only about the organisms themselves, but also their growth and preservation requirements, properties, and potential applications. Where a culture collection offers other services, such as identification and taxonomic expertise, additional specialized staff are required. Experience in culture collection management in addition to taxonomy is a fundamental prerequisite to expertise, and it is therefore important to attract and retain staff of a high caliber.

Preservation and Maintenance

There are a number of techniques available for the preservation of microorganisms, each of which has particular advantages and disadvantages. In addition, different organisms require special preservation techniques in order to ensure optimal storage and maintain purity. The choice of technique should be determined by relating the features of each method to the needs of the user. A number of considerations must be taken into account when deciding which method is most appropriate.

Purity of samples is of paramount importance, and the preservation method should minimize the chance of contamination. The process of preservation and storage should be fully validated. Because preservation could result in loss of viability and hence cell death, a process causing minimum damage should be chosen. Viability should be ascertained immediately after preservation and reassessed during storage.

Preservation may introduce the possibility of a change in the characteristics of a culture. This can happen either as a result of a significant number of cells dying during the preservation process, thus selecting for a resistant population of surviving cells, or by genetic drift as a result of mutation or, for example, loss of plasmids.

The expense of preserving and storing cultures may also influence the choice of method, and this may be further influenced by the number of cultures and factors, including staffing, equipment, materials, and storage space. The operation time for the preservation and maintenance of a small collection may be too labor intensive for a large collection. The high capital cost for methods such as freeze-drying might be considered unsuitable for a small collection, although once preserved the samples require little maintenance. Storage in liquid nitrogen vapor may be cheaper, but for large collections, a sizeable storage space is required. Considerable attention must therefore be taken when the choice of preservation is being made, particularly with regard to the long-term objectives and size of a collection.

The consequence of loss of a culture is another factor to be considered when choosing a preservation method. Valuable samples should be preserved by methods that minimize loss and should ideally be preserved by more than one technique.

Finally, the choice of preservation technique will depend on the frequency of distribution of a sample. Samples that are regularly distributed should be preserved by a method amenable to bulk storage and that does not put them at risk to loss of viability or contamination as a result of frequent manipulation. Cultures that are to be supplied through the mail should be in a form suitable for packaging and must survive transportation time and temperature in accordance with IATA (1988) Dangerous Goods Regulations for Transport of Infectious Material (3).

There are four main storage methods briefly outlined in the next section. This subject has previously been reviewed in depth, for further information see the work by Kirsop and Doyle (4).

Subculture. Preservation of samples by continued subculture is open to the risk of complete loss on the one hand and contamination on the other. Samples are incubated on a suitable medium, at an appropriate temperature, and for a suitable time. The procedure is repeated at intervals that ensure that a fresh culture is prepared before the old one dies. Samples are therefore continually in danger through loss of viability, particularly because the survival time of a particular microorganism varies considerably. *Staphylococcus* spp. and coliforms will survive for several years, but *Neisseria* spp. require subculture within a few weeks.

The majority of media used for the storage of microorganisms are unenriched and have limited nutrients. Excess carbohydrates should not be included in the medium because the acid produced might kill the culture.

Storage periods can be extended by reducing the metabolic rate of the microorganism. This can be achieved by reducing the storage temperature to 5 °C in many instances. However, exceptions occur; *Neisseria* spp. survive better at 37 °C. Restricting the availability of air with parafilm can also limit metabolism.

Thus, repeated subculturing, together with mislabeling or transposition of cultures, can result in significant contamination of deposits. Genetic drift and instability of characters also increases with each subculture. A large inoculum reduces the risk of selection but increases the risk of contamination. Therefore, this method is not to be recommended.

Freeze-Drying. Freeze-drying has found popularity with many service culture collections primarily because it is suitable for batch production and distribution. Maintenance of viability requires very little attention, storage being relatively simple and effective and allowing preservation for 50 years for some microorganisms. The capital cost of freeze-drying equipment is initially high, and the preparation of samples is quite labor intensive, but the fact that large batches of samples can be prepared, significantly reducing labor hours per ampule, quickly recoups the initial financial outlay.

The process of freeze-drying is the removal of water from frozen samples by sublimation. Cultures are suspended in a suitable medium (generally mannitol), frozen, and exposed to a vacuum. The water vapor is trapped in phosphorous pentoxide or a refrigerated condenser. The microorganisms are stored in individual ampoules or vials in an inert gas or under vacuum. There are two types of freeze-dryers in general use; however, the centrifugal rather than shelf dryer is more popular and has more advantages.

Although shelf dryers automatically stopper the ampules, the bungs can be naturally permeable to air or water vapor or theromas simply leak. The glass ampoules produced by centrifugal freezing are completely constricted, preventing leakage, and they can be plugged with cottonwool, eliminating cross-contamination and acting as a filter to prevent scatter of microorganisms into the environment when the ampoules are opened.

Even though freeze-drying has been widely used to preserve a variety of microorganisms, the process always needs refining to optimize preservation in a particular microorganism. Factors that can be adjusted and controlled are composition of medium, growth temperature, growth phase of the culture, rate of freeze-drying, final temperature, duration of drying, and the final moisture content of the culture.

In general, freeze-drying does not cause instability of characteristics and genetic drift.

Drying. Desiccation has been used across a range of microorganisms, although the process is rather a specialized application and there is less information available on the method. Long-term viability appears to be good, when samples are stored between 5 and 10 °C, and contamination is not a major problem, although strain stability is less well documented. The simplicity of the process makes it suitable for storing large numbers of cultures.

L-Drying. The term *L-drying* originates from the term *drying from the liquid state* (5). Suspensions of microorganisms are dried in ampoules under pressure, but the vacuum is adjusted to allow rapid drying without freezing. The method is widely applied and allows a significant reduction in drying time and thus labor costs when a number of microorganisms are preserved. Samples should be stored at 5 °C.

Freezing. Preservation by freezing has been carried out for a variety of different microorganisms over temperatures ranging from −20 °C (refrigerator) to −140 °C (vapor phase nitrogen) to −196 °C (liquid nitrogen). Freezing and thawing can cause injury to cells by the formation of crystals that may damage the cells' integrity or by the concentration of electrolytes through the removal of water as ice. Eutectic mixtures form at −30 °C and above, exposing cells to high salt concentrations. Temperatures of −70 °C and below have been found to be increasingly successful, and storage in the vapor phase of liquid nitrogen is now ubiquitous as a preservation method.

Storage in the liquid phase of nitrogen results in considerable safety problems resulting from nitrogen invading vials that subsequently become explosive (Fig. 1). The freezing and thawing process itself may cause a significant loss in microorganisms. However, once stored, virtually no further loss occurs. The addition of cryoprotectants, the adjustment of growth conditions, the rate of cooling and warming, and freezing methodology can reduce losses (4).

The temperature of samples should be lowered at a controlled rate of 1 to 3 °C/min to −30 °C followed by a more rapid rate of 15 to 30 °C/min to −100 °C or lower. Ideally, this should be carried out using a commercial programmable freezer, which usually operate on a differential thermocouple principle. If such equipment is not available, many microorganisms, including bacteria and fungi, can be successfully frozen by placing the ampules in a dry chest, in a mechanical refrigerator set at below −65 °C, or

Figure 1. Storage of cells in the gaseous phase of liquid nitrogen.

in the top of a liquid nitrogen storage tank for 30 minutes, and then placing the samples at the required temperature.

Preprepared plastic ampoules containing beads and cryopreservant are now commercially available to enhance preservation (Prolab Diagnostics, Microbank). A vial of 25 beads is inoculated and agitated and the microorganisms attach to the beads. Excess cryoprotectant is then removed and the ampoule frozen. Beads can then be removed when a culture is required. The vials are placed in a cryoblock before they are removed from the storage vessel. This ensures that the remaining beads do not thaw out while one bead is removed. Thus, loss of viability does not result from repeated thawing.

The major disadvantage of freezing is refrigerator failure or interruption to the supply of liquid nitrogen. Both will result in the loss of cultures; however, this can largely be overcome by 24-h alarm systems warning of refrigerator failure or loss of liquid nitrogen supply. The initial capital cost is high, but the process is not labor intensive.

Preservation of Cell Lines

It is impractical for most laboratories to maintain cell lines in culture indefinitely; moreover, cell cultures will undergo genetic drift with continuous passage and risk losing their differentiated characteristics. Therefore, it is necessary to store adequate cell stocks for future use. It is recommended practice that a limit is set for the number of passages any cell line should undergo before replacing from cryopreserved stocks. This is of less importance for undifferentiated cell lines having infinite life span, although it still remains a consideration.

Nearly all cell lines can be cryopreserved successfully in liquid nitrogen at −196 °C. A few simple criteria must be adhered to in order to provide sufficient cells for all future needs.

The essentials for efficient cryopreservation are slow freezing (i.e., at a rate between 1 and 3 °C per minute) and fast thawing, which is achieved by placing ampoules in a water bath at the temperature required for growth. The addition of a cryoprotectant, such as glycerol or dimethyl sulphoxide (DMSO), enhances survival. Nevertheless, DMSO does have certain toxic properties. Therefore, care must be taken in handling the cryoprotectant, and some consideration may be necessary as to whether cells should be washed immediately on thawing.

Sample Safety and Security. For security, and in order to minimize the possibility of strains being lost, each culture should be maintained by at least two different procedures. Cryopreservation and lyophilization are the best methods for minimizing the risk of genetic drift, therefore, at least one of these methods should be used. Where only freezing is available, duplicates should be stored with different electrical supply. Duplicates of important or irreplaceable strains should be securely held in different buildings or ideally on a different site. The documentation of samples is equally important because loss owing to natural disaster or fire would make the corresponding cultures meaningless and worthless.

Cell Banking Procedures. When a culture is received in the laboratory, it is important that a strict accessioning scheme be followed (Fig. 2). The method described prevents substantial genetic variation from the original deposit. Starter cultures or ampoules received should be propagated according to the conditions provided by the originator, and a token freeze should be produced (usually up to 5 ampoules). Cultures derived from this token freeze should then be subjected to detailed quality control and characterization. If these results are satisfactory, then a master (or seed) stock should be made. Major authentication of the culture should then be undertaken on the master cells because it is these ampoules that will be used to provide the working or distribution stock.

One ampoule from the master stock is used to produce a working or distribution stock. The number of ampoules in each bank is dependent on how often it will be used. Because the production of these banks is both expensive and time consuming, the numbers required should not be underestimated. For some industrial processes, master banks may contain up to 200 ampoules, and working banks may contain 1,000 ampoules.

Once the working bank is depleted, a fresh one is made from an ampoule from the master bank; similarly, a new master bank is made from the frozen freeze stock. If no more frozen freeze stock remains, then one of the current master bank ampoules should be used.

The level of characterization and quality control that should be carried out at each stage will depend on what it is to be used for and what the necessary regulatory bodies require (Table 1). Quality control implications are obvi-

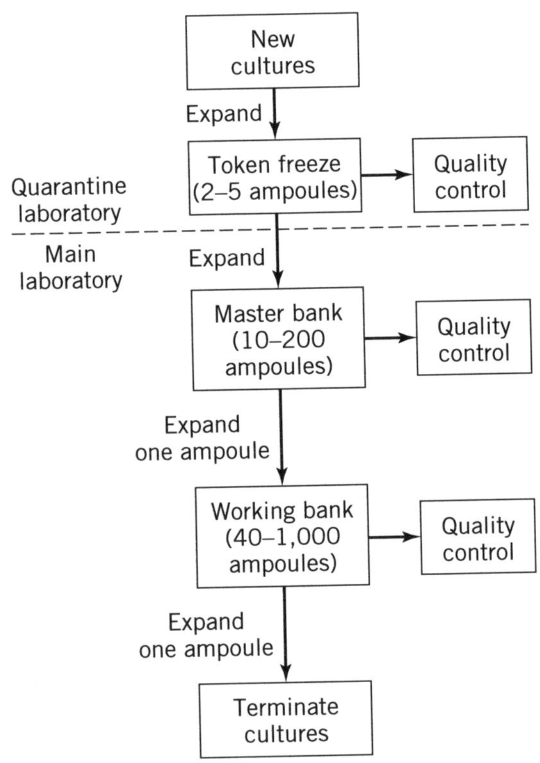

Figure 2. Cell banking procedures.

Table 1. Table of Quality Control Requirements for Cultures within Collections

Cell lines	Culture type: Bacteria	Fungi	Protozoa	Algae
Viability	Viability	Viability	Viability	Viability
Purity	Purity	Purity	Purity	Purity
Test for bacteria, yeast, and other fungi				
Test for mycoplasma				
Test for viruses				
Authenticity	Authenticity	Authenticity	Authenticity	Authenticity
Isoenzyme analysis	Gram Stain			
Cytogentic testing	Description of morphology			
DNA fingerprinting				

ously more intense for cultures used in the manufacturing industry.

CULTURE COLLECTION SERVICES

The aim of culture collections is to supply authenticated cultures to bona fide scientists, on request, promptly and without restriction on their ultimate application. The staff of culture collections have expertise and knowledge of the appropriate scientific literature and as such can advise on strain selection. In addition, they provide a range of scientific services, largely in response to the needs of the user. These include safe and patent deposit facilities and identification and preservation services. The expertise gained by culture collections and their staff enables them to run training courses and seminars for other scientists.

Culture Supply and Pricing Policy

Culture collections should be able to readily distribute cultures listed in their catalog on request. Arrangements for culture supply vary according to the financial basis and policies of the legal owner of the collection. Service-supply collections usually levy fees for supplying cultures. This charge should strike a balance between definable measurable costs and those that yield an income that is significant in relation to costs, but that do not result in a declining effective use of the collection.

Consideration should be taken as to the fact that many collections are partly government funded, unit costs would be far too high for what is a reusable commodity (i.e., they can be subcultured), and fees should not discourage the use of the collection. Therefore, culture collections operate on a nonprofit basis. It is often common practice to charge commercial companies higher fees than nonprofit organizations. Further charges may be levied for postage and packaging and any special documentation.

Great care should be given to distribution of pathogenic or toxic cultures. Originators of a request should be able to guarantee that their establishment has stipulated equipment and staff to handle pathogenic or toxic cultures. Dispatch of cultures should be in accordance with IATA (1988) Dangerous Goods Regulations for Transport of Infectious Material (3).

Restrictions to which countries cultures can be distributed also exist. Because the threat exists that microorganisms could be developed for use in biological warfare, cultures can only be issued to countries in the Australia Group of nations. The nations belonging to this group and restricted groups are detailed in *The End-Use Control* (6).

Catalogs

Catalogs are fundamental to the operation of culture collections. They describe the strains available and their salient properties, including literature citations, propagation properties, etc. Stocks are usually arranged alphabetically, in the case of bacteria by genus and species name. Increasingly catalogs are becoming available in computerized form, either on-line or on disk, because, they can can easily be updated whereas paper catalogs cannot. Such catalogs adopt a field structure and definitions that enable the data to be integrated into the international and major regional schemes that are in operation, including the Internet, on which many resource centers are accessable through the World Wide Web, Microbial Strain Data Network (MSDN), World Data Centre on Micro-organisms (WDC), Microbial Information Network Europe (MINE), and more recently CABRI (Common Access to Biotechnological Resources and Information).

CABRI is researching and developing the formation of an integrated resource center service linking databases of different organism type, genetic materials, and other biologicals in Europe so that users worldwide can access these relevant catalogs during one searching session through a common entry point and request or order products to be delivered to their place of work.

Safe Deposits

Safe deposits are held for a variety of reasons, including as a back-up to existing collections and for the retaining of ownership and confidentiality of strains. These collections are not listed in catalogs and their taxonomic status may be unclear. In some cases, the service collection may hold stocks in its collection and market them on behalf of the owner.

Patent Deposits

The rapid advance in genetic manipulation techniques and the allied boom in biotechnology over recent years has led to increased awareness of the enormous industrial potential of microorganisms. Competition is fierce in the race to exploit this potential, and large sums of money are being risked in commercial development of new microbial processes. In such circumstances, patent protection naturally becomes a major concern of any organization involved in modern industrial microbiology.

An important principle of the patent system is that an invention should be disclosed to the public. Although most kinds of invention can be disclosed simply by means of written description, those involving the use of new microorganisms cannot. No matter how detailed the written description may be, a new microbiological process cannot be tested without a culture of the organism itself. Therefore, in addition to a written disclosure, the patent offices of most industrial countries require the strain used in a microbiological invention to be deposited in a recognized public culture collection, where it will eventually be available to the public.

Patent deposits are now regulated internationally by the Budapest Treaty on the International Recognition of Deposit of Microorganisms for the purpose of patent procedures and patent depository facilities in Europe, under which certain culture collections of a generally acknowledged high scientific standard are officially recognized as international depository authorities (IDAs). The main feature of the treaty is that all countries party to it, and hence their patent offices, must recognize a deposit that is made in any one of these IDAs as being valid for the purposes of their patent procedure. This means that one deposit is all that is needed, regardless of the number of countries in which patent applications citing the strain are filed.

The Budapest Treaty contains detailed provisions on the procedure to be followed by the applicant in making a deposit and by the IDA in accepting, storing, and subsequently making the material available to the public. Broadly speaking, these provisions are designed to ensure that the applicant makes a viable deposit and provides enough information on it to enable the IDA to handle and maintain it safely and correctly. The IDA, for its part, must process the deposit promptly, issue the depositor with a receipt and certificate as proof of deposit and viability, keep the deposit for at least 30 years, and provide cultures to those parties entitled to receive them but most importantly, *to no one else.* Entitled parties in this respect are those who have the depositor's written authorization or who have an official certificate of entitlement from the appropriate patent office. The IDA is also bound by strict rules of secrecy and may not give information about a deposit to anyone not entitled to receive the patent-deposited cultures.

Microorganisms must become available to the public at some stage of the patenting procedure. To date, there are three kinds of systems in operation. In the Netherlands and Japan, patent applications are published twice, first 18 months after the filing date of priority date and before the application has been examined, and second when the application has been accepted. Once the patent office has decided to grant a patent, after the second publication date, the microorganism must be made available. The major advantage of this system to inventors is that their microorganisms do not have to be released until they has an enforceable right. In Japan, the patent owner is given a further measure of protection in that the recipients of cultures must not pass them on to a third party and must use them only for experimental purposes.

In the United States, patent applications are not published until the patent is granted, and any microorganism deposited for patent purposes need not be available until then. Thus, under the U.S. system, inventors again have an enforceable right at the time they are required to make the organism available. Additionally, under the U.S. mechanism an inventor is never put in the position of having to allow access to the organism when no legal protection is available because if a microorganism is not granted a patent, then it need never become available.

The EPC (European Patent Convention) also operates a dual publication system, but stipulates that a microorganism for which a patent is being sought must be made available at the date of first publication, before any enforceable right exists. This mechanism reflects the principle that the microorganism is an integral part of the patent application and should therefore be available at the same time as the written description. Availability at first publication can be restricted to an independant expert acting on behalf of a third party. This expert has to be selected from a list held by the EPC and is not permitted to pass the strain on to anyone else. On second publication, the strain becomes generally available, whereupon the inventor has an enforceable right.

There are 23 IDAs. A full description of these is described by Halluin (7).

Culture Authentication and Identification

Culture collections have a critical responsibility when it comes to the authentication of microorganisms and cells from their initial receipt through preservation procedures and to dispatch to external scientists. Quality control of cultures should be as extensive as possible and have three primary targets; authenticity, purity, and viability. Research using the wrong organism wastes time, is expensive, and leads to publication of invalid results.

The identification of a new accession should be ideally confirmed by a competent specialist (e.g., the depositor) or confirmed by the culture collection itself to ensure that it agrees with published descriptions. Culture collections with specialized expertise in identification can also offer an identification consultancy to the scientific community. The microorganisms identified would not necessarily enter the culture collection.

Documentation and Quality

Increasingly, many culture collections operate using a quality management system such as BS EN ISO 9001. Essentially, these regulations ensure that all procedures have proper protocols that are fully documented, thus allowing all activities to be fully traced if the need arises. All equipment has to have records of calibration and servicing, and all laboratory records, worksheets, and equipment log books are regularly audited. All staff are trained in these procedures.

In order for a culture collection to run completely successfully, it has to keep good records about the microorganisms and cells it holds, including the following categories of information, place, substrate or host, date of isolation, name or person isolating the strain, depositor, name of person identifying the strain, preservation procedure used, optimal growth media and temperature, any data on biochemical or other characteristics, and any regulatory conditions applying (e.g., contamination level).

Duplicate copies of computer files or paper records of all information should be kept to prevent unexpected loss of information.

Research

It is important that culture collections have an ongoing research program. This ensures that staff are kept abreast of current developments and are aware of the needs of the wider scientific community. Lack of adequate research programs discourages high-caliber staff from joining a culture collection and may encourage the loss of existing quality staff. As well as the obvious research and development into preservation and taxonomy, culture collections have the necessary skills to undertake a diverse spectrum of research, whether by themselves or acting as consultants. As discussed earlier, culture collections have the capacity to screen isolates for particular biochemical properties and metabolite production, evaluate and develop commercial kits and rapid detection kits for microbially produced compounds, evaluate organisms' capacity for use in bioremediation and as biocontrol agents, and select test organisms used in the evaluation of materials, media, culture vessels, and diagnostic reagents.

Table 2. The European Culture Collection Organization and the World Federation for Culture Collections

President and Secretary of ECCO and WFCC

Dr. Alan Doyle
CAMR
Porton Down
Salisbury SP4 0JG
United Kingdom
Tel: +44-1980-612512
Fax: +44-1980-611315
E-mail: *alan.doyle@camr.org.uk*

Secretary of ECCO

Dr. Maija-Liisa Suihko
VTT
Biotechnology and Food Research
PO Box 1501
FIN-02044 VTT Finland
Tel: +358-0-4565133
Fax: +358-0-4552028
E-mail: *maija-liisa.suihko@vtt.fi*

WFCC's Web Page: http://www.wdcm.riken.go.jp/wfcc.html

Safety

It is the responsibility of all culture collections to handle cultures and carry out procedures as safely as possible and within the relevant guidelines. This is particularly important with regard to microorganisms that are pathogenic to man or are genetically manipulated. Within the United Kingdom, all laboratory procedures should be performed in line with COSHH (Classification of Substances Hazardous to Health), and all matters of safety are enforced by the HSE (Health and Safety Executive), which also offers an advisory service.

All culture collections are responsible to their own staff, the delivery services, and receiving laboratories and must therefore address all matters of safety properly. All service collections have regulations for both the deposit and dispatch of cultures that have to be fully abided by before samples can be transported following the IATA (1988) Dangerous Goods Regulations for Transport of Infectious Material (3).

Within the United Kingdom, microorganisms are categorized in accordance with the Advisory Committee on Dangerous Pathogens (1995) Categorization of Pathogens According to Hazard and Categories of Containment. In the United States, a similar categorization has been devised (U.S. Department of Health, Education and Welfare, 1974; American Type Culture Collection, 1986). Both sets of guidelines have been accepted by the WHO, and other countries tend to adopt these systems. Information is given on the degree of containment and protective clothing, which should be applied during handling of such organisms in the laboratory.

The control of genetically manipulated organisms is somewhat more complicated in the United States than it is in the United Kingdom. U.S. researchers must adhere to the policy that research in this field must conform to the requirements of the coordinated framework, such as the National Institutes of Health Recombinant DNA Guidelines. In the United Kingdom, genetic manipulations are the responsibility of the Advisory Committee in Genetic Manipulations. Other countries likewise tend to adopt these systems.

All culture collections must additionally follow the regulations of their own countries and the institutes within which they work.

Training and Publications

Culture collection staff require appropriate training for all techniques and procedures they carry out. If an organization has a quality management system, such as ISO 9000, then this will be performed as a mandatory requirement. Once the staff have become skilled, they are then in an ideal position to train others in techniques relating to culture preservation, growth, and identification given that they have adequate teaching facilities and supervision.

Regular publication relating to culture collection activities and the research activities carried out within an organization is necessary to enhance its activities.

Table 3. Major Culture Collection Repositories

Country and organization	Address	Collection
	Belgium	
BCCM/IHEM	Mycology Laboratory Rue Juliette Wystman 14 B-1050 Brussel Belgium Fax: +32 2 6425519 E-mail: Dialcom Telecom Gold: 75: DB I0248	Fungi
BCCM/LMBP	Plasmid Collection University of Gent K.L. Ledeganckstraat 35 B-9000 Gent Belgium Fax: +329 2645348 E-mail: *martine@lmbl.rug.ac.be*	Plasmids
BCCM/LMG	Bacterial Collection University of Gent K.L. Ledeganckstraat 35 B-9000 Gent Belgium Fax: +32 9 2645346 E-mail: *danielle.janssens@rug.ac.be*	Bacteria
BCCM/MCLK	Faculte des Sciences Agronomiques Flace Croix du Sud 3 B-1348 Louvain-La-Neuve Belgium Fax: +3210451501 E-mail: *mucl@mbla.ucl.ac.be*	Fungi
	Brazil	
CCT	Tropical Culture Collection Rua Latino Coelho 1301 Parque Taquaral CEP 13087-010 Campinas - SP Brazil Fax: +55 (019) 242 7827 E-mail: *cct.@bdt.org.br* *World Wide Web: http://www.bdt.org.br/cct/*	Bacteria Fungi Yeasts
	France	
CIP	Institut Pasteur BP52 25 rue du Docteur Roux F-75724 Paris Cedex 15 France Fax: +33 1 40613007	Bacteria
LCP	Museum National D'Histoire Natwelle Laboratoire de Cryptogenie 12 rue Buffon F-75005 Paris France Fax: +33 1 40793594 E-mail: *roqueber@frmnhn.fr*	Fungi
	Germany	
DSMZ	Mascheroder Weg 1b D-38124 Braunschweig Germany Fax: +49 531 2616418 E-mail: dsmz@gbf.de World Wide Web: http://www.gb.de.bl/DSMZ	Bacteria Plant cells Animal and human cells Plant viruses Plasmids

Table 3. Major Culture Collection Repositories *(continued)*

Country and organization	Address	Collection
	Italy	
ICLC	CBA L.go R Benzi 10 16132 Genova Italy Fax: +39 10 5737295 E-mail: *iclc@istge.ist.unige.it* World Wide Web: *http:/arginine.umdrj.edu/home.html*	Human and animal cells
	Japan	
IFO	17-85 Juso Honmachi 2-Chome Yodogawa-Ku Osaka 532 Japan Fax: +81 6 3006555 E-mail: *cellbank@ub.infaveb.ne.ip*	Human and animal cells
Riken Gere Bank	3-1-1 Koyadai Tsukuba Science City 305 Iboraki Japan Fax: +81 2975 4 2616	
	The Netherlands	
CBS	PO Box 273 Oosterstraat 1 NL 3740AG Baarn Netherlands Fax: +31 355 416142 E-mail: *Stalpers@cbs.knaw.nl* World Wide Web: *http://www.cbs.knaw.nl*	Fungi Yeasts
	Spain	
CECT	Universidad de Valencia Edificio de Investigation Campus de Burjasot Valencia Spain Fax: +3496 3983187 E-mail: *cect@uv.es*	Bacteria Fungi
CCUG	University of Goteborg Dept of Clinical Bacteriology Guldhedsg 10 S413 46 Goteborg Sweden Fax: +46 31 825484 E-mail: *ccugef@ccug.gu.se*	Bacteria
	United Kingdom	
CCAP	Institute of Freshwater Ecology The Windermere Laboratory Far Sawrey Ambleside Cumbria LA22 OLP UK Fax: +44 15394 46914 E-mail: *ccap@ife.ac.uk*	Protozoa Freshwater algae

Table 3. Major Culture Collection Repositories *(continued)*

Country and organization	Address	Collection
CCAP	Dunstaffnage Marine Laboratory PO Box 3 OBAN Argyll PA34 4AD UK Fax: +44 1631 562244 E-mail: *ccap@dml.ac.uk*	Marine algae
NCIMB	National Collection of Industrial and Marine Bacteria 23 St. Machar Drive Aberdeen AB2 1RY UK Fax: +44 1224 487658	Bacteria
NCTC	National Collection of Type Cultures Institute of Food Research Colney Lane Norwich NR4 7UA UK Fax: +44 1603 458414 E-mail: *ian.roberts@bbsrc.ac.uk*	Bacteria
	United States	
ATCC	American Type Culture Collection PO Box 1549 Manassas Virginia 20108-1549 USA Fax: +1 703 365 2750 World Wide Web: *http://www.atcc.org/*	Bacteria Viruses Animal and human cells Plasmids Plant Cells
GIMR	Coriell Institute for Medical Research 401 Hadden Avenue Camden NJ 08103 USA Fax: +1 609 757 9737 World Wide Web: *http://locus.umdnj.edu/home*	Human cells/DNA

National and International Collaboration

Microbiologists having interests in culture collections have banded and assembled in formal or informal national federations or associations of the culture collections within them, such as UKFCC (U.K. Federation of Culture Collections). These federations tend to be further affiliated with international bodies, such as ECCO (European Culture Collections Organisation), which comprises 54 collections from 21 European countries. This organization itself is affiliated to WFCC (World Federation for Culture Collections). These federations provide excellent opportunities for discussion and exchange of information in all aspects of culture collection curation.

European Culture Collection Organization. ECCO was established in 1981. The aim of the organization is to promote collaboration and trade ideas and information about all aspects of culture collection activity. Membership is open to representatives of any resource centers that provide a professional service on demand and without restriction, that accept cultures for deposit, that provide catalogs, and that are housed in countries with microbiological societies affiliated to the Federation of the European Microbiological Society (FEMS). After 14 years of activity, ECCO comprises 54 members from 21 European countries. The total holdings of the collections are about 300,000 deposits representing a large pool of biodiversity from different sources (from humans, animals, plants, foodstuffs, environmental) as well as various sources of biotechnological interest, including GMO strains and organisms of interest for genetic engineering.

World Federation for Culture Collections. Most of the culture collections organized in national or regional federations are members of the WFCC. This international organization has been set up as a multidisciplinary commission of the International Union of Biological Science and is a federation within the International Union of Microbiological Societies (IUBS). Various committees exist within the WFCC for carrying out specific tasks in the different aspects of culture collection work. These include the publication of books, provision of training courses, collaboration with postal and patent organizations, setting up databases, providing help for endangered culture collections, and designating quality standards and guidelines for collections. One of the major activities of WFCC is to organize its quadrennial meetings (International Congress of Culture Collections).

The World Data Centre (WDC), now the World Centre on Microorganisms (WDCM), covers a central role in WFCC. Initiated in 1972 by Prof. V.B.D. Skerman and sponsored by UNEP and UNESCO, the database pioneered the organization of microbial culture collections worldwide. Relocated to the RIKEN Institute in Japan in 1986, the WDC, under the directorship of Dr. H. Sugawara, developed into a valuable information resource at the microbial level. Several databases are available as hard copies and are also accessible through the Internet. The core databases, CCINFO and STRAIN, are regularly updated and published as the WDC Directory. CCINFO covers information on nearly 500 microbial resource centers in the world, their organization, the kinds of cultures held, the expertise available, and the services offered. Strain lists all microbial species held by registered collections. The WDCM activity of continual data collection and dissemination plays an important role in the improvement of data quality of collections and for making the resource centers known to the scientific community.

Any individual with an interest in culture collection activities is eligible for membership. Culture collections as institutions may become affiliate members if they meet the standards established by the federation. Individuals or organizations may be accepted as sustaining members on the basis of extraordinary support of the objectives and activities of the Federation. Further information can be obtained from the WFCC (Table 2).

Access to Resources

Hard copy catalogs get out-of-date rapidly; the best and most up-to-date information can be obtained from most resource centers on the World Wide Web. Table 3 gives a comprehensive list of contact information for the major repositories.

BIBLIOGRAPHY

1. M. Kocur, *Proc. of the Kral Symposium*, Institute for Fermentation, Osaka, Japan, 1990, pp. 4–12
2. WFCC, *Guidelines for the Establishment and Operation of Collections of Microorganisms*. Simworth Press, Richmond, Surrey, U.K., 1990.
3. G. Stacey, A. Doyle, and P. Hambleton eds., *Safety in Tissue Culture*, Kluwer, London, 1998; IATA, 33 route de l'Aeroport, PO Box 672, CH-1216 Geneva 15, Switzerland.
4. B.E. Kirsop and A. Doyle, *Maintenance of Micro-Organisms and Cultured Cells. A Manual of Laboratory Methods*, 2nd ed., Academic Press, London, 1991.
5. D.I. Annear, *Aust. J. Exp. Biol.* **36**, 211–222 (1958).
6. DTI, *The End-Use Control*, Department of Trade and Industry, U.K. 1995, pp. 1–16.
7. A.P. Halluin, in J.C. Hunter-Cevera, and A. Belt eds., *Maintaining Cultures for Biotechnology and Industry*, Academic Press, London, 1996, pp. 1–14.

See also ASPERGILLUS; CELL CYCLE; CELL CYCLE, EUKARYOTES; FREEZE DRYING, PHARMACEUTICALS; SECONDARY METABOLITE PRODUCTION, ACTINOMYCETES, OTHER THAN STREPTOMYCES.

CULTURE MEDIA, ANIMAL CELLS, LARGE SCALE PRODUCTION

JENNIE P. MATHER
Rauen Biotechnologies, Inc.
Mountain View, California

ALISON MOORE
Amgen, Inc.
Newbury Park, California

KEY WORDS

Cell line
Fermenter
Large scale
Nutrient optimization
Product titers
Serum-free
Suspension

OUTLINE

INTRODUCTION

The process of choosing and optimizing a cell culture medium for a large-scale production process shares many biological issues and experimental approaches with the process of optimizing medium for any in vitro model system at the laboratory scale. However, there are some significant differences. The cell culture aspect of any large-scale process should fit in seamlessly with the upstream process of the transfection and selection of the protein of interest in the desired parental cell line, the equipment and scale available and desirable to run the process, and the downstream purification process. Issues surrounding the speed of development and the cost, safety, and reproducibility of the process are of major importance in developing a commercial process.

The importance of seeing the entire process—from selection of the host cell line through transfection and selection of the production line, to the growth of the seed train cells and the actual production culture, harvest, and purification of the protein—as an integrated whole cannot be overemphasized. Failure to do so will result in an unworkable process, at worst, and missed opportunities for maximizing the reliability and minimizing the cost, at best. If the end goal is kept in mind from the first, many opportunities for trade-offs in various aspects of the process will arise that allow decisions to be made in a timely manner. For example, if two candidate production lines exist, one with high specific productivity but slow doubling time and low saturation density and the other with lower specific productivity but better growth characteristics, the choice between the lines should be made taking into consideration the titer from each line and the total tank and equipment usage (seed train, production, and purification) required to obtain a given final yield of protein. It might be surprising to some how taking this broader view will change decisions made early in the process that are much harder to change later. In another case, one medium formulation was found to increase the stability of a cell line even though the titer was reduced compared to the standard medium formulation. The advantages of the increased stability were seen to far outweigh the minor loss of product in the short term.

This chapter will concentrate on specific aspects of choosing and optimizing the cell culture medium for large-scale processes. However, as will be emphasized again in the section at the end, this cannot be achieved optimally without taking into consideration the product (as well as the cell line) to be secreted and the purification process and equipment to be used. Some of the basic parameters that influence protein yield from a mammalian cell culture process are outlined in Table 1 and are discussed in detail subsequently.

OPTIMIZING CELL GROWTH AND SURVIVAL

The aim of optimizing the growth and survival of cells in production culture is to obtain the maximum average viable cell number per volume of medium over the duration of the culture (viable cell days [VCD]). Characteristics of the cell that will influence the total VCD are listed in Table 2. When selecting the production line to be used for large-scale culture, these parameters and specific productivity should be taken into consideration. A transfected cell line with high specific productivity is useless as a production line if it cannot be grown in the conditions necessary for large-scale culture. For example, a tetraploid line may have a high specific productivity but be too fragile to do well in a fermenter. Or, a low-cell inoculum may be acceptable with a rapidly growing cell line, but not with a slow growing line producing a labile product. Continuous culture systems strive to produce a steady-state situation where the cell number is constant and the growth rate and death rate are balanced. Ideally, one would wish to achieve something like a contact-inhibited G_0-arrested state in which the cells are neither growing or dying. This is based on the as-yet-unproved assumption that suspension cells producing recombinant proteins will have a specific productivity in G_0 equal to or higher than that in log phase growth. In our experience, however, it is difficult to achieve a true G_0 growth arrest state in suspension-cultured cells (1).

Although a great deal is known about the optimal growth conditions for specific types of cells, including Chinese hamster ovary (CHO), baby hamster kidney (BHK), and transformed human embryonic kidney (293) cells, it is still impossible to exactly predict the optimal medium formulation for every cell type in every culture configuration. In addition, the process of obtaining a stably transfected cell line expressing a recombinant protein can change the growth phenotype of the cells. Thus, the empirical approach is still the only method of ensuring that the medium being used is optimal for a particular production line. The optimal medium components or concentrations may vary from one CHO production line to the next. This is not surprising given that these lines are transfected with foreign DNA, selected using rather stringent conditions, and frequently amplified in the presence of a mutagen such as methotrexate. Additionally, the lines are engineered to secrete biologically active molecules that may act on the cell

Table 1. Parameters Influencing Final Product Yield in a Production Cell Line

Medium or culture configuration
Optimize nutrients
Optimize growth supplements
Minimize inhibitors
Inducers
Protein production
Specific productivity
Protein quality
Compatibility with purification process

Table 2. Characteristics to Be Considered When Choosing a Production Cell Line

Cell growth
Inoculum density
Saturation density
Doubling time
Death rate
Fragility
Suspension growth
Growth in serum-free medium
Other cell characteristics
Free of adventitious agents
Normal glycosylation
Has appropriate processing of proteins
Protease secretion low

or the medium components, altering the environment of the culture. Generally speaking, however, two media optimal for the growth of two CHO-derived cell lines will be similar to each other and differ from the medium optimized for the growth of a dissimilar cell line such as a 293-derived line. For this reason, it is best to concentrate on one or a few cell lines for large-scale production efforts. Currently, most recombinant proteins produced commercially in mammalian cells use CHO as the host cell line.

Measuring Viable Cell Number

Accurate measurement of cell viability or characterization of cell death is key to the optimization of any production cell line. Viability can be readily measured by a variety of well-known techniques; however, it has more recently been established that many cell lines grown under batch culture conditions undergo cell death via apoptosis (1,2). Apoptosis is an active programmed cell death that exhibits specific morphological traits, such as nuclear condensation and fragmentation, maintenance of organelle and membrane integrity, and gross cell shrinkage (3). This contrasts with necrotic cell death, which is passive in nature and recognized by the degradation of cellular membranes and organelles and cell swelling (4). These different forms of cell death are distinct in morphology, mechanisms, and the way in which they can be identified. An understanding of such is required for accurate reciprocal viability quantification, because the observation of simple loss of membrane integrity (e.g., using trypan blue exclusion) may be insufficient to account for all cell death caused by apoptosis. This may also be important in the accurate calculation of specific productivity. Finally, if one desires to improve the yield in an industrial process by decreasing cell death in the culture, it is crucial that one understand the mechanism of cell death in the cultures and the extent to which death decreases total VCD.

Viability can be assessed most simply by the exclusion of charged dyes such as trypan blue or propidium iodide (PI) that move freely into dead cells and exhibit minimal uptake into live cells. The extent of cell death can then be quantified by microscopy, or in the case of PI, by flow cytometry (1,5). Loss of plasma membrane integrity also results in the release of intracellular contents into the media that may then be quantified, such as the enzyme lactate dehydrogenase (LDH) (6). LDH content can be assayed in both the cellular and supernatant fractions, allowing relative quantification of cell death.

Apoptotic morphology specifically can be readily identified by staining fixed cells with a fluorescent dye such as acridine orange, which fluoresces green when bound to double-stranded nucleic acids (7). The classic apoptotic morphology of condensed or fragmented nuclear material can then be observed. An assessment of the percent apoptotic CHO cells during the course of a production run in a 2-L fermenter is shown in Figure 1. From day 7 there is an abrupt increase in the percentage of cells that show the characteristics of apoptotic cell death even though the total cell number in the culture does not change appreciably over this time period. However, the percentage of apoptotic cells increased from day 7 to a value of 50% apoptotic cells on day 11. Clearly, this large proportion of apoptotic cells will adversely affect the titers late in the run.

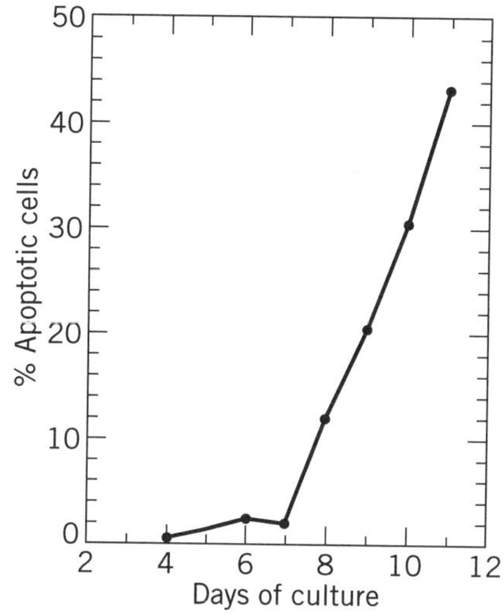

Figure 1. Apoptotic cell death in fermenters. Apoptosis was identified by light microscopy after acridine orange staining. Values are expressed as percent of total cell number that are apoptotic on days 4 to 11 of a 2-L minifermenter run using CHO cells.

An additional characteristic of apoptotic cell death that can be used for quantification is that of DNA strand breaks, which are introduced at the internucleosomal linker sections of DNA in apoptotic cells (8). The ultimate loss of short DNA fragments, coupled with reduced accessibility to intercalating fluorochromes such as PI because of the condensation of chromatin, often results in the congregation of apoptotic cells within a fixed cell population to a hypodiploid peak, when analyzed by flow cytometry (5). The percentage of cells localized to this peak can be quantified if not obscured by excessive amounts of cellular debris.

The free 3′-hydroxyl groups at the ends of the apoptotic DNA strand breaks may also be labeled with a fluorescent nucleotide using terminal deoxynucleotidyl transferase (9) or annexin (10). These techniques allow the fluorescelnation of apoptotic cells with only a low level (or no) labeling of nonapoptotic cells. Cells labeled in such a way may be counterstained with PI, allowing concurrent cell cycle analysis, so that exit from the cell cycle can be tracked by flow cytometry (1). These methods can be used to trace the effects of manipulating the culture parameters on the amount of cell death in the cultures (11). Such experiments are an invaluable adjunct to direct empirical determination of conditions that optimize titer because they can help in deciding which experimental manipulations, among the many possibilities, might yield benefits.

Considerations of Scale

Virtually all industrial productions of recombinant proteins or antibodies in mammalian cell culture are done on a large scale (100 to 12,000 L). The few exceptions may be production of antibodies or proteins for research use only. Although even these small- and intermediate-scale processes will benefit from using an optimized medium, any medium optimization will obviously be aimed at use in large-scale processes. We have found that suspension-adapted cells and suspension processes scale well from plates to fermenters. Spinners are actually less predictive, in some cases, than static plate cultures using suspension cells. (Suspension-adapted cells grown in tissue culture plates in medium containing additives such as F-68 frequently will not attach but remain in suspension.) The experiments described in this article can be performed, preferably using the suspension-adapted cells, in plates, including 96-well plates for high throughput screening. However, larger (e.g., 60 mm) plates allow multiple sampling from the same plates over time. The type of careful empirical derivation of optimal process parameters described herein really cannot be performed entirely in fermenters because too many conditions are required. However, these experiments can help narrow the number of parameters to be tested on a larger scale. Figure 2 presents data on yield of identical cells in identical medium cultured in small- and large-scale vessels. The cells in 60-mm plates, roller bottles, and spinners all performed similarly to those in the 1,000-L fermenter. The static shake flask cells that performed very poorly may have been suffering from oxygen deprivation. A few experiments, such as adjustments in medium volume, might allow closer congruence between the small-scale model selected and the performance in large-scale fermenters. All major changes or improvements to a process based on results from small-scale work should be confirmed using fermenters. The use of 2-L minifermenters allows replication of conditions and good statistical analyses of results. If these titers are then confirmed in larger-scale fermenters, one can be confident

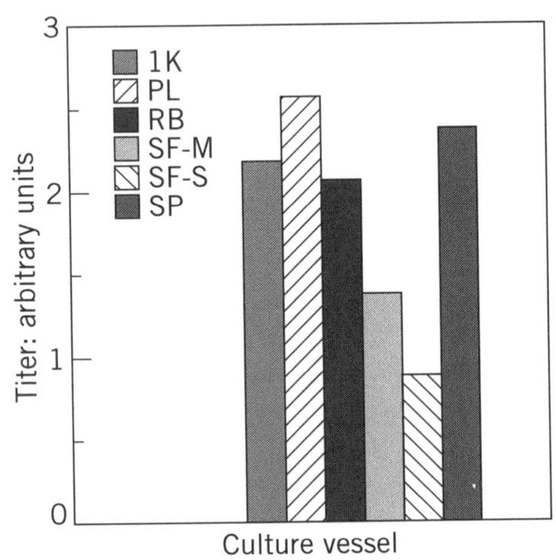

Figure 2. Cells were inoculated into a 1,000-L production fermenter. An aliquot of these cells was removed and dispensed into the various culture vessels on day 0. All vessels were sampled and the titers assayed throughout the culture period. The data shown compare the titers of product produced by cells in the different vessels on day 5 of culture. 1K, 1,000-L fermenter; SP, 1-L spinner; RB, roller bottle; PL, 60-mm tissue culture plate in CO_2 incubator; SF-S, static shake flask in incubator; SF-M, shake flask mixed using a stir bar in incubator.

that a rational approach using small-scale cultures, which allows complete analysis of the interacting variables in the medium, has led to a robust and optimal process.

Although attached cell culture systems can be scaled up via roller bottle microcarrier culture, hollow fiber systems, and so forth, these devices have generally been used for medium-scale rather than very-large-scale production. Each introduces a different set of variables that should be understood to design a small-scale equivalent system for optimizing medium, as described later. Again, any improvements should be compared directly to the standard conditions before adoption.

Some parameters change, of necessity, when going from static cultures to fermenters. The method of pH control differs because base and acid addition are used to regulate pH in fermenters in addition to the bicarbonate buffer. This results in a more stable pH over the course of a run in fermenters, compared to plates, but at the cost of a greater increase in osmolarity in the fermenters because of the added salts. Shear and foaming are also factors in fermenters, but not plates. Components of the medium that are added to fermenters to prevent shear and foaming should also be added to plate experiments when optimizing medium to allow for better concurrence between large- and small-scale results. Finally, the control of P_{O_2} is quite different in plates (diffusion) and fermenters (air or oxygen sparging). Adjusting the total surface area to volume in plates and spinners can have a significant effect on protein production and cell number and help in creating closer agreement between small- and large-scale data.

Some parameters such as pH, temperature, and P_{O_2} can be adjusted or controlled more easily in a fermenter. These parameters are best optimized directly in experiments performed in fermenters after the nutrient and growth supplements have been optimized at small scale and tested at large scale. It is important to optimize these parameters both for improved cell growth and, on occasion, to improve product quality. With a few experiments to directly compare small- and large-scale culture, and an understanding of the principle underlying differences in plate and fermenter cultures, the former can be an invaluable tool in the process of obtaining optimized media for large-scale industrial processes.

Nutrient Component of the Medium

For convenience, the culture medium can be divided into two components: the nutrient portion and the added growth factors and growth-promoting substances. The nutrient components are generally part of the commercially available powdered nutrient mixtures, such as MEM (minimal essential medium), DMEM (Dulbecco's Modified Eagle's Medium), Ham's F12 nutrient mixture, RPMI 1640, or F12/DME. These media were all empirically derived for very specific cell types and growth conditions (12), none of which are very similar to a suspension culture of genetically modified cells in a fermenter. However, they still make a good starting point for optimizing the culture medium for the specific production line of choice. An initial screen should compare several commercially available or proprietary media to determine which is optimal for the production line. If the medium is being optimized before the cells are transfected, cell number can be used as the end point. If the line is already producing the protein of interest, protein titers should be measured. If the conditions for obtaining maximal titers and maximal VCD are the same, this is a good starting point for optimizing the nutrient portion of the medium. If different media give optimal VCD and titer, then more studies should be undertaken to understand the biology behind this phenomenon. This will prove useful in designing the optimal production system.

Once a medium is found that works well, there is a temptation to move on the assumption that this adequate. However, large-scale commercial culture is one situation in which it is almost always worth the time and effort required to obtain a truly optimal medium. Optimizing the nutrient portion of the medium is important not only for the increased yields it will afford directly but because optimizing nutrients will lead to a decreased requirement for parts of the growth supplements such as serum or growth factors (13). Additionally, optimizing VCD will lead to the ability to prolong culture time while minimizing damage to the product caused by enzymes released from dying cells. Even a 10% increase in VCD can be significant, and improvements of 2-fold to 10-fold in titer are frequently seen with optimized media. Because it is very difficult to change a process after clinical trials have begun, these early improvements will result in accrued benefits extending over many years that are often worth millions of dollars during the life of a production process.

The available nutrient mixtures generally contain amino acids, a carbohydrate source (usually glucose), inorganic salts, vitamins, some type of buffering system (usually CO_2, bicarbonate, and phosphate), fatty acids or more complex lipids, and other components that may vary from medium to medium (e.g., nucleosides, nucleotides, coenzymes, detergents). These components interact with each other so that the optimal concentration will depend to some extent on concentrations of other components in the mixture. Many producers of cell culture media and components have developed their own proprietary cell culture media for specific cell types used in production of recombinant products (e.g., CHO-S-SFM, #12050 or Hybridoma-SFM, #12045 Gibco/BRL, and products from other companies such as JRH, Hana Biologics, and Sigma). These media may offer a shortcut when proteins must be produced in short order; however, incomplete disclosure of the medium components can be a cause for concern, and costs are frequently higher than those of media formulated in-house.

Optimization of the medium starts with choosing the best existing commercial or proprietary nutrient mixture. Then, media components are made up individually and in groups (12,13), and individual media are prepared missing only one component. That component is then titrated using the desired end point of viable cell number or product titer, if a good high throughput assay is available. Figure 3 shows the response of CHO cells to three media components: inositol, KCl, and isoleucine. It is obvious that the omission of KCl or isoleucine results in cell death over the production period and consequently very low protein yields. However, the omission of inositol had, if anything, a positive effect. Conversely, increasing the inositol concentration 20-fold had little detrimental effect, whereas increasing isoleucine was beneficial and high levels of KCl were toxic. Using these data, one might then adjust the inositol concentration downward by 10-fold, the KCl downward by 2-fold, and the isoleucine upward. By using the data from all the titrations, one can then determine the optimal concentrations for each component and make up the best medium. This titration can then be performed again, using the best medium as a starting point. Each successive round of optimization would be expected to yield less benefit as the optimum is approached. A comparison of titers from a production cell line in the starting (control) and round-one optimized medium is shown in Figure 4. This one round of optimization more than doubled the titers during the 8-day production period. This benefit was achieved in large part by increasing the VCD in the culture. Note that significant changes in the culture process, such as changing cell lines, changing inoculum density, changing the supplements, or going from attached to suspension culture, might well require reoptimization of the medium. With experience, it is possible to identify a few medium components that are most important and check only these after the initial optimization. Conversely, once an optimal medium is designed for a given cell line (e.g., CHO), the cells can be grown in that medium during transfection and amplification in order to select for cells that grow optimally in the medium designed.

Growth Factor Additions

The nutrient mixtures discussed in the previous section are generally supplemented with either undefined mixtures, such as serum, that promote the growth of cells or defined mixtures of growth factors, transport proteins, and hormones. The choice of whether one will attempt to reduce or eliminate serum and which supplements to use is based on the characteristics of the product and the cell line as well as cost and safety considerations. Most established cell lines have been carried in serum for years and are therefore well adapted to growth in serum. It may be difficult to achieve equivalent growth rates or titers in a serum-free medium. Table 3 illustrates the effect of medium supplementation on tissue plasminogen activator (tPA) production by Bowes melanoma cells. Even though higher specific productivities can be reached in defined medium, when compared to serum-containing medium, the total yield is lower because of the much-reduced cell growth. Note also that some of the additions primarily affect cell number, whereas others (e.g., nerve growth factor) increase specific productivity with little effect on cell yield. The cells in Table 3 are not recombinant; however, similar effects can be seen using recombinant proteins and 293, CHO, or other host cell lines. In this case, a subset of factors will increase cell number (via decreased cell death or increased cell division), and another set of factors might increase titer via increases in specific productivity of the cells or increase protein stability. The former factors might be useful for all CHO-derived production lines, whereas the latter may be product specific.

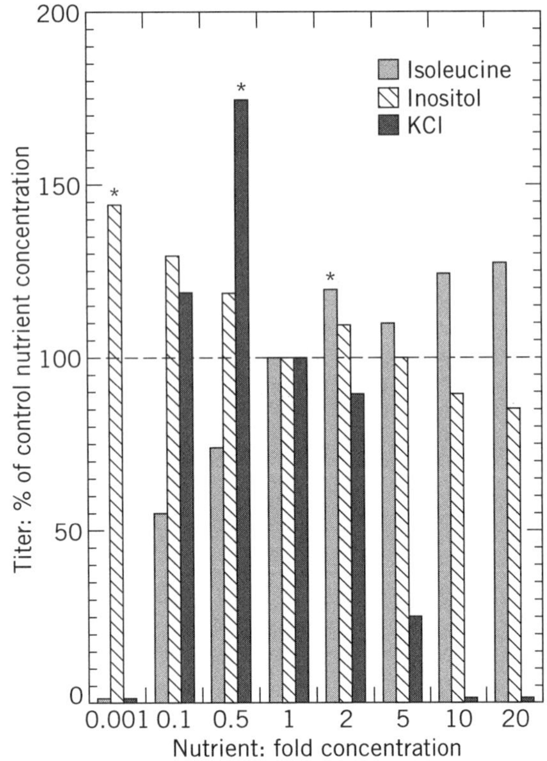

Figure 3. Titration of individual medium nutrients. CHO cells transfected to produce a recombinant protein were grown in production medium from which each of the indicated components had been omitted. These components were then titrated at the indicated concentrations. Cells were plated at standard inoculum and media collected to assay for product on day 6. All cultures were in 60-mm plates at 37 °C in triplicate. Values are expressed as percent of control where the control is the normal concentration (1×, *dotted line*) of each component in the medium. The ordinate values are the fold increase or decrease of the single component compared to control. The asterisks indicate the optimal concentration for each component that will then be chosen for use in the next round of medium optimization.

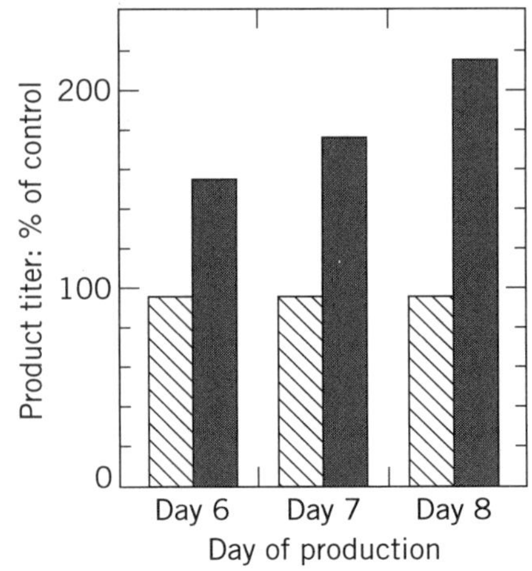

Figure 4. Comparison of titers obtained using the same cells before (control, *hatched bars*) and after (*filled bars*) optimization of the nutrient portion of the medium. These differences result entirely from changes in concentrations of the nutrients. Results are expressed as percentage of control where the control is the starting F12/DME medium.

Table 3. Cell Number, Titer, and Specific Productivity of Cells Producing tPA

		Plasminogen activator	
Supplement	Cell no. × 10^5	ng/10^5	ng/mL
Serum free	0.69	<30	Lts
+I	0.86	<30	Lts
+I+TF	1.08	<30	Lts
+I+TF+NGF	1.06	54	57
+I+TF+NGF+P	1.09	55	59
7.5% FBS	6.00	30	178

Note: Bowes melanoma cells were grown in F12/DME supplemented as indicated. Medium was collected after 5 days in culture. Lts, less than standard; I, insulin; TF, transferrin; NGF, nerve growth factor; P, progesterone; FBS, fetal bovine serum.

However, there are several reasons to eliminate serum, where possible, even if this results in a decrease in titer. Serum is expensive, contains high levels of many unknown proteins (about 40 mg/mL) that must then be purified away from the recombinant product, and contains enzymes and binding proteins that may affect product quality. For example, the production of tPA in the presence of serum will yield predominantly two-chain cleaved tPA rather than single-chain material. This is because the recombinant plasminogen activator produces plasmin from the plasminogen in the serum, which in turn cleaves the tPA. Thus, the tPA produced from the cells in Table 3 in the serum and serum-free conditions would have different characteristics. If single-chain material is desired, it is best to try and improve the cell yield by adding other supplements or adjusting the nutrient mixture.

One straightforward method for reducing contaminating proteins from serum is to simply reduce the serum concentration. This makes the medium less expensive in many cases, simplifies purification by eliminating serum proteins, and often improves product quality or cell growth. The general practice is to wean the cells by reducing the serum concentration over several passages (e.g., from 10% to 5%, 2%, 1%, 0.5%, 0.1%, etc.) allowing two or more passages at each concentration if a reduction in growth rate is observed. This method often works, but it is exceptionally time consuming. Eliminating serum altogether over one to two passages is often a reasonable solution for cells that do not have very fastidious growth requirements, although the resultant titers might not be optimal. A better solution is to derive a set of supplements that substitute for serum without any adaptation being required. In our experience, one can usually obtain titers in serum-free medium similar to, or even better than, those obtained in serum, although some thought must be taken as to the parental cell lines used for expression (see "The Cell Line").

With the advent of bovine spongiform encephalitis (BSE), it is likely that more processes will be driven to become serum free for safety considerations. Because testing for BSE on each lot of serum is time consuming and expensive, elimination of serum from the process is an easy way to avoid this problem. However, the entire process must be carried out serum free if one is to avoid the chance of infection. This is a more stringent growth condition than growing cells for only one passage serum free and requires more complete supplementation.

The question of which supplements to add is again one that can only be answered in the context of the desired end point. It should be emphasized in this context that optimizing the nutrient portion of the medium can reduce growth factor (and serum [13]) requirements. Optimizing the nutrient portion of the medium is generally less expensive than adding supplements such as serum or growth factors. For production, minimizing expense and optimizing product quality may be more important than maximizing titer. It is the final yield of purified product as a function of fermentation plus purification cost that is the determining factor in deciding what the best medium will be. In an optimized serum-free culture process, the recombinant product may comprise 10 to 80% of the total protein in the harvest medium. This makes purification easier and less costly. In contrast, if one is growing up cells for the seed train, minimizing the doubling time to have cells ready for inoculation may be of more importance than the cost of the medium because cells are grown in smaller volumes. The quality of the protein produced is also not an issue in the early passages after thaw. However, it is generally best to not make too many major changes in media while bringing cells from the thawed vial to the production tank in order to avoid lag time as the cells adjust to new conditions.

Much has been written concerning optimizing the supplements for serum-free media. Our preference is to use small-scale culture experiments to obtain defined supplements that will provide cell numbers and product titers equivalent to or higher than those obtained in serum-supplemented medium. The elimination of serum necessitates the addition of hormones, growth factors, trace elements, and lipids (12,14–16). Because of the low concentrations of these components, the absence of binding and carrier proteins provided by serum, and the instability of the components in culture medium, these must be added to the media after filtration and just before use. The most commonly required additives are insulin (1 to 10 μg/mL), transferrin (1 to 100 μg/mL), and selenium (10 to 30 nM); however, some cell lines have an added requirement for lipids. One can then determine, based on experimental data, whether the increased product yield obtained by adding the most expensive components of the medium (for example, an epidermal growth factor requirement) are worth the expense. Using such data, one can then determine a minimum required medium that contains the most cost-effective supplements. A more complete description of deriving the appropriate hormone supplement in order to eliminate serum is given by Barnes and Sato (15) and Hewlett (17). For cell lines that are particularly fragile, Pluronic F-68 (18) or Pluronic F-127 can be added to the medium to prevent shear-associated cell lysis. Finally, if the product is sensitive to proteolysis, one might wish to add protease inhibitors, as discussed more fully later.

The empirical derivation of medium formulations optimizing both the nutrient portion of the medium and the supplements is time and labor consuming. It is seldom undertaken in a thorough fashion. One often-expressed wish is the desire to be able to predict what the optimal medium will be for a given cell line. Unfortunately, at this time, it is not possible to do this with any degree of assurance. We can make good guesses, largely based on experience gained by performing previous experiments on similar cells, but this will probably result in a good enough, not an optimal, medium. The only cases in which a full optimization of the medium is undoubtedly justified are (*1*) those research laboratories interested in understanding the nutritional and hormonal requirements of cells in vitro and (*2*) in the development of a commercial process where small gains in titer, viability, and stability would, over many years of production, be expected to repay the initial effort expended in the optimization. Most large-scale culture processes fall into this category.

OPTIMIZING SPECIFIC PRODUCTIVITY

Specific productivity is the amount of recombinant protein produced per cell per specified time period. Given that the cells are growing in an optimal medium, this will be a function of the expression system (promoter, enhancer, etc.) chosen and how well that particular system works in the host cell chosen, the integration site and orientation of the recombinant plasmid, the copy number of the gene to be expressed, and any specific or nonspecific inducers used. A discussion of the many expression systems available (19) is beyond the scope of this article. However, the method used to select and clone transfected cells and amplify gene copy number frequently impacts the performance of the cells in the production culture medium and will be discussed briefly.

The process of obtaining a stable cell line expressing a foreign protein involves a method of permeabilizing the cells to allow the plasmid containing the foreign DNA to enter the cell, a method of selecting for the minority of cells that have integrated the DNA into the nucleus, cloning to select a clonal cell line producing high levels of protein, and frequently several rounds of amplification of the DNA, each followed by cloning. This process allows ample opportunity for both genetic changes in the host line and the selection of a subpopulation from the initial population. Usually, the primary, if not the sole, concern is to maintain the gene containing the recombinant protein in the cells. However, the phenotype of a good production cell line also includes the ability to grow rapidly and to high saturation density while maintaining a high viability. If one has made the effort to optimize the medium for the growth of the parental cell line, it is desirable to maintain the ability to grow well in this medium, including serum-free and suspension growth. To maintain the desired phenotype, selective pressure should be maintained for these growth characteristics during the process of transfection, amplification, and cloning. If the cells are grown in the process medium whenever possible, they will be much easier to handle when they reach the production setting. To the extent that this is not possible, the production cell line selected might benefit from a period of adaptation to the process medium and growth conditions before being banked.

OPTIMIZING PRODUCT STABILITY

The production medium is designed to provide the maximum number of cells secreting protein over the maximum amount of time. The longer the productive run time, the longer the product can accumulate in the medium and the higher the resultant titer. However, this medium, along with the metabolites and proteins being secreted into it by the cells, is also the storage medium for the proteins being produced and secreted. An environment that is 37 °C, neutral pH, fully oxygenated and stirred and where proteases and other enzymes are separated from the proteins by, at best, a layer or two of cell membrane would hardly be considered optimal for protein storage by any biochemist. This is, however, what recombinant proteins produced in the first days of the production run have to withstand for the duration of the run. Therefore, some thought should be given to protein stability in the production medium. This will be in large part dictated by the characteristics of the protein being produced. Figure 5 illustrates the accumulation of titer during a production run with two different proteins. One is stable and accumulates linearly during the run; the other is easily proteolyzed and begins to degrade faster than it is produced when the cells in the culture begin to die. Continuation of the culture at this point leads to loss of product. The most important change in this process would be to optimize the medium components to prolong cell viability. One might then adjust the inoculum density so that the number of viable cell days is optimized. Too high an initial inoculum can result in early cell lysis and degraded product, whereas too low an inoculum results in too long a production period to reach the desired titer, resulting in overly long product residence time in the medium. One might also adjust the temperature to minimize cell death (11), shorten the run (including using processes that shorten the product residence time, such as perfusion), or add protease inhibitors to the medium. Protease inhibitors can be nonspecific (e.g., serum albumin) or specific (e.g., aproteinin or small molecular weight inhibitors). The small molecular weight inhibitors are frequently toxic to cells in culture, whereas inhibitors such as aproteinin are well tolerated but very expensive for production use.

THE MARRIAGE OF THE MEDIUM AND THE PROCESS

As discussed in the introduction, medium formulation and optimization should not proceed in a vacuum but should be an integral part of the overall process design. From the initial choice of cell line through to the final purification, there should be consideration of the entire final process that will result.

The Cell Line

The production cell line should be chosen not only for its protein secretion rate measured under laboratory conditions but also for its ability to perform well in the special conditions that occur in large-scale culture (Table 2). The cells may be adapted to these conditions after transfection, as discussed earlier, or the parental cell line may be engineered to optimize its performance as a production cell line before transfection. Cells can thus be adapted to grow in suspension or serum-free medium and then be banked for use as parental cell lines. If they are maintained in these conditions during transfection and selection, then the recombinant protein producing the cell lines will have these important characteristics. Even if they cannot be handled in suspension or serum-free conditions at every step of the

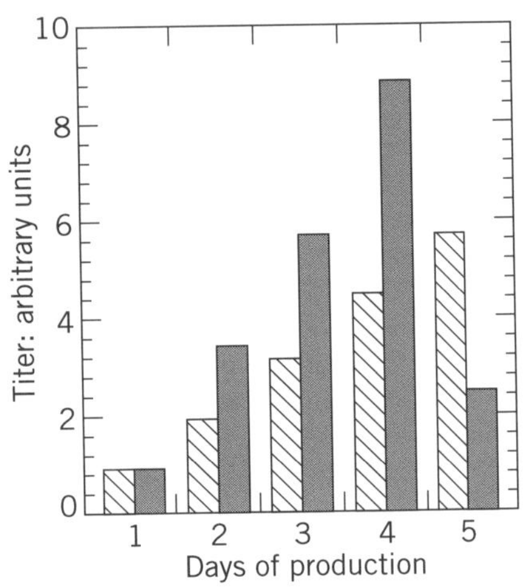

Figure 5. Titer as a function of run length. The figure shows the cumulative titers during two runs of recombinant cells, each expressing a different protein (*open bars* and *hatched bars*). The assays used only measure intact protein. Data are expressed as fold increase over control, where the control is the titer on day 1 of the culture.

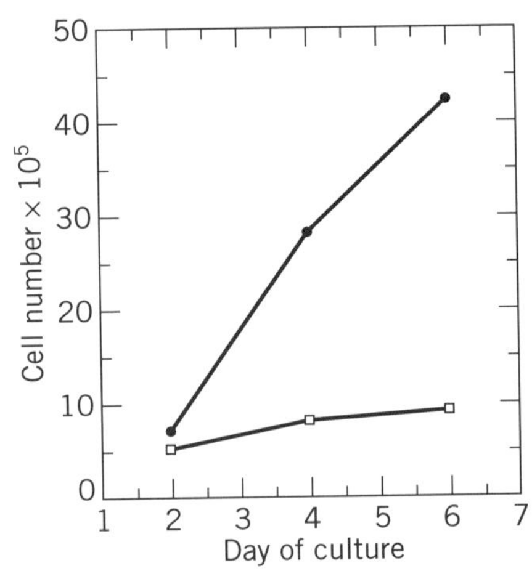

Figure 6. Insulin-independent growth of CHO cells transfected with the proinsulin gene. These cells can grow as well in the absence of insulin as their parent line can grow in medium supplemented with 10 μg/mL insulin. CHO-parent line (*squares*) and dp12-proinsulin transfected clone (*circles*) were grown in the absence of insulin.

transfection process, they will be more likely to retain the desired characteristics.

Additionally, one can genetically engineer the parental cell line to have traits that are optimal for a production cell line. One example of this is to transfect the production cell line with a required growth factor, allowing autocrine growth at large scale without exogenous addition of the factor. Figure 6 demonstrates that CHO cells that have been transfected with the proinsulin gene have become insulin independent for growth. This will, of course, alter the composition of the medium used to grow these cells.

Equipment

It is obvious that the optimal medium for a process will depend, in part, on the equipment used and the configuration of the process. Clearly, suspension cultures and attached cells might require different media. The optimal nutrient mixtures for batch cultures and for perfusion cultures would be quite different. The inoculum density and amount of growth that will be required of the cells at each stage of the scale-up process will depend on the tank sizes used to inoculate each stage. The equipment to be used for the final process should be considered from the first stages of medium optimization so that one is sure that the medium designed does, in fact, support the cell growth and viability required in the process.

Downstream Purification

Because the final goal of the process is to obtain a highly purified active protein, the downstream purification methods to be used should be considered from the beginning. Some protein characterization should be performed throughout the medium development process, in addition to a simple measurement of titer. Medium changes can alter protein characteristics as well as titers. It is therefore important to make certain that the desired product characteristics are maintained when medium changes are made to improve titer.

Finally, everything that is put into the medium by the formulation or by the cells in the culture must be removed from the final protein product. Wherever possible, it is always best to use supplements that will not raise safety considerations. Additionally, it does not make sense to add supplements to increase titer if the resulting product will be lost because of the extra purification steps that must be added. There is not as much room to alter pH, salt concentration, and so forth in a cell culture medium as there is in protein purification processes, but some understanding of the constraints and problems of both ends of the process will help make for a smooth transition and a reproducible process.

In summary, the time and effort required to optimize all components of the production medium will pay off many times over in a mammalian cell culture process. Optimal medium not only optimizes protein titers but can lead to improvements in protein quality and stability, more stable cell lines, more reproducibility from run to run, and improved use of facilities. To obtain maximal benefit from medium design it must be done in concert with the transfection and selection of the production cell line, the configuration of the manufacturing plant, and the design of the downstream purification process. All these factors working together have made large-scale mammalian cell culture an economically important part of the modern biotech industry.

ACKNOWLEDGMENTS

We would like to acknowledge the invaluable interactions with other members of Genentech's process sciences division, especially the cell culture R&D department, the cell genetics department, and assay services group. All these people contributed to the experiments and experiences described in this article.

BIBLIOGRAPHY

1. A. Moore, C.J. Donahue, J. Hooley, D.L. Stocks, K.D. Bauer, and J.P. Mather, *Cytotechnology* **17**, 1–11 (1995).
2. F. Franek, T. Vomastek, and J. Doinikova, *Cytotechnology* **9**, 117–123 (1992).
3. M.J. Arends and A.H. Wyllie, *Int. Rev. Exp. Pathol.* **32**, 223–254 (1991).
4. J.V. Moore, in C.S. Potten ed., *Perspectives in Mammalian Cell Death*. Oxford Univ. Press, Oxford, U.K., 1987, pp. 295–325.
5. W.G. Telford, L.E. King, and P.J. Fraker, *Cytometry* **13**, 137–143 (1992).
6. J.L. Goergen, A. Marc, and J.M. Engasser, *Cytotechnology* **11**, 189–195 (1993).
7. C. Dive and A.H. Wyllie, in J.A. Hickman and T. Tritton eds., *Frontiers in Pharmacology: Cancer Chemotherapy*, Oxford Univ. Press, Oxford, U.K., 1993, pp. 21–56.
8. R. Duke, R. Chervenak, and J. Cohen, *Proc. Natl. Acad. Sci. U.S.A.* **80**, 6361–6365 (1983).
9. W. Gorczyca, S. Bruno, R.J. Darzynkiewicz, J. Gong, and Z. Darzynkiewicz, *Int. J. Oncol.* **1**, 639–648 (1992).
10. R. Pitti, S. Marsters, S. Ruppert, C. Donahue, A. Moore, and A. Ashkenazi, *J. Biol. Chem.* **271**, 12687–12690 (1996).
11. A. Moore, J. Mercer, G. Dutina, C. Donahue, K. Bauer, J. Mather, T. Etcheverry, and T. Ryll, *Cytotechnology* **27**, 47–54 (1997).
12. R.G. Ham and W.L. McKeehan, *Methods Enzymol.* **58**, 44–93 (1979).
13. W.L. McKeehan and R.G. Ham, *In Vitro* **13**, 399–416 (1979).
14. J. Bottenstein, I. Hayashi, S.H. Hutchings, H. Masul, J.P. Mather, D.B. McClure, S. Okasa, A. Rizzino, G.H. Sato, G. Serroro, R. Wolfe, and R. Wu, *Methods Enzymol.* **58**, 94–109 (1979).
15. D. Barnes and G.H. Sato, *Anal. Biochem.* **102**, 255–270 (1980).
16. J.P. Mather ed., *Mammalian Cell Culture: The Use of Serum Free and Hormone Supplemented Media*, Plenum Press, New York, 1984.
17. G. Hewlett, *Cytotechnology* **5**, 3–14 (1991).
18. J.E. Swim and R.F. Parker, *Proc. Sci. Exp. Biol. Med.* **103**, 252–254 (1960).
19. D.V. Goeddel, *Methods Enzymol.* **185**, 567–577 (1991).

See also ADSORPTION, PROTEIN, BATCH; ANTIBODY PURIFICATION; MEDIUM FORMULATION AND DESIGN E. COLI AND BACILLUS SPP.; FREEZE DRYING, PHARMACEUTICALS; ROLLER BOTTLE CULTURE, MIXING.

CULTURE PRESERVATION, BACTERIA, FUNGI, YEAST, AND CELL LINES

Robert L. Gherna
American Type Culture Collection
Rockville, Maryland

KEY WORDS

Commercial freeze-drying
Component freeze-drying
Control Freeze
Dimethyl sulfoxide (DMSO)
Freeze-drying
Glycerol
Lyophilization
Manifold freeze-drying
Sublimation
Ultrafreezing
Manifold freeze-drying

OUTLINE

INTRODUCTION

The ubiquitous presence of microorganisms, as well as their metabolic diversity, has made them important factors in industrial applications and product development. Bacteria, fungi, and yeasts have been used for the production of food, medicinals, solvents, enzymes, and numerous other products. They have also been used as standards for bioassay and bioremediation applications. In addition, advances in bioengineering have enabled the incorporation of genes into microorganisms for the production of mammalian proteins and peptides.

During the past 20 years, cell line technology grew dramatically to enable the isolation and maintenance of a wide variety of tissue lines from any species of interest. The initial use of cell lines for vaccine production was expanded by developments in cell hybridization, transduction and transfection, and genetic manipulation to include the production of monoclonal antibodies, cytokines, and other pharmaceutical products. Lines that have been genetically engineered have been successfully expanded and grown in fermenters for large-scale production.

The development of microbial and cell cultures represents an enormous investment in time and money and is an important asset to industry. These cultures must be protected from accidental loss by means of standardized preservation techniques. The primary aim of culture preservation is to maintain the culture such that it is as close as possible to the original strain or line exhibiting the desired phenotypic or genetic traits. Often, industrial cultures are not analogous to taxonomic strains in that they have been modified to have special properties, either rapid growth, more active metabolic rates, or other features.

Many methods have been used to preserve bacteria and fungi, but not all species respond in a similar way to a given method. Unfortunately there is no universal method that can be applied to all microorganisms. It should be emphasized, however, that the success or failure of any preservation technique also depends on the use of the proper growth medium and cultivation procedures and on the age of the culture at the time of preservation. This is particularly true when working with microorganisms that contain plasmids or recombinant DNA, or that exhibit growth phases such as morphogenesis or spore formation.

The following provides practical information for the maintenance and preservation of bacteria, fungi, yeast, and cell lines. It should be noted, however, that not all microorganisms will be preserved by these methods, and some experimentation with cryoprotective agents and other factors may be necessary for success. Additional information on bacterial, fungal, and cell-line preservation may be found elsewhere (1–3).

CULTURE AND PRESERVATION OF BACTERIA

Short-Term Preservation

Subculture. Most bacteria can be preserved by periodic serial transfer to fresh medium. The period between transfers varies with the organism, medium employed, and the external conditions. In general, minimal media are preferred for subculturing because they lower the metabolic rate of the organism and thus prolong the interval between transfers. In rich media, growth is often much faster, with the accumulation of metabolic products. These compounds may alter environmental conditions such as pH and gas phase and shorten the interval between transfers. Some bacteria such as pathogens need complex media (4) for growth, or their retention of specific physiological properties require the addition of complex compounds to the medium. The subculturing interval for these organisms must be determined by experience.

When preserving bacteria by serial transfer, it is important to use containers that can be securely sealed. Glass test tubes or 6-mL bottles with rubber-lined screw caps are widely used. The bacteria can be grown with slightly loose caps until adequate growth is obtained and then tighten to prevent dehydration. Solid media are preferred to broth because contaminants can be readily seen. Duplicate tubes should be maintained as a precaution against loss, at least until the subculturing interval is determined. The cultures should be examined for purity after each transfer, and abbreviated characterization tests must be run periodically to ensure retention of desired traits. Storage in a refrigerator is the preferred method for subcultures. Many bacteria can be kept for 3 to 5 months between transfers if proper precautions are taken to avoid dehydration of the medium.

It should be noted, however, that the maintenance of cultures by serial transfer is precarious, time-consuming, and expensive. The frequent transfer of cultures can result in contamination, mislabeling, and the selection of strain variants.

Other methods involve storing cultures at refrigeration temperatures, drying in sterile soil, glass beads, filter paper, and other substrates. Detailed descriptions of these methods can be found in Refs. 5–7.

Short-term maintenance and preservation procedures are not recommended for industrial strains because a number of problems can occur using these methods. The main hazards are contamination, population changes through selection resulting in the loss of key genetic and physiological properties, and the expense of maintaining cultures under these conditions.

Ordinary Freezing. The use of the freezer compartment of a refrigerator or an ordinary freezer with a temperature range of 0 to −20 °C to preserve bacteria is not recommended because the freezing process (see "Long-Term Methods", Deep Freeze) damages the cells. In addition, most refrigerators that have a freezer compartment in the range of 0 to −20 °C are frost free. The frost-free state is achieved by an alternating warming and cooling cycle, which can result in further damage to the cells and a loss of viability.

Long-Term Methods

Freeze-Drying, or Lyophilization. Freeze-drying, or lyophilization, is one of the most economical and widely used methods for long-term preservation of bacterial cultures and other microorganisms. Many metabolically diverse bacterial species have been successfully preserved by this technique and have remained viable and unaltered for more than 60 years.

Freeze-drying is a complex process that involves the removal of water and other solvents from a frozen product by sublimation (8,9). Sublimation occurs when a frozen liquid goes directly to a gaseous phase without passing through the liquid phase. There are three stages in the freeze-drying process: *(1)* prefreezing of the suspension to ensure a solid frozen starting material, *(2)* primary drying during which most of the water is removed through sublimation, and *(3)* secondary drying to remove the bound water.

The rate of cooling during the prefreezing step and the final temperature of the frozen material can affect the freeze-drying process. The removal of water during the primary drying stage requires an environment that allows the water molecules to migrate from the frozen material. This is achieved by using a vacuum pump. A large-capacity pump must be used to prevent the water vapor load from compromising the efficiency of the system. The procedure also requires a moisture trap or condenser to collect and remove the water molecules before they enter the vacuum system. The condenser temperature must be lower than the frozen material for sublimation to occur. The sublimation rate is dependent on the differential vapor pressure between the frozen material and the condenser. Because the vapor pressure varies with the temperature, heat can be carefully applied to the frozen cell suspension to increase the vapor pressure and promote a faster primary drying rate. Precautions must be taken during the application of heat to prevent the frozen cell suspension from thawing. Most commercial instruments have sophisticated controllers that allow the input of heat to the frozen material to achieve maximum drying rates without damaging the product.

When the primary drying is complete, some bound water remains. This water cannot be eliminated by sublimation and requires the application of external heat to the material under low pressure and low condenser temperature. The moisture remaining in freezed-dried cells is termed residual moisture. A residual moisture content of 1–3% seems to be required for the long-term shelf life of bacteria. For an in-depth review of the complex freeze-drying process, the reader is referred to several excellent reviews (10–12).

Centrifugal freeze-drying and prefreezing are two of the most commonly used methods of lyophilization of bacteria. In centrifugal freeze-drying, the suspensions are initially frozen by evaporative freezing under a vacuum while centrifuging to prevent frothing due to removal of dissolved gases. After primary freeze-drying, the vials are constricted using a narrow flame and then placed on a manifold for secondary freeze-drying. A detailed discussion of centrifugal freeze-drying can be found in articles by Rudge (13) and Lapage (14). The prefreezing method employs a controlled rate of cooling to freeze the cell suspension and then a primary drying phase, as discussed earlier.

The simplest system consists of a vacuum pump capable of an ultimate pressure of less than 10 μmHg (a pump rated at 35–50 L/min is usually adequate), a small stainless steel condenser for cooling with dry ice, a thermocouple vacuum gauge to monitor the vacuum system, and heavy-walled vacuum/pressure tubing to connect the components. The connection of these modules into a single system is often termed a component freeze-dryer. The main components are the vacuum pump, thermocouple vacuum gauge, and a condenser that can be attached to several manifolds, to which glass vials can be attached by means of vacuum-tubing nipples, or the manifolds can be connected to a stainless steel pan containing the vials. The manifold system is relatively simple, and it has been used successfully by many laboratories to preserve cultures. A small amount of cell suspension (0.2 mL) containing a cryoprotectant is dispensed into sterile ampules, which are plugged with sterile cotton. A 1-inch piece of nonpowdered amber latex IV tubing is attached to the rim of each ampule (by using a tube stretcher) and then to the manifold. The ampules are then immersed in an ethylene glycol (50%) dry-ice bath at −40 °C to freeze the cell suspension. The freeze-drying cycle is initiated by starting the vacuum pump and allowing the temperature of the bath to rise of its own accord to ambient temperature. In general, runs can be started early in the afternoon and allowed to proceed overnight. At the end of the drying cycle, the ampules are sealed using a double-flame air/gas torch and stored at 2–8 °C. For specific details on the equipment and methods see articles by Gherna (15) and Simione and Brown (16). Details of other procedures and equipment are given by Haynes et al. (17).

The American Type Culture Collection (ATCC) has successfully lyophilized many of the bacteria described in the literature. It currently uses three methods:

1. *Component freeze-dryer*. Samples are freeze-dried in cotton-plugged inner vials, which are then sealed in outer vials under vacuum.
2. *Commercial freeze-dryer*. Cultures are freezed-dried in the double-vial system in a commercial freeze-dryer.
3. *Preceptrol cultures*. Cultures are lyophilized in a commercial freeze-dryer using glass serum vials that are sealed with a rubber stopper and metal cap.

It should be noted that after the drying cycle is complete in the commercial freeze-dryer and Preceptrol methods, the vacuum in the drying chamber is replaced with cooled nitrogen gas (2–8 °C) that has been passed through a 0.2-μm Pall filter. The condenser temperature must not be allowed to rise above −50 °C. The nitrogen gas may have to be passed through a copper coil immersed in liquid nitrogen.

Culture Preparation. Successful freeze-drying depends on using healthy cells grown under optimum conditions on the medium of choice for each strain, which ensures the retention of the desired features of the bacterium. Bacteria can be grown on agar or in shaken or static broth cultures. The cells are harvested at maximum stability and viability, in the late logarithmic or early stationary phase, and suspended in a medium. The cell suspension should contain at least 10^7–10^{10} cells/mL in order to achieve optimal results. Anaerobic cultures must be grown, harvested, and dispensed under anaerobic conditions (18).

The viability of freeze-dried cultures is greatly influenced by the suspending medium. The choice of suspending medium for freeze-drying depends on the type of bacteria and on the method used.

The ATCC uses two suspending media for the commercial freeze-drying method:

1. Reagent 18
   0.75 g trypticase soy broth
   10.0 g sucrose
   5.0 g bovine serum albumin fraction V
   100.0 mL distilled water
   Filter sterilized through a 0.2-μm filter
2. Reagent 20
   10.0 g bovine serum albumin fraction V
   20.0 g sucrose
   100 mL distilled water
   Filter sterilized through a 0.20-μm filter

Reagent 18 is more commonly used for freeze-drying bacteria and fungi. Reagent 20 is employed for bacteria that are adversely affected by trypticase soy broth and is added to an equal volume of cell suspension in the broth.

In the component freeze-drying method (double-vial), the additive is a 20% (wt/vol) sterile solution of skim milk, prepared by autoclaving a 20% skim milk solution at 116 °C for 20 min in 10-mL tubes. For cultures grown on agar surfaces, the cells are washed off with the 20% skim milk solution. Broth cultures are centrifuged, and the cell pellet is resuspended with the sterile skim milk to give a suspension of 10^8 cells/mL. Skim milk should not be used for bacteria that are inhibited by milk, instead use a 24% (wt/vol) sterile sucrose solution diluted equally with growth medium to yield a 12% (wt/vol) sucrose solution (final concentration) for the single-vial manifold or commercial freeze-dryer method. As soon as the cell suspension is prepared, it should be dispensed into the ampules. The interval between dispensing and freeze-drying should be kept to a minimum to avoid possible alteration of the culture. Other investigators have used 10% (wt/vol) dextran, horse serum, inositol, raffinose, trehalose, and other cryoprotective chemicals in the suspending medium (19–21).

Most bacteria can be lyophilized using these methods; for those cultures that require special conditions, the reader is referred to Ref. 16. The same methods have been employed successfully to freeze-dry most bacteriophages; exceptions are stored in liquid nitrogen. In general, bacteriophages can be grown in a soft agar layer or in broth. The phages are harvested by scraping off the soft agar with a sterile glass rod or rubber policeman, macerating and dispensing into a sterile centrifuge tube, and centrifuging at low speed to sediment the agar and most of the unlysed bacteria. Broth-gown cultures are centrifuged at low speed to remove the unlysed bacteria. The supernatant is then filtered through a 0.45-μm and then through a 0.2-μm Millipore filter. The filtrate should be titrated to determine the concentration of phages; a titer of 10^8 pfu/mL is desirable. The phage can be freeze-dried using 20% skim milk and the component freeze-dryer method.

Freeze-dried bacteria and bacteriophages can be stored at 2–8 °C. A longer shelf life has been obtained when cultures are stored at −30 or −70 °C in a mechanical freezer. Freezed-dried cultures should not be stored in a freezer compartment of a frost-free refrigerator because of the alternate heating and freezing cycles, nor should they be stored at room temperature.

Although lyophilization has facilitated the long-term preservation of bacteria, viability checks must be done before and after lyophilization to determine the effectiveness of the process. Quality control procedures must be in place to check for purity, retention of essential characteristics and periodic viability to determine the shelf life of the culture.

Deep Freeze. The long-term preservation of bacterial species not amenable to freeze-drying has been achieved through storage in the frozen state at a temperature of −70 °C or, more advisable, at the temperature of liquid nitrogen (−196 °C) or liquid vapor phase (−150 °C).

The freezing process results in several events that can be destructive to the cells. As the bacterial suspension begins to cool, the liquid water turns to ice at 0 °C (the exact temperature depending on the nature of the bacterial suspension). Ice formation occurs first in the external aqueous environment, resulting in an increased concentration of solutes outside the cells. The differential osmotic pressure causes water to migrate from the cells. The rate and extent of the water loss are dependent on the rate of cooling and the permeability of the cells. Removal of too much water leads to a concentration of solutes within the cells that may be harmful. If too much water is left in the cells, internal ice crystals may form, which would result in intracellular damage. It is important to maintain the critical balance between the two events by controlling the rate of cooling while freezing the cells, and warming as rapidly as possible when the cells are thawed. Chemical compounds such as glycerol and dimethyl sulfoxide (DMSO) can be added to the cells to minimize the damage caused by freezing. These compounds, labeled cryoprotective agents (see "Cryoprotectants"), exhibit certain properties that are essential for good cryoprotection (22,23). They must be nontoxic to the cells, show good permeability, and bind either the electrolytes that accumulate or the water molecule to delay freezing.

The physiological condition of the cells plays an important role in survival of freezing. In general, actively growing cultures harvested at the mid to late logarithmic phase of growth will survive the freezing process better than those harvested at an earlier or later phase. The bacteria should be grown under optimum conditions in the appropriate medium that best retains the prominent characteristics of the strain. Bacteria can be grown on broth that is static or shaken, or on agar. Shaken cultures generally reach optimal cell density faster than agar-grown cultures. Some strict anaerobes require prereduced media and anaerobic procedures for harvesting and dispensing.

Cells grown in broth cultures are harvested by aseptic centrifugation, and the resultant pellet is resuspended in sterile broth containing 30–50 % (vol/vol) sterile glycerol. Bacteria grown on agar are harvested by aseptically washing the growth with sterile broth containing the cryoprotectant. Dispense 0.4 ml of the cell suspension, containing at least 10^8 cell/mL into each sterile, prelabeled vial. Plastic presterilized screw-capped vials (such as Nunc) are recommended for bacteria. Place the filled vials into a mechanical freezer set at −70 °C. The cells are recovered by rapidly thawing the frozen cell suspensions in a 37 °C water bath.

A wide variety of bacterial have been successfully preserved at −70 °C using glass beads and 15 % glycerol as the cryoprotectant (24). The glass beads are prepared as described in Ref. 15, and 30 beads are placed in 2-mL glass screw-cap vials and sterilized for 15 min at 121 °C. The vials must be prelabeled with ink that will not come off at ultralow temperatures. Cell suspensions are prepared containing approximately 10^8 organisms/mL in sterile 15% glycerol, and 0.5 mL is dispensed into each vial. The vials are shaken gently to ensure that all the beads are wetted with the bacterial suspension, and then the excess liquid is aseptically removed with a sterile pasteur pipette. The vials are placed into a mechanical freezer at −70 °C. The cells are recovered by aseptically removing a bead from the vial with a sterile spatula or forceps and placing the bead in sterile broth. It should be noted, however, that the contents of the vial should not be allowed to thaw; this can be prevented by placing the vial in crushed dry ice and removing the bead after ensuring the beads will remain frozen. This technique has been used to preserve bacteria obtained from extreme environments such as hypersaline pools.

Precautions must be taken against electrical shutdowns and compressor malfunctions. Adequate back-up systems (such as alarms, back-up freezers, or an electrical generator) will help to prevent the loss of a valuable collection. However, additional alarm systems are recommended, such as the sound/off power-temperature monitor (Arthur H. Thomas Co.), which has a temperature range of −75 to + 200 °C. The unit is battery powered and can be mounted on a wall.

Ultrafreezing, or Cryopreservation. Although the long-term maintenance of bacteria in liquid nitrogen has been considered expensive, the successful preservation of physiologically diverse bacteria by this method, along with minimal handling and lower labor cost, makes this method feasible to use. The ATCC has successfully preserved many bacteria and bacteriophages, including fastidious ones, in liquid nitrogen for over 36 years without the loss of phenotypic properties.

A large variety of liquid-nitrogen refrigerators are now commercially available, with a wide assortment of features and storage capacities. The sizes range from 10 to 1,000 L, allowing storage of 300–40,000 ampules.

The physiological condition of the cultures plays an important role in the survival of bacteria under liquid-nitrogen freezing. Bacteria are grown in the appropriate medium and harvested at mid-late logarithmic phase of growth (25). Cultures grown in broth are harvested by aseptic centrifugation, and the resultant pellet is resuspended with sterile fresh medium containing either glycerol or DMSO. For agar-grown cells, the growth is washed off from the agar surface with sterile broth containing the appropriate cryoprotectant. Cryoprotective agents must be added to the suspending medium to protect cells from freeze damage. Cryoprotectants fall into two main classes: compounds such as glycerol and DMSO, which are permeable and appear to provide both intracellular and extracellular protection against freezing; and agents such as dextran, glucose, lactose, mannitol, polyglycol, polyvinylpyrrolidone, and sucrose, which seem to exert their protective effect external to the cell membrane. The first class of cryoprotectants has proven to be the most effective, and in general, glycerol and DMSO seem to be equally effective in protecting a wide range of bacteria. It is advisable to conduct a tolerance test on new species to ascertain whether the cryoprotectant is toxic or beneficial.

DMSO and glycerol are routinely used at concentrations of 5 % (vol/vol) and 10% (vol/vol), respectively, in the proper suspending medium. Glycerol is normally prepared as a double-strength (20 %, vol/vol) solution and then mixed with an equal amount of the cell suspension. Glycerol is sterilized by autoclaving at 121 °C for 15 min and stored in 6-mL volumes at 2–8 °C. DMSO is filter sterilized with a 0.22-μm-pore-size Teflon (polytetrafluoroethylene) membrane (Gelman or Millipore) that has been prewashed with methanol and DMSO, collected in 10 to 15-mL quantities in sterile test tubes, and stored in the frozen state at 5 °C (DMSO freezes at 18 °C) and protected from light. An opened bottle of DMSO should not be used for more than 1 month because of the accumulation of oxidative breakdown products.

Cell suspensions containing at least 10^8 cells/mL are dispensed into prelabeled sterile vials. There are a variety of vials available commercially, ranging from prescored glass to sterile plastic. With glass vials, care must be taken when sealing and storing because improperly sealed glass vials can explode when retrieved from the liquid nitrogen as a result of the rapid expansion of the liquid nitrogen that enters the vials through microscopic holes. Plastic screw-cap vials that are not properly sealed can fill with liquid nitrogen if stored in liquid phase; retrieval of the vials to a warmer temperature can result in liquid nitrogen expansion, causing the contents of the plastic vials to spray into the laboratory environment. Plastic screw-cap vials should always be stored in the vapor phase. Protective gloves and face shields should be worn when handling frozen liquid-nitrogen vials.

The rate of cooling is important in liquid-nitrogen storage. In general, the best results for preservation of bacteria have been achieved with slow cooling. This step can be controlled using programmable controllers such as the Linde BF 3-2 (Union Carbide Corp.), which has an adjustable cooling rate. The cells are frozen at a controlled rate of 1–3 °C/min to −40 °C and then at a more rapid rate of 10 °C/min to −90 °C. After this temperature is reached, the vials are transferred to a liquid-nitrogen tank and stored in the liquid phase at −196 °C or above in the vapor phase at −150 °C.

If a programmable freezer is not available, slow cooling can be obtained by placing the filled glass or plastic vials in a stainless steel pan at the bottom of a mechanical freezer at −60 °C for 1 h and then plunging the vials into a liquid-nitrogen bath for 5 min. The rate of cooling to −60 °C is approximately 1.5 °C/min. The frozen vials can then be stored in the liquid-nitrogen tank in the liquid or vapor phase.

It has been reported that the best recovery of cultures from liquid-nitrogen storage is obtained by rapid thawing (26,27). This has also been the experience with cultures frozen at the ATCC. The events during thawing are complex, and slow warming rates can lead to ice crystal formation; a rapid warming rate of approximately 100 °C/min will minimize the damage due to ice formation. A rapid rate of warming can be achieved by quickly immersing the frozen vials in a 37 °C water bath with moderate agitation until all the ice melts. This usually takes about 50 s for glass vials and 90 s for polypropylene ones.

Cryopreservation of strict anaerobic bacteria must be performed under anaerobic conditions. Technical details of the procedures involved have been described by Hill et al. (28). The reader should consult various sections of *Bergey's Manual of Systematic Bacteriology*, volumes 1 through 4 (29–32), for information on the maintenance and preservation of representative bacterial genera.

The successful cryopreservation of plasmid-containing bacteria depends on the stability of their plasmids (33). These organisms are less genetically stable and more likely to lose their foreign genetic material (34). In general, bacteria containing unstable plasmids should be grown on antibiotic-containing liquid medium at 30 °C, frozen with 10 % (vol/vol) glycerol, and stored in the vapor over liquid nitrogen.

CULTURE AND PRESERVATION OF FUNGI AND YEAST

Short-Term Preservation

Subculture. Many filamentous fungi can be successfully maintained by serial transfer on suitable media. It is important, however, to use optimal media and growth conditions to ensure healthy cells that retain their morphological and physiological characteristics. A list of suitable media and incubation temperatures for a variety of fungi is provided by Onions and Pitt (35). In general, a medium that promotes good sporulation is considered to be the most desirable. When possible, it is best to cultivate the fungi on agar slants in test tubes or culture bottles. Transfers should be made only from the growing edges of the culture, taking precautions to ensure that contamination is prevented. Some fungi degenerate when maintained on the same medium for extended periods, thus media should be alternated from time to time. A minimal medium such as potato/carrot agar can often be used to induce sporulation in a culture that has deteriorated (36). Also, the continual transfer of spores from some aging fungal cultures such as *Mucorales* can result in a deterioration of the culture; the transfer of mycelium and spores seems to minimize this problem.

The interval between transfers varies from fungus to fungus; some require transfer every 2–4 weeks, the majority every 2–4 months, while others may survive for 12 months without transfer. Cultures grown on agar slants can be stored at room temperature. Storage in the refrigerator or cold room at 5–8 °C can extend the transfer period to 4–6 months. However, there are some fungi, such as the thermophiles, that are sensitive to storage at these temperatures.

The maintenance of fungi by serial transfer has many disadvantages. The most notable are contamination, selection of variants, mislabeling of cultures, infestation with mites, and labor intensity.

Yeasts have been successfully maintained by serial transfer on either solid or liquid media. In general, many yeasts will survive for longer periods when grown on solid media rather than in broth, especially nonfermentative yeast. Fresh medium is inoculated with actively growing cells and incubated at the appropriate temperature. After sufficient growth occurs, the cells are stored at refrigerated temperature (+4 °C). The majority of yeast species will remain viable for a least 6 months. The interval between transfers must be determined by experimentation. Serial transfer is not recommended for extended maintenance because genetic drift has been observed. In addition, ascosporogenous strains may sporulate on agar slant media stored for prolonged periods, resulting in strain variability.

Many fungi and yeast can be preserved for years by drying on a suitable menstruum such as soil, silica gel, and filter paper (37), and drying under vacuum from the liquid state (L-drying) (38).

Silica Gel. Sterile anhydrous silica gel (6–22 mesh, nonindicator) is prepared by dispensing the gel into growth bottles so that they are one-quarter filled, and sterilizing in an oven at 180 °C for 2–3 h. The bottles containing the sterile gel are cooled by placing in a pan containing ethylene gycol and dry ice to the depth of the gel layer or by placing the bottles in water in a mechanical freezer at a temperature of −17 to −24 °C. Fungal spores are suspended in a sterile 5% skim milk solution that has been cooled to 4 °C. The spore suspension is added to the cooled silica gel until three-quarters of the gel has been moistened. It is important to avoid saturating the gel. The growth bottles are left in the ice bath for approximately 20–30 min and then agitated to ensure thorough dispersion of the suspension. The bottles are held at 25 °C for approximately 1–2 weeks until the crystals easily separate when shaken. Inoculated bottles are sealed with screw caps and stored over indicator silica gel in an airtight container at 4 °C. To recover the fungal spores, a few crystals are sprinkled on an appropriate medium and incubated. This method has been used successfully for a variety of fungi (39). This method has produced variable results with yeast; some are not viable after 3 months of storage, whereas other survive after 2–5 years (40).

Soil. A variety of fungi have been maintained in soil (41). The method involves the inoculation of approximately 1 mL of a spore suspension into soil that has been sterilized by autoclaving twice, 24 h apart, at 121 °C for 15 min. The soil is incubated at room temperature for 5–10 days and then stored in a refrigerator at 4–8 °C. Cultures are recovered by inoculating a suitable medium with a few grains of soil.

Paper. Yeasts have been preserved by drying on paper discs or squares and storage over the desicant silica gel (42). Whatman No. 4 filter paper is cut into small squares or discs about 10 mm across and placed on a small aluminum foil, which is folded into a packet and autoclaved at 121 °C for 15 minutes. The sterile discs or squares are inoculated by immersing into drops of a heavy yeast suspension prepared in 5% skim milk. The packets are dried in a desiccator for 2–3 weeks at 4 °C and then placed in an airtight container and stored at 4 °C. Cultures are recovered by aseptically removing a piece of filter paper and inoculating the appropriate solid or liquid medium. Yeast maintained in this manner have been viable after 2–3 years.

L-Drying. L-drying refers to liquid-state drying in such a way as to prevent freezing. Drying occurs under vacuum at temperatures generally no colder than 5–10 °C. The method has been improved and simplified by Malik (43) for the preservation of sensitive microorganisms that are damaged by freezing or freeze-drying. The technique has been used to successfully preserve some yeasts and filamentous fungal spores.

Long-Term Methods

Freeze-Drying, or Lyophilization. The equipment, cryoprotectants, and theoretical aspects of freeze-drying have already been discussed. Fungi are grown under optimum conditions using media that will produce maximum sporulation. Mature spores are harvested by flooding the agar cultures with approximately 2 ml of a 20% sterile skim milk solution. The skim milk must be cold when used, and thus it is stored at 2–8 °C until required. The spores are gently scraped from the surface of the culture to yield a suspension containing at least 10^6 spores/mL. The spore suspension is pipetted back into the tube containing the remaining milk and mixed thoroughly. If more than one tube is used, repeat the procedure and pool the suspension in one tube to yield a concentration of at least 10^6 spores/mL. The spores of cultures grown in broth medium are centrifuged and resuspended in sterile 20% skim solution. Approximately 0.2 mL of the spore suspension is dispensed into sterile vials for freeze-drying. The interval between harvesting the spores and dispensing must be minimized because many spores begin to germinate when suspended in liquid. Spores should not be in the skim milk for more than 2 h before being processed. Filled vials should be refrigerated while awaiting further processing.

The fungi are freeze-dried using one of the ATCC methods already described. Lyophilized spores can be stored at 2–8 °C. Extended shelf life can be achieved by storing the vials at −70 °C in a mechanical freezer.

Freeze-dried fungal cultures are rehydrated with 0.5–0.9 mL of sterile water, then transferred to tubes containing 6 mL sterile water and allowed to soak for 50–60 min before transferring to solid medium.

At the ATCC, yeasts are processed in the same manner as bacteria for the component freeze-drying method. Cultures are grown on the appropriate solid medium and suspended in sterile 20% skim milk or reagent 18 to a concentration of at least 10^6 cells/mL. The suspension is mixed thoroughly, and 0.2 mL is dispensed into each vial. The cultures are freezed-dried using the component freeze-drying method described earlier.

Ultrafreezing, or Cryopreservation. The preservation of fungi at ultralow temperatures of liquid nitrogen (−196 °C for liquid phase and −150 °C for vapor phase) is presently considered the best method of storage. The method can be applied to both sporulating and nonsporulating strains, as well as those fungi that have been genetically altered or harbor plasmids. The freezing rates and cryoprotective agents described for bacteria are also applicable to the freezing of fungi. Spores or mycelial fragments are harvested by flooding slants or plates with 10% glycerol or 5% DMSO and gently scraping the surface of the cultures. The fungal suspension is dispensed in 0.5-mL amounts into sterile plastic freezing vials and frozen at a controlled rate. Fungi that produce sticky mycelia or whose mycelia grow embedded in the agar can be prepared for freezing by cutting agar plugs containing new growth (hyphyl tips) with a sterile cork borer (5 mm). Three or four plugs are place into each sterile plastic vial with 0.4 mL of 10% glycerol and frozen at a slow controlled rate.

Cultures grown in broth are fragmented in a sterile Waring blender in a biological hood and suspended in equal parts of 20% glycerol and growth medium or equal parts of 10% DMSO and growth medium to give a final concentration of 10% glycerol (vol/vol) or 5% (vol/vol) DMSO, respectively. Some strains must be concentrated by centrifugation in order to obtain sufficient material for freezing. Pathogens must not be macerated in a mechanical blender because of the hazard of aerosol dispersion.

Some cultures, such as *Agaricus* strains, can consist of seeds, grains, or pollen and be frozen without the presence of a cryoprotectant. Slime molds can be preserved by freeze-drying or freezing. Spores, microcysts, and spherules are preserved by a controlled freeze in 10% glycerol and stored in the liquid-nitrogen vapor phase. Thawing of the frozen culture should be done rapidly in a 37 °C water bath with moderate agitation. Once the ice melts, the culture can be transferred to a suitable medium.

Yeast cells have been successfully preserved by controlled freezing to liquid-nitrogen temperature (−196 °C liquid, −150 °C vapor phase). The yeast cultures are handled in the same manner as bacteria and grown as described in the freeze-drying section for yeast. Cell suspensions are prepared in 10% glycerol to a concentration of at least 10^6 cells/mL. The yeast are frozen under controlled conditions as described in the bacterial section.

Several reports have described the loss of plasmid expression or the induction of respiratory-deficient mutants in yeast as a result of freeze-drying or improper freezing (44). Nierman and Feldblyum (33) reported that when plasmid-containing *Saccharomyces cerevisiae* were preserved by freezing in liquid nitrogen, no loss of viability was observed, and the fraction of plasmid-containing cells was the same before and after freezing. The preservation of plasmid-containing yeast is superior in liquid nitrogen with a controlled freezing rate than by freeze-drying.

CULTURE AND PRESERVATION OF CELL CULTURES

Long-Term Preservation

Ultrafreezing, or Cryopreservation. Storage in liquid nitrogen is currently the best method of preserving cultured cells. Cryopreservation methods for a variety of cell lines are well developed and used extensively for culture storage. For a detailed description of standard cryopreservation protocols see reports by Freshney (45) and Hay (46).

The general events that occur during freezing and the effect of the cooling rate, cryoprotective agents, and a controlled rate of cooling have been discussed in the bacterial section.

Cell suspensions are prepared in the same manner used for routine subcultivation; trypsin is added if necessary to produce a uniform suspension. The cell suspension is centrifuged at approximately 100 *g* for 10 min, and the resultant pellet is resuspended in an appropriate amount of fresh medium containing a cryoprotectant. Normally, 5–10% DMSO or glycerol is employed. The freeze medium should be prepared just prior to use by mixing fresh growth medium and the cryoprotective agent and kept at room temperature. The remaining cryoprotectant is discarded because oxidative contaminants can accumulate. A cell suspension containing approximately 10^6–10^7 cells/mL is satisfactory. The cell suspension is dispensed in 1-mL amounts into each glass vial or plastic ampule and frozen at a controlled rate of 1 °C/min to about −50 °C. The vials are immediately transferred into liquid nitrogen for storage in the liquid or vapor phase. Rapid thawing is essential for recovery of frozen cells, thus the frozen vials are immersed directly into a 37 °C water bath and agitated until the suspension is completely thawed. The thawed cell suspension is transferred to 10 mL of sterile growth medium and centrifuged at 100 *g* for 10 min. The supernatant containing the cryoprotectant is discarded, and the cell pellet is resuspended in fresh medium and propagated using standard procedures.

Quality Control. It is important that all preservation methods include standard operating procedures for the determination of culture purity, viability, retention of morphological and physiological characteristics, production of desirable products, and other key attributes. Purity checks must be conducted before and after processing microorganisms for freeze-drying and microorganisms and cell lines for freezing. Generally, the cell suspension is diluted and inoculated into suitable medium and incubated at optimum temperature. After growth appears, the cultures are examined for purity. It is also important to conduct pre- and postpreservation viability checks for freeze-drying and freezing to determine the effectiveness of the process. In addition, periodic viability assays should be performed to ascertain the shelf life of the culture.

Cell lines should also be examined for *Mycoplasma* contamination, because it has been estimated that about 10% of cell lines are contaminated (47). It is also important to conduct tests to verify cell-line identity; several rapid techniques have been developed that use DNA probes to identify the cell lines (48).

Documentation. Documentation plays an important role in the maintenance and preservation of cultures. Important data include the identity of the culture, isolation source and methods, the individual who isolated it, geographical location, morphological and physiological characteristics, maintenance data, and other salient features. If the culture was obtained from another investigator, the documentation should include the name of the investigator, culture history if available, and date of acquisition. The greater the amount of data recorded and regularly updated, the greater the value of the culture.

BIBLIOGRAPHY

1. R.J. Heckly, *Adv. Appl. Microbiol.* **24**, 1–53 (1978).
2. D. Smith and J. Kolkowski, in J.C. Hunter-Cevera and A. Belt eds., *Maintaining Cultures for Biotechnology and Industry*, Academic, New York, 1996, pp. 101–132.
3. R.J. Hay, in I. Freshney ed., *Animal Cell culture: A Practical Approach*, IRL Press, Oxford, England, 1986, pp. 71–112.

4. F. Kauffman, *The Bacteriology of Enterobacteriaceae*, Williams & Wilkins, Baltimore, 1966, pp. 1–400.
5. L.R. Hill and B.E. Kirsop eds., *Living Resources for Biotechnology*, Cambridge University Press, Cambridge, 1991.
6. B.E. Kirsop and J.J.S. Snell eds., *Maintenance of Microorganisms: A Manual of Laboratory Methods*, Academic, London, 1984.
7. R.E. Miller and L.A. Simmons, *J. Bacteriol.* **84**, 1111–1114 (1962).
8. L.R. Rey, in V.J. Cabasso and R.H. Regamey eds., *Development in Biological Standardization. International Symposium of Freeze Drying of Biological Products*, S. Karger, Basel, 1977, 19–27.
9. T.W.G. Rowe and J.W. Snowman, *Edwards Freeze-drying Handbook*, Edwards High Vacuum, Inc., Grand Island, New York, 1976.
10. R.J. Heckly, *Adv. Appl. Microbiol.* **3**, 1–76 (1961).
11. R.J. Heckly, *Dev. Ind. Microbiol.* **26**, 379–395 (1985).
12. H.T. Merryman, in H.T. Merryman ed., *Cryobiology*, Academic, New York, 1966, pp. 609–663.
13. R.H. Rudge, in B.E. Kirsop and A. Doyle eds., *Maintenance of Microorganisms and Cultured Cells*, 2nd ed., Academic, London, 1991, pp. 31–44.
14. S.P. Lapage, J.E. Shelton, T.G. Mitchell, and A.R. Mackenzie, in J.R. Norris and D.W. Ribbons eds., *Methods in Microbiol*, vol. 3A, Academic, London, 1970, pp. 135–228.
15. R.L. Gherna, in P. Gerhardt, R.G.E. Murray, W.A. Wood, and N.R. Krieg eds., *Methods for General and Molecular Bateriology*, American Society for Microbiology, Washington, D.C., 1994, pp. 278–292.
16. F.P. Simione and E.M. Brown eds., *ATCC Preservation Methods: Freezing and Freeze-drying*, American Type Culture Collection, Rockville, Md., 1991.
17. W.C. Haynes, L.J. Wickerham, and C.W. Hesseltine, *Appl. Microbiol.* **3**, 361–368 (1955).
18. H. Hippe, in B.E. Kirsop and A. Doyle eds., *Maintenance of Microorganisms and Cultured Cells*, 2nd ed., Academic, London, 1991, pp. 101–113.
19. V.J. Cabasso and R.H. Regamy eds., *International Symposium on Freeze-Drying of Biological Products*, Dev. Biol. Stand. vol. 36, 1977.
20. H.T. Merryman, in H.T. Merryman ed., *Cryobiology*, Academic, New York, 1966, pp. 1–114.
21. K.F. Redway and S.P. Lapage, *Cryobiology* **11**, 73–75 (1974).
22. G.M. Fahy, *Cryobiology* **23**, 1–13 (1986).
23. H.T. Merryman, *Cryobiology* **8**, 173–183 (1971).
24. R.K.A. Feltham, A.K. Power, P.A. Pell, and P.H.A. Sneath, *J. Appl. Bacteriol.* **44**, 313–316 (1978).
25. M.L. Speck and R.A. Cowan, in H. Iizuka and T. Hasegawa eds., *Proceedings of the First International Conference on Culture Collections*, Univ. of Tokyo Press, Tokyo, 1970, pp. 241–250.
26. A.P. Mackenzie, in V.J. Cabasso and R.H. Regamey eds., *International Symposium on Freeze-Drying of Biological Products*, S. Kargel AG, Basel, 1977, pp. 263–277.
27. P. Mazur, in H.T. Merryman ed., *Cryobiology*, Academic, New York, 1966, pp. 213–315.
28. L.R. Hill, M. Kocur, and K.A. Malik, in L.R. Hill and B.E. Kirsop eds., *Living Resources for Biotechnology*, Cambridge University Press, Cambridge, 1991, pp. 62–80.
29. N.R. Krieg and J.G. Holt eds., *Bergey's Manual of Systematic Bacteriology*, vol. 1, Williams & Wilkins, Baltimore, 1984.
30. P.H.A. Sneath, N.S. Mair, M.E. Sharpe, and J.G. Holt eds., *Bergey's Manual of Systematic Bacteriology*, vol. 2, Williams & Wilkins, Baltimore, 1986.
31. J.T. Staley, M.P. Bryant, N. Pfennig, and J.G. Holt eds., *Bergey's Manual of Systematic Bacteriology*, vol. 3, Williams & Wilkins, Baltimore, 1989.
32. S.T. Williams, M.E. Sharpe, and J.G. Holt eds., *Bergey's Manual of Systematic Bacteriology*, vol. 4, Williams & Wilkins, Baltimore, 1989.
33. W.C. Nierman and T. Feldblyum, *Dev. Ind. Microbiol.* **26**, 423–434 (1985).
34. T. Imanaka and S. Aiba, *Ann. N.Y. Acad. Sci.* **36**, 1–14 (1981).
35. A.H.S. Onions and J.I. Pitt, in D.L. Hawksworth and B.E. Kirsop eds., *Living Resources for Biotechnology*, Cambridge University Press, Cambridge, 1988, pp. 188–193.
36. D. Smith and J. Kolkowski, in J.C. Hunter-Cevera and A. Belt eds., *Maintaining Cultures for Biotechnology and Industry*, Academic, San Diego, 1996, pp. 101–132.
37. D. Smith, in D.L. Hawsworth and B.E. Kirsop eds., *Living Resources for Biotechnology*, Cambridge University Press, Cambridge, 1988, pp. 75–99.
38. I.C. Tommerup and D.K. Kidby, *Appl. Environ. Microbiol.* **37**, 831–835 (1979).
39. D. Smith, in B.E. Kirsop and J.J.S. Snell eds., *Maintenance of Microorganisms: A Manual of Laboratory Methods*, Academic, London, 1984, pp. 83–107.
40. B.E. Kirsop, in B.E. Kirsop and J.J.S. Snell eds., *Maintenance of Microorganisms: A Manual of Laboratory Methods*, Academic, London, 1984, pp. 109–130.
41. R.G. Atkinson, *Can. J. Bot.* **32**, 542–547 (1953).
42. R.G. Atkinson, *Can. J. Bot.* **32**, 673–678 (1954).
43. K.A. Malik, *J. Microbiol. Methods* **12**, 125–132 (1990).
44. B.M. Pearson, P.J.H. Jackman, K.A. Painting, and G.J. Morris, *Cryo-Letters* **11**, 205–210 (1990).
45. R.I. Freshney, *Culture of Animal Cells: A Manual of Basic Technique*, 3rd ed., Wily-Liss, New York, 1994.
46. R.J. Hay, in R.I. Freshney ed., *Animal Cell Culture: A practical Approach*, IRL Press, Oxford, England, 1986, pp. 71–112.
47. R.J. Hay, in J.C. Hunter-Cevera and A. Belt eds., *Maintaining Cultures for Biotechnology and Industry*, Academic, San Diego, 1996, pp. 161–178.
48. D.A. Gilbert, Y.A. Reid, M.H. Gail, D. Pee, C. White, R.J. Hay, S.J. O'Brien, *Am J. Hum. Genet.* **47**, 499–514.

D

DENATURATION, AGGREGATION. See

DENATURATION, PROTEINS, SOLVENT MEDIATED; PROTEIN AGGREGATION, DENATURATION; THERMAL UNFOLDING, PROTEINS.

DENATURATION, PROTEINS, SOLVENT MEDIATED

VADIM V. MOZHAEV
University of Iowa
Iowa City, Iowa

KEY WORDS

Catalytic activity
Conformation
Denaturation
Enzymes
Inactivation
Organic solvents
Proteins
Stability
Stabilization

OUTLINE

INTRODUCTION

The use of enzymes in organic solvents for biotransformations is now commonplace. The nonaqueous milieu permits the advantages of organic solvents as the reaction media, and enzymes provide the opportunity for high selectivity and catalytic rates under mild reaction conditions (1). Indeed, enzymes are fantastic catalysts in terms of effectiveness—enzyme-catalyzed reactions provide up to 10^{15}-fold rate accelerations as compared with uncatalyzed processes—and reaction specificity (e.g., enantio-, regio-, and chemioselectivity). Unfortunately, exploiting such advantages is often limited by the low stability of the biocatalysts. Even in aqueous environments, enzymes undergo denaturation that is, they loose their unique three-dimensional structure and catalytic activity. The risk of denaturation is even greater when proteins are placed into conditions much less "natural" than the conventional aqueous environment. The goal of this review is to consider the general mechanisms behind the often low stabilities of proteins in the presence of organic solvents, and to evaluate more specific mechanisms responsible for denaturation of proteins in systems with organic solvents. Special attention is given to enzymes as the most important class of proteins in biotransformations. Successful approaches elaborated to increase protein stability are also discussed.

GENERAL DEFINITIONS

Protein Denaturation

Mechanisms of protein denaturation in aqueous media have been studied for decades and are mostly understood now. Denaturation of proteins is caused by various factors such as high and low temperatures, high pressures, ultrasound, high-intensity irradiation (including microwaves), addition of organic solvents, certain salts, and specific compounds such as urea. Despite the high structural diversity of proteins, their denaturation is described by very similar features. Denaturation, or more specifically inactivation of enzymes (i.e., loss of their catalytic function) can be either reversible or irreversible. For reversible denaturation, the enzyme completely recovers the native structure and catalytic activity after the denaturing action is eliminated, for example, by reestablishing room temperature after heating or by removing chemical denaturants during dialysis. Reversible denaturation is often a two-state transition between the folded native state, N, and the unfolded denatured state, U (2). Because structural differences between N and U are significant, concentration of both forms can be followed in the transition zone quite easily by circular dichroism or fluorescence spectroscopy, or other physical methods. This ability to follow concentrations of both protein forms during denaturation enables researchers to obtain the thermodynamic characteristics of protein stability and quantitatively compare the stability under a variety of external conditions including organic solvents.

Often after incubation under denaturing conditions (especially at temperatures higher than 50–60 °C), the enzyme does not recover its native structure and catalytic activity; this is irreversible denaturation (3,4). The reason may be purely kinetic, for example, heat-denatured enzyme remains in a catalytically inactive conformation after cooling because at room temperature the protein molecule is not flexible enough to surpass the high kinetic barrier that exists between the denatured and the native states. On the other hand, in the unfolded state, the protein becomes more susceptible to aggregation and chemical changes in the structure that also lead to irreversible inactivation. Table 1 lists inactivation mechanisms.

Studying Reversible and Irreversible Denaturation. When protein is reversibly denatured, the two protein forms, N and U, are in equilibrium, which is governed by the equilibrium constant K:

$$N \overset{K}{\leftrightarrow} U, \ K = [U]/[N]$$

where [N] and [U] represent concentrations of the native and unfolded proteins, respectively, and are determined experimentally under denaturing conditions by measuring catalytic activity or physical characteristics of the protein. Other thermodynamic parameters of denaturation are determined from the value of K by studying the temperature dependence of denaturation:

$$\Delta G_D = -RT\ (\ln K); \qquad \Delta H_D/RT^2 = d(\ln K)/dT$$
$$\Delta S_D = (\Delta G_D - \Delta H_D)/T; \qquad \Delta c_p = d(\Delta H_D)/dT$$

where ΔG_D is the free energy of denaturation, R is the gas constant, and T is the temperature in degrees kelvin. ΔH_D is the enthalpy of denaturation, ΔS_D is the change in entropy of denaturation, which is a characteristic of the degree of protein unfolding, and Δc_p is the change in heat capacity on denaturation and is a measure of the hydration effects in a protein. This formalism also applies to denaturation of proteins in organic solvents, at least in cases when the denaturation is not complicated by irreversible processes such as aggregation.

Irreversible inactivation is studied by following the rate, v_{in}, of disappearance of the native form of the protein:

$$N \overset{k_{in}}{\rightarrow} I; \ v_{in} = -d[N]/dt = k_{in}[N]$$

where I is the inactivated enzyme and k_{in} is the rate constant of inactivation. Aliquots of solution, or suspensions with enzyme in organic solvents are sampled at time intervals, and the residual enzymatic activity is assayed in aqueous solution (or in organic solvent).

Protein Stability

Closely related to low stability is the problem of stabilization of enzymes. In the last two decades, numerous approaches have been elaborated to increase stability of proteins in aqueous media; see Refs. 3–6 for reviews. These approaches employ immobilization and cross-linking of proteins, the methods of site-directed and random mutagenesis, and covalent modification of proteins; also, stabilization is achieved by addition of low molecular weight compounds. Some of these approaches have been successfully used in organic solvents and are discussed in this review.

Despite the variety of organic solvent systems containing enzymes, two of them are most important in the practical sense and for understanding protein denaturation by organic solvents. First, enzymes are often dissolved in binary mixtures of water and polar organic solvents with concentrations of the solvent as high as several dozens volume percent. Second, lyophilized or immobilized enzymes are studied when suspended in neat organic solvent or in the solvent containing as little as a few volume percent of water. These two types of systems are the focus of this review.

Table 1. Inactivation Mechanisms and Experimental Methods for Their Study

Inactivation mechanism	Methods for identifying inactivation mechanism
1. Aggregation (sometimes followed by formation of intermolecular S-S bonds)	Formation of precipitates visible by eye; smaller aggregates are detected by electrophoresis or sedimentation
2. Changes in primary structure	
a. Hydrolysis of peptide bonds	Detection of low molecular weight protein fragments by SDS electrophoresis
b. Oxidation of Cys, Trp, Met	Deceleration of protein inactivation by inert gases
c. Reduction of S-S bonds by OH^-, followed by β-elimination and reactions of formed SH groups with Cys or other amino acids	Detection of emerging SH groups by Ellman's reagent; disappearance of Cys, Lys, and Arg, and appearance of lysinoalanine, dehydroalanine, and ornithinoalanine, detected by amino acid analysis
d. Modification of SH groups by metal ions to form mercaptides	Decrease in free SH groups, detected by Ellman's reagent
e. Deamidation of Asn and Gln	Formation of proteins with smaller isoelectric points; detection of ammonia with glutamate dehydrogenase
f. Racemization of amino acids	Only above 100 °C
3. Elmination of coenzyme from the active site	Disappearance of spectral bands characteristic of coenzyme attached to proteins
4. Dissociation of oligomeric proteins into monomers	Detection of dissociation products by chromatography or electrophoresis
5. Adsorption on the vessel surface	Sometimes desorption by added denaturants
6. "Incorrect refolding," with production of mismatched forms	Reactivation of mismatched protein forms by unfolding and subsequent refolding

Source: Ref. 3.

ENZYMES IN MONOPHASIC WATER–COSOLVENT MIXTURES

Reversible Denaturation

Exploiting enzymes in peptide synthesis and other biotransformations demands the use of mixtures of water with polar organic solvents (i.e., alcohols and formamides, among others) as the reaction media. However, almost all organic solvents in binary mixtures with water have unfavorable denaturing effects on enzymes (7). Even polyhydric alcohols (glycerol and ethylene glycol, among the others), which at low concentrations are known to stabilize proteins against heat treatment, induce denaturation at concentrations higher than 50 vol % (8). Catalytic activity of enzymes dissolved in mixtures of water with a polar organic solvent are described by a characteristic profile, shown in Figure 1. Addition of cosolvent to aqueous solution of enzyme up to some critical concentration does not change or increase enzymatic activity. This activation could often be explained by more favorable solvation of the polar transition state of enzymatic reaction by water–cosolvent mixture than by water. However, beginning with this critical concentration of a cosolvent, enzymatic activity drops (often reversibly) in a threshold manner, reaching a very low or even zero value. This threshold decrease in activity is accompanied by an abrupt change in spectroscopic or other physical properties of the dissolved enzyme, and this fact has been explained by a conformational transition in the protein molecule. The fact that the plummet in enzymatic activity occurs at the same concentration as the spectral changes do, implies that it is the denaturation that causes the threshold inactivation of the dissolved enzymes.

Mechanistic description of this inactivation is based on the important prerequisite that the protein molecule in aqueous solution is surrounded by a shell formed by water molecules that are bound by hydrogen bonds to the protein surface. According to a widely accepted notion (9), this hydration shell, or at least a portion of it, represents an integral part of the protein and is essential for protein structure and function. Consequently, displacement of bound water molecules by an organic solvent results in a dramatic change of the whole structure and leads to protein denaturation. In line with these ideas, reversible inactivation of enzymes in water–cosolvent mixtures consists of three major steps, described in Figure 2: *(1)* dehydration of the protein molecule, that is, removal of a certain critical amount of water molecules from the protein hydration shell; *(2)* binding of the organic cosolvent by the partially dehydrated protein molecule; and *(3)* conformational transition in the protein molecule, resulting in the formation of a denatured form, D, with concomitant loss of catalytic activity. Thermodynamic consideration of this model the details of which are beyond the scope of this article, leads to the following expression for the free energy of denaturation (ΔG_D) of proteins by organic solvents (8):

$$\Delta G_D = B_0 + B_1 n + B_2 E_T(30) + B_3 n \log P$$

where B_0, B_1, B_2, and B_3 are numerical coefficients, which remain unchanged for a given protein at constant temperature, regardless of the nature of the solvent; n is a numerical parameter that depends on the size and shape of the molecules of organic solvents and gives an idea of how many molecules of water can be displaced from the protein surface by one molecule of the organic solvent; $E_T(30)$ is the Dimroth–Reichardt parameter, which is directly related to the free energy of the solvation process of the organic cosolvent; P, the partition coefficient of the solvent in a water–octanol biphasic system, is a measure of organic solvent hydrophobicity. The values of $E_T(30)$ and P for different solvents are well documented, and the parameter n can be calculated from the information on the organic solvents contained in comprehensive handbooks on organic and physical chemistry.

Equation 1 predicts that cosolvents having high values of $E_T(30)$ and $\log P$ possess greater denaturing ability. In

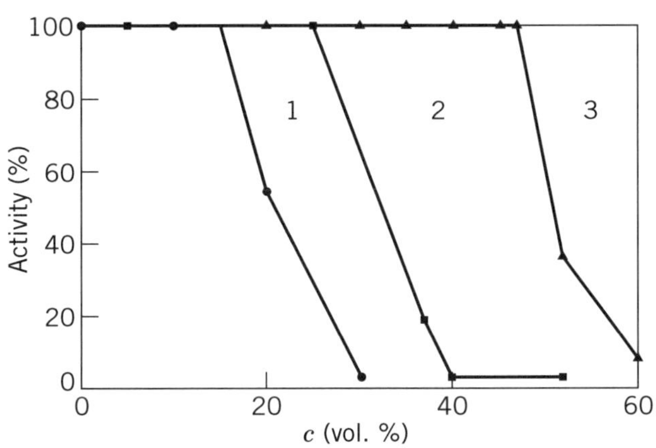

Figure 1. Dependence of relative catalytic activity of dissolved enzymes on the concentration of organic solvents in binary mixtures with water. (1) laccase in water–acetone mixtures; (2) α-chymotrypsin in water–ethanol mixtures; (3) trypsin in water–acetonitrile mixtures.

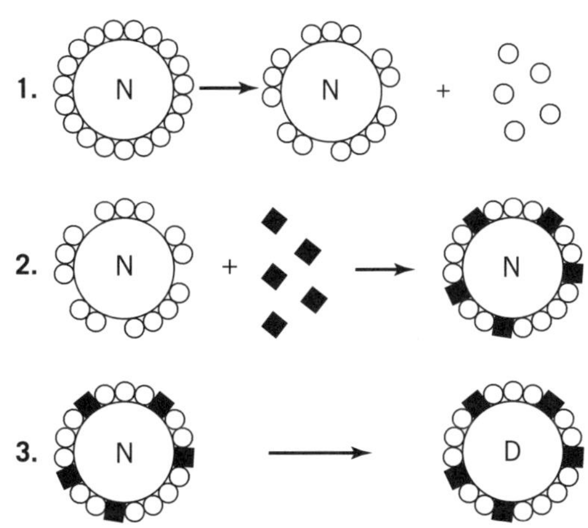

Figure 2. Schematic representation of major molecular steps involved in reversible protein denaturation by organic solvents. ○, water molecule; ■, organic cosolvent molecule; *N* states for the native protein, *D* for the denatured protein.

fact, the solvents such as 1,4-dioxane, tetrahydrofuran, and higher alcohols (e.g., isomers of butanol), which are both hydrophobic and have a high solvation capacity, appear to be strong denaturants and cause enzyme inactivation at concentrations as low as 10–30% (v/v). Inversely, Equation 1 predicts that hydrophilic solvents, such as glycerol, ethylene glycol, and formamide, have a low denaturing capacity. Indeed, it has been experimentally confirmed that many enzymes are not inactivated by these solvents at concentrations as high as 50–60 vol % (8). Thus, solutions with high concentrations of these solvents are convenient as a media for enzyme-catalyzed reactions carried out in binary water–cosolvent mixtures.

Stabilization of Enzymes against Denaturation by Organic Cosolvents

Molecular events behind enzyme inactivation in water–cosolvent mixtures also prompt ideas of how to stabilize enzymes against this denaturation. Because protein unfolding is an indispensable step in the inactivation by solvents (Fig. 2), one could try the approaches recommended as best for deceleration of protein unfolding in aqueous solutions. By using the method of covalent immobilization, proteolytic enzymes were stabilized against denaturation by alcohols and dimethylformamide (10,11); for other examples, see the literature cited in Ref. 11. The stabilization effect expressed as an increase in the threshold concentration at which enzymatic activity dropped abruptly, rose with the number of covalent bonds between the enzyme and support. Stabilization against denaturation by polar organic solvents has been also obtained when noncovalent complexes of enzymes with polyelectrolytes were formed due to multiple electrostatic interactions between oppositely charged groups in the interacting macromolecules (12). The delay in activity decrease in this case was also accompanied by a shift in the concentration of organic solvents, at which point significant structural changes in the protein molecule (unfolding) occurred. Probably the same reason (i.e., increase in rigidity of the protein molecule in water–cosolvent mixtures) underlies stabilization of enzymes by chemical cross-linking with bifunctional reagents, as was the case with cross-linked crystals of thermolysin in dimethylformamide (13). Equilibrium between the native and denatured states of a protein (step 3 in Fig. 2) can be efficiently shifted toward state N by addition of substrates and specific ligands that are preferentially bound with the native enzyme. Due to this reason, enzymes are often more stable in organic solvents in the presence of saturating concentrations of their substrates.

Another productive idea for stabilization of enzymes in water–cosolvent mixtures is to strengthen the binding of the hydration shell to the protein surface. As seen from Figure 2, protein denaturation includes the step of dehydration, namely, dissociation of essential water molecules from the hydration shell. It has been assumed that if water molecules are more tightly bound with the protein, denaturation could be decelerated. Indeed, introduction of polar and charged groups into the protein by covalent modification led to a significant increase in the concentration of organic solvents that caused denaturation of α-chymotrypsin. Stabilization was observed when enzymes were modified with low molecular weight compounds (pyromellitic anhydride, glyceraldehyde) (14) and poly(ethylene glycol) (PEG) (15). This stabilization was explained by the ability of the newly attached hydrophilic (carboxylic and hydroxylic) groups to hold water, thus preventing protein dehydration by organic cosolvent molecules. Tighter binding of the hydration shell could also be achieved due to replacements of the proteins' amino acid residues by site-directed or random mutagenesis. Recent studies from Arnold's and Wong's groups showed that such replacements increased the stability of proteases in the presence of a strong denaturant, dimethylformamide (16,17). The stabilization was explained by numerous reasons, such as enhancement of internal hydrogen bonds and ionic forces, optimization of internal hydrophobic interactions, reduction of solvation energy, and introduction of conformational restrictions to the protein molecule. Also it is possible that chemical modification or replacement of amino acid residues on the protein surface both change the protein's affinity to organic solvent molecules and, according to the schemes in Figure 2, decrease the protein's sensitivity to denaturation. A stabilizing approach recently developed by Arnold (18) and Wong (19) suggests increasing the rigidity of a protein by introducing amino acid residues (e.g., histidines) that are able to specifically bind with metal ions. In fact, in organic solvents, catalytic activity and stability of such metal-chelated proteases was significantly higher than in their nonmodified analogs.

Irreversible Inactivation: Precipitation

It has been commonly noticed that the initially reversible denaturation of proteins by organic solvents is further accompanied by irreversible inactivation during long (a dozen minutes to hours) incubation. This irreversible inactivation becomes especially important when the concentration of cosolvent is higher than the denaturation threshold. Proteins may even precipitate if their concentration in water–cosolvent solution is of the order of micromoles per liter or lower; normally these concentrations are used to detect catalytic activities. The general features of this inactivation were described by Singer (7), who found that proteins lose only a part of their native structure in organic solvents—the denatured protein retains a large portion of secondary structure, although the tertiary structure is destroyed. This conformation, analogous to the recently discovered molten globule state, turns out to be very "sticky," and proteins have a strong tendency to form inactive aggregates due to formation of interprotein hydrogen bonds and hydrophobic interactions. However, even in the insoluble form, enzymes may show a small, though detectable, activity in the presence of high concentrations (60–80 %, v/v) of polar solvents. For example, such an activity was detected when solid α-chymotrypsin was added to a water–cosolvent mixture (20). On the other hand, the enzyme does not show any activity when the precipitate is formed after an aqueous solution of the enzyme is added to the organic solvent. This difference does not seem too surprising in view of the recent fact (21) that the enzyme directly suspended in water–cosolvent mixture contains a higher percentage of the native secondary structure than the enzyme first dissolved in water and then mixed with the solvent to form a precipitate. Most likely, in the former case, only the enzyme molecules forming the surface layer of particles are inactivated due to direct contact with the organic solvent, whereas the enzyme molecules inside the particles are protected from denaturation by interprotein contacts. Oppositely, in the latter system, enzyme molecules are first denatured and then aggregate, giving rise to inactive precipitates.

This inactivation caused by aggregation has been successfully combated by protein immobilization or by modification with polymers such as PEG; both are highly efficient in preventing interprotein contacts. Aggregate formation and protein precipitation can also be delayed by chemical modification of proteins with low molecular weight compounds (for example, charged reagents) or by replacement of amino acid residues on the protein surface through mutagenesis. Sometimes these modified enzymes remain catalytically active without noticeable precipitation in the significantly wider range of cosolvent concentrations (by 30–40 vol %) than their nonmodified counterparts (V.V. Mozhaev and E.V. Kudryashova, 1996).

SUSPENSIONS OF ENZYMES IN ANHYDROUS MEDIA

Neat organic solvents or solvents containing only a few volume percent of water are very convenient media for many enzymatic reactions (Table 2). Most proteins are insoluble or sparingly soluble in neat organic solvents and most often are used as suspensions of freeze-dried protein powders. These systems have found numerous applications in chemical processing, food-related conversions, and analyses (1).

Table 2. Potential Advantages of Enzymatic Processes in Nonaqueous Media

1. Increased solubility of apolar substrates.
2. Enhanced thermostability.
3. Easier product recovery from low-boiling sovents.
4. Immobilization is often unnecessary because the enzyme is insoluble.
5. Easier recovery of the enzyme by simple filtration or centrifugation.
6. If immobilization is required for better performance, simple adsorption onto nonporous surfaces may be sufficient. Enzymes do not desorb from these surfaces in nonaqueous media.
7. Alteration in substrate specificity.
8. Suppression of water-dependent side reactions, e.g., nonenzymatic blank reaction in enzymatic hydrocyanations of aldehydes, which leads to racemic product.
9. Shifting thermodynamic equilibria to favor synthesis over hydrolysis, e.g., esterification and peptide formation.
10. The possibility of carrying out reactions with other neucleophiles, which would not be feasible in water, e.g., transesterification and ester aminolysis.
11. Easier integration of enzymatic with conventional chemical steps in a multistep synthesis (chemoenzymatic processes).

Source: Ref. 22.

The main problem that still remains with nonaqueous enzymology is a rather low catalytic activity of enzymes in anhydrous media in comparison with activity in water. Many factors such as diffusional (mass transfer) limitations, sterical shielding of the enzyme's active site from the access by a substrate, and unfavorable energetics of substrate desolvation may contribute to this decrease in activity (23). Here we focus on aspects of the problem that are related to the stability of enzymes.

Protein Denaturation During Lyophilization

It has been established that many proteins are denatured during lyophilization (24–26). This denaturation can be either reversible (lyophilized proteins regain their native three-dimensional structure and function when placed in aqueous solution) or irreversible. Especially sensitive to irreversible inactivation during lyophilization are oligomeric proteins, which often dissociate into individual subunits (25). Carpenter and coauthors have found that enzymes can lose their activity both at freezing and at drying, which constitute the lyophilization protocol, and that denaturation at these steps follows essentially different mechanisms.

Freezing. This process includes a number of "stresses" such as protein exposure to low temperatures and ice, contact with the test tube surface, concentration of the dissolved compounds, and eutectic crystallization of solution components, and each of these stresses may decrease enzyme activity. For example, eutectic crystallization from a buffer-containing solution may result in dramatic alterations in pH, for example, up to 3 pH units (27). Such changes largely influence the catalytic activity of enzymes, which is very sensitive to the ionization state of the enzyme's functional groups participating in catalysis. As shown for subtilisin Carlsberg in anhydrous media, the enzyme activity increased more than 300 times upon changing the pH of the aqueous solution (from which the enzyme was lyophilized), from 5 to 11 (28). However, addition of compounds such as polyvinylpyrrolidone and bovine serum albumin (BSA) during freezing is known to prevent such large shifts in pH (25) and may be helpful to increase the activity of enzymes in organic solvents.

Inactivation of enzymes in the freezing process is dominated mainly by unfolding of the protein molecule. Protection of labile proteins during freezing may be achieved by adding a variety of solutes, such as polyols, sugars, amino acids (normally at concentrations of a few hundred millimolar), synthetic polymer PEG, and a protein BSA, effective at concentrations of 1% (w/w) or smaller (25). This cryoprotection is explained by the mechanism of preferential hydration of proteins (6). Sugars, polyols, and some other compounds are preferentially excluded from contact with the surface of a protein because their interaction with a protein's functional groups is unfavorable in comparison with interactions of the same compounds with water molecules. Consequently, in the presence of these compounds, the protein is preferentially hydrated, that is, the hydration shell becomes more tightly bound to the protein molecule. This leads to an energetically unfavorable situation as chemical potentials of both the protein and the additive increase. As a result, the compact native structure is stabilized by the additives because denaturation (unfolding) would lead to a greater surface area of contact between the protein and the solvent and would increase this thermodynamically unfavorable tendency.

Drying. The frozen protein sample is already significantly dehydrated due to formation of ice from the bulk water. However, an essential amount of water comprising the protein's hydration shell (several hundreds of water molecules per protein of molecular weight of 20–25 kDa) remains unfrozen and is removed only by sublimation under high vacuum (i.e., by lyophilization). How much water remains bound to a protein after lyophilization is a question that is not resolved. Most likely, only a few water molecules are bound by several hydrogen bonds to polar groups in the protein's interior, whereas almost all water on the protein's surface is stripped. This latter fact leads to the appearance of a great number of polar groups with the noncompensated ability to form hydrogen bonds, which is satisfied by the formation (mainly intermolecularly) of β-sheet structures because the β-structure allows for the highest degree of hydrogen bonding in proteins (25,29). Due to low molecular flexibility in anhydrous media, these structural changes in freeze-dried enzymes are conserved, and the catalytically active conformation is not completely recovered after the enzyme is placed in organic solvent. Actually, active site titration of chymotrypsin in organic solvents has shown that about 40% of the active enzyme was destroyed at the drying step (26).

Effect of Excipients. Stabilization during the lyophilization procedure demands the protection of proteins against inactivation at both the freezing and drying steps. As discussed in the previous section, many additives are effective as protectors during freezing. Fewer compounds are known to protect enzymes against dehydration stress during drying; sugars are one example. Due to several hydroxyl groups, sugars substitute for molecules of water to form multiple hydrogen bonds with polar groups on the protein surface, thus preventing alterations in the protein's secondary structure (29). Especially successful are combinations of several additives. For example, PEG was shown to protect lactate dehydrogenase and phosphofructokinase from inactivation during freezing, but not to stabilize these proteins during drying (24). This is not surprising because PEG crystallizes during freezing and is then not able to form the hydrogen bonds with proteins that are necessary for protein protection at the drying step. However, in the presence of 1% PEG and small amounts of glucose or trehalose, which alone provide only minimal protection, the activity of both lyophilized enzymes recovers almost completely after rehydration in aqueous solution. In this case, PEG stabilizes the proteins by the mechanism of preferential hydration during freezing, while sugars act as water substitutes during drying because of their hydrogen bonding with the proteins. However, protection during freeze-drying is not the sole mechanism responsible for higher catalytic activity in organic solvents of enzymes lyophilized with excipients. For example, several proteases were shown to be activated a thousandfold in anhydrous media when they were lyophilized with an excess of KCl or other salts (30). Although certain salts are known to be good cryoprotectants, the activation of enzymes in that case was shown to be due to some other effect than protection during freeze-drying. The most plausible explanation, suggested by Khmelnitsky et al. (30), is that the activation was due to stabilization of the reaction transition state by a polar matrix microenvironment.

Inactivation at the steps of freezing and drying can, however, be avoided if a method other than lyophilization is used for preparing enzyme catalysts. Vacuum evaporation of aqueous solutions of enzymes or precipitation by acetone did not lead to better catalytic activities in anhydrous media. Much higher activities have been obtained with crystals of enzymes (31). In a crystalline microenvironment, additional polar and hydrophobic contacts between neighboring protein molecules significantly increase stability by preventing unfolding, aggregation, or dissociation in the case of oligomeric proteins. "Mild" cross-linking with bifunctional chemical reagents makes the crystals more solid (cross-linked enzyme crystals) and also increases protein stability. One more way to avoid lyophilization is to obtain noncovalent complexes of enzymes with surfactants, which exist due to ion-pairing between the molecules of surfactant and enzyme (32). Such complexes are quite soluble in nonpolar solvents (up to concentrations of several milligrams per milliliter) and possess very high catalytic activities. For example, the value of k_{cat}/K_m for transesterification of N-acetyl-L-phenylalanine ethyl ester catalyzed by the complex of subtilisin with Aerosol OT dissolved in octane was equal to 70% of the value for the hydrolysis reaction in water.

Effect of Organic Solvents on Structure and Activity of Enzymes

In a dry state almost all enzymes are inactive; they need a small quantity of water to display catalytic activity. This water provides enzymes with conformational mobility, which is necessary for catalytic function. Because of this, lyophilized (dry) enzymes are often preequilibrated with small amounts of water before starting reactions in organic solvents. However, contact with anhydrous solvents may lead to significant dehydration of enzymes ("stripping" essential water of the protein) and inactivate them (33). This tendency is especially strong with polar solvents. Some of them were shown to dehydrate enzymes on a very rapid time scale—5 min or faster (34). The driving force for dehydration is the higher affinity of water to molecules of polar solvent than to the protein's functional groups. The interaction energy for a hydrogen bond between water and dimethyl sulfoxide was calculated as equal to 8.8 kcal/mol; the solvent interaction with an amide H has the same energy, and both are much stronger than hydrogen bonds between water and a carbonyl O or an amide H and between a backbone carbonyl O and amide H (35). These data explain why dimethyl sulfoxide, as well as other polar solvents, has the tendency to strip water molecules from the protein surface and form new hydrogen bonds with protein functional groups. On the other hand, contact with nonpolar solvents leaves the major portion of the protein molecule hydrated and causes only small rearrangements of side chains on the protein surface (36).

Effects of polar solvents on the structure of suspended enzymes depend mainly on the nature of the solvent (37) and on enzyme stability (38). For example, acetonitrile, butanol, ethyl acetate, and tetrahydrofurane, each containing 1% (v/v) of water, produced few changes in the structure of bovine pancreatic trypsin inhibitor in addition to those already brought about by lyophilization (37). On the other hand, this protein was strongly denatured by methanol, dimethyl formamide, and dimethyl sulfoxide—solvents known for their ability to dissolve proteins. Contact with polar solvents is accompanied by significant changes in the protein's hydration. For example, 4% (v/v) isopropanol added to crystalline chymotrypsin in hexane nearly doubled the number of water molecules observable in the protein's crystal structure (36). These additional water molecules did not contact directly with the protein's surface groups but were integrated within the second hydration sphere of the protein. Structural changes in proteins caused by organic solvents have been successfully inhibited by the simple addition of low molecular weight compounds during lyophilization. For example, lyoprotectors such as ammonium sulfate or sucrose did not only stabilize enzymes during lyophilization but were quite efficient in protecting enzymes from inactivation by organic solvents (33), probably by maintaining water in the enzyme's active site.

Polar solvents may significantly increase molecular mobility of proteins in nonaqueous media by eliminating flexibility constraints imposed by protein–protein contacts. It has been shown that the rates of interconversion between several conformations of a peptide, gramicidin A, in an anhydrous milieu were increased in protic solvents due to a faster exchange between hydrogen bonds (39). Molecular dynamic simulations also showed that subtilisin is much more flexible in polar versus nonpolar solvents (35). Probably, such an increase in molecular flexibility by added polar cosolvents (dimethyl sulfoxide and formamide) is responsible for the more than hundredfold acceleration of the reactions catalyzed by subtilisin in neat nonpolar solvents (40), although this explanation has been disputed (36).

Recent studies with Fourrier-transform infrared spectroscopy have pointed out that enzymes conserve a much more nativelike secondary structure in neat solvents than they do in water–cosolvent solutions with a solvent concentration of 60 vol % (21). The explanation is that protein denaturation (unfolding) is significantly inhibited by the more severe conformational restrictions imparted on the enzyme molecules in anhydrous versus water–cosolvent media. This may explain why enzymes are often more catalytically active in neat organic solvents than in water–cosolvent mixtures, that is, why the addition of water is detrimental rather than beneficial for stability of the native enzyme's conformation (21).

High Thermal Stability of Enzymes in Anhydrous Media

The fact of the enormously high stability of lipase suspended in organic solvents was mentioned in one of the first works in nonaqueous enzymology: this enzyme retained high activity after several hours of incubation in anhydrous media at 100 °C (41). Further studies confirmed that high thermal stability is a hallmark of proteins in nonpolar solvents, as opposed to their low stability in aqueous media; for such comparison, see the recent review by Zaks (42). For example, the melting temperature for proteins suspended in nonpolar solvents is 110–145 °C, which is 50–70 °C higher than in aqueous solutions (43). Thermal stability of ribonuclease A was nearly the same when measured in either a free-of-solvent state or hydrophobic solvents (44). The explanation was that the conformational flexibility of proteins is significantly retarded in systems with a small water content (45), as the latter acts as a lubricant in all macromolecular motions, owing to its high propensity to form multiple hydrogen bonds with a protein. Water removal also increases the strength of intramolecular hydrogen bonds and salt bridges, due to the higher energy of electrostatic interactions in the media with low dielectrics. All this creates a high kinetic barrier between the native and unfolded states and stabilizes native conformations of proteins (42). Also molecular mobility is suppressed by protein–protein interactions, which exist in multitude in lyophilized and crystalline protein samples, where the packing density of macromolecules is very high. Elimination of water decelerates almost all detrimental chemical reactions (Table 1), which are also responsible for irreversible inactivation of proteins in nonaqueous media. For example, a very slow inactivation of proteins in nonpolar solvents at temperatures higher than 100 °C is caused by the formation of aggregates cross-linked by intermolecular reactions of transamidation and disulfide exchange (43).

On the other hand, proteins are significantly less thermostable in polar versus nonpolar solvents (46). The reason may be that in hydrophobic solvents a protein has more intraprotein hydrogen bonds, which make the protein molecule more rigid and more resistant to high-temperature unfolding. Thermal stability of proteins in organic solvents often declines on addition of even small amounts of water, which are quite sufficient to revive molecular dynamics in proteins. An almost linear decrease in the melting temperature has been observed as a function of the amount water in anhydrous media (44). This stability decrease was explained by stabilization of the unfolded protein state, rather than destabilization of the folded native state. High-temperature inactivation in anhydrous media is accompanied by migration into the solvent of water bound to protein; it was suggested that this dehydration could be the rate-limiting step of the whole inactivation process (47). The tendency to dehydrate proteins essentially depends on the hydrophobicity of organic solvents. Molecular dynamic simulation has shown that more water molecules leave the surface of a protein placed in dimethyl sulfoxide than in the acetonitrile simulation, whereas no water at all leaves the protein in carbon tetrachloride (35). Finally, protein thermal stability in anhydrous solvents significantly depends on conditions of lyophilization, such as pH of the buffer or presence of excipients (48). For example, the complex of chymotrypsin with sorbitol denatures in organic solvent at much higher temperatures than the enzyme in the absence of sorbitol (49).

CONCLUSION

The problem of low activity and stability of enzymes in organic solvents led biochemists and biochemical engineers to the necessity of protein stabilization, and some recent advances were discussed in this review. A variety of methods have been applied, including familiar immobilization and chemical modification, and recently designed genetic procedures, which have become routine in the last decade. Essentially, not only protein stability is increased by methods of enzyme engineering, but also a variety of other important enzyme properties are changed, such as pH profiles of catalytic activity and stability, substrate specificity, including chemio-, regio- and enantioselectivity, and others. It is a task of the next decade to learn how to tailor in the most rational way the whole combination of enzyme properties that are necessary for successful utilization of enzymes in biotransformations.

BIBLIOGRAPHY

1. A.M.P. Koskinen and A.M. Klibanov, eds., *Enzymatic Reactions in Organic Media*, Blackie Academic & Professional, London, 1986.
2. P.L. Privalov and S.J. Gill, *Adv. Prot. Chem.* **39**, 191–234 (1988).
3. V.V. Mozhaev, *Trends Biotechnol.* **11**, 88–95 (1993).
4. A.M. Klibanov, *Adv. Appl. Microbiol.* **29**, 1–28 (1983).
5. B.W. Matthews, *Annu. Rev. Biochem.* **62**, 139–160 (1993).
6. S.N. Timasheff, *Annu. Rev. Biophys. Biomol. Struct.* **22**, 67–97 (1993).
7. S.J. Singer, *Adv. Prot. Chem.* **17**, 1–69 (1962).
8. Yu.L. Khmelnitsky, V.V. Mozhaev, A.B. Belova, M.V. Sergeeva, and K. Martinek, *Eur. J. Biochem.* **198**, 31–41 (1991).
9. J.A. Rupley, E. Gratton, and G. Careri, *Trends Biochem. Sci.* **8**, 18–22 (1983).
10. D.S. Clark and J.E. Bailey, *Biochim. Biophys. Acta* **788**, 181–188 (1984).
11. V.V. Mozhaev, M.V. Sergeeva, A.B. Belova, and Yu. L. Khmelnitsky, *Biotechnol. Bioeng.* **35**, 653–659 (1990).
12. A.K. Gladilin, E.V. Kudryashova, A.V. Vakurov, V.A. Izumrudov, V.V. Mozhaev, and A.V. Levashov, *Biotechnol. Lett.* **17**, 1329–1334 (1995).
13. N.L. St Clair and M. Navia, *J. Am. Chem. Soc.* **114**, 7314–7316 (1992).
14. Yu.L. Khmelnitsky, V.V. Mozhaev, A.B. Belova, and A.V. Levashov, *FEBS Lett.* **284**, 267–269 (1991).
15. V.V. Mozhaev, E.V. Kudryashova, N.V. Efremova, and I.N. Topchieva, *Biotechnol. Tech.* **10**, 849–854 (1996).
16. P. Martinez and F.H. Arnold, *J. Am. Chem. Soc.* **113**, 6336–6337 (1991).
17. C.-H. Wong, *Trends Biotechnol.* **10**, 378–381 (1992).
18. F.H. Arnold and J.-H. Zhang, *Trends Biotechnol.* **12**, 189–192 (1994).
19. R.D. Kidd, H.P. Yennawar, P. Sears, C.-H. Wong, and G.K. Farber, *J. Am. Chem. Soc.* **118**, 1645–1650 (1996).
20. Y. Tomiuchi, K. Ohshima, and H. Kise, *Bull. Chem. Soc. Jpn.* **65**, 2599–2603 (1992).
21. K. Griebenow and A.M. Klibanov, *J. Am. Chem. Soc.* **118**, 11695–11698 (1996).
22. R.A. Sheldon, in *Enzymatic Reactions in Organic Media*, A.M.P. Koskinen and A.M. Klibanov eds., Blackie Academic & Professional, London, 1996, pp. 266–307.
23. J.L. Schmitke, C.R. Wescott, and A.M. Klibanov, *J. Am. Chem. Soc.* **118**, 3360–3365 (1996).
24. J.F. Carpenter, S.J. Prestrelski, and T. Arakawa, *Arch. Biochem. Biophys.* **303**, 456–464 (1993).
25. J.F. Carpenter, S.J. Prestrelski, T.J. Anchordoguy, and T. Arakawa, in *Formulation and Delivery of Proteins and Peptides*, American Chemical Society Symposium Series 567, J.L. Cleland and R. Langer, eds., American Chemical Society, Washington, D.C., 1994, pp. 134–147.
26. U.R. Desai, J.J. Osterhout, and A.M. Klibanov, *J. Am. Chem. Soc.* **116**, 9420–9422 (1994).
27. T.J. Anchordoguy and J.F. Carpenter, *Arch. Biochem. Biophys.* **332**, 231–238 (1996).
28. K. Xu and A.M. Klibanov, *J. Am. Chem. Soc.* **118**, 9815–9819 (1996).
29. K. Griebenow and A.M. Klibanov, *Proc. Nat. Acad. Sci. U.S.A.* **92**, 10969–10976 (1995).
30. Yu.L. Khmelnitsky, S.H. Welch, D.S. Clark, and J.S. Dordick, *J. Am. Chem. Soc.* **116**, 2647–2648 (1994).
31. A.L. Margolin, *Trends Biotechnol.* **14**, 223–230 (1996).
32. V.M. Paradkar and J.S. Dordick, *J. Am. Chem. Soc.* **116**, 5009–5010 (1994).
33. P.A. Burke, R.G. Griffin, and A.M. Klibanov, *J. Biol. Chem.* **267**, 20057–20064 (1992).
34. L.A.S. Gorman and J.S. Dordick, *Biotechnol. Bioeng.* **39**, 392–397 (1992).
35. Y.-J. Zheng and R.L. Ornstein, *J. Am. Chem. Soc.* **118**, 4175–4180 (1996).
36. G.K. Farber, *Ann. N. Y. Acad. Sci.* **799**, 85–89 (1996).
37. U.R. Desai and A.M. Klibanov, *J. Am. Chem. Soc.* **117**, 3940–3945 (1995).
38. A. Dong, J.D. Meyer, B.S. Kendrick, M.C. Manning, and J.F. Carpenter, *Arch. Biochem. Biophys.* **334**, 406–411 (1996).
39. F. Xu, A. Wang, J.B. Vaughn, and T.A. Cross, *J. Am. Chem. Soc.* **118**, 9176–9177 (1996).
40. O. Almarsson and A.M. Klibanov, *Biotechnol. Bioeng.* **49**, 87–92 (1996).
41. A. Zaks and A.M. Klibanov, *Science* **224**, 1249–1251 (1984).
42. A. Zaks, in *Enzymatic Reactions in Organic Media*, A.M.P. Koskinen and A.M. Klibanov eds., Blackie Academic & Professional, London, 1996, pp. 70–93.
43. D.B. Volkin, A. Staubli, R. Langer, and A.M. Klibanov, *Biotechnol. Bioeng.* **37**, 843–853 (1991).
44. N.A. Turner, D.B. Duchateau, and E.N. Vulfson, *Biotechnol. Lett.* **17**, 371–376 (1995).
45. R. Affleck, Z.-F. Xu, V. Suzawa, K. Focht, D.S. Clark, and J.S. Dordick, *Proc. Nat. Acad. Sci. U.S.A.* **89**, 1100–1104 (1992).
46. P.P. Wangikar, P.C. Michels, D.S. Clark, and J.S. Dordick, *J. Am. Chem. Soc.* **119**, 70–76 (1997).
47. G. Greco, Jr., D. Pirozzi, G. Toscano, and M. Maremonti, *Ann. N. Y. Acad. Sci.* **799**, 108–114 (1996).
48. B. Schulze and A.M. Klibanov, *Biotechnol. Bioeng.* **38**, 1001–1006 (1991).

49. A.O. Triantafyllou, E. Wehtje, P. Adlercreutz, and B. Mattiasson, *Biotechnol. Bioeng.* **45**, 406–414 (1995).

See also ENZYMES, DETERGENT; ENZYMES, DIRECTED EVOLUTION; ENZYMES, EXTREMELY THERMOSTABLE; ENZYMES, PULP AND PAPER; PEPTIDE; PROTEIN AGGREGATION, DENATURATION.

DENATURATION, THERMAL. See THERMAL UNFOLDING, PROTEINS.

DEXTRAN, MICROBIAL PRODUCTION METHODS

R.S. KARTHIKEYAN
T. SWAMINATHAN
A. BARADARAJAN
Indian Institute of Technology–Madras
Chennai, India

KEY WORDS

Biosynthesis
Dextransucrase
Exopolysaccharide
Immobilization
Leuconostoc mesenteroides
Rheology

OUTLINE

INTRODUCTION

Over the past 25 years a new class of microbial products, the microbial exopolysaccharides, has grown in industrial importance. These products can be used as alternatives to other synthetic and natural water-soluble polymers that are used as emulsifiers, stabilizers, binders, coagulants, lubricants, film formers, and thickening, gelling, and suspending agents. Despite their costly collection and extraction, plant and algal polysaccharides such as gum guar, locust bean gum, starch, alginate, carageenan, and agar are widely used in the food and pharmaceutical industries. These polymers, although valuable, have certain drawbacks such as rheological properties that may not match exactly with a specific requirement, production that is vulnerable to crop failure, climatic conditions and marine pollution that cause the quality, supply, and cost to fluctuate (1,2).

The production of microbial exopolysaccharides provides a valid alternative to plant and algal polysaccharides. The exopolysaccharides, produced by a wide variety of microorganisms, show considerable diversity in their composition, structure, and unique physical properties. Their industrial production has been stimulated by the opportunity to provide a guaranteed supply, constant quality, superior properties, and stable price. Among all of the exopolysaccharides, dextran exemplifies attainment of these objectives.

Dextran produced from a bacterial strain of *Leuconostoc mesenteroides* NRRL-B512F has a unique value in medical applications as a blood plasma substitute and a blood flow improver, properties that have provided the impetus for initiating much work on its production and further applications (3). Dextran is a microbial exopolysaccharide made up of α-D-glucose units linked mainly by α, 1 → 6 bonds. Traditionally, dextran is produced by fermenting dextran-producing bacteria in sucrose-rich media (4). The prime bacterial product of fermentation in the presence of substrate sucrose is not dextran itself but the enzyme dextransucrase. The enzyme is secreted by growing cells into the medium, where it converts the sucrose to dextran according to the following equation:

$$n\ \text{Sucrose} \xrightarrow{\text{Dextransucrase}} \begin{matrix}(\text{Glucose})_n \\ \text{or} \\ \text{Dextran}\end{matrix} + n\ \text{Fructose} \qquad (1)$$

Therefore, dextran can be produced from sucrose in cell-free environments by using the enzyme dextransucrase (5). The main dextransucrase-producing bacteria belongs to the genera *Leuconostoc* and *Streptococcus*. The enzyme dextransucrase uses sucrose as a high-energy glucopyranosyl donor for dextran synthesis, and the properties of the enzyme and the dextran produced are specific to the microorganisms used (6).

In addition to sucrose, α-D-glucopyransosyl, α-L-sorbofuranoside (7), lactulosucrose (8), α-D-glucopyranosyl fluoride (9), and *P*-nirophenyl α-D-glucopyranoside (10) also serve as glucopyranosyl donors, with sucrose having the lower Michaelis constant. These substrates have glucosidic bond energies of 5.0–5.5 kcal/mol. Hehre and Hamilton (11) and Barker et al. (12) reported that *Acetobactor capsulatum* and *A. viscosum* can transform amylodextrins to dextran. A chemical synthesis of dextran has been reported by Ruckel and Schuerch (13).

Dextran produced from microbial fermentation is called native dextran and typically has a molecular weight of 5×10^7 to 5×10^9. Native dextran is usually separated from the producing cells and precipitated by alcohol. This native dextran may then be subjected to controlled acid hydrolysis to obtain the desired molecular weight for specific applications (14,15).

HISTORICAL BACKGROUND

Dextran first came under close examination more than 100 years ago. Dextrans develop naturally in sucrose-containing solutions that have become inoculated with dextran-producing bacteria from air, plants, or soil. The resulting transformation of sucrose solution to syrupy, viscous, or ropey fluids or even gelled masses doubtlessly has plagued humankind since the inception of accumulating and storing sucrose-containing food and beverages and has caused much trouble in sugar beet industries.

In 1861 Louis Pasteur (16) identified the viscous polymer as a polysaccharide. In 1874 Scheubler (17) determined its empirical formula to be $C_6H_{10}O_5$ and named it dextran because it was closely related to starch and dextrin. Van Tieghem (18) named the causative bacteria *Leuconostoc mesenterides*. Cell-free enzymatic synthesis of dextran was first demonstrated by Hehre and Sugg (5,19). The enzyme involved in this synthesis was given the name dextransucrase by Hestrin and Shapiro-Avineri (20), and it was soon found to be a member of the general class of transglycosylases (21–23). In 1937 Flower and coworkers made the first systematic investigation of the chemical structure of dextran (24). A large body of work carried out between 1950 and 1970 on dextran production and its structure, properties, and uses is reviewed by Neely (25), Jeanes (4,26), Murphy and Whistlkr (27), and and Sidebotham (28).

CLASSIFICATION AND STRUCTURE OF DEXTRANS

The term *dextran* collectively describes a large class of bacterial extracellular hydrocolloid homopolysaccharides consisting of α-D-glucopyranosyl units polymerized predominantly in an α, 1 → 6 linkage. The diversity within the class of dextran molecules derives from the existence of non–α, 1 → 6 linkages that lead to branching from the main polymer chain. Branching mainly occurs by α, 1 → 3, α, 1 → 4, and α, 1 → 2 linkages. The type, amount, length, and arrangements of branches define the differences in dextran structure and influence the water solubility and other rheological properties of dextran (29).

Hucker and Pederson (30) classified the dextran-producing microorganisms into family, tribe, genus, and species. In 1954 Jeanes et al. (6) reported the classification and characterization of dextrans by examining 96 strains of bacteria. First they classified dextrans into three groups: water-soluble dextrans, water-insoluble dextrans, and heterogeneous dextrans. Water-soluble dextrans are further divided into less-soluble dextrans (can be precipitated by 34–37% of alcohol) and more-soluble dextrans (can be precipitated by 40–44% of alcohol). Different kinds of dextrans produced by the same strain of bacteria are called heterogeneous dextrans, and these mixtures of heterogeneous dextrans can be separated by differential alcohol precipitation (31). Further, Jeanes et al. classified the dextrans into class A, class B, and class C dextrans on the basis of the percentage of their α, 1 → 3 linkages. They characterized the dextrans by optical relation, viscosity, periodate oxidation, and physical appearance after alcohol precipitation. Dextran structures have two common features:

1. Main chain and branching chain (side chain) are made by α-D-glucose units linked together by α, 1 → 6 bonds.
2. Branching chains (or branching units) are connected to the main chain by α, 1 → 3, α, 1 → 4, or α, 1 → 2 bonds, called branching linkages.

Examples of α, 1 → 3 Branch–Linked Dextrans

Commercially produced *L. mesenteroides* B-512F dextran has 95% of the total glucouspyranosyl linkages as α, 1 → 6 linkages and 5% as α, 1 → 3 linkages (32,33) (structure A in Fig. 1). The α, 1 → 3 branching linkages occur in two ways:

1. A single α-D-glucose unit attached to the main chain by an α, 1 → 3 bond
2. A branching chain attached to the main chain by an α, 1 → 3 bond (see Fig. 1)

This dextran appears as a translucent gel. Senti et al. reported that 77% of the α, (1 → 3) linkages are single glucose residues and the remaining branch chains have 50 or more glucose residues (34). Walker and Pulkownik, studying the hydrolysis of B512F dextran by dextranglucosidase, also concluded that it has 85% single glucose branches and 33 or more glucose residues in the remaining branch chains to give the yield of glucose that they observed (35).

Streptococcus mutans 6715 produces water-soluble and water-insoluble dextrans. The water-soluble fraction has 35% α, 1 → 3 linkages and 65% α, 1 → 6 linkages. Branching single glucose units are uniformly connected to the main chain and to extended side chains (structure B in Fig. 1). The resulting structure was named an alternating bifur-

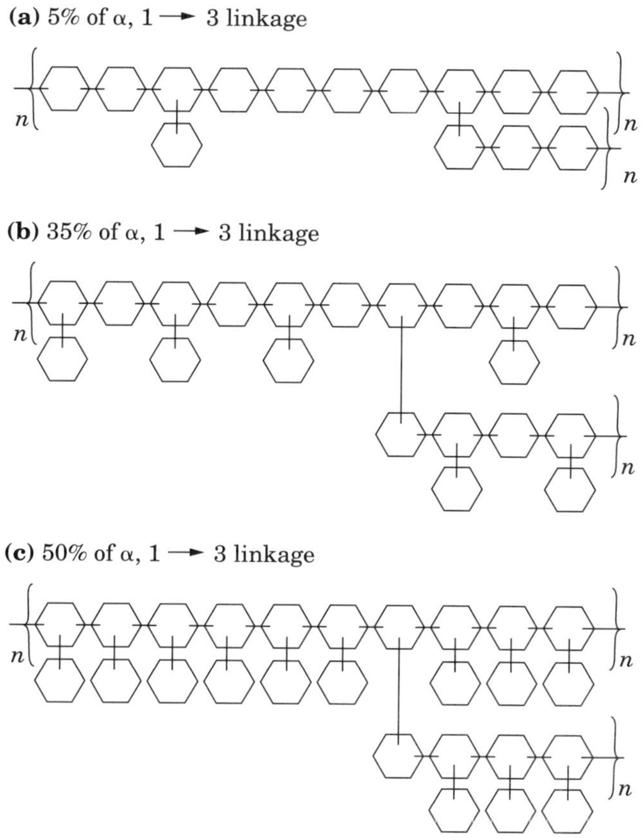

Figure 1. Structures of dextrans: Hexagonal units represent α-D-glucose units, horizontal lines joining glucose units are α, 1 → 6 bonds, vertical lines joining glucose units are α, 1 → 3 bonds: (**a**) clinical dextran, (**b**) alternating bifurcated comb structure, and (**c**) bifurcated comb structure. (Not drawn to statistically accurate scale.)

cated comb polymer by Robyt et al. (36,37), is resistant to hydrolysis by endodextranase, and appears as a heavy, opaque material (38).

L. mesenteroides B-742 strain produced more-soluble and less-soluble dextrans. The more-soluble fraction of dextrans has 50% α, 1 → 6 linkages and 50% α, 1 → 3 branching linkages. Some branching chains are attached to the main chain by α, 1 → 3 branching linkages, but single glucose units are attached to all other glucose units in the main chain by α, 1 → 3 bonds. This structure results in the most highly branched dextrans. Robyt et al. (36,37) named this structure a (structure C in Fig. 1), in which each single α, 1 → 3 linked glucose branch is like a tooth of a comb and the α, 1 → 6 linked chains are the backbone of the comb, bifurcated comb polymer (39,40). This structure is slightly resistant to hydrolysis by endodextranase. This dextran appears as a finely divided material. The α, 1 → 4 and α, 1 → 2 branched dextrans are less common (39–44).

CHARACTERISTICS OF DEXTRANSUCRASE

The unit of dextransucrase (EC 2.4.1.5, α, 1 → 6-D-glucan: D-fructose 2-glucosyltransferase) activity, DSU, is defined as the amount of enzyme that will convert 1 mg of sucrose to dextran in 1 h at 30 °C in acetate buffer solution at pH 5.2. Another unit, U, is defined as the amount of enzyme that liberates 1 μmol of reducing sugar from sucrose in 1 min at pH 5.2 and 30 °C. The relation between these two units is 1 U/mL = 20.27 DSU/mL (14).

The extensive purification of dextransucrase was first reported by Ebert and Schenk, and the molecular weight of the enzymc was determined to be 280,000 by ultracentrifugation methods (45). In 1980 Kobayashi and Matsuda separated two components of dextransucrase having different affinities for dextran gel by a column of Sephadex G-100. The major component of the purified enzyme had a molecular weight of 64,000–65,000, an isoelectric point (pI) of 4.1, and was 17% carbohydrates (46). Most of the carbohydrates in the purified enzyme were found to be endogenous dextran, covalently bound to the enzyme molecule and released from it by mild treatment with endodextranase. They concluded that dextransucrase exists in multiple forms and that endogenous dextran molecules might be responsible for the multiple forms (47). Miller et al. (48) and Fu and Robyt (49) found that the molecular weight of dextransucrase produced from a mutant strain of *Leuconostoc mesenteroides* B-512 FM initially secreted into the culture medium was 177,000; later, however, the enzyme exhibited a drop in molecular weight to 158,000 without any loss in activity.

Mode of Dextransucrase Action

The irreversible reaction of dextransucrase with sucrose is simply formulated in equation 1. Dextransucrase is one of a very few synthesizing enzymes that do not require the presence of cofactors or high-energy phosphorylated intermediates. The energy required for the condensation of two glucosyl units is provided by the hydrolysis of sucrose, distinguishing it from other mechanisms of polysaccharide synthesis (36,37).

Mechanism of Main Chain Synthesis. When dextransucrase was first described in 1941 by Hehre (5), Cori and Cori were studying the action of muscle phosphorylase (50) and Hanes was studying potato phosphorylase (51). These investigators observed that phosphorylase could elongate glycogen and starch chains by the transfer of glucose from α-D-glucopyranosyl phosphate (Glc-1-P) to the nonreducing end (C_6 moiety of glucose) of the polysaccharide chain (52) according to the following equation:

$$\begin{bmatrix}\text{Polysaccharide}\\ \text{chain (primer)}\end{bmatrix} + \begin{bmatrix}\alpha\text{-D-glucopyranosyl}\\ \text{phosphate (Glc-1-P)}\end{bmatrix} \xrightarrow{\text{Phosphorylase}} \begin{bmatrix}\text{One glucose unit added}\\ \text{to the polysacchardie chain}\end{bmatrix} \quad (2)$$

It was believed that the enzyme phosphorylase that catalyzed the preceding reaction requires a preformed polysaccharide chain, called a primer, and a high-energy monosaccharide donor, Glc-1-P. Later it was proven that the enzyme phosphorylase was a degradative enzyme that catalyzed the reaction of inorganic phosphate with the nonreducing end of the polysaccharide chain to give Glc-1-P according to the following equation:

$$\begin{bmatrix}\text{Polysaccharide}\\ \text{chain}\end{bmatrix} + \begin{bmatrix}\text{Inorganic}\\ \text{phosphate}\end{bmatrix} \xrightarrow{\text{Phosphorylase}} [\text{Glc-1-P}] + \begin{bmatrix}\text{One glucose unit reduced in the}\\ \text{polysaccharide chain (primer)}\end{bmatrix} \quad (3)$$

The reverse of the preceding reaction, which requires a primer (a polysaccharide chain), produces a polysaccharide chain by adding one glucose monomer to the primer at the nonreducing end (C_6 moiety). Thus in the 1940s and 1950s the primer mechanism was assumed for dextran synthesis (53,54).

Evidence for the primer-dependent mechanism for dextran synthesis was strengthened by Germaine and coworkers in the 1970s. They found that addition of dextran (as a primer) to dextransucrase increased the rate of dextran synthesis and that the rate reached a maximum at an average size of the added dextran of 30 glucose residues (55–57). Kobayashi and Matsuda also reported that purified dextransucrase was stimulated by dextran (as a primer), although dextransucrase could synthesize dextran without the addition of primer (46,47). Based on these results, Kobayashi et al. (58) proposed a primer-dependent mechanism for dextran synthesis. They proposed that the active site has two types of substrate binding sites, one for sucrose and the other for primer dextran. The elongation of the primer was hypothesized to take place when an enzyme nucleophile (presumably a carboxylate anion) attacks C_1 of the glucose moiety to give the glucosyl–enzyme covalent intermediate. This glucose is then transferred to the 6-hydroxyl group of the nonreducing end of the primer chain attached to the primer binding site of the enzyme, forming the next α, $1 \rightarrow 6$ linkage. For the synthesis to continue, the primer, with newly added glucose residue, must dissociate from the primer binding site and a new primer molecule must bind in the primer binding site with its nonreducing glucose residue in position to accept the next glucose unit. The synthesis in this scheme is thus a discontinuous process.

However, Robyt and Corrigan found that a dextran (as a primer) modified by a blocking group (triisoproprylbenzenesulfonyl, tripsyl) on the C_6 position of the nonreducing end of the glucose residue was as effective in increasing the rate of dextran synthesis by dextransucrase as the unmodified dextran primer. The modified dextran could not act as a primer because its sixth position was blocked, proving that the added modified dextran stimulated synthesis by some other mechanism (59). Robyt et al. have shown that the activation of dextransucrase by addition of dextran is by an allosteric mechanism rather than by an extending primer mechanism (37) as distinguished from the chain elongation of glycogen and starch by phosphorylase.

In 1962 Ebert and Patat proposed another mechanism for dextran synthesis, insertion type polymerization, in which the addition of glucose units takes place at the reducing end (C_1 of glucose moiety) of the growing dextran polymer (60). In their mechanism they assumed that the growing chain and the enzyme remain in contact in the form of a complex throughout the chain elongation cycle. This enzyme–polymer complex (EP_n) reacts in the first step with substrate to form a substrate complex $(ESP_n)_2$, which is transformed in a second step in such a way that the polymer chain in the reaction complex is extended by one unit. This can only happen if the new monomer unit is inserted into the covalent bond between the enzyme and the polymer chain in the EP_n complex.

$$EP_n + S \underset{k_{-1}}{\overset{k_1}{\rightleftarrows}} ESP_n \quad (4)$$

$$ESP_n \xrightarrow{k_2} EP_{n+1} + F \quad (5)$$
(F: fructose)

The bond between the enzyme and polymer chain does not dissociate during the reaction but must be flexible enough so that the new unit can be inserted into it. From their kinetic studies, Ebert and Schenk (45) assumed that the substrate undergoes a second reaction with the EP_n complex in which it reacts as an acceptor.

$$ESP_n + S^* \underset{k_{-3}}{\overset{k_3}{\rightleftarrows}} S^*ESP_n \rightarrow ES + S^*P_n \quad (6)$$

In the preceding reaction, the acceptor (sucrose molecule acting as an acceptor identified by the asterisk) does not compete with the substrate for the enzyme but reacts with the intermediate ESP_n complex, terminating polymer extension. This would explain why the molecular weights of the products decrease with increasing substrate concentration. From the preceding mechanism Ebert and Patat proposed the following reaction rate expressions:

$$v = \frac{V_m[S]}{K_m + [S]} \cdot \frac{K_a}{K_a + [S]} \quad (7)$$

where V_m is the maximum reaction rate (k_2E_0), K_m is the Michaelis constant of the chain extension reaction, and K_a is the corresponding constant of acceptor reaction. Ebert and Schenk conclude that there is strong evidence for their proposed mechanism because the results of their kinetic studies on dextran formation match the proposed rate in equation 7 quite well. They further conclude that primers are not necessary to initiate polymer growth, and sucrose alone can initiate polymerization (45).

The early theories of the mechanisms of dextran synthesis were described by Neely (61). Ebert and Schenk propose the sequential addition of monomer units to the reducing end by insertion between a carrier (the enzyme) and growing polysaccharide chain (45). Sidebotham (28) continued the discussion in 1974, and a series of papers by Robyt and coworkers appears to have conclusively resolved the mechanism (59,62,63).

In 1974 Robyt and coworkers proposed a two-site insertion mechanism for dextran synthesis in which the glucose unit is added to the growing polymer chain at the reducing end (C_1 moiety of glucose). They used pulse and chase techniques with ^{14}C-labeled sucrose and dextransucrase immobilized in Bio-Gel-P2 (64). In their two-site insertion mechanism (see Fig. 2), there are two sucrose binding sites (A and B) that attack two sucrose molecules to give two covalently attached glucosyl–enzyme intermediates (C). The C_6 hydroxyl of one of these glucosyl–enzyme intermediates makes a nucleophilic attack on C_1 of the other glucosyl–enzyme intermediate with the formation of an α, $1 \rightarrow 6$ glucosidic linkage and results in an isomaltosyl–enzyme intermediate (D). The vacant sucrose binding site (A) attacks another sucrose molecule to form a new glucosyl–enzyme intermediate to give a second α, $1 \rightarrow 6$ linkage and the formation of the isomaltotriosyl–enzyme intermediate (E). This process continues in a similar fashion by inserting a glucose residue between the enzyme and growing chain, which results in a dextranyl–enzyme complex (F). In this mechanism the growing dextran chain is transferred from one site to another during the synthesis. Robyt and Eklund (65,66) considered the stereochemistry of the reaction and concluded that the linkages of glucosyl and dextranyl units to the enzyme must be β to retain the configuration of the glucose residue in growing from sucrose to dextran.

Mechanism of Branch Linkage Formation. In 1959 Bovey investigated the formation of branching in dextrans using data on molecular weights obtained from light-scattering measurements. He proposed that a separate branching enzyme (acting in a way similar to Q enzyme on amylose) is responsible for the formation of branches (67). The branching enzyme, however, has never been found. In 1967 Ebert and Brosche proposed a reaction mechanism for the formation of branches in which a dextran chain itself acts as an acceptor, attacking an enzyme–dextran complex (ESP_n) (see equations 4–6). The acceptor dextran then becomes the main chain, and the dextran chain from the enzyme becomes the side-branched chain. Using a ^{3}H-labeled acceptor dextran of low molecular weight and assuming an average molecular weight of 4×10^5 for synthesized dextran, they concluded from the specific activity of synthesized product that there was only one labeled acceptor dextran molecule in each synthesized dextran (68). Although this seemed to be proof for the proposed mechanism, some doubt was cast on the mechanism because of assumptions and the circular arguments that were made.

In 1976 Robyt and Taniguchi (63) proposed a mechanism based on their study using ^{14}C-labeled sucrose and dextransucrase immobilized in Bio-Gel-P2. In their mechanism (see Fig. 3), a C_3 hydroxyl group of an interior glucose unit of an acceptor dextran (D) makes a nucleophilic attack at C_1 of either the glucosyl–enzyme complex (C) or at C_1 of the dextranyl–enzyme complex (E) to form an α, $1 \rightarrow 3$ branch linkage by displacing glucose or dextran from the enzyme. Thus, branching can take place without a separate enzyme by the action of an acceptor dextran on a glucosyl–enzyme complex (C) to give single glucose unit branch linkage (F) and on a dextranyl–enzyme complex (E) to give a long side chain branch linkage (G) (36,66).

Acceptor Reactions and Acceptor Products

In addition to catalyzing the synthesis of dextran from sucrose, dextransucrase also catalyzes the transfer of the glucose moiety from sucrose to other carbohydrates that are present or added to the reaction mixture (69). These other carbohydrates are acceptors (see Table 1). Actually, there are two classes of acceptors—those that give a homologous series of oligosaccharides, each differing one from the other by one glucose residue, and those that form only a single acceptor product that contains one glucose residue more than the acceptor (36,66).

Koepsell et al. (53) and Tsuchiya et al. (54) first observed that the presence of low molecular weight carbohydrates (acceptors) shifted the course of dextransucrase reaction from synthesizing a high molecular weight dextran to synthesizing of a low molecular weight dextran. These acceptors not only terminate the dextran synthesis but also produce a novel series of oligosaccharides. Su and Robyt systematically studied maltose acceptor reactions with dextransucrase (70). In this maltose acceptor reaction the D-glucose moiety of sucrose is diverted from synthesis of dextran but transferred to the nonreducing end of maltose (a disaccharide) to form panose (a trisaccharide). This first acceptor product, panose, can itself act as an acceptor to give a tetrasaccharide. This tetrasaccharide can again act as an acceptor to give a pentasaccharide and so forth. This series of homologous acceptor products is produced by acceptor reactions.

Mechanism of Acceptor Reaction. As described earlier, dextransucrase catalyzes the synthesis of dextran by initially forming two covalent glucosyl–enzyme complexes (Fig. 2) in which the C_6 hydroxyl moiety of one glucose attacks the C_1 of the other to form an α, $1 \rightarrow 6$ glucosidic bond to produce an isomaltosyl–enzyme complex. This is the beginning of the dextran main chain growth; the dextran chain is covalently attached to the enzyme active site and elongated by the insertion mechanism (36,66). In the presence of an acceptor, such as maltose, the transfer of glucose into a growing dextran chain is interrupted as the maltose binds to an acceptor binding site.

Robyt and Walseth proposed a mechanism for the acceptor reaction (see Fig. 4). In this mechanism, the acceptor bound at an acceptor binding site, and if the maltose molecule is an acceptor, its C_6 hydroxyl group at the nonreducing end attacks C_1 of the glucosyl–enzyme (C) or dextranyl–enzyme (E) complex to give an oligosaccharide (F) or a dextran acceptor product (G), respectively (62). Su and Robyt, using maltose as an acceptor in an equilibrium dialysis experiment, showed that there was one acceptor binding site and two sucrose binding sites (A and B) (70).

Effect of Acceptor on Dextran Production. The acceptors affect dextran production in two ways:

1. They terminate the growth of the dextran chain by reacting with the dextranyl–enzyme complex, which results in the low molecular weight dextran (product G in Fig. 4).
2. They interfere with the transfer of the glucose moiety of sucrose to the growing dextran chain by reacting with the glucosyl–enzyme complex, which results in production of a homologous series of acceptor

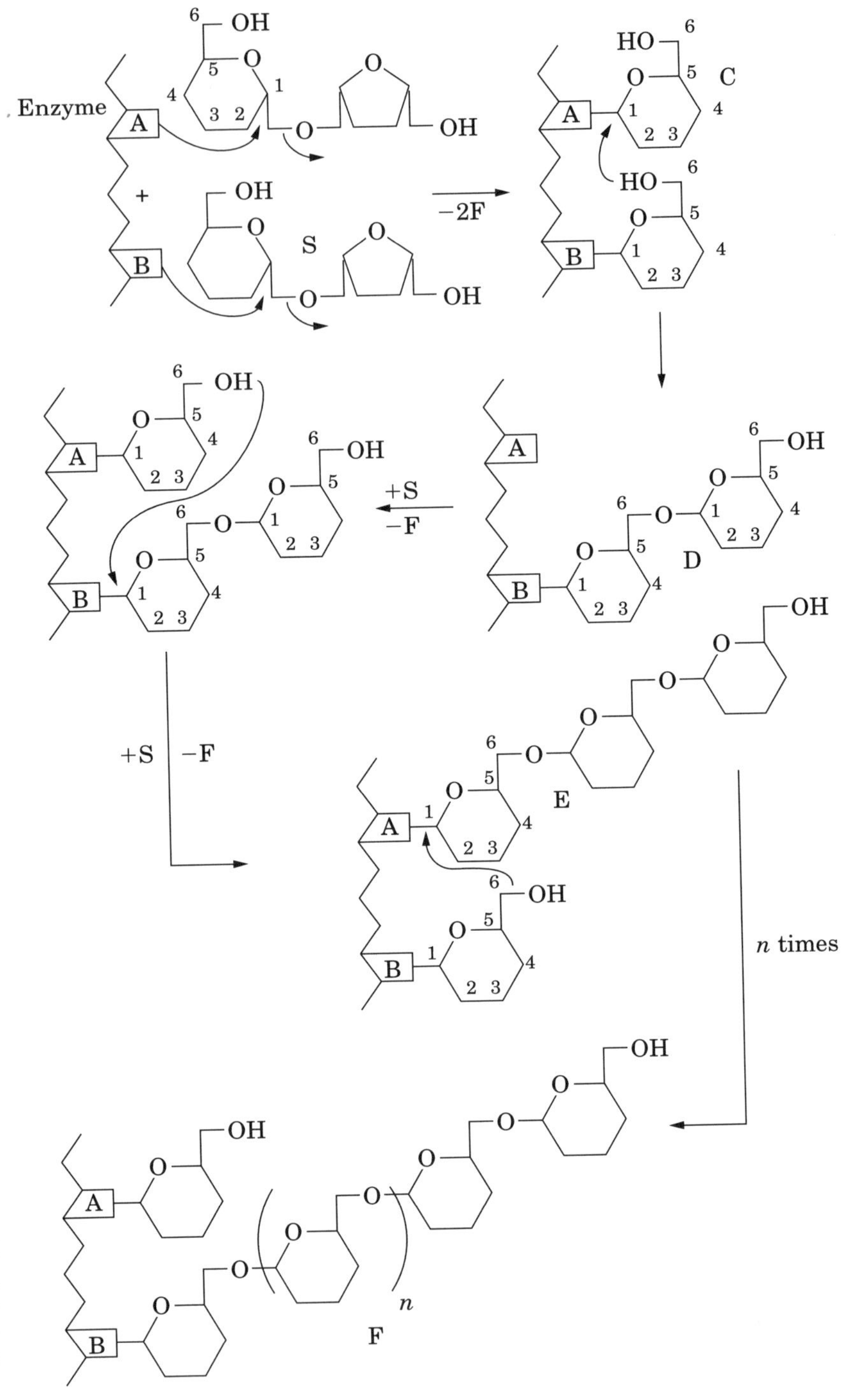

Figure 2. Two-site insertion mechanism: A and B, enzyme active sites; S, sucrose; F, fructose; C, glucosyl–enzyme complex; D, isomaltosyl–enzyme complex; E, isomaltotriosyl–enzyme complex; F, dextranyl–enzyme complex (—OH groups are omitted for clarity.)

products and reduction in the amount of dextran produced (product F in Fig. 4).

Paul et al. demonstrated the effect of the sucrose-to-acceptor ratio in the synthesis of controlled molecular weight dextrans (71). Robyt and Eklund showed that the amount of dextran synthesized decreased as the molecular ratio of maltose to sucrose increased (69). Not all of the acceptors reacted with equal efficiency. In a series of reactions with different acceptors at a 1:1 ratio of acceptor concentration to sucrose concentration, they found that the most effective acceptor for decreasing the amount of dextran synthesized was maltose and that the amount of glucose residues diverted from dextran synthesis into acceptor products depended on the ratio of maltose to sucrose used. The efficiency of other acceptors is given in Table 1, where maltose efficiency is defind as 100% (66). Su and Robyt found that an increase in the ratio of maltose to sucrose increased the amount (concentration) of acceptor products, with a concomitant decrease in dextran synthesis. Further, as the ratio was increased, the number of acceptor products (e.g., tri-, tetra-, and pentasaccharides) de-

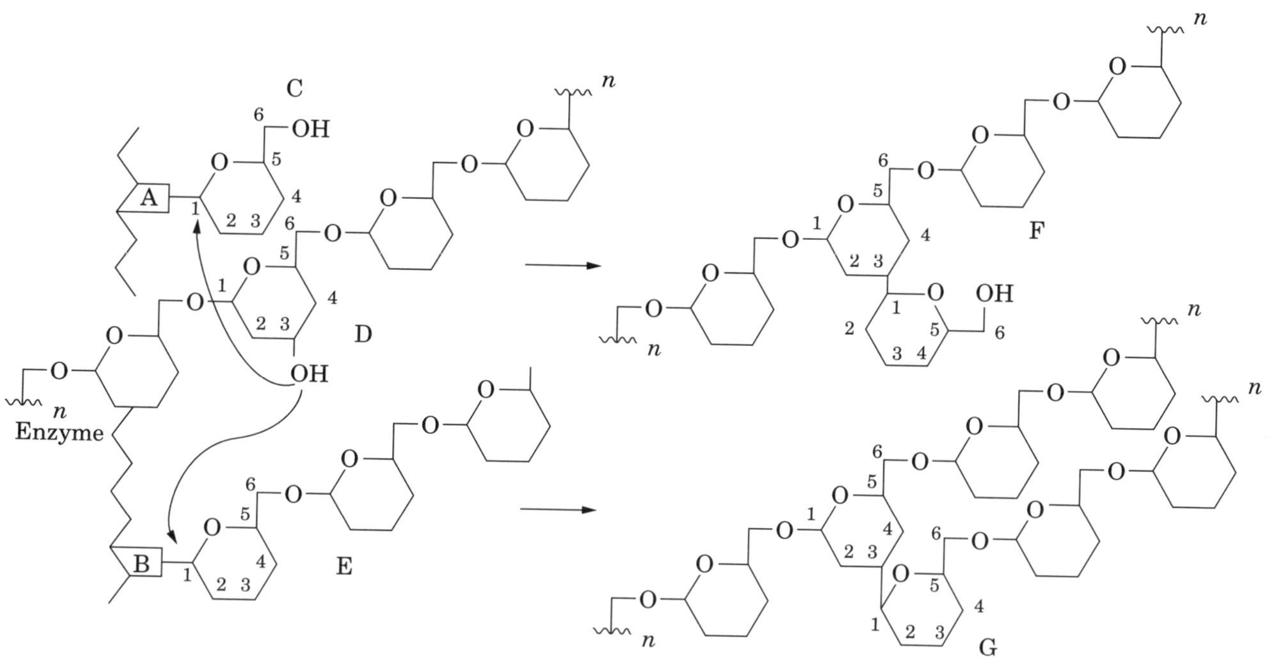

Figure 3. Mechanism of α, 1 → 3 branch linkage synthesis: D, acceptor dextran main chain; C, glucosyl–enzyme complex; E, dextranyl–enzyme complex; F, resulting α, 1 → 3 branch–linked product when C and D are reacted together; G, resulting α, 1 → 3 branch–linked product when E and D are reacted together. (—OH groups are omitted for clarity.)

Table 1. Acceptors and Their Efficiency on Dextransucrase

Acceptors[a]	Efficiency (%)
Maltose	100
Isomaltose	89
Nigerose	58
Methyl α-D-glucopyranoside	52
D-Glucose	17
Lactose	11
Cellobiose	9
D-Fructose	6.4
Maltooctaose	6.2

[a]The products from these acceptors can also act as acceptors to produce a homologous series of acceptor products.

creased. When the concentration of maltose and sucrose increased and the ratio was maintained at 1:1, there also a decrease in the amount of dextran and an increase in the amount (concentration) of acceptor products observed. In addition, the amount of dextran decreased and the amount and number of acceptor products increased when the amount of enzyme increased. From these studies, they reported that the first acceptor product (panose) can be exclusively obtained without formation of dextran by using a specific ratio and concentration of maltose and sucrose and a specific amount of enzyme (70).

Part et al. reported on the effect of borate ions on the acceptor reaction products of dextransucrase. They explained that the fructose released during dextran synthesis reacts as an acceptor with a glucosyl–enzyme complex to form fructose acceptor products (e.g., leucrose and isomaltulose), which in turn reduce the dextran yield. This fructose can be complexed with borate ions, and its acceptor effects can be minimized to increase dextran yield. They concluded that with 0.115 M sodium tetraborate in the reaction mixture, leucrose and isomaltulose formation can be minimized up to 75% with consequent increase in dextran yield (72). Therefore, the acceptor reaction between fructose and the dextranyl–enzyme complex, which leads to diminished low molecular weight dextran, can be overcome by adding borate ions to the reaction medium.

Lee et al. (73) used the idea of producing low molecular weight dextran by adding an acceptor to the reaction medium to directly produce clinical dextran (MW 70,000). They studied the effect of dextran T-10, dextran T-40, and maltose as acceptors in the batch and continuous reactors. They reported that increasing the concentration of dextran T-10 reduces the production of high molecular weight dextran and increases the production of clinical dextran, whereas increasing the dextran T-40 concentration decreases the production of clinical dextran. Maltose reduces high molecular weight dextran production but increases the production of acceptor products, thereby reducing clinical dextran production (73).

Activation, Inhibition, and Stability of Dextransucrase

Germaine et al. observed that addition of dextran to the dextransucrase (produced from the *Streptococcus mutans* reaction mixture) increased the rate of dextran synthesis. They found that the rate depended on the size of the dextran chain and reached a maximum when the average size

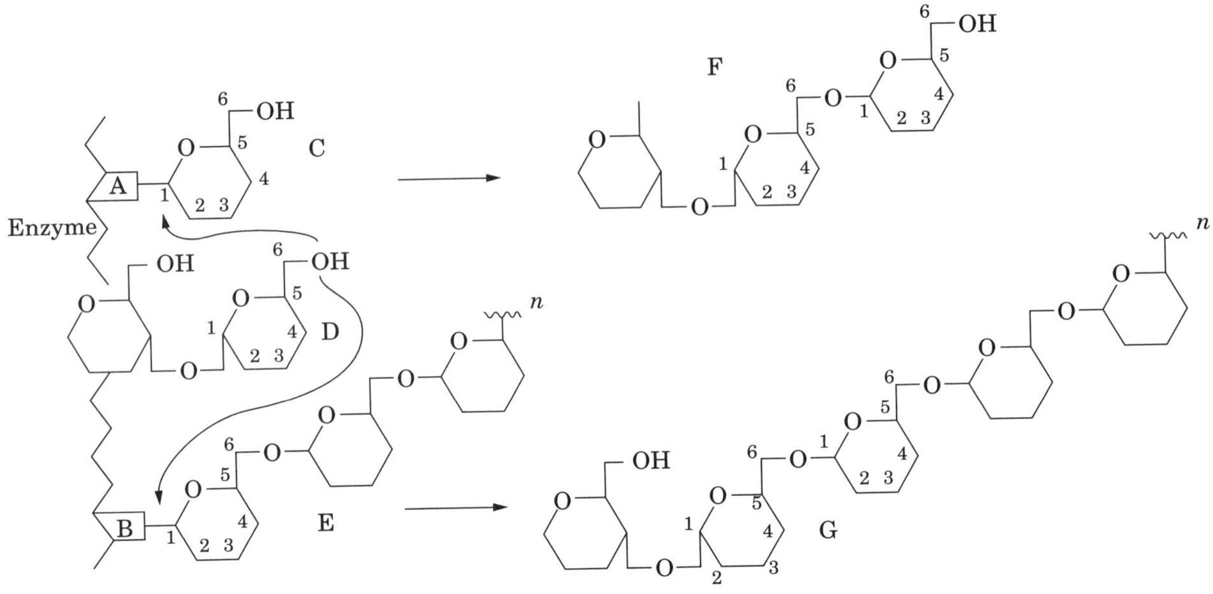

Figure 4. Mechanism of the acceptor reactions: D, maltose molecule bound at acceptor binding site; C, glucosyl–enzyme complex; E, dextranyl–enzyme complex; F, resulting acceptor product when C and D are reacted together; G, resulting acceptor product when E and D are reacted together. (—OH groups are omitted for clarity.)

of the added dextran was 30 glucose residues (55–57). Kobayashi and Matsuda also observed that purified dextransucrase was stimulated by addition of dextran, although dextransucrase could synthesize dextran without addition of dextran to the reaction mixture. The rate of dextran synthesis in dextran-free reaction mixtures was accompanied by a lag period that could be eliminated by addition of exogenous dextran (58). Robyt et al. reported that the addition of dextran to dextransucrase reaction mixtures affects the initial velocity of dextran synthesis in a sigmoidal manner characteristic of an allosteric binding of dextran to the enzyme. Further, they proposed that the allosteric activation of dextransucrase occurs when the dextran binds to the dextran binding site, where it acts as a matrix to join or induce the subunits to associate into an active form (37). Kobayashi and Matsuda reported that purified B-512 F dextransucrase contains 17% carbohydrates, mostly as endogenous dextran covalently bound to the enzyme, which can be eliminated by the mild action of endodextranase. They suggested that exogenously added dextran could stabilize the enzyme not only during storage but also during long-term incubation (46,47).

Miller and Robyt (74), from their rate kinetics of dextran synthesis, showed that concentrations of Ca^{2+} ion below 1 mM activate the enzyme by increasing V_m and decreasing the K_m for the sucrose. Above 1 mM Ca^{2+}, the ions act as weak competitive inhibitors. Although Ca^{2+} is an activator at low concentrations, it is not required for dextran synthesis. Miller and Robyt proposed a model for Ca^{2+} ion activation and inhibition behavior in which the enzyme has two Ca^{2+} binding sites. One Ca^{2+} site is an activating site and the other is an inhibitory site at which Ca^{2+} ions prevent the binding of sucrose to the sucrose binding site (48,74). Kobayashi and Matsuda reported that the purified enzyme can be activated by Ca^{2+} ions and strongly inhibited by EDTA. The inhibitory effect of EDTA can be reduced by adding Ca^{2+} ions (75,76). However, they further observed that the purified enzyme was strongly inhibited by Cu^{2+}, Fe^{2+}, and Mn^{2+} ions in this order (46). Contrary to the crude enzyme, which contains endogenous dextran, the purified enzyme shows optimum activity at pH 6.0 (46). Miller and Robyt found that the dextransucrase could be stabilized against the activity losses caused by heating or by dilution through the addition of low concentrations nonionic polymers (e.g., dextran, PEG 20000, and methyl cellulose) or of nonionic detergents at or slightly below their critical micelle concentrations (77). The effect of pH and temperature on dextransucrase is discussed in the later section "Effect of pH and Temperature".

Effect of Enzyme Concentration

In 1962 the effect of enzyme concentration on dextran synthesis was reported by Braswell et al. (78). They observed that at higher concentrations of enzyme the molecular weight of dextran synthesized was decreased and the intrinsic viscosity and branching ratio (α, $1 \rightarrow 6$ linkages to α, $1 \rightarrow 3$ linkages) was increased. Tsuchiya et al. made a related observation that increasing the enzyme concentration decreased the amount of high molecular weight dextran synthesis (54).

At the time, neither Braswell and coworkers nor Tsuchiya and coworkers put forward an explanation. However, in 1993 Su and Robyt explained in their report that the results are due to the increased acceptor reactions at higher enzyme concentrations (70) (see the earlier section "Acceptor Reactions and Acceptor Product"). Another important point is that improving the yield of enzyme from a given fermentation does not necessarily produce an improved yield of dextran because the conditions under which the enzyme is allowed to act on sucrose solution determine the dextran yield (14).

PRODUCTION OF DEXTRANSUCRASE AND DEXTRAN

The dextran fermentation process involves three phases: *(1)* the cell growth phase, *(2)* the dextransucrase production phase, and *(3)* the enzyme action phase. The enzyme action phase that produces dextran is the third phase (i.e., it occurs after cell and enzyme production). Thus, the rate of dextran production under given conditions is a function of the enzyme production rate, which in turn is a function of the cell production rate. Moreover, dextransucrase stability and its activity under given conditions influence dextran synthesis. To optimize the dextran fermentation process, one must consider the following rates of reactions: *(1)* cell growth rate, *(2)* enzyme production rate, *(3)* enzyme deactivation rate (stability), and *(4)* enzyme action rate.

The interdependency of these reaction rates under specific fermentation conditions is a key factor for in situ dextran production. Because each rate of reaction has its own optimum level, a set of conditions given for a fermentation may not be optimal for all four rates of reaction. For example, a temperature of 30 °C is optimal for cell production and enzyme action but not for enzyme production and its stability. The three phases of dextran fermentation can be separated and the optimum level of each phase can be used one by one for maximum dextran yield. Therefore, each phase and its reaction rates must be understood separately and in great detail to employ process control techniques, to find optimal conditions, and to design a high-performance reactor.

Factors That Influence Dextran Fermentation

Bacterial Strains. The genera of bacteria recognized to produce the enzyme dextransucrase, which is capable of synthesizing dextran from sucrose, are primarily *Leuconostoc* and *Streptococcus*. These two genera are gram-positive, facultative anaerobic cocci that are closely related. The notable difference between these two genera is that *Leuconostoc* species require sucrose in the culture medium to induce the formation of dextransucrase, whereas the *Streptococcus* species do not require sucrose in the culture medium to induce enzyme formation (79,80). Thus, the *Leuconostoc* species are normally inducible for the formation of dextransucrase and the *Streptococcus* species are normally constitutive for their formation. However, Kim and Robyt have obtained mutant strains of *Leuconostoc mesenteroides* B-512 FM, B-742, B-1299, and B-1355 that are constitutive for dextransucrase formation (81).

Effect of Sucrose Concentration. Sucrose is an essential substrate for dextransucrase and dextran synthesis. Although *L. mesenteroides* can grow on various carbohydrates such as glucose, fructose, and maltose, it prefers sucrose over other carbohydrates as a carbon and energy source (82,83). The carbohydrates are fermented through the hexose monophosphate pathway, giving lactic acid, carbon dioxide, and ethanol under anaerobic conditions. An aerobic oxidative metabolism that yields carbon dioxide, lactic acid, and acetic acids also exists (84).

Tsuchiya et al. found in batch culture that increasing sucrose concentration from 5 to 50 g/L resulted in an increase in enzyme productivity from 6 to 120 DSU/mL. However, they recommended 20 g/L sucrose for optimum dextransucrase production because higher sucrose concentrations increase the viscosity of the broth due to dextran synthesis, which makes cell removal difficult. The enzyme activity that they achieved at 20 g/L sucrose is 86 DSU/mL (85). It is essential to maintain the lower level of sucrose concentration (5–10 g/L) throughout fermentation to produce a higher activity of dextransucrase (86). This is possible in semicontinuous or continuous operational mode. Lopaz and Monsan achieved the enzyme activity of 180 DSU/mL by addition of sucrose at the rate of 20 g/L/h over the cell growth phase (87,88). Lawford reported that the enzyme can be synthesized only after a critical concentration of sucrose, the inducer, has been reached or surpassed (82). The highest activities of dextransucrase reported in the literature are by Schneider et al. (86) and Barker and Ajongwen (89). The former obtained 400 DSU/mL in laboratory scale and 289 DSU/mL at pilot plant scale. Barker and Ajongwen obtained activity of 450 DSU/mL under nonaerated fed-batch fermentation and 350 DSU/mL under aerated fed-batch fermentation.

At higher sucrose concentrations, the enzyme is inhibited and the conversion of sucrose to dextran decreases (see Fig. 5). Moreover, the molecular weight of synthesized dextran decreases at higher sucrose concentrations, where sucrose itself acts as an acceptor (see the earlier section "Acceptor Reactions and Acceptor Products") to terminate polymer growth. Braswell et al. observed that increasing the sucrose concentration from 50 to 300 g/L increases the intrinsic viscosity and the branching ratio of dextran produced while decreasing the molecular weight of dextran and the rate of dextran formation (78,91).

Alsop reported that higher dextran yields have been obtained at lower initial sucrose concentrations, reducing costs as a function of sucrose alone. However, the net amount (and so the concentration) of dextran produced

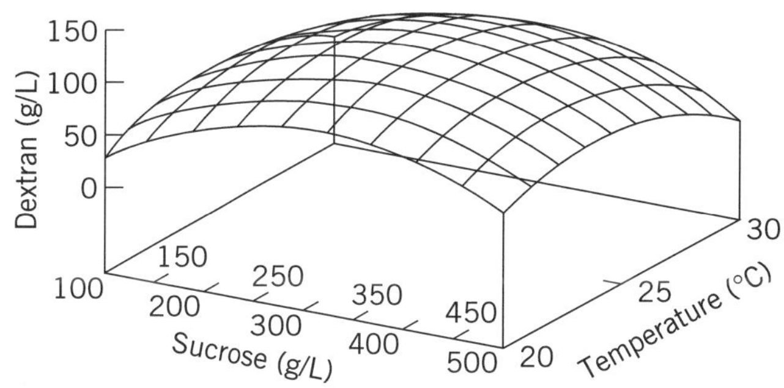

Figure 5. A calculated response surface plot of dextran yield (g/L) as a function of sucrose and temperature with a constant level of yeast extract (10 g/L), (30 g/L) K_2HPO_4, and initial pH 8.3. *Source*: From Ref. 90.

was lower, reducing plant productivity and substantially increasing ethanol and labor costs in downstream processing. Alsop calculated the optimum value on the basis of ethanol precipitation costs (which depend on the molecular weight of the dextran synthesized; high molecular weight dextrans can be precipitated with lower concentrations of alcohol) and reported that the optimum value is 170 g/L of initial sucrose concentration. However, the same author reported another optimal value, 125 g/L of sucrose, based on the operating costs of the plant (14). Martinez-Espindola and Lopez-Munguina reported that in cell-free enzymatic synthesis of dextran, an initial sucrose concentration of 150 g/L is optimum for 99% conversion of sucrose to dextran. The optimum value was based on reaction kinetics of dextransucrase and a specific design for a batch reactor (92).

In our work, we found the optimum value (based on yield) of sucrose to be 300 g/L for batch dextran production in whole cell culture based on response surface methodology (90). We observed a yield of 135 g/L dextran at this optimum level. Further, we observed substrate inhibition kinetics for cell growth at more that 100 g/L sucrose and that at more than 300 g/L sucrose the dextran yield decreases considerably (93,94) (see Fig. 5).

Effect of Nitrogen Source. The nitrogen sources used to grow *L. mesenteroides* and produce enzyme are corn steep liquor, yeast extract, soya meal preparations, and peptone. These complex nutrients also provide essential growth factors and trace elements. Ammonium ions used as the nitrogen source appear to have an adverse effect on cell growth (85). Tsuchiya et al. reported that an increase in the concentration of corn steep liquor as a nitrogen source from 2.5 to 20 g/L increases the enzyme yield but only when higher amounts of phosphate are present in the medium (85).

Barker and Ajongwen, who obtained the highest enzyme activity, used 40 g/L yeast extract as a nitrogen source in fed-batch culture. However, they observed variable performance with different yeast extract types, because different yeast extracts contain varying levels of vital nutrient components (89). In our work, we optimized yeast extract concentration to 10 g/L for conversion of 300 g/L sucrose to dextran (90) and further observed that below 10 g/L yeast extract the growth of cells and enzyme production were affected considerably.

Effect of Phosphate Concentration. Phosphates such as K_2HPO_4 and KH_2PO_4 act as buffers in the fermentation medium in addition to being nutrients. Tsuchiya et al. reported that 35 g/L KH_2PO_4 is optimum for enzyme production (85). We observed that 30 g/L K_2HPO_4 is optimum for batch dextran production to convert 300 g/L sucrose to dextran (90). Lawford et al. (82) reported that larger amounts of phosphate can remove the Ca^{2+} ions in the medium by forming a complex with Ca^{2+} ions. Calcium is an essential element for dextransucrase stability (see the earlier section "Activation, Inhibition, and Stability of Dextransucrase"). They observed loss of activity of dextransucrase during batch culture at 10 g/L potassium phosphate, whereas at 1 g/L phosphate the produced activity remained stable for a long time.

London and Webb optimized the concentration of Na_2HPO_4 to 20 g/L (95,96), and Barker and Ajongwen used K_2HPO_4 at 20 g/L in their fed-batch fermentation for higher enzyme production (89). In batch fermentation without controlled pH the pH of the medium starts to fall as the cell starts to grow. However, the rate of fall can be controlled by using higher phosphate concentrations. Cell growth, enzyme production, and dextran production are faster at pH 7.0, 6.7, and 5.2, respectively (14). The interaction between initial pH and phosphate can be seen in Figure 6 (90). However, in pH-controlled fermentations, the higher amounts of phosphates are not required and the adverse effect of Ca^{2+} ion removal can be avoided. pH control is recommended over a higher phosphate concentration in the culture.

Effect of Growth Factors and Trace Elements. Carlson and Carlson reported that the growth factors nicotinic acid, thiamine, pantothenic acid, and biotin and the amino acids valine and glutamic acid are required for *L. mesenteroides* growth (97–99). However, the complex nitrogen sources (e.g., yeast extract, corn steep liquor, and peptone) are expected to supply those vital nutrients and necessary trace elements. In 1952 a salt solution called R salts that contained 4.0% $MgSO_4 \cdot 7H_2O$, 0.2% NaCl, 0.2% $FeSO_4 \cdot 7H_2O$, and 0.2% $MnSO_4 \cdot H_2O$, was first introduced by Koepsell and Tsuchiya (100). London reported that the trace inorganic salts generally used were not found to be necessary (95,96), suggesting that nutrient requirements are culture dependent. The effect of Ca^{2+} ions on enzyme stability was discussed earlier.

Effect of pH and Temperature. The optimum levels of pH and temperature for cell, enzyme, and dextran production are given in Table 2, and the enzyme dextransucrase is stable at pH 5.2 and low temperature. The effects of pH and temperature on enzyme production and its action and stability are interrelated. Although pH 6.7 is optimum for enzyme production in this example, this value is applicable only at the optimum temperature of 23 °C. At higher temperatures, pH 6.7 inactivates the enzyme rapidly (85).

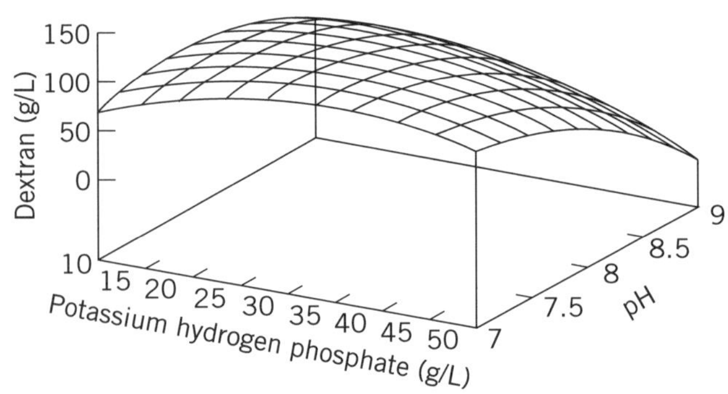

Figure 6. A calculated response surface plot of dextran yield (g/L) as function of initial pH and K_2HPO_4 with a constant level of sucrose (300 g/L) 10 g/L yeast extract, and 23 °C. *Source*: From Ref. 90.

Table 2. Optimum pH and Temperature for Cell Growth, Enzyme Production, and Dextran Production by *Leuconostoc mesenteroides*

Parameter	Cell growth	Enzyme production	Enzyme action
pH	7.0	6.7	5.2
Temperature (°C)	30	23	30

Similarly, the optimal temperature of 30 °C for cell growth and enzyme activity is too high for enzyme production and stability. The enzyme activity in culture filtrate is retained at pH 5.2 for at least 24 h at 25 °C, but 35% of the activity is lost at pH 6.7 within 1 h at 25 °C. Moreover, at higher temperatures the deleterious effect of pH is accentuated markedly. Kobali and Reilly (101) reported that the half-lives of enzyme at pH 4.9 are 13.75, 7.02, and 3.28 min at 30, 32, and 35 °C, respectively. London and Webb observed that the maximum enzyme production rate was achievable at pH 6.1, although the maximum activity was observed at pH 6.7. They further stated that at pH 6.0 and below, the enzyme was relatively stable, retaining 100% of activity for 5 h at pH 5.4 and 26 °C. Above pH 6.0 the enzyme was far less stable, retaining only 10% of its original activity at pH 6.6 and 26 °C (95,96).

The effect of temperature as the velocity of the reaction corresponds to an Arrhenius activation energy of 10.6 kcal/mol (78). However, Kobali and Reilly reported a lesser value of 8.6 kcal/mol of activation energy (101). Dextran synthesis temperature also plays an important role in the properties of the dextran produced. Braswell et al. reported that decreasing the reaction temperature from 30 to 0 °C lowers the molecular weight of dextran and increases the intrinsic viscosity. One would normally expect lesser intrinsic viscosity for lower molecular weight polymers, but a more linear polymer is produced at lower temperatures, increasing the intrinsic viscosity observed (78). Sabatie and Choplin (102) also reported that dextran produced at 5 °C was more linear than dextran produced at 30 °C. They observed 5% α, $1 \rightarrow 3$ branch linkages at 5 °C of synthesis temperature compared with 4.8% at 30 °C (78).

Effect of Aeration and Agitation. The maximum productivity and product concentration achievable in industrial microbial processes for production of biomass as product are typically limited by the rate of oxygen transfer. *L. mesenteroides* is a facultative anaerobe. However, some researchers used the aerobic fermentation for dextransucrase production. Schneider and coworkers observed a higher enzyme activity of 400 DSU/mL under highly aerated conditions in which the oxygen level was maintained at 40–80% of the initial saturated value throughout the course of fermentation (86). However, Barker and Ajongwen observed the higher activity of 450 DSU/mL under nonaerated fed-batch fermentation but achieved only 350 DSU/mL of activity under aerated fermentation (89). Foster also found that aeration gives a lower yield of enzyme (103,104).

Koepsell and Tsuchiya (104) and Tsuchiya et al. (86) noticed that enzyme yields were higher in shaken than in still flasks, and Johnson and McCleskey (105) also observed similar results. This may be due to the mixing requirements for limiting nutrients. London and Webb compared the stirred and unstirred cultures for enzyme production. They observed a lower cell growth rate and higher enzyme production in unstirred culture than in stirred culture. This is due to the limiting substrate concentration around the cells being lower under poor mixing conditions (i.e., the transfer from the liquid to the cells is insufficient to meet their demand, which results in poor cell growth rate). But as the cells produce the enzyme and acid, the enzyme and acid are expected to collect in pockets around the cells in nonagitated conditions. This lowers the pH around the enzyme to its stability optimum (95,96).

As the dextran concentration increases in the fermenter during its synthesis, the viscosity of the broth also increases, which requires proper mixing conditions for better conversion of substrate to polysaccharide and scale-up of polysaccharide fermentation (106). Margaritis and Bokkel studied the rheological behavior of dextrans (low and high molecular weight) in different concentrations (107). Their data fit well the following power law equation, which represents the pseudoplastic behavior of fluid:

$$\log\tau = \log K + n \log\gamma \qquad (8)$$

where τ is the shear stress applied on the fluid (N/m^2), K is the consistency index, n is the flow behavior index, and γ is the shear rate (s^{-1}).

From Margaritis and Bokkel's observations, the flow index n of the high molecular weight dextran solution was much less than the flow index for the low molecular weight solution. The consistency index K of high molecular weight dextran solutions was found to be higher than the K values of low molecular weight solutions for the same concentrations (1). These results show that high molecular weight dextran shows a higher degree of pseudoplastic behavior than low molecular weight dextran.

Batch Culture

Traditionally, dextran is produced by anaerobic, unstirred batch culture using dextransucrase-producing bacteria in sucrose-rich media (14). Sugar solutions are fortified with a nitrogen source such as yeast extract or corn steep liquor, a phosphate source such as K_2HPO_4 or KH_2PO_4, and trace metal salts. Little further process control is employed. The pH, which was neutral at the start of the fermentation process, falls as cells grow. Dextransucrase is not produced until near the beginning of the stationary phase. At this point the enzyme is secreted rapidly into the culture medium and reaches a maximum after a few hours (76).

Dextransucrase kinetics have generally been described by the Michaelis–Menten model and applied for sucrose levels below 100–150 g/L, where the saturation of the enzyme is achieved. Martinez-Espindola and Lopez-Munguina proposed a simple substrate inhibition model for enzyme action at higher sucrose levels as follows:

$$V_i = [V_{max} + S]/[K_m + S + (S^2/K_i)] \qquad (9)$$

For enzymatic synthesis of dextran in a batch reactor,

Martinez-Espindola and Lopez-Munguina derived the following design equation:

$$V_{max}t = S_0X - K_m \ln(1 - X) + (K_iS_0^2)(2X - X^2)/2 \quad (10)$$

where S_0 is the initial sucrose concentration, X is the conversion, and K_i is the inhibitor constant. They predicted that 150 g/L of sucrose concentration is optimum for 99% conversion of sucrose to dextran in terms of productivity in a batch reactor (92).

London and Webb (96) optimized both the medium and inoculum for batch production of dextransucrase in an unaerated and unstirred culture. The optimum conditions are given in Table 3. They achieved a final enzyme activity of 142 DSU/mL. Although it is lower than the fed-batch achievement, it was the best reported value for batch operation. However, London and Webb's scaled-up process to 1,000 L yielded only 45 DSU/mL of activity (see Table 3).

We reported the optimum conditions for batch dextran production by using response surface methodology (RSM) (109) and multistage Monte Carlo optimization (MSMCO) (110) optimization techniques with a predicted yield of 154 g/L dextran. The optimum values are given in Table 3; under these conditions we observed 135 g/L dextran yield (90). Typical batch dextran fermentation profiles can be seen in Figure 7. The optimum initial pH of 8.3, which was set before inoculation and sterilization, dropped to 7.5 at the start of fermentation after adding 10% v/v acidic inoculum (see Figs. 5 and 6). Batch operational mode is a suitable method to produce dextran through either whole cell fermentation or cell-free enzymatic synthesis due to the possibility of a high final dextran concentration (hence the productivity), which minimizes the alcohol precipitation costs in recovery of the product. However, if our aim is to produce maximum enzyme activity, fed-batch operation that keeps lower levels of sucrose concentrations in the fermenter is the best operational mode.

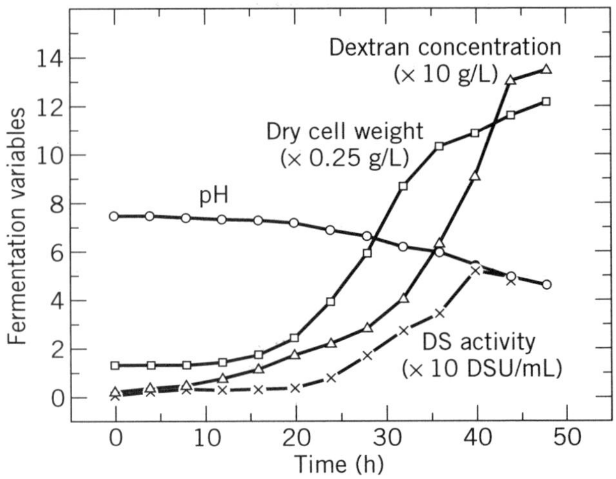

Figure 7. Typical batch dextran fermentation profiles. Fermentation conditions are given in Table 3. *Source*: From Ref. 90.

Semicontinuous Culture

Because dextran fermentation shows substrate inhibition kinetics, the substrate level can be maintained at the desired level by using fed-batch operational mode. This is shown by the various authors who produced higher enzyme activity than in batch cultures (87,89,111). Lopez and Monsan (87,88) and Monsan et al. (111,112) achieved 180 DSU/mL enzyme activity by fed-batch fermentation. They fed sucrose solution at a rate of 20 g/L/h and achieved maximum cell density and enzyme activity. Barker and Ajongwen (89) achieved the highest enzyme activity (450 DSU/mL) under low agitated, nonaerated fed-batch fermentation (see Table 3). El-Sayed and coworkers observed higher enzyme activity in fed-batch operation than batch mode in both immobilized-cell and free-cell cultures. They observed that when the sucrose feed rate is increased from 5 to 10 g/L/h, enzyme activity increases from 73 to 118 DSU/mL in free-cell culture and 110 to 223 DSU/mL

Table 3. Optimal Conditions for Production of Dextransucrase Used by Various Authors

Variables	El-Sayed et al. (108)	London and Webb (96)	Barker and Ajongwen (89)	Karthikeyan et al. (90)
Sucrose	20 g/L	40 g/L	10 g/L	300 g/L
Yeast extract	20 g/L	20 g/L	40 g/L	10 g/L
Tween 80	2 g/L	—	—	—
KH_2PO_4	2 g/L	—	—	—
K_2HPO_4	—	—	20 g/L	30 g/L
Na_2HPO_4	—	16 g/L	—	—
$CaCO_3$	10 g/L	—	—	—
R salts[a]	5 mL	—	0.5% v/v	—
pH	6.7[b]	7.1[c]	6.7[b]	8.3[c,d]
Temperature	26 °C	26 °C	23 °C	23 °C
Inoculum	—	1% v/v	—	10% v/v
Mode	Fed-batch	Batch	Fed-batch	Batch
Sucrose feed	10 g/L/h	—	Connected with alkali addition	—
Product	Dextransucrase	Dextransucrase	Dextransucrase	Dextran

[a]R salts contain 40 g $MgSO_4 \cdot 7H_2O$, 2 g NaCl, 2 g $FeSO_4 \cdot 7H_2O$, and 2 g $MnSO_4 \cdot H_2O$, dissolved in 1 L.
[b]pH-controlled fermentation.
[c]Indicates initial pH, which is uncontrolled during the fermentation.
[d]pH will become 7.5 after adding 10% v/v of acidic inoculum.

in immobilized-cell culture (113–115). Although fed-batch culture (used to maintain a low substrate level) is suitable for enzyme production, it may not be suitable for dextran production because of lower plant productivity and increased alcohol precipitation costs in downstream processing.

Continuous Culture

Ringpfeil and Selenia (116) theoretically calculated from batch culture experiments the maximum yield of dextransucrase that would be obtained from sucrose-limited continuous culture at low dilution rate. The maximum enzyme activity achieved at a dilution rate of 0.53 h^{-1} was 16 DSU/mL. However, Lawford et al. (82) reported that the enzyme can be synthesize only after a critical concentration of sucrose has been reached because they could not synthesize dextransucrase at dilution rates less than 0.2 h^{-1}. They calculated the theoretical dilution rate for maximum cell growth to be 0.67 h^{-1}. However, they achieved maximum cell growth rate at 0.35 h^{-1}, concluding that either some essential metabolite other than the energy source limited the growth or some components of the growth medium inhibited the growth. Lee et al. (73) produced low molecular weight clinical dextran in a two-stage continuous reactor using acceptors such as maltose, dextran T-40, and dextran T-10. Cells cultivated in the first reactor were transferred to the second reactor, where sucrose and acceptor were added for clinical dextran production.

Immobilized Cells and Enzymes

Use of immobilized dextransucrase, if it were of sufficient stability, would allow the production of dextran without the requirement for repetitive fermentation. Earlier reports of dextransucrase immobilization were issued from several groups. The enzyme has been immobilized ionically to DEAE sephadex A-50 (117), covalently to Bio-Gel-P2 (63,64), and covalently to polyacrylamide gel, cellulose acetate membranes, and polysulfone hollow fibers (118). In 1980 Kobali and Reilly (101) examined 13 carriers (e.g., Enacryls, DEAE cellulose, DEAE sephadex, SP sephadex, and porous silica) for dextransucrase immobilization. Among the 13 carriers surveyed, only alkaline amine porous silica retained more than 10% of activity on immobilization (10 DSU/g carrier). The optimum pH (5.2) and temperature (30 °C) of enzyme activity are not changed on immobilization, but the activation energy below 30 °C was 1.7 kcal/mol for immobilized enzyme, which is lower than for soluble enzyme (8.6 kcal/mol).

Monsan and Lopez immobilized dextransucrase on an aminoporous silica support activated with 2% glutaraldehyde by covalent coupling (111). They examined the specific surface area of the carrier silica for immobilization efficiency and observed that *(1)* lowering the specific surface area increased the coupling efficiency for dextransucrase and that *(2)* adding maltose to the dextransucrase preparation during enzyme coupling greatly increased the coupling efficiency. The production of high molecular weight dextran with immobilized dextransucrase results in extremely viscous solutions, which makes recovery of the support difficult at the end of the synthesis. They also observed from their kinetic study diffusional limitation in the immobilized catalyst because of high molecular weight dextran synthesis. Chang et al. (119) successfully immobilized the enzyme on porous phenoxy acetyl cellulose beads by hydrophobic adsorption with high efficiency. The enzyme was immobilized at more than 100 times the activity attained on porous silica beads by Kobali and Reilly (101) and at nearly double the activity found by Monsan and Lopez (111).

El-Sayed and coworkers published a series of papers (113–115) in which they studied the effect of different carriers (i.e., calcium alginate, three types of celites, celites coated with calcium alginate, stainless steel, and stainless steel coated with calcium alginate) for immobilization, mode of operation (batch and fed-batch), and the performance of different reactors (bubble column and airlift bioreactors). They further compared all the results with whole cell fermentation under the same conditions. The medium they used for dextransucrase production is given in Table 3.

El-Sayed and coworkers used the immobilized cells in calcium alginate beads for sequential production of dextransucrase followed by dextran production in three cycles. Each cycle consisted of fed-batch enzyme production for 24 h followed by batch dextran production for 24 h. The maximum dextran yields reached 40%, 44%, and 46% based on the added sucrose and were achieved 15, 7, and 3.5 h from the beginning of the first, second, and third dextran production periods, respectively. Increasing the sucrose feed rate from 5 to 10 g/L/h led to an increase in the total enzyme activity of about 88%. El-Sayed et al. compared the results with free-cell fermentation where only one cycle was possible. They observed that immobilized cells produced more than three times as much dextran as free cells during one cycle (113). When comparing different carriers they observed that the immobilized cell cultures were significantly affected by pore and particle size. Enzyme production with cells immobilized in celite R 648 resulted in the highest total enzyme activity followed by celite R 633, alginate-coated celite R 630, celite R 630, and the calcium alginate beads. A layer of calcium alginate on porous celite R 630 particles improved their mechanical stability, increased the amount of soluble dextransucrase activity, and decreased the cell leakage from the highly porous support (114). Immobilizing *L. mesenteroides* on a porous stainless steel support also increased dextransucrase and dextran production. Coating calcium alginate layers on stainless steel supports enhanced the cell loading capacity of the support and increased dextransucrase and dextran production (115).

Other Operational Strategies

Dextran fermentations consist of three phases—cell, enzyme, and dextran production—and each phase shows different optima for pH, temperature, and sucrose concentration. A major advantage in dextran fermentation is that these individual phases can be performed separately and sequentially in a reactor system that allows flexibility in operational strategy (120,121). Their separation could be achieved by either of the following simple techniques:

1. Single reactor system with multistage operation: a stepwise operation of the individual phases with their individual optimum condition.
2. Multistage reactor systems in series: simultaneous operation of individual phases (i.e., each individual phase performed with unique optimum levels in separate reactors in series).

In Figure 8, reactors a, b, and c represent single reactor–multistage operational systems with batch, fed-batch, and continuous mode of operation, respectively. In these systems, for example, if we divide the total fermentation time into three stages so as to perform cell, enzyme, and dextran production in a stepwise fashion, we can control the pH at 7 and temperature at 30 °C for the first stage to maximize cell growth. In the second stage, after reaching a high biomass concentration, we set the pH and temperature to 6.7 and 23 °C, respectively, for enzyme production. Similarly, in the third stage we can set the pH at 5.2 and temperature at 30 °C for dextran production. Further, we can control sucrose concentration in fed-batch and continuous cultures for these three stages. El-Sayed and coworkers attempted the sequential production of dextransucrase followed by dextran production using immobilized cells in an airlift bioreactor (113–115). If we divide the fermentation time for individual phases in the stepwise process the questions to be answered are as follows:

1. What is the required amount of biomass and enzyme activity to be produced to get a maximum conversion of given sucrose to dextran in the shortest time?
2. How many stages are required? Are three stages necessary, because dextran fermentation consists of three phases?
3. What is the duration required for each stage?

The term *required amount* in question 1 is important because the aim of fermentation is to produce maximum dextran, not cells or enzyme. If we produce more cells and enzymes than the required amount, it will result in low yield from the fermentation medium. Therefore, in the first and second stages we need not produce maximum cells and enzyme but a required amount to optimally convert sucrose to dextran in the overall process. Optimization can be solved by employing mathematical techniques such as dynamic programming and the continuous maximum principle (122–125). These techniques require the mathematical models for cell growth, enzyme formation and deactivation, and dextran synthesis. The model parameters should explain the effect of pH and temperature. In the stepwise process, we need not necessarily divide the total time into only the three stages suggested earlier. In fact, more than three stages will prove advantageous. The number of stages required for maximum production can also be obtained through the previously mentioned mathematical techniques.

Multistage reactors in series allow simultaneous operations. In all simultaneous reactor systems in series shown in Figure 9, the symbol F_a represents additional feed of sucrose to the respective reactors. This feed can be utilized to control sucrose concentration in the respective reactor. The optimal operational conditions for simultaneous operation can also be obtained by the previously mentioned mathematical techniques. In reactor system f in Figure 9, three reactors labeled A, B, and C are connected in series. Reactor A is kept at optimum conditions for cell growth and, similarly, B and C are kept at optimum conditions for

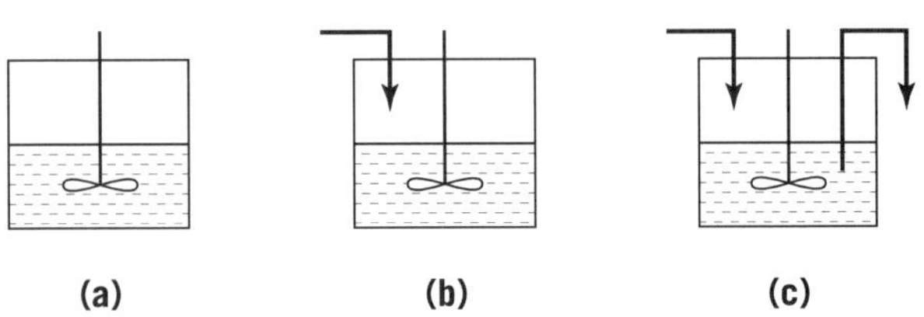

Figure 8. Reactor designs for stepwise process: (**a**) batch reactor, (**b**) fed-batch reactor, (**c**) continuous reactor.

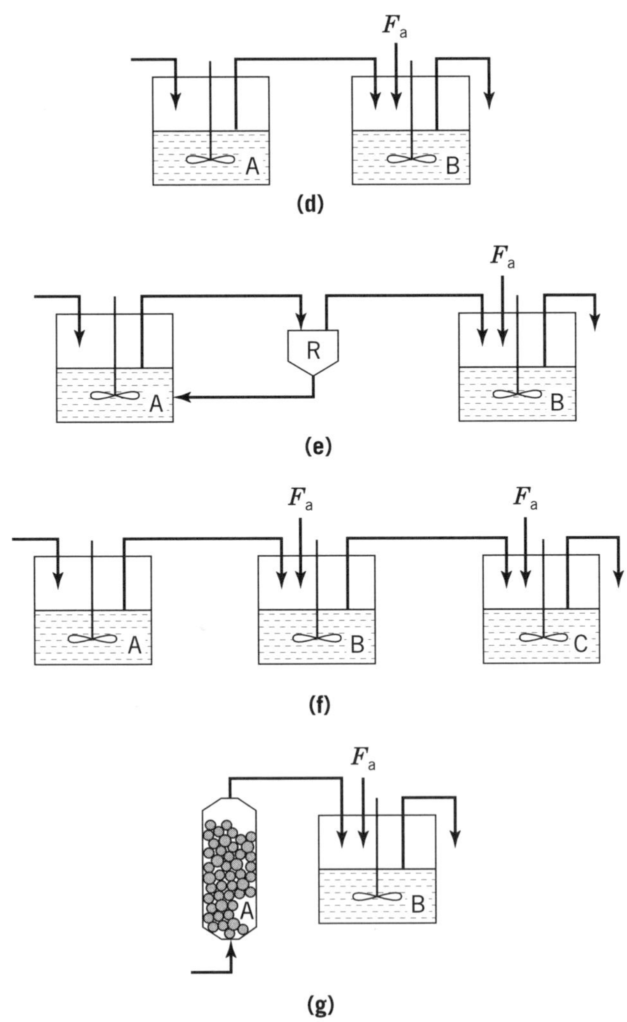

Figure 9. Reactor designs for simultaneous process (see text for description of reactors d–g).

enzyme and dextran production, respectively. Another reactor system shown in Figure 9d has only two reactors, labeled A and B. Here, reactor A supports both cell growth and enzyme production and reactor B performs the dextran synthesis. Separate sucrose feed (F_a) can also be employed in reactor B. Lee et al. (73) reported the performance of a two-stage continuous reactor for dextran synthesis. In the d reactor system in Figure 9, reactor A can also be utilized with immobilized cell systems to produce enzyme, which can be supplied to reactor B for dextran production. The reactor system shown in Figure 9e has a cell recycle unit R that separates the cells in the medium before it goes to reactor B and maximizes the cell concentration in reactor A. Another interesting system shown in Figure 9g consists of a packed-bed reactor (A) connected to a stirred tank reactor (B). Here immobilized cells are packed and are controlled at the optimum conditions for enzyme production. Using the enzyme from reactor A, reactor B supports dextran synthesis under optimum conditions. The stirred tank reactors shown in Figure 9 can be replaced by other reactors such as airlift, bubble column, and so forth.

A major advantage in reactor systems e and g in Figure 9 is that dilution of the final viscous broth for cell removal by centrifugation, is not needed, thereby saving on alcohol precipitation costs. The additional investment in terms of equipment and control systems and the operational difficulties are major disadvantages in systems with reactors in series. The overall economics of such a process should be worked out after taking into consideration the product value for such a process.

USES AND APPLICATIONS OF DEXTRANS

Depending on the microbial species employed and the fermentation conditions, a wide variety of dextrans can be produced. For example, dextrans commercially produced from *L. mesenteroides* NRRL-B512F consist of 95% α, $1 \rightarrow 6$ linkages and 5% α, $1 \rightarrow 3$ linkages. Many types of applications of this particular dextran and its derivatives have been developed (4,14,27,126–128). The important applications of dextran and its derivatives are described in the following sections.

Clinical Dextran

Dextran fractions of average molecular weight 70,000 can be used as blood plasma substitutes (or blood volume expanders). This single most important application initiated extensive research in dextran production. Clinical dextran (6% w/v) in isotonic saline provides viscous, colloidal, and osmotic properties equivalent to those of plasma proteins. Clinical dextran solutions can be administered intravenously for treatment of shock and burns, especially when blood is not available, to restore blood pressure and circulatory volume. The advantages of clinical dextran solution are its stability to sterilization by heat and in storage, independence of the blood types of recipients, low incidence of adverse reactions, and freedom from transmissible diseases (128,129). Clinical dextran (10–12% w/v) in aqueous or dextrose solutions administered intravenously provides high colloidal and osmotic activity and is used for treatment of hypoproteinemia and nephrotic and cerebral edema (128).

Rheomacrodex or Low-Viscosity Dextran

Dextran fractions of average molecular weight 40,000 can be used as blood flow improvers. Dextran (10% w/v) in sodium chloride solution (0.9% w/v) shows hematorheological properties that counteract capillary stagnation and improve the suspension stability of blood. When intravenously administered, it increases the fluidity of erythrocytes (130).

Cross-Linked Dextran Gel

Dextran is used in the manufacture of molecular sieves. It can be cross-linked with reagents such as epichlorohydrin in the presence of sodium hydroxide. The degree of cross-linking of dextran determines the porosity of the gel structure, water regain value, and molecular sieveing properties. These cross-linked gels, having three-dimensional structure, revolutionized the purification and separation, of proteins, nucleic acids, and polysaccharides based on differences in molecular size. These gels have well-defined exclusion limits in the molecular weight range of 5,000–200,000 (131). Commercial cross-linked dextran was introduced in the 1960s by Pharmacia Fine Chemical Limited under the tradename Sephadex. Hydroxyl propyl derivatives of Sephadex have been made to separate materials according to molecular size in organic solvents and have had use in separation of lipids, fatty acids, steroids, and so forth (132).

Diethylaminoethyl and carboxymethyl derivatives of cross-linked dextran gels provide adsorbants with different porosities and fractionation ranges used in the fractionation of proteins, nucleic acids, and polysaccharides on the basis of different adsorption characteristics. Cross-linked dextrans and their derivatives have found wide application in the separation and purification of biological molecules based on charge, molecular size, and so forth on both laboratory and preparative scales. Proteins are not usually denatured by hydrophilic polymer networks, and nonspecific adsorption is usually very low.

Dextran Sulfate

Dextran sulfate can be prepared by esterification of polysaccharide of molecular weight 5×10^5 with chlorosulfonic acid. This product, a polyelectrolyte containing 17% sulfur (an average 2.3 sulfate groups on each glucosyl residue), has various applications. It shows reversible interaction with plasma proteins, and under some conditions selective precipitation, and thus permits isolation of substances such as β-lipoproteins and human antihemophilic globulin (133).

Dextran of sufficiently low molecular weight (5,000–10,000), substituted with sulfate, has anticoagulant activity (128). It mimics the action of heparin but with 15% of the activity of heparin. Dextran with a molecular weight of 7,000–8,000, substituted with sulfate, is used for selective stabilization of γ-globulin and fibrinogen to heat (134).

The wide spectrum of biological properties of dextran sulfate has produced studies involving enzyme inhibition, activation, and release; adjuvant effects on immune responses; cellular interactions; and decreased virus infectivity and multiplication. It has shown some efficacy as a drug in the treatment of AIDS and tumors.

Dextran and Polyethylene Glycol in Aqueous Two-Phase Systems

The need for an extracting phase that does not harm labile enzymes, proteins, cellular organelles, and macromolecules has motivated consideration of liquid–liquid two-phase extraction systems in which each phase is primarily aqueous. The production of separate aqueous phases is achieved by dissolution of two incompatible polymers such as dextran and polyethylene glycol (135). Dextran may be replaced by its derivatives, including sodium dextran sulfate, sodium carboxymethyl dextran, or diethylaminoethyl dextran, in this two-phase system. These systems offer a new approach for extraction that is different from conventional techniques (136–138). Pullulanase and phosphorylase are extracted from *Klebsiella pneumoniae* using the dextran–polyethylene glycol (PEG) system (139).

Activated Dextrans

Several attributes (119) can be achieved by coupling ligands to dextran, including the following:

1. Stability and lifetime of ligands can be increased.
2. Because of the general compatibility of the dextran fraction in vivo, drugs that require a longer circulatory lifetime can be covalently coupled with dextran. Insulin has been coupled with dextran, and the hypoglycemic action of the complex was reported to be superior to native insulin. The complex also had a longer lifetime in the bloodstream than free insulin (140,141).
3. The antigenicity of protein ligands can be decreased.
4. Enzymes can be immobilized for enzyme reactors to produce various products (142).
5. Covalent linking of ligands for affinity chromotography can be used in purification processes (143).

Because dextran, in a well-defined molecular size range, is an established blood plasma substitute, the coupling with hemaglobin may provide a blood substitute. The preparation of hemaglobin covalently coupled with dextran was studied (144). These preparations were cleared slowly by the kidney, and it was found that the oxygen molecule bound more tightly than to free hemaglobin. Wong and associates have published a number of papers in which the preparation and experimental use of these dextran–hemaglobin products are described (145–149).

Industrial Uses

Dilute aqueous solutions of dextrans with molecular weights 18,000, 74,000, 228,000, and 7,400,000 are used as superior media for suspension polymerization of unsaturated monomers (150). Dextran (5–50% w/w; MW 10,000) incorporated into the total solids in gelatin-silver halide emulsions has significant use in the photography industry. It improves film surface coverage, maximum density contrast, and speed of gelatin-silver halide emulsions for radiologic and lithographic films (151,152).

Native Dextrans

Linear-type, water-soluble native dextrans are effective for a number of other potential uses. Dextran is superior to many polymers tested in viscous water flooding for secondary recovery of petroleum. The dextran used for this purpose has the requisite rheological properties—thermal, chemical, and solution stability and innertness to ionic effects; it does not plug earth formations and is not adsorbed on rock surfaces (153–155). Dextran is superior to carboxymethylcellulose and other common colloidal agents when used in high-calcium drilling muds even at elevated temperatures. Muds containing dextran or aldehyde-modified dextran show superior stability and performance at high pH and in saturated brines (156–159). Dextran shows relatively high resistance to deterioration in soil aggregates (160). It also makes effective protective coatings for seeds (161) and deflocculants for production of fine papers (162). Certain film-forming dextrans have value for protecting selected areas from deposition of metals in gas plating processes while remaining water soluble for ease of subsequent removal (163).

Other Uses

An iron–dextran complex has achieved significant commercial success in the treatment of iron-deficiency anemia (164). This complex is prepared by mixing a hot alkaline solution of dextran with $FeCl_3$ solution followed by neutralization. This complex containing 5% iron and 20% dextran is used for intramuscular or intravenous injections for treating iron deficiency. It is currently being used to treat anemia in newborn piglets. Dextran is used as a stabilizer in freeze-drying sensitive materials such as vaccines and bacteria (165). As recently reported, it is one of the best preservative and stabilizing agents for transplant organs (119). Compatibility of linear-type dextrans with living tissues and their ability to be metabolized permits their use in binding multiple collagen fibers into strands for surgical suture (166). Dextran's use in foods is based on its ability to stabilize syrups, confections, and the like against crystallization; to stabilize and provide viscosity to liquid sugars (167); to act as humectants and adhesives; and to show high tolerance for a low pH. Acting alone, dextran is superior for stabilizing texture in ice creams and sherbets (15).

BIBLIOGRAPHY

1. R. Margaritis and G.W. Pace, in M. Moo-Young ed., *Comprehensive Biotechnology*, vol. 3, Pergamon Press, London, 1985, pp. 1005–1044.
2. I.W. Sutherland, in H.J. Rehm and G. Reed eds., *Biotechnology*, vol. 3, Verlag Chemie, Weinheim, Germany, 1983, pp. 531–574.

3. I.W. Sutherland, *Biotechnology of Microbial Polysaccharides*, Cambridge Univ. Press, Cambridge, U.K., 1990.
4. A. Jeanes, *Encyclopedia of Polymer Science and Technology*, vol. 4, Wiley, New York, 1966, pp. 805–824.
5. E.J. Hehre, *Science* **93**, 237–238 (1941).
6. A. Jeanes, et al., *J. Am. Chem. Soc.* **76**, 5041–5052 (1954).
7. J.C. Mazza, et al., *Carbohydr. Res.* **40**, 402–406 (1975).
8. E.J. Hehre and H. Suzuki, *Arch. Biochem. Biophys.* **113**, 675–683 (1966).
9. D.S. Genghof and E.J. Hehre, *Proc. Sco. Expt. Biol. Med.* **140**, 1298–1301 (1972).
10. T.P. Binder and J.F. Robyt, *Carbohydr. Res.* **124**, 287–299 (1983).
11. E.J. Hehre and D. Hamilton, *Proc. Soc. Exp. Biol. Med.* **71**, 336 (1949).
12. S.A. Barker, et al., *J. Chem. Soc.* 4414 (1958).
13. E.R. Ruckel and C. Schuerch, *Biopolymers* **5**, 515 (1967).
14. R.M. Alsop, *Prog. Industrial Microbiol.* **18**, 1–44 (1983).
15. P.J. Barker, Jr., in R.L. Whister and J.N. Be Miller eds., *Industrial Gums: Polysaccharides and Their Derivatives*, Academic Press, New York, 1959, pp. 531–563.
16. L. Pasteur, *Bull. Soc. Chim. Fr.* 30–31 (1861).
17. C. Scheubler, *Z. Ver. Dtsch. Zuker-Ind.* **24**, 309–317 (1874).
18. P. Van Tieghem, *Ann. Sci. Nat. Bot. Biol. Veg.* **7**, 180–203 (1878).
19. E.J. Hehre and J.Y. Sugg, *J. Exp. Med.* **75**, 339 (1942).
20. S. Hestrin and S. Shapiro-Avineri, *Biochem. J.* **38**, 2 (1944).
21. P.E. Barker and E.J. Bourne, *Q. Rev.* **7**, 56 (1953).
22. E.J. Hehre, *Adv. Enzymol.* **11**, 297 (1951).
23. J. Edelman, *Adv. Enzymol.* **17**, 189 (1956).
24. F.L. Flower, et al., *Can. J. Res.* **B15**, 486 (1937).
25. W.B. Neely, *Adv. Carbohydr. Chem.* **15**, 341–364 (1960).
26. A. Jeanes, *A.C.S. Symp. Ser.* **45**, 284–298 (1977).
27. P.T. Murphy and R.L. Whistlkr, *Dextran in Industrial Gums: Polysaccharides and Their Derivatives*, 2nd ed., Academic Press, New York, 1973, pp. 513–542.
28. R.L. Sidebotham, *Adv. Carbohydr. Chem. Biochem.* **30**, 371–444 (1974).
29. M. Wales, et al., *J. Polym. Sci.* **10**, 229 (1953).
30. G.J. Hucker and C.S. Pederson, *N.Y. State Agr. Expt. Sta. Tech. Bull.* **167**, 3 (1930).
31. C.A. Wilham, et al., *Arch. Biochem. Biophys.* **59**, 61 (1955).
32. J.W. Van Cleve, et al., *J. Am. Chem. Soc.* **78**, 4435–4440 (1956).
33. A. Jeanes and F.R. Seymour, *Carbohydr. Res.* **74**, 31–40 (1979).
34. F.R. Senti, et al., *J. Polym. Sci.* **17**, 527–546 (1955).
35. G.J. Walker and A. Pulkownik, *Carbohydr. Res.* **29**, 1–4 (1973).
36. J.F. Robyt, in S.B. Peterson, B. Svensson, and S. Pedersen eds., *Carbohydrate Bioengineering*, Elsevier, Amsterdam, 1995.
37. J.F. Robyt, et al., *Carbohydr. Res.* **266**, 293–299 (1995).
38. A. Shimamura, et al., *Biochem. Biophys. Acta* **702**, 72–80 (1982).
39. F.R. Seymour, et al., *Carbohydr. Res.* **68**, 1130–121 (1979).
40. F.R. Seymour, et al., *Carbohydr. Res.* **68**, 123–140 (1979).
41. F.R. Seymour, et al., *Carbohydr. Res.* **74**, 41–47 (1979).
42. F.R. Seymour, et al., *Carbohydr. Res.* **75**, 275–294 (1979).
43. F.R. Seymono and R.D. Knapp, *Carbohydr. Res.* **81**, 105–129 (1980).
44. M.D. Hare, et al., *Carbohydr. Res.* **66**, 245–255 (1978).
45. K.H. Ebert and G. Schenk, *Adv. Enzymol.* **30**, 179–221 (1968).
46. M. Kobayashi and K. Matsuda, *Biochim. Biophys. Acta* **614**, 46–62 (1980).
47. M. Kobayashi and K. Matsuda, *J. Biochem.* **100**, 615–621 (1986).
48. A.W. Miller, et al., *Carbohydr. Res.* **147**, 119–133 (1986).
49. D. Fu and J.F. Robyt, *Prep. Biochem.* **20**, 93–106 (1990).
50. G.T. Cori and C.F. Cori, *J. Biol. Chem.* **131**, 397–398 (1939).
51. C.S. Hanes, *Proc. R. Soc. London, Ser. B* **129**, 174–208 (1940).
52. M.A. Swanson and C.F. Cori, *J. Biol. Chem.* **172**, 815–829 (1948).
53. H.J. Koepsell, et al., *J. Biol. Chem.* **200**, 793–801 (1953).
54. H.M. Tsuchiya, et al., *J. Am. Chem. Soc.* **77**, 2412–2419 (1955).
55. G.R. Germaine, et al., *J. Bacteriol.* **120**, 287–294 (1974).
56. G.R. Germaine, et al., *Infect. Immun.* **16**, 637–248 (1977).
57. A.M. Chludzinski, et al., *J. Dent. Res.* **55**, C75–C86 (1976).
58. M. Kobayashi, et al., *Agric. Biol. Chem.* **50**, 2585–2590 (1986).
59. J.F. Robyt and A.J. Corrigan, *Arch. Biochem. Biophys.* **174**, 129–135 (1976).
60. K.H. Ebert and F. Patat, *Z. Naturforsch.* **17b**, 738–745 (1962).
61. W.B. Neely, *Adv. Carbohydr. Chem.* **15**, 341–364 (1960).
62. J.F. Robyt and T.F. Walseth, *Carbohydr. Res.* **61**, 433–445 (1978).
63. J.F. Robyt and H. Taniguchi, *Arch. Biochem. Biophys.* **174**, 129–135 (1976).
64. J.F. Robyt, B.K. Kimble, and J.F. Walseth, *Arch. Biochem. Biophys.* **165**, 634–640 (1974).
65. J.F. Robyt and S.H. Eklund, *Bioorg. Chem.* **11**, 115 (1982).
66. J.F. Robyt, *Adv. Carbohydr. Chem. Biochem.* **51**, 133–168 (1995).
67. F.A. Bovey, *J. Polym. Sci.* **35**, 423 (1967).
68. K.H. Ebert and M. Brosche, *Biopolymers* **5**, 423 (1967).
69. J.F. Robyt and S.H. Eklund, *Carbohydr. Res.* **121**, 279–286 (1983).
70. D. Su and J.F. Robyt, *Carbohydr. Res.* **248**, 339–348 (1993).
71. P. Paul, et al., *Carbohydr. Res.* **149**, 433–441 (1986).
72. D. Part, et al., *Biotech. Lett.* **9**, 1–6 (1987).
73. S. Lee, et al., *Biotech. Lett.* **14**, 379–384 (1992).
74. A.W. Miller and J.F. Robyt, *Biochim. Biophys. Acta* **880**, 32–39 (1976).
75. M. Kobayashi and K. Matsuda, *J. Biochem.* **79**, 1301–1308 (1976).
76. J.F. Robyt and T.F. Walseth, *Carbohydr. Res.* **68**, 95–111 (1979).
77. A.W. Miller and J.F. Robyt, *Biochim. Biophys. Acta* **785**, 89–46 (1984).
78. E. Braswell, et al., *J. Polym. Sci.* **61**, 143–151 (1962).
79. A. Jeanes, *Methods Carbohydr. Chem.* **5**, 118–127 (1965).
80. J.E. Ciardi, et al., *Infect. Immun.* **18**, 237–241 (1977).
81. D. Kim and J.F. Robyt, *Enzyme Microb. Technol.* **16**, 659–664 (1994).
82. G.R. Lawford, *Biotech. Bioeng.* **21**, 1121–1131 (1979).

83. D.H. Bergey, *Manual of Determinative Bacteriology*, 8th ed., 1974.
84. M.K. Johnson and C.S. McCleskey, *J. Bacteriol.* **74**, 22–25 (1957).
85. H.M. Tsuchiya, et al., *J. Bacteriol.* **64**, 521–526 (1952).
86. U.K. Pat. 2079292 B (May 16, 1984), M. Schneider, et al.
87. A. Lopez and P. Monsan, *Biochemie* **62**, 325 (1980).
88. A. Lopez and P. Monsan, *Adv. Biotechnol.* **1**, 679 (1981).
89. P.E. Barker and N.J. Ajongwen, *Biotech. Bioeng.* **37**, 703–707 (1991).
90. R.S. Karthikeyan, et al., *Bioprocess Eng.* **15**, 247–251 (1995).
91. E. Braswell, et al., *J. Polym. Sci.* **41**, 467 (1959).
92. J.P. Martinez-Espindola and C.A. Lopez-Munguina, *Biotech. Lett.* **7**, 483–486 (1985).
93. R.S. Karthikeyan, et al., *Chemical Engineering Congr.*, Bombay, India, December 1993.
94. R.S. Karthikeyan, Ph.D. Thesis, I.I.T-Madras, India, 1999.
95. R.S. London, Ph.D. Thesis, UMIST, Manchester, U.K., 1982.
96. R.S. London and C. Webb, *Proc. Biochem.* 19–23 (February 1990).
97. V.W. Carlson and W.W. Carlson, *J. Bacteriol.* **58**, 135–140 (1949).
98. V.W. Carlson and W.W. Carlson, *J. Bacteriol.* **58**, 143–149 (1949).
99. V.W. Carlson and W.W. Carlson, *J. Bacteriol.* **62**, 583–589 (1951).
100. H.J. Koepsell and H.M. Tsuchiya, *J. Bacteriol.* **63**, 293–295 (1952).
101. H. Kobali and P.J. Reilly, *Biotech. Bioeng.* **22**, 1055–1069 (1980).
102. J. Sabatie and L. Choplin, *Biotech. Lett.* **8**, 425–430 (1986).
103. F.H. Foster, *Process Biotech.* **3**, 15–19 (1968).
104. F.H. Foster, *Process Biochem.* **3**, 55–62 (1968).
105. M.K. Johnson and C.S. McCleskey, *J. Bacteriol.* **74**, 22–25 (1957).
106. A. Margaritis and J.E. Zajic, *Biotech. Bioeng.* **20**, 939–1001 (1978).
107. A. Margaritis and D. Bokkel, *187th Annu. Meeting American Chemical Society*, Philadelphia, Pa., August 26–31, 1984.
108. A.M.M. El-Sayed, et al., *Biotech. Bioeng.* **36**, 338–345 (1990).
109. S. Akhnazarova and V. Kafarov, *Experiment Optimization in Chemistry and Chemical Engineering*, Moscow, Mir Publishers, 1982.
110. W.C. Conley, *Computer Optimization Techniques*, rev. ed., Princeton, N.J., 1984.
111. P. Monsan and A. Lopez, *Biotech. Bioeng.* **23**, 2027–2037 (1981).
112. P. Monsan, et al., *Ann. N.Y. Acad. Sci.* **434**, 276 (1983).
113. A.M.M. El-Sayed, et al., *Biotech. Bioeng.* **36**, 346–353 (1990).
114. A.M.M. El-Sayed, et al., *Biotech. Bioeng.* **36**, 83–91 (1990).
115. A.M.M. El-Sayed, et al., *Biotech. Bioeng.* **40**, 617–624 (1992).
116. M. Ringpfeil and M. Selenia, in I. Meleik, K. Beran, and J. Hospedka eds., *Continuous Culture of Microorganisms*, Pub. House of Zech. Acad. Sci., Prague, 1964, pp. 145–150.
117. S. Ognio, *Agric. Biol. Chem.* **34**, 1268 (1970).
118. C.R. Edwards and S.W. Drew, *Abstr. Ann. Meet. Am. Soc. Microb.* **77**, 251 (1977).
119. H.N. Chang, et al., *Biotech. Bioeng.* **23**, 2647–2653 (1981).
120. D.J. Widde, *Optimum Seeking Methods*, Prentice-Hall, Englewood Cliffs, N.J., 1964.
121. H. Ohno, E. Nakanishi, and T. Takamatsu, *Biotech. Bioeng.* **20**, 625 (1978).
122. G.S. Beveridge and R.S. Schechter, *Optimization Theory and Practice*, McGraw-Hill, New York, 1969.
123. F.S. Hiller and G.L. Liebermann, *Introduction to Operations Research*, McGraw-Hill, New York, 1989.
124. R.E. Bellman and S.E. Dreyfus, *Applied Dynamic Programming*, Princeton Univ. Press, Princeton, N.J., 1962.
125. L.T. Fan, *The Continuous Maximum Principle*, Wiley, New York, 1966.
126. J.F. Robyt, *Proc. Am. Chem. Soc.* **72**, 139–140 (1955).
127. G.H. Bixler, et al., *Ind. Eng. Chem.* **45**, 682 (1953).
128. A.J.T. Gronwall, *Dextran and Its Uses in Colloidal Infusion Solutions*, Academic Press, New York, 1957, p. 156.
129. A.J.T. Gronwall and B.G.A. Inglelman, *Acta Phys. Scand.* **7**, 97 (1945).
130. L.E. Gelin, et al., *Acta Chir. Scand.* **122**, 294–302 (1961).
131. P. Flodin, Meijels Bokindustri, Halmstad, Uppsala, 1962, p. 85.
132. Sephadex, Pharmacia Fine Chemicals Inc., New York, 1964.
133. J.L. Oneley, et al., *J. Am. Chem. Soc.* **76**, 4666–4671 (1957).
134. K.W. Walton and R.B. Pennell, *Arch. Biochem. Biophys.* **56**, 553–554 (1955).
135. P.A. Albertsson, in D. Glick ed., *Methods of Biochemical Analysis*, vol. 10, Wiley, New York, 1962, pp. 229–262.
136. M.R. Kula, *Appl. Biochem. Bioeng.* **2**, 71–94 (1979).
137. M.R. Kula, et al., *Enzyme Eng.* **4**, 47 (1978).
138. U. Matsumoto, et al., *J. Chromotogr.* **285**, 69 (1984).
139. H. Walter, et al., eds., *Partitioning in Aqueous Two-Phase Systems: Theory, Methods, Uses, and Applications in Biotechnology*, Academic Press, New York, 1985.
140. K. Armstrong, et al., *Biochem. Biophys. Acta* **498**, 102 (1977).
141. Y. Sakamoto, et al., *Biochem. Biophys. Acta* **498**, 102 (1977).
142. J. Woodward, *Immobilized Cells and Enzymes: A Practical Approach*, IRL Press, Oxford, U.K., 1996.
143. C.M. Yang and G.T. Tsao, *Adv. Biochem. Eng.* **15**, 19–42 (1982).
144. J. Blumenstein, et al., *Prog. Clin. Biol. Res.* **19**, 205 (1978).
145. J.E. Chang and J.T. Wong, *Can. J. Biochem.* **55**, 388–403 (1977).
146. P.G. Cunnington, et al., *Biochem. J.* **193**, 361–366 (1981).
147. S. Tam, et al., *Can. J. Biochem.* **56**, 981–984 (1978).
148. S. Tam, et al., *Can. J. Biochem.* **58**, 732–736 (1980).
149. S. Tam, et al., *Proc. Natl. Acad. Sci. U.S.A.* **73**, 2128–2131 (1976).
150. U.S. Pat. 2,587,367 (1958), J.J. Kearney.
151. U.S. Pat. 3,063,838 (1962), A.B. Jennings.
152. U.S. Pat. 3,203,804 (1965), A.B. Cohen and C.D. Schecklett.
153. U.S. Pat. 3,042,611 (1962), J.T. Patton.
154. U.S. Pat. 3,053,765 (1962), W.J. Sparks.
155. U.S. Pat. 3,084,122 (1963), J.C. Cybert and J.T. Patlon.
156. P.H. Monaghan and J.L. Gidley, *Oil Gas J.* **57**, 100–103 (1959).
157. U.S. Pat. 3,065,170 (1962), P.H. Monaghan and G.K. Dumbauld.
158. E.P. Mueller and Z. Angew, *Geol.* **9**, 213–217 (1963).
157. U.S. Pat. 2,602,002 (1952), W.L. Owen.
160. L.J. Novak, et al., *Agric. Food Chem.* **3**, 1028–1033 (1955).
161. U.S. Pat. 2,764,843 (1956), P.G. Peake.
162. A. De Waele, *Chem. Ind. (London)* **32**, 253 (1945).

163. U.S. Pat. 2,970,064 (1961), W.M. Bolton.
164. U.S. Pat. 2,820,740 (1958), E. London and G.D. Twigg.
165. U.S. Pat. 2,908,614 (1959), P.W. Muggleton and J. Ungar.
166. U.S. Pat. 2,748,774 (1956), L.J. Novak.
167. U.S. Pat. 2,409,816 (1946), D.V. Wardsworth and H.F. Hughes.

See also XANTHAN GUM.

DICARBOXYLIC ACID. See CITRIC ACID, PROCESSES; ENZYMES, FRUIT JUICE PROCESSING; OPINE DEHYDROGENASE, SECONDARY AMINE DICARBOXYLIC ACIDS.

DIHYDROXYPHENYLALANINE (4), PRODUCED BY MICROORGANISMS

HIDEHIKO KUMAGAI
Kyoto University
Kyoto, Japan

KEY WORDS

Ammonia
Erwinia herbicola
L-DOPA
L-Tyrosine
Pyrocatechol
Pyruvate
Tyrosine phenol-lyase (TPL)

OUTLINE

INTRODUCTION

Tyrosine phenol-lyase (TPL) (deaminating) (EC 4.1.99.2) catalyzes 3,4-dihydroxyphenyl-L-alanine (L-DOPA) synthesis from pyrocatechol, pyruvate, and ammonia in a one-step reaction. In the investigation to produce L-DOPA by this enzyme reaction, a bacterial strain (*Erwinia herbicola*) has been chosen as the best producer of TPL out of more than 1,000 others tested. Culture conditions for the preparation of cells containing high TPL activity and reaction condition for the synthesis of L-DOPA were optimized with this bacterium to support L-DOPA, appearing as the crystals in the reaction mixture, at final amounts reaching 110 g/L.

L-DOPA is the precursor of the neurotransmitter dopamine and is an effective agent in the treatment of Parkinson's disease. In this disease, the amount of dopamine in the brain of the patient is insufficient. L-DOPA had been produced mainly by a synthetic method that contains many chemical reaction steps, including a step for optical resolution. TPL is a pyridoxal phosphate dependent multifunctional enzyme obtained from bacteria that catalyzes the synthesis of L-DOPA from pyruvate, ammonia, and pyrocatechol in a one-step reaction. The L-DOPA from *E. herbicola*, is produced in cells containing high amount of TPL. This biocatalytic method is one of the most economical processes to date for the production of L-DOPA.

SELECTION OF MICROORGANISMS HAVING HIGH TPL ACTIVITY

The activity of TPL in microorganisms was investigated with 646 strains of bacteria, 140 strains of yeast, 138 strains of fungi, and 117 strains of actinomycetes (1). TPL activity was determined by measuring the amount of L-tyrosine or L-DOPA synthesized through the β-replacement reaction between L-serine and phenol or pyrocatechol, respectively. Cells of microorganisms were directly added to the reaction mixture as the enzyme for practical applications. TPL activity occurred widely in a variety of bacteria, most of which belong to the Enterobacteriaceae, especially to the genera *Escherichia, Aerobacter, Proteus*, and *Erwinia*. No activity was observed in yeasts, fungi, and actinomycetes. A large number of bacterial strains isolated from rice and corn plants showed high TPL activity. Cells of *E. herbicola* ATCC 21434, which showed the highest activity, were selected as a promising source of enzyme for the preparation of L-DOPA.

Aeration, pH, and temperature affect the formation of TPL. Cell growth requires oxygen transfer, but the condition for TPL synthesis may have a different optimum. For example, the optimum volume of the medium for the formation of TPL was 40 to 60 mL in a 500-mL shaking flask, but cell growth was rather repressed in this volume. The optimum pH for enzyme formation in this example was in the range of 7.0 to 7.5, and the optimum temperature for enzyme formation was in the range of 27 to 32 °C.

TPL activity occurs during cultivation of *E. herbicola*, and continues beyond the active growth phase. Cells were grown at 31.5 °C for 32 h with reciprocal shaking. The medium contained 0.2% tyrosine, 0.2% KH_2PO_4, 0.1% $MgSO_4 \cdot 7H_2O$, 1.0% yeast extract, 0.5% polypeptone, and 0.5% meat extract (pH 7.5) and supported maximum formation of the enzyme during the early stationary phase (cultivated for 26 to 30 h); thereafter, the enzyme gradually disappeared with the consumption of L-tyrosine added to the medium. Additions of yeast extract, meat extract, polypeptone, and the hydrolyzate of soybean protein to the basal medium enhance cell growth as well as the formation of

TPL. DL-Methionine, DL-alanine, and glycine stimulate enzyme formation. However, L-serine, L-ornithine, and L-cystine inhibit enzyme formation. Pyridoxine is added as a required precursor of coenzyme for enzyme formation.

Catabolite repression of formation of TPL by *E. herbicola* has been observed on adding glucose, pyruvate, and α-ketoglutarate to the medium at high concentrations. Glycerol is a suitable carbon source for cell growth as well as for the accumulation of the enzyme in growing cells. A marked increase in enzyme formation has been seen when glycerol is added together with succinate, fumarate or malate. Among the metallic ions tested, Fe^{2+} ion was the only one effective for the formation of TPL.

TPL is an inducible enzyme, and the addition of L-tyrosine to the medium is essential for formation of the enzyme. However, when large amounts of L-tyrosine were added, inhibition of enzyme formation and repression of cell growth were observed. The phenomena seem to be caused by the phenol that is liberated from the L-tyrosine.

Various analogs of L-tyrosine that were not decomposed during cultivation were investigated in the search for an inducer of TPL, but no effective, nondecomposable inducer was found among the L-tyrosine analogs tested. L-Phenylalanine, D-phenylalanine, and phenylpyruvate did not induce this enzyme by themselves, but they show a synergistic effect on the induction of the enzyme by L-tyrosine. L- and D-phenylalanine are competitive inhibitors of TPL.

Cells of *E. herbicola* with high TPL activity are typically prepared by growing them at 28 °C for 28 h in a medium containing 0.2% KH_2PO_4, 0.1% $MgSO_4$ $7H_2O$, 2 ppm Fe^{2+} ($FeSO_4 \cdot 7H_2O$), 0.01% pyridoxine · HCl, 0.6% glycerol, 0.5% succinic acid, 0.1% DL-methionine, 0.2% DL-alanine, 0.05% glycine, 0.1% L-phenylalanine, and 12 mL of hydrolyzed soybean protein in 100 mL of tap water, with the pH controlled at 7.5 throughout cultivation. Under these conditions, TPL is efficiently accumulated in the cells of *E. herbicola* and makes up about 10% of the total soluble cellular protein.

REACTION CONDITIONS FOR THE SYNTHESIS OF L-DOPA

Synthesis of L-DOPA from Pyruvate, Ammonia, and Pyrocatechol

The optimum pH for this reaction is around 8.0, and the optimum temperature is 16 °C. Because pyruvate is

Figure 1. The reaction mixture for the L-DOPA synthesis by TPL. Crystals of L-DOPA are precipitated on the bottom of flask.

Figure 2. Microphotograph of crystals of L-DOPA.

Table 1. Comparison of Chemical and Enzymatic Methods of L-DOPA Production

	Chemical Method	Enzymatic Method
Starting materials	Vanillin, hydantoin, acetic anhydride, hydrogen	Pyrocatechol, pyruvate, ammonia
Total number of unit-reaction	8	1
By-products	Ammonia, CO_2, acetate	H_2O
Necessity of optical resolution	Yes	No
Equipment	Specific plant	Commonly usable fermenter
Period (day)	15	3
Contaminating amino acids	L-Tyrosine, 3-methoxytyrosine, trihydroxy-phenylalanine	L-Tyrosine

unstable in the reaction mixture at a high temperature, low temperature is favorable for the synthesis of L-DOPA. The reaction is carried out at 16 °C for 48 h in a reaction mixture containing various amounts of sodium pyruvate, 5 g of ammonium acetate, 0.6 g of pyrocatechol, 0.2 g of sodium sulfite, 0.1 g of EDTA, and cells harvested from 100 mL of broth, in a total volume of 100 mL. The pH was adjusted to 8.0 by the addition of ammonia. At 2-h intervals, sodium pyruvate and pyrocatechol were added to the reaction mixture to maintain the initial concentrations. The maximum synthesis of L-DOPA was obtained when the concentration of sodium pyruvate was kept at 0.5%. L-DOPA appears as the crystals in the reaction mixture, as shown in Figures 1 and 2.

When a higher concentration of pyruvate was added to the reaction mixture during the synthesis of L-DOPA, two by-products, 1-methyl-1,3-dicarboxy-6,7-dihydroxy-1,2,3,4-tetrahydroisoquinoline and 1-methyl-3-carboxy-6,7-dihydroxy-3,4-dihydroisoquinoline, were formed. The by-products have been shown to be formed through a spontaneous condensation reaction.

TPL catalyzes a variety of α,β-elimination, β-replacement, and racemization reactions and the synthesis of L-tyrosine and its related amino acids. The mechanism for these reactions has been studied in some detail, by adopting the general mechanism proposed for pyridoxal-dependent reactions (1).

An enzymatic method for preparing L-DOPA was developed using bacterial cells with high TPL activity. This method is simple and is one of the most economical processes to date (Table 1) for preparing L-DOPA. L-DOPA is currently produced by this method with 60-m^3 scale fermenters. The yield of L-DOPA compared to pyrocatechol is 98% and to pyruvate, 90%, with the final accumulation of L-DOPA reaching 110 g/L (2). Certain L-tyrosine-related amino acids can be synthesized from the starting materials of pyruvic acid and appropriate derivatives of phenol, such as pyrocatechol (1).

BIBLIOGRAPHY

1. H. Yamada and H. Kumagai, in D. Perlman, ed. *Advances in Applied Microbiology*, vol. 19, Academic Press, New York, 1975, pp. 249–288.
2. H. Kumagai, H. Enei, H. Nakazawa, T. Tsuchida, and T. Namekawa, *Biosci. Industry* **54**, 9–15 (1996).

DILTIAZEM SYNTHESIS

HIROAKI MATSUMAE
HIROYUKI AKATSUKA
TAKEJI SHIBATANI
Tanabe Seiyaku Co., Ltd.
Osaka, Japan

KEY WORDS

Asymmetric hydrolysis
Biocatalyst
Diltiazem
Lipase
Membrane bioreactor
Methyl(2*R*,3*S*)-3-(4-methoxyphenyl)glycidate
Recombinant

OUTLINE

INTRODUCTION

Diltiazem, (2S,3*S*)-3-acetoxy-5-[2-(dimethylamino)ethyl]-2,3-dihydro-2-(4-methoxyphenyl)-1,5-benzothiazepin-4(5*H*)-one hydrochloride, is a calcium channel blocker and has been clinically used since 1974 as an antianginal and antihypertensive agent in more than 100 countries (1). Because diltiazem has two stereo centers at positions 2 and 3 in the structure shown in Figure 1, four isomers are possible. Among them, only the (2*S*,3*S*)-isomer exhibits the desired coronary vasodilating activity. The conventional production has been a nine-step process starting with *p*-anisaldehyde and methyl chloroacetate (1). The fourth step is a chemical resolution of the advanced intermediate 6. Resolution at an earlier step should save raw materials and reduce the amount of waste. (±)-MPGM is the first intermediate containing stereo centers. Two of the four isomers are excluded through the chemical synthesis of (±)-MPGM, which yields only *trans* type MPGM. Furthermore, no racemization occurs on the subsequent steps from (−)-

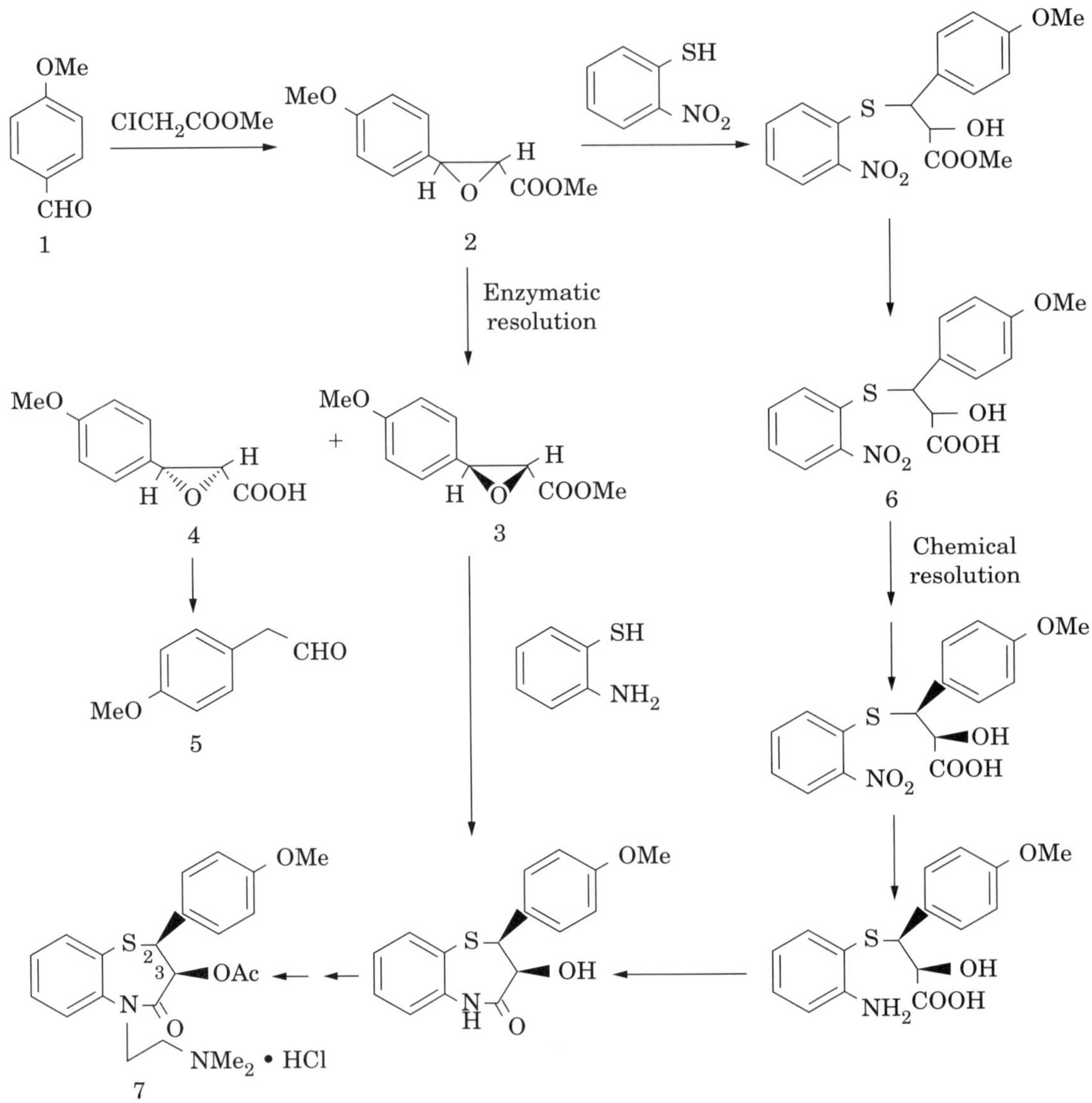

Figure 1. The synthetic process for diltiazem hydrochloride. Numerals: 1, *p*-anisaldehyde; 2, (±)-(2*RS*,3*SR*)-3-(4-methoxyphenyl)glycidic acid methyl ester [(±)-MPGM]; 3, (−)-(2*R*,3*S*)-3-(4-methoxyphenyl)glycidic acid methyl ester [(−)-MPGM]; 4, (+)-(2*S*,3*R*)-3-(4-methoxyphenyl)glycidic acid; 5, 4-methoxyphenylacetaldehyde; 6, *threo*-2-hydroxy-3-(4-methoxyphenyl)-3-(2-nitrophenylthio)propanoic acid; 7, (+)-(2*S*,3*S*)-3-acetoxy-5-[2-(dimethylamino)ethyl]-2,3-dihydro-2-(4-methoxyphenyl)-1,5-benzothiazepin-4(5*H*)-one hydrochloride (diltiazem).

MPGM to diltiazem. Therefore, the asymmetric syntheses of (−)-MPGM (2–5) and resolutions of (±)-MPGM (6,7) have been investigated for the synthesis of diltiazem. We focus on the resolution of (±)-MPGM as the more effective method of (−)-MPGM production (8,9).

Previous workers resolved (±)-MPGM by first hydrolyzing to the carboxylic acid, resolving the carboxylic acid with an optically active amine, and then esterifying (10,11). This method involves many steps and yields oily MPGM with low purity. To solve these problems, we resolved (±)-MPGM using an enzyme. Enzymes are now widely recognized as practical catalysts for asymmetric synthesis (12). Previous examples of enzyme-catalyzed enantioselective reactions of glycidate compounds include enantioselective hydrolysis of one ester group in *meso-cis*-oxirane dicarboxylic acid diester (13) and of one enantiomer of racemic glycidyl esters (14). Although these glycidate compounds are stable in aqueous solution, our substrate, (±)-MPGM, is not. The 4-methoxyphenyl group increases the reactivity of the oxirane so that (±)-MPGM decomposes easily in aqueous solution under neutral to acidic conditions. (±)-MPGM is more stable in an organic–aqueous biphasic system. In addition, the two-phase mixtures also simplify separation of reactants and products. Enzyme recovery is also very important for cost reduction in industrial enzyme reactions, but effective recovery is difficult from two-phase mixtures due to the formation of stable emulsions. We used the Sepracor membrane reactor based on the principle that permits immobilization of the enzyme and contact between organic and aqueous phases for both efficient hydrolysis and simple separation of products. Furthermore, we overproduced the enzyme using mutation, optimization of culture medium, and recombinant DNA techniques.

Our objectives were production of (−)-MPGM without chemical degradation of (±)-MPGM, construction of a lipase-hyperproducing strain, an effective isolation of (−)-MPGM from an organic–aqueous biphasic system, and a repeated use of enzyme. The screening of hydrolytic enzymes, overproduction of the enzyme, optimization of the enzyme reaction, and application of membrane reactor are reported herein.

SCREENING FOR ENANTIOSELECTIVE HYDROLASES

Screening Method

The instability of (±)-MPGM in an aqueous solution is a serious problem in the enzymatic resolution of (±)-MPGM. However, as shown in Table 1, (±)-MPGM is both more stable and more soluble in a mixture of toluene–water. Chemical hydrolysis of the oxirane ring in (±)-MPGM increases at a pH lower than 8, yielding 2,3-dihydroxy-3-(4-methoxyphenyl)propanoic acid methyl ester. We used the toluene–water biphasic system at pH 8 to screen for hydrolase with high activity and enantioselectivity toward (±)-MPGM (15). Because the reaction system contains an organic solvent, we focused on lipases that are more stable in organic solvents and can act on insoluble substrates such as oils by adsorbing to the oil droplet (16). (+)-MPGM hydrolyzing activity is assayed using the emulsion method (17) shown in Figure 2. One unit of esterase activity is defined as the amount of enzyme that hydrolyzes 1 μmol (+)-MPGM/min. One unit of lipase activity is defined as the amount of enzyme that hydrolyzes 1 μmol of ester group from 2,3-dimercaptopropan-1-ol tributyrate, as measured with the Lipase Kit S (18). The enzyme activity in "Overproduction of Enzyme" is shown by the lipase activity. *E*-values are calculated from the enantiomeric excess of the recovered (−)-MPGM and extent of conversion (19) and are not corrected for chemical hydrolysis of the methyl ester.

Table 1. The Solubility and Stability of (±)-MPGM in Organic Solvents

Organic solvent (organic solvent:water = 1:1)	Solubility (mg/mL)	Stability (%)
Water only (suspension)	<0.1	10
N,N-Dimethylformamide	>20	29
Ethanol	>20	21
Acetone	>20	70
Tetrahydrofuran	>20	85
Methyl isobutyl ketone	219	90
Ethyl acetate	560	90
n-Butyl acetate	267	90
tert-Butyl methyl ether	79	95
Toluene	251	95
Carbon tetrachloride	205	95
p-Xylene	171	95

Note: Four milligrams of (±)-MPGM is dissolved in 2 mL of various organic solvent–water (1:1) biphasic systems, and the biphase is incubated at 30 °C for 24 h. The stability of (±)-MPGM is expressed as a percentage of remaining (±)-MPGM. The solubility of (±)-MPGM is expressed as the concentration of dissolved (±)-MPGM in organic solvent at 30 °C.

Asymmetric Hydrolysis of (±)-MPGM by Various Microorganisms

Among the enzymes from 700 type strains and 30 commercially available enzymes, 453 enzymes hydrolyzed the ester of (±)-MPGM. Hydrolytic enzymes for (±)-MPGM were widely distributed in bacteria, yeasts, and molds. Many favored the (+) enantiomer of MPGM. The *E*-values of 12 enzymes were more than 10, and *Serratia marcescens* lipase has the highest enantioselectivity ($E = 135$) and hydrolyzing activity, as shown in Table 2. The ratio of the esterase activity to the hydrolyzing activity on olive oil is the highest for *S. marcescens* lipase. This indicates that *S. marcescens* lipase is the most effective enzyme for hydrolysis of (+)-MPGM. In addition, since the *E*-value includes chemical degradation of (±)-MPGM, it shows a lower value than the true value of the enzyme. *S. marcescens* lipase does not enzymatically hydrolyze (−)-MPGM; the *E*-value of this lipase itself is considered to be infinity (17).

Purification and Characterization of *S. marcescens* Lipase

Three main protein bands exist in the supernatant after centrifugation of the culture broth. The molecular masses of these bands are 62,000, 51,000, and 40,000 Da on SDS-PAGE. The hydrolytic activity toward (+)-MPGM exists in the 62,000-Da band. The esterase can be purified from culture supernatant (about 21-fold) at a yield of 20%, and its (±)-MPGM hydrolyzing activity is 570 units/mg protein. Esterase corresponds to 5% of the extracellular proteins in cells grown with dextrin as the carbon source and Meast S (beer yeast extract) as the nitrogen source. The activity of the purified enzyme is 3,100 units/mg protein for olive oil, and 520,000 units/mg protein for 2,3-dimercaptopropan-1-ol tributyrate. Because this esterase selectively hydrolyzes the 1,3-positional ester of triglyceride, the esterase is a lipase. The protein exhibiting a molecular mass of 51,000 on SDS-PAGE is a protease with hydrolytic activity toward casein. The protease does not have (±)-MPGM hydrolyzing activity.

(±)-MPGM as well as other esters having different alcohol moieties may be used as a key intermediate in the synthesis of diltiazem. The enantioselectivity of the lipase did not change upon altering the chain length of the alcohol moiety, except for the acid ethyl ester. (±)-MPGM crystallizes more easily from toluene and is less expensive than other esters, so it was selected for preparing enantiomerically pure 3-(4-methoxyphenyl)glycidic acid ester (17).

OVERPRODUCTION OF ENZYME

Cloning of *S. marcescens lipA* Gene

S. marcescens produces the lipase at a level of 3×10^3 units/mL when grown in LB medium. The amount of lipase increased 50-fold by optimizing the culture condition (15),

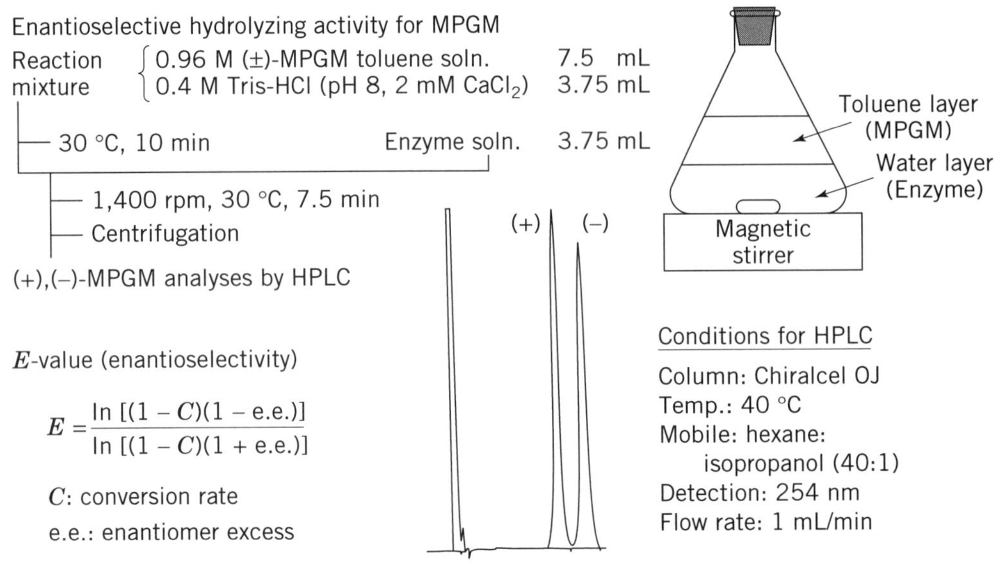

Figure 2. Analytical methods of esterase activity and enantioselectivity.

Table 2. Asymmetric Hydrolysis of (±)-MPGM by Enzymes from Various Microorganisms

Enzyme source[a]	Enzyme used[b] (mg)	Esterase activity (units/mg protein)	Conversion (%)	e.e. (%)	E^c
Serratia marcescens Sr41 8000[d]	1.4 mL	36	48.2	89.0	135
Pseudomonas putida ATCC 17426[d]	3.75 mL	18	44.2	75.1	85
Corynebacterium primorioxydans ATCC 31015[d]	3.75 mL	16	44.6	75.6	73
Pseudomonas mutabilis ATCC 31014[d]	3.75 mL	13	44.7	74.7	57
Corynebacterium alkanolyticum ATCC 21511[d]	3.75 mL	18	45.8	77.4	53
Candida cylindracea (Lipase OF)	4.0	34	50.0	82.2	26
Mucor javanicus (Lipase M)	150	0.82	48.0	75.0	22
Bacillus licheniformis (Alcalase)	1.9 mL	0.91	41.2	57.0	17
Rhizopus japonicus (Lipase saiken 100)	100	2.3	42.8	58.4	15
Bacillus subtilis (Protease 27)	300	–	45.1	62.6	14
Mucor miehei (Palatase M)	0.25 mL	27	45.3	60.3	12
Pseudomonas sp. (Lipase P)	300	0.20	49.8	68.6	11

[a]Trade name of the commercially available enzyme is shown in parentheses.
[b]The extent of hydrolysis (%) of (±)-MPGM and the optical purity (%) of the recovered (–)-MPGM are examined using this amount of enzymes.
[c]*E*-value {$E = \ln[(1 - C)(1 - \text{e.e.})]/\ln[(1 - C)(1 + \text{e.e.})]$} is calculated on the basis of the enantiomeric excess of the recovered (–)-MPGM (e.e.) and extent of conversion (C).
[d]Cells are grown in the shaking flasks (500 mL) with 2% Meast S, 1% dextrin, 0.2% $(NH_4)_2SO_4$, 0.1% kH_2PO_4, 0.05% $MgSO_4 \cdot 7H_2O$, 0.01% $CaCl_2 \cdot 2H_2O$, 0.001% $FeSO_4 \cdot 7H_2O$, 0.5% Tween 80, and 0.1% Colorin 102 (pH7) at 30 °C for 18 h (140 rpm). The esterase activity in the culture supernatant is measured.

and 2.5-fold as a result of mutation using *N*-methyl-*N*-nitro-*N*-nitrosoguanidin. Extracellular lipase activity reaches 400 × 10^3 units/mL of culture medium when the lipase-hyperproducing mutant is cultured in a fed-batch culture on proline. To further increase production of the lipase, the *lipA* gene, encoding the lipase, was cloned from *S. marcescens*. The nucleotide sequence of the *lipA* gene showed a major open reading frame encoding a 64.9-kDa protein of 613 amino acid residues; the deduced amino acid sequence contained a lipase consensus sequence, GXSXG. The *S. marcescens* lipase has 66 and 56% homologies with the lipases of *Pseudomonas fluorescens* B52 (20) and *P. fluorescens* SIK W1 (21), respectively, but does not show any overall homology with other lipases. The *Escherichia coli* cells carrying the *lipA* gene do not secrete the lipase into the medium. The *S. marcescens* lipase lacks a conventional N-terminal signal sequence and is not subjected to any processing at either the N-terminal or C-terminal regions. A specific short region, GGXGXDXXX, which is often found in secretory proteins having no N-terminal signal peptide, is observed in the amino acid sequence. The region is characterized as a glycine- and aspartic acid–rich region. Thus, these results reveal that *S. marcescens* lipase is secreted by the N-terminal-signal-peptide-independent system (22).

Secretion Mechanism of *S. marcescens* Lipase

The 20-kb *Sac*I DNA fragment allowing the lipase secretion was cloned from *S. marcescens* by lipase selection phenotype based on the clear halo formation by recombinant *E. coli* cells, as shown in Figure 3. The subcloned 6.5-kb *Eco*RV fragment contains three open reading frames, *lipB, lipC*, and *lipD*, which are composed of 588, 433, and 437 amino acid residues and constitute an operon (Fig. 3). The *E. coli* cells carrying the *lipA, lipB, lipC*, and *lipD* genes secreted the *S. marcescens* lipase, but the cells carrying only the *lipA* gene did not. Deletion mutation analysis shows that all three genes are essential for the extracellular secretion of the *S. marcescens* lipase. Thus, the secretion of the lipase requires three proteins: LipB, LipC, and LipD (23). Recent papers reported that the genes encoding the secretion device for the extracellular proteins lacking an N-terminal signal sequence are found in several Gram-negative bacteria and has been categorized as being in the ABC (ATP binding cassette) transporter family: the *hlyB, hlyD*, and *tolC* genes (24–27) for *E. coli* hemolysin; the $prtD_{EC}$, and $prtE_{EC}$, and $prtF_{EC}$ genes (28) for *Erwinia chrysanthemi* metalloprotease; the *aprD, aprE*, and *aprF* genes (29) for *Pseudomonas aeruginosa* alkaline protease; the *cyaB, cyaD*, and *cyaE* genes (30) for *Bordetella pertussis* cyclolysin; the *lktB* and *lktD* genes (31) for *P. haemolytica* leukotoxin; and the *hasD, hasE*, and *hasF* genes (32,33) for *S. marcescens* heme-binding protein and metalloprotease. The deduced amino acid sequences of the *lipB* and *lipC* genes showed high homology to those of the *E. chrysanthemi* $prtD_{EC}$ and $prtE_{EC}$ genes (56 and 46% identity, respectively), and also the *S. marcescens hasD* and *hasE* genes (54 and 45% identity, respectively). The LipB protein belongs to a member of the ABC protein family, and its putative ATP-binding site was found in the central region of the C-terminal half. No N-terminal signal sequence was identified in the LipB and LipC proteins. The LipB and LipC proteins were revealed to be inner membrane components involved in transport across the membrane. The outer membrane component corresponding to the *E. chrysanthemi* $prtF_{EC}$ gene product was encoded by the *lipD* gene in *S. marcescens* (42% identity). The products of the *lipB, lipC*, and *lipD* genes also showed 26, 20, and 22% identity with the *hlyB, hlyD*, and *tolC* gene products of *E. coli*, respectively. These results indicates that LipB-LipC-LipD is included in the ABC transporter family and is the secretion device of the *S. marcescens* lipase, as shown in Figure 4.

Overproduction of Lipase by a Recombinant Strain of *S. marcescens*

The *S. marcescens lipA* gene was cloned into the high-copy-number vector pUC19, resulting in pLIPE121 (22). This plasmid was introduced into the cells of *S. marcescens* 8000, a wild-type strain of Sr41. The extracellular lipase activity of the recombinant strain (34) reached 220 × 10^3 units/mL in the lipase production medium and is about nine fold greater than that of the strain carrying pUC19, as shown in Table 3. However, we expected an even greater increase, based on the high copy number of the plasmid. The *S. marcescens lipB, lipC*, and *lipD* genes were cloned into the low-copy-number vector pMW219 (35), resulting in pMWBCD10 (23). The *S. marcescens lipA* gene was cloned into the high-copy-number vector pBR322, resulting in pBRE121 (34). The lipase productivity of *S. marcescens* cells carrying pMWBCD10 and pLIPE121 (or pBRE121) is higher than that carrying the *lipA* plasmid only. The extracellular lipase activity of the hyperproducing strain carrying pMWBCD10 and pBRE121 reaches 420

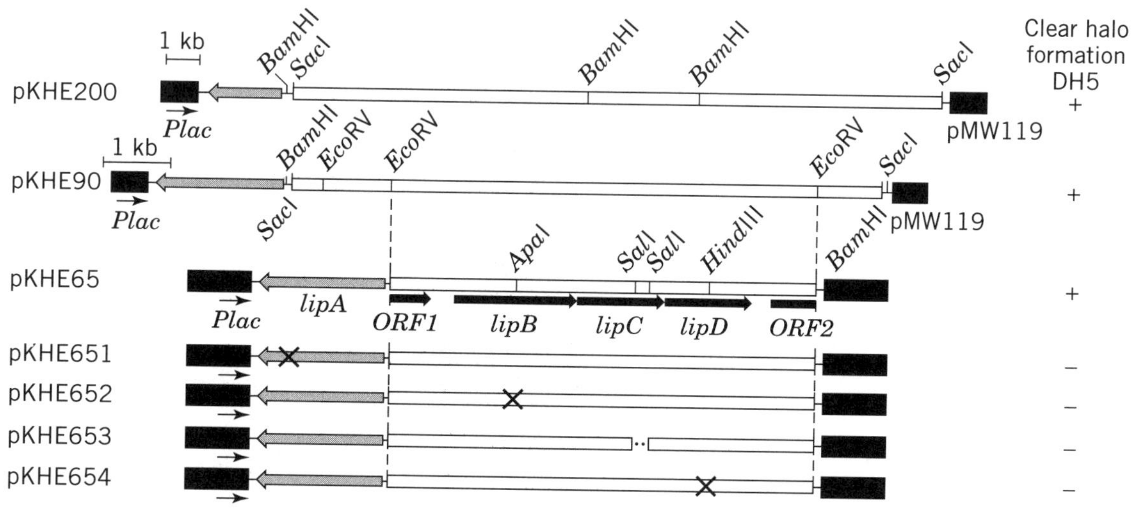

Figure 3. Restriction endonuclease maps of the inserted DNA of pKHE200, deletion derivatives, and the ability of the recombinant plasmids to give a lipase phenotype to the *E. coli* $tolC^+$ strain DH5 (22). The lipase phenotype was detected by formation of a clear halo on a tributyrin agar plate. The thick arrows indicate the open reading frame determined by sequencing. The shaded arrows represent the lipase gene. Open boxes and black bars represent the chromosomal DNA inserts and vector pMW119, respectively. The *lac* promoter carried on the plasmid pMW119 is shown by thin arrows. Only the restriction sites used for subsequent experiments are shown. The crosses symbolize the location of the mutations introduced into the inserted DNA.

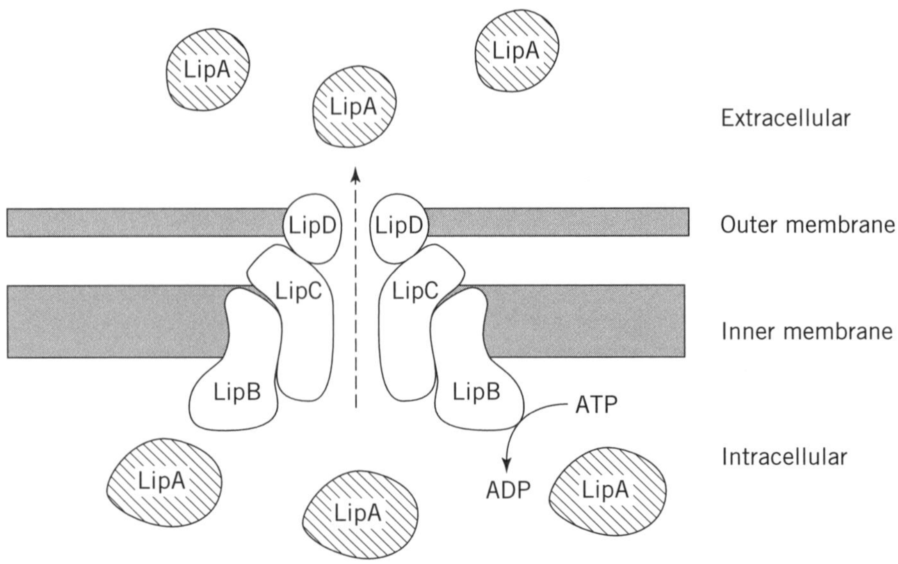

Figure 4. The secretion mechanism for the *S. marcescens* lipase and metalloprotease. LipA indicates the *S. marcescens* lipase. PrtA indicates the *S. marcescens* metalloprotease. LipB, LipC, and LipD are the gene products of the *lipB, lipC*, and *lipD* genes, which encode the ABC transporter for *S. marcescens* lipase and metalloprotease.

Table 3. Effect of *lipA* and *lipBCD* Expression on Lipase Production

Plasmid		Growth[a]	Lipase activity[b]
lipA	*lipBCD*	(OD_{660})	($\times 10^{-3}$ units/mL)
–	–	12.2	31.8
pBRE121	–	11.9	258
pBRE121	pMWBCD10	11.2	421
pLIPE121	–	12.2	220
pLIPE121	pMWBCD10	10.9	820

[a]The recombinant strains of *S. marcescens* Sr41 8000 carrying various plasmids are cultivated at 30 °C for 25.5 h in lipase production medium. Antibiotics are added at specified concentrations: ampicillin, 500 μg/mL; kanamycin, 100 μg/mL.

[b]The lipase activity in the culture supernatant is measured.

Table 4. Esterase Activity of *S. marcescens* Lipase in Various Organic Solvent–Water (1:1) Biphasic Systems

Organic solvent (organic solvent:water = 1:1)	Dielectric constant of solvent (ϵ)	Esterase activity (units/mg protein)
Carbon tetrachloride	2.2	85
Toluene	2.2	36
p-Xylene	2.2	31
n-Butyl acetate	5.0	26
Ethyl acetate	6.1	19
Methyl isobutyl ketone	13.1	11
Trichloroethane	7.5	3
Ethanol	24.6	0

$\times 10^3$ units/mL of the culture medium and is 13-fold greater than that of the wild-type strains. The extracellular lipase activity of the hyperproducing strain carrying pMWBCD10 and pLIPE121 reaches 820 $\times 10^3$ units/mL of the culture medium and is 25-fold greater than that of the wild-type strains. We are now investigating large-scale lipase production using this recombinant strain.

OPTIMIZATION OF ENZYMATIC REACTION

Conditions for Enzymatic Reaction and Asymmetric Hydrolyzing Activity

Screening for hydrolases, as already described, used the toluene–water biphasic system, which suppresses chemical degradation of (±)-MPGM and readily dissolves (±)-MPGM. In this section, (±)-MPGM hydrolyzing activities of the *S. marcescens* lipase are compared in various biphasic systems (Table 4). The hydrolyzing activities differ with the kind of organic solvent. The lower the dielectric constant of the solvent, the higher the enzyme activity obtained. Among the biphasic systems tested, the carbon tetrachloride–water biphasic system gives the highest hydrolyzing activity. However, carbon tetrachloride is not suitable for industrial application because of its toxicity. Consequently, the toluene–water biphasic system, which gives the second highest hydrolytic rate, was selected as the reaction solvent.

The lipase shows maximum activity at about pH 8. When the lipase was kept at 30 °C for 24 h at different pHs, we found it to be stable between pH 6 and 8. The apparent enantioselectivity of hydrolysis of (±)-MPGM by this lipase in a toluene–water biphasic system decreases at pH values below 8 because the chemical degradation of the oxirane ring in (±)-MPGM increases below pH 8. An Arrhenius plot indicates that the apparent activation energy of the enzymatic hydrolysis is 26 kJ/mol. The thermal inactivation of the enzyme occurs above 50 °C. On the other hand, the apparent activation energy of chemical degradation of the oxirane ring in (−)-MPGM or (+)-MPGM is 66 kJ/mol. These findings show that reaction at a lower temperature minimizes the relative amount of chemical degradation of (±)-MPGM. The best conditions were a toluene–water emulsion at pH8 and 30 °C.

As the toluene to water ratio in the reaction mixture decreases, the rate of enzymatic hydrolysis also decreases. At a toluene to water ratio lower than 1:1, the chemical degradation of (±)-MPGM increases. Therefore, we use the condition of a toluene to water ratio of 1:1 as a typical reaction for the asymmetric hydrolysis of (±)-MPGM (15).

Effect of Substrates and Solvent on the Stability of the Enzyme

Attempted reuse of the *S. marcescens* lipase showed much slower reaction rates, indicating deactivation of the lipase. When fresh lipase is incubated with thc toluene–water biphase without substrate before the enzymatic reaction is started, the progress of reaction is nearly equal that of untreated enzyme, as shown in Figure 5. Therefore, the esterase activity of the reaction decreased only slightly. However, when the first incubations are carried out in the toluene–water biphase containing (±)-MPGM or (−)-MPGM, the progress of reaction shown in Figure 5 is slowly toward that of untreated enzyme. The esterase activity of the second incubation decreases from 36 units/mg protein to 6 and 12 units/mg protein, respectively. Inactivation of the lipase is greatly influenced by MPGM. When the first incubation is carried out in the toluene–water biphase containing (±)-MPGM, the degree of inactivation of the lipase is higher than that of the lipase treated with the toluene–water biphase containing (−)-MPGM. This is because *p*-methoxyphenylacetaldehyde (aldehyde 5) derived

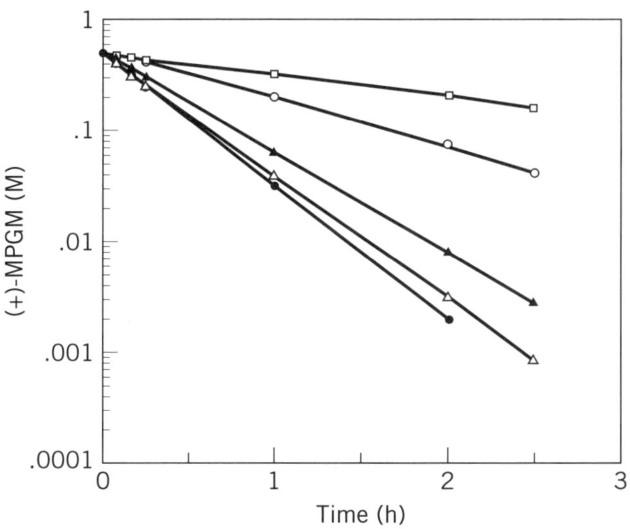

Figure 5. Repeated use of *S. marcescens* lipase for the asymmetric hydrolysis of (±)-MPGM. The toluene solution (50 mL) of 0.96 M (±)-MPGM, 0.48 M (−)-MPGM, or 20 mM *p*-methoxyphenylacetaldehyde is added to 50 mL of 0.2 M Tris-HCl buffer (pH 8, 1 mM $CaCl_2$) containing the lipase. The first incubation is carried out at 30 °C for 3 h with vigorous stirring (900 rpm). After the incubation, the mixture is centrifuged, and the water layer containing the lipase is recovered. The second incubation is carried out with the above water layer (50 mL) containing recovered lipase and a fresh toluene layer (50 mL) containing 0.96 M (±)-MPGM. The concentration of residual (+)-MPGM in the second incubation is measured by the HPLC method described in the analytical methods. Symbols: ●, without first incubation; △, toluene-water (1:1); ▲, aldehyde 5 and toluene–water (1:1); ○, (−)-MPGM and toluene–water (1:1); □, (±)-MPGM and toluene–water (1:1).

from chemical degradation of the hydrolyzed product of (+)-MPGM probably inactivates the lipase. Treatment of the lipase with the toluene–water biphase containing 20 mM and 100 mM aldehyde 5 lowers the esterase activity of the second incubation to 28 and 18 units/mg protein, respectively. Although the lipase is highly stable in a toluene–water biphase, it is unstable under the biphasic system containing (±)-MPGM and aldehyde 5. Hence, the repeated use of the lipase is difficult for this batch process (15).

Destruction of the Emulsion and Isolation of (−)-MPGM

In the organic–aqueous biphasic systems, the reaction rate is influenced by the outer diffusion and contact area on the biphase, and the highest rate is obtained by the formation of fine emulsion. However, the fine emulsion was unfavorable for industrial purposes because the destruction of the emulsion after the enzymatic reaction and subsequent phase separation are difficult, and the desired product cannot be recovered in high yields. The destruction of the emulsion after the enzymatic reaction and subsequent phase separation is a serious problem for industrial applications of the emulsion bioreactor. Several means involving an electrical method such as sonication, a physical method such as filtration, centrifugation, or heating, and a chemical method such as addition of surfactant or solvent have already been proposed to overcome this problem. Among them, the chemical method using surfactants would be the most favorable for industrial purposes because of the ease of operation. When a 0.01% (w/v) sodium dodecyl sulfate is added to the reaction mixture, the destruction of emulsion and phase separation are accomplished. In the hydrolysis of (±)-MPGM with the lipase, aldehyde 5 is accumulated in the toluene phase, with the desired (−)-MPGM. Aldehyde 5 suppresses crystallization of (−)-MPGM. To isolate (−)-MPGM efficiently, it is necessary to remove the aldehyde 5 from the toluene solution. It is well known that aldehyde compounds generally form a water-soluble adduct through the reaction with bisulfite anion (36). The aldehyde 5 can be removed by transfer to the aqueous phase by the formation of an adduct (RCHOH · SO_3Na) with the addition of sodium bisulfite (pH 6). After the phase separation and removal of the aldehyde, the toluene phase is neutralized with sodium hydrogen carbonate, washed with water, dried over anhydrous sodium sulfate, and the (−)-MPGM is crystallized. The yield of (−)-MPGM isolated from (±)-MPGM is 40–43%, and the optical purity of (−)-MPGM is nearly 100% e.e.

USE OF A MEMBRANE REACTOR

Production of (−)-MPGM Using a Membrane Reactor

The enzymatic hydrolysis of (±)-MPGM proceeds efficiently in a conventional emulsion reactor with a toluene–aqueous biphasic system (15). However, destruction of the emulsion and separation of the biphase are required for effective isolation of the product. Enzyme stability and recovery are also required for its repeated use. In particular, the recovery of enzyme is very important for cost reduction in industrial processes involving enzyme reactions. However, efficient recovery is difficult in the case of a stable emulsion. We proposed an immobilized enzyme system to solve these disadvantages of the emulsion reactor.

With respect to the immobilization of lipase, various methods have already been proposed (37–40). When these lipase-immobilized reactors are utilized, lipase can be efficiently immobilized. However, it is difficult to achieve reaction and product separation simultaneously. Furthermore, because the hydrolysis of oily fat by lipase is an interfacial reaction involving the adsorption of enzyme to the surface of oil droplets (16), the activities of most immobilized lipases are low compared with other immobilized enzymes (39). A membrane reactor permits both physical adsorption of enzyme and contact between the organic and aqueous phases; this configuration enables efficient hydrolysis and simplifies product separation. Since a pioneering work on the membrane reactor with liquid–liquid contact mode by Hoq et al. (41), several membrane bioreactors with industrial promise have been developed (42,43). We used the ultrafiltration membrane reactor technology commercialized by Sepracor, Inc., in Massachusetts, on the basis of Matson's investigation (44).

Structure of the Membrane and Reaction Conditions

The membrane reactor used in this study is a commercially available Sepracor membrane reactor model MBR 500-tm (Sepracor Inc., U.S.A.), Figure 6. The membrane module is a hydrophilic hollow-fiber ultrafiltration membrane with a molecular weight cutoff of 50,000. The membrane module contains approximately 8,500 fibers with an internal diameter of 0.21 mm and a wall thickness of 0.05 mm. The effective membrane area is 0.75 m^2. The saturated adsorption amount of proteins is 3.0 g/m^2. The membrane, composed of polyacrylonitrile, is an asymmetric membrane having a tight skin layer on the lumen side and a loose spongy layer on the shell side. Immobilization of enzyme is carried out by ultrafiltration of the lipase solution from the shell side to the lumen side. The lipase is physically immobilized on the spongy layer of the shell side. As shown in Figure 6, the reaction using a membrane reactor is achieved by circulating both toluene solution containing (±)-MPGM in the shell loop and an equimolar solution (pH 8.5) of sodium sulfite in the lumen loop. Flow rates in both loops are maintained at 750 mL/min and 500 mL/min, respectively; these are sufficient to keep mass transfer resistances at a minimum. The pressure in the shell loop is maintained at overpressure at 0.4–0.5 kg/cm^2 over the pressure in the lumen loop, to avoid the filtration of aqueous-phase solution into the shell loop through the hydrophilic membrane. The enzyme reaction is carried out at 20 °C and pH 8.5 by controlling the pH of the aqueous phase. As discussed earlier, 30 °C was selected for the optimum temperature of reaction. As for the lipase-catalyzed asymmetric hydrolysis on a membrane reactor, a long period of reaction was required, and 20 °C was selected in order to suppress chemical degradation of (±)-MPGM. If sodium bisulfite was not added to the aqueous phase in the lumen loop, the aldehyde 5 accumulated in the toluene phase, as shown in Figure 7, and deactivated the lipase. The stability of the lipase is not influenced when the sodium hydrogen sulfite concentration is below 0.1 M at pH 8.5 and 22 °C. From these results, 0.0935 M sodium hydrogen sulfite solution (pH 8.5) was applied to the aqueous phase to prevent the lowering of enzyme activity. Another by-product, methanol, is transferred from the toluene phase to the aqueous phase. Because the concentration of methanol in the aqueous phase was below 0.2% (w/v), the hydrolysis of (±)-MPGM was not influenced (45).

Rate Analysis and Optimization

Time course for hydrolysis of (±)-MPGM in a membrane reactor with *S. marcescens* lipase is shown in Figure 8. The residual concentration of (+)-MPGM decreases with re-

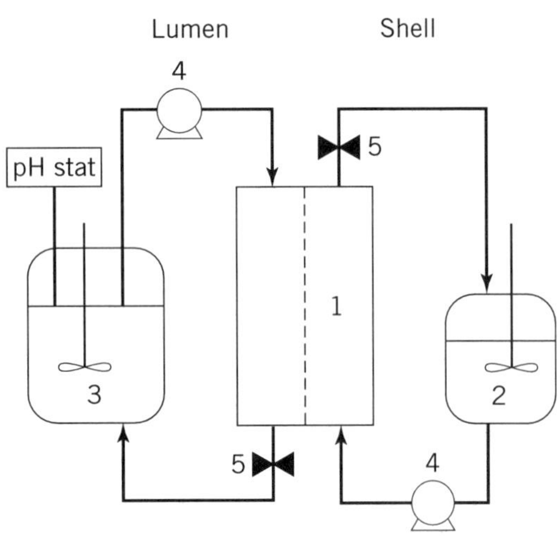

Figure 6. Flow diagram of a membrane reactor. Numerals: 1, hydrophilic membrane; 2, 1.15 L of toluene phase containing 1.15 mol of (±)-MPGM; 3, 12.3 L of aqueous phase containing 1.15 mol of sodium hydrogen sulfite; 4, recycle pump; 5, throttle valve.

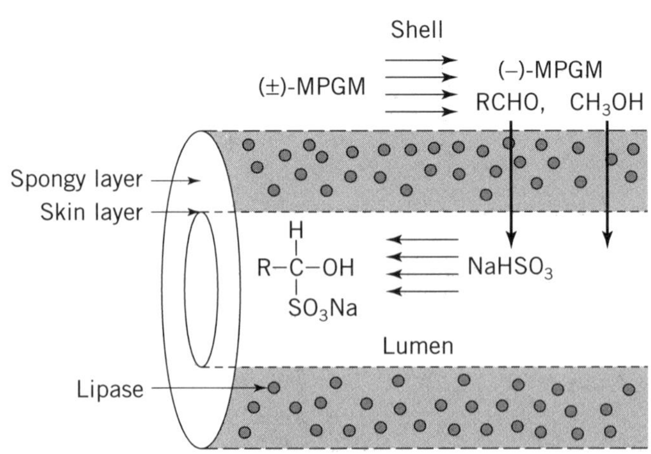

Figure 7. Behavior of products in membrane reactor. Pressure in the shell side is kept at excess compared with pressure in the lumen side. RCHO, *p*-methoxyphenylacetaldehyde; RCHOH · SO_3Na, an adduct formed between sodium hydrogen sulfite and *p*-methoxyphenylacetaldehyde.

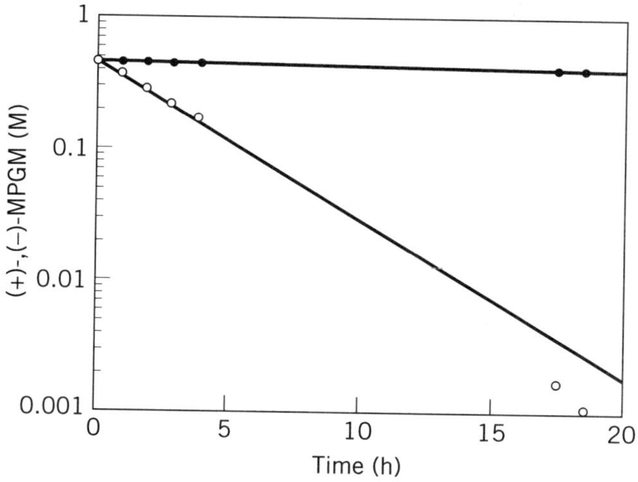

Figure 8. Time course for hydrolysis of (±)-MPGM in a membrane bioreactor with *S. marcescens* lipase. 0.6×10^5 units of lipase is loaded onto the membrane. The reaction is carried out at 22 °C using 1.15 L of 1 M (±)-MPGM toluene solution and 12.3 L of 0.094 M sodium hydrogen sulfite solution.

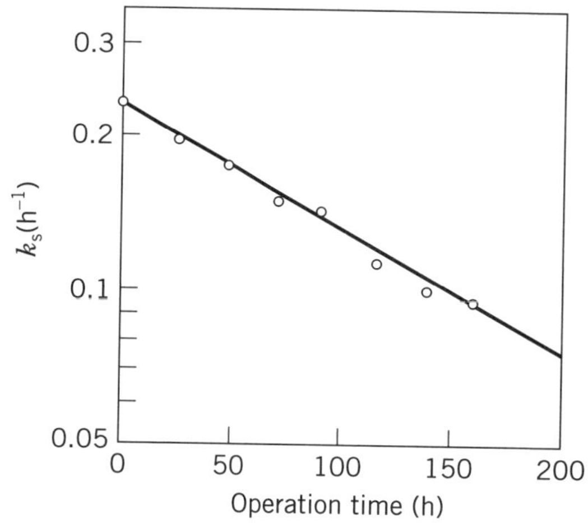

Figure 9. Operational stability of lipase immobilized on a membrane. k_s is calculated from the results of repeated batch reaction; loaded enzyme, 0.8×10^5 units/m^2; (±)-MPGM, 1 M; reaction time in individual batch, 23 h; temp., 22 °C.

action time and reaches less than 1% of initial concentration of (+)-MPGM during the first 15 h. The degradation of (−)-MPGM in merely trace amounts is observed. Thus this hydrolyzing rate of the lipase obeys the following first-order reaction kinetics:

$$r_S = -dC_S/dt = k_S C_S$$
$$r_R = -dC_R/dt = k_R C_R$$

where r_S and r_R are reaction rates for (+)- and (−)-MPGMs (M/h); C_S and C_R are (+)- and (−)-MPGM concentrations (M); k_S and k_R are reaction rate constants (h^{-1}); and t is reaction time (h). k_S and k_R are calculated to be 0.23 h^{-1} and 0.007 h^{-1}, respectively. The ratio of k_S to k_R is about 33, indicating that the lipase from *S. marcescens* has a high enantioselectivity on hydrolysis of (+)-MPGM.

Optimization of the asymmetric hydrolysis reaction with the following parameters was investigated using the reaction rate constant k (h^{-1}) as the indication of reaction. This hydrolyzing reaction is influenced by pH, temperature, phase ratio, and immobilized enzyme amount. The optimized conditions are pH 8.5; temperature 20 °C; phase ratio (=aqueous:toluene phase) 2:1; immobilized enzyme amount 0.8–1.6 $\times 10^5$ unit/m^2 (based on esterase activity). In the industrial application of the membrane bioreactor, one of the most important factors is the stabilization of enzyme. The half-life of enzyme activity was examined in a long-term continuous experiment, as shown in Figure 9. The reaction rate constants for the asymmetric hydrolysis of (±)-MPGM exponentially decrease with time, and the half-life of enzyme activity is 127 h. This half-life is much lower than for enzymes employed for the industrial production of various amino acids (46–48). However, one is about 30 times higher than that in the emulsion reactor. In this membrane bioreactor, one is able to remove inactivated enzymes by a simple back flash. When the enzyme activity in the membrane reactor lowers, one can immobilize new enzyme without regeneration of the membrane. As a result, it has become possible to achieve repeated runs of the enzyme reaction for periods up to a month. When the regeneration of the membrane after the every eight successive batch runs is repeated 10 times over a period of 4 months, the water flux rate of the old membrane is the same as that of new membrane. Thus the production method of (−)-MPGM using the membrane reactor can be industrialized (45).

Scale-Up

Advantages of the production process of (−)-MPGM using the membrane bioreactor are as follows: *(1)* the reaction and products separation are achieved simultaneously, and crystalline (−)-MPGM with a high yield of 40–43% and optical purity of 100% e.e. is obtained by just concentrating the toluene solution; *(2)* the working time taken to isolate crystalline (−)-MPGM is reduced in comparison with that of the emulsion bioreactor, *(3)* stabilization and recovery of the enzyme are achieved in laboratory-scale experiments.

Table 5. Comparison of Reactor Performances for Laboratory- and Pilot-Scale Membrane Reactors

Reactor membrane area	Laboratory scale 0.75 m^2	Pilot scale 7.8 m^2	57 m^2
L/D^a (−)	3.9	5.5	4.4
Loaded enzyme^b (units/m^2)	0.8×10^5	1.1×10^5	0.9×10^5
k_S value (h^{-1})	0.23	0.23	0.22
Half-life of enzyme activity (h)	127	137	144

[a] L, effective length of the membrane module; D, diameter of the membrane module.
[b] The amount of loaded enzyme is shown as the esterase activity.

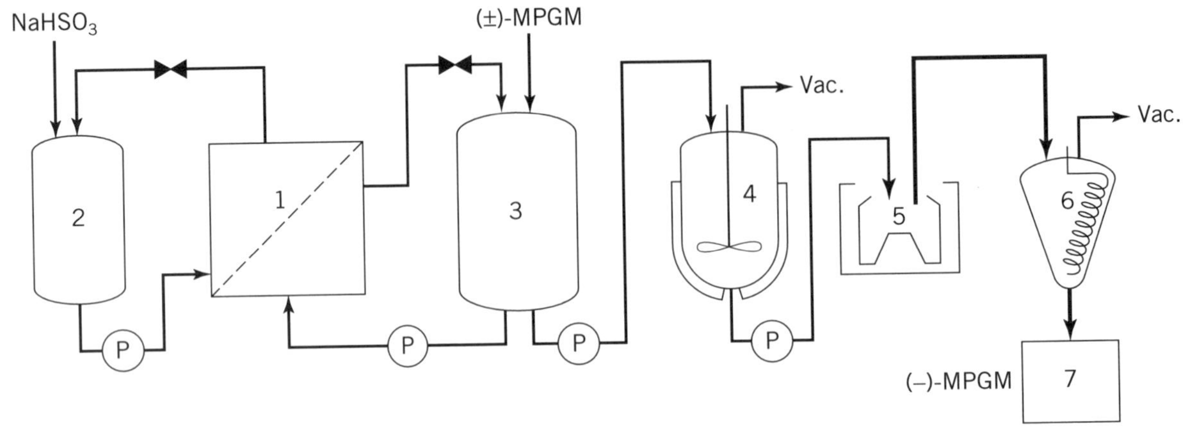

Figure 10. Schematic diagram of industrial production system of (−)-MPGM. Numerals: 1, membrane bioreactor skid; 2, aqueous solution tank; 3, organic solution tank; 4, crysytallizer; 5, centrifuge; 6, dryer; 7, product container.

On the basis of these results, the scale-up experiments were started using membrane modules having an effective membrane area of 7.8 m^2 and 57 m^2. Both membranes were magnified from that of the laboratory scale on the basis of geometrical analogy. The basic strategy of scale-up of the membrane reactor is as follows: *(1)* the flow rate of supplied fluids to the membrane area is kept constant, *(2)* when a lot of modules are used in series, a pressure gradient with respect to flow direction among the modules is not allowed to form. We have tried the reaction in pilot-scale experiments under a liquid line rate comparable to that in the laboratory-scale experiments. The functions of pilot-scale membrane reactors are identical with those of laboratory-scale membrane reactors, as shown in Table 5. From the result, it is believed that scale-up should proceed very smoothly without any trouble. On the basis of the obtained data, we made a design for an apparatus producing (−)-MPGM (Fig. 10). It is estimated that the membrane bioreactor using membrane modules having the effective membrane area of 57 m^2 produces about 50 kg (−)-MPGM m^{-2} $year^{-1}$. The pilot-plant experiments were carried out in collaboration with Sepracor, Inc., and this technique was commercialized in 1993.

BIBLIOGRAPHY

1. K. Abe, H. Inoue, and T. Nagao, *Yakugaku Zasshi* **108**, 716–732 (1988).
2. K.G. Watson, Y.M. Fung, M. Gredley, G.J. Bird, W.R. Jackson, H. Gountzos, and B.R. Matthews, *J. Chem. Soc. Chem. Commun.* 1018–1019 (1990).
3. O. Miyata, T. Shinada, I. Ninomiya, and T. Naito, *Tetrahedron Lett.* **32**, 3519–3522 (1991).
4. A. Schwartz, P.B. Madan, E. Mohacsi, J.P. O'Brien, L.J. Todaro, and D.L. Coffen, *J. Org. Chem.* **57**, 851–856 (1992).
5. K. Matsuki, M. Sobukawa, A. Kawai, H. Inoue, and M. Takeda, *Chem. Pharm. Bull.* **41**, 643–648 (1993).
6. M. Yamamoto, M. Hayashi, M. Masaki, and H. Nohira, *Tetrahedron Asymmetry* **2**, 403–406 (1991).
7. L.T. Kanerva and O. Sundholm, *J. Chem. Soc. Perkin Trans. 1*, 1385–1389 (1993).
8. Eur. Pat. 0362556 A1 (Apr 11, 1990), T. Shibatani, K. Nakamichi, and H. Matsumae (to the Regents of Tanabe Seiyaku).
9. Eur. Pat. 0343714 A1 (May 17, 1989), L.A. Hulshof and J.H. Roskam (to the Regents of Stamicarbon B. Y.).
10. Japanese Pat. Publication 60-13775 (January 24, 1985), H. Sawai, J. Yoshinaga, Y. Ikuta, and K. Nakamura (to the Regents of Sawai Seiyaku).
11. Japanese Pat. Publication 60-13776 (January 24, 1985), H. Sawai, J. Yoshinaga, and T. Masagaki (to the Regents of Sawai Seiyaku).
12. J.B. Jones, *Tetrahedron* **42**, 3351–3403 (1986).
13. G. Sabbioni and J.B. Jones, *J. Org. Chem.* **52**, 4565–4570 (1987).
14. U.S. Pat. 4,732,853 (March 22, 1988), G.M. Whitesides and W. Ladner (to the Regents of the Harvard University).
15. H. Matsumae, M. Furui, and T. Shibatani, *J. Ferment. Bioeng.* **75**, 93–98 (1993).
16. Y.J. Wang, J.Y. Sheu, F.F. Wang, and J.-F. Shaw, *Biotechnol. Bioeng.* **31**, 628–633 (1988).
17. H. Matsumae and T. Shibatani, *J. Ferment. Bioeng.* **77**, 152–158 (1994).
18. S. Kurooka, S. Okamoto, and M. Hashimoto, *J. Biochem.* **81**, 361–369 (1977).
19. C.-S. Chen, Y. Fujimoto, G. Girdaukas, and C.J. Sih, *J. Am. Chem. Soc.* **104**, 7294–7299 (1982).
20. K. Tan and K.J. Miller, *Appl. Environ. Microbiol.* **58**, 1402–1407 (1994).
21. G.H. Chung, Y.D. Lee, G.H. Jeohn, O.J. Yoo, and J.S. Rhee, *Agric. Biol. Chem.* **55**, 2359–2365 (1991).
22. H. Akatsuka, E. Kawai, K. Omori, S. Komatsubara, T. Shibatani, and T. Tosa, *J. Bacteriol.* **176**, 1949–1956 (1994).
23. H. Akatsuka, E. Kawai, K. Omori, and T. Shibatani, *J. Bacteriol.* **177**, 6381–6389 (1995).
24. M.A. Blight and I.B. Holland, *Mol. Microbiol.* **4**, 873–880 (1990).
25. N. Mackman, J.-M. Nicaud, L. Gray, and I.B. Holland, *Curr. Top. Microbiol. Immunol.* **125**, 159–181 (1986).

26. W. Wagner, M. Vogel, and W. Goebel, *J. Bacteriol.* **154**, 200–210 (1983).
27. C. Wandersman and P. Delepelaire, *Proc. Natl. Acad. Sci. U.S.A.* **87**, 4776–4780 (1983).
28. S. Letoffe, P. Delepelaire, and C. Wandersman, *EMBO J.* **9**, 1375–1382 (1990).
29. J. Guzzo, F. Duong, C. Wandersman, M. Murgier, and A. Lazdunski, *Mol. Microbiol.* **5**, 447–453 (1991).
30. L. Gilson, K. Mahanty, and R. Kolter, *EMBO J.* **9**, 3875–3884 (1990).
31. C.A. Strathdee and R.Y.C. Lo, *J. Bacteriol.* **171**, 916–928 (1989).
32. S. Letoffe, J.M. Ghigo, and C. Wandersman, *J. Bacteriol.* **176**, 5372–5377 (1994).
33. R. Binet and C. Wandersman, *Mol. Microbiol.* **22**, 265–273 (1996).
34. H. Akatsuka, E. Kawai, K. Omori, S. Komatsubara, and T. Shibatani, *J. Ferment. Bioeng.* **81**, 115–120 (1996).
35. K. Yamaguchi and Y. Masamune, *Mol. Gen. Genet.* **200**, 362–367 (1985).
36. T. Furutani, M. Furui, T. Mori, and T. Shibatani, *Appl. Biochem. Biotechnol.* **59**, 319–328 (1996).
37. R.B. Lieberman and D.F. Ollis, *Biotechnol. Bioeng.* **17**, 1401–1419 (1975).
38. Y. Kimura, A. Tanaka, K. Sonomoto, T. Nihira, and S. Fukui, *Eur. J. Appl. Microbiol. Biotechnol.* **17**, 107–112 (1983).
39. J.-F. Shaw, R.-C. Chang, F.F. Wang, and Y.J. Wang, *Biotechnol. Bioeng.* **35**, 132–137 (1990).
40. C.O. Ibrahim, H. Saeki, N. Nishio, and S. Nagai, *Agric. Biol. Chem.* **52**, 99–105 (1988).
41. M.M. Hoq, T. Yamane, and S. Shimizu, *J. Am. Oil Chem. Soc.* **61**, 776–781 (1984).
42. F. Taylor, C.C. Panzer, J.C. Craig, Jr., and D.J. O'Brien, *Biotechnol. Bioeng.* **28**, 1318–1322 (1986).
43. W. Ponk, P.J.A.M. Kerkhof, C. van Helden, and K. van't Riet, *Biotechnol. Bioeng.* **32**, 512–518 (1988).
44. S.L. Matson and J.A. Quinn, *Ann. N.Y. Acad. Sci.* **469**, 152–165 (1986).
45. H. Matsumae, M. Furui, T. Shibatani, and T. Tosa, *J. Ferment. Bioeng.* **78**, 59–63 (1994).
46. T. Sato, T. Mori, T. Tosa, I. Chibata, M. Furui, K. Yamashita, and A. Sumi, *Biotechnol. Bioeng.* **17**, 1797–1804 (1975).
47. M. Furui and K. Yamashita, *J. Ferment. Technol.* **61**, 587–591 (1983).
48. M. Senuma, O. Otsuki, N. Sakata, M. Furui, and T. Tosa, *J. Ferment. Bioeng.* **67**, 233–237 (1989).

DILTIAZEM SYNTHESIS, MICROBIAL ASYMMETRIC REDUCTION

TAKUO NISHIDA
TAKEJI SHIBATANI
Tanabe Seiyaku Co., Ltd.
Osaka, Japan

KEY WORDS

Asymmetric reduction
Baker's yeast
Biocatalysis
(2*S*,3*S*)-2,3-Dihydro-3-hydroxy-2-(4-methoxyphenyl)-1,5-benzothiazepin-4(5*H*)-one
Diltiazem
Methyl (2*R*,3*S*)-3-(4-methoxyphenyl)glycidate
Methyl (2*RS*,3*S*)-2-chloro-3-hydroxy-3-(4-methoxyphenyl)propionate
Microbial conversion
Microbial reduction

OUTLINE

INTRODUCTION

Diltiazem hydrochloride, (2*S*,3*S*)-3-acetoxy-5-[2-(dimethylamino)ethyl]-2,3-dihydro-2-(4-methoxyphenyl)-1,5-benzothiazepin-4(5*H*)-one hydrochloride (**7**; reaction products are described in Figs. 1 and 2), is a typical calcium channel blocker and is used clinically as an effective antianginal and antihypertensive agent in more than 100 countries (1). Many methods have been investigated for the synthesis of this optically active medicine (2–11). On a commercial basis, the standard process of nine chemical synthetic steps had been used for a long time. Later, lipase-catalyzed asymmetric resolution of a key intermediate, methyl *trans*-3-(4-methoxyphenyl)glycidate [(±)-MPGM] was developed, and the synthetic process was reduced to five steps (Fig. 1) (12,13). As a result, diltiazem has been produced on a commercial basis by this method since 1993. However, one drawback of the method is that the yield of the desired optically active intermediate, methyl (2*R*,3*S*)-3-(4-methoxyphenyl)glycidate [(−)-MPGM], does not exceed essentially 50% of the starting material, which means that half of the substrate goes to waste.

The microbial reduction of carbonyl compounds is a useful method in which 100% theoretical yield can be expected (14), and the procedure for the reaction is very simple. Its use has been limited to preparation of chiral alcohols on a large scale because of the following disadvantages: *(1)* a relatively low concentration of substrate in the reaction medium, *(2)* a tedious procedure for isolation and purifi-

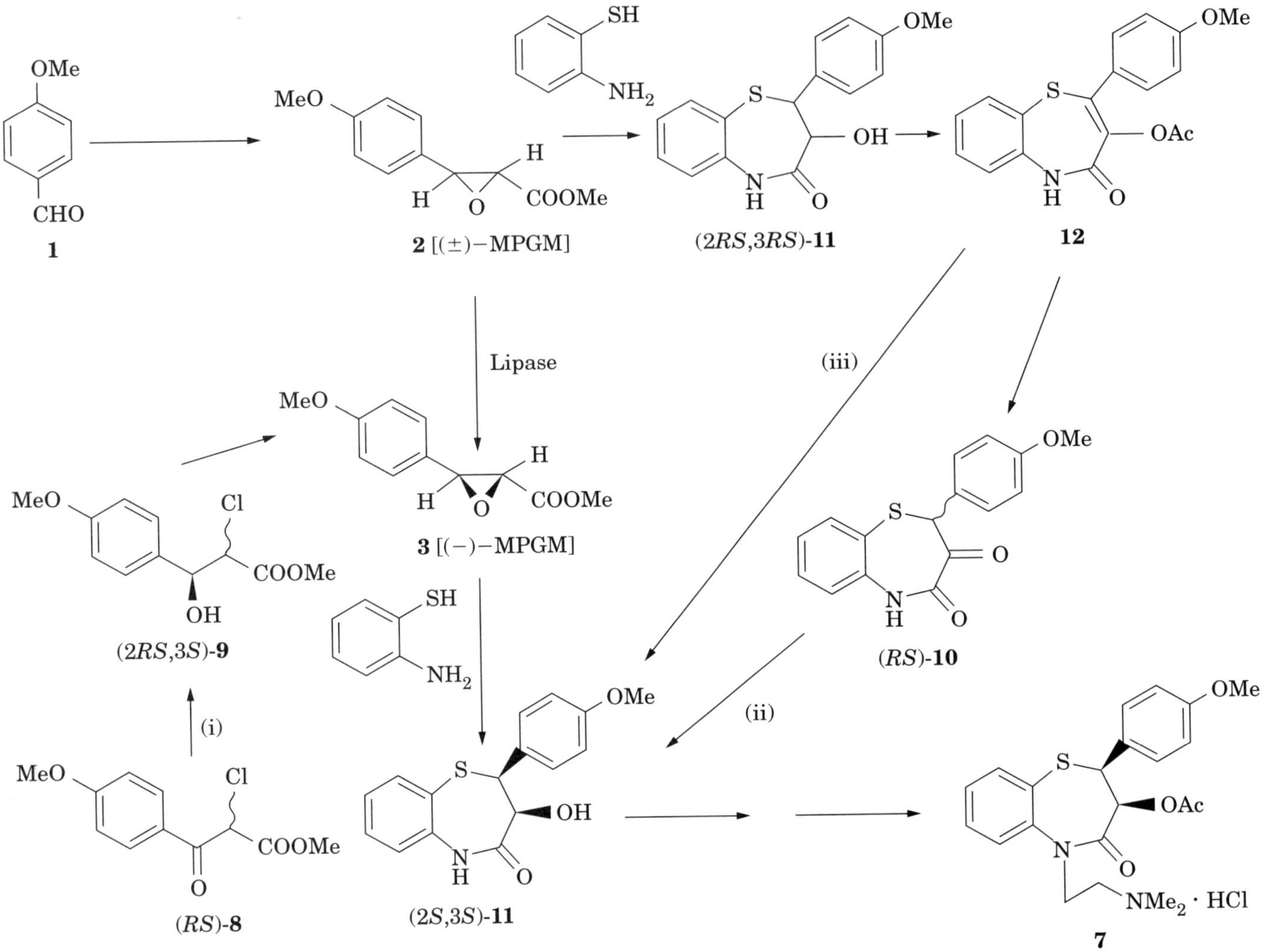

Figure 1. Synthetic process for diltiazem hydrochloride.

cation of the product, and sometimes *(3)* a relatively low optical purity of the product (15).

In order to eliminate these disadvantages, microorganisms having a high enzymatic activity and a high enantioselectivity were sought to produce the optically active diltiazem intermediates for three production methods: the reduction of methyl (*RS*)-2-chloro-3-(4-methoxyphenyl)-3-oxopropionate [(*RS*)-**8**] to methyl (2*RS*,3*S*)-2-chloro-3-hydroxy-3-(4-methoxyphenyl) propionate [(2*RS*,3*S*)-**9**] [Fig. 1(i)], the reduction of (*RS*)-2-(4-methoxyphenyl)-1,5-benzothiazepin-3,4(2*H*,5*H*)-dione [(*RS*)-**10**] to (2*S*,3*S*)-2,3-dihydro-3-hydroxy-2-(4-methoxyphenyl)-1,5-benzothiazepin-4(5*H*)-one [(2*S*,3*S*)-**11**] [Fig. 1(ii)]; and the microbial conversion of 3-acetoxy-2-(4-methoxyphenyl)-1,5-benzothiazepin-4(5*H*)-one (**12**) to (2*S*,3*S*)-**11** [Fig. 1(iii)].

Among these three routes, the second is the most promising; (2*S*,3*S*)-**11** was obtained from (*RS*)-**10** in a high yield with high enantioselectivity from baker's yeast, and was able to be easily separated from the reaction mixture. This route was also introduced to the production of (2*S*,3*S*)-**11** in a high concentration for industrial application.

SCREENING OF MICROORGANISMS FOR THE PRODUCTION OF OPTICALLY ACTIVE DILTIAZEM INTERMEDIATE

To successfully apply microbial reduction to industrial scale production it is important to have microorganisms with a high enzymatic activity and a high enantioselectivity. Three production routes for optically active diltiazem intermediates were examined (Fig. 1).

Microbial Reduction of Methyl (*RS*)-2-Chloro-3-(4-methoxyphenyl)-3-oxopropionate [(*RS*)-8]

Methyl (2*R*,3*S*)-3-(4-methoxyphenyl)glycidate [(−)-MPGM] is an important optically active intermediate for the industrial synthesis of diltiazem (12,13). (−)-MPGM can be also obtained selectively from methyl (2*RS*,3*S*)-2-chloro-3-hydroxy-3-(4-methoxyphenyl)propionate [(2*RS*, 3*S*)-**9**] by treating it with NaOMe in methanol; there is epimerization in the second position of **9** (9,16) (Fig. 1).

Table 1. Asymmetric Reduction of (*RS*)-8 by Various Microorganisms

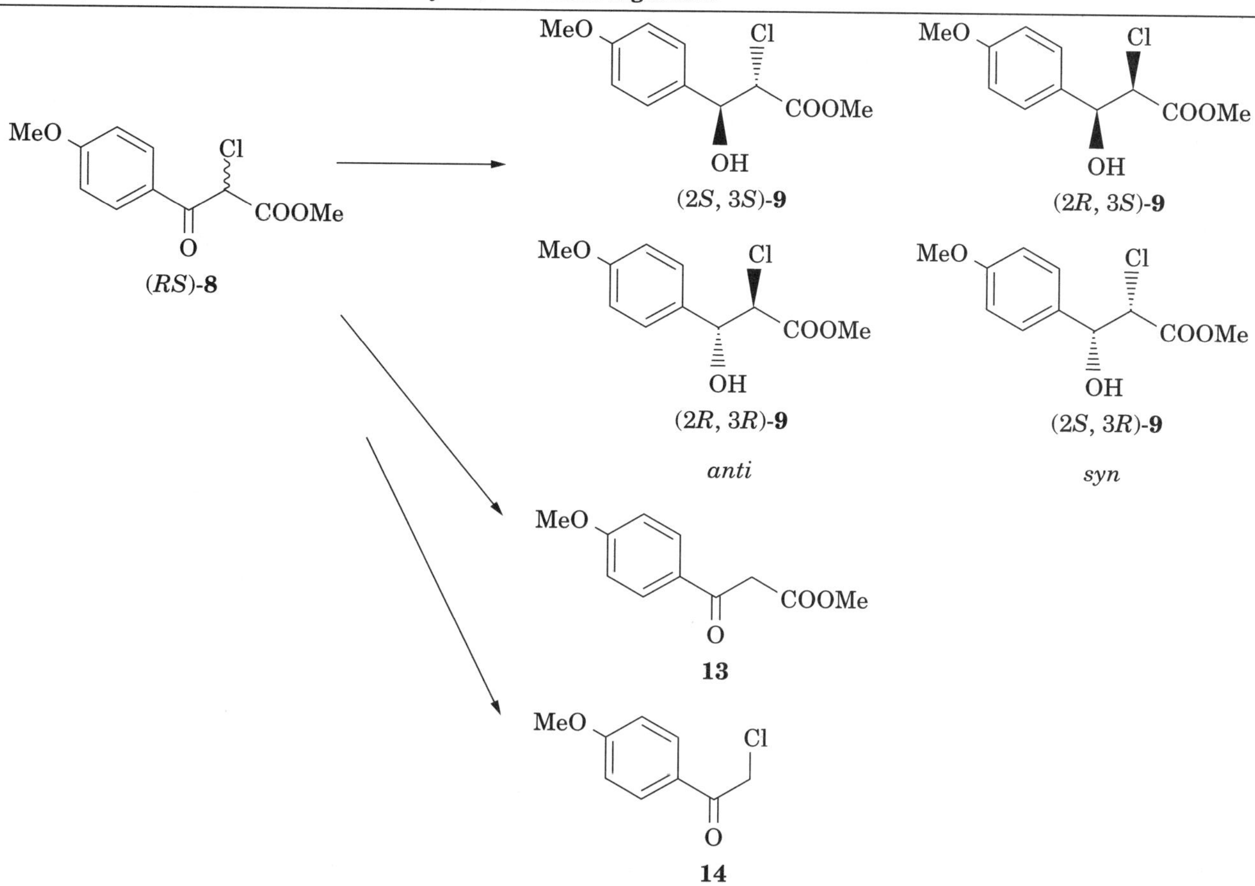

Microorganism	Yield[a] (%)	Product 9: *anti*/*syn*	Product 9: Absolute configuration	Product 9: % e.e.[a]
Absidia corymbifera IFO 4009	18	1/>99	*syn*-2*S*,3*R*	>99
Cryptococcus laurentii OUT 6027	43	80/20	*anti*-2*R*,3*R*	>99
			syn-2*S*,3*R*	99
Mucor ambiguus IFO 6742	58	1/>99	*syn*-2*S*,3*R*	>99
Mucor hiemalis OUT 1045	20	12/88	*anti*-2*R*,3*R*	68
			syn-2*S*,3*R*	>99
Mucor janssenii OUT 1050	23	5/95	*anti*-2*R*,3*R*	>99
			syn-2*S*,3*R*	>99
Mucor javanicus IFO 4570	40	7/93	*anti*-2*R*,3*R*	>99
			syn-2*S*,3*R*	>99
Trichoderma viride OUT 4642	7[b]	1/2	*anti*-2*S*,3*S*	87
			syn-2*R*,3*S*	96

The microorganisms were inoculated into test tubes (15 mm diameter × 12 cm) containing 3 mL of medium A[c] or B[d] and cultured at 30 °C for 24 h (yeast and bacteria) or for 72 h (fungi and actinomycetes) with continuous shaking (300 strokes/min). The cells were collected by centrifugation and suspended in 3 mL McIlvaine buffer (pH 5) containing 5% glucose. Then 6 μL of *N,N*-dimethylformamide (DMF) solution containing 6 mg of (*RS*)-**8** was added to the test tube. After the reaction was carried out at 30 °C for 24 h with continuous shaking (300 strokes/min), the reaction products were extracted from the reaction mixture with 2 mL of ethyl acetate. The yield of **9** was calculated by using HPLC equipped with Zorbax CN™ column (Du Pont Company). The e.e. of **9**, which was isolated by silica gel TLC (solvent: *n*-hexane: chloroform: ethyl acetate, 4:1:1) was also calculated by HPLC equipped with Chiralcel OJ™ column (Daicel Chemical Industries, Ltd.).

Source: Ref. 17.

[a]Yield and e.e. of **9** were determined by HPLC.

[b]Isolated yield.

[c]Medium A comprised 5% glucose, 0.1% KH_2PO_4, 0.05% $MgSO_4 \cdot 7H_2O$, 0.05% $CaCl_2 \cdot 2H_2O$, 0.1% yeast extract, 0.1% $(NH_4)_2SO_4$, 0.05% urea, 0.0002% $FeSO_4 \cdot 7H_2O$, 0.0002% $MnCl_2 \cdot 4H_2O$, and 0.0002% $ZnSO_4 \cdot 7H_2O$ (yeast and fungi, pH 6.5; actinomycetes, pH 7.3).

[d]Medium B comprised 1% glucose, 1% Polypepton® (Nikon Seiyaku Co. Ltd.), 0.1% KH_2PO_4, 0.05% $MgSO_4 \cdot 7H_2O$, 0.05% $CaCl_2 \cdot 2H_2O$, 0.1% yeast extract, 0.1% $(NH_4)_2SO_4$, 0.05% urea, 0.0002% $FeSO_4 \cdot 7H_2O$, 0.0002% $MnCl_2 \cdot 4H_2O$, and 0.0002% $ZnSO_4 \cdot 7H_2O$ (bacteria, pH 7).

Table 2. Asymmetric Reduction of (*RS*)-10 by Various Microorganisms

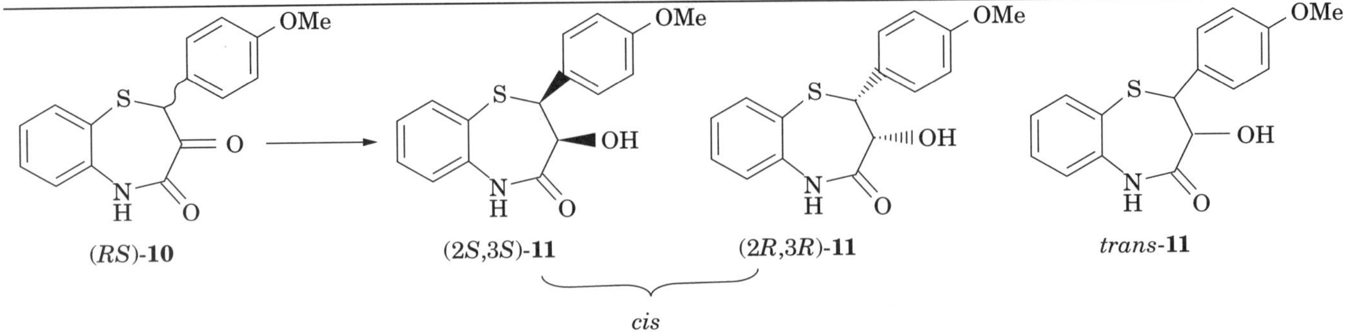

Microorganism	Product 11: Yield (%)	*cis*/*trans*	Absolute configuration	% e.e.
Brevibacterium ketoglutamicum ATCC 15588	33	98/2	2*S*,3*S*	96
Candida melinii IFO 0747	50	97/3	2*S*,3*S*	99
Candida saitoana IFO 0380	49	99/1	2*S*,3*S*	>99
Debaryomyces polymorphus IFO 1166	51	99/1	2*S*,3*S*	>99
Pichia capsulata IFO 0721	56	99/1	2*S*,3*S*	>99
Pichia pinus IFO 1793	48	>99/1	2*S*,3*S*	>99
Pseudomonas putida ATCC 17484	27	98/2	2*S*,3*S*	97
Rhizopus arrhizus IFO 5780	22	>99/1	2*R*,3*R*	99
Rhizopus japonicus IFO 4758	11	>99/1	2*R*,3*R*	>99
Rhodosporidium toruloides IFO 1638	36	99/1	2*S*,3*S*	99
Saccharomyces cerevisiae (baker's yeast)	67	95/5	2*S*,3*S*	>99
Saccharomyces cerevisiae IFO 1346	54	98/2	2*S*,3*S*	>99
Streptomyces lavendulae IFO 3145	29	98/2	2*S*,3*S*	>99

The microorganisms were inoculated into test tubes (15 mm diameter × 12 cm) containing 3 mL of medium A or B (same as those described in Table 1) and cultured at 30 °C for 24 h with continuous shaking (300 strokes/min). The cells were collected by centrifugation, and the reaction mixture contained 0.6 mmol McIlvaine buffer (pH 5.0), 0.83 mmol glucose, and cells (150 mg baker's yeast) in a total volume of 3.0 mL. Six microliters of DMF solution containing 0.01 mmol (*RS*)-**10** was added. The reaction was carried out at 30 °C for 24 h with continuous shaking (300 strokes/min); the reaction products were extracted from the reaction mixture with 4 mL of ethyl acetate. Yield and e.e. of **11** were analyzed by HPLCs equipped with ChemcoPak™ column (Develosil ODS-7) (Chemco Scientific Co., Ltd.) and Chiralcel OD™ column (Daicel Chemical Industries, Ltd.), respectively.

Source: Ref. 19.

Microbial reduction of (*RS*)-**8** has been studied for the production of an optically active **9** (17). The screening of microorganisms was carried out in a buffer adjusted to pH 5, which contained 5% glucose, substrate **8** and product **9** are unstable above pH 6. Table 1 summarizes the microorganisms that have the ability to reduce (*RS*)-**8**.

The yield and e.e. of **9** were calculated with HPLC, where the yield stands for the total yield of four optically active isomers of **9**, and the e.e. of **9** was determined for each diastereomer of *anti*-**9** and *syn*-**9**. The terms *syn* and *anti* were defined according to Masamune et al. (18). The microorganisms that produced alcohol **9** with the 3*R* configuration, which can be used for the synthesis of the enantiomer of diltiazem with high enantioselectivity, are in the genera *Absidia, Cryptococcus*, and *Mucor*. *Mucor ambiguus* IFO 6742 (IFO: Institute for Fermentation, Osaka, Japan) gives only (2*S*,3*R*)-**9** in 58% yield (>99% e.e.) among the four isomers, through keto–enol tautomerization of substrate (*RS*)-**8**. With reaction conditions of pH 4 and 30 °C, the yield of (2*S*,3*R*)-**9** increases up to 68% (isolated yield), and the e.e. of (2*S*,3*R*)-**9** is over 99%.

Furthermore, screening of microorganisms for the ability to conduct asymmetric reduction of (*RS*)-**8**, revealed that *Trichoderma viride* OUT 4642 (OUT: Department of Biotechnology, Faculty of Engineering, Osaka University, Japan) gives **9** with the 3*S* configuration, which is a key diltiazem intermediate with high enantioselectivity. The production ratio of *syn*-**9** to *anti*-**9** was 2:1, and the e.e. of the (2*R*,3*S*)-**9** and (2*S*,3*S*)-**9** is 96% and 87%, respectively. However, the total yield of (2*R*,3*S*)-**9** and (2*S*,3*S*)-**9** is low (7%) due to the formation of the by-products methyl 3-(4-methoxyphenyl)-3-oxopropionate (**13**) and 2-chloro-1-(4-methoxyphenyl)ethanone (**14**) (Table 1).

When whole cells are used as a catalyst, there are often unwanted side reactions. In this case, biotransformation by the extracted free enzyme is thought to be more effective than whole-cell biotransformation. However, if microbial reduction is performed with whole cell, an industrial device for cofactor recycling must be introduced. Desiring microbial reduction with high enzymatic activity and high enantioselectivity, two whole-cell reactions were investigated.

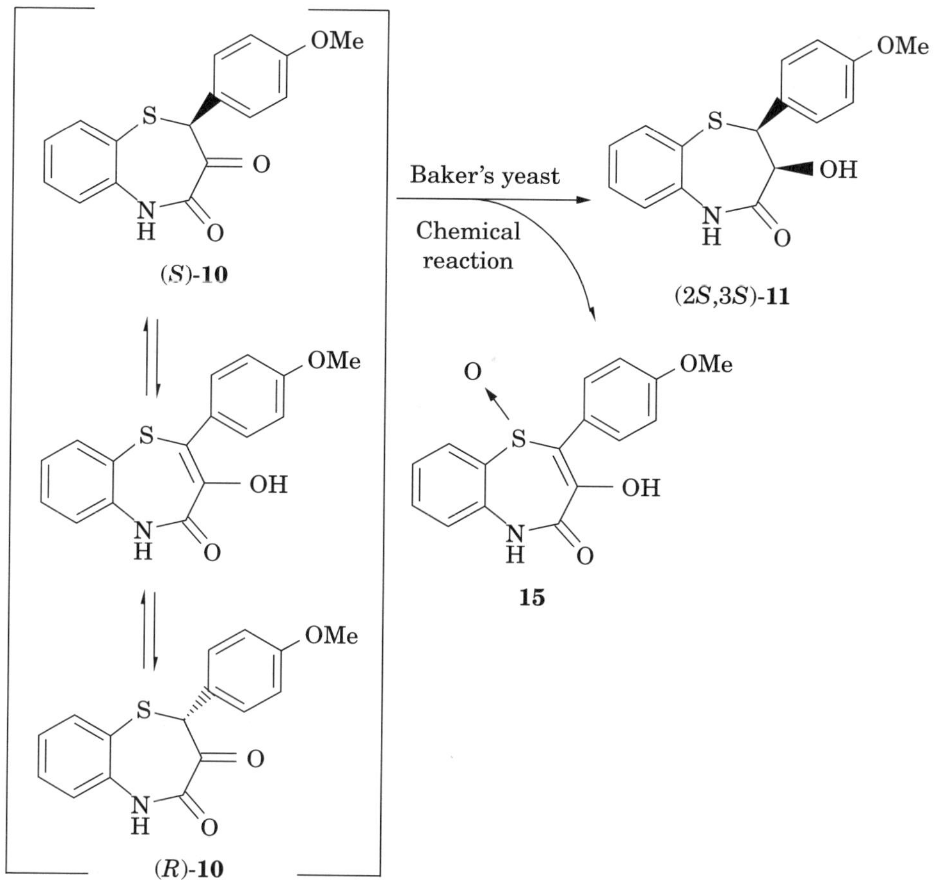

Figure 2. Reduction of (*RS*)-**10** by baker's yeast. *Source*: Ref. 19.

Microbial Reduction of (*RS*)-2-(4-Methoxyphenyl)-1,5-benzothiazepin-3,4(2*H*,5*H*)-dione [(*RS*)-10]

The microbial asymmetric reduction of (*RS*)-2-(4-methoxyphenyl)-1,5-benzothiazepin-3,4(2*H*,5*H*)-dione [(*RS*)-**10**] is a useful method for producing another optically active key intermediate, (2*S*,3*S*)-2,3-dihydro-3-hydroxy-2-(4-methoxyphenyl)-1,5-benzothiazepin-4(5*H*)-one [(2*S*,3*S*)-**11**] (19). Substrate (*RS*)-**10** is easily obtained from (±)-MPGM through three steps of chemical ring closure with aminothiophenol, chemical oxidation, and chemical hydrolysis (Fig. 1) (20).

Table 2 shows microorganisms that have the ability for the asymmetric reduction of (*RS*)-**10** (19). The e.e. of *cis*-**11** was determined by using HPLC.

Microorganisms that produce the key intermediate (2*S*,3*S*)-**11** are widely distributed in the genera *Brevibacterium, Candida, Debaryomyces, Pichia, Pseudomonas, Rhodosporidium, Saccharomyces*, and *Streptomyces*. On the other hand, only *Rhizopus* reduces (*RS*)-**10** to (2*R*,3*R*)-**11**. Among these, baker's yeast (*Saccharomyces cerevisiae*) has the highest ability for the formation of (2*S*,3*S*)-**11**, which is useful for the synthesis of diltiazem. Baker's yeast itself has been widely used as a reagent in organic synthesis (21,22) and the large-scale preparation of, for example, ethyl (*S*)-3-hydroxybutanoate using baker's yeast has already been reported (15).

In the microbial asymmetric reduction of (*RS*)-**10**, baker's yeast strictly recognizes (*S*)-**10**, and (2*S*,3*S*)-**11** forms selectively through keto–enol tautomerization of substrate (Fig. 2). A very small amount of 3-hydroxy-2-(4-methoxyphenyl)-1,5-benzothiazepin-4(5*H*)-one *S*-oxide (**15**) is formed as a by-product of the chemical reaction. In the baker's yeast reduction system, (2*S*,3*S*)-**11** is accumulated with 94% conversion under the following the low substrate concentration conditions: substrate concentration 1 g/L at pH 7 and 42 °C for 24 h. Sometimes, it is difficult to separate a product from the enzymatic reaction mixture. In this case, separation is carried out with the ease as follows: At the end of the reaction, acetone was added as a volume of three times the reaction mixture. The cells in acetone were dried to powder form and easily removed by filtration through Celite™. Then, acetone in the filtrate was evaporated under reduced pressure. The deposited precipitate in the aqueous layer was recovered by filtration and washed with water in the amount of four times the volume of the reaction mixture. The precipitate was redissolved in acetone, and then the solution was dehydrated by the addition of anhydrous magnesium sulfate. After filtration, crude crystals were obtained by evaporation of the organic layer to dryness under reduced pressure. The residue was recrystallized from methanol (10 mL methanol/1 g residue). The isolated yield was 85% based on (*RS*)-**10**, and the e.e. was over 99%.

Microbial Conversion of 3-Acetoxy-2-(4-methoxyphenyl)-1,5-benzothiazepin-4(5*H*)-one (12)

An optically active key intermediate, (2*S*,3*S*)-**11**, can be also obtained by the microbial conversion of **12** (Fig. 1) (23).

For this type biotransformation, the microbial conversion of enol ester using *Pichia farinosa* has been reported (24). The microorganisms that have the ability to catalyze the conversion of **12** into (2*S*,3*S*)-**11**, with excellent enantioselectivity, are summarized in Table 3.

The conversion yield of **12** to **11**, and the e.e. of **11** are calculated on the basis of HPLC analysis, and the isolated yield is based on **12**. No accumulation of **15** (Fig. 2) was observed in this reaction. By employing this microbial conversion, diltiazem can be synthesized in only six steps with a high yield (Fig. 1).

Figure 3 shows the time courses of the microbial conversion of (*RS*)-**10** and **12** using *Bacillus sphaericus* FERM P-6746 (FERM: Fermentation Research Institute, Agency of Industrial Science and Technology, Japan). In both cases, the products are optically active (2*S*,3*S*)-**11** (>99% e.e.) and isolated easily by the same procedure already described. With (*RS*)-**10** as the substrate, the reduction proceeds directly from the beginning of the reaction, whereas with **12**, an obvious lag phase is observed at the beginning of the reaction. Later, however, the yield of (2*S*,3*S*)-**11** from **12** increases to the same level as that from (*RS*)-**10**. Therefore, in the reaction pathway of microbial conversion of **12** to (2*S*,3*S*)-**11**, it is thought that (*RS*)-**10** is an intermediate, that hydrolysis and reduction proceed in turn, and that the hydrolysis of **12** is a rate-determining step. Furthermore, it is also confirmed that the hydrolytic enzyme is induced by **12** and that asymmetry is introduced during the (*RS*)-**10** reduction step. Because all the other microorganisms shown in Table 3 also reduce (*RS*)-**10** to (2*S*,3*S*)-**11** (79–88% conversion, >99% e.e.), the microbial conversion of **12** into (2*S*,3*S*)-**11** is thought to proceed through a similar pathway in these microorganisms.

Another approach to the synthesis of (2*S*,3*S*)-**11** from **12** in the presence of the hydrolytic enzyme from *B. sphaericus* and baker's yeast gives (2*S*,3*S*)-**11**, with a conversion of 65% (99% e.e.).

Table 3. Microorganisms with the Ability to Produce (2*S*,3*S*)-11 from 12

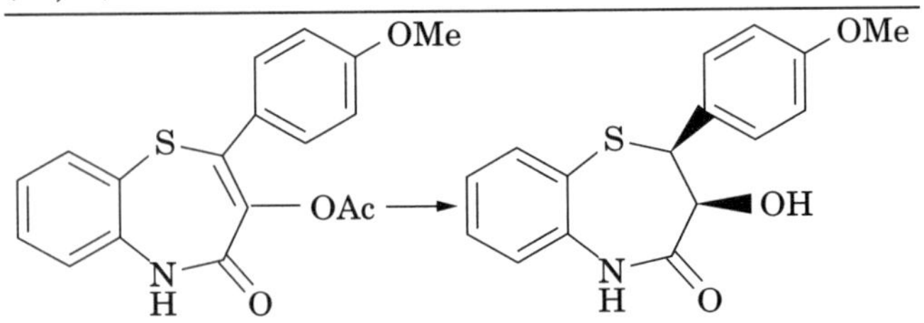

12 (2*S*,3*S*)-**11**

Microorganisms	Product (2*S*,3*S*)-**11** Conversion yield (%)	% e.e.	Isolated yield (%)
Alcaligenes faecalis FERM P-6745	76	99	72
Arthrobacter paraffineus ATCC 21219	76	>99	66
Bacillus sphaericus FERM P-6746	77	>99	74
Rhodococcus sp. ATCC 15592	67	>99	65

The microorganism were cultivated in 500-mL shaking flasks with 100 mL of medium, which comprised 1% glucose, 0.7% Polypepton™ (Nikon Seikayu Co. Ltd.), 0.5% K_2HPO_4, and 0.5% yeast extract (pH 7.2), at 30 °C for 24 h with continuous shaking (140 strokes/min). To each shaking flask was added 1 mL of DMF solution containing 100 mg **12** and 10 mL 1 M potassium phosphate buffer (pH 6.5). The mixtures were shaken at 30 °C (140 strokes/min). After 40 h, the conversion yield and e.e. of (2*S*,3*S*)-**11** were checked by HPLC equipped with Chiralcel OF™ column (Daicel Chemical Industries, Ltd.). The mixtures were shaken for an additional 8 h. According to the method already described, (2*S*,3*S*)-**11** was isolated.

Source: Ref. 23.

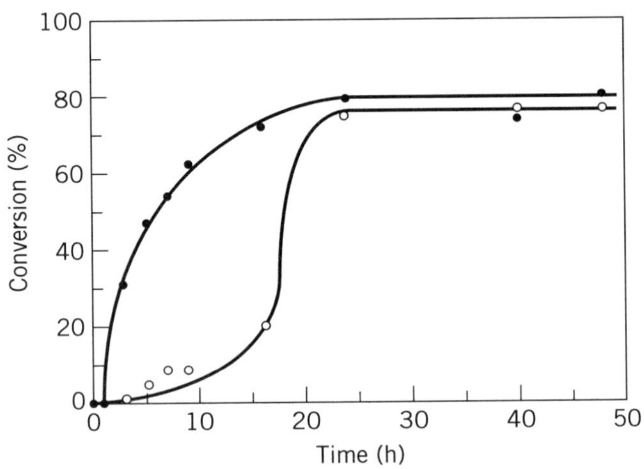

Figure 3. Time courses of (2*S*,3*S*)-**11** production from **12** (○) and (*RS*)-**10** (●) by *B. sphaericus*. *Source*: Ref. 23.

INVESTIGATION FOR INDUSTRIAL APPLICATION

In general, industrial application of enzymatic reduction is difficult due to the low concentration of the substrate in the reaction medium. Among the three reactions catalyzed by microorganisms mentioned earlier, baker's yeast–mediated reduction of (*RS*)-**10** is the most promising, for industrial application if the substrate concentration is increased, because of the high optical purity of the product, easy product isolation, and a simple reaction system (not a plural enzyme system) (25).

Because the substrate (*RS*)-**10**, obtained as crystals by recrystallization from organic solvent, is almost insoluble in an aqueous medium, the reaction was carried out by adding an *N*,*N*-dimethylformamide (DMF) solution of (*RS*)-**10**, and by using ethanol as an energy source under aerobic conditions. Although it is widely accepted that sugar is essential as an energy source, the use of ethanol as the energy source has also been considered because it is clean and efficient compared with sugar (15,26–30).

In the reduction of (*RS*)-**10**, the reduction activity of baker's yeast decreased when DMF was added at a concentration of above 10% (v/v) in order to increase the substrate concentration. By-product, *S*-oxide of the substrate (**15**), also appeared at these high DMF concentrations. Therefore, the large-scale production of (2*S*,3*S*)-**11** by a fed-batch operation on a laboratory scale was examined. A suspension of 11.2 g of baker's yeast in 50 mL of 200 mM ethanol in a 300-mL Erlenmeyer flask was shaken at 30

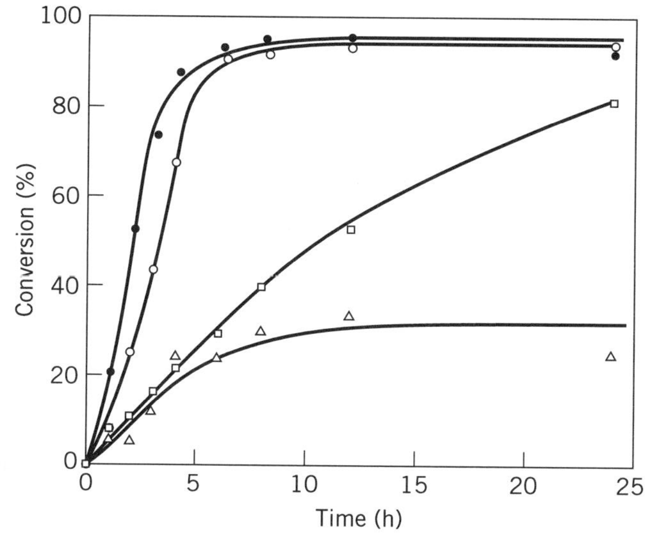

Figure 4. Conversion of (*RS*)-**10** to (2*S*,3*S*)-**11** by baker's yeast. *Source*: Ref. 25. A suspension of baker's yeast (11.2 g) and substrate (*RS*)-**10** in 200 mM ethanol (50 mL) was aerobically shaken at 30 °C at 80 rpm. Substrates using the reaction were as follows: ●, DMF solution (600 μL) containing 100 mg of (*RS*)-**10**; ○, 100 mg of the dry aggregate; □, 100 mg of the fine powder (average grain size of 2.3 μm); △, 100 mg of the crystalline form (average grain size of 21.3 μm). An aggregate of (*RS*)-**10** was prepared by the addition of 6 mL of 0.56 M (*RS*)-**10** in DMF [1.0 g of (*RS*)-**10**] to 100 mL of water for 5 min followed by stirring for 1 h at room temperature. The aggregate was collected by centrifugation at 3,000 rpm for 3 min and washed with water.

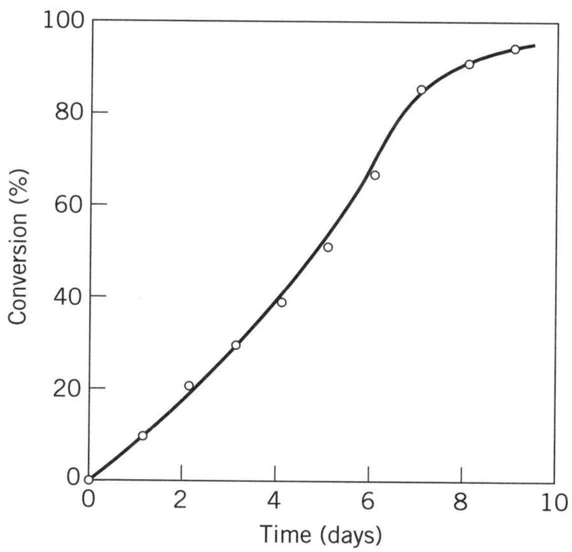

Figure 5. Changes in conversion ratio of (*RS*)-**10** to (2*S*,3*S*)-**11** in a high-concentration reaction. *Source*: Ref. 25.

°C, and 6 mL of DMF solution containing 1.0 g of (*RS*)-**10** (0.56 M) was added in 0.15 mL aliquots every 1 h. Since ethanol was consumed within 12 h, 0.575 mL of ethanol was added three times. After 42 h, the reaction was stopped and the product was isolated and purified. The yield of (2*S*,3*S*)-**11** was 0.83 g (83%), and e.e. was over 99%. By fed-batch operation, substrate concentration increased to 20.0 g/L.

In order to increase the substrate concentration further, as a yeast-mediated reduction would be inhibited by a high concentration of DMF, the reactions were also carried out in an aqueous solution without DMF. As shown in Figure 4, however, in the case of crystalline (*RS*)-**10** (an average grain size of 21.3 μm) in an aqueous solution directly, reduction did not proceed successively. Increasing the surface area of the crystalline (*RS*)-**10** by grinding (an average grain size of 2.3 μm) was not very effective in increasing the reduction rate. On the other hand, when a DMF solution of (*RS*)-**10** was added to water with stirring, a dispersed colloid of (*RS*)-**10**, an emulsion was formed, and this emulsion gradually changed to a suspension, producing small precipitates. These precipitates of (*RS*)-**10**, called aggregates, were easily collected by centrifugation (an average grain size of 13.0 μm) and were reduced with a rate comparable to that of a DMF solution of (*RS*)-**10** under low concentration of substrate, as shown in Figure 4.

The aggregate was soluble in an aqueous medium at about 20 mg/L, and DMF did not need to be added to the reaction medium. By using the aggregate as the substrate, substrate concentration was increased up to an industrial production level (100 g/L, yield 80%) by means of the following procedure: A suspension of 11.2 g of baker's yeast and the aggregate, prepared from 5.0 g of the crystalline (*RS*)-**10**, in 50 mL of 200 mM ethanol solution was shaken at 30 °C for 9 days. Ethanol (0.46 g) was added every 12 h. The reduction of (*RS*)-**10** proceeded successively as shown in Figure 5. After processing as described earlier, over 99% e.e. of (2*S*,3*S*)-**11** was obtained with 80% yield.

The substrate concentration was improved by applying this procedure. The disadvantages of *(1)* a relatively low concentration of substrate, *(2)* a tedious procedure for isolation and purification of the product, and *(3)* a relatively low optical purity of the product in the microbial reduction were resolved. This method of using baker's yeast, which is commercially available and cheap, is expected to be developed for industrial application.

BIBLIOGRAPHY

1. K. Abe, H. Inoue, and T. Nagao, *Yakugaku Zasshi* **108**, 716–732 (1988).
2. H. Inoue, S. Takeo, M. Kawazu, and H. Kugita, *Yakugaku Zasshi* **93**, 729–732 (1973).
3. M. Senuma, M. Shibazaki, S. Nishimoto, K. Shibata, K. Okamura, and T. Date, *Chem. Pharm. Bull.* **37**, 3204–3208 (1989).
4. E.N. Jacobsen, L. Deng, Y. Furukawa, and L.E. Martínez, *Tetrahedron* **50**, 4323–4334 (1994).
5. K.G. Watson, Y.M. Fung, M. Gredley, G.J. Bird, W.R. Jackson, H. Gountzos, and B.R. Matthews, *J. Chem. Soc. Chem. Commun.* 1018–1019 (1990).
6. B.R. Matthews, H. Gountzos, W.R. Jackson, and K.G. Watson, *Tetrahedron Lett.* **30**, 5157–5158 (1989).
7. A. Schwartz, P.B. Madan, E. Mohacsi, J.P. O'Brien, L.J. Todaro, and D.L. Coffen, *J. Org. Chem.* **57**, 851–856 (1992).
8. O. Miyata, T. Shinada, I. Ninomiya, and T. Naito, *Tetrahedron Lett.* **32**, 3519–3522 (1991).

9. K. Matsuki, M. Sobukawa, A. Kawai, H. Inoue, and M. Takeda, *Chem. Pharm. Bull.* **41**, 643–648 (1993).
10. H. Akita, Y. Enoki, H. Yamada, and T. Oishi, *Chem. Pharm. Bull.* **37**, 2876–2878 (1989).
11. H. Akita, I. Umezawa, H. Matsukura, and T. Oishi, *Chem. Pharm. Bull.* **39**, 1632–1633 (1991).
12. H. Matsumae, M. Furui, and T. Shibatani, *J. Ferment. Bioeng.* **75**, 93–98 (1993).
13. H. Matsumae, M. Furui, T. Shibatani, and T. Tosa, *J. Ferment. Bioeng.* **78**, 59–63 (1994).
14. C.J. Shi and C.-S. Chen, *Angew. Chem. Int. Ed. Engl.* **23**, 570–578 (1984).
15. T. Kometani, H. Yoshii, E. Kitatsuji, H. Nishimura, and R. Matsuno, *J. Ferment. Bioeng.* **76**, 33–37 (1993).
16. T. Mukaiyama and M. Murakami, *Chem. Lett.* 1129–1132 (1981).
17. T. Nishida, H. Matsumae, I. Machida, and T. Shibatani, *Biocatal. and Biotransf.* **12**, 205–214 (1995).
18. S. Masamune, S.A. Ali, D.L. Snitman, and D.S. Garvey, *Angew. Chem. Int. Ed. Engl.* **19**, 557–558 (1980).
19. H. Matsumae, H. Douno, S. Yamada, T. Nishida, Y. Ozaki, T. Shibatani, and T. Tosa, *J. Ferment. Bioeng.* **79**, 28–32 (1995).
20. S. Yamada, Y. Mori, K. Morimatsu, Y. Ishizu, Y. Ozaki, R. Yoshioka, T. Nakatani, and H. Seko, *J. Org. Chem.* **61**, 8586–8590 (1996).
21. S. Servi, *Synthesis* 1–25 (1990).
22. R. Csuk and B.I. Glänzer, *Chem. Rev.* **91**, 49–97 (1991).
23. T. Nishida, I. Machida, H. Matsumae, S. Yamada, Y. Ozaki, and T. Shibatani, *J. Ferment. Bioeng.* **81**, 534–537 (1996).
24. T. Sugai, D. Sakuma, N. Kobayashi, and H. Ohta, *Tetrahedron* **47**, 7237–7244 (1991).
25. T. Kometani, Y. Sakai, H. Matsumae, T. Shibatani, and R. Matsuno, *J. Ferment. Bioeng.* **84**, 195–199 (1997).
26. T. Kometani, E. Kitatsuji, and R. Matsuno, *Chem. Lett.* 1465–1466 (1989).
27. T. Kometani, E. Kitatsuji, and R. Matsuno, *J. Ferment. Bioeng.* **71**, 197–199 (1991).
28. T. Kometani, H. Yoshii, Y. Takeuchi, and R. Matsuno, *J. Ferment. Bioeng.* **76**, 414–415 (1993).
29. T. Kometani, Y. Morita, H. Furui, H. Yoshii, and R. Matsuno, *J. Ferment. Bioeng.* **77**, 13–16 (1994).
30. T. Kometani, Y. Morita, Y. Kiyama, H. Yoshii, and R. Matsuno, *J. Ferment. Bioeng.* **80**, 208–210 (1995).

DIMENSIONAL ANALYSIS, SCALE-UP

MARKO ZLOKARNIK
Graz, Austria

KEY WORDS

Complete and partial similarity
Dimensional analysis
Dimensionless representation
Dimensions and physical quantities
Intermediate quantities
Non-Newtonian liquids
Relevance list
Scale-up
Theory of models
Viscolastic liquids

OUTLINE

INTRODUCTION

A chemical engineer is generally concerned with the industrial implementation of processes in which chemical or microbiological conversion of material takes place in conjunction with the transfer of mass, heat, and momentum. These processes are scale dependent; that is, they behave differently on a small scale (in laboratories or pilot plants) and a large scale (in production). They include heterogeneous chemical reactions and most unit operations. Understandably, chemical engineers have always wanted to find ways of simulating these processes in models to gain insights that will assist them in designing new industrial plants. Occasionally, they are faced with the same problem for another reason: an industrial facility already exists but will not function properly, if at all, and suitable measures have to be carried out to discover the cause of the difficulties and provide a solution.

Irrespective of whether the model involved represents a "scale-up" or a "scale-down," certain important questions always apply:

1. How small can the model be? Is one model sufficient, or should tests be carried out in models of different sizes?
2. When must or when can physical properties differ? When must the measurements be carried out on the model with the original system of materials?
3. Which rules govern the adaptation of the process parameters in the model measurements to those of the full-scale plant?
4. Is it possible to achieve complete similarity between the processes in the model and those in its full-scale counterpart? If not, how should one proceed?

These questions touch on the fundamentals of the theory of models, which are based on dimensional analysis. Although they have been used in the field of fluid dynamics and heat transfer for more than a century (cars, aircraft, vessels, and heat exchangers were scaled up according to these principles), these methods have gained only modest acceptance in chemical engineering. University graduates are usually not skilled enough to deal with such problems at all. On the other hand, there is no motivation for this type of research at universities, which generally are not confronted with scale-up tasks and are not equipped with the necessary apparatus at the bench scale. This gives the totally wrong impression that the methods are, at most, of marginal importance in practical chemical engineering. It is most unfortunate that these methods are not taught and dealt with in greater depth.

DIMENSIONAL ANALYSIS

The Fundamental Principle

Dimensional analysis is based on the recognition that a mathematical formulation of a physicotechnological problem can be of general validity only when the process equation is dimensionally homogeneous.

What Is a Dimension?

A dimension is a purely qualitative description of a perception of a physical entity or a natural appearance. A length can be experienced as a height, a depth, a breadth. A mass presents itself as a light or heavy body, time as a short moment or a long period. The dimension of a length is L, the dimension of a mass is M, and so on.

What Is a Physical Quantity?

Unlike a dimension, a physical quantity represents a quantitative description of a physical quality (e.g., a mass of 5 kg). It consists of a measuring unit and a numerical value. The measuring unit of length can be a meter, a foot, a cubit, a yardstick, a nautical mile, a light-year, and so on. The measuring unit of temperature can be °C (centigrade), °R, °F, or K (kelvin). (It is therefore necessary to establish the measuring units in an appropriate measuring system.)

Basic and Derived Quantities and Dimensional Constants

We distinguish between basic and secondary quantities, the latter being often referred to as derived quantities. The basic quantities are based on standards and are quantified by comparison with them. The secondary units are derived from the primary ones according to physical laws (e.g., velocity = length/time). The borderline separating basic and derived quantities is largely arbitrary: 40 years ago a measuring system was used in which force was a primary dimension instead of mass!

All secondary units must be coherent with the basic units (Table 1); for example, the measuring unit of velocity must not be miles per hour or kilometers per hour, but meters per second. A secondary unit that has been established by a physical law may sometimes contradict another one. For example, according to Newton's second law of motion, the force F is expressed as a product of mass m and acceleration a: $F = ma$, having the measuring unit of kg m/s^2 ≡ N. According to Newton's law of gravitation, the force is defined by $F \propto m_1 m_2/r^2$, thus leading to a completely different measuring unit (kg^2/m^2). To remedy this, the gravitational constant G—a dimensional constant—had to be introduced to ensure the dimensional homogeneity of this equation: $F = G\, m_1 m_2/r^2$. Another example entails the universal gas constant R, the introduction of which ensures that in the perfect gas equation of state $pV = nRT$ the secondary unit for work $W = pV[M\, L^2/T^2]$ is not offended.

Another class of derived quantities is represented by the coefficients in diverse physical equations (e.g., transfer equations). They are established by the respective equations and determined via measurement of their constituents, (heat transfer coefficients etc.; see examples 7–10).

Table 1. Base Quantities: Their Dimensions and SI Units

Basis quantity	Basic dimension	Basic unit	
Length	L	m	meter
Mass	M	kg	kilogram
Time	T	s	second
Thermodynamic temperature	Θ	K	kelvin
Amount of substance	N	mol	mole
Electric current	I	A	ampere
Luminous intensity	I_v	cd	candela

Dimensional Systems

A dimensional system consists of all the primary and secondary dimensions and corresponding measuring units. The currently used International System of Dimensions (*Système International d'Unités*; SI) is based on seven basic dimensions. They are presented in Table 1, together with their corresponding basic units. For some of them a few explanatory remarks may be necessary.

Temperature expresses the thermal level of a system, not its energetic contents. (A fivefold mass of matter has a fivefold thermal energy at the same temperature!) The thermal energy of a system can indeed be converted into mechanical energy (base unit joule). Moles are amount of matter and must not be confused with the quantity of mass. The molecules react as individual entities regardless of their mass: one mole of hydrogen (2 g/mol) reacts with one mole of chlorine (71 g/mol) to produce 2 moles of hydrochloric acid, HCl.

Table 2 shows the most important secondary dimensions. Table 3 refers to some very frequently used secondary units that are named after famous researchers.

Dimensional Homogeneity of a Physical Content

The aim of dimensional analysis is to check whether the physical content of the case under examination can be formulated in a dimensionally homogeneous manner. The procedure necessary to accomplish this consists of two parts:

1. All physical parameters necessary to describe the problem are listed. This "relevance list" of the problem consists of the quantity in question and all the parameters that influence it. In each case only *one* target quantity must be considered; it is the only dependent variable. On the other hand, all the influencing parameters must be primarily independent of each other.
2. The dimensional homogeneity of the physical content is checked by transferring it in a dimensionless form. (*Note*: A physical content that can be transformed in dimensionless expressions is dimensionally homogeneous.)

The information given thus far can be made clear by a famous example (1).

Table 2. Often-Used Physical Quantities and Their Dimensions According to Currently Used SI Formulas in Mechanical and Thermal Problems

Physical quantity	Dimension
Angular velocity	$1/T$
Shear rate, frequency	
Mass transfer coefficient, $K_L a$	
Velocity	L/T
Acceleration	L/T^2
Kinematic viscosity	L^2/T
Diffusion coefficient	
Thermal diffusivity	
Density	M/L^3
Surface tension	M/T^2
Dynamic viscosity	$M/(LT)$
Momentum	$M\,L/T$
Force	$M\,L/T^2$
Pressure, stress	$M/(LT^2)$
Angular momentum	$M\,L^2/T$
Energy, work, torque	$M\,L^2/T^2$
Power	$M\,L^2/T^3$
Heat capacity	$L^2/T^2\Theta$
Thermal conductivity	$M\,L/T^3\Theta$
Heat transfer coefficient	$M/T^3\Theta$

Table 3. Important Secondary Measuring Units in Mechanics, Named After Famous Researchers

Secondary quantity	Dimension	Measuring unit	Eponymous researcher
Force	$M\,L/T^2$	kg m/s^2 ≡ N	Newton
Pressure	$M\,L/T^2$	kg (ms^2) ≡ Pa	Pascal
Energy	$M\,L^2/T^2$	kg m^2/s^2 ≡ J	Joule
Power	$M\,L^2/T^3$	kg m^2/s^3 ≡ W	Watt

EXAMPLE 1

What determines the period of oscillation of a pendulum?

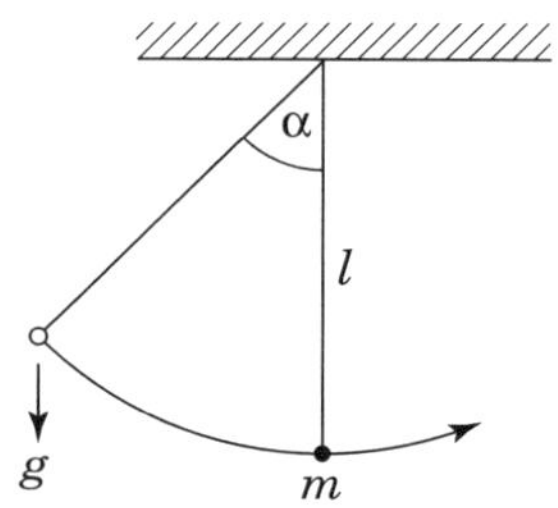

Physical quantity	Symbol	Dimension
Period of oscillation	t	T
Length of pendulum	l	L
Mass of pendulum	m	M
Gravitational acceleration	g	L/T^2
Amplitude (angle)	α	—

We first sketch a pendulum and write down all the quantities that could be involved in this question. It may be assumed that the period of oscillation of a pendulum depends on the length and mass of the pendulum, the gravitational acceleration, and the amplitude of swing.

Our aim is to express t as a function of l, m, g, and α:

$$t = f(l, m, g, \alpha)$$

Can this dependency be dimensionally homogeneous? No! The first thing that now becomes clear is that the basic unit for mass occurs only in the mass, m, itself. Changing this basic unit (e.g., from kilograms to pounds) would change the numerical value of the function. This is unacceptable. Either our list should have included a further variable containing M, or mass is not a relevant variable. If we assume the latter, the relationship is reduced to:

$$t = f(l, g, \alpha)$$

Both l and g incorporate the basic unit of length. When combined as a ratio (l/g), they become dimensionless with regard to L and thus independent of changes in the basic unit of length:

$$t = f(l/g, \alpha)$$

Because the angle α has no dimension, we are left with the dimension T on the left-hand side of the equation, and T^2 on the right. To remedy this, we will have to write $\sqrt{l/g}$. This expression will keep its dimension T only if it remains unchanged, so we have to take it out of the function f, and we obtain the following:

$$t = \sqrt{l/g}\, f(\alpha) \qquad \text{or} \qquad t\sqrt{g/l} = f(\alpha)$$

The dependency between four dimensional quantities, containing two basic dimensions (L and T) in their dimensions, is reduced to $4 - 2 = 2$ parametric relationship between dimensionless expressions ("numbers").

This equation is the only statement that dimensional analysis can offer in this case. It is not capable of producing information on the form of f. The integration of Newton's equation of motion for small amplitudes leads to $f = 2\pi$ and is independent of α. The relationship can now be expressed as

$$t\sqrt{g/l} = 2\pi$$

The elegant solution of this first example should not tempt the reader to believe that dimensional analysis can be used to solve every problem at hand. To treat this example by dimensional analysis, the law of free fall had to be known. This knowledge was gained empirically by G. Galilei in 1604 through his experiments with the inclined plane. Bridgman's (1, p. 12) comment on this situation is particularly appropriate:

> "The problem cannot be solved by the philosopher in his armchair, but the knowledge involved was gathered only by someone at some time soiling his hands with direct contact."

This transparent and easy example clearly shows how dimensional analysis deals with specific problems and what conclusions it allows. It should now be easier to understand the sarcastic comment with which Lord Rayleigh began his short essay on "The Principle of Similitude" (2):

> I have often been impressed by the scanty attention paid even by original workers in physics to the great principle of similitude. It happens not infrequently that results in the form of "laws" are put forward as novelties on the basis of elaborate experiments, which might have been predicted a priori after a few minutes' consideration.

From example 1 we also learn that a transformation of a physical dependency from a dimensional into a dimensionless form is automatically accompanied by an essential compression of the statement: the set of the dimensionless numbers is smaller than the set of the quantities contained in them, but it describes the problem equally comprehensively. In the example 1, the dependency between four dimensional parameters is reduced to a dependency between only two dimensionless numbers. This is the proof of the so-called pi theorem (pi after Π, the sign used for products), which states the following: Every physical relationship between n physical quantities can be reduced to a relationship between $m = n - r$ mutually independent dimensionless groups, whereby r stands for the rank of the dimensional matrix, made up of the physical quantities in question and generally equal to the number of the basic quantities contained in them. (*Note*: The pi theorem is often associated with E. Buckingham, who introduced this term in 1914, but the proof of it had been accomplished in 1911 in course of a mathematical analysis of partial differential equations by A. Federmann already [3, p. 7].)

THE DETERMINATION OF A PI SET BY MATRIX CALCULATION

The Establishment of a Relevance List for a Problem

As a rule, more than two dimensionless numbers are necessary to describe a physicotechnological problem, and therefore they cannot be derived by the method just described. In this case the easy and transparent matrix calculation introduced by J. Pawlowski (4) is increasingly used. Example 2, which demonstrates this calculation, treats a problem that is important in industrial chemistry and biotechnology because the mixing vessel belongs to the most important production equipment.

EXAMPLE 2

Which is the pi set for the mixing power of a stirrer?

In this case, too, we first sketch the apparatus. Then we facilitate our work by proceeding systematically toward our target quantity, the mixing power P, which depends on the following parameters:

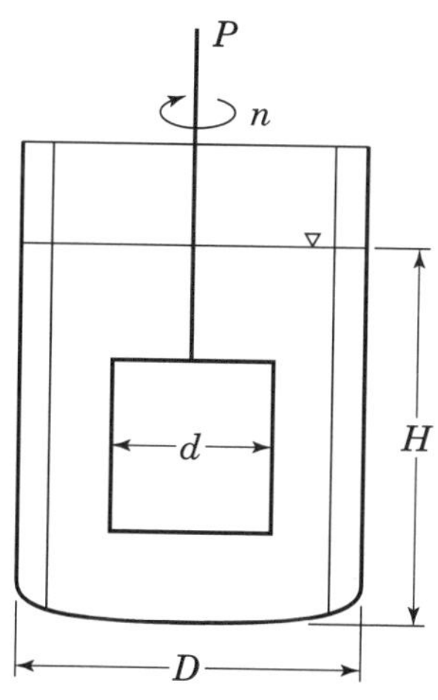

Geometrical	Stirrer diameter	d
Physical properties	Fluid density	ρ
	Kinematic viscosity	ν
Process-related	Stirrer speed	n

We obtain the following relevance list:

$$\{P;\ d;\ \rho,\ \nu;\ n\} \tag{1}$$

We interrupt the procedure by asking an important question concerning the determination of the characteristic geometric parameter.

Determination of the Characteristic Geometric Parameter

Obviously, we could name all the geometric parameters indicated in the sketch. They were all the geometric parameters of the stirrer and of the vessel, especially its diameter D and the liquid height H. In case of complex geometry, such a procedure would deflect from the problem. It is therefore advisable to introduce only one (or a few) characteristic geometric parameters, knowing that all the others can be transformed into dimensionless geometric numbers by division with this one of a few of them.

In the case of example 2, we introduced the stirrer diameter as the characteristic geometric parameter. This is reasonable. One could only think about how the mixing power would react to an increase of the vessel diameter D: it is obvious that above a certain size, D would have no influence, but a small change of the stirrer diameter d would always have an impact.

We now set forth a procedure for solving example 2.

Constructing and Solving the Dimensional Matrix

In transforming the relevance list, equation 1, of the five physical quantities into a dimensional matrix, we can minimize the calculations required by recalling the following points.

1. The dimensional matrix consists of a square core matrix and a residual matrix.
2. The rows of the matrix are formed of basic dimensions, contained in the dimensions of the quantities, and they will determine the rank r of the matrix. The columns of the matrix are presented by the physical quantities or parameters.
3. Quantities of the square core matrix may appear later on in all the dimensionless numbers as "fillers," whereas each element of the residual matrix will appear only in one dimensionless number. For this reason the residual matrix should be loaded with essential variables such as the target quantity and the most important physical properties and process-related parameters.
4. By the (extremely easy) matrix calculation (linear transformations), the core matrix is transformed into a matrix of unity: the main diagonal consists only of ones, and the remaining elements are all zero. One should therefore arrange the quantities in the core matrix in a way that will facilitate this procedure.
5. After the matrix of unity has been generated, the dimensionless numbers are created as follows: each element of the residual matrix forms the numerator of a fraction, while the denominator consists of the fillers from the matrix of unity with the exponents indicated in the residual matrix.

Let us return to example 2. The dimensional matrix reads as follows:

	Core matrix			**Residual matrix**	
	ρ	d	n	P	ν
Mass M	1	0	0	1	0
Length L	−3	1	0	2	2
Time T	0	0	−1	−3	−1

Only one linear transformation is necessary to transform −3 in the cell formed by the L row and the ρ-column into zero. The subsequent multiplication of the T row by −1 transfers −1 to 1:

	Unity matrix			**Residual matrix**	
	ρ	d	n	P	ν
M	**1**	0	0	1	0
$3M + L$	0	**1**	0	5	2
$-T$	0	0	**1**	3	1

The residual matrix contains only two parameters, therefore only two Π numbers result:

$$\Pi_1 \equiv \frac{P}{\rho^1 n^3 d^5} = \frac{P}{\rho n^3 d^5} \equiv Ne \text{ (Newton number)}$$

$$\Pi_2 \equiv \frac{\nu}{\rho^0 n^1 d^2} = \frac{\nu}{n d^2} \equiv Re^{-1} \text{ (Reynolds number)}$$

The interdependence of five dimensional quantities of the relevance list, equation 1, reduces to a set of only 5 − 3 = 2 dimensionless numbers:

$$\{Ne, Re\} \quad \text{or} \quad f(Ne, Re) = 0 \tag{2}$$

thus again confirming the pi theorem.

Determination of the Process Characteristics

The functional dependency, equation 2, is the maximum that dimensional analysis can offer here. It cannot provide any information about the form of the function f. This can be accomplished solely by experiments. But such experiments can now most certainly be conducted far more easily because it is the Reynolds number that must be varied, and this can be accomplished by altering the characteristic velocity v (here $v \propto nd$), or a characteristic length l (here d), or the kinematic viscosity ϑ. By choosing appropriate model fluids, the viscosity can be very easily altered by several orders of magnitude. This way measurements were conducted to determine the power characteristic $Ne(Re)$ of a blade stirrer of fixed geometry in a vessel. From Figure 1 it follows:

1. In the range $Re < 20$, the proportionality $Ne \propto Re^{-1}$ is found, thus resulting in the expression $NeRe \equiv P/(\eta n^2 d^3) =$ const. Density is irrelevant here—we are dealing with the laminar flow region.
2. In the range $Re > 50$ (vessel with baffles) or $Re > 5 \times 10^4$ (unbaffled vessel), a constant Newton number $Ne \equiv P/(\rho\, n^3\, d^5)$ is found. In this case, viscosity is irrelevant—we are dealing with a turbulent flow region.
3. Understandably, the baffles do not influence the power characteristic in the laminar flow region. However, their influence is extremely strong at $Re > 5 \times 10^4$. Here, the installation of baffles under oth-

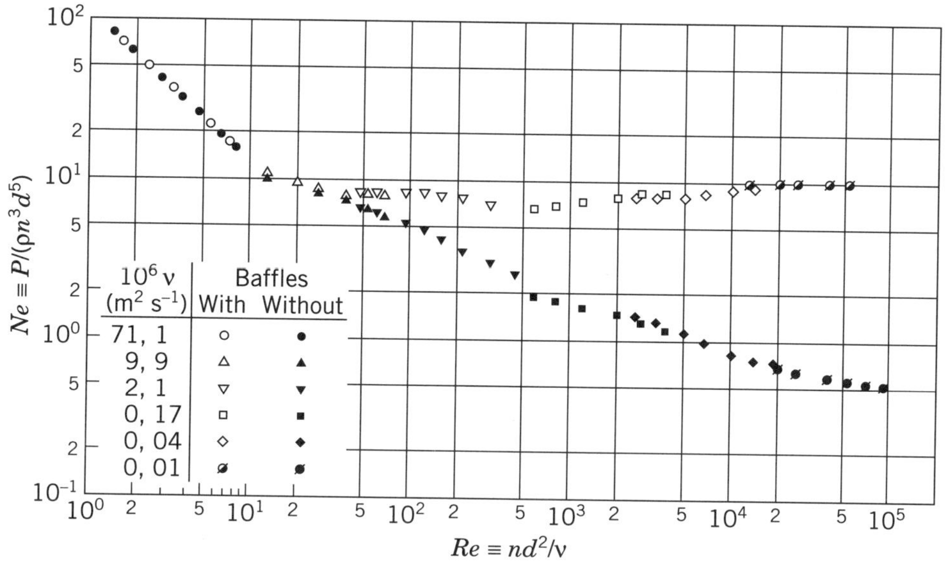

Figure 1. Power characteristics of a leaf stirrer of a given geometry and given installation conditions. *Source*: From Ref. 3.

erwise unchanged operating conditions increases the power consumption of the stirrer by a factor of 20.

4. When liquid is stirred in an unbaffled vessel, it begins to circulate, and a vortex is formed. Gravitational acceleration g, and hence the Froude number $Fr \equiv n^2 d/g$, plays no role under these circumstances as determined experimentally (Fig. 1): For confirmation, one need only look at the points on the lower $Ne(Re)$ curve where the *same* Re value was set for fluids with *different* viscosities. This was possible only by a proportional alteration of the rotational speed of the stirrer. Where Re = idem (= identical value), Fr was clearly not idem, but this has no influence on Ne: g is therefore irrelevant.

FUNDAMENTALS OF THE THEORY OF MODELS AND SCALE-UP

Theory of Models

The results in Figure 1 also could have been acquired by changing the stirrer diameter: the means by which a relevant number is changed does not matter. This is clear evidence that the representation of a physicotechnological problem in a dimensionless form is independent of scale ("scale-invariant"), and this presents the basis for a reliable scale-up: two processes may be considered completely similar if they take place in similar geometrical space and if all the dimensionless numbers necessary to describe them have the same numerical value (Π_i = identical or idem).

Clearly, the scale-up of a desired process condition from a model to industrial scale can be accomplished reliably only if the problem was formulated and dealt with according to the dimensional analysis.

Model Experiments and Scale-Up

In example 2 we evaluated process characteristics that presented a comprehensive description of the process. This often expensive and time-consuming method is certainly not necessary if one has only to scale up a given process condition from the model to the industrial plant (or vice versa). With example 2, and assuming that the $Ne(Re)$ characteristics like those in Figure 1 are not known, the task is to predict the power consumption of a leaf stirrer of d = 2 m, installed in a baffled vesser of D = 4 m (D/d = 2) and rotating with n = 10 min^{-1}. The kinematic viscosity of the liquid is $\nu = 1 \times 10^{-4}$ m^2/s.

One need know only—and this is essential—that the hydrodynamics in this case are governed solely by the Reynolds number and that the process is described by an unknown dependency $Ne(Re)$. Then one can calculate Re of the industrial plant:

$$Re \equiv nd^2/\nu = 6.7 \times 10^3$$

Let us assume that we have a geometrically similar laboratory device of D = 0.2 m and d = 0.1 m and that the rotational speed of the stirrer can be arbitrarily chosen. Which must be the rotational speed to obtain Re = idem using water $\nu = 1 \times 10^{-6}$ m^2/s as model liquid? The answer is

$$nd^2/\nu = 6.7 \times 10^3 \rightarrow n = 40 \text{ min}^{-1}$$

Under these conditions the stirrer power is measured and the power number $Ne \equiv P/(\rho n^3 d^5)$ calculated. We find Ne = 8.0. Because Re = idem results in Ne = idem, the power consumption P_T of the industrial stirrer can be obtained:

$$Ne = \text{idem} \rightarrow Ne_T = Ne_M \rightarrow \left(\frac{P}{\rho n^3 d^5}\right)_T = \left(\frac{P}{\rho n^3 d^5}\right)_M$$

From Ne = 8.0 found in the laboratory measurement, it follows for the leaf stirrer having d = 2 m and rotational speed n = 10 min^{-1}:

$$P = Ne\ \rho\ n^3 d^5 = 8.0 \times 1 \times 10^3 \times (10/60)^3 \times 2^5$$
$$= 1{,}185\ \text{W} \cong 1.2\ \text{kW}$$

We realize that in scale-up, knowledge of the functional dependency $f(\Pi_i) = 0$ is not necessary. All we need is to know which pi space describes the process.

FURTHER PROCEDURES TO ESTABLISH A RELEVANCE LIST

Consideration of the Acceleration Due to Gravity *g*

If a natural or universal physical constant has an impact on the process, it must to be incorporated into the relevant list, regardless of whether it will be altered. In this context the greatest mistakes are made with regard to the gravitational constant g. Lord Rayleigh (2) complained bitterly, saying:

> I refer to the manner in which gravity is treated. When the question under consideration depends essentially upon gravity, the symbol of gravity (g) makes no appearance, but when gravity does not enter the question at all, g obtrudes itself conspicuously.

This is all the more surprising in as much as the relevance of this quantity is easy enough to recognize if one asks the following question: would the process function differently if it took place on the moon instead of on Earth? If the answer to this question is yes, g is a relevant variable.

The gravitational acceleration g can be effective solely in connection with the density as gravity $g\rho$. When inertial forces play a role, the density ρ must be listed additionally. Thus it follows:

1. In cases involving the ballistic movement of bodies, the formation of vortices in stirring, the bow wave of a ship, the movement of a pendulum, and other processes affected by the Earth's gravity, the relevance list comprises $g\rho$ *and* ρ.
2. Creeping flow in a gravitational field is governed by the gravity $g\rho$ alone.
3. In heterogeneous physical systems with density differences (sedimentation or buoyancy), the difference in gravity $g\Delta\rho$ *and* ρ plays a decisive role.

Example 3 treats a problem in which the gravitational constant is of prime importance owing to extreme difference in densities in the gas–liquid system. This example offers opportunities to improve our skills in solving dimensional matrices and to gain deeper insights into the possibilities made available by dimensional analysis.

EXAMPLE 3

Process characteristics of a self-aspirating hollow stirrer and the determination of its optimum running conditions

As a result of their form, hollow stirrers utilize the suction generated behind their edges (Bernoulli effect) to suck in gas from the head space above the liquid. As "rotating ejectors," they are stirrers and gas pumps in one and are therefore particularly suitable for laboratory use (especially in high-pressure autoclaves) because they achieve intensive gas liquid-contact via internal gas recycling without a separate gas pump (3,5). A particularly effective stirrer of this type—the pipe stirrer—is depicted here.

In operating a hollow stirrer, the target quantities gas throughput q and stirrer power P adapt themselves simultaneously. Both target quantities depend on the following parameters:

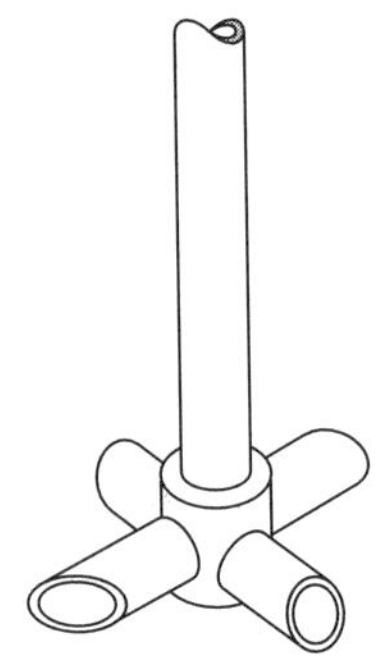

Geometry	Stirrer diameter	d
Physical properties	Density	ρ
	Dynamic viscosity	μ
Process parameters	Rotational speed	n
	Gravitational constant	g

Thus we obtain two separate relevance lists:

$$\{q;\ d,\ \rho,\ \mu;\ n,\ g\} \tag{3}$$

$$\{P;\ d,\ \rho,\ \mu;\ n,\ g\} \tag{4}$$

The problem is of a mechanical nature, and dimensions of the quantities describing it will contain only three basic quantities (M, L, T). It is to be expected that each process characteristics will therefore involve $6 - 3 = 3$ numbers.

The first relevance list, equation 3, leads to the following dimensional matrix:

	ρ	d	n	q	μ	g
Mass M	1	0	0	0	1	0
Length L	−3	1	0	3	−1	1
Time T	0	0	−1	−1	−1	−2
M	1	0	0	0	1	0
$3M + L$	0	1	0	3	2	1
$-T$	0	0	1	1	1	2

From this the following three numbers result:

$$\Pi_1 \equiv \frac{q}{\rho^0 d^3 n^1} = \frac{q}{nd^3} \equiv Q$$

$$\Pi_2 \equiv \frac{\mu}{\rho^1 d^2 n^1}\ \frac{\mu}{\rho d^2 n} \equiv \text{Re}^{-1}$$

$$\Pi_3 \equiv \frac{g}{\rho^0 d^1 n^2} = \frac{g}{dn^2} \equiv \text{Fr}^{-1}$$

where Q is the gas throughput number, Re the Reynolds number, and Fr the Froude number.

The gas throughput characteristic of a hollow stirrer thus reads as follows:

$$\frac{q}{nd^3} = f_1\left(\frac{nd^2\rho}{\mu}, \frac{n^2 d}{g}\right) \rightarrow Q = f_1(\text{Re, Fr}) \tag{5}$$

The power characteristics—obtained from the relevance list equation 4 by a similar dimensional matrix—containing P instead of q—results in the following:

$$\frac{P}{\rho n^3 d^5} = f_2\left(\frac{nd^2\rho}{\mu}, \frac{n^2 d}{g}\right) \rightarrow Ne = f_2(Re, Fr) \qquad (6)$$

Model experiments with another type of hollow stirrer (three-edged stirrer, see sketch in Fig. 2) were performed in a water–air system under defined experimental conditions, and the scale factor d_T/d_M changed in the range of 1:2:3:4:5. The results (Figs. 2 and 3) demonstrate that the Reynolds number is irrelevant. In the turbulent range of motion ($Re > 10^4$) the process is governed exclusively by the Froude number.

These process characteristics present the reliable basis for a scale-up of this hollow stirrer under given geometric conditions. But they also allow a further optimization of this process, as will be demonstrated by the following treatment. The question must be answered regarding process conditions under which this stirrer will achieve a given gas throughput with minimum power, P/q = min. The answer can easily be given by the combination of the characteristics presented in equations 5 and 6 in such a way as to produce a dimensionless expression for P/q. This is the combination

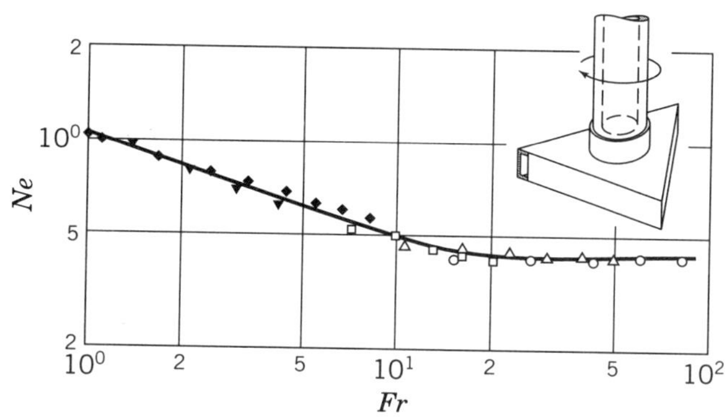

Figure 2. Power characteristics Ne (Fr) of the three-edged stirrer depicted in the inset. *Source*: From Ref. 3.

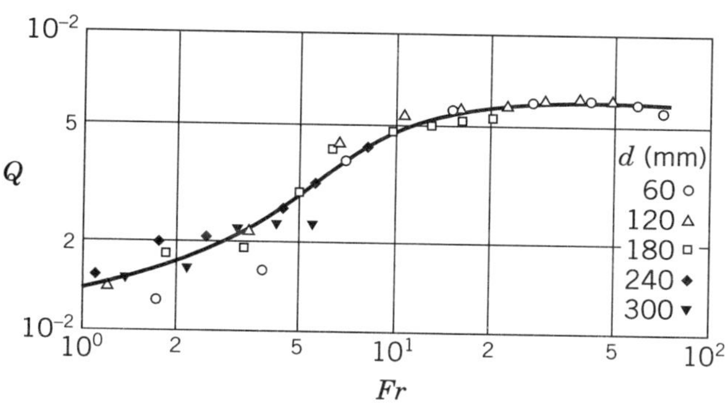

Figure 3. Gas throughput characteristics $Q(Fr)$ of a three-edged stirrer (see the sketch in Fig. 2). *Source*: From Ref. 3.

$$\frac{NeFr}{Q} \equiv \frac{P}{qd\rho g} = f_3(\text{Fr}) \qquad (7)$$

which is, as are their constituents, also a function of the Froude number. The dependency, equation 7, is plotted in Figure 4. Two sections can be distinguished:

$$Fr \leq 10\text{: } NeFr/Q = \text{const} \rightarrow P/q \propto d$$

$$Fr \geq 10\text{: } NeFr/Q \sim Fr \rightarrow P/q \propto d^2$$

In the first section P/q increases with direct proportionality to the stirrer diameter. This means that in an industrial plant having a scale-up factor, say, d_T/d_M of 10 (T, industrial scale; M, model scale), the same gas throughput requires a power consumption ten-fold higher than the model scale. At $Fr \geq 10$ the hollow stirrer would be still less efficient.

This procedure clearly shows that hollow stirrers, despite their excellent performance on the laboratory scale, do not perform economically on an industrial scale. In this case the gas throughput and the stirrer power have to be separated by choosing an appropriate stirrer (a Rushton turbine is a good choice) and providing the gas from below the stirrer by a blower.

Introduction of Intermediate Quantities

Many engineering problems involve several parameters, a condition that impedes the elaboration of the pi space. Fortunately, in some cases a closer look at a problem (or previous experience) facilitates reduction of the number of physical quantities in the relevance list. This is the case when some relevant variables affect the process by way of a so-called intermediate quantity. Assuming that this intermediate variable can be measured experimentally, it should be included in the problem relevance list if this would facilitate the removal of more than one variable from the list.

The fluid velocity v in pipes—or the superficial gas velocity v_G in mixing vessels or in bubble columns—present

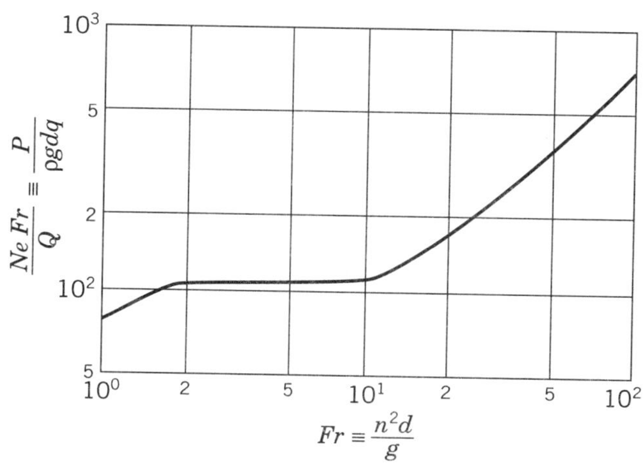

Figure 4. Optimization of a hollow stirrer according to P/q = min. The result reads: Small is beautiful.

a well-known intermediate quantity. The introduction of any such velocity into the relevance list removes two other quantities (throughput q and the diameter D) because of $v \propto q/D^2$ and $v_G \propto q_G/D^2$, respectively. The potential impact of the introduction of intermediate quantities on the relevance list is demonstrated in the following elegant examples.

EXAMPLE 4

Mixing characteristics for liquid mixtures with differences in density and viscosity

Mixing time θ necessary to achieve a molecular homogeneity of a liquid mixture (normally measured by decolorization methods) depends in material systems without differences in density and viscosity on only four parameters: stirrer diameter d, density ρ, dynamic viscosity η, rotational speed n:

$$\{\theta;\ d;\ \rho,\ \nu;\ n\} \tag{8}$$

From this, the mixing time characteristics are found to be

$$n\theta = f(Re) \qquad Re \equiv nd^2/\nu \tag{9}$$

In material systems with differences in density and viscosity, the relevance list, equation 8, enlarges by the physical properties of the second mixing component, by the volume ratio of both phases $\varphi = V_2/V_1$ and, because of the density differences, inevitably by the gravity difference $g\Delta\rho$ to nine parameters:

$$\{\theta;\ d;\ \rho_1,\ \nu_1,\ \rho_2,\ \nu_2,\ \varphi;\ g\Delta\rho,\ n\} \tag{10}$$

This results in mixing time characteristics incorporating six numbers:

$$n\theta = f(Re,\ Ar,\ \rho_2/\rho_1,\ \nu_2\nu_1,\ \varphi) \tag{11}$$

where $Re \equiv nd^2/\nu_1$ is the Reynolds number and $Ar \equiv g\Delta\rho d^3/(\rho_1\nu_1^2)$ is the Archimedes number.

Meticulous observation of this mixing process (the slow disappearance of the Schlieren patterns as result of the disappearance of density differences), reveals that macromixing is quickly accomplished compared to micromixing. This time-consuming process already takes place in a material system that can be fully described by the physical properties of the mixture:

$$\nu^* = f(\nu_1,\ \nu_2,\ \varphi) \qquad \text{and} \qquad \rho^* = f(\rho_1,\ \rho_2,\ \varphi)$$

By introducing these intermediate quantities ν^* and ρ^*, we can reduce the relevance list, equation 10, by three parameters to

$$\{\theta;\ d;\ \rho^*,\ \nu^*;\ g\Delta\rho,\ n\} \tag{12}$$

and get a mixing characteristic of only three numbers:

$$n\theta = f(\mathrm{Re},\ \mathrm{Ar}) \tag{13}$$

(In this case, Re and Ar have to be formed by ρ^* and ν^*)

The process characteristics of a cross-beam stirrer was established in this pi space by evaluation of corresponding measurements in two differently sized mixing vessels (D = 0.3 and 0.6 m) using different liquid mixtures ($\Delta\rho/\rho^* = 0.01 - 0.29$ and $\nu_2/\nu_1 = 1 - 5{,}300$). It reads (6) as follows:

$$\sqrt{n\theta} = 51.6\ Re^{-1}(Ar^{1/3} + 3) \qquad Re = 10^1\text{–}10^5;\ Ar = 10^2\text{–}10^{11}$$

Example 4, which clearly shows the great advantages achieved by the introduction of intermediate quantities, is amplified by example 5.

EXAMPLE 5

Description of the dissolved air flotation process

In the flotation process, fine gas bubbles adhere to the hydrophobic surface of solid particles and float them to the surface of the liquid, from which they are removed as foam. Dissolved air flotation has been introduced to the treatment of wastewater for the removal and subsequent concentration of biological sludge. It utilizes air that has been dissolved under a higher pressure and is released by decompression to atmospheric pressure, thus producing the finest possible gas bubbles. The appliances in use are either troughs exhibiting longitudinal flow, or flotation cells. The latter are advantageously divided by a concentric wall into two spaces (3, p. 87; 7). In the inner chamber, gas bubbles are released and brought into intimate contact with solids in suspension. Here the principal part of the flotation process occurs. The annular space surrounding the inner chamber exhibits vertical flow from top to bottom. It serves for calming down the flow, thus facilitating the final clarification (removal of the residual particles).

Flotation, a depletion process of first-order kinetics, is described by the flotation rate constant k_F $[T^{-1}]$ as the target quantity. A continuous process without back-mixing (plug flow) of the liquid is described in Ref. 7:

$$-\ln\frac{\varphi_t - \varphi_\infty}{\varphi_0 - \varphi_\infty} = k_F\tau$$

where φ is the solids content and τ is the mean residence time of the liquid throughput in the inner chamber or in the annulus, respectively.

In case of full back-mixing in the flotation unit, solids discharge A (which corresponds to the conversion factor X in first-order chemical reactions) is utilized (3, p. 87):

$$A = 1 - \frac{\varphi_{out}}{\varphi_{in}} = 1 - \frac{1}{1 + k_F\tau}$$

In constructing the relevance list of this unit operation, it is easy to list the process parameters. They are liquid feed q_{in}, which leaves the flotation cell and splits into floated output q_{fl} and "clear" output: $q_{out} = q_{in} - q_{fl}$; releasable gas content of the liquid feed q_G/q_{in}; gravitational acceleration g: $\{q_{in},\ q_{out},\ q_G/q_{in};\ g\}$.

The problem lies is the listing of all physical properties of the system, since these properties fluctuate strongly during the process (very much in contrast to the mineral flotation) and are only partially known. There is a degree of hydrophobicity (wettability) of the particle surface, described by the contact angle Θ. Another important parameter is the electrical charge of the particles, which greatly depends on the pH. Furthermore, finer particles float to the surface only after they have been agglomerated to flocs with the aid of a polyelectrolyte. Therefore the surfactant (flocculant) concentration c_f and its chemical origin (type) must be incorporated in the relevance list, too. Further material parameters will be the gravity difference $g\Delta\rho$, the solids content φ in the liquid, and the average gas bubble size (i.e., all liquid properties S_i describing bubble coalescence, which is still unknown by number and kind): $\{\Theta, \text{pH}, c_f, \text{type}, g\Delta\rho, \varphi, S_i\}$

The tentative relevance list will therefore read as follows:

$$\{A;\ D;\ \Theta, \text{pH}, c_f, \text{type}, \varphi, S_i;\ q_{in}, q_{out}, q_G/q_{in}, g\Delta\rho\} \qquad (14)$$

This relevance list can be streamlined significantly by introducing two intermediate quantities: the superficial velocity $v \propto q_{in}/D^2$ in the flotation cell and the rising velocity w of the flocs.

The rising velocity w of the flocs as an intermediate variable should allow *all* relevant physical properties to be taken into account; however, separate measurements of this intermediate quantity in suitable equipment are necessary.

The complete relevance list is then

$$\{A;\ w, v, q_{in}, q_{out}\} \qquad (15)$$

and from it we have the following pi set of three parameters:

$$\{A;\ w/v,\ q_{out}/q_{in}\} \qquad (16)$$

In case of $q_{out}/_{in}$ = const (this parameter governs the flotate concentration), the solids discharge A will depend only on w/v:

$$A = f(w/v) \qquad (17)$$

To prove this, measurements were performed in a bench-scale flotation cell (D = 1.6 m); the sample consisted of biologically clarified wastewater from a large chemical factory (160 production plants). It turned out that at $q_{out}/q_{in} = 0.9$, maintaining $w/v \approx 2$ was sufficient to ascertain a complete removal of the biosludge; see Example A 6 in Reference 3.

In the preceding example one need only to compare the tentative relevance list (equation 14) with the final five-parameter relevance list (equation 15) to fully realize the advantages of introducing intermediate quantities in complex interrelations.

Material Systems of Unknown Physical Properties

With foams, sludges, and slimes often encountered in biotechnology, we are confronted with the problem of not being able to list the physical properties because they are still unknown and therefore cannot be quantified. This situation often leads to the opinion that dimensional analysis would fail in such cases.

It is obvious that this conclusion is not valid: dimensional analysis is a *method* based on logical and mathematical fundamentals (1,3,4). If relevant parameters cannot be listed because they are unknown, one cannot blame the method. The only solution is to perform the model measurements with the same material system and to change the model scales.

EXAMPLE 6

Scale-up of a mechanical foam breaker

To obtain reliable information on dimensioning and then scale-up of a given type of mechanical foam breaker (foam centrifuge, see Fig. 5), we wish to ascertain the mode of performing and to evaluate the model measurements of the device. Preliminary experiments have shown that for each foam emergence—proportional to the gas throughput q_G—for each foam breaker of diameter d, a minimum rotational speed n_{min} exists that is necessary to control it. The dynamic properties of the foam (density and viscosity, elasticity of the foam lamella, etc.) cannot be named or measured. We will have to content ourselves by listing them wholesale as material properties S_i. In our model experiments we will of course be able to replace S_i by the known type of surfactant (foamer) and its concentration c_f (ppm).

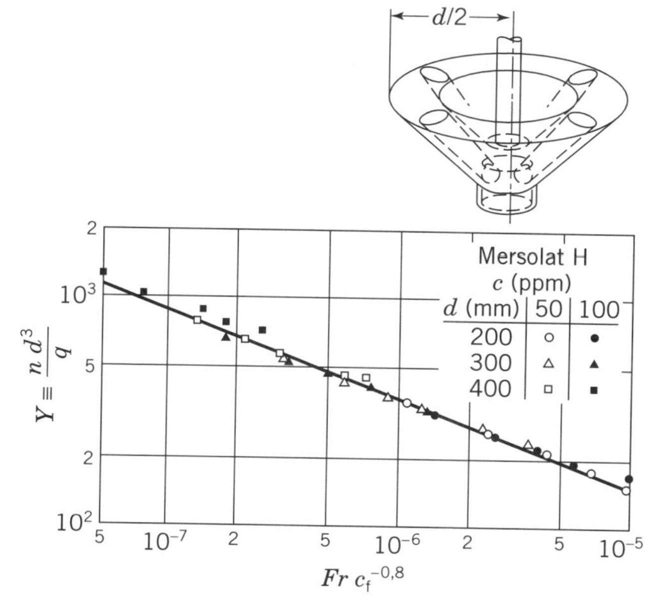

Figure 5. Process characteristics of the foam centrifuge (inset) for a particular foamer (Mersolat H, Bayer AG, Germany). *Source*: From Ref. 8.

In discerning the process parameters, we realize that the gravitational acceleration g has no impact on the foam breaking *within* the foam centrifuge: the centrifugal acceleration n^2d exceeds the gravitational one (g) by far. However, the water content of the foam entering the centrifuge depends very much on the gravitational acceleration: on the moon, the water drainage would be far less effective. In contrast to the dimensional analysis presented elsewhere (8), we are well advised to add g to the relevance list:

$$\{n_{min};\ d;\ \text{type of foamer}, c_f;\ q_G, g\} \qquad (18)$$

For the sake of simplicity, in the following, n_{min} is replaced by n, and q_G by q. For each type of foamer we obtain the following pi space:

$$\left\{\frac{nd^3}{q}, \frac{n^2d}{g}, c_f\right\} \quad \text{or, abbreviated} \quad \{Y, Fr, c_f\} \qquad (19)$$

To prove this pi space, measurements must be made in model equipment of different sizes, to produce a reliable process characteristics. Figure 5 gives results for a commercial foamer (Mersolat H, Bayer AG, Germany); the pi space of equation 19 is seen to be fully satisfied.

The straight line in Figure 5 corresponds to the analytic expression

$$Y = Fr^{-0.4} c_f^{0.32} \qquad (20)$$

which can be reduced to

$$nd = \text{const } q^{0.2} f(c_f) \quad (21)$$

Here, the foam breaker will be scaled up according to its tip speed $u = \pi nd$ in model experiments, which will also moderately depend on the foam yield (q).

In all the other foamers examined (9), the correspondence $Y \propto Fr^{-0.45}$ was found. If—by chance—the correlation

$$Y \propto \text{Fr}^{-0.5} f(c_f)$$

proved to be true, then this one can be deduced to

$$n^2 d/g = \text{const}(c_f)$$

In this case the centrifugal acceleration ($n^2 d$) would present the scale-up criterion and would depend only on the foamer concentration, not on foam yield (q).

Short Summary of the Essentials of the Dimensional Analysis and Scale-Up

The advantages made possible by the correct and timely use of dimensional analysis can be stated briefly as follows.

1. *Reduction of the number of parameters required to define the problem.* The Π theorem states that a physical problem can always be described in dimensionless terms. Thus the number of dimensionless groups that fully describe it is much smaller than the number of dimensional physical quantities and is generally equal to the number of physical quantities minus the number of basic units contained within them.
2. *Reliable scale-up of the desired operating conditions from the model to the full-scale plant.* According to the theory of models, two processes may be considered to be similar if they take place under geometrically similar conditions and all dimensionless numbers that describe the processes have the same numerical value.
3. *A deeper insight into the physical nature of the process.* When experimental data are presented in a dimensionless form, distinct physical states can be isolated from one another (e.g., turbulent or laminar flow regions), and the effect of individual physical variables can be identified.
4. *Flexibility in the choice of parameters and their reliable extrapolation within the range covered by the dimensionless numbers.* These advantages become clear if one considers the well-known Reynolds number, $Re = vl/\nu$, which can be varied by altering the characteristic velocity v, or a characteristic length l, or the kinematic viscosity ν. By choosing appropriate model fluids, the viscosity can be very easily altered by several orders of magnitude. Once the effect of the Reynolds number is known, extrapolation of both v and l is allowed within the examined range of Re.

Area of Applicability of Dimensional Analysis

The application of dimensional analysis is indeed heavily dependent on the available knowledge. The following five steps (see Fig. 6) can be outlined.

1. The physics of the basic phenomenon is unknown: dimensional analysis cannot be applied.
2. Enough is known about the physics of the basic phenomenon to compile a first, tentative relevance list: the resultant pi set is unreliable.
3. All the relevant physical variables describing the problem are known: the application of dimensional analysis is unproblematic.
4. The problem can be expressed in terms of a mathematical equation: a closer insight into the pi relationship is feasible and may facilitate a reduction of the set of dimensionless numbers.
5. A mathematical solution of the problem exists: the application of dimensional analysis is superfluous.

It must, of course, be said that approaching a problem from the point of view of dimensional analysis also remains useful even if all the variables relevant to the problem are not yet known: the timely application of dimensional analysis often lead to the discovery of forgotten variables or the exclusion of artifacts.

Experimental Methods for Scale-Up

A number of questions posed at the beginning of this article are often asked in connection with model experiments.

How small can a model be? The size of a model depends on the scale factor L_T/L_M and on the experimental precision of measurement. Where $L_T/L_M = 10$, a $\pm$ 10 % margin of error may already be excessive. A larger scale

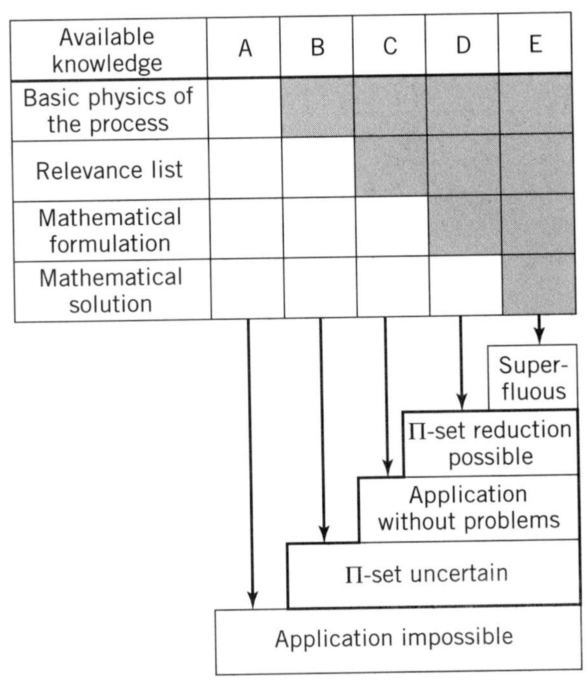

Figure 6. Graphical representation of the four levels of knowledge and their impact on the treatment of the problem by dimensional analysis. *Source*: After J. Pawlowski, Ref. 3.

for the model will therefore have to be chosen to reduce the error.

Is one model scale sufficient, or should tests be carried out in models of different sizes? One model scale is sufficient if the relevant numerical values of the dimensionless numbers necessary to describe the problem (the "process point" in the pi space describing the operational condition of the technical plant) can be adjusted by choosing the appropriate process parameters or physical properties of the model material system. If this is not possible, the process characteristics must be determined in a series of models of different sizes, or the process point must be extrapolated from experiments in technical plants of different sizes.

When must model experiments be carried out exclusively with the original material system? Where the material model system is unavailable (e.g., in the case of non-Newtonian fluids) or where the relevant physical properties are unknown (e.g., foams, sludges, slimes), model experiments must be carried out with the original material system. In such cases, measurements must be performed in models of various sizes (cf. example 6).

Partial Similarity

In the scale-up from a model to industrial scale, not only must the geometric similarity be ensured, but also all dimensionless numbers describing the problem must retain the same numerical values (Π_i = idem). This means, for example, that in the scale-up of boats or ships the dimensionless numbers governing the hydrodynamics

$$Fr \equiv \frac{v^2}{lg} \quad \text{and} \quad Re \equiv \frac{vl}{\nu}$$

must retain their numerical values: $Fr_T = Fr_M$ and $Re_T = Re_M$. It can easily be shown that this requirement cannot be fulfilled here.

Since the gravitational acceleration g cannot be varied on Earth, the Froude number, Fr, of the model can be adjusted to that of the full-scale vessel only by its velocity v_M. Subsequently, Re = idem can be achieved only by adjusting the viscosity of the model fluid. If the model size is only 10% of the full size (scale factor $L_T/L_M = 10$), Fr = idem is achieved in the model at $v_M = 0.32\ v_T$. To fulfill Re = idem for the kinematic viscosity of the model fluid ν_M, it follows

$$\frac{\nu_M}{\nu_T} = \frac{v_M}{v_T}\frac{L_M}{L_T} = 0.32 \times 0.1 = 0.032$$

No liquid exists whose viscosity would be only 3% of that of water!

Some requirements concerning physical properties of model materials cannot be implemented. In such cases only a partial similarity can be realized. For this, essentially only two procedures are available (for details, see Refs. 3 and 10). One consists of a well-planned experimental strategy in which the process is divided into parts, which are then investigated separately under conditions of complete similarity. This approach was first applied by William Froude (1810–1879) in his efforts to scale up the drag resistance of a ship's hull.

The second approach consists in deliberately abandoning certain similarity criteria and checking the effect on the entire process. This technique was used by Gerhard Damköhler (1908–1944) in his attempts to treat a chemical reaction in a catalytic fixed-bed reactor by means of dimensional analysis. Here the problem of a simultaneous mass and heat transfer arises—they are two processes that obey completely different fundamental principles.

It is seldom realized that many rules of thumb utilized for scale-up of different types of equipment are represented by quantities fulfill only a partial similarity. For example, only the volume-related mixing power P/V (widely used for scaling up mixing vessels) and the superficial velocity v (normally used for scale-up of bubble columns) should be mentioned here.

The volume-related mixing power P/V presents an adequate scale-up criterion only in L/L dispersion processes and can be deduced from the pertinent process characteristics $\delta/d \propto We^{-0.6}$ (δ, droplet diameter; We, Weber number). In the most common mixing operation, the homogenization of miscible liquids, where a macro- and back-mixing are required, this criterion fails completely (10).

Similarly, the superficial velocity v_G of the gas throughput as an intensity quantity is a reliable scale-up criterion only in mass transfer in gas–liquid systems in bubble columns. In mixing operations in bubble columns, requiring the whole liquid content be back-mixed (e.g., in homogenization), this criterion completely loses its validity (10).

Thus we must conclude that a particular scale-up criterion that is valid in a given type of apparatus for a particular process is not necessarily applicable to other processes occurring in the same device.

TREATMENT OF VARIABLE PHYSICAL PROPERTIES BY DIMENSIONAL ANALYSIS

Biotechnology deals with nonhomogeneous material systems exhibiting mostly non-Newtonian viscosity behaviors that often change their properties under shear stress. Thus a more comprehensive treatment of variable physical properties by dimensional analysis is essential.

It is generally assumed that the physical properties of the material system remain unaltered in the course of the process. Process equations (e.g., the heat characteristics of a mixing vessel as in Example 9 below, or a smooth straight pipe)

$$Nu = f(Re, Pr) \tag{22}$$

are valid for any material system with Newtonian viscosity and for any constant process temperature (i.e., for any one-time *constant* physical properties).

However, constancy of physical properties cannot be assumed in every physical process. A temperature field may well generate a viscosity field or even a density field in the material system treated. In non-Newtonian (structurally viscous or viscoelastic) liquids, a shear rate can also produce a viscosity field.

Although most physical properties (e.g., viscosity, density, heat conductivity and capacity, surface tension) must be regarded as variable, the value of viscosity in particular can be varied by many orders of magnitude under certain process conditions. In the following, we apply dimensional analysis solely to describe the variability of this one physical property. However, the approach can be adapted for any other physical property.

Dimensionless Representation of the Material Function

Similar behavior of a certain physical property common to different material systems can only be visualized by dimensionless representation of the material function of that property. It is furthermore desirable to formulate this function as uniformly as possible. This can be achieved by the "standard representation" of the material function (4,11) in which a standardized transformation of the material function $\eta(\vartheta)$ is defined such that the expression produced

$$\mu/\mu_0 = \varphi\{-\gamma_0(\vartheta - \vartheta_0)\} \tag{23}$$

meets the requirement

$$\varphi(0) = \varphi'(0) = 1$$

where $\gamma_0 \equiv [(1/\mu)(\delta\mu/\delta\vartheta)]_{\vartheta_0}$ is the temperature coefficient of the viscosity, $\mu_0 \equiv \mu(\vartheta_0)$, and ϑ_0 is *any* reference temperature. Figure 7 depicts the dramatic effect of such standard transformations.

Engineers prefer the representation

$$\mu/\mu_0 = \exp\{-\gamma_0(\vartheta - \vartheta_0)\} \tag{24}$$

which is a special case of equation 23, although this is not the best possible approximation (see dashed line in Fig. 7b). A slightly better approximation of the material function $\mu(\vartheta)$ is provided by the well-known Arrhenius relationship:

$$\frac{\mu}{\mu_0} = \exp\left\{-\frac{E_0}{R\,\vartheta_0}\left(\frac{\vartheta_0}{\vartheta} - 1\right)\right\} \tag{25}$$

To describe the process by dimensional analysis, it is advisable to formulate the reference temperature ϑ_0 in a process-related manner, using a characteristic, possibly mean process temperature, as a reference. A class of functions exists, however, that are independent (invariant) of the reference point ϑ_0. The solid line in Figure 7b shows a function of this type. (*Note*: Function equations 23 and 24 are special cases of invariant functions.)

Pi Set for Variable Physical Properties

The type of dimensionless representation of the material function affects the (extended) pi set within which the process relationship is formulated (for more information see Refs. 3 and 11). When the standard representation is used, the relevance list must include the reference viscosity η_0 instead of η and incorporate two additional parameters γ_0, ϑ_0. This leads to *two* additional dimensionless numbers in the process characteristic. With regard to the heat transfer characteristic, equation 22 of a mixing vessel, it now follows that

$$Nu = f(Re_0,\ Pr_0,\ \gamma_0\Delta\vartheta,\ \Delta\vartheta/\vartheta_0) \tag{26}$$

where the subscript to *Re* and *Pr* denotes that these two dimensionless numbers are to be formed with μ_0 (which is the numerical value of μ at ϑ_0).

If we consider that the standard transformation of the material function can be expressed invariantly with regard to the reference temperature ϑ_0 (Fig. 7b), the relevance list is extended by only *one* additional parameter, γ_0. This, in turn, leads to only *one* additional dimensionless number. For our problem, it now follows that

$$Nu = f(Re_0,\ Pr_0,\ \gamma_0\Delta\vartheta) \tag{27}$$

Non-Newtonian Liquids

The main characteristic of Newtonian liquids is that simple shear flow (e.g., Couette flow) generates shear stress τ, which is proportional to the shear rate $\dot{\gamma} \equiv dv/dy$, ($s^{-1}$). The proportionality constant, the dynamic viscosity μ, is the only material constant in the Newton's law of motion:

$$\tau = \mu\dot{\gamma} \tag{28}$$

where μ depends only on pressure and temperature.

In the case of non-Newtonian liquids, μ depends on $\dot{\gamma}$ as well. These liquids can be classified in various categories of materials depending on their flow behavior: $\eta(\dot{\gamma})$, flow curve; $\mu(\tau)$, viscosity curve.

J. Pawlowski (12) points out that the rheological properties of many non-Newtonian liquids can be described by material parameters whose dimensional matrix has a rank of 2. These physical properties can be usefully linked to produce two dimensional constants: a characteristic viscosity constant H and a characteristic time constant θ, and possibly a set of dimensionless material numbers Π_{rheol}. The relevance list is thus formed by

$$\{H,\ \theta,\ \Pi_{rheol}\} \tag{29}$$

In a change from Newtonian to non-Newtonian liquids, this interrelation has the following consequences for the complete pi set:

1. Every dimensionless number incorporating η must now be formulated with the dimensional equivalent H (e.g., μ_0, η_∞).
2. A single process-related number containing θ is added (e.g., $\dot{\gamma}$).
3. The pure material numbers are increased by Π_{rheol}.

We will demonstrate these rules on the heat transfer characteristic of a smooth, straight pipe. In the following table, line *A* stands for temperature-independent viscosity, and line *B* for a temperature-dependent viscosity.

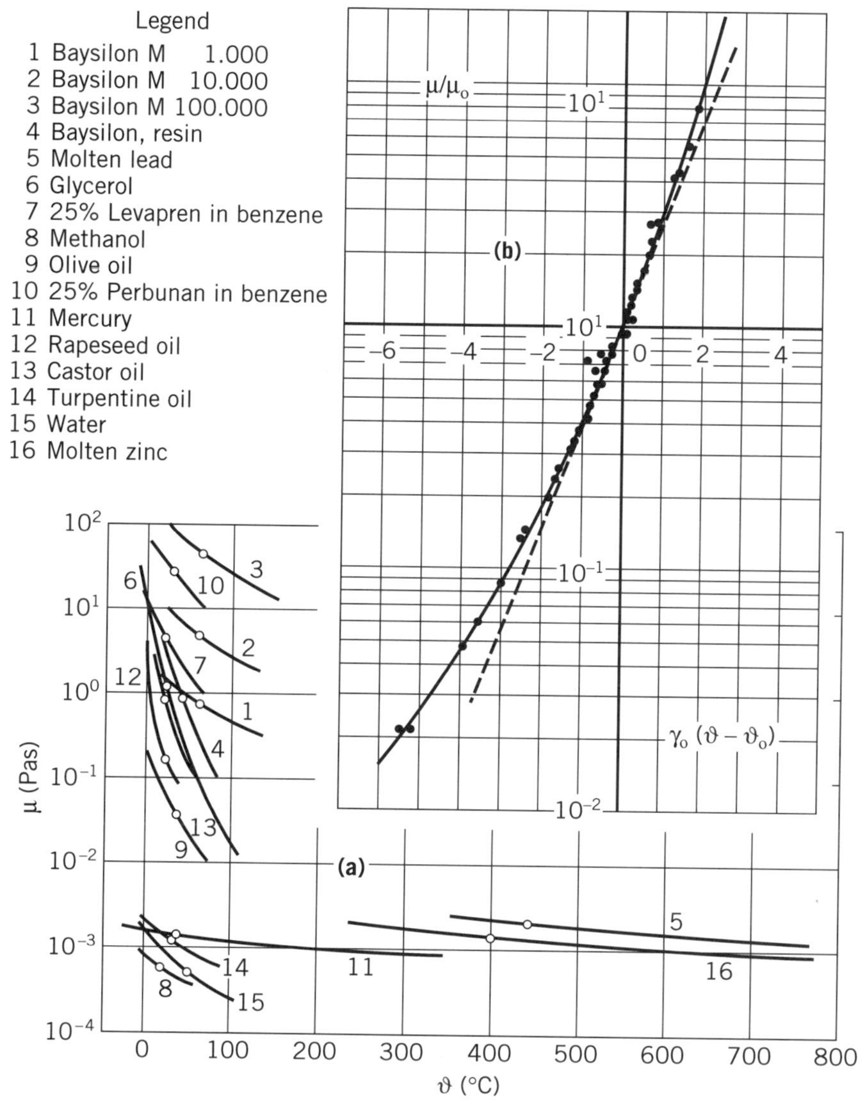

Figure 7. Temperature dependency of the viscosity $\mu(\vartheta)$ and (**b**) its standard representation, which leads to the reference-invariant approximation of the material function. Data points on the curves in (**a**) mark the reference temperatures ϑ_o chosen. From (4, 11). *Source*: From Ref. 3, and Ref. 11.

	Newtonian liquid	**Non-Newtonian liquid**
A	*Nu, Re, Pr*	$Nu, Re_H, Pr_H, v\theta/L$
B	$Nu, Re, Pr, \gamma_0 \Delta\vartheta$	$Nu, Re_H, Pr_H, v\theta/L, \gamma_H\,\Delta\vartheta, \gamma_\theta/\gamma_H$

In *B*, $\gamma_H \equiv \delta \ln H/\delta\vartheta$ and $\gamma_\theta \equiv \delta \ln \theta/\delta\vartheta$ must be added; $\Delta\vartheta$ denotes the temperature difference between the bulk of the liquid and the wall. The conventionally used expression μ/μ_w (where subscript w indicates wall) and $\gamma\Delta\vartheta$ are related by

$$\mu/\mu_w = \exp[-\gamma(\vartheta - \vartheta_w)]$$

Since little is known about the rheological properties of material systems, model experiments must be carried out with the same substance to be used in the full-scale plant. Since

$$\Pi_{\text{material}}(\text{here } Pr_H) = \text{idem} \quad \text{and} \quad \Pi_{\text{rheol}} = \text{idem}$$

the process will then take place within a Π range with only one more dimensionless number ($v\theta/L$) than in the case of Newtonian liquids.

However, when scaling up from the model to the full-scale plant, it is impossible to attain complete similarity using the same material system. In the example just described, it is obvious that in keeping ρ, H, θ = idem, we cannot ensure that $Re_H \equiv \rho vL/H$ and $v\theta/L$ also remain identical. It is therefore advisable to retain one and the same substance but to alter the scale of the model in the experiments. Pure fluid mechanical processes involving creeping (ρ being irrelevant), steady-state, and isothermal flow are exceptions. In these cases mechanical similarity can be obtained in spite of the constancy of physical properties.

Non-Newtonian liquids with flow curves obeying the so-called power law

$$\tau = K\dot{\gamma}^m$$

are known as Ostwald–de Waele fluids. The dimension of K depends on the numerical value of the exponent m and is therefore not a consistent physical quantity. To produce a material constant X having the dimension of viscosity, K must be combined with a characteristic velocity, v, and a characteristic length, l:

$$X \equiv K(v/l)^{m-1}$$

With this sort of liquid, the extension of the Π set has the following consequences:

- X replaces μ in every dimensionless number in which it appears.
- Pure material numbers are extended by m (m belongs to Π_{rheol}).

In the case of an Ostwald–de Waele fluid, the heat transfer characteristic of a smooth straight pipe is as follows:

$$Nu = f(Re_X, Pr_X, m)$$

where $Re_X = \rho v^{2-m} l^m / X$.

Viscoelastic Liquids

Almost every biological solution of low viscosity (but also viscous biopolymers such as xanthan and dilute solutions of long chain polymers—carboxymethyl cellulose CMC, polyacrylamide, polyacrylonitrile, etc.) displays not only viscous but also viscoelastic flow behavior. These liquids are capable of storing a part of the deformation energy elastically and reversibly. They evade mechanical stress by contracting like rubber bands. This behavior causes secondary flow, which often runs contrary to the flow produced by mass forces (e.g., the liquid "climbs" the shaft of a stirrer—the so-called Weissenberg effect).

Elastic behavior of liquids is characterized mainly by the ratio of first differences in normal stress, N_1, to the shear stress, τ. This ratio, the Weissenberg number $Wi = N_1/\tau$, is usually represented as a function of the rate of shear $\dot{\gamma}$.

Henzler (13) takes an approach similar to that of the Ostwald–de Waele law

$$Wi = A\dot{\gamma}^a \quad \text{or preferably} \quad Wi = A\dot{\gamma}^a + B\dot{\gamma}^b$$

To transform these material functions into dimensionless forms, a reference Weissenberg number Wi_o is chosen. This leads to the generation of the characteristic rates of shear $\dot{\gamma}_1$ and $\dot{\gamma}_2$. It thus follows that

$$\frac{Wi}{Wi_o} = \left(\frac{\dot{\gamma}}{\dot{\gamma}_1}\right)^a + \left(\frac{\dot{\gamma}}{\dot{\gamma}_2}\right)^b$$

In this case the following constituent parts of the dimensionless material function must be incorporated in the relevance list:

$$Wi_o,\ \dot{\gamma}_1,\ \dot{\gamma}_2,\ a,\ b$$

In the case of a viscoelastic liquid, the material functions of both the viscous and the elastic behaviors must be considered. Assuming that the viscous behavior of the liquid concerned is described as $\mu/\mu_o = f(\dot{\gamma}/\dot{\gamma}_1)$ and the Wi/Wi_o function is valid for its elastic behavior (Fig. 8), the Π set must be modified as follows:

- μ_o replaces μ in every dimensionless number in which it appears.
- The pure material-related numbers are extended by the following six purely rheological numbers: $\Pi_{\text{rheol}} = Wi_o,\ \dot{\gamma}/\dot{\gamma}_1,\ \dot{\gamma}/\dot{\gamma}_2,\ m,\ a,\ b$.

Examples of the Use of Dimensional Analysis in Mass and Heat Transfer in Gas–Liquid Systems

Examples 7–10 are presented for two reasons. The first lies in the fact that the compilation of the relevance list (i.e., listing all the relevant parameters of the problem) represents the only real difficulty in using this method, which can be overcome by practice. The other concerns their contents (mass and heat transfer in gas–liquid systems), which are of eminent importance to biotechnology and at the same time show the large potential of information contained in their dimensionless presentation. Clearly, information of this kind cannot be obtained by computational methods.

EXAMPLE 7

Mass transfer in a gas–liquid system in a mixing vessel

According to the so-called two-film theory (Lewis and Whiteman, 1923–1924), the mass transfer in a gas–liquid system (absorption, desorption: "sorption") is described by the overall transfer equation

$$G = k_L A \Delta c \tag{30}$$

G is mass transfer rate through the interface (M/T), k_L is liquid-side mass transfer coefficient (L/T), A is interfacial area (sum of the surfaces of all gas bubbles) (L^2), and Δc is characteristic concentration difference (M/L^3) of the dissolved gas between the interface and the liquid bulk. In the case of surface aeration the difference Δc is given by $c_s - c$. It is assumed that the interface is always saturated with the gas to be dissolved (c_s) according to the Henry's law.

Equation 30 and the foregoing definition of Δc fit well the surface aeration situation. In the commonly used volume aeration systems, it is established practice to assume a uniform distribution of gas bubbles in the liquid and to relate the mass transfer to the volume of the liquid:

$$G/V = k_L(A/V)\Delta c_m = k_L a \Delta c_m \tag{31}$$

The liquid-side mass transfer coefficient k_L as well as the volume-related interface $a \equiv A/V$ can be measured separately only with difficulty. Furthermore, both depend on same physical properties and process parameters. Therefore they are combined to form an overall mass transfer coefficient:

$$k_L a \equiv \frac{G}{V \Delta c_m} \tag{32}$$

The gas composition of bubbles rising in the column of liquid

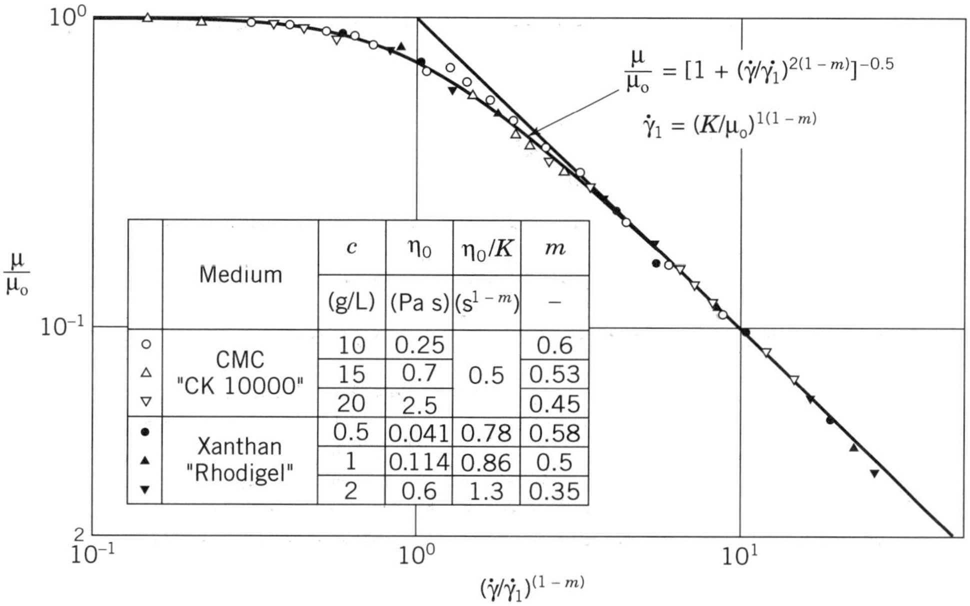

Figure 8. Dimensionless material function of some viscoelastic chemical and biological polymers. *Source*: Ref. 13.

changes as a result of the sorption processes (e.g., depletion of O_2 and absorption of CO_2). Beside this, the system pressure changes along the liquid height. These differences in partial pressures are taken into account by the logarithmic mean concentration difference Δc_m:

$$\Delta c_m = \frac{c_1 - c_2}{\ln[(c_1 - c)/(c_2 - c)]} \tag{33}$$

where c_1 and c_2 are the saturation concentrations under the (p, ϑ, x) conditions at the gas inlet and outlet, respectively, c is the concentration of the dissolved gas in the bulk of the liquid, x is a mole fraction of the dissolved gas.

The use of the volume-related and thus intensively formulated quantity k_La as the target quantity of the mass transfer implies the following consequences:

- k_La should not depend on geometric parameters, since a quasi-uniform material system is assumed.
- k_La must be independent of the physical properties of the gas phase, given that $k_G \gg k_L$.
- Because the target quantity k_La is an intensity variable, the process parameters must also be formulated intensively.

The relevance list must thus be formed with the following parameters:

1. *Target quantity*: k_La
2. *Physical properties*: density ρ, kinematic viscosity ν, diffusivity $\mathcal{D}$, and the coalescence parameters S_i of the liquid phase. (The latter are still unknown by the nature and number for dynamic processes such as aeration in mixing vessels and bubble columns.)
3. *Process parameters*: volume-related mixing power P/V, gas throughput q_G related to the volume of liquid $q_G/V_L \propto q_G/(D^2 H)$ *or* to the cross-sectional area of the vessel (superficial velocity) $v_G \propto q_G/D^2$, and—owing to extreme differences in density $\Delta\rho$—acceleration due to gravity g or $g\Delta\rho$.

We have to come to terms with two problems.

The first one is of a physical nature (S_i unknown). The nature of the liquid treated must be well known (details such as ρ, ν, $\mathcal{D}$ are not sufficient), and model measurements must be performed with the same liquid system used on the large scale.

The second problem concerns the pertinent formulation of the gas throughput. An accurate answer requires measurements with the same material system in differently scaled equipment. The decision in favor of $v_G \propto q_G/D^2$ instead of q_G/V_L was brought about by Judat (14) by comparing extensive research results of the last decade (Fig. 9). Anticipating this work, we compose the relevance list for volume aeration as follows:

$$\{k_La;\ \rho,\ \nu,\ \mathcal{D},\ S_i;\ P/V,\ v_G,\ g\} \tag{34}$$

The resulting pi set reads

$$\{(k_La)^*,\ (P/V)^*,\ v_G^*,\ Sc,\ S_i^*\} \tag{35}$$

where

$$(k_La)^* \equiv k_La(\nu/g^2)^{1/3}$$

$$(P/V)^* \equiv \frac{P/V}{\rho(\nu g^4)^{1/3}}$$

$$v^* \equiv \frac{v_G}{(\nu g)^{1/3}}$$

$$Sc \equiv \nu/\mathcal{D}$$

Quantities S_i are unknown by number and nature. Nevertheless, with the help of known physical properties (ρ, ν, $\mathcal{D}$), we can easily transform them into dimensionless numbers S_i^*.

The extent of influence of the coalescence phenomena

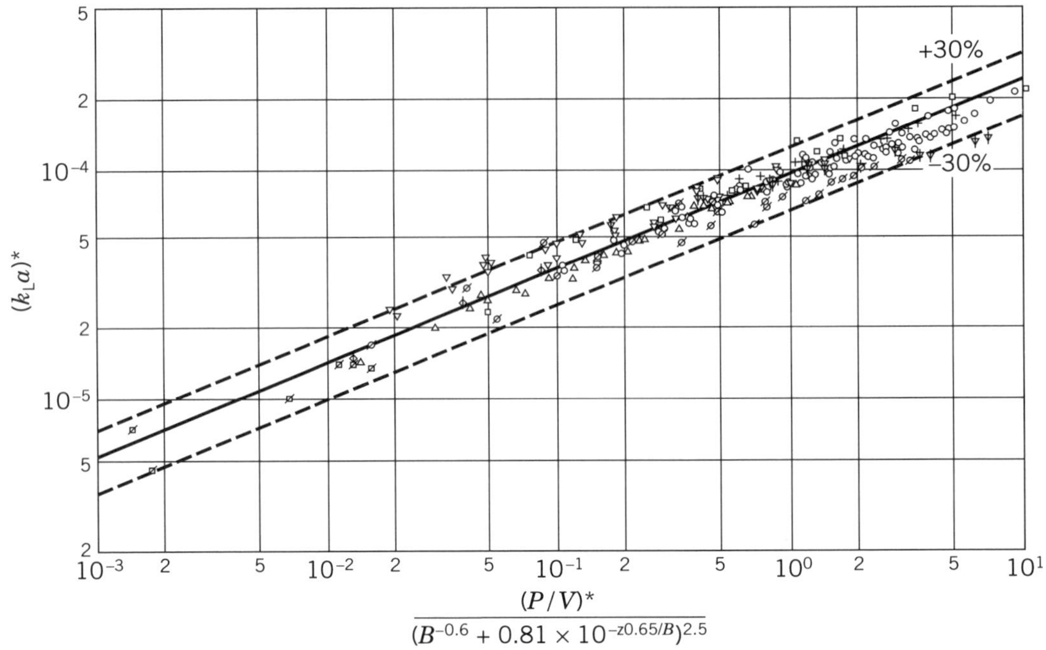

Figure 9. Sorption characteristics for a mixing vessel with turbine stirrer (Rushton turbine) for a fully coalescent water–air system. Comparative evaluation of several investigations covering water volumes of $V = 0.0025 - 906$ m^3, $B \equiv (\pi/4)\, v^*$. *Source*: From Ref. 14.

on the mass transfer in the gas–liquid system is clearly visible upon comparison of the mass transfer characteristics for the water–air system (unrestricted coalescence; Fig. 9) and for a system consisting of a 1 N solution of sodium sulfite in air (70 g salt/L; fully suppressed coalescence, Fig. 10) (15). Here self-aspirating stirrers (cf. sketches in example 3 and Fig. 2) have been used in which the mixing power and the gas throughput are coupled by the rotational speed of the stirrer. Therefore v_G is not an independent parameter here, and the pertinent pi space is given by

$$(k_L a)^* \equiv k_L a (\nu/g^2)^{1/3}$$

$$(P/V)^* \equiv \frac{P/V}{\rho(\nu g^4)^{1/3}}$$

$$Sc \equiv \nu/\mathcal{D}$$

For scale-up of surface aerators see Refs. 3 and 16.

EXAMPLE 8

Mass transfer in a gas–liquid system in bubble columns

Bubble columns are preferable in microbiological conversions under aerobic conditions. Perforated coils or sintered plates serving as gas distributors and the superficial velocity v_G present the only process parameters (besides the acceleration due to gravity, g).

When the O_2 uptake in this gas–liquid contacting method is not sufficient, stirrers must provide additional mixing. The turbine stirrers are mounted on the same shaft. The bubble column is thus transformed into a mixing vessel with $H/D \gg 1$ (mostly $H/D = 3$–5), and P/V is an additional process parameter.

In the particularly large bubble columns (tower-shaped bioreactors) increasingly used today for aerobic wastewater treatment (17), the intensification of mass transfer by stirrers would be costly: the stirrer must ensure that the entire contents of the vessel is in motion to generate in the liquid the shear rate needed for gas dispersion. In this case, injectors are the aeration devices of choice: these two-component nozzles utilize the kinetic energy of the liquid propulsion jet to disperse the gas phase into very fine gas bubbles and to distribute them into the liquid. The power of the propulsion jet P_L needed for dispersion of the gas throughput q_G in fine bubbles, P_L/q and the superficial velocity v_G are the process parameters.

Interestingly, in laboratory measurements with bubble columns, the proportionality $k_L a \propto v_G$ has been found. This connection is immediately understandable if one considers the constituents of which these quantities are built:

$$\frac{k_L a}{v_G} = \frac{G}{V \Delta c_m} \frac{S}{q_G} = \frac{G}{q_G H \Delta c_m} = \text{const}$$

where S is the surface area of the tank, H is the liquid height, and $V = SH$.

It turns out that the absorption velocity G is directly proportional to the gas throughput q_G, to the liquid height H (which determines the residence time of the gas bubbles in the liquid), and to the logarithmic mean concentration difference Δc_m (cf. equation 33).

The pi space in the three mentioned technological implementations is

1. Bubble column with a gas distributor (9): {Y; Sc, S_i^*; geometry of the distributor, aeration density}
2. Bubble column with additional stirrers (9): {Y; $(P/V)^*$ Sc, S_i^*; H/D; type of stirrer}
3. Bubble column with injectors (17, 18): {Y; $(P_L/q)^*$; Sc, S_i^*; H/D; type of injector}

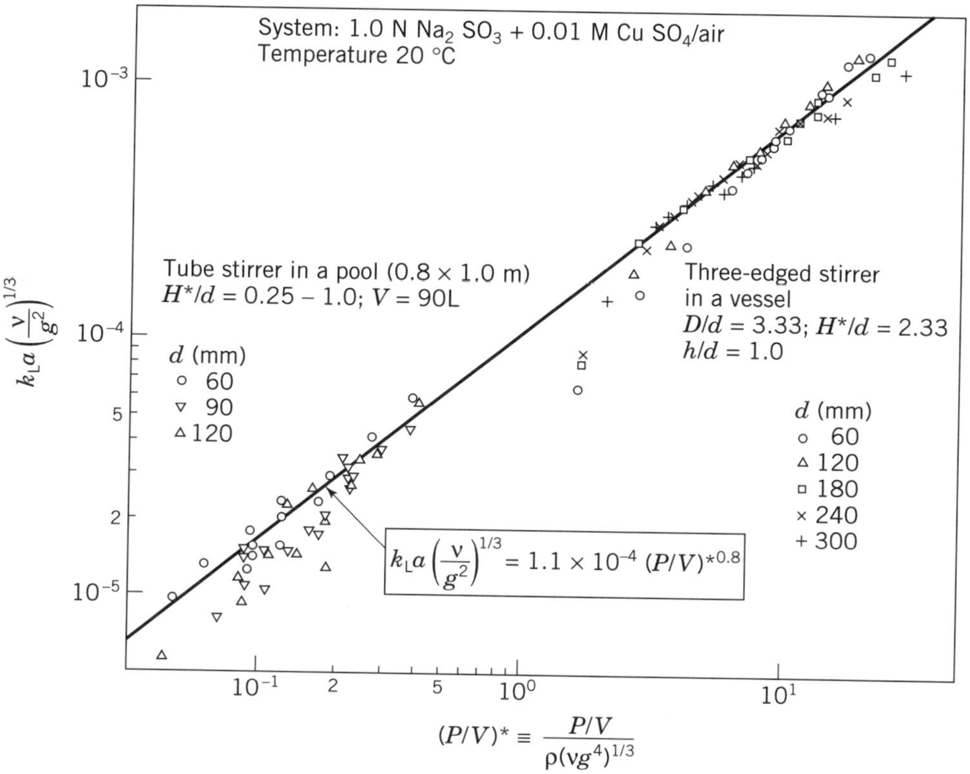

Figure 10. Sorption characteristics for a mixing vessel with two different hollow stirrers (scale factor 1:5) for a fully noncoalescing system. 1 N sodium sulfite solution in air. *Source*: From Ref. 15.

	$10^{-6}\nu$ (m²/s)		Rotational speed n (min⁻¹)									
Test liquid	75 °C	25 °C	5	6	9	12	17	24	32	48	60	87
Water	0.4	0.9									◪	◓
Glycerine, 47%	1.1	4				◐				▼		
Glycerine, 80%	5	35		▲		◑		◨		▼		
Glycerine, pure	33	750					◆					
Glucose syrup	74	2180	▽		△			□	◇			

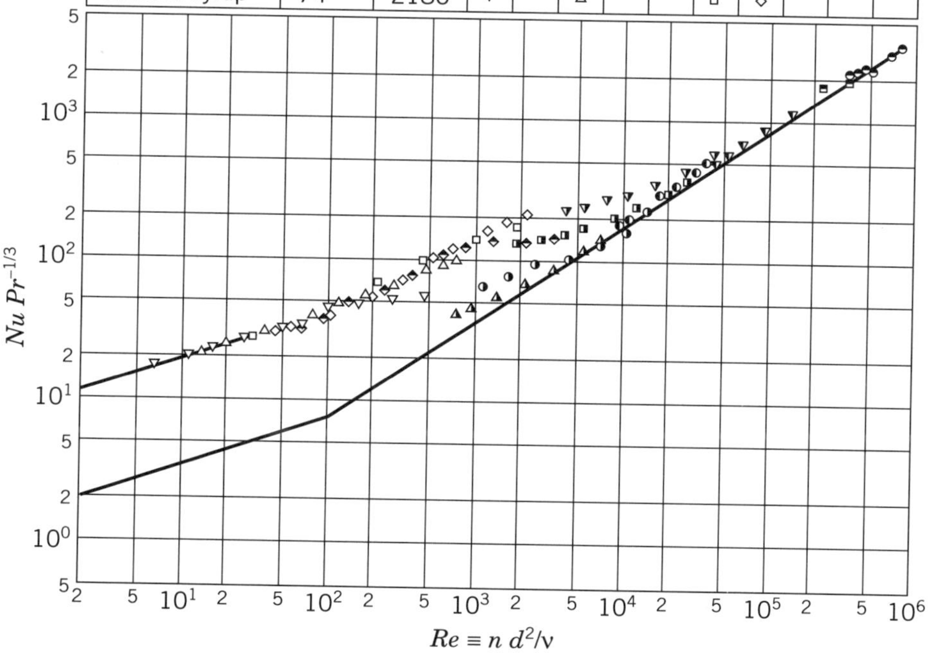

Figure 11. Enhancement of heat transfer at cooling in the laminar flow range ($Re < 10^3$) by wiping the heat transfer surface. Measurements: anchor stirrer with four arms equipped with wipers. Solid curve: anchor stirrer with two arms, D/d = 1.02 (clearance ≈ 2%). *Source*: From Ref. 20.

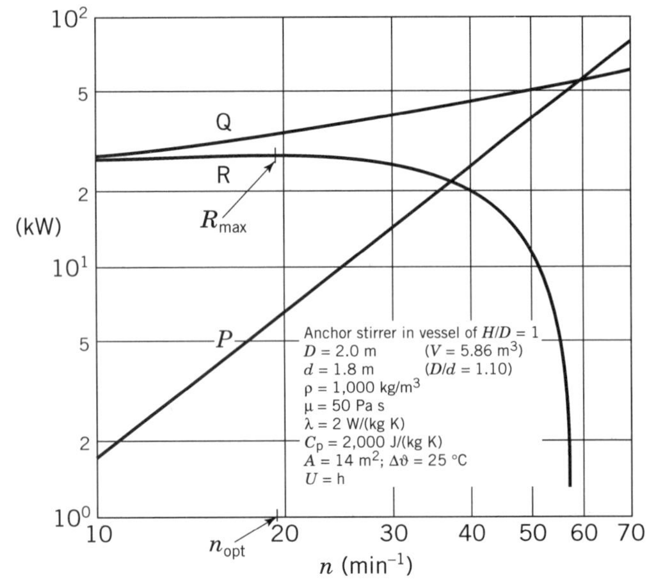

Figure 12. Determination of optimum stirring conditions (n_{opt}) for maximum removal of the heat of reaction R_{max}. *Source*: Refs. 3 and 21.

where

$$Y \equiv \frac{k_L a}{v}\left(\frac{v^2}{g}\right)^{1/3}$$

$$(P/V)^* \equiv \frac{P/V}{\rho (vg^4)^{1/3}}$$

$$(P_L/q_G)^* \equiv \frac{P_L/q_G}{\rho (vg)^{2/3}}$$

EXAMPLE 9

Heat transfer in mixing vessels

Contrary to mass transfer in fluid–fluid systems, the exchange area A in heat transfer is given and constant, which presents an advantage in dealing with mixing vessels. Here one considers three subsequent steps of heat transfer: namely, the convective heat transport to the outer side of the wall, the heat conduction through the solid wall, and the heat transport from the inner side of the wall to the fluid in the vessel that is to be cooled or heated.

The overall heat transfer equation

$$Q = UA\Delta\vartheta \quad \text{with} \quad 1/U = \frac{1}{h_o} + \frac{\delta}{\lambda} + \frac{1}{h_i} \tag{36}$$

contains thus the overall heat transfer coefficient U as a proportionality constant, which incorporates convective heat transfer coefficients h on both sides of the exchange area and the thermal conductivity λ of the wall of the thickness δ. Each of the heat transfer steps obeys different process conditions and must be described and dealt with separately. For heat transfer at the inner side of the wall, the heat transfer equation reads

$$Q_i = h_i A \Delta\vartheta_i \tag{37}$$

The heat transfer coefficient h_i as the proportionality constant in this heat transfer step is defined by this equation and can be determined only by measurement of its constituents:

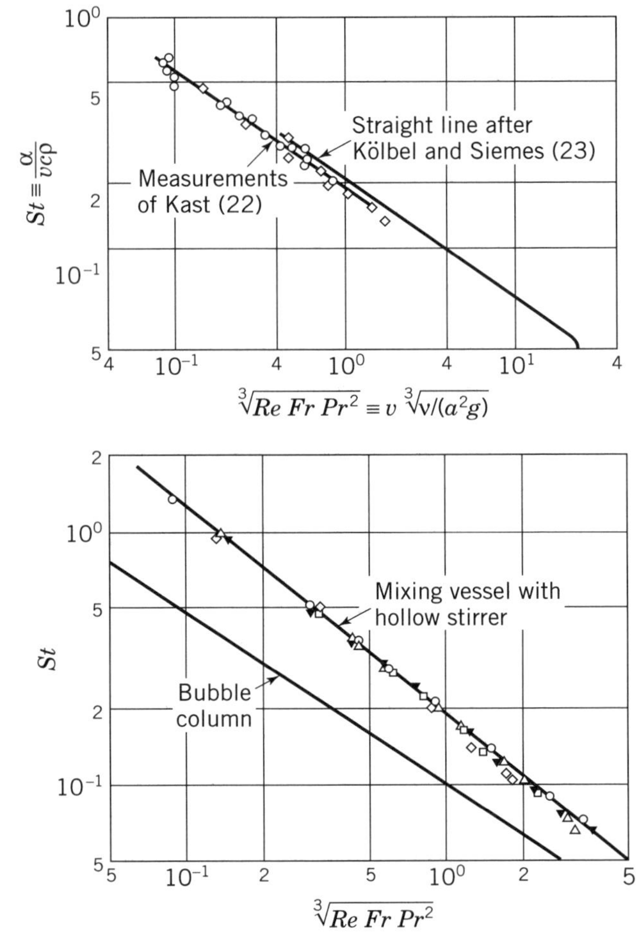

Figure 13. *Top*: Heat transfer characteristic of a bubble column. *Source*: from Ref. 22. Bottom: Comparison of bubble column heat transfer characteristic to that of a mixing vessel equipped with a hollow stirrer *Source*: Ref. 3.

$$h_i \equiv \frac{Q_i}{(A\Delta\vartheta_i)} \tag{38}$$

The heat transfer coefficient h_i in a mixing vessel of diameter D, equipped with a stirrer of a given geometry and of diameter d, will certainly depend on the physical properties of the liquid ρ, μ, μ_w, λ, C_p, (λ, heat conductivity; C_p, heat capacity, μ and μ_w, viscosity in the bulk of liquid and at the wall) and on the rotational speed n of the stirrer. (It must not depend on the temperature difference $\Delta\vartheta$, because it is defined by this value: equation 38.) It is assumed that the liquid in the vessel displays Newtonian viscosity behavior.

Then the relevance list is as follows:

$$\{h_i;\ D,\ d;\ \rho,\ \mu,\ \mu_w,\ \lambda,\ C_p;\ n\} \tag{39}$$

In connection with four basic dimensions (M, L, T, Θ) contained in the dimensions of the quantities of this relevance list, we will have to deal with $8 - 4 = 4$ dimensionless numbers:

Table 4. Important Named Dimensionless Numbers

Name	Symbol	Group	Remarks
		Mechanical unit operations	
Archimedes	Ar	$g\Delta\rho l^3/\nu^2\rho$	$\equiv Ga\ (\Delta\rho/\rho)$
Euler	Eu	$\Delta p/(\rho v^2)$	
Froude	Fr	$v^2/(lg)$	
	Fr^*	$v^2\rho/(lg\ \Delta\rho)$	$\equiv Fr\ (\rho/\Delta\rho)$
Galilei	Ga	gl^3/ν^2	$\equiv Re^2/Fr$
Knudsen	Kn	λ_m/l	
Mach	Ma	v/v_s	
Newton	Ne	$F/(\rho v^2 l^2)$	Force
		$P/(\rho v^3 l^2)$	Power
Ohnesorg	Oh	$\eta/(\rho\sigma l)^{1/2}$	$\equiv We^{1/2}/Re$
Reynolds	Re	vl/ν	$\nu \equiv \mu/\rho$
Weber	We	$\rho v^2 l/\sigma$	
		Thermal unit operations (heat transfer)	
Fourier	Fo	at/l^2	
Grashof	Gr	$\beta\Delta\vartheta g l^3/\nu^2$	$\equiv \beta\Delta\vartheta Ga$
Nusselt	Nu	hl/λ	
Péclet	Pe	vl/a	$\equiv RePr$
Prandtl	Pr	ν/a	$a \equiv \lambda/(\rho C_p)$
Rayleigh	Ra	$\beta\Delta\vartheta g l^3/(a\nu)$	$\equiv GrPr$
Stanton	St	$h/(v\rho C_p)$	$\equiv Nu/(RePr)$
		Thermal unit operations (mass transfer)	
Bodenstein	Bo	$vl/\mathcal{D}_{ax}$	
Lewis	Le	$a/\mathcal{D}$	$\equiv Sc/Pr$
Sherwood	Sh	$kl/\mathcal{D}$	
Schmidt	Sc	$\nu/\mathcal{D}$	
Stanton	St	k/v	$\equiv Sh/(Re\ Sc)$
		Chemical reaction engineering	
Arrhenius	Arr	$E/(R\vartheta)$	
Damköhler	Da	$\dfrac{c\Delta H_r}{C_p\rho\vartheta_0}$	Genuine (cf. Ref. 3, p. 148)
	Da_I	$k_1\tau$	
	Da_{II}	$k_1 l^2/\mathcal{D}$	$\equiv Da_I Bo$
			$\equiv Da_I ReSc$
	Da_{III}	$k_1\tau\left(\dfrac{c\Delta H_r}{C_p\rho\vartheta_0}\right)$	$\equiv Da_I\left(\dfrac{c\Delta H_r}{c_p\rho\vartheta_0}\right)$
	Da_{IV}	$\dfrac{k_1 c\Delta H_r l^2}{\lambda\vartheta_0}$	$\equiv Da_I RePr\left(\dfrac{c\Delta H_r}{c_p\rho\vartheta_0}\right)$
Hatta	Hat_1	$(k_1 D)^{1/2}/k_L$	First-order reaction
	Hat_2	$(k_2 c_2 \mathcal{D})^{1/2}/k_L$	Second-order reaction

$$\{Nu,\ Re,\ Pr,\ D/d,\ \mu_w/\mu\} \tag{40}$$

where

$Nu \equiv h_i(D/\lambda)$ Nusselt number

$Re \equiv nd_2(\rho/\mu)$ Reynolds number

$Pr \equiv C_p(\mu/\lambda)$ Prandtl number

The heat transfer characteristics $Nu = f(Re, Pr, D/d, \mu_w/\mu)$ are known today for almost all the stirrer types in use. They can be summarized by following process equations:

$$Re = 10^2 - 10^6 \qquad Nu = \text{const}\ Re^{2/3}Pr^{1/3}(\mu_w/\mu)^{-a}$$

$$Re < 10^2 \qquad Nu = \text{const}(RePr)^{1/3}(\mu_w/\mu)^{-a}$$

The constant in these equations depends on the geometry of the vessel and on the type of stirrer. The exponent a of the viscosity term is mostly given as = 0, 14, but in fact it is a function of the μ_w/μ term itself (9, 19):

$$\mu_w/\mu = 10^1 \quad a = 0.18$$

$$\mu_w/\mu = 10^3 \quad a = 0.12$$

Two additional aspects of great interest in biotechnology take into account the requirement, in most fermentations, that the upper temperature may not exceed normal physiological temperatures (30–37 °C). This results in a very low $\Delta\vartheta$ when a cryostat should be avoided on an industrial scale.

1. In the turbulent region, h_i can be essentially increased by implementing cooling coils in the vessel and bubble column, respectively. In the laminar range the increase of heat flow is achieved by fastening wipers on arms of an anchor stirrer (20). By this latter relatively simple implementation, the *Nu* number can be increased by a factor of 10, a demonstrated in Figure 11.
2. Increasing the rotational speed of a stirrer results in an steeper increase of the mixing power P (corresponding to heat flux into the vessel) than that of h_i (corresponding to the heat removal Q from the vessel). Therefore in each case a critical rotational speed exists at which the total heat removal equals the heat flux introduced by the stirrer minus that transferred via the vessel wall. The removable process heat is given by $R = Q - P$ (Fig. 12). The dimensional analytical treatment of this problem is given elsewhere (21).

EXAMPLE 10

Heat transfer in a bubble column

In bubble columns the pertinent process variable is the superficial velocity v_G. This intermediate quantity replaces the gas throughput and the column diameter. Therefore the relevance list reads here:

$$\{h_i;\ \rho,\ \mu,\ \lambda,\ C_p;\ v_G,\ g\} \tag{40}$$

providing 7 − 4 = 3 dimensionless numbers (22):

$$\{St,\ Pr,\ ReFr\} \tag{41}$$

where

$St\ [\equiv h_i/(v_G C_p \rho)]$	Stanton number
$Pr \equiv C_p \mu/\lambda$	Prandtl number
$ReFr \equiv v_G^3/(\nu\, g)$	Product of the Reynolds and Froude numbers

Figure 13 (22,23) depicts the dimensionless presentation of the heat transfer in bubble columns as well as in mixing vessels in which the liquid is aerated by hollow stirrers. It is shown that additional mixing roughly doubles the heat transfer. (*Note*: Such comparisons are possible solely by the use of dimensional analysis.)

Table 4 presents a list of often-used, named dimensionless numbers.

NOMENCLATURE

a	Thermal diffusivity ($\equiv \lambda/\rho C_p$)
A	Area
c, c_1, c_2	Concentrations
$\Delta c, \Delta c_m, c_f$	Concentration difference, logarithmic mean difference, and concentration of foamer and flocculant, respectively
C_p	Heat capacity at constant pressure
d	Stirrer diameter
D	Vessel diameter
$\mathcal{D}$	Diffusivity
$\mathcal{D}_{ax}$	Effective axial dispersion coefficient
E	Activation energy
F	Force
g	Gravitational acceleration
G	Mass flow, gravitational constant
h	Convective heat transfer coefficient
ΔH_R	Heat of reaction
k_L	Mass transfer coefficient (subscript L, liquid side)
k_1, k_2	Reaction rate constant (reaction order)
k_F	Flotation rate constant
l	Characteristic length
L	Dimension of length
m	Mass
M	Dimension of mass
n	Rotational speed; number of moles
$p, \Delta p$	Pressure, pressure difference
P	Power
q	Throughput
Q_i	Heat flow on the inner side of the wall
R	Universal gas constant
S	Surface area
t	Time
T	Dimension of time
U	Overall heat transver coefficient, equation 35
v	Velocity, superficial velocity
v_s	Velocity of sound
V	Liquid volume
w	Rising velocity (of gas bubbles or flocs)

Greek characters

β	Temperature coefficient of density
γ	Temperature coefficient of dynamic viscosity
$\dot{\gamma}$	Shear rate
$\vartheta, \Delta\vartheta$	Temperature, temperature difference
Θ	Dimension of temperature
λ	Thermal conductivity
λ_m	Molecular free path length
μ	Dynamic viscosity
ν	Kinematic viscosity
Π	Dimensionless product
$\rho, \Delta\rho$	Density, density difference
σ	(Interfacial) surface tension
τ	Residence time; shear stress

Subscripts

G	Gas
L	Liquid

M	Model, laboratory scale
T	Technological, industrial scale

BIBLIOGRAPHY

1. P.W. Bridgman, *Dimensional Analysis*, Yale University Press, New Haven, Conn., 1922, 1931, 1951. Reprint by AMS Press, New York, 1978.
2. Lord Rayleigh, *Nature* **95**, 2368 66–68 (1915).
3. M. Zlokarnik, *Dimensional Analysis and Scale-Up in Chemical Engineering*, Springer-Verlag, Berlin, 1991.
4. J. Pawlowski, *Die Ähnlichkeitstheorie in der physikalisch technischen Forschung* (Theory of Similarity in Physical and Technological Research; in German), Springer-Verlag, Berlin, 1971.
5. M. Zlokarnik and H. Judat, in *Ullmann's Encyclopedia of Industrial Chemistry*, vol. B2, VCH, Weinheim, Germany, 1988, pp. 1–33.
6. M. Zlokarnik, *Chem. Ing. Tech.* **42**, 1009–1011 (1970).
7. M. Zlokarnik, *Water Res.* **32**, 1095–1102 (1998).
8. M. Zlokarnik, *Ger. Chem. Eng.* **9**, 314–320 (1986).
9. M. Zlokarnik, *Rührtechnik—Theorie und Praxis* (Stirring Technology—Theory and Practice; in German), Springer-Verlag, Berlin, 1999.
10. M. Zlokarnik, *Int. Chem. Eng.* **27**, 1–9 (1987).
11. J. Pawlowski, *Veränderliche Stoffgrössen in der Ähnlichkeitstheorie* (Variable Physical Properties in the Theory of Similarity; in German), Salle Sauerländer, Aarau and Frankfurt am Main, 1991.
12. J. Pawlowski, *AIChE J.* **15**, 303–305 (1969).
13. H.-J. Henzler, *Chem. Ing. Tech.* **60**, 1–8 (1988).
14. H. Judat, *Ger. Chem. Eng.* **5**, 357–363 (1982).
15. M. Zlokarnik, *Adv. Biochem. Eng.* **8**, 133–151 (1978).
16. M. Zlokarnik, *Adv. Biochem. Eng.* **11**, 158–180 (1979).
17. M. Zlokarnik, in H.-J. Rehm and G. Reed eds., *Biotechnology*, vol. 2, VCH, Weinheim, Germany, 1985, pp. 538–569.
18. M. Zlokarnik, *Chem. Eng. Sci.* **34**, 1265/1271 (1979).
19. M. Hruby, *Int. Chem. Eng.* **7**, 86–90 (1967).
20. H. Judat and U. Johenneken, *GVC Ann. Convention*, Stuttgart, Oct. 3–5, 1990.
21. J. Pawlowski and M. Zlokarnik, *Chem. Ing. Tech.* **44**, 982–986 (1972).
22. W. Kast, *Chem. Ing. Tech.* **35**, 785–788 (1963).
23. H. Kölbel and W. Siemes, *Chem. Ing. Tech.* **30**, 400–404, 729–734 (1958).

See also BIOREACTORS, AIR-LIFT REACTORS; BIOREACTORS, FLUIDIZED-BED; BIOREACTORS, CONTINUOUS STIRRED-TANK REACTORS; MASS TRANSFER; SCALE-UP, STIRRED TANK REACTORS.

DISSOLVED OXYGEN. See FERMENTATION MONITORING, DESIGN AND OPTIMIZATION; FERMENTER DESIGN; HYPOXIA, EFFECTS ON ANIMAL CELLS; MASS TRANSFER.

E

ECONOMICS

HAROLD REISMAN
Biotechnology Results
Weston, Connecticut

KEY WORDS

Aseptic processing costs
Capital investments
Compensation issues
Contract manufacturing
Cost-of-goods
Economics
Equipment cost
Regulatory issues and validation
Scale-up
Start-up
Technology transfer

OUTLINE

INTRODUCTION

The biotechnology industry requires large investment of funds. The market capitalization of U.S. biotech companies was estimated at $93 billion in 1997 (1). Even if the profit motive were to vanish, allocation and expenditure of resources would assume great importance. Assembly and distribution of economic resources in biotechnology are required in at least six general areas. First, there is the need for discovery. Next comes the need for development, scale-up, and pilot production. Following (or concurrently) come testing, evaluation, bioactivity determination, and clinical efficacy. Manufacturing facilities must be funded. Because regulatory and validation issues are becoming more important and more complex, more resources are needed. Finally, compensation issues—which were once included in the relatively minor category of marketing and sales—are now encompassed within new study and negotiation matters that have spawned a subindustry of its own. As third-party payers become larger and as government involvement and regulations grow, economic issues and cost–benefit analyses will grow accordingly.

The current health care system is market driven rather than value driven. It is apparent that economic matters, while always present in biotechnology, continue to grow in importance and are intruding more and more into what was a less complex world of discovery, followed by production and sale of bioactive molecules. Coming decades will emphasize gene-based methods of drug discovery, targeted diagnostic applications, and speedier drug development, and more powerful and more precise treatments can be expected. It seems clear that creation of this new pipeline of pharmaceuticals and devices will be costly for the producer and for the user-payer. The profusion of small biotech companies (their creation—and in some cases demise—can be expected to continue) only adds to the complexity of funding and resource allocation for each of the following phases:

Discovery	Manufacturing
Development and scale-up	Validation and regulatory issues
Bioactivity and clinical issues	Payment and compensation

DISCOVERY

In 1993 there were believed to be some 1,100 specialty biotechnology firms worldwide (with an additional 625–725 firms with an interest in biotech) (2). According to a more recent estimate, there are now 1,274 biotech companies in the United States alone (1). Sales of biotech pharmaceuticals reached $5.9 billion in 1992. Since 1980, biotech companies have raised $20 billion in public markets. Another estimate is more sobering, if not frightening: for each U.S. public biotech company to bring one product to market, a further $40 billion would be needed. These dollar figures point to either excellent discovery or superb programs that will result in discovery. In recent years, the knowledge has circulated that even with superb drug discovery, drug development and drug creation will take much longer than first thought. Venture dollars are apparently shifting from therapeutics to devices and diagnostics (3).

The *Forbes* article (1) gives estimates of total investment (initial public offerings, secondary offerings, private capital, partnering, and venture capital) in 1994 ($3.3 billion) and in 1995 ($6.3 billion). The drug discovery process (including approval) continues to take a longer time. With a lengthening time scale, expenditures rise. The complexity of the clinical testing process coupled with the ever-problematic review procedure means that costs often grow exponentially rather than linearly. A recent time and cost estimate points to an average time for drug discovery and approval of 15 years and a total cost that may reach $400 million (4). More recently, J.L. LaMattina, vice president for discovery research at Pfizer, broached the half-billion-dollar mark by stating, "The investment, from start to finish, is on the order of $500 million (per new product)" (5). It has been estimated that for the year 1996, expenditures for research and development reached $7.9 billion for the U.S. biotech industry as a whole (1). The entire drug development and approval process is shown in Figure 1 for the general case of a novel drug.

In a decade, the cost estimate for discovery, development, and approval processes has risen by a factor of 3–4. There is no immediate expectation that this ever-increasing cost will be capped, much less reversed. It is clear, then, that expected return (dependent on sales price, cost of production, and market volume) entails greater hurdles and more complexities. The chances for failure exist at any step along the way, and financial loss can also be expected to be greater if any of the hurdles are not overcome.

Discovery does not spawn only technical articles; full-length books have been published on the process of tying science to dollars. One such view is given of the birth of Vertex (6). Even allowing for some melodrama in the narration, this book is a useful guide to the various threads that connect research ideas, academia, and Wall Street. Interesting questions of program selection, methods of concurrent negotiation, and relations with a major academic institution are detailed. As a follow-on, it is reported that Vertex has three candidates in clinical trials and has added a number of research agreements (7).

It has always been clear that genetic information is valuable. Quantifying the value has been difficult. A relatively recent sale of the genetic sequence of *Helicobacter pylori* (by Genome Therapeutics to Astra of Sweden) was made for $22 million (8). Since the fully loaded cost of a research scientist may fall in the range of $150,000 to $250,000 annually, it is incumbent on corporate managers, as well as research managers, to not only make the correct program selections but also to follow up with active direction and necessary alterations to the programs. The *burn rate*—or the dollar outflow in initial stages of a company's history during the period of no sales revenue—is followed by insiders and outsiders alike as a measure of both how much science is being performed and for how long this rate can continue. Twenty-five people in a start-up, while considered quite small, might have a burn rate of $5 million annually.

There are good sources of ideas on how to make appropriate decisions (9) and also on how to manage research and development (R&D) more effectively (10). These references describe results of extensive surveys of industrial (including biotech) organizations and give useful directions for improvement and appropriate management of R&D and technology transfer.

Current methods for discovery of leads are discussed by Norrington (11). He estimates that discovery and preclinical work could cost $10 million. Many candidates will fail at this point and others will continue in development, cost more money, and fail at a later date. The estimates given for cost of phase I trials is $2.3–7.5 million, for phase II, $7.5–15 million, and for phase III, $125–250 million. High-throughout screening (HTS) is discussed; devices for screening 1,500 samples/h with one or more readouts (spectrometry, radiometry, fluorometry/luminometry) are described. Instruments are now being developed to handle 864 well plates; concurrent testing using multiple screens is strongly suggested. Bevan et al. (12) also stress HTS screening (23 references) but provide, as well, comparisons with rational design and combinatorial synthetic methods.

A very detailed article on resource allocation at SmithKline Beecham should be read (13). Although there

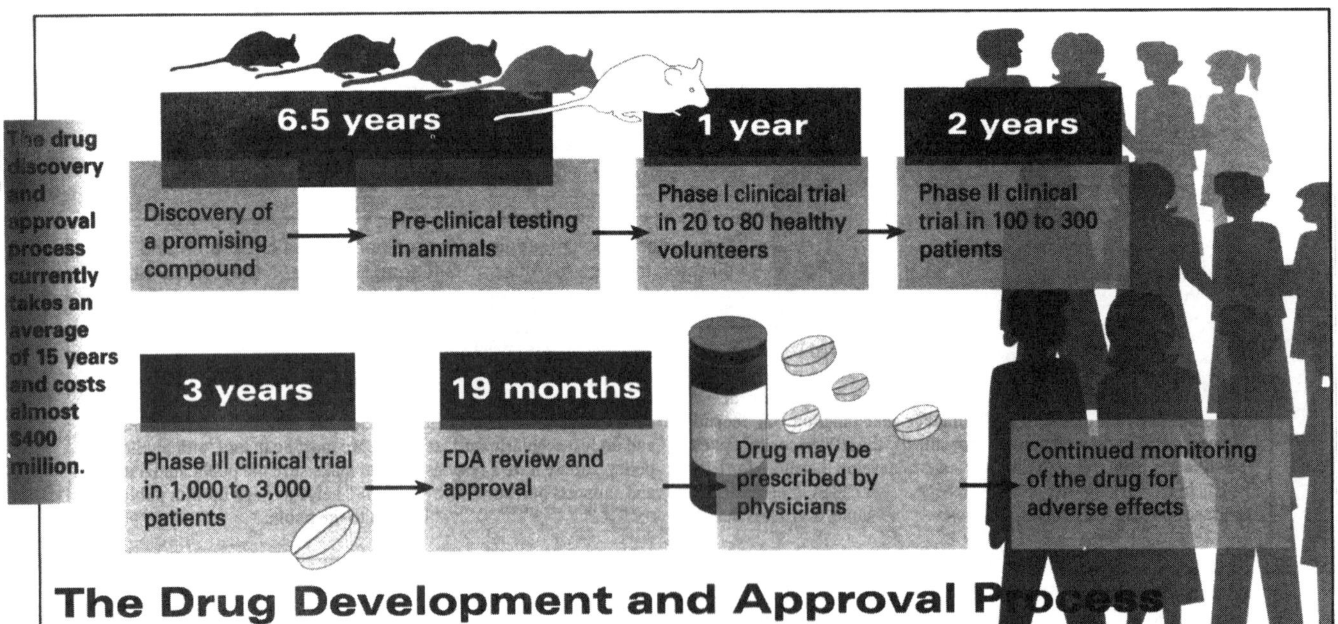

Figure 1. The drug development and approval process. *Source*: Pharmaceutical Research and Manufacturers of America, Washington, D.C.

are alternate scenarios for resource allocation, the authors carefully describe the thought processes in designing a better method for decision making. The basic steps are generation of alternatives, valuation of alternatives, and, finally, creation of a portfolio and allocation of available resources. The very important soft issues (e.g., information quality, credibility, trust) are faced and, so far as possible, defined and controlled. The effort to reach the improved allocation process took about 2 years and was implemented in late 1995; its success can only be evaluated by future results.

Although every potential area of biodiscovery cannot be covered here, two examples point to the need for continual updating of existing technology. The antibiotic market worldwide is estimated at $2.3 billion (14). In the decades following World War II, there were predictions that microbial infection could be halted, or suppressed to the extent that further work on finding novel antibiotics would be unnecessary. The facts are otherwise; drug-resistant strains have evolved in diseases such as tuberculosis, malaria, cholera, and pneumonia. The 160 antibiotics on the market may have been overprescribed or improperly used, or both. The result is a growing research effort in areas of new drug discovery, genome sequencing, and development of new vaccines. Large companies have formed partnerships (Pfizer with Microcide and Schering-Plough with Genome Therapeutics) to speed discovery.

Current antimicrobial therapies and newer agreements are described in a short review (15). The regrowth of chiral chemistry in biotechnology has been related to greater regulatory stress, tighter environmental controls, and cost control (16). Both resolution and asymmetric synthesis are being pursued. Atenolol (for hypertension) and albuterol (for asthma) are enantiomers, each of which exceeds the billion-dollar mark in annual sales. Patent protection is another reason for moving toward the single isomer (17). Drug information derived from use of the racemate may be used, in many cases, to speed clearance of the active (or more active) isomer. Schuster and Menke report that of 1,327 therapeutic compounds, 528 were chiral; of the 528, only 61 were used therapeutically as the single isomer (17).

It should be recognized that the recent emergence of "fully integrated discovery organizations" is also altering the discovery landscape (18). A company (it can be very large or very small) acts as a coordinator of discovery and/or development that takes place outside its boundaries. External companies (there can be one or more working on the same program) supply the research expertise that was once the prerogative of the pharmaceutical company itself. Turner et al. stress biocatalytic processes and report on annual production of various bioactive compounds (18). A cautionary note should be included. Discovery is not easy, technology transfer is no less complex, and scale-up has its own set of major hazards. The culture chasm (6,19) between scientists and business managers in biotech firms was vast a decade or so ago and, in many ways, bridging the gap has not been wholly successful. The scientific "set" and the business "set" tended to run their own shows, had different time scales, had few "common points of reference," and experienced trouble communicating and cooperating. Unless these real and potential difficulties are recognized, confronted, and overcome, no amount of money coupled with "good science" will result in a useful and usable device, drug, or therapeutic agent.

DEVELOPMENT AND SCALE-UP

In 1980, 2% of all prescriptions were generic; by 1993, half of all prescriptions involved generic drugs (20). In the 1993–1999 interval, patents will expire covering branded drugs with annual sales totaling $20 billion. Pisano and Wheelright give the current estimate of $359 million (as of 1992) to develop a new drug (20). This is about threefold higher than the 1987 estimate ($120 million). In the 1990s, manufacturing cost has risen to 20% of sales. All these drivers point to the need for more speed and more cleverness in process R&D. A general review is useful to cover rules of successful development (21).

Mukesh gives 15 steps for improving the success rate of process development (22). Although the overview is general, two points require special emphasis in biotech development. The first concerns the need to collect information on raw materials' sources and purities (and, one might add, impurities as well). The second emphasizes the desirability of having scientists and engineers who develop the process involved (or at least available) at various stages of advancement (or scale-up) until the process is fully commercialized. These two points are often neglected or glossed over in biotech scale-up.

One specific and useful example of a complex and successful scale-up of a bioprocess is available. Prior et al. discuss process development in the manufacture of inactivated HIV-1 (23). A process flow diagram (PFD) is given; all operations occurred within a BSL-3 facility. Cost effectiveness in scale-up is emphasized. A scalable chromatography process is described.

Another review (with 81 references) covers newer concepts in bioprocess R&D (24). Certain operations are covered in depth (mixing, foam control, use of membranes), and other subjects are reviewed and appropriate references given. Subjects include on-line analysis, estimation and control of the physiological state, automation, and error reduction. No developmental program can be expedited and relied on to generate meaningful data without a modern pilot plant. A dozen authors contribute to a review that provides 20 detailed tips on how to keep a pilot plant on the fast track (25).

Although more mundane, an article by Merck workers (26) should prove useful to anyone involved in a cell culture or fermentation pilot plant. The Merck facility was first commissioned in the early 1980s, and significant improvements have been made in hardware, maintenance, equipment replacement, instrumentation, and cleaning. Examples are given of selected upgrades. The facility has fermentors ranging from 0.28 to 19 kL in scale. The message highlighting the need for continuous improvement is clearly conveyed. In biotech operations, what was voluntary but sensible has now come under regulatory review. That is, appropriate documentation (even at the pilot scale) is more essential than ever.

Abate et al., who give careful detail on how and why effective documentation of pilot plant operations is to be achieved and offer pointers that are directly applicable to production-scale operations as well (27). The process operating log should not only include methods involved in change control, it should give the project background and list equipment to be used, as well as auxiliary systems and services. The log also should describe the process itself, identifying materials, safety issues, and process preparations. An equipment log, data acquisition and detailed process operations, and start-up and shutdown checklists should be included. Examples of forms are given (27).

In any aseptic culturing, agitation issues are invariably involved. Change is commonplace, and in recent years there has been a shift to the use of hydrofoils for improved mixing with reduced (or constant) power consumption. McFarlane and Nienow present an excellent review on this subject and follow with air–water experimental studies (28). Dream has reviewed the various bioreactors available for bioprocessing (29). Not only is there a discussion of various fermentor designs, there is further analysis of problems related to the use of serum, controls to use (including ORP), sterilization, effluent handling, surface polishing details, and relief devices. Economic factors, as well as safety issues, are discussed under the overall heading of continuous mammalian cell culture (30).

The use of batch or continuous processing is based on capital investment (usually lower for continuous processing), quantity of material to be processed, genetic stability, reliability of asepsis, and degree of flexibility needed in the production plant. Design and scale-up of monoclonal antibody production is also reviewed (31,32). Reference 32 has an exhaustive bibliography with 156 references. Reactor designs are reviewed and scale-up considerations discussed in light of the following inputs: government regulation, market requirements, supplier capabilities, capital budget limitations, cell line idiosyncrasies, existing plant layout, and alternate process systems. Actual cost data (using dimensionless factors) in an industrial setting are given by Lonza workers (33). Both fermentation and isolation are considered, and optimization covers the total process rather than one part or another. A fine chemical, L-carnitine, is the subject, and costs of raw material, personnel, overhead, and depreciation are included. A fed-batch operation gives the lowest relative cost. Since this nutrient is actually being synthesized at production scale, the article by the Lonza workers is not only fairly unique but interesting as well.

A very fine article by Middelberg (34), incorporating economic analysis in production of a recombinant protein, once again emphasizes that maximum yield in one step (biosynthesis) will not always give the optimal global solution; in the case study, maximum yield will lead to a suboptimal solution. Protein refolding introduces significant complications. The analysis incorporates direct fixed capital, consumables, waste treatment, and other costs (e.g., labor, utilities, and R&D). Many plots that show interaction of parameter estimates are given. Another paper includes the very important concept of process controllability as a major input in conceptual design (35). Conventional steady-state economic design does not include operability and control capability of each design, that is, the usual tendency is to merely select a minimum cost alternative, assuming that the plans will give the same degree of process controllability. This is a hazardous assumption. The methodology presented incorporates dynamic controllability into the overall assessment and so is a unique contribution. There are two additional examples of preliminary process design including some economic analysis. The enzyme polygalacturonase is manufactured with a 90% recovery from whole broth (36). A continuous lactulose process (including separation and purification) is described (37).

In design and scale-up of bioprocesses, one relatively new development should be considered whenever any aseptic operation is planned. In the past, filling or bottling was the only thing that came to mind when the word *asepsis* was noted. Now, however, novel cell therapies are in use and medical products are incorporating living cells. Products, for one reason or another, cannot undergo terminal sterilization. Clean rooms—or better, controlled environments—are now often considered processing space. It has been estimated that worldwide sales of clean rooms and related equipment reached $2.7 billion in 1994 (38). By the turn of the century, this figure is expected to exceed $4 billion.

The newer concept of a minienvironment or microenvironment (under the general heading of barrier technology) can result in lower investment, lower operating costs, higher product yield, and increased operator comfort. Costs for clean-room space escalate dramatically as particle size control becomes more stringent. Table 1 gives estimated cost data. Currently, class 1 and class 10 areas are found in the electronics industry; in general, class 100 is found to be acceptable for aseptic operations. Given these costs, it is obvious that overdesign would be prohibitively expensive, and efforts are under way to limit the working volumes for class 100 (or better) spaces. Regulatory requirements worldwide are being harmonized, but one should strive to satisfy current regulations of the various governing bodies; it is possible to meet the U.S. and EC guidelines without incurring onerous costs (39).

Issues in design of production facilities for cell and gene therapy are reviewed in detail (40). A range of $250–350/ft^2 is given as total installed cost for a class 10,000 space, but this includes provision and commissioning of walls, ceilings, HEPA filters, low wall returns, lights, fire protection, roll cove flooring, HVAC, controls, ductwork, and minor process support services. The requirements (such as 100% once-through air) for cell and gene therapy (same

Table 1. Cleanliness Is Costly

Class	Cost per square foot
100,000	$100
10,000	$200
1,000	$300
100	$400
10	500^+$
1	$1,000–4,000$^+$

Source: Excerpted by special permission from *Chemical Engineering*, May 1995. Copyright © 1995, by McGraw-Hill, Inc., New York.

space classification) would increase costs to \$400–450/ft^2. There are a number of useful references concerning design and cost of barrier isolation systems (41–43) and the use of isolator systems for sterility testing (44). A good overview by Akers (45) discusses numerous applications and gives comparisons of start-up costs and operating expenses for conventional clear rooms and isolator designs. Table 2 gives details of this comparison and shows the significant operating savings attainable with use of isolator technology.

The final major product of a development program is successful technology transfer. (This is equally important during the transfer of information from the research to the development organization.) A number of reports will be included, and there should be continuous interfacing between the groups involved in the transfer of the technology. An appropriate schedule should be established. Each organization has its own methodology; however, Table 3 represents an attempt to list all the information and factors that should be documented and communicated before and during the technology transfer. Any reader who has been involved in such a transfer will no doubt be able to insert additional factors.

It is surprising that insufficient emphasis is often placed on the process of technology transfer. Large, even enormous, sums are expended in research and discovery, even more might be expended in building a new facility, and an overarching goal is established to complete the technology transfer in the shortest possible time with minimum planning input. One might even view some cases of technology transfer as a sort of necessary evil to be completed as quickly as possibly so that the product can be made. It is as though time spent on technology transfer were somehow wasted or unnecessary. The results of such an approach are often painful to see, and the costs incurred (although speeding the process) sometimes dwarf other, more carefully planned costs.

Technology transfer is one of the most important steps in bringing a product to market and at minimum, the planning effort expended here should equal that devoted to all aspects of research, development, scale-up, and construction of a production plant. It is suggested that a budget (time and manpower) be created as part of the technology transfer process and the entire process receive the same high level managerial attention as other parts of the project cycle. An interesting case study is available for a real-life manufacturing development project (46). The product is a recombinant β-interferon. Subjects discussed include team creation, project scope, and timing and feedback. The plan covered some 2 years. Not surprisingly, not everything went according to plan. The difficulties noted are not unique to Biogen, and the lessons learned are probably applicable to many other biotech and regulated companies. There is also a useful appendix that summarizes the U.S. regulatory process for pharmaceuticals.

BIOACTIVITY AND CLINICAL ISSUES

This section could as easily be located after "Manufacturing" (later) but in the normal course of events, bioactivity

Table 2. Operating Expense Comparison: Isolator System versus Conventional Clean Room

	Isolator system	Clean room
Direct operating expenses (annual):	\$10,300	\$23,040
	Replacement suits, sleeves, gloves, gaskets, etc.	Full gown set (\$12)
		2 operators/day, 240 day/yr
		4 gown sets/operator/day
Gown-changing time:	\$530	\$5,330
	1 min/operator/change	10 min/operator/change
	4 min/day/operator	40 min/day/operator
	53 h/yr @ \$10/h wage	533 h/year @ \$10/h wage
Maintenance/decontamination/labor:	\$4,420	\$14,050
	3 h/week = 156 h/year	16 h/week = 832 h/yr
	\$10/h wage	\$10/h wage
	1 VHP cartridge/week	52 disinfections/yr
	\$55/cartridge	4 gown sets/week
Utilities cost (kW h/yr):	\$405 (3,372/yr)	\$2,775 (23,126/yr)
@ \$0.12/kilowatt hour	7 blowers @ 55 W	8 blowers @ 330 W
	24 h/day	24 h/day
	365 days/yr	365 days/yr
DOP testing:	\$60	\$480
	Every 6 months	Every 6 months
	1 operator/3 h, \$10/h wage	2 operators/12 h each, \$10/h wage
	Expense summary	
Start-up expenses	\$265,100	\$284,140
Annual direct expenses	15,715	45,675
Annual indirect expenses	None	300,000
	\$280,815	\$629,815

Source: This table originally appeared in J.E. Akers, *BioPharm* **7**, 43–47 (1994), with the permission of Advanstar Communications, Cleveland, Ohio. This is the corrected version (October 1994 issue, vol. 7, no. 8, p. 10).

Table 3. Documentation and Factors Involved in Bioprocess Technology Transfer

1. *Raw materials and medium composition*	
Specifications and QC tests	Shelf life
Order of addition	Storage conditions
2. *Cell strains*	
Validation and QC tests	Growth curves
In-process tests	Preparation of banks
3. *Process flow diagram (PFD)*	
Cycle times	Operating conditions
Labor use	Record (data) keeping
Sterilization sequences	Shipping studies
Compounding/packaging	
4. *Documentation*	
Written procedures	Pictures (or videos)
Process response data	In-process tests (and frequency)
Emergency situations	Start-up and shutdown
5. *Training requirements and schedule*	
6. *Acceptable performance*	
Critical variables	Inventory min/max
Acceptance criteria	Operating parameters/ranges
Finished goods variation/testing	Troubleshooting
7. *Process validation*	
Document review	Time per step
Data comparison (various scales)	Batching
QC tests	Stability and shipping studies
Master batch record	Permissible recycle
8. *Instructions for use*	
User manual	Package insert
Labels	Storage by customer
Accessories	Warranty
9. *Safety and sterility*	
Hazards	Environmental testing
Asepsis (precautions)	Other precautions
Special equipment	Cleaning (and test) procedures
Record of asepsis achieved	Off-spec material handling

determination and preclinical activities occur prior to start-up of a manufacturing plant. Initial quantities (which could amount to grams but might be hundreds of kilograms) are usually produced at large-laboratory scale or in a pilot plant. If preliminary data are promising, production plant design could be accelerated and orders might be placed for long-delivery items, but that would also mean creation of greater financial exposure. Early clinical trials (even phase I studies) might involve a few to a few dozen subjects; material to be evaluated normally is pilot plant product. Of course, many steps occur concurrently; we consider this subject at this point for convenience, with the recognition that clinical trials and clinical issues continue through full-scale plant construction and operation and for months or years afterward.

One reference already noted (43) gives the needs (including minimal capital outlay) for creation of validatable clinical production facilities. In a 1994 review, a time and cost flowchart is presented (47). Preliminary development is said to take 1–3 years, and cost is in the range of $0.5–5 million. The success rate is stated to be 5–20%. This presumably means that from 1 in 5 to 1 in 20 candidates is deemed worthy of proceeding to the next hurdle. The next step is called investigational development, and the period of this second stage (of three) is anywhere from 3 to 6 years; cost is $5–100 million. The success rate here is given as total of approximately 20%, or 1 in 5. The final stage is review by the U.S. Food and Drug Administration (FDA); cost range is $3–100 million, and time frame is 1–3 years. Success is 1–5%. It would seem that this, too, is an overall rate (of 1 in 100 to 1 in 20). There might be drugs or devices that would achieve clearance in 5 years (from discovery) and incur a total cost of $8.5 million, but these are either very simple devices or molecules, or they represent a vanishingly small segment of materials or substances that weather the difficulties described. Fitzpatrick and Mackler posit that before 1990, biotech companies could measure or expect success based on a technology portfolio, an intellectual property position, and access to capital (43). In the 1990s, regulatory management has apparently become the primary element of success. There are three reasons for the failure of the FDA to approve a given therapeutic after a clinical trial:

1. Failure in clinical design
2. Improper (or failure in) protocol compliance
3. Deficient GCP documentation

Since moving along the success curve (discovery to investigation to FDA review) means an exponentially rising dollar outlay, it makes sense to

- Plan from end to beginning (i.e., understand and define the end product, and all its requirements, and plan backward from there).
- Integrate all elements of the process.
- Conduct more phase I and II trials or include more variables; speed in starting phase III is not an unalloyed blessing.
- Be responsible and flexible (i.e., have alternatives for every potential holdup or problem).
- Seek the right people within and without the organization.
- Validate and reality-check *often* (i.e., perform many in-house reviews, studies, and audits). Find problems and defects before others do.

The clinical supply process requires a dedicated team and demands dedicated protocols. Ehrich et al. cover manufacturing, packaging, and labeling (48). The preapproval inspection of the FDA now includes an evaluation of the development history of the product. This invariably includes an evaluation of the clinical supply operation. Ehrich et al. stress the need for a manufacturing, packaging, and labeling protocol (called an MPL) and provide a sample table of contents for a typical MPL protocol. Required details are listed for the test drug, any placebo, and any positive control drug in the evaluation. It should be clear that if all this effort is properly applied, the next-generation clinical formulations will be more rapidly created (in a documented manner), and technology transfer to full-scale manufacturing and concurrent phase III trials will be speeded and, in fact, simplified.

Another report describes a semiautomatic system for R&D and clinical use for liquid-filled hard gelatin encapsulation (49). The authors and client believe that the device is well suited for small-scale batches and phase I and II studies. This reference is useful not only for the specific technology but also for the necessary procedures, testing, and validation, which can be extrapolated to any such device used for small test quantities.

It is also not uncommon to use a contract research organization (CRO) to assist in performing a clinical trial. The CRO's contribution may include formulation, dose development, protocol development, shipping, report gathering, and even analyzing results. Of course, the CRO may perform one or more of the functions involved in the overall scheme. For smaller companies, it probably is more efficient in use of time and money to hire a CRO than to do the work in house. Many of the issues relating to CROs are covered by Vogel et al. (50). The contract between the sponsor and the CRO is very complex, and its preparation is very time-consuming. Elements of the contract are given in a very detailed fashion, and relevant regulatory documents and references are shown. In 1995, *The Biomedical Engineering Handbook* was published (51). This work contains a wealth of information (relating to medical products, in the main), but two sections are relevant in the context of bioactivity (or utility as far as a medical product is concerned). A section by Oberg covers assessments for effective product development and another segment by Holohan covers health technology assessment (51). Payment and compensation issues are covered in the following, but a reasonable model of value of a proposed product should be in hand even before early clinical testing begins. Although the model certainly will be modified, it is well understood that product approval without subsequent sales is not a successful outcome.

MANUFACTURING

Starting some 15 years ago, there was a marked increase in both quantity and quality of articles and other publications involving economic issues in biotechnology. The number of these publications remains quite low compared to the burgeoning growth of all biotechnology literature. Still, the improvement noted is a positive development and speaks for maturity of the field. It is likely that before 1980, there was a tendency to protect real or perceived economic secrets. With growing interest in the field (including academia and the government) and with the understanding that many of the so-called proprietary matters were becoming common knowledge, there was an opening that continues to evolve. One well-known handbook (52) has a chapter entitled "Principles for Costing and Economic Evaluation for Bioprocesses." The first edition of this handbook (1983) had no such chapter, and very little economic information was included in the chapter entitled "Processes." The latest edition (1991) includes definitions of economic terms, various economic evaluators, indices for estimation, equipment, and operating costs (mostly 1980s), and useful information on downstream processing. Economic aspects of some traditional processes (ethanol, citric acid, single-cell protein, monosodium glutamate) are included.

A recent book (1995) on principles of fermentation technology, which could easily be used as a text, has a good chapter on fermentation economics (53). Containment issues are covered, and the point is once again made that alternative processes in extraction and purification should be evaluated concurrently in the development phase before commitment is made to a single design. The importance (and cost impact) of effluent treatment and the need for control of waste streams are emphasized. Economic matters in biotechnology were either poorly covered or not mentioned in many of the earlier introductory reference books. This was due more to the lack of information in the literature than to any disinterest on the part of the authors.

Cost estimation for biotech purification is covered by Dwyer (54), in a chapter in a book on protein biochemistry. Economic issues and the processes covered relate to high-value peptides and proteins. There is a review of available equipment and various potential processes; scale-up and related costs of chromatographic purification processes are given. Overall process design is discussed, and both equipment and operating costs are given. In an entire book devoted to economic analysis of fermentation processes, this author takes as his theme the project cycle and how economic matters impinge at every step from discovery (research planning) through plant commissioning and commercial product shipment (55). The specific example followed is a citric acid process and plant. Profitability matters are covered. One or more of these references (52–55) should be consulted so that terms are understood and available sources for further estimation are known. The many examples provided will be helpful for economic evaluation and estimation of any novel biotechnology process.

Related publications should be consulted for an understanding of where manufacturing fits and for an appreciation of facility planning including estimation of costs. Durkin (56) reiterates the point that cost of goods (COG) as a percentage of pharmaceutical sales has doubled in the last 20 years. Both cost pressure and selling price pressure demand that manufacturing strategy be developed and incorporated into the business plan. In the past, manufacturing was more or less remote from mainstream business activity; organizational inefficiency is now being remedied. The results should be speedier transition from R&D and more facile product launch. Durkin presents an interesting play on words: a biotech organization can either be "leading edge" or "bleeding edge."

Part of the determinant is the success with which manufacturing strategy is incorporated as a key element in the overall plan. Biotech facility planning has a complex set of unique requirements that must be incorporated into the more mundane and better known set of needs. Hubbard (57) gives key steps in preparing a design brief, defining project goals, and stating needs prior to preparation of a design. One step involves preparation of individual room data sheets with appropriate details; another covers the adjacency matrix and a spatial relationship diagram. A tremendous effort is needed prior to final design and commitment of major capital. A few of the issues to be mutually determined (between the biotech firm and the engineering/architectural firm) are plant room size, room location, horizontal services, service chases, vertical services distribution, and maintenance access. Many firms continue to face the need for conversion or adaptation of existing facilities.

A real-life example involving a European firm, Boehringer–Ingelheim, is available (58). A dedicated facility (single-product use) was built in the period 1985–1987 and totaled 110,000 ft^2. It was decided to change to a multiuse facility, where three different products could be made concurrently. Recognized limitations were noted in inoculum development, protein purification, and formulations areas; plans were developed to remedy the situations and also convert the HVAC system to give once-through air. Some 50,000 ft^2 of space was added. The project took 3 years to complete; successful start-up occurred in 1995. Planning and execution included a strategic capacity capability (equal to 30% of overall throughput). This is a good point to remember regardless of whether the plant being built from the ground up or converted. Added capacity for developmental runs or unforeseen new products should always be considered; the added investment may prove itself many times over.

Another case study in facility design concerns a Fujisawa product, FK 506 (tacrolimus) (59). Investment in the plant, located in Ireland, was £17 million. Conceptual and basic engineering occurred in late 1990, and the grassroots facility involved 6,000 m^2 of space. The pure bulk material was made in Japan and shipped to the site. Isolators and full air suits were installed for certain steps. Detailed design started in January 1991, and 15 months later the plant was handed over to the owner. The plant underwent a successful FDA inspection in 1993.

At some point in the developmental cycle, usually after positive bioactivity and/or clinical data are available, it becomes necessary to determine capital needs for a production facility. Cost estimating factors have been presented specifically for biopharmaceutical process equipment (60). Exponential scaling factors are given for 58 different types and sizes of equipment. Use of indices is explained. A cost scaling factor of 0.63 was the average value for all equipment covered, but the standard deviation was 0.21. Although the historic 0.6–0.7 scale factor can be used when no other information is available, overuse or inappropriate use might result in introduction of large errors in final dollar cost estimates.

Other articles, while not specific for biotech processes, are useful for newer insights in estimating costs. Use of a single scale factor can lead to problems unless there is very careful definition of auxiliary items and off-sites (61). Biotech plants often involve extensive investment in items usually grouped as outside battery limits. It is best, therefore, to calculate capital for auxiliaries in terms of product capacity in order to determine economic significance.

Life-cycle costing estimates the total cost of ownership over the entire life of the system (62). The system with the lower total cost (not lowest initial cost, necessarily) over the plant life is the optimal solution. Life-cycle costing includes process control alternatives, failure event and associated costs, and, perhaps most important, costs of lost production. Normally this technique requires use of a computer spreadsheet. Costs are usually divided into six categories: initial acquisition, initial installation, operating, maintenance, lost production, and decommissioning.

A general review on biotech manufacturing economics is available (63). Building and retaining inefficient or expensive-to-operate manufacturing facilities (often due to the need for speed to market) will cause serious problems in the long run. Multiple-track (or at least two-track) development strategy is stressed to counter such an occurrence. McKown et al. provide no hard numbers, but the information, especially as it relates to purification, should be studied for insights into the early design phase (64).

Cost modeling as a management tool is explained by Busch (64). This development tool starts with a defined process and a PFD. The assumptions used in creating the cost algorithm (materials, tooling, labor, space, energy, maintenance, capital equipment, time value of money) must be defined and understood. Comparative cost estimates and sensitivity analyses can be performed; the economic bottlenecks can be identified. The model and underlying assumptions can be followed and modified as needed in the course of development. This modeling method can and should be continued into full manufacturing, to facilitate the improvement of subsequent models as more realistic assumptions (based on experience) become available.

For those wishing to prepare an early cost estimate (without initiating a full-fledged engineering project), it is often difficult to find costs and installation factors. This is especially true for small projects (investment of less than $5 million) or for projects involving retrofitting or internal alterations. One engineering firm that is involved in biotechnology projects has supplied certain cost data, including installation, that will be of use for initial estimation purposes. The information will also be of use in negotiating details with in-house or external architectural and engineering firms. The inevitable differences between estimators notwithstanding, the sponsor should be aware of why costs are far from either published information or any other up-to-date cost estimate. Table 4 gives costs for the northeastern United States as of the middle of 1996. Escalation factors are available (noted earlier) for changes due to inflation.

New tools continue to be refined and expanded. Steinberger discusses reengineering the capital investment process (65). The key point is to produce a detailed evaluation at a very early phase (2–5% of completion) of design engineering. Computer programs are now available to unify the stages of design; process simulation software can be used in conjunction with detailed capital evaluation software. Rapid changes can be made, and many alternatives can be evaluated in a short time. This is important because about 80% of a project's capital investment is determined when conceptual design is completed. Screening multiple potential designs, rather than one or two, is clearly advantageous in selecting an optimum configuration. This capability is of great utility in discarding certain potential pathways (reasons might be high capital, excessive waste treatment cost, or scheduling problems). Furthermore, selected optima can drive further development or pilot-plant experimentation to determine real-time data and agreement to the model. Any promising model can be verified by design rather than by search. Concurrent engineering and integrated process and product development are more than concepts and become readily feasible, since the software models can be shared among different groups. A few firms have software usable in biotech design; some have the capacity to connect design information to capital estimation software. Some firms can do both under one roof. Companies that have one or more capabilities in these areas include the following:

Intelligen, Inc.
Scotch Plains, N.J.
908-654-0088

BioPro designer (plus superset and other subsets for special cases—also see Ref. 66)

- Process simulation with economic evaluation
- Models more than 70 unit operations with database of process equipment
- Waste stream characterization
- Process scheduling
- Rapid economic evaluation of alternate manufacturing processes
- Sensitivity analysis and scenario evaluation for process optimization
- Example given: β-galactosidase production by *Escherichia coli*
- Profitability analysis
- demo discs available

Icarus Corp.
Rockville, MD
301-881-9350

Process evaluator

- Evaluates process economics and design alternatives
- Capital and operating costs, schedules, investment analyses
- Rapid estimates of feasibility and profitability of alternatives
- CPM-based planning schedules (general overview, engineering, construction)
- Works from process simulator software (e.g., Aspen Plus, Chemcad III, Design II)
- Oil, chemical, petrochemical, specialty chemical, pulp and paper, and mining industries
- Expert system knowledge bases (does not use indices)

Aspen Technology
Cambridge, MA
617-577-0100

AspenPlus

- Complete library of physical properties and unit operation models
- Demo disk available
- Creation of process flowsheet and simulation model—can create PFD-style drawings
- Sensitivity analyses, optimization, and data fit (owner's or databases)
- Costing and economic analyses
- Runs on networks or stand-alone hardware (workstations and PCs)
- BioProcess Simulator (BPS) module simulates performance of bioprocesses, such as enzymes, vitamins, and antibiotics (equipment includes batch and/or continuous fermentors, sterilizer, membrane and/or depth filtration, cell disruption, disc centrifuge, ion exchange, chromatography)

Batch Process Technology
West Lafayette, IN
317-463-6473

- Simulation system to manage multiproduct batch and semicontinuous processes in biochemical, food, pharmaceutical, and specialty chemical industries
- Optimize number and size of equipment, retrofit, debottleneck, and schedule processes
- Plans for increased throughput by evaluating process capacities
- Inputs: equipment network, recipe network, and sequencing information

Table 4. Construction and Equipment Costs: Ranges Based on Mid-1996 Pricing for the Northeastern United States

Item number	Description	Size or unit of measure	Unit cost range	Comments
		Selected construction tasks		
1.	Slab demolition and removal	6 in. thick	$8–12/ft^2	
2.	Underslab piping			
	Cast iron	4 in. diameter	$26–32/linear ft	
	Polypro (single wall)	4 in. diameter	$38–55/linear ft	
	Polypro (double wall)	6 in./4 in. diameter	$75–90/linear ft	
	Stainless steel	4 in. diameter	$60–65/linear ft	Type 304L
3.	Concrete slab			
	>1,000 ft^2	6 in. thick	$3–5/ft^2	Dowel into adjacent slab
4.	Concrete slab			
	Trench infills	6 in. thick	$6–8/ft^2	Dowel into adjacent slab
5.	Masonry—interior	8 in. thick	$5–7/ft^2	
6.	Masonry—exterior	12 in. thick	$7–9/ft^2	
7.	Carpentry			
	Hollow metal door, frame and hardware	Each	$600–800	3 ft × 7 ft nominal
8.	Exterior windows			
	Strip	Square foot	$20–28	
9.	Roof, 4 ft high			
	Membrane	45 mil.	$3.50–5.50/ft^2	Approx. 1,000 ft^2
	Built up	3 ply	$3.50–5.50/ft^2	Approx. 1,000 ft^2
10.	Insulation			
	Building—walls (6 in. thick)	Square foot	$0.50–0.75	
	Piping insulation (6 in. diameter)	Linear foot	$8–12	
11.	Drywall partition			
	10 ft high	Linear foot	$32–36	
12.	Acoustic ceiling	Square foot	$1.75–2.00	2 × 4 lay-in
13.	Painting			
	Standard latex	3 coats	$0.35–0.55/ft^2	
	Exterior	3 coats	$0.55–0.75/ft^2	
	Epoxy	2 coats	$0.75–0.95/ft^2	
14.	Flooring			
	Vinyl composition tile	Square foot	$1.25–1.50/ft^2	
	Wood	Square foot	$7.50–9.50/ft^2	
	Seamless epoxy	Square foot	$7–10/ft^2	Includes integral coves
	Carpet	Square yard	$20–28	
	Ceramic tile	Square foot	$6–8	
	Seamless vinyl	Square foot	$5–7	Includes integral coves
15.	Laboratory casework	Linear foot	$250–350/linear foot	Includes "Chemsurf" countertop and approx. 40% base cabinet
16.	Process piping			
	Cast iron	2 1/2 in.	$28–36/linear ft	
	Carbon steel	6 in.	$32–42/linear ft	
	Stainless steel	1 in.	$90–225/linear ft	316L, includes valves and fittings

		Selected systems (includes installation)		
1.	Clean rooms (packaged rooms, ≤1,500 ft^2)			
	Class 100,000	Square foot	$150–175	Sidewall return
	Class 10,000	Square foot	$165–200	Sidewall return
	Class 1,000	Square foot	$185–225	Sidewall return
	Class 100	Square foot	$215–250	Sidewall return
2.	Alarm system	Per point	$750–1,000	
3.	Building demolition (interior)	Cubic foot	$0.25–0.75	
	Building demolition (including structure)	Cubic foot	$0.50–1.00	
4.	Fire protection	Square foot	$1.25–2.50	Water system
5.	Electrical system	Square foot	$10–15	Base building/no process
		Selected equipment and utilities (includes placement and installation)		
1.	Fume hoods 4 ft	Each	$6,000–7,000	Stainless steel ducted w/SS work station
2.	Culture vessels			Agitated, 30 psig, full vacuum
	100 L	Each	37,000–40,000	
	1,000 L	Each	$55,000–60,000	
	10,000 L	Each	$530,000–600,000	
3.	Air compressor	Lump sum	$78,000–85,000	600 scfm
4.	Chiller	Ton	$210–260	
5.	Freezer room	Square foot	$160–200	Excluding furnishings
6.	Emergency generator			
	100 kW	Lump sum	$30,000–32,000	Diesel, outside
	300 kW	Lump sum	$41,000–44,000	Diesel, outside
	500 kW	Lump sum	$70,000–75,000	Diesel, outside
7.	Tankage			
	Carbon steel	Gallon	$5.00–7.00	<1,000 gal
	Stainless steel	Gallon	$8.50–10.00	<1,000 gal
8.	Environmental rooms (including cold rooms)	Square foot	$120–160	Temperature controlled, excluding furnishings
9.	Steam boilers			150 psig design
	100 hp	Lump sum	$38,000–46,000	
	250 hp	Lump sum	$70,000–80,000	
	300 hp	Lump sum	$80,000–90,000	
10.	Heat exchangers			
	Carbon steel	gal/min	$100–150	
	Stainless	gal/min	$300–350	
11.	Vacuum pumps	Scfm	$900–1,100	Liquid ring
12.	RO/DI water,			
	Package system	gal/min	$1,200–1,500	Generation equipment only
13.	WFI system, package system	L/h	$110–120	Generation equipment only

Source: Data supplied by Marshall Contractors, Inc., Rumford, R.I.

- Outputs: mass balance, equipment and resource utilization, cycle time, and waiting time
- Examples: expansion of an agricultural products plant, capacity expansion of a pharmaceutical process, and optimize operating conditions for an esterification

Richardson Engineering Services
Mesa, AZ
602-497-2062

Rapid Access Cost Estimating (RACE) Spreadsheet System

- Material and manpower estimation for general and process plant construction industries
- Extensive databases are updated annually (>100,000 line items)
- Estimate new construction costs and maintenance project costs
- Create unit cost assemblies rapidly
- Multiple database access (5) while in use
- Can operate with more than one currency at a time, also English and metric interchange
- User-defined estimate spreadsheet
- Import and export: ASCII and dBASE II

Because biotech process plants are expensive and most (or all) biotech companies understand the need for a multicandidate pipeline, it would be useful to incorporate flexibility in plant design. Young presents an approach for selection of the right capacity for multiproduct plants (67). Given the unknowns of product mix, sales forecasts, and manufacturing inefficiencies, it is still better to apply some logic to the process. Young gives an example of three novel products (although greater numbers can be introduced), with the expectation that one of them will move into manufacturing; however, there is a finite chance that all three will succeed. Risks are quantified and a decision tree for probable events is created. Both dedicated and campaigned (mixed operation) programs are established. The solution gives a best estimate for long-lead items, construction cost, time of construction, and start-up. Before moving to specific examples of economic analysis, an interesting phenomenon should be noted (Fig. 2). For many diverse input variables that have a cost component, there is an economic minimum somewhere in the study region. Each set of variables requires its own specific calculations (based on theoretical formulations or empirical data); the point here is that left to chance, one will not be operating at the minimum cost point in all likelihood. One example will be given, and Table 5 lists other possible variables to study.

Suppose one is designing a plant. Let the quality of the facility be the abscissa (moving from poor at the left to perfection at the right). The left ordinate would be recurring cost. This value would be highest at the poorest design and lowest closest to the perfect design. Then label the right ordinate "initial cost": The lowest value would correspond to the poorest design, and cost would increase as one move from good to perfection in quality of design.

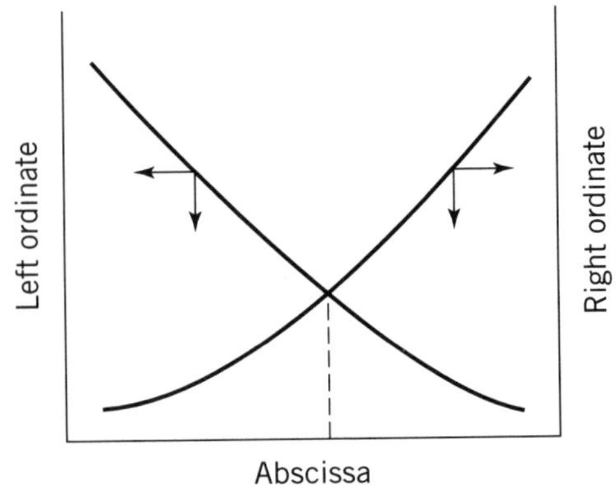

Figure 2. Determination of economic minimum.

Somewhere in the region (near or at the intersection of the curves), there is an overall cost minimum. There may be reasons for not selecting this point, but costs still must be reviewed as a function of one or more input variables. Other typical inputs are given in Table 5.

Specific cases involving economic analysis and costing can be found in the literature. The contributions of Datar are noteworthy (68,69). The earlier paper (1986) presents economics of enzyme recovery from an *E. coli*–based fermentation and extends the treatment to recombinant products. There is a breakdown of production costs for each step, capital and operating costs are developed, approaches are given for reducing manufacturing costs, and there is a clear delineation of assumptions and required off-sites. The second paper (1993) focuses on tissue plasminogen activator (tPA). Here, different processes are compared and detailed economic analyses are performed for alternative processes. Both papers are highly recommended.

A useful review exists on processing to prepare recombinant human insulin (70). The entire process is detailed, but emphasis is placed on large-scale purification. The high cost of purification is noted, but no dollar value or economic analysis is given. Another paper gives process analysis and economic evaluation for production of biosynthetic human insulin (66). Here, *E. coli* is used for expression of a precursor of proinsulin (intracellular) as a Trp-E fusion protein. The process results in 1,500 kg/yr of purified biosynthetic human insulin. Equipment is thoroughly detailed and specifications are shown. The fixed capital estimate (including contingency) comes to $142 million, using 1994 costs. Annual operating cost (including depreciation) is $56 million. Selling price is set at $110/g; a cash flow analysis is given and payback time is less than 2 years. BioPro Designer (a product of Intelligen, see preceding listing) was employed.

Economic analyses and publications continue to be generated on ethanol production from various substrates (71,72). The interesting historical phenomenon for ethanol biosynthesis (for fuel or an intermediate, not as food) is that just a little more R&D will be needed to make the specific (or any) process economically viable. This has been true for a number of years. The ethanol references noted are useful for methodology employed and probably will not result in construction of a facility in the short term.

Table 5. Determination of Diverse Cost Minima

Left ordinate	Abscissa	Right ordinate
Cost of mechanical failure	Frequency of preventive maintenance	Cost of preventive maintenance
Cost of imperfect product	Number of tests or inspections	Cost of inspection (testing)
Cost of production	Number or extent of capital items	Cost of capital
Cost of accidents	Number of safety inspections	Cost of safety programs
Cost of contamination	Number of cleaning cycles	Cost of cleaning
Cost of estimated investment	Number of planning iterations (cases)	Cost of planning or engineering
Cost of aeration	Cost of power	Cost of agitation
Cost of unplanned shutdown	Number of devices	Cost of backup or redundant equipment
Cost of energy	Extent (size) of heat exchangers or insulation	Cost of heat exchanges or insulation
Cost of product failure	Number of tests	Cost of testing (QC)
Project cost	Number of iterations	Cost of planning/engineering
Recurring cost	Quality level (poor to perfect)	Initial capital cost

Any review of economics should include some of the recent contributions to selection of the best process using payout parameters. Ward (73) stresses discounted rate of return while including consideration of investment flows over several years. Woinsky (74) presents a few different measures and seems to prefer payout time. Derivations are given and various plots are presented showing payout time as a function of pretax internal rate of return, plant economic life, and the engineering–procurement–construction period. Other methodologies are given in the reviews and books noted earlier in this section.

Another early stage technological development that may have the greatest economic impact on the biotechnology industry includes the areas of design, cost estimation, scheduling, and process control. Rapid change in computer control and especially in knowledge-based systems (KBS) offers the promise of significant cost savings by means of adaptive or predictive control of the process. These new control methods are especially useful when reaction–process mechanisms are understood poorly or not at all. The KBS spectrum includes expert systems, fuzzy logic, and neural networks.

An introductory book (75) is useful for its glossary of terms and for giving early applications in industry. Each of the separate concepts has strengths and weaknesses, and these are discussed. Software programs already exist for each of the classifications, and newer concepts are being presented with increasing frequency. For reasons of space, only a few references are given here; these works can be consulted for more publications and background. The first thing needed in consideration of computer systems is a careful definition of what the user wants. Furthermore, for biotech systems, there is the requirement for validation. The validation program should be considered in parallel with computer (and software) selection. There is, not surprisingly, a semantics and communication problem between hardware and software companies and users.

Chapman (76) gives very useful guidelines in discussing user requirements and gives a specific example for a multiproduct bulk pharmaceutical chemical plant. Documents needed for system validation are included. A paper on the use of neural networks in fermentation monitoring and control discusses optimization of inoculum development in a real case (77). Neural networks can model highly nonlinear, multivariable processes. Artificial intelligence (covering the three general classes of KBS noted above) is explained by other authors (78). Stress is placed on process control and support systems for operators. Quality improvement is reviewed. Examples are given in diagnosis and repair and alarm analysis (predictive control). One or more subgroups of KBS seem to be well suited to bioprocessing, since the programs can be used in production, staff and batch scheduling, engineering design, and project management. A Chiron computer control system (for a 1,500-L culture vessel producing a recombinant product) is discussed and explained (79). Design, building, validation, and start-up are reviewed. A supervisory control and data acquisition system above a control PC (with appropriate backup) was installed in 1992 for about $400,000.

Guidelines for computer control systems have been published (see Table 6). A large system might be found in a production facility or a pilot plant with multiple culture vessels or many purification trains and packaging capabilities. Medium scale would probably cover most pilot plants. If one is fortunate enough to have justification for a distributed-control system, provision should be made for incorporation of both the pilot plant and production equipment in a single system. Data collection and comparison are simplified, and scale-up is performed much more efficiently.

Another relevant point in this part of the review concerns contract manufacture. Just as it is sometimes efficient and prudent to use a contract research organization in clinical trials, it may be more effective to use a contract manufacturer (CMO) when product is needed. A useful starting point in this determination is an article on points to consider and methods to employ in evaluating a CMO (80). There are 24 references as well. The information relates to initial interviews, meetings, and inspections, and many issues to be incorporated into a contractual agreement are covered. In another article, describing an actual contact manufacturer in Puerto Rico, the author relates attributes to be sought in a contact partner (81). Seaver gives considerable insight into working with contract manufacturers (82). His suggestion is that a company's first protein-based product be made by a CMO. The estimate given for producing material for phase I and II trials in-

Table 6. Guidelines for Computer Control Systems

Process size	Input/output points	Number of units		Control option
		Continuous	Batch	
Large	>500	>3	>10	Distributed-control system
Medium	100–500	3	2–10	Programmable logic controller
Small	≤100	1	1–2	Personal computer, front-end system

Source: Excerpted by special permission from *Chemical Engineering*, November 1993. Copyright © 1993, by McGraw-Hill, Inc., New York.

house is 12–24 months in time and $0.5–1.5 million in cost (this latter range may be low).

Use of a CMO is not simple and requires numerous joint meetings and much effort on everyone's part to develop an agreement. More joint effort is expended during the production phase. A good suggestion is to plan on making 5–10 times more product than is needed for all the initial trials. Also, perform a test (or scale-up) run prior to the production run and save a considerable amount of product at multiple locations to serve as a long-term reference standard.

Other points related to manufacturing should be emphasized. Complete characterization of a master cell bank or a master working cell bank takes 4–6 months and costs "at least $50 thousand" (82). With newer pathogens appearing and the desirability of showing viral absence or inactivation by ever more complicated means, complete characterization might cost 1.5–2 times the figure given. It should be remembered that pooling of sublots (some of which have failed testing) to achieve an average "pass" is frowned on by regulatory bodies and is undesirable. In Table 7 the author attempts to list all issues and concerns that should be reviewed, considered, and resolved (in writing) in developing a contract with a CMO. It goes without saying that legal assistance is required in drafting or reviewing documents either during the negotiation of all conditions and terms or afterward.

VALIDATION AND REGULATORY ISSUES

Validation and regulatory issues are understood by almost all persons intimately involved in biotechnology; however, the time and cost impacts of these essential functions are woefully underestimated. The rapid growth of firms (or divisions within engineering/contractor firms) in the area of validation is not an anomaly; it is the result of ever-more intensive and in-depth requirements by regulatory agencies. Even though many large companies have in-house validation groups, variations in need (created by a new facility, a major retrofit, or a new process introduction) create a demand for outside services and consultants. Smaller companies rarely have the money to maintain a fully functional validation group that could satisfy even a majority of the company's needs. If one considers a validation technician or writer at a cost of $40–60/h, an experienced validation supervisor or technical expert at $70–120/h, a senior expert at $150–300/h, and a project that could consume several thousand man-hours for merely writing the master plan and protocols, it is easy to see that merely having an acceptable set of validation plans could cost many hundreds of thousands of dollars.

With more complicated control schemes (including programmable logic controllers or PLCs) requiring special (and often proprietary) software, there is a greater need for specialists to write the protocols and oversee their execution. Validation is the process by which documented evidence is accumulated to provide a high degree of assurance that a specific process (or part of a process) will consistently produce a product (or an intermediate) meeting its predetermined specifications and quality characteristics. Therefore, it is no wonder that this important feature comes under such careful and extensive regulatory review.

Validation cannot begin when a plant is up and running. Commissioning of a plant should include extensive written documentation, not the least part of which must include installation qualification (IQ) and operational qualification (OQ). Finally, as part of the commissioning, or hopefully soon thereafter, comes performance qualification (PQ). PQ provides the rigorous and documented testing that demonstrates the effectiveness, reproducibility, and allowable performance ranges of the process. Goswami and Lorenz discuss facility commissioning (83). A main point is that major savings are realizable if commissioning protocols are properly integrated into the overall validation plan. During design and construction, terse narrative system descriptions and single line drawings or indicators should be prepared to show cGMP control and compliance. These will prove invaluable during the preapproval inspection by the FDA.

A series of articles has been collected in a single bound issue; all articles refer to validation issues (84). This compendium will prove very useful to anyone new to the subject and also to anyone heavily involved in an ongoing validation project. Although no formula exists for cost determination, these articles can serve as background and as the framework for preparing an estimate. Furthermore, a number of firms offer to draft time and cost estimates for preparing protocols, for assisting in execution, or for completion of the entire validation process. The latter obviously would be the most costly, but there is another disadvantage. Regulatory bodies prefer some (but appreciable) direct involvement of the owner in the validation process, as opposed to providing only money and ultimate oversight.

Another area of ongoing cost (also often seriously underestimated) is that for environmental control. Even if the production process were not aseptic, many in-process steps are now performed in controlled environments. In many if not most cases, biologics, vaccines, and parenterals

Table 7. Contract Manufacturing Issues and Concerns: Audit and Contract Preparation

1. *Specifications*		
Raw materials	Cleaning	Formulation/packaging
Specialized technology	Documentation	Validation
Cell strains	Product specifications	Certificate of analysis
Testing	Safety requirements	Personnel protection
Hazard analysis	Sample retention	Labels
2. *Quantities and timing*		
Forecasting		Current planning program
Current capacity utilization		Geographic location (movement/shipping)
3. *Personnel*		
Qualifications		Training programs
Degree of automation		Training documentation
4. *Inspection* (by manufacturing, QC, QA groups of sponsor)		
Contractor's validation master plan		FDA inspection reports and status
Materials of construction		cGMP inspection (by sponsor)
5. *Regulatory review*		
Outside audits		Inside audits
Accreditation		Consultants used (plus qualifications)
6. *Experience*		
Need for specialized technology		Segregation of processes/products
History (experience in similar area)		Management capabilities
7. *Data analysis*		
Computer systems and software		Backup systems
LIMS and PLCs		Extent of (plans for) process development
8. *Sponsor supervision or participation*		
Emergencies		Timing of coverage (and access)
Planned/unplanned checks		Reports (How often? How many?)
9. *Scale-up schedule*		
Technology transfer		Additional capital needs (also ownership)
Ownership of process improvements		Program (plus explanation of deviations)
10. *Patent rights*		
11. *Payment schedule*		
Additional efforts		Criteria for payment and terms
Transfer to third party (if needed)		Change orders/changeover procedures
Termination clauses		Approvals needed

call for sterile packaging or bottling. The owner is responsible not only for design, control, and maintenance of environmental quality; the processor is responsible for evaluation of environmental quality. Alert and action levels for each controlled environment must be established; monitoring devices must be checked and qualified. It is not enough to monitor and control particulates, temperature, and humidity; additional monitoring must be in place for airborne bioburden, surface bioburden, supply and exhaust air volumes, pressure differentials between environments, and (for laminar flow only) airflow patterns and airflow uniformity. These special facilities are costly to install, costly to maintain, and costly to monitor. Cleaning methods must be verified and recertification periods must be established and followed. Generally, systems and components of the environmental control system (in aseptic processing) must be validated and requalified at these times:

Every time there is a significant system or control component repair or change

Before commissioning

Every time there is a process change or a new process is introduced

Every time out-of-control limits are noted in the operational parameters

Tedesco and Titus discuss other hidden costs that should be considered in planning, building, and operating biopharmaceutical facilities (85). They include engineering and environmental assessment fees and the cost of any required cleanup, permits, and change orders. The same article also presents generalized estimates that are useful for very early estimating:

Turnkey cost of GMP biopharmaceutical facility	$800–1,200/ft^2
Turnkey cost for pilot facilities	$400–600/ft^2

Another area of insufficient planning input is the warehouse. Appropriate inventory space (under appropriate conditions) is needed, along with quarantine and release areas. One should add space for reject materials and the various kinds of waste streams generated (chemical, biological, radioactive). Ideally, there should be no chance for mix-up or cross-contamination; proper plant layout and material flows should be carefully planned and provided.

As noted earlier, the physical plant itself, even if superbly designed and built, makes up only one part of the essential preapproval inspection by the FDA. Table 8 lists

Table 8. Preapproval Inspection Checklist

Organizational chart(s)	Integrity policy
cGMP audits	Integrity audits
Operating procedures (SOPs)	Validation documents (plus master plan)
Material specifications	Product/process flowcharts
Facility diagrams (esp. HVAC) and site plan	Computer hardware and software validation (PLCs)
Cleaning and disinfection	Product stability
Utilities	Release criteria
Employees (résumés, training)	QC review
Change control	Drug development information
Equipment list	In-process controls
Failure investigations	Labels and labeling
R&D laboratories	Internal audit summaries
Batch record(s) review	Development reports
Clinical lots	Scale-up (pilot vs. full-scale)
Documentation	Observed practices versus SOPs
Trend(s) analysis	Vendor qualification and analysis
Storage	Sampling plan(s)
Maintenance logs	Technology transfer

Review: All submissions relating to this inspection
For an existing firm: FDA-483, seizures, warning letters and responses, recalls, complaints, NDA deficiencies

a host of factors to be planned, organized, and developed prior to the inspection. A large amount of documentation (and other paperwork, in general) is required. Remember that there is a cost associated with each factor shown. The earlier the tasks and costs are recognized, planned, and budgeted, the greater the potential for a successful inspection accomplished in minimum time, at minimal cost, and with minimal aggravation.

REGULATORY ISSUES

Persons involved in biotechnology (in whatever discipline) know more or less about regulatory issues, but few come into direct contact with regulatory personnel. Therefore, those with a more academic or somewhat displaced knowledge of regulatory matters cannot appreciate the enormous amounts of time and effort that must be expended in the following:

Planning a biotech facility with compliance always an issue
Building such a facility (including records, pictures, and videotapes to show compliance)
Performing animal experiments within or outside this facility
Negotiating the many hurdles in planning a drug test
Achieving facility clearance by all appropriate regulatory bodies
Receiving product clearance (ability to sell)
Postapproval monitoring for minor and major side effects

The amount of money (and energy) needed to achieve regulatory compliance is invariably underestimated. No attempt will be made to cover all programs and costs related to regulatory clearances, but some measure of the complexity is suggested. Although everyone in a commercial biotech setting knows (or has heard about) the EPA, the FDA, and the USDA, other federal agencies have either an interest or a jurisdiction in biotechnology matters. From the simplest issues of syringe storage and the control of undenatured ethanol to animal housing and animal studies to transport of biomaterials, there are federal agencies involved (Bureau of Alcohol, Tobacco, and Firearms [BATF], National Institutes of Health, Centers for Disease Control, Department of Transportation, Occupational Safety and Health Administration, and Department of Labor). One should not forget that each state has its own regulatory bodies.

Consider the reviews needed for plant production: local, state, and federal agencies are involved. Not all the following issues apply in every case, but many will need review and approval: coastal zones, wetlands, tidelands, waterways, and historic preservation. Further, approvals will be needed for or from planning and zoning boards, boards of health, regional planning boards, discharge permitting, hazardous waste, infectious or biowaste, use and disposal of radioactive materials, transport of one or more of the foregoing, and laboratory licensing. Essentially, there is broad regulatory oversight in these areas:

- Licensing of *all* biotechnology products (biologics, drugs, and devices)
- Use of recombinant DNA and cell hybridization techniques
- Any alterations of the human gene (and related human testing and therapy)
- Planned releases of *any* genetically engineered plant or animal
- Worker protection

The cost of regulation, in its broadest sense, depends on the location of the laboratory and production plant, the process, the product and its intended use, and postmarket surveillance. Therefore, it is not possible to arrive at a total cost by formulaic means. At the lowest end (production of devices that have been in use for many decades), a minor change might involve regulatory costs of tens of thousands of dollars. For a new, simple, and noninvasive device, costs might be estimated at from a few thousand to many hundreds of thousands of dollars. Some relatively simple devices cannot be marketed because the cost of required double-blind studies to prove efficacy (there is no safety issue) is well beyond the capacity of the developer or exceeds any foreseeable financial return. For novel biologics or pharmaceuticals, regulatory costs could easily start in the tens of millions of dollars and could readily rise to the hundreds of millions of dollars, depending on test group size and overall clinical test requirements. Postmarket surveillance must be added to the total; note that this phase of the overall study is becoming ever more complex, lengthy, and costly.

Although change in regulations is common, some relatively recent reviews are useful for the state of the art at least to that point in time. An encyclopedia published in

1996 contains an enormous amount of information on the entire spectrum of pharmaceutical technology (86); however, two sections in Volume 13 are directly applicable to validation and regulatory issues. Even though the emphasis is on drugs, the tendency in U.S. regulatory affairs has been to blur the distinctions between devices and drugs. Therefore, while not every regulation covers both areas, the articles mentioned can serve as guides for the manufacture of medical products as well as for pharmaceutical development and production. The materials on registration of drugs and quality assurance as applied to drug products are very good starting points in their respective areas.

PAYMENT AND COMPENSATION

Although every phase in the foregoing discussion has elements of a more or less quantitative nature, and reasonable cost and time estimates can be made for each stage in biotech development and processing, the area of return on investment or payment for a product or service presents problems in economic analysis. It would appear that the bench scientist or the pilot-plant engineer (just to name two of many individuals involved) should have little to no interest in rate of return or, indeed, in how this rate is to be established. If that were ever true, it is no longer. From selection of a physiologic target to process design to scale-up and to all elements of operations, there is a continuing need for feedback on what range of compensation is attainable, assuming project success. Day-to-day decisions cannot be based on compensation issues; however, periodic review is essential to compare cost of good estimates with payment potential.

Given the global costs of health care and the ever-increasing involvement of third-party payers as well as government bodies, the situation is becoming critical. To assess the dollar magnitude, total 1997 sales for only five major drug firms (Merck, Johnson and Johnson, Novartis, Bristol-Myers Squibb, and Rhône-Poulenc) approaches $100 billion. Each of these firms achieved a double-digit percentage increase in sales compared to the prior year (5). Thayer also details revenues for 20 of the new major biotech firms for 1997; the total is $6.6 billion (5). With the advent of high-volume screening, genetic analysis, and the exponential growth of life-enhancement drugs, one can expect continued (and rapid) growth in sales of bioproducts and medical devices.

The complex issues of payment for biotech products or services have already elicited volumes giving fact and opinion. There is no attempt here to cover the topic completely, but some measure of the problem is needed by all workers in the field, not only those who attempt to set pricing. Fortunately, some references give both a reasoned introduction to the interplay of factors and groups (very often in conflict) and more detailed exposition of medical cost problems and potential solutions. At the least, ideas for alleviation are given.

A short article by K. Keller discusses cost of care with special attention to ethical choices sometimes overlooked (87). Keller recognizes that while government is a major factor in forcing changes in health care, government's views and approaches are hardly the same as those of either health care givers and suppliers or those who need the products and processes. Government's attention is primarily regulatory: The focus is on the process (rather than the outcome), and in general government bodies are concerned with short time scales relative to biotechnology development. Keller indicates the strong need to move beyond the "technology push" mentality; cost has now become as important as performance. Hence the creation of novel measurements (e.g., quality-adjusted life years). The point is made that all persons involved in the development of the technology must have some grasp of the complex interactions coming to the fore; these knowledgeable individuals should attempt to work constructively within the evolving framework to assist members of society in dealing with the interplay of factors and associated dislocations.

Another question has arisen that was either an insignificant problem or a nonexistent one a decade ago: Who pays for clinical trials, and just how are they to be funded? (88). Managed care organizations (MCOs) are claiming that clinical trial expenses are not part of their responsibilities. Therefore, coverage is often denied, hospital stays are shortened, and additional diagnostic tests (very often essential in the course of a trial) are denied. MCOs are taking the position that therapy in phase I, II, and III trials is unproven and that others (preferably the sponsor) should assume all associated (and additional) costs incurred during these phases. Some MCOs relent on so-called phase IV studies, which are outcome studies involving cost-effectiveness and quality of life. Not unexpectedly, elected officials (responding to biotech product suppliers, patients, or both) are becoming heavily involved with legislative ideas, proposals, and even threats. The outcome is far from clear; the problem is relatively new, and whatever the solutions, biotechnology companies will be affected.

Two separate volumes should be consulted for an overview of the cost situation in health care as well as various analyses and proposals to avoid further difficulties. Each volume is a compilation of presentations; the same sponsor arranged the meetings—one in April 1994 and the other in May 1995. The contents of the first set of proceedings (89) can be gauged by a selection of titles (the overall meeting topic is "The Role of Technology in the Cost of Healthcare"):

"Technology: Quality at What Cost?"

"How Do We Evaluate the Cost of Health Care Technology?"

"Crisis of the Uncontrolled and Misunderstood Numbers of Health Care Costs"

"Outcomes Research: Economics and Quality of Care"

The many articles in the collection cover the evolving situation, ideas on resource allocation, and the ever-increasing cost of health care. Of special interest is the review of outcomes analysis; no longer are morbidity and mortality the only issues to study and review. One must add patterns of recurrence, rehospitalization, relapse rate, procedural results, and patient satisfaction. From a number of presenters comes the unsurprising finding that the perspectives of those making cost–benefit analyses have a large impact on the findings and results. There are other articles on trial design, ethics, and current and proposed means of analysis. The second volume continues the discussion with the addition to the title of the phrase "Providing the Solutions" (90). Needless to say, solutions were offered but the problems persist. Of special note is a review by A.M. Frendrick entitled "Outcomes Research in Evaluating the Benefit of Health Care" (pp. 68–73). There is a large section with a number of papers on the impact of regulation on health care costs. Near the end of the proceedings, there were a few papers on a subject that is rather emotional. One had the provocative title "Is There a Need for Health Care Rationing?" The conclusions were rather mild, but at least the organizers had been willing to broach the subject. The authors conclude that rationing has always existed and will continue to exist. It is our job to establish appropriate limits, inasmuch as there is a definite need for setting such bounds.

A review article on disease management has been written by Merck–Medco personnel (91), who define *disease management* as "a systematic population-based approach to identifying those at risk, intervening using information from the growing field of evidence-based medicine, and measuring patient outcomes once an intervention is in effect." Economic outcomes weigh heavily in the perspective of almost every group involved in disease management. (It is odd that the only outcome listed for the perspective of the physician is clinical outcomes. This is at best idealistic and at worst naive.) The perspectives of each party to the process are discussed and analyzed. Not surprisingly, not all outlooks are congruent. Components (guidelines, patient identification, intervention methods, and outcomes assessment) are reviewed and various possible alternatives given. Disease management has as its primary goal population-wide improvements in health; an unstated subordinate goal is improved productivity—and possibly cost control—in delivery of health care.

These various issues are raised here because solutions—whether they come from HMOs, MCOs, the government, hospital conglomerates, or a combination of these bodies—will have an enormous impact on the other topics discussed earlier. Any significant change in compensation (which relates to rate of return) must have an impact on discovery, development, and plant investment. This, in turn, means that either the quality or the quantity (or likely both) of biotechnology research and product and service delivery must change. At this writing, it appears that limits may be placed on compensation. At the least, the rate of growth of health care costs will be reduced. It does not take a leap of intuition to project the impact on the biotechnology product cycle. The concluding message is that biotechnology industry participants must become involved in resource allocation in every stage from discovery to payment. The involvement and input should be reasoned; if background and understanding of the full spectrum of economic needs and impacts are clean and logical, the resulting positions will be heard.

BIBLIOGRAPHY

1. Biotechnology Industry Organization, *Forbes*, May 4, 1998, pp. 97–120.
2. J. Hodgson and K. Barlow, *CHEMTECH* **23**, 45–47 (1993).
3. Anon., *Genet. Eng. News* **16**, 3 (1996).
4. Biotechnology Medicines in Development, 1996 Survey, Pharmaceutical Research and Manufacturers of America, Washington, DC.
5. A.M. Thayer, *Chem. Eng. News* **76**, 25–37 (1998).
6. B. Werth, *The Billion-Dollar Molecule*, Simon & Schuster, New York, 1994.
7. V. Glaser, *Genet. Eng. News* **12**, 33 (1996).
8. The biotech century, *Business Week*, March 10, 1997, p. 84.
9. D. Matheson, J.E. Matheson, and M.M. Menke, *Res. Technol. Manage.* **37**, 21–24 (1994).
10. R. Szakonyi, *Res. Technol. Manage*, **33**, 31–36 (1990); **33**, 41–46 (1990).
11. I. Norrington, *Pharm. Manuf. Int.* 37–40 (1996).
12. P. Bevan, H. Ryder, and I. Shaw, *Trends Biotechnol.* **13**, 115–121 (1995).
13. P. Sharpe and T. Keelin, *Harvard Business Rev.* **76**, 45–57 (1998).
14. K.S. Brown, *Scientist*, **10**, 1, 8–9 (1996).
15. N. Pfeiffer, *Genet. Eng. News* **16**, 1, 18, 19 (1996).
16. G. Samdani and G. Ondrey, *Chem. Eng.* **100**, 35–39 (1993).
17. O. Schuster and G. Menke, *Pharm. Manuf. Int.* 85–86 (1996).
18. C. Turner, M. Turner, and K. O'Toole, *Chem. Eng.* May 16, 1996, pp. 25–27.
19. F.A. Dubinskas, *Technol. Rev.* **74**, 24–30, (1985).
20. G.P. Pisano and S.C. Wheelright, *Genet. Eng. News* **16**, 4, 32, 33 (1996).
21. R. Frumerman, *Res. Technol. Manage.* **33**, 10–11 (1990).
22. D. Mukesh, *Chem. Eng. Prog.* **92**, 110–112 (1996).
23. C. Prior, P. Bay, B. Ebert, R. Gore, J. Holt, T. Irish, F. Jensen, C. Leone, J. Mitschelen, M. Stiglitz, C. Tarr, R. Trauger, D. Weber, and M. Hrinda, *Pharm. Technol.* **19**, 30–52 (1995).
24. B. Sonnleitner, *Adv. Biochem. Engineer. Biotechnol.* **54**, 155–188 (1996).
25. J. Jones, W. Asher, J. Bomben, D. Bomberger, C. Marynowski, R. Murray, R. Phillips, D. Roberts, K. Semrau, R. Swidler, C. Witham, and J. Zuegel, *Chem. Eng.* **100**, 98–106 (1993).
26. B.H. Junker, J. Lynch, J. Leporati, J. Schmitt, J. Gieger, T. Garah, M. Stober, and P. Salmon, *Bioprocess Eng.* **13**, 279–287 (1995).
27. J. Abate, A. Chen, W.J. Dale, and D. McNeil, *Chem. Eng. Prog.* **91**, 82–87 (1995).
28. C.M. McFarlane and A.W. Nienow, *Biotechnol. Prog.* **11**, 601–607 (1995).
29. R.F. Dream, *Pharm. Eng.* **13**, 56–66 (1993).
30. R.G. Werner, F. Walz, W. Noe, and A. Konrad, *J. Bacteriol.* **22**, 51–68 (1992).
31. R.L. Bruno and P.E. Scott, *Pharm. Eng.* **13**, 42–46 (1993).
32. T.A. Bibila and D.K. Robinson, *Biotechnol. Prog.* **11**, 1–13 (1995).
33. F.W.J.M.M. Hoeks, J. Muhle, L. Bohlen, and I. Psenicka, *Chem. Eng. J.* **61**, 53–61 (1996).
34. A.P.J. Middelberg, *Chem. Eng. J.* **61**, 41–52 (1996).
35. T.R. Elliott and W.L. Luyben, *Ind. Eng. Chem. Res.* **34**, 3907–3915 (1995).

36. S. Harsa, C.A. Zaror, and D.L. Pyle, *Proc. Biochem.* **28**, 187–195 (1993).
37. M.F. Kozempel, M.J. Kurantz, and J.C. Craig, Jr., *Biotechnol. Prog.* **11**, 592–595 (1995).
38. S. Moore, G. Parkinson, and G. Ondrey, *Chem. Eng.* **102**, 33–37 (1995).
39. M.A. del Valle, *Pharm. Eng.* **15**, 14–22 (1995).
40. S.E. Mackler, *Genet. Eng. News* **16**, 3 (1996).
41. P.J. Haas, *Pharm. Technol.* **19**, 26–38 (1995).
42. W.R. Brader, R.T. Hsu, and P.E. Lorenz, *Pharm. Eng.* **15**, 30–42 (1995).
43. S.W. Fitzpatrick and S.E. Mackler, *Pharm. Technol.* **19**, 118–126 (1995).
44. J.E. Akers, J.P. Agalloco, and C.M. Kennedy, *J. Pharm. Sci. Technol.* **49**, 140–144 (1995).
45. J. Akers, *BioPharm* **7**, 43–47 (1994).
46. Harvard Business School Case Study 9-696-083, Biogen, Inc.: rBeta Interferon Manufacturing Process Development, January 29, 1996.
47. G.E. Gamerman and B.F. Mackler, *CHEMTECH* **24**, 37–41 (1994).
48. J.A. Ehrich, D.F. Bernstein, and K.H. Sills, *Pharm. Technol.* **19**, 98–112 (1995).
49. G.J. Wiley, I. Ullah, and S.N. Agharkar, *Pharm. Technol.* **19**, 72–76 (1995).
50. J.R. Vogel, R.A. Schober, and R.B. Olson, *Pharm. Technol.* **19**, 40–52 (1995).
51. J.D. Bronzino ed., *The Biomedical Engineering Handbook*, CRC Press, Boca Raton, Fla., 1995; P.A. Oberg, in J.D. Bronzino ed., *The Biomedical Engineering Handbook*, CRC Press, Boca Raton, Fla., 1995, pp. 2549–2556; T.V. Holohan, in J.D. Bronzino ed., *The Biomedical Engineering Handbook*, CRC Press, Boca Raton, Fla., 1995, pp. 2794–2801.
52. B. Atkinson and F. Mavituna, *Biochemical Engineering and Biotechnology Handbook*, 2nd ed., Stockton Press, New York, 1991. See especially Chapter 19, pp. 1059–1109.
53. P.F. Stanbury, A. Whitaker, and S.J. Hall eds., *Principles of Fermentation Technology*, Elsevier Science, Oxford, U.K., 1995, pp. 331–349.
54. J.L. Dwyer, in F. Franks ed. *Protein Biotechnology*, Humana Press, Totowa, N.J., 1993, pp. 533–572.
55. H.B. Reisman, *Economic Analysis of Fermentation Processes*, CRC Press, Boca Raton, Fla., 1988.
56. J. Durkin, *Pharm. Manuf. Int.* 103–108 (1995).
57. S. Hubbard, *Pharm. Manuf. Int.* 109–112 (1995).
58. H. Allgaier, *Pharm. Biotech. Int.* 47–48 (1996).
59. R. Maycock, *Pharm. Manuf. Int.* 113–118 (1995).
60. D.S. Remer and J.H. Idrovo, *BioPharm* **3**, 36–42 (1990).
61. D.M. Haseltine, *Chem. Eng. Prog.* **92**, 26–32 (1996).
62. W.M. Goble and B.O. Paul, *Chem. Proc.* **58**, 35–39 (1995).
63. R. McKown, R. Teutonico, and S. Fox, *Genet. Eng. News* **16**, 6, 7, 28 (1996).
64. J. Busch, *Res. Technol. Manage.* **37**, 50–56 (1994).
65. R.L. Steinberger, *Chem Eng.* **101**, 114–117 (1994).
66. D. Petrides, E. Sapidou, and J. Calandranis, *Biotech. Bioeng.* **48**, 529–541.
67. R.A. Young, *Chem. Eng. Prog.* **91**, 64–59 (1995).
68. R. Datar, *Proc. Biochem.* **21**, 19–26 (1986).
69. R.V. Datar and T. Cartwright, *Bio / Technology* **11**, 349–357 (1993).
70. M.R. Ladisch and K.L. Kohlmann, *Biotechnol. Prog.* **8**, 469–478 (1992).
71. N.D. Hinman, D.J. Schell, C.J. Riley, P.W. Bergeron, and P.J. Walter, *Appl. Biochem. Biotechnol*, **34**, 639–649 (1992).
72. M. von Sivers, G. Zacchi, L. Olson, and B. Hahn-Hagerdal, *Biotechnol. Prog.* **10**, 555–560 (1994).
73. T.J. Ward, *Chem. Eng.* **101**, 102–107 (1994).
74. S.G. Woinsky, *Chem. Eng. Prog.* **92**, 33–37 (1996).
75. W.H. Ver Duin, *Better Products Faster: A Practical Guide to Knowledge-Based Systems for Manufacturers*, Irwin Professional Publishing, Burr Ridge, Ill., 1995.
76. K.G. Chapman, *Pharm. Eng.* **16**, 8–18 (1996).
77. M.S. Slusher, R.L. Bennett, and D.L. Deitz, *Pharm. Eng.* **14**, 8–12 (1994).
78. E.R. Crowe and C.A. Vassiliadis, *Chem. Eng. Prog.* **91**, 22–31 (1995).
79. D. Mosher and P. McArthur, *Pharm. Eng.* **14**, 86–92 (1994).
80. J.L. Tedesco, T. Olson, S. Tse, and J.M. Patrick, *Pharm. Technol.* **18**, 174–187 (1994).
81. M. Borrero, *Pharm. Technol.* **19**, 113–115 (1995).
82. S.S. Seaver, *BioPharm* **8**, 20–25 (1995).
83. D. Goswami and M.R. Lorenz, *Pharm. Eng.* **16**, 14–18 (1996).
84. G. Sofer, J. McEntire, and J. Akers, *BioPharm* **8**, 60–63 (1995).
85. J.L. Tedesco and M.J. Titus, *Pharm. Eng.* **15**, 22–28 (1995).
86. J. Swarbrick and J.C. Boylan eds., *Encyclopedia of Pharmaceutical Technology*, Marcel Dekker, New York, 1996; W.J.C. Currie and R.W.J. Currie, in J. Swarbrick and J.C. Boylan eds., *Encyclopedia of Pharmaceutical Technology*, Marcel Dekker, New York, 1996, pp. 353–370; R. Kaplan, in J. Swarbrick and J.C. Boylan eds., *Encyclopedia of Pharmaceutical Technology*, Marcel Dekker, New York, 1996, pp. 207–231.
87. K.H. Keller, *CHEMTECH* **25**, 7–11 (1995).
88. M.E. Watanabe, *Scientist* June 24, 1996, pp. 1, 4–5.
89. W.S. Grundfest, ed., *Health Care Technology Policy I: The Role of Technology in the Cost of Healthcare*, International Society for Optical Engineering, Bellingham, Wash., 1994.
90. W.S. Grundfest, ed., *Health Care Technology Policy II: The Role of Technology in the Cost of Healthcare: Providing the Solutions*, International Society for Optical Engineering, Bellingham, Wash., 1995.
91. S. Epstein and M.G. McGlynn, *Dis. Manage. Health Outcomes* **1**, 3–10 (1997).

ADDITIONAL READING

A. Kornberg, *The Golden Helix: Inside Biotech Ventures*, University Science Books, Sausalito, Calif., 1995. Relates the story of DNAX and its relations with Schering-Plough as seen by a key scientific adviser. Also comments on other companies and biotech future.

V. Moses and R.E. Cape eds., *Biotechnology: The Science and the Business*, Harwood Academic Publishers, Chur, Switzerland, 1991. Covers business matters, underlying technology, and some economics; various industries (including biomass, waste treatment, foods, and pesticides).

V. Moses and S. Moses, *Exploiting Biotechnology*, Harwood Academic Publishers, Chur, Switzerland, 1995. Discusses the technology base, organizational framework, and biotech businesses; contains an extensive glossary.

A. Prokop, in S. Neidleman and A.I. Laskin eds., *Advances in Applied Microbiology*, vol. 40, Academic Press, San Diego, Calif., 1995, pp. 95–154, 155–236. An excellent and up-to-date summary of the project cycle from discovery through development to marketing. Articles combine the technical and business aspects of biotechnology, touching on many key areas and needs.

ELECTRON TRANSPORT

DAVID WHITE
Indiana University
Bloomington, Indiana

KEY WORDS

ATP synthesis
Bacteria
Chemiosmosis
Coupling sites
Electrode potentials
Electron carriers
Mitochondria
Protonmotive force
Respiration
Uncouplers

OUTLINE

INTRODUCTION

There is a current of electrons in the cell membranes of bacteria and in mitochondrial and chloroplast membranes. The electrons enter the membrane from an electron donor usually metabolically generated or supplied in the growth medium, travel through the membrane via a series of electron carriers, and leave the membrane upon being transferred to a terminal electron acceptor such as oxygen. (Photosynthetic bacteria carry out a light-driven cyclic electron flow where the electrons do not leave the membrane.) According to the chemiosmotic theory, the current of electrons, called electron transport, produces a vectoral flow of positive charges, that is, protons (H^+) to the outside of cells and organelles (1) (Fig. 1). In bacteria, the protons are translocated to the outside of the cell membrane, in mitochondria the protons are translocated from the mitochondrial matrix across the inner membrane to the cytosolic side, and in chloroplasts the protons are translocated across the membrane from the stroma to the inner thylakoid space. In all three cases, a membrane potential, outside positive, develops as does a ΔpH, outside acidic (more concentrated in protons). The return of the protons through appropriate proteins toward the negative inner side and higher pH (lower proton concentration) drives the synthesis of ATP as well as other membrane-associated processes such as solute transport. The resultant ATP is the major source of biochemical energy for the synthesis of cell material. Thus, electron transport fuels cell metabolism. This chapter describes the general routes that the electrons take in various nonphotosynthetic membranes on their way to a terminal electron acceptor such as oxygen, as well as the differences among organisms, and how the the proton current is produced. Archaeal* electron transport chains are less well studied, but they appear to be similar to bacterial electron transport chains (2).

Some Terminology and Definitions

Redox, Reactions, Electron Transport, and Respiration. The basic reaction in electron transport is the transfer of electrons from one molecule to another. The transfer of electrons from one molecule to another is called an *oxidation-reduction reaction* (or redox reaction), where the molecule that gives up the electron becomes oxidized, and the molecule that accepts the electron becomes reduced. Electron transport is actually a series of such redox reactions in membranes where the electron is passed from one molecule to another. The molecules that transfer the electrons are called *electron carriers*, and these are described later. Electron transport takes place in cell membranes of prokaryotes (organisms such as bacteria that have no organelles such as nuclei or mitochondria) and in mitochondrial and chloroplast membranes of eukaryotes (organisms with organelles). Electron transport in chloroplast membranes is called photosynthetic electron transport, whereas electron transport in other membranes is called

*There are two phylogenetic lines of prokaryotes (microorganisms without organelles such as nuclei). These are the archaea and the bacteria.

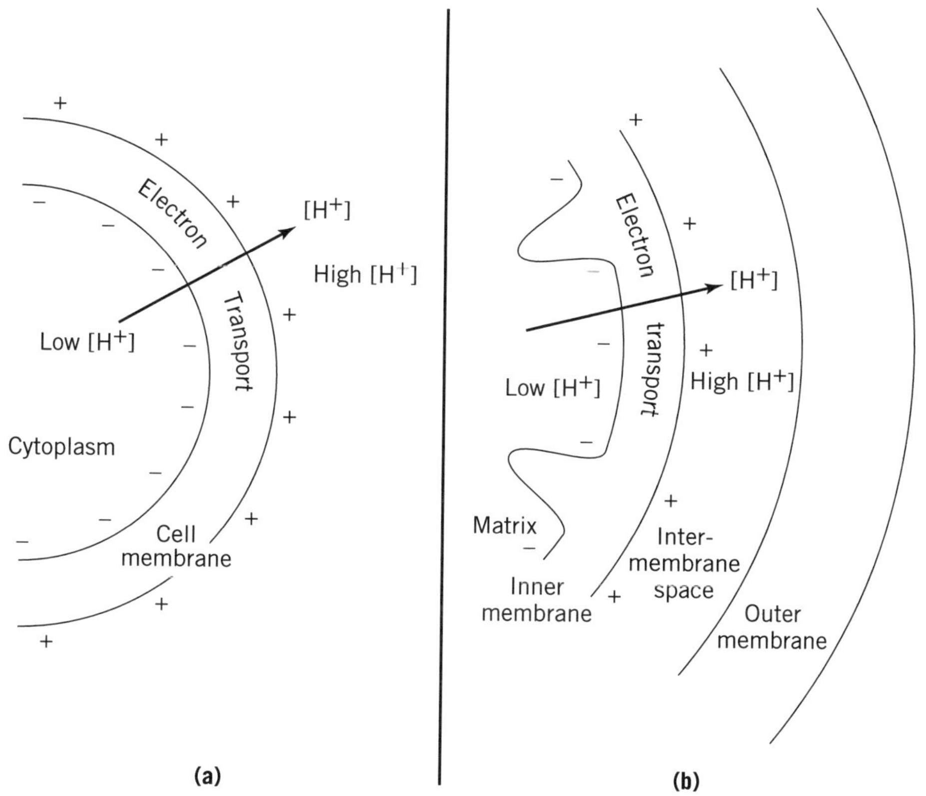

Figure 1. Electron transport in bacterial cell membranes (**a**) and mitochondrial membranes (**b**) result in proton translocation accompanied by a membrane potential (outside positive) and a pH gradient (outside acidic). The molecular mechanisms and the intermediates are not shown.

respiration. If the terminal electron acceptor in respiration is oxygen, then electron transport is called *aerobic respiration*. If it is not oxygen, then it is called *anaerobic respiration*. Mitochondria carry out only aerobic respiration, but many prokaryotes are capable of anaerobic respiration and use organic compounds such as fumaric acid or inorganic compounds such as nitrate or sulfate as terminal electron acceptors.

Electron Energy and the Redox Potential Difference. During electron transport, electrons spontaneously flow from donor to acceptor down a potential gradient called the ΔE, and it is the energy released during this process that is used to generate the vectoral flow of protons from inside to outside that is ultimately used to make ATP. By convention in biology, electron *donors* in a redox reaction that releases energy are said to have a *a more negative* redox value (E) than the molecules that accept the electron. That is to say, electrons always flow spontaneously toward the molecule with the higher redox value. The actual redox potential of a molecule depends on the ratio of oxidized to reduced forms.* The difference in electrical potential between the acceptor of the electron and the donor of the electron ($E_{acceptor} - E_{donor}$) is called the *redox potential* and is written as ΔE. The ΔE is always positive when work can be done by the moving electrons, that is, when the electrons are moving down an energy gradient. The ΔE is generally spoken of as an *electrical force* (also called the *electromotive force*) that maintains the current of electrons, and it is expressed in units of volts. The *work* that can be done by n moles of moving electrons depends on the total charge carried by the electrons and is given by n$F\Delta E$ joules where F is the charge carried by 1 mol of electrons or monovalent ion. (F is the *faraday* constant, which is equal to 96,485 coulombs per mole.) The unit of work is called the joule. In summary, then, electrons flowing over a potential gradient of ΔE volts can do a maximum of n$F\Delta E$ joules of work or exert a force of nΔE volts. As described next, the work that is done is the creation of a protonmotive force that is, a Δp, by translocating protons to the outside to a higher electrochemical potential.

*E is sometimes called the electrode potential or the reduction potential. The value of E at any particular temperature depends on the pH and relative concentrations of oxidized and reduced forms of the compound. E_o is the redox potential in the standard state, where the oxidized and reduced forms of the compound both have an activity of 1, as does the hydrogen ion concentration. E_o' refers to the standard state at pH 7 and is usually called the midpoint potential ($E_{m,7}$). The actual redox potential of a compound depends on the ratio of oxidized to reduced forms in the cell and the pH and is called E_h when the activities are not 1, that is, $E_h = E_o + RT \ln[ox]/[red]$.

The Protonmotive Force. The protonmotive force, Δp, refers to the electrochemical potential difference of the proton across bacterial, mitochondrial, and chloroplast membranes. Its units are volts. In order to understand what this means, it is necessary to realize that a membrane potential, $\Delta \Psi$, exists across these membranes such that they are positively charged on the outer surface and negatively charged on the inner surface (Fig. 1.) (How the membrane potential is created will be explained later.) The size of the $\Delta \Psi$ across bacterial cell membranes varies from about -60 mV to -200 mV, depending on the growth conditions and the bacterium. (Note that by convention the $\Delta\Psi$ is negative when the inside surface is negative with respect to the outside surface.) Because protons themselves are positively charged, it requires energy to move them to the positive side. That is to say, they are being pushed away by the $\Delta\Psi$ from the positive side toward the negative side. Another factor that will influence the direction that protons will spontaneously move is the concentration difference (ΔpH). That is to say, energy is required to move protons from the less concentrated to the more concentrated side of the membrane, and protons give up energy when they move from the more concentrated to the less concentrated side. The sum of the energies involved in the movement of protons along an electrical and concentration gradient is called the *electrochemical energy*, and it takes $+yF\Delta p$ joules to move y moles of protons against their electrochemical energy gradient from inside to outside. Accordingly, $-yF\Delta p$ joules is released and work can be done when y moles of protons return to the inside of the cell or organelle. This can be expressed as volts (a force) by dividing by the faraday ($y\Delta p$ volts). As discussed later, the energy to move protons against their electrochemical gradient is provided by respiration. In summary, then, the protonmotive force consists of contributions from the membrane potential and the ΔpH and is given in equation 1. Note that the Δp is analogous to the redox potential, ΔE, which describes the electrical force for moving electrons, except the Δp applies to moving protons and is due to both electrical and chemical energy.

$$\Delta p = \Delta\Psi - 60\ \Delta\text{pH mV at 30 °C} \qquad (1)$$

For most bacteria, the ΔpH is positive and thus contributes to the Δp. An exception are bacteria that live in environments more alkaline than the cytoplasm. Under these circumstances, the ΔpH is negative and lowers the energy available from the Δp.

The Relationship between ΔE and Δp

As mentioned, in prokaryotic cell membranes and in mitochondria (and chloroplasts), the work done by the current of electrons is to translocate protons (H^+) across the membrane to a higher electrochemical potential to create the Δp (Fig. 1). Proton translocation is done at certain redox reactions called coupling sites in the respiratory chain. The means by which this is done will be described later ("Q Loops, Q Cycles, and Proton Pumps"). The numerical relationship between the ΔE of the redox reaction at the coupling site and the Δp is given in equation 2.

$$-n\Delta E = y\Delta p \qquad (2)$$

In equation 2, n is the number of electrons transferred from the electron donor to the acceptor during the redox reaction, ΔE is the redox potential difference between the electron donor and acceptor, y is the number of protons translocated across the membrane when n electrons are transferred, and Δp is the protonmotive force. The ΔE is always positive when electrons can do work, and the Δp is always negative when protons can do work, hence the negative sign in equation 2.

Work Done by the Δp

When protons reenter the cell (or cross mitochondrial or chloroplast membranes) down their electrochemical gradient, work can be done driven by the Δp. The type of work done at the expense of the Δp depends on the specific route taken by the protons as they return across the membrane toward their lower electrochemical potential. ATP synthesis occurs when the protons cross the membrane via the membrane-bound ATP synthase, which is a complex of proteins that synthesizes ATP (see Fig. 10). Solute transport occurs when protons traverse the membrane via solute transporters, which are proteins that transport protons down their electrochemical gradient at the same time as solutes are transported against their concentration gradient, and flagellar rotation in prokaryotes occurs when the protons return to the inside of the cell via the flagellar motors, which consist of a complex system of proteins. (Flagellar motors in eukaryotes are driven by ATP rather than the Δp.) Thus, prokaryotic cell membranes as well as mitochondrial and chloroplast membranes can be viewed as containing electric machines that are powered by currents of protons, which themselves are maintained by electron transport.*

THE ELECTRON CARRIERS

Before describing the electron transport chains and the generation of the Δp, it is necessary to introduce the electron carriers. During electron transport, the electrons flow through protein and lipid electron carriers in the membranes. Some of these carry electrons only, whereas others carry both electrons and hydrogen atoms in combination. They are

1. Flavoproteins (hydrogen and electron carrier)
2. Quinones (hydrogen and electron carrier)
3. Iron–sulfur proteins (electron carrier)
4. Cytochromes (electron carrier)

The quinones are lipids (molecules that have a relatively poor solubility in water but a good solubility in lipid environments such as the interior of membranes), whereas the other electron carriers are proteins, which exist in multiprotein enzyme complexes called *oxidoreductases*.† The electrons and hydrogen are not carried in the protein per se, but in a nonprotein molecule tightly bound to the protein. The nonprotein portion that carries the electron and hydrogen is called a *prosthetic group*. The prosthetic group

*The sodium ion current also drives solute transport and in some cases flagellar rotation. Usually the Δp provides the energy to create the sodium gradient, but some marine bacteria couple electron transport to the creation of a sodium ion gradient rather than a proton gradient.

†For example, the enzyme complex that oxidizes NADH and reduces quinone is an NADH-quinone oxidoreductase. It consists of a flavoprotein and iron–sulfur proteins. (Reduced quinone is called quinol.) An enzyme complex that oxidizes quinol and reduces cytochrome c is called a quinol-cytochrome c oxidoreductase. It consists of cytochromes and an iron–sulfur protein.

in flavoproteins (fp) is a flavin that can be either flavin adenine dinucleotide (FAD) or flavin mononucleotide (FMN). The prosthetic group in iron–sulfur proteins is a cluster of iron–sulfide, which is abbreviated FeS. The prosthetic group in cytochromes is heme.

Each of the electron carriers has a different redox potential, and the electrons are transferred sequentially to a carrier of a higher redox potential. The standard redox potentials at pH 7 (E'_o) of the electron carriers and some electron donors and acceptors are listed in Table 1. Notice that redox couples are generally written in the form oxidized/reduced for biologists and biochemists. Several of the oxidation–reduction reactions in the electron transport chain can be reversed using excess energy provided by the Δp. During reversal of electron transport, the protons *enter* the cells driven by the Δp and the electrons travel to the more *negative* potential creating a ΔE. (See equation 2.)* For example, in Figure 7, the Δp can drive electrons from succinate through ubiquinone (UQ) to NAD^+. (NAD^+ is the oxidized form of NADH and will accept two electrons and a proton to become NADH.) This allows succinate to be used as an electron donor to reduce NAD^+. The NADH that is formed is used for biosynthesis. The reversal of electron transport is discussed later ("Coupling Sites").

*This means that the ratio of oxidation to reduction for several of the electron carriers (fp, cytochromes, quinones, FeS proteins) must be close to 1. Thus, for these reactions the E_h (actual potential at pH 7) of the redox couples are close to their midpoint potentials, E'_m, which is the potential at pH 7 when the couple is 50% reduced (i.e., [ox] = [red]).

Table 1. Standard Electrode Potentials (pH 7)

Couple	E'_o (mV)
Fd_{ox}/Fd_{red} (spinach)	−432
CO_2/formate	−432
H^+/H_2	−410
Fd_{ox}/Fd_{red} (*Clostridium*)	−410
NAD^+/NADH	−320
FeS(ox/red) in mitoch.	−305
Lipoic/dihydrolipoic	−290
S^o/H_2S	−270
$FAD/FADH_2$	−220
Acetaldehyde/ethanol	−197
$FMN/FMNH_2$	−190
Pyruvate/lactate	−185
OAA/malate	−170
Menaquinone(ox/red)	−74
$Cytb_{558}$(ox/red)	−75 to −43
Fum/succ	+33
Ubiquinone(ox/red)	+100
$Cytb_{556}$(ox/red)	+46 to +129
$Cytb_{562}$(ox/red)	+125 to +260
Cytd(ox/red)	+260 to +280
Cytc(ox/red)	+250
FeS(ox/red) in mitoch.	+280
Cyta(ox/red)	+290
$Cytc_{555}$(ox/red)	+355
$Cyta_3$(ox/red) in mitoch.	+385
NO_{3^-}/NO_{2^-}	+421
Fe^{3+}/Fe^{2+}	+771
O_2 (1 atm)/H_2O	+815

Note: Fd = ferredoxin; OAA = oxaloacetate; Fum = fumarate; succ = succinate.
Sources: Refs. 3 and 4.

Flavoproteins

Flavoproteins (fp) are electron carriers that have as their prosthetic group an organic molecule called flavin. There are two flavins, FMN and FAD (Fig. 2). The flavins are made from the vitamin riboflavin. The addition of phosphate to riboflavin yields FMN, and the addition of ADP to FMN yields FAD. When flavins are reduced they carry 2[H] (two electrons and two hydrogens), one on each of two ring nitrogens (Fig. 2). There are many different fp, and they differ with respect to their redox potentials and where they are situated in the electron transport chain. The differences in the redox potentials result from differences in the protein component of the enzyme, not in the flavin itself.

Quinones

Quinones are fat soluble, that is, they are soluble in the interior (lipid phase) of the membrane, which is nonaqueous. They are small and mobile and shuttle electrons between the complexes of protein electron carriers. Their structure and oxidation–reduction reactions are shown in Figure 3. Bacteria make two types of quinones that function during respiration: ubiquinone (UQ), a quinone also found in mitochondria; and menaquinone (MQ or sometimes MK). Menaquinones have a much lower electrode potential than ubiquinones and are used predominantly during anaerobic respiration where the electron acceptor has a low potential (e.g., during fumarate respiration). A third type of quinone, plastoquinone, occurs in chloroplasts

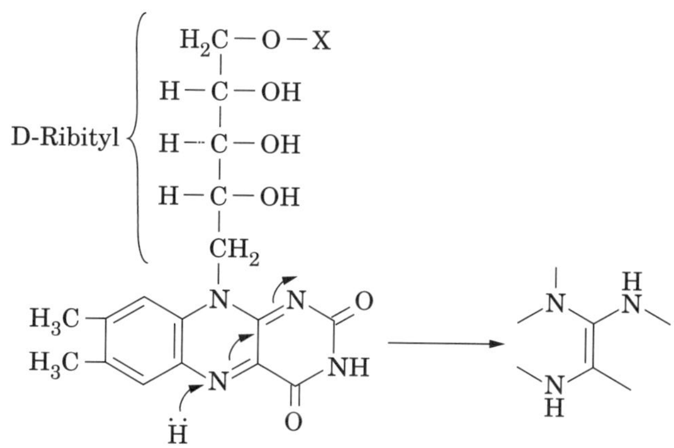

Figure 2. Structures of riboflavin, FMN, and FAD. The electronic shift upon reduction is shown. Riboflavin, X = H; FMN, X = PO_3H_2; FAD, X = ADP. For the sake of convenience, the reduction reaction is drawn as proceeding via a hydride ion even though this need not be the actual mechanism in all flavin reductions. *Source*: Ref. 5.

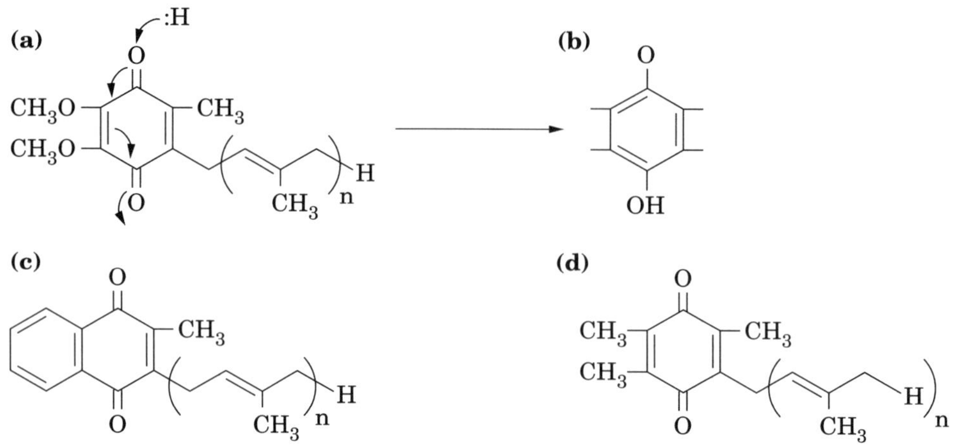

Figure 3. The structure of quinones. (**a**) Oxidized ubiquinone; (**b**) a portion of reduced ubiquinone; (**c**) oxidized menaquinone; (**d**) oxidized plastoquinone. The value of n can be 4–10 and is 8 for both quinones in *E. coli*. In *E. coli* ubiquinone plays a major role in aerobic and nitrate respiration, whereas menaquinone is dominant during fumarate respiration. One reason for this is that ubiquinone has a potential (E_o') of $+100$ mV compared to $+$ 30 mV for fumarate. It is therefore at too high of a potential to deliver electrons to fumarate. Menaquinone has a low potential, -74 mV, and is thus able to deliver electrons to fumarate. Plastoquinone is used in chloroplast and cyanobacterial photosynthetic electron transport. *Source*: Ref. 5.

and cyanobacteria, and functions in photosynthetic electron transport.

Iron–Sulfur Proteins

Iron–sulfur proteins contain nonheme iron and usually acid-labile sulfur (Fig. 4). The term *acid-labile* sulfur means that H_2S is released from the protein when the pH is lowered to approximately 1. This is because there is sulfide attached to iron by bonds that are ruptured in acid. Generally, the proteins contain clusters in which iron and acid-labile sulfur are present in a ratio of 1:1. However, there may be more than 1 FeS cluster per protein. For example, in mitochondria the enzyme complex that oxidizes NADH has at least four FeS clusters (see Fig. 9). The FeS clusters have different E_h values, and the electron travels from one FeS cluster to the next toward the higher E_h. (The E_h is the redox potential at physiological concentrations of oxidized and reduced forms.) It appears that the electron

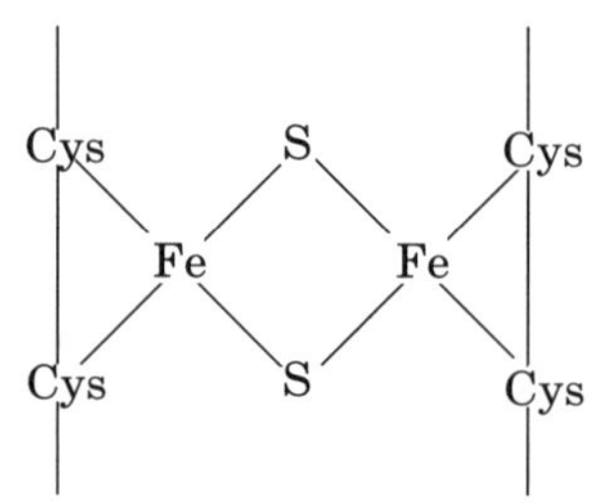

Figure 4. FeS cluster. This is an Fe_2S_2 cluster. More than one cluster may be present per protein. The sulfur atoms held only by the iron are acid labile. The iron is bonded to the protein via sulfur in cysteine residues. *Source*: Ref. 5.

may not be localized on any particular iron atom, and the entire FeS cluster should be thought of as carrying one electron, regardless of the number of Fe atoms. These proteins also contain cysteine sulfur, which is not acid labile and which bonds the iron to the protein. There are several different types of FeS proteins, and these catalyze numerous oxidation–reduction reactions in the cytoplasm as well as in the membranes. The FeS proteins cover a very wide range of potentials, from approximately -400 mV to $+350$ mV. They therefore can carry out oxidation–reduction reactions at both the low potential end and the high potential end of the electron transport chain, and indeed they are found in several locations. An example of an FeS cluster is shown in Figure 4. Note that each Fe is bound to two acid-labile S and two cysteine S. This would be called an Fe_2S_2 cluster.

Cytochromes

Cytochromes are electron carriers that have heme as the prosthetic group. Heme consists of four pyrrole rings attached to each other by methene bridges (Fig. 5). Because hemes have four pyrroles, they are called *tetrapyrroles*. Each of the pyrrole rings is substituted by a side chain. Substituted tetrapyrroles are called *porphyrins*. Therefore, hemes are also called porphyrins. (An unsubstituted tetrapyrrole is called a porphin.) Hemes are placed in different classes, described next, on the basis of the side chains attached to the pyrrole rings. In the center of each heme there is an iron atom that is bound to the nitrogen of the pyrrole rings. The iron is the electron carrier and is oxidized to ferric or reduced to ferrous ion during electron transport.* Cytochromes are therefore one-electron carri-

*The electron carried by the iron is partly delocalized over the π-electrons of the heme.

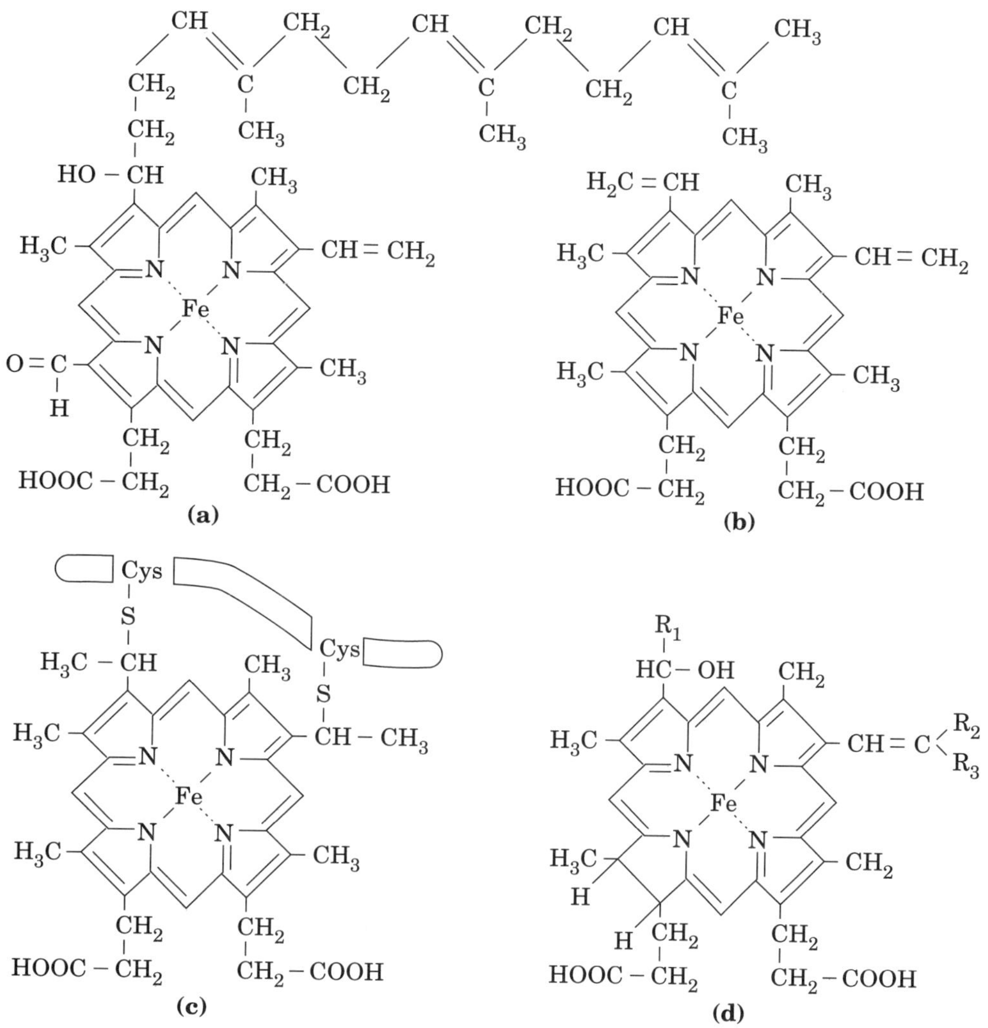

Figure 5. The prosthetic groups of the different classes of cytochromes. The hemes vary according to their side groups. Heme c is covalently bound to the protein via a sulfur bridge to a cysteine residue on the protein. The structures of R_1, R_2, and R_3 are not known. *Source*: Ref. 6.

ers. The E_h of the different cytochromes differ depending on the protein and the molecular interactions with surrounding molecules.

Classes of Cytochromes. There are four classes of cytochromes (a, b, c, and d) that have hemes a, b, c, and d. (Cytochrome o in bacteria is a b-type cytochrome.) As mentioned previously, the hemes can be distinguished according to the side groups that they possess. Hemes can also be distinguished spectrophotometrically. When cytochromes are in the reduced state they have characteristic light absorption bands in the visible range because of absorption by the heme. These are the α, the β, and the γ bands. The α bands absorb light between 500 and 600 nm, the β absorb at a lower wavelength, and the γ bands are in the blue region of the spectrum. The spectrum for a cytochrome c is shown in Fig. 6. Cytochromes are distinguished, in part, by the position of the maximum in the α band. For example, cyt. b_{556} and cyt. b_{558} differ because the former has a peak at 556 nm and the latter a peak at 558 nm.

ORGANIZATION OF THE ELECTRON CARRIERS IN MITOCHONDRIA

The electron carriers are organized as an electron transport chain in the inner mitochondrial membrane. The scheme is shown in Figure 7. The protein electron carriers (with the exception of cytochrome c) are organized in the membrane as individual complexes (complexes I–IV) that can be isolated from each other by appropriate separation techniques after mild detergent extraction. Four complexes can be recognized in mitochondria. They are complex I (NADH–ubiquinone oxidoreductase), complex II (succinate dehydrogenase), complex III (ubiquinol–cytochrome c oxidoreductase, also called the bc_1 complex), and complex IV (cytochrome c oxidase, which is cytochrome

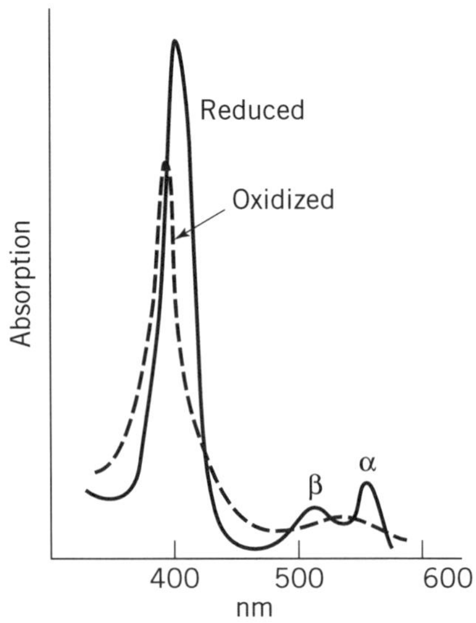

Figure 6. Absorption spectra of oxidized and reduced cytochrome c. The α band in the reduced form is used to identify cytochromes. *Source*: Ref. 5.

aa_3). Cytochrome c exists as a peripheral membrane protein that is easily extracted with water from the inner membrane. Complexes I, III, and IV are coupling sites, a Δp is generated by these complexes (see "Coupling Sites"). The electron donors are either NADH or succinic acid. NADH is generated during metabolism by various oxidation reactions in the cytoplasm. It must be reoxidized to NAD^+ in order that the cytoplasmic oxidations continue, and this is done through electron transport in respiring organisms. Succinic acid is generated from various organic compounds such as carbohydrates, amino acids, and fatty acids. It is actually an intermediate stage in the oxidation of these compounds to CO_2 (in the citric acid cycle). Succinic acid can also serve as an exogenous source of carbon and energy for growth for many microorganisms. The electrons and hydrogen are removed from the NADH or succinic acid by enzyme complexes called *dehydrogenases* (complexes I and II). The dehydrogenases are named after the substrate that donates the electrons. For example, NADH donates electrons and hydrogen to NADH dehydrogenase (complex I), and succinic acid donates electrons and hydrogen to succinate dehydrogenase (complex II). The dehydrogenases then transfer the electrons to UQ, which then transfers the electrons to cytochromes (complex III). There is a pattern here that is worth emphasizing because it is seen in bacteria as well. Dehydrogenases remove electrons and pass these to quinones which in turn pass the electrons to cytochrome oxidase complexes that transfer the electrons to oxygen. The value of the quinone is that it is a small diffusible molecule that connects the dehydrogenase complexes with the cytochrome complexes.

Each complex can have several proteins. The most intricate is complex I from mammalian mitochondria, which has about 40 polypeptide subunits, at least four FeS centers, one FMN, and one or two bound UQs. Analogous complexes have been isolated from bacteria, but in some cases (e.g., NADH–ubiquinone oxidoreductase and the bc_1 complex), they have fewer protein components (7–10).

Inhibitors of Electron Transport

There are several well-known inhibitors of the electron transport enzymes. These include rotenone and amytal, which inhibit complex I, antimycin A, stigmatellin, and myxothiazol, which inhibit complex III, and cyanide, azide, and carbon monoxide, all of which inhibit complex IV.

ORGANIZATION OF THE ELECTRON CARRIERS IN BACTERIA

Bacterial electron transport chains vary among the different bacteria and also according to the growth conditions.

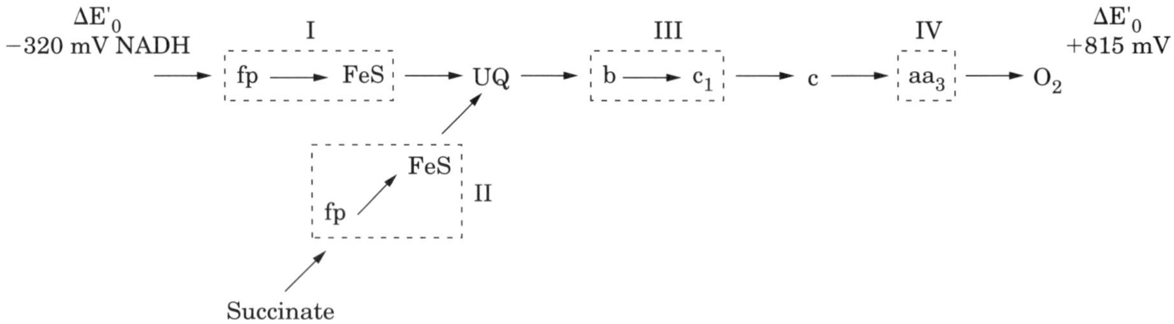

Figure 7. Electron transport scheme in mitochondria. Electrons travel in the electron transport chain from a low to a high electrode potential. Complexes I to IV are bracketed by dotted lines. Complex I is NADH dehydrogenase, also called NADH-ubiquinone oxidoreductase. Complex II is succinate dehydrogenase, also called succinate–ubiquinone oxidoreductase. Complex III is the bc_1, complex, also called ubiquinol–cytochrome c oxidoreductase. Complex IV is the cytochrome aa_3 oxidase, also called cytochrome c oxidase. There are several FeS clusters in complexes I and II and an FeS protein in complex III. Complex II also has a cytochrome b. fp, flavoprotein; FeS, iron–sulfur protein; UQ, ubiquinone; b, cytochrome b; c_1, cytochrome c_1; c, cytochrome c; aa_3, cytochrome aa_3. *Source*: Ref. 5.

However, there are certain features that bacterial electron transport chains have in common with each other and with mitochondria. As with the mitochondrial electron transport chain, the bacterial chains are organized into dehydrogenase and oxidase complexes connected by quinones (Fig. 8). The quinones accept electrons from dehydrogenases and transfer these to oxidase complexes via cytochromes, for example, via cytochromes b and c to cytochrome o or cytochrome aa_3 oxidase. As with mitochondria, the oxidase complexes reduce oxygen.

As opposed to mitochondria, bacteria are capable of using electron acceptors other than oxygen, denoted as Y in Figure 8b, during *anaerobic* respiration. A commonly used electron acceptor other than oxygen is nitrate, which can be reduced to nitrite, ammonia, or nitrogen gas. In fact, the depletion of nitrate from anaerobic soils is due to a large extent to bacteria carrying out anaerobic respiration and reducing the nitrate to nitrogen gas that escapes into the atmosphere. Sulfate is also used as an electron acceptor by certain anaerobic bacteria that are responsible for the production of hydrogen sulfide in various ecological niches, including sulfur springs. The enzyme complexes that reduce electron acceptors other than oxygen are called *reductases,* rather than oxidases.

The dehydrogenase complexes are more varied than found in mitochondria. Some of the dehydrogenase complexes are NADH and succinate dehydrogenase complexes analogous to complex I and II in mitochondria. However, in addition to these dehydrogenases, there are several others that reflect the diversity of substrates oxidized by the bacteria as sources of carbon, energy and electrons. For example, there are formic acid dehydrogenase, lactic acid dehydrogenase, methanol dehydrogenase, methylamine dehydrogenase, and so on. Many bacteria also have hydrogen gas (H_2) dehydrogenases (called hydrogenases) and use hydrogen gas as a source of energy and electrons.

Some bacteria grow on inorganic compounds as a source of electrons and energy and therefore have enzymes to oxidize the inorganic compounds and transfer the electrons to the electron transport scheme. Inorganic compounds that can be used for this purpose by certain bacteria include ammonia, nitrite, ferrous ion, hydrogen sulfide, sulfur, and (as mentioned) hydrogen gas. These bacteria have an enormous ecological impact and are crucial in the recycling of inorganic materials in the biosphere.

Depending on the source of electrons and electron acceptors, bacteria can synthesize and substitute one dehydrogenase complex for another, or reductase complexes for oxidase complexes. For example, when growing anaerobically, *Escherichia coli* makes the reductase complexes instead of the oxidase complexes, represses the synthesis of some dehydrogenases, and stimulates the synthesis of others. For this reason, the electron carrier complexes in bacteria are sometimes referred to as modules because they can be synthesized and plugged into the respiratory chain when needed.

Another important difference between mitochondrial and bacterial electron transport chains is that the routes to oxygen in the bacteria are branched, the branch point being at the quinone or one of the cytochromes. Under aerobic conditions there are often two or three branches leading to different oxidases. For example, a two-branched

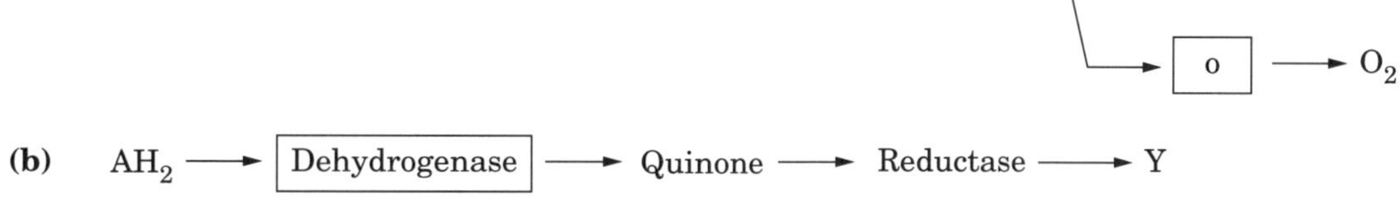

Figure 8. Generalized electron transport pathways found in bacteria. The details will vary depending on the bacterium and the growth conditions. (**a**) Aerobic respiration. A dehydrogenase complex removes electrons from an electron donor and transfers these to a quinone. The electrons are transferred to an oxidase complex via a branched pathway. Depending on the bacterium, the pathway may branch at the quinone or at cytochrome. Many bacteria have bc_1, cytochrome c, and cytochrome aa_3 in one of the branches and in this way resemble mitochondria. Other bacteria do not have a bc_1 complex, and may or may not have cytochrome aa_3. (**b**) anaerobic respiration. Under anaerobic conditions the electrons are transferred to reductase complexes that are synthesized anaerobically. Several reductases exist, each one specific for the electron acceptor. Y represents either an inorganic electron acceptor other than oxygen, such as nitrate, or an organic electron acceptor, such as fumarate. More than one reductase can simultaneously exist in a bacterium. *Source*: Ref. 5.

electron transport chain might contain a branch leading to cytochrome o oxidase and a branch leading to cytochrome aa_3 oxidase (cytochrome c oxidase). Other bacteria (e.g., *E. coli*) have cytochrome o and d oxidase branches but do not have the cytochrome aa_3 branch. The ability to synthesize branched electron transport pathways to oxygen confers flexibility on the bacteria, because the branches may differ not only in the Δp that can be generated (because they may differ in the number of coupling sites) but their terminal oxidases may also differ with respect to affinities for oxygen. Switching to an oxidase with a higher affinity for oxygen allows the cells to continue to respire even when oxygen tensions fall to very low values. This is important to ensure the reoxidation of electron carriers such as NADH in order that the various oxidative steps in metabolism continue at an acceptable rate.

COUPLING SITES

Sites in the electron transport pathway where redox reactions are coupled to proton translocation creating a Δp are called *coupling sites* (11). In mitochondria there are three coupling sites, and they are called sites 1, 2, and 3. These are shown in Figure 7. Site 1 is the NADH dehydrogenase complex (complex I), site 2 is the bc_1 complex (complex III), and site 3 is the cytochrome aa_3 complex (complex IV). The succinate dehydrogenase complex (complex II) is not a coupling site. The ratios of protons translocated per two electrons vary, depending on the complex. The bc_1 complex translocates four protons per two electrons, and depending on the reported value, complexes I and IV translocate from two to four protons per two electrons. (The consensus value for mitochondrial complex I is $4H^+/2e^-$.) Parts of the electron transport chain are reversible, a necessary situation if the source of electrons for biosynthesis is a molecule with a redox potential higher than that of the cell material being synthesized. For example, bacteria growing on ammonia or nitrite as a source of electrons must reverse electron transport in order to synthesize NADH, which is used as a source of electrons for biosynthesis. Electron transport must also be reversed when succinate is the electron donor for biosynthesis. During reversed-electron flow, protons enter through coupling sites 1 and 2, driven by the Δp, and the electrons are driven toward the lower redox potential. This creates a positive ΔE at the expense of the Δp (see equation 1.) Coupling site 3 is not physiologically reversible. Thus, water cannot serve as a source of electrons for NAD^+ reduction using reversed electron flow. However, during oxygenic photosynthesis, light energy can drive electrons from water to $NADP^+$, producing NADPH and O_2. The mechanism of photoreduction of $NADP^+$ is different from reversed electron flow.

P/O And P/2e Ratios

The synthesis of ATP during respiration is called *oxidative phosphorylation* or *respiratory phosphorylation*. The number of ATPs made per 2 e^- transfer to oxygen is called the *P/O ratio*. It is equal to the number of ATP molecules formed per atom of oxygen taken up when the Δp is used solely for ATP synthesis. When an electron acceptor other than oxygen is used, then $P/2e^-$ is substituted for P/O. In mitochondria, the oxidation of NADH results in a P/O ratio of about 3 reflecting the three coupling sites that exist between NADH and O_2. These are complexes I, III, and IV. The use of succinate as an electron donor results in a P/O ratio of approximately 2 because electrons from succinate enter at the UQ level, bypassing coupling site 1 (complex I) that occurs between NADH and UQ (i.e., the NADH-ubiquinone oxidoreductase reaction [Fig. 7]). The other two coupling sites occur between UQ and oxygen. When electrons enter the respiratory chain after the bc_1 complex, the P/O ratio is reduced to 1, indicating that the bc_1 complex (complex III) is the second coupling site and that site 3 is cytochrome aa_3 oxidase (complex IV). Each of these sites is characterized by a drop in midpoint potential of about 200 mV, which is sufficient for generating the Δp required for ATP synthesis (Fig. 9). The size of the Δp that can be generated with respect to the ΔE is given in equation 1.

Number of ATPs Made According to Chemiosmotic Theory

Although it is usually stated that the number of ATPs made per coupling site is 1, the number of ATPs made at a coupling site need not be a whole number. This is because the amount of ATP that can be made per coupling site is equal to the ratio of protons translocated at the coupling site to protons that reenter via the ATP synthase, and this ratio need not be 1 (Fig. 10). For example, if 2 mol of protons is translocated at a coupling site and 3 mol of protons enters through the ATP synthase, then 2/3 mol of ATP can be made when 2 mol of electrons passes through the coupling site. The ratio of protons translocated per two electrons traveling through all the coupling sites to O_2 is called the H^+/O ratio. It can be measured by administering a pulse of a known amount of oxygen to an anaerobic suspension of mitochondria or bacteria and measuring the initial efflux of protons with a pH electrode as the small amount of oxygen is used up. A permeable anion such as SCN^- must be present so that proton extrusion is electroneutral, thereby allowing protons to accumulate in the bulk phase. The reported values for H^+/O for NADH oxidation vary. However, there is a consensus that the true value is probably around 10. The ratio of protons entering via the ATP synthase to ATP made is called the H^+/ATP ratio. Values of H^+/ATP from 2 to 4 have been reported, and a consensus value of 3 can be used for calculations. For intact mitochondria, an additional H^+ is required to bring P_i electroneutrally from the cytosol into the mitochondrial matrix in symport with H^+, so H^+/ATP would be 4. A value of 10 for H^+/O predicts a maximum P/O ratio of 2.5 (10/4) for mitochondria. This means that the often-stated value of 3 for a P/O ratio for NADH oxidation by mitochondria may be a little too high. The number of protons translocated per 2 *e*-traveling between succinate and oxygen (couplings sites 2 and 3) in mitochondria is 6. Therefore, the maximal P/O ratio for this segment of the electron transport chain in mitochondria may be 6/4 or 1.5, rather than 2.

Uncouplers

Uncouplers of oxidative phosphorylation are toxic compounds that short-circuit the proton current so that res-

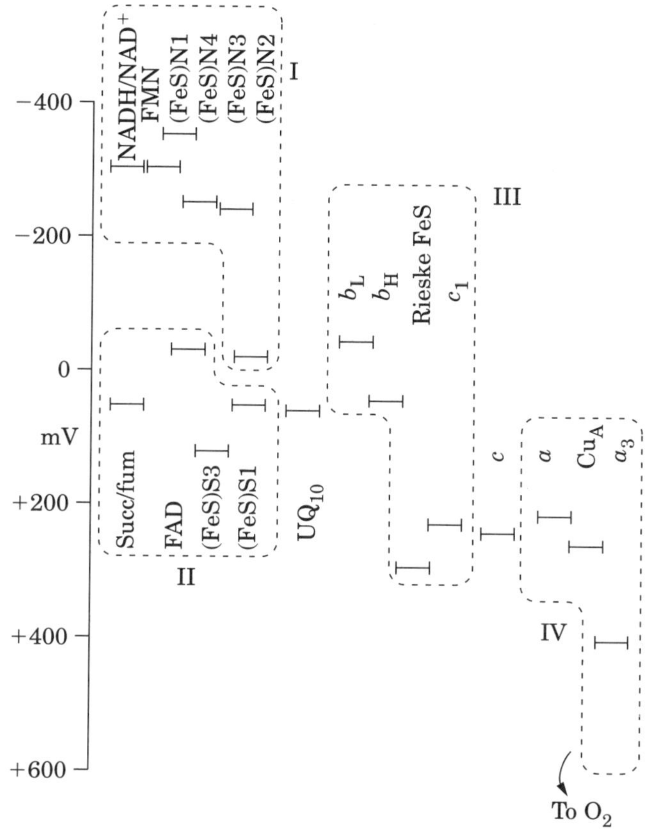

Figure 9. Average midpoint potentials, E'_m, for components of the mitochondrial respiratory chain. The complexes are in dotted boxes. The actual potentials (E_h) for most of the components is not very different from the midpoint potentials. An exception is cytochrome aa_3 whose E_h is much more positive than its midpoint potential. There are three sites where there are changes in potential of 200 mV or more. These drive proton translocation. One site is within complex I, a second within complex III, and the third between complex IV and oxygen. *Source*: Ref. 12.

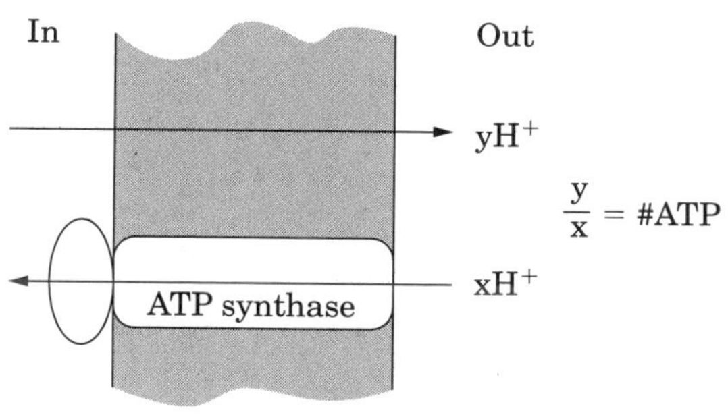

Figure 10. The ratio of protons extruded to protons translocated through the ATPase determines the amount of ATP made. The depiction of the ATP synthase does not attempt to describe the known complexity of its structure. *Source*: Ref. 5.

piration continues (usually at an increased rate) but ATP synthesis stops. The reason that ATP synthesis stops is that the protons enter via the uncoupler rather than through the ATP synthase. Two examples of uncouplers are dinitrophenol (DNP) and carbonyl cyanide-*p*-trifluoromethylhydrazone (FCCP). They are organic compounds that equilibrate protons across the membrane. As a consequence, the Δp is collapsed.

Q Loops, Q Cycles, and Proton Pumps

The previous discussion points out that proton translocation takes place at coupling sites when electrons travel downhill over a potential gradient of at least 200 mV (Fig. 9.) However, the mechanism by which the redox reaction is actually coupled to proton translocation was not explained. There are two ways in which this is thought to occur: (*1*) a Q loop or Q cycle, and (*2*) a proton pump. A Δp is established in a similar way in both the Q loop and Q cycle, as described next. However, some important differences do exist between these two mechanisms, and these are discussed separately.

In the Q loop or Q cycle, reduced quinone carries hydrogen atoms and electrons (the equivalent of H_2) across the membrane and becomes oxidized, releasing only protons on the external face of the membrane because the electrons return electrogenically via electron transport carriers to the inner membrane surface. When this occurs a $\Delta\Psi$ is established, outside positive because a positive charge (the proton) has been left on the outer surface while a negative charge (the electron) moves across the membrane toward the inside surface. (Electrogenic means that a membrane potential is established.) On the cytoplasmic side, the same number of protons are taken up as were released on the outside as the terminal oxidant (e.g., oxygen) is reduced by the electrons. The result is the net translocation of protons from the inside to the outside, although protons per se do not actually traverse the membrane, as well as the generation of a $\Delta\Psi$. A ΔpH can be established if the protons accumulate in the bulk phase outside of the cell. This can occur only if an equal number of anions (negatively charged ions) also accumulate to balance the positive charge of the protons, or if the protons exchange with cations in the medium. For most bacteria, the major portion of the Δp is due to the $\Delta\Psi$, although a small ΔpH is also created. An exception are bacteria living in environments that are much more acidic than their cytoplasm. For these bacteria, the Δp is due mostly to the ΔpH.

The Q Loop. The essential feature of the Q loop model is that the electron carriers alternate between those that carry both hydrogen and electrons (flavoproteins and quinones) and those that carry only electrons (FeS proteins and cytochromes). This is illustrated in Figure 11. The electron carriers and their sequence are flavoprotein (H carrier), FeS protein (e^- carrier), quinone (H carrier), and cytochromes (e^- carriers). The flavoprotein and FeS protein comprise the NADH dehydrogenase, which can be a coupling site, although the mechanism of proton translocation by the NADH dehydrogenase is not understood. Perhaps it is a proton pump. When electrons are transferred

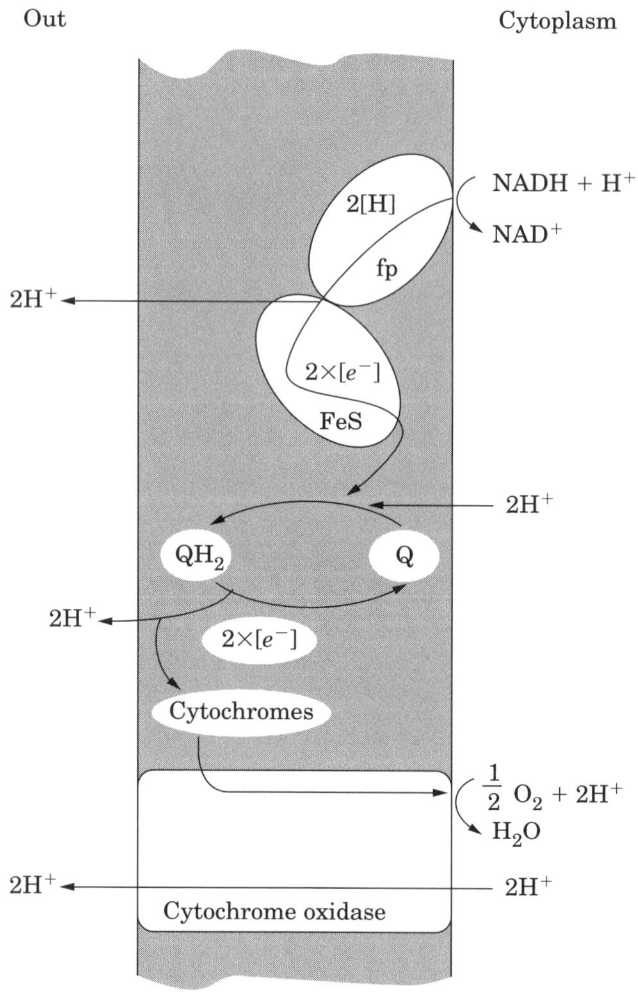

Figure 11. Proton translocation showing a Q loop and a proton pump. The diagram does not attempt to show the actual locations of the electron carriers within the membrane. It is proposed that the electron carriers exist in an alternating sequence of hydrogen (H) and electron (e-) carriers. This would be (H)flavoprotein (fp), (e-) iron–sulfur protein (FeS), (H)quinone (Q), and (e-)cytochromes. Oxidation of the fp deposits two protons on the outer membrane surface. The electrons return to the inner membrane surface where a quinone is reduced, taking up two protons from the cytoplasm. The reduced quinone diffuses to the outer surface of the membrane where it is oxidized, depositing two more protons on the surface. The electrons return to the cytoplasmic surface via cytochromes where they reduce oxygen in a reaction that consumes protons. Some cytochrome oxidases function as proton pumps. During anaerobic respiration the cytochrome oxidase is replaced by a reductase, and the electrons reduce some other electron acceptor, such as nitrate or fumarate. It should be noted that electrons can also enter at the level of quinone, such as from succinate dehydrogenase. *Source*: Ref. 5.

from the FeS protein to quinone (Q) on the inner side of the membrane, two protons are acquired from the cytoplasm. According to the model, the reduced quinone (QH_2), called quinol, then diffuses to the outer membrane surface and becomes oxidized, releasing the two protons. The electrons then return electrogenically via cytochromes to the inner membrane surface where they reduce oxygen. This would create a $\Delta\Psi$ because the protons are left on the outer surface of the membrane as the electrons move electrogenically to the inner surface. Thus, quinol oxidation is a second coupling site. As mentioned previously, the energy to create the membrane potential is derived from the ΔE between the oxidant and the reductant. One way to view this is that the energy from the ΔE pushes the electron to the negative membrane potential on the inside surface. Note that the role of the quinone is to ferry the hydrogens across the membrane, presumably by diffusing from a reduction site on the cytoplasmic side of the membrane to an oxidation site on the outer side.

During reduction of the terminal electron acceptor, cytoplasmic protons are consumed. For example, during reduction of $1/2O_2$ to H_2O, two protons are taken up from the cytoplasm. Thus, the result is a net translocation of two protons to the outside per QH_2 oxidation, even though there was no transmembrane movement of protons, per se. For the purpose of calculating the expected Δp generated at a coupling site, it makes no difference whether one postulates the transmembrane movement of protons in one direction or electrons in the opposite direction, because they both carry the same charge, and equation 1 can be used.

The Q Cycle. Although the linear Q loop as described above for the oxidation of quinol may accurately describe quinol oxidation in *E. coli* and some other bacteria, it is inconsistent with experimental observations of electron transport in mitochondria, chloroplasts, and many bacteria. For example, the Q loop predicts that the ratio of H^+ released per QH_2 oxidized is 2, whereas the measured ratio in mitochondria and many bacteria is actually 4. In order to account for the extra two protons, Mitchell suggested a new pathway for the oxidation of quinol called the *Q cycle*, which is more complicated than the Q loop (13) (Fig. 12.) The Q cycle operates in an enzyme complex called the bc_1 complex (complex III). The bc_1 complex from bacteria contains three polypeptides. These are cytochrome b with two b-type hemes, an FeS protein containing a single 2Fe-2S cluster (the Rieske protein), and cytochrome c_1 with one heme. The complex spans the membrane and has a site for binding reduced ubiquinone, UQH_2, on the outer surface of the membrane called site P (for positive); and a second site on the inner surface for binding UQ, site N (for negative) (Fig. 12). At site P, UQH_2 is oxidized to the semiquinone anion, UQ^- by the removal of one electron (Fig. 12**a**), and the two protons are released to the membrane surface. The electron travels to the FeS protein, and from there to cytochrome c_1 on its way to oxygen. The UQ^- is then oxidized to UQ by the removal of the second electron, which is transferred to b_{556}, also called b_L because of its relatively low E. The electron is then transferred across the membrane to b_{560}, also called b_H because of its relatively high E. Heme b_H transfers the electron to UQ bound at site N, reducing it to the semiquinone anion, UQ: *Electron flow from the Q_P site to the Q_N site is transmembrane and creates a membrane potential.* A second UQH_2 is oxidized at the P site and the UQ_N^- is reduced to UQH_2, picking up two protons from the cytoplasm (Fig. 12**b**). The UQH_2 enters the

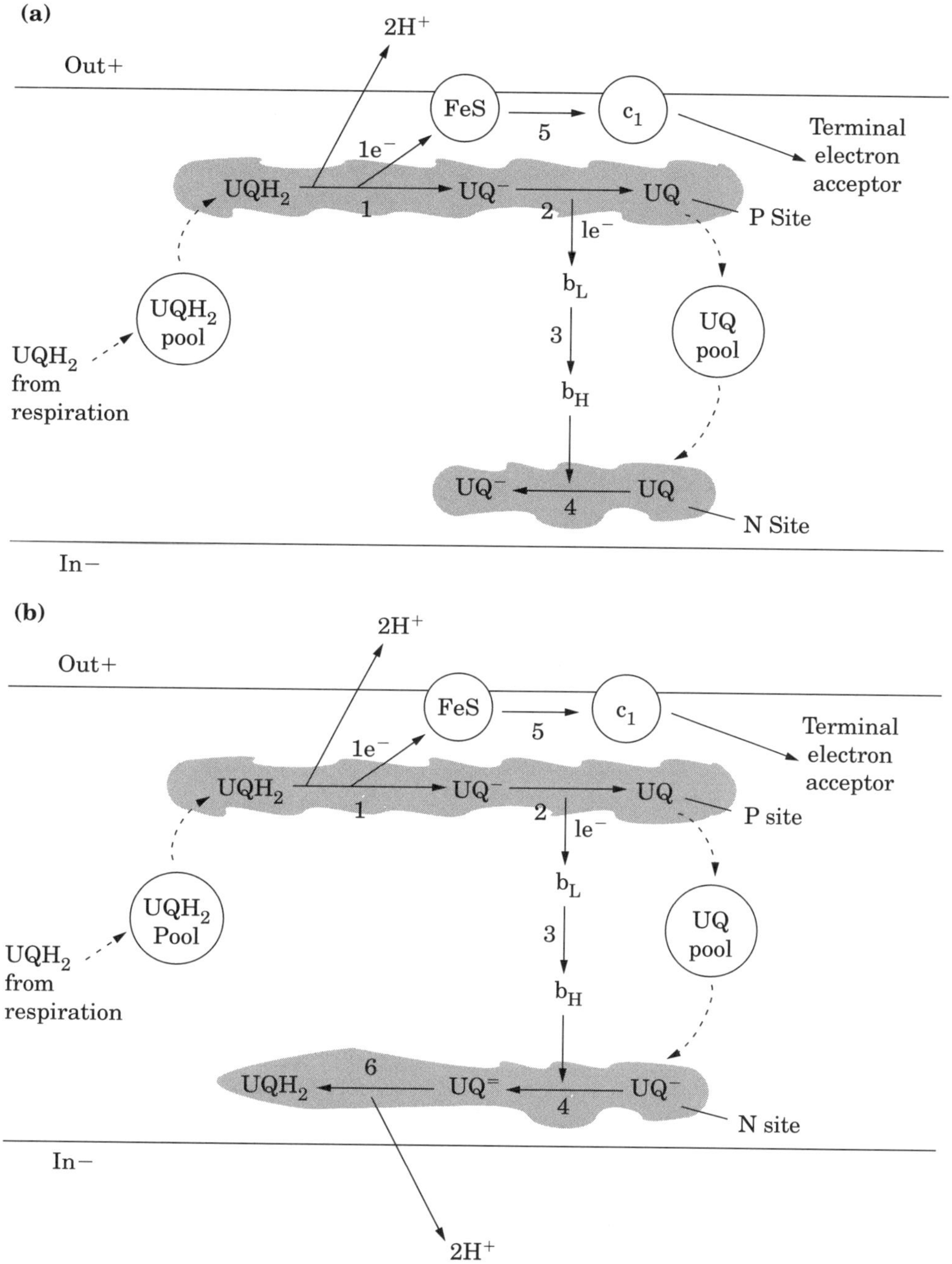

Figure 12. The bc_1 complex. The bc_1 complex isolated from bacteria contains three polypeptides, which are a cytochrome b containing two heme b groups, b_H and b_L, an iron-sulfur protein, FeS (the Rieske protein), and cytochrome c_1. It is widely distributed among the bacteria, including the photosynthetic bacteria. (Mitochondrial bc_1 complexes are similar but contain an additional 6 to 8 polypeptides without prosthetic groups.) The FeS protein and cytochrome c_1 are thought to be located on the outside (positive or p) surface of the membrane and the cytochrome b is believed to span the membrane acting as an electron conductor. On the outer surface there is a binding site in the bc_1 complex for ubiquinol, UQH_2 (the P site). On the inner surface (negative or n) there is a binding site in the bc_1 complex for ubiquinone, UQ (the N site). (**a**) Reduced ubiquinone binds to the P site and one electron is removed forming the semiquinone anion, UQ^- (reaction 1). At this time two protons are released on the outer membrane surface. The electron that is removed is transferred to the FeS protein and from there to cytochrome c1 (step 5). (Cytochrome c_1 transfers the electron to the terminal electron acceptor via a series of electron carriers in other reactions.) In reaction 2 the second electron is removed from the semiquinone anion producing the fully oxidized quinone, UQ. The second electron is transferred transmembrane via cytochrome b to ubiquinone at site N (steps 3 and 4). Because the electron travels transmembrane a membrane potential is created, outside positive. (**B**). A second reduced ubiquinone is oxidized at the P site, releasing two more protons, and the sequence of electron transfers is repeated. The UQ^- at site N becomes reduced to UQH_2, having acquired two protons from the cytoplasm. Note that for every two UQH_2 molecules that are oxidized, four protons are released, and one UQH_2 is regenerated. Therefore, four protons are released per one UQH_2 oxidized. Another way of saying this is that the ratio, H^+/e^-, is 2. *Source*: Ref. 5.

quinone pool. Thus, for every two UQH_2 that are oxidized releasing four protons, one UQH_2 is regenerated. The net result is the oxidation of one UQH_2 to UQ with the release of four protons, two from UQH_2 and two from the cytoplasm. The situation can be summarized as the following:

$$\text{P site:}\quad 2UQH_2 + 2cyt.c_{ox} \rightarrow 2UQ + 2cyt.c_{red} + 4H^+_{out} + 2e^-$$
$$\text{N site:}\quad UQ + 2e^- + 2H^+_{in} \rightarrow UQH_2$$
$$UQH_2 + 2cyt.c_{ox} + 2H^+_{in} \rightarrow UQ + 2cyt.c_{red} + 4H^+_{out}$$

In the pre–steady state, UQH_2 can be oxidized at the N site (by reversal) as well as the P site. The P site is inhibited by myxathiozol and by stigmatellin, whereas the N site is inhibited by antimycin. Therefore, a combination of inhibitors is required to completely inhibit the oxidation of UQH_2 in the pre–steady state.

Advantage of the Q Cycle Over the Q loop. Because the Q cycle translocates two protons per electron that flows to the terminal electron acceptor, it generates a larger proton current than the Q loop that translocates only one proton per electron. This can result in more ATP synthesis. Consider the situation where an ATP synthase requires the influx of three protons to make one ATP. If the transfer of an electron through the electron transport pathway resulted in the translocation of one proton, then one-third of an ATP could be made per electron. On the other hand, if electron transport resulted in the translocation of two protons per electron, then two-thirds of an ATP could be made per electron. In other words, the size of the proton current generated by respiration determines the upper value of the amount of ATP that can be made.

Pumps. Proton pumps are electron carrier proteins that couple electron transfer to the electrogenic translocation of protons (rather than electrons) through the membrane. It is not understood exactly how these pumps work at the molecular level, but perhaps the proton is passed from one amino acid side group that can be reversably protonated to another. A model for proton pumping is the light-driven proton pump bacteriorhodopsin (14). Proton extrusion accompanies the cytochrome aa_3 oxidase reaction when cytochrome c is oxidized by oxygen in mitochondria and some bacteria (Fig. 11). This can be observed by feeding electrons into the respiratory chain at the level of cytochrome c, from where they travel directly to the cytochrome aa_3 oxidase. The experimental procedure is to incubate the cells in lightly buffered anaerobic media with a reductant for cytochrome c and a permeant anion (e.g., SCN^-) or an exchangeable cation such as valinomycin plus K^+, so that protons can exit electroneutrally and accumulate in the medium. Changes in pH are measured with a pH electrode. Upon the addition of a pulse of oxygen, given as an air-saturated salt solution, a sharp, transient acidification of the medium occurs, and the H^+/O can be calculated. Because the only enzymatic reaction is the cytochrome aa_3 oxidase reaction, it can be concluded that the cytochrome aa_3 oxidase pumps protons out of the cell during the redox reaction. When similar experiments were done with *Paracoccus denitrificans* it was found that the *P. denitrificans* cytochrome aa_3 oxidase pumps protons with a stoichiometry of $1H^+/1e$ (15). Cytochrome oxidase pumping activity can also be demonstrated in proteoliposomes* made with purified cytochrome aa_3. Because the proton pumps move a positive charge across the membrane leaving behind a negative charge, a membrane potential, outside positive, develops. The membrane potential should be the same as when an electron moves inward since the proton and the electron carry the same charge (i.e., 1.6×10^{-19} C).

The mitochondrial NADH dehydrogenase complex (NADH ubiquinol oxidoreductase) translocates four protons per NADH oxidized (16). Two types of NADH ubiquinol oxidoreductases exist in bacteria (9–11,17). One of these, called NDH-1, is similar to the mitochondrial complex I in that it is a multisubunit enzyme complex (approximately 14 polypeptide subunits) consisting of FMN and FeS clusters and translocates protons across the membrane during NADH oxidation ($4H^+/2e^-$). The mechanism of proton translocation is not known. (The marine bacterium, *Vibrio alginolyticus,* has a sodium-translocating NDH [Na-NADH]). A second NADH/ubiquinol oxidoreductase, called NDH-2, is also present in bacteria. NDH-2 differs from NDH-1 in consisting of a single polypeptide and FAD and not being an energy-coupling site. In *E. coli*, NDH-1 and NDH-2 are simultaneously present (18). How *E. coli* regulates the partitioning of electrons between the two NAD dehydrogenases is not known, but clearly has important energetic consequences.

PATTERNS OF ELECTRON FLOW IN INDIVIDUAL BACTERIAL SPECIES

Although the major principles of electron transport as outlined above apply to nonphotosynthetic bacteria in general, several different patterns of electron flow exist in particular bacteria, often within the same bacterium grown under different conditions (19). The patterns of electron flow reflect the different sources of electrons and electron acceptors that are used by the bacteria. For example, bacteria may synthesize two or three different oxidases in the presence of air and several reductases anaerobically. In addition, certain dehydrogenases are made only anaerobically because they are part of an anaerobic respiratory chain. Furthermore, whereas electron donors such as NADH and $FADH_2$ generated in the cytoplasm are oxidized inside the cell, there are many instances of oxidations of other electron donors in the periplasm in Gram-negative bacteria. (Gram-negative bacteria are called thus because they fail to stain with the Gram's stain. They all possess a cell wall with common features and a periplasm. The periplasm is a compartment that exists external to the cytoplasm between the cell membrane and the cell wall. It is filled with a proteinacious liquid called periplasm, and the cell wall's rigid layer is called murein.) Examples of substances found in the growth medium or the environment that are oxidized in the periplasm include hydrogen gas, methane, methanol, methylamine, formate, perhaps ferrous ion, reduced inorganic sulfur, and elemental sulfur. In most of these instances periplasmic cytochromes c accept the electrons from the

*Proteoliposomes are lipoprotein vesicles made in the laboratory with purified proteins and phospholipids.

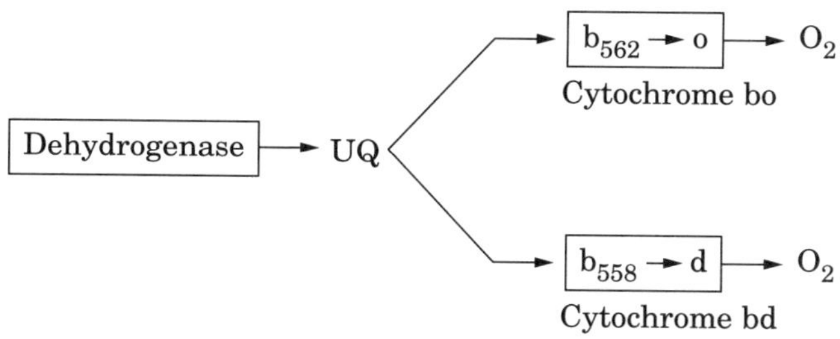

Figure 13. Aerobic respiratory chain in *E. coli.* A variety of dehydrogenase complexes are possible, such as NADH dehydrogenase, succinate dehydrogenase, lactate dehydrogenase, and glycerol-3-phosphate dehydrogenase. The chain branches at the level of ubiquinone (UQ) to two alternate quinol oxidases, cytochrome bo and bd. Cytochrome bd complex has a higher affinity for oxygen and is synthesized under low oxygen tensions where it becomes the major route to oxygen. Coupling sites, where protons are translocated to the outer surface, occur during NADH oxidation, catalyzed by NADH ubiquinone oxidoreductase (also called NDH-1), and quinol oxidation, catalyzed by cytochrome bo complex or cytochrome bd complex. Additionally, cytochrome bo is a proton pump. *E. coli* has a second NADH dehydrogenase, called NADH-2, that does not translocate protons. Therefore, the number of protons translocated per NADH oxidized can vary from two (NADH-2 and bd complex) to eight (NADH-1 and bo complex). *Source*: Ref. 5.

electron donor and transfer them to electron carriers in the cell membrane. The discussion that follows is not complete, but is meant to convey the diversity of electron transport systems found in bacteria.

Escherichia coli

E. coli is a Gram-negative heterotrophic facultative anaerobe. Heterotrophic means that it grows on organic carbon as a source of carbon. (Autotrophic means that it grows on carbon dioxide as its major source of carbon.) Facultative means that it can be grown aerobically using oxygen as an electron acceptor, anaerobically (e.g., using nitrate or fumarate as the electron acceptor), or anaerobically via fermentation. Fermentation means that it does not respire but instead reoxidizes the NADH produced during metabolism in the cytoplasm using organic compounds provided by metabolism as electron acceptors, and synthesizes ATP in the cytoplasm via a mechanism called substrate-level phosphorylation, which is different from oxidative phosphorylation. The bacteria adapt to their surroundings, and the electron transport system that the cells assemble reflects the electron acceptor that is available. When none is available, they ferment.

Aerobic Respiratory Chains. When grown aerobically, *E. coli* makes two different cytochrome oxidase complexes, cytochrome o complex and cytochrome d complex, resulting in a branched respiratory chain to oxygen (Fig. 13) (20). In these pathways, electrons flow from ubiquinol to the ter-

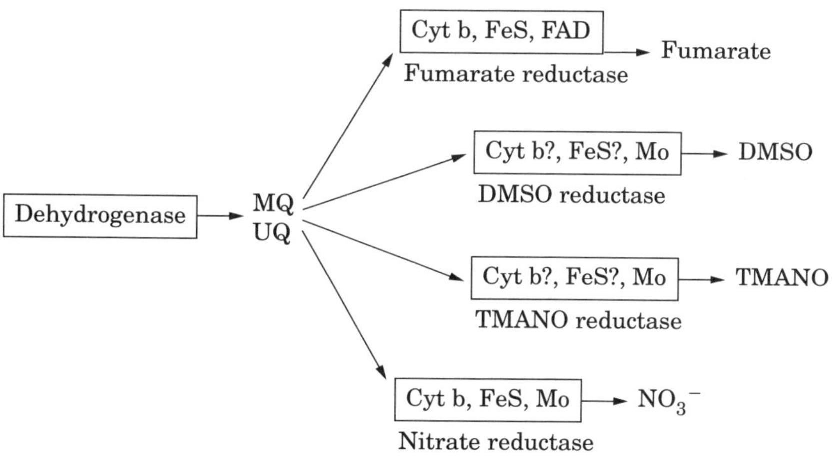

Figure 14. Anaerobic respiratory chains in *E. coli*. When oxygen is absent, *E. coli* synthesizes any one of several membrane-bound reductase complexes depending on the presence of the electron acceptors. Nitrate induces the synthesis of nitrate reductase and represses the synthesis of the other reductases. Menaquinone must be used to reduce some of the reductases, such as fumarate reductase, because it has a sufficiently low midpoint potential (−74 mV). Ubiquinol can reduce nitrate reductase. Each reductase may be a complex of several proteins and prosthetic groups through which the electrons travel to the terminal electron acceptor. The transfer of electrons from the dehydrogenases to the reductases results in the establishment of a proton potential. If the dehydrogenase has site one activity, there can theoretically be two coupling sites, one at the dehydrogenase step and one linked to quinol oxidation at the reductase step. Cyt b, cytochrome b; FeS, nonheme iron–sulfur protein; FAD, flavoprotein with flavin adenine dinucleotide as the prosthetic group; Mo, molybdenum; TMANO, trimethylamine N-oxide; DMSO, dimethylsulfoxide; MQ, menaquinone; UQ, ubiquinone. *Source*: Ref. 5.

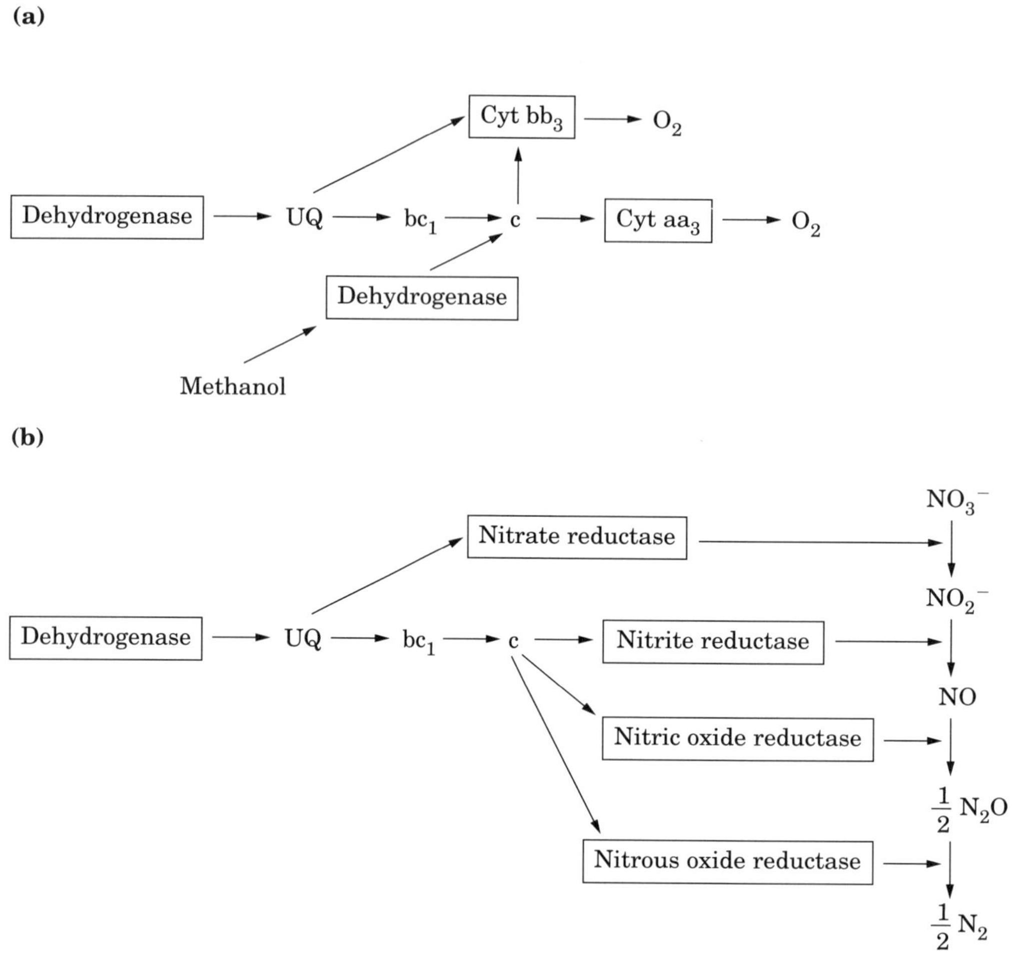

Figure 15. A model for electron transport pathways in *Paracoccus denitrificans*. (**a**) Aerobic. The pathway has two branch points. One branch is at the level of ubiquinone leading to one of two ubiquinol oxidases, that is, the cyt bc_1 complex or cyt bb_3. Both of these quinol oxidases are coupling sites. The bc_1 complex extrudes two protons per electron via the Q cycle. Cyt bb_3 is a proton pump and extrudes one proton per electron vectorially, whereas the second proton is extruded via a Q loop. A second branch point occurs at the level of the bc_1 complex. Electrons can flow either to cyt aa_3, which is a proton pump, or to cyt bb_3, which has been reported to pump protons under certain experimental conditions. NDH-1 is also a coupling site. (**b**) Anaerobic. When the bacteria are grown anaerobically using nitrate as the electron acceptor, the cytochrome aa_3 levels are very low, and the electrons travel from ubiquinone to nitrate reductase and also through the bc_1 complex to nitrite reductase, nitric oxide reductase, and nitrous oxide reductase. *Source*: Ref. 5.

minal oxidase complex, and therefore the oxidases are also called ubiquinol oxidases. The cytochrome o complex from *E. coli* is a proton pump, with a stoichiometry of $1H^+/e^-$. Cytochrome o complex is the predominant cytochrome oxidase when the oxygen levels are high. When the oxygen tensions are lowered, *E. coli* makes cytochrome oxidase d. The cytochrome d oxidase complex has a higher affinity for oxygen than does the cytochrome o complex, suggesting a rationale for why it is the dominant oxidase during growth at low oxygen tensions. Cytochrome d is not a proton pump.

The cytochrome o pathway is more energetically efficient (21). For every two electrons traveling from ubiquinol to oxygen via cytochrome o, four protons are translocated (i.e., two in the Q loop and two by the cytochrome o pump). Assuming a H^+/ATP of 3 (i.e., the ATP synthase translocates inwardly three protons per ATP made), then four-thirds or 1.33 ATPs can be made. Because cytochrome d is not a proton pump, only two-thirds or 0.67 ATPs can be made per two electrons during quinol oxidation via this route.

Physiological Significance of Alternate Electron Routes That Differ in the Number of Coupling Sites. Because the NDH-1 dehydrogenase may translocate as many as four protons per two e^-, whereas the NDH-2 dehydrogenase translocates 0, the number of protons translocated per NADH oxidized can theoretically vary from two (NDH-2 and cytochrome d complex) to eight (NDH-1 and cytochrome o complex). This means that the ATP yields per

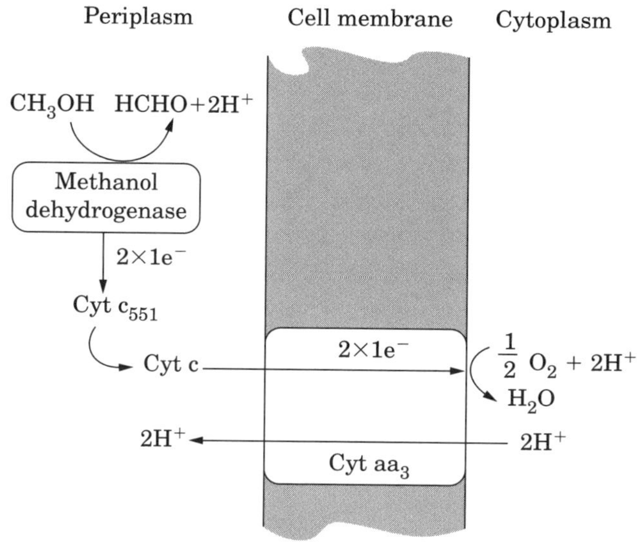

Figure 16. Oxidation of methanol by *P. denitrificans*. Methanol is oxidized to formaldehyde by a periplasmic methanol dehydrogenase. The electrons are transferred to periplasmic cytochromes c and to a membrane bound cytochrome aa_3, which is also a proton pump. A Δp is created due to the electrogenic influx of electrons and the electrogenic efflux of protons, accompanied by the release of protons in the periplasm (methanol oxidation) and uptake in the cytoplasm (oxygen reduction). *Source*: Ref. 5.

NADH oxidized can vary fourfold, from two-thirds or 0.67 to eight-thirds or 2.7. It also means that *E. coli* has great latitude in adjusting the Δp generated during respiration. Because a large Δp can drive electron transport in reverse and thus slow down oxidation of NADH, it may be an advantage to be able to direct electrons along alternate routes that bypass coupling sites and translocate fewer protons. This could ensure adequate rates of reoxidation of NADH so that the oxidative steps in metabolism can continue at a reasonable rate.

Anaerobic Respiratory Chains. In the absence of oxygen, *E. coli* can use either nitrate or fumarate as an electron acceptor. The nitrate is reduced to nitrite or further reduced to ammonia, and the fumarate is reduced to succinate, all of which are excreted into the medium* (22). The anaerobic respiratory chains consists of a dehydrogenase, a reductase, and a diffusible quinone to mediate the transfer of electrons between the dehydrogenase and the reductase (Fig. 14). There can be two coupling sites, one at the dehydrogenase step (site 1) and one at the quinone (i.e., a quinone loop). Anaerobic respiration occurs only in the absence of oxygen because oxygen represses the synthesis of the nitrate and fumarate reductases. When presented with nitrate anaerobically, only the nitrate reductase is made because nitrate represses the synthesis of the fumarate reductase. Thus, there exists a hierarchy of electron acceptors for *E. coli* (i.e., oxygen > nitrate > fumarate). This may reflect the energy yields of the electron transport pathways, because oxygen has the highest electrode potential and fumarate the lowest (Table 1).

*Fumarate is reduced to succinate using the membrane-bound enzyme fumarate reductase. Nitrate is reduced to nitrite using the membrane-bound nitrate reductase, which resembles the dissimilatory nitrate reductase in *P. denitrificans*. Both fumarate and nitrate reductases are coupled to the generation of a Δp, probably via a quinone loop. Nitrite is reduced to ammonia using an NADH-linked cytoplasmic enzyme (NADH-nitrite oxidoreductase). The ammonia is not assimilated, nor is energy conserved in the reaction. Nitrite is toxic to bacteria and probably the reduction to ammonia functions to reduce toxic levels.

Paracoccus denitrificans

P. denitrificans is a nonfermenting Gram-negative facultative anaerobe that can obtain energy from either aerobic respiration or nitrate respiration (13,23). The bacterium can grow heterotrophically on a wide variety of organic carbon sources or autotrophically on H_2 and CO_2 under anaerobic conditions using nitrate as the electron acceptor. It can be isolated from soil by anaerobic enrichment with media containing H_2 as the source of energy and electrons, Na_2CO_3 as the source of carbon, and nitrate as the electron acceptor. Electron transport in *P. denitrificans* receives a great deal of research attention because certain features closely resemble electron transport in mitochondria.

Aerobic Pathway. *P. denitrificans* differs from *E. coli* in that it has a bc_1 complex and a cytochrome aa_3 oxidase (cytochrome c oxidase), and in this way resembles mitochondria (Fig. 15). The composition of the electron transport chain varies with the growth conditions and other oxidases can be present.* The aerobic pathway is thought by some to branch at UQ or perhaps at cytochrome c.† It is not known to what extent the cytochrome bb_3 branch functions during aerobic growth. It has been suggested that the cytochrome bb_3 pathway serves as a Δp release valve. The idea is that under conditions where ATP consumption diminishes, electron flow through the two coupling sites, the bc_1 complex and cytochrome aa_3 complex, generates a Δp that can build to a point of retarding electron flow through the bc_1 complex. (Recall that electron flow can be driven in the reverse direction by the Δp.) Under these conditions, the electrons would be shunted to cytochrome o. In this way, the rates of oxidation of NADH and ubiquinol could be maintained.

P. denitrificans has three coupling sites. Site 1 is the NADH dehydrogenase (NADH-ubiquinone oxidoreductase) and this translocates two to three protons per two electrons. Site 2 is the bc_1 complex that translocates four protons per two electrons due to the Q cycle. Site 3 is the cytochrome aa_3 oxidase that pumps two protons per two electrons across the membrane. As described later, *P. denitrificans* also oxidizes methanol and in this case the electrons enter at cytochrome c, thus bypassing the bc_1 site.

*For example, when the bacteria are grown under low oxygen tensions, cytochrome aa_3 is decreased, and cytochrome o, cytochrome cd (nitrite reductase with oxidase activity), and cytochrome a_1 are increased.

†There are several cytochromes c (at least eight), some of which are in the membrane, and some in the periplasm.

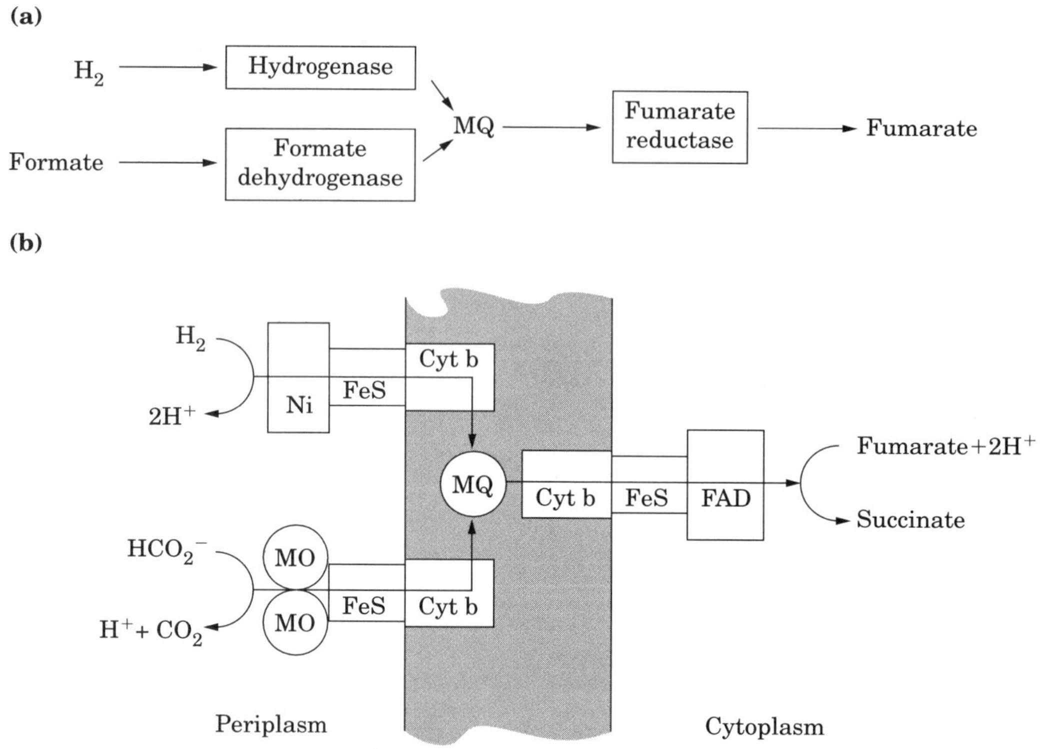

Figure 17. A model for the electron transport system of *W. succinogenes*. (**a**) Electrons flow from H_2 and formate through menaquinone (MQ) to fumarate reductase. (**b**) Illustration showing that the catalytic portions of hydrogenase and formate dehydrogenase are periplasmic, whereas fumarate reductase reduces fumarate on the cytoplasmic side. Electrons flow electrogenically to fumarate. A Δp is created because of electrogenic influx of electrons together with the release of protons in the periplasm, and their consumption in the cytoplasm. *Source*: Ref. 24.

Anaerobic Pathway. *P. denitrificans* can also grow anaerobically using nitrate as an electron acceptor, reducing it to nitrogen gas in a process called *denitrification* (Fig. 15) (25,26). During anaerobic growth on nitrate the electron transport chain is very different than during aerobic growth. The cells contain cytochrome o, nitrate reductase, nitrite reductase, nitric oxide reductase, and nitrous oxide reductase.* The levels of cytochrome aa_3 are very low. The nitrate (NO_3^-) is reduced to nitrite (NO_2^-) in a two-electron transfer via a membrane-bound nitrate reductase. The NO_2^- is reduced to nitric oxide (NO) in a one-electron transfer via a periplasmic* nitrite reductase. The NO is reduced to ½ mol of nitrous acid (½ N_2O) in a one-electron step via a membrane-bound nitric oxide reductase. And the 1/2 N_2O is reduced to 1/2 mol of nitrogen gas (1/2 N_2) in a one-electron step by a periplasmic nitrous oxide reductase. Thus, a total of five electrons flow in and out of the cell membrane through membraneous electron and periplasmic electron carriers from ubiquinol to the various reductases in order to reduce 1 mol of NO_3^- to 1/2 N_2. As shown in Figure 15, the electron transport pathway includes several branches to the individual reductases. The first branch site is at UQH_2, where electrons can flow either to nitrate reductase or to the bc_1 complex. Then there are three branches to the three other reductases after the bc_1 complex at the level of cytochrome c. In agreement with the model, electron flow to nitrate reductase is not sensitive to inhibitors of the bc_1 complex, whereas electron flow to the other reductases is sensitive to the inhibitors.

In *P. denitrificans, Pseudomonas aeruginosa*, and many other facultative anaerobes, both the synthesis and activ-

*Nitrate reductase is a membrane-bound molybdenum protein with three subunits, α, β, and γ. The γ subunit is a cytochrome b. A proposed route for electrons is from ubiquinol to the FeS centres of the β subunit, and from there to the molybdenum center at the active site in the α subunit. The α subunit then reduces the nitrate. The molybdenum is part of a cofactor (prosthetic group) called the molybdenum cofactor (Moco), which consists of molybdenum bound to a pterin. The Moco prosthetic group is in all molybdenum-containing proteins, except nitrogenase. (The molybdenum cofactor from nitrogenase contains nonheme iron and is called FeMo cofactor or FeMoco.) Nitrite reductase is a periplasmic enzyme that reduces nitrite to nitric oxide. Nitrite reductase from *P. denitrificans* has two identical subunits. These have hemes c and d_1. Nitric oxide reductase is a membrane-bound enzyme that reduces nitric oxide to nitrous oxide. It has heme b and heme c. Nitrous oxide reductase is a periplasmic enzyme that reduces nitrous oxide to nitrogen gas. It has two identical subunits containing Cu.

*The periplasm is an area in Gram-negative bacteria that exists between the cell membrane and the cell wall.

ity of the denitrifying enzymes are prevented by oxygen. However, in certain other facultative anaerobes, including *Comamonas* sp. and certain species of *Pseudomonas*, *Thiosphaera pantotropha*, and *Alcaligenes faecalis*, denitrifying enzymes are made and are active in the presence of oxygen (27). In these systems, both oxygen and nitrate are used simultaneously as electron acceptors, although aeration can significantly decrease the rate of nitrate reduction. The advantage of corespiration using both oxygen and nitrate is not obvious.

Periplasmic Oxidation of Methanol. Many Gram-negative bacteria oxidize substances in the periplasm and transfer the electrons to membrane-bound electron carriers, often via periplasmic cytochromes c. An example is *P. denitrificans*, which can grow aerobically on methanol (CH_3OH) by oxidizing it to formaldehyde (HCHO) and $2H^+$ with a periplasmic dehydrogenase, methanol dehydrogenase (Fig. 16).

$$CH_3OH \rightarrow HCHO + 2H^+$$

The electrons are transferred from the dehydrogenase to c-type cytochromes in the periplasm. The cytochromes c transfer the electrons to cytochrome aa_3 oxidase in the membrane, which reduces oxygen to water on the cytoplasmic surface. A Δp is established due to the inward flow of electrons and the outward pumping of protons by the cytochrome aa_3 oxidase, as well as the release of protons in the periplasm during methanol oxidation and consumption in the cytoplasm during oxygen reduction. Because the oxidation of methanol bypasses the bc_1 coupling site, the ATP yields are lower. In addition to methanol, the bacteria can grow on methylamine ($CH_3NH_3^+$), which is oxidized by methylamine dehydrogenase to formaldehyde, NH_4^+, and $2H^+$. The formaldehyde is eventually oxidized to CO_2, which is assimilated into cell material via the Calvin cycle. (The Calvin cycle is the pathway that green plants, cyanobacteria, and many bacteria use to grow on CO_2.)

$$CH_3NH_3^+ + H_2O \rightarrow HCHO + NH_4^+ + 2H^+$$

Methylamine dehydrogenase is also located in the periplasm and donates electrons via cytochromes c to cytochrome aa_3, bypassing the bc_1 complex.

Wolinella succinogenes

Fumarate respiration occurs in a wide range of bacteria growing anaerobically (24). This is probably because fumarate itself is formed from carbohydrates and protein during growth. *W. succinogenes* is a Gram-negative anaerobe isolated from the rumen. It can grow at the expense of H_2 or formate, both produced in the rumen by other bacteria. The electron transport pathway is shown in Figure 17. The active sites for both the hydrogenase and formate dehydrogenase are periplasmic, whereas the active site for the fumarate reductase is cytoplasmic. A Δp is produced when electrons travel electrogenically across the membrane from the electron donor on the outside to fumarate on the inside.

BIBLIOGRAPHY

1. P. Mitchell, *Nature (London)* **191**, 144–148 (1961).
2. M. Lubben, *Biochim. Biophys. Acta* **1229**, 1–22 (1995).
3. R.K. Thauer, K. Jungerman, and K. Decker, *Bacteriol. Rev.* **41**, 100–180 (1977).
4. D.E. Metzler, *Biochemistry: The Chemical Reactions of Living Cells*, Academic Press, New York, 1977.
5. D. White, *The Physiology and Biochemistry of Prokaryotes*, Oxford University Press, New York, 1995.
6. G. Gottschalk, *Bacterial Metabolism*, Springer-Verlag, New York, 1986.
7. T. Yagi, X. Xu, and A. Matsuno-Yagi, *Biochim. Biophys. Acta* **1101**, 181–183 (1992).
8. M. Finel, *J. Bioenerg. Biomemb.* **25**, 357–366 (1993).
9. T. Ohnishi, *J. Bioenerg. Biomemb.* **25**, 325–329 (1993).
10. T. Yagi, T. Yano, and A. Matsuno-Yagi, *J. Bioenerg. Biomemb.* **25**, 339–345 (1993).
11. M. Finel, *J. Bioenerg. Biomemb.* **25**, 357–366 (1993).
12. D.G. Nicholls and S.J. Ferguson, *Bioenergetics 2*, Academic Press, London, 1992.
13. B.L. Trumpower, *Microbiol. Rev.* **54**, 101–129 (1990).
14. M.P. Krebs and H.G. Khorana, *J. Bacteriol.* **175**, 1555–1560 (1993).
15. H.W. Van Verseveld, K. Krab, and A.H. Stouthamer, *Biochem. Biophys. Acta* **635**, 525–534 (1981).
16. P.C. Hinkle, M.A. Kumar, A. Resetar, and D.L. Harrs, *Biochemistry* **30**, 3576–3582 (1991).
17. T. Yagi, X. Xu, and A. Matsuno-Yagi, *Biochim. Biophys. Acta* **1101**, 181–183 (1992).
18. K. Matsushita, T. Ohnishi, and H.R. Kaback, *Biochemistry* **26**, 7732–7737 (1987).
19. V.D. Sled, T. Freidrich, H. Leif, H. Weiss, S.W. Meinhardt, Y. Fukumori, M.W. Calhoun, R.B. Gennis, and T. Ohnishi, *J. Bioenerg. Biomemb.* **25**, 347–356 (1993).
20. R.K. Poole and W.J. Ingledew, in F.C. Neidhardt, J.L. Ingraham, K.B. Low, B. Magasanik, M. Schaechter, and H.E. Umbarger eds., *Cellular and Molecular Biology*, vol. 1, ASM Press, Washington, D.C., 1987.
21. M.W. Calhoun, K.L. Oden, R.B. Gennis, M.J. Teixeira de Mattos, and O.M. Neijssel, *J. Bacteriol.* **175**, 3020–3025 (1993).
22. B.C. Berks, S.J. Ferguson, J.W.B. Moir, and D.J. Richardson, *Biochem. Biophys. Acta* **1232**, 97–173 (1995).
23. H.W. Van Verseveld and A.H. Stouthamer, in A. Balows, H.G. Truper, M. Dworkin, W. Harder, and K.-H. Schleifer eds., *The Prokaryotes*, vol. III, 2nd ed., Springer-Verlag, Berlin, 1992.
24. A. Kroger, V. Geisler, E. Lemma, F. Theis, and R. Lenger, *Microbiology* **158**, 311–314 (1992).
25. S.J. Ferguson, *TIBS* **12**, 354–357 (1987).
26. B.C. Berks, S.J. Ferguson, J.W.B. Moir, and D.J. Richardson. *Biochem. Biophys. Acta* **1232**, 97–173 (1995).
27. D. Patureau, N. Bernet, and R. Moletta. *Curr. Microbiol.* **32**, 25–32 (1996).

See also ANAEROBES; BIOENERGETICS OF MICROBIAL GROWTH; ENERGY METABOLISM, MICROBIAL AND ANIMAL CELLS.

ELECTROPHORESIS OF PROTEINS AND NUCLEIC ACIDS

SEAN R. GALLAGHER
Motorola Phoenix Corporate Research Laboratories
Tempe, Arizona

KEY WORDS

Electrophoresis
Genomics
Isoelectric focusing
Microelectrophoresis
Polyacrylamide gel electrophoresis (PAGE)
Proteomics
Pulsed-field electrophoresis
SDS PAGE
Southern blotting
Zone electrophoresis

OUTLINE

INTRODUCTION

Electrophoresis is the movement in a liquid of a charged particle under the influence of an electric field. Because of its unique ability to separate charged biological macromolecules such as proteins and nucleic acids, electrophoresis is a mainstay in all fields of academic and industrial biological sciences. Over the last 50 years, a large body of work has been generated detailing the development of various electrophoretic analysis techniques. Among the applications currently in use, one-dimensional SDS polyacrylamide gel electrophoresis (SDS PAGE) is by far the most common for protein separations. Using the negatively charged detergent sodium dodecyl sulfate (SDS) to denature and give a constant negative charged-to-mass ratio to the proteins, SDS PAGE separates according to size and is frequently used to check purity or complexity of a sample. Isoelectric focusing (IEF), which separates either native or denatured proteins by their isoelectric point, provides a different criteria for assessing purity or complexity of a sample and is the critical first step or dimension in two-dimensional SDS PAGE (2D SDS PAGE). The second dimension is SDS PAGE, giving an IEF- and size-based separation for the protein sample. 2D SDS PAGE is a mainstay in the field of proteomics, which studies gene expression at the protein level.

RNA and DNA separations use both agarose for larger molecules and acrylamide for smaller-sized nucleic acid fragments. In addition to the purification, isolation, and sizing of nucleic acids, applications relying on electrophoresis include DNA sequencing and gene analysis. DNA sequencing, which uses thin rectangular acrylamide separation gels to separate short (<500 bases) DNA oligonucleotides, is performed with both manual and automated sequencing systems and is central to the numerous genome initiatives aimed at determining the DNA sequence of an organism's entire genome.

BACKGROUND

Electrophoresis has been studied for nearly two centuries (1). Starting in the early 1800s, researchers noted that colloidal clay particles are negatively charged and will move toward the positively charged electrode (anode). This observation was expanded by many to include positively charged particles moving toward negatively charged electrodes. Furthermore, the number of charges on the particle influenced the speed of movement. The foundation of modern electrophoresis started with the Ph.D. thesis of Arne Tiselius in 1930 (University of Uppsala, Uppsala, Sweden), in which he demonstrated moving zones of serum proteins in free solution. For the first time, protein components in serum could be separated out, giving the current designation of α-, β-, and γ-globulin (2,3). His work on electrophoresis and adsorption led to the Nobel Prize in Chemistry in 1948.

By the 1950s, several variations of electrophoresis were in use. The movement of bacteria, protozoa, blood cells, and even protein-coated quartz particles could be observed directly and studied in detail with a microscope. Although useful, the study of electrophoresis using the microscope was rapidly replaced by the technique developed by Tiselius, referred to as the moving boundary procedure. The moving boundary technique, instead of observing a single particle with a microscope, monitored the bulk movement of a protein-containing solution during electrophoresis. By placing the protein solution in a U-shaped tube and then overlaying with the appropriate buffer solution, the original protein solution will begin to migrate as a boundary into the buffer solution when subjected to an electric field. This boundary represents the leading or trailing edge of the charged proteins as they move to the positive or the negative electrode.

The moving boundary procedure greatly expanded our knowledge of protein behavior in solution and permitted charged-based separation of complex mixtures and recovery of purified proteins. It is interesting to note that when the first separation of human serum was achieved with the moving boundary technique, the terminology for α-, β-, and γ-globulin was given by Tiselius to indicate the fastest, next fastest, and slowest globulin seen. By altering the pH of the buffered solutions, proteins would migrate either toward the positive or the negative electrode. By studying the movement as influenced by the pH of the buffer, the isoelectric points (or the pH where the net charge on the protein is 0) of many proteins were determined. The moving boundary procedure also brought other needed technology to bear on the problem of separations of proteins, including image analysis. Using schlieren scanning, the profile of the separated proteins could be visualized on a screen and then recorded.

The work by Tiselius was a triumph over the many physical factors that interfere with electrophoretic separations. Heat dissipation in particular continues to be a major problem in modern electrophoresis. Convective mixing is a severe problem in free solution and is caused by differential heating of the buffer or sample. When warmed, the solution expands and becomes more buoyant, leading to mixing and a loss of resolution. To minimize heat build up in the Tiselius apparatus, rectangular tubes were used to maximize the heat exchange from the U-shaped tube during electrophoresis, and the tube was further bathed in a water bath at 4 °C. At this temperature, water is most dense and is less prone to the effects of heat fluctuation.

However, moving molecules in free solution have a number of drawbacks, including mixing of zones that limit resolution. The addition of a solid support, such as agarose or acrylamide, stabilizes the separation against convective mixing, allowing higher-resolution separations. Additionally, the porosity and chemistry of the gel separation medium influences molecule migration and plays a significant role in defining separations. Electrophoresis on solid supports was originally referred to as zone electrophoresis, simply meaning that in a solid medium, proteins separate into individual component bands, or zones, rather than moving as a boundary in free solution. The zone technique in its original form used filter paper or silica gel as a solid support and was the earliest example of the modern techniques of electrophoresis that are currently being used today in laboratories for separating proteins and nucleic acids. Currently polyacrylamide or agarose is the solid support of choice for electrophoresis.

ELECTROPHORESIS THEORY

The force driving the charged molecule through the separation medium is the product of the charge Q on the protein or nucleic macro-ion and the potential gradient E across the electrophoresis chamber:

$$QE$$

Resistance from the surrounding medium counteracts the movement, with Newton's law indicating that the movement will experience an equal but opposite force. As a starting point to understand electrophoretic behavior of a molecule, Stokes's law, which approximates particle movement in a free solution, defines the resistance to electrophoresis as

$$f = 6\pi r v \eta$$

where f is the resistance of the medium to electrophoresis of the molecule, r represents radius of the particle (assumed to be a sphere), v is velocity of the particle, and η is viscosity of the liquid.

These two basic equations indicate that molecule charge, size, and shape, solution viscosity, and applied voltage gradient all play a role in electrophoresis. Although these equations are idealized, a number of general conclusions can be drawn. For a given size, more charge on the molecule will increase the separation speed, and for a given charge a larger molecule will migrate more slowly. Increasing the applied voltage gradient will speed separation as well. Several other factors also affect separation. Proteins and nucleic acid fragments are typically not spheres and can have complex structures. For example, a protein with a compact native structure will become extended upon full denaturation and reduction of internal disulfide bonds. The extended protein interacts more with the surrounding gel and buffer, experiencing more frictional resistance, and moves more slowly. Furthermore, separations usually take place in porous gels that sieve and interact with the molecules during electrophoresis. From a practical point of view, heat generation is among the most difficult problems to contend with. Increased temperatures decrease viscosity and the electrical resistance of the buffer and if not controlled will lead to localized heating, causing poor results and limited reproducibility. Typically buffer concentration is determined as a compromise between providing enough buffering capability to keep the pH of the solution stable and a low enough ionic strength to minimize heat generation at a voltage high enough to result in a rapid separation.

ELECTRICAL PARAMETERS, SAFETY, AND ELECTROPHORESIS

Ohm's law is important for understanding the practical day-to-day use of electrophoresis and can be stated in a variety of forms:

$$V = IR$$

$$I = V/R$$

$$P = VI$$

$$P = I^2R$$

where V is voltage in volts, I is current in amps, R is in ohms, and P is power in watts. A full appreciation of how the basic electrical parameters of electrophoresis interact is required to ensure safe operation and day-to-day reproducibility. A gel can be viewed as a resistor, and the power supply as the voltage and current source. Most power supplies deliver constant voltage or constant current, and more specialized power supplies give constant power. The multiple outlets on the power supply are wired in parallel, so current is additive. That is, doubling the number of separation tanks will require doubling the current output of the power supply to give the same starting voltage.

Typical separation conditions for proteins and DNA in acrylamide and agarose require <500 V and <200 mA. Exceptions are found in isoelectric focusing where voltages up to 5,000 V are used, and DNA sequencing where up to 2,000 V and 80 W are used. Keeping careful records of the starting voltage, current, and power will quickly identify problems. For example, excess heating will cause the current to go up dramatically, while atypical electrical readings at the beginning indicate an incorrect buffer concentration or pH.

When assembling and using electrophoretic instrumentation, care must be exercised to avoid a potentially lethal shock. To start, the power supply should be turned off. Note that in all manipulations connecting and disconnecting the power supply, use one hand only. Using both hands will provide a path through the chest should a shock occur. Typically the red lead from the separation chamber is connected to the red input on the power supply, and the black lead is connected to the black input. Note that proteins can be positively charged under native conditions and that the polarity of the jacks may need to be reversed in order to have electrophoresis proceed in the correct direction. This is simply because manufactures have based their standards on separation of negatively charged proteins found in SDS PAGE (see later) and nucleic acids, which are intrinsically negative due to their phosphate backbone. Once connected, turn the power supply on and program or dial in the appropriate current or voltage. When the run is complete, turn the power supply down to zero before turning off, to ensure no excess charge is left in the capacitors in the power supply. Once the display reads zero, turn off and disconnect.

PORE SIZE AND SEPARATION MATERIAL

The choice of acrylamide or agarose gel depends on the application and the size the molecule and is governed in large part by the porosity of the two matrices. Acrylamide is composed of long chains of acrylamide monomers cross-linked with the bifunctional reagent N,N'-methylene bisacrylamide to form a mesh of pores (4–6) and the sieving network that separates larger from smaller molecules and provides size discrimination. Polyacrylamide is formed by polymerizing the acrylamide monomers and the cross-linking agent N,N'-methylene bisacrylamide with the addition of the initiator ammonium persulfate and the accelerator N,N,N',N'-tetramethylethylenediamine (TEMED). Polyacrylamide is typically used at concentrations ranging from 5 to 20 % total monomer (Table 1). Acrylamide concentrations below 3–4 % T at 2–3 % bisacrylamide produce gels that are very difficult to handle and still have pores too small for many nucleic acid applications.

Table 1. Separation Size Range for Proteins in Acrylamide

% Acrylamide	Size separation range of protein standard (Da)
5	25,000–200,000
10	14,000–200,000
15	14,000–66,000
20	14,000–45,000
5–20% Gradient	14,000–300,000

Source: Refs. 7 and 8.

In contrast, agarose is a galactan hydrocolloid from agar derived from marine algae (5). To prepare a separation gel, powdered agarose is dissolved by heat in the electrophoresis buffer and then allowed to solidify into a gel by cooling. Typical concentrations of agarose (0.75–1.5%) have very large pores that can fractionate much larger molecules than those in acrylamide. Agarose is typically used for nucleic acid separations and to a lesser extent isoelectric focusing (see later).

STANDARDIZATION AND CALIBRATION

To simplify gel-to-gel comparisons, common definitions are used to describe acrylamide gel concentration and molecule mobility (4–7):

R_r = Distance migrated by band/distance migrated by reference marker

%T = Total monomer (acrylamide + bisacrylamide) in g/100 mL final gel solution

%C = Bisacrylamide g/ (acrylamide + bisacrylamide) g × 100

Relative mobility, R_f, is a useful way to compare gels because it places the migration relative to an internal reference marker or tracking dye, usually bromophenol blue. Tracking dyes are also a useful visual indicator of gel performance, and any distortions in the separation can be seen before the separation is complete.

To determine molecule size, molecular weight or size standards are required for both protein and nucleic acid separations (Table 2 and Fig. 1). These are added to a lane on the gel along with the sample for comparison after separation and visualization, and are available from a variety of commercial companies.

VISUALIZATION

Separated proteins are typically visualized through equilibration with a protein-binding blue dye, Coomassie blue R 250, or by silver staining (7,10). Alternative techniques include prelabeling the proteins with a radioactive amino acid, and then analysis by exposure to X-ray film in a procedure called autoradiography. Postseparation analysis and labeling of the proteins with a fluorescent tag and then finally analyzing the proteins through fluorescent detec-

Table 2. Protein Molecular Weight Standards

Protein	Molecular weight (Da)
α-Lactalbumin	14,200
β-Lactoglobulin	18,400
Carbonic anhydrase	29,000
Ovalbumin	45,000
Bovine serum albumin	66,000
Phosphorylase b (rabbit muscle)	97,400
β-Galactosidase	116,000
RNA polymerase, *E. coli*	155,000
	165,000
Myosin, heavy chain (rabbit muscle)	205,000

Source: Ref. 7.

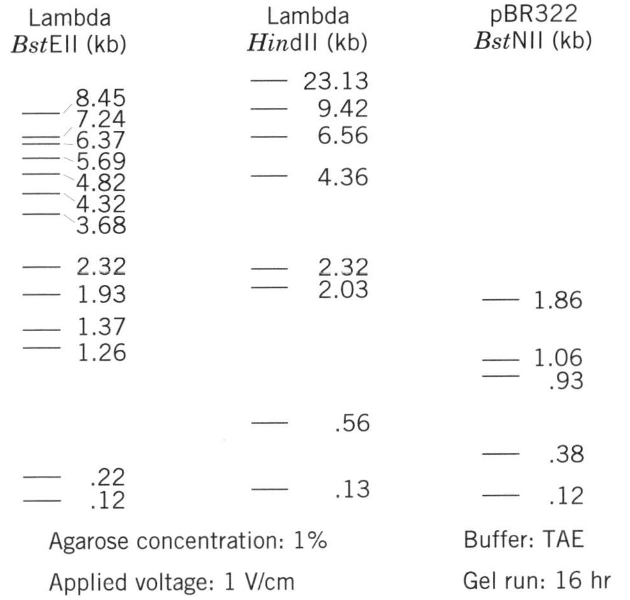

Figure 1. Migration pattern and fragment sizes of common DNA molecular weight markers. *Source*: Ref. 9.

tion is also used. Nucleic acids are in general visualized with fluorescence after soaking the gel with the intercalating dye ethidium bromide (11), but they can also be stained with silver (12). Other procedures with nucleic acids make use of radioactive labeling and autoradiography.

ELECTROPHORESIS APPLICATIONS

Over the last 50 years, a large body of work has been generated detailing the development of various electrophoretic analysis techniques. Among the applications currently in use, one-dimensional SDS polyacrylamide gel electrophoresis (SDS PAGE) is by far the most common for protein separations. Using the negatively charged detergent sodium dodecyl sulfate (SDS) to denature and give a constant negative charge-to-mass ratio to the proteins, SDS PAGE separates according to size and is frequently used to check purity or complexity of a sample (Table 3). Isoelectric focusing (IEF), which separates either native or denatured proteins by their isoelectric point, provides a different criteria for assessing purity or complexity of a sample and is the critical first step, or dimension, in two-dimensional SDS PAGE (2D SDS PAGE). The second dimension is SDS PAGE, giving an IEF- and size-based separation for the protein sample. 2D SDS PAGE is a mainstay in the field of proteomics, which studies gene expression at the protein level.

RNA and DNA separations use both agarose for larger molecules and acrylamide for smaller-sized nucleic acid fragments. In addition to the purification, isolation, and sizing of nucleic acids, applications relying on electrophoresis include DNA sequencing and gene analysis. DNA sequencing, which uses thin rectangular acrylamide separation gels to separate short (<500 bases) DNA oligonucleotides, is performed with both manual and automated sequencing systems and is central to the numerous genome initiatives aimed at determining the DNA sequence of an organism's entire genome.

NUCLEIC ACID ELECTROPHORESIS

RNA and DNA electrophoretic separations are fundamental to modern molecular biology and biotechnology (Table 4). Depending on the application, the electrophoresis is performed in either a vertical or horizontal format. Polyacrylamide is used for smaller fragments of DNA in vertical (13) or horizontal (12) slab gels, whereas agarose is used for separating larger fragments of DNA in horizontal submarine separation chambers (9) (Tables 5 and 6). Due to the phosphate backbone, nucleic acids are negatively charged, with a constant charge-to-mass ratio. Although nucleic acid mobility is largely size dependent up to 30 kb,

Table 3. Key Applications of Protein Electrophoresis

Technique	Separation principle	Application
Native PAGE	Native charge, size, shape	Purification, size of native protein and protein complex
SDS PAGE	Size dependent; SDS imparts a negative charge to proteins, giving a constant charge-to-mass ratio	Size estimation, purity, purification, subunit composition, protein expression
IEF	Intrinsic charge with both native and denatured proteins	Purification, purity check, isoenzyme analysis
2D SDS PAGE	Isoelectric point in the first dimension, and size in the second	Protein expression, purification, posttranslational analysis, proteomics

Table 4. Key Applications of Nucleic Acid Electrophoresis

Technique	Separation principle	Application
Large fragment DNA and RNA separation	Size; due to the negatively charged phosphate backbone, nucleic acids have a constant charge-to-mass ratio	Purification, determination of purity and size, genetic analysis, preparation of DNA probes
Small fragment DNA and RNA separation	Size; see above	Purification, determination of purity and size, genetic analysis, preparation of DNA probes
DNA sequencing	Size of single-stranded oligonucleotide; denaturing gels and high temperature keep DNA single stranded	DNA sequence determination, mutation, and polymorphism analysis

Table 5. Separation Size Range for DNA in Agarose

Agarose (%)	Separation range of linear DNA (kbp)
0.5	30–1
1.0	10–0.5
1.5	3–0.2

Source: Ref. 9.

Table 6. Separation Size Range for DNA in Acrylamide

Acrylamide (%)	Separation range of linear DNA (bp)
3.5	100–1,000
5.0	100–500
8.0	60–400
12	50–200
20	5–100

Source: Ref. 13.

larger fragments of DNA require the use of pulsed-field gel electrophoresis for separation (see later). Furthermore, conformation of both DNA and RNA will influence mobility. The circular DNA plasmid pBR322, for example, migrates differently depending on its form (closed circular supercoiled Type I, nicked or open circular Type II, or straight linear Type III) even though all have the same molecular weight (9) (Fig. 2). The supercoiled form has a smaller hydrodynamic radius than the linear and will move more quickly through the gel.

Pulsed-Field Electrophoresis

DNA molecules larger than 20–30 kb migrate as a single species, or band, during separation and cannot be resolved with traditional agarose gel electrophoresis. Separation of very large DNA is, however, possible through the use of pulsed-field electrophoresis (14–18). DNA molecules are forced by the alternating electric fields to change their direction through the gel, and depending on the size of the DNA molecule, the percentage of agarose, and the frequency of the alternating electric field, separation of megabase-sized DNA is possible. Pulsed-field electrophoresis is dramatically affected by temperature and field strength, particularly for large molecules. A low field strength of <2 V/cm for DNA >5 Mb is required. Cool temperatures tend to enhance resolution but also prolong the length of the separation, and a compromise of room temperature is typically used. DNA larger than 50 kb is very difficult to handle in liquid without breaking during isolation and loading and is prepared in agarose plugs. The agarose plug is loaded into the sample well prior to separation.

There are two basic methods in common use for achieving pulse-field electrophoresis. One, field inversion gel electrophoresis (FIGE) works by simply inverting the polarity on the standard electrode found in most gel electrophoresis units. Specialized electronics are either incorporated into the power supply or placed between the power supply and the gel box to achieve the switching. Thus the orientation of the DNA during separation changes by 180 degrees (14). Separation with pulsed-field electrophoresis using an angle shallower than 180 degrees is typified by the technique called (contour clamp homogenous electric field) (CHEF), allowing a separation up to 6 Mb (16,19).

Southern Blotting

Once separated, DNA is frequently transferred out of the gel and onto a flat nitrocellulose or nylon membrane for further analysis. The technique is known as Southern blotting, after E. M. Southern (20,21), or capillary blotting. Briefly, DNA is denatured and nicked to enable it to migrate from the gel. Typically capillary blotting involves loading the gel on top of paper wicks that are suspended in a tray filled with transfer buffer. On top of the gel is the nitrocellulose or nylon membrane, a buffer-saturated sheet of blotter paper, dried blotter paper, and a weight, which draws the liquid from the wicks through the gel and onto the blotter paper. This pulls the DNA out of the gel and onto the membrane where it attaches. The DNA can then be analyzed with a radiolabeled or fluorescently tagged probe containing a complementary sequence to determine the presence or absence of the sequence in the sample.

Sequencing

Determination of the DNA sequence is a prerequisite to a number of key areas of research in the biological sciences (22–25). This includes development of genetic detection technologies, identification of disease-related genes, identification of genes that indicate a individual's predisposition toward either sensitivity or insensitivity to drugs, and the precise determination of similarities among genes from different organisms.

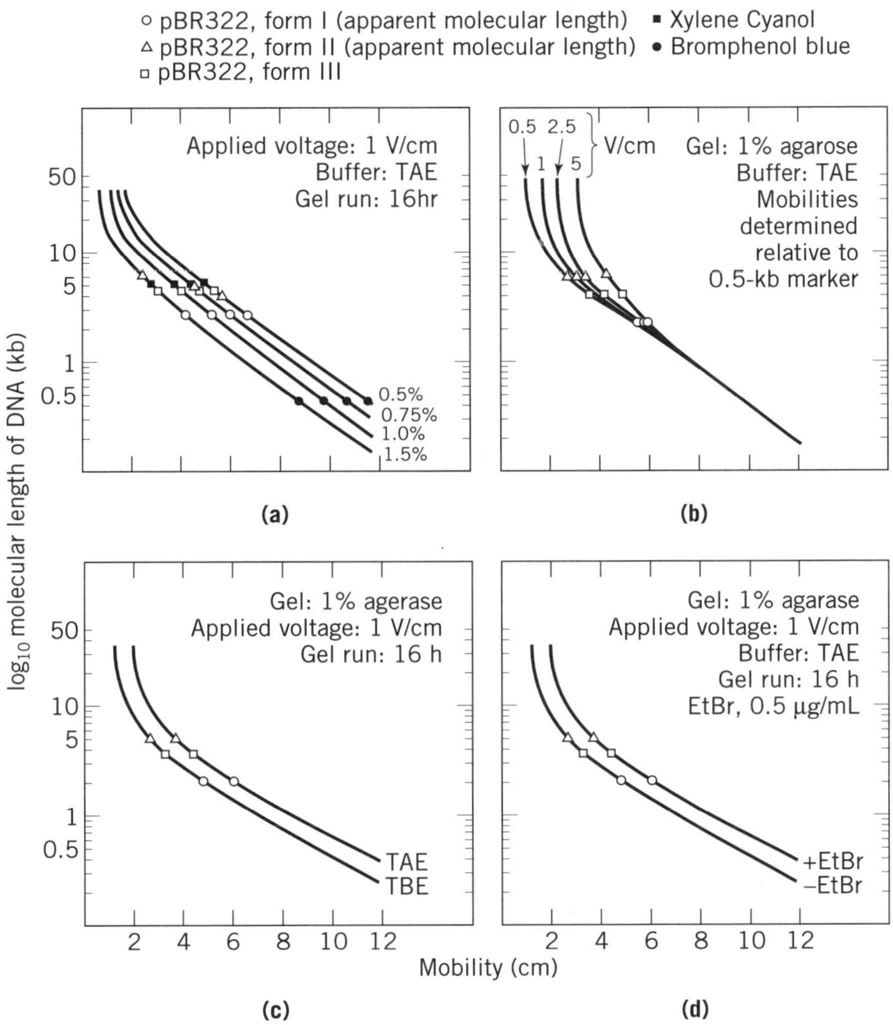

Figure 2. Effect of (**a**) agarose concentration, (**b**) applied voltage, (**c**) electrophoresis buffer, and (**d**) ethidium bromide on migration of DNA through agarose gels. *Source*: Used with Permission from Wiley.

DNA sequencing relies on electrophoretic separation of DNA fragments, oligonucleotides, that range from 10 to 500 bp in length. The separation, which is performed in thin (0.1- to 1.5-mm-thick) acrylamide gels at high temperature under denaturing conditions, yields 1-bp resolution. For most applications, dideoxyribonucleoside triphosphates (ddNTPs) chain termination sequencing is currently used (22,24). Briefly, the DNA sequence to be analyzed is annealed with a primer, which serves as the initiation point for the added DNA polymerase. With the appropriate fluorescently tagged or ^{35}S-labeled deoxyribonucleoside triphosphates (dNTPs) in the reaction, the polymerase proceeds to copy the template DNA. In each of the four separate dideoxy chain termination sequencing reactions (GATC), modified GATC nucleotides or chain terminators are added along with the unmodified nucleotides. As the chain terminators are incorporated into the growing chain, the reaction stops at A, T, C, or G. Using four separate enzymatic reactions, a set of DNA fragments with nucleotides that end with either an A, T, C, or G are placed in lanes labeled A, T, C, or G and separated by electrophoresis. The gel is then dried and placed against X-ray film to detect the radioactive nucleotides. The sequence is simply read off the resulting autoradiogram (Fig. 3). Alternatively the DNA fragments can be labeled with a fluorescently tagged nucleotide and then automatically read in an automated DNA sequencer that uses a laser to excite the fluorophore and a detector to read the fluorescence as the DNA fragment moves past a detector. For DNA larger than 300–500 bp, the DNA is generally broken into smaller fragments that can be individually sequenced.

PROTEIN ELECTROPHORESIS

Proteins are polyampholytes composed of amino acids containing a variety of ionizable acidic and basic groups that give the protein either a positive or a negative charge, depending on the pH. The acidic amino acids (aspartate and glutamate) and basic amino acids (lysine, histidine, and arginine) contribute along with other factors, such as the native structure, to the overall charge on the protein at a particular pH. Thus in acidic solutions amino groups be-

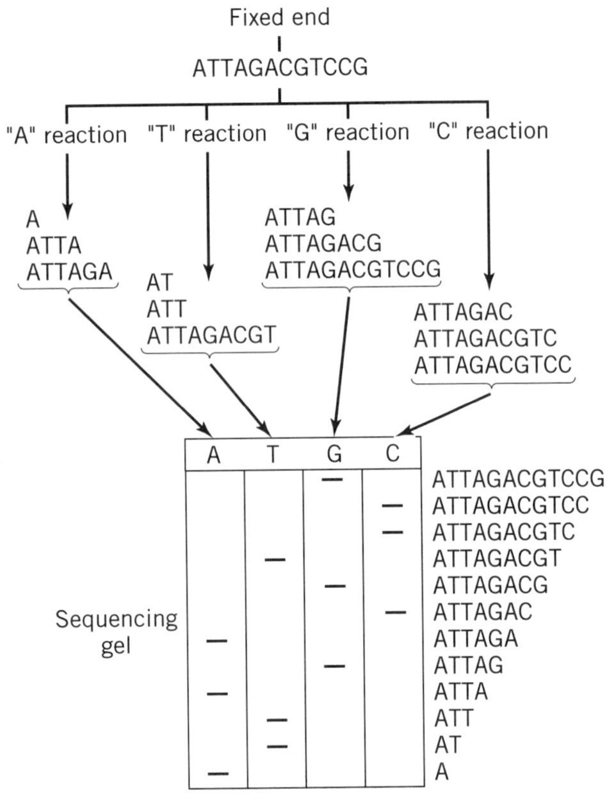

Figure 3. General strategy for DNA sequencing. To sequence a fragment of DNA, a set of radiolabeled single-stranded oligonucleotides is generated in four separate reactions. In each of the four reactions, the oligonucleotides have one fixed end and one end that terminates sequentially at each A, T, G, or C, respectively. The products of each reaction are fractionated by electrophoresis on adjacent lanes of a high-resolution polyacrylamide gel. After autoradiography, the DNA sequence can be "read" directly from the gel. *Source*: Ref. 22.

come ionized but carboxyl groups remain neutral, resulting in overall positive charge. In basic solutions, the carboxyl groups become ionized and the amino groups uncharged, creating an overall negative charge. The protein thus has an isoelectric point (pI), which is the pH at which the net charge on the protein is zero, reflecting the balance between the ionized positive and negative groups on the protein. The pI also depends on the conformation of the protein and any charged groups such as phosphorus that were added as a posttranslational modification. In the presence of denaturants or reductants that break disulfide linkages, proteins unfold and expose buried charged groups, causing an altered pI. When a protein is in its fully reduced and denatured state, both size and charge separation are reasonably accurate. SDS binds to a protein at a constant 1.4 g/g protein, completely masking the native charge, giving the protein (when fully denatured and reduced) a constant charged-to-mass ratio and making size-dependent separation possible (26).

Native Gel Electrophoresis

Under native or nondenaturing conditions, the native size, number of subunits, and isoforms can be studied in detail. Factors influencing native gel electrophoresis include native charge and shape and size of the protein (4,7,27,28). Ferguson plots (27), first applied to starch gels and later to polyacrylamide, gels provide a simple way to evaluate size and other characteristics of a protein under native conditions. First the protein sample is run under varied gel concentrations (typically 5, 10, 15, and 20%) and the slope of the log relative mobility (log R_f) versus acrylamide concentration (%T), called K_r, or the retardation coefficient, is plotted against the size of a series of standard proteins. By comparison to the standards, the approximate size of the unknown is determined.

Isoelectric focusing

Isoelectric focusing (IEF) separates proteins based on the intrinsic charge of the protein (6) and is performed by adding the protein sample to a gel containing a pH gradient. Chemically synthesized additives to the electrophoresis gel (ampholytes) generate a stable pH gradient from the basic (cathode or negative electrode) to the acidic (anode or positive electrode) end of the gel with an applied voltage. Ampholytes not only carry charge, but also are good buffers, containing a mix of acidic (carboxyl) and basic (amino) groups. Under the influence of an electric field, ampholytes migrate either as anions or cations toward the positive or negative electrode, respectively. Ampholytes gain or lose protons, and thus charge, until the isoelectric point is reached, where the molecule has a net charge of zero. Commercially available ampholytes have many species with slightly different isoelectric points that produce relatively smooth transitions in the pH gradient over a variety of ranges. The sample proteins move either toward the acid or the basic end, depending on the native charge of the protein. When the protein gets to the pH that represents its isoelectric point, or the point where the positive and negative charges on the protein balance to give a net charge of zero, the protein stops migrating and is considered focused. Initially, IEF used mobile ampholytes, but it is now possible to create a pH gradient grafted directly onto the acrylamide matrix by pouring a gradient of ionizable groups covalently attached to the acrylamide using Immobiline chemistry (12,29,30).

SDS PAGE

The most commonly used electrophoresis technique for protein separation is based on discontinuous SDS PAGE (7,31). Discontinuous or multiphasic systems separate proteins through a gel with two sections, the stacking and resolving regions. The stacking gel contains a pH 6.8 Tris-HCl buffer as well as a nonrestrictive large-pore acrylamide gel through which the sample moves. The chloride ions (called the leading ions) have an electrophoretic mobility greater than the sample proteins. Glycine ions (called the trailing ions) in the Tris-glycine electrode, or reservoir, buffer are less mobile than the sample proteins. The high mobility chloride ions leave a zone of lower conductivity between themselves and the migrating protein, causing a zone with a high voltage gradient. The proteins are then forced to "stack" in the zone between the leading and trailing ions. The proteins "destack" upon entering the resolving, or separating, gel. In the Tris-HCl pH 8.8 separating-gel buffer, glycine has a faster mobility and migrates past the proteins. The proteins then separate according to either molecular size in a denaturing SDS gel, or molecular shape, size, and charge in a nondenaturing gel.

To fully disassociate the subunits and denature the protein, the sample is mixed with SDS and a reductant, such as dithiothreitol, and heated to 100°C. SDS binds at 1.4 g/g protein, giving a constant charge-to-mass-ratio, leading to identical charge densities for the denatured proteins. Thus, the SDS-protein complexes separate according to size, not charge.

Optimal separation depends on the polyacrylamide concentration, with most proteins separated on gels containing from 5 to 15% acrylamide and 0.2 to 0.5% bisacrylamide (Table 1). The relationship between the relative mobility and log molecular weight is not linear over the entire separation range (Fig. 4). However, by using the linear portion of the curve, the molecular weight of an unknown protein is determined by comparison with known protein standards (Table 2). Gradient gels expand the linear range of separation over a much wider size range than single-concentration gels and are prepared by casting a gradient from high to low concentration of acrylamide. Proteins move into a constantly increasing concentration of acrylamide and smaller pore size, with the effect of sharpening the band and giving unparalleled resolution (Fig. 5) (5,7).

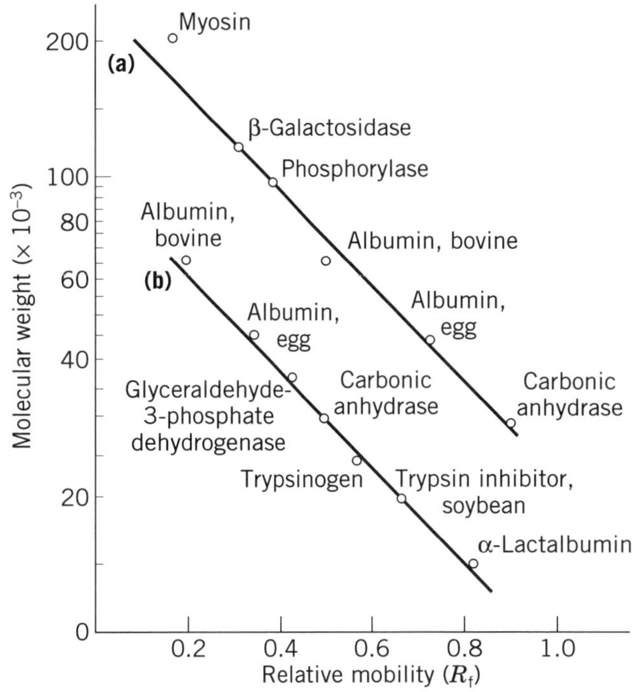

Figure 4. Typical calibration curves obtained with standard proteins separated by nongradient denaturing (SDS) discontinuous gel electrophoresis based on the method of Laemmli (31). (**a**) Gel with 7% polyacrylamide. (**b**) Gel with 11% polyacrylamide. *Source*: Redrawn with permission from Sigma.

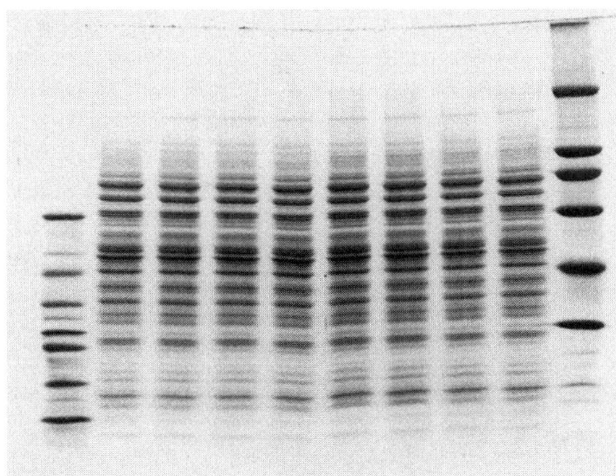

Figure 5. Separation of soluble proteins by 5.1 to 20.5%T polyacrylamide gradient SDS PAGE. Approximately 30 μL of 1 × SDS sample buffer containing 30 μg of Alaskan pea (*Pisum sativum*) soluble proteins was loaded in wells of a 14 × 14-cm, 0.75-mm-thick gel. Standard proteins were included in the outside lanes. The gel was run at 4 mA for ~15 h. *Source*: Used with permission from Wiley.

2D SDS PAGE

Two-dimensional SDS polyacrylamide gel electrophoresis (2D SDS PAGE) is a fundamental technology for many areas of basic and applied biological research including molecular biology, protein chemistry, clinical chemistry, and toxicology (32–38). In 2D SDS PAGE, proteins are first separated by isoelectric point via IEF and then by size via SDS PAGE (Fig. 6). 2D SDS PAGE simultaneously identifies the properties such as size, isoelectric point, and post-translational modifications of thousands proteins. The separation yields a high-purity product that is amenable to mass spectroscopy analysis, providing further details about the protein structure.

With the publication of work by O'Farrell (37), the basic form of current two-dimensional electrophoresis took shape. There are many applications of 2D SDS PAGE analysis in biology (32), including understanding complex interactions between expressed proteins in a cell and the environment; identifying proteins that are immunoactive; toxicology testing (where exposure to a particular drug will increase or decrease the level of one or many proteins), which is a particularly critical area where gene analysis alone does not suffice; and quality control separation of proteins in various products for both agriculture and human consumption.

2D SDS PAGE is performed by first solubilizing and denaturing the protein using nonionic detergents, chaotropic agents, and a reductant. Typically these are NP40, urea, and DTT. This basic sample buffer, however, does not work for all samples, and many variations on solubilization have been developed to handle bacterial proteins, serum samples, membrane proteins, and plant samples. The proteins are separated in the first dimension by focusing either in thin strips horizontally, or in vertical glass tubes. After the first-dimension separation, the proteins are then separated with discontinuous SDS PAGE in vertical or horizontal (12,29) systems. In the first dimension, the acrylamide matrix is porous to keep sieving effects to a minimum,

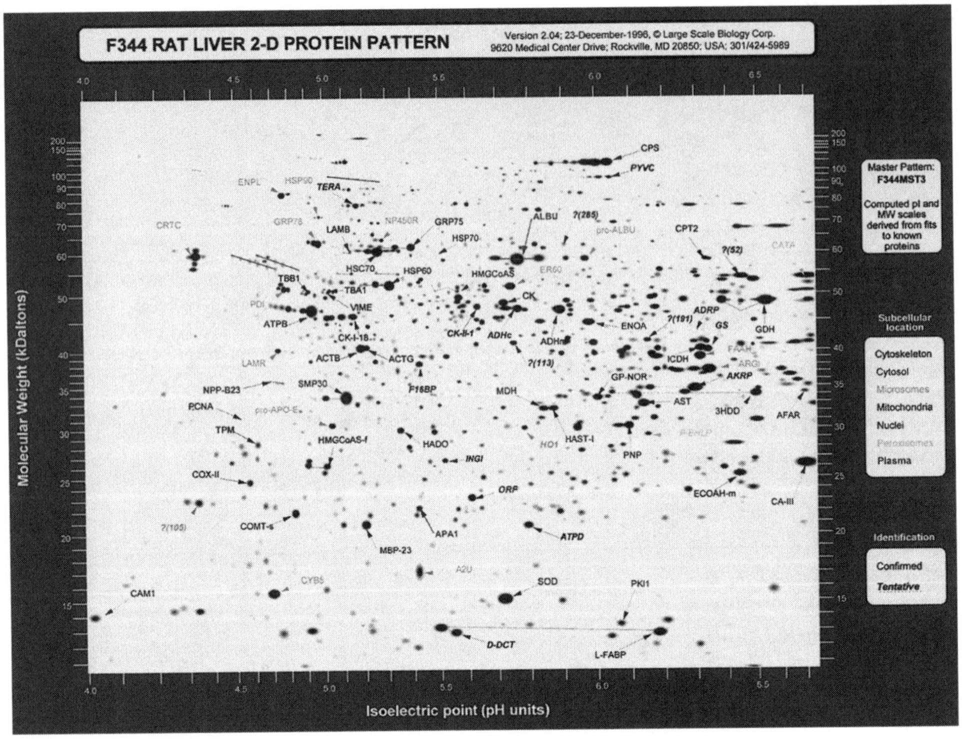

Figure 6. 2D protein pattern of rat liver. The proteins are arrayed from acidic to basic isoelectric points (right to left) and from large to small size (top to bottom). *Source*: Used with permission from Large Scale Biology, Rockville, Maryland.

whereas the second dimension is performed under conditions that emphasize the size separation by using single concentration or pore-gradient gel electrophoresis. 2D SDS PAGE is the only technique currently in use that is capable of parallel separation of literally thousands of proteins in a single separation run. The resulting gel represents an array of proteins being currently expressed in one particular cell type at that particular moment.

Although still popular, mobile carrier ampholytes present a number of challenges for researchers. Although the buffering capacity of ampholytes exceeds that of proteins, large quantities of proteins tend to interfere with the generation of the pH gradient. In addition, synthesis of the ampholytes is complex and not as reproducible as one would like from batch to batch. Furthermore, carrier ampholytes have a tendency to drift, shifting the pH gradient during separation and contributing to uncertainty about the protein position in the final separation. A new technology, immobilized pH gradient (IPG) gels (12,29,30, 36,39), helped considerably to enhance the technology of 2D SDS PAGE and make it more reproducible. With IPG gels, the pH gradient is covalently immobilized in the gel matrix during the preparation of the gel itself. When separation of the proteins occurs under electrophoresis, the pH gradient is stable. Another tremendous advantage is that, in contrast to carrier ampholytes, large quantities of proteins can be loaded without significant disruption of the separation pattern, permitting detailed analysis of the spots subsequent to separation by techniques such as mass spectroscopy. The extreme resolution capable with 2D SDS PAGE permits the identification of proteins not only by their position on the map, but by analysis of co- and posttranslational modifications (40), including phosphorylation, glycosylation, proteolysis, and prenylation.

NEW APPLICATIONS

Genomics and Proteomics

First used in 1994, the term *proteome* is defined as the study of the expressed proteins in a cell and is derived from *prote*ins expressed by the gen*ome* (32). Currently, the only way to separate and visualize all proteins in a cell is with 2D SDS PAGE. Proteomics is critical to modern cell and molecular biology because, in a single separation, number, abundance, and posttranslational modifications are apparent. Furthermore, determination of a partial sequence with mass spectrometry is now a standard technique that gives an accurate identification of the protein, or if no record of the protein is apparent, a DNA probe for identifying and isolating the gene can be developed.

The connection between genomics, drug discovery, and proteomics derives from the observation that mRNA expression and protein abundance representing mRNA do not show a good correspondence. Furthermore, because most drug targets are proteins, the analysis of drug effects must start with the expressed protein complement. However, to fully understand the protein target and associated gene family, the gene must also be identified.

Microelectrophoresis and Microfluidics

Current research in electrophoresis is directed toward automating manually intensive steps such as sample preparation, and shrinking the size of electrophoretic analysis devices to increase throughput, lower use of reagents, and generally automate electrophoretic analysis (41).

With the development of a large body of genetic analysis information and techniques, a number of diseases have been distinguished by DNA analysis. Developing assays for the diagnosis of genetic disease is a primary driver in instrumentation development. Current electrophoretic instrumentation found in research laboratories is largely manual, requiring operators to run gels either a few at a time in parallel or in a serial operation. The exception is automated DNA sequencing apparatus. However, this is also large and requires a high level of operator expertise. Recently, there has been a strong drive to not only simplify electrophoretic analysis, but also lower the cost and increase the number of samples that can be run in a given day. In order to accomplish this, researchers have turned toward microminiaturization using semiconductor microfabrication technology. Although not commercially available, microminiaturized devices for DNA and protein separation are available at the research level and have a number of attractive features. Sometimes referred to as lab-on-chip devices (LOC), these microminiaturized laboratories and electrophoresis chambers have the capabilities of not only preparing the sample but mixing the sample, amplifying the genetic material through PCR if necessary, and then providing electrophoretic separation and subsequent analysis. In fact, separation of DNA for sequencing purposes on chips has been demonstrated. As advances in microminiaturization continue, the use of microfluidics and the integration of the movement of small amounts of liquid with electrophoretic separation and detection will become increasingly important for diagnostics.

BIBLIOGRAPHY

1. G.W. Gray, *Sci. Am.* **185**, 45–53 (1951).
2. A. Tiselius, *Biochem. J.* **31**, 1464–1477 (1937).
3. A. Tiselius, *Trans. Faraday Soc.* **33**, 524 (1937).
4. A.T. Andrews, *Electrophoresis: Theory, Techniques and Biochemical and Clinical Applications*, 2nd ed., Oxford University Press, New York, 1986.
5. B.D. Hames, in B.D. Hames, D. Rickwood eds., *Gel Electrophoresis of Proteins*, Oxford University Press, Oxford, 1990, pp. 1–147.
6. P.G. Righetti, *Isoelectric Focusing: Theory, Methodology and Applications*, Elsevier Biomedical, Amsterdam, 1983.
7. S.R. Gallagher, in F.M. Ausubel, R. Brent, R.E. Kingston, D.D. Moore, J.G. Seidman, J.A. Smith and K. Struhl eds., *Current Protocols in Molecular Biology*, Wiley, New York, 1997, Unit 10.2.
8. G. Smejkal, S. Gallagher, *Biotechniques* **16**, 196–198, 200–202 (1994).
9. D. Voytas, in F.M. Ausubel, R. Brent, R.E. Kingston, D.D. Moore, J.G. Seidman, J.A. Smith and K. Struhl eds., *Current Protocols in Molecular Biology*, Wiley, New York, 1997, Unit 2.5.
10. J.H. Morrissey, *Anal. Biochem.* **117**, 307–310 (1981).
11. F.M. Ausubel, R. Brent, R.E. Kingston, D.D. Moore, J.G. Seidman, J.A. Smith and K. Struhl eds., *Current Protocols in Molecular Biology*, Wiley, New York, 1997.
12. R. Westermeier, *Electrophoresis in Practice*, VCH, Weinheim, 1997.
13. J. Chory, A.S. Baldwin, Jr., in F.M. Ausubel, R. Brent, R.E. Kingston, D.D. Moore, J.G. Seidman, J.A. Smith and K. Struhl eds., *Current Protocols in Molecular Biology*, Wiley, New York, 1997, Unit 2.7.
14. G.F. Carle, M. Frank, M.V. Olson, *Science* **232**, 65–68 (1986).
15. B.W. Birren, E. Lai, S.M. Clark, L. Hood, M.I. Simon, *Nucleic Acids Res.* **16**, 7563–7582 (1988).
16. G. Chu, *Electrophoresis* **10**, 290–295 (1989).
17. S.M. Clark, E. Lai, B.W. Birren, L. Hood, *Science* **241**, 1203–1205 (1988).
18. D.C. Schwartz, C.R. Cantor, *Cell* **37**, 67–75 (1984).
19. G. Chu, D. Vollrath, R.W. Davis, *Science* **234**, 1582–1585 (1986).
20. E.M. Southern, *J. Mol. Biol.* **98**, 503–517 (1975).
21. T. Brown, in F.M. Ausubel, R. Brent, R.E. Kingston, D.D. Moore, J.G. Seidman, J.A. Smith and K. Struhl eds., *Current Protocols in Molecular Biology*, Wiley, New York, 1997, Unit 2.9.
22. B.E. Slatko, R.L. Eckert, L.M. Albright, F.M. Ausubel, in F.M. Ausubel, R. Brent, R.E. Kingston, D.D. Moore, J.G. Seidman, J.A. Smith and K. Struhl eds., *Current Protocols in Molecular Biology*, Wiley, New York, 1997, Unit 7.1.
23. A.M. Maxam, W. Gilbert, *Methods Enzymol.* **65**, 499–560 (1980).
24. F. Sanger, S. Nicklen, A.R. Coulson, *Proc. Natl. Acad. Sci. U.S.A.* **74**, 5463–5467 (1977).
25. F. Sanger, A.R. Coulson, B.G. Barrell, A.J. Smith, B.A. Roe, *J. Mol. Biol.* **143**, 161–178 (1980).
26. W.W. Fish, J.A. Reynolds, C. Tanford, *J. Biol. Chem.* **245**, 5166–5168 (1970).
27. K.A. Ferguson, *Metabolism* **13**, 985–1002 (1964).
28. J.L. Hedrick, A.J. Smith, *Arch. Biochem. Biophys.* **126**, 155–164 (1968).
29. A. Gorg, *Biochem. Soc. Trans.* **21**, 130–132 (1993).
30. B. Bjellqvist, K. Ek, P.G. Righetti, E. Gianazza, A. Gorg, R. Westermeier, and W. Postel, *J. Biochem. Biophys. Methods* **6**, 317–339 (1982).
31. U.K. Laemmli, *Nature* **227**, 680–685 (1970).
32. M.R. Wilkins, K.L. Williams, R.D. Appel, D.F. Hochstrasser eds., *Proteome Research: New Frontiers in Functional Genomics*, Springer-Verlag, Heidelberg, 1997.
33. N.L. Anderson, N.G. Anderson, *Anal. Biochem.* **85**, 341–354 (1978).
34. N.G. Anderson, N.L. Anderson, *Anal. Biochem.* **85**, 331–340 (1978).
35. N.G. Anderson, N.L. Anderson, *Electrophoresis* **17**, 443–453 (1996).
36. A. Gorg, G. Boguth, C. Obermaier, A. Posch, W. Weiss, *Electrophoresis* **16**, 1079–1086 (1995).
37. P.H. O'Farrell, *J. Biol. Chem.* **250**, 4007–4021 (1975).
38. P.Z. O'Farrell, H.M. Goodman, P.H. O'Farrell, *Cell* **12**, 1133–1141 (1977).
39. A. Gorg, G. Boguth, C. Obermaier, W. Weiss, *Electrophoresis* **19**, 1516–1519 (1998).
40. A.A. Gooley, N.H. Packer, in M.R. Wilkins, K.L. Williams, R.D. Appel, D.F. Hochstrasser eds., *Proteome Research: New Frontiers in Functional Genomics*, Springer-Verlag, Heidelberg, 1997, pp. 65–91.
41. C.H. Mastrangero, M.A. Burns, D.T. Burke, *Proc. IEEE* **86**, 1769–1787 (1998).

ELECTROPHORESIS, PROTEINS, BATCH AND CONTINUOUS

CORNELIUS F. IVORY
Washington State University
Pullman, Washington

KEY WORDS

Analytical
Batch
Capillary
Continuous
Dispersion
Electrophoresis
PAGE
Preparative

OUTLINE

INTRODUCTION

This article is intended as a guide for scientists and engineers who are considering using electrophoresis to process proteins or other biologicals for the biotechnology industry at bench or larger scales. An overview of several types of equipment, a description of what each piece of equipment is designed for, and the problems the end user should anticipate before operating that equipment are discussed.

The discussion of each class of apparatus focuses on design principles rather than experimental protocols, emphasizing the engineering aspects of each chamber rather than the details of operation. This type of treatment provides insight into the general operating characteristics of each device and is intended to aid the reader in *(1)* choosing an appropriate instrument for a given separation and/or *(2)* setting operating conditions (e.g., column length, sample volume, or power levels) that avoid excessive dispersion (e.g., band smearing).

For the purposes of illustration, it is assumed that the buffer has the kinematic and electrical properties of pure water at 10 °C, as given in Table 1, with the exception of its electrical conductivity, σ_e, which is determined from electrolyte composition. Most of the values used are sufficiently accurate for purposes of illustration. However, the mobilities and diffusion coefficients found in practice may be quite different from those listed, since they depend primarily on the solutes processed.

The free-solution values of the mobility and diffusion coefficient correspond roughly to those of serum albumin and tend to be high compared with other proteins. The values of the mobility and diffusion coefficient given for gels are reduced from the free-solution values by a factor of 5 in order to exaggerate the retardation effects found in gels. In practice, the retardation factor depends on the composition and degree of cross-linking in the gel, and can be substantially larger or smaller than 5. The ratio of the electrophoretic mobility to the diffusion coefficient in the gel and in free solution, a parameter that frequently appears in theoretical models of electrophoresis, has been kept constant in these calculations.

Band Broadening (Dispersive) Effects

Diffusion. Before describing the various apparatuses, it is worthwhile to briefly review those phenomena that can lead to deterioration of a separation, that is, band broadening or dispersion (which is a general term for these types of effects). The most ubiquitous dispersive phenomenon is diffusion; one should always take precautions to minimize the effects of diffusion, especially when other band-broadening effects are small.

An important characteristic of diffusion-based dispersion is that it increases band width in proportion to the square root of the residence time according to the formula

$$\chi^2_{\text{diffusion}} = 2D_m t \text{ or } \sigma^2_{\text{diffusion}} = \frac{2D_m Z}{\langle u \rangle^3} \quad (1)$$

where the spatial variance χ^2 represents band spreading in space, as would be observed by scanning bands on a gel that had been run for time t. The temporal variance σ^2

Table 1. Physical Constants Used in Illustrations

Thermal conductivity	k_{gel} (W/cm °C)	0.00578
Thermal conductivity	k_{glass} (W/cm °C)	0.011
Thermal conductivity coefficient	λ[equation 4] (°C^{-1})	0.02874
Thermal viscosity coefficient	α[equation 30] (°C^{-1})	−0.02874
Thermal expansion coefficient	β[equation 40] (°C^{-1})	0.000088
Electrophoretic mobility in gel	μ (cm^2/V s) (gel)	0.00002
Electrophoretic mobility in free solution	μ (cm^2/V s) (buffer)	0.0001
Diffusion coefficient in gel	D_{gel} (cm^2/s) (gel)	1×10^{-7}
Diffusion coefficient in free solution	D_m (cm^2/s) (buffer)	5×10^{-7}

represents band spreading in time, as would be observed as bands migrate at a mean electrophoretic velocity $\langle u \rangle$ past a fixed detector located at mean position $z = Z$ relative to a starting point $z = 0$. The width of a solute band in space can be approximated as 4χ and its duration in time as 4σ. When several different mechanisms contribute to dispersion, their net effect is usually taken into account by summing the appropriate variances, in either space or time. This assumes that the individual effects are independent and is often a reasonable approximation.

Thermal Dispersion. The next major source of dispersion comes from the effect of spatial temperature variations on the electrophoretic mobility. All forms of electrophoresis require that a current be passed through a resistive medium, so electrical energy is necessarily converted to thermal energy in the medium. To efficiently remove the joule heat generated by the current, one spatial dimension of the chamber is generally kept thin, and heat is removed through the bounding surfaces along this axis.

The mobilities of charged solutes and aqueous electrolytes usually vary inversely with the buffer viscosity, which is itself quite similar to water in its temperature dependence. The viscosity of water decreases by about 25% with a 10 °C rise in temperature, so the warmer parts of a solute band migrate faster than the cooler parts and often take on parabolic or crescent shapes. This deformation results in a broadening of the fractionated sample, which usually overwhelms diffusive spreading, especially at the field strengths used in high-performance equipment (e.g., >200 V/cm).

Electroosmosis. Electroosmosis can be a severe problem in free-flow systems (i.e., those with no anticonvective medium). It arises when mobile counterions in the double layer adjacent to a charged wall are moved parallel to the wall surface by an electric field (1). These ions drag solvent with them, setting up a bulk fluid motion that can smear solute bands. Although it is difficult to eliminate electroosmosis entirely, for example, by using surface coatings or additives, steps may be taken to control it or even to take advantage of it (2). In homogeneous gels, electroosmosis plays a lesser role in band distortion but may act to displace purified bands, thereby giving inaccurate estimates of solute mobilities. In capillary zone electrophoresis at alkaline pH, electroosmosis typically provides the convective flow used to transport solutes through the capillary and does not contribute significantly to dispersion.

Other problems associated with Kohlrausch (3) effects and conductivity gradients (4,5) can have a major impact on band broadening. The former gives rise to stable band broadening or narrowing due to differences in mobilities between the target proteins and the electrolyte ions, but without inducing bulk flows. The latter may generate stable or unstable flows that can deform solute bands in a number of ways. The instability associated with conductivity gradients is particularly deleterious. As a rule of thumb, to avoid these problems with native zone electrophoresis, the electrolyte composition and conductivity should closely match those of the sample and should provide adequate buffering; the sample proteins should not be too heavily concentrated and the electric power applied should not exceed the effective cooling capacity of the column.

Natural Convection. Although problems with thermally induced natural convection (6) have long hampered scale-up of the thin-film continuous-flow electrophoresis device, these problems are largely mitigated in recycling electrophoresis equipment (7,8) and the membrane-based chambers developed by Faupel et al. (9) and Horvath et al. (10). This is not meant to imply that natural convection is a trivial problem but rather that dispersion due to natural convection can be ameliorated by innovative and careful design.

Conductive-Dielectric Flows. Although flows induced by gradients in the conductivity and dielectric constant are still largely a curiosity in electrophoretic separations, they will become major barriers to device development as electric fields are increased to boost chamber performance. Figure 1a shows the cusplike distortion of conductive dye streaks containing 100 mM NaCl in a clear solution containing 10 mM NaCl with the flow stopped and the streamlines exposed to a nominal electric field of about 26 V/cm. When the salt is replaced with different concentrations of ethanol, the induced flows have essentially the same morphology, suggesting a common mechanism for both flows. Figure 1b shows both cusps and putative roll cells in a similar experiment, but with vertical flow from top to bottom. One important feature of this instability is that conductive dispersion can be explosive, smearing finely resolved protein bands over a distance of several centimeters along the electric field lines in a few seconds.

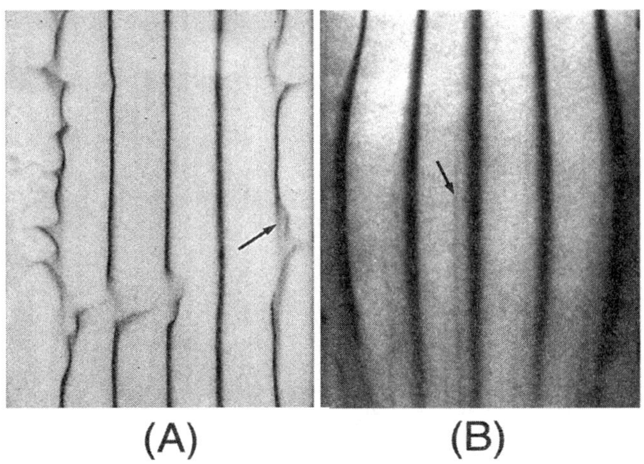

Figure 1. **(a)** Conductive–dielectric instability in a stop-flow experiment in a Plexiglas slit, 2 mm thick by 10 cm wide. Five streaks of water containing 10 mM NaCl and traces of bromophenol blue dye were injected into a moving stream of distilled water flowing between a pair of electrodes. The flow was stopped and a nominal field of 26 V/cm applied between the electrodes. The cusplike streaks are a hallmark of this instability, but roll cells *(arrow)* are also seen. **(b)** In water flowing at 1 cm/s the instability appears at a higher voltage and produces primarily roll cells *(arrow)*. Some cusps are visible in the streakline on the left.

A stable form of these flows has been described and analyzed by Roberts and coworkers (4,11) and the unstable form recently analyzed by Baygents and Baldessari (12). These papers support the empirical rule of thumb that bands in zone electrophoresis or isoelectric focusing should have conductivities as close to those of the surrounding buffer as possible to avoid conductivity gradients. These flows can be stabilized by applied shear stresses, but at fields above 400 V/cm even shear rates in excess of about 100 s^{-1} are only marginally effective at damping these flows.

Performance Metrics

Number of Theoretical Plates. In order to facilitate comparison of chamber designs and separation techniques, it is useful to have a yardstick or metric that can be universally applied not only to electrophoresis but also to other techniques. The simplest and most commonly used performance metric is the plate number N:

$$N_{\text{spatial}} = \frac{Z^2}{\chi^2} \text{ or } N_{\text{temporal}} = \frac{t^2}{\sigma^2} \tag{2}$$

for spatial or temporal analysis, respectively, where the various mean displacements and variances are taken from a single peak.

Strictly speaking, the plate efficiency is only applicable to rate-based separations, for example, zone electrophoresis or isocratic chromatography, as opposed to (pseudo)equilibrium-based separations, for example, isoelectric focusing or gradient-elution chromatography. Typical values of the plate number in zone electrophoresis range from below 1,000 for the high-capacity BioStream® apparatus (13,14) to above 10^7 for PAGE in gel-filled capillaries (15,16).

A closely related concept to plate number is plate height (17), HETP $= Z/N_{\text{spatial}}$ or $= (\langle u \rangle t)/N_{\text{temporal}}$. Although neither plate height nor plate number can be considered a truly universal metric for comparing different separation columns, they are both useful ways of checking performance within an individual column, for example, after a packing has been changed in an electrochromatography column.

Resolution A more universal metric for separation columns that can be applied to both rate-based and equilibrium-based techniques is resolution. An approximate form for the resolution that is sufficiently accurate for most applications is (17)

$$R_{\text{spatial}} = \frac{|Z_1 - Z_2\rangle}{\frac{1}{2}(W_1 + W_2)} \approx \frac{|Z_1 - Z_2|}{2(\chi_1 + \chi_2)} \quad \text{or}$$

$$R_{\text{temporal}} = \frac{|t_1 - t_2|}{\frac{1}{2}(W_1 + W_2)} \approx \frac{|t_1 - t_2|}{2(\sigma_1 + \sigma_2)} \tag{3}$$

for spatial or temporal analysis, respectively, where the relationship between the peak width at its base and the variance is $W = 4\chi$ (spatial) or 4σ (temporal). Two key components with an R equal to or slightly greater than 1 can be considered to be baseline separated, so R can be used either to compare data from different separation devices or to modify the column so that complete separation can be achieved.

BATCH ANALYTICAL TOOLS

Capillary Electrophoresis

Capillary electrophoresis (CE) offers important advantages over conventional gel electrophoresis (GE) in the characterization of proteins and nucleic acids. CE is much faster than GE, and can be fully automated and instrumented to include sample injection, detection, and collection. In addition, CE can separate solutes at a resolution significantly greater than that of HPLC. Very small sample loads (e.g., <10 ngs/run) make CE suitable primarily for analytical work. Automated equipment for CE is commercially available from several vendors, including units that allow collection of fractionated samples (18,19) or direct coupling to a secondary separation device (e.g., HPLC or mass spectrometer).

Basic Operation. In principle CE is quite simple. It can be carried out at very high power densities because the large surface area to volume ratio of the capillary lumen allows it to shed joule heat efficiently. However, in practice it has not yet reached the level of maturity of automated chromatography (i.e., where stand-alone columns can be operated by individual staff on an as-needed basis). CE is best operated by a trained technician who retains responsibility for developing protocols. The basic apparatus consists of a fused-silica capillary with a lumen diameter typically ranging from 25 to 150 μm, an outer diameter of several hundred microns (e.g., 400 μm) and a thin coat of polyimide that makes the otherwise brittle capillary quite flexible. A small window is burned in the polyimide near one end of the capillary, and this window is then placed in the beam of a spectrophotometer. To avoid damaging the sensitive electronics in the detector, this end of the capillary is dipped into a vial filled with running buffer and the electrode placed in this vial is set to electric ground (20).

Before loading a sample into a bare capillary the lumen is usually precleaned with 0.1–0.5 M sodium hydroxide and then rinsed sequentially with pure water and running buffer. The capillary is loaded by dipping the high-voltage end into a vial containing the sample and using a pressure head or electric field to load 10–100 nL of sample, which is roughly 0.01–0.1 μg at a concentration of 1 mg/mL. The high-voltage end of the capillary is then moved to a vial containing running buffer, the hot electrode is dipped into this vial, and the voltage is increased until the electric field falls to a range of 200–400 V/cm. These large fields drive solutes through the capillary at electrophoretic velocities on the order of 1–4 cm/min, yielding run times of a few minutes with plate efficiencies approaching 10^6 plates per meter of capillary in free solution.

Fundamental Principles. Two important points that must be considered when discussing CE performance are the temperature profile in the capillary and dispersion due to temperature variations. The temperature profile can be calculated using the steady-state form of the thermal energy equation in cylindrical coordinates

$$\frac{k_{\text{buffer}}}{r}\frac{d}{dr}r\frac{dT}{dr} + \sigma_{\text{buffer}}T_{\text{ref}}E^2[1 + \lambda(T - T_{\text{ref}})] = 0$$
$$\text{where } \lambda \equiv \left[\frac{d\log_e(\sigma_{\text{buffer}})}{dT}\right]_{T=T_{\text{ref}}} \quad (4)$$

where the first term accounts for radial conduction and the second for generation of joule heat. The second term in brackets accounts for the autothermal effect (21,22), that is, the tendency of warm buffer to conduct more current than cold buffer. The thermal energy equation, equation 4, must be solved together with the boundary conditions

$$\text{at } r = 0 \quad T \text{ is finite}$$
$$\text{at } r = a \quad -k_{\text{buffer}}\frac{dT}{dr} = \frac{1}{2}h_{\text{CE}}(T - T_{\text{ref}})$$
$$= \frac{1}{2a}\frac{T - T_{\text{ref}}}{\left[\frac{1}{k_{\text{g}}}\log(a_{\text{g}}/a) + \frac{1}{k_{\text{p}}}\log(a_{\text{P}}/a_{\text{g}}) + \frac{1}{a_{\text{p}}h_{\text{fluid}}}\right]} \quad (5)$$

The boundary condition at $r = 0$ states that the temperature must remain finite at the centerline, and the wall condition at $r = a$ indicates that heat transfer from the capillary buffer to the environment outside the capillary is controlled by the serial resistances of the glass capillary, the polyimide coat, and the boundary layer in the external cooling fluid surrounding the capillary. These three resistances correspond respectively to the three terms in the denominator of the boundary condition at $r = a$. The overall heat transfer coefficient between the lumen OD and the coolant is h_{CE}.

Equation 4 with the boundary conditions in equation 5 has the solution

$$\Theta_{\text{buffer}}\eta = \frac{1}{\kappa^2}\left(\frac{BiJ_0(\kappa\eta)}{BiJ_0(\kappa) - 2\kappa J_1(\kappa)} - 1\right) \quad (6)$$

with dimensionless parameters defined as

$$\Theta_{\text{buffer}} = \frac{k_{\text{buffer}}(T - T_{\text{ref}})}{\sigma_{\text{buffer}}E^2a^2} \quad \eta = \frac{r}{a}$$
$$Bi = \frac{2h_{\text{CE}}a}{k_{\text{buffer}}} \quad \kappa^2 = \lambda\frac{\sigma_{\text{buffer}}E^2a^2}{k_{\text{buffer}}}$$

where Bi is the Biot number. J_0 and J_1 are Bessel functions of the first kind of order 0 and 1, respectively. The temperature profile in the capillary lumen is roughly parabolic with its maximum at the centerline. As seen in Figure 2, the temperature in the lumen can vary markedly with applied voltage, as well as with the overall heat transfer coefficient h_{CE}. However, the variation across the radius of the lumen is usually less than 1 °C, and even this small variation can have a marked impact on dispersion in CE. This occurs because the electrophoretic mobility varies inversely with viscosity. Viscosity, in turn, decreases with increasing temperature, so solutes migrate faster in the hot center of the lumen than near the lumen wall, distorting the solute band into a parabolic shape. When this is taken into account, the effective dispersion coefficient in the capillary becomes (23,24)

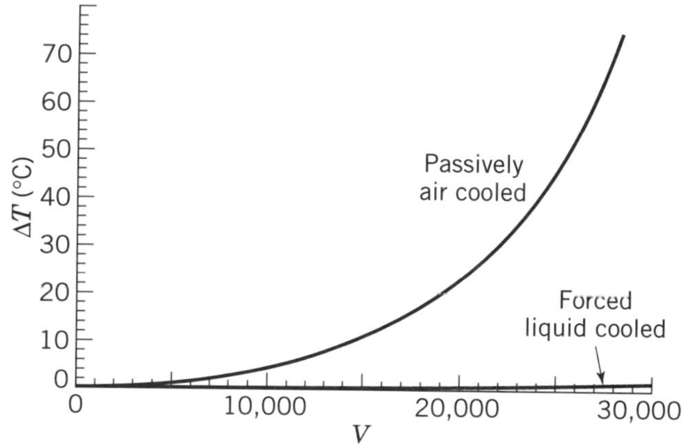

Figure 2. Temperature rise in the CZE lumen as a function of applied voltage V. Both the centerline and wall temperatures are shown but, since they differ by less than 1 °C, only a single line is apparent in this graph. The buffer is assumed to have a conductivity, $\sigma_e = 0.01$ mho/cm.

$$D_{\text{CE}} = D_{\text{m}} + \frac{a^2(\mu(T_{\text{ref}})E)^2}{3072D_{\text{m}}}\frac{\kappa^4 Bi}{(BiJ_0(\kappa) - 2\kappa J_1(\kappa))} \quad (7)$$

The first term on the right accounts for pure axial diffusion, and if only this term is included it predicts a temporal variance equal to that given in equation 1. Under normal operating conditions the second term is roughly the same order of magnitude as the first, indicating that thermal band broadening is generally significant in CE. More detailed information about the range of the Biot number, Bi, and the autothermal coefficient, κ, is available in Ref. 24.

Performance Benchmarks. Modifying the expression for the number of theoretical plates to include sample volume, wall electroosmosis, and the extended dispersion coefficient, yields the formula

$$\frac{1}{N_{\text{CE}}} = \frac{2D_{\text{CE}}}{|(\mu + \mu_{\text{eo}})\Delta\Phi|} + \frac{1}{12}\frac{(q/\pi a^2)^2}{Z^2} \quad (8)$$

where μ_{eo} is the electroosmotic mobility and q is the volume of sample loaded into the capillary. The first term on the right gives the contribution from distortion due to thermal effects and the second term gives the contribution from the variance of the initial sample injection volume q. When hydrostatic loading is used, the injection volume can be calculated as

$$q = \frac{\rho g H}{8\nu Z}\pi a^4 t_{\text{load}} \quad (9)$$

where ρ and ν are the density and viscosity of the buffer, respectively, and H is the difference in height between the liquid levels in the vials into which the two ends of the capillary are immersed, assuming both vials are at atmospheric pressure. Alternatively, if a pressure differential is used to load the capillary, then ρgH can be replaced with ΔP.

As illustrated in Figure 3, for capillaries whose lumen ID is less than 100 μm and for a loading of about 1% of the column volume, the number of theoretical plates levels out near 1×10^5. This result is approximately true regardless of how the capillaries are cooled, for example, by air in natural convection or by forced liquid flow. In small capillaries, the sample injection volume sets an upper limit on capillary performance because the sample volume can be reduced only to the point where the signal is still detectable. After that, to improve performance, the sample must be preconcentrated or stacked or the sensitivity of the detection system somehow improved. Clearly CE will benefit from the development of protocols for stacking or focusing injected solutes in free solution before starting a run (25,26).

The main advantages of forced cooling are that it *(1)* dramatically reduces the temperature rise in the capillary lumen by reducing the heat transfer resistance in the external fluid and *(2)* reduces the thermal fluctuations associated with natural convection. Using conventional sample-loading protocols, there is little benefit in plate efficiency gained by forced liquid cooling. However, as stacking procedures become more common, liquid-cooled CE systems should eventually outperform both passively and actively air-cooled units.

Polyacrylamide Gel Electrophoresis

For several decades now, polymer gel electrophoresis has been a standard method used for laboratory analysis of proteins, nucleic acids, and other biological materials (27,28). The primary advantages of GE over other applicable techniques are *(1)* moderately high resolution, that is, native PAGE routinely provides more than 10,000 theoretical plates and *(2)* the protocols are well established and widely accepted in the field (29). All engineers and scientists who enter the biotechnology industry will, at some point in their careers, find themselves running or reading results from electrophoresis gels.

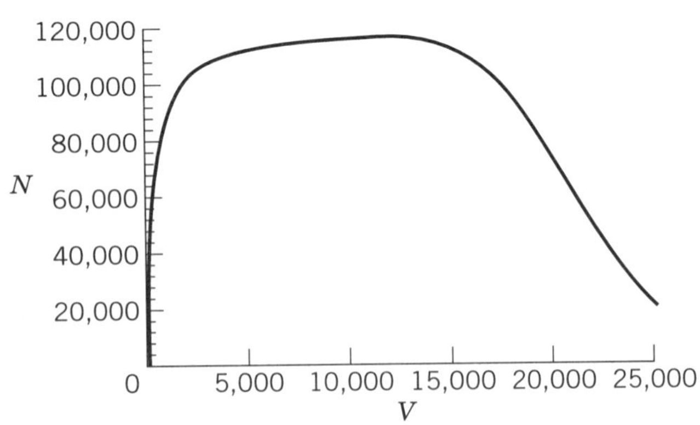

Figure 3. Predicted plate number as a function of applied voltage for a 75-μm ID × 350-μm OD × 100-cm capillary in air using an injection volume, q = 50 nL. In this example the excessive volume of the hydrostatic load limits the number of theoretical plates that can be obtained for this system. Operating at voltages near 15 kV will give the best combination of plate efficiency and speed.

The primary disadvantages of conventional GE are that it is *(1)* labor intensive; *(2)* slow, often requiring 24 h between initial buffer preparation and final destain; *(3)* difficult to recover purified solutes for subsequent steps; and *(4)* difficult to scale beyond a few milligrams of sample. My experience with PAGE is that it takes several weeks of practice to be able to produce publishable gels, and even then users may experience problems with gel-to-gel reproducibility, especially between departments within a company.

Given these limitations, it is not surprising that an effort would be made to develop instrumentation that retains the advantages of GE while minimizing its disadvantages. One can reduce the size and thickness of the gel, actively remove joule heat from one or both sides of the gel, and automate the stain and destain cycle. This greatly reduces labor requirements, turnaround time, and many of the problems associated with gel reproducibility. Also, a variety of prefabricated electrophoresis gels are now commercially available from a number of vendors for about \$10 each.

Basic Operation. Detailed protocols for preparing, running, and analyzing gels are available in any number of references (e.g., Ref. 29) and are not discussed further here. In all likelihood, the end user will run some combination of SDS or native PAGE with the sample in reduced or nonreduced form, in Tris or tricine buffer, with or without a gradient in cross-linking agent, using one of several stacking procedures and Coomassie blue or silver stain to visualize the fractionated bands. Within a biotech company, it is also likely that the gel protocols, including stain and destain procedures, will be specified by a quality assurance or analytical chemistry department or their equivalent. These protocols will almost certainly include a ladder of molecular weight markers in one lane and additional lanes containing controls appropriate to the separation.

IEF-PAGE gels or capillary IEF should be run in conjunction with the molecular weight gels to obtain pIs for the various protein isoforms and glycoforms that inevitably accompany production of recombinant therapeutics. Not only does this give a clearer picture of both contamination and heterogeneity of the purified protein, but it may also help in preparing a rational design for downstream processing.

Fundamental Principles. Although some progress has been made in understanding how proteins migrate through tight gels (30–32) from a practical standpoint it is not necessary to understand the underlying theory to produce publishable gels. Once an appropriate gel and running buffer have been selected for the molecular weight range of interest, it is best to run the gel as quickly as possible to minimize diffusive band spreading. As a rule of thumb, an electrical power density of 0.1 W/cm^3 per gel should be reasonably low for use with air-cooled, glass-walled, vertical slab PAGE. (*Note*: To calculate watts, multiply amps times voltage. If the gel is set up correctly, nearly all of this power will be dissipated in the working portion of the gel slab.) However, if the apparatus comes with a run protocol, these directions should be followed closely, at least during the initial stacking phase of the run. Once the proteins have entered the separating gel, the user has a bit more flexibility with the applied voltage, depending on how well the gel box is cooled, but the voltage should probably not exceed 0.2 W/cm^3 per gel in an air-cooled box.

Performance Benchmarks. Once the stacked proteins have entered the separating gel, the conventional wisdom is that band spreading in near-isothermal gels is controlled by diffusion. Because noneluting gels are analyzed in space, the number of theoretical plates is given by equation 2 as

$$N_{\text{SDS-PAGE}} = \frac{Z^2}{\chi^2} = \frac{Z^2}{2D_{\text{eff}}t} = \frac{(\mu_{\text{eff}}Et)Z}{2D_{\text{eff}}t} = \frac{(\mu_{\text{eff}}\Delta\Phi)Z}{2D_{\text{eff}}L_{\text{gel}}} = \frac{\mu_{\text{eff}}\Delta\Phi}{2D_{\text{eff}}}\frac{Z}{L_{\text{gel}}} \approx 10{,}000 \qquad (10)$$

where the approximation is for a 5-cm displacement. The penultimate term, which is a classical result for electrophoresis, shows how applied voltage, gel length, and so on, affect performance.

This formula also implies that high performance can be maintained as the gel is scaled down to about the size of a glass slide (33–36) if the voltage drop is fixed during miniaturization. The higher-power densities generated can be accommodated by replacing the passive air-cooled, glass-encased gel by an actively cooled (e.g., Peltier system), ceramic-encased gel and by making the gel as thin as is reasonable for handling and detection. Actively cooling both sides of the gel would allow the user to apply about twice the voltage and would therefore be expected to double performance, that is, double the run speed and increase resolution by 41%.

BATCH PREPARATIVE TOOLS

Preparative Gel Electrophoresis

Continuous elution preparative gel electrophoresis (pGE) is a relatively new entry in the field of preparative bioprocessing and is intended to address the limitations of slab GE. The essence of this technique is that a pulse of multicomponent sample can be loaded onto a column and fractionated using well-known protocols on a standard electrophoresis gel and then continuously recovered in free solution as it migrates off the end of the gel. This technique may be viewed as a melding of conventional gel electrophoresis with electroelution. In combining these two steps, it is possible to relieve several of the disadvantages of GE mentioned earlier, especially by automating and accelerating the otherwise labor-intensive sample-recovery step.

Basic Operation. A generic pGE apparatus consists of three basic parts: (*1*) a tube- or annular-shaped electrophoresis gel, (*2*) an elution chamber at the base of the gel, and (*3*) a product offtake line leading to a detector and/or fraction collector. In practice, a sample is loaded into the liquid layer immediately adjacent to the top of a standard gel and then electrophoresed, at low power to avoid mixing due to natural convection, into the top of the gel. Once inside the gel, electrophoresis of the solutes is carried out at power densities at or below 1 W/cm^3. The solutes migrate through the gel at different velocities, and as each solute band reaches the end of the gel it elutes into free solution in the elution chamber and is carried to a fraction collector by the elution buffer. Once the effluent has been aliquoted by the fraction collector, the recovered solutes, which are suspended in the elution buffer, are ready for biochemical analysis or for further processing (37).

Although only a single batch sample can be processed in each run, the gel can be reused several times before it must be replaced. A typical column load ranges from less than 100 μg to more than 10 mg per run depending on the cross-sectional area of the gel; typical run times are somewhat less than 30 min per centimeter of column.

The Electrophoresis Gel. In the electrophoresis gel, the central issue is resolution. This is the heart of the apparatus, and once the purified solutes have eluted from the gel, whatever resolution was achieved there can only be degraded in subsequent processing steps. In this part of the system the major contributors to peak variance are sample volume, molecular diffusion, and thermally induced dispersion. Sample bandwidth is directly related to the volume of the sample band as it begins electromigration through the gel. However, if a stacking gel is used, sample volume is usually not an important contributor to peak variance.

Assuming isothermal conditions and ignoring the contribution to the variance due to sample injection volume, molecular diffusion dominates band spreading by contributing a temporal variance

$$\sigma^2_{\text{diffusion}} = \frac{2D_{\text{gel}}Z}{\langle u_{\text{gel}}\rangle^3} \qquad (11)$$

as measured by a detector located at a fixed position Z. Here D_{gel} is the diffusion coefficient of the solute in the gel and $\langle u_{\text{gel}}\rangle = \langle \mu_{\text{gel}}\rangle E$ is the average electrophoretic velocity of the solute in the gel. Noting that the average holdup time for a gel of length Z is $Z/\langle u_{\text{gel}}\rangle$, the number of theoretical plates is

$$N_{\text{diffusion}} = \frac{t^2}{\sigma^2_{\text{diffusion}}} = \frac{u_{\text{gel}}Z}{2D_{\text{gel}}} \qquad (12)$$

For native PAGE, the number of theoretical plates routinely falls in the range $10^3 < N < 10^4$.

The Elution Chamber. As solutes leave the bottom of the gel tube, they are rapidly swept away from the gel, out of the electric field, and into an elution stream. In a properly designed elution chamber, dispersion is not an important issue, but dilution is. If the elution stream is too slow, some or all of each solute will be entrained by the electric field and permanently removed from the elution stream. If the elution stream flow rate is too high, the solute will be diluted, making it difficult to detect or use in subsequent processing steps.

Many different strategies could be used to reduce dilution in the elution chamber, and this part of the system offers the greatest potential for innovation. Here only two generic approaches to elution, cross-flow and counterflow, are considered to illustrate how an elution chamber is designed.

In cross-flow elution (Fig. 4) a hydrodynamic flow is directed perpendicular to the electric field at the base of the gel tube and flows out of the elution chamber through an opening on the right. The minimum elution flow rate is determined by finding the conditions under which a molecule exiting the gel tube and entering the elution stream at the farthest point from the elution channel just avoids being captured by the electric field. In Figure 4, the lower trajectory is captured on the electrode membrane, and the upper trajectory is cleared into the elution stream.

In counterflow elution, the hydrodynamic flow is directed opposite to the electrophoretic migration in the elution chamber. The minimum elution flow is found when a molecule's electrophoretic velocity in the elution buffer exactly balances the counterflow. As in Figure 4a, the lower trajectory is captured on the electrode membrane and the upper trajectory is cleared into the elution stream.

It turns out that in both of these cases the minimum elution flow rate that discourages capture of the solute by the electrode is

$$Q_{\min} = \langle u_{\text{buffer}} \rangle A_c \tag{13}$$

where A_c is the cross-sectional area of the elution chamber through which the current flows and $\langle u_{\text{buffer}} \rangle$ is the electrophoretic velocity in free solution. This is the free solution electrophoresis velocity in the elution buffer and may be substantially different from the velocity in the gel. In practice, the elution flow rate is often several times higher than this minimum value, leading to unnecessary dilution of the sample.

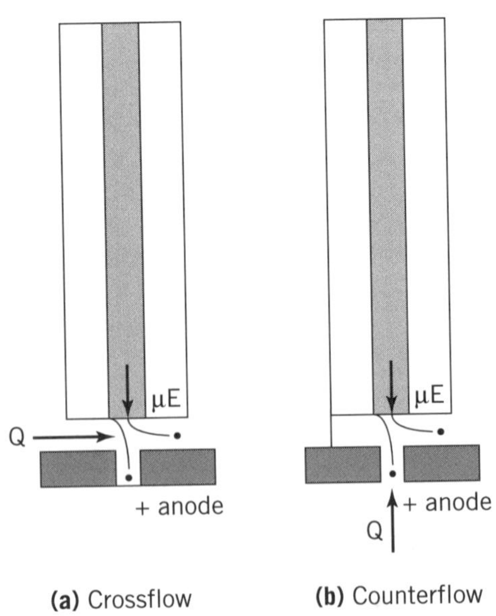

Figure 4. In cross-flow elution **(a)** the sample is swept away from the electrode port by a perpendicular flow of elution buffer. In counterflow elution **(b)** the elution buffer flow rate is directly opposed to the electrophoresis of the solute toward the lower electrode. In both cases if the elution flow rate is too low, the solute may be "captured" by the electrode. This is represented by the particle that follows the lower trajectory in this figure.

The Product Offtake Line. After being flushed from the elution chamber, the solute is carried by the elution buffer through a length of tubing to a detector and/or fraction collector. In this stage the main consideration is loss of resolution, since the purified solute bands are now subject to Taylor (diffusive–convective) dispersion (38) as they travel through the tubing. This occurs because the flow in the system is laminar, parabolic with a maximum velocity at the tube centerline. Solutes near the wall move slowly compared with those near the centerline so the solute band is elongated by the flow as it travels through the tube.

Aris–Taylor theory (39), which is strictly valid only for long tubes, predicts a dispersion coefficient that goes as

$$D_{\text{Taylor}} = D_{\text{m}} + \frac{\langle v \rangle^2 d_t^2}{192 D_{\text{m}}} = D_{\text{m}} + \frac{Q^2}{12\pi^2 d_t^2 D_{\text{m}}} \tag{14}$$

where $\langle v \rangle$ is the average hydrodynamic velocity of the buffer in the tube. Because of Taylor dispersion the offtake line can generate significant band spreading if it is improperly designed. The variance predicted by Taylor's theory for dispersion in a tube of length L is

$$\sigma^2_{\text{Taylor}} = \frac{D_{\text{m}} L}{32} \frac{\pi^3 d_t^6}{Q^3} + C \frac{\pi d_t^4 L}{384 Q D_{\text{m}}} \tag{15}$$

which indicates that Taylor dispersion may be reduced by using short, thin tubes. For tubes of length $L < Q/\pi D_{\text{m}}$, Aris–Taylor theory overpredicts dispersion and must be adjusted before it can be applied to the lengths of elution channel typically used with these instruments. To put this in perspective, for Aris–Taylor theory to apply in commercial pGE equipment, one would need $L > 1$ m when $Q = 10\ \mu\text{L/min}$ and $L > 100$ m for $Q = 1$ mL/min. The constant C, which corrects for short tubes, can be estimated from Figure 5, where $\zeta = D_{\text{m}} L / \langle v \rangle a_{\text{t}}^2$.

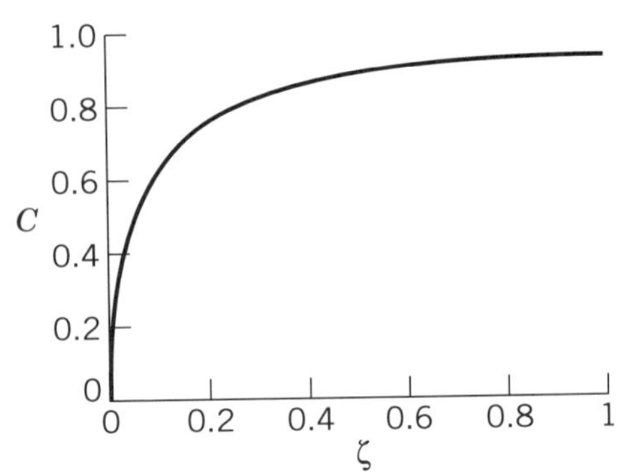

Figure 5. Ratio of the dispersion coefficient in tubes of arbitrary length to the dispersion coefficient in long tubes as a function of the dimensionless distance ζ.

For example, a 1-m-long tube with a flow rate of 1 mL/min has an adjusted variance of 8,214 s^2 at a dimensionless distance of $\zeta = 0.009425$ ($C = 0.185803$). A pulse of solute placed at the entrance of the tube will take about 6.1 min to flush through the tube entirely. The implication here is that solute bands that come off the gel less than 6.1 min apart will tend to overlap as they come out of the elution stream tube. If this becomes a problem, it is best addressed by decreasing the length of the tube.

The effect of Taylor dispersion in short tubes is summarized in Figure 6 for the two cases, $Q \approx 1$ mL/min, which corresponds roughly to the annular device, and $Q \approx 10\ \mu$L/min, which corresponds roughly to the tubular device. The temporal bandwidth, that is, the time it takes for a pulse of solute to completely elute from the tube, is approximately equal to 4σ, so this has been plotted against the tube length for each of these flow rates. Note that the temporal dispersion is independent of tube diameter (a counterintuitive result) for a fixed volumetric flow and that dispersion can be decreased by reducing Q and L.

Although existing continuous elution systems do not meet the long-tube condition required for Taylor's analysis to apply rigorously, the variance can still be significantly decreased by reducing the inner diameter and length of the elution stream tube.

Fundamental Principles. As with CE, temperature profiles in pGE can be calculated using equation 4, except that the electrical parameters used are those that pertain to the gel. The steady-state form of the thermal energy equation in cylindrical coordinates is

$$\frac{k_{\text{gel}}}{r}\frac{d}{dr}r\frac{dT}{dr} + \sigma_{\text{gel}}(T_{\text{ref}})E^2[1 + \lambda_{\text{gel}}(T - T_{\text{ref}})] = 0 \quad (16)$$

For a tubular gel, this equation is put in dimensionless form using the transformations

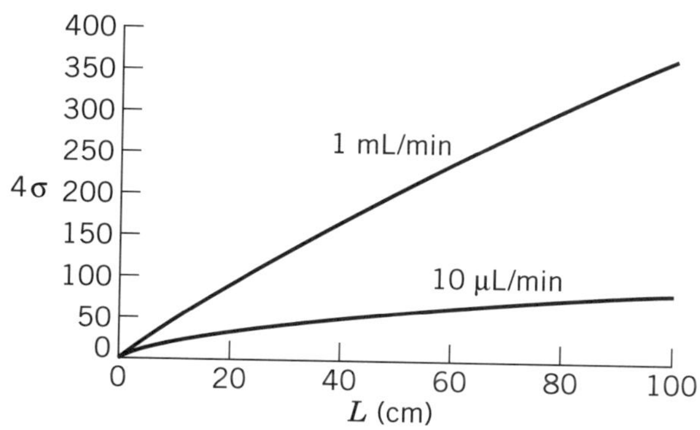

Figure 6. Contribution of the variance computed to the baseline width, in seconds, for flow in the elution stream tube. Although this is generally not a significant contribution to the overall dispersion, it can smear bands that elute within a few minutes of one another. The dispersion is alleviated by reducing the elution flowrate Q, and keeping the elution tube as short as possible.

$$\Theta_{\text{gel}}(\eta) = \frac{k_{\text{gel}}(T - T_{\text{ref}})}{\sigma_{\text{gel}}E^2a_{\text{gel}}^2} \quad Bi = \frac{2h_{\text{pGE}}a^{\text{gel}}}{k_{\text{gel}}} \approx \frac{2k_{\text{glass}}}{k_{\text{gel}}\log\left[\frac{a_{\text{gel}}}{a_{\text{glass}}}\right]}$$

$$\kappa^2 = \frac{\sigma_{\text{gel}}E^2a_{\text{gel}}^2}{k_{\text{gel}}}\left[\frac{d\log(\sigma)}{dT}\right]_{T_{\text{ref}}}$$

where a_{gel} is the radius of the gel and $\eta = r/a_{\text{gel}}$. The heat transfer coefficient h_{pGE} for the micropreparative apparatus includes resistances due to the glass wall of the tube and the external cooling fluid. If cooling is by forced air and the glass wall is thick, the heat transfer resistance is typically controlled by the glass. This is reflected in the approximate term on the far right in the relation for Bi.

These transformations yield a second-order ordinary differential equation that, when solved with the boundary conditions from equation 5, has a solution of the form

$$\Theta_{\text{gel}}(\eta) = \frac{1}{\kappa^2}\left(\frac{BiJ_0(\kappa\eta)}{BiJ_0(\kappa) - 2\kappa J_1(\kappa)} - 1\right) \quad (17)$$

which is identical to equation 6 except that the definitions of the dimensionless parameters have been changed.

Table 2 contains the physical data used to predict temperature profiles in pGE at a base power dissipation rate of 1 W/cm^3. As is clear in Figure 7, at this power setting the centerline temperature rise is only about 1 °C greater than the coolant temperature. The 1 °C variation in temperature implies that, for practical purposes, thermal excursions do not have a significant impact on dispersion or resolution when compared with contributions from other sources. Consequently, it is acceptable to ignore temperature effects when optimizing pGE performance at power densities below 1 W/cm^3.

Annular Gels. In cases where larger amounts of material must be processed in a single run, the basic tube gel used can be significantly scaled by switching to an annular gel. The gel in the annulus should be kept thin to avoid excessive temperature excursions, but the annulus perimeter can be enlarged to increase capacity. For example, an annulus with a 2-cm inner diameter and a 3-cm outer diameter would have a typical column load of roughly 20 mg/run. An annulus with a 4-cm outer diameter and a 2-cm inner diameter could process about 50 mg/run.

For a thin annulus with a large diameter and an annular gap equal to the diameter of a tubular column, the relative increase in scale of the annulus over the tube is approximately $4a_o/a_{\text{gel}}$, where a_o is the outer diameter of the annulus. In the more general case where the annulus

Table 2. Physical Constants Used in the Tubular pGE Illustrations

	Gel tube
$\sigma(T_{\text{ref}})E^2$ (W/cm^3)	1.0
Z (cm)	5.0
$2a_{\text{gel}}$ (cm)	0.25
d_i (cm)	0.8
Run time (h)	3–5

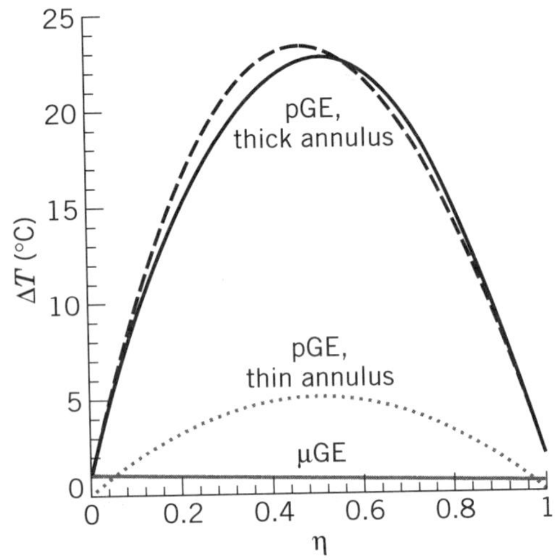

Figure 7. Temperature rise (°C) in continuous elution gel electrophoresis at a base power density, $P_w = 1$ W/cm^3 in a tube gel (———), a thin annulus (– – – –), and a thick annulus (· · · ·). Note the slight asymmetry in the annular temperature profiles. The line next to each curve is the corresponding parabolic Galerkin approximation used in the dispersion calculations.

is not thin, the relative increase in scale is given by the ratio S of the cross-sectional areas of the two columns

$$S = \frac{a_o^2 - a_i^2}{a_{gel}^2} \tag{18}$$

The capacity can be increased further by increasing the thickness of the annulus and by lengthening the column, but the former admits some degradation of performance due to thermal dispersion and the latter requires longer run times.

For an annulus, equation 16 is put in dimensionless form by using different transformations than in the previous cases. These are

$$\Theta(\eta) = \frac{k_{gel}(T - T_{ref})}{\sigma(T_{ref})E^2(a_o - a_i)^2} \qquad \eta = \frac{r - a_i}{a_o - a_i}$$

$$\kappa^2 = \frac{\sigma(T_{ref})E^2(a_o - a_i)^2}{k_{gel}}\left[\frac{d\log(\sigma)}{dT}\right]_{T=T_{ref}} \qquad \Lambda = \frac{a_i}{a_o - a_i}$$

where a_i is the radius of the inner surface of the gel and a_o is the radius of the outer surface. Note that *(1)* the dimensionless radial coordinate η is defined so that the domain extends over $0 < \eta < 1$ and *(2)* a new geometric parameter, Λ, appears in the governing equations.

These transformations yield a second-order ordinary differential equation

$$\frac{d^2\Theta}{d\eta^2} + \frac{1}{\eta + \Lambda}\frac{d\Theta}{d\eta} + \kappa^2\Theta = -1 \tag{19}$$

that has solutions of the form

$$\Theta(\eta) = AJ_0[\kappa(\eta + \Lambda)] + BY_0[\kappa(\eta + \Lambda)] - \frac{1}{\kappa^2} \tag{20}$$

where A and B are constants that must be fit using appropriate boundary conditions. J_o is a Bessel function of the first kind of order 0, and Y_o is a Bessel function of the second kind of order 0.

Appropriate boundary conditions in this case are

$$\text{at } \eta = 0 \quad +\frac{\partial\Theta(\eta)}{\partial\eta} = Bi_i\Theta(\eta) \quad \text{where } Bi_i = \frac{k_i}{k_{gel}}\frac{2}{\log(2a_i/d_i)}$$

$$\text{at } \eta = 1 \quad -\frac{\partial\Theta(\eta)}{\partial\eta} = Bi_o\Theta(\eta) \quad \text{where } Bi_o = \frac{k_o}{k_{gel}}\frac{2}{\log(d_o/2a_o)} \tag{21}$$

where d_i is the inside diameter of the cooling finger and d_o is the outer diameter of the glass tube. For the annular device, the constants of integration, A and B, are obtained by solution of the matrix equation

$$\begin{bmatrix} -\kappa J_1(\kappa(1 + \Lambda)) + Bi_oJ_0(\kappa(1 + \Lambda)) & -\kappa Y_1(\kappa(1 + \Lambda)) + Bi_oY_0(\kappa(1 + \Lambda)) \\ +\kappa J_1(\kappa\Lambda) + Bi_iJ_0(\kappa\Lambda) & +\kappa Y_1(\kappa\Lambda) + Bi_iY_0(\kappa\Lambda) \end{bmatrix}\begin{bmatrix} A \\ B \end{bmatrix} = \begin{bmatrix} \frac{Bi_o}{\kappa^2} \\ \frac{Bi_i}{\kappa^2} \end{bmatrix}$$

and autothermal runaway is predicted at values of κ for which the determinate of the 2×2 matrix on the left vanishes.

Table 3 contains the physical data used to predict temperature profiles in the preparative chamber at a base power dissipation rate of 1 W/cm^3. For purposes of illustration, two different cases are shown. The first is an instrument with a thin annular gap and the second has a thick annular gap. The predicted temperature rise in each instrument under these conditions is shown in Figure 7. Since the temperature rise in each of these devices is roughly proportional to $\sigma(T_{ref})E^2(a_o - a_i)^2/k_{gel}$, these results can be readily extrapolated to other operating conditions and gel geometries. To maintain performance levels while scaling up, the product of the electric field times the gap thickness, $E(a_o - a_i)$, should be held constant in order to limit temperature variations across the transverse axis, thereby discouraging excessive thermal dispersion.

Table 3. Physical Constants Used in the Preparative GE Illustrations

	Thin annulus	Thick annulus
L (cm)	10.0	10.0
2 a_i (cm)	1.905	1.905
2 a_o (cm)	2.8	3.7
d_o (cm)	1.565	1.565
d_i (cm)	3.14	4.04
k_{gel} (W/cm °C)	0.00578	0.00578
k_{glass} (W/cm °C)	0.011	0.011
$k_{ceramic}$ (W/cm °C)	0.0578	0.0578
Run time (h)	5–15	5–15

This implies that, while the scale may be doubled by doubling the annular gap thickness, the field must be decreased by half so the run takes twice as long. Consequently, overall productivity in preparative GE remains roughly independent of gap thickness, with the major savings coming from the reductions in the time and effort associated with reloading and purging the gel. As stated earlier, the gel apparatus is most effectively scaled by increasing the perimeter of the annulus while keeping the annular gap thickness constant.

Solute Concentration Profiles. In the absence of thermal dispersion, the concentration profiles of solutes exiting the gel can be calculated by solving the diffusion–convection equation

$$\frac{\partial C}{\partial t} + u_{\text{gel}}\frac{\partial C}{\partial z} = D_{\text{gel}}\frac{\partial^2 C}{\partial z^2} \tag{22}$$

subject to the boundary and initial conditions

$$\begin{array}{lll} \text{at } t = 0 & \text{all } z & C = 0 \\ \text{at } z = 0 & \text{all } t & C = H(0, \tau)C_o \\ \text{as } z \to \infty & \text{all } t & C \text{ is bounded} \end{array} \tag{23}$$

where the distribution function $H(0, t) = 1$ if $0 < t < \tau$ and $H(0, t) = 0$ if $t > \tau$. This problem is most conveniently solved using Laplace transforms to obtain

$$\begin{aligned} C = \frac{C_o}{2}\Bigg[e^{Pe_{\text{gel}}Z/a_{\text{gel}}\zeta}\Bigg(\text{Erfc}\left[\sqrt{\frac{Pe_{\text{gel}}Z}{4a_{\text{gel}}}}\left(\frac{\zeta}{\sqrt{\theta}} + \sqrt{\theta}\right)\right] \\ - \text{Erfc}\left[\sqrt{\frac{Pe_{\text{gel}}Z}{4a_{\text{gel}}}}\left(\frac{\zeta}{\sqrt{\theta - \Theta}} + \sqrt{\theta - \Theta}\right)\right]\Bigg) \\ + \text{Erfc}\left[\sqrt{\frac{Pe_{\text{gel}}Z}{4a_{\text{gel}}}}\left(\frac{\zeta}{\sqrt{\theta}} - \sqrt{\theta}\right)\right] \\ - \text{Erfc}\left[\sqrt{\frac{Pe_{\text{gel}}Z}{4a_{\text{gel}}}}\left(\frac{\zeta}{\sqrt{\theta - \Theta}} - \sqrt{\theta - \Theta}\right)\right]\Bigg] \end{aligned} \tag{24}$$

where the dimensionless parameters in equation 24 are defined as

$$\theta = \frac{\langle u_{\text{gel}}\rangle t}{Z} \quad \zeta = \frac{z}{Z} \quad \Theta = \frac{q}{\pi a_{\text{gel}}^2 Z}\frac{\langle u_{\text{gel}}\rangle}{u_{\text{buffer}}} \quad Pe_{\text{gel}} = \frac{\langle u_{\text{gel}}\rangle a_{\text{gel}}}{D_{\text{gel}}}$$

Here θ is the dimensionless time, ζ is the dimensionless distance, Pe_{gel} is the gel Peclet number, and Θ is a dimensionless loading factor. Note that the leading fraction in the definition of Θ is the ratio of the load volume to the gel volume and is normally kept below 5%. The trailing fraction is the ratio of the electrophoretic velocities in the gel and in the buffer. If a stacking gel is used, the estimated volume of the sample as it leaves the stacking gel should be used in calculating Θ.

Figure 8 illustrates the effect of decreasing the injection volume from 50 μl to 5 μL. At the lower injection volumes (leading peak) the peak concentration drops below the value in the feed, its breadth decreases, and the center of the peak approaches a symmetric distribution about the mean elution time, $\theta = 1$, that is, $t = Z/\langle u_{\text{gel}}\rangle$. At higher loadings the peak width expands to accommodate the added mass of solute, thus making it more difficult to resolve closely spaced peaks. To minimize the influence of the injection volume, q should be adjusted so that Θ ;< 1/300. This is most conveniently done by using a stacking procedure.

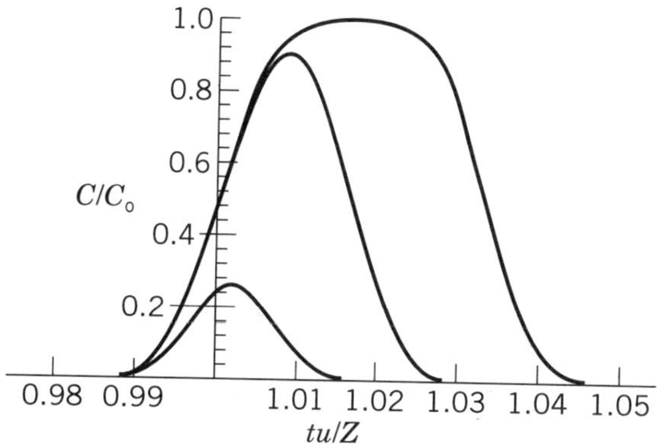

Figure 8. The effect of injection volume q, on effluent concentration profiles from the micropreparative gel electrophoresis chamber. The ordinate is dimensionless concentration C/C_o and the abscissa is dimensionless time θ. This simulation was carried out with $Pe_{\text{gel}} = 2{,}000$, $Z = 5$ cm, $a_{\text{gel}} = 0.125$ cm, and $Q = 1/30$, 1/60, and 1/300 ($q \approx 50$, 25, and 5 μL, respectively) and clearly shows the reduction in bandwidth and effluent concentration that accompanies decreasing q. The smallest injection volume is shown in the leading curve.

Thermal Dispersion. To estimate the effect of radial temperature variations on axial dispersion, it is necessary to solve the extended form of the convective–diffusion equation

$$\frac{\partial C}{\partial t} + u(r)\frac{\partial C}{\partial z} = \frac{D_{\text{gel}}}{r}\frac{\partial}{\partial r}r\frac{\partial C}{\partial r} + D_{\text{gel}}\frac{\partial^2 C}{\partial z^2} \tag{25}$$

where $u(r)$ is the axial electrophoretic velocity distribution in the gel. We can estimate the contribution of axial velocity distribution to the variance by dropping the diffusion terms, assuming that the injection volume is infinitesimally small and then using moment analysis to compute the variance. On eliminating the diffusive terms, the mass transport equation reduces to

$$\frac{\partial C}{\partial t} + u(r)\frac{\partial C}{\partial z} = 0 \tag{26}$$

A very thin volume of concentrated solute injected onto the column can be represented by the Dirac distribution (impulse function) $C_o\delta(z)$ as an initial condition and the resulting system readily solved (e.g., using Laplace transforms) to obtain

$$C(t, r, z) = \begin{bmatrix} C_o\delta\left(t - \dfrac{z}{u(r)}\right) & \text{when } \dfrac{z}{u_{\max}} < t < \dfrac{z}{u_{\min}} \\ 0 & \text{when } \dfrac{z}{u_{\max}} > t > \dfrac{z}{u_{\min}} \end{bmatrix} \tag{27}$$

These solutions are then used to calculate the radially averaged flux J and the temporal moments M_n at the outlet of the gel. The formulas for these are

Radially averaged flux

$$J(t, z) = \int_{a_\mathrm{t}}^{a_\mathrm{o}} u(r)C(t, r, z)r dr \Big/ \int_{a_\mathrm{t}}^{a_\mathrm{o}} r dr \qquad (28)$$

$$\text{Temporal moments} \quad M_n(z) = \int_0^\infty J(t, z)t^n dt \qquad (29)$$

The electrophoretic velocity distribution $u(r)$ varies with the buffer viscosity, which is a function of the temperature. For relatively small temperature variations (i.e., less than 10 °C), the velocity may be calculated by using a Taylor expansion in the temperature rise to obtain the formula

$$\begin{aligned} u(r) &= \frac{\nu(T_\mathrm{ref})}{\nu(T)}\mu(T_\mathrm{ref})E \\ &= \mu(T_\mathrm{ref})E\left[1 - (T - T_\mathrm{ref})\left(\frac{d \log(\nu)}{dT}\right)_{T_\mathrm{ref}}\right] \\ &= \mu(T_\mathrm{ref})E[1 - \alpha(T - T_\mathrm{ref})] \end{aligned} \qquad (30)$$

where μ is the electrophoretic mobility and ν is the viscosity of the buffer.

Strictly speaking, the temperature profiles are given by equation 20 for annular gels. However, to generate a relatively simple analytical form for the fluxes, Galerkin's method has been used to find an approximate solution of equation 19 in the form of a parabolic polynomial in η

$$\Theta(\eta) \approx A_0[Bi_\mathrm{b}(\eta - \eta^2) + 1] \qquad (31)$$

where

$$\begin{aligned} A_0 = & -7[10(\Lambda^2 + \Lambda)(Bi_\mathrm{b} + 6) + 3Bi_\mathrm{b} + 20] \\ & \div \{14(\Lambda^2 + \Lambda)[Bi_\mathrm{b}^2(\kappa^2 - 10) + 10Bi_\mathrm{b}(\kappa^2 - 6) + 30\kappa^2] \\ & + Bi_\mathrm{b}^2(4\kappa^2 - 49) + 14Bi_\mathrm{b}(3\kappa^2 - 25) + 140\kappa^2\} \end{aligned}$$

To simplify the analysis somewhat, only the symmetric form of the solution for the temperature profile is considered here. Although this introduces a small error in the variance computed for thick-gap devices, as seen in Figure 7, it greatly simplifies the analysis, allows closed-form solution of the resulting equations, and gives the correct qualitative tendency as the power density is increased.

For an electrophoretic velocity profile that is symmetric about the centerline of the annulus and that has the parabolic form $u(\eta) = A + B\eta + C\eta^2$, the radially averaged concentration and flux measured at position Z are

$$C(t, Z) = \left[\begin{array}{l} \dfrac{\left(B\sqrt{t} + \sqrt{B^2t - 4C(At - Z)}\right)Z}{Ct^2\sqrt{B^2t - 4C(At - Z)}(a_\mathrm{o}^2 - a_\mathrm{i}^2)} C_\mathrm{o} \\ 0 \end{array}\right. \left.\begin{array}{l} \text{when } t_\mathrm{min} = \dfrac{Z}{u_\mathrm{max}} < t < \dfrac{Z}{u_\mathrm{min}} = t_\mathrm{max} \\ \text{when } t_\mathrm{min} = \dfrac{Z}{u_\mathrm{max}} > t > \dfrac{Z}{u_\mathrm{min}} = t_\mathrm{max} \end{array}\right] \qquad (32)$$

$$J(t, Z) = \left[\begin{array}{l} \dfrac{\left(B\sqrt{t} + \sqrt{B^2t - 4C(At - Z)}\right)Z^2}{Ct^3\sqrt{B^2t - 4C(At - Z)}(a_\mathrm{o}^2 - a_\mathrm{i}^2)} C_\mathrm{o} \\ 0 \end{array}\right. \left.\begin{array}{l} \text{when } t_\mathrm{min} = \dfrac{Z}{u_\mathrm{max}} < t < \dfrac{Z}{u_\mathrm{min}} = t_\mathrm{max} \\ \text{when } t_\mathrm{min} = \dfrac{Z}{u_\mathrm{max}} > t > \dfrac{Z}{u_\mathrm{min}} = t_\mathrm{max} \end{array}\right] \qquad (33)$$

where the parameters, A, B, and C, calculated for the annular columns, are given in Table 4.

Equation 28 for the radially averaged flux is used to compute the first three temporal moments, and these are then used to compute the mean elution time and the temporal variance. The results of these calculations are summarized in Figure 9, which shows the reduction in plate efficiency due to thermal dispersion in pGE as a function of the power density.

The most interesting aspect of the thermal dispersion is its dependence on column length. As shown in Table 5, the variance associated with thermal dispersion varies with the square of the column length rather than the first power, as is common with convective–diffusive dispersive

Table 4. Parameters Used in the Preparative Electrophoresis Moment Calculations

Annular gel ($Bi_\mathrm{a} = Bi_\mathrm{b}$)
$A = +\mu(T_\mathrm{ref})E\left[1 - \alpha(T_\mathrm{ref})A_0 \dfrac{\sigma(T_\mathrm{ref})E^2b^2}{k}\right]$
$B = -C = -\mu(T_\mathrm{ref})E\left[\alpha(T_\mathrm{ref})A_0 \dfrac{\sigma(T_\mathrm{ref})E^2b^2}{k} Bi_\mathrm{b}\right]$
$t_\mathrm{min} = Z/(A + B/2 + C/4)$
$t_\mathrm{max} = Z/(A + B + C)$

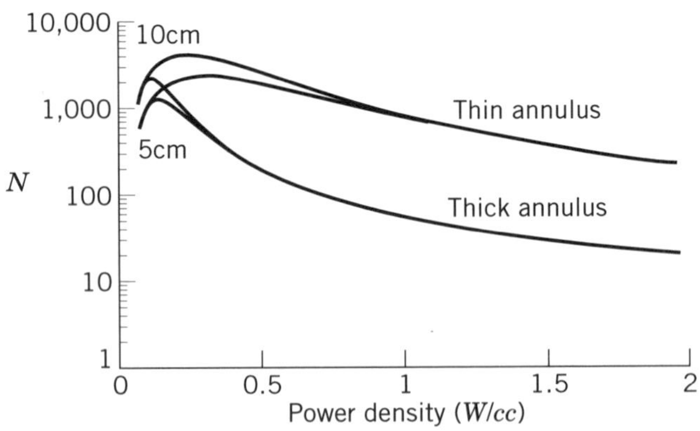

Figure 9. Plate efficiency as a function of power density in the preparative GE. The curves that meet on the upper left are calculated for $Z = 10$ cm; those on the lower left are for $Z = 5$ cm. The curves that meet on the lower right are calculated for the thick annulus; those that meet on the upper right are for the thin annulus. The effect of sample injection volume is not included in these calculations, but would tend to substantially reduce the plate efficiency of the shorter column.

Table 5. Mean Residence Time and Variance Due to Thermal Dispersion in pGE

$$t_{\text{mean}}(Z) = Z\frac{M_1(Z)}{M_o} = \frac{72Z}{72A + 11B}$$

$$\sigma^2(Z) = \frac{M_2(Z)}{M_0} - \frac{M_1(Z)^2}{M_0^2} = 192Z^2\left(\frac{\log\left[\sqrt{\frac{4A+B}{4A}}\right]}{B(72A+11B)} + \frac{\log\left[\sqrt{\frac{4A+B}{4A}} + \sqrt{\frac{B}{4A}}\right]}{\sqrt{B(4A+B)}(72A+11B)} - \frac{27}{(72A+11B)^2}\right)$$

processes. As a result, thermal dispersion dominates other sources of dispersion in long columns and at high power densities, as shown in Figure 9. As a rule of thumb, the optimum power density in existing commercial equipment for pGE should occur at a characteristic temperature rise, $\sigma E^2(a_o - a_i)^2/k_{\text{gel}} \approx 8$ °C.

Although the formulas given are approximate to the extent that molecular diffusion is ignored and only symmetric parabolic temperature distributions are admitted, they provide an excellent qualitative description of the available pGE columns and a rational quantitative basis for designing new units or upgrading existing columns. Figure 9 shows that the thick annular chamber must operate in a very narrow power dissipation range to obtain reasonable plate efficiencies. If higher resolution is needed, the user has a choice of reloading the thin-annulus device several times or building a new instrument with a larger perimeter.

Solute Dilution in the Elution Chamber. As solute leaves the gel phase, it enters the elution chamber and is then flushed into the elution stream. There is some loss of resolution at this point, but the main consideration here is dilution of the fractionated samples as they come off the gel. The dilution factor F_{dilution} is calculated by taking the ratio of the solute velocity J/C as it comes off the gel to the elution stream velocity Q/A_c. This is

$$F_{\text{dilution}} = \frac{J(Z,t)}{C(Z,t)}\frac{A_c}{Q} = \frac{Z}{t_{\text{elution}}}\frac{A_c}{Q} \gtrsim \frac{\langle u_{\text{gel}}\rangle}{\langle u_{\text{buffer}}\rangle} \qquad (34)$$

from which it is seen that the dilution factor is a function of elution time.

The approximate inequality on the far right has been included to give perspective to the dilution factor. This approximation is based on the assumption that one would like to operate under conditions in which no solute is captured by the electrodes in the elution chamber. Roughly speaking, the largest attainable dilution factor (i.e., that giving least dilution) is equal to the ratio of the electrophoretic mobility in the gel near the outlet to the electrophoretic mobility in the elution buffer. The utility of this expression is that it may be rearranged to compute the lowest admissible value of the elution stream flow rate Q

$$Q \gtrsim Q_{\min} = \frac{ZA_c}{t_{\text{elution}}}\frac{\langle u_{\text{buffer}}\rangle}{\langle u_{\text{gel}}\rangle} \approx A_c\langle u_{\text{buffer}}\rangle \qquad (35)$$

Then, if the electrophoretic velocity of a solute in free solution is known, the elution buffer flow rate can be set very near this minimum value. If a collection of solutes is being processed at a constant elution flow rate, then Q should be calculated using the mobility of the fastest solute (i.e., the first one off the gel) to avoid the loss of fast solutes to the electrode.

Note that the dilution factor depends not only on the tightness (i.e., the reduction of the solute's diffusion coefficient) of the gel being run but also on the individual electrophoretic mobilities of the solutes being processed. This implies that, to keep the dilution factor constant during a run, the elution buffer flow rate would need to be decreased in inverse proportion to time. This latter type of correction could be important in applications where peak width or height must be accurately measured, for example, so that a direct comparison of bands may be made with those on a standard gel. It may also be important in minimizing dilution of the slower solutes.

Remarks. The continuous elution pGE annular apparatus is capable of processing several hundred times more solute in a single pass than the tubular apparatus described earlier. To accomplish this, some resolution is sacrificed for capacity, and because the optimal run conditions in the annulus are found at a fraction of the power density used in the tube the runs generally take longer to complete. It is up to the user to decide for a given separation whether this trade-off between time and resolution is worthwhile.

PREPARATIVE CONTINUOUS-FLOW ELECTROPHORESIS

Thin-Film Apparatus

When still greater amounts of material must be processed, the apparatus that has the next largest capacity is a thin-film continuous electrophoresis device, such as the VaP 22, which was until recently available through Bender-Hobein, FRG. This type of apparatus, which dates back to the early 1960s, consists of a thin film of buffer that flows down between two closely spaced (e.g., 0.5-mm transverse) vertical glass plates about 20 cm across (lateral) and 50 cm long (axial), as illustrated in Figure 10. The CFE is usually run in free-flow mode, that is, without anticonvectant media.

Multicomponent sample is injected into the carrier buffer at the top of the column through a thin capillary. The solutes are convected down with the buffer, flow between electrodes that are located at the lateral edges of the plates, and are deflected by the electric field according to their electrophoretic mobilities. As the fractionated solutes

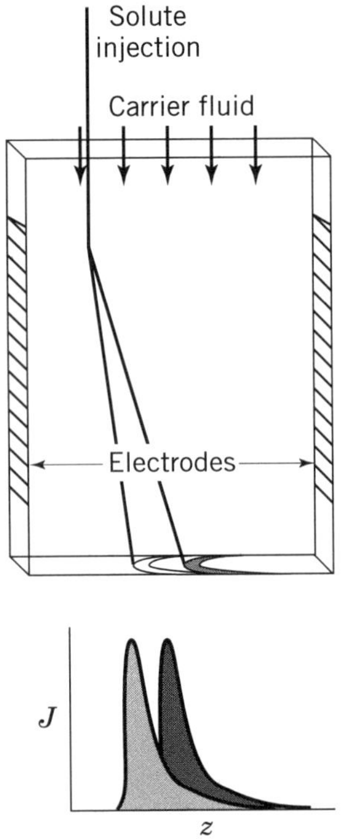

Figure 10. Schematic of the CFE showing electrode placement, multicomponent sample injection, carrier fluid flow, and product distribution. Typical elution bands are shown beneath the chamber corresponding to the two crescent-shaped bands on the lower surface of the chamber. The skewing is a deformation of the sample bands referred to as the crescent phenomena (2,40), which is caused by the interaction of hydrodynamic and electrokinetic effects.

exit the electric field, they are swept into discrete effluent ports by the carrier buffer and collected in separate vials (41).

A major advantage of the CFE over equipment described earlier is that it operates in true continuous mode, whereas the others operate in semibatch mode (i.e., samples are loaded batchwise, but product fractions are eluted continuously). Processing rates can approach 50 mg/h, and theoretical plate efficiencies can in theory exceed $N = 10{,}000$, although in practice they are generally somewhat lower.

Fundamental Principles. The first step in analyzing the performance of the CFE is to calculate the temperature profile along its thin (transverse) axis. The governing differential equation is

$$k_{\mathrm{buffer}} \frac{d}{dy}\left(\frac{dT}{dy}\right) + \sigma_{\mathrm{buffer}}(T_{\mathrm{ref}})E^2[1 + \lambda(T - T_{\mathrm{ref}})] = 0 \quad (36)$$

with the boundary conditions

$$\text{at } y = \pm b$$
$$-k_{\mathrm{buffer}} \frac{dT}{dy} = \pm h_{\mathrm{CFE}}(T - T_{\mathrm{ref}}) \approx \pm \frac{k_{\mathrm{glass}}}{d_{\mathrm{glass}}}(T - T_{\mathrm{ref}}) \quad (37)$$

Here it is assumed that both transverse surfaces are cooled, that the major resistance to heat transfer is the thickness of the transverse surfaces, and that the heat transfer coefficients are equal on both sides of the unit. Note that asymmetric conditions could significantly impair CFE performance, so every effort should be made to match the electrokinetic and thermal properties of these surfaces.

Equation 36 with symmetric boundary conditions as in equation 37 has the solution

$$\Theta_{\mathrm{CFE}}(\eta) = \frac{1}{\kappa^2}\left[\frac{Bi\cos(\kappa\eta)}{Bi\cos(\kappa) - \kappa\sin(\kappa)} - 1\right] \quad (38)$$

where the dimensionless parameters used are

$$\Theta_{\mathrm{CFE}}(\eta) = \frac{k_{\mathrm{buffer}}(T - T_{\mathrm{ref}})}{\sigma_{\mathrm{buffer}}E^2b^2} \quad \eta = \frac{y}{b} \quad Bi = \frac{h_{\mathrm{CFE}}b}{k_{\mathrm{buffer}}}$$
$$\kappa^2 = \lambda \frac{\sigma_{\mathrm{buffer}}E^2b^2}{k_{\mathrm{buffer}}}$$

and autothermal runaway occurs when $Bi\cos(\kappa) - \kappa\sin(\kappa) = 0$. The parabolic approximation to equation 38 obtained, using Galerkin's method is

$$\Theta_{\mathrm{CFE}}(\eta) \approx \frac{5(Bi + 2)(Bi + 3)}{2[5Bi(Bi + 3) - \kappa^2(2Bi^2 + 10Bi + 15)]} \cdot \left(1 - \frac{Bi\eta^2}{Bi + 2}\right) \quad (39)$$

and the predictions of both of these formulas are shown in Figure 11 for typical operating conditions. Because of the symmetry inherent in this problem, the exact and approximate solutions differ by only a few percent over most of their useful range. The agreement between equations 38 and 39 breaks down as κ approaches the critical value for autothermal runaway.

Axial Velocity Profile. Because the CFE does not normally use an anticonvective packing, the velocity profiles

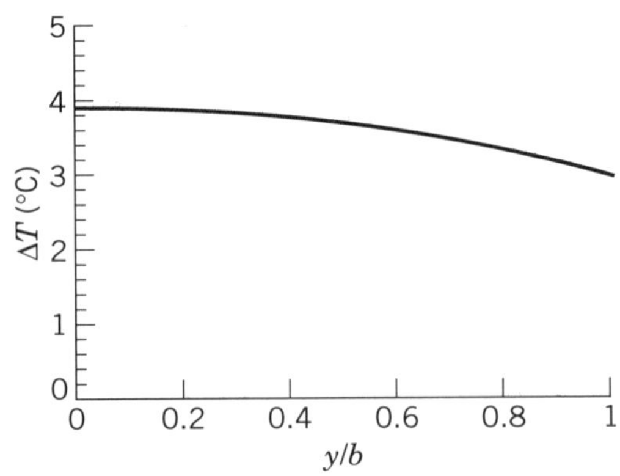

Figure 11. Predicted CFE temperature rise, in degrees centigrade from the center line at $y = 0$ to the wall at $y = b$. Since the bounding transverse walls dominate heat transfer resistance, there is a 3 °C rise above the coolant temperature at the transverse wall. The temperature difference across the slit is less than 1 °C.

in the axial and lateral directions have a significant impact on performance. Along the x axis, the velocity profile between the front and back plates is, to a first approximation, laminar and parabolic. However, because of the small temperature difference between the walls and the centerline, there is a slight distortion of the flow due to stable natural convection (42,43), which can either improve or degrade performance. Assuming that the base flow is parabolic, the perturbation to the flow is found by solving the Boussinesq (44,45) approximation to the equations of motion

$$\nu_{\text{buffer}} \frac{d}{dy}\left(\frac{dv}{dy}\right) + \rho_{\text{buffer}} g\beta(T - T_{\text{ref}}) = \frac{\Delta P'}{L} \tag{40}$$

where

$$\beta = \left[-\frac{d \log(\rho_{\text{buffer}})}{dT}\right)_{T=T_{\text{ref}}}$$

with no-slip conditions on the bounding surfaces

$$\text{at } y = \pm b \quad v(y) = 0 \tag{41}$$

and subject to the integral constraint that there is no net increase in the average velocity associated with natural convection, that is,

$$\int_{-b}^{b} v(y)dy = 0 \tag{42}$$

Note that P' is a perturbation pressure that determined by the integral constraint of equation 42. When the exact form for the temperature profile (see equation 46) is used, this system of equations yields the solution

$$v(y) = \frac{3}{2}\langle v\rangle\left((1 - \eta^2) + \frac{Gr}{Re} \cdot \frac{Bi\{3(\eta^2 - 1)[\sin(\kappa) - \kappa\cos(\kappa)] + 2\kappa[\cos(\kappa\eta) - \cos(\kappa)]\}}{3\kappa^5[Bi\cos(\kappa) - \kappa\sin(\kappa)]}\right) \tag{43}$$

while the polynomial approximation yields

$$v(y) = \frac{3}{2}\langle v\rangle\left\{(1 - \eta^2) + \frac{Gr}{Re} \cdot \frac{Bi(Bi + 3)[5(1 + \eta^2) - 6](1 - \eta^2)}{36[\kappa^2(2Bi^2 + 10Bi + 15) - 5Bi(Bi + 3)]}\right\} \tag{44}$$

where

$$Re = \frac{\rho_{\text{buffer}}\langle v\rangle b}{\nu_{\text{buffer}}} \quad \text{and} \quad Gr = \frac{\rho^2_{\text{buffer}} g\beta}{\nu^2_{\text{buffer}}}\frac{\sigma_{\text{buffer}} E^2 b^5}{k_{\text{buffer}}}$$

Under normal CFE operating conditions the ratio of the Grashof number to the Reynolds number in equations 43 and 44 is somewhat less than 1, and, as illustrated in Figure 12, the functional portion of the perturbation velocity never exceeds ± 0.01, so the contribution of natural convection in commercial CFEs, $b < 1$ mm, can be neglected

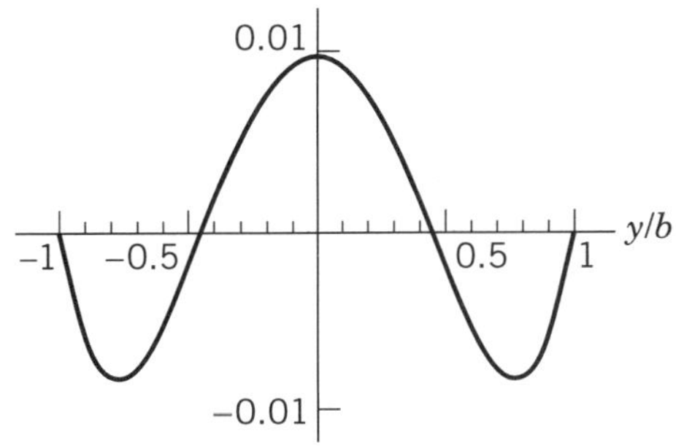

Figure 12. Predicted perturbation velocity profile due to stable natural convection in a slit. Both the exact solution, equation 43, and the polynomial approximation, equation 44, are shown here. The warmer fluid in the center of the slit rises while the cooler fluid near the wall falls.

when $Gr/Re < 1$. Note, however, that $Gr/Re \propto b^4$, so this term rapidly becomes important as the slit thickness is increased.

Lateral Velocity Profile. Because the transverse surfaces of the CFE carry a fixed charge, the electric field induces wall electroosmosis, which drives a bulk recirculating flow along the lateral axis. The shape and magnitude of this flow can be calculated from the equation of motion

$$\nu_{\text{buffer}} \frac{d}{dy}\left(\frac{dw}{dy}\right) = \frac{\Delta P''}{L} \tag{45}$$

together with the integral constraint

$$\int_{-b}^{b} w(y)dy = 0 \tag{46}$$

Assuming that the induced electroosmosis is identical on both transverse surfaces, perhaps due to application of a surface coating, a suitable set of first-kind boundary conditions is

$$\text{at} \quad y = \pm b \quad w(y) = w_{\text{eo}} \tag{47}$$

Solving the differential equation 45 subject to boundary conditions as in equation 47 and constraint from equation 46 yields the polynomial

$$w(\eta) = -\frac{3}{2} w_{\text{eo}}(1 - \eta^2) + (w_{\text{eo}} + \langle u\rangle) \tag{48}$$

where the electrophoretic velocity of the solute $\langle u\rangle$ has been tacked on to account for displacement of the solute relative to the electroosmotic flow. Note that when $w_{\text{eo}} = -u$, the term on the far right vanishes and the lateral flow becomes parabolic.

Solute Dispersion in the CFE. In practice, a multicomponent feed is injected into the CFE carrier flow through a capillary that extends over one-third to one-half of the transverse thickness of the chamber. In most instances it is possible to neglect the effects of molecular diffusion on dispersion because convective distortion of the sample band dominates band spreading. The criterion for neglecting molecular diffusion is that the time required for a molecule to diffuse from the centerline to the transverse will be much greater than the solutes' average holdup time in the chamber. This ratio is characterized by the dimensionless Graetz number

$$Gz = \frac{t_{\text{diffusion}}}{t_{\text{convection}}} = \frac{b^2/D_m}{L/\langle v\rangle} = \frac{\langle v\rangle b}{D_m}\cdot\frac{b}{L} \geq 10$$

which for proteins typically has values well in excess of 10.

Because this apparatus operates continuously and solutes are separated in space along the lateral axis rather than in time, it is appropriate to use spatial moments rather than temporal moments to characterize performance. If diffusive dispersion is ignored, the spatial moments can be found using the formula

$$M_n(L) = \left(\frac{L}{b}\right)^n C_o \int_{-\eta^*}^{+\eta^*} \frac{w(\eta)^n}{v(\eta)^{n-1}}\, d\eta \tag{49}$$

where η^* is the fraction of the transverse thickness over which sample injection extends. In this calculation the feed band is assumed to be infinitesimally thin along the lateral axis. Neglecting the effects of natural convection in equation 44, we can readily generate analytical expressions for the mean lateral displacement and spatial variance.

Table 6 gives analytical formulas for the mean lateral displacement, the spatial variance, the theoretical plate number, and the resolution in the CFE, taking into account only pure convective dispersion. Several points are brought to light in the moment analysis that are crucial to CFE operation. First, as is clear from the premultiplicative terms in the variance and the plate number, performance is optimized when $\langle u\rangle = -w_{eo}$, a condition that can be approximated by coating the transverse surfaces with a material that has a zeta potential similar to that of the key solutes being processed (2).

Second, the plate number N_{CFE} is a function neither of column length nor of the carrier fluid velocity, and is therefore not a function of residence time. This occurs because, like thermal dispersion, convective band spreading due to electroosmosis is faster than a diffusion-like process; that is, CFE variance increases with the first power of the residence time rather than the square root.

Third, as $\eta^* \to 1$, the variance increases rapidly, and, from a purely theoretical standpoint, if diffusive effects are ignored it becomes infinite in this limit. Performance degrades rapidly as η^* approaches 1, so that there is a stark trade-off between capacity and resolution. Although plate efficiencies in excess of $N = 20{,}000$ can be obtained in continuous operation, this is only possible under rather stringent conditions. However, for $\eta^* < 3/4$ and $\text{Gz} > 100$, the analysis given earlier should be sufficiently accurate to allow the user to optimize a given separation.

The contribution of the breadth of the sample injection stream to the variance is

$$\chi^2_{\text{injection}} = \frac{W^2}{12} \tag{50}$$

where W is the lateral breadth of the injected stream. This expression is readily incorporated into the calculation for the plate efficiency N by adding this contribution to the spatial variance in Table 6.

Figure 13 is a three-dimensional plot of N versus $\langle u\rangle$ and η^* for $w_{eo} = -0.01$, which shows how CFE performance varies with operating conditions. Here we see that N is greatest along the line $w_{eo} = -\langle u\rangle$ and for small values of η^*, but for reasonable injection capacities ($\eta^* > 1/3$) rarely exceeds 20,000 plates. In practice one should choose conditions that give at least baseline resolution, $R_{\text{CFE}} > 1$, of key components and differences in the mean lateral displacements z, which are at least as broad as the effluent ports.

Solute Dilution in CFE. For the CFE, the dilution factor that corresponds to that discussed with the GE systems is determined from the size of the sample injection stream as $F_{\text{dilution}} = \eta^{*-1}$, and typically ranges from 2 to 4 depending on the injection flow rate. However dilution by band smearing along the lateral axis in the CFE is often an important consideration and is relatively easy to estimate by using the formula

$$F_{\text{gross}} = \frac{1}{\eta^*}\,\frac{\sqrt{\chi^2_{\text{injection}} + \chi^2}}{\chi_{\text{injection}}} \tag{51}$$

where $\chi_{\text{injection}}$ is given by equation 50 and χ comes from Table 6. The numerator in equation 51 is proportional to the breadth of the product in the effluent stream, and the

Table 6. Statistical Expressions for the CFE Subject to Pure Convective Dispersion

Mean displacement	$z_{\text{mean}}(x) = x\,\dfrac{w_{eo}(\eta^{*3} - \eta^*) + 2\eta^*\langle u\rangle}{(3\eta^* - \eta^{*3})\langle v\rangle}$
Spatial variance	$\chi^2(x) = x^2\,\dfrac{(w_{eo} + \langle u\rangle)^2}{\langle v\rangle^2}\left(\dfrac{2}{3\eta^*(3 - \eta^{*2})}\log\left[\dfrac{1+\eta^*}{1-\eta^*}\right] - \dfrac{4}{(3-\eta^{*2})^2}\right)$
Plate number	$N_{\text{CFE}} = \dfrac{z^2_{\text{mean}}}{\chi^2} = \dfrac{3\eta^*(w_{eo}(\eta^{*2} - 1) + 2\langle u\rangle)^2}{2(w_{eo} + \langle u\rangle)^2\left[(3 - \eta^{*2})\log\left(\dfrac{1+\eta^*}{1-\eta^*}\right) - 6\eta^*\right]}$
Resolution	$R_{\text{CFE}} = \dfrac{\Delta z}{2\chi} \approx \sqrt{N_{\text{CFE}}}\left\lvert\dfrac{\Delta\langle u\rangle}{w_{eo}(\eta^{*2} - 1) + 2\langle u\rangle}\right\rvert$

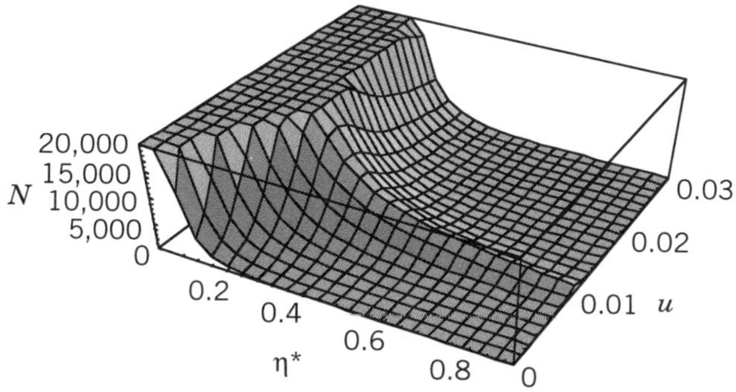

Figure 13. Isothermal plate efficiency N as a function of injection cross-section η^* and electrophoretic velocity u. N was computed for $w_{eo} = -0.01$ cm/s, $v = 1$ cm/s, $L = 100$ cm, and $2b = 0.1$ cm. In this figure it is assumed that a square injection capillary is used. This shows an optimal plate efficiency when $u = -w_{eo}$ but illustrates how small the operating regime is where plate efficiencies are greater than about 20,000 in the CFE. Note that this figure has been capped at $N = 20{,}000$.

denominator is proportional to the breadth of the sample injection stream. The premultiplicative factor $1/\eta^*$ is simply the dilution factor given earlier. As shown in Figure 14, minimum dilution occurs when $\langle u \rangle = -w_{eo}$, with the broadest operating region centered near $\eta^* = 1/3$. In practice, plate efficiency and dilution must be weighed against capacity.

Continuous-Flow Electrophoresis with Recycle (RCFE)

Effluent recycle was introduced for isoelectric focusing by Bier and Egen in 1979 (46). At that time it was believed that recycle could not be adapted to zone electrophoresis because of fundamental differences in the transport processes governing these two techniques. Specifically, isoelectric focusing is an equilibration process in which solutes placed in a pH gradient migrate to the point in the gradient where they have zero net charge. Separation depends primarily on the steepness of the pH gradient and the difference in isoelectric points (pIs) of the solutes. Zone electrophoresis, on the other hand, is a rate process in which solutes migrate at fixed velocities, so separation depends primarily on the length of the run, the electric field strength, and the difference in electrophoretic mobilities.

Recycle was first applied to zone electrophoresis in 1983 (47,48). The major obstacle that had to be overcome to accomplish this was to find a way to get two solutes with like charges but different electrophoretic mobilities to move in opposite directions. Our solution to this problem works in the following way: Two proteins with similar electrophoretic mobilities can be made to migrate in opposite directions by opposing the electrophoretic motion with a counterflow velocity set close to the average of the electrophoretic velocities of two key components.

Solutes migrate at their effective velocities (i.e., electrophoretic plus counterflow) until they reach the "regenerators" (see Figure 15) on either side of the recycle section. In the regenerators the counterflow of solvent is altered by means of the recycle lines, to keep solutes from migrating past the outlet port. At the regenerator on the left, the cross-flow rate is decreased; it is increased at the regenerator on the right and adjusted so that solutes cannot escape the offtake ports. With proper design, complete separation can be achieved with virtually no loss of biological activity.

The RCFE was set up so that on each cycle through the unit the solutes would separate slightly. However, after recycling through the chamber 100 or more times, the solutes are completely separated and can be continuously drawn off at either end of the slit with very low losses. The ultimate resolution of solutes in the RCFE depends largely on the number of recycle stages designed into the chamber. Because the design engineer has complete control of this variable, any degree of product purity can, in principle, be attained.

The primary advantage of this configuration is that it can process feeds in excess of 1 g/h and operates in true

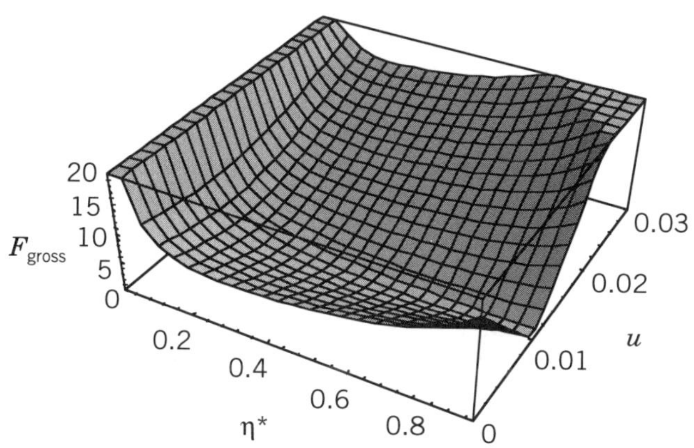

Figure 14. Gross dilution factor vs. η^* and u. This function gives the degree of dilution a solute will undergo during processing in the CFE. As seen in the center of the surface plot, under near-optimal operating conditions it is possible to obtain gross dilution factors below 4.

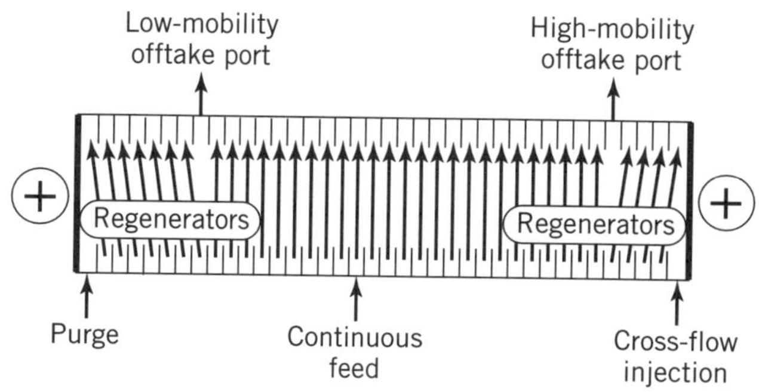

Figure 15. The RCFE with regenerators. The regenerators keep solute from migrating past the outlet ports, and this helps keep solute losses to a minimum. Note that opening up offtake ports automatically creates the regenerators and that the magnitude of the shift and the strength of the countercurrent flow is altered at each regenerator. The arrows within the chamber indicate the connection between the inlet and outlet ports. Note the change in plumbing in the right and lefthand regenerators.

continuous mode. Its major drawback is that it will split a multicomponent feed into just two products—one that contains components that are faster than the cross-flow and the other one having components that are slower than the cross-flow. Consequently, complex feeds will generally require a minimum of two passes to isolate the desired product: The first pass could be used to remove faster contaminants and the second pass to remove slower contaminants.

Separation is carried out in native buffer with background conductivities near 0.01 mho/cm (50 mM buffer), so low protein solubility is not generally a problem. Power requirements are kept low by using small electric fields (less than 40 V/cm), so the temperature rise in the chamber on a single pass is typically less than 5 °C. A multichannel heat exchanger removes joule heat from the recycle streams and sets the base temperature of the separation. In addition, the RCFE is hydrodynamically stable because the chamber is operated adiabatically with upflow.

The RCFE is not commercially available. However, commercial recycle IEF chambers, such as Rainin's MinipHor, can be readily modified to operate in RCFE mode.

Remarks. When the RCFE was first conceived, it was believed that zone electrophoresis, which can provide fine resolution of proteins in analytical equipment, would find its niche in the final polishing stages of industrial purification processes. In practice, this apparatus appears to be more useful in the early stages of downstream processing, where complex mixtures must be handled and emphasis is placed on removing solutes that are difficult to handle by other means, for example due to fouling or precipitation, or removing bulk proteins (such as serum albumin from blood plasma) inexpensively and without organic solvents or large amounts of precipitating salts.

DISCUSSION

Since 1980, the field of applied electrophoresis has undergone revolutionary changes, spurred in large part by the development and automation of capillary electrophoresis. Instrumentation is available for application of zone electrophoresis at scales up to pilot, so it is now possible to decide, on a rational basis, whether to incorporate electrophoresis as part of a purification train.

One could imagine, for instance, applying CE (along with analytical HPLC) early in process development to determine which contaminants could be removed from a product stream by electrophoresis and the optimum pH and buffer conditions at which this might be carried out. The results of this inexpensive and rapid preliminary screening step would allow the user to decide whether zone electrophoresis should be considered at all. As a rule of thumb, if the mobility difference between key components, typically the desired product and a major contaminant, measured by CE is greater than 25%, then zone electrophoresis may be considered further.

Following CE, one could scale up to native pGE in order to recover enough purified protein to check for product heterogeneity, activity, and toxicity, as well as to determine the amino acid sequence. When carried out in PAGE, the resolution achieved is boosted by the filtering power of the gel; however, separations conducted in native agarose gels can give insight to the presence (or absence) of problems with zone electrophoresis at larger scales. As before, one should look for a 25% difference in mobilities of key components before considering continuous zone electrophoresis as a possible process step. Mobility differences greater than 25%, as measured in native agarose gels, should yield a straightforward and clean separation by zone electrophoresis at pilot scale. This preparative step is somewhat more expensive and time-consuming than CE but is a better predictor of success or failure at larger scales. If an automated GE unit is not available, the user can always resort to vertical or horizontal slab gels.

Assuming the preparative step is successful, further scale-up requires a return to free-solution zone electrophoresis in either the CFE or the RCFE. (*Note*: In this step one can also use an RF3 or MiniPhor that has been outfitted for isotachophoresis. However, both the equipment itself and the protocols used must be modified for zone electrophoresis. The advantage of using this equipment is that it is commercially available and relatively inexpensive.) This phase of the study is again more expensive and time-consuming than previous ones, and should be carried out at processing rates between 10 and 100 mg/h. In free solution, the user is forced to confront potential problems that might arise as a result of heat generation or electroosmotic dispersion, hopefully for the last time during scale-up. As the conductivity of the buffer is reduced, the electric power and chamber cooling requirements decrease, and throughput generally increases because electrophoretic mobilities increase. However, as the conductivity is reduced, wall electroosmosis eventually increases to the point where the separation is ruined; an optimum conductivity should thus be sought.

In our experience, it is best to begin this search with a buffer that has a conductivity near 500 μmho/cm, which corresponds roughly to 10 mM Tris-acetate at pH 8.0, and raise the electrolyte concentration if it is necessary to reduce electroosmotic dispersion or to discourage protein aggregation. If the separation is successful at 500 μmho/cm, then the conductivity should be halved and the experiment repeated until a suitable electrolyte concentration is found. Keep in mind that whether Tris or more exotic electrolytes or additives are used, these materials are likely to constitute the major electrophoretic processing costs in zone recycle electrophoresis rather than, say, electric power or cooling.

Closing Remarks. Over the past two decades there have been a number of important developments in the area of analytical electrophoresis, the most important and most active of these centering on the commercialization of capillary electrophoresis. In addition, gel electrophoresis has metamorphosed from a labor-intensive laboratory technique to one where the instrumentation is automated and the gels are mass-produced, changes that significantly reduce operator training and improve reproducibility.

During this same period, preparative and larger-scale zone electrophoresis has remained largely stagnant, with few innovations surviving to commercialization. There is no agreement over the precise reasons for the lack of interest in electrokinetic processing, although it is clear that the biotech industry is convinced that the tools it already has, specifically the chromatographies, are superior to those currently available via electrophoresis. It will be nearly impossible to move this technology forward without the cooperation and support of the industry it will serve.

What will it take to change this mindset? In my opinion it will take at least one case study where electrophoresis shows itself to be clearly superior to existing separations technologies and offers a straightforward route to scale-up.

A number of different electrophoretic devices have not been discussed in this article. For example, neither Hoefer's Isoprime® (49–52) nor BioRad's Rotofor® (53,54) are discussed because they are designed strictly for isoelectric focusing, a topic treated elsewhere in this book. Margolis' Gradiflo® (10,55,56) is not treated both because it is not yet commercially available in this country and because I am not sufficiently familiar with that device to give it due justice.

CACE (57) and various other forms of electrochromatography (58–61) have been omitted here because they could not be adequately treated in the space allotted to this article. Other recent innovations by Shiue and Pearlstein (62), Lochmüller and Ronsick (63), and Shea et al. (64) were omitted from this article because they are still in the early stages of development.

NOMENCLATURE

Notation

a	CE capillary lumen radius
a_g	Glass outer radius
a_{gel}	Gel radius
a_i	Annulus inner radius
a_o	Annulus outer radius
a_p	Polyimide outer radius
A_c	Gel cross-sectional area
b	CFE slit half-thickness
Bi	Biot number
C	Solute concentration
C_o	Solute concentration in sample
$C(Z, t)$	Solute concentration at Z
d_{glass}	CFE glass wall thickness
d_i	Inner diameter of annulus cooling finger
d_o	Outer diameter of annulus glass tube
d_t	Diameter of elution tube
D_m	Molecular diffusivity in free solution
D_{gel}	Molecular diffusivity in gel
D_{CE}	CE dispersion coefficient
E	Electric field strength
$F_{dilution}$	Dilution factor
Gr	Grashof number
Gz	Graetz number
g	Gravitational acceleration
h	Slit thickness
h_{CE}	CE overall heat transfer coefficient
h_{CFE}	CFE overall heat transfer coefficient
h_{fluid}	Cooling fluid heat transfer coefficient
h_{pGE}	pGE heat transfer coefficient
$HETP$	Height equivalent to a theoretical plate
$hetp$	Reduced $HETP$
$J(Z, t)$	Solute flux at Z
J_0	Bessel function of first kind, order 0
J_1	Bessel function of first kind, order 1
k	Thermal conductivity
L	Elution tube length
N	Theoretical plate number
M_n	nth moment, averaged across thin axis
P	Pressure
Pe	Peclet number
q	Sample injection volume
Q	Elution flow rate
r	Radial coordinate
R	Resolution
Re	Reynolds number
S	pGE scale factor
T	Temperature
t	Time
t_{load}	CE sample injection time
u	Electrophoretic velocity
u_{max}	Maximum electrophoretic velocity in r
u_{min}	Minimum electrophoretic velocity in r
$u(r)$	Electrophoretic velocity profile in r
v	Hydrodynamic velocity
$v(r)$	Hydrodynamic velocity profile in r
w	CFE lateral velocity
w_{eo}	CFE electroosmotic wall velocity
W	Sample load width in CFE
x	Electrophoretic velocity
y	Transverse axis
Y_0	Bessel function of second kind, order 0
Y_1	Bessel function of second kind, order 1
z	Lateral or electrophoresis axis
Z	Gel column length

Greek Symbols

α	Thermal viscosity coefficient
β	Thermal volumetric expansion coefficient
δ	Dirac impulse distribution
χ^2	Spatial variance
κ	Autothermal parameter
λ	Eigenvalue or thermal conductivity coefficient
Λ	pGE geometric factor
θ	Dimensionless time

Θ	Dimensionless temperature
η	Dimensionless transverse or radial coordinate
μ	Electrophoretic mobility
μ_{eo}	Electroosmotic mobility
ν	Molecular viscosity
Φ	Electric potential
σ^2	Temporal variance
σ	Electrical conductivity
ρ	Fluid density
τ	Dimensionless time
ζ	Dimensionless distance along z

BIBLIOGRAPHY

1. R.J. Hunter, *Zeta Potential in Colloid Science*, Academic Press, New York, 1981.
2. A. Strickler and T. Sacks, *Ann. N.Y. Acad. Sci.* **209**, 497–514 (1973).
3. F.M. Everaerts, F.E.P. Mikkers, and Th.P.M. Verheggen, *Sep. Purif. Methods* **6**, 287–351 (1977).
4. P.H. Rhodes, R.S. Snyder, and G.O. Roberts, *J. Colloid Interface Sci.* **129**, 78–90 (1989).
5. J.F. Hoburg and J.R. Melcher, *Phys. Fluids* **20**, 903–911 (1977).
6. C.F. Ivory, W.A. Gobie, J.B. Beckwith, R. Hergenrother, and M. Malec, *Science* **238**, 58–61 (1987).
7. M. Bier, G.E. Twitty, and J.E. Sloan, *J. Chromatogr.* **470**, 369–376 (1989).
8. C.F. Ivory, *Electrophoresis* **11**, 919–926 (1990).
9. M. Faupel, B. Barazaghi, C. Gelfi, and P.G. Righetti, *J. Biochem. Biophys. Methods* **15**, 147–162 (1987).
10. Z.S. Horvath, G. Corthals, C.W. Wrigley, and J. Margolis, *Electrophoresis* **15**, 968–971 (1994).
11. D.A. Saville, *Phys. Rev. Lett.* **71**, 2907–2910 (1993).
12. J.C. Baygents and F. Baldessari, *Phys. Fluids* **10**, 301–311 (1998).
13. P.T. Noble, *Biotech. Progr.* **1**, 237–241 (1985).
14. J.B. Beckwith and C.F. Ivory, *Chem. Eng. Commun.* **54**, 301–331 (1987).
15. A.S. Cohen, D.R. Najarian, A. Paulus, A. Guttman, J.A. Smith, and B.L. Karger, *Proc. Nat. Acad. Sci. U.S.A.* **85**, 9660–9663 (1989).
16. A. Guttman, A.S. Cohen, D.N. Heiger, and B.L. Karger, *Anal. Chem.* **62**, 137–141 (1990).
17. J.C. Giddings, *Unified Separation Science*, Wiley, New York, 1991.
18. W.G. Kuhr and C.A. Monnig, *Anal. Chem.* **64**, 389–407R (1992).
19. Y. Xu, *Anal. Chem.* **65**(12), 426–433R (1993).
20. J.W. Jorgenson, in J.W. Jorgenson and M. Phillips eds., *New Directions in Electrophoretic Methods*, ACS Symposium Series 335, American Chemical Society, Washington, D.C., 1987, pp. 182–198.
21. E.D. Lynch and D.A. Saville, *Chem. Eng. Commun.* **9**, 201–211 (1981).
22. J.O.N. Hinckley, *J. Chromatogr.* **109**, 209–217 (1975).
23. A.E. Jones and E. Grushka, *J. Chromatogr.* **466**, 219–225 (1989).
24. W.A. Gobie and C.F. Ivory, *J. Chromatogr.* **516**, 191–210 (1990).
25. M. Albin, P.D. Grossman, and S. E. Moring, *Anal. Chem.* **65**, 489–497A (1993).
26. C. Schwer and F. Lottspeich, *J. Chromatogr.* **623**, 345–355 (1992).
27. M.P. Deutscher ed., *Guide to Protein Purification*, Academic Press, New York, 1990.
28. A.T. Andrews, *Electrophoresis: Theory, Techniques, and Biochemical and Clinical Applications*, Oxford, U.K., Clarendon Press, 1986.
29. B.D. Hames and D. Rickwood eds., *Gel Electrophoresis of Proteins: A Practical Approach*, Oxford University Press, New York, 1990.
30. J.R. Boyack and J.C. Giddings, *Arch. Biochem. Biophys.* **100**, 16 (1963).
31. A.G. Ogston, B.N. Preston, J.D. Wells, and J.McK. Snowden, *Proc. R. Soc. Lond.* **333**, 297–316 (1973).
32. P.G. De Gennes, *Scaling Concepts in Polymer Physics*, Cornell University Press, Ithaca, New York, 1979.
33. D.J. Harrison, A. Manz, Z. Fan, H., Ludi, and H.M. Widmer, *Am. Chem. Soc.* **64**, 1926–1932 (1992).
34. D.J. Harrison, K. Fluri, K. Seiler, Z. Fan, C.S. Effenhauser, and A. Manz, *Science* **261**, 895–897 (1993).
35. S.C. Jacobson, R. Hergenroder, L.B. Koutny, and J.M. Ramsey, *Anal. Chem.* **66**, 1114–1118 (1994).
36. S.C. Jacobson, R. Hergenroder, L.B. Koutny, and J.M. Ramsey, *Anal. Chem.* **66**, 2369–2373 (1994).
37. B.B. Rosenblum, *J. Liquid Chromatogr.* **14**, 1017–1024 (1991).
38. G.I. Taylor, *Proc. R. Soc. Lond.* **A219**, 186–203 (1953).
39. R. Aris, *Proc. R. Soc. A* **235**, 67–77 (1956).
40. C.F. Ivory, *J. Chromatography* **195**, 165–179 (1980).
41. K. Hannig, in J.R. Norris and D.W. Ribbons eds., *Methods in Microbiology*, vol. 5B Academic Press, New York, 1971, pp. 513–548.
42. S. Ostrach, *J. Chromatogr.* **140**, 187–195 (1977).
43. D.A. Saville and S. Ostrach, TR NASA-1 Final Report on Contract NAS-8-31349 Code 361, Dept. Chemical Engineering, Princeton University, Princeton, New Jersey, 1978.
44. S. Chandrasekhar, *Hydrodynamic and Hydromagnetic Stability*, Dover Publications, New York, 1961.
45. J.M. Mihaljan, *Astrophys. J.* **136**, 1126–1133 (1962).
46. M. Bier and N. Egen, in H. Haglund, J.G. Westerfield, and J.T. Ball eds., *Electrofocus / 78*, Elsevier/North Holland, New York, 1979, pp. 35–48.
47. C.F. Ivory, W.A. Gobie, and R.S. Turk, in H. Hirai ed., *Electrophoresis '83*, Walter de Gruyter, Berlin, 1984, pp.
48. W.A. Gobie, J.B. Beckwith, and C.F. Ivory, *Biotech. Progr.* **1**, 60–68 (1985).
49. P.G. Righetti, B. Barazaghi, M. Luzenna, G. Manfredi, and M. Faupel, *J. Biochem. Biophys. Methods* **15**, 199–206 (1987).
50. P.G. Righetti, E. Wenisch, and M. Faupel, *J. Chromatogr.* **475**, 293–309 (1989).
51. P.G. Righetti, M. Faupel, and E. Wenish, in A. Chrambach, M.J. Dunn, and B.J. Radola eds., *Advances in Electrophoresis, vol. 5*, VCH, New York, 1991, pp. 159–200.
52. P.G. Righetti and A. Bossi, *Anal. Biochem.* **247**, 1–10 (1997).
53. N.B. Egen, W. Thormann, G.E. Twitty, and M. Bier, in H. Hirai ed., *Electrophoresis '83*, Walter de Gruyter, Berlin, 1984, pp. 547–550.
54. U.S. Pat. 4,588,492 (May 13, 1986), M. Bier (to University Patents, Inc., Westport, Conn.).

55. J. Margolis, G. Corthals, and Z.S. Horvath, *Electrophoresis* **16**, 98–100 (1995).
56. G. Corthals, J. Margolis, K.L. Williams, and A.A. Gooley, *Electrophoresis* **17**, 9–11 (1996).
57. P.H. O'Farrell, *Science* **227**, 1586–1589 (1985).
58. S.R. Rudge and M.R. Ladisch, in J.A. Asenjo and J. Hong eds., *Separation, Recovery, and Purification in Biotechnology*, ACS Symposium Series 314, American Chemical Society, Washington, D.C., 1986, pp. 122–152.
59. S.R. Rudge, S.K. Basak, and M.R. Ladisch, *AIChE J.* **39**, 797–808 (1993).
60. S.K. Basak and M.R. Ladisch, *AICHE J.* **41**, 2499–2507 (1995).
61. S.K. Basak, A. Velayudhan, K. Kohlmann, and M. Ladisch, *Chromatography* **707** 69–76 (1995).
62. M.-P. Shiue and A.J. Pearlstein, *J. Chromatogr.* **707**, 87–103 (1995).
63. C.H. Lochmüller and C. Ronsick, *Anal. Chim. Acta* **249**, 297–302 (1991).
64. L.D. Shea, D.L. Feke, and U. Landau, *Biotechnol Progr.* **10**, 246–252 (1994).

ENERGY METABOLISM, MICROBIAL AND ANIMAL CELLS

MICHAEL BUTLER
RICHARD SPARLING
University of Manitoba
Winnipeg, Canada
XINFA XIAO
Apotex Fermentation, Inc.
Winnipeg, Canada

KEY WORDS

Energy charge
Glycolysis
Glutaminolysis
Heterolactic fermentation
Metabolic engineering
Methanogenesis
Oxidative phosphorylation
Pentose phosphate pathway
Secondary metabolites
TCA cycle

OUTLINE

INTRODUCTION

Every cell has a requirement for energy metabolism to maintain growth as well as to synthesize macromolecules or metabolic by-products. In the development of an industrial fermentation process it is important to be able to channel the metabolic capabilities of a selected cell to the production of a desired product. In this quest it is necessary to have a full understanding of the energy metabolism of the cell in order to be able to control the fermentation parameters to meet the cellular requirements and maximize the yield of the end product.

In this article we consider the energy metabolism of three quite distinct microbial cells—bacterial, fungal, and animal—all of which are used in large-scale commercial fermentation processes. The inclusion of animal cells as a category of microbial cells may be surprising to some, but in terms of their behavior as independent cells in homogeneous culture operations they do not differ from bacterial or fungal cells. Of course it has to be recognized that all cells used in industrial processes are likely to be selected and genetically altered in a way that marks their genotype as quite different from that found in wild strains or of cells contained within a multicellular organism.

In this article we point out the similarities and differences in energy metabolism of these cells and the aspects of the metabolism that are important for control in production processes of selected commercial products.

Common Pathways of Energy Metabolism for Bacterial, Fungal, and Animal Cells

Catabolic pathways provide energy for biosynthesis and in most instances also provide reducing equivalents for carbon reduction reactions. There are three major groups of compounds that link anabolic and catabolic pathways: adenosine phosphates (AMP, ADP, ATP), which link energy yielding and energy requiring reactions; nicotinamide adenine dinucleotide (NAD, NADH); and nicotinamide adenine dinucleotide phosphate (NADP, NADPH). The latter two link oxidative reactions of catabolic pathways to reductive reactions. The adenylate energy charge is an indicator of the pool of high-energy phosphate bonds carried on adenosyl residues:

$$\frac{[\text{ATP}] + (1/2)[\text{ADP}]}{[\text{ATP}] + [\text{ADP}] + [\text{AMP}]}$$

This index could theoretically vary between 0 (100% AMP) and 1 (100% ATP). Measurements of this charge have been shown to correlate well with the cells' physiological state (1). The energy charge of actively growing cells ranges from 0.80 to 0.95 but drops significantly when energy substrates are depleted.

From a biotechnological point of view, when the product of interest is derived from biosynthetic pathways, the adenylate energy charge is of specific importance because it participates in the regulation of metabolic pathways. There is a relationship between the energy charge and reaction rates of regulated biosynthetic (energy-requiring) and catabolic (energy-regenerating) enzymes and pathways (2). The activity of biosynthetic enzymes is low and the activity of catabolic enzymes high, at low energy charges (<0.8). The rate of protein synthesis declines in response to decreases in energy charge (3).

Because of the involvement of a variety of oxidative and reductive pathways using nicotinamides as the electron carrier, NADH/NAD and NADPH/NADP ratios, also known as the catabolic and anabolic reduction charges, respectively, have also been measured for their potential as a means of measuring the physiological state of cells (1). In *Escherichia coli*, the catabolic energy charge does not vary significantly in response to the various growth conditions tested; however, the anabolic reduction charge decreases under carbon and energy source starvation and decreases transiently under conditions of O_2 starvation (4). Changes in anabolic reduction charge may have an impact on rates of biosynthesis as well.

The Embden–Meyerhof–Parnas (EM) pathway, the pentose-phosphate (PP) pathway, and the Entner–Doudoroff (ED) pathway are three routes for the utilization of hexoses such as glucose. The EM and PP pathways are widespread in fungi and animal cells. The ED pathway is common in bacteria and in some fungi such as *Tilletia caries* and *Caldariomyces fumago* (5,6).

The EM pathway takes place in the cytoplasm and is the most common means of energy production (Fig. 1). Two molecules of ATP are generated in this pathway, whereas only one is generated the ED pathway (6). Three enzymes are involved in the initial glucose phosphorylation step of the EM pathway: hexokinase P-I, hexokinase P-II, and glucokinase. Hexokinase acts on fructose as well as glucose and mannose. These enzymes are generally present in fungal, bacterial, and animal cells. However, doubt has been expressed as to the existence of the EM pathway in some fungi such as *Candida* 107 and *Rhodotorula glutinis* because of the low activity of phosphofructokinase (7).

The ED pathway was first discovered in *Pseudomonas* and serves as a mechanism of glucose breakdown in the absence of phosphofructokinase (Fig. 2). In this pathway, glucose-6-phosphate is oxidized to 6-phosphogluconate

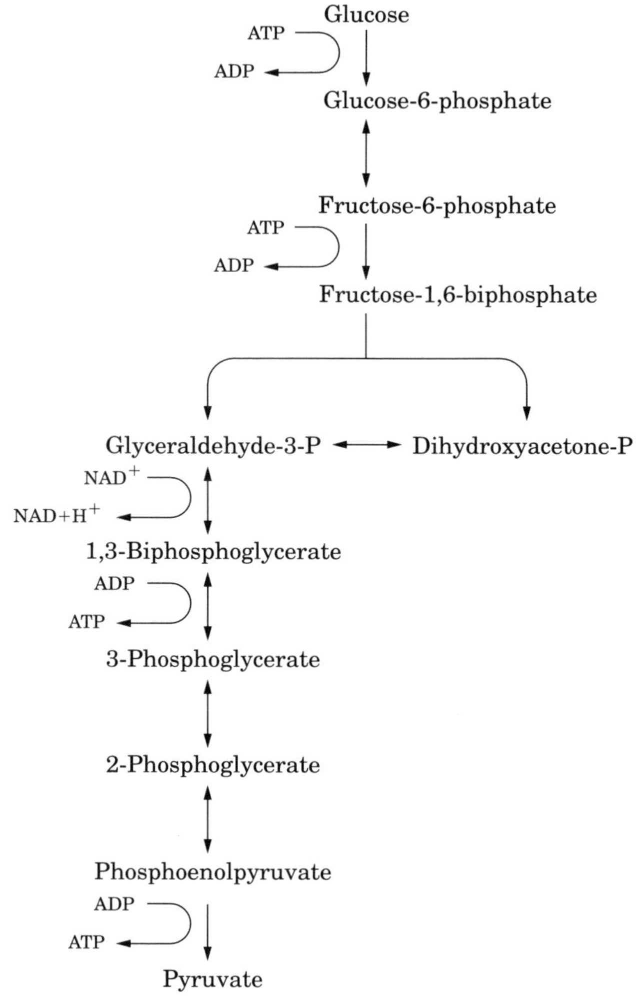

Figure 1. Embden–Meyerhof–Parnas pathway. This is the glycolysis pathway in which 1 mol glucose is degraded to 2 mol pyruvate. The net energy gain is of 2 mol ATP. The pyruvate is either reduced to lactate under anaerobic conditions or converted into acetyl-CoA for entry into the TCA cycle.

with the enzyme glucose-6-phosphate dehydrogenase. The presence of two other key enzymes (6-phosphogluconate dehydratase and 2-oxo-3-deoxy-6-phosphogluconate aldolase) allow further degradation to the three-carbon metabolites, pyruvate and glyceraldehyde-3-phosphate.

The PP pathway can be considered a side branch of the EM pathway that exists in fungal, bacterial, and animal cells (Fig. 3). Its principle role is the production of reducing power for biosynthetic reactions including nucleic acid and cell wall biosynthesis. In reductive biosynthesis, such as in fungal sporulation, the PP pathway is very important. Mutations of the PP pathway enzymes cause drastic morphological effects (6). Ribose-5-phosphate and erythrose-4-phosphate are produced from the PP pathway and are precursors for the synthesis of nucleotides and aromatic compounds, respectively (5).

The phosphoketolase (PK) pathway is an alternative pathway for sugar breakdown in fungal and bacterial cells (Fig. 4) (6). The key enzyme of the pathway (phosphoke-

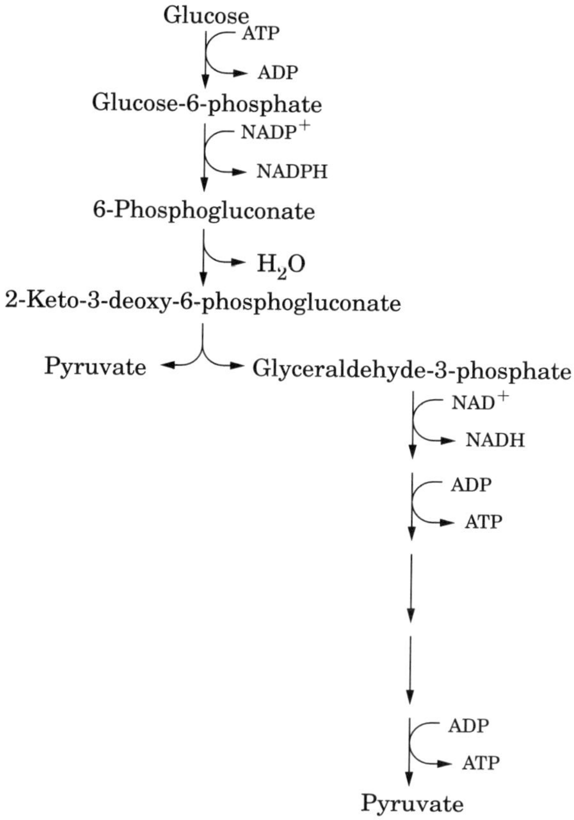

Figure 2. Entner–Duordoff pathway. Glucose is degraded in the absence of phosphofructokinase to pyruvate and glyceraldehyde-3-phosphate. The latter can be degraded further to pyruvate by the enzymes of glycolysis. This pathway is less efficient that glycolysis, yielding only 1 mol ATP/mol glucose.

tolase) converts a pentose such as xylulose-5-phosphate into acetyl-phosphate and glyceraldehyde-3-phosphate. The pathway is characteristically observed in heterolactic bacteria such as *Leuconostoc*, in which glucose may be converted into lactate, ethanol, and carbon dioxide.

Respiration in eukaryotes involves the complete oxidation of carbon-containing molecules to CO_2 and H_2O (Fig. 5). This includes three interrelated processes: the tricarboxylic acid (TCA) cycle, electron transport, and oxidative phosphorylation. These aerobic processes occur in the mitochondrion. The TCA cycle is the most efficient metabolic pathway for ATP generation (9,10) and is much more efficient than anaerobic metabolism. Using the EM pathway together with the TCA cycle under aerobic conditions, 30 to 38 molecules of ATP can be generated from one molecule of glucose, depending upon the degree of coupling of phosphorylation to the electron transport chain. By comparison, the EM pathway generates only 2 molecules of ATP from one molecule of glucose. Inorganic phosphate (P_i) supply is essential for ATP production. According to the energy requirements of cells, a mitochondrial inorganic phosphate carrier (P_ic) delivers P_i to the mitochondrial ATP synthase complex very rapidly, with a rate much higher than the rates of most mitochondrial reactions (10).

The TCA cycle is initiated when the pyruvate resulting from glycolysis is transported into the mitochondria and is converted to acetyl-CoA. Acetyl-CoA may also be formed from acetate by acetyl-CoA synthetase, and from the β-oxidation of fatty acids. When the acetyl group is transferred to oxaloacetate, the initial compound of the TCA cycle (citrate) results and is metabolized by a series of enzymatic reactions leading to the regeneration of oxaloacetate and production of CO_2, with the release of metabolic energy (5,8).

The TCA cycle is a central hub of both catabolic and anabolic metabolism. In addition to its role in the oxidation of substrates to generate energy, the TCA cycle provides important intermediates for biosynthesis. Glutamic acid, aspartic acid, and glycine are all directly derived from the TCA cycle and glyoxylate cycles, and they are important precursors for biosynthesis of other amino acids and nucleotides. The intermediates of the TCA cycle also contribute to the biosynthesis of secondary metabolites. The result of these biosynthetic processes is the depletion of intermediates from the cycle, and the TCA cycle would soon come to a halt without replenishment of intermediary metabolites. This may be accomplished through the fixation of CO_2 into oxaloacetate with phosphoenolpyruvate from glycolysis (8).

The glyoxylate cycle forms an apparent shortcut across the TCA cycle in fungi and some bacteria, linking isocitrate, succinate, and malate (Fig. 6). The net effect of this

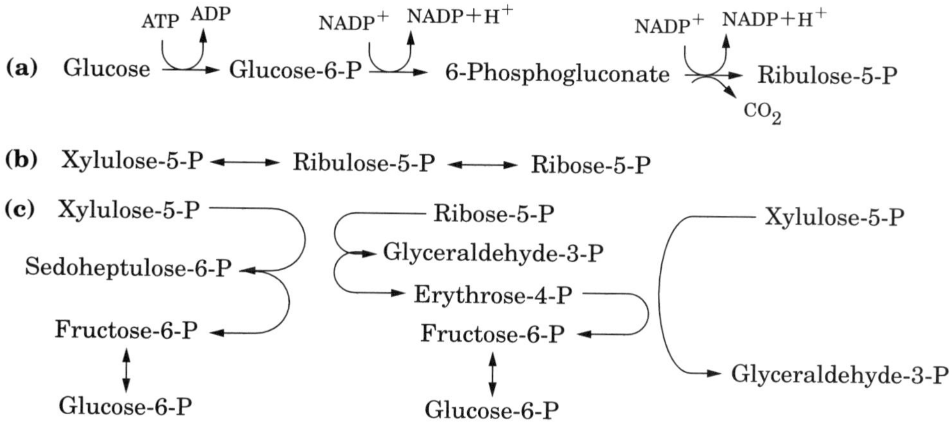

Figure 3. Pentose-phosphate pathway. This pathway is divided into three parts: (**a**) An oxidative sequence of reactions in which glucose is converted to the five-carbon, ribulose-5-phosphate. (**b**) Epimerization and isomerization reactions in which the important nucleic acid precursor, ribose-5-phosphate is formed. (**c**) A nonoxidative sequence of transaldolase and transketolase reactions.

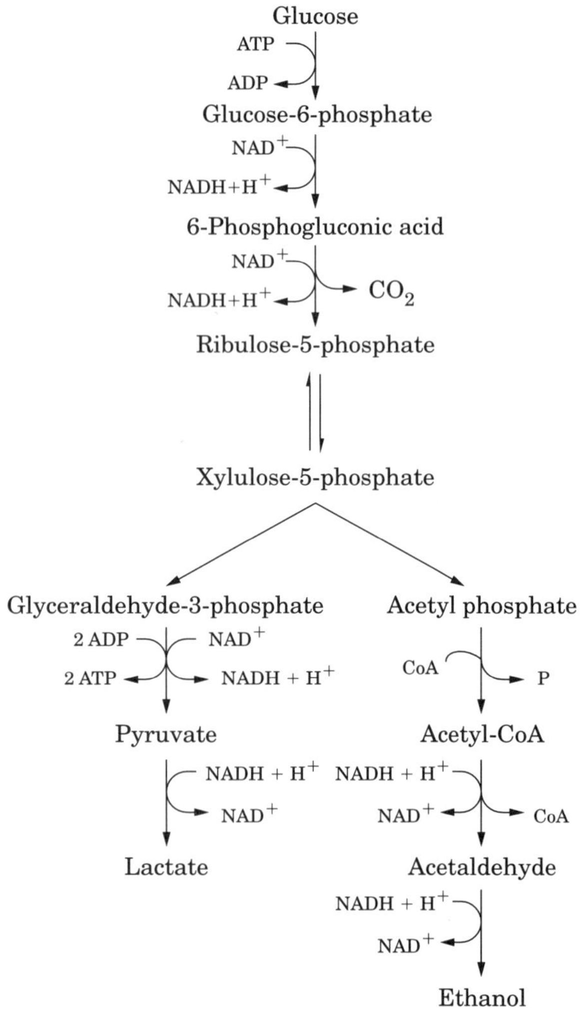

Figure 4. Phosphoketolase pathway. This pathway allows heterolactic fermentation in certain bacteria and fungi. The key enzyme is phosphoketolase, which degrades the five-carbon xylulose into a two-carbon, acetyl and three-carbon, glyceraldehyde. Lactate and ethanol are carbon products of the pathway. The net energy gain is 1 mol ATP/mol glucose utilized.

pathway is to produce one mole of succinate from two moles of acetate. The two enzymes required for this are isocitrate lyase and malate synthase. The pathway is called anaplerotic (from the Greek for "filling up") and serves to replenish the intermediates of the TCA cycle. In turn this may allow the TCA cycle intermediates to be used as precursors for biosynthetic reactions.

Mitochondrial reactions are essentially similar in plant, animal, and fungal cells. Electron transport and oxidative phosphorylation represent the final stages in the flow of electrons from organic substrates to oxygen (Fig. 7). Hydrogen atom pairs resulting from enzymatic dehydrogenation of certain intermediates of carbohydrate, fat, and protein catabolism, together with those from the TCA cycle, enter the electron transport chain (ETC), located in the inner mitochondrial membrane. The electrons of the hydrogen atom pairs are transferred to carrier components of the chain while the accompanying hydrogen ions enter the aqueous environment. The electrons are transferred by a series of oxidation–reduction reactions along the chain of redox components as far as cytochrome oxidase (cytochrome aa_3), which catalyzes electron transfer to oxygen. Each oxygen atom accepts two electrons and takes up two hydrogen ions from the environment to form water. At three sites in the ETC, oxidative phosphorylation may occur. That means ADP is phosphorylated, resulting in the formation of up to three ATPs from each electron pair transported to oxygen. For each NADH that transfers its electrons to the ETC, a maximum of three molecules of ATP result; for each $FADH_2$, a maximum of two ATP molecules result (5,8).

Besides the ubiquitous cytochrome-dependent and ATP-generating path for the oxidation of NADH + H^+, there are alternative respiration pathways. An alternative pathway, which branches from ubiquinone and transfers electrons via an unknown oxidase toward oxygen, has been described in all major classes of fungi (11). This alternative pathway is cyanide- and azide-insensitive but can be inhibited by salycilhydroxamate (SHAM). It transports electrons to oxygen without proton transport, and therefore without phosphorylation of ADP to form ATP; only heat is generated. Another alternative pathway is cyanide-insensitive and azide-sensitive and is also believed to be mitochondrial and to lack proton transport capability. This pathway accepts electrons from external NADH but not from succinate, implying lack of access to the ubiquinol pool, but whether it has access to internal mitochondrial NADH is not known (6).

In bacteria and fungi, respiration may function with an electron acceptor other than oxygen. For example, nitrate may be used as an electron acceptor and reduced to nitrite by nitrate reductase. This process, which is often called dissimilatory nitrate reduction, is coupled to the generation of ATP. Denitrification involves the complete reduction of nitrate to nitrogen and prevents the accumulation of potentially toxic levels of nitrite. Denitrification is carried out by *Pseudomonas* or *Bacillus* and also in the fungi, *Fusarium oxysporum* and *Cylindrocarpon tonkenense*. The respiratory substrates such as malate, pyruvate, succinate, and formate are effective donors of electrons. Rotenone, antimycin A, and thenoyltrifluoroacetone inhibit the reductase activity (12).

The range of energy substrates and catabolic pathways in bacteria is extremely wide. These pathways may lead to the generation of ATP or to primary ion gradients that are easily interconverted through the ubiquitous presence of an ion-coupled ATPase or synthase (13). The large variety of metabolic pathways available in prokaryotic organisms has led to their use in the production of a range of metabolic end products and enzymes.

NUTRIENTS REQUIRED FOR GROWTH

Bacteria and Fungi

Although some bacteria are autotrophic and may use carbon dioxide as a carbon source, many are heterotrophic and require some organic carbon for growth in culture. It has

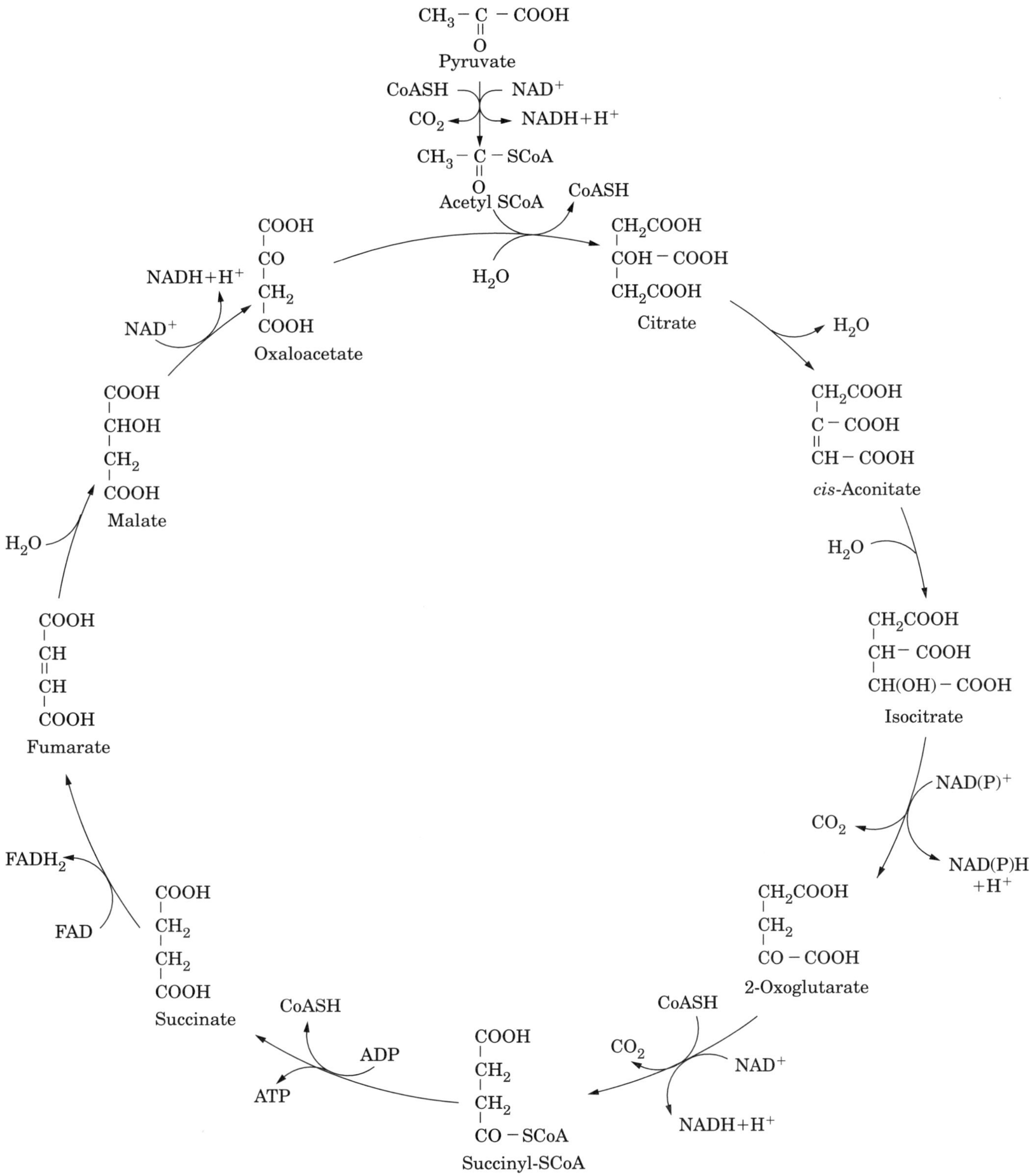

Figure 5. Tricarboxylic acid cycle. The TCA pathway operates in the mitochondria of eukaryotes. It allows the complete oxidation of pyruvate, which enters the cycle from glycolysis. The reduced coenzymes are reoxidized via the electron transport chain. The complete aerobic metabolism of glucose via glycolysis and the TCA cycle yields between 30 and 38 mol ATP/mol glucose, depending on the extent of the coupling of phosphorylation in the electron transport chain.

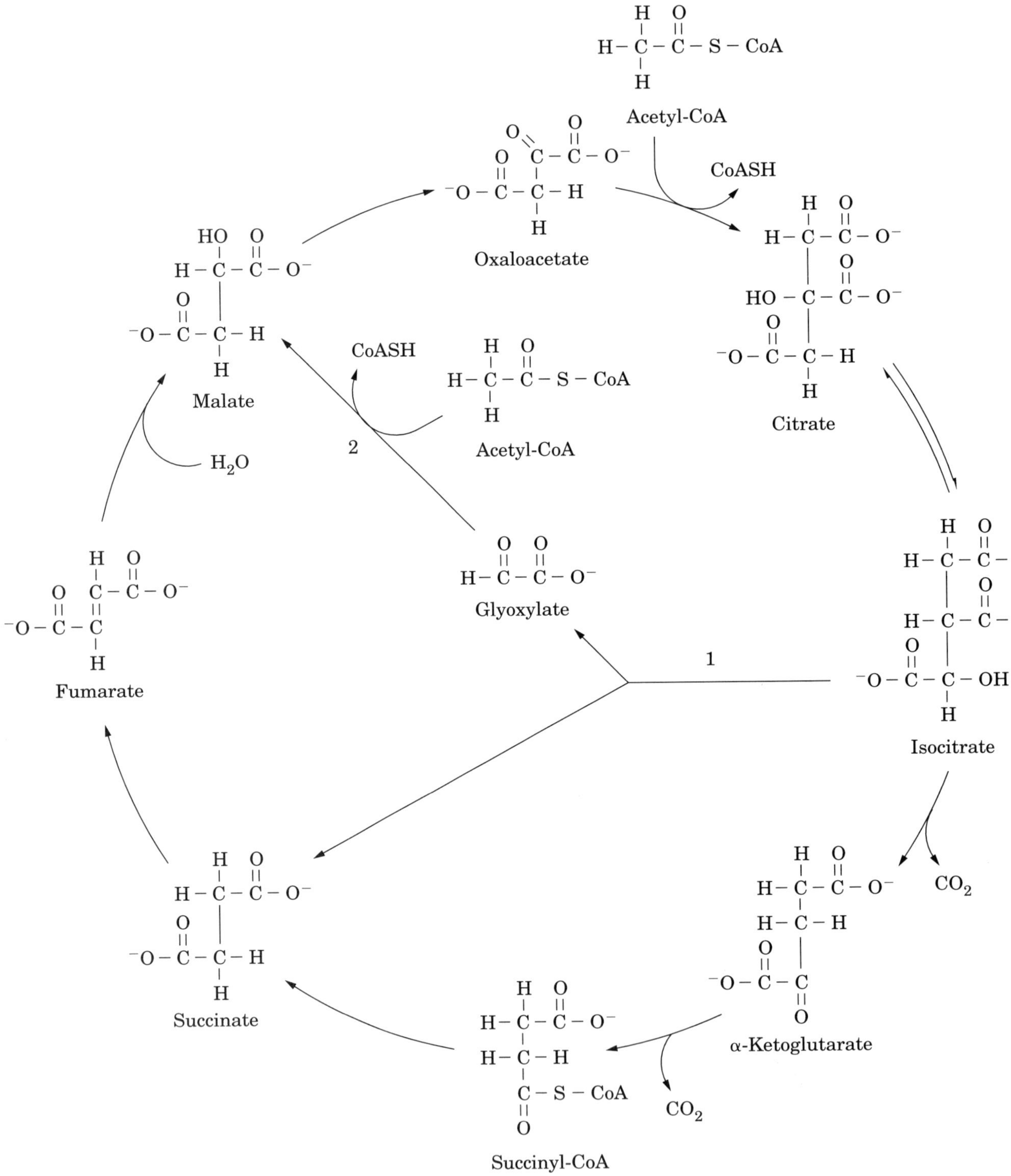

Figure 6. Glyoxalate pathway. The presence of the two enzymes, isocitrate lyase (1) and malate synthase (2) in some bacteria and fungi enables the replenishment of TCA cycle intermediates, which may be used as precursors for biosynthetic pathways. This is termed an anaplerotic pathway.

been common in laboratory cultures to use non–chemically defined sources of organic carbon such as peptones, meat extract, or yeast extract. However, to maintain consistency in large-scale cultures it is important to use a chemically defined nutrient broth consisting of an energy source in an isotonic salt solution.

Glucose and glucose-yielding carbohydrates are the most commonly used carbon energy sources for bacterial or fungal cell growth. With the breakdown of glucose through the pathways of glycolysis, fermentation, or respiration, energy and a series of small compounds are generated to support growth and biosynthesis. ATP is the energy carrier mediating between the energy donor and energy-requiring reactions. Hexoses can be converted to glucose, and pentose can be metabolized through alternative glycolytic pathways. Fatty acids, amino acids, organic acids, and alcohols enter the major catabolic pathways at various points (6,8).

Because of the extensive functional biosynthetic pathways in bacteria and fungi, the energy source and culture medium are relatively simple. Some examples of single energy sources for fungal producer cells are shown in Table 1.

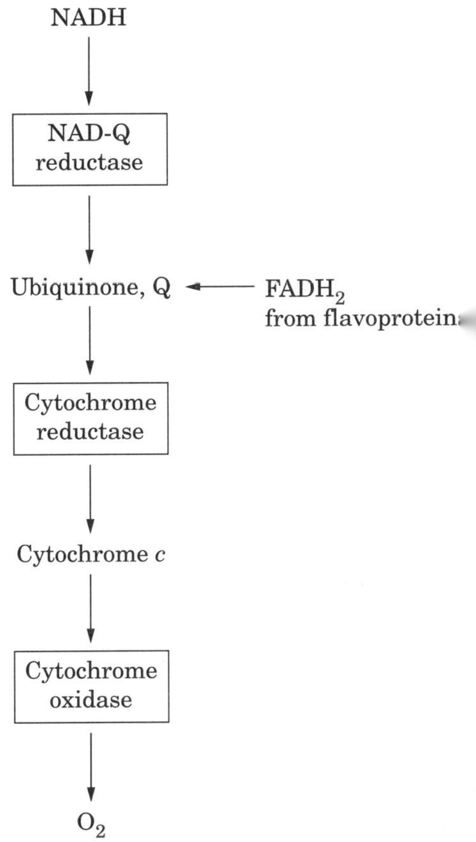

Figure 7. Electron transport chain. In this pathway electrons are transferred from NADH to oxygen through a chain of three large protein complexes (NAD-Q reductase, cytochrome reductase, and cytochrome oxidase). Ubiquinone and cytochrome *c* are relatively small molecules that shuttle between these complexes.

Unusual Energy Sources

Methanol or alkanes can be used as a sole source of carbon and energy by some species of bacteria and fungi. Methanol is oxidized to formaldehyde, which is further metabolized by two pathways. The first path oxidizes formaldehyde to formic acid and then to CO_2, with the generation of ATP and NADH. The second path combines formaldehyde with xylulose-5-phosphate to form dihydroxyacetone (DHA) and glyceraldehyde-3-phosphate. This path requires ATP to phosphorylate the DHA and provides intermediates for the pentose cycle or the EM pathway. The enzymes DHA synthase and DHA kinase are unique to this pathway and are formed only with methanol utilization. The *n*-alkanes are oxidized to corresponding alkanols by a monooxygenase enzyme complex containing cytochrome P450 and NADPH cytochrome P450 reductase. Peroxisomal enzymes carry out the remaining degradation to intermediates that are transferrable to the mitochondria. NAD-long-chain alcohol and aldehyde dehydrogenase further oxidize the alkanols to fatty acids (9). *Candida tenuis* was reported to be able to metabolize the whole range of fuel hydrocarbons. In a mixture of them, the relatively short-chain-length *n*-alkanes were used first, and then progressively longer *n*-alkanes. When all *n*-alkanes were used up, methylalkanes were exploited, but only relatively slowly (19).

Some strains of *Candida tropicalis, Rhodotorula* sp., *Pullularia pullulans*, and *Oidium* can use phenol at low concentrations as the single source of carbon and energy. *Aspergillus fumigatus* can grow on *p*-cresol or 4-ethylphenol as its sole carbon and energy source. *p*-Cresol is metabolized by two different routes, both of which converge on protocatechuate as the ring-fission substrate. One of these proceeds by initial hydroxylation of the methyl group of *p*-cresol and its further oxidation to 4-hydroxybenzoic acid, followed by ring hydroxylation; the other involves ring hydroxylation of *p*-cresol to form 4-methylcatechol as the first step, followed by oxidation of the methyl group. The protocatechuate is cleaved by *ortho*-fission, leading to formation of 3-oxoadipate. In contrast, the ring-fission substrate in the pathway for 4-ethylphenol degradation is hydroxyquinol (1,2,4-trihydroxybenzene), which is cleaved by *ortho*-fission, leading to maleylacetate. The pathway involves hydroxylation of the methylene group of 4-ethylphenol, followed by oxidation to give 4-hydroxyacetophenone. This undergoes a Baeyer–Villiger type of oxygenation to form the ester, 4-hydroxyphenyl acetate. Hydrolysis of the ester gives quinol (hydroquinone), which is converted to the ring-fission substrate by hydroxylation (20).

Toluene can be used as a sole source of carbon and energy by *Cladosporium sphaerospermum*, which is the first eukaryotic organism found to do so (21). This fungus was isolated from a biofilter used to remove toluene from waste gases. The oxygen consumption rates, as well as the measured enzyme activities, of *C. sphaerospermum* indicate that toluene is degraded by an initial attack on the methyl group.

The Composition of Animal Cell Culture Media

The substrate requirements for growth and production from animal cells is far more complex than for bacteria or

Table 1. Energy Sources for Producer Fungal Cells

Product	Producer	Source of energy and carbon	g/L	Reference
Citric acid	*Aspergillus niger*	Sugar	125–200	14
Itaconic acid	*Aspergillus terreus*	Glucose	50–150	14
Gluconic acid	*Aspergillus niger*	Dextrose	150–200	14
Penicillin G/V	*Penicillium chrysogenum*	Glucose	0–10 or 5–15	
		Lactose	20–50 or 15–60	15
Lovastatin	*Aspergillus terreus*	Dextrose	45	16
Compactin	*Penicillium citrinum*	Glucose	50	
		Glucose/maltose	10/40	17
Cyclosporin A	*Tolypocladium inflatum*	Maltose, glucose, etc.	30–50	18

fungi. Prior to the 1950s animal cells were grown in culture in undefined biological fluids, with varying degrees of success. Subsequently, chemically defined culture medium was based on the analysis of plasma and other growth-supporting biological fluids (22). The reduction of the defined components to minimum concentrations then led to "minimal essential media" such as that developed by Eagle in 1955 (23). Similar media formulations such as Dulbecco's modification of Eagle's medium (DMEM), Ham's F-12, and RPMI were later developed to promote the growth of specific cell lines. Table 2 shows the composition of DMEM as an example of a typical culture medium suitable for the growth of a wide range of animal cell lines.

The nature and concentration of the carbon and nitrogen substrates in the culture medium affect both the energy status and metabolism of the cells (Fig. 8). Medium generally contains both a carbohydrate source and a range of amino acids to satisfy cellular growth requirements. Most standard media formulations would be expected to support cell growth to a maximum density of $1–2 \times 10^6$ cells/ml in batch culture. The limitation to growth is normally a reflection of the deterioration of the culture medium due to nutrient depletion or the accumulation of toxic metabolites. The specific factor affecting the final cell yield may vary with cell line, media, and growth conditions. However, the identification of such factors and an understanding of the interaction between the cells and the media components is essential for designing a fully optimized and controllable culture system.

Carbohydrate Source. Glucose is the most commonly used carbohydrate source for animal cells and is normally included in culture at a concentration between 5 and 25 mM. At low glucose concentration (<25 μM) its primary function is to provide ribose for nucleic acid synthesis (24). It has also been shown that glucose may not be necessary if a source of pentose sugar is provided in the culture. Uridine and inosine have been found suitable, the carbohydrate requirement being derived from the ribose moiety of these nucleosides (25).

The specific glucose consumption rate of cells increases with the glucose concentration of the medium up to a maximum rate that occurs at around 5 mM glucose (24). The specific growth rate and antibody production rate of hybridomas are independent of glucose concentration over a wide range. The specific glucose consumption rate may vary with growth phase, culture pH, or cell type. During

Table 2. Dulbecco's Modification of Eagle's Medium (DMEM): An Example of a Basal Medium Commonly Used for the Growth of Animal Cells in Culture

Inorganic salts (mM)	
NaCl	110.30
KCl	5.40
$CaCl_2 \cdot 2H_2O$	1.80
$MgSO_4 \cdot 7H_2O$	0.80
$NaH_2PO_4 \cdot 2H_2O$	0.91
$NaHCO_3$	44.00
L-Amino acids (mM)	
Arginine·HCl	0.40
Cystine·diNa	0.20
Glutamine	4.00
Glycine	0.40
$Histidine \cdot HCl \cdot H_2O$	0.20
Isoleucine	0.80
Leucine	0.80
Lysine·HCl	0.80
Methionine	0.20
Phenylalanine	0.40
Threonine	0.80
Tryptophan	0.08
Tyrosine·diNa	0.40
Valine	0.80
Trace elements (μM)	
$Fe(NO_3)_3 \cdot 9H_2O$	0.25
Vitamins / cofactors (μM)	
Choline·Cl	28.60
Folic acid	9.10
Inositol	38.90
Nicotinamide	32.80
Pantothenate·Ca	8.40
Pyridoxal·HCl	19.60
Riboflavin	1.10
Thiamine·HCl	11.90
Other components	
Phenol red (uM)	26.55
Glucose (mM)	25.00
Pyruvate·Na (mM)	1.00
CO_2 (gas phase)	10%

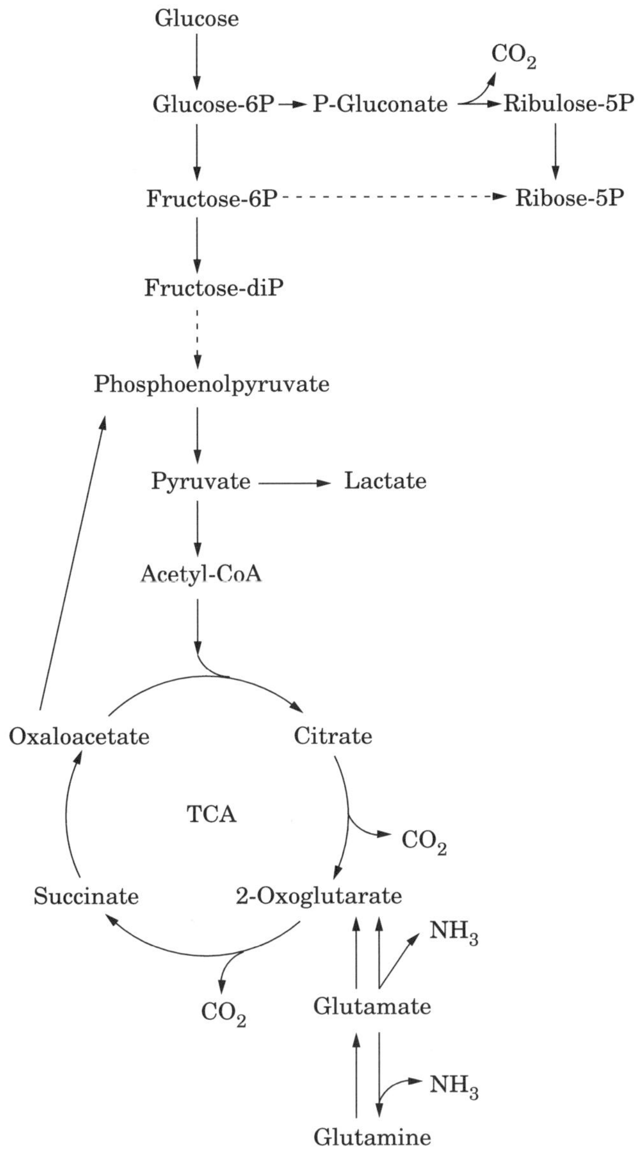

Figure 8. Interrelated pathways of energy metabolism. The two main substrates for animal cell growth in culture are glucose and glutamine. Glucose may be metabolized via glycolysis, the pentose-phosphate pathway, or the TCA cycle. Glutamine enters the TCA cycle via 2-oxoglutarate (α-ketoglutarate), which may lead to complete oxidation. Alternatively, glutaminolysis leads to incomplete oxidation of glutamine and the formation of a range of carbon by-products including lactate, alanine, pyruvate, or aspartate.

the culture of many animal cells, glucose is not completely depleted from the medium, suggesting that the glucose concentration is not the limiting factor.

Glucose utilization in culture leads to a concomitant increase in lactate concentration in the medium, which may lead to a decrease in pH if adequate buffering is not provided. The molar ratio of lactate produced to glucose consumed is often >1.8 for growing cells and is an indication of the anaerobic metabolism of glucose (26). If glucose is substituted by either fructose, galactose, or maltose, cell growth will continue, but the glycolytic rate and hence the production of lactate will decrease (27). In this situation cell growth will depend more heavily on other available substrates such as glutamine.

Glutamine. Next to carbohydrates, amino acids are the most abundant source of reduced carbon available for energy metabolism. Glutamine is normally included at a concentration higher than the other amino acids (0.7–6 mM). It is recognized as an important precursor for the synthesis of purines, pyrimidines, amino sugars, and asparagine, as well as a substrate for oxidative metabolism (28).

Glucose and glutamine utilization has been shown to be coordinately controlled in a number of cell lines (24,29). Glutamine was shown to provide 40–50% of the intracellular energy in a variety of cell lines including HeLa, human fibroblasts, and hybridoma cells. When fructose or galactose is the carbohydrate source, the proportion of cellular energy provided by glutamine is even higher.

One of the major end products of glutamine utilization is ammonia. This can increase in concentration in the medium as a result of the spontaneous chemical degradation of glutamine, as well as from its metabolic deamination. The molar ratio of ammonia produced to glutamine consumed is typically >0.75 (26). The accumulation of ammonia can be inhibitory to cell growth at >3 mM, but the sensitivity of cell lines to ammonia is variable.

There are various strategies that may be used to overcome the inhibitory effect of ammonia, including removal from the culture medium by ion exchange resins or by gas-permeable membranes (30). Alternatively, the substitution of glutamine by equivalent molar concentrations of glutamate (26) or glutamine-based dipeptides (31) can lead to reduced ammonia accumulation and, under appropriate culture conditions, to higher cell yields.

Amino Acids Other Than Glutamine. A range of amino acids is normally provided in culture medium, each at a concentration of 0.1–1.0 mM. The amino acids act primarily as precursors for protein synthesis, but some may be deaminated and degraded via the tricarboxylic acid cycle, particularly if the primary energy substrates (glucose and glutamine) have been utilized.

The branched-chain amino acids are consumed particularly rapidly by a number of cell lines including MDCK cells, human fibroblasts, mouse myeloma cells, and BHK cells. Hybridomas have been shown to metabolize serine and branched-chain amino acids when the glutamine level of the culture is low. During cell growth some amino acids may be released into the culture medium, particularly alanine and glycine. These may be products of a detoxification mechanism that prevents an excessive accumulation of ammonia in the cell.

Lipids. The inclusion of a lipid fraction or defined fatty acids in serum-free culture medium has been shown to improve cell proliferation (32). The most likely role of such fatty acids, which are normally supplemented at micromolar concentrations, is a source of precursors for the structural components of cell membranes. However, it has been shown that the availability of unsaturated fatty acids such as oleic and linoleic acids can significantly alter the metabolic profile of hybridoma cells by causing changes in the specific uptake rate of glutamine (33).

PATTERNS OF ENERGY METABOLISM

Patterns of Fungal Energy Metabolism

According to their different patterns of energy procurement, fungi can be classified into several metabolic groups. Many yeasts and most filamentous fungi belong to the group of obligate respirers. Chitridiomycetes rumen fungi belong to the group of obligate anaerobes. Between these two extreme groups are aerobic respirers, aerobic fermenters, facultative aerobic fermenters, and aeroneutrial fermenters. Few yeasts and filamentous fungi belong to the group of aerobic respirers. There is no anaerobic growth in this group, and anaerobic fermentation is possible only in pregrown cells. Aerobic fermenters and facultative aerobic fermenters have capabilities for both aerobic and anaerobic fermentation, but their respiratory capabilities are limited. Fungi in the group of aerobic fermenters, such as *Schizosaccharomyces pombe*, do not have the capability for anaerobic growth. Fungi in the group of facultative aerobic fermenters, such as *Saccharomyces*, have the capability for anaerobic growth. Some Chytridiomycetes and Oomycetes (e.g., *Blastocladia* and *Aqualinderella*) belong to the group of aeroneutrial fermenters, which lack respiration and have the capability of both aerobic and anaerobic fermentation or growth (6).

Several fungi can perform anaerobic fermentation in the presence of O_2. There are three principal types of fungal fermentation: alcoholic, lactic acid, and mixed acid fermentation. Lactic acid fermenters are found primarily among the Chytridiomycetes, Oomycetes, and Zygomycetes. Some of these fungi are obligate aerobes (e.g., *Allomyces, Sapromyces*, and *Apodachlya*), whereas others grow equally well anaerobically (e.g., *Blastocladia* and *Mindiniella*). *Rhizopus*, a member of the Zygomycetes, carries out lactic acid fermentation together with a simultaneous alcoholic fermentation. Mixed acid fermentation is found in a small group of obligately anaerobic Chytridiomycetes (6).

For most industrial fungi, oxygen is essential for both growth and production of metabolites. In an industrial aerobic fermentation process, an adequate oxygen supply is usually obtained by sparging sterile air into the fermentation broth. Because of the low solubility of oxygen it is important to maximise the mass transfer between the gas and liquid phase. The mass transfer is measured by the volumetric mass transfer coefficient, k_La, where k_L is the mass transfer coefficient in the liquid film (cm/h), and a is the specific interfacial area for air–liquid mass transfer (cm^2/cm^3). The mass transfer coefficient is related to the volumetric power input and superficial gas velocity. In large-scale aerobic fermentation, oxygen enrichment of the inlet air stream may be advantageous to meet the oxygen requirement. The profile of the oxygen uptake rate throughout the fermentation process is an important parameter for scale-up (34). The oxygen uptake rate (OUR) may be calculated by

$$\text{OUR} = k_La(C^* - C)$$

where C^* is the equilibrium oxygen concentration, and C is the dissolved oxygen in the culture broth.

Pattern of Energy Metabolism in *E. coli:* A Model Bacterium for Commercial Production

E. coli has been one of the most intensely studied species of bacteria over the last several decades and has been used extensively as the bacterium of choice for the expression of a wide variety of foreign proteins including many of industrial importance. Numerous human proteins have been expressed in *E. coli* by means of recombinant DNA technology, and some are currently in commercial production, including the examples shown in Table 3 (35). This bacterium's importance in the field of biotechnology requires that its energy conservation pathways for ATP generation be described in more detail. *E. coli* is a heterotrophic facultative anaerobe with diverse strategies for generating ATP from pathways leading to substrate level phosphorylation and proton-motive-force-coupled ATP synthesis. The anaerobic fermentative pathways of *E. coli* yield 3 ATP/mol glucose using a combination of the EM pathway of glycolysis to pyruvate followed by various pathways of pyruvate reduction leading to the formation of acetate, ethanol, formate, and lactate as the major end products (35). *E. coli* also possesses different electron transport chains coupled to a variety of electron acceptors, including O_2, NO_3^-, fumarate, and dimethylsulfoxide, for the generation of a primary proton motive force. While this versatility is of great ecological importance for the survival of this organism in the environment, most of it is of little use from a biotechnological perspective in the expression of foreign proteins, because the anaerobic pathways provide less ATP per mole of substrate used, and therefore, less cell material or product per mole of substrate.

Most of the applications of *E. coli* involve aerobic growth of the cells with O_2 as terminal electron acceptor of the respiratory chain. When using glucose as the primary energy source, oxidation of the glucose molecule to $6CO_2$ proceeds via glycolysis and the Krebs cycle, yielding 10 NADH and 2 $FADH_2$ (36). From the perspective of substrate-level phosphorylation, 4 ATP is generated within glycolysis and the Krebs cycle. Further ATP is generated in association

Table 3. Examples of Currently Licensed Therapeutic Protein Products from Animal Cell and Bacterial Cultures

Animal cell cultures	*E. coli* cultures
Interferon alpha	Insulin
OKT3 monoclonal antibody	Interferon alpha, beta, and gamma
Tissue plasminogen activator	
Erythropoietin	Human and bovine growth hormone
Human growth hormone	
Hepatitis B surface antigen	Rennin
Granulocyte colony stimulating factor	Interleukin 2
Clotting factor VIII	
DNase I	

with the electron transport chain and the generation of a primary proton motive force. The electron transport chain of *E. coli* differs from that of mitochondria in eukaryotes in that cytochrome *c* is not present (35). The chain is divided into two steps. The first step is the transfer of electrons from NADH to ubiquinone (a membrane-soluble cofactor carrying two electrons). This electron transfer is performed via the enzyme NDH1 (NADH-ubiquinone oxidoreductase I), a complex multisubunit transmembrane enzyme capable of proton translocation. The product of this reaction is ubiquinol. There are other electron transfer reactions that yield ubiquinol but are not associated with proton translocation, for example, the transfer of electrons from succinate to ubiquinone via the FAD-containing succinate dehydrogenase. In contrast to the pathway in mitochondria, where the electrons from ubiquinol are donated to cytochrome *c*, the two electrons of ubiquinol are donated directly to O_2, the reduction of which takes place on a quinol oxidase. *E. coli* can synthesize two different quinol oxidases, depending on the concentration of oxygen in the growth medium. At high concentrations of O_2, a cytochrome bo_3–containing quinol oxidase is produced, whereas a cytochrome *bd*–containing quinol oxidase, which has a higher affinity for oxygen, is produced at low O_2 concentrations. This difference is significant in that although both enzymes are involved in proton translocation, the first oxidase translocates 4 $H^+/2e^-$, whereas the second translocates only 2. Under optimal conditions (when cytochrome bo_3 quinol oxidase is used) a maximum 8 $H^+/2e^-$ has been estimated, as compared with 6 when the oxygen tension is low (35). By comparison the electron transport chain in mitochondria releases between 10 and 12 $H^+/2e^-$. These extruded protons can be used for several functions including active transport, motility, and the synthesis of ATP via the F_1F_0 ATPase. The F_1F_0 transmembrane ATPase of *E. coli* uses 3 H^+ per ATP synthesized.

The number of protons translocated per $2e^-$, and therefore also the number of ATPs that can be synthesized per $2e^-$, can be said to be dependent on both the substrate utilized and the O_2 concentration. Furthermore, the selection of bioenergetic substrate does not only have an impact on the amount of ATP generated per mole of substrate but also on the amount of ATP needed for cell biosynthesis, as demonstrated when experimentally measuring maximal cell yields per millimole of ATP (Y_{ATP}^{max}) when cells are grown on different carbon and energy sources (37). The Y_{ATP}^{max} for cells grown on glucose is twice that of cells grown on acetate. This corresponds well with theoretical calculations of the requirement for ATP for the synthesis of macromolecules (proteins, lipids, nucleotides) on the same substrates. For example the ATP requirement for protein synthesis doubles when cells use acetate as their carbon and energy source rather than glucose (42.7 vs. 20.5 mmol ATP/g cells). Thus, the choice of carbon and energy substrate not only has an impact on the amount of ATP synthesized but also on the amount of ATP needed for biosynthesis.

Patterns of Energy Metabolism in Animal Cell Culture

The metabolic pathways for the utilization of glucose in animal cells include glycolysis, the PP pathway, and the TCA cycle. The complete oxidation of glucose via glycolysis and the TCA cycle generates the greatest energy yield. However, the rate of aerobic metabolism of glucose is extremely low during the growth of many cell lines (38). By metabolic flux analysis, the proportion of glucose metabolized via the TCA cycle in cultured cells has been determined to be less than 5%, whereas over 90% is utilized via glycolysis to lactate, pyruvate, or alanine (39). The essential synthetic role of glucose is to feed the PP pathway, which leads to the formation of ribose, which is essential to maintain high rates of DNA synthesis during cell growth.

The addition of glucose to cells in culture may cause a decrease in oxygen consumption, a phenomenon commonly referred to as the Crabtree effect (40) and widely observed for many cell lines (41). This indicates a limited metabolic capacity for conversion to TCA cycle intermediates as the flux through glycolysis increases; this may be a result of a high cytoplasmic NADH/NAD ratio. The aerobic metabolism of cells in culture has also been shown to decrease at elevated dissolved oxygen concentrations (42). This may be related to the inhibition of pyruvate dehydrogenase, which is an enzyme complex that is required for linking glycolysis with the TCA cycle.

Glutamine provides a significant proportion of cellular energy by means of a metabolic network of up to eight individual metabolic pathways collectively described as glutaminolysis (43). This network is initiated by the deamination of glutamine by phosphate-activated glutaminase, which may regulate the activity of this extensive metabolic pathway. Metabolic analysis of HeLa and CHO cells by ^{15}N-NMR has shown that all the ammonia produced from glutamine originates from the amide nitrogen rather than the α-amino group (44). Because the activity of glutamate dehydrogenase is low in most cells, subsequent conversion of glutamate into α-ketoglutarate takes place by aminotransferases, notably aspartate transaminase (45). There follows a pathway of complete or partial oxidation leading to the end products of glutaminolysis, which include carbon dioxide, ammonia, alanine, aspartate, and lactate. The complete breakdown of the carbon skeleton of glutamine via the TCA cycle is the most efficient energy-releasing pathway (27 mol ATP/mol glutamine). However, it is probable that a large proportion of glutamine is incompletely oxidized to pyruvate, lactate, or alanine, resulting in a lower energy yield (9 mol ATP/mol glutamine) (43). A study of the metabolism of a murine hybridoma shows that alanine is the major end product of glutaminolysis (39). From this work it can be estimated that glutamine accounts for over 40% of the ATP generated in energy metabolism.

THE IMPORTANCE OF ENERGY METABOLISM TO SPECIFIC CELL PRODUCTS

Catabolic End Products of Fungi

Since L. Pasteur published his work on alcoholic fermentation in 1860, ethanol production from pyruvate has been best known from the studies on *Saccharomyces* (46). Many obligate aerobic fungi, such as common molds of the genera *Aspergillus, Fusarium*, and *Mucor*, are also well known for their ability to make ethanol (6).

There are excellent reviews about fermentation production of organic acids, such as citric acid, itaconic acid, gluconic acid, and fumaric acid. In economic terms, citric acid is the most important organic acid produced by fungi. Commercial citric acid production from *Aspergillus niger* was first demonstrated as feasible in 1917 (47). Strains of *Candida lipolytica* with reduced aconitase activity (which converts citrate to aconitate) are also used for citric acid production (48). It is believed that citric acid production depends on a disturbance of the metabolism of the TCA cycle. An important factor in the high citric acid productivities obtained is the existence of a powerful supplementary route to oxaloacetic acid by the carboxylation of phosphoenolpyruvate (49).

Secondary Metabolites of Fungi

The metabolic precursors of most secondary metabolites are products or intermediates of energy metabolism pathways (6). The shikimic acid pathway depends on the biosynthesis of aromatic amino acids, the polyketide biosynthetic pathway on acetyl CoA, and the mevalonic acid pathway on acetyl-CoA, polysaccharides, and peptidopolysaccharides.

Three amino acids (L-α-aminoadipic acid, L-cysteine, and L-valine) are precursors for biosynthesis of the β-lactam antibiotics penicillin G or V and cephalosporin C. The fungi *Penecillium chrysogenum* and *Acremonium chrysogenum* (also known as *Cephalosporium acremonium*) produce these two antibiotics, respectively, and use a multifunctional enzyme, δ-(L-α-aminoadipyl)-L-cysteinyl-D-valine synthetase (ACV synthetase), to convert the three amino acids to a tripeptide that is a common intermediate of β-lactam antibiotics. Cyclosporin A, an important immunosuppressant, is produced by the fungus *Tolypocladium inflatum* (also called *Beauveria nivea*) from 11 amino acids.

In recent years, the cholesterol-lowering agent lovastatin (also called mevinolin, monacolin K, or MB-530B) has been widely used for treatment of hypercholesterolemia. The fungi *Aspergillus terreus* and *Monascus ruber* produce lovastatin through a polyketide pathway from acetyl-CoA. Nine intact acetate units are incorporated into the lovastatin molecule, and the methyl group of lovastatin is derived from methionine. Acetyl-CoA is a precursor for the biosynthesis of lovastatin and fatty acids, although these two pathways occur at different times during the course of the fungal culture. Fatty acid biosynthesis is performed actively with growth at an early period of fermentation and reaches maximum on day 1 of growth, while lovastatin biosynthesis increases at a later time, with a maximum between days 3 and 5. Fatty acid biosynthesis is minimal on day 3 of growth and is associated with a decrease in the activity of fatty acid synthetase. This allows an increase in lovastatin biosynthesis, although this may be limited by a concomitant decrease in activity of the citrate cleavage enzyme that generates acetyl-CoA from citrate (50).

ATP Requirements for Fungal Secondary Metabolites. All secondary processes need energy (ATP) support. For instance, in the biosynthesis of penicillin G by *Penicillium chrosygenum*, the three precursor amino acids have to be activated by ATP at the cost of two high-energy phosphate bonds for each amino acid. Two high-energy phosphate bonds are also needed to activate phenylacetic acid (PAA) by CoA-SH. Not including energy consumption due to intracellular compartmentation and transport of penicillin from the cell to the fermentation broth, 3 ATP and 1 ADP are consumed to synthesize one molecule of penicillin G from the three amino acids (51). Cyclosporin A biosynthesis by the fungus *Tolypocladium inflatum* has a high rate of ATP consumption. The ATP concentration can be increased in a cell-free reaction mixture in order to obtain a higher cyclosporin A yield (52). A single multifunctional cyclosporin synthetase catalyses at least 40 reaction steps to complete the synthesis, and 11 amino acids need to be activated by using ATP before building up the peptide chain.

There is evidence that too much ATP production or accumulation in cells causes inhibition of the biosynthesis of secondary metabolites. Regulation of secondary metabolism by readily utilized carbon and energy sources, by high concentration of phosphate, or by the high phosphate concentration combined with an increase of consumption of the carbon source has been well reviewed, and it has been speculated that ATP could be the intracellular effector controlling the synthesis of secondary metabolism (53). A high fusidic acid–producing mutant of *Fusidium cocineum* was found to contain less intracellular ATP during production and less polyphosphate throughout the fermentation than the low producer. Similar phenomena were observed in prokaryotic antibiotic producers. Within 5 min after adding phosphate to candicidin-producing cells, the concentration of ATP doubles or triples before inhibition of antibiotic synthesis. Improved tetracyclin-producing strains have lower intracellular ATP levels than their poor-producing ancestral strains. The difference of ATP levels between the high- and low-producing strains does not mean a difference of the adenylate energy charge. In fact, the energy charge is similar between the strains with higher or lower productivity (54). In research on relationships among cellular energy status and the induction and initiation of aflatoxin synthesis in *Aspergillus parasiticus*, there was no evidence for the regulation of aflatoxin synthesis by the overall energy status of the fungal cell. However, electron microscopy indicates that aflatoxin synthesis occurs in association with a glucose-mediated inactivation of mitochondria. This suggests a possible regulation of aflatoxin synthesis by the energy status of specific subcellular compartments (55).

Fungicides and Drug Resistance Relating to Energy Metabolism

Inhibition of energy generation is one of the important mechanisms of fungicides (56). Membrane-active substances such as Ag^+, Hg^+, Cu^{2+}, Cu^+, uranyl ions, bis(tributyltin) oxide, and other organic tin compounds or polyenic antibiotics (nystatin, candicidin, amphotericin, fungimycin, etc.) indirectly inhibit fungal glycolysis by causing sublethal damage to the cytoplasmic membrane. Fluoride ions are potent inhibitors of glycolysis because of their ability to form stable complexes with magnesium and phosphate in certain enzymes involved in glycolysis. Fungicides showdomycin, gramicidin, 2,4-dinitrophenol, $CuSO_4$, and oleic acid inhibit respiration by uncoupling of oxidative phosphorylation. The fungicides carboxin, dexon, tridemorph, and antimycin A interfere with the function of the electron transport system in mitochondria. These inhibitory effects by fungicides can be attributed to a disturbance of membrane permeability and chemical reaction with sulfhydryl, carboxyl, phosphate, or amino groups of sensitive enzymes in fermentation or respiration metabolism. The fungicides terrazole and chloroneb disturb the respiratory process by causing lysis of mitochondria (56).

A mechanism of fungal resistance to the ergosterol-biosynthesis-inhibiting fungicides is energy-dependent efflux activity. For instance, the relatively high activity of energy-dependent efflux of fenarimol in resistant mutants of *Aspergillus nidulans* might be connected with constitutive membrane transport activities, which may be "fueled" with energy at the cost of other cell processes (e.g., germination, growth, and sporulation) (57). According to this principle, one might be able to manipulate fenarimol uptake by inhibiting efflux activity with other chemicals, possibly enhancing fungal toxicity of fenarimol against pathogenic fungi in agriculture. Recently, the energy-dependent drug efflux was identified as the mechanism of fluconazole resistance in *Candida glabrata* from a case of infection in which pre- and posttreatment isolates were available for comparison (58).

Catabolic End Products of Bacteria

Products of Sugar Fermentation. Typically, the EM pathway of glycolysis is used as the primary means of energy supply, yielding pyruvate, NADH, and 2 ATP from glucose (59). For the most part, species-specific variations in end products will depend on the selection of the means of recycling the electron carrying cofactor NAD and the selected substrate as an electron sink. The reductive steps are not associated with the supply of energy.

Lactic Acid. A familiar fermentation product is lactic acid (60), which is derived from the reduction of pyruvate with NADH to lactate by lactate dehydrogenase. A further means of energy conservation observed in several homolactic fermenting organisms, but which is also found in other organisms, is the generation of a proton motive force in association with the excretion of lactic acid out of the cells via a secondary transport mechanism (61). Another similar example is found in organisms performing malolactic fermentation, for example, *Lactobacillus lactis*. In this organism a combination of electrogenic malate uptake and malate/lactate antiport leads to the net efflux of 1 H^+ per malate molecule metabolized.

Ethanol. The formation of ethanol as the sole organic fermentation product is a two-step process. The first step is the decarboxylation of pyruvate to acetaldehyde plus CO_2 by pyruvate decarboxylase. NAD is recycled in the next step, the NADH-dependent reduction of acetaldehyde by alcohol dehydrogenase. Only organisms possessing pyruvate decarboxylase are capable of producing ethanol as a major fermentation product. In bacteria this enzyme is rare, and the sole group of organisms of industrial interest performing this fermentation comprises members of the genus *Zymomonas* (62), which use the E–D pathway of glucose oxidation to pyruvate, and therefore make only 1 ATP per glucose. This is in contrast to the production of ethanol in yeasts, where the E–M pathway is used for the synthesis of 2 ATP per glucose. Although *Zymomonas mobilis* is highly efficient in ethanol production, its range of fermentable substrates is very limited. In order to widen the range of usable fermentative substrates for ethanol synthesis, the genes for pyruvate decarboxylase and alcohol dehydrogenase have been transfected into *E. coli* (63).

Alcohols, Ketones, and Organic Acids. Traditionally, the biological generation of industrially useful short-chain fatty acids and solvents (64) has involved the anaerobic degradation of sugars by organisms dependent only on substrate-level phosphorylation to form ATP. From pyruvate, a further ATP can be generated via the synthesis of acetyl-CoA and acetyl phosphate, yielding acetate plus CO_2 as the end products. This reaction, however, does not recycle the NAD reduced during glycolysis. The electrons are used to reduce protons via a hydrogenase, yielding gaseous hydrogen (H_2) as an end product. An added benefit of this reaction is that the consumption of H^+ from inside the cell increases the internal pH and thus participates in the maintenance of the transmembrane ΔpH. Hydrogen may be generated from the fermentation of organic wastes (65). The buildup of hydrogen in the growth environment, however, causes a decrease in the thermodynamic efficiency of acetate production such that other electron sinks must be sought, but these various alternate end products are not necessarily associated with ATP synthesis. In the case of organisms capable of butyric acid synthesis, two molecules of acetyl-CoA can condense, forming acetoacetyl-CoA, which can then be reduced to butyryl-CoA, and finally butyrate, with the production of 1 ATP per 2 acetyl groups. Other products derived from the reduction of acetyl-CoA (ethanol, acetone, propanol, and butanol) are not associated with ATP synthesis (Fig. 9).

This shift in fermentation products, depending on hydrogen concentration, has been described in *Ruminococcus albus*, where cellulose fermentation normally yields acetate, ethanol, and hydrogen in a ratio of 1:1:2, but acetate and ethanol in a ratio of 7:1 when hydrogen is continuously removed (66). Conversely, a metabolic shift from acid production to enhanced solvent production has been observed in *Clostridium acetobutylicum* when hydrogen production is blocked by carbon monoxide, a hydrogenase inhibitor. It has also been noted that the addition of glycerol to glucose-fermenting cultures of *Clostridium acetobutylicum* also inhibits hydrogen and acid generation but enhances butanol production (67).

High concentrations of these various products (undissociated volatile fatty acids, ketones, and alcohols), because of their amphipathic character, dissolve into membrane lipids, affecting membrane fluidity and increasing membrane permeability. This results in a decrease of the transmembrane pH gradient and a reduction of the intracellular level of ATP (68) and thus limits the concentration of these products the cells can tolerate.

In certain species capable of solvent formation (alcohols, ketones), for example, *Clostridium acetobutylicum*, the

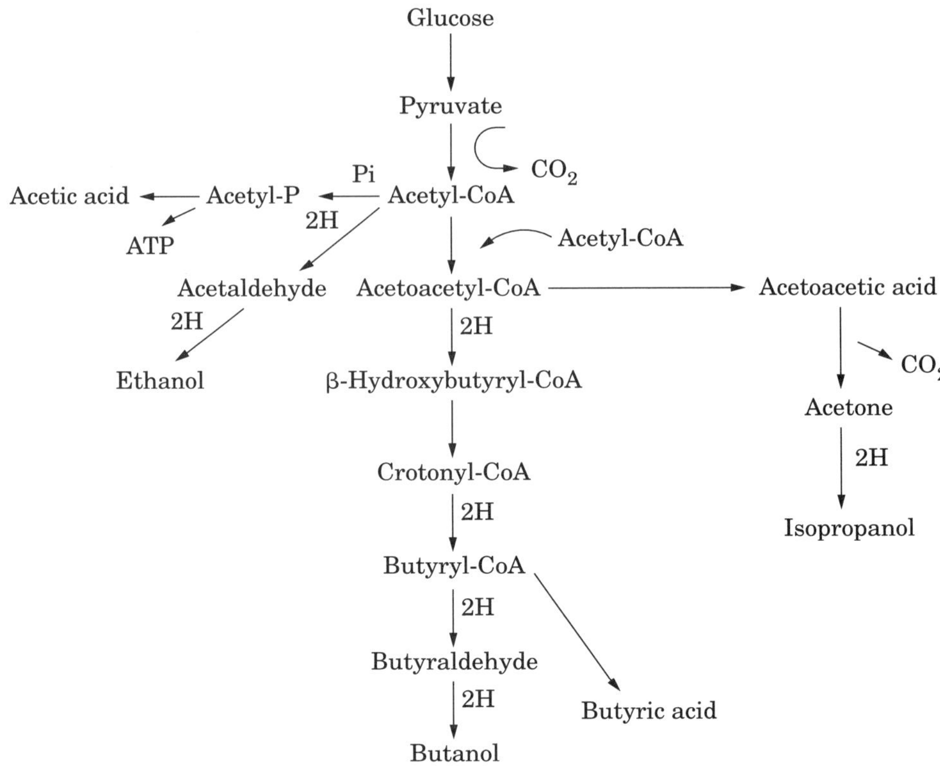

Figure 9. Butanol/acetone fermentation. This anaerobic metabolism is typical of the genus *Clostridium*. Various fermentation products are formed by reduction using NADH derived from glycolysis. The proportion of each product formed is dependent on the fermentation conditions.

drop in internal pH caused by the initial acidogenic phase can act as part of the triggering mechanism for solvent production (69). This reduction in proton motive force causes a drop in internal ATP concentration, which has been shown to act as one of the regulatory factors in turning on solvent production (70). Nonetheless, excessive butanol formation eventually leads to inability of maintaining a transmembrane pH gradient, lowers intracellular ATP levels, and prevents glucose uptake. Higher concentrations of alcohols in general have a negative impact on membrane integrity and membrane functions (71).

Acetic Acid. Acetic acid bacteria are obligate aerobes and members of the genera *Acetobacter* and *Gluconobacter*. They are capable of the conversion of sugars or other carbon substrates by fermentation to acetate with a maximum conversion of 66%, with the remaining carbon being evolved as CO_2 and the electrons serving to reduce protons to H_2 (72). These organisms have been used primarily for the synthesis of acetic acid from ethanol, whereby the oxidation of ethanol to acetate provides the electrons for O_2 reduction. In *Gluconobacter* the cells lack succinate dehydrogenase and therefore have an incomplete TCA cycle. Although *Acetobacter* does have a full TCA cycle and therefore can oxidize acetate to CO_2, its function is repressed in the presence of ethanol.

The oxidation of ethanol to acetate is a four-electron oxidation and involves two membrane-bound enzymes: alcohol dehydrogenase and acetaldehyde dehydrogenase. In *Gluconobacter suboxydans* and *Acetobacter aceti* both enzymes are localized on the outer surface of the cytoplasmic membrane and contain cytochrome *c* as well as a novel electron-transferring prosthetic group, pyrroloquinoline quinone (PQQ), with ubiquinone as the ultimate electron acceptor for these reactions (73). In view of the location of these enzymes on the outer surface of the membrane, two protons are left behind during each oxidation reaction participating in the formation of a proton gradient.

Electrons from reduced ubiquinone are then transferred to terminal ubiquinol oxidases. Depending on the growth conditions, the terminal oxidase of *Acetobacter aceti* contains a cytochrome a_1–type chromophore, which has been shown in proteoliposomes to translocate protons or a cytochrome *o* (74). In *Gluconobacter suboxydans* a cytochrome *o*–type ubiquinol oxidase can translocate protons in proteoliposomes in the presence of O_2 and ubiquinol (75).

In the overall synthesis of ascorbic acid, one step, the conversion of D-sorbitol to D-sorbose, is generally performed by *Gluconobacter suboxydans*, which possesses a D-sorbitol dehydrogenase that is bound to the outer surface of the cytoplasmic membrane. This is a cytochrome *c*–and flavin-containing protein that uses ubiquinone as an electron acceptor, which is reoxidized by terminal oxidases.

Homoacetogens. Homoacetogens are a set of anaerobes capable of stoichiometric conversion of glucose to three acetates (76). These organisms are also capable of acetate synthesis and growth as autotrophs on H_2/CO_2. The E–M pathway is used to yield the first two acetates from glucose, and the third acetate is derived from the reduction by hydrogen of the evolved CO_2 to methyl or formyl derivatives. In the reduction of CO_2 to the methyl level, ATP is hydrolyzed for the generation of formyl tetrahydrofolate, which is subsequently reduced to the methyl level. This methyl group is then transferred to a correnoid protein. The CO_2 in the carboxyl position is reduced to carbon monoxide by a carbon monoxide dehydrogenase (CODH). This same enzyme then catalyzes the formation of acetyl-CoA from this CO, the methyl group from the correnoid protein and HS-CoA. The release of acetate from acetyl-CoA via acetyl phosphate generates 1 ATP. Because of the consumption of ATP in the transfer of formate onto tetrahydrofolate, this pathway does not yield any net ATP via substrate-level phosphorylation.

Net energy is conserved in the form of a primary transmembrane ion gradient. In *Clostridium thermoaceticum*, acetogenesis from H_2 plus CO_2 generates a proton motive force. The $\Delta\mu_{H^+}$ is generated via a hydrogenase located on the outside surface of the membrane and a rudimentary electron transport chain consisting of two *b*-type cytochromes and menaquinone. The hydrogenase releases protons from the oxidation of H_2 on the outside of the membrane, and the electrons are carried through the membrane to the inner side where they are used for the generation of the methyl group of the acetate. The proton motive force thus generated can be used to drive ATP synthesis via an F_1F_0 ATPase.

The production of acetic acid may have a toxic effect. Although there is little diffusion through the cell membrane of the ion form (i.e., R-COO^-), weak acids can act as protonophores when they are present in their nondissociated form. In *C. thermoaceticum*, for example, it has been shown that, at the lower limits of the physiological pH (~pH 5), the concentration of undissociated acetic acid is sufficiently elevated to act as an effective protonophore, bringing protons into the cell and thus eliminating the proton motive force and decreasing the ATP pool size. This in turn stops growth and fermentation because ATP is necessary in the activation steps of glycolysis (77).

In *Acetobacterium woodii*, acetogenesis from H_2 plus CO_2 generates a primary transmembrane Na^+ gradient, which is used directly for ATP generation via a sodium-dependent ATPase (78). Na^+ translocation is probably associated with the transfer of the methyl group from tetrahydrofolate to the correnoid protein. The $\Delta\mu_{Na^+}$ is also necessary for flagellar motility in this organism. The capacity of acetogenic bacteria to use CO as well as CO_2 and H_2 can be used in the synthesis of organic carbon from syngas (synthesis gas) generated from the pyrolysis of coal.

Methane. Methane is one of the major end products of anaerobic biodegradation of organic material, including municipal and industrial organic wastes. Organisms that generate methane, the methanogens, are important in these environments because they allow mineralization of organic materials and generate an inflammable gas (79). Although they act on a limited range of substrates (H_2/CO_2, formate, acetate, methanol, methylamines), these represent the major products of organic polymer biodegradation. The consumption of hydrogen by the methanogens maintains a low environmental concentration, thereby reducing the amounts of alcohols and ketones produced by fermentative bacteria.

Methane is generated by H_2-dependent CO_2 reduction or conversion from methanol (80). The pathway of CO_2 reduction involves four two-electron reduction steps similar to those of the acetogenic pathway (Fig. 10). The major difference is the use of novel C_1-carrying cofactors. The reduction of CO_2 to the formyl level is catalyzed by formylmethanofuran dehydrogenase, which releases formyl methanofuran; the C_1 is then transferred to tetrahydromethanopterin, an analog of tetrahydrofolate, which becomes methylated. The methyl group is transferred to thioethanesulfonate methyl-CoM, which is then reduced to release methane.

The first step of methanogenesis from H_2 plus CO_2 involves a membrane-associated enzyme and may require energy in the form of an ion motive force to drive the reaction. The sites of energy conservation associated with the methanogenic pathway, as well as the protonophore sensitivity and the requirement for sodium, have been elucidated for *Methanosarcina*. There are two sites of ion translocation: the heterodisulfide reductase, which is associated with proton translocation, and the methyl tetrahydromethanopterin, HS-CoM methyltransferase, which is associated with sodium translocation. Thus both a proton and a sodium motive force are generated. Two different ion motive force–dependent ATP synthases have also been uncovered: a $\Delta\mu_{H^+}$-dependent A_0A_1 ATP synthase and a $\Delta\mu_{Na^+}$-dependent F_0F_1 ATP synthase (81). Proton translocation is now proposed to be associated with electron transport via a membrane-bound carrier called methanophenazine (82).

Poly-β-Hydroxyalkanoates

Poly-β-hydroxyalkanoates (PHA) serve as metabolic carbon reserves in various microorganisms and have found commercial value as biodegradable substitutes for petroleum-based plastics (83). The polymers are produced under conditions of metabolic stress in the presence of excess carbon or other energy source when, for example, nitrogen, phosphate, or oxygen limitation prevents balanced growth and acetyl-CoA accumulates in the cell. The condensation of acetyl groups leads to the production of β-hydroxybutyrate, which is esterified into long chains. Under such conditions, PHA can accumulate in the cells, forming up to 90% of the cell mass. Depending on the organism used, there can be simultaneous growth and PHA formation, or a serial process whereby the organism is allowed to buildup biomass, followed by a second stage of PHA accumulation. Although nitrogen limitation is commonly used as trigger for PHA production, the continued addition of a complex organic nitrogen source (e.g., tryptone or casein) during PHA synthesis from glucose was shown to enhance PHA synthesis without significant effect on residual cell mass yield. It has been suggested that this enhanced yield is due to the prevention of use of acetyl-CoA for amino acid and protein synthesis, thus permitting a greater proportion of the acetyl-CoA produced from glucose to form PHA rather than other cell products. Varying substrates or growth conditions can also lead to an increased yield of product and variation in the type of alkanoates formed. This permits the formation of polyesters with different physical properties. For example, β-OH-valerate is formed from the condensation of propionyl-CoA and acetyl-CoA when propionate is added to the growth medium.

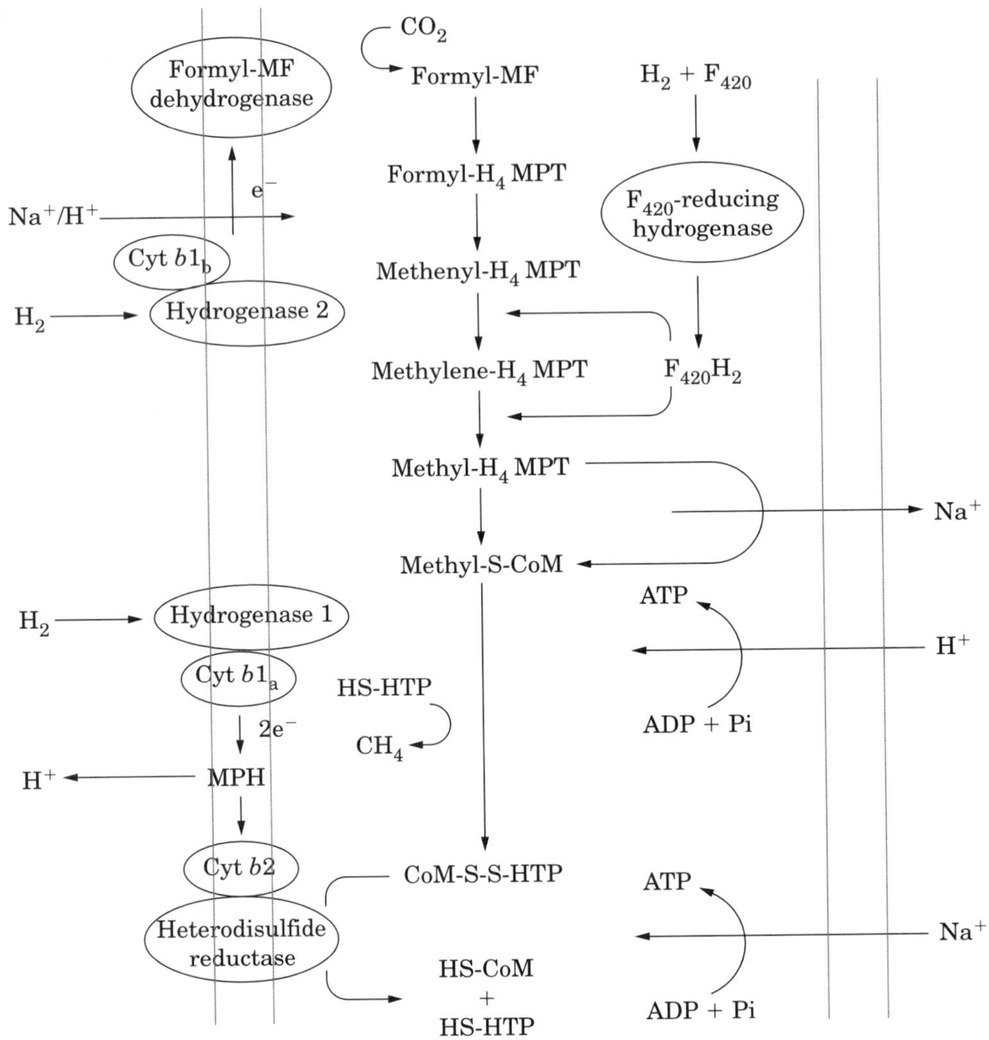

Figure 10. Scheme of methanogenesis. Methane is formed from hydrogen and carbon dioxide in a metabolic pathway present in *Methanosarcina*. The dashed lines represent the cytoplasmic membrane. A similar scheme is expected for other methanogens. H_4MPT, tetrahydromethanopterin; HS-CoM, coenzyme M (2-mercaptoethanesulfonate); HS-HTP, 7-mercaptohectanoyl threonine phosphate; MPH, methanophenazine; MF, methanofuran.

Enzyme Production in Bacteria

Aside from the proteins and enzymes expressed and produced in *E. coli* by recombinant DNA technology, there are various enzymes produced by a wide variety of bacteria that are used commercially (α-amylases, xylanases, lipases, DNA polymerases, etc.). Many of these are degradative enzymes used to break down growth substrates that cannot enter directly into the cells. Such enzymes are produced in large quantities and exported out of the cell. The need for highly stable and robust proteins capable of functioning at high temperatures and/or pH has led to the isolation and characterization of novel extremophilic organisms capable of degrading organic polymers (84,85).

A representative example of such organisms is *Pyrococcus furiosus*, a hyperthermophilic archaean that grows optimally at 100 °C and pH 6.4 and produces a thermostable α-amylase of potential use in the starch hydrolysis industry, as well as a high-fidelity DNA polymerase (86). The need for optimization of the production of these enzymes has required the characterization of the associated fermentation pathway (87). A modified EM pathway of glycolysis that employs ADP-dependent glucokinase and phosphofructokinase is used rather than the conventional ATP-utilizing enzymes. A further modification is that the oxidation of glyceraldehyde-3-phosphate to 3-phosphoglycerate occurs directly, using a glyceraldehyde-3-phosphate: ferredoxin oxidoreductase and does not involve the synthesis of 1,3-bisphosphoglycerate and the subsequent synthesis of ATP. ATP is generated from phosphoenolpyruvate via pyruvate kinase, the 2 ATP formed per glucose in this reaction balance the two high-energy phosphate bonds used to activate glucose. The only net ATP produced is from the oxidation of pyruvate to CO_2 and acetate via acetyl-CoA and acetyl phosphate. The flux of carbon through acetate, however, is dependent on the presence or absence of electron sinks. In the presence of sulfur (S^0), the electrons released during glycolysis are passed on to sulfur via a sulfhydrogenase, generating H_2S (88). In the absence of sulfur, the same enzyme is capable of reducing protons to hydrogen. However, as the hydrogen concentration in the medium increases, alanine accumulates in the growth medium and is associated with a decreased growth rate (89).

A major difficulty in batch cultures of this organism in the absence of S^0, is the accumulation of H_2, the production of alanine, and the accumulation of acetate, which can be toxic at higher concentrations. Fermenter design strategies have been developed to enhance growth yields and increase cell densities. The use of a dialysis membrane reactor permits the cells to reach 10-fold higher cell densities and a lower alanine level (89).

Production from Animal Cells

Commercial products derived from animal cells in culture include viral vaccines, monoclonal antibodies (Mabs), and recombinant proteins. Table 3 shows examples of the protein products of animal cell cultures licensed for therapeutic use. In all cases, productivity is dependent upon protein biosynthesis, which in turn must be supported by an adequate provision of intracellular energy. From data based on the macromolecular composition of a cell, it can be estimated that 60% of cellular energy is utilized for protein synthesis, ATP being required for the condensation of amino acids (90). Although some of these proteins can be synthesized in bacteria, one of the major advantages of using animal cells is that the resulting proteins are glycosylated (i.e., have attached carbohydrate structures). This is often important for the therapeutic use of the proteins.

During culture of a cell line that has been genetically engineered to secrete a specific protein, the energy demands of cell growth must be balanced with the demand for synthesis of the selected protein. There are many reports that show an inverse relationship between growth and antibody production (91). Typically, the specific rate of Mab production (q_{Mab}) occurs at 0.05 pg/cell min during the growth of a hybridoma, and this amounts to about 20% of the total cellular protein synthesis. If the growth rate is decreased, for example, in continuous culture, then the q_{Mab} may double in value as more of the cellular energy is diverted from growth to synthesis of the single secreted protein. However, there is a limit to the reduction in the growth rate of lymphocyte hybridomas beyond which apoptosis (programmed cell death) causes the rapid loss of the cell population (92).

In recombinant protein synthesis there may be an energetic burden placed on the metabolism of a cell following gene amplification. A balance must be established in the cell between the requirements of growth and protein productivity. Gu et al. (93) studied a recombinant CHO cell line that showed high expression of β-galactosidase by coamplification of the *lacZ* and *dhfr* genes. The genes were amplified 100 times by selective pressure from increasing concentrations of methotrexate, which is an inhibitor of the essential enzyme dihydrofolate reductase, the product of the *dhfr* gene. However, as a result of gene amplification, the specific cell growth rate was reduced by 62% because of the metabolic burden placed upon the cell line.

ENHANCEMENT OF PRODUCTIVITY: METABOLIC MANAGEMENT OR ENGINEERING

In the development of an industrial bioprocess, there is a strong incentive to maximize the synthetic capacity of the producer cell line by a form of metabolic management. This may be achieved by an understanding and control of the limiting factors for growth or cell yield. For example, the availability of oxygen may be critical in the profile of the energy pathways of a microorganism and is an important parameter for the production of metabolic end products. In the case of animal cell cultures, controlling the availability of glutamine is essential in attempting to limit the accumulation of ammonia, which is a major parameter in realizing a respectable cell yield. This process of metabolic management may be put to greatest effect by the development of mathematical programming models based upon metabolic flux data obtained by radioactive tracers or ^{13}C-NMR (94).

Further enhancement of productivity of a bioprocess may be gained by strain improvement, which is often achieved by genetic mutation. For example, high-lysine-producer strains of *Corynebacterium glutamicum* can be obtained by selective mutation resulting from feedback resistance of the aspartate kinase enzyme (95). Further improvement may result from mutations in the enzymes of energy metabolism. Often the metabolic basis of a synthetic enhancement is unknown but may be analyzed by changes in the pattern of metabolic flux.

A more selective form of productivity enhancement may be achieved by metabolic engineering as a means to maximize the flow of carbon and energy to the specific pathway of interest (96,97). This involves transfection of the producer cell line with a gene to manipulate the enzymatic, transport, or regulatory function of the cell. A particularly successful application of this technique is in the ability to grow a Chinese hamster ovary (CHO) cell line in a protein-free medium following transfection with genes capable of expressing essential factors such as insulin-like growth factor, transferrin, or cyclin E (98,99). Another application involves the expression of the gene for hemoglobin isolated from *Vitreoscilla* in a variety of aerobic industrial microorganisms (99). Because of the central importance of oxygen to energy metabolism, this technique has been successful in causing a significant productivity enhancement in a long list of cells including bacteria, fungi, and animal cells. With an increased understanding of the energy metabolism of industrially important producer cell lines, there comes a greater likelihood of success in designing the required phenotypic characteristics by metabolic engineering.

BIBLIOGRAPHY

1. D.M. Karl, *Microbiol. Rev.* **44**, 739–796 (1980).
2. D.E. Atkinson, *Biochemistry* **7**, 4030–4034 (1968).

3. J.S. Swedes, R.J. Sedo, and D.E. Atkinson, *J. Biol. Chem.* **250**, 6930–6938 (1975).
4. K.B. Anderson and K. von Meyenburg, *J. Biol. Chem.* **252**, 4151–4156 (1977).
5. G.M. Gadd, in D.R. Berry ed., *Physiology of Industrial Fungi*, Blackwell Scientific Publications, Oxford, 1988, pp. 21–57.
6. D.H. Griffin, *Fungal Physiology*, 2nd ed., Wiley-Liss, New York, 1994.
7. P.A. Botham and C. Ratledge, *J. Gen. Microbiol.* **114**, 361–375 (1979).
8. E.A. Dawes, *Microbial Energetics*, Blackie & Son, Glasgow, 1986, pp. 21–51.
9. T.G. Cartledge, in D.R. Berry, I. Russell, and G.G. Stewart eds., *Yeast Biotechnology*, Allen and Unwin, London, 1987, pp. 311–342.
10. G.C. Ferreira and P.L. Pedersen, *J. Bioenerg. Biomembranes* **25**, 483–491 (1993)
11. N.A.R. Gow and G.M. Gadd, *The Growing Fungus*, Chapman & Hall, London, 1995, p. 225.
12. M. Kobayashi, Y. Matsuo, A. Takimoto, S. Suzuki, F. Maruo, and H. Shoun, *J. Biol. Chem.* **271**, 16262–16267 (1996).
13. V.P. Skulachev, *Eur. J. Biochem.* **208**, 203–209 (1992).
14. M.J. Zidwick, in D.B. Finkelstein and C. Ball eds., *Biotechnology of Filamentous Fungi*, Butterworth-Heinemann, Boston, 1991, pp. 303–334.
15. G.J.M. Hersbach, C.P. Van der Beek, and P.W.M. Van Dijck, in E.J. Vandamme ed., *Biotechnology of Industrial Antibiotics*, Marcel Dekker, New York, 1984, pp. 45–140.
16. B. Buckland, K. Gbewonyo, T. Hallada, L. Kaplan, and P. Masurekar, in A.L. Demain, G.A. Somkuti, J.C. Hunter-Cevera, and H.W. Rossmoore eds., *Novel Microbial Products for Medicine and Agriculture*, Elsevier on behalf of Society for Industrial Microbiology, 1989, pp. 161–169.
17. M. Hosobuchi, T. Shioiri, J. Ohyama, M. Arai, S. Iwado, and H. Yoshikawa, *Biosci. Biotech. Biochem.* **57**, 1414–1419 (1993).
18. S.N. Agathos, C. Madhosingh, J.W. Marshall, and J. Lee, in *Annals New York Academy of Sciences*, 1987, pp. 657–662.
19. N.D. Lindley, J.F. Pedley, S.P. Kay, and M.T. Heydeman, *Int. Biodeterior.* **22**, 281–287 (1986.).
20. K.H. Jones, P.W. Trudgill, and D.J. Hopper, in A. Bousher, M. Chandra, and R. Edyvean eds., *Biodeterior. Biodegrad. 9. Int. Biodeterior. Biodegrad. Symp., [Proc.], 9th* (meeting date 1993), Institution of Chemical Engineers, Rugby, U.K., 1995, pp. 470–475.
21. F.J. Weber, K.C. Hage, and J.A. de Bont, *Appl. Environ. Microbiol.* **61**, 3562–3566 (1995).
22. J.F. Morgan, H.J. Morton, and R.C. Parker, *Proc. Soc. Exp. Biol. Med.* **73**, 1–8 (1950).
23. H. Eagle, *Science* **122**, 501 (1995).
24. H.R. Zielke, P.T. Ozand, J.T. Tildon, D.A. Sevdalian, and M. Cornblath, *J. Cell Physiol.* **95**, 41–48 (1978).
25. B.M. Wice and D. Kennell, *J. Biol. Chem.* **258**, 13134–13140 (1983).
26. T.E. Hassell and M. Butler, *J. Cell Sci.* **96**, 501–508 (1990).
27. T. Imamura, C.L. Crespi, W.G. Thilly, and H. Brunengraber, *Anal. Biochem.* **124**, 353–358 (1982).
28. W.L. McKeehan, in Morgan, M.J. ed. *Carbohydrate Metabolism in Cultured Cells*, Plenum, New York, 1986, pp. 111–150.
29. K.W. Lanks and P-W. Li, *J. Cell Physiol.* **135**, 151–155 (1988).
30. M. Schneider, I.W. Marison, U. von Stockar, *J. Biotechnol.* **46**, 161–185 (1996).
31. M. Butler and A. Christie, *Cytotechnology* **15**, 87–94 (1994).
32. M.C. Glassy, J.P. Tharakan, and P.C. Chau, *Biotech. Bioeng.* **32**, 1015–1028 (1988).
33. M. Butler, N. Huzel, and N. Barnabé, *Biochem J.* **322**, 615–623 (1997).
34. R.L. Greasham, in D.B. Finkelstein and C. Ball eds., *Biotechnology of Filamentous Fungi*, Butterworth-Heinemann, Boston, 1992, pp. 65–87.
35. F.C. Neidhardt ed., Escherichia coli *and* Salmonella *Cellular and Molecular Biology*, ASM Press, Washington D.C., 1996.
36. G. Gottschalk, *Bacterial Metabolism*, Springer-Verlag, New York, 1986.
37. A.H. Stouthamer, *Int. Rev. Biochem.* **21**, 1–47 (1979).
38. M.W. Glacken, R.J. Fleischaker, and A.J. Sinskey, *Batik. Bioeng.* **28**, 1376–1389 (1986).
39. D. Petch and M. Butler, *J. Cell Physiol.* **161**, 71–76 1994.
40. H.D. Crabtree, *Biochem. J.* **23**, 536–545 (1929).
41. M. Guppy, E. Greiner, and K. Brand, *Eur. J. Biochem.* **212**, 95–99 (1993).
42. D.C.H. Jan, D. Petch, N. Huzel, and M. Butler, *Biotech. Bioeng.* **54**, 153–164 (1997).
43. L. Haggstrom, in R.E. Spier, J.B. Griffiths, and B. Meignier eds., *Production of Biologicals from Animal Cells in Culture* (ESACT 10th Mtg.) Butterworth-Heinmann Oxford, 1991, pp. 79–81.
44. J.C. Street, A.M. Delort, P.S.H. Braddock, and K.M. Brindle, *Biochem. J.* **291**, 485–492 (1993).
45. H.A. Jenkins, M. Butler, and A.J. Dickson, *J. Biotechnol.* **23**, 167–182 (1992).
46. H.W. Doelle, *Microbial Metabolism*, Dowden, Hutchinson & Ross, Stroudsburg, Pa., 1974, pp. 8–79.
47. M.J. Zidwick, in D.B. Finkelstein and C. Ball eds., *Biotechnology of Filamentous Fungi*, Butterworth-Heinemann, Boston, 1991, pp. 303–334.
48. J.W. Bennett, in A.L. Demain and N.A. Solomon eds., *Biology of Industrial Microorganisms*, Butterworths, Boston, 1985, pp. 359–406.
49. H.L. Kornberg and A.H. Romano, *Process Biochem.* **3**, 40–42 (1968).
50. M.D. Greenspan and J.B. Yudkovitz, *J. Bacteriol.* **162**, 704–707 (1985).
51. G.J.M. Hersbach, C.P. Van der Beek, and P.W.M. Van Dijck, in E.J. Vandamme ed., *Biotechnology of Industrial Antibiotics*, Marcel Dekker, New York, 1984, pp. 45–140.
52. A. Lawen, R. Traber, D. Geyl, R. Zocher, and H. Kleinkauf, *J. Antibiot.* **42**, 1283–1289 (1989).
53. A.L. Demain, Y. Aharonowitz, and J.F. Martin, in L.C. Vining ed., *Biochemistry and Genetic Regulation of Commercially Important Antibiotics*, Addison-Wesley, Advanced Book Program/World Science Division, Reading, Mass., 1983, pp. 49–72.
54. S.W. Drew and D.A. Wallis, in J.W. Bennett and A. Ciegler eds., *Secondary Metabolism and Differentiation in Fungi*, Marcel Dekker, New York, 1983, pp. 35–54.
55. R.J. Buchanan, S.B. Jones, W.V. Gerasimowicz, L.L. Zaika, H.G. Stahl, and L.A. Ocker, *Appl. Environ. Microbiol.* **53**, 1224–1231.
56. H. Lyr, in M.R. Siegel and H.D. Sisler eds., *Antifungal Compounds*, Marcel Dekker, New York, 1977, pp. 301–332.

57. M.A. de Waard and A. Fuchs, in J. Dekker and S.G. Georgopoulos eds., *Fungicide Resistance in Crop Protection*, Centre for Agricultural Publishing and Documentation, Wageningen, The Netherlands, 1982, pp. 87–100.
58. T. Parkinson, D.J. Falconer, and C.A. Hitchcock, *Antimicrob. Agents Chemother.* **39**, 1696–1699 (1995).
59. G. Gottschalk, *Bacterial Metabolism*, 2nd ed., Springer-Verlag, New York, 1986.
60. J.H. Litchfield, *Adv. Appl. Microbiol.* **42**, 4595 (1996)
61. R. Otto, A.S.M. Sonnenberg, H. Veldkamp, and W.N. Konings, *Proc. Natl. Acad. Sci. U.S.A.* **77**, 5502–5506 (1980).
62. J. Swings and J. de Ley, *Bacteriol. Rev.* **41**, 1–46 (1977).
63. L.O. Ingram, F. Alterthum, K. Ohta, and D.S. Beall, *Develop. Ind. Microbiol.* **31**, 21–30 (1990).
64. J.G. Zeikus, *Ann. Rev. Microbiol.* **34**, 423–464 (1980).
65. S. Roychowdhury, D. Cox, and M. Levandowsky, *Int. J. Hydrogen Energy* **13**, 407–410 (1988).
66. S.G. Pavlostathis, T.L. Miller, and M. Wolin, *Appl. Microbiol. Biotechnol.* **33**, 109–116 (1990).
67. I. Vasconcelos, L. Girbal, and P. Soucaille, *J. Bacteriol.* **176**, 1443–1450 (1994).
68. P. Dürre, H. Bahl, and G. Gottschalk, in L.E. Erickson and D.Y. Fung eds., *Handbook of Anaerobic Fermentations*, Marcel Dekker, New York, 1988, pp. 187–206.
69. B. McNeil and B. Kristiansen, *Adv. Appl. Microbiol.* **31**, 61–92 (1986).
70. H. Grupe and G. Gottschalk, *Appl. Environ. Microbiol.* **58**, 3896–3902 (1992).
71. L.O. Ingram and T. Buttke, *Adv. Microbiol. Phys.* **25**, 253–300 (1984).
72. T. Asai, *Acetic Acid Bacteria Classification and Biochemical Activities*, University Park Press, Baltimore, 1968.
73. K. Matsushita, H. Ebisuya, M. Ameyama, and O. Adachi, *J. Bacteriol.* **174**, 122–129 (1992).
74. K. Matsushita, E. Shinagawa, O. Adachi, and M. Ameyama, *Proc. Natl. Acad. Sci. U.S.A.* **87**, 9863–9867 (1990)
75. K. Matsushita, E. Shinagawa, O. Adachi, and M. Ameyama, *Biochim. Biophys. Acta* **894**, 304–312 (1987).
76. H.L. Drake ed., *Acetogenesis*, Chapman & Hall, New York, 1994.
77. G. Wang and D.I.C. Wang, *Appl. Environ. Microbiol.* **47**, 294–298 (1984).
78. V. Muller and S. Bowien, *Arch. Microbiol.* **164**, 363–369 (1995).
79. L. Daniels, *Trends Biotechnol.* **2**, 91–98 (1984).
80. J.G. Ferry ed., *Methanogenesis*, Chapman & Hall, New York, 1993.
81. U. Deppenmeier, V. Müller, and G. Gottschalk, *Arch. Microbiol.* **165**, 149–163 (1996).
82. H.-J. Abken, M. Tietze, J. Brodensen, S. Baumer, U. Beifuss, and U. Deppenmeier, *J. Bacteriol.* **180**, 2027–2032 (1998).
83. C. Sasikala and C.V. Ramana, *Adv. Appl. Microbiol.* **42**, 97–218 (1996).
84. R.A. Herbert, *Trends Biotechnol.* **10**, 395–402 (1992).
85. S.E. Lowe, M.K. Jain, and J.G. Zeikus, *Microbiol. Rev.* **57**, 451–509 (1993).
86. K.S. Lundberg, D.D. Shoemaker, M.W.W. Adams, J.M. Short, J.A. Sorge, and E.J. Mathur, *Gene* **108**, 1–6 (1991).
87. S.W.M. Kengen, A.J.M. Stams, and W.M. de Vos, *FEMS Microbiol. Rev.* **18**, 119–137 (1996).
88. R.N. Schicho, K. Ma, M.W.W. Adams, and R.M. Kelly, *J. Bacteriol.* **175**, 1823–1830 (1993).
89. M. Krahe, G. Antranikian, and H. Märkl, *FEMS Microbiol. Rev.* **18**, 271–285 (1996).
90. A.H. Stouthamer, in J.R. Quayle ed., *International Review in Biochemistry, Microbial Biochemistry*, University Park Press, Baltimore, 1979, pp. 1–47.
91. M. Leno, O-W. Merten, and J. Hache, *Biotechnol. Bioeng.* **39**, 596–606 (1992).
92. R.P. Singh, M. Al-Rubeai, C.D. Gregory, and A.N. Emery, *Biotech. Bioeng.* **44**, 720–726 (1994).
93. M.B. Gu, P. Todd, and D.S. Kompala, *Cytotechnology* **18**, 159–166 (1996).
94. S.M. See, J.P. Dean and G. Dervakos, *Appl. Biochem. Biotechnol.* **60**, 251–301 (1996).
95. J.J. Vallino and G. Stephanopolous, *Biotech. Bioeng.* **41**, 633–646 (1993).
96. J.E. Bailey, *Science* **252**, 1668–1675 (1991).
97. G. Stephanopoulos and J.J. Vallino, *Science* **252**, 1675–1681 (1991).
98. S.C.O. Pak, A.M.N. Hunt, M.W. Bridges, M.J. Sleigh, and P.P. Gray *Cytotechnology* **22**, 139–146 (1996).
99. J.E. Bailey, A. Sburlati, V. Hatzimanikatis, K. Lee, W.A. Renner, and P.S. Tsai *Biotech. Bioeng.* **52**, 109–121 (1996).

See also AMMONIA TOXICITY, ANIMAL CELLS; BIOENERGETICS OF MICROBIAL GROWTH; CELL CYCLE; CELL CYCLE, EUKARYOTES; ELECTRON TRANSPORT; PYRUVATE, PRODUCTION USING DEFECTIVE ATPASE ACTIVITY.

ENZYMES, BAKING, BREAD MAKING

JOAN QI SI
Novo Nordisk Ferment Ltd.
Dittingen, Switzerland

KEY WORDS

Antistaling
Crumb elasticity
Dough
Enzyme
Firmness
Gluten strengthening
Lipase
Maltogenic α-amylase
Rheology
Xylanase

OUTLINE

INTRODUCTION

Bread is one of the most common and low-cost foods and is associated with traditions in many different countries. Bread is also closely related to *biotechnology*, which is synonymous with *high technology*. The term that relates these two subjects is *enzyme*.

Consumers have certain criteria for bread, including appearance, freshness, taste, flavor, variety, and a consistent quality. It is a great challenge for the baking industry to fulfill these criteria for several reasons. Firstly, the main ingredient of the bread—flour—varies due to varied wheat qualities, weather, and milling technology, although millers attempt to blend the wheat sorts to produce a good and consistent baking flour quality. It often proves difficult to satisfy both high-quality and low-cost standards at the same time. Secondly, because bread traditions differ, the baking industry needs various qualities of ingredients and procedures. For instance, an English toast bread with fine crumb structure and very soft texture is not popular with the French, who want baguettes with crispy crusts, large holes, and good chewiness in the crumb. Thirdly, healthier products are becoming more popular due to more intensive cultural exchanges and changes in consumer preferences. Some new variety breads can be made by simple adjustment of the formulation or baking procedure. However, others require the bakers to develop new techniques. Therefore, both millers and bakers need agents or process aids such as chemical oxidants, emulsifiers, and enzymes to standardize the quality of the products and diversify the product range.

For decades enzymes such as malt and microbial α-amylases have been used for bread making. Due to the changes in the baking industry and the demand for more varied and natural products, enzymes have gained more and more importance in bread formulations. Through new and rapid developments in biotechnology, a number of new enzymes have recently been made available to the baking industry. One example is pure xylanase, with single activity instead of traditional hemicellulase preparations, which improves the dough machinability. A lipase has a gluten-strengthening effect that results in more stable dough and improved bread quality, and a maltogenic α-amylase has a unique antistaling effect.

This article reviews the effects of these enzymes. It also presents synergistic effects when these enzymes are combined with either each other or traditional enzymes such as fungal α-amylase.

FUNGAL α-AMYLASES FOR BREAD MAKING

Wheat and thus wheat flour contain endogenous and indigenous enzymes, mainly amylases. However, the level of amylase activity varies from one type of wheat to another. The amount of α-amylases in most sound, ungerminated wheat or rye flours is negligible (1). Therefore, most bread flours must be supplemented with α-amylases, added in the form of malt flour or fungal enzymes.

Many methods are available for the determination of amylase activities, as reviewed by Kruger et al. (2). Other methods such as falling number (FN) and Brabender amylograph are used by the baking industry and millers to determine the amylase content correlating to bread making. Although FN is excellent for measuring the activity of cereal amylases including added malt flour, it is not suitable for measuring the activity of fungal α-amylases. Because fungal α-amylases are generally less thermostable, they are inactivated at temperatures near 65 °C. Therefore, fungal α-amylases cannot be detected by the standard FN method, which is conducted at 100 °C. Perten et al. (3) reported a modified FN method for measuring the fungal α-amylase activity in flour. However, in practice (B. Allvin, personal communication, January 1996), the method is less suitable for routine analysis because it is necessary to define the measuring conditions for each type of flour (because the indigenous and endogenous amylases vary in different flours) to determine the modified FN with added fungal α-amylases.

Fungal α-amylases act on the damaged starch, which varies depending on wheat sort and milling condition. Generally, flour made from hard winter wheat contains more damaged starch than soft wheat. The α-amylases widely used in the baking industry can hydrolyze amylose and amylopectin to release soluble intermediate-size dextrins of DP2–DP12 (4). α-Amylases provide fermentable sugar, which results in an increased volume, better crust color, and improved flavor. Due to hydrolysis of the damaged starch, a suitable dosage of α-amylases results in a desirable dough softening. However, extensive degradation of the damaged starch due to an overdose of α-amylases leads to sticky dough. Fungal glucoamylases are a possibility but less common as baking enzymes. Literature on glucoamylases for baking is scarce (5). Figure 1 illustrates the effect of a fungal amylase on bread quality in terms of bread volume, crumb structure, and dough characteristics.

The volume and crumb structure improve with increasing dosage of fungal α-amylases. Although a high dosage can provide a larger volume increase, the dough would be too sticky to work with. The optimum dosage is thus defined as the dosage with maximum reachable volume without a sticky dough. For the examples in Figure 1, the optimum dosage for both flours is 15 FAU/kg flour.

MALTOGENIC α-AMYLASE FOR ANTISTALING

Bread staling is responsible for significant financial losses. An estimated 3–5% of all baked goods produced in the United States are discarded every year, representing a value of \$1 billion (6). A maltogenic α-amylase has been

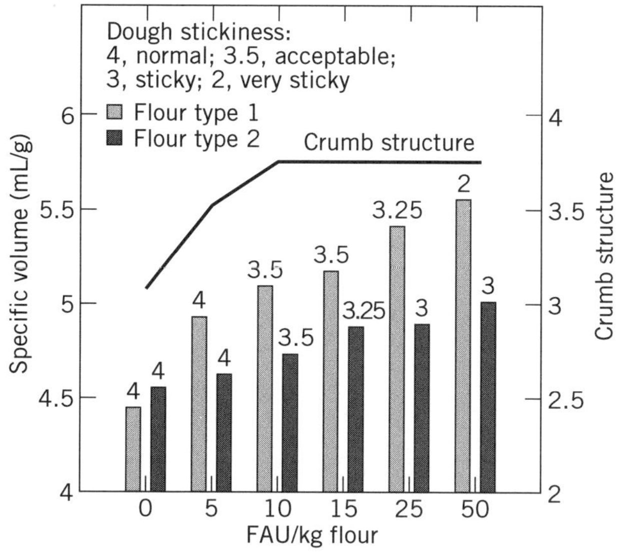

Figure 1. Dosage response of fungal amylase in different flours. FAU, units of fungal α-amylase. *Source*: Novo Nordisk Ferment.

found to have unique antistaling effects (7,8). It can degrade both amylose and amylopectin at the gelatinization temperature and produces mainly α-maltoses and a small amount of dextrins of DP1–DP12.

A study of the mechanism of antistaling effects of various amylases compared with monoglycerides (4) showed that this maltogenic α-amylase not only has a substantially improved antistaling effect compared with fungal α-amylases but also improves the elasticity of the bread crumb, which is an important factor that influences the palatability of bread. Both crumb softness and crumb elasticity are important characteristics for the description of crumb freshness perceived by consumers. Softness indicates the force needed to compress the crumbs, and elasticity indicates the resiliency or resistance given by the crumb while being pressed. The two texture characteristics may not necessarily correspond with each other; that is, the softest bread may not necessarily have the most elastic crumb texture or vice versa. Bread staling results in reduction of crumb elasticity and increase of crumb firmness.

Commercial amylases used in the baking industry are generally α-amylases that specifically hydrolyze the α-1,4 glycosidic linkages of amylose and amylopectin molecules in starch. Most known products are cereal enzymes derived from barley malt and the fungal α-amylase from *Aspergillus oryzae*. These amylases are effective in partially hydrolyzing damaged starch and are often added to flour as flour correction to develop desirable properties, such as oven spring and brown color to the crust. These amylases have limited antistaling effect due to their limited thermostability and are for the most part inactivated before the onset of starch gelatinization during baking. It was reported by Si and Simonsen (4) that fungal α-amylase, although it reduces the crumb firmness as shown in Figure 2, has no effect on reducing starch retrogradation.

Pullulanase, which specifically hydrolyzes the α-1,6 glycosidic linkages of amylopectin, does not have a positive effect and may even have a negative effect on antistaling.

The maltogenic α-amylase is a unique α-amylase that produces a significantly softer bread and maintains a high level of crumb elasticity during storage, as shown by Figure 2. Addition of maltogenic α-amylase in sponge and dough breads can prolong shelf life for at least 4 days longer compared with 0.5% powdered distilled monoglycerides, as shown by Figure 3. The crumb softness of the loaf with maltogenic α-amylase at day 7 was as soft as the loaf with 0.5% distilled monoglycerides at day 3, whereas the elasticity of the loaf with maltogenic α-amylase at day 7 is even higher than the loaf with distilled monoglycerides at day 3.

Maltogenic α-amylase has a thermostability between that of fungal α-amylase and thermostable bacterial α-amylase. Therefore, it is able to hydrolyze the glycosidic linkages during the gelatinization of starch during the baking process, but it does not excessively degrade the starch because it is inactivated during the later stage of baking.

Another major advantage of maltogenic α-amylase is its tolerance of overdosing during the bread-making process in the bakery. Too much fungal α-amylase can cause sticky dough or overbrowning of the crust. The bacterial α-amylase can easily be overdosed. Its pure endo action excessively degrades the starch during baking, causing collapse of the bread immediately after baking and sticky crumb during retail storage. Maltogenic α-amylase does not affect the dough rheological property because of its low activity at a temperature less than 35 °C. It is highly active only at the temperature during starch gelatinization and does not excessively degrade the starch but mainly produces small, soluble dextrins. The overdosing risk is thus much lower than with the other two types of amylases.

The bacterial α-amylases from *Bacillus subtilis* are able to inhibit staling by hydrolyzing glycosidic linkages within the amorphous areas of gelatinized starch. However, because of the high degree of thermostability, the enzymes can persist throughout baking and produce an excessive level of soluble dextrins. As a result the final product is often unacceptable, with a gummy or even sticky crumb texture that causes problems during slicing and retail storage. Figure 3 shows that the bacterial α-amylase gave the softest crumb but also a very gummy crumb with extremely low elasticity. The excessive degradation of starch during baking and storage creates a sticky crumb.

Soy β-amylase has been reported to have an antistaling effect. Our results, based on both sponge and dough and straight dough procedures, show that 3,000 BANU/kg flour, which is a quite high dosage, has a similar antistaling effect as 0.5% powdered distilled monoglycerides. The acid fungal α-amylase at a dosage of 100–200 AFAU/kg flour has an effect similar to 0.5% distilled monoglycerides (Fig. 3).

To understand the mode of action of this maltogenic α-amylase on bread staling, the effects of enzymes on the starch retrogradation were measured by differential scanning colorimetry (DSC), as shown in Figure 4. The endotherm area indicates the retrogradation of wheat starch. The endotherm area per gram starch gel containing monoglycerides was lower than the control, indicating its effect on retarding starch retrogradation. The endotherm area per gram of the starch treated with maltogenic α-amylase

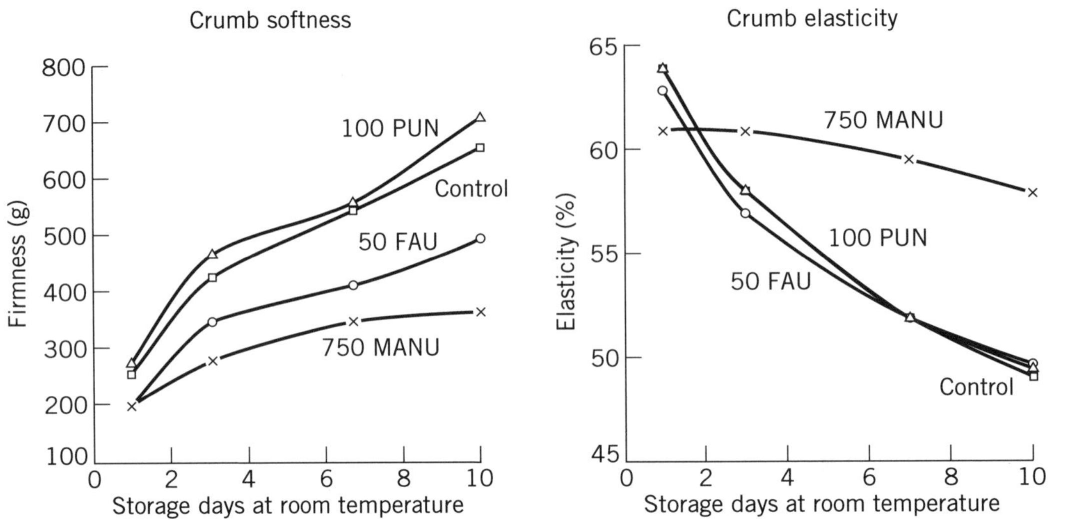

Figure 2. Effect of maltogenic α-amylase compared with fungal α-amylase and pullulanase on crumb softness and elasticity in toast bread. Straight dough procedure, controlled volume. FAU, units of fungal α-amylase/kg flour; MANU, units of maltogenic α-amylase/kg flour; PUN, units of pullulanase/kg flour.

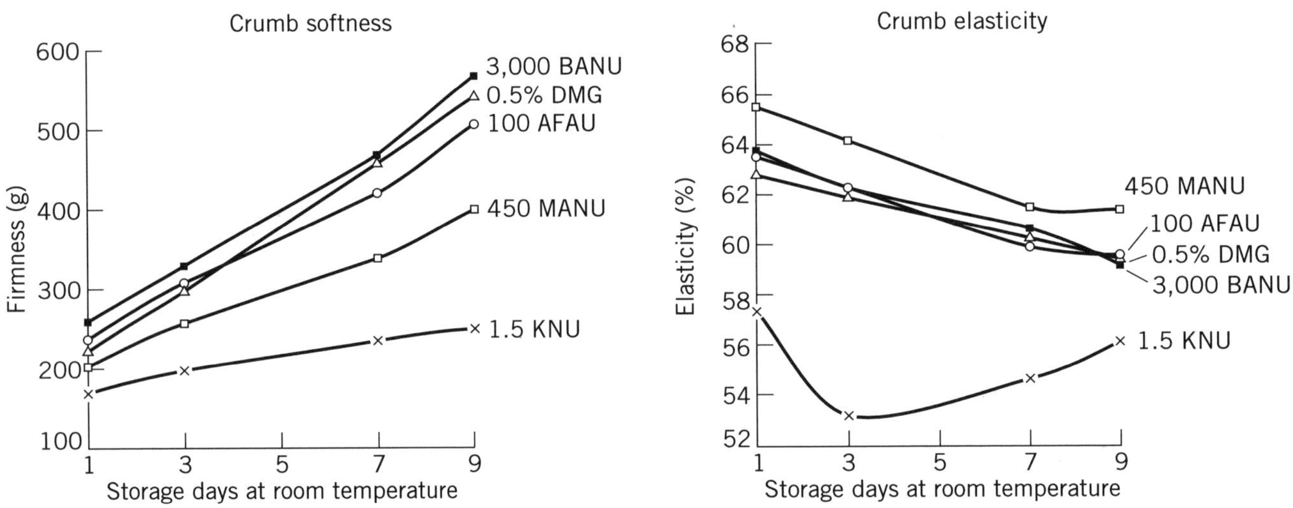

Figure 3. Effect of different amylases on crumb softness and elasticity. Sponge and dough procedure using American flour. Volume difference: maximum +/−3%, DMG, fine powdered distilled monoglycerides; MANU, units of maltogenic α-amylase/kg flour; KNU, units of bacterial α-amylase/kg flour; BANU, units of soy β-amylase/kg flour; AFAU, units of acid fungal α-amylase/kg flour. *Source*: Ref. 9.

was significantly lower than with the other samples; the difference increased after 4 days of storage and was even more pronounced after 9 days. This indicated a significantly slower rate of starch retrogradation. The fungal α-amylase had no effect on retarding starch retrogradation. The results indicate that an enzyme can reduce the crumb firmness by means of volume or crumb structure improvement, but it has no effect on retarding the retrogradation of starch during storage. Pullulanase had no effect on retarding starch retrogradation. This means that reducing the number of branches of amylopectin does not reduce the rate of bread staling and may even increase the staling rate, as reported earlier (10).

XYLANASES AND PENTOSANASES AS DOUGH CONDITIONERS

Xylanase and pentosanase, also being called hemicellulases, have long been used as dough-conditioning enzymes, especially in European-type bread. Pentosanases improve dough machinability and oven spring, resulting in a volume increase.

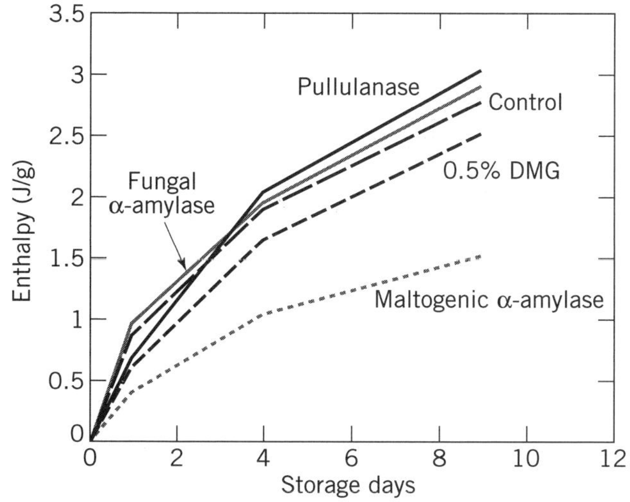

Figure 4. Effect of amylases on starch retrogradation compared with distilled monoglycerides measured by differential scanning colorimetry. *Source*: From Ref. 4.

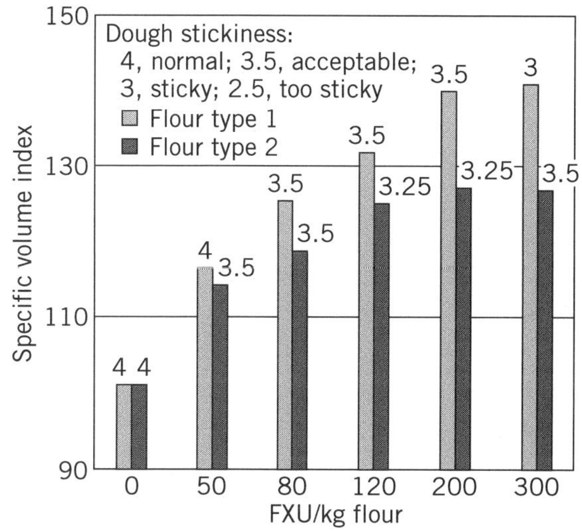

Figure 5. Dosage response of xylanase in different flours. *Source*: Novo Nordisk.

Through recombinant gene technology, some xylanases made from genetically modified organisms (GMO) are coming on the market (11). The benefit of using a xylanase instead of a traditional pentosanase preparation is that there are far fewer side activities in the xylanase product. Consequently, a lower dosage is needed to achieve the same effect with less risk of possible interference from side activities. The enzyme quality is also more consistent.

Pentosanases and xylanases can be overdosed due to overdegradation of wheat pentosans, thereby destroying the water-binding capability of the wheat pentosans. The result of overdosing is dough stickiness. A suitable dosage of these enzymes results in a desirable softening of dough, thereby improving the machinability. An optimum dosage is therefore defined as the dosage that gives the maximum improvement of the bread properties without causing dough stickiness. The optimum dosage of pentosanases or xylanases varies according to the flour, as illustrated in Figure 5. The optimum dosage for type 1 flour is 120–200 and for type 2 flour is 80 fungal xylanase units (FXU).

The true mechanism of xylanase in bread making has not been clearly elucidated, although a number of publications have attempted different approaches. The composition of pentosans varies depending on the flour type (12). The interaction between pentosans and gluten plays an important role that has not yet been elucidated. Most commercially available enzyme preparations used for studies consist of several enzyme activities including amylase, protease, and several hemicellulases. Works by Hamer (5,13) indicate that the use of pentosanases increased gluten coagulation in a diluted dough system. Si and Goddik (14) reported that a good baking xylanase increases the gluten strength measured by dynamic rheological methods on a Bohlin rheometer VOR. By measuring the gluten extracted from a dough system, the good baking xylanase increased storage modulus G′ and G″. At the same time the gluten became more elastic because the phase angle δ decreased, whereas the two other xylanases, which did not show good baking performance, had no significant effect on gluten strengthening.

Figures 6 and 7 show the effect of a commercially available baking xylanase on the rheological properties of gluten extracted from a wheat flour dough measured on Bohlin rheometer VOR (11). At the dosages that give good baking performance, 50–100 FXU/kg flour, the gluten was strengthened and exhibited a more elastic property. These improved rheological properties of the gluten explain the positive effect of the xylanase on bread making (i.e. increased oven spring, resulting in larger volume and improved dough stability). At an overdosed level of 400 FXU/kg flour, which provided a very sticky dough, the G′ is at the same level as the control and the phase angle shows a more viscous property of the gluten.

Different xylanase or pentosanase preparations have different effects on arabinoxylans in terms of their scission points and reaction products; therefore, they have different effects on bread making (15). Although many analytical methods are available for assaying pentosanases or xylanases using a wide range of substrates, it is still not possible to have an analytical method of which the enzyme units can correlate to the baking performance when comparing different types of xylanases or pentosanases. It has been reported (15) that no correlation to the baking performance was found when different enzymes from different sources were submitted to the same standardized assay and baking tests using purified substrate. One possible explanation is that once the wheat pentosans, both soluble and insoluble, were extracted from the flour, the mechanism of the interactions between wheat pentosans and gluten could no longer be assayed. Xylanase influences the gluten, as discussed earlier.

A possible mechanism of pentosanases mentioned by Hamer (5) is that the enzyme can offset the negative effects of insoluble pentosans present in the flour, because insoluble pentosans are regarded as having a negative effect on loaf volume and crumb structure. A study conducted by

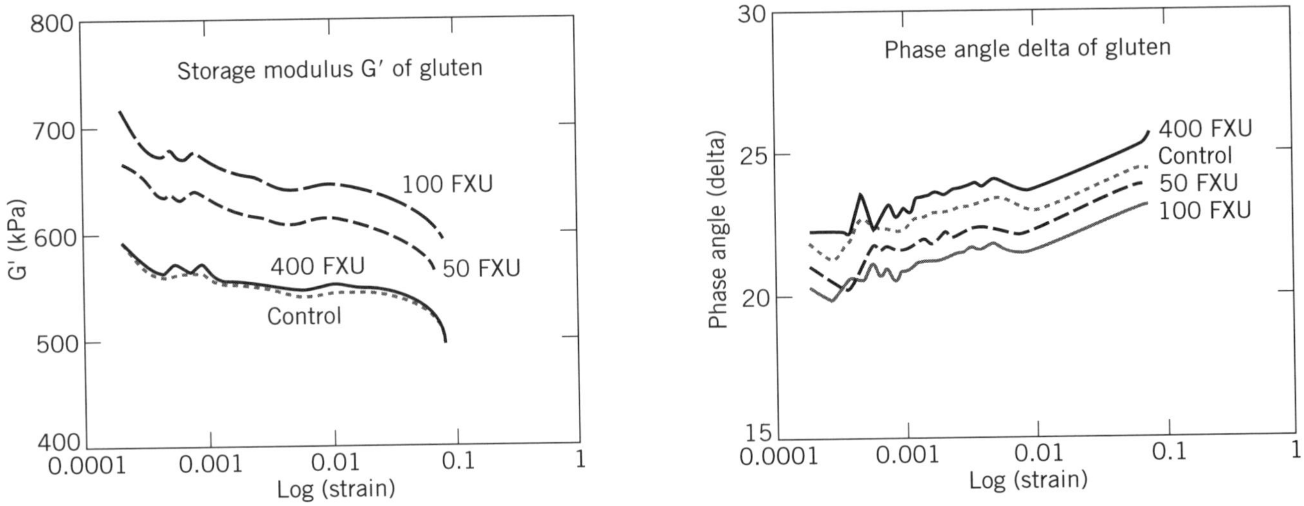

Figure 6. The effect of xylanase on the rheological properties of gluten.

Jakobsen and Si (15) using four different pure xylanases concluded that the best enzyme in terms of baking performance is the xylanase that has a certain level of activities toward both soluble and insoluble wheat arabinoxylans in a dough system. A xylanase having the major activity toward the insoluble arabinoxylans gave a dough too sticky to be accepted.

LIPASE FOR DOUGH CONDITIONING

The use of lipase for bread making was almost unknown just a few years ago, although it was reported as early as 1968 that the use of certain lipase preparations in a bread dough significantly retards the tendency of bread to become stale (16). Until recently, some 1,3-specific lipases were found to have a good dough-conditioning effect, in terms of improved dough rheological properties, increased dough stability upon overfermentation, increased oven spring resulting in a larger volume, and improved crumb structure in dough systems without added shortening (17,18).

Like most other enzymes, the effect of lipase also depends on the type of flour and the baking formulation. All three types of flour shown in Figure 7 are European baking flours with similar characteristics in terms of lipids, gluten content, and water absorption. For Meneba flour the optimum dosage was 2,000 LU/kg flour, and for Manitoba and Baguepi flours it was 1,000 LU/kg flour. An addition of 500–2,000 LU/kg flour results in a volume increase of 20–30%. At too high a dosage of lipase the dough becomes dry and stiff with a reduced volume (11).

Figure 8 shows the effect of the lipase in formulations with added fat or oil. Whereas lipase significantly improves bread quality in fat-free formulations or those with added oil, it has no beneficial effect when the formulation contains added hydrogenated shortening in terms of volume increase.

Because of its effect on volume increase and especially on the improved, more uniform crumb structure, the lipase

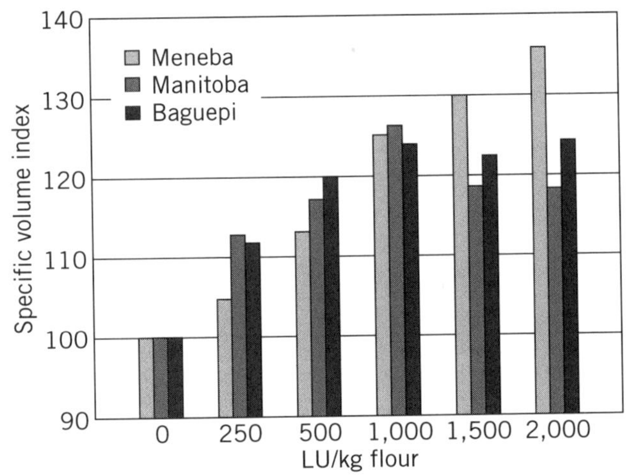

Figure 7. Dosage response of lipase in different flours.

improves crumb softness during storage. As discussed earlier, larger volume and more uniform crumb structure contribute to a softer bread. As illustrated in Figure 9, lipase reduces the crumb firmness compared with the control. It can provide a crumb softness as good as 3–6% shortening. However, it has no significant effect on improving crumb elasticity.

The mode of action of the lipase is still unknown. The postulation about producing monoglycerides in situ cannot explain the effects of the lipase demonstrated earlier. Monoglycerides are known to have an antistaling effect but have limited dough-conditioning effects compared with lipase. Furthermore, the contents of total lipids in most wheat flours is in the range of 1–1.5%. Regarding the limited hydrolysis degree (20) and limited quantity of triglycerides of saturated fatty acids in dough systems, an insufficient amount of monoglycerides of saturated fatty acids is produced to exhibit the effects described earlier. Besides, the monoglycerides produced by this 1,3-specific lipase

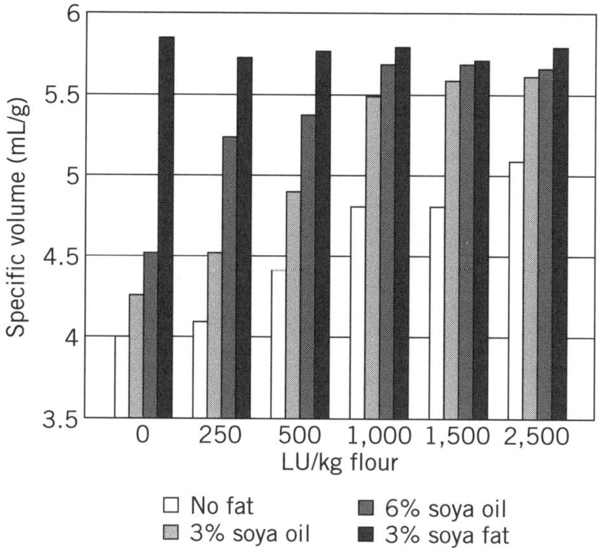

Figure 8. Effect of lipase in formulation with added lipids.

would be the second-position monoglycerides. It was reported (21) that first- and third-position monoglycerides are more able to form complexes with starch and thus having a retarding effect on starch staling.

Addition of lipase to a flour dough does not change the rheological properties of the dough measured by both Farinograph and Extensograph (9) with optimum water absorption. This is a good characteristic for bread making, because most bakers do not like major changes in the dough system. However, when we measured the rheological property of the gluten treated by the lipase using the dynamic rheological method, we found that the reason for the good dough-conditioning effect of this lipase is the increase of gluten strength, as shown in Figure 10. The gluten taken from lipase-treated wheat flour dough is significantly stronger, with an increased G′ and more elasticity, a lower δ than the control. An overdosing of lipase results in too strong of a gluten complex with a very high G′, which gives a dough that is too stiff and a smaller volume increase, as mentioned earlier. The gluten treated with lipase has a high G filt and increased relaxation time, indicating an increase of cross-links in the gluten networks measured by the dynamic stress relaxation method (20). It is not yet possible to elucidate the nature of these extra cross-links.

The knowledge of the chemistry of wheat lipids in bread making is still fragmentary. It was assumed that glycolipids, through hydrogen bonds and hydrophobic interactions, form a linkage between gliadin and glutenin (22,23). Bekes et al. (24) reported that there is a strong positive correlation between the lipid-mediated aggregates and loaf volume. However, overaggregates of the gluten may be responsible for the deterioration of bread-making quality as reported by Weegels and Hamer (25). A possible explanation of why this lipase has the positive effective is that it degrades some of the lipids, namely triglycerides, thereby reducing the overaggregation and forming a more stable gluten network.

A recent study conducted at Lund University (26) indicated that lipase increased the thermostability of the reversed hexagonal phase of the liquid–crystalline phase during heating up to 100 °C. This observation is suggested (27) as a possibility for one of the mechanisms of lipase in bread making.

Because lipases are gaining more and more attention in the baking industry, studies in the near future should give us a better understanding of the mode of action of lipase for bread making.

OXIDASES FOR BREAD MAKING: REPLACING CHEMICAL OXIDANTS

Oxidants, such as ascorbic acid, bromate, and azodicarbonamide (ADA), are widely used for bread making and have

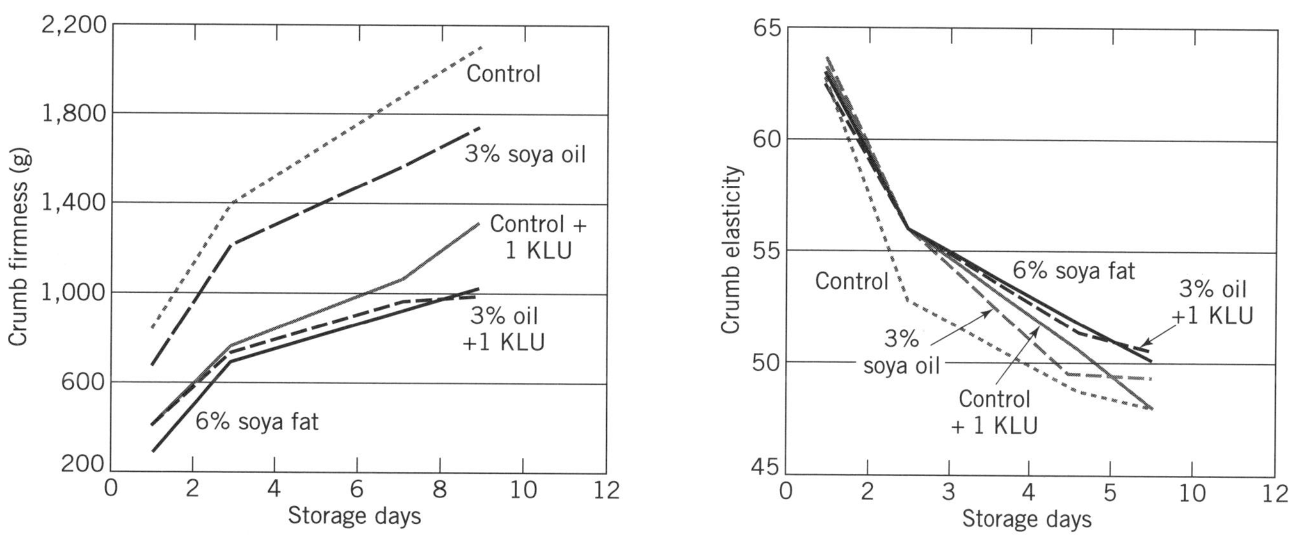

Figure 9. The effect of lipase on crumb softness compared with shortening. 1 KLU = 1,000 LU/kg flour of lipase. *Source*: From Ref. 19.

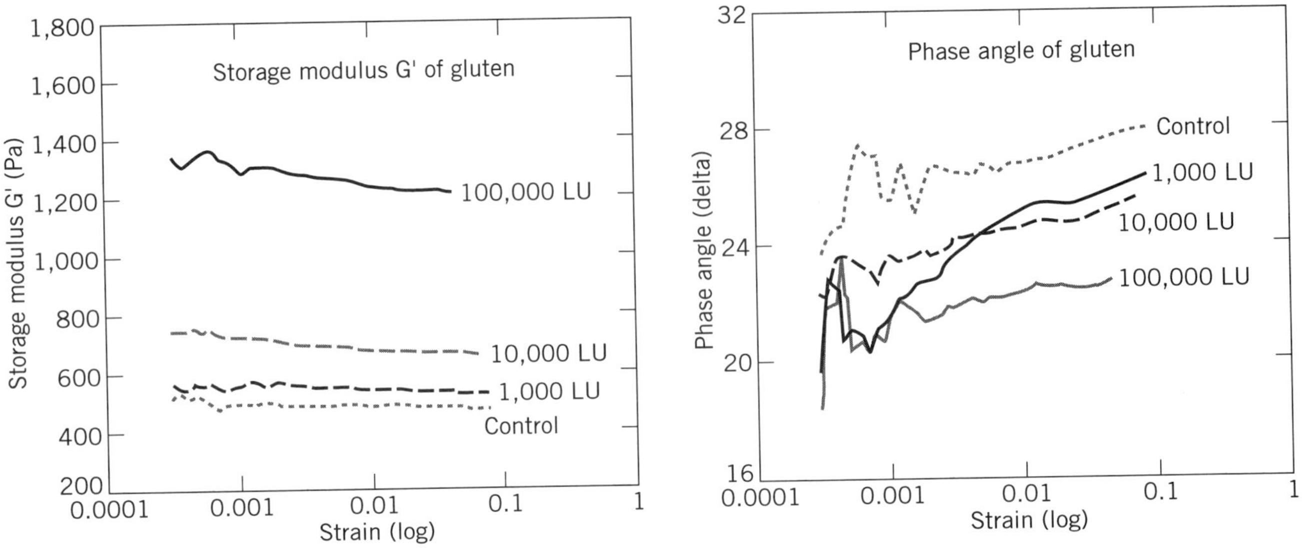

Figure 10. The effect of lipase on the rheological properties of gluten.

been well studied and reported on. However, the true mechanism of the oxidants in bread making has not been established, although a number of hypotheses were offered by numerous studies. It is generally known that bromate is a slow-acting oxidant and becomes active at high temperatures. In relation to baking, it has the maximum effect in the later stages of proofing and in the early stages of baking, whereas ascorbic acid and other oxidants are fast-acting oxidants and having the maximum effect during mixing and proofing.

Increasing demands by consumers for more natural products with fewer chemicals and especially the possible risks with bromate in food have created the need for bromate replacements. Therefore, oxidases are gaining more and more attention within the baking industry. Although glucose oxidase for bread making has been known since 1957 (28) and lipoxygenase, lysyl oxidase, sulfhydryl oxidase (29), peroxidase (30), laccase (31), and transglutaminase have been reported to have good oxidizing effects, the knowledge about oxidases for bread making is scarce and only glucose oxidase is currently available commercially.

Glucose oxidase has good oxidizing effects that result in a stronger dough. It can be used to replace oxidants such as bromate and ascorbic acid in some baking formulations and procedures. In other formulations it is an excellent dough strengthener along with ascorbic acid. Examples of applications of glucose oxidase for bread making are given elsewhere in this volume.

SYNERGISTIC EFFECTS OF ENZYMES

Enzymes for Dough Conditioning

Using enzyme combinations for bread making is not new. It is well known that hemicellulase or xylanase used in combination with fungal α-amylase has synergistic effects.

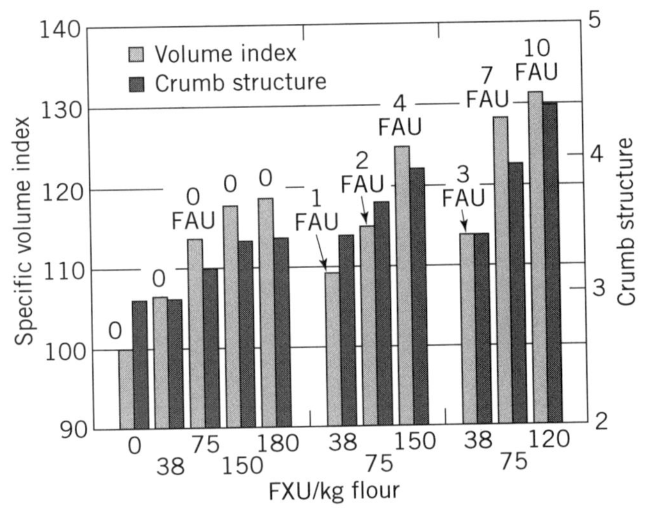

Figure 11. Synergistic effect of amylase and xylanase. FAU, fungal α-amylase units; FXU, fungal xylanase units. *Source*: Novo Nordisk, Pentopan Mono.

As shown in Figure 11, a high dosage of a pure xylanase may give some volume increase, but, the dough with this dosage of xylanase would be too sticky to be handled in practice. When the xylanase is combined with even a very small amount of fungal α-amylase, a lower dosage of xylanase with the α-amylase provides a larger volume increase and better overall score without the dough stickiness problem. The combination of xylanase and fungal amylase still has its limitations because of the dough stickiness easily caused by these enzymes.

Because of its effect on gluten strengthening, lipase improves dough stability against overfermentation. Figure 12 shows the combination of the enzymes for a French ba-

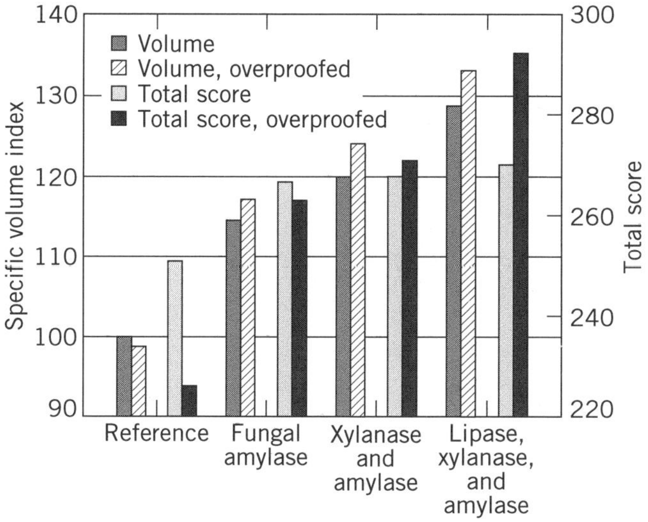

Figure 12. The synergistic effects of enzymes for the French baguette. *Source*: Novo Nordisk, Biotimes.

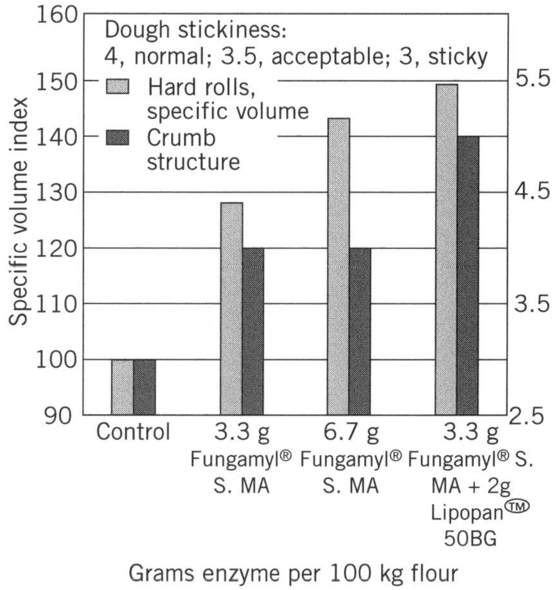

Figure 13. The combination of amylase or xylanase with lipase for hard rolls and pan bread. *Source*: Novo Nordisk Ferment.

guette (trials conducted at ENSMIC Baking Laboratory, Paris). Without any enzyme, the baguette had poor stability on overfermentation. With the addition of α-amylase alone or in combination with xylanase, the volume and overall scores of the bread improved. When α-amylase, xylanase, and lipase were used together, the overall scores were further improved. The dough exhibited especially good stability on overfermentation, which resulted in significantly better baguettes.

Because lipase does not make the dough sticky and significantly improves dough stability and crumb structure, the synergistic effects between xylanase or amylase and lipase provide many possibilities for improved bread quality. Figure 13 shows the results of hard rolls from Meneba flour. A dosage of 3.3 g Fungamyl® Super MA consisting of fungal α-amylase and xylanase gave satisfactory dough consistency but a limited volume increase. A doubling of this dosage gave a larger volume increase but a dough too sticky to be handled. When combining the low dosage of Fungamyl® Super MA with the lipase, the volume increase was significantly higher without any dough stickiness. Figure 14 shows that the combination of lipase with α-amylase or xylanase can provide a fine, silky, and uniform crumb structure to loaves based on a straight dough process.

Enzymes for Extending Shelf Life and Improving Crumb Softness and Elasticity

As mentioned earlier, the maltogenic α-amylase is a true antistaling enzyme that affects neither bread volume nor

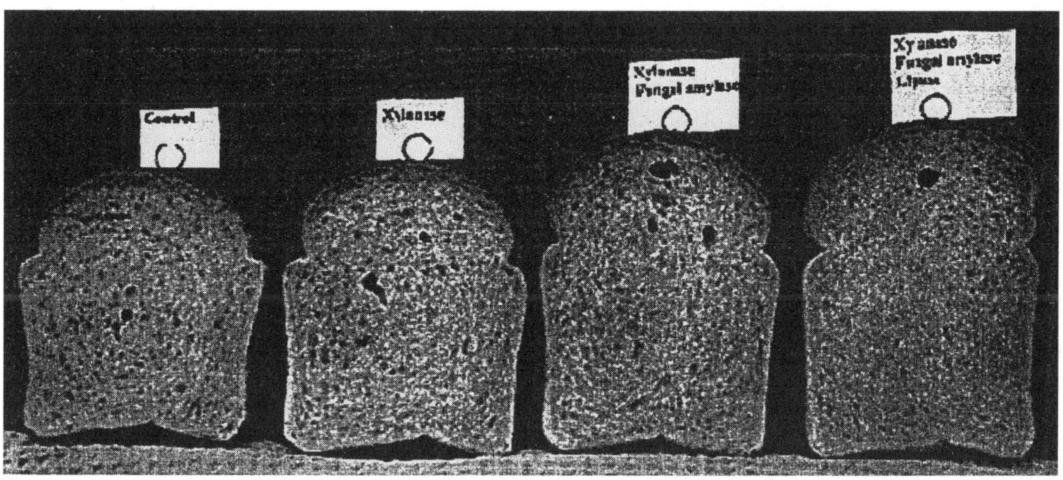

Figure 14. Synergistic effect of lipase in combination with α-amylase and xylanase.

crumb structure. Therefore, it is more feasible to use this enzyme in combination with enzymes such as fungal α-amylase, xylanase, and lipase to ensure the other bread quality parameters such as volume, dough stability, and crumb structure.

Most enzymes or other bread-improving ingredients including emulsifiers can improve the crumb softness due to the effects on bread volume and/or crumb structure, but they have a limited effect on crumb elasticity. The exception is the maltogenic α-amylase. Although the fungal α-amylase and xylanase or that in combination with thermostable bacteria α-amylase can reduce the crumb firmness during storage, as shown in Figure 15, the bread with the maltogenic α-amylase was significantly softer and had more elastic crumb than the other samples.

The preceding example also illustrates that the crumb firmness and elasticity may not necessarily correspond with each other. Many ingredients can reduce crumb firmness without affecting crumb elasticity, such as the bacterial α-amylase shown earlier. Our studies have also found that emulsifiers such as sodium stearoyl lactylate (SSL) may also have a negative effect on crumb elasticity.

Figure 16 shows different enzyme combinations compared with distilled monoglycerides in a sponge and dough system using a strong Canadian flour. The maltogenic α-amylase has the synergistic effect together with lipase, xylanase, and fungal α-amylase: 30 ppm of the maltogenic α-amylase together with the other enzymes gave a softer crumb during storage than using 45 or 75 ppm of the maltogenic α-amylase alone. The combination of the four enzymes (fungal α-amylase, xylanase, lipase, and maltogenic α-amylase) gave the softest crumb throughout the entire storage period of 9 days. The softness of the bread with these four enzymes at day 9 was at the same level as the loaf with 0.5% distilled monoglycerides at day 3. However, because only the maltogenic α-amylase has an effect on improving crumb elasticity, the more maltogenic α-amylase added, the more elastic is the bread crumb.

Enzymes for Dough Strengthening

Glucose oxidase has good oxidizing effects that result in stronger dough. It can be used to replace oxidants such as bromate and ADA in some baking formulations and procedures. In other formulations such as with ascorbic acid, it is an excellent dough strengthener (Fig. 17).

Use of glucose oxidase creates a dry and strong dough. A high dosage of fungal α-amylase gives the dough extensibility. The combination of these two enzymes can therefore achieve a synergistic effect. Figure 18 shows the combination of glucose oxidase and fungal α-amylase compared with ascorbic acid. When the two enzymes are used together with a smaller amount of ascorbic acid, the dough is not only very stable but also absorbs 1–2% more water, resulting in a greater volume increase and a crispier crust.

Glucose oxidase combined with fungal α-amylase can replace the bromate in some bread formulations. Figure 19 shows a type of South American bread, Marraqueta. This procedure requires a very stable dough because after two series of 50 min of proofing the dough is turned around before baking. A dough without strong dough stability will collapse when turned. The basic formulation contains both ascorbic acid and K-bromate. By adding glucose oxidase and fungal α-amylase instead of bromate, the final bread has a much improved appearance and a volume increase of approximately 40%.

In the previous sections each of the most commonly used enzymes for bread making has been reviewed, because of the rapid development of biochemistry and enzyme technology and the increasing interest in enzyme applications for the baking industry, more and improved enzyme products will be introduced in the near future. There will also be more reports and studies that will bring

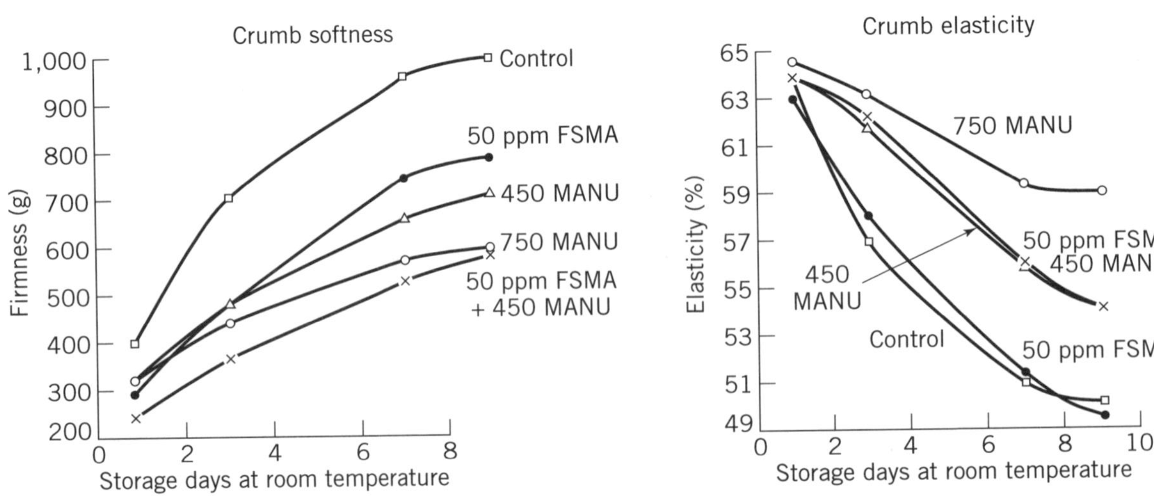

Figure 15. Antistaling effect of maltogenic α-amylase compared with thermostable bacteria amylase. 50 ppm FSMA (Novo Nordisk Fungamyl® Super MA) corresponds to 12.5 FAU and 112.5 FXU/kg flour; MANU, units of maltogenic α-amylase/kg flour.

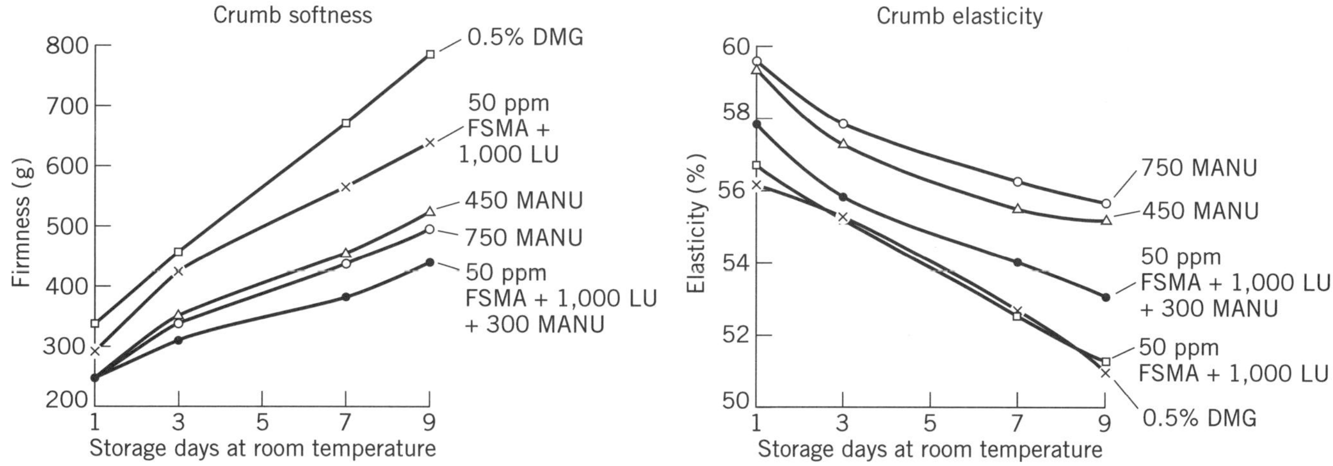

Figure 16. The synergistic effect of maltogenic α-amylase in combination with fungal α-amylase, xylanase, and lipase compared with distilled monoglycerides in American types of sponge and dough bread. FSMA (Novo Nordisk Fungamyl® Super MA) corresponds to 12.5 FAU and 112.5 FXU/kg flour; LU, units of lipase (Lipopan™)/kg flour; MANU, units of maltogenic α-amylase/kg flour; DMG, fine powder distilled monoglycerides. *Source*: Novo Nordisk Ferment.

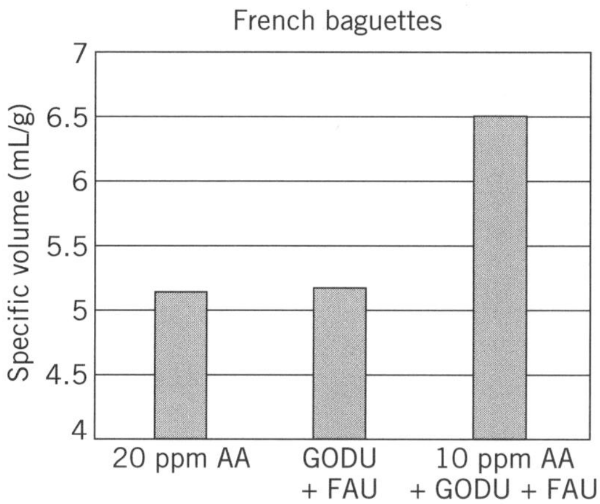

Figure 17. The combination of glucose oxidase and fungal amylase compared with ascorbic acid. The dosage of glucose oxidase is 300 GODU/kg flour plus 50 FAU. *Source*: Novo Nordisk Ferment.

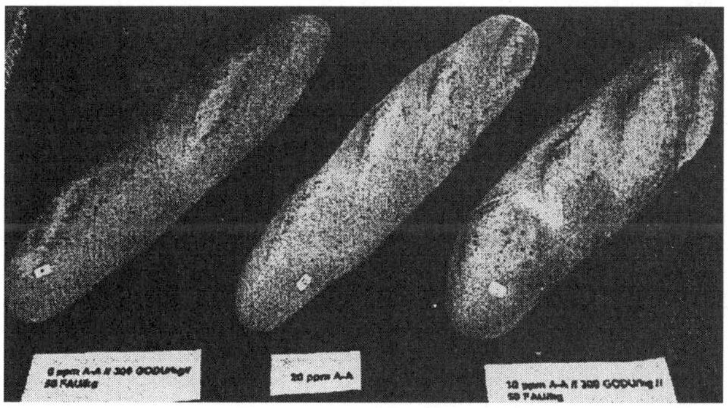

Figure 18. Glucose oxidase and fungal α-amylase for French baguettes.

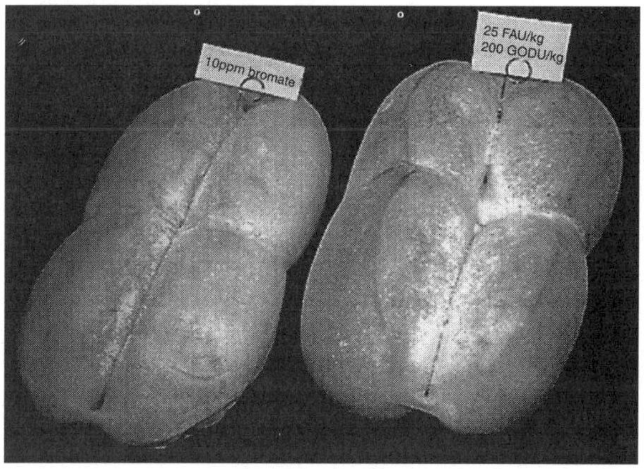

Figure 19. Glucose oxidase and fungal amylase as bromate replacer in Marraqueta. *Source*: Novo Nordisk.

us a better understanding of the mechanism of baking enzymes. In turn this understanding will lead to further improved products. An exciting future for baking enzymes lies ahead.

BIBLIOGRAPHY

1. H.W. van Dam and J.D.R. Hille, *Cereal Foods World* **37**, 245 (1992).
2. J.E. Kruger, D. Lineback, and C.E. Stauffer, in *Enzymes and Their Role in Cereal Technology*, American Association of Cereal Chemists, 1987, pp. 117–130.
3. H. Perten, *Cereal Chem.* **61**, (1984).
4. J.Q. Si and R. Simonsen, *Proc. Int. Symp. AACC / ICC / CCOA*, Beijing, November 16–19, 1994.
5. R.J. Hamer, in G. Tucker and L.F.J. Woods eds., *Enzymes in Food Process*, 1991.

6. R.E. Hebeda, et al. *Cereal Foods World* **35**, 453 (1990).
7. H. Outtrup and B.E. Norman, *Starch* **36**, 405 (1984).
8. European Pat. EP494233 (September 26, 1990), T. Olsen.
9. J.Q. Si, *Annual Meeting of the American Association of Cereal Chemists*, Baltimore, Md., September 15–19, 1996.
10. M.L. Martin and R.C. Hoseney, *Cereal Chem.* **68**, 503 (1991).
11. J.Q. Si, *Proc. 6th Mtg. Industrial Applications Enzymes*, Barcelona, November 29–30, 1995.
12. G. Cleemput, S.P. Roels, M. Van Oort, P.J. Grobet, and J.A. Delcour, *Cereal Chem.* **70**, 324 (1993).
13. R.J. Hamer, P.L. Weegels, J.P. Marseille, and M. Kelfkens, in Y. Pomeranz ed., *Wheat Is Unique*, American Association of Cereal Chemists, 1989, p. 467.
14. J.Q. Si and Goddik, *Annual Meeting of the American Association of Cereal Chemists*, Miami, Fla., October 1993.
15. T.S. Jakobsen and J.Q. Si, *Proc. Congr. Wheat Structure Biochemistry and Functions*, Reading, U.K.
16. U.S. Pat. 3,512,992 (April 2, 1968), A. Cooke and H. Johannson.
17. PCT Pat. WO 94/04035 (August 21, 1992), T. Olsen and J.Q. Si.
18. J.Q. Si, *Proc. Int. Symp. AACC / ICC / CCOA*, Beijing, November 16–19, 1994.
19. J.Q. Si, *Food Tech. Eur.* **3**, 60–64 (1996).
20. J.Q. Si and T.T. Hansen, *Proc. Int. Symp. AACC / ICC / CCOA*, Beijing, November 16–19, 1994.
21. N. Krog, *AOCS*, Hawaii.
22. R.C. Hoseney, et al. *Cereal Chem.* **47**, 135 (1970).
23. W. Bushuk, F. Bekes, G.J. MacMaster, and U. Zawistonwska, in A. Graveland and J.H. Moonen eds., *Gluten Protein: The Proceedings of the International Workshop on Gluten Protein*, 2nd ed., Wagenigen, 1984, p. 101.
24. F. Bekes, et al., *J. Cereal Sci.* **16**, 129 (1992).
25. P.L. Weegels and R.J. Hamer, *Cereal Food World* **37**, 379 (1992).
26. J. Borné, M.S. Thesis, Lund University, Sweden, 1996.
27. L. Dybdal, J.Q. Si, J. Born, and A. Eliasson, *Proc. First Eur. Symp. Enzymes Grain Processing*, Noordwijkerhout, The Netherlands, December 2–4, 1996.
28. U.S. Pat. 2,783,150 (February 26, 1957), H.G. Luther.
29. G. Matheis and J.R. Whitaker, *J. Food Biochem.* **11**, 309 (1987).
30. PCT Pat. WO 94/28729 (June 11, 1993), J.Q. Si.
31. PCT Pat. WO 94/28728 (June 13, 1994), J.Q. Si.

ADDITIONAL READING

V. Bolnedi, and C. Hoseney, Ph.D. Dissertation, Kansas State Univ., Manhattan, Kans., 1997.

J. Maat, et al. *Xylanases and Their Application in Bakery*, 1991.

Novo Nordisk, *Biotimes* (September 1996).

See also ENZYMES, PROTEIN HYDROLYSIS; FOOD PROCESS ENGINEERING; YEAST, BAKER'S.

ENZYME KINETICS. See KINETICS, ENZYMES.

ENZYMES, DETERGENT

M. S. SHOWELL
Procter & Gamble
Cincinnati, Ohio

KEY WORDS

Amylase
Cellulase
Detergents
Granulates
Lipase
Peroxidase
Protease
Stability

OUTLINE

INTRODUCTION

The use of enzymes as purposefully added ingredients to laundry products dates back to 1913 when Otto Röhm patented the use of crude pancreatic extracts in laundry presoak compositions to improve the removal of biological stains (1). It wasn't until the early 1960s, however, with the isolation and commercial production of alkaline active proteases, that enzymes became widely used in laundry products. By 1969, at least 50% of European detergents contained a protease (2). Within 6 years of the introduction of alkaline proteases in detergents, 80% of all European laundry detergents contained enzymes (3). Today, the penetration of enzymes into laundry detergents across the globe is quite high. Most premium and lower-tier laundry products contain at least one enzyme, usually a protease, and many are formulated with up to four different enzyme activities—protease, amylase, lipase, and cellulase. Enzymes in laundry and cleaning products account for about 40% of the more than 1 billion dollar industrial enzyme market, with detergent proteases accounting for about 50% of the volume (4). In the United States, the detergent sector is the largest market for enzymes, accounting for about 20% of total enzyme demand and more than 25% of industrial enzyme sales (5). The move to compact detergents, beginning in 1987, helped raise the demand for detergent enzymes, and global enzyme sales doubled between 1989 and 1994.

Four factors are responsible for the increased reliance of detergent manufacturers on enzyme technology. First, they provide consumer-recognizable cleaning benefits. Historically, proteases and amylase have been used to improve the removal of proteinaceous and starch-based soils, respectively. More recently, detergent-compatible lipases have been introduced to help clean greasy and oily stains arising from foods and body soils. Second, some enzymes enable the detergent formulator to add completely new performance benefits. Cellulases are the best and most recent example where the development of cost-effective, alkaline-active cellulases has enabled the delivery of consumer-valued benefits such as color maintenance and fabric restoration. Third, enzymes are highly weight efficient. The ability to act catalytically imparts a space efficiency to enzymes that is highly desirable to the formulator of compact detergents. The fourth factor contributing to the increased use of enzymes in detergents is that in recent years the performance/cost ratio for enzymes has improved significantly as a result of the availability of more efficient enzymes and the industry trend toward reduced pricing.

Current market trends and consumer needs are influencing the development of enzymes for detergent applications, with the emphasis on enzymes that have improved performance/cost ratios, increased cold water activity, and improved compatibility to other detergent ingredients. In addition, enzyme suppliers and detergent manufacturers are actively pursuing the development of new enzyme activities that address the consumer-expressed need for improved cleaning, fabric care, and antimicrobial benefits. For applications beyond laundry and cleaning products, enzymes with low allergenicity are being developed to minimize worker and consumer safety concerns.

THE DETERGENT ENZYME MARKET

As noted in the "Introduction", enzymes have long been of interest to the detergent industry for their ability to aid in the removal of soils and stains as well as to deliver unique benefits that cannot otherwise be obtained with conventional detergent technologies. As a result, the detergent enzyme market has grown nearly 10-fold during the past 20 years. In 1997, sales of detergent enzymes totaled about \$550 million. U.S. demand accounted for about \$160 million and is expected to grow to \$260 million by 2001 (6). Through the 1980s and early 1990s the market was split between five major suppliers: Novo-Nordisk A/S of Denmark with the dominant market share (>55%); Gist-Brocades N.V. in The Netherlands; Genencor International in the United States; Solvay in Belgium; and; Showa-Denko in Japan. These suppliers marketed a full range enzymes for liquid and powder detergents. Beginning in 1995, however, there was considerable rationalization in the detergent enzyme market owing to the relatively high cost of manufacturing coupled with increased pressure from detergent manufacturers to drive down raw material costs. Genencor International purchased the detergent enzyme businesses of Gist-Brocades and Solvay, and Novo-Nordisk acquired Showa-Denko's detergent enzyme business. Today, Novo and Genencor are the main suppliers of detergent enzymes, supplying more than 95% of the global market. Table 1 provides examples of the types of enzymes available from these two suppliers.

CURRENT DETERGENT ENZYMES

The majority of enzymes used in detergents today are hydrolases; the catalytic event involves addition of water across a chemical bond. In addition, most detergent enzymes exhibit *endo* specificity, that is, they preferentially attack internal bonds as opposed to terminal linkages.

Before the conversion of the detergent market to compact product forms, the use of enzymes in detergents was limited primarily to proteases. The market conversion to compacts, which require high levels of cleaning performance from low product dosages, fueled the interest in formulating new enzyme activities that can provide a more robust cleaning profile. This, coupled with recent advances in genetic and protein engineering, has led to new classes of enzymes being available for detergents, including amylases, lipases, cellulases, and peroxidases (Table 1). Each of these general enzyme classes is discussed in detail in the following sections.

Proteases

Most powder and liquid laundry detergents on the market today, both low density and compact, contain a protease (7). Protease enzymes enhance the cleaning of protein-based soils, such as grass and blood, by catalyzing the breakdown of the constituent proteins in these soils through hydrolysis of the amide bonds between individual amino acids. Proteases serve a multifunctional role in the overall cleaning process. They boost the efficacy of surfactant and dispersant systems by degrading proteinaceous soils into smaller, more dispersible fragments, thereby directly facilitating the soil removal process. In addition, they decrease the redeposition of proteins, especially hydrophobic proteins such as those found in blood, resulting in improved overall whiteness (8).

Figure 1 shows the dose-response curves for removal of grass stain with typical detergent proteases in standard North American liquid and powder detergent formulations. The plateau observed at higher enzyme concentrations is typical of most detergency evaluations of enzymes. Typically, detergent formulators dose enzymes at the breaking point of such curves in order to obtain optimum cleaning performance at lowest possible cost.

Figure 2 further illustrates the cleaning benefits of protease. Figure 2**a** shows an electron micrograph of blood-stained fibers from a polyester swatch washed under North American conditions in a liquid detergent without protease. The hemoglobin in the blood stain has been imaged with a colloidal-gold-labeled antibody marker to help vi-

Table 1. Detergent Enzyme Types and Suppliers: 1997

	Protease			Amylase			Cellulase		
Supplier	Alkaline	High alkaline	Bleach stable	Conventional	Bleach stable	Lipase	Multi-component	Mono-component	Peroxidase
Genencor	Optimase	Opticlean Purafect Properase	Purafect OxP	Maxamyl Rapidase	Purafect OxAm Purafect HpAm	Lumafast Lipomax		Puradax	
Novo	Alcalase	Savinase Esperase Kannase	Everlase	Termamyl BAN	Duramyl	Lipolase Lipolase Ultra LipoPrime	Celluzyme	Carezyme Endolase	Guardzyme

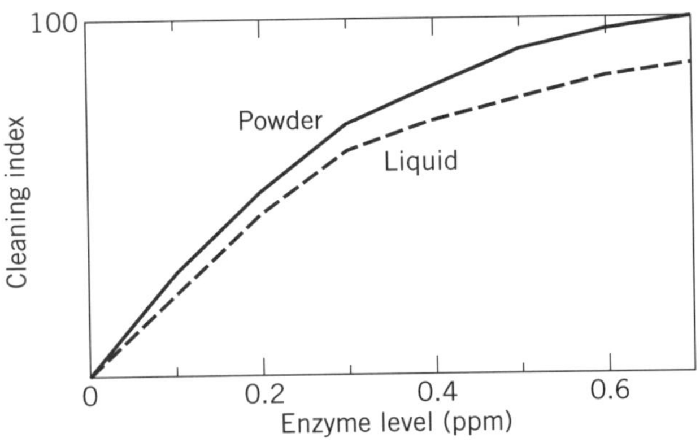

Figure 1. Protease dose response on grass. Detergent tests run at 20 °C with 1 mM mixed Ca/Mg hardness and manufacturer's recommended detergent dosage. Removal of grass on cotton based on visual grades from expert graders.

sualize remaining protein. There is a considerable amount of residual blood protein left on the fibers after washing. Figure 2**b** shows similarly stained and treated fibers that have been washed under the same conditions, in the same liquid detergent, but with 0.5 ppm protease added. There is a marked reduction in the amount of residual blood protein consistent with the visual observation of enhanced stain removal.

As shown in Table 1, a number of different commercial detergent proteases are currently being marketed. However, they are all closely related, nonspecific endopeptidases, with molecular weights around 27 kDa and highly homologous primary amino acid sequences (Fig. 3). Current commercial detergent proteases are derived from various species of *Bacillus subtilis* and are characterized by a common catalytic triad consisting of serine, aspartic acid, and histidine. The serine hydroxyl group functions as a nucleophile, attacking the peptide amide bond to form a tetrahedral intermediate that undergoes hydrogen trans-

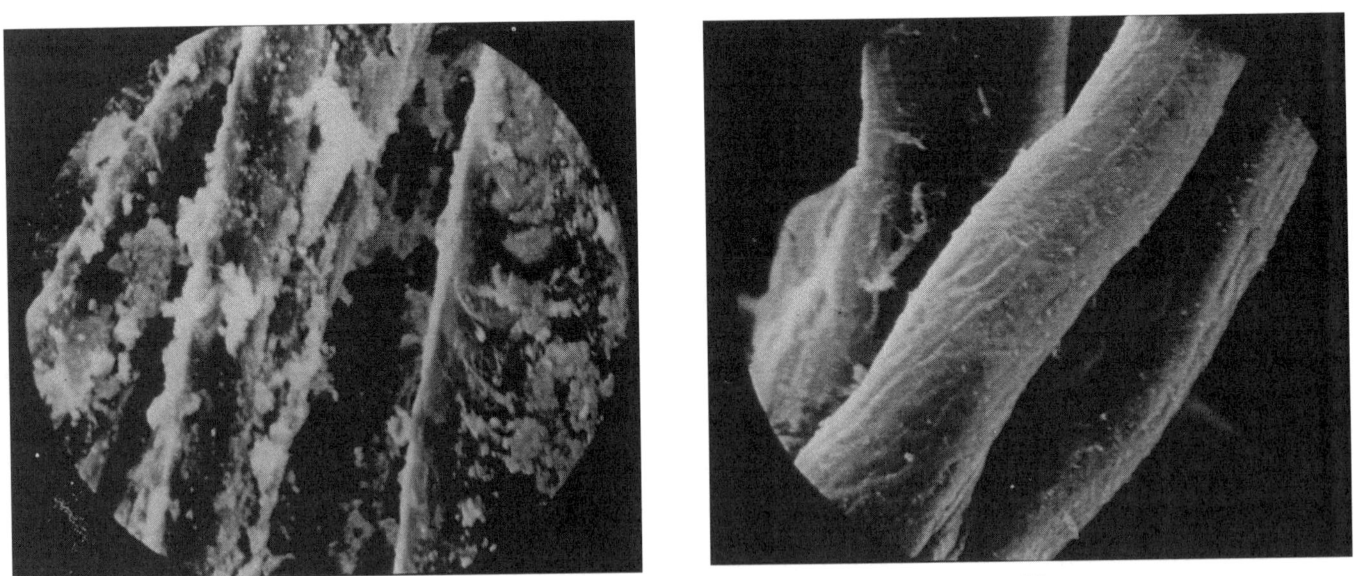

(a) **(b)**

Figure 2. (**a**) Transmission electron micrograph of blood-stained fibers after washing with a nil-protease liquid detergent. The residual hemoglobin has been marked using colloidal-gold-labeled antibodies. (**b**) Transmission electron micrograph of blood-stained fibers after washing with a protease-containing liquid detergent. The residual hemoglobin has been marked using colloidal-gold-labeled antibodies.

1

1 AQSVPWGISRVQAPAAHNRGLTGSGVKVAVL

2 AQSVPYGVSQIKAPALHSQGYTGSNVKVAVI 31

3 GQTVPYGIPLIKADKVQAQGFKGANVKVAVL

1 DTGISTHPDLNIRGGASFVPGEPST QDGNGHGTHVAGTIAALNNSIGVLGVAPSAELYAVKVLGAS

2 DSGIDSSHPDLKVAGGASMVPSETNPFQDNNSHGTHVAGTVAALNNSIGVLGVAPSASLYAVKVLGAD 99

3 DTGIQASHPDLNVVGGASFVAGEAYN TDGNGHGTHVAGTVAALDNTTGVLGVAPSVSLYAVKVLNSS

1 GSGSVSSIAQGLEWAGNNGMHVANLSLGSPSPSATLEQAVNSATSRGVLVVAASGNSG AGSISYPAR

2 GSGQYSWIINGIEWAIANNMDVINMSLGGPSGSAALKAAVDKAVASGVVVVAAAGNEGTSGSSTVGY 167

3 GSGSYSGIVSGIEWVTTNGMDVINMSLGGASGSTAMKQAVDNAYARGVVVVAAAGNSGSSGNTNTIGY

1 YANAMAVGATDQNN NRASFSQYGAGLDIVAPGVNVQSTYPGSTYASLNGTSMATPHVAGAAALVK

2 PGKYPSVIAVGAVDSSNQRASFSSVGPELDVMAPGVSIQSTLPGNKYGAYNGTSMASPHVAGAAALIL 235

3 PAKCDSVIPVGGEDSNSNRSSFSSVGAELEVMAPVSGVYSTYPTNTYTTLNGTSMASPHVAGTSALIL

1 QKNPSWSNVQIRNHLKNTATSLGSTNLYGSGLVNAEAATR

2 SKHPNWTNTQVRSSLENTTTKLGDSFYYGKGLINVQAAAQ 275

3 SKHPNLSASQVRNRLSRTATYLGSSFYYGKGLINVEAAAQ

Figure 3. Sequence alignments of common detergent proteases: Based on BPN′ numbering. 1 = protease from *Bacillus lentus* (Savinase); 2 = protease from *Bacillus amyloliquefaciens* (BPN′); 3 = protease from *Bacillus licheniformis* (Alcalase).

fer facilitated by the aspartyl and histidine functional groups acting as general base catalysts. The result is addition of water across the amide bond, generating two peptide fragments (9). These so-called *serine* proteases (EC 3.4.21.62) are particularly well suited for laundry applications owing to their alkaline pH optima (Table 2), stability to detergents, and broad substrate specificity.

Table 2. pH Profiles of Common Detergent Enzymes

Enzyme	Source (species)	pH Optimum	Range (>80% maximal activity)
Alcalase®	*Bacillus licheniformis*	8.5	8.0–10.0
BPN′	*Bacillus amyloliquefaciens*	8.5	8.0–9.5
Savinase®	*Bacillus lentus*	10.0	8.0–11.0
Esperase®	*Bacillus alcalophilus*	10.0	9.0–12.0

Source: Internal data from Proctor & Gamble; Novo-Nordisk A/S, Denmark.

Transmission electron microscopy-immunocytochemical (TEM-ICC) analysis of residual soils remaining on fabrics after washing provides more detailed clues on the mechanisms of detergent protease action. Table 3 shows the frequency and mean particle sizes (MPS) of two types of protein—chlorophyll-binding protein (CBP) and β-lipoprotein (BLP)—and four types of carbohydrate—glucose/mannose, glucosamine, fucose, and galactose/galactosamine—remaining on residual grass stains on cotton and polycotton fabrics after washing in a liquid detergent with and without protease. The general trend is for the protease to reduce the frequency but not the size of the stain components. For example, with addition of protease, the frequency of occurrence of CBP on cotton drops from 2 of every 5 fibers to 1 in 10, and on polycotton, it drops from 4 markers in every 5 fibers to 1 in 6, with little or no change in particle size. Similar effects are observed for BLP and the carbohydrates.

Although the subtilisin proteases are stable to most of the surfactants used in typical laundry detergents, they

Table 3. TEM-ICC Data for Grass Removal

		Without protease		With protease	
Component	Fabric	Frequency[a]	MPS (μm)	Frequency[a]	MPS (μm)
Chlorophyll binding protein	Cotton	2/5	0.5	1/10	0.3
	Polycotton	4/5	0.3	1/6	0.3
β-Lipoprotein	Cotton	1/2	0.2	1/5	0.3
	Polycotton	3/4	0.3	1/3	0.3
Glucose/mannose	Cotton	1/7	0.2	1/10	0.2
	Polycotton	1/4	0.2	1/15	0.2
Glucosamine	Cotton	1/7	0.5	1/10	0.2
	Polycotton	1/5	0.4	1/10	0.3
Fucose	Cotton	1/25	0.2	1/30	0.3
	Polycotton	1/20	0.3	0/10	—
Galactose/galactosamine	Cotton	1/10	0.5	1/8	0.3
	Polycotton	1/10	0.3	0/10	—

[a]Frequency is determined from the number of fibers displaying the market out of all those examined (typically 100 fibers per sample).

are susceptible to oxygen bleaches and calcium sequesterants. Oxidative attack by peroxide or peracids on the methionine residue adjacent to the catalytic serine results in a nearly 90% loss of enzymatic activity (10). Replacement of the methionine with an oxidatively stable amino acid such as alanine significantly improves the stability of the enzyme toward oxygen bleach (11). In addition, the subtilisin proteases strongly bind at least one calcium ion, which contributes to maintaining the overall three-dimensional configuration of the enzyme. The calcium sequestering agents used in many laundry products to control water hardness can remove this calcium, resulting in decreased thermal and autolytic stability. Introduction of negatively charged residues near the calcium binding site can increase the binding affinity of the enzyme for the calcium, resulting in improved stability toward calcium sequesterants (12).

Because proteases degrade proteins, there is some risk associated with washing garments constructed from natural protein fibers such as wool and silk in protease-containing detergents. Proteolytic attack against such fibers leads to increased wear and reduced tensile strength. The degree of damage is dictated by the type and level of protease as well as the wash process. Accordingly, many detergents advise against use on wool and silk garments.

In powder detergents, proteases are added in an encapsulated or granulated form (see "Powders"), which protects them from other detergent ingredients and eliminates the potential for them to degrade each other (autolysis) or other enzymes (proteolysis) in the formulation. However, in aqueous-based liquid detergents, autolysis and proteolysis are key problems requiring the formulation of reversible protease inhibitors that can shut down the activity of the enzyme in product, but are diluted away from the enzyme when the detergent is added to the wash. Boric acid and polyols, such as propylene glycol, are commonly used to inhibit protease activity in liquid detergents (3).

Detergent proteases can be generally classified into two main groups, alkaline and high alkaline. Alkaline proteases, such as those derived from *Bacillus licheniformis*, show optimum activity around pH 8, making them ideal for liquid laundry products, which typically have wash pH's between 7.0 and 8.5. Examples include Alcalase® from Novo-Nordisk and Optimase® from Genencor International.

Several high alkaline proteases are available and are typically characterized by a pH-activity optimum around 10.0. These enzymes find widespread use in powder laundry detergents and automatic dishwashing formulations. Examples include Savinase® from Novo-Nordisk and Purafect® from Genencor.

A number of analytical methods are available for assaying proteolytic activity in product and in the wash. Most methods use colormetric detection of the hydrolysis of a specific reagent. Typical reagents include dimethylcasein, azocasein, azocoll, or hemoglobin. By far the most common assay method is based on hydrolysis of small synthetic peptides functionalized with the *p*-nitroanilide chromophore. For most detergent proteases, the synthetic tetrapeptide Suc-AAPF-pNA, or succinylalanylalanylprolyl(phenylalanine)-*p*-nitroanilide (PNA), is used. Hydrolysis of the anilide bond liberates the *p*-nitroaniline group, resulting in the appearance of a yellow color that can be followed spectrophotometrically. This assay is particularly useful for establishing pH and temperature profiles of proteases and following their stability through the wash or during in-product storage (13). However, in general, PNA activity does not correlate with wash performance. Activity against larger substrates, such as casein, which more closely resemble the soil proteins found in common laundry stains, shows a better correlation with protease wash performance.

Amylases

Amylases facilitate the removal of starch-based soils, particularly food soils, by catalyzing the hydrolysis of the glycosidic linkages in starch polymers. This general class of enzymes encompasses several enzymatic activities responsible for starch hydrolysis, including α-amylase, β-amylase, pullulanase, isoamylase, and glucoamylase. The α-amylases (1,4-α-D-glucan-glucanohydrolase, EC 3.2.1.1) are most commonly used in detergents. They catalyze enhohydrolysis of 1,4-α-D-glucosidic bonds in high molecular weight starches of the type commonly found in food soils. As such, amylases find utility in both laundry and dish detergent formulations. Figure 4 shows the dose-dependent response of starch soil removal using a commercial α-amylase in a typical automatic dishwashing detergent formulation.

Current commercial detergent α-amylases are derived from various species of *Bacillus subtilis*. The α-amylase from *B. amyloliquiefacians*, is particularly well suited for detergent applications because of its thermal stability, broad pH activity, and resistance to proteolytic attack. Examples include Maxamyl® from Genencor and Termamyl® from Novo (14).

The detergent α-amylases in use today have highly homologous primary amino acid sequences and a tertiary structure comprised of three domains, A, B, and C (Fig. 5). The active site is typically located in a cleft between domains A and B and is usually comprised of acidic amino acids such as aspartic and glutamic acids. Like the subtilisin proteases, α-amylases contain an essential calcium ion

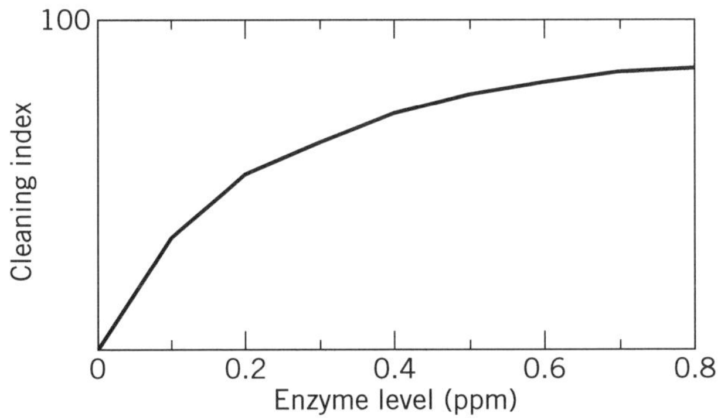

Figure 4. Dose response of α-amylase on starch soil in automatic dishwashing application. Cleaning tests run at 55 °C with 1 mM mixed Ca/Mg hardness and manufacturer's recommended detergent dosage. Starch soil removal measured gravimetrically.

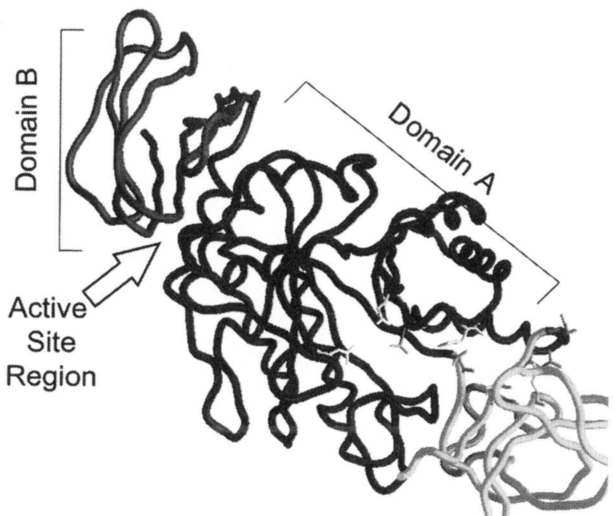

Figure 5. Three-dimensional structure of α-amylase.

that is important in maintaining the tertiary structure. Accordingly, the α-amylases show some degree of sensitivity to calcium sequesterants, with solution stability severely compromised in well built (low calcium) environments. In addition, most wild-type α-amylases are sensitive to oxygen bleach. Recently, protein engineering has been used to improve bleach stability of amylases (15). Both Novo and Genencor offer bleach-stable versions of their α-amylase (Table 1).

The use of amylase and protease combinations for improved stain removal and general cleaning was fairly well established in the mid-1980s and practiced in a number of European and U.S. brands. The use of these two enzymes in compact detergents is now fairly standard, especially in Europe. Figure 6 shows the benefit of combining an α-amylase with a serine protease for removal of food stains from cotton in a typical North American powder laundry detergent.

As with proteases, there are a number of methods available for analyzing amylase activity (16). Probably the simplest, and most common, involves monitoring the color change during hydrolysis of iodine-stained starch. Iodine staining of a standard starch substrate results in a blue color that changes to reddish brown upon hydrolysis with α-amylase. Additional methods include hydrolysis of synthetic substrates like *p*-nitrophenylmaltoheptoaside (pNP-G7). Amylolytic attack on pNP-G7 produces pNP-G3 and pNP-G4. Reacting the pNP-G3 with an excess of α-glucosidase generates glucose and the characteristic yellow color of *p*-nitrophenyl. Alternative methods involve detection of the reducing sugars formed during starch hydrolysis using 3,5-dinitrosalicylate (17), alkaline ferricyanide (18), or alkaline copper (19).

Lipases

Lipases are fairly recent entries into the detergent enzyme market. Their use in detergents is illustrated in several patents from the 1970s (20). However, the lack of a viable commercial source of alkaline-active lipases with stability

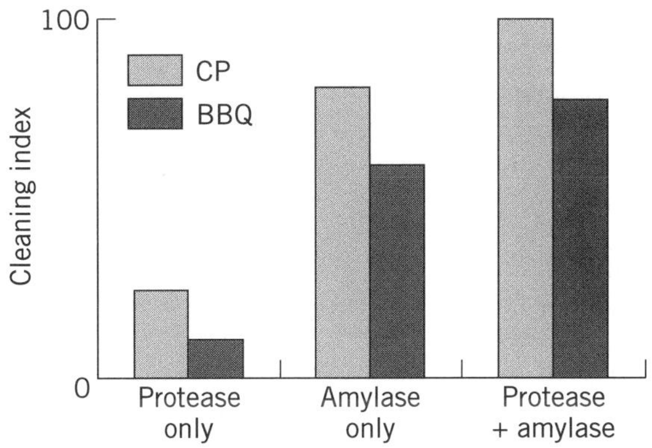

Figure 6. Cleaning benefit for protease + amylase on chocolate pudding and barbecue stains on cotton. Detergent tests run at 20 °C with 1 mM mixed Ca/Mg hardness and manufacturer's recommended detergent dosage. Enzyme dosages were 0.1 ppm each. Removal of chocolate pudding (CP) and barbecue (BBQ) stains from cotton measured by image analysis.

toward anionic surfactants essentially prohibited practical application for almost 10 years until the late 1980s when Novo Industries announced the commercial sale of the first detergent-compatable lipase, Lipolase® (21), obtained by cloning the lipase gene from *Humicola lanuginosa* into a high-expression fungal host. Lipolase® first appeared in compact powders in Japan and has since found fairly broad application in the global detergent market.

Lipases (triacylglycerol acylhydrolases, EC 3.1.1.3) degrade triglycerides by catalyzing the hydrolysis of the ester linkage between the glycerol moeity and the constituent fatty acids, generating glycerol, free fatty acids, mono- and diglycerides. Triglycerides are common components of some of the most-difficult-to-remove soils and stains, including greasy foods, cosmetics, and body soils. Owing to their generally hydrophobic nature, triglycerides are difficult to remove from fabric and may be present even after several wash cycles. As the residual triglycerides age they can oxidize, resulting in the generation of a rancid malodor as well as an overall yellowing or discoloration of the fabric. Lipolytic attack increases the water solubility of the original triglyceride and decreases the melting point of the fat, allowing for increased soil removal during the washing process.

Lipases are most active at the water–substrate interface. This phenomenon is referred to as interfacial activation (22). Lipase activity does not follow classical enzyme kinetics, which generally are only valid for reactions that occur in a homogeneous phase. Rather, lipase activity appears to follow a two-step process (23). In the first step, lipase physically binds to the surface. This is believed to involve activation of the enzyme via a structural change that exposes the hydrophobic binding domain to the substrate. The second step involves formation of the enzyme–substrate complex, leading to hydrolysis of the ester bond. Because physical adsorption to the interface is key to lipase activity, anything that interferes with this process, such as surfactants, can dramatically affect substrate hydrolysis. Exclusion from the soil–water interface is believed to be one of the key mechanisms involved in surfactant inhibition of lipase activity. Given the tendency of surfactants to accumulate at an oil and water interface, lipase can be denied access to the substrate by electrostatic repulsion and steric hindrance of the aggregated surfactant (24). Accordingly, when formulating lipase into detergents it is important to select a surfactant system that minimizes this effect. Generally, nonionic surfactants are less inhibiting than anionics.

Most commercial detergent lipases require more than one wash cycle to show a benefit. Figure 7 shows this multicycle performance effect for Lipolase Ultra®. It is generally thought this traces to deposition of the lipase onto the hydrophobic stain during the wash such that, during the drying cycle, when the water content is decreased, the enzyme is activated and can hydrolyze triglycerides in the stain (25), facilitating removal in the second wash cycle. For Lipolase®, maximum enzymatic activity against a standard lipid soil such as olive oil is obtained when the moisture content of the fabric is 20 to 30% by weight (26), suggesting that significant decomposition of the oily soil occurs during the drying process; hence the need for a second wash cycle to remove the hydrolytic fragments.

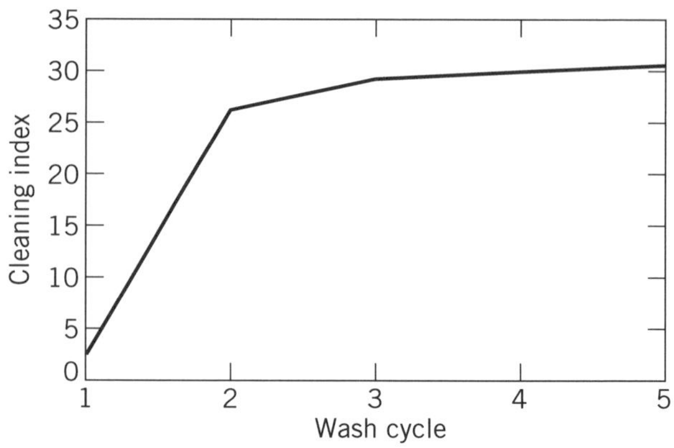

Figure 7. Multicycle stain removal of hamburger grease from cotton fabric with Lipolase® under typical North American wash conditions. Cleaning tests run at 30 °C with 1 mM mixed Ca/Mg hardness and manufacturer's recommended dosage of powder detergent. Lipolase Ultra ® dosed at 3000 LU/l. Removal of hamburger grease from cotton was measured by image analysis.

The most commonly used method to measure lipase activity is a titrimetric assay in which the fatty acid liberated during hydrolysis of an emulsified substrate, such as Triolein, is titrated with a standardized base solution. The milliequivalents of base required to maintain the solution pH at a given value is directly related to the amount of fatty acid generated. Alternate methods rely on colormetric detection of the hydrolytic products generated from synthetic substrates such as *p*-nitrophenylbutyrate.

Cellulases

Cellulases are unique among the hydrolase enzymes used in detergency. Rather than attacking substrates found in soils and stains, cellulases hydrolyze the cellulose in cotton and cotton blends to deliver cleaning benefits such as whitening and dingy cleanup as well as fabric care benefits such as color maintenance and fabric restoration. Like lipases, cellulases have only recently been introduced broadly into detergents. The first use of cellulases in detergents was described by Murata in 1981 (27). Commercial application began in 1987 with the introduction of the alkaline cellulase from *Bacillus* sp. KSM-635 in Kao's compact detergent Attack® (28). Since 1991, a number of European and North American detergents have incorporated cellulases. In 1993, Procter & Gamble introduced Novo-Nordisk's Carezyme®, an enzyme reported to help keep cotton and cotton blends looking newer longer by removing the pills and fuzz that build up on cotton garments over time, making them look worn and faded (29).

Cellulases deliver a number of benefits including cleaning (via removal of particulate and oily soils), fabric softeness, color clarification, and depilling. The cleaning benefits are believed to derive from the removal of surface fibrils that serve as anchor points for soil. Softness benefits are believed to result from the ability of cellulases to loosen surface fibrils, preventing them from intertwining and forming a rigid network with negative hand feel. Color clarification and depilling derive from the same mechanism. Repeated washing and wearing of cotton garments generates fuzz and pill buildup as surface fibrils loosen and become entwined. The fuzz and pills scatter light, making garments look worn and faded. Cellulases weaken the attachment points of the fuzz and pills, which subsequently break off in later wash cycles. The result is brighter, more colorful garments. Figure 8 shows this benefit on cotton socks after 30 wash cycles in a North American powder detergent formulation with and without Carezyme®.

Cellulases (1,4-β-D-glucan-4-glucanohydrolases, EC 3.2.1.4) catalyze the endohydrolysis of 1,4-β-D-glucosidic linkages in cellulose. Cellulose is a linear, unbranched polymer of anhydroglucose units linked through β-1,4 bonds. Each polymer contains between 10,000 and 15,000 glucose units. Up to 60 polymers are arranged in parallel bundles, called elementary fibrils, and stabilized through extensive hydrogen bond and van der Waals interactions, which promotes crystallinity of the bundle. The elementary fibrils can further aggregate to form microfibrils that, in turn, aggregate into macrofibrils of up to 500 nm in diameter. Cotton fibers are formed by the accumulation and aggregation of macrofibrils (30). Disruption of the hydrogen bonding network within these structures, either through chemical or mechanical means, leads to the formation of amorphous regions. In new cotton garments, the ratio of crystalline to amorphous regions is about 70:30. In regenerated cellulose fabrics such as rayon, the structure is much more disordered, leading to a shift in the crystalline amorphous ratio to about 30:70. The amorphous regions are the site of cellulytic attack owing to the greater accessibility and flexibility of the substrate. As a result, aged garments that have experienced a number of wash and wear cycles and, hence, have more amorphous regions, tend to show greater sensitivity to cellulase than new garments.

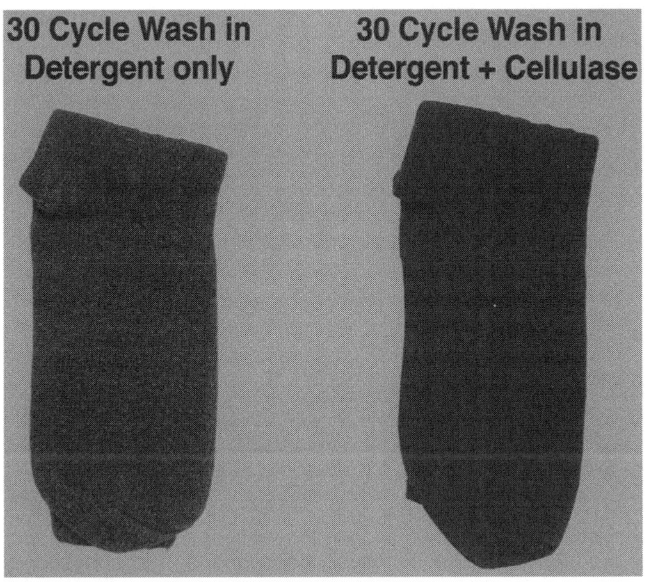

Figure 8. Effect of Carezyme® after 30 wash cycles. (**a**) 30-cycle wash in detergent only; (**b**) 30-cycle wash in detergent plus cellulase.

Cellulose hydrolysis is an acid-catalyzed process. The O-glycosidic bond is cleaved by donation of a proton from the carboxyl group of a glutamic acid residue in the active site of the enzyme, with the resulting cabonium-ion intermediate stabilized by an aspartic acid or other nucleophilic residue (31). The formation of a carbonium-ion intermediate opens the possibility of stereochemistry, with the configuration of the anomeric carbon atom of the product either being retained or inverted. The configuration of the active site of the particular cellulase enzyme determines the sterochemistry of the hydrolysis products, leading to a classification of cellulases as being either retaining or inverting types of enzymes.

Current detergent cellulases are derived from fungal sources, such as *Humicola insolens*, or bacterial sources, such as the *Bacillus* species. The fungal enzymes typically represent a mixture of *exo*- and *endo*-glucanase activities with pH optima in the acidic to neutral range. This mix of activities facilitates the degradation of crystalline cellulose. Celluzyme® from Novo-Nordisk (Table 1) is derived from *H. insolens* strain DSM 1800 and consists of a mixture of *endo*-glucanases, cellobiohydrolases, and β-glucosidases. Carezyme®, on the other hand, is a monocomponent alkaline-active *endo*-glucanase with low affinity to crystalline cellulose.

The detergent cellulases from alkalophilic *Bacillus* species are generally monocomponent *endo*-glucanases with good stability against proteases. Relative to the fungal cellulases, bacterial cellulases tend to show somewhat reduced fabric care benefits and increased detergency effects on pigmented dirt and sebum (30).

Generally, cellulases are multidomain structures comprised of a core catalytic domain, containing the active site, and a cellulose binding domain, or CBD, which is responsible for anchoring the enzyme to the cellulose substrate. Typically, the CBD is attached to the catalytic domain through a short peptide linker. In the absence of the CBD, the core will show catalytic activity against soluble substrates but drastically reduced activity against insoluble cellulose. Because the benefits derived from cellulase clearly rely on the enzyme binding to fabric surfaces, the CBD is a critical structural component. Like lipases, adsorption of cellulase onto fabric is the first step of the hydrolytic process, corresponding to a phase transfer of enzyme from solution to the fabric surface (32). Those enzymes that adsorb tightly are generally the most efficient with respect to cellulose hydrolysis (33). Thus, the CBD is an important structural feature controlling the efficacy of a particular cellulase with respect to detergency benefits. Table 4 shows activity on soluble and insoluble substrates and the depilling benefit for cellulases with (EG V, Carezyme), and without (EG 1, Endolase) a CBD as well as the catalytic core of an EG V enzyme. The importance of the CBD both to hydrolysis of insoluble substrate and in delivering a depilling benefit is apparent in the higher activity of the CBD-containing Carezyme on crystalline cellulose (Avicel) and the more than fivefold increase in depilling benefit versus the Carezyme core.

A number of analytical methods are available for assaying cellulase activity (26,34). The two most common are the viscometric method and measurement of reducing sug-

Table 4. Activity and Performance of Various Cellulases: Role of the Cellulase Binding Domain

Enzyme	CMCU/mg (soluble substrate)	Red-*Avicel*/mg (insoluble substrate)	Relative depilling benefit[a]
EG I (Endolase)	80	0.2	4
EG V (Carezyme)	100	22	100
EG V core	180	8	19

[a]Depilling measured by visual inspection after 25 wash cycles.

ars. Because of their tendency to cleave internal glycosidic linkages, detergent cellulases cause a rapid decrease in chain length of soluble cellulose polymers, resulting in a drop of solution viscosity. This change in viscosity can be measured via standard viscometeric methods and, when compared to a standard sample, can be used as a relative measure of cellulytic activity. Substituted, water-soluble cellulose substrates are commonly used in this assay, such as carboxymethylcellulose (CMC) or hydroxyethylcellulose.

Measurement of the reducing sugars formed upon hydrolysis of a cellulose substrate is often used in conjunction with the viscometric method to assess cellulase activity. Reducing sugars can be detected colorimetrically by ferricyanide.

Peroxidase

This represents the newest class of enzyme developed for detergent. There is currently only one commercially available peroxidase for detergents, Guardzyme® from Novo-Nordisk A/S (Table 1). Guardzyme® was isolated from the inkcap mushroom, *Coprinus cinereus*, and has been cloned into a production host for large-scale production (35). It is a heme-containing protein that, in the presence of hydrogen peroxide, can mediate the oxidation of fugitive dyes in solution, thereby providing a dye-transfer inhibition benefit.

Peroxidases (EC 1.11.1.7) are a subclass of the general class of enzymes known as oxidoreductases. Oxidoreductases catalyze oxidation–reduction reactions on a broad range of substrates, including alcohols, carboxyl groups, alkenes, and amines. Peroxidases specifically catalyze the reaction:

$$\text{Donor} + H_2O_2 \rightarrow \text{Oxidized donor} + 2H_2O$$

where the donor molecule can be anything with an oxidizable group. Specifically, if the donor molecule is a fugitive dye that has bled off a colored garment, peroxidase can be used to bleach the dye in solution, effectively eliminating the transfer of the colored dye species to other fabrics.

For Guardzyme®, the dye-transfer inhibition system consists of three components: the enzyme, a mediator, and hydrogen peroxide. The mediator is generally a substituted phenolic compound that shuttles electrons from the substrate to peroxidase. Through selection of the appropriate mediator, a catalytic oxidation cycle can be established as follows:

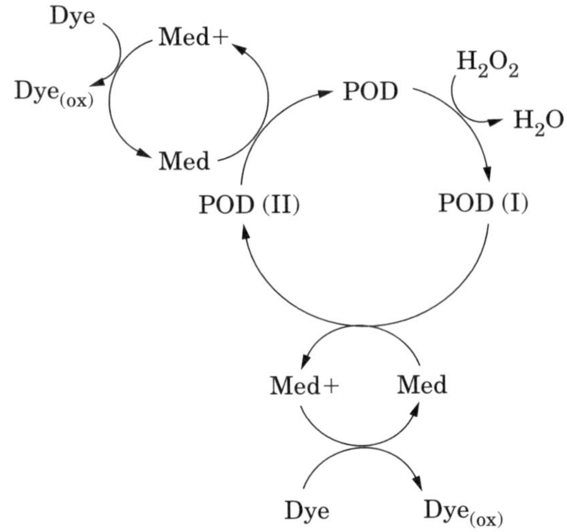

where POD (I) and POD (II) are oxidized forms of the peroxidase. Figure 9 shows the dye-transfer inhibition benefit for a Guardzyme®-mediator system, where the mediator is phenothiazine-10-propionic acid. The peroxidase-based dye-transfer inhibition system essentially eliminates transfer of dye from a blue bleeder (Direct Blue 1 on cotton) to a white cotton tracer.

FORMULATION OF DETERGENT ENZYMES

The key challenge to the use of enzymes in detergents is stability. This is particularly true in liquid detergents, where the enzyme is in intimate contact with the formulation ingredients. For powder detergents, enzymes are typically supplied as a coated granulate, thereby separating the enzyme from other detergent ingredients and generally delivering a stable composition. Liquid detergents, however, must be formulated to minimize damage to the enzyme. Denaturation by surfactants and builders, unfavorable interactions with perfume ingredients, and deg-

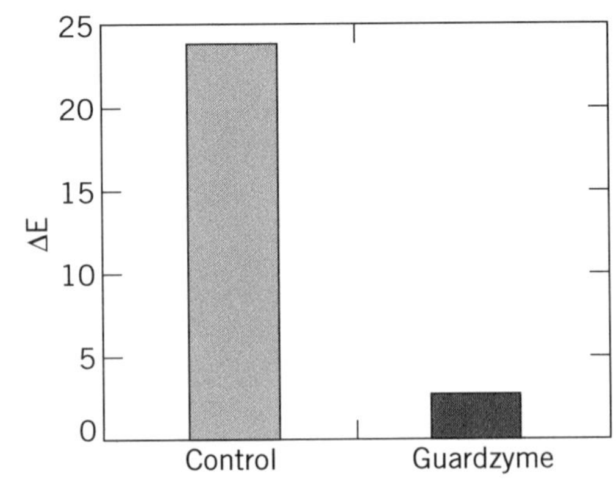

Figure 9. Dye transfer inhibition benefit for Guardzyme®/mediator under North American wash conditions. Dye transfer was measured by recording the difference in reflected light (reported as ΔE from a Hunter colorimeter) from a white cotton tracer swatch before and after washing with a bleeder fabric (Direct Blue 1 on cotton) in a North American powder detergent at 35 °C with 1-mM water hardness.

radation by protease are the principal processes responsible for loss of enzyme activity upon storage.

Liquids

Most modern liquid detergents contain fairly high concentrations of anionic surfactants for cleaning. These surfactants are particularly good at denaturing proteins, including enzymes. In addition, most heavy-duty liquids contain multiple enzyme systems, including proteases, amylases, lipases, and cellulases. Proteolytic degradation of these enzymes by the added protease poses a significant long-term stability issue.

In order to achieve acceptable shelf stability and ensure delivery of a high level of cleaning performance over time, detergent manufacturers and enzyme suppliers have taken three basic approaches to improve the stability of enzymes in liquid detergents:

1. *Formulation of stabilizing additives*. A number of ingredients can be added to the liquid detergent to improve enzyme stability. Calcium is known to stabilize both proteases and amylases. The added calcium serves to tie up detergent actives that could, otherwise remove the essential calcium from the enzyme, resulting in a loss of the catalytic three-dimensional structure. Figure 10 shows the effect added calcium has on the stability of a protease in a heavy-duty liquid detergent. The half-life of the enzyme under stressed storage conditions increases as calcium concentration is increased.

   In addition, short-chain carboxylates have been shown to promote protease stability (36) through a noncompetitive-inhibition mechanism. Accordingly, calcium and sodium formates are commonly added stabilizing agents in enzyme-containing liquid detergents. Reversible protease inhibitors can also be added to shut down the activity of the protease in product. The combination of a polyol with vicinal hydroxyl groups, such as propylene glycol, and a boron compound, such as boric acid, results in formation of a borate-diol complex that effectively inhibits proteolytic activity. Upon dilution of the product into the wash, the inhibitor is diluted away from the protease yielding a fully active enzyme. Substituted boronic acids (37) and peptide derivatives (38) have also been shown to be effective reversible inhibitors.
2. *Selection of more stable wild-type enzymes*. Enzymes with enhanced stability to detergent ingredients can be obtained by classical screening methods. For example, the Kao Corporation in Japan has reported isolation of protease-resistant cellulases from cultures of *Streptomyces* strains (39). Kawai et al. (40) identified a number of new cellulases from *Bacillus* species with good stability to surfactants, proteases, and chelants. Similarly, new α-amylases have been found that do not appear to be substantially affected by anionic surfactants (41). Natural isolate screening for new wild-type lipases with improved detergent compatibility has met with limited success to date, although lipases from *Fusarium culmorium* have been isolated and reportedly show improved activity versus Lipolase® in the presence of alcohol ethoxylate surfactants (42).
3. *Protein engineering*. Another way to improve stability of enzymes for detergents is to engineer in specific stabilizing mutations. For example, replacing a neutral or positively charged amino acid in the calcium-binding domain of an enzyme with a negatively charged amino acid would be expected to increase electrostatic interactions, thereby minimizing stability loss by removal of the essential calcium ion. Aehle et al. (43) used this approach to modify the calcium binding site of an α-amylase. The resulting variant showed a 2.5-fold increase in stability versus the native enzyme in low calcium environments. Similarly, amino acid substitutions in the vicinity of a hydrophobic domain in the alkaline *Bacillus lentus* protease are reported to improve storage stability of the resulting variants in both liquid detergents and soap bars (44). Schülein et al. (45) reported a number of modifications to Carezyme® that improved its alkaline activity and detergent compatibility.

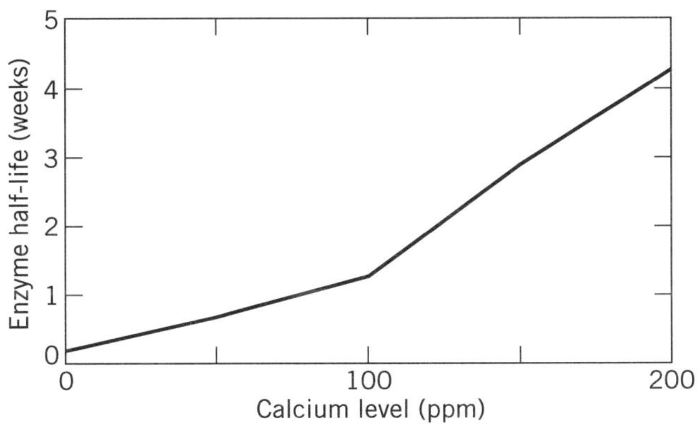

Figure 10. Half-life of protease in a heavy-duty liquid detergent as a function of calcium level. Liquid detergent stored at 35 °C. Enzyme activity measured by PNA assay.

In addition to stability, aesthetics are another major consideration for the formulation of enzymes into liquid products (46). Odor, color, and clarity are key aesthetic signals for consumers. Enzyme raw materials, being products of microbial fermentation, often contain a number of smelly by-products, including volatile amines and low molecular weight carboxylic acids, that must be removed to avoid consumer malodor complaints. Likewise, fermentation-derived by-products may cause turbidity, flocculation, or color changes in liquid detergent formulas. Therefore, the formulation of enzymes into liquid detergents requires extensive evaluations of odor, color, and clarity in order to select enzyme formulations that minimize aesthetic negatives.

Powders

Many of the same considerations for the formulation of enzymes in liquids apply to powders. Odor, color, and stability are important considerations although to a somewhat lesser extent owing to the physical separation of the enzyme from the bulk detergent via granulation. In addition, particle size, shape, flowability, and bulk density become important when considering mixing and segregation of enzyme particles with powdered detergents. Probably the most critical factor in formulating solid enzymes is dust generation during detergent manufacture. This was clearly evident in the experiences of the detergent industry in the late 1960s and early 1970s, with the raw powdered enzyme formulations originally supplied for detergent use. These formulations led to high airborne dust levels in the manufacturing plants and an increase in worker sensitization. As a result, enzyme manufactures replaced these early powder and agglomerated enzymes with various granulate formulations (13,46), the most important of which are summarized below:

1. *Enzyme prills.* In this process, dry enzyme is dispersed into a molten wax, nonionic surfactant, or polymer matrix, then sprayed into a cooling tower to form solid, spherical particles. The molten phase is typically comprised of a water-soluble or water-dispersable material with a melting point above 50 °C. The key advantages of prilling are high throughput and the ability to recycle particles that fall outside the desired size range. The key disadvantage is the relatively poor physical strength of the particle, leading to break up and high dust generation in subsequent processing. Overcoating with an inert, mechanically tough material such as polyethylene glycol can be used to improve physical strength and reduce dustiness. However, this typically is not sufficient to ensure no break up of the particles during detergent processing and, as a result, enzyme prills are not in widespread use today.
2. *Extruded granulates.* By combining an enzyme powder or liquid concentrate with binders, such as clay, sugar, or starch, an extrudable dough can be produced that can then be pressed through a perforated metal plate. Upon drying, the extrudates are reasonably durable. In order to produce spherical particles of the appropriate size the wet extrudate is fed into a Marumerizer, where a rotating textured plate breaks up the noodles into smaller particles and rounds the corners to yield a more spherical shape. After sieving, the particles are coated with pigments, such as TiO_2, and a protective outer layer to achieve the desired appearance and further improve granulate integrity. Key disadvantages of this method are the high capital investment in a multistep process and the sensitivity of the process to variation in feedstock moisture and composition.
3. *High shear granulation.* This method is used by Novo-Nordisk to produce their T-granulate products (47). In this process, enzyme is mixed with controlled amounts of water, binders, such as polyethylene glycol, ethoxylated fatty alcohols, fatty acids, bentonite, kaoline, or other clays, and other granulation ingredients to form a low-moisture agglomerate. The agglomerate is fed into a high-shear mixer that breaks it up into smaller particles. The particles are dried in a fluidized bed and coated with a final protective layer of polymers and pigments. By incorporating cellulose fibers into the original agglomerate, particles with high mechanical stability and physical strength can be produced (48). Enzyme T-granulates are widely used in dry laundry products and are among the lowest-dusting enzyme raw materials available. The process is reasonably flexible, especially with respect to enzyme feedstock concentration, allowing granulation of a given enzyme, such as Savinase®, across a wide range of enzyme payload.
4. *Fluid bed coating.* In this process, liquid enzyme feedstock is sprayed onto a solid, inert support, or core, such as NaCl or a sucrose-starch nonpareil. Once the enzyme layer has dried, additional coatings can be applied, including stabilizers, chelants, antioxidants, and pigments. The outer coating typically consists of a film-forming polymer, such as polyvinyl alcohol, and a pigment, such as TiO_2. Genencor International uses this granulation process to deliver their Enzoguard line of dry enzyme formulations. The key advantage of fluid bed coating is that a single piece of equipment, the fluid bed coater, can be used to produce the final product.

Figure 11 shows scanning electron micrographs of typical granulates produced by the high shear granulation process (**a**) and fluid bed coating (**b**). Both processes produce spherical granulates with fairly uniform particle size.

Industrial Hygiene

Controlling protein dust from enzyme raw materials is a key consideration in the manufacture of enzyme-containing laundry products because enzymes are proteins, like animal dander or pollen, with the potential to act as respiratory allergens. Enzyme formulations, such as the granulate compositions described above, that minimize protein dust or reduce the tendency of enzyme protein to form aerosols are critical to protein dust control. Just as important is the need for appropriate industrial hygiene practices and in-plant exposure control technology to further minimize the potential for exposure to enzyme protein. This involves compliance to operational guidelines that require no visible dust in the manufacturing environment, no recurring enzyme spills or spills of enzyme-containing detergent, no gross skin contact with raw material enzyme, and cleanup processes that minimize dust and aerosol generation. Design of dust handling equipment should follow current best practices for enclosure and local exhaust ventilation. Spill-containment systems should be in place for those areas where either spills of raw material enzyme or enzyme-containing product are likely. Additionally, an appropriate preventive maintenance program should be implemented to ensure the equipment performs as intended. Strict adherence to guidelines such as these has resulted in a drop in airborne enzyme dust levels in North American and Western European detergent manufacturing plants by more than three orders of magnitude versus 1970 values (48).

Figure 12 shows a simplified schematic for safe handling of bulk solid enzymes in a detergent manufacturing

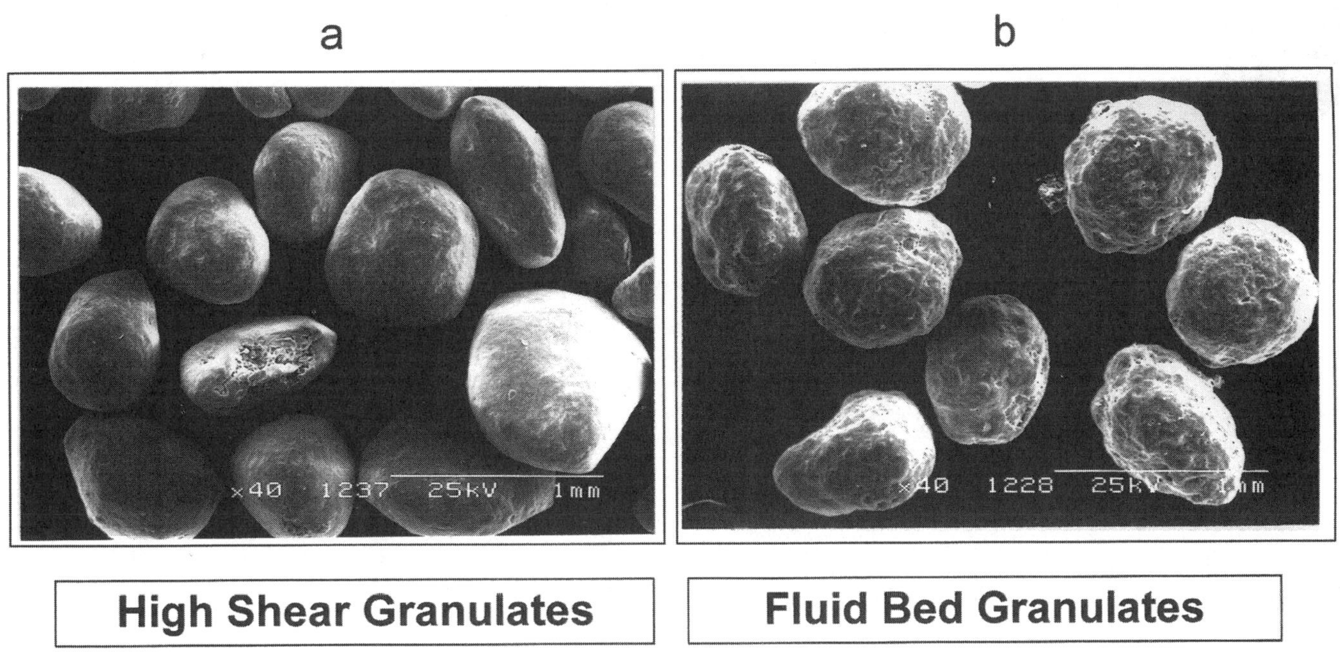

Figure 11. Electron micrographs (40×) of enzyme granulates. (**a**) High shear granulates; (**b**) Fluid bed granulates.

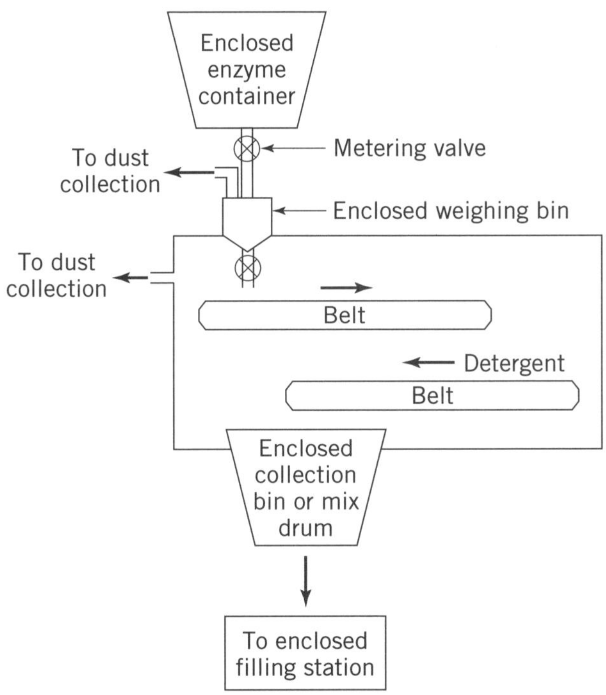

Figure 12. Simplified schematic of dry enzyme handling station.

environment. Enzymes are received from the supplier in an enclosed container that can be coupled to a metering system for accurate dosing. From the weigh station, enzyme is mixed with a stream of detergent base granules in a dust-proof enclosure and sent to either a mix drum or holding bin in preparation for pack-out to cartons. The entire process, including packaging of the final product into boxes or cartons, is housed in dust-proof enclosures to minimize airborne protein dust. Systems such as these ensure low risk for enzyme handling.

MANUFACTURE OF DETERGENT ENZYMES

Nearly all detergent enzymes in use today are produced through large-scale fermentation of microorganisms. As discussed in "Current Detergent Enzymes", detergent enzymes are derived from bacterial and fungal sources. Large-scale fermentation is typically carried out with optimized strains of appropriate bacterial (*Bacillus* or *Pseudomonas* strains) or fungal species (*Aspergillus*) that have been genetically engineered to express high levels of the target enzyme, excrete that enzyme into the bulk fermentation medium for ease of recovery, and minimize production of other, unwanted proteins. After fermentation, the enzyme is recovered from the broth by removal of the whole cells and other particulates through centrifugation or filtration, concentrated via ultrafiltration and diafiltration techniques, and, if necessary, further purified by selective precipitation of nonactive proteins (13).

Because multiton quantities of low-cost enzymes are needed to support the requirements of the global detergent business, enzyme manufacturers closely adhere to the following principles (49):

1. The production organism must be capable of secreting the enzyme extracellularly into the bulk fermentation medium. This greatly facilitates recovery because the cost, both in time and money, to recover intracellular enzymes is to great to support the needs of the detergent industry. Enzyme secretion

can be maximized by appropriate optimization of fermentation conditions.

2. The production organisms should be able to be tuned to give the highest possible yields. Strain optimization can be accomplished either through classical mutagenesis and screening methods or via genetic engineering, where multiple copies of the gene coding for the enzyme of interest are inserted into the host DNA so that each microbe in the strain can produce several copies of the enzyme of interest.
3. The number of steps in the downstream recovery process should be minimized to avoid yield losses and keep overall processing costs down.
4. The production organisms should produce the target enzyme in relatively high purity with no contaminating side activities or proteins. This ensures that only basic recovery methods, such as filtration and centrifugation, are needed to obtain the final product and eliminates the need for high-cost process steps like solvent precipitation. By deleting the genes that code for unwanted enzymes and proteins, strains can be prepared that secrete only the enzyme of interest into the bulk media.

Herrmann et al. (49) provide a thorough review of the principles and practices of detergent enzyme manufacture.

TRENDS IN DETERGENT ENZYMES

A number of market trends and consumer needs are currently driving research on new enzyme technologies for detergent application. Two critical changes occurring in the marketplace today that are expected to have a major impact on future enzyme technologies are the continued move to cooler wash temperatures and fundamental changes in washing machine design leading to reduced water consumption. In the United States, there has been a slow but steady decline in wash temperatures since the early 1980s. This trend has been matched in Europe where more loads are being washed at 40 °C. Asia and Latin America have historically been cold-water-wash geographies. In order to deliver consumer valued benefits under these conditions, enzymes with improved activity at low temperatures will be required, particularly for those laundry stains and soils that rely on thermal energy for removal, such as mechanical or food grease and oil.

In response to anticipated legislation on energy consumption, U.S. washing machines will become more energy efficient, work at lower temperatures, and require less water per load. The net effect will be a much higher ratio of fabric and soil to wash liquor. These new machine designs will likely require enzymes with increased stability to other detergent ingredients in low water environments. In addition, changes in enzyme raw material formulation may be necessary for improved solubility.

There is a trend in the detergent industry toward development of technologies that can deliver fabric care benefits. This ranges from enzymes that can maintain the color of a garment over successive wash cycles to systems capable of reducing the wear associated with the laundry process. As a result, there is increased emphasis on identification and development of enzymes that can extend the useful wear life of fabrics. Cellulases are probably the best example of this for cotton garments. Recently, proteases have been considered as a way to deliver comparable benefits to washable wools and silks.

Finally, there is growing concern among consumers about communicable disease and the ability of laundry and cleaning products to eliminate this risk. The trend toward colder wash temperatures is expected to decrease the efficacy of traditional detergent ingredients, especially oxygen bleaches, toward microbial kill. Recent patent literature has cited the use of *endo*-glycosidases alone for removal of bacteria from fabric and, in combination with traditional antimicrobial actives, for improved microbial kill during the wash process (50). Antimicrobial detergent compositions in which the active is a peroxidase or laccase have also been disclosed (51).

Current trends in the development of new detergent enzymes are focused on improved activity in cold water, increased stability and compatibility with other detergent actives, and the identification of new enzymes that can deliver new benefits such as fabric care and santization.

ACKNOWLEDGMENTS

The author wishes to thank the following for their help in preparing this chapter: Chanchal Ghosh, Genevieve Kipte, Yong Zhu, Eric Wells, Paul Correa, Michael Fedak, Stanton Boyer, and Christian Thoen.

BIBLIOGRAPHY

1. German Pat. DE 28923 (1913), Otto Röhm.
2. J.H. Houston, in J.H. van Ee, O. Misset, E.J. Baas eds., *Enzymes in Detergency*, vol. 69, Surfactant Science Series, Dekker, New York, 1997, pp. 11–21.
3. T. Krawczyk, *INFORM*, **8**, 6–14, (1997).
4. P. Van Arnum, *Chem. Market. Reporter*, **251** (suppl.), 5–6 (Jan. 27, 1997).
5. Business Communications Company, Inc., Study No. C-147NA: *Enzyme for Industrial Applications: Products, Uses, Technologies and Economic Values*, P. Rotheim (1998).
6. E.M. Kirschner, *Chem. Eng. News*, p. 46 (Jan. 26, 1998).
7. *Chem. Market. Reporter*, SR18 (Jan. 16, 1995).
8. M.G. Venegas, in M.S. Showell ed., *Powdered Detergents*, vol. 71, Surfactant Science Series, Dekker, New York 1997, pp. 295.
9. E.J. Bass, M.M.P. Bollier, P.F. Plank, and D.S. Winetzky, in J.H. van Ee, O. Misset, E.J. Baas eds., *Enzymes in Detergency*, vol. 69, Surfactant Science Series, Dekker, New York, 1997, pp. 23–59.
10. C.E. Stauffer and D. Etson, *J. Biol. Chem.* **244**, 5333–5338 (1969).
11. J.A. Wells, D.B. Powers, R.R. Bott, B.A. Katz, M.H. Ultsch, A.A. Kossiakoff, S.D. Power, R.M. Adams, H.H. Heyneker, B.C. Cunningham, J.V. Miller, T.P. Graycar, and D.A. Estell, in *Protein Engineering*, Alan R. Liss, 1987, pp. 279–287.
12. M. Egmond, in J.H. van Ee, O. Misset, E.J. Baas, eds., *Enzymes in Detergency*, vol. 69, Surfactant Science Series, Dekker, New York, 1997, pp. 61–74.

13. E. Gormsen, E. Marcussen, and T. Damhus, in *Enzymes in Detergency*, J.H. van Ee, O. Misset, E.J. Baas eds., vol. 69, Surfactant Science Series, Dekker, New York, 1997, pp. 137–163.
14. W. Aehle, in J.H. van Ee, O. Misset, E.J. Baas eds., *Enzymes in Detergency*, vol. 69, Surfactant Science Series, Dekker, New York, 1997, pp. 213–229.
15. PCT Pat. Application WO 95/10603 (1995) H. Bisgaard-Frantzen, T. Borchert, A. Svendsen, M.H. Thellersen, and P. Van Der Zee (to Novo-Nordisk A/S).
16. T. Takagi, H. Toda, and T. Isemura, in P.D. Boyer ed., *The Enzymes*, vol. 5, Academic Press, New York, 1971, pp. 265–266.
17. E.H. Fischer and E. Stein, *Biochem. Prep.* **8**, 27 (1961).
18. J.F. Robyt and S. Bemis, *Anal. Biochem.* **19**, 56 (1967).
19. S. Dygert, L.H. Li, D. Florida, and J.A. Thoma, *Anal. Biochem.* **13**, 367 (1965).
20. East German Pat. 14,296 (1958), F. Leidholdt. French Pat. 2,011,578 (1970), C.M. Brooke and R.T. Littleton. British Pat. 1,442,418 (1976), R.E. Montgomery.
21. *Chem. Eng. News*, p. 6 (Feb. 8, 1988).
22. O. Misset, in J.H. van Ee, O. Misset, E.J. Baas, eds., *Enzymes in Detergency*, vol. 69, Surfactant Science Series, Dekker, New York, 1997, pp. 107–131.
23. R. Verger and G.H. de Haas, *Annu. Rev. Biophys. Bioeng.* **5**, 77 (1976).
24. A. Svendsen et al., in *Protein Engineering of Microbial Lipases*, Academic Press, New York, 1997.
25. P.N. Christensen and E. Gormsen, *Chim. Oggi.* **11**, 15 (1993).
26. N. Eriksen, in T. Godfrey and S. West eds., *Industrial Enzymology*, 2nd ed. Macmillan London, 1996.
27. Japanese Pat. 5,631,674 (1981), M. Murata.
28. E. Hoshino and S. Ito, in J.H. van Ee, O. Misset, E.J. Baas eds., *Enzymes in Detergency*, vol. 69, Surfactant Science Series, Dekker, New York, 1997, pp. 149–174.
29. S.M. Colwell, *Soap Cosmetics Chem. Specialties*, **70**, 28 (1994).
30. K.-H. Maurer, in J.H. van Ee, O. Misset, E.J. Baas eds., *Enzymes in Detergency*, vol. 69, Surfactant Science Series, Dekker, New York, 1997, pp. 175–202.
31. J. Knowles, P. Lehtovaara, and T. Teeri, *Trends Biotechnol.* **5**, 255 (1987).
32. B. Henrissat, *Cellulose* **1**, 169 (1994).
33. A.A. Klyosov, *Biochemistry* **29**, 10577 (1990).
34. K.R. Sharrock, *J. Biochem. Biophys. Methods* **17**, 81 (1988).
35. E. Gormsen, *JAOCS / AOCS Conference*, Kuala Lumpur, Malaysia, September 1997.
36. M.C. Crossin, *JAOCS*, **66**, 1010–1014 (1989).
37. U.S. Pat. 5,442,100 (1995), D.W. Bjorkquist and R.K. Panandiker (to Procter & Gamble Co.) U.S. Pat. 5,472,628 (1995, R.K. Panandiker et al. (to Procter & Gamble Co.) U.S. Pat. 5,488,157 (1996), D.W. Bjorkquist and R.K. Panandiker, (to Procter & Gamble Co.).
38. WO 98/13461 A1 (1998) (to Procter & Gamble Co.).
39. JP0629575 A (1994) (to Kao Corp).
40. EP 270974 and EP 269977 (1988), S. Kawai et al. (to Kao Corp.).
41. JP09206073 (1998) (to Kao Corp.).
42. WO 96/19570 A1 (1994) (to Novo-Nordisk A/S).
43. W. Aehle, *American Oil Chemists Society 86th Annual Meeting*, San Antonio, 1995.
44. WO 96/34935 (1996) (to Unilever).
45. WO 94/07998 (1994), M. Schülein, H. Fredholm, C. Hjort, G. Rasmussen, E. Nielsen, and P. Rosholm (to Novo-Nordisk A/S).
46. T. Becker, G. Park, and A. Gaertner, in J.H. van Ee, O. Misset, E.J. Baas eds., *Enzymes in Detergency*, vol. 69, Surfactant Science Series, Dekker, New York, 1997.
47. U.S. Pat. 4,106,991 (1978), E.K. Markusseen and A.W. Schmidt, (to Novo Industries A/S).
48. G. Peters and D.P. Mackenzie, in J.H. van Ee, O. Misset, E.J. Baas, eds., *Enzymes in Detergency*, vol. 69, Surfactant Science Series, Dekker, New York, 1997.
49. H.A. Herrmann, I. Good, and A. Läufer, in J.H. van Ee, O. Misset, E.J. Baas eds., *Enzymes in Detergency*, vol. 69, Surfactant Science Series, Dekker, New York, 1997.
50. U.S. Pat. 5,238,843 (1993), R.S. Carpenter, I.J. Goldstein, and P.J. Lad (to Genencor Int. and Procter & Gamble Co.); U.S. Pat. 5,258,304 (1993), R.S. Carpenter, P.J. Lad, and A.M. Wolff (to Genencor Int. and Procter & Gamble Co.).
51. WO 97/43383 A-1 and WO 97/42825—A1 (1997) (to The Procter & Gamble Co.).

See also PROTEIN EXPRESSION, SOLUBLE; PROTEIN PURIFICATION, AQUEOUS LIQUID EXTRACTION; PROTEIN SECRETION, *SACCHAROMYCES CEREVISIAE*; PROTEIN ULTRAFILTRATION.

ENZYMES, DIRECTED EVOLUTION

FRANCES H. ARNOLD
PATRICK L. WINTRODE
California Institute of Technology
Pasadena, California

KEY WORDS

Biocatalysis
Catalyst improvement
DNA shuffling
High-throughput screening
Protein engineering
Random mutagenesis
Recombination
Selection

OUTLINE

INTRODUCTION

Staunch Darwinists attribute all the complexity of living things to an algorithm of mutation and natural selection. The exquisite products of this evolution algorithm are apparent at all levels, from the amazing diversity of life all the way down to individual protein molecules. Scientists and engineers who wish to redesign these same molecules are now implementing their own versions of the algorithm. Directed evolution allows us to explore enzyme functions never required in the natural environment and for which the molecular basis is poorly understood. This bottom-up design approach contrasts with the more conventional, top-down one in which proteins are tamed "rationally" using computers and site-directed mutagenesis. We will describe how molecular evolution can be directed in the test tube in order to produce useful biocatalysts. It is not possible to provide a complete account of all the methods proposed for in vitro evolution; this article therefore introduces methods and strategies used successfully in our laboratory for directing the evolution of enzymes. Some alternative methods for biocatalyst evolution are also be discussed. The reader should also be aware that there is a rather substantial and largely separate literature on combinatorial approaches to engineering binding molecules. Recent advances in the ability to create genetic diversity and to screen or select for improved functions in large libraries of enzyme variants are being combined in a robust approach to solving difficult molecular design problems. With directed evolution we now have the ability to tailor individual proteins as well as whole biosynthetic and biodegradation pathways for biotechnology applications.

WHY DIRECTED EVOLUTION?

The Limitations of Nature's Biocatalysts Present Insurmountable Challenges to Rational Design

When natural enzymes are recruited for industrial applications—from serving as catalysts in chemicals synthesis to additives for laundry detergents—we discover that they are often not well suited to these tasks. Due to poor substrate solubility, breakdown of unstable products, or competing chemical reactions, the conditions for an enzyme reaction may be unsuitable for large-scale applications. Reflecting their participation in complex biochemical networks inside living cells, enzymes are often inhibited by their own substrates or products, either of which may severely limit the productivity of a biocatalytic process. Evolution is usually the culprit: enzymes are optimized and often highly specialized for specific biological functions within the context of a living organism. Biotechnology, in contrast, needs enzymes that are stable and active over long periods of time (a feature that might clash with the need for rapid protein turnover inside a cell), are active in nonaqueous solvents (a feature probably not required in most biological milieu), and can accept different substrates (substrates not present in nature).

It is possible to produce new enzymes in recombinant organisms, altering the amino acid sequence and therefore the properties through appropriate modifications at the DNA level. We are hobbled, however, by near complete ignorance of how the amino acid sequence affects every aspect of enzyme performance, from its ability to be expressed in a heterologous host to its catalytic activity in nonnatural environments. Numerous protein engineering experiments have demonstrated that changes in protein properties are brought about by the cumulative effects of many small adjustments, many of which are distributed or propagated over significant distances. Furthermore, proteins are usually teetering on the brink of instability, with folded structures that are more stable than unfolded—and therefore inactive—ones by the equivalent of a few hydrogen bonds out of the hundreds that form. Superimposing on the need to retain this relatively fragile folded state, the additional requirements of having to fold in the first place, and the need of maintaining or even reengineering a catalytic site that is affected at some level by virtually any modification yields a design problem of such complexity that any rational design effort will require enormous inputs of structural, mechanistic, and dynamic information. Information that is available for but a tiny fraction of interesting catalysts. Even if one trait is successfully designed (e.g., enhanced stability), it is virtually impossible to predict the cost to another (e.g., catalytic activity or expression level). The relatively few examples where rational design has yielded *useful* enzymes do not negate the view that rational enzyme design is often a fruitless exercise.

Extending Natural Diversity by Laboratory Evolution

All these hurdles to the rational design of enzymes are bypassed by evolution. The power of evolution as an algorithm for molecular design is perhaps best appreciated by studying its products. By constructing the evolutionary histories of today's proteins we have learned that they are highly adaptable molecules, at least on evolutionary time scales. Well illustrated by the panoply of α/β-barrel enzymes (1), enzymes catalyzing very different reactions have evolved divergently from a common ancestral protein of the same general structure, acquiring diverse capabilities by processes of random mutation, recombination, and natural selection. We also know that enzymes sharing a common function (for example, all catalyzing a particular step in a metabolic pathway) in addition to three-dimensional structure can exhibit widely different properties (stability, solubility, tolerance to pH, etc.), depending on where they are found.

Enzymes evolve, and adapt, at the molecular level. The structures of protein modules are conserved (although the modules themselves are often shuffled to create new, multifunctional proteins). Function, however, can vary. Specific features such as substrate specificity or thermostability vary significantly. Amino acid sequences can vary to such an extent that evolutionary relationships may no longer be apparent from sequences alone (1).

Evolution is a powerful algorithm with proven ability to alter enzyme function and especially to "tune" enzyme properties. It is also an algorithm that can be implemented in the laboratory for redesign. The challenge is to collapse the time scale to months, or even weeks.

CHOOSING AN EVOLUTIONARY STRATEGY

Evolutionary mechanisms at work in nature assure adaptability to ever-changing environments. Evolution does not work toward any particular direction, nor is there a goal; the underlying processes occur spontaneously during reproduction and survival of the whole organism. In contrast, a directed molecular evolution experiment has a defined goal, and the key processes (mutation, recombination, and screening or selection) are controlled by the experimenter. Although there may be multiple ways to reach a defined goal (i.e., a desired enzyme function), the approach that minimizes the effort is preferred.

The major steps in a typical directed enzyme evolution experiment are outlined in Figure 1. The genetic diversity for evolution is created by mutagenesis and/or recombination of one or more parent sequences. These altered genes are inserted into a plasmid for expression in a suitable host organism (bacteria or yeast). Clones expressing improved enzymes are identified in a high-throughput screen or by selection, and the gene(s) encoding those improved enzymes are isolated and recycled to the next round

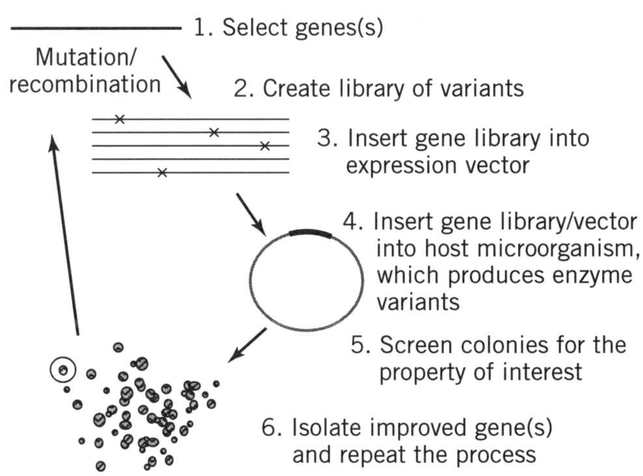

Figure 1. Key steps of a typical directed enzyme evolution experiment.

of directed evolution. Approaches for carrying out these key steps are discussed in some detail in the "Methods of Directed Evolution" section. Here we focus on more fundamental considerations that help to define workable strategies for directed evolution.

To appreciate the challenge of designing and carrying out a successful directed protein evolution experiment, it is important to underscore the powerful combinatorial features of this system. A typical enzyme is a linear chain of N amino acids (N is usually several hundred), and there are 20 possible amino acids at each position in the chain. Thus the "sequence space" of possible proteins is huge beyond the imagination (N^{20}). Even in 4 billion years, nature has had a chance to explore but a tiny fraction of these possibilities. A laboratory exploration of this vast space of sequences and their corresponding functions must obviously be severely limited and carefully guided (2). Because much of sequence space will be devoid of the desired function, and probably even of folded proteins, it is best to direct the evolution of an existing enzyme rather than look for function in random peptide libraries.

Evolution is often referred to as a hill-climbing exercise in the fitness landscape of sequence space (3,4). The fitnesses (performance, for laboratory evolution) of the proteins in sequence space make up this landscape, whose most basic features are still quite unknown. The landscape for laboratory evolution will be different for each property or collection of properties undergoing evolution. An uphill climb in a protein landscape is more likely to be successful if it can take place in small steps (one or two amino acid substitutions). The high dimensionality of the surface (there are $19N$ one-mutant neighbors of any given sequence) offers many opportunities to find improved mutants. Although we may never reach the "global optimum," the improvements achieved by taking even a simple random up-hill walk via single amino acid mutations often yield useful results. A widely effective evolutionary strategy, illustrated in Figure 2, is one in which the steps are small (preferably one or two amino acid substitutions in each generation), and multiple such mutations are accumulated either sequentially (5) or by recombination (6,7) to acquire the desired function. Such an approach is compatible with a low level of random mutagenesis over the entire gene. An alternative approach is to direct a much higher level of random mutation to a relatively small region of the gene (8). Both approaches have their advantages. Mutagenesis over the entire gene allows discovery of unanticipated solutions (a common experience). More intense, directed mutation, however, may yield novel combinations of amino acid substitutions, combinations that would be inaccessible by single-step walks because the intermediates are unfavorable. Details of the evolutionary exploration are ultimately dictated by a combination of *(1)* the power of the search tool, *(2)* the frequency of beneficial mutations (usually small!), and *(3)* the choice of starting point(s).

A different approach to creating diversity for directed evolution is the in vitro shuffling of homologous genes, or "family shuffling" (9), illustrated in Figure 3. Here, recombination of two or more parent genes yields a chimeric gene library for evolution of the desired features. Because the recombined sequences are related through divergent evo-

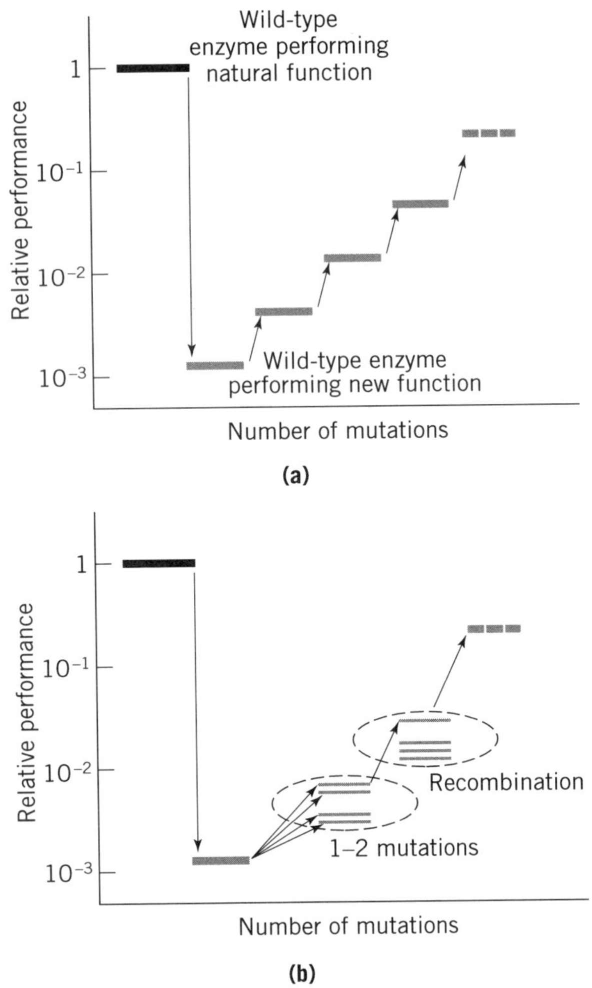

Figure 2. An effective strategy for directed enzyme evolution is one in which small changes associated with one or two amino acid substitutions are accumulated sequentially (**a**) or by recombination of improved genes (**b**).

lution from a common ancestor of similar structure and function (and therefore the sequence differences are to some extent neutral with respect to structure and function), it appears that very large jumps in sequence space can yield functional proteins (9,10). In vivo recombination can also yield interesting new chimeric enzymes (11–14).

Good Problems for Directed Enzyme Evolution

There are four requirements for successful directed evolution. *(1)* The desired function must be physically possible. *(2)* The function must also be biologically, or evolutionarily, feasible. In practice, this means that there exists a mutational pathway to get from here to there through ever-improving variants (see earlier discussion). Although we cannot know a priori that the path exists, a good experiment will maximize the likelihood. *(3)* One must be able to make libraries of mutants complex enough to contain rare beneficial mutations. This usually means functional expression in a suitable microorganism such as *Escherichia coli* or *Saccharomyces cerevisiae*. *(4)* One must have a rapid screen or selection that reflects the desired function. Just how rapid the screen must be depends on how rare mutations leading to the desired property are and how many must be accumulated to achieve the desired result.

Whether directed evolution will solve a particular problem depends to some extent on how hard natural evolution has already worked at it. If a particular trait is already under selective pressure (e.g., catalytic activity), it is unlikely that further improvements can be obtained in the laboratory by small mutational steps. However, if biological function has imposed additional constraints, for example the trait is coupled to another trait that is also under selective pressure (e.g., high thermostability), then this balance can be altered during laboratory evolution (15). While selected traits are often difficult to improve, they should be relatively easy to remove (e.g., product inhibition). Many traits are not under selective pressure; they may be changing as a result of random genetic drift,

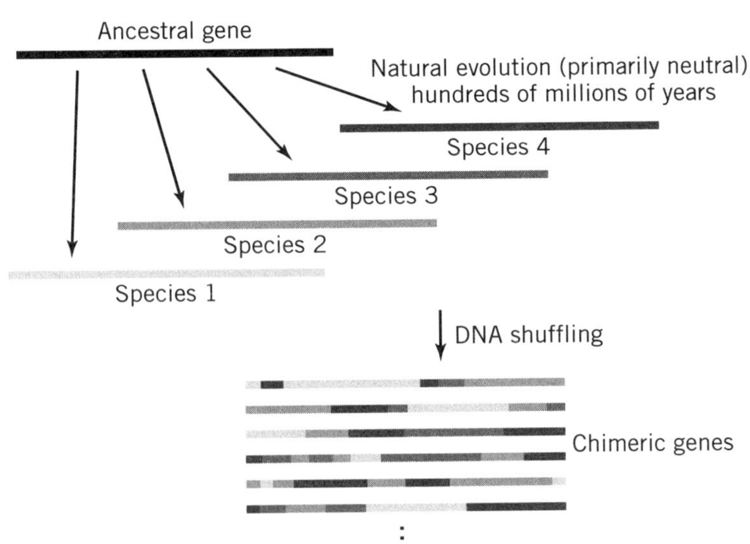

Figure 3. Homologous proteins are descended by divergent evolution from a common ancestral protein and share its overall three-dimensional structure. Recombining homologous genes creates chimeras, some fraction of which should also fold into that structure. Such family-shuffled libraries could be rich in novel function.

they may be vestigial—reflecting the enzyme's history—or they may be coupled to selected traits (16). In general, nonselected traits are easier to improve, but may be more difficult to remove. It is especially easy to improve traits never required for biological function, such as stability or activity in a nonnatural environment or activity toward a new substrate; there is often much room for improvement, and small changes in sequence and function can be accumulated. As expected for any hill-climbing exercise, the number of pathways leading uphill diminishes as a peak is approached. Thus the ease with which improved mutants are identified (the frequency of improved clones) should eventually decrease as the sequences move closer to an optimum.

The preceding discussion focuses only on altering existing enzyme traits. Evolving a completely new function is a risky venture because we rarely know how far in sequence space we have to go in order to create the new function, or how frequent the solutions will be. If there is good reason to believe that a new function, for example, activity toward a substrate not accepted at all by the wild-type enzyme, can be obtained (at a level measurable during a high-throughput screen) by making one or two amino acid substitutions, then evolution makes sense. However, if the new function requires the simultaneous placement of multiple new amino acid residues, it is unlikely to appear in a random library of mutants. Such a problem is probably a better candidate for a combination of rational design and combinatorial tuning (17).

Table 1 summarizes the key features of selected directed enzyme evolution experiments. In nearly all cases, the desired trait(s) was at a measurable, albeit low, level in the starting enzyme(s). The problems can be roughly divided into a few major categories: improving function in nonnatural or extreme environments (where activities or stabilities are low), improving activity toward a new substrate, tuning specificity (enantioselectivity), and increasing functional expression in a heterologous host.

METHODS OF DIRECTED EVOLUTION

Methods for Generating Genetic Diversity

The first step in a directed evolution experiment is the creation of a mutant library containing an appropriate degree of molecular diversity. For this, the mutation rate must be tuned to the power of the sorting method. Useful, reasonably sized libraries can be created by introducing multiple mutations in a particular region or across a limited number of positions (e.g., by combinatorial mutagenesis using oligonucleotide cassettes [42]). However, such an approach excludes the many useful solutions found in unexpected places. We therefore usually try to evolve the entire gene, rather than target particular positions. Whole-gene evolution requires a correspondingly lower mutation rate (see later).

Diversity is created by two operations: point mutagenesis and recombination (Fig. 4). Most random point mutagenesis methods create mutations in single bases. Due to the degeneracy of the genetic code, this provides access to only about 6 amino acid substitutions instead of 19, significantly reducing the potential diversity. In addition, many mutagenesis methods are not really random, further limiting the number of amino acid substitutions actually accessible in a given experiment. For example, a commonly used method, error-prone PCR, shows a strong bias for transitions over transversions (43).

Recombination methods combine sequences from multiple parent genes. In vitro recombination methods include DNA shuffling (6,7,44), random-priming recombination (45), and the staggered extension process (StEP) (46). In yeast, in vivo recombination is particularly simple (11,14). All these methods have a (controllable) level of associated point mutagenesis. Directed evolution can begin from multiple, closely related starting points rather than a single sequence. Recombination of existing functional sequences (i.e., homologous enzymes) will create another level of diversity that point mutagenesis cannot generate (9).

Point Mutagenesis. The most important factors in random mutagenesis are the mutation frequency and mutation bias. Mutation frequency is the average number of mutations per gene and is usually reported as a percentage. The optimal target mutation frequency can be calculated from the length of the DNA coding sequence and an estimate of the desired number of mutations per sequence. In deciding on the optimal number of mutations, one must balance the cost of combinatorially searching large libraries against the likelihood of finding mutants with improved properties. The number of possible variants increases drastically with the number of mutations introduced. The number of possible variants that can be produced by simultaneously substituting M amino acids in a protein of length N is $19M[N!/(N-M)!M!]$ (47). For example there are only 3,800 possible single mutants, but 7,183,900 double mutants, and 9,008,610,600 triple mutants of a protein of only 200 amino acids. The single-mutant library of an enzyme is generally within our ability to screen exhaustively. Double-mutant libraries are already on the outer limit of current screening technology, and triple-mutant libraries are well beyond our capability for the foreseeable future. To these numerical considerations we add the biochemical fact that beneficial mutations are rare, and combinations of beneficial mutations are extremely rare. In light of these considerations it appears that, at present, single-mutant libraries usually represent the best compromise between search effort and diversity: searching single-mutant libraries maximizes the chance of finding beneficial mutations in a reasonable amount of time. If high-throughput screening technology is available, or if significant fractions of possible mutations are neutral, then double-mutant libraries may be the optimal choice. An average of one amino acid substitution per sequence corresponds to approximately three base substitutions per gene. Thus for a 1-kb sequence, the mutation frequency should be ~0.3–0.5%.

Although numerous methods for making random DNA mutations exist, error-prone PCR is often preferable because the procedure is simple, rapid, robust, and most importantly, the mutation frequency can be precisely controlled. Error-prone PCR does have some intrinsic bias as to the location of mutations (mutations at AT base pairs

Table 1. Selected Examples of Directed Enzyme Evolution

Target enzyme	Target function	Change effected	Approach	Organism	Reference
Kanamycin nucleotidyltransferase	Thermostability	>200-fold increase in half-life at 60–65 °C	Mutator strain + selection	*B. stearothermophilus*	18
Subtilisin E	Activity in organic solvents	~170-fold increase in −60% dimethylformamide	Error-prone PCR + screening	*B. subtilis*	5,19
β-Lactamase	Activity toward new substrate	32,000-fold greater resistance to cefotaxime	DNA shuffling + selection	*E. coli*	6
Subtilisin BPN′	Stability in the absence of Ca^{2+}	1,000-fold increase in half-life	Loop removed + casette mutagenesis + screening	*B. subtilis*	20
para-Nitrobenzyl esterase	Activity toward pNB esters; activity in organic solvent	60 to 150-fold increase	Error-prone PCR and DNA shuffling + screening	*E. coli*	21,22,23
Thymidine kinase	Substrate specificity (gene therapy)	43-fold increase in sensitivity to gancyclovir in hamster cells	Cassette mutagenesis + selection and screening	*E. coli*	8
β-Galactosidase	Activity toward new substrate; substrate specificity	66-fold increased activity; 1,000-fold increase in substrate specificity	DNA shuffling + screening	*E. coli*	24
Subtilisin E	Expression level; activity in organic solvents	500-fold increase in total activity	Error-prone PCR + screening	*B. subtilis*	25
*O*6-Alkylguanine-DNA alkyltransferase	Protection against alkylating agents (gene therapy)	10-fold increased protection against toxic methylating agent	Cassette mutagenesis + selection	*E. coli*	26
Arsenate detoxification pathway	Arsenic resistance	12-fold increased rate of arsenate reduction	DNA shuffling + screening	*E. coli*	27
Aminoacyl-tRNA synthetase	Aminoacylation of a modified tRNA	55-fold increase in activity	DNA shuffling + selection	*E. coli*	28
Aspartate aminotransferase	Activity toward β-branched amino and 2-oxo-acids	10^5 increase	DNA shuffling + selection	*E. coli*	29
Lipase	Wash performance	Improved performance in one-cycle wash	Mutagenesis and in vivo recombination + screening	*S. cerevisiae*	30
Lipase	Enantioselectivity in hydrolysis of *p*-nitrophenyl 2-methyldecanoate	Increase in enantiomeric excess from 2% to 81%	Error-prone PCR + screening	*Pseudomonas aeruginosa* PAOI	31
Lipases	Activity toward long chain *p*-nitrophenyl esters	3-fold increase	In vivo recombination of homologous genes + screening	*E. coli*	13

pNB esterase	Thermostability	14 °C increase in T_m + increased activity at all temperatures	Error-prone PCR, DNA shuffling + screening	*E. coli*	15
Esterase	Enantioselectivity of hydrolysis of a sterically hindered 3-hydroxy ester	Increase in enantiomeric excess from 0% to 25%	Mutator strain + selection	*E. coli*	32
Subtilisin E	Thermostability	17 °C increase in T_m + increased activity at all temperatures	Error-prone PCR, DNA shuffling + screening	*B. subtilis*	33
Subtilisin E	Thermostability	50-fold increase in half-life at 65 °C	DNA shuffling + screening	*B. subtilis*	34
B. lentus subtilisin	Expression level (total activity of secreted enzyme)	50% increase	Error-prone PCR + enrichment in hollow fibers	*B. subtilis*	35
Subtilisin BPN′	Activity at 10 °C	2-fold increase	Chemical mutagenesis + screening	*B. subtilis E. coli*	36
3-Isopropylmalate dehydrogenase	Thermostability	3.4-fold increase in activity at 70 °C	Spontaneous mutations + selection	*T. thermophilus*	37
Cephalosporinases	Activity toward moxalactam	270 to 540-fold increased resistance	DNA shuffling of homologous genes + selection	*E. coli*	9
Chorismate mutase	Conversion to monomeric enzyme (solubility)	Functional monomeric enzyme	Oligonucleotide-directed codon mutagenesis + selection	*E. coli*	17
Biphenyl dioxygenases	Degradation of polychlorinated biphenyls (PCBs)	Gained activity toward substrates, poorly degraded by native enzymes, improved activity toward various substrates	DNA shuffling of homologous genes + screening	*E. coli*	38
FLP recombinase	In vivo recombination efficiency at elevated temperatures in *E. coli* and mammalian cells; in vitro thermostability	Improved recombination efficiency in *E. coli* and mammalian cells	Error-prone PCR and DNA shuffling + screening	*E. coli*	39
Echinocandin B deacylase	Activity	3-fold increase	Error-prone PCR, random-priming recombination and screening	*Streptomyces lividans*	40
Horseradish peroxidase	Functional expression in *E. coli*	11-fold increase in expressed activity	Error-prone PCR and screening	*E. coli*	41

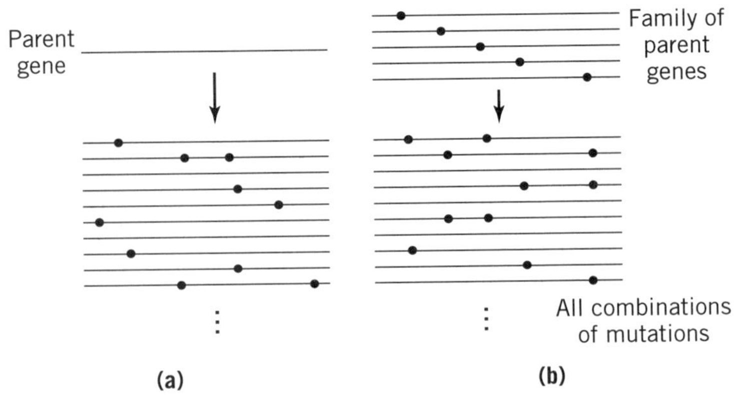

Figure 4. Two general approaches for generating molecular diversity: random point mutagenesis (**a**) and in vitro recombination (**b**).

occur much more frequently than mutations at GC base pairs) as well as their type (A is more frequently substituted with G). PCR modifications reduce bias, but do not eliminate it (48).

The error-prone PCR method routinely used was originally outlined by Leung et al. (49) and further examined by Cadwell and Joyce (43) and Shafikhani et al. (50). Mutation frequencies ranging from 0.11 to 2% have been obtained under different reaction conditions. In particular, the mutation frequency can be controlled simply by adjusting the concentration of manganese ions in the reaction mixture (48).

Because polymerase fidelity also depends on the nature of the target sequences, different mutation rates will be observed on different sequences, even when the exact same reaction conditions are used. A straightforward way to assess the overall mutation frequency and the nature of mutations is to sequence a few random clones from the amplified population. However, sequencing is time-consuming and expensive. A simple and efficient alternative is to estimate the mutation frequency from the fraction of active clones from the amplified population (51). Activity profiles of clones sampled from libraries of subtilisin E variants produced by error-prone PCR with different $MnCl_2$ concentrations are plotted in Figure 5. The clones are sorted and plotted in descending order to create local profiles, or landscapes, of the enzyme's fitness. The higher the Mn^{2+} concentration, the higher the mutation rate, which manifests itself in a higher proportion of inactivated clones. Even though different enzymes show different activity profiles for different mutation rates, the fraction of active clones is a convenient index of mutation frequency and can be used as a diagnostic check for the successful creation of the desired randomly mutated library.

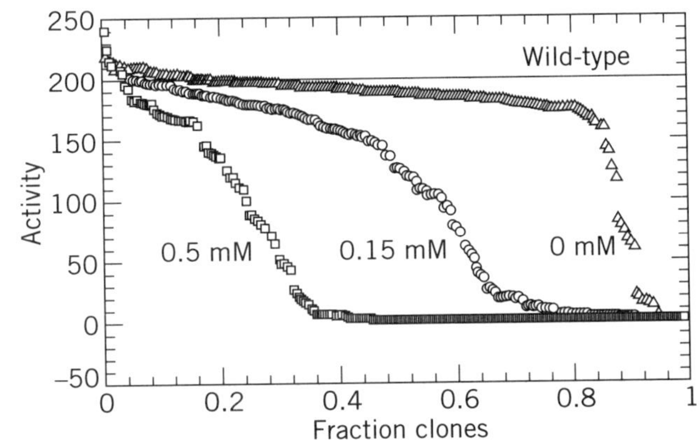

Figure 5. Activity profiles of subtilisin E mutant libraries generated by error-prone PCR with different concentrations of $MnCl_2$. Mutants are sorted according to activity in descending order. Horizontal line indicates the activity of the wild-type enzyme. *Source*: Ref. 51.

Recombination. When an enzyme is evolved by sequential generations of random mutagenesis and screening, only the best variant identified in each generation is used to parent the next generation. Other improved variants are set aside and must be rediscovered in subsequent generations in order to become incorporated. This is wasteful because screening is time-consuming. If the mutation rate is high, this approach can also accumulate deleterious mutations, possibly limiting the fitness that can be reached. "DNA shuffling" (6,7,44) and related in vitro recombination methods (45,46,48) can overcome these limitations of the sequential approach. Recombining parental genes to produce libraries of different mutation combinations, and screening for improved variants quickly accumulates the beneficial mutations, while removing deleterious ones.

The goal is to create gene libraries containing novel combinations of mutations present in the parent genes (Fig. 4). This library can be screened in order to find the combinations of mutations giving rise to the best enzymes. Screening eliminates unfavorable combinations of mutations. For example, if two point mutations that are beneficial by themselves become harmful when combined, the variant carrying this combination will be removed during screening. We now discuss some general considerations for recombination experiments before describing different recombination methods.

Statistics of Random Recombination. When recombining genes from multiple improved variants it is important to consider the recombined library size. The various in vitro recombination methods can recombine any number of parent genes. However, the resulting libraries may be so impossibly large that the probability of finding large improvements in function is effectively zero.

For the recombination of N sequences and M total mutations, the probability of generating progeny sequences containing μ mutations is equal to the number of ways a μ mutation sequence can be generated (C_{μ}^{M}) multiplied by the probability of generating any single μ-mutation sequence (22)

$$P_{\mu} = C_{\mu}^{M}\left(\frac{1}{N}\right)^{\mu}\left(\frac{N-1}{N}\right)^{M-\mu} = \frac{M!}{(M-\mu)!\mu!}\left(\frac{1}{N}\right)^{\mu}\left(\frac{N-1}{N}\right)^{M-\mu}$$

For recombination of N single mutants the probability of generating single-mutant parents or wild-type grandparents is approximately 75%, and only 25% of the library consists of new sequences. The probability of generating individual sequences decreases precipitously with increasing numbers of parents. The least-frequent sequences (often the most desirable) are those containing the majority of mutations from the parent population. The rarest sequence will be the one containing all mutations ($\mu = M$). The probability P_M of generating this sequence is $1/N^m$.

Some degree of oversampling is required in practice to maximize the chance of discovering a given variant. The sampling S required in order to achieve a given level of confidence of having sampled the rarest variant in a library is given by (22)

$$(1 - P_M)^S < 1 - (\text{confidence limit})$$

Generally, for 95% certainty that a specific clone has been sampled, the oversampling is between 2.6 and 3.0.

Recombination Methods. Stemmer described the first method for in vitro recombination of DNA sequences, called DNA shuffling (6,44). This method, illustrated in Figure 6, involves enzymatic digestion of the parental DNA into short fragments, followed by reassembly of the fragments into full-length genes. Because the fragments are free to associate with complementary fragments from other, similar genes, mutations from one parent can be combined with mutations from other parent(s) to generate novel combinations. The fidelity of this process is not perfect, and new point mutations not present in any of the parents will occur at a finite rate. The original method has been modified to simplify it and to yield better control over the associated mutagenic rate (48,52). The mutagenic frequency can be controlled over a wide range, from about 0.05 to 0.7%, by the inclusion of Mn^{2+} or Mg^{2+}, by the choice of DNA polymerase, and/or by using restriction enzyme digestion to prepare the starting DNA (48). The finite error frequency associated with DNA shuffling has been used by Stemmer and coworkers to supply point mutations for recombination and evolution. When starting with a single sequence, we prefer to use error-prone PCR under controlled conditions to generate libraries of variants. Improved sequences identified during screening can then be recombined.

An alternative method for recombination was developed by Shao et al. (45). In this method the template DNA sequences are primed with random-sequence primers, which are then extended by DNA polymerase to generate a pool of short DNA fragments. After removal of the template, these fragments can be reassembled, as with the Stemmer shuffling method, to give a library of full-length sequences containing information from multiple parents. Yet another method for in vitro recombination, the staggered extension process (StEP), does not require fragmentation and reassembly of the parent sequences (34). This method is illustrated in Figure 7 for recombination of two parent genes. In StEP recombination the template genes are primed and very briefly extended before denaturation and reannealing. The growing fragments can reanneal to different templates and therefore pick up sequence information from different parents as they grow into full-length sequences over hundreds of very brief cycles. StEP (in principle) involves only a single relatively short protocol as opposed to the multiple steps of fragmentation, isolation, and annealing required in DNA shuffling.

"Family Shuffling": Recombination of Homologous Genes. The approaches already described (point mutation and recombination of improved variants) allow a local exploration of sequence space. Only sequences that are quite close to the parent sequence are sampled. One would like to sample distant regions of sequence space to search for new functions, but this is not readily accomplished using these methods. The simultaneous introduction of more than a few random mutations into a wild-type sequence will almost always result in variants that are inactive or unfolded. The probability that an improved enzyme will be found is usually much smaller than the screening capacity. An alternative approach to making large jumps in sequence space is by "family shuffling" (9) (see Fig. 3). In this approach, homologous genes from different organisms are recombined to create a library of chimeric molecules. Family shuffling exploits the fact that the sequence differences between two homologous proteins are *not* random. Consider the two proteases subtilisin E and subtilisin thermitase. These two enzymes differ at 157 of their amino acid positions, and yet they fold to the same overall three-dimensional structure and catalyze the same proteolytic reaction. The amino acid substitutions in these and other naturally occurring subtilisins have been preselected to be largely neutral with respect to folding and gross function. Mutations resulting in a misfolded or inactive enzyme have been eliminated by natural selection. Accordingly, recombination of this pool of largely neutral mutations will be far more likely to result in folded, active proteins than recombination of purely random mutations. The use of family shuffling to access remote unexplored regions of sequence space may yield proteins with a variety of desirable, but not yet found, functions.

Searching for Improved Enzymes

In general, the most time-consuming and expensive part of directed enzyme evolution is setting up and implementing the search for improved variants. Developing an optimal search strategy is crucial for success. A fundamental rule of directed evolution, "you get what you screen for," specifies that the screen (or selection) should reflect the desired result as closely as possible.

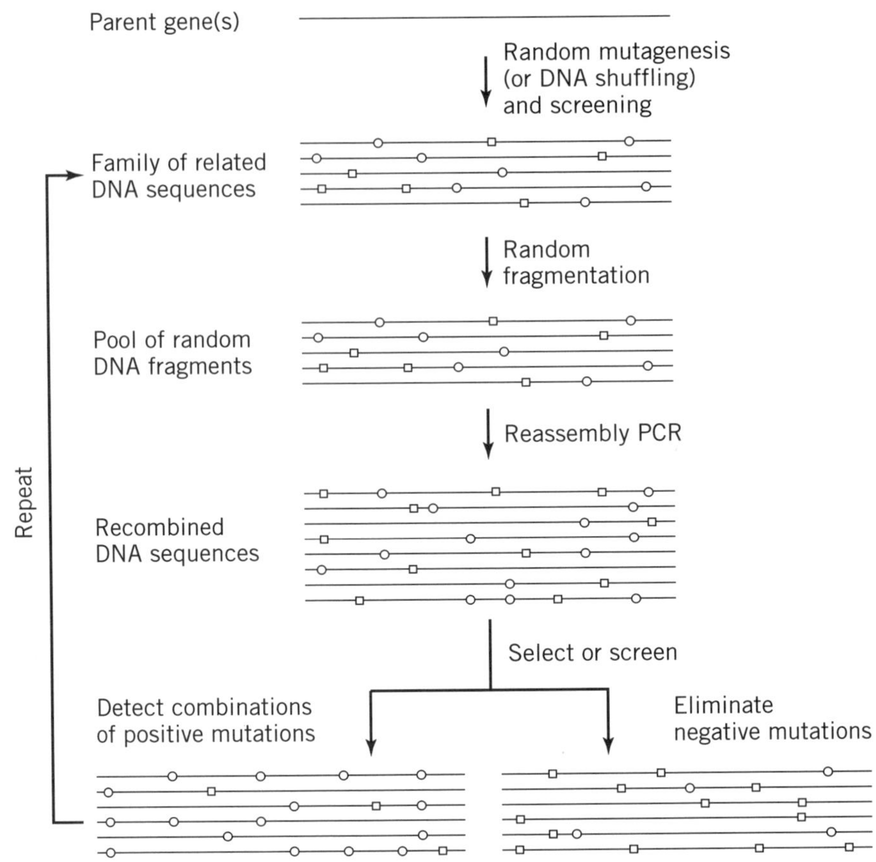

Figure 6. Recombination of parent genes by DNA shuffling. *Source*: Ref. 44.

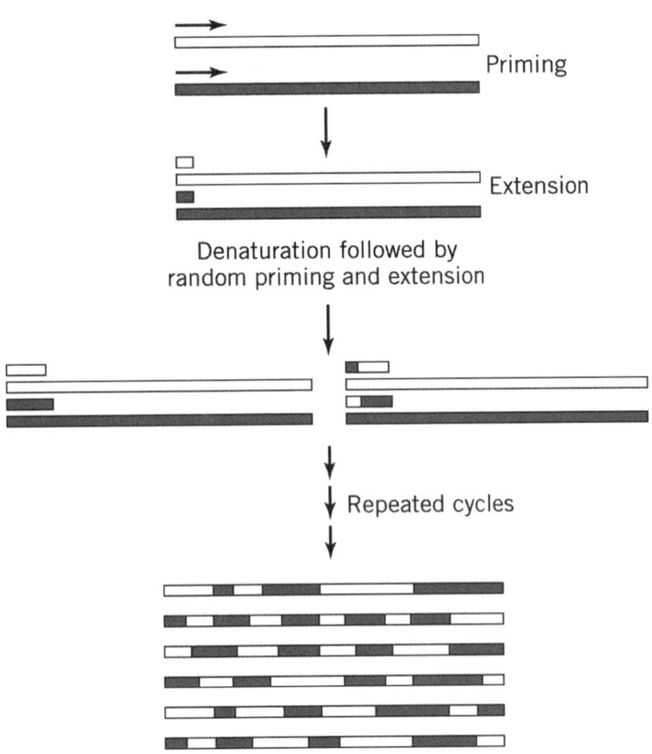

Figure 7. Recombination of two parent genes by StEP. *Source*: Ref. 34.

Mutant libraries are searched by screening or selection. In a genetic selection, the ability of an enzyme to perform the desired function is directly coupled to the survival and reproduction of the host organism. A straightforward example is selection for drug resistance: clones expressing an enzyme capable of degrading an antibiotic are selected for their ability to grow on antibiotic-containing plates (53,54). Another example is the selection of functional mutants of a human methyltransferase. Active mutants were selected based on their ability to protect alkyltransferase-deficient *E. coli* cells from a methylating agent (26). Such approaches are attractive in principle because they allow one to search larger libraries than does screening (10^6–10^7 for selection, as opposed to 10^4–10^5 for screening). Kast and Hilvert (55) recently reviewed genetic selections in directed evolution.

However, selections suffer serious drawbacks. The types of new properties that can be searched for are limited. For example, enzymes tolerate a number of environments that cannot sustain life (e.g., organic solvents). It becomes extremely difficult to tie the desired function to the survival or growth of an organism when the organism cannot survive. The main drawback, however, is the complexity of the organisms themselves and the surprising solutions that spontaneously arise during adaptation. Given our limited understanding of even simple, well-studied organisms such as *E. coli*, it is difficult to ensure that improving a particular enzyme is the *only* way for an organism to adapt to the selective pressure applied. One may go through several rounds of selection only to find that the organism has found some alternate way to solve the problem.

Various in vitro selections have been proposed based on column binding or "panning" (56). In these methods, the proteins are displayed on the surface of a phage or a cell. Very large libraries ($>10^9$) can be searched based on the ability of the displayed protein to bind to a substrate immobilized on a column. The greatest limitation to this approach is that it is not generally applicable to screening for properties other than binding.

For many problems, and especially those of practical interest, libraries of variants must be screened rather than selected. A typical screening strategy involves the construction of an arrayed enzyme library and application of a rapid assay of the desired property. The screen can be more or less sensitive, depending on the willingness of the researcher to accept false positives (and to apply additional tests) (57).

Prerequisites for an effective screening strategy include the following:

1. The screen should be sufficiently rapid that large numbers of clones can be searched in a reasonable amount of time. It should be feasible to screen at least a few hundred, and preferably a few thousand, in a day. For obvious reasons, the screen should not require purification of the enzyme. A second-level assay can eliminate false positives.
2. Only rarely do very large improvements result from a single amino acid substitution. Directed evolution most often succeeds by accumulating a number of modest improvements over several generations. Therefore a screen must be accurate enough to allow modest improvements to be identified against the background variation.
3. The phenotype (the enzyme property that is being screened) must be physically coupled to the genotype (the DNA or RNA sequence coding for the enzyme). This is most often accomplished trivially by the fact that proteins are expressed in host organisms: the same cell expressing the enzyme contains the gene coding for its sequence. Coupling can also be accomplished by display on the surface of a virus containing the genetic information in the form of DNA or RNA (58). Recently developed in vitro translation systems such as ribosome display (59) or RNA–peptide fusions (60) link the enzyme directly to the genetic information in the form of RNA without using cells. Tawfik and Griffiths (61) have encapsulated the in vitro translation apparatus and genes within reverse micelles.

Simple visual screens are widely used when the function of interest can generate a visible signal. Various oxidases, for example, produce hydrogen peroxide as a byproduct of their oxidation reaction. As illustrated in Figure 8, coupling this to a second enzyme that is easily assayed colorimetrically (horseradish peroxidase) provides a visual screen for the oxidase activity. Clones secreting active pro-

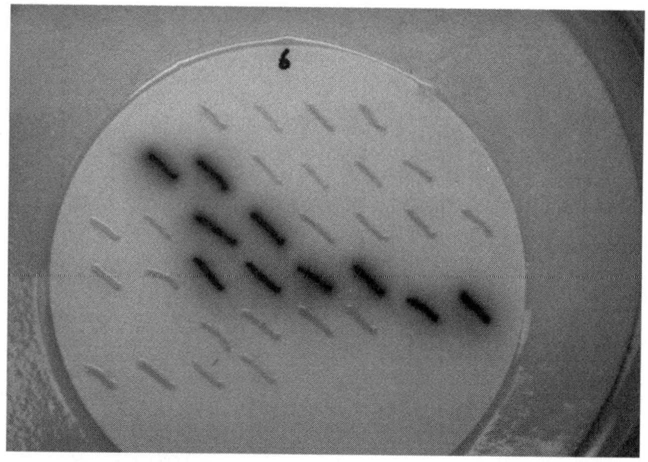

Figure 8. Example data from a visual screen for galactose oxidase activity toward D-galactose. Colonies are grown on agar plates and transferred to a membrane. Active colonies generate a purple color when incubated with 2,2′-azinobis(3-ethylbenzothiazline-6-sulfonic acid), horseradish peroxidase, and D-galactose. *Source*: M. Yagasaki, unpublished results, 1998.

teases produce a zone of clearing or "halo," the size of which is proportional to the hydrolytic activity when the organism is grown on agar plates containing casein or skim milk proteins. This simple method has been used extensively in screening protease libraries (25,62).

Although visual screening based on color or halo formation is rapid and efficient, it is also nonquantitative and often relatively insensitive. Digital imaging spectroscopy has been developed to increase the sensitivity and throughput of filter- and agar-plate-based screens (63). However, the 96-well microtiter plate remains the standard format for automated, high-throughput screening. Screening automation and quantification of the results are highly desirable; 96- and 384-well plates appear to be the format most compatible with currently available robotic arms, liquid handling systems, and plate readers. Data collection and analysis are greatly facilitated by computerized data acquisition (computerized data are easily sorted into local enzyme fitness landscapes such as those shown in Fig. 5). Computerized data acquisition will certainly facilitate exploitation of the enormous amounts of information available in protein libraries. Many conventional enzyme assays can be readily converted into automated formats.

If activity against the desired substrate does not result in an easily detectable signal (such as a change in absorbance or fluorescence), it is sometimes possible to screen against a surrogate substrate that does. Naturally, this approach requires caution. It must be verified that activity against the replacement substrate reflects activity against the real substrate under the conditions of interest. A successful application of this approach is described in the example provided later of the evolution of a *p*-nitrobenzyl esterase (21).

Enzyme properties other than activity are often targeted for improvement. A convenient approach to screening for thermostable variants of irreversibly inactivated enzymes is shown in Figure 9. Duplicate 96-well plates are made from a single master plate. One is incubated at high temperature for a fixed period. Clones in both are then assayed for activity, and the ratio of activity after heating to activity without heating provides a measure of enzyme stability at the temperature of incubation.

An attractive approach to screening large libraries is to couple functional complementation with screening. By requiring that at least some of the biological function of the protein is retained, functional complementation can greatly reduce the subsequent screening requirements. The initial selection asks an essentially binary question (is there any activity? yes or no), and only in the later screening stage are differences in the level of activity detected. Loeb and coworkers have applied this strategy to several enzymes of interest in gene therapy and other applications (8,26). However, the major limitation is finding or constructing an appropriate complementation system. Furthermore, retention of biological function may preclude acquisition of other functions.

Extracting Useful Information from the Results of Evolution

Two major types of information can be extracted from the results of directed evolution experiments. The first and most obvious is the information contained in the sequences of the evolved proteins. The second consists of the overall activity profiles, or local fitness landscapes, that can be generated by analysis of the entire screened library.

Clearly, we would like to identify the specific amino acid substitutions that are responsible for conferring new properties. In contrast to comparisons of evolutionarily related proteins found in nature, this identification can be accomplished relatively easily in directed evolution experiments. In nature, related enzymes that are separated by millions of years of evolution will generally have accumulated a large number of neutral mutations in addition to those that are responsible for changes in specific properties. In contrast, in laboratory evolution we strive to allow only functional mutations to survive and carry on to the next generation. A striking example of the difference between natural and laboratory evolution can be seen in a recent study in which subtilisin E was converted into a functional equivalent of the thermophilic enzyme thermitase (33). Subtilisin E differs from thermitase at 157 positions, but only eight substitutions were required to turn it into a thermitase-like enzyme, with an 18 °C increase in its temperature optimum and a 250-fold increase in its half life at 65 °C.

However, it is not always possible to identify functional mutations simply by examining the sequences of the evolved enzymes. In the experimental procedures for introducing random point mutations, only the *average* error rate can be controlled. Thus, some variants will accumulate multiple mutations in a single generation. If a given variant has multiple mutations, only one of which is functional, the remaining neutral mutations can "hitchhike" their way into subsequent generations and complicate analysis of the final evolved sequences. A method for rapidly identifying functional mutations by back-crossing with wild type has been described (64).

It is hoped that detailed structural analysis of evolved enzymes will provide insight into the underlying physical–chemical mechanisms responsible for protein adaptation for new functions or environments. However, the results of single point mutations can be very subtle, and the analysis is not straightforward even when the three-dimensional structure is known. For example, both subtilisin E and pNB esterase have been successfully evolved for activity in aqueous organic solvents, but functional mutations in

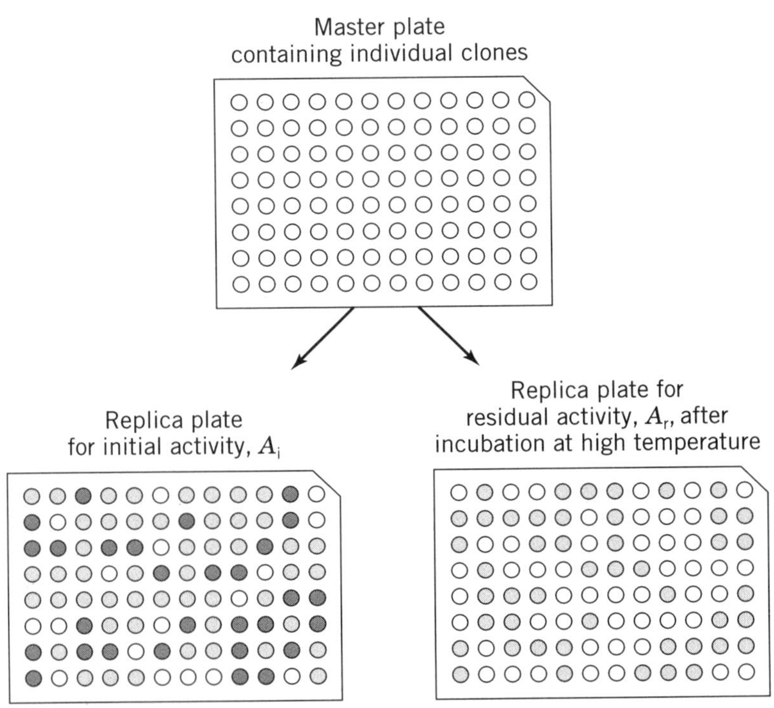

Figure 9. A rapid screen for enzyme thermostability based on catalytic activity before and after incubation at high temperature. *Source*: Ref. 34.

the evolved enzymes follow no discernible pattern: large residues are replaced by small, small by large, charged by uncharged, uncharged by charged (5,21,25). Further, mutations are found both close to and far from the enzyme active sites. Numerous studies now attest to the subtlety of the effects of amino acid substitutions on enzyme structure and properties (8,24,38,39).

The distribution of fitnesses (defined in terms of activity, stability, etc.) in a mutant library can provide useful information for designing evolutionary strategies. One way to access this information is by presenting the data in a local fitness landscape such as those shown in Figure 5. As mentioned previously, the mutation rate can be estimated from the fraction of inactive mutants. More generally, such a plot represents (at least for single- or double-mutant libraries) the distribution of fitness in the region of sequence space immediately surrounding the wild-type sequence. Most mutants will be equally or less fit than the wild type; a small number may be better. The fraction of improved mutants can be used to estimate the evolvability of the property of interest. When two or more properties are being evolved simultaneously, plots of one versus the other can be informative. Typically, single mutations that simultaneously improve two independent properties are extremely rare. However, individuals may be recombined to yield an enzyme with multiple improved properties (Fig. 10) (65).

EXAMPLES OF DIRECTED EVOLUTION APPLICATIONS

Evolution of Lipases with High Activity in Detergent Solutions

New lipase variants that give greatly improved residual lipid removal after one wash cycle have been developed by directed evolution (30). Lipases are used as detergent enzymes to remove lipid or fatty stains from clothes and other textiles. A drawback of the known detergent lipases is that they exert the best lipid-removing effect after more than one wash cycle, presumably because they are more active during a certain period of the drying process than during the wash itself. Therefore at least two wash cycles, separated by a sufficiently long drying period, are required to remove fatty stains.

An intense protein engineering effort involving both site-directed and random mutagenesis yielded a large number of variants of *Thermomyces lanuginosa* lipase that were strongly improved in multicycle wash performance, but not significantly improved in one-cycle wash tests. Random mutant libraries were screened using filter assays containing detergent and low calcium concentrations, to mimic wash conditions. Consecutive rounds of mutagenesis and screening with increasing amounts of detergent were performed.

New variants with a strong one-cycle wash effect were obtained by recombining 20 variants with the best performance in multicycle wash tests and screening. The recombination was performed using a simple in vivo method in *S. cerevisiae* in which PCR fragments from the 20 variants mixed with the opened vector are used to transform competent *S. cerevisiae* cells. Upon transformation, the vector and fragments recombine and shuffle the variants (14). Screening yielded seven variants with a strong one-cycle wash effect; the best of these is now commercially available.

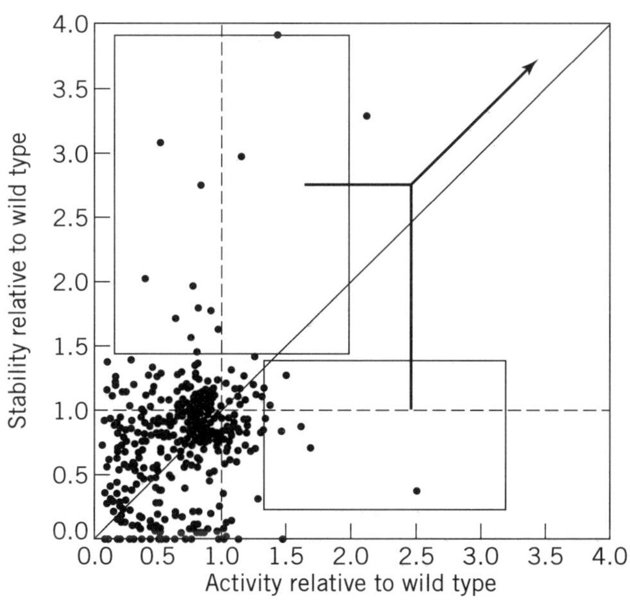

Figure 10. Activity and stability to hydrogen peroxide of a randomly mutagenized enzyme library. The individual populations shown in boxes can be evolved in parallel and recombined to obtain enzymes that are both more active and more stable. *Source*: Reproduced with permission from Ref. 65.

Evolution of an Esterase for Deprotection of *p*-Nitrobenzyl Esters and Enhanced Thermostability

By screening microbial cultures, scientists at Eli Lilly identified an enzyme in *Bacillus subtilis* that could remove the *p*-nitrobenzyl ester protecting group used during the large-scale synthesis of certain β-lactam antibiotics. However, the enzyme's relatively poor activity, particularly in the solvents required to solubilize the ester substrate, made it a poor competitor to the chemical catalyst for deprotection of a loracarbef synthetic intermediate. The natural function of *B. subtilis* pNB esterase is unknown.

Moore and Arnold were able to evolve highly active pNB esterases that also function well in mixed aqueous–organic solvents (21,22). A surrogate, chromogenic substrate (the antibiotic *p*-nitrophenyl ester) was used in a rapid screen to identify potential positives. Variants identified during the rapid screen could be verified during a second-level screen on the *p*-nitrobenzyl ester using HPLC. Four generations of PCR mutagenesis and two rounds of recombination by DNA shuffling yielded a clone with more than 100 times the total activity of wild type in 15–20% dimethylformamide (DMF) toward loracarbef-*p*-nitrobenzyl ester. The enzyme's catalytic efficiency increased more than 50-fold (Fig. 11). The total activity toward the screening substrate (loracarbef-*p*-nitrophenyl ester) increased more than 150-fold. Although the contributions of individual effective amino acid substitutions to enhanced activity were small (usually approximately two-fold increases in activity), the accumulation of multiple mutations over a num-

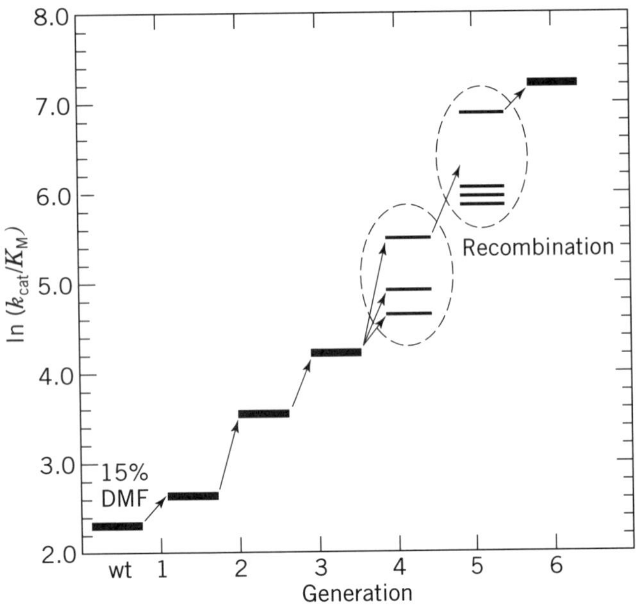

Figure 11. Evolutionary progression of catalytic efficiency of pNB esterase (toward LCN-pNB) in 15% DMF through four generations of random mutagenesis, followed by two rounds of recombination of circled populations. *Source*: Ref. 66.

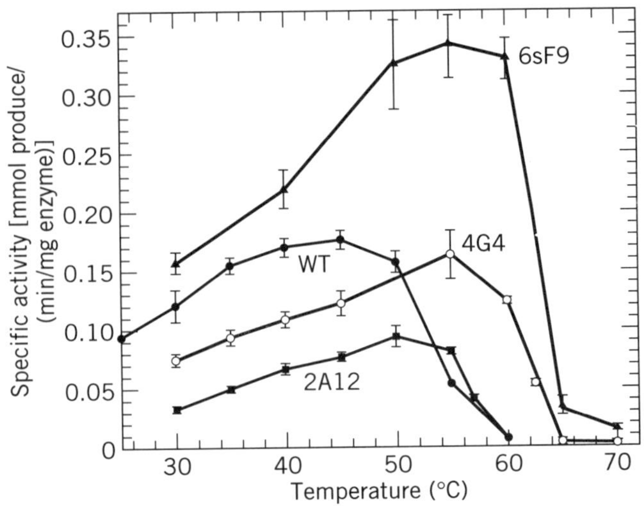

Figure 12. Activities of wild-type and evolved pNB esterases as a function of temperature. *Source*: Ref. 15 and A. Gershenson, unpublished results.

ber of generations significantly improved the biocatalyst for this nonnatural reaction. Moore and Arnold concluded that none of the mutations accumulated in the evolved enzyme are in direct contact with the substrate. Some are as far away as 20 Å. Limiting the search for possible solutions to residues that line the substrate binding site would have overlooked important beneficial mutations (and may not have succeeded).

Directed evolution has also significantly increased the thermostability of the *B. subtilis* pNB esterase (15). A simple screen was developed based on retention of catalytic activity after incubation at high temperature. Positives were then subjected to differential scanning calorimetry (DSC) to verify an increase in melting temperature (T_m) relative to the parent enzyme. Accumulating mutations over eight generations of random mutagenesis and DNA shuffling yielded an increase in T_m of 17 °C (15, A. Gershenson, unpublished results). This large increase in thermostability is equivalent to the difference between proteins from mesophiles and many thermophilic organisms. Furthermore, because only those variants that retained their catalytic activity as well were chosen, the increased thermostability was accompanied by a very significant increase in enzyme activity at elevated temperatures (Fig. 12). Reflecting the fact that the thermostability screen involved activity toward *p*-nitrophenyl acetate, the resulting enzymes are highly active toward this substrate and less active toward the antibiotic substrate and *p*-nitrobenzyl esters. The most thermostable pNB esterase variant has 13 amino acid substitutions (out of 490).

Evolution of Herpes Simplex Virus Thymidine Kinase for Gene Therapy

The Loeb group has pioneered the use of directed evolution for developing improved enzymes for cancer gene therapy. Here, greater activity toward nonnatural substrates is desired, with the goal of selectively protecting bone marrow cells (26) or sensitizing tumor cells (8) to toxic agents. This group consistently directs intense mutagenesis of a limited region of the gene and genetic selection to identify the sequences that are still biologically functional. Screening this much smaller subset of sequences identifies those with the desired properties (altered substrate specificity, ability to protect or sensitize cells).

In one study (8), directed evolution was used to create herpes simplex virus thymidine kinase (HSV TK) variants that enhance cell killing by the guanine analogs ganciclovir and acyclovir, which terminate DNA synthesis and prevent viral and host cell replication. Mutant HSV TKs that selectively phosphorylate gancyclovir and/or acyclovir were obtained after complete randomization of a limited set of amino acid positions (six) adjacent to the putative nucleoside binding site and previously identified to tolerate some substitution. Of the approximately 10^6 *E. coli* transformants, 426 expressed active TK, as determined by genetic complementation. These were then assayed for their ability to enhance the sensitivity of the *E. coli* to the guanine analogs. Hamster cells transfected with one of the improved variants were more than 43 times more sensitive to gancyclovir and 20 times more sensitive to acyclovir than cells expressing the wild-type HSV TK. This mutant contained four amino acid substitutions, all within the TK active site pocket.

Evolution of Biphenyl Dioxygenases for Bioremediation of PCBs

Biphenyl dioxygenases (BPDO) are responsible for the initial oxidation of polychlorinated biphenyls (PCBs) during their biodegradation by various organisms. In an effort to create BPDOs with enhanced ability to oxidize a wider range of PCB congeners, Furukawa and coworkers used DNA shuffling to create a chimeric library from two genes

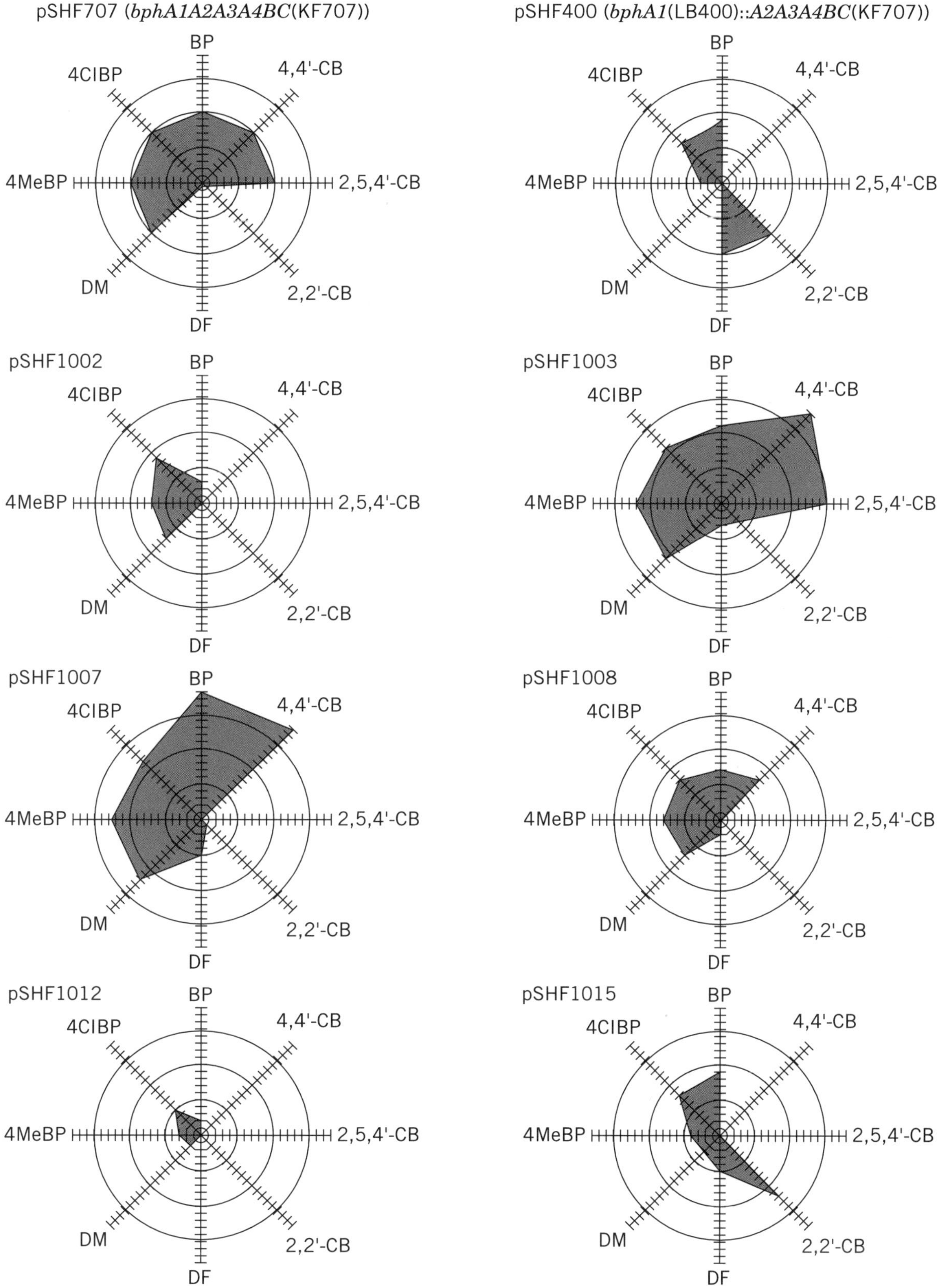

Figure 13. Formation of *meta*-cleavage yellow products from a variety of biphenyl compounds by *E. coli* expressing chimeric biphenyl dioxygenases. BP, biphenyl; 4ClBP, 4-chlorobiphenyl; 2.2′-CB, 2,2′-dichlorobiphenyl; 4,4′-CB, 4,4′-dichlorobiphenyl; 2,5,4′-CB, 2,5,4′-trichlorobiphenyl; MeBP, 4-methylbiphenyl; DM, diphenylmethane; DF, dibenzofuran. The relative values of yellow products formed from each compound are plotted for *E. coli* strains carrying the evolved enzymes. *Source*: Reprinted with permission from Ref. 38.

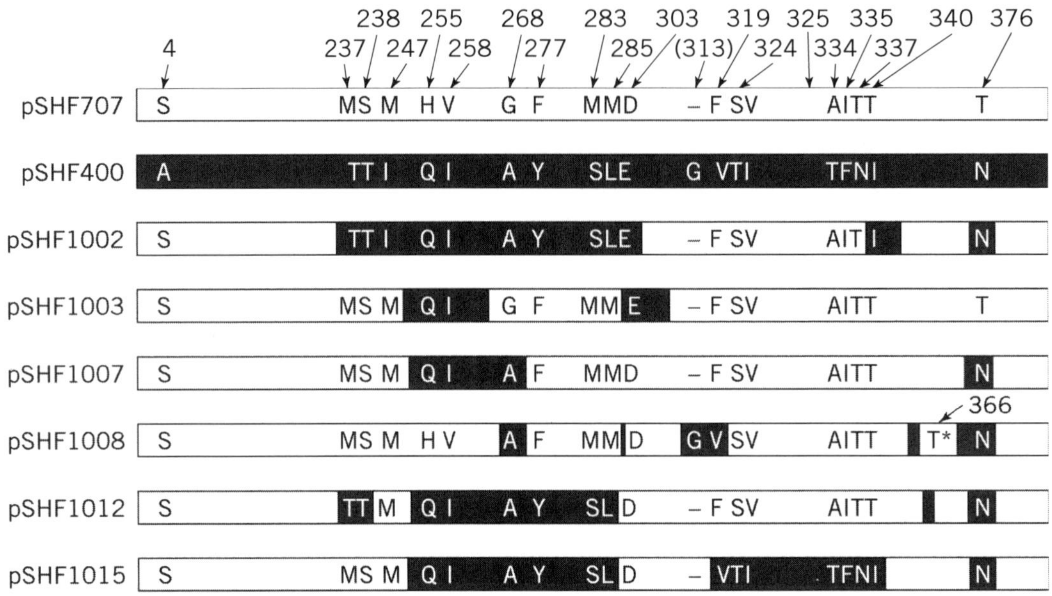

Figure 14. Sequence analysis of the shuffled *bphA1* genes. Twenty different amino acids between KF707-BphA1 and LB400-BphA1 are shown. *, Indicates the mutational change (I366T) that occurred during PCR; –, indicates an amino acid lacking in KF707 relative to the LB400 sequence. *Source*: Reprinted with permission from Ref. 38.

for the terminal dioxygenases of BPDO (*bphA1*) from *Pseudomonas pseudoalcaligenes* KF707 and *Burkholderia cepacia* LB400 (38). A high-fidelity DNA shuffling protocol (48) was found to increase the frequency of clones expressing active PBDO. The two BphA1 protein sequences are extremely similar, differing only at 20 amino acids. The genes are 96.4% identical.

E. coli containing the chimeric genes were screened for their ability to produce yellow *meta*-cleavage products from a variety of biphenyl compounds in order to identify functional chimeras with enhanced activities and altered activity profiles. Enzymes with greater activity toward various biphenyl compounds than either parent were identified (Fig. 13), as were enzymes with relaxed specificity and enzymes active toward smaller aromatic hydrocarbons such as benzene and toluene. These latter compounds are poor substrates for the two parent enzymes.

Sequence analysis of six shuffled functional *bphA1* genes demonstrated how the parent sequences recombined to yield these new enzymes (Fig. 14). A very low level of point mutagenesis was observed. The specific roles of the amino acids in determining substrate specificity are difficult to pinpoint in the absence of a three-dimensional structure of the enzyme. However, the results are generally in agreement with previous studies of BPDO chimeras and site-directed mutagenesis.

ACKNOWLEDGMENTS

F.H.A. wishes to thank the many talented undergraduate, graduate, and postdoctoral students who have contributed to directed evolution research at Caltech. The financial support of the Office of Naval Research, the Department of Energy, and the Army Research Office is gratefully acknowledged.

BIBLIOGRAPHY

1. L. Holm and C. Sander, *Proteins: Struct. Funct. Genet.* **28**, 72–82 (1997).
2. F.H. Arnold, *Acct. Chem. Res.* **31**, 125–131 (1998).
3. M. Eigen, *Chemica Scripta* **26B**, 13–26 (1986).
4. S.A. Kauffman, *The Origins of Order*, Oxford University Press, Oxford, 1993.
5. K. Chen and F.H. Arnold, *Proc. Natl. Acad. Sci. U.S.A.* **90**, 5618–5622 (1993).
6. W.P.C. Stemmer, *Nature* **370**, 389–391 (1994).
7. U.S. Pat. 5,605,793 (February 25, 1997) W.P.C. Stemmer (to Affymax Technologies N.V.).
8. M.E. Black, T.G. Newcomb, H.M.P. Wilson, and L.A. Loeb, *Proc. Natl. Acad. Sci. U.S.A.* **93**, 3525–3529 (1996).
9. A. Crameri, S.A. Raillard, E. Bermudez, and W.P.C. Stemmer, *Nature* **391**, 288–291 (1998).
10. F.H. Arnold, *Nat. Biotechnol.* **16**, 617–618 (1998).
11. D. Pompon and A. Nicolas, *Gene* **83**, 15–24 (1989).
12. U.S. Pat. 5,093,257 (March 3, 1992) G.L. Gray (to Genencor International, Inc.).
13. M.D. van Kampen, N. Dekker, M. Egmond, and H. Verheij, *Biochemistry* **37**, 3459–3466 (1998).
14. PCT Pat. WO 97/07205 (August 12, 1996), J.S. Okkels (to Novo Nordisk A/S).
15. L. Giver, A. Gershenson, P.O. Freskgard, and F.H. Arnold, *Proc. Natl. Acad. Sci. U.S.A.* **95**, 12809–12813 (1998).
16. S.A. Benner, *Chem Rev.* **89**, 789–806 (1989).
17. G. MacBeath, P. Kast, and D. Hilvert, *Science* **279**, 1958–1961 (1998).
18. H. Liao, T. McKenzie, and R. Hageman, *Proc. Natl. Acad. Sci. U.S.A.* **83**, 576–580 (1986).

19. U.S. Pat. 5,316,935 (May 31, 1994), F.H. Arnold and K. Chen (to California Institute of Technology).
20. S.L. Strausberg, P.A. Alexander, D.T. Gallagher, G.L. Gilliland, B.L. Barnett, and P.N. Bryan, *Biotechnology* **13**, 669–673 (1995).
21. J.C. Moore and F.H. Arnold, *Nat. Biotechnol.* **14**, 458–467 (1996).
22. J.C. Moore, H.M. Jin, O. Kuchner, and F.H. Arnold, *J. Mol. Biol.* **272**, 336–347 (1997).
23. U.S. Pat. 5,741,691 (April 21, 1998), F.H. Arnold and J.C. Moore (to California Institute of Technology).
24. J-H. Zhang, G. Dawes, and W.P.C. Stemmer, *Proc. Natl. Acad. Sci. U.S.A.* **94**, 4504–4509 (1997).
25. L. You and F.H. Arnold, *Protein Eng.* **9**, 77–83 (1996).
26. F.C. Christians and L.A. Loeb, *Proc. Natl. Acad. Sci. U.S.A.* **93**, 6124–6128 (1996).
27. A. Crameri, G. Dawes, E. Rodrigues, Jr., S. Silver, and W.P.C. Stemmer, *Nat. Biotechnol.* **15**, 436–438 (1997).
28. D.R. Liu, T.J. Magliery, M. Pasternk, and P.G. Schultz, *Proc. Natl. Acad. Sci. U.S.A.* **94**, 10092–10097 (1997).
29. T. Yano, S. Oue, and H. Kagamiyama, *Proc. Natl. Acad. Sci. U.S.A.* **95**, 5511–5515 (1997).
30. PCT Pat. WO 97/07202 (February 27, 1997), J.S. Okkels, M. Thellersen, D.A. Petersen, A. Svendsen, S.A. Patkar, and K. Borch (to Novo Nordisk A/S).
31. M.T. Reetz, A. Zonta, K. Schimossek, K. Liebeton, and K.E. Jaeger, *Angew. Chem. Int. Ed. Engl.* **36**, 2830–2832 (1997).
32. U.T. Bornscheuer, J. Altenbuchner, H.H. Meyer, *Biotechnol. Bioeng.* **58**, 554–555 (1998).
33. H. Zhao and F.H. Arnold, *Protein Eng.* **12**, 47–53 (1999).
34. H. Zhao, L. Giver, Z. Shao, J.A. Affholter, and F.H. Arnold, *Nat. Biotechnol.* **16**, 258–235 (1998).
35. D. Naki, C. Paech, G. Ganshaw, and V. Schellenberger, *Appl. Microbiol. Biotechnol.* **49**, 290–294 (1998).
36. S. Taguchi, A. Ozaki, and H. Momose, *Appl. Environ. Microbiol.* **64**, 492–495 (1998).
37. S. Akanuma, A. Yamagushi, N. Tanaka, and T. Ooshima, *Protein Sci.* **7**, 698–705 (1998).
38. T. Kumamaru, H. Suenaga, M. Mitsuoka, T. Watanabe, and K. Furukawa, *Nat. Biotechnol.* **16**, 663–666 (1998).
39. F. Buchholz, P.O. Angrand, and A.F. Steward, *Nat. Biotechnol.* **16**, 657–662 (1998).
40. Z. Shao, M. Callahan, and F.H. Arnold, *Appl. Environ. Microbiol.* (in press).
41. Z. Lin, T. Thorsen, and F.H. Arnold, *Biotechnol. Prog.* (in press).
42. J. Reidhaar-Olson and R. Sauer, *Science* **241**, 53–57 (1988).
43. R. Cadwell, and G. Joyce, *PCR Methods Appl.* **3**, S136–S140 (1994).
44. W.P.C. Stemmer, *Proc. Natl. Acad. Sci. U.S.A.* **91**, 10747–10751 (1994).
45. X. Shao, H. Zhao, L. Giver, and F.H. Arnold, *Nucleic Acids Res.* **26**, 681–683 (1998).
46. H. Zhao, L. Giver, X. Shao, J. Affholter, and F.H. Arnold, *Nat. Biotechnol.* **16**, 258–261 (1998).
47. F.H. Arnold, *Chem. Eng. Sci.* **51**, 5091–5102 (1996).
48. H. Zhao and F.H. Arnold, *Nucleic Acids Res.* **25**, 1307–1308 (1997).
49. D. Leung, E. Chen, and D. Goeddel, *Technique* **1**, 11–15 (1989).
50. S. Shafikhani, R. Siegel, E. Ferrari, and V. Schellenberger, *Biotechniques* **23**, 304–310 (1997).
51. H. Zhao, J. Moore, and F.H. Arnold, in *ASM Manual of Industrial Microbiology and Biotechnology*, ASM Press, Washington D.C., 1999, pp. 597–604.
52. I.A. Lorimer and I. Pastsn, *Nucleic Acids Res.* **23**, 3067–3068 (1995).
53. J. Petrosino and T. Palzkill, *J. Bacteriol.* **178**, 1821–1828 (1996).
54. W. Huang, J. Petrosino, M. Hirsch, P. Shenkin, and Y. Palzkill, *J. Mol. Biol.* **258**, 688–703 (1996).
55. P. Kast and D. Hilvert, *Curr. Opin. Struct. Biol.* **7**, 470–479 (1997).
56. C. Wang, Q. Yang, and C. Craik, *Methods Enzymol.* **267**, 52–68 (1996).
57. H. Zhao and F.H. Arnold, *Curr. Opin. Struct. Biol.* **7**, 480–485 (1996).
58. K. O'Neil and R. Hoess, *Curr. Opin. Struct. Biol.* **5**, 443–449 (1995).
59. J. Hanes and A. Pluckthun, *Proc. Natl. Acad. Sci. U.S.A.* **94**, 4937–4949 (1997).
60. R. Roberts and J. Szostak, *Proc. Natl. Acad. Sci. U.S.A.* **94**, 12297–12302 (1997).
61. D. Tawfik and A. Griffiths, *Nat. Biotechnol.* **16**, 652–656 (1998).
62. S. Sidhu and T. Borgford, *J. Mol. Biol.* **257**, 233–245 (1996).
63. D. Youvan, E. Goldman, S. Elgrave, and M. Yang, *Methods Enzymol.* **246**, 732–749 (1995).
64. H. Zhao, and F.H. Arnold, *Proc. Natl. Acad. Sci. U.S.A.* **94**, 7997–8000 (1997).
65. O. Kuchner and F.H. Arnold, *TIBTECH* **15**, 523–530 (1997).
66. F.H. Arnold and J. Moore, *Adv. Biochem. Eng. Biotechnol.* **58**, 2–14 (1997).

See also ENZYMES, BAKING, BREAD MAKING; ENZYMES, DETERGENT; GOOD MANUFACTURING PRACTICE (GMP) AND GOOD INDUSTRIAL; PROTEIN EXPRESSION, SOLUBLE; PROTEIN SECRETION, *SACCHAROMYCES CEREVISIAE*.

ENZYMES, EXTREMELY THERMOSTABLE

PAULA M. HICKS
North Carolina State University
Raleigh, North Carolina

MICHAEL W. W. ADAMS
University of Georgia
Athens, Georgia

ROBERT M. KELLY
North Carolina State University
Raleigh, North Carolina

KEY WORDS

Archaea
Biomass
Enzymes
Hyperthermophilic
Proteases
Protein stability
Thermostable

OUTLINE

INTRODUCTION

The influence of temperature on the structure and function of proteins has been extensively studied. However, it is still not possible to determine in any definitive sense which factors underlie the vast differences in thermostability and thermoactivity between seemingly similar proteins (1,2). The discovery of hyperthermophilic microorganisms (those with growth temperatures of 90 °C and above) and the subsequent purification and characterization of their extremely thermostable and thermoactive enzymes have underscored the complex nature of protein stability. It is clear now that differences of 100 °C or more may separate the temperature optima of enzymes carrying out identical biocatalytic functions. Information available on amino acid sequences of enzymes with widely varying temperature optima, and even three-dimensional structural data, may not provide much insight into how seemingly similar enzymes can respond to thermal energy so differently (3). Thus, enhancing the thermostability of a given enzyme to a significant extent remains an elusive objective left to fortuitous improvements through random mutagenesis, or to extensive screening efforts from natural high temperature biotopes (4).

In cellular environments, protein stability is a property that has important physiological implications. Protein architecture must be consistent with in vivo requirements. For example, the free energy of stabilization of an enzyme (ΔG), a balance between large but opposing entropic ($T\Delta S$) and enthalpic forces (ΔH), is equivalent to only a few weak bonds (5), presumably so that it can be regulated in response to changing cellular conditions. Once an enzyme has performed a needed cellular function that is no longer necessary or becomes detrimental, it may denature, be deattenuated through a regulatory response, or be proteolytically inactivated.

Although enzymes are desirable in industry because of their specificity and availability, in order to be useful for commercial purposes, some degree of stability is required to make economic sense. In fact, proteins may need to not only withstand thermal forces, but also be resistant to nonaqueous solvents, high ionic strengths, detergents, and extremes of pH as well. In addition to stability, it is important that the enzyme be functional in the specific extreme environment in which it is to be used. Thus, a highly thermostable and thermoactive enzyme may not be useful in applications at very low temperatures, even though it may be very stable under these conditions. Clearly, the maintenance of an enzyme's structural integrity and its catalytic efficiency need both be considered for possible application of these biocatalysts in an industrial setting.

Even though site-directed and random mutagenesis have yet to become a routine approach for conferring thermostability on an enzyme, nature has fortunately provided a source of thermostable biocatalysts in extremely thermophilic microorganisms (optimal growth temperatures above 75 °C). An ever-increasing list of such enzymes shows that most significant activities from mesophiles (optimal growth around 37 °C) have high-temperature, and presumably thermostable, counterparts. The biochemical and biophysical characteristics of many of these enzymes are currently being examined for both scientific and technological purposes (4,6). In this article, the methodology being used to discover extremely thermostable enzymes will be discussed in addition to information on their function and use. It is expected that the extraordinary stability of these enzymes will make them candidates for replacing many currently used biocatalysts in addition to opening up many new applications.

EXTREMELY THERMOSTABLE ENZYME DISCOVERY

The methodology for identifying enzymes with technological importance has gone through considerable changes in recent years with the ever-expanding use of molecular biological tools. Concomitantly, the search for enzymes from microorganisms in extreme environments has also expanded and has made use of new discovery techniques (7). Gone are the days when industrial enzymes were exclusively mined from microorganisms isolated from soil samples and grown in pure culture. Now, only the gene encoding an enzyme of interest is, in principle, necessary if it can be overexpressed in an appropriate host. Another development that is sure to impact enzyme discovery is the increasing availability of genome sequences for a variety of mesophilic and thermophilic microorganisms (8). Various approaches for the identification of extremely thermostable enzymes are discussed in the following sections and summarized in Figure 1.

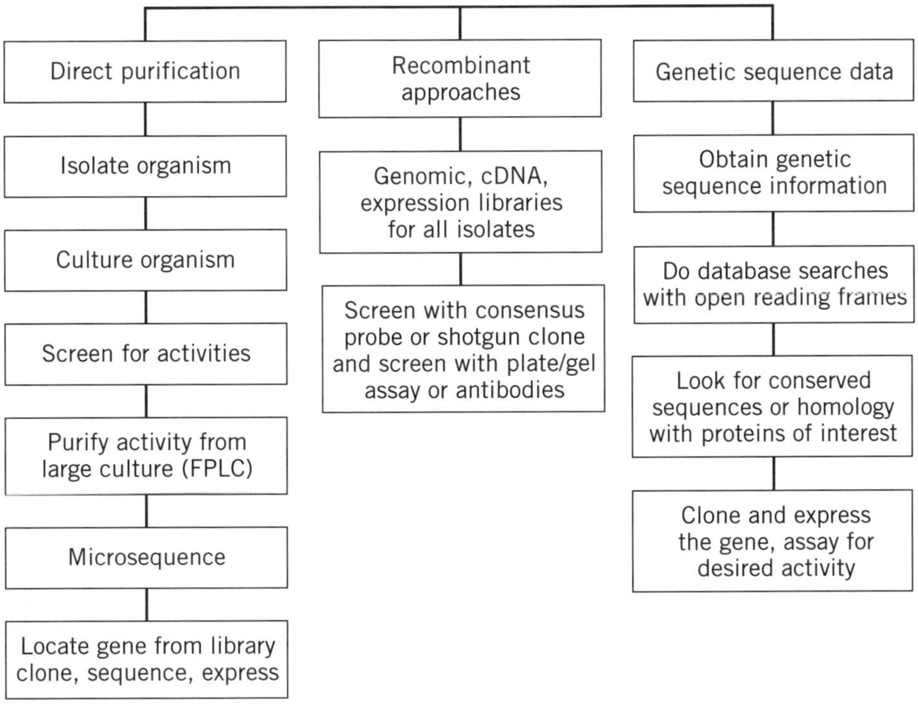

Figure 1. Screening for extremely thermostabile enzymes.

Direct Purification From Extreme Thermophile Biomass

Expeditions to geothermal habitats, as accessible as hot springs in Yellowstone National Park (9) and as inaccessible as deep sea geothermal vents (10), have led to the isolation of an array of high-temperature microorganisms covering a diversity of physiological types (11). Typically, isolates are obtained through serial dilution techniques because high temperatures preclude cultivation on solid media in many cases. Most microorganisms classified as extreme thermophiles ($T_{opt} \geq 75$ °C) are anaerobic, which is not surprising considering the low solubility of O_2 at elevated temperatures. Notable exceptions are members of the order Sulfolobales, many of which are thermoacidophiles (11). A subset of the extreme thermophiles, the hyperthermophiles (capable of growth at 90 °C or higher) are almost exclusively anaerobic, and many reduce sulfur to sulfide during growth (11). Additional details on these microorganisms can be found elsewhere in this volume (THERMOPHILIC MICROORGANISMS).

Cultivation of extreme thermophiles on a scale sufficient to provide biomass for enzyme purification efforts presents many challenges (12,13). In short, lack of knowledge of microbial physiology, low biomass yields, generation of copious amounts of hydrogen sulfide, explosive substrate and product gases, salt-laden media, anaerobic conditions, and cell sensitivity to shear are all potential obstacles. A few heterotrophic hyperthermophiles, such as *Pyrococcus furiosus* (14), *Thermococcus litoralis* (15), and *Thermotoga maritima* (16), can be cultivated without sulfur and have, thus, been the primary focus of direct purification efforts. These organisms can be grown to biomass yields of 1 kg (wet) or higher in 600-L fermentation systems (17). From such efforts, more than 20 enzymes have been purified from the biomass generated from a single fermentation (13).

Once sufficient biomass has been generated, enzyme purification can proceed along conventional lines using multistep protocols involving various types of liquid chromatography. Cells can be broken in a variety of ways including freeze-thaw cycles, French pressing, or sonication. Because many extreme thermophiles are members of the domain archaea, which have unusual cell membrane features, including a lack of peptidoglycan, lysozyme treatments are typically not effective. Most types of prep-scale chromatography have been used in the purification of these enzymes. These include ion exchange, hydrophobic interaction, gel filtration, and affinity methods. The extreme thermostability of these enzymes does provide a convenience: Purification can be carried out at ambient temperatures rather than in refrigerated environments. On the other hand, thermostability does present some difficulties in assessing the homogeneity of products of the purification process. Conventional sodium dodecyl sulfate-polyacrylamide gel electrophoresis (SDS-PAGE) can be misleading because extremely thermostable enzymes may not be sufficiently denatured in this process. The result is that both the native and partially denatured forms of multimeric proteins may appear on SDS-PAGE, making assessment of M_r (relative molecular mass) and homogeneity difficult. Unfortunately, extended periods of heating to ensure complete denaturation may also lead to covalent modifications of labile amino acid residues.

Because it is economically impractical to produce large amounts of extremely thermostable enzymes in their native hosts, recombinant approaches must be used. The purified enzyme may be N-terminally or C-terminally (or both) sequenced, and conventional methods for identifying the corresponding gene in genomic preparations by nucleic acid probe hybridization or polymerase chain reaction (PCR) techniques can be used. Overexpression of genes encoding extremely thermophilic enzymes is still in its earliest stages, and most efforts to date have used a variety of vectors in *Escherichia coli* (4). Although many efforts have yielded a recombinant enzyme seemingly identical to the native at low expression levels (18,19), it is not known whether existing expression systems will be suitable to produce the large amounts of enzyme needed for certain industrial uses. This is a critical issue for the development of applications making use of these enzymes.

Expression Cloning

Although not specific to extremely thermophilic enzymes, molecular biology has provided an array of techniques that circumvent the tedious process of obtaining biomass from large fermentations done in pure culture. If an assay is available for a particular enzyme activity, expression libraries containing DNA from an organism of interest may be screened to isolate the gene encoding this activity. This assumes that the gene manifests itself as an expression product. This same approach has been used to mine enzymes directly from environmental samples for which no effort is made to sort out individual organisms in pure culture (7). For extreme thermophiles, the intrinsic thermostability of the enzyme made as the expression product facilitates partial purification. A heating step will serve to denature the less-stable enzymes produced by a mesophilic host so that the thermostable enzyme is enriched in the resulting mixture. Unfortunately, expression cloning will not always work because some genes may not be expressed as an active enzyme or a suitable screening assay may not be easily used. However, this method can be a particularly attractive alternative to large-scale production of extreme thermophile biomass and can also be used as the basis for high throughput screening and directed evolution (7).

Sequence Analysis of Genomic DNA

Recent efforts to sequence entire genomes for microorganisms are sure to have an impact on the enzyme discovery process (8). Because many of the initial microbial sequencing projects involve extreme thermophiles, this bank of information can be used to identity putative proteins of industrial importance. The *Methanococcus jannaschii* genome has been sequenced (20), and many other extreme thermophiles are currently being sequenced, such as *Archaeoglobus fulgidus, Aquifex aeolicus, Methanobacterium thermoautotrophicum, Pyrobaculum aerophilum, Pyrococcus furiosus, Pyrococcus horikoshi, Sulfolobus solfataricus*, and *Thermotoga maritima* (21). Computer databases can be searched using amino acid or nucleotide sequences corresponding to known motifs, and resulting homologous sequences can be the basis for cloning the gene of interest. At this time, however, 50% or more of the genome sequence obtained from specific microorganisms contain open reading frames (ORFs) that are not attributable to a protein whose function is known. Furthermore, there are relatively little sequence data available for many hydrolytic enzyme types of potential commercial significance so that identification of homologous versions of certain enzymes can be an imprecise exercise. Nevertheless, as entire genome sequence data expand, this approach should not be ruled out because it may be difficult to purify a particular enzyme directly from the native source.

ENZYMES FROM EXTREME THERMOPHILES

General Characteristics

Enzymes of all known classes have been purified from extremely thermophilic microorganisms and characterized biochemically. Many of these proteins are intrinsically thermostable (Table 1). In many cases, the genes encoding these enzymes have been sequenced and in some cases expressed in active form in a foreign host, usually *E. coli* (Table 2). With the availability of genome sequence data (20), we also now know that these microorganisms contain thermostable counterparts to many proteins previously characterized from mesophilic sources, many of which are linked to central metabolic pathways (162). Also, because most existing enzyme applications make use of hydrolytic enzymes, these enzyme types have been sought in extreme thermophiles (163–165).

Enzymes from Extreme Thermophiles Involved in Intermediate Metabolism

Because of the evolutionary placement of extreme thermophiles, based on 16S rRNA phylogeny (8,166), interest in the origins of metabolic pathways has led to the isolation and characterization of enzymes directly involved in intermediate metabolism and bioenergetics (162). These include a number of oxidoreductases (167), hydrogenases (71,72,74), dehydrogenases (60–65), and redox proteins (167). Many such enzymes from the most studied hyperthermophile *P. furiosus* (14) appear to be intrinsically thermostable and function in vitro at the organism's optimal growth temperature of 98 to 100 °C (4). The study of these enzymes in vitro presents certain challenges because assays to detect activity may involve thermally labile substrates, anaerobic conditions, and coenzymes. For example, even though *P. furiosus* contains several dehydrogenases, these catalyze reactions involving NADH as a cofactor, which itself is not very thermostable. This organism is also known to contain a modified Emden-Myerhoff-Parnos pathway for glycolysis, which implies that glucose is processed at high temperatures through several intermediate steps (168). The mechanisms by which simple sugars, which may be sensitive to thermal modifications, can be processed at high temperatures are not known.

Proteases from Extreme Thermophiles

Proteolysis (hydrolysis of peptide bonds) is essential to cellular function and is implicated in nutrition, heat shock response, immune response, and certain housekeeping roles (169). Historically, proteases have found important industrial uses in detergents, leather bating, food processing, and much more, collectively comprising the largest volume use of commercial enzymes (170). But in the past 10 years, these enzymes have also been shown to be active in organic solvents for peptide synthesis and are often selective enough to produce pharmaceuticals (171,172). Recently, protease inhibitors have received much attention in medical applications for combatting the HIV virus and some cancerous tumors (173,174).

There are many potential benefits to be derived from the use of proteases from extreme thermophiles. If mesophilic proteins are to be hydrolyzed, it is very likely that at elevated temperatures their denatured forms will facilitate access of thermostable proteases to specific peptide bonds. Also, at higher temperatures, many substrates are more soluble, solution viscosity is reduced, and diffusion rates are higher, which can lead to higher reaction rates (175). Because these enzymes appear to also be more stable to denaturing forces other than heat, proteases from extreme thermophiles should be amenable to use in organic solvents for synthetic reactions.

Many proteases have been purified and characterized from extreme thermophiles (163). From an evolutionary viewpoint, these proteases are of interest because they presumably predate those found in less thermophilic cells and organisms in terms of structure and function. Some of these proteases appear to be novel with no previously known mesophilic counterparts, whereas others represent more stable forms of known proteases.

Proteolytic activities produced by the hyperthermophilic archaeon *P. furiosus* (14) have been the most extensively studied among the extreme thermophiles. This organism appears to produce a number of proteases (176), which is consistent with its excellent growth on peptide-based media (177). The properties of several proteases from this organism are discussed below.

***P. furiosus* Protease I (PfpI).** *P. furiosus* protease I (PfpI) is a major intracellular proteases in *P. furiosus*. Previously, PfpI was named S66 because, after heating at 95 °C in 1% SDS for 24 hours, it migrates to a molecular mass corresponding to 66 kDa on a denaturing gel (25). It was subsequently found to occur in at least two active forms (hexamer and trimer composed of 18.8-kDa subunits) in vitro (26) and appears to have at least one active larger form (Hicks and Kelly, unpublished data). The calculated pI (from amino acid sequence data) of the 18.8-kDa subunit is 6.1, whereas the experimentally determined pIs of the hexamer and trimer are 6.1 and 3.7 to 3.9, respectively. If all asparagine and glutamine residues are changed to their corresponding acidic residues, the pI of the monomer is calculated to be 4.8. Thus, at high temperatures, it is possible that the protein is deamidated, especially in its trimeric form, explaining in part the difference from the measured and predicted pI values (26). Upon long exposures to heat (24 to 48 hours), the predominant band on an overlay gel containing gelatin is in fact the smaller form (26). PfpI is specific toward basic and bulky, hydrophobic P1 amino acid residues in peptide substrates (26) and most active toward *N*-succinyl-alanine-alanine-phenylalanine-7-amido-4-methylcoumarin (AAF-MCA) (26). Large proteins, such as azocasein and gelatin, are also degraded by PfpI, although specific cleavage sites in these proteins have not yet been identified. The gene for PfpI (*pfpI*) has been expressed in *E. coli* using the T7 promoter and infection with λ phage CE6 (107). Vector constructs are often unstable, and toxicity has been a problem in producing recombinant PfpI. Expression levels were highest (1 mg per 100 mL culture) when a histidine tag was used to make a fusion protein, which could then purified using an affinity column (107). The amino acid sequence of PfpI was found to be highly homologous to putative proteins of unknown function contained in representatives of all three domains of life (e.g., *E. coli* (bacteria), *M. jannaschii* (archaea), *Arabidopsis thaliana* (eukarya)) (178). However, the role of this protease in *P. furiosus* or in other cells and organisms is not clear at this time.

Pyrolysin. Pyrolysin is a cell-envelope-associated protease that can be purified from *P. furiosus* membrane fractions (27,179). Two versions of this enzyme, one based on a 150-kDa subunit and one that was apparently cleaved by carboxy-terminal autoproteolysis, have been reported. The *pls* gene encoding the subunit has been cloned and sequenced (27). The gene corresponds to a 4194-bp open reading frame, containing a prepro sequence, whereas the mature protein is encoded by a 1249-bp region. The N-terminus shows high homology to subtilisin-like serine proteases, whereas multiple glycosylation sites are proposed at the carboxy terminus. Pyrolysin shows highest homology at the amino acid level with eucaryl tripeptidyl peptidases (TPP), although it has different cleavage specificities; TPPs are typically exopeptidases (27). Pyrolysin appears to be a version of the surface protein assembly (designated tetrabrachion) identified in another hyperthermophile, *Staphylothermus marinus*, which contains a protease with trypsin- and chymotrypsin-like activities (28).

Archaeal Proteasome. Several thermophilic archaea contain a version of the eukaryotic proteasome (29,33,180). *P. furiosus* produces a version of the archaeal proteasome that appears to be composed of α (25 kDa) and β (22 kDa) subunits with an overall M_r of 640 kDa ($\alpha_{14}\beta_{14}$) (33). The *P. furiosus* proteasome has a temperature optimum of 95 °C, and preliminary biochemical analysis showed that methionine was preferred in the P1 position for hydrolysis of short peptides, although aromatic residues were also cleaved (33). Because the purification fold of the proteasome from *P. furiosus* cell extracts was approximately 16, it appears that it is a major cytosolic protease. Studies to assess its physiological role are under way. Unlike PfpI, no evidence for smaller active forms of the proteasome in *P. furiosus* was detected.

Other Pyrococcal Proteases. There have been reports of other proteases produced by pyrococcal species. A prolyl endopeptidase was identified in *P. furiosus*, the gene for which was cloned and expressed in *E. coli* (108). The recombinant enzyme was found to be optimally active between 85 and 90 °C. The gene encoding the enzyme corresponds to a polypeptide of 70 kDa, which was consistent with SDS-PAGE analysis of the recombinant version. A thiol protease from *Pyrococcus* strain KOD1 was isolated and characterized (181). This protease, one of at least three extracellular proteases produced by this strain, had a molecular mass of 44 kDa and was optimally active at pH 7.0 and 110 °C. Its threefold activation in the presence of 8 mM cysteine led to its classification as a thiol protease.

Table 1. Types of Enzymes Purified from Extreme Thermophiles

Enzyme	M_r^a	T_{opt}^b	$t_{50\%}^c$	Function	Source[d]	Reference
		Hydrolases				
Aminopeptidase	320 (α_4)	75°	NA	Peptide hyrolysis	Ss	22
Carboxypeptidase	170 (α_4)	85°	<15 min 95°	Peptide hydrolysis	Ss	23
Archaelysin	52 (α)	98°	1.5 h/95°	Peptide hydrolysis	DM	24
Protease PfpI-H	66 (α_3)	105°	33 h/98°	Peptide hydrolysis	Pf	25
Protease PfpI-Cl	113 (α_6)	85°	<0.5 h/98°	Peptide hydrolysis	Pf	26
Protease pyrolysin	150 (α)	115°	9 h/95°	Peptide hydrolysis	Pf	27
Tetrabrachion	130 (α)	>90°	NA	Structural/proteolytic	Sm	28
Thiol protease	44 (α)	110°	1 h/100°	Peptide hydrolysis	P KOD1	29
Proteinase	68 (α)	85°	22 h/90°	Peptide hydrolysis	Ts	30
Thermopsin	46 (α)	75°	90 min/90°	Peptide hydrolysis	Sa	31
Proteinase I	108 (α_2)	90°	5.7 h/92°	Peptide hydrolysis	Ss	32
Proteasome	640 ($\alpha_{14}\beta_{14}$)	95°	>24 h/95°	Protein turnover	Pf	33
α-Glucosidase	125 (α)	110°	48 h/98°	Maltose hydrolysis	Pf	34
α-Glucosidase	80 (α)	105°	3.0 h/95°	Maltose hydrolysis	Ss	35
Amylase	132 (α_2)	100°	2 h/120°	Starch hydrolysis	Pf	36,37
Amylase	70 (α)	100°	6 h/100°	Starch hydrolysis	Pw	38
Amylopullulanase	140 (α)	118°	20 h/98°	Starch degradation	ES	39
β-Glucosidase	230 (α_4)	105°	13 h/110°	Cellobiose hydrolysis	Pf	40
Sucrose α-glucohydrolase	114 (α)	105°	48 h/95°	Unknown	Pf	41
β-Mannosidase	240 (α_4)	105°	1.1 h/98°	Mannobiose hydrolysis	Pf	42
β-Mannosidase	200 (α_2)	87°	17 h/85°	Mannobiose hydrolysis	Tn	43
β-Mannanase	55 (α)	92°	34 h/85°	Mannan hydrolysis	Tn	43
α-Galactosidase	66 (α)	105°	9 h/85°	Galactomannan hydrolysis	Tn	43
Endo-β-1,4-glucanase	27 (α)	95°	>6 h/80°	Glucan hydrolysis	Tm	44
Exo-β-1,4-glucanase	29 (α)	95°	>6 h/80°	Glucan hydrolysis	Tm	44
Xylanase A	120 (α)	92°	45 min/90°	Xylan/xylose hydrolysis	Tm	45
Xylanase B	40 (α)	105°	125 min/95°	Xylan/xylose hydrolysis	Tm	45
Exo-1,4-β-cellobiohydrolase	36 (α)	105°	70 min/108°	Cellulose hydrolysis	TF	46
β-Xylosidase	174 (α_2)	80°	2 h/80°	Xylobiose hydrolysis	TF	47
α-Arabinofuranosidase	92 (α)	80°	4 h/98°	Arabinoxylan hydrolysis	TF	48
Fumarate hydratase	170 (α_4)	85°	15 min/90°	Sugar metabolism	Ss	49
N^5N^{10}-methenyltetrahydro-methanopterin cyclohydrolase	41.5 (α)	95°	>1 h/90°	Methanogenesis	Mk	50

			Oxidoreductases			
Aldehyde OR	85 (α)	>95°	6 h/80°	Glycolytic enzyme	Pf	51
Pyruvate OR	120 ($\alpha\beta\gamma\delta$)	>95°	0.3 h/90°	Pyruvate oxidation	Pf	52
Formaldehyde OR	280 (α_4)	>95°	2 h/80°	Unknown	Tl	53
Indolepyruvate OR	180 ($\alpha_2\beta_2$)	>95°	6 h/80°	Peptide fermentation	Pf	54
Gly. 3-phosphate OR	63 (α)	>95°	NA	Glycolytic enzyme	Pf	55
Isovalerate OR	220 ($(\alpha\beta\gamma\delta)_2$)	>95°	0.6 h/90°	Peptide fermentation	Pf	56
2-Ketoglutarate	220 ($(\alpha\beta\gamma\delta)_2$)	>95°	0.3 h/90°	Peptide fermentation	Pf	57
Pyruvate OR	200 ($(\alpha\beta\gamma\delta)_2$)	>95°	15 h/80°	Pyruvate oxidation	Tm	58
Quinone OR	280 ($\alpha\beta\gamma\delta\epsilon\zeta\eta$)	65°	5 min/75°	Electron donor	Af	59
Glutamate dehydrogenase	270 (α_6)	95°	10 h/100°	Glutamate oxidation	ES	60
GAPDH	150 (α_4)	NA	0.7 h/100°	Glycolysis (?)	Pw	61
GAPDH (NAD)	196 (α_4)	NA	0.3 h/100°	Unknown	Tt	62
GAPDH (NADP)	156 (α_4)	NA	0.5 h/100°	Unknown	Tt	62
Sulfide dehydrogenase	92 ($\alpha\beta$)	>95°	12 h/95°	Sulfur reduction	Pf	63
Alcohol dehydrogenase	200 (α_4)	80°	2.0 h/85°	Unknown	Tl	64
Isocitrate dehydrogenase	90 (α_2)	NA	NA	Sugar metabolism	Ss	65
Quinol oxidase	149 ($\alpha\beta\gamma\delta\epsilon$)	NA	NA	Proton gradient	Aa	66
Methyl coenzyme M reductase	270 ($(\alpha\beta\gamma)_2$)	NA	NA	Methanogenesis	Mk	67
N^5N^{10}-methylenetretrahydromethanopterin reductase	300 (α_2)	90°	>1 h/100°	Methanogenesis	Mk	68
N^5N^{10}-methylenetetrahydromethanopterin dehydrogenase	170 (α_4)	>90°	Very labile	Methanogenesis	Mk	69
Adenylylsulphate reductase	160 (α_2)	85°	NA	Electron donor	Af	70
			Hydrogenases / redox proteins			
Hydrogenase	118 ($\alpha\beta$)	>90°	1 h/98°	H_2 production	Pf	71,72,73
Hydrogenase	280 (α_4)	>95°	1 h/90°	H_2 production	Tm	74
Ferredoxin	7 (α)	>95°	>24 h/95°	Electron transfer	Pf, Tl	75
Rubredoxin	5 (α)	>95°	>24 h/95°	Unknown	Pf	76
Thioredoxin (glutaredoxin)	24.8 (α)	NA	>3 h/90°	Reduces disulfides	Ss, Pf	77,78
			Isomerases / invertases			
Xylose isomerase	204 (α_4)	97°	2.3 h/95°	Sugar metabolism	Tm	79
Triose phosphate isomerase	100 (α_4)	NA	4.7 h/104°	Unknown	Pw	80
Prolyl *cis/trans* isomerase	18 (α)	NA	NA	Unknown	Ss	81

Table 1. Types of Enzymes Purified from Extreme Thermophiles *(continued)*

Enzyme	M_r^a	T_{opt}^b	$t_{50\%}^c$	Function	Source[d]	Reference
		Nucleic-acid modifying enzymes				
DNA polymerase	93 (α)	>75°	20 h/95°	Replication	Pf	82,83
Endonuclease SuaI	NA	65°	30 min 80°	Restriction	Sa	84
Reverse gyrase	200 ($\alpha\beta$)	NA	NA	Supercoiling	Mk	85
DNA topoisomerase II	230 ($\alpha_2\beta_2$)	80°	35 m/80°	DNA relaxation	Ssh	86
DNA topoisomerase III	108 (α)	NA	NA	DNA relaxation	Damy	87
DNA toposisomerase V	110 (α)	75°	NA	DNA relaxation	Mk	88,89
RNAse p2	13 (α_2)	NA	<15 min/90°	RNA hydrolysis	Ss	90
		Synthetases				
ACS I	140 ($\alpha_2\beta_2$)	>95°	18 h/80°	Acetyl-CoA formation	Pf	91
ACS II	140 ($\alpha_2\beta_2$)	>95°	8 h/80°	Aryl-CoA formation	Pf	91
Carbamoyl-P synthetase	46.6 (α)	>90°	3 h/95°	Pyrimidine pathway	Paby	92
PEP-synthase	2000 (α_7)	NA	NA	Metabolism	Sm	93
		Transferases				
Aromatic aminotransferase	90 (α_2)	>95°	NA	Peptide fermentation	Pf	94
Aspartate transcarbamoylase	NA	70°	>6 h/90°	Nucleotide synthesis	Paby	95
Propylamine transferase	110 (α_3)	90°	>1 h/100°	Amino acid metabolism	Ss	96
Formyltransferase	35 (α)	90°	>1 h/90°	Methanogenesis	Mk	97
		Other				
Glucokinase	93 (α_2)	105	3.7 h/100°	Sugar metabolism	Pf	98
Enolase	180 (α_4)	>95°	0.7 h/100°	Glycolysis	Pf	99
Chaperonin	920 (α_{16})	80°	NA	Protein refolding	Ss	100
PEP carboxylase	240 (α_4)	88°	>2 h/80°	Metabolism	Ms	101
5′Methylthioadenosine phosphorylase	160 (α_6)	120°	15 min/130°	Nucleoside catabolysis	Ss	102

Note: NA = not available (not determined or not reported); GADPH = glyceraldehyde-3-phosphate dehydrogenase; ACS = acetyl CoA synthetase. PEP = phophoenolypyruvate.

[a]Molecular weight in kDa (the number of subunits is given in parentheses).

[b]Optimum temperature (°C) for catalytic activity in vitro.

[c]Time required to lose 50% of catalytic activity after incubation at the indicated temperature.

[d]The sources of the enzymes are Aa = *Acidianus ambivalens*; Af = *Archaeoglobus fulgidus*; Damy = *Desulfurococcus amylolyticus*; Dm = *Desulfurococcus mucosus*; ES = Es-4; Mk = *Methanopyrus kandleri*, Ms = *Methanothermus sociabilis*; Paby = *Pyrococcus abyssi*, Pf = *Pyrococcus furiosus*; P KOD1 = *Pyrococcus* KOD1; Pw = *P. woesei*; Sa = *Sulfolobus acidocaldarius*; Sm = *Staphylothermus marinus*; Ss = *Sulfolobus solfataricus*; Ssh = *Sulfolobus shibatae*; Tf = *Thermotoga* FjSS3-B.1; Tl = *Thermococcus litoralis*; Tm = *Thermotoga maritima*, Tn = *Thermotoga neapolitana*; Ts = *Thermococcus stetteri*; Tt = *Thermoproteus tenax*.

Table 2. Proteins from Extreme Thermophiles Whose Genes Have Been Cloned and Expressed

Organism/Enzyme (host)	Reference
Pyrococcus furiosus	
Ferredoxin	103
Glutaredoxin	78
DNA polymerase	104
α-Amylase	105
Glutamate dehydrogenase	106
Protease I	107
Prolyl endopeptidase	108
Citrate synthase	109
β-Glucosidase	42,110
β-Mannosidase	42
Pyrococcus woeseii	
Glyceraldehyde-3-phosphate dehydrogenase	61
3-Phosphoglycerate kinase	111
Pullulanase	112
GTP-binding protein (expression library)	113
Transcription factor	114
Pyrococcus strain ES-4	
Glutamate dehydrogenase	60
Pyrococcus sp. KOD1	
Aspartyl-tRNA synthetase	115
TATA-binding protein	116
α-Amylase	117
RecA/RAD51 protein	118
Molecular chaperonin beta subunit	119
Pyrococcus sp. strain ST700	
Adenylosuccinate	120
Pyrodictium occultum	
DNA polymerase 1	121
DNA polymerase 2	121
Thermococcus litoralis	
DNA polymerase	122
Thermococcus sp. 9 degree N-7	
DNA polymerase	123
Thermotoga neapolitana⁺	
Phosphoglycerate kinase	124
Triosephosphate isomerase	124
Enolase	124
Xylose isomerase	125
Endoxylanase A	126
Thermotoga maritima⁺	
Dihydrofolate reductase	127
L-Lactate dehydrogenase	128
D-Glyceraldehyde-3-phosphate dehydrogenase	129
Chemotaxis protein Y	130
Chemotaxis protein W	130
Ferredoxin	131
RuvB	132
trp Gene products (tryptophan pathway)	133

Table 2. Proteins from Extreme Thermophiles Whose Genes Have Been Cloned and Expressed (*continued*)

Organism/Enzyme (host)	Reference
4-α-Glucanotransferase	134
α-Amylase	135
3-Phosphoglycerate kinase	136
Triosephosphate isomerase	136
β-glucosidase	137
β-galactosidase	137
β-glucanases (2)	138
Histidine protein kinase	139
Response regulator	139
Elongation factor Tu	140
Thermotoga FjSS3-B.1	
Endoxylanase A (expression library)	141
Desulfurococcus SY	
Aspartate racemase	142
Sulfolobus acidocaldarius	
Sac 7d gene product	143
Genanylgeranyl-diphosphate synthase	144
Adenylate kinase	145
Sulfolobus sp. strain 7	
Ferredoxin	146
Sulfolobus solfataricus	
Anthranilate synthase	147
β-Glycosidase (Ec), (Sc), (mammalian)	148–150
Phosphoglycerate kinase	151
Glyceraldehyde-3 phosphate dehydrogenase	151
Elongation factor I	152
Glycosyltransferase	153
α-Amylase	153
Aspartate aminotransferase	154
3-Hydroxy-3-methylglutaryl CoA reductase	155
Sulfolobus shibatae	
ATP(CTP):tRNA nucleotidyl transferase	156
Methanothermus fervidus	
2-Phosphoglycerate kinase	157
Methanobacterium thermoautotrophicum	
Methylene-5,6,7,8-tetrahydromethanopterin dehydrogenase	158
HMt 1 and 2 (histone proteins)	159
Methanopyrus kandleri	
Formylmethanofuran:tetrahydromethanopterin Formyl transferase (Ftr)	160
Caldococcus noboribetus	
Isocitrate dehydrogenase	161

Note: All genes expressed in *E. coli* unless otherwise indicated.
⁺ = denote hyperthermophilic bacterium; Ec = *E. coli*; Sc = *Saccharomyces cerevisiae*.

Glycosyl Hydrolases from Extreme Thermophiles

The hydrolysis of natural polymers, such as cellulose, hemicellulose, and starch, by glycosyl hydrolases is the focus of a significant portion of all current commercial uses of enzymes as biocatalysts. It is desirable to hydrolyze natural polymers with these enzymes at elevated temperatures because of increased substrate solubility and lowered solution viscosity, which improves enzyme access to susceptible bonds (165). It is somewhat surprising that the source of these thermostable enzymes, extreme thermophiles, are thought to be primitive life forms and presumably predate the emergence of higher life forms, such as plants, which produce most naturally occurring polysaccharides. Nonetheless, these organisms have proved to be a fertile source of an array of glycosyl hydrolases, many of which have potential as commercial enzymes (163,165). Classification schemes for glycosyl hydrolases have advanced considerably in recent years. Because these enzymes are often capable of hydrolyzing a range of glycosidic bonds, classification in terms of substrate specificity can be confusing. Amino acid sequence information has been used to fortify traditional classifications schemes so that glycosyl hydrolases have now been arranged in over 60 families (182). As three-dimensional structures and detailed kinetic mechanisms are resolved for enzymes within these families, a comprehensive basis for distinguishing among various glycosyl hydrolases can be developed.

Hydrolysis of α-Specific Glycosidic Bonds. Starch, glycogen, pullulan, and maltose have proved to be good growth substrates for a variety of heterotrophic extreme thermophiles. These microorganisms contain a variety of *exo*-acting and *endo*-acting α-specific glycosyl hydrolases that are apparently used to hydrolyze complex carbohydrates to glucose. *P. furiosus*, for example, produces an extracellular amylopullulanase that can hydrolyze either α-1,4 or α-1,6 linkages (183), both intracellular (36) and extracellular (165) α-amylases, and an α-glucosidase (34). Versions of these enzymes have been found in many other heterotrophic extreme thermophiles (163,165). In fact, glycosyl hydrolases may be more widely distributed among the extreme thermophiles than previously thought. The genome sequence of *M. jannaschii* indicates that this methanogen also produces an intracellular α-amylase and glucoamylase, although this organism has not been reported to grow on starch-based carbon sources (20).

Hydrolysis of β-Specific Glycosidic Bonds. Because of the natural abundance of cellulose and hemicellulose, enzyme systems that are capable of hydrolyzing these polysaccharides to simple sugars have long been sought to facilitate recycling of renewable resources (184). The structure and molecular composition of β-linked natural biopolymers are highly variable, often necessitating the use mixtures of enzymes to get substantial hydrolysis. On the other hand, unsubstituted cellulose and hemicellulose (e.g., xylan and mannan) are sparingly soluble in water, and enzymes capable of hydrolyzing these compounds often contain binding domains used to pry open the structure for enzymatic access (184).

Exo-acting, β-specific glycosyl hydrolases have been found among extremely thermophilic bacteria and archaea (163,165). *P. furiosus* produces an intracellular β-glucosidase (40,185) and an intracellular β-mannosidase (42); although the former appears to be induced by cellobiose in the growth media and has broad substrate specificity, the function of the latter is not clear. β-glucosidases, β-galactosidases, β-xylosidases, and cellobiohydrolases have also been identified in other extreme thermophiles (163,165), which suggests that biopolymeric substrates associated with these specificities are present in geothermal environments.

The wide occurrence of *exo*-acting, β-specific glycosyl hydrolases among the extreme thermophilic archaea suggests that *endo*-acting enzymes capable of hydrolyzing β-linked polysaccharides are also present. However, no such *endo*-acting, β-specific glycosyl hydrolases have been identified to this point in the extremely thermophilic archaea. Although some evidence for *endo*-acting hemicellulase activity toward xylan-based polysaccharides among the extremely thermophilic archaea has been mentioned (165), a xylanase corresponding to this activity has not been isolated. The hyperthermophilic bacterial genus *Thermotoga*, however, has been a rich source of *endo*-acting, β-specific enzymes. Glucanases (44,186), xylanases (45,141,186–189), and mannanases (43,190) have all been purified and characterized from various *Thermotoga* species.

MECHANISMS OF THERMOSTABILITY

Although there has been much progress in isolating and characterizing the biochemical features of extremely thermophilic enzymes (4), little is yet known about the intrinsic basis for their astounding thermostability. It was only a few years ago that the amino acid sequences of these enzymes were eagerly awaited, because it was anticipated that they could be used as templates for the design of stable forms of industrially important, but thermally labile, enzymes. However, as more and more genes encoding hyperthermophilic enzymes have become available and their amino acid translations scrutinized, it is evident that the present sophistication of protein chemistry provides little encouragement that strategies for protein stabilization in a generic sense are at hand (7,191). In fact, three-dimensional structures are now available for four hyperthermophilic proteins (192–195) but these have not yielded specific mechanisms for stabilization (6). At present, it appears that the most fruitful approach to obtaining extremely stable proteins is to isolate them from extreme thermophiles or to express the genes encoding these proteins in mesophilic hosts. But even in these cases, it is important to examine both intrinsic and extrinsic factors to ensure that the recombinant protein so produced resembles its natural counterpart.

Much has been done to examine intrinsic factors that lead to the thermostabilization of hyperthermophilic enzymes. However, much less has been done to examine extrinsic factors, such as small molecules and other proteins present in the microenvironment in which the protein folds. By mimicking the intracellular environment encountered by these enzymes, their in vitro stability and function might be optimized. In addition, such conditions may be useful in improving the stability of less thermophilic proteins, particularly those that are used industrially where their stability is a limiting factor.

Intrinsic Factors Influencing Protein Stability

A number of thermostable enzymes from extremely thermophilic microorganisms have been discovered during the past decade (Table 1). In most cases, it is not unusual to find the $t_{1/2}$ of these enzymes to be on the order of a day or more at temperatures near or even above 100 °C. Just as is the case with mesophilic organisms, enzymes from a particular hyperthermophile vary in their catalytic efficiency (k_{cat}/K_m), T_{opt}, and thermostability because of factors presumably related to their physiological roles. However, the intrinsic bases for protein thermostabilization at high temperatures are no better understood than the bases for stabilization at moderate conditions. No one feature seems to underlie the fact that proteins exist in stable conformations as a result of a delicate balance of large, opposing enthalpic and entropic forces. Additional stabilization arises from the efficient packing of amino acid residues, optimization of charged interactions within the structure, and minimization of exposure of hydrophobic elements to the solvent (5). This all implies a degree of cooperativity within the protein structure (196), which is difficult to attribute in a quantitative way to its component parts.

Extrinsic Factors Influencing Protein Stability

Extrinsic factors clearly play an important role in protein stabilization. Environmental conditions such as temperature and pressure have a universal effect on the collection of proteins within a given cell, but other extrinsic factors that influence protein thermostability are more specific. Some proteins benefit from the presence of critical concentrations of certain cations, whereas others function best and are the most stable in concert with coenzymes and cofactors, such as ATP. It is difficult to be definitive about how the in vitro stability of an intracellular protein relates to its in vivo stability, because the latter can be influenced by so many intracellular factors. This assessment may be especially difficult for extremely thermophilic proteins because in vitro, they are subject to an array of potentially damaging chemical reactions, involving covalent modifications to thermally labile amino acid residues, such as deamidation and oxidation. Protection against such destabilizing forces is presumably provided in the intracellular environment and may or may not be important in vitro.

There are ways that protein stability can be influenced in vitro. Of course, some conditions that optimize enzyme activity can be readily determined, such as pH, ionic strength, temperature and enzyme concentration. Other factors that may be important in vivo are less easily reproduced. These include the presence of compatible solutes (197,198) or molecular chaperones (199). Each of these will be discussed with regard to extremely thermophilic proteins.

Compatible Solutes. Although enzymes purified from extreme thermophiles are intrinsically more thermostable than those purified from their less thermophilic counterparts, the intracellular environments of extreme thermophiles thus far examined contain high levels of certain unusual solutes. These compounds are thought to play a role in the thermoprotection of at least some of their constituent biomolecules in vivo (200). High concentrations of a novel phosphodiester, di-*myo*-inositol-1,1′-phosphate (DIP), have been found in a number of hyperthermophiles, and the potassium salt of this compound has been shown to offer a significant degree of thermoprotection to at least one hyperthermophilic enzyme in vitro (201). The mechanisms for protein stabilization and the pathways used to synthesize DIP have not been determined, nor have the factors that influence its intracellular production. These studies are necessary to understand the intrinsic basis for the function of organisms at extremely high temperatures as well as to determine factors responsible for the stabilization of proteins in a general sense.

Scholz et al. (202) reported that the S°-dependent hyperthermophile *Pyrococcus woesii* (T_{opt} ~ 100 °C) contains high concentrations (~0.6 M) of DIP, and Ciulla et al. (203) subsequently found that the hyperthermophilic methanogen *Methanococcus igneus* (T_{opt} ~ 80 °C) also produced DIP when grown at temperatures above 80 °C or when exposed to high external salt concentrations. They also showed that DIP is not used as a metabolic intermediate. Martins and Santos (204) determined the intracellular concentration of DIP to be ~60 mM when *Pyrococcus furiosus* (T_{opt} 100 °C) is grown at 95 °C, but this increases to above 1.0 M at supraoptimal temperatures. In addition, they found that at high salinity, the organism accumulated the unusual sugar derivative β-mannosyl glycerate (~30 mM). In contrast to DIP, the latter compound had previously been found in some moderately thermophilic bacteria (205), suggesting that it may play a role in both osmolysis and thermal protection in these organisms. Moreover, the extraordinary increase in DIP concentrations in *P. furiosus* at the higher growth temperatures strongly suggests that it plays some role in thermoprotection (204). Accordingly, Hensel and Jakob (201) showed that a crude preparation of DIP (~115 mM) dramatically increased the thermal stability of glyceraldehyde-3-phosphate dehydrogenase (61) from *P. woesii* (the time for a 50% activity loss at 105 °C increased by 10-fold). They also showed that compounds related in structure to DIP, such as citrate and phosphate, stabilized two hyperthermophilic enzymes (201).

Molecular Chaperones in Extreme Thermophiles and Protein Aggregation. Although many proteins have been shown to fold spontaneously into their native "correct" structures in vitro, it is also clear that molecular chaperones often assist in the folding process in vitro and, presumably, in vivo. In one sense, the temperatures at which extreme thermophiles grow would seem to create a perpetual situation of thermal stress, yet these organisms respond to heat shock in many of the same ways as less thermophilic microorganisms (206). Members of the extremely thermophilic archaea have been shown to produce heat shock proteins (HSPs) at supraoptimal temperatures (207–210). Kagawa et al. (211) showed that *Sulfolobus shibatae* produces an HSP60-like protein, similar to the one identified in bacteria (199), that is a double-ring complex formed from two subunits. Similar structures have been isolated from another hyperthermophilic archaeon, *Pyrodictium occultum* (212,213) and from the moderately thermophilic archaeon *Thermoplasma acidophilum* (214). All these proteins appear to have ATPase activity and have been shown to be involved in the folding process in vitro. The genes for the archaeal HSP60s that have been sequenced show that they are closely related to a family of eukaryotic proteins termed TCP1s (215).

Although the response to thermal stress in extreme thermophiles thus far appears to be similar to the stress response of more conventional organisms at lower temperatures, the degree to which molecular chaperones are involved in protein folding in vivo is not clear. One general characteristic that has been proposed that differentiates extremely thermophilic proteins from others is increased hydrophobicity (4). If so, this may mean that the tendency for extremely thermophilic proteins to aggregate during the folding process is enhanced. Thus, molecular chaperones that can bind to the nascent protein as it folds may take on added importance relative to mesophilic proteins. Aggregation will be especially problematic when high protein concentrations are present and no means to prevent collisions between interactive protein surfaces exist (5). Hence, approaches that would prevent aggregation of extremely thermophilic proteins through the addition of extrinsic factors are highly desirable, especially in systems where high levels of overexpression are needed.

Recombinant Extremely Thermophilic Enzymes

There have been several literature reports of the successful cloning and expression of the genes encoding extremely thermophilic proteins in foreign hosts (4), a summary of which is provided in Table 2. In addition, expression cloning of genomic DNA from hyperthermophiles, coupled with robotic screening for enzyme activities of interest, has led to the discovery of a number of thermostable enzymes (7). This approach, however, says little about what genes were not successfully cloned and expressed. Moreover, of the proteins listed in Table 2, there have been only a few reports on the comparative properties of the native and recombinant forms. In fact, only in a limited number of cases has the native protein been characterized and the recombinant protein purified. For example, only DNA polymerase, glyceraldehyde-3-phosphate dehydrogenase, glutaredoxin, and β-glucosidase from *P. furiosus* have been shown, thus far, to have comparable thermal stability and molecular and catalytic properties in their native and recombinant forms. In contrast, native glutamate dehydrogenase (GDH) from *P. furiosus* has a half-life at 100 °C of about 10 h (216), whereas the recombinant form has a half-life of minutes at 100 °C, suggesting that the protein is incompletely folded or not assembled (106). The latter was supported by the fact that the native enzyme is a homohexamer (α_6), whereas the recombinant form was comprised of inactive monomers. Interestingly, some conversion of the latter into the active hexameric form was accomplished by heating (85 °C for 15 min). The opposite situation appears to exist with the α-amylase of *P. furiosus* (105), as the recombinant form of this enzyme, although not purified, was of higher molecular weight than the native form, suggesting some type of aggregation had occurred. Last, the recombinant form of *P. furiosus* citrate synthase was reported to have a half-life of only 1 min at 100 °C (109). Unfortunately, the native enzyme has not been purified but one could anticipate that it would be much more stable than that produced by *E. coli*. Clearly, some recombinant extremely thermophilic enzymes are as stable as their native counterparts, but it is also apparent that this is not a general phenomenon.

USES OF EXTREMELY THERMOSTABLE ENZYMES

General Considerations

The increasing variety of enzymes purified from extreme thermophiles (Table 1) supports the prospect of their future use in industrial applications. In some situations, existing enzyme-based approaches could potentially be improved by the use of substantially more thermostable versions of the same enzyme type. This presumes that the increase in enzyme stability achieved with an extremely thermophilic enzyme is not offset by a loss of enzyme activity at the desired operating temperature; this may or may not be the case. The fact that extremely thermophilic enzymes tend to be less active at ambient temperatures may be desirable in some instances, especially in cases where it is necessary to have strict control over the timing or extent of the reaction. In other instances, strategic opportunities that make use of improved biocatalyst stability and also advantageously use higher operating temperatures are desirable. Probably the most intriguing aspect is the use of extremely thermophilic enzymes to catalyze reactions that otherwise were not amenable to biocatalysis. Examples of these cases are presented below.

Polymerase Chain Reaction

The revolution in molecular biology has to a significant extent relied on the ability to amplify small amounts of DNA to levels that facilitate cloning, sequencing, and expression of particular genes of interest. This can be accomplished through the use of thermostable DNA polymerases that are produced by thermophilic microorganisms (122). To be effective in catalyzing the DNA polymerization reaction, the polymerase must be able to withstand repeated thermal swings from low to high temperatures so that the DNA template can denature to be copied and then reanneal before the next cycle. Although the DNA polymerase from a moderately thermophilic bacterium, *Thermus aquaticus*, was initially used in PCR applications, a host of polymerases from bacterial and archaeal sources are now available (122). Recent advances include the ability to replicate long sequences (20 to 40 kb) by adding to the reaction mixture an additional DNA polymerase with proofreading (3′ to 5′ *exo*-nuclease activity) capability to repair mistakes that might otherwise terminate the replication process (217). PCR is an example of the case in which protein thermostability is used strategically because thermally labile polymerases would not survive the repeated exposures to high temperatures, thus requiring addition of enzyme at each cycle.

Replacements of Existing Industrial Enzymes

Glucose Isomerase. The biocatalyzed conversion of glucose to fructose for the production of high-fructose corn syrup (HFCS) by xylose (glucose) isomerase represents the largest existing industrial application of an immobilized enzyme. Because of enzyme thermostability limitations, this process is currently operated at approximately 60 °C, at which temperature the equilibrium conditions limit the HFCS concentration of fructose to a maximum of 40 to 42% (218). Because 55% fructose concentration is desired for sweetening properties, a chromatographic separation step is subsequently used to enrich the mixture to this level by the removal of unreacted glucose. However, because the equilibrium yield of fructose increases with temperature (55% fructose estimated at 95 °C) (218), thermostable versions of glucose isomerase (GI) that could be used at higher temperatures are desirable to minimize the need for the glucose removal step.

Brown et al. (79) reported the presence of xylose isomerase in members of the hyperthermophilic eubacterial genus *Thermotoga*. For example, *T. neapolitana* produces a version of GI with an optimal temperature of at least 95 °C, which was also found to be highly active over a broad pH range (125). Compared to many other GIs that have been used commercially, the *T. neapolitana* GI possesses a 50 amino acid insert at the N-terminus, the function of which is unknown. The catalytic efficiency of this enzyme was superior to those determined for other GIs from less thermophilic sources at their respective optimal temperature ranges (125).

The GI from *T. neapolitana* again illustrates the potential strategic benefits that arise from the availability of thermostable versions of enzyme used in current industrial applications. By operating at elevated temperatures, the separation step needed for fructose enrichment can be minimized or eliminated because of the improved reaction yield. However, many other issues must be resolved before this enzyme can be considered as a replacement for the one used in existing processes. For example, consideration must be given to the levels of overexpression that can be attained in a foreign host, the impact of and approach used for immobilization, problems with potential side reactions of sugars at high temperatures, and regulatory approvals that might be required in the production of a substance for human consumption.

Hydrolysis of Natural Polymers. Interest in renewable resources has driven efforts to identify enzyme systems capable of hydrolyzing natural biopolymers, such as cellulose and hemicelluse, to more readily usable saccharides. This may be done in a nonspecific way so that the biopolymer is extensively hydrolyzed to simple sugars or in a specific way so that the properties of the biopolymer are systematically modified (219). As mentioned earlier, solutions containing natural polymers are less viscous at high temperatures compared to ambient conditions so that enzyme access to specific sites for hydrolysis is facilitated. Therefore, natural polymers may be hydrolyzed more readily by extremely thermophilic enzymes. Additionally, because these enzymes are typically stable at temperatures much lower than their optimal temperatures for hydrolysis, these hydrolases may retain their function at suboptimal temperatures for extended periods of time.

In the stimulation of oil and gas wells by hydraulic fracturing, aqueous solutions of natural polymers such as guar gum (a molecule consisting of a mannan backbone substituted with galactose through α-1,6 linkages) are used to transport particles to the site of the fracture (220). The objective is to prop open fissures in the bedrock perpendicular to wellbore and, hence, promote the flow of oil or gas to the point where it is readily recovered. The viscous properties of the guar gum solution used advantageously to transport the particles become problematic once the particles are in place. Thus, various methods, including the use of hemicellulases, are used to hydrolyze the guar to reduce viscosity. The high temperatures encountered in deep wells reduce the effectiveness of conventional enzyme systems but create an opportunity for extremely thermophilic enzymes that can withstand temperatures above 100 °C (43). An additional advantage of using thermostable enzymes is their lower activity at ambient conditions that otherwise can lead to premature hydrolysis of the guar (190). Such an enzyme system, consisting of β-mannanase, α-galactosidase, and β-mannosidase from the hyperthermophile *T. neapolitana*, has been shown to work to hydrolyze guar at high temperatures but much less so at moderate temperatures (190).

New Opportunities

To date, extremely thermophilic enzymes have been considered mostly as replacement enzymes to improve existing bioprocesses. However, their high degree of stability to thermal and other denaturing forces makes them candidates for catalyzing reactions not previously amenable to biocatalysis (6). This includes reactions conducted in nonaqueous environments such as organic solvents and supercritical fluids. The fact that their operational ranges are consistent with conditions used in chemical processing environments presents some intriguing possibilities for biocatalysis. It remains to be seen if these enzymes will be used for these purposes. Certainly, any insights gained into the intrinsic or extrinsic basis for enzyme stability that can be translated into bioprocessing improvements for enzymes from any source would be a welcome development.

SUMMARY

Extremely thermophilic microorganisms have proved to be a rich source of thermostable enzymes. The diversity of such enzymes and the associated biochemical properties are only beginning to be understood. The discovery process has made good use of the tools of molecular biology, and these will also be critical for producing sufficient enzymes for various applications. The promise that these enzymes hold for the expansion of uses of biocatalysis is considerable but will require input from those outside of this field to find the most suitable applications. Nonetheless, biocatalysis has added a new dimension with the increased availability of extremely thermostable enzymes, which bodes well for the development of more efficient and effective bioprocesses.

BIBLIOGRAPHY

1. R. Jaenicke, *Eur. J. Biochem.* **202**, 715–728 (1991).
2. R. Jaenicke, H. Schurig, N. Beaucamp, and R. Ostendorp, in M.W.W. Adams ed., *Advances in Protein Chemistry* 48, Academic Press, San Diego, 1996, pp. 181–269.
3. R. Jaenicke, *FASEB J.* **10**, 84–92 (January 1996).
4. M.W.W. Adams, F.B. Perler, and R.M. Kelly, *Bio / Technol.* **13**, 662–668 (1995).
5. R. Jaenicke, *FEMS Microbiol. Rev.* **18**, 215–224 (1996).
6. M.W.W. Adams and R.M. Kelly, *Chem. Eng. News* **73**, 32–42 (1995).
7. D.E. Robertson, E.J. Mathur, R.V. Swanson, B.L. Marrs, and J.M. Short, *SIM News* **46**, (1996).
8. E.J. Strauss and S. Falkow, *Science* **276**, 707–712 (1997).
9. T.D. Brock, *Thermophilic Microorganisms and Life at High Temperatures*, Springer-Verlag, New York, 1978.
10. J.A. Baross and J.W. Deming, in D.M. Karl ed., *Deep-Sea Hydrothermal Vents*, CRC Press, Boca Raton, 1995, pp. 169–217.
11. K.O. Stetter, *FEMS Microbiol. Rev.* **18**, 149–158 (1996).
12. R.M. Kelly, S.H. Brown, I.I. Blumentals, and M.W.W. Adams, in M.W.W. Adams and R.M. Kelly ed., *Biocatalysis at Extreme Temperatures*, American Chemical Society, Washington, D.C., 1992, pp. 23–41.
13. K.D. Rinker, C.J. Han, M.W.W. Adams, and R.M. Kelly, in *ASM Manual of Industrial Microbiology and Biotechnology*, in press.
14. G. Fiala and K.O. Stetter, *Arch. Microbiol.* **145**, 56–61 (1986).
15. A. Neuner, H.W. Jannasch, S. Belkin, and K.O. Stetter, *Arch. Microbiol.* **153**, 205–207, (1990).
16. R. Huber, T.A. Langworthy, H. Konig, M. Thomm, C.R. Woese, U.B. Sleytr, and K.O. Stetter, *Arch. Microbiol.* **144**, 324–333 (1986).
17. M.W.W. Adams, in F.T. Robb, A.R. Place, K.R. Sowers, H.J. Schreier, S. DasSarma, and E.M. Fleischmann ed., *Archaea: A Laboratory Manual*, Cold Spring Harbor Laboratory Press, New York, 1995, pp. 47–49.
18. C.-C. Chen, M.W.W. Adams, and J. Westpheling, *4th Annual Hyperthermophile Symposium*, Del Mar, Calif., May 14–16, 1997.
19. D. Hess and R. Hensel, *Gene* **172**, 121–124 (1996).
20. C.J. Bult, O. White, G.J. Olsen, L. Zhou, R.D. Fleischmann, G.G. Sutton, J.A. Blake, L.M. Fitzgerald, R.A. Clayton, J.D. Gocayne, A.R. Kerlavage, B.A. Dougherty, J.F. Tomb, M.D. Adams, C.I. Reich, R. Overbeek, E.F. Kirkness, K.G. Weinstock, J.M. Merrick, A. Glodek, J.L. Scott, N.S.M. Geoghagen, and J.C. Venter, *Science* **273**, 1058–1073 (1996).
21. The Institute for Genomic Research (TIGR), 1996, 1997, (*http: / / www.tigr.org /*).
22. M. Hanner, B. Redl, and G. Stoffler, *Biochim. Biophys. Acta 1033*, 148–153 (1990).
23. S. Colombo, S. D'Auria, P. Fusi, L. Zecca, C.A. Raia, and P. Tortora, *Eur. J. Biochem.* **206**, 349–357 (1992).
24. D.A. Cowan, K.A. Smolenski, R.M. Daniel, and H.W. Morgan, *J. Biochem.* **247**, 121–133 (1987).
25. I.I. Blumentals, A.S. Robinson, and R.M. Kelly, *Appl. Environ. Microbiol.* **56**, 1992–1998 (1990).
26. S.B. Halio, M.W. Bauer, S. Mukund, M.W.W. Adams, and R.M. Kelly, *Appl. Environ. Microbiol.* **63**, 289–295 (1997).
27. W.G.B. Voorhorst, R.I.L. Eggen, A.C.M. Geerling, C. Platteeuw, R.J. Siezen, and W.M. de Vos, *J. Biol. Chem.* **271**, 20426–20431 (1996).
28. J. Peters, M. Nitsch, B. Kuhlmorgen, R. Golbik, A. Lupas, J. Kellerman, H. Engelhardt, J.P. Pfander, S. Muller, K. Goldie, A. Engel, K.O. Stetter, and W. Baumeister, *J. Mol. Biol.* **245**, 385–401 (1995).
29. J.A. Maupin-Furlow and J.G. Ferry, *J. Biol. Chem.* **270**, 28617–28622 (1995).
30. M. Klingeberg, B. Galunsky, C. Sjoholm, V. Kasche, and G. Antranikian, *Appl. Environ. Microbiol.* **61**, 3098–3104 (1995).
31. M. Fusek, X. Lin, and J. Tang, *J. Biol. Chem.* **265**, 1496–1501 (1990).
32. N. Burlini, P. Magnani, A. Villa, F. Macchi, P. Tortora, and A. Guerritore, *Biochim. Biophys. Acta* **1122**, 283–292 (1992).
33. M.W. Bauer, S.H. Bauer, and R.M. Kelly, *Appl. Environ. Microbiol.* **63**, 1160–1164 (1997).
34. H.R. Costantino, S.H. Brown, and R.M. Kelly, *J. Bacteriol.* **172**, 3654–3660 (1990).
35. M. Rolfsmeier and P. Blum, *J. Bacteriol.* **177**, 482–485 (1995).
36. K.A. Laderman, B.R. Davis, H.C. Krutzsch, M.S. Lewis, Y.V. Griko, P.L. Privalov, and C.B. Anfinsen, *J. Biol. Chem.* **268**, 24394–24401 (1993).
37. R. Koch, P. Zablowski, A. Spreinat, and G. Antranikian, *FEMS Mirobiol. Lett.* **71**, 21–26 (1990).
38. R. Koch, A. Spreinat, K. Lemke, and G. Antranikian, *Arch. Microbiol.* **155**, 572–578, (1991).
39. J.W. Schuliger, S.H. Brown, J.A. Baross, and R.M. Kelly, *Mol. Mar. Biol. Biotechnol.* **2**, 76–87 (1993).
40. S.W.M. Kengen, E.J. Luesink, A.J.M. Stams, and A.J.B. Zehnder, *Eur. J. Biochem.* **213**, 305–312 (1993).
41. H.R. Badr, K.A. Sims, and M.W.W. Adams, *Syst. Appl. Microbiol.* **17**, 1–6 (1994).
42. M.W. Bauer, E.J. Bylina, R.V. Swanson, and R.M. Kelly, *J. Biol. Chem.* **271**, 23749–23755 (1996).
43. G.D. Duffaud, C.M. McCutchen, P. Leduc, K.N. Parker, and R.M. Kelly, *Appl. Environ. Microbiol.* **63**, 169–177 (1997).
44. K. Bronnenmeier, A. Kern, W. Liebl, and W.L. Staudenbauer, *Appl. Environ. Microbiol.* **61**, 1399–1407 (1995).
45. C. Winterhalter and W. Liebl, *Appl. Environ. Microbiol.* **61**, 1810–1815 (1995).
46. L.D. Ruttersmith and R.M. Daniel, *J. Biochem.* **277**, 887–890 (1991).
47. L.D. Ruttersmith and R.M. Daniel, *Biochim. Biophys. Acta* **1156**, 167–172 (1993).
48. L.D. Ruttersmith, R.M. Daniel, and H.D. Simpson, *Ann. N. Y. Acad. Sci.* **672**, 137–141 (1992).
49. S. Puchegger, B. Redl, and G. Stoffler, *J. Gen. Microbiol.* **136**, 1537–1541 (1990).
50. J. Breitung, R.A. Schmitz, K.O. Stetter, and R.K. Thauer, *Arch. Microbiol.* **156**, 517–524 (1991).
51. S. Mukund and M.W.W. Adams, *J. Biol. Chem.* **266**, 14208–14216 (1991).
52. E.T. Smith, J.M. Blamey, and M.W.W. Adams, *Biochemistry* **33**, 1008–1016 (1994).

53. S. Mukund and M.W. Adams, *J. Biol. Chem.* **268**, 13592–13600 (1993).
54. X. Mai and M.W.W. Adams, *J. Biol. Chem.* **269**, 16726–16732 (1994).
55. S. Mukund and M.W.W. Adams, *J. Biol. Chem.* **270**, 8389–8405 (1995).
56. J. Heider, X. Mai, and M.W. Adams, *J. Bacteriol.* **178**, 780–787 (1996).
57. X. Mai and M.W. Adams, *J. Bacteriol.* **178**, 5890–5896 (1996).
58. J.M. Blamey and M.W. Adams, *Biochemistry*, **33**, 1000–1007 (1994).
59. J. Kunow, D. Linder, K.O. Stetter, and R.K. Thauer, *Eur. J. Biochem.* **223**, 503–511 (1994).
60. J. DiRuggiero, F.T. Robb, R. Jagus, H.K. Klump, K.M. Borges, X. Mai, M. Kessel, and M.W.W. Adams, *J. Biol. Chem.* **268**, 17767–17774 (1993).
61. P. Zwickl, S. Fabry, C. Bogedain, A. Haas, and R. Hensel, *J. Bacteriol.* **172**, 4329–4338 (1990).
62. R. Hensel, S. Laumann, J. Lang, H. Heumann, and F. Lottspeich, *Eur. J. Biochem.* **170**, 325–333 (1987).
63. K. Ma and M.W.W. Adams, *J. Bacteriol.* **176**, 6509–6517 (1994).
64. K. Ma, F.T. Robb, and M.W. Adams, *Appl. Environ. Microbiol.* **60**, 562–568 (1994).
65. M.L. Camacho, R.A. Brown, M.-J. Bonete, M.J. Danson, and D.W. Hough, *FEMS Microbiol. Lett.* **134**, 85–90 (1995).
66. W.G. Purschke, C.L. Schmidt, A. Peterson, and G. Schafer, *J. Bacteriol.* **179**, 1344–1353 (1997).
67. S. Rospert, J. Breitung, K. Ma, B. Schwoerer, C. Zirngibl, R. K. Thauer, D. Linder, R. Huber, and K.O. Stetter, *Arch. Microbiol.* **156**, 49–55 (1991).
68. K. Ma, D. Linder, K.O. Stetter, and R.K. Thauer, *Arch. Microbiol.* **155**, 593–600 (1991).
69. K. Ma, C. Zirngibl, D. Linder, K.O. Stetter, and R.K. Thauer, *Arch. Microbiol.* **156**, 43–48 (1991).
70. N. Speich and H.G. Truper, *J. Gen. Microbiol.* **134**, 1419–1425 (1988).
71. F.O. Bryant and M.W.W. Adams, *J. Biol. Chem.* **264**, 5070–5079 (1989).
72. K. Ma, R.N. Schicho, R.M. Kelly, and M.W.W. Adams, *Proc. Natl. Acad. Sci. USA* **90**, 5341–5344 (1993).
73. A.F. Arendsen, P.T.M. Veenhuizern, and W.R. Hagen, *FEBS Lett.* **368**, 117–121 (1995).
74. A. Juszczak, S. Aono, and M.W. Adams, *J. Biol. Chem.* **266**, 13834–13841 (1991).
75. S. Aono, F.O. Bryant, and M.W.W. Adams, *J. Bacteriol.* **171**, 3433–3439 (1989).
76. P.R. Blake, J.B. Park, F.O. Bryant, S. Aono, J.K. Magnuson, E. Eccleston, J.B. Howard, M.F. Summers, and M.W.W. Adams, *Biochemistry* **30**, 10885–10895 (1991).
77. A. Guagliardi, V. Nobile, S. Bartolucci, and M. Rossi, *Int. J. Biochem.* **26**, 375–380 (1994).
78. A. Guagliardi, D. de Pascale, R. Cannio, V. Nobile, S. Bartolucci, and M. Rossi, *J. Biol. Chem.* **270**, 5748–5755 (1995).
79. S.H. Brown, C. Sjøholm, and R.M. Kelly, *Biotechnol. Bioeng.* **41**, 878–886 (1993).
80. M. Kohlhoff, A. Dahm, and R. Hensel, *FEBS Lett.* **383**, 245–250 (1996).
81. A. Guagliardi, L. Cerchia, S. Bartolucci, M. Rossi, P. Gallo, and C. Pedone, *Prot. Peptide Lett.* **1**, 5–8 (1994).
82. T. Uemori, Y. Ishino, H. Toh, K. Asada, and I. Kato, *Nucleic Acids Res.* **21, 25–265** (1993).
83. M. Imamura, T. Uemori, I. Kato, and Y. Ishino, *Biol. Pharm. Bull.* **18**, 1647–1652 (1995).
84. D.A. Prangishvili, R.P. Vashakidze, M.G. Chelidze, and I.Y. Gabriadze, *FEBS Lett.* **192**, 57–60 (1985).
85. S.A. Kozyavkin, R. Krah, M. Gellert, K.O. Stetter, J.A. Lake, and A.I. Slesarev, *J. Biol. Chem.* **269**, 11081–11089 (1994).
86. A. Bergerat, D. Gadelle, and P. Forterre, *J. Biol. Chem.* **269**, 27663–27669 (1994).
87. A.I. Slesarev, D.A. Zaitzev, V.M. Kopylov, K.O. Stetter, and S.A. Kozyavkin, *J. Biol. Chem.* **266**, 12321–12328 (1991).
88. A.I. Slesarev, K.O. Stetter, J.A. Lake, M. Gellert, R. Krah, and S.A. Kozyavkin, *Nature* **364**, 735–736 (1993).
89. A.I. Slesarev, J.A. Lake, K.O. Stetter, M. Gellert, and S.A. Kozyavkin, *J. Biol. Chem.* **269**, 3295–3303 (1994).
90. P. Fusi, G. Tedeschi, A. Aliverti, S. Ronchi, P. Tortora, and A. Guerritore, *Eur. J. Biochem.* **211**, 305–310 (1993).
91. X. Mai and M.W.W. Adams, *J. Bacteriol.* **178**, 5897–5903 (1996).
92. C. Purcarea, V. Simon, D. Prieur, and G. Herve, *Eur. J. Biochem.* **236**, 189–199 (1996).
93. C. Cicicopol, J. Peters, J. Kellerman, and W. Baumeister, *FEBS Lett.* **356**, 345–350 (1994).
94. G. Andreotti, M.V. Cubellis, G. Nitti, G. Sannia, X. Mai, M.W. Adams, and G. Marino, *Biochim. Biophys. Acta* **1247**, 90–96 (1995).
95. C. Purcarea, G. Erauso, D. Prieur, and G. Herve, *Microbiology* **140**, 1967–1975 (1994).
96. G. Cacciapuoti, M. Porcelli, M. Carteni-Farina, A. Gambacorta, and V. Zappia, *Eur. J. Biochem.* **161**, 263–271 (1986).
97. J. Breitung, G. Borner, S. Scholz, D. Linder, K.O. Stetter, and R.K. Thauer, *Eur. J. Biochem.* **210**, 971–981 (1992).
98. S.W.M. Kengen, J.E. Tuininga, F.A.M. de Bok, A.J.M. Stams, and W.M. de Vos, *J. Biol. Chem.* **270**, 30453–30457 (1995).
99. M.J. Peak, J.G. Peak, F.J. Stevens, J. Blamey, X. Mai, Z.H. Zhou, and M.W. Adams, *Arch. Biochem. Biophys.* **313**, 280–286 (1994).
100. A. Guagliardi, L. Cerchia, S. Bartolucci, and M. Rossi, *Protein Sci.* **3**, 1436–1443 (1994).
101. Y. Sako, K. Takai, A. Uchida, Y. Ishida, *FEBS Lett.* **392**, 148–152 (1996).
102. G. Cacciapuoti, M. Porcelli, C. Bertoldo, M. De Rosa, and V. Zappia, *J. Biol. Chem.* **269**, 24762–24769 (1994).
103. A. Heltzel, E. T. Smith, Z.H. Zhou, J.M. Blamey, and M.W.W. Adams, *J. Bacteriol.* **176**, 4790–4793 (1994).
104. B.S. Mroczkowski, A. Huvar, W. Lernhardt, K. Misono, K. Nielson, and B. Scott, *J. Biol. Chem.* **269**, 13522–13528 (1994).
105. K.A. Laderman, K. Asada, T. Uemori, H. Mukai, Y. Taguchi, I. Kato, and C.B. Anfinsen, *J. Biol. Chem.* **268**, 24402–24407 (1993).
106. J. DiRuggiero and F.T. Robb, *Appl. Environ. Microbiol.* **61**, 159–164 (1995).
107. S.B. Halio, I.I. Blumentals, S.A. Short, B.M. Merrill, and R.M. Kelly, *J. Bacteriol.* **178**, 2605–2612 (1996).
108. V.J. Harwood, J.D. Denson, K.A. Robinson-Bidle, and H.J. Schreier, *J. Bacteriol.* **179**, 3613–3618 (1997).
109. J.M. Muir, R.J. Russell, D.W. Hough, and M.J. Danson, *Protein Eng.* **8**, 583–592 (1995).
110. W.G. Voorhorst, R.I. Eggen, E.J. Luesink, and W.M. de Vos, *J. Bacteriol.* **177**, 7105–7111 (1995).

111. D. Hess, K. Kruger, A. Knappik, P. Palm, and R. Hensel, *Eur. J. Biochem.* **233**, 227–237 (1995).
112. A. Rudiger, P.L. Jorgensen, and G. Antranikian, *Appl. Environ. Microbiol.* **61**, 567–575 (1995).
113. P. Baumann and S. Jackson, *Proc. Natl. Acad. Sci. USA* **933**, 6726–6730 (1996).
114. T. Rowlands, P. Baumann, and S.P. Jackson, *Science* **264**, 1326–1328 (1994).
115. T. Imanaka, S. Lee, M. Takagi, and S. Fujiwara, *Gene* **164**, 153–156 (1995).
116. N. Rashid, M. Morikawa, and T. Imanaka, *Gene* **166**, 139–143 (1995).
117. Y. Tachibana, M.M. Leclere, S. Fujiwara, M. Takagi, and T. Imanaka, *J. Ferment. Bioeng.* **82**, 224–232 (1996).
118. N. Rashid, M. Morikawa, and T. Imanaka, *Mol. Gen. Genet.* **253**, 397–400 (1996).
119. Z. Yan, S. Fujuwara, K. Kohda, M. Takagi, and T. Imanaka, *Appl. Environ. Microbiol.* **63**, 785–789 (1997).
120. A. Bouyoub, G. Barbier, P. Forterre, and B. Labedan, *J. Mol. Biol.* **261**, 144–154 (1996).
121. T. Uemori, Y. Ishino, H. Doi, and I. Kato, *J. Bacteriol.* **177**, 2164–2177 (1995).
122. F.B. Perler, D.G. Comb, W.E. Jack, L.S. Moran, B. Qiang, R.B. Kucera, J. Benner, B.E. Slatko, D.O. Nwankwo, S.K. Hempstead, C.K.S. Carlow, and H. Jannasch, *Proc. Natl. Acad. Sci. USA* **89**, 5577–5581 (1992).
123. M.W. Southworth, H. Kong, R.B. Kucera, J. Ware, H.W. Jannasch, and F. Perler, *Proc. Natl. Acad. Sci. USA* **93**, 5281–5285 (1996).
124. J.S. Yu and K.M. Noll, *FEMS Microbiol. Lett.* **131**, 307–312 (1995).
125. C. Vieille, J.M. Hess, R.M. Kelly, and J.G. Zeikus, *Appl. Environ. Microbiol.* **61**, 1867–1875 (1995).
126. V. Zverlov, K. Piotukh, O. Dakhova, G. Velikodvorskaya, and R. Borris, *Microbiol. Biotechnol.* **45**, 245–247 (1996).
127. M. Van de Casteele, C. Legrain, V. Wilquet, and N. Glansdorff, *Gene* **158**, 101–105 (1995).
128. R. Ostendorp, W. Liebl, H. Schurig, and R. Jaenicke, *Eur. J. Biochem.* **216**, 709–715 (1993).
129. A. Tomschy, R. Glockshuber, and R. Jaenicke, *Eur. J. Biochem.* **214**, 43–50 (1993).
130. R.V. Swanson, M.G. Sanna, and M.I. Simon, *J. Bacteriol.* **178**, 484–489 (1996).
131. B. Darimont and R. Sterner, *EMBO J.* **13**, 1772–1781 (1994).
132. J. Tong and J.G. Wetmur, *J. Bacteriol.* **178**, 2695–2700 (1996).
133. R. Sterner, A. Dahm, B. Darimont, A. Ivens, W. Liebl, and K. Kirshner, *EMBO J.* **14**, 4395–4402 (1995).
134. W. Liebl, R. Feil, J. Gabelsberger, J. Kellermann, and K.H. Schleifer, *Eur. J. Biochem.* **207**, 81–88 (1992).
135. W. Liebl, I. Stemplinger, and P. Ruile, *J. Bacteriol.* **179**, 941–948 (1997).
136. N. Beaucamp, R. Ostendorp, H. Schurig, and R. Jaenicke, *Prot. Peptide Lett.* **2**, 281–286 (1995).
137. J. Gabelsberger, W. Liebl, and K.H. Schleifer, *FEMS Microbiol. Lett.* **109**, 131–138 (1993).
138. W. Liebl, P. Ruile, K. Bronnenmeier, K. Riedel, F. Lottspeich, and I. Greif, *Microbiology* **142**, 2533–2542 (1996).
139. P.-J. Lee and A.M. Stock, *J. Bacteriol.* **178**, 5579–5585 (1996).
140. A.M. Sanangelantoni, P. Cammarano, and O. Tiboni, *Microbiology* **142**, 2525–2532 (1996).
141. D.J. Saul, L.C. Williams, R.A. Reeves, M.D. Gibbs, and P.L. Bergquist, *Appl. Environ. Microbiol.* **61**, 4110–4113 (1995).
142. M. Yohda, I. Endo, Y. Abe, T. Ohta, T. Iida, T. Maruyama, and Y. Kagawa, *J. Biol. Chem.* **271**, 22017–22021 (1996).
143. J.G. McAfee, S.P. Edmondson, P.K. Datta, J.W. Shriver, and R. Gupta, *Biochem.* **34**, 10063–10077 (1995).
144. S. Ohnuma, M. Suzuki, and T. Nishino, *J. Biol. Chem.* **269**, 14792–14797 (1994).
145. T. Kath, R. Schmid, and G. Schafer, *Arch. Biochem. Biophys.* **307**, 405–410 (1993).
146. T. Wakagi, T. Fujii, and T. Oshima, *Biochem. Biophys. Res. Comm.* **225**, 489–493 (1996).
147. M.L. Tutino, A. Tosco, G. Marino, and G. Sannia, *Biochem. Biophys. Res. Comm.* **230**, 306–310 (1997).
148. M. Moracci, A. La Volpe, J.F. Pulitzer, M. Rossi, and M. Ciaramella, *J. Bacteriol.* **174**, 873–882 (1992).
149. M. Moracci, M. Ciaramella, R. Nucci, L.H. Pearl, I. Sanderson, A. Tincone, and M. Rossi, *Biocatalysis* **11**, 89–103 (1994).
150. R. Cannio, D. De Pacale, M. Rossi, and S. Bartolucci, *Biotechnol. Appl. Biochem.* **19**, 233–244 (1994).
151. C.E. Jones, T.M. Fleming, D.A. Cowan, J.A. Littlechild, and P.W. Piper, *Eur. J. Biochem.* **233**, 800–808 (1995).
152. G. Ianniciello, M. Masullo, M. Gallo, P. Arcari, and V. Bocchini, *Biotechnol. Appl. Biochem.* **23**, 41–45 (1996).
153. K. Kobayashi, M. Kato, Y. Miura, M. Kettoku, T. Komeda, and A. Iwamatsu, *Biosci. Biotech. Biochem.* **60**, 1882–1885 (1996).
154. M.I. Arnone, L. Birolo, M.V. Cubellis, G. Nitti, G. Marino, and G. Sannia, *Biochim. Biophys. Acta* **1160**, 206–212 (1992).
155. D.A. Bochar, J.R. Brown, W.F. Doolittle, H.-P. Klenk, W. Lam, M.E. Schenk, C.V. Stauffacher, and V.W. Rodwell, *J. Bacteriol.* **179**, 3632–3638 (1997).
156. D. Yue, N. Maizels, and A.M. Weiner, *RNA* **2**, 895–908 (1996).
157. A. Lehmacher and R. Hensel, *Mol. Gen. Genet.* **242**, 163–168 (1994).
158. B. Mukhopadhyay, E. Purwantini, T.D. Pihl, J.N. Reeve, and L. Daniels, *J. Biol. Chem.* **270**, 2827–2832 (1995).
159. R. Tabassum, K.M. Sandman, and J.N. Reeve, *J. Bacteriol.* **174**, 7890–7895 (1992).
160. S. Shima, D.S. Weiss, and R.K. Thauer, *Eur. J. Biochem.* **230**, 906–913 (1995).
161. M. Aoshima, A. Yamagishi, and T. Oshima, *Arch. Biochem. Biophys.* **336**, 77–85 (1996).
162. R.M. Kelly and M.W.W. Adams, *Antonie van Leeuwenhoek* **66**, 247–270 (1994).
163. M.W. Bauer, S.B. Halio, and R.M. Kelly, in M.W.W. Adams ed., *Advances in Protein Chemistry* 48, Academic Press, San Diego, 1996, pp. 271–310.
164. C. Leuschner and G. Antranikian, *W. J. Microbiol. Biotechnol.* **11**, 95–114 (1995).
165. A. Sunna, M. Moracci, M. Rossi, and G. Antranikian, *Extremophiles* **1**, 2–13 (1997).
166. C.R. Woese, O. Kandler, and M.L. Wheelis, *Proc. Natl. Acad. Sci. USA* **87**, 4576–4579 (1990).
167. M.W.W. Adams and A. Kletzin, in M.W.W. Adams ed., *Advances in Protein Chemistry* 48, Academic Press, San Diego, 1996, pp. 101–173.

168. S.W.M. Kengen, A.J.M. Stams, and W.M. de Vos, *FEMS Microbiol. Rev.* **18**, 119–137 (1996).

169. C.G. Miller, in F. Neidhardt ed., E. coli and Salmonella *typhimurium. Cellular and Molecular Biology*, vol. 1, American Society for Microbiology, Washington D.C., 1987, pp. 680–691.

170. *Enzymes at Work*, product literature from Novo Nordisk A/S, Bagsvaerd, Denmark, 1995.

171. O. Almarsson and A.M. Klibanov, *Biotechnol. Bioeng.* **49**, 87–92 (1996).

172. H. Kise, A. Hayakawa, and H. Noritomi, *J. Biotechnol.* **14**, 239–254 (1990).

173. H.L. Sham, C. Zhao, K.D. Stewart, D.A. Betebenner, S. Lin, C.H. Park, X.P. Kong, W. Rosenbrook, T. Herrin, D. Madigan, S. Vasavanonda, N. Lyons, A. Molla, A. Saldivar, K.C. Marsh, E. McDonald, N.E. Wideburg, J.F. Denissen, T. Robins, D.J. Kempf, J.J. Plattner, and D.W. Norbeck, *J. Med. Chem.* **39**, 392–397 (1996).

174. B.F. Sloane, *Nat. Biotechnol.* **14**, 826–827 (1996).

175. K. Peek, S.A. Wilson, M. Prescott, and R.M. Daniel, *Ann. N. Y. Acad. Sci.* **672**, 471–477 (1992).

176. H. Connaris, D.A. Cowan, and R.J. Sharp, *J. Gen. Microbiol.* **137**, 1193–1199 (1991).

177. L.J. Snowden, I.I. Blumentals, and R.M. Kelly, *Appl. Environ. Microbiol.* **58**, 1134–1140 (1992).

178. P.M. Hicks and R.M. Kelly, in F. Woessner, N. Rawlings, and A. Barret ed., *Handbook of Proteolytic Enzymes*, Academic Press, London, 1999, pp. 1579–1581.

179. R. Eggen, A. Geerling, J. Watts, and W.M. de Vos, *FEMS Microbiol. Lett.* **71**, 17–20 (1990).

180. J. Lowe, D. Stock, B. Jap, P. Zwickl, W. Baumeister, and R. Huber, *Science* **268**, 533–539 (1995).

181. M. Morikawa, Y. Izawa, N. Rashid, T. Hoaki, and T. Imanaka, *Appl. Environ. Microbiol.* **60**, 4559–4566 (1994).

182. B. Henrissat, *Biochem. J.* **280**, 309–316 (1991).

183. S.H. Brown and R.M. Kelly, *Appl. Environ. Microbiol.* **59**, 2614–2621 (1993).

184. G.P. Hazlewood and H.J. Gilbert, in M.P. Coughlan and G.P. Hazlewood ed., *Hemicellulose and Hemicellulases*, Portland Press, Chapel Hill, N.C., 1993, pp. 103–126.

185. S.W.M. Kengen and A.J.M. Stams, *Biocatalysis* **11**, 79–88 (1994).

186. J.D. Bok, S.K. Goers, and D.E. Eveleigh, *ACS Symp. Ser.* **566**, 54–65 (1994).

187. H.D. Simpson, U.R. Haufler, and R.M. Daniel, *J. Biochem.* **277**, 413–417 (1991).

188. C. Winterhalter, P. Heinrich, A. Candussio, G. Wich, and W. Liebl, *Mol. Microbiol.* **15**, 431–444 (1995).

189. A. Sunna, J. Puls, and G. Antranikian, *Biotechnol. Appl. Biochem.* **24**, 177 (1996).

190. C.M. McCutchen, G.D. Duffaud, P. Leduc, A.R.H. Petersen, A. Tayal, S.A. Khan, and R.M. Kelly, *Biotechnol. Bioeng.* **52**, 332–339 (1996).

191. M.W.W. Adams, *Annu. Rev. Microbiol.* **47**, 627–658 (1993).

192. M.W. Day, B.T. Hsu, L. Joshua-Tor, J.-B. Park, Z.H. Zhou, M.W.W. Adams, and D.C. Rees, *Prot. Sci.* **1**, 1494–1507 (1992).

193. M.K. Chan, S. Mukund, A. Kletzin, M.W.W. Adams, and D.C. Rees, *Science* **267**, 1463–1469 (1995).

194. K.S.P. Yip, T.J. Stillman, K.L. Britton, P.J. Artymiuk, P.J. Baker, S.E. Sedelnikova, P.C. Engel, A. Pasquo, R. Chiaraluce, V. Consalvi, R. Scandurra, and D.W. Rice, *Structure* **3**, 1147–1158 (1995).

195. I. Korndorfer, B. Steipe, R. Huber, A. Tomschy, and R. Jaenicke, *J. Mol. Biol.* **246**, 511–521 (1995).

196. K.A. Dill, *Biochemistry* **29**, 7133–7155 (1990).

197. J.F. Carpenter, J.S. Clegg, J.H. Crowe, and C.N. Somero, *Cryobiology* **30**, 2001–2041 (1993).

198. R.J. Ellis and S.M. van der Vies, *Annu. Rev. Biochem.* **60**, 321–347 (1991).

199. D.A. Parsell and S. Lindquist, in R.I. Morimoto, A. Tissieres, and C. Georgopoulos eds., *The Biology of Heat Shock Proteins and Molecular Chaperones*. Cold Spring Harbor Laboratory Press, Plainview, N.Y., 1994, pp. 457–494.

200. G. Zellner and H. Kneifel, *Arch. Microbiol.* **159**, 472–476 (1993).

201. R. Hensel and I. Jakob, *System. Appl. Microbiol.* **16**, 742–745 (1994).

202. S. Scholz, J. Sonnenbichler, W. Schafer, and R. Hensel, *FEBS Lett.* **306**, 239–242 (1992).

203. R.A. Ciulla, S. Burggraf, K.O. Stetter, and M.F. Roberts, *Appl. Environ. Microbiol.* **60**, 3660–3664 (1994).

204. L.O. Martins and H. Santos, *Appl. Environ. Microbiol.* **61**, 3299–3303 (1995).

205. O.C. Nunes, C.M. Manaia, M.S. DaCosta, and H. Santos, *Appl. Environ. Microbiol.* **61**, 2351–2357 (1995).

206. J.D. Trent, *FEMS microbiol. Rev.* **18**, 249–258 (1996).

207. J.D. Trent, J. Osipiuk, and T. Pinkau, *J. Bacteriol.* **172**, 1478–1484 (1990).

208. J.F. Holden and J.A. Baross, *J. Bacteriol.* **175**, 2389–2843 (1993).

209. J.D. Trent, M. Gabrielsen, B. Jensen, J. Neuhard, and J. Olsen, *J. Bacteriol.* **176**, 6148–6152 (1994).

210. C.J. Han, S.H. Park, and R.M. Kelly, *Appl. Environ. Microbiol.* **63**, 2391–2396 (1997).

211. H.K. Kagawa, J. Osipiuk, N. Maltsev, R. Overbeek, E. Quaite-Randall, A. Joachimisk, and J.D. Trent, *J. Mol. Biol.* **253**, 712–725 (1995).

212. B.M. Phipps, A. Hoffmann, K.O. Stetter, and W. Baumeister, *EMBO J.* **10**, 1711–1722 (1991).

213. B.M. Phipps, D. Typke, R. Hegerl, S. Volker, A. Hoffmann, K.O. Stetter, and W. Baumeister, *Nature* **361**, 475–477 (1993).

214. T. Waldmann, E. Nimmesgrern, M. Nitsch, J. Peters, G. Pfeifer, S. Müller, J. Kellermann, A. Engel, F.-U. Hartl, and W. Baumeister, *Eur. J. Biochem.* **227**, 848–856 (1995).

215. J.D. Trent, E. Nimmesgern, J.S. Wall, F.-U. Hartl, and A. Horwich, *Nature* **354**, 490–493 (1991).

216. F.T. Robb, J.B. Park, and M.W.W. Adams, *Biochim. Biophys. Acta* **1120**, 267–272 (1992).

217. W.M. Barnes, *Proc. Natl. Acad. Sci. USA* **91**, 2216–2220 (1994).

218. Y.B. Tewari and R.N. Goldberg, *Appl. Biochem. Biotechnol.* **11**, 17–24 (1985).

219. E.J. VanDamme and W. Soetaert, *FEMS Microbiol. Rev.* **16**, 163–186 (1995).

220. A. Tayal, R.M. Kelly, and S.A. Khan, *J. Soc. Petr. Eng.*, SPE 38432 (1997).

See also ENZYMES, DIRECTED EVOLUTION; PROTEIN SECRETION, *SACCHAROMYCES CEREVISIAE*; THERMOPHILIC MICROORGANISMS.

ENZYMES, FOR FLAVOR PRODUCTION

PETER S. J. CHEETHAM
Zylepsis, Ltd.
Ashford, United Kingdom

KEY WORDS

Bioactive molecules
Biocatalysis
Downstream processing
Fermentation
Isolated enzymes
Natural flavoring substances
Nature-identical flavoring substances
Receptor

OUTLINE

INTRODUCTION

Flavor and aroma chemicals are exceptionally bioactive molecules that exert their very characteristic taste and smell effects even when present at very low concentrations. This is because of their highly selective interactions with receptors in the mouth and nose. It is this high activity, in combination with the usually very pleasant, evocative, and identity-conferring properties that make flavor and aroma chemicals so valuable. In particular their tastes and aromas are very characteristic of particular foods, beverages, flowers, and so forth. This is made possible because of the many combinations and permutations of different flavor and aroma chemical molecules, with different characteristic mixtures found in each particular natural source. Hence complex mixtures of flavor chemicals are generally much more valuable because of their more rounded tastes. These organoleptic characteristics are generally very precisely determined by the exact molecular structures of the chemicals involved. As a result, even slight modifications, such as another hydroxyl group, a *trans* rather than a *cis* double bond, or an *R* rather than an *S* isomer, can completely change or abolish a molecule's taste and smell properties, both in terms of its character and identity, and also the threshold concentration at which it can be detected.

Many flavor and aroma chemicals are produced naturally in foods by the action of endogenous enzymes and/or by naturally occurring microorganisms such as those contributing to the tastes and smell of bread, cheese, wine, beer, and many oriental products such as soy. The need to make processed foods with taste and smell characteristics redolent of traditional products creates a derived demand for natural flavors, which in turn, creates a need for processes to make them. This is particularly the case for many highly processed foods that lose many of their volatile or heat-unstable flavor chemicals during processing. It is very appropriate to use modern enzyme and microbial technologies to manufacture flavor and fragrance materials, particularly as biocatalysis techniques provide the opportunity to carry out new reactions, and especially more selective reactions not possible with conventional organic chemical synthesis.

Use of biocatalysis methods is especially advantageous for the production of flavors because, when carried out rigorously, such techniques allow them to be described as "natural," and so command the premium prices conferred on them as a result of the insistent consumer demand for natural or "green" food ingredients. This advantage is, however, less strong for aroma chemicals because the consumer demand for natural fragrance materials is, so far, less well established. Biocatalysis processes are also very useful when alternative methods such as chemical synthesis or agricultural production are either not possible or not cost-effective. Perhaps the biggest challenge is in developing processes that can reliably and reproducibly deliver products at cost-effective prices.

The normal range of isolated enzyme, fermentation, and bioconversion processes are used. However, biocatalytic processes for flavors and fragrances are different from those in other areas of biotechnology in that they generally do not utilize genetic engineering techniques. This is because the products of genetic engineering don't fully meet the stringent consumer criteria for natural products. However, in all other respects this article on the use of bioprocessing to make flavors has been written in the context of, and with reference to, other successful bioprocesses used in the drug, agrochemical, food, and chemicals industries.

Ultimately the goal is products that are of good quality, are cost-effective and safe to use, and that meet customer demands. So far this combination of market pull, the consumer demand for natural flavors, and the technical push of enzyme and microbial technology has been successful in bringing a wide range of natural flavours to market, produced by enzyme, bioconversion, and fermentation techniques—whichever is most effective for each particular flavor material. Notable exceptions include a microbially produced natural vanilla flavor, and indeed any product using plant cell culture as a manufacturing technology. For reviews see Refs. 1–3.

The world market for flavor sales has been estimated at $4.5 billion in 1994 (4,5); this figure excludes the very considerable in-house production and use of flavors by some major food and beverage manufacturers. Europe still constitutes 40% of these sales (ca. $1.8 billion per year), followed by the United States, with sales in Asia having expanded most rapidly and with eastern European markets offering potential for the future. The major uses of flavors are in soft drinks (29%), followed by savory flavors such as meat and snack products (26%), and then applications in many other categories of products such as alcoholic beverages.

Major commercial trends in the flavor industry include globalization, consolidation, compliance, coping with the pressure of flavor costs, and customer service. Globalization trends are especially aimed at gaining a commercial presence in emerging markets such as China. Consolidation is proceeding apace as companies join together and are acquired so as to create the critical mass in research and development, purchasing, sales, and so forth in order to compete worldwide and against larger competitors. Globalization and consolidation trends combine as the smaller local companies with particular skills and niche markets are purchased by multinational flavor companies. This often results in a rationalization of manufacturing and fewer but much larger flavor-producing plants.

Compliance with regulatory, legislative, and customer requirements is increasing rapidly, and the resources that must be devoted to compliance issues are increasing much more rapidly than the growth in sales revenues. This is despite growth in flavor sales of 6–7% per year over the last decade, which is significantly higher than for the economy in general.

Due to the pressure on large food and beverage manufactures to reduce the contribution of flavor ingredients to the overall cost of their final products, and increasing competition between different flavor manufacturers, there has been an interesting trend to provide a more valuable service, for instance, in the form of more highly formulated flavors that are easier for the end-manufacturer to use. Other trends are to focus on the absolutely key "character-conferring" flavor chemicals, especially when they are sufficiently powerful to allow use at only low concentrations in the final product; to search for generic manufacturing processes whereby a small range of different flavors, such as C6 alcohols and aldehydes, can all be made using the same manufacturing plant; and to anticipate emerging consumer demands, such as for ethnic food influences or particular types of healthy ingredients.

Despite recent consolidation, the flavor ingredient supply market is still very fragmented, with the top 10 flavor companies having together still only 54% of the 1994 market of $4.5 billion in sales, and with the 5 leading companies, IFF, Quest, Givaudan, Haarmann & Reimer, and Firmenich, having 14, 10, 9, and 7% market shares, respectively (Table 1). This means that even these market leader companies have turnovers that are much smaller than their biggest customers such as Nestlé, Unilever, and Sara Lee, who are among the largest companies in the world. Another indication of economic importance is that European Community estimates show that eventually the impact of biotechnology in food and related businesses could be greater than in pharmaceuticals.

Any successful product requires a strong market demand and also a cost-effective manufacturing process to meet that demand. This article concentrates on the latter, but it is important to touch on some aspects of market demand first. Any bioprocess to make a natural flavor or aroma chemical exists only because of the market need for that flavor or aroma chemical. In turn the need for the flavor or aroma chemical arises as a result of a derived demand from the consumer's choice of products that contain the flavor or aroma chemical in a fully formulated and often branded product. The current world consumption of flavor and fragrance products (6) is summarized in Figure 1, and Hausler and Munch (7) report more than 65% of all

Table 1. Estimated Flavor and Fragrance Product Sales of Leading Companies

Company	Millions of dollars[a]	% of total industry sales
International Flavours and Fragrances (U.S.)	1,315	13.6
Givaudan-Roure (Switzerland)	980	10.1
Quest International (U.K./Netherlands)	930	9.6
Bayer (Haarmann & Reimer, & Florasynth, U.S. and Germany)	850	8.8
Firmenich (Switzerland)	708	7.3
Takasago International (Japan)	500	5.2
Bush Boken Allen (U.S.)	375	3.9
T. Hasegawa Co. (Japan)	356	3.7
Dragoco Gerberding and Co. (Germany)	260	2.7
Tastemaker (U.S.) (now acquired by Givaudan-Roure)	210	2.2
Universal Flavours (U.S.)	180	1.9
V. Mane Fils (France)	160	1.7
Ogawa & Co. (Japan)	158	1.6
Robertet (France)	130	1.3
Subtotal	7,112	73.4
Other companies	2,575	26.6
Total	9,687	100

Source: Ref. 6.

[a]Flavor and fragrance sales only; does not include sales of nonflavor and fragrance products such as food additives, speciality chemicals, juices, ice cream, and beverage bases, into which many of these companies have expanded as a logical diversification of their original product ranges.

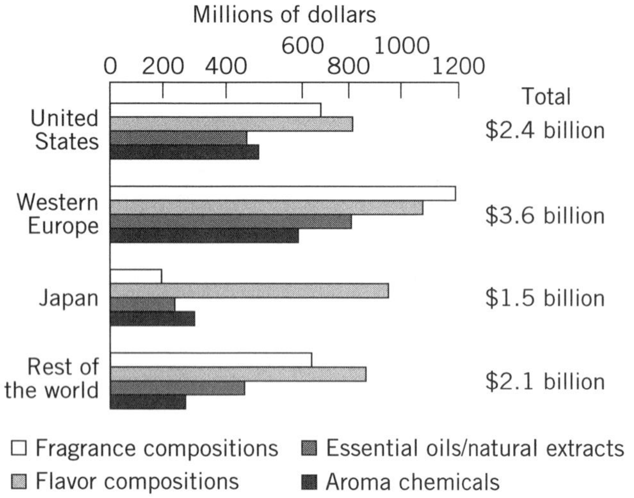

Figure 1. World consumption. Estimated consumption of flavor and aroma products, 1994.

flavoring ingredients used commercially in the United States are labeled as natural and have a food market potential exceeding $9 billion. This estimates the worldwide market for flavors alone at approaching $4 billion per year, and the total for all flavor, aroma, and essential oil ingredients at almost $10 billion per year. Another illustration is the high percentage of naturally flavored foods sold in Europe and the United States that contain natural flavor chemicals (8) (Table 2).

The economic value of biocatalytic processes is not just in terms of the materials produced, but also in the very much higher economic value of the flavored end products, which include branded consumer goods that depend on availability of flavors of the right quality and cost.

HISTORY

Flavors have a long history. The first products involving the bioproduction of flavors included beer, wine, and bread; yogurt made using lactobacilli; cheeses matured using a variety of microorganism including fungi. Also in Asia soy, miso, tempeh, and other products made using *Aspergillus oryzae* and other strains. Very often the success of these processes depends on the use of fast-growing anaerobes that accumulate products that discourage the growth of contaminants, such as ethanol or lactic acid, and also supplementation with additives such as hops and spices. This first phase of products was refined over many millennia.

Table 2. Proportions of Natural Aromas (%) in Flavored Foods

Product group	Europe	U.S.
Nonalcoholic beverages	90	83
Dairy products	50	75
Savory products	80	80

Source: Ref. 8.
Note: It has been estimated that, in 1990, 30 metric tons of aroma compounds was produced by biotechnological processes, sufficient to flavor several million tons of food products.

Many of the traditional flavors and flavored products still set the pattern for modern products because of deeply ingrained consumer preferences. For instance, beer and cheese were first made at least 6,000 years ago, that is some 250 generations ago. So there has been plenty of time for humans to develop inate flavor preferences, as well as preferences for other qualities such as color, clarity, foaming behavior, and mouthfeel properties. Some of these additional properties are the direct result of microbial biochemistry, for instance, glycerol is important in conferring mouthfeel, and it is produced from dihyroxyacetone by yeast.

It was not until the nineteenth century and the rise of a technological culture that second-phase products arrived in the form of much more chemically defined flavor products such as monosodium glutamate, citric acid, lactic acid, acetic acid in the form of improved vinegar producing processes, and also the products of a range of microbial enzymes. These processes required strain selection, development of special growth media, and the maintenance of hygienic or sterile conditions.

In 1837 Liebig and Wöhler identified the first flavor compound, benzaldehyde. The isolation and synthesis of vanillin by Tiemann, Haarmann, and Reimer in 1874–1876 began the modern flavor industry, and microbially produced flavors were first described by Omelianski in 1923. Also, once the technology for producing enzymes and cells had become widespread, the use of these enzymes and cells as biocatalysts in biotransformation processes became possible.

Some other indications of the high sensory effects of flavors and the impact that biomanufacturing has already had are the estimates that only about 400 aroma chemicals are manufactured in quantities of greater than 1 tonne per year, and only 68 materials are reported to be used at a rate greater than 3 tonne per year, but that some 50–100 flavor materials are reported to be made by bioprocesses (9). Several key considerations need to be recognized because they strongly influence the type of processes used (see following).

SOME CHARACTERISTIC FEATURES OF FOOD BIOTECHNOLOGY

Raw materials often contain endogenous enzyme activities and are often physically and chemically complex and diverse. They include oils and fats, proteins, and carbohydrates. Many are biopolymers that are usually produced from substainable agricultural sources, rather than petrochemicals, and contain many chiral molecules.

Products have the following characteristics:

- Have a wide range of value, from £ several 000/kg for some flavors to less than £100/tonne for animal feed (compare the prices of citric acid and vitamin B_{12}).
- Product value depends on consumer satisfaction (flavor, mouthfeel, etc.), which can vary with age, culture, and so forth.

- Genetic engineering, especially genetically modified organisms, have poor consumer acceptability.
- Flavor/aroma chemicals with comparatively similar chemical structures usually have quite different characteristics and intensities. For example, both diacetyl (2,3-butadione), which gives a buttery taste to dairy products, and hop-α-iso acids, which gives beer its bitter taste, are both vicinal diketones.
- It is not possible to record taste and smell sensations in the same way as sight and hearing sensations, and so the flavor and aroma quality of a material cannot be accurately assessed on-line, or even easily off-line. In a process, only the proportions of the various chemicals that contribute to the overall taste and smell can be analyzed. True flavor quality can be properly assessed only by an expert flavorist or trained taste panel, or on a gross basis by the reaction of consumers to a particular flavored product. However, some progress toward a taste–aroma sensor is being made, with olfactory neurones being shown to respond to physiological levels of volatiles in culture media (10) and a bovine binding protein being coupled to a resin (11). A functional odor receptor has been expressed in the noses of rats and shown to exhibit specificity toward just four C7–C10 aliphatic aldehydes and with the greatest response being to octanal (12). This will hopefully lead to a proper working knowledge of how the 500–1,000 odor receptors can recognize the thousands of distinct aromas, probably by acting in combinational mechanisms, with any one receptor responding to several different aroma chemicals and any one such aroma chemical capable of triggering several receptors.
- Flavor and aroma chemicals are often very potent materials, well suited to small-scale processing, and are usually diluted considerably when added to the final product. Their flavor value is defined as the concentration of flavor used, divided by its flavor threshold (that is, the concentration at which it can be just perceived). Because of the potent activities of flavors, care must be taken in multipurpose processing operations that produce a range of flavor products to ensure that there is no cross-contamination of one particular flavor by residues of another flavor produced in the preceding run.
- The actual perceived flavor of a molecule depends to some degree on the physical form in which it is eaten, thus the same flavor molecule formulated in products based on fat or oil, carbohydrate, water (as in carbonated beverage), and ethanol (as in an alcoholic beverage) can have quite different tastes, for instance due to solubility effects and the influence of mouthfeel and other factors on taste perception.
- Often one single flavor chemical will substantially define the characteristic of a particular flavor or aroma chemical, such as benzaldehyde for almond taste and vanillin for vanilla. These character-impact flavor chemicals can often command particularly high prices because of their high flavor value.
- Even quite closely related molecules can have very different flavor sensations, for instance geraniol (flower aroma) can be microbially converted into 6-methyl heptene-2-one, which is a component of tomato flavor.
- The natural or nature-identical status of flavors is important and deserves a more detailed consideration because of the impact of such definitions on the design and operation of flavor bioprocesses.
- Flavor and aroma chemicals must meet both industry and government standards for safety such as those provided by the International Organization of the Flavor Industry and the U.S. Food and Drug Administration, respectively. Increasingly, kosher and halal standards are also being used as criteria of good quality. For a more detailed discussion of the issues involved see work by Manley (9).
- Very often different isomeric forms of the same molecule have quite different flavor and aroma sensations. For instance (*S*)-(+)- and (*R*)-(−)-linalool are described as sweet and lavender, respectively, *R* and (−)-*trans*-α-ionone are violet-fruity and woody, (*S*)-(+)- and (*R*)-(−)-carvone are caraway and spearmint, (+)- and (−)-*cis*-rose oxide are sweet and fruity, (+)- and (−)-nootkatone are grapefruit and woody, (+)- and (−)-limonene are orange and turpentine, and (+)- and (−)-*p*-menthene-8-thiol are fruity and grapefruit tasting, respectively. In addition marked differences in flavor threshold often occur. Thus the (+)- and (−)-nootkatones have odor thresholds of 0.6–1.0 ppm and 400–800 ppm, respectively, and the (*R*) and (*S*)-isomers of δ-decalactone have odor thresholds of 1.5 and 5.6 ppb, respectively (13).

Natural and nature-identical flavors are defined as follows. These definitions obviously very much influence the types of processing operations that can be used.

Natural flavoring substances. These are materials obtained by appropriate physical processes (including distillation and solvent extraction) or enzymatic or microbiological processes from material of vegetable or animal origin, either in the raw state or after processing for human consumption by traditional food-preparation processes (including drying, torrefaction, and fermentation).

Nature-identical flavoring substances. These are materials obtained by chemical synthesis or isolated by chemical processes and are chemically identical to a substance naturally present in material of vegetable or animal origin.

Market Need

In order to succeed, a product made by any process needs to meet market demand factors such as the following:

- The declining availability and quality of many traditional materials, especially when they cannot meet the demand of modern high-volume-, high-quality-specification mass-market products, for instance, some plant spices and colors

- Changes in the technologies used for cooking and food processing (e.g., microwaved foods, extruded products, and liposome delivery of cosmetic ingredients)
- Increasing size of low-cost mass markets, especially in branded and own-brand consumer products, detergents, air fresheners, washing powders, and so on
- Changes in consumer demands, for example, healthy eating trends that include preferences for low saturated fat, salt, sugar, azo dyes, high fiber, and vitamins
- Increased legislative concerns that amplify the drive toward safety and healthiness

These factors create opportunities for technical innovations such as the following:

- Making available alternative raw materials
- New reaction and processing techniques, as well as processes adapted to use new raw materials
- Improved analytical techniques including those useful for process monitoring and control, and improved formulation for better performance (e.g., encapsulation and knowledge of synergy effects)

If technical innovations are successful, the flavor will then be used in formulated products recommended by flavorists on the basis of availability and cost of raw materials, the organoleptic value of the product, and the product's physical properties that affect ease of formulation and so on.

A more detailed business analysis will include a consideration of the following factors:

- Applications
- Customer needs and product segmentation
- Volume of demand and likely longevity of demand, price, and especially likely profit margins
- Specifications for different grades of product
- Competitors, both current and prospective
- Raw materials prices and availabilities
- Processing costs
- Patent position, including freedom to patent and non-infringements position
- Likely time and investment required to reach first sales
- Return on investment, especially allowing for risk factors, regulatory and safety considerations, and strategic benefits such as other uses for the technology
- Manufacturing capacity required, including downstream processing and any special requirements such as continuous operation

The major flavor and fragrance companies, most of which have significant in-house bioresearch and bioprocess development capabilities, are as given in Table 1.

The dynamic nature of the industry is illustrated by recent changes. The Tastemaker company has recently been acquired by Hoffman LaRoche (Givaudan-Roure), Quest International has now been sold by Unilever to ICI, Bayer has integrated Haarmann and Reimer and Florasynth, and Danisco has purchased both Borthwicks and Becks, is an interesting example of a major food ingredients company expanding its flavor businesses, rather than the more usual diversification of flavor companies into food ingredients that will move them into the top dozen flavor companies (Table 1).

Bioprocesses to Make Flavor and Aroma Chemicals

The necessary bioprocesses to make flavor and aroma chemicals fall into four broad categories: *(1)* traditional processes, *(2)* processes for low-intensity flavors, *(3)* processes for high-intensity flavors, and *(4)* production of taste enhancers.

Traditional Processes. Traditional processes are used for products such as beer, wine, cheese, yogurt, and soy sauce. These generally involve microbial action on chemically complex plant- or sometimes animal-derived raw materials, with the formation of relatively low concentrations of flavor chemicals that may very well have good preservative properties as well. These can be described mostly as flavored foods produced by naturally occurring microorganisms and/or enzymes derived from them. For instance, in the beer-brewing process, flavor is derived both from the ingredients used, such as hops and malt, and from the processing: the action of the yeasts and also the heating that forms the butter-tasting α-iso acids from precursors present in the hops (Fig. 2). The traditional brewing process is typical of more recently developed bioprocesses that consist of three main stages; work-up of raw materials, the microbial reaction step, and product processing, which puts it into the form required by the customer. This last step may involve formulation with other flavor materials and/or bulking agents.

In many cases traditional processes have been industrialized and adapted. For instance, intense cheese flavors can be manufactured by the hydrolysis of casein with proprietary proteases and peptidases combined with the hydrolysis of butter fat by lipases and esterases. *Aspergillus niger* and *Rhizomuor miehei* are the preferred sources for lipases for cheese making.

Other bioprocesses for the dairy flavor chemicals diacetyl, acetaldehyde, and various lactones, as well as for accelerated milk souring, have been developed. From Asian countries there are also many excellent solid-state fermentation products (Table 3).

Low-Intensity Flavors. Manufacturing has already been established for a range of low-intensity flavoring materials such as high-fructose corn syrups (10 billion tons/year produced worldwide), citric and acetic acids (acidulants), protein hydrolysates, and cheeses and cheese flavors (14). The bioprocesses that have been developed for their large-scale manufacture are well described elsewhere: citric acid (15), milk souring (16), protein hydrolysates (17). More recently processes involving the sequential use of enzymes to modified cheese flavors have been developed (Fig. 3).

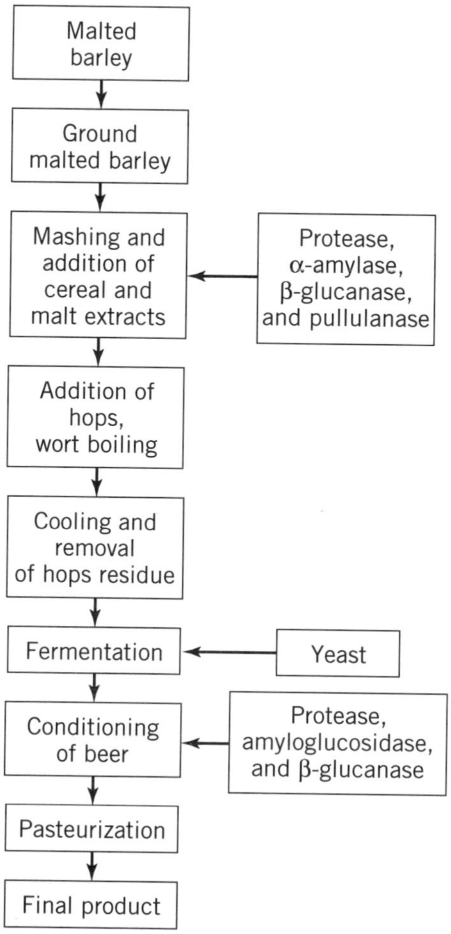

Figure 2. Beer manufacture. A flow diagram of the main operations. Note that pasteurization is omitted in the brewing of live beers. *Source*: Ref. 2.

High-Intensity Flavor Chemicals. More recent developments include the production of high-intensity flavors, such as methylketones (blue cheese flavor) and γ-decalactone (peach and other fruit flavors), in the form of pure natural chemicals by fermentation and bioconversion processes.

Taste Enhancers. Taste enhancers include materials such as monosodium glutamate (MSG) and mononucleotide flavor enhancers such as guanosine monophosphate (GMP) and inosine monophosphate (IMP), which are also produced as pure chemicals.

Important Considerations in Developing a Successful Process

It is important to appreciate that scaling up a process to meet cost and product specification targets can be at least as challenging as the original discovery research necessary to invent the process.

Much lower flavor concentrations need to be achieved for a flavor chemical when it is actually created in the food, in the way that flavor chemicals are produced in traditional products such as beer, cheese, and bread, rather than as a pure chemical, which is then added to the food or beverage at the required level. This does, of course, presume that any microorganisms involved are completely safe and acceptable, and also that flavor production will occur under the very different conditions existing in the food. Where processed foods are concerned, production in as pure and concentrated a form as possible is the most economic. The flavor can then be diluted as required for each end-use product and also combined with other flavors and flavor chemicals to give improved flavor quality. Some indication of the relationship between the concentration at which a flavor is produced and the concentration at which it is used in the final food or beverage product is given by the production-to-use ratio (Table 4), which can be illustrated by the following. If a material is produced at a concentration of 10 g L^{-1} and then used at 0.5 g L^{-1} (500 mg L^{-1}) in the product, or is present in a natural food or beverage at a concentration of 0.5 g L^{-1}, then its production-to-use ratio will be 20:1.

Key variables include the type of biocatalyst to be used, its performance, and in particular the medium used to carry out the reaction. The key features of the main methods are summarized as follows:

- *Fermentation production*. Fermentation is very suitable for multistep reactions. When using unstable enzymes and when cofactor regeneration is needed, it requires expensive fermenters and associated equipment. Operation is relatively expensive, because of the requirement for sterile conditions. The product is usually dilute and heavily contaminated with cells and by-products, and product quality can be variable due to batch-to-batch variations.
- *Use of isolated enzymes*. This method is especially suitable when extracellular stable enzymes are available, when single-step reactions are required that do not require cofactors (up to 10^3- to 10^4-fold regeneration of cofactors is necessary to enable them to be used cost-effectively, making synthetic reactions,

Table 3. Examples of Flavored Foods Obtained by Solid-State Fermentation

Product	Primary genus	Substrate	Further processing required?	Countries of origin
Soy sauce	*Aspergillus*	Soybean, wheat	Yes	China, Japan, southeast Asia
Sake	*Aspergillus*	Rice	Yes	Japan
Miso	*Aspergilus*	Rice, soybean	Yes	Japan, China, Taiwan, Indonesia
Tempeh	*Rhizopus*	Soybean	No	Indonesia, Surinam, New Guinea
Sufu	*Actinomucor*	Tofu	Yes	China
Ontjom	*Neurospora*	Peanut, coconut cakes	No	Indonesia

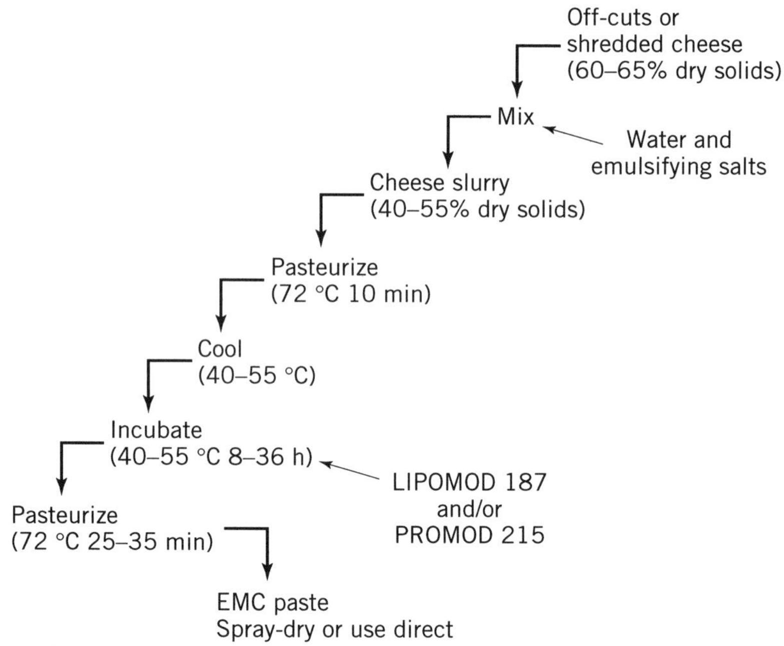

Figure 3. Outline of scheme for enzyme-modified cheese production. *Source*: Ref. 14.

Table 4. Production-to-Use Ratio

Traditional food products, e.g., beer, wine, bread, cheese	1.0 or greater if used as an ingredient in a more complex food
Industrialized forms of traditional products, e.g., enzyme-modified cheeses (EMCs) have flavor intensities about 20-fold greater than traditional cheeses	ca. 20
Traditional condiments, e.g., vinegar	ca. < 1.0
High-impact flavor chemicals, e.g., if γ-decalactone is made at 5 gL^{-1} and used at 100 ppm	50
trans-2-Hexenol	3,000 as compared with the concentration present in free-living plants

among others, very difficult to carry out using isolated enzyme preparations), and when hygienic but not sterile processing conditions can be used (although, of course, fermentation is necessary to produce the original cells and/or enzymes).

- Screening. In both cases, screening is a vitally important technique for discovering new and proprietary biocatalysts with the required activities and other properties required to make the needed molecules.

Key Performance Parameters

Cost-effective productivities can be met by biocatalysts that have the required activities in terms of their stability, specificity, and activity (grams of product per gram of cells per hour, or gram of product per liter reactor per hour), and with the right combination of the following characteristics (18,19):

- Cost
- Regulatory approval (safe and acceptable sources)
- Easy fit into process, including downstream processing and process tolerance (process integration)
- Availability of raw materials, including access to new materials
- Easy scale-up
- Broad specificity
- Activity on concentrated reactants
- Equilibrium point (high degree of conversion)
- Good stability
- High selectivity (few side products)
- Sterile conditions *not* required (or fermentation not susceptible to contamination)
- Ease of strain improvement by mutation or genetic engineering
- Rapid growth rates and nonfastidious nutritional requirements
- Robust and easily reproducible performance

The complexities involved in selecting biocatalysts with the optimal performance is illustrated by the cascade of enzymes used to create cheese flavors (20) (Fig. 3), and by a comparison of the rates of reaction of some common lipases when used either to hydrolyze triglycerides, or in the reverse reaction, to synthesis a flavor ester (Table 5).

Similarly there is a very large variation in the ability of different lipases to make a range of esters, as shown by the variabilities in degrees of conversion shown in Table 6 (21). These differences in reactivity are less surprising. It

Table 5. A Comparison of Lipolytic and Esterolytic Activities of Some Microbial Lipases

Microbial strain	Lipase activity ($U\ g^{-1}$)	Esterase activity ($U\ g^{-1}$)	Ratio of esterase: lipase activities
Chromobacter viscosium	144,000	2,650,000	18.4
Mucor miehei	420	22,000	5.24
Mucor miehi[a]	8,100	150	54
Pseudomonas fluorescens	8,500	26,300	3.1
Pseudomonas fluorescens[a]	15,000	150	0.01
Penicillium roquerfortii	290	625	2.2
Rhizopus japonicus	8,750	13,300	1.5
Rhizopus japonicus	36,000	39,400	1.09
Rhizopus japonicus[a]	47,000	0	∞
Candida cylindracae	85,700	53,600	0.6
Candida cylindracae[a]	95,000	370	257
Aspergillus niger	1,580	940	0.6
Geotrichum candidum	4,650	170	0.04
Candida lipolytica	10,500	150	0.01

[a]Data used is from Ref. 13 together with independant data for four of the enzymes in which esterase activity was assayed using geranyl proprionate in hexane and hydrolysis using olive oil emulsion as the substrate.

Table 6. Ester Formation Using Lipases

Lipase	Acetate	Propionate	Butyrate	Acetate	Propionate	Butyrate
Alcaligenes (Meito-Sangyo)	0.5	3	18	<0.5	4	24
Aspergillus (Rohm)	23[a]	33	40[a]	6[a]	15	6[a]
Fungal (Rohm)	0.5	40	65[a]	<0.5	48	61
A. niger (Palatase)	0.5[a]	22	98[a]	<0.5[a]	5	40[a]
C. rugosa (cylindracea) (MY)	3.5	95	98	<0.5	95	90
G. candidum (SNEA)	<0.5	2	18[a]	<0.5	2	5
M. miehei (Gist-Brocades)	80[a]	92	97	14	96	96
M. miehi (Lipozyme)	15	42	98	<0.5	30	93
P. cyclopium A (SNEA)	0.5	4	48	<0.5	2.5	38
P. cyclopium B (SNEA)	18	3.5	45	14	3.5	38
Porcine pancreas (Rohm)	0.5	55	95	<0.5	94	85
Porcine pancreas (Kochlight)	0.5	0.5	35	0.5	<0.5	12
R. arrhizus (SNEA)	5	70	97	3.5	58	97

Note: Data was obtained using 0.2 g lipase with analysis at 24 h, unless otherwise indicated.
[a]0.5 g of lipase was used instead of 0.2 g.

is noted that the catalytic structure of various lipases are different. Thus, whereas the catalytic triad in the active site of *Homo sapiens* and *Mucor miehei* lipases is Asp, His, Ser, in *Humicola lanuginosa* it is Asp, His, Try, and in *Geotrichum candidum* it is Glu, His, Ser.

Fermentation medium requirements include the following important considerations when developing a production medium that will be as inexpensive and reliable as possible (19):

- Selective advantages for the production strain, allowing fast growth, high yields, and no remaining by-products (a big challenge, not least because most wild-type strains are surface growers, and considerable effort may be necessary to get them to adapt to efficient production in submerged fermentations)
- Defined and balanced medium (no excess of any medium component)
- No by-products, color, odor, or substrates, remaining at the end of biotransformation
- Substrate(s) stable, soluble, and of constant quality
- Reduced risk of contamination due to medium composition
- Product not degraded by use as C- or N-source for cell growth in any circumstances
- Oxygen requirements (this can become rate-limiting for rapidly growing cultures; provision of air for growth is expensive because of filtration costs to ensure sterility and mixing power costs necessary to ensure good gas transfer throughout the fermenter)
- Little or no foaming problem
- Short lag period

Overall the bioprocess will be quantified and costed on the basis of the following seven factors:

- Specific growth rate
- Biomass concentration (g dry wt/L)
- Oxygen transfer rates and power consumption rates (usually turbine impellors are used so as to minimize power input costs and maximize mixing)

- Grams of product produced per liter of reactor per hour
- Concentration of product achieved (g/L)

Yields are very dependant on the type of metabolite being formed by the cells. Estimates of rates vary from 100 to 500 nmol/g $cells^{-1}$ h^{-1} for amino acid formation, to 1 to 5 nmol/g $cells^{-1}$ for vitamin and coenzyme synthesis (Table 7). Degree of conversion of precursor(s) into product (%), and purity of product (especially with respect to other chemically similar materials that could make its purification difficult) are also important. This can be also described as the selectivity of the reaction: grams of product required per gram of all products formed.

Problems in bioprocessing are well illustrated by the terpenes. Whereas terpenes are very common flavor and aroma materials with a very wide range of tastes, smells, and applications, production to date is entirely by extraction from plant raw materials. It appears that no true bioproduction process for a terpene has yet been established. This is probably because monoterpenes have low water solubilities, combined with high cytotoxicities, together with high volatilities and often poor chemical stabilities. By contrast, sesquiterpenes can be biotransformed much more successfully. For instance, the processes for producing Ambrox and sclareolide are described elsewhere in this article. Various bioengineering aspects of flavor process are summarized in Table 8.

Of course, each product and process can be expected to have some individual features. These include the optimization of fermentation media to maximize productivity and minimize cost, and to allow for particular features, for instance, often the product concentration achieved can vary with the specific growth rate of the fermentation. Some indication of the challenges encountered in developing a bioprocess are illustrated by the examples given in this article.

Technicocommercial Factors

In addition to the scientific factors just discussed, the following technicocommercial factors also have a big influence in many instances.

Flavors and fragrances are manufactured only when required by end-user companies, who are usually the manufacturers of consumer end products, that is, they are subject to a derived demand. This is accomplished following the issuing of a competitive brief to a number of companies. Thus, the requirement for the individual flavor chemicals and materials required to create the flavor is subject to a derived demand, and as the actual flavor is usually tailored specifically to the needs of one end-user company, the amounts of any one flavor chemical or material required can vary considerably from customer to customer.

Compounding and formulation of flavor chemicals is very often the last step in the manufacturing process and is a comparatively simple operation that adds significant value because of the improved complexity and character that can be created. This step depends on the creative input of skilled and experienced flavorists who create the formulations that dictate the types and amounts of flavor chemicals that are used. High flavor/aroma impact and character- or identity-conferring chemicals are very important because of their high value, even when they are expensive to make.

Control of organoleptic purity involves far more than just maintenance of chemical purity, as a trace impurity with a high flavor impact present at instrumentally undetectable levels can still have a very detrimental effect on flavor and aroma quality. This factor is very important when considering the batch-to-batch variations in quality because very few flavor chemicals are required in large enough volumes to justify production by continuous processes. Control of organoleptic quality also extends to the storage of ingredient chemicals and formulated flavors because cross-contamination can occur from ingredient to ingredient (vanillin is especially prone to this) and also from materials leached out of containers and packaging.

Economies of scale of manufacture are difficult to achieve for all but the largest-volume flavor chemicals. Instead, the small scale of manufacture creates the opportunity for generic manufacture using the same multipurpose equipment to make a range of ingredient chemicals on a campaign basis in direct response to market demands. This requires careful scheduling so as to properly match demand with supply, and scrupulous cleaning of equipment between runs.

Patent protection is important, especially for new flavor materials, but the enforcement of process patents is sometimes difficult, which is a particular problem for key producer microbial strains. Because of the individual nature of the flavor chemicals, including extraction from natural raw materials and complex formulation, special manufacturing know-how is often involved that cannot be protected by patents.

Table 7. A Comparison of the Relative Rates of Metabolic Processes Producing Different Classes of Metabolites as Bioprocess Products

Substrate family	Reaction type	Turnover rate (mmol/h/g cells)
Carbon source	Uptake, oxidation	10
Monomers, polymers	Synthesis, polymerization	1
Amino acids	Uptake, synthesis, incorporation in protein	0.1–0.5
Bases, nucleotides	Uptake, synthesis	
	Incorporation in RNA	0.15
	Incorporation in DNA	0.01
Vitamins, coenzymes	Uptake, synthesis	0.001–0.005

Table 8. Bioengineering Aspects of Bioflavor Production

γ-Decalactone	The precursor is an oil, and so water-miscibility issues had to be solved. Supply of the precursor in the form of an ester reduces foaming problems in the fermentor.
Benzaldehyde	Toxic HCN is produced when using cyanogenic nitrile glycoside precursors appropriate for microbial benzaldehyde production. Product inhibition reduces the rates of formation and final product concentrations, in situ product recovery by perevaporation has been successfully trialed.
Vanillin	A complex pathway is involved. When eugenol is used as the precursor then precursor toxicity to the cells must be overcome. Easy recovery of the product as a Schiff's base is incompatible with natural processing.
Soy sauce	Several strains of microorganism have been used, separately immobilized and operated in series.
Milk souring	In order to achieve balanced productivity, two fermentors have to be used in series. Because this is a rare example of a continuous fermentation process and because a food product is being made, special care has to be taken to ensure aseptic operation.
Hexenals and other flavor aldehydes	Use of an alcohol oxidase in two-phase water–hexanol reaction has been demonstrated to be high yielding.
Ethylisovalerate	Recovery of product from the exit gases from the reactor by adsorption to charcoal has been reported as successful.
Monosodium glutamate	Rod-shaped β-crystals of monosodium glutamate are easier to filter and wash, but the α-form crystalizes preferentially and so is redissolved and then recrystallized in the β-form.
Methylbutyric acids	Fed-batch reactions are used because of the toxicity of the methylbutanol precursors.
Methylketones (blue cheese flavor)	Extractive fermentation can be employed in which product is removed during the reaction to encourage continued production of the methylketones.
Offensive odors in exit gases	Mixed microbial culture biofilters used to degrade off-odors in exit gas streams.
Enzyme-modified cheese flavors	A number of different types of enzyme have to be added in the right sequence in order to obtain the desired flavor.
δ-Decalactone	Flavor quality was significantly reduced by trace impurities carried over from the fermentation media. Therefore before the bioconversion step, the 11-hydroxypalmitic acid precursor had to be purified.

The very extensive process validation requirements for biopharmaceutical processes are not generally required for flavor chemicals. The standards that are required are consistent with best practice in the food and fine chemicals industries, such as food hygiene standards and the sterility necessary to perform uncontaminated fermentions.

Manufacture of high-volume flavor and aroma chemicals is increasingly becoming attractive to chemical companies, rather than just flavour and fragrance companies, and internationalization is occurring fast, as indicated by the entry of Chinese manufacturers.

Because of the volatile nature of flavor and aroma chemicals, distillation is a dominant downstream processing operation in their manufacture. Usually, sophisticated equipment is used, controlled by computers, and upgraded to minimize oxidation, overheating, and so forth that can so easily produce off-flavors. Solvent extraction is common, and new methods are gaining credibility, for example, the use of supercritical carbon dioxide as an extraction technique; it has a polarity similar to that of hexane, and because of its volatility, leaves no residue in the product. Solvent extraction is limited to a short list of solvents that are classed as being consistent with natural processing (Table 9). Formulation of flavors often begins with spray-drying, often onto a carrier such as maltodextrins, which also serve to stabilize the flavor.

Table 9. Comparison of Various Solvents for the Extraction of 2-Phenylethanol Produced by Bioconversion

Solvent	K_d
Ethylacetate	0.95
Butylacetate	34
Butanol-1	11
Butanol-2	10
Diethylether	12
Hexane	0.5
Methyl propanol-1	22

Source: Ref. 22.
Note: K_d is the ratio of the dissolved substance in the extracting phase to that in the aqueous phase. Data assumes 20 °C with an aqueous solution containing 1 g/L 2-phenylethyl alcohol.

Raw materials are usually purchased from a great variety of sources, rather than produced in-house. Sometimes the availability of a raw material is a key driving force in a company's decision to produce a flavor. One example is that a good supply of by-product yeast is very useful for making savory flavors. Variabilities in the supply, price, and quality of natural raw materials, such as vanilla beans, can frequently pose big problems. In other cases it is the availability of the natural precursor molecule for a bioprocess that is the problem that can prevent the commercialization of that process. Examples include the 11-hydroxypalmitic acid required to make δ-decalactone, the methionine required for methional and related flavor chemicals, and also precursors for raspberry ketone (Fig. 5).

Effluent disposal from manufacturing plants can create some novel problems because of the objectionable smells that can be created for neighbors. One widely adopted process to combat this problem is to route all exit gases through a bed of supported microorganisms that metabolize much of the objectionable materials, thereby valorizing the exit gases to acceptable emission levels. One such design of biofilter is the Bioton, supplied by Monsanto Enviro-Chem Systems.

In any process, its performance, and therefore the cost and quantity of the product, is dependent on a complex interrelationship between a large number of process variables, as illustrated for the use of glucose isomerase (GI) to make high-fructose corn syrup bulk sweeteners (23). This is a splendid example of the high degree of process integration that must be achieved in order for a mature bioprocess to continue to be successful (Fig. 4). The process involves a semicontinuous process with three enzyme reaction steps together with raw material processing, downstreaming, and sophisticated chromatographic separation of glucose and fructose sugars. Since GI is now a mature product, new versions of this enzyme demonstrate the big improvements necessary for second-generation processes to succeed. GI from *Streptomyces murinus* has about a 60-fold greater productivity in terms of tons of product produced/kg enzyme, which results in a 30-fold lower cost of product.

Protection of Technological Advantages

Another important factor is the protection of the technology by patenting (24). The key objectives are to avoid infringement of existing patents and to gain patent protection oneself. In this respect biochemical and microbiological inventions create some particular problems:

- Functionally identical biological materials (e.g., enzymes) can have significantly different structures and amino acid sequences.

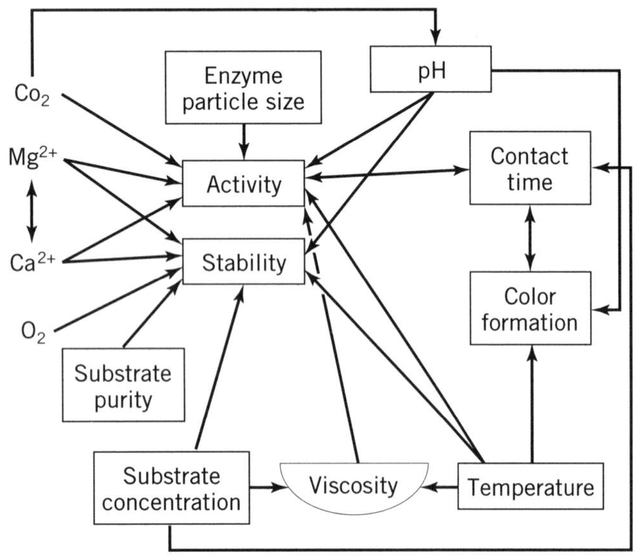

Figure 4. The interactions between the various process variables that can influence the productivity of glucose isomerase when used for the industrial production of high-fructose syrups from hydrolyzed starch.

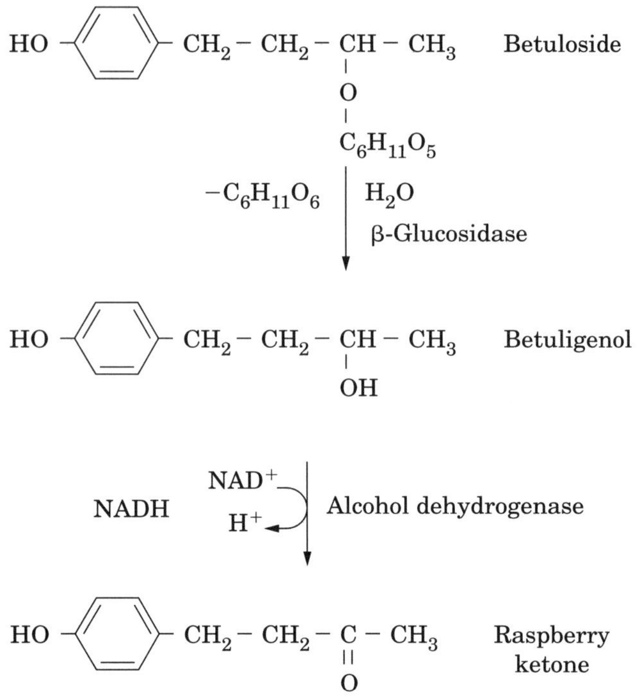

Figure 5. Scheme for making raspberry ketone flavor chemical from the plant-derived precursor betuloside.

- Functionally identical biological materials can be obtained from very different biological sources (e.g., enzymes with the same substrate specificity from different microbial sources).
- The requirement to deposit samples of microorganisms is a substantial overdisclosure as compared to patenting of other inventions. This often leads to some processes being kept as in-house secrets.
- Biological materials, especially living organisms, are too complex to be described in sufficient detail and with enough precision to easily satisfy the requirement of patent law. For instance, existing microbial classifications are not always directly applicable to newly isolated microorganisms.
- Varying criteria and definitions are used to distinguish biocatalysts, for example, substrate and inhibitor specificity and immunochemical cross-reactivity in the case of enzymes, and nutrient requirements and fatty acid composition and so on for microorganisms.

Categorization of Bioflavor Processes

Depending on the complexity of the biochemical reactions required to produce the desired products, bioflavor processes can be categorized along the lines of the following examples:

Single-Enzyme Reactions. Examples of single-enzyme reactions include glutaminase, 3′–5′ nucleotide diesterase (e.g., from *Penicillium citrinium*), adenylic deaminase (e.g., from *Brevibacterium ammoniagenes*), glycosidases for the release of terpene flavors from grape and other flavor sources, lipases and esterases for flavor ester synthesis and for resolution of racemic mixtures (such as obtaining *l*-menthol from *dl*-menthol).

Mixed-Enzyme Reactions. Examples of mixed-enzyme reactions include use of mixed proteases for producing protein hydrolysate flavors; β-glucosidase and nitrile lyase to make benzaldehyde from amygdalin; lipoxygenase, hydroperoxide lyase, and oxidoreductase to make, for instance, *cis*-3-hexenol from linoleic acid via *cis*-3-hexenal), and proteases, peptidases, lipases, and esterases to make enzyme-modified cheese flavors (EMCs) (Fig. 3). Recently, mixed-enzyme preparations have become widely used in the extraction of fruit flavors.

Use of Mixed Enzymes to Enhance Fruit Flavors. Recently it has been found that very many fruit flavor chemicals such as the 13-norisoprenoids and terpenes (e.g., linalool and geraniol) occur in plants as glycosides (24). Geraniol was the first glycosidically bound aroma chemical to be identified as such in rose extracts, with a structure of geranyl-xylose-glucose, and the C13-norisoprenoids (e.g., ionones, raspberry ketones, and others such as *trans*-3-hexenol are present in the fruit of origin, predominantly in glycosidially bound forms; see Ref. 25). Usually the proportions of these glycosidically bound flavor chemicals to free flavor chemicals are in the ratio of 2:1 to 5:1. This is possibly because the glycosides are more water soluble, more resistant to acid hydrolysis, and effectively detoxify these terpenes, phenols, and similar compounds.

The glycosides have similar structures with the aglycone bound first to glucose and then to a another sugar such as α-L-arabinose, α-L-rhamnose, β-D-xylose, β-D-apiose, or another β-D-glucose unit, depending on the plant source. Hydrolysis of such glycosides liberates the aglycose, thereby enhancing the flavor content of a conventional fruit product and even allowing the generation of flavor from waste materials such as peel and stems. However, the enzymes required to achieve these advantages must have some special characteristic such as good activity at low pH (about pH 3.5), resistance to product inhibition by glucose, and even resistance to moderate concentrations of ethanol in the case of fermented products. Most importantly, flavor chemicals are released, and flavor is enhanced only when two enzyme activities are present: a β-glucosidase and also another glycosidase such as α-L-arabinosidase or α-L-rhamnosidase, depending on the plant material being treated. Because of these requirements, glycosidically bound flavors are released only rarely in existing processes, such as fermentation, because yeast β-glucosidase is insufficiently active. However, one process that absolutely depends on endogenous glycosidase action is in flowers, where it appears that aroma materials are stored as glycosidases and then, upon the flower opening, glycosidases are produced that hydrolyze the glycosides, releasing the aroma chemicals that create the fragrance of the flower.

Use of such special glycosidase combinations have now been widely adapted, for instance, in grapefruit juice extraction, and the cloning of the appropriate enzymes into suitable host strains is underway. An example is the *A. niger* enzymes used by Gist-brocades (26). A very good example of the improved yields of flavor that can be achieved by the use of glycosidase is the improved yields of vanilla flavor that can be obtained by treating vanilla beans with emulsin, which is a mixed enzyme preparation especially rich in β-glucosidase that is obtained as a crude extract of almond meal.

Use of Single Microbial Strains. Single microbial strains are used to produce the following: methylketones with spores of *A. niger*, γ-decalactone with *S. cerevisiae* strains, δ-decalactone with *Cladosporiun suaveolens*, diacetyl with *Streptococcus cremoris* and *Streptococcus diacetylactis*, soured milk for margarine production with *S. cremoris*, methylbutyric acids with *Acetobacter acetii*, and pyrazines with *Corynebacterium glutamicum*.

Combined use of Several Microorganisms. A combination of several microorganisms is used for such processes as the Unilever two-stage continuous process for manufacturing soured milk, which makes use of *Lactobacillus* and *Streptococcus* strains (27) (Fig. 6), and the accelerated process for soy sauce manufacture using *Pediococcus halophilus, Saccharomyces rouxii*, and *Torulopsis / Candida versatilis*. Also several microbial strains are used to achieve a synergistic effect in cheese fermentations. For instance, propionic acid is an important flavor component of Emmental and some other Swiss cheeses. It can be produced by first forming lactic acid with strains such as *Lactobacillus lactis* or *Streptococcus thermophillus* and then metabolizing the lactic acid into propionic acid by *Propionibacter* strains. Using this two-stage method, concentrations of propionic acid of 40 g L^{-1} can be made. Propionic acid is also useful to form flavor esters such as ethyl, benzyl, citronelyl, and geranyl propionates, as is butyric acid, also produced by fermentation in concentrations of 20–30 g L^{-1}, which is used to make ethyl, isobutyl, and amyl butyrate flavor esters.

Combined use of Enzymes and Microorganisms

Savory Flavor. Production of savory flavors using proteases, lipases, ribonucleases, and polysaccharidases to break down plant raw materials, and then microorganisms to convert the products of hydrolysis into flavor chemicals are used in processes such as the Biosol process (28). Improved flavors using this approach could be possible by genetically engineering yeast strains to contain higher concentrations of the savory flavor enhancers GMP and IMP, and also of the flavorsome octapeptide "beefy meaty peptide."

Fruit and Dairy Flavor. The traditional use of the acetic acid bacterium *A. acetii* to make vinegar by oxidizing the ethanol in wine into acetic acid has been redeployed by modern flavor biotechnologists. This time *A. acetii* is used for the bioconversion of isobutyric alcohols present in fusil oil into isobutyric acids. Thus both the 2- and 3-methylbutanols are oxidized into 2- and 3-methylbutyric acids. This oxidation takes place with very little racemization.

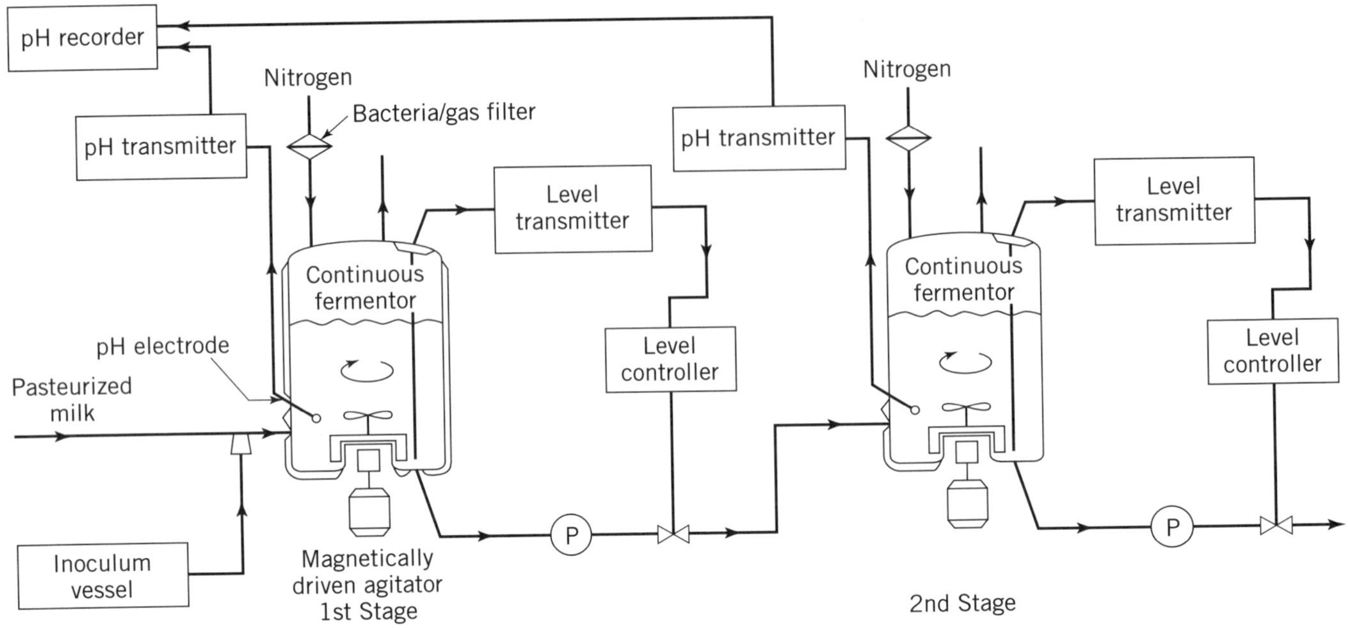

Figure 6. Two-stage continuous microbial souring of milk. *Source*: Ref. 16.

The overall reaction requires two sequential reactions: a dehydrogenation into the aldehyde intermediate than can tautomerise easily, and then oxidation to the acid. These isobutyric acids are good flavor chemicals in their own right, but in addition their value and range of uses can be increased and expanded by esterification either by using lipases or esterases "in reverse," or by azeotropic distillation (29) (Fig. 7).

Plant-Derived Biocatalysts. Plant-derived biocatalysts include lipoxygenase for the production of β-ionone, *cis*-3-hexenol, and so forth. Soy lioxygenase produces 13-hydroperoxides that break down to C6 aldehydes, whereas tomato lipoxygenase has a different substrate specificity and produces 9-hydroperoxides from linolenic acid, rather than the 13-hydroperoxides, with the 9-hydroperoxides breaking down to form C9 aldehydes such as *cis*-3-nonenal. By contrast, Fusarium lipoxygenase forms 9- and 13-hydroperoxides. *Morchella* sp. are used for 1-octen-3-ol production and hydroperoxide lyase obtained from plant sources such as green pepper. So, in general there would appear to be a historical trend from traditional fermented products to the use of isolated enzymes, submerged fermentation, and now bioconversion methods. Obviously the complexity of the biocatalytic system used has a direct effect on the ease of operation and costs of the process, and so a more technically sophisticated process can be justified only when the resulting product is sufficiently valuable, both in terms of profit margins created and markets that can be gained, to justify its use.

Examples of Actual Bioflavor Processes

Isolated enzymes are used to make a variety of flavor products, as described in the following sections.

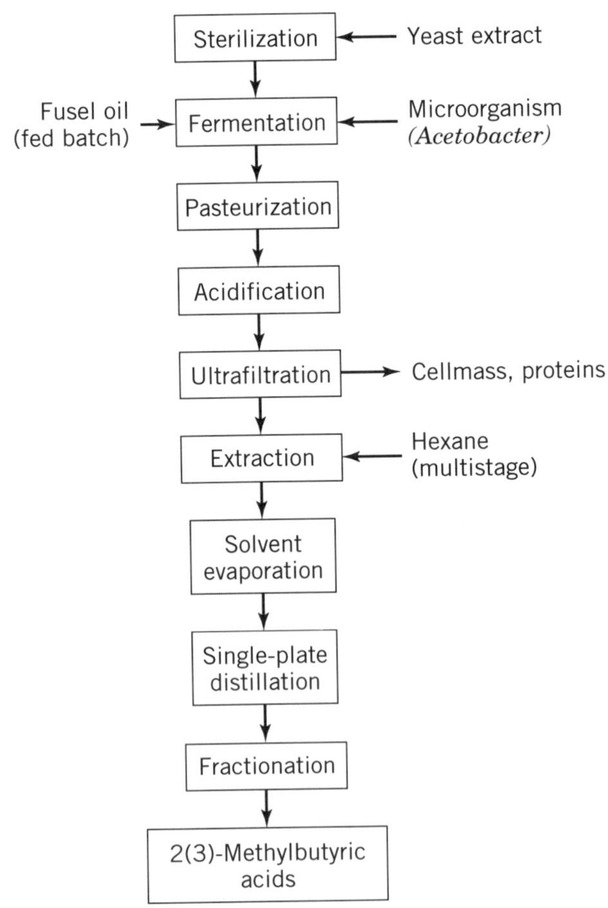

Figure 7. Large-scale process for the manufacture of 2- and 3-methylbutyric acids for flavor chemicals and flavor precursors, using fusel oil as the source of the 2- and 3-methylbutanol precursors. *Source*: Ref. 29.

Protein Hydrolysates. Different mixtures of microorganisms and plant and enzyme proteases are used to produce protein hydrolysates, depending on the source of the protein, the process used, and the product needed, especially because different end-use applications require different degrees of hydrolysis of the proteins. A particular challenge is to minimize the bitter taste of the hydrolysate, which is caused by peptides that have N-terminal hydrophobic amino acids. This can now be achieved using specially developed peptidases, especially aminopeptidases, which are usually obtained from lactic acid bacteria such as *Leuconostoc lactis* and from *Aspergillus oryzae*. These were discovered as enzymes from some dairy starter cultures that were found to have debittering activity (30) and are being cloned to make them available more cheaply and in greater quantities. Some carboxypeptidases are also used. Note that the process lends itself to the creating of a range of savory products, depending on whether the material is dried, concentrated, or supplied as a dilute liquor.

Monosodium Glutamate. Over 500,000 tonnes of monosodium glutamate the taste enhancer is made worldwide each year, mostly by fermentation using specially selected strains of *Corynebacterium glutamicum*. These strains have a low α-ketoglutarate dehydrogenase activity that prevents the formation of succinate, and so they instead form glutamate by reductive amidation of the α-ketoglutarate. In addition, successful and high-yielding fermentations depend on the use of media that have a very low biotin content (less than 5 $\mu g\ L^{-1}$). This makes the cell walls of the microorganism leaky so that the glutamate passes out of the cell and accumulates in the medium. After fermentation the cells are removed, and the glutamate is precipitated (by the addition of HCl) as α-form crystals. These are heated in water to convert them to β-form rod-shaped crystals that are easier to filter and wash. This is followed by neutralization, decolorization, and recrystallization (17).

Flavor Nucleotides. Hydrolysis of yeast RNA into flavor-potentiating nucleotides such as guanosine-5′-monophosphate is carried out by processes similar to the following (17). Yeast is grown such that it has a high RNA content. It is then harvested in late log phase with a hot alkaline solution (Fig. 8). Since inosine monophosphate (IMP) is a potent flavor enhancer, the activity of the hydrolysate is frequently enhanced by treatment with an adenylic deaminase, which converts 5′-AMP into 5′-IMP, as shown in Figure 9.

Use of Enzymes in the Extraction of Plant Materials. Plant materials are a traditional source of flavoring materials. Recently it has been found that many plant flavor chemicals, especially terpenols such as linalol, nerol, and geraniol, occur glycosidically bound to the plant tissue. As a result, specialized glycosidase enzyme preparations have been developed to augment traditional processing by enabling increased rates of extraction and higher yields of products. This approach has been increasingly employed in wine making; enzyme preparations such as those from *A. niger* are added to the must.

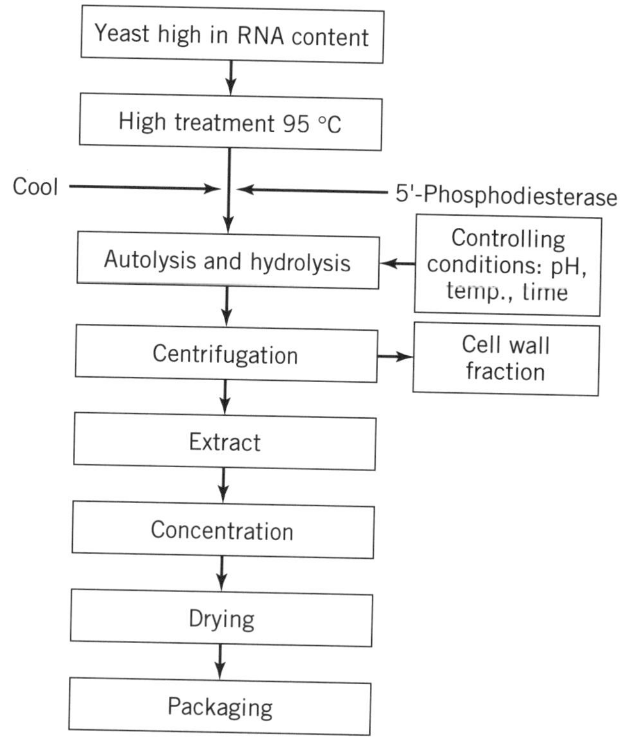

Figure 8. Flow diagram of the manufacturing process for AMP flavor-enhancer production.

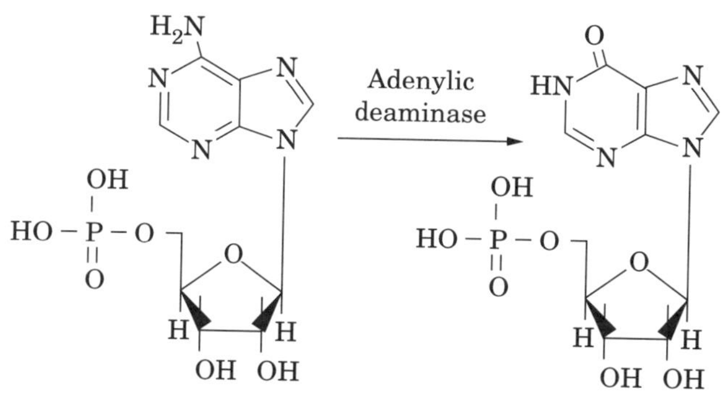

Figure 9. Conversion of 5′-AMP to 5′-IMP by the action of adenylic deaminase.

Curiously, endogenous plant glycosidases and the glycosidases of wine and beer yeasts are barely active against these glycosidically bound flavors, which explains the low yields obtained during traditional extraction and fermentation. This is because the glycosidases present in plants and yeast tend to be active at a moderate pH (around pH 5), are inhibited by glucose and ethanol, and have only a narrow aglycone substrate specificity. Such specialized enzymes have now been made available by enzyme supplier companies such as Gist-brocades.

One good example of extraction is the preparation of L-rhamnose, which is the key precursor molecule for the very important flavor Furaneol® (2,5,-dimethyl-4-hydroxy-(2*H*)-furan-3-one), which is prepared by heating rhamnose with an amine source such as proline. L-Rhamnose can be isolated by the enzyme hydrolysis of glycosides such as naringin, neohesperidin, and rutin present in citrus peel and other plant sources. These molecules are flavanone glycosides with rhamnose, in the terminal position. Enzymes are preferred that exclusively or, more usually, preferentially hydrolyse rhamnose, while minimizing the release of other sugars. This is to avoid an off-taste that develops during the reaction of glucose with the proline. Therefore any glucose released is removed by fermentation to ethanol and CO_2 by yeast, or into 5-ketogluconic acid, with its removal by precipitation as its calcium salt (31). Taking this approach a stage further, by coupling hydrolysis of a precursor glycoside with its bioconversion into a useful flavor chemical, raspberry ketone (4-(4′-hydroxyphenyl)-butan-2-one) has been synthesized (32) (Fig. 5).

Enzyme treatment can also be used to reduce an undesirable flavor of a plant juice, rather than to enhance a pleasant taste. For instance, the bitterness of citrus juices is due to the flavanones limonin, naringin, and neohesperidin, which are often present in fruit juice as glycosides. Bitterness is a particular problem for grapefruit, in which naringin predominates and has a taste threshold of just 20 ppm. Commercially available naringinase, which is a mixture of glucosidase and rhamnosidase, acts sequentially by converting the naringin to prunin and to then the much less bitter naringenin, has been proved in trials to provide an effective solution to this problem.

Benzaldehyde. Another source of an enzymatically treated plant-derived flavor is natural benzaldehyde (33). A number of companies have operated a bioprocess to make natural benzaldehyde, based on the combined use of β-glucosidase and mandelonitrile lyase, which are both conveniently present in almond meal (emulsin). This is used to treat the mandelonitrile (a nitrile glycoside) present in certain plant materials. A preferred industrial source is the cherry stones remaining after processing to remove the cherry flesh. However, the liberation of poisonous hydrogen cyanide by the lyase creates safety problems and has led to a search for alternative processes. L-Phenylalanine is a suitable precursor, especially since it is now available cheaply from a fermentation process developed to make phenylalanine as a component of the high-intensity sweetner Aspartame. The basidiomycete *Ischnoderma benzoinum*, combined with effective in situ product recovery, produces 1 g benzaldehyde/L, together with some 3-phenylpropanol.

Methylketone Production. Methylketone provides the characteristic flavor of blue cheeses such as Roquefort, Camembert, and Stilton (34) and consists of mixtures of C5 to C11 2-alkanones, especially 2-pentanone, 2-heptanone, 2-nonanone, and 2-undecanone. This is because only medium-chain-length fatty acids are converted into methyl ketones. This may be because only such medium-chain fatty acids, and not longer-chain fatty acids, can enter the cells as free acids and seem to cause significant morphological changes to the cells that take them up. Such flavor chemicals are commercially important in the manufacture of consumer products, such as salad dressings, soups, and pizza toppings, in which a cheese flavor can predominate, and also as a component of other flavors, especially savory flavors.

Methylketones are manufactured using the spores of microorganisms, such as *Penicillium roquefortii*, that still retain the required enzyme activities despite being dormant. This makes the production-scale use of spores very easy, as once grown they are easy to store until required and then require no growth media. They are eminently suitable for use in a logistically convenient and cost-effective biotransformation process. This process is much simpler than a conventional fermentation because of, for instance, very moderate oxygen transfer requirements. Using this method, lipolyzed fats or oils can be transformed into cheese flavors with 10 times the methylketone content of normal blue cheeses (Fig. 10).

The details of the biochemical pathway whereby microorganisms convert fatty acids into methylketones have been elucidated and used as the basis for large-scale fermentation processes, although the detailed physiologically control processes are still not properly understood. Commercially successful microbial processes for making methylketones have been used for some time by companies such as Unilever and Bush Boake Allen (35). The way microorganisms such as *P. roquefortii* produce methyl ketones from medium-chain fatty acids is of considerable scientific interest because it resembles conventional β-oxidation until the final step, at which a decarboxylase produces the methyl ketone, replacing the thioesterase found in β-oxidation. Methylketone formation is favored under con-

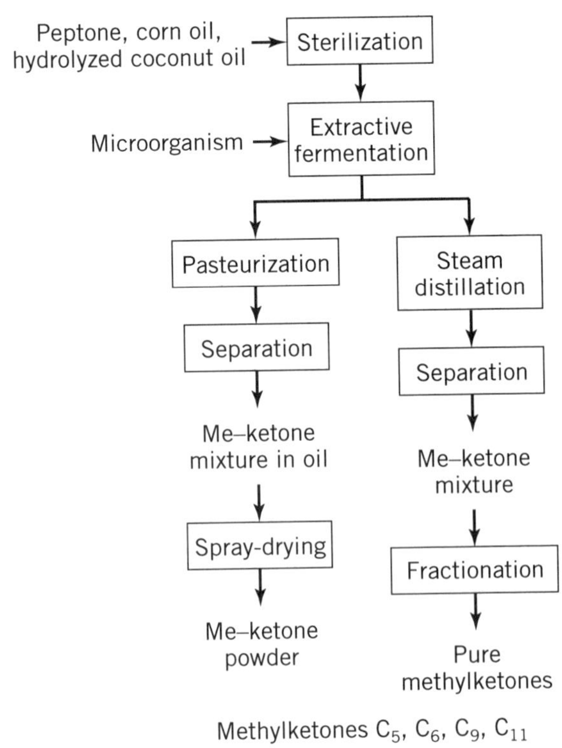

Figure 10. The unit operations involved in manufacturing processes for the methyl ketone blue cheese flavoring materials.

ditions in which cell growth is constrained. Spores are more active in producing methyl ketones than mycelium; use of immobilized cells have been described, and solid-state fermentations are often used. For instance, Bush Boake Allen's process involves mixing cellulose powder with coconut oil, which contains a high concentration of octanoic acid, and inoculating with *A. niger* spores, which converts the octanoic acid into 2-heptanone, which has strong blue cheese aroma and flavor (20). The yield is reported to be about 40%, and 2-undecanone to be 60% of the total products. By comparison, using spores of *P. roqueforti* in solid-state fermentation, Larroche and Gross (36) achieved product concentrations of 75 g 2-heptanone L^{-1}, with a yield of almost 80%. This appears to be the only manufacturing process for flavor chemicals that employs solid-state fermentation, despite the advantageous features of this technique. This may be because the "culture" of applied microbiology at the manufacturing level is dominated by the stirred-tank approach, developed with such success for antibiotic production, except for traditional processes such as cheese ripening and silage making.

γ- and δ-Decalactones. The γ- and δ-decalactones are commercially very important flavor chemicals for which successful industrial bioconversion processes have been established. This is despite the fact that de novo production by fermentation of glucose and other carbon sources is very low yielding. An outline of the process (29) for making γ-decalactone, useful for peach and other flavors, and a typical fermentation profile (37) are shown in Figures 11 and 12.

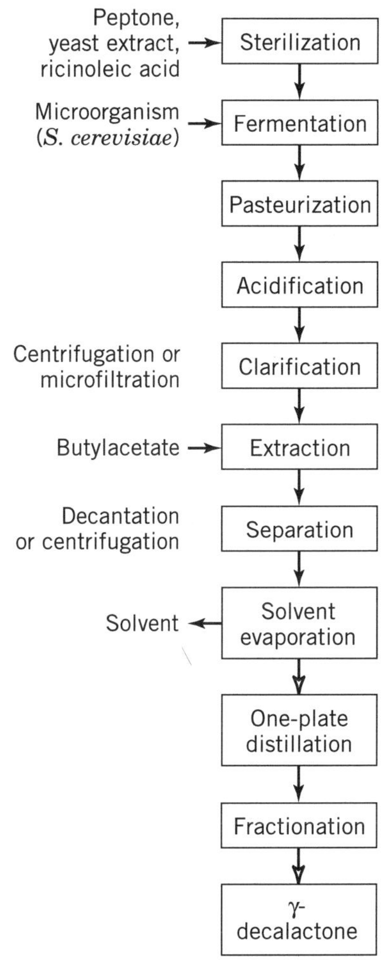

Figure 11. Outline of the process unit operations required for large-scale production of the flavor chemical γ-decalactone.

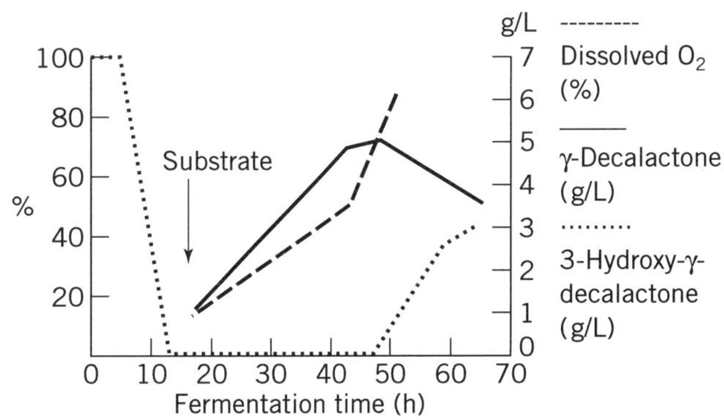

Figure 12. Fermentation profile for γ-decalactone formation from caster oil substrate.

The yields of lactones obtained by this process are greatly enhanced by acidification during product recovery. Only about 25% of the 4-hydroxydecanoic acid is lactonized during the fermentation at pH 6.0–7.0, but full yields are reported when extraction by steam distillation and solvent extraction are performed at pH 2 (38).

By-product formation can also be an important consideration affecting the yield of a product and the ease of its isolation. For example, 3-hydroxy-γ-decalactone continues to be formed even after production of the desired product, γ-decalactone, has ceased. It is associated with the fermentation becoming more aerobic (Fig. 11), although some 3,4-dihydroxydecanoic acid can be cyclized into 3-hydroxy-γ-decalactone, which has no positive flavor value, or it can be dehydrated into a mixture of 2-decen-4-olide and some 3-decen-4-olide, which together make up some 5% of the microbial metabolites. Upon reduction by yeast, these metabolites give δ-decalactone and the *R*-γ-decalactone, respectively. In addition, minor amounts of other lactones are also formed from hydroxylated fatty acid intermediates in the β-oxidation pathway; 3-dodecen-6-olide is formed from the C12 fatty acid intermediate.

Yields of the desired γ-decalactone product can be maximized by terminating the fermentation at the right point, or the 3,4-dihydroxy-γ-decalactone can be converted into 3,4-unsaturated-γ-decalactones by distillation and then converted into some additional γ-decalactone by a stereoselective *S. cerevisiae* reduction.

In all of the processes developed, yeasts are used as the biocatalyst. This is in part because yeast allows fatty acids to diffuse freely across their cell membranes, making them readily available for reaction. By contrast most bacteria find fatty acids to be toxic, particularly in the undissociated forms that prevail at low pH, and cellular uptake is deliberately limited and requires specific uptake proteins.

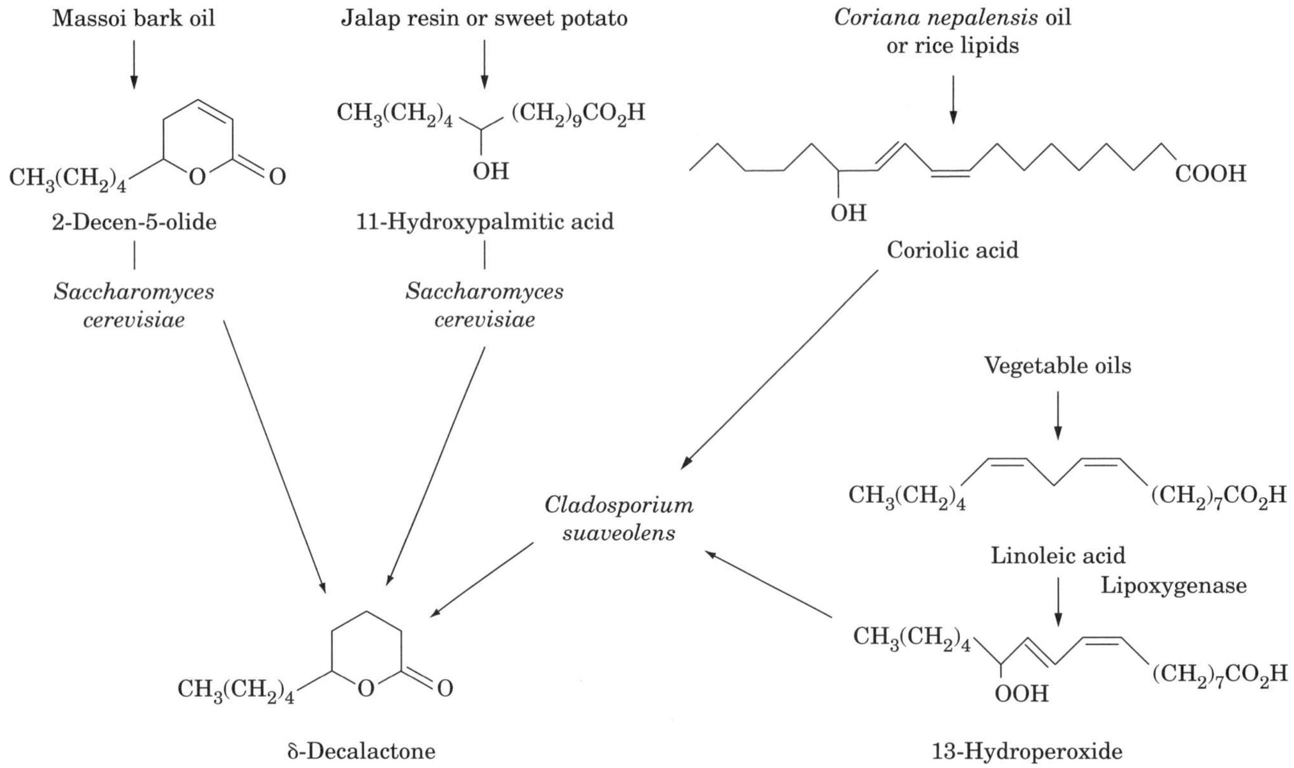

Figure 13. Various precursor molecules and biotransformation routes used for making δ-decalactone flavor.

It should also be remembered that the process from the microorganisms' viewpoint consists of diffusion of the precursor molecule across the cell membrane and into the cytosol, thence into the peroxisome, where reaction takes place, with subsequent movement of the lactone out of the peroxisome and then out of the cell. Hence, the concentration of fatty acid precursor supplied is important: too little and it is all consumed to meet microbial growth and maintenance requirements, too much and the excessive fatty acid concentration is toxic, which also reduces yields of the desired product. Considerable research has been devoted to finding the best raw material for γ-decalactone production. Castor oil is a cheap and readily available source of *R*-ricinoleic acid, but plant sources of oleic acid have also been converted into ricinoleic acid by microbial hydration. In addition, alkyl ricinoleate has been shown to have the advantage of reducing the amount of foaming that takes place in the fermenter. Even more ingenuity has been displayed in developing sources of the precursors and biosynthetic pathways to δ-decalactone, as demonstrated in Figure 13. Bioconversations have also been carried out using water-immiscible extracting solvents separated from the aqueous phase by a semipermeable membrane to allow selective passage of the δ-decalactone into extraction solvent (39).

These routes to δ-decalactone are also of interest because *S. cerevisiae* is used to carry out two different reactions: reduction of the 2-decen-5-olide and lactonization of the 11-hydroxypalmitic acid. Jalap resin also contains some 3,11-dihydroxymyristic acid, and this is converted by *S. cerevisiae* during the process into its corresponding lactone δ-decalactone.

Diacetyl. Shukunobe and Takato (40) described the production of diacetyl, useful as a component of dairy flavors, from citric acid using *Streptococcus cremoris* and *S. diacetylactis*, achieving yields of 14 g L^{-1} of product. A feature of this process was the requirement of oxidizing agents for the effective recovery of the diacetyl by distillation.

1-Octen-3-ol. Schindler and Seipenbusch (41) reported an industrial process for making 1-octen-3-ol, which is a character impact flavor chemical of mushrooms, by fermenting mycelia of morchella and then adding linoleic acid. Under high shear conditions, about 2.5 g product/kg dry w mycelia was formed, presumably by lipoxygenase action on the linoleic acid.

Methylbutyric Acids. 2-Methylbutyric acid and 3-methylbutyric acid are useful flavor components, especially when esterified. They are manufactured from their corresponding 2- and 3-methylbutanols, which are extracted from fusel oils by oxidation using *A. acetii* (vinegar bacteria) by the process shown in Figure 7 (29).

C6 "Green" Flavor Aldehydes and Alcohols As an example of a more recently developed bioprocess, Firmenich claims to have developed a flexible and viable industrial-scale biotechnological process for the production of natural

C6 aldehydes and alcohols. These include *trans*-2-hexenol, which has a leafy green and fruity taste, and *cis*-3-hexen-1-al, which has a grasslike green and fresh taste (42). The current world market for these materials is estimated to be high, at up to \$50M per year. Another stimulus to develop a bioprocess is that although demand for C6 "green" flavors is growing, the traditional sources such as mint oil continue to be in short supply compared with demand. However, there are no good microbial sources of the key hydroperoxide lyase enzymes required; plant sources have hitherto been too low in activity, and cloning of these plant enzyme genes to provide more active enzyme sources is difficult because of the reluctance of consumers to accept gene modification as natural.

The Firmenich process simulates the metabolic path whereby these flavor molecules are made in plants: from polyunsaturated fatty acids by the sequential action of lipoxygenase, hydroperoxide lyase, and oxidoreductase enzymes. Firmenich uses soy bean lipoxygenase to make the hydroperoxide fatty acid derivatives, with the reaction carried out in aqueous medium at pH 9.5, and supplying pure oxygen to ensure that the reaction is completed within 1 h. The next step, cleavage of the hydroperoxides, was carried out at pH 8.5 using a plant extract specially selected to have a high lyase activity. For instance, for producing *cis*-3- and *trans*-2-hexenals, green pepper and cucumber extracts had the highest activities.

If desired these flavor aldehydes could be converted to their corresponding alcohols, such as *trans*-2-hexenol, using the alcohol dehydrogenase activity of *S. cerevisiae*. Using this process, typical results include the production of *cis*-3-hexenol at 4.2 g L^{-1}, with a yield of 42% based on the fatty acid supplied, and *trans*-2-hexenal at 1.8 g L^{-1} at a yield of 24%. These values compare with typical concentrations of these flavor chemicals in fruits and vegetables of only about 2.1 mg kg^{-1} and 6.2 mg kg^{-1}, respectively; in this respect, the bioprocesses are almost 2,000- and 200-fold more effective, respectively (Table 4). However, this method is not so easily applicable to *cis*-3-hexenol because, although *trans*-2-hexenal and *cis*-3-hexenal are initially produced as a mixture, the *cis*-3-hexenal is then subject to rapid nonenzymic conversion into *trans*-2-hexenal, and then the former is easily reduced by the yeast first into *trans*-2-hexenol and then into *n*-hexanol. In addition, *trans*-2-hexenal can react with acetaldehyde to form 4-octen-2,3-diol (37). Therefore, in order to make *cis*-3-hexenol, the 13-hydroperoxide, plant lyase, and yeast must be incubated simultaneously so that *cis*-3-hexenal is immediately reduced into *cis*-3-hexenol without any isomerization.

A similar approach is used in the manufacture of 1-octen-3-ol and other related C8 alcohols and aldehydes that are very characteristic of mushroom flavors from *Agaricus bisporus* and related species. This fungus has a lipoxygenase with a special substrate specificity that converts linoleic acid into the 10-hydroperoxide, which is then cleaned to give the 1-octen-3-ol. This flavor is manufactured using unwanted mushroom stems as the source of the biocatalyst.

β-Ionone. β-Ionone production is a good example of an emerging bioprocess that also uses lipoxygenase, but requires the development of novel reaction conditions. Both α- and β-ionones are widely used in the perfume and flavor industries, and are probably produced in nature by oxidation of carotenes and fatty acids. In this process soya lipoxygenase is used to convert carotenes, from carrots or algae, in a very concentrated reaction together with fatty acids. The reaction is carried out in a kneading trough under pressure and with air renewal. Products include α- and β-ionones and aldehydes such as hexanal and 2,4-decadienal, produced presumably from the carotene and fatty acids, respectively (43).

Vanillin. An example of a new and emerging approach, but using a different technical approach is the Harmaan and Reimer process for vanillin, the key flavor constituent of vanilla. Screening allowed the isolation of a novel *Pseudomonas* strain capable of converting eugenol, which is readily extracted from clary sage, into a number of products, several of which could be isolated in good yields, the choice of product being dictated by the fermentation conditions employed (44) (Fig. 14). From a biochemical perspective, there is an interesting parallel between phenylpropenoid degradation, such as of ferulic acid and fatty acid β-oxidation. A recent advance has been achieved by Gasson et al. (45). They have characterized the pathway of ferulic acid degradation in a strain of a soil bacterium, *Pseudomonas fluorescens*, which they isolated by its ability to grow on ferulic acid as sole carbon source. A key feature of the pathway is the hydration and retroaldol cleavage of feruloyl-CoA to produce vanillin and acetyl-CoA, catalyzed

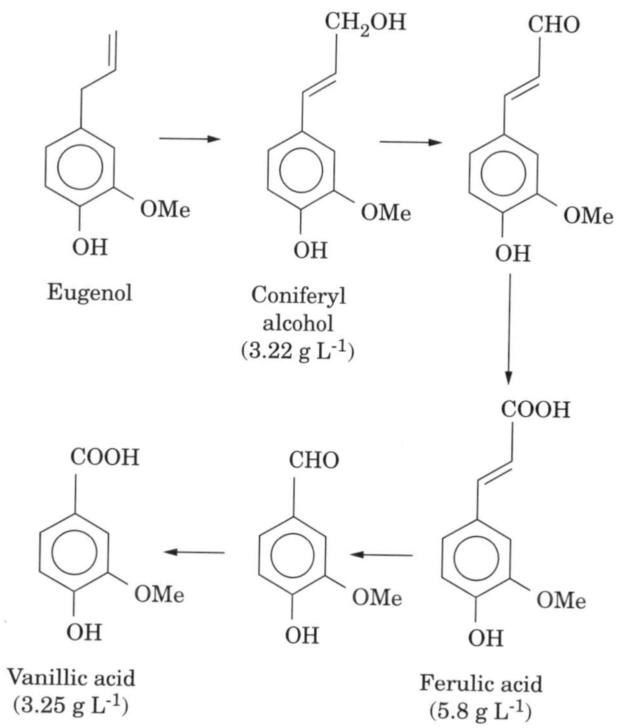

Figure 14. Eugenol metabolism using newly isolated *Pseudomonas* sp. and using different fermentation media to produce the various metabolites as isolated products. *Source*: Ref. 44.

by a newly discovered enzyme, (+)hydroxycinnamoyl-CoA hydratase/lyase. Comparison of the amino acid sequence of this enzyme with other known enzymes reveals that it is a member of the enoyl-CoA hydratase/isomerase superfamily (45).

Pyrazines. Pyrazines contribute substantially to the characteristic tastes of roasted and toasted foods and also contribute to many cheese, vegetable, nut, and fermented food flavors. *Bacillus subtilis* and *Brevibacterium linens* have been used to produce 2,5-dimethylpyrazine and tetramethylpyrazine at 4 g L^{-1} and 1 g L^{-1}, respectively, by fermentation of soy beans or flour enriched with the precursors L-threonine, acetoin, and ammonia and then supplied in a fed-batch mode, followed by extraction and distillation. The highest reported yield of a pyrazine as produced by a bioprocess is 12.5 g L^{-1} of 2-isobutyl-3-methoxypyrazine, produced by a mutant strain of *Pseudomonas perolens*. This pyrazine has a characteristic bell pepper flavor. The report provides a good example of strain-to-strain variations in flavor production by microorganisms, as the parent strain of *P. perolens* produced only 46 mg L^{-1} of this pyrazine when grown on the same medium (46).

Irones. Irones are degradation products of carotenoids that are valuable because of their very high flavor impact and low flavor thresholds. Traditionally they have been isolated from stored iris rhizomes, but now accelerated microbial processes have been developed, for instance, by Givaudan-Roure using microbial strains isolated from rhizomes such as *Serratia liquifaciens*. Using this method, 1 g mixed irones/kg rhizome could be obtained after 8 days, compared with 0.4 g/kg after 3 years for the traditional process (a 340-fold increase in productivity), and with the proportions of *trans*-α-, *cis*-α-, *cis*-γ-, and β-irones being very similar in both products (47).

Terpenes. Despite terpenes comprising a class of molecules that is very important in flavor and aroma products, very little real progress toward developing robust industrial processes appears to have been made. This is in part because of the water insolubility and cytotoxic characteristics of terpenes due to their effect on cell membranes. Toxicity seems to be related to their log *P* (partition coefficient) values, and terpenoids are often more toxic than hydrocarbon terpenes. However, a couple of promising attempts to develop productive process have been made. (*R*)-(+)-limonene used at 20 g/L has been converted on a 70-L scale into 900 g of pure 1*S*,2*S*4(*R*)-*p*-menth-8-ene-1,2-diol, with very few side products, using *Corynespora cassiicola* (Fig. 15). So as to avoid toxicity problems, the limonene was fed continuously; glucose, which is required to maintain enzyme activities, was also fed continuously in response to the CO_2 level in the exit gases (an indicator of cellular respiration rates). (Fig. 16). Using a second microorganism, *Penicillium digitatum*, the (*R*)-(+)-limonene component of racemic limonene was transformed by the same workers into α-terpineol in a 46% yield (48) (Fig. 15). Another promising terpene biotransformation is of the sesquiterpene patchoulol into 10-hydroxypatchoulol, a key intermediate to nor-patchoulenol, which is the major odiferous component of patchouli oil; 2–4 g L^{-1} of the 10-hydroxypatchoulol was obtained using strains of *Pithomyces* obtained by a selective screening exercise (49). Another promising terpene biotransformation is the use of immobilized 3-β-hydroxysteroid dehydrogenase from *Cel-*

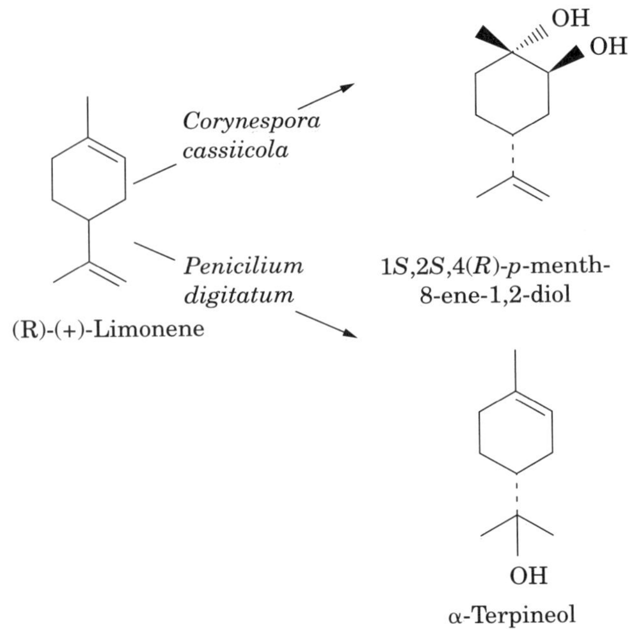

Figure 15. Microbial biotransformation of the abundant monoterpene (*R*)-+-limonene.

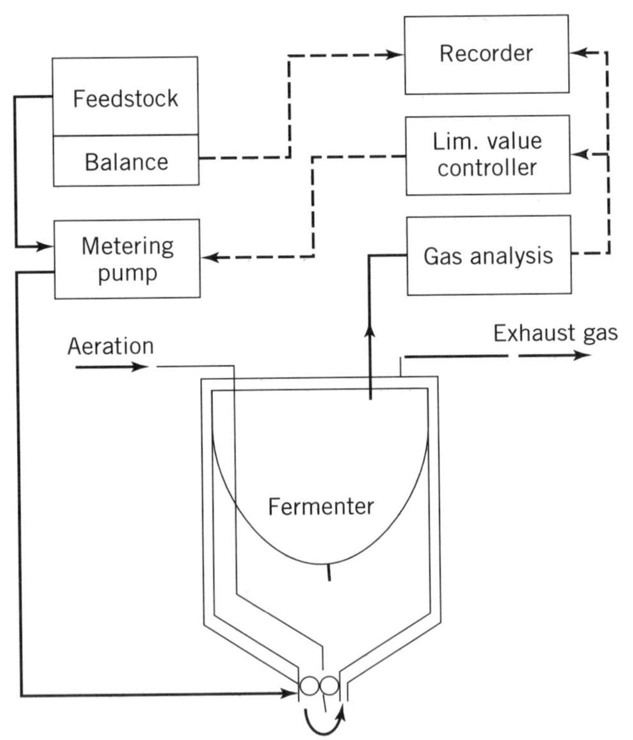

Figure 16. Pilot-scale reactor developed for the biotransformation of terpenes.

lulomonas turubata to convert L-menthone into L-menthol (50). In this reaction, NADH was recycled by oxidation of L-methylisobutyl carbinol by the same enzyme. Although this dehydrogenase was active only on L-menthone at 0.3% of its activity toward its natural steroid substrate, it was stable, with a half-life for the immobilized enzyme of 25 days, and produced 25 g L^{-1} of L-menthol product. Reactor productivity was improved by continuous in situ product removal by adding solvent to extract the L-menthol, with selective passage of solvent out of the reactor through a hydrophobic membrane. This process is potentially very useful for industrial use because immature peppermint plants produce high yields of an essential oil that consists mostly of L-menthone, only 40% of which is converted into the L-menthol and is normally harvested, which opens up the possibility of a higher-yielding process involving harvesting the immature peppermint oil and then producing L-menthol by the sort of dehydrogenase reaction described earlier.

Accelerated Soy Sauce Production. Traditional soy sauce takes 6 months to make, which adds to the costs of the process. For a description of the traditional process, see Springham (51). Osaki et al. (52) succeeded in making soy sauce of acceptable flavor with a system of sequential immobilized cell columns. The system shown in Figure 17 has been operated on a 280-L scale. Best results were obtained using *P. halophilus, Zygosaccharomyces rouxii*, and *Candida versatilis* cells immobilized on ceramic supports and run in series, which required just a 6-day reaction period.

Dairy Flavors. The variety of different yogurt, cheese, and other flavors is well known, each being due to a different mixture of flavor chemicals, most of which are produced by microbial action (53). These include proprionic and butyric acids, peptides and amino acids, δ- and γ-lactones, methylketones, pyrazines, diacetyl and acetoin, acetaldehyde, phenylethanol, methanediol, dimethylsulfide and even phenol. Another important factor is control of the degree of bitterness caused by peptides containing hydrophobic amino acids by use of the most appropriate mixture of proteases and peptidases.

The original objective of bioprocesses was to reduce costs by accelerating the natural fermentation and thereby reducing storage times and costs. The next objective has now become a quest for improved flavor quality, and even the generation of novel flavors. Since the actual flavor development process remains the same, the goal is to develop superior starter cultures. Microbial improvement of cheese making is a challenging task for several interrelated reasons. First, mixed cultures are common, so there is often sequential growth of different strains, with growth of the second strain taking place after the pH has been changed and using metabolites produced by the first strain. Then there is the solid nature of the cheese as the fermentation medium, with both surface growth of strains and growth within the cheese taking place. Consequently both microbial growth and flavor production can be very slow compared with submerged fermentation. The type of microbial growth that takes place also seems to influence the flavor of the cheese produced. For surface-ripened cheeses, such as Camembert and Brie, flavor development is dominated by proteolysis and the resulting peptides formed, whereas for cheeses ripened by internal mold growth, such as Roquefort and Danish Blue, the flavor is dominated by methyl ketones and fatty acids, respectively. A well-established innovation is the so-called enzyme modified cheese flavors available from a wide range of suppliers. These are made by treating cheeses with selected lipases and proteases and result in well-balanced cheese flavors with intensities up to 20 times those of the conventional mature cheeses.

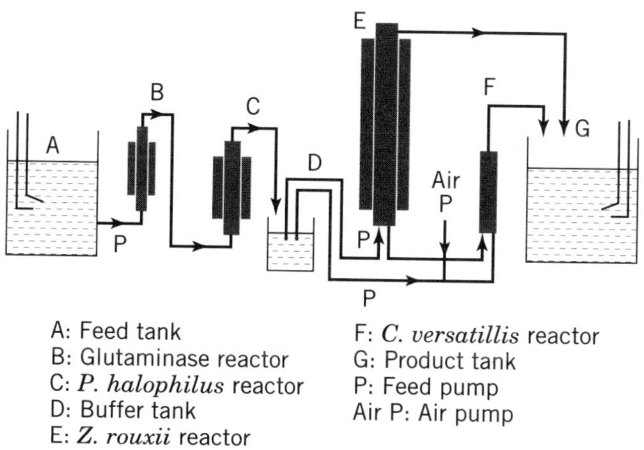

Figure 17. Pilot-scale process for the accelerated production of soy sauce using three different immobilized cell reactors in series. Preceded by immobilized glutaminase reactor.

Savory Flavor Processes. *Gluconobacter suboxydans* is used to manufacture 5-ketogluconic acid from glucose as the key precursor for the important meat flavor 4-hydroxy-5-methyl-2,3-(2*H*)-furanone. The precursor is easily recovered from the bioconversion as its calcium salt and is then converted into the furanone by heating. A related sulfur-containing furanone is formed by reacting in the presence of cysteine (54).

The process described in Figure 18 has been developed by Quest International. It can be seen as a more advanced form of yeast flavor built up from traditional yeast extracts (Fig. 7). By varying the types of enzyme and microorganisms used, the flavor profile of the product can be modified to best match the needs of particular applications (28). Additional flavor character and range can also be created by a subsequent reaction flavor step.

THE APPLICATION OF GENETIC ENGINEERING TECHNIQUES TO THE PRODUCTION OF BIOFLAVORS

The public acceptance of genetically engineered products in the food industry is not very well developed. Also there is considerable opposition to being able to assign the label "natural" to flavors produced by genetic engineering. Hence, there are still not very many good examples of successful processes using this technology. However, there are

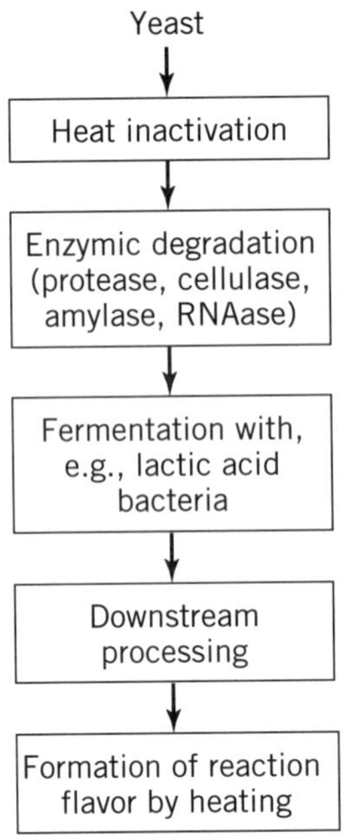

Figure 18. The major operations involved in the manufacture of savory flavors by combination of enzyme treatment of yeast and subsequent reaction flavor formation.

many instances of well-researched processes that are waiting in the wings to be used once consumer perceptions have changed. What is quite well established is the use of enzymes produced on an industrial scale using engineered producer strains in which the gene for the required enzyme has been cloned into a host strain with good fermentation characteristics. These include lipases, and chymosin for milk clotting. A comprehensive review by Muheim et al. (55) also describes the use of genetic engineering technology to make debittering proteases, to produce a flavorsome beefy meaty octapeptide, and to make *cis*-3-hexenol and methional.

One very effective use of genetic engineering is to remove the off-flavor caused in beers by the accumulation of diacetyl and thereby reduce the time required for the slow metabolism of diacetyl by the maturing yeast, which converts it into acetoin. The diacetyl is formed from α-acetolactate, and it can either be prevented from forming, or be rapidly removed before it can be converted into diacetyl. In order to prevent its formation, acetolactate decarboxylase (ALDC) can be cloned into the yeast. Yeasts transformed with the *Acetobacter aceti* ssp. xylinum ALDC gene have been used for beer production at the pilot scale and produce much less diacetyl during fermentation than their untransformed parents, resulting in a significant saving in maturation time and therefore production costs (56).

Diacetyl formation can be reduced by modifying genes in the amino acid biosynthetic pathway that lead to α-acetolactate, especially the IL V2 gene that codes for acetolactate synthetase, followed by mating of these strains to generate hybrids that produce both little diacetyl and good quality beer (57).

Another success is the development of *B. subtilis* strains that produce the flavor enhancer IMP in enhanced yields because of an absence of IMP dehydrogenase that is responsible for its metabolism into guanosine synthesis. Alternatively guanosine production can be enhanced by increasing the number of gene copies of this enzyme (58).

DOWNSTREAM PROCESSING

Biochemical engineering has the key role of transforming promising laboratory discoveries into cost-effective and reproducible large-scale manufacturing processes in a timely fashion and with minimal expenditure. This usually involves real-time analysis of pH and dissolved oxygen and carbon dioxide levels and sophisticated data logging and analysis as well as a careful consideration of the chemical and physical properties of the materials involved, for instance, the separation characteristics of biological materials are very often pH dependent. Also of critical importance is maintaining good sterile or hygenic conditions (e.g., by using cleaning-in-place methods), product specifications and cost targets, by-products and their disposal costs, the intended scale of production and any economies of scale possible, effects on the sizing of process equipment, operating regime (e.g., batch vs. continuous reactors, distillation vs. solvent extraction of product), and safety and good management practice constraints including process validation requirements. Of particular consideration for flavor and aroma products is that they often consist of a mixture of different chemicals, many of which are necessary to give the product its real value. As well as downstream processing, upstream processing is also important. This is because so many flavor and aroma products are manufactured from natural raw materials extracted from plants, so upstream processing operations for such intrinsically variable raw materials also pose special challenges, especially for successful process integration.

The main problems that are usually associated with bioprocessing are the low concentrations of product(s) that are formed, and consequently the need to remove large amounts of water, and the instability of these biochemical products with respect to heat and other extreme conditions that may occur during their purification, concentration, and recovery. Consequently the costs of downstream processing can be greater than those of the fermentation or bioconversion unit operation and can be as great as 50% of the overall processing costs.

Because the value of a flavor or aroma chemical is so dependent on its taste and smell qualities, the effects on these of trace impurities carried over from fermentation media and other contaminants can be very significant. For example, in a process to produce δ-decalactone flavor, considerable efforts were made to purify the 11-hydroxypalmitic acid precursor to an acceptable purity so as not to compromise the flavor quality of the product (29). This involved extraction of the botanical source (sweet potato residue or jalap roots) followed by hydrolysis and purification.

Obviously it is the performance of the overall process that determines the success of the process and the product it produces. Therefore the integration of all the individual unit operations, whether bioproduction steps or product isolation operations, is vitally important. Also important is the reproducibility of the process; an important factor in many processes for natural flavors is the variability in the quality and prices of raw materials.

As compared with chemical processing, precursor and product inhibition effects often dominate bioreactions, necessitating specialized techniques including fed-batch reactions, continuous supply of precursors, use of slow-release forms of products, two-phase reactions (in which the second phase or resin acts as a product reservoir for product sequestered out of the aqueous phase), loop reactors (in which product is recovered by an in-line resin column or membrane extraction unit), and the recovery of product by its adsorption from the exit gases of a reactor. The potential for such sophisticated bioengineering is illustrated by the good productivities already achieved for some nonflavor products such as L-tertiary leucine and *N*-acetylneuraminic acid, which have productivities of 638 and 470 g product $(\text{L day})^{-1}$, respectively, using membrane reactors in which the product is continuously removed and the cofactor is regenerated in situ, allowing the enzymes to work at as high a rate as possible (59).

Other important considerations are the requirements of good manufacturing practice, including process validation and safety audit procedures and for meeting "natural flavor" and kosher criteria. For instance, a European Community directive on natural flavors exists, and the list of extraction solvents that may be used during the processing of raw materials for food stuffs, of food components, or of food ingredients is given in Table 9 (60).

On occasions the maximum permitted amounts of residue solvents are defined, and multiple extractions are required to obtain good yields. It may also be necessary to use more sophisticated extraction equipment, such as rotating disc extractors. In general the solvents used should be nonvolatile (for safety and to prevent contamination of the aroma of the product), not prone to emulsion formation, and immiscible with water to facilitate solvent recovery and reuse and minimize environmental problems, and cells should be dried prior to solvent extraction so as to prevent emulsion formation with residual water.

In any consideration of the use of solvents, the importance of good quality water is vital. An early illustration of this in the history of biotechnology is that the best beers were brewed using the special, high-quality water obtained at Burton-on-Trent, U.K. The subsequent introduction of ion exchange for the purification of water for brewing allowed the production of superior quality beers to become widespread.

A good account of the biochemical engineering aspects of process development is given by Chisti and Moo-Young (61). These include the following unit operations.

Chromatography

Chromatography is very often expensive and difficult to fit into processes, and so it is used only for high-value products. Reverse-phase chromatography is very useful because the desired material can be absorbed straight from the aqueous phase in which it has been produced. Chromatography is usually used after clarification to remove any impurities that would foul the column, such as proteins released by any cell lysis, and then product is recovered by washing with ethanol, in which form the flavor can often be used directly. An example is the malt flavor produced by fermentation using *Streptococcus lactis* var. maltigenes, which (following ultrafiltration to remove cell mass and protein remaining from the fermentation media) can either be subject to reverse osmosis and then spray-dried with a carrier to make a powder form of the flavor, or steam distilled and the reverse-phase chromatography carried out with ethanol as the eluent to make a liquid flavor (29).

Cell Disruption

Cell disruption is necessary for the liberation of intracellular products. This imposes two constraints on the process. First, the amount of product that can be made is limited to that which can be accumulated within the cell, whether the intracellular product is an enzyme or a small molecule. Second, cell disruption liberates most of the contents of the cell together with the desired product, and so considerable purification is often required. It is also interesting to note that cell disruption is a unit operation that has had to be developed de novo in response to the need from bioprocesses. Therefore, unlike many other operations such as filtration, centrifugation, and crystallization, cell disruption has no equivalent in conventional chemical engineering.

Centrifugation

Centrifugation is required variously to recover cells, for the clarification of liquids prior to extraction, chromatography etc., for breaking emulsions (such as are often formed by cell disruption), or for washing cells prior to another operation such as disruption or immobilization.

OTHER CONSIDERATIONS

Many other biological factors are also important in the production and processing of flavor chemicals. Trace contaminant strains can introduce traces of undesirable flavor component that make the product less valuable or even unusable. The BOD of fermentor wastes needs to be minimized. Exit gases from fermentation need to be sterilized often by microfiltration. Noxious smells should be removed prior to emission into the local environment, although flavor factories are often sited with respect to prevailing winds so that emissions are less likely to reach major sites of population (62). In biotransformation processes the precursors will ideally be appreciably water soluble and with minimal cytotoxicity. As in any bioprocess, the media will be formulated so as to achieve a number of aims, including minimization of the lag phase, the inoculum size required, and the duration of the fermentation. Other objectives are to minimize power input costs and to maximize mixing so as to maximize the utilization of expensive filtered air.

Perhaps the newest innovation in flavor recovery technology are the use of in situ solvent extraction (Table 10), and of perevaporation, which has been demonstrated to be effective for the in situ recovery of benzaldehyde from fer-

Table 10. List of Solvents Accepted for Use in the Manufacturing of Natural Flavors

Name
Propane
Butane
Butyl acetate
Ethyl acetate
Ethanol
Carbon dioxide
Acetone
Nitrous oxide

mentation broth using *Bjerkandera adusta* (63). Another continuous in situ product recovery method is the continuous addition of an extracting solvent to the reaction vessel, with its selective removal together with the product using a hydrophobic membrane (50). In addition pervaporation has been used to recover 6-pentyl-α-pyrone from *Trichoderma viride* cultures, obtaining an 85% yield of product and a 20-fold enrichment after 2 weeks of fermentation (64).

COMPARISON OF DIFFERENT PROCESSES

In comparing various processes, the productivity of the biocatalyst is very important. As a general guide, Table 10 provides an indication of the typical rates of enzyme-catalyzed bioorganic reactions of various substrate classes in bacteria.

Production of flavors by plant cells is a very desirable method that still requires major advances to make it both practicable and cost-effective. These advances may be developed for pharmaceutical products and then transferred to flavor processes. For instance, Phyton has made big advances in developing plant cell fermentations to make paclitaxel, the active ingredient in Bristol-Myers Squibb's taxol anticancer drug. Therefore, it is interesting to note that, by comparison with the microbial systems already described, it is reported that the rates of capsaicin biosynthesis in ripening pepper plants is ca. 0.5 mg g dry wt^{-1} day^{-1}, and in pepper plant cell culture it is ca. 0.1 mg g dry wt^{-1} day^{-1}, which is equivalent to a turnover rate of 0.00006 mmol g^{-1} h^{-1}. To make progress it is usually necessary to elucidate the biochemical pathway the plant uses to make the flavor. For instance, the biosynthesis of Furaneol by strawberries (2,5-dimethyl-4-hydroxy-(2*H*)-furan-3-one) has been recently described (65), as has the biosynthesis of raspberry ketone by coupling coumaroyl-CoA and malonyl-CoA, followed by reduction, by raspberry fruit (66).

If sufficient research and development resources are available, biocatalysts with very high productivities can be developed, and these have created vast industries. For example, consider the commodity products made by biocatalysts, as shown in Table 11. Because flavor and aroma chemicals are much higher-value products, the bioprocesses can afford to be less productive but still need to achieve good productivities because the volume of demand is not as great as for other products. For instance ca. 10^7 tons/year of high fructose corn syrup are made, but processes for flavors and aroma chemicals do not benefit from

Table 11. Examples of Productivities That Have Been Achieved in Bioprocesses for Flavor and Aroma Chemicals

Products	Concentration of product in reactor ($g\ L^{-1}$)	Scale of production ($t\ year^{-1}$)
Diol precursor of Ambrox	150	
Citric acid	120–150	500,000
Acetic acid	130	
C13 dicarboxylic acid precursor of musk	140	
ω-Hydroxy hexadecanoic acid (hexadecanolide precursor)	>100	
Glutamic acid, especially as monosodium glutamate	80	300,000
Methylketone (2-heptanone)	75	
Coriolic acid (precursor of δ lactones)	25	
Esters	30	
Acetaldehyde	28	
Butyric acid	20–30	
L-Phenylalanine[a]	25	
Fructose in high-fructose corn syrup[a]	17	7,000,000
Diacetyl	14	
Pyrazines	12.5	
Penicillin[a]	10–15	5,000
β-Cyclohomogeraniol	9	
Rosmarinic acid[a]	6.7	
Proprionic acid	6.3	
Gamma decalactone	5	
Trans-2-hexenol	4.2	
Tetramethylpyrazine	3	
Lactic acid	2.8	
Ionones	1	
6-Pentyl-2-pyrone	0.03	

[a]Nonflavor/aroma benchmark products.

such big economies of scale. The concentration at which a product can be produced in an enzyme or microbial reaction is a good guide to how cost-effectively it can be manufactured. Table 11 is illustrated by examples of high productivities that biocatalysts must reach to achieve a successful and widespread industrial use (Table 12). High reaction productivity must be complemented by effective downstream processing and by good process intensity in each unit operation of the manufacturing process. This is because product concentration reflects the size of equipment, especially of reactors, and the hydraulic throughput rates that are required to make a given amount of product, and thus influences both capital costs and costs of mixing, drying, pumping, and so forth. Also the product concentration indicates the volumes of fluid that will have to be processed to obtain the required amount of product and the cost of concentrating and isolating the product. The relationship between production concentration and sale price (Table 11) is well established for a wide range of bioproducts ranging from very high value biopharmaceuticals to low-cost food ingredients such as citric acid. Thus, if a flavor material can be produced only at quite low concentrations, it is very likely that the only market opportunity for it will be in prestige and/or high-value niche markets, if they exist. A major constraint on the product concentrations that can be achieved is product inhibition. Therefore, a determination of inhibition constants is as important as determining the rate of reaction parameters, and work on developing effective in situ product recovery methods that will remove product as it is made are very important. For flavor molecules, such in situ product recovery techniques can include the recovery of volatile flavors from the exit gases of bioreactors. Product inhibition is, of course, exacerbated if it is difficult for the product to diffuse out of the cell in which it has been produced.

FLAVOR APPLICATIONS CONSTRAINTS

Once a flavor has been produced it must always be remembered that its value will vary with its applications. A graphic example is provided by vanilla flavor when used in low-fat ice creams. Reformulation of the flavor is required because the less oil soluble flavor chemicals are less well perceived in the low-fat products and so need to be present in relatively higher concentration than does the more oil soluble flavor component in traditional high-fat ice creams (67).

Table 12. Comparison of the Productivities of Three of the Most Economically Important Biotransformation Manufacturing Processes for Chemical, Pharmaceutical, and Food Applications

Nitrile hydratase (Nitto Chemicals)	50 g Acrylamide/L reactor/h
Penicillin G amidase (SmithKline Beecham and others)	1–2 tonnes 6-APA/kg of immobilized enzyme
Glucose isomerase (Novo Nordisk and others)	20 tonnes HFCS/kg of immobilized enzyme

Big advantages can be obtained if positive applications results can be obtained, for example, in creating reductions in fat and salt contents of products, improved flavor stability and controlled release, emulsion stabilisation, as well in as the growing area of health benefits (nutraceuticals).

Changes can also occur during food processing and cooking, especially when extreme conditions are used. For instance, so-called reaction flavors can be formed from flavor components during heating, particularly via the Mailard reaction that occurs between reducing sugars and the amino groups of amino acids. Condensate flavors can also arise upon cooling, especially when local high concentrations of reactants occur. An example is the reaction of acetaldehyde and hydrogen sulfide to form 3,5-dimethyl-1,2,4-trithiolene.

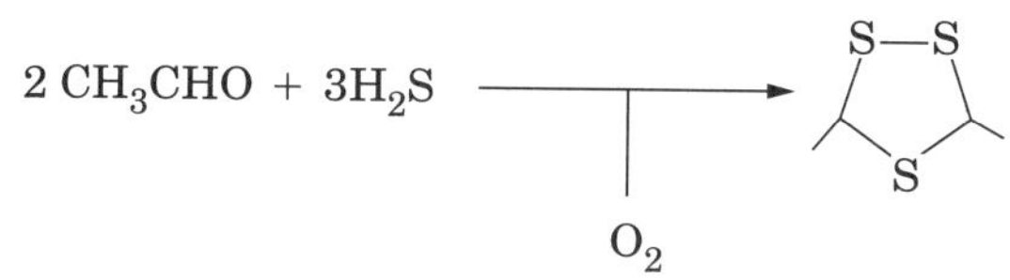

Product Approval and Registration

The necessary tests and procedures for product approval and registration can be both lengthy and expensive, and different countries often have different requirements. The most difficult task is to gain approval for a new chemical; approval for new processes to make existing materials with a proven record of safety is easier to gain. The ideal is "generally regarded as safe" approval, which indicates that long-term and significant consumption in the diet results in no evidence of any adverse effects.

Consumer Approval

However novel the underlying science and however efficient the performance of the manufacturing process, the product will be successful only if it achieves both regulatory approval and consumer acceptability, and indeed, generates a marked consumer preference because of the perceived value of the products to individual consumers and interest groups. This is a difficult area not least because of the marked difference between lay and expert perceptions of risk, with the consumer tending to put more weight on the risk of new uncontrolled factors and the effects directly on themselves and their families, rather than relying on the statistical risk factors based on broad demographics. In particular individual consumers fear loss of control, particularly to organizations that they find difficult to trust because of their vested interests that may act counter to their own interests, such as companies' profit motive. So in this respect there is a very big difference between the use of bioprocesses for flavor and for pharmaceuticals. Whereas, both types of product have very high biological activities and are discovered and then developed along similar lines, the consumer needs and decisions for their uses are quite different. There is a strong "cure" imperative for drugs, which are usually used only an occasionally, whereas the use of food flavors is much more an everyday pleasure-based activity, leading to an ethical concern expressed as a preference for natural-flavored products. For instance, Abraham et al. (66) reports that about 70% of the food flavors used in Germany in 1990 were natural.

FUTURE OPPORTUNITIES

One of the recurrent themes in flavor and fragrance biotechnology is the need to oxygenate molecules, such as terpenes, in order to convert them into molecules with higher-intensity flavors. This can be extended to the introduction of sulfur, as not only is sulfur close to oxygen in the periodic table, but many important flavor molecules contain sulfur. For instance, an important component of blackcurrant flavour has been biosynthesized as follows (68):

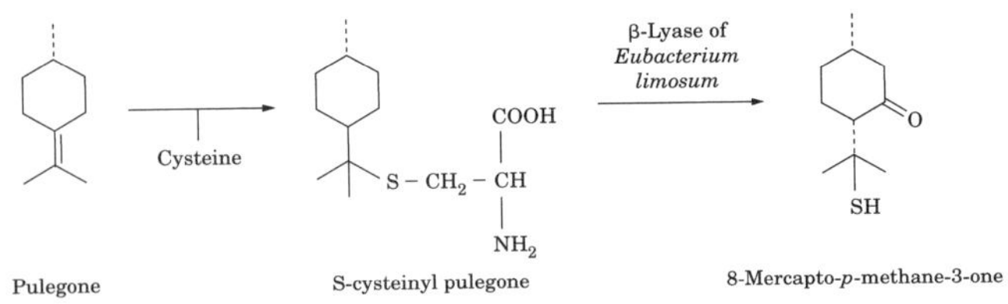

Similarly there is the biosynthesis of furfurythiol using the β-lyase activity of an *Enterobacter* strain (69):

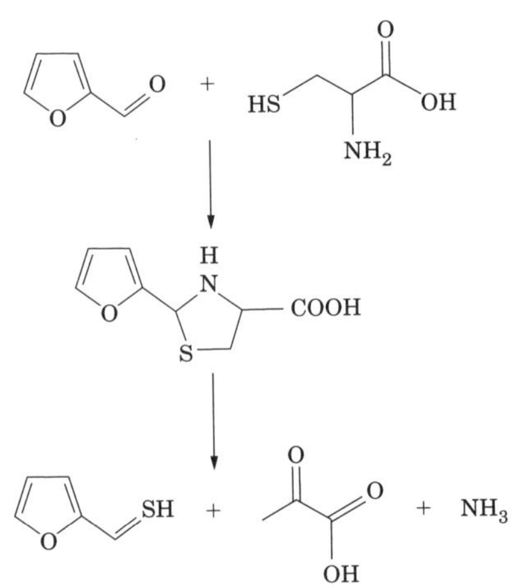

Healthy, indeed so-called nutraceutical food ingredients, are gaining increasing consumer recognition and commercial momentum. Natural flavors with some health benefits may be possible. Allicin is an important precursor of flavor compounds in garlic and onion; allicin has some antibacterial properties but also forms molecules with marked antithrombolytic properties.

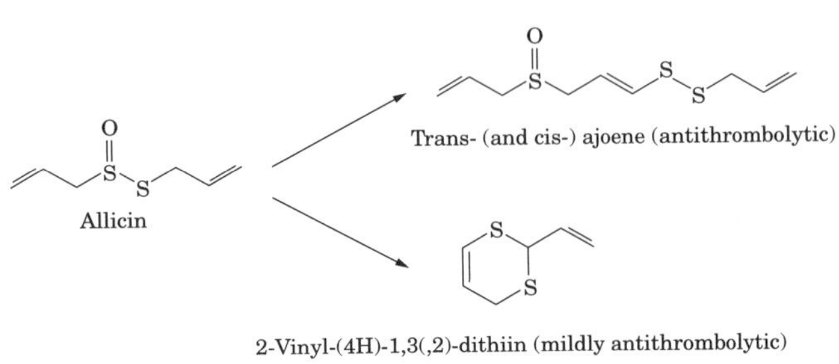

CONCLUSIONS

Despite biopharmaceuticals being widely regarded as the cutting edge of modern biotechnology, a surprisingly high proportion of current flavors and food ingredients are manufactured using bioprocesses. These include not only refined and developed versions of traditional processes, such as for cheese, yogurt, and protein hydrolysates, but also many new and sophisticated bioprocesses for key chemically defined materials such as the flavor chemicals methyl ketones, and γ-decalactone, and flavor enhancers such as GMP and IMP. This article has attempted to explain the technical basis for these products as used in industry and indicate exciting future developments in this field that are based on the pull of market demand forces but still constrained by cost, patents, labeling, and regulatory considerations. However, a comprehensive account is not yet possible because full details are not available concerning the bioprocesses now known to be operating. In many cases it is even uncertain at what scale processes are operating.

In conclusion, some important commercial factors that must be considered including the following:

- Significant market demand (especially for label-friendly natural ingredients and for products with superior performance such as fat reduction, health benefits, or improved convenience)
- Substantial profit margins
- Economies of scale of production
- Availability of low-cost raw materials
- Patentable (proprietary) technologies
- Weak competition
- A range of products with different market uses from a single process
- Ease of regulatory approval

As regards cost, Welsh (Ref. 3) has suggested a relationship between selling price and market size, which has been reinterpreted in Figure 19. This shows that although the highest-priced products, up to $1,250 per kg, appear superficially most attractive, the largest markets are for the

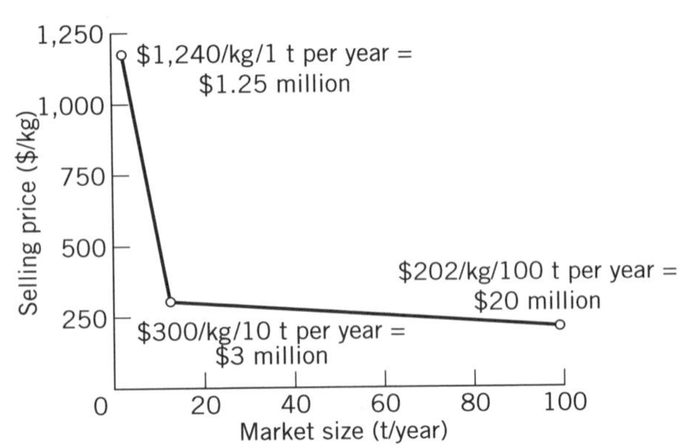

Figure 19. A comparison between the selling price and market size for flavor chemicals, with the market values of the products rising from $1.25 million, to $3 million, to $20 million as the tons sold per year increases and the sale price decreases.

lower-priced products, with an approximate market size of $20 million, as compared with only $3 million and $1.25 million for the higher-priced products. These values of natural flavors do not compare unfavourably with the price of gold, currently about $12,000 per kilogram. Much exciting research is taking place, and the pace of change in the food and beverage industries is intense, which is creating many new opportunities for bioflavors, especially considering the sustained demand from consumers for products containing natural ingredients. However, these opportunities must be balanced against constraints, including the availability of natural raw materials in sufficient quantity, quality and low price, the difficulty in developing bioprocesses that give high conversions of concentrated substrate streams into products with no by-products, and the difficulties in easily concentrating and purifying the required bioproducts. Perhaps in this last respect more efforts should be made to obtain the benefits of biomolecular selectivity for product isolation and recovery.

The objective of bioprocesses in the flavor and fragrance industry is to produce the highest quality product at the minimum processing costs. I hope this chapter illustrates the complex interrelationships between what is needed (consumer demands), how to make it (the development of bioprocesses), and also the important conditioning factors, such as what it is possible to patent, safe operation, and the need to meet other demands, such as for naturalness. Perhaps this chapter can in some small way help others in this exciting area of technology to succeed in developing more new and improved bioprocesses for flavors.

BIBLIOGRAPHY

1. T. Godfrey and S. West eds., *Industrial Enzymology*, 2nd ed., Macmillan, New York, 1996.
2. T. Scheper managing ed. and R.G. Berger volume ed., *Advances in Biochemical Engineering*, vol. 55, Springer, Berlin, 1997.
3. A. Gabelman ed., *Bioprocess Production of Flavour, Fragrance, and Color Ingredients*, Wiley Interscience, New York, 1994.
4. H. Hartmann, Perfumer and Flavourist **20**, 35–42 (1995).
5. H. Hartmann, Perfumer and Flavourist **22**, 21–23 (1996).
6. L.P. Somogyi, *Chem. Ind.* 170–173 (March 1996).
7. A. Hausler and T. Munch, *ASM News* **63**, 551–559 (1997).
8. J. Hariel, data of 1988–1990 quoted in R.G. Berger, *Aroma Biotechnology*, Springer, Berlin, 1995, pp. 170.
9. C.H. Manley, in A. Gabelman ed., *Bioprocess Production of Flavour, Fragrance, and Color Ingredients*, Wiley Interscience, New York, 1994, pp. 19–40.
10. P.P.C. Graziadei, *Cytotechnology* **11**, 3 (1993).
11. L. Bussolati, R. Ramoni, S. Grolli, G. Donotrio, and E.J. Bignetti, *Biotechnology* **30**, 225 (1993).
12. H. Zhao and S. Firestein, *Science* **279**, 237 (1998).
13. R.G. Berger, *Aroma Biotechnology*, Springer-Verlag, Berlin, 1995.
14. S. West and D. Pawlett, in T. Godfrey and S. West eds., *Industrial Enzymology*, 2nd ed., Macmillan, New York, 1996, pp. 209–225.
15. P.E. Milson, in R.D. King and P.S.D. Cheetham, eds., *Food Biotechnology*, vol. 1, Elsevier Applied Science, London, 1987, pp. 273–308.
16. H.L.M. Lelieveld, *Process Biochem.* **19**, 112–113 (1982).
17. T.W. Nagodawithana, in A. Gabelman ed., *Bioprocess Production of Flavour, Fragrance, and Color Ingredients*, Wiley Interscience, New York, 1994, pp. 135–168.
18. P.S.J. Cheetham, in A. Wiseman ed., *Handbook of Enzyme Biotechnology*, 3rd ed., Ellis Horwood, Chichester, 1995, pp. 83–234.
19. H.-P. Meyer, *Biotech. Forum Eur.* **8**, 602–606 (1991).
20. S. West, in T. Godfrey and S. West eds., *Industrial Enzymology*, 2nd ed., Macmillan, New York, 1996, pp. 155–175.
21. D.W. Armstrong and L.A. Brown, in A. Gabelman ed., *Bioprocess Production of Flavour, Fragrance, and Color Ingredients*, Wiley Interscience, New York, 1994, pp. 41–94.
22. C.E. Fare, P.J. Blanc, A. Marty, G. Goma, I. Souchen, and A. Voilley, *Perfumer and Flavourist* **21**, 27–39 (1996).
23. P. Fulbrook, in T. Godfrey and J. Reichelt eds., *Industrial Enzymology*, 1st ed., Nature Press, London, 1983, pp. 8–110.
24. P.S.J. Cheetham, *Biocatalysis* **6**, 1–30 (1992).
25. P. Winterhalter and P. Schreirer, *Flavour and Fragrance* **J9**, 281–287 (1994).
26. European Pat. EP 416713 A1 (1991), Gist-brocades.
27. H.M. Lelieveld, *Proc. Biochem.* **19**, 112 (1984).
28. European Pat. 850245 (1986), J.F.M. de Rooij and M.J.J. Hakkaart.
29. H.J. Takken, N.M. Groesbeck, and R. Roos, in R.L.S. Patterson, B.V. Charlwood, G. Macleod, and A.A. Williams eds., *Biotrans of Flavours*, Royal Society of Chemists, Cambridge, 1992, pp. 186–200.
30. N.J. Izawa, *Agric. Food* **45**, 543 (1997).
31. U.S. Pat. 5,077,206 (1991), P.S.J. Cheetham and M.A. Quail.
32. P. Hugueny, B. Dumont, F. Ropert, and J.M. Belin, in P. Etievant and P. Schreier eds., *Bioflavour '95*, INRA, Paris, 1995, p. 269.
33. U. Krings, M. Hinz, and R.G. Berger, *J. Biotechnol.* **51**, 123–129 (1996).
34. M. Starkle, *Biochemistry* **2**, 151, 371 (1924).
35. U.S. Pat. 5,185,252 (1993), A. Humphrey et al.
36. C. Larroche and J.B. Gross, in R.G. Berger, ed., *Advances in Biochemical Engineering*, 1997, pp. 174–220.
37. I. Gatfield, *Perfumer and Flavourist* **20**, 5–14 (1995).
38. I.L. Gatfield, M. Guntert, H. Sommer, and P. Werkhoff, *Chem. Mikrobiol. Technol. Lebensm.* **15**, 165–170 (1993).
39. French Pat. 2 705 971 (1993), H.E. Spinnier, L. Dufosse, I. Souchen, A. Latrasse, C. Piffaut-Juffard, C. Voilley, and A. Philippe.
40. Japanese Pat. 01 168 256 (1989), Y. Shukunobe and S. Takato.
41. F. Schindler and R. Seipenbusch, *Food Biotech.* **4**, 77 (1990).
42. B.L. Muller, C. Dean, and I.M. Whitehead, in P. Etievant and P. Schreier eds., *Bioflavour '95*, INRA, Paris, 1995, pp. 339–344.
43. F. Ropert, B. Dumont, and J.M. Belin, in P. Etievant and P. Schreier eds., *Bioflavour '95*, INRA, Paris, 1995, pp. 275–278.
44. J. Rabbenhorst, *Appl. Microbiol. Biotechnol.* **46**, 470–474 (1996).
45. M.J. Gasson, Y. Kitamura, W.R. McLauchan, A. Narbad, J. Parr, E.L. Parsons, J. Payne, M.J.C. Rhodes, and N.J. Walton, *J. Biol. Chem.* **273**, 4163–4170 (1998).
46. R.C. McIver and G.A. Reineccius, *ACS Symp. Ser.* **317**, 266–274 (1986).
47. U.S. Pat. 4,963,480 (1990), B. Belcour, D. Coutois, and C. Ehret.

48. W.R. Abraham, H.M.R. Hoffmann, K. Kieslich, G. Reng, and B. Stumpf, in *Enzs in Orig. Syn. CIBA Foundation Symposium* III, 1985, pp. 146–160.
49. Y. Suhara, S. Itoh, M. Ogawa, K. Yokose, T. Sawada, T. Sano, R. Ninomiya, and B. Maruyama, *Appl. Environ. Microbiol.* **42**, 187–191 (1981).
50. S. Kise and M.J. Hayashida, *Biotechnology* **14**, 221 (1990).
51. D.G. Springham, in V. Moses and R.E. Cape eds., *Biotechnology: The Science and the Business*, Harwood Academic, 1991, pp. 249–296.
52. K. Osaki, T. Ishiyama, and H. Motai, *Appl. Microbiol. Biotechnol.* **31**, 346 (1989).
53. D.C. Eaton, in A. Gabelman ed., *Bioprocess Production of Flavour, Fragrance, and Color Ingredients*, Wiley Interscience, New York, 1994, pp. 169–203.
54. European Pat. 58870 (1982) (to Unilever).
55. A. Muheim, A. Hausler, B. Scilling, and K. Lerch, *Perfumer and Flavourist* **23**, 21–27 (1998).
56. S. Tada, T. Takeuchi, H. Sone, S. Yamono, M.A. Schofield, J.R.M. Hammond, and T. Inoune, *Proc. Eur. Brewery Conv. Cong.*, Brussels, 1995, p. 369.
57. M.C. Kielland-Brandt, C. Gjermansen, T. Nilsson-Tilgren, and S. Holmberg, *Proc. Eur Brewing Conv. Cong.*, Zurich, 1987, p. 37.
58. K. Miyagawa, et al. *Biotechnology* **7**, 821–825 (1989).
59. U. Kragl, M. Kittelmann, O. Ghisalba, and C. Wandrey, *Ann. N.Y. Acad. Sci.* **750**, 300–305 (1995).
60. EEC Council Directive 13.6.88 (88/334/EEC).
61. Y. Christi and M. Moo-Young, in V. Moses and R.E. Cape eds., *Biotechnology: The Science and the Business*, Harwood Academic Publishers, 1991, pp. 167–209.
62. T.O. Williams and F.C. Miller, *Biocycle* **33**, 72 (1992).
63. T. Lamer, H.E. Spinnler, I. Souchon, A. Voilley, *Process Biochem.* **31**, 533–542 (1996).
64. W. Bonejsza-Wysocki and G. Hrazdina, *Phytochemistry* **35**, 623 (1994).
65. I. Zabetakis, J.W. Gramshaw, and D.S. Robinson, in H. Maarse and D.G. vd Heij eds., *Trends Flav. Res. Proc. 7th Weurman Symp.*, Elsevier, Amsterdam, pp. 91–93.
66. B. Abraham, U. Krings, and R.G. Berger, *GIT Fachz Lab.* **38**, 370–375 (1994).
67. E. Graf and K.B. de Roos, in R.I. McGorrin and I.V. Leland, eds., *Flavour–Food Interactions*, American Chemical Society, Washington, D.C., 1996.
68. U.S. Pat. 5,182,194 (1988), A. Kerkenaar, D.J.M. Schmedding, and J. Berg (to Int. Flavors and Fragrances).
69. P. Van der Schaft, I. van Gee, G. de Jong, and N. ter Berg, in H. Maarse and D.G. van de Heil eds., *Trends in Flavour Research*, Elsevier, Amsterdam, 1994, p. 437.

ENZYMES, FRUIT JUICE PROCESSING

Catherine Grassin
Pierre Fauguenbergue
Gist-brocades
Seclin Cedex, France

KEY WORDS

Cell wall
Enzyme
Fruit
Juice
Pectin
Process

OUTLINE

INTRODUCTION

The fruit juice industry has to deal with many intricate problems in dealing with material whose composition varies widely depending on the variety of the fruit, the year, and the fruit ripeness. The main objective of the fruit juice industry is to process as much fruit as possible at the lowest cost, while maintaining, or if possible improving, the organoleptic quality and the stability of the finished product. Because it is very viscous and forms a gel, fruit pectin is responsible for numerous problems encountered during processing. The first application of pectinases for apple juice clarification started in the 1930s (1) (Table 1).

The induction of the juice clarification after the pectin breakdown by pectinases and the decrease of the juice viscosity have resulted in a shorter process, improving greatly the quality of industrial apple juice. Enzymes have been applied to fruits other than apples, such as pear, black currant, strawberry, raspberry, cherry, cranberry, and grape. In the production of clear and stable red berry juice or concentrate, the hot maceration with enzymes ensures a higher juice yield and a better color extraction. The use of pectinases and amylases, for apple starch degradation, during the hot clarification stage has prevented postbottling haze formation, and for the first time, the concentration of the juice by a factor of six has been possible. This has resulted in a smaller storage volume and cheaper transportation together with a better concentrate stability by preventing spoilage. Apple pulp enzyming with pectinases and hemicellulases appeared later (2): by lowering the viscosity of the pulp, the press capacity and the yield were significantly improved at the same time. Juice depectinization with pectinases (and arabanases) has improved the product quality even more by preventing araban haze formation after juice concentration.

Table 1. History of Enzymes in Fruit Processing

	Product innovation	Enzyme innnovation
1930	Clear apple juice	Introduction of pectinases
1953	Berry processing/hot maceration	Heat-stable pectinases
1960	Apple juice concentrate	Amylases to prevent starch haze
1976	Hot clarification process	Pectin and starch degradation at 50 °C
1980	Apple pulp enzyming for yield increase	Pectinases + hemicellulases
1983	Stable apple juice concentrate	Arabanases to prevent araban haze
1985	Advanced maceration with decanter	Pectinases + hemicellulases + cellulases
1994	Apple liquefaction	Pectinases + rhamnogalacturonase + hemicellulases + cellulases
1995	Enzymes from genetically modified organisms	Single enzyme activity

The addition of cellulases to the fruit pulp has resulted in a liquefaction in which solids are easily separated from the juice with decanters instead of presses. The maximal degradation of apple tissue with a broad range of enzymes has been obtained with a total liquefaction process and decanters. Today, enzyme suppliers can offer to fruit juice producers tailor-made enzyme preparations, optimally blended on the base of the fruit structure for every fruit process—citrus, tropical fruit, or even vegetable. Numerous new enzymes have recently been discovered based on fruit cell wall biochemical structure. The improvement in the knowledge of fruit composition, cell wall chemistry, and enzymology in parallel with better control of large-scale enzyme production has made possible the supply of almost pure enzymes from genetically modified organisms (GMOs) blended in optimal composition based on the fruit composition. Fruit juice technologies now require more accurate transformations than the simple mechanical pressing or physical extractions. Application of exogenous enzymes leads to the degradation of fruits or the selective extraction of some of their components, allowing the creation of new types of finished products and fruit derivatives. They improve quality and stability of finished products together with a shorter process duration and a larger plant capacity. Associated with new equipments and processing technologies, industrial enzymes allow processors to add value to raw material for food and feed and to reduce waste quantity.

In this article, fruit juice processes are described nonexhaustively; in fact, processes are very variable and numerous. We will describe the most common of them, but variations exist in fruit processing companies and factories. Processing of apple, pear, red berry, stone fruit, tropical fruit, and citrus fruit is detailed. The aspect of juice authentification is discussed.

BIOCHEMISTRY OF FRUIT CELL WALL

The pulp of fruits and vegetables is composed of cells. They are surrounded by a cell wall. This is what separates and maintains the integrity of the cells in the pulp and gives them their rigidity. This cell wall stands up to internal pressure and external shocks. Polysaccharides constitute 90 to 100% of the structural polymers of the walls of growing plant cells called *primary cell walls*. *Secondary cell walls* develop from primary cell walls during cell growth. Cell wall composition varies among fruit species and evolves depending on agronomic and climatic conditions, fruit ripeness, and the method and duration of fruit storage (cell growth and cell senescence). Plant cell wall composition has been widely studied, and numerous models of three-dimensional structure have been proposed (3,4). Three major independent domains are distinguished: the xyloglucan network, the pectin matrix, and the structural proteins. The cellulose–xyloglucan network is embedded in the pectin matrix. Pectin is the major structural polysaccharide component of the fruit lamella and cell wall. Three pectic polysaccharides are present in all primary cell walls that have been described: homogalacturonan, rhamnogalacturonan I (RG-I), and rhamnogalacturonan II (RG-II) (5). Recent models divide pectin in smooth regions as unbranched homogalacturonan and hairy regions as highly branched RG-I (6,7) (Fig. 1). Smooth region or homogalacturonan is a homopolymer of (1→4)-α-D-galactosyluronic acid residues, renowned for its ability to form gels (Fig. 1**a**). Carboxyl groups of the galactosyluronic acid residues of primary cell wall homogalacturonan can be methyl esterified at C6 position and acetyl esterified at C2 or C3 position (i.e., sugar beet). Helical chains of homogalacturonan that are less than 50% methyl esterified can form a gel-like structure and condense by cross-linking with calcium ions, which are present in primary cell walls, to form junction zones.

Methyl and acetyl esterification degrees and molecular weight of pectin are specific of fruit species (Table 2). Both decrease during fruit ripeness. With time, the protopectin in primary cell wall is slowly transformed into soluble pectin by the action of endogenous pectinases present in the fruit. The decrease of molecular weight is due to pectin depolymerases, and the decrease of esterification degree is due to pectinmethylesterase and pectinacetylesterase. However, these activities are very low. Homogalacturonan contains about 200 galacturonic acid unit chains 100 nm long, with some rhamnose making a kink in the backbone. A xylogalacturonan subunit (XGa1A) has been identified as part of galacturonan backbone substituted with xylose at the C3 position of the galacturonic acid residue. It can be methylesterified. In apples, XGa1A content is high. Hairy regions are composed of RGI. RG-I has a backbone

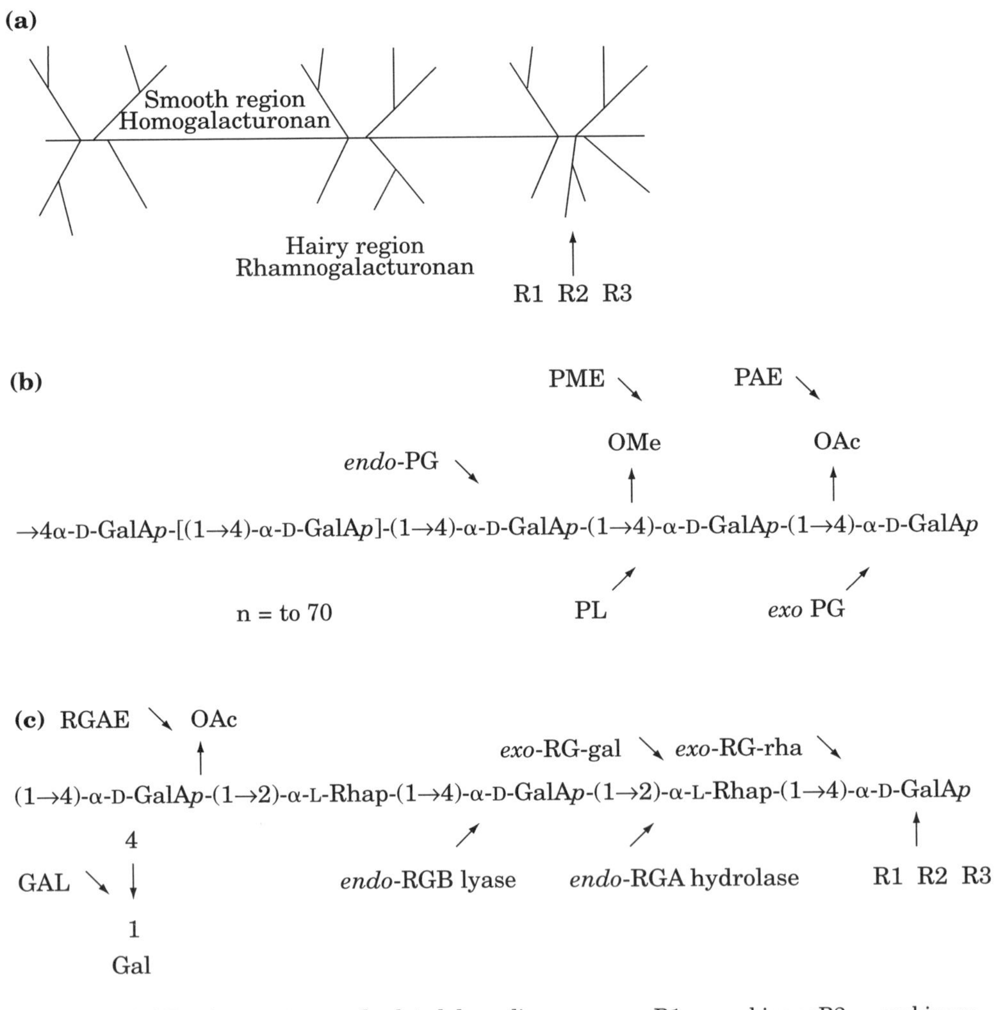

Figure 1. (**a**) Pectin structure and related degrading enzymes. R1 = arabinan; R2 = arabinogalactan type I; R3 = arabinogalactan type II. (**b**) Smooth region, homogalacturonan; PL = pectinlyase; PME = pectinmethylesterase; PAE = pectinacetylesterase; PG = polygalacturonase. (**c**) Hairy region, rhamnogalacturonan type I. RGAE = rhamnogalacturonase acetylesterase; RG-gal = rhamnogalacturonase galacturonohydrolase; GAL = galactosidase; RG-rha = rhamnogalacturonase rhamnohydrolase; RGB = rhamnogalacturonase lyase; RGA = rhamnogalacturonase hydrolase; R1 = arabinan; R2 = arabinogalactan type 1; R3 = arabinogalactan type II. (**d**) Hairy region, arabinan (R1). ARA = arabanase; ARFA = arabinofuranosidase A; ARFB = arabinofuranosidase B. (**e**) Hairy region arabinogalactan type I (R2). AGal = $\beta(1 \rightarrow 4)$ arabinogalactanase. (**f**) Hairy region, arabinogalactan type II (R3). AGal = $\beta(1 \rightarrow 3)$, $\beta(1 \rightarrow 6)$ arabinogalactanase.

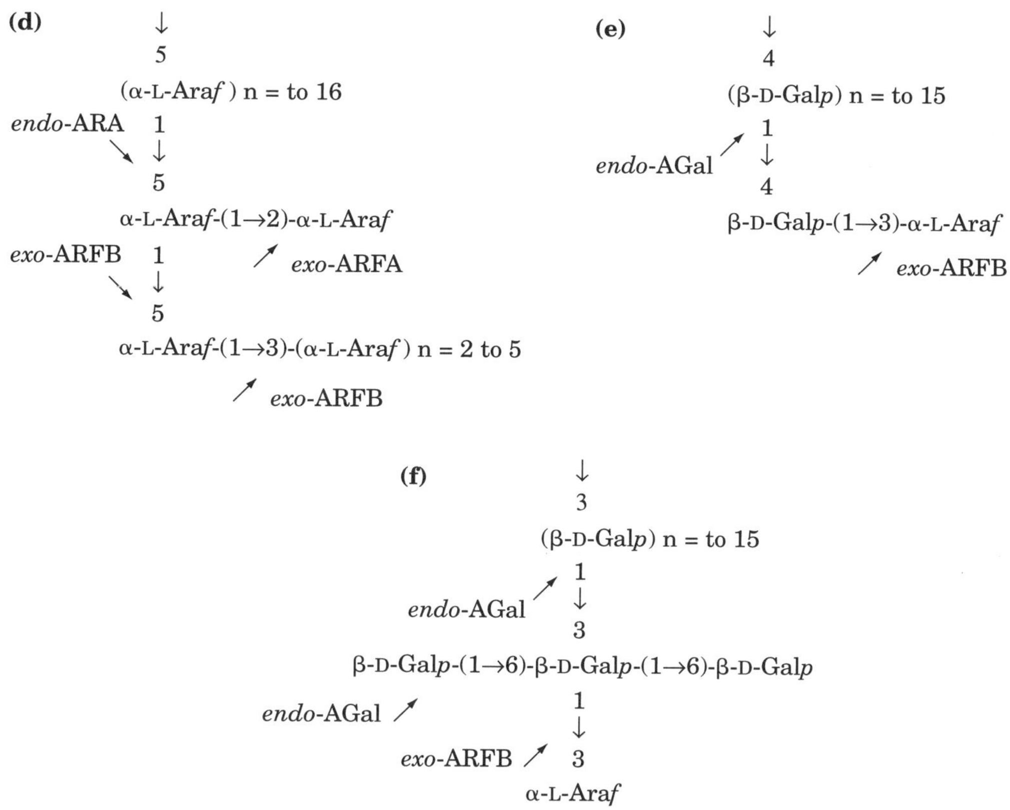

Figure 1. *(continued).*

of as many as 100 repeats of the disaccharide [→2)-*α*-L-rhamnosyl-(1→4)-*α*-D-galactosyluronic acid-(1] (5) (Fig. **1b**). Arabinosyl- and galactosyl-rich side chains are attached to the *O*-4 of rhamnosyl residues. The side chain can vary in size from a single glycosyl residue to 50 or more glycosyl residues (5). In general, about half of the rhamnosyl units of RGI have side chains, but this can vary with cell type and physiological state. Arabinans are mostly five-linked arabinofuranosyl units forming helical chains, but arabinosyl units can be connected to each other at every free position *O*-2, *O*-3, and *O*-5, forming a diverse group of branched arabinans (8) (Fig. **1c**). RG-I of primary cell walls is branched with (1→5)-*α*L-arabinan (1→3) or (1→2)-*α*-L-arabinosyl residues (Fig. **1c**), (1→4)-*β*-D-galactan, arabinogalactan (AG) type I with a (1→4)-*β*-D-galactan backbone (Fig. **1d**), and AG type II with a (1→3),(1→6)-*β*-D-galactan backbone; both AGs have (1→3)*α*-L-arabinosyl residues (9) (Fig. **1e**).

Arabinan and galactan of primary cell walls are present as covalently attached side chains of RGI (5) (Fig. 1). But AG would be unattached to cellulose or xyloglucan, and arabinogalactan-proteins (AGPs) are associated with the plasma membrane (10). RG-II is a low molecular weight (about 4.8 kDa), complex polysaccharide composed of 12 different sugars (Fig. 2). These different sugars are apiose, 2-*O*-methyl-1-fucose, 2-*O*-methyl-D-xylose, aceric acid (3-*C*-carboxy-5-deoxy-1-xylose), Kdo (3-deoxy-D-manno-octulosonic acid), Dha (3-deoxy-D-lyxo-heptulosaric acid) in more than 20 different linkages. RG-II has a homogalacturonan backbone composed of about nine (1→4)-*α*-D-galactosyluronic acid residues (11). Four different complex side chains are attached to *O*-2 or *O*-3 of four of the backbone residues. Two main hemicelluloses of all primary cell walls are xyloglucan and arabinoxylan. Minor components such as glucomannan or galactoglucomannan have also been described (5). Hemicelluloses bind tightly via hydrogen bonds to the surface of cellulose-linking or cross-linking microfibrils, creating a cellulose-hemicellulose network. Interconnections with the pectic polysaccharides are of first importance concerning the integrity of the pectin network. Cellulose accounts for about 20 to 30% of the dry matter of most primary cell walls and is particularly abundant in secondary cell walls. This is a (1→4)-*β*-D-glucan polysaccharide component (Fig. 3). Cellulose microfibrils give the shape and tensile strength to primary cell wall and move relative to another for cell growth but form intractable aggregates in secondary cell walls (5). Microfibrils are coated with hemicelluloses that bind tightly to their surface (12). In the primary cell wall, glucan chains are randomly arranged; the cellulose is said to be amorphous (Fig. 3). This makes the primary cell wall probably more susceptible to enzymatic degradation.

The cellulose of secondary cell walls has a rather high degree of crystallinity, with the chains running in the same direction in a parallel arrangement. In 1993–1994, Vincken et al. demonstrated that xyloglucan was a key structure of apple cell wall for the degradation of cell wall embedded cellulose (13). The cellulose–xyloglucan network

Table 2. Fruit Composition

Fruit	Pectic substances (% wet weight)	Methylation (degree %)	Endogenous enzymes	Main neutral sugars	Pectin[a]	Hemicellulose[a]	Cellulose[a]	Lignin[a]	Proteins[a]	Total[a]	Titrable acidity (% as tartaric)
Apple	0.5–1.6	65–92	PE, PG, PPO	Glu, Ara, Gal	272	169	349	2	76	868	0.5–1.4
Banana	0.7–1.2		PE, PG, PPO								0.3–0.4
Black currant	1.1	50–80	PE								3.0–4.0
Cherry	0.2–0.3	40	PE, PG, CEL	Glu, Ara, Gal	396	49	130	169	244	988	0.4–0.6
Grape	0.1–0.4	50–65	PE, PG, PPO	Glu, Gal, Ara							0.4–1.3
Guava	0.8–1.0	54–62	PE, PG, CEL, PPO, AMY	Glu, Xyl							0.9–1.3
Lemon pulp	2.5–4.0		PE								4.0–4.5
Mango	0.3–0.4	78–85	PE, PG, CEL	Glu, Xyl, Ara	408	91	236	27	127	889	0.2–1.2
Orange peel	3.5–5.5	65	PE								0.8–1.1
Papaya	0.7–1.0		PE, PG, CEL, PPO	Glu, Gal, Man	364	165	124	4	127	784	0.5–0.8
Peach	0.1–0.9	60–80	PG, CEL, PPO								0.8–1.1
Pear	0.7–0.9	50–70	PE, PG	Glu, Xyl, Ara	281	148	267	69	82	847	0.2–0.4
Pineapple	0.04–0.1	22–40	PG	Glu, Xyl, Ara	163	267	210	85	94	819	0.8–1.3
Plum	0.7–0.9	75	PE, PPO								1.4–1.7
Raspberry	0.4–0.5	20	PE, PG, CEL, PPO	Glu, Xyl, Ara	168	89	177	73	277	784	1.4–1.6
Strawberry	0.5–0.7	20–60	PE, PG, CEL, PPO	Glu, Ara, Gal	411	66	232	11	253	975	0.6–1.5

Note: PE = pectinesterase; PG = polygalacturonase; CEL = cellulase; PPO = polyphenoloxidase; AMY = amylase; Glu = glucose (from cellulose); Ara = Arabinose; Xyl = Xylose; Gal = galactose; Man = mannose.

[a]Components in g/kg of ethanol insoluble residue from fresh matter.

determines the strength of the cell wall and is embedded in an independent pectin matrix, hemicelluloses, and proteins. The backbone is (1→4)β-D-glucopyranose bound to α-D-xylopyranose residues attached to the 6-position of glucose. The apple xyloglucan fraction makes up about 24% of the total amount of sugar. The cellulose–xyloglucan complex accounts for approximatively 57% of the apple cell wall matrix. In fruit liquefaction process, efficient enzymic degradation of this complex is crucial. This involves at least endoglucanase and cellobiohydrolase (13). About 5% of the mass of some primary cell walls is made up of a structural glycoprotein hydroxyprolin-rich *extensin* binding some polysaccharides together. Nature and localization of polysaccharides and macromolecules within the fruit cell wall are summarized in the Table 3. Starch is present in unripe apple in amyloplasts. This is the largest biological molecule, with 10^5 to 10^9 Da molecular weight. Starch is composed of two components: amylose and amylopectin (Fig. 4). Amylose is the minor component of the two and has a linear structure. This is a 1→4-α-D-glucose polymer. Various degrees of polymerization have been ascribed to this fraction, with chain lengths in the range of 100 to 1,000 glucose units. Amylopectin contains both α-1→6 plus α-1→4 glucose linkages. Amylopectin exhibits branching at the 1→6 position, and the degree of polymerization is far greater in amylopectin than in amylose. The ratio of amylose to amylopectin can vary in natural starches in the general range of 1:3 to 1:4.

CELL WALL DEGRADING ENZYMES

Pectinases

Recent progress in enzymology has been so fast that certain activities are not yet described in enzyme classification (Table 4). Many microorganisms produce enzymes that degrade the fruit cell wall. Commercial pectinases for the fruit juice industry come from selected strains of *Aspergillus* sp. Enzymes are purified and concentrated. Pectinases are defined and classified on the basis of their action toward the pectin. Two main groups are distinguished: pectinesterases and pectin depolymerases (14). The pectin lyase (PL) is a pectin depolymerase of endo type that has a great affinity for long methylated chains and acts by β-elimination with the formation of C4-C5 unsaturated oligouronides (15) (Fig. **1a**). *Aspergillus* produces several endo-PL isoenzymes with an average pH optimum of 5.0 to 5.5. Acting on the same substrate as the PL, the pectinmethylesterase (PME) removes methoxyl groups from methylated pectin, decreasing at the same time the affinity of PL for this substrate. This results in the formation of methanol and lower methylated pectin, but PME has very low effect on pectin viscosity (Fig. **1a**). PME from *Aspergillus* has a pH optimum around 4.0 to 4.5 and has a strong affinity for high methoxylated pectin such as apple pectin. Fruit PME from orange, for example, acts according to a blockwise mechanism, but *Aspergillus* PME acts according to a multichain mechanism (16). Demethylation with PME and deacetylation with pectinacetylesterase (PAE) reveals free carboxylic acid groups and the pectin becomes negatively charged. Polygalacturonase (PG) exists in two forms:

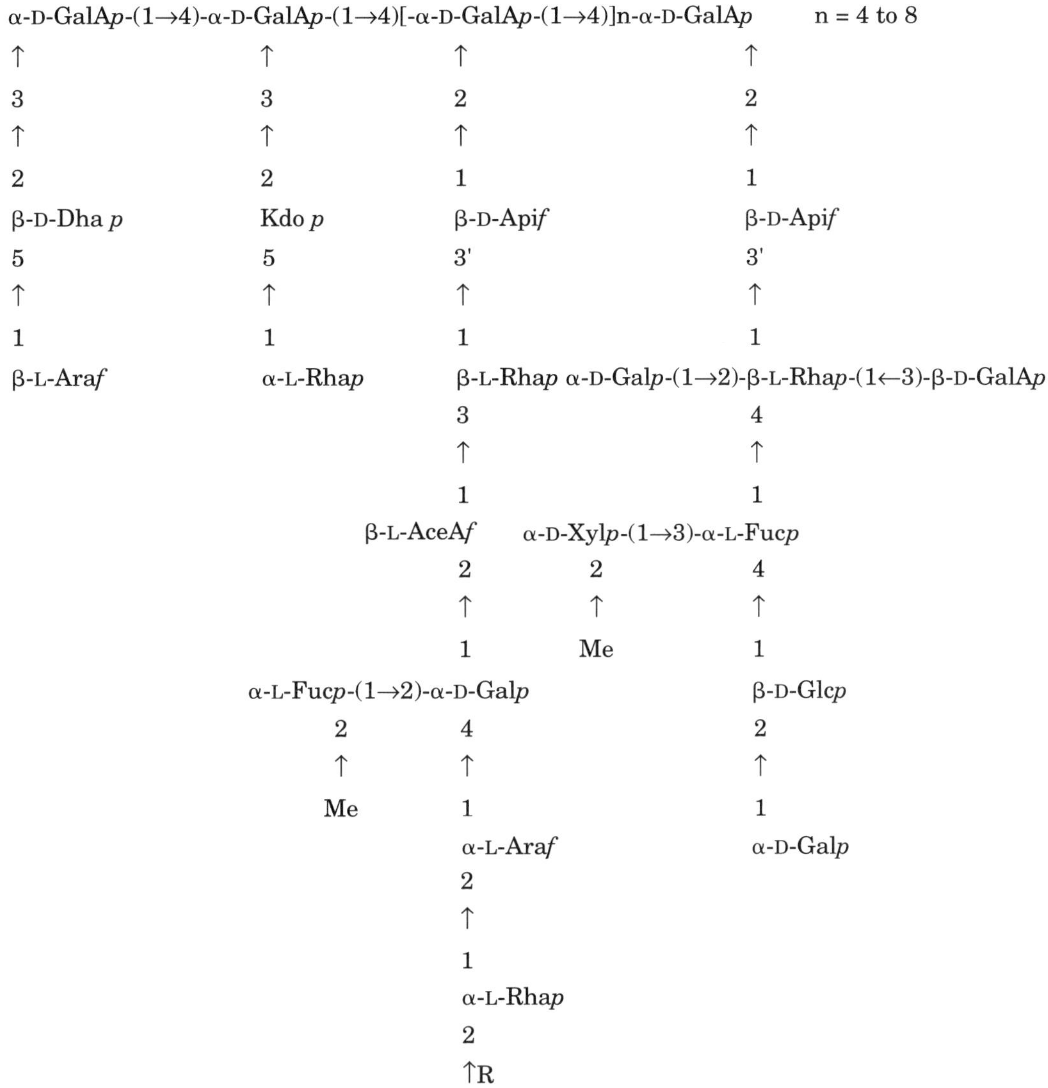

Figure 2. Rhamnogalacturonan type II.

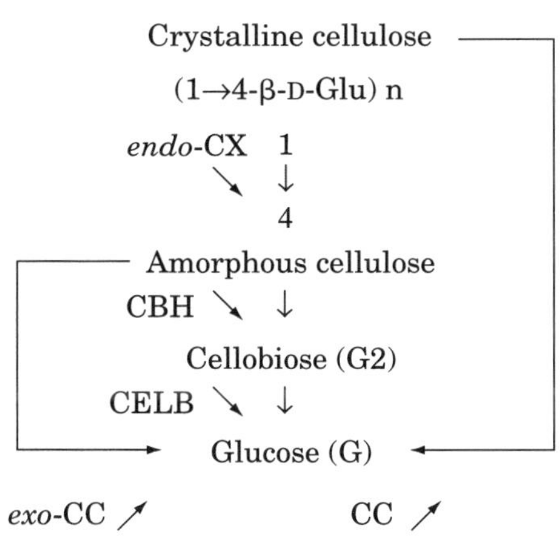

Figure 3. Cellulose structure and related degrading enzymes. CX = β(1→4) *endo*-glucanase; cc = glucohydrolase; CBH = cellobiohydrolase; CELB = cellobiase.

endo-PG and *exo*-PG. Both enzymes are pectin depolymerases and act only on low-esterified pectin (below 50 to 60%) (Fig. **1a**).

endo-PG acts randomly on the polygalacturonic backbone and *exo*-PG acts on the relevant bond at the nonreducing end of the chain. *exo*-PG releases small fragments from the chain and does not significantly reduce the viscosity. Five *endo*-PG isoenzymes and one *exo*-PG isoenzymes have been described from *A. niger* (17). *endo*-PGs from *Aspergillus* have a pH optimum around 4.0 to 4.5, and *exo*-PG has a pH optimum of 5.0. *endo*-PG rapidly induces pectin depolymerization and reduces its viscosity. Different enzymes active on RG-I were identified and purified from *Aspergillus aculeatus* using apple-pectin-modified hairy regions as substrate (18) (Fig. **1b**). RGase A was identified as a hydrolase, splitting the α-D-GalA*p*-(1→2)-α-L-Rha*p* linkage of RG-I, while RGase B appeared to be a lyase splitting the α-L-Rha*p*-(1→4)-α-D-GalA*p* linkage by β-elimination. Two novel enzymes were also identified: a rhamnogalacturonan rhamnohydrolase and a rhamnoga-

Table 3. Nature and Localization of Fruit Cell Wall Polysaccharides

Polysaccharides	Class	Middle lamella	Primary cell wall	Secondary cell wall
Cellulose	β1,4-Glucan		+	+
Pectins	Galacturonan	+	+	
	Rhamnogalacturonan	+	+	
	Arabinans	+	+	
	Galactans	+	+	
	Arabinogalactans	+	+	
Hemicelluloses	Xyloglucans		+	
	Arabinoxylans		+	+
	Xylans			+
	Glucomannans			+
	Galactoglucomannans			+
Glycoproteins		+	+	
Lignin		+	+	+

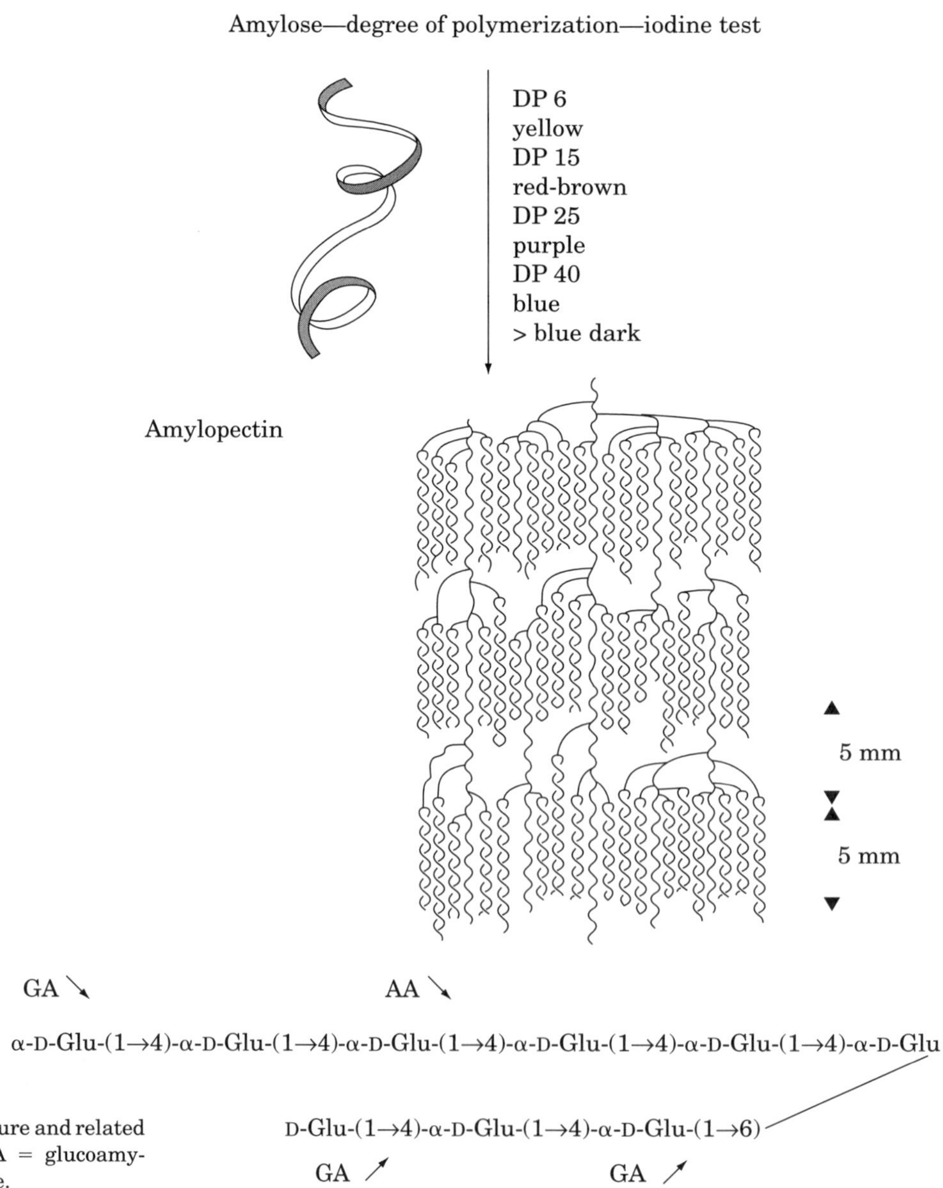

Figure 4. Starch structure and related degrading amylases. GA = glucoamylase; AA = acid amylase.

Table 4. Enzyme Classification

Enzyme	Symbol	EC No.	Substrate	Action mode
Pectin lyase	PL	4.2.2.10	Highly methylated homogalacturonan	*endo*-$\alpha 1 \rightarrow 4$ β-elmination
Pectinmethylesterase	PME	3.1.1.11	Highly methylated homogalacturonan	Deesterification multichain
Pectinacetylesterase	PAE	3.1.1.6	Acetylated homogalacturonan	Deacetylation
Polygalacturonases	PG	3.2.1.15	Deesterified homogalacturonan	*endo*-$\alpha 1 \rightarrow 4$ depolymerization
	PG	3.2.1.67	Polygalacturonic acid	*exo*-α-$1 \rightarrow 4$
Rhamnogalacturonases	RHG A		Rhamnogalacturonan	Gal α-$1 \rightarrow 2$ Rha hydrolysis
	RHG B		Rhamnogalacturonan	Rha α-$1 \rightarrow 4$ Gal β-elmination
			Rhamnogalacturonan	*exo*-Rhan-Rha
			Rhamnogalacturonan	*exo*-Rhan-GalA
	RGAE		Rhamnogalacturonan	Deacetylation
Arabinosidase	ARA	3.2.1.99	Pectin araban	*endo*-α $1 \rightarrow 5$
	ARF A	3.2.1.55	Pectin side chains	*exo*-α $1 \rightarrow 2$, $1 \rightarrow 3$
	ARF B	3.2.1.55	Pectin side chains	*exo*-α $1 \rightarrow$, $1 \rightarrow 5$
Galactanase	GAL	3.2.1.89	Pectin side chains	*endo*-$\beta 1 \rightarrow 4$
Galactosidase	Gal	3.2.1.23	Pectin side chains	*exp*
Arabinogalactanase	AGal	3.2.1.90	RG side chains	*endo*-$\beta 1 \rightarrow 3$
	AGal	3.2.1.90	RG side chains	*endo*-$\beta 1 \rightarrow 6$
Xylanase	XYL	3.2.1.8	Xyloglucan	*endo*-$\beta 1 \rightarrow 4$
Xylosidase	Xyl	3.2.1.37	Xyloglucan	*exp*
Amylase	AMY	3.2.1.1	Amylose	*endo*-α $1 \rightarrow 4$
Amyloglucosidase	AG	3.2.1.3	Amylose amylopectin	Nonreducing end *exo*-α $1 \rightarrow 4$, $1 \rightarrow 6$
Mannanase	MAN	3.2.1.78	Mannan	*endo* $\beta 1 \rightarrow 4$
Galactomannanase	GMan		Galactomannan	*endo*
Cellulase	CC	3.2.1.4	Amorphous cellulose	*endo*-$\beta 1 \rightarrow 4$
	CX	3.2.1.74	Crystalline cellulose	*exo*-$\beta 1 \rightarrow 4$
Cellobiohydrolase	CBH	3.2.1.91	Crystalline cellulose	Dimer
Cellobiase	CELB	3.2.1.21	Cellobiose	*exo*

lacturonan galacturonohydrolase (Fig. **1b**). As an accessory enzyme for the RGases, rhamnogalacturonan acetyl esterase (RGAE) was discovered in the same preparation. Although a hypothetical structure of RG-II has been described, RG-II degrading enzymes are still to be discovered. Arabanases are pectinases because they remove arabinose covalently bound to the homogalacturonan backbone. Three enzymes have been described: one *endo*-arabinanase (α-1→5) (*endo*-ARA) and two arabinofuranosidases, among which one *exo*-arabinofuranosidase A (α-1→2; α-1→3) (ARFA) and one *exo*-arabinofuranosidase B (α-1→3; α-1→5) (ARFB) (Fig. **1c**). All three are produced by *Aspergillus* sp. (19). Activity level and ratio must vary in enzyme preparations, depending on the fruit species processed (i.e., high activities are required for apple and pear processing). Highly branched arabinans are not a substrate for *endo*-ARA. However, when *exo*-ARFA activity is too high, the linearization of arabinan chains leads to a reassociation of these chains at low water activity, forming a haze in apple and pear concentrates (20). *A. niger* ARAs have pH optimum at 4.0.

Hemicellulases

Hemicellulases are enzymes hydrolyzing arabinogalactans, galactans, xyloglucans, and xylans. Arabinanases are active on arabinans, and also on arabinogalactans or arabinoxylans. *Aspergillus* produces enzymes able to hydrolyze arabino (1→4)-β-D-galactans type I (Fig. **1d**) and arabino (1→3)-(1→6)-β-D-galactans type II (Fig. **1e**) (9,21,22). The *exo*-(1→3)-β-D-galactanase is able to release galactose and (1→6)-β-D-galactobiose. *Aspergillus* enzyme is able to bypass a branch point in a β-(1→3) backbone. The action of this enzyme is enhanced by the presence of ARAB. Xylanases hydrolyze the (1→4)-β-D-xylans in synergy with ARAs, glucuronidase, and acetylesterase (23). They are important enzymes for pear cell wall breakdown. Other hemicellulases such as mannanases hydrolyze the β-(1→4) mannans, galactomannans, and glucomannans present in coconut or pineapple.

Cellulases

Commercial cellulases come from selected strains of *Trichoderma* sp. Beldman has widely described cellulases from *Trichoderma* (24). Cellulase is a multienzyme system composed of several enzymes with numerous isoenzymes that act in synergy (Fig. 3). Nevertheless, products of reaction can retroinhibit the enzymes (i.e., glucose and cellobiose toward *exo*-glucanases and cellobiohydrolases).

endo-Glucanase isoenzymes (CX) hydrolyze the (1→4)-β-D-glucan chain of cellulose at random (Fig. 3). Vincken et al. in 1994 demonstrated that removal of the xyloglucan coating from cellulose of isolated apple cell walls by a xyloglucanase (XGase) purified from *Trichoderma viride* or *endo*-IV glucanase (CX) enhanced the enzymic degradation of cell wall embedded cellulose (5). XGase can work in synergy with pectic enzymes such as PL and is a key activity for the enzymatic liquefaction of apple (25). *exo*-Cellulase (CC) hydrolyzes the β-D-(1→4) glucan chain of cellulose at its reducing extremity with an *exo* behavior. After CX action, cellobiohydrolases (CBHs) hydrolyze bonds at the nonreducing end of the crystalline cellulosic chain, producing cellobiose. Cellobiase (CELB) or β-glucosidase splits the cellobiose to glucose. Many models of cellulose hydrolysis have been proposed (26). Native crystalline cellulose is opened by *endo*-glucanase, which acts on regions of low crystallinity to form amorphous cellulose. *endo*-CX plus CBH can degrade amorphous cellulose to cellobiose easily. This cellulolytic complex is of first importance in every process of fruit liquefaction.

Amylases

In the fruit juice industry, fungal acid amylases are used to process fruits that contain starch. This is the case of unripe apples just after the crop period. *A. niger* produces acid α-amylase and amyloglucosidase (27). These exogenous enzymes are applied after starch gelatinization (occurring above 75 °C (167 °F) for preventing postbottling haze formation. Acid *endo*-amylase (AA) is active on (1→4)-α-D-glucose chain or amylose (Fig. 4). AA from *A. niger* has an optimal pH of 3.0 to 4.0, and the optimal temperature is 70 to 75 °C (158 to 167 °F).

After action of AA that gives dextrins, glucoamylase or AG (which is an *exo*-hydrolase) acts on gelatinized starch by releasing glucose from the nonreducing end of the α-(1→4) bonds of amylose chains but can also hydrolyze α-(1,6) bonds of amylopectin chains (Table 4). AG from *A. niger* has an optimal pH of 4.0 to 4.5, and the optimal temperature is 55 to 60 °C (131 to 140 °F).

INDUSTRIAL ENZYMES

Standardization of Enzyme Production

Commercial enzymes are produced and sold by several companies, such as Gist-Brocades, Novo Nordisk, Solvay, Röhm, Quest, and Danisco. The enzymes most frequently offered to fruit juice producers are pectinases coming from *Aspergillus* species, the most common from *A. niger*. Next to pectinases, numerous side-activities-type hemicellulases or glycosidases are secreted, all working in synergy with the pectinases on the cell wall substrate. The spectrum of enzymes and their activities are natural to the strain (constitutive) and to the culture conditions (inducible). This is why the enzyme spectrum differs from one commercial product to another. Pectinases are exocellular enzymes and are the main activities produced on carbohydrate substrates. These enzymes are concentrated and sold mainly as liquids, but also as powders or granulates. In the past few years, some juices producers have chosen the new technology of fruit liquefaction. In addition to pectinases, hemicellulases and cellulases are needed for complete cell wall degradation. These enzymes are produced from *Aspergillus* or *Trichoderma* spp. It is possible to blend cell wall degrading enzymes coming from different cultures optimally, depending on the application and the fruit composition (Table 5).

Various microorganisms are used by enzyme producers as enzyme source. All the microorganisms used to produce food enzymes bear the GRAS status (generally recognized as safe). Enzymes are produced under extremely well defined and reproducible conditions (Gist-Brocades has ISO 9002 certification and applies HACCP). Quality assurance and standardization of industrial enzymes are strictly defined concerning the enzyme activity, its concentration, the presence of chemical and microbiological contaminents (heavy metals, toxins, yeasts, bacteria, molds) depending on legislature requirements. Enzymes are proteins considered as processing aids; they are inactivated in fruit processes and are not present or active in the final product. Good manufacturing practices (GMP) in the microbial food enzyme industry ensures that enzyme preparations are produced, packed, and handled in a hygienic way. The key issues in GMP are the microbiological control of the microorganism selected for enzyme production, the control and monitoring systems ensuring pure culture and optimum enzyme productivity conditions during fermentation, and the control of the hygienic conditions throughout recovery and finishing of the enzyme preparations. GMP will ensure that contamination of the product during recovery is minimized. The final product has to meet the criteria formulated by the WHO/FAO Joint Experts Committee on Food Additives (JECFA) and the Food Chemicals Codex (FCC). Gist-Brocades and other enzyme suppliers are members of AMFEP (Association of Manufacturers of Fermentation Enzyme Products). AMFEP makes sure that regulatory aspects of enzyme applications in the food industry are respected by its members.

Genetically Modified Organisms

Enzymes produced by recombinant DNA (rDNA) technology will increasingly find application in fruit juice production. Application of so-called rDNA enzymes has been slow for a number of reasons, mainly because of the technical complexity of the processes involved, but it can be expected to increase because knowledge about enzyme mixtures is increasing rapidly. The application of rDNA technology to produce enzymes for fruit juices involves the following steps: (*1*) The genes coding for these enzymes have to be cloned. This is not a small task, because many enzymes are involved and for many enzymatic activities more than one gene can be found in the traditional strains that are being used. Today, more than 25 genes encoding so-called plant cell wall degrading enzymes that are important for this process are known. (*2*) The genes have to be overexpressed in a suitable safe host strain. This strain is preferably belonging to the same genus as the production strain, for instance *A. niger*. The technology for introducing these genes in production strains resulting in overexpression of the relevant enzymes are well established now. It results in a genetically modified organism (GMO) that has a small change in its genetic makeup, thereby overproducing one of the enzymes of interest, for instance, a pectin-lyase. If no DNA of other species is introduced in this process, it is called homologous recombination, which can be compared to the classical genetic techniques. If foreign DNA is introduced in the receptor strain as well, the process is called heterologous recombination. (*3*) The enzyme produced by the GMO has to be tested in application studies. This is, of course, a very critical process because a balanced mixture of enzymes is required to obtain a juice of the right quality and yield.

Table 5. Fruit, Enzymes, Process

Fruit/process	Juice extraction	Juice depectinization concentration	Fruit liquefaction
Apple	PL or PME + PGs + RGs + ARFs + ARA	PL + PME + PGs + RGs + ARFs + ARA + AGals + AMY + AG	PL + PME + PGs + RGs + ARFs + ARA + CX + C1 + CBH + CELB
Pear	PL or PME + PGs + RGs + ARFs + ARA	PL + PME + PGs + RGs + ARFs + ARA + AGals + AMY + AG	PL + PME + PGS + RGS + ARFs + ARA + XYL + CX + C1 + CBH + CELB
Red Berry	PME + PGs + ARA + ARFs + AGals + CX	PME + PGs + ARAs + AGals + XYL	
Grape	PG + ARFs + ARA + AGals	PL + PME + PGs + RGs + ARFs + ARA + AGals	
Pineapple		PG + GMan + AGals + XYL	
Stone fruit			PL + PME + PGS + RGS + ARFS + ARA + XYL + CX + C1 + CBH + CELB
Citrus		PL	

Note: PL = pectin lyase; PME = pectinmethylesterase; PG = polygalacturonase; RG = rhamnogalacturonase; ARF = arabinofuranosidase; ARA = arabanase; CX = endoglucanase; C1 = exoglucanase; CBH = cellobiohydrolase; CELB = cellobiase; XYL = xylanase; GMan = galactomannanase; AGal = arabinogalactanase; AMY = amylase; AG = amyloglucosidase.

The genes that have been cloned from *Aspergillus* species that are currently relevant in this respect include enzymes that degrade pectin (*endo*- and *exo*-polygalacturonase, pectinlyase, pectinmethylesterase, rhamnogalacturonase). Other examples are arabinan-degrading enzymes (*endo* and *exo*), cellulases (*endo*-glucanases and β-glucosidases) and galactanases. Starch-degrading enzymes and proteases are relevant as well. Despite the impressive list of known enzymes, a number of others, for instance those degrading the so-called hairy regions of pectin, still have to be identified and their genes cloned. But, in theory it will be possible in the future to make optimal mixtures of enzymes produced by GMOs for every application (Table 5). Fruit juice enzymes are usually produced from *Aspergillus* and *Trichoderma* strains. The GMO technology guarantees that enzyme proteins produced from GMOs are identical to those produced from non-GMOs and allows the production of single pure enzyme activity. GMOs are absent from final products. Various GMO-derived food enzymes have already been approved and used for more than 15 years in Europe and the United States in the starch, dairy, and beverages industries. In summary, it can be predicted that enzymes produced by GMOs will start a new era in fruit juice manufacture, provided that it will be possible to further unravel the complex mixtures that are used today. Another important prerequisite is that the products will present an advantage for both the fruit juice manufacturer and the consumer.

ENZYMES AND APPLE JUICE PROCESSES

Apple is by far the most important raw material worldwide for the production of clear juice and clear concentrate. Annual production is around 54 million tons of apples, from which about 20% are processed mainly as clear concentrate, but a small part is processed as cloudy juice and apple sauce. The major producers of apple juice concentrate are located in the United States, Poland, Argentina, Italy, Chile, China, and Germany. In the traditional process, enzymes are used at two different stages. Pectinase preparations containing numerous side activities are added to the fruit pulp for a maceration during which soluble pectin and cell walls are partially hydrolyzed, resulting in a dramatic decrease of the viscosity, increasing the yield because the free-run juice volume is larger and the pressing easier. Mash enzyming increases also the productivity of the plant. Pectinases with amylases are added to the juice after pressing and juice pasteurization. These enzymes decrease the juice viscosity, increase its filterability, making the clarification and the concentration easier. Both enzyme treatments allow production of clear and stable concentrate. At the same time, enzyme treatments prevent postbottling haze formation (starch haze or araban haze) in concentrates. A few producers in Germany and the United States make natural or cloudy apple juice. Until now, they have not added enzymes because the existing commercial pectinases induce a fast clarification of the juice. Gist-Brocades sells a purified pectinmethylesterase without pectin depolymerase activity. It can be used in the cloudy juice process and also in the French cider process with flotation. Finally, the new technology of total liquefaction, achieved by using pectinases, hemicellulases, and cellulases in blend, is performed with only one enzyme addition. Solids are removed from the liquefied pulp with decanter or press. The yield is maximal.

Clear Apple Juice Concentrate

Process Description. The traditional process (Fig. 5) (28) consists of the following stages: apples are selected, washed, and crushed. Macerating enzymes are added continuously during crushing after their dilution in water through a dosing pump. The enzymed pulp is then pumped into the maceration tank where the incubation takes place during 30 to 60 minutes at ambient temperature. Pectinases such as Pectinex UltraSP (Novo Nordisk) or Rapidase® Press (Gist-Brocades) are added at 40 to 75 g/

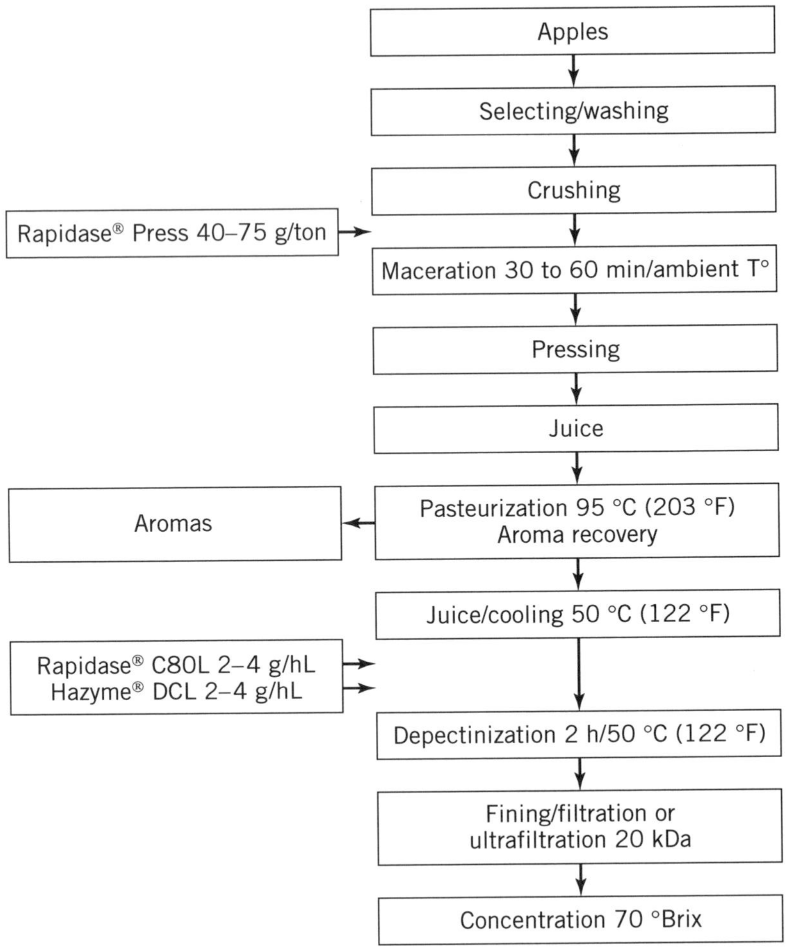

Figure 5. Clear apple juice concentrate process.

ton of apples depending on fruit ripeness. In the apple pulp, macerating pectinases from *Aspergillus* are stable at about 75% of their optimal activity at pH 3.5 to 4.0. After this enzymatic maceration, the mash is pumped to the pressing equipment. After pressing, the juice is flash pasteurized for aroma recovery and cooled down. Pectinases are added to the juice at 45 to 50 °C (113 to 122 °F) at 2 to 4 g/hL. Pectinases Pectinex 3XL (Novo Nordisk) or Rapidase® C80L (Gist-Brocades) and amylases AMG (Novo Nordisk) or Hazyme®, DCL (Gist-Brocades) are added at the same dosage at the beginning of the season when apples contain starch. After 2 h, the juice no longer contains any pectin or starch. In recent years, the development of tangential ultrafiltration with membranes of 20 to 30 kDa cutoff has enabled the fast clarification of the juice by removal of insoluble particles and macromolecules. The brilliant and clear ultrafiltered juice is then concentrated at 72 ° Brix and stored in the dark at low temperature.

Apple Pulp Maceration

Fruit juice producers look for an optimization of the yield and of the plant capacity to make more profit from raw material. Press efficacy is very much higher after pulp enzyming. The application of exogenous pectinases from *Aspergillus* sp. on apple mash is a necessity because endogenous enzymes present in apple such as pectinmethylesterase and polygalacturonase have too low activity to cause immediate visible effect (29,30). Because apple pectin is approximately 90% methylated, commercial enzyme preparation must contain a high level of pectinylase or pectinmethylesterase in association with polygalacturonase and arabanases (Table 5). In 1990, Schols et al. described a new pectinase rhamnogalacturonase produced by *Aspergillus* sp. and its important role on the maceration of apple tissue (7). In 1996, new enzymes are active toward pectic structures, among which different rhamnogalacturonases have been described (18). Rapidase® Press (Gist-Brocades), Pectinex UltraSP (Novo Nordisk) or Rohapect MA plus (Röhm), sold as, mash enzyme preparations, contain activities that are necessary to obtain high yields throughout the entire season. Commercial products contain more or less these enzyme activities but with a different ratio. About 70% of the producers use horizontal presses; 15%, belt presses; and others, decanter or screw presses. After pulp enzyming, extraction of the juice can increase the yield up to 90% with hydraulic press (with pomace leaching), compared to a maximum 80% yield without enzyme treatment. Apple varieties processed just at the crop period (fresh apples) are easily pressed with a relatively high yield. The current trend is to process apple juice from table varieties (golden delicious, granny smith, jonagold, red delicious) with quality defects. These apples are rich in pectin and poor in tannins in contrast to old apple varieties that have almost disappeared (cider-type apples). These apple varieties are often stored at low temperature for several months in controlled atmosphere (stored apples), and overripe apples are processed in juice throughout the year. During storage, bonds between different polysaccharide fractions such as pectin, hemicelluloses, and cellulose are modified. The insoluble protopectin is slowly transformed into soluble pectin through the endogenous pectinases of the apple, and starch is degraded by endogenous amylases (Fig. 6). In the apple, the elasticity and the cohesion of the cells decrease with time. The water in the fruit moves to the crystalline cellulose in the periphery, linked to lignin in a more organized and thick structure (31). Moreover, as apples become overripe, especially after being stored for a long time at a low temperature and in a controlled atmosphere, they give a decreased juice yield (Fig. 6). The mash pressability is bad, and the pomace sticks to the press cloth. Apples are impossible to press without enzymatic maceration because of the high soluble pectin content. This is why the dosage of enzymes, added to decrease the viscosity of the pulp, must be higher at the end of the season.

Apple Juice Depectinization

Pectin is the main cause of turbidity. One liter of juice with a dry matter of 13% can contain 2 to 5 g of pectin after pressing, depending on fruit ripeness. After aroma recovery and pasteurization, pectinases and amylases (only at the beginning of the processing season) are added simultaneously to the juice during 1 to 2 h at 45 to 60 °C (113 to 140 °F). *Aspergillus* pectinases are stable at 50 to 55 °C (122 to 131 °F) in apple juice at pH 3.0 to 4.5, and some of them are stable until 60 °C (140 °F). Pectinase dosage is determined at the laboratory by performing the alcohol test. PL, or pectinmethylesterase plus polygalacturonases, and arabanases are the most important activities.

Arabinose represents 55% of neutral sugars in apple pectin hairy regions. Arabanases prevent the haze formation after the juice concentration. The juice clarification is actually a secondary effect of the enzymatic depectinization (32). It is caused by enzymatic and physicochemical mechanisms. The first stage consists of the destabilization of the cloud first because of PL, without visible effect but that results in a strong decrease of the viscosity of the juice (Fig. 7). The PL cuts the pectin at random, and cutting 1 to 2% linkages is enough to reduce apple juice viscosity by 50%. PL affinity for apple pectin is high because of the high methylation degree. PGs become active only after the action of PME because they cannot depolymerize the pectin when the methylation degree is higher than 50 to 60% and have a greater affinity for the active site after its partial demethylation (steric hindrance). At this stage, the PL is no longer active and the pectin hydrolysis results from the system PME and PGs. The second stage is flocculation and sedimentation. The sedimentation necessary to lead to the clarification of an apple juice occurs only after these enzymatic degradations of pectin and starch. The cloud is composed of proteins with positive charges when the pH of the juice is 3.5 to 4.0 because their isoelectric point is in-between 4.0 and 5.0. These proteins are bound to hemicelluloses surrounded by the pectin as a protective, negatively charged colloid layer. Pectinases partially hydrolyse the pectin gel, resulting in the aggregation of oppositely charged particles, such as tannins, and in the flocculation of the cloud and then clarification of the juice. The optimal pH for this mechanism is 3.6. The clarification occurs after the solubilization of the insoluble protopectin by pulp enzymes, the decrease of the juice's viscosity (in fact, in a first stage, the viscosity increases first because of the progressive solubilization on the part of protopectin and then decreases after pectin hydrolysis), and the partial hydrolysis of the colloid protector.

The sedimentation occurs quickly during the neutralization electrostatic phase inducing the clarification. In the past, this step occurred after a very long time and the juice was sensitive to contamination. The use of pectinases saves time. The alcohol test shows if residual pectin is still present in the juice. We recommend using acidified ethanol (2 vol of ethanol containing 0.5% HCl + 1 vol of apple juice) to precipitate the residual pectin with a molecular weight equal or lower than 3,000 Da. It is important to use acidified ethanol; if not, the precipitation of organic acids or calcium pectate occurs because of the modification of the pH after mixing of alcohol with the juice, giving a false-positive answer. At the beginning of the processing season, unripe apples contain starch, 5 to 7 g/L of juice. At this time, the iodine test will give a blue dark color, and the starch will precipitate with iodine. Starch is in a granular

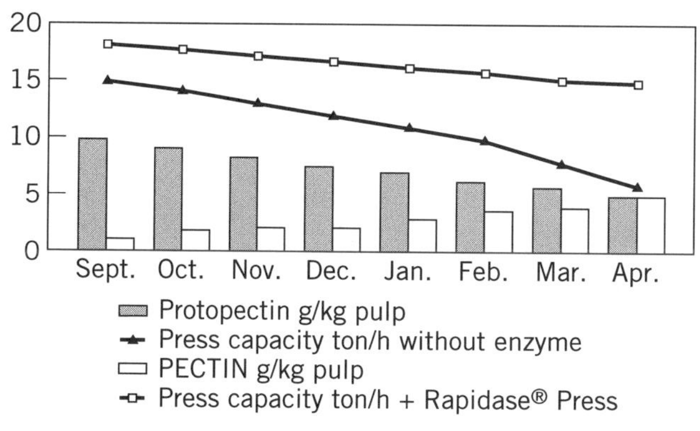

Figure 6. Apple ripeness, protopectin, pectin, and pressability (in northern hemisphere).

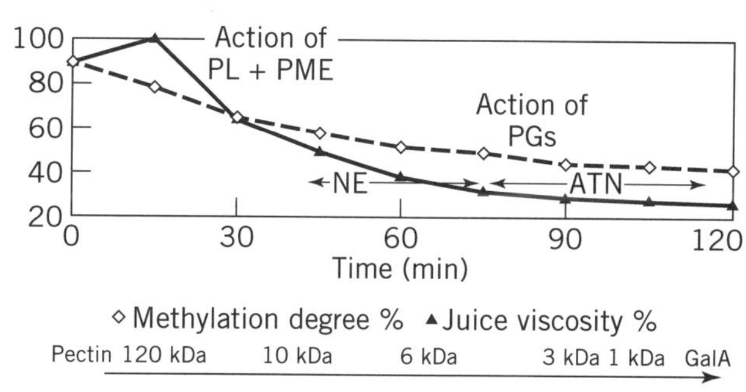

Figure 7. Apple juice depectinization. NE = neutralization of electrostatic charges; ATN = alcohol test negative.

form, ranging from 2 to 13 μm and is composed of 30% amylose and 70% amylopectin chains. Amylopectin can fix 20% its weight in iodine (Fig. 4). Apples contain endogenous amylases; however, the process is too short and activities are too low to degrade the starch. When the juice is heated up to 75 to 80 °C (67 to 176 °F), starch moves from an insoluble form to a soluble form (gelatinization with water). If no amylase is added, these molecules can retrograde later and make big aggregates that are difficult to hydrolyze and clott the filters. The addition of fungal amylase results in the hydrolysis of soluble starch until glucose monomer. Retrogradation and postbottling haze formation are prevented. Clarification and filterability are improved. Preparations such as Hazyme® DCL (Gist-Brocades) or AMG (Novo Nordisk) contain α-amylase and amyloglucosidase activities from *A. niger*. The dosage is determined by the iodine test.

Amylases are added during the depectinization stage at 2 to 4 g/hL simultaneously with pectinases, for 2 h at 50 °C (122 °F). At the end of the hydrolysis, the iodine test and alcohol tests are negative. The third stage is the clarification with fining agents (gelatin and bentonite), filtration, or ultrafiltration (UF). UF membranes made of inorganic substances have an average cutoff of 20,000 Da. The juice goes through the membrane as filtrate. Insoluble particles and soluble colloidal material of molecular weight higher than the pore size (nondegraded part of RG-I, RG-II, proteins such as enzymes, polyphenols) are removed as retentate. Recently, enzymes active toward the rhamnogalacturonan part of pectin have gained attention, because it was found that in juice processing, rhamnogalacturonan parts can foul UF membranes. For this reason, the UF stage was sometimes a bottleneck in the process, especially when the extraction of cell wall components was too strong. We know that a more complete pectin degradation is required to increase the UF flux rate. This is why enzyme preparations proposed for apple juice depectinization must contain a broad range of pectinases, including rhamnogalacturonases, hemicellulases, and amylases (Table 5). With the all these enzymes, the juice clarification is optimal, and its concentration up to 70 °Brix is easy, without risk of gel or haze formation.

Apple Total Liquefaction

The enzymatic liquefaction of apple consists of the hydrolysis of cell wall polymers by exogenous enzymes into soluble components, mainly acidic and neutral sugars, thus increasing the extraction level and the yield. In relation to the cell wall composition and in comparison with maceration, more activities are needed for cell wall liquefaction (33).

The apple cell wall can be solubilized more than 80% by the combined action of pectinases, cellulases, and hemicellulases (Table 5) (34). Recently, new enzymes such as rhamnogalacturonases and xyloglucanases have been found to be very important for apple liquefaction yield and UF flux rate. They also strongly decrease the viscosity of the mash, and solids are discarded from the juice by centrifugation more easily. The new trend is a process including a hot break stage.

Process Description. Apples are selected, washed, and crushed, and then the pulp is heated up to 85 °C (185 °F) for 4 to 5 min (pectinases work more easily after the physical degradation of pulp cells) (Fig. 8). During cooling, the pulp goes to a finisher for seed peel and core removal. The liquefaction is then carried out for 2 to 3 h at 50 °C (122 °F) with a continuous slow stirring at 40 to 60 rpm. During the tank filling, commercial pectinases such as AFPL (Novo Nordisk), Grindamyl (Danisco) or Rapidase® LIQ EXTRA (Gist-Brocades) are added, from 100 to 250 g/ton of pulp. In this process, no cellulase is needed, and there is no cellulose degradation to cellobiose and glucose in the juice. The yield is still very high. Enzymes are stable in the process conditions, and pectolytic and hemicellulolytic activities are high enough to depectinize the juice during the liquefaction time. After about 3 h, the liquefied pulp, which contains 25 to 30% solids (in volume), is easily centrifuged with a decanter, giving a first juice. The pomace is mixed with hot water and pumped to a second decanter or press, giving a second juice. Both juices are mixed and contain less than 1.5% solids. After pasteurization, active carbon and gelatin are often added to decolorize the juice and to improve UF flux rate. Fining agents are removed by vacuum filtration. Then the juice is ultrafiltered and concentrated up to 70 °Brix. The total liquefaction process results in a high yield increase, with an average of 95 to 96% along the whole processing season and according to the process described in Figure 8. The yield is calculated as follows:

$$[(\text{juice n°1 weight} \times \text{°brix}) + (\text{juice n°2 weight} \times \text{°Brix})] \div [(\text{apples weight} \times \text{°Brix of apple})]$$

The titrable acidity of the final juice is high because of the galacturonic acid content. Liquefied apple juice can contain methanol, up to 300 to 500 mg/L of juice. However, its major part is removed during the concentration stage. The yield after apple total liquefaction process is compared with the traditional process in Table 6. The quantity of residual waste is very low. Production cost is decreased (higher yield, cheaper equipment, less labor in a continuous process). This process could be applied to extract the juice from fruits such as mango or guava for which no equipment has been developed.

Apple Pomace Liquefaction

Process Description. Gist-Brocades is one of the first enzyme suppliers that has proposed enzymes dedicated to apple pomace liquefaction (35) (Fig. 9). Depending on the equipment, the first extraction yield can be adjusted precisely by using macerating pectinases, and we recommend 65 to 70% yield. After addition of water (Fig. 10), the pomace is liquefied with enzymes. Rapidase® Pomaliq Extra is used at 150 to 250 g/ton calculated according to the weight of apples at the start of the process. The pomace is stirred at 40 to 60 rpm for about 3 h at 50 °C (122 °F). The liquefied pomace is pumped to the press or the decanter for obtaining a second juice. During the pomace liquefaction, the Brix degree of the juice increases from 6.0 to about 7.5, mainly because of the degradation of the pectin.

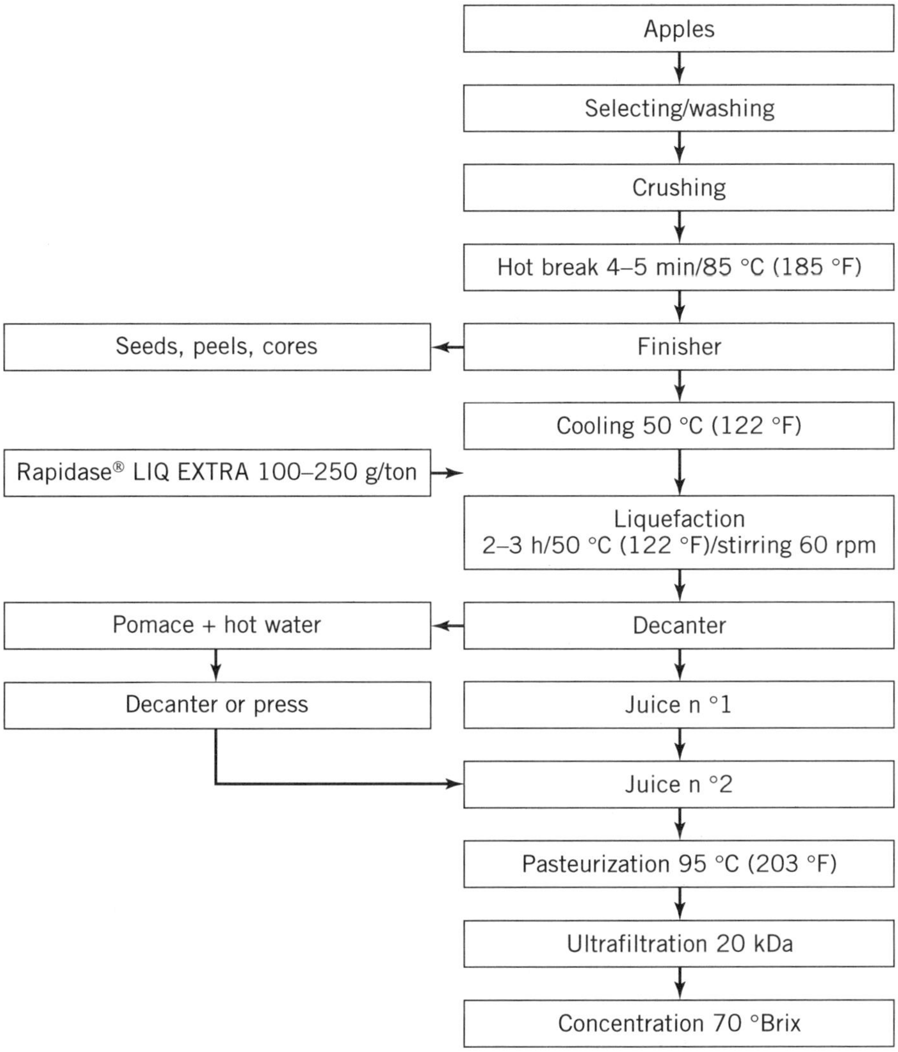

Figure 8. Apple total liquefaction process.

This juice can be blended to the first pressing juice that has been depectinized in the case of clear juice production. The second juice, which is very acidic, can be used separately for other applications (to increase the acidity, cider, vinegar). Further stages are pasteurization, UF, and concentration. In this process, the quantity of waste pomace is dramatically reduced (around 60 kg pomace per ton of apples using horizontal press). The final yield is around 97 to 98% (Table 6). This technology is very flexible because the producer can produce cloudy juice, clear juice, or clear concentrate, gaining greater value from the raw material. Pomace liquefaction can be performed with the existing pressing equipment (only a heating system and a tank with a stirrer are needed). This results in a dramatic increase of the output of the plant with an important additional weight of concentrate.

French Apple Cider

French cider is produced from cider varieties of apples, bitter, bittersweet, sweet, or sour, with high tannin content (Bedan, Bisquet, Domaine, Douce Coetligné). In the traditional French process, the fermentation of the apple juice is very slow (a few weeks to 3 months). According to the French government, residual sugar content depends on the type of cider (brut, <28 g/L; demi-sec, 28 to 42 g/L; doux, >35 g/L with less than 3% ethanol). Because of the variability of raw material and the lack of hygiene, quality was difficult to manage. The use of continuous pressing and the diffusion system, the improvement of filtration equipment, and the control of bacteria and yeast strains resulted in improving the quality of French cider (36). In 1909, Warcollier described cider production, and the technology is almost the same today. The major part of French cider is produced by the technology of juice "defecation."

Traditional French Apple Cider Process. The existing process can be divided in three stages (Fig. 11). The first stage consists of the preparation of the apple must. Apples are crushed and pressed, and then a second extraction is performed by diffusion (countercurrent of cold water in the pomace) or second pressing. The second stage is the traditional clarification by "defecation." It consists of the pectin demethylation by the apple endogenous PME, giving pectic acid, which binds to apple calcium ions and becomes

Table 6. Yield and Technology

For 1 ton apples at 12.5 °C Brix	Pressing depectinization	Maceration depectinization	Total liquefactor	Pomace liquefaction
Enzyme	No Rapidase® C8OL	Rapidase® Press Rapidase® C80L	Rapidase® LIQ EXTRA	Rapidase® POMALIQ EXTRA
Dose	0 3 g/hL	50 g/ton 3 g/hL	150 g/ton apples	150 g/ton apples
Juice (kg)	750	850	860	750 + 455
Brix °	12.5	12.5	13.5	12.5 + 7.0
Concentrate (kg) 70 ° brix	134.0	151.8	165.9	179.5
Increase	–	+17.8 kg	+3.1 kg	+45.5 kg

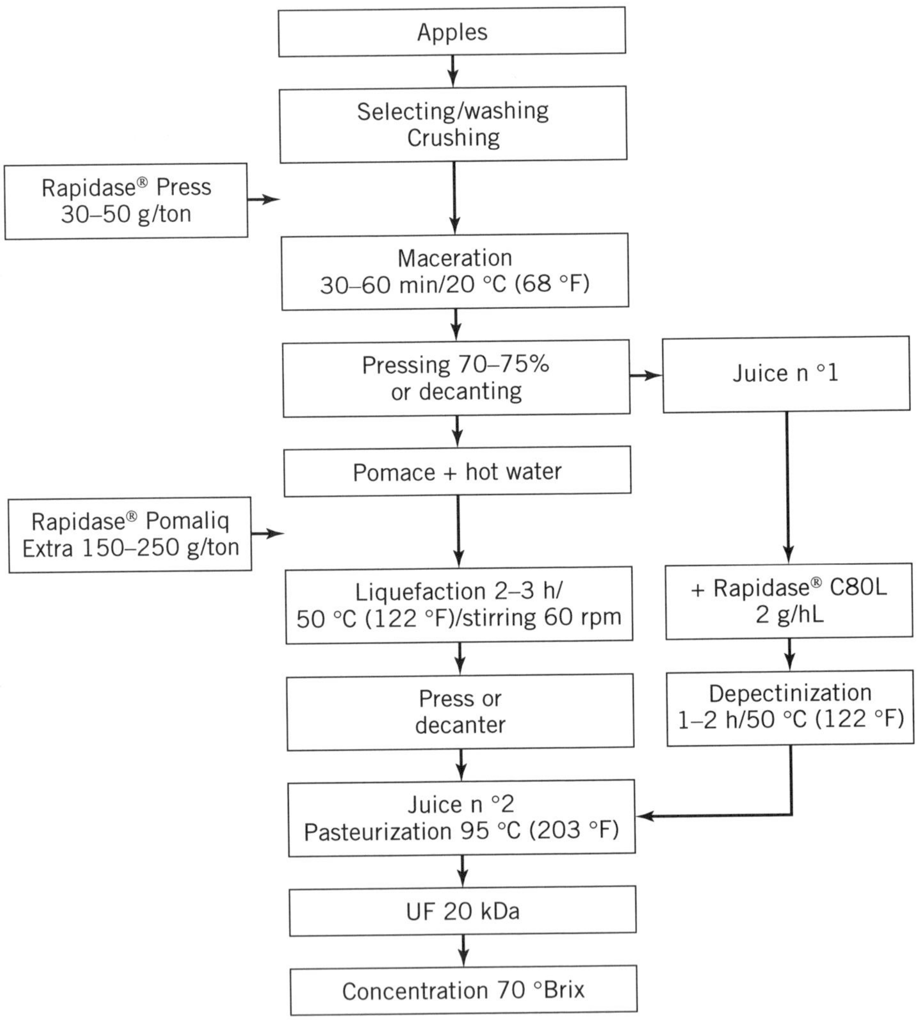

Figure 9. Apple pomace liquefaction process.

insoluble in the form of calcium pectate. The so-called coagulum, which entraps pectate, proteins, and tannins, rises to the surface because of CO_2 bubbles formed by the alcoholic fermentation that has started by this time. This coagulum is then discarded. This type of clarification lowers the nitrogen content, resulting in a slower fermentation that gives a more aromatic cider. This process, which was unreliable because of the variability of endogenous PME activity present in the juice, is now better controlled by the use of commercial fungal PME. Gist-Brocades and INRA Le Rheu have developed a new process of defecation using a purified pectin methylesterase Rapidase® CPE from *A. niger* that enables the cider producer to control and to accelerate the mechanism of coagulum formation. Rapidase® CPE is added at 7 to 10 mL/hL, plus anhydrous calcium chloride, 44 g/hL. A gel of insoluble calcium pectate is formed after about 36 h at 15 °C (59 °F) or a few days at low temperature that rises to the top of the tank on the surface of the must. This coagulum is then discarded. Before the fermentation, sulphur dioxide (50 mg/

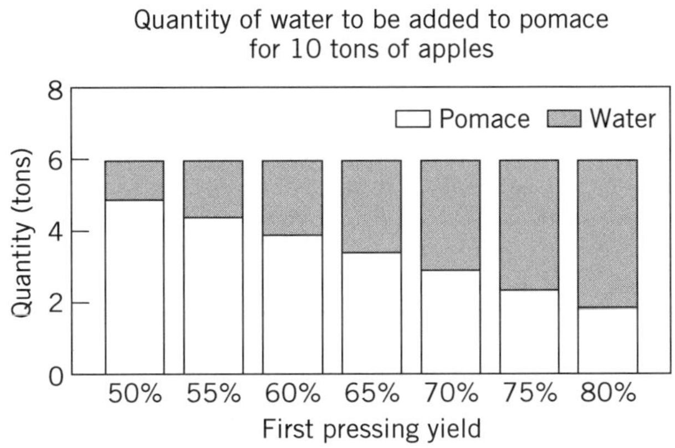

Figure 10. Water addition for apple pomace liquefaction.

L) can be added to the must to inhibit undesired microorganisms. The alcoholic fermentation takes place over 1 to 2 weeks at cellar temperature or up to 3 months in cold storage conditions.

The third stage consists of post-treatments after alcoholic and possibly malolactic fermentation. The cider is centrifuged to discard yeasts at a gravity selected according to the type of cider. It is filtered, and carbon dioxide (standard is above 5 g/L) is added just before bottling. Most of the ciders are pasteurized after bottling to ensure microbiological stability, but traditional ciders are not. In conclusion, the use of a purified PME like Rapidase® CPE (Gist-Brocades) results in producing French cider of high and constant quality whatever the raw material and process conditions. Recently, a new enzyme has been introduced in the cider industry that can be used as a macerating enzyme, improving the juice yield and at the same time making the clarification with flotation possible. This is a highly purified PME without side activity. Because of its high purity, this enzyme does not hydrolyze apple pectin, but the pectin can be extracted from the dry pomace after cider processing by pectin manufacturers.

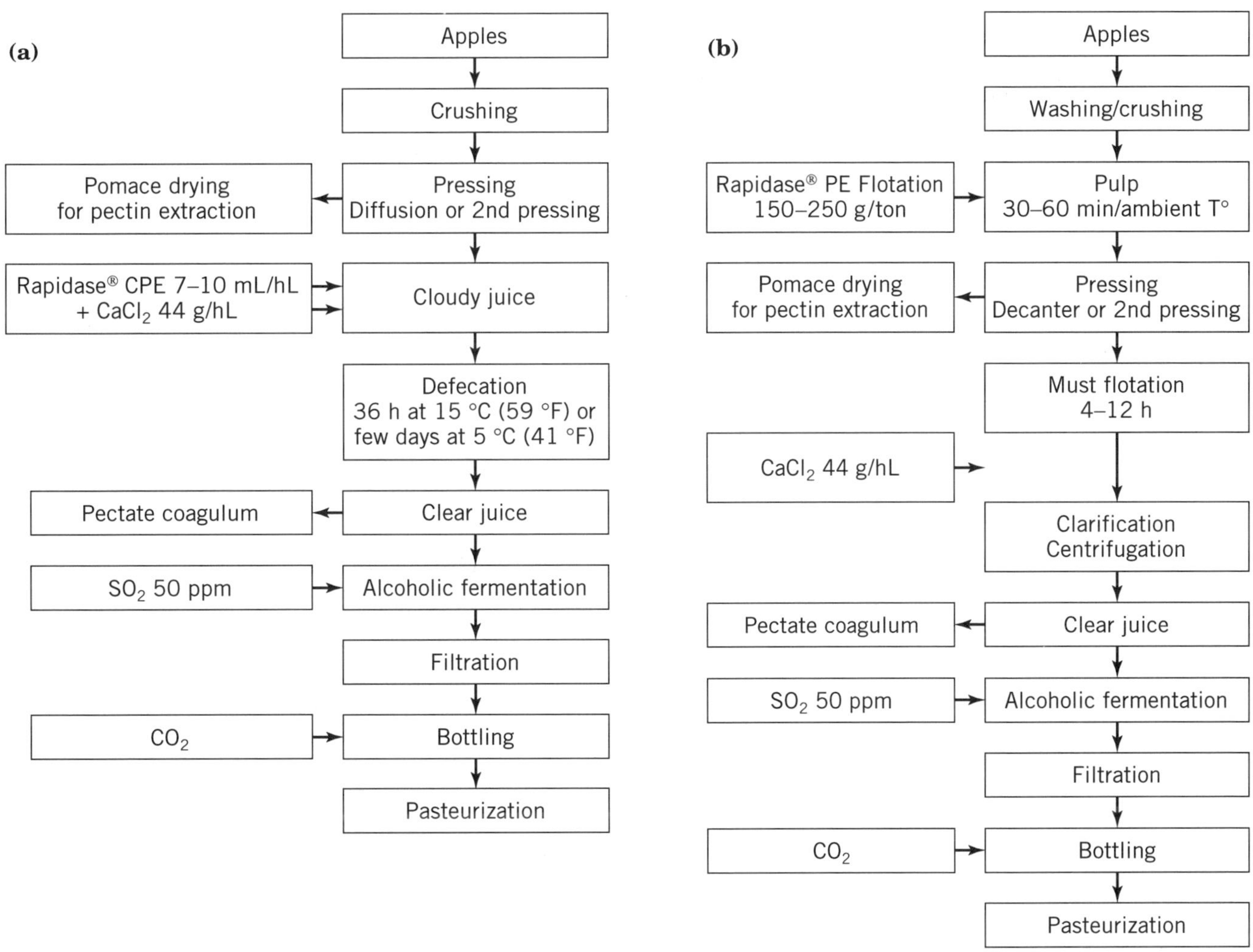

Figure 11. (**a**) Traditional French apple cider process. (**b**) French apple cider process with pulp enzyming.

French Apple Cider Process with Pulp Enzyming. The process using pulp enzymes is shown in Figure 11*b*. Apples are washed and crushed. The pulp is macerated with 150 to 250 g/ton Rapidase® PE FLOTATION (Gist-Brocades) at ambient temperature for 30 to 60 minutes. The juice extraction is performed with press or decanter, depending on available machinery and expected yield. After pressing and leaching, the pomace is dried and sold to pectin manufacturers. The juice is demethylated in 4 to 12 hours after extraction, depending on the juice temperature and the enzyme dosage. The demethylated juice is then clarified by flotation. Anhydrous $CaCl_2$ is added at 44 g/hL. The clarified juice is centrifuged and can be fermented at low temperature or stored. Fermentation is stopped at the required sugar gravity by centrifugation and earth filtration or microfiltration.

Cloudy Apple Juice

Stable cloudy juice is difficult to process. Because of the natural clarifying action of actual commercial pectinases, cloudy juice production with enzymes has not been possible until now. Prevention of juice oxidation is also important to avoid problems after bottling. To conciliate quality preservation together with a good extraction yield, the right enzyme and process must be chosen. The following process has been designed to improve the production of highly cloudy apple juice (about 1,500 NTU) with improved yield and reduced cloud loss risk.

Process Description. The process for making cloudy apple juice is shown in Figure 12*a*. Apples are washed sorted out and crushed. Pulp is added with 400 ppm of ascorbic acid to prevent oxidation. Maceration is carried out with 25 g/ton Rapidase® CLOUDY (Gist-Brocades) for 15 to 30 min at ambient temperature. Stirring must be avoided to prevent oxidation. Juice extraction can be performed with decanters, belt presses, or horizontal presses, depending on available equipment and required yield. Juice must be pasteurized as soon as possible after extraction (20 min is a maximum delay between pressing and pasteurization) to prevent any clarification. The juice is centrifuged for cloudiness standardization. It is then stored in aseptic tanks or immediately bottled. Some processors want to valorize apple pomace for pectin production. The process described next has been designed to improve juice yield by enzymatic treatment of the pulp without degradation of the protopectin of the pomace (37). Enzymes used in this process for pulp maceration do not contain pectin depolymerases but only a purified PME. In the mash, the pectin demethylation is followed by the formation of insoluble calcium pectate, which remains within the pomace after pressing.

The stronger firmness of the pulp containing pectate results in an increase of free-run juice and an easier release of the juice when high pressure is applied, such as with a hydraulic press. The partially demethylated pectin binds to other cloud components such as proteins. It remains in suspension in the juice after pressing, providing a better cloud stability. The extracted juice can be processed either to slightly cloudy juice (200 NTU) or to clear juice or concentrate.

Cloudy Apple Juice Process with Pomace Recovery. Apples are washed, sorted, and crushed (Fig. 12**b**). Pulp is macerated with 100 to 300 g/ton during crushing to apple mash, simultaneously with ascorbic acid at 200 to 400 ppm Rapidase® PE MASH for 30 to 60 min at ambient temperature. If the temperature is too low, the pulp can be heated up to 20 °C (68 °F). Stirring should be avoided to prevent oxidation. Juice extraction can be performed with decanters or presses. The juice is pasteurized and can be later processed to cloudy single-strength juice or clear juice. It must be centrifuged if processed as cloudy juice. It is then bottled or stored in aseptic tanks. It is depectinized if processed as clear juice with 2 to 4 g/hL Rapidase® PRO (Gist-Brocades) for 1 to 2 h at 50 °C (122 °F). The depectinized juice is clarified with traditional fining and filtration or UF and then concentrated.

ENZYMES AND PEAR JUICE PROCESSES

Commercial enzymes sold for maceration, depectinization, and liquefaction of apples work well to degrade pear pulp. However, juice producers sometimes have problems in processing such fruit. The pear is very fragile, and it ripens fast, resulting in a decrease of pressability and yield. Changes during ripening and postharvest storage are similar to those described for apples. The soluble pectin content increases, and there is a pectin de-esterification (endogenous PME) and partial pectinolysis to lower molecular weight uronides (endogenous PG) (Table 2). Pectate content increases. Phenolic compounds are oxidized during ripening and form insoluble compounds (lignin). In pear, the presence of scleroids or stone cells in the pulp results in a lower textural quality. These stone cells are lignocellulosic and contain approximatively 18% lignin and 82% carbohydrates, which makes filtration or UF difficult (38). Araban, hemicellulose, and cellulose contents are higher than in apples. Linear arabans can also reassociate after concentration, producing araban haze. This is why it is important to use pectinases with strong arabanase and arabinosidase activities. Enzymes such as Rapidase® Pear (Gist-Brocades), Pectinex AR (Novo Nordisk) or Pearex (Solvay) prevent haze formation and improve the stability of the concentrate.

Clear Pear Juice Concentrate

Pear Pulp Maceration and Juice Depectinization. The process is similar to that for apples (Fig. 5). Pears are crushed and enzymes, for example, Rapidase® Press (Gist-Brocades), are added to the pulp at about 75 g/ton for 30 to 60 min at ambient temperature.

Pectinases PL, PME, and PG and arabanases contained in Rapidase® Press improve the extraction and the juice yield. After pressing, the juice is depectinized with pectinases containing high *endo-* and *exo-*arabanase activities (e.g., Rapidase® Pear from Gist-Brocades) for about 1 h at 45 to 50 °C (113 to 122 °F). The juice is then fined and filtered or ultrafiltered and concentrated up to 70 °Brix. Pears can also be processed to purees or nectars.

Pear Total Liquefaction. Next to pectinases, hemicellulases are of first importance for pear liquefaction (Table 5).

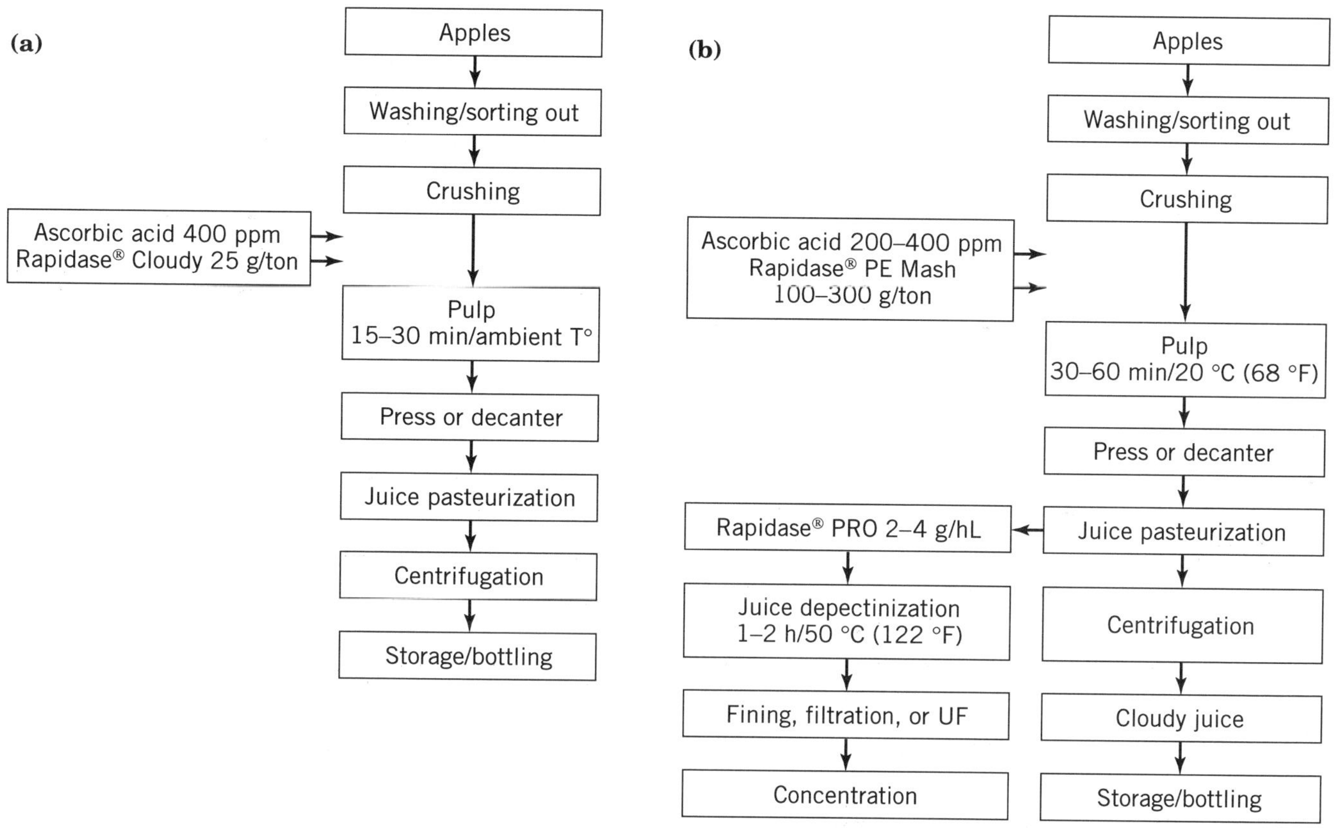

Figure 12. (**a**) Cloudy apple juice process. (**b**) Cloudy apple juice process with pomace recovery.

The pear total liquefaction process is similar to the apple total liquefaction process, including a hot break before enzyme addition (Fig. 8). Pears are crushed and heated up to 85 °C (185 °F) for 4 to 5 min to destroy the endogenous PPO and to open the cell wall. Cooled down at 50 °C (122 °F), the pulp goes through a finisher that discards seeds, stems, and stone cells. The liquefaction is carried out during 2.5 h at 52 °C (125.6 °F), with enzymes such as Rapidase® Pear Liq (Gist-Brocades) or Pearex (Solvay) added from 200 to 250 g/ton with continuous stirring at 60 rpm. The juice is centrifuged with a decanter and pasteurized. Then, aroma recovery is performed. The juice is either fined to remove proteins and tannins and filtered or ultrafiltered and then concentrated up to 70 °Brix. Pear total liquefaction is carried out in only one step; no depectinization is required.

ENZYMES AND RED BERRY JUICE PROCESSES

The production of clear juices and concentrates from strawberries or blackberries requires enzymatic depectinization. The clarification, filtration, and concentration remains difficult because these juices contain high amounts of pectic substances (39). It is always difficult to obtain a complete depectinization of the juices, and it is normal to find a residual low molecular weight pectin level of 7 g/L, compared with about 0.5 g/L for apple juice. The assumption is that hairy regions of the strawberry pectin remain as colloidally dissolved residual pectins in the juice, whereas rhamnogalaturonans are the main group of residual pectins in raspberry juice. Neutral sugars are also important in the depectinization. With typical molecular weights from 10,000 to 1 million Da, polysaccharides, arabinose, and galactose are the main sugar constituents, probably in arabinogalactans. Moreover, residual pectins and hemicelluloses tend to bind to phenolics and proteins during processing and storage. The result is the formation of irreversibly linked complexes that enzymes cannot break down anymore. An additional problem is related to the frequent contamination of red berries, mainly strawberries and raspberries, by *Botrytis cinerea.* This parasitic fungus, which in many cases cannot be seen from inspection of the fruit, secretes a β1,3–1,6-linked glucan into the berries with molecular weight around 10^6 Da. This gum reduces the filterability and the clarity of the juice. It is possible to hydrolyze this glucan with β-glucanases such as Glucanex (Novo Nordisk). Berry juices must have a good flavor and are attractive in color. For this reason, they have a long history of use in the preparation of fruit derivatives and other fruit beverages. With the exception of blackberry and cranberry juices, none is commonly used as a table beverage. Common berries have low sugar content (from 4 to 8%).

The average acidity is about 1% calculated as citric acid. Cranberry is an exception, with about 1.9% or more. It is for this reason that most berry juices are diluted with a half volume of light sugar syrup. Pectinases are used to improve the pressing stage and to extract the red color, while retaining the organoleptic properties of the fruit. The color is located in the skin of the berries. It can be released by heat, alcohol, or enzymes. However, it is not advantageous to extract the color if it is then destroyed by oxidation. Oxidation can be chemical, caused by atmospheric oxygen, or enzymatic, from the endogenous polyphenoloxidase (PPO) and peroxidase in the fruit. These enzymes can be inhibited with sulfur dioxide addition or with heating. It is therefore recommended that the pulp be heated to over 85 °C (185 °F) to inhibit these endogenous enzymes. This action not only improves the release of color but also the protopectin solubilization, and this is why pectinases are required. Black currants are very acid (pH of 2.6 to 2.8) and contain high levels of phenolics and anthocyanins. Both acids and polyphenolics inhibit fungal pectinases. This results in high doses of enzymes that are added and emphasizes the importance of enzyme stability in the juice.

Red Berries Single Stage Process. The single-stage process (Fig. 13**a**) is recommended for berries such as blackberry, bilberry, elderberry, gooseberry, raspberry, and black currant. This single-stage process allows the performance of maceration and depectinization at the same time. Yield is maximized, and good quality is preserved. This process is recommended when decanters are used for juice extraction. Fruits are inspected and sorted out, then thawed if necessary and crushed. The pulp is heated up to 90 °C (194 °F) for a thermal break of 2 min minimum to increase color extraction and to destroy endogenous polyphenoloxidase (PPO). Then the pulp is cooled down to 50 °C (122 °F) for enzymatic maceration. Enzymatic maceration is carried out with 200 to 400 g/ton Rapidase® BE Super (Gist-Brocades) or Pectinex BE (Novo Nordsik) at 50 °C (122 °F) for 1 to 2 h. If the pulp pH is below 3.2, we recommend the use of Klerzyme® 120 (Gist-Brocades) at same dosage instead of Rapidase® BE Super. Juice separation can be performed either with press or with decanter. The extracted juice should be pectin free (acidified alcohol test or acetone test should be negative). No further enzymatic treatment is required. Juice is pasteurized after extraction for protein denaturation. A minimal temperature of 90 °C (194 °F) is necessary. Cooling down below 50 °C (122 °F) leads to protein insolubilization. Pasteurized juice can be clarified either with UF or with traditional fining with gelatin, bentonite, and silicasol. In the case of UF being used, membrane cutoff should not be below 30,000 Da to prevent color loss. The juice is concentrated.

Red Berry Two-Stage Process. Applied to the same berries as described for the single-stage process, the two stages process consists of an enzymatic maceration of fruit pulp followed by a second enzymatic treatment of the extracted juice for its depectinization (Fig. 13**b**). The low processing temperature after the hot break prevents aroma losses, and the best quality juices and concentrates can be produced. Fruits are inspected and sorted, thawed if necessary, and crushed. Fruit pulp is heated up to 90 °C (194 °F) for a thermal break of 2 min minimum to increase color extraction and to destroy endogenous PPO. Then the pulp is cooled down to 20 to 25 °C (68 to 75 °F) for enzymatic maceration, which is carried out with Pectinex BE or Rapidase® BE Super at 100 to 200 g/ton at 20 °C (68 °F) for 30 to 60 mins. If the pulp pH is below 3.2, we recommend the use of Klerzyme® 120 at same dosage instead of Rapidase® BE Super.

Juice separation can be performed either with the horizontal or belt press. After extraction, juice is pasteurized at minimum 90 °C (194 °F) for protein denaturation. Cooling down to 20 °C (68 °F) leads to protein insolubilization. Depectinization is carried out with Rapidase® BE Super or Pectinex BE at 4 to 8 g/hL at 20 °C (68 °F) for 1 to 2 h. If the pulp pH is below 3.2, we recommend the use of Klerzyme® 120 at same dosage instead of Rapidase® BE Super. The acidified ethanol or acetone test should be negative at the end of the treatment. The juice can be clarified either with UF or with traditional fining with gelatin, bentonite, and silicasol. In the case of UF being used, membrane cutoff should not be below 30,000 Da to prevent color loss. The juice is concentrated.

Strawberry

In contrast to black currant or red currant, it is impossible to heat the strawberry pulp because it would create an unpressable puree. Furthermore, many aromas would be lost during heating, oxidation, destruction, or evaporation. Because it is not recommended to heat such pulp, endogenous PPO cannot be destroyed, which can result in condensation reactions between oxidized products and anthocyanins and proteins and pectin, leading to great loss of color. Additionally, anthocyanin pigments are temperature and oxidation sensitive. Thus, it is essential to work at low temperatures to process delicate fruit such as strawberry. Special care is required to prevent oxidation and to keep aromas. Quality in term of the sanitary state of processed fruit is of first importance. Contaminated strawberries could contain very viscous β-glucan from *B. cinerea*. Prevention of oxidation with quick processing is also of key importance.

Process Description. Strawberries are inspected and sorted, thawed, and crushed with a hammer mill fitted with a 0.635-cm screen (Fig. 14). Pectinex BE (Novo Nordisk) or Rapidase® BE Super (Gist-Brocades) can be added at 100 to 150 g/ton to the pulp at 20 °C (68 °F) for 30 to 60 min. For the processing of frozen fruits, enzymes are added after heating when the temperature has decreased to about 20 °C. The juice is extracted with a horizontal press, belt press, or decanter. Extraction yield is around 70 to 80%. The juice is flash pasteurized at minimum 90 °C (194 °F) for protein denaturation and a simultaneous aroma stripping. The juice is then cooled down at 30 °C (86 °F) to be depectinized with the same enzymes at 2 to 5 g/hL over 1 to 2 h. In countries where glucanase treatment is authorized, Cytolase CL (Gist-Brocades) or Glucanex (Novo Nordisk) glucanase-containing products can be added for glucan hydrolysis at 2 to 5 g/hL together with pectinases. To control depectinization and glucan hydrolysis, two specific tests are performed. The specific alcohol test is per-

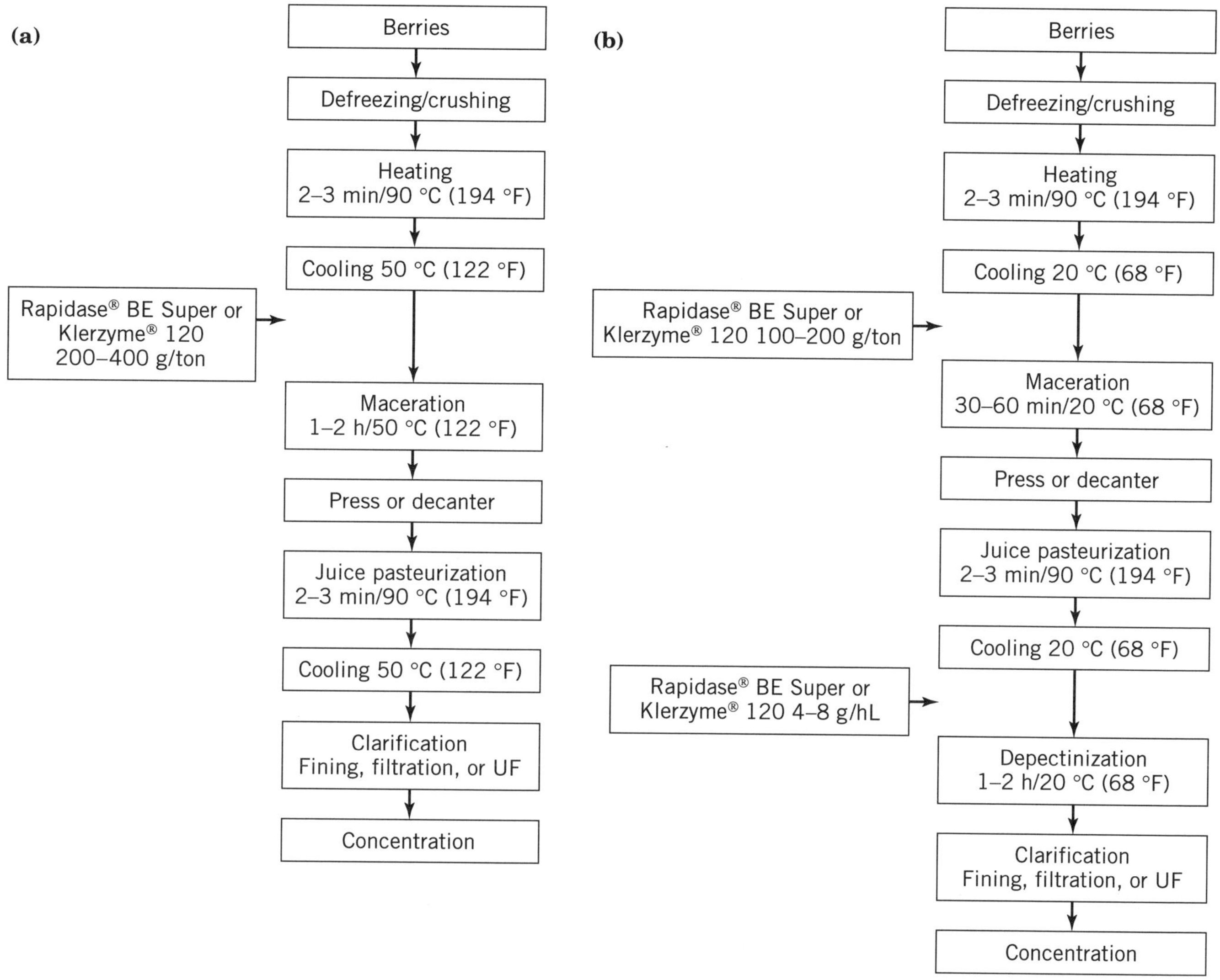

Figure 13. (**a**) Red berry juice, single-stage process. (**b**) Red berry juice, Two-stage process.

formed to check glucan degradation. One vol ethanol plus 1 vol juice are gently mixed. White filaments appear when glucanes are present. The specific acetone test is performed to check pectin degradation. The alcohol test used to check the depectinization of apple juice is not suitable for red berries because anthocyanins and other compounds can also precipitate together with pectin. We recommend mixing 1.5 vol juice with 1 vol acetone. The test is negative when the sediment is less than 1:10 of the control or when the amount of precipitated colloids is less than 1 g/L (when strawberry concentrates are blended for the liquor industry, no precipitation or haze formation should occur with this test). Strawberry juice is clarified either with UF or with traditional fining treatment with gelatin, bentonite, and silicasol. In UF, the membrane cutoff should not be below 30,000 Da to prevent any color loss. The clear juice is then concentrated. *Note*: It is important to have stainless steel equipment to avoid oxidation catalyzed by metallic ions such as iron, copper, or aluminium. A final UF step before concentration is recommended.

Cherry

The global production of cherries, supplied by more than 55 countries, varies from year to year between 2 and 3 million tons, more than 60% of which is sour cherries. Fruit processing involves an average of 40% of sour cherries and 15% of sweet cherries. The most important world producers of cherries (in thousands of tons) are Germany (280), the United States (170), Italy (140), and the former Soviet Union (130). Production in China is growing fast. Cherry pectin has smooth and hairy regions, with arabinose and galactose as main neutral sugars.

Process Description. After washing and sorting, cherries are destoned and crushed (Fig. 15). The pulp is heated up to 85 °C (185 °F) for 3 min to destroy endogenous PPO and to increase color extraction, and then cooled down to 50 °C (122 °F) for enzymatic maceration. Maceration is carried out with Klerzyme® 120 at 300 to 400 g/ton for 2 h at 50 °C (122 °F). The juice is extracted with a horizontal

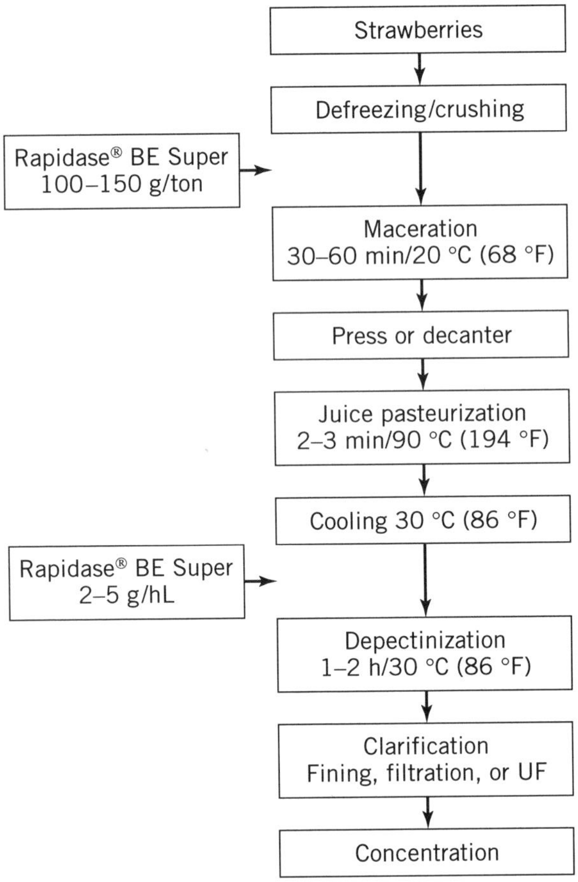

Figure 14. Strawberry juice process.

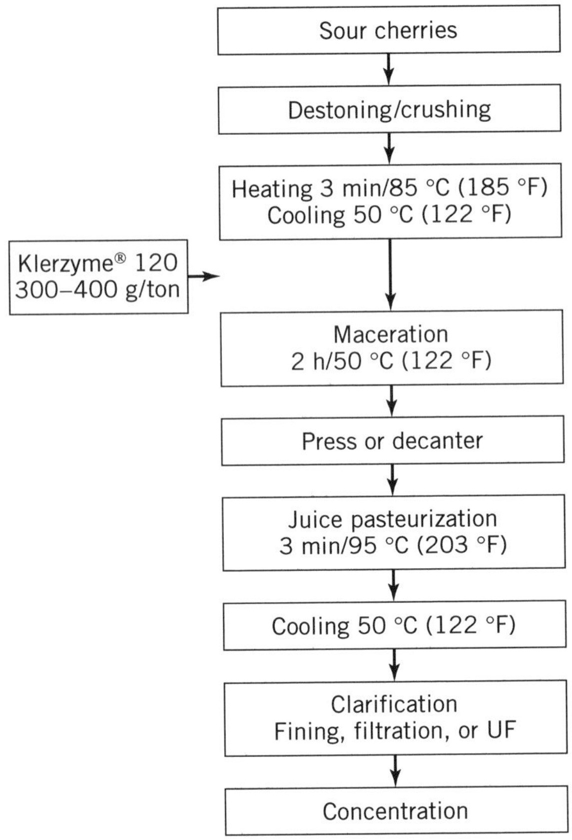

Figure 15. Cherry juice process.

press or decanter. Extracted juice should be pectin free (alcohol test negative). It must be pasteurized to insolubilize proteins at minimum 95 °C (203 °F). The juice is then cooled down to 50 °C (122 °F) for clarification either by traditional fining and earth filtering or by UF, with the membrane cutoff above 30,000 Da. The juice is then concentrated.

Cranberry

Cranberry juice is very acidic, with an acid content exceeding 2% w/w, and a pH of 2.6 or lower. The juice is very astringent because of a high natural level of tannins and leukoanthocyannins. Because the single-strength juice, with a Brix-to-acid ratio of 3.75, is unpalatable (the ratio for grapefruit is 10), formulated cranberry juice blends have been developed by companies such as Ocean Spray. Many other cranberry-based juice drink blends have been marketed. They have a fresh taste and attractive pink color. Cranberry is reported to be the third favorite flavor in the United States after orange and apple (40). Cranberry pectin exhibits different properties than citrus or apple pectin. Cranberry sauce is manufactured with a solid level of only 40%, whereas other fruits need 65% solids. In addition to citric and malic acid, the fruit contains also quinic and benzoic acids. With a low carbohydrate content and significant quantities of vitamin C, potassium, and phenolic glycosides, this juice is considered healthy.

Process Description. Fresh or frozen fruits are milled and heated up to 85 °C (185 °F) for 5 min, and then pulp is cooled down to 50 °C (122 °F) for enzymatic maceration (Fig. 16). Biopectinase BP300 (Quest) or Rapidase® Cranberry (Gist-Brocades) pectinase is added to the pulp at 200 to 400 g/ton for 1 to 2 h. The juice is extacted with a decanter or horizontal press, pasteurized, ultrafiltered, and concentrated up to 50 °Brix.

Concord Grape

Grape juice is not consumed in large amounts because it is too sweet (about 200 g/L sugars) or too acidic (up to 10 g/L tartaric acid). It is mixed with other fruit juices such as apple. Pederson in 1980 reviewed the technology of grape juice processing (41). Red grapes, especially Concord variety and its hybrids, are the principal grapes for the U.S. industry. These grapes are processed to juice. The Concord variety is rich in pectin and anthocyanins and has a unique flavor due to methyl anthranilate. The acidity is 13 g/L as tartaric acid. This juice is often recommended in special diets. Because of the high pectin content (5 to 10 g/L), berries are difficult to crush and to press.

Process Description. Grapes are destemmed, crushed, and heated up to 85 °C (185 °F) for 3 min to release red color pigments from the skins and to destroy PPO (Fig. 17). After cooling down to 60 °C (140 °F), enzymes such as Cytolase® PCL5 (Gist-Brocades) or Ultrazym (Novo Nor-

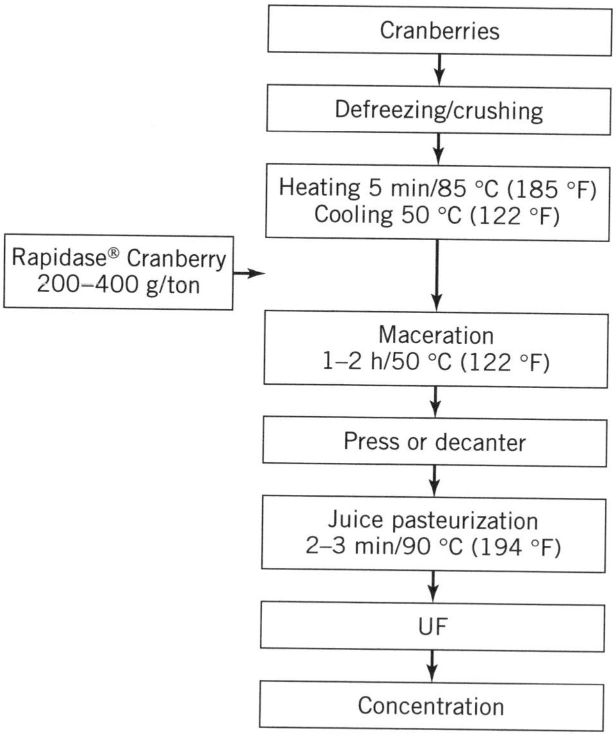

Figure 16. Cranberry juice process.

disk) are added at 50 g/ton for 1 h. Precise monitoring of temperature is important; it should not be below 55 °C (131 °F). After 1 h, juice extraction can be carried out with decanters, a screw press, or a horizontal press, after addition of cellulose paper for absorbing particles. The juice is pasteurized at 88 °C (190 °F) for 30 sec and then cooled down to − 1 °C (30 °F) to prevent start of alcoholic fermentation. The juice is clarified at low temperature for 7 to 14 days with Klerzyme® 200 (Gist-Brocades) at 200 to 400 ppm. Enzymatic clarification of grape juice as well as other juices is based on the partial hydrolysis of soluble pectin that acts as a protective colloid. Pectinases, mainly PG and hemicellulases-like arabinogalactanases, help insoluble particles to flocculate.

At the same time, cold temperature induces the insolubilization and sedimentation of tartaric acid salt, reducing the total acidity to acceptable levels. The juice is then filtered with diatomaceous earth on vacuum filter and then on pressure filter and concentrated or pasteurized and bottled. Concord grapes can also be processed with enzymatic liquefaction. A thermal break is necessary. Pulp is heated up to 85 °C (185 °F) for 3 min and then cooled down to 60 °C (140 °F) for enzymatic maceration, which is carried out with 300 g/ton Klerzyme® LIQ. Temperature monitoring is important and should not be below 55 °C (131 °F) nor above 60 °C (140 °F). Juice extraction is carried out with decanters. It is pasteurized at 88 °C (190 °F) for 30 sec. Clarification takes place at cold temperature, − 1 °C (30 °F) to 2 °C (36 °F) for 3 to 7 days using 10 to 20 g/hL Klerzyme® 200. Tartaric stabilization occurs during this period. Clarified juice is usually vacuum filtered and then pressure filtered and concentrated.

White Grape

Different red and white grape varieties such as Niagara, Thomson seedless, Sultana, and Muscat are processed to juice and concentrate. Because grape sugar content is very high and because the berry has endogenous PPO activity, it is necessary to process this type of fruit very quickly to prevent start of alcoholic fermentation and juice oxidation.

Process Description. White grapes are destemmed and crushed (Fig. 18**a**). Special care should be taken to avoid rotten fruits, frequent contamination by *B. cinerea* results in the presence of glucan and laccase in the fruit. Glucans cause pressing problems, and laccase is a fungal polyphenoloxidase causing juice browning. Grape pulp is macerated at ambient temperature for 1 to 2 h with 40 to 80 g/ton Klerzyme® 120. If the temperature is too warm, cooling down to 15 to 20 °C (59 to 68 °F) is required to prevent fermentation. Pressing can be carried out with a horizontal press, pneumatic press, or screw press. The juice is quickly pasteurized. After cooling at 50 °C (122 °F), depectinization is carried out with 2 g/hL Klerzyme® 120 or Ultrazym (Novo Nordisk) for 1 to 2 h. Enzyme is added at the beginning of tank filling. No stirring is required. Depectinized juice can be filtered or ultrafiltered. It is then concentrated.

Red Grape

A main concern for red grape juice producers is color extraction. Process conditions, the sanitary state of grapes, and enzymes are key factors for good color stabilization in the final juice. Anthocyanidins, responsible for the red color, are glycosylated with glucose. These pigments turn from red to brown depending on light, pH, temperature, enzyme oxidases (naturally present in the grape), and β-glucosidases (present in grape and a side activity of pectinase products) that destabilize anthocyanins after hydrolysis of their glycosidic part.

Process Description. Red grapes are sorted out before crushing (Fig. 18**b**). Berries are heated up to 90 to 95 °C (194 to 203 °F) for 3 min to increase color extraction and to destroy endogenous PPO. Pulp is cooled down to 50 °C (122 °F) for the enzymatic maceration. Rohapect VR (Röhm) or Klerzyme® Color is added at 40 to 100 g/ton for 2 to 5 h. Pulp should be carefully mixed with enzymes during tank filling and just before pressing. No stirring is required during maceration. Color extraction is monitored, and the juice should be pectin free (alcohol test). Juice extraction can be carried out with horizontal, pneumatic, screw presses, or decanters. Juice should be pasteurized quickly after extraction for a better color stability. Juice is vacuum filtered, and then pressure filtered or ultrafiltered (membrane cutoff higher than 50,000 Da) and concentrated.

ENZYMES, STONE FRUIT AND TROPICAL FRUIT PROCESSES

The production of fruit puree, nectar, pulpy juices, and fruit blends are of great commercial importance. Many

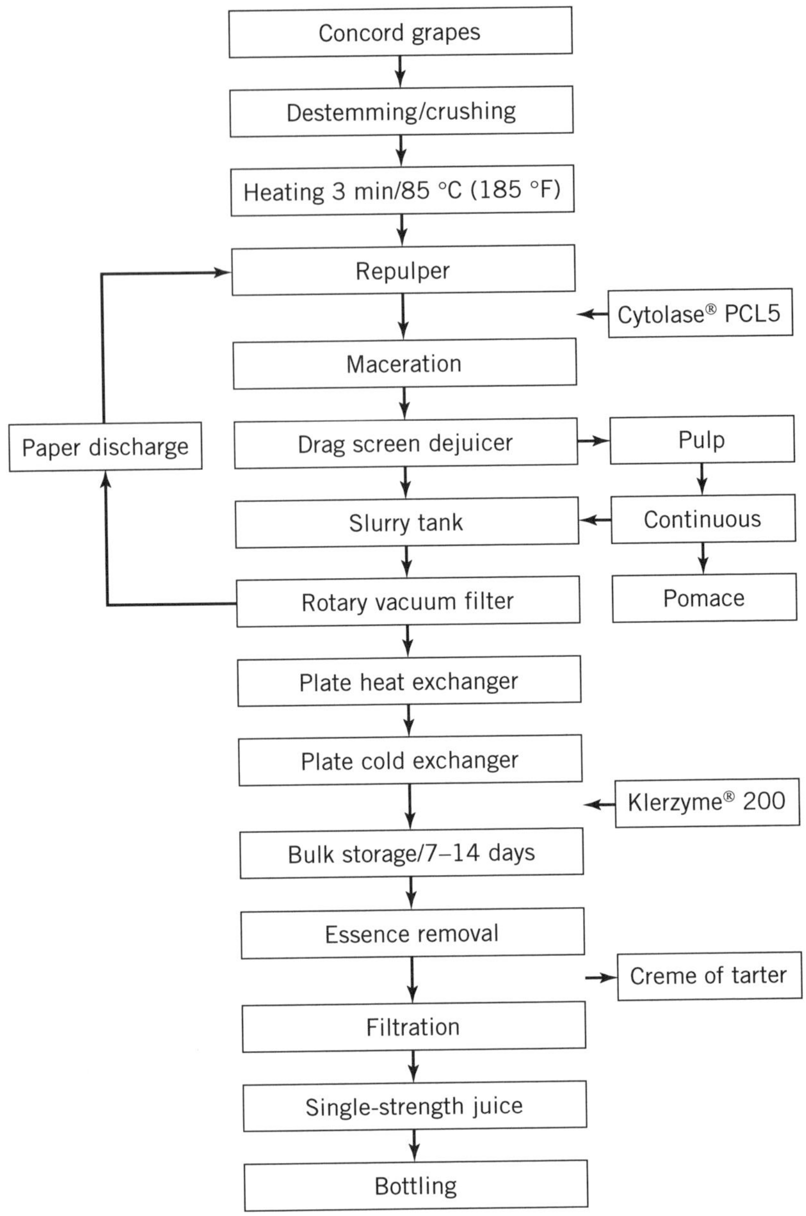

Figure 17. Concord juice process.

fruits are not pressable, or they are too acidic or too strongly flavored to be pleasant beverages without dilution or blending or both. These juices are delicious after diluting with thin syrup or blending with other juices; good examples of this are apricot, peach, pear, and tropical fruit such as mango, guava, papaya, and banana. Other fruits may not have enough flavor, and they are greatly improved if the entire fruit is processed into a smooth pulpy beverage. Fruit nectars are pulpy fruit juices blended with sugar syrup water and citric acid to produce a ready-to-drink beverage. Their composition is strictly defined in the Code of Federal Regulations (1984). They come from diluted purees or pulpy juice process. Cloudy beverages have a high content of fruit pulp (mango and guava have more than 35% pulp). The most important quality factor of these nectars is that the cloud be stable. After mixing with water, sugar, and acid, the cloud particles tend to settle down and sometimes form a gel with a clear supernatant layer. This separation is a serious defect that reduces the attractiveness of the product. Fruit polysaccharide structures change during ripening because of endogenous enzymes (Table 2). The fruit texture is modified and has an effect on the texture, viscosity, and stability of further-processed nectars. Prevention of cloud loss or instability in nectars has focused on the use of homogenization as a mechanical means to reduce the particle size. The use of exogenous enzymes can improve the cloud stability in nectars or make puree concentration easier. The general process is used to produce puree, nectar, cloudy, or clear juice from pear, peach, apricot, mango, or guava.

Process Description. Fruits are washed, sorted, halved, and destoned (Fig. 19). Destoned fruits are heated with

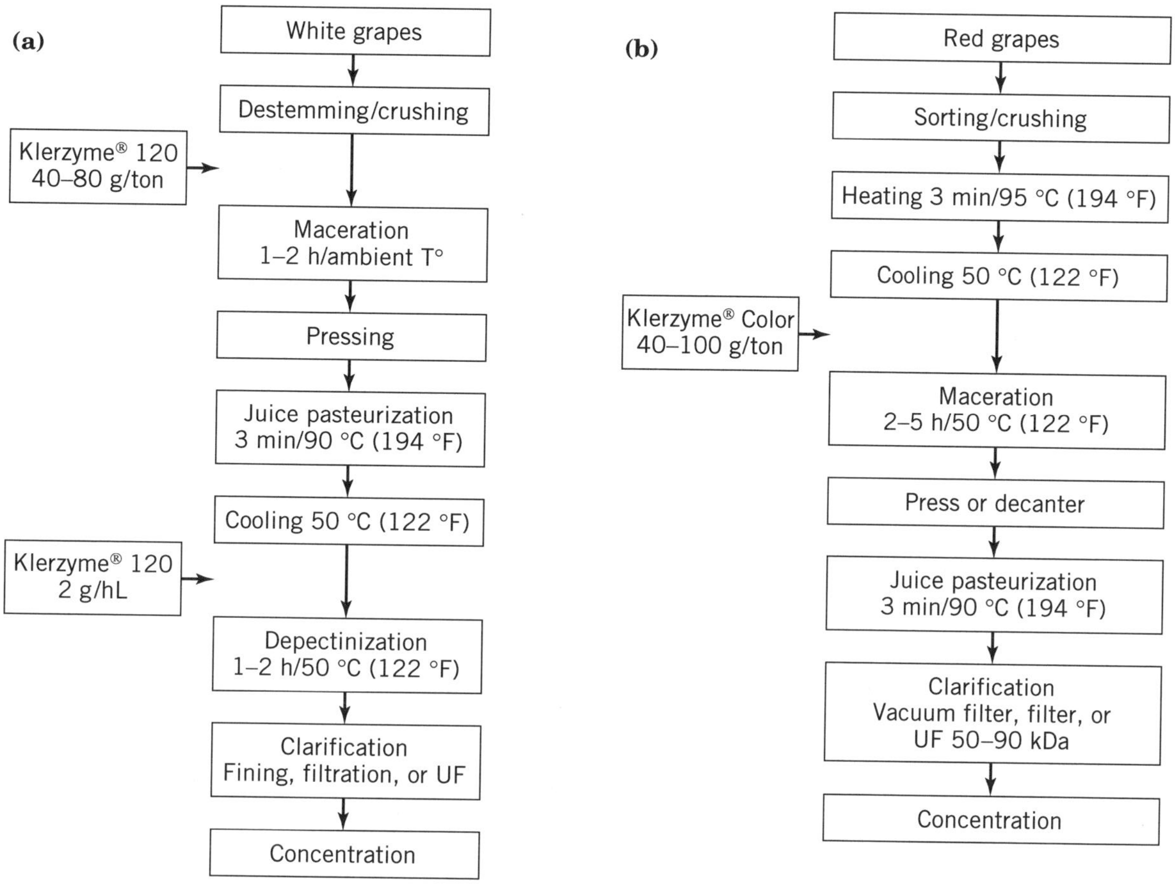

Figure 18. (**a**) White grape juice process. (**b**) Red grape juice process.

steam in a continuous blanching step at about 85 °C (185 °F) for 5 min to inactivate endogenous PPO, to avoid puree browning, and to kill microorganisms. Enzyme preparations should have high PG or PL activities combined with cellulases and hemicellulases. For puree processing, maceration is carried out with Pectinex UltraSP (Novo Nordisk), Rohapect TF (Röhm), or Rapidase® TF (Gist-Brocades) at 50 to 100 g/ton for 1 to 2 h at 50 °C (122 °F). These enzymes have been found to decrease the viscosity of puree or cloudy juice without destabilizing the cloud and to improve the ability to concentrate the product. Puree extraction is performed with a finisher (with a 0.6 to 0.8-mm screen). The resulting puree is concentrated or deaerated and passed through a steam-heated tubular heat exchanger where it is quickly heated to pasteurization temperature, 88 to 95 °C (190.4 to 203 °F), before canning. For the production of nectar, the puree is sweetened with sugar syrup and acidified with citric acid to maintain a constant ratio of sugars to acid. The nectar is then canned or bottled, sealed and processed at 100 °C (212 °F) for 2 to 3 min. The Tetra-Pak system is now widely used for aseptic packing of nectars. Cloud instability in nectars depends mainly on the presence of large pulp particles associating together, and it is also related to the endogenous PME. This enzyme leads to the formation of calcium pectates that form sediment and result in clarification. It is necessary to inactivate this activity as soon as possible in the process, because even a low residual activity after processing would destabilize the nectar during storage after bottling. It has been found that PME from guava or mango (at pH greater than 4) had a high heat stability, and it was necessary to heat the nectar for at least 1 min at 97 °C for its inactivation.

For clear juice processing, 150 to 200 g/ton Rapidase® TF for 2 to 2.5 h at 50 °C (122 °F) is added to the puree. The juice is extracted with a horizontal press or decanter and then pasteurized at 95 °C (203 °F) to denaturate proteins. Clarification is carried out with diatomaceous earth filter or UF. The clear juice can be concentrated.

Peach and Apricot

The role of various polysaccharide degrading enzymes for cloud stabilization in peach and apricot nectar has been demonstrated. Fruit primary cell walls contain around 90% polysaccharide and 10% protein. Apricot nectar exhibits a phenomenon of cloud loss. This separation can be avoided by grinding the pulp very finely (42). The addition of pectolytic enzymes separately or in combination with homogenization results in a cloud-stable apricot nectar, but homogenization alone fails to achieve cloud stability. The cellulose microfibrils liberated by mechanical action become more hydrated and interact with constituents in the serum, increasing the yield and the cloud stability. En-

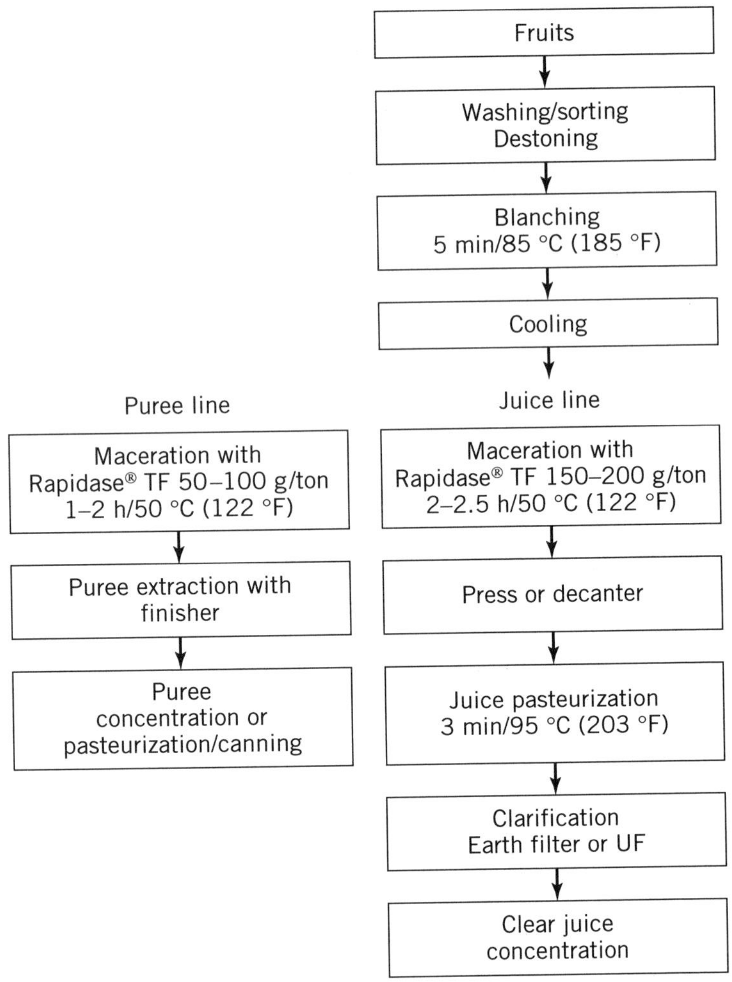

Figure 19. Stone fruit and tropical fruit juice process.

zymes such as Rapidase®TF are able to degrade the arabinan-rich pectin, open cells in the pulp, and assist in the production of clear peach and apricot juice concentrate.

Mango

Mango is the most important tropical fruit. India is the largest producer, but it is also produced in Mexico, China, Brazil, and the Philippines. The annual world production is about 20 million metric tons. The skin color of ripe fruits is yellow-red, under which the flesh is white, pink, green, or orange-yellow and juicy. The taste varies from sweet to sour. It is a source of vitamin C (100 mg per 100 g) and nicotinic acid. Mango is processed as a puree or nectar, juice, and various juice drink blends. A typical mango nectar is one with 20 to 30% puree, 12 to 18 ° Brix, pH around 3.5, and titratable acidity 0.2 to 0.3%. Exogenous enzymes are used to increase the concentration of mango puree or to produce clear juice and concentrate. Enzymatic treatment of mango destoned flesh provides a yield of around 80% at 15 °Brix. Pectinases and hemicellulases are important activities to degrade mango pulp (Tables 2 and 5).

Guava

Guava is a popular fruit throughout the tropics, eaten fresh or made into juice, nectar, puree, and jam or jelly after removal of the seeds. The major producers are South Africa, India, and Hawai. Guava puree is reprocessed into nectars, juice drink blends, syrups, jams, or jellies. Fruit pulp is increasingly treated with exogenous enzymes to improve juice extraction. Guava pericarp has two main cellular types: parenchyma and stone cells, which are also encountered in pear, plum, and olive mesocarps (43). Stone cells are strongly lignified woody material. Cell walls from stone cells represent the major proportion of total cell wall material found to be insoluble in alcohol (74% of insoluble substances). The mesocarp portion contains 90% of total cell wall material of guava pulp, 77% of which are stone cells. Half of stone cells are made up of neutral polysaccharides, 94% of which are cellulosic glucose and xylose (Table 2). Uronic acids are also present and *O*-acetyl groups are important and could be related to xylose-containing polymers. The dominant structural feature in parenchyma cell walls is cellulose, with side chains of glucans and xyloglucans, highly branched arabinans, and type I arabinogalactans; the rhamnose comes from rhamnogalacturonans.

Guava stone cells contain large amounts of β-(1→4) xylan as the main constituent. Compared to other fruit juices, an additional centrifugation is required to remove stone cells.

Yellow Banana

Banana production is a key activity in a lot of equatorial countries. A considerable amount of bananas are rejected because they do not meet quality standards; this waste can be substantial (160,000 tons of rejected bananas in Costa Rica in 1978). The quantity of processed bananas is very small because of the competition with fresh fruits throughout the year. Many banana products have been described, such as powder, flour, chips, puree, canned slices, jams, beverages, and alcohol. The banana is widely appreciated for its flavor and aroma (isoamylacetate) either as banana juice or a mixture with other juices. However, banana juices are too pulpy and too viscous to yield beverage juices easily, even if they contain endogenous pectinases such as PME and PG (44) (Table 2). Bananas contain 3 to 4% fiber, 1% pectin, and 2% hemicellulose (45). The composition of banana polysaccharides changes considerably during ripening, with a large decrease in hemicellulose, cellulose, pectin, and starch. Bananas can be processed at the green stage, mainly for alcohol production, or at the ripe stage, for the production of clear juice, requiring different enzyme treatments.

Process Description. Yellow bananas are peeled and ground (Fig. 20). Banana PPO, which is very active, must be inhibited by heating or sulfite treatment (100 mg/L of potassium metabisulfite) to prevent browning. Addition of ascorbic acid is also possible. After heating at 85 °C (185 °F) for 3 min and then cooling at 45 °C (113 °F), Rapidase® Press (Gist-Brocades) can be added at 75 to 100 g/ton over 1 to 2 h. After decanting, the clear juice at about 85% yield is pasteurized, filtered, and concentrated.

Clear Pineapple Juice

Annual pineapple production worldwide is 5 million tons in Thailand, the Philippines, Indonesia, and Kenya. Smooth Cayenne is the main variety. Thailand was the world's leading producer of commercial canned pineapple (420,500 metric tons) and pineapple juice concentrate (87,000 metric tons) in 1991. There are two types of pineapple juices: beverage juice (cloudy) from eradicated fruit and mild juice, the clear pineapple juice from wastes. The processing of pineapple mainly involves the fruit being sliced or cubed and then canned. A clear concentrate can be processed. The latter is used to cover the slices and cubes in the cans.

Process Description. Pineapple juice results from the pressing of the following by-products of the canning factory: the surplus of juice drained from the eradicator fruit, juice drained during crushed pineapple preparation, fruits that are too small for slicing, and fruit shell peel cores, trimmings, or end cuts (Fig. 21). All are blended, crushed, and pressed with a screw press to obtain a 8 to 10 °Brix

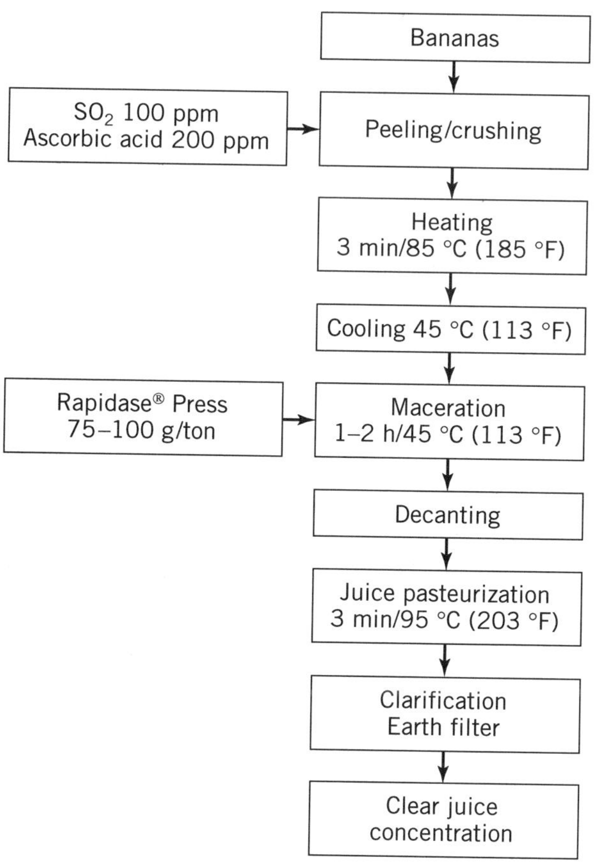

Figure 20. Banana juice process.

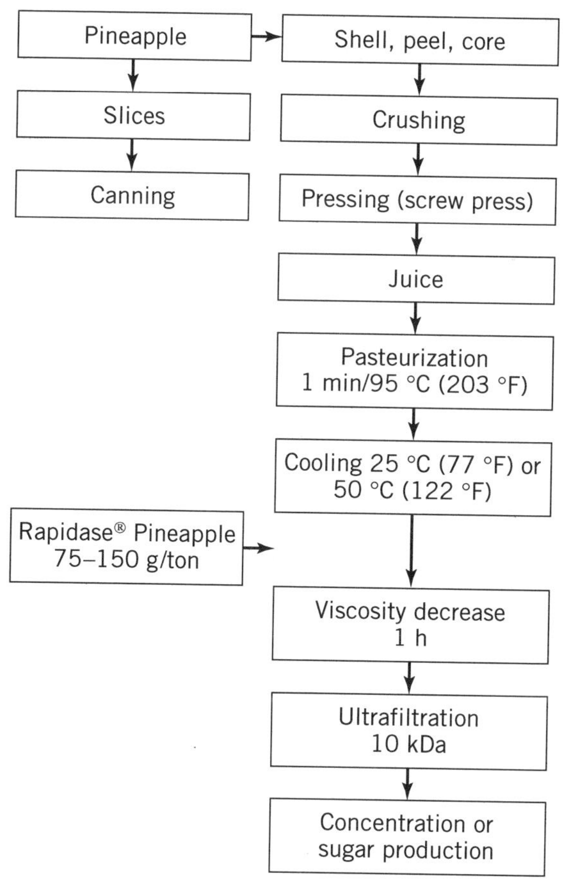

Figure 21. Clear pineapple juice processing.

juice with a of pH 3.5 to 4.0. The juice must be centrifuged to reduce its fiber content; the percentage of insoluble solids (spin test at 3,000 rpm for 5 min) should be below 1.0 to 1.5%. The juice is pasteurized to insolubilize bromelain protein at 95 °C (203 °F) for 1 min.

The juice is cooled down and depectinized at 25 °C (77 °F) for 1 h. Enzymes such as Rapidase® Pineapple (Gist-Brocades) or Rohapect B1L (Röhm) are added for the depectinization from 7.5 to 15 g/hL. Rapidase® Pineapple contains pectinases and specific hemicellulases that hydrolyze pineapple gum. This pineapple juice contains a small amount of pectin but has a high hemicellulose content (galactomannans, arabinogalactans, and galactoglucomannans), whereas the insoluble parts are rich in arabinoxylans (Table 5). The natural gum in pineapple juice was found to decrease the UF capabilities and increase pulp suspension and foaming properties of the juice (46). This gum is a neutral polysaccharide containing 70% sugars, which are predominantly galactomannans (mannose-to-galactose ratio of 2.25:1). Because of this gum, the UF flux rate quickly drops and becomes a bottleneck in the process when using UF membranes with a cutoff of 10,000 Da. Enzymes in Rapidase® Pineapple improve the UF flux rate because the xylanase and other hemicellulase activities reduce the polysaccharides' molecular weight. The juice viscosity is decreased, the UF flux rate is improved, and the juice is more easily concentrated at 62 °Brix.

Kiwi

The processing of kiwi puree requires no enzyme application. The processing of kiwi juice and concentrate is a new demand, especially in Korea, New Zealand, and Chile. This fruit is very rich in ascorbic acid (vitamin C) and contains more than 0.5% proteins. The fruit pulp is very viscous, and pectinases are required for processing because of a high pectin content of about 260 mg%.

Process Description. Kiwis are washed and sorted out then crushed with a hammer mill (Fig. 22). SO_2 is added to prevent pulp browning at 30 to 50 ppm. Pectinase products such as Rohapect D5L (Röhm) or Rapidase® Press (Gist-Brocades) at 100 to 150 g/ton are added to the pulp to improve the juice extraction. Maceration is carried out for 1 to 2 h at ambient temperature. After pressing or decanting, the juice is pasteurized at 95 to 98 °C (203 to 210 °F) for 1 min for protein denaturation. Depectinization is carried out with Cytolase® PCL5 at 2 to 4 g/hL for 1 to 2 h at 50 °C (122 °F). Centrifugation may be useful to remove heat-stable proteins. Clarification is obtained after fining with bentonite (100 to 200 g/hL) or with bentonite plus silicasol (75 g each) followed by earth filtration or UF. The clear juice is concentrated up to 65 °Brix.

Cloudy Passion Fruit Juice

Most of the passion fruit juice is sold as cloudy concentrate juice. Cloud stability juice yield, and concentrability are the main concerns of juice producers. Enzymes, pectinases, hemicellulases, and cellulases can increase the extraction juice yield and improve the concentration after decreasing the viscosity of the juice.

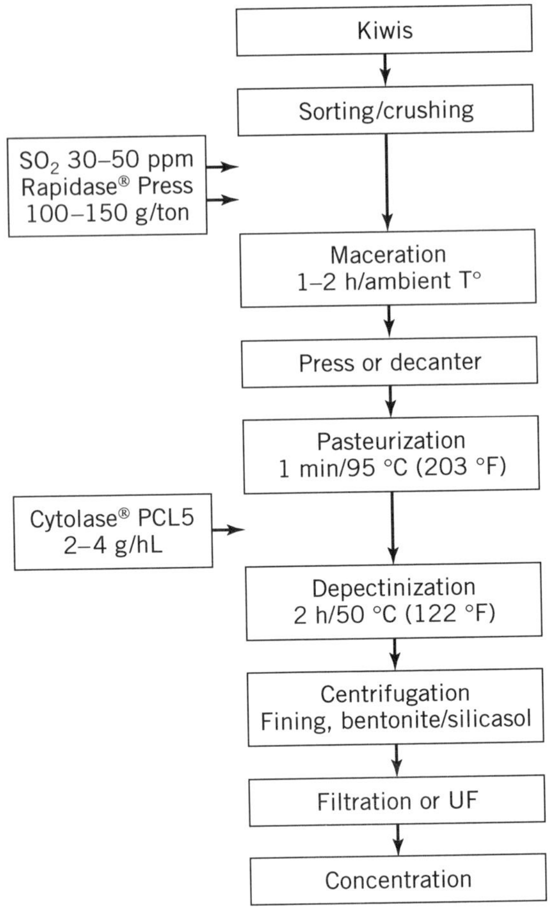

Figure 22. Kiwi juice process.

Process Description. Fruits are inspected, washed, brushed, and rinsed before crushing (Fig. 23). Pulp extraction is carried out on specific extractors with a desteming system. The pulp is heated up to 95 °C (203 °F) for 2 min to inactivate endogenous PPO and is cooled down at 50 °C (122 °C). Enzymatic maceration is carried out with 100 to 200 g/ton Rapidase® Tropical Cloud (Gist-brocades) for 1 to 2 h. Juice is extracted with a pulp finisher, centrifuged, and pasteurized at 85 °C minimum (185 °F) for 30 to 45 sec and concentrated up to 55 °Brix.

Orange, Grapefruit, and Tangerine

The commercial production of citrus for processing into fruit sections and juice began in the 1920s in Florida. The development of frozen concentrated orange juice (FCOJ) started in the 1940s and now exceeds all other citrus products combined, and it is the largest single frozen food item. Although citrus juice is the most important product, the by-products, of which pectin and essential oils are the most significant, provide added value for producers. Although some countries forbid the use of enzymes in the production of premium citrus juices from orange, tangerine, and grapefruit, enzymes can be used to increase the pulp wash yield, to make the concentration easier, to improve essential oil recovery, and to debitter or clarify the juice (lemon

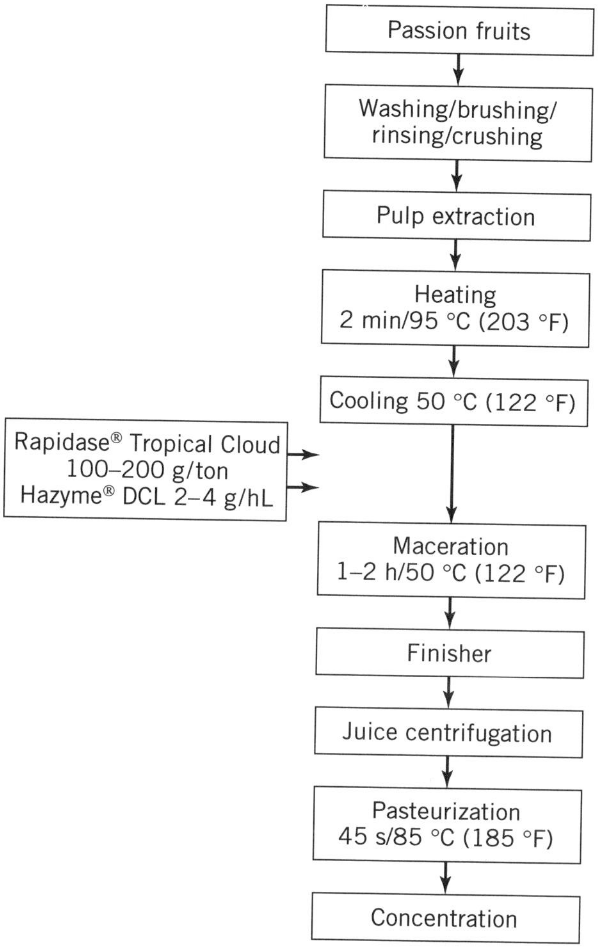

Figure 23. Passion fruit juice process.

juice). After incoming fruits are sampled and tested (the ripeness standard is the ratio of Brix to acid, defined in the 1985 Florida Citrus Laws), oranges enter the extractor. A schematic diagram of extraction and finishing in citrus juice plant is shown in Figure 24. Methods of extraction and finishing operations affect the yield, quality, and characteristics of the citrus juice produced. To maintain high quality, the yields obtainable by processors in Florida are highly regulated by citrus laws (e.g., size and type of extractors). Officials decide what processors are allowed to recover as pure juice. There are two main types of extractors and finishers, manufactured by FMC and Brown Companies in the United States. The expressed juice from the extractors contains pulp and cell membrane residue, and processing in a finisher (rotating screw principle) is required to reduce the pulp content. Raw juice enters at one end, pulp is squeezed out at the other end, and the juice passes radially through sieving cylindrical-shaped screens. Standard finishing reduces the pulp content of juice to about 12%. The excess pulp and solid by-products of the extracted juice from the primary finishers are processed through a pulp wash system (47).

Pulp Wash

The pulp wash system consists of a series of five to seven stages of secondary finishers. Water is added in a multistage countercurrent washing system. This water, along with a high finisher screen pressure, removes the remaining soluble solids containing a high level of pectin from the fruit pulp. About 5 to 7% of the additional solids can be recovered. After removal of the soluble solids, the product is concentrated. Rapidase® Citrus 2000 is added before or after pulp wash extraction at 2.0 g per 100 kg pulp (or per hL) to reduce the viscosity (mainly from the pectin) for achieving the desired 65 °Brix concentrate (Fig. 24). The result is a better extraction of sugars and soluble solids, resulting in a higher yield and therefore a lower viscosity. The pulp wash product can be easily concentrated to 65 °Brix without risk of gelling and keeping a good cloud stability. The pulp wash product is added back to sodas as the cloud source, stable after centrifugation and pasteurization. After this treatment, the pectin degradation is limited to lowering the viscosity without attacking the insoluble pectin that maintains the stability of the cloud. It is very important to use pectinases with the lowest PME activity as possible, to avoid clarification of the pulp wash. The ideal enzyme is a pure PL. Next to PL, Rapidase® Citrus 2000 contains glycosidases that decrease the pulp wash bitterness and improve its taste, mouthfeel, and flavor level.

Oil Recovery

The essential oil content is 2 to 5 kg/ton of processed raw material, but the oil is mainly located in the fruit peel. Essential oils of citrus are located in special cells of the flavedo. The flavedo contains hydrocarbons (terpenes and sesquiterpenes), oxygenated compounds (aldehydes, esters, alcohols, ketones, phenols), and nonvolatile residues (waxes, coumarins, flavonoids, fatty acids). Although the Brown system requires a separate machine to extract oil before juice extraction, the FMC extractor is capable of extracting juice and peel oil in the same operation.

Process Description. After extraction, the juice is sent to the finisher for juice production, and the peel particles containing the oil are washed in a separate stream. After juice extraction, the peel particles and the oil–water emulsion are separated through a finisher. The proportion of oil in the emulsion is between 1 and 3% (9 lb/ton fruit). Enzymes can speed the process time and increase the yield of oil from the emulsion (cream) and the quality of the final product. Enzymes needed for a better separation of oil and water from the emulsion are pectinases and acid proteases. By hydrolyzing the complex pectin proteins, oil is released more easily from the aqueous phase. Enzymes preparations such as Citrozym CEO (Novo), Rohapect VR (Röhm), or Rapidase® Citrus Oil (Gist-Brocades) are added at 10 to 20 g/hL of emulsion before the desludger centrifugation for 30 to 60 min at room temperature. Oil yield can be increased by 10 to 15%.

The finished emulsion passes through a sand cyclone and is fed to a desludger centrifuge to produce an oil-rich emulsion that is sent to a polisher that recovers the clear oil. After the desludger stage, the oil is concentrated up to 60 to 80%. In the second stage, the polisher removes the remaining water and the very fine solid particles. A cold storage allows dewaxing of the oil for better purification.

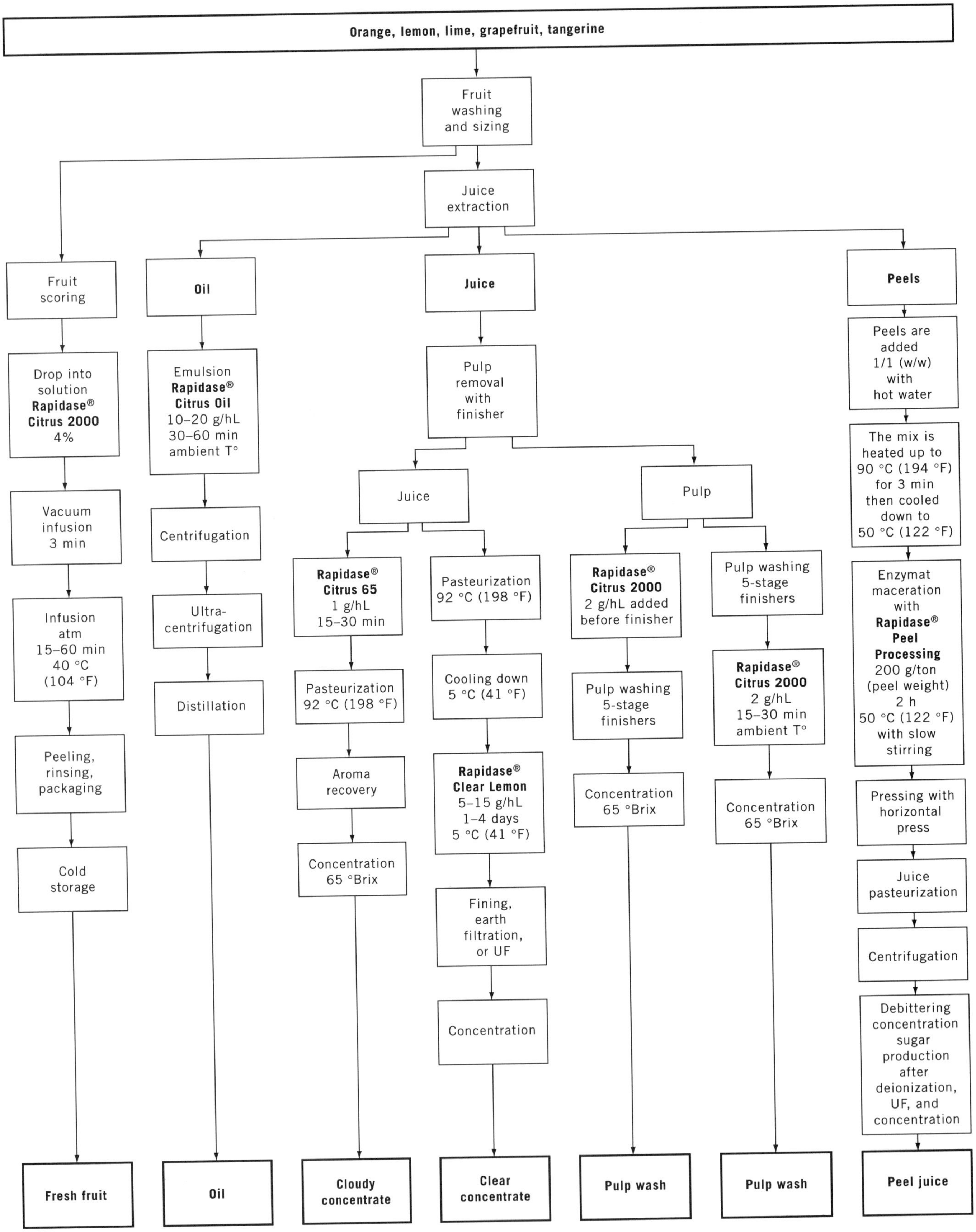

Figure 24. Citrus processing.

Highly Concentrated Citrus Base

Extracted finished juice is sensitive to microbiological contaminations, enzymatic activities, and chemical reactions, all leading to a quick quality degradation. Problems with unpasteurized juices are cloud loss and juice separation caused by citrus PME, gelation of the concentrate, fermentation, and spoilage, even if the pH is low (3.0 to 4.0). These problems are prevented by fast pasteurization, although it causes undesirable browning reactions and flavor and taste changes in the juices before aseptically packaging. Inactivation of citrus PME by heating has been determined to occur at 90 °C (194°F) for 1 min (48). The juice is then concentrated with multieffect evaporators at 65 °Brix. Volatiles recovered during juice concentration are referred to as essences. Concentrate is chilled immediately after evaporation to maintain high flavor quality. Citrus juices can be concentrated after decrease of the juice viscosity by special pectinases in countries where it is authorized, such as Italy. The soluble pectin, which is responsible for many problems in processing, must be slightly degraded to avoid the risk of gelatinization in the concentrator.

Process Description. Added after the first finisher, enzymes such as Rohapect SK1 (Röhm) or Rapidase® Citrus 65 (Gist-Brocades) at 1 g/hL for 15 to 30 min make the concentration of the premium juice easier without clarification and enable the production of concentrated citrus base above 65 °Brix (Fig. 24).

Prepeeling

Citrus fruits are sometimes treated with enzymes that digest white albedo, which binds the flavedo peel to fruit segments. After treatment, hand peeling is easier and faster. Fruit peel is first scored, and fruits are then dropped into an enzyme solution with 4% Peelzym (Novo Nordisk) or Rapidase® Citrus 2000 (Gist-Brocades). Fruits are vacuum infused for 2 to 3 min in this enzyme solution (Fig. 24). After vacuum break, fruits are maintained at 40 °C (104 °F) for 15 to 60 min for albedo digestion inside the fruit (49). Fruits will be later be very easily hand peeled. Fruits are then rinsed and cooled in a water bath at 5 °C (41 °F) and packed.

Peel Pressing

Orange juice yield is low, around 40%. This is why peel valorization is of high interest. After juice extraction, peels are added with same weight of hot water (Fig. 24). The blend is heated up to 90 °C (194 °F) for 3 min and then cooled down to 50 °C (122 °F). Enzymes such as Pectinex UltraSP (Novo Nordisk) or Rapidase® Peel Pressing (Gist-Brocades) are added at 200 g/ton for 2 h at 50 °C (112 °F) with slow stirring. Pressing is carried out with a horizontal press or decanter. The extracted juice is pasteurized, centrifuged, possibly debittered, and concentrated.

Lemon

Lemon juice is extracted with FMC or Brown extractors. Before concentration, seeds, bits of peel, and pulp must be removed (Fig. 25). Lemon pulp, which is added back to lemonade, is recovered by hydrocyclone or cylindrical perforated screen. Primary oil emulsion discharged from extractors still contains peel fragments that must be discarded by finishers before essential oil is recovered from the emulsion by centrifugation. Cold-pressed lemon oil has a large range of applications, and the market value of lemon oil is very high. There is a strong demand for lemon oil as flavoring in foods and soft drinks and in cosmetics, pharmaceuticals, and household products. Citral is the predominant aldehyde in lemon oil and has long been considered the chief flavor constituent of the product. Lemon juice contains a complex colloidal system of particles composing the cloud. These particles, remaining in suspension after the largest ones have been removed, play a major role in the organoleptic qualities of the juice. Lemon cloud is composed of approximatively equal amounts of proteins and pectin with flavonoids and phospholipids (50). Lemons can be processed as cloudy juice. In this case, it is necessary to destroy the endogenous PME, which could lead to a cloud destabilization. The temperature of lemon PME inactivation is 74 °C (165.2 °F) for 14 s. Heating the pulpy lemon juice for too long may result in browning. The juice is then concentrated by evaporation or reverse osmosis at 55 to 65 °Brix. Demand for clarified lemon juice concentrate has increased in recent years. Clarification of single-strength lemon juice may be carried out directly by UF or addition of polygalacturonic acid or commercial pectinases active at low pH. Lemon juice is not easily clarified and concentrated.

In the past, juice clarification took as long as 4 to 16 weeks and occurred only after the addition of huge amounts of sulfur dioxide (about 1,000 to 2,000 mg/L) to

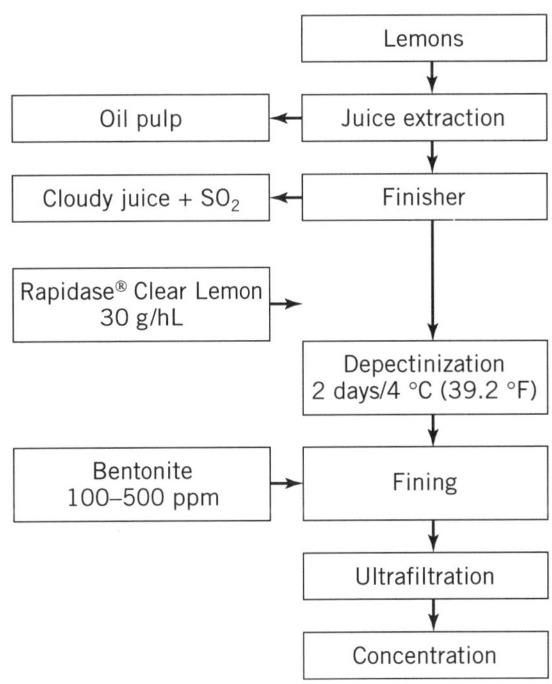

Figure 25. Lemon juice process.

aid the action of lemon PME. Depectinization requires pectinases to be stable at low pH (2.2 to 2.8) with high citric acid concentration. Moreover, pectinases alone did not allow juice clarification because the cloud is caused by pectin, hemicelluloses, and cellulose, with a high protein content.

Process Description. Enzymes proposed for lemon juice production are Rapidase® Clear Lemon (Gist-Brocades), Rohapect C (Röhm), or Citrozym (Novo). After lemon juice extraction, the juice goes through a finisher and is then depectinized (Fig. 25). Rapidase® Clear Lemon is added at 30 g/hL to the juice at low temperature, 4 to 8 °C (39.2 to 46.4 °F), to avoid juice oxidation. Lemon albedo contains about 33% of its dry weight as pectin. The complete degradation of pectin that is required is finished after about 2 days at low temperature or a few hours at 40 °C (104 °F). However, even if the juice is depectinized, it is not clarified. At this stage, it is still possible to produce a cloudy concentrate or a clear juice and concentrate (Fig. 24). A fining step is necessary for the clarification, with high dosis of bentonite (about 100 g/hL or 500 g/hL without enzyme treatment) or silicasol. Then, flocculation and sedimentation of the cloud can occur. After a few hours, the juice is centrifuged, ultrafiltered, and concentrated easily up to 52 °Brix.

FRUIT JUICE AUTHENTIFICATION

Today, quality, authenticity, and detection of adulteration of fruit juices are absolute requirements from juice purchasers and consumers. The evolution of technology and processes using enzymes in parallel with more sensitive analytical techniques, especially in the field of liquid and gas chromatography, and new analytical methods such as SNIF-NMR reveal a broad spectrum of components in marketed fruit juices. It is sometimes difficult to identify these different compounds and precisely determine their origin. Most of them result from the physical and enzymatic degradation of natural substrates released during the fruit processing, but deviations from the standard ranges may indicate adulteration, and expert interpretation is required. In Europe, a code of practice has been developed to maintain quality and authenticity standards. Members of Association of the Industry of Juices and Nectars (AIJN) have established reference guidelines to judge the quality and authenticity of fruit juices (51). The values should reflect the commercial standards. The most widely used of these are the German Fruit Juice Industry guide values known as RSK values, introduced in the 1970s. Fruit juice authenticity is assessed by taking into account numerous chemical analyses. Apple has been extensively studied, but products coming from the degradation of cell wall polymers by enzymes are not yet fully described, which creates problems in juice authenticity: what is a "standard" apple juice and what is the reference process? At the same time, the development of enzymatic liquefaction technology with pectinases, hemicellulases, and cellulases and the development of new analytical techniques with a very high sensitivity show new and atypical analytical profiles of juices, revealing more numerous chromatography peaks caused by sugar oligomers that are not easily identified and from which the origin is not obvious.

For example, starch, pectin, and cellulose are polysaccharides naturally present in most of fruits. Granular starch is mainly present at the crop period in immature apples. It is hydrolyzed during fruit ripening by endogenous amylases. Apple starch is composed of 70% amylopectin and 30% amylose. Starch degradation results in the formation of simple sugars such as glucose, maltose, and maltitol. Because endogenous enzyme activity is very low and the process duration is short, it is necessary to add high exogenous fungal amylases that release glucose, maltose, and other isomers such as isomaltose and pannose. We assume that side enzyme activities, which are present in commercial amylase preparations such as maltase and α-glucosidase, can synthesize these compounds. Pectin is the main substrate being degraded by exogenous enzymes during processing. Pectinases from *A. niger* release all sizes of oligomers and monomers of galacturonic acid, arabinose, and galactose and also parts of nondegraded RG-I and RG-II. Analyses of juice composition with new techniques can now reveal all these different compounds. With the purpose of optimizing fruit sugar extraction, new technologies such as fruit liquefaction (total or pomace) have appeared. The related new processes increase the juice yield highly. The yield is high because of the almost total hydrolysis of protopectin in oligomers and monomers of galacturonic acid and neutral sugars, but also because of partial hydrolysis of cellulose into oligomers and monomers of glucose, including cellobiose. Cellulases, by decreasing the mash viscosity, improve pectinases efficiency. The sugar profile depends also on physical conditions of the process submitted to the fruit. One sugar molecule can easily epimerize at a slightly alkaline pH of 8.0–8.5. This can be the case during fruit pulp or pomace heating with steam condensate water being at alkaline pH. The contact between glucose in the fruit pulp and alkaline water can form fructose, thus changing the ratio of glucose to fructose that is a criterion of fruit juice authenticity. A high temperature could also change the sugar profile in the final juice.

In conclusion, numerous laboratories of authentification have recently appeared that can detect the adulteration of juices with cheap sugars coming from cereals. However, it is urgent now that analysts at these laboratories enlarge the description of standard juice composition, including the following parameters: the fruit variety, the geographical area where raw material originates, climatic conditions of the year, fruit ripeness (endogenous enzymes), and all process conditions. Communication among analysts, enzyme producers, fruit juice producer organisations, and legislators is necessary, and studies are ongoing. The role of exogenous enzymes that are necessary to produce clear and stable concentrate is crucial in determining the final composition of the juice. All these parameters should be integrated in the description of new standards in order to avoid confusion in the interpretation of results.

BIBLIOGRAPHY

1. A. Mehlitz, *Biochem. Z.* **221**, 217–231 (1930).
2. J.G. Faigh, *Food Technol.* **49**(9), 79–83 (1995).
3. M.C. MacCann, and K. Roberts, *Agro-Food Industry Hi-Tech.* 43–46 (March–April 1994).
4. N.C. Carpita and D.M. Gibeaut, *Plant J.* **3**, 1–30 (1993).
5. P. Albersheim, A.G. Darvill, M.A. O'Neill, H.A. Schols, and A.G.J. Voragen, in J. Visser and A.G.J. Voragen ed., *Pectin and Pectinases*, Elsevier Science, Amsterdam, 1996, pp. 47–55.
6. J. De Vries, A. Voragen, F. Rombouts, and W. Pilnik, in M. Fishman and J. Jen ed., *Chemistry and Function of Pectins*, ACS Symposium Series 310, Am. Chem. Soc., Washington, D.C., 1986, pp. 38–48.
7. H. Schols, M.A. Posthumus, and A. Voragen, *Carbohydr. Res.* **206**, 117–129 (1990).
8. P. Dey and K. Brinson, *Adv. Carbohydr. Chem. Biochem.* **42**, 265–282 (1984).
9. F. Will, Ph.D. Thesis, University of Giessen, Germany, 1990.
10. P. Pellerin, S. Vidal, P. Williams, and J.M. Brillouet, *Carbohydrate Res.* **277**, 135–143 (1995).
11. T. Doco, P. Williams, S. Vidal, and P. Pellerin, *Carbohydr. Res.* **297**, 181–186 (1997).
12. F. Barnoud, in B. Monties ed., *Les Polymères Végétaux*, Gauthier-Villars, Paris, 1980, pp. 66–86.
13. J.P. Vincken, G. Beldman, and A. Voragen, *Plant Physiol.* **104**, 99–107 (1994).
14. L. Rexova-Benkova and O. Markovic, *Adv. Carbohydr. Chem. Biochem.* **33**, 323–385 (1976).
15. A. Voragen, Ph.D. Thesis, Wageningen University, The Netherlands, 1972.
16. C. Versteeg, Ph.D. Thesis, Wageningen University, The Netherlands, 1979.
17. J.A.E. Benen, H.C.M. Kester, Lucie Parenicova, and J. Visser, in J. Visser and A.G.J. Voragen ed., *Pectin and Pectinases*, Elsevier Science, Amsterdam, 1996, pp. 221–230.
18. G. Beldman, G. Mutter, M.J.F. Searle-van-leeuwen, L.A.M. Van den Broek, H.A. Schols, and A.G.J. Voragen, in J. Visser and A.G.J. Voragen ed., *Pectin and Pectinases*, Elsevier Science, Amsterdam, 1996, pp. 231–245.
19. F.M. Rombouts, A. Voragen, M. Searle-van-leeuwen, C. Gereads, H. Schols, and W. Pilnik, *Carbohydr. Polym.* **9**, 5–47 (1988).
20. P. Ducroo, *Flüssiges Obst.* **5**, 265–269 (1987).
21. H. Schols, P. Veld, W. Van Deelen, and A. Voragen, *Z. Lebensm.-Unters.-Forsch.* **192**, 142–148 (1991).
22. L. Saulnier, J.M. Brillouet, M. Moutounet, C. Herve du Penhoat, and V. Michon, *Carbohydr. Res.* **224**, 19–235 (1992).
23. F. Kormelink, M. Searle-van-Leeuwen, T. Wood, and A. Voragen, *J. Biotechnol.* **27**, 249–265 (1993).
24. G. Beldman, M. Searle-van-Leeuwen, F.M. Rombouts, and A. Voragen, *Eur. J. Biochem.* **146**, 301–308 (1985).
25. A. Kliosov and M. Rabinovitch, in L. Wingard, J. Berezin, and A. Kliosov ed. *Enzyme Engineering: Future Directions*, Plenum Press, New York, 1980, pp. 82–165.
26. C. Renard, M. Searle-van-Leeuwen, A. Voragen, J.F. Thibault, and W. Pilnik, *Carbohydr. Polym.* **14**, 295–314 (1991).
27. W. Fogarty and C. Kelly, in A. Rose ed., *Economic Microbiology*, vol. 5, Academic Press, New York, 1980, pp. 115–170.
28. C. Grassin and P. Fauquembergue, in Godfrey T. and S. West ed., *Industrial Enzymology*, 2nd ed., MacMillan, London, 1995, pp. 227–264.
29. M. Yamasaki, T. Yasui, and K. Arima, *Agric. Biol. Chem.* **28**, 779–787 (1964).
30. Q. Wu, M. Szabacs-Dobozi, M. Hemmat, and G. Hrazdina, *Plant Physiol.* **102**, 219–225 (1993).
31. I. Bartley and M. Knee, *Food Chem.* **9**, 47–58 (1982).
32. C. Grassin, *Flüssiges Obst.* **57**, 519–522 (1992).
33. C. Grassin and P. Fauquemberque, *Fruit Process.* **7**, 242–245 (1993).
34. A.G.J. Voragen, R. Heuting, and W. Pilnik, *J. Appl. Biochem.* 452–468 (1980).
35. C. Grassin and P. Fauquembergue, *Fruit Process.* **12**, 2–7 (1996).
36. J.F. Drilleau, *IAA* 47–58 (Sept. 1985).
37. European Pat. Specification EP 0547,648,B1 (November 6, 1992), C. Grassin and P. Fauquembergue (to Gist-Brocades).
38. A. Ranadive and N. Haard, *J. Food Sci.* **38**, 331–333 (1973).
39. F. Will, G. Hasselbeck, and H. Dietrich, *Flüssiges Obst.* **59**, 352–353 (1992).
40. A. Holmes and M. Starr, in S. Nagy, C.S. Chen, and P.E. Shaw eds., *Fruit Juice Processing Technology*, Agscience Inc., Auburndale, Fla., 1993, pp. 515–531.
41. C. Pederson, in P. Nelson and D. Tresslier ed., *Fruit and Vegetable Juice Processing Technology*, AVI Publishing, Westport, Conn., 1980, pp. 268–309.
42. H. Siliha, Ph.D. Thesis, Wageningen University, The Netherlands, 1985.
43. O. Marcellin, P. Williams, and J.M. Brillouet, *Carbohydr. Res.* **240**, 233–243 (1993).
44. O. Markovic, K. Heinrichova, and B. Lenkey, *Collect. Czech. Chem. Commun.* **40**, 769 (1976).
45. E. Koffi, C. Sims, and R. Bates, *J. Food Quality* **14**, 209–218 (1991).
46. K. Chen Chin and H. Yamamoto, *J. Food Sci.* **43**, 1261–1263 (1978).
47. S. Chen, P. Shaw, and M. Parish, in Nagy, Chen, and Shaw ed., *Fruit Juice Processing Technology*, Agscience, Auburndale, Fla., 1993, pp. 110–165.
48. B. Eagerman and A. Rouse, *J. Food Sci.* **41**, 1396–1397 (1976).
49. U.S. Pat. 5,200,217 (April 6, 1993), R. Elliot and J. Tinibel (to Sunkist Growers).
50. B. Carter, in Nagy, Chen, and Shaw ed., *Fruit Juice Processing Technology*, Agscience, Auburndale, Fla., 1993, pp. 215–270.
51. G.G. Martin, E. Jamin, J. Gonzales, G. Remaud, V. Hanote, P. Ströber, and N. Naulet, *Fruit Process.* **9**, 344–349 1997.

See also ENZYMES, PROTEIN HYDROLYSIS; FOOD PROCESS ENGINEERING.

ENZYMES, IMMOBILIZATION METHODS

TADASHI SATO
TETSUYA TOSA
Tanabe Seiyaku Co., Ltd.
Osaka, Japan

KEY WORDS

Carrier binding
Cross-linking
Entrapping enzyme
Immobilization method
Immobilized enzyme

OUTLINE

Bibliography

In 1916, Nelson and Griffin (1) found that an enzyme in water-insoluble form showed catalytic activity. They reported that invertase extracted from yeast was adsorbed on charcoal, and the adsorbed enzyme showed the same activity as native enzyme. In 1948, Sumner (2) found that urease from jack bean became water insoluble on standing in 30% alcohol and sodium chloride for 1 to 2 days at room temperature and that the water-insoluble urease was active. Thus, it has been known for a long time that enzymes in water-insoluble form show catalytic activity. However, the first attempt to immobilize an enzyme to improve its properties for a particular application was not made until 1953, when Grubhofer and Schleith (3) immobilized enzymes such as carboxypeptidase, diastase, pepsin, and ribonuclease by using diazotized polyaminopolystyrene resin. On the other hand, before this (1949), immobilization of physiologically active protein was carried out by Micheel and Ewers (4). Further, in 1951, Campbell et al. (5) prepared immobilized antigen by binding albumin to diazotized *p*-aminobenzylcellulose. Subsequently, a number of reports on the preparation and application of immobilized antigens and antibodies appeared. Such studies may also be applicable to immobilized enzymes.

In the 1960s, many papers appeared; in particular, Professor Katzir-Katchalski and his co-workers at the Weizmann Institute of Science, Israel, carried out extensive studies on new immobilization techniques, and on the enzymatic, physical, and chemical properties of immobilized enzymes.

In 1969, Dr. Chibata et al. at Tanabe Seiyaku Co., Ltd., in Japan succeeded in the industrialization of a continuous process for the optical resolution of DL-amino acids using immobilized aminoacylase (6). This was the first industrial application of immobilized enzymes in the world.

In 1971, the first Enzyme Engineering Conference was held at Henniker, New Hampshire, and the predominant theme of this conference was immobilized enzymes. A definition and classification of immobilized enzymes were proposed at the conference. This conference is biannual, and the main topics have continued to be immobilized enzymes. Some books on immobilized enzymes have been published (7,8) as well.

Immobilized enzymes are defined as "enzymes physically confined or localized in a certain defined region of space with retention of their catalytic activities, and which can be used repeatedly and continuously." Accordingly, enzymes modified to a water-insoluble form by suitable techniques satisfy this definition of immobilized enzymes. In addition, when an enzyme reaction using a substrate of high molecular weight is carried out in a reactor equipped with a semipermeable ultrafiltration membrane without leakage of the enzyme from the reactor, this can also be considered as a kind of immobilized enzyme system.

The term *immobilized enzyme* was recommended at the First Enzyme Engineering Conference in 1971. Before that time, various terms such as water-insoluble enzyme, trapped enzyme, fixed enzyme, and matrix-supported enzyme had been used.

As functional groups involved in enzyme immobilization, free, amino and carboxyl groups, the sulfhydryl group of cystein, the imidazole group of histidine, phenolic groups, and hydroxyl groups of serine and threonine may be considered. For immobilization of an enzyme, it is necessary that functional groups in the active center should not be involved in the reaction leading to immobilization of the enzyme. Further, because the tertiary structure of enzyme protein is maintained by relatively weak binding forces, such as hydrogen, hydrophobic, and ionic bonds, it is necessary to carry out the immobilization reaction under mild conditions.

Methods for enzyme immobilization can be classified into three basic categories as follows.

1. Carrier-binding method: the binding of enzymes to water-insoluble carriers such as polysaccharide derivatives, synthetic polymers, and porous glass, etc.
2. Cross-linking method: intermolecular cross-linking of enzymes by means of bifunctional or multifunctional reagents such as glutaraldehyde, bisdiazobenzidine, and hexamethylene diisocyanate, etc.
3. Entrapping method: incorporating enzymes into the lattice of a semipermeable gel on enclosing the enzymes in a semipermeable polymer membrane, such as collagen, gelatin cellulose triacetate polyacrylamide, and κ-carrageenan, etc.

These methods are shown schematically in Figure 1.

The carrier-binding method is the oldest immobilization method for enzymes, and many papers have been published on this method. When enzymes are immobilized in this way, care is required regarding the selection of carriers as well as in binding techniques. The carrier-binding method can be further divided into three categories according to the binding mode of the enzyme, that is, physical adsorption, ionic binding, and covalent binding.

The cross-linking method is based on the formation of chemical bonds, but water-insoluble carriers are not used in this method. The immobilization of enzymes is performed by the formation of intermolecular cross-linkages between the enzyme molecules by means of bifunctional or multifunctional reagents.

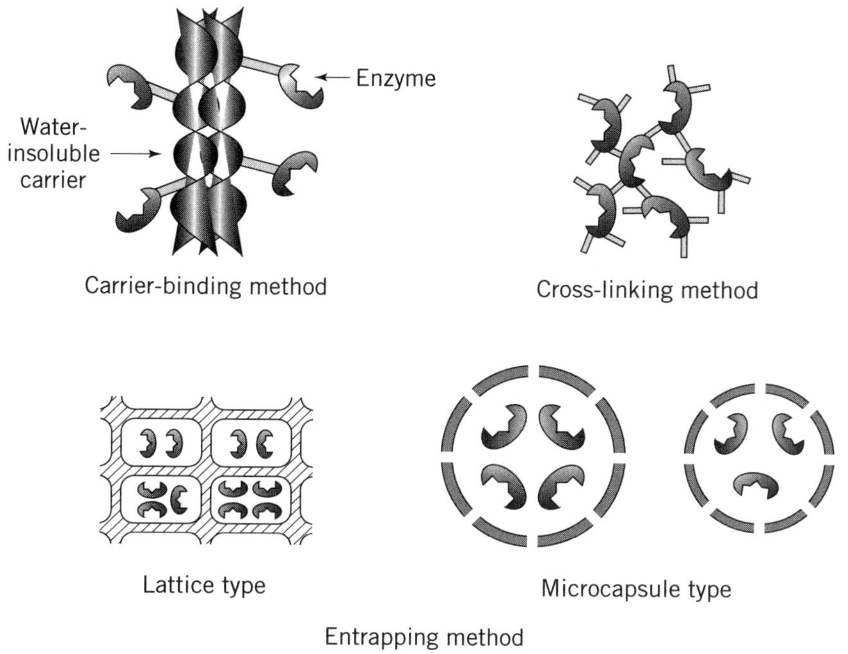

Figure 1. Schematic diagrams of immobilized enzymes.

The entrapping method is based on confining enzymes in the lattice of a polymer matrix or enclosing enzymes in semipermeable membranes, and it can be classified into the lattice and microcapsule types. This method differs from the covalent binding and cross-linking method in that the enzyme itself does not bind to the gel matrix or membrane. Thus, this method may have wide applicability. However, if a chemical polymerization reaction is used for entrapping, relatively severe conditions are required and lose of enzyme activity occurs in some cases. Therefore, it is necessary to select the most suitable conditions for the immobilization of various enzymes.

Preparation methods and characteristics of immobilized enzymes are summarized broadly in Table 1, though there are many exceptions.

Immobilization of enzymes by covalent binding or cross-linking is carried out under relatively severe conditions in comparison with those of physical adsorption or ionic binding. Thus, in the former cases, conformational change of the enzyme structure and partial destruction of the active center may occur. Accordingly, unless immobilization of an enzyme by covalent binding is carried out under well-controlled conditions, immobilized enzyme having high activity cannot be obtained. However, the binding forces between the enzyme and carrier are strong, and the enzyme cannot easily be lost from carriers even in the presence of substrates or salts at high concentrations. However, when the activity of enzymes immobilized by covalent binding decreases on long-term operation, regeneration is impossible.

On the other hand, immobilization of enzymes by the ionic-binding method can be achieved simply under mild conditions. Accordingly, in this case, preparations having relatively high activity are obtained. However, the binding forces between enzyme and carrier are weak in comparison with those in the covalent-binding methods. Therefore, leakage of the enzyme from the carrier may occur after changes in the ionic strength, pH of the substrate, or product solution. This type of immobilized enzyme can be regenerated when the enzyme activity decreases after prolonged operation. Thus, the ionic-binding method is advantageous in comparison with the covalent-binding method, particularly when expensive carriers or enzymes are used.

Unlike the ionic-binding, covalent-binding, and cross-linking methods in the entrapping method, no binding

Table 1. Preparation and Characteristics of Immobilized Enzyme

	Carrier-binding method				
Characteristic	Physical adsorption	Ionic binding	Covalent binding	Cross-linking method	Entrapping method
Preparation	Easy	Easy	Difficult	Difficult	Difficult
Enzyme activity	Low	High	High	Moderate	High
Substrate specificity	Unchangeable	Unchangeable	Changeable	Changeable	Unchangeable
Binding force	Weak	Moderate	Strong	Strong	Strong
Regeneration	Possible	Possible	Impossible	Impossible	Impossible
General applicability	Low	Moderate	Moderate	Low	High
Cost of immobilization	Low	Low	High	Moderate	Low

between enzyme and carrier should occur in theory. Therefore, in many cases, preparations having high activity are obtained. However, in this method, regeneration of activity losses is impossible, as in the case of the covalent-binding method. A major disadvantage of the entrapping method is that its application is limited to small molecular substrate and product molecules; the entrapped enzyme shows little or no activity toward macromolecular substrates.

Although a number of enzyme immobilization methods have been studied, no ideal general methods applicable for the immobilization of many enzymes have yet been developed. Each method has specific disadvantages. Therefore, in practice, it is necessary to find a suitable method and conditions for the immobilization of a particular enzyme in the light of the intended application.

Because of the special characteristics of immobilized enzymes, a considerable amount of work has been carried out on application of immobilized enzymes as solid catalysts having high substrate specificity. Immobilized enzymes are used in various fields, such as chemical processes, analysis, medical treatment, food processing, affinity chromatography, and so forth. For the efficient utilization of such immobilized systems, chemical engineering studies are necessary.

BIBLIOGRAPHY

1. J.M. Nelson and E.G. Griffin, *J. Am. Chem. Soc.* **38**, 1109–1115 (1916).
2. J.B. Sumner, *Science* **108**, 410–413 (1948).
3. N. Grubhofer and L. Schleith, *Naturwissenschaften* **40**, 508 (1953).
4. F. Micheel and J. Ewers, *Makromol. Chem.* **3**, 200–209 (1949).
5. D.H. Campbell, E. Luescher, and L.S. Lerman, *Proc. Nat. Acad. Sci.* **37**, 575–581 (1951).
6. I. Chibata, T. Tosa, T. Sato, T. Mori, and Y. Matuo, *Proc. of the 4th Int. Fermentation Symp.: Fermentation Technology Today*, 1972, p. 383–389.
7. L.B. Wingard Jr. ed., *Enzyme Engineering*, Wiley, New York, 1972.
8. O.R. Zaborsky, *Immobilized Enzymes*, Chemical Rubber Co. Press, Ohio, 1973.

See also BIOFILMS, MICROBIAL; BIOREACTORS, AIR-LIFT REACTORS; BIOREACTORS, FLUIDIZED-BED; BIOREMEDIATION.

ENZYMES, IMMOBILIZED, REACTORS

UDO KRAGL
LASSE GREINER
CHRISTIAN WANDREY
Forschungszentrum Jülich GmbH
Jülich, Germany

KEY WORDS

Amino acylase
Enzyme, immobilized
Glucose-fructose isomerase
Glycerol kinase
Lipase
Mass transport limitation
Membrane
Nitrile hydratase
PenG acylase
Reactor

OUTLINE

INTRODUCTION

The use of enzymes in food as well as in pharmaceutical and chemical industries has increased steadily since the early 1970s. Applications range from synthesis in the milligram scale for natural or nonnatural oligosaccharides up to industrial synthesis for the production of bulk chemicals such as acrylamide. Some examples are listed in Table 1. This has been well documented by a large number of textbooks, reviews, and original papers (1,5–20). This article highlights some of the most important aspects for the application of immobilized enzymes and presents selected examples.

The most common definition for immobilized enzymes is that proposed by Katchalski-Katzir in the 1960s (16): "Enzymes physically confined or localized in a certain defined region of space with retention of their catalytic activities, which can be used repeatedly and continuously." According to this definition, three types of immobilized enzymes can be distinguished:

Table 1. Examples of the Industrial Use of Immobilized Enzymes

Enzyme (ref.)	Substrate	Product	Scale (ton a^{-1})
Penicillin amidase (1,2,3)	Penicillin G	6-Amino penicillanic acid	6,000
Cephalosporin amidase (1)	Glutaryl-7-ACA	7-ACA	?
Amino acylase (1)	Acyl-D-L-amino acid	L-amino acid	approx. 300
Aspartase (1)	Fumaric acid	L-aspartic acid	1,200
Fumarase (1)	Fumaric acid	L-malic acid	360
Lactase (1) (galactosidase)	Lactose GalGlc	Low lactose milk Gal + Glc	?
Aspartase β-decarboxylase (1)	Aspartic acid	L-alanine	120
Nitrile hydratase (H. Yamada, personal communication)	Acrylonitrile	Acrylamide	30,000
Glucose isomerase (1)	Glucose	High fructose corn syrup	8,000,000
Lipase (4) (expected)	Rac-1-phenylethyl-amine	(*S*)-1-phenylethyl-amine	200

1. Heterogenization of the soluble enzyme by coupling to an insoluble support by adsorption or covalent binding, by cross-linking of the enzyme or entrapment in a lattice or in microcapsules such as alginate beads
2. Retention of the enzyme by means of ultrafiltration membranes
3. Use of whole cells for biotransformations using their enzyme apparatus

Because of the scope of this article, only the first type is considered. Membrane reactors have been reviewed quite recently (21–28), as have applications of whole cell processes (1,5,13,14,17,29–33). Retention of the cells within the reactor may be achieved by membrane separation or by the same immobilization methods that are used for isolated enzymes (34). In principle, the cell itself can be regarded as a form of native immobilization of enzymes. Biosensors are a very special form of carrier-fixed biocatalysts, for which the reader is referred to the literature (35–37).

The major goal behind immobilization (in its most common definition) is the recovery of the biocatalysts, separation from products and reactants, and subsequent reuse in batch or continuous processes. The latter is especially important for a reduction of the catalyst costs. Enzymes immobilized on a support often show enhanced stability when compared with the soluble form (38–41). When considering the use of soluble or carrier-fixed enzymes, the following topics have to be addressed:

1. Additional costs for support and chemicals performing the immobilization have to be balanced against the increase of stability.
2. Loss of activity during the immobilization step.
3. When the catalyst is immobilized only by adsorption or entrapment without covalent attachment, its leakage from the carrier support has to be examined and compared with the overall deactivation rate (42).
4. Mass transfer limitations for enzymes on a support may cause problems when adjusting of the pH is necessary during the reaction.
5. With soluble enzymes, higher volumetric activities at high catalyst concentrations are possible, enabling conversion of poor substrates at reasonable rates.
6. Whereas membrane reactors can be easily sterilized before use, this is not possible for reactors with carrier-fixed enzymes. To prevent microbial contamination, these processes are quite often operated at higher temperatures. These are discussed in more detail later.
7. When enzymes are to be used together with organic solvents to increase reactant or product solubility or to alter their kinetics, it may become necessary to immobilize them on a support (8,18,43–48). The support will at the same time act as a water pool to maintain the enzymatic activity (49,50). In such systems, water-insoluble organic solvents have less effect on the enzyme stability than water-soluble solvents.

Table 2 summarizes some arguments in favor or against immobilization of biocatalysts, namely enzymes. The final decision for a certain reactor design should be based on an optimization process covering all relevant factors contributing to the overall costs, including investment, catalyst consumption, or productivity.

Reactors for Immobilized Enzymes

The methods for the heterogenisation (or localization) of enzymes by coupling them to insoluble supports or by entrapment have been discussed in the literature

Table 2. Comparison of Processes Using Soluble or Carrier-Fixed Enzymes

Enzyme fixed on a carrier or membrane	Soluble enzyme retained by a membrane
Continuous or batch process	Continuous or batch process
Improved stability	Lower enzyme stability
Loss of activity during immobilization	No loss of activity during immobilization
Immobilization costs	No costs for immobilization
Mass transfer limitations (heterogeneous reaction)	No mass transfer limitation (homogeneous reaction)
Established technology[a]	New technology[b]
	Sterile process possible
Higher investment costs	Easy supply of new enzyme

[a]Immobilization protocol may be different for each enzyme.
[b]Can be used for different enzymes without modification.

(5,15,34,51,52). The types of reactors used for immobilized enzymes are summarized in Figure 1. The principles developed for general heterogeneous catalysis in synthetic chemistry are valid, resulting in well-known reactor configurations. Differences between enzyme catalysis and other systems result from the nature of the biocatalyst and reaction medium. For example, soft particles containing the biocatalyst, such as alginate beads, may limit the pressure drop in fixed-bed reactors.

The decision as to specific reactor design will be based on a careful analysis of the kinetic properties of the reaction system (53–55). For example, if the enzyme shows a strong substrate-surplus inhibition, a continuously operated reactor with complete backmixing working at high conversion is advantageous. A reaction with strong product inhibition may utilize a batch reactor or a plug flow reactor to achieve higher volume and catalyst specific productivities. An extractive bioreactor may be used if substrates and products show different solubilities (Fig. 1**i** and **j**) (see "Selected Examples" section). By using this reactor configuration, the destabilizing effect of organic solvents may also be overcome, because the enzyme is separated from the organic phase, which is used to extract the insoluble product (56,57). The aqueous phase containing the enzyme will be saturated until the maximum solubility of with the substrate is reached.

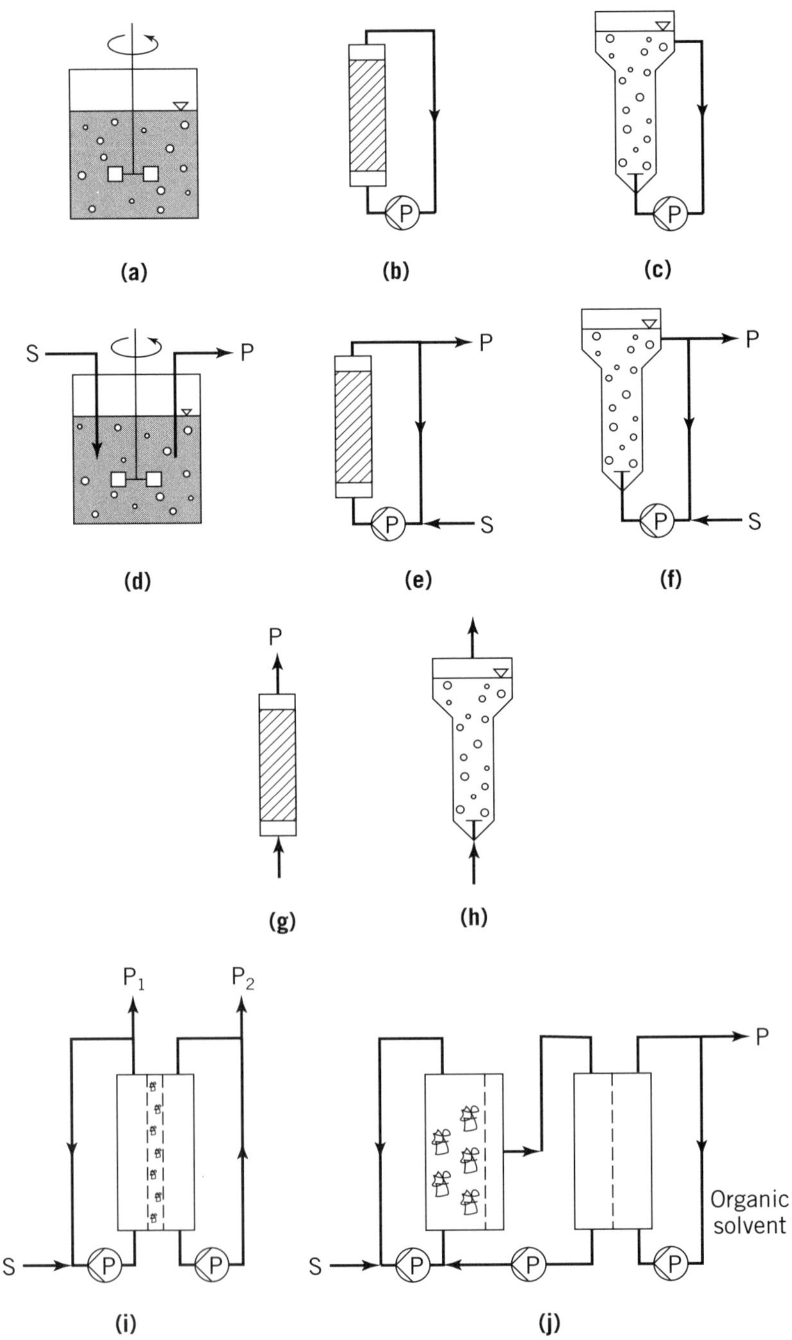

Figure 1. Reactors for immobilized enzymes. (**a–c**) Batch reactors with complete backmixing; (**a**) Stirred-tank reactor; (**b**) Fixed-bed reactor; (**c**) Fluidized-bed reactor. (**d–f**) Continuously operated reactors with complete backmixing. (**g–h**) Continuously operated reactors with plug-flow behavior. (**i**) Reactor with the enzyme immobilized in or on a membrane that may at the same time separate two phases such as water and organic solvent. (**j**) reactor with physically separated enzyme and organic solvent in order to prevent denaturation of the protein (the membrane reactor shown may be replaced by one of the types for immobilized enzymes).

Reactions using biocatalysts are normally performed in aqueous solution at temperatures between 10 and 80 °C and at ambient pressure. Due to the inhibition of some enzymes by heavy metals, the materials of construction must not release these elements. Reactors are operated under conditions that prevent microbial contamination. The reactor itself as well as the substrate may be sterilized prior to reaction by using chemical agents (ethanol, formaldehyde, ethylenoxide, Velcorin™) or steam. Ultraviolet rays or γ-rays may be used to sterilize the immobilized enzyme on its support (58). Alternatively, the immobilization may be performed under sterile conditions. Antibacterial agents may be added to the reaction mixture to prevent microbial growth while the reactor is running. In some cases, the reactants may act as sterilants or inhibitors of microbial growth, such as ketones or alcohols. At higher concentrations (more than 500 mmol/L), solutions may become autosterile because of osmotic pressure effects. Industrial processes are often performed at elevated temperatures, above 55 °C, reducing the danger of microbial contamination.

For a constant product quality and reproducibility of downstream processing, the reactor should be operated at constant conversion. To overcome the deactivation per unit of time that shows all biocatalysts as a result of denaturation processes, either the residence time has to be increased or fresh enzyme has to be supplied. The latter is especially easy for soluble enzymes. For carrier-fixed enzymes, a combination of both methods is used, as discussed later.

CHARACTERIZATION OF IMMOBILIZED ENZYMES

An exhaustive treatment of all aspects of immobilized enzymes, including a complete mathematical description, is beyond the scope of this article. This information may be found elsewhere (22,25,53,59–70).

The well-known Michaelis-Menten equation (equation 1) describes the enzymatic activity as function of the substrate concentration for an irreversible one-substrate reaction without inhibition. This idealized situation is rarely found in nature. Equation 2 gives the rate for a reversible reaction A⇔B. Using this form, product inhibition is considered as a competition of A and B for the active site as expressed by the two K_M values. This form might be extended to cover substrate surplus inhibition as well as multisubstrate reactions by adding additional terms in the numerator (53–55). The form of equation 2 resembles the kinetics of heterogeneously catalyzed reactions (Hougen-Watson kinetics) (71).

$$v = c_E \times \frac{A_{max} \times c_A}{K_M + c_A} \quad (1)$$

$$v = c_E \times \frac{\dfrac{A_{max,A} \times c_A}{K_{M,A}} - \dfrac{A_{max,B} \times c_B}{K_{M,B}}}{1 + \dfrac{c_A}{K_{M,A}} + \dfrac{c_B}{K_{M,B}}} \quad (2)$$

Upon immobilization of the enzyme on a support, the intrinsic properties expressed as kinetic constants A_{max} and K_M may be altered. This may be because of to conformational changes of the protein by the attachment resulting in a variation of the active site. More often, however, there might be a direct influence of the support or the microenvironment (59). For example, the polymer matrix may cause a pH difference between the microenvironment of the immobilized enzyme and the bulk fluid. Thus, the activity as a function of pH is often different for the soluble and the immobilized enzyme. In a similar way, the apparent K_M value of an immobilized enzyme may be altered compared to that of the soluble form. Additional attracting or repulsing interactions of the support matrix with the substrates or products may lead to different values when compared with the native enzyme. A similar problem arises when lipases are used in a heterogeneous reaction system. Because of their natural properties, these enzymes enrich at hydrophopic/hydrophilic interfaces as found in oil-in-water emulsions. If the inhibiting product enriches there, too, the enzyme will experience a very high local product concentration that will drastically alter its activity (72).

In heterogeneous catalysis, mass transport affects control kinetics at the solid–liquid interphase (Fig. 2). The external mass transfer or film diffusion is determined by the velocity difference between the particle and the solution. The influence of this type of mass transfer resistance can be minimized by rapid flow through a fixed bed or fluidized bed reactor or vigorous mixing in a stirred tank reactor. The limiting property will be the mechanical stability of the support used.

The example given in Scheme 1 illustrates the influence of film diffusion (73,74). The enzyme fructose-1,6-bisphosphate aldolase from rabbit muscle catalyzes the reaction between dihydroxyacetone phosphate (**1**) and various aldehydes (**2**) (10). The enzyme has been immobilized on an insoluble support, thereby enhancing its stability and enabling the reaction in a continuously operated slurry reactor (Fig. **1d**). If the mass transport is only diffusion controlled (relative velocity between particle and medium = 0), the Sherwood number describing the mass transport with spheres is defined as

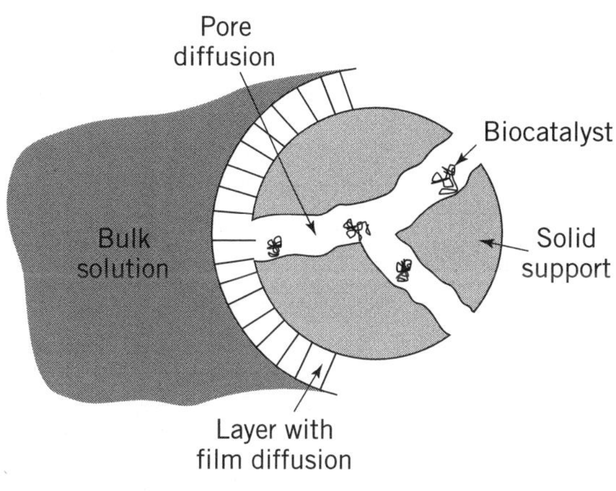

Figure 2. Scheme of a particle with immobilized biocatalyst.

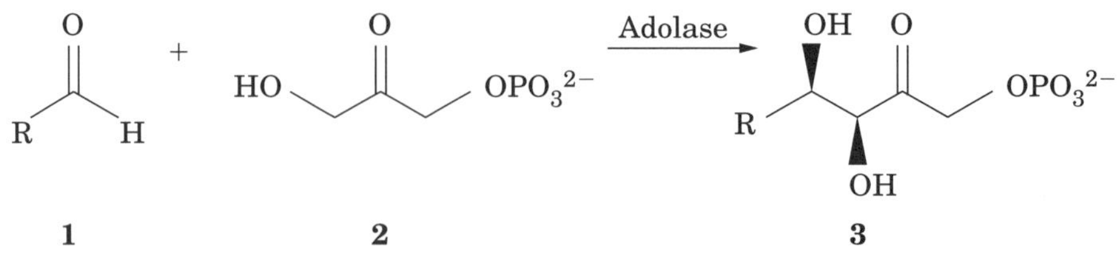

Scheme 1.

$$Sh = \frac{k \times d}{D} = 2 \tag{3}$$

The molar flux of mass transport is given by

$$v_{mt} = k \times (c_0 - c) \tag{4}$$

Assuming Michaelis-Menten-type kinetics, the enzymatic reaction will be rate limiting as long as

$$k > \frac{A_{max,A}}{K_M} \tag{5}$$

For rabbit muscle aldolase bound to Eupergit™ C1Z (nonporous carrier, diameter 1 μm), typical values for $A_{max,A}$ are 50 μmol m^{-2} s^{-1} corresponding to 405.9 nkat g^{-1} (24.3 U g^{-1}). K_M values range between 0.1 and 10 mmol L^{-1}. From equation 3, typical k values of 2×10^{-3} m s^{-1} result. Thus, the mass transport is much faster than the enzymatic reaction, since

$$5 \cdot 10^{-6}\ ms^{-1} < \frac{A_{max,A}}{K_M} < 500 \cdot 10^{-6}\ ms^{-1}$$

If, however, particles with a diameter of 100 μm were used for immobilization, film diffusion would be rate limiting (Fig. 3). The mass-transport resistance might be reduced to a certain extent by establishing a high velocity between particle and solution. The influence of internal mass transfer or pore diffusion becomes more dominant with increasing particle size. This may reduce the effectiveness by substrate limitation or product inhibition in the inner part of the particle, and when the reaction causes a strong pH-shift, local high values may deactivate the enzyme. This may be avoided by higher buffer concentrations. An example is the production of 6-aminopenicillanic acid (see "Selected Examples" section) (75, 76).

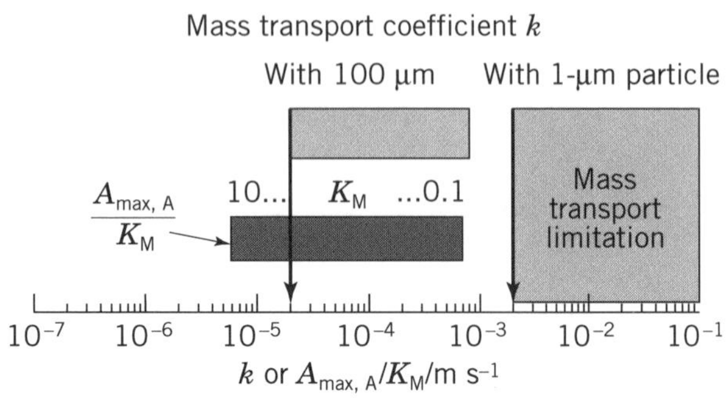

Figure 3. Influence of particle diameter on mass transport in comparison with enzymatic reaction.

In a very simple way the effect of internal mass transfer may be considered as an efficiency coefficient η, which is a function of size, shape, and porosity of the support matrix and the diffusion of the compounds involved:

$$\eta = \frac{\text{activity of immobilized catalyst}}{\text{activity of catalyst without mass transfer}} \tag{6}$$

$$v_{app} = \eta \times v \tag{7}$$

Pore diffusional resistance is indicated when apparent activity rises after grinding the catalyst particles (homogeneous catalyst distribution). If no increase of activity is observed, film diffusion is already limiting. These effects, together with the kinetics of the immobilized catalyst, can be easily investigated in a differential recirculation reactor (62,66). The basic setup is the same as in Figure **1b**. Only a small amount of catalyst is used to minimize conversion per pass through the bed, as measured in the bypass. By adjusting the flow rate, film diffusion and pore diffusion may be easily distinguished.

For a more elaborate description of pore diffusion, the Thiele modulus, interpreted as the reaction rate relative to the diffusion rate within the particle (59,68–70), has to be used.

The number of catalytically active reaction sites per unit mass of catalyst can be easily determined for immobilized enzymes. For this active-site test, a suitable substrate, such as an ester, is converted by the enzyme and the product is determined analytically (77,78). This test is independent from mass transport phenomena because only the additional amount of liberated product after addition of the catalyst is used for calculation. According to the Michaelis-Menten kinetics of an enzymatic reaction with varying reaction order, the efficiency coefficient η is a function of conversion. This has been demonstrated for immobilized α-chymotrypsin in the cleavage of amino acid esters (Fig. 4) (79). The use of immobilized enzymes in processes with precipitated substrates or products has been addressed recently (80).

SELECTED EXAMPLES OF REACTIONS USING IMMOBILIZED ENZYMES

One of the first enzymes shown to be active after immobilization by adsorption on charcoal was invertase (12,81). This enzyme is still used for the large-scale production of invert sugar by hydrolysis of sucrose. One kilogram of im-

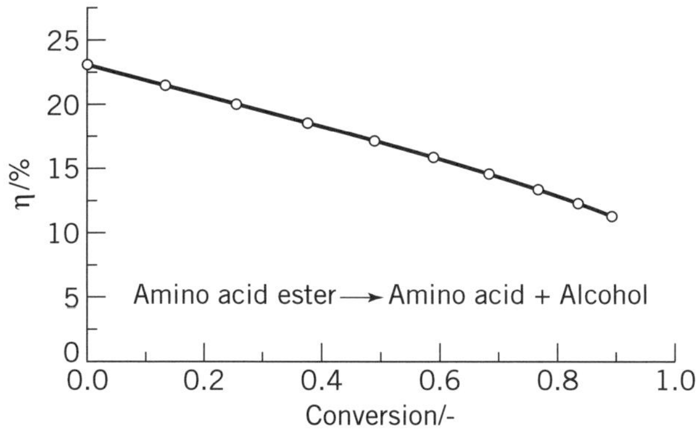

Figure 4. Efficiency coefficient η as function of conversion; initial ester concentration 200 mmol L^{-1}; Thiele modulus 10.1.

mobilized enzyme supports production of approximately 10,000 kg of invert sugar (12).

Production of L-Amino Acids Using Amino Acylase

Enantiomerically pure L-amino acids are required for use in parenteral nutrition, as food additives or as chiral synthons. In addition to isolation from natural sources and direct fermentation, racemic mixtures of L- and D-amino acid can be resolved by enzymatic reaction with one of the two enantiomers. Scheme 2 illustrates this for a racemic amino acid (4) using amino acylase I from *Aspergillus* (EC 3.5.1.14). (*Note*: the EC number, in parentheses, is the internationally recognized Enzyme Commission number (82). The six main classes are oxidoreductases, transferases, hydrolases, lyases, isomerases, and lipases). The unreacted D-enantiomer (**6**) can be racemized easily and recycled to increase the overall yield. The first industrial process of this type was established by Tanabe Seiyaku Co. in 1969 using enzyme immobilized on an insoluble support (1,83).

One of the best methods to obtain a stable catalyst in their example has been to immobilize by ionic binding to DEAE-Sephadex™. To compensate for the loss of activity during operation (20% after 1 month), fresh enzyme has been passed through the column periodically to establish the former activity again. The main products of this process are L-methionine, L-phenylalanine, and L-valine (9). Typically, a 0.2 mol L^{-1} solution of D, L-acetyl amino acids is feed into a 1-m^3 fixed-bed reactor with various flow rates. For L-methionine, the space-time yield reported is about 715 g L^{-1} d^{-1}.

Production of 6-Amino Penicillanic Acid and 7-Amino Cephalosporanic Acid Using Amidases

6-Amino penicillanic acid (6-APA) (**8**) and 7-amino cephalosporanic acid (7-ACA) (**11**) are important intermediates for semisynthetic β-lactam antibiotics. Both compounds can be obtained readily by amidase-catalyzed cleavage of the corresponding amides (**7** and **10**) that are accessible by fermentation (Scheme 3).

Whereas penicillin amidase (EC 3.5.1.11) has been used for production of 6-APA since the early 1970s, cephalosporin C amidase (EC 3.5.1.3) has been used just recently in a chemoenzymatic synthesis of 7-APA (1,84). Amidases have been prepared from several different sources, such as *Pseudomonas, Bacillus*, and (recombinant) *Escherichia coli* (1,85,86). In general, the enzyme is immobilized on and in various supports such as polyacrylonitrile fibers, polyacrylamid beads, or Sephadex™. For 6-APA batch production, typically a 10% solution of penicillin G is circulated through the bioreactor or processed in a stirred-tank reactor. One cycle lasts about 3 h. The life time for the enzyme is more than 360 cycles. The space-time yield for this process is about 800 g L^{-1} d^{-1} (1). One of the main problems for both processes is the liberation of a strong acid. To avoid large pH shift, an effective method to control and maintain the desired pH is needed. High recirculation rates are used for the enzyme immobilized on hollow fibers. For the enzyme immobilized in beads, both a small bead diameter (typically around 100 to 600 μm) and good control of film and pore diffusion by high fluid velocities are needed.

Production of High-Fructose Corn Syrup Using Glucose Isomerase

The production of high-fructose corn syrup, a mixture of glucose (**13**) and fructose (**14**), is the largest enzymatic process in commercial practice so far utilizing immobilized biocatalysts (Scheme 4). The process using glucose isomerase (EC 5.3.1.5) was first commercialized in 1970 (1,12). The enzyme is available from different bacterial sources. By site-directed mutagenesis, more thermostable mutants have been obtained. Most of the enzyme preparations commercially available are cross-linked with glutaraldehyde (1). The enzyme is used in fixed-bed reactors, and the feed flow is adjusted to overcome enzyme deactivation and en-

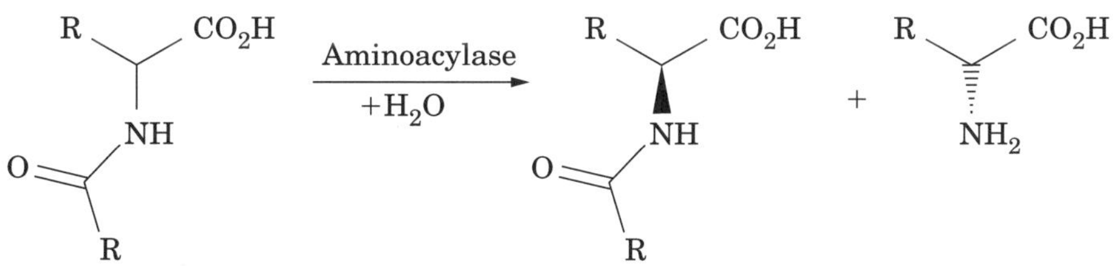

4 5 6

Scheme 2.

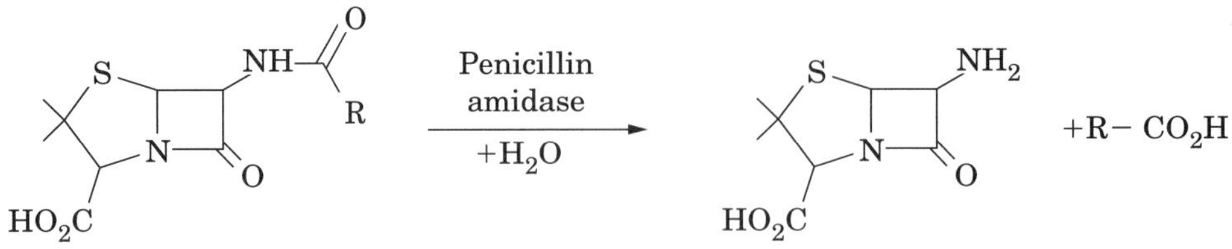

7 **8** **9**

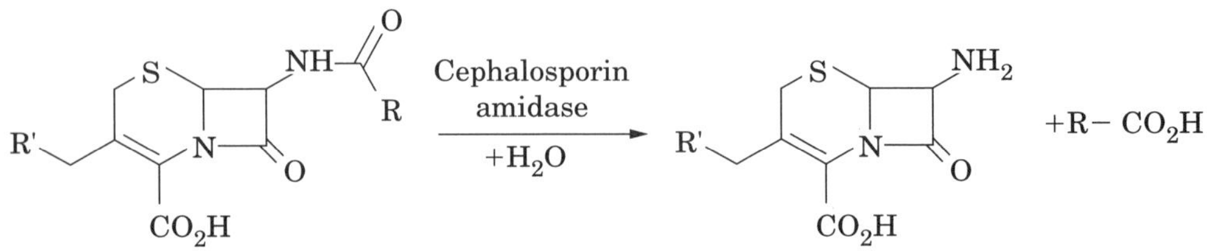

10 **11** **12**

Scheme 3.

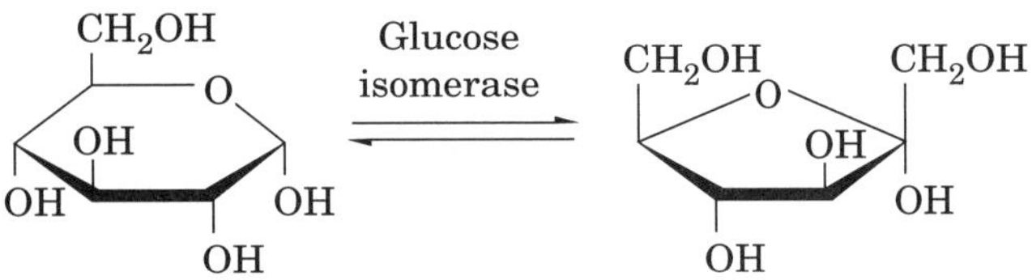

13 **14**

Scheme 4.

sure a constant product quality (conversion). Different columns of different enzyme age are operated in parallel, resulting in an average conversion of 45% in a process of uniform throughput. The equilibrium conversion is about 50%. If the residual activity has dropped to about 10% of the initial activity, the enzyme is replaced. For the most-stable commercially available enzyme preparations, the half-life is more than 100 days. Typical temperatures range from 50 to 60 °C with 50% substrate solution.

Production of Acrylamide Using Nitrile Hydratase

Biotransformations have been also established for the production of bulk chemicals. The enzymatic production of acrylamide (**16**) from acrylonitrile (**15**) using nitrile hydratase (EC 4.2.1.84) from *Rhodococcus rhodochrous* has replaced the copper-catalyzed process in Japan (Scheme 5) (1,87). For this process, whole cells were immobilized in a cationic acrylamide-based polymer gel. Typical reaction conditions range from 0 to 15 °C and pH 6.5 to 8.0. Acrylonitrile is added gradually in order to achieve a final concentration of acrylamide of 400 g L^{-1}. The conversion of acrylonitrile is more than 99%, with a selectivity for acrylamide of more than 99%. Because of its broad substrate specificity, the nitrile hydratase may be also used for production of nicotinamide, isonicotinamide, or pyrazinamide (87).

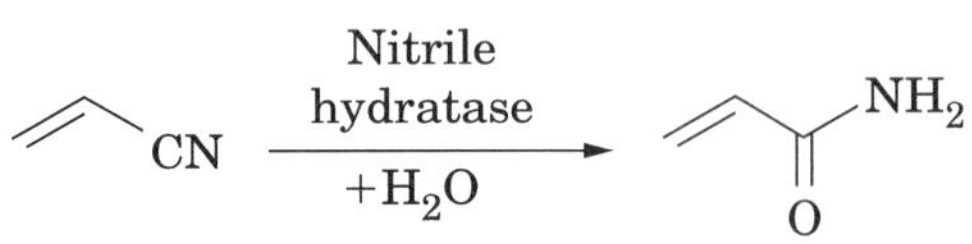

15 **16**

Scheme 5.

Production of Glycerol-3-Phosphate Using Glycerol Kinase

Glycerol-3-phosphate (G3P) (**18**) is a useful starting material for the synthesis of phospholipids. These compounds are of importance for numerous medicinal, diagnostic, and research applications. Their potential applications include microemulsion and liposomal drug delivery systems (80). The enzymic synthesis of G3P using glycerol kinase (EC 2.7.1.30) has been commercialized by Genzyme (88–90). Adenosine triphosphate (ATP) is used as the phosphorylating agent (Scheme 6), and regeneration of ATP is achieved by pyruvate kinase (EC 2.7.1.40) using phosphoenol pyruvate (PEP) (**22**).

The enzymes used are immobilized on a solid support. Especially for the glycerol kinase the stability was increased dramatically when compared with the native form. Its half-life in aqueous solution is about 15 min, because of dissociation of the subunits. The reaction is typically performed in a batch reactor containing the immobilized enzyme in a fixed bed. Multikilogram-scale production typically uses 1 mol L^{-1} glycerol and 1 mol L^{-1} phosphoenol pyruvate. For this process, space-time yield is about 81 g L^{-1} d^{-1}. Catalytic amounts of ATP (1 to 10 mol%) are used, resulting in total turnover numbers up to 100. These and even higher values are attainable without requiring coimmobilization of the cofactors. NADH values of about 1,000 may be reached easily (27,56). These processes are economically feasible because of the low cost of the cofactors. (*Note*: Cofactors in bulk quantities are available from Oriental Yeast, Japan.)

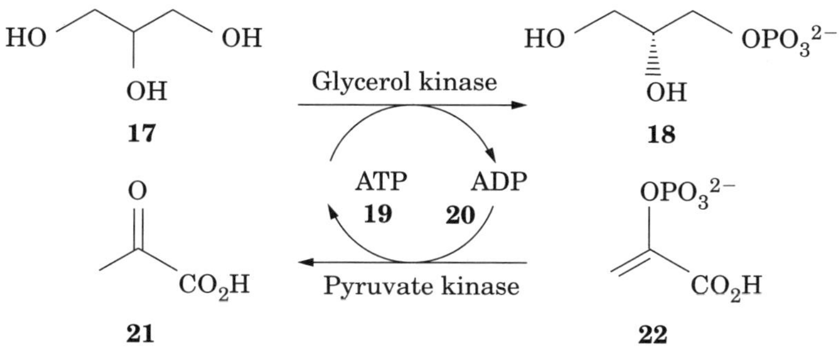

Scheme 6.

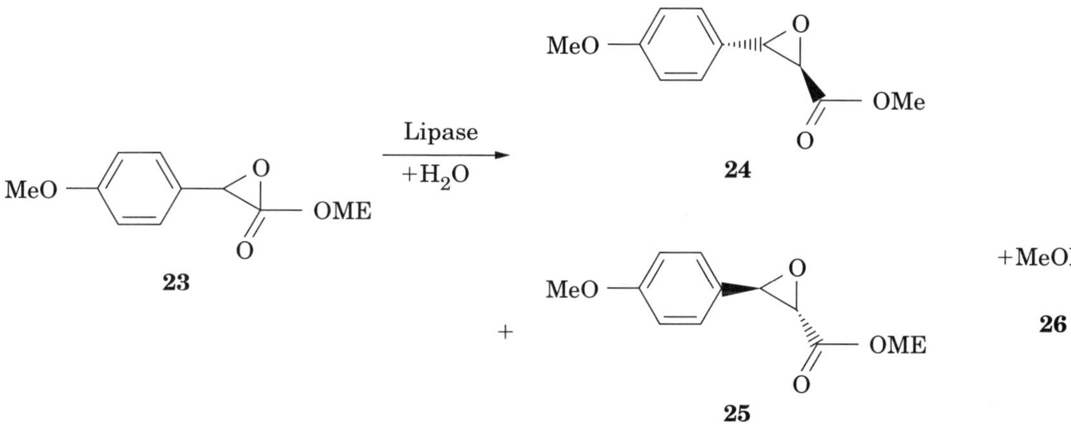

Scheme 7.

Table 3. Further Examples for Application of Immobilized Enzymes

Enzyme	Substrate	Product	Ref.
Glucose kinase	Glc, ATP	Glucose-6-phosphate	97
Oxynitrilase	Aldehydes, HCN	Cyanohydrines	44
Transaminase	Ketone	Amine	98,99
Pyranose oxidase	Sugar	Keto-sugars	100
Hydantoinase	Hydantoin	Amino acid	101
Oxidoreductase	Benzoylformate	Mandelate	102
Sialidase	Neu5Ac, Gal	Neu5Ac-Gal	103
Glucosidase	Glc, octanol	Octylglycoside	104
Chymotrypsin	Amino acids	Oligopeptides	105
Glycosyltransferase	Carbohydrate	Oligosaccharide	10,106

Production of (2R,3S)-Methyl-*p*-Methoxyphenylglycidate Using a Lipase

Optically pure (2R,3S)-methyl-*p*-methoxyphenylglycidate (**24**) is used as chiral building block for the synthesis of diltiazem. Sepracor has developed a process for enantioselective hydrolysis of the racemic ester using a lipase, which has reached a 50 t a^{-1} scale (91,92) (Scheme 7). The reaction is performed in a multiphase membrane reactor with the enzyme reversibly entraped within the porous matrix of an asymmetric hollow fiber ultrafiltration membrane (92–94). This reactor is shown in Figure 1**i**. An aqueous buffer is circulated on one side of the hydrophilic membrane, providing an aqueous environment for the enzyme. A 17% solution of the racemic ester (**23**) dissolved in toluene is circulated on the other side of the membrane. Despite its lower enantioselectivity, *Candida cylindracea* lipase (EC 3.1.1.3) is being used; lipases from other sources were not retained sufficiently within the membrane. The reactor is used in a batch mode. The space-time yield, based on the membrane area, is about 125 g m^{-2} d^{-1}.

This type of reactor has also been used for the hydrolysis of butter oil (95) or the production of peroxycarboxylic acids (96).

Miscellaneous Processes Using Immobilized Enzymes

Table 3 lists some reactions that have been performed using immobilized enzymes to give an impression of the range of enzymatic biotransformations. For a comprehensive overview the reader is refered to reviews and textbooks (1,5–20). Some of the examples are already used on industrial level. Application of soluble enzymes retained by ultrafiltration membranes have been successful as well (21–28).

NOMENCLATURE

Symbols and Abbreviations

Note:	1 kat = 1 mol s^{-1}, 1 U = 1 μmol min^{-1}
	1 kat = 60 * 106 U
A_{max}	Maximum mass specific activity, U g^{-1}, kat g^{-1}, or mol s^{-1} g^{-1}
$A_{max,A}$	Maximum area specific activity, mol s^{-1} m^{-2}
ATP	Adenosine triphosphate
c	Concentration, mol L^{-1}, g L^{-1}
D	Diffusion coefficient, m^2 s^{-1}
d	Particle size, m
Gal	Galactose
Glc	Glucose
k	Stoff transport coefficient, m s^{-1}
K_m	Michaelis-Menten constant, mol L^{-1}
η	Efficiency coefficient, –
Neu5Ac	*N*-acetylneuraminic acid
Sh	Sherwood number, –
v	Reaction rate, kat L^{-1}
v_{mt}	Mass transport, mol m^{-2} s^{-1}

Indices

0	Time zero or in bulk phase
A	Substance A
app	Apparent
B	Substance B
E	Enzyme

BIBLIOGRAPHY

This is a modified version of the article first published as U. Kragl and C. Wandrey, "Biocatalytic Reactions with Immobilized Enzymes" in G. Ertl, H. Knözinger, J. Weitkamp eds., *Handbook of Heterogeneous Catalysis*, Wiley-VCH Verlag GmbH, D-69451 Weinheim, Germany, 1997, pp. 2436–2447.

1. A. Tanaka, T. Tosa, and T. Kobayashi eds., *Industrial Application of Immobilized Biocatalysts*, Decker, New York, 1993.
2. J.G. Shewale, B.S. Deshpande, V.K. Sudhakran, and S.S. Ambedkar, *Process. Biochem. Int.* **25**, 97–103 (1990).
3. V. Kasche, W. Hass, C. Niebuhr-Redder, and E. Piotratscke, *Bioengineering* **2**, 69–77 (1992).
4. F. Balkenhol, K. Ditrich, B. Hauer, and W. Ladner, *J. Prakt. Chemie* **339**, 381–384 (1997).
5. I. Chibata ed., *Immobilized Enzymes, Research and Development*, Kodansha, Tokyo, 1978.
6. H.G. Davies and R.H. Green, *Biotransformations in Preparative Organic Chemistry: The Use of Isolated Enzymes and Whole Cell Systems in Synthesis*, Academic Press, London, 1989.
7. M.W. Blanch and D.S. Clark eds., *Applied Biocatalysis*, vol. 1, Decker, New York, 1991.
8. K. Faber, *Biotransformations in Organic Chemistry*, Springer-Verlag, Berlin, 1995.
9. S.M. Roberts, K. Wiggins, and G. Casy, *Whole Cells and Isolated Enzymes in Organic Chemistry*, Wiley, Chichester, U.K., 1992.
10. C.H. Wong and G.M. Whitesides, *Enzymes in Synthetic Organic Chemistry*, Pergamon Press, Oxford, U.K., 1994.
11. W. Gerhartz, Y.S. Yamamoto, L. Kandy, J.I. Rounsaville, and G. Schulz eds., *Ullman's Encyclopedia Industrial Chemistry*, 5th ed., vol. 9, Wiley-VCH, Weinheim, 1987, pp. 341–530.
12. W. Gerhartz, *Enzymes in Industry*, VCH, Weinheim, 1990.
13. A.N. Collins, G.N. Sheldrake, and J. Crosby eds., *Chirality in Industry*, Wiley, Chichester, U.K., 1992.
14. D. Rozzell and F. Wagner, *Biocatalytic Production of Amino Acids & Derivatives*, Hanser, Munich, 1992.
15. K. Mosbach ed., *Methods Enzymol.* **135B**, **136C**, **137D**, 1987.
16. E. Katchalski-Katzir, *Trends Biotechnol.* **11**, 471–478 (1993).
17. R.A. Sheldon, *Chirotechnology: Industrial Synthesis of Optically Active Compounds*, Decker, New York, 1993.
18. K. Drauz and H. Waldmann eds., *Enzyme Catalysis in Organic Synthesis*, VCH, Weinheim, Germany, 1995.
19. B.P. Sharma, L.F. Bailey, and R.A. Messing, *Angew. Chem. Int. Ed. Engl.* **21**, 837 (1982).
20. A. Rosevar, *J. Chem. Tech. Biotechnol.* **34B**, 127–150 (1984).
21. M.-R. Kula and C. Wandrey, *Methods Enzymol.* **136**, 9–21 (1987).
22. U. Kragl, D. Vasic-Racki, and C. Wandrey, *Chem. Ing. Tech.* **64**, 499–509 (1992).
23. U. Kragl, D. Vasic-Racki, and C. Wandrey, *Ind. J. Chem.* **32B**, 103–117 (1993).
24. A.S. Bommarius, K. Drauz,U. Groeger, and C. Wandrey, in A.N. Collins, G.N. Sheldrake, J. Crosby eds., *Chirality in Industry*, Wiley, Chichester, U.K., 1992, pp. 371–397.
25. A.S. Bommarius, in H.-J. Rehm, G. Reed, A. Pöhler, P. Stadtler, G. Stephanopoulus eds., *Biotechnology* vol. 3, VCH, Weinheim, 1993, pp. 427–466.
26. D.M.F. Prazeres and J.M.S. Cabral, *Enzyme Microb. Technol.* **16**, 738–750 (1994).
27. U. Kragl, in T. Godfrey, M. Daniels, S. West eds.,*Industrial Enzymology*, Macmillan, London, 1996, pp. 271–283.
28. U. Kragl, A. Gödde, C. Wandrey, W. Kinzy, J.J. Cappon, and J. Lugtenburg, *Tetrahedron Asymmetry* **4**, 1193–1202 (1993).
29. P. Brodelius and E.J. Vandamme, in H.J. Rehm, G. Reed, J.F. Kennedy eds., *Biotechnology*, vol. 7a, VCH, Weinheim, 1987, pp. 405–464.
30. A. Grobiollot, D.K. Boadi, D. Poncelet, and R. Neufeld, *Crit. Rev. Biotechnol.* **14**, 75–107 (1994).
31. S. Norton and J.C. Vuillemard, *Crit. Rev. Biotechnol.* **14**, 193–224 (1994).
32. J.A.M. de Bont, J. Visser, B. Mattiasson, and J. Tramper eds., *Physiology of Immobilized Cells*, Elsevier, Amsterdam, 1990.
33. J.F. Kennedy and J.M.S. Cabral, in H.J. Rehm, G. Reed, J.F. Kennedy eds., *Biotechnology*, vol. 7a, VCH, Weinheim, 1987, pp. 347–404.

34. W. Keim and B. Dreißen-Hölscher, in G. Ertl, H. Knözinger, J. Weitkamp eds., *Handbook of Heterogeneous Catalysis*, Wiley-VCH, Weinheim, 1997, pp. 231–240.
35. L.J. Blum and P.R. Coulet eds., *Biosensor Principles and Applications*, Decker, New York, 1991.
36. R.D. Schmid and F. Scheller eds., *Biosensors: Fundamentals, Technologies and Applications*, VCH, Weinheim, 1992.
37. J.E. Zull, J. Reed-Mundell, Y.W. Lee, D. Vezenow, N.P. Ziats, J.M. Anderson, and C.N. Sukenik, *J. Industr. Microbiol.* **13**, 137–143 (1994).
38. P. Wang, T.G. Hill, C.A. Wartchow, M.E. Huston, L.M. Oehler, M.B. Smith, M.D. Bednarski, and M.R. Callstrom, *J. Am. Chem. Soc.* **114**, 378–380 (1992).
39. D. Leckband and R. Langer, *Biotechnol. Bioeng.* **37**, 227–237 (1991).
40. V.V. Mozhaev, N.S. Melik-Nubarov, M.V. Sergeeva, V. Siksnis, and K. Martinek, *Biocatalysis* **3**, 179–187 (1990).
41. R.D. Schmidt, in T.K. Gose, A. Fiechter, N. Blakebrough eds., *Advances in Biochemical Engineering*, vol. 12, Springer-Verlag, Berlin, 1979, pp. 41–118.
42. R. Ulbrich, R. Golbik, and A. Schellenberger, *Biotechnol. Bioeng.* **37**, 280–287 (1991).
43. C. Laane, J. Tramper, and M.D. Lilly eds., *Biocatalysis in Organic Media*, Elsevier, Amsterdam, 1987.
44. F. Effenberger, *Angew. Chem. Int. Ed. Engl.* **33**, 1555–1564 (1994).
45. L.E.S. Brink, J. Tramper, K.C.A.M. Luyben, and K. Van't Riet, *Enzyme Microb. Technol.* **10**, 736–743 (1988).
46. M.N. Gupta, *J. Biochem.* **203**, 25–32 (1992).
47. A.M. Klibanov, *Acc. Chem. Res.* **23**, 114–120 (1990).
48. W.F.M. Stöcklein and F.W. Scheller, *Chem.-Ing. Techn.* **67**, 69–77 (1995).
49. E. Wehtje, P. Adlerkreutz, and B. Mattiasson, *Biotechnol. Bioeng.* **41**, 171–178 (1993).
50. P. Adlerkreutz, *Eur. J. Biochem.* **199**, 609–614 (1991).
51. I. Chibata, T. Tosa, and T. Sato, in H.-J. Rehm, G. Reed, and J.F. Kennedy eds., *Biotechnology*, vol. 7a, VCH, Weinheim, 1987, pp. 653–684.
52. W. Gerhartz, Y.S. Yamamoto, L. Kandy, J.F. Rounsaville, and G. Schulz eds., *Ullman's Encyclopedia Industrial Chemistry*, 5th ed., vol. A9, Wiley-VCH, Weinheim, 1987, pp. 382–388.
53. M. Biselli, U. Kragl, and C. Wandrey, in K. Drauz and H. Waldmann eds., *Enzyme Catalysis in Organic Synthesis*, VCH, Weinheim, 1995.
54. I.H. Segel, *Enzyme Kinetics*, Wiley, New York, 1975.
55. A. Cornish-Bowden, *Fundamentals of Enzyme Kinetics*, Butterworths, London, 1981.
56. W. Kruse, W. Hummel, and U. Kragl, *Recl. Trac. Chim. Pays-Bas* **115**, 239–243 (1996).
57. A. Liese, M. Karutz, J. Kamphuis, C. Wandrey, and U. Kragl, *Biotechnol. Bioeng.* **51**, 544–550 (1996).
58. P. Surmann and E. Nürnberg eds., *Hagers Handbuch der Pharmazeutischen Praxis*, vol. 2, Springer-Verlag, Berlin, 1991, pp. 426–428.
59. D.S. Clark, *Trends Biotechnol.* **12**, 439–443 (1994).
60. J.E. Prenosil, I.J. Dunn, and E. Heinzle, in H.J. Rehm, G. Reed, J.F. Kennedy eds., *Biotechnology*, Vol. 7a, VCH, Weinheim, 1987, pp. 489–545.
61. B. Elvers, S. Hawkins, M. Ravenscraft, and G. Schulz eds., *Ullmann's Encyclopedia Industrial Chemistry*, 5th ed., vol. 14, Wiley-VCH, Weinheim, 1989, pp. 1–48.
62. C. Wandrey and E. Flaschel, in T.K. Ghose, A. Fiechter, N. Blakebrough, and A. Fiechter eds., *Advances in Biochemical Engineering*, Vol. 12, Springer-Verlag, Berlin, 1979, pp. 147–218.
63. O. Levenspiel, *Chemical Reaction Engineering*, Wiley, New York, 1972.
64. J.E. Bailey and D.F. Ollis, *Biochemical Reaction Fundamentals*, McGraw-Hill, New York, 1986.
65. H. Chmiel ed., *Bioprozeßtechnik Band 1 und 2*, Gustav Fischer, Stuttgart, 1991.
66. K. Buchholz, *Characterisation of Immobilized Biocatalysis*, vol. 84, VCH, Weinheim, 1979.
67. F. Shadman-Yazdi and E.E. Petersen, *Chem. Eng. Sci.* **27**, 227–237 (1972).
68. B.J. Rovito and J.R. Kittrell, *Biotechnol. Bioeng.* **15**, 143–161 (1973).
69. R. Lortie and D. Thomas, *Biotechnol. Bioeng.* **28**, 1256–1260 (1986).
70. Y.H. Park, M.H. Han, and H.-K. Rhee, *J. Chem. Tech. Biotechnol.* **34B**, 57–69 (1984).
71. M. Baerns, H. Hoffmann, and A. Renken, *Chemische Reaktionstechnik*, Georg Thieme, Stuttgart, 1992.
72. J.G.T. Kierkels, L.F.W. Vleugels, E.T.F. Gelad, D.P. Vermeulen, J. Kamphuis, and C. Wandrey, *Enzyme Microb. Technol.* **16**, 513–521 (1994).
73. B. Bossow-Berke, Ph.D. Thesis, Bonn University, Bonn, 1989.
74. B. Bossow-Berke, W. Berke, and C. Wandrey, *Bioengineering* **8**, 12–19 (1992).
75. J. Danzig, W. Tischer, and C. Wandrey, *Ind. J. Chem* **32B**, 40–43 (1993).
76. J. Danzig, W. Tischer, and C. Wandrey, *Chem. Eng. Technol.* **18**, 256–259 (1995).
77. M.L. Bender, M.L. Begecanton, R.L. Blakely, L.J. Brubacher, J. Feder, C.R. Gunter, F.J. Kezdy, J.V. Killhefter, T.H. Marshall, C.G. Miller, R.W. Roeske, and J.K. Stoops, *J. Am. Chem. Soc.* **88**, 24 (1966).
78. J.R. Ford, R.P. Chambers, and W. Cohen, *Biochim. Biophys. Acta* **309**, 175–180 (1973).
79. W. Halwachs, C. Wandrey, and K. Schügerl, *Biotechnol. Bioeng.* **20**, 541–554 (1978).
80. V. Kasche and B. Galunsky, *Biotechnol. Bioeng.* **45**, 261–267 (1995).
81. J.M. Nelson and E.G. Griffin, *J. Am. Chem. Soc.* **83**, 1109–1115 (1916).
82. Int. Union of Biochemistry and Molecular Biology, *Enzyme Nomenclature*, Academic Press, San Diego, 1992.
83. T. Tosa, T. Mori, N. Fuse, and I. Chibata, *Agric. Biol. Chem.* **33**, 1047–1052 (1969).
84. W. Tischer, U. Gieseke, G. Lang, A. Röder, and F. Wedekind, *Ann. N.Y. Acad. Sci.* **672**, 502–509 (1992).
85. J.G. Shewale and H. Sivaram, *Process Biochem.* **24**, 146–154, (1989).
86. *United Nations Industrial Development Organization, Enzyme Engineering at the Industrial Level—Present Status and Future Prospects*, Wien, 1989, pp. 39–47.
87. T. Nagasawa and H. Yamada, *Trends Biotechnol.* **7**, 153–158 (1989).
88. A.E. Walts, D.G. Shena, E.M. Fox, J.T. Davis, and M.R. Mische, in A.N. Collins, G.N. Sheldrake, J. Crosby eds., *Chirality in Industry*, Wiley, New York, 1992, pp. 223–235.
89. D.C. Crans and G.M. Whitesides, *J. Am. Chem. Soc.* **107**, 7019–7027 (1985).

90. A.E. Walts, D.G. Shena, E.M. Fox, J.T. Davies, and M.R. Mischke, *Chim. Oggi.* **9**, 29–33 (1991).
91. J.L. Lopez, *First Int. Symposium on Technologies for the Production of Enantiomerically Pure Compounds*, Amelia, Florida, 1993.
92. M. Furui, T. Furutani, T. Shibatani, Y. Nakamoto, and T. Mori, *J. Ferment. Bioeng.* **81**, 21–25 (1996).
93. WO 88/07582 (Oct. 6, 1988), S.L. Matson (to the regents of Sepracor Inc.).
94. J.L. Lopez, S.A. Wald, S.L. Matson, and J.A. Quinn, *Ann. N.Y. Acad. Sci.* **613**, 155–166 (1990).
95. F.X. Malcata, C.G. Hill Jr., and C.H. Amundsen, *Biotechnol. Bioeng.* **39**, 984–1001 (1992).
96. F.P. Cuperus, S.T. Bouver, G.F.H. Kramer, and J.T.P. Derksen, *Biocatalysis* **9**, 89–96 (1994).
97. H. Nakajima, H. Kondo, R. Tsurutani, M. Dombou, I. Tomioka, and K. Tomita, *ACS Symp. Ser.* **466**, 111–120 (1991).
98. G. Matcham, A. Bowen, S. Lee, P. Pienkos, and A. Zeitlin, *Proc. Chiral 94 USA Symposium*, Spring Innovations, Reston, Va., 1994, pp. 309–315.
99. D.I. Stirling, in A.N. Collins, G.N. Sheldrake, J. Crosby eds., *Chirality in Industry*, Wiley, New York, 1992, pp. 209–222.
100. A. Huwig, H.J. Daneel, and F. Griffhorn, *J. Biotechnol.* **32**, 309–315 (1994).
101. E.U. Pat. 92,118,795 (1992), F. Wagner, M. Pietsch, and C. Syldatk.
102. K. Hosono, S. Kajiwara, Y. Yamazaki, and H. Maeda, *J. Biotechnol.* **14**, 149–156 (1990).
103. J. Thiem and B. Sauerbrei, *Angew. Chem. Int. Ed. Engl.* **30**, 1503 (1991).
104. G. Ljunger, P. Adlerkreutz, and B. Mattiason, *Enzyme Microb. Technol.* 16, 751–755 (1994).
105. A.O. Richards, I.S. Gill, and E.N. Vulfson, *Enzyme Microb. Technol.* **15**, 928–935 (1993).
106. J. Hodgson, *Biotechnology* **13**, 38–41 (1995).

ENZYMES, PROTEIN HYDROLYSIS

HANS SEJR OLSEN
Novo Nordisk
Bagsvaerd, Denmark

KEY WORDS

- Dairy products
- Detergents
- Enzymes
- Food processing
- Hydrolysis
- Industrial cleaning
- Leather
- Osmometry
- pH stat
- Whipping agents

OUTLINE

INTRODUCTION

Industrial enzymology is an important branch of bioprocess technology. Enzymatic processes permit natural raw materials to be upgraded and finished. Using enzymes as a functional ingredient in detergents, laundry and dishes, for example, are cleaned in an efficient, environmentally mild, and energy-saving way. The most used industrial enzymes are proteases for hydrolysis reactions within the detergent, dairy, protein, pet food, feed, and leather tanning industries.

In detergents proteases are the most widely used enzymes. In laundry detergents, protein stains such as grass, blood, egg, and human sweat are removed through proteolysis. In automatic dishwashing detergents, proteases secure the removal of proteinaceous food films, which are a particular problem with glassware and cutlery.

Enzymes offer alternative ways of making products previously made using conventional chemical processes. Proteases, in particular, offer attractive ways of performing hydrolysis reactions needed for most tasks within the mentioned industries. Historically, proteases were used in ancient Greece for the production of cheese, as referenced in Greek epic poems dating from about 800 B.C.

The development of the submerged fermentation technique has resulted in tremendous progress in the field of industrial enzymology. Due to their efficiency, specific action, ability to work under mild conditions, high purification, and standardization, proteases are ideal catalysts for the food industry. Simple equipment can be used, and because moderate temperature and pH values are required for the reactions, only a few by-products influencing taste and color are formed. Furthermore, reactions using proteases are easily controlled and can be stopped when the desired degree of conversion is reached.

World Consumption of Industrial Proteases

Worldwide consumption of proteases amounted to approximately $600 million in 1996. The total growth (including nonfood applications) in volume of the enzyme business from 1974 to 1986 was 10–15% per year, although no major new applications emerged in this period. From 1980 to 1990 the growth was estimated as 5–10% per year, and from 1992 to 1996 as 8–10%.

The principal producers of industrial proteases are Novo Nordisk A/S (headquartered in Denmark) and Genencor International, Inc. (headquartered in the United States). These companies provide more than 90% of the total volume of enzyme products to the detergent enzyme market.

Regulatory Aspects and Quality Assurance of Detergent Enzymes

In most countries, regulatory status classification, and labeling of enzymes are determined in accordance with existing schemes for chemicals. Many enzyme types are listed on chemical inventories, for example, European Inventory of Existing Commercial Chemical Substances (EINECS) in the European Union and Toxic Substances Control Act (TSCA) in the United States. In some cases, enzymes are considered natural substances exempt from listing. In other cases enzymes are regulated by specific biotechnology products legislation.

The Association of Manufacturers of Fermentation Enzyme Products (AMFEP) has defined good manufacturing practice (GMP) for microbial food enzymes. This practice is generally followed for detergent enzymes as well. The most important element is to ensure a pure culture of the production organism.

When an enzyme is used for a nonfood and nonfeed industrial technical application, its regulatory status is determined by its properties as a naturally occurring substance. These properties determine the classification and consequent labeling in accordance with existing schemes for chemicals.

Regulatory Aspects and Quality Assurance of Food Enzymes

The application of enzymes in food processing is governed by food laws. Within the European Union large parts of food laws have been harmonized by directives and regulations. For general purposes, the Joint FAO/WHO Expert Committee on Food Additives (JECFA) and the Food Chemicals Codex (FCC) have made guidelines available for the application of enzymes as food additives. AMFEP (in Europe) and the Enzyme Technical Association (ETA) (in the United States) work nationally, as well as internationally, for harmonization of regulations with regard to enzymes.

AMFEP members ensure that the enzymes used in food processing are obtained from nonpathogenic and nontoxicogenic microorganisms, that is, microorganisms that have clean safety records, with no reported cases of pathogenicity or toxicosis attributed to the species in question. When the production strain contains recombinant DNA, the characteristics and safety record of each donor organism contributing genetic information to the production strain are assessed.

The majority of food enzymes are used as processing aids and have no function in the final food. In this case they need not be declared on the label, because they will not be present in the final food in any significant amount. Some enzymes are used as both processing aids and food additives. When used as additives, they must be declared on the food label.

GMP is used in the microbial food enzyme industry. The key issues in GMP are the microbial control of the microorganism selected for enzyme production, the control and monitoring systems ensuring pure culture and optimum enzyme productivity conditions during fermentation, and control of the hygienic conditions throughout recovery and finishing of the enzyme preparation.

Commercial enzyme products are usually formulated in aqueous solutions or processed to nondusty dry products (granulates or microgranulates). Both types of preparations must be formulated to be suitable for the final application.

Requirements mainly regarding the storage stability (e.g., enzyme activity stability, microbial stability and physical stability) and formulations as significant quality parameters of industrial enzyme products must be taken into consideration by both the producer and the user.

Safety

Like many other proteins foreign to the human body, enzymes are potential inhalation allergens. Inhalation of even small concentrations of a foreign protein in the form of dust or aerosol can stimulate the body's immune system to produce antibodies. In some individuals, increased concentrations of antibody–enzyme protein complexes can trigger increased concentrations of histamine. The latter compounds can cause hay fever–like symptoms, such as watery eyes, runny nose, and sore throat. When exposure ceases, these effects cease also.

Enzymes must be inhaled to present a risk of causing sensitization, which may lead to an allergic reaction. Working environments in which enzymes are used are therefore subjected to extensive monitoring to confirm that threshold limit values (TLV) for atmospheric enzymes are not exceeded. In many countries, the TLV for enzymes is based on the proteolytic enzyme subtilisin, and is stated as 0.00006 mg/m^3 of pure crystalline subtilisin in air (1).

Successful experience in controlling enzyme exposure in detergent manufacturing and in protecting workers' health has been reported by Procter & Gamble (2). As early as 1971, a National Research Council (NRC) report concluded that consumers of enzyme-containing laundry products did not develop respiratory allergies (3). Further studies over the years have confirmed that enzyme-containing laundry and dishwashing detergents are safe for consumer use.

Use of proteases as aids in food processing virtually never represent safety problems for consumers. Because liquid enzyme preparations are normally used in processing plants in very small concentrations, dust- or aerosol-containing enzymes are usually not released from process equipment based on closed systems.

INDUSTRIAL PROTEASES

It is difficult to tell which applications spurred the development of new proteases in the early days. Besides for production of foods such as cheese, beer, and shoyu (soy sauce), proteases have been developed for the tanning industry, and in the past the detergent industry has been the driving force in development. Scientific interest in proteolytic action on different food proteins has resulted in the development of new proteases for both extraction processes and production of functional ingredients for the food industry. By far the largest tonnage production of protease is based on microbial sources. Recombinant DNA methods and protein engineering are today's means to develop important proteases for both food and nonfood uses.

Proteases are classified according to their source of origin (animal, plant, microbial), their catalytic action (endopeptidase or exopeptidase), and the nature of the catalytic site (active site). They are characterized by common names and trade names, typical pH range, and preferential specificity. Based on a comparison of active sites, catalytic residues, and three-dimensional structures, four major protease families are recognized: serine, thiol, aspartic, and metalloprotease. The serine protease family contains two subgroups: the chymotrypsin-like and the subtilisin-like proteases—the most important industrial products. Many commercial proteases are mixtures of the different types. This is especially the case for pancreatin, papain (crude), and some proteases from *Bacillus amyloliquefaciens, Aspergillus oryzae, Streptomyces*, and *Penicillium duponti*, mainly used within the food area.

Some commercial proteases have the characteristics shown in Table 1. The enzymes Alcalase, Esperase, Savinase, and Trypsin are serine proteases, while the enzyme Neutrase is a metalloprotease having Zn^{++} in its active site. Furthermore, this enzyme is stabilized by Ca^{++}. Durazym and Everlase are protein-engineered variants of Savinase. The milk-clotting enzyme product Rennilase or Fromase is an aspartic protease. Examples of thiol (or cysteine) proteases are plant proteases from papaya latex (papain), from pineapple stem (bromelin), and from fig latex (ficin).

Table 2 shows the status of today's merchant market for detergents: Practical selections of proteases are made on the basis of performance and stability tests carried out under certain standard conditions chosen in conjunction with detergent formulators.

Protein Hydrolysis in Enzyme Applications

Proteases catalyze the hydrolytic degradation of the peptide chain, as shown for reactants and products in Figure 1. In aqueous solutions and suspensions of protein the equilibrium lies so far to the right that degradation rather than synthesis of larger molecules is thermodynamically favored.

When a protease acts on a protein substrate (see Fig. 1) a catalytic reaction actually consists of three consecutive reactions. This reaction mechanism is simplified by the scheme shown in Figure 2.

1. Formation of the Michaelis complex between the original peptide chain (the substrate) and the enzyme (referred to as ES)
2. Cleavage of the peptide bond to titrate one of the two resulting peptides
3. A nucleophilic attack on the remains of the complex to split off the other peptide and to reconstitute the free enzyme

The rate-determining step is the acylation step characterized by the reaction velocity constant k_{+2}. The ratio k_{-1}/k_{+1} is equal to the Michaelis constant k_w for serine proteases. In other words classical Michaelis–Menten kinetics are valid for protein hydrolysis.

The principal difference of the hydrolysis conditions of food and feed proteins and of protein stains in detergent applications is shown in Table 3. In detergent applications the substrate and enzyme concentrations are very small. Strangely enough, the $E/S\%$ ratio is of the same order of magnitude.

Controlling the Hydrolysis Reaction

The technical discipline of enzymatic hydrolysis of food proteins has been developed over the past 20 years to become a significant area of modern food processing. The number of literature references has grown enormously. Descriptions of the principles, methods of controlling the hydrolysis reaction, ways of carrying out industrial processing, and characterization of functional properties of proteins treated with proteases have been intensively researched and developed at Novo Nordisk A/S. A great deal of this work was introduced by Adler-Nissen (4). Examples and highlights are described in the following.

Table 1. Some Commercial Proteolytic Enzymes

Product names	Microorganism or other origin	State of product	Practical application range pH	Practical application range temperature (°C)
Alcalase®	*Bacillus spp.*	Liquid or granulate	6–10	10–80
Esperase®	*Bacillus spp.*	Liquid or granulate	7–12	10–80
Everlase™	*Bacillus GMO* (engineered)	Liquid or granulate	8–11	15–30
Savinase®	*Bacillus GMO*	Liquid or granulate	8–11	15–75
Durazym®	*Bacillus* spp. *GMO* (engineered)	Liquid or granulate	8–11	15–70
Neutrase®	*Bacillus* spp.	Liquid or granulate	6–8	10–65
Protamex™	*Bacillus* spp.	Microgranulate	6–8	10–65
Flavourzyme™	*Aspergillus* spp.	Liquid or granulate	4–8	10–55
Rennilase® or Fromase®	*Rhizomucor meihei*	Liquid or granulate	3–6	10–50
Trypsin	Pancreas	Granulate	7–9	10–55

Table 2. Merchant Market Detergent Proteases

Producer	Alkaline	High alkaline	High bleach stability	Cold water
Genencor International	Maxatase®	Purafect®	Maxapem®	Properase®
	Optimase®	Maxacal®	Purafect® OxP	
		Opticlean®	Opticlean® plus	
Novo Nordisk A/S	Alcalase	Esperase	Durazym	New developments
		Savinase	Everlase	

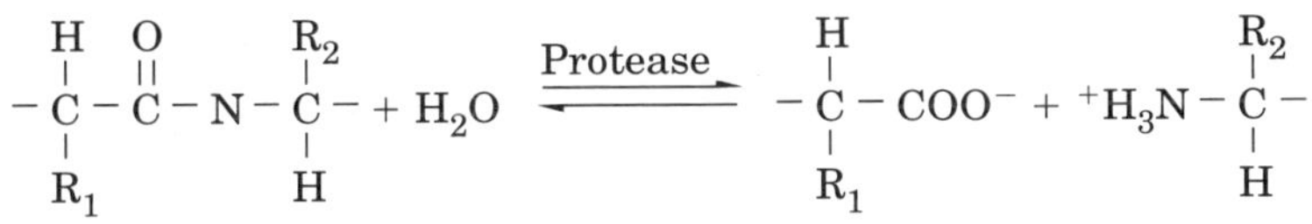

Figure 1. The hydrolysis reaction: Enzymatic cleavage of a peptide bond.

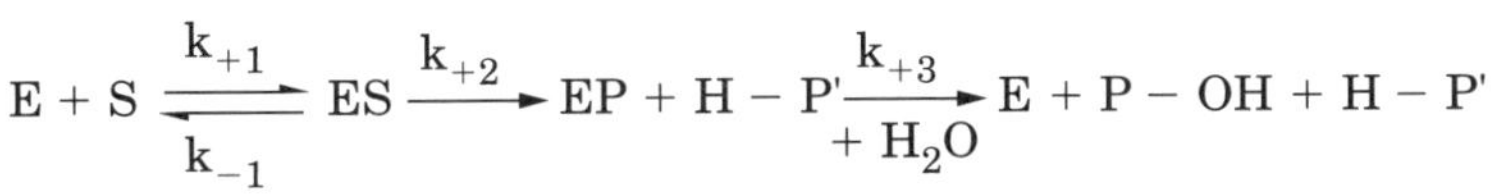

Figure 2. A catalytic mechanism of proteolysis: E, enzyme, S, substrate, P and P′, resulting peptide, k, reaction velocity constant.

The most important parameters of hydrolysis reactions are S (percent protein in the reaction mixture), E/S (enzyme–substrate ratio in activity units per kilogram protein), pH, and temperature. Together with the specificity and properties of the enzyme itself, these parameters are responsible for the course of reaction on a given protein raw material. The quantitative criterion of a proteolytic reaction is the degree of hydrolysis, which is calculated from determination of the number of peptide bonds cleaved and the total number of peptide bonds in the intact protein.

As a quantitative measure for the hydrolytic reaction, hydrolysis equivalents are applied, termed h. The unity of h is equivalent peptide bonds cleaved per kilogram protein. The term h is used for kinetic investigations of protein hydrolysis, but in other usages the derivative quantity DH (degree of hydrolysis) is preferred. DH is calculated from the following equation:

$$DH = \frac{h}{h_{tot}} \times 100\% \qquad (1)$$

h_{tot} is estimated on the basis of the amino acid composition of the protein. For most food proteins, the average molecular weight of amino acids is about 125 g/mol, making h_{tot} about 8 g equiv/kg protein (calculated as $6.25 \times N$). Table 4 gives more exact figures for Kjeldahl conversion factors f_N and for h_{tot} for some common proteins.

The degree of hydrolysis of enzyme-treated proteins determines the properties of relevance to food applications. Therefore, it is of the utmost importance that the degree of hydrolysis can be measured while the reaction is going on. Only in this way is it possible to stop the reactions at a well-defined stage, when the desired property of the product has been obtained. Within detergent applications (laundry or dishwashing), no estimates of this parameter have been reported, presumably because the substrate concentration is considerably low and therefore difficult to measure.

When sufficient substrate is available, relatively simple analytical tools that can be applied directly during the reactions are often used. Some of them are the pH-stat technique, pH drops, osmometry, viscosimetry, and chemical determination of free amino groups, using, for example, a measurement of liberated α-amino groups by reaction with

Table 3. Properties and Parameters Used during Protein Hydrolyses

Property or parameter	In food or feed hydrolysis processes	In detergent applications
Type of proteases in use	endoprotease and/or exoprotease activities	Endoprotease
Substrate concentration	Usually high (>6% protein)	Usually low (<0.1% protein)
Substrate state	Native or denature	Denatured
Enzyme concentration during application.	e.g., 0.025 g enzyme protein/L	e.g., 0.0005 g enzyme protein/L
$E/S\%$ ratio	0.03% (enzyme protein to substrate)	0.05% (enzyme protein to substrate)
Temperature	45–80 °C	15–95 °C
pH	4–8	10–12
Reaction time	1–4 hr	15–45 min

Table 4. Kjeidahl Conversion Factors and Content of Peptide Bonds for Various Food

Protein	Kjeldahl conversion factor, f_N	h_{tot} equiv/kg ($N \times f_N$)
Casein	6.38	8.2
Whey protein isolate	6.38	8.8
Meat	6.25	7.6
Fish muscle	6.25	8.6
Egg white	6.25	approx. 8
Soy meal, concentrate and isolate	6.25	7.8
Cotton seed	6.25	7.8
Red blood cells	6.25	7.8
Wheat protein	5.7	8.3
Gelatin	5.55	11.1

trinitrobenzenesulfonic acid (TNBS), whereby a colored derivative of the peptide is formed. The TNBS method has been adapted for use for hydrolysates of food proteins, which usually are partly soluble (i.e., the sample should be dispersed in sodium dodecyl sulphate (SDS)) (6). The *O*-phthaldialdehyde and the ninhydrin reaction with liberated α-amino groups may also be used, but with some limitations (7). The pH-stat technique and the osmometry technique are quicker and less labor-demanding than the "chemical" methods. When simultaneous solubilization of protein occurs as an effect of the protein hydrolysis measurements of soluble nitrogen, for example, by Kjeldahl analysis or a simple detection of soluble dry matter by measuring of the increase of refraction index (°BRIX) can be used to describe the hydrolysis reaction.

Calculation of Degree of Hydrolysis of Proteins Using the pH-Stat Technique

Originally, the pH-stat method was developed at the Carlsberg Laboratory (8). It is based on the principle that pH is kept constant during hydrolysis by means of automatic titration with a base, when the hydrolyses are carried out under neutral to alkaline conditions. When the hydrolysis is carried out under acidic conditions, (e.g., by use of pepsin at pH 3) the titration must be made with acid. The high buffer capacity of proteins means that at extremely pH values (i.e., pH > 11 or pH < 3) the pH stat is inoperable (8).

The free carboxyl and free amino groups found after hydrolyses will be more or less ionized depending on the pH of the hydrolysis reaction. Working in the pH range of 6–9.5, the amino groups will be partially protonated. This means that the hydrolysis of proteins in this pH region is accompanied by a release of H^+, as already noted by Sörensen (9). Consequently, at pH values above 7.5–7.8 (the p*K* values at 25 °C), the amino group will be less than half protonated, but the carboxyl groups will be fully dissociated. This leads to a net release of 0.5–1 mol H^+ for each mole of peptide bonds cleaved. In the pH range of 4–6.5, amino groups and carboxyl groups will be approximately equally dissociated. Therefore, in this pH range the pH state is not accurate. *DH* is calculated on the basis of the titration equations as follows:

$$DH = \frac{h}{h_{tot}} \times 100\% \tag{1}$$

$$DH = B \times N_b \times \frac{1}{\alpha} \times \frac{1}{MP} \times \frac{1}{h_{tot}} \times 100\% \tag{2}$$

where B is base consumption in mL (or L), N_b is normality of base, α is average degree of dissociation of the α-NH_2 groups (see following), MP is mass of protein ($N \times f_N$) in g (or kg), h is hydrolysis equivalents in mequiv/g protein (or equiv/kg protein), and h_{tot} is total number of peptide bonds in the protein substrate (mequiv/g protein or equiv/kg) (see Table 4). The degree of dissociation is

$$\alpha = \frac{10^{pH - pK}}{1 + 10^{pH - pK}} \tag{3}$$

p*K* is the average p*K* value of the α-amino groups liberated during the hydrolysis, and it can be determined from a direct assay of these amino groups by use of the TNBS methods. p*K* also varies significantly with temperature because the ionization enthalpy of the amino group is considerable (4). Inserting the ionization enthalpy (+45 kJ/mol) in the Gibbs–Helmholtz equation, it was found that p*K* changed about 0.23 pH units for a change of 10 °C in the hydrolysis temperature (10). Table 5 gives results of calculations of the degrees of dissociation α from equation 3 for various pH values as a function of the temperature using the previously described method.

Calculation of Degree of Hydrolysis of Proteins Using the Osmometry Technique

The osmometry technique can be used to calculate the degree of hydrolysis of proteins when no soluble components are added during the reaction (i.e., in ranges of the pH

Table 5. Degree of Dissociation (α Values)

	T(°C)					
	40	50	60	70	75	80
pK	7.3	7.1	6.9	6.7	6.6	6.5
pH						
6.5	—	0.20	0.29	0.39	0.44	0.50
7.0	0.33	0.44	0.55	0.67	0.71	0.76
7.5	0.61	0.71	0.80	0.86	0.89	0.91
8.0	0.83	0.89	0.93	0.95	0.96	0.97
8.5	0.94	0.96	0.97	0.98	0.99	0.99

Source: From Ref. 5

scale where the pH-stat principle is not functioning) (11). A freezing-point osmometer can record the osmolality of a sample in a few minutes on a sample taken directly from the hydrolysis mixture. By drawing samples as a function of time during a hydrolysis reaction, a good reaction curve can be constructed. With the osmometer ΔT is measured and converted to osmole though the simple proportionality $\Delta T = K_f \times$ osmolality.

From the increase in osmolality ΔC, DH is calculated by use of the following equation:

$$DH = \frac{\Delta C}{S\% \times f_{osm}} \times \frac{1}{\omega} \times \frac{1}{h_{tot}} \times 100\% \qquad (4)$$

where ΔC is the increase in osmolality measured in milliosmol/kg H_2O, $S\%$ is the protein substrate concentration ($N \times 6.25$), w/w%, ω is the osmotic coefficient ($\omega = 0.96$ for most actual concentrations of protein [4]), and f_{osm} is the factor to convert percent to grams per kilogram H_2O, which can be calculated using the following equation:

$$f_{osm} = \frac{1000}{100 - D\%} \qquad (5)$$

where $D\%$ is percent dry matter present in the reaction mixture, and h_{tot} is total number of peptide bonds in the protein substrate (mequiv/g protein or equiv/kg).

Some Kinetic Aspects of Protein Hydrolysis

The pH-stat technique has been utilized at Novo Nordisk to study the proteolytic degradation of hemoglobin, for example, and to understand the reaction mechanism.

In discussions of initial proteolysis, Linderstrøm-Lang (12) referred to the reaction of a protease on native hemoglobin molecules as one-by-one, indicating that a particular protease molecule degraded one substrate molecule at a time. No appreciable amounts of intermediary products will be present. The reaction mixture will consist of native proteins and end products only.

In another reaction mechanism discussed, the native protein molecules were rapidly converted into intermediary forms, which degraded more slowly to end products—the zipper reaction.

As seen on the hydrolysis curves made by using Alcalase in Figure 3 (see Table 1) at 50 and at 55 °C in the pH stat,

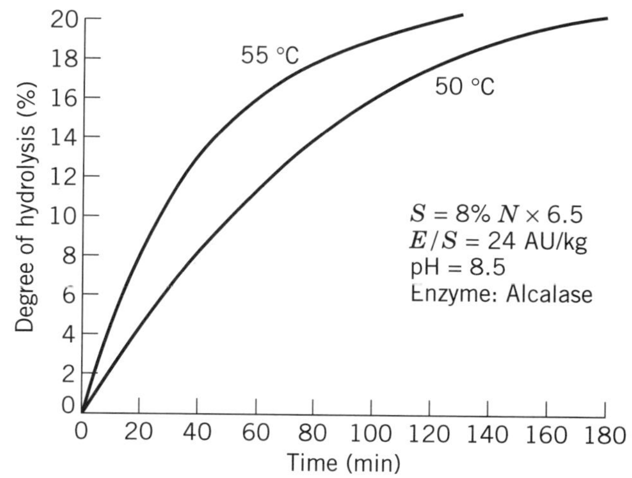

Figure 3. Hydrolysis curves for a hemolyzed red blood cell fraction.

the slight bending at low DH values of the blood-cell-fraction hydrolysis curves indicates a thorough initial degradation to small peptides (a zero-order reaction). This degradation is close to the ideal one-by-one reaction. The hydrolysis proceeds rapidly with little decrease in velocity until the substrate concentration is so low that first-order kinetics take over.

A gel chromatogram of the soluble part of the hemoglobin hydrolysate at DH 10% and DH 20% (Fig. 4) fully confirms the degradation pattern of the hemoglobin substrate. At DH 10%, the hydrolysate consists of high molecular weight material and small peptides in accordance with the one-by-one mechanism. At DH 20% the high molecular weight material has completely disappeared, as predicted by Linderstrøm-Lang as early as 1952 (12).

This is an important factor behind the application of ultrafiltration in the industrial process for separation of the hydrolysate from precipitated hemin, called the sludge (13). If the hydrolysis was not of the one-by-one type, large

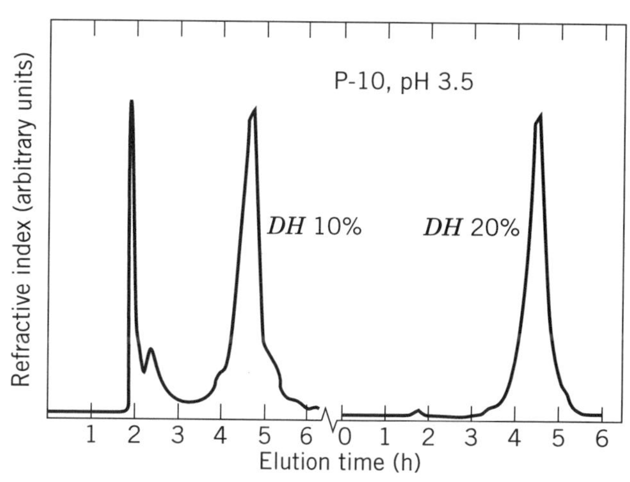

Figure 4. Gelchromatography of hemoglobin hydrolysate. *Source*: From Ref. 4.

peptides would be present also, and consequently the yields and fluxes would have been much lower. One should be careful not to draw too simple conclusions about kinetic models on enzymatic hydrolysis on even a fairly simple and well-defined substrate as hemoglobin. The HPLC spectrum of blood cell hydrolysates (BCH) is shown in Figure 5. At least 50 different peptide compounds are present under the single peak in the gel chromatogram of *DH* 20%.

The Effect of Proteolysis

When the protein structure is modified by proteolysis, the various effects are considered either positive or negative, depending on whether the effect is wanted or unwanted. The developments within enzymatic hydrolysis have often been within both these criteria. The uses and importance of wanted effects including increased solubility, reduced surface tension, increased emulsification, increased foaming properties, and detergency of textile fibers and hard surfaces have been studied. Unwanted effects, such as increased bitterness and off-flavor developments, have been key problems to solve within the food area.

During an enzymatic hydrolysis of proteins, the size character of the soluble peptides produced is changed. Key indices such as %*DH*, protein solubility index (PSI), and protein solubility in 0.8 M trichloroacetic acid (TCA index) increase, whereas peptide chain length (PCL) is reduced, as shown in Figure 6.

Protease Inhibitors

Some proteins are natural inhibitors of proteases. Trypsin and chymotrypsin inhibitors of legume and cereal grains have thus been reviewed and studied (14). Protease inhibitors for bacterial proteases, such as serine proteases (e.g., subtilisin) have been purified and characterized from legume seeds (15) and potatoes (16).

The probably best-known and most-studied protease inhibitors are the trypsin inhibitors of soybeans. The nutritional significance of the antitryptic activity of soybeans, soybean foods, and feed materials has been known for a long time. The way to inactivate these inhibitors has been an inherent part of most manufacturing processes for processes based on soybean material (17).

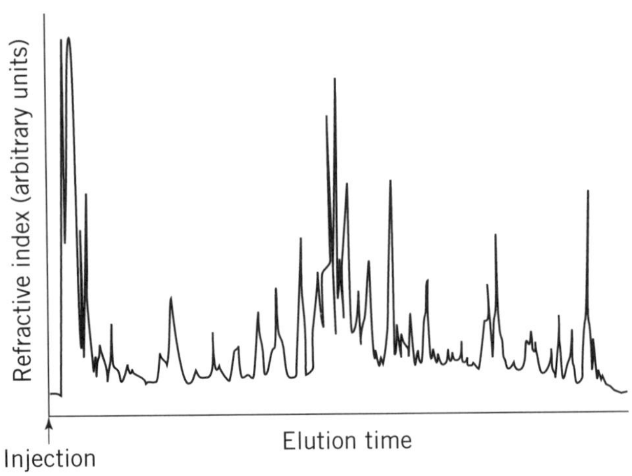

Figure 5. HPLC chromatogram of blood cell hydrolysates, *DH* 20%. *Source*: From Ref. 4.

Addition of protease inhibitors is known to protect proteases in detergents during storage. In powder detergents, where proteases are granulated and encapsulated, all enzymes are stable in the absence of water. In liquid detergents (more than 40% of the U.S. market [18]), the aqueous environment is challenging enzyme stability, and inhibitors are necessary to prevent autodigestion of the proteases themselves or to prevent other enzymes from being degraded. Polyols such as glycerin and boric acid have been used in combination to inhibit protease activity and reduce autodigestion. Borate works as a reversible protease inhibitor by interacting with groups in the active site of the enzymes. Once in the wash water, these inhibitors are diluted and enzyme activity increases (18). Protein hydrolysate products and amino acids are also known to be able to stabilize proteases by binding to the active site of the enzyme (19).

In liquid formulations used in the food industry, the way to stabilize proteases against loss of catalytic activity by denaturation and against microbial growth is by combining low water activity and a reduced pH.

PROTEASES FOR DETERGENTS

Over the past 30 years, the importance of enzymes in detergents has changed from being minor additives to being key ingredients. The main areas where the use of proteases has grown are household laundry, automatic dishwashing, and industrial and institutional cleaning (I&I). I&I covers laundry, dishwashing, hard-surface cleaning, cleaning-in-place, and membrane cleaning.

All commercial proteases from the early 1990s are serine endoproteases produced by *Bacillus* strains (Tables 1 and 2). Some have a maximum activity in the high pH range (pH 8–12) and others have a maximum activity in the pH range 7–10. Protease performance is influenced by factors such as detergent pH, ionic strength, wash temperature, wash time, detergent composition, and mechanical handling. Both performance and stability are influenced by some detergent surfactants and some bleach systems.

To illustrate the reaction conditions under which enzymes must operate, Table 6 shows the components of household detergents in broad terms. Heavy-duty powdered detergent formulations may vary across geographic regions, with a tendency for higher content of builders in North America than in Europe and Japan (20). The ratio of anionic to nonionic surfactant usage also varies from region to region. Furthermore, the content of surfactants is higher in Japan than elsewhere (20). The remarkable difference between detergent formulations for laundry and dishwashing is a much lower content of surfactants in dishwashing powder.

The variable compositions in different regions should be seen in terms of the laundry practices common in North America, Europe, and Japan. Table 7 shows some general laundry washing conditions and washing machine types in

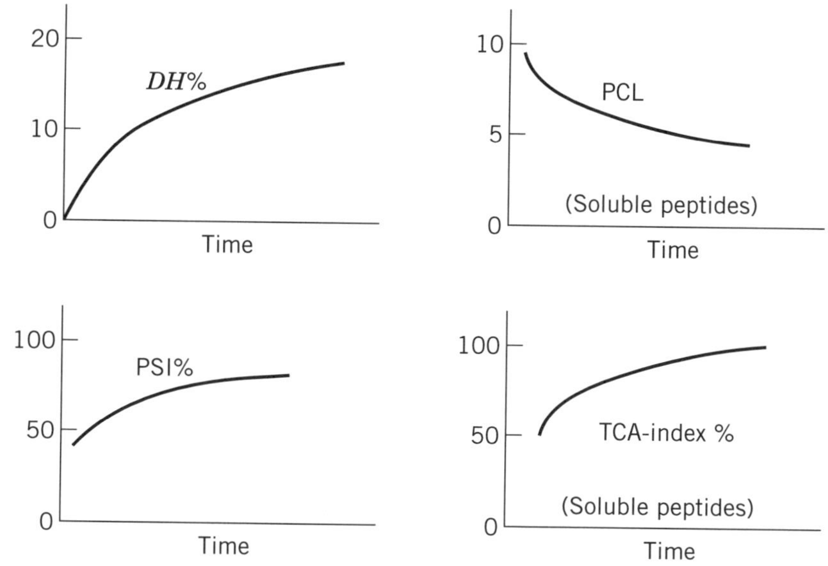

Figure 6. Qualitative illustration showing the time dependence of key indices during enzymatic hydrolysis of protein. *Source*: From Ref 4.

the three principal geographical regions. The ratio of wash liquor to fabric load, temperatures, the usual dosage of detergents, and water hardness around the world are significantly different, as are detergent compositions. Most of these parameters may influence the detergency of proteases in a washing machine or automatic dishwasher. Particularly in laundry, the water hardness (i.e., the calcium concentration) is of concern. Calcium-ions are known to stabilize some enzymes and destabilize others. Usually, sufficient amounts of enzymes are present in the powders to secure the desired detergency power.

Washing Performance Evaluations Laundry

Proteases are usually evaluated by wash trials of test pieces soiled with milk, blood, or grass. Commercial presoiled test pieces are available from a number of research and test institutes. In laboratories, washing performance studies are carried out at relevant temperatures using Terg-O-tometers, which simulate the U.S. top-loading type of washing machine, and Launder-O-meters, which simulate European drum-type machines. Launder-O-meters to some degree simulate washing mechanics and temperature profiles, including initial temperatures, heating rate, and maximum temperatures. Washing results depend on a number of factors, including detergents (Table 6), water, type of soils, textiles, washing machine (temperature, time, mechanics, and number of rinses), and the conditions mentioned in Table 7. Dose-response trials may be made in ordinary washing machines with ballast laundry and artificially soiled fabrics. A sufficient amount of standard swatches may be fixed onto the ballast cloth. Evaluation of the effects on the test pieces may be made visually or by measuring the reflectance of light remitted at 460 nm when illuminating the test pieces. The intensity of reflected light % R (% remission) can thus be measured. The value $\Delta R = R_{\text{washed}} - R_{\text{unwashed}}$ is a measure of the total detergency. $\Delta R_{\text{Enz}} = R_{\text{washed}} - R_{\text{washed without enzyme}}$ reflects the enzyme's effect.

Figure 7 is an example of a dose-response curve of Savinase (Table 1) in a European machine at 25 °C (21). Using a particular standard test textile swatch, EMPA 117 (cotton and polyester), containing blood, milk, and carbon black, Figure 7 shows that the optimal effect of the enzyme was almost reached at a dosage of about 0.2% of the Savinase 6.0 T formulation. In practice, a dosage of about 1% is used to compensate for the detergency factors mentioned earlier.

Figure 8 is an example of a dose curve made in a Launder-O-meter. At $\Delta R < 22$ the fabric was clean. This means that the maximum measured R value was almost identical to that of an unsoiled test piece. Figure 8 shows that the performance is better at 60 °C than at 40 °C until a dosage of about 3% Savinase 6.0 T (granulate) in the tested detergent was used.

Stored detergents in open packages may become humid. At storage conditions of about 55% relative humidity, the bleaches (percarbonate or perborate) become activated. After some weeks of storage the residual enzyme activity may become reduced (22). However, stability of proteases in activated bleach systems has been improved significantly in the past few years by the use of genetic and protein engineering. Using these techniques, oxidation-sensitive regions of amino acids in protease molecules have been replaced by more stable amino acids (22).

Washing Performance Evaluations: Automatic Dishwashing

A standard laboratory procedure for evaluating the cleaning efficacy of enzymes, mainly proteases and amylases, has been published (23). To test the efficiency of proteases, stainless steel plates are soiled with a baked egg–milk film. After washing light reflectance values, R, are measured directly on the plates. Calculation of percentage of removed, protein film, *RPF%*, is shown by the formula

$$RPF\% = \frac{(R_{\text{after wash}} - R_{\text{before wash}})}{(R_{\text{clean plate}} - R_{\text{before wash}})} \times 100\% \qquad (6)$$

The soil-removal values are affected by the temperature at which the protein film has been baked. The effect of the baking temperature is shown in Figure 9 using a citrate-

Table 6. General Composition of Detergents for Laundry and Dishwashing

Raw materials	General functions	Type of compounds	Contents (w/w%) Laundry	Dishwashing
Builders	Washing alkalis Water softening Buffer action Stabilizers Corrosion protection	Phosphates (sodium triphosphate) Phosphonates Na citrates Zeolites Soda Silicates Complex formers (e.g., nitrilotriacetate [NTA])	30–60	10–40
Surfactants	Emulsify particles Interfacial activity Reduce surface tension	Surfactants Anionics Nonionics Cationics	10–30	1–4
Soap	Prevents overfoaming, in the machine	Soap	1–5	0
Bleach as peroxygen or activator	Oxidize polyphenolics from fruits, coffee, tea, wine, and vegetables	Na perborates Na percarbonates Activators (e.g., tetraacetylethylenediamine [TAED])	0–25	4–15
Enzymes	Removal of starch-based stains	Amylases	0.4–1	1–3
	Removal of protein complexes	Proteases	0.4–2	0.5–2
	Removal of fat stains	Lipases	0.2–1	0.5
	Textile color brightening, softening, soil removal, whiteness maintenance	Cellulases	1–3	0
Salt	Process aid to promote flow and loading properties (powders)	Na sulfate	Balance to 100	Balance to 100
Perfume	Adds pleasant smell to fabrics	Various esters	0.1–0.5	0–0.1
Other organics	Graying inhibitors Optical brighteners Solvents	Carboxymethylcellulose (CMC), polycarboxylates Alcohols	0.5–5	0–1

Note: Data from general brochures, package information, encyclopedias, and other general literature, including Ref. 20.

Table 7. Laundry Washing Conditions and Procedures by Region

Conditions	United States and Canada	Japan	Europe
Machine type	Top loading Vertical axis Agitator	Top loading Vertical axis Impeller/pulsator	Front loading Horizontal axis Drum type
Fabric load (kg)	2–3	1–1.5	3–5
Wash liquor (L)	35–80	30–45	15–25
Wash temperature (°C)	20–50	10–30	30–95
Washing time (min)	10–15	15–35	40–60
Water hardness (ppm $CaCO_3$) (simplified)	100	50	250
Detergent dosage levels (g/L)			
Powder	1–5	0.5–1.5	5–10
Liquid	2–4	0.75–1.0	7.5–10

Note: Information from various general brochures, package information, encyclopedias, and other general literature.

based model detergent. In Figure 10, a dosage response (activity basis) for Savinase is shown for removal of protein film baked at 120 °C for 60 min.

Protease Effects for Cleaning-in-Place and Membrane Cleaning

Proteinaceous soils often contain complexes formed between proteins, fats, and carbohydrates. Thus it may be advantageous to combine enzymes to obtain a synergism during the hydrolysis of the protein. This has been demonstrated by the addition of lipase to protease in a model hydrolysis experiment.

A homogeneous freeze-dried model substrate called burned whole milk was prepared from whole milk stirred in a pressure cooker at 120 °C for 30 min. This substrate was tested for protein degradation using combinations of Esperase (Table 1) and Lipolase™ (a lipase produced by Novo Nordisk A/S). Surprisingly, it was found that when Esperase was combined with Lipolase, a synergistic effect on the release of free amino groups was found (TNBS method). Esperase used without Lipolase resulted in a

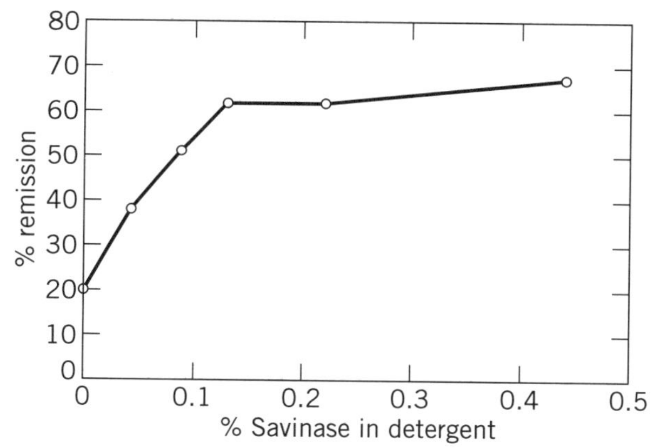

Figure 7. Dose-response curve of Savinase® in a European machine at 25 °C. Test system: test fabric, EMPA117; 4 g/L European powder detergent; 18 German degree of hardness (°dH) water hardness (equivalent to 320.4 ppm $CaCO_3$). *Source*: From Ref. 21.

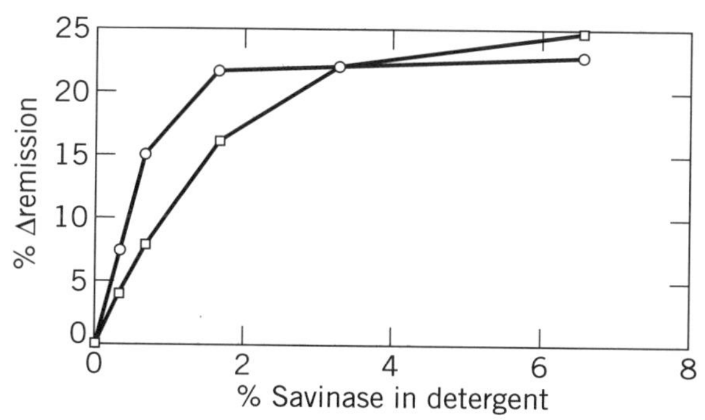

Figure 8. Launder-O-meter data of Savinase 6.0 T. Test system: EMPA 117; 4 g/L European powder detergent: —□—, 40 °C for 47 min; —○—, 60 °C for 40 min.

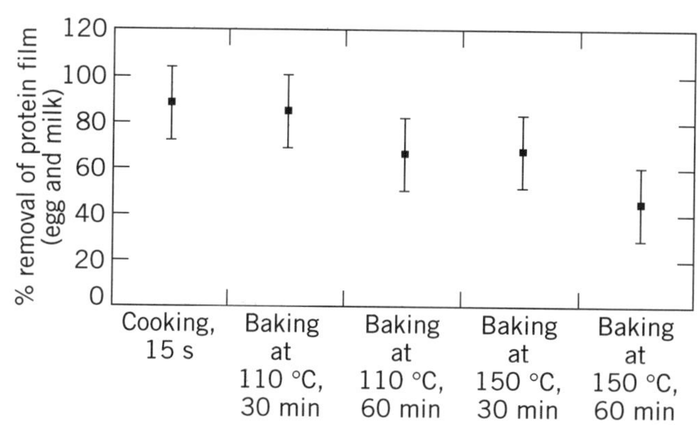

Figure 9. Effect of baking temperature on removal of protein film in automatic dishwashing. Enzyme: 0.5% Savinase 16L added to a citrate-based model automatic dishwashing detergent (ADD); 95% confidence intervals are shown based on five plates/machine; Each plate was measured at six points.

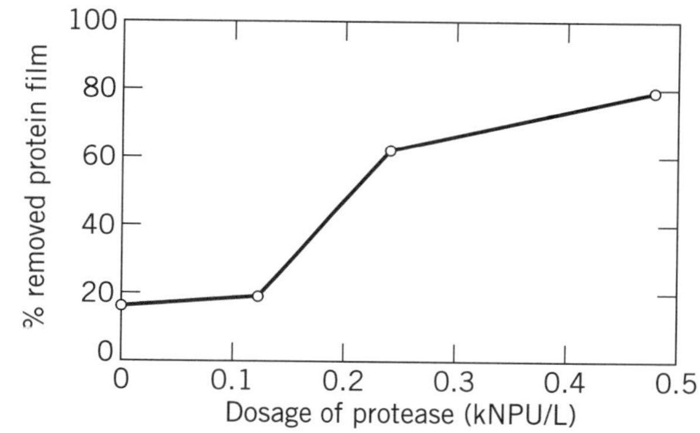

Figure 10. Dosage response (activity basis) of Savinase® at 55 °C for automatic dishwashing. Soiling system: Film of egg and milk on stainless steel plates baked at 120 °C for 60 min; detergent: European ADD–powder; enzyme: Savinase 6.0 T. kNPU is the proteolytic activity in kilo Novo protease units measured as the hydrolysis rate of dimethyl casein and compared with a standard reference protease preparation.

lower solubilization of the burned whole milk (Fig. 11). Using fresh milk, high levels of Lipolase did not greatly improve the degradation of the protein (Fig. 11).

A practical utilization of this hydrolysis synergism was demonstrated for removing burned-on soil on a heat ex-

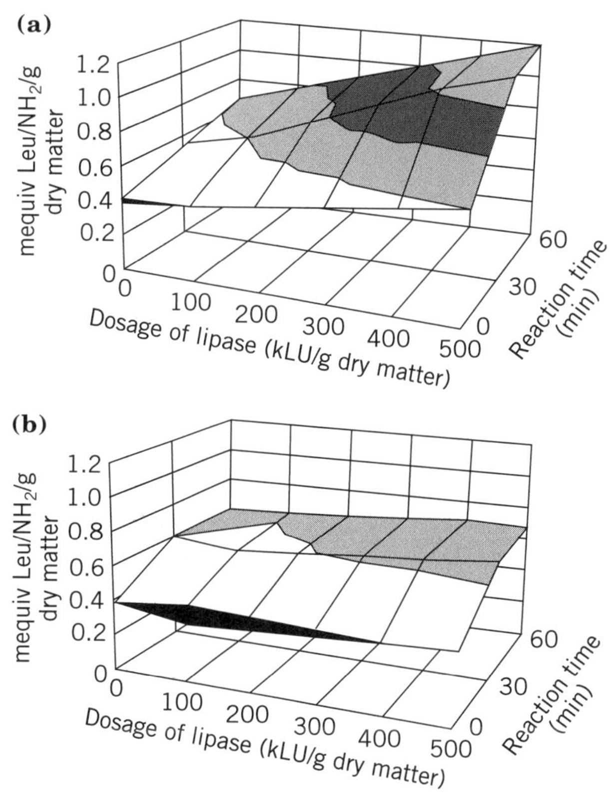

Figure 11. (**a**) Protein hydrolysis of burned whole milk assisted by use of lipase. (**b**) Protein hydrolyses of fresh whole milk. Enzymes used: Esperase and Lipolase; substrate concentration 0.5% dry matter; T, 50 °C; pH, 8.0; protease activity: kNPU/g; lipase activity: kLU/g.

changer used for high pasteurization of whole milk. A 2- to 3-mm layer of burned milk was removed using a protease and lipase at pH 8 (phosphate buffer), 50 °C for 2 h (24). Scale was removed using 0.5% nitric acid after the enzyme treatment.

During the enzyme cleaning, surfactants (soaps) were produced in situ from the fat, and emulsifiers and foaming agents were formed from the proteins. This was demonstrated by measuring the surface tension of a reaction mixture as a function of the consumed amount of base in a pH-stat trial using lipase and protease (Fig. 12). A significant reduction of the surface tension was shown.

A complete cleaning-in-place (CIP) program using enzymatic cleaning of heat exchangers is estimated to be economical and useful when there are problems with conventional chemical cleaning processes.

Membrane cleaning was tested in the laboratory and pilot plant using a pure enzyme–containing medium in dairy applications. Spiral-wound modules are very sensitive to harsh chemicals and in general not easy to clean by use of traditional cleaning products. Protease and lipase used at 40–60 °C, pH 7–10, for 10–120 min under recirculation at high speed seem to solve the problems. In dairy plants enzymatic CIP are usually carried out according to the following scheme: rinsing → enzymes → rinsing → acid → rinsing.

Understanding Enzymatic Cleaning Mechanisms

To understand the mechanism of performance of proteases during washing of cloth it is important to know the molecular binding properties of the surfaces. Issues such as adsorption and desorption of stains to textile, composition of dirt substrate, specificity of the enzyme, net charge, and isoelectric properties of the enzyme in relation to the pH value of the detergent solution must be considered (22,25). Research within this field may involve activity tests on soluble and insoluble substrates and wash trials using test textile swatches with common stains, such as milk, blood, grass, and egg. Detergent proteases hydrolyze the insoluble proteins in stains into fragments that can be easily removed or dissolved when assisted by the detergent. Larger peptide-protein fragments from hydrolysis with specific proteases could be removed more efficiently from the surface than small hydrophobic peptides (22). The simple analytical tools mentioned earlier may be used for evaluation of the protein hydrolysis reaction during a washing process (26). The one-by-one reaction, described earlier, may, together with the absorption theory, be part of the explanation of why the surfaces may be cleaned in a short time (15–45 min).

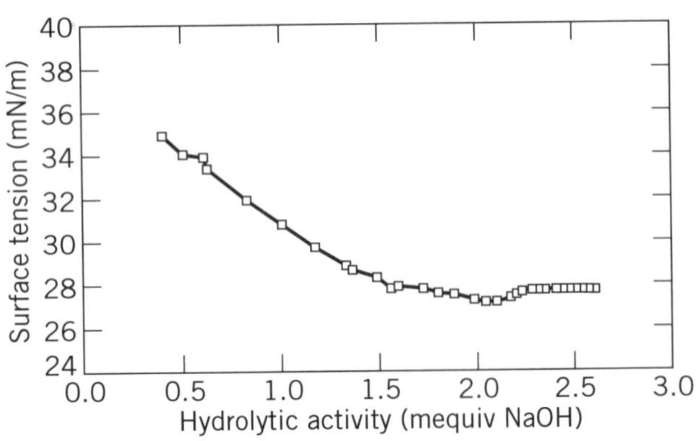

Figure 12. Surface tension versus hydrolytic activity using protease and lipase on whole milk (0.4% dry matter) at 50 °C and pH 8.0.

Protein Hydrolysates for Detergents

Enzymatically produced protein hydrolysates made by processes similar to those described in the following can, in combination with anionic surfactants, modify viscosity, foaming ability, foam stability, and cleaning efficiency for fat on textile fabric (27). Such protein hydrolysates have been acylated (condensated) with fatty acids of various lengths (28,29) and have been found to be powerful amphiphiles in the sense that they can lower surface and interfacial tension, enhance wetting and foaming ability, enhance skin compatibility, and reduce viscosity (28). Such products are produced from wheat gluten and are used in shampoos and cosmetics (30). Enhanced emulsifying properties were demonstrated in laundry and cosmetic products, such as bath soap, conditioning shampoo, and cream hand cleanser, when enzymatic soy protein hydrolysates were used to partly replace detergent components (31,32).

N-acyl protein hydrolysates made from enzymatic hydrolysates of keratin by acylation with fatty acid chlorides may be used as detergents during scouring of raw wool or other wool-cleaning operations or as fiber-protective agents during wool dyeing and bleaching (33).

HYDROLYSIS OF PROTEINS FOR FOOD PROCESSING

Hydrolysis of proteins with enzymes is an attractive means of obtaining better functional and nutritional properties of food proteins from vegetable origin or from by-products (e.g., from slaughterhouses). Some important properties of proteins and their applications in foods are shown in Table 8. Industrial baking of biscuits, conversion of milk to cheese, and fermentation of soy sauce are examples of applications of proteases to produce food.

Proteases in Baking

The proteins of wheat flour form a network with disulfide bridges. The structure of gluten becomes stronger as more

Table 8. Functional Properties of Proteins in Foods and Their Applications

Property	Application
Emulsification	Meats, coffee whiteners, salad dressing
Hydration	Dough, meats
Viscosity	Beverages, doughs
Gelation	Sausages, gel desserts
Foaming	Toppings, meringues, angel food cakes
Cohesion binding	Textured products, dough
Textural properties	Textured foods
Solubility	Beverages

sulfur bridges are present. Oxidizing agents, such as bromate and dehydroascorbic acid, strengthen the dough in this manner. By hydrolysis of the gluten fraction of wheat, neutral bacterial endoproteases may weaken a flour that is too strong or may provide the plastic properties required in a dough to be used for biscuits. Figure 13 shows a simplified model of the gluten network with disulfide bridges and the possibilities for chemical and enzymatic modification, which have been further discussed elsewhere (34). In biscuit manufacturing, cither sodium meta-bisulfite or a neutral bacterial protease, such as Neutrase® (Table 1) is used to weaken the gluten, so that the biscuits, cookies, or crackers keep their shape and size.

Dairy Products

Milk is clotted to cheese by the application of proteases. Proteases of various kinds are used for acceleration of cheese ripening, for modification of the functional properties, and for modification of milk proteins to reduce allergic properties of cow milk products.

Production of Cheese. For industrial cheese production, the milk must be standardized with respect to protein and fat content. This is normally achieved by blending milk batches, skimming (by centrifugation) to remove fat, or adding fat. The standardized milk is normally pasteurized at 72 °C for 15 s. Subsequently, the milk is cooled to 30 °C before being transferred to the cheese tank. Starter culture is added to the pasteurized milk to initiate fermentation. Rennet (the milk-clotting enzyme) and calcium chloride are added to promote the milk protein clotting reaction.

The milk-clotting effect of rennet is due to a limited hydrolysis of the κ-casein surrounding the protein micelles. Thereby, the micelles become electrostatically discharged and able to aggregate with the help of calcium and phosphate ions and can form a network entrapping the fat micelles. A gel structure is thus formed.

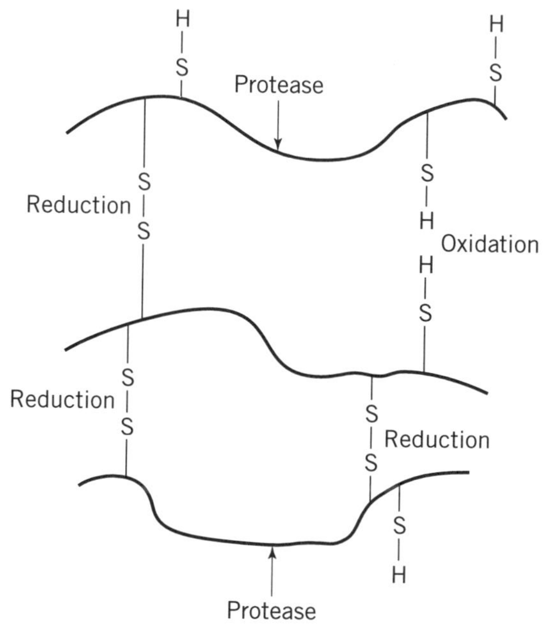

Figure 13. Model of gluten network and chemical and enzymatic modifications.

After about 25 min at 30 °C, the gelled milk (the coagulum) is cut or stirred to promote syneresis, or exudation of the whey. Syneresis is further promoted by heating the curd–whey mixture. Specialized equipment for drainage of the whey is used. After drainage, the cheese is pressed, and lactic acid fermentation continues during pressing and subsequent storage. The cheeses are then steeped in brine. The fermentation of lactose continues until all lactose has been fermented, so that after removal of the brine the cheese is free of lactose. The cheese is then ripened during an appropriate period of storage.

Chymous Enzymes. Rennet, also called rennin, is a mixture of chymosin and pepsin. Rennet is extracted from the gastric mucosa of the abomasum of young mammals (e.g., calves and lambs). The pepsin-to-chymosin ratios differ in different rennet preparations because chymosin is present only in the stomachs of unweaned mammals and is later replaced by pepsin. Thus, the content of pure chymosin depends on the age and species of the animal. Chymosin rapidly splits κ-casein into a glycomacropeptide and *para*-κκ-casein by selectively cleaving the 105–106 bond between phenylalanine and methionine. Pepsin and other proteolytic enzymes are less specific and can give rise to a number of degradation products, which tend to taste bitter (35).

Rennet Substitutes. Microbial rennets from a number of producers are marketed under various trade names (e.g., Fromase®, Hannilase®, Marzyme®, and Rennilase®). The enzyme costs for treating the same volume of milk are considerably lower than using standard rennet (36).

Microbial rennets are produced by submerged fermentation of selected strains of fungi, often *Rhizomucor miehei*. The properties of microbial rennets have proven similar to those of chymosin, and only slight modifications of the cheese-making technique are made in practice (37).

The gene for calf chymosin has been cloned into selected bacteria, yeasts, and molds. Chymosins made with recombinant DNA techniques on *Kluyveromyces lactis* (Gist Brochades), *Escherichia coli* (Pfizer), and *Aspergillus nidulans* (Hansen) are now commercially available (38).

Extraction or Modification of Protein Substrates

Protein structure can be modified to improve solubility, emulsification, and foaming properties. Chemical modification is not desirable for food applications because of harsh reaction conditions, nonspecific chemical reagents, and difficulties of removing residual reagents from the final product. On the other hand, enzymes provide several advantages, including fast reaction rates, mild conditions, and, most importantly, high specificity. There is also a need for gentle methods that modify the food itself, in order to limit the use of additives. Over the years many different protein raw materials have been used, with different objectives. Examples of extraction processes with enhanced yields include production of soya milk, recovery of scrap meat, cleaning of bone from slaughterhouses, recovery of gelatin, and production of meat (flavor) and yeast extracts. Furthermore, processes such as evaporation of fish or meat stickwater, rendering of fat, deskinning of fish roe, and tenderizing of meat are assisted by proteases. Functional food ingredients from protein sources are made by use of proteases, including soluble protein hydrolysates used for nutrition, and foaming and emulsifying ingredients. Examples of such products are isoelectric soluble soy protein (ISSPH), egg white substitute of soy protein, emulsifier of soy protein, soluble wheat gluten, foaming wheat gluten, blood cell hydrolysate, whey protein hydrolysates, casein hydrolysates, soluble meat proteins, and gelatin hydrolysates.

Functional Protein Hydrolysates

A limited hydrolysis of soy protein isolates was made at a very early stage with *Aspergillus oryzae* protease at varying ratios of enzyme to substrate (39). The *DH* was not directly measured, but from the data given by Puski in 1975 (39) the maximum *DH* value obtained was about 5%. It was shown that whipping and emulsifying capacities were improved.

The first application of the pH-stat technique for a controlled production of enzymatically hydrolyzed soy proteins was by use of Alcalase at pH 8 and Neutrase at pH 7 (Table 1) (40). The isoelectric solubility, the foaming capacity, and the emulsifying capacity were all increased. If the reactions were continued to *DH* values higher than 5%, the foaming and emulsifying capacities again decreased. Also, a significant difference of the functional capacities was observed, depending on which of the two different proteases was applied. Alcalase had a much better performance, in general. The reason for this effect could be due to the fact that foams and oil emulsions are stabilized better by peptides having a higher content of nonterminal hydrophobic amino acids than by peptides having terminal hydrophobic amino acids. Figure 14 shows the whipping expansion versus *DH* from the previously mentioned early demonstration.

In further research on functional ingredients based on enzymatic modification, improved products were developed with other raw soy materials or when physical treatments, such as ultrafiltration, were used during purification of the active components responsible for the specific effect. When a soy protein concentrate was hydrolyzed to *DH* 3–5%, with Alcalase, a functional protein hydrolysate with high foam expansion and remarkably good foam stability was obtained (41). On the other hand, the emulsifying capacity was considerably lower than if soy isolate was modified the same way.

Using a native soy protein isolate produced by ultrafiltration of an aqueous extract of defatted soybean meal (42,43), the shape of the hydrolysis curve was remarkably different compared to that obtained for acid-precipitated protein, as seen in Figure 15. The hydrolysis curves are shown for acid-precipitated soy protein isolate (denatured and nearly insoluble in water) and ultrafiltered soy protein isolate (native and soluble in water). From Figure 15 it appears that the ultrafiltered protein isolate is hydrolyzed

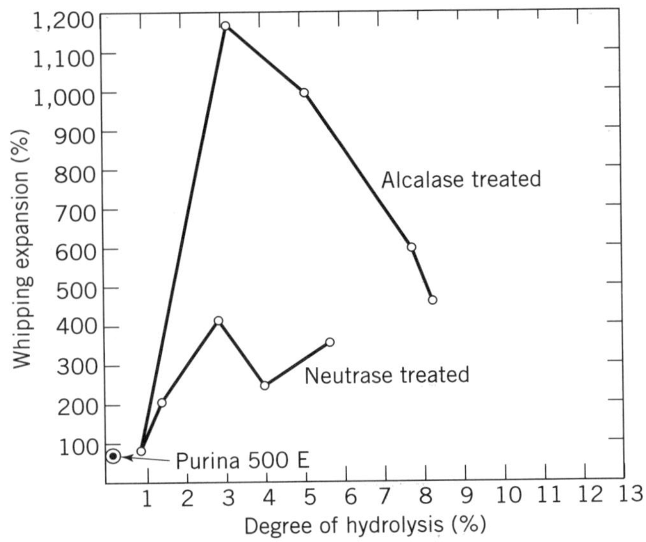

Figure 14. Whipping expansion as a function of the degree of hydrolysis (DH) for soy protein hydrolysates. *Source*: From Ref. 40.

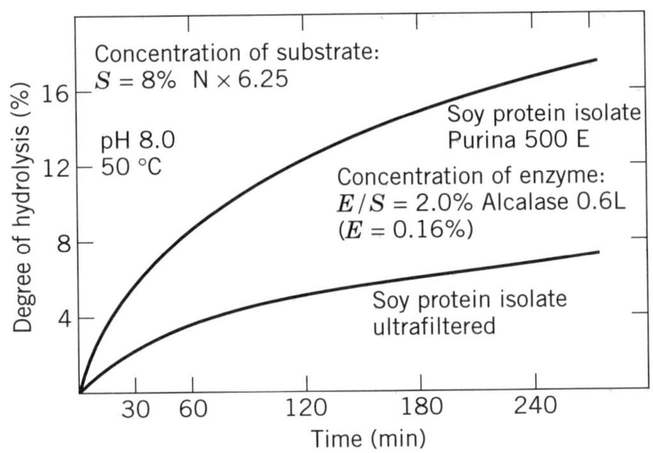

Figure 15. Hydrolysis curves for soy protein isolates. *Source*: From Ref. 43.

more slowly than the acid-precipitated protein. This is due to the compact molecular structure of the ultrafiltered protein, which is still in its native stage. The degree of denaturation of a protein substrate has a profound influence on the kinetics of proteolysis, as has been known for a long time (44). Using this soy protein isolate, functional protein hydrolysate with better foam expansion and better foam stability was obtained compared to those from the acid-precipitated protein (Fig. 14). If the small peptides in this hydrolysate were removed by ultrafiltration, the foam stability was improved even further. The use of ultrafiltration before or after the enzymatic hydrolysis was investigated in pilot-plant scale in a number of process combinations (43,45). The high molecular weight hydrolysates had excellent whipping properties; furthermore, they were also able to heat-coagulate, so that they could substitute for egg white 100% on a protein basis in a meringue batter.

Enzymatic Hydrolysis of Soy Protein Concentrate in Alcohol. Soya protein concentrate is produced by leach of defatted soybean grits using a water and alcohol mixture of 20–40% H_2O in ethanol. By this process the protein denatures so that the protein solubility is reduced and does not function adequately. A way to increase the solubility might be to enzymatically hydrolyze the separated protein before the alcohol is stripped off. To test if a protease such as Alcalase® could operate in relatively high ethanolic environments, a series of hydrolysis curves were made under various conditions. In Figure 16, pH-stat hydrolysis curves are shown for a soy protein concentrate at various concentrations of ethanol. Figure 17 shows the effects on the protein solubility index (PSI%) of percent ethanol, temperature, and percent DH during hydrolysis. For Figure 17, freeze-dried samples were produced to the percent DH as indicated. The alcohol reduced the enzyme's effect considerably, but even when the percent DH was low, an improvement could be obtained on the protein solubility index. The practical utilization of these data could be to add water to an alcohol-precipitated protein sediment from a soy concentrate factory and to perform a protein hydrolysis before drying. A product having improved functional properties may thereby be produced.

Practical Aspects of Enzymatic Processing

Enzymatic processing methods have become an important area for new processes used by the food industry to produce a large and diversified range of products based on protein raw materials. The principal advantages offered by enzymatic processes are the enzymes' specific mode of attack and the mild temperature and pH conditions required. Gentle treatment of the raw material is achieved, and only

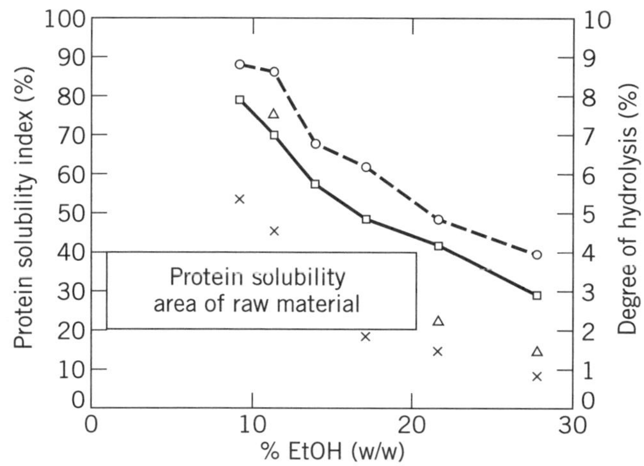

Figure 17. Enzymatic hydrolysis of soya protein concentrate in alcoholic solution. Protein solubility index and degree of hydrolysis at two different hydrolysis temperatures. —○—, 57 °C; —□—, 62 °C; △, % DH at 57 °C; x, % DH at 62 °C.

a few by-products that may affect the taste and smell of the final products are produced.

Considering a typical enzyme application in details, several effects influence the enzyme reaction and the final product. The scheme shown in Figure 18 shows a selection of parameters that can be varied when optimizing and developing new enzyme applications.

Inactivation of the enzymes is of particular concern. In most enzymatically treated food products, active enzyme is undesirable because of changes that could appear in the final product. Therefore an efficient method for irreversible inactivation of the enzyme must be developed. Usually an efficient heat treatment is used, but neutral and alkaline

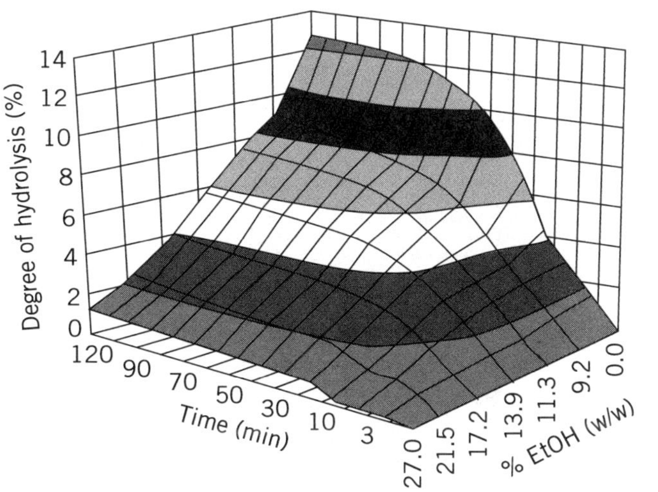

Figure 16. Percentage DH measured in pH stat by use of Alcalase® on soy protein concentrate versus alcohol concentration. Enzyme, Alcalase 2.4 L; E/S, 2.0%; pH 8.0 (pH stat); temperature, 55–60 °C; substrate, soya concentrate; danpro H; S = 8.0% protein; EtOH concentrations, w/w: 0, 9.2, 11.3, 13.9, 17.2, 21.5, 27.7.

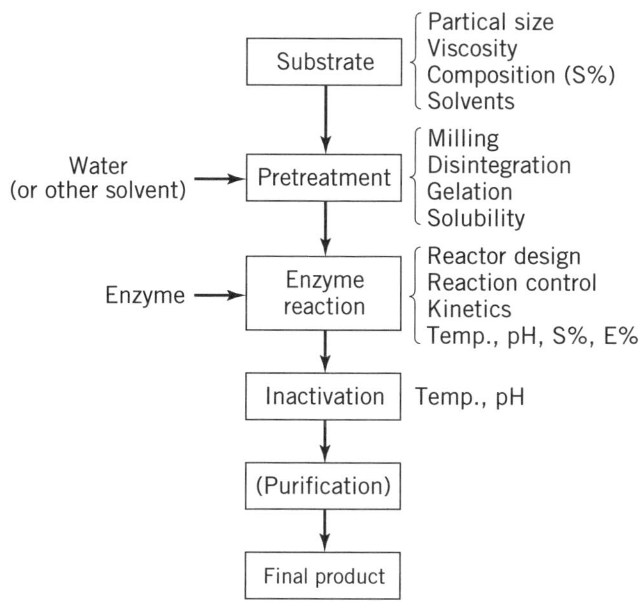

Figure 18. Practical aspects of enzyme applications.

proteases are also effectively inactivated when pH is adjusted to 4 or below.

Engineering aspects with regard to the up-scaling of inactivation must be taken into account for enzyme treatments carried out in production scale. Examples of inactivation procedures are discussed in the following. Purification of the products made enzymatically in which the enzymatic method does not produce the product in situ may in most cases be wanted. The selection of unit operations and the understanding of the performance of downstream processes are therefore also important parts of the application technology.

EXAMPLES OF APPLIED PROCESSING USING PROTEASES

Hydrolysis Processes and Simultaneous Inactivation of the Protease

If the reaction is carried out at a temperature above the enzyme's denaturation temperature, the protease is inactivated simultaneously during the hydrolysis (46). Examples of processes using this principle are shown for modification of wheat gluten and, as an alternative to rendering, for upgrading fresh bones from slaughterhouses to valuable products. The meat protein hydrolysates from the latter process are used as functional ingredients in meat products or as a clear, soluble meat extract for soups and seasonings. The cleaned bones make an excellent raw material for production of gelatin (46).

Modification of Wheat Gluten. Figure 19 shows examples of hydrolysis curves of wheat gluten when using a pH stat at 72.5 °C. After a certain time, which seems to be dependent on the enzyme dosage used, the curves end at a plateau, showing that no further hydrolysis is occurring. Two product types from wheat gluten are considered.

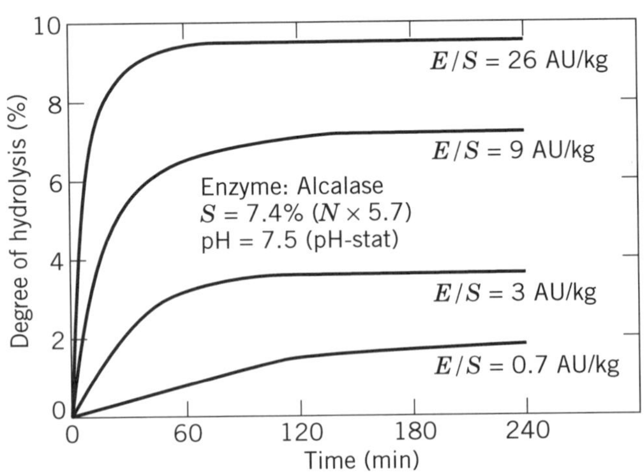

Figure 19. Enzymatic hydrolysis of wheat gluten at 72.5 °C. One Anson unit (AU) is the amount of enzyme that under standard conditions digests hemoglobin at an initial rate, liberating per minute an amount of TCA soluble products that give the same color with phenol reagent as 1 mequiv of tyrosine. *Source*: From Ref. 47.

The first is soluble wheat gluten. High protein solubility may be obtained at higher degrees of hydrolysis, as shown in Figure 20. A *DH* value of about 10% results in solubilization of more than 90% of the gluten. A 100% soluble, bland-tasting wheat gluten hydrolysate with high yield may be recovered by centrifugation and concentration.

A completely soluble wheat gluten hydrolysate is desirable for application as protein enrichment of food (e.g., soft drinks). Also, such protein hydrolysates may be excellent raw material for production of enzymatically produced hydrolyzed vegetable protein (HVP) for flavoring using, for example, exopeptidases. Nonfood applications were mentioned earlier.

The second type of wheat gluten is whipping wheat gluten. In Figure 20 whipping expansions are shown for wheat gluten hydrolysates made at different *DH* percentages. The optimal whipping properties have been found at a degree of hydrolysis of 2–3%. The active whipping protein is recovered by centrifugation and drying. A whipping protein product is required for baked goods and for different types of candy.

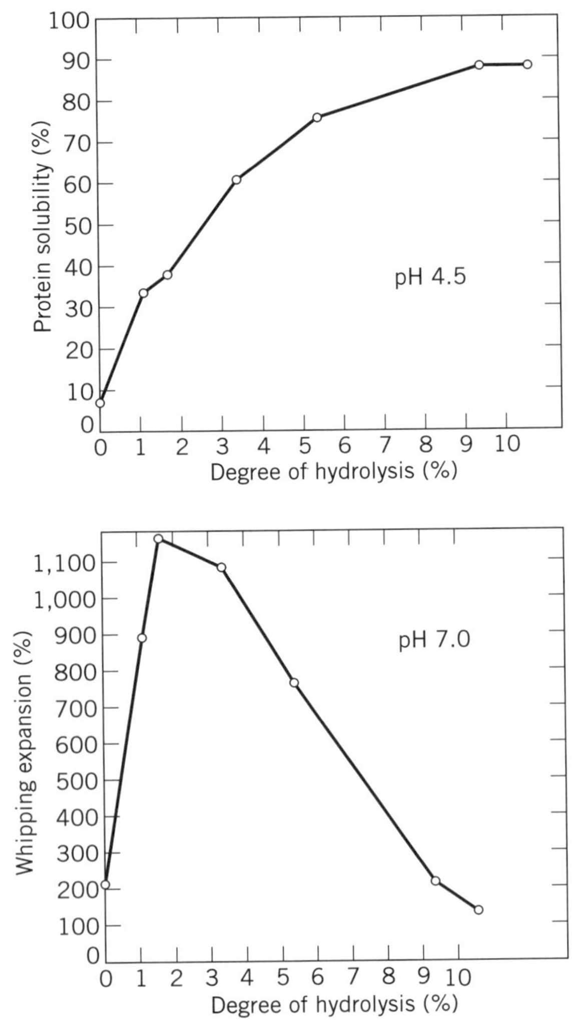

Figure 20. Protein solubility (top curve) and whipping expansion (bottom curve) for high-temperature enzyme-hydrolyzed wheat gluten.

Bone Cleaning. A two-step enzyme process in which functional meat extracts are produced by the first enzymatic extraction and cleaned bones by the second enzymatic extraction can be performed (46).

The fresh bone material (beef or pork) is crushed and mixed with hot water. The enzymes Neutrase, Alcalase, Esperase, and Flavourzyme, used alone or in combination, are added, and the slurry is agitated and extracted batchwise or continuously at 60–75 °C at neutral pH (6.5–7). Poultry material may also be used, but the bone material is not in general used for gelatin production.

Bone cleaning trials have been carried out in Novo Nordisk's pilot plant for process development. Using a bone crusher, 250 kg of beef bones were milled to a particle size of approximately 20 mm. This material was poured into 250 kg of 80 °C water and the temperature was adjusted to 65–70 °C. Hereafter the enzymatic hydrolysis reaction was carried out under powerful stirring. Samples, drawn as a function of time, were centrifuged for 3 min before determination of °BRIX (dry matter), osmolality, and pH. The total reaction mixture was sieved and the liquid phase was pasteurized, defatted by centrifugation, concentrated, and dried. The bone material was washed and dried. Hydrolysis data recorded from such a trial using Esperase are shown in Figure 21

Within the 20 min of reaction the °BRIX value and the osmolality were still increasing. This indicates that the enzyme was not completely inactivated at this temperature (−70 °C). The value mOSM/°BRIX is a measurement of the molecular size of the soluble phase. In the beginning, when the content of soluble protein is low, mOSM/°BRIX is high because of a high content of low molecular weight compounds. Later, this value stabilizes because no further degradation occurs when the protein is solubilized from the surface of the bone material (Fig. 21). The one-by-one reaction system seems to be confirmed.

Inactivation of the enzyme Esperase may simultaneously be achieved at a higher temperature (approximately 75 °C). Centrifuging at 90 °C during defatting of the hydrolysate inactivates the enzyme and pasteurizes the product.

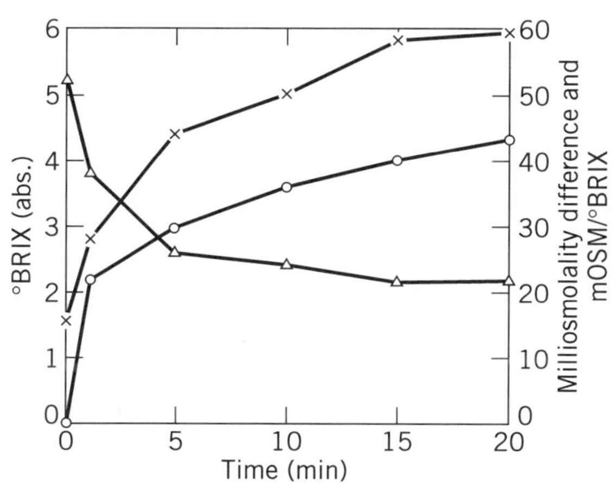

Figure 21. Data from a pilot-plant bone-cleaning trial. Enzyme, Esperase 7.5 FG; dosage, 2.0 kg/t bone —x—, °BRIX; —o—, mOSM (diff.); —△—, mOSM/°Brix; T, 70 °C; pH, 6.7–7.1.

Hydrolysis of Meat Protein in Relation to Bone Cleaning. Based on the osmolality difference, ΔC in equation 4, *DH* can be calculated and compared with those values from hydrolyses of lean beef. Hydrolyses using Esperase were carried out at 55, 65, and 75 °C on minced beef (raw or cooked). Both pH-stat trials and hydrolyses followed by osmometer were done at temperatures in the high region. The pH was adjusted to 7.5 so there was no appreciable hydrolysis of triglycerides. Figure 22a shows the hydrolyses curves and corresponding protein solubility for the osmometer trials. The percent *DH* was higher for raw meat, and the highest *DH* value for raw meat was obtained at 55 °C. For both the raw and cooked meat the highest solubilization was found at 65 °C (Fig. 22b). This was also seen in pH-stat trials. In Table 9 the hydrolysis data are compared for the bone cleaning trial (Fig. 21) and for hydrolysis of raw meat. The enzyme dosage based on protein was doubled for the bone trial as compared to the meat trial, yet a high protein-solubilizing effect of Esperase was seen in both cases. But, surprisingly, a lower percent *DH* secured a high solubility for the raw meat than for the meat on bones.

Use of Membranes in Protein Hydrolysis Processes

Membrane processes are used within many areas of industrial enzymatic hydrolysis of proteins. Table 10 shows membrane processes can be applied for pretreatment of proteins, for both the reaction step and the purification step.

Functional Protein Hydrolysates. Examples of the application of ultrafiltration processes in various process combinations for production of highly functional protein hydrolysates of soy protein were mentioned earlier. Principally, the technique of removing small peptides from total hydrolysates may be used on many substrates other than soy. Thereby, excellent foaming and coagulating protein products may be isolated (purified). The first process step after inactivation is normally a centrifugation carried out to remove insoluble material, which may reduce the whipping expansion and the foam stability. To purify the supernatant as effectively as possible, ultrafiltration can be carried out in three steps. In the first step, the concentrate reaches a protein content of about 10%. Then, a diafiltration process is performed for washing out small, low-function peptides. A final ultrafiltration can bring the protein content up to approximately 20%. Finally, the concentrate can be spray-dried. The efficiency of these processes and the overall yields are dependent on the actual ultrafiltration equipment used, the membrane cutoff value, and the character of the hydrolysate (enzyme used, *DH*%, pH, protein type) (43,45).

Low Molecular Weight Hydrolysates. Low molecular weight hydrolysates are produced from raw materials that are hydrolyzed to a high degree of hydrolysis. Examples of such products are the mild-tasting decolorized BCH developed from the hygienically collected red cell fraction of slaughterhouse blood (13) and the isoelectric soluble soy protein hydrolysate (ISSPH) (48). Flowcharts for these

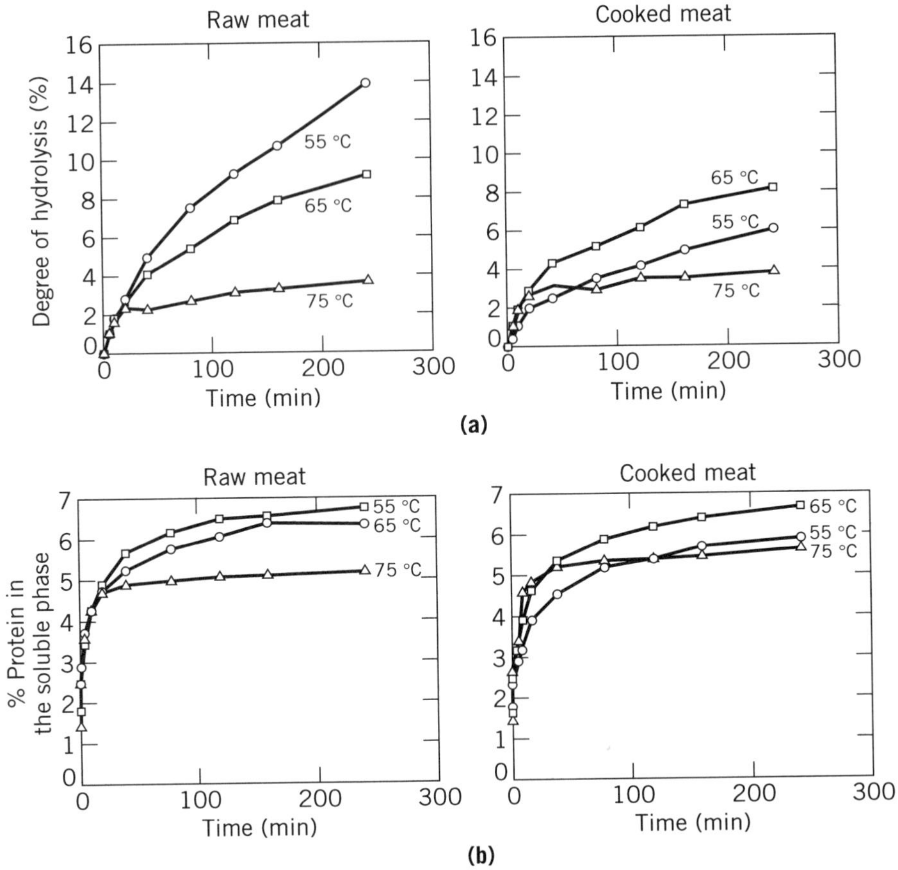

Figure 22. (**a**) Hydrolyses curves (osmometer) at 8% beef, protein. enzyme, Esperase 7.5 FG, dosage 0.5% of protein, —○—, 55 °C; —□—, 65 °C; —△—, 75 °C. (**b**) Solubilization of protein by hydrolyses of beef; meat, enzyme, Esperase 7.5 FG; dosage, 2.0 kg/t bone.

Table 9. Bone or Meat Treated with Esperase at 65–70 °C

	Bone			Raw meat		
Time (min)	% *DH*	pH	% Soluble protein	% *DH*	pH	% Soluble protein
0	0	7.14	1.2	0	7.5	1.8
1	5.7	6.82	2	0.1	7.33	2.5
5	7.7	6.81	3.2	1	7	3.4
10	9.3	6.67	3.7	2.1	6.85	4.1
15	10.3	6.67	4.2	2.5	6.77	4.5
20	11.1	6.68	4.3	2.8	6.71	4.8

hydrolysates are shown in Figure 23 (BCH) and Figure 24 (ISSPH). These two examples demonstrate how the membrane processes can be used for different purposes.

In the BCH process, the pH is 4.2 and insoluble hem product may be removed by centrifugation or ultrafiltration. Comparison of the centrifugation process and the ultrafiltration process shows that the ultrafiltration process was most economical when new plants were installed (13,49). Ultrafiltration equipment from various suppliers has been tried with technical success.

Application Trials Using Blood Cell Hydrolysate. The earliest application trials were for pumping of meat cuts. Us-

Table 10. Application of Membrane Processes during Enzymatic Modification of Proteins

Types of enzymatic modified proteins	Pretreatment	Enzymatic reaction step	Posttreatment
Highly functional proteins	Production of native protein isolate by ultrafiltration		Molecular separation by ultrafiltration
Low molecular weight protein hyrolysates		Membrane reactors by ultrafiltration	Concentration or desalination by hyperfiltration or nanofiltration

Source: From Ref. 43.

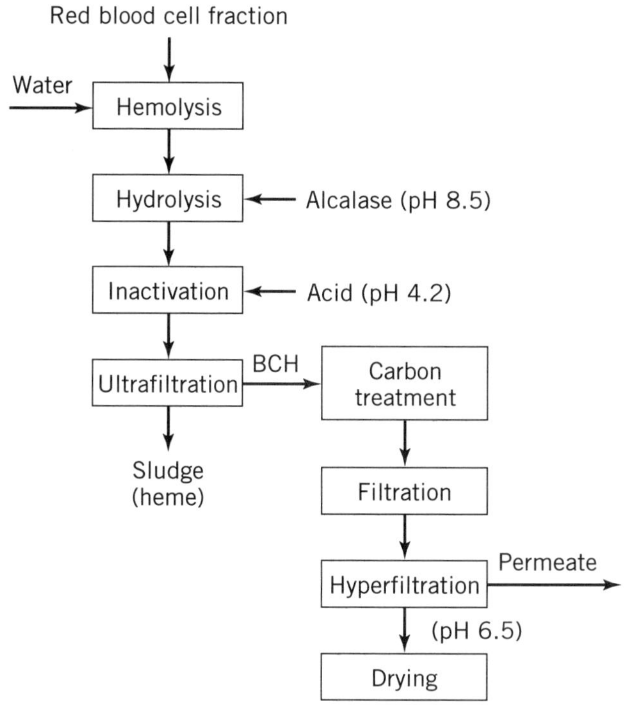

Figure 23. Flowchart for production of blood cell hydrolysate.

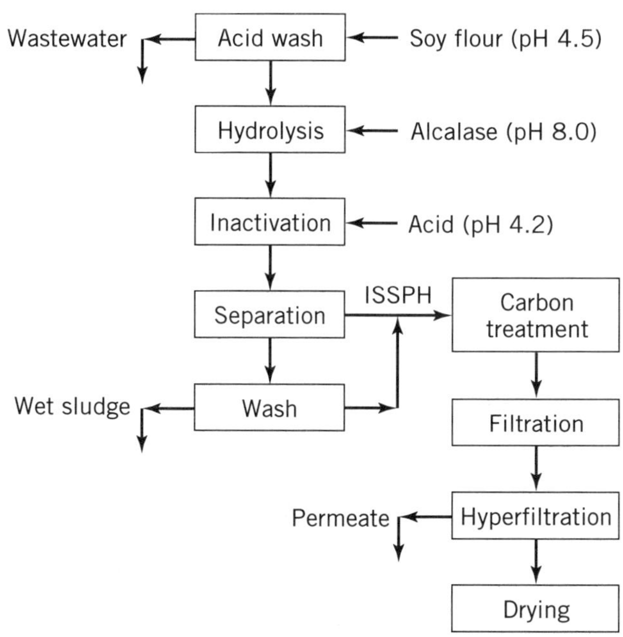

Figure 24. Flowchart for production of isoelectric soluble soy protein hydrolysate (ISSPH).

ing up to 3% BCH on a final meat product basis, nutritional protein equivalence was obtained, yields were increased, costs were reduced, toughness of meat was reduced, and the fat to protein ratio was reduced.

In a study using BCH in low concentrations in meat, it was demonstrated that BCH has a polyphosphate-like character in water-added meat systems. BCH has also been compared to sodium triphosphate in luncheon meat heat-treated at 75 °C and 108 °C, respectively (50).

During production of the soy protein hydrolysate (Fig. 24), the clear filtrate after carbon treatment is concentrated by hyperfiltration as an alternative to evaporation. A falling film evaporator can concentrate the hydrolysate to above 50% dry matter. Hyperfiltration can only be used efficiently up to about 25% dry matter but allows the simultaneous removal of salt if nanofiltration membranes are used for the concentration. Thus, about 75% of the salt in ISSPH was removed by direct concentration and 93% if a diafiltration was included in a pilot-plant trial on a selected cellulose acetate membrane (43).

Membrane Reactor. The membrane reactor is an ultrafiltration system in which a high concentration of hydrolytic enzyme is confined. High molecular weight substrate is fed continuously to the reactor, and the low molecular weight products are removed simultaneously as they permeate through the membrane. Ideally, a steady state is reached in which the degradation of substrate is carried out indefinitely with high efficiency and negligible loss of enzyme. A considerable number of publications on the use of membrane reactors for enzymatic hydrolysis of proteins appeared 10–15 years ago. However practical use of the system has not been widespread. One of the reasons may be that the economical advantage of the membrane reactor depends critically on the need for purging of the reactor. A small part of the substrate is always nondegradable. Therefore inert material will rapidly build up in the reactor, causing mechanical problems. A considerable purge is thus necessary if the concentration of this material is to be kept at a reasonably low level. This has a drastic, negative influence on the yield of product. Also, the enzyme loss during purging is considerable. Kinetic investigations of protein hydrolysis in the membrane reactor have been presented and a mathematical model established (43). Based on these calculations, it was found that the concomitant loss of enzyme in certain cases may lead to an overall enzyme consumption close to that of the batch hydrolysis process.

The Bitterness Problem

Unpleasant bitterness has been observed for some protein hydrolysates. This problem can be managed by proper selection of the hydrolysis parameters and the enzymes used. Bitterness is a complex problem that can be influenced by the following variables (4,51):

1. Hydrophobicity of the substrate, hydrolysates become bitter if amino acid side chains containing hydrophobic groups become exposed due to hydrolysis of the protein (52)
2. High *DH* and, therefore, the concentration of soluble hydrophobic amino acids and their peptide chain lengths
3. The specificity of the protease, to some extent, determines the bitterness

4. Separation steps applied in downstream processing, for example, centrifuging at isoelectric pH and carbon treatment
5. Masking effects from other components in the protein hydrolysate; organic acids such as malic acid and citric acid exert a significant masking effect in soy protein hydrolysates (53) that can be utilized in the final formulation of the food, where the hydrolysate is incorporated

In addition, recent recognition of the following effects points out other solutions to solve bitterness problems in production of protein hydrolysates with high *DH* values. A debittering of protein hydrolysates can be performed by a selective extraction or removal of hydrophobic peptides to a 2-butanol phase (4,51). This knowledge can be used to explain why it was surprisingly possible to produce a nonbitter *DH* 20% soy protein hydrolysate from full-fat soybean meal using Alcalase (54). This *DH* 20% hydrolysate was fully acceptable from an organoleptic point of view. The oil served as the hydrophobic extraction medium and eliminated the bitter peptides. The oil was simultaneously recovered by centrifugation.

Reaction parameters such as pH and temperature, by which the specificity of the proteases are changed, also play an important role in managing bitterness. Furthermore, combinations of enzymes can change the overall mode of degradation of a protein so that bitterness is reduced. The application of exopeptidases is a generally recognized way to remove bitterness of high-*DH* hydrolysates.

Other Off-Flavor Problems in Relation to Bitterness

Animal raw materials considered useful for protein hydrolysis usually do not contain many in situ bitter compounds of significance. Bitterness can appear from rancid traces of fat or from extended hydrolysis of proteins.

Oilseeds contain a variety of phenolic compounds, in particular, bitter-tasting phenolic acids (55). Some volatile compounds are hydrophobic and bind to the protein phase, especially to denatured protein. Off-flavors may be released by enzymatic hydrolysis of such proteins, some of which, but usually not all of which can be removed. A solution to such problems has been described for industrial production of ISSPH (48). The white flakes were washed at pH 4.5 with water in a four-step extraction process to produce a soy concentrate with a low level of beany off-flavor. This soy concentrate was immediately hydrolyzed with Alcalase to a specific degree of hydrolysis. The hydrolysis reaction was terminated by lowering the pH, and the supernatant was recovered by centrifugation at the isoelectric point. The hydrolysate was carbon treated, concentrated (Fig. 24), and used for protein-fortified fruit juices and for extension of cured, whole meat products. This was possible because the ISSPH had a bland taste and no bitterness.

Enzymatic Tenderization of Meat

The use of the vegetable proteinases papain, ficin, and bromelain for enzymatic tenderization of meat has been known for many years. These enzymes have a relatively strong potency on muscle tissue components such as collagen and elastin. Microbial proteinases have also shown the ability to tenderize meat, but not to the same extent as the vegetable proteinases.

The most successful meat, tenderizing enzymes are those having the ability to hydrolyze the connective tissue proteins as well as the proteins of the muscle fibers. The enzymes are applied to the meat by sprinkling the enzyme powder on a thin slice of meat, dipping the meat in an enzyme solution, or injecting an enzyme solution into the meat. A method has been developed in which the enzyme is introduced directly into the circulatory system of the animal shortly before slaughter (56) or after stunning the animal to cause brain death (57).

Whatever method is used, the difficulty in using external enzymes to obtain an even distribution of the enzyme throughout the tissues seems to be in the control of the hydrolysis reaction.

Production of Flavor Enhancers

Proteins in their natural state do not contribute chemically to formation of flavor in foods. The products of the hydrolysis of proteins, peptides, and amino acids are much more reactive, as are other breakdown products in general.

Hydrolyzed protein, mainly produced by the use of hydrochloric acid, is one of the largest contributors to flavoring produced from protein. Using various sources, many savory products are on the market. The hydrolysis with hydrochloric acid usually takes place in the presence of triglyceride fat material. It is well known that chloro- and dichloropropanols are formed during acid hydrolysis under harsh conditions, although they are found in small quantities in the final food products. Concern for the safety of the products of hydrochloric acid hydrolysis of proteins has led to investigations into the safety of these chlorocompounds (58).

Glutamic acid, as monosodium glutamate (MSG), is by far the most consumed single flavor enhancer of proteinogenic origin. About 270,000 tons are produced per year (59). Glutamates are known as the fifth basic taste (in addition to sweet, sour, salty, and bitter). This taste is called *umami* by Japanese and savoriness by Americans. MSG is used at concentrations of 0.2–0.8% in a variety of foods, such as soups, broths, sauces, gravies, flavoring and spice blends, canned and frozen meats, poultry, vegetables, and finished dishes. To be able to reduce the intake of sodium and to produce in situ flavor enhancers in protein hydrolysates, peptidoglutaminases are of interest for industrial purposes (60–62).

Through their reactions with taste sites, peptides may give flavors that are bitter, sweet, salty, or *umami*. Also, sour taste and astringency have been attributed to peptides isolated from protein hydrolysates or by synthesis. Many studies have been made of the effects of flavor in dairy products, meat and fish products, and yeast extracts. The direct production of flavor products can be made with proteases used alone, or in combination with fermentation processes. Chemical hydrolysis of protein and carbohydrates combined with heat treatments form Maillard reactions that allow many flavors to be produced. Soy and wheat hydrolysates are used for flavoring and flavor enhancement of soups and seasonings; meat hydrolysates for addition of flavor to meat products, soups, bouillon, and sauces; and milk and cheese hydrolysates for production of cheese flavors (62).

Flavors from Hydrolyzed Vegetable Proteins

Enzymatic hydrolysis usually leads to products that are light in color and have a much less pronounced meaty or savory flavor than, for example, yeast extracts (34). Meat flavor and similar condiments are often produced from many sources, and their production often involves a natural microbial fermentation. Among these can be included the Asian products based on soy protein. Wheat and maize gluten are also hydrolyzed to flavor intense peptides.

The content of free glutamic acid is important for the *umami* flavor. A way to exhibit flavor enhancement characteristics by using mild acid hydrolysis of the protein, followed by enzymatic hydrolysis, has been described (63). The preferred protein for producing savory flavors is wheat gluten. The mildly acidic hydrolysis is done to obtain a deamidation in the first step. This step was carried out for 1 h at about 95 °C at a hydrogen ion concentration of about 1.0 M. The enzymatic hydrolysis was preferably carried out by use of an *Aspergillus oryzae* protease (Prozyme 6 from Amano) having both endoproteases and exoproteases. The overall degree of hydrolysis was 50–70%. The conditions during the acid treatment were mild enough to avoid the formation of substantial amounts of monochloropropanols.

Flavourzyme™ used in combination with Alcalase may obtain a degree of hydrolysis of 60–70% for hydrolysis of soy protein isolate (62). Except for the deamidation step, the same degree of hydrolysis could thus be obtained without the acid treatment.

An enzymatically hydrolyzed soy protein, so-called eHVP, has been developed by Miwon C., Ltd. (64). This product is suitable for special uses, such as infant diets, sports nutrition, and medical diets. eHVP gave no limitations of dosage in formulation as a flavor enhancer.

Flavors from Hydrolyzed Animal Proteins

Products that have a high flavor of meat extract are used in soups, sauces, and prepared meals. Proteinaceous material is recovered with proteases from coarse and fine scrap-bone residues from mechanical fleshing of beef, pig, turkey, or chicken bones. The flavor intensity depends on the content of free amino acids and peptides and their reaction products. Reactions that develop flavor are, for example, Maillard reactions between reducing sugars and amino acids, thermal degradation due to the Maillard second stage, and deamination, decarboxylation, or degradation of cysteine and glutathione. The latter reaction can give rise to a large number of volatile compounds of importance to aroma and taste (58).

For the production of protein hydrolysates from meat, the first step is an efficient solubilization of the product. For this purpose the bacterial serine proteases and metalloproteases are very efficient. Unless certain treatments are undertaken, meat hydrolysates are usually found to be bitter when the degree of hydrolysis is above 10%, which is needed for sufficient solubilization (65). A method comprising a series of steps using raw meat hydrolyzed with a specific combination of neutral and alkaline proteases has been developed (66). This meat hydrolysate exhibits excellent organoleptic properties and can be used as a meat-flavored additive in soup concentrates. The degree of hydrolysis above 20% did not show bitterness when this specific combination of enzymes was used. This effect may be due to the preferential proteolytic specificity being favorable when metalloprotease and serine protease are used simultaneously.

A method for further enhancement of the flavor of a natural beef-juice concentrate is hydrolysis using endoproteases (metalloprotease, serine protease, or combinations) followed by treatment with an exoprotease such as Flavourzyme™ or by incubation with a culture of a food-grade microorganism that is capable of producing exoproteases (e.g., aminopeptidase) (67).

PROTEASES FOR LEATHER MANUFACTURING

The processing of skin and hides for leather has been based on enzymes since 1908, when Otto Röhm patented the first standardized bate based on pancreatic enzymes (68). Leather chemistry research helped to improve understanding of the bating process and at the same time spurred on developments to improve leather processing (69).

The stages in the processing of hides to leather include curing, soaking, liming or dehairing, deliming, bating, pickling, and tanning. The benefits of using enzymes in the different stages of leather manufacturing are mainly reduction of process time, increased opening of the fibrous structure, improved clarity of the surface, increased softness, increased uptake of water, improvement of area yield, and reduction of the need for chemicals.

The animal hide is composed of three main layers. The grain is the outer layer in which the hairs are embedded. This layer also contains the small glands that produce oil and sweat. The corium represents the principal part of the hide. To convert raw hide into leather a great deal of interfibrillar material must be removed from this layer. Collagen fibers, elastin, and proteoglycans, such as dermatan sulfate, can be regarded as a kind of cement, that binds collagen fibrils. The connective tissue of the underside of the hide contains the blood vessels, fat tissue, and some meat proteins.

Raw or cured hides must be properly soaked to obtain a satisfactory hydration and removal of unwanted materials, including meat and fat. Bacterial or pancreatic alkaline proteases are normally preferred for hydrolysis of interfibrillar proteins.

Enzymes for Dehairing, Liming, and Bating

During liming, the highly charged dermatan sulfate may be removed from nearly each collagen fibril by use of alkaline-stable proteolytic enzyme preparations such as NUE 0.6 MPX from Novo Nordisk A/S (69,70). An accelerated dehairing process and an up to 40% reduction of the requirement for sulfide have been measured (69). Furthermore, increased strength of grain leather and increased area yield are useful results.

Before the bating process, the hides are delimed with ammonium sulfate or ammonium chloride. When proteases are applied a thickness similar to that of hairless pelt is reestablished. The early preparations proposed by Röhm were pancreatic trypsins. Some of the microbial proteases shown in Table 1 are used today as bating enzymes to make the hides soft and supple before tanning.

OTHER BIOPROCESSES BASED ON ENZYMES AND PROTEIN HYDROLYSIS

A number of applications of proteases have been mentioned. The emphasis of this article has been on my research over the years and on the most significant industrial applications, as seen from an enzyme producer's point of view. Some applications of proteases have not been mentioned in detail. This is the case, for example, with regard to the use of enzymes in fish processing, where proteases are used for production of fish sauce, for salted matjes herring, for deskinning of fish, for production of protein hydrolysate as a cryoprotective agent, and for roe caviar purification (71).

Using a glutamic acid–specific protease from *Bacillus licheniformis*, the viscosity of a solution of whey protein concentrate could increase dramatically during hydrolysis due to production of plastein aggregates (7). An increased water-binding capacity and increased viscosity during hydrolysis of mechanically deboned meat was also reported (7). Extending hydrolysis to smaller peptides reduced viscosity again. A practical application of these findings remains to be further developed.

Use of proteases as processing aids for reduction of viscosity of stickwater before and during evaporation in a fishmeal plant or a meat-rendering plant could reduce the fouling on heat surfaces. The capacity of the evaporators was then increased due to the reduction of the overall heat transfer coefficients and because a greater amount of dry solids could be obtained. A direct economic advantage was shown (72,73).

Evaporation of raw fruit water from potato production represented an insurmountable problem, because gelling due to the protein components occurs at about 15% dry matter. It was obviously necessary to add proteolytic enzymes, as mentioned earlier, for concentration of stickwater. However, it was surprisingly found that no effects of the enzymes could be observed until heat treatment, preferably jet cooking to at least 125 °C using a holding time of about 3 min, had inactivated a powerful protease inhibitor in the raw fruit water (74).

CONCLUSION

Proteases are by far the most researched enzymes for industrial bioprocessing. The major uses of these enzymes are divided among many applications, all based on hydrolysis as the catalytic concept. Similarities in the catalytic mechanism within the widespread uses of proteases for detergents, for modification of food proteins, for better utilization of waste fractions of the food and feed industry, and for production of basic food have been shown. No doubt more development within proteolysis, as broadly defined, will be seen in the future. To meet challenges within the detergent industry and the food–feed industry, discovery of new protease molecules must be made through protein engineering and recombinant DNA techniques.

BIBLIOGRAPHY

1. American Conference of Governmental Industrial Hygienists, *Documentation of the Threshold Limit Values*, 5th ed., ACGIH, Cincinnati, Ohio, 1986, pp. 540–541.
2. G. Peters and D.P. Mackenzie, in J.H. van Ee, O. Misset, and E.J. Baas eds., *Enzymes in Detergency*, Marcel Dekker, New York, 1997, pp. 327–340.
3. National Research Council (NRC), *Enzyme-Containing Laundering Compounds and Consumer Health*, Report no. PB 204 118 of the ad hoc Committee on Enzyme Detergents, Division of Medical Sciences, National Academy of Science—National Research Council, Washington, D.C., 1971, pp. 1–31.
4. J. Adler-Nissen, *Enzymic Hydrolysis of Food Proteins*, Applied Science Publishers, London, 1986.
5. Novo Nordisk, *Enzymatic Modification of Proteins Using Novo Proteinases*, Bagsvaerd, Denmark, 1989.
6. J. Adler-Nissen, *J. Agric. Food Chem.* **27**, 1256–1262 (1979).
7. P.M. Nielsen, in S. Damodaran and A. Paraf eds., *Food Proteins and Their Applications*, Marcel Dekker, New York, 1997, pp. 443–472.
8. C.F. Jacobsen, J. Léonis, K. Linderstrøm-Lang, and M. Ottesen, *Methods Biochem. Anal.* **4**, 171–210 (1957).
9. S.P.L. Sörensen, *Biochem. Z.* **7**, 45–101 (1908).
10. J. Adler-Nissen, *J. Chem. Technol. Biotechnol.* **32**, 138–156 (1982).
11. J. Adler-Nissen, *J. Chem. Technol. Biotechnol.* **34B**, 215–222 (1984).
12. K. Linderstrøm-Lang, *Proteins and Enzymes, III: The Initial Stages in the Breakdown of Proteins by Enzymes,* vol. 6, Stanford Univ. Press, Stanford, Calif., 1952, pp. 53–72.
13. H.S. Olsen. *ZFL: International Zeitschrift für Lebensmittel-Technologie und Verfahrenstechnik*, **34**, 406–412 (1983).
14. Y. Birk, *Methods Enzymol.* **45**, 695–739 (1976).
15. J.K. Chavan and J. Hejgaard, *J. Sci. Food Agric.* **32**, 857–862 (1981).
16. A. Shinzaki, Y. Aizono, H. Yamagata, T. Iwasaki, and M. Sawano, *Agric. Biol. Chem.* **55**, 1521–1530 (1991).
17. I.E. Liener, in J. Adler-Nissen, B.O. Eggum, L. Munck, and H. Sejr Olsen eds., *Proceedings of the 11th FEBS Meeting, vol. 44. Biochemical Aspects of New Protein Food*, Pergamon Press, Oxford, U.K. 1978, pp. 129–138.
18. T. Krawczyk, *INFORM* **8**, 7–14 (1997).
19. U.S. Pat. 4,497,897 (February 5, 1985), J.H. Eilertsen, A.D. Fog, and K. Gibson (to Novo Nordisk).
20. M.S. Showell, in M.S. Showell and E.J. Baas eds., *Powdered Detergents*, Marcel Dekker, New York, 1998, pp. 1–19.
21. R. Fleuren, P. Bauditz, S. Schløsler and H. Bodenhoff, *88th AOCS Annual Meeting*, Seattle, Wash., May 14–18, 1997.
22. N. Eriksen, in T. Godfrey and S. West eds., *Industrial Enzymology, 2nd ed.*, Stockton Press, New York, 1996, pp. 189–200.
23. G. Jensen, *SÖFW J.* **123**, 723–731 (1997).
24. Pat. WO 97/02753 (January 30, 1997), H.S. Olsen (to Novo Nordisk A/S).

25. D. Aaslyng, E. Gormsen, and H. Malmos. *J. Chem. Tech. Biotechnol.* **50**, 321–330 (1991).
26. H.S. Olsen and P. Falholt, *J. Surfact. Deterg.* **1**, 555–567 (1998).
27. H.-P. Welzel, R. Neumann, and K. Haage. *Tenside Surf. Det.* **31**, 286–293 (1994).
28. K. Haage, B. Weiland, C. Wedler, and H.-P. Welzel. *Tenside Surf. Det.* **32**, 45–54 (1995).
29. Pat. DE 44 10 000 C1 (March 2, 1995), E.V. Kries, E. Eberhard, and A. Sander (to Henkel KGaA).
30. Anonymous, *SÖFW J.*, **121**, 383 (1995).
31. W. Wu and N.S. Hettiarachchy, *J. Surfact. Deterg.* **1**, 241–246 (1998).
32. W. Wu, N.S. Hettiarachchy, and K.C. Rhee. *J. Surfact. Deterg.* **1**, No. 3. 387–392 (1998).
33. K. Schäfer and H. Höcker, *Proceedings of the 38th International Detergency Conference*, wfk-Forschungsinstitut für Reinigungstechnologie, e.V., D-47798, Krefeld, Germany, 1998, pp. 116–121.
34. H. Sejr Olsen, in G. Reed and T.W. Nagodawithana eds., *Biotechnology*, vol. 9: *Enzymes, Biomass, Food, and Feed*, VCH, 2nd ed., Weinheim, Germany, 1995, pp. 666–736.
35. P.F. Fox, *Neth. Milk Dairy J.* **35**, 233–253 (1981).
36. K. Burgess and M. Shaw, in T. Godfrey and J. Reichelt eds., *Industrial Enzymology*, Macmillan, Surrey, England, 1983, pp. 260–283.
37. I.A. Fedrick and S.C. Fuller, *Aust. J. Dairy Tech.* **43**, 12–15 (1988).
38. P.F. Fox, *J. Food Biochem.* **17**, 173–199 (1993).
39. G. Puski, *Cereal Chem.* **52**, 655–666 (1975).
40. J. Adler-Nissen and H.S. Olsen, *Am. Chem. Soc. Symp. Ser.* **92**, 125–146 (1979).
41. J. Adler-Nissen, S. Eriksen, and H.S. Olsen, *Qual. Plant. Plant Foods Hum. Nutr.* **32**, 411–423 (1983).
42. H.S. Olsen, *International Zeitschrift für Lebensmittel—Technologie und Verfahrenstechnik* **11**, 57–64 (1978).
43. H.S. Olsen and J. Adler-Nissen, *ACS Symp. Ser.* **154**, 133–169 (1981).
44. L.K. Christensen, *Comptes Rendus des Travaux du Laboratoire Carlsberg—Série Chimique*, **28**, 39–169 (1952).
45. U.S. Pat. 4,431,629 (February 14, 1984), H.S. Olsen (to Novo Industri A/S).
46. H.S. Olsen, *Food Technology International Europe 1988*, Sterling Publications Limited, London, 1988 pp. 245–250.
47. L. Anson, *J. Gen. Physiol.* **22**, 79–89 (1939).
48. H.S. Olsen and J. Adler-Nissen, *Process Biochem.* **14**, 6–11 (1979).
49. S. Kristensen, A/S De Danske Sukkerfabrikker File 1972-GB-1182, 1985.
50. H.S. Olsen, *Meat Symposium,* Zalaergerszeg, Hungary, April 3, 1991 (Novo Nordisk file no. A-06143, 1991.)
51. J. Adler-Nissen, in G. Charalambous ed., *Frontiers of Flavor*, Elsevier, Amsterdam, 1988, pp. 63–77.
52. K.H. Ney, *Z. Lebensm Untersuch. Forsch.* **147**, 64–71 (1971).
53. J. Adler-Nissen and H.S. Olsen, in G. Charalambous and G. Inglett eds., *Chemistry of Foods and Beverages:—Recent Developments*, Academic Press, New York, 1982, pp. 149–169.
54. U.S. Pat. 4,324,805 (April 13, 1982), H.S. Olsen (to Novo Industri A/S).
55. F. Sosulski, *J. Am. Oil Chem. Soc.* **56**, 711–715 (1979).
56. H.F. Bernholdt, in G. Reed ed., *Enzymes in Food Processing*, 2nd ed., Academic Press, New York, 1975, pp. 473–492.
57. European Pat. EP 0471 470 A2 (February 19, 1992), S.J. Warren (to British Beef Company Ltd.).
58. G.S.D. Weir, in B.J.F. Hudson ed., *Biochemistry of Food Proteins*, Elsevier, London, 1992, pp. 363–408.
59. BIOTOL, *Biotechnological Innovations in Food Processing*, Oxford, U.K., Butterworth-Heinemann, 1991, pp. 211–252.
60. J.S. Hamada, in B.J.F. Hudson ed., *Biochemistry of Food Proteins*, Elsevier, New York, 1992, pp. 249–270.
61. J.S. Hamada, *Crit. Rev. Food Sci. Nut.* **34**, 283–292 (1994).
62. P.M. Nielsen, *Food Ingredients Europe '94*, Earls Court, London, October 4–6, 1994.
63. European Pat. EP 0495390 AI (January 9, 1992), D.J. Hamm (to CPC International).
64. H.J. Chae, M. In, and M.H. Kim, *Agric. Chem. Biotechnol.* **40**, 404–408 (1997).
65. G.M. O'Meara and P.A. Monro, *Meat Sci.* **11**, 227–238 (1984).
66. Pat. WO94/01003 (January 20, 1994), H.H. Pedersen,, H.S. Olsen, and P.M. Nielsen (to Novo Nordisk A/S).
67. European Pat. EP 0 505 733 A1 (February 21, 1992), S.S-Y. Kwon, M.A. Marsico, and D.V. Vadehra (to Societe des Produits Nestle S.A.).
68. U.S. Pat. 886,411 (May 5, 1908), O. Röhm (to Otto Röhm of Esslingen, Germany).
69. K.T.W. Alexander, *JALCA*, **83**, 287–316 (1988).
70. Application of NUE 0.6 MPX for liming and unhairing of hides, Application sheet, Novo Nordisk A/S, 1991.
71. G. Stefansson, in T. Nagodawithana and G. Reed eds., *Enzymes in Food Processing*, 3rd ed., Academic Press, New York, 1993, pp. 459–470.
72. F. Jacobsen, *Process Biochem.* **20**, 103–108 (1985).
73. F.M. Christensen, *Biotechnol. Appl. Biochem.* **11**, 249 (1989).
74. U.S. Pat. 5,573,795 (November 12, 1996), H.S. Olsen (to Novo Nordisk A/S).

See also ENZYMES, DETERGENTS; ENZYMES, FRUIT JUICE PROCESSING; ENZYMES, PULP AND PAPER PROCESSING; FOOD PROCESS ENGINEERING.

ENZYMES, PULP AND PAPER PROCESSING

J.S. TOLAN
Iogen Corporation
Ottawa, Canada

KEY WORDS

Bleaching
Deinking
Dewatering
Enzymes
Ligninase
Lipase
Papermaking
Pulp
Refining
Xylanase

OUTLINE

INTRODUCTION

The use of enzymes in the pulp and paper industry started in the late 1980s and has grown rapidly since then. This growth has been fueled by several factors, including (*1*) an improved understanding of the interactions between enzymes and the constituents of pulp and paper processing, (*2*) an increased need for the industry to adopt environmentally benign technology, and (*3*) development of cost-effective production technology for the relevant enzymes. The pulp and paper industry processes wood and other cellulose-containing fiber, such as straw. Typically, pulp is made by processing the fiber feedstock, and paper is made by combining one or more types of pulp with various non-pulp additives designed to achieve the desired properties, such as smoothness, opacity, or color.

Enzymes are used to aid pulp and paper manufacturing by modifying the following types of substrates:

1. The cellulose fiber
2. Noncellulosic constituents of the fiber, such as hemicellulose and lipids
3. Paper additives, such as starch
4. Contaminants, such as slime

This article provides general background on pulp and paper processing and describes the use of enzymes in each of the processing areas.

PULPING PROCESSES

Generalized flow sheets for three major pulp production processes are shown in Figure 1. Although these processes have common elements, the differences are sufficiently large that enzymes are used for different purposes.

Mechanical Pulps

Newsprint and other bulk papers are made primarily by mechanical means, that is, the raw fiber is processed primarily by mechanical action. First, the fiber feedstock is cleaned and chipped to uniform size. The raw fiber is then torn apart by mechanical means in the pulping. Closely associated with the pulping is refining, which improves the strength of the pulp. Enzymes are used to improve the dewatering and refining of mechanical pulps. After refining, the pulp is mildly bleached, usually with peroxide. Enzymes are not used to aid in the bleaching of mechanical pulps, although enzymes are used in chemical pulp bleaching. After bleaching, the pulp is ready for papermaking. Mechanical pulp is produced at lower cost than chemical pulp, but the strength and surface properties are not suitable for more demanding applications, such as writing paper or wrapping paper.

Chemical Pulps

These higher performance papers are made from chemical pulps, most notably by cooking in alkali (kraft pulp) or in acid (sulfite pulp). For chemical pulps, the fiber source and initial handling are similar as for mechanical pulps. However, the chemical pulping processes remove impurities, principally lignin, from the fiber, which improves the strength and makes the surface more uniform. After pulping, bleaching is carried out at harsher conditions and with more aggressive oxidizing chemicals than used for bleaching mechanical pulps; the resulting product is very white. Enzymes are widely used to decrease the usage of these bleaching chemicals, particularly chlorine gas and other chlorine-based oxidizing chemicals. After bleaching, the pulp is refined (the reverse of the order for mechanical pulp). Data suggest that enzymes can enhance the refining of chemical pulps, but this is not as widely reported as with mechanical pulps.

Secondary Fiber

A third source of pulp that is becoming more prominent involves the use of recycled paper as the feedstock. This is known as secondary fiber processing. After mechanical shredding and repulping, the ink is removed from the pulp, a process known as deinking. There are laboratory data and limited mill results that show benefits of enzyme use in deinking. The deinked stock is then lightly refined and blended with pure mechanical pulps for paper production. Enzymes are also reported to be beneficial in the refining of secondary fiber.

Papermaking

Regardless of the pulping process, the finished pulp is combined with inorganic fillers, paper additives to improve properties, and process additives to improve the performance of the paper machine. The pulp and additives are formed into sheets and dried on a paper machine. Enzymes are used in papermaking to modify the starches used in paper sizing, to remove pitch, and to prevent slime accumulation. Although an important benefit of enzyme use in the industry is a decreased amount and toxicity of effluent

Operation	Mechanical pulp	Chemical pulp	Recycled pulp
Feedstock	Virgin fiber	Virgin fiber	Secondary fiber
Handling	Cleaning Chip sizing	Cleaning Chip sizing	Shredding
Pulping	Neutral pH Mechanical action on fiber	Alkali (kraft) or acid (sulfite)	Repulp
Refining, dewatering	Integrated closely with pulping†	Usually after bleaching*	Often required†
Bleaching	Mild	Severe with strong oxidants†	Mild, after deinking*
Papermaking Enzyme uses†: Starch Pitch removal Slime removal	Newsprint	Fine paper	Blend with mechanical pulp

Figure 1. Pulping processes. *Widely reported enzyme effect. †Commercial enzyme process.

discharges, there is little use of enzymes for direct treatment of waste streams. The environmental benefits of enzyme use in the industry are mentioned in the following relevant process sections, and each processing area and the associated use of enzymes is described in more detail.

ENZYMES IN PULP PRODUCTION

Fiber Handling

Historically, wood has been by far the most prominent cellulose-containing feedstock for the industry. Numerous species of hardwood and softwood are used around the world. Nonwoody fibers such as bamboo, kenaf, and straw are also widely used, particularly in China and tropical areas. Regardless of the fiber source, a certain amount of physical handling is carried out to obtain cleanliness and particle size suitable for pulping. This processing takes the form of debarking, washing, and screening, to remove the foreign matter, and mechanical chipping for most pulping processes. Nonwoody fibers require removal of seeds and other nonfiber constituents. In virtually all cases, the intact fiber structure is not very susceptible to enzyme attack. Consequently, there have been few laboratory scale, let alone commercial, enzyme processes involving treating intact fiber.

There has been much research focused on the use of whole cells to digest wood chips. This is targeted particularly to remove lignin to improve the subsequent pulping and bleaching (1) and to remove pitch by digesting lipids, of which at least one commercial product is available (2). Whole cell applications on intact fiber are plagued by the long (several day) reaction times required, the extreme variability in day-to-day chip storage times, and the vulnerability of the microbes to subfreezing temperatures. Whole cell applications will not be discussed further in this article.

Pulping

Pulping is the process by which the macroscopic structure of raw wood fiber is broken apart, rendering the fiber much more pliable and thereby suitable for papermaking. Two general classes of pulping operations are chemical pulping and mechanical pulping. These operations will be discussed in terms of wood processing, although the same operations apply to the processing of nonwoody fibers.

Chemical Pulping. In chemical pulping, the fiber is heated in the presence of chemicals that dissolve a portion of the fiber, particularly the lignin. Today, most of the chemical pulp is from the alkali or kraft process, in which wood chips are cooked at 160 to 190 °C (320–374 °F) for about 3 h in a concentrated liquor of sodium hydroxide and sodium sulfide. After the alkali cook, the pulp is a slurry with the liquor; the cooking liquor is reused through an extensive recovery process. The kraft cooking treatment greatly increases the cellulose content of the fibers (Table 1).

The other significant chemical pulping process is acid (sulfite) pulping. Sulfite pulping is particularly prevalent in central Europe. In sulfite pulping, the wood chips are cooked in sulfurous acid analogous to the alkali liquor of kraft pulp. The cellulose content is higher than that of kraft pulp, but the resulting pulp is not as strong. Sulfite pulp is used primarily as a cellulose chemical feedstock. The conditions of chemical pulping are too harsh to be tol-

Table 1. Typical Chemical Pulp Composition

Constituent	Feedstock	Pulp
Cellulose	60	85
Hemicellulose	15	10
Lignin	20	4
Miscellaneous: resin, ash, etc.	5	1

erated by commercial enzymes, so there has not been significant attention paid to enzymes in this area. The most significant use of enzymes in chemical pulp has been in conjunction with bleaching rather than pulping.

Mechanical Pulping. In mechanical pulping, the fiber is refined by rotating plates; today, this is usually carried out at temperatures of about 140 °C (284 °F) and designated as thermomechanical pulping. The power input onto the fiber and the spacing between the plates that the pulp is forced through determine the severity of the pulping. Refining is a series of compressions and decompressions that gently separate and fibrillate individual fibers. This enhances pulp strength relative to the older stone groundwood process, which involved coarse grinding on a large stone. As with chemical pulp, the conditions of mechanical pulping are too harsh for enzyme treatment. However, enzyme treatment of mechanical pulp has important implications in the pulping and refining, as described in the next section.

Pulp Dewatering and Refining

One process parameter that is of primary interest in mechanical pulping and in secondary fiber processing is the rate of drainage of water from the pulp. This is important because the subsequent rate of water removal on the paper machine is often rate limiting in mill production throughput. The presence of fine particles in the pulp markedly decreases the rate of drainage. Typically, pulps consist of about 12% to 15% fines that pass through a 200-mesh screen; particle size is usually based on a Bauer-McNett classifier. Secondary fiber tends to have particularly high levels of fines, and the level of fines increases with repeated recycling of stock. Refining, which is required in many cases to increase the strength of the pulp, also creates fines that decrease the rate of drainage.

Mixtures of cellulase enzymes and hemicellulase enzymes are used to increase the drainage rate of mechanical pulp and secondary fiber. The speed of the paper machine can then be increased. Alternatively, additional refining or pulping can be carried out to obtain pulp of increased strength with the original rate of drainage. This is shown in Table 2.

The fines consist largely of cellulose and hemicellulose and, relative to the remainder of the pulp, have a high surface area for enzyme attack. There is some evidence for three distinct actions by the enzymes: (*1*) a portion of the fines are solubilized, (*2*) there is a flocculation of the particles, and (*3*) there is a cleaning of the small fibrils on the fiber surface. All these would tend to increase the drainage of the pulp.

Table 2. Pulp Dewatering by Enzyme (3)

Parameter	Control[a]	Enzyme-Treated[b]	Refine + Enzyme[b]
Freeness (mL CSF)	192	274	156
Burst (kPa m^2/g)	1.83	1.84	2.07
Corrugation (newtons force)	129	124	141

[a]Corrugated container board refined on Sprout Waldron disc refiner.
[b]Enzyme treatment 3% consistency, 50 °C (122 °F,) pH 4.8, 30 min with 0.1% enzyme.

A commercial product of Nalco (Naperville, Ill.) that is used to aid pulp dewatering is a mixture of cellulase and hemicellulases from the fungus *Trichoderma*. The enzyme is supplied in liquid solution and is pumped into the storage chest before the pulp machine headbox or, in the case of secondary fiber, just after repulping (Figure 2). The pulp is diluted in water (2 to 4% solids consistency) in a stock chest, and the enzyme is simply added to this slurry; the flow and mixing are sufficient to disperse the enzyme into the pulp. The process pH of 5 to 7 is in the optimal range for the enzyme; the process temperature of 35 to 50 °C (95 to 122 °F) is adequate. The retention time is at least 30 mins, and the enzyme dosage is 1.5 L (~1/2 U.S. gallon) per ton of pulp. The drying or refining of the sheet destroys the enzyme, preventing further action on the pulp. Careful control of the enzyme dosage is important, excessive enzyme treatment can weaken the pulp because of cellulose hydrolysis. In addition, excessive fines creation or destruction can affect the opacity of the sheet.

Bleaching

The process by which pulp is brightened or made completely white with oxidizing chemicals is bleaching. En-

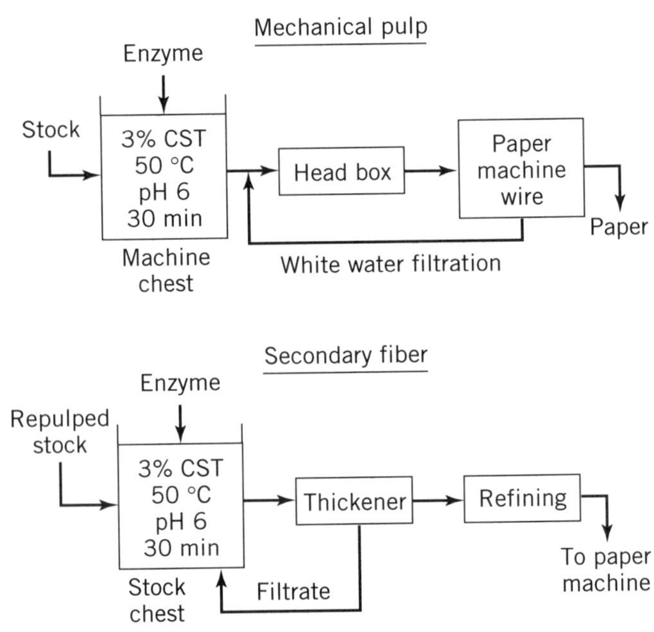

Figure 2. Enzymes for pulp dewatering.

zymes are used to aid in the bleaching of kraft pulp. In most cases, mechanical pulp is only mildly bleached with hydrogen peroxide and sodium hydrosulfite. There are few effluent problems with these bleaching chemicals. The major usage of bleaching chemicals and associated effluent problems are with kraft pulp.

In bleaching kraft pulp to produce bright white finished pulp, the residual lignin is virtually removed. Bleaching is carried out as a continuous process. The pulp is reacted with oxidizing compounds that preferentially attack the aromatic rings on the lignin while damaging the cellulose as little as possible. Bleaching is carried out in 3 to 6 discrete stages, which taken together are known as a bleaching sequence. A classical three-stage sequence uses chlorine gas in the first stage, followed by an alkali extraction, then finally reaction with chlorine dioxide, a chemical that is particularly suited to brightening kraft pulp.

In recent years, the bleached pulp industry has been motivated to decrease the use of chlorine-based bleaching compounds for environmental, regulatory, and market reasons. There are several options for mills to do this, but most require the expenditure of large sums of capital. Enzymes can be used to eliminate some of the chlorine-based compounds with little capital investment and are thus attractive to the industry.

Xylanase. Xylanase enzymes, which act on the xylan in the pulp, have been used commercially by the industry and are the example of the growing use of enzymes to aid in the bleaching of kraft pulp. At this writing (late 1997), the percentage of the bleached kraft pulp that is treated with xylanase enzymes, by country, is: Canada, 12%; Finland, 6%; Sweden, 3%; United States, 2%. These enzymes are marketed to mills in North America by Iogen Corporation of Ottawa, Canada; ICI of Toronto, Canada; and Clariant of Charlotte, N.C.

The beneficial effect of xylanase enzymes in bleaching was reported by Viikari et al. (5). The commercial enzymes used for enhancing the bleaching of kraft pulp all contain β-1,4 endo-xylanase activity that performs most of the action on the pulp. These enzymes may also contain other hemicellulase enzymes. Xylanase enzymes act on the xylan in the pulp; this is a portion of the hemicellulose with a β-1,4-linked xylose backbone, with arabinosyl and other side chains. The enzymes do not bleach the pulp, but rather change the pulp structure such that the oxidizing bleaching chemicals (whether chlorine-based or otherwise) work more effectively on the pulp. Xylanase enzymes can decrease the use of oxidizing chemicals by 15 to 20%; more typically, a decrease of 10 to 15% is observed. The enzyme treatment can also increase the brightness of the bleached pulp and decrease the quantity of off-white fiber particles, known variously as shives, sclereids, and vessels.

Figure 3 shows the process streams at the point of xylanase enzyme treatment in a kraft mill. The unbleached pulp, known as brownstock, is pumped from the final washer into a storage tank before it enters the bleach plant. The pulp is a mat about 2 inches thick, of 8 to 14% solids consistency. Although the pulp is well washed, it has sufficient residual alkali soda that its pH is 9 to 12. This is higher than the effective pH range of the commercial enzymes, so the pulp must be acidified, usually with sulfuric acid, to the appropriate pH. The temperature of the pulp is typically 50 to 70 °C (122 to 158 °F). The capacity in the storage tank usually is sufficient for at least 1 to 2 hours of stock retention, which is sufficient time for the enzyme to act on the pulp. The pulp is pumped from the storage tank into the bleach plant, where contact with the bleaching chemicals destroys the enzyme. The enzyme-treated pulp then proceeds as normal through the bleach plant, except with a decreased requirement for bleaching chemicals. The decrease in demand for bleaching chemicals requires changes to be made in the bleach plant control strategy. The details of the how this is done are covered elsewhere (6).

The enzymes are supplied in liquid solution, often in 1,000-L (275 gallon) totes. At the typical enzyme usage of 0.2 to 0.6 L per ton of dry pulp, a mill that processes 1,000 tons of pulp per day requires 2 to 5 totes per week of enzyme. The temperature on the operating floor can reach 35 °C (95 °F), so the enzyme must withstand this temperature during storage or be maintained in a refrigerated tank. It is imperative that the enzyme, which is added to pulp in such small quantities, be evenly dispersed into the pulp mat to obtain uniform treatment and bleaching response. The enzyme solution is usually added to the pulp with some dilution water to increase the bulk liquid dispersion. The combined stream is added before a stock pump to take advantage of the mixing provided by these pumps. The equipment that is used to deliver this stream to the pulp varies from mill to mill but has included simple hoses, shower bars, and other devices.

The pH of the pulp must be decreased for effective enzyme treatment. This is accomplished by adding acid to the pulp, usually well in advance of the point of enzyme addition. The most common acid used is sulfuric, but sulfurous acid, carbon dioxide, and waste acid from various mill process streams have been used. When sulfuric acid is used, it is stored as a 93 to 98% concentrate. The typical acid requirement is 3 to 5 kg/ton (6.6 to 11 lb/ton) pulp, which is an acid flow of 1 L of concentrate per minute on a 1,000 ton per day bleaching line. As with the enzyme, the acid must be dispersed uniformly throughout the pulp. The acid stream is pumped through a hose or shower bar and onto the pulp. The choice of location and equipment to handle the acid and add it to the pulp takes into account the mill's equipment configuration and desires for safe acid handling and avoidance of equipment corrosion.

Several properties of xylanase enzymes are important in bleaching applications. These include:

1. *The bleaching benefit*. The precise mechanism of the enzyme's action on pulp is not known. Hence, no single enzyme activity has been identified that characterizes the enhanced bleaching of pulp. The bleaching benefit for a given enzyme is measured by treating pulp and bleaching. That having been said, xylanase activity is present in all the commercial enzymes and is important in enhancing pulp bleaching. However, it is by no means the only active enzyme present; enzymes of the same xylanase activity can have very different effects in bleaching (7).

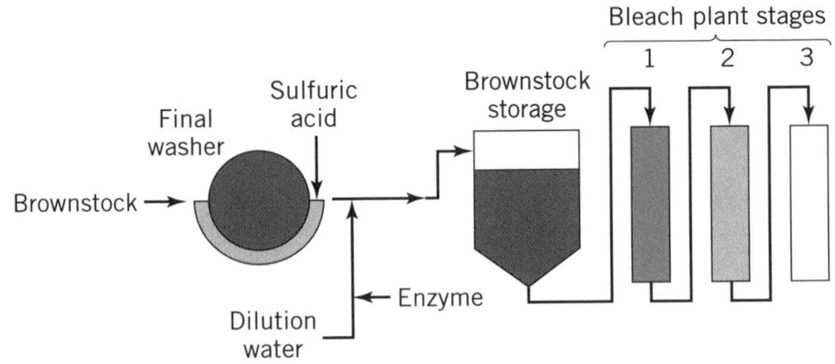

Figure 3. Enzyme treatment in pulp bleaching.

2. *The conditions of use.* All the commercial xylanases are characterized by a range of pH and temperature that must be maintained for full effectiveness. The pH range is typically 1 unit broad, and the temperature range spans about 10 °C (18 °F). Figure 4 shows the implications of various pH and temperature ranges on mill operation; different enzymes are preferred for different mills. Enzymes that can operate at pH 10 to 12 and at temperatures up to 70 °C (158 °F) are the subject of much research.
3. *The absence of deleterious effects on the pulp.* This includes degradation of the pulp by cellulase; the enzymes must have little or no cellulase activity.

Enzyme manufacturers have used several different technologies to produce xylanase enzymes that comply with these requirements. These technologies are as follows:

1. *Produce xylanase using a natural microbe that does not make cellulase.* This seemingly straightforward approach has not generally been successful, because most such microbes make only a limited amount of protein. Examples of such microbes are species of *Bacillus* and *Streptomyces* bacteria and *Aspergillus* fungi.
2. *Clone and express xylanase in a microbe that does not make cellulase.* This approach has been somewhat more successful than the first, with host organisms including *Aspergillus oryzae, Bacillus pumilis*, and *Streptomyces lividans*.
3. *Produce xylanase along with cellulase and remove the cellulase using downstream processing.* This approach has worked well using *Trichoderma* fungi, which are well known as high protein-producing microbes. Methods of cellulase removal include size-differentiated filtration (8), polyethylene glycol precipitation (9), precipitation with metal ions (10), ion exchange chromatography (11), and salting out (12).
4. *Delete cellulase genes from* Trichoderma. This approach has also worked well for commercial products (13).

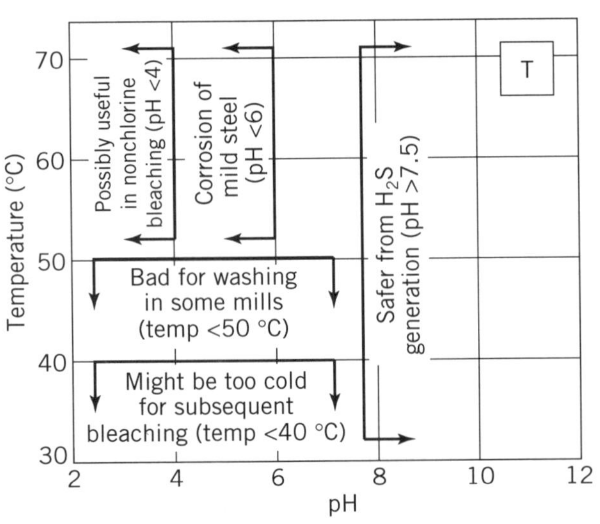

Figure 4. pH and temperature for enzymes in pulp bleaching: Effects on mill operations. The box labeled T indicates typical brownstock conditions.

In addition to xylanase, many other hemicellulase enzymes have been evaluated for their ability to enhance the bleaching of pulp. Of these, only mannanase appears to deliver benefits comparable to xylanase (14). Mannanase enzymes are not used commercially for bleaching at present.

Ligninase. Research over 3 decades focused on the use of ligninase enzymes to oxidize and remove the lignin from the pulp. The motivation for this research was straightforward: lignin removal is the objective of bleaching. However, for many years ligninase technology was not suitable for mill applications because reaction times of several days were required. A breakthrough discovery (15) decreased the time required to bleach pulp with ligninase enzymes to 1 hour, which has motivated much recent effort in the area.

Four proteins have been identified that participate in enzymatic delignification: lignin peroxidase, cellobiose dehydrogenase, manganese peroxidase, and laccase. Lignin peroxidase is a manganese-containing enzyme that requires hydrogen peroxide to oxidize the aromatic rings and side chains of lignin. Cellobiose dehydrogenase is a multifunctional enzyme that has recently been shown to be able to participate in enzymatic delignification (16).

From the point of view of commercial-scale use, laccase is the furthest along in development of the ligninase enzymes. Laccase is a copper-containing enzyme that catalyzes the oxidation of phenolic groups on lignin. The most prominent microbe for laccase production is the white-rot fungus *Trametes versicolor*, and production of the enzyme is induced by phenolic compounds. The copper that forms the redox-active center of laccase is part of the functional enzyme as it is produced and does not need to be added to the system for lignin degradation to occur.

Call (15) determined that inclusion of a mediator compound dramatically increases the rate of lignin oxidation. The mediator reported by Call is 1-hydroxybenzotriazole; many other effective mediators have since been reported (17–19). Laccase and mediator are proposed to be used in a mill as follows. Pulp at an intermediate brownstock washer is adjusted to pH 4.5 and 45 °C (113 °F), then exposed to oxygen at moderate pressure. The enzyme and mediator are then added and act on the pulp for 1 hour. After treatment, the pulp is washed to remove the soluble lignin. The laccase-mediator process is reported to remove 50% of the lignin in a single-stage reaction and a further 25% in a second reaction. This magnitude of effect is much greater than that possible with xylanase and would be an extremely important step forward for mills to eliminate chlorine-based oxidizing chemicals entirely. The laccase-mediator bleaching has been demonstrated in pilot plant trials, but it has not been used commercially because of high costs of production of the enzyme and the mediator.

The fourth ligninase enzyme, manganese peroxidase, offers the potential advantage over laccase in that it does not require a mediator. Manganese peroxidase is an iron-containing enzyme that uses hydrogen peroxide to oxidize chelated manganese, which then acts on lignin. The enzyme is most effective at 45 °C (113 °F), pH 4.5. The rate and extent of lignin removal of manganese peroxidase is not as great as with a laccase-mediator system, but it is still very impressive (i.e., 40% delignification in 4 h). The areas for development of this technology are the need for careful control of the levels of peroxide, manganese, chelating agent, and enzyme and the costs.

Deinking

The removal of ink is an important part of secondary fiber processing. Large-scale deinking is new and evolving rapidly. Deinking is carried out by repulping at 3 to 4% solids consistency, then diluting to 1% consistency and aerating the slurry while adding flocculating surfactants and ink solvents. The ink particles float to the surface and are collected and removed.

Several reports indicate that cellulase enzymes can increase the efficiency of deinking processes (20,21). It appears that cellulase can release ink particles, both bound to fiber fines and separately from fiber, and thereby enhance the removal of ink by flotation. This in turn decreases the ink count, allows a higher brightness to be achieved, or alternatively decreases the usage of the deinking chemicals. There is some question, however, as to whether the enzyme treatment decreases the strength of the pulp (22). Some reports state that the enzymatic deinking process is best carried out under alkaline conditions (23). There has been one published report of a mill scale trial using this enzymatic deinking; the results confirmed the laboratory data (24). EDT of Norcross, Ga., has marketed cellulase and xylanase enzymes for mill-scale deinking applications.

ENZYMES IN PAPERMAKING

Most of the content of paper is finished pulp (which can be blends of both mechanical and chemical pulps) and inorganic fillers such as clay. The remainder of the paper, perhaps up to 10% of the total weight, consists of a variety of additives designed to achieve certain properties in the paper and other additives to improve the operation of the paper machine in the mill. Among the additives to influence paper properties is modified starch for coated paper. Amylase enzymes are used in the starch modification. Enzymes are also used to remove pitch from the paper and to eliminate slime buildup on the paper machine.

Starch Modification

The oldest use of enzymes in the pulp and paper industry is in the modification of starches for coated paper. This practice dates back to the early 1970s and is now carried out more frequently within the starch processing plants than in the paper mills. Starch imparts many beneficial properties to paper, including strength, stiffness, and erasability. The starches used in papermaking are often modifications of the natural starch polymers found in the locale of the mill, such as corn starch in the United States, potato starch in Britain, and rice starch in Japan. Of particular interest in enzyme use is starch for sizing. Sizing refers to an increased stiffness of the paper that is useful as a coating or for further paper processing.

The starch is added to the paper partway through the drying section of a paper machine. The starch is added at a size press, which is two rollers through which the paper picks up starch solution while passing. The presses run best at 45 to 60 °C (113 to 140 °F), so the viscosity of the starch solution must be in an acceptable range at this temperature. The standard starch products for the paper industry are suitable for some grades of paper, but in other cases the viscosity is too high for good spreading through the press. In other cases, a mill desires the flexibility to vary the starch viscosity for different paper grades. More recent work has suggested that amylase enzymes can form unique starch products for the paper industry (25).

The generic process for decreasing starch viscosity is to decrease its molecular weight by hydrolyzing the polymer to shorter chains. The first method to do this involved a dilute acid cooking of the starch; this is still practiced in some mills today. The second method, hydrolysis with α-amylase enzymes, was originally carried out in the late 1960s. These enzymes catalyze the addition of a water molecule across the linkages between two glucose units in the starch polymer. The desired reactions for paper starch is to shorten the starch polymer. Complete conversion to glucose, which would be achieved using glucoamylase and pullulanase enzymes, is undesirable.

The two types of reactor systems used for enzymatic starch hydrolysis in paper mills are batch and continuous.

In batch systems (Fig. 5), roughly 1,000 kg (2,200 lb) of starch is diluted in 4,000 L (1,000 gallons) of water in an agitated tank. The temperature is maintained at, depending on the enzyme used, 25 to 90 °C (77 to 194 °F). The pH is controlled to 4.5 to 7. The treatment time is typically 30 minutes, after which the system is heated to a suitable temperature to destroy the enzyme's activity, or, for the thermostable enzymes, chemical denaturants are added. Batch systems offer the advantages of simplicity of control and the ability to run at solids levels of up to 40 %.

In continuous systems (Fig. 6), the starch and water are mixed in a slurry tank and pumped through a steam-heated tube. Enzyme is injected at the mouth of the tube. The running temperature is 125 °C (257 °F) for 2 min at pH 6. At this point, the target viscosity is reached, and the enzyme, which is added to a concentration of 0.5 g/L, is fully denatured. This system offers advantages of easy automation and high volumetric productivity. The continuous systems are generally for the larger starch users.

Pitch Removal

Pitch is the sticky substance that is prevalent in softwoods, especially pine. Pitch is a collection of compounds known as lipids, which are long-chain fatty acids connected by glycerol linkages. These compounds are analogous to the fatty tissue of mammals and have both hydrophobic and hydrophilic sites on the molecule. This character contributes to the processing difficulties caused by pitch; the compounds are slightly soluble in some points of the process, and they can form a difficult precipitate in other areas. The most troublesome pitch formation is from Japanese Red Pine. Pitch comprises typically less than 1% of the weight of wood, but can reach 3% of the total on this species.

The removal of pitch by chemical pulping and bleaching is not particularly efficient, but it is usually sufficient to prevent severe deposition. In mechanical pulping, the pitch problems are more severe, particularly with pitch exposed on the paper machine. At this point, pitch interacts with other additives, decreasing their effectiveness as well as decreasing the rate of drainage of the sheet. Chemical dispersants have been used for pitch control, and the pitch content is decreased by longer storage periods in the woodyard, which is limited by fungal growth and chip deterioration.

A third method of decreasing pitch deposition is by using lipase enzymes to digest the pitch. Lipases hydrolyze the lipids to their constituent fatty acids and substituted polyol (Fig. 7). The polyol (glycerol for triglycerides) is water soluble, and the fatty acids can be bound with talc or other charged compounds. Lipase for pitch removal has been particularly successful in Japan, where 120,000 to 140,000 tons of groundwood pulp are treated annually with lipase, representing 20% of the total groundwood pulp produced in Japan (26).

Lipase has been reported to be added to the groundwood pulping line and to the storage chest before the paper machine (26). When added to the pulping line (just before the post refiner), a mixture of two lipases added at levels of about 100 ppm and about 600 ppm reduced the triglyceride content of the pulp by 70% over a 1-month mill trial, with an enzyme reaction time of 3 to 3.5 h. This corresponded to a decease in pitch deposits on the paper and equipment and defects in the resulting paper sheets. However, one disadvantage of this location for lipase treatment is the frequent use of alum, particularly in American mills, which can inactivate the enzyme.

In another mill trial, lipase was added at 1 to 5 ppm in the storage chest before the paper machine, for a retention time of at least 1 h. The process temperature and pH of 40 °C (104 °C) and 6.0, respectively, were compatible with the enzyme treatment. The enzyme was added to the dilute slurry in the chest without additional mixing, although in at least one mill lack of good enzyme dispersion caused substandard results (27). Results of a mill trial using lipase for pitch removal are shown in Table 3. Lipase does not affect the strength or brightness of the pulp. The yeasts *Candida cylindracae* and *Candida rugosa* and *Aspergillus* fungi have been reported to make lipase in high productivity and are used commercially to make lipase.

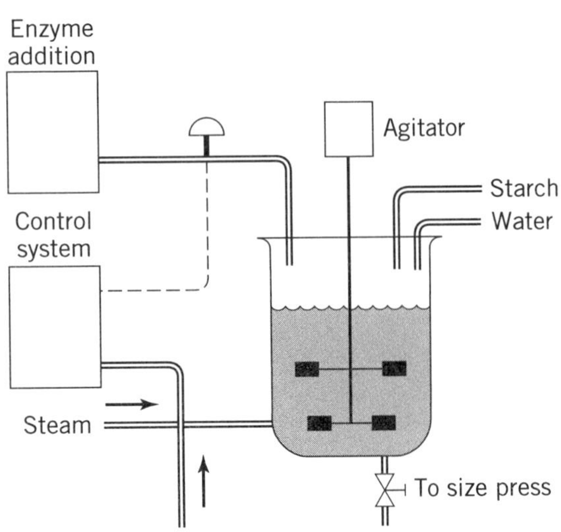

Figure 5. Batch starch conversion system.

Slime Removal

Like pitch, slime deposition causes significant operating problems around a paper machine. Slime is the generic name for deposits of a microbial origin in a paper mill. It is impractical to run a paper mill as a sterile system. As a result, a vast array of microbes contaminate the mills, and many of the resulting slime compounds have not been characterized. When confronted with a slime, often the strategy is to try every available biocide until one is found that targets the microbe and destroys the source of the slime.

In some cases, however, specific slime compounds have been characterized, and efficient strategies for their removal can be used (28). One such case is with levan, which is a β-2,6-linked polymer of fructose that forms a slime film. This compound is secreted by several species of *Bacillus* and *Pseudomonas* bacteria that can grow in the recirculated water around the paper machine, especially for fine paper, where the level of inhibiting compounds is low. The

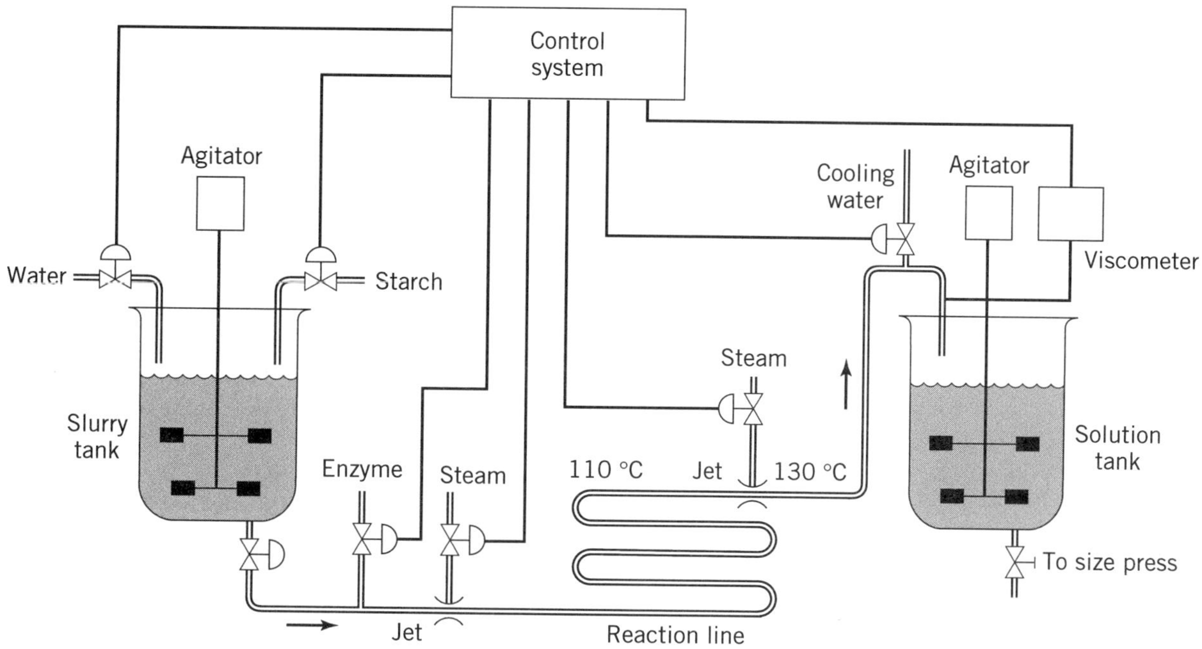

Figure 6. Continuous starch conversion system.

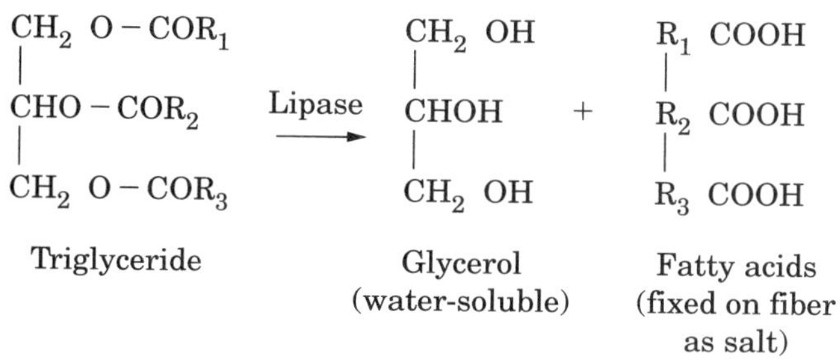

Figure 7. Hydrolysis of pitch by lipase.

Table 3. Lipase for Pitch Removal

Parameter	Newsprint Control[a]	Newsprint Lipase[b]	Telephone book Control	Telephone book Lipase
Pitch, g/day	242.6	32.6	110.2	46.1
Paper web, holes/day	11.0	1.4	4.5	3.4

Source: Data from Irie et al. (27).
[a]Red pine groundwood pulp, 200 ton/day, paper machine speed 830 m/min.
[b]Enzyme added to mixing chest, retention time 60–120 min, lipase 3 ppm.

enzyme levan hydrolase can hydrolyze this polymer to low molecular weight polymers that are water soluble, thereby cleaning the slime out of the system. Another advantage of the enzymatic treatment is that it destroys the ability of the microbe to bind to the surface, thereby increasing the effectiveness of chemical biocides.

Commercial levan hydrolase is supplied as the product EDC-1 by Henkel Corporation (Morristown, N.J.). The enzyme does not hydrolyze cellulose, so is not harmful to the paper. The enzyme is usually added at the headbox of the paper machine, although in some cases it has been added at the dryer discharge. The enzyme is effective at pH 4 to 8 and runs best at 50 °C. The enzyme can be used alone or in combination with chemical biocides.

Other enzymes reported to be effective for slime removal are proteases, cellulase, amylase, and pentosanase (28). These enzymes are used more typically to treat the process waters of the mill to prevent slime buildup in cooling towers, heat exchanger surfaces, and other surfaces exposed to water.

BIBLIOGRAPHY

1. T.K. Kirk, H.M. Chang, *4th Int. Conf. on Biotechnology in the Pulp and Paper Industry Proc.*, Raleigh, N.C., May 16–19, 1989.
2. R.L. Farrell, R.A. Blanchette, T.S. Brush, B. Gysin, Y. Hadar, J.J. Perollaz, P.A. Wendler, W. Zimmerman, in *Proc. of the Fifth Int. Conf. of Biotechnology in the Pulp and Paper Industry*, Kyoto, Japan, May 27–30, 1992, pp. 27–32.
3. J.C. Pommier, J.L. Fuentes, G. Goma, *TAPPI* 187–191 (June 1989).
4. L.S. Jackson, J.A. Heitmann, T.W. Joyce, TAPPI **76**, 147–154 (1993).
5. L. Viikari, M. Ranua, A. Kantelinen, J. Sundquist, M. Linko, *Proc. Int. Conf. Biotechnol. Pulp and Paper Ind.*, Stockholm, 1986, pp. 67–69.
6. J.S. Tolan, *Proc. Non-Chlorine Bleaching Conf.*, Hilton Head, S.C. March 12–17, 1993.
7. J.S. Tolan, *TAPPI Pulping Conference Proc.*, Boston, November 6–10, 1992.
8. L.U.L. Tan, K.K.Y. Wong, J.N. Saddler, *Enz. Microb. Technol.* **7**, 425–430 (1985).
9. U.S. Pat. 5,320,960 (June 14, 1994), B. Bower (to Genencor International).

10. J.M. Chauvet, J. Comtat, P. Noe, *Paris Symposium Wood and Pulping Chemistry*, 1987, pp. 325–327.
11. A. Lappalainen, *VTT Publication* **50**, 22 (1988).
12. K. Poutanen, *VTT Publication* **47**, 13–20 (1988).
13. U.S. Pat. 5,298,405 (March 29, 1994), H. Nevalainen, J. Knowles, P. Suominen, M. Pentilla, A. Mantyla (to Alko Ltd.).
14. A. Suurnakki, M. Tenkanen, J. Buchert, L. Viikari, L., *Adv. Biochem. Eng. Biotechn.* **57**, 261–287 (1997).
15. H.P. Call, Proc. of the 1994 TAPPI Pulping Conference, San Diego.
16. P. Ander, G. Daniel, H. Pettersson, U. Westermark, in T. Jeffries and L. Viikari eds., *Enzymes for Pulp and Paper Processing*, ACS Symposium Series 655, 1997, pp. 297–307.
17. Eur. Pat. Appl. 717,143 (June 20, 1996), H.P. Call (to Lignozym GmbH).
18. PCT Pat. Appl. WO 95/01426 (January 12, 1995), P. Schneider, A. Pedersen (to Novo Nordisk).
19. M.G. Paice, R. Bourbonnais, I. Reid, F.S. Archibald, 9th *Int. Symposium on Wood and Pulping Chemistry Proc.* Montreal, 1997, pp. 1–4.
20. J.A. Heitmann, T.W. Joyce, D.Y. Prasad, in *Proc. of the Fifth Int. Conf. of Biotechnology in the Pulp and Paper Industry*, Kyoto, Japan, May 27–30, 1992, pp. 175–180.
21. K. Rutledge-Cropsey, T. Jeffries, J.H. Klungness, M. Sykes, *TAPPI 1994 Recycling Symposium*, Boston, May 1994, pp. 103–105.
22. G. Stork, H. Preira, T.M. Wood, E.M. Dusterhoft, A. Toft, J. Puls, *TAPPI 1994 Recycling Symposium*, Boston, pp. 103–105.
23. U.S. Pat. 5,364,501 (November 15, 1994), J.A.G. Baret, M. Leclerc, J.P. Lamort (to Novo Nordisk).
24. T.J. Kim, S.S.K. Ow, T.J. Eom, in *Pulping Conference 1991 TAPPI Proc.*, October 1991, pp. 1023–1027.
25. S.E. Nachtergale, F.J. van der Meeren, *Stark* **39**, 135–141 (1987).
26. K. Hata, M. Matsukura, H. Taneda, Y. Fujita, in T. Jeffries and L. Viikari eds., *Enzymes for Pulp and Paper Processing*, ACS Symposium Series 655, 1997, pp. 280–296.
27. Y. Irie, M. Matsukura, M. Usui, K. Hata, *Papermakers Conference / TAPPI Proc.* 1990, p. 10.
28. S.C. Johnsrud, *Adv. Biochem. Eng. Biotechn.* **57**, 311–328 (1997).

See also BIOREMEDIATION; ENZYMES, EXTREMELY THERMOSTABLE; ENZYMES, PROTEINS HYDROLYSIS; HEMICELLULASES; HEMICELLULOSE CONVERSION.

ENZYMES, STARCH CONVERSION

I.S. BENTLEY
ABM Brewing and Enzymes Group/Rhone-Poulenc
Stockport, U.K.

KEY WORDS

Corn
Enzymes
Fructose
Glucose
Maize
Maltodextrins
Maltose
Starch
Syrups
Wheat

OUTLINE

INTRODUCTION

Starch is the main storage carbohydrate of plants and a primary source of energy in the human diet. It can be found in cereals such as wheat or barley, in the seeds of the maize plant, and in the tuberous roots of potato and cassava. Industrially, starch and starch-derived sweeteners are mainly produced from maize, wheat, potato, and tapioca. The typical starch contents of these raw materials are given in Table 1.

The use of carbohydrate hydrolyzing enzymes to convert starch to nutritive sweeteners is not a new technology. For centuries malt has been used to produce a sweet syrup from starch, although it was not known at the time that the enzymes in malt were responsible for the reaction. It was the development of amylases from fungal and bacterial sources at the turn of the twentieth century that paved the way for the use of enzymes in starch hydrolysis. Even so, it was not until the 1960s that enzymes began to replace mineral acids in the production of dextrose from starch.

The term *glucose syrup* is often used to describe the refined, concentrated, aqueous solution of saccharides obtained by the controlled partial hydrolysis of edible starch.

Table 1. Starch Content of Basic Raw Materials

Raw material	Starch content (%)
Maize	72
Wheat	64
Potato	15
Tapioca	20–30

This conversion may be catalyzed by food-grade acids and enzymes. Depending on the degree of hydrolysis, the resultant syrup contains varying amounts of D-glucose (1). Syrup produced from maize starch is often called corn syrup rather than glucose syrup.

The term *glucose syrup* is ambiguous. In the glucose syrup industry, glucose is used to describe any starch-derived syrup; however, from a chemical viewpoint, it is the monomeric saccharide α-D-glucose. To distinguish the monomer from the syrup, the term *dextrose* is used to describe the monomer. Because starch is composed of D-glucose units, dextrose is the ultimate product of nondestructive starch hydrolysis (2).

The extent to which the starch molecules are broken down is described by the term *dextrose equivalent* (DE). Starch has a DE of 0, whereas the ultimate product, dextrose, has a DE of 100. In general, sweetness will increase with increasing DE value, although the nutritive value will remain the same. It is the DE value that is important in describing a glucose syrup, since the characteristics of the syrup will vary with the DE.

Products with a DE of less than 20 are referred to as maltodextrins, those with a DE of between 20 and 80 as glucose syrups, and those with a DE greater than 80 as hydrolysates. At a DE lower than 20, the functionality of the maltodextrin is often related to the botanical source of the starch. This is related to the ratio of amylose and amylopectin within the particular starch structure.

The Structure of Starch

In its native state, starch is present in plant cells as microscopic granules arranged in layers. These granules vary in size and shape—from the relatively large granules found in potato (50–150 μm) to the smaller granules of maize and wheat (2–40 μm). Starch is made up of two types of high molecular weight polymers of glucose, amylose and amylopectin. Amylose and amylopectin are found concomitantly within the starch granule, and the ratio varies depending on the source of the starch. Table 2 shows typical levels of amylose and amylopectin in some common starches.

Starches from tubers, such as potato and tapioca, tend to have lower levels of amylose compared to starches from grain. The ratio of amylose to amylopectin also varies with the variety of the crop. For example, some maize varieties yield up to 99% amylopectin (waxy maize), whereas other starches have a high amylose content (high amylose starches). These are specialist starches, and unlikely to be used in the production of glucose syrup.

The basic building block of both amylose and amylopectin is α-D-glucose monomer linked through an α-1,4 glucosidic bond. In amylose, the glucose monomers are linked solely through the α-1,4 glucosidic bonds to yield linear chains of several hundred glucose units. In solution, the amylose chains may be coiled in the form of a helix. In amylopectin, the basic structure is still linear chains of glucose monomers linked through α-1,4 glucosidic bonds, but these are linked to other chains through α-1,6 glucosidic bonds. These branch points occur every 10–12 glucose units. This gives amylopectin a branched structure that may contain many thousands of glucose units per molecule.

Table 2. Level of Amylose and Amylopectin in Common Starches

	Amylose (%)	Amylopectin (%)
Maize	28	72
Wheat	28	72
Potato	20	80
Tapioca	17	83

In a starch granule, amylose and amylopectin are linked together through hydrogen bonding to form structures that contain both crystalline and amorphous regions. It is the intra- and intermolecular hydrogen bonding that gives starch its insolubility in water. On heating an aqueous suspension of starch, the hydrogen bonds are disrupted and the starch disperses. On cooling, the hydrogen bonds reform and the starch precipitates. The rate at which these hydrogen bonds form is related to the concentration of starch in the dispersion and the temperature. This precipitation is called retrogradation and can cause low yields and processing difficulties if allowed to occur in the production of glucose syrups.

USE OF ENZYMES IN THE CONVERSION OF STARCH

The manufacture of glucose syrup and nutritive sweeteners from starch-containing raw materials can be divided into a number of individual processes:

1. Starch separation
2. Gelatinization and liquefaction
3. Dextrinization
4. Saccharification
5. Isomerization

Within each of these defined stages, the use of microbial enzymes facilitates the process to give a product that in turn becomes the raw material for the subsequent stage. The first of the stages that makes use of enzymes is the separation of the starch from the starch-containing raw material.

Starch Separation

The separation of starch is primarily based on physical processes. With tubers such as potatoes and tapioca, the starch is separated using a process of chopping, disintegration (rasping), and sieving. It is unusual to use enzymes to facilitate the separation of starch from tuberous raw materials.

The processes of starch separation for maize and wheat are much more complex than those for potato and tapioca. In recent years enzymes have come to be used extensively to improve the starch separation from wheat. Enzymes can be used to improve the yield of starch from maize (3). In both processes the enzymes employed act on hemicellulose compounds such as arabinoxylans and β-glucans. The particular enzyme depends on the type of hemicellulose present in the raw material. Typical values for arabinoxylans and β-glucans in maize and wheat are given in Table 3.

In the case of maize, whole maize kernels are steeped in an aqueous environment under controlled conditions of temperature, time, and concentrations of sulfur dioxide and lactic acid. The main purpose of the steeping process is to soften the maize kernel for milling while inhibiting microbial growth to enable optimum starch recovery (4). It has been suggested that the addition of enzymes to the steep tank improves the separation of the starch from the fiber by degrading the arabinoxylans that bind the starch to the fiber.

Studies have shown that the use of an enzyme preparation in the steeping of maize can result in increased starch yield, increased grind capacity, and decreased steep times (5). The effectiveness of enzyme preparations when added to the steeping of maize is a subject of some debate due to the increase in starch yield versus the amount of enzyme added. It has also been suggested that enzymes are not absorbed along with the water into the maize kernel because the enzyme is a large molecule that could not penetrate the kernel through the tip cap.

This problem could be overcome by adding the enzyme during the first grind (6), after the initial cracking of the maize, allowing the enzyme access to the arabinoxylan gums that bind the starch. Many maize starch processors do not have the facilities to have a holding time between the first and second grind to allow enzyme action. It is up to the individual maize starch processor, therefore, to determine the feasibility of using enzymes.

In contrast to the separation of corn starch, wheat starch is produced by at least five different commercial processes. In all of these, wheat is ground and combined with water to form either a dough or a batter. The fractionation is achieved by physical means, and vital wheat gluten and starch are products of high value.

Wheat flour contains a high level of arabinoxylans (Table 3), a large proportion of which are soluble in water. These high molecular weight gums can bind significant amounts of water and hence give rise to high viscosities in doughs or batters. Addition of an enzyme (xylanase) to the dough or batter can reduce the viscosity by degrading the arabinoxylan and, hence, improve the separation process. The enzymes used not only improve the separation of gluten and starch but also enhance gluten coagulation.

In all enzymes used to aid starch separation, the presence of proteinase or amylase side activities is detrimental to the process, since they will break down protein and starch, cause material loss, and cause deterioration in the quality of the end products, gluten and starch.

Once separated, the starch may be dried as a product in itself or further processed to nutritive sweeteners. The conversion of starch to nutritive sweeteners is the main use of enzymes in the starch industry.

Table 3. Percentage of β-Glucan and Arabinoxylan in Maize and Wheat

	β-glucan (%)	Arabinoxylan (%)
Maize	0.3	3.5
Wheat	0.4	8.5

Gelatinization

At ambient temperatures starch granules are insoluble in water, however, if an aqueous starch suspension is heated to above 60 °C the granules begin to swell and disrupt. Eventually the starch molecules are dispersed into solution. This process is known as gelatinization.

The temperature required for complete dispersion or gelatinization depends on the source of the starch. For most starches, a temperature of 105–110 °C is sufficient. At 25–40% dry solids, the viscosity of gelatinized starch is extremely high, and so a thinning agent is required. Viscosity reduction may be achieved by a combination of the shear forces exerted on the slurry as it is continuously pumped through a jet cooker (mechanical thinning) and the action of acid or enzyme. The gelatinization temperatures of starches from the most commonly used raw materials are given in Table 4.

During the process of gelatinization and thinning of a starch slurry, proteins, lipids, and hemicellulases that complex with the amylose and amylopectin are released and coagulate to some extent. The lipids that are complexed mainly with the amylose are dissociated at higher temperatures. Subsequent cooling may again lead to the precipitation of the amylose–lipid complexes. These precipitates cause hazes and flocculation later in the process and are difficult to remove. They are often the cause of a positive iodine reaction to starch in saccharification reactors (7).

To prevent retrogradation and reprecipitation of starch–amylose complexes, it is necessary to continue the thinning process until partial hydrolysis occurs. This process is called liquefaction.

Liquefaction

Liquefaction is the first primary action in hydrolysis or thinning of the starch slurry whereby viscosity is reduced and first-stage degradation of starch molecules takes place (Fig. 1).

Acid Liquefaction Process. Traditionally, the process of thinning, liquefaction, and dextrinization of starch was

Table 4. Gelatinization Temperature for Some Common Starches

Starch type	Gelatinization temperature (°C)
Maize	72–76
Wheat	60–64
Potato	65–70
Tapioca	70–75

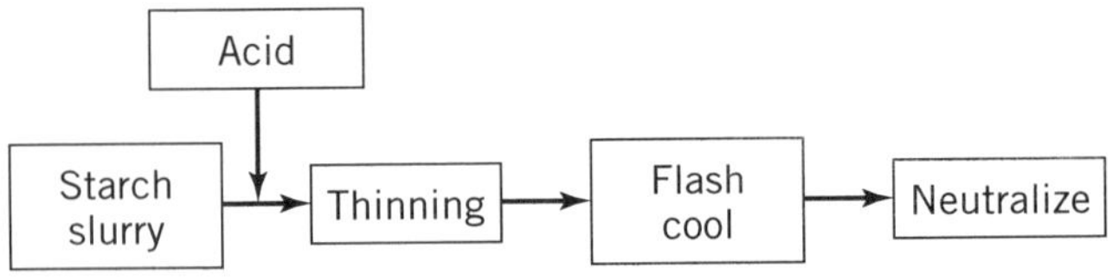

Figure 1. Scheme for the acid liquefaction of starch; thinning conditions: pH 1.5–2.0, 140–150 °C, 5–8 min.

carried out using acid. In acid liquefaction, a starch slurry of between 30 and 40% dry solids is adjusted to a pH of 1.5–2 with addition of an acid, usually hydrochloric acid.

The acidified slurry is then heated to 140–150 °C by either direct steam injection in a jet cooker or indirect heating in a reactor for approximately 5–8 min. The hydrolysate is then flash-cooled to atmospheric pressure and neutralized (Fig. 1).

Acid liquefaction results in complete gelatinization, thinning, and a liquefact that can be easily filtered. However, the process also results in a product with a high color and ash content, and significant purification costs. The action of the acid on the starch molecules is nonspecific, and by-products: such as furfuraldehyde derivatives are formed, resulting in a lower yield of desired products.

The use of a heat-stable endo-α-amylase allows for less rigorous conditions to be employed in the liquefaction process. By-product formation becomes less of a problem and refining costs are reduced.

Two-Stage Enzyme Liquefaction Process. The first enzyme to be employed in starch liquefaction was the endo-α-amylase produced by *Bacillus subtilis* var. *amyloliquefaciens*. The enzyme is utilized in a two-stage enzyme liquefaction process. A starch slurry at 30–40% dry solids is thinned at a temperature of 85 °C for 1 h at pH 6.0–6.5. This is followed by a short heat treatment at 140 °C, flash-cooled to 85 °C (at which point a second enzyme addition is made), and the reaction continued until liquefaction is complete. For optimum enzyme stability a calcium concentration of 150 ppm is required; this is usually added as calcium hydroxide or calcium chloride into the starch slurry (Fig. 2).

If a thermostable α-amylase is employed, such as that from *Bacillus licheniformis* or *Bacillus stearothermophilus*, the heat treatment at 140 °C may be eliminated. The use of thermostable α-amylases has largely replaced acid liquefaction and processing using the less heat-stable α-amylase from *B. subtilis*.

Single-Stage Enzyme Liquefaction Process. A starch slurry at 30–40% dry solids is prepared, and the pH is adjusted to 5.8–6.2 by the addition of sodium of calcium hydroxide. Calcium salts are added, if necessary, to give a level of 75–100 ppm of free calcium ions. The slurry is pumped continuously through a jet cooker operating at 103–107 °C, enzyme having already been added to the slurry stream. The higher shear produced by the action of pumping through the jet cooker gives a degree of mechanical thinning, and this, in addition to the action of the α-amylase, means that peak viscosities are avoided. The slurry is then transferred to a holding coil, which maintains the temperature for 5–10 min before the liquefied starch is flash-cooled to 95 °C, where dextrinization continues.

In the single-stage liquefaction process, the control of the jet cooker is critical (Fig. 3). Control of the temperature of the slurry through the jet cooker and into the holding coil must be accurate. If the temperature is allowed to drop much below 105 °C, incomplete gelatinization will occur and filtration problems may be encountered later in the process. However, at temperatures much above 105 °C, the enzyme will be inactivated and the viscosity of the slurry is not effectively reduced. Operating pH should be between 5.8 and 6.2, although lower pH values (down to 5.5) have been reported with the use of *B. stearothermophilus* heat-stable α-amylase. At lower pH values the enzymes are less stable and, in combination with high temperatures, lead to rapid inactivation; on the other hand, higher pH levels may cause color formation and production of by-products, resulting in lower end product yields.

Other factors that affect the performance of the liquefaction process are dry substance and calcium ion concentration. Higher levels of dry substance have a stabilizing effect on the enzyme but may also lead to incomplete gelatinization and subsequent filtration problems. A dry substance level of 30–35% gives adequate enzyme stability, and at the same time ensures complete gelatinization. Calcium is required to give the heat-stable α-amylases stability at higher temperatures; for *B. licheniformis* enzyme the requirement is 75–100 ppm free calcium ions, and for *B. stearothermophilus* it is 100–150 ppm. The level of calcium ions is usually adjusted by addition of calcium hydroxide or calcium chloride to adjust the pH of the slurry.

The final parameter that requires consideration is enzyme dose rate: Too little enzyme leads to incomplete liquefaction and starch positive reaction after dextrinization; too much enzyme results in a large enzyme carryover into dextrinization and a dextrose equivalent higher than desired. The DE is used to monitor the degree of starch breakdown. It is calculated by expressing the reducing sugars, such as dextrose, as a percentage of the dry substance. Typical enzyme dose rates for the single-stage liquefaction process range from 0.6 to 1.0 L of a heat-stable α-amylase per ton of dry substance.

Enzymes produced from *B. licheniformis* give different starch breakdown products to heat-stable α-amylases produced using *B. stearothermophilus*. The difference shows in the measurement of DE for the same degree of starch degradation. The DE given by *B. stearothermophilus* enzyme can be as much as two percentage points higher than *B. licheniformis* enzyme.

Because of their different breakdown products, mixtures of heat-stable α-amylases from *B. licheniformis* and *B. stearothermophilus* give a more efficient breakdown of starch during liquefaction. A further potential use of the

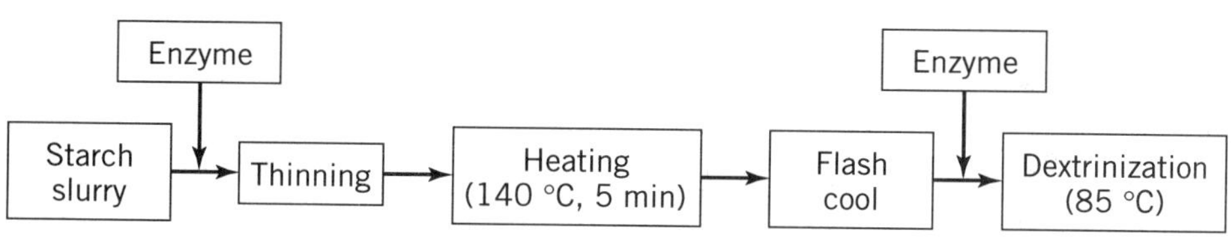

Figure 2. Liquefaction using a non-heat-stable α-amylase (*Bacillus subtilis*); thinning conditions: pH 6.0–6.5, 80–90 °C, 5–8 min.

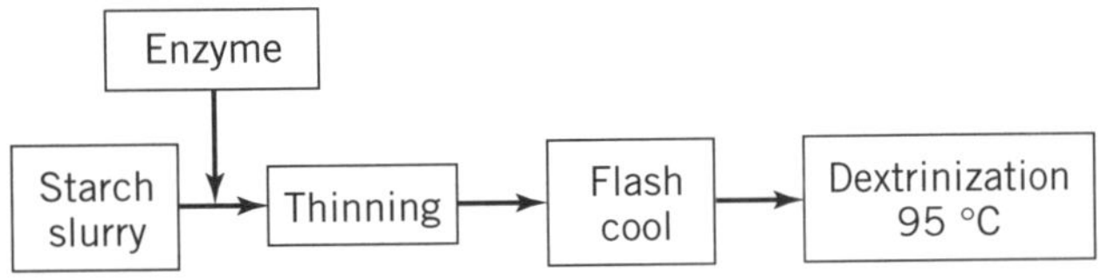

Figure 3. Single-stage liquefaction process; thinning conditions: pH 5.8–6.2, 105–107 °C, 5–11 min.

B. stearothermophilus heat-stable α-amylase is using lower pH liquefaction; the enzyme is more stable than *B. licheniformis* enzymes at pH values down to 5.5 (8). The advantage of lower-pH liquefaction is reduced by-products and color formation.

Acid or Enzyme Liquefaction. A process that takes advantage of acid thinning (which can improve filtration downstream due to better separation of fat and protein) and enzyme liquefaction is the acid or enzyme liquefaction process.

A starch slurry containing between 30 and 40% dry substance is adjusted to give a pH of 2–5. The slurry is jet cooked at 140 °C for about 5 min, where mechanical thinning due to both shear and acid thinning takes place. The pH is very important, if it is too low, color and by-product formation occurs, but if it is too high, insufficient thinning results.

After this initial cooling, the thinned starch is cooled to 95–100 °C, the pH is adjusted to 5.8–6.2, and enzyme is added. The major disadvantages of this process are increased energy consumption and potential by-product formation.

Dextrinization

After liquefaction of the starch slurry, the action of the heat-stable α-amylase on the starch is allowed to continue in a process called dextrinization. This process involves further starch degradation until a syrup with the required dextrose equivalent is obtained. Dextrinization may involve the further addition of α-amylase, as in the case of maltodextrin production, or there may be sufficient enzyme remaining after the cooling process.

The degree to which dextrinization is allowed to continue is determined by the type of product manufactured. In the production of high-dextrose syrups (i.e., syrups containing 94–97% dextrose), dextrinization is allowed to continue to a DE of 15–18. However, in the production of very-high-maltose syrups (68–75% maltose) the DE must be controlled to a maximum of 12. Addition of a further quantity of bacterial α-amylase allows maltodextrin with intermediate DEs up to 30 to be prepared.

Once the desired DE has been reached during dextrinization, the α-amylase may require inactivation. This can be achieved by a second jet-cooking step at temperatures of 120 °C and above with a low pH, preferably between 3.8 and 4.5. The resulting hydrolysate may then be filtered, clarified with carbon, concentrated, and dried or spray-dried to give a maltodextrin.

The process of liquefaction and dextrinization produces a hydrolyzed starch suspension of low viscosity that is stable to retrogradation and suitable for further enzyme processing to high-dextrose, glucose, or high-maltose syrups.

Saccharification

The process in which the dextrinized starch is converted to nutritive sweeteners is called saccharification (Fig. 4). Further hydrolysis of the oligosaccharides produced by the process of liquefaction and dextrinization may be carried out by a number of different exoamylases and debranching enzymes, depending on the required product (Table 5).

Dextrose (Glucose) Production. During the production of glucose, the enzyme amyloglucosidase is used in saccharification (Fig. 5). Amyloglucosidase is an exo-α-amylase that catalyzes the hydrolysis of α-glucosidic bonds by successively recovering glucose from the nonreducing end of oligosaccharides in a stepwise manner. The enzyme is able to hydrolyze both α-1,4 and α-1,6 glucosidic bonds; however, the rate of hydrolysis of α-1,6 bonds is much slower than the rate of hydrolysis of α-1,4 bonds. It has been reported that amyloglucosidase hydrolyzes α-1,4 bonds between 20 and 30 times faster than α-1,6 bonds.

Amyloglucosidase preparations are usually available at activities of either 200 or 300 U mL^{-1}. After liquefaction and dextrinization the solution is cooled to 60 °C and the pHs adjusted to 4.2–4.6 with acid. After dextrinization, the dry substance should be between 25 and 37%. Amyloglu-

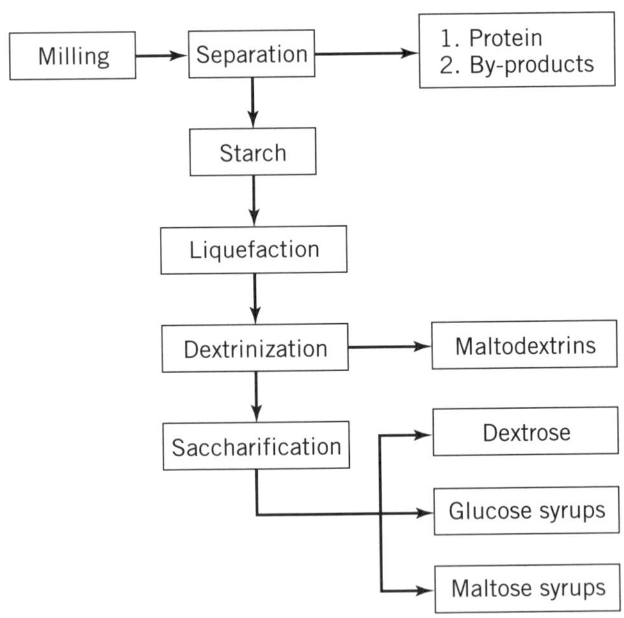

Figure 4. Production of nutritive sweeteners from starch.

Table 5. Listing of the Main Saccharification Enzyme Required for Various Desired Products

Product required	Main saccharification enzyme
Glucose	Amyloglucosidase
Maltose (50%)	Fungal α-amylase
Maltose (≥50%)	Cereal β-amylase

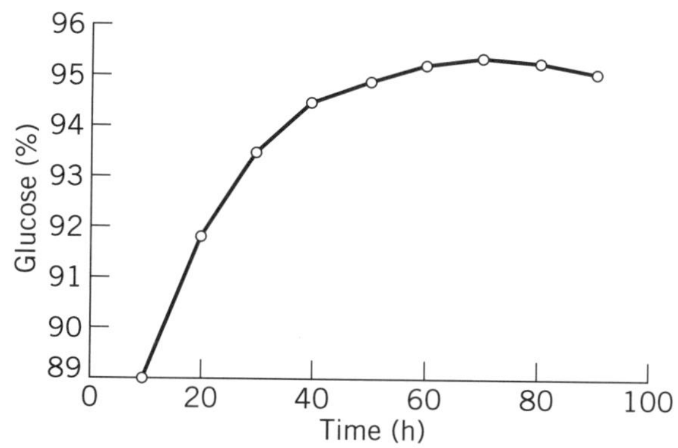

Figure 5. Typical saccharification profile with amyloglucosidase.

cosidase (300 U mL^{-1}) is added, typically at dose rates of between 0.5 and 1.1 L ton^{-1} dry substance, depending on the required saccharification time. The saccharification reaction takes place in stirred reactors and is allowed to continue until peak dextrose is reached, typically 40–72 h.

Given that the enzyme is able to hydrolyze both types of bonds present in starch, it would be reasonable to expect that amyloglucosidase is able to produce quantitative amounts of glucose from liquefied starch. At very low solid levels, glucose levels approaching 100% are possible, but at the higher solid levels used in industrial saccharification this is not the case.

Amyloglucosidase can polymerize glucose by the reverse reaction of hydrolysis. This reversion is a condensation reaction and is accompanied by the elimination of water. The products of this reversion are maltose and isomaltose. The reversion reaction takes place at a much slower rate than the hydrolysis reaction and is dependent on the dry substance level and the residual enzyme activity at the end of saccharification. At lower dry substance the rate of reversion is slower and higher levels of dextrose can be attained (Fig. 6).

Peak dextrose levels may also be reduced by the presence of a secondary enzyme activity within amyloglucosidase preparations, namely transglucosidase. This enzyme is a glucose transferase and catalyzes the formation of α-1,6 branched oligosaccharides such as isomaltose and panose from glucose and maltose. The buildup of isomaltose and panose during saccharification leads to an overall reduction in final dextrose levels. Transglucosidase activities have recently been essentially removed from amyloglucosidase preparations either after the fermentation process or by selection of an *Aspergillus niger* strain that does not produce the activity.

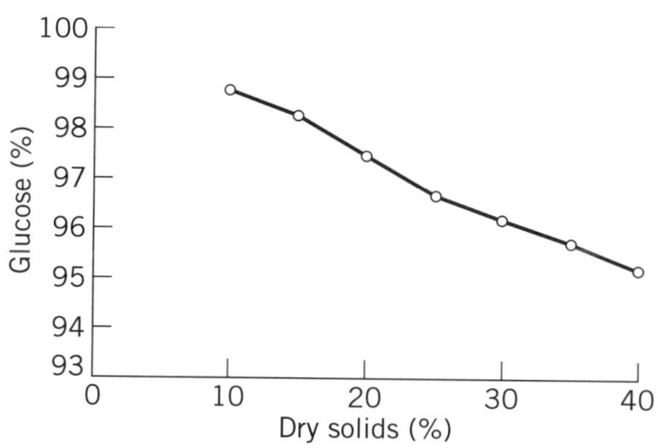

Figure 6. The effect of solids on maximum dextrose yield.

Amyloglucosidase itself is very inefficient at hydrolyzing the products of the reverse reaction, such as isomaltose, isomaltotriose, and panose. The formation of these products is the major factor in governing the dextrose level after saccharification. It is important, therefore, to limit the rate of the reversion reaction as much as possible. One of the ways to achieve this is to lower the level of amyloglucosidase in saccharification.

The relative inefficiency of amyloglucosidase in hydrolyzing α-1,6 glucosidic linkages in the branched oligosaccharides is a major obstacle to dextrose production. An excess of enzyme is therefore required to achieve the desired saccharification rates. This excess results in a higher rate of reversion and ultimately lower dextrose levels.

It is possible to reduce the level of amyloglucosidase and maintain the reaction rate by the addition of an enzyme that can specifically hydrolyze the α-1,6 links of the branched oligosaccharides. This enzyme is called pullulanase, and blends of pullulanase and amyloglucosidase are now available. The addition of a specific α-1,6 debranching enzyme augments the α-1,6 activity of amyloglucosidase and, therefore, the addition rate of the glucose-producing enzyme may be lowered. The reduction in amyloglucosidase in the saccharification tank decreases the rate of the reversion reaction. The overall effect is an increase in the peak dextrose level. The dose rate and conditions of use of the blended products should be the same as the dose rate for amyloglucosidase alone.

The choice of an amyloglucosidase or a blend of pullulanase and amyloglucosidase depends on the individual saccharification requirements. A fast saccharification can be achieved by a high addition of amyloglucosidase, but this in turn will mean quicker reversion rates; therefore, if dextrose levels are important, the enzyme blend is the preferred route. Often the residual oligosaccharides referred to as DP_4+ are considered undesirable, and again the enzyme blend has been shown to lower the levels of this oligosaccharide fraction. The use of an amyloglucosidase–pullulanase blend gives greater flexibility and control over the saccharification process.

Further alternatives are available to achieve specific saccharification requirements. For example, all amyloglucosidase preparations contain acidic α-amylase activity. The presence of this activity enhances the saccharification rate by degrading the higher molecular weight oligosaccharides into lower molecular weight fractions, which also results in a reduction of DP_4+ oligosaccharides in the final product.

Higher dextrose levels may also be obtained by addition of an α-1,4-glucosyl transferase to amyloglucosidase (9). Unlike transglucosidase, which catalyzes the formation of α-1,6 glucosyl linkages, α-1,4 glucosyl transferase produces α-1,4 glucosyl linkages. Such an enzyme is produced by the fermentation of *Bacillus megaterium*. Substrates for the enzyme include isomaltose and isomaltotriose; these are combined with glucose to form molecules that can then be hydrolyzed by amyloglucosidase.

A combination of amyloglucosidase and α-1,4 glucosyl transferase results in higher dextrose levels by hydrolysis of the products of the reversion reaction. The enzyme may be added at the start of saccharification along with amyloglucosidase (10), however, the pH optimum of Megadex is above that of amyloglucosidase and the enzyme may be added to the process after a pH adjustment.

Maltose Syrups. Syrups containing high levels of maltose have different properties than those containing high dextrose levels. Maltose has a different sweetness characteristic, and its syrups have a resistance to crystallization. Unlike the production of high-dextrose syrups, it is difficult to achieve maltose levels in excess of 80–85% at high dry substance levels.

There are two main maltose-producing enzymes. The first is a maltogenic α-amylase produced by *Aspergillus oryzae* and is often referred to simply as fungal α-amylase. The second is a β-amylase, which is found in cereals such as barley, wheat, rye, oats, and sorghum, and also in soya beans and sweet potato. Recently a microbial β-amylase has been produced from a *Bacillus* species.

The production of high-maltose-containing syrups using *A. oryzae* (fungal) maltogenic α-amylase follows a similar process as the production of high-dextrose syrups. After dextrinization using a thermostable α-amylase the solution is cooled to 55–60 °C and the pH adjusted to 4.7–5.0. Fungal α-amylase is added to the stirred reactor at a level of between 0.3 and 0.7 L ton^{-1} dry solids. The reaction is allowed to continue for 24 to 40 h.

After saccharification the crude syrup may be filtered and ion exchanged to remove ash. The syrup can then be decolorized by activated charcoal and finally evaporated to the desired solids level. Syrups produced by fungal α-amylase can contain up to 60% maltose based on dry substance. Dextrose levels may be controlled by limiting dextrinization with the thermostable α-amylase. One disadvantage of maltose syrup production using a fungal α-amylase is the high level of maltotriose obtained.

Maltose syrups produced by cereal β-amylases have lower levels of maltotriose than those made with fungal α-amylases. After dextrinization, the starch solution is cooled to 54–58 °C and the pH lowered to 5.2–5.6. Cereal β-amylase is added between 0.25 and 0.75 L ton^{-1} dry solids and the reaction is allowed to continue for 36–48 h, depending on enzyme dose rate. The level of maltose produced by cereal β-amylases is limited to between 50 and 55% on dry substance.

β-Amylases produce maltose sequentially from the nonreducing ends of the oligosaccharides formed by the action of the thermostable α-amylase. Once the α-1,6 branch points are reached, the enzyme is unable to continue. The major products of the reaction are limit dextrins and maltose itself. Table 6 shows a typical carbohydrate spectrum of maltose syrups produced using fungal α-amylase and cereal β-amylase.

Syrups containing between 50 and 60% of maltose on a dry solids basis are referred to as high-maltose syrups.

Table 6. Carbohydrate Spectrum of Maltose Syrups Produced Using Fungal α-Amylase and Cereal β-Amylase

	Fungal α-amylase	Cereal β-amylase
Glucose	3	1
Maltose	50	60
Maltotriose	26	8
Higher dextrins	21	31

Higher levels of maltose may be achieved by addition of a debranching enzyme (pullulanase). The pullulanase hydrolyzes the α-1,6 glucosidic linkages in branched dextrins and amylopectin and allows the β-amylase to continue producing maltose. With careful control of the process, up to 75% maltose may be achieved. Such syrups are called very-high-maltose syrups (VHMS).

Careful control of dextrinization is imperative to produce maltose in excess of 70%. After liquefaction, dextrinization is controlled to give a DE of between 7 and 10, the solution is cooled to 54–58 °C, and the pH adjusted to 5.2–5.6. Cereal β-amylase is added at 0.5 and 0.75 L ton^{-1} dry solids and the pullulanase is added in a similar amount. The reaction is allowed to continue for 36–58 h and the product refined and evaporated. Maltose-producing enzymes do not catalyze the reverse reaction, so there is no reduction in maltose level if the reaction is allowed to proceed.

Maltotriose is a barrier to higher levels of maltose being achieved. β-Amylases are unable to hydrolyze maltotriose, so it is important to keep the level of this oligosaccharide as low as possible. One of the ways this can be achieved is by controlling the DE to a low level during dextrinization.

An alternative is to use an enzyme that has some action toward maltotriose, unfortunately, these enzymes, in addition to producing maltose, also produce glucose. It is often a requirement of maltose syrups that the glucose level be kept low (<3.0% on dry substance).

Filtration Enzymes

Maize starch hydrolyzates generally exhibit good filtration characteristics. High filtration rates make it easy to remove undesirable, insoluble components from the hydrolysates in a relatively short time. However, wheat starch hydrolysates have very poor filtration characteristics due to low filtration rates and obstruction of the filter cake. To overcome this problem so-called filtration enzymes are commonly used in the industry.

The major cause of the poor filtration has been identified as a monoacyl lipid compound that is thought to occur in inclusion complexes with the amylose fraction. In wheat the native starch contains about 0.8% of this internal lipid, the principal component of which is lysophosphatidylcholine (lysolecithin). Amylose molecules are helical, with a hydrophobic interior to the helix, which can accommodate certain lipophilic molecules, such as lysophosphatidylcholine.

On hydrolysis of the amylose fraction of wheat starch, the lysophosphatidylcholine is released into the solution. Lysophospholipids are water soluble and excellent emulsifiers. It has been reported that the lipid forms micelles above a concentration of 0.025 g kg^{-1}. When the critical micelle concentration is exceeded, the filtration rate of the hydrolysate is decreased (11). The critical micelle concentration is 40 times lower than the amount of lysophosphalipids in wheat starch hydrolysates. Therefore, the concentration of lysophosphalipids needs to be reduced to less than 2.5% of the original amount in wheat starch hydrolysates to achieve acceptable filtration rates.

The reduction in lysophosphalipids can be achieved by the addition to the hydrolysate of the enzyme lysophospholipase, which cleaves the remaining fatty acid to give a free fatty acid and a glycerophosphatide. This cleavage destroys the emulsifying ability of the phospholipid. The molecule no longer contains a hydrophobic part and a hydrophilic part, but instead a water-insoluble molecule (the fatty acid) and a water-soluble molecule (the glycerophosphatide) are formed.

Lysophospholipase preparations are produced by the fermentation of *A. niger*. The enzymes are very low in amyloglucosidase activity and are therefore suitable for use in the production of maltose syrups. Manufacturers of glucose from wheat starch should choose an amyloglucosidase enzyme preparation that has an appreciable amount of lysophospholipase as a side activity. There should be no need to add further quantities of lysophospholipase to aid filtration of glucose syrups produced using amyloglucosidase.

A second cause of poor filtration in wheat starch hydrolysates has been identified as the gummy substances from wheat starch, the arabinoxylans, often referred to as pentosans. Pentosans are important macromolecules of wheat flour because of their ability to bind water. Because of the high water-binding capacity, the soluble pentosans form highly viscous solutions with intrinsic viscosities that are about 15–20 times higher than those of soluble proteins. The addition of an enzyme with the ability to break down the water-soluble pentosans will reduce the viscosity of a wheat starch hydrolysate, improving filtration performance.

A second class of pentosans can be separated by means of centrifugation of a flour–water suspension in the so-called tailings, which form a layer above the primary starch. These are the water insoluble pentosans of wheat flour.

The structure of the water-insoluble pentosans is similar to that of the soluble pentosans, but unlike the soluble pentosans, on hydration the water-insoluble pentosans swell without going into solution. On contact with the filter bed these water-insoluble pentosans impede flow through the filter. A second type of pentosan-degrading enzyme can be added to degrade the insoluble pentosan fraction and improve the flow through the filter. The preferred pentosan-degrading enzyme for wheat starch hydrolysate combines action toward both soluble and insoluble pentosans.

Isomerization

The worldwide demand for fructose syrups has increased in recent years due to the inclusion of enriched fructose corn syrup (EFCS) containing 55% fructose in popular soft drinks. Fructose is a naturally occurring sugar that is approximately 1.5 times as sweet as sucrose on an equal weight basis.

Although fructose can be produced by the process of inversion of sucrose followed by separation of the fructose and glucose by chromatography, it is also produced by the enzymatic hydrolysis of starch. The production of fructose from starch initially involves the production of glucose, as already outlined.

Glucose is converted into fructose by a process called isomerization. The enzyme used to catalyze this conversion is glucose isomerase and is usually used in an immobilized form suitable for a fixed-bed continuous column process. The enzyme-catalyzed reaction is reversible, and under the process conditions the equilibrium constant is approximately 1.

Glucose solution from the saccharification is refined using filtration, carbon treatment, and ion exchange and is evaporated to 40–50% dry matter. Calcium ions inhibit the activity of the enzyme, and their levels should be kept as low as possible. The effect of residual calcium ions can be counteracted by the presence of magnesium. Magnesium sulfate is added to give a magnesium ion concentration above 50 ppm and 20 times greater than the calcium ion concentration.

The pH is adjusted to 8.1–8.3 and the temperature to between 55 and 58 °C. The solution is then pumped through a fixed-bed column containing immobilized glucose isomerase. The flow rate through the column should be adjusted in proportion to the enzyme activity in such a way that the fructose content of the product remains constant. The enzyme activity in the column decreases during use and the flow rate must be increased over the lifetime of the column to maintain the fructose level. The typical carbohydrate profile for the products of isomerization is given in Table 7.

The product from the isomerization columns may be refined and concentrated to approximately 70% solids to give a high-fructose syrup at 42% fructose based on dry solids. The 55% fructose syrup can be achieved by concentrating this 42% high-fructose syrup to 90% fructose by a separation process, such as chromatographic adsorption. This concentrated 90% fructose syrup can then be blended with 42% high-fructose syrup to produce the required 55% enriched fructose syrup. A scheme for the production of 55% fructose is shown in Figure 7.

The stability of the enzyme and its ability to retain its activity under process conditions are essential factors in the productivity of the immobilized enzyme. The operating conditions given in Table 8 illustrate the feedstock parameters that have an influence on enzyme performance and therefore productivity.

Table 8 is not an exhaustive list of parameters that effect enzyme performance, but it gives an indication of the conditions that will give optimal enzyme performance.

Table 7. Carbohydrate Profile for Products of Isomerization

Type of carbohydrate	% w/w on dry solids
Fructose	42
Glucose	50–55
Others	3–8

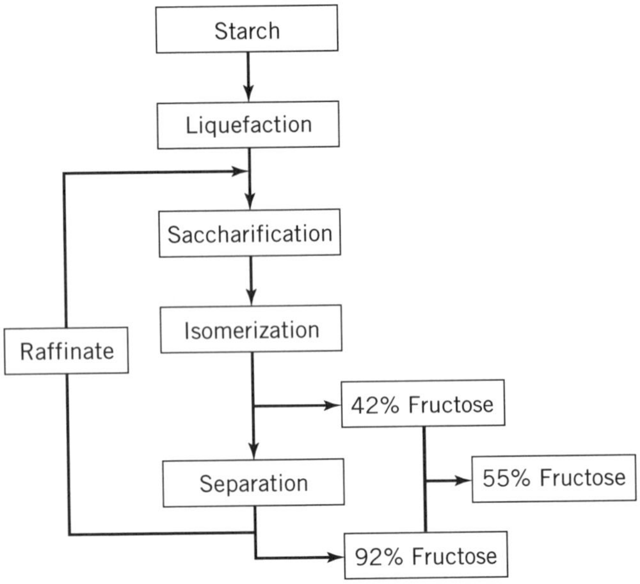

Figure 7. The production of 55% fructose syrup.

Table 8. Conditions Required by Certain Parameters to Promote Optimal Enzyme Performance

Parameter	Range
Dry Substance (%)	40–50
Glucose (% dsb)[a]	≥94
Fructose (% dsb)[a]	≤3
Temperature (°C)	55–58
pH	7.8–8.3
Magnesium (ppm)	≥50
Sulfite (ppm)	≥125

Secondary Product Processing

The processing of the secondary products of the starch industry can be improved with the addition of enzymes. The major use of enzymes in this area is in wheat starch processing.

The yield of quality starch from wheat that is suitable for processing into maltodextrins and sweeteners is lower than the total starch content of the wheat itself. Between 10 and 15% of the total starch of wheat is not separated along with the large particle starch (A starch) during the separation process. This starch, depending on the separation process, is called the B starch or C starch fraction and is usually sold as a low-value animal feed ingredient.

The B or C starch fraction contains high levels of protein and gums (pentosans) that make this fraction unsuitable for processing into sweeteners. These high levels of protein and gums lead to problems with color formation and evaporation of the syrup.

As with other starches, B starch can be liquefied and dextrinized using a heat-stable α-amylase. A saccharification enzyme, such as amyloglucosidase, can be used to produce sugars. The value of the B starch in animal feed is often related to the DE of the syrup. The conditions of liquefaction, dextrinization, and saccharification are no different from those used in the production of nutritive sweeteners.

To aid evaporation, a pentosanase enzyme is also added during saccharification to reduce the viscosity of the syrup. The level of protein can be reduced by introducing a heating step that allows the protein to coagulate. This protein can be dried and sold as a further product in animal feedstuffs.

The enzyme used in the reduction of viscosity can be heat stable, such as that produced by *Penicillium emersonii*, which allows a continued action in the evaporators. Alternatively, there may be no need for a heat-stable enzyme if the saccharification time is long enough to allow the viscosity to be reduced sufficiently.

Using enzymes to reduce the viscosity of B starch syrups has the following benefits.

1. Higher solids content of the final evaporated syrup
2. Improved heat transfer during evaporation and less energy consumption
3. Less fouling of the evaporators, leading to more efficient cleaning
4. Higher evaporation capacity

Hemicellulases (xylanases) may also be used to aid the filtration of maize gluten. The enzyme reduces the viscosity of the gluten suspension and thereby improves the rate of filtration.

CONCLUSION

The development of the enzyme-catalyzed conversion of starch has led to great advancements in the variety and quality of starch-derived sweeteners. The production and use of enzymes is still a relatively new technology, and with the help of modern biotechnology further improvements are not far in the future.

BIBLIOGRAPHY

1. Committee on Food Institute of Medicine ed., *The Food Chemicals Codex*, 4th ed., National Academy Press 1997.
2. E.R. Kooi and F.C. Armbruster vol. 2, in R.L. Whistler ed., *Starch: Chemistry and Technology*, Academic Press, New York, 1984, pp. 553–568.
3. J.D. Steinke and L.A. Johnson, *Cereal Chem.* **68**, 7–12 (1991).
4. D.S. Jackson and D.L. Shandera, Jr., *Adv. Food Nutr. Res.*, **38**, 271–300 (19xx).
5. J. Steinke and L. Johnson, *Cereal Chem.* **68**, 7–17 (1991).
6. A. Caransa, M. Simell, A. Lehmussaari, M. Vaara, and T. Vaara, *Starch* **40**, 409–411 (1988).
7. P.J. Brumm, R.E. Hebeda, and W.M. Teague, *Starch* **41**, 343–348 (1989).
8. P.J. Brumm, R.E. Hebeda, and W.M. Teague, *Food Biotechnol.* **2**, 67–80 (1988).
9. M.-H. David, H. Gunther, and H. Röper, *Starch* **39**, 436–440 (1987).
10. R.E. Hebeda, C.R. Strylund, and W.M. Teague, *Starch* **40**, 33–36 (1988).
11. A.M. Matser and P.A.M. Steeneken, *Detmold Starch Conference*, Germany. 1997.

ERYTHROPOIETIN

Seiji Masuda
Ryuzo Sasaki
Kyoto University
Kyoto, Japan

KEY WORDS

Carbohydrate structure
Erythropoietin
Monoclonal antibody
Recombinant protein

OUTLINE

INTRODUCTION

Erythropoietin (EPO) stimulates red blood cell production by promoting both the growth of late erythroid precursor cells and their maturation into proerythroblasts, in which globin synthesis starts. It is widely accepted that EPO is a major physiological regulator of the process of erythroid differentiation and that it controls the maintenance of physiological levels of the circulating erythrocyte mass. EPO is a heavily glycosylated protein and produced mainly by the kidney in human adults. Anemia associated with renal failure often results from a decreased level of EPO, and EPO administration results in a dramatic improvement with few side reactions in these anemic patients. Carbohydrates attached to EPO play a critical role in expression of the in vivo activity. In this paper, we mainly focus on the production of recombinant human EPO by mammalian cells and structure and function of carbohydrate chains, comparing with urinary EPO and recombinant product. Basic aspects of EPO, including a novel function of EPO in brain, are also described.

BIOCHEMISTRY AND PHYSIOLOGY OF EPO

It has taken a long time to isolate and characterize EPO. More than 100 years ago, it had been suggested that hypoxia stimulated red blood cell formation. In 1977, human EPO was purified from patients with aplastic anemia (1). The human gene for EPO was cloned in 1985 (2,3), and recombinant EPO (rEPO) is now available. One unit (U) of EPO activity was defined as an erythropoiesis-stimulating activity elicited when 5 μmol of $CoCl_2$ are injected into a starved Sprague Dawler male rat (4). Cobaltous ion promotes erythropoiesis through stimulation of EPO production, because this ion mimics hypoxia in terms of EPO production. The latest standard human rEPO gives 130,000 U/mg protein when it is fully glycosylated. The values of the serum concentration in healthy adult humane range 10 to 30 mU/mL. In 1989, EPO receptor was cloned from erythroleukemia cell line (5), and the signal transduction pathway of EPO has been analyzed in recent decades.

Chemistry

Human EPO gene has been mapped on chromosome 7q11-q22. The mouse EPO gene is localized on chromosome 5. There is a single copy with a length of 4 kb that covers a minimum promoter, five exons, four introns, and hypoxia-inducible enhancer (6,7). The promoter, first intron, and hypoxia-inducible enhancer are highly homologous between human and mouse EPO. The amino acid sequences of EPO in human, monkey, mouse, and rat are presented in Figure 1. Deduction from cDNA indicated 193 amino acid residues in human EPO. The mature EPO isolated from human urine consists of 165 residues and is formed by removal of N-terminal 27 signal peptide and C-terminal Arg residue through the post-translational processing. Two disulfide bond are formed between positions 7 and 161 and 29 and 33. The disulfide linkage between 7 and 161 is essential for EPO bioactivity. Because there is 80% protein homology between mouse and human EPO, human EPO crossbinds with mouse EPO receptor and vice versa. EPO protein has a significant homology to mature thrombopoietin, which is a primary regulator for megakaryopoiesis (8).

A tertiary conformation for EPO has been predicted based on the protein structure modeling. Despite little homology in the amino acid sequence, EPO has a similarity to the three-dimensional structures of growth hormone, prolactin, interleukin-6, and granulocyte-colony stimulating factor (9). By analogy with growth hormone that has been shown to contain four antiparallel α-helices (10), EPO is also thought to contain four antiparallel α-helices and three loops, which is supported by the analysis of site-directed mutation within EPO molecule. Mutations of buried portion of the helix A (first helix from the N-terminal

```
          -27                         +1
Human:    MGVHECPAWLWLLLSLLSLPLGLPVLGAPPRLICDSRVLERYLLEAKEAE
Monkey:                    V        P
Rat:          P R TL -      LI       C               I
Mouse:        P R TL -      LI       C               I

          *             *          50
          NITTGCAEHCSLNENITVPDTKVNFYAWKRMEVGQQAVEVWQGLALLSEA
           V M  S S
           V M    GPR S                  K EE          S
           V M    GPR S                    EE  I       S

                    *              100
          VLRGQALLVNSSQPWEPLQLHVDKAVSGLRSLTTLLRALRAQKEAISPPD
                 V A     F      M   I     I          -    L
          I QA    QA     P  S   I   I       S   V  G   LM
          I QA     A     P  T   I   I       S   V  G   LM

             ↓                     150
          AASAAPLRTITADTFRKLFRVYSNFLRGKLKLYTGEACRTGDR
                         C                       R
           TQ      L     C                       R
          TTPP     L V   C      A             V  R
```

Figure 1. The amino acid sequences of human, monkey, mouse, and rat EPO. Numbering starts from N-terminus of mature human EPO. N-linked glycosylation sites are shown by *asterisks*, and an O-linked glycosylation site is indicated by *arrows*. The disulfide bonds are drawn as bridges.

of EPO) and D (last helix) show little effect in vitro activity, whereas those of surface portion of the helix A and D reduced the activity (11). The mutational analysis suggests that EPO has two separate receptor binding sites, which is consistent with the ligand-induced receptor homodimerization, an initial step for the signal transmission (12).

EPO Biosynthesis

As far as kidney cells responsible for EPO production are normal, the relationship between concentrations of EPO and hemoglobin in blood is normally inverse. A decrease in the hemoglobin content, for instance from blood loss, is detected by the EPO-producing cells in kidney as a reduction of oxygen supply (Figure 2). Low oxygen stimulates transcription of EPO gene, and the resulting increase in blood EPO promotes the proliferation and differentiation of colony-forming unit erythroids (CFU-Es), erythroid precursor cells on which EPO acts, into matured red blood cells. Recovery of the red blood cell mass and the concomitant oxygen supply suppresses EPO production. The molecular mechanism of hypoxic activation of EPO gene transcription has been extensively studied, although the molecule that senses changes of oxygen concentration is poorly understood (13). An important *cis*-acting sequence required for the hypoxic induction of Epo gene was defined in the 3′-flanking region. This enhancer is a 50-bp element consisting of three important segments. The highly conserved sequence near the 5′ end of the enhancer is a hypoxia-inducible factor-1 (HIF-1) binding site. HIF-1 is a transcription factor consisting of two basic helix-loop-helix-PAS proteins (HIF-1α and HIF-1β). The middle segment containing CA repeats in human gene is not well conserved between human and mouse, but the mutation of this region abolished the inducible behavior of both the human and murine enhancers. Thus far the specific protein that binds to this region has not been demonstrated. The third element is the 3′ segment, which is again highly conserved and amplifies the hypoxic signal. Binding of hepatocyte nuclear factor-4 (HNF-4), which belongs a member of the steroid hormone receptor family, to this element appears to increase the hypoxic inducibility.

EPO Production Sites

The kidney is the primary site of EPO production in adults, whereas the liver is the production site in the fetal stage. Hypoxia activates expression of the EPO gene in kidney

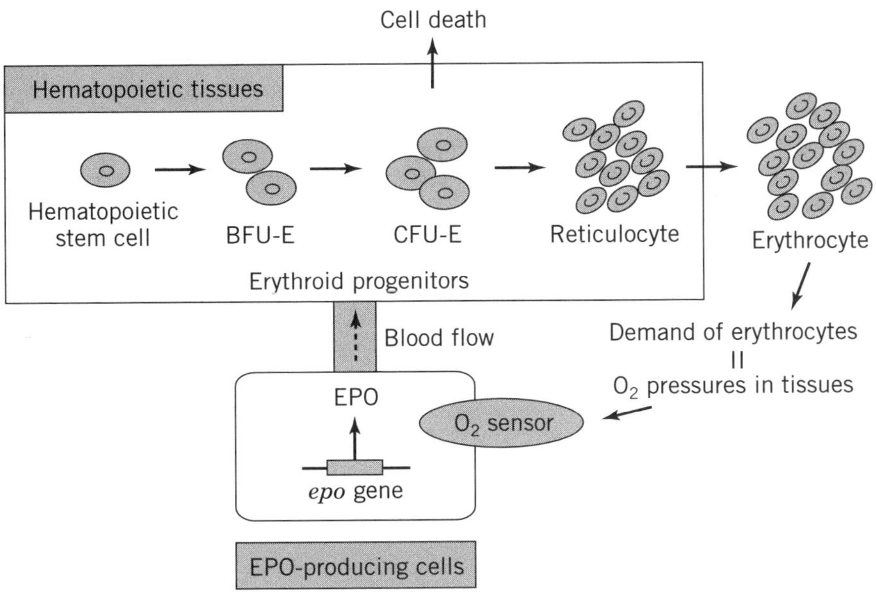

Figure 2. Regulation of erythropoiesis by EPO.

and liver (6,7,12–15). The-EPO producing cells in the kidney have been controversial but recent studies indicate that EPO is produced by peritubular interstitial fibroblasts. The hepatic EPO production is essential for fetal erythropoiesis, because homozygous mice carrying null mutations of EPO gene die around embryonic day 13 due to deficiency in erythropoiesis (16). Although EPO production in liver is hypoxia inducible, the failure of renal EPO production cannot be substituted by the liver, because uremic patients with severe anemia show low levels of serum EPO. EPO has been believed to act exclusively on erythroid precursor cells, but several recent findings indicate that EPO acts on neurons in the central nervous system (CNS) (17–19). Astrocytes produce brain EPO, and low oxygen tension stimulates the production of EPO as it does in liver and kidney. EPO in CNS functions as a neurotrophic factor as described in the next section.

EPO Action

Hematopoiesis is governed by several hematopoietic growth factors. EPO is a glycoprotein that plays a essential role in the regulation of erythropoiesis by stimulating the proliferation and differentiation of erythroid precursor cells (6,7,12–15). All the terminally differentiated blood cells have been derived from pluripotent stem cells with self-renewing activity. Hematopoiesis occurs in bone marrow, and the spleen is also an active hematopoietic site in murine. Pluripotent stem cells are committed to differentiate to multipotent stem cells with differentiation activity to several lineage-specific progenitors that have lost the self-renewality. Each lineage-specific progenitors differentiate to mature blood cells such as erythrocytes, monocytes, granulocytes, and megakaryocytes. The most primitive erythroid-specific progenitor cells derived from multipotent stem cells are burst-forming unit-erythroids (BFU-Es). Although BFU-Es express EPO receptor, their response to EPO is low, and they require additional cytokines such as interleukin 3 or granulocyte-macrophage–colony-stimulating factor for their proliferation and differentiation into CFU-Es that show the highest sensitivity to EPO. Homozygous mice carrying null mutation in EPO receptor gene die around embryonic day 13, because mature erythrocytes are not formed, although CFU-Es abundantly exist (16). Thus, EPO is essential for differentiation and proliferation of CFU-E. Recently, a novel function in the CNS has been proposed. Neuronal cell lines such as PC12 and SN6 express EPO receptor, and binding of EPO to PC12 cells increases the intracellular concentration of monoamines (20). EPO protects primary cultured hippocampal and cerebral cortical neurons from NMDA receptor-mediated glutamate toxicity (19), which is believed to be a major cause of neuron death by ischemia. These findings suggest that EPO plays a neuroprotective role when brain is exposed to hypoxia or ischemia.

PRODUCTION AND PURIFICATION OF RECOMBINANT EPO

In some cases, carbohydrates attached to the proteins play a key role in expression of their biological activities. The carbohydrates may be required for the in vitro activity as well as the in vivo activity. Even when the carbohydrates have nothing to do with the in vitro activity, they may be very important in vivo; carbohydrates protect glycoproteins from heat inactivation and proteolytic degradation, resulting in an increased lifetime in circulating blood. Eukaryotic cells are used for production of recombinant glycoproteins. For clinical applications, not only glycosylation but also the structure of attached carbohydrates are critical; a glycoprotein whose terminal carbohydrate residues are recognized by hepatic receptors would be rapidly eliminated from the circulation. The kind of carbohydrate structure may also affect the antigenicity of the recombinant product. When recombinant glycoproteins are produced in heterologous animal cells, their amino acid sequences govern the positions onto which carbohydrate chains are attached, but the maturation of the chains may entirely depend on the host cells. It is important, therefore, to compare the properties of recombinant glycoproteins produced in different cell lines with those of naturally occurring ones. The glycoprotein may also be dependent on the culture conditions.

Isolation and Expression of EPO Gene

Peptides derived from urinary EPO (u-EPO) (21) were sequenced to prepare synthetic DNA probes for screening a human fetal liver genomic library for EPO gene clones. Restriction endonuclease mapping and Southern analysis showed the gene fragments contained in three phage clones to be identical. One, designated λEPO41, was used for sequence analysis. The entire amino acid sequence of EPO was determined. The clone λEPO41 contained a whole EPO coding region, although half of 3′-untranslated region was missing. A vector for the expression of EPO in animal cells was constructed by the procedures diagrammed in Figure 3. The EPO expression vector, pZIP-NeoSV(X) 1-EPO, contained the EPO gene preceded by a long terminal repeat (LTR) functioning as transcriptional promoter and followed by DNA sequences derived from the transposon Tn5 (Neo fragment in Fig. 3) that confers G418 resistance upon mammalian cells. The expression vector was introduced into two mammalian cells, BHK cells established from baby hamster kidney, and ψ2 cells derived from NIH/3T3. Both cells were transfected with pZIP-NeoSV(X) 1-EPO, using calcium phosphate method. EPO-producing cells were isolated by selection in G418 media and by two rounds of limiting dilution. The EPO levels in culture supernatants of EPO-producing cells assayed by radioimmunoassay (RIA) were 150 U/mL for BHK cells and 300 U/mL for ψ 2 cells.

Cell Culture and EPO Production

BHK cells were cultured in Eagle basal medium supplemented with 10% calf serum and 10% bacto tryptose phosphate broth. The ψ 2 cells were cultured in Dulbecco's modified Eagle's medium supplemented with 10% calf serum. Cell cultures for EPO production were started at a cell density of 2×10^5 cells/mL in plastic culture trays (Cell Factory, 6,000 cm^2, NUNC) containing 2 L of the medium. After 4 days of culture, the cells grew to be nearly confluent. The culture medium was replaced with the low-serum medium (1.5% calf serum) and culture supernatants were harvested every 3 days.

Purification of Human rEPO

Human urinary EPO directed MAb (21) was effective in the purification of EPO from human urine and culture supernatants of BHK cells that produced rEPO (22), but pretreatment of the EPO with SDS was needed for EPO to bind with the immobilized MAb. To develop a more rapid purification procedure, the new MAbs, R2, (see next section) were used for rEPO purification (23).

The culture supernatant (100 L) was filtered with suction and concentrated to 2 L by ultrafiltration on a hollow-fiber device (Amicon DC-10, Danvers, Mass.) with a nominal molecular cutoff of 10,000. The concentrate was put on an immunosorbent column (4.4 × 6.6 cm) containing EPO-directed MAb R2 fixed on Affi-Gel 10 equilibrated with phosphate buffered saline (PBS). The column was extensively washed with PBS, 10 mM NaPi, pH 7.4, containing 0.5 M NaCl and 0.15 M, in this order. EPO was eluted by 0.2 M acetate. pH 2.5, containing 0.15 M NaCl. After neutralization of the eluted fraction, the protein was concentrated by ultrafiltration with a hollow-fiber device (Mini-Module NM-3. Asahikasei. Tokyo) and was precipitated with 90% ethanol. The precipitate was dissolved in a small amount of PBS and put on a Sephadex G-100 column (2.2 × 94 cm) equilibrated with PBS. The EPO in the fractions was concentrated and precipitated with 90% ethanol. The precipitate was dissolved in 10 mM NaPi, pH 6.8, containing 10 μM $CaCl_2$ and put on a hydroxyapatite column (100 × 7.8 mm: Bio-Gel HPHT, Bio-Rad) equilibrated with the same buffer. The adsorbed EPOs were eluted by increasing concentration of NaPi. The rEPO-B appeared in the flowthrough of a column equilibrated with 10 mM NaPi, whereas rEPO-ψ was eluted by 50 mM NaPi.

The large-scale purification of rEPO-B in the supernatant of BHK cells was attempted. Purification with the immunosorbent column was effective (Table 1); most of the protein and little EPO activity in the concentrate of culture supernatants emerged in the flowthrough fractions without being adsorbed, and EPO was eluted sharply at pH 2.5. About 2,800-fold purification was achieved by this single step, and 84% of the activity was recovered. Mab R2 was suitable for purification of EPO with a high yield. Subsequent chromatography with Sephadex G-100 and then hydroxyapatite yielded purified EPO with 3,260-fold purification and 52% recovery of the starting activity. The purified EPO had the specific activity of 137,000 U/$A_{280\,nm}$, which was higher than that of the EPO purified with an earlier preparation of an MAb (21). The final EPO preparation was pure on SDS-polyacrylamide gel electrophoresis and high-pressure liquid chromatography. The immobilized R2 could be used 30 times without any deterioration.

Figure 4 shows the electrophoretic patterns of rEPOs and uEPO on SDS-polyacrylamide gels. Purified uEPO showed two bands (lane 1). The minor one with the lower molecular weight is the asialylated form of uEPO (21). Both rEPOs migrated as a single band, but they differed in electrophoretic mobility (lanes 2 and 3). Recombinant EPO from ψ 2 cells migrated slightly faster than uEPO, and the rEPO-B band was broader than those of rEPO-ψ and uEPO. Thirty amino acids in the N-terminal portion of rEPO-ψ and rEPO-B were consistent with those of uEPO and also with those deduced from the cDNA sequence (2,3). The amino acid sequence in the C-terminal region of both rEPOs and uEPO determined by the carboxypeptidase method was identical to that of uEPO, indicating that the recombinant products had the same primary structure as uEPO. Therefore, differences between purified rEPOs in their electrophoretic mobility on SDS-polyacrylamide gel, therefore, are attributable to the difference in the carbohydrate chains.

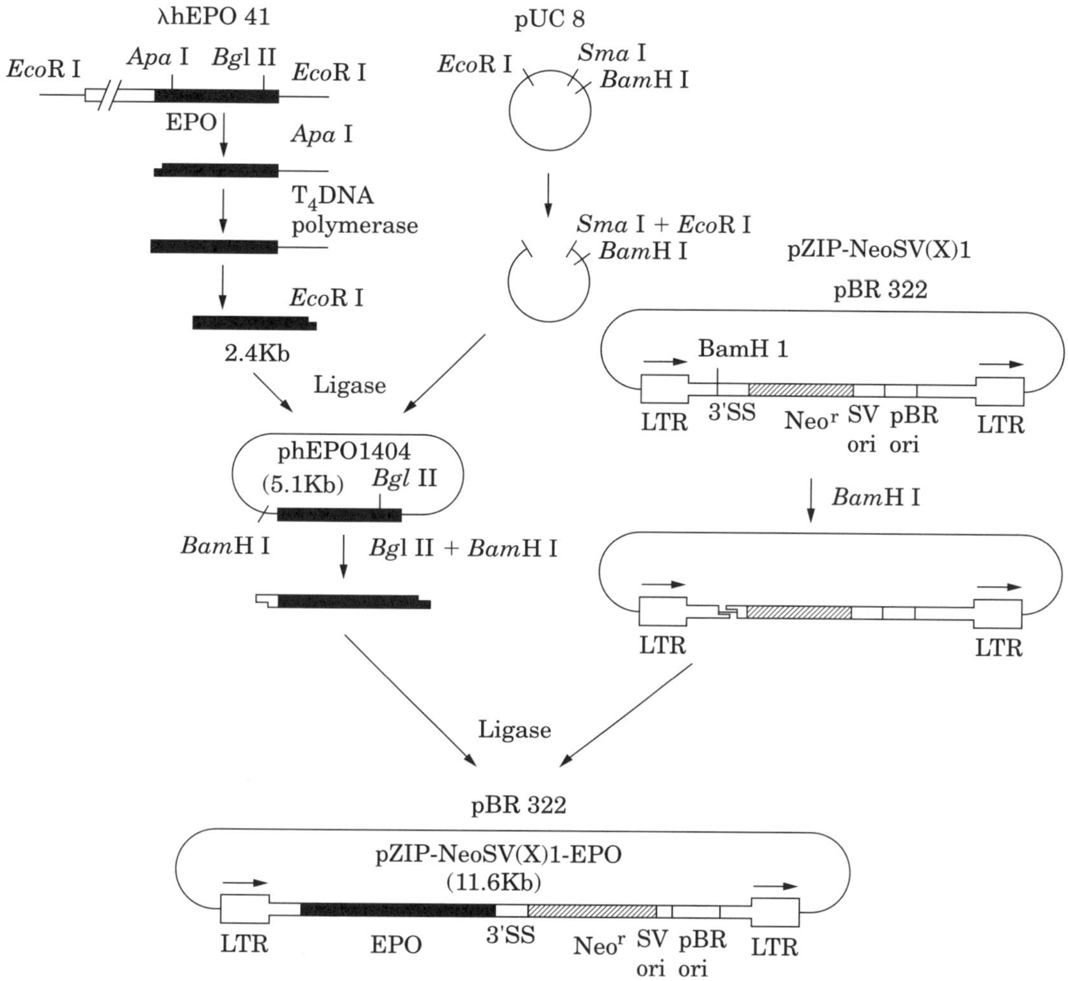

Figure 3. Construction of EPO expression plasmid. The *Apa*I site of the EPO gene in λhEPO41 is a unique site located at 62 bp upstream from the translation initiation codon, and the *Eco*RI site is for inserting human DNA fragments into the bacteriophage cloning vector, Charon 4A. The *Bgl*II site of the EPO gene in phEPO1404 is 185 bp downstream from the translation stop codon. The details in pZIP-NeoSV(X)1 have been described elsewhere (22). LTR = long terminal repeat of Molony murine leukemia virus; 3′SS = 3′ splicing signal; Neo^r = sequence derived from the transposon Tn5; SV ori = sequence of SV40 origin; pBR ori = sequence of pBR322 origin. *Source*: Goto et al. (22).

Table 1. Purification of Recombinant Human EPO from BHK Cells

Procedures	Protein ($A_{280\ nm}$)	Activity ($U \times 10^{-4}$)	Specific activity ($U/A_{280\ nm}$)	Yield (%)	Purification (fold)
Concentrated culture supernatant	473,000	1,990	42	100	1
Immunosorbent column	141	1,680	119,000	84	2,830
Sephadex G-100	97	1,310	135,000	66	3,210
Hydroxyapatite	75	1,030	137,000	52	3,260

Source: Goto et al. (23).

Biological Activity of Isolated EPOs

The Second International Reference Preparation for human EPO obtained from the WHO International Laboratory for Biological Standards, National Institute for Medical Research, London (24), was used as the standard. The in vivo and in vitro activities of three purified EPOs were assayed (Table 2). The stimulatory effect of EPO on the incorporation of ^{3}H-thymidine into DNA of cultured mouse fetal liver cells was used to assay the in vitro activity, and the in vivo activity was assayed with the use of starved rats (4). The stimulatory effect of the EPO preparations were completely repressed by incubation with antiserum against EPO, indicating that the mitogenic effect of a con-

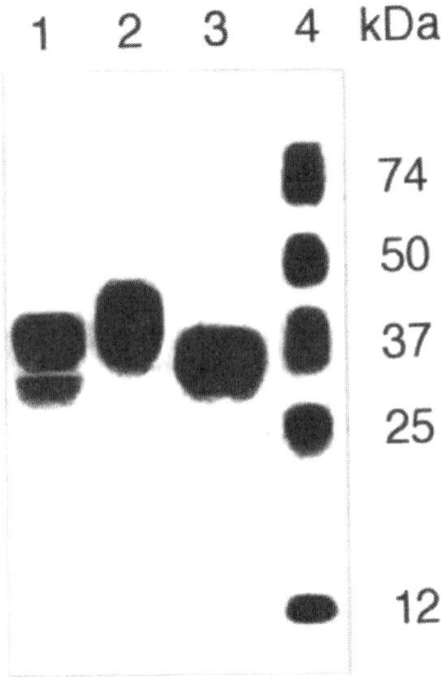

Figure 4. Analysis of human urinary and recombinant EPOs with SDS-polyacrylamide gel electrophoresis. Lanes 1, 2, and 3 are EPOs from urine, the culture supernatant from BHK, and ψ2 cells, respectively. Lane 4 shows the molecular standards. *Source*: Goto et al. (22).

Table 2. Biological Activities of Human EPO

EPO	Treatment	EPO activity (U/A_{280}) In vitro	EPO activity (U/A_{280}) In vivo
u-EPO	Untreated	77,000	72,000
	Sialidase treated	99,000	0
r-EPO-B	Untreated	96,000	72,000
	Sialidase treated	109,000	0
r-EPO-ψ	Untreated	153,000	18,000
	Sialidase treated	173,000	0

Source: Goto et al. (22).

taminant did not contribute to the in vitro activity. The in vivo activity of rEPO-B was equivalent to that of uEPO, and its in vitro activity was slightly higher than that of uEPO. The in vitro activity of rEPO-ψ was 1.6-fold that of uEPO and of rEPO-B, whereas, the in vivo activity was only 25% that of other EPOs. The in vitro activities of the three EPO preparations slightly increased after sialidase treatment, but their in vivo activities were abolished. It seems to be that the lower the amount of sialic acid, the higher the in vitro activity. The high activity of the native rEPO-ψ as compared with other native EPOs may be due to the low content of sialic acid, because the asialo EPOs have an increased affinity to EPO receptor (25,26). However, the low in vivo activity of rEPO-ψ is probably due to the small amount of sialic acid, because the removal of sialic acid from EPO by sialidase treatment abolished in vivo activity. Loss of in vivo activity is caused by the removal of asialoglycoproteins from the circulation by the liver (20). Thus, it is likely that the in vivo lifetime of rEPO-B is longer than that of rEPO-ψ.

ISOLATION OF EPO-DIRECTED MONOCLONAL ANTIBODY AND APPLICATION TO ELISA

We reported an EPO-directed monoclonal antibody (MAb) that binds to the antigen after it is treated with SDS (21). It appears that this antibody binds to the epitope exposed by a conformation change of EPO caused by the SDS. This antibody worked efficiently in the purification of EPO from human urine and human recombinant EPO from the supernatant of cultured cells, but treatment of preparations containing EPO with SDS is time consuming. We have prepared MAbs capable of binding with the native EPO and thereby to achieve the following purposes: (*1*) establishment of a more rapid procedure for the purification of EPO by immobilized MAb; (*2*) development of a sandwich-type ELISA, which is easier to handle and more rapid and sensitive than radioimmunoassay (RIA), by use of MAbs that bind to EPO at different antigenic determinants.

Isolation of EPO-Directed MAb-Producing Hybridomas

Female BALB-c mice 8 weeks old were immunized intraperitoneally with 90 μg of pure rEPO emulsified 1:1 in Freund's complete adjuvant. After three times of immunization, hybridomas were prepared by procedures described previously (21). Hybridoma cells were maintained in RPMI 1640 supplemented with 10% fetal calf serum. Two hybridoma clones (R2 and R6) that were stable in their rapid growth and their production of MAbs to EPO were isolated. Both antibodies were produced in the ascitic fluid of BALB/c mice and purified by the ammonium sulfate precipitation and diethylaminoethyl (DEAE)-cellulose column chromatography. The subclass of both antibodies was IgG1. The apparent dissociation constant, k_d was calculated by the binding to ^{125}I-labeled EPO. Both MAbs bound to EPO with high affinities, with k_d = 1–2 nM, respectively. The fully deglycosylated rEPO bound to the MAbs, indicating that they recognized peptide sequences of the antigen but not the carbohydrates attached to the antigen. Moreover, each antibody recognizes a different epitope.

Enzyme-Linked Immunosorbent Assay

The finding that the antigen site recognized by MAb R2 differed from the sites recognized by R6 prompted us to develop a sandwich-type ELISA for the measurement of EPO levels. EPO was sandwiched between R6 fixed in microtiter wells and the enzyme-linked R2. R2 was linked covalently with alkarine phosphatase with 0.2% glutaraldehyde. The MAb R6 solution (10 μg/mL) was added to the wells of microtiter plates with 96 flat-bottomed wells and incubated for 2 h. After washing, 100 μL/well of the sample solution containing 0.05% Tween 20 was added to the wells, which were incubated for 2 h to form the fixed R6 and EPO complexes. After washing, 100 μL/well of the alkaline phosphatase-linked R2 solution was added and the wells were incubated for 2 h more to form ternary complexes (fixed R6, EPO, and enzyme-linked R2). The wells were washed, and then 100 μL of 2 mg/mL *p*-nitrophenyl phosphate disodium was added. After incubation for 30 min, the reaction was stopped by the addition of 3 M

NaOH. The optical density at 405 nm was measured with a microtiter reader. This procedure is referred to as a two-step reaction. In the two-step reaction, EPO in samples was bound to the fixed R6 and then alkaline phosphatase-linked R2 was added to the wells to form ternary complexes of R6, EPO, and the enzyme-linked R2. In the alternative procedure, the one-step reaction mixtures of EPO samples and enzyme-linked MAb R2 were added to the wells coated with MAb R6.

The relationship of the amounts of rEPO and the enzyme activity was linear regardless of different procedures (Fig. 5); use of the one-step reaction shortened the time needed for completion of the process to 7 h. Once the microtiter wells in which the first antibody, R6, have been fixed, results can be obtained within 3 h. To check the reproducibility of the ELISA, intra-assay and interassay studies were done. In both studies, the coefficients of variation were satisfactory (<20%). The assay measured plasma levels of EPO as low as 2 mU/mL within several hours.

We have developed RIA for EPO by the use of antisera against human u-EPO and measured EPO levels in plasma samples from normal subjects and anemic or polycythemic patients (27). A sample diluted to about 1 U/mL EPO was mixed with 10,000 cpm of ^{125}I-rEPO, rabbit anti-rEPO antiserum. The mixture was incubated at 4 °C overnight and sheep antirabbit γ-globulin (Cappel) was added. The immunocomplexes were precipitated by centrifugation. The radioactivity in the precipitate was counted with a γ-counter. The amounts of EPO in the sample were estimated from a standard curve (2 to 128 mU/assay).

From total 38 samples, EPO levels of normal subjects and patients were measured with ELISA to compare with RIA. The results are shown in Figure 6 and indicate a high

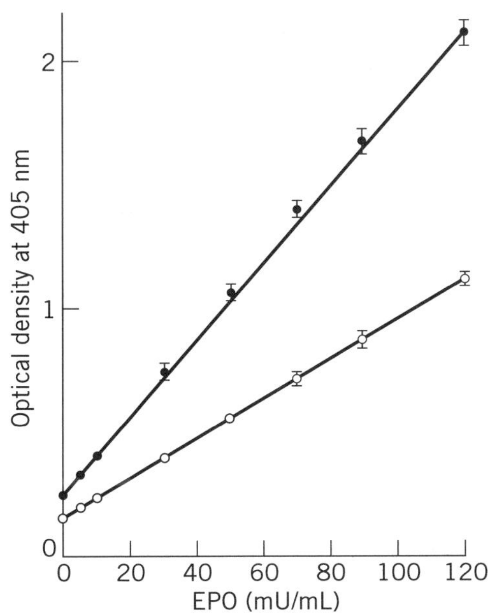

Figure 5. EPO titration by ELISA. ○ indicates one-step reaction; ● indicates two-step reaction. Each point is the mean of triplicate samples. *Source*: Goto et al. (23).

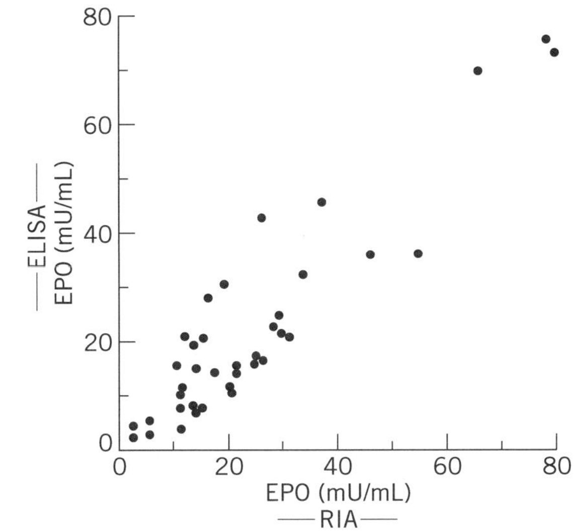

Figure 6. Comparison of RIA and ELISA in human plasma. The ordinate shows EPO in the plasma samples assayed with ELISA by two-step reaction, and the abscissa indicates EPO estimated with RIA by using polyclonal antibody raised in a rabbit. Each point is the mean of triplicate samples. *Source*: Goto et al. (23).

degree of correlation between these two methods, with a correlation coefficient of 0.965. The ELISA was comparable with the RIAs in sensitivity. One can assay EPO for a short time without using any radioactive compounds. Moreover, the assay conditions, once established, can be used semi-permanently, because MAbs with the constant properties can be supplied continually. The sandwich-type ELISA developed by the use of MAbs thus had advantages over the RIAs in practicality. The findings that the MAbs R6 and R2 bound to EPO with high affinities and recognized different antigenic determinants made it possible to develop a sensitive and easier-to-handle method for EPO assay. The sandwich-type ELISA should be useful to assay EPO levels in human plasma for investigating an etiology of defective erythropoiesis and also to monitor rEPO given to anemic patients.

FUNCTION OF CARBOHYDRATES ATTACHED TO EPO IN ERYTHROPOIESIS

The following biological functions can be assigned to carbohydrates attached to proteins such as EPO that act in the endocrine system. First, the carbohydrates of glycoproteins are needed for their biosynthesis and secretion, including folding of the protein during translation and protection of the protein from intracellular degradation. Second, carbohydrates confer stability to the protein so that it is not eliminated from the circulation during its travel from the production site to its target cells. Mammalian cells, particularly liver cells, express proteins (lectins) that bind to carbohydrates in glycoproteins. Glycosylation of proteins thus can be a signal determining their lifetime in the circulation. Third, carbohydrates are involved in the expression of biological activity, including the steps of the binding of the protein to its target cells and intracellular signal transduction.

To examine these possible functions of carbohydrates, the following strategies can be used. First, there are some mammalian cell lines lacking an enzyme involved in glycosylation. One can use these cells to produce a recombinant glycoprotein and compare with the natural product or the fully glycosylated recombinant product. Second, one may produce the protein in procaryotes incapable of attaching carbohydrates, and if the product has the biological activity, one can get some information of functions of carbohydrates. Third, one can examine the effects of chemical or enzymatic removal of carbohydrate residues from a glycoprotein on its biological properties. Discovery of enzymes that remove carbohydrate residues from a glycoprotein is particularly useful, because the enzymatic treatment is milder by far than the chemical treatment. Fourth, glycosylation sites can be removed by site-directed mutagenesis and the biological properties of the glycosylation-defective mutant glycoproteins are compared with those of wild type.

Enzymatic Removal of Carbohydrates from EPO

Figure 7 shows representative structures of N-linked and O-linked oligosaccharides of recombinant human EPO produced by BHK cells (22,25) and cleavage sites of the enzymes that we used for removal of carbohydrate residues. The native EPO is represented as EPO-0, and deglycosylated forms are as EPO-1~5 in Table 3. Sialidase removes terminal sialic acids, producing asialo-EPO (EPO-1). *β*-Galactosidase cleaves off the exposed Gal residues of N-linked oligosaccharides in EPO-1 but does not split the b1→3 linkage between Gal and GalNAc in the O-linked oligosaccharide, giving EPO-2. Double digestion of EPO-1 with *β*-galactosidase and *β*-*N*-acetylhexosaminidase removes all the exposed Gal and GlcNAc residues from N-linked oligosaccharides, and the core structure is exposed (EPO-3). N-Glycanase completely eliminates N-linked oligosaccharides from EPO-0, yielding EPO-4. *Endo*-*α*-*N*-acetylgalactosaminidase removes O-linked oligosaccharides from asialo-EPO. Therefore, successive digestion of EPO with sialidase, *endo*-*α*-*N*-acetylgalactosaminidase, and *N*-glycanase yields the fully deglycosylated EPO,

Table 3. Biological Activities of Recombinant EPO Deglycosylated by Glycosidases

EPO	In vitro activity (%)	Affinity to EPO-R [I_{50} (nM)]	In vivo activity (%)
EPO-	100	1.2	100
EPO-1	252	0.3	0
EPO-2	252	0.3	0
EPO-3	266	0.3	0
EPO-4	81	0.6	0
EPO-5	106	0.3	0

Source: Tsuda et al. (25).
Note: EPO-0-native EPO; EPO-1-sialidase-digested EPO-0; EPO-2-sialidase/*β*-galactosidase-digested; EPO-3-sialidase/*β*-galactosidase/*β*-N-acetylhexosaminidase-digested; EPO-4-*N*-glycanase-digested; EPO-5-fully deglycosylated (sialidase/*endo*-*α*-*N*-acetylgalactosaminidase/*N*-glycanase-digested) (see also Fig. 7). Affinity to EPO receptor (EPO-R) does not represent a dissociation constant but a relative affinity. The lower the value, the higher the affinity.

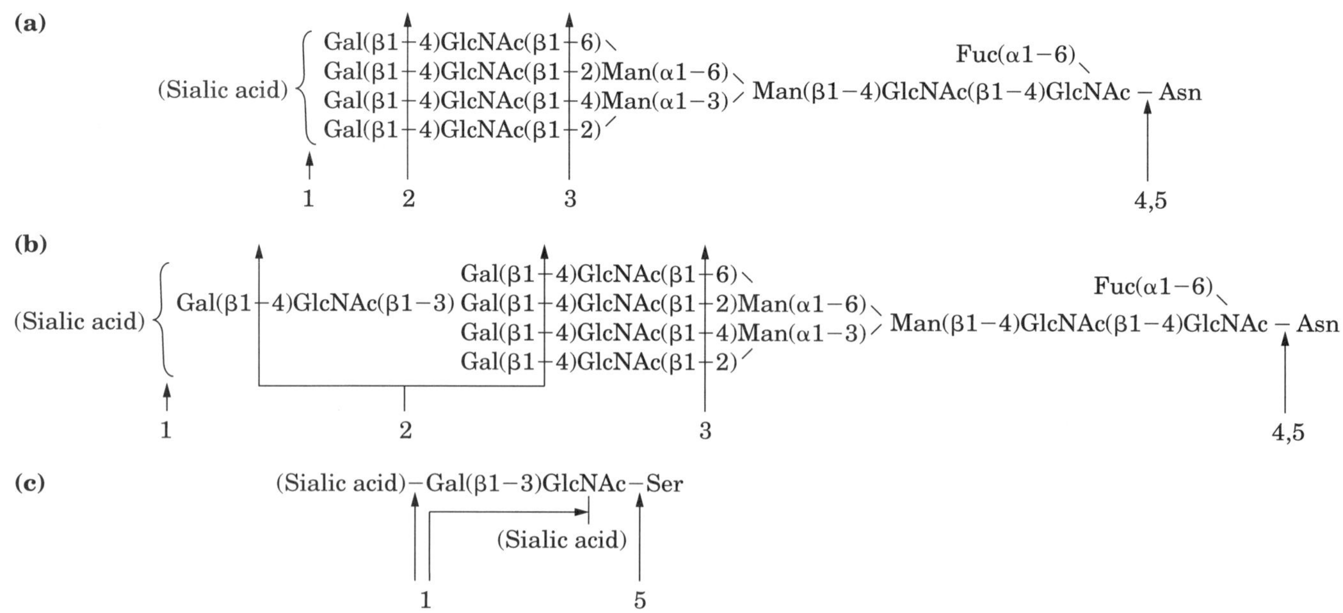

Figure 7. Enzymatic removal of sugars from recombinant human EPO. Cleavage site of sugars in EPO by glycosidases are shown. (**a** and **b**) Structures of representative complex-type N-linked sugars (**c**) O-linked sugar. The lines with arrowheads numbered 1–5 indicate the cleavage sites by glycosidases: 1-sialidase; 2-*β*-galactosidase after sialidase digestion; 3-*β*-galactosidase/*β*-*N*-acetylhexosaminidase after sialidase-digestion; 4-*N*-glycanase; 5-sialidase/*endo*-*α*-*N*-acetylgalactosaminidase/*N*-glycanase. *Source*: Tsuda et al. (25).

EPO-5. Completion of the enzyme reactions were confirmed by SDS-polyacrylamide gel electrophoresis of the products.

Effects of these enzymatic removals on the biological properties are summarized in Table 3. Affinities to EPO receptors do not represent the true dissociation constants but the relative values; the lower value, the higher affinity. Carbohydrates with terminal sialic acids are essential for expression of the in vivo activity. The carbohydrates are not important for expression of the in vitro activity. Deglycosylation increased the in vitro activity by increasing the affinity to EPO receptor. The presence of terminal sialic acids is responsible for an decreased affinity. EPO-4 and 5 have in vitro activity similar to that of the native EPO (EPO-0), although their affinity to EPO receptor is higher than EPO-0. They may be partially inactivated during assay thorough aggregate formation. The removal of terminal sialic acids from EPO has been shown to destroy its in vivo activity (21,22,28–31). Desialylated EPO is cleared from the circulation faster by far than the native EPO (32,33). In contrast, the in vitro activity of EPO increases with desialylation (21,29). All these results are consistent with those in Table 3.

A possible function of carbohydrates in the in vitro activity of EPO was controversial. It has been reported that *N*-glycanase digestion of EPO produced in Chinese hamster ovary (CHO) cells caused almost complete loss of activity (34), which was contradictory to our conclusion that carbohydrates were not important in expression of the in vitro activity. However, all the results published later support our conclusion (26,35–37). We have to be very careful to assay the activity of EPO that has lost N-linked sugars, because the deglycosylated EPO is temperature-sensitive and sticks to experimental vessels (25).

Site-Directed Mutagenesis for Production of Glycosylation-Defective Mutant EPO

We constructed EPO genes mutated by site-directed mutagenesis so that EPOs lacking N-linked carbohydrates in every possible combination of the three N-glycosylation sites were produced (26). Because the codon of Asn (N) at the N-glycosylation sites is replaced with that of Gln (Q) in these mutants, we referred to them as NQ0, NQ1, and so on (Table 4). In this nomenclature, for example, NQ13 means the mutant at positions 24 and 83. NQ0 corresponds to the wild type and NQ123 is the mutant with no N-glycosylation sites. Transient production of mutant EPOs by BHK cells was examined to find roles of sugars in biosynthesis of EPO. The mutant EPOs produced by stably transfected BHK cells were partially purified and their properties, including the biological activity both in vivo and in vitro, were investigated. In some mutants, the in vitro activity was higher than that of wild-type, and the high activity is probably due to an increased affinity to EPO receptor. The in vitro activity of EPO with no glycosylation sites was comparable with that of the wild-type EPO, whereas the in vivo activity severely decreased. The elimination of all three N-glycosylation sites decreased EPO production to 10% of that of the wild-type EPO. These results indicate that N-linked sugars of EPO have two major functions: N-glycosylation is important for (*1*) proper biosynthesis and/or secretion and (*2*) expression of the in vivo activity. N-Linked sugars of EPO affect binding affinity of the ligand to the receptor but do not play a key role in expression of the in vitro activity. It has been reported that replacement of Asn to Gln at position 24 or 38 gives EPOs with the specific activity in vitro of about 20% that of the wild type (38), which is contradictory to our results (26). A possible cause has been discussed (26), and the results in a paper published later (36) are consistent with our data.

Preventing O-linked glycosylation at position 126 by site-directed mutagenesis (38) severely reduced the production into the culture medium, suggesting that O-linked sugar is important for proper biosynthesis and/or secretion. But the result from other laboratories showed the normal production of the mutant (36). The latter was supported by the result obtained using the mutant cells that are deficient in an enzyme involved in the carbohydrate synthesis (39). Involvement of O-linked sugar in expression of the biological activity is unlikely (25,36,38,39). O-linked sugar is present only in primate EPOs but not in rodents (6,7,14).

Others

Interestingly, fractionation of recombinant EPO containing biantennary and tetraantennary N-linked sugars revealed that EPO containing more extensively branched chains has greater in vivo activity (40). This suggests a possible unique role of the N-linked sugar in vivo. Attempt

Table 4. Biological Properties of N-Glycosylation Site-Defective EPO

Mutant EPO	Mutation points (Asn → Gln)	In vitro activity (%)	Affinity to EPO-R [I_{50}(nM)]	In vivo activity (%)	Secreted EPO (%)
NQ0	Wild type	100	0.50	100	100
NQ1	24	180	0.25	129	70
NQ2	38	107	0.28	60	6
NQ3	83	130	0.40	106	78
NQ12	24,38	140	0.15	42	31
NQ13	24,83	170	0.18	63	58
NQ23	38,83	66	0.31	51	41
NQ123	24,38,83	100	0.14	8	10

Source: Yamaguchi et al. (26).
Note: Affinity to EPO receptor (EPO-R) represents a relative value but not a dissociation constant. The lower the value, the higher the affinity.

to produce EPO in *Escherichia coli* has been made (36). They were unable to compare the biological activity of EPO produced in *E. coli* with that of EPO produced in mammalian cells because of the poor stability of *E. coli* EPO, but they confirmed that *E. coli* EPO had the in vitro activity. *Bacillus brevis* cells secrete recombinant human EPO, which has in vitro activity similar to that of the fully glycosylated EPO (42). These results support our conclusion that the N-linked sugars in EPO are not required for expression of the in vitro activity. When human wild-type EPO was expressed in the polarized Madin-Darby canine kidney (MDCK) epithelial cells, EPO was preferentially secreted from the apical domain (43). This polarized secretion was perturbed by the treatment of the cells with tunicamycin, suggesting the involvement of N-linked sugars in the apical sorting mechanism in MDCK cells. N-Linked sugar chain at position 38 is critical for the polarized secretion, but the biological significance of this finding remains unknown.

BIBLIOGRAPHY

1. T. Miyake, C.K.-H. Kung, and E. Goldwasser, *J. Biol. Chem.* **252**, 5558–5564 (1977).
2. K. Jacobs, C. Shoemaker, R. Rudersdorf, S.D. Neill, R.J. Kaufman, A. Mufson, J. Seehra, S.S. Jones, R. Hewick, E.F. Fritsch, M. Kawakita, T. Shimizu, and T. Miyake, *Nature* **313**, 806–810 (1985).
3. F.-K. Lin, S. Suggs, C.-H. Lin, J.K. Browne, R. Smalling, J.C. Egrie, K.K. Chen, G.M. Fox, F. Martin, Z. Stabinsky, S.M. Badrawi, P.-H. Lai, and E. Goldwasser, *Proc. Natl. Acad. Sci. USA* **82**, 7580–7584 (1985).
4. J.F. Garcia, in J.W. Fischer ed., *Kidney Hormones*, vol. 2, Academic Press, London, 1977, p. 7.
5. A.D. D'Andrea, H.F. Lodish, G.G. Wong, *Cell* **57**, 277–285 (1989).
6. W. Jelkmamnn, *Phys. Rev.* **72**, 449–489 (1992).
7. D. Roberts and D.J. Smith, *J. Mol. Endocrinol.* **12**, 131–148 (1993).
8. S. Lok, K. Kaushansky, R.D. Holly, J.L. Kuijper, C.E. Lofton-Day, P.J. Oort, F.J. Grant, M.D. Heipel, S.K. Burkhead, J.M. Kramer, L.A. Bell, C.A. Sprecher, H. Blumberg, R. Johnson, D. Prunkard, A.E.T. Ching, S.L. Mathewes, M.C. Bailey, J.W. Forstrom, M.M. Buddle, S.G. Osborn, S.J. Evans, P.O. Sheppard, S.R. Presnell, P.J. O'Hara, F.S. Hagen, G.J. Roth, and D.C. Foster, *Nature* **369**, 565–568 (1994).
9. J.F. Bazan, *Biochem. Biophys. Res. Commun.* **147**, 392–398 (1989).
10. M.V. Abraham, M. Ultsch, and A.A. Kossiakoff, *Science* **255**, 306–312 (1992).
11. S. Elliott, T. Lorenzini, D. Chang, J. Barzilay, and E. Delorme, *Blood* **89**, 493–502 (1997).
12. J.W. Fisher, *Proc. Soc. Exp. Biol. Med.* **216**, 358–369 (1997).
13. H.F. Bunn and R.O. Poyton, *Physiol. Rev.* **76**, 839–885 (1996).
14. S.B. Krantz, *Blood* **77**, 419–434 (1991).
15. P.J. Ratcliffe, *Kid. Int.* **44**, 887–904 (1993).
16. H. Wu, X. Liu, R. Jaenisch, and H.F. Lodish, *Cell* **83**, 59–67 (1995).
17. S. Masuda, M. Okano, K. Yamagishi, M. Nagao, M. Ueda, and R. Sasaki, *J. Biol. Chem.* **269**, 19488–19493 (1994).
18. M. Digicaylioglu, S. Bichet, H.H. Marti, R.H. Wenger, L.A. Rivas, C. Bauer, and M. Gassmann, *Proc. Natl. Acad. Sci. USA.* **92**, 3717–3720 (1995).
19. E. Morishita, M. Masuda, M. Nagao, Y. Yasuda, and R. Sasaki, *Neuroscience* **76**, 105–116 (1997).
20. S. Masuda, M. Nagao, K. Takahata, Y. Konishi, F. Gallyas Jr., T. Tabira, and R. Sasaki, *J. Biol. Chem.* **268**, 11208–11216 (1993).
21. S. Yanagawa, K. Hirade, H. Ohnota, R. Sasaki, H. Chiba, M. Ueda, and M. Goto, *J. Biol. Chem.* **259**, 2707–2710 (1984).
22. M. Goto, K. Akai, A. Murakami, C. Hashimoto, E. Tsuda, M. Ueda, G. Kawanishi, N. Takahashi, A. Ishimoto, H. Chiba, and R. Sasaki, *Bio / Technology* **6**, 67–71 (1988).
23. M. Goto, A. Murakami, K. Akai, G. Kawanishi, M. Ueda, H. Chiba, and R. Sasaki, *Blood* **74**, 1415–1423 (1989).
24. L.M. Annable, P.M. Cotes, and M.V. Mussett, *Bull. World Health Org.* **47**, 99–112 (1972).
25. E. Tsuda, G. Kawanishi, M. Ueda, S. Masuda, and R. Sasaki, *Eur. J. Biochem.* **188**, 405–411 (1990).
26. K. Yamaguchi, K. Akai, G. Kawanishi, M. Ueda, S. Masuda, and R. Sasaki, *J. Biol. Chem.* **266**, 20434–20439 (1991).
27. H. Mizoguchi, K. Ohta, T. Suzuki, A. Murakami, M. Ueda, R. Sasaki, and H. Chiba, *Acta Hematol. Jpn.* **50**, 15–24 (1987).
28. W.A. Lukowsky and R.H. Painter, *Can. J. Biochem.* **50**, 909–917 (1972).
29. E. Goldwasser, K.H. Kung, and J. Eliasson, *J. Biol. Chem.* **249**, 4202–4206 (1974).
30. M.S. Dordal, F.F. Wang, and E. Goldwasser, *Endocrinology* **116**, 2293–2299 (1985).
31. R. Sasaki, S. Yanagawa, and H. Chiba, *Methods Enzymol.* **147**, 328–340 (1987).
32. M.N. Fukuda, H. Sasaki, L. Lopez, and M. Fukuda, *Blood* **73**, 84–89 (1989).
33. J.L. Spivak and B.B. Hogans, *Blood* **73**, 90–99 (1989).
34. M. Takeuchi, S. Takasaki, M. Shimada, and A. Kobata, *J. Biol. Chem.* **265**, 12127–12130 (1990).
35. A.J. Sytkowski, L. Feldman, and D.J. Zurbuch, *Biochem. Biophys. Res. Commun.* **176**, 698–704 (1991).
36. E. Delmore, T. Lorenzini, J. Giffin, F. Martin, F. Jacobsen, T. Boone, and S. Elliot, *Biochemistry* **31**, 9871–9876 (1992).
37. M. Higuchi, M. Oh-eda, H. Kuboniwa, K. Tomonoh, Y. Shimonaka, and N. Ochi, *J. Biol. Chem.* **267**, 7703–7709 (1992).
38. S. Dube, J.W. Fisher, and J.S. Powell, *J. Biol. Chem.* **263**, 17516–17521 (1988).
39. L.C. Wasley, G. Timony, P. Murtha, J. Stoudemire, A.J. Dorner, J. Caro, M. Krieger, and R.J. Kaufman, *Blood* **77**, 2624–2632 (1991).
40. M. Takeuchi, N. Inoue, T.W. Strickland, M. Kubota, M. Wada, R. Shimizu, S. Hoshi, H. Kozutumi, S. Takasaki, and A. Kobata, *Proc. Natl. Acad. Sci. USA* **86**, 7819–7822 (1989).
42. M. Nagao, K. Inoue, S.K. Moon, S. Masuda, H. Hakagi, S. Udaka, and R. Sasaki, *Biosci. Biotech. Biochem.* **61**, 670–674 (1997).
43. Y. Kitagawa, Y. Sano, M. Ueda, K. Higashio, H. Narita, M. Okano, S. Mastumoto, and R. Sasaki, *Exp. Cell Res.* **213**, 449–457 (1994).

EXPRESSION SYSTEMS. See TRANSIENT EXPRESSION SYSTEMS.

See TRANSIENT EXPRESSION SYSTEMS.

EXPRESSION SYSTEMS, *E. COLI*

SAVVAS C. MAKRIDES
PRAECIS Pharmaceuticals, Inc.
Cambridge, Massachusetts

KEY WORDS

Chaperone
E. coli
Fusion protein
Gene expression
Vector

OUTLINE

INTRODUCTION

Escherichia coli is a widely used organism for the production of recombinant proteins for both basic research and therapeutic applications. In spite of the extensive knowledge of the genetics and molecular biology of *E. coli*, the robust expression of genes in this prokaryote is not necessarily a routine matter. This may be due to the unique structural features of the gene sequence, the stability and translational efficiency of mRNA, the ease of protein folding, degradation of the protein by host cell proteases, major differences in codon usage between the foreign gene and native *E. coli*, and the potential toxicity of the protein to the host. Progress in our understanding of many of these processes is certain to broaden the usefulness of *E. coli* as a tool for gene expression. The major drawbacks of *E. coli* as an expression system are the lack of a secretion mechanism for the efficient release of protein into the culture medium, and the limited ability to facilitate extensive disulfide bond formation. In addition, although bacteria are capable of carrying out a variety of posttranslational modifications, including some types of glycosylation (1), these are not as complex as those performed by eukaryotic cells. On the other hand, many eukaryotic proteins retain their full biological activity in a nonglycosylated form and, therefore, can be produced in *E. coli*.

The objectives here are to focus on expression systems and methodological approaches useful for the high-level production of proteins in *E. coli*. Nutrient composition and fermentation variables are not discussed here, as these have been reviewed in detail (2). Due to space limitations many original references are not cited. The reader is referred to recent reviews for detailed discussion of specific topics (3–7).

EXPRESSION VECTORS

The construction of an expression plasmid requires several elements whose configuration must be carefully considered in order to ensure the highest levels of protein synthesis (7). The promoter is positioned approximately 10–100 bp upstream of the ribosome binding site (RBS) and is under the control of a regulatory gene that may be present on the vector itself or integrated in the host chromosome. Downstream of the promoter is the Shine and Dalgarno (SD) site, which interacts with the 3′ end of 16S rRNA during translation initiation. The distance between the SD region and the start codon ranges from 5 to 13 bases, and its sequence should eliminate the potential for secondary structure formation in the mRNA transcript, which can reduce the efficiency of translation initiation. Both 5′ and 3′ regions of the RBS exhibit a bias toward a high adenine content. The transcription terminator is located downstream of the coding sequence and serves both as a signal to terminate transcription as well as a protective element comprising stem–loop structures thereby protecting the mRNA from exonucleolytic degradation. In addition, vectors contain a gene that confers antibiotic resistance on the host to aid in plasmid selection and propagation. Finally, the copy number of plasmids is determined by the origin of replication. Although the use of runaway replicons has been reported to result in higher yields of plasmid-encoded protein, in general, higher copy number vectors are not advantageous (7).

Transcription

Promoters. There are many promoters available for gene expression in *E. coli*, including those derived from Gram-positive bacteria and bacteriophages (Table 1). A promoter for use in *E. coli* should have certain characteristics to render it suitable for high level protein synthesis. First, it must be strong, resulting in the accumulation of protein comprising 10–30% or more of the total cellular protein. Second, it should exhibit a minimum level of basal transcriptional activity. Large-scale gene expression preferably employs cell growth to high density and minimal promoter activity, followed by induction or derepression of the promoter. The tight regulation of a promoter is essential for the synthesis of proteins that may be detrimental to the host cell. Furthermore, incompletely repressed expression systems can cause plasmid instability, a decrease

Table 1. Promoters Used for the High-Level Expression of Genes in *E. coli*

Promoter (source)	Regulation	Induction
lac (E. coli)	*lacI, lacI*^{Q}	IPTG[a]
	*lacI*ts[b], *lacI*^{Q}ts[b]	Thermal
	*lacI*ts[c]	Thermal
lpp (E. coli)		IPTG, lactose[d]
trp (E. coli)		Tryptophan depletion, indole acrylic acid
tnaA (E. coli)[e]		Tryptophan, glucose depletion
Mn-SOD *(E. coli)*[f]		Oxygen, paraquat
phoA (E. coli)	*phoB* (positive)	Phosphate depletion
	phoR (negative)	
recA (E. coli)	*lexA*	Nalidixic acid
araBAD (E. coli)	AraC	L-Arabinose
proU (E. coli)		Osmolarity
cst-1 (E. coli)		Glucose depletion
tetA (E. coli)		Tetracycline
cadA (E. coli)	*cadR*	pH
nar (E. coli)	*fnr*	Anaerobic conditions, nitrate ion
nirB (E. coli)[g]		Anaerobic conditions
tac (hybrid) *(E. coli)*	*lacI, lacI*^{Q}	IPTG
	lacI[h]	Thermal
trc (hybrid) *(E. coli)*	*lacI, lacI*^{Q}	IPTG
	*lacI*ts, *lacI*^{Q}ts	Thermal
lpp-lac (hybrid) *(E. coli)*	*lacI*	IPTG
$P_{syn(synthetic)}$ *(E. coli)*	*lacI, lacI*^{Q}	IPTG
$P_{LtetO\text{-}1}$ *(E. coli)*[i]		Anhydrotetracycline
$P_{lac/ara\text{-}1}$ *(E. coli)*[i]		IPTG, arabinose
Starvation promoters *(E. coli)*		
P_L (bacteriophage λ)	*lacI*ts857	Thermal
P_L-9G-50 (mutant) (λ)		Reduced temperature (<20 °C)
cspA (E. coli)		Reduced temperature (<20 °C)
P_R, P_L (tandem) (λ)	*lacI*ts857	Thermal
T7 (T7)	cascaded system[j]	IPTG
T7-*lac* operator (T7)	*lacI*^{Q}	IPTG
λP_L, P_{T7} (tandem) (λ, T7)	*lacI*ts857, *lacI*^{Q}	Thermal, IPTG
T3-*lac* operator (T3)	*lacI*^{Q}	IPTG
T5-*lac* operator (T5)	*lacI*^{Q}, *lacI*	IPTG
T4 gene 32 (T4)		T4 infection
nprM-lac operator (Bacillus)	*lacI*^{Q}	IPTG
vhb (Vitreoscilla)	*fnr*	Oxygen, cAMP-CAP[k]
protein A (*Staphylococcus aureus*)		
Pu (Pseudomonas putida)[l]	*xylR*	*m*-Toluic acid
Pm (P. putida)[l]	*xylS*	3-Methylbenzylacohol

Source: Updated from Ref. 7 with permission from the publisher.
Note: Except where indicated, original references may be found in Ref. 7.
[a]Isopropyl-β-D-thiogalactopyranoside.
[b]*lacI* gene with single mutation, Gly187Ser.
[c]*lacI* gene with three mutations, Ala241Thr, Gly265Asp, Ser300Asn.
[d]The constitutive *lpp* promoter (P_{lpp}) was converted into an inducible promoter by insertion of the *lac*UV5 promoter/operator region downstream of the P_{lpp}. Expression occurs only in the presence of a *lac* inducer.
[e]Ref. 8.
[f]Ref. 9.
[g]Ref. 10.
[h]Wild-type *lacI* gene.
[i]Ref. 11.
[j]Expression of T7 RNA polymerase is controlled by the *lacUV5* promoter, which is regulated by the *lac* repressor. Production of T7 RNA polymerase causes transcription of the recombinant gene, which is under the control of the φ10 promoter.
[k]Cyclic AMP-catabolite activator protein.
[l]Ref. 12.

in cell growth rate, and loss of recombinant protein production (13). A third important characteristic of a promoter is its inducibility in a simple and cost-effective manner. The most widely used promoters for large-scale protein production use thermal induction (λP_L) or chemical inducers (tac, trp) (Table 1). IPTG-inducible promoters are widely used in basic research. However, the use of IPTG for the large-scale production of human therapeutic proteins is undesirable due to its toxicity and cost. Moreover, IPTG is detrimental to the physiology of *E. coli* (14). The availability of a mutant *lacI*ts gene that encodes a thermosensitive *lac* repressor now permits the thermal induction of these promoters (15,16).

Cold-responsive promoters have been shown to facilitate efficient gene expression at reduced temperatures. The rationale behind the use of cold-responsive promoters for gene expression is based on the proposition that the rate of protein folding will be only slightly affected at about 15–20 °C, whereas the rates of transcription and translation, being biochemical reactions, will be substantially decreased. This, in turn, will provide sufficient time for protein folding, yielding active proteins and avoiding the formation of inactive protein aggregates (inclusion bodies) without reducing the final yield of the target protein (17).

Transcription Terminators. Transcription through a promoter may inhibit its function, a phenomenon known as promoter occlusion (18). This interference can be prevented by the proper placement of a transcription terminator downstream of the coding sequence in order to prevent continued transcription through another promoter. Efficient transcription terminators serve several important functions: they minimize background transcription, stabilize plasmids, enhance mRNA stability, and can increase substantially the level of protein production (7).

The advantages of tightly regulated promoters have led to the design of completely repressible expression systems (19) that are particularly useful for the expression of genes whose products are detrimental to host growth. Most of these systems have not yet been used for the high-level production of proteins on a large scale.

Translation

Translational Initiation. The wide range of efficiencies in the translation of different mRNAs is predominantly due to the unique structural features at the 5′ end of each mRNA species. To date no universal sequence for the efficient initiation of translation has been devised. The translational initiation region of most sequenced *E. coli* genes contains the initiation codon AUG (91%). GUG is used by about 8% of the genes, and UUG (1%) is rarely used as a start site (20). The spacing between the SD site and the initiating AUG codon can vary from 5 to 13 nucleotides, and it influences the efficiency of translational initiation. The SD sequence UAAGGAGG is optimal for protein production (21). The secondary structure at the translation initiation region of mRNA plays a crucial role in the efficiency of gene expression. It is believed that the occlusion of the SD region and/or the AUG codon by a stem–loop structure prevents accessibility to the 30S ribosomal subunits and inhibits translation. Several different strategies have been devised to minimize mRNA secondary structure and enhance the expression of genes. These include the enrichment of the RBS with adenine and thymidine residues, the mutation of specific nucleotides upstream or downstream of the SD region, and the use of translational coupling or dicistronic systems (7). The insertion of a single amino acid residue (preferably arginine or lysine) between the N-terminal methionine and the protein of interest was reported to increase expression levels from 0.4% to 22% of the total cell protein (22). This strategy was effective for several types of proteins and different promoters. The recombinant proteins were converted to their native sequences by removal of dipeptides with cathepsin C. The precise mechanism for the observed increase in protein yields is unclear.

Translational Termination. The design of expression vectors frequently includes the insertion of all three stop codons in order to prevent possible ribosome "skipping." In *E. coli* there is a preference for the UAA stop codon (Table 2). A statistical analysis of more than 2,000 *E. coli* genes revealed local nonrandomness both in the stop codon and in the nucleotide immediately following the triplet and concluded that UAAU is the most efficient translational termination sequence in *E. coli* (23).

Codon Usage. Genes in both prokaryotes and eukaryotes show a nonrandom usage of synonymous codons (24). These observations imply that heterologous genes enriched with codons that are rarely used by *E. coli* (Table 2) may not be expressed efficiently in *E. coli*. The minor arginine $tRNA^{Arg(AGG/AGA)}$ has been shown to be a limiting factor in the bacterial expression of several mammalian genes because the codons AGA and AGG are infrequently used in *E. coli*. Many studies have demonstrated that the substitution of high-usage for low-usage codons or the coexpression of the cognate tRNA genes can ameliorate the detrimental effect of the rare codons on gene expression levels (25). The location of low-usage codons in the transcript may adversely impact translation. Thus, the production of β-galactosidase decreased when AGG codons were inserted before the 10th codon from the initiation codon of the *lacZ* gene (26). Similarly, translational inhibition of a test mRNA was much stronger when consecutive low-usage arginine codons were located near the 5′ end of the mRNA (27,28). Moreover, 5′ translational blockage was also observed with a different sequence of consecutive low-usage leucine codons (CUA), making it unlikely that struc-

Table 2. Codon Usage by *E. coli*

UUU	F	21.7	UCU	S	9.1	UAU	Y	16.2	UGU	C	5.1
UUC	F	16.7	UCC	S	8.9	UAC	Y	12.4	UGC	C	6.4
UUA	L	13.4	UCA	S	7.6	UAA	*	2.0	UGA	*	0.9
UUG	L	13.2	UCG	S	8.7	UAG	*	0.3	UGG	W	14.4
CUU	L	11.1	CCU	P	7.0	CAu	H	12.6	CGU	R	21.1
CUC	L	10.7	CCC	P	5.3	CAC	H	9.8	CGC	R	21.5
CUA	L	3.9	CCA	P	8.4	CAA	Q	14.7	CGA	R	3.6
CUG	L	51.9	CCG	P	22.8	CAG	Q	29.0	CGG	R	5.5
AUU	I	29.8	ACU	T	9.4	AAU	N	18.0	AGU	S	8.8
AUC	I	25.1	ACC	T	23.1	AAC	N	21.8	AGC	S	15.7
AUA	I	4.9	ACA	T	7.5	AAA	K	34.4	AGA	R	2.5
AUG	M	27.4	ACG	T	14.0	AAG	K	11.2	AGG	R	1.5
GUU	V	19.0	GCU	A	16.1	GAU	D	32.1	GGU	G	25.3
GUC	V	14.9	GCC	A	25.2	GAC	D	19.5	GGC	G	29.2
GUA	V	11.2	GCA	A	20.4	GAA	E	39.9	GGA	G	8.3
GUG	V	25.8	GCG	A	32.8	GAG	E	18.4	GGG	G	11.0

Source: Taken from GeneBank (*http: / /www.dna.affrc.go.jp /~nakamura / codon.html*).

Note: Codon usage is expressed as frequency per 1,000 codons, based on a total of 4,143,606 codons.

tural effects on the mRNA were responsible (28). In another case tandem AGG triplets caused a substantial inhibition of gene expression independently of their localization in mRNA, a phenomenon that was attributed to competition of the tandem AGGAGG codons with the natural SD sequence (29). In contrast, other workers reported efficient expression of genes that contained low-usage codons (7).

To date, it has not been possible to formulate precise "rules" to predict whether the content of low-usage codons in a specific gene might adversely affect the efficiency of its expression in *E. coli*. The experimental results may possibly be confounded by several variables such as positional effects, the clustering or interspersion of the rarely used codons, secondary structure of the message, and other effects (25). Nevertheless, it is clear that the codon context of specific genes *can* have adverse effects on both the quantity and quality of the expressed protein. Usually, this problem can be rectified by the alteration of the codons in question, or by the coexpression of the cognate tRNA genes.

mRNA Stability. mRNA degradation is not effected randomly by nonspecific endonucleolytic cleavage, since there is no inverse correlation between mRNA length and half-life (30). Two classes of protective elements are known to stabilize mRNAs in *E. coli*. One class consists of sequences in the 5′ untranslated regions (UTRs) of mRNAs, and the other class includes stem–loop structures from the 3′ UTRs and intercistronic regions. Some of these elements act as stabilizers when fused to heterologous mRNAs, but only under restricted conditions (7). In contrast, the 5′ UTR of the *E. coli ompA* transcript prolongs the half-life of a number of heterologous mRNAs in *E. coli* under normal conditions of rapid cell growth (31). Belasco and colleagues showed that the presence of a stem–loop at or very near the extreme 5′ terminus of the *ompA* 5′ UTR is essential for its stabilizing effect. Furthermore, the half-life of a normally labile mRNA was prolonged in *E. coli* by adding a hairpin structure at its 5′ terminus, and it was proposed that *E. coli* mRNAs beginning with a long single-stranded segment are preferentially targeted by a ribonuclease that interacts with the 5′ terminus before cleaving the mRNA at internal sites (31). It appears, therefore, that the addition of the *ompA* 5′ stabilizer to heterologous genes might enhance gene expression in *E. coli*. However, it is possible that mRNAs that contain internal RNA processing sites may not be protected by the presence of the 5′ stabilizer.

The other class of mRNA protective elements consists of 3′ UTR sequences that can form stem–loop structures, thereby blocking exonucleolytic degradation of the transcript from the 3′ terminus. Such an element was identified within the transcription terminator of the crystal protein gene of *Bacillus thuringiensis*. Its fusion to the 3′ termini of the penicillinase *(penP)* gene of *B. licheniformis* and the human interleukin 2 cDNA increased the half-life of the mRNAs and enhanced the production of the corresponding polypeptides in both *B. subtilis* and *E. coli* (32). However, as in the case of some of the 5′ stabilizers, this 3′ "retroregulator" is unlikely to act as a universal mRNA stabilizer. For example, the replacement of the 3′-terminal hairpins of labile mRNAs with those from stable messages did not enhance the expression of the labile transcripts (32–34). It has been suggested that gene expression might be enhanced by the use of host strains that are deficient in specific ribonucleases, such as RNase II or PNPase. This too is unlikely to be an effective strategy, because the absence of RNase II or PNPase, as well as the overproduction of RNase II, had no effect on the average half-life of *E. coli* bulk mRNA (35). Moreover, strains that were deficient in both RNase II and PNPase were inviable (36).

The deletion of 110 bp in the 3′ untranslated region of a cDNA encoding the human TNF-α was reported to enhance the expression level of this cytokine (37). The authors suggested that a UUUAUUA sequence at the 3′ end of the mRNA may have an inhibitory effect on the expression of TNF-α in *E. coli*. It is well known that 3′ AU-rich sequences in cytokines, growth factors, and oncoproteins have an inhibitory effect on translation and may destabilize the mRNA (reviewed in Ref. 38). However, AU-binding proteins (39) are not known to exist in bacteria (Gary Brewer, personal communication, 1999). Therefore, the observed increase in TNF-α yield (37) is probably due to some other mechanism.

CELLULAR COMPARTMENTS

Cytoplasm

Each of the four "compartments" of *E. coli*, i.e. the cytoplasm, periplasm, membranes/cell surface, and the extracellular medium, offers advantages as well as disadvantages for protein production (Table 3). The formation of inclusion bodies remains a significant barrier to gene expression in the cytosol. Inclusion bodies do offer several advantages (Table 3), particularly the production of proteins that are toxic to *E. coli*. For example, the highly insoluble bovine prochymosin has recently been used as a fusion partner for the production of antimicrobial peptides as inclusion bodies in *E. coli* (40). On the other hand, the advantages of inclusion bodies are counterbalanced by the arduous task of refolding the aggregated protein, the uncertainty of whether the refolded protein retained its biological activity, and the reduction in yield of the refolded and purified protein. To date the precise physicochemical parameters that contribute to the formation of inclusion bodies remain unclear (41). A statistical analysis of the composition of 81 proteins that do and do not form inclusion bodies in *E. coli* concluded that charge average and turn-forming residue fraction are strongly correlated with inclusion body formation (42). These observations led to the development of a model to predict the probability of inclusion body formation solely based on the amino acid composition of a protein (42). The model was used to predict accurately the insolubility of the human T-cell receptor Vβ5.3 in *E. coli* (15), as well as to enhance the solubility of a protein by adding negatively charged residues (43).

Several experimental approaches have been utilized to minimize the formation of inclusion bodies and improve protein folding. These include the growth of bacterial cultures at lower temperatures, selection of different *E. coli* strains, substitution of selected amino acid residues, co-

Table 3. Relative Merits of Different Compartments for Gene Expression in *E. coli*

Advantages	Disadvantages
Cytoplasm	
Inclusion bodies. Facile isolation of protein in high purity and concentration, target protein protected from proteases, desirable for production of proteins that, if active, are lethal to host cell.	*Inclusion bodies.* Protein insolubility, refolding to regain protein activity, refolded protein may not regain its biological activity, reduction in final protein yield, increase in cost of goods.
Higher protein yields	Reducing environment: does not facilitate disulfide bond formation.
Simpler plasmid constructs.	N-terminus authenticity.
	Proteolysis.
	Purification is more complex (more protein types).
Periplasm	
Purification is simpler (fewer protein types).	Signal peptide does not always facilitate transport.
Proteolysis is less extensive.	Inclusion bodies may form.
Improved disulfide bond formation/folding.	
N-terminus authenticity.	
Membranes / cell surface	
Surface display may facilitate drug screening, biocatalysis and other applications. Some progress made in overexpression of membrane proteins.	
Extracellular	
Least level of proteolysis.	No secretion, usually.
Purification is simpler (fewest protein types).	Protein diluted, more difficult to purify.
Improved protein folding.	
N-terminus authenticity.	

production of chaperones, use of *E. coli* DsbA or thioredoxin either as fusion partners or coproduced with the protein of interest, growth and induction of the cells under osmotic stress in the presence of sorbitol and glycyl betaine, addition of nonmetabolizable sugars to the growth medium, alteration of the pH of the culture medium, and the use of strains deficient in thioredoxin reductase (7,43).

The reducing potential of the cytoplasmic redox state (44) presents still another problem. Bacterial cytoplasmic proteins contain few cysteine residues and few disulfide bonds (44). Most proteins that contain stable disulfide bonds are exported from the cytoplasm (45). Thus, mammalian proteins whose complex tertiary structure depends in part on disulfide bond formation may not be produced in their correct conformation in the bacterial cytoplasm. Mutant *E. coli* strains were isolated that allow the formation of disulfide bonds in the cytoplasm (46). These mutations inactivate the *trxB* gene that encodes thioredoxin reductase and contributes to the sulfhydryl reducing potential of the cytoplasm. Thioredoxin itself was unnecessary for disulfide bond formation (46). The authors suggested that the cytoplasm may contain another thioredoxin-like protein that can be reduced by thioredoxin reductase; in the absence of thioredoxin reductase, the oxidized form of this protein facilitates formation of disulfide bonds in the cytoplasm. Other workers have utilized *E. coli* strains carrying null mutations in the *trxB* gene, and observed significant amounts of functional disulfide-containing protein in the cytoplasm (47).

Proteins produced in the cytoplasm may not possess an authentic N-terminus. The cytoplasmic expression of a gene without a leader requires the presence of an initiation codon, the most common one encoding methionine. Although this amino acid might have no adverse effect on the protein synthesized, there are specific cases in which the extraneous methionine residue may have profound consequences (7). Bacterial translation is initiated by *N*-formylmethionine, which is deformylated during synthesis (48) but not necessarily removed. The N-terminal methionine might be cleaved off by an endogenous methionine aminopeptidase (MAP) (49), depending on the side-chain length of the second amino acid residue (50,51). Thus, residues with small side chains such as Gly, Ala, Pro, Ser, Thr, Val, Cys, and to a lesser degree, Asn, Asp, Leu, and Ile facilitate the MAP-catalyzed removal of the N-terminal methionine. Several strategies have been successfully used to remove the extra methionine residue from recombinant proteins in vivo, including the coexpression of the *E. coli* methionine aminopeptidase gene and the use of exopeptidases (7).

Protein degradation in the cytosol and in other compartments is another concern, discussed in the section "Protein Stability". Finally, cytosolic gene expression necessitates purification of the target protein from the large pool of intracellular proteins.

Periplasm

The periplasm offers several advantages for protein targeting. In contrast to the cytosolic compartment, the periplasm contains only 4% of the total cell protein (52). The target protein is thus effectively concentrated, and its purification is considerably less onerous. The oxidizing environment of the periplasm facilitates the proper folding of proteins, and the cleaving in vivo of the signal peptide during translocation to the periplasm is more likely to yield the authentic N-terminus of the target protein. Protein degradation in the periplasm may be less extensive than in the cytosol, but it is still a significant issue.

The transport of a protein through the inner membrane to the periplasm normally requires a signal sequence. A wide variety of signal peptides have been employed successfully in *E. coli* for protein translocation to the periplasm. These include prokaryotic signal sequences, such as the *E. coli* PhoA signal, OmpA, OmpT, LamB, OmpF, β-lactamase, enterotoxins ST-II, LT-A, LT-B, protein A from *Staphylococcus aureus*, endoglucanase from *B. subtilis*, PelB from *Erwinia carotovora*, a degenerate PelB signal sequence, the murine ribonuclease, and the human growth hormone signal (7). However, the presence of a signal peptide does not always ensure efficient protein translocation through the inner membrane. It is clear that besides the signal peptide, other structural features in proteins are involved in membrane transport. Strategies for the improved translocation of proteins to the periplasm include the supply of components involved in protein transport and processing, such as overproduction of signal peptidase I, use of *prlF* mutant strains, coexpression of the *prlA4* and *secE* genes, coexpression of the *prlF* gene, and expression of the *pspA* gene (7).

The intracellular accumulation of unprocessed proteins has often been interpreted as a possible saturation of the *E. coli* translocation system (e.g., Ref. 5). For example, Hsiung et al. (53) observed the intracellular accumulation of a greater amount of human growth hormone precursor after IPTG induction, but no increase in the amount of translocated growth hormone. Similarly, the *Chlamydia trachomatis* major outer membrane protein (MOMP) was successfully targeted to the *E. coli* surface, but upon prolonged induction, the unprocessed higher molecular weight MOMP accumulated (54). Although this may have been caused by saturation of the transport mechanisms, the authors point out that other explanations are also possible, such as differences in the carboxyl terminus of the chlamydial leader peptide, or formation of inclusion bodies, making a large proportion of MOMP unavailable for processing and resulting in the accumulation of unprocessed MOMP (54). Other workers dismiss the possibility of jamming of the protein export machinery, pointing out that under proper conditions correctly processed and fully active proteins can accumulate in the periplasm at a level 20–50% of the total cell protein (6). In agreement with this, insulin-like growth factor I (IGF-I) was expressed in the periplasm at up to 30% of the total cell protein (55). Moreover, there was no unprocessed IGF-I detected in the cytosol (John Joly, personal communication, 1998). However, IGF-I expression was achieved in the presence of the thiol oxidant proteins DsbA or DsbC, which might have increased protein yields by possibly assisting translocation of IGF-I. Interestingly, the overexpression of the disulfide oxidoreductases increased the fraction of aggregated IGF-I for reasons not understood (55).

The basic protein Skp has been identified as a periplasmic molecular chaperone (56). Overexpression of Skp was shown to increase the amount of fusion proteins incorporated into phage (57). The effect was most notable on scFv fragments that are aggregation prone, suggesting that Skp acts by increasing the amount of correctly folded proteins, which can then get incorporated into the phage coat. The chaperone composition of the periplasm is not as well characterized as that of the cytoplasm, and the mechanism of action of Skp is unclear (57). Further studies should demonstrate whether Skp and other periplasmic molecular chaperones have wide applicability in periplasmic protein expression.

Membranes and Cell Surface

The cell surface of *E. coli* may facilitate several applications including vaccine development, biocatalysis, protein–protein interactions, and drug screening (58). Progress in the high-level expression of membrane proteins in *E. coli* has been reviewed in detail (5). In many cases production of membrane proteins is limited by toxicity and bacterial cell death. A procedure has recently been described for the selection of *E. coli* strains that facilitate the overproduction of proteins and their incorporation into the membrane (59).

Extracellular Medium

The targeting of synthesized proteins for secretion to the culture medium presents significant advantages (Table 3). Unfortunately, *E. coli* normally secretes very few proteins, and the manipulation of the various transport pathways to facilitate secretion of foreign proteins remains a formidable task. The methodological approaches to protein secretion in *E. coli* may be conveniently divided into two categories: (*1*) the exploitation of existing pathways for "truly" secreted proteins (60); and (*2*) the use of signal sequences, fusion partners, permeabilizing proteins, nutrients, or other agents that effect protein secretion as a result of "leakage" or selective and limited permeability of the outer membrane. The first approach offers the advantage of specific secretion of the protein of interest, hence minimum contamination by nontarget proteins. For example, the hemolysin gene has been used for construction of secreted hybrid proteins (60). Secretion, however, is not a particularly efficient process. The second approach relies on the induction of limited leakage of the outer membrane to render protein secretion, albeit with modest yields. Examples are the use of the *pelB* leader, *ompA* leader, protein A leader, the coexpression of bacteriocin release protein along with the addition of glycine to the culture medium, and the coexpression of the *kil* gene for membrane permeabilization (7). A mutated bacteriocin release protein was reported to provide more efficient extracellular production of proteins by *E. coli* (61).

PROTEIN STABILITY

Proteolysis is a selective, highly regulated process that plays an important role in cellular physiology. *E. coli* contains at least 40 enzymes that catalyze the hydrolysis of peptide bonds (62). These enzymes participate in a wide range of metabolic processes including the removal of N- or C-terminal amino acid residues, catabolism of dipeptides, oligopeptides, and proteins, the selective removal of abnormal proteins, and the inactivation of regulatory proteins (62,63). Protein damage or alteration may result from a variety of conditions, such as incomplete polypeptides, mutations caused by amino acid substitutions, excessive synthesis of subunits from multimeric complexes, posttranslational damage through oxidation or free radical attack, and genetic engineering. Such abnormal proteins are efficiently removed by the bacterial proteolytic machine. *E. coli* proteases are localized in the cytoplasm, the periplasm, and the inner and outer membranes (Table 4).

Some determinants of protein instability have been elucidated. The "N-end rule" relates the metabolic stability of a protein to its amino-terminal residue (64). Thus, in *E.*

Table 4. Localization of Proteases in *E. coli*

Cytoplasm	Inner membrane	Periplasm	Outer membrane
La (Lon)	SppA (protease IV)	Deg P (HtrA, Do)	OmpT (protease VII)
ClpP	DegS (HhoB)	DegQ (HhoA)	OmpP
ClpX	Protease VI	Ptr (Pi, III, pitrilysin)	OmpX
ClpA (Ti)	LepB (signal peptidase I)	Prc (Tsp, Re)	
ClpQ	LspA (signal peptidase II)	Mi	
Clp Y	SohB		
HflB (FtsH, TolZ)	Protease V		
Protease II			
Ci			
Fa			
In			
So			

Source: Adapted from Refs. 62 and 63.

Note: The nomenclature for *E. coli* proteolytic activities is confusing, and some of these proteases are found under multiple names. A few proteins originally identified as proteases turned out to have other primary activities. In addition to the listed proteases, there are at least 12 peptidases in the cytoplasm.

coli, N-terminal Arg, Lys, Leu, Phe, Tyr, and Trp conferred 2-min half-lives on a test protein, whereas all the other amino acids, except proline, conferred greater than 10-h half-lives on the same protein. As already discussed in "Cellular Compartments, Cytoplasm", amino acids with small side chains in the second position of the polypeptide facilitate the MAP-catalyzed removal of the N-terminal methionine. Therefore, these studies suggest that Leu in the second position would probably be exposed by the removal of the methionine residue and destabilize the protein. The second determinant of protein instability is a specific internal lysine residue located near the amino terminus (65). This residue is the acceptor of a multiubiquitin chain that facilitates protein degradation by a ubiquitin-dependent protease in eukaryotes. Another correlation between amino acid content and protein instability is presented in the PEST hypothesis, which proposes that proteins are destabilized by regions enriched in Pro, Glu, Ser, and Thr, flanked by certain amino acid residues (66). It was suggested that PEST-rich proteins may be produced efficiently in *E. coli*, which apparently lacks the PEST proteolytic system.

Strategies for minimizing proteolysis of recombinant proteins in *E. coli* have been reviewed (7,67,68). These include protein targeting to the periplasm or the culture medium, use of protease-deficient host strains, growth of the host cells at low temperature, construction of N- and/or C-terminal fusion proteins, tandem fusion of multiple copies of the target gene, coexpression of molecular chaperones, coexpression of the T4 *pin* gene, replacement of specific amino acid residues in order to eliminate protease cleavage sites, modification of the hydrophobicity of the target protein, and optimization of fermentation conditions. The usefulness of some of these methods may be limited, depending on the intended use of the recombinant protein. Thus, for example, the presence of fusion moieties on the target protein may possibly interfere with functional or structural properties or therapeutic applications of the product. The engineering of enzymatic or chemical cleavage sites for the subsequent removal of the fusion partners involves many considerations: the accessibility of the cleavage sites to enzyme digestion, the purity, specificity and cost of the commercially available enzymes, the authenticity of the N- or C-termini upon enzymatic digestion, the possible modification of the target protein upon chemical treatment, and so forth (e.g., Refs. 69, 70). For the large-scale production of fusion proteins, some of these difficulties are amplified. Similarly, the fusion of multiple copies of the target gene to effect multidomain polypeptides requires the subsequent conversion to monomeric protein units by cyanogen bromide cleavage. In this case, the target protein must not contain internal methionine residues and must be able to withstand harsh reaction conditions. Moreover, a limited extend of amino acid side-chain modification may occur, and the toxicity of CNBr presents a significant issue for large-scale cleavage reactions. Similarly, the "rational" modification of a protein sequence requires structural information that may not be available. Molecular chaperones (see later) have been used successfully to stabilize specific proteins (68), but this approach to date remains a hit or miss affair (71).

The cytoplasm of *E. coli* appears to contain a greater number of proteases than the periplasm. Therefore, proteins located in the periplasm are less likely to be degraded (72). However, proteolytic activity in the periplasm is substantial (62). Secretion into the culture medium would provide a better alternative in terms of protein stability. Unfortunately, the technology for secretion of proteins from *E. coli* into the culture medium is still in its infancy. A major catalyst of protein degradation in bacteria is the induction of heat shock proteins in response to a variety of stress conditions, such as the thermal induction of gene expression or the accumulation of abnormal or heterologous proteins in the cytoplasm (73). Under these conditions the production of the *lon* gene product, protease La, and other proteases, is enhanced. This problem is minimized by the use of host strains deficient in the *rpoH (htpR)* locus (74). The *rpoH* gene encodes the RNA polymerase σ^{32} subunit, which regulates several proteolytic activities in *E. coli* Hosts that carry the *rpoH* mutation have been demonstrated to increase substantially the production of foreign proteins in *E. coli* (7). A large number of protease-deficient hosts exists (e.g., Refs. 74, 75), including those that are deficient in all known protease loci that affect the stability of secreted proteins (76).

PROTEIN FOLDING

The efficient posttranslational folding of proteins, the assembly of polypeptides into oligomeric structures, and the localization of proteins are mediated by specialized proteins termed molecular chaperones. The demonstration that the production and assembly of prokaryotic ribulose bisphosphate carboxylase in *E. coli* requires both groES and groEL proteins, led to an increasing interest in the use of molecular chaperones for high-level gene expression in *E. coli*. The experimental results from the use of chaperones, however, have been inconsistent, and thus far the effects of chaperone coproduction on gene expression in *E. coli* appear to be protein specific (71, 77). It is also unclear whether the in vivo levels of different chaperone species are limiting under conditions of gene overexpression (78).

Contradictory results may be due to several reasons. Proteins that are drastically destabilized will probably fold incorrectly, even in the presence of chaperones. Thus, the truncation of polypeptides, the production of single domains from multisubunit protein complexes, lack of formation of disulfide bonds (which ordinarily contribute to protein structure), or absence of posttranslational modifications such as glycosylation may make it impossible to attain thermodynamic stability. It was shown that calmodulin coexpression was essential to generate active inducible nitric oxide synthase (iNOS) in *E. coli*, and it was suggested that calmodulin binds to iNOS after translation and aids in folding and stabilizing the enzyme (79). It is now clear that different types of chaperones normally act in concert (80). Therefore, the overproduction of a single chaperone may be ineffective (81). Another variable to consider is growth temperature. For example, GroES/EL coexpression increased the production of β-galactosidase at 30 °C, but not at 37 °C or 42 °C, whereas DnaK and DnaJ were effective at all temperatures tested (82). Finally, the overexpression of chaperones can lead to phenotypic changes, such as cell filamentation, that can be detrimental to cell viability and protein production (81).

The coexpression of the human (83) or rat (84) protein disulfide isomerase (PDI) enhanced the yield of correctly folded protein in the *E. coli* periplasm. Disulfide bond formation in the *E. coli* periplasm is facilitated by a group of proteins that maintain the correct redox potential (85). It is thought that DsbA, a soluble periplasmic protein, directly catalyzes disulfide bond formation in proteins, whereas DsbB, an inner membrane protein, is involved in the reoxidation of DsbA. Eukaryotic PDI was capable of complementing the phenotypes of *dsbA* null mutants (83,84), but its function was virtually abolished in *dsbB* mutants (84). In addition, the ability of PDI to enhance the yield of target proteins was increased in the presence of exogenously added glutathione (84). These observations suggest that PDI depends on the presence of bacterial redox proteins for its reoxidation. The coexpression of rat PDI has also been reported to enhance the correct folding of tissue plasminogen activator (84).

Protein misfolding can be attributed to the intracellular concentration of aggregation-prone intermediates, suggesting that reducing the rate of protein synthesis should disfavor protein misfolding. Indeed, the use of weaker promoters or conditions of partial induction from stronger promoters can result in higher amounts of soluble protein (82). The ability of *E. coli* to secrete a large amount of alkaline phosphatase into the periplasm was due to a lower synthetic rate of the *phoA* gene product (86), and a reduction in promoter strength enhanced the extracellular production of exoglucanase by *E. coli* (87).

FUSION PROTEINS

In recent years there has been a remarkable increase in the variety of fusion proteins for biological research (70,88). Table 5 includes most of the published fusion moieties. Other studies have addressed the design and engineering of excision sites for the chemical or enzymatic cleavage and removal of fusion partners (69,70). This section briefly summarizes the use of selected fusion systems that have a direct impact on high-level production and, in some cases, secretion of target proteins.

Uhlén and colleagues developed a multifunctional fusion partner based on staphylococcal protein A or synthetic derivatives thereof. In addition to its utility as an affinity tag for purification, the protein A moiety acts as a solubilizing partner to improve folding (92), and the presence of the protein A signal peptide causes the secretion of the gene product to the culture medium. An alternative fusion partner is derived from streptococcal protein G (SPG), a bacterial cell wall protein that has separate binding regions for albumin within the amino-terminal domains, and for IgG within the carboxyl-terminal domain. The combination of both protein A and SPG domains formed a tripartite fusion protein, thus providing an additional purification option and further protecting the target protein from proteolytic degradation (93).

Another fusion partner, thioredoxin, has been shown to dramatically increase the solubility of fusion proteins produced in the *E. coli* cytoplasm, and to prevent the formation of inclusion bodies (88). Similarly, the thioredoxin homolog DsbA has been used as a fusion partner to direct the transport of proteins to the periplasm (94). More recently, a mutant form of DsbA ($DsbA^{mut}$) was used as a fusion partner to confer higher solubility to 10 different eukaryotic proteins expressed in the cytoplasm at 37 °C (43). The consistent performance of $DsbA^{mut}$ led the authors to conclude that this protein may be used as a generic fusion partner for enhancing the solubility of a variety of heterologous eukaryotic proteins. An additional advantage of $DsbA^{mut}$ may be its ability to exert its solubilizing effect at 37 °C rather than at the lower growth temperatures often employed for enhancing solubility of proteins: at reduced temperatures the cell growth cycle becomes longer, with a significant impact on fermentation cycle time and a reduced protein yield (43).

The fusion of genes to the ubiquitin sequence significantly increased the yield of proteins (95). The remarkable

Table 5. Fusion Partners

Fusion partner	Ligand/matrix	Purification conditions
Flag®	Anti-Flag monoclonal antibodies M1, M2	Low calcium, EDTA, glycine
His_6	Ni^{2+}-nitrilotriacetic acid	Imidazole
Glutathione-S-transferase	Glutathione-Sepharose	Reduced glutathione
Staphylococcal protein A	IgG-Sepharose	Low pH, IgG-affinity ligand
Streptococcal protein G	Albumin	Low pH, albumin-affinity ligand
Calmodulin	Organic or peptide ligands, DEAE-Sephadex	Low calcium, EGTA
Calmodulin-binding peptides[a]	Calmodulin	Low calcium, EGTA
Thioredoxin	ThioBond™ resin	Ion exchange
Flagellin *(E. coli)*[b]		
β-Galactosidase	TPEG[c] Sepharose	Borate
Ubiquitin		
Chloramphenicol acetyltransferase	Chloramphenicol-Sepharose	Chloramphenicol
S-peptide (RNAse A, residues 1-20)	S-protein (RNAse A, residues 21–124)	Denaturing or nondenaturing conditions
Myosin heavy chain		Differential solubility in low/high salt
DsbA		
Biotin subunit (in vivo biotinylation)		
Avidin	Biotin	Denaturation (urea, heat)
Streptavidin	Biotin	Denaturation (urea, heat)
Strep-tag	Streptavidin	2-Iminobiotin, diaminobiotin
c-*myc*	anti-*myc* antibody	
Dihydrofolate reductase	Methotrexate-agarose	Folate buffer
CKS[d]		
Polyarginine	S-Sepharose	NaCl
Polycysteine	Thiopropyl-Sepharose	Dithiothreitol
Polyphenylalaline	Phenyl-Superose	Ethylene glycol
lac repressor	*lac* operator	Lactose analogue, DNAse, restriction endonuclease
T4 gp55		
Growth hormone, N-terminus		
Maltose-binding protein	Amylose resin	Maltose
Galactose-binding protein	Galactose-Sepharose	Galactose
Cyclomaltodextrin glucanotransferase	α-Cyclodextrin-agarose	α-Cyclodextrin
Cellulose-binding domain	Cellulose	Water
Hemolysin A *(E. coli)*		
λ cII protein		
TrpE or TrpLE		
Protein kinase site(s)		
$(AlaTrpTrpPro)_n$		Aqueous two-phase extraction
HAI[e]-epitope		
BTag (VP7 protein region of Bluetongue virus)	Anti-BTag antibodies	
Green fluorescent protein		
Prochymosin[f]		
TNFb[g]		

Source: Updated from Ref. 7 with permission from the publisher.
Note: Except where indicated, original references may be found in Ref. 7.
[a]Ref. 89.
[b]Ref. 90.
[c]*p*-Amino-phenyl-β-D-thiogalactoside.
[d]CTP:CMP-3-deoxy-D-manno-octulosonate cytidyltransferase.
[e]Influenza virus hemagglutinin; "reverse epitope tagging": tagging of the chromosomal rather than the plasmid-encoded protein, to avoid the need to remove the fusion partner.
[f]Highly insoluble, facilitates formation of inclusion bodies (40).
[g]Mutated tumor necrosis factor gene (91).

increase in protein yield was thought to be due to protection of the target protein from proteolysis, improved folding, and efficient mRNA translation. Ubiquitin or the ubiquitin metabolic pathway is absent in prokaryotic organisms. In order to remove the ubiquitin moiety from fusion proteins, Baker et al. (96) coexpressed the ubiquitin-specific protease Ubp2 in *E. coli*, thus effecting the cotranslational cleavage of ubiquitin from the fusion protein.

SUMMARY

An efficient prokaryotic expression vector contains a strong and tightly regulated promoter, a Shine–Dalgarno site that is positioned approximately 9 bp 5′ to the translation initiation codon and is complementary to the 3′ end of 16S rRNA, and a transcription terminator positioned 3′ to the protein-coding sequence. In addition, the vector requires an origin of replication, a selection marker, and a gene that facilitates the stringent regulation of promoter activity. The translational initiation region of a gene must be free of secondary structures that may occlude the initiation codon and/or block ribosome binding. UAAU is the most efficient translation termination sequence in *E. coli*.

Each of the four "compartments" for targeted protein production, namely, the cytoplasm, periplasm, membranes/cell surface, and the growth medium, offers advantages and disadvantages for gene expression. The formation of inclusion bodies can be minimized using a variety of techniques, but it remains a significant barrier to high-level protein production in the cytoplasm. The effectiveness of molecular chaperones has been protein specific so far.

Codon usage can have adverse effects on the synthesis and yield of recombinant proteins. Positional effects appear to play an important role in protein synthesis. Thus, the presence of rare codons near the 5′ end of a transcript probably affects translational efficiency. This problem may be rectified by the alteration of the rare codons, and/or the coexpression of the cognate tRNA genes.

To date there is no generally applicable strategy to prevent the degradation of a wide variety of mRNA species in *E. coli*. Although certain 5′ and 3′ stem–loop structures have been shown to block mRNA degradation, these seem to stabilize only specific mRNAs, under restricted conditions. One exception appears to be the 5′ UTR of the *E. coli ompA* transcript, which prolongs the half-life of a number of heterologous mRNAs in *E. coli*.

The wide variety of existing fusion partners have utility in the production, detection, and purification of recombinant proteins. Specific fusion moieties can increase the folding, solubility, resistance to proteolysis, and secretion of recombinant proteins into the growth medium. Protein misfolding, attributed to the intracellular concentration of aggregation-prone intermediates, may be minimized by a combination of experimental approaches: replacement of amino acid residues that cause aggregation, coexpression of molecular chaperones and foldases, reduction of the rate of protein synthesis, the use of solubilizing fusion partners, and the careful optimization of growth conditions.

Proteolysis in *E. coli* may be minimized by the use of protease-deficient host strains, the construction of fusion proteins, the coexpression of molecular chaperones, the elimination of protease cleavage sites through genetic engineering, and the optimization of fermentation conditions. Host strains that are deficient in the *rpoH (htpR)* locus are effective, particularly for thermally induced expression systems.

BIBLIOGRAPHY

1. E.I. Tuomanen, *J. Clin. Invest.* **98**, 2659–2660 (1996).
2. S.Y. Lee, *Trends Biotechnol.* **14**, 98–105 (1996).
3. D.V. Goeddel, *Methods Enzymol.* **185**, 3–7 (1990).
4. A.R. Shatzman, *Curr. Opin. Biotechnol.* **6**, 491–493 (1995).
5. R. Grisshammer and C.G. Tate, *Q. Rev. Biophys.* **28**, 315–422 (1995).
6. G. Georgiou, in J.L. Cleland and C.S. Craik eds., *Protein Engineering: Principles and Practice*, Wiley-Liss, New York, 1996, pp. 101–127.
7. S.C. Makrides, *Microbiol. Rev.* **60**, 512–538 (1996).
8. K.C. Sitney, M.B. Mann, G.W. Stearns, A.D. Menjares, J.L. Stevenson, M.D. Snavely, J.C. Fieschko, C. Curless, and L.B. Tsai, *Ann. NY Acad. Sci.* **782**, 297–310 (1996).
9. B.F. Gao, S.C. Flores, S.K. Bose, and J.M. McCord, *Gene* **176**, 269–272 (1996).
10. S.M.C. Newton, P.E. Klebba, M. Hofnung, and A. Charbit, *Res. Microbiol.* **146**, 193–202 (1995).
11. R. Lutz and H. Bujard, *Nucleic Acids Res.* **25**, 1203–1210 (1997).
12. J.M. Blatny, T. Brautaset, H.C. Winther-Larsen, P. Karunakaran, and S. Valla, *Plasmid* **38**, 35–51 (1997).
13. W.E. Bentley, N. Mirjalili, D.C. Andersen, R.H. Davis, and D.S. Kompala, *Biotechnol. Bioeng.* **35**, 668–681 (1990).
14. M.J. Kosinski, U. Rinas, and J.E. Bailey, *Appl. Microbiol. Biotechnol.* **36**, 782–784 (1992).
15. B. Andrews, H. Adari, G. Hannig, E. Lahue, M. Gosselin, S. Martin, A. Ahmed, P.J. Ford, E.G. Hayman, and S.C. Makrides, *Gene* **182**, 101–109 (1996).
16. N. Hasan and W. Szybalski, *Gene* **163**, 35–40 (1995).
17. Patent WO 96/03521, A.B. Oppenheim, H. Giladi, D. Goldenberg, S. Kobi, and I. Azar.
18. S. Adhya and M. Gottesman, *Cell* **29**, 939–944 (1982).
19. N. Hasan and W. Szybalski, *Gene* **56**, 145–151 (1987).
20. C.O. Gualerzi and C.L. Pon, *Biochemistry* **29**, 5881–5889 (1990).
21. S. Ringquist, S. Shinedling, D. Barrick, L. Green, J. Binkley, G.D. Stormo, and L. Gold, *Mol. Microbiol.* **6**, 1219–1229 (1992).
22. R.M. Belagaje, S.G. Reams, S.C. Ly, and W.F. Prouty, *Protein Sci.* **6**, 1953–1962 (1997).
23. E.S. Poole, C.M. Brown, and W.P. Tate, *EMBO J.* **14**, 151–158 (1995).
24. T. Ikemura, *Mol. Biol. Evol.* **2**, 13–34 (1985).
25. J.F. Kane, *Curr. Opin. Biotechnol.* **6**, 494–500 (1995).
26. G.-F.T. Chen and M. Inouye, *Nucleic Acids Res.* **18**, 1465–1473 (1990).
27. E. Goldman, A.H. Rosenberg, G. Zubay, and F.W. Studier, *J. Mol. Biol.* **245**, 467–473 (1995).
28. W. Gao, S. Tyagi, F.R. Kramer, and E. Goldman, *Mol. Microbiol.* **25**, 707–716 (1997).
29. I. Ivanov, R. Alexandrova, B. Dragulev, A. Saraffova, and M.G. AbouHaidar, *FEBS Lett.* **307**, 173–176 (1992).
30. C.-Y.A. Chen and J.G. Belasco, *J. Bacteriol.* **172**, 4578–4586 (1990).
31. S.A. Emory, P. Bouvet, and J.G. Belasco, *Genes Dev.* **6**, 135–148 (1992).
32. H.C. Wong and S. Chang, *Proc. Natl. Acad. Sci. U.S.A.* **83**, 3233–3237 (1986).
33. J.G. Belasco, G. Nilsson, A. von Gabain, and S.N. Cohen, *Cell* **46**, 245–251 (1986).
34. C.-Y.A. Chen, J.T. Beatty, S.N. Cohen, and J.G. Belasco, *Cell* **52**, 609–619 (1988).
35. W.P. Donovan and S.R. Kushner, *Proc. Natl. Acad. Sci. U.S.A.* **83**, 120–124 (1986).

36. W.P. Donovan and S.R. Kushner, *Nucleic Acids Res.* **11**, 265–275 (1983).
37. J. Chang, W. Cai, L. Xu, C. Li, and S. Zhao, *Chin. J. Biotechnol.* **11**, 213–219 (1995).
38. V. Kruys and G. Huez, *Biochimie* **76**, 862–866 (1994).
39. G.M. Wilson and G. Brewer, in K. Moldave ed., *Progress in Nucleic Acid Research and Molecular Biology*, vol. 62, Academic, San Diego, 1999, pp. 257–291.
40. C. Haught, G.D. Davis, R. Subramanian, K.W. Jackson, and R.G. Harrison, *Biotechnol. Bioeng.* **57**, 55–61 (1998).
41. R. Rudolph and H. Lilie, *FASEB J.* **10**, 49–56 (1996).
42. D.L. Wilkinson and R.G. Harrison, *Bio / Technology* **9**, 443–448 (1991).
43. Y. Zhang, D.R. Olsen, K.B. Nguyen, P.S. Olson, E.T. Rhodes, and D. Mascarenhas, *Protein Expr. Purif.* **12**, 159–165 (1998).
44. R.C. Fahey, J.S. Hunt, and G.C. Windham, *J. Mol. Evol.* **10**, 155–160 (1977).
45. J.M. Thornton, *J. Mol. Biol.* **151**, 261–287 (1981).
46. A.I. Derman, W.A. Prinz, D. Belin, and J. Beckwith, *Science* **262**, 1744–1747 (1993).
47. K. Proba, L.M. Ge, and A. Plückthun, *Gene* **159**, 203–207 (1995).
48. J.M. Adams, *J. Mol. Biol.* **33**, 571–589 (1968).
49. A. Ben-Bassat, K. Bauer, S.-Y. Chang, K. Myambo, A. Boosman, and S. Chang, *J. Bacteriol.* **169**, 751–757 (1987).
50. F. Sherman, J.W. Stewart, and S. Tsunasawa, *BioEssays* **3**, 27–31 (1985).
51. P.-H. Hirel, J.-M. Schmitter, P. Dessen, G. Fayat, and S. Blanquet, *Proc. Natl. Acad. Sci. U.S.A.* **86**, 8247–8251 (1989).
52. N.G. Nossal and L.A. Heppel, *J. Biol. Chem.* **241**, 3055–3062 (1966).
53. H.M. Hsiung, N.G. Mayne, and G.W. Becker, *Bio / Technology* **4**, 991–995 (1986).
54. J.E. Koehler, S. Birkelund, and R.S. Stephens, *Mol. Microbiol.* **6**, 1087–1094 (1992).
55. J.C. Joly, W.S. Leung, and J.R. Swartz, *Proc. Natl. Acad. Sci. U.S.A.* **95**, 2773–2777 (1998).
56. R. Chen and U. Henning, *Mol. Microbiol.* **19**, 1287–1294 (1996).
57. H. Bothmann and A. Plückthun, *Nat. Biotechnol.* **16**, 376–380 (1998).
58. G. Georgiou, H.L. Poetschke, C. Stathopoulos, and J.A. Francisco, *Trends Biotechnol.* **11**, 6–10 (1993).
59. B. Miroux and J.E. Walker, *J. Mol. Biol.* **260**, 289–298 (1996).
60. J.A. Stader and T.J. Silhavy, *Methods Enzymol.* **185**, 166–187 (1990).
61. F.J. van der Wal, G. Koningstein, C.M. ten Hagen, B. Oudega, and J. Luirink, *Appl. Environ. Microbiol.* **64**, 392–398 (1998).
62. C.G. Miller, in F.C. Neidhardt, R. Curtiss III, J.L. Ingraham, E.C.C. Lin, K.B. Low, B. Magasanik, W.S. Reznikoff, M. Riley, M. Schaechter, and H.E. Umbarger eds., Escherichia Coli and Salmonella. *Cellular and Molecular Biology,* vol. 1, American Society for Microbiology Press, Washington, D.C., 1996, pp. 938–954.
63. S. Gottesman, *Annu. Rev. Genet.* **30**, 465–506 (1996).
64. A. Varshavsky, *Cell* **69**, 725–735 (1992).
65. A. Bachmair and A. Varshavsky, *Cell* **56**, 1019–1032 (1989).
66. S. Rogers, R. Wells, and M. Rechsteiner, *Science* **234**, 364–368 (1986).
67. F. Baneyx and G. Georgiou, in T.J. Ahern, and M.C. Manning eds., *Stability of Protein Pharmaceuticals. Part A: Chemical and Physical Pathways of Protein Degradation*, Plenum, New York, 1992, pp. 69–108.
68. M. Murby, M. Uhlén, and S. Ståhl, *Protein Expr. Purif.* **7**, 129–136 (1996).
69. P. Carter, in M.R. Ladisch, R.C. Willson, C.C. Painton, and S.E. Builder eds., *Protein Purification: From Molecular Mechanisms to Large-Scale Processes*, vol. 427, American Chemical Society, Washington, D.C., 1990, pp. 181–193.
70. P.-Å. Nygren, S. Ståhl, and M. Uhlén, *Trends Biotechnol.* **12**, 184–188 (1994).
71. J.G. Wall and A. Plückthun, *Curr. Opin. Biotechnol.* **6**, 507–516 (1995).
72. K. Talmadge and W. Gilbert, *Proc. Natl. Acad. Sci. U.S.A.* **79**, 1830–1833 (1982).
73. S.A. Goff and A.L. Goldberg, *Cell* **41**, 587–595 (1985).
74. S. Gottesman, *Methods Enzymol.* **185**, 119–129 (1990).
75. F. Baneyx and G. Georgiou, *J. Bacteriol.* **173**, 2696–2703 (1991).
76. H.J. Meerman and G. Georgiou, *Bio / Technology* **12**, 1107–1110 (1994).
77. T. Yasukawa, C. Kaneiishii, T. Maekawa, J. Fujimoto, T. Yamamoto, and S. Ishii, *J. Biol. Chem.* **270**, 25328–25331 (1995).
78. A. Knappik, C. Krebber, and A. Plückthun, *Bio / Technology* **11**, 77–83 (1993).
79. C. Wu, J. Zhang, H. Abu-Soud, D.K. Ghosh, and D.J. Stuehr, *Biochem. Biophys. Res. Commun.* **222**, 439–444 (1996).
80. J. Buchner, *FASEB J.* **10**, 10–19 (1996).
81. P. Blum, J. Ory, J. Bauernfeind, and J. Krska, *J. Bacteriol.* **174**, 7436–7444 (1992).
82. G. Georgiou and P. Valax, *Curr. Opin. Biotechnol.* **7**, 190–197 (1996).
83. D.P. Humphreys, N. Weir, A. Mountain, and P.A. Lund, *J. Biol. Chem.* **270**, 28210–28215 (1995).
84. M. Ostermeier, K. De Sutter, and G. Georgiou, *J. Biol. Chem.* **271**, 10616–10622 (1996).
85. J.C.A. Bardwell, *Mol. Microbiol.* **14**, 199–205 (1994).
86. H. Kadokura, K. Yoda, S. Watanabe, Y. Kikuchi, G. Tamura, and M. Yamasaki, *Appl. Microbiol. Biotechnol.* **41**, 163–169 (1994).
87. T.-L. Lam, R.S.C. Wong, and W.-K.R. Wong, *Enzyme Microb. Technol.* **20**, 482–488 (1997).
88. E.R. LaVallie and J.M. McCoy, *Curr. Opin. Biotechnol.* **6**, 501–506 (1995).
89. C.-F. Zheng, T. Simcox, L. Xu, and P. Vaillancourt, *Gene* **186**, 55–60 (1997).
90. B. Westerlund-Wikström, J. Tanskanen, R. Virkola, J. Hacker, M. Lindberg, M. Skurnik, and T.K. Korhonen, *Protein Eng.* **10**, 1319–1326 (1997).
91. L. Wang, H. Wu, F. Dou, W. Xie, and X. Xu, *Biochem. Mol. Biol. Int.* **41**, 1051–1056 (1997).
92. E. Samuelsson, T. Moks, B. Nilsson, and M. Uhlén, *Biochemistry* **33**, 4207–4211 (1994).
93. M. Murby, L. Cedergren, J. Nilsson, P.-Å. Nygren, B. Hammarberg, B. Nilsson, S.-O. Enfors, and M. Uhlén, *Biotechnol. Appl. Biochem.* **14**, 336–346 (1991).
94. L.A. Collins-Racie, J.M. McColgan, K.L. Grant, E.A. DiBlasio-Smith, J.M. McCoy, and E.R. LaVallie, *Bio / Technology* **13**, 982–987 (1995).
95. A.L. Pilon, P. Yost, T.E. Chase, G.L. Lohnas, and W.E. Bentley, *Biotechnol. Prog.* **12**, 331–337 (1996).

96. R.T. Baker, S.A. Smith, R. Marano, J. McKee, and P.G. Board, *J. Biol. Chem.* **269**, 25381–25386 (1994).

See also CHINESE HAMSTER OVARY CELLS, RECOMBINANT PROTEIN PRODUCTION; EXPRESSION SYSTEMS, MAMMALIAN CELLS; INSECT CELL CULTURE, PROTEIN EXPRESSION; INSECT CELLS AND LARVAE, GENE EXPRESSION SYSTEMS; PICHIA, OPTIMIZATION OF PROTEIN EXPRESSION.

EXPRESSION SYSTEMS, MAMMALIAN CELLS

SUBINAY GANGULY
ALLAN R. SHATZMAN
SmithKline Beecham Pharmaceuticals
King of Prussia, Pennsylvania

KEY WORDS

293 cells
Amplification
CHO cells
COS cells
mAbs
Regulated expression
Selectable markers
Stable expression
Transient expression

OUTLINE

INTRODUCTION

Gene expression systems offer a way to produce a heterologous protein encoded by a specific gene of interest through its introduction into a specific host cell. Heterologous gene expression of recombinant proteins has many applications, for example, in the biochemical and structure and function analysis of complex proteins, promoter analysis of genes of interest, and development of cell/membrane-based assays. One of the most crucial applications of mammalian cell expression systems is the production of therapeutic proteins such as growth factors and recombinant monoclonal antibodies (mAbs). Mammalian gene expression systems are also being used in gene-therapy-related experiments for the treatment of human diseases. Moreover, various expression systems utilizing mammalian cells can be used to determine functions for novel genes that are being identified through genomics. Dozens (if not hundreds) of mammalian cell expression systems have been described in the literature. The efficiency of expression of heterologous genes in mammalian cells is determined by the choice of host cells, method of gene transfer, sites at which the heterologous genes are integrated into the host genome (in the case of stable expression), and *cis*-acting DNA elements (e.g., enhancers) present in the vector carrying the gene of interest. In addition, efficiency of transcription, mRNA transport and stability, efficiency of translation, and secretion of the expressed protein into the media (for secreted proteins) may also play a role in the successful expression of recombinant proteins. Steps such as posttranslational modification, glycosylation, phosphorylation, and proteolytic processing must also be considered because they can affect the quality of the protein made in or secreted by recombinant host cells. This review summarizes various host–vector and expression systems commonly available for transient and stable expression of heterologous proteins in mammalian cells (for review see Refs. 1–5) and aims to illustrate the use of some of the most common ones.

GENERAL CONSIDERATIONS AND STRATEGIES FOR OPTIMIZING EXPRESSION

Host Cell Types

Commonly Used Cell Types. Mammalian cells such as chinese hamster ovary (CHO) cells, myeloma cells, human embryonic kidney (HEK)293 cells, baby hamster kidney (BHK) cells, mouse Ltk$^-$ cells, and NIH3T3 cells have been frequently used for stable expression of heterologous genes. In contrast, cell lines such as Cos and HEK293 (see later) are routinely used for transient expression of recombinant proteins.

CHO-K1 and DHFR$^-$ CHO cells DG44 (6) and DUK-B11 (7) have been extensively used for high-level protein production (see later) because these cells enable the amplification of genes of interest via selection with the amplifiable marker DHFR (8). A wide variety of proteins have been expressed in CHO cells, including enzymes, recombinant mAbs, blood coagulation factors, 7TM-receptors, and other cell surface markers, chemokines, adhesion molecules, growth factors, and their receptors (see later for examples).

Myeloma cells such as Sp2/0, YB2/0, NSO, and P3X63.Ag8.653 have also been commonly employed for high-level production of mAbs (9–13). Several examples of this application will be described later in this review.

HEK293 cells, which were derived after transformation of primary human kidney cells by adenovirus (Ad virus) early-region-encoded oncoproteins, have been widely used to stably express many 7TM receptors that can functionally couple to G-proteins. It has been shown that the human cytomegalovirus (hCMV) enhancer/promoter in an expression plasmid is highly active in these cells because Ad virus E1A protein expressed in HEK293 cells transactivates the hCMV enhancer/promoter (14). Examples of 7TM receptor expression in these cells are the *Drosophila* dopamine D1 receptor (15), the human PTH receptor (16), the human muscarinic cholinergic receptor Hm1 (17), the human CGRP type 1 receptor (18), and the rat bradykinin B2 receptor (19). In addition to these receptors, a number of chemokine receptors such as the human CC CKR5 receptor (20), the mouse mCCR2 receptor (21), the mouse eotaxin receptor (22), the human MCP3 receptor (23), and the human MCP-1 receptor (24) have been successfully expressed in HEK293 cells. High-level expression on the cell surface and functional coupling of most the 7TM receptors expressed in HEK293 cells have been achieved. HEK293 cells have also been used for expression of various enzymes, as exemplified by the bovine constitutive nitric oxide synthase (eNOS) (25), and the human prostacyclin synthase (26). HEK293 cells have been used for heterologous expression of a wide variety of proteins, for example, interleukin 1 receptor–IgG fusion protein (immunoadhesion) (27), the human insulin growth factor binding protein 3 (28), platelet integrin $\alpha_{IIb}\beta_3$ (an integral membrane protein that mediates adhesion of cells to the extracellular matrix) (29), and the human tissue factor (a procoagulant factor that serves as a cofactor for the blood coagulation factor VII) (30).

Cos cells have been used to transiently express a large number of heterologous proteins. Cos cells constitutively express simian virus 40 (SV40) T-antigen and therefore support replication of expression plasmids containing the SV40 origin of replication. This results in high copy number (~40,000 copies cell^{-1}) of the transcribed coding sequences (31). Cos cells are primarily used for three reasons: *(1)* to determine if the expression plasmid produces the protein of interest, *(2)* for rapid expression of recombinant proteins to perform initial functional studies, and *(3)* expression cloning. Hundreds (if not thousands) of proteins have been expressed transiently in Cos cells. For example, Patel and Grundy (32) used a Cos cell expression system to study biogenesis of apolipoprotein B–containing lipoproteins. Functional expression of 7TM-receptors, for example, the guniea pig α2A, α2B, and α2C-adrenoreceptors (33), and the rabbit bradykinin B1 receptor (34), have been successfully achieved in Cos cells. Cos cells have been used commonly to express small amounts of functional enzymes, such as the rat liver microsomal glutathione transferase (35), the human 11β-hydroxylase and aldosterone synthase (36). Functionally active recombinant mAbs have also been expressed transiently in Cos cells (for a review see Ref. 4). It should be remembered however, that the Cos expression system is a transient expression system, and therefore, multiple repeat transfection experiments, which are time-consuming and labor intensive, are needed to produce milligram quantities of proteins for initial studies. More importantly, data obtained from a Cos expression system cannot necessarily be extrapolated to the level of expression expected from stable expression systems.

Modified Hosts. Some of the host cell lines just described have been engineered to produce milligram quantities of recombinant proteins in a rapid manner. In most cases this has been achieved by expressing (constitutively) a transcriptional transactivator that stimulates transcription from the promoter linked to the gene of interest. Following are descriptions of some of these modified hosts.

CHOE1a. To quickly express a heterologous protein, Cockett et al. (37) have established a CHO cell line constitutively expressing the Ad virus E1A, which transactivates the CMV promoter in the expression plasmid. Transfection of a human procollagenase gene into this cell line produced a 13-fold higher level in stable expression compared with that in CHO-K1 cells. This system has been used to express mAbs (13,38) and single-chain Fvs (39).

CHO/Pit-1. CHO cells were engineered to express the rat Pit-1 transactivator (a pituitary specific transcription factor). Stable transfection of a reporter gene (tPA) under the control of the rat prolactin promoter showed an enhancement of expression levels (~20- to 60-fold) compared with the same promoter in parallel in CHO cells (40).

BHK/VP16. Hippenmeyer and Highkin (41) have established a BHK cell line that stably expresses the herpes simplex (HSV) VP16, a transcriptional transactivator. Using this cell line, tPA and bGH were expressed at high levels (1–20 μg 10^6 cells^{-1} day^{-1}) as compared with that in the parental BHK cell line (<0.2 μg 10^6 cells^{-1} day^{-1}). Since then, a wide variety of proteins have been stably expressed in the BHK/VP16 cell line (42).

293/EBNA. HEK293 cell lines have been established after stably transfecting the Epstein–Barr virus (EBV) nuclear antigen 1 (EBNA-1) protein, which is required for the maintenance of a plasmid carrying the EBV origin of replication extrachromosomally (43). Such cell lines have been used for heterologous protein expression. For example, using a secreted form of human placental alkaline phosphatase (SEAP) as a reporter gene, Cachianes et al. (44) obtained 10-fold higher expression in 293 cells expressing EBNA-1 as compared with that in wild-type 293 cells. A number of other proteins (e.g., tPA, vascular endothelial growth factor, hepatocyte growth factor, soluble tumor necrosis factor) have been produced in this cell line at levels ranging from 0.2 to 10 μg ml^{-1} (44).

Mammalian Cells Expressing T7 RNA Polymerase Mouse Ltk$^-$ cells, CHO cells, and mouse myeloma Sp2/0 cells expressing T7 RNA polymerase have been established (45).

In these cells, T7 RNA polymerase has been modified by fusion of a nuclear localization signal to the N-terminus of the polymerase so that it can function in the cell nuclei. Transfection of a heterologous gene under the control of a T7 promoter into these modified host cell lines enables high-level expression of the gene of interest. For example, using this system the human growth hormone has been stably expressed at levels of 20–30 μg 10^6 $cells^{-1}$ day^{-1} (45).

Vector Systems

A typical mammalian expression vector contains a cDNA copy of the gene of interest placed downstream of a promoter sequence to ensure efficient transcription; placed upstream is a polyA addition signal, which is required for efficient transcription termination. Enhancer sequences, which stimulate transcription, are usually placed upstream of the linked promoter, whereas a selectable marker (e.g., amp^r gene) and a plasmid origin of replication are included for propagation of the expression construct in *E. Coli*. In some instances, a drug selectable/amplifiable marker expression cassette (see later) for use in eukaryotic cells is linked to the heterologous gene expression unit. As mentioned, for transient expression in Cos cells, the SV40 origin of replication is included in the expression plasmid. See Figure 1 for examples of mammalian expression vectors.

A wide variety of promoters and enhancers and a combination thereof have been used for heterologous expression in mammalian cells and are included in vectors that are commercially available. Among the more popular promoters are those from *(1)* animal viral genomes such as SV40, human CMV, Ad virus, Rous sarcoma virus (RSV), murine leukemia virus (MuLV), HSV, and so forth; and *(2)* cellular genes such as β-globin, immunoglobulin, heat shock, and metallothionein genes (2,3,46,47). SV40 early and late promoters, hCMV IE promoter, RSV long-terminal repeat (LTR), and SRα promoter (a composite promoter consisting of the SV40 early promoter and a part of the LTR of human T-cell leukemia virus) (48) have all been successfully utilized in CHO-K1 and $DHFR^-$ CHO cells for expression of a variety of proteins. However, the choice of promoter is not usually critical when high-level expression is desired in $DHFR^-$ CHO cells where gene amplification methodology will be employed (see later). Plasmids containing the CMV promoter have been shown to produce high-level expression in HEK293 cells (49). Most of the expression vectors used for mAb expression in myeloma cells contain both H- and L-chain cDNAs under the transcription control of IgG promoter, although the hCMV promoter has been used successfully for mAb expression in NSO myeloma cells (10). A polyA addition signal (AA/TTAAA) and downstream sequences that influence transcription termination can be derived from many sources including the SV40 genome, the bovine growth hormone gene, and Ig genes.

The expression level of the gene of interest can sometimes be increased several fold through the use of an intronic sequence, for example, β-globin (50) or chimeric J/Ck intron (51). It is advisable to include the intron at the 5′ end of the coding sequence because, in some examples (52), aberrant processing of transcript was found when the intron was placed at the 3′ end of the cDNA. However, it should be pointed out that the effect of introns in enhancing expression levels is not a general phenomenon and in most cases does not significantly affect expression levels.

Kozak Sequence

In most vertebrate mRNAs examined to date, a sequence related to GCCACC*AUG*G, called a Kozak sequence (the underlined AUG is the initiator methionine), has been identified at the 5′ end, which can promote efficient translation of a linked open reading frame (ORF) (53). Therefore, many investigators include an optimal Kozak sequence in their expression vector. For example, Ho et al. (54) studied translation initiation and assembly of rat peripherin, an intermediate filament protein in transfected cells. A full-length peripherin cDNA (designated as p61-11), when expressed in an adrenocarcinoma cell line (SW13 c1.2 vim^-) without any detectable cytoplasmic intermediate filament, did not yield a functional filamentous network because of low-level expression of p61-11. When the first ATG was placed in the context of a better Kozak consensus sequence, the translational efficiency of p61-11 was increased significantly, and the resulting 57 kDa protein was able to form a functional filamentous network in the vim^- cells.

Translational Enhancers

Regulatory sequences that stimulate translation, hence called translational enhancers, have been identified at the 5′ end of many eukaryotic viral RNAs and some cellular mRNAs (55–57). These sequences bind to translation initiation factors with higher affinity than their cellular counterparts. The Ad virus tripartite leader sequence is such a translational enhancer and has been used in mammalian expression vectors. For example, Sheay et al. (58) compared expression of a reporter CAT gene from expression vectors containing the Ad virus tripartite leader sequence placed downstream of various promoters such as CMV, RSV, and MuLV. In transient transfection assays, expression vectors containing the Ad virus tripartite leader sequence gave 5- to 20-fold higher CAT activity when compared with that from vectors without the leader sequence. A translational enhancer has also been identified in the 5′ UTR of a cellular mRNA (e.g., the the human β-kinesin mRNA encoding the first exon) (59). High levels of reporter gene expression have been achieved from transfection of plasmids carrying the kinesin enhancer sequences when compared with that obtained without the enhancer. For example, in Cos transient transfections, the kinesin enhancer placed downstream of the HSV tK promoter gave a four-fold increase in reporter gene (CAT) activity when compared with that from a plasmid without the enhancer.

A 34-bp-long translational enhancer has been described in the *Xenopus* β-globin mRNA. When placed upstream of a reporter gene, up to 300-fold increases in translational efficiency have been demonstrated in in vitro systems (60).

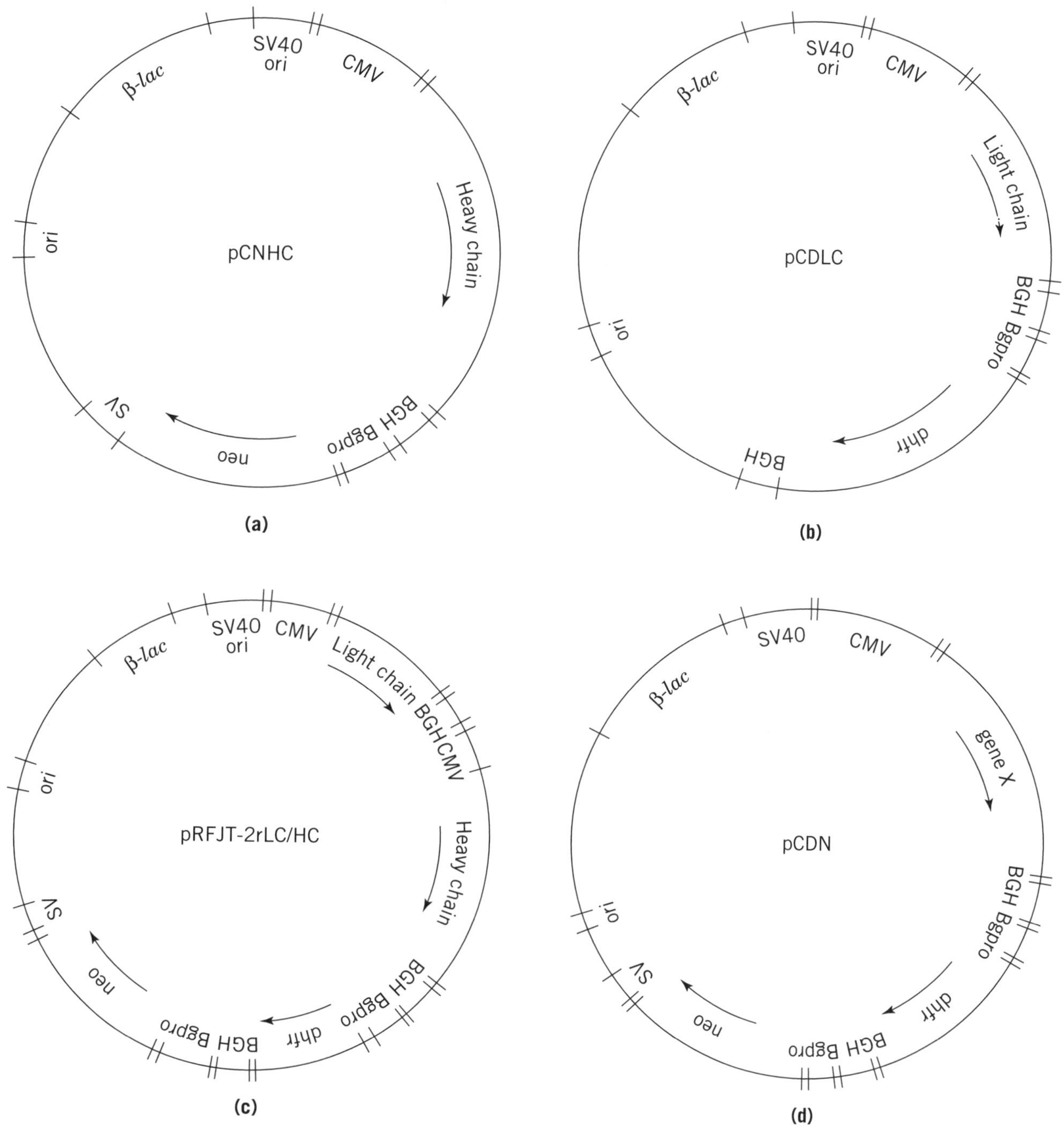

Figure 1. Mammalian expression vectors. pCNHC **(a)** and pCDLC **(b)** are used for mAb expression after subcloning Ig heavy- and light-chain cDNAs separately. **(c)** Single-vector pRFJT-2rLC/HC for mAB expression. **(d)** pCDNX for expression of gene of interest (X) in CHO, Cos, and 293 cells. CMV, cytomegalovirus promoter; BGH, bovine growth hormone polyA addition signal; Bgpro, mouse *β*-globin promoter; dhfr, dihydrofolate reductase cDNA; neo, neomycin phosphotransferase gene; SV, SV40 polyA additional signal; ori, plasmid origin of replication; *β-lac*, *β*-lactamase gene; SV ori, SV40 origin of replication. Orientation of the genes is shown by arrows.

Translational enhancers have also been identified in many plant viral genomes (55), with the tobacco mosaic virus (TMV) enhancer receiving the most attention. However, the use of this and other plant translation enhancers for improving heterologous mammalian expression systems has not yet been explored.

Selectable/Amplifiable Markers

A number of dominant selectable markers such as neomycin resistance (Neo^r), hygromycin resistance (Hyg^r), *E. Coli.* xanthine-guanylyl phosphotransferase (Ecogpt), and glutamine synthetase (GS) have been used for initial selection of clones obtained after transfection of the expression plasmids into appropriate host cells (2,8,61,62). A mouse DHFR cDNA has been used for both initial selection of DHFR resistant CHO cells, as well as gene amplification (8). The selectable/amplifiable marker expression unit (under the control of a weak promoter and polyA addition signal) can either be linked to the heterologous gene expression unit or can be on a separate plasmid that is usually cotransfected into the host cells. With the use of an expression vector containing a linked DHFR gene, the gene copy number in the initially selected $DHFR^+$ clones can be amplified by selecting the clones further in the presence of stepwise increasing concentrations of methotrexate (MTX), a competitive inhibitor of DHFR.

Glutamine synthetase can be used as a dominant selectable/amplifiable marker when cells are grown in the presence of methionine sulfoxamine (MSX), an inhibitor of GS (62) (see later). Other gene amplification markers that have been used to a lesser extent are encoded by adenosine deaminase (ADA) and multiple drug resistance (MDR) genes (8,61).

Transfection Methods

A variety of methods can be used to introduce expression plasmids into mammalian host cells (2,63,64). These include $CaPO_4$, DEAE-dextran, electroporation, and lipofection. Receptor-mediated gene transfer has also been described in the literature (65). Interestingly, a single, standard, optimized method has yet to be described in the literature for establishing mammalian cells expressing heterologous gene products. In a typical stable expression experiment, the plasmid carrying the gene of interest and a drug selectable/amplifiable marker is transfected into host cells using any of the gene transfer methods just mentioned. The transfected cells are selected in the presence of a drug until the drug-resistant colonies are obtained (usually within 2–3 weeks). The resultant colonies are then screened using a variety of methods to determine which are producing the desired heterologous gene product. The detection method is dependent on the nature of the expressed protein, availability of antibodies, and so forth. For secreted proteins, conditioned media can easily be screened by ELISA, western blotting, and so forth. On average, one-third to one-half of the drug-selected colonies produce the protein of interest at a detectable level. The other clones are either expressing the desired gene product at very low levels or not at all, presumably due to inactivation or rearrangements of the heterologous gene.

Growth Media

Growth media and process development play an important role in the successful expression of a heterologous gene product in mammalian cells. Serum-free media has been developed for large-scale growth of CHO and myeloma cells. Serum-free or protein-free media offers the following advantages: *(1)* they are less expensive compared with media containing serum; *(2)* downstream purification of heterologous gene products becomes easier, especially for proteins that are secreted into the media; and *(3)* contaminating agents present in serum are avoided (66–69).

Increases in expression levels of heterologous gene products can normally be achieved through media manipulation and process development. For example, Bebbington et al. (10) have reported a three- to fourfold increase in mAb expression in NSO cells amplified with the GS system when cells were grown in a defined serum-free medium. When an amino acid feed was applied to these cells, an additional increase in expression level up to 560 mg L^{-1} was achieved. Using NSO myeloma cells and the GS system, expression level of a number of mAbs in the range of 0.5–2 g L^{-1} have been reported through medium development, nutrient supplementation, and other process development considerations (70). Similar results have also been obtained with mAbs produced in $DHFR^-$ CHO cells (4,70,71).

Glycosylation patterns of recombinant proteins can vary widely when expressed in different mammalian host cell types. For example, Parekh et al. (72,73) reported different glycosylation patterns for recombinant human tPA expressed in four different cell types: CHO cells, murine epithelial (C127) cells, the Bowes melanoma cells, and human colon fibroblast (CCD-18Co) cells. Cell culture conditions also determine glycosylation pattern of recombinant proteins (for review see Ref. 74). Borys et al. (75) found that culture pH affects expression rate and glycosylation of recombinant mouse placental lactogen proteins produced in CHO cells. The optimum expression rate was found to be in pH range of 7.6 to 8.0. The extent of glycosylation was decreased at low pH (below 6.9) and high pH (above 8.2). N-linked oligosaccarides were found to be different in composition when recombinant human tissue kallikrein was produced by CHO cells grown in serum-free suspension culture or on microcarrier beads (76).

Regulated Expression

Inducible expression systems have been proven to be valuable for elucidating the function of a number of genes in *E. Coli*, yeast, and *Drosophila*. In mammalian cells a tightly regulated expression system may also offer an advantage for expression of those proteins that are potentially cytotoxic. One way to achieve regulated expression of a heterologous gene is through the use of an inducible promoter. A number of inducible promoters have been described in the literature, such as IFN-γ, hsp70, metallothionein (Mtn), and MMTV LTR (for review see Refs. 46 and 77). Following are some popular methods for achieving inducible expression of heterologous genes in mammalian cells.

The Lacswitch System. Bacterial transcription control systems such as the *lac* operator/repressor have been adapted for regulated expression of heterologous genes in mammalian cells (78). The *lac* repressor, made from a *lacI* expression vector, blocks transcription of a reporter gene placed downstream of *lac* operator sequences. Addition of the inducer IPTG results in derepression of reporter gene transcription by up to 10- to 60-fold. Labow et al. (79) modified the *lac* repressor/operator system by converting the *lac* repressor into a transactivator by fusing it to the HSV

VP16 transactivation domain. In the absence of IPTG, this *lac* transactivator stimulates expression from a minimal promoter containing *lac* operator sequences up to 1,000-fold.

The *tet* System. Gossen and Bujard (80) described a regulated expression system based on the control elements of the *E. coli* tetracycline (Tc) resistance operon. In this system (for review see Ref. 81) the *tet* repressor has been fused to the HSV VP16 transactivator domain, and the resultant transactivator, called tTA, stimulates expression from a hCMV minimal promoter linked to *tet* operator sequences. In the presence of Tc (1 μg ml^{-1}), the tTA cannot bind to the operator sequence, and transcription from the minimal promoter shuts off, whereas removal of Tc leads to activation of transcription. In stable HeLa clones expression of a reporter gene has been shown to be modulated up to 10^5-fold. Derivatives of Tc, such as anhydrotetracycline (82) and doxycyclin (83), have also been used to regulate expression using this system.

An expression vector with the *tet*-responsive promoter, tTA expression unit, origin of replication of EBV, EBNA-1 expression unit, and a selectable marker has been described for the use in transient and stable expression of heterologous genes (84). Using this vector, luciferase activity in transiently transfected cells can be regulated by several orders of magnitude by varying both the Tc concentration and incubation time in the presence of Tc.

The *tet*-responsive system has been shown to produce different responses in different cell lines. Howe et al. (85) used the *tet* system to regulate *(1)* expression of dopamine D3 receptor in stable GH3 cells and *(2)* GluR6 (a glutamate receptor subunit) in stable HEK293 cells. In GH3 cells, a high-level induction of dopamine D3 receptor expression was obtained in the absence of Tc, whereas only a modest level induction of GluR6 was seen in HEK293 cells.

Recently, a mutant form of the *tet* repressor has been used to create a transactivator after fusion with the HSV VP16 (83). This transactivator, called rtTA, has reversed DNA binding activity as compared with that of wild-type *tet* repressor or tTA. In the presence doxycycline or anhydrotetracycline, more than 1,000-fold induction has been achieved in stable HeLa clones.

Ecdysone-Inducible System. The *Drosophila* ecdysone (Ec) receptor, a member of the steroid hormone receptor superfamily, can function as a ligand-dependent transcription factor in animal cells (86). Functional EcR is a heterodimer of EcR and the *Drosophila* ultraspiracle gene product (USP). The activity of the EcR is regulated by certain ecdysteroids such as ecdysone (Ec) and not by mammalian hormones such as dexamethasone and estradiol. EcR by itself is incapable of high-affinity DNA binding and transcriptional activation. In the presence of Ec, physical association between EcR and USP is greatly enhanced (87). Structural analysis shows that steroid hormone receptors contain a DNA-binding domain, ligand-binding domain, and transactivation domain. Mammalian retinoic acid receptor (R-X-R) can replace USP to form a functional heterodimer.

Recently, No et al. (88) described an ecdysone-inducible gene expression system for mammalian cells. Mammalian cells cotransfected with EcR and USP gave only a threefold induction upon treatment with a micromolar concentration of muristerone (MurA), a synthetic analog of Ec. Modifications of EcR were made such that the N-terminal transactivation domain in EcR was replaced by the corresponding domain of the glucocorticoid receptor (GR). This hybrid receptor gave a 34-fold induction in a transient reporter gene assay. A more potent heterodimer was made by combining either R-X-R and VpEcR (an N-terminus of EcR was truncated and fused to the VP16 transactivation domain) or R-X-R and VgEcR (which recognizes a hybrid Ec/GRE). Stable lines generated with these regulatory elements and induced with 100 nM MurA gave 20,000-fold induction of reporter gene (β-galactosidase) expression.

All three of the inducible systems are commercially available. For system it is important to determine the level of leaky expression in the absence of inducer (for the *lac* and Ec system) and the presence of Tc (for the *tet* system). The optimal concentration of the inducer and the kinetics of induction need to be determined. More importantly, one needs to determine if the high-fold induction obtained through any inducible system correlates with the desired high-level expression of a recombinant protein.

Dicistronic Vectors

In prokaryotic systems, many mRNAs encode two or more gene products in polycistronic messages. Expression of the distal coding sequences are often regulated by the translation of the upstream coding sequence in these messages. In contrast, eukaryotic mRNAs are not generally polycistronic. However, translation of polycistronic message in eukaryotic systems has now been achieved through the use of an internal ribosome entry site (IRES). IRESs have been identified in the single-stranded RNA genome of picornaviruses such as encephalomyocarditis (EMC) virus, foot and mouth disease virus, and polio virus. Ribosomes bind to a 450- to 600-nt-long secondary structure (IRES) present in a 600- to 1,200-nt-long 5′ UTR followed by an ORF in the viral genome. Kaufman (89) first reported the use of an IRES placed in between the gene of interest (a cDNA-encoding eukaryotic initiation factor 2α [eIF-2α]) and a selectable marker (DHFR). The resultant expression vector is called a dicistronic vector. When transfected into DHFR$^-$ CHO cells, efficient internal translation initiation of DHFR yielded increased transfection efficiency. High-level expression of eIF-2α was observed when the DHFR$^+$ clones were grown in presence of increasing concentration of MTX.

A number of dicistronic vectors have been described in the literature for heterologous gene expression (90,91). In order to increase the expression level of a gene of interest (5′ cistron), vectors have been designed such that the translation initiation of the selectable marker gene (3′ cistron) is reduced. The resultant high-level expression of the heterologous gene in the marker-selected (drug-selected) clones can be due to the integration of the expression plasmid in a transcriptionally active region of the host genome (92). For example, Kobayashi et al. (93) constructed a dicistronic vector in which transcription is under the control of the CMV enhancer/chicken β-actin promoter. The vector contains an IRES of EMC virus downstream of a reporter β-galactosidase gene, followed by a neomycin resistance gene. The translation of *neo*^{r} gene was weakened by placing spacers of various lengths in between AUG-11 (located at the 3′ end of IRES) and the AUG of *neo*^{r}. The number of Neo^r 3T3 clones obtained after transfection of the dicistronic vector with the weakened *neo*^{r} was 4- to 7-fold less compared with that from a conventional dicistronic vector, but these clones gave about 10-fold higher levels of *β-gal* activity. A similar strategy of weakening the selectable marker gene expression from a dicistronic vector was described by Rees et al. (94). Expression levels of serotonin receptors in 10 independent stable HEK293 clones varied from 1.5 to 11.9 pmol mg^{-1} cellular protein, which was significantly higher than the reported expression levels (98 fmol mg^{-1} in HeLa to 4 pmol mg^{-1} in Lmtk$^-$ cells).

Lucas et al. (95) reported a dicistronic intron vector (or DI vector) in which the mouse DHFR cDNA is placed into the intron of the pRK vector (96) and is under the transcriptional control of the CMV enhancer/promoter. A cDNA encoding a gene of interest (e.g., tPA or TPO) was placed downstream of the intron. When transfected into DHFR-CHO cells, full-length mRNA produced by DI vectors was translated into DHFR, while spliced mRNA produced the desired recombinant protein. DI-vector-derived CHO clones grown for 5 days gave 11-fold more tPA (~1 mg L^{-1}) and 13-fold more TPO than were produced by conventional vector-derived cells. The DI vector system has also been used to produce antibodies at levels ~4–5 mg L^{-1} 10 day^{-1}. The weakening of DHFR expression and/or its linkage to product expression gave fewer clones, but they had more productivity.

APPLICATIONS OF MAMMALIAN CELL EXPRESSION TECHNOLOGY

Specific Examples of CHO Cell Expression

Growth Factors. The following examples describe the use of DHFR$^-$ CHO cells for expression of various growth factors. The expression vectors were either cotransfected with a DHFR expression plasmid to select the DHFR$^+$ clones or contained their own DHFR expression cassette. Expression levels of the growth factors were increased via amplification in the presence of increasing concentrations of MTX, as already described.

Madisen et al. (97) achieved high-level expression of TGF-β2 and a precursor TGF-β2 (414) in CHO cells using an expression vector similar to that described by Gentry et al. (98) for TGF-β1. A pSV2 expression vector containing the genes encoding TGF-β2 (414) and DHFR was utilized to obtain DHFR$^+$ clones. These clones were subsequently adapted to 0.5, 2.5, 10.0, and 50 μM MTX. Two clones grown at 10 and 50 μM MTX were analyzed for expression of recombinant mature TGF-β2 (12 kDa under reducing conditions). In addition to the mature protein, proteins of 85, 105, and 130 kDa, consisting of both mature and proregion sequences were also produced, as determined by immunoblotting. The pro-TGF-β2 was found to be both glycosylated and phosphorylated.

Bourdrel et al. (99) expressed recombinant human TGF-β1 after subcloning the cDNA into the pDSVα2 expression vector, a derivative of the pCD vector (100) in which the SV40 polyA addition signal was replaced by the polyA addition signal of the α-subunit of bovine follicular stimulating hormone and a mouse DHFR minigene (downstream of the expression cassette). CHO cells were transfected using the $CaPO_4$ method, and a single DHFR$^+$ clone expressing human TGF-β1 at 7 mg L^{-1} in the conditioned media was identified.

Ferrara et al. (101) used CHO cells to produce the full-length vascular endothelial growth factor (VEGF) after subcloning the cDNA encoding the full-length human VEGF cDNA into a derivative of the vector pCISh (49), which carries a DHFR gene. The transcription of VEGF cDNA and that of the DHFR are both under the control of the SV40 promoter. The initial selected clones, which released the highest mitogenic activity, were amplified in the presence of increasing concentration of MTX, resulting in clones that were resistant to 1 μM MTX and produced 0.5 μg ml^{-1} day^{-1} of VEGF.

The human neurotrophin 3 (NT-3) gene product was purified and characterized after expression in CHO cells (102). A construct that contained a chimera encoding the human nerve growth factor pre-pro region and NT-3 mature region under the control of MuLV-LTR was transfected into CHO cells. DHFR$^+$ clones were serially subcultured in media containing 0.1, 1.0, and 10 μM MTX, yielding clones expressing as much as 1 mg recombinant NT-3 L^{-1}. The purified recombinant NT-3 was biologically active, as shown through the induction of neurite outgrowth in neurons prepared from an 8-day-old chicks.

Burgess et al. (103) used CHO cells to study biosynthetic processing of the rat *neu* differentiation factors (NDFs). pDSRα2-derived vectors (104) containing NDF-β2c and NDF-β4a were transfected using the $CaPO_4$ method, and the selected clones were screened by western blotting. High-expressing clones for NDF-α2c and NDF-β4a were identified. Expression of NDF-β4a was gradually amplified with MTX (70 nM). The precursor of NDF undergoes typical glycosylation, and a portion of the molecule is proteolytically cleaved, leading to secretion of soluble mature NDF. A portion of the newly synthesized precursors escape processing and are retained on the cell surface.

Monoclonal Antibodies. In recent years DHFR$^-$ CHO cells have been successfully used to produce a number of recombinant mAbs at a high level (4,51,71,105–107). For example, Page and Sydenham (105) used DUK-B11 CHO cells to obtain expression of the humanized mAb Campath-1 at levels of 100 μg 10^6 $cells^{-1}$ day^{-1}. Using the same host cells, Fouser et al. (51) reported expression of a chimeric ganglioside GD2 antibody at 80–110 μg 10^6 $cells^{-1}$ day^{-1}. Recently, Reff et al. (106) expressed a chimeric antibody against CD20 in DG44 CHO cells at levels of 30 μg 10^6 $cells^{-1}$ day^{-1}, using only one round of gene amplification with MTX.

A variety of promoters have been used to obtain expression of H- and L-chains in these CHO cells, for example, the β-actin promoter (105), the human CMV MIE promoter (106), the Ad virus major late promoter (MLP) (108), the RSV promoter (4), and a murine leukemia virus LTR (109). A number of vectors for mAb expression are described in the literature in which the two Ig chains are carried by two different plasmids with an independent selectable/amplifiable marker. For example, Fouser et al. (51) linked the Ig H-chain with the ADA gene, and the L-chain with the DHFR gene. Using these vectors, an expression level of 80–110 μg of mAb 10^6 cells^{-1} day^{-1} was achieved. Trill et al. (4) have used vectors containing H-chain linked to a DHFR marker, and an L-chain expression cassette with the Neo^r marker and vice versa to obtain ~180 mg of a humanized mAb L^{-1} 7 day^{-1} in spinner flasks. The methods used for initial selection and subsequent amplification vary from laboratory to laboratory. In general, high-level mAb expression can be obtained using the following steps: initial selection and subsequent amplification of candidate clones, coselection (e.g., in cases where both H-chain and L-chain expression vectors carry DHFR expression unit) and amplification, coamplification using different amplifiable markers, and initial selection and amplification in mass culture, followed by dilution cloning to identify individual high-expressing clones.

Because integration sites may influence the efficiency of H-chain and L-chain expression and overall mAb expression, single vectors have been created in which the two Ig-chain expression units are placed in tandem. These vectors also carry a dominant selectable marker such as Neo^r and the DHFR expression cassette (4,105,107,109). Using the single-vector system and two rounds of amplification, Newman et al. (107) achieved high-level expression (>200 μg ml^{-1}) of a primatized anti-CD4 mAb. Similar high-level expression of two different humanized mAbs has recently been reported by Trill et al. (4).

Enzymes. Recently, Laubach et al. (110) used the GS system to achieve high-level expression of the human inducible nitric oxide synthase (iNOS) in CHO-K1 cells. The iNOS expression plasmid, a derivative of pEE14 vector (Celltech Ltd., U.K.) was transfected into CHO cells using Lipofectamine™ and selected with 25 μM MSX in the presence of ethylisothiourea (EITU), an inhibitor of iNOS (to reduce toxicity due to NO production). Further amplification of initially selected clones in the presence of 400 μM MSX gave stable clones with activity > 3 nmole min^{-1} mg^{-1}. Sodium butyrate treatment of cells increased the level of expression by another 1.5-fold.

Receptors. A coding sequence for the soluble form of CD4 (111) was constructed through the introduction of a translation termination codon at the extracellular boundary of the transmembrane domain, and subcloning the cDNA downstream of the SV40 early promoter and upstream of the bovine growth hormone poly A addition signal. This expression cassette was linked to a mouse DHFR expression cassette, and the resultant plasmid was transfected into DHFR$^-$ CHO DXB-11 cells. The initially selected DHFR$^+$ clones were exposed to increasing concentration of MTX, and the expression level in several clones was >3pg cell^{-1} day^{-1}. This specific productivity translated into an expression level of ~40 mg of sCD4 L^{-1} under high-density suspension culture conditions.

Newman-Tancredi et al. (112) reported high-level expression of recombinant 5-HT1A receptors in CHO cells using a derivative of pSVL (Pharmacia). DHFR-CHO cells were cotransfected with pSV2dhfr (113) using the $CaPO_4$ method. Out of 35 DHFR$^+$ clones tested for expression, 33 had receptor expression at 20–3,000 fmol mg^{-1}, as determined by ^{3}H-8OH-DPAT binding assays. The highest-expressing clone (expressing at 2.8 pmol receptor mg protein^{-1}) was identified, and functional coupling was demonstrated through the action of a series of 5-hydroxytryptamine agonists that inhibited forskoline-stimulated adenyl cyclase (AC) activity. The receptor density and receptor regulation are maintained in suspension grown cells, but the inhibition of AC by the receptor is lost gradually (probably due to a change in G-protein–AC interaction and not due to a change in receptor–G-protein coupling).

The GS system has also been utilized to express a 7TM receptor for human thyrotropin (TSH) in CHO cells (114). Cell lines expressing up to 78,000 receptors, cell^{-1}, which efficiently couple to AC, have been identified.

Adehesion Molecules. The soluble form of human E-selectin (sE-selectin) has been expressed at a high level in DHFR$^-$ CHO cells (115). The sE-selectin cDNA was subcloned into the RLDN expression vector (in which expression of sE-selectin was under the control of RSV promoter) and transfected into CHO cells. The G418^r clones expressing sE-selectin were further selected stepwise in media containing 10 nM and 100 nM MTX. A 100 nM MTX^r clone was identified. In spinner flasks it produced 100 mg L^{-1} of sE-selectin over a 5-day period.

Chemokines. Liao et al. (116) expressed recombinant human Mig, a CXC chemokine, after subcloning the Mig cDNA into the vector pMSXND. Expression of Mig was under the control of the Mtn I promoter. The expression vector contained the SV40 small t-antigen intron and poly A addition signal, a DHFR expression cassette under the control of the SV40 early promoter, and a *neo*^{r} under the control of the tk promoter. The construct was transfected into CHO cells (proline$^-$), and the *neo*^{r} clones were subjected to amplification using 0.2 μM MTX. The MTX^r clones were induced with $CdSO_4$, and the conditioned media screened by SDS-PAGE showed a collection of immunoreactive polypeptides (due to protein processing of the high molecular weight Mig), which were similar to that secreted from IFN-treated peripheral blood monocytes and THP-1 cells.

Other Examples. Cockett et al. (117) developed the GS system for high-level expression of heterologous genes in CHO cells. Transfection of an expression vector containing a tissue inhibitor of metalloproteinases (TIMP) cDNA (under the transcriptional control of the hCMV promoter) and a GS mini gene (under the control of the SV40 late promoter) into CHO-K1 cells (followed by selection with 20 μM

MSX) yielded clones expressing up to 9 μg TIMP 10^6 cells^{-1} day^{-1}. Further selection with higher concentration of MSX (500 μM) produced a clone expressing TIMP at 110 μg 10^6 cells^{-1} day^{-1}.

Yonemura et al. (118) have developed a high-level expression system for blood coagulation factor VIII (consisting of a 90-kDa H-chain and an 80-kDa L-chain). The expression plasmids contained a chicken β-actin/rabbit β-globin hybrid promoter (119) and a DHFR transcription unit under the control of the SV40 early promoter. DHFR$^-$ CHO cells (DG44) were cotransfected with H- and L-chain expression vectors. After selecting for DHFR$^+$ clones (using nucleoside-free media), cells that tested positive in a precoagulation assay were cultured in increasing MTX concentrations (upto 1,000 nM) for gene amplification. A clone resistant to 1,000 nM MTX was established that produced 15 IU 10^6 cells^{-1} day^{-1} of factor VIII activity.

Specific Examples of Myeloma Cell Expression

Monoclonal Antibodies. High-level production of mAbs have been reported by several groups (9–12). For example, Gillies et al. (9) expressed two chimeric mAbs, Ch14.18 and ChB72.3, in Sp2/0 cells. H-chain V-region cDNA and a hybrid human–mouse L-chain V-region cDNA were fused to the their respective genomic constant regions. The H-chain expression unit (under the control of the mouse H-chain enhancer/metallothionein MT I promoter) was linked to a DHFR expression cassette. The L-chain expression unit (under the control of a synthetic murine K-promoter) was subcloned into the H-chain expression plasmid to generate a single vector. The expression vector was transfected into Sp2/0 cells, and the MTX^r clones secreting antibody were adapted to increasing concentration of MTX. Subclones of 5 μM MTX^r clones had expression levels of 120–180 μg ml^{-1}. Shitara et al. (12) constructed a "tandem vector" in which both H- and L-chain cDNA expression units (both under the control of Moloney virus LTR), DHFR, and Neo^r expression units were linked. Expression from this vector was compared in three host cells: mouse Sp2/0, P3X63Ag8U.1, and rat YB2/0 cells. YB2/0 cells gave higher numbers of G418^r clones and higher levels of mAb expression. Subsequent amplifications with 50–200 nM MTX yielded several clones secreting 70–100 μg of mAb 10^6 cells^{-1} day^{-1}.

The GS system has been widely utilized for antibody expression in myeloma cells. Bebbington et al. (10) reported high-level expression of a chimeric antibody in NSO myeloma cells. An expression vector was created after placing in tandem a GS expression cassette, the H-chain transcription unit, and the L-chain transcription unit. The expression of the GS gene was under the control of the SV40 early promoter, while the expression of both H-chain and L-chain was under the transcriptional control of hCMV promoter. Transfection of this expression plasmid into NSO myeloma cells, followed by selection in glutamine free medium, gave a pool of clones expressing up to ~4 pg antibody cell^{-1} day^{-1}. Further selection with 20–100 μM MSX gave rise to the highest-expressing clone, with a specific productivity of about 10 pg cell^{-1} day^{-1}. Southern blot analysis showed that this cell line had approximately four copies of the vector per cell.

CONCLUSIONS

As described in this review, a large number of expression systems are available for heterologous expression of a wide variety of proteins in mammalian cells. The choice of a particular host/vector system depends primarily on the desired expression level, nature of expression (transient vs. stable), intracellular versus extracellular expression, constitutive versus inducibile expression, and so forth. Using the expression systems described, it is now possible to produce therapeutic proteins such as mAbs at very high levels (~1–2 gm L^{-1}). However, the challenge still remains to produce such therapeutic proteins in an even more economical manner. Therefore, alternate expression systems still need to be developed to *(1)* achieve even higher levels of expression and *(3)* reduce cycle time (the time interval between construction of an expression vector to production of the protein of interest in large scale). Further understanding of cellular processes such as transcription, translation, and secretion will be essential to achieve such a goal.

NOMENCLATURE

AC	Adenyl cyclase
ADA	adenosine deaminase
BHK	Baby hamster kidney
$CaPO_4$	Calcium phosphate
CAT	Chloramphenicol acetyl transferase
$CdSO_4$	Cadmium sulfate
CGRP	Calcitonin-gene-related peptide
CHO	Chinese hamster ovary
DHFR	Dihydrofolate reductase
DI	Dicistronic intron
EBV	Epstein–Barr virus
EcR	Ecdysone receptor
EMC	Encephalomyocarditis
GH	Growth hormone
GR	Glucocorticoid receptor
GRE	Glucocorticoid response element
hCMV	Human cytomegalo virus
HEK	Human embryonic kidney
hsp	Heat shock protein
HSV	Herpes simplex virus
IE	Immediate early
IFN	Interferon
IgG	Immunoglobulin
IPTG	Isopropyl β-thiogalactopyranoside
IRES	Internal ribosome entry site
L	Light
LTR	Long terminal repeat
mAB	Monoclonal antibody
MCP	Monocyte chemotactic protein
MLP	Major late promoter
MMTV	Mouse mammary tumor virus

MSX	Methionine sulfoxamine
MTN	Metallothionein
MTX	methotrexate
MuLV	Murine leukemia virus
MurA	Muristerone
NOS	Nitric oxide synthase
8-OH-DPAT	2-(*N,N*-dipropylamino)-8-hydroxy-1,2,3,4-tetrahydronapthalate
ORF	Open reading frame
PTH	Parathyroid hormone
RSV	Rous sarcoma virus
SV	Simian virus
TGF	Transforming growth factor
TIMP	Tissue inhibitor of metalloproteinase
TM	Transmembrane
tPA	Tissue plasminogen activator
TPO	Thrombopoietin
UTR	Untranslated region
VEGF	Vascular endothelial growth factor

BIBLIOGRAPHY

1. A.D. Levinson, *Methods Enzymol.* **185**, 485–487 (1990).
2. M. Kriegler, *Gene Transfer and Expression: A Laboratory Manual*, Stockton Press, New York, 1990.
3. S.E. Kane, in J.K. Setlow ed., *Genetic Engineering*, vol. 13, Plenum, New York, 1991, pp. 167–182.
4. J.J. Trill, A.R. Shatzman, and S. Ganguly, *Curr. Opin. Biotechnol.* **6**, 553–560 (1995).
5. G. Gross and H. Hauser. *J. Biotechnol.* **41**, 91–110 (1995).
6. G. Urlaub, E. Kas, A.M. Carothers, and L.A. Chasin. *Cell* **33**, 405–412 (1983).
7. G. Urlaub and L.A. Chasin, *Proc. Natl. Acad. Sci. U.S.A.* **77**, 4216–4220 (1980).
8. R.J. Kaufman, *Methods Enzymol.* **185**, 537–566 (1990).
9. S.D. Gillies, K-M. Lo, and J. Wesolowski, *J. Immunol. Methods* **125**, 191–202 (1989).
10. C.R. Bebbington, G. Renner, S. Thomson, D. King, D. Abrams, and G.T. Yarronton, *Bio/Technology* **10**, 169–175 (1992).
11. M.A. Walls, K.-c. Hsiao, and L.J. Harris, *Nucleic Acids Res.* **21**, 2921–2929 (1993).
12. K. Shitara, K. Nakamura, Y. Tokutake-Tanaka, M. Fukushima, and N. Hanai, *J. Immunol. Methods* **167**, 271–278 (1994).
13. R.J. Owens and R.J. Young, *J. Immunol. Methods* **168**, 149–165 (1994).
14. C.M. Gorman, D. Gies, G. McCray, and M. Huang, *Virology* **171**, 377–385, (1989).
15. F. Gotzes and A. Baumann, *Biochem. Biophys. Res. Commun.* **222**, 121–126 (1996).
16. M. Pines, S. Fukayama, K. Costas, E. Meurer, P.K. Goldsmith, X. Xu, S. Muallem, V. Behar, M. Chorev, M. Rosenblatt, A.H. Tashijan, Jr., and L.J. Suva, *Bone* **18**, 381–389 (1996).
17. L.M. Tolbert and J. Lameh, *J. Biol. Chem.* **271**, 17335–17342 (1996).
18. N. Aiyar, K. Rand, N.A. Elshourbagy, Z. Zeng, J.E. Adamou, D.J. Bergsma, and Y.A. Li, *J. Biol. Chem.* **271**, 11325–11329 (1996).
19. S. Tippmer, B. Bossenmaier, and H. Haring, *Eur. J. Biochem.* **236**, 953–959 (1996).
20. C. Combadiere, S.K. Ahuja, H.L. Tiffany, and P.M. Murphy, *J. Leukoc. Biol.* **60**, 147–152 (1996).
21. T. Kurihara and R. Bravo, *J. Biol. Chem.* **271**, 11603–11606 (1996).
22. J-L. Gao, A.I. Sen, M. Kitaura, O. Yoshie, M.E. Rothenberg, P.M. Murphy, and A.D. Luster, *Biochem. Biophys. Res. Commun.* **223**, 679–684 (1996).
23. A. Ben-Baruch, L. Xu, P.R. Young, K. Bengali, J.J. Oppenheim, and J.M. Wang, *J. Biol. Chem.* **270**, 22123–22128 (1995).
24. S. Yamagami, Y. Tokuda, K. Ishii, H. Tanaka, and N. Endo. *Biochem. Biophys. Res. Commun.* **202**, 1156–1162 (1994).
25. P. Martasek, Q. Liu, J. Liu, L.J. Roman, S.S. Gross, W.C. Sessa, and B.S.S. Masters, *Biochem. Biophys. Res. Commun.* **219**, 359–365 (1996).
26. T. Hatae, S. Hara, C. Yokoyama, T. Yabuki, H. Inoue, V. Ullrich, and T. Tanabe, *FEBS Lett.* **389**, 268–272 (1996).
27. R.M. Pitti, S.A. Marsters, M. Haak-Frendscho, G.C. Osaka, J. Mordenti, S.M. Chamow, and A. Ashkenazi, *Mol. Immunol.* **31**, 1345–1351 (1994)
28. R.G. Clark, D. Mortensen, D. Reifsynder, M. Mohler, T. Etcheverry, and V. Mukku, *Growth Regul.* **3**, 50–52 (1993)
29. A.J. Pelletier, S.C. Bodary, and A.D. Levinson, *Mol. Biol. Cell.* **3**, 989–998 (1992)
30. L.R. Poborsky, B.M. Fendly, K.L. Fisher, R.M. Lawn, B.J. Marks, G. McCray, K.M. Tate, G.A. Vehar, and C.M. Gorman, *Protein Eng.* **3**, 547–553 (1990).
31. P. Mellon, V. Parker, Y. Gluzman, and T. Maniatis, *Cell* **27**, 279–288 (1981).
32. S.B. Patel and S.M. Grundy, *J. Lipid Res.* **36**, 2090–2103 (1995).
33. S.P.S. Svensson, T.J. Bailey, A.C. Porter, J.G. Richman, and J.W. Regan, *Biochem. Pharmacol.* **51**, 291–300 (1996).
34. T. MacNeil, K.K. Bierilo, J.G. Menke, and J.F. Hess, *Biochim. Biophys. Acta* **1264**, 223–228 (1995).
35. R. Weinander, E. Mosialou, J. DeJong, C-P.D. Tu, J. Dypbukt, T. Bergman, H.J. Barnes, J-O. Hoog, and R. Morgenstern, *Biochem J.* **311**, 861–866 (1995).
36. K. Denner, J. Doehmer, and R. Bernhardt, *Endocrinol. Res.* **21**, 443–448 (1995).
37. M.I. Cockett, C.R. Bebbington, and G.T. Yarronton, *Nucleic Acids Res.* **19**, 319–325 (1991).
38. E. Bender, J.M. Woof, J.D. Atkin, M.D. Barker, C.R. Bebbington, and D.R. Burton, *Hum. Antibod. Hybridomas* **4**, 74–79 (1993).
39. H. Dorai, J.E. McCartney, R.M. Hudziak, M-S. Tai, A.A. Laminet, L.L. Houston, J.S. Huston, and H. Oppermann, *Bio/Technology* **12**, 890–897 (1994).
40. P.J. Hippenmeyer, A.M. Rankin, B.A. Reitz, D.R. McWilliams, B.B. Brightwell, R.A. Wolfe and T.G. Warren, *Mol. Cell. Endocrinol.* **107**, 155–164 (1995).
41. P.J. Hippenmeyer and M. Highkin, *Bio/Technology* **11**, 1037–1041 (1993).
42. P.J. Hippenmeyer and L.E. Pegg, *Curr. Opin. Biotechnol.* **6**, 548–552 (1995).
43. R.F. Margolskee, P. Kavathas, and P. Berg, *Mol. Cell. Biol.* **8**, 2837–2847 (1988).

44. G. Cachianes, C. Ho, R.F. Weber, S.R. Williams, D.V. Goeddel, and D.W. Leung, *Biotechniques* **15**, 255–259 (1993).
45. A. Lieber, V. Sandig, W. Sommer, S. Bahring, and M. Strauss, *Methods Enzymol*. **217**, 47–66 (1993).
46. R.J. Kaufman, *Methods Enzymol.* **185**, 487–511 (1990).
47. M. Kriegler, *Methods Enzymol.* **185**, 512–527 (1990).
48. Y. Takebe, M. Seiki, J-I. Fujisawa, P. Hoy, K. Yokota, K-I. Arai, M. Yoshida, and N. Arai, *Mol. Cell. Biol.* **8**, 466–472 (1988).
49. C.M. Gorman, D.R. Gies, and G. McCray, *DNA Protein Eng. Techn.* **2**, 3–10 (1990).
50. A.R. Buchman and P. Berg, *Mol. Cell. Biol.* **8**, 4395–4405 (1988).
51. L.A. Fouser, S.L. Swanberg, B-Y. Lin, M. Benedict, K. Kelleher, D.A. Cumming, and G.E. Riedel, *Bio / Technology* **10**, 1121–1127 (1992).
52. M.T.F. Huang and C.M. Gorman, *Mol. Cell. Biol.* **10**, 1805–1810 (1990).
53. M. Kozak, *J. Cell Biol.* **115**, 887–903 (1991).
54. C-L. Ho, S.S.M. Chin, K. Carnevale, and R.K.H. Liem, *Eur. J. Cell Biol.* **68**, 103–112 (1995).
55. R. Turner and G.D. Foster, *Mol. Biotecnol.* **3**, 225–236 (1995).
56. W.C. Merrick, *Microbiol. Rev.* **56**, 291–315 (1992).
57. V.M. Pain, *Eur. J. Biochem.* **236**, 747–771 (1996).
58. W. Sheay, S. Nelson, I. Martinez, T.-H.T. Chu, S. Bhatia, and R. Dornburg, *Biotechniques*, **15**, 856–862 (1993).
59. Br. Pat. GB2284210-A (November 25, 1994), British Technology Group, Ltd.
60. D. Falcone and D.W. Andrews, *Mol. Cell. Biol.* **11**, 2656–2664 (1991).
61. R.E. Kellems, *Curr. Opin. Biotechnol.* **2**, 723–729 (1991).
62. C.R. Bebbington and C.C.G. Hentschel, in D. Glover ed., *DNA Cloning*, vol. III, Academic, New York, 1987, pp. 163–188.
63. W.A. Keown, C.R. Campbell, and R.S. Kucherlapati, *Methods Enzymol.* **185**, 527–537 (1990).
64. J. Barsoum, in J.A. Nickoloff ed., *Methods in Molecular Biology*, vol. 48 (*Animal Cell Electroporation and Electrofusion Protocols*), Humana Press, Totowa, New Jersey, 1995, pp. 225–237.
65. E. Wagner, M. Cotten, R. Foisner, and M.L. Birnstiel, *Proc. Natl. Acad. Sci. U.S.A.* **88**, 4255–4259 (1991).
66. M. Ogata, K-I. Wakita, K. Kimura, Y. Marumoto, K. Oh.-i., and S. Shimizu, *Appl. Microbiol. Biotechnol.* **38**, 520–525 (1993).
67. C. Gandor, C. Leist, A. Fiechter, and F.A.M. Asselbergs, *FEBS Lett.* **377**, 290–294 (1995).
68. A.S. Lubiniecki, K. Anumula, J. Callaway, J. L'Italien, M. Oka, B. Okita, G. Wasserman, D. Zabriskie, R. Arathoon, S. Builder, R. Garnick, M. Wiebe, and J. Browne, *Dev. Biol. Stand.* **76**, 105–115 (1992).
69. P.G. Zaworski and G.S. Gill, *Biotechniques*, **15**, 863–866 (1993).
70. T.A. Bibila and D.K. Robinson, *Biotechnol. Prog.* **11**, 1–13 (1995).
71. M. Reff, *Curr. Opin. Biotechnol*, **4**, 573–576 (1993).
72. R.B. Parekh, R.A. Dwek, J.R. Thomas, G. Opdenakker, T.W. Rademacher, A.J. Wittwer, S.C. Howard, R. Nelson, N.R. Siegel, M.G. Jennings, N.K. Harakas, and J. Feder, *Biochemistry* **28**, 7644–7662 (1989).
73. R.B. Parekh, R.A. Dwek, P.M. Rudd, J.R. Thomas, T.W. Rademacher, T. Warren, T.-C. Wun, B. Hebert, B. Reitz, M. Palmier, T. Ramabhadran, and D.C. Tiemeier, *Biochemistry* **28**, 7670–7679 (1989).
74. C.F. Goochee, M.J. Gramer, D.C. Andersen, J.B. Bahr, and J.R. Rasmussen, *Bio / Technology* **9**, 1347–1355 (1991).
75. M. Borys, D.I.H. Linzer, and E.T. Papoutsakis, *Bio / Technology*, **11**, 720–724 (1993).
76. E. Watson, B. Shah, L. Leiderman, Y-R. Hsu, S. Karkare, H.S. Lu, and F-K. Lin, *Biotechnol Prog.* **10**, 39–44 (1994).
77. G.T. Yarronton, *Curr. Opin. Biotechnol.* **3**, 506–511 (1992).
78. M.C.-T. Hu and N. Davidson , *Cell* **48**, 555–566 (1987).
79. M.A. Labow, S.B. Baim, T. Shenk, and A.J. Levine, *Mol. Cell. Biol.* **10**, 3343–3356 (1990).
80. M. Gossen and H. Bujard, *Proc. Natl. Acad. Sci. U.S.A.* **89**, 5547–5551 (1992).
81. M. Gossen, A.L. Bonin, S. Freundlieb, and H. Bujard, *Curr. Opin. Biotechnol.* **5**, 516–520 (1994).
82. M. Gossen and H. Bujard, *Nucleic Acids Res.* **21**, 4411–4412 (1993).
83. M. Gossen, S. Freundlieb, G. Bender, G. Muller, W. Hillen, and H. Bujard, *Science*, **268**, 1766–1769 (1995).
84. Z. Lang and J.M. Feingold, *Gene*, **168**, 169–171 (1996).
85. J.R. Howe, B.V. Skryabin, S.M. Belcher, C.A. Zerillo, and C. Schmauss, *J. Biol. Chem.* **270**, 14168–14174 (1995).
86. K.S. Christopherson, M.R. Mark, V. Bajaj, and P.J. Godowski, *Proc. Natl. Acad. Sci. U.S.A.* **89**, 6314–6318 (1992).
87. T-P. Yao, B.M. Forman, Z. Jiang, L. Cherbas, J.-D. Chen, M. McKeown, P. Cherbas, and R.M. Evans, *Nature* **366**, 476–479 (1993)
88. D. No, T-P. Yao, and R.M. Evans, *Proc. Natl. Acad. Sci. U.S.A.* **93**, 3346–3351 (1996).
89. R.J. Kaufman, M.V. Davies, L.C. Wasley, and D. Michnick, *Nucleic Acids Res.* **19**, 4485–4490 (1991).
90. W. Dirks, M. Wirth, and H. Hauser, *Gene* **128**, 247–249 (1993).
91. D.L. Cadena and G.N. Gill, *Protein Expr. Purif.* **4**, 177–186 (1993).
92. M.V. Davies and R.J. Kaufman, *J. Virol.* **66**, 1924–1932 (1992).
93. M. Kobayashi, Y. Yamauchi, A. Tanaka, and S. Shimamura, *Biotechniques* **21**, 398–402 (1996).
94. S. Rees, J. Coote, J. Stables, S. Goodson, S. Harris, and M.G. Lee *Biotechniques,* **20**, 102–110 (1996).
95. B.K. Lucas, L.M. Giere, R.A. DeMarco, A. Shen, V. Chisholm, and C.W. Crowley. *Nucleic Acids. Res.* **24**, 1774–1779 (1996).
96. L.J. Suva, G.A. Winslow, R.E.H. Wettenhall, R.G. Hammonds, J.M. Moseley, H. Diefenbach-Jagger, C.P. Rodda, B.E. Kemp, H. Rodriguez, E.Y. Chen, P.J. Hudson, T.J. Martin, and W.I. Wood, *Science* **237**, 893–896 (1987).
97. L. Madisen, M.N. Lioubin, H. Marquardt, and A.F. Purchio. *Growth Factors* **3**, 129–138 (1990).
98. L.E. Gentry, N.R. Webb, G.J. Lim, A.M. Brunner, J.E. Ranchalis, D.R. Twardzik, M.N. Lioubin, H. Marquardt, and A.F. Purchio, *Mol. Cell. Biol.* **7**, 3418–3427 (1987).
99. L. Bourdrel, C-H. Lin, S.L. Lauren, R.H. Elmore, B.J. Sugarman, S. Hu, and K.R. Westcott. *Protein Expr. Purif.* **4**, 130–140 (1993).
100. H. Okayama and P. Berg, *Mol. Cell. Biol.* **3**, 280–289 (1983).
101. N. Ferrara, J. Winer, T. Burton, A. Rowland, M. Siegel, H.S. Phillips, T. Terrell, G.A. Keller, and A.D. Levinson, *J. Clin. Invest.* **91**, 160–170 (1993).

102. M. Iwane, T. Watanabe, A. Shintani, Y. Kaisho, S. Matsumoto, R. Sasada, and K. Igarahasi, *Appl. Microbiol. Biotechnol.* **41**, 225–232 (1994).
103. T.L. Burgess, S.L. Ross, Y.-x. Qian, D. Brankow, and S. Hu. *J. Biol. Chem.* **270**, 19188–19196 (1995).
104. Y.A. DeClerck, T-D. Yean, H.S. Lu, J. Ting, and K.E. Langley, *J. Biol. Chem.* **266**, 3893–3899 (1991).
105. M.J. Page and M.A. Sydenham, *Bio/Technology* **9**, 64–68 (1991).
106. M.E. Reff, K. Carner, K.S. Chambers, P.C. Chinn, J.E. Leonard, R. Raab, R.A. Newman, N. Hanna, and D.R. Anderson, *Blood* **83**, 435–445 (1994).
107. R. Newman, J. Alberts, D. Anderson, K. Carner, C. Heard, F. Norton, R. Raab, M. Reff, S. Shuey, and N. Hanna, *Bio/Technology* **10**, 1455–1460 (1992).
108. C.R. Wood, A.J. Dorner, G.E. Morris, E.M. Alderman, D. Wilson, R.M. O'Hara, and R.J. Kaufman, *J. Inmmunol.* **145**, 3011–3016 (1990).
109. H. Tada, T. Kurokawa, T. Seita, T. Watanabe, and S. Iwasa, *J. Biotechnol.* **33**, 157–174 (1994).
110. V.E. Laubach, E.P. Garvey, and P.A. Sherman, *Biochem. Biophys. Res. Commun.* **218**, 802–807 (1996).
111. K.C. Deen, J.S. McDougal, R. Inacker, G. Folena-Wasserman, J. Arthos, J. Rosenberg, P.J. Maddon, R. Axel, and R.W. Sweet, *Nature* **331**, 82–84 (1988).
112. A. Newman-Tancredi, R. Wootton, and P.G. Strange, *Biochem J.* **285**, 933–938 (1992).
113. S. Subramani, R. Mulligan, and P. Berg, *Mol. Cell. Biol.* **1**, 854–864 (1981).
114. E. Harfst and A.P. Johnstone, *Anal. Biochem.* **207**, 80–84 (1992).
115. P. Hensley, P.J. McDevitt, I. Brooks, J.J. Trill, J.A. Feild, D.E. McNulty, J.R. Connor, D.E. Griswold, V. Kumar, K.D. Kopple, S.A. Carr, B.J. Dalton, and K. Johanson. *J. Biol. Chem.* **269**, 1–10 (1994).
116. F. Liao, R.L. Rabin, J.R. Yannelli, L.G. Koniaris, P. Vanguri, and J.M. Farber, *J. Exp. Med.* **182**, 1301–1314 (1995).
117. M.I. Cockett, C.R. Bebbington, and G.T. Yarranton. *Bio/Technology* **8**, 662–667 (1990).
118. H. Yonemura, K. Sugawara, K. Nakashima, Y. Nakahara, T. Hamamoto, I. Mimaki, K. Yokomizo, Y. Tajima, K.-i. Masuda, and A. Imaizumi, A. Funatsu, and J.-i. Miyazaki, *Protein Eng.* **6**, 669–674 (1993).
119. I.-i. Miyazaki, S. Takaki, K. Araki, F. Tashiro, A. Tominaga, K. Takatsu, and K.-i. Yamamura, *Gene* **79**, 269–277 (1989).

See also CHINESE HAMSTER OVARY CELLS, RECOMBINANT PROTEIN PRODUCTION; EXPRESSION SYSTEMS, *E. COLI*; PICHIA, OPTIMIZATION OF PROTEIN EXPRESSION.

F

FERMENTATION MONITORING, DESIGN AND OPTIMIZATION

Jens Nielsen
Technical University of Denmark
Lyngby, Denmark

KEY WORDS

Fermentation
High-performance bioreactors
Metabolic control analysis
Metabolic engineering
Metabolic flux analysis
On-line monitoring
Quantitative physiology

OUTLINE

INTRODUCTION

The term *fermentation* is derived from Latin *fermentum* to ferment, and it has been used to describe the metabolism of sugars by microorganisms since ancient times. Thus, fermentation of fruits is so old that ancient Greeks attributed its discovery to one of their gods, Dionysos. Among the classical fermentation processes are beer brewing, which has documented from as early as 3000 B.C. in Babylonia, soya sauce production in Japan and China, and fermented milk beverages in the Balkans and in Central Asia (1). Before World War II, however, fermentation processes mainly found their application in the food area, and it was with the introduction of the penicillin production that large-scale fermentation was first used in the production of pharmaceuticals. Today fermentation processes are used in the production of many different products and these processes can be divided into nine categories, according to the product (Table 1).

Development of fermentation processes can roughly be divided into four phases (see Fig. 1). First the product is identified. In the case of a pharmaceutical this may be a result of random screening for different therapeutic effects by microbial metabolites (e.g., by high throughput screening of secondary metabolites from *Actinomycetes*) or it may be the result of a targeted identification of a novel product (e.g., a peptide hormone with known function). Outside the pharmaceutical sector the product may also be chosen after a random screening procedure (e.g., screening for a novel enzyme to be used in detergents) or it may be chosen in a more rational fashion. With the rapid progress in the different genomic sequencing programs (the yeast genome was completely sequenced in 1996 and the genome of several bacteria have been completely sequenced), there is much focus on identification of novel products, and today topics such as bioinformatics and gene function are frontline research topics (2).

The next step in the development phase is to choose a strain. In the past this choice was normally obvious after the product had been identified; for example, *Penicillium chrysogenum* was chosen for penicillin production because it was the first organism identified to produce penicillin.

Table 1. Categories and Typical Application of Fermentation Processes

Category	Typical processes
Production of microbial cells	Baker's yeast, single cell protein, lactic acid bacteria
Production of primary metabolites	Ethanol (both in beer and wine production and for technical use), lactic acid, citric acid, acetic acid, amino acids, vitamins
Production of secondary metabolites	Antibiotics (penicillins, cephalosporins, clavulanic acid, tetracycline), skikonin
Production of microbial enzymes	Lipases, proteases, amylases, glucose isomerase
Production of pharmaceutical proteins	tPA, human insulin, erythropoietin, growth hormone, hepatitis B vaccine, monoclonal antibodies
Production of polysaccharides	Xanthan gum, dextran
Production of DNA	Genes for vaccines or gene therapy
Biotransformations	Steroids, L-sorbose (from L-sorbitol)
Production of tissue	Bone marrow, skin

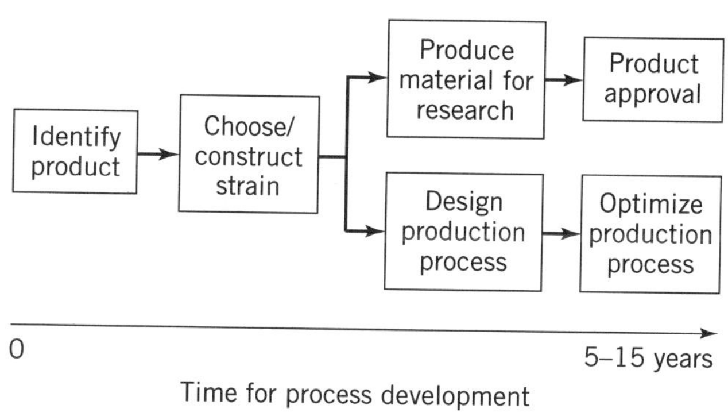

Figure 1. Different phases in the development of a fermentation process.

With the introduction of recombinant DNA (rDNA) technology, it is now possible to choose almost any host for the production. Thus, a strain of *Escherichia coli* has been constructed that can produce ethanol at a high yield (3), and a recombinant strain of *P. chrysogenum* can now be used to produce 7-ADCA (a precursor used for synthesis of cephalosporins) directly by fermentation (4). The choice of strain does, however, often depend much on tradition within the company, and most of the fermentation industries have a set of favorite organisms that are used in the production of many different products. Thus the major players on the industrial enzyme market (Novo Nordisk, Genencor, and Gist-brocades) have chosen a few organisms (typically one to two strains of filamentous fungi and one to two species of bacteria) as production vehicles for a wide range of enzymes. With the production of a heterologous protein, that is, expression of a foreign gene in a given organism, it is also necessary to consider many other aspects (e.g., whether the protein is correctly folded and glycosylated) Table 2 gives an overview of the advantages and disadvantages of different cellular systems for the production of recombinant proteins.

After the strain has been constructed, one of the first aims is to produce sufficient materials for further research, and this is typically done in pilot-plant facilities. For a pharmaceutical compound sufficient material must be produced for clinical trials. For other products it may be necessary to carry out tests of the product and examine any possible toxic effects. In parallel to the continuing research in the application of the product, the production process is also designed. The final steps in development are product approval by the proper authorities and construction of the production facility (which in some cases may be through retrofitting of an existing plant). After the process has been developed, there is continuous optimization of the process, and in some cases better strains are introduced. However, for pharmaceutical products new strains are, only introduced when there is a significant improvement in the process because it is costly to obtain approval from the authorities for redesign of the process.

CRITERIA FOR DESIGN AND OPTIMIZATION

The criteria used for design and optimization of fermentation processes depends on the product. Thus, the criteria used for a high-volume–low-value-added product are normally completely different from the criteria used for a low-volume–high-value-added product. For products belonging to the first category (which includes most whole cell products, most primary metabolites, many secondary metabolites, most industrial enzymes, and most polysaccharides) the three most important design parameters are

- Yield of product on the substrate
- Productivity
- Final titer

A high yield of product on the substrate is for these processes very important, since this gives a good utilization of the raw materials, which often accounts for a significant part of the total costs. Thus, in penicillin production the costs of glucose alone may account for up to 15% of the total production costs (5). Productivity is important because it ensures an efficient utilization of the production capacity (i.e., the bioreactors). In an increasing market it is especially important to increase productivity, because this may prevent new capital investments. Final titer is of importance for the further treatment of the fermentation medium (e.g., purification of the product). Thus, if the product is present in a very low concentration at the end of the fermentation, it may be very expensive to perform purification of the product.

In the production of novel pharmaceuticals, which typically belong to the category of low-volume–high-value-added products, the three previously mentioned design parameters are normally not very important. For these processes quickness to market is generally much more important, and change of the process after implementation is often complicated due to the required U.S. Food and Drug Administration (FDA) approval. In the initial design phase

Table 2. Advantages and Disadvantages of Different Hosts for Production of Recombinant

Host	Advantages	Disadvantages
Bacteria (*E. coli*)	Wide choice of cloning vectors	Posttranslational modifications lacking
	Gene expression easy to control	High endotoxin content
	Large yields possible	Protein aggregation (inclusion bodies)
	Good protein secretion	
Yeast (*S. cerevisiae*)	Generally regarded as safe (GRAS)	Less cloning vectors available
	No pathogens for humans	Glycosylation not identical to mammalian glycosylation
	Large-scale production established	Genetics less understood
	Some posttranslational modifications possible	
Filamentous fungi	Experience with large-scale production	High level of heterologous protein expression has not been achieved
	Source of many industrial enzymes	Genetics not well characterized
	Excellent protein secreters	
Mammalian cells	Same biological activity as natural protein	Cells difficult to grow in bioreactors
	Expression vectors available	Expensive
		Slow growth
		Low productivity
Cultured insect cells	High level of gene expression possible	Not always 100% active proteins
	Posttranlational modification possible	Mechanisms largely unknown

it is, however, still important to keep these three design criteria in mind. The requirement for a high final titer is especially important since the cost for purification (or downstream processing) often accounts for more than 90% of the total production costs.

STRAIN CONSTRUCTION AND STRAIN IMPROVEMENT

Today high-performance production strains to be used in fermentation processes are often constructed using rDNA technology. However, the concept of metabolic pathway manipulation is by no means new. Thus, for decades better strains of *Saccharomyces* to be used for beer fermentation have been obtained through classical breeding and crossing of different strains, and in the production of penicillin productivity has increased by more than 500 times through repeated rounds of mutation and selection of new strains of *P. chrysogenum*. With the introduction of rDNA technology it has become possible to apply a more rational approach to strain improvement–namely by the introduction of targeted genetic changes resulting in strains with a phenotype that gives a better process. This rational approach has been named cellular and metabolic engineering, and different definitions of the approach have been given (6). The term *metabolic engineering* was first introduced by Bailey (7), who defined it as "improvement of cellular activities by manipulation of enzymatic transport and regulatory functions of the cell with the use of recombinant DNA technology." His definition of metabolic engineering includes the following:

1. Inserting new pathways in microorganisms with the aim of producing novel metabolites (e.g., production of polyketides by *Streptomyces* [8,9]), with the aim of degrading toxic compounds (e.g., in bioremediation), or with the aim of constructing a novel biotransformation system
2. Production of heterologous peptides, such as production of human insulin, erythropoitin, and tPA, or industrial enzymes, such as lipases and cellulases
3. Improvement of pathway fluxes leading to higher yields of metabolites (e.g., increasing the flux toward antibiotics or primary metabolites) or yields of biomass (e.g., increasing the cell mass yield in baker's yeast production using industrial media containing mixed sugars [10])

What characterizes metabolic engineering is the rational approach to performing genetic changes, and as with all other fields of engineering it consists of two steps: analysis and synthesis (11). As a consequence of the difficulties in performing detailed analysis of cellular metabolism there has mainly been focus on synthesis in the past, such as expression of new genes in various host cells, amplification of endogenous enzymes, and deletion of genes or modulation of enzyme activities. With modern experimental techniques it has, become possible, however, to perform detailed analysis of cellular function through both in vivo and in vitro measurements, and in the following some of these techniques are described.

MODERN EXPERIMENTAL TECHNIQUES

Whereas the development in rDNA technology has enabled the design of novel high-performance strains that are tailor-made to specific needs, the revolution in computer technology and in analytical chemistry has allowed a much more fundamental characterization of cellular function than previously possible. Thus, flow cytometry allows quantification of population dynamics, while advances in microscopy and image analysis have allowed studies of cell growth and function at the single cell level (12). In studies of fermentation processes it is important to study the function of whole cultures, and here the development of high-performance bioreactors has been an important contribution. When microbial or cell cultures are grown in these bioreactors, it is ensured that all the cells experience the same conditions, and it is therefore possible to perform quantitative physiological studies, such as study of growth and production kinetics, study of gene expression under well-controlled conditions, quantification of metabolic fluxes, and study of the control of flux through the different cellular pathways.

High-Performance Bioreactors

High-performance bioreactors are characterized as being practically ideal bioreactors with a very low mixing time and a very high gas–liquid mass transfer (13,14). In these bioreactors the cells are subjected to an unchanging environment when they are circulated throughout the liquid medium, and the response to imposed variations in the environment is therefore a consequence of the cellular behavior only. High-performance bioreactors are equipped with a large number of in situ sensors, and to control operating variables a flexible direct digital control (DDC) is used, rather than classical single-purpose controllers. This allows precise control of many operating variables, including temperature, pH, dilution rate, and dissolved oxygen concentration, which therefore can be taken to be culture parameters (Table 3) (15). Many commercially available laboratory bioreactors (volumes less than 10 L) normally fulfill the requirements for being high-performance bioreactors. Several bioreactor companies offer systems with flexible DDCs that enable precise control of the culture variables, and when these bioreactors are designed with

Table 3. In Situ Sensors Used for Measuring Culture Parameters

Culture parameter	Sensor	Range	Accuracy
Temperature	Pt-100	0–150 °C	0.1 °C
Pressure	Piezoresistor	0–2 bar	20 mbar
Gas flow	Thermal mass flow meter	0–20 L min^{-1}	20 mL min^{-1}
pH	pH electrode	2–12	0.02
pO_2	Polarografic Clark electrode	0–400 mbar	2 mbar
pCO_2	Membrane-covered pH electrode	0–100 mbar	2 mbar

the proper stirrer and sparger it is possible to ensure a very low mixing time (less than 1 s) and a high gas–liquid mass transfer (k_1a values above 0.1 s^{-1}). In addition to sensors for the culture parameters, modern high-performance bioreactors are normally equipped with in situ sensors and on-line analyzers that allow frequent monitoring of key culture variables. Figure 2 illustrates a typical high-performance bioreactor system. The bioreactor is equipped with on-line flow injection analyzers for monitoring of important culture variables, exhaust gas analysis, and an automatic sampling system. Hereby the most important culture variables can be measured at a high frequency (30–1,000 h^{-1}), whereas other culture parameters can be measured in the automatically withdrawn samples at a lower frequency (1–3 h^{-1}). The bioreactor system can be operated as a constant mass, continuous culture with measurement of the feed flow. It is therefore possible to vary the dilution rate (with a precision of 0.005 h^{-1}), and through proper design of the feed medium it is possible to operate the bioreactor such that there is a single limiting substrate component, such as glucose or ammonia. Finally, the bioreactor is equipped with gas blending, which allows variations in the composition of the inlet gas to the bioreactor, so the influence of, for example, the oxygen concentration on the growth and production kinetics can be studied at the same specific growth rate (16).

With advanced computer control of culture parameters it is possible to perform very reproducible experiment, which is of paramount importance for a fundamental understanding of the underlying cellular reactions. This is illustrated in Figure 3, where the carbon dioxide evolution rate is shown for two different batch cultivations of the filamentous fungi *P. chrysogenum*. With remarkable reproducibility the CER signal can obviously be used to map the different phases of the batch fermentation, which were carried out with sucrose as the carbon source. In the first phase sucrose is hydrolyzed to glucose and fructose and glucose is metabolized. When glucose is exhausted after 45 h there is a distinct shift in the metabolism, and when fructose is exhausted after 60 h there is another shift in the metabolism (i.e., the CER decreases rapidly). Even when both sugars are exhausted there is still some cellular activity due to metabolism of gluconic acid, which was formed in the initial phase during growth on glucose (17).

In situ Sensors. An in situ sensor can be inserted directly into the bioreactor and can be sterilized in place. Table 3 gives the characteristics of in situ sensors used to monitor culture parameters in a typical high-performance bioreactor. Most of these sensors are extremely reliable and are used routinely not only in research laboratories but also in industrial plants.

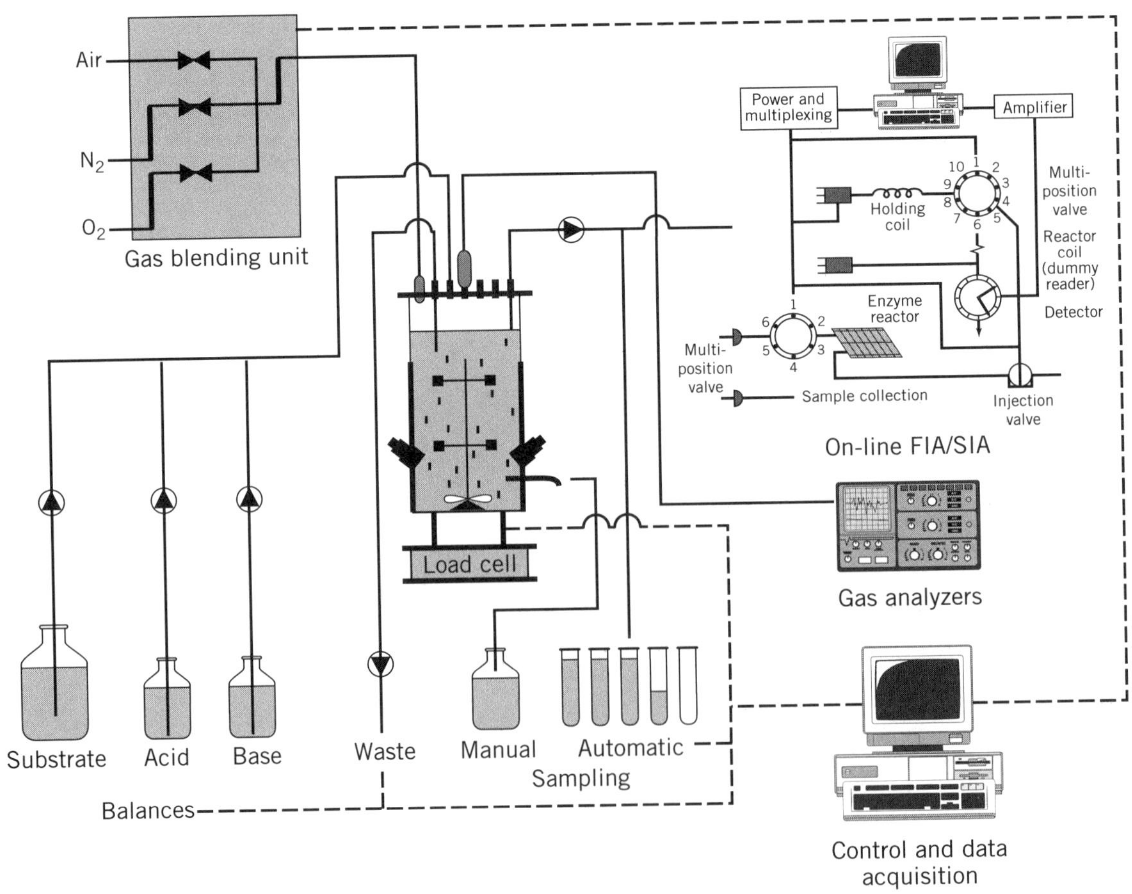

Figure 2. Typical high-performance bioreactor system.

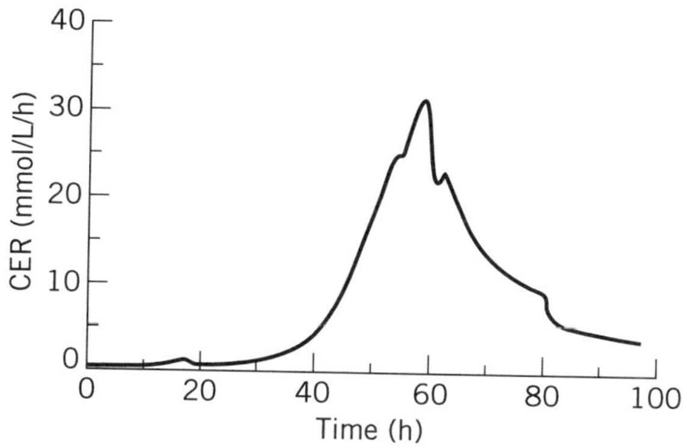

Figure 3. On-line measurement of the carbon dioxide evolution rate (CER) during two consecutive batch cultivations of *P. chrysogenum*. The batch cultures were inoculated with spores at time zero. *Source*: From Ref. 5.

Table 4. Advantages and Disadvantages of Different Measurement Principles for In Situ Biomass Sensors

Principle	Advantages	Disadvantages
Optical density	Wide linear range	Some interference
Culture fluorescence	High sensitivity	Much interference
	Measurement of cellular activity	Signal interpretation difficult
Capacitance	Wide measurement range	Interferance from aeration and agitation
	Measurement of cellular activity	Signal interpretation difficult
Ultrasonic	Wide linear range	Interference from aeration and agitation
	Self-cleaning	Temperature sensitive

For monitoring of culture variables there are, however, only a few commercially available in situ sensors available, and besides a few sensors based on enzymatic analysis of medium components (e.g., for glucose analysis [18]), these sensors measure biomass using different measurement principles (see Table 4) (19). There are several commercially available sensors for in situ monitoring of the optical density, which is linearly correlated with the biomass concentration. These sensors have a wide linear range, but there is interference from gas bubbles and solids in the medium. In some sensors special designs are used to reduce this interference such as by passing the sample into an internal measurement chamber (19). Sensors measuring the culture fluorescence (normally the NAD(P)H-related fluorescence) were first introduced as biomass sensors, but despite high sensitivity and correlation of the culture fluorescence with the total cellular activity, the signal from these sensors is difficult to interpret, and there is generally a poor correlation with the biomass concentration (17,20). Measurement of dielectric properties (or capacitance) is one of the later contributions to measurement of biomass, and it has the advantage of a wide measurement range. However, the signal depends on the cellular activity, which although it is an interesting variable, may not necessarily correlate with the biomass concentration. Even though none of the commercially available in situ biomass sensors stands out as the best biomass sensor, several of the sensors are very useful additions to the sensors normally applied in high-performance bioreactors.

On-Line Analyzers. With the lack of reliable in situ sensors for measurement of important culture variables, including glucose and other medium components, on-line analysis is frequently used. Here a sample is automatically withdrawn and analyzed. For this purpose flow injection analysis (FIA) is well suited due to its high speed, good precision, and good reliability (21). The major drawback of FIA is that it is a single component analysis, and even though a novel technique termed *sequential injection analysis* (SIA) offers the possibility to measure several components using the same hardware (22), other analytical systems such as mass spectrometry (MS), high-performance liquid chromatography (HPLC), and gas chromatography (GC) may sometimes be preferred. These analytical systems offer the ability to measure many components in a single run, but, in general, it is difficult to match the high measurement frequency and reliability of FIA.

A prerequisite for on-line monitoring is automatic sampling, and the sampling system normally serves as the sterile barrier between the bioreactor and the analytical system. Automatic sampling can be performed either by withdrawing a sample directly or by using a membrane module. In direct sample withdrawal, a nonfiltered sample is pumped directly from the bioreactor through a syringe or catheter. Direct sample withdrawal is a very simple solution that allows for on-line measurement of the biomass concentration using FIA (23) and measurement of intracellular components (24). However, a construction with a syringe is not sufficiently robust to be used in an industrial environment. For sampling via a membrane module there are several commercially available designs (25), which can be divided into two groups:

1. *Membrane modules placed in a recycle loop connected to the bioreactor*. The advantage of these modules is that it is possible to change the membrane while the process is running. However, the continuous pumping of the medium through the membrane loop may have a significant effect on the cellular behavior (especially for shear-sensitive organisms).
2. In situ *membrane modules, which are inserted into the bioreactor and sterilized in place*. The advantage of these modules is that the modules do not influence the bioreactor operation and the withdrawn sample is a good representation of the medium in the bioreactor. However, to ensure mechanical breakage of the membrane it is necessary to apply rather thick membranes, which influences the response time (26).

With a reliable membrane module for automatic sampling it is possible to perform routine on-line monitoring of important medium components, and if FIA is used it is especially possible to obtain high frequency measurements, which allows precise quantification of the fluxes in and out of the cells (e.g., the substrate uptake rates and the product formation rates). These fluxes form the basis for quantitative physiological research, as discussed later, and high-performance bioreactors equipped with on-line analyzers for the most important culture variables are therefore indispensable tools in such research.

Measurement of Intracellular Variables. For quantitative physiological studies, the measurement of intracellular variables is extremely important. These variables should ideally be performed in vivo, while cellular function is maintained, but presently there are no techniques that allow for precise in vivo measurements of intracellular variables; therefore they are taken in vitro. To ensure that the in vitro measurements are good representatives of the in vivo conditions, it is critical to have techniques that allow for rapid sampling and quenching of the cellular activity. For the measurement of intracellular metabolites and cofactors, such as glycolytic intermediates, ATP, and NADH, which have a turnover time of a few seconds, it is extremely important to have an almost instantaneous quenching of the cellular metabolism. Using a specially designed sampling module and rapid cooling of the sample in precooled test tubes containing small glass beads, Theobald et al. (27) were able to measure the intracellular level of many metabolites of the EMP pathway and the cofactors ATP, ADP, and AMP in *S. cerevisiae*. Furthermore, using a freeze–quench method, where the sample is mixed with a reagent and then quenched by pumping it into a reservoir of liquid methanol at −40 °C, de Koning and van Dam studied the uptake kinetics of glucose by *S. cerevisiae* at the scale of 5 ms to 1 s (28). Using these sampling techniques, together with HPLC and enzymatic assays, it is possible to obtain a fundamental understanding of the metabolite levels inside the cells at different growth conditions, which, together with knowledge of the metabolic fluxes (discussed later), will give the ultimate physiological characterization.

Quantification of Morphology. For filamentous microorganisms there is in many cases a relation between morphology and product formation. In the past, detailed studies of the growth mechanisms of these organisms have been hampered by the lack of good experimental techniques. However with the rapid development in image analysis it has become possible to study the growth of these organisms in much detail (29). Yet, because of the many different processes that influence morphology, it is important to combine experimental work and mathematical modeling to extract information about the underlying mechanisms. When the morphology is measured for a given number of hyphal elements in a sample taken from a fermentation with filamentous microorganisms, there will be a certain distribution of properties, governed by the underlying mechanisms for growth and hyphal breakage (30). It is difficult to extract information about the different mechanisms from these measurements directly. However, using small growth chambers, where the developing morphology can be monitored on-line, it is possible to obtain fundamental information about the growth mechanisms (12). When this information is combined with computer simulations of hyphal element populations, it is possible to obtain some insight into the mechanisms behind hyphal breakage during submerged fermentations (31).

Quantitative Physiological Studies

Despite many years of extensive analysis of glycolysis, all the details in regulation of glycolytic flux have not been elucidated. Even in *S. cerevisiae*, where studies of a large number of mutants in the glycolysis have resulted in a meticulous mapping of regulation patterns, all the control structures have probably not been revealed. Thus, it was only a few years ago that fructose-2,6-bisphosphate was discovered not only as an important regulator of phosphofructokinase but of other enzymes as well. Further physiological studies are therefore extremely important to identify yet-unknown regulatory patterns and regulatory compounds. With the recent advances in genome characterization it is now possible to rapidly construct many different mutants by introduction of gene disruptions or overexpression of key enzymes, and this is a good basis for further studies. However, it is important to combine genetic work with quantitative physiological studies in which gene regulation is studied under well-controlled growth conditions, and metabolic fluxes and their control are quantified.

Study of Gene Regulation at Well-Controlled Growth Conditions. To illustrate how gene regulation can be studied in submerged fermentations carried out in high-performance bioreactors, two examples are considered. The first example is regulation of the strong TAKA (or α-amlyase) promoter of *Aspergillus oryzae*, which is often used as expression promoter for production of industrial enzymes by filamentous fungi. In the past there has been much speculation on the regulation of this promoter, but most studies were based on shake-flask experiments, in which it is difficult to distinguish between repression and induction. However, through the use of continuous cultures it can clearly be demonstrated that there is both glucose repression and maltose induction (32,33). Thus, by adding a glucose pulse to a steady-state chemostat operated at a low dilution rate, when the glucose concentration is very low (about 5 mg L^{-1}; i.e., derepressed conditions), it was found that there is an immediate decrease in the α-amylase synthesis (Fig. 4). This clearly indicates that there is glucose repression. Furthermore, by shifting from a glucose-based medium to a maltose-based medium when the cells were grown under derepressed conditions (i.e., at a low glucose concentration) it was possible to show that there is maltose induction (33). Besides allowing qualitative conclusions to be drawn, measurements of the glucose concentration at the milligram per liter level even allowed quantification of the degree of glucose repression (32). These findings are consistent with mapping of the TAKA promoter, which indicates the presence of several *creA* binding sites (creAp is a regulatory protein associated with glucose repression in filamentous fungi) and a binding site for a protein that is induced by starch (or maltose) (34). In a classical approach to analyze the promoter further, one would construct promoter fragments and identify the function of the individual binding sites. In this analysis shake-

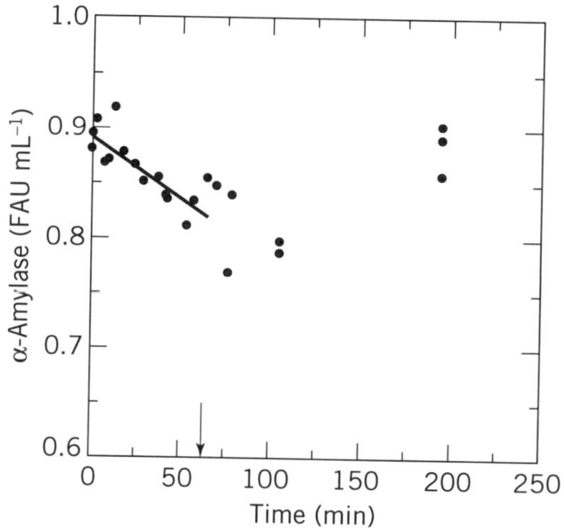

Figure 4. Glucose repression of the TAKA promoter in *A. oryzae*. At time zero a glucose pulse (resulting in a concentration of 600 mg L^{-1}) is added to a steady-state chemostat where the glucose concentration is low (i.e., derepressed conditions). Right after the glucose pulse is added the production of α-amylase decreases instantaneously to the level observed at high glucose concentrations. The α-amylase concentration (given as units per milliliter) therefore decreases due to washout from the bioreactor. When the glucose pulse is exhausted (indicated by the arrow) there is derepression and α-amylase production increases. *Source*: From Ref. 32.

flask cultures may be useful in drawing qualitative conclusions (i.e., whether glucose repression has been eliminated). However, to draw quantitative conclusions these experiments should be followed by another set of continuous fermentation experiments.

The other example is on glucose control in the yeast *S. cerevisiae*. In the production and application of baker's yeast, glucose repression is a serious problem. An often-used carbon source is molasses, which consists of glucose, fructose, sucrose, and raffinose (the trisaccharide galactose–glucose–fructose), and in the uptake and metabolism of the different sugars glucose repression results in lag phases. Furthermore, in the application of baker's yeast in break making, maltose is the typical carbon source and glucose repression of maltose uptake and metabolism results in a reduction in the gassing power. One approach to alleviate glucose repression is to delete the genes coding for key regulatory proteins, for example, *MIG1*, a general repressor that influences the transcription of a number of genes, including the *SUC*, the *GAL*, and the *MAL* genes. Analysis of the effect of deletion of *MIG1* can only be done properly using well-controlled fermentation experiments, and the effect of *MIG1* deletion on sucrose and maltose utilization has been studied using batch cultures (10,35). Two different approaches were applied to silence *MIG1*: *(1)* gene disruption and *(2)* introduction of high copy number plasmids with antisense fragments of *MIG1*. Use of antisense is especially interesting in silencing genes in polyploid industrial yeast strains, but even though the antisense fragments were expressed, there was no change in the degree of glucose repression (10) (Table 5). However,

Table 5. Invertase Activity in Different Haploid Laboratory Strains of *Saccharomyces cerevisiae* during Growth on Glucose[a]

Strain	Genetic construct	Invertase activity [U (g DW)$^{-1}$]
X2180-1A	Wild-type	61
TA8	Antisense, plasmid	55
TA25	Antisense, chromosomal	55
T301	Disruption	263

Note: The antisense fragments were inserted either in a high-copy plasmid or integrated into the chromosome.
Source: From Ref. 10.

with gene disruption of *MIG1* a significantly higher activity of invertase is obtained during glucose consumption (Table 5), resulting in an almost parallel uptake of glucose and sucrose (10), as seen in Figure 5. Disruption of *MIG1* was also found to partly alleviate glucose repression on maltose uptake and metabolism in laboratory strains; there was an almost parallel uptake of glucose and maltose, whereas there was no effect of *MIG1* disruption on maltose uptake in a polyploid industrial strain (35). This was explained to be a consequence of strong catabolite inactivation of the maltose permease in the industrial strain (35). Based on the physiological characterization of the recombinant strains it was concluded that silencing of *MIG1* may alleviate glucose repression, but other effects including catabolite inactivation, also have an important influence. Only through careful physiological characterizations was it possible to identify the different effects.

Quantification of Metabolic Fluxes. In connection with optimization of metabolite production, where the aim is to direct as much carbon as possible from the substrate into the metabolic product to ensure a high yield, it is of prime importance to quantify how the carbon fluxes are distributed through the various cellular pathways. A very powerful technique for calculation of the fluxes through various pathways is the so-called metabolic flux analysis, where the intracellular fluxes are calculated using a stoichiometric model for all the major intracellular reactions and applying mass balances around the intracellular metabolites. A set of measured fluxes as input in the calculations is used, and these fluxes typically are the uptake rates of substrates and secretion rates of metabolites (36–38). If one performs experiments with ^{13}C-enriched carbon sources and measures the fractional enrichment of ^{13}C in intracellular metabolites, it is possible to apply an additional set of constraints in the model, and a better estimation of the intracellular fluxes may be obtained (39). Finally, if one applies information about the isotopomer distributions for the intracellular metabolites much more information is supplied, and it may even be possible to quantify the net fluxes through reactions that are practically reversible (40).

Besides quantification of the various pathway fluxes, and therefore determination of the carbon flows inside the cell, metabolic flux analysis is useful for the following:

- *Identification of possible rigid branch points (or nodes) in the cellular pathways*. Through comparison

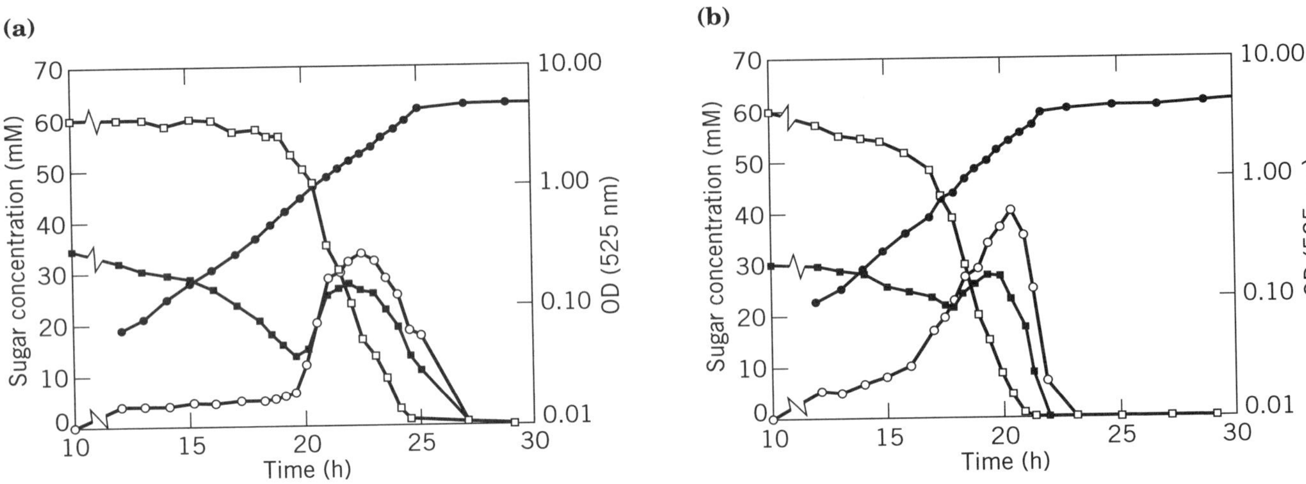

Figure 5. Alleviation of glucose repression by disruption of *MIG1* in *S. cerevisiae*. Results are shown for two batch fermentations carried out on a glucose–sucrose mixture: (**a**) with a wild-type haploid strain and (**b**) with the same strain where *MIG1* has been disrupted. Obviously disruption of *MIG1* results in metabolism of sucrose at a higher glucose concentration as compared with the wild-type strain (i.e., the sucrose concentration begins to decrease right at the start of the fermentation). □, sucrose concentration (mM); ●, biomass concentration (OD); ■, glucose concentration (mM); ○, fructose concentration (mM). *Source*: From Ref. 10.

of the distribution of fluxes at different operating conditions and in different mutants it is possible to identify whether a pathway node is rigid or flexible (41). Thus through metabolic flux analysis of various mutants of *C. glutamicum*, Vallino and Stephanopoulos (42,43) concluded that in lysine production the glucose-6-phosphate node is flexible and the pyruvate node is weakly rigid.

- *Identification of the existence of different pathways.* Formulation of the reaction stoichiometry, which is the basis for metabolic flux analysis, requires detailed information of the biochemistry. However, for many microorganisms certain details in the pathway stoichiometries are unknown, and it may not be known whether a given pathway is active. Furthermore, in many cells there are several isoenzymes whose functions are not known in complete details. By calculating the metabolic fluxes with a different set of cellular pathways it may be possible to identify which pathway is most likely to exist or to obtain indications of the functions of different isoenzymes and/or pathways. Thus, in analysis of anaerobic growth of *S. cerevisiae*, we found that alcohol dehydrogenase III, a mitochondrial enzyme whose physiological function has not been identified, plays an important role in maintaining the redox level inside the mitochondria (38). However, this type of analysis should always be followed by enzyme assays in which the presence of enzyme activity in cell extracts is tested.
- *Calculation of nonmeasured extracellular fluxes.* Normally the number of fluxes that can be measured is larger than what is needed for calculation of the intracellular fluxes. In this case it is possible to calculate some of the extracellular fluxes, such as the rate of production of various by-products, by use of the stoichiometric model and the measured fluxes.
- *Examination of the influence of alternative pathways on the distribution of fluxes.* In connection with optimization of metabolite production it may be possible to identify one or several constraints for increasing the yield of a metabolite on the substrate or the flux leading to the desired metabolite. In these cases it is possible to envision various scenarios, such as examining whether insertion of a new pathway or an isoenzyme (or perhaps deletion of an isoenzyme) can have a positive effect on removing this constraint, leading to an increased flux toward the desired metabolite. In a study of penicillin production we calculated that the yield of penicillin on glucose is likely to be higher if cysteine, one of the precursors for penicillin production, is synthesized by direct sulfhydrylation rather than by the transsulfuration pathway claimed to exist in *P. chrysogenum* (37).
- *Calculation of maximum theoretical yields.* Based on a stoichiometric model it is normally possible to calculate the maximum theoretical yield of a given metabolite if a given set of constraints is specified. This value is of interest because it specifies an upper limit for the yield in the process. This has been illustrated both for lysine production by *C. glutamicum* (36) and for penicillin production by *P. chrysogenum* (37). In these calculations it is necessary to introduce a constraint, for example, that reactions do not run in thermodynamically unfavorable directions.

Control of Metabolic Fluxes. One of the most important aspects of quantitative physiology is control of flux. As discussed, the concept of metabolic flux analysis is useful for studies of interactions between different pathways and for quantification of flux distributions around branch points, but it does not allow for evaluation of how the fluxes are controlled (i.e., how the rates of synthesis and conversion of metabolites are kept in close balance over a very wide range of external conditions without catastrophic rises or falls in the metabolite concentrations). The discoveries in the 1950s of feedback inhibition, cooperativity, covalent modification of enzymes, and control of enzyme synthesis introduced a number of molecular effects that may play a role in control of fluxes. With these many different mechanisms it is not surprising that disputes often arises over how the flux through a given pathway is controlled. Since the many reports on enzyme regulation typically are verbal and qualitative (e.g., phosphofructokinase is the major flux-controlling enzyme of glycolysis in muscle), it is difficult to discriminate between different findings, and in many cases different findings may seem conflicting. Furthermore, one often finds terms such as *rate-limiting steps* and *bottleneck enzyme* when control of flux is discussed. Thus, the early findings that it is the first enzyme in a pathway or the first enzyme after a branch that is under some type of control (e.g., by feedback inhibition) often results in statements such as "the first step in a pathway is the rate-limiting step."

The concept of metabolic control analysis (MCA) tells us that these kinds of qualitative statements have no meaning, since flux control is distributed over all the steps in a pathway, but that some steps may exert a higher degree of flux control than others. The concept of MCA was developed from the landmark papers of Kacser and Burns (44) and Heinrich and Rapoport (45). Its basis is a set of parameters, called elasticity coefficients and control coefficients, that quantify the control within a reaction network. The elasticity coefficients are given by

$$\epsilon^{\mathrm{i}}_{x_{\mathrm{j}}} = \frac{X_{\mathrm{j}}}{v_{\mathrm{i}}}\frac{\partial v_{\mathrm{i}}}{\partial X_{\mathrm{j}}} \qquad (1)$$

and they express the sensitivity of the rates of the enzyme-catalyzed reactions (v_i) toward the metabolite concentrations (X_j). The most often used control coefficients are the flux control coefficients (FCCs), which are given by

$$C^{J_{\mathrm{j}}}_{\mathrm{i}} = \frac{v_{\mathrm{i}}}{J_{\mathrm{j}}}\frac{\partial J_{\mathrm{i}}}{\partial v_{\mathrm{i}}} \qquad (2)$$

The FCCs express the fractional change in the steady-state flux through the pathway (J_j) that results from an infinitesimal change in the activity of the enzymes (or reaction rates). The FCCs and the elasticity coefficients are related to each other via the summation theorem, which states that the sum of all the FCCs is 1, and the connectivity theorem, which states that the sum of the product of the elasticity coefficients and the FCCs is zero.

Three different groups of experimental methods are available for determination of the FCCs:

- Direct methods, where the control coefficients are determined directly
- Indirect methods, where the elasticity coefficients are determined and the control coefficients are calculated from the theorems of MCA
- Transient metabolite measurements, where the metabolite concentrations are measured during a tran-

Table 6. Overview of Methods for Determination of Flux Control Coefficients

Method	Procedure	Advantages and disadvantages
	Direct	
Genetic manipulations	Alternate the expressed enzyme activity through genetic manipulation, such as by inserting inducible promoters	Robust method that gives direct answers, but is very laborious
Enzyme titration	Vary the enzyme activity through titration with purified enzymes	Simple and straightforward procedure, but it can only be applied for pathway segments that are completely decoupled from the rest of the cell
Inhibitor titration	Vary the enzyme activity through titration with specific inhibitors	Simple and easy to apply, but requires the existence of specific inhibitors
	Indirect	
Double modulation	Measure the metabolite levels at different environmental conditions and determine the elasticity coefficients by calculation of differentials	Elegant approach, but it requires two independent changes in the metabolite levels, which are difficult to obtain due to the high degree of coupling between intracellular reactions
Single modulation	Similar to double modulation, but it is based on knowledge of one of the elasticity coefficients	More robust than double modulation, but it requires knowledge of one elasticity coefficient
Top down approach	Based on grouping of reactions and then using double modulation, for example	Very useful, but do not directly give all the FCCs of the system
Kinetic models	Direct calculation of the elasticity coefficients from a kinetic model	Robust, but relies on the availability of a reliable kinetic model for the individual enzymes in the pathway

sient and this information is used to determine the control coefficients (46)

Table 6 gives an overview of the different direct and indirect methods.

Whereas the elasticity coefficients are properties of the individual enzymes, the FCCs are properties of the system. The FCCs are therefore not fixed but change with the environmental conditions. This has been illustrated in analysis of the penicillin biosynthetic pathway (47,48). Based on a kinetic model for the enzymes in this pathway, the FCCs were calculated at different stages of fed-batch fermentation, and a drastic shift in the flux control was found. During the first part of the cultivation the flux control was mainly at the first step in the pathway, the formation of the tripeptide LLD-ACV by ACV synthetase (ACVS). Later in the cultivation flux control shifted to the second step in the pathway, the conversion of LLD-ACV to isopenicillin N by isopenicillin N synthetase (IPNS). This shift in flux control is due to intracellular accumulation of LLD-ACV, which is an inhibitor of ACVS. Obviously, it makes no sense to talk about a rate-limiting step or a bottleneck enzyme in this process. Besides the shift in flux control, it is interesting to note that most of the flux control is by the first two steps in the pathway (48). Furthermore, through analysis of the kinetic model it was found that the value of the FCCs depends on the dissolved oxygen concentration, which is a substrate in the IPNS catalyzed reaction (48).

BIBLIOGRAPHY

1. B. Sikyta, *Prog. Ind. Microbiol.* **31**, 13–32 (1995).
2. A. Goffeau et al., *Science* **274**, 546–567 (1996).
3. K. Ohta, D.S. Beall, J.P. Mejia, K.T. Shanmugam, and L.O. Ingram, *Appl. Environ. Microbiol.* **57**, 893–900 (1991).
4. L. Crawford, A.M. Stepan, P.C. McAda, J.A. Rambosek, M.J. Conder, V.A. Vinci, and C.D. Reeves, *Biotechnology* **13**, 58–62 (1995).
5. J. Nielsen, *Physiological Engineering Aspects* of Penicillium chrysogenum, World Scientific Publishing, Singapore, 1997.
6. D.C. Cameron and I.-T. Tong, *Appl. Biochem. Biotechnol.* **38**, 105–140 (1993).
7. J.E. Bailey, *Science* **252**, 1668–1674 (1991).
8. R. McDaniel, S. Ebert-Khosla, D.A. Hopwood, and C. Khosla, *Science* **262**, 1546–1550 (1993).
9. C.R. Hutchinson, *Biotechnology* **12**, 375–380 (1994).
10. L. Olsson, M.E. Larsen, B. Rønnov, J.D. Mikkelsen, and J. Nielsen, *Appl. Environ. Microbiol.* **63**, 2366–2371 (1997).
11. G. Stephanopoulos, *Curr. Opinion Biotechnol.* **5**, 196–200 (1994).
12. A.B. Spohr, C.D. Mikkelsen, M. Carlsen, J. Nielsen, and J. Villadsen, *Biotechnol. Bioeng.* **58**, 541–553 (1998).
13. B. Sonnleitner and A. Fiechter, *Anal. Chim. Acta* **213**, 199–205 (1988).
14. J. Nielsen and J. Villadsen, in G. Stephanopoulos ed., *Biotechnology*, vol. 3, 2nd ed., VCH, Weinheim, Germany, 1993, pp. 77–104.
15. B. Sonnleitner and A. Fiechter, *Adv. Biochem. Eng. / Biotechnol.* **46**, 141–159 (1992).
16. C.M. Henriksen, J. Nielsen, and J. Villadsen, *Biotechnol. Prog.*, submitted (1997).
17. J. Nielsen, C.L. Johansen, and J. Villadsen, *J. Biotechnol.* **38**, 51–62 (1994).
18. J. Bradley, W. Stöcklein, and R.D. Schmid, *Proc. Control Qual.* **1**, 157–183 (1989).
19. L. Olsson and J. Nielsen, *Trends Biotechnol.*, submitted (1997).
20. G. Lidén, *Pure Appl. Chem.* **65**, 1927–1932 (1993).
21. J. Nielsen, *Proc. Control Qual.* **2**, 371–384 (1992).
22. R.W. Min, J. Nielsen, and J. Villadsen, *Anal. Chim. Acta* **320**, 199–205 (1996).
23. S. Benthin, J. Nielsen, and J. Villadsen, *Anal. Chim. Acta.* **261**, 145–153 (1992).
24. H.-A . Kracke-Helm, L. Brandes, B. Hitzmann, U. Rinas, and K. Schügerl, *J. Biotechnol.* **20**, 95–104 (1991).
25. K. Schügerl, in K. Schügerl ed., *Biotechnology*, vol. 4, 2nd, ed., VCH, Weinheim, Germany, 1991, pp. 149–178.
26. L.H. Christensen, J. Nielsen, and J. Villadsen, *Chem. Eng. Sci.* **46**, 3304–3307 (1991).
27. U. Theobald, W. Mailinger, M. Reuss, and M. Rizzi, *Anal. Biochem.* **214**, 31–37 (1993).
28. W. de Koning and K. van Dam, *Anal. Biochem.* **204**, 118–123 (1992).
29. C.R. Thomas and G.C. Paul, *Curr. Opin. Biotechnol.* **7**, 35–45 (1996).
30. J. Nielsen, *Trends Biotechnol.* **14**, 438–443 (1996).
31. P. Krabben, J. Nielsen, and M. Michelsen, *Chem. Eng. Sci.* **52**, 2641–2652 (1997).
32. M. Carlsen, J. Nielsen, and J. Villadsen, *J. Biotechnol.* **45**, 81–93 (1996).
33. R. Mørkeberg, M. Carlsen, and J. Nielsen, *Microbiol.* **141**, 2449–2454 (1995).
34. N. Tsukagoshi, M. Kato, A. Aoyama, F. Naruse, T. Kawaguchi, and T. Kobayashi, *presented at 3 European Conference on Fungal Genetics*, Münster, March 27–30, 1996.
35. C.J.L. Klein, L. Olsson, B. Rønnov, J.D. Mikkelsen, and J. Nielsen, *Appl. Environ. Microbiol.* **62**, 4441–4449 (1996).
36. J.J. Vallino and G. Stephanopoulos, *Biotechnol. Bioeng.* **41**, 633–646 (1993).
37. H.S. Jørgensen, J. Nielsen, J. Villadsen, and H. Møllgaard, *Biotechnol. Bioeng.* **46**, 117–131 (1995).
38. T.L. Nissen, U. Schulze, J. Nielsen, and J. Villadsen, *Microbiology* **143**, 203–214 (1997).
39. A. Marx, A.A. de Graaf, W. Wiechert, L. Eggeling, and H. Sahm, *Biotechnol. Bioeng.* **49**, 111–129 (1996).
40. K. Schmidt, M. Carlsen, J. Nielsen, and J. Villadsen, *Biotechnol. Bioeng.* **58**, 831–840 (1997).
41. G. Stephanopoulos and J.J. Vallino, *Science* **252**, 1675–1681 (1991).
42. J.J. Vallino and G. Stephanopoulos, *Biotechnol. Prog.* **10**, 320–326 (1994).
43. J.J. Vallino and G. Stephanopoulos, *Biotechnol. Prog.* **10**, 327–334 (1994).
44. H. Kacser and J.A. Burns, *Symp. Soc. Exp. Biol.* **27**, 65–104 (1973).
45. R. Heinrich and T.A. Rapoport, *Eur. J. Biochem.* **42**, 89–95 (1974).
46. J. Delgado and J.C. Liao, *Biochem. J.* **282**, 919–927 (1992).
47. J. Nielsen and H.S. Jørgensen, *Biotechnol. Prog.* **11**, 299–305 (1995).
48. P.N. Pissarra, J. Nielsen, and M.J. Bazin, *Biotechnol. Bioeng.* **51**, 168–176 (1996).

See also AMINO ACIDS, PRODUCTION PROCESSES; AMMONIA TOXICITY, ANIMAL CELLS; BIOREACTORS, AIR-LIFT REACTORS; BIOREACTORS, CONTINUOUS STIRRED-TANK REACTORS; BIOREACTORS, FLUIDIZED-BED; SECONDARY METABOLITE PRODUCTION, ACTINOMYCETES, OTHER THAN STREPTOMYCES.

FERMENTER DESIGN

MARVIN CHARLES
Lehigh University
Bethlehem, Pennsylvania
JACK WILSON
ABEC, Inc.
Allentown, Pennsylvania

KEY WORDS

Bulk mixing
Clean-in-place
Heat transfer
Impellers
Oxygen transfer
Power input
Sampling systems
Seals
Sterilization
Stirred tank

OUTLINE

INTRODUCTION

Scope

The scope of this article is limited to design of agitated vessels used for aerobic, single-cell (bacteria and yeast), and filamentous (bacteria and fungi) fermentations. It presents general design principles, combining basic concepts with practical matters including regulatory compliance, safety, maintenance, cleanability, ease of use, and cost, all of which are strongly interrelated. It focuses primarily on fermenters used to produce human and animal health products or their precursors, but the general approaches discussed are applicable to other types of products (e.g., industrial enzymes, food products); the primary differences are emphasized.

Included in the presentation are fermenter vessels; agitation; aeration; heat transfer; sterilization; cleaning; and piping systems. Control and data acquisition are discussed only to the extent that they influence the items already noted.

It is assumed that the reader is familiar with the basics of fermentation, general fermenter construction, and nomenclature. Those not having such background are advised to consult basic references (1–3).

Design Philosophy

Our basic philosophy comprises the following simple principles:

1. Each design case has unique requirements and calls for individualized application of the basic design concepts and practices discussed herein. The one-size-fits-all approach usually results in sleeves that drape into your food, arm holes that cut off circulation, pant legs that are made to trip over, or some combination of these features and others. The magic number approach—all problems have simple solutions based on codified numbers such as fixed geometric ratios (4)—leads to the same place.
2. Successful design requires a systems approach. A fermenter is a system that is part of a process system, and the process is part of a plant system. All these systems interact with each other, with real people, with control systems, and with other systems related to regulatory compliance, safety, documentation (including protocols, SOPs, etc.), change control, maintenance, and so forth. Failure to take these

interactions into account during fermenter design usually results in considerable pain, not only for those guilty of the omission, but also for innocents who were never asked for planning and/or design input, but who must live with the result. Note that appropriate documentation, protocols, SOPs, change control, and so on should be considered important elements of design and operation of any plant, licensed or not.

3. Compromise is always necessary. Nothing in any project is 100% right, and nothing has to be. He or she who looks for 100% of anything will hold up a project needlessly and will generate a lot of animus. (The words actually used in such cases cannot be used here.)
4. The time to be thinking about all of these points is prior to and during design, not when you're standing on the plant floor trying to validate a continuous mixer that has only one port and no instrumentation.

We discuss next the bases for safety and regulatory considerations, primarily to sensitize readers to these issues early on. They are all too often overlooked during early stages of design, when the focus is on satisfying process requirements (e.g., oxygen transfer). Unfortunately, this frequently leads to important constraints being ignored and hence to designs that require expensive and often cumbersome "fixes."

SAFETY AND REGULATORY COMPLIANCE

All fermenter designs are influenced by safety and regulatory factors, the extent depends on *(1)* the nature of the product and its intended use, *(2)* the process, and *(3)* the nature and location of the facility. For example, even fermentation products not regulated by the U.S. Food and Drug Administration (FDA) must adhere to safety requirements, which in some cases are very strict. Furthermore, no professional design can escape the effects of a long list of code requirements, for example, the ASME pressure vessel code (5). By the same token, not all products that are FDA regulated are equally affected by safety, regulatory, and code requirements. We must consider each case on the basis of its own unique requirements.

Our objective in the rest of this section is to introduce some of the requirements and to provide some rational bases for their implementation, details of which are discussed later. Much of the discussion is directly applicable to fermenters used for production of human or animal health products or for their precursors, but also applies to other cases to varying extents.

Containment: Worker and Community Biosafety

A fermenter must be an integral part of a system designed to insure safety at three levels: product, worker, and community. Many of the methods used to protect the product also serve to protect the plant personnel and the community (e.g., use of closed systems and HEPA filters); however, there are potential points of conflict (6). For example, some containment practices that call for completely welded hard piping to a contained drain line for any condensate that could be exposed to culture fluid could expose the product to drain line contaminants. One resolution of this problem is to use steam locks in all such lines. All such conflicts encountered to date have been resolved, but not always easily or relatively inexpensively.

Insurance of product safety is tied primarily to compliance with FDA (or similar organizations) regulations. This will be considered in the subsection "Product Safety." Here we deal with worker and community safety.

Protection must be provided against potential ill effects of fermentation products (e.g., cytotoxins, allergens), organisms (e.g., pathogens, whether recombinant or not), fermentation byproducts (e.g., pharmacologically active precursors of the active product), or some combination. The practice of providing such protection is called *containment*, which is defined as insuring that deleterious fermentation components can't be transported to any area inside or outside the plant before they have been rendered harmless. To accomplish this, containment must be exercised at several levels (7). We consider in this article only direct or "primary" containment of fermenters; however, one should not lose sight of the fact that the other levels must be considered during fermenter design, if for no other reason than to ensure that the fermenter design and operation will be consistent with the overall containment strategy.

Containment levels have been defined by the National Institutes of Health (NIH) (8), based on the potential danger of organisms. These levels, in increasing order of potential danger are GILSP, BL1-LS, BL2-LS, BL3-LS, and BL4-LS. (BL4-LS must be handled on a case-by-case basis in consultation with the Centers for Disease Control and others. It is, thank goodness, well beyond our scope.) There are other classifications that have been developed in the United Kingdom (9), the European Union (10), and elsewhere, but these do not differ markedly from the U.S. scheme. It is important to know that *(1)* all of these guidelines address what must be accomplished but not how they should be accomplished, and *(2)* there is not complete agreement on what is actually required to implement the guidelines. In addition, the guidelines do not address containment of products. Although many of the considerations for product containment are the same or very similar to those for organism containment, there are some very important differences. Here, we confine our attention to organisms; the interested reader should consult the Refs. 11–14 for more specific information concerning hazardous products. One also should consult U.S. Environmental Protection Agency (EPA) regulations applicable to microorganisms and their products.

Most existing processes use organisms that require either GILSP or BL1-LS containment; nevertheless, many plants are built to satisfy requirements for BL2-LS (actually, somewhere between BL2-LS and BL3-LS). The reasons for this include satisfying FDA requirements (e.g., safety of the product) and the fact that the additional cost is not large and is relatively cheap insurance. There also is the oft-stated reason that such a route insures greater flexibility for future operation (15). One might add to this the fact that the rules are not well established and that requirements could change substantially between the time of design and the time production begins. It also is worth noting that there is growing interest in using organisms that do, in fact, require BL2-LS and even BL3-LS containment, which will probably focus greater attention on further development of acceptable implementation practices as well as changes in the guidelines. Note that equipment requirements and cost for BL2-LS and BL3-LS are currently about the same (this probably will change); BL3-LS facilities requirements and cost are very much greater than for BL2-LS.

Finally, it is important to note that containment is nothing new. Highly pathogenic organisms have long been used to produce therapeutics and biological warfare components; hence, there is a considerable body of experience dealing with the subject (11,16). There also is considerable guidance to be had from the nuclear industry; nevertheless, the natures, sizes, and large number of new commercial processes, along with new methods, materials, and so on, will force heightened awareness, scrutinity, review, and modernization or change.

Physical Safety

Most physical safety issues are addressed by U.S. Occupational Safety and Health Administration (OSHA) requirements (17) and by various national codes such as the ASME Pressure Vessel Code (5) and the National Electrical Code. There also are many local construction codes that must be satisfied. Among these are various earthquake-resistant construction codes that are also applicable to containment considerations. And then there are the requirements of the final arbitors: the insurance companies.

Product Safety

We focus here on FDA requirements, but the reader is urged to keep in mind that he or she also must consider those of other regulatory agencies and be aware of the unsettling facts that it is not always clear which agency has jurisdiction, and there often are conflicting requirements.

The primary purpose of FDA regulations is to insure product safety, purity, and efficacy. The legal bases for the regulations and their enforcement can be found in the *Code of Federal Regulations* (CFR) (18) and a number of other FDA documents such as *Points to Consider* (19,20), *Inspection Guides* (21,22), and *Guides to Industry* (23,24). Industrial implementation of all the regulations, along with ongoing improvements, is referred to as current good manufacturing practices (cGMP).

There is an endless stream of writings in professional journals, short courses, and such concerning the regulations and cGMP practice. Unfortunately, there is a great deal that is not covered clearly in writing, and there is a constant proliferation of new and/or revised guidelines followed closely by lots of (frequently conflicting) interpretations and opinions. In addition there has been a problem of nonuniform application. For example, regulatory scrutiny has been far stricter for fermenters used to produce active molecules directly than for those used to produce precursors; there has been considerable variation from product to product within a given class; and manufacture of human drugs has been regulated much more rigidly than manufacture of animal drugs. While understandable to some extent, such nonuniformity has caused considerable confusion. These gaps are beginning to be closed, and this may have considerable influence on the design of fermenters—particularly those used to produce precursors and animal drugs.

In simplest terms, cGMP requires that the combination of fermentation process equipment and operating protocols must consistently yield product that meets acceptable specifications and is capable of being converted/purified consistently to final product, meeting approved product specifications. From this general sense of intent it can be argued that failure to have control over the fermentation puts the final product at risk. Few would argue with this general concept; however, there is considerable debate as to interpretation and extent.

Requirements that flow from the need to have control are based on relatively simple ideas:

1. If an organism is subject to environmental, medium, or other conditions outside the range in which it is known to yield product meeting acceptable specifications and capable of being converted/purified consistently to final product meeting approved product specifications, then we have no guarantee that it does not produce other products with which the recovery system can not cope and that can escape detection by the analytical methods in place.
2. If a fermenter becomes contaminated with other microbes, said microbes could produce toxins that could be carried undetected to the final product. It is possible that this could occur without altering the behavior of the process organism. It also is possible that products of the contaminant could cause the process organism to make toxins that could go undetected into the final product. Given these possibilities, plus the fact that it is not possible to prove that the contaminant will always be the same, evidence of contamination is evidence for lack of control.
3. If a fermenter is not cleaned properly, deleterious microbial or nonmicrobial products could remain to contaminate the next batch in such a way that impurities could be carried undetected to the final product. Similar statements can be made about other contaminants introduced via other routes as a result of poor cleaning practices.

Again, it is unlikely that many would argue with these points, but again there is considerable debate as to interpretation and extent.

Requirements 1–3 can be translated to specific requirements for, among other things, controllability and reliability of fermentation conditions, sterilization, and aseptic and cleaning operations. Controllability requires that the fermenter and its subsystems be designed and constructed in ways that make possible control of environmental variables within the operational ranges required. This obviously means that mechanical design must be in harmony with control system design.

Sterilization/aseptic operation translates to *(1)* destroying any microbial contaminants that may be present in any part of the equipment that might contact process fluids and *(2)* insuring that no microbes (other than the production organism) can enter after the equipment has been sterilized. The latter embodies the concept of the "sterile barrier." To these ends, the following apply:

- The fermenter and all its ports and direct attachments must be sterilizable.
- All piping that will contact process fluids (including additives) and/or provide paths into the system (e.g., the air exhaust line) must be sterilizable initially. Some (e.g., sampling lines, addition lines) must also be sterilizable at any time during a fermentation.
- Inlet gases (e.g., air, oxygen) and all additives (e.g., medium components, acid, base, antifoam) must be sterilized before they contact any sterile process piping.
- All penetrations (e.g., drive shaft, probe ports) must be sterilizable.

The preferred sterilization method is automatic sterilization-in-place (SIP) with steam. This is discussed at length later.

Clean operation requires, among other things, the following:

- The system must be designed such that any surface that can contact a process stream can be cleaned consistently to a level that insures that the product will be free of soils resulting from a fermentation.
- Nothing in the system that comes in contact with process fluid can introduce unacceptable and unidentifiable materials.

Note (for example) that 21 CFR 211(Part D) (guidelines for design of equipment for licensed facilities) does not have any specific requirements for fermenter sterilization or cleaning; however, the preceding discussion coupled with the intended use and cGMP concepts makes it difficult to to argue the point. But again, there remain the questions of rational interpretation and extent of application.

Riding along with cGMP is the practice of validation. This is an extensive, expensive, and controversial area for which and about which reams of reams have written. Following is the formal definition of validation:

> Establishing documented evidence that provides a high degree of assurance that a specific process will consistently produce a product meeting its predetermined specifications and quality attributes. (25)

What this means is that not only does the equipment have to be designed, installed, and operated so as to run consistently within well-defined ranges of process variables to consistently yield product that meets acceptable specifications and is capable of being converted/purified consistently to final product, meeting approved product specifications, but that there must be adequate documentation to prove this before product can be sold. In addition, one must demonstrate that this goal can be achieved under so called worst-case conditions; there is often considerable debate as to what this really means (26). It is beyond our scope to consider validation in detail. Suffice it to say that it can translate to checking and testing every component of the equipment and documenting not only its proper function but also its history. The brave of heart are referred to the bulging literature on the subject; Refs. 27–30 provide a start.

All of this means we must take great care to consider *all* the details, such as materials, corrosion issues, machining methods, vendors, welds, and types of steam and water, keeping in mind that all must be considered as part of a system that must interface "seamlessly" with the systems of cGMP and validation. Failure to consider this before and during design is almost always very costly. Design should be done so as to facilitate validation, and subsequently, routine cGMP operation, maintenance, and so forth. "Simple" things such as judicious placement of access valves and piping isolation can go a very long way to ensuring minimum pain and maximum operability. Having said all this, we reiterate that not all fermenters are subject to all these requirements, and even those that are, are subject to them in varying degrees (at least in practice).

DESIGN BASIS AND OTHER GENERAL CONSIDERATIONS

A design basis for fermentation equipment should derive from a facility/process design basis. The latter should include (among other items) the general nature of the facility (e.g., research and development vs. production, single product vs. multiuse); product(s) specifications; regulatory, containment, and other requirements; level of automatic operation; general processing scheme (e.g., batch); nature of individual process steps; productivity, concentrations, and so forth; cleaning requirements; special considerations (e.g., earthquake-proof construction); critical valving and instrumentation; staffing requirements and constraints; architectural and general floor plan constraints; and utilities. Most of these will have some influence on fermenter design—some in more subtle ways than others.

It is difficult to overstate the importance of the initial definition that will derive from the design basis. Obviously, a rational design basis for a fermenter must also be based on fermentation characteristics as well as operating cycle and productivity required (which should derive from the overall process design basis). From these will flow the sizes and number of vessels, and definitions of oxygen transfer, heat transfer, power, and bulk mixing requirements. The design must satisfy these but must also satisfy requirements for regulatory compliance, safety, cleaning, facile operation, and maintenance. All of these factors are highly interactive (e.g., design for oxygen transfer affects design for cleaning); hence, responsible design will almost always require several iterations to ensure the greatest probability of success and to minimize lost time caused by installation, operation, and various problems. The iterative nature of the design (and, unfortunately, construction) process makes most important the existence of a well-crafted, well-implemented, and well-documented change control process. Finally, as with any engineering project, failure to have a solid basis of design, a complete scope, accurate process flow diagrams and piping and instrumentation diagrams (P&IDs), and accurate process timing will almost certainly result in added cost, lost time, and worse.

PROCESS REQUIREMENTS: BASICS

One of the first steps in fermenter design is translation of process demands to oxygen transfer, heat transfer, bulk mixing, and power requirements—the so-called transport processes (TPs). Sound translation is not straightforward because the TPs interact not only with each other, but also with vessel geometry and other design factors. The nature of each TP and the relationships among them vary with the process requirements. They also vary significantly with the basic nature of the organism and the culture broth. The major factors here are the rheological nature of the broth and the sensitivity of the organism to fluid mechanical forces. In some cases, sensitivity of the organism to temperature and other environmental factors, such as pH, impose tighter constraints (e.g., wall temperature control, uniformity of mixing).

Single-cell organisms (most bacteria and yeasts) tend to tolerate fluid forces very well (there are a few exceptions) and tend to have low-viscosity, Newtonian broths. Mycelial organisms (fungi and some mycelial bacteria such as streptomycetes) tend to be more prone to damage by fluid mechanical forces than are single-cell organisms and tend to have high-viscosity, non-Newtonian broths. Detailed discussion of all these factors is beyond the scope of this article; however, we now discuss a few points concerning rheological behavior, and point to fluid effects wherever appropriate in the remainder of the article.

Broth Rheology and Its Effects on Transport Processes

Newtonian viscosity depends only on composition and temperature; the nature of the fluid motion does not affect it. Non-Newtonian viscosity does depend on the nature of the fluid motion. This dependency is usually expressed in terms of fluid shear rate (a measure of how rapidly fluid velocity changes from one point to another point close by).

There are several types of non-Newtonian behavior (e.g., pseudopalstic, Bingham) that can be described quantitatively by a host of mathematical models. One model used frequently to describe non-Newtonian fermentation broths (31–33) is the so-called power law:

$$\eta = K\gamma^{(n-1)} \qquad (1)$$

where η is viscosity, γ is shear rate, and n and K are constants. Such relationships can be useful if one has rheological data for the subject broth and if the design correlations used incorporate rheological properties in a meaningful and accurate way. It is seldom that either of these conditions is satisfied, and almost never that both are satisfied. Also, there is a considerable amount of misleading and/or inaccurate information in the literature concerning broth rheology, its measurement, and its use; the reader is cautioned to tread carefully in this area (4,34). If accurate viscosity information is available, it can be a valuable qualitative and sometimes semiquantitative guide, if interpreted and used properly.

Mass transfer (oxygen), heat transfer, and bulk fluid motion all depend strongly on rheological characteristics. Generally, all are poorer in non-Newtonian than in Newtonian broths; indeed, the rheological characteristics of mycelial broths can and do impose severe constraints. Two of many examples are the following:

1. Heat and oxygen transfer rates tend to be anywhere from 50 to 5% of what they would be for typical (e.g., *Escherichia coli*) bacterial fermentations.
2. Bulk mixing quality (homogeneity) is much poorer than in typical Newtonian broths; therefore, accurate monitoring and control are much more difficult.

Difficulties associated with viscous, non-Newtonian rheology also extend to other areas such as cleaning.

Oxygen Transfer/Aeration

Please note that in this and subsequent sections we include design and operating rules of thumb based on our experience and the experiences of others. They are intended to be helpful guides, not edicts.

Requirements. Oxygen transfer rate (OTR) requirements are usually dictated by conditions during the most active part of the growth phase (other phases require oxygen, but supply rate usually is not as high). The requirement is calculated from

$$\text{OTR} = (Y_{x/o})^{-1}(\mu X)_{\max} \qquad (2)$$

where μ is the specific growth rate (h^{-1}), X is the cell mass concentration (g/L dw), and $Y_{x/o}$ is the cell yield coefficient on oxygen (g cells dw/g O_2). Note that the maximum growth rate and maximum cell mass are not always reached simultaneously.

Satisfying the Oxygen Material Balance: Gas Flow and Linear Velocity. The very first thing we must do in satisfying an OTR requirement is to insure that we have a closed oxygen material balance. For most practical operating conditions, the rate of change of oxygen inventory in a fermenter is very small compared to oxygen flows in the inlet and outlet and to the OTR. It also can be demonstrated easily that the total molar flow rate of gas does not change enough to cause any loss of sleep. Given these practical realities, the oxygen material balance is

$$\text{OTR} = (1{,}000 \times 60)\text{F}(y_{\text{in}} - y_{\text{out}})/V_{\text{L}} \qquad (3)$$

where, F is the gas flow rate (mol/min), y is the mole fraction of oxygen in the gas, and V_L is the liquid volume (L). Note that the usual units for OTR are millimoles per liter per hour and that the factors of 1,000 and 60 in equation 3 are conversion factors needed to keep unit consistency.

The material balance also can be expressed as

$$F = \text{OTR} \times V_L/(1{,}000 \times 60\ \varepsilon y_{in}) \quad (4)$$

where ε is the oxygen transfer efficiency and y_{in} is the oxygen mole fraction in the inlet gas. Equation 4 is a more useful form because we know that under practical conditions ε will be between 0.15 and 0.35 for typical Newtonian broths, and between 0.05 and 0.15 for typical non-Newtonian broths.

As a general rule, heroic efforts will be required to get OTRs over 300 mmol/(L h) in large fermenters (>5,000 L). Even if the effort is expended and it is successful, it will usually create heat transfer problems (see later) that will require even greater effort to overcome. Our advice is to consider other options very seriously.

The material balance provides information about the gas volumetric flow rate, which often is expressed as the standard flow, v_{std}, in units of standard liters per minute (SLPM):

$$v_{std} = 22.4F \quad (5)$$

This is useful because most flow meters are calibrated in terms of standard flow; however, actual flow is more useful for fermenter design purposes. One problem encountered in calculating the actual flow is that it increases from the bottom to the top of the broth because of pressure change. As a practical matter, however, it is usually adequate to use the average pressure in the tank:

$$v_{act} = v_{std} \times (1/P_{avg})(T_f/298) \quad (6)$$

where v_{act} is the gas flow at average pressure (L/min), P_{avg} is the average pressure (atm, abs), and T_f is the fermentation temperature (K). One might want to revisit this for very tall vessels.

The standard gas flow rate also is expressed in standard gas volumes per liquid volume per minute (VVM).

Among the important information that can be obtained from v_{act} is the gas linear velocity, V_S (cm/min):

$$V_S = 4 \times 1000\ v_{act}/(\pi \times D_t^2) \quad (7)$$

where D_t is the inside diameter of the fermenter. V_S affects mass transfer, deliverable power via the impellers, foaming, gas holdup, and aerosol formation. As a rule of thumb, V_S should be held below 200 cm/min to avoid problems associated with excessive gas holdup, foaming, and aerosol formation. These effects are discussed later.

Mass Transfer. The rate at which oxygen can be transferred is dictated by three factors: the driving force, the resistance to transfer, and the contact area between the gas and liquid phases. This is usually expressed as

$$\text{OTR} = K_g a (\Delta P)_{LM} \quad (8)$$

where K_g is the mass transfer coefficient ([mmole M]/[L h atm]), and a is the interfacial area per unit volume (M^2/M^3). $(\Delta P)_{LM}$ is the log mean pressure driving force, defined as

$$(\Delta P)_{LM} = (P_{O,in} - P_{O,out})/LN[(P_{O,in} - P_O^*)/(P_{O,out} - P_O^*)] \quad (9)$$

where $P_{O,in}$ is the partial pressure of oxygen in the gas inlet (atm, abs), $P_{O,out}$ is the partial pressure of oxygen in the gas outlet (atm, abs), andP_O^* is the partial pressure of oxygen that would be in equilibrium with the dissolved oxygen concentration in the existing liquid (atm, abs). Note that use of the log mean driving force assumes that the liquid is mixed perfectly (homogeneous liquid phase) and that the gas moves in plug flow through the fermenter. These assumptions are acceptable for most low-viscosity Newtonian broths, but they can be quite poor for high-viscosity, non-Newtonian broths. Unfortunately, no useful alternatives have yet been proposed; therefore, extra caution should be exercised in interpreting calculated results for non-Newtonian broths.

Other mass transfer coefficients are used for driving forces other than $(\Delta P)_{LM}$. The most common among these is $k_l a$ (min^{-1}), which is used in conjunction with a dissolved oxygen concentration driving force. The reader is cautioned that $K_g a$ and $k_l a$ are sometimes confused with each other—even in the literature.

A given OTR requirement can be satisfied by various combinations of driving force and mass transfer coefficient. It is important to note, however, the following:

- Factors that influence driving force also can influence $K_g a$.
- Each combination will have different effects on the ultimate design of the vessel, piping system, and support systems, as well as on operational factors related to regulatory compliance cleaning, maintenance, and so on. Such effects are discussed as we proceed.

Driving force is affected by total pressure, gas inlet oxygen mole fraction, total gas flow rate, and the dissolved oxygen concentration. Total pressure and oxygen mole fraction affect the oxygen partial pressure directly. The effect of total gas flow is a bit more subtle. For a given oxygen transfer rate, increasing the total gas (fixed inlet conditions, pressure, etc.) increases the mole fraction of the outlet gas flow, thereby increasing the average mole fraction of oxygen in the gas; hence, the average driving force is increased. This can be seen quantitatively via an oxygen material balance which gives

$$P_{O,out} = P_{O,in} - \text{OTR} \times P/(1{,}000F) \quad (10)$$

where P is the total absolute pressure (atm) and F is the total molar flow (mmol/h) of gas.

The characteristics of K_g and a are complex and not completely understood. Our first clue to this is that $K_g a$ is expressed as a product in terms of empirical correlations (we don't understand enough about them to get reliable quantitative guidance from first principles). Typical correlations have the form

$$K_ga = \delta \times (P_g/V_L)^{\varphi}(V_S)^{\kappa} \quad (11)$$

where P_g is the power delivered to aerated broth (HP), V_L is the unaerated liquid volume (L), and δ, φ, and κ are usually advertised as constants. There also a several variations on this form. A discussion of all the fine points of such correlations is beyond our scope; however, the the reader should take note of the fact that published correlations usually cannot be relied upon to provide accurate predictions. They should be used as qualitative or semiquantitative guides only. Reasons for this caveat include the following:

- They are usually applicable only under conditions at or near the ones used to develop them. Unfortunately, most are developed at scales and under conditions that are not realistic for production.
- They can have pronounced dependence on fermenter geometry.
- They do not scale up well, or at all. The "constants" in equation 11 and its relatives usually aren't constant.

These caveats should be made more emphatically for viscous non-Newtonian broths than for Newtonian broths because non-Newtonian rheological characteristics can have profound effects on K_ga (as well as the other transport properties). Most K_ga correlations do not account for broth rheology. Those that do, usually do so inadequately or in ways valid only for a particular broth. It also is important to note that the broth rheology can vary dramatically during a fermentation and that the rheological characteristics can be affected by the fermentation conditions, ranging from the nature of the inoculum to the history of the agitation speed.

As noted earlier, we can get some qualitative guidance from the form of equation 11. We'll do that, but please keep in mind that the relationships are highly nonlinear, the factors that influence K_ga are dependent on each other in practical operation, and it is not always possible to control variables at the levels you would like. For example, equation 11 predicts that increasing power input per unit liquid volume will increase K_ga. This usually turns out to be true in practice if the power can actually be delivered to the fluid and delivered in a manner that will contribute to K_ga. Such might not be true in any given case.

The fact that a drive's power rating is 100 hp doesn't mean that 100 hp can be delivered to the broth. Deliverability depends on, among other things, geometry (vessel and agitator), rheology, agitator speed, and gas linear velocity (see later). The effects of all are complex and interactive; the interrelations are more complex for non-Newtonian broths than for Newtonian broths. The manner in which power is delivered depends largely on impeller type and broth rheology. For example, oxygen transfer rates to highly viscous, non-Newtonian mycelial broths have been shown to be larger for A-315 (hydrofoil) impellers than for Rushton (disk turbine) impellers delivering the same power (35). There appears to be no meaningful difference for Newtonian broths.

The bottom line is that one should try to obtain experimentally determined correlations for the subject organism and broth at meaningful scales. Failing that (which usually is the case because of time and resource constraints), one should try to use correlations developed for systems as similar as possible to the one at hand. Considerable art is involved here.

The reader can find more about K_ga in the literature (35–38).

Power

Power delivered to the broth is used for micromixing and gas dispersion, which are related to mass transfer, and macromixing, which provides overall homogeneity (discussed later). Agitated vessel power is delivered via two mechanisms: direct mechanical power from the impellers and gas expansion. The bulk (>90%) of the power comes from the impellers as long as the fluid motion is under their control, a condition that prevails so long as the impellers are not flooded. There are reasonably reliable correlations available to determine flooding conditions for Newtonian broths (39); however, flooding usually is not a problem under typical conditions used in most Newtonian broth fermentation. Flooding is more likely in highly viscous, pseudoplastic broths typical of mycelial and polysaccharide fermentations because the viscosity near the impeller is much lower than in the rest of the broth. As a result (assuming air is introduced under the impeller), air tends to channel toward the impeller, thereby enshrouding it and decreasing the deliverable power (4). This can dramatically decrease overall mass transfer (and heat transfer) rates and quality of bulk mixing quality (see later). Reliable flooding correlations for non-Newtonian broths have not been published, but some companies have accumulated a considerable amount of data for the fermentations they practice.

Calculation of power input relies on empirical correlations. As with K_ga, many correlations have been published (see for example Refs. 40–42), but they are not particularly reliable for many of the same reasons given for the unreliability of K_ga correlations; therefore, we give the same advice here as we did for K_ga.

Among the more popular approaches used is the one that relies on the aeration number, N_A (dimensionless), defined as

$$N_A = v/ND_i^3 \quad (12)$$

where v is the gas volumetric flow rate (m^3/min), N is the agitation speed (min^{-1}), and D_i is the impeller diameter (m). Published correlations give the ratio of gassed power to ungassed power as empirical functions of N_A:

$$P_g/P_{ug} = \mathrm{Func}(N_A) \quad (13)$$

where P_{ug} is the power delivered to the same broth agitated under the same conditions but unaerated.

Each such correlation pertains to a specific impeller type, a single fermenter geometry, and a fairly narrow range of operating conditions. In some such correlations

(40), a unique function is presented for each impeller type (Fig. 1). In others the relationship is shown not to be unique (43). In most cases, the correlations are for a single aerated impeller. The situation is more confusing for non-Newtonian broths.

Obviously, to use equation 13 one must be able to calculate the ungassed power. This can be done via other empirical correlations that give ungassed power in terms of the dimensionless power number (N_P), defined as

$$N_P = P_{ug} \times g_c/N^3 \times D_i^5 \times \rho \quad (14)$$

where g_c is Newton's law conversion factor, and ρ is the ungassed broth density (lb/ft^3). There are published correlations that give N_P as a function of Reynolds number for a wide range of single impellers (44). Each is specific to a particular vessel geometry, liquid height, and so forth. Interestingly enough, there are fairly reliable correlations for Newtonian and non-Newtonian fluids so long as the constraints of the correlations are observed.

It is important to note that some impellers that have high power numbers under unaerated conditions can "unload" considerably under aerated conditions. For example, it is fairly typical for a Rushton turbine to deliver, under aerated conditions, only 40% of the power it can deliver when unaerated. On the other hand, the SCABA 6SRGT (45), which is a curved-blade disc turbine, does not tend to unload at all. Some hydrofoils also exhibit very little unloading. This is an important consideration when one is trying to rank impellers in terms of how much power they can deliver for oxygen transfer purposes. (Often the ranking tends to be done qualitatively or "semiquantitatively.")

Multiimpeller Systems. Most fermenters are equipped with more than one impeller. The reasons for this are to improve bulk mixing (see later) and power distribution and to avoid the need for very large impellers and/or very high agitator speeds. Impeller size is limited practically not only by vessel internals but also by the following:

- *Torque transmitted to the drive shaft.* The larger the torque, the stronger and thicker the shaft must be. This also translates to higher torque and more expensive gear boxes.
- *The size of the vessel manway.* This is particularly important if the impellers must be single piece for better cleaning/sterilization characteristics.

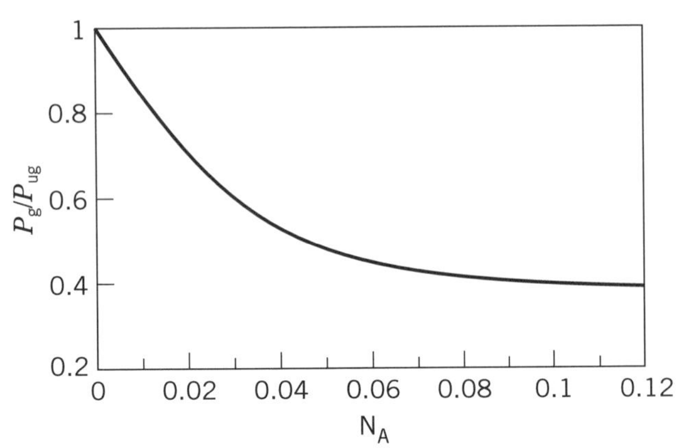

Figure 1. Power ratio as a function of aeration number for a Rushton turbine.

Agitator speed is limited by the natural frequency of the agitation system (4). This is because severe and potentially dangerous vibrations will occur if the rotational speed of the agitator approaches the natural frequency. We recommend that the maximum shaft speed not exceed 70% of the natural frequency.

Having said all this, we are left with the problem of doing the power calculations. This usually involves an iterative calculation in which the parameters are (assuming we have selected impeller types) number of impellers, impeller diameter(s), and shaft speed. (There are other considerations we discuss later.) To do this we use the power correlations already noted. But these usually apply only to single impellers. In most cases, they apply to the lowest impeller, which is the one under which inlet air is introduced. Several approaches have been suggested. The simplest is to treat all the impellers as though each is the only one present. Among other things, this assumes that the impellers do not interact with each other. There is empirical evidence (46pp.264–266) that Rushton turbines will not interact substantively in Newtonian broths if they are spaced at least an impeller diameter apart. Obviously, the assumption of noninteraction is not meaningful for impellers that produce significant axial flow (see later).

Another approach used frequently, but not recommended by the authors, is to use the correlations for the lowest impeller but to use them for the upper impellers:

$$P_g/P_{ug} = (1 + \text{HU})^{-1} \quad (15)$$

where HU is the gas holdup. You guessed it: there are empirical correlations for HU. One example for Newtonian broths is (47)

$$\text{HU} = 1.8(P_g/V_L)^{0.14}V_S^{0.75} \quad (16)$$

The usual caveats apply.

Typical values for gas holdup range from about 0.1 to 0.3; therefore, equation 15 predicts about 10–20% unloading for the upper impellers. Aeration number correlations predict around 50–60% unloading. Given the uncertainty in all this, and assuming that no other (reliable) information is available, we suggest the more conservative approach of applying the aeration number correlation to each impeller independently.

The added cost for the larger drive will not be a major factor for vessels up to about 5,000 L; this is fairly cheap insurance. One should do pilot agitation studies at a meaningful scale for much larger vessels.

Organism Sensitivity to Fluid Mechanical Forces. Finally, an additional constraint must be imposed if the organism is sensitive to fluid mechanical forces. This is seldom true of unicellular organisms. There are, however, some mycelial organisms that are sensitive. The extent of sensitivity and the nature of the forces that cause damage should be determined experimentally. One also should keep in mind that the character of fluid forces changes significantly with scale. For example, some turbulent forces that are negligible at small scale can be large and potentially destructive at large scale. Reliable guidance in this area is almost nonexistent. There is, however, a rough rule of thumb that states that damage will occur if impeller tip speed exceeds 1,500 ft/min. Bear in mind that this rule was developed almost 50 years ago and was based on information for organisms used at that time in vessels that had working volumes in the 30,000 to 50,000-gal range. It is clear that many factors (including broth rheology) will affect the potential for damage. Unfortunately, most of the information is seldom available when needed. In such cases, the "any port in a storm" approach to design usually becomes operable. Fortunately, we seldom encounter a circumstance in which the rule of thumb must be violated.

Bulk Mixing

Good bulk mixing is needed to insure homogeneity, which is required for reliable data acquisition and control. There are several important factors that affect mixing quality.

Broth rheology. This is the primary factor affecting our ability to provide good mixing quality, but we seldom have any control over it. We simply must recognize that we will have to go to greater and greater lengths as the broth becomes more viscous and non-Newtonian. Also, one must bear in mind that the nonlinear behavior of most non-Newtonian fluids often results in nonobvious responses to various actions. For example, increasing agitator speed in a highly non-Newtonian, pseudoplastic broth often leads to a decrease in power input and a decrease in bulk mixing quality.

Type(s) of impeller(s). Impeller types fall into three basic categories: radial flow (e.g., Rushton turbine), axial flow (e.g., marine propellor), and mixed axial and radial flow (e.g., Lightnin' A-315). Radial flow impellers tend to deliver the high power required to enhance micromixing and mass transfer but do not promote top-to-bottom mixing; therefore, they are not the best choice for promoting homogeneity. Multiple radial flow impellers often are used in an attempt to compensate for poor bulk mixing. This can work fairly well for low-viscosity Newtonian broths when the ratio of liquid height to tank diameter does not exceed 1.7–2, but it is a very poor solution when the broth is very viscous and highly non-Newtonian. Anyone who has observed classical streptomycete or xanthan gum fermentations can attest to this fact. Pure axial flow impellers do not deliver much power but do tend to promote good top-to-bottom mixing and hence contribute significantly to bulk homogeneity. They do not, however, contribute much to mass transfer. Some mixed flow types appear to provide a good balance between bulk mixing and mass transfer requirements, particularly for non-Newtonian broths. We recommend that they be considered seriously for such cases. We also have had considerable success with combination systems, for example, those with turbines as the lower impellers and a hydrofoil on top. This appears to work well for Newtonian and non-Newtonian broths.

Impeller size(s). In most cases, large-diameter impellers distribute power better and promote bulk mixing better than do small-diameter impellers. As a rough rule of thumb, based on considerable empirical evidence, we recommend D_i/D_t ratios of between 0.4 and 0.5 unless some other constraint (e.g., torque) is violated. Exceptions to this are some cases in which an axial flow impeller is the top impeller. In such cases, the ratio should not exceed 0.35 to avoid the risk of vortexing. This is particularly true for low-viscosity Newtonian broths.

Vessel geometry. We suggest that the ratio of the aerated liquid height to the tank diameter be kept below 2.0 to minimize bulk mixing problems. This must be reconciled with impeller diameter and impeller spacing requirements (see later) and with architectural, shipping, and other constraints.

Impeller spacing. As noted earlier, there is a body of empirical evidence (48) that supports the spacing of Rushton turbines 1–1.5 impeller diameters apart in Newtonian broths. This spacing appears to provide a balance between preventing impellers from interfering with their neighbors and at the same time avoiding dead zones between them. The same sources of empirical data support placing the lowest impeller one impeller diameter from the bottom of the vessel (assuming a standard dished head) and placing the uppermost impeller one impeller diameter below the liquid surface. We have found these rules to work reasonably well for Newtonian broths and a few non-Newtonian broths, and for radial flow impellers other than Rushtons. Please note that these conclusions are based on limited observations. Finally, keep in mind that one cannot avoid interaction among impellers when at least one of them is an axial flow type or has a significant axial flow component, and that spacing effects in highly non-Newtonian broths are not well understood.

Baffling. Baffles are placed in vessels to minimize fluid swirling and vortex formation (see Ref. 48 for flow pattern diagrams). The usual approach is to use four baffles on 90° centers, each baffle having a width equal to 0.1 of the D_t. Baffling tends to increase transmittable power and to improve mixing (except for the dead spots, which tend to form behind the baffles). Elimination of significant vortexing is important for safety as well as for improved bulk mixing. A vortex can reach down to the impeller in an unbaffled vessel. When this happens, a sizable portion of the impeller becomes air enshrouded, thereby decreasing resistance on the impeller. This tends to drive up the impeller speed, at least briefly (or the speed may be increased intentionally in an attempt to increase the power input). Unfortunately, vortices are unstable and can collapse shortly after formation, resulting in very large and potentially catastrophic stresses (impulse) on the impeller, particularly if the impeller speed is increased significantly while it is enshrouded.

Gas flow. Gas flow has a complex effect on bulk mixing. The flow alone does tends to promote bulk mixing (as in bubble tanks), but it also tends to decrease the effect of the impellers, particularly at high values of gas linear velocity. Fortunately, this is not a major problem in most cases, but it can be for very large fermenters and for highly viscous non-Newtonian broths.

Gas Holdup, Foaming, and Aerosol

There are several problems that can interfere seriously with a fermentation. Gas holdup (discussed earlier) decreases the effective volume of a fermenter. Foaming and aerosol formation can constrain operation, cause major cleaning and asepsis problems, and in the extreme cause termination of a fermentation. All three are dependent on broth characteristics, power input, and gas flow rate. There is some information in the literature, but these points have not been given the attention that even begins to reflect their importance. In general, all one can say is that all three become bigger problems for a given fermentation as gas flow and power input increase. Beyond that, each case is a new adventure. More will be said about the practical aspects later.

Heat Transfer

Heat Loads. Heat transfer is required during fermentation to maintain constant temperature conditions, and at other times (i.e., sterilization, induction) to increase or decrease broth temperature. In most cases, cooling is required during most of an active, aerobic fermentation. A good approximation to the total heat load, Q_{tot}, during such fermentations in which the primary carbon source is glucose or a similar carbohydrate is (49)

$$Q_{tot} = Q_{metab} + Q_{mech}$$
$$Q_{tot} = 0.48(\mathrm{OTR})V_L/Y_{x/o} + 2545P_g \tag{17}$$

where Q_{metab} is the heat generated by metabolism (Btu/h), Q_{mech} is the heat generated by power input (Btu/h), and $Y_{x/o}$ is the yield of cells on oxygen (g dry wt cells per g oxygen consumed). Heat transfer can be more of a limiting factor in aerobic fermentations than is oxygen transfer—particularly in large fermenters. The major heat transfer demand is usually during the maximum growth period (i.e., when μX is greatest). It is always wise, however, to check other loads. For example, rapid cooldown after an induction phase could put enormous demands on the cooling system. Cooling a 10,000-L (wv) fermenter from 38 °C (104 °F) to 10 °C (50 °F) in 30 min would require an average heat transfer rate of approximately 2.4×10^6 Btu/h. The same heat transfer rate would support an oxygen transfer rate of approximately 495 mmol/L(h), which is very high by almost anyone's standards. Similarly, one should check other potentially high loads such as cooldown after sterilization (see later). In any case, careful thought and solid, empirical facts (e.g., product thermal degradation rates) should form the basis of any specification that will require extraordinary transfer rates.

Heat Transfer Rate. The rate at which heat can be transferred is dependent on *(1)* the driving force for heat transfer, *(2)* the area across which transfer must occur and, *(3)* the resistance to heat transfer:

$$Q = (1/\text{resistance}) \times (\text{area}) \times (\text{driving force}) \tag{18}$$

The driving force for agitated vessels is usually taken to be the log mean temperature difference, ΔT_{LM}, defined as

$$\Delta T_{LM} = (T_{c,o} - T_{c,i})/LN[(T_f - T_{c,i})/(T_f - T_{c,o})] \tag{19}$$

where $T_{c,o}$ is the coolant outlet temperature (°F), $T_{c,i}$ is the coolant inlet temperature (°F), and T_f is the fermentation temperature (°F). Note that use of equation 19 assumes that the broth is homogeneous and can be characterized by a single temperature.

The transfer area, A_{JA}, is the actual contact area (not necessarily the same as the available area) between the broth and the heat exchange surface, which is usually a jacket and/or an internal coil. The actual area can be found if the height of aerated liquid in the tank is known, along with the geometry of the jacket and any internal exchange area (e.g., a coil). The aerated liquid height measured along the centerline of the vessel, h_{LA} (ft), is

$$h_{LA} = [(1 + \mathrm{HU})V_{LU} - V_D]/(0.785D_t^2) + \mathrm{IDD} \tag{20}$$

where V_{LU} is the unaerated liquid volume (ft^3), V_D is the lower dish volume (ft^3), IDD is the inside depth of lower dish (ft), and D_t is the tank inside diameter (ft).

The resistance to heat transfer is a little more complicated. It is composed of five resistances in series and is expressed as

$$1/U = 1/h_i + 1/h_o + 1/h_{fi} + 1/h_{fo} + t/k \tag{21}$$

where U is the overall heat transfer coefficient (Btu/[h °F ft^2]), h_i is the broth film heat transfer coefficient (Btu/[h °F ft^2]), h_o is the jacket fluid heat transfer coefficient (Btu/[h °F ft^2]), h_{fi} is the broth fouling factor (Btu/[h °F ft^2]), h_{fo} is the jacket fluid fouling factor (Btu/[h °F ft^2]), t is the fermenter wall thickness (ft), and k is the fermenter wall thermal conductivity (Btu/[h °F ft]). The resistance of the tank wall can be predicted accurately because both k and wall thickness are known accurately. It should be noted that the k for stainless steel is very low (about 10 Btu/[°F h ft]). This means that the tank wall should be kept as thin as possible to minimize its contribution to heat transfer resistance (see later).

Correlations for h_o can be found in the literature (46pp. 282–283) but are seldom as reliable as those available from vendors for the specific jackets they fabricate.

The problem in predicting U is in predicting reliable values of h_i, h_{fi}, and h_{fo} Published correlations (50) for h_i values for aerated, Newtonian fermentation broths have been published but are not very reliable. Correlations for non-Newtonian broths have also been published (51), but as was the case for mass transfer coefficients, h_i is strongly dependent on broth rheological characteristics. Unfortunately, existing correlations do not do a very good job of accounting adequately for broth rheology, a problem exacerbated by the fact that good rheological information is seldom available. Finally, the fouling factors are essentially impossible to predict. The best one can do is is keep fouling to a minimum by means of good cleaning practices and proper filtration of coolants. Effects of individual resistances on U are illustrated in Figure 2.

The following rough guides (based on many years of experience) are offered because the authors are unable to offer any collection of reliable prediction tools:

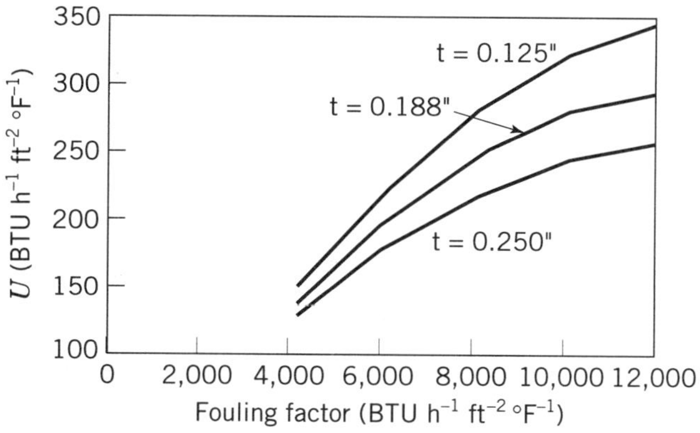

Figure 2. Effects of metal thickness and fouling resistances on the overall heat transfer coefficient.

1. U values for simple, low viscosity Newtonian broths usually range from about 120 to 150 Btu/(h °F ft^2) for new, clean stainless steel tanks.
2. U for viscous, non-Newtonian broths ranges from about 30 to 60 Btu/(h °F ft^2)

It also should be noted that U's for heating usually are a bit higher than U's for cooling.

The heat transfer rate can now be expressed as

$$Q = UA\Delta T_{\mathrm{LM}} \tag{22}$$

Similar equations can be developed for heat transfer via an internal coil. The only difference is that the heat transfer coefficients for coils are usually a bit higher than those for jackets. That having been said, we will try everything possible to discourage you from using internal coils. The reasons relate primarily to mechanical and cleaning considerations and are discussed later.

In applying the preceding, we suggest that the reader consider the following suggestions:

- Avoid the use of internal cooling surfaces (e.g., coils). They make cleaning difficult to impossible, are potential sources of contamination (leaking of coolant), put a lot of mechanical strain on the vessel, and can cause deterioration of bulk mixing (particularly for viscous non-Newtonian broths).
- Avoid subfreezing coolant. Sooner or later, subfreezing coolant will cause severe valve freeze-up and worse.
- Try to keep coolant flow rates low enough to avoid pipe diameters greater than 3 in.; cost increases considerably and availability of fittings and so forth decreases as size increases.

Finally, there are some cases in which high wall temperatures have been cited as causing problems with temperature-sensitive organisms. The wall temperature, T_w, depends on the temperatures of the broth and of the jacket fluid, as well as on the relative values of the heat transfer resistances. A simple energy balance across the heat transfer path yields

$$T_w = (T_f + \beta T_j)/(1 + \beta) \tag{23}$$

where

$$\beta = k/h_i/(t + k/h_o) \tag{24}$$

Note that this does not account for locally high wall temperatures that would be caused by steam applied to steam locks during fermentation.

Sterilization

Introduction. Sterilization and aseptic operation taken together (as they must be) have a much greater influence on fermenter design and operation than does any other requirement.

In theory *(1)* sterilization destroys or removes all foreign organisms in all process equipment (including piping, seals, etc.) that might come into contact with the process fluid, and *(2)* aseptic operation insures that *no* contaminating organisms enter the fermenter after sterilization. These ideals are not attainable in practice because (among other reasons) *(1)* there is a finite probability that a very low concentration of contaminating organisms will not be captured in a sample used to test for contamination, and *(2)* there is a finite probability of false positives due to sample contamination and so forth. It also is important to note that there is a continuing debate concerning the definitions of pure culture and sterility (6,51).

The driving forces for sterilization and aseptic operation range from minimizing product losses to insuring strict compliance with regulatory requirements. Among the many specific problems that contamination can cause are the following:

- Production of a toxin that can't be removed by the purification system
- Production of an enzyme that degrades the product
- Decreased product yield due to use of substrate by contaminants
- Production of toxins that inhibit the producer strain
- Production of compounds (e.g., polysaccharides) that interfere with the operation of recovery and purification equipment

Which specific problems will exist and where in the spectrum a particular case will lie depend primarily on the product, its economic value, whether it is regulated, and how it will be used. Given all these variables and the fact that absolute sterility is an unachievable abstraction, the extents to which one should go should be considered on a case-by-case basis. As a practical matter, sterilization has to be interpreted as "effective sterilization," meaning that the design and procedures (including sampling and detection) are suitable for the specific case. For example, sterilization of fermenters used to produce parenterals should be held to a much higher standard than sterilization of fermenters used to produce amylases for starch hydrolysis. Unfortunately, there remains a lot of room for disagreement among well-intentioned people (see also the section "Product Safety").

Quantification of Sterilization. Mechanical details and other practical considerations of sterilization and aseptic operation are discussed in the subsection "Sterile Piping Systems."

Several methods have been developed to quantify sterilization. The simplest is based on the empirical observation that the death kinetics of many vegetative organisms and spores can be described by a simple first-order expression

$$dN/dt = -kN \tag{25}$$

where N is the number of viable organisms (or spores) at any time, t, and k is the thermal death constant. k is a function of most environmental variables but is most strongly affected by temperature. This dependence is very often expressible as an Arrhenius relationship

$$k = A \exp(-E_A/RT) \tag{26}$$

where A is the preexponential factor (min^{-1}), E_A is the thermal death activation energy (cal/mol), T is the absolute temperature (K), and R is the gas constant. The effect of temperature on k is pronounced for most organisms because their E_A's are so high (e.g., 65,000 cal/mol for *Bacillus stearothermophilus* spores).

Among other methods of quantifying sterilization concepts are the following:

- *Probability-based theories* (52). These are conceptually different from the first-order model, but give essentially the same results for the levels of kill that must be achieved practically.
- *Methods based on the "decimal reduction factor," D*. These are used widely (53). It is important to note, however, that the nature of temperature dependence has no rational, physical basis and differs considerably from an Arrhenius relationship: it is based primarily on an empirical observation for relatively small temperature ranges.

We use the first-order model for the rest of the discussion.

Medium Sterilization. The primary design basis for sterilization systems and procedures is usually the level of sterility that must be achieved. This is most frequently expressed in terms of the probability of sterilization failure (i.e., the probability that a single organism will survive). In most instances probabilities of 10^{-3} to 10^{-4} are quite adequate. This means 1 failure in 1,000 or 10,000 sterilization operations, respectively. It also is important to note that the probability of failure is numerically very close to the value of N in the first-order model.

The reader should also take note of the fact that in some quarters, sterilization criteria are stated in terms of "logs of kill." That is to say, a fixed number of logs (e.g., 12) is taken to be adequate. This is meaningless in that it does not take into account the initial contaminant loading.

Now we consider prediction of requirements for batch sterilizing a fermentation medium as part of the fermenter design process to ensure that we'll have adequate heating and cooling capacity. It should also be done as part of the process design to ensure adequate timing. Batch sterilization involves heating the medium in the fermenter to sterilization temperature (T_{ster}), holding the medium temperature constant at T_{ster} for an adequate time, and then cooling the medium down to fermentation temperature (see the section "Mechanical Design" for additional details). The medium is kept in the fermenter throughout.

It is assumed in what follows that *(1)* heating and cooling are via the jacket only, and *(2)* the energy and condensate contributed by the small amount of steam that flows into the vessel as a result of sterilizing piping, air filters, and so on is negligible for energy balance purposes.

The rate of kill at any instant during the sterilization process is

$$dN/dt = -kN = -A \exp(-E_A/RT)N \tag{27}$$

therefore,

$$N/N_o = A \int \exp(-E_A/RT)dt \tag{28}$$

where N_o is the number of contaminating organisms initially present. The integration of equation 26 must be performed over heating, hold, and cooling phases of sterilization. This is done by numerical integration of equation 28, coupled with the equations that describe broth temperature as a function of time.

For heating

$$T_f = T_{ST} - (T_{ST} - T_{F,O}) \exp(-U * A_{JA} * t/M_F * C_{P,m}) \tag{29}$$

where T_f is the medium temperature (°F), T_{ST} is the steam temperature (°F), $T_{F,O}$ is the initial medium temperature (°F), A_{JA} is the active jacket area (ft^2), t is the time since beginning of heating (h), M_F is the medium mass (lb), and $C_{P,m}$ is the medium heat capacity (Btu/[lb °F]).

For cooling

$$T_f = T_{C,i} + (T_{ster} - T_{C,i}) \exp(t\ (W_J C_{P,c}/(M_F C_{P,m})) \cdot (\exp(UA_{JA}/(W_J C_{P,c})) - 1) \tag{30}$$

where $T_{C,i}$ is the inlet coolant temperature (°F), T_{ster} is the sterilization temperature (°F), t is the time from beginning of cooling (h), W_J is the coolant flow (lb/h), and $C_{P,c}$ is the coolant heat capacity (Btu/[lb °F]). Equations 29 and 30 can be derived via energy balances.

One chooses times and temperatures such that the desired kill, N/N_o, is achieved. Unfortunately, one usually does not know the level of contamination or the nature of the contaminants. Indeed, it is very unlikely that contamination will remain constant from one fermentation to the next. The usual practice, therefore, is to design for what is thought to be the "worst case." This usually gives results that are quite satisfactory for design specifications. A "typical" worst case is a loading of 5×10^6 spores/mL of the organism *B. stearothermophilus*. Such spores are highly resistant and have the added advantage of being used widely to validate sterilization operations (see later). It is important to note that the purpose of the calculations is to provide a rational basis for obtaining reasonable design information and ensuring a high probability that the desired level of sterilization can be achieved over the whole range of anticipated operating conditions.

Figure 3 illustrates the results of such calculations for the following case:

Medium volume = 15,000 L
Initial spore load = 5×10^6/mL
Initial medium temperature = 70 °F
Steam temperature = 300 °F
Coolant temperature = 35 °F
Coolant flow rate = 100 gpm
Fermentation temperature = 86 °F
Active jacket area = 225 ft^2
U = 150 BTU/(°F h ft^2)
$A = 2.1 \times 10^{36}$/min
E_A = 65,000 cal/mol
N (probability of failure) = 0.0001

As seen in this example, heating and cooling phases usually contribute only a small fraction of the total sterilization kill, but they do affect turnaround time significantly. Long heating and cooling times can also have other effects:

- Cause damage to the fermentation medium to the extent that the fermentation can be compromised. The seriousness of this depends on some combination of economics and regulatory compliance.
- Cause medium changes that have negative effects on recovery and purification without causing fermentation problems. This includes the possibility of introducing foreign materials that may pass undetected into the final product.

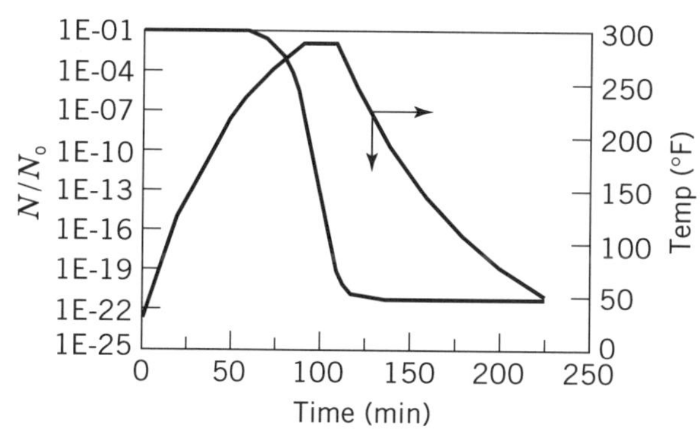

Figure 3. Temperature profile and kill ratio for a batch sterilization.

The example also shows that cooling time is considerably longer than heating time. This results primarily from the lower temperature-driving forces during cooling. The problem is exacerbated considerably when the heat transfer coefficient is very low as is the case for viscous non-Newtonian broths. Another method used for batch sterilization is direct steam injection: live steam is injected directly into the medium as the main source of thermal energy. This decreases heating time as well as the overall steam requirement. It also increases the medium volume by about 20% (as a result of steam condensation), which can cause some problems, including the following:

- *Increased cooling time.*
- *Medium dilution.* This will be a significant problem if the initial medium cannot be made concentrated enough to account for the dilution. Some reasons include low solubility of medium components, increased viscosity, and increased reaction rates among medium components at elevated temperatures.
- *Introduction of impurities.* This depends primarily on the quality of the injected steam. In some cases plant steam is acceptable if boiler cleaning agents do not cause problems. At the other extreme is the requirement to use clean steam (WFI quality). It should be noted with regard to this point that some steam will be injected directly even when jacket heating is used as the main energy source; therefore, one will always be in the position of having to evaluate the effects of contaminants carried by the steam.

For cases in which sterilization times are too long or medium alterations cause too many problems, one might consider continuous sterilization. The interested reader is directed to the literature for additional information (54pp. 159–176).

Piping System Sterilization. Heating and cooling times are not usually significant issues for sterilizing piping; however, there can be some serious heat transfer problems related to piping length, diameter, and orientation. Some of these are discussed in the subsection "Vessel Design." The reader is also referred to literature reports of some practical experiments and theoretical analyses (55,56).

Air Sterilization. Theoretical aspects of air sterilization are not discussed here. Suffice it to say that modern filters, if sized, installed, and maintained properly, will provide very reliable results. We'll discuss some of these practical aspects in the section "Mechanical Design." See Refs. 54pp. 176–183 and 57 for additional information.

Cleaning

Scrupulous cleaning is necessary to decrease nonbiological contamination and prevent cross-contamination of batches. Experience also has shown that reliable sterilization is difficult or impossible to achieve in the absence of rigorous cleaning.

As noted earlier, fermenter cleaning is not mentioned specifically in 21 CFR, but the basis is provided in section

211.67, which requires equipment be cleaned as per a defined plan and by trained people. Furthermore, the FDA has made clear the importance it attaches to cleaning not only in various agency publications (58), but also in more stringent enforcement. This is particularly true in multiproduct facilities (a key issue for most modern biotech plants). This agency initiative and the desire to generally improve operation have caused the industry to develop more reliable cleaning systems and protocols and to move to fully automated cleaning-in-place (CIP) systems, which provide greater assurance of successful validation and long-term compliance.

Developing and designing reliable fermenter cleaning systems is not as straightforward as it might appear, and considerable controversy continues. Among the reasons for this is that while much is known about the basic science of cleaning in general, very little is known about the basic science of cleaning fermenters and little has been published. The approach taken is based primarily on experience derived from the dairy, food, and beverage industries. Such information is useful but is not directly applicable in general to pharmaceutical and biotech processes where soils are different and the cleaning requirements are far more stringent. What meager information there is concerning pharmaceutical and biotech soils has been obtained from experiments done on single soil components and/or studies done under conditions not truly representative of the process conditions (59,60). There have been no significant attempts to develop systematic analyses based on experiments done with complex mixtures typical of real fermentation soils under conditions found in real processes. As a result there are no reliable general methods, tools, or correlations on which to base the development of cleaning agents, protocols and CIP system design; decisions tend to be made based on arbitrary criteria (e.g., coupon bake on studies). In keeping with our general design philosophy, we think it important that such arbitrariness be avoided and that cleaning protocols be developed along with the fermentation. It is clear that this will help to ensure not only proper design but will also minimize cleaning validation studies and any questions concerning the presence of contaminants and/or cleaning residues in the commercial product that could not have been present in clinical trial material.

The selection of fermenter cleaning agents and protocols should be based not only on the specific soil but also on the materials of construction, surface finishes, and so on. In addition, one should consider the issues of *(1)* compatibility of each material with the cleaning agents and protocols, and *(2)* potential materials interaction during cleaning. These are decisions that usually are made during the design phase, but really should be defined much earlier. These issues are discussed in the subsection "Agitation Systems."

Fermenter cleaning validation is beyond the scope of this article; however, the design must be done so as to minimize validation problems and ensure maximum ease in ongoing compliance. The design team must therefore be aware of the procedures that will be used for validation and recognize that this is an issue that has not yet been resolved in the general community.

Putting It Together: Preliminary Design Calculations

We now begin to see some of the interactions that can arise among the transport processes, and between them and the operating variables. This is a necessary step in developing sound designs, and a precursor to preliminary design calculations. As an example, consider a case in which we are trying to find ways to increase the OTR above some base value. One approach would be to increase the gas flow rate (v) keeping other variables constant:

The overall driving force tends to increase (equation 10).

V_s increases (equation 7).

Increased V_s tends to increase K_ga (equation 11).

Increased v also increases N_A (equation 12).

Increased N_A tends to decrease power input (Fig. 1).

Decreased power input tends to decrease K_ga (equation 11).

As another example, consider increasing the vessel pressure while keeping other variables constant:

The overall driving force tends to increase (equation 10).

The actual gas volumetric flow rate (v) tends to decrease (assuming constant molar flow rate) (equation 6).

V_s tends to decrease (equation 7).

The decrease in V_s tends to decrease K_ga (equation 11).

The decrease in v also tends to decrease N_A (equation 12).

This tends to increase power input (Fig. 1).

Increased power input tends to increase K_ga (equation 11).

Such complex interrelationships develop regardless of what steps are taken to increase OTR. The nature of these will change depending on the approach taken and the system being considered. There will also be additional pluses and minuses related to factors such as cost, safety, cleaning, and regulatory compliance. For example, using pure oxygen to increase the driving force carries with it very distinct cost and safety problems. Many of these more practical implications are be discussed in the section "Mechanical Design."

Now let's carry the second approach a little further. We do this because increasing pressure is usually the easiest and seemingly least expensive way to increase OTR. The decrease in V_s caused by the increased pressure has the positive effects of decreasing foaming, aerosol formation, and gas holdup. But it has the negative effect of requiring thicker tank walls, thereby increasing heat transfer resistance in the face of an increased heat load due to increased metabolic activity. This will become a significant problem when the pressure desired is greater than that required for vessel sterilization. The larger the tank diameter, the greater the possible problem. It should also be noted that thicker walls can make it more difficult to get a good surface finish and therefore increases the possibility of cleaning problems. In addition, the higher the pressure and thicker the walls, the shorter the list of qualified vendors.

This type of reasoning process is applicable to any of the many permutations possible. We extend it after we consider some additional basic concepts, mechanical design practices, and a few more important practical constraints.

We now consider two examples of preliminary design calculations, the objectives of which are to, first, define vessel geometry to ensure that it will not violate architectural constraints and be able to satisfy general rules of thumb for good bulk mixing, impeller spacing, and so forth. Some of this will require checking calculations done later (see later) and may require some iteration.

The second objective is to establish ranges of air flow, pressure, oxygen enrichment, and power that will accomplish the following:

- Satisfy the design basis maximum OTR requirement.
- Keep V_S low enough to avoid foaming, holdup, and aerosol problems. We have found a limit of 200 cm/min to be a safe guide; however, there are cases (Newtonian broths) in which considerably higher values up to 300–350 cm/min tolerable and others (very viscous, highly non-Newtonian broths) in which it is not wise to go beyond 75–100 cm/min.
- Not violate compressor and other component constraints.
- Not require heroic efforts (e.g., extremes in pressure, power, etc.).

Please note that we are trying to establish reasonably broad ranges for each of the variables within which we can satisfy the requirements with a reasonable degree of comfort (based on experience). This is important because the correlations usually available have questionable accuracy, and it allows some breathing room for process improvements and so forth.

The third objective is to define impeller type(s), sizes, and speed that will do the following:

- Provide the power required
- Provide good bulk mixing
- Satisfy impeller spacing rules of thumb
- Not violate constraints related to the sensitivity of the organism
- Make possible single-piece impeller construction and so forth (see the section "Mechanical Design" for consideration of mechanical constraints)
- Not violate any other constraints specified

The fourth objective is to define heat exchange area and coolant flow and temperature that achieve the following:

- Satisfy peak heat transfer requirements
- Not violate coolant temperature and flow constraints and such
- Avoid internal heat exchange area
- Not require subfreezing coolant or flow that will require very large pipe diameters

(*Note*: One should check to make sure that other heat loads [e.g., cooling after fermentation] are not greater than the load during maximum growth activity. We assume that to be the case here.)

The fifth objective is to use the results of the calculations not only to define mechanical details (see the section "Mechanical Design"), but also as a basis for recommendations for rational changes in design basis.

Please note that experimentally derived correlations for the subject cases are used in the examples (this is a design gift as rare as hen's teeth). All were obtained at 500-L scale, and we are just within our comfort range concerning their scalability to 15,000 L (design scale). The reader is cautioned, however, that there is no basis to believe that the correlations would be applicable to other cases, nor would we recommend extrapolation to much larger scales. It also is worth noting that there are some good examples in the literature that might be applicable to the problems at hand or to others; however, one must exercise considerable caution if the conditions specified in the reference differ significantly from those for the case being analyzed. Unfortunately, it is often not possible to determine if significant differences exist.

Case 1.

Product	Intracellular protein
Organism	*E. coli*
Final broth volume	15,000 L
Maximum OTR	234 mmol/(L h)
Dissolved oxygen requirement	10% relative to atmospheric conditions
Broth rheological type	Newtonian
Maximum broth viscosity	5 cp
Regulation	CBER
Containment	BL2-LS
Maximum coolant flow	250 gpm
Minimum fluid temperature	35 °F
Internal coils	No
Compressor max press/max flow	50 psig/25,000 SLPM
Minimum heat transfer coefficient	150 Btu/(h °F ft^2)
Ceiling height	336 in.
Floor space	1000 ft^2, unobstructed

The following experimental correlations are applicable:

$$K_g a = 108(P_g/V_L)^{0.6} V_S^{0.2}$$

$$P_g/P_{ug} = 1.0 - 18.56N_A + 215.64N_A^2 - 1082.66N_A^3 + 1859.68N_A^4$$

$$\text{HU} = 0.0133(P_g/V_L)^{0.4}(60V_S/^{0.5}) - 0.0333$$

These were developed under the following conditions for single Rushton turbines:

$0.8 < P_g/V_L < 2.9$ hp/100 gal
$100 < V_S < 200$ cm/min
$0 < N_A < 0.1$
$0.33 < D_i/D_t < 0.45$

Vessel geometry calculations are the same for both cases. We start by assuming a vessel diameter and wall thickness. (Note that the thickness will probably be changed a bit later to satisfy code requirements. This will not have a significant effect on height and volume calculations to follow.) In this case we assume a 96-in. outside diameter and a 0.25-in. wall thickness. By straightforward geometry and using tabulated values of volumes and inside depth of dish (IDD) for standard ASME heads, we get the following:

Liquid height	134.2 in.
(Liquid height)/tank diameter	1.41
Total tank height	192.5 in.
Total volume	21.1 L
Percent fill (unaerated)	71.1%

This satisfies rules of thumb concerning liquid height-to-diameter ratio and fill percent. If we add about 72 in. for legs and 30 in. for opening the manway, we get a total height requirement of 294.5 in. We should be able to satisfy the maximum height limit.

The results of oxygen transfer calculations for several operating conditions are given in Table 1. From these results we conclude that the following will satisfy the OTR required and give us a reasonable amount of "breathing room."

Gassed power	100 hp (includes transmission efficiency)
Maximum operating pressure	40 psig
Maximum air flow	10,000 SLPM

We do not recommend oxygen enrichment at this point, but we do suggest that the "holes be punched" to facilitate future installation of an oxygen system. The agitation system calculations are based on the assumption of using two 40-in. diameter Rushton turbines. The results are given in Table 2. A maximum speed of 150 rpm should be adequate. Note also that the impellers can be spaced to satisfy the rules of thumb for bulk mixing rules discussed previously. Although this should give satisfactory results for such a low viscosity broth, we recommend consideration of a slightly larger turbine as the lower impeller and a hydrofoil (e.g., A-315) as the top impeller.

The results of the heat transfer calculations are given in Table 3. A minimum coolant temperature of 35 °F was used in all cases. The results show that the heat loads can be handled, but the conditions are uncomfortably tight. For example, in no case will increasing the coolant flow to the maximum make up for a 10% decrease of the heat transfer coefficient. The potential consequences of inadequate heat transfer in this case should be considered seriously before final design commitments are made. Alternatives (e.g., lower coolant temperature) should be explored.

Case 2.

Product	Antibiotic
Organism	*Streptomyces* sp.
Final broth volume	15,000 L
Maximum OTR	50 mmol/(L h)
Dissolved oxygen requirement	10% relative to atmospheric conditions
Broth rheological type	Non-Newtonian (pseudoplastic)
Maximum broth viscosity	1,000–1,500 cp
Regulation	CDER
Containment	BL1-LS
Max coolant flow	250 gpm
Min fluid temperature	35 °F
Internal coils	No
Compressor max press/max flow	50 psig/25,000 SLPM
Min heat transfer coefficient	40 btu/(h °F ft^2)
Ceiling height	336 in.
Floor space	1000 ft^2, unobstructed

The following empirical equations are applicable:

$$K_g a = 8.0(P_g/V_L)^{0.5}V_S^{0.4}$$

$$P_g/P_{ug} = 1.0 - 26.99N_A + 417.37N_A^2 - 2789.43N_A^3 + 6643.30N_A^4$$

Table 1. Oxygen Transfer Calculation Results (Single Cell)

Pressure (PSIG)	35	20	20	2
O_2 trans eff. (%)	21	30	28	28
O_2 (MOL % in)	21	21	40	40
Gas flow (in, SLPM)	29,762	20,833	11,719	11,719
VVM	1.98	1.39	0.78	0.78
Average V_s (cm/min)	186	183	103	200
hp/100 gal	1.3	2.7	1.0	2.6
Theor. P_g	52.8	105.6	41.1	103.8

Table 2. Citation System Calculation Results (Single Cell)

Shaft speed (rpm)	114	141	88	141
Turb imp dia (in)	40.00	40.00	40.00	40.00
Turb D_i/D_t ratio	0.42	0.42	0.42	0.42
Ungassed hp (turb)	123.1	233.0	56.6	233.0
N_a, aeration numb	0.071	0.057	0.051	0.062
P_g/P_o ratio	0.43	0.46	0.48	0.45
Gassed hp	53	106	42	104

Table 3. Heat Transfer Calculation Results (Single Cell)

Heat load (MMBtu/h)	1.808	1.943	1.779	1.938
Hold-up	0.199	0.271	0.123	0.282
Jacket area (ft^2)	297	317	276	320
UJKT (BTU/ft^2 h °F)	150	150	150	150
Coolant temp in (°F)	35	35	35	35
Coolant flow (GPM)	190	210	235	200

$$HU = 0.012(P_g/V_L)^{0.3}(60V_S/100)^{0.52} - 0.029$$

These were developed under the following conditions for single Rushton turbines:

$1.0 < P_g/V_L < 2.3$ hp/100 gal
$60 < V_S < 130$ cm/min
$0 < N_A < 0.1$
$0.36 < D_i/D_t < 0.42$

The results of oxygen transfer calculations for several operating conditions are given in Table 4. From these results we conclude that the following will satisfy the OTR required.

Gassed power	100 hp (includes transmission efficiency)
Maximum operating pressure	40 psig
Maximum air flow	25,000 SLPM

Oxygen enrichment should be considered in this case despite the fact that the calculations show that it is not necessary. The reason is that the gas flow rate, and hence V_S, is quite high without enrichment. This results primarily from the very low oxygen transfer efficiency typical of very viscous non-Newtonian broths. Although it is true that we haven't broken the rule of thumb for holdup, foaming, and aerosol we should take note of the fact that high air flows in pseudoplastic broths can exacerbate impeller unloading and cause flooding more readily than for non-Newtonian broths. It also is generally true that our correlations become less reliable as conditions become more extreme and further removed from the conditions under which they were developed. Note in particular that our correlations were not developed for the high values of V_S found in the calculations. It also is clear that we should specify higher pressures than shown in the results in Table 4. Additional calculations show that 40–45 psig would bring the V_S into a comfortable range. We suggest, therefore, the following:

Gassed power	100 hp (includes transmission efficiency)
Maximum operating pressure	45 psig
Maximum air flow	25,000 SLPM
Oxygen enrichment	40% oxygen in inlet stream

The agitation system calculations are based on the assumption of using two 40-in.-diameter Rushton turbines. The results are given in Table 5. A maximum speed of 140 rpm appears to be adequate. Note also that the impellers can be spaced to satisfy the bulk mixing rules of thumb discussed previously.

That's very nice, but we have significant reservations about the results. An alternative should be given serious consideration because of the rheological nature of the broth, the profound effect this can have on mixing, and the issues of scalability for such broths. At the very least, the lower turbine should be replaced by one having a D_i/D_t approaching 0.5, and the upper turbine should be replaced by a hydrofoil. The reader should understand that this is based on experience only: we have no solid correlations for this case. The best route to take would be to do a few pilot mixing studies in as large a vessel as possible with the real broth.

The results of the heat transfer calculations are given in Table 6. The minimum coolant temperature of 35 °F was used in all cases. The results show that the heat loads can be handled. It also appears that we are well below the maximum available coolant flow. This may be comforting, but the reader should realize that there is not a lot of breathing room for process improvement. In particular, an increase in OTR up to 60–65 mmol/(L h) would put us over the top even if there were no deterioration in heat transfer coefficient. Again, one should give this serious consideration before committing to a final design.

We trust that the discussion and examples in this section have helped the reader to understand not only the many interactions that must be considered, but also the nature of preliminary design calculations and their limi-

Table 4. Oxygen Transfer Calculation Results (Mycelial)

Pressure (PSIG)	35	25	20	10
Oxygen transfer eff (%)	6	6	6	6
O_2 mol % in	21	21	40	40
Gas flow IN (SLPM)	22,222	22,222	11,667	11,667
VVM	1.48	1.48	0.78	0.78
Average V_s (cm/min)	154	172	102	140
hp/100 gal required	1.7	2.0	1.0	1.6
Theoretical Pg	60.9	79.9	41.7	61.8

Table 5. Agitation System Calculation Results (Mycelial)

Shaft speed (rpm)	124	137	107	124
Ungassed hp	159	214	102	159
Aeration number	0.049	0.055	0.042	0.049
Gassed hp	61	80	42	62

Table 6. Heat Transfer Calculation Results (Mycelial)

Heat load (MMBtu/h)	0.533	0.560	0.463	0.515
Gas hold-up	0.202	0.230	0.124	0.182
Jacket area (ft^2)	298	306	276	292
UJKT [(Btu/ft^2 h °F)]	40	40	40	40
Coolant temp in (°F)	35	35	35	35
Coolant flow (GPM)	190	210	235	200

tations. The results obtained here must now be translated into a mechanical design that must also take into consideration the items to be discussed in the next section.

MECHANICAL DESIGN

Mechanical Design Basis

Much of the mechanical design follows from the results of calculations discussed in the previous section 4; however, these must be tempered by other considerations that should be included in the process and/or facility design basis. Some of these (e.g., regulatory compliance, containment) have already been discussed and are revisited in greater detail in this section. Other items that should be included and will affect fermenter design include the following:

- *Extent of automation, nature of plant wide control system, etc.* Such items affect not only the instrumentation (beyond our scope) and similar factors, but also valving, piping, vessel ports, and so forth. Among some of the major issues here are identification of critical (as defined by cGMP requirements) valves and other components. It must be kept in mind that failure to identify critical instrumentation and control components frequently leads to very complex valving systems in which the failure of a single switch can bring everything to a grinding halt. Unfortunately, there are too many people who do not think this through (as a system) before final design begins. Much of this results from the human tendency to rationalize and/or push "problems" downstream, the kind of thinking that results in the thought "Why worry? We'll get it whether or not we need it." Such thinking is in no one's best interest (buyer or seller).
- *Staff requirements and the anticipated nature of the operating staff.* This can influence the complexity and physical layout of (among other things) the piping system(s).
- *Plant location.* This will have some affect on overall design and component selection if for no other than service issues.
- *Available utilities.* This can influence such things as the designs of the cooling and aeration systems.
- *Architectural constraints.* These can have profound effects on the ways in which process requirements will be satisfied. Floor space and ceiling height constraints often require that a fermenter be designed in such a way that it violates some or most or even all of the rules of thumb mentioned in the previous section. This is not good but is better than building a vessel that does not fit into the plant.
- *Transportation constraints.* The fermenter has to be moved from the fabricator's shop to the plant—at a cost not greater than the total project budget. This may require some thumb bending or breaking. In extreme cases it may be better to fabricate the fermenter on site; however, this can carry large penalties, particularly in licensed facilities.
- *Maintenance.* Many design decisions can have major effects on the ease of maintenance. It is also often true that some of the design features that ease maintenance are more costly than those that do not. In most cases, the savings in capital expenditure will not come near paying for losses that will result later because of maintenance problems. This is particularly true in licensed facilities where regulators view good, facile maintenance as an integral, indispensible part of cGMP operation.
- *Definition of standards.* Standards for welds, finishes, and so on should be standardized for all parts of the equipment that will contact process fluid. It does not, for example, make sense to call for a high-quality vessel finish (e.g., 320 grit, EP) and at the same time accept unpolished tubing and valves in inoculation and medium addition lines.

Finally, there are a few recurring themes that are encountered in fermenter design:

- *Building-in "versatility."* The types of versatility desired range from wanting a fermenter that can operate well over a very wide range of conditions but with a narrow range of organisms, to wanting a convertible bioreactor that can handle microbes, mammalian cells, and maybe (someday) transgenic animals. Most people understand the value of versatility, but many do not understand the attendant problems and costs. As a general rule, converting laboratory glassware directly into large, stainless steel equivalents usually costs more than any rational person should be willing to pay. Also as a general rule, versatility should decrease as a process goes from the lab to the production floor. Versatility in the laboratory is almost a requirement because change is in the nature of laboratory work. Versatility on the plant floor, however, leads to more procedures, more paper work, more testing, and more confusion, particularly when equipment modifications are required to achieve the versatility; change is not in the basic nature of most plant work. Equipment capital savings can justify the added costs of all the preceding for many cases in which the equipment is designed to handle multiple, similar fermentations on a campaign basis and without significant equipment modification. This statement becomes less true as the fermentations become less similar, as more modifications are necessary, and as regulatory scrutiny increases. Bottom line: analyze very carefully any inclination to want versatility built into plant equipment (or even pilot equipment, in some cases).

- *Retrofitting existing equipment*. The usual thinking is that capital and time savings can be had by refurbishing "old faithful." Just how true this is depends not only on the condition of the existing equipment but also on the intended application(s) of the reborn version. Comments similar to the ones made for versatility apply here. The chances of success decrease as one goes from a lab to licensed production facility. There are ample, expensive corpses to prove this point.

Vessel Design

Please note that Figures 4 and 5 are simplified vessel and piping diagrams intended to assist the reader in the discussions to follow.

Materials of Construction. The major choices that must be made are *(1)* the type of metal to be used for the vessel and nozzles, and *(2)* the type of elastomer to be used for static seals (see later for rotating seals). The selections should be based on compatibility with the organism, compatibility with the product, corrosion resistance, cleanability (also related to finish), welding characteristics, and cost and durability. All of these should be determined during process development but seldom are. In almost all cases, the metal selected will be some grade of stainless steel (usually SS304, SS304L, SS316, or SS316L). The choice is usually associated with the nature and value of the product, although some of the other factors noted earlier may be considered. SS304 is usually good enough for lower-value, unlicensed products, whereas SS316L is the material of choice for high-value, licensed products. L-grade is selected when better corrosion resistance and good multipass welding characteristics are required; it adds about 15% to the cost of the vessel.

It is important in making metal selection to keep in mind that clean steam, purified water, and WFI are all highly corrosive. The reader should note that there is considerable controversy concerning metal selection and is directed to Refs. 61–63 for additional information and opinion.

Static seal (O-rings and gaskets) materials are usually limited to silicone, EPDM, Viton, or Teflon. Silicone and EPDM are the materials of choice for headplates and elsewhere, respectively. (The reader should be alert to the fact that there are several grades of both, not all of which are FDA approved.) Teflon has much better temperature resistance than either silicone or EPDM, but it doesn't stretch, and it cold flows. Viton hardens on use.

Finally, one should be aware of that some O-ring/gasket-forming processes can leave very small quantities of metals in the seals. These may not be enough to cause fermentation or product problems per se, but they can be the cause of considerable corrosion.

Finish. In most instances, the need for a high-quality finish is closely related to the need for high-quality cleaning: there is little question that CIP systems function better when surfaces are smoother. Smoothness is expressed on several scales, two of the more common being grit number and surface R_A. Grit number is related directly to the abrasive used to achieve the finish (mechanical); surface R_A is the actual surface roughness as measured by means of a profilometer and expressed in microns:

Grit	R_A (μm)
120	3.2
220	0.8
320	0.4

The two most common ways of achieving smoothness are mechanical polishing and electropolishing. Mechanical polishing is done by means of a series of continually decreasing abrasive particle sizes (grits). Electropolishing is achieved by electrochemical removal of metal from high points on the surface. One is usually best advised to electropolish only after getting a good mechanical polish (240-grit minimum). It has long been held that the very good finishes required for vessels where cleaning is critical could be achieved only by electropolishing a 320-grit mechanical finish (although 240-grit EP is the most common practice). There is some evidence in the literature (62,63) to support this, but it is not overwhelmingly convincing. It also is important to note that recent advances in mechanical finishes have produced surfaces that easily rival the appearance of any electropolished surface (Lee Industries, personal communication, 1996).

It should be (but is not always) evident that cost increases with surface smoothness. It is also important to note that just because a surface is almost perfectly smooth does not guarantee that it will be easy to clean. Cleanability will also depend on the nature of the interaction between foulants and the surface material. The only way to know for sure is to try your broth on various surfaces and

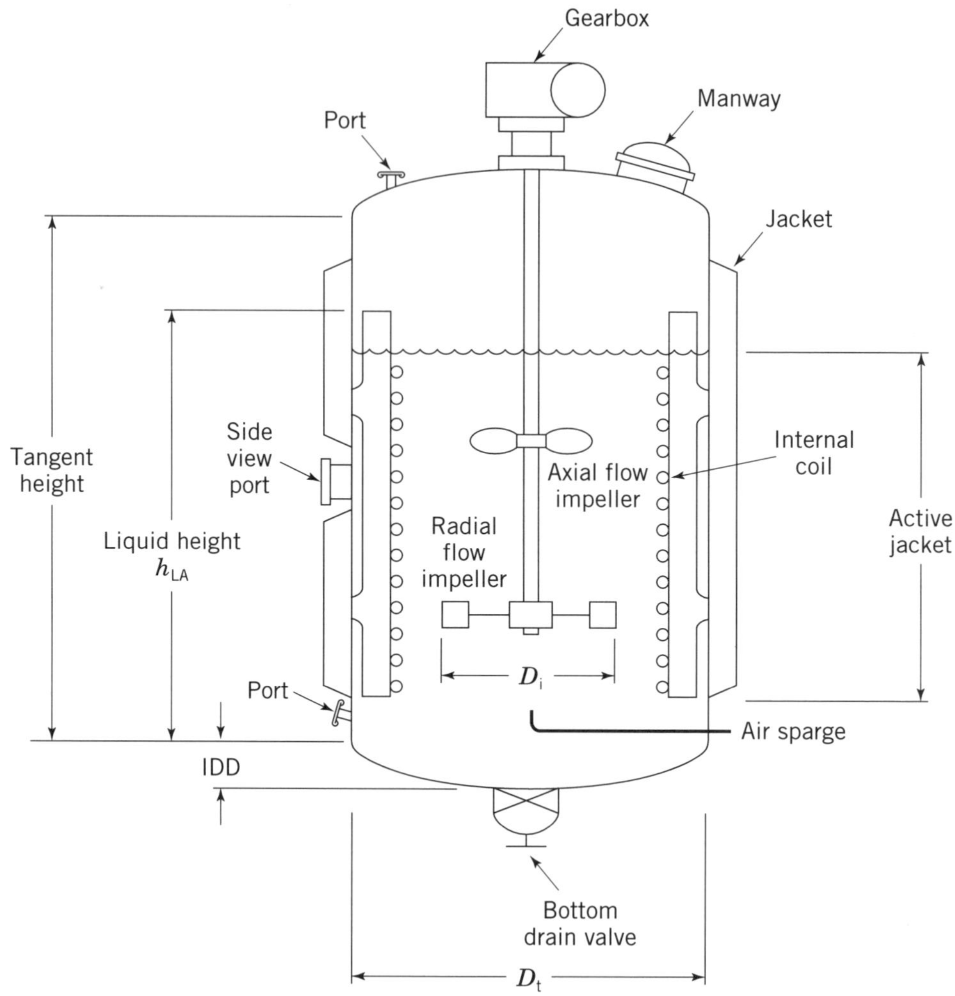

Figure 4. Fermenter vessel schematic and terminology.

then make a specification based on rational evaluation of the experimental information. Surface material and finish, cleaning agents, and cleaning protocols must be considered together—preferably before design.

Passivation must also be considered. This is a treatment that restores oxides that protect stainless steel from corrosion but can be removed during vessel fabrication (primarily welding). The process is rather simple, but the surface chemistry is not and is not completely understood (64,65); hence, there is controversy as to which method is best and whether it is really as necessary for fermenters as it is for high-purity water systems where rouge formation is a serious problem (66). Two methods are in common use: *(1)* treatment with sodium hydroxide, citric acid, and nitric acid; and *(2)* treatment with chelating agents. The second is much milder and does not generate the noxious waste of the first. Both appear to be equally effective; hence, the chelating method is becoming more popular.

Nozzles. Ports for additions and probes must be designed to be sterilizable and cleanable. Among other things, this means they must have reliable seals and be free draining. The major debate here usually focuses on the choice between Ingold ports and those designed for sanitary clamp connection (Fig. 6). There have been sterility problems with Ingolds, most of which have been traced to the ports being out of round as a result of distortions caused by welding. Such distortion causes obvious problems with the O-ring seals. This problem can be corrected quite easily by making the nozzle thick enough to avoid distortion. Our experience has been that such a correction yields nozzles that are just as reliable and more easily cleaned than are sanitary fittings when both are mounted at the usual 15°.

Side View Ports. The first word that comes to mind is "don't." Side view ports let the bugs see out more than they let you see in. The little that's gained by being able to see less than 1% of the total action does not seem to justify the expense, the jacket coving, and the cleaning and sterility problems caused by these little gems. If you must have one, we suggest the circular Metaglas™ type (Fig. 7) attached to the vessel with sanitary clamps; the glass is bonded directly to the metal, which avoids the sealing problems encountered with other types. It does, however, have the disadvantage of having a smaller viewing area than does a standard, circular view port, and it is more expensive. We

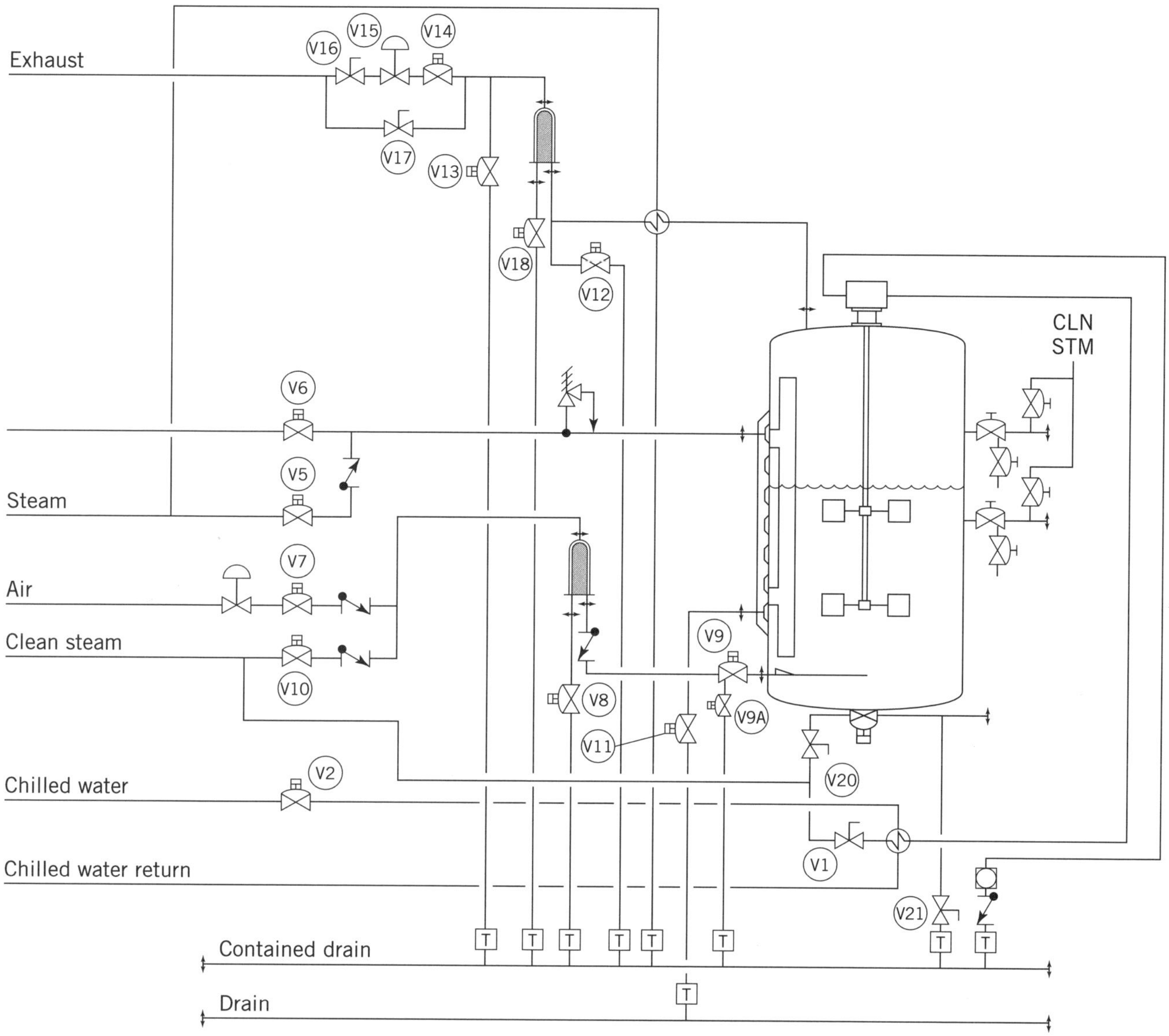

Figure 5. Simplified fermenter piping diagram.

recommend avoiding long rectangular view ports, which are very difficult to keep clean and free draining, and types in which the glass has to be sealed with gaskets and/or O-rings on both sides (Fig. 8). Keep in mind that tolerances for glass are measured in fractions of an inch, not thousandths. That, coupled with the fact that the gaskets will compress means that the probability of repeatedly achieving a good seal and good cleaning is lower than your odds of winning more than once at Atlantic City.

Baffles. Baffles are usually required in high-power systems to prevent swirling and vortexing, thereby increasing the power that can be delivered to the fluid. The usual practice is to use four baffles on 90° centers welded directly to the wall. Each baffle should have a width equal to 10% of the tank diameter and should have long slots cut out of the edge facing the wall so as to prevent solids build up and to make cleaning easier. Removable baffles are used in some cases; however, we discourage this practice, particularly in cases where cleaning is a critical issue, simply because unsealed joints resulting from baffle removal make cleaning more difficult.

Jackets. Several types (67) of jacket are used on fermenters; the choice of type is usually not critical for heat transfer purposes and is best left to the vessel fabricater. In some cases, however, a jacket type (e.g., half pipe) may be chosen to increase the vessel pressure rating. The following have proven to be useful practices:

- The jacket should extend from the probe ring (about 2 in. above the bottom tangent line) to the top tangent line and should be zoned. This allows additional active surface area to come in contact with the broth as the volume increase due to additions and to increas-

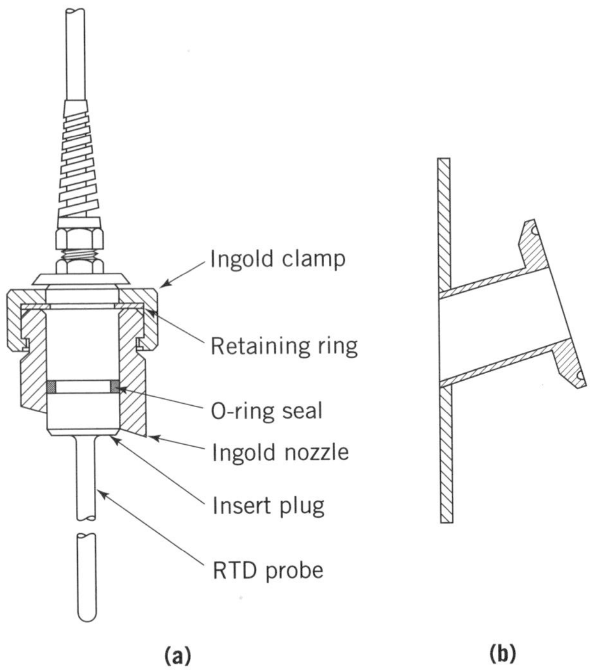

Figure 6. Ingold and sanitary clamp fittings.

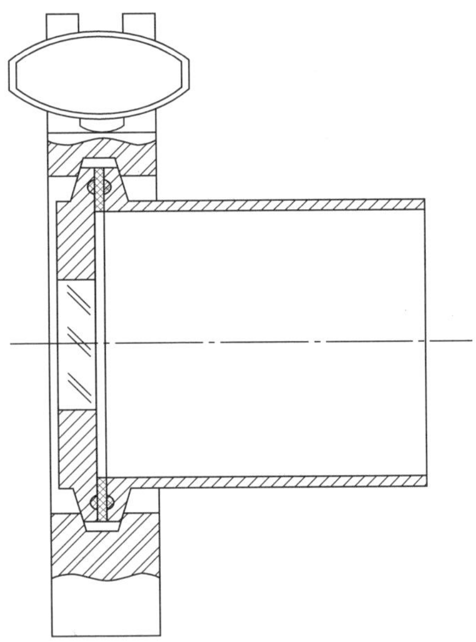

Figure 7. Metaglas™ side viewport.

ing gas holdup, minimizing cooling-loop pressure drop, and minimizing medium bake-on.

- Connections to the jacket should be via sanitary clamps or flanges, not by screwed fittings. This minimizes possible damage to jacket welds when cooling lines are connected or disconnected.
- Coolant should be filtered to avoid solids buildup and/or jacket fouling.
- Bottom jackets should be considered only for vessels larger than 5,000 L. The additional area will be about 10% of the straight side jacket area (allowing for bottom drain, etc. and will add about 5% to the cost of the vessel). In cases where there is an internal coil, bottom jacketing isn't necessary or desirable.
- Jacket coving to accommodate view ports, decreases jacket area and effectiveness and increases cost considerably. It should be avoided.

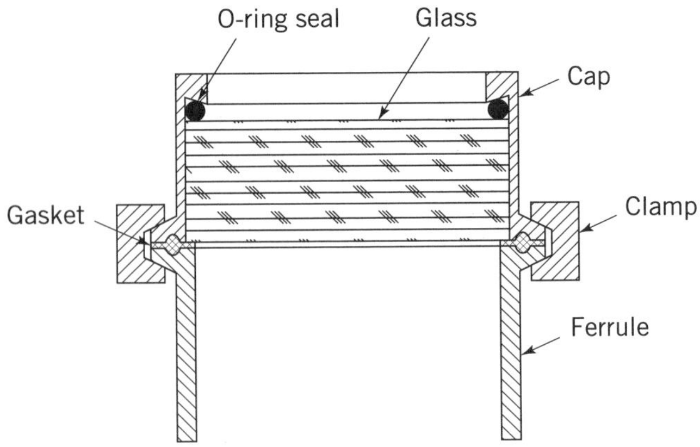

Figure 8. Side viewport with separate glass piece sealed by two gaskets.

Internal Cooling Surfaces. Coils are the most common internal cooling surfaces, although there are other types (68). They can easily double the available heat transfer area and tend to be more effective than jackets; however, they can have big disadvantages:

- They make cleaning very difficult.
- They can cause additional bulk mixing problems, particularly for very viscous non-Newtonian broths.
- They will eventually leak nonsterile coolant into the broth.
- They can add up to 25% to the cost of the vessel.

We recommend that every other alternative (including slight decreases in growth rate and/or cell mass) be considered before yielding to the temptation to use coils. Keep in mind that, once installed, they will be there for the life of the vessel. If you absolutely, positively must use a coil, there are several points to remember:

- Mount it in a way that will insure minimum stress during heat-up and cooldown.
- Weld cladding over the butt welds used to join the coil pipe sections.
- Leak test (69) after construction and build in design features that will simplify leak testing on a regular basis thereafter.
- Space coil turns at least 3 in. apart. Anything closer will insure major cleaning problems.

Spargers. There has been a lot of discussion concerning the pros and cons of ring and single-orifice spargers, and the details of design of each. We have seen both work well and have not found any evidence for the validity of claims concerning the importance of hole size (for example) for oxygen transfer per se. The primary focus should be on gas distribution (which will depend on other aspects of the agitation system) and on aseptic operation. We usually favor single-orifice spargers.

There also have been advocates of porous (frits) spargers. The rationale presented has focused primarily on the small bubble size such spargers produce. There is some basis for these claims in cases where little mechanical energy is available for bubble breakup (e.g., in mammalian cell systems). There might also be some value in cases for which bubble coalescence is not a problem (far and few between in practical systems). They can be extremely difficult to clean, particularly for mycelial organisms. We usually suggest they not be used except in very special cases, as just noted.

Finally, it is important that the sparger be made and mounted in such a way that it can be removed easily for cleaning.

Piping and valving

Design and construction details of the piping system depend on process, sterility, cleaning, and containment requirements, taken together.

Service Lines for Non-Process-Contacting Fluids. All lines providing fluids that cannot contact process fluid surfaces (e.g., coolant lines, plant steam lines for heating only) fall into this category. Satisfactory service is provided by either copper or stainless steel piping along with a rational combination of welded and compression fittings. Durability, servicability, cost, and corrosion resistance are major considerations. Ball valves are satisfactory on lines not turned on and off frequently (diaphragm valves tend to withstand a greater number of on–off cycles prior to failure).

Sterile Piping Systems. Before we discuss design of sterile piping systems, we reiterate the importance of a systems approach to integrating vessel and piping design. Independent design almost always results in a lot of aggravation, as well as higher cost and lost time.

A sterile piping system can usually be divided into five major subsystems: process, steam/condensate, air, harvest, and seal lubrication. Each has specific requirements dependent on its unique functions as well as the specific demands of the process. None of these systems is completely independent of the others, and all share common components. All also interact with the CIP piping system, which is discussed later.

Details of each of these subsystems follow a simplified description of sterilization procedures (Fig. 5). The reader is advised that the procedure is one of several commonly practiced; there are, for example, different opinions as to when live steam to the air inlet line should be turned on.

Phase 1. After the vessel is filled with set medium, steam flow is started through the jacket. The agitator is run at low speed, and particulate free steam lubricates the shaft seal. The exhaust line is open for venting to the atmosphere. All other valves are closed. During this time not only are the vessel and the set medium being heated, but air is being purged from the medium. If the air is not purged, the vessel pressure during sterilization can become dangerously high, and there will not be a reliable correlation between pressure and temperature. Phase 1 continues until the medium temperature reaches 100 °C.

Phase 2. Steam continues to flow through the jacket. The air exhaust line is closed; the only path out is via steam traps. Steam is now admitted through the air inlet line to sterilize the inlet filter; steam leaving the vessel through the air exhaust line sterilizes the exhaust line piping and the exhaust filter. Steam flow is started through subsurface ports via steam lock assemblies (Fig. 9). (See later for further discussion of steam locks.) Ports above the liquid surface are sterilized by steam escaping through them from the vessel and then to condensate lines. Steam flow also is started through the bottom drain valve. All steam flows continue until the cooldown begins. Phase 2 continues until the medium temperature reaches 121 °C.

Phase 3. Steam flow to the jacket is throttled to hold sterilization temperature constant until cooldown begins.

Phase 4. Cooldown. This is not discussed here.

General Principles. There are several general principles that should be applied to all sterile piping:

- Make all piping system components free draining. This requires special attention to the details of pipe pitches, valve orientations, and so forth
- Design all components and the piping layout so as to eliminate nooks and crannies where contamination (biological and nonbiological) can hide so as to escape cleaning and/or sterilization. Frequently overlooked problems include such things as deadlegs and ridges formed by welding operations.
- Make steam and condensate piping at least 3/8 in. diameter to insure proper steam flow and condensate draining.

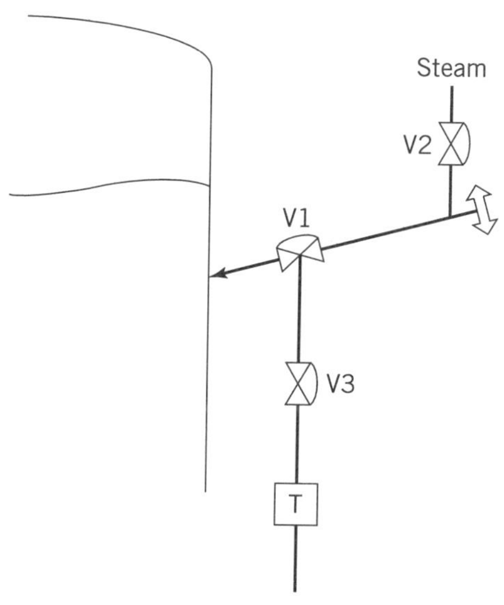

Figure 9. Sterilization of subsurface ports.

- Make all piping lengths as short as possible without interfering with good fabrication practices (GFP), operability, and maintenance.
- Use check valves only when no other solution is possible (e.g., in overpressurized lines).
- Do not use sight glasses in condensate drain lines except in the seal lubricant drain line.
- Pay careful attention to piping orientation to avoid the possibility of air traps and inadequate heating (55,56).
- Make sure that bottom drain valves are flush-mounted diaphragm valves capable of being steamed in the closed position (Fig. 10). They must be free draining.
- Avoid dip tubes unless there is no other way (there almost always is).

Finally, the design should take into consideration the methods that will be used for validation of sterilization operations. These will vary considerably with the product and the organism. At one extreme is virtually no validation; at the other extreme is temperature mapping of the vessel and all the trap lines, as well as the use of spore strips and/or spore ampules in the vessel and in the trap lines. In some cases, tee sections have been included in drain lines for insertion of spore strips.

Welds. Orbital welding of the piping system produces the most reliable results with regard to aseptic operation and cleanability; it is essentially a requirement for many classes of regulated products, particularly those regulated by Center for Biologics Evaluation and Research (CBER). It also is the most expensive to build and to maintain. There are, however, many processes for which an economic combination of welded and compression fittings is acceptable. These can be operated quite satisfactorily if proper attention is paid to cleaning, sterilization, and maintenance protocols. The fact is that such systems have been used for many years to manufacture regulated products safely and efficaciously.

Valves. Diaphragm valves have become essentially required for most classes of regulated fermentation products. There is little argument that their design provides the greatest reliability for clean, aseptic operation. They have the added advantage of integral sterile access ports, which permit very short piping runs, particularly in valve clusters. Historically, diaphragm valves have been more expensive than have ball valves, but this has changed recently, particularly for some materials (e.g., SS316L). It therefore makes sense to use diaphragms for most applications. The primary exceptions are steam lines or other applications in which diaphragm life is a problem; in those cases one should consider ball valves or plug valves, as appropriate.

Another important consideration is the extent to which one uses valve actuators and position indicators. Obviously, this will depend in large measure on the nature of the control systems employed. In any event, one should consider very seriously the problems (e.g., space and orientation constraints, reliability) associated with valve "extras" while control system decisions are being made.

Steamlock Assemblies. The primary purpose of steamlocks (also known as block and bleed) is to allow sterilization of ports and process piping at any time during a fermentation in such a way that connections can be made and broken without risk of breaching the sterile barrier and/or the containment barrier. A basic steamlock assembly is illustrated in Figure 11. In this case, vessel T1 contains presterilized nutrient that must be added at some time during the fermentation. T1 is connected to the fermenter by means of sanitary clamps (other means are possible). Steam flow is then started (open V1, V3, and V5) so as to sterilize the connecting hose and all the valving not previously sterilized. Condensate goes to drain (sanitary, if necessary) via V3 and V5. The steam and condensate valves are closed after sterilization is complete; process valves V2 and V4 can be opened anytime thereafter to make the sterile transfer. Note that the diaphragm valve sterile access ports make it much easier to sterilize the

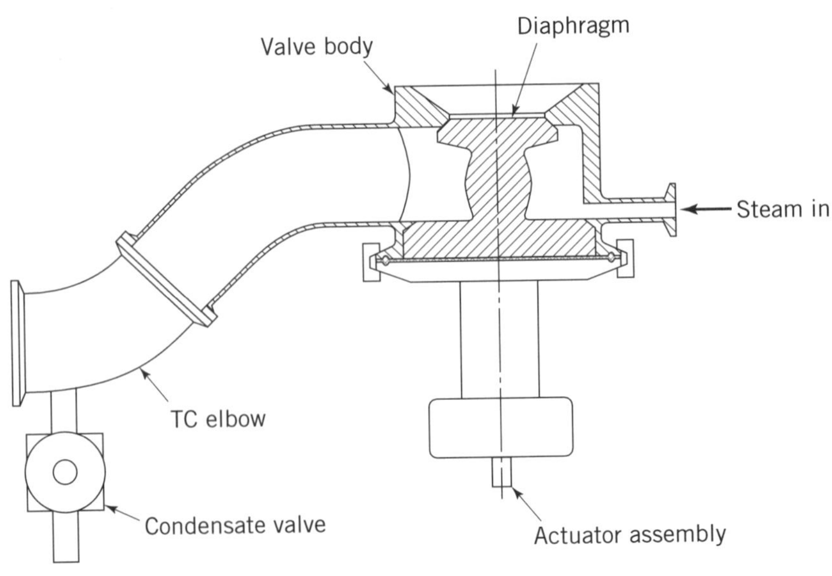

Figure 10. Bottom drain valve.

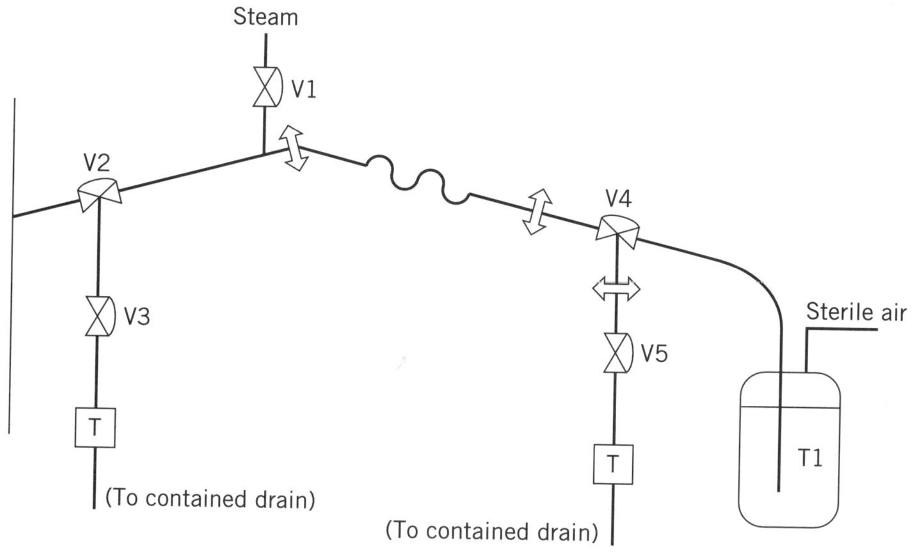

Figure 11. Steam lock assemblies for transfering sterile medium.

steam side of the valves than is possible with other valve types. Variations on this theme are better suited to cases involving long transfer lines or having other special requirements.

Air Inlet and Outlet. Figure 12 is a simplified diagram of a "typical" air inlet/outlet system. The major problems for this system are related mainly to the filters. Chief among these (other than outright failure) is wetting of both filters caused by condensate during sterilization, and wetting and clogging of the exhaust filter by condensate and/or aerosol during fermentation. These problems are not all easy to resolve and should be dealt with most seriously during design. Wetting and/or clogging of the inlet filter by condensate and/or compressor oil during fermentation also can be a problem but is usually the result of not using proper quality air, having no or improper prefilters in the air inlet line, and/or poor maintenance (e.g., of the compressor). All of these are easily remedied.

The extent of condensate problems during sterilization can be remedied best by ensuring that whatever condensate does form can drain freely. This can be complicated by the fact that it is undesirable to have the sterile side of the filter connected directly to the drain line. Another approach is to steam heat the filter housing such that condensation cannot occur. This is effective but has the disadvantages of adding cost and decreasing filter life.

Other factors that can affect condensate problems are the positioning and orientation of the housing. There are differing opinions concerning these; one is best advised to consider the advice the filter and the fermenter vendors for specific cases.

Plugging of the exhaust filter by condensate and aerosols during fermentation require special consideration. Air leaving the fermenter will be essentially saturated with water vapor at fermentation temperature. The exhaust line, filter, and so on usually are colder than the fermenter; therefore, condensation is inevitable. The amount of condensation will depend on temperature differences, air flow rate, and the nature of the surfaces of the components in the exhaust line (the maximum potential is easily calculable). Approaches that have been used to deal with these problems include (70):

- Condensers
- Heat exchangers; used before the exhaust filter to avoid condensation from the fermenter off gases and after the pressure control valve to prevent reflux from the exterior exhaust line
- Steam-heated filter housings
- Heated exhaust lines
- Coalescers
- All of the above

None of these is completely successful, and each has its proponents and detractors. Suffice it to say that each case should be considered on its own merits (e.g., fermentation temperature, duration, and air flow) and that combinations of the approaches should not be ignored.

The condensate problem is exacerbated by the aerosol problem. Aerosols will form in any aerobic fermentation and will carry liquid along with dissolved solids and particulates (including organisms) to the exhaust line. The only questions are how much and what problems they will cause. Among the problems are deposition of dissolved and particulate solids, not only on the filter, but also on heated surfaces or condenser surfaces. Among the problems deposition causes is reduction in heat exchange effectiveness, which reduces the capabilities of the previously mentioned devices to eliminate condensation. In severe cases, exhaust line blockage can lead to other problems related to pressure buildup and/or higher linear velocities in the exhaust line. The problem should be dealt with on a case-by-case basis during process development and fermenter design.

Provisions should be made for in situ integrity testing of sterile filters. This can be done by means of any of several commercial electronic testing devices that are based

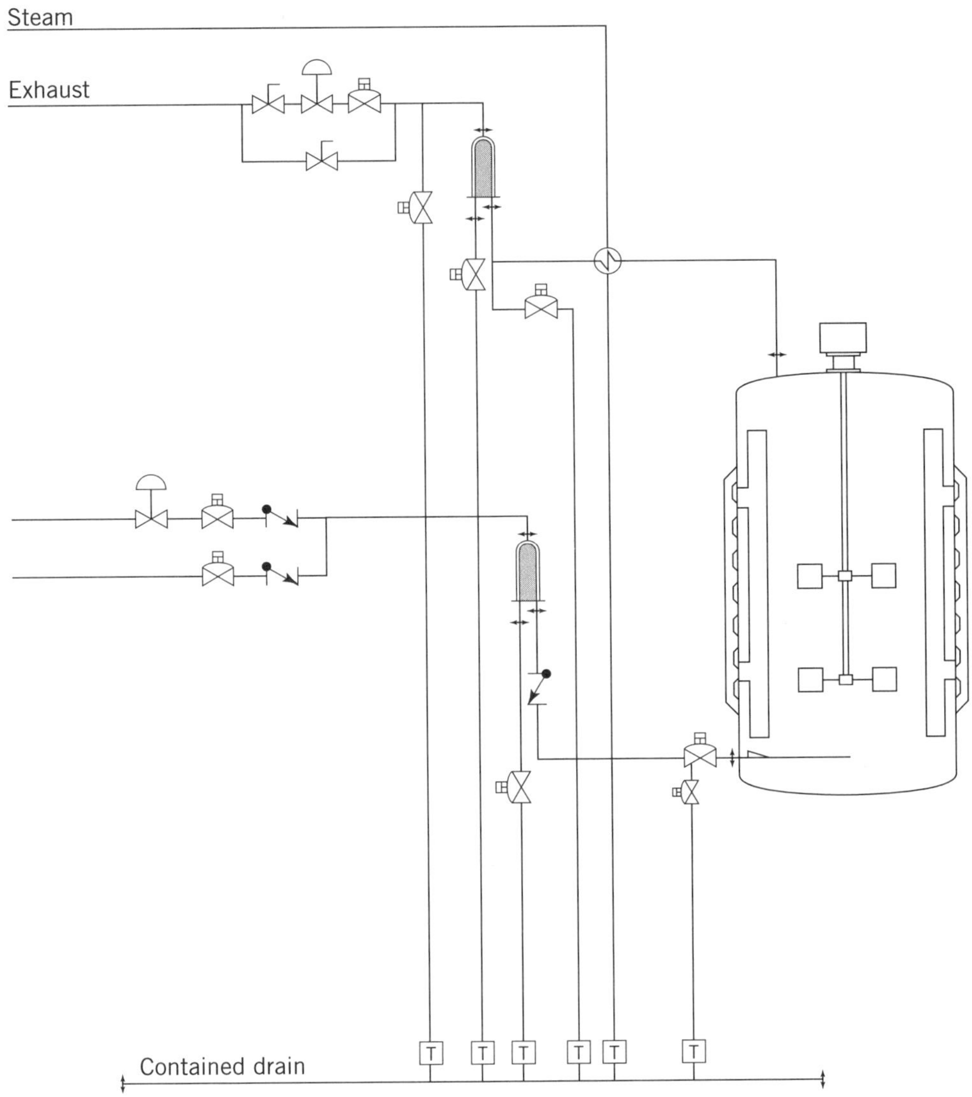

Figure 12. Simplified piping diagram illustrating air inlet and outlet systems.

on liquid intrusion, diffusion, or some other well-established method for testing membrane filter integrity (71–73). One example of the piping necessary to perform the test on an exhaust filter is shown in Figure 13.

Finally, two pressure control valves are required in the case of constant pressure sterilization (continuous flow of steam into the vessel and out to atmosphere). The reason is that the steam flow will be very much lower than the air flow during fermentation; hence, the valve C_v required to maintain steam pressure accurately will be far too small to control air pressure accurately.

Sampling Systems. One example of a simple aseptic sampling system is illustrated in Figure 14. The sample vial, vent filter, and valve V1 are sterilized as a unit in an autoclave. The unit then is connected to the fermenter piping via a sanitary clamp. After the connection is made, steam is introduced via V3 and passes across the nonsterile sides of V2 and V1 and then to trap via V4. Once the valves and line are sterile, the steam and condensate valves are closed. Samples may be taken any time after the sampling system cools. Samples are taken via V1 and V2. Lines and valves are resterilized before the sanitary connection is broken.

The valves selected for this system will depend on the class of service required. We suggest that vessel sampling valve V3 be a flush-mounted diaphragm valve (e.g., Asepco(TM), NovaSeptic(TM)) designed in such a way as to minimize dead volume and piping lengths; it is an excellent choice for service requiring very high levels of asepsis and cleanability. Less expensive valves can be used for less-demanding service, but we think that the initial extra cost will more than pay for itself. Diaphragm valves are suggested for V1, V3, and V4 in cases of demanding cGMP or containment requirements. This is not required for all classes of service; indeed, it is usually satisfactory to use plug valves for steam and condensate lines even in many demanding circumstances. It should be noted, however, that you may encounter a perception problem if you use them for applications deemed to be sensitive.

There are several other designs that have been used successfully. A few have been designed specifically for systems requiring BL3-LS containment or higher. These are

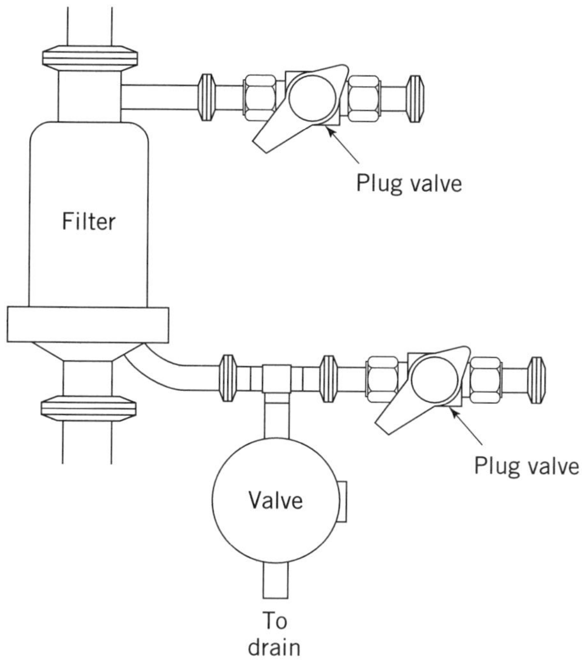

Figure 13. Piping diagram for in situ air outlet integrity test.

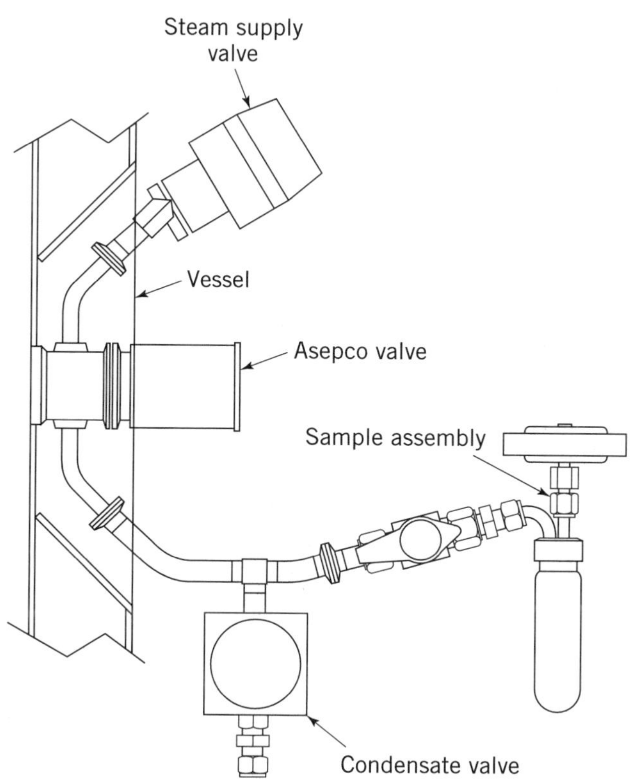

Figure 14. Aseptic sampling system.

discussed later. Finally, please note that addition system design and operation are very similar to those for sample systems and are not discussed here.

Agitation Systems

Agitation can be done by direct mechanical coupling of the shaft to the drive or by magnetic coupling. The latter has been recommended for cases where high levels of containment are required (74). There also is the perception that magnetic drives are more suitable for maintaining more stringent levels of asepsis and cleanliness than is possible with direct drive. Neither claim has any substantive basis; indeed, there is good reason to believe that existing magnetic drives may present greater cleaning difficulties because of the manner in which the driven magnet must be mounted at the bottom of the vessel. In addition, power transfer by magnetic drive is quite low.

Direct drive is the predominant current choice at almost any scale of operation. The major mechanical decisions that must be made in this case (other than power) have been discussed at length elsewhere (4,75). The major arguments for bottom drive are ease of maintenance, shorter shafts, less support structure, and lower overall height. The arguments against bottom (and for top drive) focus primarily on the potential of catastrophic spills resulting from bottom seal failure, seal grinding as a result of broth particulates working into a bottom seal, and greater cleaning difficulties. If the seals are designed and maintained properly, none of these is a problem. We have seldom seen any real difference in aseptic operability between top and bottom drives, and we are reasonably certain that the organisms don't care. The choice probably will continue to be driven primarily by personal preference.

There are very few cases in which double mechanical seals are not (or should not) be used in fermenters. The major debate focuses on seal orientation and the details of individual seal designs. There are basically two orientations used: inline (Fig. 15) and back-to-back (Fig. 16). The details are discussed in Refs. 4 and 75. We recommend inline design because we have found it to operate more cleanly and require a simpler sterilizing/lubricant system (4,75).

Seal lubrication is usually provided by means of sterile steam condensate. It is extremely important that this condensate be free of particulates: their presence guarantees rapid seal failure, contaminated fermentations, and a hyperactive maintenance program. It is also important to note that during sterilization live steam flows through the seal housing. There are some who insist on keeping the steam flowing throughout the fermentation. (Obviously, they have no faith in the seals.) The one thing this will guarantee is much more rapid wearing of the seals (perhaps supporting the lack of faith in the seals).

One must also decide on the means for controlling lubricant flow rate. Most use a valve for this purpose. We suggest an orifice sized to deliver the proper flow. This avoids the cost and maintenance of a valve and insures fiddle-proof operation. It does, however, require the use of particulate-free condensate, but then so does proper operation of the seals. We also recommend including a sight

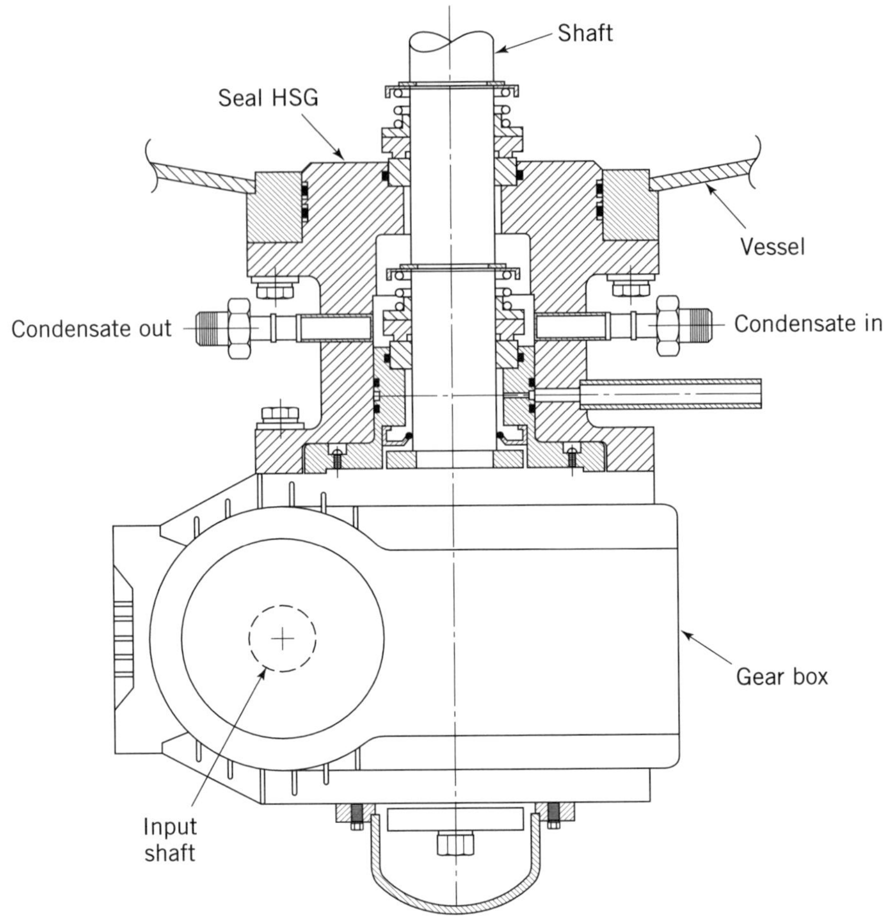

Figure 15. In-line double mechanical fermenter shaft seal.

glass in the lubricant drain line as well as a seal leak detector (Fig. 17). Other more complex detection systems might be considered for specific circumstances (e.g., severe containment requirements).

Finally, preventing shaft vibration is another important factor in agitation system design. Shaft vibration is a safety hazard. It will also cause premature seal failure and other costly mechanical damage. Details have been presented elsewhere (4).

Cleaning Systems

As noted earlier, specific requirements for cleaning depend on several factors including the nature of the fermentation broth. There is a major difference between most microbial broths and most mycelial broths. Single-cell microbial broths, with the exception of those containing a lot of undissolved particulates and high viscosity components (e.g., xanthan broths), tend to be free draining and readily amenable to cleaning based primarily on the physicochemical action of the cleaning agents. Many fungi and mycelial bacteria, on the other hand, tend to cling to fermenter internals and may require mechanical action (e.g., high-velocity jets) in addition to cleaning agents.

The following guidelines are applicable for most systems. These principles must be applied in light of the actual cleaning agents and protocols to be used:

- Eliminate internals and nooks and crannies to the greatest extent possible in the vessel and throughout the piping system. This practice is consistent with design for aseptic operation.
- Drill and position sprayballs to insure complete coverage of all surfaces inside the vessel. This usually requires an empirical approach. Coverage can be tested by means of the riboflavin test. The reader is cautioned, however, that complete coverage is a necessary but not sufficient condition for cleaning. Also, please note the following:
  - Spray balls designed for sanitary operation are self-draining and self-cleaning. They can be sterilized in situ. There is, however, considerable debate concerning whether they should be removed prior to fermentation.
  - High-velocity, rotating devices that may be necessary when large clumps of sticky residue must be removed (e.g., as with a fungus) are not designed to be inherently self-cleaning, self-draining, or sterilizable. They should not be mounted permanently.
- Eliminate deadlegs in piping and deadspaces in valves, fittings, and other system components. This is also consistent with design for aseptic operation.
- Specify the same materials and finishes for process and CIP piping as are specified for the fermenter. Ob-

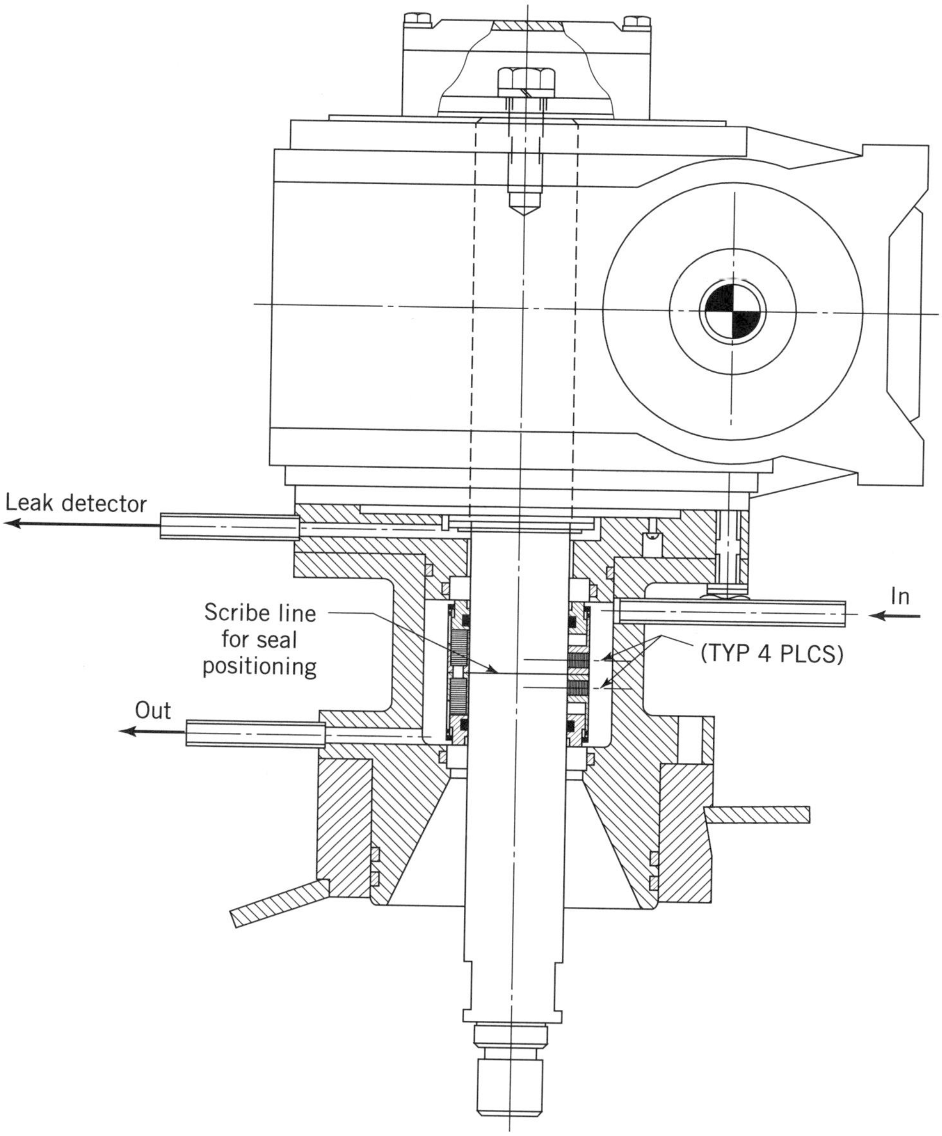

Figure 16. Back-to-back double mechanical fermenter shaft seal.

viously, these must be compatible with the cleansers and conditions used.

- Avoid threaded joints. A completely welded system is best, but compression fittings can be satisfactory when cost is a major consideration, particularly for nonregulated products.
- Make CIP piping as simple as possible. For example, make dual use of process and other piping as much as possible.
- Avoid complex, expensive transfer panels wherever possible. Use swing elbows wherever practical.
- Design and construct the system to facilitate validation and on-going testing for removal of contaminants, including the cleaning agents. This design should provide for swabbing, obtaining rinse samples, or whatever else the cleaning protocol requires.

A portion of one approach to integral CIP piping is illustrated in Figure 18.

There are many cases for which a portable CIP system is preferable to an integral system. For such cases, the fermenter and the portable unit should be designed to allow simple mechanical attachment of the necessary hoses between the two units and with utilities, and a straightforward means for interfacing the instrumentation, logging data, and controlling the systems of the two units.

Finally, we note that classical CIP systems rely on continuous cleanser flow and the maintenance of a shallow puddle in the bottom of the fermenter. To achieve this they rely on special pumping devices such as eductors (74). Aside from the design, control, and other operating problems this causes, it has been observed in some facilities that the stable pool leads to the formation of "cleaning rings" at the bottom of the fermenter. These rings may not be real problems (additional evidence is still required), but they do cause perception problems. We suggest the use of a pulsed-flow system to overcome not only the need for special pumping devices but also the "cleaning ring" problem.

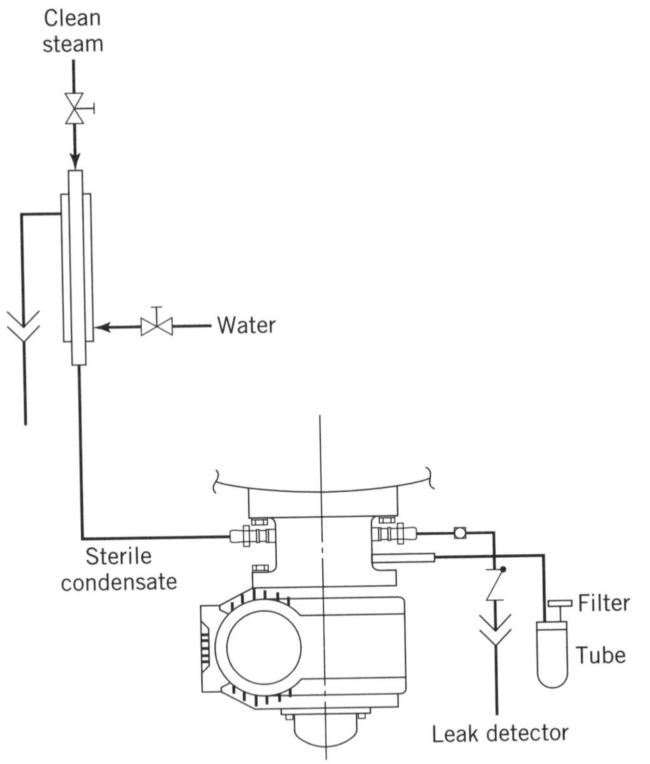

Figure 17. Piping diagram illustrating lubricant flow sight glass and seal leak detector.

Such pulsed systems have been found to accomplish both objectives in practice.

Additional details concerning fluid velocity, vortex elimination, and many other aspects of CIP system design and operation are described elsewhere (59).

Containment

Our scope is limited to containment of fermenters; however, we reiterate here the importance of making certain that the equipment, the facility, and the operating protocols be considered simultaneously to ensure compatibility with regard to containment (just as in the cases of sterility and cleaning). As noted earlier, most new facilities for regulated products are being built to satisfy BL2-LS facility requirements and (in most cases) BL3-LS fermenter containment requirements. With few exceptions (about which there is no agreement) there is no difference with regard to the equipment containment requirements at the two levels—at present.

The guidelines require that BL2-LS/BL3-LS fermenters be designed and constructed to prevent organism release (primary containment) from the vessel and from any of the subsystems noted earlier; hence, they must be built as closed systems. (Note that it also is possible to enclose a fermenter in a biosafety "cabinet," but this is not usually practical or necessary.) Some areas that have been subject to greater scrutiny than has been required for cGMP compliance are as follows. In evaluating the discussion, the reader should bear in mind that there is not much information readily available concerning containment reliability data. Some progress is being made via the Industrial Biosafety Project in the United Kingdom.

- *Static seals.* It has been suggested that double static seals (e.g., in headplates) be used for BL2-LS and that double static seals with a steam trace between the seals be used for BL3-LS (76). There has been little to demonstrate that these are either needed or desirable. Indeed, there have been substantive arguments made against using such devices (76). There is as yet no consensus.
- *Rotating seals.* Arguments have been made in favor of using magnetic drives at anything higher than BL2-LS. This was discussed earlier. There are many who argue that only top drive be considered for BL2-LS and above. Their reasons have mainly to do with avoiding the possibility of large spills in the event of catastrophic seal failure. Such a circumstance is possible but highly unlikely. Leak detectors (which would normally be installed for cGMP requirements) would almost certainly indicate a leak long before there was the remotest possibility of catastrophic failure. One also should consider that a leak in top drive seal might be more difficult to detect and could lead to release of a more insidious nature. Again there is no consensus. Finally, it has been suggested that a low-pressure sensor or flow sensor be used to detect lubricant flow failure. This could help to avoid seal failure (77).
- *Sampling systems.* One system recommended (78) for contained sampling is illustrated in Figure 19. Note that it is similar to the system in Figure 14 recommended for cGMP compliance. It is a bit more complicated and may provide some incremental benefit; however, this might be outweighed by the greater likelihood of operator error. There have also been more complex systems suggested that involve more intricate valving and/or biosafety cabinets. It is apparent that this is an area that needs considerably more development work.
- *Air exhaust system.* It has been suggested that two in-series sterile filters or a sterile filter followed by an incinerator be used at BL2-LS and higher. All other aspects are covered by the earlier discussion of exhaust lines.
- *Pressure relief.* This is an area of some controversy because of the need to satisfy physical and biological safety requirements simultaneously. It has been suggested by many that relief venting be done via a large kill tank protected by a HEPA filter; however, this could be in conflict with physical safety codes that require there be no devices in the relief path (79). We do not have anything approaching universal agreement here; however, it does seem fairly apparent that limiting the supply pressure would help to minimize the risk. There does appear to be agreement on the use of rupture disks rather than pressure relief valves; this is consistent with accepted practice for

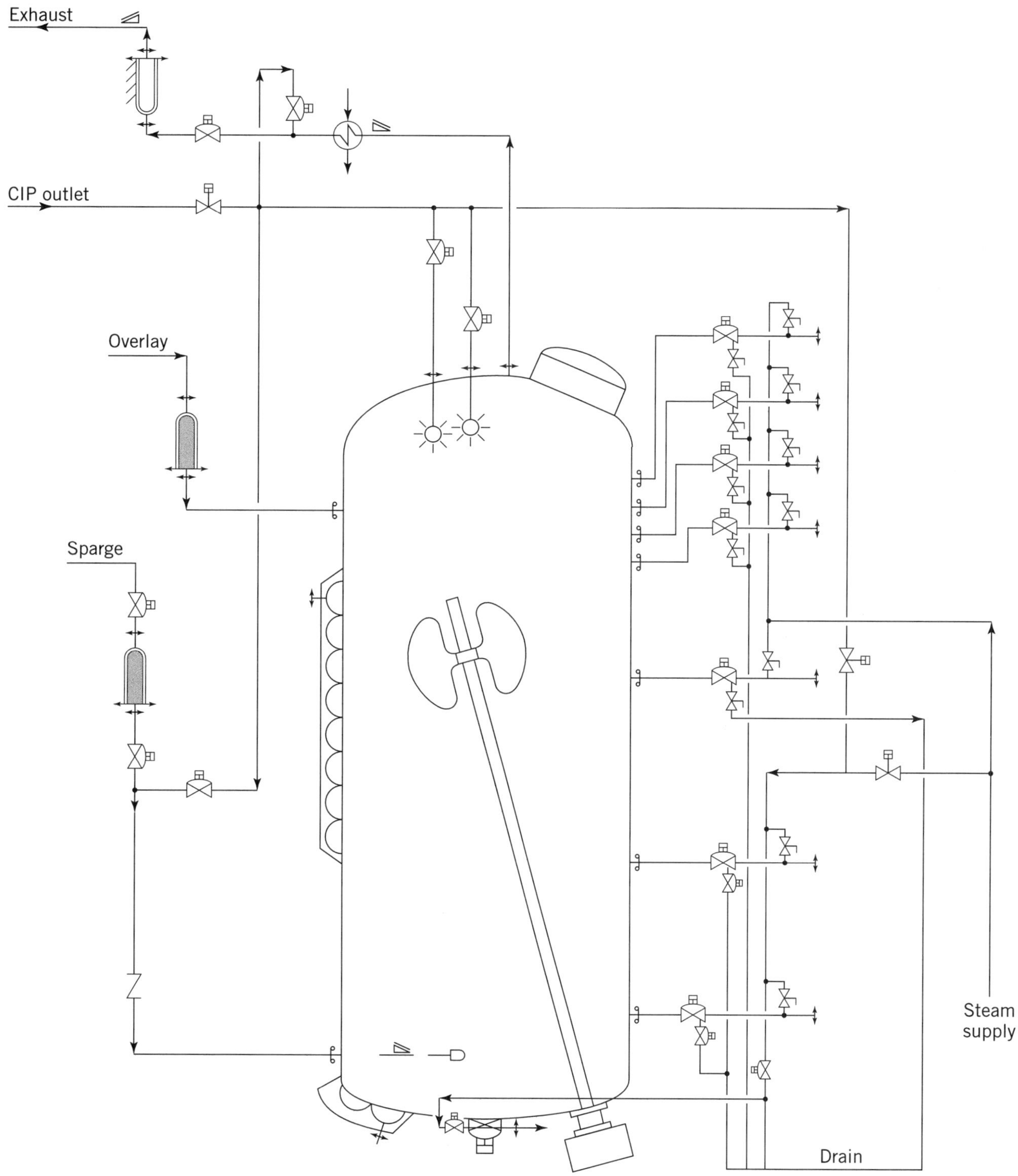

Figure 18. Simplified CIP piping diagram.

cGMP compliance. Rupture disks are cleaner and are not prone to sticking.

Finally, HAZOP evaluations for fermenter containment have been recommended. Such analyses would evaluate potential hazards that might be caused by a wide range of incidents (e.g., fire) not limited to normal operation.

CONCLUSION

Our intent has been to present a balanced view of stirred-tank fermenter design and construction for microbial and and mycelial organisms. There are, however, several important aspects that space and scope considerations prevent us from considering in the depth they warrant. In

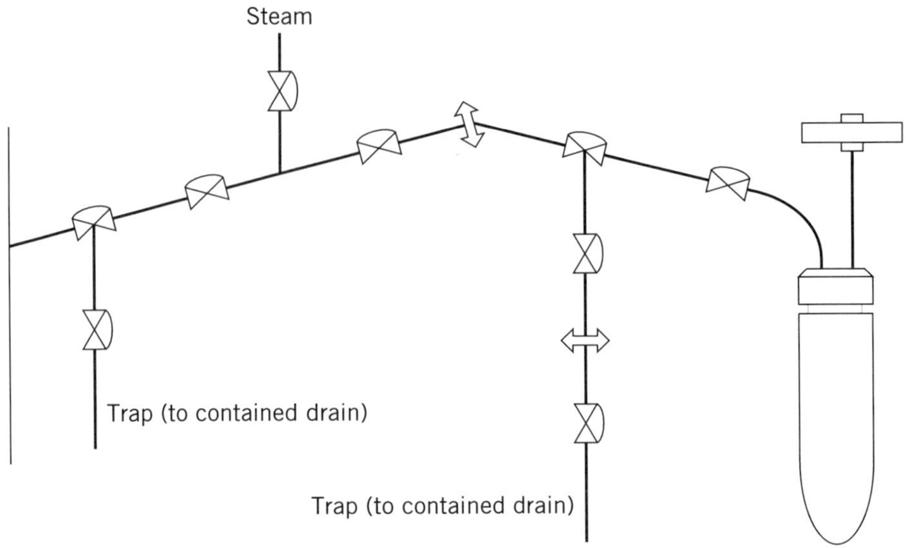

Figure 19. Contained sampling system.
Source: Adapted from Van Houten (78).

particular, we would like to have included more information related to overall process considerations, instrumentation, and control; however, we believe that these are addressed elsewhere in this volume. We trust that we have conveyed the sense that fermenter design is very much an art and that we have convinced you to consider each case separately.

BIBLIOGRAPHY

1. D.I.C. Wang, C.L. Cooney, A.L. Demain, P. Dunill, and A.E. Humphrey, *Fermentation and Enzyme Technology*, Wiley, New York, 1979.
2. M.L. Schuler and F. Kargi, *Bioprocess Engineering: Basic Concepts*, Prentice-Hall, Englewood Cliffs, N.J., 1992.
3. H.W. Blanch and D.S. Clark, *Biochemical Engineering*, Marcel Dekker, New York, 1996.
4. M. Charles and J. Wilson, in B.K. Lydersen, N.A. D'Elia, and K.L. Nelson eds., *Bioprocess Engineering*, Wiley, New York, 1994.
5. S. Yokell, *Chem. Eng.* **93**, 75–83 (1986).
6. S.P. Vranch, in W.C. Hyer ed., *Bioprocessing Safety, Worker and Community Safety and Health Considerations*, STP 1951, ASTM, Philadelphia, 1990, pp. 39–57.
7. C.H. Collins, in C.H. Collins and A.J. Beale eds., *Safety in Industrial Microbiology and Biotechnology*, Butterworth Heineman, Oxford, U.K. 1992, pp. 23–33.
8. Guidelines for Research Involving Recombinant DNA Molecules, *Fed. Regist.* **59**(127), 34496–34547 (July 5 1994).
9. *Management of Health and Safety at Work Regulations*, HMSO Publication Centre, London, 1992.
10. Council Directives on the Protection of Workers from Risks Related to Exposure to Biological Agents at Work (90/679/EEC), *Off. J. Eur. Commun.* **33**, 1–12 (1990).
11. M.C. Flickinger, E.B. Sansone, *Biotech. Bioeng.* **26**, 860–870 (1984).
12. *Design Criteria for Viral Oncology Research Facilities*, DHEW 78-891, Department of Health, Education, and Welfare, Washington, D.C., 1978.
13. M. Dick and E. Hanel, *Design Criteria for Microbiological Facilities, Fort Dietrick, Maryland*, vols. 1 and 2; Dept. of the Army, Technical Engineering Division Project 1B662706A072, 1970.
14. R.S. Runkle and G.B. Phillips, *Microbial Contamination Control Facilities*, Van Nostrand Reinhold, New York, 1969.
15. C.A. Perkowski, in B.K. Lydersen, N.A. D'Elia, and K.L. Nelson eds., *Bioprocess Engineering*, Wiley, New York, 1994, pp. 730–743.
16. E.L. Paul, in W.C. Hyer ed., *Bioprocessing Safety, Worker and Community Safety and Health Considerations*, STP 1951, ASTM, Philadelphia, 1990, pp. 65–73.
17. *Code of Federal Regulations*, Title 29, Washington, D.C. 1996.
18. *Code of Federal Regulations* Title 21, Washington, D.C. 1996.
19. *Points to Consider in the Production and Testing of New Drugs Produced by Recombinant DNA Technology*, Food and Drug Administration Washington, D.C., 1993.
20. *Points to Consider in the Manufacture and Testing of Monoclonal Antibody Products for Human Use*, Food and Drug Administration, Washington, D.C., 1991.
21. *Biotechnology Inspection Guide*, Food and Drug Administration, Division of Field Investigations, Washington, D.C., 1991.
22. *Guide to Inspection of Bulk Pharmaceutical Chemicals*, Food and Drug Administration, Washington, D.C., 1991.
23. *Guidance for Industry for the Submission of Chemistry, Manufacturing, and Controls Information for a Therapeutic Recombinant DNA-Derived Product or a Monoclonal Antibody Product for In Vivo Use*, Food and Drug Administration, Washington, D.C. 1996.
24. *FDA Guidance Document Concerning Use of Pilot Plant Manufacturing Facilities for the Development and Manufacture of Biological Products*, Food and Drug Administration, Washington, D.C., 1995.
25. *Guidelines on the General Principles of Process Validation*, Food and Drug Administration, Washington, D.C., 1987.

26. J.Y. Lee, in I.R. Berry and R.A. Nash eds., *Pharmaceutical Process Validation*, Marcel Dekker, New York, 1993, pp. 573–597.
27. T.J. Naglak, M.G. Keith, and D.R. Omstead, *BioPharm* **7(6)**, 28–36 (1994).
28. R. Baird and P. De Santis, in B.K. Lydersen, N.A. D'Elia, and K.L. Nelson eds., *Bioprocess Engineering*, Wiley, New York, 1994.
29. E.K. White and D.M. Marks, in J.K. Shillenn ed., *Validation Practices for Biotechnology Products*, ASTM STP 1260, ASTM, Philadelphia, 1995.
30. M.G. Beatrice, in G. Stephanopoulos ed., *Biotechnology: Bioprocessing*, vol. 3, VCH Weinheim, 1993.
31. L. Tuanchi and D. Yu, *Biotechnol. Bioeng.* **42**, 777–784 (1993).
32. L. Leong-Poi and D.G. Allen, *Biotechnol. Bioeng.* **40**, 403–412 (1992).
33. Y. Kawase and T. Kumagai, *Bioprocess Eng.* **7**(1), 25–28 (1991).
34. M. Charles, *Adv. Biochem. Eng.* **8**, 1–62 (1978).
35. B.C. Buckland, K. Gwebonyo, D. DiMasi, G. Hunt, G. Westerfield, and A.W. Nienow, *Biotechnol. Bioeng.* **31**, 737–742 (1988).
36. M. Reuss, in G. Stephanopoulos ed., *Biotechnology: Bioprocessing*, vol. 3, VCH, Weinheim, 1993.
37. Y. Kawase, B. Halard, and M. Moo-Young, *Biotechnol. Bioeng.* **39**, 1133–1140 (1992).
38. J.M.T. Vasconcelos and S.S. Alves, *Chem. Eng. J.* **47**, B35–B44 (1991).
39. D.S. Dickey, *72nd AIChE Ann. Meeting*, San Francisco, November 25–29, 1979.
40. A. Bakker, J.M. Smith, and K.J. Myers, *Chem. Eng.* **101**(12), 98–104 (1994).
41. C.M. McFarlane, X.-M. Zhao, and A.W. Nienow, *Biotechnol. Prog.* **11**, 608–618 (1995).
42. G.J. Balmer, I.P.T. Moore, and A.W. Nienow, in C.S. Ho and J.Y. Oldshue eds., *Biotechnology Processes, Scale-Up and Mixing*, AIChE, New York, 1987, pp. 117–127.
43. N.M.G. Oosterhuis and N.W.F. Kossen, *Biotechnol. Lett.* **3**(11), 645–650 (1981).
44. C.J. Geankopolis, *Transport Processes and Unit Operations*, Allyn and Bacon, Boston, Mass., 1983, pp. 154–157, 169–170.
45. A.W. Nienow, *Trends Biotechnol.*, **8**, 224–233 (1990).
46. J.Y. Oldshue, *Fluid Mixing Technology*, McGraw-Hill, New York, 1983.
47. M. Yasukawa, M. Onodero, K. Yamagiwa, and K. Ohkawa, *Biotechnol. Bioeng.* **39**, 629–636 (1991).
48. D.G. Cronin, A.W. Nienow, and G.W. Moody, *Trans. IChemE.* **72** (part C), 35–40 (1994).
49. C.L. Cooney, D.I.C. Wang, and R.I. Mateles, *Biotechnol. Bioeng.* **11**, 269–281 (1968).
50. R. Pollard and H.H. Topiwala, *Biotechnol. Bioeng.* **18**, 1517–1535 (1976).
51. P. Mohan, A.N. Emery, and T. Al-Hassan, *Trans. IChemE.* **70** (part C), 200–204 (1992).
52. S. Aiba and K. Toda, *J. Ferm. Tech. (Japan)* **43**, 527 (1965).
53. U. Pflug and R.G. Holcomb, in S.S. Block ed., *Disinfection, Sterilization and Preservation*, Lea and Feibiger, Philadelphia, 1991.
54. G.K. Raju and C.L. Cooney, in G. Stephanopoulos ed., *Biotechnology: Bioprocessing*, vol. 3, VCH, Weinheim, 1993.
55. J.H. Young, *Biotechnol. Bioeng.* **42**, 125–132 (1993).
56. J.H. Young, W.C. Lasher, and R.P. Gaber, *Bioprocess. Eng.* **12**, 293–304 (1995).
57. T. Oakley, in B.K. Lydersen, N.A. D'Elia, and K.L. Nelson eds., *Bioprocess Engineering*, Wiley, New York, 1994, pp. 500–521.
58. *Guide to Inspection of Validation of Cleaning Processes*, Food and Drug Administration, Division of Field Investigations, Washington, D.C., 1993.
59. R. Brunkow, D. DeLucia, S. Haft, J. Hyde, J. Lidsay, J. McEntire, R. Murphy, J. Myers, K. Nichols, B. Terranova, J. Voss, and E. White, in *Cleaning and Cleaning Validation: A Biotechnology Perspective*, PDA, Bethesda, Md., 1996, pp. 41–62.
60. S. Lombardo, P. Inampudi, A. Scotton, G. Ruezinsky, R. Rupp, and S. Nigam, *Biotechnol. Bioeng.* **48**, 513–519 (1995).
61. C.P. Dillon, D.W. Rahoi, and A.H. Tuthill, *BioProcessing* **8**(5) 32–35, **8**(7), 31–33 (1992).
62. J. Villafranca and E.M. Zambrano, *Pharm. Eng.* **5**(6), 28–30 (1985).
63. J. Butters and A. Reynolds, *Process Eng.* **71**(2), 65–79 (1990).
64. K.B. Balmer and M. Larter, *Pharm. Eng.* **13**(3), 21–28 (1993).
65. D.C. Coleman and R.W. Evans, *Pharm. Eng.* **10**(2), 43–49 (1990).
66. D.C. Coleman and R.W. Evans, *Pharm. Eng.* **11**(4), 75–79 (1991).
67. R.E. Markovitz, *Chem. Eng.* **78**(11), 156–163 (1971).
68. U.S. Pat. 4,670,397 (July 2, 1987), E.H. Wegner and H.R. Hunt (to Phillips Petroleum).
69. C.A. Perkowski, *Biotechnol. Bioeng.* **26**, 857 (1984).
70. M.A. Fogglesong, in W.C. Hyer ed., *Bioprocessing Safety, Worker and Community Safety and Health Considerations*, STP 1951, ASTM, Philadelphia, 1990, pp. 14–19.
71. T.H. Meltzer, M. Jornitz, P.J. Waibel, *Pharm. Technol.* **18**(9), 76–84 (1994).
72. S.F. Emory, *Pharm. Technol.* **13**(9), 68–77 (1989).
73. J.J. Errico, in F.J. Carleton and J.P. Agalloco eds., *Validation of Aseptic Pharmaceutical Process*, Marcel Dekker, New York, 1990, pp. 427–471.
74. P.D. Walker, T.J. Narendranathan, D.C. Brown, F. Woolhouse, and S.P. Vranch, in M.S. Verall and M.J. Hudson, eds., *Separations for Biotechnology*, Ellis Horwood, Chichester, U.K., 1987, pp. 469–479.
75. J.D. Wilson and T.E. Andrews, *Biotechnol. Bioeng.* **25**, 1205–1214 (1983).
76. N.J. Titchener-Hooker, P.A. Sinclair, M. Hoare, S.P. Vranch, A. Cottman, and M.K. Turner, *BioPharm* **6**(8), 32–37 (1993).
77. S.R. Miller and D. Bergmann, *J. Ind. Micro.* **11**, 223–234 (1993).
78. J. Van Houten, in W.C. Hyer ed., *Bioprocessing Safety, Worker and Community Safety and Health Considerations*, STP 1951, ASTM, Philadelphia, 1990, pp. 91–100.
79. G. Leaver, in P. Hambleton, J. Melling, and T.T. Salusbury, eds., *Biosafety in Industrial Biotechnology*, Blackie Academic and Professional, London, 1994, pp. 213–239.

FILTER AIDS

TONY HUNT
Advanced Minerals Corporation
Santa Barbara, California

KEY WORDS

Celpure
Diatomite
Extractables
Microparticulate removal
Solid–liquid separation

OUTLINE

INTRODUCTION

Among the most capable and attractive means of microparticulate (greater than 0.1 μ) separation used in biopharmaceutical processing today is the use of highly porous powdered media in dynamic systems. Powdered media, often called filter aids (Fig. 1), provide versatility, high solids-loading capacity, high product recovery, low cost, and ease of scale-up in any filtration process. Moreover, recent technical advances have stimulated a much greater breadth of applications that use this media filtration in biopharmaceutical processes, with a concomitant exponential increase in product innovation.

PROCESS CLARIFICATION

The first unit operation in downstream processing is clarification. It involves the removal of cells, cell debris, or precipitated components from a fermentation broth or process supernatant. Four types of solid–liquid separations are common to the industry: (*1*) The removal of whole cells in fermentation broths where the product is expressed into the supernatant; (*2*) The removal of cell debris from process broths after cell disruption and extraction to release the product of interest; (*3*) The removal of precipitated contaminants (usually proteins) from a process fluid, for example, in the removal of misfolded forms of recombinant proteins; and (*4*) The selective precipitation of the target protein and the capture of this precipitate from the process supernatant.

Most processes use either centrifugation or filtration for solid–liquid separations. Centrifugation, although widely accepted, has some disadvantages in biopharmaceutical applications. The challenge in many applications is to remove both cellular and subcellular debris without applying excessive shear forces on the solids. Centrifuges depend on differences in density and on the centrifugal force applied for solid–liquid separation. However, centrifugation becomes less practical when the product and waste possess similar densities. In addition, centrifuges can be difficult to maintain, clean, and sterilize. Finally, it is important to avoid the generation of aerosols when dealing with biological fluids, and this is often difficult to avoid when using centrifuges for this unit operation.

Depth filters containing diatomite, perlite, and cellulose are suited to low solids applications, because these products tend to blind quickly when subjected to moderate to high levels of solids. Their capacity can be increased by the addition of filter aid (as body feed) into the unfiltered broth, thus extending the lifetime of the pad.

The use of powdered media offers many advantages in biopharmaceutical applications. These systems are much more flexible and dynamic because powdered media addition can be adjusted based on changes in incoming solids level. In addition, they are extremely efficient at removing suspended solids and are particularly effective at clarifying supernatants heavily contaminated with colloids and small particles. The media can be used in all four applications described above, including the selective precipitation of the target protein and the capture of this precipitate from the process supernatant. Once captured on the filter media, the protein or product of interest can be redissolved by adjusting the pH or salt concentration in the wash buffer and recirculating this solution through the filter media bed. After resolubilization, the product can be eluted from the media with approximately two bed volumes of buffer.

Capital costs for powdered-media-based systems are relatively low when compared with centrifugation or other filtration systems. Because the media are disposable, media cleaning and lifetime studies do not need to be performed, hence the cost of validating the filtration process is substantially reduced. In addition, the ability to postwash the filter cake helps to maximize product recovery, a process that is extremely expensive or impractical with centrifuges. Finally, with improvements in both powder

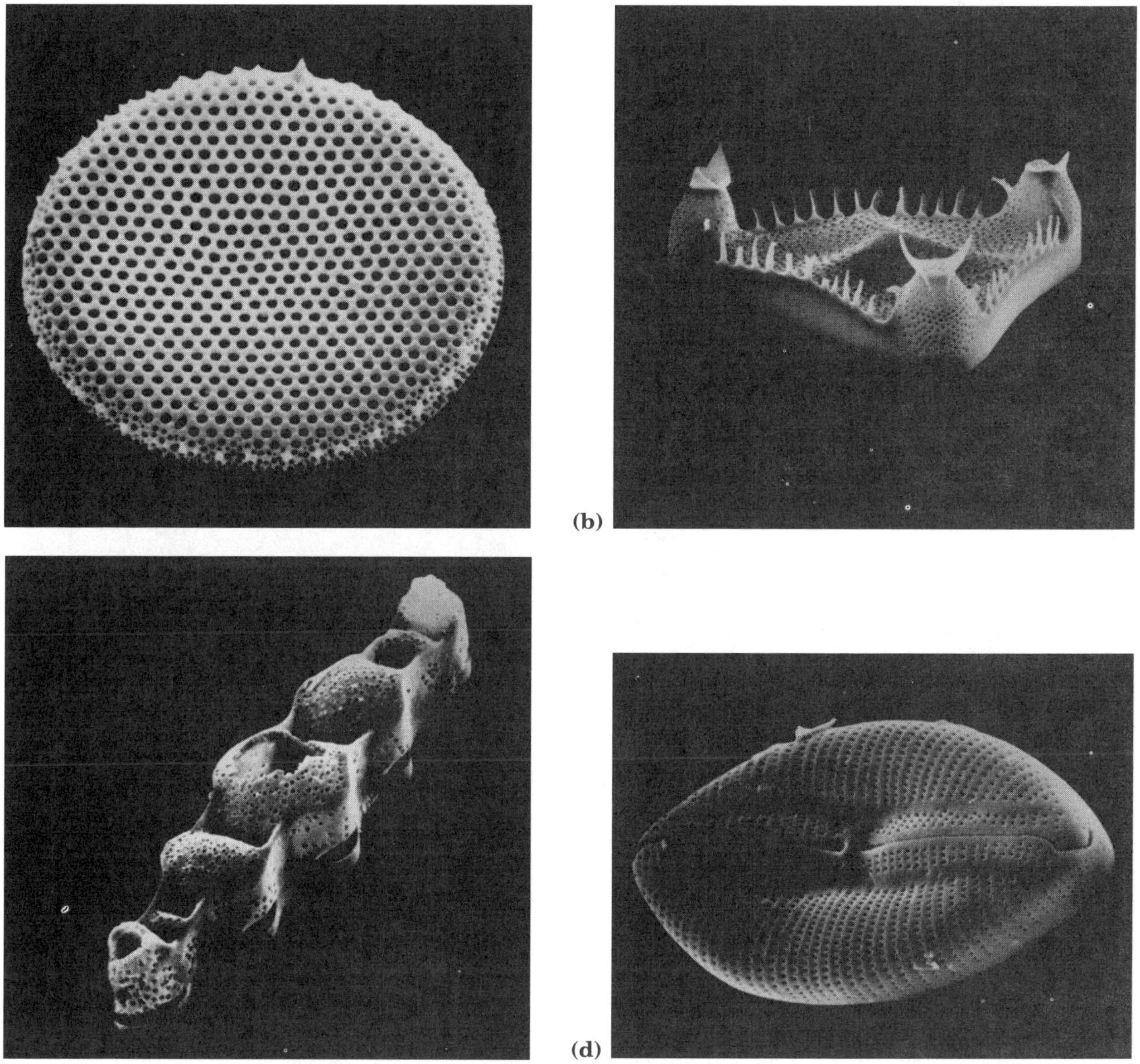

Figure 1. Size and shape of diatoms commonly found in diatomaceous earth deposits.

containment and filter equipment design, complete containment is now easily achieved both in the media preparation area and in GMP suites.

The combination of these improvements in powder containment, filter design, and powdered media technology position filter aids as a very cost-effective and economical approach to clarification.

POROUS MEDIA IN DYNAMIC PROCESS FILTRATIONS

Diatomite, perlite, and cellulose are the most widely used porous media in dynamic process filtrations, with a high percentage of applications using diatomite.

Diatomite

Diatomite is often used in combination with cellulose; however, diatomite is often the key ingredient of static or fixed-bed filters such as filter sheets and filter pads. Diatomite products are especially characterized by an inherently intricate and highly porous structure composed primarily of silica. These products are obtained from diatomaceous earth, a sediment greatly enriched in biogenic silica in the form of siliceous frustules of diatoms, a diverse array of microscopic, single-celled golden-brown algae of the class Bacillariophyceae. Surprisingly, these frustules are sufficiently durable to retain much of their ultrastructure nearly intact through long periods of geologic time when preserved in conditions that maintain chemical equilibrium.

Perlite

Perlite is a naturally occurring volcanic glass that thermally expands upon processing. After milling, porous, complicated structures result (Fig. 2) that are also useful in dynamic process filtrations of biopharmaceuticals. Because its structure is not as intricate as that of diatomite, perlite is better suited to the separation of coarse microparticulates from liquids having high solids loading. Like diatomite, perlite is also useful as a functional filtration component of filter sheets and filter pads.

Cellulose

Cellulose, like perlite, possesses a less intricate structure than diatomite. As a result, the use of cellulose is generally limited to coarse filtrations or in specialized applications where a fibrous precoat on the septum is required.

The rest of this chapter will focus attention on the use of diatomite in solid–liquid separations.

FUNDAMENTAL PRINCIPLES OF DIATOMITE FILTRATION

A discussion on solid–liquid separations can cover a wide range of techniques and applications, but each method is often defined by the relative size of solids being removed or collected from a given process stream. This chapter will

Figure 2. Scanning electron micrograph of perlite.

focus on microparticulate filtration, where the solids are typically greater than 0.1 μ. The key objective is to remove unwanted solids from a process stream, so that the refined solution (filtrate) is adequately clarified and qualifies for further downstream processing.

Solids Composition

In general, solids can either be rigid or compressible in nature. Solids that form rigid cakes have some degree of inherent permeability, dictated by particle size distribution and packing arrangement, but most solids in biopharmaceutical applications are typically compressible. These often gelatinous, highly compressible solids retain a degree of permeability if low flow rates and low differential pressures are used for filtration. Unfortunately, these characteristics are contrary to typical processing requirements where this unit operation needs to be completed quickly and economically.

Dynamic Filtration

The problem in static fixed-bed filtration is that unwanted solids have very limited permeability when they collect or accumulate onto the filter septum (pad, paper, or fabric or metallic or plastic woven wire screen). If these solids build up on a septum, then eventually these solids will lack sufficient permeability for fluid drainage and filtration terminates. This behavior highlights a mistake frequently applied in filtration practice; often, higher differential pressure does not guarantee faster filtration rates precisely because these solids collapse to form an impermeable cake.

Introduction of a filter medium changes the composition of the accumulated cake, and therefore its filtration behavior. Powdered media (filter aids) are essential to this technology because they provide two functions: (*1*) as a precoat applied before the start of a filtration cycle and (*2*) as body feed added to the unfiltered feed throughout the filtration cycle. The precoat layer protects the filter septum, preventing penetration and blinding by the unwanted solids. This layer also facilitates septum cleaning and provides for a high-quality filtrate from the beginning of the filtration cycle.

The addition of powdered media (as body feed) to the process feed increases permeability in the accumulating filter cake, restricts solids movement, provides channels for filtrate recovery, and extends cycle length (Fig. 3). The extended permeability afforded by these media slows the rise in differential pressure for constant flow processes and retards the drop in flow for constant pressure operations. At the conclusion of a filtration cycle, product recovery can be maximized by rinsing or purging the accumulated cake in place. It is extremely important to maximize the body feed in a process. By adding too little, the solids quickly blind the filter cake and the filtration is terminated. Adding too much is not only wasteful but leads to a rapid increase in filter-cake thickness, with a corresponding increase in resistance to flow and a subsequent reduction in cycle length (1).

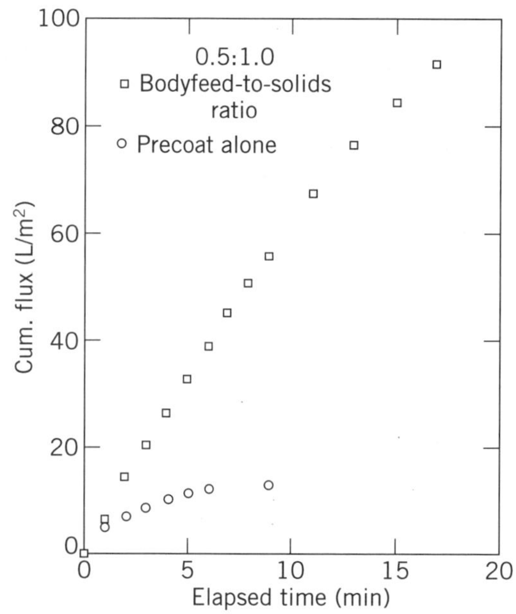

Figure 3. Addition of body feed to a filtration process extends the filtration cycle length.

Filtration Theory Overview

The following section is not intended to be a comprehensive treatment of filtration theory, but rather an overview to describe how the permeability of an accumulated filter cake (solids and filter aid) ultimately affects filtration performance (rate). A thorough discussion of filtration mechanics can be found in numerous reviews (2–4).

Any discussion on porous media filtration can start with a simplified form of the Darcy equation (equation 1), which is used to describe laminar (nonturbulent) fluid flow through a homogeneous, porous medium.

$$k = \frac{Q \times \Delta x \times \mu}{A \times \Delta P} \tag{1}$$

The relative permeability (k) is measured in darcy units, and it is described in terms of the instantaneous volumetric flow rate (Q) and viscosity (μ) of the fluid, the thickness (Δx) and cross-sectional area (A) of the porous medium (accumulating filter cake), and the pressure drop or differential pressure (ΔP) associated with flow. The flow rate (Q) can also be represented by the differential change in volume over time (dV/dt). A value of 1 darcy corresponds to the permeability through a filter media 1-cm thick that allows 1 cm^3 of fluid with a viscosity of 1 cP to pass through an area of 1 cm^2 in 1 s under a pressure differential of 1 atm (i.e., 101.325 kPa). Although fluid viscosity varies with temperature and shear forces, values for it and the other variables can be obtained from tables and experimental observations. Consequently, the permeability (k) is an empirical value (constant) used to characterize the filter cake composition and not a property that can be readily altered by changing the process flow rate or differential pressure. In practice, permeability may appear to be influenced by flow rate or pressure, but those responses are strictly consequences of composition, and they are often inconsistent with the behavior described by equation 1. The type and relative usage rate of filter aid coupled with the nature and concentration of solids define composition, which in turn controls permeability and filtration performance.

The constancy of k explains why (*1*) the differential pressure increases as filter cake accumulates (Δx) for constant flow (Q_0) processes (equation 2)

$$k = \frac{Q_0 \times \mu}{A} \text{ (constant) } \times \frac{\Delta x}{\Delta P} \tag{2}$$

and (*2*) flow decreases as cake accumulates for constant pressure (ΔP_0) processes (equation 3)

$$k = \frac{\mu}{A \times \Delta P_0} \text{ (constant) } \times Q \times \Delta x \tag{3}$$

Permeability measurements are not essential in the application of dynamic filtration technology. In fact, once the unfiltered feed is introduced to the filter, the overall permeability (as measured in darcies) of the filter cake can be reduced by orders of magnitude. Ultimately, the manipulation of permeability (k), through filter aid selection and usage, describes the optimization of this solid–liquid separation process.

As an example, initial (laboratory scale) testing develops enough information to determine the permeability (k_{app}) in a constant pressure application. In fact, a key laboratory test is to determine the elapsed time (t_s) it takes to produce the maximum cake thickness (Δx_f; limited by the physical capacity of the filtration equipment, typically 3 cm) at optimum filter aid usage and terminal pressure (ΔP_f). Hence, the Darcy equation can be put to better use for process scale-up (equation 4):

$$[k_{app}] = \frac{Q_{0\ lab} \times \Delta x_f \times \mu}{A_{lab} \times \Delta P_f} = \frac{Q_{0\ plant} \times \Delta x_f \times \mu}{A_{plant} \times \Delta P_f} \tag{4}$$

Notice that differential pressure (ΔP_f) and filter cake thickness (Δx_f) need not be affected by the scale of operations. Differences in allowable cake thickness (Δx_f) must be accommodated by adjusting filtration area (A_{plant}) requirements. A physical capacity test also determines the total volume of process feed that can be filtered on the given area. In the simplest case of equal cake thickness, scale-up becomes a matter of sizing filter area to accommodate production rate (equation 5).

$$\frac{Q_{0\ lab}}{A_{lab}} = \frac{Q_{0\ plant}}{A_{plant}} \tag{5}$$

GRADE SELECTION AND OPTIMIZATION

An optimized filtration process maximizes throughput and product recovery and minimizes pressure drop while maintaining the desired filtrate clarity, all in a reasonable time. There are five main parameters one needs to consider when selecting an appropriate grade of powder media for a given process.

Product Stability

Careful attention should be paid to soluble metals when selecting a grade. High concentrations of soluble metals in reagents can lead to contamination of biopharmaceutical products, for example, aluminum contamination of albumin (5–7). Also, high concentrations of soluble metals are known to oxidize proteins and enzymes and can activate proteases in fermentation broths (8–10). Therefore, the use of high-purity reagents for filtration will help reduce or eliminate product contamination and degradation issues further downstream.

The industry demand for high purity and high performance reagents has led to the development of a new generation of powdered media. These products, sold under the trade name Celpure™ (Advanced Minerals Corporation, Calif.) offer greater filtration capacity with a corresponding reduction in powdered media consumption and reduction in disposal costs (Fig. 4).

Filtration processes with Celpure grades typically use less media because of a combination of the higher solids loading capacity and improved flow properties of the media. This results in a reduction in overall processing times compared with conventional grades of diatomaceous earth, such as the widely used Celite® grades (Fig. 5). Finally, the products are extremely pure and have very low levels of extractable metals such as aluminum and iron (Table 1) and correspondingly low electrical conductivities.

Filtrate Clarity

Most processes have a clarity specification that needs to be met or exceeded. Achieving stringent clarity specifications can extend the life of downstream filters and protect chromatography columns. The filtrate clarity achieved is dictated by the grade selected and the nature of the turbidity removed. Once the grade is chosen, its level of usage (body feed addition) combined with the available differential pressure to induce flow will control the volume of unfiltered feed that can be processed by a given filtration area.

Product Throughput

Product throughput and filtrate clarity are tightly linked when it comes to grade selection. The goal is to select a grade that achieves the desired clarity and maximizes throughput. By selecting a grade that is too fine, the clarity specification can be exceeded, but the throughput rate may be extremely low with correspondingly high differential pressures.

Operating Pressure

Actual operating pressure can be limited by shear sensitivity of the product or feedstream by equipment constraints or by overall plant design. Many processes have mechanical pressure limits of approximately 30 to 45 psi or 2 to 3 bar.

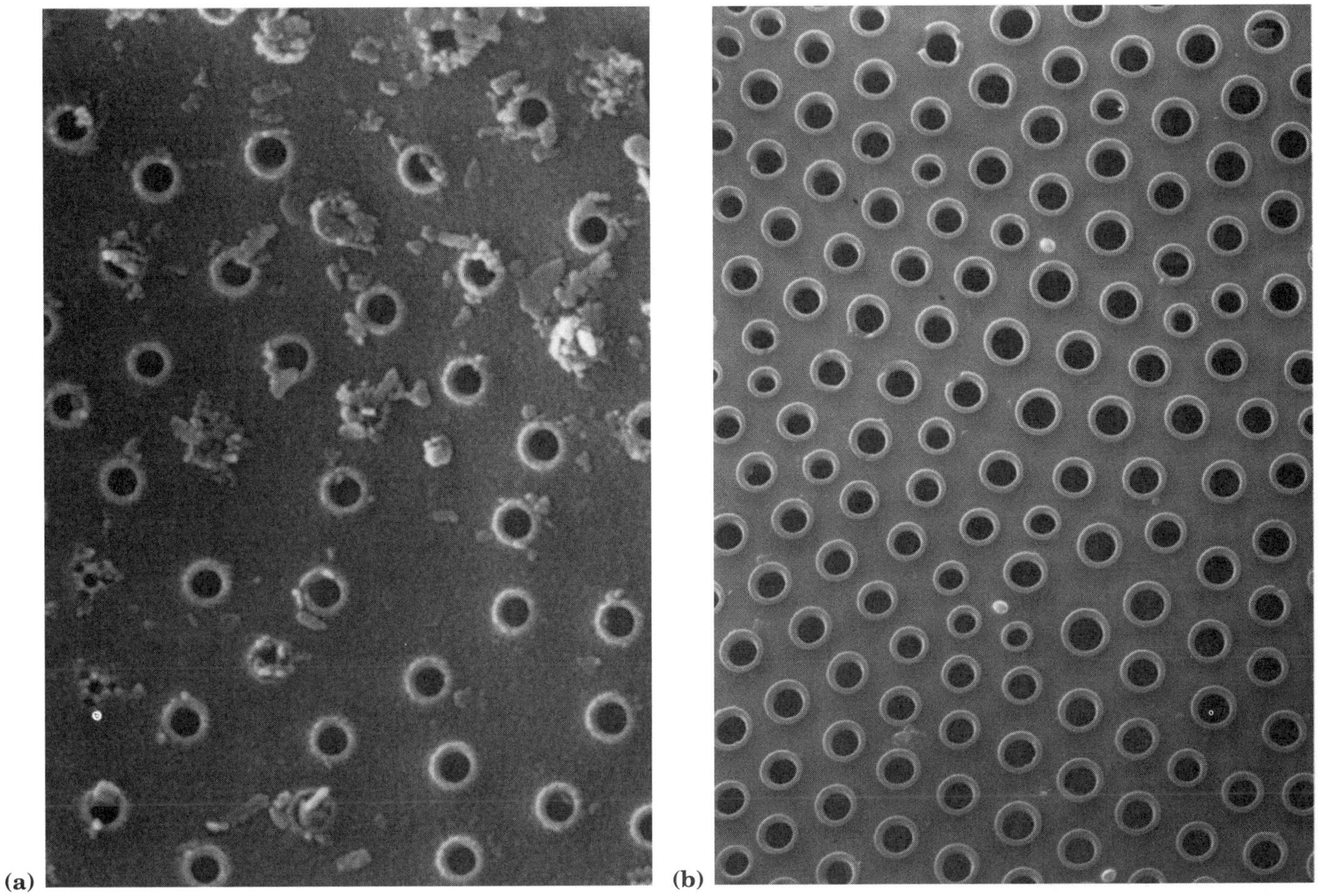

Figure 4. Scanning electron micrograph of the surface of an individual diatom from a conventional Celite grade (**a**) and a Celpure grade (**b**).

Product Recovery

This can be sometimes overlooked in initial filtration studies. At the conclusion of any filtration cycle, the accumulated solids should be washed with a product-compatible buffer to maximize product recovery. Upon completion of the filtration cycle (including rinse), it is important to confirm that the product of interest does not interact with the filtration media. Product recovery issues can often be improved by selecting a more permeable grade of filter aid. As you increase the porosity of a filter aid, the available surface area decreases and any nonspecific interaction decreases. All these factors are tightly linked; therefore, we recommend a systematic approach to grade selection and optimization.

SYSTEMATIC METHODS DEVELOPMENT APPROACH TO GRADE SELECTION

As stated above, the overall goal is to achieve optimum clarity and maximum throughput for a process while minimizing pressure and product losses. The following is a systematic approach to grade selection. Figure 6 shows a typical experimental setup for laboratory-scale constant rate filtration experiments. To perform constant pressure filtrations, a pressurized vessel is required to deliver unfiltered feed to the filtration housing.

Determine the Percent of Solids in Your Feedstream

Any information on distribution and size of feed solids is useful in grade selection. If this information is not available, a recommended procedure is outlined below.

Determine percent solids by weight in your fermentation broth. Filter a small aliquot onto a preweighed filter disk, dry in an oven, and reweigh to determine percent solids in unfiltered feed.

1. Use a Whatman 934-AH filter paper as a septum with the filtration housing. This is an open-pore septum and enables you to precoat without having to custom make a wire screen.
2. Resuspend 1.5 g of media in 100 to 150 mL of buffer (similar in pH to your sample solution; 1.5 g of most grades gives a precoat of approximately 2 mm on 20 cm^2 of surface area). Start with the coarsest grade, Celpure™ 3000 or Celite® 545 AW. By starting with the coarsest grade, you will always achieve maxi-

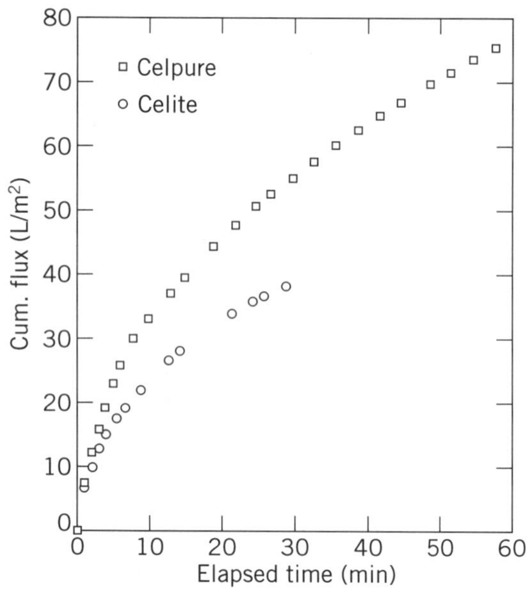

Figure 5. Equal weight comparison of Celpure and Celite grades in the clarification of a fermentation broth with 0.5:1.0 body-feed-to-solids ratio. Runs were terminated at minimum (equal) flow for reasonable production.

Table 1. Typical Celpure™ Properties (Applicable to All Grades)

Property	Value
Soluble aluminum (mg Al/kg)	
Extraction in fermentation broth, pH 4.5.	<3
Soluble iron (mg Fe/kg)	
Extraction in fermentation broth, pH 4.5	<3
Conductivity in H_2O (μS-cm)	<12.5

mum throughput and optimal clarity for a given process.

3. Recirculate the solution until filtrate is clear (typically 2 to 3 mins). The recommended flux rate is 40 L/m^2/min. On a 20-cm^2 surface area, this is equivalent to 80 mL/min.
4. Once the precoat has been formed, reduce the volume of buffer in the chamber until 2 to 3 mm above surface of precoat.
5. Add media to the unfiltered feed solution. As a starting point match the percent of solids in your feed with powdered media addition (2% solids by weight, add 2% by weight of powdered media). Add product/media solution to filter chamber at a flux rate of 1 to 5 L/m^2/min. On a 20-cm^2 surface area, this is equivalent to 2 to 10 mL/min.
6. Measure initial flow rate and monitor both flow rate and pressure throughout the coarse of the filtration. Allow the pressure to rise to 30 psi or 2 bar (or maximum system pressure), and maintain this pressure for the duration of the filtration by reducing flow or until flow rate reaches 20% of initial flow.
7. At the end of the filtration cycle, wash the filter bed with two bed volumes of compatible wash buffer. This will maximize the product recovery for a process. Finally, measure the total volume processed, the time of filtration, filtrate clarity, and the bed height in the filter housing. Save a sample for product recovery analysis.

If clarity is not acceptable, then repeat steps 1 to 7 with a less permeable grade of filter aid. Powdered media are rated based on their Darcy permeability. Celpure 3000 and Celite 545 AW are considered the coarsest grades and Celpure 65 and Celite Filter-Cel® the finest.

For constant pressure filtrations, follow above procedure through step 4. Then switch to a pressurized vessel containing unfiltered broth mixed with appropriate grade of powdered media. Ramp the pressure up to the desired output and monitor decay in flow rate. Repeat experiments with different grades as outlined above.

Body Feed Optimization

Once grade selection is completed, the next step is to optimize the body feed addition. The level of body feed addition controls the permeability and volume of accumulating cake. Higher pressure does not mean greater throughput, especially when body feed usage coupled with the handling characteristics of the solids dictate filter cake

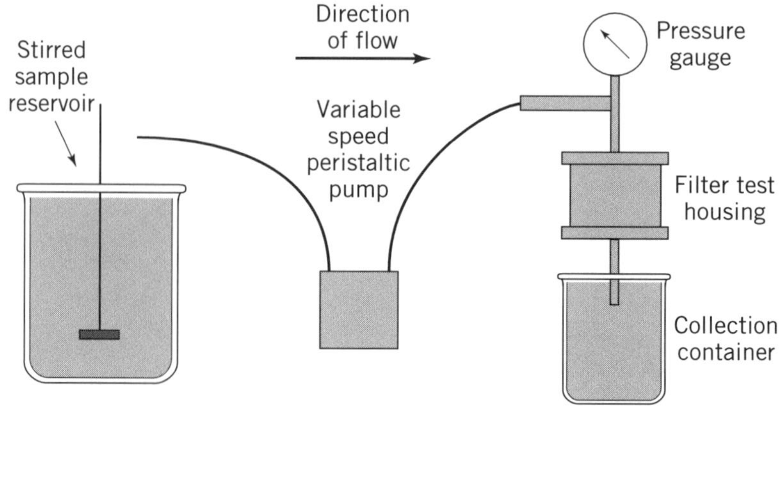

Figure 6. Experimental setup for constant rate experiments.

compressibility. The dilatant nature of diatomite resists cake compression and thereby retains permeability. Feed properties (pH, viscosity, etc.) also affect filtration performance; therefore, these factors must be considered in process development and optimization studies.

Past work with biological solutions at high solids levels (greater than 5% by weight) suggests that constant flux (filtration) rates above 10 L/m^2/min exhibited steep rises in differential pressure because of compression and subsequent loss of cake permeability. Modest flux rates (5.0 L/m^2/min) extended filtration cycles, often producing better economy. Although dilution using a compatible solution can reduce shear and also make filtration easier, the goal to expedite processing implies that it will be used only after others options have failed.

As described earlier, an excellent starting point is to match the percent solids in the process stream on a weight basis with percent media also on a weight basis. If this ratio of 1:1 is not sufficient, i.e., the filtration terminates quickly because of a rapid rise in pressure and subsequent blinding of filter cake with solids, then consider increasing your body feed addition. Repeat the above steps using body feed additions of 1.5:1 or 2:1.

If a body feed addition of 1:1 allows you to process your batch with little or no rise in differential pressure, then you can optimize the body feed by reducing the ratio of powdered media to solids. Reduce the body feed until an acceptable throughput rate is achieved (Fig 7). This approach will allow you to size an appropriate filter and minimize the size of filter needed to process a given batch.

Process Scale-Up and Filter Selection

Candle elements, pressure filters, and filter presses are all used in biopharmaceutical applications. Your choice of filter will depend on a number of factors, including plant design, equipment compatibility with existing systems, clean-in-place and steam-in-place considerations, scale of operation, and overall capital costs. All these factors need to be considered in selecting an appropriate filter. Once the filter design has been selected, the data generated in the methods development stage can be used to predict filter size. A smaller filter can be considered if the cycle time is short, and multiple cycles (including time for cleaning and setup) can be performed, in a typical 8 hour shift, to clarify the batch.

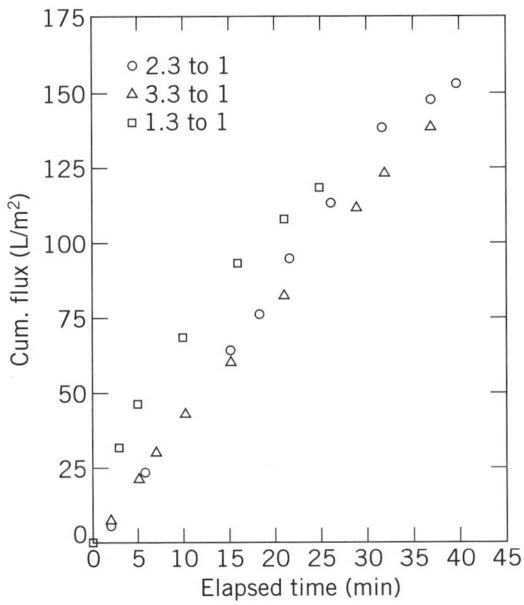

Figure 7. Body feed optimization study for a biological broth with body-feed-to-solids ratios.

SUMMARY

The use of advanced filtration media in dynamic filtration systems is a flexible, efficient, and attractive means for the solid–liquid separation of microparticulates from various process streams. Powdered media provide versatility, high solids-loading capacity, high product recovery, low cost, and ease of scale-up in any filtration process. These systems are cost effective and extremely flexible and efficient at removing suspended solids and clarifying supernatants rich in colloids and small particles. Newer products such as Celpure™ demonstrate improvements in purity and performance and further address the needs of the biopharmaceutical industry.

In addition to the advances that have been made in powdered media, improved systems for media handling and containment are now available from vendors that readily meet the needs of the industry. Finally, a wide variety of economical filter housings of various designs, including candle elements, pressure filters, and filter presses, are manufactured to biopharmaceutical specifications.

BIBLIOGRAPHY

1. R.H. Rees and C.W. Cain, *Chem. Eng.* 72–74 (August 1990).
2. J. Kiefer, *Brauwelt International IV*, 1991, pp. 300–309.
3. J. Bear, *Dynamics of Fluids in Porous Media*, Dover Publications, New York, 1988, pp. 119–195.
4. J. Hermia and S. Brocheton, *Filtration Separation* 721–725 (November 1994).
5. W.P. Olson and R.S. Kent, *Transfusion*, **29**, 86–87 (1989).
6. S.W. King, M.R. Wills, and J. Savory, *Res. Commun. Chem. Pathol. Pharmacol.* **26**, 161–169 (1979).
7. M. Inoue, Y. Gion, H. Itoh, K. Ikariya, K. Takechi, S. Tomioka, K. Furuta, and S. Yabushita, *Vox Sang.* **66**, 249–252 (1994).
8. W.A. Banks, L.M. Maness, M.F. Banks, and A.J. Kastin, *Neurotoxicol. Teratol.* **18**, 671–677 (1996).
9. S.A. Hulea, E. Wasowicz, F.A. Kummerow, *Biochim. Biophys. Acta* **1259**, 29–38 (1995).
10. N. Ahmed, J. Liggins, A.J. Furth, *Biochem Soc. Trans. (England)* **21**, 93S (1993).

See also CELL SEPARATION, SEDIMENTATION; MICROENCAPSULATION.

FILTRATION, AIR

HOLLY HAUGHNEY
Pall Corporation
Long Island, New York

KEY WORDS

Air
Fermentation
Filtration
Membrane
Sterilization
Vent

OUTLINE

INTRODUCTION

In the bioprocessing industry there exist a number of applications for air (or gas) filtration (1–3). The type of filter used for the application will depend on the specific removal requirements. The filtration required can be to the 0.2-μ (sterilizing) level or to a courser filtration level for particulate removal. Some air applications for which sterile filtration can be a requirement are fermenter inlet air, fermenter vent gas, vents on water-for-injection tanks, and vacuum break filters during lyophilization. Courser filtration can 90be used for particulate removal. In many cases, the course filter acts as a prefilter to the sterilizing grade filter. Other applications use coalescers as prefilters for the removal of liquid droplets, such as oil or water.

TYPES OF AIR FILTERS

Packed towers were the first air filters used by the industry for air sterilization. Packed towers consist of beds or pads of a fibrous material, such as paper, cotton wool, glass wool, or mineral slag wool. Membrane filters typically are constructed of a porous, hydrophobic membrane material, such as PVDF (polyvinylidienefluoride) or PTFE (polytetrafluoroethylene).

When filters are used to produce sterile air, the filters can be referred to as sterilizing grade, 0.2-μ filters. Microbial retention tests can be used to verify that the filters produce sterile air. The standard test organism is *Brevundimonas (Pseudomonas) diminuta* (ATCC 19146).

The removal efficiency of filtration media used in packed towers or membrane filters can be dependent on the following mechanisms:

1. Direct interception by the fibers
2. Inertial impaction
3. Brownian motion or diffusional interception
4. Electrostatic attraction between the fibers and particles

Brownian motion (also referred to as diffusional interception) applies to small particles at low face velocities. When air molecules are in a state of random motion, small particles suspended in the particles can be struck by moving air molecules and displaced. The movement of particles resulting from molecular collisions is known as Brownian motion. This phenomenon can increase the probability of capture of the particle by the fibers in a packed tower or by the diffusional interception in a membrane.

The filtration mechanism for *direct interception* is a sieving action that mechanically retains the particles. The filter material acts as a screen that stops particles that are larger than the pores. Direct interception is independent of face velocity and mostly involves particles with large diameters.

Inertial impaction refers to the deviation of a particle from its streamline during flow because of inertia. This deviation can result in the retention of the particle by the fibers in the packed tower or by the membrane in a membrane cartridge filter. As the face velocity increases, the probability of inertial impaction increases.

An *electrostatic attraction* can exist between the fibrous material in the depth filter, or the membrane material, and the particles in the air stream. This attraction can enhance the ability of the filter to remove particles.

A general description of packed towers as well as membrane filters used for air filtration is provided in the next section.

Packed Towers

Description of Packed Towers. Packed towers for air filtration are comprised of beds of fibrous material (1). The diameter of the fibrous material is between 0.5 and 15 μ, and the space between the fibers can be many times this range, depending on how the tower is packed.

Figure 1 illustrates the basic design of a packed tower air filter (5). The filter consists of a steel structure filled with loose fibrous packing. The air inlet is on the bottom and the outlet is on the top of the filter. The packing is supported by a grid or perforated plate. For proper opera-

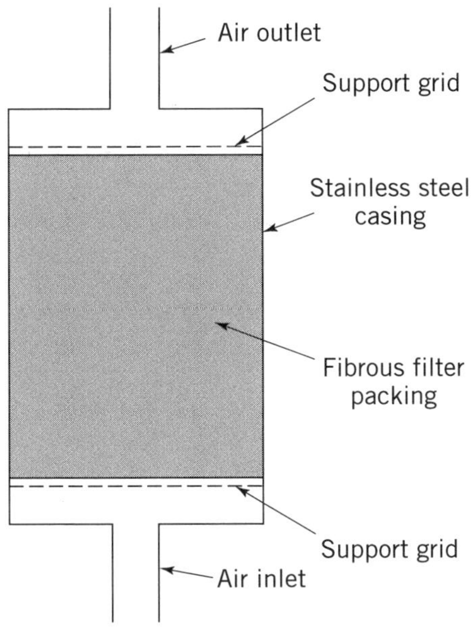

Figure 1. Packed air tower design.

tion, it is necessary to ensure that the appropriate packing density for the application is obtained. If the tower has been packed properly, there will not be movement of the fibers during use. Fiber movement during use can lead to channeling of air, which will lead to an inefficient packed tower, because only a portion of the bed will be acting as a filter. Fiber repositioning can also lead to the dislodging of trapped microorganisms.

Once the filter has been packed with the appropriate fibrous material, a support grid or plate is fitted to the top of the bed to ensure that the bed remains compressed. After the first steam sterilization of a packed tower air filter, the packing will tend to settle further. Additional fibers can be added to the bed after the first steam sterilization to maintain the correct packing density.

Bonded fiber mats have been developed to use in place of loose fibers. Resins are used to bond the fibers, and the mats must be tested to determine if they are compatible with steam sterilization. When mats are used, it is necessary to have a good seal between the mat and the tower wall, so that channeling does not occur. Fiber pads often contain mineral wool with an average fiber diameter of 4 to 5 μ.

Thin sheets of small diameter fibers are another option for packed towers. The sheets are placed on top of each other with a mesh or grid in between each sheet for support. The edges of the sheets are sealed between flanges.

Sterilization of Packed Tower Air Filters. Before using a packed tower for the sterilization of fermenter inlet air, the filter itself must be sterilized. There are two techniques that can be used: steam sterilization and dry heat sterilization (6).

Passing steam at a pressure of 15 psig through a packed tower for 2 h is typically adequate for sterilization. Actual conditions required for a specific installation may vary. The presence of air in the packed tower during the sterilization can prevent complete sterilization of the packing. A drain at the bottom of the packed tower is used to purge the steam and to drain any residual condensate in the filter. It is necessary to remove condensate from the bed because wetted fibers are less efficient for microorganism removal and may also decrease the retention efficiency of the packed tower well below its design value, especially if channeling through the wetted media occurs.

Some fiber material as well as material used to bond fibers can be degraded by steam sterilization. An alternative to steam sterilization is dry heat sterilization, which will avoid the possibility of steam degradation and fiber wetting. This can be accomplished by using a heating device at the inlet of the tower and passing air at a temperature of 160 to 200 °C through the bed for 2 h. During dry heat sterilization, the filter is isolated from the rest of the process. Fiber material can be stable up to 800 °C, thus there is no chance of damage to the bed from this sterilization technique.

Operating Considerations for Air Sterilization by a Packed Tower. The microorganism retention efficiency of packed towers is dependent on the inlet velocity of the air. The relationship between air velocity and filtration efficiency has been determined experimentally in a number of studies (7). In the example shown in Figure 2 (7), the particle retention efficiency may change by a factor of 10 as a result of a relatively small change in the inlet air velocity. The results also show that the most difficult particles to remove are in the size range of small microorganisms. Therefore, it is quite possible to encounter air flow conditions in a packed tower that can reduce the statistical probability for the complete retention of all microorganisms.

The hydrophilic nature of the fibrous material used in packed towers (e.g., glass wool) can contribute to a reduction of the microorganism removal efficiency of the packed tower. Water vapor can enter the system with the air discharge of the compressor. The air and water vapor mixture is initially at an elevated temperature. As the gas stream is cooled, water droplets may condense and wet the hydrophilic glass fibers. The wetted fibers are less efficient for microorganism removal and may decrease the retention

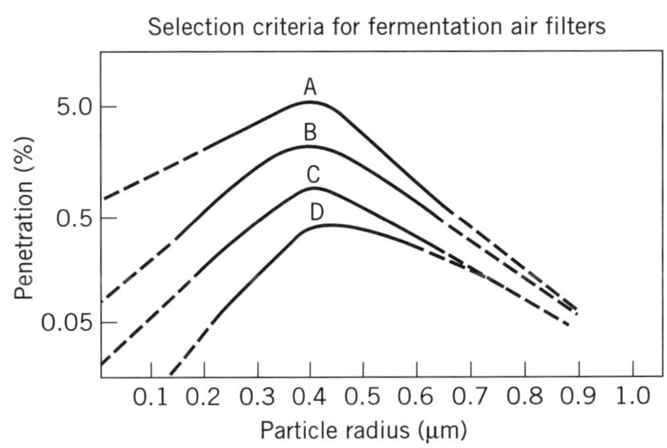

Figure 2. Particle retention efficiency.

efficiency of the packed tower well below its design value, if channeling through the wetted media occurs. Also, organic components present in the compressor exhaust can provide a nutrient source for the retained microorganisms and increase the possibility of bacterial growth and eventual penetration through the depth filter medium. Although a number of approaches have been tried to overcome the wetting problem, such as heat tracing to maintain an elevated temperature, these are expensive to operate and have not consistently resolved the problem. The pressure differential across the tower may also increase significantly with wetting.

Although it is obvious that operating deviations such as these do reduce the reliability of a packed tower for air sterilization, there is no quantitative procedure to determine if the filtration efficiency is adequate to ensure a sterile inlet air condition. The lack of such a quantitative procedure adds an element of uncertainty that affects both the operation and maintenance decisions. Skilled operators are necessary to monitor the packed tower to ensure that it is packed and operating efficiently. The actual source of contamination of a production batch is often difficult to identify, and the air filtration system is always suspect during a contamination outbreak. Without a technique for testing the efficiency of a packed tower, it is difficult to ascertain whether the packed tower is the cause of the contamination outbreak.

Design of Fibrous Air Filters. In order to model a fibrous air filter, several assumptions must be made. It is assumed that (*1*) Once a particle is trapped by a fiber, it then will remain trapped; (*2*) At a particular depth across the filter, the particle concentration does not vary; and (*3*) The removal efficiency at a given depth is equivalent across the filter.

The following equation describes how the concentration of particles varies with the depth position in the filter (6):

$$\frac{dN}{dt} = -Kx \tag{1}$$

where N is the particle concentration, x is the depth, and K is a constant.

Solving the equation between a depth of 0 and x and a particle concentration of N_0 particles entering the filter and particle concentration of N particles leaving the filter yields:

$$\ln\left(\frac{N}{N_0}\right) = -Kx \tag{2}$$

The relationship between depth and the logarithm of the ratio of particles removed to particles incident is known as the log–penetration relationship (6). This relationship has been used in sizing depth filters.

The constant K in equation 2 will vary with the type of packing and is dependent on linear velocity through the packed bed. If the relationship between the constant K and linear velocity through the bed is known, equation 2 can be used to size a packed bed for a given log reduction of particles.

Another consideration for the design of a packed bed is the pressure drop across the bed. Typically, the ΔP is linear with the linear air velocity for a given depth (6). The pressure drop across the packed tower can be dependent on the type of medium, the packing density, the air density, and the linear air velocity through the filter. As an example of an equation for the pressure drop across the packed bed is given in Richards (6):

$$\Delta P = \frac{2\rho v^2 \alpha x C}{\pi D_f} \tag{3}$$

where ΔP is the pressure drop, v is the linear air velocity, α is the ratio of filter density over fiber density, x is the filter bed depth, C is the drag coefficient, D_f is the fiber diameter, and ρ is the air density. The relationship in equation 3 indicates that the pressure drop is proportional to the square of the linear air velocity. At relatively low linear air velocities (2 to 3 ft/s), the relationship is linear.

Membrane Filter Cartridges

Membrane filter cartridges are available as either prefilter (particulate contaminant rating) or sterilizing filter (bacterial contaminant rating) configurations (1–3). Prefilters in air service can be used for particulate removal or aerosol removal. Prefilters are positioned upstream of the final (sterilizing) filters to protect the final filter from premature plugging, thereby prolonging the life of the final filter. The following are brief descriptions of the membrane filters (sterilizing and prefilter types) that can be used for air filtration.

Sterilizing Grade Membrane Filters for Air Service. Membrane filters used for fermenter and bioreactor sterile inlet air and exhaust gas vents, sterile pressure gas, sterile nitrogen blankets, storage tank sterile vents, formulation tank sterile vents, and sterile air for aseptic packaging usually contain a membrane made of hydrophobic materials such as PVDF or PTFE.

Hydrophobic membrane filters are desired in these sterile gas filtration applications because hydrophobic filters do not spontaneously wet with water. When a hydrophilic filter is wetted with water, it will not pass air until the water-wet bubble point of the filter is exceeded, and this water-wet bubble point can be greater than 50 psi. The inherent hydrophobicity of membrane filters used for fermenter air sterilization allows these filters to be able to remove bacteria completely from inlet air, even when exposed to moisture (8). The filter cartridge is expected to remove bacteria and bacteriophage from air streams with 100% efficiency.

The hydrophobic membrane filter material is pleated together with a layer of material (for example, nonwoven polypropylene) on the upstream and downstream side of the membrane. These layers provide mechanical support to the membrane and proper drainage of the fluid. The pleated membrane pack is formed into a cylinder. The longitudinal side seal of the pleated membrane filter pack should be an integral, homogeneous melt seal without any additional or extraneous materials, glues, or resins.

A rigid, perforated inner core is present to provide support against operating pressure. An outer cage placed on the upstream side of the membrane filter pack is provided for additional support and protection during handling. The cage provides retention of structural integrity against accidental reverse pressure (usually up to about 50 psi). Polypropylene and PTFE are examples of materials that can be used for the cage, core, and support material in a membrane filter.

End caps are attached by melt sealing to imbed the medium in the plastic. All hardware should be specified as produced from virgin materials.

Membrane filters are available in a variety of shapes and sizes. The most typical configuration for pharmaceutical air applications are 10-in. elements. These elements can be flame welded together end to end to form multilength configurations (typically up to 40 in.). Multiple elements can be used in a filter housing, as required. A membrane filter is illustrated in Figure 3.

Membrane filter elements may be integrity tested to ensure that the air is sterile. The integrity tests have been correlated with bacteria removal efficiency by direct microbial challenge testing.

Sterilization of Membrane Filters. Hydrophobic membrane process filter cartridges can be subjected to multiple sterilization cycles and must be designed to be repeatedly steam sterilized in either direction of flow or repeatedly autoclaved. The vendor should provide information on filter cartridge sterilization procedures and operating data on process limitations (time and temperature). These filters are typically capable of withstanding multiple in situ

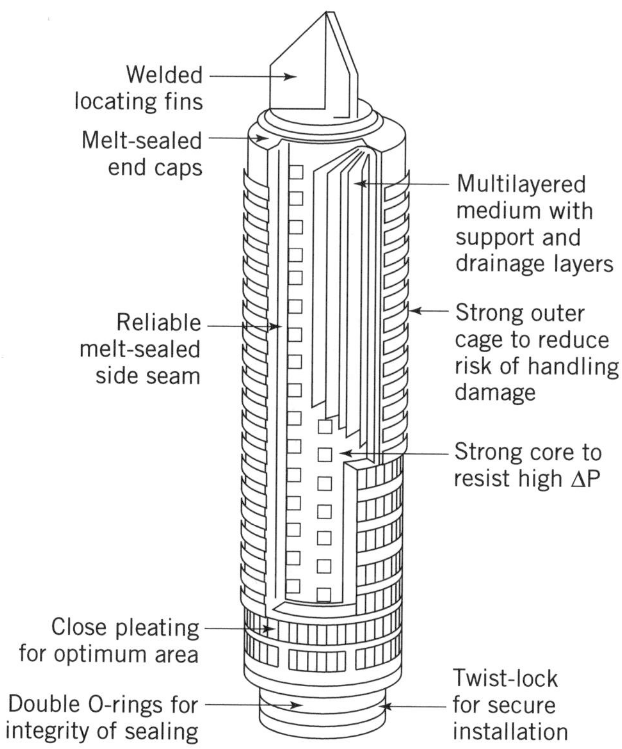

Figure 3. Membrane filter.

steam sterilizations (for example, at least 165 cumulative hours at up to 142 °C). If needed, the membrane filter can be autoclaved and aseptically installed into the process application.

Hydrophobic membrane filters can be in situ steam sterilized in 30 min of exposure time (or longer if the process requires a longer time) at a temperature of 125 °C. Because the filters are hydrophobic, no additional drying time is required. It is necessary for membrane filters as well as for packed towers to drain the entrained condensate on the inlet side of the filter. A membrane filter housing can be isolated under pressure at the end of an aeration period and therefore does not require sterilization for each cycle.

Prefilters. For some applications, it is necessary to use a prefilter to remove particulate material or liquid aerosols. It is often desirable to use a prefilter to protect the sterilizing grade filter. Typically, the use of a prefilter will reduce the overall filtration economics.

Particulate Removal. Membrane filters composed of materials such as polypropylene or cellulose can effectively be used to remove particulate material from an air stream. The filter micron ratings range from the order of 1 to the order of 100 μ. The appropriate filter can be selected for the particulate contaminant removal. The following are descriptions of examples of membrane filters that can be used as prefilters in pharmaceutical air filtration, including porous stainless steel filters, cellulose pleated filters, and polypropylene pleated filters.

1. *Porous stainless steel filters cartridges.* Porous stainless steel medium is made by sintering small particles of stainless steel or other high alloy powder together to form a porous metal medium. Porous stainless steel can be formed as a flat sheet or, when used in filter elements, as a seamless cylinder. This special manufacturing process produces a high dirt-capacity medium that is temperature and corrosion resistant. The recommended alloy is type 316LB, which has a higher silicon content than type 316L and provides a stronger, more ductile product with better flow properties. Elements have absolute ratings of 0.4 to 11 μ in gas service applications. Porous stainless steel filter cartridges are chemically or mechanically cleanable, offering economy of reuse. Porous stainless steel filters are used for steam filtration and for air sparging.
2. *Cellulose pleated filter cartridges.* Pleated cellulose filter cartridges (8 μ rating) are applicable as prefilters for inlet air for fermentors and bioreactors. These filter cartridges can be constructed of pure cellulose medium, without resin binders. The cellulose membrane can be pleated into a high area cylinder and has a longitudinal side seal with an appropriate polypropylene (approved for food contact usage by the FDA). Cellulose media cartridges are assembled with hardware components consisting of a perforated inner support core, an outer support cage, and end caps melt sealed to imbed the medium in the plastic. All hardware components should be of pure

polypropylene, without filler or reinforcement, to ensure a minimum of soluble extractables.

3. *Polypropylene pleated filter cartridges.* Polypropylene pleated depth filters (8-μ rating) are applicable as prefilters for prefiltration of exhaust gases. These process filter cartridges are constructed using nonmigrating continuous strands of nonwoven polypropylene filaments. The medium should have a constant pore size section (downstream) for absolute rated filtration and a continuously graded upstream section for effective prefiltration. The thin sheet of polypropylene media is pleated and formed into a cylinder with a longitudinal side seal of melt seal polypropylene. The cylinder is then melt sealed to injection molded polypropylene end caps to ensure no fluid bypass. Polypropylene hardware components consisting of an inner support core and an external protective outer cage are incorporated.

Liquid Aerosol Removal. A coalescer can be used for the removal of liquid aerosols containing water or oil droplets. This is desirable as a prefiltration for a sterilizing air filter, because the liquid aerosol could prevent the flow of air through the sterilizing grade filter.

Coalesers operate efficiently if they are able to separate the liquid and the gas in the liquid aerosol. The three basic steps that are required are (*1*) aerosol capture (*2*) unloading or draining of the liquid, and (*3*) separation of the liquid and gas.

Figure 4 is an illustration of a liquid–gas coalescer. The coalescer has a gravity separator, which allows for the removal of large liquid aerosols (typically >300 μm). The coalescer flow direction is in to out to prevent reentrainment. The liquid is captured through the coalescence of fine aerosols (0.1 to 300 μm) to large droplets (1 to 2 mm). The coalescence of the droplets is illustrated in Figure 5. The large droplets flow downward from a drainage layer. The separation is completed when the liquid is drained from the system. The air leaves the system from the top of the assembly.

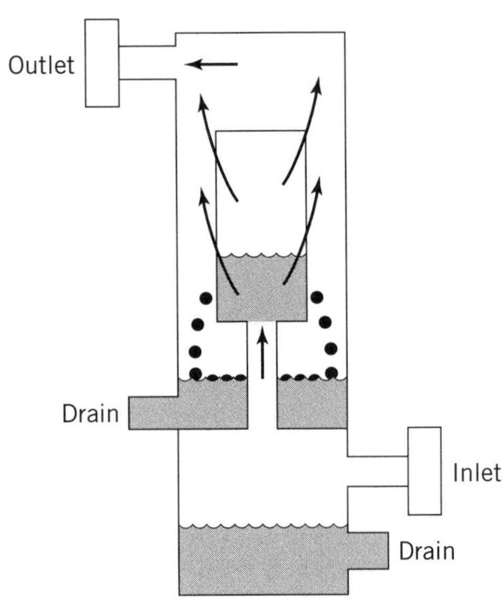

Figure 4. Liquid–gas coalescer.

Filter Housings. Cartridge filters used for air filtration are placed in sanitary, air or gas service filter housings. Cartridge filter housings usually consists of a head and a bowl. The filter is attached to the head, and the bowl is clamped to the head to provide a complete enclosure of the filter.

The size of the housing should be adequate for the flow and differential pressure requirements. Filter housings for bioprocessing applications are typically constructed of stainless steel (e.g., 304, 316, 316L, etc.) or carbon steel, with 316-series stainless steel internal hardware and cartridge seating surfaces. Housings typically have quick-release mechanisms such as V-band clamps or fast-action swing bolts to facilitate filter change-outs.

Design operating pressure of all filter housings should be specified as minimum psig and rated for full vacuum service. Design maximum operating temperature of the housing should also be specified. The housings or pressure vessels that are within the scope of the ASME Boiler and Pressure Vessel Code, Section VIII, Division 1, should be designed and U stamped per the code. TIG weld construction should be used in sanitary-style housings to minimize weld porosity and ensure high-quality, clean joints, with all internal welds ground smooth and flush. All weld procedures and welders should be qualified to ASME/BPVC Section IX.

Housings should be capable of in situ steam sterilization in accordance with the manufacturer's recommended procedures, and housing or system design should provide for condensate drainage. Gasket material and O-ring elastomers must also be capable of withstanding repeated steam sterilization cycles, along with being compatible with process fluids.

Industrial style housings (used for prefilters) provide cartridge mounting on a tie rod and sealing to the tie rod assembly by use of a seal nut at the top of the assembly. Tube sheet adapters should be seal welded to the tube sheet to prevent fluid bypass. Filter cartridges are thereby sealed in the housing independent of any cover assembly, ensuring positive sealing and no fluid bypass. Filter cartridges should be seated on the tube sheet adaptor assemblies above the tube sheet to ensure complete drainage of nonfiltered fluid before cartridge replacement. This prevents potential contamination of downstream surfaces during change-out of filter elements.

QUALIFICATION TESTS FOR MEMBRANE FILTERS USED FOR AIR FILTRATION

Sterilizing Grade Membrane Filters

Organism Retention Tests. Microorganism retention tests can be conducted to verify that membrane filters produce sterile air (1,2). Liquid challenge tests with *Brevundimonas (Pseudomonas) diminuta* (ATCC 19146), measuring 0.3 × 0.6 to 0.8 μm, is a standard challenge test for the validation of sterilizing grade filters (0.2 μm) in the pharmaceutical industry. Aerosol challenge tests with *P.*

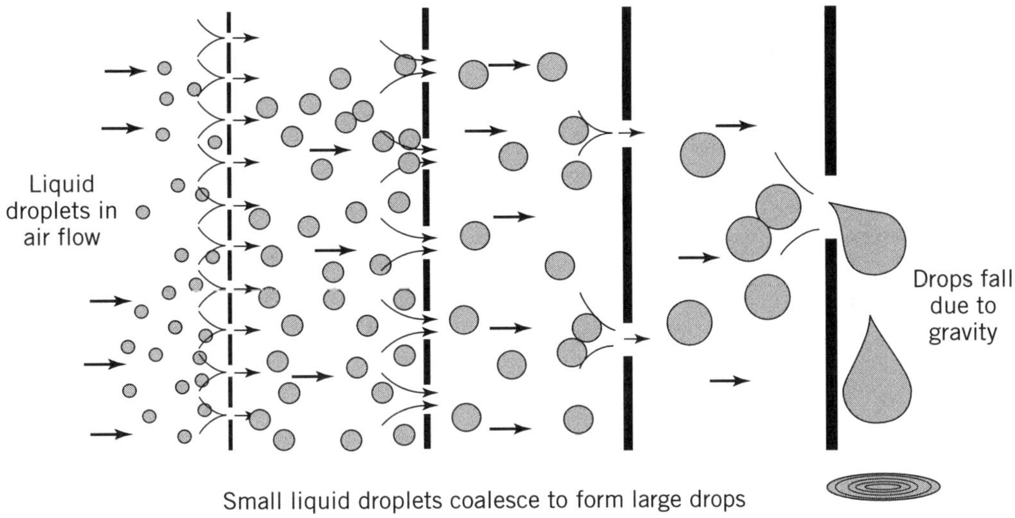

Figure 5. Coalescence of droplets.

diminuta should approximate extreme air flow conditions. Aerosol challenges with T_1 bacteriophage ($0.05 \times 0.1\,\mu m$), for example, can provide a test of a filter's retention efficiency of extremely small organisms, so small that the bacteriophage in a liquid suspension will penetrate a 0.2-μm sterilizing grade filter.

The retention efficiency of a given filter is less when a liquid challenge is used instead of an aerosol challenge. Thus, a liquid challenge test is a more stringent test of a filter's retention capability. A liquid challenge test can also provide retention information for process conditions such as extreme moisture after sterilization or air entrained with water drops. An example of the technique used to perform a liquid bacterial challenge on sterilizing grade membrane cartridges was published by Pall Corporation (9).

The liquid test involves challenging a test filter with a known quantity of *P. diminuta*, no less than 1×10^7 organisms per 2 square centimeter of filter area. The challenge sample is passed through the filter suspended in sterile water at a defined flow rate and time. All of the effluent from the test filter is passed through an analysis membrane. After the challenge is completed, the system is flushed for 5 to 10 minutes with sterile water. The analysis membranes are removed and placed on Mueller Hinton Agar at 32 °C for 48 h. After incubation, the plate is examined for the presence or absence of microbial colonies. Figure 6 illustrates a liquid bacterial challenge test stand.

The aerosol challenge test system can consist of a nebulizer loaded with the challenge microorganism suspension, a separate line for dry air makeup, and split-stream impingers to sample the aerosol challenge with and without the test filter. A schematic of a test stand that can be used for the aerosol challenge procedure described is given in Figure 7. Detailed descriptions of the aerosol test protocols for *P. diminuta* and T_1 bacteriophage Ref. 9.

During the aerosol challenge, an aerosol is generated with a nebulizer. The aerosol is introduced into the test filter at a given flow rate. The filter effluent is collected in dual liquid impingers. Controls are performed simultaneously via a split stream by using a two-channel timer to direct air flow, on an alternating basis, from the test side filter impingers to the unfiltered control side impingers for recovery.

The impingers contain sterile buffer and after the challenge is completed, the buffer can be analyzed for the test organism. If *P. diminuta* is the test organism, then the buffer is analyzed by putting the buffer solutions through an analysis membrane and placing the membrane on Mueller Hinton Agar for 48 h before titering. If T_1 bacteriophage is the test organism, then samples of the buffer are diluted with nutrient broth and mixed with liquid nutrient agar (1.5% agar concentration; 48 °C) and *Escherichia coli* in the log phase of growth. After vortex mixing all three components, the mix is poured over nutrient agar plates and incubated for 3 to 5 h at 37 °C, so that the plaques can be counted.

Integrity Tests for Membrane Filter Cartridges. During the sterilization of fermenter air, it is necessary to achieve the highest possible assurance of filter integrity and removal efficiency. The installation of integrity-testable filters and the performance of routine integrity testing by the user are essential to demonstrate that the system is performing to specification. Tests that qualify the retention characteristics of a membrane filter can be defined as destructive or nondestructive tests.

Destructive tests are performed using an appropriate contaminant to meet a specific claim for retention of the contaminant. The test procedure must be sensitive enough to detect the passage of contaminants of interest. For sterilizing grade 0.2-μm membrane filters, the industry standard test organism (i.e., contaminant) is *P. diminuta* (ATCC 19146). The organism and minimum challenge level (10^7 CFU/cm^2 filter area) are specified in the ASTM standard F383,-83 (10) and referenced in the FDA guideline on sterile drug process produced by aseptic processing (11).

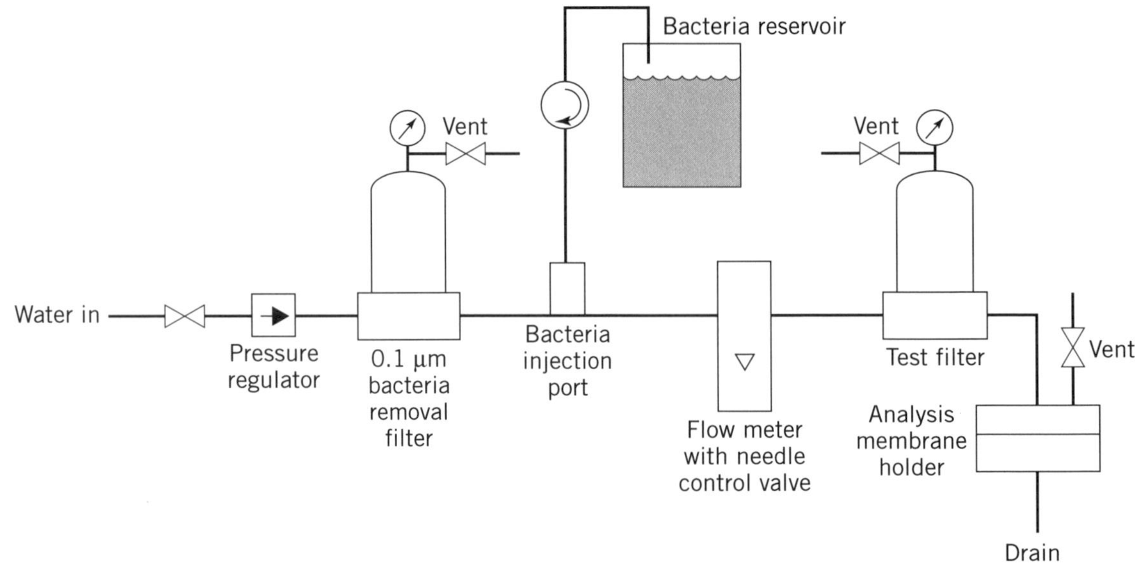

Figure 6. Liquid bacterial challenge test stand.

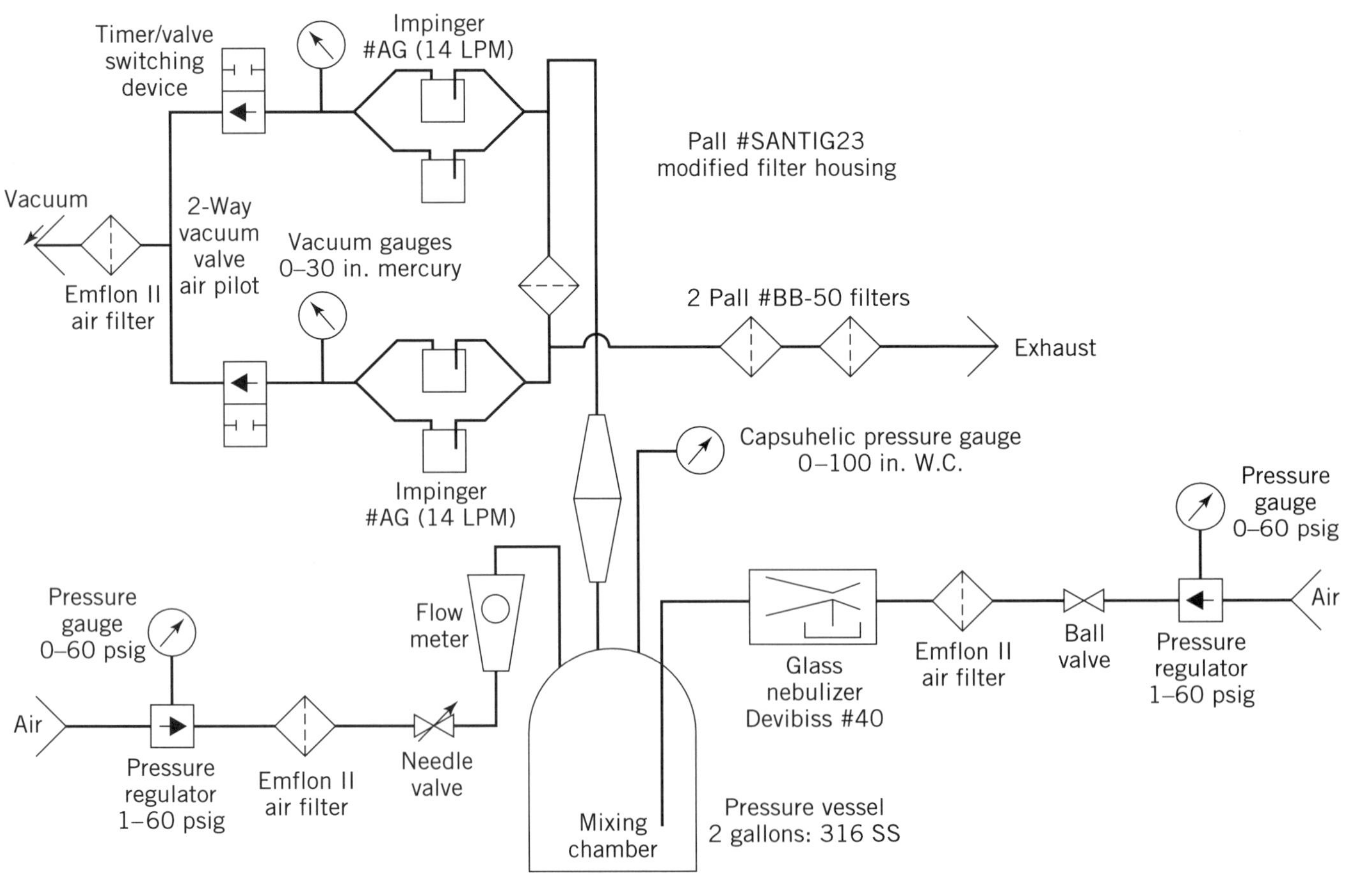

Figure 7. Aerosol challenge test stand.

Because most filter users would not want to perform a destructive test in a process environment and because the filter is not useable after the destructive test, nondestructive tests related to the retention results of the destructive test are used instead. From the relationship developed between a nondestructive and a destructive test, membrane filter performance can be safely and conveniently verified in the production environment. The relationship between a nondestructive integrity test and the assurance of bacterial retention constitutes a filter validation study and is extremely important for microbial retentive filters used in critical fluid processes.

There are four nondestructive integrity tests that can be used for sterilizing grade hydrophobic membrane filters used for air filtration: (*1*) bubble point test, (*2*) forward flow test, (*3*) pressure hold test, and (*4*) water intrusion test. These tests are described in FILTRATION, CARTRIDGE.

For a nondestructive test to be useful, the results of the test must be able to predict the ability of the filter to remove bacterial. The manufacturing quality control of the membrane filters tested to obtain this prediction of bacterial removal for the filters must be used to establish quality control procedures that are consistently maintained in the manufacturing of the filter. The combination of the nondestructive integrity test, the correlated bacterial challenge, and manufacturing quality control allows membrane filters to be used reliably for sterile filtration.

Tests for Prefilters

Many filter manufacturers use a nominal micron rating for removal efficiency. A nominal rating is an arbitrary micron rating assigned by the filter manufacturer. Such ratings are subject to a lack of reproducibility.

An alternative method for rating filters is the Oklahoma State University (OSU) F-2 Test. This rating method (ISO 4572, ANSI B93.31) has received wide acceptance for use on lubricating and hydraulic fluids. Pall Corporation, for example, uses this method for oils extensively and has adapted it for use in water with contaminants ranging from 0.5 to 25 μm.

The test is based on continuous on-line particle counts of different particle sizes, both in the influent and the effluent. The β ratio at a specific particle size is defined as β_x: the number of particles of a given size (X) and larger in the influent, divided by the number of particles of the same size (X) and larger in the effluent, where X is the particle size in microns. The percent removal efficiency can be calculated from the β-value. The percent removal efficiency is $[(\beta_x - 1)/\beta_x)]$ 100 (12).

As an example, Pall prefilters are given a micron rating that corresponds to a 100% removal efficiency or the value in microns at which the OSU F-2 Test gives a β-value of more than 5000.

AIR FILTRATION APPLICATIONS

Fermentation and downstream processing are two major operation categories in bioprocessing. During both fermentation and downstream processing, sterilizing-grade 0.2-μm hydrophobic membrane filters can be required for processing air (or gas) streams.

Sterilizing-grade 0.2-μm hydrophobic membrane filters are used during fermentation for the sterilization of fermentation inlet air and for the filtration of fermentation exhaust gas. During downstream processing, sterilizing grade filters can be found in use as sterile tank vents. In a final purification, membrane filters can be used for vacuum breaking in processes such as lyophilization.

As discussed above, hydrophobic membrane filters are desired in sterile gas filtration applications because hydrophobic filters do not spontaneously wet with water.

Sterile air filtration applications discussed in this section include fermenter inlet air, exhaust gas (vent), high temperature air filtration, and lyophilizer vacuum break filtration. Special attention will be given to the specific requirements for a sterilizing filter when used for these applications. After the discussion of the applications is a general discussion of recommendations for sterilizing-grade filter usage.

Fermentation

During a fermentation process a specific cell (yeast, bacteria, or mammalian) is grown to provide a desired product (1–3). Products can include cells, antibiotics, amino acids, or recombinant proteins. There can be a variety of sizes for the fermenter, ranging from very small (100 L or less) cell culture reactor to very large scale antibiotic production (100,000 L). In these applications, there is often the need to maintain sterility in both liquid and gas (air or nitrogen) feeds to support growth of the desired cells.

Air filtration applications for fermentation are illustrated in Figure 8. Those applications specific to fermentation are described here, and filtration of utilities used in fermentation such as steam, air, and water are discussed in the "Utilities" section.

Prefiltration of Fermentation Air. Compressors are often used to generate air flow for the manufacturing facility. There are two types of compressors, oil free and oil lubricated. In older facilities where oil-lubricated air compressors are commonly used, prefiltration of inlet air is necessary for removal of oil droplets. A coalescing filter can provide greater than 99.9% removal of oil and water droplets in the 0.01- to 0.5-μm range and larger. This also acts as an excellent prefilter for the hydrophobic membrane pleated filters that are commonly used for sterilizing the inlet air to the fermenter. The typical gas flow rate per 10-in. filter module is 200 to 400 standard cubic feet per minute (SCFM).

For oil-free compressors, a prefilter acts to remove dirt in the air system, extending the service life of the final filter. For use with fermentation air, a cellulose pleated filter with an absolute rating of 8.0 μm is normally the filter of choice. Alternately, polypropylene (2.5-μm rated) pleated filters also serve as excellent prefilters for this application. The typical gas flow rate per 10″ filter module is 75 SCFM.

Sterile Air Filtration for Pilot and Production Fermenters and Feed Tanks. One of the largest applications for sterile

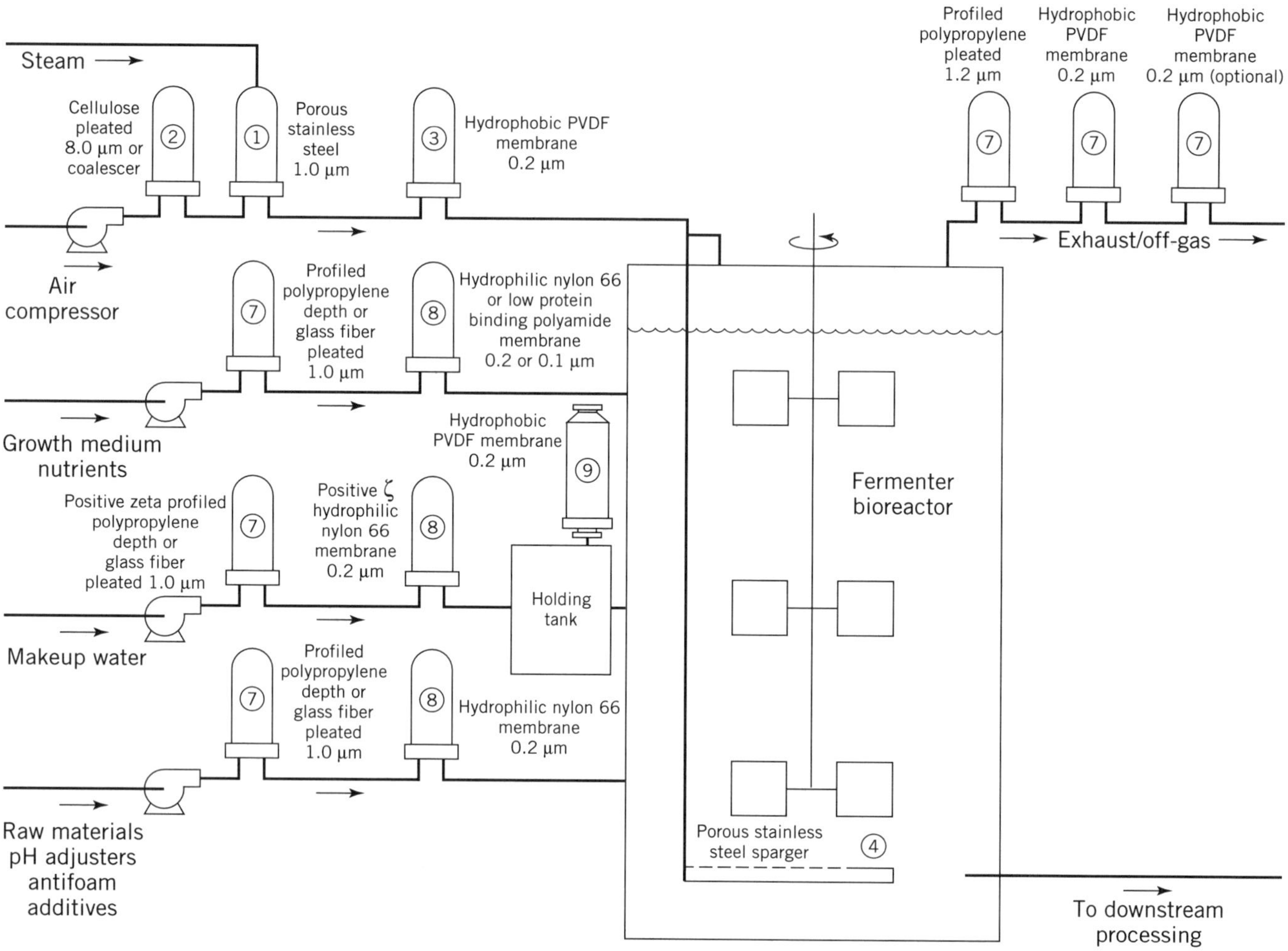

Figure 8. Fermentation air applications.

air filtration is the sterile air used for an aerobic fermenter during a typical production cycle. Typically, 1 vol of air per volume of broth per minute is used. Thus, for a 100,000 L fermenter on line for 48 h a total of 1.01×10^9 cubic feet of air requires sterilization.

The contaminants present in compressed air can include dust, lubricating oil, hydrocarbons, water, rust, and microorganisms including molds, bacteria, and viruses. Microorganisms in air are often associated with carrier particles, such as dust. Water and oil can be present as bulk liquid, vapor, or an aerosol. The air distribution system can give rise to contaminants such as rust and water. The concentration and size distribution of particles in compressed air are variable. The size range is generally between 0.001 and 30 μ, with a concentration between 10^{-2} to 10^{-4} g/m^3 (4,6).

Bacteria and bacteriophage, when present in air feeds, can enter fermentation tanks or bioreactors and contaminate the product. Bacteriophage or other viruses can destroy the producing cells and reduce yields.

The process requirements to supply this sterile air can be quite restrictive. The air sterilization process must (*1*) process a large volume of compressed air, (*2*) provide a high degree of reliability, and (*3*) operate economically. Several methods have been considered for the sterilization of fermenter inlet air. These include filtration, heat, irradiation, washing with sterilizing chemicals, and electrostatic precipitation. Washing and electrostatic precipitation are not effective for the removal of microorganisms. Heat and irradiation are not economical. Filtration is the only technique that meets all the requirements for sterilizing fermenter inlet air.

Packed towers, an early filtration approach, were used widely in the industry. Since the early 1980s, filtration technology has advanced (13), and there has been an ongoing trend since the early 1980s to replace depth filters with hydrophobic membrane cartridge filters (e.g., Hoffman-La Roche [14] and E. R. Squibb and Sons [8]).

The recommended filters for sterilization of air feeds to fermenters and bioreactors are the hydrophobic membrane pleated filters. The hydrophobic (water repelling) nature of these membranes can provide for bacteria and bacteriophage removal with 100% efficiency under moist or dry operating conditions. This is an important benefit over fiberglass towers and cartridges. Filters for sterile air feeds should have a 0.2-μm absolute bacterial rating in liquids and a 0.01-μm particulate rating in air service. The typical gas flow rate per 10-in. filter module is 75 to 100 SCFM.

For some fermentations, the requirement may be for the filtration of fermenter air at an elevated temperature. If an application involves hot air and a longer service life is desired, then a filter that can withstand the elevated temperature is required. High Temperature Emflon® filters manufactured by Pall Corporation can be used in continuous service at a temperature up to 120 °C. These filters have a 0.2-μm microbial rating in liquid service and a particulate removal rating of 0.01-μm in gas service. The filter membrane is made of inherently hydrophobic PTFE, and the cage, core, and end caps are specifically designed for high temperature applications.

Sparging. Sparging acts to disperse air evenly in the fermenter or bioreactor containing the growth media and product. The product of choice for this application is a porous stainless steel sparging element, which can provide an exceptionally uniform and fine aeration gas dispersion. These elements are fabricated with one face of porous metal and one face of solid metal. If both surfaces of the sparging elements were porous, bubbles from the under surface may coalesce with bubbles from the top surface.

Porous stainless steel sparging elements should be positioned horizontally in the fermentation tank, with the porous stainless steel facing upward. Fine grades of porous stainless steel (e.g., 3.0 μm absolute liquid rated) are ideal even for sheer-sensitive mammalian cell cultures because of their high gas transfer and low shear aeration capability. Elements are typically available in standard and custom designs.

Exhaust Air–Off Gas Filtration. The purpose of a vent filter on a sterile fermentation tank is twofold: to prevent contamination of the tank and to provide containment of the material inside the tank. Prevention of contamination in the tank is desirable for processes that involve long fermentation cycles or require a sensitive fermentation medium (e.g., tissue culture medium). Genetic engineering techniques as well as fermentation of pathogenic organisms (such as organisms used for the manufacture of vaccines) have made it necessary to protect the environment and prevent the escape of microorganisms from the fermentation tank. The exhaust filtration system for a recombinant or mammalian cell fermenter or bioreactor must yield sterile air to the environment and provide a sterile barrier to prevent ingression of contaminants. Additionally, it must be in situ steam sterilizable and typically has a clean differential pressure less than 1 psid.

The removal efficiencies for simple depth filters (as described earlier) are typically poor under wet or variable flow conditions. Therefore, membrane filters are recommended for vent filtration applications.

The fermenter or bioreactor exhaust gas line can be contaminated with microorganisms or cells, growth medium components expelled from the fermenter or bioreactor as droplets or as solid particles, and aerosol condensate droplets formed during cooling of the gas in the exhaust system. These aerosol droplets, when present, can potentially block the final filter and must be removed before reaching the final filter. Mechanical separation devices, such as cyclones, condensers, and demisters, may not achieve effective aerosol removal below 5 μm (15). Removal efficiencies and pressure drops also vary significantly with flow rate in such equipment. Recent studies have shown that aerosols in exhaust lines are predominantly in the very fine 1- to 5-μm range (16).

The contaminants present will depend on the fermentation conditions, the growth medium, and the design of the exhaust gas system. The basic requirements for a vent filter are ability to provide sterility, a low pressure drop, and in situ steam sterilizable.

The recommended exhaust filtration system design entails two stages using a polypropylene pleated depth-filter cartridge as a prefilter to a 0.2-μm absolute rated hydrophobic membrane pleated filter cartridge. The purpose of the prefilter (typically 1.2-μm absolute rated) is to remove aerosolized particles and liquid droplets containing cells or growth media from the fermentation off-gas or exhaust air. This serves to extend the service life of the final sterilizing filter. If the medium contains only fully dissolved components, such as with a sterile filtered cell culture medium, and if the fermentation is run at low temperatures (<30 °C) and low aeration rate (1 to 1.5 volume of air per volume of media per minute [VVM]), the prefilter may be optional. The typical gas flow rate per 10-in. filter module is 40 SCFM.

Like the final sterilizing filter, the pleated polypropylene prefilter should be multiple steam sterilizable. As additional benefit of the prefilter is to retard foam-outs from reaching the final sterilizing filter.

Sterilizing-grade 0.2-μm absolute rated hydrophobic membrane pleated final filters, with PVDF or PTFE membranes, can prevent organisms from entering or leaving the controlled reaction zone, even in the presence of water droplets and saturated gas (8,17). Steam sterilizability and integrity test values correlated to microbial retention studies under worst-case liquid challenge conditions provide the highest degree of assurance performance. Redundant systems using a second 0.2-μm rated sterilizing filter in series are recommended for high-risk recombinant organisms.

Condensate control is usually the most critical consideration for this application. In cases in which there is condensate accumulation and if the fermenter is operated with overpressure in the fermenter head, the amount of condensate accumulation can be reduced if a pressure control valve is placed at the fermenter exit, upstream of the exhaust gas filter. An alternative technique for the prevention of condensate accumulation is to use a heating section in the exhaust gas pipe upstream of the filter installation. This can be also be accomplished by specifying steam jacketing on exhaust filter housings. In this case, the exhaust gas temperature at the terminal filter must lie above the temperature of the exhaust gas at the fermenter exit. The heater must be properly sized based on the process parameters.

Downstream Processing

Starting with the cells and conditioned broth medium from the fermenter or bioreactor, the objective of downstream processing can be to produce a highly purified, biologically active protein product, free of contaminants such as endotoxins, bacteria, particles or other biologically active molecules. This phase of bioprocessing typically comprises a series of unit operations including cell and cell debris separation, fluid clarification and polishing, concentration and purification, and membrane filtration sterilization of the purified product.

Cartridge filters are used in many stages of downstream processing, which involves filtration of both the harvest fluid and product intermediates as well as filtration of air and gases required throughout the process. Air filtration applications include vacuum break filters for lyophilizers, sterile nitrogen blankets, tank vents, and sterile air for container cleaning.

Absolute rated cartridge filters eliminate contaminants and impurities from air, nitrogen, and other gases used in downstream processing to prevent contamination of product and further protect concentration and purification equipment. Vent filtration ensures containment and freedom from product contamination during fluid transfer operations and protects processing equipment during sterilization cycles.

In fermentation, cartridge filters are used to maintain the sterility of the makeup water, feeds, additives, media in holding tanks, and in fermenter or bioreactor exhaust. Cartridge filters are typically used in downstream processing for the filtration of air, gases, and venting applications when it is necessary to vent tanks during fluid transfers; pressurize tanks using inert gases such as nitrogen and argon; protect vacuum lines, sterile vent holding tanks, and lyophilizers; and for gas purging, blanketing, drying, and when sterilizing equipment by in situ steaming or autoclaving.

The recommended filters for nonsterile particulate removal applications are polypropylene pleated filters. Hydrophobic membrane pleated filters such as PVDF or PTFE are recommended for aseptic processing. The absolute removal rating for the latter filters should be 0.2 μm determined under liquid flow conditions. The typical gas flow rate per 10-in. filter module is 75 to 100 SCFM.

In downstream processing applications, Cell and cell debris separation and clarification processes can be broken into a primary separation, secondary separation, and a cell concentrate section. During some primary separations, a cyclone can be used for particulate removal; a sterilizing air filter can be used as a vent on the cyclone. A variety of holding and receiving tanks can be used during the separation and clarification process; these tanks can be fitted with sterile vent filters.

During secondary separation, a nitrogen blanket may be needed; the nitrogen gas can be sterile filtered with a hydrophobic filter. Downstream processing can also involve the concentration and purification of clarified harvest fluid. Applications for air filtration include sterilizing-grade vent filters for solvent or buffer tanks and for holding or buffer tanks needed for ultrafiltration and chromatography.

The final pharmaceutical product will often need to be packaged. During filling processes, a sterile nitrogen blanket and thus a sterilizing grade hydrophobic filter may be needed. The final product is can be placed into a container; sterile air or nitrogen may be needed for container cleaning. A vent filter can be required on holding tanks. The typical gas flow rate per 10-in. filter module for tank vent applications is 75 to 100 SCFM.

Several specific applications for final processing are addressed in the next section.

Vacuum Break Filters. Sterilizing grade filters are used in freeze dryer installations to filter the gases used to maintain the chamber pressure and to break vacuum during operation and in sterilizers for vacuum break purposes.

Blow-Fill-Seal Equipment. Blow-fill-seal equipment can be used for the aseptic filling of pharmaceutical products. The container is formed and sealed aseptically. Air filtration is required to ensure sterility in this unit operation. A typical arrangement is illustrated in Figure 9.

For buffer-tank air hydrophobic membrane filters are used to supply sterile air to a buffer tank on the blow-fill-seal machine. This blanket air is used to drive the sterile solution through a pneumatically controlled dosing system. The air used in the buffer tank is referred to as the gas cushion or buffer tank air.

Hydrophobic membrane filters provide sterile air used to form the hot moldable plastic tube (parison). The air used to form the parison is known as the parison support air. The parison is subsequently blow molded into the shape of a ampule strip or a bottle. Typical requirements for the filters used include that the filters be steam sterilizable, integrity testable, and the proper size to prevent restriction of gas flow. Hydrophobic membranes are used to prevent wetting out and to maintain high flow rates even in moist conditions.

Utilities

There are a number of peripheral unit operations required during a sterile process. The air filtration applications for these applications are described in this section.

Water-for-Injection Tank Vent. Sterilizing grade membrane filters can be used in vent applications in which the fluid in the tank is at an elevated temperature. One such application is the vent used to prevent contamination in a water-for-injection tank. The water is at 80 °C or higher. When a sterilizing-grade air filter is used for this type of vent service, a steam-jacketed housing is typically used. It is only necessary to maintain the temperature of the filter cartridge at a temperature *slightly above* the dew point of the vapor. The steam introduced into the jacket should be at ambient pressure. Continuous operation of the jacket at a significantly higher steam pressure and temperature can reduce the service life of the filter caused by accelerated aging of the hardware by oxidation.

Steam Filtration. Process equipment and final filters are frequently sterilized by direct steam flow in situ, during the normal line sterilizing cycle. This eliminates the need for making aseptic connections and risking recontamination. Filtered steam is required for this sterilizing-in-place

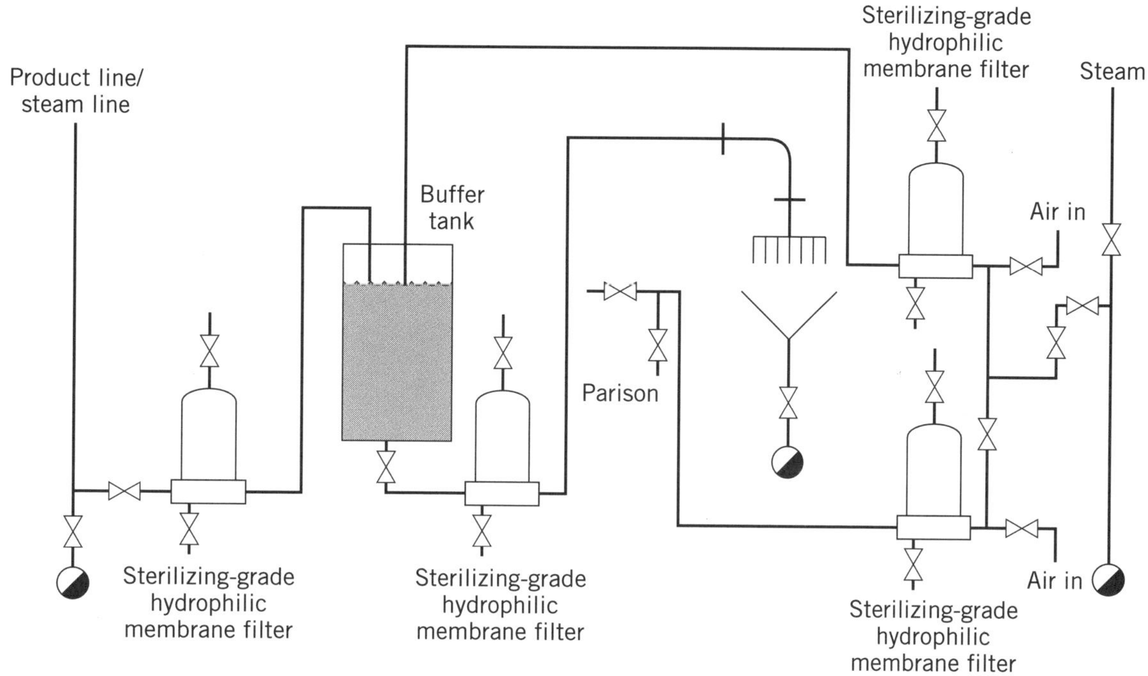

Figure 9. Air filtration applications in a Blow-Fill-Seal operation.

of filters, piping, vessels, and filling equipment. Steam is also required for general equipment cleaning and sterilizing. The steam often contains significant amounts of pipe scale and other corrosion products. This particulate material should be removed in the interest of overall cleanliness and to avoid burdening the prefilters and final filters.

Particulate contamination in process steam is efficiently retained by porous stainless-steel filters with an absolute gas rating of 1.2 μm. Porous stainless-steel filter assemblies are typically sized at steam flow rates of 30 to 40 ACFM per square foot of filter medium.

GENERAL CONSIDERATIONS FOR OPERATION

Integrity Test Considerations and Guidelines

Field experience with the integrity testing of sterilizing grade filters shows that various combinations of integrity test procedures are in use and that different testing schedules, both pre- and post-use, are followed. Perhaps the best approach is to perform the test or combination of tests that provides the highest degree of accuracy commensurate with the economics and the practicalities of the process.

Since the requirements for the use and testing of sterile gas filters can depend on location and application, several different types of filter tests can be used. Filter users typically use the forward flow or pressure decay tests, which require the use of a low surface-tension solvent, such as 60:40 isopropyl alcohol/water or 25:75 *t*-butyl alcohol/water. This is a well-proved, widely accepted approach directly correlatable to bacterial challenge (9).

Air filters are typically integrity tested when they are installed and retested periodically based on service conditions and operating requirements (e.g., once a month). Change-out schedules based on filter life studies have also been used in conjunction with integrity testing. Because filters in these applications are often sterilized in situ and can be damaged during in situ steam sterilization by reverse pressurization if the sterilization procedure is not properly controlled, filter life study data or a periodic integrity test regimen can be used only if the filter sterilization procedure is validated and in control.

In certain cases (such as in facilities in which the use of a nonflammable fluid is required, or when disposal of organic solvents is a concern), some filter users in aseptic processes have recently considered the use of water-based tests for hydrophobic filters.

There are a number of integrity tests possible for a hydrophobic air filter. The selection of the appropriate test and the appropriate test schedule depends on the specific application.

After an integrity test has been completed, it is typically desirable to remove the wetting fluid from the filter. This can be accomplished by blowing clean, dry (-40 °C dew point) air or nitrogen through the filter. It is necessary to qualify this procedure, because every system is different.

Filter Service Life

Filter cartridge change-out is usually based on actual experience, with a safety factor. Filters should be inspected on a monthly basis for oxidation. This should be supplemented by monitoring the pressure drop across the filters during operation to determine if the filters are plugging and routine integrity testing to confirm filter integrity during the service life of the filter. Alternatively, filter life studies, with an appropriate safety factor, could also be used to set a change-out schedule. Actual conditions for each application should be used during filter life studies.

For elevated temperature applications, conditions leading to oxidation need to be considered. Aging of the membrane filters by oxidation depends on the status of the system. Oxidation does not occur when the cartridge is being steamed, because there should be no air present in a properly operating steam-in-place system. If a cartridge is exposed to air at an elevated temperature, oxidation of the material in the filter, such as polypropylene hardware, will be accelerated. Oxidation will also occur when the filter is in a stagnant situation, that is, it has no air flow going through it. The flow of air through a filter can moderate the temperature environment, whereas under stagnant conditions the temperature of the filter will rise to the temperature of the housing. Stagnant conditions can exist when the tank is not being used or when the tank is empty. To prolong service life, the steam jacket should be turned off when there is no air flow through the filter for extended periods of time, when operating conditions permit.

Steam Sterilization Guidelines for Sterilizing Grade Membrane Filters

Membrane filters can be sterilized by chemical sterilants (such as ethylene oxide, hydrogen peroxide in vapor form, propylene oxide, formaldehyde, and glutaraldehyde), radiant energy sterilization (such as γ-irradiation) or steam sterilization. The most common method of sterilization is steam sterilization.

Steam sterilization of a membrane filter can be accomplished either by an autoclave or by in situ steam sterilization. In situ steam sterilization can be effectively accomplished by a variety of different process arrangements. Steam sterilization is often the most critical portion of the process, and it is important that the procedures followed lead to sterilization of the system and do not impart any damage to the membrane filters. Some general procedural recommendations are included for prevention of either forward or reverse pressurization damage.

Reverse pressurization conditions can be prevented by using a noncondensing gas, such as air or nitrogen, at the end of a steam cycle. If this step is not followed, and the steam valve is shut off without the introduction of air or nitrogen, the filter housing will act like an isolated system. As such, the temperature will be different on the upstream (outside of the housing) and downstream sides of the filter. Due to the temperature difference, steam will condense at different rates on the upstream and downstream sides of the membrane. This can lead to pressure differences of 5 to 15 psi in the reverse direction. Even if this reverse pressure condition exists for a short period of time, the filter can suffer permanent damage.

Condensation during the steam cycle can lead to long cycle times and filter damage by excessive forward pressurization. If during steam sterilization the filter membrane is wet with condensate and in order to overcome the resistance of the wetted membrane the differential pressure in the forward direction is greater than required maximum (typically 5 psi), then the filter can be damaged by excessive forward pressurization. In this case, the filter will appear crushed. Precautions should be taken to minimize condensate accumulation. Drains for condensate removal should be strategically placed and should be cracked during the steam cycle.

CONCLUSION

This chapter has provided information on a variety of air filtration applications: fermenter inlet air, fermenter vent gas, vents on water-for-injection tanks, and vacuum break filters during lyophilization. In addition to a description of the air filtration applications, guidelines were provided on the usage of the filters in these applications.

BIBLIOGRAPHY

This is a modified version of the article first published as H. Haughney, "Air Filtration" in T. Meltzer and M. Jornitz eds. *Filtration in the Biopharmaceutical Industry*, Marcel Dekker, Inc., New York, 1998, Chapter 22. Reprinted by courtesy of Marcel Dekker, Inc.

1. H. Haughney, in T. Meltzer and M. Jornitz eds., *Filtration in the Biopharmaceutical Industry*, Marcel Dekker, Inc., New York, 1998.
2. H. Haughney, in K.E. Avis ed., *Sterile Pharmaceutical Products: Process Engineering Applications*, Interpharm Press, Buffalo Grove, Ill., 1995, pp. 305–355.
3. J.M. Martin, A.M. Trotter, P. Schubert, and H. Katz, in B.K. Lydersen, N.A. D'Elia, K.L. Nelson ed., *Bioprocess Engineering: Systems, Equipment, Facilities*, Wiley Interscience, New York, 1994, pp. 317–370.
4. N.B. Bruckshaw, *Filtr. Sep.* **296** (May/June 1973).
5. P.F. Stanbury and A. Whitaker, *Principles of Fermentation Technology*, Pergamon Press, Oxford, U.K., 1984.
6. J.W. Richards, *Introduction to Industrial Sterilization*, Academic Press, London, 1968.
7. N.A. Fuchs, in C.N. Davies ed., *The Mechanics of Aerosols*, Macmillan, New York, 1964.
8. C.F. Bruno and L.A. Szabo, *Biotechnol. Bioeng.* **25**, 1223 (1983).
9. Pall Corporation, *Validation Guide for Pall Emflon®II Membrane Cartridges for Air and Gas Filtration*, Pub. TR-870b, 1994.
10. American Society for Testing and Materials (ASTM), *Standard Test Method for Determining Bacterial Retention of Membrane Filters Utilized for Liquid Filtration*, ASTM Standard F838-83, 1983.
11. Food and Drug Administration, Center for Drugs and Biologics, Rockville, Md., *Guideline on Sterile Drug Products Produced by Aseptic Processing*, 1987.
12. T. Uberoi, *Chem. Eng.* 75 (March 1992).
13. R.S. Conway, *Biotechnol. Bioeng.* **26**, 844 (1984).
14. C.A. Perkowski, *Biotechnol. Bioeng.* **25**, 1215 (1983).
15. Porter, in C.H. Chilton and R.H. Perry, *Chemical Engineers Handbook*, 5th ed., vol. 26, McGraw-Hill, New York, vol. 26, 1973.
16. Pall Corporation, *Process Parameters on an Economic Use of Membrane Filters in Fermentor Exhaust Gas Filtration*, Scientific and Technical Report (STR) 15, Pall Corporation, East Hills, N.Y., 1992.

17. R.S. Conway, in C. Cooney ed., *Comprehensive Biotechnology*, Academic Press, New York, 1984.

See also STERILIZATION-IN-PLACE.

FILTRATION, CARTRIDGE

HOLLY HAUGHNEY
Pall Corporation
Long Island, New York

KEY WORDS

Cartridge
Filtration
Fluid
Membrane
Sterilization

OUTLINE

INTRODUCTION

Bioprocess technology operations have included membrane cartridge filtration for more than 20 years. Cartridge filters have been used for the removal of submicron particles, such as bacteria, and for the removal of particulate material, usually greater than 1 μ. Many applications involve the removal of bacteria in sterile filtration applications; particulate removal cartridge filters can be used as prefilters to protect the sterilizing grade filters. Some examples of applications for liquid filtration include fermentation media, deionized water, sera, and sanitizing chemicals. The purpose of this article is to provide information on operational considerations for cartridge filters and to explain the usage of cartridge filters in bioprocessing. (1–3)

DESCRIPTION OF MEMBRANE CARTRIDGE FILTRATION

Cartridge filters used in bioprocessing applications are used for the removal of bacteria (sterilizing grade filtration) or for the removal of particulate material (prefiltration). The mechanisms at work during filtration are provided, as well as, a general description of sterilizing grade filters and prefilters. Prefilters are often used upstream of the final (sterilizing) filters to protect the final filter from premature plugging, thereby prolonging the life of the final filter.

Filtration Mechanisms

There are four mechanisms that can govern the particle removal efficiency of membrane filters: (*1*) direct interception, (*2*) inertial impaction, (*3*) Brownian motion or diffusional interception, and (*4*) electrostatic attraction between the membrane and particles (4).

Direct interception involves a sieving action that mechanically retains the particles on the filter surface. The filter acts as a screen that stops particles that are larger than the pores (openings) in the membrane. Direct interception is independent of face velocity and mostly involves particles with relatively large diameters.

Inertial impaction refers to the deviation of a particle from its streamline during flow because of inertia, resulting in the retention of the particle by the membrane. As the face velocity increases, the probability of inertial impaction increases.

Brownian motion or diffusional interception applies to small particles at low face velocities in a gas stream. When air molecules are in a state of random motion, small particles suspended in the air can be struck by moving air molecules and displaced. The movement of particles resulting from molecular collisions is known as Brownian motion. This phenomena can increase the probability of capture of particles by diffusional interception within the membrane.

An electrostatic attraction can exist or be induced between the membrane material and particles in the fluid stream. This attraction can enhance the removal of particles and their retention by the filter membrane.

The pores (open area) in most membrane filter materials are typically not straight through the membrane. In membrane filters, the membrane can be considered to be constructed of multiple screens that provide a tortuous path for the fluid through the membrane.

In bioprocessing applications, there are many processes that require bacterial removal (sterile filtration). The filters used to provide bacterial removal are generally 0.2-μ sterilizing grade filters. (The tests used to ensure that the filters do remove the bacteria are described below)

Removal of bacteria by a 0.2-μ sterilizing grade filter is not only by a sieving action on the surface, but also by size-exclusion entrapment in the membrane structure. As shown in Figure 1, a membrane filter structure is sponge-like. The ability of such a membrane to remove bacterial contaminants is related to its thickness as well as to the size of its pores. For example, it has been demonstrated that whereas a single 150-μ thick layer of 0.8-μ cellulose ester membrane can provide only a titer reduction of 70 for *Brevundimonas (Pseudomonas) diminuta*, 10 layers can provide a titer reduction of 2.4×10^8 (5).

Because membrane filters are constructed of porous materials, they will have a certain amount of open area. The open area allows for fluid flow through the membrane; this area, expressed as a percent, is known as the porosity of the membrane.

Sterilizing Grade Membrane Filters

Construction. Membrane filters used for sterile liquid filtration are typically constructed of polymeric microporous hydrophilic materials, such as nylon, cellulose acetate, modified polyvinylidenefluoride (PVDF), or other polymers. Most liquids that require sterile filtration are water based, and hydrophilic membranes will spontaneously become wet with the fluid. Absolute rated hydrophilic membrane filter cartridges have been used for sterile filtration of parenterals, diagnostic reagents, purified water and water-for-injection, dry gases, organic solvents, buffers, and biological fluids such as serum, plasma, tissue culture media, nondilute protein solutions, and fermentation harvest fluids (1–3).

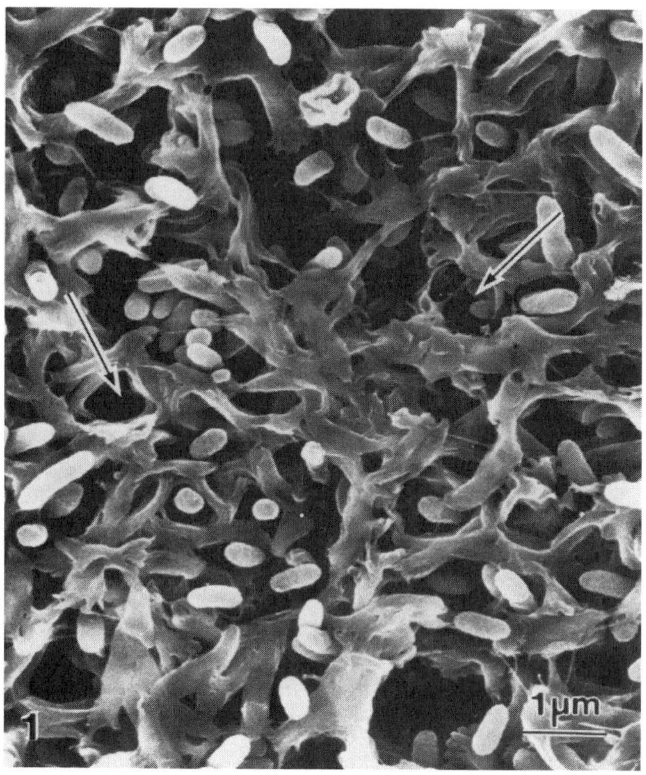

Figure 1. Membrane filter structure. Courtesy of Pall Corporation (20).

Membrane filters used for the sterile filtration of gas streams, such as nitrogen blankets, storage tank sterile vents, formulation tank sterile vents, sterile air for aseptic packaging, sterile filtration of fermenter air, or vent gases, typically contain a membrane made of hydrophobic materials such as PVDF or PTFE (polytetrafluoroethylene). Hydrophobic membrane filters are desired in these sterile gas filtration applications because hydrophobic filters do not spontaneously wet with water. When a hydrophilic filter is wetted with water, it will not pass air until the water-wet bubble point of the filter is exceeded. This water-wet bubble point can be greater than 50 psi (see FILTRATION, AIR).

Membrane materials can be modified to optimize their performance for certain types of applications. Positive-charged hydrophilic membranes contain cationic (positive charged) functional groups that impart a positive ζ-potential when immersed in an aqueous solution and provide enhanced retention of particles smaller than the absolute rating. Most particles have a negative charge, and the positive charge provides a mechanism, in addition to the tortuous path of the membrane, by which the particles are retained. Applications involving charged membrane filters are discussed in the Refs. 6, 7, and 8.

Membrane surfaces can also be modified to expose hydroxyl groups, which tend to prevent the retention of proteins. This is a benefit for dilute protein solutions (typically less than 1 mg/mL), in which the protein is the desired product. Such fluids include serum-free tissue culture media and protein additives, protein-based therapeutics and diagnostics, dilute protein containing diluents and buffers, recombinant proteins, hormones and growth factors, protein chromatography feeds and eluates, vaccines, and other dilute biologicals. Filtration applications involving low protein-binding membrane filters are described in Refs. 9 and 10.

The membranes used for sterile filtration bioprocessing applications are usually pleated and formed into a cylinder. Figure 2 illustrates the pleated cartridge construction.

The filter membrane is typically cast onto a support material (e.g., nonwoven polyester or polypropylene substrate) that provides high tensile strength while retaining flexibility. Layers of the membrane are corrugated, or pleated, along with upstream and downstream layers of courser material (e.g., nonwoven polyester or polypropylene). These layers provide support and drainage for the membrane. The sides of the corrugated membrane pack are sealed, often by a heat seal.

The corrugated filter membrane pack is fitted around the core. The purpose of the rigid, inner core is to provide support for the filter element against pressure in the forward direction. An external cage is provided for additional support and protection during handling. The flow path for the cartridge filter is from the outside (cage) to the inside (core). The material used for the core, cage, and support layers will depend on the intended application for the filter. Polypropylene, polyester, and PTFE are examples of ma-

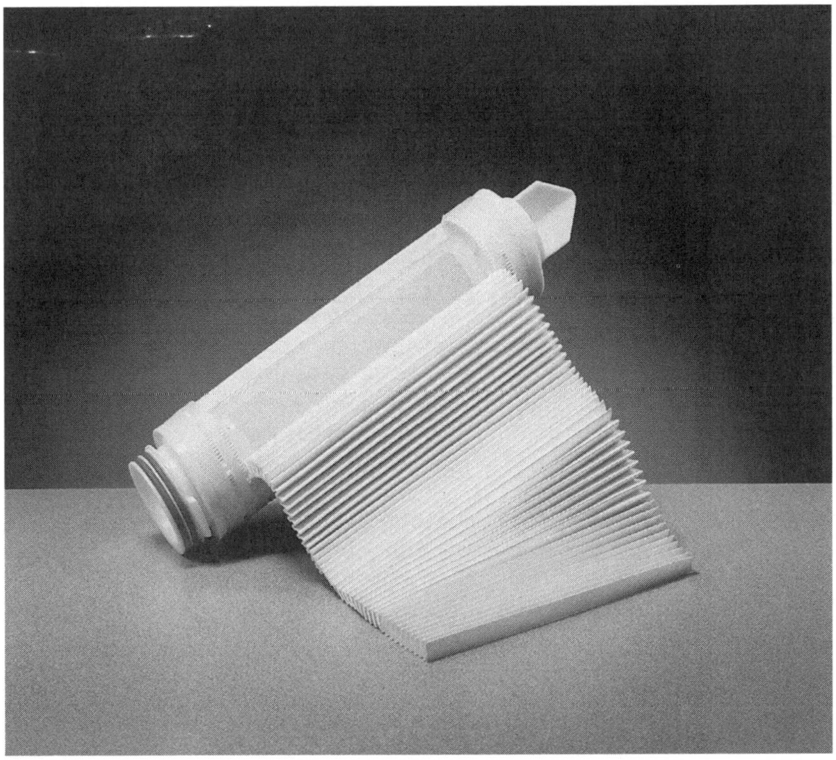

Figure 2. Pleated cartridge configuration. Courtesy of Pall Corporation (29).

terials that can be used for the cage, core, and support material in a membrane filter. The membrane pack ends are sealed, usually by heat, with end caps. The end caps are constructed of the same types of materials as the core and cage. The final step in the construction of a membrane filter is to attach the proper end closure and O-ring adaptor by an appropriate welding technique.

The end closure is used for placement of the filter in a filter housing. A finned or flat blind end cap on one end is generally attached on one end and an open outlet at the other end. Sealing into the housing is accomplished with a double O-ring seal at the open end to ensure no fluid bypass. Filter elements can have locking tabs at the double O-ring adaptor base for bayonet locking of the element to ensure positive sealing in the housing base.

Most single filter elements are 10 inches in length. For the construction of a multiple length cartridge (20, 30, or 40 inches), the sections are attached by welding 10-in. subassemblies end to end. Figure 3 shows the construction types.

Requirements. For sterile filtration processes, the biological safety of the membrane filter or filter cartridge should be demonstrated by the performance of the USP ⟨88⟩ Class VI (121 °C) *Plastics Test for Biological Reactivity* (11). Another typical qualification used to select filter materials of construction is a listing for food contact in CFR Title 21 Part 177 (12).

For sterile aseptic processes, a typical requirement is sterilization by passage through a 0.2-μm sterilizing grade membrane filter, as defined by ASTM Standard F838-83 (13).

Figure 3. Cartridge filter construction. Courtesy of Pall Corporation (30).

A membrane filter as well as all system components can contribute extractables to the process stream. Many filter manufacturers document the extractables level of a particular filter in an appropriate solvent as a nonvolatile residue (NVR). These extractables are typically composed of the oligimers or additives of the plastic materials present in the filter element. It is important to note that the amount of extractables from a filter element as well as any extractable from the rest of the process system can be reduced by flushing the system before filtration.

Many filter manufacturers address the issue of effluent quality requirements by the performance of appropriate tests on filter samples from manufacturing lots. These include tests for

- *Cleanliness*: Per current USP limits under *Particulate Matter in Injections ⟨788⟩* and conformance with requirements for a non-fiber-releasing filter per Title 21 of the U.S. CFR Title 21, Part 211.72 and Part 210.3(b)(6) (14)
- *Oxidizable substances*: Per current USP requirements under *Purified Water* after flushing (15)
- *pH*: Per current USP requirements under *Purified Water* after flushing (15)
- *Pyrogens*: Per current USP requirements under *Bacterial Endotoxins Test* as determined using the limulus ameobocyte lysate (LAL) reagent with an aliquot from a soak solution (16)

Prefiltration

Cartridge filters used for prefiltration can be either pleated or nonpleated (depth) filters. Standard design depth filter cartridges are open cylinders of thick filter material (double open-ended). Prefiltration filter cartridges are typically available in 2-1/2 or 2-3/4-in. o.d., in multiple lengths of 10 in. up to 40 in. long.

Prefiltration encompasses a wide variety of fluid types and removal requirements. The following is a discussion of some important characteristic of prefilter construction and a sampling of some of the types available.

The particles or fibers in the prefilter membrane should not become dislodged, slough off, shed, or in some other way contaminate the filtrate (media migration). The filter medium should also not unload or release contaminants retained by the filter medium into the effluent as the differential pressure across the membrane increases. Data that indicate filtration efficiency as a function of increasing differential pressure should be available from the vendor to support the claim of nonunloading.

Polypropylene Depth Type Filter Cartridges. Polypropylene depth filters are made using nonmigrating continuous filaments of polypropylene filament medium without resin binders. Absolute rated polypropylene depth filter cartridges are applicable as prefilters for serum, vaccines, diagnostics, tissue culture products, deionized water, container washing, and final product. These filters are used in fermentation for liquid feeds, makeup water, solvents, and antifoam. In downstream processing, they are used for cell and cell debris removal, buffers, cleaning agents, and sanitizing solutions. Positive charged versions of these filters are possible and have the added ability to remove organisms and particles such as bacterial endotoxins (pyrogens) that are much smaller than the absolute pore size rating.

The membrane for these depth filtration cartridges has an inner (downstream) section in which the pore diameter is constant (this section provides absolute rated filtration) and an outer section in which the pore diameter varies continuously from that of the absolute rated section to as much as 90 μm or more. Pore size variation of the filter cartridge is achieved by varying the fiber diameter while maintaining constant void volume throughout the medium. The constant void volume provides increased dirt capacity and a low clean pressure drop of the filter cartridge.

Polypropylene Pleated Filter Cartridges. Polypropylene pleated filters are applicable as prefilters for makeup and rinse water, for serum for culture media, and in downstream processing for cell debris removal and prefiltration of solvents and buffers. These process filter cartridges can be constructed of the same type of membrane material as depth filters.

The thin sheet of polypropylene media is pleated and formed into a cylinder with a longitudinal side seal of melt seal polypropylene. The cylinder is then melt sealed to injection molded polypropylene end caps to ensure no fluid bypass. Polypropylene hardware components consisting of an inner support core and an external protective outer cage are incorporated. The process filter cartridge with polypropylene hardware should be rated to withstand differential pressure of 80 psi up to 50 °C (122 °F) and 60 psi up to 80 °C (176 °F).

Resin-Free Cellulose Pleated Filter Cartridges. Resin-free pleated cellulose filter cartridges are applicable as prefilters for makeup and rinse water, reverse osmosis membranes, deionized water, and inlet air for fermenters and bioreactors. These filter cartridges are constructed of pure cellulose medium without resin binders, which is pleated into a high area cylinder. The longitudinal side seal of the process filter cartridge should be a polypropylene (approved for food contact usage).

Cellulose media cartridges are assembled with hardware components consisting of a perforated inner support core, an outer support cage, and end caps melt sealed to imbed the medium in the plastic. All hardware components should be of pure polypropylene, without filler or reinforcement, to ensure a high quality filtrate with a minimum of soluble extractables. The process filter cartridge with polypropylene hardware should be rated to withstand differential pressure of 80 psi up to 50 °C (122 °F) and 60 psi up to 80 °C (176 °F).

Resin-Impregnated Fiberglass Pleated Filter Cartridges. Resin-impregnated fiberglass filters are applicable as prefilters for deionized water, solvents, serums, diagnostic reagents, and cell debris removal from harvest fluids. The resin impregnation makes available a choice of filters with either a positive or negative ζ-potential when used in aqueous service. Filter cartridges with negative ζ-potential are constructed of glass fiber that has a natural negative ζ potential and reinforced with a resin binder that is also naturally negative when immersed in water. Positive ζ potential filter cartridges are constructed using a resin binder that coats the glass fibers and imparts a positive ζ potential to the medium.

The filter cartridges are available with hardware components consisting of end caps and an internal perforated support core made of either polypropylene or stainless steel and in an external polypropylene outer cage or protective net. Stainless-steel end caps are attached to the filter by an inert synthetic resin. Polypropylene end caps are melt sealed to imbed the medium in the plastic. Filter cartridges with polypropylene hardware should be rated to withstand differential pressures of 80 psi up to 50 °C (122 °F) and 60 psi up to 80 °C (176 °F). With stainless-steel hardware, the filter can be rated to 75 psi up to 135 °C (275 °F).

Filter Assemblies Used in Processes

During filtration, the membrane filter cartridges are placed in housings that provide a means of fluid contact with the filter cartridge. The housings are typically composed of 304, 316, and 316L stainless steel; plastic housings are also available. The housing design is typically composed of a head and a bowl. The head is the portion of the housing in which the filter is attached. The bowl is clamped to the head. A housing is illustrated in Figure 4.

Membrane filters can also be obtained in complete assemblies. These disposable assemblies consist of a membrane filter cartridge welded into a plastic filter housing. Disposable assembles or capsule filters are typically used in small-scale applications (e.g., 100-L batches or less, depending on the fluid) and are particularly useful in applications in which operator contact with the fluid is not desirable, because the entire assembly can be removed from the system and discarded without operator contact with the product-wet cartridge. Figure 5 shows disposable assemblies.

MEMBRANE FILTER RATINGS AND TESTS

Many filter manufacturers use a nominal micron (μm) rating for particle removal efficiency. This is defined by the American National Standards Institute (ANSI) as an "arbitrary micrometer value indicated by the filter manufacturer. Due to lack of reproducibility this rating is deprecated." Further, nominal rating standards are arbitrary and a comparison of nominally rated filters is imprecise. In addition, nominal ratings can be misleading because the filter can allow passage of particles larger than the rating indicates.

In order to establish a more meaningful filter rating, a test is performed on the filter using an appropriate contaminant to meet a specific claim for retention of the contaminant. The test procedure must be sensitive enough to detect the passage of contaminants of interest. The testing should be performed under carefully controlled conditions using industry-accepted reference standards. These reference standards include silica suspensions, latex beads, or microorganisms.

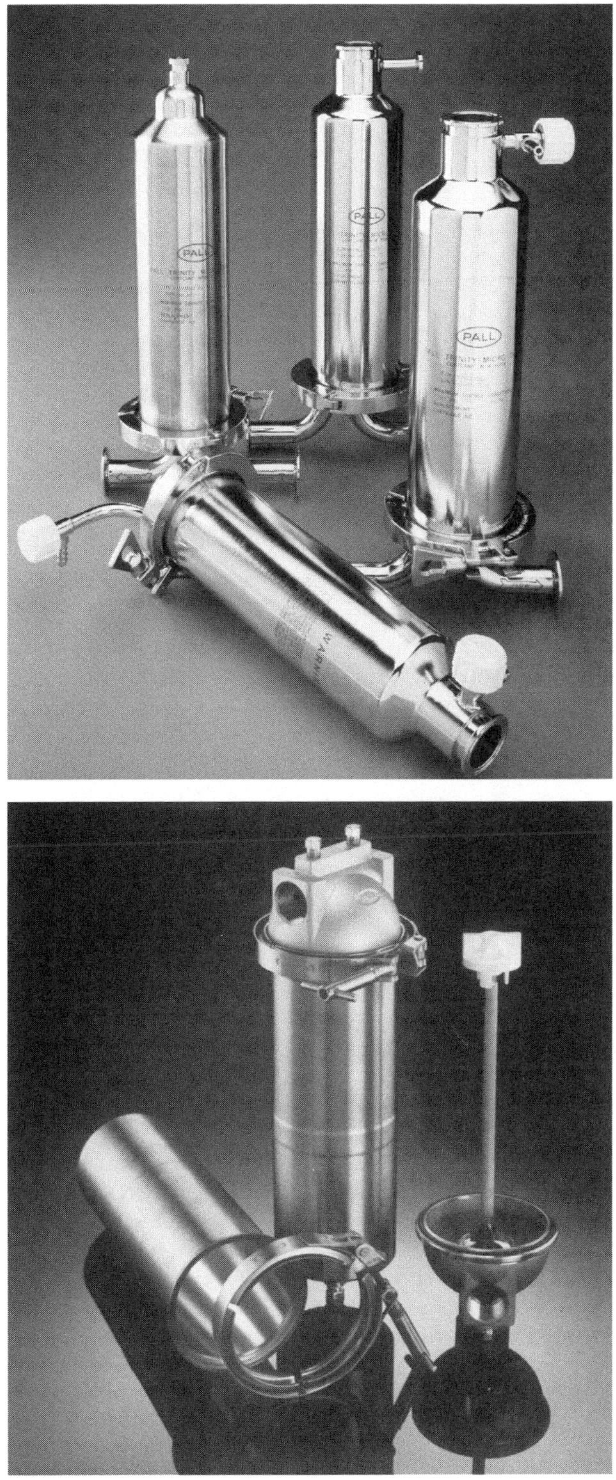

Figure 4. Housing configurations. Courtesy of Pall Corporation (31,32).

Particulate Removal Filters

The removal ratings for particulate removal filters (typically used as prefilters in sterile filtration processes) can

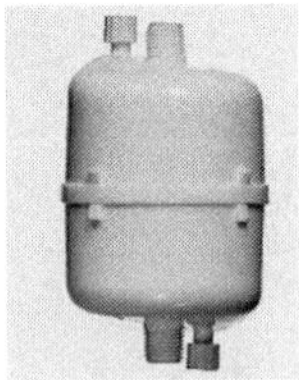

Figure 5. Disposable cartridge assembly. Courtesy of Pall Corporation (29).

be established using the Oklahoma State University (OSU) F-2 Test. This rating method (ISO 4572, ANSI B93.31) has received wide acceptance for use on lubricating and hydraulic fluids. The technique has been used for oils extensively and has been adapted for use in water with contaminants ranging from 0.5 to 25 μm.

The test is based on continuous on-line particle counts of different particle sizes, both in the influent and the effluent stream. The β-ratio at a specific particle size is defined as β_x: the number of particles of a given size (X) and larger in the influent, divided by the number of particles of the same size (X) and larger in the effluent, where X is the particle size in micrometers. The percent removal efficiency can be calculated from the β-value. The percent removal efficiency is $[(\beta_x - 1)/\beta_x]\ 100$ (17).

Sterilizing-Grade Membrane Filter Ratings and Integrity Tests

Sterilizing-grade hydrophilic membrane filters require testing using a liquid bacterial suspension challenge that is sensitive enough to detect the passage of any microorganisms, for the establishment of the micron rating of the filter. A sterilizing-grade filter is defined by the FDA's *Guideline on Sterile Drug Products by Aseptic Processing* (18) as one that will produce sterile effluent when challenged with the test organism *Brevundimons (Pseudomonas) diminuta* (ATCC 19146) to the level of greater than 10^7 CFU/cm^2 filter area. Such microorganism retention tests are conducted with *P. diminuta* (ATCC 19146), measuring 0.3×0.6 to 0.8 μm, to verify that membrane filters produce sterile fluid filtrates. This is a standard test for the validation of sterilizing-grade filters (0.2-μm pore size rating) in the bioprocessing industry.

In practical terms, there is a limit to how many bacteria a filter element can be challenged with before it becomes plugged. When a filter has been challenged at the level of 5×10^9 CFU/cm^2 it becomes effectively clogged. This challenge level corresponds to about 2×10^{13} CFU for a 5-ft^2 membrane filter (17).

The liquid test involves challenging a test filter with a known quantity of *P. diminuta*, no less than 1×10^7 organisms per cm^2 of filter area. The challenge sample is suspended, for example, in sterile water and then filtered at a defined flow rate and time through a test filter, followed by an analysis membrane. After the challenge is complete, the system is flushed for 5 to 10 min with sterile water. The analysis membranes are removed and placed on Mueller Hinton agar and incubated at 32 °C for 48 h. After incubation, the plate is examined for the presence or absence of microbial colonies. Figure 6 illustrates a liquid bacterial challenge test stand.

For some applications (e.g., mycoplasma control or removal) a membrane with a more stringent rating (0.1 μm) is required. These applications typically involve sera and tissue culture media. One organism that has been used extensively in validating 0.1-μm rated filters is *Acholeplasma laidlawii* (19).

A liquid bacterial challenge is a destructive test; the filter cannot be used after the challenge. Therefore, a correlation is made to a nondestructive test. It is the correlation, not the nondestructive test alone, that provides assurance that the filter performs as required. During the production of a sterile product, the filter should be tested before and after the filtration to ensure that the filter meets the specification, is properly installed and intact, and confirms the filter rating.

The three major tests used to determine the integrity of a membrane filter are the bubble point, forward flow, and pressure-hold integrity tests. These tests are based on the flow of a gas (air or nitrogen gas) through a liquid wetted membrane under applied gas pressure. A fourth test, the water intrusion test, is sometimes used to determine the integrity of hydrophobic membrane filters.

For a nondestructive test to be useful, the results of the test must be able to predict the ability of the filter to remove bacteria. Such a test should also be easy to perform and highly reproducible. The manufacturing quality control of the membrane filters tested to obtain this prediction must be used to establish quality control specifications that are consistently maintained in filter manufacturing. It is the combination of the nondestructive integrity test, the correlated bacteria challenge test, and manufacturing quality control that allows membrane filters to be reliably used for sterile filtration of bioprocessing products.

Forward Flow, Pressure Hold, and Bubble Point Tests. The industry-accepted, nondestructive tests used to verify sterilizing grade filter integrity are the forward flow, pressure hold, and the bubble point tests (20). These tests are performed by applying a preset air (or nitrogen gas) pressure to a filter wet with an appropriate fluid. For hydrophilic filters (filters that wet readily with water) the wetting fluid is typically water or a water-based fluid; for hydrophobic membrane filters (filters that do not spontaneously wet with water and are used in air service) 60/40 isopropyl alcohol/water is a typical wetting fluid.

The wetting fluid fills the voids in the membrane. For this to be occur uniformly, the wetting must be complete. The result of the wetting can be considered as a layer of water supported by the membrane. When gas pressure is applied to one side of the membrane, the test gas will dissolve into the liquid layer to an extent determined by the solubility and pressure of the gas (as described by Henry's law [21]). Downstream of the membrane, the pressure is lower, and the gas in the liquid is driven out of solution. The result is a net flow of gas through the membrane. The forward flow, pressure hold, and the bubble point tests all involve the observation or measurement of a gas flow through a wetted membrane. The measured flow is a com-

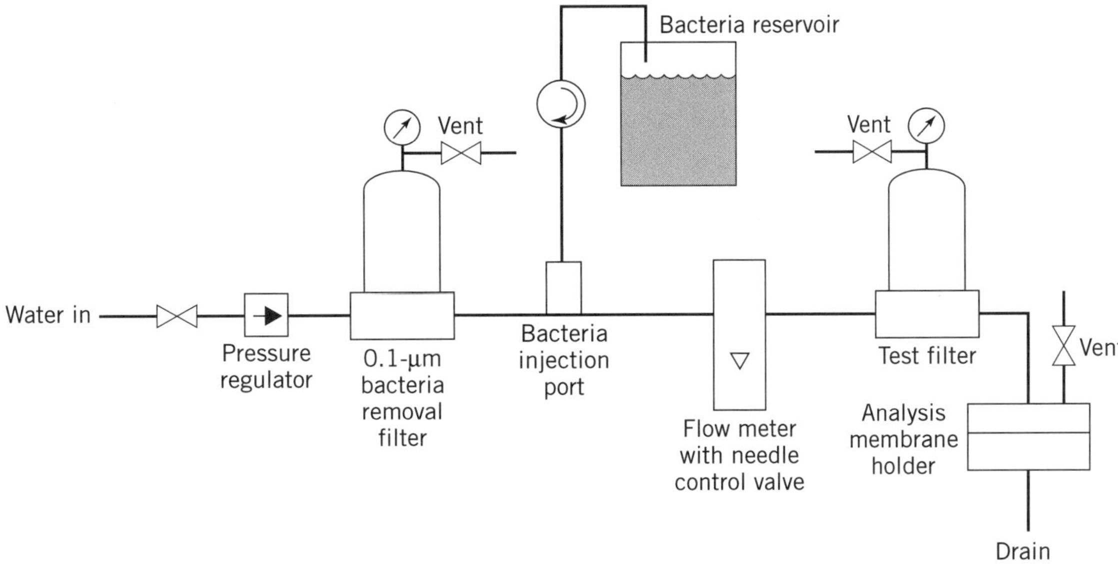

Figure 6. Liquid bacterial challenge. Adapted from "Engineering Considerations in Sterile Filtration Processes" by Holly Haughney in *Sterile Pharmaceutical Products: Process Engineering Applications*, edited by Kenneth E. Avis. Copyright © 1995 by Interpharm Press, Inc. All rights reserved. Adapted by permission.

bination of the diffusion of the gas through the liquid layer and bulk gas flow through any open passageways.

The rate of diffusive flow is a function of the test pressure, the diffusivity of the gas in the liquid, the solubility of the gas in the liquid, the membrane void volume, and the membrane thickness. This diffusive flow is described by Fick's first law (13), which is presented below in an integrated form:

$$N = \frac{DH\Delta P\rho}{L} \qquad (1)$$

where N is the permeation rate, D is the diffusivity of the gas in the liquid, H is the solubility coefficient of the gas, L is the thickness of the liquid in the membrane, ΔP is the differential pressure, and ρ is the void volume.

As the applied upstream gas pressure is increased, the diffusive flow increases proportionally. At some point, the pressure becomes great enough to expel the fluid from a passageway and establish a path for bulk flow of gas. As a result, the gas flow through the wetted filter begins to increase in a nonlinear manner due to the initiation of bulk gas flow through these open paths. The pressure at which this occurs can be modeled by a modified capillary rise equation:

$$P = F\frac{4\gamma \cos\theta}{D} \qquad (2)$$

where P is the pressure (dynes/cm^2), γ is the surface tension (dynes/cm), θ is the contact or wetting angle, D is the diameter (cm), and F is the shape factor (to account for the structure of the membrane). As the pressure is further increased, more passages are opened and the proportion of the total gas flow due to the bulk air flow increases. If the pore size distribution is narrow, then the possible passages through the membrane will provide similar resistance to flow.

For a wetted high-area pleated membrane filter, a plot of differential gas flow pressure versus gas flow will show a gradual transition from diffusive to bulk flow. An alternate plot of differential pressure versus gas flow or differential pressure will reveal a rapid transition from diffusive flow to bulk flow, as seen in Figure 7. This graph is referred to as a diffusive flow spectrum or a K_L (knee location) curve. The K_L pressure represents the transition from the diffusive flow region to the bulk gas flow region.

The bubble point test is a test designed to determine the pressure at which a continuous stream of bubbles is initially seen downstream of a wetted filter under gas pressure. There are several ways of performing this test. The first of these is the visual bubble point test. Generally, the

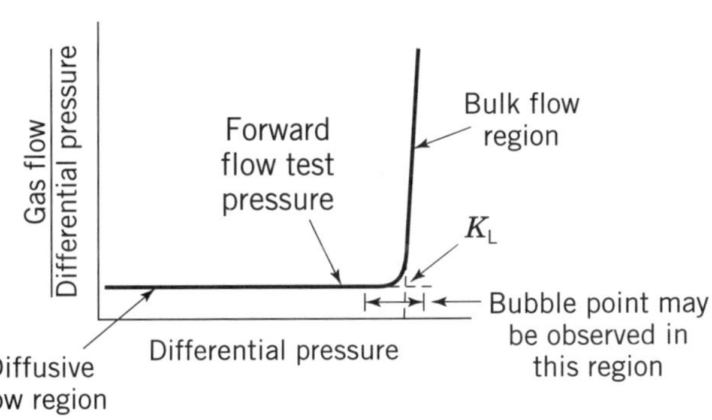

Figure 7. Diffusive flow spectrum. Courtesy of Pall Corporation (20).

visual bubble point test is not used except on small area disks where the whole membrane surface may be examined. To perform a visual bubble point test, gas pressure is applied to one side of a wetted membrane, with a layer of the wetting fluid maintained on the other side. The gas pressure is slowly increased. The rate of increase will affect the measured bubble point value. At some point, the pressure becomes high enough to displace the liquid from one or more passageways. This establishes the beginning of bulk flow of air through the membrane, and a thin stream of bubbles can be observed on the downstream side of the membrane. The pressure at which this stream of bubbles is observed is referred to as the first bubble point.

A manual bubble point is measured with a filter installed in a housing, with tubing running from the housing outlet to a beaker of water as shown in Figure 8. The geometry and position of the filter housing, and the diameter and length of the tubing will affect the measured bubble point value. In this case, the bubble point is considered to be the pressure at which a steady stream of bubbles is seen coming from the tubing; however, the determination of a steady stream is often subjective.

The manual bubble point (Fig. 9) will generally occur at a higher pressure than the first bubble point determined by a visual examination of the filter surface, because greater flow is necessary to obtain a clear stream of bubbles from the volume of the tubing. When determining the manual bubble point on a high area cartridge, the diffusion of air through the wetted filter at pressures below the bubble point can be significant. The complications resulting from the length and diameter of the tubing, the housing, and operator interpretation render bubble points determined in this manner not easily reproducible or comparable to bubble points determined by other techniques.

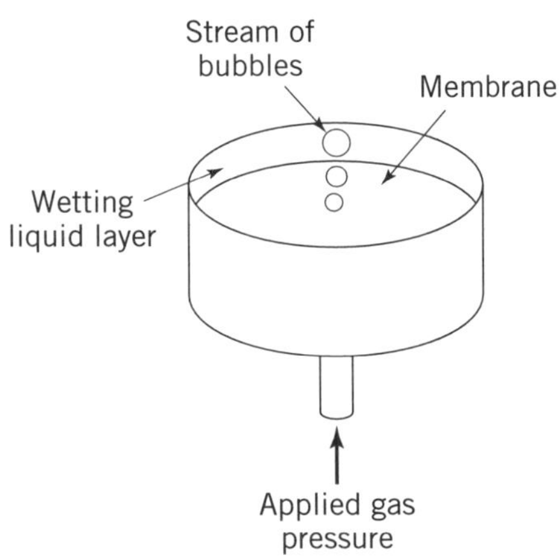

Figure 8. Bubble point test. Courtesy of Pall Corporation (20).

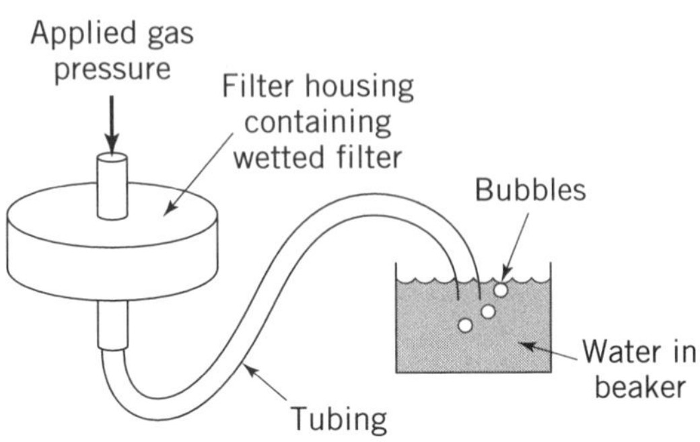

Figure 9. Manual bubble point test. Courtesy of Pall Corporation (20).

The forward flow test and the related pressure hold test were first introduced by Pall Corporation in 1973 (22). During these tests (Fig. 10), a wetted membrane is subjected to a predetermined gas pressure on the upstream side of the membrane. The gas will diffuse through the wetted membrane at a measurable rate. During the forward flow test, the diffusion of the gas and flow through any open pores is measured at the specified test pressure.

The pressure hold test (Fig. 11) is a modification of the forward flow test. The upstream side of filter housing is pressurized at the test pressure and then the filter is isolated from the pressure source. The diffusion of gas through the wetted membrane and flow through any open pore is measured as a decay in pressure over a specified period of time.

The pressure chosen as the test pressure for the forward flow or pressure hold test must be close enough to the K_L to allow differentiation between different grades of membrane and high enough to present a measurable flow. It is also necessary that the test pressure chosen be such so that the background diffusive flow does not interfere with the sensitive detection of defects in the filter. The exact pressure chosen for the forward flow test must be below the K_L. The test pressure is set by filter manufacturers and should be used during tests performed to establish the correlation with bacterial retention.

Integrity tests can be performed with automated equipment. Automated filter integrity test instruments (e.g., TruFlow™, Integritest®, Palltronic®, and Sartocheck®) have been developed to provide accurate and reproducible filter integrity test values. The instruments traditionally were designed to perform a pressure hold test. The forward flow value can be calculated from the pressure hold value. Newer equipment designs use direct upstream air flow measurements.

The tests are controlled and monitored by a built-in microprocessor, and the equipment often can be used with a programmable logic controller (PLC) for full system automation. An example of an automated integrity test application is provided in Ref. 23.

Automated devices cannot perform a true bubble point test. Instead, a modified bubble point test is performed, which essentially consists of a series of pressure hold tests. In the automated bubble point test, upstream air pressure is increased in steps. Between each pressure increase, the instrument isolates the upstream volume and monitors the pressure decay upstream of the filter. As the pressure decay exceeds a certain value, the software uses a specific algorithm for calculating the bubble point from relative values of this series pressure decays. The instrument uses a series of pressure hold tests as the basis for the calculation of the bubble point rather than measuring the bubble point directly. Figure 12 illustrates how pressure hold

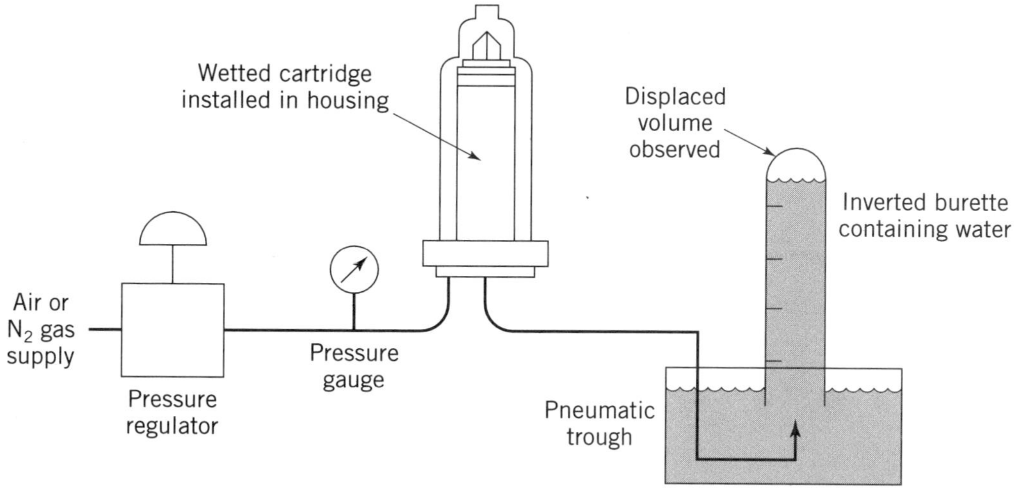

Figure 10. Forward-flow test. Adapted from "Engineering Considerations in Sterile Filtration Processes" by Holly Haughney in *Sterile Pharmaceutical Products: Process Engineering Applications*, edited by Kenneth E. Avis. Copyright © 1995 by Interpharm Press, Inc. All rights reserved. Adapted by permission.

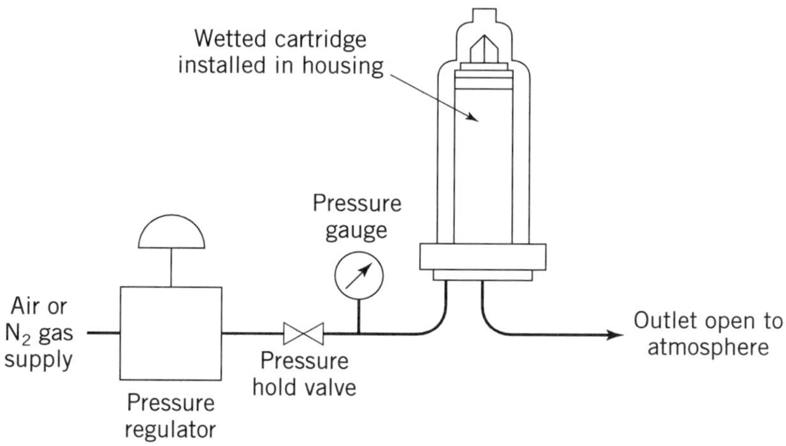

Figure 11. Pressure-hold test. Adapted from "Engineering Considerations in Sterile Filtration Processes" by Holly Haughney in *Sterile Pharmaceutical Products: Process Engineering Applications*, edited by Kenneth E. Avis. Copyright © 1995 by Interpharm Press, Inc. All rights reserved. Adapted by permission.

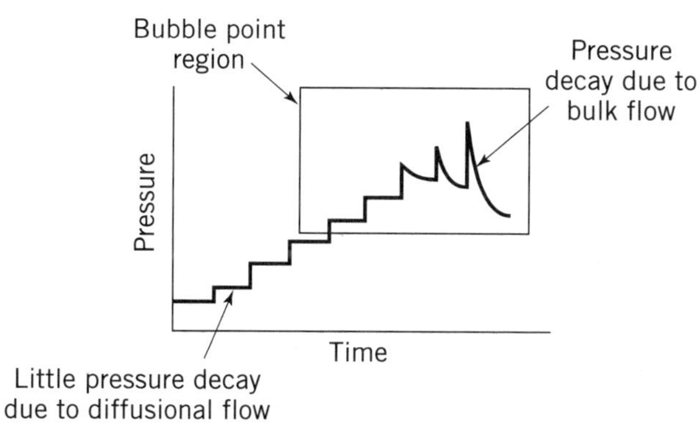

Figure 12. Pressure-hold test for bubble point determination with an integrity test device. Courtesy of Pall Corporation (20).

tests can be used to determine a bubble point. The actual algorithms used in the automated test devices may differ between automated filter integrity test instrument manufacturers.

Water Intrusion Test for Hydrophobic Filters. The water intrusion test can be used to test hydrophobic membrane filters (24–27). To perform a forward flow, pressure hold, or bubble point test on a hydrophobic filter, the membrane must be wet with an alcohol solution (e.g., 60/40 IPA/water). For an in situ integrity test, the wetting agent has to be added to the system to perform the test and then removed to restore air flow through the filter. The potential flammability of the alcohol is a consideration, and for some applications it can be necessary to validate the complete removal of the solvent to avoid contamination of the product. For these reasons, integrity testing of hydrophobic filters in situ has not been widely used, and off-line testing has been more common. The increasing industry demand and growing regulatory requirement for in situ testing of sterilizing-grade air filters has led to the development of water-based tests.

If the upstream side of a dry hydrophobic filter assembly is filled with water and pressurized, the hydrophobic nature of the porous membrane will prevent the bulk flow of water through the filter until the intrusion pressure is reached. At pressure below the intrusion pressure, a small but measurable flow of water through the membrane occurs.

The presence of larger pores in the filter will be detected by an increased flow due to bulk water flow through these pores (Fig. 13). This principle is the basis of the water intrusion test.

At the start of the test, the upstream side of the filter assembly is filled with water and a predetermined gas pressure is supplied. This causes a fall in the water level within the filter housing because of factors such as pleat compression and the passage of trapped air through the membrane. It is important that the changes in the upstream volume caused by these effects have stabilized so that the very small flows of water through the membrane can be accurately measured. A typical test arrangement is illustrated in Figure 14.

Adequate stabilization time (e.g., 10 mins) is also required to ensure thermal equilibrium. Water flow can be measured by determining the upstream flow of gas required to keep the pressure constant. The use of a direct flow measurement integrity test instrument can be used for water intrusion test measurements. After the test is over, the water can be drained and in many cases the filter is ready for use. Where dry air or gas is required, a drying procedure can be used.

FILTRATION SYSTEM DESIGN AND OPERATIONAL CONSIDERATIONS

Filter Selection

As detailed earlier, there are a number of different types of cartridge filter types for a wide variety of applications (1–3). In order to properly design a filtration system, all the requirements for the system and fluid must be considered. As a starting point, the removal efficiency must be defined. Other considerations can include process temperature, flow rate, pressure drop restrictions, integrity test regimen, and sterilization.

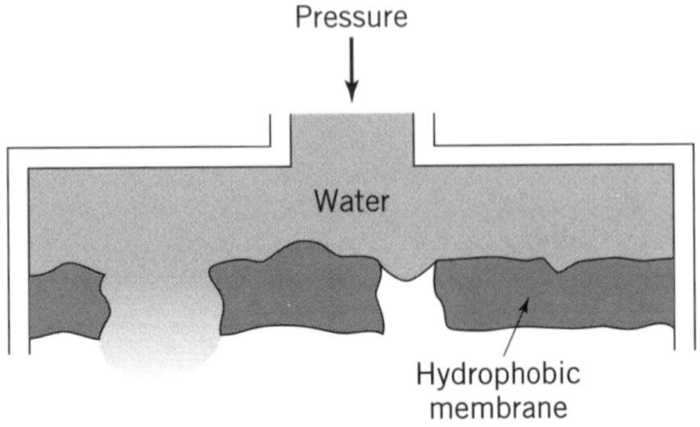

Figure 13. Bulk flow of water through pores. Courtesy of Pall Corporation (27).

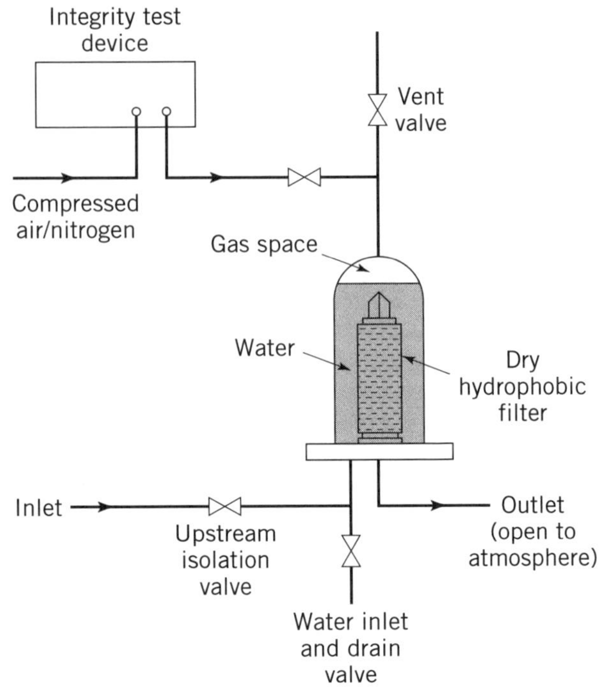

Figure 14. Water intrusion test. Courtesy of Pall Corporation (33).

All cartridge filter materials (membrane, cage, core, support material and housing) should be compatible with the process fluid over the temperature and pressure range indicated for a given process application. In general, most bioprocessing fluids are water based and are compatible with most membrane materials, such as polyamides (nylon), PVDF, polysulfone, and cellulose acetate.

For many filtration processes, it is desirable to have a staged filtration system. A staged system will have one or more prefilters of higher micron rating protecting the final filter. The final filter in the stage is the filter that provides the desired effluent quality. The prefilters are typically courser filters that remove larger contaminants in the fluid and reduce the loading of the final filter. For a sterile filtration process, the final filter is typically a 0.2-μm membrane filter. In a staged system, the life of the final filter is enhanced by the use of the prefilters. This will usually reduce the cost of the system, because the cost of the final filter is generally greater than the cost of the prefilters.

The size of the filtration system will depend upon the volume of the fluid being filtered and the ease with which the fluid is filtered or its filterability. For production applications, the membrane filters are typically provided as 10-in. segments or filter elements. These segments can be welded together to form 20, 30, and 40-in. filter elements. The diameter of the cartridge is approximately 3 in. Filter housings can hold a single cartridge (10- to 40-in. elements) or multiple filter cartridges. For applications that involve relatively small volumes (for example 10 L), disposable filter assemblies (capsule filters) with smaller membrane areas are available. If prefiltration is required, then the prefilters will also be a part of the filter train. For many applications, there is a one-to-one correspondence of prefilters to final filters. Optimization for a particular process may indicate that fewer, or more, prefilters are needed.

In some production environments, it is desirable to have a dual filtration system (filters arranged in parallel) to allow for a changeover from one system to the other. While one system is being used, the other system can be prepared for use (e.g., it can be cleaned and tested and the filters can be changed out). The decision to use a dual filtration system requires a balancing of the ease of operation with the additional space required for the additional system. For some sterile applications, redundant final 0.2-μm filters (filters in series) are used to provide additional assurance of sterility. If one of the filters fails, because of processing conditions, then the other filter is in place as a backup. If the backup filter is integral, then the fluid has been sterile filtered. The decision point is whether further assurance of maintaining sterility throughout the entire filtration processing time can justify a redundant system.

Flow Rate and Pressure Considerations

All components of a system will cause a resistance to flow, which results in a pressure drop (ΔP). Pressure drops, or pressure losses in a system, can be caused by piping, connections, valves, and filling heads and by the filter and its assembly. The pressure drop through a clean filter assembly is due to the sum of component pressure drops, including the filter housing, the filter hardware, and the filter membrane. The filter membrane will cause a resistance to flow, or a pressure drop, in a filtration system. As the filtration progresses and the filter membrane becomes plugged with the particulate contaminants in the fluid, an increasing pressure drop will occur. The pressure drop will be different, depending on the fluid hydrodynamics (e.g., viscosity) and the filter membrane material characteristics (such as porosity).

An adequate pressure source is needed to overcome the pressure drop across the system, including the filter assembly and to allow for the maintenance of a constant flow rate. If the initial clean pressure drop across a filtration system is close to the maximum available pressure, the flow rate will not be maintained because as the filter plugs the pressure drop will increase and there will not be enough pressure to overcome the increased pressure drop throughout the filtration. This will cause a decrease in flow. If possible, where constant flow is required, pumping capacity should be increased. An alternative remedy would be to increase the membrane filtration area so that the initial pressure drop is reduced.

Ultimately, the filter will become plugged with contaminants. This will depend on the amount of the contaminant in the fluid that is being filtered. For most pharmaceutical sterile filtration applications the load (percent solids) is typically low. Filters upstream of the sterile filter are more likely to become plugged in a properly designed filter train. As plugging occurs, the differential pressure will increase. System design should include a change-out schedule so that the differential pressure constraints for the filtration system are not exceeded. For example, in a batch system, the filter change-out should occur after the batch filtration has been completed. If multiple in-line stages of filtration are needed, then the differential pressure for each stage must be added, and the sum must be less than the maximum differential pressure allowed for the filtration process to allow for increased pressure caused by membrane fouling.

At no time during a process should the pressure drop exceed the recommendations for pressure drop (usually specified for a temperature range of the individual filter unit). At elevated temperatures the recommended pressure drop is typically lower than the recommended pressure drop for ambient conditions. A combination of elevated temperature and a high pressure drop can lead to structural filter damage. Structural damage to the filter is possible under these conditions because at elevated temperatures the polymeric components of the filter can become softer and are more likely to be deformed if the pressure drop is too high. For example, under extreme conditions, the core of the filter as well as the cage can be crushed.

Required flow rate, maximum pressure drop, and available pressure for the filtration system must be considered when the system is designed. Flow rate is typically reported as either gallons per minute (GPM) or liters per minute (LPM). The flow rate achievable through a filtration system is directly related to the applied pressure and inversely related to the resistance to flow. Thus, if the applied pressure is increased, the flow rate will increase. If the resistance to flow is increased, such as by membrane plugging, then the flow rate will decrease. It is important to note that an initial higher pressure (and flow rate) can lead to premature membrane plugging with many products.

The nature of microporous membrane filters is that they will tend to plug rapidly if they are subjected to a relatively high flow rate (directly related to a high applied pressure) during the start-up of a filtration. For fluids with a significant particle loading, under conditions of initial high flow rate, the microporous membrane can become rapidly plugged, or fouled, and the pressure drop will increase and the throughput, or filter life, will be reduced. This is especially true for products that may contain gels, such as biological products. Filter life, or throughput, can be increased if the initial flow rate is reduced. If the initial flow rate is decreased, the pressure drop will increase at a slower rate and the throughput will be increased. Another approach is to increase filter surface area for the given flow rate. Additionally, the pressure drop across a filter assembly can be reduced by increasing the filtration area. Increasing filtration area can often be an economical approach because the increase in throughput is often greater than linear for an increase in filtration area. Figure 15 illustrates the concept of a lower flow rate and differential pressure providing a higher throughput (total volume filtered).

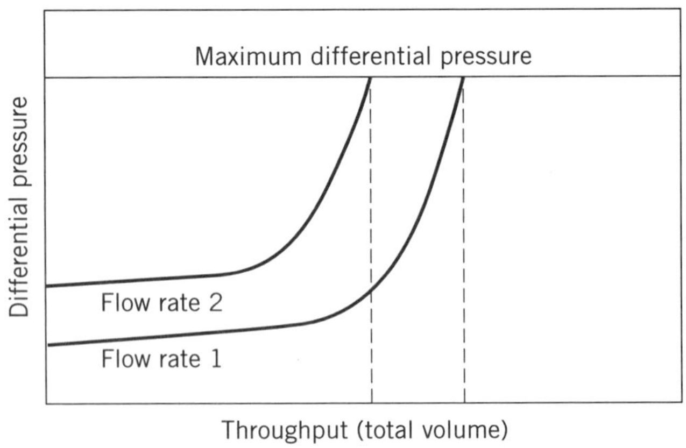

Figure 15. Comparison of a filtration performed with different flow rates. Adapted from "Engineering Considerations in Sterile Filtration Processes" by Holly Haughney in *Sterile Pharmaceutical Products: Process Engineering Applications*, edited by Kenneth E. Avis. Copyright © 1995 by Interpharm Press, Inc. All rights reserved. Adapted by permission.

Fluids with higher viscosities will be more resistant to flow and the pressure drop across the filter will be greater. For fluids with high viscosities, a greater applied pressure will be required to maintain the process flow rate. The smaller the pore size rating of a filter, for a fluid of a given viscosity, the greater the resistance to flow.

Scale-up

Although a particle size distribution evaluation can often provide useful information about the contaminant material in a process stream, the filtration system usually cannot be selected without an actual scaled-down filterability test. Actual testing is required because the particle size distribution fails to provide information on the quantitative particle load, its effect on membrane life, and how the membrane filters will perform with the process fluid under process conditions. In addition, other factors such as gels and high molecular weight biopolymers present in some products will also affect filter life and can only be evaluated by actual test.

The scaled-down, or filterability, test will require a relatively small representative sample of the process fluid (typically 2 to 4 L). Membrane filter disks with a 47-mm diameter are typically used. If a process stream is subject to variable composition, then the worst case fluid should be defined and tested. Only membrane materials that are compatible with the fluid are tested during the filterability test, and test conditions (e.g., temperature and flow rate) should match the process conditions as nearly as feasible. Several filtration schemes, including staged filtration schemes, are usually evaluated to optimize the filtration. The filterability tests can be performed at either a constant pressure or a constant flow rate. The use of constant pressure or constant flow rate must be considered because the system size, system life, process time, and throughput can depend on these process parameters. The flow rate can be scaled linearly based on filtration area. Optimization parameters must be defined and can include effluent quality, time (or throughput) to reach terminal (maximum allowable) differential pressure at a given flow rate, or flow rate for a given applied pressure.

System scaling can be based on the following ratios for the final filter in the system, where it is necessary to solve for full scale system area:

$$\frac{\text{Full scale system area}}{\text{Scaled down system area}} = \frac{\text{Full scale system flow rate}}{\text{Scaled down system flow rate}}$$

$$\frac{\text{Full scale system area}}{\text{Scaled down system area}} = \frac{\text{Full scale system throughput}}{\text{Scaled down system throughput}}$$

Throughput (total fluid filtration volume required), flow rate, and process time requirements must be met by the filtration system.

If the system involves prefiltration stages, the prefilter area will usually be equivalent to the final filter area; however, the size of each stage in a multistage system will depend on the overall requirements for practical and economical filtration of the product. Further optimization is possible by performing a filterability test with the prefilters alone. The issues associated with the sizing of a filter system with pre- and final filters are described next.

Figure 16 illustrates the behavior of a typical optimized (ideal) filtration system. The differential pressure (ΔP) is plotted as a function of volume filtered. The limiting or maximum differential pressure and the total batch volume are indicated on the graph. Ideally, the batch volume (throughput requirement) is reached before the limiting ΔP is reached. This allows a safety factor in the filtration and permits complete processing of the batch without change-out of the filter.

Figure 17 shows the effect of limiting ΔP versus volume for a filtration system in which the prefilter is too coarse and is not able to remove a sufficient number of particles in the fluid (the particles go through the prefilter instead

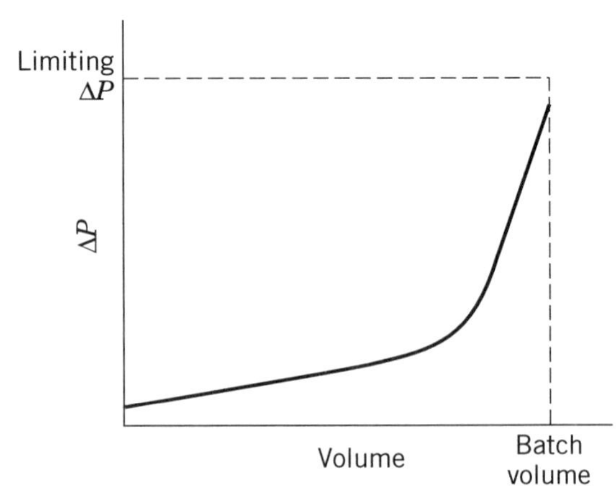

Figure 16. Example of an ideal filtration system. Adapted from "Engineering Considerations in Sterile Filtration Processes" by Holly Haughney in *Sterile Pharmaceutical Products: Process Engineering Applications*, edited by Kenneth E. Avis. Copyright © 1995 by Interpharm Press, Inc. All rights reserved. Adapted by permission.

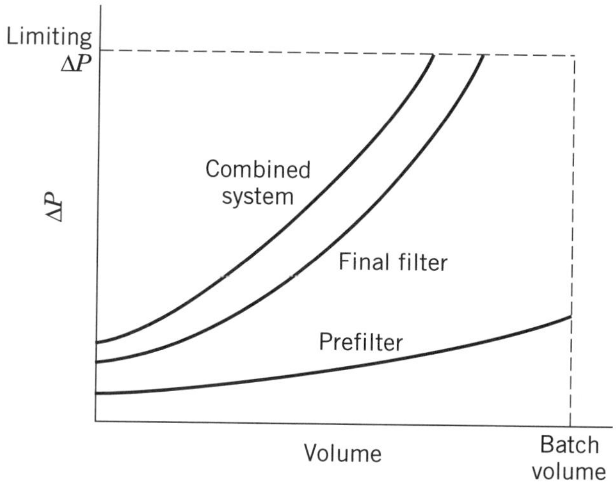

Figure 17. Example of a filtration system with the prefilter too course. Adapted from "Engineering Considerations in Sterile Filtration Processes" by Holly Haughney in *Sterile Pharmaceutical Products: Process Engineering Applications*, edited by Kenneth E. Avis. Copyright © 1995 by Interpharm Press, Inc. All rights reserved. Adapted by permission.

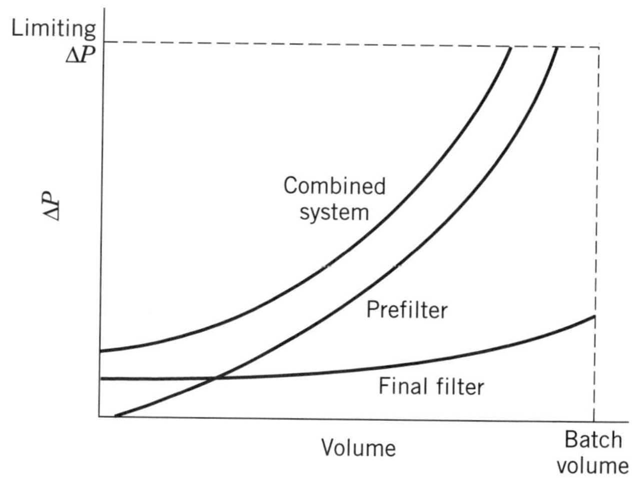

Figure 18. Example of a filtration system with the prefilter too small. Adapted from "Engineering Considerations in Sterile Filtration Processes" by Holly Haughney in *Sterile Pharmaceutical Products: Process Engineering Applications*, edited by Kenneth E. Avis. Copyright © 1995 by Interpharm Press, Inc. All rights reserved. Adapted by permission.

of being retained by the prefilter). The prefilter is unable to adequately protect the final filter. Representative curves are provided for the prefilter, the final filter, and for the combined system. The combined system curve is a sum of the differential pressures for the final filter and the prefilter. The combined system reaches the limiting differential pressure (plugging) before the complete batch is filtered. Thus, a finer prefilter or additional final filter surface area would be required for this batch volume to be filtered at differential pressure less than the limiting differential pressure.

Figure 18 illustrates what occurs in a system in which the prefilter area is too small, but performs its protective function. In this case, the final filter did not reach the limiting ΔP, but the prefilter did. Additional prefilter area would allow the filtration to be optimized. Figure 19 contains representative curves for an optimally staged filtration system. In the optimized system, the combined system is able to completely process the batch at a differential pressure lower that the limiting (plugging) differential pressure. The differential pressures across the final filter are slightly higher that the differential pressure of the prefilter. This is the desired situation for a system that requires prefiltration.

Depending on the knowledge of the filtration process for a particular fluid, it may be desirable to perform a side-stream or a pilot-scale test under process conditions before installing the system at full scale. Further optimization of the system may be possible based on the results of the side-stream test. A side-stream test is recommended when the scaled-down test was limited in its ability to match the process conditions for any stage of the filter system.

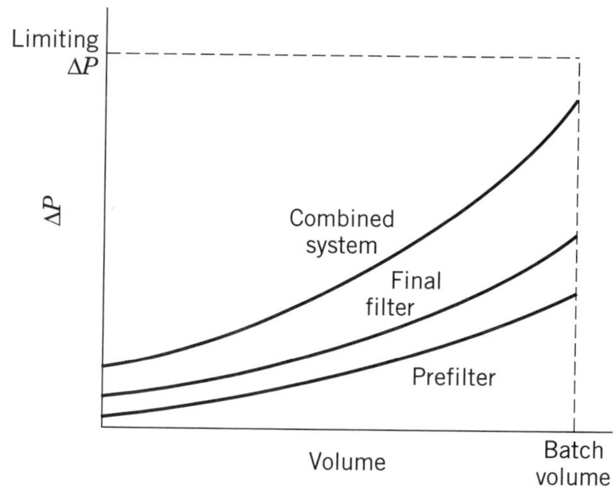

Figure 19. Example of a filtration with optimal prefiltration. Adapted from "Engineering Considerations in Sterile Filtration Processes" by Holly Haughney in *Sterile Pharmaceutical Products: Process Engineering Applications*, edited by Kcnneth E. Avis. Copyright © 1995 by Interpharm Press, Inc. All rights reserved. Adapted by permission.

System Design

Once the filters have been specified, the system components must be considered. A filtration system will typically contain the following filters, filter housings, gaskets or O-rings, pressure gauges, thermocouples, pipe or tubing, connections for the piping, and a pump (or gas pressure source). Some pharmaceutical systems may require an inert gas for product stability as well as pressure. The system components must be compatible with the process fluid, the full-scale process fluid temperature, the process pressure, and the sterilization method.

Most sterile filtration applications require a sanitary design. In a sanitary design the components (e.g., sanitary valves) and materials of construction are selected for their ability to prevent the buildup of contaminants in the system. For a sanitary process, 316L stainless steel is typically used for portions of the system that will have product contact, provided it is compatible with the process fluid. Internal surfaces should be finished to an appropriate microinch R_s specification (e.g., a minimum of 20 to 25 μin.); external surfaces should be mechanically polished to a high quality sanitary finish. Both internal and external surfaces should be electropolished for a smooth adhesion and corrosion resistant surface.

System welds should be constructed in such a way that the weld porosity is minimized and the joints are of high quality and are clean. Any internal welds that have product contact should be ground smooth and flush to reduce the potential for the buildup of material from the process stream. Proper weld procedures need to be followed.

Only sanitary fittings should be used in the system to ensure sterilizibility of the seals and cleanliness. Any ports in the system, such as those required for vents, drains, pressure gauges, and thermocouples, should have a sanitary stem or sanitary valve design. Absolutely no threaded connections should be used in the portions of a sanitary system where there is a potential for contact with the process fluid. Threaded connections can lead to system contamination because of a potential buildup of contaminants.

There are a number of considerations for the filter housing. Housings are essentially pressure vessels, and as such, may be subject to the appropriate ASME boiler and pressure code. A housing should be rated for the appropriate pressure and temperature with a safety margin for the process. A full vacuum rating may also need to be considered, especially for housings used in vent service or lyophilizers. The design of the housing as well as the physical placement of the housing in the production facility, must facilitate filter change-out. The size of the housing must meet the requirements for the flow rate and differential pressure process specifications. Inlet and outlet ports should not lead to excessive pressure losses.

The system and housing design should not contain any portions through which there is no flow (dead legs). All product contact areas need to be accessible for cleaning purposes. Drains should be designed and positioned to minimize holdup, or retention, of the process fluid in the system.

The filter assembly should be optimized for use in a sanitary system, especially where aseptic processing is required. For example, the filter should have an O-ring sealing mechanism in the filter housing instead of a gasket sealing mechanism. For some small batch applications, disposable filter assemblies (in effect filter plus plastic housing) can be used. These assemblies should have proper sanitary fittings.

For most sterile filtration systems a appropriate pump (e.g., a positive displacement pump or a centrifugal pump) is used to provide pumping capacity. The components of the pump head should be of a sanitary design and compatible with the process fluid.

Pressure gauges should be installed upstream and downstream of each filter assembly in the system so that the differential pressure across the filter can be monitored. The ability to monitor ΔP can be used to indicate filter life and that filter change-out is required. For some applications it is desirable to have temperature probes or thermocouples installed in the system. During an in situ steam sterilization of a system, the ability to monitor the temperature is needed to ensure that the temperature conditions for steam sterilization are met. For a process that requires a temperature other than ambient, the temperature of the fluid should be monitored. Temperature measurement should be performed immediately downstream of the filter installation to ensure that the filter installation is at the correct temperature.

Throughout the system, sample ports may be required to allow for sample collection during a process. It is important to ensure that the collection of a sample does not cause contamination in the system. This can be attained by using sample ports that are fitted with septa. For systems that are not hard piped (plumbed with permanent connections), the ability to make sterile (or aseptic) connections must be available.

Sterilization

In sterile filtration processes, the downstream side of the filter must be sterilized and must remain sterile during the process. The sterilization process must be properly validated to ensure that the sterile condition is met for a given system. Filters can be sterilized by a number of techniques, including in situ steam sterilization, autoclave sterilization, irradiation, and chemical treatment (hydrogen peroxide vapors or ethylene oxide). For any method, it is important to ensure that the sterilization is adequate for the system and that the technique does not damage the filter. The Parenteral Drug Association (PDA) published a technical monograph on the validation of steam sterilization cycles (28). The majority of sterile processes employ in situ steam sterilization or autoclave sterilization and will be the subject of this discussion.

Methods that involve steam are validated through the use of thermocouples or biological indicators to ensure that the system has been properly sterilized. The maximum steam temperature for most sterilizing grade filter cartridges is 140 °C; the maximum conditions for temperature and pressure should be obtained from the filter manufacturer. Typically, the steam pressure is held at a minimum of 15 psi at 121 °C for a minimum of 30 min. Although steam temperatures up to 140 °C can be used, the service life of the filter may be shortened, especially where multiple uses require repeat sterilization.

A decision must be made on the appropriate sterilization technique for the process. Larger systems tend to be in situ steam sterilized, while small volume systems tend to use autoclave sterilization. A system that utilizes in situ steam sterilization can be automated, which is becoming a common practice for a new system.

Integrity Testing Design and Operation Considerations

The issues that must be considered for integrity testing of a filtration assembly include the technique (forward flow, pressure hold, bubble point, or water intrusion for hydrophobic filters), the wetting fluid (water or product wet), and the test time (after installation, after sterilization, after filtration) and the test location (in situ or off line). Once a plan for integrity testing is established, the system must be designed so that the integrity test regimen can be properly carried out.

Field experience with the integrity testing of sterilizing grade filters shows that various combinations of integrity test procedures are in use and that different testing schedules, both pre- and postuse, are followed. In some sterile processes, redundant filters are used for convenience in integrity testing or for increasing assurance of sterility. For others, filter usage is based on filter life studies. A common approach is to perform the test or combination of tests that meet regulatory guidelines and that provide the highest degree of accuracy commensurate with the economics and the practicalities of the process.

The highest degree of confidence in the maintenance of filter integrity can be attained by routinely integrity testing the filter system, either by the forward-flow or pressure-hold integrity test, before sterilization, after sterilization, and after filter use.

Pre- and poststerilization integrity tests are recommended, because both tests can provide information about the process conditions. However, if only one preuse integrity test is to be performed, then an integrity test after sterilization will provide more useful information, because improper sterilization procedures can lead to filter damage. Regardless of any preuse testing performed, a postuse test is always recommended. The assurance of bacterial retention during the filtration of critical fluids, such as parenterals, biological liquids, and media for fermentations, is extremely important and, at the very least, filters are integrity tested after use.

It is often desirable for a filter user to perform a product–wet integrity test after the filter has been used to filter the product. Product–wet testing allows the filter user to perform the integrity test under actual process conditions and does not require a flush step to remove the product from the filter membrane. Flushing out the product and testing with water can lead to a false failure if there is incomplete removal of the product that may have a lower surface tension than water and thus reduce the bubble point to below the test pressure required for a water-wet integrity test.

Appropriate integrity test parameters for specific applications should be obtained from the filter manufacturer in writing because the filter manufacturer can provide the appropriate test parameters for a fluid and can ensure the proper relationship to the claims for the specific filter. The filter user should confirm performance of the tests before incorporating the test parameters into their standard operating procedures. If any anomalies are noted during confirmation, then these must be resolved.

Integrity test parameters can be provided with either air or nitrogen (or other inert gas) pressurization. Carbon dioxide cannot be used as a test gas, because the diffusive flow through the filter will be extremely high and the gas is reactive. Pure oxygen is typically not acceptable as a test gas, because it is a strong oxidizer. Process conditions, including temperature and gas used for pressurization, must be considered for the issuance of integrity test values.

BIBLIOGRAPHY

1. H. Haughney, in T. Meltzer and M. Jornitz ed., *Filtration in the Pharmaceutical Industry*, Dekker, New York, in press.
2. H. Haughney, in K.E. Avis ed., *Sterile Pharmaceutical Products: Process Engineering Applications*, Interpharm Press, Buffalo Grove, Ill., 1995, pp. 305–355.
3. J.M. Martin, A.M. Trotter, P. Schubert, and H. Katz, in B.K. Lydersen, N.A. D'Elia, K.L. Nelson ed., *Bioprocess Engineering: Systems, Equipment, Facilities*, Wiley Interscience, New York, 1994, pp. 317–370.
4. N.B. Bruckshaw, *Filtr. Sep.* **296** (May/June 1973).
5. D.B. Pall and E.A. Kirnbauer, *52nd Colloidal and Surface Science Symposium*, University of Tennessee, Knoxville, Tenn., 1978.
6. H. van Doorne, *J. of Parenter. Sci. Technol.* 192–198 (Sept./Oct. 1993).
7. K. Wickert, *Pharm. Eng.* **11** (1991).
8. S. Brown and A. Fuller, *J. Parenter. Sci. Technol.* 285–288 (Nov./Dec. 1993).
9. J. Martin and R. Manteuffel, *BioPharm* (Nov./Dec. 1988).
10. R. Datar, J. Martin, and R. Manteuffel, *J. Parenter. Sci. Technol.* **46**, 35–42 (1988).
11. United States Pharmacopoeia (USP) XXIII, *Biological Reactivity Tests, In Vivo ⟨88⟩*, The United States Pharmacopoeial Convention, Rockville, Md., 1994, pp. 1699–1703.
12. Code of Federal Regulations, *Indirect Food Additives Subpart B: Substances for Use as Basic Components of Single and Repeat Use Food Contact Surfaces*, Title 21, Part 177, U.S. Government Printing Office, Washington, D.C., 1994.
13. American Society for Testing and Materials (ASTM) *Standard Test Method for Determining Bacterial Retention of Membrane Filters Utilized for Liquid Filtration*, ASTM Standard F838-83, 1983.
14. United States Pharmacopoeia (USP) XXIII, *Particulate Matter in Injections ⟨788⟩*, The United States Pharmacopoeial Convention, Rockville, Md., 1994, 1813–1819.
15. United States Pharmacopoeia (USP) XXIII, *Purified Water*, The United States Pharmacopoeial Convention, Rockville, Md., 1994, p. 1637.
16. United States Pharmacopoeia (USP) XXIII, *Bacterial Endotoxins Test ⟨85⟩*, The United States Pharmacopoeial Convention, Rockville, Md., 1994, p. 1696–1697.
17. T. Uberoi, *Chem. Eng. Prog.* 75–80 (March 1992).
18. Food and Drug Administration, *Guideline on Sterile Drug Products Produced by Aseptic Processing*, Center for Drugs and Biologics, Rockville, Md., June 1987.
19. J. Meeker, E. Hickey, J. Martin, and G. Howard Jr., *BioPharm*, March 1992, Tech Note.
20. J.D. Brantley and J.M. Martin, *Integrity Testing Sterilizing Grade Filters*, Pall Corporation Publication PBB-STR-28, 1997.
21. F.M.H. Villars and G.B. Benedek, *Physics with Illustrative Examples From Medicine and Biology*, 2nd ed., Addison-Wesley, Reading, Mass., 1974.
22. D.B. Pall, *Annual Meeting of Parenteral Drug Association*, New York, November 1973.
23. T. Takegoshi, *Proc. of the Int. Congress: Advanced Technologies for Manufacturing of Aseptic and Terminally Sterilized Pharmaceuticals and Biopharmaceuticals*, Basel, Switzerland, February 17–19, 1993.

24. M.P. Dosmar, K. Wolber, H. Bracht, H. Troger, and P. Waiber, *J. Parenter. Sci. Technol.* (July/August 1992).
25. M.P. Dosmar, K. Wolber, H. Bracht, H. Troger, and P. Waiber, *J. Parenter. Sci. Technol.* (July/August 1992), *Filtr. Sep.* June 1993.
26. R. Jaenchen and J. Malsey, *Analysis of Water Penetration Through Hydrophobic Filters as a Reliable Method for Integrity Testing.*
27. *The Pall Water Intrusion Test for Integrity Testing Sterile Gas Filters* Pall Corporation Publication PBB-STR-29, 1997.
28. Parenteral Drug Association (PDA), Inc. *Validation of Steam Sterilization Cycles*, Technical Monograph No. 1, 1978.
29. Pall corporation, *Filtration Products for the Food and Beverage Industry*, Pub. FBB-100, 1990.
30. Pall Corporation, Pub. E-33, out of print.
31. Pall Corporation, *Selection Guide CLL Style Filter Housings*, Housing Data Sheet H26a, 1987.
32. Pall Corporation, *Selection Guide V-Series Filter Housings*, Housing Data Sheet H28, 1990.
33. Pall Corporation, *Water Intrusion Test for Integrity Testing Hydrophobic Sterile Gas Filters*, Pub. PBB-STR-30, 1996.

FLOW CYTOMETRY

SUSANN MÜLLER
University of Leipzig
Leipzig, Germany

THOMAS BLEY
Dresden University of Technology
Dresden, Germany

KEY WORDS

Cell cycle
Cell sorting
Cytometry
Fluorescence
Immunolabeling
Microbial performances
Population dynamics
Process monitoring
rRNA probes
Viability

OUTLINE

INTRODUCTION

Cytometry is a technique for measuring physical or chemical characteristics of single cells (or other biological particles). Such measurements are usually performed with the aid of a microscope, but there are severe limitations both in the number of cells that can be investigated microscopically and in the number of features that can be observed simultaneously.

In flow cytometry (which avoids these limitations), the cells pass the measuring volume in a fluid stream, preferably in a single file, and in a short time. The first flow cytometers, developed in the 1940s, were based on the Coulter principle, that is, the detection of changes in the electrical conductivity or impedance of a small saline-filled orifice as cells pass through.

The history involved leads most people to think of flow cytometers as instruments that measure optical signals from cells. These optical signals result from light being scattered by the cells (differentially, according to cell type and status) and the fluorescent light emitted by different cell compartments when excited by light of a certain spectral range. The cellular parameters detected can be characterized as intrinsic or extrinsic, depending on whether their measurement requires use of reagents or probes.

A cytometer consists of five main components:

1. *Illumination optics*. Usually consisting of an arc lamp (high-pressure mercury) or a laser (argon ion) combined with crossed cylindrical lenses to focus the light.
2. *Flow cell*. A device for focusing the cells in a linear file by hydrodynamics. The core stream with the cells travels in a laminar flow to the illumination point with the aid of a sheath stream of clean water. The main types are stream-in-air systems and variations thereof passing through a (flat-sided) quartz cuvette.
3. *Collection optics*. Usually consisting of a combination of optical filters for spectral separation, generally set perpendicular to the illuminating beam and in four spectral ranges (green, 510–540 nm; yellow, 560–580 nm; orange, 605–635 nm; and red, 650 nm and above).
4. *Detector electronics*. These consist of photodiodes or photomultiplier tubes (PMTs). The particles of inter-

est in biotechnology, which are frequently very small, (e.g., bacteria), emit only a small number of photons, typically about 1,000 in passage through the detection chamber. Therefore, an appropriate combination of high-power illumination optics and sensitive detectors is necessary for the application of flow cytometry in biotechnology.

5. *Data analysis*. The first flow cytometers displayed and stored the data in multichannel pulse height analyzers. Now the data are stored as standard list mode files in PCs, ready for statistical analysis and sophisticated visualization.

The result of a flow cytometric investigation is a frequency distribution, that is, a histogram. In the channels of this histogram, cells are counted according to the light intensity detected, which corresponds to the relative size or quantity of a specific cellular compartment or component, such as the DNA content. Measuring rates up to 10,000 cells s^{-1} are possible, although about 100 cells s^{-1} is more usual.

Flow cytometry is not recommended for measuring the absolute content or concentration of cell compartments. However, it is the best method for classifying cell types with different physiological properties and for selecting and quantifying subpopulations in a cell ensemble.

If subpopulations of (vital) cells are to be chemically analyzed or cultivated further, the information provided by flow cytometric measurement can be used for selecting appropriate subpopulations by an electrical or mechanical sorting process. Depending on the type of flow cell and the type of illumination involved, three basic constructions are used:

- Stream-in-air
- Contained flow chamber
- Dark-field illumination

The devices range from large constructions for research purposes to small and inexpensive laptop machines on the way to an online approach. The development of new flow cytometric methods is often accomplished using special modular constructions ("cyto mutt" comp. [1]).

A further development is the slit-scanning procedure, which allows the shape of the signal generated by the cell passing through the illumination beam to be used for gathering additional information. In this way, morphological features and the distribution of cellular compartments can be recorded.

Special demands have led to appropriate modifications. The distribution of aquatic microorganisms can be monitored in the size range of 0.1 to 2000 μm by such a device, for instance (2).

Flow cytometers are best used to answer questions concerning the dynamics of subpopulations in a cell population, for example, cells in a mammalian tissue that have a higher DNA content than that of G1-phase cells. Thus, in medicine, flow cytometry was first applied in oncology, where it was quickly (and frequently) used in cancer diagnosis. These results generated a solid basis for mathematical modeling of cell-cycle regulation in eukaryotic organisms.

Using immunofluorescence and multiparameter analysis, flow cytometry became a successful, and standard, method in laboratory hematology and clinical immunology in the 1980s. Several commercial devices are now available, constructed specifically for medical purposes. An excellent and comprehensive documentation of the method of flow cytometry is given by Shapiro (1).

The use of flow cytometry in biotechnology, and especially in bacteriology was for a long time not as successful as it was in medicine. After early, inspiring investigations by Bailey et al. (3), Slater et al. (4), and Hutter and Eipel (5) further development seems to have been delayed. There are probably three major reasons for this:

- The high cost of the method
- The small size of the cells investigated (an *Escherichia coli* cell has only 10^{-3} to 10^{-4} times the size or DNA content, and consequently the potential fluorescence energy gain, of a mammalian cell)
- The belief of most biotechnologists that analysis of physiological state distributions do not have an important role to play in understanding the dynamics of bioprocesses or for running them safely and economically

The first two handicaps have now been overcome because powerful and inexpensive microcomputers have considerably reduced the cost of the technique. Flow cytometry has also benefited from new methods of data analysis and mining, such as neural network analysis, which allow a comprehensive evaluation of the rich and complex information gained by flow cytometry (6). Furthermore, new staining methods combined with improved illumination, detection, and amplification have made flow cytometric analysis highly reliable, even for bacteria and microplankton as small as 1 μm in diameter.

Nevertheless, the key to more widespread use of flow cytometry in biotechnology is more general acknowledgment of the importance of a segregated model concept. Toward this end, Fredrickson et al. (7) proved that the statistics, the distribution of cell populations, and the dynamics of bioprocesses are inseparably connected. Moreover, Munch et al. (8) demonstrated, with a sophisticated approach to flow cytometry, that knowledge concerning the distribution of states in the cell cycle is essential for deeper understanding of the growth dynamics of yeast and the metabolism of a biomass.

BACTERIAL CELLS

Bacteria exist as dynamic and diverse populations in nature. Even the cellular heterogeneity within a pure bacterial culture is far greater than previously assumed. To understand bacterial ecology in nature and in bioprocesses, new techniques are required for answering questions that could not be solved by traditional methods alone. Now more than ever, flow cytometry has been recognized as a valuable asset for microbiologists and is perceived to be an important and indispensable tool for investigations on the microenvironmental level.

Bacterial flow cytometry enables visualization of cell states and allows the analyst to follow growth, death, replication, cell division, metabolism, and surface phenomena, greatly enhancing the ability to understand and control cell physiology. A great assortment of fluorescent stains is commercially available; their numbers have rapidly increased recently. It is not possible to give an overview of their application in this article, but Lloyd (9) provides a fine introduction to this topic. Summaries are given in Refs. 10 and 11, too.

Generally, however, it should be realized that these techniques need to be handled more carefully for use with bacteria than with eukaryotes to avoid unclear results from artifacts arising either from sampling or from analytical errors.

Growth and the Cell Cycle

All survival strategies should be programmed, above all, to safeguard the genome for the future. Results obtained by flow cytometry allow conclusions to be drawn and tested about how the physiological state of cells is connected with survival strategies under changing environmental conditions. Consequently, the most obvious targets for bacterial staining are the nucleic acids. Steen and Boye (12) and Skarstad et al. (13) presented a model for calculating the relative rates of DNA synthesis per cell from studies based on *E. coli* strains. By means of this model and flow cytometric evaluation of the DNA distribution, the generation time (τ) and the initiation and termination of the replication and postreplication periods (C and D, respectively) can be estimated. Features of replication (e.g., the number of initiation origins) seem to be dependent both on the genetically determined disposition of the bacterial strain analyzed and the growth conditions leading to the initial cell mass. In unperturbed bacterial cultures with high growth rates, several replication forks may be initiated in each cell, if the growth conditions are optimal. Then, after treatment with antibiotics, *E. coli* cells will contain two, four, or eight fully replicated chromosomes. In contrast, bacteria growing slowly, at doubling rates of less than 1 per hour, exhibit an increase in the prereplication period as well as in the C and D periods. This effect can be amplified by further decreasing the growth rate using chemostat cultures under limited growth conditions (12,13).

Besides the considerations just mentioned, it must be stressed that during limited, and even unrestricted, growth, the behavior of individual cells differs. As pointed out by Åkerlund et al. (14), each cell goes through different discontinuous processes in its life (e.g., the cell cycle). They are therefore different both in size and metabolic activity. This is generally true, except in steady-state conditions, to some extent, where only the distribution of the cells and the cell size are time invariant).

Considering these results, it is clear that key events in cell physiology, including initiation of replication and cell division, are tightly bound to the microenvironmental conditions in the surroundings of the bacterial cell. Obtaining information about the replication behavior of biotechnologically useful bacterial strains is, therefore, a crucial prerequisite for controlling industrial bioprocesses.

This knowledge could also be used for investigations into the population dynamics of marine bacteria. Because of fluctuating climatic influences, these organisms must constantly adapt to changing conditions and move between states of growth and starvation. In such cases, major alterations in the level of other cellular components, such as lipids and proteins, may be observed in addition to changes in the nucleic acid contents.

Bioprocessing

Progress in bioprocess engineering depends ultimately on the level of understanding and control of the physiological state of the bacterial population being exploited. Process efficiency is strongly sensitive to changes in the cell state. Characterization of these states via fluorescence monitoring, evaluation of the data obtained, and subsequent regulation of the process regime by controlling the surrounding natural or artificial conditions is one of the most ambitious tasks in the near future for biotechnology. Flow cytometry appears ideally suited for determination of such parameters as cell concentration and size as well as for visualizing intracellular performance in industrial processes promoted by bacteria. Using this technique, the state of the cellular system can be precisely monitored, allowing optimization of process efficiency in production or degradation of the target substances (15).

Although bioprocess control on the cellular level is not being widely practiced as yet, some processes are now being adjusted by flow cytometry. One main limitation seems to be the high content of particulate constituents in most industrial media, because discriminating between microorganisms and noise signals from media is not a trivial undertaking.

Microorganisms respond to an unbalanced supply of nutrients or deviations from optimum physical factors by increasing the synthesis of intermediates. This phenomenon is called overflow metabolism. The metabolites overproduced can be excreted or accumulated intracellularly. Flow cytometric methods have been established for detecting intracellular accumulation of polyhydroxybutyrate (PHB), a typical overflow metabolite in many bacterial strains (16). PHB serves as an energy and carbon reserve and has received attention as a thermoplastic and biologically degradable polymer. Furthermore, researchers are searching for novel bacterial strains that may accumulate novel types of polyhydroxyacid (PHA) possessing better physical and mechanical qualities. Some bacterial yield up to 90% of their dry weight in this material. The inclusion bodies so formed alter both the size of cells and their light-scattering behavior. The effect of different cultivation conditions on PHB formation has been investigated in *Ralstonia eutropha* and recombinant *E. coli* transformed with PHA-synthesis genes. In these cases, cell-sorting technologies can be used to isolate highly efficient strains at a high speed. Flow cytometry can also be used to quantify heterogeneity of PHB production and accumulation, as found in *Methylobacterium rhodesianum* cultures. These investigations are based on the idea that growth and product formation (overflow metabolism) are coupled to specific cell states, which are an expression of maturation of individuals and are connected to stages in the proliferation cycle. It has been found that under growth-limiting conditions, the cells first go though the DNA replication program, thereby safeguarding the genetic information by doubling the chromosome content. Doing this, the organisms maintain the chance of restarting multiplication as a forward strategy of survival if better conditions arise. Cells only lay down PHB as an energy reserve in this kind of situation (17).

Introduction of foreign genes into bacterial cells enables synthesis of desirable products such as proteins, lipids, and a wide variety of other biologically active compounds. Normally, the expression of these genes alters the normal pattern of interaction and synthetic activities in the host cell. It may be observed that highly active cloned gene expression or extremely large plasmid content leads to a markedly slower growth rate in recombinant cells. Generally, however, some of the recombinant cells revert during prolonged cultivation, either (in the case of plasmids), through defective partitioning during cell division, or through changes in plasmid structure. By this process, a subpopulation of nonrecombinant cells develops that has a higher specific growth rate. In chemostat experiments, the recombinant cells are usually replaced by this phenomenon, and the plasmid-free population comes to dominate the dynamics of the reactor (18). It is generally believed that the production of heterologous products is metabolically harmful to the cell. Occasionally, it has been proposed that the decrease in the growth rate seen is more likely to be a result of deleterious and injurious interaction between the expressed product and some component of the cell. Flow cytometry can be used to detect segregational instability of a plasmid-containing expression system as well as to estimate the physiological state of the cells involved.

There have only been a few biotechnological studies investigating the stability of recombinant cells on near-industrial scales. By way of illustration, however, the method has been used to examine recombinant *E. coli* cells forming protein inclusion bodies via wide-angle light-scattering measurements (19).

Fluorescence-activated cell sorting has also been found to be capable of isolating gramicidin S hyperproducing mutant cells of *Bacillus brevis* from wild-type cells (20). Furthermore, flow cytometry can be used to study the progress of gene expression in a wide range of bacterial cell systems. The expression of foreign genes in bacterial cells or the stability of transformed plasmids can generally be determined by the use of bioreporter genes. These systems allow flow cytometric monitoring of the gene expression even when the gene product is difficult to assay. The promoterless reporter gene is placed under the expression control of the promoter of the foreign gene. A typical example is use of the *lacZ* gene from *E. coli*, one of the most widely used bioreporter systems. The *lacZ* gene encodes the enzyme β-galactosidase, which cleaves a fluorogenic product from specially designed substrates. The emission of the resulting fluorescence is proportional to the amount of the enzyme in the cell, corresponding to the expression of the target gene. Consequently, the β-galactosidase-positive cells exhibit a level of fluorescence that is measurable for each cell by flow cytometry (21). However, a primary problem in single-cell β-galactosidase assays is leakage of the intracellular fluorescent marker, although a number of methods have been developed to prevent this movement (22). Other bioreporter systems encode detectable markers such as green fluorescent proteins or surface antigens for monitoring microbial performance.

There have also been attempts to use polymerase chain reaction (PCR) in DNA amplification methods for sensitive detection of specific innate or foreign DNA or RNA sequences inside intact cells (23). Some of the population seem to remain unbroken after such treatment, allowing subsequent flow cytometric assessment. For example, PCR can be used to detect plasmid-encoded gene sequences or poorly expressed mRNA in individual bacterial cells (24).

Vitality

In contrast to biotechnologically processes, bacterial cells in nature live neither in pure cultures nor in surroundings of constant temperature, pH, or nutrient availability. Many of the (microscopically) visible bacterial cells are inactive or dying. Every change in the habitat is potentially a source of stress, which results in a change in growth rates. The sum of the cellular responses to stress is a factor associated with the survival strategy.

Responses to stress take place at two distinct cellular levels: first, activation of existing enzymes and transport processes and, second, gene expression. The biochemical bacterial stress response can be precisely determined using flow cytometry, by analyzing changes in the type of duplication, cellular pH, membrane potential, and the amount and kind of storage products. The physiological links between the environment and the genetic systems with respect to survival of stress situations are often mediated by expression of proteins involved in a range of tolerance mechanisms. It is commonly accepted that specific genetic programs exist for prolonging the survival of non-growing bacteria exposed to starvation or stress. Bacterial programmed cell death seems to be mostly related to population development, because the lysed cells are usually essential as sources of nutrient compounds, allowing completion of the developmental cycle of the remaining, living population (25).

To date, knowledge of the connections between bacterial processes and bacterial community structure in nature is inadequate and restricted by procedural flaws. Most commonly used methods for the detection and enumeration of bacteria have serious limitations because of their incapacity to allow certain bacteria to grow. The reason may be that the bacteria concerned need symbiotic partners, anaerobic microenvironments, other uncommon conditions, or unknown growth factors.

However, for a variety of reasons, quantifying the bacteria present in the investigated system, independently of growth, is a fundamental requirement for almost all bacteriological studies. Living cells must be distinguished from dormant cells (induced by starvation) as well as from viable (but not culturable) and truly dead bacteria. To characterize the degree of viability of bacteria, flow cytometric detection of basic cell functions such as reproductive activity, metabolic activity, and membrane integrity is an invaluable tool (26,27). Reliable flow cytometric methods for viability assessment that circumvent the need for culture were largely developed by Porter et al. (28). Investigations associated with the state of dormancy in bacteria, in which the cells survived for extended periods of time without growth or multiplication, have been performed by Kell et al. (29). Cationic lipophilic viability dyes should be handled with care in such studies, because they are very often extruded in an energy-dependent manner. There are several indications that a transport system is involved. Excellent correlations can regularly be observed between viability of the cell and the ability to translocate the dye to the cell's surroundings.

A wide variety of bacteria are able to metabolize xenobiotica as sources of carbon and energy for growth and to decontaminate polluted ecosystems in this way. However, the effects these toxic substances have on the artificially introduced bacterial species should be considered in such biotechnologically forced processes. Above certain, critical concentrations the chemicals become toxic, the bacteria are poisoned, the bioremediation process slows down and finally stops. It is well known that most of such substances are membrane active and burden the cell energetically. Consequently, cellular changes in the energetic state and the form of the membrane potential must be analyzed with the solute transport, ATP synthesis, and pH homeostasis. To control and optimize such processes, information is necessary about death characteristics, which depend on the physiological state of the cells (30). Flow cytometric assessment of the membrane potential allows a rapid and differentiated analysis of the ecotoxic potential of pollutants toward bacterial cells and detection of the rapidly changing physiological behavior of the cell in growth-phase-dependent variations of the membrane potential (Fig. 1). Very little information exists regarding the members of bacterial communities with respect to morphology and taxonomy that are responsible for the decontamination process. There is also little information about differences in the physiological state of specific strains under certain conditions. For this reason, methods must be developed that relate specific cellular parameters to taxonomic classes (Fig. 2). The first such investigations of this type were done by Herrmann et al. (31).

In medical science, situations also arise in which living and dead cells must be distinguished. In this way, flow cytometry can be used in the assessment of biocidal drugs, offering rapid assays for prokaryote drug susceptibility. The aim is to use the technique in the clinic for rapid susceptibility testing as an aid for choice of therapy. Common methods for investigating resistance to antibiotics and disinfectants and their effects on bacterial strains are based on estimating the minimum growth inhibitory concentration (MIC) of the compound or the decrease in colony-forming units after exposure. These conventional techniques need at least 24 to 48 h to get information about the action of antibiotics.

Flow cytometry is much faster and, furthermore, gives considerable advantages in drug research for investigating the mode of action of novel antibiotics. The susceptibility of bacterial cells to antimicrobial agents may be so profoundly influenced by age, growth, and metabolism to be heterogeneous within a population. The permeability of the bacterial membrane changes during the life cycle, for instance. Recently, it has become apparent that the permeation of lipophilic substances through the envelope of Gram-negative bacteria has great potential in the development of new chemotherapeutic agents. Generally, using flow cytometry, key drug-induced changes in light-scattering behavior, uptake of vitality stains, DNA ploidy, and protein patterns can be estimated (32,33).

Mixed Populations

Bacterial ecology requires the application of new techniques to help address problems that cannot be solved by traditional methods alone. Efforts to demonstrate the value of flow cytometry in identifying different species within complex populations have relied on immunofluorescence-based methods such as the application of antibodies (34), lectins (35), or rRNA-targeted oligonucleotide probes (36). The methods chosen must be applicable to as wide a range as possible of different organisms. These techniques have also found uses in areas such as quality control of water and foodstuffs as well as in soil and water ecology.

Today, fluorescently labeled oligonucleotide probes are most commonly used for monitoring specific strains. The probes are complementary to group-specific regions of the highly conserved multicopy 16S or 23S ribosomal RNA molecules and are bound using in situ hybridization. In this manner, the bacterial strain of interest can be detected without separating it from the surrounding microflora and contaminating debris. Furthermore, comparative sequencing and the design of nucleic acid hybridization probes have served to establish an alterated phylogenetic framework.

Differentiation between artificial cultures is possible if the members have contrasting proportions or absolute amounts of guanine–cytosine or adenine–thymine. Estimations of guanine–cytosine content vary from nearly 20% up to 80% (37). Using dyes with different binding preferences to DNA and double fluorescence excitation, composition analysis seems to be workable.

Another way of differentiating mixed populations is the genetic insertion of a fluorescent system to split populations with special physiological capacities.

Food

The use of fluorescent labels is accepted as a successful and sensitive approach for detecting bacteria in food. Lactic acid bacteria are fermentative Gram-positive organisms that are widely used as starter cultures for manufacturing a great assortment of foods and drinks, such as cheese, yogurt, and sausages. The quality of a starter culture is highly important. It must be taken into account that industrial fermentations may involve stressful conditions that can affect bacterial gene expression, arrest cell multiplication, interfere with metabolic activity and survival potential, and lead, possibly, to cell death. Studying responses of lactic bacteria to stresses such as freezing and

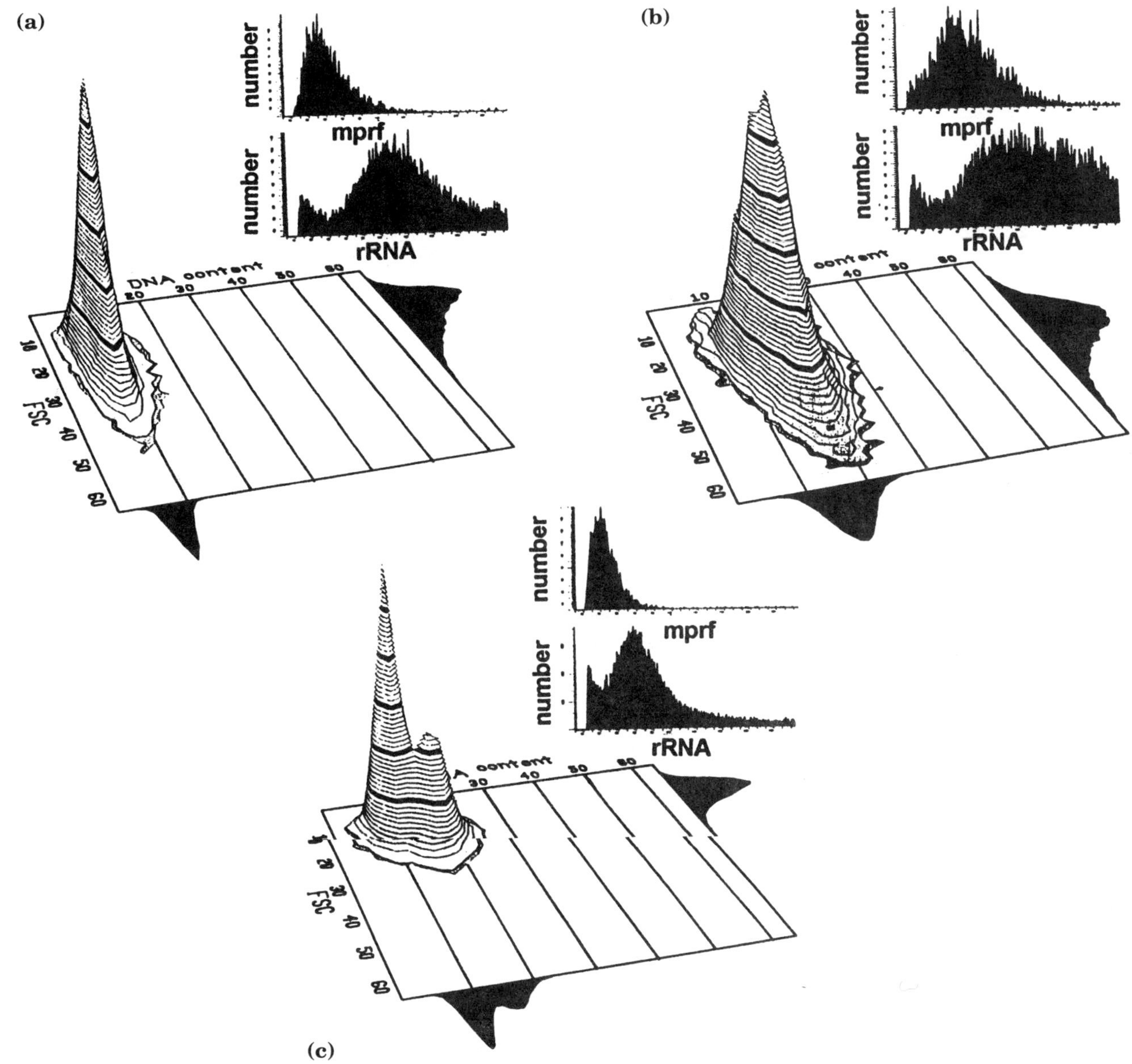

Figure 1. Analysis of the cellular DNA content, membrane-potential-related fluorescence (MPRF), and rRNA content of *Ralstonia eutropha* JMP 134 in relation to the growth rate during batch cultivation on phenol. Samples were harvested at the lag-phase (**a**), early log-phase (**b**), and stationary phase (**c**). The cells were stained with the carbocyanine dye di-OC_6 (3) for MPF, with oligonucleotide probes for analyzing the rRNA, and with 4′,6-diamidinophenylindole (DAPI) for the DNA. The proportions of cells having two chromosomes is estimated to be 3.88% (**a**), 87.80% (**b**), and 38.96% (**c**).

drying, with the aim of monitoring, controlling, and promoting appropriate responses in starter cultures is a typical requirement of flow cytometric investigations in food industries. The method enables determination of cell counts, cell viability, and the effects of stress on the bacterial physiological state. The method can suffer from interference by particulate materials, but multiparameter counting based both on light scattering and fluorescence measurements generally enables separation of bacteria from background material (38).

There are many situations where the total bacterial concentration is the only measurement that is important. Bacteria are a principal cause of food spoilage. Contamination by harmful organisms can cause severe human food-borne diseases, either caused by the bacteria themselves or the toxins released by the bacteria. The time required for conventional tests (plate-count techniques) can lead to substantial delays (of 24 to 72 h), which is a serious disadvantage. Direct epifluorescent filter enumeration is another commonly used technique that allows microscopic

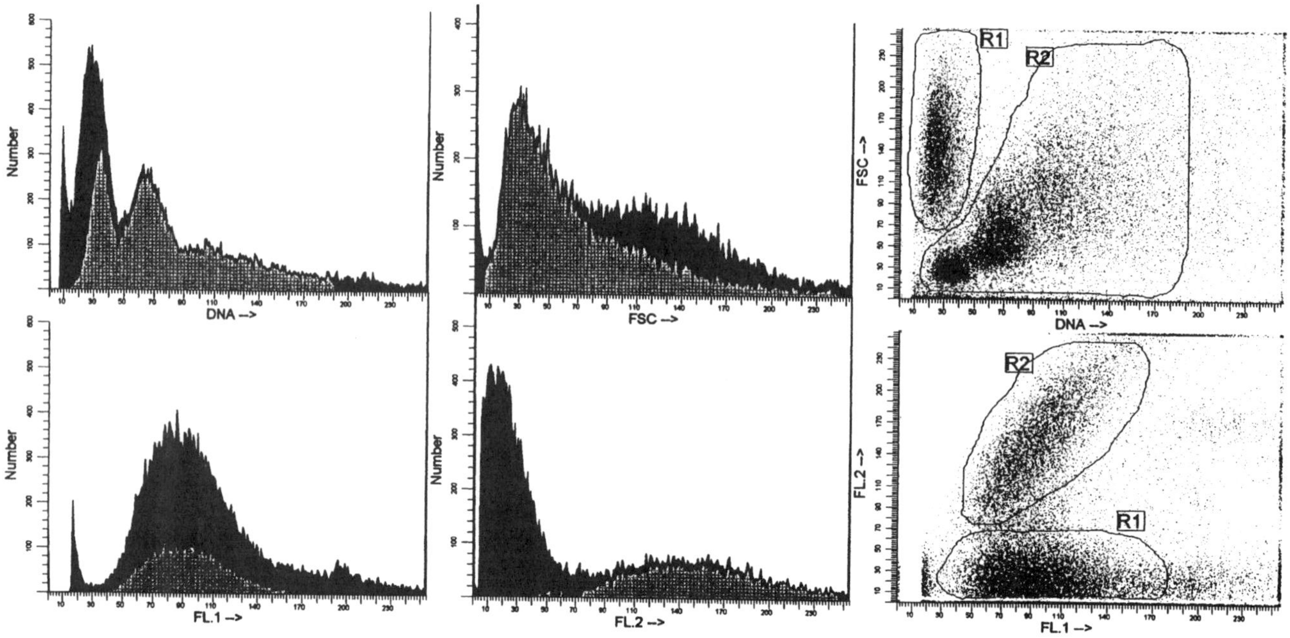

Figure 2. Analysis of the DNA content using DAPI and cell size differentiation via forward light scatter measurements of a mixed population of *Ralstonia eutropha* JMP 134 (R1) and *Acinetobacter calcoaceticus* 69-V (R2) (upper three histograms). Discrimination of the two strains was achieved after in situ hybridization with differently labeled groups and strain-specific oligonucleotide probes using the same sample (lower three histograms). Both strains were harvested during the middle log-phase in mixed batch culture.

counting of bacteria retained on a filter. Over the past 10 to 15 years, this method has been used extensively for studies on the survival of bacteria in food and for estimating biomass in drinks (39).

However, results suggest that flow cytometry also has significant potential for the detection of pathogenic microorganisms in the food industry. Using this method, pathogenic bacterial cells can be detected by applying fluorescently labeled monoclonal antibodies, for rapidly detecting *Salmonella* or *Listeria*, for example. Accurate detection has been demonstrated down to levels of below 10^4 cells/mL or even 1 cell/mL after preenrichment (40,41).

Further Applications

The function of bacterial strains in certain human or animal diseases is increasingly being assessed using flow cytometry. For instance, specific cellular characteristics of bacteria involved in skin diseases are being studied by these techniques (42). Furthermore, flow cytometry offers a rapid method for characterizing anaerobic bacteria present in human fecal suspensions using specific physical and biochemical features (43). Attempts have been made to detect bacterial toxins via fluorescently labeled antibodies bound to polystyrene microbeads in stool specimens (44). Moreover, flow cytometric investigations have been used to identify antisera produced against bacteria capable of specifically binding to surfaces of the target bacteria (45). The resurgence of tuberculosis has caused considerable effort to be focused on developing rapid methods for inhibiting growth of the causal organism by a variety of agents (46). Also, in animal disease investigations, there is growing interest in the capacity of fish-pathogenic bacteria to survive long-term starvation (in a nonculturable mode) in seawater. Knowledge of how long pathogenic bacteria preserve their infective capacity after being released to the environment is clearly of great practical importance (47).

In conjunction with other biochemical and genetical methods, flow cytometry has been used as a rapid method for testing the effects of shear stress, extreme temperatures, action of chemical compounds, sonication, and electroporation on bacterial vitality (48). Bioluminescent *E. coli* harboring lux genes have been used for detecting environmental pollutants, for example. Furthermore, fluorescence-activated cell sorting is often the best available method for obtaining purified samples from natural environments in order to apply molecular techniques or for extracting or enriching samples for detecting low numbers of specific bacteria, such as pathogenic or genetically engineered microorganisms.

The function of bacterial cell surfaces has been investigated in various ways, such as through the interaction of bacterial cells with electrically charged surfaces, macrophages, or other cells through electrostatic interaction. The investigated phenomena may be mediated by bacterial polysaccharides, which are habitually associated with the outer surface of the bacterium. These polysaccharides form a unique class of polyelectrolytes, possessing antigenic properties, for which a lot of fluorescent monoclonal or polyclonal antibodies are available (49).

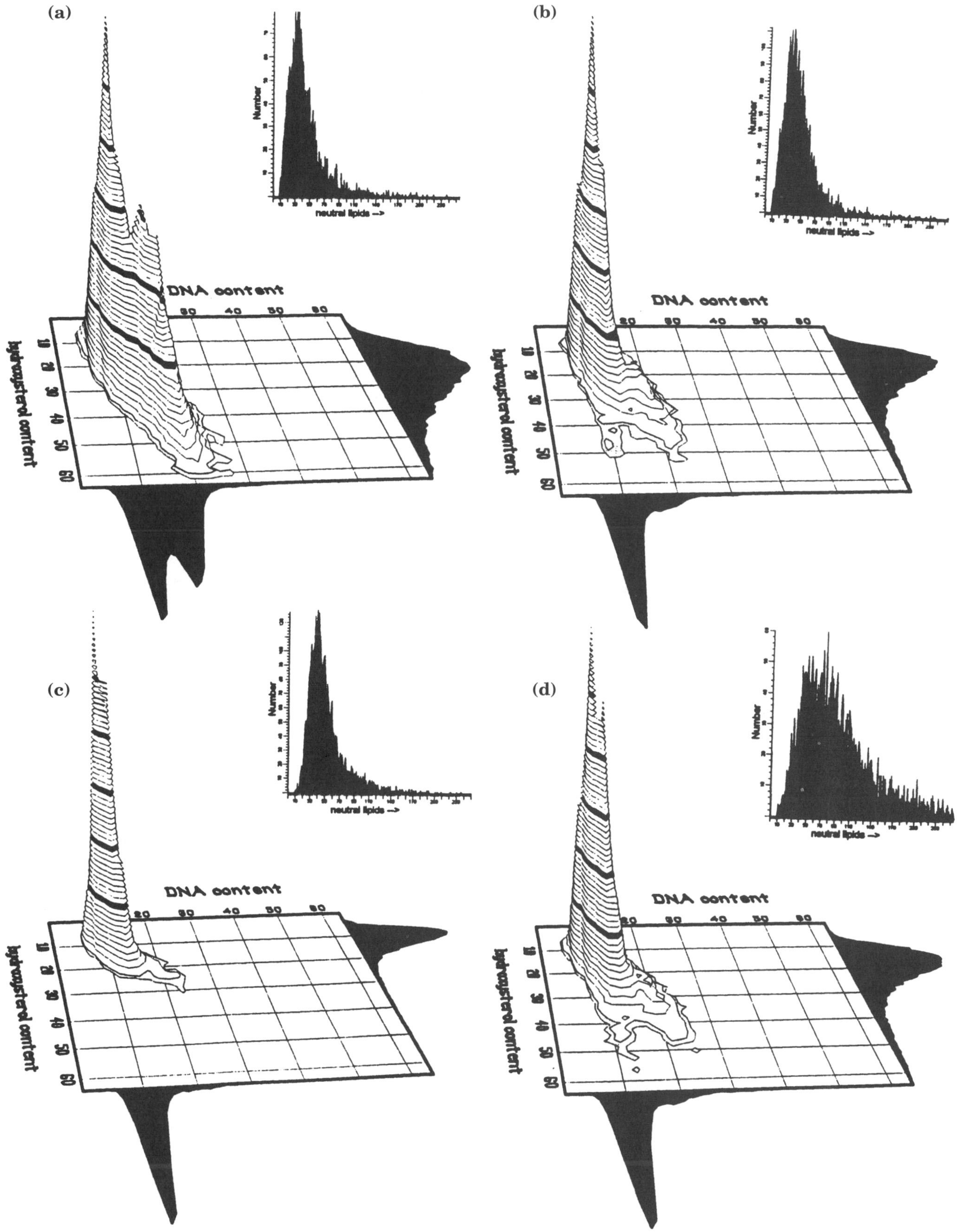

Figure 3. Flow cytometric analysis via double fluorescent measurement of the DNA and 3-β-hydroxysterol content of *Saccharomyces carlsbergensis* during fermentation. The first sample (**a**) was harvested 20 after inoculation of the brewing reactor and is characterized by a high rate of proliferation. The second sample (**b**) was harvested the following day, the third (**c**) after 4 days, and the fourth (**d**) after 6 days. The fermentation process was finished on the 7th day by carbon limitation. As shown in the last three samples, most of the cells (especially in the case of **c**) were in the G1 phase of the cell cycle and possessed a limited amount of 3-β-hydroxysterols. The insets show the distribution of the neutral lipid content. A large increase in Nile red fluorescence could be observed after 6 days (**d**).

YEASTS

Of all the biotechnologically important microorganisms, the yeast *Saccharomyces cerevisiae* has been the most extensively investigated using flow cytometry to date. As baker's yeast and producer (along with certain related species) of beer and wine, this yeast is of great economic significance. Nevertheless, the chief reasons for the exceptional interest for flow cytometry in this yeast is that it is a widely accepted model organism for eukaryotic cells, because of its ease of cultivation and its well-developed genetics. Investigations of other yeast genera have been very rare and have not been done systematically.

Growth and the Cell Cycle

Basic research into yeast growth and cell cycling by flow cytometry began in the late 1970s with one-dimensional analysis of cell protein content (by FITC-staining) and cell DNA content (using mithramycin staining) by Gilbert et al. (50), Slater (4), and Hutter (51). Both common batch-growth phenomena and the typical diauxic growth pattern of *S. cerevisiae* were investigated. These authors elucidated basic rules concerning the timing of DNA synthesis in cells growing at different rates. The first experimental evidence supporting the hypothesis that the eukaryotic G0/G1 phase is the most variable phase of the cell cycle, and that the duration of the S phase is nearly constant and is not greatly influenced by the cell's environmental conditions, was generated by these flow cytometrically monitored experiments.

In the 1980s, Alberghina et al. (52) and others introduced two-dimensional approaches, involving a light-scattering signal and a fluorescence signal. In particular, methods for the determination of fluorescently stained single-cell protein were added by Scheper et al. (53). Furthermore, sharper detection of the DNA content and, therefore, of the cell cycle state distribution was also made possible by the introduction of more specific staining methods such as propidium iodide (excitation at 488 nm) and DAPI (excitation in the near UV at 351 nm). The theory that DNA replication of *S. cerevisiae* is initiated when a cell has reached a certain critical size was proved experimentally by these investigations.

Asymmetric cell division, subcellular characteristics, and numerous features of the intracellular distribution of cellular components were also shown to be detectable by the application of slit-scanning data acquisition (54). A double flow cytometric tag allows the progress of the cell cycle to be tracked in the different cohorts of mother and newborn daughter cells during balanced exponential growth (55).

Bioprocessing

Substantial progress also came from bioprocess type investigations. Because *S. cerevisiae* populations tend toward autosynchronization under certain process conditions in a continuously stirred tank reactor (CSTR), the changing distribution of states in the cell cycle can be determined by flow cytometry and correlated with the physiological performance of the population. It was demonstrated in this way that the type and rate of substrate utilization and cell compartment synthesis are dependent on the stage of the cell proliferation cycle (53).

In a key paper from 1992, Munch et al. (8) proved on the basis of a comprehensive use of flow cytometric methods that understanding cell cycle behavior is essential for the characterization of growth dynamics in bioprocesses. Using Calcofluor White M2R, the chromosomal and the mitochondrial DNA content in the yeast cells could be separately monitored, and the aging process in the mother cells of the budding yeast could be recorded.

Changing respiration coefficients, combined with changes in substrate fluxes and cell processes, such as the storage and mobilization of carbohydrates and the excretion of ethanol, can now be correlated with different stages in the cell cycle.

A further important cell component, with cell-cycle sparking and bulk membrane functions (3-β-hydroxysterol), became recordable by staining with FITC covalently coupled to the amino group containing polyene-macrolide antibiotic, Nystatin A1 (56,57).

Brewing

Demands of the brewing industry for ever-faster production, without prejudicing the quality of the product, have made flow cytometry more and more significant in this classical branch of biotechnology. The use of cylindrical cone-shaped brewing reactors with volumes of at least 160 m^2 requires increasingly abundant information concerning yeast physiology. Close monitoring of changes in the distribution of DNA, neutral lipid, and 3-β-hydroxysterol contents in *Saccharomyces* cells during propagation, fermentation, and storage enables time-saving process control (58).

DNA distribution monitoring has shown that only cells with a single chromosome content produce ethanol and CO_2 efficiently. On the basis of flow cytometrical investigations, the optimum point for controlling the wort and oxygen supply can be accurately determined. Double-staining techniques prove that cells with a high content of neutral lipids and 3-β-hydroxysterols have a high survival capacity. Only cells with such features are able to generate vigorous subsequent fermentation. However, additional measurements for determining the proportions of living cells or detecting bacterial infections are only sporadically performed in the brewing industry by flow cytometry.

Further Applications

Wittrup and Bailey (59) use flow cytometry to monitor the β-galactosidase activity in *S. cerevisiae* by measuring the fluorescence decay of the fluorogenic substrate resorufin-β-D-galactopyranoside. With this method, the expression of foreign protein, coupled to the β-galactosidase gene, could be measured, and a segregated model mapping β-galactosidase activity as a function of plasmid copy number was presented (60).

Another method for detecting this protein was a quantitative immunofluorescence procedure developed by Eitzman et al. (61). These authors demonstrated that the activity of the enzyme is correlated with the stage in the cell cycle that individual cells have reached.

Peroxisomes are inducible organelles that may occupy large proportions of the cell volume when yeasts are growing on methanol-containing media. *Hansenula polymorpha*, a methylotrophic yeast, was investigated via side-scattered light (which was found to depend on cell volume, morphology, and structure) and FITC retention (which was largely dependent on the vacuole.) A wild strain and a strain partially repressed by glucose were compared (62). Furthermore, it has been shown that the viability of the pathogenic *Candida* yeasts can be determined by measuring the membrane potential (63) and that of viable blastospores using tetrabromofluorescein (64).

CELLS OF OTHER ORGANISMS

Aquatic Organisms

In aquatic science, flow cytometry can make an important contribution to ecological and physiological studies of natural populations of microbial plankton. It has been increasingly used to analyze natural communities of aquatic microorganisms because of its sensitivity and quantification capacity. To understand the processes that influence the behavior of any planktonic species, information is required on both the variability between individuals and groups and the major environmental factors.

Flow cytometry was introduced in freshwater and marine sciences at the beginning of the 1980s. However, natural samples of aquatic particles are generally very heterogeneous; there are often immense differences in concentrations and cell sizes (from 0.2 to more than 20,000 μm in diameter) as well as in cell types. Additionally, they may be contaminated by inorganic particles, and artifacts may be generated that can make the data difficult to interpret.

However, the application of flow cytometry to planktology is no longer hampered by shortcomings of commercial instruments. High-resolution and high-speed sorters are available for rapid and precise estimation of cell numbers, differentiation of cell types, and other pertinent problems. Neural network analysis provides an innovative approach for examining taxonomic properties and the state of the marine food web, including its involvement in the surrounding environment.

The prokaryotic fraction of natural marine communities is composed of both heterotrophic and autotrophic organisms. In spite of their very small sizes (0.2 to 2 μm), even the oceanic picoplanktonic cells are now easy to identify. Major groups of photosynthetic prokaryotes can be discriminated by flow cytometry because of their different and, compared to other algae, unusual pigment compositions and their forward light-scatter signal. The heterotrophic organisms are mostly bacteria. Their lack of pigments makes them less easy than autotrophs to analyze by fluorescence methods. Nevertheless, flow cytometry has been used to evaluate kinetic constants of nutrient uptake in relation to cell quantity and, additionally, to characterize individual parameters of the marine bacteria (65). Frequently, total counts of these bacteria include a large fraction of non-nucleoid-containing bacteria (ghosts), which may be cell residues of virus-lysed bacteria or remains of protozoan grazing. Nevertheless, simultaneously measured DNA content allows further discrimination between autotrophic and heterotrophic components of the picoplankton. In some cases, it has been demonstrated that the cell cycle of autotrophic planktonic prokaryotes progresses in phase with the daily light cycle (66), in contrast to cyanobacterial populations, which apparently do not adjust their circadian timekeeping.

Eukaryotic algae exhibit a strong natural autofluorescence emanating from chlorophyll or other pigments such as phycobiliproteins. The simultaneous measurement of DNA and chlorophyll fluorescence, together with cell size, is necessary to distinguish subpopulations of phytoplankton. Flow cytometry allows the quantification of changes in cellular pigmentation (e.g., chlorophyll and phycoerythrin concentrations) of divergent phytoplankton subpopulations. Such alterations arise in response to nonsteady light or nutrient regimes, such as may be encountered along a depth profile.

Information on the physiological status or response induced by different, artificial treatments can be easily assessed by flow cytometry. Biomass production in phytoplankton and its dependence on environmental conditions can be detected via measurement of the activities of various enzymes such as esterases (FDA) or nitrate reductase (by immunolabeling). The recent development of taxon-specific probes should increase the applicability of flow cytometry for rapid identification of cultured phytopico- and nanoplanktonic strains, especially those that lack taxonomically useful morphological features. Moreover, rRNA probes may provide new information on species that are not amenable to culture in the laboratory (67).

Furthermore, contemporary research has shown that protozoa are important in the dynamics of the aquatic system. Grazing by heterotrophic protozoa has now been accepted as the most important factor responsible for limiting the numbers of bacterial cells in aquatic ecosystems and maintaining nutrient availability. There is evidence suggesting that bioremediation processes involving bacteria may be strongly inhibited in the presence of phagotrophic protozoa, for instance. Sometimes, microbial population growth rates are difficult to ascertain because ubiquitious grazers remove cells as quickly as they are produced. These taxonomically heterogeneous organisms (including ciliates, amoebae, and a variety of flagellates) can graze voraciously on phytoplanktonic and bacterial cells that form the foundation of the marine food web. Discrimination has been achieved between prey cells and a dinoflagellate through DNA staining and isolation of the eukaryotic nuclei. Furthermore, differentiation of phagocytosic activity within populations of protozoa is measurable using fluorescently labeled microbeads (68).

Some investigations have been performed on viruses, regarded as ubiquitious, and biologically active members of marine and freshwater microbial communities, in which the viruses may be responsible for the destruction of a considerable proportion of the bacterial and cyanobacterial populations. In some cases, virus and host support each other in a symbiotic manner, in contrast to the observed annihilation of alternative host populations within the microbial community.

Plants

Application of flow cytometry and cell sorting procedures to study higher plant cells required methods for the production of single-cell suspensions. The need to probe the performance of protoplasts was the main motivation for developing methods for analyzing plant cells. This also required a series of modifications to the flow cytometric instrumentation, because plant cells were sometimes found to be much larger than those of other living organisms.

Flow cytometry has been accepted as a rapid and trustworthy technique for genome analysis; for obtaining information about the size, ploidy, and aneuploidy changes during plant evolution; and for the differentiation and state of replication of plant cells. It provides an accurate method for determining the proportions of cells in the G1, S, and G2/M stages of the cell cycle and for calculating cell cycle times. Consequently, cell-cycle-dependent events such as the expression of distinct proteins can be observed in synchronized cell cultures. For example, the expression of a *Nicotiana tabacum* cyclin, a homologue to the *cdc2* gene, disappears in the G1 and S phases of the cell cycle, but is present in the G2/M phase (69).

Flow karyotyping from mitotic cells and chromosome sorting provide opportunities for gene mapping and the construction of chromosome-enriched DNA libraries. Thus, the analysis and sorting of plant chromosomes is of considerable economic interest. Also, the occurrence and extent of polyploidization, which often accompanies differentiation of plant cells, is important for understanding the regulation of gene expression in differentiated tissues. Furthermore, in situ hybridization methods allow accurate quantification of copy numbers of repetitive DNA sequences on different chromosomes within a species.

Examination of intra- and interspecific variation of DNA content can be important in plant hybridization and genetic manipulation programs. Therefore, the method is widely used in molecular cytogenetic research. For example, it can be used to search for foreign chromosomes added to a species, as in the introduction of resistance genes into wheat from the germplasm of rye. Moreover, one of the major problems of using wheat–rye addition lines is their genetic instability. The absence of the easily lost rye chromosomes must be detected, but counting the chromosomes by methods other than flow cytometry is tiresome and time consuming.

For some properties it is necessary to obtain data concerning distinct compartments within a plant cell, and a fundamental goal is appraisal of the physiological behavior of plant cell protoplasts. For this, there are a number of flow cytometric methods for assessing different intracellular components, such as reserve metabolites as well as enzyme activities. Changes in the cytosolic pH using common pH-specific probes have also been shown using these methods. Indeed, it has been proved that analysis of the quantity and quality of cellular fluorescence can yield important information concerning tissue status and derivation.

Furthermore, plant flow cytometry can be a powerful tool in the examination of symbiotic systems. For instance, *Azolla filiculoides* harbors phototrophic, aerobically grown cyanobacteria in vegetative cells and nitrogen-fixing anaerobic cyanobacteria in heterocysts. Some of these differentiated tissues have been investigated regarding the internal distribution of the blue-green algae and the capacities of the various tissues for nitrogen fixation using DNA estimation (70).

Recently, the green fluorescent protein (GFP) from the jellyfish *Aequora victoria* has been used as a powerful new vital reporter in diverse plant cells. In higher plant protoplast studies, it can be used to monitor gene expression, signal transduction, cotransfection, transformation, protein trafficking and localization, protein–protein interactions, and cell separation and purification. In plant cell systems, the bright green fluorescence of GFP is visible even in the presence of the red chlorophyll autofluorescence. GFP is very stable in plant cells and shows little photobleaching. Viable cells, marked with GFP, can be recovered after fluorescence-activated cell sorting (71).

Animal Cells

Flow cytometry is widely exploited in the study of spermatogenesis and routinely applied to estimate viability. It is also used to determine the chromosome content and sex of the offspring. It is a valuable and adaptable tool for research into the genesis of germ cells in both normal and perturbed situations, and it can be used in the control of reproduction, domestic animal breeding, and wildlife conservation. Furthermore, cytogenetic investigation of eukaryotic organisms is important for understanding evolutionary processes and essential for testing evolutionary models. Improved knowledge of the morphology of germ cells, and the motility and fertility of spermatozoa enhance the efficiency of preinsemination technology. Also, functional heterogeneity can be easily detected, as in the assessment of mitochondrial lesions, which are connected to subfertility. The abundance of live cells, altered sperm morphology, and chromatin structure and the resistance of membrane-bound enzymes and membrane permeability to environmental stress factors can also be calculated from flow cytometric data.

In contrast to other methods, the productivity of the flow cytometric separation of the X or Y chromosome-containing germ cells (using highly efficient staining and sorting methods) yields highly purified samples that can be used for sex control of domestic animals (72). More accurate still is slit-scan technology, which allows the whole X or Y germ cell to be spatially visualized along the direction of flow. This is important, because the intensity of fluorescence detected is affected by the flat ovoid head shape and the compactness of the chromatin. Thus, the orientation of the germ cell during passage through the laser beam, detectable by slit-scan flow cytometry, is critical for a high efficiency sorting procedure.

Another range of applications in flow cytometry lies in improving the control of antibody production. Generally, hybridoma cell lines serve as sufficient carriers. Optimizing the productivity of hybridoma cell lines depends partly on developing suitable methods for screening and selecting highly productive cultures and on understanding the regulation of antibody production. Surface-antibody immunofluorescence can be used to select high yielding cells by sorting and to gauge the efficacy of the whole population by cytometric cell analysis. Furthermore, information can be obtained about the stability of expression, the yield, and the integrity of the antibodies obtained. Antigen production can be improved using cytostatically growth-arrested cells, because they do not need to consume cellular resources for biomass production (73). Besides the hybridoma lines, other cell lines (often from insects) have been systematically developed for somatic cell genetics, expression of transfected genes, and synthesis of hormone-inducible proteins. Toward this end, flow cytometric-based studies on animal cell systems can be used for developing convenient and widely applicable procedures for gene cloning. There are, for instance, protocols for DNA transfection and subsequent selection of rare transfectants by sorting. For such investigations, the green fluorescent protein (GFP) may be inserted. Several new GFP variants have been generated that are brighter or have altered excitation spectra, facilitating the monitoring of the marked cell lines.

Viruses have a particularly important biotechnological role in the expression of both foreign prokaryotic and eukaryotic genes in host cells. However, only a few virus-cell systems have been analyzed by flow cytometry until now, though the method is competent for broad applications in detecting and quantifying virus-infected cells.

Viruses can be assayed in animal cell cultures by in situ hybridization using fluorescently marked oligonucleotides and quantitatively estimated by flow cytometry. Characterization of the accessible virus host systems offers information about their capability for providing adequate quantities of virus outgrowth. The process of productive and nonproductive virus infection in animal cell lines can also be investigated, for example, by in vivo analysis of promoter activity regarding transcription of the virus gene. The staging of the virus can be estimated using a virus reporter gene, whose product is, for instance, immunofluorescently stainable. Another possibility is to insert an insect luciferase gene, encoding a protein with characteristic light-emission properties that are measurable by flow cytometry. High sensitivity and rapidity of the reaction (in combination with the luminescence, which does not naturally occur in most of the cellular systems used) makes the method valuable for flow cytometric estimation of virus production. It is also possible to get information about the reproductive capabilities of the virus-infected and uninfected host systems. Differences can be analyzed between highly susceptible and less susceptible cells. Furthermore, virus-induced DNA synthesis may be monitored by assisted determination of BrdUrd incorporation, using monoclonal antibodies. An overview is given by McSherry (74).

Another biological field where the prospects offered by flow cytometry seem extremely encouraging, besides clinical diagnostics, is pharmacology. Before the application of a drug is allowed for human use, the mode of action at cellular and subcellular levels must be clear. The experimental conditions must be as close as possible to physiological drug–cell interactions within living target cells. Model systems have been developed to study the operation of curative drug systems. On-line flow cytometry is the most sensitive and fastest method available, and it causes the least possible harm to living cells.

Additionally, the method is appropriate for investigating the mode of action of other biologically active substances. In phytopathological sciences, for instance, insect cell lines are very suitable target systems for studying the cytotoxic potential of microbe-produced cytostatic agents.

BIBLIOGRAPHY

1. H.M. Shapiro, *Practical Flow Cytometry*, Wiley-Liss, New York, 1995.
2. R. Hüller, E. Glossner, S. Schaub, J. Weingärtner, and V. Kachel, *Cytometry* **17**, 109–118 (1994).
3. J.E. Bailey, J. Fazel-Madjlessi, D.N. McQuitty, L.Y. Lee, J.C. Allred, and J.A. Oro, *Science* **198**, 1175–1176 (1977).
4. M.L. Slater, S.O. Sharrow, and J.J. Gart, *Proc. Natl. Acad. Sci. USA* **74**, 3850–3854 (1977).
5. K.J. Hutter and H.E. Eipel, *Eur. J. Appl. Microbiol. Biotechnol.* **6**, 223–231 (1979).
6. L. Boddy, C.W. Morris, M.F. Wilkins, G.A. Tarran, and P.H. Burkill, *Cytometry* **15**, 283–293 (1994).
7. A.G. Fredrickson, D. Ramkrishna, and H.N. Tsuchiya, *Math. Biosci.* **1**, 327–374 (1967).
8. Th. Münch, B. Sonnleitner, and A.J. Fiechter, *Biotechnology* **22**, 329–352 (1992).
9. D. Lloyd, *Flow Cytometry in Microbiology*, Springer, Berlin, 1992.
10. A. Degelau, R. Freitag, F. Linz, C. Middendorf, Th. Scheper, Th. Bley, S. Müller, P. Stoll, and K.F. Reardon, *J. Biotechnol.* **25**, 115–144 (1992).
11. P. Fouchet, C. Jayat, Y. Hechard, M.-H. Ratinaud, and G. Frelat, *Biol. Cell* **78**, 95–109 (1993).
12. H.B. Steen and E. Boye, *Cytometry* **1**, 32–36 (1980).
13. K. Skarstad, H.B. Steen, and E. Boye, *J. Bacteriol.* **163**, 661–668 (1985).
14. T. Åkerlund, K. Nordström, and R. Bernander, *J. Bacteriol.* **177**, 6791–6797 (1995).
15. Th. Scheper, B. Hitzmann, U. Rinas, and K. Schügerl, *J. Biotechnol.* **5**, 139–148 (1987).
16. S. Müller, A. Lösche, Th. Bley, and T. Scheper, *Appl. Microbiol. Biotechnol.* **43**, 93–101 (1995).
17. J.-U. Ackermann, S. Müller, A. Lösche, Th. Bley, and W. Babel, *J. Biotechnol.* **39**, 9–20 (1995).
18. J.-H. Seo and J.E. Bailey, *Biotechnol. Bioeng.* **27**, 156–165 (1985).
19. P. Fouchet, C. Manin, H. Richard, G. Frelat, and J.N. Barbotin, *Appl. Microbiol. Biotechnol.* **41**, 584–590 (1994).
20. T. Azuma, G.I. Harrison, and A.L. Demain, *Appl. Microbiol. Biotechnol.* **38**, 173–178 (1992).
21. J.D. Chung, S. Conner, and G. Stephanopoulos, *Cytometry* **20**, 324–333 (1995).
22. M. Poot and S. Arttamangkul, *Cytometry* **28**, 36–41.
23. J. Porter, R. Pickup, and C. Edwards, *FEMS Microbiol. Lett.* **134**, 51–56 (1995).
24. T. Tolker-Nielsen, K. Holmstrom, and S. Molin, *Appl. Environ. Microbiol.* **63**, 4196–4203 (1997).

25. J. Chaloupka and V. Vinter, *Folia Microbiol.* **41**, 451–464 (1996).
26. G. Nebe-von Caron and R.A. Badley, in *Flow Cytometry Applications in Cell Culture*, Dekker, New York, 1996.
27. D. Lloyd and A.J. Hayes, *FEMS Microbiol. Lett.*, **133**, 1–7 (1995).
28. J. Porter, D. Deere, R. Pickup, and C. Edwards, *Cytometry* **23**, 91–96 (1996).
29. D.B. Kell, H.M. Ryder, A.S. Kaprelyants, and H.V. Westerhoff, *Antonie van Leeuwenhoeck* **60**, 145–158 (1991).
30. S. Müller, N. Loffhagen, Th. Bley, and W. Babel, *Microbiol. Res.* **151**, 127–131 (1996).
31. C. Herrmann, A. Lösche, S. Müller, Th. Bley, and W. Babel, *Acta Biotechnol.* **17**, 19–38 (1997).
32. J.V. Ordóñez and N.M. Wehmann, *Cytometry* **14**, 811–818 (1993).
33. D.J. Masen, R. Lopez-Amoros, R. Allmann, J.M. Stark, and D. Lloyd, *J. Appl. Bacteriol.* **78**, 309–315 (1995).
34. E. Sahar, R. Nir, and R. Lamed, *Cytometry* **15**, 213–221 (1994).
35. J. Yagoda-Shagam, L.L. Barton, W.P. Reed, and R. Chiovetti, *Appl. Environ. Microbiol.* 1831–1837 (July 1988).
36. G. Wallner, R. Amann, and W. Beisker, *Cytometry* **14**, 136–143 (1993).
37. C.A. Sanders, D.M. Yajko, W. Hyun, R.G. Langlois, P.S. Nassos, M.J. Fulwyler, and W.K. Hadley, *J. Gen. Microbiol.* **136**, 359–365 (1990).
38. S.B. Riis, H.M. Pedersen, N.K. Sørensen, and M. Jakobsen, *Food Microbiol.* **12**, 245–250 (1995).
39. C. Laplace-Builhé, K. Hahne, W. Hunger, Y. Tirilly, and J.L. Drocourt, *Biol. Cell.* **78**, 123–128 (1993).
40. R.A. Patchett, J.P. Back, A.C. Pinder, and R.G. Kroll, *Food Microbiol.* **8**, 119–125 (1991).
41. R.G. McClelland and A.C. Pinder, *J. Appl. Bacteriol.* **77**, 440–447 (1994).
42. C.M. Futsaether and A. Johnsson, *J. Microbiol.* **40**, 439–445 (1994).
43. L.A. van der Waaij, G. Mesander, P.C. Limburg, and D. van der Waaij, *Cytometry* **16**, 270–279 (1994).
44. E.D. Renner, *Cytometry* **18**, 103–108 (1994).
45. E.E. Hughes, J.M. Matthews-Greer, and H.E. Gilleland Jr., *Can. J. Microbiol.* **42**, 859–862 (1996).
46. M.A. Norden, T.A. Kurzynski, S.E. Bownds, S.M. Callister, and R.F. Schell, *J. Clin. Microbiol.* **33**, 1231–1237 (1995).
47. B. Husevåg, *FEMS Microbiol. Ecol.* **16**, 25–32 (1995).
48. S. Langsrud and G. Sundheim, *J. Appl. Bacteriol.* **81**, 411–418 (1996).
49. I.S. Roberts, *Annu. Rev. Microbiol.* **50**, 285–315 (1996).
50. M.F. Gilbert, D. McQuitty, and J.E. Bailey, *Appl. Env. Microbiol.* **36**, 615–617 (1978).
51. K.-J. Hutter, *FEMS Microbiol. Lett.* **3**, 35–38 (1978).
52. Alberghina, M. Mariani, E. Martegani, and M. Vanoni, *Biotechnol. Bioeng.* **25**, 1295–1310 (1983).
53. Th. Scheper, H. Hoffmann, K. Schügerl, *Enzyme Microb. Technol.* **9**, 399–405 (1987).
54. D.E. Block, P.D. Eitzmann, J.D. Wangesteen, and F. Srienc, *Biotechnol. Prog.* **6**, 504–512 (1990).
55. D. Porro, B.M. Ranzi, C. Smeraldi, E. Martegani, and L. Alberghina, *Yeast* **11**, 1157–1169 (1995).
56. S. Müller and A. Schmidt, *Acta Biotechnol.* **9**, 71–75 (1989).
57. S. Müller, A. Lösche, and Th. Bley, *Acta Biotechnol.* **12**, 365–375 (1992).
58. S. Müller, K.J. Hutter, Th. Bley, L. Petzold and W. Babel, *Bioproc. Eng.* **17**, 287–293 (1997).
59. K.D. Wittrup and J.E. Bailey, *Cytometry* **9**, 394–404 (1988).
60. K.D. Wittrup, J.E. Bailey, B. Ratzkin, and A. Patel, *Biotechnol. Bioeng.* **35**, 565–577 (1990).
61. P.D. Eitzman, J.L. Hendrick, and F. Srienc, *Cytometry* **10**, 175–483 (1989).
62. C. Smeraldi, E. Berardi, and D. Porro, *Microbiology* **140**, 3161–3166 (1994).
63. J.V. Ordonez and N.M. Wehman, *Cytometry* **22**, 154–157 (1995).
64. P.J. Constantino, D.E. Budd, and N.F. Gare, *Cytometry* **19**, 370–375 (1995).
65. D.K. Button and B.R. Robertson, *Cytometry*, **10**, 558–563 (1989).
66. D. Vaulot, D. Marie, R.J. Olsen, and S.W. Chisholm, *Science* **9**, 1480–1482 (1995).
67. N. Simon, N. LeBot, D. Marie, F. Partensky, and D. Vaulot, *Appl. Environ. Microbiol.* **61**, 2506–2513 (1995).
68. V.S. Avery, J.L. Harwood, and D. Lloyd, *Appl. Environ. Microbiol.* **61**, 1124–1132 (1995).
69. L.X. Qin, C. Perennes, L. Richard, M. Bouvier-Durand, C. Trehin, D. Inze, and C. Bergounioux, *Plant. Mol. Biol.* **32**, 1093–1101 (1996).
70. A. Canini, M.G. Caiola, and L. Ferrucci, *Cytometry* **13**, 299–306 (1992).
71. J. Sheen, S. Hwang, Y. Niwa, H. Kobayashi, and D.W. Galbraith, *Plant J.* **8**, 777–784 (1995).
72. S. Damjanovich, R. Gaspar Jr., L. Tron, and A. Aszalos, *Am. Biotech. Lab.* **3**, 11–21 (1985).
73. M. Fussenegger, X. Mazur, and J.E. Bailey, *Biotechnol. Bioeng.* **55**, 927–939 (1997).
74. J.J. McSherry, *Clin. Microbiol. Rev.* **7**, 576–604 (1994).

FLUORESCENCE TECHNIQUES FOR BIOPROCESS MONITORING

CARSTEN LINDEMANN
STEFAN MAROSE
THOMAS SCHEPER
University of Hannover
Hannover, Germany

HANS OLE NIELSEN
DELTA Light&Optics
Lyngby, Denmark

KENNETH F. REARDON
Colorado State University
Fort Collins, Colorado

KEY WORDS

Bioprocess monitoring
Fluorescence
On-line
Optical sensor

OUTLINE

INTRODUCTION

In recent years, various bioanalytical systems for improved bioprocess monitoring have been developed and investigated. The application of these systems in pilot facilities and production plants can provide the information needed to better understand all types of bioprocesses, including cultivations, biotransformations, and downstream operations. In addition, better documentation of the processes, especially important for the production of pharmaceuticals, can be obtained. However, most of these sensors are invasive, and thus there is the chance of impacting the cultivation process (e.g., contamination) and causing complications in the processes. For this reason, more and more research groups are involved in the development and application of noninvasive bioanalytical tools, which can be interfaced to the bioprocess without causing problems such as contamination or changes in cell metabolism.

Optical sensors offer tremendous advantages for bioprocess monitoring. They can be interfaced noninvasively to the bioprocess via glass fiberoptics. The information can be transported via fiberoptics without interferences caused by electromagnetic fields over long distances, and signals can be obtained instantaneously, because spectro-optical systems have extremely short response times.

Fluorescence sensors have been investigated intensively during the past 15 years for different applications in biotechnology, especially biomass estimation, reactor characterization, metabolic studies, and general bioprocess monitoring. These so-called fluorosensors were developed mainly for monitoring of the reduced form of nicotine adenine dinucleotide (NADH) and its phosphorylated form (NADPH). This coenzyme is present in all living cells and is involved in different metabolic reactions; its level is extremely sensitive to environmental conditions.

PRINCIPLES OF SINGLE-WAVELENGTH FLUOROSENSORS

The NAD(P)H pool gives information about the metabolic state of the cells. In 1957, Duysens and Amesz first showed that relative NAD(P)H measurements could be performed in living cells using spectrofluorometric techniques (1). UV light of 340 to 360 nm was used to excite fluorescence of NAD(P)H in cells in a suspension. The NAD(P)H fluorescence was monitored at 460 nm while metabolic experiments were performed. The first spectrometer device for biotechnological application was designed by Harrison and Chance in 1970 (2). It could be used in glass bioreactors. The excitation light was guided through a quartz window into the bioreactor, and the fluorescence was collected at an angle of 60° through a second quartz window. The penetration depth of the light was only a few centimeters at low cell densities. The sensitivity decreased as the cell density increased. However, it was possible to monitor aerobic and anaerobic transitions in *Klebsiella aerogenes*. Zabriskie and Humphrey used this technique in 1978 for biomass estimation in different cultivation experiments (e.g., *Saccharomyces cerevisiae, Streptomyces* spp., and *Thermoactinomyces* spp.) (3).

Based on these experiments, different smaller miniaturized process fluorometers were developed that could be interfaced to a bioreactor via a standard port (Fig. 1). This port is used to guide the excitation light into the bioreactor and to collect the backward (180°) fluorescence. In 1981, Beyeler et al. (4) developed a miniaturized fluorosensor in an open-end detection mode. This design has the advantage of eliminating the inner filter effect by monitoring the surface fluorescence at the observation window. Three commercial versions of the open-end spectrofluorometer were produced by BioChem Technology/USA (FluoroMeasure), Mettler Toledo AG/Switzerland (Fluorosensor), and BioBalance A/S, Denmark (Biobalance).

All these sensors shared two disadvantages. First, they were limited to detection of NAD(P)H fluorescence, it was only possible to excite fluorophores in the range of 340 to 360 nm and to detect the fluorescence intensities at 450 to 460 nm. And they also lacked the ability to distinguish between different fluorophores that might interfere in this region.

In order to obtain more information, a flexible two-channel fluorometer was developed by Scheper and Schügerl (5). In this device, the light of the UV lamp was focused out the optical filters and lenses onto a quartz fiber bundle. This fiberoptic was used to guide the light into the bioreactor. The backward fluorescence light was collected via the same fiber bundle and split into two bundles, passing

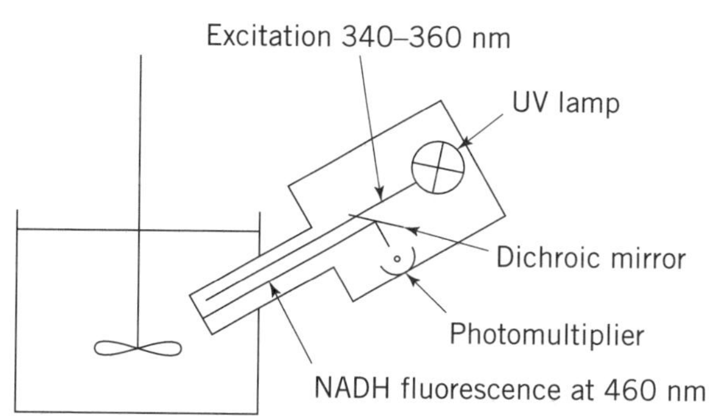

Figure 1. Principle setup of miniaturized spectrofluorometer.

interference filters that selected for different wavelength bands. Thus, it was possible to monitor simultaneously different fluorescence wavelengths.

APPLICATION OF SINGLE-WAVELENGTH FLUOROSENSORS

Single-wavelength fluorosensor devices have been used to monitor different cultivation processes, including suspended as well as immobilized organisms (Table 1). Investigations in which the NAD(P)H fluorescence of these cultivations was monitored had one of two goals: on-line determination of biomass concentration or on-line detection of metabolic changes. A significant advantage of using fluorescence for biomass concentration monitoring is that only living cells are detected. To use culture fluorescence as an indicator for biomass concentration, it is necessary that the culture fluorescence per cell is constant during the cultivation process. In this case, a linear correlation between fluorescence and the biomass concentration could be expected. However, the intracellular NAD(P)H pool changes in a cell during the batch experiment. Thus, it is necessary to linearize the corresponding biomass concentration and fluorescence data to obtain a calibration plot. One problem that must be overcome is that all process conditions affecting the culture fluorescence intensity cause problems with the biomass determination. The two most common interferences are gas bubbles and the presence of compounds with fluorescence spectra that overlap that of NAD(P)H. Interestingly, fouling of observation windows was not observed.

Many investigators have reported efficient biomass concentration determination via culture fluorescence monitoring. Examples include cultivations of *Methylomonas mucosa, Zymomonas moblis* in synthetic medium, and *Pseudomonas putida*. In all these applications, suspended cell cultivations were studied, and it was shown that on-line biomass estimation is also possible in technical media under certain process conditions (for detailed information, see Ref. 6).

Table 1. Application of Fluorescence Monitoring for Different Cultivation Processes and Organisms

Suspended
Saccharomyces cerevisiae
Penicillium chrysogenum
Cephalosporium acremonium
Escherichia coli (incl. recombinant)
Zymomonas mobilis
Bacillus licheniformis
Clostridium acetobutolyticum
Spodoptera frugiperda
Baby hamster kidney cells
Wastewater treatment
Immobilized
Cephalosporium acremonium
Saccharomyces cerevisiae
Zymomonas mobilis

Source: See Refs. 6 and 7.

A wide range of metabolic changes have also been studied using NAD(P)H culture fluorescence (Table 2). For example, Ristroph et al. (8) studied the cultivation of *Candida utilis* under different conditions and developed a feeding strategy for fed-batch cultivation of this organism on the basis of NAD(P)H-dependent culture fluorescence. Einsele (9) reported on different applications of culture fluorescence monitoring and determined the mixing-time behavior of different bioreactors and the substrate uptake rate of different microorganisms during cultivation. Reardon et al. (6) studied immobilized *Clostridium acebutylium* in a fixed-bed reactor. The cells were immobilized in calcium alginate beads, and the cultivation medium was circulated through the reactor. The growth and productivity of the cells could be monitored under different cultivation conditions. Scheper et al. reported (10) on the application of this technique for the monitoring of immobilized yeast cells.

NEW DEVELOPMENTS IN FLUOROSENSOR TECHNOLOGY AND APPLICATION

Several approaches have been taken in the past 8 years to circumvent the restrictions of single-wavelength NAD(P)H-dependent culture fluorescence monitoring. In 1991, Li et al. (11) reported on a special fluorosensor with five different wavelengths for the monitoring of *Saccharomyces cerevisiae* and *Candida utilis* cultivations. They showed that other wavelengths could be used for biomass concentration estimation and for metabolic studies. Horvath et al. (12) used a fluorescence spectrometer coupled to a bioreactor via quartz glass fibers for the determination of *S. cerevisiae* biomass concentrations. Based on his results, the tryptophan fluorescence was better correlated to the cell mass concentration than was the NAD(P)H fluo-

Table 2. Different Phenomena Studied via Fluorescence Monitoring

Phenomen studied	Organism
Aerobic–anaerobic transition	*Klebsiella aerogenes* *Saccharomyces cerevisiae* *Candida tropicalis* *Escherichia coli* *C. guilliermondii*
Aeration rate	*Penicillium chrysogenum*
Addition of carbon source to starved cells	*Saccharomyces cerevisiae* *Candida tropicalis* *Escherichia coli*
Diauxic growth	*Saccharomyces cerevisiae*
Dilution rate changes	*Saccharomyces cerevisiae* *Escherichia coli* *Pseudomonas putida*
Glycolytic oscillation	*Saccharomyces cerevisiae*
Culture synchrony	*Saccharomyces cerevisiae*
Metabolic shifts	*Thermoactinomyces* sp. *Clostridium acetobutylicum*
Addition of metabolic uncouplers	*Saccharomyces cerevisiae*

Source: Based on Refs. 3 and 4.

rescence. A more recent development was described by Chenina in 1993 (13) for the cultivation of yeasts (*Cyathus striatus, Eurotium cristatum*, and *Crinipellis stipitaria*). A special fluorescence spectrometer was developed that can also be interfaced to a bioreactor via light guide cables. Monochromators were used to monitor the fluorescence of wide excitation and emission ranges.

TWO-DIMENSIONAL FLUORESCENCE

These fluorescence spectrometers can be used to produce a two-dimensional (2-D) fluorescence spectrum, which represents the emission spectra at different excitation wavelengths. Such a spectrum is described by the three parameters of excitation wavelength, emission wavelength, and fluorescence intensity, which can be shown as an excitation-emission-data matrix. In this matrix, each row represents an emission scan and each column an excitation scan. Different compounds can be identified by their position in the matrix. The most illustrative way to show a 2-D fluorescence spectrum is to use an isometric plot, but contour plots are the most useful because no fluorescence peaks are hidden (Fig. 2). Dark shadings in the fluorescence plots indicate high fluorescence intensities.

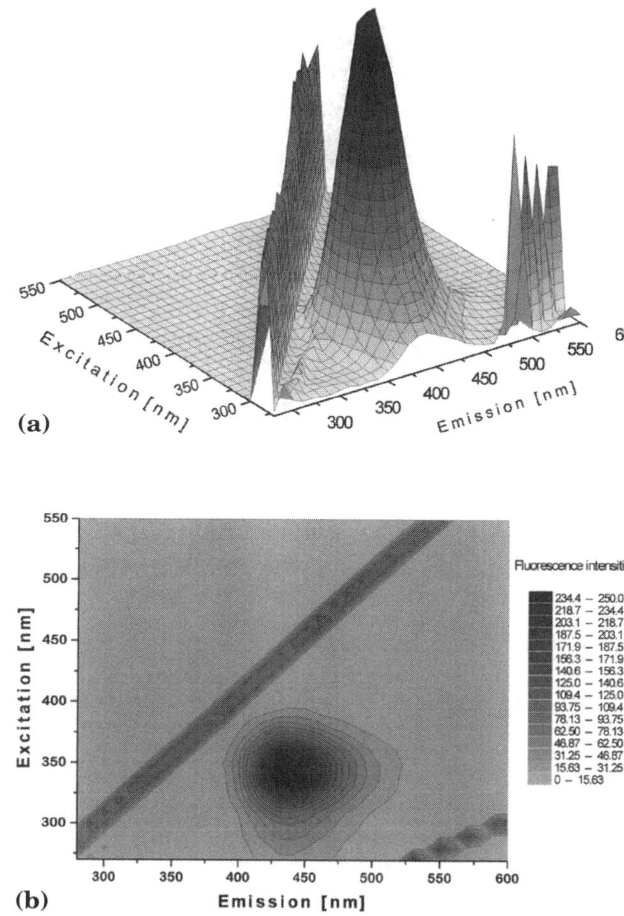

Figure 2. (**a**) Isometric projection of a spectrum of 7-amino-4-methylcoumarin. (**b**) Contour plot of a spectrum of 7-amino-4-methylcoumarin.

BIOGENIC FLUOROPHORS

The spectral ranges of important biogenic fluorophores, which occur inside the cell and can be detected simultaneously by 2-D fluorescence spectroscopy, are shown in Figure 3. Almost all biological processes depend on proteins, which fluoresce because of aromatic amino acids such as tryptophan, tyrosine, and phenylalanine. In another region of the 2-D fluorescence spectrum (λ_{Ex} = 310–480 nm, λ_{Em} = 350–550 nm), vitamins (e.g., pyridoxine [vitamin B-6]), riboflavin [vitamin B-2]), and coenzymes (e.g., NADH, NADPH, FMN, FAD) can be found.

INSTRUMENTATION

For on-line monitoring and control of biotechnological processes using 2-D fluorescence, the instrument must be capable of measuring a complete 2-D fluorescence spectrum in a short time (about 1 min), and the time between two measurements should not be longer than 5 to 15 min. This is to ensure that a complete 2-D fluorescence spectrum represents the fluorescence at a certain time and that changes during a cultivation can be monitored in real time.

HITACHI F4500 SPECTROFLUOROMETER

For 2-D fluorescence measurements, fluorescence spectrometers for use in laboratories can be used. One suitable apparatus is the fluorescence spectrophotometer model F4500 (Hitachi, Japan). This spectrometer allows a very fast scan speed (30,000 nm/min) and an entire 2-D spectrum with a wavelength range of 250 to 550 nm for excitation (10-nm step size) and 260 to 600 nm for emission (20-nm step size) can be measured within 1 min. For selection of excitation and emission wavelength, grating monochromators are used. A 150-W Xe lamp is used for excitation, and a photomultiplier at a voltage of 700 V is used for detection of fluorescence.

The spectrometer is connected by a 1-m bifurcated liquid light conductor (Lumatec, Germany) to a quartz window in a 25-mm port of a bioreactor (Fig. 4). The excitation light is guided via the fiberoptic into the reactor, and the backward fluorescent light is guided via the second fiberoptic to the emission detection unit. With this setup, in situ measurements are possible without interference from light outside the bioreactor, and there is no risk of contamination because the sensor is not in contact with the medium. Furthermore, sterilization is possible without connecting the sensor to the reactor.

BIOVIEW

The BioView sensor (DELTA Light & Optics, Denmark) is a fluorescence sensor optimized for fully automated industrial measurements and applications to monitoring and control of bioprocesses. It is very robust and suitable for harsh environments (e.g., high temperature or moisture)

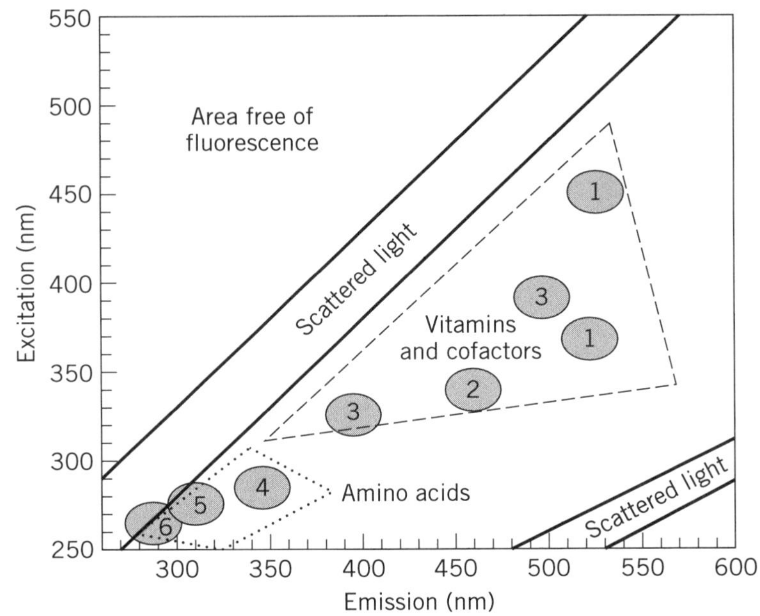

Figure 3. Biogenic fluorophores in a 2-D fluorescence spectrum. 1 = riboflavin, FAD, FMN; 2 = NAD(P)H; 3 = pyridoxine, pyridoxamine, pyridoxal-5′ = phosphate; 4 = tryptophan; 5 = tyrosine; 6 = phenylalanine). *Source*: Adapted from reference 9.

and electromagnetic interference. For data transfer, a single-fiber modem is used, which allows a distance between sensor and computer up to several hundred meters. The optical parts of the BioView are optimized for use in biotechnological processes in that interference filters (DELTA Light & Optics, Denmark), rather than spectrometer gratings, are used by the BioView sensor for selection of excitation and emission wavelengths. Filter wheels (one for excitation and another for emission) with 16 filters each are used (Fig. 5). These filters can be designed and chosen individually. Usually, measurements in steps of 20 nm are performed in a wavelength range of 270 to 550 nm for excitation and 290 to 590 nm for emission. For the measurement of a complete 2-D fluorescence spectrum, the excitation filter wheel is set to the first filter position. With this excitation filter, the fluorescence spectrum is monitored via the emission filters that switch from filter to filter until a cycle is completed. Afterward, the excitation filter switches to the next position and the next fluorescence spectrum is measured. This procedure is continued until the excitation filter wheel has performed a complete cycle. The complete measurement takes about 1 min, depending on the number of different filter positions and measurements for each filter combination, which can be chosen individually. The connection to the bioreactor is the same as that described for the Hitachi instrument, except that a 2-m long liquid light conductor is used.

APPLICATIONS

Two-dimensional fluorescence spectroscopy offers the possibility of simultaneous analysis of several fluorophores. The location of the peak maximum in a 2-D spectrum characterizes each fluorophore, and the relative fluorescence intensity correlates with its concentration. Despite this conceptual simplicity, the interpretation of 2-D fluorescence spectra is not trivial. Interactions of the fluorophores with one another and with other nonfluorescent compounds, bubbles, and other physical influences such as pH, temperature, and viscosity, quenching effect, and changes in morphology of the cells all cause problems in the interpretation of the 2-D spectrum.

To make the system user friendly, automated interpretation by artificial intelligence was developed for 2-D fluorescence spectra. The intelligent BioView sensor is able to predict the course of a cultivation if it is taught with a training cultivation under comparable cultivation conditions. Modern chemometric systems like principle component analysis (PCA) or neural networks are able to consider all influences on the fluorescence signal and lead to a sensor with high capacity, able to measure various cultivation variables.

The changes in fluorescence intensities during cultivation processes are shown in difference spectra, which are obtained by subtraction of a 2-D spectrum taken during the cultivation from the one in the beginning. Through comparison to the spectrum of a mixture of pure biogenic fluorophores, such as riboflavin, NADH, pyridoxine, pyridoxamine, pyridoxal-5′-phosphate, and tryptophan, the peaks could be assigned to the expected fluorophores. All changes have a biological origin and offer information about the cell growth and metabolism. Even small changes in the concentration of a single fluorophore can be seen in a complex mixture of several fluorophores. The regions of changes are regarded as the areas of interest.

Measurements made nearly continuously allow us to construct the course of various culture variables during a bioprocess. One of the most important is the biomass concentration. Off-line analysis of counting cells, measuring optical density or weighting dry mass are time consuming and fail if filamenteous fungi are used. On-line tools such as a turbidimeter often cause problems. The search for a reliable automated technique for on-line monitoring of biomass still continues. In a 2-D fluorescence spectrum, vari-

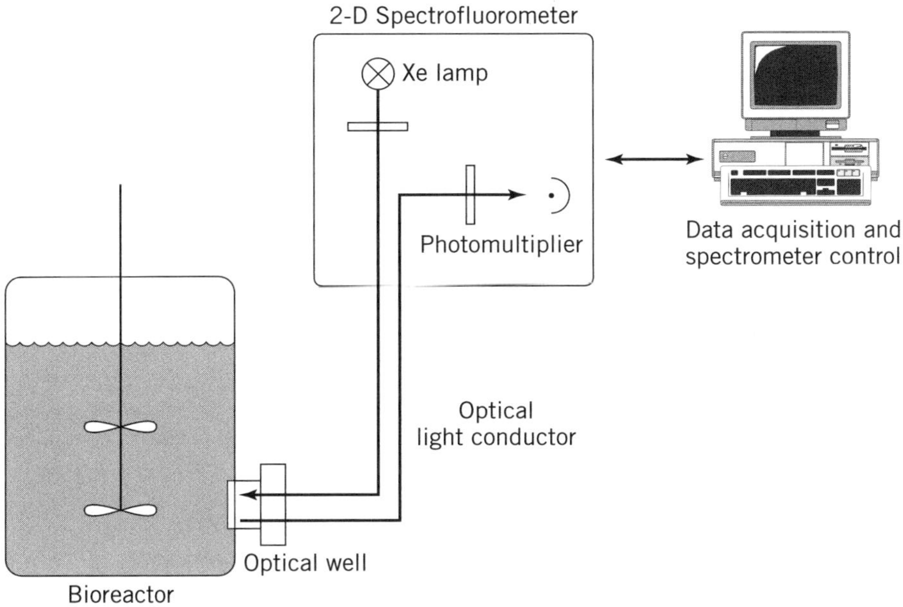

Figure 4. Measuring setup.

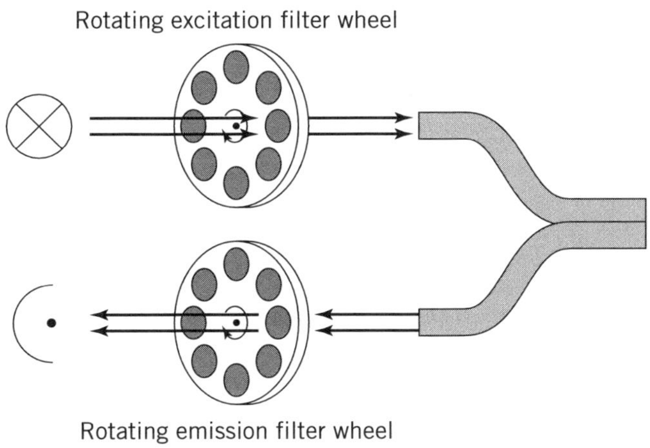

Figure 5. Measuring principle of the BioView sensor.

ous excitation–emission regions can be correlated to biomass concentration. This offers the advantage to change to another region if one cannot be used because of overlapping or other influences. Fluorescence of proteins (because of their high quantum yield, especially in the case of tryptophan), pyridoxine, the area of NAD(P)H, and the area of riboflavine and its derivatives (FAD or FMN) show good correlations with biomass for several organisms (e.g., *Acremonium crysogenum, Claviceps purpurea, Sphingomonas yanoikuyae, Bacillus licheniformis*, different strains of *Escherichia coli, Enterobacter aerogenes, Saccharomyces cerevisiae, Schizosaccharomyces pombe*, mammalian cells or *Tetrahymena thermophila* (9,14). For the filamenteous fungi, *Claviceps purpurea* fluorescence intensity in the region of riboflavin and its derivatives (λ_{Ex} = 450 nm, λ_{Em} = 530 nm) allow on-line monitoring of biomass concentration, which is not possible by optical density or cell count because of the mycelial growth. Scattered light is also measured in a 2-D fluorescence spectrum. These data correlate well with data derived from conventional turbidimetric sensors.

Along with the ability to estimate biomass concentration, 2-D fluorescence sensors such as the BioView system offer on-line information about intracellular variables without interference into the process. This is achieved by monitoring changes in the excitation–emission regions associated with the redox-sensitive coenzymes NADH, FAD, and FMN. This approach can be used to study aerobic–anaerobic transitions, the tricarboxylic cycle, effects of uncoupling agents such as 2,4-dinitrophenol (DNP) on oxidative phosphorylation, diauxies, the effects of pulse additions of substrate, and oscillations of a culture.

Fluorescent products can be observed by 2-D fluorescence. The productivity of ergot alkaloids could be estimated by 2-D fluorescence spectroscopy (9). Degradation of fluorescent substrates such as polycyclic aromatic hydrocarbons can be observed. Other extracellular or physical variables such as the microenvironment of the cells are captured by 2-D fluorescence.

Two-dimensional fluorescence spectroscopy is ideal for a wide variety of organisms and installation uses and can therefore be seen as a multianalyzer that is capable of replacing several other sensors in the future.

BIBLIOGRAPHY

1. L.N.M. Duysens and J. Amesz, *Biochim. Biophys. Acta* **24**, 19–26 (1957).
2. D.E.F. Harrison and B. Chance, *Appl. Microbiol.* **19**, 446–450 (1970).
3. D.W. Zabriskie and A.E. Humphrey, *Appl. Eur. Microbiol.* **35** 337–343 (1978).
4. W. Beyeler, K. Gschwend, and A. Fiechter, *Chem. Ing. Technol.* **55**, 869–871 (1983).

5. T. Scheper and K. Schügerl, *J. Biotechnol.* **3**, 221–229 (1986).
6. T. Scheper and K.F. Reardon, in W. Göpel, J. Hesse, and J.N. Memel eds., *Sensors*, vol 2, VCH, Weinheim, Germany, 1991, pp. 1024–1026.
7. K. Reardon and T. Scheper, in H.J. Rehm and G. Reeds eds., *Biotechnology*, 2nd ed., VCH, Weinheim, Germany, 1991, pp. 179–233.
8. D.L. Ristroph, C.M. Watteeuw, W.B. Armiger, and A.E. Humphrey, *J. Ferment. Technol.* **55**, 559–608 (1977).
9. A. Einsele, D.L. Ristroph, and A.E. Humphrey, *Biotechnol. Bioeng.* **20**, 1487–1492 (1978).
10. T. Scheper, K.-D. Anders, M. Busch, W. Müller, and K.F. Reardon, in J.A.M. de Bont, J. Visser, B. Mattiasson, and J. Tramper eds., *Physiology of Immobilized Cells*, Elsevier, Amsterdam, 1990, pp. 625–636.
11. J.-K. Li, E.C. Asali, and A.E. Humphrey, *Biotechnol. Prog.* **7**, 21–27 (1991).
12. J.J. Horvath, S.A. Glazier, and C.J. Spangler, *Biotechnol. Prog.* **9**, 666–670 (1993).
13. R. Chenina, A. Dochnahl, T. Huff, H.-G. Kuball, B. Lange, H. Anke, and T. Anke, *DECHEMA Monographien* **129**, 171–182 (1993).
14. S. Marose, C. Lindemann, and T. Scheper, *Biotechnol. Prog.* **14**, 63–74 (1998).

FOOD PROCESS ENGINEERING

Henry Schwartzberg
University of Massachusetts
Amherst, Massachusetts

KEY WORDS

Concentration
Drying
Expression
Extraction
Extrusion
Heat transfer
Membrane processes
Preservation
Separations
Thermal processing

OUTLINE

INTRODUCTION

Foods and beverages are processed to increase their safety and shelf life, to favorably modify their sensory and nutritional quality and composition, and to improve ease and reliability of preparation by consumers. Foods must be protected from contamination and stored and transported safely and without adversely affecting quality. Food process engineers employ engineering principles and methods to identify, select, design, develop, and "debug" processes, equipment, systems, and plants used to accomplish these tasks. Food process engineers also supervise the start-up and operation of food processing systems and plants; devise related control, maintenance, and sanitation systems; devise ways to eliminate hazards and minimize waste; and help develop new food products and ways to upgrade existing processes. In carrying out these tasks, engineers must account for the biological activity and chemical instability of foods and for the ability of microorganisms and pests to invade and grow in or on foods.

Many different types of food products are produced. The processes used are extremely diverse (1) and too numerous to cover in detail. Therefore, this article examines only the engineering aspects of a limited number of processes or operations. Many of these are used to preserve foods. Others are used to separate food components, and still others are used to change the shape, size, and texture of foods. Food processing frequently involves heat transfer. This article briefly examines such transfer.

Individual operations may produce effects that fall into several processing categories. Thus drying, which separates water from other food ingredients, is used mainly to preserve foods. Drying often changes the texture of foods and involves transfer of heat from hot air to moist solids. It also involves transfer of moisture (mass) within solids and from solids to air and transfers of momentum, which affects airflow.

HEAT TRANSFER IN FOOD PROCESSING

Food process engineers are frequently asked to model heat transfer, devise ways to control or improve such transfer, and calculate heating and cooling loads and sizes of equipment needed to provide desired amounts of heating or cooling. These tasks are accomplished by means of methods widely employed by other engineers. However, in using these methods, food process engineers must account for the physical, heat transfer, and sensory properties of foods, and for their temperature sensitivity. Food components rapidly decompose in undesirable ways at high temperatures. The highest product temperature reached during food processing is probably 250 °C, the end-of-roast temperature for very darkly roasted coffee. Maximum tolerable temperatures for many foods are much lower. Higher processing media temperatures may be used (e.g., gas temperatures ranging up to 500 °C are used in roasting coffee). For economic reasons, foods are rarely cooled to temperatures lower than −40 °C.

Foods usually are either solid or complex composites (e.g., moist multicellular tissue, porous materials, stiff doughs) that act like solids. Heat transfer in such foods is modeled by means of partial differential equations based on Fourier's laws of conduction. Modeling also requires specification of boundary conditions based on surface temperatures, surface heat transfer coefficients, or surface rates of heat transfer. Solutions to partial differential equations for conductive heat transfer and many boundary conditions of practical interest can be found in the work of Carslaw and Jaeger (2) and other standard texts. Similar solutions for partial differential equations and boundary conditions describing diffusive mass transfer equations are provided by Crank (3). Many pairs of heat transfer and mass transfer equations and boundary conditions are formally equivalent. Therefore equation solutions for diffusive mass transfer often can be readily converted into solutions that apply for conductive heat transfer and vice versa.

Heat transfer calculations for solid foods may be complicated by a need to account for heats generated by ongoing biological processes, surface evaporation, anisotropic conduction, and internal vaporization and condensation. Often boundary conditions are spatially nonuniform, or they may vary with time during food processing. The thermal conductivities of foods may be difficult to predict. Nevertheless, most problems relating to heat transfer in solid foods can usually be solved fairly easily by means of computer-based numerical methods.

Food engineers also deal with heat transfer that is due to natural convection in liquid-filled containers, in pore spaces in stacks of heat-generating, respiring produce, and in the air that surrounds solid foods discharged from heating devices. In addition to widely encountered problems involving forced convection of heat in heat exchangers, evaporators, and condensers, they deal with convective heat exchange between streams of air or water and solid foods lined up one behind another in rows on shelves. These problems are often solved empirically. Food engineers working on aseptic preservation and ohmic heating processes are now attempting to precisely predict heat transfer to individual solids in groups of solids suspended in heated, flowing fluid.

Food engineers also deal with prediction of temperature changes during microwave and dielectric heating. In doing so they must account for attenuation and focusing of energy as electromagnetic radiation passes through foods, refraction and reflection of such radiation at food surfaces, energy distribution patterns in food-containing microwave cavities, and temperature-induced changes in the energy absorption characteristics of foods.

PRESERVATION OF FOODS

Drying, chilling, freezing, heating-induced inactivation of microorganisms and enzymes, application of bactericides, storage in controlled atmospheres, solute-induced water activity or pH reduction, and combinations of some of these processes are used to preserve foods.

Drying

Drying, the evaporative removal of water from moist solids or solidifiable solutions, may well be the earliest method of preserving foods and still serves that purpose today. It is also used to harden foods that have been shaped in a moist state, to reduce food weight, and volume, and to reduce moisture contents to facilitate processing.

Drying preserves foods because biological activity (e.g., microbial growth) radically slows or stops and chemical reactants become immobile when little water is present. Fresh fruits and vegetables usually contain 5–10 kg water/kg dry solids; stable, dried vegetables, 0.03–0.1 kg water/kg dry solids; and stable, sugar-rich dried fruits, 0.20–0.39 kg water/kg dry solids (4). The water content of freshly harvested grains is up to 0.22–0.43 kg water/kg dry solids; to provide safe storage, this must be reduced to 0.11–0.15 kg water/kg dry solids.

Hot Air Drying. Food drying most frequently is carried out by transferring heat to moist food particles from flowing hot air. This causes food moisture to evaporate and move away with the air. Consequently, the air moisture content (humidity) rises, and the air temperature drops proportionately. Relative humidity (RH) is the partial pressure of water in air divided by the vapor pressure of pure water at the same temperature. Water activity (a_w) is a measure of the chemical activity of water. It is also the equilibrium RH for a food at a given moisture content. Food moisture sorption isotherms, such as those shown for apples and potatoes in Figure 1, are constant-temperature plots of a_w versus a food's dry-basis moisture content, X. Moisture sorption isotherms shift downward moderately as temperature rises. The minimum X obtainable by drying is X_E, the sorption isotherm X at which a_w at the air temperature equals RH.

After start-up transients have decayed, food temperatures during drying depend on a_w at the food's surface. As long as the surface a_w is greater than 0.9, the food temperature will be no more than 1 °C higher than the air wet-bulb temperature; if air conditions are almost constant, the drying rate will be almost constant. Constant-rate drying does not last very long for foods. Surface a_w depends on X_S, X at the surface. As drying removes water from the surface, water diffusing to the surface from the food's interior partially replaces it. Because water diffuses slowly in foods, X_S very rapidly drops enough to drive a_w below 0.9. After this, a_w drops sharply as X_S decreases, the food temperature rises and progressively approaches the air tempera-

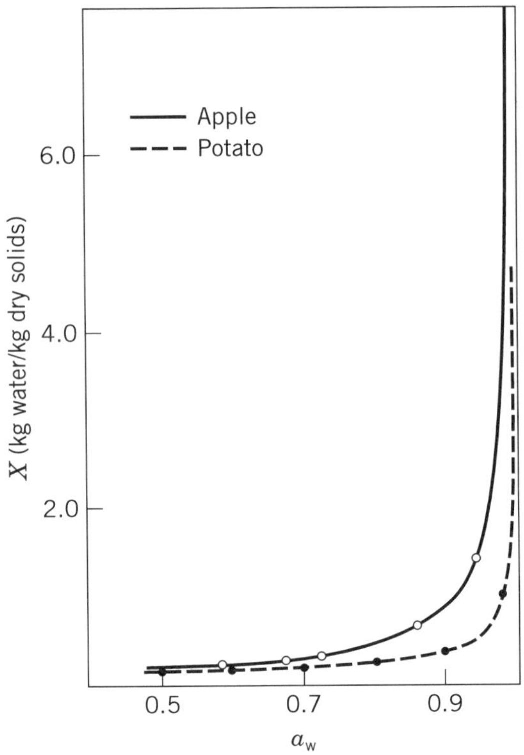

Figure 1. Moisture sorption isotherms for apples and potatoes.

ture, and the drying rate progressively decreases. Drying rates are strongly influenced by heat transfer and airflow rates as long as the air temperature is significantly higher than the food temperature. As food temperature approaches the air temperature, drying is more and more controlled by water diffusion in the food.

Near the end of drying, diffusion completely controls drying rates. The air temperature sets the food temperature. Hence, it affects diffusion and drying rates. Enough airflow must be provided to remove evaporated moisture fast enough to prevent X_E from rising excessively. Other than that, airflow no longer affects drying.

Drying Rates. Water diffusivity in foods often decreases markedly as drying proceeds. Drying also causes shrinkage. Theoretical analysis of water diffusion in drying foods is difficult because of these changes. Though special computational techniques can be used to deal with these complications, in most cases, drying rates are correlated empirically rather than theoretically predicted. After constant-rate drying stops, drying rates at constant air conditions are often proportional to $(X_A - X_E)^n$, where X_A is the average X and n is an empirical constant, which ranges between 0.7 and 2.4.

Drying air temperature and humidity affect food quality in addition to affecting drying rates and thermal efficiency. Therefore, air conditions are adjusted to minimize quality loss and to prevent in-process microbial growth. Dried products may not rehydrate well. This can be partially remedied by vapor-induced puffing, elicited by short-term use of high air temperatures late in the drying process.

Wet particles may stick together during drying. This property can be used to provide desired agglomeration. When sticking occurs and is unwelcome, particles are broken apart by passing them between rubber-faced rolls after drying.

Shallow-Bed Dryers. Foods are air-dried in many different types of equipment. In conveyor dryers (Fig. 2), moist food particles are uniformly spread as a shallow bed on a perforated belt and conveyed through a series of stages where heated air is blown through the bed and belt. Different air temperatures, humidities, and flow rates are often used in the stages. To improve uniformity of drying, upflow is used in some stages and downflow in others.

In conveyor-based drying of vegetables, 60–85% of the food's moisture content is removed in the first 1.5–2 h of drying. An additional 2–8 h usually is needed to complete the process. Air temperatures between 66 and 120 °C are used in early stages of drying. Lower air temperatures (32–52 °C) are used later, at the end of drying. Quite different conditions are used for some foods. Low temperatures are used during first stages of gelatin strip drying to prevent melting. Quick-cooking rice is dried in less than 15 min using temperatures up to 200 °C. Temperature around 175 °C may be used to puff diced potatoes at late stages of drying.

In tunnel dryers, hot air passes over shallow beds of food particles conveyed through a rectangular tunnel on racked trays mounted on carts. Airflows in the direction of cart movement or counter to that direction. Clearances between the carts and the tunnel are small, to minimize bypassing of air. Periodically, a freshly loaded cart is pushed into the tunnel and a cart containing dried material simultaneously leaves. Drying tunnels often are used in series. Cocurrent drying with high inlet-temperature air is followed by countercurrent drying with lower temperature air. Conveyor dryers are used instead of tunnel dryers wherever feasible because tray loading is burdensome and conveyor dryers permit automatic loading and discharging. In addition, drying conditions are more readily controlled in conveyor dryers.

Deep-Bed Drying. Bin dryers in which air passes through a deep bed of particles may be used to complete the drying of partly dried vegetables discharged from shallow-bed dryers. Grain is often directly dried in deep beds in bins, where air is introduced through a false floor or ducts near the floor (5). Ambient air may be used if its relative humidity is less than 0.55, or it may be heated to 11 °C above ambient. In such dryers, moisture desorption waves (Fig. 3) and accompanying temperature-change waves move upward through the grain bed. Most of the time, the exit air temperature equals the initial grain temperature and the exit air RH equals the initial grain a_w. Design engineers select airflow rates that provide drying rapid enough to prevent mold growth in moist region: 0.007 m^3 air/(m^3 grain s) may be used for relatively dry feed; 0.08 m^3 air/(m^3 grain s) may be needed for very moist feed (5).

In other grain dryers, air flows across descending beds of grain that are 0.2–0.4 m thick (5,6). In these cases, air-

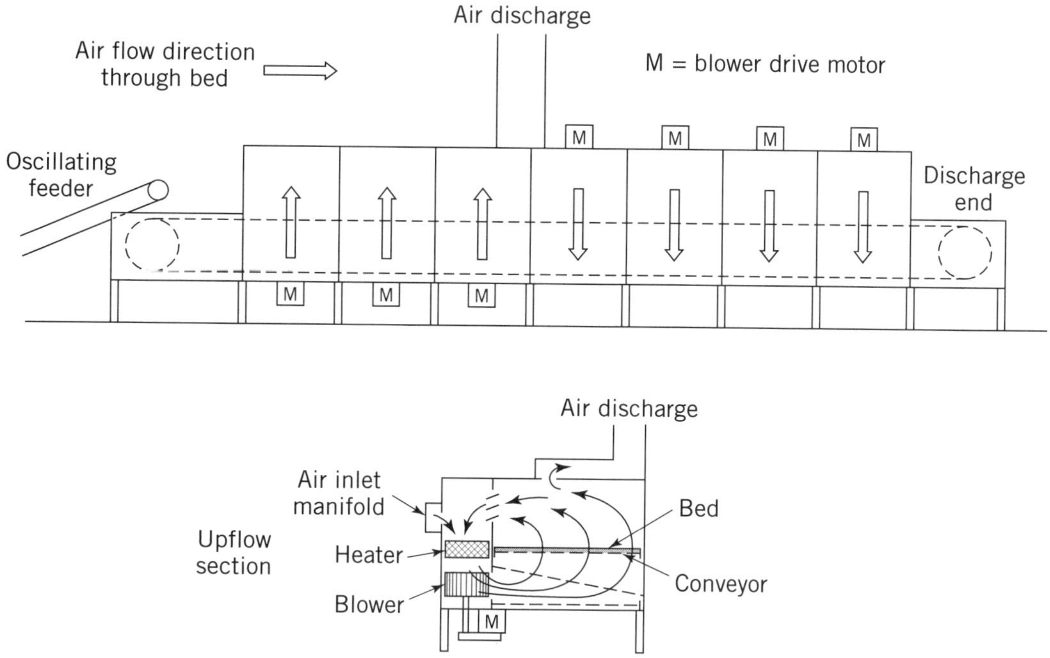

Figure 2. Conveyor dryer.

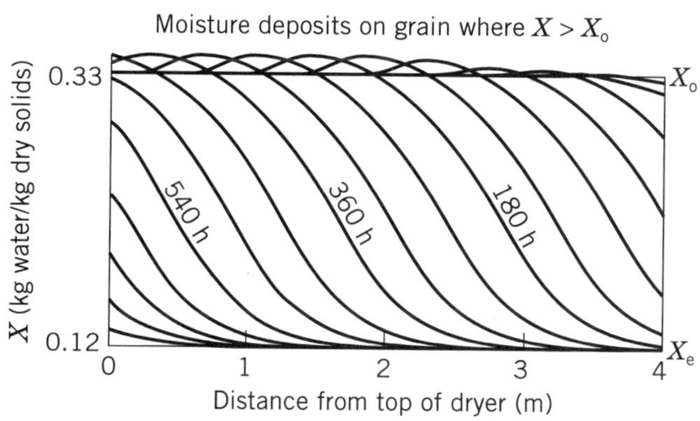

Figure 3. Moisture migration profiles for grain dryer (X vs. position at different times).

flow rates of 0.7–2.7 m^3 air/(m^3 grain s) are used. Inlet air temperatures up to 82 °C are used for animal feeds. Less than 54 °C is used for grain that is to be milled and processed, and less than 45 °C is used for seed grain. Grain is cooled with cool, dry air after such drying. Grain may be partly dried with high temperature air and then transferred to a deep bin where local moisture contents equilibrate without airflow. Final drying and cooling are accomplished by blowing cool, dry air through the grain.

Pasta Dryers. Special dryers are used to dry pasta. Freshly extruded pasta contains roughly 0.45 kg water/kg dry solids. That moisture content is reduced roughly 0.13 kg water/kg dry solids to set pasta's shape and preserve it. Long goods (e.g., spaghetti) are draped over moving rods and carried through the dryer system. Short goods (e.g., macaroni) are conveyed through dryers by simpler means. To prevent adjacent pieces of pasta from sticking together, the pieces are predried using 63 °C, 65% RH air. They leave the predryer and enter a large, final dryer where very humid air is used. Zones in which active drying occurs are followed by zones in which internal equilibration of moisture can take place. To prevent cracking (checking) due to stress buildup, relatively low temperatures (e.g., 50 °C) and very long drying times (up to 18 h) were and often still are used for long goods. Now, based on better modeling of moisture diffusion and stress development, temperatures up to 110 °C and drying times of 5 h are frequently used for long goods. Short goods are less susceptible to cracking and can now be dried in 2 h. Use of higher temperatures also reduces microbial growth.

Spray Dryers. Spray dryers are used to dry liquid foods and pumpable food slurries (7). Such foods are converted into drops with mean diameters ranging between 50 and 250 μm by nozzles or rotary atomizers and sprayed into a large chamber. The drops enter at high velocity but soon decelerate. Hot air is blown in and flows in the same direction as the descending drops. One type of spray dryer is shown in Figure 4. Inlet air temperatures as high as 270 °C may be used; often they range between 160 and 200 °C. Outlet air temperatures usually range between 80 and 110 °C. Drop temperatures quickly rise to the wet-bulb temperature of the inlet air and then progressively approach the local air temperature as drying proceeds.

Because air cools as it moves through the dryer, cocurrent air-drop flow prevents drops from overheating. Dried particles leave the dryer with temperatures 10–20 °C below the exit air temperature. Large particles drop out of the airstream. Smaller particles are collected in cyclones.

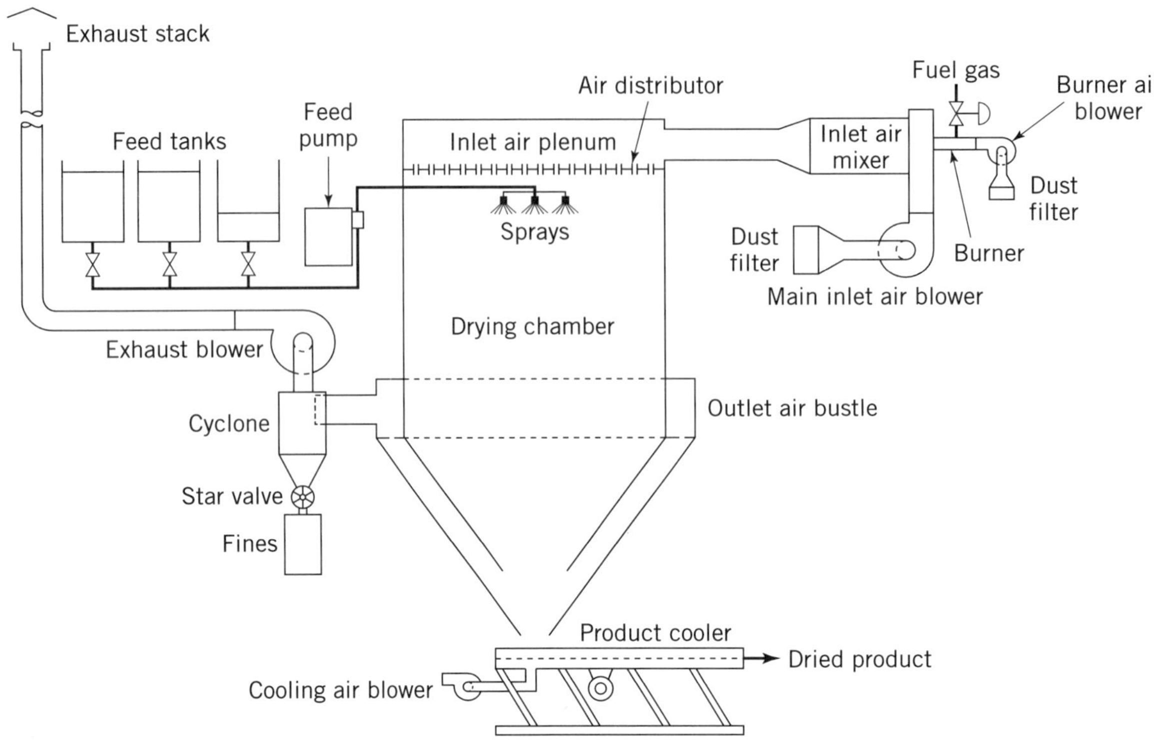

Figure 4. Spray dryer in which large particles drop out of the drying air by gravity and small particles are collected in a cyclone.

Drying times range from 5 s for small drops to 50 s for large drops. Spray-drying involves exchanges of heat, momentum, and mass between the air and the drops, nonlinear diffusion of moisture within drops, and changes in drop size and shape. Complex computer programs have been developed to account for these processes (8). Spray-dried foods are rapidly cooled after drying by sucking cool, dry air into solids discharge lines or by drawing cool, dry air through a screen-supported product bed on a vibrating conveyor.

Freeze-Dryers. Freeze-dryers are used to dry beverage extracts, shrimp, soup ingredients, and military and camper rations. In these dryers, moisture sublimes from frozen food in a vacuum chamber and condenses on very cold refrigerated coils. Heat transfers radiantly from heated panels to the surface of the food and then passes by conduction through a porous dry food layer to a sublimation interface. The interface recedes into the food as drying proceeds (9,10). Vapor created by sublimation escapes from the food through pores produced by prior sublimation.

Temperatures at the product surface and sublimation interface depend on the radiant panel temperature and on the thermal conductivity, vapor permeability, and thickness of the dry layer. Depending on whether heat conduction or vapor permeation controls the drying rate, radiant panel temperatures are programmed either to prevent product surface overheating or to prevent melting at the sublimation interface (11). Small, frozen particles are sometimes used to circumvent drying rate retardation caused by low dried-layer permeability.

Noncondensibles must be removed before and during freeze-drying. If adequate noncondensibles removal is achieved, chamber pressures in the dryer approach ice's vapor pressure at the condenser temperature. Condenser temperatures of -40 °C are frequently used, resulting in absolute pressures around 100 μm Hg. Local chamber pressures in plant-scale freeze-dryers are effected by pressure drops due to high-velocity vapor flows between dryer trays and heating panels (12).

Other Dryers. Foods are also dried in air-lift systems, in fluidized beds, on the surface of rotating, steam-heated drums, by contact with hot air in rotating, hollow cylinders of various types, by passage under or through arrays of high-velocity air jets, and by exposure to the sun in hot dry regions. Sun-dried foods may be treated with SO_2 to prevent browning and fumigated to prevent insect infestation.

a_w Reduction

Microorganism growth can be prevented by sufficiently reducing a_w, the chemical activity of water (13). Thus, *Clostridium botulinum* will not grow at $a_w < 0.94$, most bacteria will not grow at $a_w < 0.905$, most yeasts at $a_w < 0.88$, and most molds at $a_w < 0.8$; even osmiophilic molds and yeasts will not grow at $a_w < 0.6$. Drying preserves foods in part by reducing a_w. Moderate a_w reduction can be used in combination with other steps (e.g., pH reduction, refrigeration or freezing, and use of antimicrobial agents and fungistats to provide storage stability). Water activity values are also reduced by evaporation or by adding solutes that reduce the mole fraction of water in solution. Different combinations of salting and drying are used to preserve cod. Lightly salted cod is stable when its moisture content is 12.3%; heavily salted cod is stable with a moisture content of 52.4%.

Chilling

Chilled, nonfrozen storage is widely used to extend the storage life of fresh produce, meats, fish, and dairy products (14). Recommended temperatures and humidities and approximate storage lives for the commercial storage of fruits and vegetables are listed in U.S. government publications (15) and by the American Society of Heating, Refrigeration, and Air-Conditioning Engineers (ASHRAE) (16). These sources also provide heat capacities and heats of respiration for many of the commodities listed. Typical rates of heat influx are also given to permit computation of refrigeration loads for storage facilities. Storage temperatures close to 0 °C and humidities ranging between 85 and 95% are frequently recommended; cold-sensitive produce is stored at temperatures between 7 and 13 °C. Storage life at these conditions varies widely, ranging from as short as 2 days to as long as 12 months; storage life is less than 1 month for roughly 60% of the foods surveyed. Design and operating considerations for food storage refrigeration systems are treated by ASHRAE (16). Prechilling before storage is carried out by blowing chilled air on products, by immersing products in chilled water containing chlorine or an approved phenol compound or spraying similar chilled water on them, or by mixing products with ice or slush. Leafy products that conduct heat poorly are often cooled by evaporating a small part of their water content in a vacuum chamber. To minimize weight loss, such products may be prewetted before vacuum cooling.

Controlled Atmospheres. Chilling is combined with use of controlled atmospheres that contain as little as 2–5% O_2 and up to 5% CO_2 to increase the storage life of apples, pears, and cabbage (17). CO_2-rich controlled atmospheres are also used during long-distance shipment of meat. The rooms or shipping containers used must be sealed and airtight. Exceeding tolerable CO_2 levels or falling below tolerable O_2 levels damages apples. These levels significantly differ for different varieties of apples. Respiration can be used to reduce O_2 and raise CO_2 levels. Any excess CO_2 produced is absorbed by hydrated lime, ethanolamine (which can be regenerated by heating), water, or less frequently activated charcoal or molecular sieves. To more rapidly adjust O_2 and CO_2 levels, catalytic or noncatalytic combustion of fuel and flushing with nitrogen are used. Multilocale O_2 and CO_2 sensors and loggers have been used to ensure that improper local environments do not develop in controlled-atmosphere storage rooms. The use of selectively permeable wrapping can generate modified atmospheres in packages and improve shelf life during food distribution. Vacuum packing is also used.

Freezing

Freezing, the cooling-induced conversion of most of a food's water content into ice, is used to preserve virtually all types of moist food (18,19). It is also used as a step in freeze-concentration and freeze-drying and to produce ice cream and sherbet. Freezing preserves foods by reducing rate constants for chemical reactions, radically reducing reactant mobility, and reducing a_w. The water activity of a frozen food depends solely on its temperature and is independent of the nature of the food involved, as long as it is suitably moist; $a_w = 0.908$ at -10 °C, 0.824 at -20 °C, and 0.748 at -30 °C.

Solid foods are frequently frozen by exposing them to suitably cold streams of refrigerated air. This may be done by passing cold air through a shallow, fluidized bed of small particles or by blowing cold air over larger pieces of food carried on perforated conveyors or mounted on trays in racks. Meat carcasses are frozen by passing cold air around them while they hang by the hind legs on hooks. Foods have also been frozen by direct contacting with brine–ice mixtures or by contact with an inert, evaporating refrigerant or cryogenic gas. Packaged foods are often frozen by compressing them between refrigerated plates.

Ice cream and sherbets are partially frozen in scraped-surface freezers. Finish freezing (hardening) is carried out in plate freezers. Ice cream pops are frozen in molds conveyed through baths of refrigerated brine. Descriptions of most types of commercial freezing equipment, their specifications, and operating conditions are provided by Postolski and Gruda (20).

Most water in moist foods will freeze, but part (0.17–0.31 kg water/kg solids, depending on the particular food) is bound to food solutes or solids in nonsolvent form and will not freeze, even at very low temperatures. Food solutes depress food freezing points. Equilibrium initial freezing points for moist foods range between -0.4 and -3.2 °C, and usually lie between -1 and -2 °C. Foods frequently supercool 5 or 6 centrigrade degrees below their equilibrium initial freezing point before ice nucleates and freezing starts. Then the temperature rapidly rises to the initial freezing point. As freezing proceeds, solutes concentrate in residual unfrozen solution, further depressing the freezing point. Thus, food freezing occurs over a range of temperatures rather than at a single temperature.

Fractional conversions of freezable water to ice, fractions of water that remain unfrozen, and thermal properties of partly frozen foods can be predicted with the aid of the freezing point depression equation (21). These predicted values can be used in calculating freezing times for foods and temperature profiles for foods undergoing slow to moderately rapid freezing (i.e., freezing in which local phase equilibrium can be assumed without serious error) (22–24). Freezing involves incorporation of water in ice crystals and diffusion of water to active growth surfaces on ice crystals. These processes occur at rates slow enough to cause significant local deviations from phase equilibrium during very fast freezing. They also largely determine ice crystal size and shape (but not the total amount of ice formed) during slower freezing.

Ice usually grows as dendrites (treelike crystals) when foods freeze. Spacings between ice dendrites tend to be inversely proportional to the square root of the initial freezing rate. Specific surface areas and surface free energies of dendrites are much larger than those of rounded crystals.

Therefore ice dendrites tend to convert into rounded ice. Conversion occurs rapidly in agitated slushes but is quite slow in unagitated frozen food, particularly food that is kept well below its initial freezing point. Specific surface energies are larger for small crystals than for large crystals. Therefore small ice crystals tend to sacrificially melt and cause growth of larger ice crystals. If storage temperatures are not low enough or are allowed to cycle excessively, ice crystal size will increase appreciably after long storage, affecting texture adversely.

When unblanched fruits and vegetables are frozen, ice initially nucleates outside cells. If freezing is slow, much water transfers outward from cell interiors and forms extracellular ice before nucleation occurs inside cells. This affects texture adversely. Good practice calls for rapid freezing. Then, intracellular nucleation occurs quickly, most water freezes inside cells, and adverse textural effects are minimized. Preferential growth of ice in extracellular space usually does not occur during slow freezing of blanched vegetables. Fast freezing of meat and fish helps minimize losses of water-holding power and decreases drip following thawing.

Thermal Processing

Sterilization, pasteurization, and blanching are processes in which foods are heated and held at suitable temperatures long enough to kill the microorganisms and inactivate the enzymes they contain. Commercial sterilization is designed to kill substantially all microorganisms that can grow or spores that can germinate at storage conditions. Because sterilized foods are packed in sealed containers that contain virtually no air, anaerobic organisms are of greatest concern. Destruction of *C. botulinum* spores is often used to evaluate adequacy of sterilization for low-acid foods (i.e., foods with $pH > 4.5$). Fractional survival levels must be less than 1×10^{-12} for such spores. Adequacy of sterilization for low-acid foods may also be based on destruction of a hardier putrefactive anaerobe called (PA) 3679. Adequacy of sterilization for foods with pH values less than 4.5 is based on destruction of one or another of the facultative anaerobes, *Bacillus coagulans, B. mascerans*, or *B. polymaxa*. Fractional survival levels for these organisms should be less than 1×10^{-5} or 1×10^{-6}. Sterilization of low-acid foods involves maximum temperatures in the 116–149 °C range. Maximum temperatures up to 110 °C are used for more acidic foods.

Sterilization of canned products is carried out batchwise by steam heating in vertical or horizontal retorts. The retorts must be vented for at least 5 min with steam flow to remove air before steam pressurization and starting a heating cycle. Foods in jars are often sterilized in air-pressurized hot water in retorts. Sterilization based on steam heating is also carried out continuously in retorts with sections separated by pressure locks, in rotary sterilizers (where a rotor causes cans to move through the unit along spiral guide rails), and in hydrostatic sterilizers (where entering and leaving cans are conveyed through a long, steam-heated zone confined between vertical water legs that balance the steam pressure). Figure 5 shows an example of a hydrostatic sterilizer.

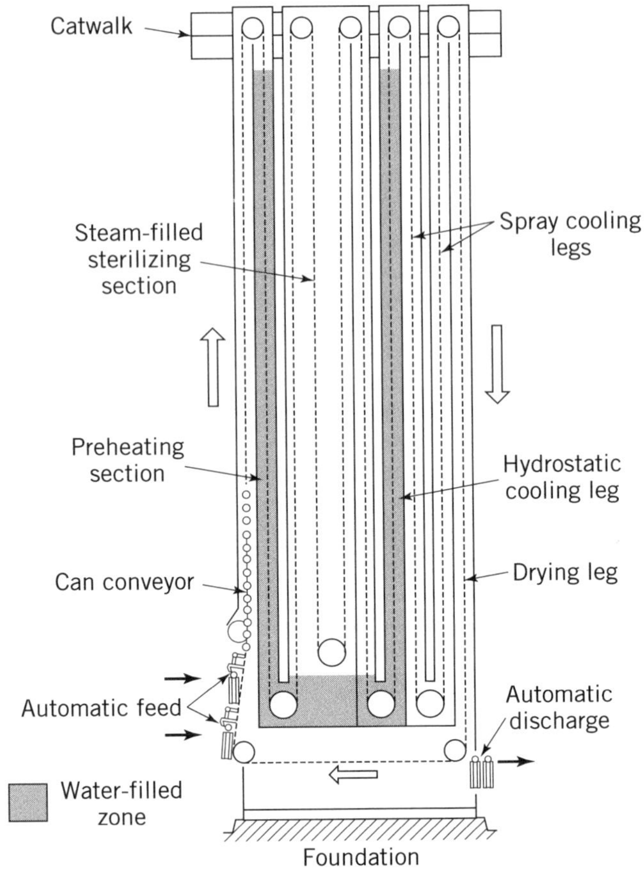

Figure 5. Hydrostatic sterilizer. Reprinted with permission of the Chemetron Corporation.

Headspace is left when food containers are filled. The container walls or tops have elements that flex. These elements and headspace prevent thermal expansion from damaging containers during sterilization. Headspace air and dissolved gases are partly removed from cans, jars, and their contents before sterilization by steam flushing or vacuum evacuation before capping. Thermal exhaustion (i.e., preheating material to expel dissolved gases) is sometimes used instead of or in conjunction with these processes.

Thermally processed products are cooled under air pressure with chlorinated cold water immediately after heating. Air pressure is used to prevent bulging of containers or unseating of caps caused by excess residual pressures inside containers. Cooling is continued until products are just warm enough to fully evaporate water clinging to the container surfaces. Sterilization, container filling, and capping procedures for specific products differ in detail but are described in standard references (25).

Pasteurization. Pasteurization, the thermal treatment at temperatures less than 100 °C, is used to inactivate enzymes, virtually all pathogenic microbes, and many food spoilage organisms. It is used in combination with protective packaging and other treatments (e.g., chilling, pH, and a_w reduction) that greatly retard subsequent microbial growth. The target organism or enzyme on which adequate inactivation is based depends on the product involved. For milk it is *Coxiella burnetti*, which causes Q fever. For wine or beer it is wild yeasts. For high-acid fruits it is molds and yeasts. For citrus juices it is pectinesterase. Inactivation of pectinesterase also provides more than adequate microbial inactivation. Acceptable holding time-temperature combinations may be specified by government regulations. In the United States, at least 30 min at 62.8 °C or at least 15 s at 71.7 °C is required for pasteurization of milk. Most frequently, the latter conditions are used. Other, equally effective time–temperature combinations can be approved. Ultrahigh temperature pasteurization carried out at 88.3 °C for 1 s or 100 °C for 0.01 s is used sometimes. Liquid mixes used to produce ice cream and frozen desserts are pasteurized for at least 24 s at 79.4 °C.

Milk is pasteurized before it is put in containers. This is done by regeneratively and directly heating milk to a specified temperature in a plate heat exchanger, holding it at that temperature for a required time and then cooling it in the same exchanger. Integrated systems that also provide cream removal by centrifugation and homogenization are often used (see Fig. 6). Other arrangements for pasteurizing milk are described by Kessler (26), who also provides correlations for overall heat transfer coefficients versus fluid velocities or Reynolds numbers for different plate–heat exchanger flow arrangements. Simpler plate–heat exchanger setups are used to pasteurize fruit juices.

Other foods and beverages are pasteurized in containers. Products in glass jars are conveyed on metal-link belts, heated gradually, held at temperature, and cooled gradually in stages to prevent thermal shock to the glass. Long hot water baths or sprays followed by similar cold water baths or sprays are used for jarred pickles, canned fruit, and bottled beer. Heating by passage through steam at atmospheric pressure is sometimes used in pasteurizing products packed in glass.

Blanching. Blanching, heating in water somewhat below it boiling point or in steam at atmospheric pressure, is used to inactivate enzymes that adversely affect food storability and quality. Sterilization and pasteurization processes usually inactivate enzymes as well as microorganisms; blanching is used primarily before drying and freezing for vegetables and some fruits. Blanching also causes expulsion of gases from foods, softens foods, facilitates cutting and peeling, sets some colors, and provides some microbial inactivation. Unfortunately it also leaches nutrients from foods, which in turn increases the biological oxygen demand (BOD) of effluent blanch water. Less leaching occurs when steam blanching is used and still less when very humid hot air is used.

Aseptic Processing. Aseptic processing involves heating liquids, semiliquid products, or liquids containing small particles to sterilization or pasteurization temperatures in heat exchangers or scraped-surface heaters or by injection or infusion of culinary steam and then passing the products through a holding tube. They are then cooled in heat exchangers or, for products heated by steam addition, by vacuum evaporation, and put in sterilized containers. The containers may be cans, jars, or drums that have been heated with superheated culinary steam. They are capped with steam-sterilized covers in chambers filled with super-

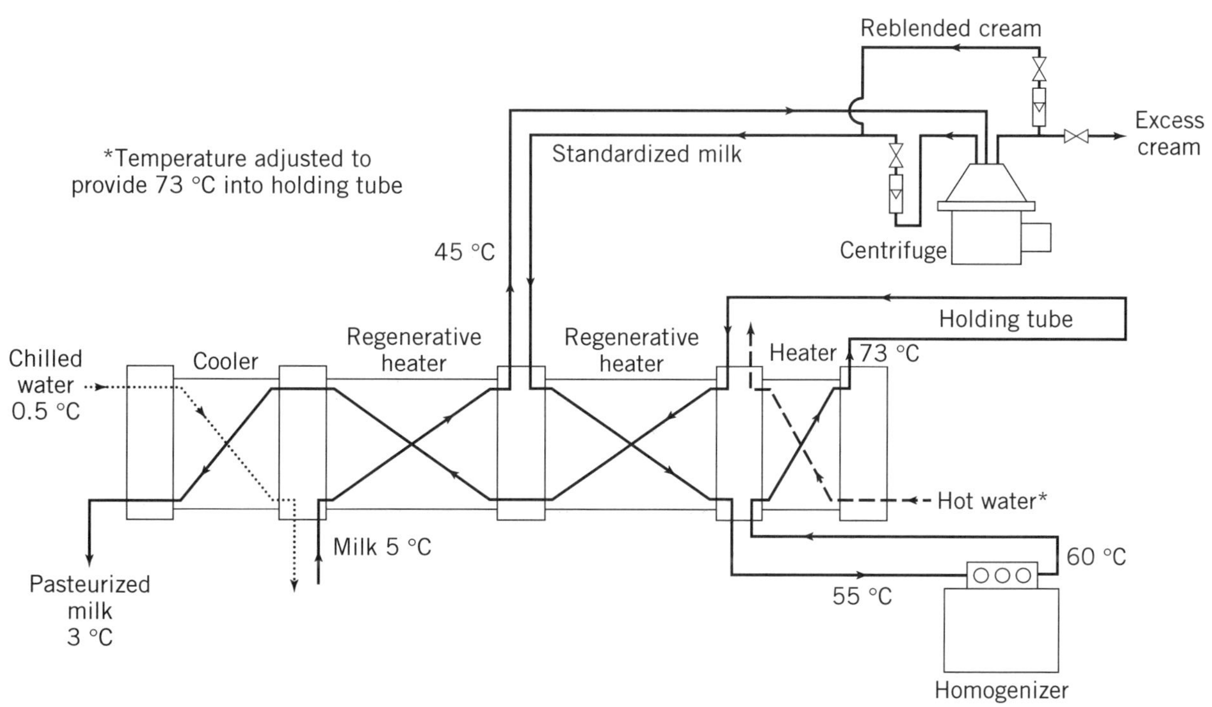

Figure 6. System for standardizing, homogenizing, and pasteurizing milk. Reprinted with permission of Kessler Verlag.

heated culinary steam. Continuously formed laminated containers or plastic cups are also used. Before they are filled, they are bathed in H_2O_2 and dried. These containers are heat-sealed by fusing together narrow bands of a meltable plastic. Heated product temperatures range from slightly less than 100 °C for acidic products to 149 °C for low-acid products.

Inactivation Kinetics. Thermally based microbial destruction and enzyme inactivation are treated as first-order reactions; that is, destruction or inactivation rates are proportional to the residual concentrations of living microbes or active enzymes involved. Inactivation rates are strongly affected by pH and food composition. If $k(T_p)$ is the destruction rate constant for a particular microbe or enzyme in a food expressed as a function of product temperature T_p and T_p is a function of time t, we can write

$$\ln\left(\frac{N}{N_o}\right) = -\int_{t=0}^{t} k(T_p)dt \quad (1)$$

where N is the residual number of microbes per unit volume or enzyme concentration at t and the spot where T_p and $k(T_p)$ are evaluated and N_o is the initial value of N. Figure 7 depicts how $k(T_p)$, T_p, and N/N_o change during thermally based sterilization. To achieve acceptable destruction or inactivation, the highest N/N_o at any locale in

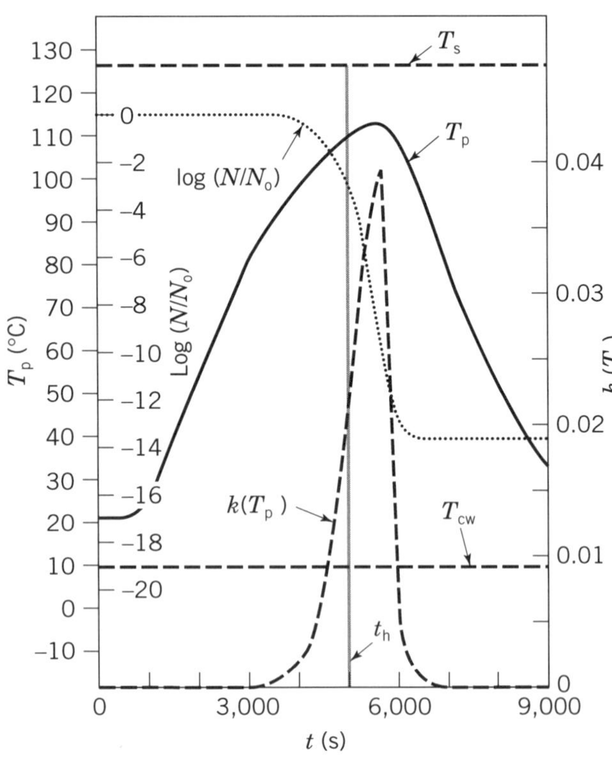

Figure 7. T_p, $k(T_p)$, and N/N_o versus t during sterilization of a canned solid food by thermal processing: T_s and T_{cw} are, respectively, the steam temperature and the cooling water temperature in the retort, T_p is the temperature at the center of the food, and t_h is the time at which heating ends and cooling begins. Reprinted with permission of John Wiley & Sons.

a product for the most critical microbe or enzyme in question must be equal to or less than a required value. For low-acid foods, N/N_o must be less than 10^{-12} for *C. botulinum* spores. For nonmobile canned foods, the greatest N/N_o is assumed to occur at the center of a can's contents. Convection occurs in foods containing free liquid, and the greatest N/N_o is assumed to occur at the slowest heating point. This usually lies along a can's axis and near its bottom, but its locus should be experimentally determined.

In the food industry, adequacy of thermal processing (27) is often evaluated by determining whether inactivation is equal to or greater than that provided by treatment for a specified time at standard temperature, 121.1 °C (250 °F). Temperature-induced changes in thermal inactivation rates are accounted for in terms of Z, the temperature reduction that will cause $k(T_p)$ to decrease by a factor of 10. A standard Z of 10 °C (18 °F) is often used for *C. botulinum*. Depending on the food and corresponding Z involved, acceptable treatment of low-acid foods should provide *C. botulinum* destruction equal to that provided by 1.2–2.4 min isothermal exposure at 121.1 °C.

In evaluating adequacy of thermal processing, one must account for changes in product temperature that occur during both heating and cooling. Product temperatures predicted from heat transfer analysis are often used in trial-and-error fashion to determine suitable heating and cooling times. These calculations are useful for estimating required processing time, but determination of adequacy of processing is usually based on measured product temperature–time data. This is particularly true for new products, where adequacy of sterilization must be further verified by microbial inactivation tests in inoculated packs.

Formula methods and nomographs based on characteristics of semilog plots of unaccomplished temperature change versus time were once widely used to determine adequacy of sterilization and needed thermal processing time. Now, simple computer programs are used for the same purpose.

Sterilization causes partial destruction of some nutrients and often adversely affects food flavor and color. Destructions of nutrients and some colors and flavors are first-order or pseudo–first-order processes for which kinetic parameters are sometimes available (28). New flavors, colors, and textures that affect product quality also develop by processes with more complex kinetics. Effects of product temperature on nutrient, flavor, and color destruction often are framed in terms of activation energies and the Arrhenius equation instead of Z. In evaluating nutrient destruction and quality reduction, fractional retention of the nutrient or attribute being evaluated is averaged over the whole volume of the container. Sterilization methods and temperature–time conditions that provide adequate pathogen or spoilage agent inactivation but minimize adverse effects should be used. Rates for reactions that destroy nutrients and adversely affect sensory properties increase less sharply than microbial destruction rates as temperature increases. Therefore high-temperature, short-time (HTST) processing often improves nutrient and flavor retention. Heat-resistant enzymes may not be adequately inactivated when microbial destruction is adequate if very high temperatures are used for very short times.

Outer regions of containers are often overprocessed and suffer excessive quality degradation when the points that heat most slowly are just adequately processed. Overprocessing can be reduced by providing more uniform heating (e.g., by using aseptic processing and microwave and ohmic heating). Other preservation processes that avoid or greatly limit heating (e.g., radiation preservation, microfiltration, and high-pressure treatment) have been extensively investigated or, in some cases, are already in use. Their advantages and limitations were examined in a recent review (29).

Concentration Processes

Evaporators. Water and solvents can be removed from liquid foods by means of evaporators. In multiple-effect evaporators, the most commonly used type, vapor generated by steam or vapor heating in an effect provides heating that causes evaporation in a subsequent effect operating at a lower pressure. In *N*-effect evaporators, roughly *N* kg of vapor is generated for each kilogram of steam used. In some cases, a steam ejector is used to compress part of the vapor generated in a second or third effect, and the compressed vapor and motive steam are used to heat the first effect, providing more than *N* kg of vapor per kilogram of steam used (30). Vapor recompression may also be carried out by large centrifugal compressors. Electrical energy is used to drive the compressor; virtually no steam is used directly. Evaporation, particularly multiple-effect and vapor recompression evaporation, is more thermally efficient than drying. Therefore evaporators are often used to concentrate liquid foods before drying. Evaporation is also used to concentrate juices, syrups, sauces, fermentation residues, stillage bottoms, and absorption refrigeration solutions. It is also used to induce crystallization and produce glassy supersaturated sugar solutions used to make hard candies.

Vacuum evaporation, often combined with single-pass operation in each effect to provide low product holdup, is used to minimize thermal damage to foods (31). In single-pass evaporators, steam condenses on the outer walls of vertical heat transfer tubes 6–12 m long mounted in a steam chest. Water evaporates from liquid flowing in film form down the inner walls of the tubes. Concentrated liquid and vapor leave through the bottom of the tubes. Entrained liquid is separated from the vapor in a cyclone. A single-pass, multieffect system is depicted in Figure 8. High-velocity recirculation in liquid-filled tubes is used to improve heat transfer in evaporators used to concentrate viscous materials such as tomato paste and gelatin and pectin solutions. Liquid superheats as it passes through tubes in such evaporators and flashes as it leaves the tubes. Cyclones are used to separate entrained liquid from the vapor produced by flashing. Scraped-surface, thin-film evaporators are used for very viscous products (e.g., hard candy melts) and for heat-sensitive products.

Natural convection and boiling cause circulation of liquid in evaporators containing short, steam-heated tubes mounted in a wide-diameter assembly (calandria). Vapor separates from the foaming, boiling liquid in a large, vapor-filled chamber above the calandria. Heat transfer coefficients in these evaporators peak at a certain liquid level in the calandria. Multieffect systems containing such evaporators are frequently used to concentrate sugar solutions in sugar refineries. Part of the vapor produced may be withdrawn from effects and used for process heating elsewhere in the refinery.

Most food aromas have high relative volatility with respect to water and almost completely vaporize during initial stages of evaporation. Such aromas can be recovered in concentrated form from first-stage vapor streams and used for add-back purposes.

Membrane Processes. Pressure-driven flow through permselective membranes is used to gently remove water or aqueous solutions of low molecular weight solutes from liquid food products. Reverse osmosis (i.e., selective removal of water) is used to raise the solids content of maple sap from roughly 2% to 10–12%, to recover water and solutes from waste processing streams, and to concentrate dilute aqueous caffeine extracts. Ultrafiltration (selective removal of water and low molecular weight solutes) is used to concentrate proteins in milk, milk whey, and soy whey and to recover protein from potato fruit water. Protein-rich concentrates obtained from heated milk by ultrafiltration are used to make cheese with reduced loss of milk solutes and reduced use of rennet and starter cultures. Diafiltration (ultrafiltration accompanied by water addition) is used to de-ash gelatin. Pervaporation has been used experimentally to produce aroma-rich concentrates by selectively evaporating water through permselective membranes. Mass balances for retained species and equations governing permeation rates are used in designing membrane-based processes. Design calculations must not neglect the effects of concentration polarization on permeation rates and selectivity.

Freeze Concentration. Freezing is sometimes used to gently and selectively remove water from fruit juices, beverage extracts, beer, and wine. Scraped-surface coolers partially freeze the solution being processed, producing small ice crystals. Ostwald ripening (i.e., growth of large crystals induced by sacrificial melting of smaller crystals) is then used to produce large ice crystals, which are separated from residual solution in wash columns (32) or centrifuges. Thermal efficiency and upper concentration limits have been increased, and costs reduced, by means of countercurrent freeze concentration (33), but the use of freeze concentration remains limited because of its high cost.

Solid–Liquid Extraction

Solid–liquid extraction is used to recover sugar from sugar beets and sugarcane, oil from oilseeds, juice solutes from fruits, beverage extracts from coffee and tea, protein-rich materials from soybeans, fermentable solutes from grains, and hydrocolloids, gums, and natural food colors from various sources. It is also used to remove unwanted constituents (e.g., caffeine from coffee, bitter compounds from olives, cyanogenic compounds from cassava, sarcoplasmic

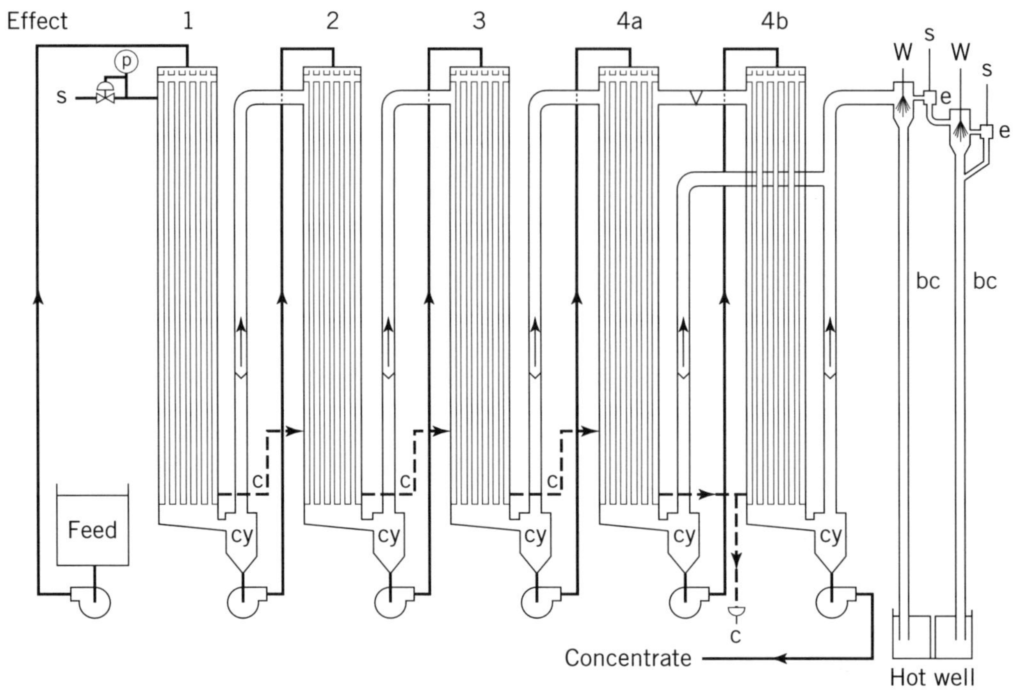

Figure 8. Single-pass, falling-film, multieffect evaporator. The fourth and fifth stages in terms of liquid flow together make up the fourth effect in terms of vapor utilization. bc, barometric condenser; e, ejector; cy, cyclone; s, steam; w, cold water; c, condensate; v, vapor.

protein, and undesired flavors from minced fish, and salt from pickles) (34,35). Extraction also occurs when flavoring agents leach out of charred barrels during the aging of wines and distilled spirits, when solutes leach out of vegetables and fruits during blanching and fluming, and when plasticizers leach out of plastic containers into liquid foods and beverages.

Vegetable or animal matter from which solutes are to be extracted is cut or otherwise divided into small particles. Oilseed grits are flaked or extruded as steam-puffed cylindrical particles. This coalesces isolated oil bodies and creates fissures or pores that allow solvent or extract to reach the oil. During extraction, solutes diffuse out of the solid through internal liquid-filled paths and pass into solvent or extract that contracts the solid. Dry materials (e.g., coffee and tea) must imbibe solvent or extract before diffusion can proceed. In rare cases (e.g., extraction of vanilla from vanilla beans with alcohol), imbibition is very slow and is a rate-controlling step. Plasma membranes are denatured to facilitate water-based extraction of solutes from cellular material. Then, pores in cell walls become the main diffusion bottleneck. In-particle diffusivities for water-soluble extractibles are 0.1–0.5 times as large as in water alone. Required extraction times are inversely proportional to the in-solid solute diffusivity and proportional to the particle size squared.

The extract carries away the solute, following a contacting path that promotes high solute recovery. Extraction efficiency and extract concentration depend, in part, on the stripping factor, the solute distribution coefficient multiplied by the ratio of extract to moist solids. High concentration extracts can be obtained by using a low stripping factor (i.e., low solvent flow rates or small amounts of solvent), but efficient extraction can be obtained only if the stripping factor is greater than 1.0. Stripping factors of 1.2–1.4 are often used in water-based extractions. Higher stripping factors (e.g., 2.4) are used for hexane-based extraction of oil from oilseeds.

Small-sized particles favor fast extraction but cause excessive flow pressure drop and displacement instability. Large particles extract slowly. Particle thicknesses or diameters in the range of 3–5 mm are a good compromise. Where possible, particle shapes that prevent blinding of surfaces by interparticle contact are used.

Commercial extraction is often carried out in very large, continuous or semicontinuous countercurrent extractors where extract percolates downward through a rising bed of solids (see Fig. 9) or in units where a bed of solids moves horizontally while extract repeatedly percolates through it in a flow arrangement that effectively provides countercurrent contact. In other cases, a cyclically loaded and discharged set of interconnected columns is used to provide nearly countercurrent contacting. Most of these processes can be analyzed and designed by means of solutions of partial differential equations describing diffusive mass transfer in solids (3,34) or mass transfer analogs of partial differential equations describing conductive heat transfer in solids (2). Solutes are sometimes produced from insoluble precursors by reactions whose kinetics must be accounted for in designing the extraction process. Extraction processes involving foods often are subject to imperfect contacting (e.g., channeling, unsteady displacement, and axial

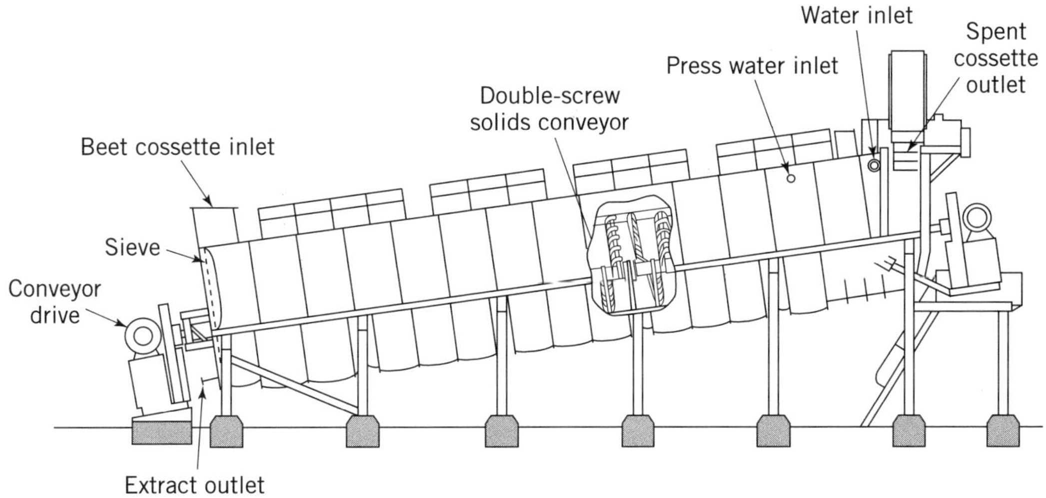

Figure 9. DDS extractor, a type often used to extract sugar from sugar beets. Reprinted with permission of De Danske Sukkerfabrikker.

dispersion). These imperfections must be accounted for in design calculations.

Supercritical carbon dioxide is now used as a solvent in decaffeinating coffee and tea and in recovering solutes from hops. Supercritical carbon dioxide is a particularly attractive solvent because it is nontoxic and can be readily removed from foods. Further, its dissolving capacity and selectivity for solutes such as caffeine can be adjusted by adjusting temperature and pressure. Figure 10 shows a schematic of a large supercritical extraction system used to decaffeinate green coffee beans.

Solid–Liquid Separations

Filtration. Filtration is used to remove solid particles from or to clarify juices, extracts, vegetable and fish oils, fermented beverages, recirculated, cooking oil, flume water, milk, and soy milk. It is also used to separate potato starch from potato fruit water, high-melting fats from vegetable oils in fractionation processes, crystals from mother liquors, and chemically precipitated impurities from sugar juice.

The engineering principles that govern filtrations involving foods are very similar to those that apply to the filtration of chemicals (36), but superior sanitation must be maintained. A fluid that contains particles is passed through a cloth or other porous medium that retains particles. The particles deposit on the medium, forming a porous cake that thickens as deposition continues. As long as total pressure drop remains constant, total flow resistance increases linearly as the cake thickens. Depending on the feed pumping system used, flow resistance increases cause either decreases in flow rates and solids deposition rates, increased fluid pressure drop, or a combined increase in pressure drop and decrease in flow rate. Filtration is stopped when these changes become excessive, and the filter cake is manually or mechanically removed.

Extents of these changes and optimal scheduling of cake removal can be predicted by means of calculations that relate cake depth and flow pressure drop to the initial solids content of the fluid, the amount of filtrate discharged, and the experimentally determined relationships between applied pressure, cake density, specific filtration resistance, and applied pressure. Some food-based filter cakes are highly compressible, and their specific filtration resistance increases markedly as pressure increases. Filter aids (inert, highly porous solids) are sometimes added to filter feeds to reduce cake filtration resistance and compressibility and to improve solids retention. Cake filtration resistance tends to be inversely proportional to the particle size squared. Therefore chemical agents that cause particles to flocculate (stick together in clumps) are often added to filter feeds.

Deep porous media that capture and retain particles within their pores are used for some feeds. Media flow resistance increases as filtration proceeds. Consequently flow pressure drops increase and/or flow rates decrease. The medium is replaced or cleaned by backwashing when these changes become excessive. Unlike cake filtration, where total filtration resistances at constant pressure increase linearly as solids removal progresses, filtration resistance for in-media solids capture is more than linearly proportional to the amount of solid removed. Therefore in-media capture is used only for fluids with light solids loads.

Sterile microfiltration based on use of membrane filters with 0.2-μm pores is used to remove bacteria from beers, thereby eliminating or reducing needs for pasteurization. Ceramic microfilters that can be sterilized and cleaned by back-flushing are used to clarify fruit juices. They are very expensive but will last for 10 years or more if care is taken to avoid heat shock.

Beds of particles used in solid–liquid extraction and adsorption and ion-exchange processes act like deep filter cakes. Pressure drops during flows through such beds often can be predicted with the aid of the Ergun equations or Kozeny–Carman equation. In other cases, flow pressure drops are calculated by means of empirically determined permeabilities or specific filtration resistances.

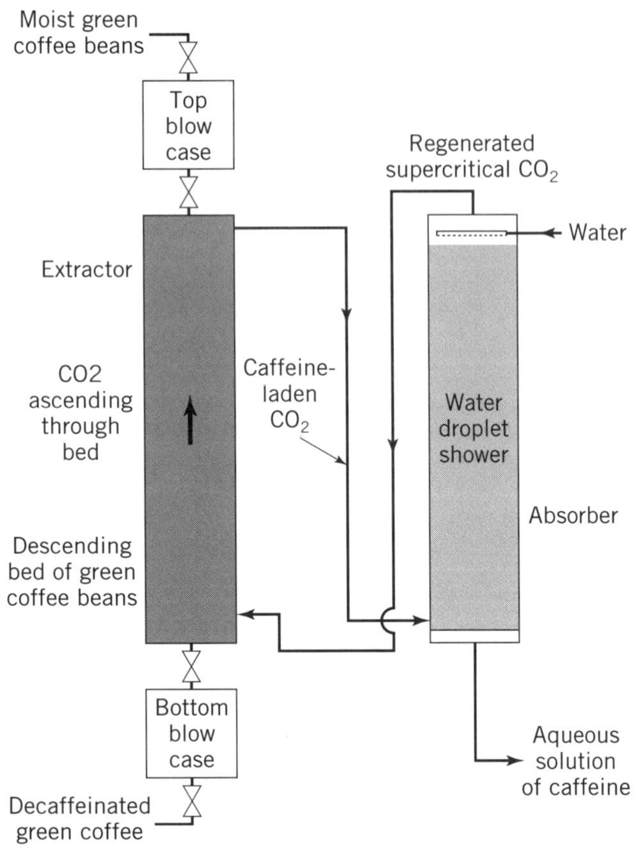

Figure 10. System for extraction of caffeine from green coffee by supercritical carbon dioxide. Reprinted with permission of Chemical Engineering Communications.

Centrifugation. Particles forced to rotate about an axis at high angular velocity in a centrifuge experience a radially acting force proportional to the radial distance from the axis, the angular velocity squared, and the density difference between the particle and surrounding fluid. At rotational speeds used in commercial centrifuges, particles denser than the fluid move outward and particles lighter than the fluid move inward thousands to tens of thousands times faster than they would sink and rise, respectively, under the influence of gravity. Therefore, centrifuges are used to separate and collect dispersed particles and droplets too small to be cleanly and quickly separated by other means. Centrifuges are used, for example, to separate dispersed oil, microorganisms, and fine biological particles and precipitated curds from aqueous solutions (23) and to remove yeasts from fermented beverages and waste treatment sludges from dischargeable water. Density differences driving centrifugal settling are usually much smaller for food processing than for chemical processing; the particles involved are more compressible, and, again higher levels of sanitation are required.

The particles or droplets involved move outward or inward until they strike a collection surface or material previously deposited on that surface. Deposited material then slides along the collection surface or is mechanically conveyed along it to a conduit or port that permits it to leave the centrifuge. Settled solids may leave through peripherally located nozzles as a dense slurry or leave in even denser form through intermittently opened peripherally located slits. Settled solids may also be knifed or augered out of centrifuges. Clarified liquid and separated oil or oil-rich creams leave through ports located closer to the rotational axis of the centrifuge. Centrifuges are sized to provide holdup time sufficient to permit complete settling or a desired level of settling. Large numbers of thin, closely spaced, concentric conical disks are mounted in centrifuges to reduce settling distances and times for extremely fine materials (e.g., butterfat droplets, and microorganisms). This greatly increases usable throughput rates for feeds containing such fine material.

Filtering centrifuges have a finely perforated wall or a perforated wall covered with filter cloth. Centrifugal force causes solids to deposit as a cake on the wall or cloth and then forces liquid through the cake. Wash liquor may also be forced through the cake. Such centrifuges are used to recover crystals (e.g., sugar and citric acid) from concentrated mother liquors, to separate freeze-concentrated vinegar and beverage extracts from ice, and to recover juice from milled fruit.

Centrifugal force generated by rotational flow is produced by tangentially injecting solids-laden liquid or gas into hydroclones or gas cyclones. These units have a cylindrical shell attached to a conical bottom, a central outflow port at their top, and a bottom port through which solids or solids-rich underflows leave. The injected liquid or gas first spirals downward in an outer vortex and then spirals upward in an inner vortex. The vortex flow tends to cause particles to move outward toward the wall. In gas cyclones virtually all the gas leaves through the top port. Gas cyclones are sized to provide enough flow residence time to permit wall impact by all particles greater than a desired size. Gas cyclones are used to separate fine spray-dried material from discharged drying air, pneumatically elutriated materials from carrier gas, chaff from coffee roaster discharge gases, and pneumatically conveyed material from conveying air.

In hydroclones, most entering liquid leaves through the upper port as overflow, but enough leaves through the lower port to provide a flowable slurry, the underflow. The centrifugal force generated by vortex flow is partly counterbalanced by drag force generated by radial flow of liquid from the outer to the inner vortex. Consequently, small particles or low-density particles may migrate inward and be carried away with the overflow. Larger or denser particles leave with the overflow. Large banks of miniature hydroclones used in parallel are employed to separate starches from protein- and solute-rich solutions during the wet milling of corn.

Expression. Pressing of fluid-rich biological matter is used to recover fruit and sugar juice, vegetable and fish oil, pectin released by cooking apple pomace, and protein-rich juice from leaves. It is also used to expel whey and to fuse curds together during cheese manufacture and to dewater spent food processing wastes, by-products, and processing intermediates (37). Expression is carried out in a wide variety of equipment (38). In screw presses, feed is transported down the length of a perforated barrel by a rotating single screw or by twin screws. The feed compacts, and fluid is expelled through barrel perforations because the screw's root diameter increases, its pitch or rotary speed decreases, or the barrel diameter decreases, or because constricted solids outflow from the barrel causes a progressive buildup of pressure.

In roll mills, a grooved, rotating roll is hydraulically pressed against two similar counterrotating rolls that lie beneath it. As the rolls rotate, they drag feed (usually shredded sugarcane) through nips between the top and bottom rolls. This compacts the feed and expels juice. Some of the juice flows back through the advancing feed; some flows laterally into the grooves and then backward. The process is analyzed by relating local backflow and flow pressure drop to the cumulative extent and rate of compaction and the local specific filtration resistance of the compacted feed.

In belt presses, feed is deposited on a moving cloth belt and pressed between that belt and an opposing, moving cloth belt driven over a series of solid or perforated rolls. In some cases the belts follow a serpentine path. Expelled fluid flows through the cloths and is collected in troughs beneath the rolls. Figure 11 shows a serpentine belt press of a type used to press juice out of milled apples. In presses of other types, feeds confined in a perforated cage or in stacked cloth pouches are pressed by a diaphragm or ram. Fluid is also expelled from cakes compacted by centrifugal force in decanter centrifuges.

Engineering treatment of expression is based on numerical solution of partial differential equations that describe how flow in press cakes is driven by fluid pressure gradients or related solid stress gradients and how rates of local cake volume change as a function of flow rate gradients (38,39). Empirically determined relationships between solid stress, specific filtration resistance, and the solids-based specific volume of the cake are used in solving these equations. Methods have been developed for determining these relationships from constant-pressure or constant-rate pressing tests carried out in suitably instrumented equipment (40).

Some press feeds (e.g., waste treatment sludges and fruit with thin cell walls and weak intercellular bonds) compact excessively as pressure increases. When such feeds are pressed, fluid pressures drop precipitously and solid stress rises sharply over very thin layers near outflow surfaces. Pressing can be improved for such feeds by adding an inert material (e.g., rice hulls) that stiffens the press cake. Other feeds (e.g., spent coffee grounds and oilseeds) have thick cell walls that adhere strongly together. Fluid pressure builds up inside the solid when such feeds are pressed. The extent of in-solids pressure buildup depends on the rate of pressing and the extent of compaction. Pressing of such materials can be improved by treatments that promote cell wall rupture.

PEELING, SKINNING, AND HULLING

Peels, skins, fatty layers, hulls, and shells frequently must be removed from foods. Skins are often manually cut from meat and fish with the assistance of special tools. Wholly mechanical systems are also used. These grab a carcass, carcass part, or fish at a selected spot and drag it past suitably positioned knives and saws. Fat-depth-sensing devices are sometimes used to guide cutting.

Moisture contents of shell-encased and hull-encased foods are often adjusted to increase shell and hull friability and separability. Hulls and shells are then cracked by impaction in rapidly rotating pin mills or by passing the food involved through pairs of rollers or plates that turn at different speeds and produce a pressing, rolling action. Thin shell and hull fragments are then pneumatically separated from more compact usable food particles.

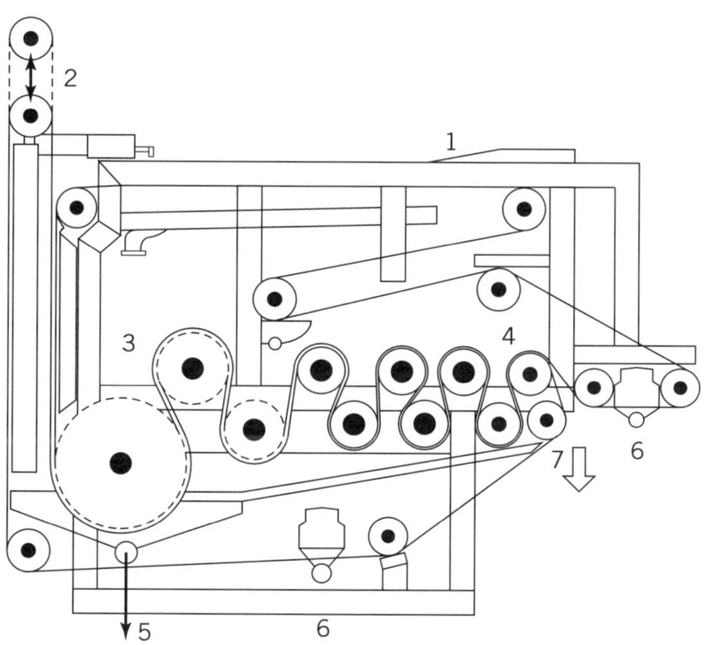

Figure 11. Serpentine belt press. 1, Mash deposition; 2, wedge area; 3, low-pressure press zone; 4, high-pressure press zone; 5, juice runoff; 6, belt cleaning; 7, press-cake discharge. Reprinted with permission of Bellmer GmbH and Co.

Manual peeling of vegetables and fruit is costly; abrasion peeling removes too much peel and is wasteful. Therefore, treatments that break down bonds between cells or decompose cells in or near skins are used. These include short time heating with high pressure steam, short-time immersion in baths of lye, and passing product through a flame. The loosened skin is then brushed off or washed off by water jets, a step that also removes residues of chemical peeling agents. The process can be analyzed in terms of short-term heat or mass transfer in surface layers and the kinetics of protopectin or cell wall breakdown. To sharply limit intercellular bond breakdown, internal temperature gradients near the product surface must be very large and exposure times very short. Even so, undesirable textural and color changes often occur in regions close to the peeled surface.

CLEANING

Washing-based cleaning of food surfaces and cleaning and disinfection of equipment are widely used food processing operations that have been analyzed from an engineering point of view (41–43). Efficient cleaning and rinsing can be achieved only when smooth, highly polished, crevice- and pocket-free equipment is used. Equipment conforming to the sanitary design standards of the Technical Committee of Dairy and Food Industry Supply Association and made of type 304 or type 316 stainless steel is frequently used. Parts may also be made of plastics and synthetic materials approved by the U.S. Food and Drug Administration or Department of Agriculture.

Equipment cleaning frequently involves an initial water rinse, a caustic detergent wash, a second water rinse, an acid wash (if calcium-rich deposits have formed), and a final water rinse. Conditions that promote corrosion and etching (e.g., use of chlorides) must be avoided. Disinfection is provided by cleaning or rinsing at high temperatures, by steam heating of equipment, or by use of chemical agents (e.g., hypochlorites). Cleaning formerly involved manual hosing down and brushing of disassembled pipelines and equipment and one-time use of cleaning agents. Manual cleaning procedures are still frequently used in meat processing plants and in food processing pilot plants. Spray wands operating at pressures up to 3.5 MPa (500 psig) are used to ensure removal of adherent material.

Clean-in-place (CIP) systems that accomplish thorough cleaning without disassembling piping and permit some reuse of cleaning agents have been developed for many plants. Because cleaning agents and rinse water run through pipelines and narrow-bore equipment in CIP systems, they must be laid out to permit hang-up–free drainage. Product contact surfaces in tanks and large-bore equipment are cleaned with the aid of sprays issuing from permanently installed spraying devices. Globelike spray heads or reciprocating sprays are used to provide complete washing and rinsing of internal surfaces. Removal of cleaning agents during rinsing has been successfully modeled so that rinsing can be accomplished without excessive use of water.

Soil deposition and fouling interfere with process heat transfer and greatly reduce permeation rates in membrane-based processes. Therefore fouling and soil deposition mechanisms and kinetics for such processes have been extensively studied. The results of these studies have been used to optimize frequencies of cleaning and to select operating conditions and surface treatments that retard fouling.

HOMOGENIZATION AND EMULSIFICATION

Fat in freshly drawn cow's milk is dispersed as butterfat globules that individually contain a fatty core coated with a lipoprotein layer roughly 10 nm thick. Depending on the type of cow involved and how long the animal has been lactating, mean volume/surface diameters of butterfat globules range between 2.5 and 4.5 μm (43). Unless treated, the globules rise, producing a layer of cream on top of milk. To prevent this, and to minimize butterfat clumping, the globules are broken into droplets whose mean volume/surface diameter ranges between 0.22 and 0.5 μm. This is usually done by passing milk through spring-loaded, narrow-clearance valves in high-pressure piston pumps, called homogenizers (26,43). Cavitation produced by rapid increases in velocity in the pump valve and flow-induced shear break up the globules. The mean size of the resulting droplets is inversely proportional to the homogenization pressure used raised to the 0.6 power. Brownian motion would very rapidly cause collision, clumping, and coalescence of incompletely coated droplets. Therefore the lipoprotein coating must redistribute itself rapidly enough to completely coat freshly created drops before they collide with each other and stick together.

Dispersions of small drops of oil in a continuous aqueous phase are often created by intensive agitation of a mixture of an oil and an aqueous solution. The agitation produces turbulent velocity fluctuations, which in turn cause local shearing intense enough to overcome the interfacial tension forces that tend to hold drops together. Mean drop sizes produced by such agitation are frequently proportional to the interfacial tension raised to the 0.6 power and inversely proportional to agitation power expenditure per unit volume raised to the 0.4 power. Simple dispersions rapidly coalesce and separate when mixing stops. To avoid this, surfactants or agents that coat drops and prevent or greatly retard drop coalescence are added. Dispersions stabilized in this way are called emulsions. Stiff emulsions that act like soft solids (e.g., mayonnaise) contain so much dispersed material that drops butt against each other. A great deal of work has been done to develop gelled low-fat food systems with textural properties resembling those of stiff, oil-rich emulsions.

SOLIDIFICATION

Gelling and Coagulation

Fluid foods often are converted into fluid-rich solids with desirable textures by means of gelling and coagulation. Acid produced by bacterial action and bond site production due to the action of the enzyme rennet serve to coagulate milk casein proteins and produce cheese (44–46). Acidification is used to precipitate and gel food proteins with low isoelectric pH values (e.g., casein, milk whey proteins, and soy proteins solubilized by base addition). Calcium salts precipitate soy proteins and cause them to gel during the production of tofu (47). Heat-induced gelation occurs during the cooking of many protein-rich solutions and slurries. Other protein solutions (e.g., gelatin) gel when cooled and melt when heated. High-methoxy pectin causes acidic sugar-rich fruit juices and slurries to gel when jams and jellies are produced. Fully hydrated (gelatinized) starches are used to produce cold-set soft puddings.

Liquid foods gel because reactions or physical chemical forces bond together sites on adjacent, long-chain food molecules or chains of globular proteins (48). Low extents of bonding produce increases in viscosity. Gelling occurs when bond junction densities become large enough to form coherent three-dimensional polymeric networks. Gelling occurs only if the gelling agent concentration C exceeds a critical value C_o. Gelling times are often inversely proportional to $(C - C_o)^n$, where n is a constant. Simple kinetic considerations indicate that n should equal 1, but it is frequently larger than 1. Gelling and coagulation rates may be controlled by the rate of bond site production or by rates of bond forming reactions or events. Cross-linking continues after gels form, and the gels progressively stiffen. This reduces the mobility of network sections, slowing and eventually stopping the bonding process. At gelling agent concentrations substantially greater than C_o, gel strength tends to increase roughly linearly as $(C/C_o)^2$ increases.

Cooling-Induced Solidification

Hard sugar candies are made by evaporating water from a concentrated solution of mixed sugars until the solution is essentially converted into a hot melt containing less than 1 % moisture. Then the melt is cooled. So viscous that crystal growth does not occur, the melt ultimately becomes a vitreous glass. Flavoring and coloring agents are blended into the melt. When the hardening melt becomes stiff enough, it is processed in devices that convert it into pieces of appropriate size and shape. Chewy candies are made by cooling, flavoring, coloring, and shaping evaporatively concentrated mixtures of sugars, corn syrup, fat, and milk solids containing 12–15% moisture. In both cases, solidification is effectively caused by large cooling-induced increases in viscosity and flow yield strength. Cooling is controlled to prevent excess stress from developing and to permit shaping (e.g., molding, rolling, and extrusion) before the completion of hardening.

Chocolate is made by molding and cooling finely milled, thoroughly worked mixtures of molten cocoa butter, cocoa, sugar, and in some cases milk solids. Cocoa butter can exist in four polymorphic forms that solidify or change form at different temperatures. Only one, the β form, is stable at room temperature. Therefore carefully controlled cooling and seeding with fine β-form crystals is used in the solidification of chocolate.

SHAPING AND TEXTURIZING OPERATIONS

Rolling and Flaking

Food processing intermediates used to produce bread, noodles, cookies, and chewing gum are often rolled into sheets before being further shaped by cutting and other forming operations. Cooked cereal grits are rolled to form flakes that are dried and toasted to make ready-to-eat breakfast cereals. Oilseed grits are rolled into flakes to prepare them for extraction. Ropes of partially hardened candy melts are pulled through sequences of pairs of rollers turning at progressively faster speeds to produce narrow-diameter ropes, which are then cut into pieces that are pressed into shape by other devices. Engineering aspects of these operations can be dealt with by methods developed to analyze the calendaring of plastics (48). These methods involve the use of partial differential equations to describe flow fields and pressure gradients that develop in plastic materials passing between rolls. These equations can be solved to determine the pressing force and torque acting on the rolls and the power required to drive them. Dimensional analysis has been applied to circumvent complications due to the complex rheological characteristics of food doughs (49).

Extrusion

Foods are frequently shaped, mixed, cooked, and texturized in screw-driven extruders (50). Flow driven by shear between the rotating screw and the stationary barrel conveys, mixes, and kneads the material being processed and forces it through dies that shape the product. Rotating or oscillating knives cut the extruded material into pieces of suitable size. Action of the screws on the processed material can generate a great deal of frictional heat, particularly when screws with shallow flights are used and the feed's moisture content is low. Supplemental heating is often provided by heated jackets. Cooling jackets are sometimes used. Initially, single-screw extruders were used; now, extruders containing corotating twin screws that contain changeable sequences of conveying, kneading, and pressurizing elements (Fig. 12) are frequently used. Flow, mixing, and power expenditure in extruders of both types have been treated by methods developed for extrusion processing of polymers (49,51). Foods processed in extruders usually contain 15–31% water so that they can be made plastic when worked on by the screw.

Extruders operating close to room temperature and containing screws with deep flights are used to form pasta by forcing semolina-based dough through large arrays of parallel holes. Die hole inserts that provide nonuniform clearance are used in the production of elbow macaroni. High-temperature extrusion, based on use of screws with shallow flights or elements that provide intense kneading, is used to produce confectionery doughs and puffed and textured foods. Temperatures between 130 and 180 °C are frequently used in making these products. Extrusion is used to convert soy protein into products that resemble meat in structure and texture. Such texturization appears to involve the formation of links between ϵ acids and ϵ amines on amino acids in neighboring protein chains. Kinetic models for the reaction indicate that its rate depends on shear rates in the extruder and die as well as the concentrations of the reacting species.

Concentrated purees or slurries containing components capable of entering into setting reactions are pumped or

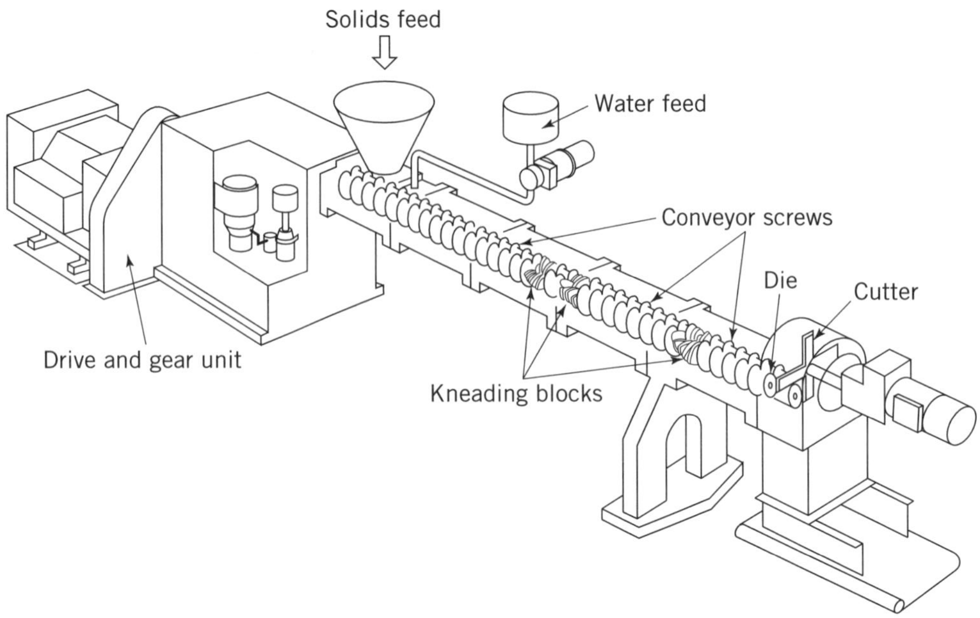

Figure 12. Twin-screw extruder. Reprinted with permission of Krupp Werner and Pfleiderer Corporation.

extruded through dies into baths or vapor-rich atmospheres containing setting agents. Examples include the use of extruded, slurried, partially dissolved collagen or viscose solutions to form sausage casings and use of pumped slurries containing alginates and pureed fruits or vegetables to form artificially shaped fruit and vegetable pieces. Pump- or piston-based extrusion is used to form products that are subsequently heat set (e.g., doughnuts, masa-based puffed products, and certain types of cookie dough). Extrusion is also used to stuff meat emulsions into sausage casings. Coextrusion is used to form fruit-filled bars.

Moist feeds used to produce shaped cereal pieces and pelletized animal feeds and pet food are pushed by rollers through die holes in thick, rotating, perforated rings. Extrudate emerging from the holes is cut into short pieces by knives.

OTHER OPERATIONS

Because of space limitations, many food processing operations have not been covered in this article. These include mixing; grinding, and milling; cutting, slicing, and dicing; screening; pneumatic separation and classification; pumping; mechanical, pneumatic, and hydraulic conveying; slaughtering and carcass disassembly; casting and molding; distillation and deodorization; liquid–liquid extraction; gas absorption; crystallization; adsorption and ion exchange; chemical modification of food ingredients (e.g., hydrogenation of vegetable oils); various types of cooking and baking; treatment of foods with ionizing radiation; pore generation and puffing; coating and enrobing; sorting and inspection; packaging and filling; fermentation, including brewing, wine making, and the production of pickled vegetables and fermented dairy products; aquaculture; shellfish depuration; automated egg production, and automated growing of plants and plant tissue in controlled environments.

BIBLIOGRAPHY

1. N.W. Desrosier, *Elements of Food Technology*, AVI Publishing, Westport, Conn., 1977.
2. H.S. Carslaw and J.C. Jaeger, *Conduction of Heat in Solids*, Oxford Univ. Press, Oxford, U.K., 1959.
3. J. Crank, *Mathematics of Diffusion*, Oxford Univ. Press, London, 1970.
4. B.K. Watt and A.L. Merrill, *Composition of Foods: Agriculture Handbook 8*, U.S. Department of Agriculture, Washington, D.C., 1975.
5. D.B. Brooker, F.W. Bakker-Arkema, and C.W. Hall, *Drying and Storage of Grains and Oil Seeds*, Chapman & Hall, New York, 1992.
6. Midwest Plans Service, *Grain Drying, Handling, and Storage Handbook*, Midwest Plans Service, Iowa State University, Ames, Iowa, 1988, pp. 3.1–3.21.
7. K. Masters, *Spray Drying Handbook*, 5th ed., Wiley, New York, 1991.
8. F.G. Kieviet and P.J.A.M. Kerkhof, in *Proceedings 5th World Congress of Chemical Engineering*, vol. 2, American Institute of Chemical Engineers, New York, 1996, pp. 87–92.
9. J.D. Mellor, *Fundamentals of Freeze Drying*, Academic Press, New York, 1979.

10. C.J. King, *Freeze Drying of Foods*, CRC Press, Cleveland, Ohio, 1971.
11. S.A. Goldblith, L. Rey, and W.W. Rothmayr, *Freeze Drying and Advanced Food Technology*, Academic Press, New York, 1975, pp. 123–480.
12. H.G. Schwartzberg, *69th Annu. American Institute of Chemical Engineers Meeting*, Chicago, Ill., December 1976, pp. 1–54.
13. M. Karel, in M. Karel, O.R. Fennema, and D.B. Lund eds., *Principles of Food Science, II: Physical Principles of Food Preservation*, Marcel Dekker, New York, 1975, pp. 237–263.
14. O.R. Fennema, in M. Karel, O.R. Fennema, and D.B. Lund eds., *Principles of Food Science, II: Physical Principles of Food Preservation*, Marcel Dekker, New York, 1975, pp. 133–171.
15. R.E. Hardenburg, A.E. Watada, and C.Y. Wang, *The Commercial Storage of Fruits, Vegetables, Florist, and Nursery Stocks*, Agriculture Handbook No. 66. U.S. Department of Agriculture, Washington, D.C., 1986, pp. 1–130.
16. *1994 ASHRAE Handbook—Refrigeration* American Society of Heating, Refrigeration, and Air-Conditioning Engineers, Atlanta, Ga., 1994.
17. D.H. Dewey, in S. Thorne, ed., *Developments in Food Preservation*, vol. 2, Applied Science Publishers, London, 1983, pp. 1–24.
18. O.R. Fennema, W.D. Powrie, and E.H. Marth, *Low Temperature Preservation of Foods and Living Matter*, Marcel Dekker, New York, 1973.
19. O.R. Fennema, in M. Karel, O.R. Fennema, and D.B. Lund eds., *Principles of Food Science, II: Physical Principles of Food Preservation*, Marcel Dekker, New York, 1975, pp. 173–215.
20. J. Postolski and Z. Gruda, *Zamrażani Żywnosci*, Wydawnictwa Naukowo-Techiczne, Warsaw, 1985.
21. H.G. Schwartzberg, in *Freezing, Frozen Storage and Freeze Drying of Biological Materials and Foodstuffs*, International Institute of Refrigeration, Paris, 1977, pp. 303–310.
22. H.G. Schwartzberg, *American Institute of Chemical Engineers Summer National Meeting*, Detroit, Mich., 1981, pp. 1–33.
23. A.C. Cleland, *Food Refrigeration Processes, Analysis, Design, and Simulation*, Elsevier Applied Science, New York, 1990.
24. R.P. Singh and J.D. Mannapperuma, in H.G. Schwartzberg and M.A. Rao eds., *Biotechnology and Food Process Engineering*, Marcel Dekker, New York, 1990, pp. 309–358.
25. D.L. Downing, *A Complete Course in Canning and Related Processes*, 13th ed., CTI Publications, Baltimore, Md., 1996.
26. H.G. Kessler, *Food Engineering and Dairy Technology*, Kessler Verlag, Freising, Germany, 1981, pp. 59–81, 119–207.
27. D.B. Lund, in M. Karel, O.R. Fennema, and D.B. Lund eds., *Principles of Food Science, II: Physical Principles of Food Preservation*, Marcel Dekker, New York, 1975, pp. 31–92.
28. R. Villota and J.G. Hawkes, in M. Okos, ed., *Physical and Chemical Properties of Foods*, American Society for Agricultural Engineering, St. Joseph, Mich., 1986, pp. 266–366.
29. H.G. Kessler, in *Proceedings 5th World Congress of Chemical Engineering*, vol. 2, American Institute of Chemical Engineers, New York, 1996, pp. 29–34.
30. H.G. Schwartzberg, *Food Technol.* **31**, 67–76 (1977).
31. H.G. Schwartzberg, in R.P. Singh and A.G. Medina eds., *Food Properties and Computer-Aided Design of Food Processing Systems*, Kluwer Academic Publishers, Boston, Mass., 1989, pp. 443–470.
32. H.A.C. Thijssen, in A. Spicer, ed., *Advances in Preconcentration and Dehydration of Foods*, Wiley, New York, 1973, pp. 115–149.
33. H.G. Schwartzberg, in H.G. Schwartzberg and M.A. Rao eds., *Biotechnology and Food Process Engineering*, Marcel Dekker, New York, 1990, pp. 127–202.
34. H.G. Schwartzberg, in R.W. Roussea, ed., *Handbook of Separation Process Technology*, Wiley, New York, 1987, pp. 540–577.
35. H.G. Schwartzberg, *Chem. Eng. Prog.* **76**, 67–85 (1980).
36. M. Shirato, T. Murase, E. Iritani, F.M. Tiller, and A.F. Alciatore, in M.J. Matteson and C. Orr eds., *Filtration Principles and Practices*, Marcel Dekker, New York, 1987, pp. 299–424.
37. H.G. Schwartzberg, in *McGraw-Hill Encyclopedia of Science and Technology*, vol. 7, McGraw-Hill, New York, 1987, pp. 254–255.
38. H.G. Schwartzberg, *Sep. Purif. Methods* **26**, 1–213 (1997).
39. H.G. Schwartzberg, in M. Peleg and E. Bagley, eds., *Physical Properties of Foods*, AVI Publishing, Westport, Conn., 1983, pp. .
40. B. Hallstrom, D.B. Lund, and C. Trägårdh, eds., *Fundamentals and Applications of Surface Phenomena Associated with Fouling and Cleaning in Food Processing*, Department Food Science, University of Wisconsin, Madison, Wisc., 1981.
41. D.B. Lund, E. Plett, and C. Sandu, eds., *Fouling and Cleaning in Food Processing*, Department of Food Science, University of Wisconsin, Madison, Wisc., 1985.
42. H.G. Kessler and D.B. Lund eds., *Fouling and Cleaning in Food Processing*, Kessler Verlag, Freising, Germany, 1989.
43. H. Mulder and P. Walstra, *The Milk Fat Globule*, Commonwealth Agricultural Bureau, Farnham Royal, England, 1974, pp. 163–194.
44. H.G. Kessler, *Food Engineering and Dairy Technology*, Kessler Verlag, Freising, Germany, 1981, pp. 426–449.
45. F. Kosikowski, *Cheese and Fermented Milk Products*, Brooktendale, New York, 1978.
46. P.F. Fox ed., *Cheese: Chemistry, Physics, and Microbiology*, vol. 1, *General Aspects*, Chapman & Hall, London, 1993.
47. W. Shurtleff and A. Aoyagi, *Tofu and Soymilk Production*, New-Age Foods Study Center, Lafayette, Calif., 1979.
48. S. Middleman, *Fundamentals of Polymer Processing*, McGraw-Hill, New York, 1977, pp. 123–187, 295–353.
49. K. Valentas, L. Levine, and J. Peter Clark, *Food Processing Operations and Scale-Up*, Marcel Dekker, New York, 1991, pp. 359–365.
50. J.M. Harper, in H.G. Schwartzberg and M.A. Rao eds., *Biotechnology and Food Process Engineering*, Marcel Dekker, New York, 1990, pp. 295–308.
51. L.P.B.M. Janssen, *Twin-Screw Extrusion*, Elsevier Science Publishers, New York, 1978.

See also ENZYMES, BAKING, BREAD MAKING; ENZYMES, FRUIT JUICE PROCESSING; FREEZE DRYING, PHARMACEUTICALS; PRODUCTION OF L-AMINO ACIDS BY AMINOACYLASE; SOLID SUBSTRATE FERMENTATIONS, ENZYME PRODUCTION, FOOD ENRICHMENT; YEAST, BAKER'S.

FORMULATION AND DELIVERY, PROTEIN PHARMACEUTICALS

XANTHE M. LAM
JEFFREY L. CLELAND
Genentech, Inc.
South San Francisco, California

KEY WORDS

Degradation
Drug delivery
Formulation
Lyophilization
pH dependence
Recombinant protein
Shelf life
Stability
Stress testing
Temperature dependence

OUTLINE

INTRODUCTION

For the last twenty years, many biotechnology-derived pharmaceuticals have been manufactured for the treatment of diseases such as diabetes, cardiovascular diseases, cancers, dwarfism, anemia, AIDS, cystic fibrosis, chronic granulomatous disease, and kidney diseases. Some of these recombinant protein drugs have already been approved by the U.S. Food and Drug Administration (FDA) for marketing, others are still being tested in clinics or are under review for approval. The success of these protein pharmaceutical drugs greatly depends on the delivery of their biologically active forms to the target site. Therefore, the development of a stable formulation, in which the protein can maintain its native conformation and bioactivity, is one of the important steps in the production of protein pharmaceuticals. In fact, the FDA usually requires the pharmaceutical manufacturers to provide real-time stability data to demonstrate that a pharmaceutical drug contains at least 90% potency throughout the entire shelf life period (usually $\geq$ 2 years) before it can be approved to be a marketed product. In addition, the manufacturers are required to demonstrate that the degradation products that may cause a loss in the potency of the drug do not have any adverse effects on the safety and efficacy of the drug.

In developing a protein drug formulation and designing a delivery system, formulation scientists must consider the physicochemical properties, clinical indication, site of action, pharmacokinetics, and toxicity of the drug. The physicochemical properties of proteins such as the amino acid composition, molecular weight, isoelectric point (pI), glycosylation, and conformation can affect the pharmacokinetics and toxicity as well as the clinical indication. The potential clinical application of a protein drug depends upon the biological function and potency of the product, and the physical and chemical properties of a protein determine its biological function and potency.

The unique physical and chemical properties of each protein also determine its in vitro and in vivo stability. To obtain in vivo information on the pharmacokinetics and toxicity of a drug, the drug must be administered with a stable formulation. In considering the best formulation for the intended applications of a protein drug, the formulation scientist must also consider the route of administration. For initial animal testing, protein drugs are usually administered systemically via an intravenous (i.v.) injection. However, some indications may require a high local drug dose that cannot be achieved by i.v. administration due to toxicity issues. In this case, an alternative route of delivery is necessary. For example, Pulmozyme® (rhDNase) is delivered to patients with cystic fibrosis via the pulmonary route of administration (1). The drug is best administered directly into the lungs of the patients as an aerosol using nebulizers in order to achieve an efficacious dose at the target site (lungs). Thus, the development of both a stable formulation and suitable delivery route or system is critical in determining the success of a final product.

During the past 10 years, several review articles and texts related to protein formulation and stability have been published (2–7). The purpose of this article is to provide a comprehensive overview of formulation and delivery aspects of protein drugs. Protein degradation pathways, analytical characterization of proteins, the process of formulation development, and drug delivery systems are reviewed.

PROTEIN DEGRADATION PATHWAYS

Chemical and physical degradation is the major cause of instability for biopharmaceutical products. There are many degradation routes for proteins. Aggregation, deamidation, and oxidation are the three most common ones, others include isomerization, cleavage, thiol disulfide exchange, and β-elimination. To develop a stable protein formulation, protein degradation in the formulation must be inhibited or minimized throughout the shelf life of the product. Fortunately, with the advances in analytical techniques for protein analysis, most of these expected degradation routes can be identified, and the degradation products can also be characterized by formulation scientists. In addition, with the continued success of biotechnology, more recombinant proteins have become available for studying degradation mechanisms, thus increasing the general knowledge in formulating new proteins.

Aggregation

Protein degradation via aggregation pathways is considered a physical reaction that does not involve the breakage or formation of covalent bonds. For a protein to retain its biological function and stability, it must maintain its conformational or tertiary structure. Protein conformation is stabilized mainly by hydrophobic interactions (8–11). Thus, monomeric globular proteins often exist in native or folded conformations such that the hydrophobic groups are not exposed on the surface (12). Loss in the tertiary globular structure or unfolding of proteins due to denaturation results in exposure of hydrophobic residues to the environment. Protein aggregation or denaturation in a formulation (liquid or lyophilized) can be caused by a change in temperature, extreme pH, ionic strength, pressure, or presence of denaturants (11,13). Usually, the native, N, and unfolded, U, states of a protein are in equilibrium during denaturation and folding in a two-state process, as shown in equation 1. The reversibility of the unfolding process depends upon the formulation conditions.

$$N \rightleftharpoons U \qquad (1)$$

$$\begin{array}{c} N \rightleftharpoons I \rightleftharpoons U \\ \downarrow \\ A \end{array} \qquad (2)$$

Sometimes, a protein denatures via a different pathway that involves the formation of at least one stable, partially unfolded intermediates. As shown in equation 2, the native protein unfolds to form an intermediate, I, that has the internal hydrophobic residues exposed to the environment. This hydrophobic intermediate on the refolding pathway may then form aggregates, A, due to intermolecular hydrophobic interactions. According to this scheme, the protein must be denatured or unfolded before nonnative aggregates can be formed. The formation of a stable intermediate during denaturation and folding has been observed for many proteins (14–18). For example, the dimeric native form of interferon-γ (IFN-γ), unfolds to form a partially denatured monomeric intermediate, below pH 4.5 and in the absence of NaCl (19,20). Upon dialysis, both native IFN-γ and aggregates were obtained. The aggregates lost most of the native tertiary structure as well as bioactivity.

Deamidation

Deamidation of protein is the acid- and base-catalyzed hydrolysis of the side-chain amide on glutamine (Gln) and asparagine (Asn) residues to form a carboxylic acid (Scheme 1). Studies of deamidation in peptides by Robinson and his coworkers suggested that Asn is more susceptible to deamidation than Gln and degrades more readily in the presence of an adjacent glycine (Gly) residue on the C-terminal side of the sequence (21,22). It was also discovered that the deamidation of Asn-Gly was accelerated at pH $\geq$ 7.0. In addition, the studies found that the amino acid sequence plays a major role in determining the rate of deamidation. For example, the presence of polar amino acid residues adjacent to an Asn or Gln enhances the rate of deamidation, whereas adjacent bulky hydrophobic residues decrease the deamidation rate. Serine (Ser) or threonine (Thr) next to an Asn or Gln can act as a general acid by providing a proton in the acid-catalyzed reaction, thereby enhancing the rate of deamidation. The major factors influencing the rate of protein deamidation include pH, temperature, ionic strength, and buffer species. Degradation of proteins via deamidation has been shown to exist in recombinant protein pharmaceuticals such as insulin (23,24) and human growth hormone (25,26) in liquid formulations.

Isomerization

Deamidation of an asparagine (Asn) residue to yield aspartate (Asp) can also lead to the formation of an isoaspartate (iso-Asp) via intermediate succinimide formation (27). Nucleophilic attack of the main-chain amide nitrogen on the carbonyl carbon of the Asn side-chain amide group results in a five-membered succinimide ring. This cyclic imide can be hydrolyzed to form either aspartate or isoaspartate (Scheme 2). An aspartate residue can also undergo deamidation via intermediate succinimide formation to yield isoaspartate. In this case, nucleophilic attack of the main-chain amide nitrogen on the carbonyl carbon of the Asp side-chain carboxylic group results in the formation of a cyclic imide (Scheme 3). The additional CH_2 group in the glutamine (Gln) side chain makes the formation of a cyclic imide intermediate by nucleophilic attack less favorable than for an Asn or Asp residue. Therefore, the deamidation rate of Gln via isomerization is very slow. When a Gly or Ser residue is next to the Asn or Asp residue, the isomerization via cyclic imide formation is accelerated in solution (28). An increase in temperature or pH also increases the rate of isomerization for peptides containing an Asn-Gly or Asp-Gly sequence (29). Cyclic imide formation was demonstrated during the storage of lyophilized recombinant methionyl human growth hormone (met-rhGH) at 45 °C for 4 months (30), and the Asp-130 residue in the protein underwent isomerization (31).

Cleavage

Cleavage of peptides at an asparagine residue can result from the nucleophilic attack of the side-chain amide nitrogen on the main-chain peptide carbonyl to form a C-terminal succinimide (Scheme 4). The reaction is spontaneous and dependent on protein sequence. Asparagine

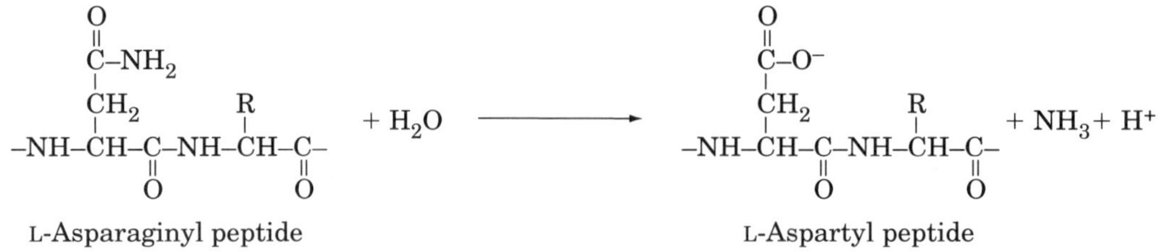

Scheme 1. Deamidation of asparagine residue in a peptide by direct hydrolysis.

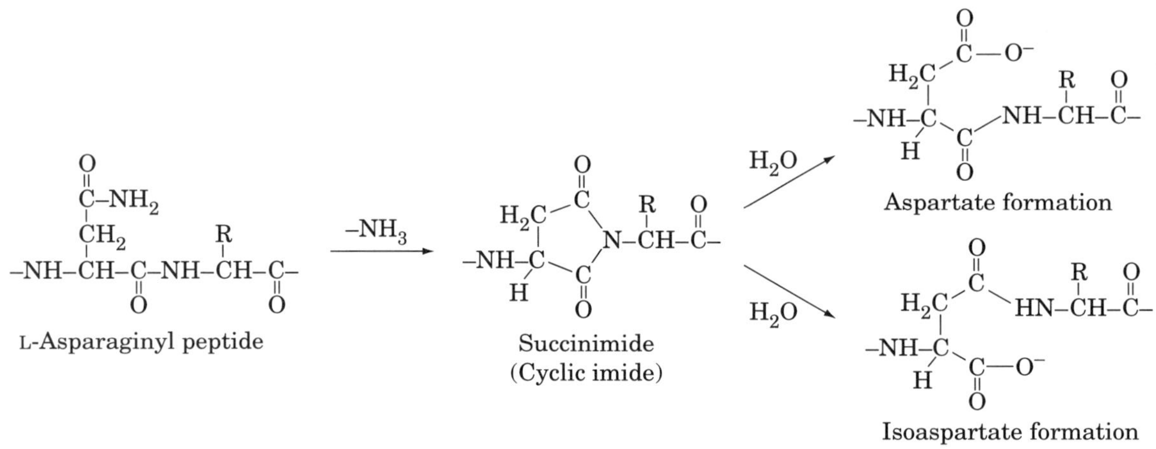

Scheme 2. Pathway for asparagine deamidation in peptides or proteins via intermediate succinimide formation.

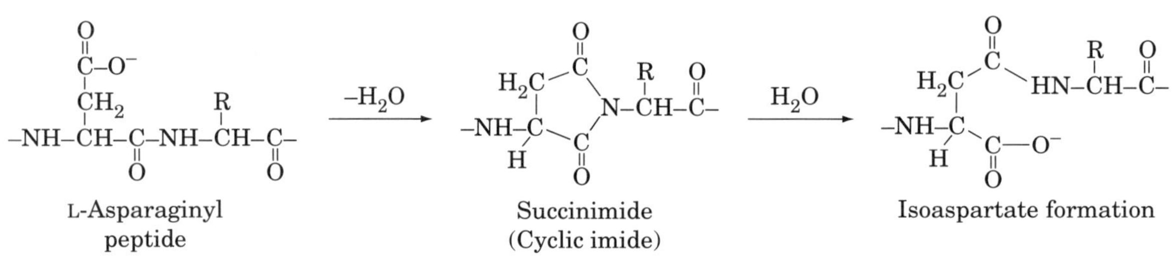

Scheme 3. Deamidation of aspartate to yield isoaspartate via succinimide formation.

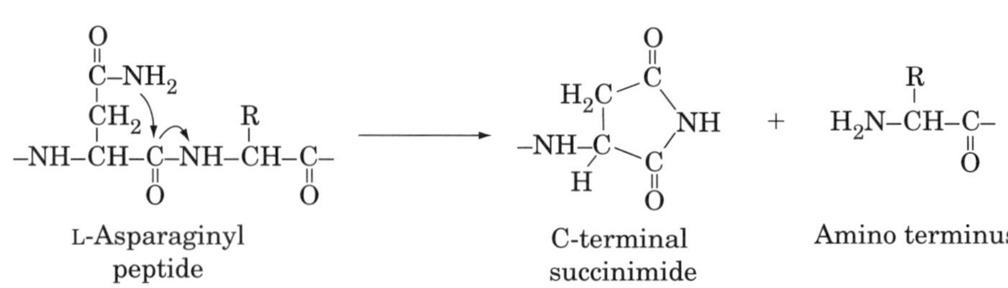

Scheme 4. Spontaneous cleavage of peptide as asparagine residue.

having an adjacent Gly or Ser residue on the C-terminal side is more labile to spontaneous cleavage. Cleavage at an Asn-Ser linkage has been demonstrated in proteins such as bovine and porcine somatotropins (32).

Oxidation

Oxidation is one of the major degradation pathways for protein pharmaceuticals. Amino acids that can undergo oxidation include methionine, cysteine, histidine, tryptophan, and tyrosine. Oxidation of methionine has been demonstrated in many recombinant proteins such as interleukin 2 (33), relaxin (34), parathyroid hormone (35), and human growth hormone (36). At low pH, the thioether group of methionine is not protonated and is susceptible to oxidation, resulting in methionine sulfoxides. Under extremely oxidative conditions, the sulfoxides can be further oxidized to sulfones. Methionine can react with a variety of oxygen-reactive species such as hydrogen peroxide, alkylhydroperoxides, molecular oxygen, and singlet oxygen generated by heat and light. Methionine oxidation can be catalyzed by the presence of transition metal ions such as Cu^{2+}, Fe^{2+} or Fe^{3+}. Although there are many oxidative pathways for methionine reported in literature (3,7,37,38), oxidation via a free radical, singlet oxygen, or nucleophilic substitution are the three common mechanisms. In contrast to methionine, oxidation of cysteine usually occurs at higher pH where the thiol is deprotonated (39). Cysteine can react with molecular oxygen to form cystine disulfide. The reaction can also be catalyzed by transition metal ions, especially Cu^{2+}. In some proteins cysteine can be oxidized to give sulfenic, sulfinic, or sulfonic acids (40). Histidine, tryptophan, and tyrosine are susceptible to photooxidation via the singlet oxygen pathway. The rate of photooxidation is pH dependent. At neutral pH, histidine and tryptophan photooxidize faster than tyrosine (41). At low pHs, tryptophan and methionine are the most photoreactive amino acids (42).

Oxidation of methionine:

$$\underset{\text{methioine}}{RSCH_3} \rightarrow \underset{\text{methionine sulfoxide}}{RS(O)CH_3} \rightarrow \underset{\text{methionine sulfone}}{RS(OO)CH_3}$$

Oxidation of cysteine by oxygen:

$$RS^- + O_2 \rightarrow RS\cdot + O_2\cdot^-$$
$$RSH + O_2\cdot^- \rightarrow RS\cdot + HOO^-$$
$$RS\cdot + RS\cdot \rightarrow RS\text{-}SR$$

Thiol Disulfide Exchange

Thiol disulfide exchange occurs when a disulfide bond is reduced to two cysteines, and one of them reacts with another cysteine to form a new disulfide (5). The incorrect linkage of two cysteines in a disulfide bond can induce loss of biological activity of the protein. For example, interleukin 2 contains three cysteines at positions 58, 105, and 125. The native protein form has a disulfide linkage between the two cysteines at 58 and 105. In the presence of copper ions, two less active isomers with disulfide linkages between the cysteines at 58 and 125 and the cysteines at 105 and 125 are formed (5). The reaction is also base catalyzed and concentration dependent. The reaction rate can be influenced by pH, temperature, buffer composition, and the presence of other metal ions.

β-Elimination

Inactivation of proteins in an alkaline solution can result from *β*-elimination of the cystine residue, with the heterolytic cleavage of the disulfide and the formation of dehydroalanine and persulfide (equation 3). The persulfide can further react with hydroxide to form hydrosulfide (HS), as shown in equation 4 (5). The rate of *β*-elimination increases with increase in temperature and pH. The presence of metal ions also enhances the reaction rate.

$$R'CH_2SSCH_2R'' + OH^- \rightarrow R'CH_2SS^- + CH_2 = CR'' + H_2O \quad (3)$$

$$R'CH_2SS^- + OH^- \rightarrow R'CH_2SO^- + HS^- \quad (4)$$

PROTEIN FORMULATION DEVELOPMENT

After gathering all the background information (physicochemical properties, application, site of action, etc.) on the protein to be formulated, the first step that a formulation scientist usually takes is to characterize the protein in the initial formulation by analytical methods. During this step, analytical methods are also developed or optimized for use in formulation and stability studies. Because protein degradation is often dependent on pH and temperature, the second step of formulation development is to determine the relationship between the major degradation pathways of the protein and these parameters. Formulation scientists usually perform short-term studies of pH and elevated temperature (e.g., 25–40 °C) on initial liquid formulations. After determining the major degradation pathways in these formulations, the formulation scientist will select the most stable formulation that can achieve an optimum balance between the different degradation pathways. Finally, long-term stability testing is performed on the final chosen formulation to obtain real-time stability data. This section reviews the process of developing a stable formulation for protein pharmaceuticals. A simplified process diagram for protein formulation development is also illustrated in Figure 1.

Analytical Methods for Assessment of Protein Formulations

The selection of formulations for protein pharmaceuticals greatly depends on the stability results obtained in the formulation screening. Therefore, it is important to have suitable and sensitive analytical methods to study the degradation of proteins in formulations. Review articles by Jones provide detailed principles and applications of many analytical methods that can be used for protein characterization (43,44). In the following section, the commonly used analytical methods for determining protein formulation stability during formulation development are discussed. A summary of the analytical methods used for identifying protein degradation is shown in Table 1.

High Performance Liquid Chromatography. High performance liquid chromatography (HPLC) methods are widely used in the pharmaceutical industry for the analysis of peptides and proteins. The common HPLC methods for use in formulation development to assess protein stability are size exclusion chromatography (SEC), ion exchange chromatography (IEC), reversed-phase chromatography

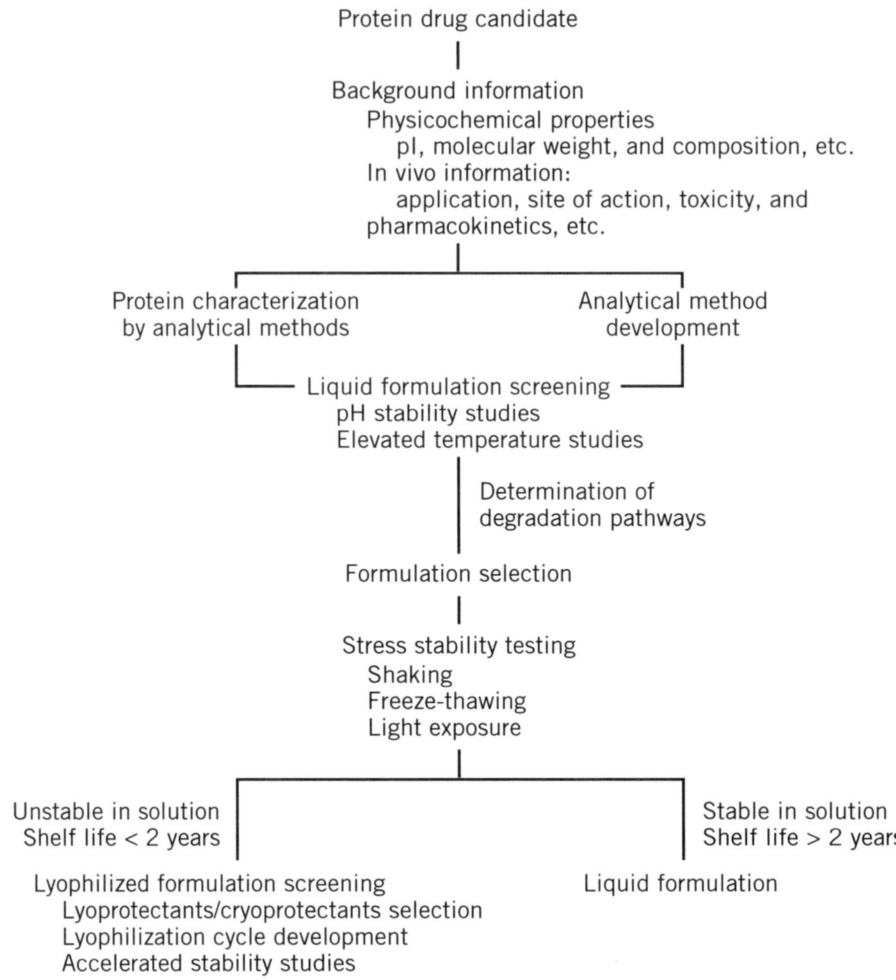

Figure 1. Simplified process diagram for protein formulation development.

(RPC), and hydrophobic interaction chromatography (HIC). Degradation of proteins via aggregation and cleavage results in the formation of soluble and insoluble aggregates (oligomers) and fragments, respectively. These soluble aggregates and fragments have different apparent molecular weights and can be separated from the intact protein on a size exclusion column that contains porous particles with different pore diameters and has been calibrated with molecular weight standards, usually globular proteins of known molecular weight.

IEC is often employed during formulation and stability studies for the separation of degraded proteins with charge heterogeneity resulting from deamidation or isomerization. Charged protein tends to bind to ion-exchanged resin (e.g., sulfopropyl, DEAE) of opposite charge by ionic interaction. Bound proteins can be eluted at their own critical salt concentration, allowing separation of the various charge forms (44). The amount of salt required to elute a charged species depends on its affinity for the resin (e.g., number of binding sites), the net charge, and the ability of the salt to displace the bound protein. By this method, deamidated forms of a protein that exhibit different net charge from each other and the native protein can be separated and quantified.

RPC methods are commonly used for the separation of small molecules and peptides, especially for peptides that result from proteolytic digestion of proteins in peptide mapping (45). A hydrophobic surface on the protein binds to the hydrophobic site of the resin (e.g., alkyl groups such as butyl [C4], hexyl [C6], or octadecyl [C18] or aromatics such as phenyl) in a reversed-phase column by hydrophobic interaction. Such hydrophobic interactions can be weakened by increasing the content of organic modifier in the mobile phase (46). Thus, proteins can be eluted in the order of their hydrophobic interaction strengths by an organic modifier such as acetonitrile. Although the application of RPC for analysis of recombinant proteins is less common than for peptides due to difficulties in resolution of large molecules, it has been used for characterization or separation, as in the case of met-rhGH and rhGH, which differ by the only one amino acid (47). This method is also useful in detecting the formation of more polar degraded proteins resulting from oxidation.

Similar to RPC, HIC is another useful technique for the separation of proteins based on their differences in surface hydrophobicity. Most proteins have some hydrophobic groups exposed to the surface. When the hydrophobic surface of a protein is in contact with the hydrophobic ligands on the beads of a column, water is released from the hydrophobic regions, and the entropy of the system is increased. The release of water increases the entropy of the

Table 1. Analytical Methods Used for Analysis of Protein Degradation During Formulation Development

Analytical method	Protein degradation/alteration
High performance liquid chromatography	
Size exclusion	Aggregation, polypetide cleavage
Ion exchange	Charge alteration (e.g., deamidation)
Reversed phase	Oxidation, disulfide alterations, charge alteration (using peptide map)
Hydrophobic interaction	Neutral alteration (e.g., Met oxidation)
Electrophoresis	
SDS-PAGE	Aggregation, polypeptide cleavage, disulfide alterations
Isoelectric focusing	Charge alteration (e.g., deamidation)
Capillary electrophoresis	
Capillary zone electrophoresis	Charge alteration (e.g., deamidation)
Capillary isoelectric focusing	Charge alteration (e.g., deamidation)
SDS-dynamic sieving capillary electrophoresis	Aggregation, polypeptide cleavage, disulfide alterations
Spectroscopy	
UV absorption	Tertiary structure alteration Change in protein concentration
Circular dichroism	Secondary structure alteration (far-UV CD) Tertiary structure alteration (near-UV CD)
Light scattering	Aggregation

system, causing the free energy to decrease on adsorption (48). Protein adsorption on an HIC column can be adjusted by changing the salt concentration. At high salt concentrations, protein adsorption is mainly due to hydrophobic interactions. By running the column with decreasing salt concentration, protein will be eluted from the column when the salt concentration is not high enough to maintain or stabilize the interaction.

Electrophoresis. Electrophoresis is a common method for the separation of proteins based on their net charge and size. Upon applying an electric field to a protein, it migrates toward the anode or cathode, depending on its net charge. There are two basic forms of electrophoresis: sodium dodecyl sulfate polyacrylamide gel electrophoresis (SDS-PAGE) and isoelectric focusing (IEF). In SDS-PAGE, proteins are complexed with the denaturing agent SDS to give similar mass-to-charge ratios so that they can have free electrophoretic mobility. When SDS-complexed proteins are electrophoresed in a polyacrylamide gel with the appropriate pore size, their migration rate will be based on the size of the proteins. Thus, a mixture of proteins with different molecular weights can be separated using this method.

Separation of proteins by IEF is based on their isoelectric points (pIs), where the mobility of each protein in an electric field is zero and the net charge of the each protein is also zero. Because the mobility of nondenatured protein is dependent on its net charge, and this charge is determined by the pH of the solution, an electric field in the presence of a pH gradient causes the protein to migrate until the pH is the same as the pI (49). This technique is usually employed by preparing a polyacrylamide gel containing ampholytes of appropriate pI and buffer capacity to produce a pH gradient that is stable in the applied electric field. This method has been used to assess protein deamidation during stability studies of many pharmaceutical products such as recombinant human growth hormone (50).

Capillary Electrophoresis. Due to recent advances in technology, capillary electrophoresis (CE) is being more widely used in the biotechnology industry for protein analysis. The sensitivity and speed of CE make it a useful technique in early protein formulation development when only a small amount of material is available. CE provides a wide range of separation modes that can be performed with a single instrument. The basic separation principles are the same as with conventional electrophoresis techniques.

Capillary zone electrophoresis (CZE) is the most common CE separation method, and it employs a single buffer system in free solution. Separation of proteins using CE is based on their differences in mass-to-charge ratio. Separation takes place in free solution inside a capillary (usually made of fused silica) filled with electrolyte. Proteins with different mass-to-charge ratios are separated into individual zones as electrical force drives them at different rates through the capillary. As the mass-to-charge ratio increases, the electrophoretic mobility decreases, resulting in a longer migration time. Coated capillaries and buffer additives are often used to prevent protein adsorption and electroendosmotic flow. CZE has been used for the detection of the heterogeneity of recombinant proteins (51) as well as the separation of the deamidation products of growth-hormone-releasing factor (52).

Capillary isoelectric focusing (cIEF) is another separation mode in CE for the separation of proteins based on their isoelectric points. There are three basic steps in cIEF: sample injection, focusing, and mobilization. Proteins are premixed with ampholytes and injected into the capillary by pressure. High voltage is applied to the capillary during focusing so that sample ions can migrate to their neutral charge point in the pH gradient formed by the ampholytes along the capillary. After focusing, the focused protein bands are moved to the detection point to generate a signal. The on-line UV detection of cIEF provides a more accurate quantitation than the slab gel technique. A coated capillary is recommended for cIEF because the proteins stay in the capillary for focusing and mobilization. cIEF has been applied to the characterization of charge heterogeneity of recombinant monoclonal antibodies as a result of deamidation (53,54). The high resolving power of the method also makes it useful for the characterization of the heterogeneity of the glycoforms of glycoproteins such as recombinant tissue plasminogen activator (55).

SDS-dynamic sieving capillary electrophoresis (SDS-DSCE) is a useful separation mode in CE for the molecular weight determination of proteins. Because all proteins have the same mass-to-charge ratio after complexation with SDS, the migration rate of proteins depends only on their molecular size. Although SDS-DSCE is equivalent to the conventional SDS-PAGE, it provides a rapid and automated system for quantitative analysis of proteins based on their molecular weight. Monoclonal antibodies degraded by the formation of aggregates and fragments can easily be separated by this method (54,56,57).

Spectroscopy. Ultraviolet (UV) absorption spectroscopic measurement is the most convenient and accurate method for the determination of protein concentration. Proteins with aromatic amino acids have a maximum in UV absorption between 275 and 282 nm at pH values below 8 (58). Thus, the protein concentration in a sample can be determined by measuring its absorbance at the wavelength maximum near 280 nm, and this absorbance is then compared with the absorbance of a solution with a known protein concentration or, if known, the molar extinction coefficient or absorptivity of the protein may be used. The absorptivity (often incorrectly described as the extinction coefficient) of a protein is defined as the absorbance of a 1 mg/mL protein solution through a 1-cm path at the wavelength maximum (e.g., near 280 nm). The molar extinction coefficient is the absorptivity of a 1.0 M solution of the protein through a 1.0-cm path. It is important to measure the protein concentration accurately for extinction coefficient determination. Quantitative amino acid analysis, nitrogen determination, or dry weight measurement are often used for this purpose.

Circular dichroism (CD) spectroscopy is a useful technique to evaluate the secondary and tertiary structures of protein (59,60). The secondary structure of a protein such as α-helices, β-sheets, and random coils can give rise to CD signals in the far-UV region (170–250 nm). CD spectroscopy is commonly employed for the study of denaturation or unfolding of proteins. The tertiary structure of a protein affects the local environment of the aromatic amino acids such as trytophan, tyrosine, and phenylalanine, as well as the disulfide bonds, and these effects contribute to CD signals in the near-UV region (240–320 nm). Because CD signals can be either positive or negative, it is not easy to obtain detail from CD spectra. However, a change in the far- or near-UV CD spectrum can represent a change in the secondary or tertiary structure of the protein, respectively. Therefore, CD is a useful tool to measure protein denaturation as well as aggregation, both of which result in structural or conformational changes.

Light scattering spectroscopy can be used to detect protein aggregation due to the fact that larger particles scatter more light per weight unit than small ones. Aggregated proteins can be either soluble or insoluble. Insoluble aggregates usually cause opalescence (turbidity) in a protein liquid formulation, yielding light scattering at wavelengths even in the visible region of the spectrum. Turbidity can also be conveniently measured in the near UV (340–360 nm) if no other chromophores are present in the same region. Thus, an increase in absorbance in this region often indicates the formation of aggregates. Correction for light scattering errors is necessary for the measurement of protein concentration made by the absorbance at the peak wavelength near 280 nm (61). Pharmaceutical final product solutions are required by the FDA to be inspected for clarity and degree of opalescence. Therefore, light-scattering determination may be substituted for the visual inspection method. Because native protein in solution also scatters light, light-scattering spectroscopy can be used to estimate the molecular weight of a protein. Dynamic laser light scattering, also known as photon correlation spectroscopy (PCS) or quasi-elastic light scattering (QELS), can be used to calculate the average hydrodynamic diameter and determine the diffusion coefficient of a protein by autocorrelation methods (62,63).

Formulation Screening Methods

The methods commonly used by formulation scientists for protein formulation screening to select the best candidate formulation include the short-term accelerated pH and temperature stability studies. Accelerated conditions are defined as the extremes of the parameter range (e.g., 40 °C and high (≥9.0) or low (≤3.0) pH).

Accelerated pH Stability Studies. Solution pH often plays an important role in the stability of the protein product because most of the major protein degradation pathways are pH dependent. Thus, the stability of a protein in different formulations can be compared by performing stability studies on the protein formulated at a range of pH using a variety of buffer solutions. Examples of buffers that are used to maintain the pH of formulations in commercially available recombinant protein drug products are listed in Table 2. Acetate, succinate, citrate, and phosphate are the most common buffers used in recombinant protein formulations. For rapid formulation screening, a pH stability study is usually performed in a short period of time and at an elevated temperature of 40 °C because the rate

Table 2. Examples of Buffers Commonly Used in Recombinant Protein Formulations for pH Maintenance

Buffers	pH	Commercial products[a]
Sodium acetate	4.0	Neupogen® (Amgen)
Sodium succinate	5.0	Actimmune® (Genentech)
Sodium citrate	6.0	Nutropin® AQ (Genentech)
	6.5	Rituxan® (IDEC, Genentech)
	6.9	Epogen® (Amgen)
L-Histindine HCl	6.0	HERCEPTIN® (Genentech)
Sodium phosphate	7.3	Intron-A® (Schering-Plough)
	7.4	Protropin® (Genentech)
	7.5	Humatrope® (Lilly)
	7.0	Orthoclone OKT®3 (Ortho)
	7.5	Proleukin® (Chiron)
	7.4	Nutropin® (Genentech)
Phosphoric acid	7.3	Activase® (Genentech)

[a]Recombinant biological products approved for marketing in the United States. *Source*: Ref. 64.

of degradation is slow at low temperature. Often, a pH stability profile for each formulation can be developed by plotting the rate constants of all degradation reactions versus pH. The most stable protein formulation selected should be the one with a pH at which the overall degradation reactions are minimal.

Accelerated Temperature Stability Studies. Because temperature also has effects on protein degradation reactions, stability testing at accelerated temperatures is often carried out by formulation scientists for formulation screening, and the degradation rate profiles in various formulations are compared using Arrhenius plots. An Arrhenius plot allows one to predict the degradation rate constants at storage temperatures other than those actually studied, assuming that the protein does not undergo physical changes at a given temperature. Formulation scientists often use the Arrhenius plot to extrapolate results obtained at higher temperatures to 5 °C for product shelf life prediction. However, at least three temperatures are required to obtain an Arrhenius plot, and extrapolation of these data should be done with understanding of the assumptions made regarding the linearity of the response and the accuracy of the rate constants. Small-molecule drugs that often degrade via a single degradation pathway usually follow Arrhenius behavior. However, due to complexity in protein structure, more than one degradation reaction can occur in a protein product. This may result in nonlinear Arrhenius behavior. Sometimes, the degradation pathways for proteins at higher temperatures are not the same as those at lower temperatures. For example, interleukin 1β undergoes deamidation at or below 30 °C and aggregation at or above 39 °C (65). In this case, the formulation selected based on accelerated stability results may not be stable enough to protect the protein from degradation at lower temperatures. However, accelerated stability testing can still be a useful tool for protein formulation screening if the protein primarily degrades via only one degradation pathway within a selected temperature range, or if the degradation pathways are not influenced by one another (e.g., deamidation rate not affected by oxidation).

Stress Testing

After a stable protein formulation is selected from the formulation screening studies, stress tests are usually performed to determine whether additives are required to stabilize the protein against stresses such as agitation, freeze-thawing, and light exposure.

Agitation Studies. Proteins are susceptible to denaturation by mechanical force such shaking or agitation. Shaking of liquid protein products can occur during filling, shipping, and handling. The mechanism of denaturation may involve an increase in area of the air–liquid interface during shaking. Proteins have a tendency to unfold and expose their hydrophobic groups at an air–liquid interface, resulting in denaturation and the formation of aggregates. Proteins also tend to concentrate at the air–liquid interface due to the presence of both polar and nonpolar side chains. Irreversible unfolding or aggregation of the protein occurs when the interface surface area increases (66). Henson explained that the turbidity observed in some protein solutions during vigorous shaking could be due to the formation of aggregates when the air–liquid interface increases (67). Formulation scientists often perform agitation studies on liquid protein formulations to determine if a surfactant is required to protect the protein from denaturation when agitation is unavoidable during manufacturing, shipping, and handling. The addition of surfactants could either reduce the interfacial tension or solubilize the concentrated protein at the interface (2). Surfactants commonly used as stabilizers in the pharmaceutical industry include polysorbate 20, polysorbate 80, pluronic 68, and brij, which are polymers composed of various combinations of ethylene oxide, propyl oxide, sorbitan, and fatty alcohols.

Freeze–Thaw Studies. Under many circumstances, such as processing, storage, shipping, and stability testing, freezing and freeze-thawing stresses may be applied to proteins. For instance, it is very common for a liquid protein bulk to be manufactured and stored as frozen bulk for long-term storage and then thawed before filling into containers for packing as a liquid protein product or for freeze-drying as a lyophilized product. Sometimes, a protein formulation is frozen as a concentrated bulk for storage and then thawed and diluted with placebo before use. The risks of accidental freezing and thawing of proteins in liquid formulations may also exist during shipping and handling. Formulation scientists often perform short-term stability testing such that samples at each timepoint are kept frozen and then thawed at the end of the study for analysis. In lyophilized formulations, proteins are subjected to freezing during the lyophilization process. Therefore, freeze-thawing stability studies (at least three cycles of freezing and thawing) are usually performed on potential formulation candidates during formulation development to ensure the final product is stable upon freeze-thawing.

Protein denaturation during freezing is mainly caused by the phase separation due to the formation of ice. When ice is formed in a liquid protein formulation, the local concentration of protein as well as all excipients increases dramatically. The concentrated excipients in the non-ice phase may become destabilizing agents and cause the protein to denature (68). Some buffer species such as sodium phosphate and sodium succinate exhibit a decrease in pH during freezing. Freezing-induced pH changes may cause instability of recombinant proteins, as observed in IFN-γ formulated in succinate buffer (69).

If a protein formulation is evaluated to be denatured during freeze-thawing studies, formulation scientists may consider adding cryoprotectants such as sugars, polyols, amino acids, and inorganic and organic salts to prevent denaturation. The preferential exclusion mechanism of protection of proteins by these cryoprotectants during freeze-thawing was proposed by Timasheff and his coworkers (70). They suggested that the cryoprotectants, acting as cosolutes in the solution, are preferentially excluded from the protein. Consequently, the protein itself is preferentially hydrated by water. Hydration of a protein plays an important role in maintaining the conformational structure. Water contributes to the hydrogen bonds and hydrophobic interactions that are responsible for the formation of the secondary structure as well as maintaining the protein in a folded or native form (tertiary structure).

Light Stability Studies. Some amino acid residues in proteins can absorb energy from UV light and become modified to form photooxidized degradation products, especially in the presence of photosensitisers such as molecular oxygen, riboflavin, dyes, and polymers containing ether linkages (e.g., polysorbates). For example, the methionine residue in recombinant human relaxin can undergo photooxidation to form methionine sulfoxide (71), and the tryptophan residue in monoclonal IgG is converted to kynurenine and *N*-formylkynurenine upon exposure to UV light (72). The degree of photolytic degradation in the protein formulation can be influenced by many factors, including the buffer species and its concentration, excipients, and the formulation pH. Light intensity, duration of exposure, and storage temperature can also affect the rate of photolytic degradation.

Light stability testing is often performed during protein formulation development. The aim of a photostability study is to accelerate the formation of photodegradation products, if any, in a particular formulation upon exposure to light. FDA guidelines call for light stability testing of finished drug formulations but do not provide specifics of how to perform the test. A recent review article by Nema et al. provides an extensive overview of the protocols currently being used to perform photostability studies in pharmaceutical industry (73). Fluorescent lamps are usually used by pharmaceutical formulation scientists to simulate sunlight in photostability studies. Finished drug products are stored in a light cabinet with light intensity of 2–180 klx. The total exposure is between 8 and 4,500 klx days. Temperature inside the light cabinet is maintained at 25 $\pm$ 2 °C. Control samples, wrapped in aluminum foil, are exposed concurrently with the test samples.

Photolytic degradation of protein can be prevented by adding antioxidants to the formulation. Antioxidants can be classified into four main categories: chelating agents, reducing agents, oxygen scavengers, and chain terminators (74). If the addition of antioxidants in a protein formulation does not inhibit oxidation, the use of light-resistant containers is an alternative way of protecting light-sensitive protein drugs.

Lyophilized Formulation

Protein products that are unstable in solution or whose most-stable liquid formulations have insufficient long-term storage stability for marketing require lyophilization. However, the lyophilized products are usually limited to reconstitution and use in a short period of time. Lyophilized products also require longer time and more labor for formulation development than the ready-to-use liquid products. Lyophilization is also expensive due to the added costs of drying. Therefore, lyophilization is usually considered as an alternative method in protein formulation development when success with a liquid formulation cannot be achieved.

Lyophilization, also known as freeze-drying, is a process usually employed in the protein pharmaceutical industry to convert protein liquid formulations into solids of sufficient stability for storage. The process involves three major steps: freezing, primary drying, and secondary drying. During the initial freezing step, solvent (usually water) from an aqueous solution of protein drug that has been filled into glass vials is frozen and crystallized into solid ice at a low temperature such as -40 °C in a freeze dryer. The solutes including the protein and excipients are usually converted into an amorphous solid phase, which consists of uncrystallized solutes and uncrystallized water below their glass transition temperature, although some excipients such as buffer salts and mannitol may crystallize (75). The second step of the freeze-drying process, in which ice is removed from the frozen product by sublimation and condensation, is called primary drying. After the entire protein solution has been solidified, vacuum is applied to the drying chamber, and the shelf temperature is increased to supply heat to sublime ice from the frozen solid. Water vapor formed as a result of sublimation is condensed on cold surfaces in the condenser chamber. Primary drying is the longest part of the freeze-drying process and may require several days. After the bulk water (ice) has been removed from the partially dried product, secondary drying begins. During the stage of secondary drying, the final step of the freeze-drying process, the shelf temperature is increased to provide an elevated product temperature to remove the unfrozen water remaining in the amorphous solid phase. Secondary drying is usually performed over several hours at a shelf temperature between 25 and 50 °C, and the product temperature is maintained at about 25–35 °C.

The successful development of a lyophilized protein product requires the development of a liquid formulation that is stable to the freeze-drying process. Proteins can undergo denaturation or aggregation during both freezing and drying. The effects of freezing and drying on protein conformation described by Arakawa and Carpenter are illustrated in Figure 2. Formulation scientists often consider adding cryoprotectants and lyoprotectants to formulations to protect the protein from damage by freezing and drying, respectively, in the lyophilization process. Often, lyoprotectants improve the stability of the lyophilized products during subsequent storage. During freezing, all solutes in a formulation are concentrated when ice is formed, and buffer salts may crystallize, resulting in a pH shift (69). These freezing stresses can induce irreversible conformational changes that lead to protein instability. Thus, the use of buffers at low concentration and with minimal pH shift during freezing is recommended for formulations to be prepared by the freeze-drying process (76). The addition of cryoprotectants such as polymers, carbohydrates, and amino acids can provide cosolutes to preserve the native structure of protein in the frozen state. The mechanism for solute-induced protein stabilization during freezing proposed by Arakawa et al. is applicable to these conditions as just described. Amorphous sugars such as trehalose, sucrose, and lactose are the most effective lyoprotectants for stabilizing proteins during drying and storage. Carpenter and his coworkers have found that the formation of an amorphous solid is required for preservation of proteins by sugars, and some sugars such as

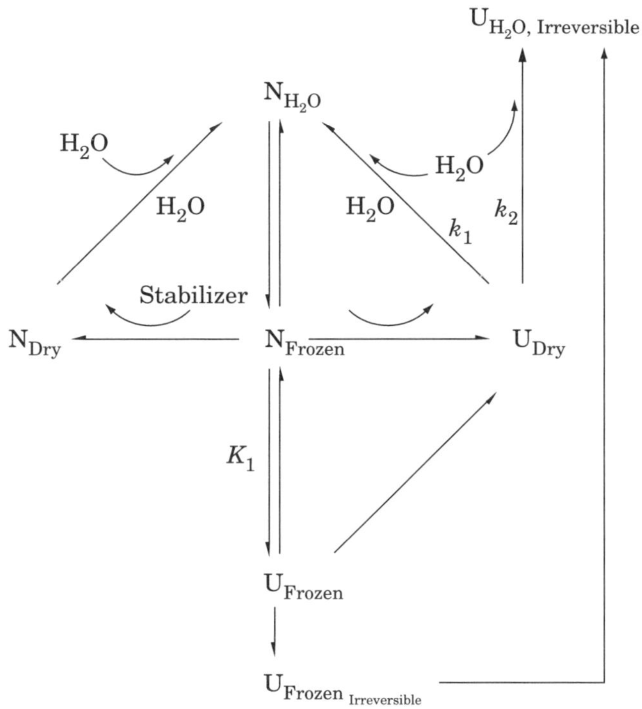

Figure 2. Schematic representation of the conformation of a protein during freezing, drying, and rehydration. N (native), U (unfolded), K_1 (conformational equilibrium formed upon freezing), k_1 (rate constant for refolding), k_2 (rate constant for formation of irreversibly denatured forms). *Source*: Reprinted with permission from Ref. 78.

mannitol fail to protect labile proteins during freeze-drying because the sugar forms a separate crystalline phase during freezing (77). It has been hypothesized that lyoprotectants stabilize proteins by acting as a water replacement through hydrogen bonding to the dried protein, preventing conformational changes during drying (78).

In developing a freeze-drying cycle for lyophilized protein product, temperature control is important in each step of the process. During freezing, the protein solution is frozen so that most of the water is converted to ice and all solutes are converted to either crystalline solids or amphorous solids below their glass transition temperature (79). The temperature at which the concentrated solute crystallizes during freezing is called the eutectic temperature. If the solute crystallizes, the shelf temperature must be reduced to well below the lowest eutectic temperature to ensure complete crystallization. For an amorphous solute, the shelf temperature must be lowered below the glass transition temperature (79). Typically, the final product temperature during freezing is about −40 °C. Usually, rapid ice growth as a result of low freezing temperature produces a high degree of supercooling (79,80). The degree of supercooling determines the number and size of ice crystals and therefore determines the drying characteristics and appearance of the dried cake. A very high degree of supercooling will result in very small pores, which means slower primary drying. Therefore, a very high degree of supercooling is undesirable. The ideal freezing method is to provide a moderate supercooling and rapid ice growth. The time required for primary drying decreases by a factor of 2 for a 5 °C increase in product temperature (79). For minimizing process time, it is important to maintain product temperature as high as possible during primary drying without affecting the product quality. Usually, the product temperature in primary drying is controlled at a safe margin (2–5 °C) below the collapse temperature. After all ice has been removed, the product temperature is usually increased to about 25–35 °C in the secondary drying step to allow the removal of unfrozen water remained in the product. To avoid collapse, the product temperature should not be raised until all the ice has been removed during primary drying.

Long-term Stability Studies.

Long-term premarket stability studies of protein formulations are designed to establish the specific storage conditions and expiration dating for the drug product. It is usually required by the FDA that a pharmaceutical product contain not less than 90% of the label claim of active ingredient at the expiration date. Although real-time stability data are required by the FDA to demonstrate that the pharmaceutical product can meet this requirement prior to approval, an extrapolated expiration date based on the stability results obtained from clinical batches is usually allowed. Thus, stability-indicating assays are important for determining the stability of drug products. Analytical methods such HPLC, electrophoresis, and spectroscopy have been widely used for this purpose.

In designing a long-term stability protocol for drug potency testing, one should consider the storage conditions and storage intervals. For refrigeration (5 °C) and room temperature (15–30 °C, by U.S.P. definition) storage, samples should be stored at 5, 15, and 25–30 °C and monitored at 0, 1, 2, 3, 6, 9, 12, 18, and 24 months, and once yearly after. For lyophilized products, samples can be assayed at longer storage intervals and at higher storage temperatures such as 40 °C. Pharmaceutical drugs are often monitored at high temperatures for the estimation of the degradation constants at low temperatures by Arrhenius plots. For high-temperature storage, samples are usually stored at 37 or 40 °C and assayed at 1, 3, and 6 months. Samples may also be stored at 45 or 50 °C and assayed at 2 weeks and 1 and 2 months. However, these elevated temperatures may approach the denaturation temperature of the protein, resulting in physical denaturation of the protein that would not be encountered in real situations, or extrapolate to lower-temperature behavior. In addition to determining the amount of active ingredient in a drug product, evaluation of the product should also include the following: pH, ionic strength, appearance, color, clarity, particulate matter, sterility, and pyrogenicity.

PROTEIN DRUG DELIVERY

Typically, proteins and many peptides must be administered by injection because of their poor bioavailability (fraction of administered drug in the circulation) by other routes. Initial clinical studies of these molecules is often conducted by dosing the drug i.v., intramuscularly (i.m.), or subcutaneously (s.c.). These clinical studies assess the safety and tolerability of systemic exposure to the new drug. The results from early clinical trials combined with preclinical studies in animal models may indicate the need for a drug delivery system. Drug delivery systems are often required when the efficacious dose is not well tolerated, indicating a requirement to localize the drug at the site of action or maintain a low systemic steady-state level (e.g., provide same total drug dose at a lower maximum serum concentration). Localization of the drug to the site of action may also be necessary due to biological barriers preventing sufficient delivery of the drug via the systemic circulation (e.g. blood–brain barrier, blood–ocular barrier). In some situations, the disease state may be more readily accessible to an alternative delivery method than traditional parenteral administration. In the case of cystic fibrosis, a disease of the lungs, aerosol delivery of rhDNase has proven to be an effective therapy (81). Another rationale for using a drug delivery system is patient compliance or ease of use. For example, if a patient is required to have multiple injections of a drug per day or per week, it will be difficult to maintain proper compliance unless the drug effects are immediately felt by the patient and/or the disease is debilitating. These issues along with the in vitro (physicochemical properties) and in vivo (pharmacokinetics, pharmacodynamics, and toxicology) characteristics of the drug must be well understood prior to embarking on the development of a drug delivery system.

Selection of a Delivery System

Several potential options are available for drug delivery, as shown in Table 3 (82). These options include both invasive methods involving injections or noninvasive approaches that may or may not provide sufficient bioavailability. In each case, a stable formulation must first be developed, as described in the previous sections. For invasive methods, drug delivery may be applied with either devices or biodegradable depots. Delivery devices such as syringe pens and needle-less injectors (injection by high-pressure gas) are gaining acceptance for many frequently administered drugs (83,84). In addition, implantable pumps are currently used with protein drugs such as insulin (85). Several injectable depot formulations made from the biodegradable polymer, poly(lactic-coglycolic acid) (PLGA), are either approved (86,87) or in clinical trials (88). These depot systems offer the advantage of eliminating multiple injections and localizing drug to the site of action, if necessary. However, a major challenge in the development of these systems is the production of sterile injectable depots, usually microspheres on the order of 20 μm or greater in diameter. Approaches to developing microsphere depot systems for clinical and commercial use have recently been reviewed (89,90).

For noninvasive administration, aerosol delivery of proteins and peptides is the most developed and accepted delivery option. Aerosol delivery offers the unique option of delivering the drug through a single inhalation instead of a single injection. The lung provides a large surface area for absorption of the drug, but the aerodynamic properties of the drug (liquid or solid) play a key role in the deposition and, ultimately, the amount of drug delivered systemically. Several key issues such as drug stability and device selection have been reviewed recently (91). This work demonstrates that if the key parameters are analyzed and the appropriate drug formulation and device are selected, then aerosol delivery of proteins may be an effective method of administration. In contrast, the bioavailability of the protein or peptide when delivered by the other noninvasive routes is usually quite low ($<1\%$), indicating that they are not an effective method for achieving significant systemic levels of the drug (for a review see Ref. 92). However, they may be very effective for local delivery of protein drugs if a low dose is efficacious.

Protein Stability in a Delivery System

After selecting a delivery system, the formulation scientist is required to develop a stable formulation that is compatible with the system and its intended use. These studies usually involve the formulation development worked described in the previous sections, as well as protein–delivery system interaction studies. The type of studies undertaken to assure delivery of biologically active and nonimmunogenic protein depend upon the system characteristics.

For injection devices, it is necessary to assess the protein stability in the container used to store the protein (e.g., cartridges) and the effect of the injection itself on the protein stability. These studies often involve storage stability studies of the protein in the device container as already described for protein filled into vials. The protein delivered from the device is also assessed by injecting into an empty sample tube or a buffer solution. Both recombinant human growth hormone (rhGH) (93) and insulin (94) are currently available in injection devices. Typically, high-pressure devices such as the needle-less injectors may cause denaturation of the protein due to the air–water interface and the high shear force applied to the liquid. Surface-induced denaturation may also occur at the device surface or at an air–water interface in the device. This degradation can be reduced and potentially eliminated by the addition of surfactants (e.g., polysorbates). In addition, implantable pump devices require studies to demonstrate maintenance of protein stability during the protein's residence time in the pump at physiological temperatures ($\sim$37 °C). Several studies have been performed with insulin in these pumps (85), indicating the difficulty in designing stable formulations for these pumps. These systems pose a formidable challenge due to the potential for accumulation of degraded protein over time with repeated use of a refillable pump.

Similar issues are also encountered in the development of the injectable depot formulations. In this case, the protein remains in a biodegradable depot at physiological conditions for prolonged periods. The protein must maintain its potency during its in vivo release from the depot. Recent studies with rhGH revealed that while it degrades during incubation at physiological pH, temperatures, and ionic strength, it maintains its biological activity and is not immunogenic (95,96). This result, however, cannot be assumed for each protein. Therefore, stability studies at physiological conditions should be performed prior to the development of a depot formulation to define the feasibility of the system and the acceptable duration of release (i.e.,

Table 3. Routes of Delivery for Proteins and Peptides

Delivery routes	Formulation and device requirements	Commercial products[a]
	Invasive	
Direct Injection	Liquid or reconstituted solid, syringe	
Intravenous		Activase®
Subcutaneous		Nutropin®
Intramuscular		Recombi Vax®
Intracerebral vein		
Depot system	Biodegradable polymes, liposomes, permeable polymers (not degradable), microspheres, implants	
	[LHRH analogs (subucutaneous or intramuscular)]	Lupron Depot® Zoladex® Decapeptyl®
	Noninvasive	
Pulmonary	Liquid or powder formulations, nebulizers, metered dose inhalers, dry powder inhalers	Pulmozyme®
Oral	Solid, emulsion, microparticulates, absorption enhancers	
Nasal	Liquid, usually requires permeation enhancers	Synarel®
Topical	Emulsion, cream or paste (liposomes)	
Transdermal	Electrophoretic (iontophoresis), electroporation, chemical permeation enhancers, prodrugs, ultrasonic	
Buccal, rectal, vaginal	Gels, suppositories, bioadhesives, particles	

Source: Ref. 8 with permission from the American Chemical Society.
[a]Nutropin® (recombinant human growth hormone), Activase® (recombinant human tissue plasminogen activator), and Pulmozyme® (recombinant human deoxyribonuclease I) are all products of Genentech, Inc. RecombiVax® (recombinant hepatitis B surface antigen) is produced by Merck & Co. Lupron Depot® (leuprolide acetate–PLGA) is a product of Takeda Pharmaceuticals. Zoladex® (goserelin acetate–PLGA) is produced by the Imperial Chemical Industries, Ltd. Decapeptyl® is a manufactured by Debiopharm. Synarel® (nafarelin acetate) is made by Syntex Corporation.

what level of degradation can be tolerated in terms of both potency and immunogenicity). These systems also pose a challenge in protein stability during the manufacture of the depot. For example, PLGA formulations require the dissolution of the polymer in organic solvents or use of high temperatures to melt the polymer, followed by dispersion of the protein in the PLGA and extraction of the organic solvent or reduction of the temperature, respectively. This exposure to harsh conditions often causes denaturation of the protein. Screening studies are performed to select excipients that will stabilize the protein against denaturation under these conditions. An example of this type of study was recently described using the model proteins rhGH and recombinant human IFN-γ (97). These studies revealed that excipients (e.g., sugars), which cause preferential hydration of the protein (cause an increased water layer at the protein surface), also increase protein stability during exposure to organic solvents. If these stability issues are overcome, an injectable depot system such as PLGA microspheres may be developed and tested in clinical trials.

Unlike depot systems, aerosol formulations for proteins do not require long-term stability at physiological conditions or stability in organic solvents. However, the aerosolization of proteins in their liquid state often leads to some protein degradation due to the air–water interface as well as the shear stress generated during the process. Several studies have been performed with traditional nebulizer devices that generate aqueous droplets of the appropriate diameter for efficient delivery to the lung. For rhGH, a surfactant was required to stabilize the protein during the process of aerosol formation (98), whereas rhDNAse was very stable in an unbuffered isotonic solution (1). In many cases, it may not be possible to develop an aqueous protein formulation that both prevents degradation and is not an irritant to the lung. In addition, devices required to generate liquid droplets are usually not easily portable, making them less attractive to patients that are not hospitalized or bed-ridden. However, dry powder protein formulations may be delivered in small hand-held devices. To produce a protein powder with the proper aerodynamic properties, spray-drying of the protein is usually performed (99). This process usually involves high temperatures and aerosolization of the protein, both of which may cause denaturation, as observed for rhGH (100). Once again, a formulation must be developed to prevent this denaturation. In addition, the dry powder protein formulation may be required to have stability at controlled room temperature (15–30 °C) and variable levels of residual humidity. The formulation scientist then carries out formulation screening experiments at different temperatures and residual humidities. If a protein powder is feasible, it offers advantages in ease of use and, perhaps, greater long-term stability.

Manufacturing Issues

The development of drug delivery systems may also require elaborate manufacturing procedures due to the complexity of both the system and the protein. When considering the manufacture of injection devices such as pens and pumps, some of the key issues are the ability to perform a sterile filling operation, the stability in the device, and the interaction between the drug, excipients, and the device or manufacturing components. The level of complexity is further increased in the case of pulmonary delivery. A nebulized liquid formulation is usually filled into ampules or other containers that allow easy addition of the liquid to the nebulizer by the patient. The filling of plastic ampules also has unique challenges in the ability to maintain sterility of the final product, the permeability of the material to gases and volatile solvents, and the physical properties of the container. In contrast, if a dry powder is used as the aerosol, the manufacturing process requires large-scale spray-drying equipment and machines for filling the dry powder into blister packs that can be used for metered-dose inhalers (99).

Manufacturing complexity is further increased when considering depot formulations such as PLGA microspheres. These microspheres are typically 20–90 μm in diameter and are therefore not readily sterilized by the traditional filtration methods (e.g., 0.22-μm filters). Terminal sterilization of these microspheres causes degradation of the polymer backbone, altering the release properties and often causing significant degradation of the protein within the microspheres. Microspheres must then be produced aseptically to avoid terminal sterilization while still maintaining the delivery of a sterile product to the patient. An aseptic process may be achieved through the use of barrier technology (isolators) for raw material input into the process and final product removal from the process. The remainder of the process operations is then designed as steam-in-place systems to assure a sterile environment throughout the production process.

If the unique challenges of a drug delivery system are overcome for a given protein, protein delivery may be successfully achieved. To reach commercial use, the drug delivery system must be well matched with the clinical and marketing requirements as well as the physiological requirements of the protein.

BIBLIOGRAPHY

1. D.C. Cipolla, I. Gonda, K.C. Meserve, S. Weck, and S.J. Shire, in J.L. Cleland and R. Langer eds., in *Formulation and Delivery of Proteins and Peptides*, American Chemical Society Symposium Series, ACS Books, New York, 1994, pp. 322–342.
2. Y.-C.J. Wang and M.A. Hanson, *J. Parenter. Sci. Technol.* **42**, S1–S26 (1988).
3. M.C. Manning, K. Patel, and R.T. Borchardt, *Pharm. Res.* **6**, 903–918 (1989).
4. T. Chen, *Drug Dev. Ind. Pharm.* **18**, 1311–1354 (1992).
5. T.J. Ahen and M.C. Manning eds., *Stability of Protein Pharmaceuticals, Part A: Chemical and Physical Pathways of Protein Degradation*, vol. 2, Plenum, New York, 1992.
6. T.J. Ahen and M.C. Manning eds., *Stability of Protein Pharmaceuticals. Part B: In vivo Pathways of Degradation and Strategies for Protein Stabilization*, vol. 3, Plenum, New York, 1992.
7. J.L. Cleland, M.F. Powell, and S.J. Shire, S. Bruck ed., in *Critical Reviews in Therapeutics Drug Carrier Systems*, CRC Press, Boca Raton, Fl., 1993, pp. 307–377.
8. R.T. Sauer and W.A. Lim, *Curr. Opin. Struct. Biol.* **2**, 46–51 (1992).
9. W.A. Lim and R.T. Sauer, *J. Mol. Biol.* **219**, 359–376 (1991).
10. A.E. Eriksson, W.A. Baase, X.-J. Zhang, D.W. Heinz, M. Blaber, E.P. Baldwin, and B.W. Matthews, *Science* **255**, 178–183 (1992).
11. K.A. Dill, *Biochemistry* **29**, 7133–7155 (1990).
12. K.A. Drill, *Biochemistry* **24**, 1501–1509 (1985).
13. C.N. Pace, *Trends Biochem. Sci.* **15**, 14–17 (1990).
14. T. Kiefhaber, R. Rudolp, H.-H. Kohler, and J. Buchner, *Biotechnology* **9**, 825–829 (1991).
15. M.R. DeFelippis, L.A. Alter, A.H. Pekar, H.A. Havel, and D.N. Brems, *Biochemistry* **32**, 1555–1562 (1993).
16. J.L. Cleland and D.I.C. Wang, *Biochemistry* **29**, 11072–11078 (1990).
17. H.A. Havel, E.W. Kauffman, S.M. Plaisted, and D.N. Brems, *Biochemistry* **25**, 6533–6538 (1986).
18. S. Tandon and P.M. Horowitz, *J. Biol. Chem.* **264**, 9856–9866 (1989).
19. T. Arakawa, N.K. Alton, and Y.-R. Hsu, *J. Biol. Chem.* **260**, 14435–14439 (1985).
20. D.A. Yphantis and T. Arakawa, *Biochemistry* **26**, 5422–5427 (1987).
21. A.B. Robinson and C.J. Rudd, in B.L. Horecker and E.R. Stadtman eds., *Current Topics in Cellular Regulations*, vol. 8, Academic, New York, 1974, pp. 247–295.
22. A.B. Robinson, J.W. Schotchler, and J.H. McKerrow, *J. Am. Chem. Soc.* **95**, 8156–8159 (1973).
23. S.A. Berson and R.S. Yalow, *Diabetes* **15**, 875–879 (1966).
24. B.V. Fisher and P.B. Porter, *J. Pharm. Pharmacol.* **33**, 203–206 (1981).
25. U.J. Lewis, E.V. Cheever, and W.C. Hopkins, *Biochim. Biophys. Acta* **214**, 498–508 (1970).
26. G.W. Becker, P.M. Tackitt, W.W. Bromer, D.S. Lefeber, and R.M. Riggin, *Biotechnol. Appl. Biochem.* **10**, 326–337 (1988).
27. E.E. Haley, B.J. Coreoran, F.E. Dorer, and D.L. Buchanan, *Biochemistry* **5**, 3229–3235 (1966).
28. T. Geiger and S. Clarke, *J. Biol. Chem.* **262**, 785–794 (1987).
29. K. Patel, C. Oliyai, R.T. Borchardt, and M.C. Manning, *J. Cell. Biochem.* **13A**, 88 (1989).
30. G. Teshima, J.T. Stults, V. Ling, and E. Canova-Davis, *J. Biol. Chem.* **266**, 13544–13547 (1991).
31. B.A. Johnson, J.M. Shirokawa, W.S. Hancock, M.W. Spellman, L.J. Basa, and D.W. Aswad, *J. Biol. Chem.* **264**, 14262–14271 (1989).
32. B.N. Violand, M.R. Schlittler, P.C. Toren, and N.R. Siegel, *J. Protein Chem.* **9**, 109–117 (1990).
33. K. Sasaoki, T. Hiroshima, S. Kusumoto, and K. Nishi, *Chem. Pharm. Bull.* **37**, 2160–2164 (1989).
34. T.H. Nguyen, J. Burnier, and W. Meng, *Pharm. Res.* **10**, 1563–1571 (1993).
35. Y. Nabuchi, E. Fujiwara, K. Ueno, H. Kuboniwa, Y. Asoh, and H. Ushio, *Pharm. Res.* **12**, 2049–2052 (1995).
36. L.C. Teh, L.J. Murphy, N.L. Huq, A.S. Surus, H.G. Friesen, L. Lazarus, and G.E. Chapman, *J. Biol. Chem.* **262**, 6472–6477 (1987).
37. J.Q. Oeswein and S.J. Shire, in V.H.L. Lee ed., *Peptide and Protein Drug Delivery*, Advances in Parenteral Sciences 4, Marcel Dekker, New York, 1990, pp. 167–202.
38. T.H. Nguyen, in J.L. Cleland and R. Langer eds., *Formulation and Delivery of Proteins and Peptides*, American Chemical Society Symposium Series, ACS Books, New York, 1994, pp. 59–71.
39. L. Philipson, *Biochim. Biophys. Acta* **56**, 375–382 (1962).
40. C. Little and P.J. O'Brien, *Eur. J. Biochem.* **10**, 533–538 (1969).

41. L.A. Holt, B. Milligan, D.E. Rivett, and F.H.C. Stewart, *Biochim. Biophys. Acta* **499**, 131–138 (1977).
42. L.A. Sluyterman, *Biochim. Biophys. Acta* **60**, 557–561 (1962).
43. A.J.S. Jones, in J.L. Cleland and R. Langer eds., *Formulation and Delivery of Proteins and Peptides*, American Chemical Society Symposium Series, ACS Books, New York, 1994, pp. 22–45.
44. A.J.S. Jones, in G. Poste and E. Tomlinson, eds., *Advanced Drug Delivery Reviews*, Elsevier Science Publishers B.V., Amsterdam, 1993, pp. 29–90.
45. C.S. Fullmer and R.H. Wasserman, *J. Biol. Chem.* **254**, 7028–7212 (1979).
46. P.H. Cowan, in R.W.A. Oliver ed., *HPLC of Macromolecules: A Practical Approach*, IRL Press, Oxford, 1989, pp. 127–156.
47. W.J. Kohr, R. Keck, and R.N. Harkins, *Anal. Biochem.* **122**, 348–359 (1982).
48. T.W. Perkins, D.S. MaK, T.W. Root, and E.N. Lightfoot, *J. Chromatogr. A* **766**, 1–14 (1997).
49. P.G. Righetti, *Isoelectric Focusing: Theory, Methodology and Applications*, Elsevier Science Publishers B.V., Amsterdam, 1993.
50. R. Pearlman and T.H. Nguyen, in *Therapeutic Peptides Proteins: Formulation, Delivery and Targeting*, D. Marshak and D. Liu eds., Cold Spring Harbor Laboratory, Cold Spring Harbor, N.Y., 1989, pp. 23–30.
51. B.L. Karger, A.S. Cohen, and A. Guttman, *J. Chromatogr.* **492**, 585–614 (1989).
52. L. Bongers, T. Lambros, A.M. Felix, and E.P. Heimer, *J. Liquid Chromatogr.* **15**, 1115–1128 (1992).
53. M.A. Costello, C. Woititz, J. DeFeo, D. Stremlo, L.-F.L. Wen, D.J. Palling, K. Iqbal, N.A. Guzman, *J. Liquid Chromatogr.* **15**, 1081–1097 (1992).
54. G. Hunt, K.G. Moorhouse, and A.B. Chen, *J. Chromatogr. A* **744**, 295–301 (1996).
55. A.B. Chen, C.A. Rickel, A. Flanigan, G. Hunt, and F.G. Moorhouse, *J. Chromatogr. A* **744**, 279–284 (1996).
56. L.E. Bennett, W.N. Charman, D.B. Williams, and S.A. Charman, *J. Pharm. Biomed. Anal.* **12**, 1103–1108 (1994).
57. D.J. Kroon, S. Goltra and B. Sharma, *J. Cap. Elec.* **2**, 34–39 (1995).
58. D.B. Wetlaufer, *Adv. Protein Chem.* **17**, 303–390 (1962).
59. R.H. Strick, *CRC Crit. Rev. Biochem.* **2**, 113–175 (1974).
60. T.A. Bewley, *Rec. Prog. Hormone Res.* **35**, 155–213 (1979).
61. A.F. Winder and W.L.G. Gent, *Biopolymers* **10**, 1243–1251 (1971).
62. V.A. Bloomfield, *Ann. Rev. Biophys. Bioeng.* **10**, 421–450 (1981).
63. G.D.J. Phillies, *Anal. Chem.* **62**, 1049A–1057A (1990).
64. *Physicians' Desk Reference (PDR)*, 52nd ed., Medical Economics Company, Montvale, N.J., 1998.
65. L.C. Gu, E.A. Erdos, H.S. Chiang, T. Calderwood, K. Tsai, G.C. Visor, J. Duffy, W.-C. Hsu, and L.C. Foster, *Pharm. Res.* **8**, 485–490 (1991).
66. A.W. Adamson, *Physical Chemistry of Surfaces*, 2nd ed., Interscience Publishers, New York, 1960.
67. A.F. Henson, J.R. Mitchell, and P.R. Musselwhite, *J. Colloid Interface Sci.* **32**, 162–165 (1970).
68. P.H. Von Hippel and T. Shleich, in S.N. Timasheff and G.D. Fasman eds., *Structure and Stability of Macromolecules*, vol. 2, Dekker, New York, 1969, pp. 417–574.
69. X.M. Lam, H.R. Costantino, D.E. Overcashier, T.H. Nguyen, and C.C. Hsu, *Int. J. Pharm.* **142**, 85–95 (1996).
70. S.N. Timasheff and T. Arakawa, in T.E. Creighton ed., *Protein Structure, A Practical Approach*, IRL Press; New York, 1988, pp. 331–345.
71. D.C. Cipolla and S.J. Shire, in J.J. Villafranca ed., *Techniques in Protein Chemistry II*, Academic, New York, 1990, pp. 543–555.
72. H. Mach, C.J. Burke, G. Sanyal, P.-K. Tsai, D.B. Volkin, and C.R. Middaugh, in J.L. Cleland and R. Langer eds., *Formulation and Delivery of Proteins and Peptides*, American Chemical Society Symposium Series, ACS Books, New York, 1994, pp. 72–84.
73. S. Nema, R.J. Washkuhn, and D.R. Beussink, *Pharm. Technol.* **19**, 170–185 (1995).
74. K.A. Connors, G.L. Amidon, and V.J. Stella, in *Chemical Stability of Pharmaceuticals: A Handbook for Pharmacists*, 2nd ed., John Wiley, New York, 1986, pp. 82–104.
75. M.J. Pikal, *BioPharmacy* **3**, 18–27 (1990).
76. M.J. Pikal, *BioPharmacy* **3**, 26–30 (1990).
77. J.F. Carpenter, S.J. Prestrelski, T.J. Anchordoguy, and T. Arakawa, in J.L. Cleland and R. Langer eds., *Formulation and Delivery of Proteins and Peptides*, American Chemical Society Symposium Series, ACS Books, New York, 1994, pp. 132–146.
78. S.J. Prestrelski, T. Arakawa, and J.F. Carpenter, in J.L. Cleland and R. Langer, eds., *Formulation and Delivery of Proteins and Peptides*, American Chemical Society Symposium Series, ACS Books, New York, 1994, pp. 148–169.
79. M.J. Pikal, in J.L. Cleland and R. Langer eds., *Formulation and Delivery of Proteins and Peptides*, American Chemical Society Symposium Series, ACS Books, New York, 1994, pp. 120–132.
80. L.A. Gatlin and S.L. Nail, *Bioprocess Technol.* **18**, 317–367 (1994).
81. R.C. Hubbaard, N.G. McElvaney, P. Birrer, S. Shak, W.W. Robinson, and C. Jolley, *New Engl. J. Med.* **326**, 812–815 (1992).
82. J.L. Cleland and R. Langer, in J.L. Cleland and R. Langer eds., *Formulation and Delivery of Proteins and Peptides*, American Chemical Society Symposium Series, ACS Books, New York, 1994, pp. 1–19.
83. P.D. Gluckman and W.S. Cutfield, *Arch. Dis. Childhood*, **66**, 686–688 (1991).
84. C. Pavia, C. Valls, M. Gallart, R. Velasco, and M.C. Gonzalez, *Acta Ped. Espanola* **53**, 25–28 (1995).
85. J.R. Brennan, S.S.P. Gebhart, and W.G. Blackard, *Diabetes* **34**, 353–359 (1985).
86. R. Sharifi, R.C. Bruskewitz, M.C. Gittleman, S.D. Graham, Jr., P.B. Hudson, and B. Stein, *Clin. Ther.* **18**, 647–657 (1996).
87. O.A. Oyesany, S.K. Teo, E. Quah, N. Abdurazak, F.Y. Lee, and W.C. Cheng, *Hum. Reprod.* **10**, 1042–1044 (1995).
88. J.L. Cleland, O.L. Johnson, S.D. Putney, and A.J.S. Jones, *Adv. Drug Delivery Rev.* **28**, 71–84 (1997).
89. J.L. Cleland, in *Protein Delivery: Physical Systems*, L. Sanders and W. Hendren eds., Plenum, New York, 1997, pp. 1–43.
90. H. Okada and H. Toguchi, *Crit. Rev. Ther. Drug Carrier Systems* **12**, 1–99 (1995).
91. R.W. Niven, *Pharm. Technol.* **17**, 72–74 (1993).
92. L.L. Wearley, *Crit. Rev. Ther. Drug Carrier Systems* **8**, 331–394 (1991).
93. J.T. Jorgensen, *J. Ped. Endocrinol.* **7**, 175–180 (1994).
94. J.O. Hornquist, A. Wikby, U. Stenstrom, P.O. Anderson, *Diabetes*, **28**, 63–72 (1995).
95. J.L. Cleland, A. Mac, B. Boyd, J. Yang, E.T. Duenas, D. Yeung, D. Brooks, C. Hsu, H. Chu, V. Mukku, and A.J.S. Jones, *Pharm. Res.* **14**, 420–425 (1997).

96. J.L. Cleland, E.T. Duenas, and A.L. Daugherty, M. Marian, J. Yang, M. Wilson, A.C. Celniker, A. Shhzamani, V. Quarmby, H. Chu, V. Mukku, A. Mac, M. Roussakis, N. Gillette, B. Boyd, D. Yeung, D. Brooks, Y.-F. Maa, C. Hsu, A.J.S. Jones, *J. Controll. Rel.* **49**, 193–205 (1997).
97. J.L. Cleland and A.J.S. Jones, *Pharm. Res.* **13**, 1462 (1996).
98. J.Q. Oeswein, A.L. Daugherty, E.J. Cornavaca, J.A. Moore, H.M. Gustafson, B.M. Eckhardt, in R.N. Dalby and R. Evans eds., *Proceedings of the Second Respiratory Drug Delivery Symposium*, Continuing Pharmacy Education, University of Kentucky, Lexington, Ky., 1991, pp. 14–49.
99. V. Naini, P.R. Byron, and R.N. Dalby, in R.N. Dalby, P.R. Byron, and S.J. Farr eds., *Respiratory Drug Delivery V*, Interpharm Press, Buffalo Grove, Ill., 1996, pp. 382–384.
100. M. Mummenthaler, C.C. Hsu, R. Pearlman, *Pharm. Res.* **11**, 12–17 (1994).

FREEZE-DRYING, PHARMACEUTICALS

R. BRUTTINI
Criofarma Freeze-drying Equipment
Turin, Italy

KEY WORDS

Amorphous solutes
Bound water
Crystalline solutes
Free water
Freeze-drying
Lyophilization
Primary drying
Secondary drying

OUTLINE

INTRODUCTION

The primary object of freeze-drying is to preserve biological materials and pharmaceuticals without injury by freezing the contained water and then removing the ice by sublimation. Thus one can expect to combine the advantages of both freezing and low-temperature dessication to obtain a more propitious state of preservation. In freeze-drying, the water or other solvent is removed as a vapor by sublimation from the frozen material in a vacuum chamber, leaving a dry porous mass of approximately the same size and shape as the original frozen mass. The resulting product is in a stable form and can be redissolved rapidly in water; the material displays what is called lyophilic behavior, which gave rise to the term *lyophilization* to describe the process of freeze-drying.

Lyophilization is employed soon after production for injectable pharmaceuticals that exhibit poor stability in solution, preventing their deactivation over a period of time and preserving their bioactivity. For example, a compound that is heat sensitive and undergoes rapid decomposition in aqueous solution can be formulated in a stable, rapidly soluble form.

THE FREEZE-DRYING PROCESS

Many pharmaceuticals and biological materials, which may not be heated even to moderate temperatures in ordinary drying, can be freeze dried (1–6). Furthermore, many pharmaceutical products in solution deactivate over a period of time; the bioactivity of such compounds can be preserved by freeze-dried soon after their production, which stabilizes their molecules (2,6). Freeze-drying can also be applied to systems in which interfacial forces during drying need to be eliminated and where homogeneous powder products (derived from multicomponent, multiphase precursors) are required (2,5,7).

As a rule, freeze-drying produces the highest-quality pharmaceutical product obtainable by any drying method. One important factor is the surface structural rigidity of the frozen substance, where sublimination occurs. This rigidity, to a large extent, prevents collapse of the solid matrix remaining after drying. The result is a porous, nonshrunken structure that facilitates rapid and almost complete rehydration when water is added at a later time. Freeze-drying of biological materials and pharmaceuticals also has the advantage of low processing temperatures, the relative absence of liquid water, and the rapid transition of any local region of the material being dried from a fully hydrated to a nearly completely dehydrated state, all of which minimize the degradative reactions that normally occur in ordinary drying processes, such as protein denaturation and enzymatic reactions. In any pharmaceutical or biological material, some nonfrozen water (bound, or sorbed, water) will almost unavoidably be present during freeze-drying, but there is very often a rather sharp transition temperature for the still-wet region during drying (5), below which the product quality improves markedly. This improvement shows that sufficient water is frozen to provide the beneficial characteristics of freeze-drying.

However, freeze-drying is an expensive form of dehydration for pharmaceuticals because of the slow drying rate and the use of vacuum. The cost of processing is offset to some extent by the absence of any need for refrigerated handling and storage.

Systematic freeze-drying is a procedure mainly applied to the following categories of material (1–18):

1. Nonliving matter, such as blood plasma, serum, hormone solutions, foodstuffs, pharmaceuticals (e.g., antibiotics), ceramics, superconducting materials, and materials of historical importance (e.g., archaeological wood)
2. Surgical transplants, which are made nonviable so that the host cells can grow on them and use them as a skeleton (e.g., artery, bone, and skin)
3. Certain living cells that must remain viable for long periods of time, such as bacteria, yeasts, and viruses

Freeze-drying requires very low pressures or high vacuum in order to produce a satisfactory drying rate. If the water were in a pure state, freeze drying at or near 0 °C at an absolute pressure of 4.58 mmHg could be performed. But because the water usually exists in a combined state or a solution, the material must be cooled below 0 °C to keep the water in the solid phase. Most pharmaceutical freeze-drying is done at −20 °C or lower at absolute pressures of about 0.3 mmHg or less.

In short, freeze drying is a multistep operation:

1. The material is frozen hard by low-temperature cooling.
2. It is dried by direct sublimation of the frozen solvent and by desorption of the sorbed or bound solvent (nonfrozen solvent), generally under reduced pressure.
3. It is then stored in the dry state under controlled conditions (free of oxygen and water vapor and usually in airtight, opaque containers, filled with inert dry gas).

If correctly processed, most products can be kept in such a state for an almost unlimited period of time while retaining all their initial physical, chemical, biological, and organoleptic properties, and remaining available at any time for immediate reconstitution. In most cases reconstitution is accomplished by addition of the exact amount of solvent that has been extracted, thus giving to the reconstituted product a structure and appearance as close as possible to the original material. However, in some instances, reconstitution can be monitored in order to yield more concentrated or diluted products by controlling the amount of solvent. Vaccines and pharmaceutical materials are very often reconstituted in physiological solutions that are quite different from the original but are best suited for intramuscular or intravenous injection.

These examples are not exhaustive; detailed presentations on the uses of the freeze-drying process and of freeze-dried products are given in Refs. 1–10, 12, and 17–22.

FREEZE-DRYING STAGES

The freeze-drying separation method (process) involves three stages: *(1)* the freezing stage, *(2)* the primary drying stage, and *(3)* the secondary drying stage.

In the freezing stage, the product or solution to be processed is cooled down to a temperature at which all the material is in a frozen state.

In the primary drying stage, the frozen solvent is removed by sublimation; this requires that the pressure of the system (freeze-dryer) where the product is being dried must be less than or near to the equilibrium vapor pressure of the frozen solvent. If, for instance, frozen pure water (ice) is processed, then sublimation of pure water at or near 0 °C and at an absolute pressure of 4.58 mmHg could occur. But because the water usually exists in a combined state (e.g., gel suspension or biological material) or a solution (e.g., pharmaceutical product), the material must be cooled below 0 °C to keep the water in the frozen state. For this reason, during the primary drying stage the temperature of the frozen layer (Fig. 1) is most often at −20 °C or lower and at absolute pressures of about 0.3 mmHg or less. As the solvent (ice) sublimes, the sublimation interface (plane of sublimation), which started at the outside surface (Fig. 1), recedes, and a porous shell of dried material remains. The heat for the latent heat of sublimation (2,840 kJ/kg ice) can be conducted through the layer of dried material and through the frozen layer, as shown in Fig. 1. The vaporized solvent (water vapor) is transported through the porous layer of dried material. During the primary drying stage, some of the sorbed water (nonfrozen water) in the dried layer may be desorbed. The desorption process in the dried layer could affect the amount of heat that arrives at the sublimation interface, and therefore, it could affect the velocity of the moving sublimation front (interface). The time at which there is no more frozen layer (that is, there is no more sublimation interface) is taken to represent the end of the primary drying stage.

The secondary drying stage involves the removal of solvent (water) that did not freeze (sorbed, or bound, water). The secondary drying stage starts at the end of the primary drying stage, and the desorbed water vapor is transported through the pores of the material being dried.

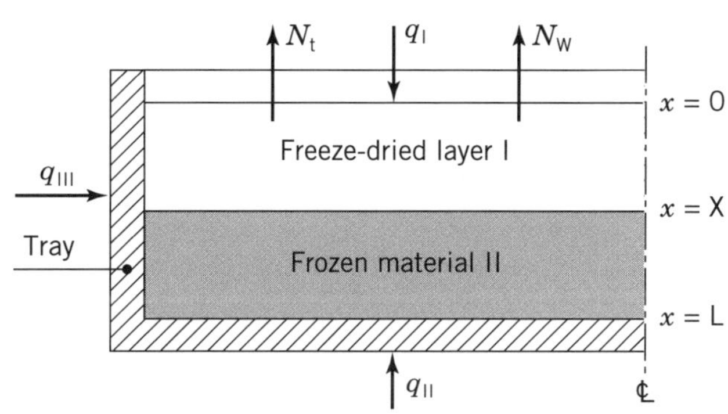

Figure 1. Diagram of a material on a tray during freeze-drying; the variable X denotes the position of the sublimation interface (front) between the freeze-dried layer (layer I) and the frozen material (layer II).

The Freezing Stage

The freezing stage represents the first separation step in the freeze-drying process, and the performance of the overall freeze-drying process depends significantly on this stage. The objective in freezing is to convert most of the water from the system by formation of ice and convert all solutes into solids, either crystalline solids or a glass. The material system to be processed (e.g., gel suspension, pharmaceutical solution, or biological material) is cooled down to a temperature (which depends on the nature of the product) that is always below the solidification temperature of the material system. For instance, if the material to be freeze-dried is a solution whose equilibrium phase diagram presents a eutectic point (e.g., the solution of NaCl and water presents a eutectic point at −21.6 °C), then the value of the final freezing temperature must be below the value of the eutectic temperature; in this case, the material becomes wholly crystalline. In practice, materials display one of two different types of freezing behavior: *(1)* the liquid phase suddenly solidifies (eutectic formation) at a temperature that depends on the nature of the solids in the sample, or *(2)* the liquid phase does not solidify (glass formation), but rather it just becomes more and more viscous until it finally takes the form of a very stiff, highly viscous liquid. In this latter case there is no eutectic temperature, but there is a minimum freezing temperature.

At the end of the freezing step, there already exists a separation between the water to be removed (frozen water in the form of ice crystals) and the solute. In many cases, at the end of the freezing stage about 65–90% of the initial (at the start of the freezing stage) water is in the frozen state, and the remaining 10–35% of the initial water is in the sorbed (nonfrozen) state. The shape of the pores, the pore-size distribution, and pore connectivity (1–3,6,9–11,13,15,16,21,23,24) of the porous network of the dried layer formed by the sublimation of the frozen water during the primary drying stage depend on the ice crystals that formed during the freezing stage; this dependence is of extreme importance because the parameters that characterize the mass and heat transfer rates in the dried layer are influenced significantly by the porous structure of the dried layer. If the ice crystals are small and discontinuous, then the mass transfer rate of the water vapor in the dried layer could be limited. On the other hand, if large dendritic ice crystals are formed and homogeneous dispersion of the preeutectic and posteutectic frozen solution can be realized, the mass transfer rate of the water vapor in the dried layer could be high, allowing the product to be dried more quickly. Thus, the method and rate of freezing, as well as the shape that contains the solution and the nature of the product, are critical to the course of lyophilization because they affect the drying rate and the quality of the product.

In industrial pharmaceutical freeze-dryers the freezing of the product is usually done in the same plant where the drying occurs. In certain nonsterile pharmaceutical freeze-dryers the freezing of the product is accomplished by spraying liquid nitrogen into the drying chamber containing the product. In sterile pharmaceutical freeze-dryers the freezing stage is realized by contact between cooled plates and product-supporting containers.

The Primary Drying Stage

After the freezing stage, the drying chamber where the product is placed is evacuated and the chamber pressure is reduced down to a value that allows the sublimation of solvent (water) to take place. When the water molecules sublime and enter the vapor phase, they also keep with them a significant amount of the latent heat of sublimation (2,840 kJ/kg ice), and thus, the temperature of the frozen product is again reduced. If there is no heat supplied to the product by a heat source, then the vapor pressure of the water at the temperature of the product reaches the same value as that of the partial pressure of the water vapor in the drying chamber; therefore, the system reaches equilibrium and no additional water sublimation from the product occurs. In order to have continuous sublimation of water from the product, the latent heat of sublimation must be provided by a heat source. The heat is usually supplied by conduction, convection, and/or radiation; conduction is realized by contact between heated plates and product-supporting containers.

The amount of heat that can be supplied to the product can not be increased freely because there are certain limiting conditions that have to be satisfied during the primary drying stage. One of the constraints has to do with the maximum temperature that the dried product (freeze-dried layer in Fig. 1) can tolerate without *(1)* loss of bioactivity, *(2)* color change, *(3)* the possibility of degradative chemical and biochemical reactions, and *(4)* structural deformation in the dried layer. The maximum temperature that the dried product can tolerate without suffering any of these deleterious effects is denoted, for a given product, by T_{scor} (T_{scor} is often called, by convention, the temperature of the scorch point of the dried product). Another constraint has to do with the maximum temperature the frozen layer can tolerate and still remain frozen. If the material has a eutectic form and if the temperature of the lowest eutectic point is exceeded during the primary drying stage, then melting in the frozen layer (Fig. 1) can occur. Melting at the sublimation interface or any melting that occurs in the frozen layer can give rise to gross material faults such as puffing, shrinking, and structural topologies filled with liquid solution. When melting has occurred at some point in the frozen layer, then the solvent at that point cannot be removed by sublimation, and therefore, there is process failure (Fig. 1). There has also been, at least, a loss in structural stability. If the material is in a glass form and if the minimum freezing temperature is exceeded during the primary drying stage, then the phenomenon of collapse can occur, with a loss of rigidity in the solid matrix. Again in this case, there is a process failure because the water can no longer be removed from the frozen layer only by sublimation, and there has also been, at least, a loss in structural stability. The structural stability of a material relates to its ability to go through the freeze-drying process without change in size, porous structure, and shape. The maximum allowable temperature in the frozen layer is determined by both structural stability and product stability (e.g., product bioactivity) factors; that is, the maximum value of the temperature in the frozen layer during the primary drying stage must be such that the drying process is conducted without loss of product properties (e.g., bioactivity) and structural stability. Sometimes the product stability factors are related to structural stability factors (as in melting). There are systems in which the product stability factors do not depend on structural stability factors, as is the case for many vaccines, viruses, and bacteria where the temperature of the frozen layer during the primary drying stage must be kept well below the melting temperature so that there is a good level of bioactivity and organism survival after drying. In general, product stability is related to the temperature of the frozen layer during the primary drying stage. The maximum allowable temperature that the frozen layer can tolerate without suffering melting, puffing, shrinking, collapse, and loss of product property or stability, is denoted, for a given product, by T_m (T_m is often called, by convention, the melting temperature of the sublimation interface or of the frozen layer).

The water vapor produced by the sublimation of the frozen water in the frozen layer and by the desorption of sorbed (nonfrozen) water in the dried layer during the primary drying stage travels by diffusion and convective flow through the porous structure of the dried layer and enters the drying chamber. (It should be noted that most of the water removed during the primary drying stage is produced by sublimation of the frozen water in the frozen layer.) The water vapor must be continuously removed from the drying chamber in order to maintain nonequilibrium conditions for the drying process in the system. This is usually accomplished by fitting a refrigerated trap (called an ice condenser) between the drying chamber and the vacuum pump; the water vapor is collected on the cooled surface of the condenser in the form of ice. The time at which there is no more frozen layer is taken to represent the end of the primary drying stage.

Secondary Drying Stage

The secondary drying stage involves the removal of water that did not freeze (sorbed or bound water). In an ideal freeze-drying process, the secondary drying stage starts at the end of the primary drying stage. The word *ideal* is used here to suggest that in an ideal freeze-drying process only frozen water should be removed during the primary drying stage, whereas the sorbed water should be removed during the secondary drying stage. But, as just discussed, in real freeze-drying systems a small amount of sorbed water could be removed by desorption from the dried layer of the product during the primary drying stage, and thus, there could be some secondary drying in the dried layer of the product during the primary drying stage. In real freeze-drying processes the secondary drying stage is considered to start when all the ice has been removed by sublimation (end of primary drying stage). It is then considered that, during the secondary drying stage, most of the water that did not freeze (bound water) is removed. The bound moisture is present due to mechanisms of *(1)* physical adsorption, *(2)* chemical adsorption, and *(3)* water of crystallization. Although the amount of bound water is about 10–35% of the total moisture content (65–90% of the total moisture could be free water that was frozen and then removed by sublimation during the primary drying stage), its effect on the drying rate and overall drying time is very significant, and the time that it takes to remove the sorbed water could be as long or longer than the time that is required for the removal of the free water.

The bound water is removed by heating the product under vacuum. But, as in the case of primary drying, the amount of heat that can be supplied to the product can not be increased freely because there are certain constraints that have to be satisfied during the secondary drying stage. The constraints have to do with the moisture content and the temperature of the product; these two variables influence the structural stability, as well as the product stability, during and after drying. For structural stability, the same phenomena, as in the case of the primary drying stage, have to be considered: collapse, melting (if the temperature is increased at constant moisture), or dissolution (if moisture is increased at constant temperature) of the solid matrix can occur. Product stability (e.g., bioactivity) is a function of both moisture content and temperature in the sample, and during secondary drying the moisture concentration and temperature in the sample could vary widely with location and time. This implies that the potential for product alteration to occur in the sample will vary with time and location. The moisture concentration profile is related to the temperature profile in the dried layer, and thus, the moisture content in the sample cannot be controlled independently. Because many products are temperature sensitive, it is usual to control product stability by limiting the temperature during the secondary drying process, and the final moisture content is checked before the end of the cycle (6,10,25).

In the secondary drying stage, the bound water is removed by heating the product under vacuum, and the heat is usually supplied to the product by conduction, convection, and/or radiation. The following product temperatures are usually employed: *(1)* below 0 °C for vaccines, *(2)* between 10 °C and 35 °C for heat-sensitive products, and *(3)* 50 °C or more for less heat sensitive products.

The residual moisture content in the dried material at the end of the secondary drying stage, as well as the temperature at which the dried material is kept in storage, are critical factors in determining product stability during storage life. Some vaccines remain stable for many years when they are stored at −20 °C, but a significant loss of titer can be found after one year if they are stored at 37 °C (26). Furthermore, certain vaccines such as live rubella and measles can be damaged by overdrying (final moisture content of about 2% is required for best titer retention), whereas other materials such as chemotherapeutics and antibiotics must be dried to a residual moisture content as low as 0.1% for best results.

FREEZE-DRYING EQUIPMENT

Freeze-drying equipment for pharmaceuticals is able to perform the three stages of the freeze-drying process: *(1)* freezing of the pharmaceutical product (usually a pharmaceutical aqueous solution), *(2)* primary drying of the product by removing the frozen solvent by sublimation, and *(3)* secondary drying of the product by removing the sorbed, or bound, water (solvent that did not freeze).

All three stages of the freeze-drying process are energy consuming. The freezing stage involves freezing of the wet product. Because this is normally considered one of the preparatory steps before the actual freeze-drying, we concentrate here on the other two steps, which take place in the freeze-drying chamber (2,5,12). The primary and secondary drying stages involve the controlled supply of heat to the product to cover requirements for the sublimation and desorption processes, respectively. The removal from the freeze-drying chamber of the vast volumes of water vapor released during the sublimation and desorption processes always consumes the largest amount of energy. The efficiency of water vapor removal (the vapor trap system) therefore has a decisive effect on the total energy consumption of the freeze-drying plant.

The vapor trap is placed in an ice condenser chamber communicating with the freeze-drying chamber. The water vapor condenses to ice on its refrigerated surfaces.

When in operation, the efficiency of the vapor trap is shown by a small total temperature difference, ΔT, between the saturation temperature for water vapor at the pressure in the freeze-drying cabinet and the evaporation temperature of the refrigerant (Fig. 2). This total temperature difference, ΔT, results mainly from the following three resistances:

1. Pressure difference, ΔP; equivalent to the pressure drop caused by the resistances to the vapor flow from the freeze-drying chamber to the cold surfaces of the vapor trap placed in the ice condenser chamber.
2. The temperature difference ΔT_{ice} through the layer of ice on the cold surface.
3. The temperature difference ΔT_{refr} between the cold surface and the evaporating refrigerant.

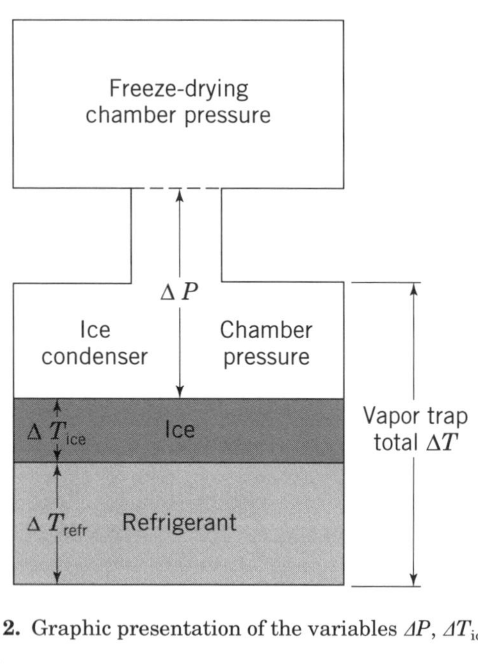

Figure 2. Graphic presentation of the variables ΔP, ΔT_{ice}, ΔT_{refr}, and ΔT.

For an efficient vapor trap it is necessary to have a combination of a large cross-sectional area for the vapor flow (low ΔP), an efficient deicing system (low ΔT_{ice}), and an efficient refrigerating system (low ΔT_{refr}). A less efficient vapor trap means a higher ΔT, thus demanding a lower evaporation temperature of the refrigerating plant to maintain the required vacuum in the freeze-drying cabinet. A lower evaporation temperature means higher operation costs. In this temperature range, an evaporation temperature 10 °C lower means a 50% increased in energy consumption.

The principal problem in the freeze-drying of pharmaceutical solutions is to operate in sterile conditions. In modern plants, the internal sterilization of the equipment is usually made with pressurized steam at 121 °C or more; in old plants, sterilization is realized with the use of certain proprietary sanitizing agents (27,28). A pharmaceutical freeze-dryer must be designed according to current good manufacturing practice (cGMP) in order to be validated before and during use.

When evaluating industrial freeze-drying plants, the following characteristics are thus of prime importance:

1. Operation reliability
2. Ease, safety, and quality of process control
3. Product losses
4. Vapor trap efficiency
5. Internal sanitization
6. cGMP requirements

In the following sections, some of the commonly used types of pilot and industrial freeze-drying equipment for pharmaceuticals are presented, and their most important technical features are discussed.

Pilot Freeze-dryers

The freeze-drying of pharmaceutical products can be performed in the freeze-dryer as: *(1)* bulk solution freeze-drying in trays, or *(2)* freeze-drying in vials.

Freeze-drying pilot units appropriate for use in the pharmaceutical industries and in the laboratory are in high demand because they are used to explore possibilities for the preservation of labile products, especially those of biological origin. These units are portable and of convenient size for developmental work on freeze-dried products in laboratories and factories around the world. A large number of designs incorporate self-contained facilities for refrigeration, heating, and vacuum pumping, and they can freeze-dry batches consisting of 1 to 10 kg of frozen product. Because of the large variety of pilot freeze-dryers that are employed in industries and laboratories, and because of the limitation space we have for description, a pilot freeze-dryer is presented here whose characteristics are very close to the characteristics of industrial large-scale lyophilizers.

A schematic diagram of the pilot unit (Criofarma model C8-2), is shown in Figure 3a. The unit consists of *(1)* a freezing fluid system (R404A) that can be routed to the heat exchanger section of the condenser or into the refrig-

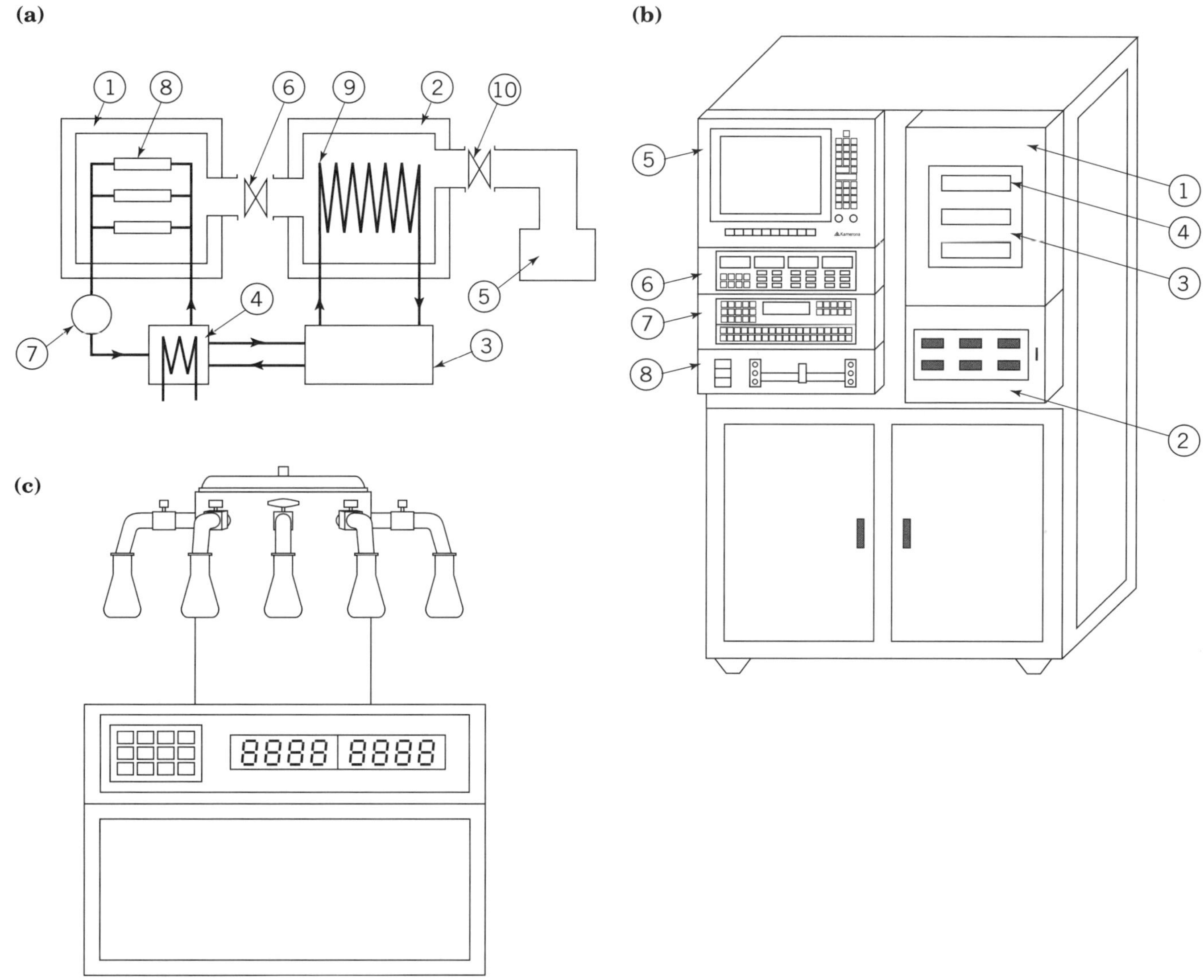

Figure 3. Pilot freeze-dryer (Criofarma model C8-2). (**a**) Diagram of a pilot freeze-dryer: 1, drying chamber; 2, ice condenser chamber; 3, refrigeration unit; 4, cooling/heating system for the plates; 5, vacuum unit; 6, isolation butterfly valve; 7, silicon oil pump; 8, cooling/heating plate; 9, refrigerated coil; 10, condenser vacuum valve. (**b**) Frontal view of a pilot freeze-dryer: 1, drying chamber; 2, ice condenser chamber; 3, cooling/heating plate; 4, inspection window; 5, computer system; 6, vacuum indicator and regulator; 7, temperature control panel; 8, printer. (**c**) Frontal view of a laboratory freeze-dryer (Criofarma model Criolab-8). *Source*: Courtesy of Criofarma.

eration coils for product freezing, *(2)* a heating circuit (silicon oil is the heating fluid) for plate heating and defrosting of the condenser, and *(3)* a vacuum system for evacuating air from the apparatus before and during drying.

The rectangular drying chamber shown in Figure 3b is mounted on top of the condenser; the dimensions are 0.45 m × 0.45 m, and 0.5 m deep. Viewing windows are incorporated in the drying and condensation sections. The refrigeration and vacuum systems are in the internal part of the apparatus, whose complete dimensions are 1.1 m × 0.8 m, and 2.0 m high.

For pilot use, the freeze-dryer offers full control of the process variables and is able to achieve conditions of pressure and temperature beyond the limits of production units. The shelf and ice condenser temperatures of the pilot unit can be −55 °C and −75 °C, respectively, and the pressure in the drying chamber can be as low as 1 Pa or less. The pilot freeze-dryer has a control panel fully accessoried with instruments that record and display *(1)* the temperature inside the product, *(2)* the temperature on the plates, *(3)* the temperature of the coils of the condenser, *(4)* the pressure in the drying chamber, *(5)* the pressure in the vacuum unit, and *(6)* the pressure in the condenser section. The use of a personal computer in this pilot unit, with programmable temperature during the freeze-drying cycle and programmable input–output logic in the different freeze-drying stages, offers a wide variety of drying cycles and the capability for extensive data acquisition, so process optimization can be examined and studied without the risk and cost of investigating a full-size production freeze-dryer. Furthermore, the use of a computer is important to make use of manometric temperature measurement methodology (29) in the primary drying, which is based upon an analysis of the rate of pressure increase in the drying chamber when the valve separating the drying chamber from the ice condenser chamber is periodically quickly closed for a brief period of time. A mathematical analysis of the pressure rise data for a given apparatus gives the average temperature of the product without placing temperature sensors in the product and, therefore, without risk of sterility compromise.

Figure 3c shows a small pilot freeze-dryer, suitable for freeze-drying a few vials (Criofarma model Criolab-8).

Industrial Freeze-Dryers

The objective in a freeze-drying process is to convert most of the water into ice in the freezing stage, remove the ice by direct sublimation in the primary drying stage, and finally remove most of the unfrozen water in the secondary drying stage by desorption. In the industrial freeze-dryer it is possible to perform all three stages of the freeze-drying process. Most of the industrial freeze-dryers in operation are of the vacuum-batch type with freeze-drying of the product in trays. There are two main subtypes, depending on the type of condenser used. In one, the condenser plates are alongside the tray-heater assembly and in the same chamber; in the other the ice condenser is in a separate chamber joined to the first, generally, by a wide butterfly valve. This latter type of plant is always used in pharmaceutical industries. Because of the wide variety and complexity of the problems associated with the production of pharmaceuticals by freeze-drying, in the following paragraphs the principal features of an industrial tray freeze-dryer for pharmaceuticals are presented.

The principal problem in the freeze-drying of pharmaceutical solutions is to operate in sterile conditions. The pharmaceutical solution may be sterile filtered immediately before introduction into a vial or a tray. Also the location of the plant must be able to guarantee sterile conditions during the filling, charging before drying, and discharging after drying of the pharmaceutical product. This is realized by facing the drying chamber door in a wall that separates the sterile room from the machine, or nonsterile, room. In the plant, this separation is accomplished with an isolation valve that separates the ice condenser from the drying chamber; this valve is also able to permit *(1)* the pressure rise test at the end of the freezing-drying cycle, *(2)* the simultaneous discharging/loading of the product and condenser defrosting, and *(3)* the reduction of cross-contamination between batches to a minimum. All the internal parts of the freeze-dryer are of stainless steel type AISI 316L with a finished surface of 0, 1–0, 4 rugosity or less. In modern plants, internal sterilization of the equipment is usually done with pressurized steam at 121 °C or more; in old plants, sterilization is realized with the use of hydrogen peroxide (27) or certain proprietary sanitizing agents (28).

The product containers (vials or bottles loaded on stainless steel trays) are usually sterilized in a separate unit before filling and charging in the freeze-dryer. These operations require the presence of people in the sterile room, with consequent handling of the containers and possible contamination of the batch. For this reason, the human presence in the sterile room is usually reduced the minimum necessary. A new freeze-dryer plant concept has been developed in order to reduce the risk of product contamination. The plant, as shown in Figure 4 (Criofarma model C292-8S), has two full doors. In the first (usually closed) there is a small door for loading the product before drying; the second full door, located opposite to the small door, is for discharging the product after drying. The condenser could be placed on the same floor, or on a floor that is below the floor where the drying chamber is located. The shelves of the freeze-dryer are lowered to the bottom of the drying chamber and are then lifted one by one to a position in line with the loading machine. The charging of the product is made under a laminar flow of sterile air; the small door is opened only for each plate loading and is then immediately closed. If the product is unstable and must be frozen within a short time after it is filled into its container, then it is possible to load trays of product onto the precooled shelves a half plate at a time. When the product container is a bottle, as shown in Figure 4, it usually has a silicon plug, which is partially inserted into the bottle; the solvent vapor leaves the container from the free space between the inserted portion of the plug and the container. After drying and before product discharge, the bottles are stoppered in the drying chamber with the plugs, which are now fully introduced into the bottles. The stoppering operation is done in vacuum conditions, or at atmospheric pressure by breaking the drying chamber vacuum with sterile nitrogen, which prevents successive oxidation of the product; the latter method is most often employed in practice. The silicon plug in the stoppered bottles provides a protection from contamination, and it may be possible to discharge the product in a less-sterile environment from the full door of the freeze-dryer in only one operation. The entire process may be fully automated as the bottles are removed from the filling machine; the disadvantage of automation is that the loading time of the freeze-dryer may become as large as the time it usually takes to complete the filling step of the operation, and this could reduce the theoretical freeze-dryer production for a large installation.

Pharmaceutical freeze-dryers are very often used to produce raw materials such as ampicillin, cloxacillin, and cefazolin (usually as sodium salt), or other specialty materials such as collagen. In these systems, the product is usually charged on stainless steel or polyethylene film trays, and the plant is usually a medium or large unit with a loading surface varying from 20 m^2 to 60 m^2. If the product to be freeze-dried is not particularly unstable (e.g., collagen) and can withstand a delay of some hours between being filled into its tray and being frozen, then one can usually accumulate the trays of product on a loading trolley. When the loading trolley is filled, it is placed in front of the freeze-dryer, and the trays are automatically pushed on the shelves without sliding contact (in order to avoid particle generation), in a single operation. This system is advantageous because it permits maximum utilization of the freeze-dryer; the trolley may be loaded ahead of time,

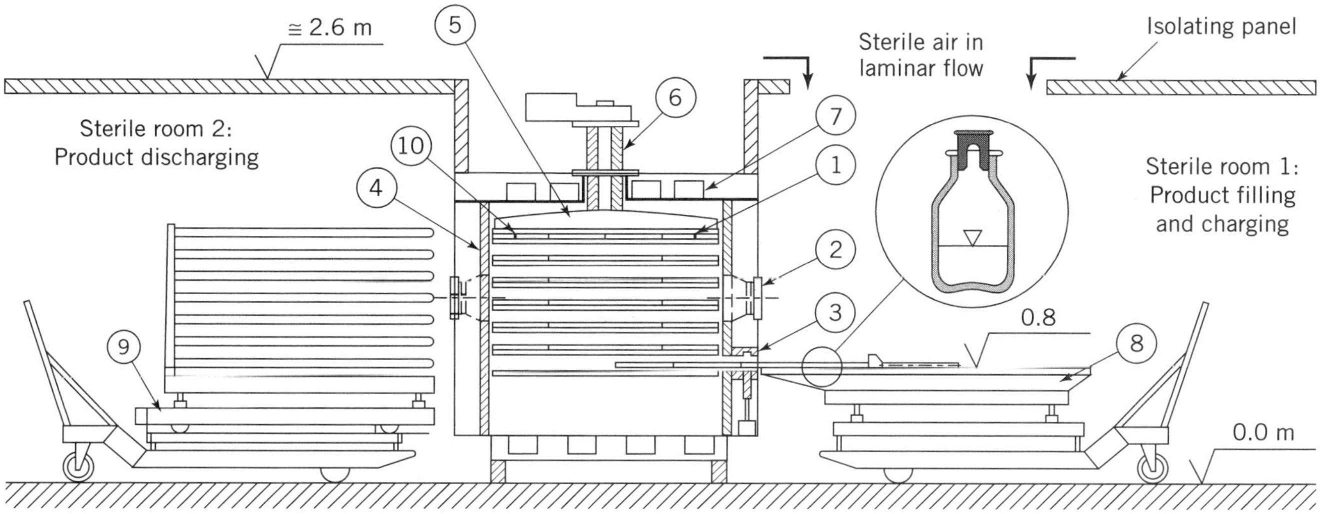

Figure 4. Layout of an industrial freeze-dryer with stoppering device (Criofarma Model C292-8S): 1, drying chamber; 2, inspection window; 3, automatic small door opening; 4, full door; 5, hydraulic press for stoppering the bottles after drying; 6, stainless steel elbows for double sterile condition inside the stoppering plug; 7, reenforcing member and cooling coils after steam sterilization; 8, loading device; 9, discharging device; 10, loaded shelves. *Source*: Courtesy of Criofarma.

so that loading into the freeze-dryer can be carried out in a few minutes. If the product is not stable in the liquid state (e.g., ampicillin sodium salt) and must be frozen within a short time after its preparation, it is common to charge the empty trays on the precooled shelves and then to fill the trays, so that the freezing step is very quick and can proceed during the whole loading operation. This approach is also advantageous because it reduces the freeze-drying cycle time; this happens because the cooling phase starts at the same time as the loading phase, with a consequent reduction in the total time of these two steps.

The freeze-drying system used in the pharmaceutical industry basically consist of the following six units:

1. *Drying chamber*. Holds the product to be dried on shelves an provides an evacuated space for freeze-drying.
2. *Ice condenser chamber*. Collects and disposes of the water vapor produced during drying stages.
3. Shelf heating and cooling unit. Permits the heat transfer required for the freeze-drying cycle.
4. *Vacuum pumping unit*. Evacuates both the drying chamber and the ice condenser chamber to operating pressure and removes noncondensables such as air.
5. *Refrigeration unit*. Provides refrigeration for the chamber shelves during the freezing stage, and ice condenser plates for the primary and secondary drying stages.
6. *Instruments and control panel*. Indicate, record, and control shelf, product, and condenser temperature, as well as drying chamber, ice condenser chamber, and vacuum unit pressure.

A typical freeze-drying equipment lay-out (Criofarma model C292-8S) is showed in Figure 5.

The user of a pharmaceutical freeze-dryer must observe current cGMP for processes and equipment to be validated before and during use. As is the case in any validation effort, the protocol includes sections for installation, operation, and process. Each section of the protocol needs to be comprehensive and sufficiently detailed. The objectives of the validation program are to adequately demonstrate the operation's ability to support the process and to develop documentation demonstrating that control of the unit's operation can be maintained. To develop and implement a protocol for lyophilization, careful consideration should be given to the installation qualification (IQ) and operational qualification (OQ) sections. The IQ section is the foundation of the protocol. Documentation of a general description, specifications, the support utility services, the equipment subsystems, and the installation of the lyophilization system should be addressed within the IQ section. This portion of the protocol is similar to that of any operation being validated. As part of the installation, an initial calibration should be performed to ensure that the instrumentation is in working condition after shipment and to ensure the accuracy of any data collected during start-up of the system. Calibration procedures and results also should be documented.

The OQ section can begin after successful completion of the installation qualification portion of the validation protocol. The objective of the operational qualification is to assure that the system performance is adequate to support the process for which the system is intended. The performance of the system needs to be compared either to the original equipment manufacturer's specification or to the process parameter requirements as noted in the acceptance criteria. Tests should be performed on shelf temperature control, condenser cool-down, vacuum pumping rates, vacuum integrity, sublimation and condensation rates, vacuum level control, steam-in-place (SIP) system,

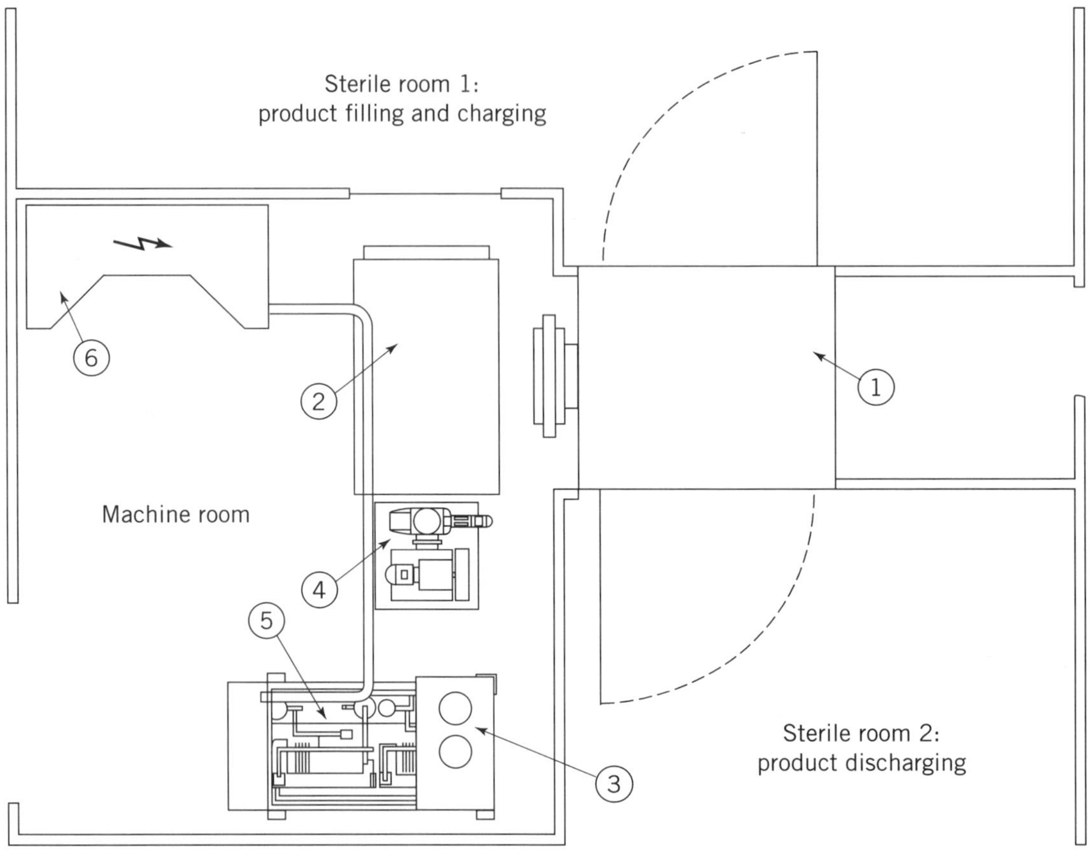

Figure 5. Lay-out of an industrial double-door freeze-dryer (Criofarma model C292-8S): 1, drying chamber with double full doors; 2, ice condenser chamber; 3, shelf heating and cooling unit; 4, vacuum pumping unit; 5, refrigeration unit; 6, instrument and control panel. Courtesy of Criofarma.

clean-in-place (CIP) system (cleaning the inside part of the plant with flushing steam or sterile water), and control hardware and software. The tests for validation of control hardware and software basically require suitable computer hardware and computer programs that perform consistently within preestablished operational limits so that analysis of the effects of possible failures can be carried out. If the tests are conducted in a manner that is in compliance with the procedures and that satisfies the acceptance criteria, the system will perform adequately at levels required to support the process.

The process validation section can be done as concurrent validation or as retrospective validation. Concurrent validation is applicable in situations in which the process data are collected and analyzed for a product that is to be marketed. Retrospective validation, which is particularly applicable to validation of lyophilization, is achieved through analysis of historical batch records. In this method, the data are gathered from records and then evaluated and analyzed. It is important to note that the validation of lyophilization processes is a sophisticated area that requires an understanding of the equipment, the process, and the product if a validation program for the process is to be implemented successfully.

Useful information for cGMP compliance, process, and computer system validation, can be found in Refs. 30–37.

FREEZE-DRYING OF PHARMACEUTICALS

Freeze-drying of pharmaceuticals can be divided according to the kind of product containers used to perform the process. It is possible to have freeze-drying of pharmaceutical products on trays, and freeze-drying of pharmaceutical products in vials. The different amount and kind of product processed (usually raw materials are freeze-dried on trays, whereas vaccines or drugs are freeze-dried in vials) and the different geometry of the container are very important relative to the process parameters and drying rates.

In Figure 1, a material being freeze-dried in a tray is shown. The thickness of the sides and bottom of the tray, as well as the material from which the tray is made, are most often in practice such that the resistance of the tray to heat transfer can be considered to be negligible (1–2,5–6,8–10, 38–39). Heat q_{I} can be supplied to the surface of the dried layer by conduction, convection, and/or radiation from the gas phase; this heat is then transferred by conduction to the frozen layer. Heat q_{II} is supplied by a heating plate and is conducted through the bottom of the tray and through the frozen material to reach the sublimation interface or plane. The magnitude of the amount of heat q_{III} in the vertical sides of the tray is much smaller (1–2,5–6,8–10,38–39) than that of q_{I} or q_{II}; q_{III} represents the amount of heat transferred between the environment in the drying chamber and the vertical sides of the tray. Because the contribution of q_{III} is rather negligible when compared with the contributions of q_I and q_{II}, the contribution of q_{III} to the drying rate will not be further considered (1–2,8–9). The terms N_W and N_t in Figure 1 represent the mass flux of water vapor and the total mass flux, respectively, in the dried layer. The total mass flux is equal to the sum of the mass fluxes of water vapor and inert gas, $N_t = N_W + N_{in}$, where N_{in} denotes the mass flux of the inert gas.

In Figure 6, a typical lyophilizing vial (40) is shown during freeze-drying. The energy of sublimation of ice and desorption of sorbed water is supplied to the vial through radiation from the two shelf surfaces above and below the vial, conduction from vial–vial and vial–shelf contacts, and gaseous transfer along the bottom and side. The thickness of the pharmaceutical product charged into the vial usually is equivalent to the diameter of the vial. In pharmaceutical freeze-drying in vials, therefore, the heat transferred between the environment in the drying chamber and the surface of the vertical sides of the vial cannot be considered negligible (as in the case of pharmaceutical freeze-drying on trays) when compared with the contribution of the heat transferred between the dryer shelf and the surface of the bottom of the vial. For this reason the sublimation interface at time $t = 0$ is planar (as in the case of pharmaceutical freeze-drying on trays), but after the lyophilization process progresses, the moving interface acquires a parabolic shape because of energy entering along the sides of the vial (3,6,11). The traditionally accepted course of the moving sublimation interface is approximately 0-1-2-3-4. Experimental results (41–42) suggest that the dried-product mass transfer resistance is the dominant resistance, accounting for 82–92% of the overall mass transfer

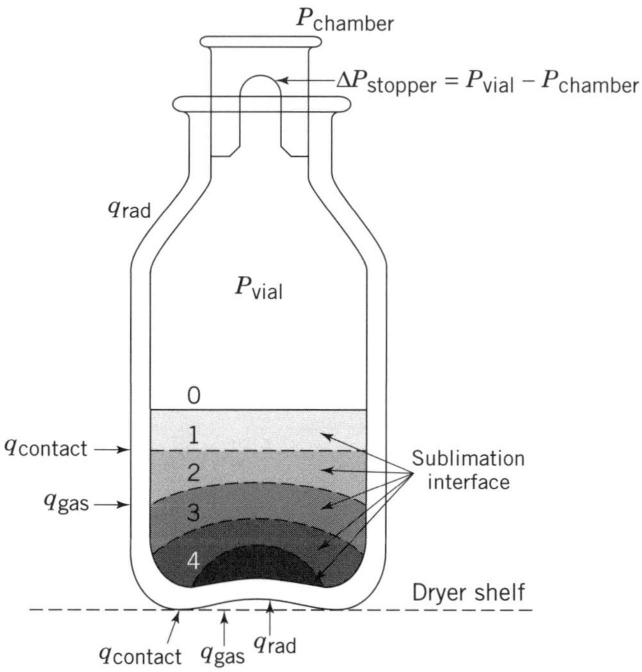

Figure 6. Modes of energy transfer and traditional view of movement of the sublimation interface in a vial during freeze-drying.

resistance, even when small (13-mm) closures are used. Closure and drying chamber mass transfer resistances are estimated (42) to be of the same order of magnitude, and in a well-designed freeze-dryer the chamber resistance (1–3,6) should be less than the resistance of the closure. A simple but satisfactory expression for estimating the resistance of the closure can be found in Ref. 42.

The goal of the pharmaceutical process designer and of the processor is to provide an economical drying process that gives reliably uniform and high product quality (1–4,18–20,43–44). A knowledge of the basic phenomena and mechanisms involved in freeze-drying is essential. A qualitative description and a mathematical model of the freeze-drying process on trays can be found in Refs. 1–3, 6, 9–10, 13, 15–16, and 39; the model can be used to analyze rates of freeze drying of pharmaceuticals on trays (1–6,8–11,13–16,19,45–48). Qualitative description and a mathematical model of the freeze drying process in vials can be found in Refs. 3, 6, 9, 11, and 49.

TEMPERATURE AND PRESSURE CONSTRAINTS IN FREEZE-DRYING

A point that should be stressed is that, in any freeze-drying process, it will be desirable to fix the design and operating conditions such that the process is not rate limited by external resistances to either heat or mass transfer. The internal heat and mass transfer resistances are characteristic of the material being dried, but the external resistances are characteristic of the freeze-dryer. The design conditions of the equipment refer to having appropriate capacities for the vacuum pump, water vapor condenser, and heaters, and that the spacings between trays are such that the external heat and mass transfer resistances are not significant.

It is important to note at this point that, because a batch of a pharmaceutical product in an industrial freeze-dryer can easily be worth significant amounts of money, it is of paramount importance that the units and the control systems always operate under conditions at which there is insignificant loss in the quality of the product being freeze-dried. For this purpose, the freeze-dryer usually has one additional refrigeration and/or vacuum unit in a standby condition, and the control instrumentation is designed in such a way that the control policies can be implemented either automatically or by manual override.

In a well-designed freeze-dryer, the external resistances should not be the controlling factor in determining the drying rate (1–4,8–16). The internal heat and mass transfer resistances control the drying rate, and therefore, the freeze-drying process should be conducted within certain temperature and pressure limits in order to avoid product damage during the process.

Two temperature constraints (limits) may possibly be reached during the primary drying stage. First, the surface temperature of the dried layer must not become too high because of the risk of thermal damage; in fact the surface temperature of the dried layer must be kept below the scorch temperature, T_{scor}, of the material being dried. Second, the temperature of the interface, T_X, must be kept well below the melting temperature, T_m. If the outer surface temperature limit (T_{scor}) is encountered first as the surface temperature of the dried layer is raised, the process is considered to be heat transfer controlled; to further increase the drying rate, the thermal conductivity, k_{Ie}, of the dried layer must be raised (50,51). Many commercial freeze-drying processes are heat transfer controlled (1–6). If the melting point temperature, T_m, is encountered first, then the process is considered to be mass transfer limited, and in order to increase the drying rate, the effective diffusivity of water vapor in the dried layer, $D_{win,e}$, and the total mass flux, N_t, must be raised (an increase in N_t implies that the convective velocity of the vapor in the pores of the dried layer is increased); the values of $D_{win,e}$ and N_t could be raised by decreasing the pressure in the drying chamber (1,2,43). Because of the risks of lost bioactivity or structural and chemical damage of the product, the frozen layer temperature must be maintained below the melting point, which may in some cases be 10 °C or more below the melting point of ice (1–6, 8–17). Typical ice temperatures existing in the freeze-drying of pharmaceuticals under conditions in which the total pressure was primarily due to water vapor and the heat transfer took place via the dried (I) and frozen (II) layers are shown in Table 1.

The end of the primary drying stage occurs (1–3,5,6,10) when the position of the moving sublimation interface is at $X = L$; this condition implies that at the end of the primary drying stage there is no frozen (II) layer, and therefore, there is no (moving) sublimation interface.

During the secondary drying stage, the temperature everywhere in the dried (I) layer must be kept below the value of the scorch temperature, T_{scor}.

The pressure constraints (limits) are more complex to define because of the pressure effect on both the heat transfer (pressure has effect on the thermal conductivity, k_{Ie}, of the dried layer) and the mass transfer (pressure has effect on the effective diffusivity of water vapor in the dried layer, $D_{win,e}$, and the total mass flux, N_t).

The effective thermal conductivity, k_{Ie}, in the dried material (51) has been found to vary significantly with the total pressure and with the type of gas present. At very low pressures the thermal conductivity reaches a lower asymptotic value independent of the surrounding gas. This asymptotic conductivity reflects the geometric structure of the solid matrix itself, with no contribution from the gas in the voids of the material because the gas pressure is so low. At high pressures the thermal conductivity levels out again at a higher asymptotic value. This higher asymptote is characteristic of the heterogeneous matrix composed of solid material and the gas in the voids. Consequently, the high-pressure thermal conductivity is dependent upon the nature of the gas present and increases as the thermal conductivity of the gas increases and, hence, as the molecular weight of the gas decreases. When the thermal conductivity attains the high-pressure asymptotic value, the mean free path of the gas molecules within the void spaces of the dried layer has become substantially less than the dimensions of the void spaces. During the transition in thermal conductivity from the low-pressure asymptote to the high-pressure asymptote, the mean free path of the gas molecules rivals the void space dimensions in magnitude, but once the mean free path is reduced to the point at which the gas phase within the solid matrix obeys simple kinetic theory, the thermal conductivity stops rising. This reflects the fact that the thermal conductivity of a gas obeying simple kinetic theory is independent of the pressure. The transition in thermal conductivity between asymptotes usually occurs between 0.1 and 100 mmHg, which includes the pressures characteristic of freeze-drying processes. The pressure range over which the transition in thermal conductivity between asymptotes occurs is characteristic of the pore-size distribution of the void spaces within the freeze-dried material (52). A smaller pore dimension means that the gas must achieve a higher pressure in order for the mean free path of the gas to become comparable to the pore spacing and, hence, means that the transition between asymptotes will occur at higher pressures. Because fast freezing before freeze-drying leads to smaller pore spacing after freeze-drying (53,54), it follows that faster freezing should lead to lower thermal conductivities at a given pressure. If a freeze-drying process is rate limited by internal heat transfer, the rate of freeze-drying for fast-frozen material should then be less than that of a slowly frozen material (45,46). Thermal conductivity data for freeze-dryed substances and pharmaceuticals are reported in Ref. 2.

The pressure effect on the mass transfer is explained by the fact that the effective diffusivity (2,6) of water vapor in the dried layer, $D_{win,e}$, and the total mass flux, N_t, are a function of the total pressure, P, in the drying chamber (43). Elaborate expressions of $D_{win,e}$ and $D_{inw,e}$ can be found in Refs. 2, 6, 9, 11, 15, and 16. In all cases the magnitude of $D_{win,e}$ and $D_{inw,e}$ decrease if the total pressure, P, is increased. If the value of $D_{win,e}$ is decreased because of increased total pressure, then (2,6) the value of the effective pore diffusivity, $D_{win,e}$, decreases as the total pressure is

Table 1. Frozen Layer and Maximum Dry Surface Temperatures in a Typical Freeze-Drying Operation Conducted with Heat Input through the Dry and Frozen Layers

Pharmaceutical material	Chamber pressure (mm Hg)	Maximum surface temperature (°C)	Frozen-layer temperature (°C)
Ampicillin sodium salt	0.15	40	−24
Cephazolin sodium salt	0.15	40	−25
Cloxacillin sodium salt	0.20	40	−20
Collagen	0.30	70	−20
Fructose-1,6-diphosphate	0.05	30	−45
Lactobacillus	0.10	20	−25
Rubella vaccine	0.05	10	−25

increased. Therefore, if the total pressure in the drying chamber is increased, then the effective diffusivity, $D_{win,e}$, decreases, and thus, the diffusional mass flux of water vapor in the dried layer decreases. Furthermore, when the total pressure in the drying chamber is increased, the gradient of the total pressure, $\partial P/\partial x$, in the dried layer is reduced, and this decreases the convective velocity, v_p, and the total mass flux, N_t. Because the freeze-drying process will become internal mass transfer controlled above a certain pressure ($D_{win,e}$ and N_t decrease with increasing pressure, and k_{Ie} increases with pressure), the highest rate under mass transfer control will occur at the pressure of transition from heat transfer control to mass transfer control, and the attainable drying rate will decrease at higher pressures.

In general, operating conditions in freeze-drying pharmaceuticals include maximum surface temperatures of 20–40 °C and chamber pressures of 0.1–1 mmHg. Freeze drying of biological specimens, vaccines, and microorganisms is usually conducted with maximum surface temperatures of -20 to $+20$ °C and chamber pressures below 0.1 mmHg.

There has been considerable interest in investigating the factors affecting the batch time of freeze-dryers because this variable is most amenable to control, and efforts have bene made to minimize the batch time (2,5,10,13,15,16,45–47).

The heat variables q_I and q_{II} from the energy sources, and the drying chamber pressure P_{ch}, $P_{ch} \cong P^0 = P_w^0 + P_{in}^0$ ($P_{ch} = P^0$ when the external mass transfer resistance is insignificant, as it would be the case with a well-designed freeze-dryer) are natural control variables. It should be emphasized at this point that (as the equations of the mathematical models in Ref. 2 indicate) the effects on the heat and mass transfer rates resulting from changes in the values of q_I, q_{II}, and P_{ch} are coupled. The variable P_w^0 is taken to represent the water vapor pressure in the drying chamber (the external mass transfer resistance is taken to be insignificant), and its value is determined by the design and the operational temperature of the ice condenser. Thus, P_{ch} may be changed by changes in P_w^0 (P_w^0 could be changed by changes in the temperature of the ice condenser), and by increasing or decreasing P_{in}^0. Therefore, changes in the temperature of the ice condenser affect the pressure, P_{ch}, (through P_w^0) in the drying chamber, and thus, the mass transfer rate in the dried layer.

Liapis and Litchfield (13) performed a quasi-steady-state analysis for a system where $q_I \neq 0$ and $q_{II} = 0$ and obtained general guidelines about the optimal control policy at the beginning of the drying process (when neither of the state constraints is active), as well as during operation, when the process may be heat and/or mass transfer limited (10,13).

Millman et al. (10) studied the freeze-drying of skim milk under various operational policies that included the case where $q_I \neq 0$ and $q_{II} \neq 0$. They found that the control policy that produced the shortest primary drying stage was also the policy that provided the shortest overall drying time. Their results show that at least 80% of the heat used during the primary drying stage was transferred through the frozen layer of the sample. They also showed that the type of criterion used in terminating the secondary drying stage is of extreme importance, especially for samples of large thicknesses because it may lead to an undesirable sorbed (bound) water profile that may deteriorate the quality of the dried product.

CONCLUSION

The evolution of freeze-drying in the past 40 years indicates that this separation process (unit operation) is a convenient method for drying those decomposable products (mostly pharmaceuticals) that cannot be stabilized in any other way (e.g., plasma, vaccines, antibiotics, sera, and growth hormones).

The economics of the process indicate that freeze-drying can be suitable for high-value products with specific biological and/or physicochemical properties. The highest cost advantages would be obtained from the processing of concentrated solutions of expensive materials; in this respect, freeze-drying could represent a viable alternative to filtration and crystallization. It is certain that freeze-drying has a future, and it is likely that this will essentially be in the fields of biological sciences and biotechnology. Its potential evolution is still great and will, of course, depend upon progress in basic research (4) and upon the level of creativity in the design and operating conditions of plants and instruments.

NOMENCLATURE

$D_{inw,e}$	Effective pore diffusivity of a binary mixture of inert gas and water vapor in the dried layer, m^2/sec
$D_{win,e}$	Effective pore diffusivity of a binary mixture of water vapor and inert gas in the dried layer, m^2/sec
k_{Ie}	Effective thermal conductivity in the dried layer, kW/(m K)
k_{II}	Thermal conductivity in the frozen layer, kW/(m K)
L	Sample thickness, m
M_{in}	Molecular weight of inert gas, kg/kg mol
M_w	Molecular weight of water vapor, kg/kg mol
N_{in}	Mass flux of inert gas in the dried layer, kg/(m^2 s)
N_t	Total mass flux in the dried layer ($N_t = N_{in} + N_w$), kg/(m^2 s)
N_w	Mass flux of water vapor in the dried layer, kg/(m^2 s)
p_{in}^0	Partial pressure of inert gas at $x = 0$, N/m^2
p_w^0	Partial pressure of water vapor at $x = 0$, N/m^2
p_{wX}	Partial pressure of water vapor in equilibrium with the sublimation front ($p_{wX} = g(T_X)$), N/m^2
P	Total pressure ($P = p_{in} + p_w$) in the dried layer, N/m^2

p^0	Total pressure at $x = 0$, N/m^2
p_{ch}	Total pressure in the drying chamber, N/m^2
$P_{chamber}$	Total pressure in the drying chamber, N/m^2
P_{vial}	Vial pressure, N/m^2
q_I	Heat flux at $x = 0$, kW/m^2
q_{II}	Heat flux at the bottom of the tray, kW/m^2
q_{III}	Heat flux at the sides of the tray, kW/m^2
$q_{contact}$	Conduction heat flux, kW/m^2
q_{gas}	Heat flux due to gaseous transfer, kW/m^2
q_{rad}	Radiation heat flux, kW/m^2
t	Time, s
T_I	Temperature in the dried layer, K
T_{II}	Temperature in the frozen layer, K
T_m	Melting temperature, K
T_{scor}	Scorch temperature, K
T_X	Temperature of the sublimation front, K
v_p	Convective velocity of the binary mixture of water vapor and inert gas in the porous dried layer, m/sec
X	Position of sublimation front (interface), m

Greek Symbols

ΔP	Pressure drop, N/m^2
ΔT	Total temperature difference in Figure 2, K
ΔT_{ice}	Temperature difference through the layer of ice on the cold surface, K
ΔT_{refr}	Temperature difference between the cold surface and the evaporating refrigerant, K

Subscripts

I	Dried layer
II	Frozen layer

BIBLIOGRAPHY

1. A.I. Liapis and R. Bruttini, *Sep. Technol.* **4**, 144–155 (1994).
2. A.I. Liapis and R. Bruttini, in A.S. Mujumdar ed., *Handbook of Industrial Drying* 2nd ed., Marcel Dekker, New York and Basel, 1995, pp. 309–343.
3. A.I. Liapis and R. Bruttini, *Drying Technol.* **13**, 43–72 (1995).
4. A.I. Liapis, M.J. Pikal, and R. Bruttini, *Drying Technol.* **14**, 1265–1300 (1996).
5. J.D. Mellor, *Fundamentals of Freeze-Drying*, Academic Press, London, 1978.
6. A.I. Liapis and R. Bruttini, in I. Turner and A.S. Mujumdar eds., *Mathematical Modelling and Numerical Techniques in Drying Technology*, Marcel Dekker, New York and Basel, 1997, pp. 481–535.
7. S.A. Goldblith, L. Rey, and W.W. Rothmayr, *Freeze-Drying and Advanced Food Technology*, Academic Press, London, 1975.
8. R. Bruttini, G. Rovero, and G. Baldi, *Chem. Eng. J.* **45**, B67–B77 (1991).
9. R. Bruttini, Ph.D. Thesis, University of Missouri, Rolla, Mo., 1994.
10. M.J. Millman, A.I. Liapis, and J.M. Marchello, *AIChE J.* **31**, 1594–1604 (1985).
11. M.M. Tang, A.I. Liapis, and J.M. Marchello, in A.S. Mujumdar ed., *Proceedings of the Fifth International Drying Symposium*, vol. 1, Hemisphere Publishing, New York, 1986, pp. 57–65.
12. C.J. King, 1971, *Freeze-Drying of Food*, Chemical Rubber Co. Press, Cleveland, Ohio, 1971.
13. A.I. Liapis and R.J. Litchfield, *Chem. Eng. Sci.* **34**, 975–981 (1979).
14. M.J. Pikal, S. Shah, M.L. Roy, and R. Putman, *Int. J. Pharm.* **60**, 203–217 (1990).
15. R.J. Litchfield and A.I. Liapis, *Chem. Eng. Sci.* **34**, 1085–1090 (1979).
16. R.J. Litchfield and A.I. Liapis, *Chem. Eng. Sci.* **37**, 45–55 (1982).
17. J.W. Snowman, in Avshalom Mizrahi ed., *Downstream Processes: Equipment and Techniques*, Alan R. Liss, New York, 1988, pp. 315–351.
18. M.J. Pikal, in L. Cleland and R. Langer ed., *Formulation and Delivery of Proteins and Peptides*, ACS Symposium Series 567, American Chemical Society, Washington, D.C., 1994, pp. 120–133.
19. C.D. Holland and A.I. Liapis, *Computer Methods for Solving Dynamic Separation Problems*, McGraw-Hill, New York, 1983.
20. P.A. Belter, E.L. Cussler, and W.S. Hu, *Bioseparations: Downstream Processing for Biotechnology*, Wiley-Interscience, New York, 1988.
21. R.J. Litchfield, F.A. Farhadpour, and A.. Liapis, *Chem. Eng. Sci.* **36**, 1233–1238 (1981).
22. R.G. Livesey and T.W.G. Rowe, *J. Parenteral Sci. Technol.* **41**, 169–171 (1987).
23. J.H. Petropoulos, J.K. Petrou, and A.I. Liapis, *Ind. Eng. Chem. Res.* **30**, 1281–1289 (1991).
24. J.H. Petropoulos, A.I. Liapis, N.P. Kolliopoulos, J.K. Petrou, and N.K. Kanellopoulos, *Bioseparation* **1**, 60–69 (1990).
25. M.L. Roy and M.J. Pikal, *J. Parenteral Sci. Technol.* **43**, 60–66 (1989).
26. C.C. Freyrichs and C.N. Herbert, *J. Biol. Stand.* **2**(1), 59 (1974).
27. J.W. Johnson, J.F. Arnold, S.L. Nail, and E. Renzi, *J. Parenteral Sci. Technol.* **46**, 215 (1992).
28. A. Bardat, R. Schmitthaeusler, and E. Renzi, *PDA J. Pharm. Sci. Technol.* **50**(2), 83–88 (1996).
29. N. Milton, S.L. Nail, M.L. Roy, and M.J. Pikal, *Pharm. Res.* **10**(S-170), 6174 (1993).
30. J.Y. Lee, *Pharmaceutical Technology*, October, 54–60 (1988).
31. U.S. Federal Drug Administration, *Lyophilization of Parenterals, Inspection Technical Guide*, U.S. Department of Health and Human Services, Washington, D.C., 1986.
32. U.S. Food and Drug Administration, *Guideline on General Principles of Process Validation*, U.S. Department of Health and Human Services, Washington D.C., 1987.
33. U.S. Food and Drug Administration, *Guide to Inspection of Bulk Pharmaceutical Chemical Manufacturing*, U.S. Department of Health and Human Services, Washington D.C., 1987.
34. E.H. Trappler, *Pharmaceutical Technology*, January, 56–60 (1989).
35. K.G. Chapman and J.R. Harris, *Pharmaceutical Technology International*, May/June, 54–58 (1989).
36. J.S. Alford, and F.L. Cline, *Pharm. Technol.* **14**(9), 88 (1990).
37. P.M. Masterson, *Pharmaceutical Technology*, January, 48–54 (1989).
38. A.I. Liapis and R. J. Litchfield, *Comput. Chem. Eng.* **3**, 615–621 (1979).

39. A.I. Liapis and J.M. Marchello, in J.C. Ashworth ed., *Proceedings of the Third International Drying Syposium*, vol. 2, Drying Research Limited, Wolverhampton, England, 1982, pp. 479–486.
40. M.D. Cise, *AIChE Summer National Meeting*, Cleveland, Ohio, 1982, paper 22d.
41. M.J. Pikal, S. Shah, D. Senior, and J.E. Lang, *J. Pharm. Sci.*, **72**, 635–650 (1983).
42. M.J. Pikal, M.L. Roy, and S. Shah, *J. Pharm. Sci.* **73**, 1224–1237 (1984).
43. C.J. Geankoplis, *Transport Processes and Unit Operations*, Allyn and Bacon, Boston, 1983.
44. A.I. Liapis, *Sep. Purif. Methods*, **19**(2), 133 (1990).
45. M.J. Millman, A.I. Liapis, and J.M. Marchello, *J. Food Technol.* **19**, 725 (1984).
46. M.J. Millman, A.I. Liapis, and J.M. Marchello, *J. Food Technol.* **20**, 541 (1985).
47. M.J. Millman, Ph.D. Thesis, University of Missouri, Rolla, Mo., 1994.
48. H. Sadikoglu and A.I. Liapis, in C. Strumillo and Z. Pakowski ed., *Drying 96*, vol. B, Marcel Dekker, New York and Basel, 1996, p. 1284.
49. J.L. Duda, M.F. Malone, R.H. Notter, and J.S. Vrentas, *Int. J. Heat Mass Transfer* **18**, 901–910 (1975).
50. A.V. Luikov, A.G. Shashkov, L.L. Vasiliev, and Y.E. Fraiman, *Int. J. Heat Mass Transfer* **11**, 117 (1968).
51. L. Marcussen, in T. Ashworth and D.R. Smith eds., *Thermal Conductivity*, vol. 18, Plenum Press, New York, 1985, pp. 585–598.
52. C.-J. King, W.K. Lam, and O.C. Sandall, *Food Technol.* **22**, 1302 (1968).
53. C.G. Haugh, C.S. Huber, W.J. Stadelman, and R.M. Peart, *Trans. ASAE* **11**, 877 (1968).
54. G. Lusk, M. Karel, and S.A. Golblith, *Food Technol.* **19**, 620 (1965).

See also CULTURE COLLECTIONS; GOOD MANUFACTURING PRACTICE (GMP) AND GOOD INDUSTRIAL LARGE SCALE PRACTICE (GLSP).